DICTIONARY
OF
BUILDING AND CIVIL ENGINEERING

S. N. KORCHOMKIN,
S. V. KURBATOV,
N. B. SHEIKHON,
G. B. VILJKOVYSKAJA

MOSCOW
RUSSKY YAZYK PUBLISHERS
1985

DICTIONARY
OF
BUILDING
AND
CIVIL ENGINEERING

ENGLISH
GERMAN
FRENCH
DUTCH
RUSSIAN

von BPH - Kollegin
Hamburg, Januar 1995
Anke Lindemann

1985

MARTINUS NIJHOFF PUBLISHERS
THE HAGUE/BOSTON/LANCASTER

BORDAS DUNOD
PARIS

KLUWER TECHNISCHE BOEKEN
DEVENTER

Distributors:

for the Benelux:
Kluwer Technical Books B. V./Libresso B. U.
P. O. Box 23
7400 GA Deventer
The Netherlands
ISBN 90 201 18633
D/1985/0108/200

for France and Canada:
Bordas Dunod
17, rue Rémi-Dumoncel
B. P. 50
75661 Paris Cedex 14
France
ISBN 2-04-015 799

for the United States:
Kluwer Boston, Inc.
190 Old Derby Street
Hingham, MA 02043
USA

for all other non-socialist countries:
Kluwer Academic Publishers Group Distribution Center
P. O. Box 322
3300 AH Dordrecht
The Netherlands
ISBN 90 201 18633

PREFACE

In the last few decades civil engineering has undergone substantial technological change which has, naturally, been reflected in the terminology employed in the industry. Efforts are now being made in many countries to bring about a systematization and unification of technical terminology in general, and that of civil engineering in particular.

The publication of a multilingual dictionary of civil engineering terms has been necessitated by the expansion of international cooperation and information exchange in this field, as well as by the lack of suitable updated bilingual dictionaries.

This Dictionary contains some 14 000 English terms together with their German, French, Dutch and Russian equivalents, which are used in the main branches of civil engineering and relate to the basic principles of structural design and calculations (the elasticity theory, strength of materials, soil mechanics and other allied technical disciplines); to buildings and installations, structures and their parts, building materials and prefabrications, civil engineering technology and practice, building and road construction machines, construction site equipment, housing equipment and fittings (including modern systems of air conditioning); as well as to hydrotechnical and irrigation constructions.

The Dictionary also includes a limited number of basic technical expressions and terms relating to allied disciplines such as architecture and town planning, as well as airfield, railway and underground construction. The Dictionary does not list trade names of building materials, parts and machines or the names of chemical compounds. Nor does it give adverbial, adjective or verbal terms.

In compiling the Dictionary the authors used a wide range of recent publications: encyclopedias, terminological dictionaries, civil engineering thesauruses, glossaries, reference books, terminological standards, monographs, textbooks, periodicals and technical documentation.

Unequal semantic value or inadequacy of certain English terms and their counterparts in the other languages of the Dictionary have in some instance compelled the authors to provide explanatory or discriptive translations. The user should also realise that in cases of multisemantic English terms only those equivalents that immediately pertain to civil engineering have been listed.

The Dictionary is intended for civil engineers, research workers, teachers, postgraduates and college students, as well as for translators of scientific and technical publications.

Being the first of its kind, the Dictionary may well have certain drawbacks or lack some useful terms. The authors would, therefore, appreciate any remarks and suggestions on its contents. These should be forwarded to Kluwer Technische Boeken B. V. Postbus 23,7400 GA Deventer, The Nederlands or to 103012 Moskva, Staropanski per., 1/5, Izdatelstvo "Russky Yazyk".

The Authors

USING THE DICTIONARY

English terms are arranged alphabetically, with compound terms being treated as single words, e.g.

deck girder
decking
deck level

Each entry comprises an English term and its German, French, Dutch and Russian equivalents. For a crossreference from an English term to its synonym, a note *see* is used. German, French, Dutch and Russian terms indicate gender (*m*, *f*, *n*) and, where necessary, plural (*pl*). Each English term is given an index number plus an index letter corresponding to the initial letter of the term so as to facilitate the search for a German, French, Dutch or Russian equivalent. This makes it possible to translate from any of these languages. A typical entry therefore, appears as:

B66 *e* baked brick
   d gebrannter Ziegel(stein) *m*
   f brique *f* cuite
  nl baksteen *m*
   r обожжённый кирпич *m*

Translations of the different meanings of a multisemantic English term are indicated by arabic numbers. Semantically close equivalents of an English term are divided by a semicolon, while a comma is used to separate its synonymous counterparts.

For the sake of printing space the interchangeable parts of the synonymous equivalents of an English term are in some cases placed in square brackets, while their optional parts are enclosed in parentheses, e.g.

rechnerische Belegung [Besetzung]	= rechnerische Belegung *or* rechnerische Besetzung
emulsion bitumineuse [de bitume]	= emulsion bitumineuse *or* emulsion de bitume
gekleurde verglaasde pan [tegel]	= gekleurde verglaasde pan *or* gekleurde verglaasde tegel
Einspann(ungs)moment	= Einspannmoment *or* Einspannungsmoment
tuile (cuite) plate	= tuile cuite plate *or* tuile plate
insteek(verdieping)	= insteek *or* insteekverdieping

Geographical area of an English term's usage is indicated as follows:

UK = United Kingdom
US = United States

To translate from German, French, Dutch or Russian the user should consult the relevant appendix where alphabetically arranged terms are supplied with letter-number references corresponding to the indexes of their English counterparts.

VORWORT

In der materiellen Produktion haben sich in den letzten Jahrzehnten weitreichende Wand-lungen vollzogen, die sich in der Terminologie ihren Niederschlag fanden. In vielen Ländern wird ständig an der Einordnung (Standardisierung und Normung) des technischen Wortgutes darunter auch für den Bereich des Bauwesens, gearbeitet.

Die sich immer mehr ausweitende Zusammenarbeit und der zunehmende Informationsaus-, tausch zwischen den verschiedenen Ländern auf dem Gebiet des Bauwesens sowie das Fehlen von zweisprachigen Wörterbüchern, die dem heutigen Entwicklungsstand Rechnung tragen, verlangten nach einem mehrsprachigen Wörterbuch für diesen Fachbereich.

Das vorliegende Wörterbuch enthält etwa 14 000 englische Begriffe und ihre Entsprechun-gen in Deutsch, Französisch, Niederländisch und Russisch. Sie erstrecken sich auf die wichtigste Bereiche des Bauwesens wie Grunglagen der Projektierung und Berechnung (Elastizitäts- und Festigkeitslehre, Baugrundmechanik sowie angrenzende technische Disziplinen); Wohnungs-und Ingenieurhochbau; Baukonstruktionen und ihre Elemente; Baustoffe und vorgefertigte Bauelemente; Organisation und Technologie des Bauwesens; Ingenieurausrüstungen der Bauwer-ke wie zum Beispiel Klimaanlagen; Wasserstau-, Kraftwerks- und Bewässerungsanlagen.

Nur begrenzt wurden in das Wörterbuch allgemeintechnische sowie an das Bauwesen angren-zende Begriffe wie z. B. aus den Bereichen Architektur und Städtebau, Flugplatzbau sowie Tief- und Eisenbahnbau aufgenommen. Nicht erfaßt sind die Bezeichnungen der Herstellfirmen von Materialien, Erzeugnissen und Ausrüstungen, die Namen von chemischen Verbindungen und ihren Komponenten sowie die Adverbien, Adjektive und Verben zu den jeweiligen Begriffen.

Bei der Zusammenstellung des Wörterbuchs wurde reichhaltiges Quellenmaterial in den jeweiligen Sprachen ausgewertet: Lexika und Fachwörterbücher, Thesaurien des Bauwesens, Glossare, Terminologiestandards, Monographien, Lehrbücher, Periodika sowie Entwurfsunter-lagen.

Erhebliche Schwierigkeiten bereitete den Autoren hin und wieder das Auffinden der exakten lexikalischen Äquivalente für die englischen Termini. Der Erfassungsbereich der Begriffe weicht in den verschiedenen Sprachen oft erheblich ab. In einigen Fällen sahen sich die Autoren deshalb zu Erläuterungen oder umschreibenden Übersetzungen gezwungen. Bei der Wahl der Entsprechungen für mehrdeutige englische Termini sind in das Wörterbuch nur diejenigen Begriffe aufgenommen worden, die unmittelbaren Bezug auf das Bauwesen haben.

Das vorliegende Wörterbuch ist für einen breiten Kreis von Fachleuten bestimmt: Bauin-genieure und -techniker, Wissenschaftler, Lehrkräfte, Aspiranten, Studenten sowie Übersetzer von technischer Literatur.

Die Autoren sind sich! völlig bewußt, daß dieser erste Versuch, ein mehrsprachiges Wör-terbuch des Bauwesens zu schaffen, das den Belangen von Theorie, Lehre und Praxis gleicher-maßen genügt, nicht frei von Lücken und Mängeln sein kann. Kritische Hinweise und Vorschläge zur Verbesserung des Wörterbuches bitten wir deshalb an folgende Adresse zu richten: Kluwer Technische Boeken B. V. Postbus 23, 7400 GA Deventer, Nederland oder 103012, Moskau, Staropanski per., 1/5, Verlag «Russkij Jazyk».

Die Autoren

HINWEISE FÜR DIE BENUTZUNG

Die englischen Stichwörter sind alphabetisch geordnet:
deck girder
decking
deck level
Eine Wortstelle besteht aus dem jeweiligen englischen Begriff und seinen Entsprechungen in Deutsch, Französisch, Niederländisch und Russisch. Für den Hinweis auf einen synonymen englischen Terminus wird der Vermerk *see* (*siehe*) verwendet. Alle deutschen, französischen, niederländischen und russischen Begriffe sind mit der Angabe des Geschlechts (*m, n, f*) und, wo es sich um die Pluralform handelt, mit dem Zeichen (*pl*) versehen.

Die englischen Begriffe sind numeriert, so daß der entsprechende deutsche, französische, niederländische oder russische Begriff in dem im Anhang befindlichen Register mühelos aufgefunden werden kann.

Die Numerierung erfolgt für jeden Buchstaben getrennt. Folglich sieht eine Wortstelle folgendermaßen aus:

B66 *e* **backed brick**
   d gebrannter Ziegel(stein) *m*
   f brique *f* cuite
  nl baksteen *m*
   r обожжённый кирпич *m*

Wenn der angeführte englische Begriff mehrdeutig ist, sind die sinnabweichenden Äquivalente in den anderen Sprachen durch Ziffern, sinnverwandte Varianten durch ein Semikolon und Synonyme durch ein Komma getrennt.

In den Übersetzungen sind die austauschbaren Teile synonymer Varianten in eckige Klammern gesetzt; der fakultative, also verzichtbare Teil des Begriffs steht in runden Klammern, z. B.:

rechnerische Belegung [Besetzung]	= rechnerische Belegung *oder* rechnerische Besetzung
emulsion bitumineuse [de bitume]	= emulsion bitumineuse *oder* emulsion de bitume
gekleurde verglaasde pan [tegel]	= gekleurde verglaasde pan *oder* gekleurde verglaasde tegel
Einspann(ungs)moment	= Einspannmoment *oder* Einspannungsmoment
tuile (cuite) plate	= tuile cuite plate *oder* tuile plate
insteek(verdieping)	= insteek *oder* insteekverdieping

Dort, wo im englischen und anglo-amerikanischen Sprachraum unterschiedliche Begriffe gebräuchlich sind, wurden diese mit den Zeichen *UK* (Vereinigtes Königreich von Großbritannien) bzw. *US* (Vereinigte Staaten) versehen.

PRÉFACE

Le bâtiment en tant que secteur économique a subi au cours des dernières décennies d'importants changements qualitatifs, ce qui trouve évidemment sa répercussion dans la terminologie employée dans ce domaine. A l'heure actuelle beaucoup de pays tâchent de normaliser la terminologie technique et notamment celle du bâtiment.

Le développement de la coopération internationale et de l'échange d'information entre divers pays en matière de bâtiment, d'une part, et l'absence des dictionnaires bilingues correspondant au niveau actuel de cette branche, d'autre part, ont nécessité la composition d'un dictionnaire du bâtiment rédigé dans les principales langues européennes.

Le présent dictionnaire comporte près de 14.000 termes anglais et leurs équivalents en allemand, français, néerlandais et russe, employés dans les principales branches du bâtiment: les principes des calculs et des plans (théorie de l'élasticité, résistance des matériaux, mécanique du sol et autres disciplines techniques appropriées); les bâtiments et les ouvrages de génie civil; les structures de construction et leurs éléments; les matériaux et produits de construction; l'organisation et la technologie des travaux du bâtiment; machines de construction et les machines servant pour l'aménagement de routes; le matériel de chantier; l'équipement technique de bâtiments (y compris les systèmes modernes d'air conditionné); les ouvrages hydrauliques et d'irrigation.

Le dictionnaire comporte également un nombre limité de termes appartenant aux secteurs techniques généraux ou proches du bâtiment ainsi qu'à l'architecture, à l'urbanisme, à la construction d'aérodromes, de communications souterraines et de chemins de fer. Les auteurs n'ont pas jugé utile d'inclure dans le dictionnaire les appellations données par les firmes à leurs matériaux, articles ou équipements, ni les noms de composés chimiques et de leurs constituants.

Les auteurs ont utilisé une vaste collection d'ouvrages modernes dans les langues étrangères ainsi que des dictionnaires techniques, des encyclopédies du bâtiment, des glossaires, des normes terminologiques, des monographies et des périodiques.

Dans les cas où les auteurs n'ont pas réussi à trouver un équivalent pour un terme anglais, ils ont eu recours aux méthodes descriptives et aux explications détaillées. Les équivalents des termes anglais polysémantiques ne figurent dans le dictionnaire que dans le sens directement lié au bâtiment. Ceci doit être pris en compte lors de l'utilisation de l'ouvrage.

Le dictionnaire est destiné à un large cercle de spécialistes du bâtiment: ingénieurs, scientifiques, enseignants, étudiants et traducteurs d'œuvres scientifiques et techniques.

Les auteurs sont parfaitement conscients du fait qu'un premier essai de création d'un dictionnaire de ce genre ne peut pas être exempt d'erreurs et de lacunes.

Prière d'envoyer toutes les remarques et suggestions visant à améliorer le présent dictionnaire à l'adresse suivante: Kluwer Technische Boeken B.V., Postbus 23, 7400 GA Deventer, Nederland ou
103012 MOSCOU, Staropanski per., 1/5, Izdatelstvo «Russkij Jazyk»

Les Auteurs.

LE BON USAGE DU DICTIONNAIRE

Les termes anglais du dictionnaire sont rangés par ordre alphabétique. Les termes composés y sont considérés comme un seul mot, par exemple:

deck girder
decking
deck level

Chaque article du dictionnaire se compose d'un terme anglais et des ses équivalents allemand, français, neérlandais et russe. Le signe «see» renvoie à un terme anglais synonyme. Les termes allemands, français, néerlandais et russes sont suivis de la mention du genre (*m*, *f*, *n*) et au besoin de celle du pluriel (*pl*). Pour faciliter la recherche des équivalents allemands, français, néerlandais ou russes dans les index correspondants et pour permettre la traduction à partir de ces langues, les termes anglais sont classés dans un système, qui les numérote à l'intérieur de la classification alphabétique. Les articles du dictionnaire sont donc présentés de la façon suivante:

B66 *e* **baked brick**
   d gebrannter Ziegel(stein) *m*
   f braque *f* cuite
   nl baksteen *m*
   r обожжённый кирпич *m*

Dans le cas d'un terme anglais polysémantique les différents sens sont séparés par des chiffres arabes de leurs équivalents dans les autres langues, les synonymes apparentés par un point-virgule et les vrais synonymes par une virgule.

Pour des raisons d'économie de place les parties interchangeables des variantes synonymiques ont été mises entre crochets et les parties facultatives entre parenthèses.
Par exemple:

rechnerische Belegung [Besetzung]   = rechnerische Belegung *ou* rechnerische Besetzung

emulsion bitumineuse [de bitume]   = emulsion bitumineuse *ou* emulsion de bitume

gekleurde verglaasde pan [tegel]   = gekleurde verglaasde pan *ou* gekleurde verglaasde tegel

Einspann(ungs)moment   = Einspannmoment *ou* Einspannungsmoment

tuile (cuite) plate   = tuile cuite plate *ou* tuile plate
insteek(verdieping)   = insteek *ou* insteekverdieping

Pour la traduction d'ouvrages allemands, français, néerlandais ou russes, ayant rapport au bâtiment, il faut utiliser les index alphabétiques à la fin du dictionnaire où tous les termes sont accompagnés de lettres et de chiffres correspondant au numérotage des termes anglais.

VOORWOORD

Het bouwwezen als bedrijfstak heeft in de laatste decennia aanzienlijke kwalitatieve veranderingen ondergaan, hetgeen vanzelfsprekend weerklank vindt in de bouwterminologie. In vele landen wordt tegenwoordig veel gedaan om de technische en in het bijzonder de bouwterminologie te ordenen, vooral deze te unificeren en te standaardiseren.

Het uitbreiden van de samenwerking op wereldschaal en van de informatie-uitwisseling tussen verschillende landen op gebied van bouwactiviteiten, alsook het gebrek aan tweetalige woordenboeken die het hedendaagse niveau van bouwontwikkelingen adequaat weergeven,— dat zijn de redenen die het uitgeven van een veeltalig voordenboek van bouwkundige termen noodzakelijk maken.

Het onderhavige Woordenboek bevat ongeveer 14.000 Engeles termen en hun equivalenten in het Duits, Frans, Nederlands en Russisch qua hoofdtakken van het bouwvakbedrijf: grondlagen van ontwerpen en berekening (taaiheidstheorie, materiaalsterkte, grondmechanica, en andere ermee verbonden technische vakgebieden); gebouwen en industriele bouwwerken; bouwconstructies en -elementen; bouwstoffen en bouwprodukten; organisatie en technologie van bouwprocessen; bouw- en wegenbouwmachines; het uitrusten van bouwterreinen; technische uitrusting van gebouwen (incluis moderne systemen voor luchtverversing); water- en irrigatiewerken.

\Algemeentechnische en aan het bouwwezen grenzende termen, alsook termen inzake architectuur en ruimtelijke ordening, het bouwen van vliegvelden, tunnels en spoorwegen komen in het Woordenboek in nogal beperkte omvang voor. Firmabenamingen van materialen, produkten en installaties, chemische verbindingen en hun componenten zijn in het Woordenboek niet opgenomen.

Bij het samenstellen van het Woordenboek werd gebruik gemaakt van talrijke moderne edities in respectievelijke talen, zoals encyclopedische en terminologische woordenboeken, bouwencyclopedieën, glossaria, naslagedities, terminologische standaarden, speciale verhandelingen, leerboeken, periodieken en technische ontwerpsdocumentatie.

Bij het kiezen van de juiste equivalenten voor Engelse termen hadden de auteurs met vele moeilijkheden te maken ten gevolge van ongelijke omvang of onvolledige semantische overeenkomst van vaktermen in verschillende talen. In sommige gevallen kon men niet anders dan gebruik maken van speciale uitleg of omschrijvende vertaling. Bij het selecteren van equivalenten voor polysemantische Engelse termen nam men in het Woordenboek alleen die betekenissen van de termen op, die in rechtstreeks verband stonden met het bouwvakbedrijf, waarmee bij het gebruik van het Woordenboek rekening gehouden moet worden.

Het Woordenboek is bestemd voor een brede kring van bouwdeskundigen: bouwingenieurs, wetenschappers, leraren, aspiranten, studenten en vertalers van technisch-wetenschappelijke literatuur.

De auteurs zien duidelijk in dat de eerste poging om een dergelijk woordenboek samen te stellen niet vrij is van tekortkomingen en leemten. Alle kritische opmerkingen en voorstellen tot verbetering van het Woordenboek kunt u sturen naar het volgende adres: Kluwer Technische Boeken B. V., Postbus 23, 7400 GA Deventer, Nederland of Moskou 103012, Staropanski per. 1/5, Izdatelstvo «Roesskij jazyk».

De auteurs

AANWIJZINGEN VOOR HET GEBRUIK

De Engelse termen in het Woordenboek zijn gerangschikt volgens het alfabet, waarbij samengestelde termen als aaneengeschreven woorden beschouwd worden, bijvoorbeeld:
deck girder
decking
deck level
Ieder artikel bevat een Engelse term met zijn equivalenten in het Duits, Frans, Nederlands en Russisch. Het teken *see (zie)* staat ter verwijzing van een synonieme Engelse term naar een andere. Het grammaticale geslacht en zonodig het meervoud bij de Duitse, Franse, Nederlandse en Russische termen worden aangegeven met tekens *m*, *f*, *n*, resp. *pl*.

De Engelse termen zijn genummereerd bij iedere letter van het alfabet, wat de nodige Duitse, Franse, Nederlandse of Russische term in de indexen gemakkelijk helpt vinden, d.w.z. de vertaling maken uit iedere taal van het Woordenboek. Een artikel in het Woordenboek ziet dus eruit als volgt:
B66 *e* baked brick
    d gebrannter Ziegel(stein) *m*
    f brigue *f* cuite
    nl baksteen *m*
    r обожжённый кирпич *m*
Is het Engelse trefwoord polysemantisch, dan worden verschillende betekenissen van zijn equivalenten in de andere talen onderscheiden met Arabische cijfers, verwante (synonieme) varianten — met een kommapunt, en echte synoniemen — met een komma.

Om plaats te bezuinigen kunnen de onderling vervangbare delen van synonieme varianten van Engelse termen binnen vierkante haken, en facultatieve delen van termen binnen ronde haken geplaatst worden, bijvoorbeeld:

rechnerische Belegung [Besetzung]	= rechnerische Belegung *of* rechnerische Besetzung
emulsion bitumineuse [de bitume]	= emulsion bitumineuse *of* emulsion de bitume
gekleurde verglaasde pan [tegel]	= gekleurde verglaasde pan *of* gekleurde verglaasde tegel
Einspann(ungs)moment	= Einspannmoment *of* Einspannungsmoment
tuile (cuite) plate	= tuile cuite plate *of* tuile plate
insteek(verdieping)	= insteek *of* insteekverdieping
Gebruiksbereik van de term:	
UK (United Kingdom)	= Verenigd Koninkrijk (van Engeland en Noord-Ierland)
US (United States)	= Verenigde Staten (VS)

Bij het vertalen van bouwkundige literatuur uit het Duits, Frans, Nederlands en Russisch dient men het achter in het Woordenboek geplaatste alfabetische register (index) te gebruiken, waarin alle termen zijn aangeduid met letters en cijfers die overeenkomen met de nummerering van de Engelse termen.

ПРЕДИСЛОВИЕ

Строительство как одна из отраслей материального производства за последние десятилетия претерпело значительные качественные изменения, что нашло свое отражение в строительной терминологии. В настоящее время во многих странах проводится большая работа по упорядочению (унификации и стандартизации) технической и, в частности, строительной терминологии.

Расширение международного сотрудничества и широкий обмен информацией между разными странами в области строительства, а также отсутствие двуязычных словарей, отражающих современный уровень развития строительства, обусловили необходимость издания многоязычного словаря по строительству.

Предлагаемый Словарь содержит около 14 000 английских терминов и их эквиваленты на немецком, французском, нидерландском и русском языках по основным разделам строительства: основы строительного проектирования и расчета (теория упругости, сопротивление материалов, механика грунтов и другие связанные с этим технические дисциплины); здания и инженерные сооружения; строительные конструкции и их элементы; строительные материалы и изделия; организация и технология строительного производства; строительные и дорожные машины, оборудование и оснащение строительных площадок; инженерное оборудование зданий (включая современные системы кондиционирования воздуха); гидротехнические и ирригационные сооружения.

Общетехнические и смежные со строительством термины, а также термины по архитектуре и градостроительству, по аэродромному, подземному и железнодорожному строительству включены в Словарь в ограниченном объеме. Не вошли в Словарь фирменные наименования материалов, изделий и оборудования, названия химических соединений и их компонентов, термины-наречия, термины-прилагательные, термины-глаголы.

При составлении Словаря использована обширная современная литература на соответствующих иностранных языках: энциклопедические и терминологические словари, тезаурусы по строительству, глоссарии, справочники, терминологические стандарты, монографии, учебники, периодические издания, проектная документация.

Авторам пришлось преодолеть немало трудностей при подборе точных лексических эквивалентов английских терминов из-за различного объема или неполного соответствия понятий на различных языках. В отдельных случаях пришлось прибегать к пояснениям или описательному переводу. При подборе эквивалентов к многозначным английским терминам в Словарь включались лишь те значения термина, которые имеют непосредственное отношение к строительству, и это надо учитывать при пользовании Словарем.

Словарь предназначен для широкого круга специалистов в области строительства: инженеров-строителей, научных работников, преподавателей, аспирантов, студентов, а также переводчиков научно-технической литературы.

Авторы отдают себе отчет в том, что первый опыт создания подобного словаря не свободен от недостатков и пробелов. Все критические замечания и предложения по улучшению Словаря просим направлять по адресу: Москва, 103012, Старопанский пер., д. 1/5, издательство «Русский язык».

Авторы

13

О ПОЛЬЗОВАНИИ СЛОВАРЕМ

Английские термины расположены в Словаре в алфавитном порядке, причем составные термины рассматриваются как слитно написанные слова, например:

deck girder
decking
deck level

Словарная статья состоит из термина на английском языке и его эквивалентов на немецком, французском, нидерландском и русском языках. Для ссылки с одного синонимичного английского термина на другой используется помета *see (смотри)*. Немецкие, французские, нидерландские и русские термины имеют указания рода *(m, f, n)* и, при необходимости, множественного числа *(pl)*. Английские термины имеют порядковую нумерацию в пределах каждой буквы алфавита, что помогает быстро отыскать по указателям соответствующий немецкий, французский, нидерландский или русский термин, то есть делать перевод с любого из языков Словаря. Таким образом, словарная статья имеет следующий вид:

B66 *e* baked brick
 d gebrannter Ziegel(stein) *m*
 f brique *f* cuite
 nl baksteen *m*
 r обожжённый кирпич *m*

Если приведенный английский термин многозначен, то разные значения его эквивалентов на других языках разделяются арабскими цифрами, близкие по смыслу синонимичные варианты — точкой с запятой, синонимы — запятой.

Для экономии места взаимозаменяемые части синонимичных вариантов эквивалентов английских терминов могут быть заключены в квадратные скобки, а факультативные (не обязательные для употребления) части терминов — в круглые, например:

rechnerische Belegung [Besetzung]	= rechnerische Belegung *или* rechnerische Besetzung
emulsion bitumineuse [de bitume]	= emulsion bitumineuse *или* emulsion de bitume
gekleurde verglaasde pan [tegel]	= gekleurde verglaasde pan *или* gekleurde verglaasde tegel
Einspann(ungs)moment	= Einspannmoment *или* Einspannungsmoment
tuile (cuite) plate	= tuile cuite plate *или* tuile plate
insteek(verdieping)	= insteek *или* insteekverdieping
Область употребления термина: UK (United Kingdom)	= Соединенное Королевство Великобритании и Северной Ирландии
US (United States)	= США

Для перевода литературы по строительству с немецкого, французского, нидерландского и русского языков следует пользоваться помещенными в конце Словаря алфавитными указателями, в которых все термины имеют буквенно-цифровые обозначения, соответствующие нумерации английских терминов.

A

A1   *e* **abatement**
d 1. Verminderung *f*; Abschwächung *f*,
Verminderung *f* der Intensität 2.
Abfälle *m pl*
f 1. abattement *m*; affaiblissement *m*
2. déchets *m pl*
nl 1. vermindering *f*; afname *f* 2. afval
n
r 1. уменьшение *n*; снижение *n*
(интенсивности); ослабление *n*
2. отходы *m pl*

A2   *e* **abat-vent**
d 1. Deflektor *m* 2. Jalousie *f*
f 1. abat-vent *m*, déflecteur *m*
2. jalousie *f*
nl 1. deflector *m*, schoorsteenkap *f (m)*
2. jaloezie *f*
r 1. дефлектор *m* 2. жалюзи *pl*

A3   *e* **A-block**
d Betonhohlstein *m*
f bloc *m* creux préfabriqué en béton
nl holle betonsteen *m*
r пустотелый бетонный блок *m*

A4   *e* **abnormal wear**
d unzulässiger Verschleiß *m*
f usure *f* inadmissible
nl ontoelaatbare slijtage *f*
r недопустимый износ *m*

A5   *e* **abode**
d Wohnung *f*
f demeure *f*, logis *m*, habitation *f*;
local *m*
nl woning *f*
r жилище *n*; помещение *n*

A7   *e* **above-grade pipeline, above-
-ground pipeline**
d oberirdische (Rohr-)Leitung *f*
f tuyauterie *f* [conduite *f*] terrestre
nl bovengrondse pijpleiding *f*
r наземный трубопровод *m*

A8   *e* **above-ground structure**
d oberirdisches Bauwerk *n*
f bâtiment *m* [ouvrage *m*] terrestre
nl bovengronds gebouw *n* [bouwwerk *n*]
r наземное сооружение *n*

A9   *e* **abrasion**
d 1. Abreiben *n*, Abrieb *m* 2. Ab-
schleifung *f*
f 1. abrasion *f*, usure *f* 2. frottement *m*,
meulage *m*
nl 1. afslijten *n*, afschuren *n*, afwrijven
n; afschaven *n*, afschrappen *n*;
slijtage *f*; slijpsel *n* 2. afslijpen *n*
r 1. истирание *n*, абразивный износ *m*
2. шлифование *n*

A10   *e* **abrasion hardness**
d Abriebhärte *f*
f dureté *f* à l'abrasion
nl slijtvastheid *f*
r твёрдость *f* на истирание

A11   *e* **abrasion resistance**
d Abriebfestigkeit *f*
f résistance *f* à l'abrasion
nl slijtvastheid *f*, slijtbestendigheid *f*
r износостойкость *f*; сопротивление *n*
истиранию

A12   *e* **abrasion value**
d Abnutzungsgrad *m*
f degré *m* d'abrasivité
nl slijtagegraad *m*
r степень *f* износа

A13   *e* **abrasive aggregates**
d griffige Zuschläge *m pl*
f agrégats *m pl* [granulats *m pl*]
abrasifs
nl stroefheidstoeslagen *m pl*
r шероховатые заполнители *m pl*

A14   *e* **abrasive disc [disk]**
d Schleifscheibe *f*
f meule *f*, disque *m* abrasif
nl slijpschijf *f(m)*
r абразивный круг *m*

A15   **abrasive hardness** *see* **abrasion
hardness**

A16   **abrasive materials** *see* **abrasives**

A17   *e* **abrasive paper**
d Schleifpapier *n*
f papier *m* abrasif [émérisé], papier-
-émeri *n*
nl schuurpapier *n*
r наждачная бумага *f*

A18   *e* **abrasives**
d Schleifmittel *n pl*
f abrasifs *m pl*, matières *f pl* abrasives
nl slijpmiddelen *n pl*
r абразивы *m pl*, абразивные
материалы *m pl*

A19   *e* **abrasive tile**
d gleitsichere Fußbodenfliese *f*
[Fußbodenbelagplatte *f*]
f carreau *m* abrasif
nl ruwe plaat *f(m)*
r шероховатая плитка *f*

A20   *e* **absolute deformation**
d Absolutverformung *f*, absolute
Verformung *f*
f déformation *f* absolue
nl absolute vervorming *f*
r абсолютная деформация *f*

A21　e absolute displacement
　　　d Absolutverschiebung *f*, absolute
　　　　Verschiebung *f*
　　　f déplacement *m* absolu
　　　nl absolute verplaatsing *f*
　　　r абсолютное перемещение *n*

A22　e absolute filter
　　　d absolutes Filter *n*
　　　f filtre *m* absolu
　　　nl absolute filter *m, n*
　　　r абсолютный фильтр *m*

A23　e absolute humidity
　　　d absolute Feuchtigkeit *f*
　　　f humidité *f* absolue
　　　nl absolute vochtigheid *f*
　　　r абсолютная влажность *f*

A24　e absolute pressure
　　　d absoluter Druck *m*
　　　f pression *f* absolue
　　　nl absolute druk *m*
　　　r абсолютное давление *n*

A25　e absolute viscosity
　　　d dynamische Viskosität *f*
　　　f viscosité *f* absolue
　　　nl absolute viscositeit *f*
　　　r динамическая вязкость *f*

A26　e absorbed water
　　　d absorbiertes Wasser *n*
　　　f eau *f* absorbée
　　　nl geabsorbeerd water *n*
　　　r абсорбированная вода *f*

A27　absorbency *see* absorptive capacity

A28　e absorbing ceiling
　　　d Schallschluckdecke *f*, akustische
　　　　Decke *f*
　　　f plafond *m* acoustique
　　　nl geluidabsorberend [akoestisch]
　　　　plafond *n*
　　　r звукопоглощающий [акустический]
　　　　потолок *m*

A29　e absorbing well
　　　d Sickerschacht *m*, Schluckbrunnen *m*,
　　　　Schluckschacht *m*, Saugbrunnen *m*,
　　　　Senkbrunnen *m*
　　　f puits *m* absorbant
　　　nl bronbuis *f(m)*, bron *f(m)*
　　　r поглощающий колодец *m*

A30　e absorption basin
　　　d Beruhigungsbecken *n*, Tosbecken *n*
　　　f bassin *m* de tranquillisation
　　　　[d'amortissement]
　　　n woelkom *f(m)*
　　　r водобойный колодец *m*;
　　　　успокоительный бассейн *m*

A31　e absorption chamber
　　　d Beruhigungskammer *f*, Toskammer *f*
　　　f chambre *f* d'amortissement
　　　nl woelkelder *m*
　　　r камера *f* гашения энергии

A32　e absorption test
　　　d Wasseraufnahmeprobe *f*.
　　　　Wasseraufnahmeprüfung *f*.
　　　　Wasseraufnahmeversuch *m*

　　　f essai *m* d'absorptivité [d'absorption]
　　　nl absorptieproef *f(m)*
　　　r испытание *n* на водопоглощение

A33　e absorptive capacity
　　　d Absorbiervermögen *n*,
　　　　Aufnahmefähigkeit *f*
　　　f pouvoir *m* absorbant
　　　nl absorptievermogen *n*
　　　r поглощающая способность *f*

A34　e absorptive forms
　　　d wasserabsorbierende Schalung *f*
　　　f coffrage *m* absorbant
　　　nl waterabsorberende bekisting *f*
　　　r абсорбирующая опалубка *f*

A35　e ABS plastic
　　　d ABC-Plast *m*
　　　f plastique *m* ABS
　　　nl ABS-plastiek *n*
　　　r акрилонитрилбутадиенстирол *m*

A36　e abutment
　　　d 1. Widerlager *n*, Wehrwange *f*,
　　　　Landpfeiler *m* 2. Kämpfer *m*
　　　　3. Anschluß *m*
　　　f 1. culée *f* 2. naissance *f* (*de voûte
　　　　ou de l'arc*) 3. aboutement *m*
　　　nl 1. landpijler *m*; landhoofd *n*
　　　　2. rechtstand *m*, geboorte *f* (*van een
　　　　gewelf*) 3. aansluiting *f*
　　　r 1. береговой устой *m* 2. пята *f*
　　　　(*арки, свода*) 3. примыкание *n*,
　　　　сопряжение *n*

A37　e abutment wall
　　　d Widerlagerflügelmauer *f*;
　　　　Widerlagerböschungsflügel *m*
　　　f mur *m* en retour [en aile]
　　　nl rechtstandmuur *m*; vleugelmuur *m*,
　　　　retourmuur *m*
　　　r открылок *m* плотины; откосное
　　　　крыло *n* (*берегового устоя*)

A38　e abutting joint
　　　d Stoßverbindung *f*, Stumpfstoß *m*;
　　　　Stoßverband *m*, Hirnfuge *f*,
　　　　Hirnverbindung *f*
　　　f joint *m* abouté [par aboutement],
　　　　assemblage *m* en about
　　　nl stompe verbinding *f*, haakse las *f* (*m*),
　　　　stootvoeg *f(m)*
　　　r стыковое соединение *n*,
　　　　соединение *n* встык, торцовое
　　　　соединение *n*; торцовое примыкание *n*

A39　e abutting tenon
　　　d Jagdzapfen *m* (*Holz*)
　　　f about *m*, tenon *m* d'about
　　　nl pen *m* (*hout*)
　　　r торцовый шип *m*

A40　e Abyssinien well
　　　d Rammbrunnen *m*
　　　f puits *m* abyssinien
　　　nl geslagen pompbuis *f(m)*
　　　r забивной трубчатый колодец *m*

A41　e accelerated curing of concrete
　　　d Schnellnachbehandlung *f* von Beton
　　　f cure *f* de béton accélérée

nl verhardingsversnellende
nabehandeling f van het beton
r уход m за бетоном, обеспечивающий
ускоренное твердение

A42 e **accelerated test**
d Schnellversuch m, Kurzzeitversuch
m, abgekürzte Probe f
f essai m accéléré
nl versnelde proef f(m), versneld
onderzoek **n**
r ускоренное испытание n

A43 e **accelerated weathering**
d künstliche Bewitterung f,
Kurzbewitterung f
f altération f accélérée
nl versnelde verwering f
r ускоренное испытание n
атмосферостойкости; ускоренное
выветривание n

A44 e **accelerating admixture**
d Abbinde(zeit)beschleuniger m,
Erstarrungsbeschleuniger m
f accélérateur m de prise
nl verhardingsversnellende
toevoeging f
r добавка f, ускоряющая
схватывание

A45 e **acceleration lane**
d Beschleunigungsspur f,
Beschleunigungsstreifen m
f voie f d'accélération
nl versnellingsstrook f(m)
r полоса f ускорения [разгона]

A46 e **accelerator**
d 1. Abbindebeschleuniger m
2. Umlaufpumpe f,
Zirkulationspumpe f
f 1. accélérateur m (p.ex. de prise)
2. pompe f de circulation
nl versnellingsmiddel n
r 1. ускоритель m (напр.
схватывания); 2. циркуляционный
насос m

A47 e **acceptance of constructional work**
d Bauabnahme f, Kollaudierung f
f réception f des travaux de
construction
nl (aanvaarding f van de) oplevering f
r приёмка-сдача f строительных работ

A48 e **acceptance of tender**
d Ausschreibungsvergebung f
f adjudication f
nl gunning f
r утверждение n заявки на подряд
(на торгах)

A49 e **acceptance test**
d Abnahmeprüfung f
f essai m [épreuve f] de réception
nl afnameproef f(m), keuring f bij afname
r приёмочное испытание n

A50 e **access**
d 1. Eintritt m; Zutritt m 2. Zugang
m; Zufahrt f

f 1. accès m, entrée f 2. accès m
nl 1. ingang m 2. toegang m
r 1. вход m; доступ m 2. подход m,
подъезд m

A51 e **access balcony**
d offener Gang m, Laubengang m
f balcon m d'accès
nl toegangsgalerij f
r входная (открытая) галерея f
(обеспечивающая доступ в квартиры)

A52 e **access door**
d Bedienungstür f
f porte f d'accès, trappe f
nl toegangsdeur f(m), inspectieluik n
r дверца f люка

A53 **access gallery** see **access balcony**

A54 e **access hole**
d Zugangsöffnung f
f regard m de visite [de service]
nl toegangsluik n, mangat n
r смотровой люк m [лаз m]

A55 e **accessible duct**
d begehbarer (Leitungs-)Kanal m
f canal m praticable [accessible]
nl toegankelijk kanaal n
r проходной канал m

A56 e **accessories**
d 1. Zubehör n, Beiwerk n 2. Fittings
n, m pl
f 1. accessoires m pl, auxiliaires m pl
2. robinetterie f
nl toebehoren n pl
r 1. вспомогательные принадлежности
f pl или изделия n pl 2. арматура f
(водопроводная)

A57 **accessory building** see **ancillary
building**

A58 e **access ramp**
d Auffahrtrampe f, Anschlußrampe f
f rampe f d'accès
nl oprit m
r съезд m; наклонный въезд m;
аппарель f; пандус m, рампа f

A59 e **access road**
d Zufahrt f, Zubringer m, Zufahrtsweg m
f route f d'accès
nl toegangsweg m
r подъездная дорога f

A60 e **access tunnel**
d Baustollen m, Zugangsstollen m
f galerie f d'accès; galerie f pilote
nl bouwtunnel m, toegangsgalerij f
r строительный туннель m;
передовая штольня f; подходная
галерея f

A61 e **accidental loading**
d unvorhergesehene Belastung f
f sollicitation f [(sur)charge f]
accidentelle
nl toevallige belasting f
r случайное нагружение n,
случайная нагрузка f

A62 *e* accident prevention
    d Unfallverhütung *f*
    f prévention *f* des accidents
    nl bedrijfsveiligheid *f*
    r меры *f pl* предупреждения
    несчастных случаев (*техника
    безопасности*)

A63 *e* accident prevention regulations
    d Unfallverhütungsvorschriften *f pl*,
    Sicherheitsschrift *f*
    f règlement *m* de sécurité
    nl (bedrijfs)veiligheidsvoorschriften *f pl*
    r правила *n pl* техники безопасности

A64 *e* accident prevention tag
    d warnendes Anhängeschildchen *n*
    f étiquette *f* de précaution
    nl etiket *n* met waarschuwing
    r бирка *f* с предупредительной
    надписью (*о неисправности*)

A65 *e* accomodations
    d Unterkünfte *f pl*
    f locaux *m pl* destinés au personnel
    nl huisvesting *f*
    r бытовые помещения *n pl*

A66 *e* accordion door
    d Akkordeontür *f*, Falttür *f*
    f porte *f* (en) accordéon [(re)pliante]
    nl vouwdeur *f(m)*
    r складная [складывающаяся] дверь *f*

A67 *e* accordion partition
    d Harmonikatrennwand *f*, Faltwand *f*
    f paroi *m* accordéon
    nl vouwwand *m*
    r складная [складывающаяся]
    перегородка *f*

A68 *e* accuracy of calculations
    d Berechnungsgenauigkeit *f*
    f précision *f* des calculs
    nl nauwkeurigheid *f* van berekeningen
    r точность *f* вычислений [расчётов]

A69 *e* accuracy of fixing
    d Genauigkeit *f* der Befestigung
    f précision *f* de fixation
    nl nauwkeurigheid *f* van de bevestiging
    r точность *f* закрепления

A70 *e* acetylene welding
    d Azetylenschweißen *n*
    f soudage *m* à l'acétylène
    [acétylènique]
    nl autogeen lassen *n*
    r ацетилено-кислородная сварка *f*

A72 *e* acid proof brick, acid-resistant brick
    d säurefester Ziegel *m*
    f brique *f* antiacide
    nl zuurvaste baksteen *m*
    r кислотостойкий [кислотоупорный]
    кирпич *m*
    acoustic... *see* **acoustical...**

A73 *e* acoustical absorptivity
    d Schallabsorptionskoeffizient *m*
    f absorptivité *f* acoustique
    nl geluidabsorptievermogen *n*
    r коэффициент *m* звукопоглощения

A74 *e* acoustical board
    d Schallschluckplatte *f*, Akustikplatte
    f, Schalldämmplatte *f*
    f panneau *m* acoustique
    nl geluidabsorberende [akoestische] plaat
    f(m)
    r акустическая [звукопоглощающая]
    панель *f* [плита *f*]

A75 *e* acoustical ceiling
    d Akustikdecke *f*, Schallschluckdecke *f*
    f plafond *m* acoustique
    nl akoestisch plafond *n*
    r акустический потолок *m*

A76 *e* acoustical fiberboard
    d Akustikfaserplatte *f*,
    Schallschluckfaserplatte *f*
    f 1. panneau *m* acoustique en fibre de
    bois 2. carton-fibre *m* acoustique
    nl akoestische vezelplaat *f(m)*
    r 1. акустическая древесно-
    -волокнистая плита *f* 2. акустический
    фибровый картон *m*

A77 *e* acoustical insulation
    d Schalldämmung *f*
    f isolation *f* acoustique [sonore,
    phonique], isolement *m* acoustique
    nl geluidisolatie *f*
    r звукоизоляция *f*

A78 *e* acoustical lining
    d Schallschluckauskleidung *f*,
    Akustikbelag *m*, Schallschluckbelag
    m
    f revêtement *m* acoustique [insonore,
    absorbant le bruit]
    nl geluiddempende bekleding *f*
    r акустическая [звукопоглощающая]
    облицовка *f*

A79 acoustical panel *see* acoustical board

A80 *e* acoustical plaster
    d Schallschluckputz *m*, Akustikputz *m*
    f enduit *m* acoustique [phonique,
    d'insonorisation]
    nl akoestisch pleister *n*
    r акустическая штукатурка *f*

A81 *e* acoustical resistance
    d akustische Resistenz *f*, akustischer
    Widerstand *m*
    f résistance *f* acoustique
    nl akoestische weerstand *m*
    r акустическое сопротивление *n*

A82 *e* acoustical screen
    d Schallschirm *m*, Schallwand *f*
    f écran *m* acoustique, barrière *f* contre
    le bruit
    nl geluiddempend scherm *n*
    r акустический экран *m*

A83 *e* acoustical tile
    d Schallschluckfliese *f*, Akustikfliese *f*,
    Schallschluckplatte *f*, Akustikplatte *f*
    f carreau *m* acoustique [absorbant le
    bruit, insonore]
    nl geluiddempende tegel *m*
    r акустическая облицовочная плитка *f*

A84 *e* **actinic glass**
  d aktinisches Glas *n*
  f verre *m* actinique
  nl actinisch glas *n*
  r актиничное стекло *n*

A85 *e* **activated alumina**
  d aktivierte Tonerde *f*
  f alumine *f* activée
  nl geactiveerde aluinaarde *f(m)*,
  geactiveerd aluminiumoxide *n*
  r активированный глинозём *m*

A86 *e* **activated carbon filter**
  d Aktivkohlefilter *n*, *m*
  f filtre *m* à charbon activé
  nl adsorptiekoolfilter *n*, *m*
  r фильтр *m* с активированным углем,
  угольный фильтр *m*

A87 *e* **activated sludge**
  d belebter [aktivierter] Schlamm *m*
  f boue *f* activée
  nl geactiveerd slib *n*
  r активный ил *m*

A88 *e* **activated sludge plant**
  d Belebungsanlage *f*
  f installation *f* de boues activées
  nl biologische afvalwaterzuiveringsinstallatie *f*
  r очистная станция *f* с аэротенками
  или окислительными каналами;
  станция *f* аэрации

A89 *e* **activated sludge process**
  d Belebungsverfahren *n*
  f procédé *m* des boues activées
  nl biochemische afvalwaterzuivering *f*
  r очистка *f* (сточных вод) активным
  илом

A90 *e* **active earth pressure**
  d aktiver Erddruck *m*
  f poussée *f* active des terres
  nl werkzame gronddruk *m*
  r активное давление *n* грунта

A91 *e* **active residential solar heating system**
  d aktive Solarheizanlage *f* für
  Wohnbauten
  f chauffage *f* activé par énergie solaire
  nl zonne-energieverwarmingsinstallatie *f*
  r энергоактивная система *f* солнечного
  отопления зданий

A92 *e* **activity**
  d Teilprozeß *m*, Aktivität *f*
  (*Netzplantechnik*)
  f activité *f*; opération *f* de travail
  nl werking *f*; werkzaamheid *f*;
  activiteit *f*
  r работа *f* (*в сетевом планировании*);
  рабочая операция *f*

A93 *e* **activity duration, activity time**
  d Dauer *f* des Teilprozesses
  (*Netzplantechnik*)
  f durée *f* d'activité
  nl duur *m* van de werking
  r продолжительность *f* работы

A95 *e* **actual construction time**
  d Istbauzeit *f*, tatsächliche Bauzeit *f*
  f délai *m* de construction réel, temps
  m de construction effectif
  nl werkelijke [eigenlijke] bouwtijd *m*
  r действительная продолжительность *f*
  строительства

A96 *e* **actual damage**
  d tatsächlicher Schaden *m*,
  tatsächliche Beschädigung *f*
  f dommage *m* [endommagement *m*] réel
  nl werkelijke schade *f*
  r действительные размеры *m pl*
  повреждения [разрушения]

A97 *e* **actual loading**
  d angreifende Last *f*, Wirklast *f*,
  Angriffslast *f*
  f charge *f* sollicitante; charge *f* active
  nl werkelijke belasting *f*
  r действительное нагружение *n*,
  фактическая нагрузка *f*

A98 *e* **actual size**
  d Istgröße *f*, tatsächliche Größe *f*
  f cote *f* [dimension *f*] effective
  nl ware grootte *f*
  r действительный размер *m*

A99 *e* **actual strength**
  d erreichte [vorhandene, tatsächliche]
  Festigkeit *f*
  f résistance *f* effective
  nl werkelijke sterkte *f*
  r действительная прочность *f*

A100 *e* **acute arch**
  d überhöhter Lanzettbogen *m*
  f arc *m* en lancettes [en ogive surélevé]
  nl spitsboog *m*
  r остроконечная [стрельчатая] арка *f*

A101 *e* **addition**
  d 1. Zusatz *m*, Beigabe *f* 2. Anbau *m*;
  Überbau *m*, Aufbau *m* 3.
  veranschlagte zusätzliche Kosten *pl*
  f 1. addition *f*, adjonction *f* 2. annexe *f*
  3. frais *m pl* supplémentaires
  nl 1. toevoeging *f*, bijslag *m* 2. aanbouw
  m, bijgebouw *m* 3. bijkomende
  kosten *m pl*
  r 1. добавка *f*; присадка *f*; примесь *f*
  2. пристройка *f*; надстройка *f*
  3. дополнительные расходы *m pl* по
  смете

A102 *e* **additional services**
  d Zusatzleistungen *f pl*, zusätzlicher
  Kundendienst *m*
  f services *f pl* addionnelles
  nl extra dienstverlening *f*
  r дополнительные услуги *f pl* (*не
  учтённые договором*)

A103 *e* **addition of forces**
  d Hinzufügung *f* von Kräften
  f composition *f* des forces
  nl krachtensamenstelling *f*
  r сложение *n* сил

A104 **additive** *see* **admixture**

2*

A105 *e* **adhesion agent**
d Haftmittel *n*, Haftanreger *m*,
adhäsionsfördernder Zusatzstoff *m*
f agent *m* d'adhésion
nl hechtmiddel *n*
r добавка *f*, повышающая адгезию
[прилипание]

A106 adhesion stress *see* bond stress

A107 *e* **adhesion-type ceramic veneer**
d Mörtelverkleidung *f* mit
Keramiktafeln
f revêtement *m* en carreaux céramiques
posés au mortier
nl in de mortel gezette tegels *m pl*
r облицовка *f* из керамических
плиток на строительном растворе
или мастике

A108 *e* **adhesive**
d Klebstoff *m*, Klebemittel *n*, Kleber *m*
f adhésif *m*, colle *f*
nl lijm *m*, kleefmiddel *n*, kleefstof
f(m)
r клей *m*, клеящее вещество *n*,
клеящий состав *m*

A109 *e* **adhesive failure**
d Zerstörung *f* einer Klebeverbindung
f rupture *f* des joints [des assemblages]
collés
nl slechte hechting *f* van de lijm
r разрушение *n* клеевого шва
[соединения]

A110 *e* **adhesive nail-on method**
d Bindemittel- und Nagelverfahren *n*
f méthode *f* de fixation par adhésive
et par clous
nl opbrengen *n* met lijm en spijkers
r гвоздеклеевое соединение *n*

A111 *e* **adhesiveness**
d Klebrigkeit *f*, Haftfestigkeit *f*,
Grenzflächenkraft *f*
f adhésivité *f*, adhésion *f*
nl lijmvermogen *n*, kleefkracht *f(m)*,
hechting *f*
r связность *f*, клейкость *f*,
прилипаемость *f*, адгезионная
способность *f*

A112 *e* **adhesive power**
d Adhäsionskraft *f*, Haftfähigkeit *f*,
Klebekraft *f*
f pouvoir *m* adhésif
nl kleefkracht *f(m)*
r адгезионная прочность *f*, клеящая
способность *f*, прилипаемость *f*

A113 adhesive strength *see* adhesiveness

A114 *e* **adiabatic cooling**
d adiabatische Kühlung *f*
f refroidissement *m* adiabatique
nl adiabatische koeling *f*
r адиабатическое охлаждение *n*

A115 *e* **adiabatic curing**
d adiabatische Betonbehandlung *f*
f traitement *m* adiabatique du béton

nl adiabatische betonbehandeling *f*
r обработка *f* бетона в адиабатическом
режиме

A116 *e* **adjustable ball hinge**
d einstellbares Scharnierband *n*
f paumelle *f* à rotule réglable
nl verstelbaar kogelscharnier *n*
r регулируемая шарнирная петля *f*
(*двери*)

A117 *e* **adjustable grille**
d Verstellgitter *n*, regelbare
Jalousieklappe *f*
f jalousie *f* ajustable [réglable]
nl verstelbaar rooster *n*
r регулируемая вентиляционная
решётка *f*

A119 *e* **adjustable spanner, adjustable wrench**
US
d Rollgabelschlüssel *m*
f clé *f* [clef *f*] à molette [à bouche
réglable]
nl verstelbare [engelse] sleutel *m*
r разводной (гаечный) ключ *m*

A120 *e* **adjusted net fill**
d Auftrag *m* [Schüttung *f*, Damm *m*]
mit Setzungszugabe
f remblai *m* avec compensation pour
le tassement
nl ophoging *f* met overmaat voor de
zetting
r насыпь *f* с припуском на осадку

A121 *e* **adjusting screw**
d Einstellschraube *f*, Justierschraube *f*
f vis *m* de réglage
nl stelschroef *f(m)*, justeerschroef *f(m)*
r установочный [регулировочный] винт
m

A122 administration building *see* office
building

A123 *e* **admixture**
d Zusatz *m*, Zusatzmittel *n*,
Beimischung *f*, Beimengung *f*
f adjuvant *m*, addition *f*
nl toevoeging *f*, bijmenging *f*, bijslag *m*
r добавка *f*

A124 *e* **admixture effects**
d Einwirkung *f* des Zusatzmittels
f effets *m pl* des adjuvants
nl uitwerking *f* van de toevoeging
r эффект *m* воздействия добавок

A125 *e* **adobe brick**
d Lehmziegel *m*, Adobe(ton)ziegel *m*
f brique *f* crue [brute]
nl in de zon gedroogde (ongebakken)
steen *m*
r кирпич-сырец *m*, саман *m*

A126 *e* **adsorbed water**
d Haftwasser *n*, Benetzungswasser *n*,
Adsorptionswasser *n*
f eau *f* adsorbée
nl geadsorbeerd water *n*
r адсорбированная [адсорбционная]
вода *f*

A127 *e* **adverse slope**
   d Gegengefälle *n*, Gegensteigung *f*,
     Gegenneigung *f*
   f contre-pente *f*, pente *f* négative
   nl tegenhelling *f*, tegengestelde helling
     f
   r обратный уклон *m*

A128 *e* **advertisement for bids**
   d Ausschreibung *f*
   f appel *m* d'offres
   nl bericht *n* van aanbesteding
   r объявление *n* о торгах

A129 *e* **adz** *US*, **adze** *UK*
   d Texel *m*, Dachsbeil *n*, Krummhaue *f*
   f essette *f*
   nl dissel *m*
   r тёсло *n*

A130   **aerated concrete** *see* **cellular concrete**

A131 *e* **aerated spillway**
   d Saugüberfall *m*, Vakuumüberfall *m*,
     Vakuumüberlauf *m*
   f déversoir *m* aéré [à vide]
   nl overlaat *m*
   r вакуумный водослив *m*

A132 *e* **aeration**
   d 1. Belüftung *f* (*des Wassers*),
     Durchlüftung *f* 2. regelbare
     natürliche Lüftung *f*, Durchlüftung *f*
     (*des Gebäudes*)
   f aération *f*
   nl 1. beluchten *n*, aëratie *f* 2. ventileren
     n, luchten *n*
   r аэрация *f*

A133   **aeration basin** *see* **aeration tank**

A134   **aeration plant** *see* **activated sludge
   plant**

A135 *e* **aeration skylight**
   d Belüftungsaufsatz *m*, Belüftungslaterne
     f
   f lucarne *f* d'aération
   nl ventilatiekap *f(m)* met bovenlicht,
     lantaarn *f(m)* met ventilatie
   r аэрационный фонарь *m*

A136 *e* **aeration tank**
   d Durchlüftungsbecken *n*,
     Belüftungsbecken *n*
   f bassin *m* de boues activées
   nl beluchtingsbak *m*
   r аэротенк *m*, резервуар-аэратор *m*

A137 *e* **aerial mast**
   d Antennenmast *m*
   f mât *m* d'antenne
   nl antennemast *m*
   r антенная мачта *f*; мачта-антенна *f*

A138 *e* **aerial pipe crossing**
   d Rohr(leitungs)brücke *f*
   f traversée *f* aérienne du pipe-line
   nl bovengronds kruisende pijpleiding *f*
   r подвесной [воздушный] трубный
     переход *m*

A139   **aerial ropeway** *see* **cableway**

A140   **aerial tramway** *see* **cableway 1.**

A141 *e* **aerobic treatment**
   d aerobe Abwasserreinigung *f*
     [Abwasserbehandlung *f*]
   f épuration *f* des eaux usées aérobie
   nl aërobe afvalwaterzuivering *f*
   r аэробная очистка *f* сточных вод

A142 *e* **aerodynamic instability**
   d aerodynamische Instabilität *f*
     [Labilität *f*]
   f instabilité *f* aérodynamique
   nl aërodynamische onstabiliteit *f*
   r аэродинамическая неустойчивость *f*
     (*напр. висячих мостов*)!

A143 *e* **aerofilter**
   d Tropfkörper *m* mit künstlicher
     Lüftung
   f aérofiltre *m*, filtre *m* d'aération
   nl druppellichaam *n* met kunstmatige
     beluchting
   r аэрофильтр *m*

A145 *e* **aerograph**
   d Farbspritzpistole *f*
   f aérographe *m*
   nl verfspuit(pistool) *f(m)* (*n*)
   r аэрограф *m*; пистолет-
     -краскораспылитель *m*

A146 *e* **aerosol paint**
   d Aerosolfarbe *f*
   f peinture *f* aérosol
   nl verf *f(m)* in spuitbuis
   r аэрозольная краска *f*

A147   **aerotank** *see* **aeration tank**

A148 *e* **afflux**
   d 1. Hebung *f* des Wasserspiegels,
     Zufluß *m* 2. Stau *m*, Stauung *f*;
     Stauhöhe *f*
   f 1. afflux *m* 2. remous *m*, différence *f*
     de niveaux entre l'amont et l'aval
   nl 1. toestroming *f* 2. stuwing *f*
   r 1. приток *m* (*воды*) 2. подпор *m*

A149 *e* **A-frame**
   d A-Rahmen *m*
   f cadre *m* (en) *A*
   nl bok *m* (*hijswerktuig*)
   r А-образная рамная конструкция *f*

A150 *e* **after air filter**
   d Sekundärfilter *n*, *m*
   f filtre *m* à air secondaire, épurateur *m*
     d'air secondaire
   nl secundair filter *n*, *m*
   r воздушный фильтр *m* второй ступени

A151 *e* **after cooler**
   d Nachkühler *m*
   f sous-refroidisseur *m*
   nl nakoeler *m*
   r доводчик-доохладитель *m* (*воздуха*)

A152 *e* **after heater**
   d Nachwärmer *m*, Nachheizgerät *n*
   f réchauffeur *m* d'air secondaire
   nl naverhitter *m*
   r воздухонагреватель *m* [калорифер *m*]
     второго подогрева

A153 *e* **after-purification**
   d Nachreinigung *f*, Nachbehandlung *f*
   f épuration *f* secondaire
   nl nareiniging *f*
   r доочистка *f* (*сточных вод*)

A154 *e* **agent**
   d Zusatzmittel *n*, Zusatzstoff *m*
   f agent *m*, produit *m*, substance *f*
   nl middel *n*, agens *n*
   r вещество *n*, средство *n*; (химический) агент *m*; активная добавка *f*

A155 *e* **age-strength relation**
   d Alter-Festigkeit-Beziehung *f*
   f rapport *m* âge-résistance, rapport *m* entre l'age et la résistance
   nl verband *n* tussen ouderdom en sterkte
   r зависимость *f* прочности от возраста

A156 *e* **agglomerate**
   d Agglomerat *n*
   f agglomérat *m*, aggloméré *m*
   nl agglomeraat *n*
   r агломерат *m*

A157 *e* **agglomeration**
   d 1. Agglomeration *f*, Sinterung *f*, Verkittung *f* 2. Agglomeration *f*, Ballungsgebiet *n*
   f agglomération *f*
   nl agglomeratie *f*
   r 1. агломерация *f*, спекание *n* 2. городская агломерация *f*

A158 *e* **aggloporite**
   d Aggloporit *m*
   f aggloporite *m*
   nl aggloporit *m*
   r аглопорит *m*

A159 *e* **aggregate**
   d Zuschlag(stoff) *m*
   f agrégat *m*, granulat *m*
   nl toeslagmateriaal *n*, aggregaat *n*
   r заполнитель *m* (*напр. для бетона*)

A160 *e* **aggregate base**
   d Schotterbettung *f*, Schottergründung *f*, Grobschlagtragschicht *f*
   f couche *f* de fondation de pierre cassée *ou* de gravier
   nl wegfundering *f* van steenslag
   r щебёночное *или* гравийное дорожное основание *n*

A161 *e* **aggregate batching plant**
   d Zuschlag-Dosieranlage *f*
   f installation *f* de dosage de granulat
   nl toeslagdoseringsinstallatie *f*
   r установка *f* для дозирования заполнителей

A162 *e* **aggregate bin**
   d Zuschlag(stoff)silo *n*
   f trémie *f* [silo *m*] à agrégat [à granulat]
   nl toeslagsilo *m*
   r бункер *m* для хранения заполнителей

A163 *e* **aggregate blending**
   d Kornmischung-Zubereitung *f*, Herstellung *f* des Zuschlagstoffgemisches, Zubereiten *n* des kornabgestuften Zuschlagstoffgemisches
   f malaxage *m* [mélangeage *m*] des granulats
   nl mengen *n* van de toeslagmaterialen
   r смешение *n* заполнителей различных фракций

A164 *e* **aggregate-cement ratio**
   d Zuschlag(stoff)-Zement-Verhältnis *n*
   f rapport *m* agrégat-ciment
   nl toeslag/cement-verhouding *f*
   r отношение *n* заполнитель — вяжущее

A165 *e* **aggregate-coated panel**
   d Furnierholzplatte *f* mit Zierverkleidung aus Körnern des Zuschlagstoffes
   f panneau *m* en contreplaqué revêtu par granulat
   nl gegranuleerde [bezande] plaat *f(m)*
   r фанерная панель *f* с фактурной облицовкой из зёрен заполнителей

A166 *e* **aggregate exposure**
   d Freilegung *f* der Zuschläge, Bloßlegung *f* der Zuschlagstoffkörner
   f mise *f* à nu des grains de granulat (*pour les faire ressortir de la surface de béton*)
   nl uitwassen *n* van het beton
   r обнажение *n* зёрен заполнителя на лицевых поверхностях бетонных конструкций (*вид декоративной обработки*)

A167 *e* **aggregate gradation**
   d Kornabstufung *f* der Zuschläge
   f granulométrie *f* [granularité *f*] d'agrégat
   nl korrelgrootteverdeling *f* [gradatie *f*] van het toeslagmateriaal
   r гранулометрический состав *m* заполнителей

A168 *e* **aggregate preparation plant, aggregate production plant**
   d Zuschlagstoff-Aufbereitungsanlage *f*
   f usine *f* à granulat; installation *f* de concassage-criblage [de préparation de granulat]
   nl toeslagvoorbereidingsinstallatie *f*
   r завод *m* по производству заполнителей; дробильно-сортировочная установка *f*

A169 *e* **aggregate weighing batcher**
   d Zuschlagstoffwaage *f*
   f doseur *m* pondéral à granulat [à agrégat]
   nl gewichtsdoseringsapparaat *n* voor de toeslagen
   r весовой дозатор *m* заполнителей

A170 *e* **aggressive water**
   d Aggressivwasser *n*, angreifendes Wasser *n*, Schadwasser *n*

f eau f agressive
nl agressief water n
r агрессивная вода f

A171 **agitating lorry** *UK* **see agitating truck**

A172 e **agitating truck**
   d (Beton-)Rührfahrzeug n, (Beton-)Nachmischer m, (Beton-)Rührwagen m
   f camion m (avec) agitateur
   nl betonvrachtwagen m met draaiende trommel
   r автобетоносмеситель m; автобетоновоз m (с лопастным валом для побуждения бетонной смеси)

A173 e **agitation**
   d Auflockern n, Rühren n
   f agitation f
   nl roeren n, bewegen n
   r побуждение n (напр. смеси), перемешивание n

A174 e **agitation tank**
   d Rührbehälter m, Mischtank m
   f réservoir m de mélange
   nl mengtrommel f(m), tank m met roerwerk
   r смесительный резервуар m

A175 e **agitator**
   d 1. Rührwerk n, Rührmischer m 2. Naßbaggerkopf m, Schleppkopf m 3. Transportbirne f, Fertigbeton-Nachmischer m 4. Zirkulationspumpe f, Umwälzpumpe f
   f 1. agitateur m, mélangeur m 2. désagrégateur m 3. agitateur m, malaxeur m 4. pompe f de circulation
   nl 1. roerwerk n 2. zuigmond m van een zandzuiger 3. roterende trommel f(m) van betonvrachtauto 4. circulatiepomp f(m)
   r 1. мешалка f, механический смеситель m 2. механический рыхлитель m (землесосного снаряда) 3. перемешивающее устройство n (напр. для бетонной смеси) 4. циркуляционный насос m

A177 e **agricultural buildings**
   d Landwirtschaftsgebäude n pl, landwirtschaftliche Betriebsgebäude n pl
   f bâtiments m pl agricoles
   nl gebouwen m pl voor de landbouw
   r сельскохозяйственные здания n pl

A178 e **air balance**
   d Zuluft- und Abluftbilanz f, Volumenstrombilanz f, Lufthaushalt m
   f bilan m d'air
   nl luchtbalans f(m)
   r воздушный баланс m

A179 e **air blast**
   d 1. Gebläsewind m, Zwangsluftstrom m 2. Gebläse n
   f 1. courant m d'air; soufflage m 2. soufflante f, soufflerie f

   nl krachtige luchtstroom m
   r 1. дутьё n 2. воздуходувка f

A180 e **airborne noise**
   d Luftlärm m, Luftschall m
   f bruit m aérien
   nl door de lucht voorgeplant geluid n
   r воздушный шум m

A181 e **air breakwater**
   d pneumatischer Wellenbrecher m
   f brise-lames m pneumatique
   nl pneumatische golfbreker m
   r пневматический волнолом m

A182 e **air brick**
   d Lüftungsziegel m
   f brique f d'aérage [aérée, de ventilation]
   nl ventilatiesteen m
   r кирпич m с вентиляционными каналами, вентилируемый кирпич m

A183 e **air brush**
   d Farbspritzpistole f
   f pinceau m de peinture par pulvérisation, pinceau m à peindre pneumatique
   nl verfspuit f(m), airbrush m
   r небольшое краскораспылительное устройство n, пистолет-краскораспылитель m

A184 e **air chamber**
   d Luftglocke f
   f chambre f à air
   nl luchtkamer f(m), luchtketel m
   r воздушный колокол m (для работы под водой)

A185 e **air change**
   d Luftwechsel m, Luftaustausch m, Lufterneuerung f
   f renouvellement m [échange m] d'air
   nl luchtverversing f
   r воздухообмен m

A186 e **air change rate**
   d Luftwechselzahl f
   f degré m de renouvellement d'air
   nl luchtuitwisselingsgetal n
   r кратность f воздухообмена

A187 e **air channel**
   d Luftkanal m, Lüftungskanal m
   f canal m de ventilation, gaine f d'air
   nl luchtkanaal n
   r вентиляционный канал m, воздуховод m

A188 e **air chimney**
   d Abluftkamin m, Abzugsschacht m, Absaugschacht m
   f cheminée f d'appel [de tirage]
   nl luchtschacht f(m), luchtkoker m
   r вытяжная [вентиляционная] труба f

A189 e **air chute**
   d Luftkanal m (Luftleitungsteil mit rechteckigem Querschnitt)
   f gaine f d'air
   nl luchtkanaal n
   r вентиляционный короб m

A190 *e* **air circulation**
   d Luftzirkulation *f*, Luftumwälzung *f*
   f circulation *f* d'air
   nl luchtcirculatie *f*
   r циркуляция *f* воздуха

A191 *e* **air cleaner**
   d Luftreiniger *m*
   f épurateur *m* d'air
   nl luchtreiniger *m*, luchtfilter *n*, *m*
   r воздухоочиститель *m*

A192 *e* **air cock**
   d 1. Lufthahn *m*, Luftventil *n*,
   Rohrentlüfter *m*, Entlüftungsventil *n*,
   Luftabscheider *m* 2. Druckluftventil
   n, Drucklufthahn *m*
   f 1. robinet *m* d'air 2. robinet *m* d'air
   comprimé
   nl 1. pijpontluchter *m*, afblaaskraan
   f(m) 2. persluchtventiel *n*
   r 1. воздуховыпускной клапан *m*,
   воздухоотводчик *m*, вантуз *m*
   2. пневмовентиль *m*

A193 *e* **air compressor**
   d (Luft-)Kompressor *m*, (Luft-)-
   Verdichter *m*
   f compresseur *m* (d'air)
   nl luchtcompressor *m*
   r (воздушный) компрессор *m*

A194 *e* **air conditioner**
   d Klimagerät *n*, Klimaaggregat *n*
   f conditionneur *m* d'air, climatiseur *m*
   nl klimaatregelingstoestel *n*
   r кондиционер *m* (воздуха)

A195 *e* **air-conditioning**
   d Klimatisierung *f*, Klimatechnik *f*
   f conditionnement *m* de l'air
   nl luchtconditionering *f*,
   klimaatregeling *f*
   r кондиционирование *n* воздуха

A196 *e* **air-conditioning convector**
   d Kühlkonvektor *m*
   f convecteur *m* de refroidissement
   nl luchtverwarmings- en
   koelingsconvector *m*
   r воздухоохладитель *m* с
   естественной конвекцией

A197 *e* **air conditioning load**
   d Klimalast *f*
   f charge *f* sur un système de
   conditionnement d'air
   nl belasting *f* van luchtconditionering
   r нагрузка *f* системы
   кондиционирования воздуха

A198 *e* **air conditioning plant**
   d Zentralklimaanlage *f*, Klimazentrale *f*
   f installation *f* de conditionnement
   d'air; climatiseur *m* central
   nl luchtbehandelingsinstallatie *f*,
   klimaatregelingsinstallatie *f*
   r установка *f* кондиционирования
   воздуха; центральный кондиционер *m*

A199 *e* **air-conditioning substation**
   d Klimastation *f*, Klimaunterzentrale *f*

   f climatiseur *m* zonal
   nl klimaatregelingsstation *n*
   r зональный доводчик *m*

A200 *e* **air-conditioning unit**
   d Klimaaggregat *n*
   f appareil *m* de conditionnement d'air
   [de climatisation]
   nl luchtbehandelingstoestel *n*
   r агрегат *m* кондиционирования
   воздуха

A201 *e* **air conditions**
   d Luftzustandsgrößen *f pl*
   f paramètres *m pl* d'état de l'air
   nl luchttoestandgrootheden *f pl*
   r параметры *m pl* состояния воздуха

A203 *e* **air content**
   d Luftanteil *m*, Luftgehalt *m*
   f teneur *m* en air
   nl luchtinhoud *m*
   r содержание *n* воздуха,
   воздухосодержание *n*

A204 *e* **air cooled condenser**
   d luftgekühlter Kondensator *m*
   f condens(at)eur *m* à refroidissement
   par air
   nl luchtgekoelde condensator *m*
   r конденсатор *m* с воздушным
   охлаждением

A205 *e* **air cooler**
   d Luftkühler *m*
   f refroidisseur *m* d'air
   nl luchtkoeler *m*
   r воздухоохладитель *m*

A206 *e* **air cooling**
   d 1. Luftkühlung *f*, Kühlung *f* von Luft
   2. Kühlung *f* durch Luft
   f 1. refroidissement *m* d'air
   2. refroidissement *m* par l'air
   nl luchtkoeling *f*
   r 1. охлаждение *n* воздуха
   2. воздушное охлаждение *n*

A207 *e* **air curtain**
   d Luftschleier *m*
   f voile *m* [rideau *m*] d'air
   nl luchtgordijn *n*, *f(m)*
   r воздушная завеса *f*

A208 *e* **air cushion construction vehicles**
   d Luftkissenfahrzeuge *n pl* für
   Bauzwecke
   f véhicules *m pl* de chantier sur
   coussin d'air
   nl luchtkussenvoertuigen *n pl* voor
   bouwwerken
   r строительные транспортные средства
   n pl на воздушной подушке

A209 *e* **air damper**
   d Luftklappe *f*
   f volet *m* d'aération, registre *m*
   nl luchtklep *f(m)*
   r воздушный клапан *m*

A210 *e* **air dehumidification**
   d Luftentfeuchtung *f*
   f déshumidification *f* [séchage *m*] d'air

nl luchtdrogen *n*
r осушение *n* воздуха

A211   **air demand** *see* **air requirement**

A212 *e* **air-detraining admixture**
   d Luftporenminderer *m*
   f désaérateur *m*, anti-mousse *f*
   nl luchtverdrijvingsmiddel *n*
   r добавка *f*, снижающая содержание воздуха

A213 *e* **air diffuser**
   d 1. Luftverteiler *m* 2. Filterplatte *f*
   f 1. diffuseur *m* d'air 2. plaque *f* diffuseuse
   nl 1. luchtinblaas- en verdeelelement *n* 2. filterplaat *f(m)*
   r 1. диффузор *m*, воздухораспределительный плафон *m*, воздухораспределитель *m* 2. фильтрос *m*

A214 *e* **air diffusion**
   d Luftdiffusion *f*, Luftverteilung *f* (*im Raum*)
   f diffusion *f* d'air
   nl luchtverdeling *f* (*in de ruimte*)
   r воздухораздача *f*, распределение *n* воздуха (*в вентилируемом помещении*)

A215 *e* **air-diffusion aeration**
   d Druckluftverfahren *n* (*Abwasserbelüftung*)
   f aération *f* [aérage *m*] par (diffusion de) l'air comprimé
   nl pneumatische afvalwaterbeluchting *f*
   r пневматическая аэрация *f* (*сточных вод*)

A216 *e* **air digger**
   d Druckluft-Spatenhammer *m*, pneumatischer Spatenhammer *m*
   f pelle *f* à air comprimé
   nl pneumatische spade *f(m)*
   r пневматическая механическая лопата *f*

A217 *e* **air discharge grille**
   d Zuluftgitter *n*
   f grille *f* d'insufflation
   nl luchtinlaatrooster *n*
   r приточная решётка *f*

A218 *e* **air distribution**
   d Luftverteilung *f*
   f distribution *f* [répartition *f*] d'air
   nl luchtverdeling *f*
   r воздухораспределение *n*

A219 *e* **air drain**
   d 1. Zuluftkanal *m* 2. Feuchtigkeitsschutzgraben *m* (*an der Außenseite von Grundmauer*)
   f 1. conduit *m* d'air 2. cour *f* anglaise
   nl 1. luchtkanaal *n* 2. luchtopening *f*
   r 1. воздушный канал *m* (*в строительных конструкциях*) 2. приямок *m*, траншея *f* (*у фундаментной стены для защиты от увлажнения*)

A220 *e* **air-dried lumber**
   d lufttrockenes Schnittholz *n*
   f bois *m* séché à l'air
   nl winddroog bouwhout *n*
   r воздушно-сухой лесоматериал *m*

A221 *e* **air drill hammer**
   d Druckluft-Bohrhammer *m*
   f perforateur *m* [perforatrice *f*] pneumatique [à air comprimé]
   nl pneumatische boorhamer *m*
   r пневматический бурильный молоток *m*, пневмоперфоратор *m*

A222 *e* **air drying**
   d Lufttrocknung *f*
   f séchage *m* à l'air
   nl drogen *n* aan de lucht
   r атмосферная [воздушная] сушка *f*

A223 *e* **air duct**
   d Luft(leit)kanal *m*, Lüftungskanal *m*
   f conduit *m* [gaine *f*] d'air
   nl luchtleiding *f*, luchtkoker *m*
   r воздуховод *m*

A225 *e* **air-entrained concrete**
   d AEA-Beton *m*, belüfteter Beton *m*, Luftporenbeton *m*
   f béton *m* à occlusion d'air, béton *m* aéré
   nl luchthoudend beton *n*
   r бетон *m* с воздуховововлекающей добавкой

A226 *e* **air-entraining admixture**
   d luftporenbildender Zusatzstoff *m*, LP-Zusatz *m*
   f entraîneur *m* d'air
   nl luchtbellenvormer *m* (*in beton*)
   r воздуховововлекающая добавка *f*

A227 *e* **air-entraining cement**
   d Luftporenzement *m*, LP-Zement *m*
   f ciment *m* à entraîneur d'air
   nl luchthoudend cement *n*
   r цемент *m* с воздуховововлекающей добавкой

A228 *e* **air entrainment**
   d Lufteinschluß *m*; Lufteinführung *f*
   f entraînement *m* d'air; aération *f* du béton
   nl luchtopname *f*
   r вовлечение *n или* подсос *m* воздуха

A229 *e* **air entrainment test**
   d (Beton-)Luftgehaltprüfung *f*
   f détermination *f* de la teneur d'air en béton frais
   nl bepalen *n* van luchtgehalte in beton
   r определение *n* содержания воздуха в бетонной смеси

A230 *e* **air exhaust**
   d Luftabsaugung *f*
   f exhaustion *f* de l'air
   nl luchtafzuiging *f*
   r вытяжка *f* воздуха

A231 *e* **air exhaust opening**
   d Abluftöffnung *f*, Entlüftungsöffnung *f*

f orifice *m* de soutirage, bouche *f* d'aspiration
nl afzuigopening *f*
r вытяжное отверстие *n*

A232 *e* air face
d Luftseite *f*, Talseite *f*
f parement *m* [talus *m*] aval
nl dalzijde *f*
r низовая грань *f*, низовой откос *m*

A233 *e* air filter
d Luftfilter *n, m*
f filtre *m* d'air
nl luchtfilter *n, m*
r воздушный фильтр *m*

A234 *e* air filtration
d Luftfilterung *f*
f filtration *f* [filtrage *m*] d'air
nl luchtfiltratie *f*
r фильтрация *f* воздуха

A235 *e* air float
d Druckluft-Glättkelle *f*, pneumatische Glättkelle *f*
f lisseur *m* [lisseuse *f*] pneumatique
nl pneumatische plakspaan *f(m)*
r пневматическая гладилка *f*

A236 *e* air flow meter
d Luftströmungsmesser *m*
f débitmètre *m* d'air
nl luchtstroommeter *m*
r расходомер *m* воздуха

A237 *e* air flow rate
d Luftmengenstrom *m*, Luftdurchfluß *m*, Luftdurchsatz *m*
f débit *m* d'air
nl hoeveelheid *f* doorstromende lucht
r расход *m* воздуха

A238 *e* air flow switch
d (Luft-)Strömungswächter *m*, (Luft-)-Strömungsüberwachungsrelais *n*, Strömungssicherung *f*
f relais *m* [interrupteur *m*] de courant d'air
nl luchtstroomregelaar *m*
r реле *n* воздушного потока

A239 *e* air gap
d 1. Schlitz *m*, Abzug *m*, Zugloch *n*, Entlüfter *m* 2. Luftspalt *m*
f 1. fente *f* de respiration 2. entrefer *m*, lame *f* [espace *f*] d'air
nl 1. luchtspleet *f(m)* 2. luchtopening *f*, ventilatieopening *f*
r 1. (вентиляционная) отдушина *f*, вентиляционная щель *f* 2. воздушный зазор *m*

A240 *e* air hammer
d Drucklufthammer *m*
f marteau *m* pneumatique [à air comprimé]
nl luchtdrukhamer *m*
r пневматический молоток *m*

A241 *e* **air-handling capacity**
d Luftleistung *f*
f capacité *f* de production d'air

nl luchtbehandelingscapaciteit *f*
r воздухопроизводительность *f*

A242 *e* **air handling unit**
d Luftbehandlungsaggregat *n*
f 1. climatiseur *m* multizone 2. groupe *m* de traitement de l'air
nl luchtbehandelingstoestel *n*
r 1. неавтономная установка *f* кондиционирования воздуха 2. приточная камера *f*; воздухоприготовительная установка *f*

A243 *e* air heater
d Lufterhitzer *m*
f réchauffeur *m* d'air
nl luchtverhitter *m*
r воздухонагреватель *m*, калорифер *m*

A244 *e* air heating
d (Warm-)Luftheizung *f*, Überdruckluftheizung *f*
f chauffage *m* à air chaud
nl verwarming *f* met hete lucht
r воздушное отопление *n*

A245 *e* air hoist
d Druckluftaufzug *m*, pneumatischer Aufzug *m*
f élévateur *m* [treuil *m*] pneumatique [à air comprimé]
nl pneumatische hijsinrichting *f*
r пневматический подъёмник *m*, пневмоподъёмник *m*

A247 *e* air humidification
d Luftbefeuchtung *f*
f humidification *f* d'air
nl luchtbevochtiging *f*
r увлажнение *n* воздуха

A248 *e* air humidifier
d Luftbefeuchtungsgerät *n*, Luftbefeuchter *m*
f humidificateur *m* d'air
nl luchtbevochtiger *m*
r увлажнитель *m* воздуха

A249 *e* air hydraulic jack
d pneumatisch-hydraulischer Hebebock *m*
f vérin *m* pneumo-hydrolique
nl pneumatisch-hydraulische vijzel *f(m)*
r пневмогидравлический домкрат *m*

A250 *e* air infiltration
d Lufteindringung *f*, Luftinfiltration *f*
f infiltration *f* d'air
nl luchtinfiltratie *f*
r инфильтрация *f* воздуха

A251 *e* air inflow
d Luftzufuhr *f*
f afflux *m* d'air
nl luchttoevoer *m*
r приток *m* воздуха

A252 air influx *see* air inflow

A253 *e* air inlet
d Lufteinlaß *m*
f bouche *f* d'entrée d'air
nl luchtinlaat *m*, luchtinlaatopening *f*
r воздуховпускное отверстие *n*

A254 *e* **air intake**
 d Lufterfassungsöffnung *f*,
 Luftaufnahmeöffnung *f*, Lufterfasser
 m, Absaugvorrichtung *f*
 f ouverture *f* d'air, bouche *f* d'entrée
 d'air, dispositif *m* de prise d'air
 nl luchtinlaat(opening) *m* (*f*)
 r воздухозаборное отверстие *n*,
 воздухоприёмник *m*,
 воздухоприёмное устройство *n*

A255 *e* **air-intake shaft**
 d Lüftungsschacht *m*, Ansaugschacht *m*
 f puits *m* d'aérage
 nl luchtinlaatschacht *f* (*m*)
 r воздухозаборная *или* аэрационная
 шахта *f*

A256 *e* **air leakage**
 d Luftleckung *f*, Luftaustritt *m*
 f fuite *f* d'air
 nl luchtlek *n*
 r утечка *f* воздуха

A257 *e* **air leakage factor**
 d Fugendurchlässigkeit *f*,
 Fugendurchlaßkoeffizient *m*
 f coefficient *m* de fuite d'air
 nl (lucht)lekkagecoëfficiënt *m*
 r коэффициент *m* утечки воздуха;
 коэффициент *m* воздухопроницаемости швов

A258 *e* **airless spraying**
 d (druck)luftloses Spritzen *n*
 f pulvérisation *f* sans air [airless]
 nl luchtloze drukverstuiving *f*
 r безвоздушное распыление *n*

A259 *e* **airless spray unit**
 d luftloser Zerstäuber *m*
 f appareil *m* de pulvérisation sans air
 [airless]
 nl luchtloze verstuiver *m*
 r безвоздушная распылительная
 установка *f*

A260 *e* **air-lift, air-lift pump**
 d Druckluftpumpe *f*,
 Druckluftwasserheber *m*,
 Mammutpumpe *f*, Luftheber *m*
 f émulseur *m* à air comprimé, pompe *f*
 à émulsion, air-lift, pompe *f*
 mammouth
 nl luchtlift *m* (*in een bronbuis*)
 r эрлифт *m*

A261 *e* **air lock**
 d 1. Luftschleuse *f*, Windfang *m*,
 Vorraum *m* 2. Luftsack *m*
 3. Zug(ab)dichtung *f*
 f 1. écluse *f* à air, sas *m* d'air 2. poche *f*
 [tampon *m*] d'air 3. latte *f* de
 recouvrement
 nl 1. luchtsluis *f*(*m*), luchtportaal *n*
 2. luchtzak *m* (*in een leiding*)
 r 1. воздушный шлюз *m* 2. воздушная
 пробка *f* 3. нащельник *m*, нащельная
 рейка *f*

A262 *e* **air mixing box**
 d Luftmischkammer *f*, Mischkasten *m*

 f boîte *f* de mélange (de l'air)
 nl luchtmengkamer *f*(*m*)
 r воздухосмесительная камера *f*

A263 *e* **air outlet**
 d Luftauslaß *m*, Luftaustrittsöffnung *f*
 f sortie *f* d'air
 nl luchtuitlaat(opening) *m* (*f*)
 r воздуховыпускное отверстие *n*

A264 *e* **air output**
 d Luftfördermenge *f*
 f rendement *m* de ventilateur
 nl ventilatorcapaciteit *f*
 r подача *f* вентилятора

A265 *e* **air permeability factor**
 d Luftdurchlaßkoeffizient *m*
 f coefficient *m* de perméabilité à l'air
 nl luchtdoorlaatbaarheidscoëfficiënt *m*
 r коэффициент *m* воздухопроницаемости

A266 *e* **air pipe line**
 d Druckluftleitung *f*
 f canalisation *f* d'air comprimé,
 conduite *f* d'air
 nl luchtleiding *f*, persluchtleiding *f*
 r трубопровод *m* сжатого воздуха,
 воздухопровод *m*

A267 *e* **air-placed concrete**
 d Spritzbeton *m*, Torkretbeton *m*
 f béton *m* projeté
 nl spuitbeton *n*
 r торкрет-бетон *m*

A269 *e* **air placer, air placing machine**
 d Druckluft-Betonförderer *m*,
 pneumatischer Betonförderer *m*
 f bétonneuse *f* pneumatique
 nl betonspuitinstallatie *f*
 r пневматический бетоноукладчик *m*

A270 *e* **air pollution control**
 d Luftreinhaltung *f*
 f protection *f* contre la pollution d'air
 nl toezicht *n* op luchtverontreiniging
 r охрана *f* атмосферы, борьба *f*
 с загрязнением воздуха

A271 *e* **airport facilities**
 d Flughafenanlagen *f pl*
 f infrastructure *f* aéronautique,
 installations *f pl* des services
 d'aérodrome
 nl luchthaveninstallaties *f pl*
 r наземное оборудование *n* аэропорта

A272 *e* **air preheater**
 d Vorwärmer *m*, Vorlufterhitzer *m*
 f préchauffeur *m* d'air
 nl luchtvoorverwarmer *m*
 r воздухонагреватель *m* [калорифер *m*]
 первого подогрева

A273 *e* **air purification**
 d Luftreinigung *f*
 f épuration *f* d'air
 nl luchtzuivering *f*
 r очистка *f* воздуха

A274 *e* **air purifier** *see* **air cleaner**

A275 *e* **air receiver**
 d Windkessel *m*, Druckluftkessel *m*

f réservoir *m* d'air, récipient *m* à air
comprimé
nl persluchtketel *m*
r ресивер *m* сжатого воздуха

A276 *e* **air register**
d Luftregler *m*, Lüftungsgitter *n* mit
Mengeneinstellung
f grille *f* à registre, registre *m* d'air
nl luchtregelaar *m*, schuif *f(m)*
r воздухораспределительная решётка
f с воздушным клапаном

A277 **air-relief valve** *see* **automatic air valve**

A278 *e* **air requirement**
d Luftbedarf *m*
f demande *f* d'air
nl luchtbehoefte *f*
r требуемое количество *n* воздуха

A279 *e* **air riveting hammer**
d Nietrevolver *m*
f riv(et)euse *f* pneumatique [à air
comprimé]
nl pneumatische klinkhamer *m*
r пневматический клепальный
молоток *m*

A280 *e* **air sampling**
d Luftprobenahme *f*
f prélèvement *m* d'air
nl luchtbemonstering *f*
r отбор *m* проб воздуха

A281 *e* **airslide conveyor**
d Luftrutsche *f*, Luftförderrohr *n*
f convoyeur *m* pneumatique
nl luchtbedtransporteur *m*
r аэрожёлоб *m*

A282 *e* **air-space ratio**
d Luftporenanteil *m*, scheinbare
Porosität *f*
f indice *m* de pourcentage d'air en
volume
nl percentage *f* holle ruimte
r коэффициент *m* воздухосодержания

A283 *e* **air supply**
d Luftzufuhr *f*, Luftförderung *f*,
Lufteinführung *f*
f approvisionnement *m* d'air
nl luchttoevoer *m*
r подача *f* воздуха

A284 *e* **air supply grille**
d Lufteintrittsgitter *n*,
Lufteinlaßgitter *n*, Zuluftgitter *n*
f grillage *m* d'admission [d'entrée]
nl lucht(inlaat)rooster *n*
r нерегулируемая приточная решётка *f*

A285 *e* **air supply plant**
d Zuluftzentrale *f*, Zuluftanlage *f*
f installation *f* d'alimentation en air
nl luchtvoorzieningsinstallatie *f*
r приточная установка *f*

A286 *e* **air supply system**
d Frischluftsystem *n*, Frischluftanlage *f*
f système *m* d'alimentation en air
nl luchttoevoersysteem *n*
r приточная система *f* вентиляции

A287 *e* **air supply unit**
d Belüftungsgerät *n*
f unité *f* [groupe *m*] d'aérage
nl beluchtingstoestel *n*
r вентиляционный агрегат *m*

A288 *e* **air-supported structure**
d luftgetragenes Bauwerk *n*,
Tragluftkonstruktion *f*
f structure *f* [ouvrage *m*] gonflable
nl opblaasbaar gebouw *n*
r пневматическое воздухоопорное
сооружение *n*

A289 *e* **air tamper**
d Druckluftramme *f*
f dam(eus)e *f* pneumatique, dameur *m*
à air comprimé
nl pneumatische stamper *m*
r пневматическая трамбовка *f*,
пневмотрамбовка *f*

A290 *e* **air terminal**
d Blitzauffangstange *f*
f partie *f* supérieure d'un paratonnerre
nl bliksemafleiderspits *f(m)*
r молниеприёмник *m*

A291 *e* **air terminal unit**
d Nachbehandlungsgerät *n*,
Endaggregat *n*
f climatiseur *m* [conditionneur *m* d'air]
terminal
nl nabehandelingsinrichting *f*
r доводчик *m* (*кондиционирования воздуха*)

A292 **air termination** *see* **air terminal**

A293 *e* **air test**
d Druckluftprobe *f*
f essai *m* à l'air comprimé
nl beproeving *f* met druklucht
r пневматическое испытание *n*,
испытание *n* на герметичность

A294 *e* **air throw**
d Wurfweite *f* eines Luftstrahles
f lancée *f* du jet d'air
nl worpwijdte *f* (van een straal)
r дальнобойность *f* приточной струи
(*воздуха*)

A295 *e* **air-tight concrete**
d Luftdichtbeton *m*,
luftundurchlässiger Beton *m*
f béton *m* imperméable à l'air
nl luchtdicht beton *n*
r воздухонепроницаемый бетон *m*

A296 *e* **air-tight door**
d abgedichtete Tür *f*
f porte *f* étanche
nl hermetisch sluitende deur *f(m)*
r герметическая [герметичная] дверь
f

A297 *e* **air-tight seal**
d luftdichter Verschluß *m*,
luftundurchlässige Abdichtung *f*
f joint *m* étanche à l'air
nl hermetische afsluiting *f*
r воздухонепроницаемое уплотнение *n*

A298 *e* **air-to-air heat recovery unit**
 d Luft-Luft-Wärmerückgewinnungsanlage *f*
 f agrégat *m* de récupération de chaleur d'air-air
 nl lucht-lucht-warmterecuperator *m*
 r теплоутилизатор *m* с воздухо-воздушным теплообменником

A299 *e* **air-to-air system**
 d Luft-Luft-Klimaanlage *f*
 f système *m* (de climatisation) d'air-air
 nl lucht-lucht-conditioneringssysteem *n*
 r воздухо-воздушная система *f* кондиционирования воздуха

A300 *e* **air tool**
 d Druckluftgerät *n*, pneumatisches Werkzeug *n*
 f outil *m* pneumatique [à air comprimé]
 nl persluchtgereedschap *n*
 r пневматический инструмент *m*

A301 *e* **air-to-water system**
 d Luft-Wasser-Klimaanlage *f*
 f système *m* (de climatisation) d'air-eau
 nl lucht-water-conditioneringssysteem *n*
 r водовоздушная система *f* кондиционирования воздуха

A302 *e* **air-to-water heat exchanger**
 d Luft-Wasser-Wärmeübertrager *m*
 f échangeur *m* thermique d'air à l'eau
 nl lucht-water-warmtewisselaar *m*
 r воздуховодяной теплообменник *m*

A303 *e* **air treatment**
 d Luftbehandlung *f*, Luftaufbereitung *f*
 f traitement *m* d'air
 nl luchtbehandeling *f*
 r обработка *f* воздуха, воздухоподготовка *f*

A304 **air trowel** *see* **air float**

A305 *e* **air valve**
 d Rohrentlüfter *m*
 f ventouse *f*, soupape *f* d'air
 nl ontluchtingsventiel *n*
 r вантуз *n*, воздушный клапан *m*

A306 *e* **air vent**
 d 1. Rohrentlüfter *m* 2. Luftloch *n*, Zugloch *n*
 f 1. ventouse *f*, soupape *f* d'air 2. aspirail *m*, évent *m*
 nl 1. luchtklep *f(m)* 2. ventilatieopening *f*, luchtuitlaat *m*
 r 1. вантуз *m*, воздушный клапан *m* 2. вентиляционная отдушина *f*, вентиляционное отверстие *n*

A307 *e* **air ventilator**
 d Belüftungsöffnung *f*
 f ouverture *f* d'aération
 nl luchtinlaat *m*
 r аэрационное отверстие *n*, аэрационный проём *m*

A308 **air vessel** *see* **air receiver**

A309 *e* **air voids**
 d Luftporen *f pl*
 f vides *m pl* d'air
 nl luchtporiën *f pl*
 r воздушные пустоты *f pl* [поры *f pl*]

A310 *e* **air volume**
 d Luft(volumen)menge *f*
 f débit *m* d'air volumétrique, débit-volume *m* d'air
 nl luchtvolume *n*
 r (объёмный) расход *m* воздуха

A311 *e* **air volume control**
 d Luftmengenregelung *f*
 f contrôle *m* de volume d'air
 nl regeling *f* van de luchthoeveelheid
 r регулирование *n* расхода воздуха

A312 *e* **air washer**
 d Luftwäscher *m*, Luftwaschanlage *f*
 f laveur *m* [nettoyeur *m*] d'air
 nl luchtreinigingsinstallatie *f*
 r контактный аппарат *m* для обработки воздуха; центральная воздухоприготовительная установка *f* с камерой орошения; камера *f* орошения

A313 *e* **air-water heat pump**
 d Luft-Wasser-Wärmepumpe *f*
 f thermo-pompe *f* [pompe *f* à chaleur] air-eau
 nl lucht-water-warmtepomp *f(m)*
 r тепловой насос *m* с передачей тепла от воздуха к воде

A314 *e* **air winch**
 d Druckluftwinde *f*, pneumatische Winde *f*
 f treuil *m* pneumatique [à air comprimé]
 nl pneumatisch windas *n*, pneumatische lier *f(m)*
 r пневматическая лебёдка *f*, лебёдка *f* с пневматическим приводом

A315 *e* **aisle**
 d Durchgang *m*
 f passage *m*, couloir *m*
 nl zijbeuk *m* (*van een kerk*), gangpad *n*, gang *m*
 r проход *m*

A316 *e* **alabaster**
 d 1. Alabaster(gips) *m* 2. Baugips *m*
 f 1. albâtre *m* 2. plâtre *m* de construction
 nl albast *n*
 r 1. алебастр *m* (*минерал*) 2. строительный гипс *m*

A317 *e* **alarm system**
 d Alarmanlage *f*
 f système *f* [installation *f*] d'alarme
 nl alarminstallatie *f*, alarmsysteem *n*
 r аварийно-сигнальная система *f*, аварийно-сигнальное устройство *n*

A318 *e* **alclad**
 d Alclad *m* (*mit Reinaluminium plattiertes Dural*)
 f produit *m* d'acier plaqué par aluminium
 nl met aluminium geplateerde duraluminiumplaat *f(m)*

r стальное изделие *n*, плакированное алюминием

A319 *e* **algae control**
d Algenverhütung *f*, Algenbekämpfung *f*
f lutte *f* [défense *f*] contre les algues
nl algenbestrijding *f*
r обработка *f* воды, противодействующая росту водорослей (*в контактных аппаратах*)

A320 *e* **aligning, alignment**
d 1. Abfluchtung *f*, Einfluchtung *f*
2. Trasse *f* 3. Absteckung *f*,
Linienführung *f* 4. Flußbettregulierung *f*
f 1. alignement *m* 2. tracé *m* 3. traçage *m* 4. correction *f* du lit
nl 1. richten *n* 2. tracé *n* 3. rooilijn *f*(*m*) 4. beddingregeling *f*
r 1. выравнивание *n*, рихтовка *f*; центрирование *n* 2. трасса *f* 3. трассирование *n*; провешивание *n* 4. выправление *n* русла

A321 *e* **A-line**
d A-Linie *f* (*Einflußlinie der Auflagerkraft*)
f ligne *f* d'influence de réactions aux appuis
nl invloedlijn *f* van de oplegdruk
r линия *f* влияния опорных реакций

A322 *e* **alite**
d Alit *m*
f alite *m*, silicate *m* tricalcique
nl alit *n*, (gemalen) portlandcementklinker *m*
r алит *m*, трёхкальциевый силикат *m*

A323 *e* **alkali-aggregate expansion**
d Betonausdehnung *f* infolge der alkalischen Reaktion der Betonzuschläge mit dem Zement
f expansion *f* du béton due à la réaction alcali-agrégat
nl betonuitzetting *f* (*als gevolg van de reactie van de alkalische toeslag met het cement*)
r расширение *n* бетона в результате реакции между щёлочью в цементе и заполнителями

A324 *e* **alkali discoloration**
d alkalische Verfärbung *f*
f décoloration *f* due à l'action d'alcali
nl verkleuring *f* door alkalische werking
r обесцвечивание *n* (*краски*) от щелочей

A325 *e* **alkaline soil**
d alkalischer Boden *m*
f sol *m* alcalin
nl alkalische grond *m*
r щелочной грунт *m*

A326 *e* **alkali resistance**
d Alkali(en)festigkeit *f*, Alkalibeständigkeit *f*, Alkali(en)widerstand *m*
f résistance *f* aux alcalis

nl alkalibestendigheid *f*
r щёлочеупорность *f*, щёлочестойкость *f*

A327 *e* **alkali-resistant glass fiber**
d alkalibeständige Glasfaser *f*
f fibre *f* de verre résistante aux alcalis
nl alkalibestendige glasvezel *f*(*m*)
r щёлочеупорное [щёлочестойкое] стекловолокно *n*

A328 *e* **alkyd paint**
d Alkydharzfarbe *f*
f peinture *f* à base de résines alkyds
nl alkydverf *f*(*m*)
r алкидная краска *f*

A329 *e* **all air heat pump**
d Luft-Luft-Wärmepumpe *f*
f thermo-pompe *f* [pompe *f* à chaleur] air-air
nl lucht-lucht-warmtepomp *f*(*m*)
r тепловой насос *m* с передачей тепла от воздуха к воздуху, воздухо-воздушный тепловой насос *m*

A330 *e* **all air system**
d Ganzluftsystem *n*, Nur-Luft-Klimaanlage *f*
f système *m* (de climatisation) entièrement à l'air
nl luchtconditioneringsinstallatie *f* met lucht als medium
r полностью воздушная система *f* кондиционирования воздуха

A331 *e* **allege**
d Brüstung(smauer) *f*
f allège *f*
nl borstwering *f*
r тонкая подоконная часть *f* стены

A332 *e* **alley**
d 1. Allee *f* 2. Durchfahrt *f*
f 1. allée *f* 2. passage *m*, ruelle *f*
nl 1. laan *f*(*m*), steeg *f*(*m*) 2. doorgang *m*, doorloop *m*
r 1. аллея *f*; дорожка *f* 2. проход *m*; проезд *m*

A333 *e* **all-flanged tee**
d T-Formstück *n*
f té *m* à trois brides, raccord *m* en T à trois brides
nl T-stuk *n* met drie flenzen
r фланцевый тройник *m*

A334 *e* **all-glass façade**
d Ganzglasfassade *f*, Voll-Glasfassade *f*
f façade *f* entièrement vitrée
nl geheel glazen gordijngevel *m*
r полностью застеклённый фасад *m*

A335 *e* **alligatoring**
d Oberflächenrißbildung *f*, Hautrißbildung *f*
f fissuration *f* superficielle, fendillement *m*
nl craquelure *f*
r поверхностное трещинообразование *n*; растрескивание *n*, сетка *f* трещин

A336 *e* **alligator shears**
d Alligatorschere *f*, Hebelschere *f*

f cisailles *f* *pl* à levier [à mâchoires]
nl hefboomschaar *f(m)*
r аллигаторные [рычажные]
ножницы *pl*

A337 *e* **alligator wrench**
d Rohrschlüssel *m*, Rohrzange *f*
f clé *f* [clef *f*] à tube, pince *f*
universelle à «crocodile»
nl pijpsleutel *m*
r аллигаторный (трубный) ключ *m*

A338 *e* **all-in-aggregate**
d 1. natürliches Sand-Kies-Gemisch *n*
2. ungesiebtes gebrochenes Material
n, ungesiebte Zuschlagstoffe *m* *pl*
f 1. mélange *m* naturel sable-gravier
2. gravier *m* non tamisé [de carrière]
nl 1. natuurlijk zand-en-grindmengsel *n*
2. ongezeefd grind *n*
r 1. природная песчано-гравийная
смесь *f* 2. несортированный щебень *m*

A339 *e* **all-metal building**
d Ganzmetallgebäude *n*
f bâtiment *m* tout (en) métal
nl metalen gebouw *n*
r цельнометаллическое здание *n*

A340 *e* **all outside air system**
d Außenluft-Klimaanlage *f*,
Primärluftklimaanlage *f*
f système *m* de climatisation à air
extérieur
nl luchtbehandelingsinstallatie *f* met
toevoer van buitenlucht
r прямоточная система *f*
кондиционирования воздуха

A341 *e* **allowable bearing pressure**
d zulässiger Auflagerdruck *m*
[Stützendruck *m*]
f pression *f* de contact admissible
nl toelaatbare oplegdruk *m*
r допускаемое опорное давление *n*

A342 *e* **allowable bearing value**
d zulässige Bodenpressung *f*
f pression *f* admissible sur le sol
nl toelaatbare gronddruk *m*
r допускаемое давление *n* на грунт

A343 *e* **allowable load**
d zulässige Last *f* [Belastung *f*]
f charge *f* admissible
nl toelaatbare belasting *f*
r допускаемая нагрузка *f*

A344 *e* **allowable noise level**
d zulässiger Geräuschpegel *m*
f niveau *m* de bruit admissible
nl toelaatbaar geluidniveau *n*
r допустимый уровень *m* шума

A345 *e* **allowable pile bearing load**
d zulässige Belastung *f* des Pfahles;
Pfahltragfähigkeit *f*
f charge *f* sur pieu admissible; pouvoir
m porteur du pieu
nl toelaatbare paalbelasting *f*;
paaldraagvermogen *n*
r допускаемая нагрузка *f* на сваю;
несущая способность *f* сваи

A346 **allowable soil pressure** *see* **allowable
bearing value**

A347 *e* **allowable stress**
d zulässige Spannung *f*
f contrainte *f* admissible [admise],
tension *f* admissible
nl toelaatbare spanning *f*
r допускаемое напряжение *n*

A348 *e* **allowable stress design**
d Berechnung *f* nach zulässigen
Spannungen
f calcul *m* d'après les contraintes
admissibles
nl berekening *f* op grond van
toelaatbare spanningen
r расчёт *m* по допускаемым
напряжениям

A349 *e* **allowance**
d 1. Toleranz *f* 2. Übermaß *n*, (Maß-)-
Zugabe *f*
f 1. tolérance *f* 2. surépaisseur *f*
nl 1. tolerantie *f* 2. overmaat *f(m)*,
extra ruimte *f*
r 1. допуск *m* 2. припуск *m*

A350 *e* **alloy steel**
d legierter Stahl *m*
f acier *m* allié
nl staallegering *f*
r легированная сталь *f*

A351 *e* **all-position electrode**
d Allpositionselektrode *f*
f électrode *f* pour soudage dans toutes
les positions
nl elektrode *f* voor het lassen in alle
houdingen
r электрод *m* для сварки во всех
положениях

A352 *e* **all-purpose excavator**
d Mehrzweckbagger *m*,
Universalbagger *m*, Vielzweckbagger
m
f excavateur *m* universel
nl universele graafmachine *f*
r универсальный экскаватор *m*

A353 *e* **all-purpose insulation**
d Allzweckdämmstoff *m*
f matériau *m* d'isolation [isolant]
universel
nl universeel isolatiemateriaal *n*
r универсальный изоляционный
материал *m*

A354 *e* **all-purpose road**
d allgemeine Verkehrsstraße *f*
f route *f* universelle
nl weg *m* voor alle verkeer
r дорога *f* универсального назначения

A355 *e* **all-socket cross**
d Kreuzstück *n* für Muffenverbindung,
Muffen-Kreuzstück *n*
f croix *f* à quatre emboîtements
nl kruisstuk *n* voor mofverbinding
r раструбный крест *m*, крест *m*
с раструбами на всех патрубках

31

A356 *e* **all-terrain vehicle**
 d Geländefahrzeug *m*, geländegängiges Fahrzeug *m*
 f véhicule *m* tous terrains [tous chemins]
 nl terreinvoertuig *n*, terreinwaardig voertuig *n*
 r вездеход *m*

A357 *e* **alluvial deposit(s), alluvium**
 d alluviale Ablagerungen *f pl*, Alluvium *n*
 f dépôts *m pl* alluviaux, alluvions *f pl*
 nl alluviale afzettingen *f pl*
 r аллювиальные отложения *n pl*, аллювий *m*

A358 *e* **all-welded steel structure**
 d Ganzschweißstahlbauwerk *n*
 f ossature *f* en acier entièrement soudée
 nl geheel gelaste stalen constructie *f*
 r цельносварная стальная конструкция *f*

A359 *e* **all-weld-metal test specimen**
 d Ganzschweißprobe *f*
 f éprouvette *f* d'essai en métal déposé
 nl proefstuk *n* van ingesmolten metaal
 r испытательный образец *m* из наплавленного металла

A360 *e* **all year air conditioning**
 d ganzjährige Klimatisierung *f*
 f conditionnement *m* d'air été-hiver
 nl permanente klimaatregeling *f*
 r круглогодичное кондиционирование *n* воздуха

A361 *e* **alteration**
 d Anbau *m*; Aufbau *m*, Überbau *m*
 f annexe *f*; superstructure *f*
 nl bijbouw *m*; opbouw *m*
 r пристройка *f*; надстройка *f*

A362 *e* **alterations**
 d 1. bauliche Veränderungen *f pl* 2. Umbau *m*
 f 1. amendements *m pl*, modifications (*du projet*) 2. reconstruction *f*, rénovation *f*
 nl 1. veranderingen *f pl* (*van ontwerp*) 2. verbouwing *f*
 r 1. изменения *n pl* (*проекта*) 2. перестройка *f*, реконструкция *f*

A363 *e* **alteration work**
 d Umbauarbeiten *f pl*
 f travaux *m pl* de reconstruction [de rénovation]
 nl verbouwingswerkzaamheden *f pl*
 r работы *f pl* по перестройке [по реконструкции]

A364 *e* **alternate bending strength**
 d Wechselbiegefestigkeit *f*
 f résistance *f* aux flexions alternées
 nl buigvastheid *f* bij heen-en-weer buigen
 r прочность *f* [предел *m* выносливости] при знакопеременном изгибе

A365 *e* **alternate depths**
 d zusammengehörige [kohärente, reziproke, konjugierte] Tiefen *f pl*
 f hauteurs *f pl* [profondeurs *f pl*] conjuguées
 nl reciproke diepten *f pl*
 r сопряжённые глубины *f pl*

A366 *e* **alternate strength**
 d Wechselfestigkeit *f*
 f résistance *f* aux efforts répétés alternés
 nl weerstand *m* tegen wisselende belasting
 r прочность *f* [предел *m* выносливости] при знакопеременной нагрузке

A367 *e* **alternating bending test**
 d Hin- und Her-Biegeversuch *m*
 f essai *m* de fatigue par flexions alternatives répétées
 nl heen-en-weerbuigproef *f(m)*
 r испытание *n* на выносливость [усталость] при знакопеременном изгибе

A368 *e* **alternating load, alternating loading**
 d Wechsellast *f*, wechselnde Belastung *f*
 f charge *f* alternée
 nl wisselende belasting *f*
 r знакопеременная нагрузка *f*

A369 *e* **alternating stress cycle**
 d schwingende Beanspruchung *f*, Wechsellastspiel *n*
 f cycle *m* des contraintes alternées
 nl periode *f* van de spanningswisseling
 r знакопеременный цикл *m* напряжений

A370 *e* **alternating stresses**
 d Wechselbeanspruchung *f*
 f contraintes *f pl* alternées
 nl wisselende spanning *f*
 r знакопеременные напряжения *n pl*

A371 *e* **altitude valve**
 d Wasserhochbehälter-Niveauregler *m*
 f régulateur *m* de niveau
 nl waterhoogteregelaar *m*
 r регулятор *m* уровня (в водонапорном баке)

A372 *e* **alumina**
 d Tonerde *f*, Alaunerde *f*
 f alumine *f*
 nl aluminiumoxyde *n*, aluinaarde *f(m)*
 r оксид *m* алюминия, глинозём *m*

A373 **alumina cement** *see* **aluminous cement**

A374 *e* **aluminium alloy plate**
 d Aluminiumgrobblech *n*
 f tôle *f* grosse d'aluminium
 nl aluminiumplaat *f(m)*
 r толстолистовой алюминий *m*

A375 *e* **aluminium alloys**
 d Aluminiumlegierungen *f pl*
 f alliages *m pl* d'aluminium
 nl aluminiumlegeringen *f pl*
 r алюминиевые сплавы *m pl*

A376 e **aluminium alloy sheet**
d Aluminiumfeinblech *n*
f tôle *f* mince d'aluminium
nl aluminiumblik *n*
r тонколистовой алюминий *m*

A377 e **aluminium-coated fabric**
d aluminiertes Stahlgewebe *n*
f grillage *m* en fils d'acier aluminés
nl gealumineerd weefsel *n*
r алюминированная [алитированная] стальная сетка *f*

A378 e **aluminium-coated sheet steel**
d dünnes aluminiertes Stahlblech *n*
f tôle *f* mince d'acier aluminée
nl gealuminiseerd staalblik *n*
r алюминированная тонколистовая сталь *f*, тонколистовая сталь *f*, плакированная алюминием

A379 e **aluminium foil**
d Aluminiumfolie *f*
f feuille *f* d'aluminium
nl aluminiumfolie *f(m)*
r алюминиевая фольга *f*

A380 **aluminium oxide** *see* **alumina**

A381 e **aluminium paint**
d Aluminiumfarbe *f*
f peinture *f* aluminium [à l'aluminium]
nl aluminiumverf *f(m)*
r алюминиевая краска *f*, краска *f*, пигментированная алюминиевой пудрой

A382 e **aluminium powder**
d Aluminiumpulver *n*
f poudre *f* d'aluminium, aluminium *m* en poudre
nl aluminiumpoeder *n*, *m*
r алюминиевая пудра *f* (*пигмент*)

A383 e **aluminium solar reflective coating**
d reflektierender Aluminiumfolie--Sonnenschutzbelag *m*
f revêtement *m* réflecteur en feuille d'aluminium
nl warmtereflecterende bekleding *f* met aluminiumfolie
r солнцеотражающая теплоизоляция *f* из алюминиевой фольги

A384 e **aluminium trim**
d Aluminiumprofil-Einrahmung *f*
f encadrement *m* en profilés d'aluminium
nl aluminium omranding *f*
r алюминиевое обрамление *n*

A385 e **aluminized coating**
d Aluminiumfarbe *f*
f peinture *f* à l'aluminium
nl aluminiumhoudende verf *f* (*m*)
r отражающее окрасочное покрытие *n* на алюминиевой пудре

A386 e **aluminizing**
d Aluminieren *n*
f aluminiage *m*, aluminure *f*
nl aluminiseren *n*
r алюминирование *n*, плакирование *n* алюминием

A387 e **aluminous cement**
d Tonerdezement *m*, Aluminozement *m*
f ciment *m* alumineux [fondu]
nl aluminiumcement *n, m*
r глинозёмистый цемент *m*

A388 e **aluminous fire brick**
d tonerdereicher Schamottestein *m*
f brique *f* d'alumine réfractaire
nl vuurvaste aluinaardesteen *m*
r глинозёмистый [алюминатный] огнеупорный кирпич *m*; шамотный кирпич *m*

aluminum... *see* **aluminium...**

A389 e **ambient conditions**
d Umweltbedingungen *f pl*, Umgebungsverhältnisse *n pl*
f conditions *f pl* d'ambiance
nl omgevingsvoorwaarden *f pl*
r окружающие условия *n pl*, условия *n pl* окружающей среды

A390 e **ambient noise**
d Umgebungsgeräusch *n*
f bruit *m* ambiant
nl omgevingsgeluid *n*
r фоновый шум *m*

A391 e **ambient temperature**
d Umgebungstemperatur *f*
f temperature *f* ambiante
nl omgevingstemperatuur *f*
r температура *f* окружающей среды; наружная температура *f*

A392 **Ambursen dam** *see* **flat-slab buttress dam**

A393 e **Ambursen dam deck slab**
d Stauplatte *f*
f dalle *f* plane du barrage Ambursen
nl stuwplaat *f(m)*
r напорная плита *f*

A394 e **amenities**
d 1. Wohnkomfort *m*, Bequemlichkeiten *f pl* 2. ingenieurtechnischer Ausbau *m*, ingenieurtechnische Erschließung *f* [Ausstattung *f*] des Baugeländes
f 1. confort *m* des intérieurs 2. aménagement *m* du territoire
nl 1. leefbaarheid *f* 2. situering *f* in het terrain
r 1. удобства *n pl* жилища 2. благоустройство *n* территории

A395 e **American basement**
d Sockelgeschoß *n*, Souterrain *n*
f sous-sol *m* habitable; rez-de-chaussée *m*
nl kelderverdieping *f*, souterrain *n*
r цокольный этаж *m*

A396 **American bond** *see* **common bond**

A397 **American caisson** *see* **box caisson**

A398 e **amorphous materials**
d amorphe Materialien *n pl*
f matériaux *m pl* [matières *f pl*] amorphes
nl amorfe materialen *n pl*
r аморфные материалы *m pl*

A399 *e* **amorphous structure**
 d amorphe Struktur *f*, amorphes Gefüge *n*
 f structure *f* amorphe
 nl amorfe structuur *f*
 r аморфная структура *f*, аморфное строение *n*

A400 *e* **amphibious shovel**
 d Wat(löffel)bagger *m*
 f excavateur *m* amphibie
 nl amfibische graafmachine *f*
 r одноковшовый экскаватор-амфибия *m*

A401 *e* **amphibious site**
 d amphibische Baustelle *f*, Land--Wasser-Baustelle *f*, Wasser-Land--Baustelle *f*
 f chantier *m* (de construction) amphibie
 nl bouwplaats *f(m)* aan het water
 r строительная площадка *f* на частично затопленной территории

A402 *e* **amplitude of ground motion**
 d Erdmassenschwingungsamplitude *f*
 f amplitude *f* du mouvement de la terre
 nl amplitude *f* van grondgolving
 r амплитуда *f* сейсмических колебаний грунта

A403 *e* **anaerobic digester**
 d Schlammfaulkammer *f*, Methankammer *f*
 f digesteur *m* de boues
 nl rottingskelder *m*
 r метантенк *m*

A404 *e* **analysis by successive approximations**
 d schrittweise Näherung *f*, Iterationsverfahren *n*
 f calcul *m* par (méthode des) approximations successives
 nl berekening *f* door opeenvolgende benaderingen
 r расчёт *m* методом последовательных приближений, итерационный метод *m* расчёта

A405 *e* **analysis of sections**
 d Dimensionierung *f*
 f dimensionnement *m* des sections, calcul *m* des dimensions de profilés
 nl doorsnedenkeuze *f* (*van staven*)
 r подбор *m* сечений (*ферм, рам*)

A407 *e* **analysis of the truss joints**
 d Knotenpunktmethode *f*, Rundschnittverfahren *n*
 f méthode *f* des nœuds
 nl knooppuntenberekening *f*, snedemethode *f*
 r метод *m* вырезания узлов

A408 *e* **analysis of trusses**
 d Binderberechnung *f*
 f calcul *m* des poutres en treillis
 nl berekening *f* van vakwerkspanten en -liggers
 r расчёт *m* ферм, определение *n* усилий в стержнях фермы

A409 *e* **anchor**
 d Anker *m*
 f ancre *m*
 nl anker *n*
 r анкер *m*, анкерное устройство *n*

A410 *e* **anchorage**
 d 1. Verankerung *f*; Festpunkt *m* 2. Ankerblock *m*; Ankerstütze *f*
 f 1. ancrage *m* 2. ancre *m*, ancrage *m*, dispositif *m* d'ancrage
 nl 1. verankering *f* 2. grondanker *n*
 r 1. анкеровка *f* 2. анкерное устройство *n*

A411 *e* **anchorage bond stress**
 d Haftspannung *f* in Verankerungszone der Bewehrung
 f contrainte *f* d'adhérence dans la zone d'ancrage
 nl hechtspanning *f* in verankeringsbereik van de wapening
 r напряжение *n* сцепления в зоне анкеровки арматуры

A412 *e* **anchorage by bond**
 d Haftkraftverankerung *f* der Spannbewehrung
 f ancrage *m* d'armature à béton par adhérence
 nl verankering *f* van de wapening door hechting aan het beton
 r анкеровка *f* за счёт сцепления арматуры с бетоном

A413 *e* **anchorage device**
 d Ankervorrichtung *f*
 f appareil *m* [dispositif *m*] d'ancrage
 nl verankeringsinrichting *f*
 r анкерное устройство *n*

A414 *e* **anchorage length**
 d Verankerungslänge *f*
 f longueur *f* d'ancrage
 nl verankeringslengte *f*
 r длина *f* зоны анкеровки

A415 *e* **anchorage of reinforcement**
 d Bewehrungsverankerung *f*, Verankerung *f* der Bewehrung
 f ancrage *m* d'armature
 nl verankering *f* van de wapening
 r анкеровка *f* арматуры

A416 *e* **anchorage pier**
 d Ankerpfeiler *m*
 f pile *f* d'ancrage
 nl ankerpijler *m*
 r анкерная опора *f* (*моста*)

A417 *e* **anchorage zone**
 d Verankerungszone *f*
 f zone *f* d'ancrage
 nl verankeringszone *f*
 r зона *f* анкеровки

A418 *e* **anchor bearing plate**
 d Ankerlagerplatte *f*
 f plaque *f* (d'appui et) d'ancrage
 nl verankerde steunplaat *f(m)*, ankerplaat *f(m)*
 r опорная анкерная плита *f*

A419 *e* **anchor block** *US*
  d 1. nagelfester Block *m* 2. Ankerblock *m*, Verankerungsblock *m*
  f 1. taquet *m* en bois (*p. ex. dans le béton*) 2. appui *m* d'ancrage; bloc *m* d'ancrage
  nl 1. spijkerklos *m* in metselwerk 2. ankerblok *n*, ankersteen *m*
  r 1. пробка *f* (*напр. в бетоне*) 2. анкерная опора *f*; анкерный блок *m*

A420 *e* **anchor bolt**
  d Ankerbolzen *m*, Verankerungsschraube *f*, Fundamentanker *m*
  f boulon *m* d'ancrage [de soutènement, de fondation]
  nl ankerbout *m*
  r анкерный [фундаментный] болт *m*

A421 *e* **anchor cone**
  d Verankerungskonus *m*, Ankerkonus *m*
  f cône *m* d'ancrage
  nl ankerkegel *m*
  r анкерный конус *m*

A422 **anchored bulkhead** *see* **anchored sheet pile wall**

A423 *e* **anchored pavement**
  d verankerte Betonfahrbahndecke *f*
  f revêtement *m* routier ancré [en dalles ancrées]
  nl verankerd betonwegdek *n*
  r дорожное покрытие *n* из заанкеренных бетонных плит

A424 *e* **anchored retaining wall**
  d Ankerstützwand *f*, Ankerstützmauer *f*
  f mur *m* de soutènement ancré
  nl verankerde keermuur *m*
  r заанкеренная подпорная стенка *f*

A425 *e* **anchored sheet pile wall**
  d verankerte Spundwand *f*, Bohlwerk *n*
  f rideau *m* de palplanches ancrées
  nl verankerde damwand *m*
  r заанкеренная шпунтовая стенка *f*

A426 *e* **anchored-type ceramic veneer**
  d Keramikverkleidung *f* mit Ankerbefestigung
  f revêtement *m* en carreaux céramiques ancrés
  nl keramische bekleding *f* met ankerbevestiging
  r керамическая облицовка *f* с анкерными креплениями

A427 *e* **anchor female cone**
  d äußerer Verankerungskonus *m*
  f cône *m* d'ancrage femelle
  nl uitwendige ankerkegel *m*
  r анкерная втулка *f* [колодка *f*]

A428 *e* **anchor ice**
  d Grundeis *n*
  f glace *f* de fond
  nl grondijs *n*
  r донный лёд *m*

A429 **anchoring** *see* **anchorage**

A430 *e* **anchor log**
  d Ankerpfahl *m*
  f poteau *m* d'ancrage
  nl ankerblok *n*
  r анкерный столб *m*, мертвяк *m*

A431 *e* **anchor male cone**
  d innerer Verankerungskonus *m*
  f cône *m* d'ancrage mâle
  nl inwendige ankerkegel *m*, ankerprop *f(m)*
  r анкерный вкладыш *m*, анкерная пробка *f*

A432 *e* **anchor nut**
  d Verankerungsmutter *f*
  f écrou *m* d'ancrage
  nl ankermoer *f(m)*
  r анкерная гайка *f*

A433 *e* **anchor pile**
  d Ankerpfahl *m*, Verankerungspfahl *m*
  f pieu *m* d'ancrage
  nl ankerpaal *m*
  r анкерная свая *f*

A434 *e* **anchor rod**
  d Verankerungsstange *f*, Zugstange *f*
  f barre *f* ancrée [d'ancrage]
  nl ankerstang *f(m)*, ankerstaaf *f(m)*
  r анкерный стержень *m*, заанкеренная подвеска *f*

A435 *e* **anchors-cramps and dowels**
  d Ankerklammern *f pl* und Dübel *m pl*
  f pièces *f pl* d'attache (*p.ex. ancres, crampons, goujons*) pour fixer des pierres de taille
  nl muurankers *m pl* en doken *f(m) pl*
  r металлические крепёжные детали *f pl* для облицовочных камней и плит (*напр. закрепы, штыри, скобы*)

A436 *e* **anchor tie**
  d 1. Ankerzugstange *f* 2. Fugenanker *m*
  f tige *f* d'ancrage
  nl ankertrekstang *f(m)*
  r 1. анкерный тяж *m*, анкерная тяга *f* 2. анкерная связь *f* каменной кладки

A437 *e* **anchor wall**
  d Ankerwand *f*, Ankervorlage *f*
  f mur *m* d'ancrage
  nl ankerwand *m*
  r анкерная стенка *f*

A438 *e* **ancillary building**
  d 1. Nebengebäude *n* 2. Anbau *m*
  f 1. bâtiment *m* auxiliaire 2. annexe *f*, bâtiment *m* [édifice *m*] annexe
  nl 1. hulpgebouw *n* 2. aanbouw *m*
  r 1. вспомогательное здание *n* 2. пристройка *f*

A440 *e* **angle**
  d 1. Winkel *m* 2. Winkelprofil *n*
  f 1. angle *m* 2. cornière *f*, équerre *f*
  nl 1. hoek *m* 2. hoekijzer *n*, hoekprofiel *n*
  r 1. угол *m* 2. уголковый профиль *m*, уголок *m*

A441 *e* **angle bead**
  d 1. Dichtungsraupe *f* (*für Eck-*

3*

verbindungen) 2. Eck(schutz)leiste
f, Kantenschützer *m*,
Kantenschutzschiene *f*
f 1. garniture *f* [bourrelet *m*]
d'étanchéité 2. éclisse *f* cornière
nl 1. hoeklas *f(m)* 2. hoeklijst *f(m)*
r 1. уплотнительный валик *m (для
угловых соединений)* 2. уголковая
накладка *f (для защиты рёбер
конструкций)*

A442 *e* angle brace
d 1. Kopfband *n*, Winkelband *n*;
Diagonalstab *m*, Strebe *f*; Bug *m*
2. Eckbohrmaschine *f*
f 1. entretoise *f* d'angle 2. perceuse *f*
d'angle
nl 1. draagband *m*, schouderband *m*,
hoekbeugel *m*, schoor *m*
2. hoekboor(machine) *m(f)*
r 1. угловая связь *f*; угловая схватка
f 2. угловая ручная сверлильная
машина *f*

A443 *e* angle bracket
d Absetzwinkel *m*, Stützwinkel *m*,
Aufsetzwinkel *m*
f console *f* angulaire, corbeau *m*
nl hoekkraagsteen *m*, hoekconsole *f*,
raamhoek *m*
r угловой кронштейн *m*

A444 *e* angle brick
d Schrägziegel *m*
f brique *f* biaise [chanfreinée]
nl schuine baksteen *m*
r кирпич *m* со скошенной боковой
гранью, клиновой кирпич *m*

A445 *e* angle buttress
d Eckpfeiler *m*
f contrefort *m* d'angle [du coin]
nl hoeksteunpijler *m*, hoekbeer *m*
r угловой контрфорс *m*

A446 *e* angle capital
d Ecksäulenkopf *m*
f chapiteau *m* d'angle
nl hoekkapiteel *n*
r угловая капитель *f*, капитель *f*
угловой колонны

A448 *e* angle cleat, angle clip
d Winkellasche *f*
f éclisse *f* cornière
nl hoekbeslag *n*
r уголковая накладка *f*

A449 *e* angle collar
d Muffenkrümmer *m*
f coude *f* à deux emboîtements
nl dubbele mof *f*
r отвод *m* с раструбами на обоих
концах

A450 *e* angle dozer, angledozer
d Seitenräumer *m*,
Schwenkschildplanierraupe *f*
f angledozer *m*, bouteur *m* biais
nl angledozer *m*
r англедозер *m*, бульдозер *m* с
поворотным отвалом

A451 *e* angle fillet
d Dreikantleiste *f*, Kehlleiste *f*,
Holzleiste *f*; Hohlkehle *f*
f baguette *f* d'angle
nl plint *f*; lijst *f(m)*
r галтель *f*; плинтус *m*; раскладка *f*

A452 *e* angle float
d Kantenlehre *f*, Eckenputzlehre *f*
f lisseuse *f* d'angle
nl hoektroffel *m*
r угловое правило *n*, угловой
полутёрок *m*

A453 angle hiptile *see* arris hip tile

A454 *e* angle iron
d Winkeleisen *n*, Winkelstahl *m*
f cornière *f* en acier fer *m* d'angle
nl hoekijzer *n*, hoekstaal *n*
r уголковая сталь *f*

A455 *e* angle joint
d Eckverbindung *f*, Eckstoß *m*
f joint *m* [assemblage *m*] d'angle
nl hoekverbinding *f*
r угловое соединение *n*

A456 *e* angle lacing
d Ausfachung *f* aus einzelnen
Winkelprofilen
f treillis *m* en cornières
nl spanstaven *f(m) pl* van hoekprofiel
r решётка *f* из одиночных уголков

A457 *e* angle-lighting luminaire
d Schrägstrahler *m*
f luminaire *m* à répartition oblique
nl scheefstraler *m* (*verlichting*)
r кососвет *m* (*светильник*)

A457a angle loads method *see* elastic weights
method

A458 *e* angle of external friction
d äußerer Reibungswinkel *m*
f angle *m* de frottement extérieur
nl uitwendige wrijvingshoek *m*
r угол *m* наружного трения

A459 *e* angle of fall
d Bruchwinkel *m*
f angle *m* d'éboulement
nl (terrein)hellingshoek *m*
r угол *m* обрушения

A460 *e* angle of friction
d Reibungswinkel *m*
f angle *m* de frottement
nl wrijvingshoek *m*
r угол *m* трения

A461 *e* angle of internal friction
d innerer Reibungswinkel *m*
f angle *m* de frottement interne
nl inwendige wrijvingshoek *m*
r угол *m* внутреннего трения

A462 *e* angle of repose
d 1. natürlicher Böschungswinkel *m*
2. Ruhewinkel *m*
f 1. angle *m* de talus naturel 2. angle
m d'équilibre limite
nl hellingshoek *m* van het natuurlijk
talud

r 1. угол *m* естественного откоса
   2. угол *m* предельного равновесия

A463 **angle of rest** *see* angle of repose 1.

A464 *e* **angle of rotation**
   d Drehwinkel *m*
   f angle *m* de rotation
  nl draaiingshoek *m*
   r угол *m* поворота

A465 *e* **angle of shear**
   d Scherwinkel *m*, Schubwinkel *m*
   f angle *m* de cisaillement
  nl schuifhoek *m*
   r угол *m* сдвига

A466 *e* **angle of slide**
   d Gleitwinkel *m*
   f angle *m* de glissement
  nl glijhoek *m*
   r угол *m* скольжения

A467 *e* **angle of slope**
   d Böschungswinkel *m*
   f angle *m* de talus
  nl hellingshoek *m*
   r угол *m* откоса

A468 *e* **angle of torsion**
   d Torsionswinkel *m*
   f angle *m* de torsion
  nl torsiehoek *m*
   r угол *m* закручивания

A469 *e* **angle post**
   d Eckpfosten *m*, Eckständer *m*
   f poteau *m* [montant *m*] d'angle
  nl hoekstijl *f(m)*
   r угловая стойка *f*

A470 *e* **angle rafter**
   d Anfallsparren *m*, Dachschifter *m*
   f arêtier *m*
  nl hoekspantbeen *n*, hoekspar *m*
   r угловая стропильная нога *f*

A471 *e* **angle section**
   d Winkelprofil *n*
   f cornière *f*, profilé *m* en L
  nl hoekprofiel *n*
   r уголковый профиль *m*, уголок *m*

A472 *e* **angle tee**
   d 45°-Abzweig *m*
   f embranchement *m* [culotte *f*] simple
  nl T-stuk *n* met spruit onder
    45°-aftakking
   r косой тройник *m*

A473 *e* **angle valve**
   d Eckventil *n*
   f robinet *m* d'équerre [d'angle]
  nl hoekklep *f(m)*
   r угловой вентиль *m*

A474 *e* **angular aggregate**
   d kantiger Zuschlag(stoff) *m*
   f agrégat *m* [granulat *m*] anguleux
  nl toeslag *m* met scherpe kanten
   r неокатанный заполнитель *m*,
    заполнитель *m* угловатой формы

A475 *e* **angular displacement**
   d Winkelverrückung *f*,
    Winkelverschiebung *f*

   f déplacement *m* [translation *f*]
    angulaire
  nl hoekverplaatsing *f*
   r угловое перемещение *n*, угловая
    деформация *f*; угол *m* поворота
    (сечения)

A476 *e* **angular restraint**
   d wink(e)lige Einspannung *f*
   f encastrement *m* contre rotation
  nl inklemming *f* tegen draaiing
   r защемление *n* (*опоры балки*) против
    углового смещения

A477 *e* **angular retaining wall**
   d Winkelstützmauer *f*
   f mur *m* de soutènement cantilever
  nl vleugelmuur *m*
   r подпорная стенка *f* с консольной
    подошвой, стенка *f* уголкового типа

A478 *e* **anhydrite, anhydrous gypsum**
   d Anhydrit *m*, wasserfreier Gips(stein)
    m
   f anhydrite *m*
  nl anhydriet *n*
   r ангидрит *m*, безводный гипс *m*

A480 *e* **animal glue**
   d Glutinleim *m*, Tierleim *m*
   f colle *f* animale
  nl dierlijke lijm *m*
   r животный (столярный) клей *m*

A481 *e* **anion bed**
   d Anionenbett *n*, Anionenfüllkörper *m*
   f lit *m* anionique
  nl anionenafzetting *f*
   r анионообменная насадка *f*
    [загрузка *f*]

A482 *e* **anionic emulsion**
   d anionaktive Emulsion *f*
   f émulsion *f* anionique
  nl anionactieve emulsie *f*
   r анионная эмульсия *f* (*дорожный*
    материал)

A483 **annexe** *see* ancillary building

A484 **announcement of bidding** *see*
    advertisement for bids

A485 *e* **annual flood**
   d Jahreshochwasser *n*
   f crue *f* annuelle
  nl jaarlijkse hoogwaterlijn *f(m)*
   r максимальный годовой паводок *m*

A486 *e* **annual range of temperature**
   d Jahresamplitude *f* der Temperatur
   f amplitude *f* de fluctuations annuelles
    de température
  nl jaarlijkse temperatuurschommeling *f*
   r годовые колебания *n pl* температуры

A487 *e* **annual runoff**
   d jährlicher Abfluß *m*
   f écoulement *m* annuel
  nl jaarlijkse waterafloop *m*
   r годовой сток *m*

A488 *e* **annual storage reservoir**
   d Jahresspeicher *m*
   f réservoir *m* annuel

nl waterreservoir *n* voor jaarlijkse opslag
r водохранилище *n* годового регулирования

A489 *e* **annular girder**
d Kreisringträger *m*
f poutre *f* annulaire
nl ringvormige balk *m*
r кольцевая балка *f*

A490 *e* **annular ringed nail**
d Kreisringnagel *m*
f clou *m* à crampon [en forme d'U]
nl ringvormige nagel *m*
r гвоздь-скоба *m*

A491 *e* **annular vault**
d zylindrisches Gewölbe *n*, Tonnengewölbe *n*
f voûte *f* cylindrique
nl cilindrisch tongewelf *n*
r цилиндрический свод *m*

A492 *e* **anodic coating**
d Eloxalschicht *f*
f revêtement *m* [enduit *m*] anodique
nl geanodiseerde laag *f(m)*
r анодированное покрытие *n*

A493 *e* **anodic protection**
d Anodenschutz *m*
f protection *f* anodique
nl anodische bescherming *f*
r анодная защита *f*

A494 *e* **anodizing**
d Anodisieren *n*, Eloxieren *n*
f anodisation *f*
nl anodiseren *n*, eloxeren *n*
r анодирование *n*

A495 *e* **anta**
d Eckpilaster *m*
f ante *f*, pilastre *m* d'angle
nl hoekpilaster *m*
r угловая пилястра *f*

A496 *e* **antechamber**
d Vorzimmer *n*; Vorraum *m*
f antichambre *f*
nl hal *f(m)*
r прихожая *f*, передняя *f*; вестибюль *m*

A497 **antenna mast** *see* **aerial mast**

A498 **anteroom** *see* **antechamber**

A499 *e* **anticorrosive coating**
d Korrosionsschutzanstrich *m*
f revêtement *m* anticorrosif
nl anticorrosieve (dek)laag *f(m)*
r антикоррозионное покрытие *n*

A501 **anti-drumming coating** *see* **sound-
-deadening coating**

A502 *e* **anti-foaming agent**
d Antischaummittel *n*, Entschäumer *m*, Schaumdämpfer *m*
f agent *m* antimousse
nl antischuimmiddel *n*
r противовспенивающая [противопенная] добавка *f*

A503 *e* **antifouling paint**
d anwuchsverhindernder Anstrich *m,* Unterwasser-Schutzanstrich *m*
f peinture *f* antiputride
nl antiseptische verf *f(m)*
r противогнилостная краска *f*

A504 *e* **anti-freezing admixture**
d Frostschutzmittel *n*
f antigel *m*, agent *m* antigel, dégivreur *m*
nl antivriesmiddel *n*
r противоморозная добавка *f,* антифриз *m*

A506 *e* **antireflective coating**
d nichtreflektierende Deckschicht *f*
f revêtement *m* antiréflecteur
nl niet-reflecterende (dek)laag *f(m)*
r поглощающее покрытие *n*, поглощающая изоляция *f*

A507 *e* **antiseepage cofferdam**
d Sickerschutzfangdamm *m*
f bâtardeau *m* étanche
nl kistdam *m*, koffer(dam) *m* tegen kwel
r противофильтрационная перемыч- ка *f*

A508 *e* **anti-skinning agent**
d Hautverhinderungsmittel *n*, Hautverhütungsmittel *n*
f agent *m* prévenant la formation de la pellicule
nl toevoeging *f* die velvorming tegengaat
r вещество *n*, предотвращающее образование плёнки (*при хранении краски в таре*)

A509 *e* **antislip paint**
d Gleitschutzfarbe *f*, rutschsichere Farbe *f*
f peinture *f* antidérapant
nl antislipverf *f(m)*
r краска *f*, предупреждающая скольжение

A510 *e* **antistatic agent**
d antistatisches Mittel *n*
f agent *m* antistatique
nl antistatische toeslag *m*
r антистатическая добавка *f*

A511 *e* **antisymmetrical load**
d antisymmetrische Belastung *f*
f charge *f* antisymétrique
nl antisymmetrische belasting *f*
r обратносимметричная нагрузка *f*

A512 *e* **anti-vibration mounting**
d schwingungsdämpfendes Maschinen- fundament
f fondation *f* anti-vibratile
nl trillingsdempende opstelling *f*
r виброизолирующее основание *n*

A513 *e* **anvil**
d 1. Rammhaube *f*, Schlaghaube *f* 2. Amboß *m*
f 1. casque *m* de pieu 2. enclume *f*
nl 1. paalmuts *f(m)* 2. aambeeld *n*
r 1. свайный наголовник *m* 2. наковальня *f*

A514 *e* **apartment**
 d Wohnung *f*
 f appartement *m,* logement *m*
 nl woning *f*
 r квартира *f*

A515 *e* **apartment block**
 d Mehrfamilienhaus *n*
 f immeuble *m* collectif
 nl flat(gebouw) *m* (*n*)
 r многоквартирный жилой дом *m*

A516 *e* **apartment hotel**
 d Apartementhaus *n*, Mittelganghaus *n*
 f immeuble *m* type hôtel
 nl apartementhotel *n*
 r дом *m* гостиничного типа

A517 **apartment house** *see* **apartment block**

A518 *e* **aperture**
 d Wandöffnung *f*; Balken(träger)tasche
 f, Balken(träger)kammer *f*
 f baie *f*; ouverture *f*, ope *m*
 nl opening *f*; balkgat *n* (*in de muur*)
 r проём *m* (*в стене*); гнездо *n* (*в стене для укладки балки*)

A519 *e* **apex**
 d Scheitel(punkt) *m*; Anfallpunkt *m*
 f sommet *m*; pointe *f*; crête *f*; faîte *m*
 nl top *m*; dakvorst *m*
 r вершина *f*; гребень *m*; конёк *m*

A520 *e* **apex block, apex stone**
 d Bogenschlußstein *m*,
 Bogenscheitelstein *m*
 f clef *f* de voûte, pierre *f* de clé [de clef]
 nl sluitsteen *m*
 r замковый [ключевой] камень *m* (*арки, свода*)

A522 *e* **apparent cohesion**
 d scheinbare Kohäsion *f* [Haftfestigkeit *f*]
 f cohésion *f* apparente
 nl schijnbare cohesie *f*
 r кажущееся сцепление *n*

A523 **apparent density** *see* **bulk density**

A524 *e* **apparent elastic limit**
 d technische Elastizitätsgrenze *f*
 f limite *f* conventionnelle d'élasticité
 nl technische elasticiteitsgrens *f(m)*
 r условный предел *m* упругости

A525 *e* **apparent elastic range**
 d scheinbarer elastischer Bereich *m*
 f domaine *m* élastique apparent
 nl schijnbare elasticiteitszone *f*
 r кажущаяся упругая зона *f* [область *f*]

A526 *e* **application consistency**
 d Verarbeitungskonsistenz *f*, Verarbeitungssteife *f* (*Beton*)
 f consistance *f* du béton mis en place
 nl verwerkingsconsistentie *f* (*beton*); lijvigheid *f* (*verf*)
 r подвижность *f* бетонной смеси в момент укладки

A527 *e* **application for payment**
 d Zwischenrechnung *f* für Abschlagszahlung
 f décomte *m* provisoire des travaux exécutés
 nl tussenrekening *f* voor betaling in gedeelten
 r счёт *m* на оплату части работ, выполненных подрядчиком

A528 *e* **application life**
 d Gebrauchsdauer *f*
 f durée *f* d'utilisation
 nl levensduur *f* (*mengsels*)
 r жизнеспособность *f* (*составов, смесей*)

A529 *e* **application of roofing felt**
 d Aufkleben *n* vom Ruberoid
 f collage *m* du carton bitumé (*sur toiture*)
 nl dakvilt aanbrengen *n*
 r наклейка *f* рубероида

A530 *e* **application techniques**
 d Auftragungsverfahren *n* pl, Aufbringungsmethoden *f* pl
 f techniques *f* pl d'application
 nl opbrengmethoden *f* pl (*verf e. d.*)
 r приёмы *m* pl нанесения (*штукатурки, краски, торкрет-бетона*)

A531 **applied column** *see* **attached column**

A533 *e* **applied hydrology**
 d angewandte Hydrologie *f*
 f hydrologie *f* appliquée
 nl toegepaste hydrologie *f*
 r прикладная гидрология *f*

A534 *e* **applied load**
 d eingetragene Belastung *f*
 f charge *f* appliquée
 nl aangebrachte belasting *f*
 r приложенная нагрузка *f*

A535 *e* **applied moment**
 d Angriffsmoment *n*
 f moment *m* appliqué [sollicitant]
 nl aangrijpingsmoment *n*
 r приложенный момент *m*, моментная нагрузка *f*

A536 *e* **approach**
 d Zugang *m*, Zutritt *m*, Zufahrt *f*, Einfahrt *f*
 f accès *m*, approche *f*
 nl toegang *m*, nadering *f*
 r подход *m*, подъезд *m*

A537 *e* **approach channel**
 d Zufahrtrinne *f*; Zugangskanal *m*
 f canal *m* d'amenée [d'accès]
 nl toegangskanaal *n*
 r подходной фарватер *m* [канал *m*]

A538 *e* **approach road**
 d Zufahrtsstraße *f*
 f route *f* [chemin *m*] d'accès
 nl toegangsweg *m*
 r подъездная дорога *f*

A539 *e* **approach span**
 d Landfeld *n*, Landöffnung *f*
 f travée *f* d'approche

nl aanbrug *f(m)*
r береговой пролёт *m*

A540 *e* **approach velocity**
d Zufahrtrinnengeschwindigkeit *f* von Wasser
f vitesse *f* d'approche
nl stroomsnelheid *f* in het toevoerkanaal
r средняя скорость *f* воды в подходном фарватере

A541 *e* **approximate analysis**
d Näherungsberechnung *f*, Überschlagsrechnung *f*
f calcul *m* approximatif
nl benaderende [voorlopige] berekening *f*
r приближённый расчёт *m*

A542 *e* **appurtenances**
d Ausbauteile *m pl*, Zubehörteile *m pl*
f auxiliaires *m pl*, accessoires *m pl*, appartenances *f pl*
nl toebehoren *pl*
r вспомогательные части *f pl* и оборудование *n* (*зданий, сооружений*)

A543 *e* **apron**
d 1. Betonvorlage *f* (*z. B. vorm Hangar*) 2. Schürze *f*, Vorlage *f* 3. Flutbett *n* und dazugehörige Teile *n pl* (*Vorboden m, Sturzboden m, Sturzbett n, Rißberme f*) 4. Abweiserblech *n* 5. Fensterbrüstung *f* 6. Waschbeckenschürze *f*
f 1. aire *f* bétonnée (*p.ex. devant les hangars*) 2. partie *f* sous-marine du revêtement des rives 3. radier *m* 4. bavette *f* 5. allège *f* 6. panneau *m* de cloison derrière un lavabo ou un évier
nl 1. betonnen platform *n* (*schuur*) 2. kraag *m* (*oeverbekleding*) 3. storte bed *n* 4. loodloket *n* 5. borstwering *f* 6. spatscherm *n* (*sanitair*)
r 1. бетонная площадка *f* (*напр. перед ангаром*) 2. подводная часть *f* береговой одежды 3. флютбет *m*; понур *m*; водобой *m*; рисберма *f* 4. защитный кровельный фартук *m*, отлив *m* из кровельной стали 5. подоконная стенка *f* 6. панель *f* позади умывальника *или* мойки

A544 *e* **apron-conveyor feeder**
d Aufgabeplattenband *n*, Plattenbandspeiser *m*, Gliederbandbeschicker *m*
f alimentateur *m* [distributeur *m*] à palettes
nl toevoer-platformtransporteur *m*
r пластинчатый питатель *m*

A545 **apron flashing** *see* **apron 4.**

A546 **apron wall** *see* **apron 5.**

A547 *e* **aptitude test**
d Eignungsprüfung *f*
f épreuve *f* [vérification *f*] de l'aptitude pour l'emploi
nl geschikheidstest *m*

r квалификацинные испытания *n pl*, квалификационная проверка *f* (*персонала*)

A548 *e* **aquatorium**
d Aquatorium *n*, Hafenwasserfläche *f*
f plan *m* d'eau, bassin *m* intérieur
nl havenbekken *n*
r акватория *f* (*порта*)

A549 *e* **aqueduct**
d Aquädukt *m*, Wasserleitung *f*
f aqueduc *m*, pont-canal *m*
nl aquaduct *n*, bovenkruisend kanaal *n*
r акведук *m*, водовод *m*

A550 *e* **aqueduct trough**
d Trog *m* des Aquädukts
f canal *m* d'aqueduc
nl aquaducttrog *m*
r несущий лоток *m* акведука

A551 *e* **aqueous wood preservative**
d wäßriges Holzschutzmittel *n*
f produit *m* antiseptique soluble dans l'eau
nl in water oplosbaar houtconserveringsmiddel *n*
r водорастворимый антисептик *m*

A552 *e* **aquiclude**
d verzögernde Gesteinsschicht *f*
f aquiclude *m*, formation *f* à microporosité
nl (doorzijpeling) vertragende gesteentelaag *f(m)*
r водоупор-водоуловитель *m* (*пористая формация, медленно адсорбирующая воду, но не обеспечивающая питание колодцев или источников*)

A553 *e* **aquifer**
d Grundwasserleiter *m*, Aquifer *m*
f nappe *f* [couche *f*] aquifère
nl (grond)watervoerende laag *f(m)*
r водоносный горизонт *m* [слой *m*, пласт *m*]

A554 *e* **aquifuge**
d wasserundurchlässige Gesteinsschicht *f*
f formation *f* imperméable
nl waterondoorlatende gesteentelaag *f(m)*
r водоупор *m*, образованный скальной породой

A555 *e* **arc**
d Bogen *m* (*Kurve*)
f arc *m*
nl boog *m*
r дуга *f*

A556 *e* **arcade**
d Arkade *f*, Bogengang *m*, Reihengewölbe *n*
f arcade *f*
nl arcade *f*, bogengaanderij *f*
r аркада *f*, сводчатая открытая галерея *f*

A557 *e* **arc-boutant**
d Strebebogen *m*, Schwibbogen *m*
f arc-boutant *m*

nl schoorboog *m*, schraagboog *m*
r аркбутан *m*

A558 *e* arc cutting
d Elektroschneiden *n*,
Lichtbogenbrennschneiden *n*
f coupage *m* à l'arc
nl snijden *n* met vlamboog
r электродуговая резка *f*

A559 *e* arch
d Bogen *m*
f arc *m*, arceau *m*
nl boog *m*, booglatei *f(m)*, toog *m*,
gewelf *n*
r арка *f*

A560 arch apex *see* arch key

A561 arch beam *see* arched beam

A562 *e* arch brick
d Keilziegel *m*, Bogenziegel *m*
f brique *f* d'arc [de voûte], claveau *m*
nl wigsteen *m*, gewelfsteen *m*
r клинчатый [клиновой] кирпич *m*

A563 *e* arch bridge
d Bogenbrücke *f*
f pont *m* en arc [arqué]
nl boogbrug *f(m)*
r арочный мост *m*

A564 arch buttress *see* arc-boutant

A565 *e* arch center, arch centre
d Bogenlehrgerüst *n*, Bogenlehre *f*,
Wölbgerüst *n*
f cintre *m*
nl boogformeel *n*, porringpunt *n*
r кружало *n*

A566 arch crown *see* arch key

A567 *e* arch dam
d Bogenstaumauer *f*, gewölbte Mauer *f*
f barrage-voûte *m*
nl (gebogen) stuwdam *m*
r арочная плотина *f*

A568 arched barrel roof *see* barrel roof

A569 *e* arched beam
d Bogenbalken *m*, Bogenträger *m*
f poutre *f* à membrure supérieure
bombée
nl boogligger *m*
r балка *f* с выгнутым верхним поясом

A570 *e* arched retaining wall
d Gewölbestützmauer *f*
f mur *m* de soutènement en arc
nl gewelfde keermuur *m*
r арочная подпорная стенка *f*

A571 *e* arched spillway
d Bogenüberlauf *m*
f déversoir *m* avec crête arrondie
nl gebogen overloop *m* (*afvoer*)
r криволинейный водослив *m*

A572 *e* arch-gravity dam
d Bogengewichtssperre *f*,
Bogengewichtsstaumauer *f*
f barrage-poids *m* à voûte
nl zware gebogen stuwdam *m*
r арочно-гравитационная плотина *f*

A573 *e* arching
d Gewölbebildung *f*, Gewölbewirkung *f*
f effet *m* de voûte [d'arc]
nl brugvorming *f* (*stortgoed in silo*)
r сводообразование *n*, образование *n*
свода (*в грунте, сыпучем материале*)

A574 *e* architect
d Architekt *m*
f architecte *m*
nl 1. architect *m* 2. bouwkundige *m*
r 1. архитектор *m* 2. фирма *f* или
лицо *n*, разрабатывающие проект
и/или ведущие контроль за
строительством

A575 *e* architect's approval
d Architektenzustimmung *f*
f approbation *f* par [de] l'architecte
nl aanbeveling *f* van de architect
r утверждение *n* архитектором
(*проектных решений, выбора материалов, методов производства работ*)

A576 *e* architectural acoustics
d Bau- und Raumakustik *f*
f acoustique *f* architectural
nl bouwakoestiek *f*
r архитектурная акустика

A577 *e* architectural coating
d Zierverkleidung *f*
f revêtement *m* architectural
[ornemental]
nl sierbekleding *f*
r архитектурная облицовка *f*

A578 *e* architectural concept
d architektonische Gestaltung *f*,
Raumgestaltung *f*
f conception *f* architecturale, parti *m*
architectural
nl architectonisch concept *n*
r архитектурное решение *n*

A579 *e* architectural concrete
d Architekturbeton *m*
f béton *m* architectural
[architectonique]
nl sierbeton *n*
r декоративный бетон *m*

A580 *e* architectural constructional materials
d Architekturbaustoffe *m pl*
f matériaux *m pl* architectoniques
nl afwerkingsmaterialen *n pl*
r материалы *m pl* для архитектурной
отделки зданий

A581 *e* architectural design
d 1. architektonische Planung *f*
2. architektonischer Entwurf *m*
f 1. élaboration *f* du projet architectural
2. projet *m* architectural, conception
f architecturale
nl bouwkundig ontwerp *n*
r 1. архитектурное проектирование *n*
2. архитектурный проект *m*

A582 *e* architectural drawing
d Architekturzeichnung *f*

f dessin *m* architectural [d'architecture]
nl bouwkundige tekening *f*
r архитектурный чертёж *m*

A583 *e* **architectural engineering**
d Baudurchformung *f*; Baugestaltung *f*;
bautechnische Projektierung *f*
f élaboration *f* des projets de
construction
nl bouwkundig ontwerpen *n*
r проектирование *n* строительных
конструкций; разработка *f*
конструктивных систем зданий;
разработка *f* строительной части
проекта

A584 *e* **architectural finish(ing)**
d Oberflächengestaltung *f*
f fini *m* architectural
nl bouwkundige afwerking *f*
r архитектурная отделка *f*

A585 **architectural hardware** *see* **builder's finish hardware**

A586 *e* **architectural planning**
d Architekturplanung *f*
f aménagement *m* architectural
nl een bouwplan maken
r разработка *f* архитектурно-
-планировочных решений;
планировка *f* зданий

A587 *e* **architectural surface**
d Zieroberfläche *f*, Ornamentoberfläche *f*
f surface *f* décorative
nl oppervlakte *f* met artistieke afwerking
r поверхность *f* с декоративной
отделкой

A588 *e* **architectural terra-cotta**
d Architekturterrakotta *f*,
Bauterrakotta *f*
f terra-cotta *f* architecturale, terre *f*
cuite architecturale [ornementale]
nl bouwterracotta *n*
r архитектурная терракота *f*

A589 *e* **architectural volume**
d umbauter Raum *m*, Rauminhalt *m*
des Gebäudes
f volume *m* construit [bâti]
nl bouwvolume *n*
r кубатура *f* [строительный объём *m*]
здания

A590 *e* **architrave**
d 1. Architrav *m* 2. Fassung *f*,
Türverkleidung *f*; Fensterverkleidung *f*
f 1. architrave *f* 2. chambranle *m*
nl 1. architraaf *f(m)* 2. dekbalk *m* van
een kozijn
r 1. архитрав *m* 2. наличник *m*

A591 *e* **arch key**
d Bogenscheitel(punkt) *m*
f clé *f* [clef *f*] d'un arc
nl gewelfkruin *f(m)*, top *m* van een boog
r замок *m* арки

A592 *e* **arch thrust**
d Bogenschub *m*

f poussée *f* de l'arc
nl horizontale oplegkracht *f(m)* van een
boog
r распор *m* арки

A593 *e* **arch truss**
d Bogenbinder *m*, Fachwerkbogen *m*
f poutre *f* (à treillis) en arc
nl boogligger *m*
r арочная ферма *f*

A594 *e* **archway**
d Bogenfeld *n*; überwölbter Durchgang
m
f baie *f* en arc; passage *m* voûté
nl boogvormige opening *f*; doorgang *m*
onder de boog
r арочный проём *m*; проход *m* под
аркой

A595 *e* **arc welding**
d Lichtbogenschweißen *n*
f soudage *m* à l'arc (électrique)
nl (vlam)booglassen *n*
r электродуговая сварка *f*

A596 *e* **area**
d 1. Fläche *f*; Flächeninhalt *m*,
Flächenraum *m*; Oberfläche *f*
2. Platz *m*, Stelle *f* 3. Innenhof *m*
4. Zone *f*
f 1. aire *f*, surface *f* 2. terrain *m* 3. cour
f intérieure 4. zone *f*
nl 1. oppervlakte *f* 2. terrein *n*,
gebied *n* 3. binnenplaats *f* 4. zone *f*
r 1. площадь *f*; поверхность *f*
2. площадка *f* 3. внутренний двор
m 4. зона *f*

A597 *e* **area light**
d Lichtöffnung *f*
f lumière *f*, baie *f* d'admission de la
lumière
nl lichtopening *f*
r световой проём *m*

A598 *e* **area method**
d Veranschlagung *f* nach
Flächeneinheitspreisen
f élaboration *f* du devis estimatif basée
sur le coût de construction d'une
unité de la surface bâtie (*d'un
bâtiment analogue*)
nl begroting *f* op basis van de grondprijs
r разработка *f* [составление *n*] сметы
по стоимости единицы площади
здания, сметно-финансовый расчёт
m по укрупнённым показателям (*по
стоимости единицы площади*)

A599 *e* **area moment**
d Flächenmoment *n*
f moment *m* (statique) d'une surface
nl oppervlaktemoment *n*
r (статический) момент *m* площади

A600 *e* **area-moment method**
d Kraftmethode *f*, Kraftverfahren *n*
f méthode *f* des forces
nl oppervlaktemomentenberekening *f*
r метод *m* сил

A601 *e* **area of building**
d bebaute Fläche *f*
f surface *f* bâtie
nl bouwplaats *f(m)*
r площадь *f* застройки

A602 *e* **area of influence**
d Absenkungsbereich *m*,
Absenkungsfläche *f*
f zone *f* d'appel
nl invloedsvlak *n*
r зона *f* влияния
(*скважины, колодца*)

A603 *e* **area of reinforcing steel**
d Bewehrungsquerschnitt *m*,
Armierungsquerschnitt *m*
f aire *f* de la section des armatures
nl doorsnede *f(m)* van de wapening
r площадь *f* поперечного сечения
арматуры

A604 *e* **area of waterway**
d Abflußquerschnitt *m*,
Durchflußquerschnitt *m*,
Ablaufquerschnitt *m*,
durchströmter Querschnitt *m*
f section *f* mouillée
nl doorstromingsprofiel *n*
r живое сечение *n* водотока

A605 *e* **area ratio of a sampler**
d Flächenverhältnis *n* der
Probeentnahmestange
f rapport *m* de surface d'un
échantillonneur
nl oppervlakteverhouding *f* van een
monster
r отношение *n* площадей стенок
грунтоноса и образца (*при оценке
качества грунтоноса*)

A606 *e* **area-volume curve**
d Inhaltslinie *f*
f courbe *f* surface-volume
nl functiekromme *f* van oppervlakte en
inhoud
r объёмная кривая *f* (*кривая зависимости объёма водохранилища от площади водной поверхности*)

A607 *e* **area wall**
d Wand *f* des Kellerlichtschachtes
f mur *m* (de soutènement) de la **cour**
anglaise
nl wand *m* van de lichtschacht
r подпорная стенка *f* приямка у подвального окна

A608 *e* **areaway**
d 1. Passage *f*, Durchgang *m*
2. Kellergraben *m*, Kellervorhof *m*
f 1. passage *m* 2. cour *f* anglaise
nl 1. doorgang *m*; doorrit *m*
2. keldertoegang *m*
r 1. проход *m*; проезд *m* 2. приямок *m*
у подвального окна

A609 *e* **argillaceous limestone**
d Tonkalk(stein) *m*
f calcaire *m* argileux
nl leemkalksteen *n*, *m*

r глинистый известняк *m*

A610 *e* **argillaceous rocks**
d Tongesteine *n pl*
f roches *f pl* argileuses
nl harde kleihoudende gesteentes *n pl*
r глинистые породы *f pl*

A611 *e* **argillite**
d Argyllit(h) *m*
f argil(l)ite *f*, argillyte *f*
nl argilliet *n*, kleilei *n*
r аргиллит *m*

A612 *e* **argillization**
d Toninjektion *f*
f injection *f* de l'argile, argilisation *f*
nl leeminjectie *f*
r глинизация *f* (*напр. русла канала*)

A613 *e* **argon-arc welding**
d Argonarc-Verfahren *n*, Wolfram-
-Inertgas-Schweißen *n*, WIG-
-Schweißen *n*
f soudage *m* à l'arc dans l'argon
nl argonbooglassen *n*
r аргоно-дуговая сварка *f*

A614 *e* **armoured cable**
d armiertes [bewehrtes] Kabel *n*,
Panzerkabel *n*
f câble *m* (électrique) armé
nl gepantserde kabel *m*
r бронированный (электро)кабель *m*

A615 *e* **armoured glass**
d Drahtglas *n*
f verre *m* armé
nl draadglas *n*
r армированное стекло *n*

A616 *e* **armoured plywood**
d gepanzertes Sperrholz *n*
f contre-plaqué *m* armé
nl gepantserd multiplex *n*
r фанера *f* с металлической обшивкой

A617 *e* **armoured wood**
d blechummanteltes Holz *n*
f bois *m* armé [renforcé] par tôles
métalliques
nl gewapend hout *n*
r лесоматериал *m*, усиленный
металлической обшивкой

A618 *e* **armour hose**
d Panzerschlauch *m*, armierter Schlauch
m
f tuyau *m* flexible armé, manche *f*
armée
nl pantserslang *f(m)*
r армированный шланг *m*

A619 *e* **armour plate**
d 1. Panzerplatte *f*, Panzerblech *n*
2. Schutzblech *n*, Stoßplatte *f*
f 1. plaque *f* de blindage 2. plaque *f*
de protection (*au bas d'une porte*)
nl 1. pantserplaat *f(m)* 2. schopplaat *f(m)*
r 1. броневая плита *f* 2. защитный
(металлический) лист *m* (*двери, ворот*)

A620 *e* **arrangement**
d 1. Anordnung *f*, Aufstellung *f* 2. Plan

m, Schema *n* 3. Einrichtung *f*,
Vorrichtung *f*
f 1. arrangement *m*; disposition *f*;
emplacement *m* 2. plan *m*; schéma *m*
3. dispositif *m*
nl 1. plaatsing *f* 2. (plaatsings)plan *n*
3. inrichting *f*
r 1. расположение *n*; расстановка *f*,
размещение *n* 2. план *m*; схема *f*
3. устройство *n*

A621 *e* **arrangement of beams**
  d Balkenanordnung *f*
  f disposition *f* [arrangement *m*] des
poutres
  nl plaatsing *f* van balken
  r расположение *n* балок

A622 *e* **arrangement of reinforcement**
  d Bewehrungsanordnung *f*, Anordnung *f*
der Stahleinlagen,
Bewehrungsverteilung *f*
  f disposition *f* de l'armature [de
ferraillage]
  nl plaatsing *f* van wapening
  r расположение *n* арматуры

A623 *e* **arrester**
  d 1. Funkenfänger *m*, Funkenlöscher *m*
2. Auffangstange *f*,
Blitzauffanganlage *f*
  f 1. étouffoir *m*, pare-étincelles *m*
2. éclateur *m*, paratonnerre *m*,
parafoudre *m*
  nl 1. vonkvanger *m* 2. bliksemafleider *m*
  r 1. искроуловитель *m* 2.
молниеприёмник *m*; разрядник *m*

A624 *e* **arrière-voussure**
  d Innenbogen *m*, Innengewölbe *n*
  f arrière-voussure *f*
  nl venstertoog *m*; deurtoog *m*
  r внутренняя арка *f*, внутренний свод
m (*в толще стены*)

A625 *e* **arris**
  d Rippe *f*; Grat *m*, Gratlinie *f*,
(scharfe) Kante *f*; Außenwinkel *m*
  f arête *f* (vive); angle *m* saillant
  nl ribbe *f(m)*; uitwendige hoek *m*
  r ребро *n*; наружный угловой выступ
m; внешний угол *m*

A626 *e* **arris hip tile**
  d Firstziegel *m*
  f tuile *f* arêtière [faîtière]
  nl nokpan *f(m)*, keperpan *f(m)*
  r коньковая черепица *f*

A627 *e* **arris rail**
  d Dreikantleiste *f*
  f latte *f* (à section) triangulaire
  nl driekantige lijst *f(m)*
  r рейка *f* [раскладка *f*] треугольного
сечения

A628 *e* **arrissing tool**
  d Kantenlehre *f*, Eckenputzlehre *f*
  f lisseuse *f* en cornière
  nl hoekpleistermal *m*
  r лузговое *или* усёночное правѝло *n*

A629 *e* **arrow diagram**
  d Netzdiagramm *n*
  f diagramme *m* à flèches, graphe *m*
par chemin critique
  nl planningsnetwerk *n*
  r сетевой график *m*

A630 *e* **arterial drain**
  d Hauptsammler *m*, Hauptdrän *m*,
Sammeldrän *m*
  f collecteur *m* principal
  nl hoofdriool *n*, verzamelriool *n*
  r главный магистральный коллектор
m

A631 *e* **arterial highway**
  d Hauptverkehrsstraße *f*,
Fernverkehrsstraße *f*, Autobahn *f*
  f autoroute *f* principale, artère *f*,
autostrade *f*
  nl hoofdverkeersweg *m*
  r магистральная автомобильная
дорога *f*, автомагистраль *f*

A632 *e* **artesian head**
  d artesische Druckhöhe *f*
  f charge *f* artésienne
  nl artesische drukhoogte *f*
  r артезианский напор *m*

A633 *e* **artesian well**
  d artesischer Brunnen *m*
  f puits *m* artésien
  nl artesische bron *f(m)*
  r артезианская скважина *f*;
артезианский колодец *m*

A634 *e* **articulated boom**
  d abgewinkelter [geknickter] Ausleger
m, Knickausleger *m*
  f flèche *f* articulée
  nl gelede giek *m*
  r шарнирно-сочленённая стрела *f*

A635 *e* **articulated boom platform**
  d Schwebearbeitsbühne *f*, die mit Hilfe
vom Knickausleger aufgehoben wird
  f nacelle *f* sur flèche articulée
  nl hoogwerker *m*
  r монтажная платформа *f* [рабочая
площадка *f*], поднимаемая
шарнирно-сочленённой стрелой

A636 **articulated connection** *see* **hinged joint**

A637 *e* **articulated drop chute**
  d angelenktes Betonschüttrohr *n*
  f goulotte *f* articulée pour le bétonnage
  nl scharnierend aangekoppelde
betongoot *f(m)*
  r шарнирно-сочленённая труба *f*
(*для гравитационной подачи бетонной
смеси*)

A638 *e* **articulated system**
  d Gelenksystem *n*
  f système *m* articulé
  nl geleed systeem *n*
  r шарнирная система *f*

A639 *e* **artificial ageing**
  d künstliche Alterung *f*
  f vieillissement *m* artificiel

nl kunstmatige veroudering *f*
r искусственное старение *n*

A640 *e* **artificial aggregate**
d künstlicher Zuschlag(stoff) *m*
f agrégat *m* [granulat *m*] artificiel
nl kunstmatige toeslag *m*
r искусственный заполнитель *m*

A641 *e* **artificial harbour**
d künstlicher Hafen *m*
f port *m* artificiel
nl kunstmatige haven *f(m)*
r искусственная гавань *f*

A642 *e* **artificial lake**
d Stausee *m*, Staubecken *n*,
Speicherbecken *n*
f lac *m* artificiel, réservoir *m*, retenue
f
nl stuwmeer *n*
r водохранилище *n*; искусственное
озеро *n*

A643 *e* **artificial lighting**
d künstliche Beleuchtung *f*
f éclairage *m* artificiel
nl (verlichting *f* met) kunstlicht *n*
r искусственное освещение *n*

A644 *e* **artificial marble**
d Kunstmarmor *m*, Marmorimitation *f*
f marbre *m* artificiel
nl kunstmarmer *n*
r искусственный мрамор *m*

A645 *e* **artificial recharge, artificial
replenishment**
d künstliche Grundwasseranreicherung *f*
f épandage *m* d'eau, alimentation *f*
artificielle
nl kunstmatige grondwateraanvulling *f*
r искусственное подпитывание *n*
подземных вод

A647 *e* **artificial seasoning**
d künstliche Trocknung *f* (*von Holz*)
f séchage *m* artificiel (*du bois*)
nl kunstmatige droging *f* (*hout*)
r искусственная сушка *f* (*древесины*)

A648 artificial stone *see* cast stone

A649 artificial weathering *see* accelerated
weathering

A650 *e* **asbestine**
d Asbestine *f*
f asbestine *f*
nl asbestverf *f(m)*
r асбестин *m*

A651 *e* **asbestos**
d Asbest *m*
f asbeste *m*, amiante *m*
nl asbest *n*
r асбест *m*

A652 *e* **asbestos-base asphalt felt, asbestos-
-base asphalt paper**
d Asbest-Bitumenpappe *f*, Bitumen-
-Asbestpappe *f*
f carton *m* d'amiante bitumé,
hydroïsol *m*

nl met bitumen geïmpregneerd
asbestvilt *n*
r гидроизол *m*

A654 *e* **asbestos blanket**
d Asbestdämmschicht *f*
f tissu *m* d'amiante
nl asbestdeken *f(m)*
r асбестовая ткань *f*

A655 *e* **asbestos board**
d Asbestkarton *m*; Asbestpappe *f*;
Asbestplatte *f*
f carton *m* d'amiante [d'asbeste],
feuille *f* d'amiante; plaque *f*
d'amiante
nl asbestplaat *f(m)* asbestboard *n*
r асбестовый картон *m*; асбестовый
лист *m*; асбестовая плита *f*

A656 *e* **asbestos cement**
d Asbestzement *m*
f amiante-ciment *m*, ciment *m*
d'asbeste, fibrociment *m*
nl asbestcement *n*, *m*
r асбестоцемент *m*

A657 *e* **asbestos-cement board**
d Asbestzementplatte *f*
f plaque *f* d'amiante-ciment
nl asbestcementplaat *f(m)*
r асбестоцементный лист *m*;
асбестоцементная плита *f*

A658 *e* **asbestos-cement cladding**
d Asbestzementwanddeckung *f*
f bardage *m* en amiante-ciment
nl bekleding *f* (op basis) van
asbestcement
r асбестоцементная стеновая
облицовка *f*; асбестоцементные
облицовочные листы *m pl*

A659 *e* **asbestos-cement form board**
d Asbestzementplatte *f* für Schalung
f plaque *f* en amiante-ciment pour
coffrage
nl asbestcementplaat *f(m)* voor bekisting
r асбестоцементный лист *m* для
опалубки

A660 *e* **asbestos-cement pipe**
d Asbestzementrohr *n*
f tuyau *m* en amiante-ciment
nl asbestcementpijp *f(m)*
r асбестоцементная труба *f*

A661 asbestos-cement sheet *see* asbestos-
-cement board

A662 asbestos cement siding *see* asbestos
cement cladding

A663 *e* **asbestos felt**
d 1. Asbestpappe *f*
2. Bitumenasbestpappe *f*
f 1. carton *m* (en) amiante 2. carton *m*
d'amiante bitumé, hydroïsol *m*
nl asbestvezelkarton *n*
r 1. асбестовый картон *m* 2. гидроизол
m

A664 *e* **asbestos fiber**
d Asbestfaser *f*

f fibre *f* d'asbeste [d'amiante]
nl asbestvezel *f(m)*
r асбестовое волокно *n*

A665 *e* **asbestos fire curtain**
d Brandschutzvorhang *m* aus
Asbestfaser,
Asbestfeuerschutzvorhang *m*
f rideau *m* coupe-feu en fibres
d'amiante
nl brandwerend gordijn van asbest
r противопожарный занавес *m* из
асбестовой ткани

A666 *e* **asbestos plaster**
d Asbestmörtelputz *m*
f enduit *m* de mortier d'asbeste
nl asbestpleister(werk) *n*
r асбестовая штукатурка *f*

Б667 *e* **as-built drawings**
d Baubestandspläne *m* pl,
Bestandszeichnungen *f* pl
f dessins *m* pl d'exécution
nl revisietekeningen *f* pl
r исполнительные [натурные]
чертежи *m* pl

A668 *e* **as-dug gravel**
d Grubenkies *m*, Sand-Kies-Gemisch
n
f gravier *m* de carrière non tamisé
nl ongesorteerd grind *n*, grindzand *n*
r карьерный несортированный гравий
m; карьерная песчано-гравийная
смесь *f*

A669 *e* **aseismic joint**
d Erdbebenfuge *f*
f joint *m* antiséismique
nl aardbevingsvoeg *f(m)*
r антисейсмический шов *m*

A670 **aseismic structures** *see* **earthquake-
-resistant structures**

A671 *e* **ashlar**
d 1. Quader *m*, Werkstein *m*,
bearbeiteter Bruchstein *m*
2. Werksteinmauerwerk *n*,
Quadermauerwerk *n*
f 1. pierre *f* taillée [façonnée]
2. maçonnerie *f* en pierres taillées
nl 1. behouwen natuursteen *m*
2. metselwerk *n* in natuursteen
r 1. тёсаный камень *m* [блок *m*], блок
m из природного камня 2. кладка *f*
из тёсаных камней [блоков]

A672 *e* **ashlar facing**
d Hausteinvorsatz *m*,
Werksteinverblendung *f*
f revêtement *m* en pierres taillées
nl natuurstenen bekleding *f*
r облицовка *f* из тёсаного камня

A673 *e* **ashlaring**
d 1. Quader *m* pl, Werksteine *m* pl
2. Dachverschalungsstützen *f* pl,
Dachausschalung *f*
f 1. pierres *f* pl taillées de construction
2. poteaux *m* pl de comble

nl 1. natuurstenen metselwerk *n*
2. verticale aftimmering *f* van de
kap
r 1. штучные тёсаные камни *m* pl
2. стойки *f* pl на чердаке между
балками и стропилами

A674 *e* **ashlar masonry**
d Hausteinmauerwerk *n*,
Werksteinmauerwerk *n*
f maçonnerie *f* de pierre(s) taillé(s),
maçonnerie *f* en pierres de taille
nl metselwerk *n* in natuursteen
r кладка *f* из тёсаных камней

A675 *e* **ashpit**
d Aschenraum *m*, Aschfall *m*,
Aschengrube *f*
f cendrier *m*
nl aslade *f*, askuil *m*
r зольник *m*

A676 *e* **ash removal**
d Entaschung *f*
f évacuation *f* des cendres
nl asafvoer *m*
r золоудаление *n*

A677 **as-mixed concrete** *see* **fresh concrete**

A678 *e* **aspect**
d 1. Gebäudeanordnung *f* nach
Himmelsrichtungen 2. orientierte
Fassade *f* 3. Hangexposition *f*
f 1. orientation *f* d'un édifice 2. façade
f orientée par rapport aux points
cardinaux 3. exposition *f* d'un versant
nl ligging *f* ten opzichte van de
windstreken *of* een helling
r 1. ориентация *f* фасадов здания по
странам света 2. ориентированный
фасад *m* здания 3. экспозиция *f*
склона

A679 *e* **aspect ratio**
d Längenverhältnis *n* des rechteckigen
Querschnittes (*z. B. eines Luftkanals*)
f rapport *m* d'allongement [de la
largeur à la hauteur] (*d'une section*)
nl lengte-breedteverhouding *f*
r соотношение *n* сторон сечения (*напр.
воздуховода*)

A680 *e* **asphalt**
d 1. Bitumen *n* 2. Asphalt *m* 3. Bitumen-
-Mineral-Gemisch *n*
f 1. bitume *m* 2. asphalte *m* 3. béton
m bitum(in)eux [asphaltique]
nl asfalt *n*; asfaltbitumen *n*
r 1. нефтяной битум *m* 2. асфальт *m*
3. асфальтобетон *m*

A681 *e* **asphalt-asbestos**
d Bitumen-Asbestfaser-Aufstrich *m*,
Dichtungskitt *m* auf Bitumen-
-Asbestfaser-Basis
f peinture *f* bitum(in)euse à filler
d'asbeste
nl bitumen *n* met asbesttoeslag
r битумная краска *f* с асбестовым
наполнителем

A682 *e* **asphalt base course**
  d Asphalttragschicht *f,*
    Bitumentragschicht *f*
  f couche *f* de base en asphalte
  nl wegfundering *f* van asfaltbeton
  r асфальтобетонное основание *n*
    дорожного покрытия

A683 *e* **asphalt binder course** *US*
  d Asphaltbinder(schicht) *m* *(f)*
  f couche *f* de liaison [binder *m*] en
    béton asphaltique
  nl hechtlaag *f(m)* [tussenlaag *f(m)*] voor
    asfaltbetonweg
  r нижний (связующий) слой *m*
    асфальтобетонного дорожного
    покрытия

A684 *e* **asphalt block**
  d Asphaltpflasterstein *m*
  f bloc *m* de béton bitum(in)eux
  nl asfaltbetonklinker *m*
  r асфальтобетонный штучный камень
    m [блок *m*]

A685 *e* **asphalt block pavement**
  d Fahrbahnbelag *m* [Straßendecke *f*]
    aus Asphaltpflastersteinen
  f revêtement *m* routier en blocs de
    béton bitum(in)eux
  nl plaveisel *n* van asfaltbetonklinker
  r дорожное покрытие *n* из
    асфальтобетонных блоков [камней]

A686 *e* **asphalt cement**
  d Straßenbaubitumen *n,* Asphaltkitt
    m, Asphaltmastix *m*
  f ciment *m* asphaltique, bitume *m*
    routier
  nl asfaltcement *n,* *m,* bitumenkit
    f(m)
  r асфальтовое вяжущее *n,*
    дорожный [вязкий] битум *m*

A687 *e* **asphalt coating compound**
  d Bitumenanstrichmasse *f,*
    Bitumenaufstrichmasse *f*
  f enduit *m* bitum(in)eux [d'asphalte]
  nl asfaltbitumenbedekking *f*
  r битумный обмазочный состав *m*

A688 *e* **asphalt concrete**
  d Asphaltbeton *m*
  f béton *m* asphaltique [bitum(in)eux]
  nl asfaltbeton *n*
  r асфальтобетон *m*

A689 *e* **asphalt concrete base**
  d Asphaltbetontragschicht *f*
  f fondation *f* en béton asphaltique
  nl fundering *f* van asfaltbeton
  r асфальтобетонное основание *n*

A690 *e* **asphalt concrete pavement**
  d Asphaltbetonbelag *m,*
    Asphaltbetondecke *f*
  f revêtement *m* routier en béton
    asphaltique
  nl wegverharding *f* van asfaltbeton
  r асфальтобетонное дорожное
    покрытие *n*

A691 *e* **asphalt cutter**
  d Asphaltmeißel *m,* Fugenschneider *m*
    für Asphaltbetondecken
  f machine *f* à couper le revêtement
    d'asphalte
  nl asfaltbetonzaag *f(m)* *(wegdek)*
  r машина *f* для нарезки швов
    в асфальтобетонном дорожном
    покрытии

A692 **asphalt emulsion** *see* **bituminous
    emulsion**

A693 *e* **asphalt emulsion slurry seal**
  d Bitumenschlämmeabsiegelung *f,*
    Bitumenschlämme-Porenschluß *m*
  f mortier *m* à émulsion de bitume
  nl geëmulgeerd bitumen *n* voor
    wegondergrond
  r асфальтовый раствор *m* на битумной
    эмульсии

A694 *e* **asphalt fog seal**
  d Bitumenemulsion- *oder*
    Flüssigbitumenanstrich *m*
  f traitement *m* superficiel à l'émulsion
    ou au bitume
  nl oppervlaktebehandeling *f* met
    geëmulgeerd bitumen
  r поверхностная обработка *f*
    эмульсией *или* жидким битумом

A695 *e* **asphalt heater**
  d Bitumenhitzer *m,*
    Bitumenschmelzkessel *m,*
    Bitumenkocher *m*
  f réchauffeur *m* de bitume
  nl asfaltketel *m,* bitumenverhitter *m*
  r разогреватель *m* [нагреватель *m*]
    битума

A696 *e* **asphalting**
  d 1. Asphaltieren *n,* Asphaltierung *f*
    2. Bituminieren *n,* Bituminierung
    f
  f 1. asphaltage *m* 2. bitumage *m*
  nl asfalteren *n*
  r 1. асфальтирование *n* 2. обмазка *f*
    битумом *(для гидроизоляции)*

A697 **asphalt intermediate course** *see*
    asphalt binder course

A698 *e* **asphalt joint filler**
  d Bitumen-Vergußmasse *f,*
    Bitumenfugenkitt *m,*
    Bitumenfugenmastix *m*
  f mastic *m* bitum(in)eux [d'asphalte]
    pour remplissage de joints
  nl bitumenvoegvulling *f*
  r битумная мастика *f* для заполнения
    швов и трещин

A699 *e* **asphalt levelling course**
  d Asphalt-Ausgleichlage *f,* Asphalt-
    -Ausgleichschicht *f*
  f couche *f* de nivellement [de
    profilage] en béton d'asphalte
  nl asfaltbetonnen egalisatielaag *f(m)*
  r асфальтобетонный выравнивающий
    слой *m*

A700 *e* **asphalt macadam**
 d Asphaltmakadam *m, n,*
 Bitumenmakadam *m, n*
 f macadam *m* bitum(in)eux [au bitume]
 nl asfaltmacadam *n,* bitumenmacadam *n*
 r щебёночное (дорожное) покрытие *n,*
 обработанное битумом

A701 *e* **asphalt mastic**
 d Asphaltmastix *m,* Bitumenmastix *m*
 f mastic *m* bitum(in)eux [d'asphalte]
 nl asfaltmastiek *n, m* bitumenmastiek *n,*
 m
 r битумная мастика *f*

A702 *e* **asphalt membrane**
 d bituminöse Abdichtungshaut *f*
 [Isolierhaut *f*]
 f membrane *f* bitum(in)euse
 nl waterdichte bitumenlaag *f(m)*
 r битумный гидроизолирующий слой
 m

A703 *e* **asphalt overlay**
 d Asphaltverschleißschicht *f,*
 Asphaltverschleißlage *f*
 f couche *f* de roulement [d'usure]
 asphaltique
 nl asfaltslijtlaag *f(m)*
 r верхний слой *m* асфальтобетонного
 покрытия; слой *m* износа

A704 *e* **asphalt paint**
 d Bitumenschutzanstrichmittel *n*
 f peinture *f* bitum(in)euse
 nl bitumenverf *f(m)*
 r битумная краска *f,* битумный
 окрасочный состав *m*

A705 *e* **asphalt paper**
 d Asphalt(dach)pappe *f,*
 Bitumen(dach)pappe *f*
 f carton *m* asphalté [bitumé]
 nl asfaltpapier *n*
 r пергамин *m*

A706 *e* **asphalt pavement recycling**
 d Wiederherstellung *f* von
 Asphaltbeton-Straßendecken
 f recyclage *m* d'un revêtement routier
 asphaltique
 nl hergebruik *n* van
 asfaltbetonverhardingsmateriaal
 r восстановление *n* асфальтобетонных
 покрытий (*путём снятия, переплавки
 и повторной укладки смеси*)

A707 *e* **asphalt pavement structure**
 d Aufbau *m* von Asphaltbeton-
 -Straßendecken
 f structure *f* d'une corps de chaussée
 asphaltique
 nl samenstelling *f* van een asfaltbetonweg
 r конструкция *f* [конструктивные
 слои *m pl*] асфальтобетонной
 дорожной одежды

A708 *e* **asphalt prime coat**
 d Bitumengrundanstrich *m*
 f couche *f* de fond en bitume
 nl bitumengrondlaag *f(m)*
 r битумный грунтовочный слой *m*

A709 *e* **asphalt primer**
 d Bitumengrund(ierung) *m (f),*
 Bitumengrundmittel *n*
 f apprêt *m* en bitume
 nl waterdichte bitumenlaag *f(m)*
 r битумный грунтовочный состав *m,*
 битумная грунтовка *f*

A710 *e* **asphalt seal coat**
 d Asphaltdichtungsschicht *f*
 f couche *f* de scellement [de fermeture]
 en bitume
 nl waterdichte bitumenlaag *f(m)*
 r верхний [защитный] битумный слой
 m

A711 *e* **asphalt soil stabilization**
 d Bitumenverfestigung *f,*
 Bodenverfestigung *f* mit Bitumen
 f stabilisation *f* de sol au bitume
 nl grondstabilisatie *f* met bitumen
 r битумизация *f* грунта (*укрепление
 грунта битумом*)

A712 *e* **asphalt surface course**
 d Asphaltdecklage *f,*
 Asphaltverschleißschicht *f*
 f couche *f* de roulement en asphalte
 nl asfaltslijtlaag *f(m)*
 r поверхностный слой *m*
 асфальтобетонного дорожного
 покрытия, слой *m* износа

A713 *e* **asphalt surface treatment**
 d Bitumenanstrich *m* der Straßendecke
 f traitement *m* superficiel au bitume
 nl behandeling *f* van het wegdek met
 bitumen
 r поверхностная обработка *f*
 дорожного покрытия битумом

A714 *e* **asphalt tack coat**
 d Bitumenbinder(schicht) *m (f)*
 f couche *f* d'accrochage [de liaison] en
 bitume
 nl bitumineuze hechtlaag *f(m)*
 r битумный связующий слой *m*

A715 *e* **asphalt tile**
 d Asphalt(fuß)bodenplatte *f*
 f carreau *m* d'asphalte [asphaltique]
 nl asfalttegel *m*
 r битумная плитка *f*

A716 *e* **asphaltum** *UK*
 d Naturbitumen *n;* natürliches Bitumen-
 -Mineralgemisch *n,* Bitumen-
 -Mineralmischung *f*
 f bitume *m* naturel [natif]; asphalte *m*
 naturel
 nl natuurlijk asfaltbitumen *n*
 r природный битум *m;* природный
 асфальт *m*

A717 **asphalt-wearing course** *see* **asphalt
 surface course**

A718 *e* **aspiration pipe**
 d Saugrohr *n*
 f conduite *f* d'aspiration
 nl aanzuigbuis *f(m)*
 r всасывающая труба *f*

A719 as-placed concrete *see* green concrete

A720 assemblage *see* assembling

A721 *e* assembling
   d Zusammenbauen *n*, Montage *f*
   f assemblage *m*, montage *m*
   nl samenstelling *n*, assemblage *f*
   r сборка *f*, монтаж *m*

A722 *e* assembling bolt
   d Heftbolzen *m*, Montagebolzen *m*
   f boulon *m* de montage
   nl montagebout *m*
   r монтажный [сборочный] болт *m*

A723 *e* assembling work
   d Montagearbeiten *f pl*
   f travaux *f pl* d'assemblage [de montage]
   nl montagewerk *n*
   r сборочные [монтажные] работы *f pl*

A724 *e* assembly
   d 1. Maschinenaggregat *n*, Baugruppe *f*
   2. Zusammenbau *m*, Montage *f*, Aufbau *m*
   f 1. groupe *m* 2. assemblage *m*, montage *m*, installation *f*
   nl montage *f*
   r 1. (машинный) агрегат *m*; узел *m* 2. сборка *f*, монтаж *m*, установка *f*

A725 *e* assembly workshop
   d Zusammenbauwerkstatt *f*, Montagehalle *f*
   f atelier *m* de montage
   nl montagewerkplaats *f(m)*
   r цех *m* укрупнительной сборки конструкций; сборочный [монтажный] цех *m*

A726 *e* assize
   d 1. Trommel *f* (*Säule*) 2. Schicht *f* des Steinmauerwerkes
   f 1. bloc *m* cylindrique en pierre (*d'une colonne*) 2. assise *f* de maçonnerie en pierres
   nl 1. cilindrisch steenblok *n* (*zuil*) 2. steenlaag *f(m)* (*metselwerk*)
   r 1. цилиндрический каменный блок *m* (*колонны*) 2. ряд *m* каменной кладки

A727 assumed load *see* design load

A728 *e* asymmetrical(ly-placed) load(s)
   d asymmetrische Belastung *f*
   f charge *f* asymétrique
   nl asymmetrische belasting *f*
   r асимметричная [несимметричная] нагрузка *f*

A729 *e* at-grade intersection
   d plangleiche [ebenerdige] Kreuzung *f*
   f croisement *m* à niveau
   nl gelijkvloerse kruising *f*
   r пересечение *n* дорог в одном уровне

A730 *e* atmospheric condenser
   d Berieselungskondensator *m*
   f condens(at)eur *m* atmosphérique [à ruissellement]
   nl bevloeiingscondensator. *m*
   r оросительный конденсатор *m* с естественным воздушным охлаждением

A731 *e* atmospheric steam curing
   d (Niederdruck-)Dampfbehandlung *f* (*von Beton*)
   f traitement *m* (*du béton*) à la vapeur sous pression atmosphérique
   nl behandeling *f* met stoom onder atmosferische druk (*beton*)
   r пропаривание *n* (*бетона*) при атмосферном давлении

A732 *e* atomic blast excavation
   d Atomsprengungsaushub *m*
   f excavation *f* par explosion nucléaire
   nl atoomontploffingsexcavatie *f*
   r ядерная экскавация *f*, выемка *f* грунта с помощью ядерных взрывов

A733 *e* atomic-hydrogen welding
   d atomare Wasserstoffschweißung *f*
   f soudage *m* à l'hydrogène atomique
   nl lassen *n* met atomaire waterstof
   r атомно-водородная сварка *f*

A734 *e* atomic reactor containment structure
   d Einschlußbauwerk *n* [Container *m*] einer Reaktoranlage
   f blindage *m* [protection *f*] du réacteur nucléaire
   nl stralingsmantel *m* (*kernreactor*)
   r противорадиационная защитная оболочка *m* атомного реактора

A735 *e* atomization
   d Zerstäubung *f*, Versprühen *n*
   f pulvérisation *f*, atomisation *f*
   nl vernevelen *n*, verstuiven *n*
   r распыление *n*

A736 *e* atomizer
   d Zerstäuber *m*, Sprühapparat *m*
   f atomiseur *m*
   nl sproeier *m*, verstuiver *m*
   r распылитель *m* (*напр. штукатурного раствора*)

A737 atomizing *see* atomization

A738 *e* atomizing application
   d Aufdüsen *n*, Aufsprühen *n*
   f application *f* par pulvérisation
   nl opstuiven *n* (*opbrengen door verstuiven*)
   r нанесение *n* методом распыления

A739 *e* at-rest earth pressure coefficient
   d Ruhedruckbeiwert *m*
   f coefficient *m* de pression du sol naturelle
   nl coëfficiënt *m* van gronddruk in ongeroerde grond
   r коэффициент *m* статического давления грунта

A740 *e* attached column
   d eingebundene Säule *f*, Halbsäule *f*
   f colonne *f* adossée [engagée]
   nl halfzuil *f(m)* (*tegen de muur*)
   r полуколонна *f*

A741 *e* **attached garage**
d Einbaugarage *f*; angebaute Garage *f*
f garage *m* encastré; garage *m* adossé
nl aangebouwde garage *f*
r гараж *m*, встроенный в здание;
пристроенный гараж *m*

A742 *e* **Atterberg test**
d Konsistenzprüfung *f*
[Konsistenzprobe *f*] nach Atterberg
f essai *m* d'Atterberg
nl bepaling *f* van Atterbergse waarden
r определение *n* пластичности грунта,
определение *n* пределов Аттерберга

A743 *e* **attic**
d 1. Dachraum *m*, Dachboden *m*
2. Dachetage *f*, Dachgeschoß *n*
3. Attika *f*
f 1. grenier *m* 2. étage *m* sous le toit,
étage *m* de combles 3. attique *m*
nl 1. zolder *m* 2. zolderkamer *f(m)*
3. vliering *f*
r 1. чердак *m* 2. мансардный этаж
m; мансардное помещение *n* 3. аттик
m, фронтон *m*

A744 **attic storey** *see* attic 2.

A745 *e* **attic tank**
d Dachboden-Wasserdruckbehälter *m*
f réservoir *m* [bac *m*] posé sous comble
nl waterreservoir *n* op de bovenste
verdieping
r водонапорный резервуар *m* на
чердаке дома

A746 *e* **attic window**
d Mansard(en)dachfenster *n*
f lucarne *f*
nl zoldervenster *n*, dakraam *n*
r слуховое [чердачное] окно *n*

A747 *e* **attracting groyne**
d Leitdamm *m*, Leitwerk *n*
f épi *m* de guidage, guideau *m*
nl leidam *m*
r струенаправляющая дамба *f*

A748 *e* **auger**
d 1. Bohrer *m* 2. Löffelbohrer *m*,
Bodenbohrer *m*, Bohrlöffel *m*
3. Verteilerschnecke *f*,
Auflockerungsschnecke *f*
(*Schwarzbelageinbau*)
f 1. mèche *f* pour bois, foret *m* à langue
d'aspic 2. cuiller *m*, tarière *f* 3. vis *f*
d'alimentation, transporteur *m* à vis
nl 1. fretboor *f(m)*; avegaar *m*
2. grondboor *m*, avegaar *m*,
3. woelworm *m*
r 1. бурав *m*; пёрка *f* 2. ложечный бур
m, почвенный бурав *m*; буровая
желонка *f* 3. распределительный
шнек *m*

A749 *e* **auger-backfiller**
d Vielzweck-Schnecke *f*,
Planierschneckenfräse *f*
f transporteur *m* à vis universel
nl sleuvendichter *m*
r универсальный шнек *m*

A750 **auger bit** *see* **auger 1., 2.**

A751 **augered pile** *see* **bored pile**

A752 **autoclave curing** *see* **autoclaving**

A753 *e* **autoclaved aerated concrete**
d Autoklav-Porenbeton *m*
f béton *m* poraux autoclavé
nl autoclaafporiënbeton *n*
r автоклавный ячеистый бетон *m*,
ячеистый бетон *m* автоклавного
твердения

A754 *e* **autoclaved concrete product**
d Autoklavbetonerzeugnis *n*
f produit *m* en béton autoclavé
nl autoclaafbetonnen produkt *n*
r изделие *n* из бетона автоклавного
твердения

A755 *e* **autoclaving**
d Autoklavbehandlung *f*,
Autoklav(is)ierung *f*
f autoclavage *m*, traitement *m* à
l'autoclave
nl behandeling *f* met de autoclaaf
r автоклавная обработка *f*

A756 *e* **autoclaving cycle**
d Autoklavbehandlungsdauer *f*
f cycle *m* d'autoclavage
nl duur *m* van de behandeling met de
autoclaaf
r цикл *m* [продолжительность *f*]
автоклавной обработки

A757 *e* **autogenous curing**
d Betonerhärtung *f* mit Aufbewahrung
von Feuchte und Hydratationswärme
f cure *f* autogène (de béton)
nl autogene verharding *f* (*beton*) bij
vochtbehouding en bewaring van
hydratatiewarmte
r твердение *n* бетона в условиях
сохранения им влажности и теплоты
гидратации

A758 *e* **autogenous cutting and welding
apparatus**
d Brennschneid- und Schweißgerät *n*
f appareil *m* de soudage et coupage
aux gaz
nl machine *f* voor autogeen lassen en
snijden
r аппарат *m* для газовой сварки
и резки

A759 *e* **autogenous cutting machine**
d Brennschneid(e)maschine *f*
f machine *f* de découpage à gaz,
machine *f* d'oxycoupage
nl machine *f* voor autogeen snijden
r машина *f* для газовой резки

A760 *e* **autogenous healing of concrete**
d Selbstheilung *f* des Betons
f scellement *m* autogène des fissures sur
la surface de béton
nl zelfsluiten *n* van de scheuren in het
beton
r самозалечивание *n* бетона,
самозакрытие *n* трещин в бетоне

A761 *e* **autogenous volume change**
    d spontanes Schwellen *n* des Betons,
      Betonaufblähung *f*
    f augmentation *f* autogène de volume
      du béton
    nl spontane uitzetting *f* van het beton
    r самопроизвольное набухание *n*
      бетона

A762 *e* **autogenous welding**
    d Gasschweißen *n*
    f soudage *m* autogène [au gaz]
    nl autogeen lassen *n*
    r газовая сварка *f*

A763 *e* **automatic air valve**
    d Rohrentlüfter *m*, automatisches
      Entlüftungsventil *n*, Luftventil *n*,
      Luftabscheider *m*
    f soupape *f* automatique de dégagement
      d'air, soupape *f* d'évacuation d'air
      automatique, purgeur *m* d'air
    nl zelfinstellende luchtklep *f(m)*,
      automatisch ventiel *n*
    r автоматический воздуховыпускной
      [деаэрационный] клапан *m*,
      вантуз *m*

A764 *e* **automatic arc welding machine**
    d Lichtbogenschweißautomat *m*
    f soudeuse *f* automatique (à l'arc)
    nl automaat *m* voor booglassen
    r дуговой сварочный автомат *m*

A765 *e* **automatic batcher**
    d Dosierautomat *m*
    f doseur *m* automatique
    nl automatisch doseringsapparaat *n*
    r автоматический дозатор *m*

A766 *e* **automatic boiler**
    d Automatikkessel *m*
    f chaudière *f* automatique
    nl thermostatische boiler *m*
    r автоматизированная котельная
      установка *f*

A767 *e* **automatic closing device**
    d automatischer Türschließer *m*
    f ferme-porte *m* automatique
    nl deurdranger *m*
    r автоматический закрыватель *m*
      двери

A768 **automatic closing door** *see* **automatic fire door**

A769 *e* **automatic control valve**
    d automatisches Regelventil *n*
    f vanne *f* de réglage automatique
    nl zelfwerkende regelklep *f(m)*
    r автоматический регулирующий
      клапан *m*

A770 *e* **automatic cutting-off table**
    d automatischer Abschneider *m*,
      Abschneideautomat *m*
      (*Ziegelindustrie*)
    f machine *f* [dispositif *m*] automatique
      pour couper les briques brutes
    nl zelfwerkender afsnijder *m*
    r автомат *m* для резки кирпича-сырца

A771 *e* **automatic detection**
    d automatischer Nachweis *m*,
      automatisches Auffinden *n*
    f détection *f* automatique
    nl automatische detectie *f*
    r автоматическое обнаружение *n*
      (*повреждений, утечек*) ;

A772 **automatic door closer** *see* **automatic closing device**

A773 *e* **automatic door operator**
    d automatische (Tür-)Öffnungsanlage *f*
    f opérateur *m* [dispositif *m*] d'ouverture
      automatique
    nl automatische deuropener *m*
    r автоматический открыватель *m*
      двери

A774 *e* **automatic damping batcher scale**
    d automatische Ausschüttwaage
    f balance *f* doseuse à déversement
      automatique
    nl zelflossende doseerautomaat *m*
    r весовой дозатор *m* с автоматическим
      разгрузочным устройством

A775 *e* **automatic elevator**
    d automatischer Aufzug *m*
    f ascenseur *m* automatique
    nl lift *m* met automatisch bediende
      deuren
    r лифт *m* с автоматически
      открывающимися дверьми

A776 *e* **automatic fire alarm system**
    d selbsttätige Feuermeldeanlage *f*
    f système *m* d'avertissement
      automatique d'incendie
    nl (automatisch) brandmeldingssysteem *n*
    r автоматическая система *f* пожарной
      сигнализации

A777 *e* **automatic fire door**
    d automatische Brand(schutz)tür *f*
    f porte *f* coupe-feu automatique
    nl automatisch sluitende branddeur *f(m)*
    r автоматическая противопожарная
      дверь *f*

A778 *e* **automatic fire protection system**
    d automatische Brandschutzanlage *f*
    f système *m* automatique de protection
      contre l'incendie
    nl automatisch brandbeveiligingsysteem *n*
    r автоматическая система *f*
      (противо)пожарной защиты

A779 *e* **automatic fire pump**
    d automatische Feuerlöschpumpe *f*
    f pompe *f* automatique à incendie
    nl automatische brandspuit *f(m)*
    r автоматический пожарный насос *m*

A780 *e* **automatic flushing cistern**
    d automatischer Spülkasten *m*
    f réservoir *m* de chasse automatique
    nl zelfwerkende stortbak *m*
    r автоматический смывной бачок *m*

A781 *e* **automatic gas-fired water heater**
    d vollautomatischer
      Gaswarmwasserbereiter *m*

f chauffe-eau *m* automatique à gaz
nl automatische gasboiler *m*
r автоматический газовый
водонагреватель *m*

A782 e **automatic gate**
d automatische Schütze *f*,
Selbstschlußschütze *f*
f vanne *f* (à fermeture) automatique
nl automatische afsluiter *m*
r автоматический затвор *m*
(*гидросооружения*)

A783 e **automatic heating**
d automatische Heizung *f*
f chauffage *m* automatique
nl automatische verwarming *f*
r автоматическое отопление *n*

A784 e **automatic opener**
d selbstöffnende Aufzugsschachttür *f*
f porte *f* automatique du cage [de la
gaine] d'ascenseur
nl zelfwerkend openingsmechanisme *n*
r автоматически открывающаяся дверь
f лифтовой шахты

A785 e **automatic oxygen cutting**
d automatisches
Sauerstoffbrennschneiden *n*
f oxycoupage *m* automatique, coupage
m aux gaz automatique
nl automatisch snijbranden *n*
r автоматическая кислородная резка *f*

A786 e **automatic return of water**
d Dampfschleife *f*
f retour *m* d'eau automatique
nl automatische waterterugvoer *m*
r паровая петля *f*
(*устройство для автоматического
возврата в котёл воды, увлекаемой
давлением пара*)

A787 e **automatic return trap**
d automatischer Rückspeiser *m*
(*Dampfheizung*)
f soupape *f* de retour automatique
nl automatische
(stoom)terugvoerinrichting *f*
r автоматический обратный клапан *m*
(*в системе парового отопления*)

A788 e **automatic siphon**
d selbsttätiger Heber *m*
f siphon *m* automatique
nl automatische hevel *m*
r автоматический сифон *m*

A789 e **automatic siphon spillway**
d selbsttätiger Heberüberlauf *m*
f évacuateur *m* à siphon automatique
nl automatische heveloverloop *m*
r автоматический сифонный водосброс
m

A790 e **automatic smoke detector**
d Rauchmelder *m*, Rauchmeldeautomat
m (*Brandschutz*)
f détecteur *m* de fumée automatique
nl automatische rookdetector *m*
r автоматический дымовой [пожарный]
извещатель *m*

A791 e **automatic sprinkler system**
d automatische Sprinkleranlage *f*
f système *m* de sprinkler automatique
nl (automatisch bediende)
sprinklerinstallatie *f*
r автоматическая спринклерная
система *f* (*пожаротушения*)

A792 e **automatic taper**
d Dichtungsband-Klebeautomat *m*,
automatische Dichtungsband-
-Klebevorrichtung *f*
f dispositif *m* automatisé à coller des
bandes d'étanchéité
nl automatische bandwikkelmachine *f*
r автоматизированное устройство *n*
для наклейки уплотняющих
[герметизирующих] лент (*на швы*)

A793 e **automatic weigh-batching**
d automatische Gewichtsabmessung *f*
[Gewichtsdosierung *f*]
f dosage *m* automatique pondéral [en
poids, par pesée]
nl automatische gewichtsdosering *f*
r автоматическое весовое дозирование
n

A794 e **automatic weir**
d selbsttätiges Wehr *n*
f déversoir *m* [évacuateur] *m*
automatique
nl zelfwerkende stuw *m*
r автоматический водосброс *m*

A795 **automatic welder** *see* **automatic
welding machine**

A796 e **automatic welding**
d automatisches Schweißen *n*
f soudage *m* automatique
nl automatisch lassen *n*
r автоматическая сварка *f*

A797 e **automatic welding machine**
d Schweißautomat *m*
f machine *f* à souder automatique
nl lasautomaat *m*
r сварочный автомат *m*

A798 e **automobile crane**
d Auto(mobil)kran *m*
f grue *f* automobile [camion]
nl kraanwagen *m*
r автомобильный кран *m*, автокран

A799 e **autonomous house**
d autonomes Wohngebäude *n* (*mit
unabhängigen Versorgungssystemen*)
f bâtiment *m* autonome (*avec
installations du comfort des
intérieurs autonomes*)
nl woonhuis *n* met eigen water- en
energievoorzieningen
r жилой дом *m* с автономными
системами инженерного
оборудования

A800 e **autosilo, autostacker**
d Turmgarage *f* mit Lift, Parkturm *m*,
Autosilo *n*, *m*
f autosilo *m*, garage-tour *m* à
élévateur(s)

nl parkeergarage f met lift
r башенный гараж m с лифтом

A802 e **auxiliary hoist**
   d Hilfshubwinde f
   f treuil m auxiliaire
   nl hulplier f(m), hulpwindas f(m)
   r лебёдка f вспомогательного подъёма
     (на кране)

A803 e **auxiliary prestressing**
   d Hilfsvorspannung f
   f précontrainte f auxiliaire
   nl hulpvoorspanning f
   r дополнительное преднапряжение n

A804 e **auxiliary rafter**
   d Hilfsdachverband m, Hilfssparren m
   f arbalétrier m auxiliaire
   nl hulpspar m
   r вспомогательные стропила n pl;
     вспомогательная стропильная нога f

A805 e **auxiliary reinforcement**
   d Hilfsbewehrung f; schlaffe Bewehrung f
   f armature f [ferraillage m] auxiliaire
   nl hulpwapening f; onbelaste wapening f
   r вспомогательная арматура f;
     ненапрягаемая арматура f

A806 e **auxiliary stairs**
   d Hilfstreppe f; Diensttreppe f
   f escalier m auxiliaire; escalier m de
     service
   nl diensttrap m
   r вспомогательная лестница f;
     служебная лестница f

A807 e **auxiliary structure**
   d Behelfskonstruktion f; behelfsmäßiges
     Bauwerk n
   f construction f auxiliaire; bâṭiment m
     secondaire
   nl hulpconstructie f
   r вспомогательная конструкция f;
     вспомогательное сооружение n

A808 e **available head**
   d verfügbares [nutzbares] Gefälle n,
     Nutzfallhöhe f
   f hauteur f de chute utile
   nl nuttige drukhoogte f, beschikbaar
     verhang n
   r полезный напор m

A809 e **available power**
   d verfügbare Leistung f
   f puissance f disponible
   nl beschikbaar vermogen n
   r располагаемая мощность f

A810 e **avalanche baffle wall**
   d Lawinen(schutz)mauer f
   f mur m de protection contre les
     avalanches
   nl tegen lawines beschermende muur m
   r лавинозащитная стена f

A811 e **avalanche brake structure**
   d Lawinenbrecher m
   f brise-avalanche m
   nl lawine-brekende constructie f
   r лавинозадерживающее сооружение n

A812 e **avalanche gallery**
   d Lawinenschutzdach n, Halbtunnel m
   f pare-avalanche m, galerie f
     antiavalanche
   nl tegen lawines beschermende
     overkapping f
   r противообвальная [лавинозащитная]
     галерея f

A813 e **avalanche load**
   d Lawinenbelastung f
   f charge f d'avalanche (de neige)
   nl lawinebelasting f
   r нагрузка f от (снежной) лавины

A814 e **avalanche protection dike**
   d Lawinenleitwerk n
   f digue f (de protection) contre les
     avalanches
   nl tegen lawines beschermende dam m
   r лавиноотбойная дамба f

A815 e **avalanche protection works**
   d Lawinenverbau m,
     Lawinenschutzwerke n pl
   f ouvrages m pl de protection contre
     les avalanches
   nl beveiligingsconstructies f pl tegen
     lawines
   r противолавинные [лавинозащитные]
     сооружения n pl

A816 e **avant-corps**
   d Vorbau m; Fassadenwandvorsprung m
   f avant-corps m
   nl uitspringend gedeelte n in de
     voorgevel
   r выступающая часть f здания; выступ
     m фасада

A817 e **avenue**
   d Allee f, Avenue f, Promenade f,
     bepflanzte Straße f; Zufahrt f,
     Zugang m
   f 1. rue f plantée d'arbres 2. avenue f
     3. voie f d'accès
   nl 1. avenue f 2. laan f(m) 3. oprijlaan
     f(m)
   r 1. обсаженная деревьями улица f
     2. авеню f; проспект m 3. подъезд
     m; подход m

A818 e **average annual rainfall**
   d mittlere Jahresniederschlagsmenge f
   f précipitations f pl annuelles moyennes
   nl (gemiddelde) jaarlijkse
     neerslag(hoeveelheid) m (f)
   r среднегодовые осадки m pl

A819 e **average bond stress**
   d mittlere Haftspannung f
   f contrainte f moyenne de l'adhérence
   nl gemiddelde hechtspanning f
   r среднее расчётное значение n
     напряжений сцепления

A820 e **average flow**
   d Abflußmenge f bei Mittelwasser, MQ
   f débit m moyen du cours d'eau
   nl gemiddelde afvoer m, gemiddeld
     debiet n
   r средний расход m водотока

A821 *e* **average velocity of groundwater**
 d mittlere Geschwindigkeit *f* des
  Grundwassers
 f vitesse *f* moyenne d'eau souterraine
 nl gemiddelde stroomsnelheid *f* van het
  grondwater
 r средняя скорость *f* грунтовых вод

A822 *e* **award of a contract**
 d Vergabe *f* (einer Arbeit);
  Auftrag(s)erteilung *f*
 f passation *f* du marché [d'un contrat]
 nl gunning *f* (bij aanbesteding);
  bouwopdracht *f(m)*
 r сдача *f* подряда *n*; заключение *n*
  подрядного договора *или* контракта

A823 *e* **awl**
 d Ahle *f*, Pfriem(en) *m*
 f alêne *f*
 nl els *f(m)*, priem *m*
 r шило *n*

A824 *e* **awning**
 d Markise *f*, Sonnendach *n*, Wetterdach *n*
 f auvent *m* de toile; toile *f*
 nl markies *f(m)*, zonnescherm *n*, luifel
  f(m)
 r мягкий (тканевый) навес *m*; тент *m*

A825 *e* **awning blind**
 d aufklappbare Blende *f*, Klappblende
  f, Anschlagladen *m*
 f volet *m* se rabattant à l'extérieur
 nl zonnescherm *m* met beweegbare
  jalouzieën
 r верхнеподвесная [откидная] ставня *f*

A826 *e* **awning window**
 d Klappflügelfenster *n*, aufklappbares
  Fenster *n*
 f fenêtre *f* poussant extérieure
 nl uitzetraam *n*
 r верхнеподвесное [откидное] окно *n*

A827 *e* **axe**
 d Axt *f*; Beil *n*
 f hache *f*; cognée *f*
 nl bijl *f(m)*, aks *f(m)*
 r топор *m*; колун *m*

A828 *e* **axed work**
 d scharriertes Mauerwerk *n*
 f smillage *m*, taille *f* smillée
 nl gepunthamerde natuursteen *m*
 r точечная фактура *f* штучного камня
  в кладке

A829 *e* **axhammer**
 d 1. Krönel *m*; Bossierhammer *m*
  2. Maurerhammer *m*
 f 1. rustique *f*, smille *f*; têtu *m*
  2. marteau *m* de maçon
 nl 1. steenhamer *m* 2. metselaarshamer
 r 1. закольник *m* 2. молоток-кирочка
  m (каменщика)

A830 *e* **axial compression**
 d axialer Druck *m*
 f compression *f* centrée [axiale]
 nl axiale druk *m*
 r осевое сжатие *n*

A831 *e* **axial deformation**
 d axiale Formänderung *f* [Verformung *f*]
 f déformation *f* axiale
 nl vervorming *f* in langsrichting
 r продольная деформация *f*

A832 *e* **axial-flow fan**
 d Axiallüfter *m*, Axialventilator *m*
 f ventilateur *m* axial
 nl axiaalventilator *m*
 r осевой вентилятор *m*

A833 *e* **axial-flow pump**
 d Axialpumpe *f*, Propellerpumpe *f*
 f pompe *f* axiale
 nl axiaalpomp *f(m)*, schroefpomp *f(m)*
 r осевой насос *m*

A834 *e* **axial force diagram**
 d Längskraftlinie *f*,
  Längskraftdiagramm *n*
 f diagramme *m* des forces axiales
 nl diagram *n* van overlangse krachten
 r эпюра *f* продольных сил

A835 *e* **axial load**
 d Axiallast *f*
 f charge *f* [sollicitation *f*] axiale
 nl axiale last *f(m)*
 r осевая нагрузка *f*

A837 *e* **axial moment of inertia**
 d axiales Trägheitsmoment *n*
 f moment *m* d'inertie axial
 nl axiaal traagheidsmoment *n*
 r осевой момент *m* инерции

A838 *e* **axial prestressing**
 d axiale Vorspannung *f*
 f précontrainte *f* axiale
 nl axiale voorspanning *f*
 r продольное [осевое] предварительное
  напряжение *n*

A839 *e* **axial restraint**
 d axiale Einspannung *f*
 f encastrement *m* axial
 nl axiale inklemming *f*
 r осевое защемление *n*, защемление *n*
  в продольном направлении

A840 *e* **axial section**
 d Axialschnitt *m*
 f section *f* longitudinale [axiale]
 nl langsdoorsnede *f*
 r продольный разрез *m*

A841 **axial strain** *see* **axial deformation**

A842 *e* **axial stress**
 d Axialspannung *f*
 f contrainte *f* axiale
 nl axiaalspanning *f*
 r осевое напряжение *n*

A843 *e* **axial tension**
 d Axialzug *m*, mittiger Zug *m*
 f tension *f* axiale
 nl axiale trek *m*
 r осевое растяжение *n*

A844 **axial tensioning** *see* **axial prestressing**

A845 *e* **axis of channel**
 d Stromstrich *m*

f axe *m* du lit
nl as *f(m)* van een kanaal
r геометрическая ось *f* русла

A846   axis of gravity *see* centroidal axis

A847 *e* axis of inertia
d Trägheitsachse *f*
f axe *f* d'inertie
nl traagheidsas *f(m)*
r ось *f* инерции

A848 *e* axis of rotation
d Drehachse *f*
f axe *m* de rotation
nl draaiingsas *f(m)*
r ось *f* вращения

A849 *e* axis of simmetry
d Symmetrieachse *f*
f axe *f* de symétrie
nl symmetrie-as *f(m)*
r ось *f* симметрии

A850   axis stress *see* axial stress

A851 *e* axisymmetrical loading
d achs(en)symmetrische
[axialsymmetrische] Last *f*
[Belastung *f*]
f charge *f* axi-simétrique
nl axiaal-symmetrische belasting *f*
r осесимметричная нагрузка *f*

A852 *e* axle load
d Achslast *f*
f charge *f* d'essieu [par essieu]
nl asbelasting *f*
r нагрузка *f* на ось

A853 *e* axonometric projection
d Parallelprojektion *f*, parallele
[axonometrische] Projektion *f*
f projection *f* axonométrique
nl axonometrische projectie *f*
r аксонометрическая проекция *f*

B

B1   *e* back
d 1. Rückseite *f* 2. Putzgrund *m*,
Putzträger *m*; Putzfläche *f* 3.
Oberkante *f*, Sichtfläche *f* (*des
stückigen Materials für
Dacheindeckung*) 4. Rippe *f*,
Oberkante *f* (*des Trägers*) 5.
Hauptbinder *m* 6. Unterschicht *f* der
Furnierung
f 1. arrière *m*; envers *m* 2. couche *f*
de base; surface *f* à enduire 3. face *f*
supérieure des plaques de toiture
4. surface *f* supérieure (*p. ex. d'une
poutre*) 5. arbalétrier *m* principal
6. couche *f* inférieure du revêtement
en contre-plaqué
nl 1. achterkant *m*, keerzijde *f*
2. ondergrond *m* voor pleisterwerk
3. bovenzijde *f* (*van dakbedekking*)
4. buitenzijde *f* (*van gewelf*) 5. kop *m*

van een wig 6. onderlaag *n* van een
fineerbekleding
r 1. тыльная [обратная] сторона *f*
2. основание *n* под штукатурку;
поверхность *f*, подлежащая
оштукатуриванию 3. верхняя
[лицевая] грань *f* (*штучных
кровельных материалов*) 4. верхняя
часть *f* [грань *f*] (*напр. балки*)
5. главное стропило *n* 6. нижний
слой *m* фанерной обшивки

B2   backactor *see* backhoe-pullshovel

B3   *e* back arch
d verborgener Bogen *m* (*im Mauerwerk*)
f arc *m* muré dans la maçonnerie
nl ontlastingsboog *m* (*in een muur*)
r арка *f*, скрытая в каменной кладке

B4   *e* backband
d Halteleiste *f*, Dreikantleiste *f*
f parclose *f*; latte *f*
triangulaire [moulurée]
nl buitenzijde *f* van een kozijn
r раскладка *f* (*погонажное изделие*)

B5   back cutting *see* borrow pit

B6   backdigger *see* backhoe-pullshovel

B7   *e* back edging
d Glasur-Steingutrohrschneiden *n* (*in
zwei Arbeitsgängen*)
f coupage *m* des tuyaux en grès
(-cérame) vernissé (*en deux procédés*)
nl geglazuurde keramische buis (*in twee
arbeidsgangen*) snijden
r резка *f* глазурованных керамических
труб (*в два этапа*)

B8   *e* backer board
d Unterlegplatten *f* pl, Unterlegtafeln
f pl
f plaque *f* de sous-couche
nl plaatmateriaal *n* voor dakbeschot,
dakplaat *f(m)*
r подшивочный *или* подкладочный
листовой материал *m*

B9   *e* backfill
d 1. Hinterfüllung *f* 2. Fachwerkaus-
füllung *f* 3. Hintermauerung *f*,
Mauerhinterfüllung *f*
f 1. remblayage *m*, remblai *m* de
comblement [de remplissage]
2. remplissage *m* du pan
3. maçonnerie *f* de remplissage
nl 1. aanaarding *f* 2. opvullen *n*
(*vakwerkconstructie*) 3. achtermetseling
f
r 1. обратная засыпка *f* 2. заполнение
n фахверка 3. забутка *f*, забутовка *f*

B10   *e* backfill compactor
d Grabenverdichter *m*;
Aufschüttungsverdichter *m*
f compacteur *m* de sol pour terrains
remblayés
nl grondverdichter *m* (*voor gestorte grond*)
r уплотнитель *m* обратной засыпки;
уплотнитель *m* насыпи

B11 **back filling** *see* **backfill**

B12 *e* **backfill rammer**
d Grabenramme *f*
f dame *f* pour (compacter) les terrains remblayés
nl stamper *m* (*voor grondverdichting*)
r трамбовка *f* для уплотнения грунта в насыпи

B13 *e* **backflow**
d 1. Rückströmung *f*, Rückfluß *m* 2. Rückstau *m*
f 1. écoulement *m* de retour 2. remous *m*
nl 1. terugstroming *f* 2. terugstuwing *f*
r 1. обратный поток *m*, обратное течение *n* 2. распространение *n* подпора вверх по течению

B14 *e* **backflow barrier**
d Rückstauverschluß *m*, Rückstauklappe *f*
f clapet-valve *m* de retenue [antiretour, de non-retour]
nl terugslagklep *f(m)*
r обратный клапан *m* во внутренней канализационной сети

B16 **back-flushing** *see* **back washing**

B17 *e* **background heating**
d Grundheizung *f*
f chauffage *m* d'appoint
nl basisverwarming *f*
r дежурное отопление *n*

B18 **background noise** *see* **ambient noise**

B19 *e* **background noise level**
d Grundgeräuschpegel *m*
f niveau *m* de bruit de fond
nl niveau *n* van het achtergrondgeluid
r уровень *m* фонового шума

B20 *e* **backhoe-pullshovel, backhoe shovel**
d Tief(löffel)bagger *m*
f pelle *f* (mécanique) rétro
nl lepelschop-graafmachine *f*
r экскаватор *m* с обратной лопатой

B22 *e* **backing**
d 1. Hintermauerung *f* 2. putztragende Belattung *f* 3. Mörtellage *f* [Mörtelschicht *f*] für Fliesenbefestigung 4. Hinterfüllung *f* (*der Stützmauer*); Entlastungsprisma *n* (*der Kaimauer*) 5. Stau *m*, Rückstau *m*
f 1. maçonnerie *f* de remplissage 2. couche *f* de base en lattes à plâtre 3. sous-couche *f* en mortier de ciment pour carrelage 4. terre-plein *m* d'un mur de soutènement; prisme *m* de décharge (*d'un quai*) 5. retenue *f*, remous *m*
nl 1. achtermetseling *f* 2. latwerk *n* 3. mortellaag *f(m)* waarin de tegels worden gezet 4. grondaanvulling *f* achter een keermuur 5. terugstroming *f*
r 1. забут(ов)ка *f* 2. драночное основание *n* под штукатурку 3.

прослойка *f* цементного раствора для крепления плиток 4. обратная засыпка *f* грунта за подпорную стенку; разгрузочная призма *f* набережной 5. подпор *m*

B23 *e* **backing brick**
d Hintermauerungsziegel *m*
f brique *f* de remplissage
nl achterwerker *m*
r кирпич *m* для забутки

B24 *e* **backing strip**
d Schweißunterlage *f*
f bande *f* de renforcement de soudure
nl voering *f* (*onder de lasnaad*)
r подкладка *f* (*под сварным швом*)

B25 **backing up** *see* **backup 1.**

B26 **back of arch** *see* **extrados**

B27 *e* **back painting**
d 1. Grundierung *f* (*von Metallkonstruktionen*) 2. Schutzanstrich *m* für unsichtbare Flächen
f 1. peinture *f* de fond (*pour charpentes métalliques*) 2. peinture *f* de protection des surfaces invisibles
nl 1. (metaalconstructies) in de grondverf zetten 2. verven van uit het zicht blijvende gedeelte
r 1. грунтовочная краска *f* (*для металлоконструкций*) 2. защитная окраска *f* нелицевых поверхностей

B28 *e* **back pressure**
d 1. Rückstau *m*, Stau *m* 2. Gegendruck *m*, Auftrieb *m* 3. Saugdruck *m*
f 1. retenue *f*; remous *m* 2. contre-pression *f* 3. pression *f* d'aspiration; sous-pression *f*
nl 1. stuwing *f* 2. tegendruk *m* 3. onderdruk *m*
r 1. подпор *m* 2. противодавление *n* (*на сооружения*) 3. давление *n* всасывания

B29 *e* **back pressure valve**
d Saugdruckventil *n*, Saugdruckregler *m*
f vanne *f* régulatrice de contre-pression; régulateur *m* de pression d'aspiration
nl terugslagklep *f(m)*
r регулирующий клапан *m* на линии всасывания, регулятор *m* противодавления; регулятор *m* давления «до себя»

B30 *e* **back prop**
d Schrägsteife *f*
f étai *m* (incliné) de blindage
nl schoorpaal *m*
r подкос *m* креплений (*стенок земляных выемок*)

B31 *e* **back saw**
d Rückensäge *f*, Bogensäge *f*
f scie *f* à main, égoïne *f*, égohine
nl toffelzaag *f(m)*, rugzaag *f(m)*
r ножовка *f*

B32 *e* **back siphonage**
d Rücksaugen *n*
f siphonnage *m* inverse
nl terugheveling *f*
r обратное сифонирование *n*,
засасывание *n* загрязнённой воды из
канализации в водопроводную сеть

B33 *e* **back surge**
d Absperrschwall *m*, Stauschwall *m*
f onde *f* de remous
nl stuwgolf *f(m)*
r волна *f* подпора

B34 *e* **back-to-back houses**
d Reihenhäuser *n pl*
f maisons *f pl* dos à dos
nl rijtjeshuizen *n pl*
r сблокированные дома *m pl* с общей
стеной; дома *m pl* рядовой застройки

B35 *e* **backup**
d 1. Hintermauerung *f* 2. elastische
Zwischenlage *f* 3. Rückstau *m* (*in
Rohrleitungen*) 4. Unterlegstreifen *m*
f 1. briques *f pl* de remplissage;
maçonnerie *f* de remplissage
2. plaque *f* intercalaire élastique
3. accroissement *m* de pression (*dans
un conduit*) 4. bande *f* de renforcement;
bande *f* d'étanchéité
nl 1. achterwerk *n* 2. elastische
tussenlaag *f(m)* 3. terugstuwing *f* (*in
pijpleidingen*) 4. onderlaag *f(m)*
r 1. забут(ов)ка *f* 2. упругая
прокладка *f* 3. подпор *m*
(*в трубопроводе*) 4. подкладка *f*
(*под швом*)

B36 *e* **back-up block**
d Hintermauerungsblock *m*
f bloc *m* de [pour] remplissage
nl metselblok *n* voor achterwerk
r блок *m* для забут(ов)ки [для
заполнения]

B37 *e* **back-up wall**
d 1. Hintermauerung *f*
2. Innenverkleidung *f* der
Vorhangwandplatte (*zur Erhöhung
des Feuerfestigkeitsgrades*)
f 1. briques *f pl* de remplissage;
maçonnerie *f* de remplissage 2.
revêtement *m* intérieur du mur-rideau
(*pour augmenter la résistance au feu*)
nl 1. achtermetseling *f* 2.
binnenbekleiding *f* van gevelpaneel
(*ter verhoging van de brandveiligheid*)
r 1. забут(ов)ка *f* 2. внутренняя
облицовка *f* навесной стеновой
панели (*повышающая огнестойкость*)

B38 *e* **backward-dumping wheel scraper**
d rückwärtskippender Radschrapper *m*
f scraper *m* à roues à vidage par
l'arrière
nl wielscraper *m* met achterwaarts
lossende laadbak
r колёсный скрепер *m* с задней
разгрузкой

B39 *e* **back washing**
d 1. Umkehrspülung *f*, indirekte
Spülung *f* 2. Rückspülung *f*
f 1. chasse *f* par reflux 2. lavage *m*
par contre-courant (*d'un filtre*)
nl 1. omkeerspoeling *f* 2. terugspoeling *f*
(*van riool of filter*)
r 1. обратная промывка *f* (*метод
освоения водоносного горизонта*)
2. обратная промывка *f* (фильтра)

B40 *e* **backwater**
d 1. Stau *m*, Wasserstau *m*;
Staubereich *m*, Staugebiet *n* 2.
Liegehafen *m* 3. Rücklaufwasser *n*
f 1. remous *m* 2. anse *f* 3. eaux *f pl*
de restitution [recyclées]
nl 1. (water)stuw *m*; stuwzone *f*
2. lighaven *f(m)* 3. teruglopend
water *n*
r 1. подпор *m*; зона *f* подпора
2. затон *m* 3. возвратные воды *f pl*

B41 **backwater area** *see* **backwater 1.**

B42 *e* **backwater curve**
d Staulinie *f*, Staukurve *f*
f courbe *f* de remous
nl stuwlijn *f(m)*, stuwkromme *f(m)*
r кривая *f* подпора

B43 *e* **backwater jump**
d eingestauter Wassersprung *m*
f ressaut *m* noyé
nl (water)sprong *m* door opstuwing
r затопленный (гидравлический)
прыжок *m*

B44 **backwater profile** *see* **backwater curve**

B45 *e* **backwater structure**
d Wasserstauanlage *f*
f ouvrage *m* de retenue
nl stuwwerk *n*, stuw *m*
r водоподпорное сооружение *n*

B46 **backwater valve** *see* **backflow barrier**

B47 *e* **bacteria bed, bacteria filter**
d Tropfkörper *m*
f lit *m* bactérien
nl bacteriologisch filter *n*, *m*
r биологический фильтр *m*, биофильтр
m

B49 *e* **badger**
d Kratzer *m*
f grattoir *m*
nl krabber *m*
r шабровка *f*; очистительный скребок
m

B50 *e* **badger plain**
d Kehlhobel *m*, Spundhobel *m*
f guillaume *m*, rabot *m* à rainurer
nl sponningschaaf *f(m)*
r зензубель *m*, шпунтгебель *m*,
пазник *m*, калёвочник *m*

B51 *e* **baffle**
d 1. Energievernichter *m*,
Energiebrecher *m* und Auflöser *m*
2. Prallblech *n*, Leitblech *n*,
Leitwand *f*, Strömungsleiteinrichtung

f, Strahlablenker *m* 3. Klappe *f*
f 1. chicane *f*, tranqui lisateur *m*
   2. déflecteur *m*, bloc *m* de
   diffusion 3. volet *m*
nl 1. stroomdemper *m* 2. keerplaat *f(m)*,
   leischot *n* 3. schuifafsluiter *m*
r 1. гаситель *m* энергии потока;
   гаситель-растекатель *m* 2.
   (струе)направляющая перегородка *f*,
   лопатка *f* (*в воздуховодах*) 3. заслонка *f*

B52 *e* **baffle beam**
   d Staubalken *m*
   f poutre *f* masque
   nl stuwbalk *m*
   r забральная балка *f*

B53 *e* **baffle block**
   d Dämpferblock *m*, Ablenkblock *m*,
     Prallblock *m*, Schikane *f*,
     Bremspfeiler *m*
   f bloc *m* chicane
   nl rempijler *m*
   r шашечный гаситель *m*

B54 *e* **baffle diffuser**
   d Prallplattenluftverteiler *m*
   f diffuseur *m* d'air à disque
   nl luchtverdeler *m* met scherm
   r дисковый воздухораспределительный
     плафон *m*, диффузор *m* с экраном

B55 *e* **baffle pier**
   d Sohlblock *m*,
     Energievernichtungspfeiler *m*,
     Störpfeiler *m*, Bremspfeiler *m*
   f bloc *m* de dissipation d'énergie, pile *f*
     de freinage
   nl stroombreker *m*, stromingsbreker *m*
   r пирсовый гаситель *m*

B56 *e* **baffle plate**
   d Prallplatte *f*, Stauscheibe *f*,
     Trennwand *f*
   f chicane *f*, baffle *m*
   nl stootplaat *f(m)*, stuwplaat *f(m)*
   r отражатель *m*, экран *m*,
     перегородка *f*

B57 *e* **baffle wall**
   d Toswand *f*, Stoßnase *f*, Gegensperre *f*;
     Prallwand *f*
   f contre-barrage *m*; mur *m* déflecteur
   nl keermuur *m*
   r водобойная стенка *f*;
     отклоняющая стенка *f*

B58 *e* **bag**
   d 1. Sack *m* 2. Schlauch *m*,
     Filterbeutel *m*
   f 1. sac *m* 2. manche *f* (*de filtre*)
   nl 1. zak *m* 2. filterzak *m*
   r 1. мешок *m* 2. рукав *m* (*фильтра*)

B59 *e* **bag-cleaning device**
   d Schlauchfilter-Regeneriereinrichtung *f*
   f dispositif *m* de nettoyage de filtres
     à manche
   nl zakfilterreinigingsinstallatie *f*
   r устройство *n* для регенерации [для
     очистки от пыли] рукавных фильтров

B60 *e* **bag conveyor**

d Sackförderer *m*
f transporteur *m* de sacs
nl zakkentransporteur *m*
r конвейер *m* для транспортирования
   сыпучих материалов в мешках

B61 *e* **bag dam**
   d Sacksperre *f*
   f digue *f* en sacs de sable *ou* de béton
   nl dam *m* van zandzakken
   r дамба *f* из мешков с песком *или*
     бетоном

B62 *e* **bag filter**
   d Schlauchfilter *n, m*
   f filtre *m* à manche(s)
   nl zakfilter *n, m*
   r рукавный фильтр *m*

B63 *e* **bagwork**
   d Sackbeton(ab)deckung *f*
   f 1. ouvrage *m* de défense du littoral
     en sacs de béton 2. mise *f* en œuvre
     du béton sec en sacs
   nl oeverbescherming *f* met zandzakken *of*
     zakken betonspecie
   r 1. берегоукрепление *n* из мешков
     с тощим бетоном 2. укладка *f* сухой
     бетонной смеси в мешках

B64 *e* **Bailey bridge**
   d Bailey-Behelfsbrücke *f*
   f pont *m* (de) Bailey
   nl Baily-brug *f(m)*
   r мост *m* Бэйли, сборно-разборный
     мост *m* с металлическими фермами

B65 *e* **bail handle**
   d Griffstange *f*
   f poignée *f* de porte en éttrier, poignée-
     barre *f*
   nl deurklink *f(m)*, sluitbalk *m* (*van een*
     schuurdeur)
   r ручка-скоба *f*

B66 *e* **baked brick**
   d gebrannter Ziegel(stein) *m*
   f brique *f* cuite
   nl baksteen *m*
   r обожжённый кирпич *m*

B67 *e* **baked enamel finish**
   d gebrannter Emailüberzug *m*
   f revêtement *m* émaillé séché au four
   nl glazuur *n*
   r эмалевая отделка *f*, закреплённая
     обжигом

B68 **baker dolphin** *see* **bell dolphin**

B69 **balance bridge** *see* **bascule bridge**

B70 *e* **balanced cantilever method**
   d Freivorbauverfahren *n* (*Montage des*
     Brückentragwerkes)
   f montage *m* en porte-à-faux [en
     encorbellement]
   nl vrije-uitbouwmethode *f* (*bruggenbouw*)
   r монтаж *m* моста методом
     (уравновешенной) навесной сборки

B71 *e* **balanced draught**
   d ausgeglichener Zug *m*
   f tirage *m* (*de cheminée*) équilibré

nl evenwichtige trek *m*
r уравновешенная тяга *f* (*в дымоходах*)

B72 *e* **balanced earthworks**
d Erdbau *m* mit Massenausgleich
f terrassements *m pl* équilibrés
nl massanivellement *n*
r сбалансированные земляные работы *f pl*

B73 *e* **balanced flue**
d Rauchkanal *m* mit ausgeglichenem Zug, kombiniertes Abgas-Frischluft--System *n*
f conduit *m* de fumée vertical équilibré
nl rookkanaal *n* met evenwichtige trek
r дымоход *m* с уравновешенной тягой

B74 *e* **balanced-flue heater**
d Luftheizgerät *n* [Luftheizaggregat *n*] mit ausgeglichenem Zug im Rauchkanal
f installation *f* de chauffage à air chaud avec le tirage équilibré
nl luchtverhitter *m* met evenwichtige trek
r агрегат *m* воздушного отопления с уравновешенной тягой в дымоходах

B75 *e* **balanced gate**
d Gleichgewichtsklappe *f*, Gleichgewichtsverschluß *m*
f clapet *m* équilibré; vanne *f* équilibrée
nl balansklep *f(m)*
r уравновешенный затвор *m*

B76 *e* **balanced load**
d Bruchbelastung *f* des Stahlbetonbauelementes
f sollicitation *f* limite équilibrée (*provoquant simultanément la rupture du béton et l'écoulement plastique de l'armature*)
nl grensbelasting *f*
r предельная нагрузка *f*, вызывающая одновременно разрушение бетона и пластическое течение арматуры

B77 *e* **balanced moment**
d 1. ausgeglichene Momentenbelastung *f* 2. Bruchmoment *n* (*Stahlbetonbau*)
f 1. moment *m* équilibré 2. moment *m* limite équilibré (*provoquant simultanément la rupture du béton et l'écoulement plastique de l'armature*)
nl samenvallend scheur- en vloeimoment *n*
r 1. уравновешенный момент *m* 2. предельный момент *m*, вызывающий одновременное разрушение бетона и пластическое течение арматуры

B78 *e* **balanced reinforcement**
d gleichfeste Bewehrung *f*
f armature *f* équilibrée (*égale en résistance au béton*)
nl uitgebalanceerde wapening *f*
r арматура *f* (*напр. балки*), равнопрочная с бетоном

B79 *e* **balanced ventilation system**
d ausgeglichene Be- und Entlüftung *f*

f système *m* de ventilation équilibré
nl uitgebalanceerd ventilatiesysteem *n*
r сбалансированная приточно--вытяжная система *f* вентиляции

B80 *e* **balance pressure**
d Gleichgewichtsdruck *m*
f pression *f* d'équilibre
nl evenwichtsdruk *m*
r равновесное давление *n*

B81 *e* **balance storage**
d Ausgleichbecken *n*
f réservoir *m* journalier
nl vereffeningsbekken *n*
r водохранилище *n* суточного регулирования

B82 *e* **balancing**
d 1. Ausgleich *m*, Auswuchtung *f* 2. Abstimmung *f*, Koordinierung *f*
f 1. balancement *m*; équilibrage *m* 2. compensation *f*
nl 1. in evenwicht brengen *n*, balancering *f* 2. vereffening *f* (van meetfouten)
r 1. уравновешивание *n*; балансировка *f* 2. увязка *f* погрешностей (*геодезической съёмки*)

B83 *e* **balancing and commissioning**
d Einregulierung *f* und Inbetriebnahme *f*
f travaux *m pl* de mise au point
nl afstellen *n* en in werking stellen *n*
r пусконаладочные работы *f pl*, пусковая наладка *f*

B84 *e* **balancing of building systems**
d Ausgleichen *n*, Abgleichung *f*, Einregulierung *f*, Einregelung *f* (von Gebäudeausrüstungsanlagen)
f réglage *m*, mise *f* au point (de l'équipement immobilier)
nl op elkaar afstemmen *n* van bouwsystemen
r наладка *f* инженерных систем здания

B85 *e* **balancing of stresses**
d Spannungsausgleich *m*
f balancement *m* [équilibrage *m*] des contraintes
nl spanningsvereffening *f*
r уравновешивание *n* [выравнивание *n*] напряжений

B86 *e* **balancing reservoir**
d Ausgleichbecken *n*, Ausgleichweiher *m*
f bassin *m* de compensation
nl compensatiebekken *n*
r компенсационное [регулирующее] водохранилище *n*; выравнивающий бассейн *m*

B87 *e* **balancing tank**
d Ausgleichspeicher *m*; Ausgleichbecken *n*
f cheminée *f* d'équilibre
nl compensatiebassin *n*
r уравнительный бассейн *m*; усреднительный резервуар *m*

B88 e **balancing valve**
  d Druckausgleichventil *n*,
    Einstellventil *n*
  f soupape *f* égalisatrice (de pression),
    clapet *m* d'équilibrage, vanne *f*
    égalisatrice
  nl instelklep *f(m)*
  r наладочный клапан *m*

B89 e **balcony exterior exit**
  d Notausgang *m* durch Balkon
  f sortie *f* de secours par balcon
  nl nooduitgang *m* over het balkon
  r аварийный выход *m* через балкон

B90 e **balcony slab**
  d Balkonplatte *f*
  f dalle *f* de balcon
  nl balkonplaat *f(m)*
  r балконная плита *f*

B91 e **balcony-window**
  d Balkonfenster *n*
  f fenêtre *f* à [de] balcon
  nl balkonvenster *n*
  r балконное окно *n*

B92 e **ballast**
  d 1. Zusatzladung *f*, tote Last *f*,
    Ausgleichslast *f* 2. Bettungsmaterial *n*
    (*Bahnbau*); Schotterbett *n*
  f 1. lest *m* 2. ballast *m*; couche *f* de
    ballast
  nl 1. ballast *m*; last *m* 2. ballast *m*,
    ballastbed *n*
  r 1. балласт *m*; груз *m* 2. балласт *m*
    (*инертный материал*); балластный
    слой *m*

B93 **ballast bed** *see* **ballast 2.**

B94 e **ballast crusher**
  d Gleisschotterbrecher *m*
  f concasseur *m* pour ballast
  nl steenbreker *m*, steenbreekmachine *f*
  r дробилка *f* для (приготовления)
    балласта

B95 e **ballast pit**
  d Schottergrube *f*
  f ballastière *f*
  nl steengroeve *f*
  r гравийный карьер *m*

B96 e **ball clay**
  d Ballton *m*, Töpferton *m*,
    Tonklumpen *m*
  f argile *f* en mottes
  nl pottenbakkersklei *f(m)*
  r комовая глина *f*

B97 e **ball joint**
  d Kugelgelenk *n*
  f articulation *f* sphérique, joint *m*
    à bille
  nl kogelscharnier *n*
  r шаровой шарнир *m*

B98 e **ball jointed rocker bearing**
  d Kugelzapfenkipp(auf)lager *n*
  f appui *m* oscillant à pivot sphérique
  nl kogeltap-pendeloplegging *f*

  r шарнирная балансирная опора *f*
    (*моста*)

B99 e **ball lock**
  d Kugelverschluß *m*
  f vanne *f* sphérique
  nl kogelafsluiter *m*
  r шаровой затвор *m*

B100 e **ball mill**
  d Kugelmühle *f*
  f broyeur *m* à boulets, moulin *m*
    à billes
  nl kogelmolen *m*
  r шаровая мельница *f*

B101 e **ball pivot**
  d Kugellager *n*
  f pivot *m* sphérique
  nl kogeltap *m*
  r шаровая пята *f*

B102 e **ball test**
  d Kugelschlagprüfung *f* nach Kelly
  f essai *m* (de) Kelly
  nl kogelproef *f(m)* (*ter bepaling van
    consistentie van beton*)
  r определение *n* консистенции бетона
    по погружению цилиндрического
    стержня с полусферическим нако-
    нечником при его падении

B103 e **ball valve**
  d 1. Schwimmerventil *n* 2. Kugel-
    -Rückschlagventil *n* 3. Kugelhahn *m*
  f 1. soupape *f* à flotteur 2. clapet *m*
    à bille 3. robinet *m* à corps sphérique
  nl 1. vlotterventiel *n* 2. kogelklep *f(m)*,
    kogelventiel *n* 3. kogelkraan *f(m)*
  r 1. поплавковый клапан *m*
    2. шаровой (обратный) клапан *m*
    3. шаровой кран *m*

B104 e **baluster**
  d Baluster *m*, Geländerpfosten *m*
  f balustre *m*
  nl baluster *m*, leuningstijl *m*
  r балясина *f*

B105 e **band**
  d 1. Zugband *n* 2. Triebriemen *m*
    3. Kranz *m*, Gesims *n* 4. flaches
    leistenförmiges Bauelement *n* 5. Lage
    f, Schicht *f*
  f 1. frette *f*, bandage *m* 2. courroie *f*
    de transmission 3. bandeau *m* 4. latte
    f moulurée 5. nappe *f*
  nl 1. trekband *m* 2. drijfriem *m* 3.
    krans *f(m)* 4. lijst *f(m)*, lint *n* 5. laag
    f(m)
  r 1. стяжной хомут *m* 2. приводной
    ремень *m* 3. поясок *m*, сандрик *m*
    4. плоская погонажная деталь *f*
    5. слой *m*

B106 **bandage** *see* **pipe shell**

B107 e **B and better lumber**
  d Schnittholz *n* der Güteklasse I
  f bois *m* de charpente de première
    qualité

nl bezaagd hout *n* van de eerste soort
r первосортный пиломатериал *m*

B108 **band conveyor** *see* **belt conveyor**

B109 *e* **band chain**
d Meßkette *f*, Vermessungskette *f*
f chaîne *f* d'arpenteur
nl meetband *m*, meetketting *m*, *f*
r мерная цепь *f* (*для геодезических работ*)

B110 *e* **band filter**
d Bandfilter *n*, *m*
f filtre *m* à bande
nl bandfilter *n*, *m*
r ленточный фильтр *m*

B111 *e* **band saw**
d Bandsäge *f*
f scie *f* à ruban
nl lintzaag *f(m)*
r ленточная пила *f*

B112 *e* **band screen**
d Bandrechen *m*, Siebband *n*
f tamis *m* à bande
nl bandzeef *f(m)*, bandrooster *n*
r ленточное сито *n*

B113 *e* **bank**
d 1. Wall *m*, Damm *m*, Deich *m*
2. Auftrag *m*, Schüttung *f*
3. Bodenböschung *f*, Erdböschung *f*
4. abzutragender Boden *m*
5. Abraummassiv *n* über dem Niveau der Grubensohle
f 1. digue *f* 2. remblai *m* 3. talus *m* de remblai 4. sol *m* à exaver
5. massif *m* de terrain au-dessus du niveau de l'excavation
nl 1. wal *m*; dijk *m* 2. ophoging *f*, aarden wal *m* 3. glooiing *f*, talud *n*
4. te ontgraven grond *m* 5. af te graven bovengrond
r 1. вал *m*; дамба *f* 2. насыпь *f*
3. спланированный [профилированный] грунтовый откос *m* 4. грунт *m*, подлежащий выемке
5. массив *m* грунта над уровнем копания

B114 *e* **banked-up water level**
d Stauspiegel *m*
f niveau *m* de retenue
nl stuwpeil *n*
r подпорный уровень *m*

B115 *e* **bankhead**
d Buhne *f*; Querdeich *m*, Querbau *m*
f épi *m*; traverse *f*
nl dwarsdijk *m*
r полузапруда *f*, струенаправляющая буна *f*; поперечная дамба *f*, траверс *m*

B116 *e* **banking**
d Eindeichung *f*
f endiguement *m*
nl indijking *f*
r обвалование *n*

B117 **bank material** *see* **bank 4.**

B118 *e* **bank measure**
d Maß *n* in fester Masse (*der Rauminhalt von Erdreich oder Fels in ungestörter Lagerung*)
f volume *m* du sol en place, cubage *m* de matériaux en place
nl volume *n* van de ongeroerde grond *of* gesteente
r объём *m* грунта в естественном залегании, кубатура *f* земляных работ

B119 *e* **bank protection dam**
d Uferschutzdamm *m*
f digue *f* parafouille [de défense des rives]
nl oeverbeschermende dam *m*
r берегозащитная дамба *f*

B120 *e* **bank protection structure**
d Uferschutzwerk *n*, Küstenschutzwerk *n*
f ouvrage *m* de défense des côtes
nl oeverbescherming *f*
r берегозащитное сооружение *n*

B121 *e* **bank protection work**
d Uferschutz *m*, Deckwerk *n*, Uferbefestigung *f*, Uferausbau *m*
f protection *f* des berges [des côtes]
nl beschoeiing *f*
r берегозащитные мероприятия *n pl*; берегоукрепительные работы *f pl*

B122 *e* **bank revetment**
d Uferbekleidung *f*, Uferbefestigung *f*, Ufersicherung *f*, Uferdeckwerk *n*
f revêtement *m* des berges
nl oeverbekleding *f*
r берегоукрепление *n*, береговая одежда *f*; одежда *f* откосов

B123 *e* **bank-run gravel**
d Naturkies *m*
f gravier *m* de carrière
nl gedolven grind *n*
r природная песчано-гравийная смесь *f*

B124 *e* **bank sloping**
d Abböschen *n*, Böschungsziehen *n*
f talutage *m*, mise *f* en talus
nl talud maken *n*
r планировка *f* откоса

B125 *e* **banksman**
d Anschläger *m*
f signaleur *m*
nl seingever *m*
r сигнальщик *m* (*на монтажных работах*)

B126 **bank stabilisation** *see* **bank protection work**

B127 *e* **banquette**
d 1. Bankett *n*, Dammbankett *n*, Sickerkörper *m* 2. Berme *f*
f 1. banquette *f* 2. berme *f*
nl 1. banket *n* 2. berm *f*
r 1. банкет *m*, дренажная призма *f*
2. берма *f*

B128 *e* **bar**
 d 1. Stab *m* 2. Bewehrungsstab *m*
 f 1. barre *f* 2. barre *f* d'armature,
 rond *m* à béton
 nl 1. staaf *f(m)*, stang *f(m)*, spijl *f(m)*
 2. wapeningsstaaf *f(m)*
 r 1. стержень *m* 2. арматурный
 стержень *m*

B129 *e* **barbed nail**
 d Dachnagel *m* mit Widerhaken
 f clou *m* barbelé
 nl daknagel *m* met weerhaken
 r заершённый гвоздь *m*

B131 *e* **bar bender, bar-bending machine**
 d Betonstahl-Biegemaschine *f*
 f coudeuse *f*, cintreuse *f* (*pour ronds*
 à béton)
 nl buigmachine *f* voor betonstaal
 r станок *m* для гибки арматуры,
 гибочный станок *m*

B132 **bar chairs** *see* **bar spacers**

B133 *e* **bar cropping machine**
 d Rundstahlschneidemaschine *f*
 f tronçonneuse *f*
 nl snijmachine *f* voor betonstaal
 r станок *m* для резки арматурной
 стали

B134 *e* **bar cross-section**
 d Stabquerschnitt *m*, Gliedquerschnitt
 m
 f section *f* d'une barre
 nl staafdoorsnede *f*
 r поперечное сечение *n* стержня

B135 *e* **bar cutter**
 d Betonstahl(hand)schneider *m*,
 Betonstahlschere *f*
 f cisailles *f pl* à barres
 nl handsnijbank *f(m)* [hefboomschaar
 f(m)] voor betonstaal
 r ручной станок *m* для резки
 арматурной стали

B136 **bar cutting shears** *see* **bar cutter**

B137 *e* **barefaced tenon, bareface tenon**
 d Seitenzapfen *m*
 f ténon *m* mi-bois
 nl eenzijdige pen *f(m)*, pen *f(m)* met
 één borst (*pen en gatverbinding*)
 r боковой шип *m* вполдерева

B138 *e* **bar list**
 d Bewehrungs(stahl)liste *f*
 f nomenclature *f* des ronds [des fers]
 à béton
 r l specificatie *f* van de wapening
 r спецификация *f* арматуры

B139 *e* **barrage**
 d Stauwehr *n*, Staudamm *m*;
 Umleitungssperre *f*; Fang(e)damm *m*
 f barrage *m* de dénivellation [en
 rivière]; barrage *m* de garde
 nl stuw *m*, stuwdam *m*
 r водоподъёмная плотина *f*;
 заградительная перемычка *f*

B140 *e* **barrage power station**
 d Stauwasserkraftwerk *n*,
 Sperrenkraftwerk *n*
 f usine *f* de retenue, usine-barrage *m*,
 centrale-barrage *m*
 nl waterkrachtcentrale *f* (*in een
 stuwdam*)
 r приплотинная ГЭС *f*

B141 *e* **barrel**
 d Heberrohr *n*, Siphonrohr *n*,
 Heberleitung *f*
 f gaine *f* de siphon [d'aspiration]
 nl pompbuis *f(m)*
 r труба *f* сифона, сифонный водовод *m*

B142 *e* **barrel roof, barrel shell roof**
 d zylindrisches Gewölbe *n*
 f coque-voûte *f*; voûte *f* cylindrique
 [en tonnelle]
 nl cilindrisch schaaldak *n*
 r цилиндрическая оболочка *f*;
 цилиндрический свод *m*

B143 *e* **barrel theory**
 d Tonnentheorie *f*
 f théorie *f* de coques-voûtes
 nl (theoretische) berekening *f* van
 schaaldaken
 r теория *f* расчёта цилиндрических
 оболочек

B144 *e* **barre vault**
 d Tonnengewölbe *n*
 f voûte *f* cylindrique [en tonnelle]
 nl tongewelf *n*
 r цилиндрический свод *m*

B145 *e* **barricade**
 d Absperrung *f*, Sperre *f*
 f barricade *f*, clôture *f* de protection
 nl versperring *f*, barricade *f*
 r заграждение *n*, ограждение *n*,
 преграда *f*

B146 *e* **barricade lantern**
 d Sperrlichtsignal *n*
 f feu *m* d'obstacle, lanterne *f*
 portative d'obstacle
 nl opstakellamp *f(m)*
 r оградительный огонь *m* (*при производ-
 стве строительных или дорожных
 работ*)

B147 *e* **barrier railings**
 d Schutzgeländer *n pl*
 f garde-corps *m*
 nl wegafzetting *f*, dranghekken *n pl*
 r перильные ограждения *n pl*

B148 *e* **barrier shielding**
 d Strahlenschutzumhüllung *f*,
 Strahlenschutzpanzer *m*
 f caisson *m* du réacteur nucléaire, écran
 m étanche
 nl afscherming *f* (*tegen radioaktieve
 straling*)
 r защитное ограждение *n* (*реактора*)

B149 *e* **barrow**
 d Schiebkarre *f*, Benne *f*

f brouette *f*
nl kruiwagen *m*
r тачка *f*

B150 *e* **barrow run**
d Karrdiele *f*
f planche *f* pour le roulement [pour la circulation] de brouettes
nl loopplank *f(m)* voor kruiwagens
r катальный ход *m*

B151 *e* **bar slope**
d Gliedablenkung *f*
f inclinaison *f* de la diagonale du treillis
nl helling *f* van de vakwerkdiagonaal
r наклон *m* раскоса (*решётки фермы*)

B152 *e* **bar spacers**
d Betonstahl-Abstandhalter *m pl*, Bewehrungsabstandhalter *m*, *pl*
f espaceurs *m pl* [distanciers *m pl*] pour ferraillage
nl afstandhouders *m pl* voor de wapeningsstaven
r фиксаторы *m pl* арматуры (*для установки в проектное положение*)

B153 *e* **bar spacing**
d Bewehrungsstababstand *m*
f espacement *m* de barres
nl afstand *m* tussen de staven
r шаг *m* арматурных стержней, осевое расстояние *n* между арматурными стержнями

B154 *e* **bar stress**
d Stabspannung *f*, Gliedspannung *f*
f contrainte *f* dans la barre (*de treillis*)
nl spanning *f* in de staaf
r усилие *n* в стержне (*решётки фермы*)

B155 **bar supports** *see* **bar spacers**

B156 *e* **bar system**
d Stabsystem *n*, Stabwerk *n*
f système *m* de barres; système *m* réticulé [à trellis]
nl staafwerk *n*
r стержневая система *f*; решётчатая система *f*

B157 *e* **bascule bridge**
d Klappbrücke *f*, Wippbrücke *f*
f pont *m* basculant [à bascule]
nl basculebrug *f(m)*
r раскрывающийся (разводной) мост *m*

B158 *e* **base**
d 1. Fundament *n*, Grundmauerwerk *n*, Gründung *f* 2. obere Tragschicht *f* (*Straße*) 3. Plinthe *f*, Fußleiste *f*, Scheuerleiste *f*
f 1. fondation *f* 2. couche *f* de base 3. antébois *m*, plinthe *f*
nl 1. fundament *n*, voetplaat *f(m)*, sokkel *m* 2. fundering *f* (*wegenbouw*) 3. plint *f(m)*
r 1. фундамент *m*; основание *n* 2. основание *n* дорожной одежды 3. плинтус *m*

B159 *e* **base bid, base bid price**
d Angebotssumme *f*, Angebotspreis *m*
f montant *m* de la soumission (*non compris les offres alternatives*)
nl inschrijvingsbedrag *n*, prijsopgave *f*
r стоимость *f* строительства, заявленная на торгах

B161 *e* **base bid specifications with alternates**
d technische Vorschriften *f pl* und Leistungsverzeichnis *n*, die Veränderungen seitens Bauauftragnehmer zulassen
f cahier *m* des charges avec alternatives
nl bestek *n* met alternatieve uitvoeringsmogelijkheden
r основные технические условия *n pl* на строительство, допускающие альтернативные предложения со стороны подрядчиков

B162 *e* **baseboard**
d Fußleiste *f*, Scheuerleiste *f*
f antébois *m*, plinthe *f*
nl plint *f(m)*
r плинтус *m*

B163 *e* **baseboard heating**
d Fußleistenheizung *f*
f chauffage *m* par plinthes chauffantes
nl plintverwarming *f*
r плинтусное отопление *n*

B164 *e* **baseboard radiator**
d Fußleisten-Radiator *m*
f radiateur *m* à plinthe, plinthe *f* chauffante
nl plint-radiator *m*
r плинтусный радиатор *m*

B165 *e* **base coat**
d 1. Rauhwerk *n*, Grobputzschicht *f*, Grundputzlage *f* 2. Grundanstrich *m*
f 1. sous-enduit *m* 2. couche *f* de fond
nl ondergrond *m* voor pleister- of verflaag
r 1. нижние слои *m pl* [нижний слой *m*] штукатурного покрытия (*напр. обрызг и грунт*) 2. грунт *m* (*под лакокрасочное покрытие*)

B166 *e* **base course**
d 1. Binder *m*; Ausgleichschicht *f* 2. Binder *m*, Unterbauauflage *f* 3. untere Mauerwerkschicht *f*
f 1. couche *f* de profilage; couche *f* de nivellement 2. couche *f* de liaison [de base] 3. assise *f* de base
nl 1. onderlaag *f(m)* 2. funderingslaag *f(m)* 3. onderste laag *f (m)* van een gemetselde muur
r 1. подстилающий [выравнивающий] слой *m* 2. нижний слой *m* дорожного покрытия (*под слоем износа*) или основания 3. нижний ряд *m* кладки

B167 *e* **base flow**
d 1. unterirdischer Zufluß *m*, Grundwasserzufluß *m* 2. Grundwasserabfluß *m*

f 1. alimentation *f* souterraine 2. débit *m* de base
nl grondwaterstroming *f*
r 1. подземное питание *n* (*приток подземных вод в водотоки и водоёмы*) 2. подземный сток *m*

B168 base-flow recession *see* depletion curve

B169 *e* base line
d Basis *f*, Grundlinie *f*
f ligne *f* de base
nl basis *f*
r базовая линия *f*

B170 *e* basement
d 1. Kellergeschoß *n*, Halbkeller *m* 2. Unterbau *m*; Sockel *m*, Säulenfuß *m*
f 1. socle *m*; sous-sol *m* 2. base *f*, fondation *f*, soubassement *m*
nl 1. kelder *m*, souterrain *n* 2. grondmuur *m*, sokkel *m*
r 1. цокольный *или* полуподвальный этаж *m*; подвал *m*; подвальный этаж *m* 2. основание *n*, фундамент *m*; цоколь *m*; база *f* колонны

B171 *e* basement extension
d außenseits der oberirdischen Gebäudeumrißlinie liegender Teil *m* des Kellergeschosses
f partie *f* du sous-sol hors le contour du bâtiment
nl buiten het huis uitstekend keldergedeelte *n*
r часть *f* подвала за пределами дома

B172 base of foundation *see* foundation base

B173 *e* base plate
d Fußplatte *f*, Sohlplatte *f*, Lagerplatte *f*; Fundamentplatte *f*
f plaque *f* d'appui
nl voetplaat *f(m)*, grondplaat *f(m)*, fundatieplaat *f(m)*
r опорная плита *f* (*напр. колонны*); плита *f* фундамента

B174 base radiator *see* baseboard radiator

B175 base runoff *see* base flow 2.

B176 *e* base sheet
d Unterschicht *f* der mehrschichtigen Dacheindeckung
f couche *f* inférieure de la couverture en carton asphalté
nl onderste laag *f(m)* van een bitumineuze dakbedekking
r нижний слой *m* многослойной мягкой кровли

B177 base slab *see* foundation slab

B178 *e* basic floor area
d Nutzfläche *f* des Gebäudes
f surface *f* utile d'un bâtiment
nl nuttige oppervlakte *f* van een gebouw
r полезная площадь *f* здания

e basic geometry
B179*d* geometrische Grundwerte *m pl* und Grundeigenschaften *f pl*
caractéristiques *f pl* géométriques *f* incipales; configuration *f*

nl hoofdmaten *f(m)* en indeling *f*
r основные геометрические характеристики *f pl* (*напр. сооружения, конструкции*); геометрическое очертание *n*; конфигурация *f*

B180 *e* basic hydrologic data
d hydrologische Grundwerte *m pl*
f données *f pl* hydrologiques de base
nl hydrologische basisgegevens *pl*
r основные [исходные] гидрологические данные *pl*

B181 *e* basic items of work
d Hauptleistungsverzeichnis *n* (*in Baukostenvoranschlag*)
f liste *f* [énumération *f*] des activités essentielles
nl hoofdonderdelen *n pl* van het werk
r перечень *m* основных видов строительно-монтажных работ (*напр. в смете, договоре*)

B182 *e* basin
d 1. Becken *n* 2. Handwaschbecken *n*
f 1. bassin *m*, réservoir *m* 2. lavabo *m*
nl 1. bekken *n*, bassin *n*, havenbekken *n* 2. wastafel *f(m)*, wasbak *m*
r 1. бассейн *m*; водоём *m* 2. умывальник *m*, кухонная раковина *f*

B183 *e* basin for shipping
d Hafenbecken *n*
f darce *f*, darse *f*, bassin *m* portuaire [deport]
nl havenbekken *n*
r портовый бассейн *m*

B184 *e* basin lock
d Sparschleuse *f*, Speicherschleuse *f*
f écluse *f* avec bassins d'épargne
nl spaarsluis *f(m)*
r шлюз *m* со сберегательными бассейнами

B185 basis of design *see* design principles

B186 *e* basket hitch
d Ringstropp *m*
f élingue *f* annulaire [en boucle]
nl ringstrop *m*
r кольцевой [охватывающий] строп *m*

B187 *e* bat
d Halbziegel *m*
f demi-brique *f*, brique *f* à deux quartiers
nl halve steen *m*
r половняк *m*, полкирпича

B188 *e* batch
d 1. Mischung *f* 2. Charge *f*; Dosis *f*; Dosierung *f*, Zuteilung *f*
f 1. gâchée *f* 2. charge *f*; dose *f*
nl 1. vulling *f* (*betonmolen*) 2. gemeten hoeveelheid *f* materiaal
r 1. замес *m* (*напр. бетона*) 2. отмеренная порция *f* (*материала*); доза *f*

B189 *e* batch box
d Abmeßkasten *m*, Dosierkasten *m*
f boîte *f* de dosage [de mesure]

nl meetkast *f(m)*
r мерный ящик *m*

B190 *e* **batch counter**
d Chargenzähler *m*, Mischungszähler *m*
f enregistreur *m* des gâchées
nl ladingenteller *m* *(betonmolen)*
r счётчик *m* (числа) замесов

B191 *e* **batched water**
d Anmach(e)wasser *n* *(für eine Mischung)*
f eau *f* de gâchage *(pour une seule gâchée)*
nl per lading betonspecie afgemeten hoeveelheid *f* materiaal
r вода *f* затворения *(на один замес)*

B192 *e* **batcher**
d Dosierer *m*, Dosierwaage *f*
f doseur *m*, doseuse *f*
nl doseringsapparaat *n*
r дозатор *m*

B193 *e* **batching**
d Dosierung *f*, Zuteilung *f*, Zumessung *f*, Abmessung *f*
f dosage *m*
nl dosering *f*
r дозирование *n*, дозировка *f*

B194 *e* **batching cycle**
d Dosierzyklus *m*, Zumeßfolge *f*
f cycle *m* de dosage
nl doseringscyclus *m*
r дозировочный цикл *m*, цикл *m* дозировки

B195 *e* **batching plant**
d Dosieranlage *f*, Zumeßanlage *f*, Zuteilanlage *f*
f poste *m* [groupe *m*] (de) dosage
nl doseringsinrichting *f*
r дозировочная установка *f*

B196 *e* **batch mixer**
d Chargenmischer *m*, Stoßmischer *m*
f malaxeur *m* discontinu, malaxeuse *f* discontinue
nl betonmolen *m*
r бетоносмеситель *m* периодического действия, цикличный бетоносмеситель *m*

B197 *e* **batch weights**
d Dosierteile *m*, *n pl*, Zumeßteile *m*, *n pl*
f doses *f pl* pondérales [en poids] des composants du béton
nl bestanddelen *n pl* in afgemeten hoeveelheden
r массовые порции *f pl* составляющих (бетонной смеси)

B198 *e* **bathroom**
d Badezimmer *n*, Bad *n*
f salle *f* de bain
nl badkamer *f(m)*
r ванная комната *f*

B199 *e* **bath trap**
d Badsiphon *m*
f siphon *m* de baignoire

nl stankafsluiter *m*
r сифон *m* ванны

B200 *e* **batt**
d Isoliermatte *f*, Isolierplatte *f*
f matelas *m* isolant, natte *f* isolante
nl isolerende mat *f(m)*
r звуко- *или* теплоизолирующий мат *m* *(напр. из войлока)*

B201 *e* **batten**
d Dachlatte *f*; Leiste *f*; Fugendeckleiste *f*
f latte *f* de toit; liteau *m*, listeau *m*; latte *f*
nl op lengte bezaagd hout *n*; lat *f(m)*; regel *m*, balk *m*; stierlijst *f(m)*, deklijst *f(m)*
r обрешетина *f*; рейка *f*; планка *f*

B202 *e* **batten cleat**
d 1. Verbindungslasche *f*, Zangenholz *n* 2. Fugendeckleiste *f*
f 1. moise *f* 2. bande *f* de recouvrement
nl 1. klamp *m*, versterkingslat *f(m)*, klos *m* 2. deklijst *f(m)*
r 1. соединительная планка *f* [схватка *f*] 2. нащельник *m*

B203 *e* **batten door**
d Brettertür *f*, Lattentür *f*
f porte *f* en planches
nl opgeklampte deur *f(m)*, spiegeldeur *f*
r дощатая дверь *f*

B204 *e* **batten plates**
d Bindebleche *n pl*
f plats *m pl* de liaison *(des poteaux composés)*
nl koppelplaten *f(m) pl*
r соединительные планки *f pl* *(ветвей стальной колонны)*

B205 *e* **batter**
d 1. Böschung *f*; Böschungsneigung *f* 2. rückläufiger Mauerabsatz *m* 3. Tonbrei *m*, gekneteter Ton *m* [Lehm *m*] 4. Mörtelhaue *f*, Mörtelkrücke *f*
f 1. talus *m*; pente *f* du talus 2. retraite *f* de maçonnerie 3. pâte *f* d'argile, barbotine *f*; argile *f* pétrie 4. racloir *m*, raclette *f*
nl 1. talud *n*, achterwaartse helling *f* 2. trapsgewijs verlopend metselwerk *n* 3. kleimortel *m* 4. mortelschop *f(m)*
r 1. откос *m*; уклон *m* откоса 2. убежный уступ *m* каменной кладки 3. глиняное тесто *n*; мятая глина *f* 4. скребок *m* для затирки раствора

B206 *e* **batter boards**
d Visiergerüst *n*, Schnurgerüst *n*
f planches *f pl* de piquetage
nl bouwplanken *f(m) pl*
r обноска *f*

B207 *e* **battering**
d Lotabweichung *f* *(der Wand)* nach innen
f fruit *m*

nl buik *m* (*in metselwerk*)
r отклонение *n* (*стены*) от вертикали внутрь

B208 *e* **batter pile**
d geneigter Pfahl *m*, Schrägpfahl *m*
f pieu *m* incliné [biais]
nl schuiningepaal *m*, steekpaal *m*
r откосная [наклонная] свая *f*

B209 *e* **battery cyclon**
d Zyklon-Batterie *f*
f batterie *f* de cyclons
nl batterijcycloon *m*
r батарейный циклон *m*

B210 *e* **battery garages**
d Reihengaragen *f pl*
f garage *m* à boxes
nl garageboxen *m pl*
r гараж *m* с отдельными боксами

B211 **battery mould** *see* **gang mould**

B212 *e* **Baudelot cooler**
d Rieselkühler *m*, Baudelot-Kühler *m*, Berieselungskühler *m*
f refroidisseur *m* de Baudelot [à ruissellement]
nl bevloeiingskoeler *m*
r поверхностный оросительный охладитель *m*

B213 *e* **Baudelot evaporator**
d Rieselverdampfer *m*
f évaporateur *m* de pulvérisation
nl bevloeiingsverdamper *m*
r оросительный испаритель *m*

B214 **bauxite cement** *see* **high alumina cement**

B215 *e* **bay**
d 1. Gebäudesektion *f*, Gebäudeabschnitt *m*; Hallenschiff *n* 2. Wehröffnung *f* 3. Betonierungsabschnitt *m*
f 1. nef *f*, travée *f*; section *f* (*d'un bâtiment*) 2. pertuis *m* 3. section *f* de bétonnage
nl 1. gedeelte *n* van een gebouw; vak *n* 2. vak *n* van een stuw 3. betonstortsectie *f*
r 1. отсек *m*; секция *f*; пролёт *m* (*здания*) 2. водосливное отверстие *n* плотины 3. блок *m* бетонирования

B216 *e* **bay bridge**
d Baibrücke *f*, Buchtbrücke *f*
f pont *m* au-dessus [à travers] d'un golfe
nl brug *f(m)* met meer dan één overspanning
r мост *m* через залив

B217 *e* **bay window**
d Erkerfenster *n*, Erker *m*
f fenêtre *f* en saillie [en tribune] (*avec murs portants*)
nl erkerraam *n*, erker *m*
r эркер *m* с несущими стенами

B218 *e* **beaching**
d Uferbefestigung *f*, Uferschutzverkleidung *f*;

Steinabdeckung *f*, Steinsatz *m*, Steinpackung *f*; Böschungsverkleidung *f*
f perré *m*, enrochement *m* de protection
nl oeverbekleding *f*; taludbekleding *f*
r берегоукрепление *n*, крепление *n* откоса камнем, одежда *f* откоса

B219 *e* **beach rehabilitation**
d Strandwiederherstellung *f*
f rétablissement *m* du littoral
nl strandherstel *n*
r восстановление *n* пляжей (*наносозадерживающими мероприятиями*)

B220 *e* **beacon**
d 1. Leuchtfeuer *n* 2. Triangulationssignal *n*, Triangulationshochbau *m*
f 1 balise *f*, phare *m* 2. signal *m*
nl 1. vuurtoren *m*, baken *n* 2. triangulatiepunt *n*
r 1. береговой маяк *m* 2. триангуляционный сигнал *m* [знак *m*]

B221 *e* **bead**
d 1. Anschlagleiste *f*, Dichtungsleiste *f* 2. Dichtungsstrick *m*; Kittdichtungsraupe *f* 3. Falzleiste *f* 4. Deckleiste *f*, Ecklasche *f* 5. Versteifungssicke *f*
f 1. battée *f* 2. baguette *f* de vitrerie 3. parclose *f*, parclause *f* 4. moulure *f* cornière [d'angle] 5. nervure *f* de rigidité
nl 1. aanslag *m* 2. stopverfzoom *m* 3. felsrand *m* 4. lasrups *f(m)* 5. ril *f(m)*, kraal *f(m)*
r 1. притвор *m*; уплотняющий брусок *m* 2. уплотняющая прокладка *f* (*жгут*); валик *m* замазки 3. штапик *m* 4. (прямая *или* угловая) металлическая накладка *f* (*для защиты штукатурки, углов*) 5. зиг *m*, валик *m* жёсткости

B222 **beading plane** *see* **bead plane**

B223 *e* **bead moulding**
d Kehlleiste *f*, konvexe Halteleiste *f*
f parclose *f*, liteau *m* à moulures (*en relief*), batonnet *m* perlé
nl kraallijst *f(m)*
r калёвка *f*, раскладка *f* выпуклого профиля

B224 *e* **bead plane**
d Rund(stab)hobel *m*
f rabot *m* à moulures
nl holschaaf *f(m)*
r калёвка *f* (*рубанок*)

B225 *e* **beam**
d Balken *m*; Träger *m*
f poutre *f*, poutrelle *f*; solive *f*, madrier *m*
nl draagbalk *m*; ligger *m*
r балка *f*; прогон *m*

B226 *e* **beam anchorage**
d Balkeneinspannung *f*

f ancrage *m* de la poutre
nl balkanker *m*
r заделка *f* концов балок

B227 *e* **beam-and-girder construction, beam and girder framing**
d Trägerrost *m*, Balkenrostwerk *n*
f poutres *f pl* croisées
nl balkconstructie *f*; draagbinten *pl* met dwarsbalken
r балочная клетка *f*; система *f* перекрёстных балок

B229 *e* **beam-and-girder slab**
d Kassettenplatte *f*
f dalle *f* en caisson
nl cassettenplaat *f(m)*
r кессонная плита *f*

B230 *e* **beam-and-slab floor**
d Rippendecke *f*
f plancher *m* nervuré [à nervures]
nl uit balken en platen bestaande systeemvloer *m*
r ребристое перекрытие *n*

B231 *e* **beam bottom**
d 1. Balkenunterkante *f* 2. untere Schaltafel *f* des Balkenschalenkastens
f 1. face *f* inférieure d'une poutre 2. panneau *m* inférieur du coffrage de poutre, fond *m* du coffrage (*d'une poutre*)
nl 1. onderkant *m* van een balk *m* 2. bodem *m* van de bekisting van een betonnen balk
r 1. нижняя грань *f* балки 2. нижний щит *m* опалубочного короба балки

B232 *e* **beam calculation**
d Balkenberechnung *f*, Trägerberechnung *f*
f calcul *m* de poutres
nl sterkteberekening *f* van een ligger
r расчёт *m* балок

B233 *e* **beam-column**
d querbelasteter Druckstab *m*
f poutre *f* comprimée et fléchie; poutre *f* soumise à la compression excentrée
nl dwarsbelaste drukstaaf *f(m)*
r элемент *m*, работающий на изгиб со сжатием, сжато-изогнутый элемент *m*

B234 *e* **beam construction**
d Balkenkonstruktion *f*, Balkensystem *n*, Trägerkonstruktion *f*, Trägersystem *n*
f construction *f* en poutres
nl balkconstructie *f*
r балочная конструкция *f*

B235 *e* **beam deflection, beam deflexion**
d Balkendurchbiegung *f*, Balkendurchsenkung *f*, Trägerdurchbiegung *f*, Trägerdurchsenkung *f*
f flèche *f* de la poutre
nl doorbuiging *f* van een balk
r прогиб *m* балки

B237 *e* **beam design formula**
d Balkenbemessungsformel *f*
f formule *f* de calcul des poutres
nl formule *f(m)* voor dragersberekening
r формула *f* для расчёта балок

B238 **beam ends out-of-square** *see* **beam flanges out-of-square**

B239 **beam fixed at both ends** *see* **fixed beam**

B240 *e* **beam flanges out-of-square**
d Verkantung *f* der Trägerflansche
f gauchissement *m* des ailes de poutres laminées
nl stalen balk *m* met scheve flenzen
r перекос *m* полок (стальных прокатных) балок

B241 *e* **beam flexural theory**
d Balkenbiegetheorie *f*, Trägerbiegetheorie *f*
f théorie *f* de flexion des poutres
nl doorbuigingstheorie *f*
r теория *f* изгиба балок

B242 *e* **beam forms**
d Balkenschalung *f*
f coffrage *m* de poutres
nl balkbekisting *f*
r опалубка *f* для (бетонирования) балок

B243 *e* **beam grid, beam grillage**
d Balkenrost *m*, Trägerrost *m*, Balkenkreuzwerk *n*, Trägerkreuzwerk *n*
f grillage *m* de poutres
nl balkenrooster(werk) *n*
r ростверк *m* из (стальных) балок; балочный ростверк *m*; балочная клетка

B244 *e* **beam hangers**
d 1. Balkenaufhänger *m pl* 2. Aufhänger *m pl* für Balkenschalung
f 1. suspentes *f pl* de poutre 2. tiges *f pl* [barres *f pl*] de suspension (*pour accrochage le coffrage de poutres*)
nl 1. balkhangers *m pl* 2. console *f* voor de balkbekisting
r 1. подвески *f pl* балок 2. подвески *f pl* для крепления опалубки балок (*к несущим конструкциям*)

B245 *e* **beam mould**
d Trägerschalung *f*, Schalung *f* für Stahlbetonfertigbalken
f moule *m* à poutres préfabriquées
nl mal *m* voor (het maken van) betonbalken
r форма *f* для изготовления сборных железобетонных балок

B246 *e* **beam on elastic foundation**
d Balken *m* auf elastischer Bettung
f poutre *f* reposant sur fondation élastique
nl elastisch opgelegde balk *m*
r балка *f* на упругом основании

B247 *e* **beams and stringers**
d Kantholz *n* (*Sortiment*)

f bois *m* carré [équarri] (*assortiment*)
nl kanthout *n*, gezaagde balken *m pl*
en platen *f(m) pl*
r деревянные брусья *m pl* (*сортамент*)

B248 *e* beam sides
   d 1. Seitenflächen *f pl* des Balkens [des
   Trägers] 2. Seitenschaltafeln *f pl* der
   Balkenschalung
   f 1. faces *f pl* latérales d'une poutre
   2. joues *f pl* du coffrage de poutre
   nl 1. zijkanten *f(m) pl* van een balk
   2. schotten *n pl* van de balkbekisting
   r 1. боковые грани *f pl* балки
   2. боковые щиты *m pl*
   опалубочного короба балки

B249 *e* beam steel
   d Balkenbewehrung *f*,
   Stahlbetonträgerbewehrung *f*
   f armature *f* [ferraillage *m*] de poutres
   nl wapeningsstaal *n* voor betonbalken
   r арматура *f* железобетонных балок

B250 beam structural system *see* beam
   construction

B251 beam theory *see* beam flexural theory

B252 *e* beam-to-beam connection
   d Balkenanschluß *m*, Trägeranschluß *m*
   f nœud *m* d'assemblage des poutres
   croisées
   nl balkenverband *n*
   r сопряжение *n* [соединение *n*] балок

B253 *e* beam-to-beam moment connection
   d steifer Balkenanschluß *m*
   [Trägeranschluß *m*]
   f assemblage *m* de poutres croisées à
   nœuds rigides
   nl stijf balkenverband *n*
   r глухое сопряжение *n* балок

B254 *e* beam-to-column connection
   d Balkenanschluß *m* an die Säule [an
   die Stütze]
   f nœud *m* d'assemblage de la poutre
   avec le poteau
   nl verbinding *f* van balken en kolommen
   r сопряжение *n* [соединение *n*] балок
   с колоннами

B255 *e* beam vibrator
   d Bohlenrüttler *m*
   f poutre *f* [règle *f*] vibrante
   nl staafvibrator *m*, trilstaaf *f(m)*
   r виброрейка *f*; вибробрус *m*

B256 *e* beam with overhangs
   d Einfeldkragbalken *m*, Kragbalken *m*,
   Konsolträger *m*
   f poutre *f* à une travée avec deux
   consoles
   nl balk *m* met overstekken [met
   uitkragingen]
   r однопролётная балка *f* с двумя
   консолями

B257 *e* bearer
   d 1. Haupttragebalken *m* 2. Unterzug
   m, Längsbalken *m*, Pfette *f*

3. Treppenpodestträger *m*
4. Querriegel *m* des Baugerüstes
f 1. poutre *f* principale 2. panne *f*
3. poutre *f* de palier, solive *f* palière
4. boulin *n*
nl 1. hoofddraagbalk *m*, moerbint *n*,
onderslagbalk *m* 2. langsbalk *m*,
draagbalk *m*, gording *f* 3. draagbalk
m met trapbordes 4. steigerbalk *m*
r 1. главная балка *f* 2. прогон *m*
3. балка *f* лестничной площадки
4. поперечина *f* лесов *или* подмостей

B258 *e* bearing
   d 1. Auflagerung *f* 2. Lager *n*
   3. Richtungswinkel *m*, Azimut *m*, *n*
   f 1. appui *m*, support *m* 2. palier *m*
   3. azimut *m*, gisement *m*
   nl 1. oplegging *f*, draagvlak *n* 2. lager *n*
   3. richting *f*, peiling *f*
   r 1. опора *f* 2. подшипник *m*
   3. азимут *m*

B259 *e* bearing bar
   d Fußschwelle *f*, Sohlschwelle *f*;
   Stützstab *m*
   f barre *f* d'appui
   nl steunstang *f(m)*
   r опорный брус *m*; опорный
   стержень *m*

B260 *e* bearing block
   d Lagerstuhl *m*; Lagerblock *m*; Knagge *f*
   f bloc *m* d'appui; selle *f* d'appui
   nl oplegstoel *m*, kussenblok *n*
   r опорный блок *m*; опорная
   подкладка *f*; подферменник *m*;
   опорная подушка *f*

B262 *e* bearing length
   d Auflagerlänge *f*; Einspannlänge *f*
   f longueur *f* d'appui; longueur *f*
   d'encastrement
   nl dragende lengte *f*; (over)spanning *f*
   r длина *f* опорной части (*напр.
   балки*); глубина *f* защемления
   (*напр. перемычки над проёмом*)

B263 bearing pad *see* bearing block

B264 *e* bearing pile
   d Gründungspfahl *m*, Druckpfahl *m*;
   Standpfahl *m*, Säulenpfahl *m*
   f pieu *m* portant [de support]; pieu-
   -colonne *f*
   nl dragende paal *m*; drukpaal *m*
   r несущая свая *f*; свая-стойка *f*, свая-
   -колонна *f*

B265 *e* bearing plate
   d Auflagerplatte *f*, Unterlagsplatte *f*,
   Lastplatte *f*
   f plaque *f* d'appui
   nl oplegplaat *f(m)*, onderlegplaat *f(m)*
   r опорная плита *f*

B266 *e* bearing pressure
   d Auflagerdruck *m*, Auflagerpressung
   f pression *f* sur les appuis [d'appui]
   nl oplegdruk *m*
   r давление *n* на опорной поверхности,
   опорное давление *n*

B267 e bearing stress
  d Auflagerspannung *f*
  f contrainte *f* (due à la force) de
    compression localisée
  nl oplegdruk *m*
  r напряжение *n* смятия

B268 e bearing support
  d Auflagerteil *n*, *m* (*des
    Brückentragwerkes*); Auflagerkonstruk-
    tion *f*
  f appareil *m* d'appui
  nl opleggingsdeel *n* (*van een
    brugligger*)
  r опорная часть *f* (*пролётного
    строения моста*); опорная
    конструкция *f*

B269 e bearing test
  d Tragfähigkeitsversuch *m*,
    Tragfähigkeitsprobe *f*,
    Belastungsversuch *m*
  f essai *m* de capacité portante du sol
  nl vaststellen *n* van het draagvermogen
    van de bodem
  r определение *n* несущей
    способности грунта

B270 e bear-trap dam
  d Bärenfallen(klappe-Dach)wehr *n*
  f barrage *m* en toit
  nl beer *m*
  r водосливная плотина *f*
    с крышевидным затвором

B271 bear-trap gate *see* roof gate

B272 e beaumontage
  d Füllkitt *m*, Füllstoff *m*
  f beaumontage *m* (*mélange de résine,
    cire et shellac pour sceller des trous
    et fissures en bois ou métal*)
  nl vulmiddel *n*, vulstof *f(m)*
  r бомонтаж *m* (*специальный уплот-
    няющий состав для заделки рако-
    вин и трещин в дереве и металле*)

B273 e bed
  d 1. Flötz *n* 2. Mörtelschicht *f*
    3. hartes Lager *n* 4. Bettschicht *f*
    5. Flußbett *n*
  f 1. couche *f* de terre 2. couche *f* [lit
    m] de mortier 3. lit *m* de pose 4.
    assise *f* de pierres *ou* de briques
    5. lit *m* de fleuve
  nl 1. (gesteente)laag *f(m)* 2. specielaag
    f(m) 3. bed *n* 4. steenlaag *f(m)* (in
    metselwerk) 5. stroombedding *f*
  r 1. пласт *m* 2. слой *m* раствора
    3. постель *f* (*строительного камня*)
    4. ряд *m* камней в кладке
    5. русло *n*, ложе *n* (*реки*)

B274 e bed-and-joint width
  d Weite *f* der Fuge zwischen den
    Verkleidungsplatten
  f largeur *f* des joints entre plaques de
    revêtement
  nl voegbreedte *f* tussen bekledingsplaten
  r ширина *f* шва между
    облицовочными плитами

B275 e bedding
  d 1. Sauberkeitsschicht *f*,
    Ausgleichschicht *f*,
    Unterbettungsschicht *f*, Unterbettung
    f, Auflager *n* 2. Einbettung *f*,
    Vermauern *n* 3. Verlegermasse *f*;
    Dichtungskitt *m*
  f 1. lit *m* (*p. ex. de béton*); sous-
    -couche *f*; semelle *f* 2. scellement *m*
    (*p. ex. d'une poutre*) 3. mastic *m*
    pour scellement
  nl 1. bed *n* van schraal beton, uitvlaklaag
    f(m), onderlaag *f(m)* 2. verzegelen *n*,
    pakking *f* 3. afdichtmiddel *n*
  r 1. постель *f*; выравнивающий слой
    m; подстилающий слой *m*;
    подготовка *f*; опорная подушка *f*
    2. заделка *f* 3. уплотняющий
    состав *m*

B276 bedding compound *see* bedding 3.
B277 bedding course *see* bed of mortar
B278 e bedding value
  d Bettungsziffer *f*
  f module *m* de réaction (du sol)
  nl beddingsmodulus *m*
  r коэффициент *m* постели

B279 e bed groin
  d Grundbuhne *f*
  f épi *m* de fond
  nl grondkrib *f(m)*
  r донная полузапруда *f*

B280 e bed joint
  d 1. Mörtelbett *n* 2. Lagerfuge *f*
  f 1. lit *m* de mortier 2. assise *f*
    horizontale (de maçonnerie)
  nl 1. strekse voeg *f(m)* 2. voeg *f(m)* in
    een strekse boog
  r 1. горизонтальный [постельный]
    слой *m* раствора 2. горизонтальный
    шов *m* кладки

B281 e bed load
  d Bodenlast *f*, Geschiebelast *f*,
    Geschiebefracht *f*, Geschiebe *n*,
    Sinkstoffe *m pl*; Geschiebedurchfluß
    m
  f charriage *m* (de fond); débit *m* de
    charriage (de fond)
  nl door een rivier meegevoerde vaste
    stof *f(m)*
  r донные наносы *m pl*; расход *m*
    донных наносов

B282 e bed load function
  d Sohlfrachtfunktion *f*
  f fonction *f* du charriage
  nl functie *f* van de afvoer van vaste stof
  r расход *m* донных наносов в функции
    диаметра частиц при данном
    расходе воды в водотоке

B283 e bed-load trap
  d Geschiebefalle *f*, Geschiebefang *m*,
    Schotterfalle *f*, Schotterfang *m*
  f trappe *f* à sédiments, collecteur *m*
    de charriage

nl bezinkselvanger *m*, sedimentvanger *m*
r ловушка *f* донных наносов

B284 *e* **bed of brick**
d Lagerfläche *f*
f lit *m* d'une brique
nl stenen onderlaag *f(m)*
r постель *f* кирпича

B285 *e* **bed of mortar**
d Mörtelbett *n*
f lit *m* de mortier
nl speciebed *n*
r слой *m* [грядка *f*] раствора

B286 *e* **bedrock**
d Felsuntergrund *m*
f assise *f* rocheuse, fond *m* rocheux
nl vast gesteente *n*
r коренная [материковая] порода *f*; скальное основание *n*

B287 *e* **bed roughness**
d Gerinnerauhigkeit *f*
f rugosité *f* du lit de cours d'eau
nl beddingruwheid *f*
r шероховатость *f* русла

B288 *e* **bed scour**
d Flußbettauswaschung *f*
f affouillement *m* du lit de cours d'eau
nl uitspoeling *f* van de rivierbedding
r размыв *m* русла

B289 *e* **bed stabilisation**
d Flußbettsicherung *f*, Flußbettbefestigung *f*
f stabilisation *f* du lit de cours d'eau
nl stabilisatie *f* van de rivierbedding
r крепление *n* русла

B290 *e* **beetle**
d Stampfer *m*; großer Holzhammer *m*
f batte *f*; maillet *m*; couperet *m* (de paveur)
nl stamper *m*; zware houten hamer *m*
r деревянная трамбовка *f* (*для мощения дорог*); молоток *m* мостовщика; киянка *f*

B291 *e* **behaviour of structures**
d Verhalten *n* der Konstruktionen
f comportement *m* des constructions
nl gedrag *n* van de constructies
r работа *f* конструкций [сооружений]

B292 *e* **bell dolphin**
d Glockendalben *m*, glockenförmige Dalbe *f*
f borne *f* d'amarrage en cloche
nl dukdalf *m*, meerstoel *m*
r колоколообразный пал *m*

B293 *e* **belled-out pile, belled up pile**
d Klumpfußpfahl *m*
f pieu *m* moulé avec base élargie
nl heipaal *m* met verzwaarde voet
r набивная свая *f* с уширенной пятой

B296 *e* **bell end, bell mouth**
d Rohrmuffe *f*
f évasement *m*, emboîtement *m*
nl pijpmof *f(m)*
r раструб *m*

B297 *e* **bellmouth intake**
d Einlauftrompete *f*
f trompe *f* d'entrée
nl inlooptrechter *m*
r входная воронка *f* (водозабора)

B298 *e* **bellows expansion piece**
d Wellrohrausdehnungsstück *n*
f compensateur *m* à soufflet
nl expansie-compenserende balg *m*
r линзовый компенсатор *m*

B299 *e* **bellows seal**
d Balgdichtung *f*
f étanchage *m* [étanchement *m*] à soufflet
nl balgafdichting *f*
r сильфонный сальник *m*

B300 *e* **below-ground masonry**
d unterirdisches Mauerwerk *n*
f maçonnerie *f* au-dessous du (niveau de) terrain
nl metselwerk *n* onder het maaiveld
r каменная кладка *f* ниже уровня земли

B301 *e* **below-ground tank**
d Tiefbehälter *m*, Erdbehälter *m*
f réservoir *m* souterrain
nl ondergronds reservoir *n*
r подземный резервуар *m*

B302 *e* **belt conveyor**
d Bandförderer *m*, Gurtförderer *m*
f convoyeur *m* [transporteur *m*] à courroie [à bande]
nl bandtransporteur *m*
r ленточный конвейер *m*

B303 *e* **belt course**
d 1. horizontale [waagerechte] Ziegelschicht *f* (*Mauerwerk*) 2. auskragende Ziegelschicht *f* 3. Rödelbohle *f*
f 1. assise *f* 2. bandeau *m* 3. listel *m*
nl uitspringende laag *f(m)* (*in metselwerk*)
r 1. горизонтальный ряд *m* кладки 2. поясок *m* (*в кирпичной кладке*) 3. поясок *m* (*из доски*)

B304 **belt elevator** *see* **belt-type bucket elevator**

B305 *e* **belt loader**
d 1. Pflugbagger *m* mit kreisförmiger Diskuspflugschar 2. Bandförderer *m*
f chargeur *m* [chargeuse *f*] à courroie
nl 1. graafmachine *f* met transportband 2. transportband *n*
r 1. экскаватор *m* с ленточным конвейером 2. ленточный погрузчик *m*, погрузчик *m* непрерывного действия

B305a **belt saw** *see* **band saw**

B306 *e* **belt stress**
d waagerechte Spannung *f*
f contrainte *f* annulaire
nl ringspanning *f*
r кольцевое напряжение *n*

B307 *e* **belt-type bucket elevator**
d Gurtbecherwerk *n*, Bandelevator *m*
f élévateur *m* à godets
nl bandelevator *m*
r ковшовый (ленточный) элеватор *m*,
нория *f*

B308 *e* **belt-type concrete placer**
d Betonbandförderer *m*
f bétonneuse *f* à courroie
nl betonstortinstallatie *f* met
transportband
r бетоноукладчик *m* ленточного типа
(*с ленточным конвейером*)

B309 *e* **belt-type water strainer**
d Bandwasserfilter *n*, *m*
f tamis *m* d'eau à bande
nl band-waterfilter *n*, *m*
r ленточный водяной фильтр *m*

B310 **bench** *see* **berm**

B311 *e* **bench grinder**
d Tischschleifmaschine *f*
f meuleuse *f*, affûteuse *f*
nl tafelslijpmachine *f*
r шлифовальный станок *m*; заточный
станок *m*

B312 *e* **bench mark**
d Pegel *m*, Höhenmarke *f*, Festpunkt
m, Bezugspunkt *m*
f repère *m*, point *m* de repère [de
référence]
nl peilmerk *n*, vast meetpunt *n*,
merkteken *n*
r репер *m*, опорная отметка *f* высоты

B313 *e* **bend**
d 1. Verbiegung *f*, Krümmung *f*, Knick
m 2. Krümmer *m*, Kniestück *n*,
Rohrbogen *m* 3. Bahnkurve *f*
4. Flußschlinge *f*, Mäander *m*
f 1. pli *m*, pliage *m*, recourbement *m*
2. coude *m* (*d'un tuyau*) 3. courbe *f*
(*d'une route*) 4. coude *m*, tournant *m*
(*d'une rivière*)
nl 1. verbuiging *f*, knik *m* 2. bocht *f(m)*
(*pijp*) 3. bocht *f(m)* (*weg, rivier*)
r 1. изгиб *m*, загиб *m*, перегиб *m*
2. колено *n* (*трубы*);
отвод *m* (криволинейного
очертания) 3. изгиб *m* дороги;
дорожная кривая 4. излучина *f* (*реки*)

B314 **bender** *see* **bending machine**

B315 *e* **bending**
d Biegung *f*
f 1. flexion *f* 2. cintrage *m*; pliage *m*
nl 1. (door)buiging *f* 2. buigen *n* (*van
hout*)
r 1. изгиб *m* 2. гнутьё *n*, гибка *f*

B316 *e* **bending analysis**
d Biegeberechnung *f*
f calcul *m* à la flexion
nl buigberekening *f*
r расчёт *m* на изгиб

B317 **bending and reinforcement assembly**
shop *see* **reinforcement assembly shop**

B318 *e* **bending bench**
d Biegetisch *m*, Betonstahlbiegemaschi-
ne *f*
f banc *m* de pliage
nl buigtafel *f(m)*, betonstaalbuigmachine
f
r (верстачный) станок *m* для гибки
арматуры

B319 **bending curve** *see* **flexure curve**

B320 *e* **bending machine**
d Biegemaschine *f*
f plieuse *f*
nl buigmachine *f*
r гибочный станок *m*

B321 *e* **bending moment coefficients**
d Koeffizienten *m pl* der Biegemomente
f coefficients *m pl* des moments
fléchissants
nl coëfficiënten *m pl* van de buigende
momenten
r коэффициенты *m pl* изгибающих
моментов

B322 *e* **bending moment diagram**
d Biegemomentenlinie *f*,
Biegungsmomentendiagramm *n*
f diagramme *m* des moments, ligne *f*
du moment fléchissant
nl buigende-momentenlijn *f(m)*
r эпюра *f* изгибающих моментов

B323 *e* **bending press**
d Biegepresse *f*, Abkantpresse *f*
f presse *f* à cintrer [à plier]
nl buigpers *f(m)*, buigmachine *f*
r гибочный пресс *m*

B324 **bending rigidity** *see* **flexural**
rigidity

B325 *e* **bending schedule**
d Biegeliste *f*, Stahlliste *f*
f tableau *m* de pliage
nl buigstaat *m*
r заготовительная спецификация *f*
арматурных стержней (*для гибки*)

B326 *e* **bending strength**
d Biegefestigkeit *f*
f résistance *f* à la flexion
nl buigsterkte *f*
r предел *m* прочности при изгибе;
временное сопротивление *n* изгибу

B327 *e* **bending stress**
d Biegespannung *f*, Biegungsspannung *f*
f contrainte *f* de flexion
nl buigspanning *f*
r напряжение *n* изгиба [при изгибе]

B328 *e* **bending tensile test**
d Biegezugversuch *m*, Biegezugprobe *f*
f essai *m* de traction par flexion
nl buigtrekproef *f(m)*
r определение *n* прочности на
растяжение при изгибе

B329 *e* **bending theory**
d 1. Biegetheorie *f* 2. Balkentheorie *f*
f théorie *f* de flexion
nl 1. buigingstheorie *f* 2. balkentheorie *f*

r 1. моментная теория *f* (расчёта оболочек) 2. теория *f* расчёта балок

B330 e **bent**
   d 1. Querrahmen *m*; Rahmenstütze *f* (*Brücke*) 2. Biegung *f*, Krümmung *f* 3. Hügelabhang *m*
   f 1. portique *m* 2. courbure *f* 3. talus *m*
   nl 1. vakwerkligger *m*, spant *n* 2. schraagbeen *n* 3. talud *n*; glooiing *f*
   r 1. поперечная рама *f*; рамная опора *f* (*напр. моста*) 2. изгиб *m*, кривизна *f* 3. склон *m*; откос *m*

B331 e **bent member**
   d gebogenes Bauelement *n*
   f élément *m* courbé, pièce *f* courbée
   nl gebogen onderdeel *f(m)* (*van constructie*)
   r изогнутый [криволинейный] элемент *m* (*конструкции*)

B332 e **bentonite**
   d Bentonit *m*
   f bentonite *f*
   nl bentoniet *n*
   r бентонит *m*, бентонитовая глина *f*

B333 e **bentonite mud, bentonit slurry**
   d Bentonitsuspension *f*
   f suspension *f* de bentonite
   nl bentonietsuspensie *f*
   r бентонитовая суспензия *f*

B334 e **bent reinforcement bar, bent-up bar**
   d aufgebogenes Eisen *n*, abgebogener Stab *m*
   f barre *f* relevée
   nl opgebogen (wapenings)staaf *f(m)*
   r отгиб *m*, арматурный стержень *m* с отгибом

B336 e **berm**
   d Berme *f*, Bankett *n*
   f berme *f*
   nl berm *m*
   r берма *f*

B337 e **berth**
   d 1. Anlegestelle *f*, Anlegeplatz *m*, Liegeplatz *m*; Kaianlage *f* 2. Stapel *m*
   f 1. quai *m* d'amarrage 2. cale *f*
   nl 1. aanlegplaats *f(m)*, ligplaats *f(m)* (*van schepen*) 2. (scheeps)kooi *f(m)*
   r 1. причал *m* 2. стапель *m*

B338 e **berthing facilities**
   d Kaianlage *f*, Anlegevorrichtung *f*
   f ouvrage *m* d'accostage
   nl aanlegfaciliteiten *f pl*, havenvoorzieningen *f pl*
   r причальное сооружение *n*

B339 e **berthing impact**
   d Schiffsstoß *m*, Auflagen *n* des Schiffes
   f choc *m* de navire
   nl schok *m* bij afmeren
   r навал *m* судна (*на причальное сооружение*)

B340 e **bevel**
   d 1. Gehrung *f*, Kantenabschrägung *f*, Schrägabschnitt *m*, Schrägkante *f*

2. Schmiege *f*, Stellwinkel *m*, Gehrmaß *n*
   f 1. chanfrein *m*, biseau *m* 2. équerre *f* mobile, souterelle *f*, fausse-équerre *f*
   nl 1. verstek *n*; schuinte *f*, schuine kant *m* 2. zwei *m*, zwaaihaak *m*
   r 1. скос *m*; скошенная кромка *f*, фаска *f* 2. малка *f*

B341 e **bevel board**
   d keilförmiges Brett *n* (*im Querschnitt*)
   f planche *f* biaise [en biais, en biseau, en coin]
   nl plank *f(m)* met schuine kanten
   r клиновая доска *f* (*скошенная в поперечном сечении*)

B342 **bevelled angle** *see* **chamfered angle**

B343 e **bevelled glass**
   d Facettenglas *n*
   f verre *m* plat chanfreiné
   nl glas *n* met schuingeslepen rand
   r листовое стекло *n* с фаской

B344 e **bevel siding**
   d Bretterbeschlag *m*, Bretterverkleidung *f* der Wände (*mit keilförmigen Beschlagbrettern*)
   f bardage *m* en planches jointées en biais; paroi *f* en planches avec joint à recouvrement; parement *m* en bordillon
   nl potdekselwerk *n*, overnaadse beschieting *f*
   r обшивка *f* внакрой клиновыми досками

B345 **B.F.B.** *see* **broad-flanged beam**

B346 e **biaxiality**
   d Zweiachsigkeit *f*
   f rapport *m* de la contrainte principale minimale à la contrainte principale maximale
   nl tweeassigheid *f*
   r отношение *n* меньшего главного напряжения к большему (*при двухосном напряжённом; состоянии*)

B347 e **biaxial stress**
   d ebener Spannungszustand *m*
   f contrainte *f* biaxiale, état *m* de contrainte biaxial [à deux dimensions]
   nl tweedimensionale spanningstoestand *m*
   r двухосное напряжённое состояние *n*

B348 e **bib cock, bib tap**
   d Zapfhahn *m*, Wasserhahn *m*, Auslaufventil *n*
   f robinet *m* de puisage [à bec-courbe]
   nl tapkraan *f(m)*
   r водопроводный [водоразборный] кран *m*

B350 e **bid**
   d Anmeldung *f*, Angebot *n*
   f soumission *f*, offre *f*
   nl bod *n*, inschrijving(sprijs) *f*
   r заявка *f* на участие в торгах, заявка *f* на подряд

B351 *e* **bid date**
  d Anmeldedatum *n*, Submissionstermin *m*
  f date *f* de l'ouverture des offres
  nl datum *m* van aanbesteding
  r дата *f* представления заявки на участие в торгах

B352 *e* **bidder**
  d Bieter *m*, Bewerber *m*, anbietende Firma *f*
  f soumissionnaire *m*
  nl bieder *m*, inschrijver *m*
  r участник *m* торгов; подрядчик *m*, подающий заявку на участие в торгах

B354 *e* **bidding documents**
  d Angebotsunterlagen *f pl*, Ausschreibungsunterlagen *f pl*, Fertigungsunterlagen *f pl*
  f documentation *f* relative à l'offre
  nl inschrijvingsbescheiden *n pl*
  r документация *f* для проведения торгов

B355 **bid time** *see* **bid date**

B356 *e* **bifold door**
  d Doppelfalttür *f*
  f porte *f* accordéon à deux battants
  nl vouwdeur *f(m)* met twee bladen
  r двупольная складчатая дверь *f*

B357 *e* **bifurcation**
  d Gabelung *f*, Verzweigung *f*, Verästelung *f*
  f bifurcation *f*
  nl bifurcatie *f*, (af)splitsing *f*
  r бифуркация *f*, разветвление *n*

B358 *e* **bi-level**
  d zweistöckiges Einfamilienhaus *n*
  f demeure *f* bi-étagée [à deux niveaux]
  nl eengezinshuis *n* met twee verdiepingen; tweeverdiepings...
  r жилой дом *m* в двух уровнях

B359 *e* **bilinear diagram**
  d bilineares [Prandtlsches] Spannungsdiagramm *n*
  f diagramme *m* de Prandtl
  nl diagram *n* van Prandtl
  r условная диаграмма *f* напряжений в форме двух прямых, диаграмма *f* Прандтля

B360 *e* **bin**
  d Trichterbunker *m*, Trichter *m*
  f trémie *f*, silo *m*
  nl (trechter)silo *m*, reservoir *n*
  r бункер *m*

B361 *e* **binder**
  d 1. Bindemittel *n*, Binder *m* 2. Bügel *m*
  f 1. liant *m* 2. attache *f*, bride *f*
  nl 1. kopse steen *m* 2. beugel *m* (*in gewapend beton*)
  r 1. вяжущее вещество *n* [средство *n*], вяжущий материал *m*, вяжущее *n* 2. хомут *m*; серьга *f*

B362 *e* **binder distributor**
  d selbstfahrender Bitumensprengwagen *m*, Auto(mobil)-Teerspritzmaschine *f*
  f goudronneuse *f* automobile
  nl bitumensproeiwagen *m*
  r автогудронатор *m*

B363 *e* **binder storage heater**
  d Bitumenwagen *m* mit Bindemittel-Erwärmungsvorrichtung
  f citerne *f* de stockage et de chauffage de liants
  nl verwarmde opslaginrichting *f* voor bitumen
  r битумная цистерна *f* с системой подогрева

B364 **binding agent** *see* **binder 1.**

B365 **binding power** *see* **bond strength**

B366 *e* **binding reinforcement**
  d Verbindungsbewehrung *f*, Anschlußbewehrung *f*
  f armature *f* de liaison
  nl verbindingswapening *f*
  r монтажная арматура *f*

B367 *e* **binding wire**
  d Rödeldraht *m*, Bindedraht *m*
  f fil *m* de ligature
  nl binddraad *m*
  r вязальная проволока *f*

B368 *e* **bin-type retaining wall**
  d Zellenstützmauer *f*
  f mur *m* de soutènement à cellules
  nl trogvormige keermuur *m*
  r ячеистая подпорная стенка *f*

B369 *e* **bin wall**
  d 1. Zellenstützmauer *f*; Zellenfangdamm *m*, Steinkastenstützmauer *f* 2. Bunkerwand *f*
  f 1. batardeau *m* cellulaire 2. paroi *m* de trémie
  nl 1. steenkist *f(m)* 2. silowand *m*
  r 1. ячеистая перемычка *f*; ряжевая стенка *f* 2. стенка *f* бункера

B370 *e* **biochemical oxygen demand**
  d biochemischer Sauerstoffbedarf *m*
  f demande *f* biochimique d'oxygène
  nl biochemische zuurstofbehoefte *f*
  r биохимическая потребность *f* в кислороде

B371 *e* **biochemical sewage treatment**
  d biochemische Abwasserreinigung *f*
  f traitement *m* biochimique des eaux résiduaires
  nl biochemische afvalwaterzuivering *f*
  r биохимическая очистка *f* сточных вод

B372 **biofilter** *see* **bacteria bed**

B373 *e* **biofilter loading**
  d Tropfkörperbelastung *f*
  f charge *f* d'un lit bactérien
  nl belasting *f* van het biologisch filter
  r нагрузка *f* биофильтра

B374 *e* **biological film**
  d biologischer Rasen *m*
  f film *m* [pellicule *f*] biologique

nl biologische film *m*
r биологическая плёнка *f* (*биофильтра*)

B375 biological filter *see* bacteria bed

B376 *e* biological protection
  d Uferbefestigung *f* durch Grasaussaat und [oder] Gesträuchanpflanzungen
  f protection *f* biologique
  nl oeverstabilisatie *f* door begroeiing
  r биологическое крепление *n*, зелёная защита *f* (*берегов, откосов*)

B377 *e* biological sewage treatment
  d biologische Abwasserreinigung *f*
  f épuration *f* des eaux d'égouts par procédés biologiques
  nl biologische afvalwaterzuivering *f*
  r биологическая очистка *f* сточных вод

B378 biological shielding concrete *see* radiation shield concrete

B379 biological shielding wall *see* radiation shield wall

B380 biological sludge *see* activated sludge

B381 *e* biparting door
  d zweiteilige Schiebetür *f*
  f porte *f* couliss(ant)e à deux vantaux
  nl dubbele schuifdeur *f(m)*
  r двупольная раздвижная дверь *f*

B382 *e* bird's mouth
  d Klaue *f*
  f assemblage *m* à chevronnage
  nl keep *f(m)* (*houtverbinding*)
  r угловая врубка *f* в стропильной ноге, стропильная врубка *f*|

B383 *e* bitumastic
  d Bitumenmastix *m*
  f bitumastic *m*, mastic *m* bitumineux
  nl bitumineuze mastiek *m*
  r битумная мастика *f*

B384 *e* bitumastic paint
  d Bitumenanstrich *m*
  f peinture *f* bitumineuse
  nl bitumineuze verf *f(m)*
  r битумный окрасочный состав *m*, битумная краска *f*

B385 *e* bitumastic sealer
  d Bitumendichtungskitt *m*
  f bitumastic *m* d'étanchéité, mastic *m* d'étanchéité bitumineux
  nl bitumineuze (afdichtings)kit *f(m)*
  r уплотняющий [герметизирующий] битумный состав *m*

B386 *e* bitumen
  d Bitumen *n*
  f bitume *m*, asphalte *m*
  nl bitumen *n*
  r битум *m*; асфальт *m*

B387 *e* bitumen boiler
  d Bitumen-Schmelzkessel *m*, Bitumenkocher *m*, Schwarzbindemittel-Kocher *m*
  f fondoir *m* à bitume
  nl bitumenverhitter *m*
  r битумный котёл *m*

B388 bitumen emulsion *see* bituminous emulsion

B389 *e* bitumen macadam
  d Bitumenmakadam *m*, Asphaltmakadam *m*, Tränk-Makadam *m*
  f macadam *m* bitumineux [au bitume], bitumacadam *m*
  nl bitumenmacadam *n*
  r щебёночное дорожное покрытие *n*, обработанное битумом

B390 *e* bitumen prime coat
  d Bitumengrundanstrich *m*
  f sous-couche *f* bitumineuse
  nl bitumineuze grond(verf)laag *f(m)*
  r битумный грунтовочный слой *m*

B391 *e* bitumen slip coating
  d Bitumengleitanstrich *m*
  f enduit *m* bitumineux antifriction
  nl anti-sliplaag *f(m)* op basis van bitumen
  r битумная обмазка *f*, снижающая трение

B392 *e* bitumen-tar binder
  d Bitumenteerbindemittel *n*
  f liant *m* mixte en goudron et bitume
  nl bindmiddel *n* op basis van teer en bitumen
  r битумно-дёгтевое вяжущее *n*

B393 *e* bituminated filler
  d bituminierter Füllstoff *m*
  f poudre *f* [charge *f*] minérale bitumée
  nl gebitumineerde vulstof *f(m)*
  r битумизированный минеральный порошок *m* (*дорожный материал*)

B394 *e* bituminizing
  d Bituminieren *n*
  f bitumage *m*
  nl bitumineren *n*
  r битуми(ни)рование *n*, пропитка *f* *или* покрытие *n* органическим вяжущим; обработка *f* битумом

B395 *e* bituminous cements
  d Kohlenwasserstoffbindemittel *n pl*, Bitumenbindemittel *n pl*
  f liants *m pl* hydrocarbonés [à base de bitume]
  nl bitumenkitten *f(m) pl*
  r битуминозные вяжущие материалы *m pl*

B396 *e* bituminous coating
  d Bitumendecke *f*, Bitumendeckschicht *f*
  f enduit *m* bitumineux; revêtement *m* bitumineux
  nl bitumenverflaag *f(m)*, bitumineuze deklaag *f(m)*
  r покрытие *n* из битуминозных материалов; асфальтовое *или* битумное покрытие *n*

B397 *e* bituminous concrete
  d Asphaltbeton *m*, Schwarzbeton *m*, bituminöser Beton *m*

f béton *m* asphaltique [bitumineux],
béton *m* (à base) de bitume
nl asfaltbeton *n*
r асфальтобетон *m*

B398 *e* **bituminous emulsion**
d Bitumenemulsion *f*
f émulsion *f* bitumineuse [de bitume]
nl bitumenemulsie *f*
r битумная эмульсия *f*

B399 *e* **bituminous felt**
d Bitumen-Rohfilzpappe *f*
f feutre *m* bitumé
nl bitumineus dakvilt *n*, ruberoid *n*
r рубероид *m*

B400 *e* **bituminous grout**
d Bitumenmörtel *m*
f coulis *m* bitumineux [au bitume]
nl bitumenvulmiddel *n*, bitumen-
-aangietmiddel *n*
r битумный раствор *m* [состав *m*]

B401 *e* **bituminous mixing plant**
d Schwarzdeckenmischanlage *f*,
bituminöse Mischanlage *f*
f poste *m* [centrale *m*, installation *f*]
d'enrobage
nl asfaltbetoninstallatie *f*
r асфальтосмесительная установка *f*

B402 **bituminous paint** *see* **bitumastic paint**

B403 *e* **bituminous plastic cement**
d Bitumenklebemittel *n*,
Bitumenklebemasse *f*
f ciment *m* asphaltique liquefié,
bitume-filler *m* liquéfié
nl bitumineus kleefmiddel *n*,
bitumineuze kleefmassa *f(m)*
r разжиженное битумное вяжущее *n*

B404 *e* **bituminous putty**
d Bitumenkitt *m*, Asphaltkitt *m*
f mastic *m* bitumineux
nl bitumenkit *f(m)*, asfaltkit *f(m)*
r битумная мастика *f* или замазка *f*

B405 *e* **bituminous road surfacing finisher**
d Schwarzbelageinbaumaschine *f*,
Schwarzdeckenfertiger *m*
f machine *f* à goudronner, goudronneuse
f automobile
nl asfaltafwerkmachine *f*
r автогудронатор *m*

B406 *e* **bituminous waterproofing**
d Bitumen(dichtungs)anstrich *m*
f enduit *m* hydrofuge bitumineux,
couche *f* hydrofuge bitumineuse
nl bitumineuze waterafdichting *f*
r битумная гидроизоляция *f*

B407 *e* **bituminous waterproofing membrane**
d Bitumen-Dichtungshaut *f*
f membrane *f* hydrofuge en polyéthylène
bitumineux
nl bitumineuze afdichtingshuid *f(m)*
r гидроизоляционная мембрана *f* из
битуминизированной
полиэтиленовой плёнки

B408 *e* **bivalent heat pump system**
d bivalente Wärmepumpenanlage *f*
f système *m* à thermopompes bivalentes
nl bivalent warmtepompsysteem *n*
r теплонасосная система *f* с двумя
взаимодополняющими источниками
тепловой энергии

B410 *e* **blacktop**
d Schwarzbelag *m*, Schwarzdecke *f*
f revêtement *m* hydrocarboné [en
asphalte]
nl geasfalteerd [bitumineus] wegdek *n*
r асфальтовое (дорожное) покрытие *n*;
битумное (дорожное) покрытие *n*

B411 **black-top paver** *see* **bituminous road
surfacing finisher**

B412 *e* **blade**
d 1. Planierschild *m*, Hobelmesser *n*,
Schälblatt *n*; Schneidscheibe *f*
2. Schaufel *f*
f 1. lame *f* 2. palette *f*, ailette *f*, aube *f*
nl 1. blad *n* (*bijl, schop e.d.*); dozerblad *n*
(*grondschaaf*); lemmet *n* 2. schoep
f(m), waaierblad *n*
r 1. отвал *m*, лезвие *n*, нож *m*
2. лопасть *f*; лопатка *f* (*лопаточной
машины*)

B413 *e* **Blaine test**
d Blaine-Probe *f*, Feinheitsprüfung *f*
nach Blaine
f essai *m* de finesse de mouture d'après
Blaine
nl fijnheidsproef *f(m)* volgens Blaine
r испытание *n* тонкости помола (*напр.
цемента*) по Блейну

B414 *e* **blanket**
d 1. Isoliermatte *f*
2. Stromsohlensicherung *f*
f 1. matelas *m* isolant 2. recharge *m*
du fond de fleuve (*avant le
creusement du tunnel sous-fluvial*)
nl 1. isolatiemat *f(m)* 2. afdichtingsmat
f(m) op rivierbodem
r 1. звуко- и теплоизолирующее
полотнище *n*; изоляционный мат *m*
2. временная пригрузка *f* дна реки
перед забоем (*при проходке под-
водного туннеля*)

B415 *e* **blanket grouting**
d Flächen-Vermörtelung *f*,
Flächenverpressung *f*
f injection *f* superficielle d'un terrain
nl oppervlakteverdichting *f* door
injecteren
r площадная цементация *f*

B416 *e* **blanket-type insulant**
d Isoliermatte *f*
f isolant *m* en matelas, matelas *m*
isolant
nl isolatiemat *f(m)*
r изоляционный мат *m*

B417 *e* **blast area**
d Sprengbereich *m*
f zone *f* des tires

nl springgebied *n*, onveilige zone *f*
r зона *f* производства взрывных работ

B418 *e* **blast coil**
d Ventilator-Konvektor *m*,
Lufterhitzer *m*
f serpentin *m* [radiateur *m*] de
chauffage ventilé
nl luchtverwarmer *m* met aanjager
r обдуваемый [вентиляторный]
теплообменник *m*,
калорифер *m*

B419 *e* **blast furnace cement**
d Hochofenzement *m*
f ciment *m* de haut fourneau
nl hoogovencement *n*
r шлакопортландцемент *m*

B420 *e* **blast-furnace slag aggregate**
d Hochofenschlacken-Zuschlag *m*
f agrégat *m* de laitier de haut fourneau
nl toeslag *m* van hoogovenslakken
r заполнитель *m* из доменных шлаков

B421 *e* **blast gate**
d Luftschieber *m*, Luftklappe *f*
f volet *m* d'air, papillon *m* à air
nl luchtklep *f(m)*
r воздушная заслонка *f*, воздушный
клапан *m*

B422 **blast heater** *see* **air heater**

B423 *e* **blast hole**
d Sprengloch *n*
f trou *m* de mine
nl boorgat *n*
r шпур *m*, взрывная скважина *f*

B424 *e* **blasting**
d 1. Sprengen *n*, Sprengung *f*,
Sprengarbeit *f* 2. Sandstrahlen *n*,
Kugelstrahlbehandlung *f*,
Putzstrahlen *n*
f 1. tir *m* 2. sablage *m*, nettoyage *m*
au jet de sable; grenaillage *m*
nl 1. (doen) springen *n*, opblazen *n*,
schieten *n* 2. zandstralen *n*;
gritstralen *n*
r 1. взрывание *n*, производство *n*
взрывных работ, взрывные работы
f pl 2. пескоструйная *или*
дробеструйная очистка *f*

B425 *e* **blasting agent**
d Sprengmittel *n*
f explosif *m*; mélange *m* explosible
nl springstof *f(m)*
r взрывчатое вещество *n*;
взрывчатая смесь *f*

B426 *e* **blasting cap**
d Sprengkapsel *f*, Zündhütchen *n*
f détonateur *m*, amorce *f*, capsule *f*
nl ontsteker *m*, slagpijpje *n*
r капсюль-детонатор *m*

B427 *e* **blasting machine**
d Zündmaschine *f*
f exploseur *m*
nl ontstekingstoestel *n*

r взрывная машинка *f*

B428 *e* **blast-resistant door**
d Luftstoßschutztür *f*
f porte *f* résistant à l'explosion
nl drukbestendige deur *f(m)*
r взрывозащитная [взрывостойкая]
дверь *f*

B429 *e* **bleaching**
d Bleichen *n*; Entfärbung *f*
f blanchiment *m*, blanchissage *m*;
décoloration *f*
nl bleken *n*; verschieten *n* (*kleuren*),
ontkleuring *f*
r отбелка *f*, беление *n*; осветление *n*
(*напр. древесины*); обесцвечивание *n*

B430 *e* **bleaching agent**
d Bleichmittel *n*; Entfärbungsmittel *n*
f agent *m* de blanchiment; décolorant *m*
nl bleekmiddel *n*
r отбеливающее *или* осветляющее
вещество *n*; отбеливатель *m*

B431 **bleeder** *US see* **bleed pipe**

B432 *e* **bleeder valve**
d 1. Entlüftungsventil *n* 2. Ablaßventil
n, Entleerungsventil *n*
f 1. soupape *f* [purgeur *m*] d'air 2. valve
f de drainage
nl 1. ontluchtingsventiel *n*
2. aflaatventiel *n*
r 1. выпускной (воздушный) клапан *m*
2. дренажный [сливной] клапан *m*

B433 *e* **bleeder well**
d Abzapfbrunnen *m*,
Entlastungsbrunnen *m*,
Entlastungsbohrung *f*
f puits *m* filtrant [de décharge, de
décompression, de drainage
ascendant]
nl aflooput *m*
r дренажный [разгружающий]
колодец *m*; разгружающая скважина *f*

B434 *e* **bleeding**
d 1. Bluten *n*, Wasserabstoßen *n*,
Wasserabsonderung *f* 2. Bluten *n*,
Ausschwitzen *n* (*Schwarzdecken*)
3. Pechausbringen *n* 4. Ausbluten *n*
5. Durchsickerung *f*, Versickern *n*
6. Ablassen *n*
f 1. bleeding *m*; ressuage *m*, laitance *f*
2. bleeding *m*, suintement *m* du
bitume 3. séparation *f* de la résine
4. exsudation *f* de la couche
inférieure de la peinture
5. infiltration *f* [suintement *m*] de
l'eau 6. échappement *m* des produits
gazeux *ou* liquides sous pression
nl 1. bloeden *n*; afstoten *n* van
cemetmelk; afstoten *n* van water
2. zweten *n* (*wegdek van asfalt*)
3. afscheiden *n* van hars (*hout*)
4. doorslaan *n* (*verf*) 5. doorzijpelen *n*
(*water*) *n* 6. aflaten *n*, afblazen *n*
r 1. водоотделение *n*; выделение *n*
воды *или* цементного молока (*на*

поверхности бетона) 2. выпотевание
n (битума на дорожном покрытии)
3. выделение *n* смолы *(на дереве)*
4. выпотевание *n* нижнего слоя
краски 5. просачивание *n (воды)*
6. выпуск *m* в атмосферу *(среды,
находящейся под давлением)*

B435 *e* **bleeding asphalt**
   d Bitumenblüte *f*, ausgeschwitzte
   Bitumenhaut *f*
   f bitume *m* exsudé du revêtement
   routier en asphalte
   nl uitgezwete bitumenfilm *m*
   r плёнка *f* выпотевшего битума

B436 *e* **bleed off**
   d 1. Kühlwasserablaß *m*
   2. Dampfentnahme *f*
   f 1. drainage *m* de l'eau (de
   refroidissement) 2. prise *f* de vapeur
   nl 1. koelwater aflaten 2. stoom afblazen
   r 1. слив [спуск] *m* охлаждающей
   воды из холодильной системы
   2. (нерегулируемый) отбор *m* пара

B437 *e* **bleed pipe**
   d Anzapfrohr *n*
   f tube *m* de soutirage
   nl aftappijp *f(m)*
   r отводная труба *f*

B438 **bleed valve** *see* **bleeder valve**

B439 *e* **blender**
   d Mischpumpe *f*
   f pompe *f* mélangeuse
   nl mengpomp *f(m)*
   r насос-смеситель *m*

B440 *e* **blending**
   d Vermischen *n*, Vermengen *n*;
   Homogenisierung *f*
   f malaxage *m*, mélangeage *m*
   nl (ver)mengen *n*; homogeniseren *n*
   r смешение *n*, смешивание *n*;
   перемешивание *n*

B441 *e* **blind drain**
   d geschlossene Dränung *f*,
   unterirdischer Drän *m*, Rigole *f*;
   Steindrän *m*, Sickerdohle *f*
   f fossé *m* couvert, drain *m* souterrain
   nl (ondergrondse) draineringsleiding *f*
   r (за)крытая дрена *f*; каменная дрена *f*

B442 **blind drainage** *see* **subsurface drainage**

B443 *e* **blind flange**
   d Blindflansch *m*, Verschlußflansch *m*,
   Deckelflansch *m*
   f bouchon *m* à bride, bride *f* aveugle
   [feinte]
   nl blinde flens *m*
   r фланцевая заглушка *f*; глухой
   фланец *m*

B444 *e* **blind hoistway**
   d Aufzugsschacht *m* ohne
   Zwischenausgänge
   f puits *m* [gaine *f*, cage *f*] d'ascenseur
   sans sorties aux étages
   intermédiaires

   nl (over de volle lengte) gesloten
   liftschacht *f(m)*
   r лифтовая шахта *f* без выходов на
   промежуточных этажах

B445 *e* **blinding**
   d 1. Unterbeton *m*; Ausgleichschicht *f*,
   Sauberkeitsschicht *f*, Estrichschicht *f*,
   Aufstrich *m* 2. Schüttung *f*,
   Absplitterung *f* 3. Aufbringen *n* der
   Ausgleichschicht, Absplittern *n*
   f 1. sous-couche *f*; chape *f* (de béton,
   gravier, sable) 2. grenailles *f pl*
   3. épandage *m* des grenailles
   nl 1. spreidlaag *f(m)* 2. slijtlaag *f(m)*
   3. aanbrengen *n* van spreid- *resp.*
   slijtlaag
   r 1. подготовка *f*, стяжка *f*,
   выравнивающий слой *m (тощего
   бетона, гравия, песка)* 2. посыпка *f*,
   расклинцовка *f* 3. нанесение *n*
   высевок, расклинцовка *f*

B446 *e* **blind nailing**
   d verdeckte Nagelung *f*
   f clouement *m* [clouage *m*] noyé
   [masqué]
   nl verdekt nagelen *n*
   r скрытое крепление *n* гвоздями,
   крепление *n* гвоздями впотай

B447 *e* **blinds**
   d Jalousie *f*, Rollvorhang *m*;
   Fensterladen *m*
   f jalousie *f*
   nl jaloezieën *pl*, vensterluiken *n pl*
   r жалюзи *pl*; ставни *m pl*

B448 *e* **blister**
   d Blase *f*, Gußblase *f*, Gasblase *f*
   f bulle *f*, gonflement *m*, bombement
   m, ondulation *f*
   nl blaas *f(m)*, gietgal *f(m)*, bladder *m*
   (verf)
   r вздутие *n*, пузырь *m (напр. на
   кровельном покрытии)*

B449 *e* **blistering**
   d Blasenbildung *f (Anstrichfehler)*
   f cloquage *m*, formation *f* des bulles
   et des cloques, boursouflure *f*
   nl blaasvorming *f (verfwerk)*
   r образование *n* пузырей и вздутий
   *(на окрашенной или оштукатуренной
   поверхности)*

B450 *e* **blister repair**
   d Blasenbeseitigung *f*
   (Pappdachausbesserung)
   f réparation *f* de l'ondulation [du
   gonflément] sur la couverture en
   carton asphalté
   nl verwijderen *n* van blazen
   (bitumineuze dakbedekking)
   r ремонт *m* вздутых мест мягкой
   кровли

B451 *e* **block**
   d 1. Wandblock *m (Beton, Stein)*
   2. Wasserbaumassiv *n*, Block *m*
   3. Holzklotz *m* 4. Abstandhalter *m*,

Spreizstück *n* 5. Seilrolle *f*,
Riemenscheibe *f* 6. Stadtviertel *n*
f 1. bloc *m* mural 2. bloc *m* artificiel
3. billot *m* 4. étrésillon *m* 5. poulie *f*
6. bloc *m* de bâtiments, quartier *m*
nl 1. steenblok *n*, muurblok *n*
2. funderingsblok *n* 3. (hout)blok *n*
4. afstandstuk *n* 5. (hijs)blok *n*
6. (huizen)blok *n*
r 1. стеновой блок *m* 2. искусственный
[бетонный] массив *m* 3. колодка *f*;
чурбан *m* 4. распорка *f* 5. блок *m*;
шкив *m* 6. квартал *m*; жилой массив
m

B452 *e* **blockboard**
d Tischlerplatte *f* mit Stabmittellage
f panneau *m* latté
nl meubelplaat *f(m)*
r столярная плита *f* с реечной основой

B453 *e* **block cutting machine**
d Blocktrennmaschine *f*,
Blockschneider *m*
f machine *f* pour scier les blocs
nl blokkensnijmachine *f*
r машина *f* для резки блоков

B454 *e* **block flooring**
d 1. Parkettafelfußboden *m* 2. Fußboden
m aus Natursteinplatten,
Steinfußboden *m*
f 1. parquet *m* d'assemblage; parquet *m*
à l'anglaise 2. pavé *m* de pierres
appareillées
nl 1. parket(vloer) *m* 2. natuurstenen
vloer(bedekking) *m* (*f*)
r 1. пол *m* из щитового паркета *или*
досок 2. пол *m* из каменных
штучных материалов

B455 *e* **block foundation**
d Blockgründung *f*
f fondation-bloc *f*
nl blokfundering *f*
r фундамент *m* из массивов, блочный
фундамент *m*

B456 *e* **block hydroelectric station**
d durchströmtes Wasserkraftwerk *n*,
Wasserkraftwerk *n* in
Verbundbauweise
f barrage-usine *m*
nl in de stuwdam ingebouwde
waterkrachtcentrale *f*
r гидроэлектростанция *f* совмещённого
типа

B457 *e* **blocking**
d 1. Maueranschluß *m* mit Verzahnung
2. Verstopfung *f*, Blockieren *n*,
Sperren *n*
f 1. liaison *f* de murs à arrachement
2. bouchage *m*, colamatage *m*
nl 1. muuraansluiting *f* met vertanding
2. verstopping *f*, versperring *f*
r 1. соединение *n* кирпичных стен
с устройством штрабы 2. засорение
n; закупорка *f*

B458 *e* **block-making machine**
d Steinfertiger *m*, Steinformmaschine *f*
f presse *f* à blocs de béton
nl steenpers *f(m)*
r пресс *m* [машина *f*] для изготовления
бетонных блоков

B459 **block masonry** *see* **blockwork**

B460 **block of flats** *see* **apartment block**

B461 *e* **block paving**
d 1. Blockpflasterung *f* 2. Blockpflaster *n*
f 1. pavage *m*, pavement *m* 2. pavé *m*
de pierre; chaussée *f* pavée
nl 1. steenbestrating *f* 2. blokkenplavei-
sel *n*
r 1. мощение *n* (блоками) 2. мостовая *f*
из блоков

B462 *e* **block quay wall**
d Blockkaimauer *f*
f mur *m* de quai en blocs lourds
nl uit betonblokken opgebouwde kade-
(muur) *f(m)* (*m*)
r причальная *или* набережная стенка *f*
из массивов

B463 **block splitter** *see* **block cutting machine**

B464 *e* **blockwork**
d Blockmauerwerk *n*
f maçonnerie *f* massive en moellons
nl blokkenmetselwerk *n*
r кладка *f* из массивов, блочная
кладка *f*

B465 **blondin** *see* **cableway**

B466 *e* **bloom**
d Ausblühung *f*, Anlauf *m*
f efflorescence *f*
nl uitslag *n*, uitbloeiing *f*
r выцвет *m*, выцветание *n*;
потускнение *n*

B467 *e* **blotchy appearance**
d fleckiges Aussehen *n* (*Fehler der
Betonoberflächen*)
f taches *f pl* sur la surface de béton
nl gevlekt oppervlak *n* (*betonfout*)
r пятнистость *f* (*дефект бетонных
поверхностей*)

B468 *e* **blow**
d 1. Stoß *m* 2. Explosion *f* 3.
Durchblasen *n*; Blasen *n* 4.
Grundbruch *m*, Fließerscheinung *f*;
Unterspülung *f*
f 1. choc *m*; percussion *f*, coup *m*
2. explosion *f* 3. purge *f*, soufflage *m*;
air *m* soufflé 4. renard *m*
nl 1. stoot *m* 2. explosie *f*
3. (door)blazen *n* 4. gasuitstroming *f*
(*kolenmijn*)
r 1. удар *m*; толчок *m* 2. взрыв *m*
3. продувка *f*; дутьё *n* 4. грифон *m*;
подмыв *m* (*сооружения*)

B469 *e* **blow-down branch**
d Durchblasstutzen *m*
f tubulure *f* d'évacuation de boue
nl aftapbuisje *n* voor bezinksel
r грязеспускной патрубок *m*

B470 *e* **blower**
  d Gebläse *n*
  f soufflerie *f*, soufflet *m*, ventilateur *m* à basse pression
  nl blazer *m*, ventilator *m*, aanjager *m*
  r вентилятор *m* низкого давления

B471 **blowing action** *see* **pumping**

B472 *e* **blown asphalt, blown bitumen**
  d geblasenes [oxydiertes] Bitumen *n*
  f bitume *m* soufflé [de soufflage]
  nl geblazen asfalt *n*
  r продутый [окислённый] битум *m*

B474 *e* **blow off** *US*
  d 1. Abblasen *n*, Kesselabschlämmung *f* 2. Ablaß *m*, Ablaßvorrichtung *f*
  f 1. purge *f* de la chaudière 2. ouverture *f* de purge [de chasse]
  nl 1. de ketel afblazen *n* 2. uitlaat-(voorziening) *m* (*f*)
  r 1. продувка *f* котла; спуск *m* воды (*из котла, трубопровода*) 2. промывное отверстие (*в трубопроводе*)

B475 *e* **blow-off valve**
  d Ablaßventil *n*, Entlüftungsventil *n*
  f robinet *m* de purge
  nl afblaasventiel *n*; ontluchtingsventiel *n*
  r спускной клапан *m*; воздушный кран *m*, воздуховыпускной клапан *m*

B476 **blow-out** *see* **boil**

B477 *e* **blow through fan system**
  d Gleichstrom(-Lüftungs)anlage *f*, Zwangsluftstromanlage *f*, Luftdruckanlage *f*
  f système *m* à passage; système *m* de pression (*de ventilation*)
  nl doorblaasventilatiesysteem *n*
  r прямоточная *или* напорная система *f* вентиляции (*с функциональными элементами на стороне нагнетания вентилятора*)

B478 *e* **blue lead**
  d Bleiglanz *m*
  f plomb *m* bleu
  nl loodglans *m* (*erst*), galeniet *n*
  r серо-голубой свинцовый пигмент *m*

B479 *e* **blush, blushing**
  d Trübung *f*, Weißwerden *n* (*von Lacken*)
  f trouble *m*; formation *f* des taches troubles
  nl witte-sluiervorming *f*
  r помутнение *n* (*лакового покрытия*), образование *n* матовых пятен

B480 **M.D.** *see* **bending moment diagram**

B481 *e* **board**
  d Brett *n*
  f planche *f*
  nl plank *f* (*m*), plaat *f*(*m*)
  r доска *f*

B482 *e* **board and batten**
  d Außenbretterverschalung *f* [Außenbretterverkleidung *f*] mit Fugendeckleisten
  f revêtement *m* en planches (*extérieur*) avec recouvrement des joints
  nl buitenbeschieting *f*
  r наружная дощатая обшивка *f* с нащельниками

B483 *e* **boarding**
  d Verbretterung *f*, Brettverkleidung *f*
  f planchéiage *m*, bardage *m* en bois, voligeage *m*
  nl beplanking *f*
  r дощатая обшивка *f*

B484 *e* **board-type insulant**
  d Plattenisoliermaterial *n*, Isolierplatten *f pl*
  f isolant *m* en panneau
  nl plaatvormig isolatiemateriaal *n*
  r изоляционные плиты *f pl*

B485 *e* **Boatswain's chair**
  d Montagekorb *m*, Montagehängerüstung *f*
  f sellette *f*, nacelle *f*
  nl hangsteiger *m*, vliegende steiger *m*
  r монтажная люлька *f*

B486 **BOD** *see* **biochemical oxygen demand**

B487 *e* **BOD loading**
  d BSB-Belastung *f* (*Abwasserreinigung*)
  f charge *f* en DBO
  nl naar behoefte zuurstof toevoeren (*biologische afvalwaterzuivering*)
  r нагрузка *f* по БПК (*очистка сточных вод*)

B488 *e* **BOD reduction**
  d BSB-Abbau *m* (*Abwasserreinigung*)
  f réduction *f* de DBO
  nl afname *f* van zuurstofbehoefte (*biologische afvalwaterzuivering*)
  r снижение *n* БПК

B489 *e* **body**
  d 1. Körper *m*; Gehäuse *n* 2. Zähigkeit *f*, Steifigkeit *f*, Konsistenz *f* 3. Farbkörper *m*
  f 1. corps *m* 2. consistance *f*, viscosité *f* (*de la peinture*) 3. corps *m* (*de la peinture*)
  nl 1. lichaam *n*; romp *m* 2. consistentie *f*, lijvigheid *f* (*verf*) 3. basisverf *f*(*m*)
  r 1. тело *n*; корпус *m* 2. вязкость *f*; консистенция *f* 3. основа *f* (*лакокрасочного материала*)

B490 *e* **body brick**
  d Ofenziegel *m*, feuerfester Ziegel *m*
  f brique *f* bien cuite
  nl metselsteen *m* voor opgaand metselwerk
  r хорошо [окончательно] обожжённый кирпич *m*

B491 *e* **body coat**
  d 1. Zwischenschicht *f* des Anstrichaufbaus 2. Deckschicht *f* des Anstrichaufbaus

f 1. couche *f* intermédiaire de la peinture 2. couche *f* de finition de la peinture
nl 1. tweede verflaag *f(m)* 2. deklaag *f(m)* (*verfwerk*)
r 1. промежуточный слой *m* окраски [окрасочного покрытия] 2. верхний отделочный слой *m* окраски

B492 *e* **bog drainage**
d Moordränage *f*
f dessèchement *m* des marais
nl droogleggen *n* van moeras, waterafvoer *m* naar zinkputten
r осушение *n* болот

B493 *e* **boil**
d Sandquelle *f*, (hydraulische) Grundbruch *m*
f renard *m*
nl wel *f(m)* (*bron*)
r грифон *m*

B494 *e* **boiled linseed oil**
d Leinölfirnis *m*, gekochtes Leinöl *n*
f huile *f* cuite
nl gekookte lijnolie *f*
r олифа *f*

B495 *e* **boiler**
d 1. Warmwasserspeichergerät *n*, Boiler *m*, Sieder *m* 2. Dampfkessel *m*, Heizkessel *m*
f 1. bouilleur *m*, chauffe-eau *m* 2. chaudière *f* à vapeur [de chauffage]
nl 1. boiler *m* 2. stoomketel *m*
r 1. бойлер *m*, подогреватель *m* сетевой воды, водоподогреватель *m*, водогрейный котёл *m* 2. парогенератор *m*, отопительный котёл *m*

B496 *e* **boiler accessories**
d Kesselzubehör *n*
f garnitures *f pl* [accessoires *m pl*] de chaudière
nl ketelappendages *f pl*
r котельная гарнитура

B497 *e* **boiler efficiency**
d Kesselwirkungsgrad *m*
f rendement *m* de la chaudière
nl ketelrendement *n*
r коэффициент *m* полезного действия котлоагрегата

B498 **boiler capacity** *see* **boiler power**

B499 *e* **boiler evaporating surface**
d Ausdampffläche *f*
f surface *f* d'évaporation
nl verdampend oppervlak *n* van de stoomketel
r испарительная поверхность *f* котла

B500 *e* **boiler flow temperature**
d Heizwassertemperatur *f*; Kesselvorlauftemperatur *f*
f température *f* d'eau de chaudière
nl circulatietemperatuur *f*

r (расчётная) температура *f* сетевой воды; температура *f* воды, поступающей в сеть из котла

B501 *e* **boiler heating surface**
d Kesselheizfläche *f*
f surface *f* de chauffe de la chaudière
nl verwarmd oppervlak *n* van de ketel
r поверхность *f* нагрева котла

B502 *e* **boiler house**
d Kesselhaus *n*
f bâtiment *m* des chaudières, chaufferie *f*
nl ketelhuis *n*
r котельная *f*, здание *n* котельной

B503 *e* **boiler plant**
d Kesselanlage *f*
f installation *f* de chaudières
nl ketelinstallatie *f*
r котельная установка *f*

B504 *e* **boiler power**
d Kesselleistung *f*, Dampfleistung *f*
f capacité *f* de la chaudière
nl ketelvermogen *n*
r мощность *f* [паропроизводительность *f*] котельной установки

B505 *e* **boiler room**
d Kesselraum *m*
f chaufferie *f*, salle *f* des chaudières
nl ketelruimte *f*
r котельная *f* (*помещение*)

B506 *e* **boiler unit**
d Kesseleinheit *f*
f groupe *m* de chaudières
nl keteleenheid *f*, ketelgroep *f(m)*
r котельный агрегат *m*, котлоагрегат *m*

B507 *e* **boiler water**
d Kesselwasser *n*
f eau *f* de chaudière
nl ketelwater *n*
r котловая вода *f*

B508 *e* **boiling water method**
d Siedewasserverfahren *n* (*Schnellprüfung der Betonfestigkeit nach ASTM-Verfahren*)
f méthode *f* d'essai accélérée de résistance du béton (*après le durcissement final dans l'eau bouillie*)
nl kookkoekproef *f(m)* (*cement*)
r ускоренное определение *n* прочности бетона (*после твердения в кипящей воде*)

B509 *e* **bollard**
d 1. Poller *m*, Haltepfahl *m*, Festmacher *m*, Dalbe *f*, Anlegepfosten *m* 2. Prellstein *m*, Prellpfosten *m*
f 1. bollard *m* 2. borne *m*
nl 1. bolder *m*, meerpaal *m* 2. schutpaal *m* (*wegenaanleg*)
r 1. причальная [швартовная] тумба *f*, кнехт *m* 2. тротуарная тумба *f*; оградительный столбик *m*

B510 *e* **bolt**
  d 1. Bolzen *m* 2. Sperriegel *m*
    3. Rundholzstammabschnitt *m*
  f 1. boulon *m* 2. verrou *m* 3. tronçon *m*
    de grume
  nl 1. bout *m* 2. grendel *m*, schoot *m*
    (*van een slot*) 3. dolk *m*, klos *m*
    (*hout*)
  r 1. болт *m* 2. засов *m*; задвижка *f*
    3. короткий отрезок *m* бревна

B511 *e* **bolted gland flexible joint**
  d bewegliche Muffen-
    -Flanschenrohrverbindung *f* mit
    Dichtungsring
  f raccord *m* flexible à emboîtement et
    à bride avec anneau élastique
  nl flexibele pijpkoppeling *f* met flenzen
    en afdichtingsring
  r подвижное раструбно-фланцевое
    трубное соединение *n* с уплотнитель-
    ным кольцом

B512 *e* **bolted joint**
  d Bolzenverbindung *f*
  f assemblage *m* [joint *m*] boulonné
  nl boutverbinding *f*
  r болтовое соединение *n*

B513 *e* **bolt sleeve**
  d Bolzenhüllrohr *n*
  f manchon *m* de protection pour faire
    passer le boulon (*à travers le mur*)
  nl bouthuls *f(m)*
  r патрубок *m* для пропуска болта
    (*в бетонной стене*)

B514 *e* **bond**
  d 1. Adhäsion *f*, Haftung *f*
    2. Mauerverband *m*, Verband *m*
    3. Aufteilung *f* 4. Verleimung *f*
  f 1. adhérence *f*, adhésion *f*
    2. appareil *m* 3. division *f* 4. collage *m*
  nl 1. hechting *f* 2. metselverband *n*
    3. (hout)verbinding *f* 4. lijmen *n*
  r 1. сцепление *n* 2. перевязка *f*
    (*каменной кладки*) 3. разрезка *f*
    (*сооружения, кладки или облицовки
    рабочими швами*) 4. склеивание *n*

B515 *e* **bond area**
  d Haftfläche *f*, Verbundfläche *f*
  f surface *f* d'adhérence
  nl hechtingsvlak *n*
  r площадь *f* или поверхность *f*
    сцепления

B516 *e* **bond course**
  d Binderschicht *f* (zur Verbindung
    zwischen Schnurschichten und
    Hintermauerung)
  f assise *f* de boutisses
  nl patijtse laag *f(m)*, koplaag *f(m)*
  r тычковый ряд *m* кладки

B517 *e* **bonded gravel screen**
  d gebundenes Polymer-
    Kiesbelagfilter *n*, *m* (*des
    Rohrbrunnens*)
  f filtre *m* (*d'un puits de forage*) à
    gravier lié par résines

  nl pompfilter *n*, *m* van gebonden grind
  r гравийный фильтр *m* (*бурового
    колодца*) на полимерном вяжущем

B518 *e* **bonded post-tensioning**
  d Vorspannung *f* mit nachträglichem
    Verbund
  f post-tension *f* [post-contrainte *f*] avec
    injection de coulis
  nl achteraf aangebrachte voorspanning *f*
  r напряжение *n* арматурных пучков
    [прядей] с последующим
    заполнением каналов раствором

B519 *e* **bonded tendons**
  d vorgepreßte [ausgepreßte]
    Spannglieder *n pl*
  f torons *m pl* [câbles *m pl*] de
    précontrainte adhérés au béton
  nl voorspanstaal *n*, voorgespannen
    wapeningsstaven *f(m) pl*
  r преднапряжённые арматурные пучки
    m pl [пряди *f pl*], сцеплённые
    с бетоном

B520 *e* **bonder, bonder header**
  d Strecker *m*, Binder(stein) *m*
  f brique *f* de liaison [d'ancrage];
    boutisse *f*
  nl bindsteen *m*, kopsteen *m*
  r соединительный [анкерный] камень
    m; тычковый кирпич *m*

B522 *e* **bond failure**
  d Haftbruch *m*, Verbundbruch *m*
  f rupture *f* d'adhérence [d'adhésion]
  nl slechte hechting *f*, loslaten *n* van de
    lijm
  r нарушение *n* сцепления;
    разрушение *n* от потери сцепления

B523 **bonding** *see* **bond**

B524 *e* **bonding adhesive**
  d Klebemischung *f*
  f adhésif *m*; composition *f* adhésive
    [d'adhésion]
  nl lijm *m*
  r клей *m*; клеевой состав *m*;
    связующее *n*

B525 *e* **bonding admixture**
  d Netzhaftmittel *n*, Haftmittel *n*,
    Haftvermittler *m*
  f adjuvant *m* adhésif
  nl hechtingsverbeterend middel *n*
  r добавка *f*, улучшающая сцепление

B526 *e* **bonding cement**
  d Klebekitt *m*, Glasfaserzement *m*
  f ciment *m* à fibres de verre
  nl hechtende kit *f(m)*
  r стеклофиброцемент *m*

B527 *e* **bond plaster**
  d Haftputz *m*
  f couche *f* d'enduit de scellement
  nl hechtend pleister *n*
  r штукатурный связующий слой *m*

B528 *e* **bond prevention**
  d Verbundvorbeugung *f*
  f prévention *f* de l'adhérence

nl verhindering *f* van de hechting
r предупреждение *n* сцепления

B529 bond stone *see* bonder

B530 *e* **bond strength**
d Verbundfestigkeit *f*, Haftfestigkeit *f*
f résistance *f* d'adhésion [d'adhérence]
nl hechtsterkte *f*, hechting *f*
r прочность *f* сцепления; адгезионная
прочность *f*

B531 *e* **bond strength of cement to coarse aggregate**
d Haftfestigkeit *f* der Zementleim-
-Grobzuschlagstoff-Verbindung
f adhésivité *f* ciment-agrégat
nl hechting *f* van cement aan de grove
vulstof
r прочность *f* сцепления цементного
камня с крупным заполнителем

B532 *e* **bond stress**
d Haftspannung *f*, Verbundspannung *f*
f contrainte *f* d'adhérence
nl hechtspanning *f*
r напряжение *n* сцепления

B533 *e* **bond zone**
d Haftfläche *f*, Verbundfläche *f*
f zone *f* d'adhérence
nl hechtvlak *n*
r зона *f* сцепления

B534 *e* **boom angle**
d Auslegerwinkel *m*
f angle *m* d'inclinaison de la flèche
nl giekhoek *m* (*kraan*)
r угол *m* наклона стрелы (*крана*)

B535 *e* **boom derrick**
d Mastauslegerkran *m*,
Derrick(ausleger)kran *m*
f grue-derrick *f*
nl torenkraan *f(m)* met giek
r мачтово-стреловой [мачтовый] кран
m, деррик-кран *m*

B536 *e* **boom guards**
d Auslegerwegbegrenzer *m pl*
f limiteurs *m pl* de mouvement de la
flèche (*d'une grue*)
nl bewegingsbegrenzers *m pl* van de giek
r ограничители *m pl* движений стрелы
(*крана*)

B537 *e* **boom harness**
d Auslegerseilflasche *f*
f palan *m* de la flèche (*d'une grue*)
nl hijstuig *n*
r стреловой полиспаст *m*

B538 *e* **boom hoist**
d Auslegerwinde *f*
f treuil *m* de flèche
nl gieklier *n*
r стреловая лебёдка *f*

B539 *e* **boom point**
d Auslegerspitze *f*
f tête *f* de flèche (*d'une grue*)
nl giekeinde *n*
r оголовок *m* стрелы (*крана,
экскаватора*)

B540 *e* **boom stop**
d Auslegerhubbegrenzer *m*
f limiteur *m* de (re)levage de la flèche
(*d'une grue*)
nl hefbegrenzer *m* van de giek
r ограничитель *m* подъёма стрелы
(*крана*)

B541 *e* **booster**
d Booster *m*, Vorschaltverdichter *m*;
Druckverstärker *m*
f moteur *m* auxiliaire; surpresseur *m*
nl aanvullende krachtbron *f(m)*;
aanjager *m*
r вспомогательный агрегат *m* (*насос,
компрессор*), повышающий
интенсивность циркуляции в системе;
аппарат *m* для повышения давления
в системе

B542 *e* **booster pump**
d Druckerhöhungspumpe *f*,
Verstärkungspumpe *f*,
Vordruckpumpe *f*
f pompe *f* d'alimentation
nl boosterpomp *f(m)*
r подкачивающий насос *m* (*для
повышения давления в сети*);
вспомогательный насос *m*

B543 *e* **booster pump station**
d Druckerhöhungsstation *f*,
Zwischenpumpwerk *n*,
Verstärkeranlage *f*
f station *f* de pompage intermédiaire
nl tussenpompstation *n*
r повысительная [промежуточная]
насосная станция *f*

B544 *e* **booth**
d 1. Kabine *f* 2. Absaughaube *f*
f 1. cabine *f* 2. capot *m* d'aspiration
nl 1. cabine *f* 2. afzuigkap *f(m)*
r 1. кабина *f* 2. вентиляционное
укрытие *n*; отсос *m* полуоткрытого
типа

B545 *e* **booth front opening**
d Ansaugquerschnitt *m*
f orifice *m* d'entrée de la hotte; orifice
m d'entrée du capot d'aspiration
nl aanzuigopening *f*
r входное отверстие *n* вытяжного
зонта; рабочий проём *m*
вентиляционного укрытия

B546 *e* **bootstrap system**
d 1. Bootstrap-Verfahren *n*
2. Bootstrap-Kälteanlage *f*
f 1. système *m* bootstrap
2. réfrigérateur *m* à air bootstrap
nl bootstrap-systeem *n*
r 1. теплоутилизационная система *f*
2. воздушная холодильная машина *f*
с дожимающим компрессором

B547 *e* **border**
d Kante *f*; Umrandung *f*; Bordkante *f*;
Fries *m*
f bord *m*; rive *f*; bordure *f*; frise *f*
nl rand *m*; omranding *f*; fries *f(m)*

r кромка *f*; окантовка *f*; бордюр *m*;
фриз *m*

B548 *e* **bordering**
d Eindämmung *f*
f endiguement *m*
nl indammen *n*
r обвалование *n*

B549 *e* **bore**
d 1. Rohrweite *f*, lichte Weite *f*
2. Bohrung *f* 3. Bohrloch *n* 4. Bore *f*,
Mascaret *n*, Flußgeschwelle *n*,
Sturzwelle *f*
f 1. calibre *m*, diamètre *m* intérieur
(*de tuyau*) 2. trou *m* alésé 3. trou *m*
de forage 4. mascaret *m*, onde *f*
à front raide
nl 1. kaliber *n*, binnendiameter *m*
2. boorgat *n*, boring *f* 3. boorgat *n*
4. vloedgolf *f(m)*, bandjir *m*
r 1. внутренний диаметр *m* (*трубы,
отверстия*) 2. цилиндрическое
отверстие *n* 3. буровая скважина *f*
4. бор *m*, маскарэ *m*, приливный
вал *m*

B550 **bored cast-in-place pile** *see* **bored pile**
B551 *e* **bored pile**
d Bohrpfahl *m*, Ortpfahl *m*
f pieu *m* coulé en place, pieu *m* moulé
(dans le sol)
nl in het boorgat gestorte betonpaal *m*
r буровая [буронабивная] свая *f*

B552 *e* **bored pile with expanded base**
d Klumpfußpfahl *m*, Bohrpfahl *m* mit
Fußverbreiterung
f pieu *m* moulé avec base élargie
nl betonpaal *m* met verzwaarde voet
r буронабивная свая *f*
с уширенным забоем,
камуфлетная свая *f*

B553 *e* **bored well**
d Bohrbrunnen *m*
f puits *m* foré
nl geboorde bron *f(m)* [pompput *m*]
r буровой колодец *m*

B554 *e* **borehole, bore hole**
d Bohrloch *n*
f trou *m* [puits *m*] de forage
nl boorgat *n*
r буровая скважина *f*

B555 *e* **borehole casing**
d Bohrlochmantelrohr *n*,
Brunnenrohr *n*, Futterrohr *n*
f tube *m* de cuvelage [de revêtement]
nl boorbuis *f(m)*
r обсадная труба *f*

B556 *e* **borehole mouth**
d Bohrlochmündung *f*, Mundloch *n*
f tête *f* de forage
nl monding *f* van het boorgat
r устье *n* скважины

B557 *e* **borehole pump**
d Tiefpumpe *f*
f pompe *f* de fond

nl dieptepomp *f(m)* zuigperspomp *f(m)*
in boorput
r глубинный насос *m*

B558 *e* **bore pit**
d Schürfung *f*, Schürfgrube *f*,
Schürfloch *n*
f fouille *f* d'essai [de recherche]
nl proefput *m*, proefschacht *f(m)*
r шурф *m*[j]

B559 *e* **borer holes**
d Wurmstichigkeit *f*, Baumfraß *m*
f vermoulure *f*, piqûre *f*
nl (hout)wormgangen *m pl*
r червоточина *f* (*порок древесины*)

B560 **bore well** *see* **borehole**
B561 *e* **bore well filter**
d Brunnenfilter *n*, *m*
f crépine *f* de puits, filtre *m* de puits
tubulaires
nl bronfilter *n*, *m*
r фильтр *m* трубчатого колодца

B562 *e* **boring**
d 1. Bohren *n* 2. Bohrloch *n*
f 1. forage *m* 2. trou *m* de forage
nl 1. boren *n* 2. boorgat *n*
r 1. бурение *n* 2. буровая скважина
f

B563 *e* **boring machine**
d Bohrmaschine *f*
f foreuse *f*, sondeuse *f*
nl boormachine *f*
r буровой станок *m*

B564 *e* **borrow**
d Entnahmegruve *f*, Entnahmezone *f*
f emprunt *m* de terre
nl groeve *f*, ontgravingsgebied *n*
r резервный карьер *m*, резерв *m*,
резервная выемка *f*

B565 *e* **borrowed light**
d Innenfenster *n*
f fenêtre *f* intérieure
nl binnenvenster *n*; indirekt licht *n*
r внутреннее окно *n*; фрамуга *f* (*в
стене между помещениями*)

B566 *e* **borrow pit**
d Entnahmegrube *f*
f ballastière *f*; emprunt *m* de terre
nl groeve *f*, ontgavingsgebied *n*
r карьер *m*, резерв *m* грунта

B567 *e* **bottle silt sampler**
d Schlammproben-Flasche *f*
f turbidisonde *f* à bouteille
nl slibmonster-flesje *n*
r бутылочный батометр *m*, батометр
m для взятия проб взвешенных
наносов

B568 *e* **bottle trap**
d Flaschensiphon *m*,
Flaschengeruchverschluß *m*
f siphon *m* à bouteille
nl bekersifon *m*, bekerstankafsluiter *m*
r бутылочный сифон *m*

B569 *e* **bottom**
d 1. Liegende *n*, Sohle *f*, Bausohle *f*
2. Flußsohle *f* 3. Unterkante *f*
f 1. pied *m*, sol *m*, radier *m* 2. fond *m*
(*de cours d'eau*) 3. face *f* inférieure
nl 1. grondvlak *n* 2. rivierbodem *m*
3. ondervlak *n*, onderzijde *f*
r 1. подошва *f* (*выработки*) 2. дно *n*
(*водотока*) 3. нижняя поверхность *f*
[грань *f*]

B570 *e* **bottom chord**
d Untergurt *m* (*Fachwerkbinder*)
f membrure *f* inférieure
nl onderrand *m* (*van vakwerk*)
r нижний пояс *m* (*фермы*)

B571 *e* **bottom-chord stress**
d Untergurtbeanspruchung *f*
f contrainte *f* [effort *m*] dans la
membrure inférieure
nl spanning *f* in de onderrand van het
vakwerk
r усилие *n* в нижнем поясе (*фермы*)

B572 **bottom discharge** *see* **bottom discharge opening**

B573 *e* **bottom discharge opening**
d Grundablaß *m*, Bodenöffnung *f*,
Sohlenöffnung *f*
f vidange *m* basse [de fond], orifice *m*
de fond
nl onderuitlaat *m*
r донный водоспуск *m* [водосброс *m*]

B574 *e* **bottom fiber**
d untere Faser *f* (*Balken*)
f fibre *f* inférieure (*de la poutre*)
nl onderste vezel *f(m)* (*balk*)
r нижнее волокно *n* (*балки*)

B575 *e* **bottom gate**
d Bodenverschluß *m*
f vanne *f* de profondeur
nl bodemschuif *f(m)*, bodemklep *f(m)*
r глубинный затвор *m*

B576 *e* **bottom-hinged ventilator**
d Kippflügel *m*,
Fensterbelüftungsklappe *f*
f volet *m* de fenêtre tombant
intérieur
nl valraam *n* (*bovenraam*)
r нижнеподвесная фрамуга *f*

B577 **bottom ice** *see* **anchor ice**

B578 *e* **bottom intake**
d Sohlenentnahme *f*
f prise *f* de fond
nl onderinvoer *m*
r донный водозабор *m*

B579 *e* **bottom layer of the reinforcing steel**
d Zugbewehrung *f*
f nappe *f* d'armature inférieure
nl trekwapening *f*
r нижний слой *m* арматуры

B580 *e* **bottom of foundation**
d Fundamentsohle *f*
f surface *f* de contact d'une fondation

nl aanleg *m* van de fundering
r подошва *f* фундамента

B581 **bottom outlet** *see* **bottom discharge opening**

B582 *e* **bottom pipe distribution**
d untere Verteilung *f*
f distribution *f* basse
nl onderverdeling *f* (*verwarmingssysteem*)
r нижняя разводка *f* труб (*в системе
отопления*)

B583 *e* **bottom rail**
d Unterfries *m* (*Fensterrahmen*);
Türsockelfries *m*
f bâti *m* dormant; pièce *f* d'appui
nl onderdorpel *m* (*raam, deur*)
r нижний брусок *m* створки окна
или обвязки дверного полотна

B584 *e* **bottom reinforcement bars**
d untere Bewehrung *f*
f armature *f* inférieure
nl trekwapening *f* (*balk*)
r арматура *f* растянутой зоны,
нижняя арматура *f*

B585 *e* **bottom sill**
d Grundschwelle *f*
f seuil *m* à la base
nl onderdorpel *m* (*kozijn*)
r донный порог *m*

B586 **bottom water outlet** *see* **bottom discharge opening**

B587 *e* **boulder**
d Feldstein *m*, Findling *m*
f gros galet *m*
nl rolsteen *m*
r валун *m*, крупный булыжник *m*

B588 *e* **boulder clay**
d Geschiebemergel *m*
f argile *f* de moraine à blocs
nl keileem *n*
r моренная *или* валунная глина

B589 *e* **boundary**
d Grenze *f*
f limite *f*, frontière *f*, borne *f*
nl grens *f(m)*, begrenzing *f*
r граница *f*; предел *m*

B590 *e* **boundary action**
d Randwirkung *f*
f effet *m* de conditions aux limites;
sollicitation *f* due aux conditions aux
limites
nl grenstoestand *m*
r влияние *n* граничных условий

B591 *e* **boundary conditions**
d Grenzbedingungen *f pl*
f conditions *f pl* aux limites
nl grensvoorwaarden *f pl*
r граничные условия *n pl*

B592 *e* **boundary force**
d Randkraft *f*
f effort *m* aux limites
nl randkracht *f(m)*
r краевое усилие *n*

B593 *e* **boundary layer**
d Grenzschicht *f*
f couche *f* limite [frontière]
nl grenslaag *f(m)*
r граничный слой *m*

B594 *e* **boundary member**
d Randglied *n*
f élément *m* de bordure [de bord, de rive]
nl randstaaf *f(m)*
r краевой элемент *m*

B595 *e* **boundary of saturation**
d Sättigungsgrenze *f*
f frontière *f* de saturation
nl verzadigingsgrens *f(m)*
r граница *f* насыщения (*поверхность раздела между насыщенным и ненасыщенным грунтом*)

B596 *e* **boundary stress**
d Randspannung *f*
f contrainte *f* due aux efforts aux limites, contrainte *f* aux limites
nl randspanning *f*
r напряжение *n*, вызванное краевыми усилиями; напряжение *n* на контуре

B597 *e* **boundary survey**
d Grenzenaufnahme *f*
f levé *m* topographique aux limites de chantier
nl (de bouwplaats) uitzetten *n*
r геодезическая съёмка *f* границ стройплощадки

B598 *e* **bow**
d 1. Bogen *m*; Krümmung *f*, Verbiegung *f* 2. Flächenverwerfung *f* 3. Längskrümmung *f*
f 1. arc *m*, cambrure *f* 2. gauchissement *m*, voilement *m* 3. courbure *f* longitudinale
nl 1. kromming *f*, (ver)buiging *f* 2. kromtrekken *n* 3. langskromming *f* (*hout*)
r 1. дуга *f*; изгиб *m*; выгиб *m* 2. продольное коробление (*напр. панели*) 3. продольная кривизна *f*

B599 *e* **bowel urinal**
d Pissoirbecken *n*
f urinal *m*
nl urinoir *n*
r писсуар *m*

B600 *e* **Bow's notation**
d Bowsche Bezeichnungsweise *f* (*für reziproken Kräfteplan des ebenen Fachwerkes*)
f notation *f* de Bow (*pour la construction d'un diagramme de Crémona*)
nl Bow-notatie *f* (*staafkrachten in Cremona-diagram*)
r обозначение *n* силовых полей фермы (*при построении диаграммы Кремоны*)

B601 *e* **bowstring truss**
d Bogenträger *m* mit Zugband, Bogenfachwerk *n*
f poutre *f* en arc à tirant, arc *m* à tirant
nl boogligger *m*
r арочная ферма *f* с затяжкой

B602 **box beam** *see* **box girder**

B603 *e* **box caisson**
d Schwimmkasten *m*, gestrandeter [amerikanischer] Senkkasten *m*
f caisson *m* géant [flottant]
nl dooscaisson *m*
r массив-гигант *m*

B604 *e* **box channel**
d 1. Rechteckkanal *m* 2. Rechteckgerinne *n*
f canal *m* (à section) rectangulaire
nl kanaal *n* met rechthoekige doorsnede
r канал *m* прямоугольного сечения

B605 *e* **box culvert**
d Kasten-Durchlaß *m*
f ponceau *m* (à section) rectangulaire; dalot *m*
nl rechthoekige duiker *m*
r водопропускная труба *f* прямоугольного сечения

B606 *e* **box dam**
d Umschließungsfang(e)damm *m*
f batardeau *m* circulaire
nl omsluitende damwand *m*
r перемычка *f*, полностью окружающая строительную площадку; замкнутая перемычка *f*

B607 *e* **box drain**
d Kastendrän *m*
f drain *m* (à section) rectangulaire
nl afvoerleiding *f* met rechthoekige doorsnede
r дрена *f* прямоугольного сечения

B608 *e* **box girder**
d Hohlkastenträger *m*
f poutre-caisson *f*
nl kokerligger *m*
r балка *f* коробчатого сечения

B609 *e* **boxing out, box out**
d Bildung *f* von Aussparungskasten im Beton
f formation *f* de baies *ou* de niches en béton
nl uitsparingen maken in beton
r образование *n* проёмов [ниш] в бетоне

B610 *e* **box pile**
d Kastenpfahl *m*
f pieu *m* caisson
nl kokerpaal *m*
r коробчатая свая *f*

B611 *e* **box-shaped module open on two sides**
d zweiseitig offener wohnungsgroßer Raumblock *m*, kastenförmige Raumzelle *f* in Wohnungsgröße

f cellule *f* tridimensionnelle préfabriquée
ouverte aux abouts
nl aan twee zijden open ruimtemoduul *n*
r объёмная блок-квартира *f*,
открытая с двух сторон, блок-труба *f*

B612 *e* **box stairs**
d wandaufgestützte Treppe *f*
f escalier *m* reposant sur murs,
escalier *m* encloisonné
nl in de muren opgelegde trap *m*
r лестница *f*, опирающаяся на стены
[на стеновые косоуры]

B613 *e* **box system**
d räumliche Plattenkonstruktion *f*
f système *m* de construction en
panneaux préfabriqués (*sans
ossature*)
nl (systeem *n* van) zelfdragende
ruimtemodulen *n pl*
r пространственная панельная
конструкция *f* (*без несущего каркаса*)

B614 *e* **brace**
d 1. Verband *m*; Strebe *f*; Spreize *f*;
Verspannung *f*; Diagonalzangenbalken
m; Versteifungsglied *n* 2. Handbohrer
m
f 1. contreventement *m*; entretoise *f*;
hauban *m*; diagonale *f*; élément *m*
de raidissement 2. perceuse *f* à main
nl 1. schoor *m*, op trek belaste
verbinding(sbalk) *f* (*m*);
verstijvingsbalk *m* 2. klamp *m*,
booromslag *m*, *n*, beugel *m*
r 1. связь *f*; подкос *m*; распорка *f*;
оттяжка *f*; диагональная схватка *f*;
элемент *m* жёсткости 2. ручная
дрель *f*; коловорот *m*

B615 *e* **braced arch**
d Fachwerkbogen *m*, Bogenfachwerk *n*
f arc *m* à treillis, poutre *f* en arc
à treillis
nl vakwerkboog *m*
r сквозная арка *f*, арочная ферма *f*

B616 *e* **braced frame**
d Rahmen *m* mit Aussteifungsverbänden,
verspannter [verstrebter] Rahmen *m*
f portique *m* [cadre *m*] contreventé
nl vakwerkskelet *n*
r рама *f*, раскреплённая связями,
жёсткая рама *f*

B617 *e* **braced structures**
d 1. Gitterkonstruktionen *f pl*,
Fachwerke *n pl*
2. Skelettkonstruktionen *f pl*
[Fachwerkbauten *m pl*] mit
Aussteifungsverbänden, verstrebte
Skelettkonstruktionen *f pl*
f 1. constructions *f pl* réticulées
[à treillis] 2. constructions *f pl* de
charpente contreventée
nl 1. vakwerkconstructies *f pl*
2. geschoord skelet *n*
r 1. решётчатые [сквозные] конструкции *f pl*

2. каркасные конструкции *f pl*,
раскреплённые связями

B618 *e* **bracing**
d 1. Steife *f*, Versteifungsglied *n*,
Aussteif(ungs)glied *n*
2. Aussteifungssystem *n*
3. Aussteifen *n*, Absteifen *n*
f 1. contreventement *m* 2. système *m*
de contreventements 3. pose *f* de
contreventement, montage *m* de
contreventements
nl 1. verstijving(selement) *f* (*n*),
verband *n*, verspanning *f* 2.
verstijvingssysteem *n* 3. montage *f*
van verstijvingselementen
r 1. связь *f* жёсткости; элемент *m*
связи 2. система *f* связей
3. постановка *f* [монтаж *m*] связей

B619 *e* **bracket**
d 1. Konsole *f*, Halterung *f*,
Krageisen *n* 2. Lasche *f*
f 1. console *f*; appui *m* 2. éclisse *f*
nl 1. console *f*(*m*), karbeel *m*; uitstek *n*
2. knoopplaat *f*(*m*), lasplaat *f*(*m*)
r 1. кронштейн *m*; консоль *f*; выступ
m 2. накладка *f*

B620 *e* **bracket scaffolds**
d Konsolegerüst *n*, Auslegerrüstung *f*,
Krag-Baugerüst *n*
f échafaudage *m* à console
nl (uit)steeksteiger *m*
r выпускные [консольные] леса *pl*

B621 *e* **bracket-type retaining wall**
d Stützmauer *f* mit Konsole
f mur *m* de soutènement à console
nl keermuur *m* met vleugels
r подпорная стенка *f* с разгружающей
консолью

B622 *e* **braided nylon rope**
d geklöppeltes Nylonseil *n*
f câble *m* en fibres de nylon tressées
nl nylonkoord *n*, koord *n* van
gevlochten nylon
r плетёный нейлоновый канат *m*

B623 *e* **branch**
d 1. Zweigleitung *f*, Anschlußleitung *f*
2. Anschlußstutzen *m*
f 1. conduite *f* secondaire; tuyau *m* de
branchement 2. tubulure *f*, raccord *m*
nl 1. aftakking *f* van een pijpleiding *f*
2. T-stuk *n*
r 1. ветвь *f* [ответвление *n*]
трубопровода 2. патрубок *m*

B624 *e* **branch duct**
d Abzweigkanal *m*, Abzweigluftleitung *f*
f conduit *m* (à air) secondaire
nl aftakking *f* van een luchtleiding
r ответвление *n* воздуховода

B625 *e* **branch fitting**
d Abzweigstück *n*, Abzweigstutzen *m*
f tubulure *f* de branchement
nl secundaire leiding *f*
r труба-ответвление *f*; труба *f*
[патрубок *m*] с отростком

B626 branch-off *see* **branch 1.**

B627 *e* **branch sewer**
  d Nebensammler *m*
  f collecteur *m* secondaire
  nl secondaire verzamelleiding *f*
  r боковой коллектор *m*;
  канализационный коллектор *m*
  второго порядка

B628 *e* **brand of cement**
  d Zementmarke *f*, Zementgüteklasse *f*
  f qualité *f* de ciment
  nl cementsoort *f(m)*, *n*
  r марка *f* цемента

B629 *e* **breakaway**
  d 1. Strahlablösung *f* von der Ebene
  2. Strahlablösung *f*, Abreißen *n*,
  Überziehen *n*, Abkippen *n*
  f 1. décollement *m* de la veine
  2. décrochement *m*
  nl 1. het losbreken van een stroming
  2. wegvallen *n*
  r 1. отрыв *m* струи от плоскости
  2. срыв *m* потока

B630 *e* **breaker**
  d 1. Brecher *m* 2. Aufbrechhammer *m*,
  Aufreißhammer *m*
  f 1. concasseur *m*, broyeur *m*
  2. perforateur *m* de béton, marteau *m*
  brise-béton
  nl 1. moker *m* 2. sloophamer *m* voor
  beton
  r 1. дробилка *f* 2. тяжёлый отбойный
  молоток *m*; бетонолом *m*

B631 *e* **breaking**
  d Bruch *m*
  f rupture *f*
  nl breuk *f(m)*
  r разрушение *n*; разрыв *m*

B632 *e* **breaking bending moment**
  d Bruchbiegemoment *n*,
  Bruchbiegungsmoment *n*
  f moment *m* fléchissant de rupture
  nl buigmoment *n* bij breuk
  r разрушающий [предельный]
  изгибающий момент *m*

B633 *e* **breaking cross-section**
  d Bruchquerschnitt *m*
  f section *f* de rupture
  nl breukdoorsnede *f*
  r поперечное сечение *n* в месте разрыва

B634 *e* **breaking elongation**
  d Bruchdehnung *f*
  f allongement *m* de rupture
  nl breukverlenging *f*
  r удлинение *n* при разрыве

B635 *e* **breaking joints**
  d Fugenversatz *m*
  f joints *m pl* alternés, appareil *m*
  alterné
  nl metselverband *n* met verspringende
  voegen
  r вертикальные швы *m pl* вразбежку
  (*напр. каменной кладки*)

B636 *e* **breaking load**
  d Bruchbelastung *f*
  f charge *f* de rupture
  nl breukbelasting *f*
  r разрывная [разрушающая]
  нагрузка *f*

B637 *e* **breaking plant**
  d Brech(er)anlage *f*
  f installation *f* de concassage
  nl steenbreker *m*
  r дробильная установка *f*

B638 *e* **breaking strength**
  d Bruchwiderstand *m*, Bruchfestigkeit *f*
  f résistance *f* à la rupture, résistance *f*
  ultime
  nl breukvastheid *f*
  r прочность *f* при разрыве, временное
  сопротивление *n* разрыву

B639 *e* **breaking stress**
  d Bruchspannung *f*
  f contrainte *f* de rupture
  nl breukspanning *f*
  r разрушающее напряжение *n*,
  напряжение *n* при разрыве

B640 *e* **breaking test**
  d Bruchprüfung *f*, Bruchversuch *m*
  f essai *m* [épreuve *f*] de rupture (à la
  traction)
  nl breukproef *f(m)*
  r испытание *n* на разрыв

B641 *e* **breaking up**
  d 1. Aufbrechen *n*, Aufbruch *m*,
  Aufreißen *n*; Zersetzung *f*, Zerfall *m*,
  Dekomposition *f* 2. Auflösung *f*,
  Verdünnung *f* 3. Auflockerung *f*
  4. Eis(auf)bruch *m*
  f 1. rupture *f*; décomposition *f*
  2. dilution *f* 3. ameublissement *m*
  du sol 4. débâcle *f*
  nl 1. (op)breken *n* 2. uiteenvallen *n*
  3. loswerken *n* van grond
  4. breken *n* van het ijs, kruien *n* (*ijs*)
  r 1. разрыв *m*; распадение *n*
  2. разбавление *n*; разжижение *n*
  3. разрыхление *n* грунта
  4. вскрытие *n* ледяного покрова

B642 *e* **breakwater**
  d Hafenaußenwerk *n*, Mole *f*,
  Hafendamm *m*, Wellenbrecher *m*
  f ouvrage *m* d'abri; brise-lames *m*
  nl havendam *m*, pier *m* (*havenwerken*)
  r внешнее оградительное сооружение
  n (*порта*); волнолом *m*, мол *m*

B644 *e* **breakwater arm, breakwater pier**
  d Mole *f*, Hafendamm *m*
  f jetée *f*, môle *m*
  nl pier *m* (*havenwerken*)
  r мол *m*

B645 *e* **breast**
  d 1. Brüstungsmauer *f* 2. vorstehender
  Mauerteil *m* 3. untere Kante *f*
  f 1. allège *f* 2. partie *f* de mur en saillie
  3. (sur)face *f* inférieure (*p.ex. de
  poutre*)

nl **1.** borstwering *f* (*venster*) **2.** boezem *m*
(*schoorsteen*) **3.** onderkant *m* van een
leuningbalk
r **1.** подоконная стенка *f*
2. выступающая часть *f* тены
3. нижняя грань *f* (*балки, поручней, стропил*)

B646 *e* **breasting dolphin**
d Dückdalbe *f*, Prallpfahl *m*
f duc *m* d'Albe, pieu *m* de défense
nl dukdalf *m*
r отбойный пал *m*

B647 *e* **breast wall**
d **1.** Tauchwand *f* **2.** Stützmauer *f*,
Stützwand *f* **3.** Sturzwand *f*,
Fallwand *f*
f **1.** masque *m* **2.** mur *m* de butée [de
soutènement] **3.** mur *m* de chute
nl **1.** borstwering *f* **2.** steunmuur *m*
3. valmuur *m*, stortmuur *m* (*sluis*)
r **1.** забральная стенка *f* **2.** подпорная
стенка *f* **3.** стенка *f* падения

B648 *e* **breathing apparatus**
d Atemschutzgerät *n*, Atmungsgerät *n*
f respirateur *m*
nl ademhalingstoestel *n*
r респиратор *m*

B649 *e* **breech fitting**
d Hosenrohr *n*
f embranchement *m*, bifurcation *f* de
tuyau
nl gaffelpijp *f(m)*
r разветвление *n* трубопровода;
штанообразный тройник *m*

B650 *e* **breeching**
d Abzugskrümmer *m*, Fuchs *m*
f carneau *m*
nl broek *f(m)*, broeking *f*, schoorsteentrek
m
r боров *m*, соединительный дымоход *m*

B651 *e* **breeze concrete**
d Kesselschlackenbeton *m*
f béton *m* de mâchefer
nl slakkenbeton *n*
r шлакобетон *m* на котельном шлаке

B652 **Breuchaud pile** *see* **pipe pile**

B653 *e* **brick**
d Ziegel *m*
f brique *f*
nl baksteen *m*
r кирпич *m*

B654 *e* **brick and brick**
d halbtrockenes Ziegelmauerwerk *n*
(*Mörtel ist nur für Ausfüllen der
Unebenheiten gebraucht*)
f maçonnerie *f* (en briques) à demi-sec
(*mortier n'est utilisé que pour remplir
les cavités en briques*)
nl halvdroog metselwerk *n*
r полусухая кирпичная кладка *f*
(*с заполнением раствором лишь не-
ровностей в кирпичах*)

B655 *e* **brick facing**
d Ziegelverkleidung *f*, Ziegelauskleidung
f, Ziegelbekleidung *f*
f revêtement *m* en briques
nl bekleding *f* met baksteen
r облицовка *f* из кирпича, кирпичная
облицовка *f*

B656 *e* **brick hammer**
d Maurerhammer *m*
f marteau *m* de maçon
nl metselaarshamer *m*, kaphamer *m*
r молоток-кирочка *m* каменщика

B657 *e* **brick-jointer**
d Fug(en)eisen *n*, Fugenkelle *f*
f tire-joint *m*
nl voegsprijker *m*
r расшивка *f* (*инструмент*)

B658 *e* **bricklayer**
d Maurer *m*
f maçon *m*, limousin *m*; briqueteur *m*
(*Canada*)
nl metselaar *m*
r каменщик *m*

B659 *e* **bricklayer's square scaffolds**
d Maurergerüst *n* mit
Rechteckstirnwänden
f échafaud(age) *m* de maçon à cadres en
planches
nl schraagsteiger *m*, juksteiger *m*
r подмости *pl* каменщика на
конвертах

B660 *e* **brick masonry**
d Ziegelmauerwerk *n*
f maçonnerie *f* de [en] briques
nl metselwerk *n*
r кирпичная кладка *f* (*конструкция*)

B661 *e* **brick nogging**
d Fachwerk-Ziegelausfüllung *f*
f remplissage *m* des pans par briques
nl opvulling *f* van vakwerk met
metselwerk
r заполнение *n* фахверка кирпичной
кладкой

B662 *e* **brick paving**
d Klinkerpflaster *n*
f pavé *m* [pavage *m*] en briques cuites
nl klinkerbestrating *f*
r клинкерное дорожное покрытие *n*;
клинкерная мостовая *f*

B663 **brick veneer** *se* **brick facing**

B664 *e* **brick wall**
d Ziegelmauer *f*
f mur *m* en briques
nl bakstenen muur *m*
r кирпичная стена *f*

B665 *e* **brick walling**
d **1.** Ziegelmauern *f pl*;
Ziegelsperrmauer *f* **2.** Errichtung *f*
der Ziegelmauern
f **1.** murs *m pl* en briques; enceinte *f*
[clôture *f*] en briques **2.** construction *f*
de murs en briques

nl 1. ommuring *f* in baksteen 2. muren
metselen *n*
r 1. кирпичные стены *f pl*; кирпичная
ограда *f* 2. возведение *n* кирпичных
стен

B666 *e* **brickwork**
d Mauerwerk *n*
f maçonnerie *f*
nl metselwerk *n*
r каменная кладка *f*

B667 *e* **brickwork casing**
d Ziegelverkleidung *f*
f revêtement *m* en brique(s)
nl bekleding *f* met metselwerk
r кирпичная облицовка *f*

B668 *e* **brick works**
d Ziegelfabrik *f*, Ziegelwerk *n*,
Ziegelbrennerei *f*
f briqueterie *f*, usine *f* à briques
nl steenfabriek *f*, steenbakkerij *f*
r кирпичный завод *m*

B669 *e* **bridge**
d 1. Brücke *f* 2. Gehwegüberdachung *f*,
Bürgersteigüberdachung *f*
f 1. pont *m* 2. auvent *m* de protection
nl 1. brug *f(m)* 2. veiligheidsluifer *f(m)*
(stijger)
r 1. мост *m* 2. защитный козырёк *m*
над тротуаром [*у стройплощадки*)

B670 *e* **bridge abutment**
d Brückenwiderlager *n*, Uferpfeiler *m*
f culée *f* de pont
nl landhoofd *n*
r береговой устой *m* моста

B671 *e* **bridge approach embankment**
d Zufahrtdamm *m* zu einer Brücke
f remblai *m* routier aux accès de pont
nl brugoprit *m*
r насыпь *f* на подходах к мосту

B672 *e* **bridgeboard**
d Treppenwange *f*, Wange *f*
f limon *m* (*d'escalier*)
nl trapboom *m*
r тетива *f*, косоур *m* (*лестницы*)

B673 *e* **bridge crossing**
d Brückenübergang *m*
f pont *m* avec ouvrages d'accès
nl brugwegviaduct *n*
r мостовой переход *m*

B674 *e* **bridge deck**
d Brückentafel *f*, Brückenüberbau *m*
f tablier *m* de pont
nl brugdek *n*
r мостовой настил *m*

B675 *e* **bridge engineering**
d Brückenbau *m*
f technique *f* de construction des ponts
nl bruggenbouw *m*
r мостовое дело *n*; мостостроение *n*;
конструирование *n* и строительство
n мостов

B676 **bridge flooring** *see* **bridge deck**

B677 *e* **bridge girder**
d Brückenträger *m*
f poutre *f* de pont
nl hoofdligger *m*
r главная балка *f* моста

B678 *e* **bridge opening**
d Brückenöffnung *f*
f débouché *m* du pont
nl doorvaartopening *f*, doorvaartwijdte *f*
r отверстие *n* моста

B679 *e* **bridge pier**
d Brückenpfeiler *m*
f pile *f* de pont, pilier *m*
nl brugpijler *m*
r бык *m*, промежуточная опора *f*
моста

B680 **bridge plank** *see* **bridge deck**

B681 *e* **bridge travel**
d Laufweg *m* des Brückenkranes
f parcours *m* du pont roulant
nl traject *n* van een brugkraan
r пробег *m* [путь *m* движения]
мостового крана

B682 *e* **bridge truss**
d Brückenfachwerkträger *m*
f poutre *f* à treillis de pont
nl hoofdligger *m* [vakwerkligger *m*] van
een brug
r главная мостовая ферма *f*

B683 *e* **bridging**
d 1. Brückenschlag *m*, Überbrückung *f*
2. Spreize *f*; Ausspreizung *f*,
Spreizwerk *n* (*Trägerverbindung*)
3. rissenüberdeckende Farbhaut *f*
4. Brückenbildung *f*, Gewölbebildung
f (*bei Schüttgut*)
f 1. pontage *m* 2. croisillon *m ou*
système *m* de croisillons 3. pellicule *f*
de peinture au dessus des fissures
4. effet *m* de voûte
nl 1. overbruggen *n*, brugslag *m*
2. schoren *m pl*
3. scheuren-bedekkende verflaag *f(m)*
4. brugvorming *f* (*bij stortgoed*)
r 1. наводка *f* моста 2. распорка *f*;
связь *f*; система *f* связей 3. плёнка *f*
краски поверх трещин
4. сводообразование *n* (*в сыпучем
материале*)

B684 *e* **bridle wire rope sling**
d mehrsträngiger Stropp *m*,
Seilgehänge *n* mit mehreren
Aufhängepunkten
f élingue *f* en câble d'acier à plusieurs
brins réunis par bride
nl dubbele hijsstrop *f(m)*
r многоветвевой канатный строп *m*
(*с общей верхней серьгой или кольцом*)

B685 *e* **brine air cooler**
d Soleluftkühler *m*
f refroidisseur *m* d'air de saumure
nl pekelluchtkoeler *m*
r рассольный воздухоохладитель *m*

B686 *e* **briquette**
  d 1. Brikett *n*; Preßziegel *m*
  2. Zugprobekörper *m*
  f 1. briquette *f* 2. éprouvette *f* de
  mortier en forme d'un huit
  nl 1. briket *f(m)* 2. proefstuk *n* voor
  het bepalen van de trekvastheid van
  mortel
  r 1. брикет *m*; прессованный кирпич
  m 2. образец *m* в виде восьмёрки
  для испытания цемента на
  растяжение

B688 *e* **brittle failure, brittle fracture**
  d Sprödbruch *m*
  f rupture *f* fragile
  nl brosheidsbreuk *f(m)*
  r хрупкое разрушение *n*; хрупкий
  разрыв *m*

B689 *e* **brittle-lacquer technique**
  d Reißlackverfahren *n*
  f méthode *f* expérimentale des vernis
  craquelants
  nl scheurlakmethode *f*
  r метод *m* лаковых покрытий
  (*для исследования напряжений*)

B690 *e* **brittleness**
  d Sprödigkeit *f*
  f fragilité *f*
  nl brosheid *f*
  r хрупкость *f*

B691 *e* **broadcrested weir**
  d breitkroniges Wehr *n*, breitkroniger
  Überfall *m*, Überfall *m* mit flacher
  Kante
  f déversoir *f* à seuil épais [à crête
  épaisse]
  nl stuw *m* met brede kruin
  r водослив *m* с широким порогом

B692 *e* **broad-flanged beam**
  d Breitflanschprofil *n*, breitflanschiger
  Doppel-T-Stahl *m*, Breitflanschträger
  m
  f poutrelle *f* H, poutrelle *f* à ailes
  larges
  nl breedflensbalk *m*; T-profiel *n*
  r широкополочная двутавровая балка
  f, широкополочный двутавр *m*

B693 *e* **broad irrigation**
  d Abwasserbewässerung *f*,
  Oberflächenberieselung *f* mit
  Abwasser
  f épandage *m* des eaux d'égout
  nl bevloeiing *f* met afvalwater
  r поверхностное орошение *n*
  сточными водами

B694 *e* **broken-colour work**
  d Anstrich *m* mit Imitation (*des
  Marmors, des Holzes*) auf echt
  f peinture *f* peignée, application *f* (*de
  peinture*) par peigne
  nl verfwerk *n* in hout- *of* marmerimitatie
  r окраска *f* с разделкой
  (*напр. под мрамор, древесину*)

B695 *e* **broken stone bed**
  d Schotterlage *f*
  f lit *m* de pier es concassées
  nl steenslaglaag *f(m)*
  r слой *m* щебня, щебёночная
  подготовка *f*

B696 *e* **broom finish**
  d Besenabzug *m*
  f fini *m* au balai
  nl bezemen *n* (*afwerking van beton*)
  r отделка *f* бетонных поверхностей
  щёткой (*для придания шероховатости*)

B697 *e* **brown coat**
  d zweite Unterputzschicht *f*
  f sous-couche *f* d'enduit
  nl grondlaag *f(m)* voor de pleisterlaag
  r грунт *m* штукатурного покрытия;
  второй слой *m* штукатурного намёта

B698 *e* **brush**
  d Pinsel *m*; Bürste *f*
  f brosse *f*; pinceau *m*
  nl penseel *n*, kwast *m*; borstel *m*
  r кисть *f*; щётка *f*

B699 *e* **brushed plywood**
  d aufgestrichenes Sperrholz *n*
  f contre-plaqué *m* brossé [traité par
  brossage]
  nl met staalborstel bewerkt multiplex *n*
  r фанера *f*, обработанная
  металлическими щётками
  (*для получения рельефной текстуры*)

B700 *e* **brushed surface**
  d mit Bürsten bearbeitete
  Betonoberfläche *f*
  f béton *m* surfacé au balai
  nl met borstels bewerkt betonoppervlak
  n
  r бетонная поверхность *f*,
  обработанная щёткой

B701 *e* **brushing**
  d 1. Pinselanstrich *m* 2. Abbürsten *n*
  f 1. peinturage *m* au pinceau
  2. brossage *m*
  nl 1. schilderwerk *n*, verfwerk *n*
  2. afborstelen *n*
  r 1. окраска *f* кистью 2. очистка *f*
  [обработка *f*] поверхности щёткой

B702 *e* **brush matting**
  d Rauhwehr *n*
  f revêtement *m* en matelas de branches
  nl rijstuin *m*, vlechtwerk *n* van rijshout
  als oeververdediging
  r укрепление *n* (*берега*) плетняком

B703 *e* **brush rake**
  d Gestrüppharke *f*, Roderechen *m*
  f débroussailleuse *f*
  nl struikenhark *f(m)*
  r корчеватель *m* кустарника

B704 *e* **brush revetment**
  d Berauhwehrung *f*
  f revêtement *m* aux [en] branches
  sèches

nl rijsbed *n*, rijsberm *m*
r хворостяная одежда *f* [выстилка *f*]

B705 *e* **brushwood**
   d Faschinenholz *n*, Reisholz *n*; Reisig *n*
   f fascines *f pl*; fagots *m pl*; branches *f pl* sèches
   nl rijsbos *m*; rijs *n*
   r фашинник *m*; хворост *m*

B706 *e* **brushwood mattress**
   d Buschpackwerk *n*, Buschmatte *f*, Reisigpackung *f*, Sinkstück *n*
   f matelas *m* en fascines [en fagots]
   nl zinkstuk *n* van rijshout
   r хворостяной тюфяк *m*

B707 *e* **brushwork**
   d 1. Rauhwehr *n* 2. Pinselanstrich *m*
   f 1. revêtement *m* en matelas de branchages 2. application *f* au pinceau, peinturage *m* [peinture *f*] au pinceau
   nl 1. vlechtwerk *n* (*oeververkleding*) 2. verfwerk *n*
   r 1. укрепление *n* (*берега*) плетняком 2. окраска *f* кистью

B708 *e* **bubble breakwater**
   d pneumatischer Wellenbrecher *m*, Druckluft-Wellenbrecher *m*
   f brise-lames *m* pneumatique
   nl pneumatische golfbreker *m*
   r пневматический волнолом *m*

B709 *e* **bucket**
   d 1. Kübel *m* 2. Sprungnase *f*, Sprungschnauze *f*, Überfallsprung *m*
   f 1. benne *f*, godet *m* auget *m* 2. auge *f* de pied, auget *m* (*de barrage déversoir*)
   nl 1. emmer *m*, bak *m* 2. vangbak *m* van een overlaat
   r 1. бадья *f*, ковш *m* 2. вогнутый носок *m*, уступ *m*, трамплин *m* (*водослива*)

B710 *e* **bucket chain**
   d Becher(werk)kette *f*, Elevatorkette *f*
   f chaîne *f* à godets
   nl emmerketting *m*, *f*
   r ковшовая цепь *f*

B711   **bucket elevator** *see* **belt-type bucket elevator**

B712 *e* **bucket-ladder dredge**
   d Eimerketten-Naßbagger *m*, Eimerketten-Schwimmbagger *m*
   f drague *f* à godets
   nl emmerbaggermolen *m*
   r многоковшовый землечерпательный снаряд *m*

B713   **bucket-loader excavator** *see* **trencher**

B714 *e* **bucket pump**
   d 1. Kolbenpumpe *f*; Eimerpumpe *f* 2. Becherwerk *n*
   f 1. pompe *f* élévatoire 2. chapelet *m* hydraulique, pompe *f* à chapelet
   nl 1. plunjerpomp *f(m)* 2. jacobsladder *m*
   r 1. водоподъёмный насос *m*

2. водоотливная нория *f*

B715 *e* **bucket wheel excavator**
   d Schaufelradbagger *m*, Becherradbagger *m*
   f excavateur *m* [excavatrice *f*] à roue à godets, excavateur *m* rotatif, excavatrice *f* rotative
   nl graafmachine *f* met emmerwiel
   r многоковшовый роторный экскаватор *m*

B716 *e* **bucket wheel suction dredger**
   d Schneidkopf-Saugbagger *m*
   f drague *f* à roue à couteaux, drague *f* suceuse à désagrégateur
   nl baggermolen *m* met inbreekschoepenrad
   r роторный землесосный снаряд *m*

B717 *e* **bucket wheel type agitator**
   d Naßbagger-Schaufelrad *n*
   f désagrégateur *m* rotatif
   nl schoepenrad-agitator *m*
   r роторный разрыхлитель *m* (*землесосного снаряда*)

B718 *e* **buckle**
   d 1. Knickung *f*, Knick *m* 2. Klammer *f*; Bügel *m*; Spannmuffe *f*
   f 1. flambage *m*, flambement *m* 2. étrier *m*, manchon *m* de serrage
   nl 1. knik *m*, deuk *f(m)* 2. gesp *m*, *f*, spanschroef *f(m)*
   r 1. продольный изгиб *m* 2. скоба *f*; хомут *m*; стяжная муфта *f*

B719 *e* **buckling**
   d Knick *m*, Knickung *f*
   f flambage *m*, flambement *m*
   nl knik *m*
   r продольный изгиб *m*

B720 *e* **buckling analysis**
   d Knickberechnung *f*
   f calcul *m* de flambage
   nl knikberekening *f*
   r расчёт *m* на продольный изгиб

B722 *e* **buckling coefficient**
   d Knickbeiwert *m*, Knickzahl *f*
   f coefficient *m* de flambage [de flambement]
   nl knikgetal *n*, knikcoëfficiënt *m*
   r коэффициент *m* продольного изгиба

B723 *e* **buckling configuration**
   d Knickfigur *f*
   f configuration *f* de la courbe de flambement
   nl knikfiguur *f(m)*
   r форма *f* (кривой) продольного изгиба

B724 *e* **buckling length**
   d Knicklänge *f*
   f longueur *f* de flambement
   nl kniklengte *f*
   r приведённая длина *f* (*элемента при продольном изгибе*)

B725 *e* **buckling load**
   d Knicklast *f*
   f charge *f* de flambage [de flambement], charge *f* d'Euler

nl knikbelasting *f*
r критическая нагрузка *f* (*при
продольном изгибе*)

B726 *e* **buckling of plate**
d Ausknicken *n* [Ausbeulen *n*] der
Platte
f voilement *m* des plaques
nl plooien *n* van de plaat
r выпучивание *n* [потеря *f* местной
устойчивости] пластинки

B727 *e* **buckling strength**
d Knickfestigkeit *f*, Knickwiderstand *m*
f résistance *f* au flambage [au
flambement]
nl kniksterkte *f*, knikvastheid *f*
r предел *m* прочности при
продольном изгибе

B728 *e* **buckling stress**
d Knickspannung *f*
f contrainte *f* de flambage [de
flambement]
nl knikspanning *f*
r напряжение *n* при продольном
изгибе

B729 *e* **budget for construction project**
d Gesamtbaukosten *pl*
f valeur *f* contractuelle, coût *m* total
des travaux de construction
nl totale bouwkosten *m pl*
r общая стоимость *f* строительства
(объекта), обусловленная
заказчиком

B730 *e* **buffing**
d Polieren *n*
f polissage *m*
nl polijsten *n*
r полирование *n*

B731 *e* **bugholes**
d Lunker *m pl*
f cavités *f pl*, trous *m pl* (*sur la surface
de béton*)
nl holten *f pl* (*in het betonoppervlak*)
r раковины *f pl* (*на поверхности
бетона*)

B732 *e* **builder**
d 1. Baufachmann *m*; Bauarbeiter *m*
2. Baufirma *f*
f 1. constructeur *m*; travailleur *m*
[ouvrier *m*] du bâtiment 2. compagnie
f de construction
nl 1. bouwvakarbeider *m* 2. bouwfirma *f*
r 1. строитель *m*; строительный
рабочий *m* 2. строительная фирма *f*

B733 *e* **builder's air heater**
d Bautrockner *m*, Bautrocknungsgerät *n*
f appareil *m* de chauffage à air de
chantier [pour chantier]
nl verwarmingsapparaat *n* voor de
bouwplaats
r воздушно-отопительный агрегат *m*
для сушки зданий

B734 *e* **builder's finish hardware**
d Beschläge *m pl*, Baubeschläge *m pl*

f quincaillerie *f* de bâtiment
nl bouwbeslag *n*
r строительные скобяные изделия *n
pl*

B735 *e* **builder's road**
d Baustellenweg *m*
f route *f* de chantier
nl weg *m* voor het bouwverkeer
r внутриплощадочная дорога *f*

B736 *e* **builder's staging**
d Baugerüst *n*
f échafaudage *m* (de construction) de
pied fixe (en bois)
nl bouwsteiger *m*
r коренные леса *m pl*

B737 **builder's workshop** *see* **site workshop**

B738 *e* **building**
d 1. Gebäude *n*; Bau *m*, Bauwerk *n*
2. Bauen *n*, Gebäudeerrichtung *f*
f 1. bâtiment *m*, édifice *m*,
immeuble *m* 2. construction *f*
nl 1. bouwwerk *n*, gebouw *n* 2. bouw *m*,
bouwen *n*
r 1. здание *n*; сооружение *n*;
постройка *f* 2. строительство *n*,
возведение *n* зданий

B739 *e* **building area**
d überbaute [bebaute] Fläche *f*
f surface *f* bâtie
nl bebouwd oppervlak *n*, bouwterrein *n*
r площадь *f* застройки

B740 *e* **building berth**
d 1. Stapel *m*; Ausrüstungskai *m*
2. Helling *f*, *m*, Helgen *m*, *f*, Helge *f*
f 1. cale *f* de construction, cale *f* de
lancement; quai *m* d'achèvement
2. cale *f* couverte
nl 1. scheepshelling *f* 2. afbouwkade
f(*m*)
r 1. стапель *m*; достроечная
набережная *f* 2. эллинг *m*

B741 *e* **building block**
d Mauerwerkblock *m*, Blockstein *m*,
Block *m*, Stein *m*
f bloc *m* de maçonnerie, aggloméré *m*,
parpaing *m*, pierre *f* de taille [taillée]
nl bouwsteen *m*, bouweenheid *f*
r строительный [стеновой] блок *m*,
камень *m* (для кладки стен)

B742 *e* **building brick**
d Mauer(werk)ziegel *m*
f brique *f* de maçonnerie
nl metselsteen *m*, baksteen *m*
r строительный кирпич *m*

B743 *e* **building climatology**
d Bauklimatologie *f*
f climatologie *f* de bâtiment
nl bouwklimatologie *f*
r строительная климатология *f*

B744 *e* **building construction**
d 1. Hochbau *m*; Bauausführung *f*,
Baudurchführung *f* 2. Baukonstruktion
f, Gebäudekonstruktion *f*

f 1. construction *f* de bâtiments
2. charpente *f*
nl 1. uitvoeren *n* van een bouwwerk
2. constructie *f* van een gebouw
r 1. жилищное строительство *n*;
строительство *n* зданий
2. строительная конструкция *f*

B745 *e* **building construction programme**
d Hochbauprogramm *n*, Bauzeitplan *m*
f programme *m* de construction
[d'avancement des travaux]
nl planningsschema *n* voor de bouw
r план-график *m* производства
строительных работ

B746 building control *see* building
supervision

B747 *e* **building core**
d Gebäudekern *m*
f noyau *m* central de bâtiment, tour-
-gaine *f* centrale
nl kern *f(m)* van het gebouw
r центральный ствол *m* [техническое
ядро *n*] здания

B748 *e* **building density**
d Baudichte *f*,
Bebauungsdichte *f*
f densité *f* de construction
nl bebouwingsdichtheid *f*
r плотность *f* застройки

B749 *e* **building design**
d 1. Projektierung *f* der Gebäude,
Gebäudeplanung *f* 2. Gebäudeprojekt *n*
f 1. étude *f* de projets des bâtiments
2. projet *m* d'un bâtiment
nl 1. bouwplan *n* 2. bouwontwerp *n*
r 1. проектирование *n* зданий
2. проект *m* здания

B750 *e* **building fiberboard**
d Holzfaserplatte *f*
f panneau *m* en fibres de bois
nl houtvezelplaat *f(m)*
r древесноволокнистая плита *f*

B751 *e* **building inspector**
d Bauaufseher *m*
f inspecteur *m* des travaux
nl bouwinspecteur *m*; bouwpolitie *f*
r должностное лицо *n* [представитель
m] государственного технадзора за
строительством; государственный
строительный инспектор *m*

B752 *e* **building line**
d Baulinie *f*, Fluchtlinie *f*
f alignement *m* des bâtiments
nl rooilijn *f(m)*
r линия *f* застройки, красная линия *f*

B753 *e* **building material machines**
d Baustoff(-Herstellungs)maschinen *f pl*
f machines *f pl* de production des
matériaux de construction
nl machine *f pl* voor de productie van
bouwmateriaal
r машины *f pl* для изготовления
строительных материалов

B754 *e* **building operations**
d Hochbauarbeiten *f pl*
f opérations *f pl* de construction,
travaux *m pl* de bâtiments; corps *m*
d'état
nl uitvoering *f* van bouwwerken
r строительные операции *f pl*;
рабочие процессы *m pl*;
отдельные виды *m pl* строительных
работ

B755 *e* **building paper**
d Baupapier *n*, Baupappe *f*
f papier *m* de construction
nl wegenpapier *n*
r (тонкий) строительный картон *m*

B756 *e* **building part**
d Gebäudeteil *n, m*
f partie *f* d'un bâtiment; travée d'un
bâtiment
nl deel *f(m)* van een gebouw
r часть *f* здания; отсек *m* здания

B757 *e* **building performance**
d Baueigenschaften *f pl* (*des Gebäudes,
des Baustoffes*)
f caractéristiques *f pl* constructives
(*d'un ouvrage, d'un matériau*)
nl bouwkarakteristieken *f pl*
(*bouwmateriaal*)
r строительные характеристики *f pl*
(*здания, материала*)

B758 *e* **building permit**
d Bauerlaubnis *f*, Baugenehmigung *f*,
Baubewilligung *f*
f permis *m* [autorisation *f*] de
construire [de bâtir]
nl bouwvergunning *f*
r разрешение *n* на производство
строительных работ

B759 *e* **building population**
d Bewohneranzahl *f*
f indice *m* d'habitat; nombre *m* moyen
d'habitants par bâtiment
nl bewonersaantal *n*
r расчётная численность *f*
пользователей [жильцов] здания

B760 *e* **building pressure**
d Innenluftaufstau *m*, Luftaufstau *m*
im Gebäude
f surpression *f* d'air dans le bâtiment
nl overdruk *m* in een gebouw
r подпор *m* воздуха в здании

B761 *e* **building products**
d Bauwaren *f pl*, Bauerzeugnisse *n pl*,
Bauartikel *m pl*
f produits *m pl* de construction
nl bouwmaterialen *n pl*
r строительные изделия *n pl*

B762 *e* **building program(me)**
d Projektierungsauftrag *m*, Vorprojekt
n
f avant-projet *m*
nl bouwplan *n*
r техническое задание *n* на
строительство

B763 *e* **building sealant**
 d Fugenverguẞmasse *f*, Verguẞmasse *f*,
  Fugenkitt *m*
 f matériau *m* de remplissage [de
  garnissage] des joints
 nl voegenkit *f(m)*, voegen(vul)massa *f(m)*
 r герметик *m* [герметизирующий
  материал *m*] для уплотнения швов
  строительных конструкций

B764 *e* **building services**
 d Hausinstallationen *f pl*, technische
  Gebäudeausrüstung *f*,
  Gebäudebetriebsanlagen *f pl*
 f èquipement *m* technique et mécanique
  des bâtiments
 nl technische installaties *f pl* van een
  gebouw
 r инженерное оборудование *n* здания;
  инженерные сети *f pl* здания

B765 *e* **building sides**
 d Fassaden *f pl*, Fronten *f pl* (*des
  Gebäudes*)
 f façades *f pl* [côtés *m pl*] du bâtiment
  (*A-principale*, *B*, *C*, *D*,-*latérales et
  arrière*)
 nl gevels *m pl*
 r стороны *f pl* здания
  (*главная фасадная сторона* — *A*,
  остальные — *B*, *C*, *D*)

B766 *e* **building site**
 d Baustelle *f*; Baugelände *n*
 f chantier *m* (de construction)
 nl bouwterrein *n*
 r строительная площадка *f*; территория строительства

B767 *e* **building slipway**
 d Slip *m*, Aufschleppe *f*; Stapel *m*
 f slip *m*, cale *f* de halage [de
  construction]; cale *f* de lancement
 nl sleephelling *f*; stapel *m*
 r слип *m*; стапель *m*

B768 *e* **building space**
 d umbauter Raum *m*
 f espace *m* bâti
 nl bouwvolume *n*
 r застроенное пространство *n*

B769 *e* **building stone**
 d (Natur-)Baustein *m*, Baugestein *n*
 f pierre *f* à bâtir [de construction];
  roche *f* de construction
 nl bouwsteen *m*
 r строительный камень *m*

B770 *e* **building supervision**
 d Bauüberwachung *f*
 f surveillance *f* des travaux
 nl toezicht *n* op de bouw
 r строительный надзор *m* [контроль *m*]

B771 *e* **building technician**
 d Bautechniker *m*
 f technicien *m* du bâtiment
 nl bouwkundige *m*
 r техник-строитель *m*

B772 *e* **building technology**
 d Bautechnik *f*, Bautechnologie *f*,

Technologie *f* der Bauproduktion
 f technique *f* du bâtiment et des
  travaux publics
 nl bouwkunde *f*
 r технология *f* строительного
  производства, строительная
  технология *f*

B773 *e* **building types**
 d Hochbauarten *f pl*
 f types *m pl* des bâtiments [des
  édifices]
 nl gebouwtypen *n pl*
 r типы *m pl* зданий

B774 *e* **building unit**
 d Hochbaufertigelement *n*
 f composant *m* [élément *m*] préfabriqué
  de construction
 nl prefab-(bouw)element *n*
 r сборный элемент *m* конструкции,
  сборный конструктивный элемент *m*

B775 **building ways** *see* **building slipway**

B776 *e* **built-in beam**
 d eingespannter Balken *m* [Träger *m*]
 f poutre *f* encastrée (aux extrémités)
 nl aan beide zijden ingeklemde balk *m*
 r балка *f* с защемлёнными концами

B777 *e* **built-ins, built-in items**
 d Einbauteile *n pl*
 f pièces *f pl* incorporées
 nl ingebouwde onderdelen *n pl*
 r закладные детали *f pl*; встроенные
  приборы *m pl* и устройства *n pl*

B778 *e* **built-up area**
 d bebautes Gebiet *n*, bebaute Zone *f*
 f surface *f* bâtie
 nl bebouwd oppervlak *n*
 r застроенная территория *f* [зона *f*];
  площадь *f* застройки

B779 *e* **built-up beam**
 d zusammengesetzter Balken *m*
  [Träger *m*]
 f poutre *f* composée [composite]
 nl samengestelde balk *m* [ligger *m*]
 r составная балка *f*

B780 *e* **built-up laminated wood**
 d zusammengesetztes Schichtholz *n*
 f bois *m* lamellé(-collé)
 nl gelamineerd houten constructiedeel *n*
 r дощато-клеёный составной элемент *m*

B781 *e* **built-up membrane**
 d mehrschichtige Feuchtigkeitssperre *f*
  (*z.B. Ruberoidbelagteppich*)
 f membrane *f* d'étanchéité à plusieurs
  couches
 nl meervoudige dampwerende laag *f(m)*
 r многослойный гидроизоляционный
  (рубероидный) ковёр *m*

B782 *e* **built-up timber**
 d zusammengesetztes
  Holz(bau)element *n*
 f pièce *f* composée en bois
 nl samengesteld houten constructie-
  -element *n*

r составной элемент *m* деревянных конструкций

B783 *e* **bulb**
d Temperaturfühler *m*
f bulbe *m*
nl thermosensor *m*
r термобаллон *m*, термочувствительный элемент *m*

B784 *e* **bulk asphaltic-bitumen distributor**
d Bitumen-Sprengwagen *m*, Drucktankwagen *m*
f goudronneuse *f*
nl bitumensproeiwagen *m*
r гудронатор *m*

B785 *e* **bulk cement**
d loser [unverpackter, ungesackter] Zement *m*
f ciment *m* en vrac
nl los gestorte cement *n*, *m*, bulkcement *n*, *m*
r цемент *m* насыпью (*без упаковки*)

B786 *e* **bulk cement lorry** *UK*, **bulk cement truck** *US*
d Zement-Lastkraftwagen *m*
f camion-citerne *m* à ciment
nl vrachtauto *m* voor bulkcement
r (автомобиль-)цементовоз *m*

B787 *e* **bulk density**
d mittlere Dichte *f*; Schüttdichte *f*, Rohdichte *f*
* *f* densité *f* moyenne; densité *f* en vrac
nl materiaaldichtheid *f* bij bulkvervoer
r средняя плотность *f*; насыпная плотность *f*

B788 *e* **bulkhead**
d 1. Schott *n*, Schotte *f*, Abschlußwand *f*; Fang(e)damm *m* 2. Stützwand *f*, Stützmauer *f*, Böschungsmauer *f*; Kaimauer *f* 3. Dränauslauf *m*; Ausmündungsbauwerk *n*
f 1. batardeau *m*, cloison *f* étanche 2. mur *m* de butée [de soutènement, de quai] 3. tête *f* du drain; embouchure *f*
nl 1. waterdicht schot *n*; vangdam *m* 2. keermuur *m*, damwand *m*, kaaimuur *m* 3. rechtstands- en vleugelmuren *m pl* van een duikersluis
r 1. диафрагма *f*; (водо)непроницаемая перемычка *f* 2. подпорная стенка *f*; набережная стенка *f* 3. устье *n* дрены; устьевое сооружение *n*

B789 *e* **bulkhead wall**
d Anlegemauer *f*
f quai *m* d'accostage
nl aanlegkade *f(m)*, kaaimuur *m*
r причальная стенка *f*

B790 *e* **bulking**
d Quellen *n*, Schwellen *n*
f foisonnement *m*, gonflement *m*
nl wellen *n*, zwelling *f*
r вспучивание *n*, пучение *n*

B791 *e* **bulking sludge**
d Blähschlamm *m*
f boue *f* gonflée
nl opwellend slib *n*
r вспухающий ил *m*

B792 *e* **bulk loading**
d sacklose Verladung *f*
f chargement *m* en vrac
nl los gestorte lading *f*, bulklading *f*
r погрузка *f* (*сыпучих материалов*) навалом [россыпью]

B793 **bulk material** *see* **bulk product**

B794 *e* **bulk modulus**
d Volumenelastizitätsmodul *m*, Raummodul *m*
f module *m* de compression volumétrique [triaxiale]
nl compressiemodulus *m*
r модуль *m* объёмной упругости [объёмного сжатия]

B795 *e* **bulk product**
d Massengut *n*; Schüttgut *n*
f chargement *m* en vrac; matériau *m* pulvérulent
nl stortgoed *n*
r навалочный груз *m*; сыпучий материал *m*

B796 *e* **bulldozer**
d 1. Bulldozer *m*, Fronträumer *m*, Planierraupe *f* 2. Querschild *m*, Geradschild *m*, Planierschild *m*, Räumschild *m* 3. Biegepresse *f*, Biegemaschine *f*
f 1. bull(dozer) *m*, bouldozer *m*, bouteur *m* 2. lame *f* du bull(dozer) 3. presse *f* à plier des profilés
nl 1. bulldozer *m* 2. (dozer)blad *n* 3. opstuikmachine *f*
r 1. бульдозер *m* (*землеройная машина*) 2. отвал *m* бульдозера 3. гибочный пресс *m*; гибочная машина *f* (*для гибки проката*)

B797 *e* **bulletproof glass, bullet-resistant glazing**
d kugelsicheres Glas *n*; kugelsichere Verglasung *f*
f verre *m* *ou* vitrage *m* résistant aux balles
nl kogelvrij glas *n*, pantserglas *n*; kogelvrije beglazing *f*
r пуленепробиваемое стекло *n*; пуленепробиваемое остекление *n*

B798 *e* **bull float**
d Betonglätter *m*, Reibebrett *n*, Reibeholz *n*
f lisseur *m* [lisseuse *f*] de béton; lisseur *m* [lisseuse *f*] mécanique
nl betonplakspaan *f(m)*
r ручная *или* механическая гладилка *f*

B799 *e* **bullfloat finishing machine**
d Längsbohlenfertiger *m*
f finisseuse *f* routière
nl betonverdichtingsmachine *f* met afrijbalk

r финишер *m*, отделочная машина *f* для дорожных покрытий

B800 *e* **bull header**
d hochkant gestellter Ziegel *m*
f brique *f* de champ
nl hoeksteen *m*, dorpelsteen *m*
r кирпич *m*, уложенный на ребро (*в кладке*)

B802 *e* **bulwark**
d Bohlwerk *n*, Bohlwand *f* (*im Hafenbau*)
f mur *m* de soutènement mince ancré
nl bolwerk *n* (*zeewering*)
r больверк *m*

B803 **bumper** *see* **fender**

B804 *e* **bund**
d senkrechter Hafenkai *m*
f mur *m* de quai vertical *ou* avec un faible fruit
nl steile kade *f(m)*
r вертикальная *или* крутонаклонная набережная стенка *f*

B805 *e* **bundle**
d 1. Stapel *m* (*z. B. Sperrholz*) 2. Bund *n*, Bündel *n*
f 1. pile *f* (*p.ex. de contre-plaqué*) 2. faisceau *m*, botte *f*, liasse *f*
nl 1. stapel *m* platen 2. bundel *m*, bos *m*
r 1. стопа *f* (*напр. фанеры*) 2. связка *f*; пучок *m*; моток *m*

B806 *e* **bundled bars, bundle of reinforcement**
d Bewehrungsbündel *n*
f faisceau *m* de ronds d'armatures
nl bundel *m* wapeningsstaven
r пучок *m* арматурных стержней

B807 *e* **bunker**
d Bunker *m*
f trémie *f*
nl bunker *m*
r бункер *m*

B808 *e* **buoyancy**
d 1. Auftrieb *m* 2. Schwimmkraft *f*
f 1. sous-pression *f*, force *f* ascensionnelle, portance *f*, poussée *f* archimédéenne [d'Archimède] 2. flottabilité *f*
nl 1. opwaartse druk *m* 2. drijfvermogen *n*
r 1. подъёмная [выталкивающая, архимедова] сила *f*; взвешивающая сила *f*; взвешивающее давление *n* 2. плавучесть *f*

B809 *e* **buoyant foundation**
d Plattenfundament *n*, Plattengründung *f*
f fondation-dalle *f*
nl drijvende fundering *f*
r плитный [сплошной] фундамент *m*

B810 **buried drain** *see* **blind drain**

B811 *e* **buried laying**
d verdeckte Verlegung *f*
f mise *f* en place en sous-sol; pose *f* dissimulée (sous crépi)
nl ingraven *n*, montage *f* uit het zicht
r скрытая прокладка *f*

B812 *e* **buried-tube irrigation system**
d Bewässerungssystem *n* mit erdverlegten Rohrleitungen
f système *m* d'irrigation clos
nl ingegraven bevloeiingssysteem *n*
r закрытая система *f* орошения

B813 **buried wiring** *see* **hidden wiring**

B814 *e* **burl**
d Maser *f*, Drehwuchs *m*, Auswuchs *m*, Beule *f* (*Holzfehler*)
f croissance *f* (*du bois*) ondulée
nl knoest *m*
r свиль *f*, свилеватость *f*; нарост *m*, наплыв *m* (*порок древесины*)

B815 *e* **burlap**
d Sackleinen *n*
f toile *f* à sac (de jute)
nl zakkenlinnen *n*, (geweven) jute *f(m)*
r джутовая мешочная ткань *f*, мешковина *f*

B816 *e* **burner**
d Brenner *m*
f brûleur *m*
nl brander *m*
r горелка *f*; (топливная) форсунка *f*

B817 *e* **burning of refractories**
d Brennen *n* der Feuerfeststoffe
f cuisson *f* des réfractaires
nl bakken *n* van vuurvaste steen
r обжиг *m* керамических изделий *или* огнеупоров

B818 *e* **burnt and ground lime**
d gebrannter und gemahlener Kalk *m*, gemahlener Branntkalk *m*
f chaux *f* anhydre [calcinée] moulue
nl gebrande en gemalen kalk *f(m)*
r молотая известь-кипелка *f*

B819 *e* **burst**
d Bruch *m*
f éclatement *m*
nl breuk *f(m)*
r разрыв *m*

B820 *e* **bush hammer**
d Stockhammer *m*
f boucharde *f*
nl bouchard(hamer) *m*
r бучарда *f*

B821 *e* **bush hammer finish**
d 1. Bearbeitung *f* mit Stockhammer 2. gestockte [aufgespitzte] Oberflächenbehandlung *f*
f 1. bouchardage *m* 2. surface *f* bouchardée
nl 1. boucharderen *n* 2. gebouchardeerd betonoppervlak *n*
r 1. обработка *f* *или* отделка *f* бучардой 2. фактурная поверхность *f* (*камня, бетона*) после обработки бучардой

B822 bush hammering *see* bush hammer finish 1.

B823 *e* butterfly damper
d Drosselklappe *f*
f registre *m* papillon
nl 1. smoorklep *f(m)* 2. rookklep *f(m)*
r 1. поворотный (дроссельный) воздушный клапан *m* 2. двустворчатый обратный воздушный клапан *m*, клапан-бабочка *m*

B824 *e* butterfly gate
d Drosselklappe *f*
f vanne *f* papillon
nl tolklep *f(m)* (*sluis*)
r дисковый затвор *m*

B825 *e* butterfly roof
d Satteldach *n* mit Gegenhängen, zweihängiges Dach *n* mit Wiederkehr, Wiederkehrdach *n*, Kehlendach *n*
f cômble *f* à papillon (*avec deux versants inclinés à l'intérieur vers la noue centrale*)
nl sheddak *n*, zaagdak *n*
r двускатная крыша *f* с обратными скатами (*наклонёнными к центральной ендове*)

B826 *e* butterfly valve
d Drosselventil *n*, Drosselklappe *f*, Doppelsitzventil *n*; Flügelventil *n*
f papillon *m*; soupape *f* à papillon
nl vlinderklep *f(m)*
r дроссельный клапан *m*; дроссельная заслонка *f*

B827 *e* buttering
d Auftragen *n* des Putzmörtels [Anwerfen *n* des Putzes] mit der Kelle
f application *f* à la truelle du mortier sur la brique
nl mortel op de baksteen leggen *n* met de troffel
r нанесение *n* раствора на кирпич кельмой

B828 *e* butt fusion
d Abschmelz(stumpf)schweißen *n*
f soudage *m* en bout par fusion
nl stomplassen *n* door afsmelten
r стыковая сварка *f* оплавлением

B829 *e* butt joint
d Stoß *m*, Stoßverbindung *f*, Stoßfuge *f*, Stumpfstoß *m*
f assemblage *m* par aboutement [bout à bout]; joint *m* abouté [d'about]
nl stompe verbinding *f* [naad *m*]
r стык *m*, стыковое соединение *n*; стыковой шов *m*

B830 *e* button-head ancorage
d Stauchkopfverankerung *f*
f ancrage *m* par boutonnage
nl knopverankering *f*
r анкеровка *f* стержневой арматуры с высаженными головками

B831 *e* buttress
d Strebepfeiler *m*; Pfeiler *m*; Gegenpfeiler *m*

f contrefort *m*; pilier *m* d'arc-boutant
nl steunpijler *m*; steunbeer *m*, conterfort *m*
r контрфорс *m* (*с лицевой стороны сооружения*); опора *f*; устой *m*; бык *m*

B832 *e* buttress bracing strut
d Aussteifungsbalken *m* des Pfeilers
f jambe *f* de force
nl verstijvingsbalk *m* van de conterfort
r балка *f* жёсткости контрфорса

B833 *e* buttress dam
d Pfeilerstaumauer *f*, Pfeilersperre *f*, aufgelöste Staumauer *f*
f barrage *m* à contreforts
nl stuwdam *m* met steunpijlers
r контрфорсная плотина *f*

B834 *e* buttress head
d Pfeilerkopf *m*
f tête *f* du contrefort
nl conterforthoofd *n*, pijlerkop *m*
r оголовок *m* контрфорса

B835 *e* buttress type power house
d Pfeilerkrafthaus *n*
f bâtiment *m* de commande d'une centrale-pile [d'une usine-pile]
nl tegen de stuwdam gebouwde waterkrachtcentrale *f*
r здание *n* ГЭС бычкового типа

B836 *e* buttress water power station
d Pfeiler-Wasserkraftwerk *n*
f centrale-pile *f*, usine-pile *f*
nl in een steunpijler opgenomen waterkrachtcentrale *f*
r бычковая гидроэлектростанция *f*

B837 *e* butt strap
d Stoßlasche *f*, Verbindungslasche *f*, Anschlußlasche *f*
f couvre-joint *m*
nl verdekte las *f(m)*
r стыковая накладка *f*

B838 *e* butt welding
d Stumpfschweißung *f*
f soudage *m* en bout (par résistance)
nl stomplassen *n*
r стыковая сварка *f* (сопротивлением)

B839 *e* butyl rubber tape
d Butylkautschuk-Isolierband *n*
f ruban *m* isolant en butylcaoutchouc
nl isolatieband *m* (op basis) van butylrubber
r бутилкаучуковая изоляционная лента *f*

B840 bye-channel *see* by-pass canal

B841 *e* bye wash
d Überlaufkanal *m*; Beipaßkanal *m*, Umleitungskanal *m*
f canal *m* de décharge; canal *m* de dérivation
nl aflaatkanaal *n*
r (боковой паводковый) водосбросный канал *m*; обводной канал *m*; деривационный канал *m*

B842 e **bypass**
d Beipaß *m*, Umführungsleitung *f*,
Umgehungsleitung *f*
f by-pass *m*, conduit *m* de by-pass
nl omloopleiding *f*
r обводная линия *f*, байпас *m*,
обводной трубопровод *m*

B843 e **by-pass canal, by-pass channel**
d Umgehungskanal *m*, Umlaufkanal *m*,
Umleitungskanal *m*,
Umführungskanal *m*
f canal *m* de décharge amont; canal *m*
latéral; canal *m* de déviation
nl lateraal kanaal *n*
r отводной канал *m*; обводной канал
m; водосбросный канал *m*

B844 e **bypassing**
d Überleitung *f*, Überlauf *m*
f by-pass *m*, dérivation *f*
nl rechtstreeks verbinden *n*
r перепуск *m*

B845 e **by-pass pipe**
d Umlaufrohr *n*, Beipaßrohr *n*
f tuyau *m* de dérivation
nl omloopleiding *f*
r перепускная труба *f*, байпас *m*

B846 e **bypass valve**
d Beipaßventil *n*, Nebenstromventil *n*;
Umgehungsventil *n*
f soupape *f* de by-pass; vanne *f* de
dérivation
nl omloopventiel *n*
r перепускной клапан *m*; обводной
клапан *m*

B847 **bywash** *see* **bye wash**

C

C1 e **cab**
d Kabine *f*, Fahrerhaus *n*
f cabine *f*; abri *m*
nl cabine *f*
r кабина *f* (*водителя*); будка *f*
(*машиниста*)

C2 e **cabin**
d 1. Bauleiter-Arbeitsraum *m* (*am
Bauplatz*) 2. Kabine *f*, Fahrerhaus *n*
f 1. abri *m* du conducteur de travaux
(*sur chantier*) 2. cabine *f* (*du
conducteur d'engin*)
nl 1. directiekeet *f(m)* 2. cabine *f*
r 1. помещение *n* [конторка *f*]
производителя работ (*на
стройплощадке*) 2. кабина *f*
(*водителя*)

C3 e **cabinet-type air conditioner**
d Klimaschrank *m*, Schrankklimagerät *n*
f climatiseur *m* armoire
nl luchtbehandelingstoestel *n*
(*kastmodel*)
r шкафной кондиционер *m*

C4 e **cable**
d 1. Seil *n*; Kabel *n* 2. vorgespanntes
Bewehrungsdrahtbündel *n*
f 1. câble *m* 2. câble *m* d'armature de
précontrainte
nl 1. kabel *m* 2. voorspankabel *m*
r 1. канат *m*, трос *m*; кабель *m*
(*электрический*); многожильный
провод *m* 2. пучок *m или* прядь *f*
(пред)напрягаемой арматуры

C5 **cable bridge** *see* **cable-suspended bridge**

C6 e **cable clamp, cable clip**
d Kabelhülse *f*, Kabelklammer *f*,
Kabelklemme *f*
f serre-câble *m*
nl kabelklem *f(m)*
r сжим *m* стального каната

C7 **cable conduit** *see* **cable duct** 1.

C8 e **cable conveyor**
d Gehängeförderer *m*
f convoyeur *m* [transporteur *m*] à
câble
nl kabelbaantransporteur *m*
r канатный конвейер *m*

C9 e **cable drill**
d Seil(schlag)bohrgerät *n*,
Seilschlaggerät *n*
f foreuse *f* à câble, engin *m* de forage
à câble
nl kabelboorinstallatie *f*
r ударно-канатный буровой станок *m*,
станок *m* ударно-канатного бурения

C10 e **cable drilling**
d Seilbohrung *f*
f forage *m* au câble
nl boren *n* aan de kabel
r канатное [ударно-канатное] бурение *n*

C11 e **cable duct**
d 1. Kabelkanal *m*, Hüllrohr *n* 2.
Spanngliederkanal *m*, Spannglied-
-Rohrhülle *f*
f 1. conduite *f* à câbles; caniveau *m*
pour câbles 2. trou *m* pour passage
des câbles d'armature
nl 1. kabelgoot *f(m)* 2. mantelbuis *f(m)*
voor de spankabel
r 1. кабельный канал *m*,
кабелепровод *m*; труба *f* кабельной
канализации 2. канал *m* для
преднапрягаемой арматуры

C12 e **cable joint**
d Kabelmuffe *f*
f boîte *f* à câble
nl kabelmof *f(m)*
r кабельная муфта *f*

C13 e **cable-laid rope**
d sechslitziges Seil *n* mit der Seele
f câble *m* à âme composé de six torons
nl kabelslagtouw *n*
r шестипрядевый стальной канат *m*
с сердечником

C14 e **cable layer**
d Kabelverlegemaschine *f*

f machine f à poser les câbles
nl kabellegmachine f
r кабелеукладчик m

C15 e cable protection pipe
d Kabelschutzrohr n
f tuyau m protecteur du câble, tuyau m à câble
nl mantelbuis f(m)
r трубный кабельный канал m, труба f для защиты кабеля

C16 e cable railway
d Seil-Standbahn f
f chemin m de fer à câble, voie f ferrée à câble
nl kabelbaan f(m)
r наземная канатная дорога f

C17 e cable sag
d Seildurchhang m
f flèche f du câble
nl kabeldoorhang m
r провес m каната

C18 e cable saw
d Kabelsäge f
f scie f à fil [à câble], fil m à scier
nl kabelzaag f(m)
r канатная пила f

C19 e cable-stayed bridge
d Schrägseilbrücke f, Zügelgurtbrücke f
f pont m à poutres haubanées
nl liggerbrug f(m) met spankabels
r балочно-вантовый мост m

C20 e cable-suspended bridge
d Seil(hänge)brücke f
f pont m suspendu à câbles
nl hangbrug f(m)
r висячий мост m с несущими кабелями

C21 e cableway
d 1. Kabelkran m 2. Luftseilbahn f, Seilschwebebahn f
f 1. blondin m, grue f à câble, câble m grue 2. téléférique m, funiculaire m
nl 1. kabelkraan f(m) 2. kabelbaan f(m)
r 1. кабель-кран m 2. канатная подвесная дорога f

C22 e cableway excavator
d Kabelbagger m
f blondin m à benne preneuse
nl kabelbaggermolen m
r грейферный кабель-кран m

C23 cableway transporter see cable conveyor

C24 e cadmium yellow
d Kadmiumgelb n
f jaune m de cadmium
nl cadmiumgeel n, cadmiumsulfide n
r кадмиевая жёлтая f (пигмент)

C25 e cage of reinforcement
d Bewehrungskorb m
f cage f d'armature
nl wapeningskooi f(m)
r пространственный арматурный каркас m

C26 e caisson
d 1. Druckluftsenkkasten m, Caisson m 2. Senkbrunnen m, Senkkasten m, Brunnenschacht m 3. Verschlußponton m, Schwimmtor n 4. Säulenpfahl m
f 1. caisson m pneumatique 2. caisson m ouvert 3. bateau-porte m 4. pieu-colone m, pieu m résistant à la pointe
nl 1. caisson m, zinkkist f(m), duikerklok f(m) 2. zinkbuis f(m) 3. sluitcaisson m 4. verzonken paneel n in een gewelf
r 1. кессон m 2. опускной колодец m 3. плавучий затвор m, батопорт m 4. свая-колонна f

C27 e caisson ceiling
d 1. Kassettendecke f 2. Senkkastendecke f, Caissondecke f
f 1. plafond m en caisson 2. plafon m de la chambre de travail du caisson
nl 1. kassettenplafond n 2. caissondak n
r 1. кессонный потолок m 2. покрытие n рабочей камеры кессона

C28 e caisson chamber
d Druck(luft)kammer f, Arbeitskammer f des Caissons
f chambre f de travail du caisson
nl werkruimte f in de duikerklok
r рабочая камера f кессона

C29 e caisson foundation
d Kastengründung f, Kastenfundation f, Caissongründung f
f fondation f sur caissons
nl caissonfundatie f
r кессонный фундамент m; фундамент m из опускных колодцев или массивов-гигантов

C30 e caisson gate
d 1. Schwimmtor n, Verschlußponton m, Schwimmponton m, Torschiff n 2. Rollponton m, Schleusenschiff n
f 1. bateau-porte m 2. porte f coulissante d'écluse à sas
nl 1. caissondeur f(m) 2. doorlaatcaisson m
r 1. плавучий затвор m, батопорт m, откатной затвор m 2. откатные ворота pl кессонного шлюза

C31 e caisson pile
d Hülsenpfahl m, Ortbetonpfahl m mit verlorenem Füllrohr, Mantelrohrpfahl m; Großbohrpfahl m, Ortbetonpfahl m mit einem Durchmesser von 0,6 bis 1,0 m und mehr
f pieu m à tube battu, pieu m tubé, pieu m tubulaire; pieu m caisson
nl holle betonpaal m, die na plaatsing wordt volgestort
r набивная свая f с обсадной трубой; трубчатая свая f; набивная свая f крупного диаметра (0,6—1 м)

C32 e cake
d (Filter-)Kuchen m, Schlammkuchen m

f gâteau *m* de filtre
nl filterkoek *m*
r кек *m*, осадок *m*, лепёшка *f*
  (*остаток на фильтре*)

C33 *e* calcareous clay
d Kalkton *m*
f argile *f* calcaire [calcarifère]
nl kalkverweringsklei *f(m)*
r известковая глина *f*

C34 *e* calcinating kiln
d Brennofen *m*
f four *m* de calcination [à calciner, de grillage]
nl kalkbranderij *f*, kalkoven *m*
r обжиговая печь *f*

C35 *e* calcined gypsum
d Branntgips *m*, gebrannter Gips *m*
f gypse *m* [plâtre *m*] semi-hydraté
nl gipskalk *m*
r низкообжиговое гипсовое вяжущее *n*; полуводный гипс *m*

C36 calcium aluminate cement *see* high--alumina cement

C37 *e* calcium chloride
d Chlorkalzium *n*, Kalziumchlorid *n*
f chlorure *f* de calcium
nl calciumchloride *n*
r хлорид *m* кальция (*ускоритель твердения бетона*)

C38 *e* calcium hardness
d Kalziumhärte *f*, Kalkhärte *f*
f dureté *f* calcaire [calcique]
nl calciumhardheid *f*
r кальциевая жёсткость *f* (*воды*)

C39 *e* calcium hydroxide
d Kalzium-Hydroxid *n*, gelöschter Kalk *m*
f hydroxyde *m* de calcium, chaux *f* éteinte [blanche]
nl calciumhydroxyde *n*
r гидроокись *f* кальция, гашёная известь *f*

C40 *e* calcium oxide
d Kalziumoxid *n*, Branntkalk *m*, ungelöschter Kalk *m*
f oxide *m* de calcium
nl calciumoxyde *n*
r окись *f* кальция, негашёная известь *f*

C41 *e* calcium silicate brick
d Kalksandstein *m*, weißer Mauerstein *m*
f brique *f* silicieuse [de silice]
nl kalkzandsteen *m*
r силикатный кирпич *m*

C42 *e* calcium silicate concrete
d Silikatbeton *m*
f béton *m* silico-calcaire
nl silicaatbeton *n*
r силикатный бетон *m*

C43 *e* calcium silicate insulation
d Silikalzit-Wärmedämmung *f*
f isolation *f* silico-calcaire

nl glaswolisolatie *f*
r теплоизоляция *f* на известково--кремнезёмистом связующем

C44 *e* calcium silicate products
d Silikalziterzeugnisse *n pl*
f produits *m pl* silico-calcaire
nl glaswo!produkten *n pl*
r изделия *n pl* на известково--кремнезёмистом связующем

C45 *e* calculated settlement
d errechnete Setzung *f*
f affaissement *m* [tassement *m*] calculé
nl berekende zetting *f*
r вычисленное значение *n* осадки

C46 *e* calculation assumption
d rechnerische Annahme *f*
f hypothèse *f* de calcul
nl aanname *f* voor de berekening
r расчётное предположение *n*, расчётная гипотеза *f*

C48 calculation hypothesis *see* calculation assumption

C49 *e* calculation of areas
d Flächenberechnung *f*
f calcul *m* de surfaces
nl oppervlakteberekening *f*
r вычисление *n* [расчёт *m*] площадей

C50 *e* calculation of heat losses
d Wärmelastberechnung *f*
f calcul *m* de pertes thermiques [calorifique, de chaleur]
nl berekening *f* van de warmteverliezen
r расчёт *m* тепловых потерь

C51 *e* calculation of stresses
d Spannungsberechnung *f*
f calcul *m* de contraintes
nl spanningsberekening *f*
r расчёт *m* напряжений

C52 caldron *see* cauldron

C53 *e* calfdozer
d Klein-Planierschlepper *m*, Klein--Bulldozer *m*, Kleinräumer *m*
f bulldozer *m* de petites dimensions
nl lichte bulldozer *m*
r малогабаритный бульдозер *m*

C54 *e* California bearing ratio
d kalifornischer Tragfähigkeitswert *m*
f indice *m* portant Californien
nl draagvermogen *n* van de grond, berekend volgens de CBR-methode
r коэффициент *m* относительной прочности (грунта) по калифорнийскому методу

C55 *e* California bearing ratio test
d CBR-Verfahren *n*, CBR-Versuch *m*
f essai *m* d'indice portant Californien
nl CBR-methode *f*
r определение *n* несущей способности грунта калифорнийским методом

C56 calked joint *see* caulked joint

C57 calking *see* caulking

C58　*e* calling for tenders
　　d Ausschreibung *f*
　　f appel *m* d'offre
　　nl aanbesteding *f*
　　r объявление *n* о торгах

C59　*e* calorifier
　　d Wasserheizer *m*, Warmwasserbereiter *m* mit Speicherung, Warmwasserspeicher *m*
　　f chauffe-eau *m* à accumulation
　　nl elektrisch heetwaterreservoir *n*
　　r ёмкостный водонагреватель *m*

C60　*e* camber
　　d 1. Aufwölbung *f*, Wölbung *f*, Überhöhung *f* 2. Stich *m*, Stichhöhe *f* 3. Straßenwölbung *f* 4. Dammüberhöhung *f* 5. Gezeitenbecken *n*, Tidebecken *n* 6. Torkammer *f*
　　f 1. contre-flèche *f* 2. flèche *f*, montée *f*, hauteur *f* sous clef 3. bombement *m* du profil de la route 4. supplément *m* de hauteur du barrage pour compenser le tassement 5. petit bassin *m* de marée 6. chambre *f* de vanne
　　nl 1. ronding *f* 2. zeeg *f(m)* 3. tonrondte 4. ophoging *f* 5. getijdebekken *n* 6. sluisdeurkamer *f(m)*, schutdeurkamer *f(m)*
　　r 1. строительный подъём *m* 2. стрела *f* подъёма 3. выпуклая часть *f* [подъём *m*] поперечного профиля дороги 4. запас *m* высоты плотины на осадку 5. малый приливный бассейн *m* 6. камера *f* затвора

C61　*e* camber beam
　　d Träger *m* mit gewölbtem Obergurt
　　f poutre *f* à membrure supérieure bombée, poutre *f* à membrure supérieure en forme d'arc
　　nl gewelfde balk *m*
　　r балка *f* с выпуклым верхним поясом

C62　*e* camber of a road *see* camber 3.

C63　*e* camp sheathing, camp shedding
　　d zweireihiger Holzspundfangdamm *m*, zweireihige Ufereinfassung--Holzspundwand *f*; Bollwerk *n*
　　f batardeau *m* [enceinte *f*] à double paroi de palplanches
　　nl houten beschoeiing *f*
　　r двухрядная деревянная шпунтовая конструкция *f* (*перемычка, набережная стенка*); больверк *m*

C64　*e* can
　　d Hüllung *f* von Uranblöcken [Uranstäben]
　　f enveloppe *f* des blocks [des barres] d'uranium
　　nl omhulling *f* van de (uranium) brandstofstaven
　　r защитная оболочка *f* топливных [урановых] стержней (*атомного реактора*)

C65　*e* canal
　　d Kanal *m*
　　f canal *m*
　　nl kanaal *n*
　　r канал *m*

C66　canal bank *see* canal slope

C67　*e* canal bridge
　　d Kanalbrücke *f*
　　f pont-canal *m*
　　nl kanaalbrug *f(m)*
　　r мост-канал *m*

C68　*e* canal cleaning
　　d Kanalreinigung *f*, Kanalräumung *f*
　　f curage *m* [nettoyage *m*] du canal
　　nl kanaalreiniging *f*
　　r очистка *f* канала

C69　*e* canal concrete paver
　　d Kanal-Betoniermaschine *f*
　　f bétonneuse *f* de canal (*du radier et des talus du canal*)
　　nl betonneermachine *f* voor kanaaloevers
　　r машина *f* [агрегат *m*] для бетонирования одежды канала

C70　*e* canal embankment
　　d Kanaldamm *m*
　　f digue *f* [remblai *m*] de canal
　　nl kanaaldijk *m*
　　r дамба *f* обвалования канала

C71　*e* canal head
　　d Kanaleinlauf *m*, Kanaleinlaßbauwerk *n*
　　f tête *f* d'un canal
　　nl kanaalinlaat *m*
　　r головной узел *m*, головное сооружение *n*, голова *f* канала

C72　*e* canal in a cut *US*, canal in cuttiug *UK*
　　d Kanal *m* im Einschnitt, Einschnittkanal *m*, Abtragkanal *m*
　　f canal *m* en déblai
　　nl kanaal *n* in een insnijding
　　r канал *m* в выемке

C73　*e* canalization
　　d 1. Kanalisierung *f*, Stauregelung *f* 2. Kanalbau *m* 3. Kanalisation *f*
　　f canalisation *f*
　　nl 1. kanaliseren *n* 2. kanaalaanleg *m*
　　r 1. канализирование *n* [шлюзование *n*] реки 2. строительство *n* каналов 3. проводка *f* (*сеть трубопроводов или коммуникационных каналов*)

C74　*e* canalization section
　　d kanalisierte Strecke *f* des Flusses
　　f tronçon *m* canalisé d'une rivière
　　nl gekanaliseerde riviersectie *f*
　　r канализированный [шлюзованный] участок *m* реки

C75　*e* canalized river, canalized stream
　　d kanalisierter Strom *m* [Fluß *m*]
　　f fleuve *m* canalisé, rivière *f* canalisée
　　nl gekanaliseerde rivier *f(m)*
　　r канализированная река *f*

C77　e **canal lift**
　　　d Trog *m*, Schiffshebewerk *n*,
　　　　Schiffshebeanlage *f*
　　　f ascenseur *m* à bateaux
　　　nl scheepslift *m*
　　　r судоподъёмник *m*

C78　e **canal lining**
　　　d Kanalverkleidung *f*,
　　　　Kanalauskleidung *f*
　　　f revêtement *m* du canal
　　　nl oeverbekleding *f*, beschoeiing *f*
　　　r одежда *f* канала

C79　e **canal lock**
　　　d einfache Kanalschleuse *f*;
　　　　Einkammerschleuse *f*
　　　f écluse *f* en canal; écluse *f* à sas
　　　　simple
　　　nl kanaalsluis *f(m)*
　　　r шлюз *m* на канале; однокамерный
　　　　шлюз *m*

C80　e **canal mouth**
　　　d Kanalmündung *f*
　　　f débouché *m* [bouche *f*] d'un canal
　　　nl kanaalmonding *f*
　　　r устье *n* канала

C81　e **canal on embankment**
　　　d Dammkanal *m*, Kanal *m* im Auftrag
　　　f canal *m* en remblai
　　　nl hooggelegen kanaalpand *n*
　　　r канал *m* в насыпи

C82　e **canal pumping station**
　　　d Kanalpumpstation *f*,
　　　　Kanalpumpwerk *n*
　　　f station *f* de pompage de canal
　　　nl gemaal *n*
　　　r насосная станция *f* канала

C83　e **canal reach**
　　　d Kanalhaltung *f*
　　　f bief *m* de canal
　　　nl kanaalpand *n*
　　　r бьеф *m* канала

C84　e **canal sealing by deposition of silt**
　　　d Schlämmdichtung *f* des Kanals
　　　f scellement *m* [étanchement *m*] d'un
　　　　canal par la vase [par le limon]
　　　nl kanaalbedding *f* van een kleilaag
　　　　voorzien
　　　r заиливание *n* канала (*обеспечение
　　　　водонепроницаемости*)

C85　e **canal silting**
　　　d Kanalverschlammung *f*,
　　　　Verlandung *f* des Kanals
　　　f envasement *m* d'un canal
　　　nl dichtslibben *n* van een kanaal
　　　r заиление *n* канала, засорение *n*
　　　　канала илом

C86　e **canal slope**
　　　d Kanalböschung *f*
　　　f talus *m* [berge *f*] de canal
　　　nl kanaal(dijk)talud *n*
　　　r откос *m* канала

C87　e **canal slope concrete paver**
　　　d Kanalböschungsbetoniermaschine *f*

　　　f bétonneuse *f* de talus
　　　nl betonneermachine *f* voor (kanaal)talud
　　　r машина *f* для бетонирования
　　　　одежды откосов канала

C88　e **canal slope trimmer, canal trimmer**
　　　d Kanalböschungsprofiliermaschine *f*
　　　f taluteuse *f* (pour canaux)
　　　nl profileermachine *f* voor (kanaal)talud
　　　r профилировщик *m* откосов канала

C89　e **canal tunnel**
　　　d Kanaltunnel *m*, Schiffstunnel *m*
　　　f tunnel *m* pour canal
　　　nl kanaaltunnel *m*
　　　r судоходный туннель *m*

C90　e **can hook**
　　　d Faßgreifer *m*
　　　f pince *m* à tonneau
　　　nl kanhaak *m*
　　　r захват *m* для подъёма бочек

C91　e **canopy**
　　　d Sonnenzelt *n*; Verdeck *n*,
　　　　Schutzdach *n*; Vordach *n*, Kragdach *n*
　　　f tente *f*; auvent *m*
　　　nl zonnetent *f(m)*; afdak *n*, zonnedak *n*
　　　r тент *m*; навес *m*; козырёк *m*

C91a　**canopy hood** *see* **exhaust hood**
C92　e **cant**
　　　d 1. geneigte Fläche *f*, Schräge *f*
　　　　2. Überhöhung *f* 3. Kantholz *n*
　　　f 1. pente *f*; fruit *m* 2. surélévation *f*;
　　　　relèvement *m*; dévers *m* 3. bois *m*
　　　　équarri; madrier *m*
　　　nl 1. schuine kant *m*; helling *f* 2.
　　　　verkanting *f*
　　　r 1. скос *m*; наклон *m* 2. наклон *m*
　　　　виража *или* наружного рельса
　　　　внутрь пути 3. деревянный брус *m*
　　　　(*полностью или частично окантован-
　　　　ный*)

C93　　**cant brick** *see* **splay brick**
C94　e **canted wall**
　　　d spitz- *oder* stumpfwinklige Mauerecke
　　　　f
　　　f mur *m* transversal en biais
　　　nl muur *m* die niet haaks op een andere
　　　　muur aansluit
　　　r стена *f*, примыкающая к другой
　　　　стене под косым углом

C95　e **cantilever**
　　　d Kragarm *m*; Auskragung *f*;
　　　　Ausladung *f*
　　　f cantilever *m*; console *f*; porte-à-faux
　　　　m
　　　nl uitgekraagde ligger *m*
　　　r консоль *f*; выступ *m*; вылет *m*; свес
　　　　m

C96　e **cantilever arm**
　　　d Kragarm *m* der Kragträgerbrücke
　　　f console *f* du pont à poutres cantilever
　　　nl uitkraging *f* van de hoofdligger van
　　　　een cantileverbrug
　　　r консоль *f* балочно-консольного
　　　　моста

C97   e **cantilever beam**
    d Kragträger *m*, Auslegerträger *m*
    f poutre *f* en porte-à-faux, poutre-
       -console *f*, poutre *f* cantilever
   nl uitgekraagde ligger *m*
    r консольная балка *f*, консоль *f*

C98   e **cantilever bridge**
    d Kragbrücke *f*, Auslegerbrücke *f*
    f pont *m* à poutres cantilever
   nl cantileverbrug *f(m)*, brug *f(m)* met naar
       het midden uitgekraagde hoofdliggers
    r балочно-консольный мост *m*

C99   e **cantilever crane**
    d Ausleger-Bockkran *m*
    f grue *f* à portique cantilever
   nl kraan *f(m)* met horizontale uitleggers
    r консольный козловой кран *m*

C100 e **cantilever deck dam**
    d Plattenpfeiler(stau)mauer *f* mit
       Pfeilervorkragungen
    f barrage *m* à contreforts à tête ronde
       (*système Noetzli*)
   nl stuwdam *m* met uitgekraagd rijdek
    r массивно-контрфорсная плотина *f*
       с консольными оголовками

C101 e **cantilevered quay wall**
    d auskragende Kaimauer *f*
    f mur *m* de quai lourd type à chaise
   nl uitkragende kaaimuur *m*
    r набережная стенка *f*
       с разгружающей консолью

C102 e **cantilever foundation**
    d auskragendes Fundament *n* (*der Säule*)
    f fondation *f* à console
   nl vooruitspringend fundament *m* (*van
       een zuil*)
    r консольный фундамент *m* (*колонны*)

C103   **cantilever girder** *see* **cantilever beam**

C104 e **cantilever retaining wall**
    d auskragende Stützmauer *f*,
       Konsolstützmauer *f*
    f mur *m* de soutènement à console, mur
       m de soutènement (à) cantilever
   nl verbrede voet *m* van een keermuur
    r консольная подпорная стенка *f*

C105 e **cantilever-type drop**
    d Sprungschanzenüberfall *m*
    f chute *f* (à) cantilever
   nl uitgekraagde overloop *m*
    r консольный перепад *m*

C106 e **cantilever walking crane**
    d Zweiradkran *m*
    f grue *f* à vélocipède
   nl rijdende torenkraan *f (m)*
    r велосипедный кран *m*

C107 e **canvas**
    d Plane *f*, Zeltleinwand *f*, Segeltuch *n*
    f bâche *f*, grosse toile *f*, toile *f* de tente
   nl dekzeil *n*; canvas *n*
    r брезент *m*; парусина *f*

C108 e **cap**
    d 1. Aufsatz *m* 2. Kapitell *n*,
       Säulenkopf *m* 3. Betonbettung *f*

[Betonausgleichschicht *f*] auf der
Grubensohle 4. Rohrverschluß *m*;
Kappe *f*, Deckel *m* 5. Zündkapsel *f*
    f 1. chaperon *m*; chapeau *m* 2. chapiteau
       m 3. lit *m* de béton au fond d'une
       fouille 4. bouchon *m*; couvercle *m*
       5. détonateur *m*
   nl 1. kap *f(m)* 2. kolomkop *m* 3.
       muurafdekking *f* 4. (afsluit)stop *m*;
       dop *m* 5. slaghoedje *n*
    r 1. венчающий элемент *m*; насадка *f*;
       оголовок *m* 2. капитель *f*
       (*колонны*) 3. бетонная подготовка *f*
       на дне котлована 4. заглушка *f*;
       крышка *f* 5. капсюль-детонатор *m*

C109 e **capacity**
    d 1. Fassungsvermögen *n*, Stauraum *m*,
       Speicherfähigkeit *f*, Speicherraum *m*
       2. Leistungsfähigkeit *f*,
       Durchlaßkapazität *f* 3. Hubfähigkeit
       f, Tragfähigkeit *f* 4. Leistung *f*,
       Leistungsvermögen *n* 5. Fördermenge
       f, Liefermenge *f* 6. Hubraum *m*
    f 1. capacité *f* 2. débit *m* 3. pouvoir *m*
       porteur 4. rendement *m*, capacité *f*
       5. débit *m* de la pompe 6. cylindrée *f*
   nl capaciteit *f*
    r 1. вместимость *f*, ёмкость *f* (*напр.
       водохранилища*) 2. пропускная
       способность *f*; расход *m*
       3. грузоподъёмность *f*
       4. производительность *f* 5. подача *f*
       (*насоса*) 6. рабочий объём *m*
       (*цилиндра*)

C110 e **capacity curve**
    d Speicherinhaltskurve *f*
    f courbe *f* de jaugeage d'un
       réservoir
   nl capaciteitskromme *f(m)* (*van een
       waterreservoir*)
    r кривая *f* объёма водохранилища

C111 e **capacity formula**
    d Durchflußgleichung *f*
    f formule *f* de débit
   nl doorstromingsvergelijking *f*
    r формула *f* расхода

C112 e **capacity level**
    d Stauziel *n*, Stauhöhe *f*
    f niveau *m* [cote *f*] de retenue
    n. benodigd stuwpeil *n*
    r нормальный подпорный уровень *m*

C113 e **capacity of stream**
    d 1. Transportvermögen *n* der Strömung
       2. Durchlaßkapazität *f*,
       Leistungsfähigkeit *f* des offenen
       Gerinnes
    f 1. capacité *f* de transport des
       sédiments 2. capacité *f* de transport
       de l'eau
   nl 1. transportvermogen *n* van een
       stroom 2. doorlaatcapaciteit *f*
    r 1. транспортирующая способность *f*
       потока 2. максимальная
       пропускная способность *f*
       открытого русла

C114 *e* **cap crimper**
  d Sprengkapselzange, Anwürgzange *f*
  f pince *f* à sertir, sertisseur *m*
  nl wurgtang *f(m)*
  r обжимные щипцы *pl (для капсюлей-*
  -детонаторов)

C115 *e* **capel**
  d Seilabschlußmuffe *f* mit einer Öse
  f tendeur *m* du câble à œillet
  nl kabeleindmof *f(m)* met oog
  r концевая муфта *f* каната
  с проушиной [с ушком]

C116 *e* **capillary cell washer**
  d Kapillarluftwäscher *m*,
  Füllkörperkammer *f*;
  Kapillarrieselwäscher *m*
  f humidificateur *m* [humecteur *m*] d'air
  capillaire; laveur *m* de gaz capillaire
  nl capillair luchtzuiveringstoestel *n*
  r оросительный воздухоувлажнитель
  m с капиллярной насадкой,
  капиллярный воздухоувлажнитель
  m; камера *f* орошения с
  капиллярными ячейками; скруббер
  m с капиллярной насадкой

C117 *e* **capillary fitting**
  d Fitting *n* mit Kapillarlötverbindungen
  f raccord *m* de tuyaux capillaire
  nl capillaire soldeerverbinding *f*
  r фитинг *m* с капиллярными паяными
  соединениями (*пайка — по капилляр-*
  ному зазору)

C118 *e* **capillary fringe water**
  d hängendes Haftwasser *n*
  f eau *f* retenue par capillarité
  nl capillair gebonden water *n*
  r подвешенные воды *f pl*

C119 *e* **capillary joint**
  d Lötverbindung *f* mit haarfeiner Fuge,
  Kapillarlötverbindung *f*
  f joint *m* capillaire
  nl capillaire soldeerverbinding *f*
  r капиллярное паяное соединение *n*

C120 *e* **capillary pressure**
  d Kapillardruck *m*
  f pression *f* capillaire
  nl capillaire druk *m*
  r капиллярное давление *n*

C121 *e* **capillary rise**
  d kapillarer Aufstieg *m*, kapillare
  Steighöhe *f*
  f ascension *f* capillaire; hauteur *f*
  capillaire
  nl capillaire stijghoogte *f*
  r капиллярное поднятие *n*; высота *f*
  капиллярного подъёма

C122   **capillary type air humidifier** *see*
  capillary cell washer

C123   **capillary washer** *see* **capillary cell**
  washer

C124 *e* **capillary water**
  d Kapillarwasser *n*, Porensaugwasser *n*
  f eau *f* capillaire

nl capillair gebonden water *n*
  r капиллярная вода *f*

C125 *e* **capstan**
  d Gangspill *n*, Auflaufhaspel *f*,
  Ankerwinde *f*
  f cabestan *m*, treuil *m* vertical
  nl kaapstander *m*, windas *n*
  r кабестан *m*

C126 *e* **captation**
  d 1. Wasserfassung *f*
  2. Wasserfassungsanlage *f*
  f 1. captage *m* 2. ouvrage *m* de
  captage des eaux
  nl 1. watervang *m* 2. waterwinning *f*
  r 1. каптаж *m* 2. каптажное сооружение *n*

C127   **caption of a drawing** *see* **title block**

C128 *e* **capture**
  d Flußanzapfung *f*, Flußenthauptung
  f
  f capture *f*
  nl aftapping *f* van rivierwater
  r перехват *m* (реки)

C129 *e* **capture velocity**
  d Auffanggeschwindigkeit *f*,
  Erfassungsgeschwindigkeit *f*
  f vitesse *f* de capture
  nl aftapsnelheid *f*
  r расчётная скорость *f* воздуха
  в вытяжном факеле местного отсоса

C130 *e* **carbon-arc welding**
  d Schweißen *n* mit Kohleelektrode,
  Kohle-Lichtbogenschweißen *n*
  f soudage *m* à l'électrode de charbon
  nl booglassen *n* met koolspitsen
  r дуговая сварка *f* угольным
  электродом

C131 *e* **carbonate hardness**
  d Karbonathärte *f*, vorübergehende
  Härte *f*
  f dureté *f* en carbonate
  nl carbonaathardheid *f*, tijdelijke
  hardheid *f*
  r карбонатная [временная] жёсткость
  f (воды)

C132 *e* **carbonation of fresh concrete**
  d Karbonisation *f* des Frischbetons
  f carbon(is)ation *f* du béton frais
  nl carbonisatie *f* van vers beton
  r карбонизация *f* (*насыщение*
  углекислым газом из воздуха)
  свежеуложенного бетона

C133 *e* **carbon black**
  d Ruß *m*, Rußflocke *f*
  f noir *m* (de carbon)
  nl roet *n*
  r сажа *f* (*пигмент*)

C134 *e* **carbon dioxide fire extinguisher**
  d Kohlendioxid-Feuerlöscher *m*, CO_2-
  -Feuerlöschapparat *m*
  f extincteur *m* à neige carbonique
  nl koolzuursneeuwblusapparaat *n*
  r углекислотный огнетушитель *m*

C135 e **carbon fibers**
  d Kohlenfaden m pl
  f fibres f pl carboniques
  nl koolstofvezels f(m) pl
  r углеродные волокна n pl

C136 e **carburising**
  d Aufkohlung f
  f carburation f
  nl carburatie f (staal)
  r науглероживание n

C137 e **carcase, carcass**
  d Rohbau m
  f carcasse f; ossature f
  nl ruwbouw m, skelet n
  r каркас m, остов m (здания); здание n
    без отделочных работ

C138 e **carcassing**
  d Rohbauarbeiten f pl
  f montage m [construction f] de la
    charpente
  nl ruwbouw m, oprichten n van het
    skelet
  r возведение n [монтаж m] каркаса
    здания

C139 e **carcassing timber**
  d Bau(nutz)holz n, Konstruktionsholz n
  f bois m de charpente [de construction]
  nl bouwhout n voor dragende
    constructies
  r лесоматериал m для несущих
    конструкций; строительный лесо-
    материал m

C140 **carcass work** see **carcassing**

C141 e **cardboard**
  d Pappe f, Baupappe f
  f carton m de construction
  nl karton n
  r строительный картон m

C142 e **cargo berth, cargo terminal**
  d Anlegestelle f für Güterverkehr
  f quai m (d'accostage) de chargement
  nl laad- en loskade f
  r грузовой причал m

C143 e **car park**
  d Parkplatz m
  f aire f de stationnement
  nl parkeerterrein n
  r зона f стоянок автомобилей;
    автостоянка f

C144 e **carpenter**
  d Zimmermann m
  f charpentier m
  nl timmerman m
  r плотник m

C145 e **carpenter's hammer**
  d Zimmermannshammer m,
    Latt(en)hammer m
  f marteau m de charpentier
  nl klauwhamer m
  r плотничный молоток m

C146 e **carpenter's wooden vise**
  d Bankschraubstock m
  f étau m d'etabli de menuisier

  nl timmermansbankschroef f(m)
  r столярные верстачные тиски pl

C147 e **carpenter's work, carpentry**
  d Zimmer(manns)arbeiten f pl
  f charpenterie f, travaux m pl de
    charpenterie en bois
  nl timmerwerk n
  r плотничные работы f pl;
    плотничное дело n

C148 e **carriage**
  d 1. Wagen m; Fahrgestell n
    2. Schlitten m
  f chariot m; bogie m
  nl 1. wagen m 2. slede f(m)
  r 1. тележка f 2. каретка f

C149 e **carriage mounting**
  d (Jumbo-)Bohrwagen m, Wagenaufbau
    m
  f chariot m de forage, jumbo m
  nl mobiele boorinstallatie f
  r буровая каретка f

C150 e **carriage piece**
  d Mittelwangenträger m
  f limon m intermédiaire
  nl middenboom m (bij brede steektrap)
  r промежуточная тетива f
    (лестницы)

C151 e **carriageway**
  d Fahrbahn f (Straße, Brücke)
  f chaussée f; tablier m (du pont)
  nl rijbaan f(m); brugdek n
  r проезжая часть f (дороги, моста)

C152 **carriageway marking** see **road marking**

C152a e **carriageway surfacing**
  d Fahrbahnbelag m
  f revêtement m de chaussée
  nl wegdek n van de rijbaan
  r верхний слой m дорожного
    покрытия

C153 **carrier canal** see **water supply canal**

C154 **carrier drain** see **collecting drain**

C155 e **carrying capacity**
  d 1. Tragfähigkeit f 2.
    Durchlaßfähigkeit f (Rohrleitung)
  f 1. capacité f portante; capacité f de
    charge [de levage] 2. rendement m
    de la tuyauterie [de la conduite]
  nl 1. draagvermogen n; laadvermogen n
    2. doorlaatvermogen n (buizen)
  r 1. несущая способность f (напр.
    грунта); грузоподъёмность f
    2. пропускная способность f
    (трубопровода)

C156 e **carry-over storage**
  d Überjahresspeicherung f;
    Mehrjahresspeicher m
  f emmagasinement m interannuel;
    reservoir m interannuel
  nl spaarbekken n, buffervoorraad m
  r водохранилище n многолетнего
    регулирования

C157 e **cartridge-operated fixing gun**
  d Bolzenschießgerät n,

Bolzensetzwerkzeug *n*
f appareil *m* de scellement à cartouches
explosives
nl spijkerpistool *n*, boutenschiethamer *m*
r строительно-монтажный пистолет *m*

C158 *e* **cascade**
d Kaskade *f*; Stufenabsturz *m*
f cascade *f*
nl getrapte uitloop *m*
r каскад *m*; ступенчатый перепад *m*

C159 **cased pile** *see* **caisson pile 1.**

C160 **cased structures** *see* **encased structures**

C161 *e* **cased well**
d Mantelrohrbrunnen *m*
f puits *m* tubé
nl geboorde bron *f(m)*
r трубчатый колодец *m*

C162 *e* **casein glue**
d Kaseinleim *m*
f colle *f* de caséin
nl caseïnelijm *m*
r казеиновый клей *m*

C163 *e* **casein paint**
d Kaseinfarbe *f*
f peinture *f* à la caséin
nl caseïneverf *f(m)*
r казеиновая краска *f*

C164 *e* **casement**
d Fenster-Flügel *m*, Flügel *m*
f battant *m* [vantail *m*] de fenêtre
nl raam *n*
r створка *f* окна

C165 **casement door** *see* **glazed door**

C166 *e* **casement window**
d Flügelfenster *n*
f fenêtre *f* à battants
nl openslaand venster *n*
r створчатое окно *n*

C167 *e* **casing**
d 1. Rahmen *m*; Tür- *oder*
Fenstereinfassung *f* 2. Gehäuse *n*,
Kasten *m*, Mantel *m* 3. Mantelrohr
n, Futterrohr *n*
f 1. chambranle *m*; encadrement *m*
2. revêtement *m*; enveloppe *f* 3. tube
f de revêtement, tube *f* fourreau
nl 1. kozijn *n* 2. huis *n* (*ommanteling*)
3. mantelbuis *f(m)*
r 1. рама *f*; обрамление *n* дверного
или оконного проёма 2. облицовка *f*;
короб *m*; кожух *m* 3. обсадная
труба *f*

C168 *e* **castellated beam**
d Lochwandstegträger *m*
f poutre *f* évidée [crénelée]; poutre *f*
Litzke
nl stalen balk *m* met geperforeerd lijf
r перфорированная балка *f*

C169 **casting area** *see* **casting yard**

C170 *e* **casting bed**
d Betonierbett *n*
f banc *m* de bétonnage

nl gietvorm *m*, mal *m*
r стенд *m* для изготовления сборных
железобетонных изделий

C171 *e* **casting of concrete without forms**
d schalungsloses Betonieren *n*,
schalungsloser Betoneinbau *m*
f betonnage *m* [coulage *m*] sans coffrage
nl betonstorten *n* zonder bekisting
r безопалубочное бетонирование *n*

C172 *e* **casting plant**
d Betonfertigteilwerk *n*
f usine *f* de préfabrication (*pour*
éléments en béton), usine *f* pour produits en béton
nl betonwerk *n*
r завод *m* сборных железобетонных
конструкций [изделий]

C173 *e* **casting yard**
d Betonierfläche *f*, Beton-
-Fertigteilplatz *m*; Beton-
-Fertigteilwerk *n*
f aire *f* de bétonnage, chantier *m* de
bétonnage des élements préfabriqués
nl bouwplaats *f(m)* voor geprefabriceerde
betonelementen
r полигон *m* [цех *m*] для
изготовления сборных
железобетонных изделий

C174 *e* **cast-in member**
d einbetoniertes Element *n*, Einlegeteil
n
f pièce *f* encastrée
nl in het beton opgenomen onderdeel *n*
r закладной элемент *m*, закладная
деталь *f*

C175 *e* **cast-in-place and precast**
construction
d Stahlbetonverbundkonstruktion *f*,
Stahlbetonfertigteilkonstruktion *f*
mit monolithischem Verbund
f construction *f* (combinée) en béton
coulé sur place et en éléments
préfabriqués
nl monolietconstructie *f* met
geprefabriceerde onderdelen
r сборно-монолитная конструкция *f*

C176 *e* **cast-in-place concrete**
d Ortbeton *m*
f béton *m* coulé sur place [en place]
nl in het werk gestort beton *n*
r монолитный бетон *m*

C177 *e* **cast-in-place pile**
d Ort(beton)pfahl *m*, Betonortpfahl *m*
f pieu *m* coulé en place, pieu *m* moulé
dans le sol
nl in de grond gevormde betonpaal *m*
r набивная (бетонная) свая *f*

C178 *e* **cast-in-place shelless pile**
d Ortpfahl *m* ohne Mantelrohr
f pieu *m* coulé en place sans tube de
cuvelage
nl in de grond gevormde betonpaal *m*
zonder mantelbuis
r набивная свая *f* без обсадной трубы

C179   cast-in-situ concrete *see* cast-in-place concrete

C180   cast-in-situ pile *see* cast-in-place pile

C181 *e* cast iron pipe
   d Guß(eisen)rohr *n*
   f tuyau *m* de fonte
  nl gietijzeren pijp *f(m)*
   r чугунная труба *f*

C182 *e* cast iron radiator
   d Gußeisenradiator *m*
   f radiateur *m* de fonte
  nl gietijzeren radiator *m*
   r чугунный радиатор *m*

C183 *e* cast iron segment
   d Gußeisentübbing *m*
   f anneau *m* de cuvelage en fonte
  nl gietijzeren segment *n*
   r чугунный тюбинг *m*

C184 *e* cast steel
   d Stahlguß *m*
   f acier *m* coulé [fondu, moulé]; moulage *m* d'acier
  nl gietstaal *n*, gegoten staal *n*; gietstuk *n*
   r литая сталь *f*; стальная отливка *f*

C185 *e* cast stone
   d Betonwerkstein *m*
   f aggloméré *m* (de béton), bloc *m* [corps *m*, parpaing *m*] en béton
  nl betonsteen *m*
   r бетонный камень *m* [блок *m*]

C186 *e* catalyzed aggregates
   d katalysierte Zuschlagstoffe *m pl*
   f agrégats *m pl* [granulats *m pl*] catalysés [activés par catalyseur]
  nl met katalysatoren behandelde toeslagstoffen *f(m) pl*
   r заполнители *m pl*, активированные катализатором (*для безусадочных бетонов*)

C187 *e* catastrophic flood
   d Katastrophenhochwasser *n*
   f crue *f* catastrophique
  nl catastrofaal hoogwater *n*
   r катастрофический паводок *m*

C188 *e* catch basin US
   d Sinkkasten *m*, Regen(wasser)einlauf *m*, Auffangbecken *n*
   f bouche *f* d'égout
  nl slijkvanger *m*, grof-vuilvanger *m*
   r дождеприёмник *m*; грязеуловитель *m*

C189 *e* catch bolt
   d Federverschluß *m*
   f verrou *m* glissant (à ressort)
  nl veergrendel *m*
   r пружинная дверная задвижка *f*

C190   catch drain *see* catch water drain

C191 *e* catch gallery
   d Brunnengalerie *f*, Sickergalerie *f*
   f galerie *f* de captage [de prise d'eau]
  nl vuilwaterkelder *m*
   r водосборная галерея *f*

C192   catchment *see* captation

C193 *e* catchment area
   d Abflußgebiet *n*, Einzugsgebiet *n*; Abflußfläche *f*
   f bassin *m* versant; surface *f* de réception, aire *f* de captage
  nl waterwingebied *n*
   r водосборный бассейн *m*; площадь *f* водосбора

C194 *e* catchment area of aquifer
   d Einzugsgebiet *n* eines Grundwasserleiters
   f bassin *m* versant [d'alimentation] d'une formation aquifère
  nl waterwingebied *n* voor grondwater
   r площадь *f* водосбора подземных вод

C195 *e* catchment area storage
   d Beckenregulierung *f*
   f régularisation *f* des débits de bassin
  nl opslag *m* in waterwingebied
   r бассейновое регулирование *n*

C196 *e* catchment management
   d Bewirtschaftung *f* eines Einzugsgebiets
   f exploitation *f* d'un bassin versant (naturel), exploitation *f* d'un bassin d'alimentation
  nl beheer *n* van het waterwingebied
   r мероприятия *n pl* по поддержанию территории водосбора в состоянии, обеспечивающем оптимальное качество стекающей воды

C197 *e* catchment yield
   d Ergiebigkeit *f* eines Einzugsgebietes
   f rendement *m* d'un bassin versant
  nl wateropbrengst *f*
   r дебит *m* водосборного бассейна

C198   catch pit *see* catch basin

C199   catch-water ditch *see* interception channel

C200 *e* catch water drain
   d Abfangsammler *m*, Sammeldrän *m*, Abfanggraben *m*, Hanggraben *m*
   f contre-fossé *m*, fossé *m* collecteur, contre-canal *m* de drainage
  nl verzameldrain *m*
   r нагорная [перехватывающая] канава *f*, нагорный канал *m*

C201 *e* catchwaters
   d Abfangsammler *m*, Sammeldrän *m*, Fangdrän *m*
   f fossés *m pl* de ceinture
  nl vangdrain *m*
   r ловчие каналы *m pl или* дрены *f pl*

C202 *e* catenary
   d Kettenlinie *f*
   f chaînette *f*
  nl kettinglijn *f(m)*
   r цепная линия *f*; линия *f* провеса каната

C203 *e* catenary suspension
   d Vielfachaufhängung *f* des Oberleitungsdrahtes

f tige *f* de suspension du fil de contact
nl kettinglijnophanging *f*
r цепная контактная подвеска *f*

C204 *e* caterpillar
d Raupenschlepper *m*
f tracteur *m* à chenille(s)
nl rupstrekker *m*, trekker *m* op rupsbanden
r гусеничный трактор *m*

C205 *e* caterpillar gate *US*
d Raupenschütz(e) *n* (*f*)
f vanne *f* à chenilles
nl rupsschuif *f(m)*
r гусеничный затвор *m*

C206 *e* cat-head sheave
d Spillseilrolle *f*
f poulie *f* de tête (*de la sonnette ou
du mât de levage*)
nl liertrommel *f(m)*
r грузоподъёмный блок *m* (*на ого-
ловке копра или мачты*)

C207 *e* cathodic protection
d Katodenschutz *m*
f protection *f* cathodique
nl kathodische bescherming *f*
r катодная защита *f* (*от коррозии*)

C208 *e* cation bed
d Kationenbett *n*, Kationenfüllkörper *m*
f lit *m* (filtrant) d'échange des cations
nl kationenbed *n* (*filter*)
r катионообменная загрузка *f* (*фильтра*)

C210 *e* cationic emulsion
d kationische Emulsion *f*
f émulsion *f* cationique
nl kationische emulsie *f*
r катионная эмульсия *f* (*для дорожных
покрытий*)

C211 *e* cattle creep
d Viehdurchlaß *m* (*unter einem
Verkehrsweg*)
f passage *m* (en dessous) pour bétail
nl veetunnel *m* (*onder de rijweg*)
r туннельный проход *m* для скота
(*под дорогой*); скотопрогон *m*

C212 *e* cattle crossing, cattle crossing bridge
d Viehdurchlaß-Kreuzung *f*
f passage *m* en dessus pour bétail
nl veebrug *f(m)*
r скотопрогон *m* мостового типа

C213 *e* cat walk
d Hängesteg *m*, Stegbrücke *f*, Laufsteg
m
f passerelle *f* suspendue
nl voetbrug *f(m)*, loopbrug *f(m)*
r подвесной переходной мостик *m*;
мостки *pl* для пешеходов

C214 *e* cauldron
d Bitumen- und Teerkocher *m*,
Bitumen- und Teerschmelzkessel *m*
f fondoir *m* [chaudière *f* de fusion] pour
liants hydrocarbonés; fondoir *m*
[chaudière *f*] à bitume

nl grote ketel *m* (*b.v. ter verhitting van
bitumen*)
r открытый котёл *m* для подогрева
чёрных вяжущих

C215 *e* caulked joint
d Stemmuffe *f*; Stemmnaht *f*,
verstemmte Fuge *f*
f emboîtement *m* [joint *m*
d'emboîtement] calfaté
nl gebreeuwde naad *m*
r зачеканенное раструбное соединение
n; зачеканенный шов *m*

C216 *e* caulking
d Verstemmen *n*; Kalfatern *n*
f calfeutrement *m*, calfatage *m*
nl kalfaten *n*; breeuwen *n*
r чеканка *f*; конопатка *f*

C217 *e* caulking hammer
d Schlegel *m*, Stemmhammer *m*,
Kalfaterhammer *m*
f matoir *m*, marteau *m* à mater
nl breeuwhamer *m*
r киянка *f* для конопатки; молоток *m*
для зачеканивания труб

C218 *e* caulking iron
d Verstemmeisen *n*, Stopfeisen *n*
f cordoir *m*
nl kookbeitel *m*
r чеканка *f* (*инструмент*)

C219 *e* causeway
d 1. Dammstraße *f* 2. gepflasterter
Wegübergang *m*
f 1. route *f* sur remblai; route-digue *f*
2. route *f* sur le fond; route-cassis
f
nl 1. opgehoogde weg *m* in laag terrein
2. bestrate weg *m*
r 1. дорога *f* по насыпи (*на заболоченной или низменной территории*)
2. мощёный переезд *m*

C220 *e* caution sign
d Warn(ungs)zeichen *n*
f signal *m* d'avertissement
nl waarschuwingsbord *n*
r предупредительный знак *m*

C221 *e* cave-in
d Gleitflächenbruch *m*,
Böschungsbruch *m*
f rupture *f* de talus [de pente]
nl afkalving *f*
r обрушение *n* откоса

C222 *e* cavern
d Kaverne *f*
f caverne *f*, cavité *f*
nl holte *f*
r каверна *f*

C223 *e* cavil
d Schlageisen *n*, Zahneisen *n*
f laye *f* [laie *f*] pointue à une extrémité
nl punthamer *m*, steenhamer *m*
r закольник *m* с одним заострённым
концом, зубатка *f*

C224 *e* **caving**
  d Unterspülung *f*, Unterkolkung *f*,
    Uferbruch *m*
  f affouillement *m* (*d'une berge, d'une
    digue*)
  nl onderspoeling *f* (*oever*); dijkval *m*
  r подмыв *m* (*берега, дамбы*)

C225 *e* **cavitation**
  d Kavitation *f*; Blasenbildung *f*,
    Hohlraumbildung *f*
  f cavitation *f*
  nl cavitatie *f*
  r кавитация *f*

C226 *e* **cavity**
  d Hohlraum *m*, Kaverne *f*;
    Luftschicht *f*
  f cavité *f*, caverne *f*; lame *f* [vide *m*]
    d'air
  nl holte *f*, caviteit *f*
  r полость *f*, каверна *f*; воздушная
    прослойка *f*

C227 *e* **cavity block, cavity concrete block**
  d (Beton-)Hohlblock *m*
  f bloc *m* (de béton) creux [évidé],
    parpaing *m* creux [vide] (en béton)
  nl holle bouwsteen *m* [betonsteen *m*]
  r пустотелый (бетонный) блок *m*

C228 *e* **cavity dam**
  d Hohl(stau)mauer *f*, Zellenwehr *n*
  f barrage *m* cellulaire
  nl holle stuwdam *m*
  r ячеистая плотина *f*

C229 *e* **cavity wall**
  d Hohlmauer *f*
  f mur *m* creux, mur *m* à double paroi
  nl spouwmuur *m*
  r пустотелая [полая] стена *f*, стена *f*
    с воздушной прослойкой

C230 *e* **ceiling**
  d Decke *f*
  f plafond *m*
  nl plafond *n*, zoldering *f*
  r потолок *m*

C231 *e* **ceiling air diffuser**
  d Deckendiffusor *m*,
    Deckenluftverteiler *m*, Deckenauslaß
    m
  f diffuseur *m* de plafond
  nl bovenluchtinlaat *m*
  r потолочный диффузор *m*; потолочный
    вентиляционный плафон *m*

C232 *e* **ceiling air distribution**
  d Deckenluftverteilung *f*
  f distribution *f* d'air par système de
    plafond
  nl luchtinlaat *m* langs het plafond
  r потолочное воздухораспределение *n*

C233 *e* **ceiling effect**
  d Anlegen *n* des Zuluftstrahles an die
    Decke, Anlehnung *f* der
    Luftströmung
  f effet *n* de plafond
  nl plafondeffect *n*

  r настилание *n* приточной струи на
    потолок

C234 *e* **ceiling fan**
  d Deckenlüfter *m*, Deckenventilator *m*
  f ventilateur *m* de plafond
  nl plafondventilator *m*
  r потолочный вентилятор *m*

C235 *e* **ceiling fitting** *UK*
  d Deckenleuchte *f*
  f luminaire *m* de plafond, plafonnier *m*
  nl plafonnier *m*
  r потолочный светильник *m*

C236 *e* **ceiling heating**
  d Decken(strahlungs)heizung *f*
  f chauffage *m* par plafond
  nl plafondverwarming *f*
  r потолочное панельное отопление *n*

C237 *e* **ceiling joist**
  d Deckenbalken *m*
  f solive *f* [poutre *f*] du plafond
    (suspendu)
  nl vloerbalk *m*, zolderbalk *m*
  r потолочная балка *f*; балка *f*,
    несущая подвесной потолок

C238 **ceiling lighting fitting** *see* **ceiling
    fitting**

C239 *e* **ceiling outlet**
  d Deckenaustrittsöffnung *f*;
    Deckenauslaß *m*
  f bouche *f* de plafond; sortie *f* de
    plafond
  nl afvoeropening *f* in het plafond
  r потолочное приточное отверстие *n*;
    потолочный воздухораспределитель
    m

C240 *e* **celerity**
  d Wellen-Schnelligkeit *f*,
    Wellenfortpflanzungsgeschwindigkeit
    f
  f célérité *f* [vitesse *f*] des ondes
  nl (voorplantings)snelheid *f*
  r скорость *f* движения волн

C241 *e* **cellar**
  d Keller *m*; technischer Kellerraum *m*
  f cave *f*
  nl kelder *m*
  r нежилой подвал *m*; техническое
    подполье *n*

C242 *e* **cellular air filter**
  d Zellenluftfilter *n*, *m* Kassettenfilter
    n, *m*
  f filtre *m* à air à cellules
  nl cellulair luchtfilter *n*, *m*
  r ячейковый воздушный фильтр *m*

C243 *e* **cellular block**
  d Zellenblock *m*
  f bloc *m* [corps *m*, parpaing *m*]
    cellulaire
  nl celblok *n*
  r ячеистый блок *m*

C244 *e* **cellular cofferdam**
  d Zellenfangdamm *m*
  f batardeau *m* cellulaire

nl celvormige kistdam *m*
r ячеистая шпунтовая перемычка *f*

C245 *e* **cellular cofferdam with semicircular cells**
d Segmentzellenfangdamm *m*
f batardeau *m* en cellules semi-circulaires
nl segmentenkistdam *m*
r сегментная перемычка *f*

C246 *e* **cellular concrete**
d Zellenbeton *m*
f béton *m* cellulaire [aéré]
nl cellenbeton *n*
r ячеистый бетон *m*

C247 *e* **cellular dam**
d Zellenstaumauer *f*, Zellensperre *f*
f barrage *m* cellulaire
nl cellenstuwdam *m*
r ячеистая [массивно-контрфорсная] плотина *f*

C248 *e* **cellular dust collector**
d Zellenluftfilter *n*, *m*, Zellenentstauber *m*
f dépoussiéreur *m* à cellules
nl cellulair stoffilter *n*, *m*
r ячейковый пылеуловитель *m* [фильтр *m*]

C249 *e* **cellular glass**
d Schaumglas *n*
f verre *m* moussé
nl schuimglas *n*
r пеностекло *n*, ячеистое стекло *n*

C250 **cellular plastics** *see* **foamed plastics**

C251 *e* **cellular rubber**
d Porengummi *m*, Schaumgummi *m*
f caoutchouc-mousse *m*
nl schuimrubber *m*, *n*
r пенорезина *f*, пористая [ячеистая] резина *f*

C252 *e* **cellular sheetpile breakwater**
d Stahlzellen-Wellenschutzanlage *f*, Stahlzellen-Wellenbrecher *m*
f brise-lames *m* [brise-mer *m*] à batardeau cellulaire
nl stalen damwand *m* als golfbreker
r оградительное сооружение *n* [волнолом *m*] ячеистой шпунтовой конструкции

C253 *e* **cellular silencer**
d Zellenschalldämpfer *m*
f amortisseur *m* de bruit cellulaire
nl gecompartimenteerde geluiddemper *m*
r сотовый шумоглушитель *m*

C254 *e* **cement**
d 1. Zement *m*; (unorganisches) Bindemittel *n* 2. Kleber *m*
f 1. ciment *m* 2. colle *f*
nl 1. cement *n* 2. lijm *m*
r 1. цемент *m*; неорганическое вяжущее *n* (*вещество*) 2. клей *m*

C255 *e* **cement additive**
d Zementzusatz(mittel) *m* (*n*)
f adjuvant *m* [additif *m*] de ciment

nl cementtoeslag *m*
r добавка *f* к цементу

C256 *e* **cement-aggregate bond**
d Bindemittelhaftfestigkeit *f*, Zementhaftung *f*
f adhésivité *f* ciment-agrégat
nl hechting *f* van het cement aan het toeslagmateriaal
r сцепление *n* цементного камня с заполнителем

C257 *e* **cementation**
d Bodenversteinerung *f*, Zementinjektion *f*, Zementeinpressung *f*, Zementverfestigung *f*, Zementierung *f*
f cimentation *f*, cimentage *m*, injection *f* de ciment, stabilisation *f* au ciment
nl cementspecie injecteren *n*
r цементация *f* (*пород*)

C258 *e* **cement batcher**
d Zementdosierer *m*
f doseur *m* de ciment
nl doseringsapparaat *n* voor cement
r дозатор *m* цемента

C259 *e* **cement bin**
d Zementaufgabesilo *n*, *m*
f trémie *f* à ciment
nl cementsilo *m*
r цементный бункер *m*, бункер *m* для цемента

C260 *e* **cement-bound macadam**
d Zementschotterdecke *f*, Mörteleingußdecke *f*
f macadam-mortier *m*, macadam--ciment *m*
nl met cement gestabiliseerde steenslag *m*
r щебёночное (дорожное) покрытие *n*, укреплённое сухой смесью цемента с песком; цементно-щебёночное (дорожное) покрытие *n*

C261 *e* **cement clinker**
d Zementklinker *m*
f clinker *m* de ciment
nl cementklinker *m*
r цементный клинкер *m*

C262 *e* **cement content**
d Zementanteil *m*, Zementfaktor *m*, Zementgehalt *m*; Zementverbrauch *m*
f teneur *m* en ciment; dosage *m* en ciment
nl gehalte *n* aan cement
r содержание *n* цемента; расход *m* цемента

C263 *e* **cement external rendering**
d Außenzementputz *m*
f enduit *m* extérieur au ciment
nl buitenpleisterwerk *n*
r наружная цементная штукатурка *f*

C264 **cement factor** *see* **cement content**

C265 *e* **cement grinding mill**
d Zementmühle *f*

f broyeur *m* à ciment
nl cementmolen *m*
r цементная мельница *f*

C266 *e* cement grout
d 1. Zementschlämme *f*
2. Zementrohschlamm *m*
f 1. mortier *m* [coulis *m*] de ciment
2. pâte *f* crue à ciment
nl cementvoegspecie *f*
r 1. жидкий цементный раствор *m*
2. цементный сырьевой шлам *m*

C267 cement grouting *see* cementation

C268 *e* cement gun
d Zementkanone *f*, Zementspritze *f*,
Zementmörtelinjektor *m*,
Betonspritzmaschine *f*
f canon *m* à ciment, injecteur *m* de
mortier de ciment, guniteuse *f*
nl cementspuitmachine *f*, cement-
-injectieapparaat *n*
r цемент-пушка *f*, торкрет-аппарат *m*

C269 *e* cement in bulk
d loser [unverpackter] Zement *m*
f ciment *m* en vrac [en masse]
nl onverpakt cement *n*, bulkcement *n*
r цемент *m* навалом

C270 cementing materials *see* cementitious
materials

C271 *e* cementing properties
d Bindeeigenschaften *f pl*
f propriétés *f pl* liantes [cohésives, de
cohésivité]
nl bindvermogen *n*
r вяжущие свойства *n pl*

C272 *e* cementitious materials
d mineralische Bindemittel *n pl*
[Binder *m pl*]
f liants *m pl* minéraux
nl minerale bindmiddelen *n pl*
r минеральные вяжущие вещества *n*
pl

C273 *e* cement lime mortar
d Kalkzementmörtel *m*,
Zementkalkmörtel *m*
f mortier *m* de chaux et de ciment
nl kalkcementmortel *m*
r известково-цементный раствор *m*

C274 *e* cement minerals
d Zementmineralien *n pl*
f minéraux *m pl* de clinker de ciment
nl minerale bestanddelen *n pl* van het
cement
r цементные минералы *m pl*

C275 *e* cement mortar
d Zementmörtel *m*
f mortier *m* de ciment
nl cementmortel *m*
r цементный раствор *m*

C276 *e* cement paint
d 1. Zementfarbe *f*, Zementanstrich *m*
2. Steinemaille *f*
f 1. peinture *f* au ciment 2. peinture *f*
à la caséine *ou* à la base d'autres

matériaux résistants à l'action
d'alcali
nl cementverf *f(m)*
r 1. цементная краска *f* 2. окрасочный
состав *m* на основе казеина и других
щёлочестойких материалов

C277 *e* cement paste
d Zementpaste *f*
f pâte *f* de ciment
nl cementbrij *f*, cementmelk *f(m)*
r цементное тесто *n* [молоко *n*]

C278 *e* cement plaster, cement rendering *see*
cement external rendering

C279 *e* cement screed
d Zementestrichboden *m*
f chape *f* (de mortier) en ciment
nl tot diktemal dienende pleisterstrook
f(m)
r цементная стяжка *f*, стяжка *f* из
цементного раствора

C280 *e* cement silo
d Zementsilo *n*, *m*
f silo *m* à ciment
nl cementsilo *m*
r силос *m* для хранения цемента,
цементный бункер *m*

C281 *e* cement slurry
d 1. Zement-Rohschlamm *m*
2. Zementschlämme *f*
f 1. pâte *f* crue à ciment 2. coulis *m*
de ciment
nl cementbrij *m*
r 1. цементный шлам *m* 2. жидкое
цементное тесто *n*

C282 *e* cement soil stabilization *see*
cementation

C283 *e* cement stabilized material
d zementstabilisierter
[zementverfestigter] Untergrund *m*
[Boden *m*]
f sol *m* stabilisé au ciment
nl met cement gestabiliseerd materiaal *n*
r грунт *m*, стабилизированный
цементным раствором, грунт *m*,
подвергнутый цементации

C284 cement testing sand *see* standard
sand

C285 *e* cement tests
d Zementprüfverfahren *n pl*
f essais *m pl* de ciment; méthodes *f pl*
d'essai de ciment
nl cementproeven *f(n) pl*
r методы *m pl* испытания цементов

C286 *e* cement-water ratio
d Zement-Wasser-Faktor *m*
f rapport *m* ciment-eau
nl water-cementfactor *m*
r цементно-водное отношение *n*

C287 *e* cement-wood floor
d Sägemehlbetonbelag *m*
f revêtement *m* de sol en béton de bois
nl houtmeelcementvloer *m*
r покрытие *n* пола из арболита

C283 *e* cement works
 d Zementwerk *n*
 f cimenterie *f*, usine *f* à ciment
 nl cementfabriek *f*
 r цементный завод *m*

C289 centering *see* centers

C290 *e* center line
 d Mittellinie *f*, Achse *f*
 f axe *m*, ligne *f* centrale
 nl hartlijn *f(m)*
 r осевая линия *f*, ось *f*

C291 center line of inertia *see* axis of inertia

C292 *e* center of curvature
 d Krümmungsmittelpunkt *m*
 f centre *m* de courbure
 nl kromtemiddelpunt *n*
 r центр *m* кривизны

C293 *e* center of gravity
 d Schwerpunkt *m*
 f centre *m* de gravité
 nl zwaartepunt *n*
 r центр *m* тяжести; центр *m* масс

C294 center of moments *see* moment center

C295 *e* center of pressure
 d Druckmittelpunkt *m*
 f centre *m* de pression
 nl drukpunt *n*
 r центр *m* давления

C296 *e* center of rotation
 d Rotationszentrum *n*, Drehpunkt *m*,
 Momentenpol *m*
 f centre *f* de rotation
 nl draaipunt *n*
 r центр *m* вращения

C297 center of twist *see* shear center

C298 *e* center pegs
 d 1. Pflöcke *m pl* auf der Fahrbahnachse
 2. Pflocken *n* auf der Fahrbahnachse
 f 1. piquets *m pl* le long du tracé
 d'une route 2. piquetage *m* du tracé
 d'une route
 nl pennen *f pl* in de as van de rijbaan
 r 1. пикеты *m pl* по продольной оси
 дороги 2. пикетаж *m* по
 продольной оси дороги

C299 *e* centers
 d Lehrgerüst *n*, Bogenlehre *f*
 f cintre *m* (de construction d'un arc),
 cintre *m* de charpente
 nl formelen *n pl*
 r кружала *n pl*

C300 *e* center span
 d Mittelfeld *n*, Mittelöffnung *f*
 f travée *f* centrale [médiane],
 ouverture *f* centrale
 nl centrale overspanning *f*
 r центральный пролёт *m*

C301 *e* center support
 d Mittelstütze *f*; Mittelstiel *m*
 f appui *m* central; pile *f* centrale;
 poteau *m* central
 nl middensteun *m*; middenstijl *m*

 r центральная [средняя] опора *f*;
 центральная стойка *f*

C302 *e* center-to-center distance
 d Mittenabstand *m*
 f entraxe *m*, entre-axe *m*, distance *f*
 d'axe en axe
 nl afstand *m* hart-op-hart
 r осевое расстояние *n*

C303 *e* central air conditioner
 d Zentralklimagerät *n*, Klimazentrale *f*
 f climatiseur *m* central, conditionneur
 m de l'air central
 nl centraal klimaatregelingstoestel *n*
 r центральный кондиционер *m*

C304 *e* central air conditioning plant
 d Zentralklimaanlage *f*
 f installation *f* centrale de conditionnement d'air, installation *f* centrale
 de climatisation
 nl centrale klimaatregelingsinstallatie *f*
 r центральная установка *f*
 кондиционирования воздуха

C305 central air-conditioning system *see*
 central plant air-conditioning system

C306 *e* central air-handling plant
 d zentrale Lüftungsanlage *f*,
 Zuluftzentrale *f*
 f installation *f* centrale de préparation
 d'air
 nl centrale luchtbehandelingsinstallatie *f*
 r центральная воздухоприготовительная [вентиляционная] установка *f*

C307 *e* central boiler house
 d Heizzentrale *f*, Zentralkesselhaus *n*,
 Kesselgebäude *n*
 f bâtiment *m* des chaudières central
 nl centraal ketelhuis *n*
 r центральная котельная *f*

C308 *e* central fan system
 d zentrales Luftfördersystem *n*
 f système *m* centralisé
 d'approvisionnement en air (*pour
 ventilation, chauffage à air chaud ou
 climatisation*)
 nl centraal ventilatiesysteem *n*
 r центральная система *f* (*вентиляции,
 воздушного отопления или кондиционирования воздуха*)

C309 *e* central heating
 d Zentralheizung *f*, Sammelheizung *f*
 f chauffage *m* central
 nl centrale verwarming *f*
 r центральное отопление *n*

C310 *e* central heating boiler
 d Zentralheizungskessel *m*
 f chaudière *f* de chauffage central
 nl centrale verwarmingsketel *m*
 r котёл *m* центрального отопления

C311 *e* central heating plant
 d 1. Heizzentrale *f* 2. Zentral-
 heizungsanlage *f*,
 Sammelheizungsanlage *f*

f 1. bâtiment *m* des chaudières central
2. chauffage *m* central
nl centrale verwarmingsinstallatie *f*
r 1. центральная котельная *f* 2. система
f центрального отопления

C312 *e* **central heating station**
d Wärmeübertragungsstation *f*
f station *f* centrale de chaleur
nl centraal ketelhuis *n*
r центральный тепловой пункт *m*

C313 *e* **central humidifier**
d zentraler Luftbefeuchter *m*,
zentrale Luftbefeuchtigungseinrichtung
f
f humidificateur *m* [humecteur *m*]
(d'air) central
nl centrale luchtbevochtigingsinrichting
f
r центральный воздухоувлажнитель *m*

C314 *e* **centralized hot-water supply system**
d Sammel-Warmwasserversorgung *f*,
zentrale Warmwasserbereitung *f*
f système *m* central de distribution
d'eau chaude
nl centraal systeem *n* voor
warmwatervoorziening
r централизованная система *f*
горячего водоснабжения

C315 *e* **centralized monitoring**
d Zentralkontrolle *f*
f contrôle *m* à distance centralisé
nl centrale regeling *f*
r централизованное телеуправление *n*;
централизованный (дистанционный)
контроль *m*

C316 *e* **central moment of inertia**
d zentrales Trägheitsmoment *n*
f moment *m* d'inertie central
nl centraal traagheidsmoment *n*
r центральный момент *m* инерции

C317 *e* **central plant air-conditioning system**
d Zentralklimaanlage *f*
f installation *f* de climatisation [de
conditionnement d'air] centralisée,
centrale *f* de climatisation [de
conditionnement d'air]
nl centraal luchtbehandelingssysteem *n*
r центральная система *f*
кондиционирования воздуха

C318 *e* **central point load**
d mittige Punktlast *f*, Einzellast *f* in
Feldmitte
f (sur)charge *f* appliquée au point
central (*p.ex. d'une poutre*)
nl centrale puntlast *m*
r сосредоточенная нагрузка *f*,
приложенная в середине пролёта
(*напр. балки*)

C319 *e* **central refrigeration station**
d zentrale Kältestation *f*
f station *f* frigorifique centrale
nl centraal koelstation *n*
r холодильный центр *m*;
центральная холодильная станция *f*

C320 *e* **central reservation**
d Mittelstreifen *m*
f terre-plein *m* central
nl middenberm *m*
r центральная разделительная
полоса *f* (*дороги*)

C321 *e* **central shaft**
d Sammelbrunnen *m*
f puits *m* de captage central
nl centrale schacht *f(m)*
r водосборный колодец *m* (*шахтный
колодец в центре группы трубчатых
колодцев*)

C322 *e* **central warm-air heating system**
d zentrale Luftheizungsanlage *f*
f système *m* central(isé) de chauffage
à air chaud
nl centraal luchtverwarmingssysteem *n*
r центральная система *f* воздушного
отопления

C323 *e* **centrifugal compressor**
d Kreiselradverdichter *m*
f compresseur *m* centrifuge
nl centrifugaalcompressor *m*
r центробежный компрессор *m*

C324 *e* **centrifugal dust collector**
d Zentrifugalentstauber *m*, Fliehkraft-
-Abscheider *m*
f dépoussiéreur *m* à (tube) cyclone,
collecteur *m* de poussières cyclone,
cyclone *m* dé poussiéreur
nl centrifugale stofvanger *m*
r центробежный пылеуловитель *m*

C325 *e* **centrifugal fan**
d Radiallüfter *m*, Radialventilator *m*,
Kreisellüfter *m*, Zentrifugalgebläse *n*
f ventilateur *m* centrifuge
nl centrifugaalventilator *m*
r радиальный вентилятор *m*

C326 *e* **centrifugally cast concrete**
d Schleuderbeton *m*
f béton *m* centrifugé
nl gecentrifugeerd beton *n*
r центрифугированный бетон *m*

C327 *e* **centrifugally-cast pipe**
d Schleuderbetonrohr *n*
f tuyau *m* (de béton) centrifugé
nl buis *f(m)* van gecentrifugeerd beton
r центрифугированная железобетонная
труба *f*

C328 *e* **centrifugal pump**
d Kreiselpumpe *f*
f pompe *f* centrifuge
nl centrifugaalpomp *f(m)*
r центробежный насос *m*

C329 *e* **centrifugal scrubber, centrifugal
washer**
d Zentrifugalwäscher *m*, Fliehkraft-
wäscher *m*
f scrubber *m* [laveur *m* de gaz]
centrifuge
nl centrifugale gasreinigingsinstallatie *f*
r центробежный скруббер *m*

C330 e **centroidal axis**
    d Zentralachse *f*, Schwerachse *f*
    f axe *m* central [de gravité]
    nl centroïdale as *f(m)*
    r центральная ось *f*

C331 e **centroid of a section**
    d Schwerpunkt *m*, Massenmittelpunkt *m*
    f centre *m* de gravité de la section
    nl zwaartepunt *n* van een sectie
    r центр *m* тяжести сечения

C332 e **ceramic drain**
    d Röhrendrän *m*
    f drain *m* céramique [en terre-cuite]
    nl keramische drain *m*
    r гончарная дрена *f*

C333 e **ceramic fibre**
    d keramische Faser *f*
    f fibre *f* céramique
    nl keramische vezel *f(m)*
    r керамическое волокно *n*

C334 e **ceramic materials**
    d Baukeramik *f*, keramische Baustoffe
      m pl
    f matériaux *m pl* céramiques (pour la
      construction)
    nl keramische (bouw)materialen *n pl*
    r строительные керамические
      материалы *m pl*

C335 e **ceramic mosaic** *US*
    d venezianischer Estrich *m*,
      Mosaikterrazzo *n*
    f carreaux *m pl* (de sol) mosaïques en
      céramique
    nl terrazzovloer *m*
    r керамические мозаичные плитки *f pl*
      для полов

C336 e **ceramics**
    d Baukeramik *f*, keramische
      Erzeugnisse *n pl*
    f céramique *f* de bâtiment
    nl keramische produkten *n pl*
    r керамические изделия *n pl*,
      строительная керамика *f*

C337 e **ceramic tile**
    d Keramikfliese *f*, Keramikplatte *f*,
      keramische Fliese *f* [Platte *f*]
    f carreau *m* (en) céramique
    nl keramische tegel *m*
    r керамическая плитка *f*

C337a **CBR** *see* **California bearing ratio**

C338 e **ceramic veneer** *US*
    d keramische Wandfliesen *f pl*
      [Verkleidungstafeln *f pl*]; Verkleidung
      f mit Keramiktafeln,
      Keramikfliesenbelag *m*
    f carreau *m* céramique; carrelage *m*
      céramique
    nl keramische wandtegels *m pl*
    r керамические облицовочные плиты *f*
      pl [плитки *f pl*]; облицовка *f*
      керамическими плит(к)ами

C339 e **ceramic wall**
    d Wand *f* aus Keramikblöcken

    f mur *m* en blocs céramiques
    nl wand *m* van keramische blokken
    r стена *f* из керамических блоков

C340 e **certificate of approval**
    d Billigungsbescheinigung *f*
    f certificat *m* (d'agrément)
    nl goedkeuringscertificaat *n*
    r сертификат *m*, удостоверяющий
      качество материалов и изделий

C441 e **cess**
    d Einschnittsböschungsfuß-Rohrdrän *m*
    f drain *m* à tuyaux au fond de la
      fouille
    nl put *m* voor bronbemaling
    r трубчатая дрена *f* у дна выемки

C342 e **cesspit, cesspool**
    d Senkgrube *f*, Fäkaliengrube *f*
    f fosse *f* d'aisance
    nl zinkput *m*
    r выгребная яма *f*

C343 e **chain**
    d 1. Kette *f* 2. Meßkette *f*,
      Vermessungskette *f*
    f 1. chaîne *f* 2. chaîne *f* d'arpenteur
    nl 1. ketting *f* 2. meetketting *f*
    r 1. цепь *f* 2. мерная цепь *f*

C345 **chain book** *see* **field book**

C346 e **chain-bucket dredge, chain-bucket**
    dredger
    d Eimerketten-Naßbagger *m*
    f drague *f* à godets
    nl emmerbaggermolen *m*
    r многоковшовый землечерпательный
      снаряд *m*

C349 e **chain hoist**
    d Schraubenflaschenzug *m*,
      Schneckenkettenzug *m*
    f palan *m* à chaîne
    nl kettingtakel *m*
    r ручная цепная таль *f*

C350 e **chain lever hoist**
    d Handkettenwinde *f*
    f treuil *m* à manivelle [à bras, à main,
      manuel] à chaîne
    nl handtakel *m*
    r цепная рычажная лебёдка *f*

C351 e **chain mortiser**
    d Kettenfräsmaschine *f (Holzbearbeitung)*
    f chaîne *f* à mortaiser, mortaiseuse *f*
      à chaîne
    nl kettingfreesmachine *f*
    r цепной долбёжник *m* (*деревообра-*
      батывающий инструмент)

C352 e **chain of locks**
    d Schleusentreppe *f*, Koppelschleuse *f*
    f écluses *f* étagées [superposées]
    nl sluizentrap *m*
    r многоступенчатый
      [многокамерный] шлюз *m*

C353 e **chain of power plants**
    d Wasserkraftwerkkette *f*,
      Ausbautreppe *f*
    f chaîne *f* de centrales hydrauliques

nl keten *f(m)* van waterkrachtcentrales
r каскад *m* гидроэлектростанций

C354 *e* **chain saw**
d Handkettengsäge *f*
f scie *f* à chaîne
nl kettingzaag *f(m)*
r цепная пила *f*

C355 *e* **chain sling**
d Kettenschlinge *f*
f élingue *f* en chaîne
nl kettingstrop *m*
r цепной строп *m*

C356 *e* **chain survey**
d Kettenvermessung *f*
f chaînage *m*
nl kettingmeting *f*
r измерение *n* [промер *m*] длин,
измерение *n* мерной цепью

C357 *e* **chain tongs**
d Ketten(rohr)zange *f*
f pince *f* (à tubes) à chaîne, clé *f*
[clef *f*] à chaîne
nl kettingpijptang *f(m)*
r цепной трубный ключ *m*

C358 *e* **chain suspension bridge**
d Ketten(hänge)brücke *f*
f pont *m* suspendu à chaînes
nl kettingbrug *f(m)*
r висячий цепной мост *m*

C359 **chain trencher** *see* **ladder trencher**

C360 *e* **chamber**
d Kammer *f*
f chambre *f*
nl 1. kamer *f(m)*; vertrek *n* 2. sluiskolk
m
r камера *f*; отсек *m*; комната *f*;
помещение *n*

C361 *e* **chambering**
d Kesselschießen *n*, Vorkesseln *n*,
Auskesseln *n* eines Bohrloches
f chambrage *m*, pochage *m* (*de trous
de mine*)
nl het boorgat doorschieten
r прострелка *f* (*скважин, шпуров*)

C362 **chamber lock** *see* **single-lift lock**

C363 *e* **chamfer**
d Abfasung *f*; Abschrägung *f*; Fase *f*;
Schräge *f*
f chanfrein *m*
nl schuinte *f*; afschuining *f*
r скос *m*; фаска *f*; срез *m* (*кромки*)
под углом

C364 *e* **chamfered angle**
d gebrochene Kante *f*
f arête *f* chanfreinée
nl afgeschuinde hoek *m*
r скошенная кромка *f*, фаска *f*

C365 *e* **change in length**
d Längenänderung *f*
f changement *m* de longueur
nl lengteverandering *f*
r изменение *n* длины

C366 *e* **change of gradient**
d Wechselpunkt *m*
f point *m* de changement (de
déclivité)
nl knik *m* in de helling
r точка *f* перелома профиля

C367 *e* **change of stress state**
d Spannungszustandsänderung *f*
f changement *m* de l'état de contrainte
[de tension]
nl verandering *f* van spanningstoestand
r изменение *n* напряжённого состояния

C368 *e* **change-over system**
d umschaltbares System *n*,
Umschaltanlage *f*
f système *m* d'air climatisé à
commutateur de fonctions
nl omschakelsysteem *n*
r система *f* кондиционирования
воздуха с переключением режимов
работы

C369 *e* **channel**
d 1. Stromrinne *f*, Gerinne *n*; Kanal
m; Wasserlauf *m*; Fahrrinne *f*,
Fahrwasser *n* 2. Rinne *f*;
Ablaufgerinne *n* 3. Leitungskanal *m*
4. U-Stahl *m*, U-Träger *m*
f 1. chenal *m*; canal *m*; cours *m* d'eau
2. caniveau *m* 3. voie *f* de
(télé)communication 4. profilé *m*
en U
nl 1. vaargeul *f(m)*; kanaal *n*; waterloop
m; stroombed *n* 2. groef *f(m)*,
kilgoot *f(m)* 3. leidinggoot *f(m)*
4. U-staal *n*
r 1. русло *n*; канал *m*; водоток *m*;
фарватер *m* 2. борозда *f*;
водосточный лоток *m*
3. коммуникационный канал *m*
4. швеллер *m*

C370 *e* **channel beam**
d U-Profil-Träger *m*
f poutre *f* en U
nl U-profiel *n*, U-balk *m*
r швеллер *m*, швеллерная балка *f*;
швеллерный прогон *m*

C371 *e* **channel capacity**
d Abflußvermögen *n*
f capacité *f* de débit d'un canal
nl afvoervermogen *n* van een waterloop
r пропускная способность *f* русла

C372 *e* **channel check**
d Querregler *m*
f régulateur *m* d'écoulement (*des eaux
en canal d'irrigation*)
nl regelschot *n* (*bevloeiing*)
r регулятор *m* (*на оросительном канале*)

C373 *e* **channel excavation**
d Kanalaushub *m*
f excavation *f* [creusement *m*] de canal
nl (een) kanaal graven
r отрывка *f* канала, выемка *f* под
канал

C374 *e* **channel flow**
  d Flußbettabfluß *m*
  f écoulement *m* en canal
  nl afvoer *m* van een waterloop
  r русловой сток *m*]

C375 *e* **channel hydraulics**
  d Gerinnenhydraulik *f*
  f hydraulique *f* des canaux découverts,
    hydraulique *f* fluviale
  nl hydraulica *f* van open leidingen
  r гидравлика *f* открытых русел
    [руслового потока]

C376 *e* **channel lines** US
  d Schiffahrtstiefengrenze *f*
  f limite *f* de la profondeur navigable
    d'un chenal
  nl bevaarbaarheidsgrens *f(m)*
  r граница *f* судоходных глубин

C377 *e* **channel of approach**
  d Zuführungskanal *m*, Zuleitungskanal
    m, Zufahrtrinne *f*
  f canal *m* d'amenée [d'adduction,
    adducteur]
  nl aanvoerkanaal *n*
  r подводящий канал *m*

C378 *e* **channel race**
  d Triebwasserkanal *m*
  f canal *m* d'amenée
  nl aan- *of* afvoerleiding *f* van een
    waterkrachtturbine
  r подводящий канал *m* ГЭС

C379 *e* **channel revetment**
  d Auskleidung *f*, Befestigung *f*,
    Böschungsverkleidung *f*
  f revêtement *m* de canal [d'étanchéité
    pour canal]
  nl kanaalbeschoeiing *f*
  r одежда *f* канала

C380 *e* **channel roughness**
  d Rauhigkeitsbeiwert *m*,
    Gerinnerauhigkeit *f*,
    Bettrauhigkeit *f*
  f rugosité *f* de lit de canal
  nl oneffenheid *f* van de bodem
  r коэффициент *m* шероховатости русла

C381 **channel runoff** *see* **channel flow**

C382 *e* **channels**
  d Fahrspuren *f* pl]
  f ornières *f* pl, frayés *m* pl, traces *f* pl
    de roues
  nl rijsporen *n* pl
  r колеи *f* pl, борозды *f* pl (*в дорожном
    покрытии*)

C383 *e* **channel scour**
  d Flußbetterosion *f*
  f affouillement *m* du lit d'un cours
    d'eau, érosion *f* par ravinement
  nl erosie *f* van het rivierbed]
  r русловая эрозия *f*

C384 *e* **channel section**
  d U-Profil *n*
  f section *f* en U; profilé *m* en U

nl U-profiel *n*
  r швеллерное сечение *n*; швеллер *m*

C385 *e* **channel type spillway**
  d Hochwasserentlastungsanlage *f*,
    Überlaufkanal *m*, Überströmkanal *m*
  f canal *m* d'évacuation
  nl ontlastkanaal *n* (*van stuwmeer*)
  r водосбросный канал *m*

C386 *e* **characteristic concrete cube strength**
  d normativer Betonwürfelwiderstand *m*
  f résistance *f* caractéristique sur cube
  nl normale betonkubusweerstand *m*
  r нормативная кубиковая прочность *f*
    бетона

C387 *e* **characteristic curve**
  d Kennlinie *f*, charakteristische Kurve *f*
  f courbe *f* caractéristique [de
    performance]
  nl karakteristiek *f*
  r характеристическая кривая *f*

C388 *e* **characteristic load**
  d Regellast *f*, Normlast *f*, normative
    Last *f*
  f charge *f* caractéristique
  nl normale belasting *f*
  r нормативная нагрузка *f*

C389 *e* **characteristic properties**
  d charakteristische Eigenschaften *f* pl;
    Klassifizierungseigenschaften *f* pl
  f propriétés *f* pl caractéristiques
  nl kenmerkende eigenschappen *f* pl
  r характерные свойства *n* pl;
    классифицирующие признаки *m* pl

C390 *e* **characteristics**
  d Charakteristik *f*, Baumerkmale *n* pl
  f caractéristiques *f* pl
  nl bedrijfskarakteristieken *f* pl
  r технико-эксплуатационные
    характеристики *f* pl (*оборудова-
    ния*)

C391 *e* **characteristic strength**
  d Nennfestigkeit *f*, normativer
    Widerstand *m*
  f résistance *f* caractéristique
  nl normale sterkte *f*
  r нормативное сопротивление *n*

C392 *e* **charge**
  d Beschickung *f*; Füllung *f*; Ladung *f*
  f chargement *m*; charge *f*
  nl charge *f(m)*; lading *f* (*installatie*)
  r порция *f*; количество *n* (*напр.
    вещества для заправки системы*);
    заряд *m*

C393 *e* **charging hopper**
  d Beschickungskübel *m*,
    Kippkübel *m*
  f benne *f* de chargement
  nl kiepbak *m*
  r загрузочный ковш *m*

C394 *e* **Charpy impact machine**
  d Charpyscher Pendelschlaghammer *m*
  f mouton-pendule *m* (de) Charpy
  nl pendelslaghamer *m* van Charpy

r копёр *m* Шарпи (*для определения ударной вязкости*), маятниковый копёр

C395 e **Charpy test**
d Kerbschlagversuch *m* nach Charpy
f essai *m* Charpy
nl kerfslagproef *f(m)* van Charpy
r определение *n* ударной вязкости по Шарпи

C396 e **chart**
d 1. Rechentafel *f*, Berechnungstafel *f* 2. Diagramm *n*, Schaubild *n*
f 1. table *f* de calcul, barème *m*, tableau *m* barème 2. abaque *f*; diagramme *m*; schéma *m*
nl 1. tabel *f(m)* 2. diagram *n*, grafiek *f*
r 1. расчётная таблица *f* 2. диаграмма *f*; схема *f*; график *m*; номограмма *f*

C397 e **chase**
d Rille *f*; Mauerschlitz *m*; eingemeißelte Rinne *f* (*zur Rohrverlegung*)
f rainure *f*; goujure *m* (*pour la pose des tuyaux de canalisation dans les murs*)
nl leidingsleuf *f(m)* (*in de muur of de groef*)
r борозда *f*; паз *m*; узкий канал *m* (*для прокладки инженерных сетей в зданиях*)

C398 e **chasing**
d Schlitzestemmen *n*
f mortaisage *m* de rainures
nl sleuven hakken *n*
r долбление *n* [выборка *f*] пазов

C399 e **check**
d 1. Kontrolle *f*, Überprüfung *f*, Nachprüfung *f* 2. Sternriß *m*, radialer Treibriß *m* (*im Holz*); durchgehender Riß *m* (*im Farbüberzug, Beton*) 3. Stauschleuse *f*, Einlaßregler *m* 4. Stauabteilung *f*, eigedeichter Teil *m* eines Bewässerungsfeldes
f 1. contrôle *m*, vérification *f* 2. gerce *f* (radiale), gerçure *f* (*à la surface de peinture, de béton*) 3. régulateur *m*; ouvrage *m* de régula(risa)tion (*sur canal*) 4. bassin *m* d'arrosage [de submersion]
nl 1. controle *f(m)* 2. hartscheur *f(m)* {*hout*}; barst *m* in de verflaag 3. aanslag(dorpel) *m* van sluisdeur 4. bevloeid perceel *n*
r 1. контроль *m*, проверка *f* 2. радиальная трещина *f* (*в бревне, брусе*); поверхностная трещина *f* (*в окрасочном покрытии, бетоне*) 3. регулятор *m* (*на канале*); шлюз--регулятор *m* 4. чек *m*, обвалованный орошаемый участок *m*

C400 e **check and drop**
d Regulierungsabsturz *m*, Kontrollabsturz *m*
f chute *f* fonctionnant comme régulateur
nl regelende overlaat *m*
r регулирующий перепад *m*

C401 e **check calculation**
d Nachrechnung *f*, Nachrechnen *n*
f calcul *m* de contrôle
nl controleberekening *f*
r поверочный расчёт *m*

C402 e **check gate**
d Regulierverschluß *m*, Schleusenkopf *m*
f vanne *f* de réglage [commandée, dirigée]
nl puntdeur *f(m)* (*sluis*)
r регулирующий затвор *m*

C403 e **checklist**
d Kontrolleliste *f*
f liste *f* de contrôle (*p.ex d'instruments, de matériaux*)
nl controlelijst *f(m)*
r контрольный перечень *m* [список *m*] (*напр. материалов, инструментов*)

C404 **check post** see **bollard** 1.

C405 e **check rail**
d 1. Zwangsschiene *f*, Leitschiene *f* 2. Führungsleiste *f* des versenkbaren Fensters
f 1. contre-rail *m*, garde-rail *m* 2. rail *m* de guidage (d'une fenêtre à guillotine)
nl 1. tegenrail *f(m)*, veiligheidsrail *f(m)* 2. geleiding *f*
r 1. контррельс *m* 2. направляющая *f* (опускного окна)

C406 e **check test**
d Gegenprobe *f*, Gegenprüfung *f*, Gegenversuch *m*
f essai *m* de contrôle, contre-essai *m*, contre-épreuve *f*
nl controleproef *f(m)*
r контрольное [проверочное] испытание *n*

C407 e **check throat**
d Wassernase *f*, Tropfbrett *n*, Tropfleiste *f*
f larmier *m*
nl waterhol *n*
r слезник *m* (*под подоконником или порогом*)

C408 e **check valve**
d Rückschlagventil *n*
f clapet *m* anti-retour [de retenue, de non-retour]
nl terugslagklep *f(m)*
r обратный клапан *m*

C409 e **cheek**
d Wange *f*; Seitenwand *f*, Seitenkante *f*
f joue *f*; paroi *f* latérale
nl wang *f(m)*; zijkant *m*
r щека *f*; боковая грань *f*

C410 e **chemical gauging**
d Fließgeschwindigkeitsmessung *f* nach dem Konzentrationsverfahren, chemische Messung *f*
f jaugeage *m* chimique [par titration]
nl chemische hydrometrie *f*
r химическая гидрометрия *f*

C411 *e* **chemical grouting**
   d chemische Injektion *f* [Verfestigung *f*]
   f injection *f* chimique, stabilisation *f*
   (de sols) chimique
   nl chemische grondstabilisatie *f*
   r химическое укрепление *n* грунта

C412 *e* **chemical oxygen demand**
   d chemischer Sauerstoffbedarf *m*
   f demande *f* chimique d'oxygène
   nl scheikundige behoefte *f* aan zuurstof
   r химическая потребность *f*
   в кислороде

C413 *e* **chemical precipitation**
   d chemische Klärung *f*
   f dépot *m* [précipitation *f*] chimique
   nl uitvlokking *f* met behulp van
   (scheikundige) reagentia
   r химическое осаждение *n*

C414   **chemical stabilization** *see* **chemical**
   grouting

C415 *e* **chemical water treatment**
   d chemische Wasserbehandlung *f*
   [Wasseraufbereitung *f*]
   f épuration *f* [traitement *m*] chimique
   (*des eaux potables*)
   nl chemische waterbehandeling *f*
   r химическая обработка *f* воды,
   химическая водоподготовка *f*

C416 *e* **chemise**
   d Futtermauer *f*, Böschungsmauer *f*
   f mur *m* de talus; revêtement *m* de
   talus
   nl bekledingsmuur *m*
   r наклонная подпорная стенка *f* для
   защиты откоса; облицовка *f*
   земляного берега

C417 *e* **chemistry of concrete**
   d Betonchemie *f*
   f chemie *f* du béton
   nl betonchemie *f*
   r химия *f* бетона

C418 *e* **chequer plate**
   d Riffel(belag)platte *f*
   f plaque *f* striée
   nl ruitplaat *f(m)*
   r рифлёная плита *f* настила

C419 *e* **cherry picker**
   d 1. Bockderrik *m* 2.
   Wagenwechselanlage *f* (*Tunnelbau*)
   f 1. petit derrick *m* de montage
   2. poutre *f* roulante dans un tunnel
   ferroviaire
   nl 1. lichte hoogwerker *m*
   2. spoorwisselkraan *f(m)*
   r 1. (небольшой) деррик-кран *m*
   с А-образной стрелой 2. мостовой
   кран *m* [кран-балка *f*] в
   железнодорожном тоннеле

C420 *e* **chevron drain**
   d Fischgrätendrän *m*
   f système *m* de drainage en arête de
   poisson
   nl visgraatdrainage *f*

   r система *f* дренажа с расположением
   дрен «ёлочкой», древовидный
   дренаж *m*

C421 *e* **chilled water**
   d gekühltes Wasser *n*
   f eau *f* réfrigérée
   nl gekoeld water *n*
   r охлаждённая вода *f*

C422 *e* **chilling**
   d 1. mechanische Kühlung *f*,
   Maschinenkühlung *f* 2. Anlaufen *n*,
   Weißanlaufen *n*, Weißwerden *n*
   (*von Lacken*)
   f 1. réfrigération *f* mécanique 2. trouble
   m; formation *f* des taches troubles
   nl 1. geforceerde koeling *f* 2. dofworden
   n van verse vernislaag
   r 1. машинное охлаждение *n* (*воды*)
   2. помутнение *n* (*лакового или кра-*
   сочного покрытия)

C423   **chilling water** *see* **cooling water**

C424 *e* **chimney**
   d Schornstein *m*, Kamin *m*;
   Rauchkanal *m*
   f cheminée *f*
   nl schoorsteen *m*
   r дымовая труба *f*; дымоход *m*

C425 *e* **chimney back**
   d Hinterboden *m* des Kamins *oder*
   Feuerraumes
   f paroi *f* arrière de la cheminée
   nl achterplaat *f(m)* van de vuurhaard
   r задняя стена *f* камина *или* топки
   печи

C426 *e* **chimney block** *US*
   d Kamin(form)stein *m*
   f boisseau *m*
   nl schoorsteenblok *n*
   r блок *m* (*керамический или*
   бетонный) дымохода

C427 *e* **chimney breast**
   d Kaminvorsprung *m*
   f chambranle *m* de cheminée
   nl schoorsteenboezem *m*
   r передняя выступающая часть *f*
   камина *или* печи с топкой

C428 *e* **chimney cap**
   d Schornsteinkopf *m*
   f chapiteau *m* de cheminée
   nl schoorsteenkap *f(m)*
   r оголовок *m* дымовой трубы

C429 *e* **chimney drain**
   d Kaminentwässerung *f*, Kamindrän *m*
   f drain *m* cheminée
   nl verticale drainage *f*
   r вертикальный дренаж *m* (*насыпи,*
   плотины), вертикальная дрена *f*

C430 *e* **chimney effect**
   d Zugwirkung *f*, Schornsteinwirkung *f*
   f tirage *m* de cheminée; effet *m* de
   tirage
   nl schoorsteentrek *m*; trek-effect *n*
   r тяга *f* в дымовой трубе; эффект *m*

тяги (*напр. в лестничной клетке многоэтажного здания*)

C431 *e* **chimney flue**
  d 1. Schornstein *m* 2. Rauchkanal *m*, Fuchs *m*
  f 1. conduit *m* de fumée vertical 2. conduit *m* de fumée; carneau *m*
  nl rookkanaal *n*
  r 1. дымовая труба *f* 2. дымовой канал *m*

C432   **chimney hood** *see* **chimney cap**

C433 *e* **chimneys brick**
  d Schornsteinziegel *m*
  f brique *f* pour cheminées
  nl baksteen *m* voor schoorstenen
  r кирпич *m* для кладки дымовых труб

C434 *e* **chimney shaft**
  d Schornsteinschaft *m*
  f fût *m* de cheminée
  nl schoorsteenschacht *f(m)*
  r ствол *m* свободностоящей дымовой трубы

C435 *e* **chimney stack**
  d Dachschornstein *m*
  f cheminée *f* de maison
  nl schoorsteen *m*
  r дымовая труба *f* дома (*часть дымовой трубы над крышей здания*)

C436 *e* **china sanitary ware**
  d Sanitärkeramik *f*
  f appareils *m pl* sanitaire en faïence
  nl keramisch sanitair *n*
  r санитарная керамика *f*

C437   **chip** *see* **chippings**

C438 *e* **chipboard**
  d Spanplatte *f*
  f panneau *m* de [en] copeaux
  nl spaanplaat *f(m)*
  r древесностружечная плита *f*

C439 *e* **chippings**
  d Splitt *m*, gebrochener Kies *m*
  f gravillon *m* concassé [de concassage]
  nl split *m*
  r (мелкий) щебень *m*; высевки *pl*; каменная мелочь *f*

C440 *e* **chippings spreader**
  d Splittstreumaschine *f*
  f répandeur *m* [répandeuse *f*] de gravillon
  nl splitstrooier *m*
  r распределитель *m* каменной мелочи [высевок]

C441 *e* **chisel**
  d Meißel *m*, Stechbeitel *m*
  f bédane *m*, bec *m* d'âne
  nl (steek)beitel *m*
  r стамеска *f*; долото *n*

C442 *e* **chlorinated rubber paints**
  d Chlorkautschuk(anstrich)farben *f pl*
  f peintures *f pl* au chlorocaoutchouc
  nl verf *f* (*m*) op basis van gechloreerde rubber
  r хлоркаучуковые окрасочные составы *m pl* [краски *f pl*]

C443 *e* **chlorination**
  d Chlor(ier)ung *f*, Chloren *n*
  f chlor(ur)ation *f*, chlorage *m*
  nl chloreren *n*
  r хлорирование *n*

C444 *e* **chlorination plant**
  d Chlorieranlage *f*, Chlorungsanlage *f*
  f installation *f* de chlorage [de chloration, de traitement au chlore]
  nl chloreerfabriek *f*
  r хлораторная установка *f*

C445 *e* **chlorinator**
  d Chlorator *m*, Chlorzusatzgerät *n*
  f chlorateur *m*
  nl chloreertoestel *n*
  r хлоратор *m*

C446   **chock vibrator** *see* **jolting vibrator**

C447 *e* **choker hitch**
  d Schlingstropp *m*, Schlingkette *f*
  f élingue *f* (*en chaîne ou en câble d'acier*) à nœud coulant
  nl kettingstrop *m*, (kabel)strop *m*
  r петлевой строп *m* (*канатный или цепной*)

C448 *e* **chord**
  d 1. Gurt *m* 2. Sehne *f*
  f 1. membrure *f* (*d'une poutre*) 2. corde *f*
  nl 1. (boven- of onder)rand *m* van een vakwerk 2. koorde *f*
  r 1. пояс *m* (*балки, фермы*) 2. хорда *f*

C449 *e* **chrome green**
  d Chromgrün *n*
  f vert *m* de chrome
  nl chromaatgroen *n*
  r зелёный хром *m* (*пигмент*)

C450 *e* **chrome yellow**
  d Chromgelb *n*
  f jaune *m* de chrome
  nl chromaatgeel *n*
  r жёлтый хром *m*, жёлтая хромовая *f* (*пигмент*)

C451 *e* **chute**
  d 1. Schußrinne *f* 2. Rinne *f*; Rutsche *f* 3. Schüttkasten *m*, Schütt-Trichter *m* 4. Gefälleleitung *f*
  f 1. chute *f*, canal *m* à forte pente 2. canal *m* jaugeur 3. trémie *f* de chargement 4. conduite *f* gravitaire
  nl 1. leiding *f* met groot verval 2. (stort)goot *f* (*m*) 3. stortkoker *m* 4. vultrechter *m*, hopper *m*
  r 1. быстроток *m* 2. лоток *m*; спускной жёлоб *m* 3. загрузочный бункер *m*; воронка *f* 4. самотёчный водовод *m*

C452   **cill** *see* **sill**

C453   **ciment fondu** *see* **high-alumina cement**

C454 *e* **cinder block** *US*
  d Schlackenstein *m*

f bloc *m* [parpaing *m*] de béton de laitier
nl slakkenbetonsteen *m*
r шлакоблок *m*, шлакобетонный камень *m*

C455 e **Cipoletti weir**
d Trapezwehr *n*, Cipoletti-Wehr: Meßwehr *n* nach Cipoletti
f déversoir *m* Cipoletti, déversoir *m* trapézoïdal
nl meetschot *n* van Cipoletti
r трапецеидальный водослив *m*, водослив *m* Чиполетти

C456 e **circle of influence**
d Absenkungskreis *m*
f périmètre *m* d'appel circulaire
nl invloedsfeer *f* (*m*)
r зона *f* влияния (*скважины, колодца*); окружность *f* зоны влияния

C457 e **circle of stress**
d Spannungskreis *m*
f cercle *m* des contraintes [d$_e$ Mohr]
nl cirkel *m* van Mohr
r круг *m* (напряжений) Мора

C458 e **circuit**
d 1. Kreisleitung *f*, Leitungskreis *m*, Zirkulationsschleife *f*; Kreislauf *m* 2. Stromkreis *m*
f 1. circuit *m* d'eau 2. circuit *m* de courant [électrique]
nl 1. kringloop *m* 2. schakeling *f*, stroomketen *f* (*m*)
r 1. трубопроводное кольцо *n*; контур *m* циркуляции 2. электрическая цепь *f*

C459 e **circular-arc method**
d Gleitkreisverfahren *n*
f méthode *f* de cercle de frottement [glissement]
nl glijcirkelmethode *f*
r метод *m* расчёта откосов на скольжение по цилиндрической поверхности

C460 e **circular cell cofferdam**
d Zylinderzellenfangdamm *m*, Kreiszellenfangdamm *m*
f batardeau *m* en cellules circulaires
nl uit cilindrische cellen opgebouwde cofferdam *m*, ringcellencofferdam *m*
r ячеистая цилиндрическая перемычка *f*, перемычка *f* с цилиндрическими ячейками

C461 e **circular cofferdam**
d Ring(fang)damm *m*, Kreis(fang)damm *m*
f batardeau *m* circulaire
nl cirkelvormige cofferdam *m*
r кольцевая (*круглая в плане*) перемычка *f* [дамба *f*]

C462 e **circular frequency**
d Kreisfrequenz *f*
f fréquence *f* circulaire

nl cirkelfrequentie *f*
r круговая [угловая] частота *f*

C463 e **circular saw**
d Kreissäge *f*
f scie *f* circulaire
nl cirkelzaag *f*(*m*)
r дисковая пила *f*

C464 e **circular settling tank**
d Rund(klar)becken *n*, radialdurchflossenes Absetzbecken *n*
f décanteur *m* radial
nl ronde bezinktank *m*
r радиальный отстойник *m*

C465 **circular stair** *see* **spiral stair**

C466 e **circular tunnel**
d Rohrtunnel *m*, Kreistunnel *m*
f tunnel *m* circulaire
nl tunnel *m* met ronde doorsnede
r туннель *m* круглого сечения; туннель-труба *m*

C467 **circular type cellular cofferdam** *see* **circular cell cofferdam**

C468 e **circulating space heater**
d Umlufterhitzer *m*, Zirkulationsluftheizgerät *n*
f installation *f* de chauffage à circulation par air chaud
nl convectieluchtverhitter *m*
r циркуляционный агрегат *m* воздушного отопления

C469 e **circulating system**
d Zirkulationssystem *n*, Umlaufsystem *n*
f système *m* de circulation
nl circulatiesysteem *n*
r циркуляционная система *f*

C470 e **circulating water**
d Umlaufwasser *n*
f eau *f* de circulation
nl circulatiewater *n*
r циркулирующая вода *f*

C471 e **circulation cooling**
d Umlaufkühlung *f*
f système *m* de refroidissement à circulation
nl omloopkoeling *f*
r циркуляционная [оборотная] система *f* охлаждения

C472 e **circulation heating**
d Umlaufheizung *f*, Umwälzheizung *f*
f chauffage *m* à circulation
nl circulatieverwarming *f*
r циркуляционная система *f* отопления

C473 e **circulation pump**
d Umlaufpumpe *f*, Zirkulationspumpe *f*, Umwälzpumpe *f*
f pompe *f* de circulation
nl circulatiepomp *f*(*m*)
r циркуляционный насос *m*

C474 e **circulation ratio**
d Umwälzzahl *f*, Umlaufzahl *f*
f rapport *m* de circulation

nl mate *f(m)* van circulatie
r кратность *f* циркуляции

C475　**circulator pump** *see* **circulation pump**

C476 *e* **circumferential stress**
d Umfangspannung *f*
f contrainte *f* circonférentielle
nl omtrekspanning *f*
r касательное напряжение *n* по окружности, тангенциальное напряжение *n*

C477 *e* **cistern**
d 1. offener Behälter *m* 2. Zisterne *f* 3. Spülkasten *m*
f 1. réservoir *m* ouvert 2. cuve *f* fermée; citerne *f* 3. reservoir *m* de chasse
nl 1. bak *m*, reservoir *n* 2. tank *m* 3. spoelkast *f(m)*
r 1. открытый резервуар *m* 2. закрытый резервуар *m*; цистерна *f* 3. смывной бачок *m*

C478　**city planning** *see* **town planning**

C479 *e* **city water**
d Leitungwasser *n*
f eau *f* de robinet
nl leidingwater *n*
r водопроводная вода *f*

C480 *e* **civil engineer**
d Bauingenieur *m*
f ingénieur *m* civil [des travaux publics]
nl civiel ingenieur *m*
r инженер-строитель *m*

C481 *e* **civil engineering**
d Bauingenieurwesen *n*, Ingenieurbau *m*
f génie *m* civil
nl civiele bouwkunde *f*
r гражданское строительство *n* (*проектирование, строительство и эксплуатация гражданских, транспортных и промышленных зданий и сооружений*)

C482 *e* **cladding**
d 1. Verkleidung *f*, Außenhaut *f*; Umhüllungskonstruktion *f* 2. Plattierungsschicht *f*
f 1. parement *m*; bardage *m*; couverture; revêtement *m* d'étanchéité 2. couche *f* de placage *ou* de rechargement; couche *f* de métal déposé
nl 1. bekleding *f*; omhulsel *n* 2. platering *f*
r 1. наружная обшивка *f*; гидроизолирующие ограждения *n pl* (*стен, кровли*); несущие ограждающие конструкции *f pl* 2. наплавленный слой *m* металла; плакировочный слой *m* металла

C483 *e* **cladding panel**
d 1. Ausfachungstafel *f*

2. Verkleidungsplatte *f*; Umschließungswandplatte *f*
f 1. panneau *m* de remplissage 2. panneau *m* (d'habillage) de façade
nl 1. ommantelingsplaat *f(m)* 2. gevelbekledingspaneel *n*
r 1. панель *f* заполнения (*каркаса, фахверка*) 2. облицовочная панель *f*; панель *f* ограждения

C484 *e* **clad steel**
d plattierter Stahl *m*
f acier *m* plaqué
nl geplateerd staal *n*
r плакированная сталь *f*

C485 *e* **clamping device**
d 1. Schraubzwinge *f* 2. Klemmvorrichtung *f*. Einspannvorrichtung *f*; Spannband *n*, Sickenband *n*, Rohrschelle *f*; Zwinge *f*
f 1. serre-joint *m* 2. bride *f*; dispositif *m* de serrage; étrier *m*; collier *m* de serrage
nl 1. opspaninrichting *f* 2. klem *f(m)*, lijmtang *f(m)*, sergeant *m*
r 1. струбцина *f* 2. зажимное приспособление *n*; бандаж *m*; хомут *m*

C486 *e* **clamping plate**
d Krallenband *n*, Krallenplatte *f*
f crampon *m* denté
nl kramplaat *f(m)*
r когтевая [зубчатая] шпонка *f*

C487 *e* **clamping screw**
d Spannschraube *f*
f vis *f* de serrage
nl klemschroef *f(m)*, spanschroef *f(m)*
r зажимной винт *m*

C488 *e* **clamshell bucket**
d Zweischalengreifkorb *m*
f benne *f* preneuse à deux mâchoires
nl grijper *m*
r двухчелюстной грейферный ковш *m*

C489 *e* **clamshell bucket dredger**
d Greifer-Naßbagger *m*, Greifer- -Schwimmbagger *m*
f drague *f* à mâchoires
nl grijperbaggermolen *m*
r грейферный земснаряд *m*

C490 *e* **clamshell grab**
d Schalengreifer *m*
f benne *f* preneuse
nl baggergrijper *m*
r грейферный ковш *m*

C491　**clapboard** *see* **weather-boarding**

C492　**clap sill** *see* **lock sill**

C493 *e* **clarification**
d Abklärung *f*, Absetzklärung *f*, (mechanische) Klärung *f*, Sedimentation *f*
f clarification *f*, décantage *m*, décantation *f* (*d'eau*)
nl vuil afscheiden *n* door bezinking, klaren *n*

r осветление *n*, механическая
очистка *f* (*воды*)

C494　clarification tank *see* clarifier

C495 *e* clarification time
d Klärzeit *f*, Absetzzeit *f*
f durée *f* de décantation
nl klaringstijd *m*
r время *f* отстаивания [осветления]

C496 *e* clarified sewage
d geklärtes [entschlammtes]
Abwasser *n*
f eaux *f pl* usées traitées [décantées]
nl geklaard afvalwater *n*
r осветлённые сточные воды *f pl*

C497 *e* clarifier
d Schwebefilter *n*, *m*; Klärbecken *n*,
Kläranlage *f*
f décanteur *m*
nl klaringsbak *m*; sedimentatiefilter *n*, *m*
r осветлитель *m*; отстойник *m*

C498 *e* classification of soils
d Bodenklassifikation *f*,
Bodeneinteilung *f*
f classification *f* des sols
nl bodemclassificatie *f*
r классификация *f* грунтов

C499 *e* classified road
d klassifizierte Straße *f* [Autobahn *f*]
f route *f* classée
nl geclassificeerde weg *m*
r автомобильная дорога *f*
определённой категории

C500 *e* classifier
d Klassifikator *m*, Klassierer *m*
f classeur *m*, classificateur *m*
nl classificeerapparaat *n*
r классификатор *m* (*обогатительный
аппарат*)

C501 *e* class of buildings
d Gebäudeklasse *f*
f catégorie *f* de bâtiments [de
constructions]
nl gebouwenklasse *f*
r класс *m* зданий

C502 *e* class of loading
d Lastart *f*
f classe *f* [catégorie *f*] de charge
nl belastingsklasse *f*
r класс *m* нагрузки

C503 *e* claw
d Nagelzieher *m*, Nageleisen *n*
f arrache-clou *m*, tire-clou *m*; pied-
-de-biche *m*
nl spijkertrekker *m*
r гвоздодёр *m*

C504　claw bar *see* pinch bar

C505　claw hammer *see* carpenter's hammer

C506 *e* clay
d Ton *m*; Tonboden *m*, Lehmboden *m*
f argile *f*; sol *m* argileux
nl klei *f(m)*
r глина *f*; глинистый грунт *m*

C507 *e* clay blanket
d Tondichtungsschürze *f*
f masque *m* étanche d'argile
nl klei-afdekking *f* (*dijk*)
r глиняный противофильтрационный
экран *m*

C508 *e* clay concrete
d Tonbeton *m*
f béton *m* d'argile
nl leembeton *n*
r глинобетон *m*

C509 *e* clay core
d Tondichtungskern *m*
f noyau *m* d'argile [argileux]
nl kleikern *f(m)* (*dijk*)
r глиняное противофильтрационное
ядро *n*

C510　clay engineering brick *see* clinker
brick

C512　clay grouting *see* clay sealing

C513 *e* clay lath
d Drahtziegelgewebe *n*
f sous-couche *f* (d'enduit) de toile
métallique avec remplissage en
plaques d'argile
nl steengaas *n*
r основание *n* под штукатурку из
металлической сетки с заполнением
из керамических пластинок

C514 *e* clay mortar
d Lehmmörtel *m*
f mortier *m* d'argile
nl leemmortel *m*
r глиняный раствор *m*

C515 *e* clay pipe
d Tonrohr *n*, Steinzeugrohr *n*
f tuyau *m* de grès (cérame)
nl gresbuis *f(m)*
r гончарная [керамическая] труба *f*

C516 *e* clay puddle
d Lehmschlag *m*, Tonschlag *m*,
Puddle *m*; Lehmmörtel *m*
f mortier *m* d'étanchéité en argile;
enduit *m* en argile
nl aangestampte kleilaag *f(m)*;
leemmortel *m*
r глинобетонная смесь *f*; глиняный
раствор *m* (*для гидроизоляции*)

C517 *e* clay sealing
d Tondichtung *f*, Toninjektion *f*
f injection *f* d'argile
nl afdichting *f* van de bedding met
klei
r глинизация *f* (*русла каналов*)

C518 *e* clay tile
d 1. Tonziegel *m*, Dachziegel *m* aus
Ton 2. keramische Fliese *f*,
Keramikfliese *f* 3. Tondränrohr *n*
f 1. tuile *f* en terre cuite 2. carreau
m céramique 3. drain *m* à tuyau
en grès (cérame), tuyau *m* de
drainage en grès (cérame)

nl 1. gebakken dakpan f(m)
2. keramische tegel m 3. keramische drainagebuis f(m)
r 1. глиняная кровельная черепица 2. керамическая облицовочная плитка f; метлахская плитка f 3. гончарная дрена f, керамическая дренажная труба f

C519 e clayware
d Steinzeug n, Keramikerzeugnis n
f grès m (cérame), produit m céramique
nl keramiek aardewerk n
r керамическое изделие n

C520 e cleandown of concrete surface
d Betonoberflächenreinigung f und -verstrich m
f nettoyage m et lissage de la surface de béton
nl reinigen en dichtstrijken n van het betonoppervlak
r зачистка f и затирка f поверхности бетона

C521 e cleaning door
d Reinigungsluke f, Reinigungstür f
f orifice m de nettoyage
nl schoonmaakluik n
r очистной люк m

C523 e cleanout, cleaning eye
d Reinigungsdeckel m
f bouchon m de nettoyage
nl reinigingsluik n
r очистной лючок m, ревизия f

C524 e clean-out hole
d Reinigungsöffnung f
f trou m de nettoyage
nl reinigingsopening f
r лючок m для прочистки воздуховода

C525 e clean room
d reiner Raum m
f local m [chambre f] à atmosphère pure
nl stofvrije ruimte f
r высокочистое помещение n

C526 e clean timber
d astreines Bauholz n
f bois m de construction sans nœuds
nl timmerhout n zonder kwasten en scheuren
r лесоматериал m или пиломатериал m без сучков

C527 e cleanup work
d Reinigungsarbeiten f pl, Baustellen(auf)räumung f
f nettoyage m de chantier
nl opruimingswerk n
r уборка f [очистка f] строительных объектов или площадок

C528 e clearance
d 1. Spielraum m 2. Außenmaß n, lichter Raum

f 1. jeu m 2. tirant m d'air; espace m libre
nl 1. tussenruimte f, speling f 2. vrije hoogte f, doorvaarthoogte f
r 1. зазор m 2. размер m в свету, габарит m

C529 e clearance diagram
d Lichtraumprofil n; Durchfahrtsprofil
f gabarits m pl de passage [d'espace libre], cotes f pl d'encombrement
nl profiel n van vrije ruimte
r габариты m pl; габарит m приближения строений

C530 e clear distance
d lichte Weite f
f écartement m intérieur
nl spanwijdte f, doorvaartwijdte f
r расстояние n в свету

C531 e clearing and grubbing
d Räumung f, Freilegung f von Baugelände, Räumungsarbeiten f pl (Entfernen aller Hindernisse pflanzlicher Natur)
f nettoyage m et défrichement m du chantier
nl de bouwplaats van begroeiing ontdoen
r расчистка f и раскорчёвка f строительной площадки

C532 e clearing hole
d Bolzenloch n mit Spielraum
f trou m (de boulon) avec jeu
nl boutgat n met speling
r болтовое отверстие n с зазором

C533 e clear overflow weir
d 1. freier [nichteingestauter, vollkommener] Überfall m 2. Überfallwehr n
f 1. déversoir m libre 2. barrage-déversoir m, barrage m déversant
nl stuw m met vrije overstort
r 1. незатопленный [совершенный] водослив m 2. водосливная плотина f

C534 e clear span
d lichte Weite f, freitragende Spannweite f
f portée f [ouverture f] libre
nl vrije overspanning f, spanwijdte f
r пролёт m в свету

C535 e clearstory window
d Ober(licht)fenster n, Hochschiffenster n
f fenêtre f du jour
nl bovenlicht(venster) n
r окно n верхнего света, окно n в верхней части стены

C536 e clear timber
d fehlerfreies Bauholz n
f bois m de construction sans défauts visibles
nl gaaf timmerhout n
r лесо- или пиломатериал m без видимых дефектов

C537 *e* **clear water**
 d Klarwasser *n*
 f eau *f* clarifiée [épurée]
 nl drinkwater *n*
 r осветлённая вода *f*

C538 *e* **clear-water reservoir**
 d Reinwasserbehälter *m*
 f réservoir *m* d'eau potable
 nl drinkwaterreservoir *n*
 r резервуар *m* чистой [очищенной]
 воды; регулирующий
 [распределительный] резервуар *m*

C539 *e* **clearway**
 d 1. Autoschnellstraße *f* 2.
 Straßenabschnitt *m* ohne ruhenden
 Verkehr 3. Fahrrinne *f*, Fahrwasser *n*
 f 1. voie-express *f*, route *f* rapide
 2. voie *f* d'arrêt interdit 3. voie *f*
 navigable; chenal *m*
 nl 1. autosnelweg *m* 2. straat *f(m)*
 zonder langzaam verkeer
 r 1. скоростная (автомобильная)
 дорога *f* 2. участок *m* дороги, где
 запрещена остановка транспорта
 3. фарватер *m*

C540 *e* **cleat**
 d 1. Knagge *f* 2. Hafter *m*
 f 1. tasseau *m*, taquet *m* 2. agrafe *f*
 nl klamp *m*, *f*, klemwig *f(m)*
 r 1. опорный брусок *m* (*вдоль
 стены*); пробка *f* (*в стене для
 крепления деталей*) 2. клямера *f*

C541 *e* **cleats**
 d Leitersprossen *f pl*
 f échelons *m pl*
 nl sporten *f(m) pl* van een ladder
 r перекладины *f pl* приставной
 лестницы *или* стремянки

C542 *e* **cleavage**
 d Aufspaltung *f*, Spaltbarkeit *f*
 f clivage *m*
 nl splijten *n*
 r кливаж *m*

C543 *e* **clenching, clench nailing**
 d Annageln *n* mit Abbiegung der
 Nagelspitze
 f clouage *m* [clouement *m*] avec le
 rabattement des pointes de clous
 nl op elkaar spijkeren *n* en daarna de
 spijkerpunt omslaan *n*
 r прибивка *f* гвоздями с загибанием
 выступающего конца с обратной
 стороны

C544 *e* **clerk of works**
 d Baukontrolleur *m*, Vertreter *m* des
 Bauauftraggebers
 f surveillant *m* des travaux
 nl opzichter *m* van de bouwheer
 r инспектор *m* технадзора;
 представитель *m* заказчика на
 стройплощадке, куратор *m*

C545 *e* **clevis**
 d Klammer *f*, Kausche *f*; Schäkel *m*;
 Gabelkopf *m*

 f boucle *m*, étrier *m*: manille *f*
 nl trekhaak *m*, lasthaak *m*,
 schluitschalm *m*
 r рым *m*, захватная скоба *f*; серьга *f*

C546 *e* **client**
 d Bauauftraggeber *m*, Auftraggeber
 m, Bauherr *m*
 f maître *m* d'ouvrage
 nl bouwheer *m*
 r заказчик *m*

C547 *e* **climatic region**
 d Klimazone *f*, Klimagebiet *n*
 f région *f* [zone *f*] climatique
 nl klimaatzone *f*
 r климатический район *m*

C548 *e* **climatic test chamber**
 d Klimakammer *f*, Klimaraum *m*
 f chambre *f* climatique
 nl klimaatruimte *f*
 r камера *f* искусственного климата,
 климатическая камера *f*

C549 *e* **climatological data**
 d klimatologische Daten *pl*
 f données *pl* climatologiques
 nl klimatologische gegevens *n pl*
 r климатологические данные *pl*

C550 *e* **climbing forms**
 d Kletterschalung *f*; zerleg- und
 umsetzbare Schalung *f*
 f coffrage *m* grimpant
 nl opschuifbare bekisting *f*; klimkist
 f(m)
 r шагающая или разборно-
 -переставная опалубка *f*

C551 *e* **climbing lane**
 d Kriechstreifen *m*, Steigstreifen *m*,
 Verzögerungsstreifen *m*
 f voie *f* de ralentissement
 nl kruipstrook *f*
 r полоса *f* замедленного движения
 (*на подъёмах*)

C552 *e* **clinker**
 d 1. Kesselschlacke *f* 2. Klinker *m*
 f 1. mâchefer *m* 2. clinker *m*
 nl 1. sintel *m*, slak *f(m)* 2. klinker *m*
 r 1. котельный шлак *m* 2. клинкер
 m

C553 **clinker block** *UK see* **cinder block**

C554 *e* **clinker brick**
 d Tiefbauklinker *m*,
 Ingenieurbauklinker *m*
 f brique *f* de clinker [de pavage]
 nl (straat)klinker *m*
 r клинкер *m*, клинкерный кирпич *m*

C555 *e* **clinometer**
 d Neigungsmesser *m*, Klinometer *n*
 f inclinomètre *m*
 nl clinometer *m*, hellingmeter *m*
 r инклинометр *m*, уклономер *m*

C556 *e* **clip**
 d Seilklemme *f*, Spannbügel *m*;
 Spannband *n*, Rohrschelle *f*,
 Sickenband *n*

f serre-câble *m*; collier *m* de
serrage; étrier *m*
nl kabelklem *f(m)*; clip *m*
r сжим *m* (*для каната*); бандаж *m*;
хомут *m*

C557 *e* clip band
d Spannband *n*, Rohrschelle *f*;
Sickenband *n*
f collier *m* de serrage; bandage *m*;
frette *f*
nl ophangband *m*
r бандаж *m*; хомут *m*

C558 clipped gable roof *US see* hipped-
-gable roof

C559 *e* clockwise moment
d rechtsdrehendes Moment *n*
f moment *m* dextrosum
nl rechtsdraaiend moment *n*
r момент *m*, вращающий по часовой
стрелке

C560 *e* clogging
d 1. Kolmation *f*, Kolmatage *f*,
Verschlammen *n* 2. Filterverstopfung
f
f 1. colmatage *m* 2. envasement *m*;
engorgement *m*
nl 1. samenklonteren *n*
2. (filter)verstopping *f*
r 1. заиление *n*, кольматация *f*
2. закупорка *f* [загрязнение *n*,
заиливание *n*] фильтра

C561 *e* clogging of air ducts
d Luftkanalverfilzung *f*,
Luftkanalverstopfung *f*
f engorgement *m* des gaines [des
conduits] d'air
nl verstopping *f* van luchtkanalen
r зарастание *n* воздуховодов

C562 *e* close boarding
d 1. dichte Dachbelattung *f* 2.
Bohlwand *f*
f 1. voligeage *m* (de toit); lattes *f pl*
de toit jointives 2. rideau *m* de
palplanches en bois
nl 1. dakbeschot *n* 2. houten wand *m*
r 1. сплошная обрешётка *f* кровли
2. шпунтовая дощатая стена *f*

C563 *e* closed conduit drop, closed
conduit fall
d geschlossener Absturz *m*,
Absturzschacht *m*
f chute *f* en conduite
nl stortleiding *f*, valleiding *f*
r закрытый перепад *m*; напорный
перепад *m*

C564 *e* closed container
d hermetisches Gefäß *n*,
hermetischer Behälter *m*
[Container *m*]
f container *m* [récipient *m*]
hermétique
nl luchtdichte container *m*
r герметичный сосуд *m* [контейнер
m]; герметичная ёмкость *f*

C565 *e* closed cornice
d Kastengesims *n*
f corniche *f* fermée (à section en
caisson)
nl rondgaande kroonlijst *f(m)*
r закрытый коробчатый карниз *m*

C566 closed drainage *see* subsurface
drainage

C567 *e* closed force polygon
d geschlossenes Kräftepolygon *n*
f polygone *m* de forces fermé
nl gesloten krachtenveelhoek *m*
r замкнутый силовой многоугольник *m*

C568 *e* closed heat-supply system
d Fernwärmeversorgungssystem¹ *n* mit
indirekter Wärmeübergabe für die
Warmwasserbereitung
f système *m* de chauffage urbain à
circuit fermé
nl stadsverwarmingssysteem *n*
r закрытая система *f* теплоснабжения

C569 *e* closed sheeting
d Ausrammung *f*, Einspundung *f*,
Spundwandumschließung *f*
f enceinte *f* de palplanches,
encloisonnement *m* en palplanches
nl gesloten damwand *m*
r замкнутое шпунтовое ограждение *n*

C570 *e* closed system
d geschlossener Kreislauf *m*, Anlage *f*
mit geschlossenem Kreislauf
f système *m* à cycle fermé
nl gesloten systeem *n*
r система *f* с замкнутым циклом,
замкнутая система *f*

C571 *e* closed traverse
d geschlossener Polygonzug *m*,
Ringpolygon *n*
f polygone *m* [cheminement *m*] fermé
nl rondgaande [gesloten] landmeting *f*
r замкнутый полигон *m*,
замкнутый полигонометрический
ход *m*

C572 *e* closed-type steam heating system
d geschlossenes Dampfheizungssystem
n
f système *m* de chauffage à vapeur
à circuit fermé
nl gesloten stoomverwarmingssysteem *n*
r замкнутая система *f* парового
отопления

C573 *e* closer
d Abschluß(spund)bohle *f*
f palplanche *f* serre-file
nl sluit(dam)plank *f(m)*, sluitsteen *m*
r замыкающая шпунтовая свая *f*

C574 *e* closet
d 1. Abort *m*, Klosett *n*, Abtritt *m*
2. Abstellkammer *f*; Einbauschrank *m*
f 1. cabinet *m* d'aisances, water-
-closet *m* 2. placard *m*
nl 1. watercloset *n* 2. bergruimte,
inbouwkast *f(m)*

r 1. уборная f 2. кладовая f;
стенной шкаф m

C575 e **close timbering**
d Bohlwand f
f planches f pl de blindage
nl dichte plankenbeschieting f
r стенка f сплошного крепления
траншеи, выполненная из пластин;
стенка f больверка

C576 e **closing device**
d Türschließer m
f ferme-porte m
nl deursluiter m, deurdranger m
r закрыватель m (двери),
закрывающее устройство n

C577 e **closing error**
d Abschlußfehler m
f écart m de fermeture
nl sluitfout f(m)
r невязка f замыкания

C578 e **closing line**
d 1. Schlußlinie f 2. Schließseil n
f 1. ligne f de fermeture 2. câble f
de fermeture (p. ex. de la benne
preneuse)
nl 1. sluitlijn f(m) 2. sluitingsdraad m
(grijper)
r 1. замыкающая (линия) f
2. закрывающий канат m (напр.
грейферного ковша)

C579 **closing rope** see **closing line** 2.

C580 e **closing stile**
d Schloßfries m
f montant m de battement
nl kozijnstijl m met scharnieren
r притворный брус m дверной рамы

C581 e **closing up of cracks**
d Schließen n von Rissen
f fermeture f [scellement m] de
fissures
nl dichting f van scheuren
r закрытие n трещин

C582 e **closure**
d Flußbettabriegelung f,
Flußbettabsperrung f
f coupure f de cours d'eau
nl (af)sluiting f, afdamming f
r перекрытие n русла реки

C583 e **closure channel**
d Durchflußrinne f
f avulsion f
nl stroomgeul f(m)
r проран m

C584 e **closure embankment**
d 1. Umwallung f 2. Abschlußdamm
m, Abschlußdeich m
f 1. endiguement m 2. digue f de
fermeture
nl 1. bedijking f 2. afsluitdijk m
r 1. обвалование n 2. замыкающая
[огораживающая, отсекающая]
дамба f

C585 e **cloth filter**
d Gewebefilter n, m
f filtre m en tissu
nl filtreerdoek n
r тканевый фильтр m

C586 e **cloth roof**
d Gewebedacheindeckung f
f toiture f en tissu
nl dakvilt n
r тканевое покрытие n (здания)

C587 **clout nails** see **felt nails**

C588 e **cloverleaf junction**
d Kleeblattkreuzung f
f croisement m en trèfle [double-huit]
nl klaverbladkruising f
r пересечение n (автомобильных
дорог) типа «клеверный лист»

C589 e **coagulant**
d Fällmittel n, Flockungsmittel n,
Koagulationsmittel n,
Koagulans n
f coagulant m
nl coaguleermiddel n
r коагулянт m

C590 e **coagulation basin**
d Koagulationsbecken n,
Flockungsbecken n
f bassin m de coagulation
nl coagulatiebak m
r камера f хлопьеобразования

C591 **coak scarf joint** see **cog scarf joint**

C592 e **coal tar enamel**
d Teerpechlack m
f vernis m au goudron de houille
nl koolteerlak n, m
r эмаль f [лак m] на
каменноугольной смоле

C593 e **Coanda effect**
d Coanda-Effekt m
f effet m Coanda
nl Coanda-effect n
r эффект m Коанды (налипание
воздушной струи на
криволинейную поверхность)

C594 e **coarse aggregate**
d Grobzuschlag(stoff) m, grobkörnige
Zuschläge m pl
f agrégats m pl gros, granulat m
gros, gros agrégat m [granulat m]
nl grove toeslag m
r крупный заполнитель m

C595 e **coarse air filter**
d Grobfilter n, m
f préfiltre m à air
nl grof luchtfilter n, m
r воздушный фильтр m грубой
очистки

C596 e **coarse-graded aggregate asphaltic
concrete**
d Asphaltgrobbeton m
f béton m bitum(in)eux [asphaltique]
gros

nl asfaltbeton *n* met grove toeslag
r крупнозернистый асфальтобетон *m*

C597 *e* **coarse-grained filter, coarse-grain filter**
d grobkörniges Filter *n*, *m*, Grobkornfilter *n*, *m*
f filtre *m* à gros grains [en gravier]
nl grofkorrelig filter *n*, *m*
r крупнозернистый фильтр *m*

C598 *e* **coarse screen**
d Grobrechen *m*
f grille *f* grosse (de protection)
nl grofkuilrooster *n*
r сороудерживающая решётка *f*

C599 *e* **coarse stuff**
d Grobputzmörtel *m*, Unterputzmörtel *m*
f mortier *m* bâtard
nl pleistermortel *m* voor grondlaag (*binnenwerk*)
r штукатурный раствор *m* для двух нижних слоёв штукатурного намёта

C600 *e* **coarse textured timber**
d grobfaseriges Holz *n*
f bois *m* (de construction) à texture grosse
nl grofvezelig hout *n*
r древесина *f* или пиломатериал *m* с крупной текстурой

C601 *e* **coastal canal**
d Küstenkanal *m*
f canal *m* côtier
nl kustkanaal *n*
r прибрежный канал *m*

C602 *e* **coast protection**
d Küstenschutz *m*, Küstensicherung *f*
f défense *f* de berge [de rive], défense *f* de (la) côte; défense *f* du littoral
nl kustverdediging *f*
r берегозащитные мероприятия *n pl*

C603 *e* **coast-protection dam**
d Küstenschutzdamm *m*
f digue *f* de protection du littoral
nl zeedijk *m*
r (морская) берегозащитная дамба *f*

C604 *e* **coast-protection works**
d Küstenschutzbauwerke *n pl*
f ouvrages *m pl* de défense de côte [de littoral]
nl zeewering *f*
r берегозащитные сооружения *n pl*

C605 *e* **coat**
d Anstrich *m*; Deckschicht *f*, Putzschicht *f*
f couche *f*; enduit *m*
nl (opgebrachte) laag *f(m)*
r слой *m* (*напр. краски, асфальта, штукатурки*); покрытие *n*

C606 *e* **coated chippings, coated grit**
d Mischsplitt *m*, umhüllter Splitt *m*
f gravillon *m* enrobé de bitume ou de goudron

nl geteerde split *n*
r чёрный мелкий щебень *m*, высевки *pl*, покрытые чёрными вяжущими

C607 *e* **coated macadam**
d Mischmakadam *m*, *n*
f macadam *m* enrobé; macadam *m* en bitume [bitum(in)eux]
nl voorgemengd macadam *n*
r щебёночное дорожное покрытие *n*, обработанное чёрными вяжущими

C608 *e* **coating**
d 1. Überzug *m*, Auftrag *m*, Anstrich *m*; Schicht *f*, Lage *f* 2. Anstreichen *n*, Überziehen *n*
f 1. enduit *m* 2. application *f* d'enduit
nl 1. bedekking *f*, bekleding *f*, laag *f(m)* 2. een laag opbrengen
r 1. покрытие *n*; слой *m* 2. нанесение *n* покрытия

C609 *e* **cob**
d 1. ungebrannter Ziegel *m* mit Strohzusatz, Gemenge *n* von Lehm, Kies und Stroh 2. Lehmwand *f*, Lehmstampfmauerwerk *n*
f 1. brique *f* crue, torchis *m* 2. pisé *m* (de terre)
nl stroleem *n* met kalk
r 1. саман *m*, саманный кирпич *m* 2. стены *f pl* из уплотнённого связного грунта или гр унтоцемента

C610 *e* **cobble, cobblestone**
d Katzenkopf *m*, Kopfstein *m*; Kies *m*
f galet *m*; caillou *m*
nl straatkei *m*, kinderhoofd *n*
r булыжник *m*; галька *f*

C612 *e* **cobble-stone masonry**
d Bruchsteinmauerwerk *n*
f maçonnerie *f* en galets [en moellons, en pierres brutes]
nl van rolstenen gemetselde muur
r бутовая кладка *f*

C613 *e* **cobwork** *US*
d 1. Blockhaus *n* 2. see cob 2.
f 1. maison *f* en rondins (de bois); murs *m pl* en plein bois 2. see cob 2.
nl 1. blokhut *f(m)* (VS) 2. vakwerkmuur *m* van leemmortel
r 1. сруб *m*; рубленный дом *m* из брёвен 2. see cob 2.

C614 *e* **cock**
d Hahn *m*
f robinet *m*
nl kraan *f(m)*
r кран *m*

C615 **cocking** 1. see cogging 2. see rising butts

C616 **cocking piece** see sprocket piece

C617 *e* **cockscomb**
d gezahntes Kratzeisen *n*, gezahnte Maurerschippe *f*

f racloir *m* [raclette *f*] à dents
nl lijmkam *m*
r зубчатый скребок *m*

C618 *e* **code of recommended practice**
d bautechnische Richtlinien *f pl*,
Baunormen *f pl* und
Bauvorschriften *f pl*,
Baubestimmungen *f pl*
f règlement *m* (de travaux) de
construction; code *m* du bâtiment
nl richtlijnen *pl*
r строительные нормы *f pl* и правила
n pl

C619 *e* **coefficient of compressibility**
d Verdichtungsbeiwert *m*
f coefficient *m* de compressibilité
nl samendrukbaarheidscoëfficiënt *m*
r коэффициент *m* сжимаемости

C620 *e* **coefficient of consolidation**
d Verfestigungsziffer *f*,
Konsolidierungsbeiwert *m*
f coefficient *m* de consolidation
nl verdichtingsgraad *m*
r коэффициент *m* уплотнения
[консолидации] грунта

C621 *e* **coefficient of contraction**
d Kontraktionskoeffizient *m*,
Einschnürungszahl *f*
f coefficient *m* de contraction
nl samentrekkingscoëfficiënt *m* (*hydr*)
r коэффициент *m* сжатия струи

C622 **coefficient of creep** *see* **creep coefficient**

C623 *e* **coefficient of discharge**
d Ausflußbeiwert *m*,
Durchflußkoeffizient *m*
f coefficient *m* de débit
nl afvoercoëfficiënt *m* (*hydr*)
r коэффициент *m* расхода

C624 *e* **coefficient of earth pressure**
d aktiver Grunddruckfaktor *m*
f coefficient *m* de pression [poussée]
du sol
nl coëfficiënt *m* van actieve gronddruk
r коэффициент *m* активного
давления грунта

C625 **coefficient of expansion** *see*
coefficient of thermal expansion

C626 *e* **coefficient of filtration**
d Durchlässigkeitsbeiwert *m*
f coefficient *m* de filtration
nl filtratiecoëfficiënt *m*
r коэффициент *m* фильтрации

C627 *e* **coefficient of friction**
d Reibungsbeiwetr *m*
f coefficient *m* de frottement
nl wrijvingscoëfficiënt *m*
r коэффициент *m* трения

C628 *e* **coefficient of heat transfer**
d Wärmeübergangszahl *f*
f coefficient *m* d'échange de chaleur
[d'échange thermique]

nl warmteoverdrachtscoëfficiënt *m*
r коэффициент *m* теплоотдачи

C629 *e* **coefficient of internal friction**
d innerer Reibungsbeiwert *m*
f coefficient *m* de frottement interne
nl inwendige wrijvingscoëfficiënt *m*
r коэффициент *m* внутреннего
трения

C630 *e* **coefficient of moisture precipitation**
d Feuchtigkeitsausscheidungskoeffizient
m
f coefficient *m* de précipitation
d'humidité
nl neerslagcoëfficiënt *m*
r коэффициент *m* влаговыпадения

C631 *e* **coefficient of overall heat
transmission**
d Wärmedurchgangszahl *f*
f coefficient *m* global de transmission
thermique
nl coëfficiënt *m* van de totale
warmteoverdracht
r коэффициент *m* теплопередачи

C632 *e* **coefficient of passive earth pressure**
d Erdwiderstandsbeiwert *m*
f coefficient *m* de pression passive
du sol
nl nuttig effect *n*
r коэффициент *m* сопротивления
[отпора, пассивного давления]
грунта

C633 *e* **coefficient of performance**
d Leistungsziffer *f*
f coefficient *m* performance
nl rendement *n*
r коэффициент *m* преобразования
энергии, холодильный коэффициент *m*

C634 *e* **coefficient of permeability**
d Durchlässigkeits(bei)wert *m*,
Koeffizient *m* des Leitungsvermögens,
Ableitungsfaktor *m* nach Darcy
f coefficient *m* de perméabilité [de
conductivité hydraulique, de
filtration], coefficient *m* de Darcy
nl permeabiliteitscoëfficiënt *m*
r коэффициент *m* водопроницаемости
[водопроводимости, фильтрации],
коэффициент *m* Дарси

C635 *e* **coefficient of radiation**
d Strahlungszahl *f*,
Strahlungskoeffizient *m*
f coefficient *m* de rayonnement
nl stralingscoëfficiënt *m*
r коэффициент *m* излучения

C636 **coefficient of resistance** *see*
resistance coefficient

C637 *e* **coefficient of roughness, coefficient
of rugosity**
d Rauhigkeitszahl *f*,
Rauhigkeitsbeiwert *m*
f coefficient *m* de rugosité
nl ruwheidscoëfficiënt *m*
r коэффициент *m* шероховатости

C638 coefficient of soil reaction *see*
coefficient of subgrade reaction

C639 *e* coefficient of storage
d Koeffizient *m* der Wasserabgabe,
Speicherungskoeffizient *m* eines
Wasserleiters
f coefficient *m* de rendement d'eau,
coefficient *m* d'emmagasinement
nl coëfficiënt *m* voor de waterberging
r коэффициент *m* водоотдачи,
коэффициент *m* водовместимости

C640 *e* coefficient of subgrade reaction
d Bettungszahl *f*, Druck-Setzungs-
-Quotient *m*
f module *m* de réaction du sol
nl coëfficiënt *m* voor de reactie van
het ballastbed
r коэффициент *m* постели

C641 *e* coefficient of thermal conductivity
d Wärmeleitzahl *f*
f coefficient *m* de conductibilité
thermique
nl warmtegeleidingscoëfficiënt *m*
r коэффициент *m* теплопроводности

C642 *e* coefficient of thermal expansion
d (Wärme-)Ausdehnungskoeffizient *m*
f coefficient *m* d'expansion
(thermique)
nl uitzettingscoëfficiënt *m*
r коэффициент *m* (температурного)
расширения

C643 *e* coefficient of transmissibility
d Koeffizient *m* des
Durchlässigkeitsvermögens,
Durchleitungskennziffer *f*
f coefficient *m* de transmissibilité
nl doorlaatbaarheidscoëfficiënt *m*,
doorlaatbaarheidscijfer *n*
r проводимость *f* подземного
бассейна

C644 coefficient of transmission *see*
coefficient of permeability

C645 *e* coefficient of uniformity
d Ungleichförmigkeitsgrad *m*
f coefficient *m* d'uniformité
nl gelijkvormigheidscoëfficiënt *m*
r коэффициент *m* однородности

C646 *e* coefficient of velocity
d Geschwindigkeitszahl *f*
f coefficient *m* de vitesse
nl snelheidscoëfficiënt *m*
r коэффициент *m* скорости

C647 *e* coefficient of viscosity
d Viskositätsbeiwert *m*,
Zähigkeitsbeiwert *m*
f coefficient *m* de viscosité
nl viscositeitscoëfficiënt *m*
r коэффициент *m* вязкости

C648 *e* coffer
d 1. Balkenfeld *n*, Deckenfeld *n*,
Kassette *f* 2. Schleusenkammer *f*
f 1. caisson *m* (*compartiment creux
de plafond*) 2. sas *m* (d'écluse)

nl 1. caisson *m* 2. sluiskolk *f(m)*
r 1. кессон *m* (потолка) 2. камера *f*
шлюза

C649 *e* cofferdam
d Fang(e)damm *m*, Vordamm *m*,
Hilfssperre *f*
f batardeau *m*
nl kistdam *m*
r перемычка *f* (*гидротехническая*)

C650 cogeneration of heat and power *see*
combined heat and power generation

C651 *e* cogging
d Aufkämmung *f*, Stirnversatz *m*
f mortaise *f* à embrèvement,
assemblage *m* avec embrèvement
nl overkeping *f* met (loef en)
voorloeven
r щековая врубка *f*, врубка *f*
в гребень

C652 *e* cog scarf joint
d schiefes Blatt *n*
f enture *f* à trait de jupiter simple,
sifflet *m* simple
nl schuine haaklas *f(m)*
r врубка *f* в полдерева со скосом

C653 *e* cohesion
d Kohäsion *f*, Haftung *f*;
Bindigkeit *f*
f cohésion *f*; adhérence *f*
nl cohesie *f*; kleef *f(m)*
r когезия *f*, сцепление *n*; связность
f (*грунта*)

C654 *e* cohesionless soil
d kohäsionsloser [nichtbindiger]
Boden *m*
f sol *m* non cohérent
nl losse grond *m*
r несвязный грунт *m*

C655 *e* cohesive soil
d bindiger Boden *m*
f sol *m* cohésif [cohérent]
nl samenhangende grond *m*
r связный грунт *m*

C656 coign *see* quoin

C657 *e* coil
d 1. Rohrschlange *f*
2. Wärmeaustauscher *m*,
Wärmeübertrager *m* 3. Bund *m*,
Coil *n*
f 1. serpentin *m* 2. échangeur *m* de
chaleur [thermique] 3. rouleau *m*;
botte *f*
nl 1. winding *f*, gewonden buis *f(m)*
2. spiraalvormige warmtewisselaar
m 3. spoel *f(m)*
r 1. змеевик *m* 2. теплообменник *m*
3. рулон *m* (*тонколистового
метала*); бухта *f* (*проволоки,
каната*)

C658 Colcrete *see* colloidal concrete

C659 *e* cold-drawn wire
d kaltgezogener Draht *m*

f fil *m* écroui [étiré, tréfilé] (à froid)
nl koudgetrokken draad *f(m)*
r холоднотянутая проволока *f*

C660 *e* **cold feed**
d Frïschwasserzulauf *m*
f alimentation *f* en eau froide
nl verswaterinlaat *m*
r подача *f* холодной воды

C661 *e* **cold generation**
d Kälteerzeugung *f*
f production *f* du froid
nl koude-opwekking *f*
r производство *n* холода

C662 *e* **cold-laid asphaltic concrete**
d kalteinbaufähiger Asphaltbeton *m*
f béton *m* bitum(in)eux [asphaltique] posé à froid
nl koudasfalt *n*
r асфальтобетон *m*, уложенный в холодном состоянии; холодный асфальтобетон *m*

C663 *e* **cold-laid asphalt surface**
d kalteingebaute Asphaltbetondecke *f*
f revêtement *m* routier en béton bitum(in)eux [asphaltique] à froid
nl wegdek *n* van koudasfalt
r дорожное покрытие *n* из холодного асфальтобетона

C664 *e* **cold source**
d Kältequelle *f*
f source *f* de réfrigération [du froid]
nl koudebron *f(m)*
r источник *m* холода

C665 *e* **cold store**
d Kühlhaus *n*
f entrepôt *m* frigorifique
nl koudhuis *n*
r склад-холодильник *m*

C666 *e* **cold supply**
d Kälteversorgung *f*
f alimentation *f* [approvisionnement *m*] en froid
nl koeling *f*
r холодоснабжение *n*

C667 *e* **cold supply system**
d Kältesystem *n*
f système *m* de refroidissement [d'alimentation en froid]
nl koelsysteem *n*
r система *f* холодоснабжения

C668 *e* **cold-twisted bar**
d kaltverwundener Bewehrungsstab *m*
f barre *f* à béton tordue
nl koudgebogen staaf *f(m)*
r кручёный [скрученный] арматурный стержень *m*

C669 *e* **cold water supply**
d Trinkwasserversorgung *f*, Kaltwasserversorgung *f*
f alimentation *f* en eau potable
nl drinkwatervoorziening *f*
r хозяйственно-питьевой водопровод *m*

C670 *e* **cold weather construction**
d Bau *m* bei Frost
f construction *f* par temps froid
nl (bij vorst) doorwerken
r строительство *n* в холодный период года

C671 *e* **cold-worked steel, cold-worked steel reinforcement**
d kaltverformter Betonstahl *m*
f armature *f* écrouie à froid
nl koudbewerkt wapeningsstaal *n*
r стальные арматурные стержни *m pl* или проволока *f*, получаемые методами холодной обработки

C672 *e* **cold working**
d 1. Kaltverfestigung *f* 2. Kaltverformung *f*
f écrouissage *m*
nl 1. koudbewerken *n* 2. koud vervormen *n*
r 1. наклёп *m* 2. деформация *f* в холодном состоянии

C673 **Colgrout** *see* **colloidal grout**

C675 **collapse load** *see* **failure load**

C676 *e* **collapse method of structural design**
d plastisches Berechnungsverfahren *n*
f calcul *m* (des constructions métalliques) compte tenu des déformations plastiques
nl vloeiberekeningsmethode *f* (staalconstructies)
r расчёт *m* (стальных конструкций) с учётом пластических деформаций [с учётом образования пластического шарнира]

C677 *e* **collapsible needle weir**
d Nadelwehr *n*
f barrage *f* à (fermettes et) aiguilles
nl naaldstuw *m*
r спицевая плотина *f*

C678 **collapsible weir** *see* **movable weir**

C679 *e* **collar**
d 1. Halsring *m* 2. Bund *m*, Wulst *m*; Hals *m*; Flansch *m* 3. Manschette *f* 4. Doppelmuffe *f*
f 1. collerette *f* d'étanchéité 2. collet *m*, collier *m*, bride *f* 3. manchon *m*, manchette *f* 4. emboîtement *m* double; manchon *m* d'assemblage
nl 1. ronde onderlegplaat *f(m)* 2. kraag *m*, flens *m* 3. manchet *f(m)* 4. (verbindings)mof *f(m)*
r 1. гидроизолирующий воротник *m* 2. буртик *m*; шейка *f*; фланец *m* 3. манжета *f* 4. двойной раструб *m*; соединительная муфта *f*

C680 *e* **collar beam**
d Kehlbalken *m*
f entrait *m*
nl trekplaat *f(m)*, hanebalk *m*
r затяжка *f* висячих стропил, стропильная затяжка *f*

681 *e* **collar beam roof**
  d Kehlbalkendach *n*
  f comble *m* à arbalétriers réunis par
    entrait retroussé [par faux entrait]
  nl zadeldak *n*
  r крыша *f* с висячими стропилами
    с повышенной затяжкой

C682 *e* **collar for flange connection**
  d Flanschverbinder *m*
  f tubulure *f* [manchon *m*] de
    raccord(ement) à brides
  nl flenskoppeling *f*
  r патрубок *m* с фланцевым
    соединением

C683 *e* **collar saddle**
  d Sattelstück *n*
  f branche *f* [branchement *m*] de
    tuyau à collier [à collet]
  nl muurplaat *f(m)* (*kraan*)
  r ответвление *n* (*трубопровода*) с
    воротником

C684 *e* **collecting drain**
  d Sammeldrän *m*
  f drain *m* collecteur [de captage]
  nl (hoofd)afvoerleiding *f*
  r собирающая дрена *f*

C685 *e* **collecting manhole, collecting well**
  d Sammelbrunnen *m*, Sammelschacht *m*
  f puits *m* collecteur
  nl verzamelput *m*
  r сборный колодец *m*

C686 **collection line** *US see* **house connection 2.**

C687 *e* **collector**
  d 1. Sammler *m*, Sammelrohr *n*
    2. Sammler *m*, Sammeldrän *m*;
    Abfangdrän *m*
  f 1. collecteur *m* d'eaux (pluviales)
    2. drain *m* collecteur; drain *m*
    d'interception
  nl 1. opvanginrichting *f*, vergaarbak
    m 2. verzameldrain *m*
  r 1. (водоприёмный) коллектор *m*
    2. дрена-собиратель *f*;
    перехватывающая дрена *f*

C688 *e* **collector drain**
  d Sammler *m*, Sammeldrän *m*;
    Sammler *m*, Entwässerungsstollen
    m
  f fossé *m* collecteur; galerie *f* de
    drainage
  nl verzameldrain *m*
  r дренажный коллектор *m*;
    дренажная галерея *f*

C689 *e* **collimation line**
  d Kollimationsachse *f*
  f ligne *f* [axe *m*] de visée
  nl vizierlijn *f(m)*
  r линия *f* визирования, визирная
    ось *f*

C690 *e* **colloidal concrete**
  d Kolloidalbeton *m*, Colcretebeton *m*
  f béton *m* colloïdal [Colcrete]

  nl kolloïdaal beton *n*
  r коллоидальный бетон *m* (*на
    коллоидном строительном растворе*)

C691 *e* **colloidal grout**
  d kolloidaler Zementmörtel *m*,
    Colgrout-Mörtel *m*,
    Einpreßmörtel *m*
  f mortier *m* colloïdal [Colgrout]
  nl kolloïdale voegspecie *f*
  r коллоидный цементный раствор *m*

C692 *e* **colloidal mixer**
  d Kolloidmörtelmischer *m*
  f malaxeur *m* pour mortier Colgrout
  nl menger *m* voor kolloïdale specie
  r (растворо)смеситель *m* для
    приготовления коллоидных
    растворов

C693 *e* **colmatage**
  d Kolmatage *f*, Kolmation *f*,
    Kolmatierung *f*
  f colmatage *m*
  nl aanslibben *n*
  r кольматаж *m*

C694 *e* **colorimetric test**
  d kolorimetrische Prüfung *f*
  f éssai *m* colorimétrique
  nl colorimetrie *f*
  r колориметрическое испытание *n*

C695 *e* **colour**
  d Farbe *f*, Farbton *m*
  f couleur *f*
  nl kleur *f(m)*
  r цвет *m*, колер *m*

C696 *e* **coloured cement**
  d farbiger Zement *m*, Farbzement *m*
  f ciment *m* coloré [à pigment]
  nl gekleurd cement *n*
  r цветной цемент *m*

C697 *e* **colouring admixture**
  d Pigment *n*, Farbstoff *m*
  f adjuvant *m* colorant, pigment *m*
  nl kleurstof *f(m)*, pigment *n*
  r окрашивающая добавка *f*,
    пигмент *m*

C698 *e* **colour stability**
  d Farbbeständigkeit *f*
  f stabilité *f* de la couleur
  nl kleurbestendigheid *f*
  r цветостойкость *f*; стабильность *f*
    окраски [цвета]

C699 **colour test** *see* **colorimetric test**

C700 *e* **column**
  d Stütze *f*, Säule *f*; Pfosten *m*,
    Ständer *m*
  f colonne *f*; poteau *m*
  nl kolom *f(m)*; stijl *m*, zuil *f(m)*
  r колонна *f*; столб *m*, стойка *f*

C701 *e* **column base**
  d Säulenfuß *m*; Stützenfuß *m*
  f base *f* de colonne [de poteau]
  nl kolomvoet *m*; zuilvoet *m*
  r база *f* [башмак *m*] колонны;
    подколонник *m*

C702 *e* **column curve**
   d Bezugskurve *f* «kritische Spannung—
      Schlankheit der Säule»
   f courbe *f* de flambage d'unecolonne
      [d'un poteau], courbe *f* repésentant
      les contraintes critiques en
      fonction de l'élancement
   nl kritische-slankheidskromme *f* (*van
      een kolom*)
   r кривая *f* зависимости
      критических напряжений
      гибкости колонны

C703 *e* **column drill**
   d Säulenbohrmaschine *f*
   f marteau *m* perforateur à colonne
   nl kolomboormachine *f*
   r колонковый бур *m*; колонковый
      бурильный молоток *m*

C704 *e* **column effective length**
   d Knicklänge *f* der Säule
   f longueur *f* effective d'une colonne
      [d'un poteau]
   nl kniklengte *f* van een kolom
   r приведённая длина *f* колонны

C705 *e* **column footing**
   d Säulengrundwerk *n*
   f pied *m* d'une colonne
   nl kolomfundering *f*
   r фундамент *m* колонны

C706 *e* **column foundation block**
   d Fundamenthülse *f*, Hülsenfunda-
      mentblock *m*
   f bloc *m*(évidé) de fondation d'une
      colonne [d'un poteau]
   nl poer *f(m)*
   r стаканный фундаментный блок *m*

C707 *e* **column head**
   d Säulenkopf *m*, Pilzkopf *m*
   f tête *f* de colonne, chapiteau *m*
   nl kolomkop *m*
   r оголовок *m* колонны

C708 *e* **column slenderness ratio**
   d Schlankheit *f* der Säule [der
      Stütze]
   f élancement *m* d'une colonne [d'un
      poteau]
   nl slankheidsfactor *m* van een kolom
   r гибкость *f* колонны [стойки]

C709 *e* **combed fascine raft**
   d Senkstück *n*
   f matelas *m* de fascin(ag)es
   nl zinkstuk *n* van rijswerk
   r фашинный тюфяк *m*

C710 *e* **combed joint**
   d Zinkeneckverbindung *f*
   f assemblage *m* (*de planches*) en
      L à queues droites
   nl meervoudige slisverbinding *f*
   r угловое соединение *n* (*досок*)
      прямыми шипами

C711   **combination plane** *see* **universal
      plane**

C712 *e* **combination railing**
   d Geländer *n* für Autos und
      Fußgänger
   f garde-corps *m* combiné (*susceptible
      d'arrêter des véhicules et assurer la
      sécurité des piétons*)
   nl gecombineerde leuning *f* en
      vangrail
   r перильные ограждения *n pl*
      комбинированного типа (*для
      автомобилей и пешеходов*)

C713   **combination sewer** *see* **combined
      sewer**

C714 *e* **combination tap assembly**
   d Zweiventilmischbatterie *f*
   f batterie *f* de robinets à mélangeur
   nl mengklepcombinatie *f*
   r двухвентильный смеситель *m*

C715   **combination-type pavement structure**
      see **composite-type pavement
      structure**

C716 *e* **combination well**
   d Brunnenreihe *f*
   f puits *m* complexe
   nl bronnencomplex *n*
   r групповой колодец *m*

C717 *e* **combination window**
   d Doppelfenster *n*
   f fenêtre *f* double
   nl dubbel venster *n*
   r окно *n* с двойными раздельными
      переплётами

C718 *e* **combined bending and axial loading**
   d Biegung *f* mit Längskraft
   f flexion *f* avec force longitudinale,
      flexion-compression *f*, flexion-
      -traction *f*
   nl buiging *f* met langskracht
   r продольно-поперечный изгиб *m*

C719 *e* **combined drainage system**
   d kombiniertes Dränagesystem *n*,
      Ramspol-Verfahren *n*
   f méthode *f* de Ramspol, système *m*
      de drainage combiné
   nl gecombineerd drainagestelsel *n*
   r комбинированная система *f*
      дренажа (*с отводом поверхностных
      и подземных вод по одним и тем
      же дренам и коллекторам*);
      комбинированная оросительно-
      дренажная система *f*

C720 *e* **combined effects of settlement and
      creep**
   d Zusammenwirkung *f* von Setzung
      und Kriechen
   f sollicitation *f* composée due au
      tassement et au fluage
   nl zamenwerking *f* van zetting en
      kruip
   r совместное действие *n* осадки и
      ползучести

C721 *e* **combined feed and expansion cistern**
   d Speise- und Ausdehnungsgefäß *n*

f réservoir *m* d'expansion et
d'alimentation
nl gecombineerd voedings- en
expansievat *n*
r расширительный бак *m* системы
отопления, одновременно
питающий систему горячего
водоснабжения

C722 e **combined heat and power generation**
d Kraft-Wärme-Kopplung *f*
f thermofication *f* (*chauffage urbain
à base d'une centrale thermique*)
nl warmte-krachtkoppeling *f*
r теплофикация *f*

C723 e **combined sewer**
d gemeinsamer Sammelkanal *m*,
Mischwassersammler *m*,
Mischwasserkanal *m*
f égout *m* combiné [unitaire]
nl gemengd riool *n*
r коллектор *m* общесплавной
канализации

C724 e **combined sewerage system, combined
sewers**
d Mischkanalisation *f*,
Schwemmkanalisation *f*,
Mischsystem *n*, Vollentwässerung *f*
f système *m* (d'assainissement)
unitaire
nl gemengd rioleringsstelsel *n*
r общесплавная канализация *f*

C726 e **combined stress, combined stresses**
d komplizierter Spannungszustand *m*,
zusammengesetzte Beanspruchung *f*
[Spannungen *f pl*]
f sollicitations *f pl* composées
nl gecombineerde spanningstoestand *m*
r сложное напряжённое состояние *n*

C727 e **combined water**
d 1. chemisch gebundenes Wasser *n*,
Konstitutionswasser *n*
2. Mischwasser *n*
f 1. eau *f* liée; eau *f* de cristallisation
2. eaux *f pl* usées évacuées par
système unitaire
nl 1. chemisch gebonden water *n*
2. met het afvalwater gemengd
hemelwater
r 1. связная *или* кристаллизационная
вода *f* 2. стоки *m pl*, отводимые
сетью общесплавной канализации

C728 e **combustible building materials**
d brennbare Baustoffe *m pl*
f matériaux *m pl* de construction
inflammables
nl brandbare bouwmaterialen *n pl*
r возгораемые строительные
материалы *m pl*

C729 e **combustion chamber**
d 1. Verbrennungsraum *m*
2. Brennkammer *f*
f chambre *f* de combustion
nl 1. vlamkast *f(m)* (*ketel*)
2. verbrandingskamer *f(m)*

r 1. топка *f*, топочная камера *f*
2. камера *f* сгорания

C730 e **comfort**
d Komfort *m*, thermische
Behaglichkeit *f*
f confort *m* thermique
nl behaaglijkheid *f*
r тепловой комфорт *m*

C731 e **comfort air-conditioning**
d Komfortklimatisierung *f*
f climatisation *f* de confort
nl luchtbehandeling *f* ten behoeve
van de behaaglijkheid
r комфортное кондиционирование *n*
воздуха

C732 e **comfort chart**
d Diagramm *n* für Komfortbedingungen
f diagramme *m* de confort (*thermique*)
nl behaaglijkheidsgrafiek *f*
r диаграмма *f* комфортных условий

C733 e **comfort conditions**
d Komfortbedingungen *f pl*,
Behaglichkeitsklima *n*
f conditions *f pl* de confort
nl behaaglijkheidsvoorwaarden *f pl*
r комфортные условия *n pl*

C734 e **comfort cooling**
d Komfortkühlung *f*
f refroidissement *m* de confort
nl koeling *f* ten behoeve van de
behaaglijkheid
r комфортное охлаждение *n*

C735 e **comfort cooling system**
d Komfortluftkühlanlage *f*
f installation *f* de refroidissement
d'air de confort
nl koelsysteem *n* ten behoeve van de
behaaglijkheid
r установка *f* [система *f*]
комфортного охлаждения воздуха

C736 e **comfort index**
d Komfortgradindex *m*,
Komfortgradkennziffer *f*
f indice *m* de confort
nl index *m* van de (mate van)
behaaglijkheid
r индекс *m* комфортных условий

C737 e **comfort zone**
d Behaglichkeitsbereich *m*,
Behaglichkeitsfeld *n*, Komfortzone *f*
f zone *f* de confort
nl behaaglijkheidsbereik *n*
r область *f* комфортных условий

C738 e **commercial building**
d Geschäftshaus *n*
f bâtiment *m* [immeuble *m*, édifice
m] commercial
nl handelsgebouw *n*
r торговое здание *n*

C739 e **commercial specifications**
d technische Liefer- und
Gütebestimmungen *f pl*
f cahier *m* des charges du fournisseur

nl technische specificaties *f pl*
r технические условия *n pl* (фирмы-
-изготовителя)

C740 *e* **comminutor**
d Rechengutzerkleinerer *m*
f dilacérateur *m*
nl vermaler *m*
r комминутор *m*, решётка-дробилка *f*

C741 *e* **commissioning**
d Einregulierung *f* und Inbetriebnahme
f
f mise *f* au point; ajustage *m*,
réglage *m*
nl inbedrijfstelling *f*
r пусконаладочные работы *f pl*

C742 *e* **commissioning of plant**
d Gerätevorhaltung *f*
f mise *f* au point [réglage *m*] d'une
installation
nl inbedrijfstelling *f* van een installatie
r предпусковые (приёмосдаточные)
испытания *n pl* системы

C743 *e* **commode step**
d Wendelstufe *f*
f marche *f* dansante [gironnée,
balancée]
nl bloktrede *f* met gebogen stootbord
r забежная ступень *f* (*лестницы*)

C744 *e* **common ashlar**
d Quader *m*
f moellon *m* taillé [de taille],
pierre *f* (en bloc) taillée [de
taille]
nl gebouchardeerd hardstenen blok *n*
r тёсаный [штучный] камень *m*

C745 *e* **common bond**
d amerikanischer (Mauerwerk-)
Verband *m*
f appareil *m* (de maçonnerie)
américaine
nl halfsteenverband *n* met om de vijf
lagen een koppenlaag
r многорядная перевязка *f*
(*каменной кладки*)

C746 *e* **common brick**
d Mauerziegel *m*, gewöhnlicher
Tonziegel *m*
f brique *m* d'argile [de terre cuite]
ordinaire
nl baksteen *m* voor binnenmuren en
achterwerkers
r обыкновенный глиняный кирпич *m*

C747 *e* **common joist**
d Holz(-Decken)unterzug *m*
f poutrelle *f* en planches alignées
mises sur arête
nl vloerbalk *m*
r деревянная балка *f* (*перекрытия*)
из досок, уложенных на ребро

C748 *e* **common partition**
d selbsttragende hölzerne
Gerippentrennwand *f*

f paroi *f* simple à ossature en bois
nl scheidingswand *m*
r самонесущая деревянная каркасная
перегородка *f*

C749 *e* **common rafter**
d Bindersparren *m*
f chevron *m* ordinaire
nl dakspar *m*
r стропильная нога *f* наклонных
стропил; стропильная нога *f*,
опирающаяся на прогоны

C750 **common wall** *US see* **party wall**

C751 *e* **communication pipe**
d 1. Endstrang *m* 2.
Schwachstromkabelrohr *n*
f 1. tuyau *m* de branchement
[d'embrachement] 2. canalisation *f*
pour câbles téléphoniques
nl 1. dienstleiding *f*
2. telefoonkabelbuis *f(m)*
r 1. тупиковая ветвь *f* трубопровода
2. кабелепровод *m* линий связи

C752 *e* **compactability**
d Verdichtungsfähigkeit *f*
f compactibilité *f*
nl verdichtbaarheid *f*
r уплотняемость *f*, способность *f* к
уплотнению

C753 *e* **compacted rock fill**
d verdichtete Steinschüttung *f*
f enrochement *m* compacté
nl verdichte steenaanvulling *f*
r уплотнённая каменная отсыпка *f*

C754 *e* **compacted yard**
d Bodenvolumen *n* im verdichteten
Zustand
f volume *m* de remblai compacté
nl grondvolume *n* in verdichte toestand
r объём *m* грунта в уплотнённом
состоянии

C755 *e* **compacting equipment**
d Verdichtungsgeräte *n pl*
f matériel *m* de compactage
nl verdichtingsmachines *f pl*
r оборудование *n* [машины *f pl*]
для уплотнения (*грунта, бетона*)

C756 *e* **compacting factor**
d Verdichtungsquotient *m*,
Verdichtungsfaktor *m*
f facteur *m* de compacité
nl verdichtingsgraad *m*
r коэффициент *m* уплотнения

C757 *e* **compacting factor test**
d Verdichtungsquotientprüfung *f*
f essai *m* de consistance du béton frais
par détermination du facteur de
compacité
nl vaststellen *n* van de
verdichtingsgraad van betonmortel
r определение *n* подвижности бетонной
смеси по степени её уплотнения
(*при падении в стандартный сосуд
с заданной высоты*)

C758 *e* **compaction**
   d Verdichtung *f*
   f compactage *m*
  nl verdichting *f*
   r уплотнение *n*

C759 *e* **compaction by rolling**
   d Verdichten *n* durch Walzen
   f compactage *m* par cylindrage [par roulage]
  nl verdichting *f* door walsen
   r уплотнение *n* катками

C760 *e* **compaction curve**
   d Verdichtungskurve *f*
   f courbe *f* de compactage
  nl verdichtingskromme *f*
   r кривая *f* уплотнения

C761 *e* **compaction test**
   d Verdichtungsversuch *m*
   f essai *m* de compactage
  nl verdichtingsproef *f(m)*
   r испытание *n* уплотняемости

C762 *e* **compact material**
   d dichter Boden *m* (*relative Bodenskelettdichte über 90 %*)
   f sol *m* compact (*à compacité relative plus de 90%*)
  nl vaste grondslag *m*
   r плотный грунт *m* (*с относительной плотностью скелета свыше 90%*)

C763 *e* **compactness**
   d Lagerungsdichte *f*
   f compacité *f*
  nl dichtheid *f*
   r плотность *f*, степень *f* уплотнения

C764 *e* **compactor**
   d Stampfgerät *n*, Verdichter *m*
   f compacteur *m*
  nl stamper *m*, verdichter *m*
   r уплотняющая машина *f*; (механическая) трамбовка *f*

C765 *e* **companion flange**
   d Gegenflansch *m*
   f contrebride *f*
  nl tegenflens *m*
   r ответный фланец *m*, контрфланец *m*

C766 *e* **compartment**
   d 1. Zelle *f* 2. Brandabschnitt *m* 3. Gewölbefeld *n*; Bogenbrückenfeld *n* 4. Gefach *n* (*Fachwerkbau*) 5. Stauabteilung *f*, Bewässerungsabschnitt *m*
   f 1. compartiment *m* 2. compartiment *m* coupe-feu [pore-feu] 3. travée *f* d'une voûte 4. pan *m* de fer 5. bassin *m*
  nl 1. afdeling *f*, cel *f(m)* 2. brandsector *m* 3. gewelfkap *f(m)* 4. brugvak *n* 5. afgescheiden gedeelte *n*
   r 1. отделение *n*, отсек *m* 2. противопожарная зона *f*, противопожарный отсек *m* 3. пролёт *m* [звено *n*] свода; арочный пролёт *m* 4. фахверк *m*, сквозная конструкция *f* 5. чек *m*

C767 *e* **compartmentation**
   d Aufteilung *f* eines Gebäudes durch Feuer(schutz)mauern
   f compartimentage *m*
  nl compartimentering *f* (*door brandmuren*)
   r членение *n* [разделение *n*] здания противопожарными ограждениями (*напр. стенами, перекрытиями*) на отсеки

C768 *e* **compass**
   d Zirkel *m*
   f compas *m*
  nl passer *m*
   r циркуль *m*

C769 *e* **compass plane**
   d Bogenhobel *m*, Rundhobel *m*, Schiffshobel *m*
   f rabot *m* cintré
  nl toogschaaf *f(m)*, ronde schaaf *f(m)*
   r горбатик *m* (*рубанок*)

C770 *e* **compass window**
   d rundes Erkerfenster *n*, Bogenfenster *n*
   f fenêtre *f* cintrée
  nl rond venster *n*
   r полукруглое эркерное окно *n*, полукруглый эркер *m*

C771 *e* **compatibility conditions**
   d Verträglichkeitsbedingungen *f pl*
   f conditions *f pl* de compatibilité
  nl tolerantievoorwaarden *f pl*
   r условия *n pl* совместности

C772 *e* **compatibility equations**
   d Kompatibilitätsgleichungen *f pl*
   f équations *f pl* de compatibilité
  nl tolerantievergelijkingen *f pl*
   r уравнения *n pl* совместности

C773 *e* **compensated control**
   d Kompensationsregelung *f*, Ausgleich(s)regelung *f*
   f réglage *m* compensateur
  nl compensatieregeling *f*
   r компенсационное регулирование *n*

C774 *e* **compensating basin**
   d Ausgleich(s)becken *n*
   f bassin *m* de compensation
  nl vereffeningsbekken *n*
   r выравнивающий [буферный] бассейн *m*

C775 **compensating reservoir** *see* **balancing reservoir**

C776 *e* **compensating rope**
   d Gegengewichtsseil *n*
   f câble *m* de contrepoids
  nl vereffeningskabel *m*
   r канат *m* противовеса; компенсирующий канат *m*

C777 *e* **competence of stream**
   d Transportvermögen *n* der Strömung
   f capacité *f* de transport des

sédiments, capacité *f* transportante
d'un cours d'eau
nl transportvermogen *n* van een stroom
r транспортирующая способность *f*
потока

C779 *e* **complete air-conditioning**
d vollständige Klimatisierung *f*
f conditionnement *m* de l'air parfait
[complet, intégral]
nl volledige luchtbehandeling *f*
r полное кондиционирование *n*
воздуха (*с независимым
регулированием температуры и
влажности*)

C780 *e* **completion**
d Fertigstellung *f*, Abschluß *m*
f achèvement *m*
nl beëindiging *f*, voltooiing *f*
r окончание *n*, завершение *n*

C781 *e* **compliance**
d Nachgiebigkeit *f*
f déformabilité *f*, compressibilité *f*,
flexibilité *f*
nl vermogen *n* tot niet-plastische
vervorming
r податливость *f*

C782 *e* **compo**
d 1. Kalkzementmörtel *m*
2. Bleilegierung *f* 3. Komposition *f*,
Zusammensetzung *f*
f 1. mortier *m* de ciment et de chaux
2. alliage *m* de plomb
3. composition *f*, composé *m*
nl 1. cementmortel *n* 2. gipskalkmortel
n
r 1. цементно-известковый раствор *m*
2. свинцовый сплав *m*
3. композиция *f*, состав *m*

C783 **compo mortar** *see* **compo 1.**

C784 *e* **component**
d 1. Bauteil *m* 2. Bestandteil *m*,
Komponente *f*
f 1. composant *m* 2. composante *f*
nl 1. bestanddeel *n*, bouwelement *n*
2. component *m* (*krachten*)
r 1. компонент *m*, элемент *m*
(*конструкции*) 2. составляющая
f, компонента *f*

C785 *e* **component of a force**
d Kraftkomponente *f*
f composante *f* de force
nl krachtcomponent *m*
r составляющая *f* силы

C786 *e* **composite action**
d Verbundwirkung *f*
f action *f* composite
nl samenwerking *f*
r совместная работа *f*; совместное
воздействие *n*

C787 *e* **composite board**
d Verbundplatte *f*, Sandwichplatte *f*
f panneau *m* multicouche, panneau *m*
[plaque *f*] sandwich

nl sandwichplaat *f*(*m*)
r многослойная плита *f*

C788 *e* **composite bridge**
d Verbundbrücke *f* (*Stahl/Beton*)
f pont *m* composite (*en acier et
béton armé*)
nl samengestelde brug *f*(*m*)
(*staal/beton*)
r сталежелезобетонный мост *m*,
мост *m* объединённой конструкции
(*из стали и бетона*)

C789 *e* **composite cofferdam**
d Verbundfangdamm *m*
f batardeau *m* composite (*à double
paroi de palplanches avec
remplissage en sable*)
nl kistdam *m*
r двухрядная шпунтовая перемычка *f* с
грунтовой засыпкой

C790 *e* **composite column**
d Verbundsäule *f*
f colonne *m* composite, poteau *m*
composé
nl samengestelde kolom *m*
r составная колонна *f*

C791 *e* **composite construction**
d Verbundbau *m*, Verbundkonstruktion *f*
f construction *f* composite
nl samengestelde constructie *f*
r комбинированная [составная]
конструкция *f*; объединённая
конструкция *f*

C792 *e* **composite insulation**
d Verbunddämmstoff *m*
f isolation *f* thermique multicouche
nl samengestelde warmte-isolatie *f*
r многослойная теплоизоляция *f*

C793 *e* **composite member**
d Verbundbauteil *m*
f élément *m* de construction
composite [composé]
nl samengesteld onderdeel *n*
r составной (конструктивный)
элемент *m*

C794 *e* **composite type pavement structure**
d Verbunddecke *f*; Asphalt(beton)decke
f mit starrem Unterbau
f corps *m* de chaussée composite;
couche *f* de surface en béton
asphaltique placée sur la couche de
fondation rigide
nl samengestelde wegverharding *f*
r дорожная одежда *f*
комбинированного типа;
асфальтобетонное покрытие *n* по
жёсткому основанию

C795 *e* **composite unit graph**
d zusammengesetzte Einheitsganglinie
f
f hydrogramme *m* unitaire composé
nl samengesteld eenheidshydrogram *n*
r композиционный [составной]
единичный гидрограф *m*

C796 *e* **composition**
  d 1. Zusammensetzung *f*,
    Komposition *f* 2. Struktur *f*,
    Aufbau *m*
  f composition *f*, composé *m*
  nl 1. samenstelling *f*, compositie *f*
    2. structuur *f*
  r 1. состав *m*, композиция *f*
    2. структура *f*; строение *n*

C799 *e* **compound**
  d Gemisch *n*; Zusammensetzung *f*
  f composé *m*; mélange *m*, pâte *f*
  nl mengsel *n*
  r смесь *f*; состав *m*, композиция *f*

C800 *e* **compound beam**
  d zusammengesetzter Holzträger *m*,
    mehrteiliger Holzbalken *m*
  f poutre *f* composite [composée]
  nl samengestelde houten ligger *m*
  r составная (деревянная) балка *f*

C801 *e* **compound compressor**
  d Verbundverdichter *m*,
    Verbundkompressor *m*
  f compresseur *m* compound
  nl meertrapscompressor *m*
  r многоступенчатый компрессор *m*

C802 *e* **compound curve**
  d Korbbogen *m*
  f anse *f* de panier
  nl overgangsboog *m*
  r сложная кривая *f*; коробовая
    кривая *f*

C803 *e* **compound gage, compound gauge**
  d Manovakuummeter *n*
  f manomètre *m* et indicateur *m* de vide
    combiné
  nl mano-vacuümmeter *m*
  r мановакуумметр *m*

C804 *e* **compound hydrograph**
  d zusammengesetzte Ganglinie *f*
  f hydrogramme *m* composé
  nl samengestelde hydrogram *n*
  r сложный гидрограф *m*

C805 *e* **compound pile**
  d Verbundpfahl *m*,
    zusammengesetzter Pfahl *m*
  f pieu *m* composé, pieu *m*
    préfabriqué assemblé de plusieurs
    segments
  nl samengestelde heipaal *m*
  r комбинированная свая *f*

C806 *e* **compound wall, compound walling**
  d Wand *f* mit Verkleidung aus
    verschiedenen Materialien
  f mur *m* composite (*mur avec
    revêtement en matériaux différents*)
  nl uit verscheidene lagen opgebouwde
    wand *m*
  r стена *f* с облицовкой из разных
    материалов

C807 *e* **compound well**
  d Verbundbrunnen *m*, Brunnen *m* mit
    Rohren verschiedenen Durchmessers
  f puits *m* tubulaire à diamètre variable
  nl samengestelde bron *f(m)*
  r трубчатый колодец *m* с обсадными
    трубами различных диаметров

C808 **compregnated wood** *US see* **high-
  -density plywood**

C809 *e* **compressed air**
  d Druckluft *f*
  f air *f* comprimé
  nl samengeperste lucht *f(m)*,
    druklucht *f(m)*
  r сжатый воздух *m*

C810 **compressed air caisson** *see* **caisson 2.**

C811 *e* **compressed-air ejector**
  d Druckluftheber *m*, pneumatischer
    Heber *m*
  f éjecteur *m* pneumatique [à air
    comprimé]
  nl persluchtstraalpijp *f(m)*
  r пневматический эжектор *m*

C812 *e* **compressed-air gun**
  d Druckluftpistole *f*
  f pistolet *m* pneumatique [à air
    comprimé]
  nl persluchtpistool *n*
  r пневмопистолет *m*, пневматический
    пистолет-краскораспылитель *m*

C813 *e* **compressed-air jack**
  d Druckluftheber *m*, Druckluftwinde *f*
  f vérin *m* pneumatique [à air
    comprimé]
  nl pneumatische vijzel *f(m)*
  r пневматический домкрат *m*

C814 *e* **compressed-air pile hammer**
  d Druckluftrammhammer *m*
  f mouton *m* pneumatique [à air
    comprimé]
  nl pneumatisch heiblok *n*
  r пневматический сваебойный молот
    m

C815 *e* **compressed-air plant, compressed-air
  plant station** *see* **compressor station**

C816 *e* **compressed-air shield driving** *see*
  pneumatic shield driving

C817 *e* **compressed flange**
  d Druckgurt *m*, Druckflansch *m*
  f membrure *f* [semelle *f*] comprimée
  nl gedrukte flens *m*
  r сжатый пояс *m*, сжатая полка *f*
    (*балки*)

C818 **compressed straw slab** *see* **strawboard**

C819 **compressed union** *see* **compression
  joint**

C820 *e* **compressibility**
  d Verdichtungsfähigkeit *f*,
    Zusammendrückbarkeit *f*
  f compressibilité *f*
  nl samendrukbaarheid *f*
  r сжимаемость *f*

C821 *e* **compressibility factor**
  d Komprimierbarkeit *f*,
    Zusammendrückbarkeitsbeiwert *m*

f coefficient *m* de compressibilité
nl samendrukbaarheidscoëfficiënt *m*
r коэффициент *m* сжимаемости

C822 *e* **compression**
d Kompression *f*, Verdichtung *f*
f compression *f*
nl compressie *f*, verdichting *f*
r сжатие *n*, компрессия *f*

C823 *e* **compression curve**
d Verdichtungskurve *f*
f courbe *f* de compression
nl verdichtingskromme *f*
r компрессионная кривая *f*; кривая *f* уплотнения [сжатия]

C824 *e* **compression efficiency**
d Kompressionswirkungsgrad *m*, Verdichtungswirkungsgrad *m*
f efficacité *f* de compression
nl compressierendement *n*
r адиабатическая эффективность *f* сжатия (*в компрессоре*)

C825 *e* **compression element**
d Druckelement *n*, Druckglied *n*
f élement *m* (*de construction*) comprimé
nl drukstaaf *f(m)*
r сжатый элемент *m* (*конструкции*)

C826 **compression flange** *see* **compressed flange**

C827 *e* **compression joint**
d Quetschkonusverbindung *f*
f joint *n* à compression
nl op druk belaste verbinding *f*
r компрессионное [обжимное] соединение *n* (*труб*)

C828 **compression member** *see* **compression element**

C829 *e* **compression refrigerating machine**
d Kompressionskälteanlage *f*
f machine *f* frigorifique à compression
nl compressiekoelmachine *f*
r компрессионная холодильная машина *f*

C830 **compression reinforcement** *see* **compressive reinforcement**

C831 **compression release valve** *see* **pressure-relief valve**

C832 **compression stress** *see* **compressive stress**

C833 *e* **compression tank**
d 1. Druckgefäß *n* 2. Luftpolster--Expansionsbehälter *m*
f 1. réservoir *m* à pression 2. réservoir *m* d'expansion à coussin d'air
nl 1. drukketel *m* 2. expansievat *n* met luchtkamer
r 1. сосуд *m* под давлением 2. замкнутый расширительный сосуд *m* с воздушной подушкой (*в системе водяного отопления замкнутого типа*)

C834 *e* **compression test**
d 1. Druckfestigkeitsprüfung *f*, Druckfestigkeitsprobe *f*
2. Kompressionsversuch *m*
f 1. essai *m* de compression; essai de résistance à l'écrasement 2. essai *m* de compressibilité (des sols)
nl beproeving *f* op druk
r 1. испытание *n* на сжатие 2. определение *n* сжимаемости грунтов

C835 *e* **compression-testing machine**
d Baustoffprüfpresse *f*
f presse *f* d'essai
nl toestel *n* voor proefbelasting op druk
r испытательный пресс *m*

C836 *e* **compression wave**
d Kompressionswelle *f*
f onde *f* de compression
nl compressiegolf *f(m)*, drukgolf *f(m)*
r волна *f* давления (*в грунте*); волна *f* сжатия (*при взрыве*)

C837 *e* **compression wood**
d Druckholz *n*
f bois *m* de compression
nl drukhout *n*
r крень *f*, эксцентричность *f* годовых колец (*порок древесины*)

C838 *e* **compressive force**
d Druckkraft *f*
f force *f* de compression
nl drukkracht *f(m)*
r сжимающее усилие *n*

C839 *e* **compressive reinforcement**
d Druckbewehrung *f*
f armature *f* comprimée
nl drukwapening *f*
r сжатая арматура *f*

C840 *e* **compressive strength**
d Druckfestigkeit *f*
f résistance *f* à la compression
nl druksterkte *f*
r предел *m* прочности при сжатии; временное сопротивление *n* сжатию

C841 *e* **compressive stress**
d Druckspannung *f*
f contrainte *f* de compression
nl drukspanning *f*
r сжимающее напряжение *n*; напряжение *n* сжатия

C842 **compressive zone**
d Druckzone *f*
f zone *f* comprimée
nl drukzone *f*
r сжатая зона *f*

C843 *e* **compressor**
d Verdichter *m*, Kompressor *m*
f compresseur *m*
nl compressor *m*
r компрессор *m*

C844 *e* **compressor station**
d Kompressorenstation *f*, Verdichterstation *f*

f poste m [centrale f] de
compresseurs, station f d'air
comprimé
nl persluchtcentrale f
r компрессорная станция f

C845 e **compressor unit**
d Kompressorsatz m
f groupe m compresseur
nl compressoraggregaat n
r компрессорный агрегат m

C846 **computation of stresses** see **stress analysis**

C847 e **concave joint**
d konkave Mauerfuge f
f joint m concave (de maçonnerie)
nl verdiepte voeg f(m)
r вогнутый шов m (кирпичной кладки)

C848 e **concealed conduit**
d Unterputzleitung f
f conduit m dissimulé
nl leidingen f pl uit (het) zicht
r скрытая проводка f

C849 **concentrated flow** see **channel flow**

C850 e **concentrated force**
d konzentrierte Kraft f, Einzelkraft f
f force f concentrée [ponctuelle]
nl geconcentreerde kracht f(m)
r сосредоточенная сила f

C851 e **concentrated load**
d Punktlast f, Einzellast f
f charge f concentrée [ponctuelle]
nl puntlast m
r сосредоточенная нагрузка f

C852 e **concentration**
d Konzentration f; Anreicherung f; Eindicken n
f concentration f
nl concentratie f; indikking f
r концентрация f; сосредоточение n; обогащение n (минерального сырья); выпаривание n; сгущение n

C853 e **concrete**
d Beton m
f béton m
nl beton n
r бетон m

C854 e **concrete accelerator**
d Abbindebeschleuniger m (für Beton)
f accélérateur m de durcissement (de béton)
nl verhardingsversneller m
r ускоритель m твердения (бетона)

C855 e **concrete additive, concrete admixture**
d Betonzusatz m
f adjuvant m [additif m] du béton
nl bijmenging f [bijslag m] voor het beton
r добавка f к бетону

C856 **concrete aggregate** see **aggregate**

C857 e **concrete antifreezer**
d Betonfrostschutzmittel n

f antigel m, agent m antigel, antigivre(ur) m (pour bétons)
nl vorstwerend middel n
r морозозащитная добавка f (для бетона)

C858 e **concrete base** see **concrete bed**

C859 e **concrete batching plant**
d Betondosieranlage f
f centrale f [poste m] de dosage du béton, centrale f doseuse du béton
nl inrichting f voor het doseren van betoncomponenten
r установка f для дозирования компонентов бетонной смеси

C860 e **concrete bed**
d Betonbett n
f lit m de béton
nl betonbed n
r бетонная подготовка f

C861 e **concrete belt placing tower**
d Betonierbandturm m
f tour f distributrice de béton à transporteur à bande
nl betontoren m met transportband
r бетонолитная башня f с ленточным конвейером

C862 e **concrete bin**
d Betonbunker m
f trémie f à béton frais
nl bunker m voor betonspecie
r приёмный бункер m для бетонной смеси

C863 e **concrete blanket**
d Betonaußendichtung f (Staudamm)
f écran m étanche [d'étanchéité] de béton (pour barrage)
nl filtratiebeschermingsscherm n
r бетонный противофильтрационный экран m (плотины)

C864 e **concrete blinding coat**
d Betondeckschicht f; Betonausgleichschicht f; Vorsatzbeton m, Sichtbetonschicht f; Sauberkeitsbetonschicht f, Betonunterschicht f
f chape f en béton; enduit m de finition en ciment; sous-couche f [lit m] de béton
nl betondeklaag f(m)
r бетонная стяжка f; бетонный отделочный [затирочный] слой m; бетонная подготовка f

C865 e **concrete block**
d Betonquader m, Betonblock m, Betonstein m
f bloc m de béton
nl betonblok n
r бетонный массив m [блок m, камень m]

C866 e **concrete block pavement**
d Betonblockpflaster n
f pavage m en blocs de béton
nl straatweg m van betonklinkers

r мостовая *f* из бетонных блоков [камней]

C867 *e* **concrete blockwork**
d Betonblockmauerwerk *n*
f maçonnerie *f* en blocs de béton
nl metselwerk *n* van betonsteen
r кладка *f* из бетонных блоков [камней]

C868 **concrete bonding plaster** *see* **bond plaster**

C869 *e* **concrete breaker**
d Beton-Aufbruchhammer *m*, Betonreißhammer *m*, Betonbrecher *m*
f marteau *m* brise-béton, brise-béton *m*
nl betonbreker *m*
r бетонолом *m*

C870 *e* **concrete brick**
d Betonbaustein *m*, Betonwerkstein *m*
f brique *f* [aggloméré *m*] en béton
nl betonsteen *m*
r бетонный камень *m*

C871 *e* **concrete bridge**
d Betonbrücke *f*
f pont *m* en béton
nl betonbrug *f(m)*
r бетонный мост *m*

C872 *e* **concrete buggy**
d Betonkarre *f*
f chariot *m* à béton
nl kipwagen *m* voor betonmortel, japanner *m*
r тележка *f* для подачи бетонной смеси

C873 *e* **concrete caisson**
d Stahlbeton-Senkkasten *m*
f caisson *m* (ouvert) en béton, massif *m* creux en béton
nl benonnen caisson *m*
r железобетонный массив-гигант *m*

C875 *e* **concrete carriageway**
d Betonfahrbahn *f*
f chaussée *f* bétonnée [en béton]
nl betonweg *m*
r проезжая часть *f* с бетонным покрытием

C876 *e* **concrete catch gutter**
d Betonablaufrinne *f*
f caniveau *m* de surface en béton
nl betonnen afvoergoot *f(m)*
r бетонный дренажный лоток *m*

C877 *e* **concrete chute**
d Betongießrinne *f*
f goulotte *f* à béton
nl stortgoot *f(m)* voor beton
r жёлоб *m* для подачи бетонной смеси; бетонолитный лоток *m*

C878 *e* **concrete chuting cable crane**
d Betonkabelkran *m*, Kabelkran *m* mit Beton-Gießvorrichtung

f blondin *m* [grue *f* à câble] avec disposition de coulage de béton par gravité
nl kraan *f(m)* voor de stortgoot
r бетонолитный кабель-кран *m*

C879 *e* **concrete compacted by jolting**
d Schockbeton *m*, Stoßbeton *m*
f béton *m* choqué [compacté par chocs]
nl schokbeton *n*
r бетон *m*, уплотнённый встряхиванием

C880 *e* **concrete compactor**
d Betonverdichter *m*
f engin *m* de serrage [de compactage] de béton, compacteur *m* de béton
nl betonverdichter *m*
r уплотнитель *m* бетонной смеси

C881 *e* **concrete composition**
d Betonaufbau *m*, Betonzusammensetzung *f*
f composition *f* du béton
nl betonsamenstelling *f*
r состав *m* бетона

C882 *e* **concrete compression strength, concrete compressive strength**
d Betondruckfestigkeit *f*
f résistance *f* de béton à la compression
nl druksterkte *f* van het beton
r прочность *f* [предел *m* прочности] бетона при сжатии

C883 *e* **concrete consistence**
d Betonkonsistenz *f*; Betonsteife *f*
f consistence *f* de béton frais
nl betonconsistentie *f*
r консистенция *f* или подвижность *f* бетонной смеси

C884 *e* **concrete construction**
d Betonbau *m*
f 1. construction *f* [ouvrage *m*] en béton 2. construction *f* des ouvrages de béton
nl 1. betonnen constructie *f* 2. bouwen *n* in beton
r 1. сооружение *n* из бетона 2. строительство *n* бетонных сооружений

C885 *e* **concrete-conveying pipe**
d Betonförderrohr *n*, Betontransportrohr *n*
f conduit *m* à béton (frais)
nl betontransportleiding *f*
r бетоно(про)вод *m*

C886 *e* **concrete core**
d Beton(bohr)kern *m*, Betonprobekern *m*
f carotte *f* (cylindrique) découpée de béton
nl betonmonsterboorkern *f(m)*
r бетонный керн *m*, цилиндрический образец *m* бетона, вырезаемый из толщи бетонной конструкции

C887 *e* **concrete core slab, concrete core unit**

d Stahlbetonhohldiele *f*,
Stahlbetonhohlplatte *f*
f dalle *f* creuse [alvéolée, vide] en
béton (armé)
nl holle vloerplaat *f(m)* van gewapend
beton
r пустотелая железобетонная плита-
-настил *f*, многопустотная
железобетонная плита *f* [панель *f*]

C888 *e* concrete cover
d Betonüberdeckung *f* der
Stahleinlagen
f couche *f* de protection; recouvrement
m des armatures
nl betonlaag *f(m)* buiten de wapening
r защитный слой *m* бетона

C889 concrete crushing strength *see*
concrete compression strength

C890 *e* concrete curing
d Betonnachbehandlung *f*
f cure *f* [curing *m*] du béton
nl nabehandeling *f* van beton
r уход *m* за бетоном; выдерживание
n бетона

C891 *e* concrete cutting machine
d Betontrennmaschine *f*
f machine *f* à scier le béton,
machine *f* pour le sciage du béton
nl betonsnijmachine *f*
r машина *f* для резки бетона

C892 *e* concrete dam
d Betonstaumauer *f*, Betontalsperre *f*,
Betondamm *m*
f barrage *m* en béton
nl betonnen stuwdam *m*
r бетонная плотина *f*

C893 concrete desintegration *see* concrete
segregation

C894 *e* concrete dispenser
d Betonverteiler *m*, Betoneinbringer
m
f distributeur *m* de béton, machine *f*
de répartition du béton
nl betonverdeler *m*
r раздатчик *m* [распределитель *m*]
бетонной смеси

C895 *e* concrete finishing machine
d Beton(fahrbahn)fertiger *m*,
Betondeckenfertiger *m*;
Betonglättmaschine *f*,
Abreibmaschine *f*
f finisseur *m* [finisseuse *f*] de
revêtement en béton; lisseur *m*
[lisseuse *f*] mécanique pour béton
nl betonafwerkmachine *f*
r бетоноотделочная машина *f*;
затирочная машина *f* (*для
бетонных покрытий*)

C896 *e* concrete foundation
d Betonfundament *n*
f fondation *f* en béton
nl betonfundering *f*
r бетонный фундамент *m*

C897 *e* concrete grade
d Betongüteklasse *f*
f classe *f* de qualité de béton
nl betonkwaliteit *f*
r марка *f* бетона

C898 *e* concrete grouter
d Betonspritzmaschine *f*
appareil *m* de projection du béton
frais
nl betonspuitmachine *f*
r бетоншприцмашина *f*

C899 *e* concrete gun
d Torkretiergerät *n*, Torkretkanone *f*
f cément-gun *m*, guniteuse *f*
nl mortelspuit *f(m)*
r цемент-пушка *f*

C900 *e* concrete hardening
d Beton(er)härtung *f*
f durcissement *m* de béton
nl verharding *f* van het beton
r твердение *n* бетона

C901 concrete injection unit *see* concrete
grouter[;]

C902 concrete-in-mass *see* mass concrete

C903 *e* concrete interlocking tile
d Beton-Dachfalzstein *m*,
Betonfalz(dach)stein *m*
f tuile *f* à emboîtement en béton [en
ciment]
nl betonnen dakpan *f(m)*
r цементная черепица *f*, черепица *f*
из цементно-песчаных растворов
(*с замковым соединением*)

C904 *e* concrete joint sealing compound
d Betonfugenvergußmasse *f*
f mastic *m* pour étanchement de
joints de béton
nl voegenvulmassa *f(m)* voor
betonplaten
r уплотняющая мастика *f* для]
заливки швов в бетонных
конструкциях

C905 *e* concrete-laying machine
d Betondeckenfertiger *m*
f bétonneuse *f* routière
nl betonverwerkingsmachine *f*
r дорожный бетоноукладчик *m*

C906 *e* concrete levelling course
d Betonausgleichschicht *f*
f couche *f* d'égalisation en béton
nl vlaklaag *f(m)* van beton
r выравнивающий слой *m* бетона

C907 *e* concrete lighting column, concrete
lighting mast
d Beton-Lichtmast *m*, Beton-
-Beleuchtmast *m*
f poteau *m* de ligne d'éclairage en
béton (préfabriqué)
nl betonnen lichtmast *m*
r бетонная осветительная мачта *f*

C908 concrete line *see* concrete conveying
pipe

C909 *e* **concrete lining**
   d Betonauskleidung *f*, Betonmantel *m*;
     Betonummantelung *f*, Betonausbau
     m
   f revêtement *m* en béton; cuvelage *m*
     en béton
  nl betonmantel *m*
   r бетонная облицовка *f*; бетонная
     обделка *f* [крепь *f*]

C910 *e* **concrete mixer**
   d Betonmischer *m*
   f bétonnière *f*, malaxeur *m*
     [malaxeuse *f*] à beton, mélangeur *m*
     à béton
  nl betonmolen *m*
   r бетоносмеситель *m*

C911 *e* **concrete mixer with vibrating
     blades**
   d Vibroschaufel-Betonmischer *m*
   f malaxeur *m* [malaxeuse *f*] à
     ailettes vibratoires
  nl betonmolen *m* met vibrerende
     schoepen
   r вибробетоносмеситель *m*

C912 *e* **concrete mixing plant**
   d 1. Betonmischanlage *f*
     2. Betonfabrik *f*, Großbetonanlage *f*
   f 1. centrale *f* à béton 2. usine *f* de
     production de béton
  nl beton(mortel)centrale *f*
   r 1. бетоносмесительная установка *f*
     2. бетонный завод *m*

C913 *e* **concrete mixture**
   d Betonmischung *f*, Betongemisch *n*
   f béton *m* frais, mélange *m* de béton
  nl betonmortel *m*
   r бетонная смесь *f*

C914 *e* **concret-mix vibration**
   d Beton(ein)rütteln *n*
   f vibrage *m* [vibration *f*] de béton
  nl betontrillen *n*
   r вибрирование *n* бетонной смеси

C915 *e* **concrete nail** *US*
   d Dübelbolzen *m*
   f clou *m* à béton
  nl betonspijker *m*
   r гвоздь *m* для бетона *или* кирпичной
     кладки, дюбель-гвоздь *m*

C916 *e* **concrete patching**
   d Betonreparatur *f*, Betonausbesserung
     f, Schlaglochausfüllung *f*
   f point *m* à temps
  nl betonreparatie *f*
   r ямочный ремонт *m* бетонного
     покрытия

C917 *e* **concrete pavement**
   d Betonfahrbahndecke *f*
   f revêtement *m* routier en béton
  nl betonnen wegdek *n*
   r бетонное дорожное покрытие *n*

C918 *e* **concrete pavement spreader**
   d Betondeckenverteiler *m*
   f bétonneuse *f* de route [routière]

  nl betonspreidmachine *f*
   r дорожный бетоноукладчик *m*

C919 *e* **concrete paver**
   d Straßenbetoniermaschine *f*
   f bétonnière *f* distributrice motorisée
  nl betonverwerkingsmachine *f*
   r дорожный бетоносмеситель-
     -бетоноукладчик *m*

C920 *e* **concrete paving block**
   d Betonpflasterstein *m*
   f bloc *m* de béton pour pavage; pavé
     m en béton
  nl beton(straat)klinker *m*
   r бетонный блок *m* или шашка *f*
     для мощения; бетонная брусчатка
     f

C921 *e* **concrete pile**
   d Betonpfahl *m*
   f pieu *m* en béton
  nl betonpaal *m*
   r бетонная свая *f*

C922 *e* **concrete-pile follower**
   d Schlagjungfer *f*
   f avant-pieu *m*
  nl heimuts *f(m)*
   r подбабок *m* (*при забивке свай*)

C923 *e* **concrete pile foundation**
   d Betonpfahlgründung *f*,
     Betonpfahlfundation *f*
   f fondation *f* sur pieux en béton
  nl betonpaalfundering *f*
   r железобетонный свайный
     фундамент *m*, фундамент *m* из
     железобетонных свай

C924 *e* **concrete piling**
   d Beton-Spundwand *f*
   f rideau *m* [cloison *m*] de
     palplanches en béton
  nl betonnen damwand *m*
   r бетонная шпунтовая стенка *f*

C925 *e* **concrete pipe**
   d Betonrohr *n*
   f tuyau *m* en béton
  nl betonnen buis *f (m)*
   r бетонная труба *f*

C926 *e* **concrete placed in the work**
   d eingebrachter Beton *m*
   f béton *m* mis en œuvre
  nl in het werk gestort beton *n*
   r бетон *m*, уложенный в сооружение

C927 *e* **concrete placement**
   d Betoneinbau *m*, Betoneinbringung *f*,
     Betonieren *n*
   f bétonnage *m*, coulage *m* [coulée *f*]
     du béton, mise *f* en place [en
     œuvre] du béton
  nl beton storten
   r укладка *f* бетонной смеси,
     бетонирование *n*

C928   **concrete placer** *see* **pneumatically
     operated concrete placer**

C929 *e* **concrete plasticiser**
   d Betonweichmacher *m*

f plastifi(c)ant *m* de béton, adjuvant *m* plastifi(c)ant de béton
nl betonplastificeerder *m*
r пластификатор *m* бетона, пластифицирующая добавка *f* к бетону

C930 concrete pouring *see* concrete placement

C931 concrete production plant *see* concrete mixing plant

C932 *e* concrete progress chart
d Betoneinbauplan *m*, Betonierungsplan *m*
f programme *m* de bétonnage
nl stortprogramma *n*
r график *m* бетонирования

C933 *e* concrete pump
d Betonpumpe *f*
f pompe *f* à béton
nl betonpomp *f(m)*
r бетононасос *m*

C934 *e* concrete pumping
d Betonieren *n* mit Pumpe, Pumpförderung *f* von Beton
f pompage *m* de béton, bétonnage *m* à la pompe
nl opvoer *m* van het beton met pomp
r подача *f* бетонной смеси бетононасосом

C935 *e* concrete radiation shield
d Betonabschirmung *f*
f enveloppe *f* [enceinte *f*] étanche en béton (*protection contre le rayonnement radioactif*)
nl betonnen stralingsscherm *n*
r бетонный биологический (противорадиационный) экран *m*

C936 *e* concrete reactor shield
d Beton-Reaktor(strahlen)schutzanlage *f*
f protection *f* en béton (*d'un réacteur nucléaire*)
nl betonnen (kern)reactorscherm *n*
r бетонная защита *f* реактора; противорадиационный экран *m* реактора

C937 *e* concrete road
d Betonstraße *f*
f route *f* en béton (de ciment), route *f* bétonnée
nl betonweg *f(m)*
r бетон(ирован)ная дорога *f*, дорога *f* с бетонным покрытием

C938 *e* concrete roof
d Betondach *n*
f couverture *f* en béton (armé)
nl betonnen dak *n*
r (железо)бетонное покрытие *n*; (железо)бетонная крыша *f*

C939 *e* concrete roof deck
d Beton(decken)belag *m*
f dalle *f* de couverture en béton (armé); plancher-toiture *m* en béton

(armé), dalle *f* de toiture en béton (armé)
nl betonnen dakplaten *f(m) pl*
r бетонный настил *m* покрытия

C940 *e* concrete sand
d Betonsand *m*
f sable *m* à béton
nl betonzand *n*
r песок *m*, годный для приготовления бетонной смеси

C941 *e* concrete saw
d Betontrennsäge *f*; Fugenschleifgerät *n*
f scie *f* à béton
nl betonzaagmachine *f*
r дисковая пила *f* для резки бетона; дисковая пила *f* для нарезки швов в бетонном дорожном покрытии

C942 *e* concrete segregation
d Betonzersetzung *f*, Entmischung *f*
f ségrégation *f* du béton frais
nl ontmengen *n* van betonmortel
r расслоение *n* [расслаивание *n*] бетонной смеси

C943 *e* concrete shrinkage
d Betonschwindung *f*
f retrait *m* du béton
nl betonkrimp *m*
r усадка *f* бетона

C944 *e* concrete slab
d Betonplatte *f*
f dalle *f* [plaque *f*] en béton
nl betonnen vloerplaat *f(m)*
r бетонная плита *f*

C945 concrete spraying *see* gunite work

C946 *e* concrete spreader
d Betonverteiler *m*
f distributeur *m* de béton, machine *f* de répartition de béton
nl betonspreidmachine *f*
r бетонораспределительная машина *f*

C947 *e* concrete strength
d Betonfestigkeit *f*; Betongüte(klasse) *f*
f résistance *f* du béton
nl betonsterkte *f*
r прочность *f* бетона; марка *f* бетона

C948 *e* concrete structure
d 1. Betonbauwerk *n*
2. Betonaufbau *m*, Struktur *f* des Betons
f 1. ouvrage *m* en béton
2. structure *f* du béton
nl 1. betonskelet *n* 2. structuur *f(m)* van het beton
r 1. бетонное сооружение *n*
2. структура *f* бетона

C949 *e* concrete technology
d Betontechnologie *f*
f technologie *f* du béton
nl betontechnologie *f*
r технология *f* бетона

C950 e concrete test cube
d Betonprobewürfel m
f cube m d'essai (en béton)
nl betonproefkubus m
r бетонный куб(ик) m (для
испытания)

C951 e concrete texture
d Betongefüge n
f texture f du béton
nl textuur f van het beton
r текстура f бетона [бетонной
поверхности]; фактурная отделка f
бетонной поверхности

C952 e concrete tile
d Betonfliese f, Betonplatte f
f tuile f en béton
nl betontegel m
r цементная черепица f

C953 e concrete truck
d Betontransportwagen m
f camion-benne m [camion-citerne m]
pour béton
nl betontransportauto n
r бетоновоз m

C954 e concrete vibrating machine
d Betonrüttelmaschine f
f machine f vibreuse pour route en
béton, machine f vibrofinisseuse,
vibrofinisseur m, vibro-surfaceur m
nl betontrilmachine f
r машина f для виброуплотнения
бетонного покрытия

C955 e concrete vibrator
d Beton-Rüttler m, Beton-Vibrator m
f vibr(at)eur m à béton
nl trilnaald f(m)
r вибратор m для бетона

C956 e concrete waffle slab
d Beton-Kassettenplatte f
f dalle f caisson de béton, dalle f
[plaque f] alvéolaire en béton
nl betoncassettenplaat f(m)
r кессонная бетонная плита f
(перекрытия)

C957 e concrete waterproofer
d Betondichter m
f reducteur m de perméabilité de
béton
nl waterdichtheid verhogende bijslag m
r гидрофобная добавка f,
обеспечивающая водонепроницае-
мость бетона

C958 e concrete works
d 1. Betonarbeiten f pl
2. Betonbauwerke n pl
f 1. travaux m pl de bétonnage
2. ouvrages m pl en béton
nl 1. betonarbeid m
2. betonbouwwerken n pl
r 1. бетонные работы f pl 2. бетонные
сооружения n pl

C959 e concreting boom
d Betongießausleger m

f flèche f (articulée) distributrice
nl betonstortgiek m
r бетонораспределительная стрела f

C960 e concreting hopper
d Betonier(ungs)bunker m
f trémie f de bétonnage
nl storttrechter m voor beton
r бункер m для бетонирования

C961 e concreting in lifts
d schichtweises Betonieren n
f bétonnage m [mise f en place du
béton] par couches
nl laagswijs betonstorten n
r послойное бетонирование n

C962 concreting operations see concrete
placement

C963 e concreting paper
d Papierunterlage f, Unterlagspapier n
f papier m utilisé dans les coffrages
à béton
nl kraftpapier n, wegenpapier n
r крафт-бумага f, картон m (для
предупреждения сцепления бетона
с опалубкой)

C964 e concreting sequence
d Betonierfolge f
f séquence f (des travaux) de
bétonnage
nl stortvolgorde f
r последовательность f [план m]
бетонирования (блоков
массивного сооружения)

C965 e concreting skip
d Betonkübel m
f benne f de bétonnage
nl betonstortbak m
r ковш m [бадья f] для подачи
бетона; скип m

C966 e concreting train
d Betonierzug m
f convoi m [colonne f] de machines
routières pour l'exécution d'une
chaussée en béton [pour le
bétonnage d'une chaussée]
nl betontrein m
r бетонопоезд m (группа машин,
осуществляющая строительство
бетонного покрытия дороги по
принципу технологического
потока)

C967 e concurrent flow
d Gleichstrom m
f écoulement m rectiligne, équicourant m
nl gelijkstroom m
r прямоток m

C968 e concurrent forces
d zusammenwirkende [sich kreuzende]
Kräfte f pl
f forces f pl intersectées [croisées]
nl samenwerkende krachten f(m) pl
r пересекающиеся силы f pl

C969 e condensate
d Kondensat n, Kondenswasser n

f condensat *m*, eau *f* de condensation
nl condensaat *n*
r конденсат *m*

C970 *e* **condensate collector, condensate header**
d Kondensatsammelrohr *n*, Kondensatsammler *m*
f collecteur *m* de condensat
nl condensaat(op)vanger *m*
r коллектор *m* для отвода попутного конденсата, конденсатосборник *m*

C971 *e* **condensate line**
d Kondens(at)leitung *f*
f conduit *m* de condensat
nl condensaatleiding *f*
r конденсатопровод *m*

C972 *e* **condensate pump**
d Kondensatpumpe *f*
f pompe *f* à condensat
nl condensaatpomp *f(m)*
r конденсатный насос *m*

C973 *e* **condensate return pipe**
d Kondensat(rückspeise)leitung *f*
f conduit *m* de retour de condensat
nl terugvoerleiding *f* voor het condensaat
r конденсатопровод *m*, линия *f* возврата конденсата

C974 **condensate trap** *see* **steam trap**

C975 *e* **condensation**
d Kondensation *f*
f condensation *f*
nl condensatie *f*
r конденсация *f*

C976 *e* **condensation heat**
d Kondensationswärme *f*
f chaleur *f* de condensation
nl condensatiewarmte *f*
r теплота *f* конденсации

C977 *e* **condensation tank**
d Kondensatbehälter *m*
f réservoir *m* de condensat
nl condensaatvat *n*
r бак *m* для конденсата

C978 *e* **condenser**
d Verflüssiger *m*, Kondensator *m*
f condensateur *m*
nl condens(at)or *m*
r конденсатор *m*

C979 *e* **condenser water**
d Kondensatorwasser *n*
f eau *f* du condenseur
nl condensatiewater *n*
r конденсаторная вода *f*

C980 *e* **condensing pressure**
d Kondensationsdruck *m*
f pression *f* de condensation
nl condensatiedruk *m*
r давление *n* конденсации

C981 *e* **condensing unit**
d Kondensatorsatz *m*
f groupe *m* compresseur-condenseur

nl condensorinstallatie *f*
r компрессорно-конденсаторный агрегат *m*

C982 *e* **conditioned space**
d klimatisierter Raum *m*
f espace *m* conditionné
nl geconditioneerde ruimte *f*
r кондиционируемое пространство *n* [помещение *n*]

C983 *e* **conditioning**
d Konditionierung *f* (*Holz*)
f conditionnement *m* (*du bois*)
nl conditioneren *n* (*hout*)
r выдерживание *n,* кондиционирование *n* (*древесины*)

C984 *e* **condition of instability**
d Instabilitätsbedingung *f*
f condition *f* (d'état) d'instabilité
nl instabiele toestand *m*
r условие *n* неустойчивости [неустойчивого состояния]

C985 *e* **conditions of exposure**
d Verhältnisse *n pl* der Umwelteinwirkung
f effets *m pl* des conditions climatiques [d'ambiance]
nl inwerking *f* van het klimaat
r характер *m* [условия *n pl*] воздействия окружающей среды

C986 *e* **conductor**
d 1. Stromleiter *m* 2. innenliegendes Regenfallrohr *n*, innere Dachentwässerung *f*
f 1. conducteur *m*; fil *m*, câble 2. descente *f* (pluviale) interne
nl 1. geleider *m* 2. ingebouwde regenpijp *f(m)*
r 1. проводник *m*; провод *m* 2.водосточная труба *f* (*внутри здания*), внутренний водосток *m*

C987 *e* **conduit**
d 1. Kanal *m*, Rohrleitung *f*; Zubringerleitung *f* 2. Kabelrohr *n*, Kabelkanal *m*, Leitungskanal *m*
f 1. canal *m*, conduit *m*, conduite *f* 2. caniveau *m* pour câbles
nl 1. kanaal *n*, leiding *f* 2. leidinggoot *f(m)*, mantelbuis *f(m)*
r 1. канал *m*, трубопровод *m*; водовод *m* 2. кабелепровод *m*; электротехнический короб *m*; кабельная канализация *f*

C988 *e* **cone of depression**
d Senkungstrichter *m*, Entnahmetrichter *m*, Depressionstrichter *m*, Depressionsspiegel *m*, Absenkkegel *m*
f cône *m* d'appel, entonnoir *m* de dépression de la nappe
nl indrukking *f* van de kegel
r воронка *f* депрессии, депрессионная поверхность *f*

C989 *e* **cone penetration test**
d Kegeleindringversuch *m*

f essai *m* de pénétration au cône
nl kegelindrukkingstest *m* (*van het draagvermogen van de bodem*)
r определение *n* несущей способности грунта методом погружения конуса

C990 e **cone penetrometer**
d Kegelgerät *n*, Kegeleindringungsapparat *m*
f pénétromètre *m* à cône [conique]
nl kegelpenetrometer *m*
r конусный пенетрометр *m*

C991 e **cone shell**
d Kegelschale *f*
f coque *f* conique
nl kegelomhulsel *n*
r коническая оболочка *f*

C992 e **confined aquifer**
d artesischer Grundwasserleiter *m*
f nappe *f* aquifère artésienne
nl artesische watervoerende laag *f(m)*
r артезианский водоносный горизонт *m*

C993 e **confined water** *US*
d artesisches Wasser *n*, gespanntes Grundwasser *n*
f eau *f* artésienne, nappe *f* captive
nl artesisch water *n*
r напорные воды *f pl*

C994 e **confining bed**
d undurchlässige Schicht *f*, Grundwassersohle *f*
f couche *f* encaissante, substratum *m* imperméable d'une nappe souterraine
nl waterkerende bodemlaag *f(m)*
r водоупор *m*, водонепроницаемый слой *m* [пласт *m*]

C995 **conical shell** *see* **cone shell**
C996 **conjugate depths** *see* **alternate depths**
C997 e **conjugation**
d Anschluß *m*, Zusammenschluß *m*, Verbindung *f*
f conjugaison *m*, (con)jonction *f*, raccordement *m*
nl verbinding *f*
r сопряжение *n*

C998 e **connecting angle**
d Anschlußwinkel *m*
f cornière *f* d'assemblage [d'attache]
nl verbindingshoekijzer *n*
r соединительный уголок *m*

C999 e **connecting reinforcement**
d Anschlußbewehrung *f*
f armature *f* de raccordement
nl verbindingswapening *f*
r соединительная арматура *f*, стыковые выпуски *m pl* арматуры

C1000 e **connecting sewer**
d Grundstückanschlußleitung *f*, Anschlußkanal *m*

f conduite *f* d'assainissement de raccordement
nl aansluitriool *n*
r соединительная ветка *f* канализации

C1001 **connection angle** *see* **connecting angle**
C1002 e **connection dimension**
d Anschlußmaß *n*
f dimension *f* de raccordement
nl aansluitmaat *f*
r присоединительный размер *m*

C1003 e **connector**
d Anschlußstück *n*, Verbinder *m*
f pièce-raccord *f*
nl koppelstuk *n*
r соединительный элемент *m*

C1004 e **conoidal shell**
d Konoidschale *f*
f coque *f* conoïde
nl kegelvormig omhulsel *n*
r коноидальная оболочка *f*

C1005 e **conservation of energy**
d Erhaltung *f* der Energie
f conservation *f* de l'énergie
nl behoud *n* van arbeidsvermogen [van energie]
r сохранение *n* энергии; экономия *f* энергии

C1006 **conservation storage reservoir** *see* **carry-over storage**
C1007 e **consistence, consistency**
d Konsistenz *f*, Steife *f*
f consistance *f*
nl consistentie *f*
r консистенция *f*

C1008 e **consistency index**
d Konsistenzzahl *f*
f indice *m* de consistance
nl consistentiewaarde *f*
r показатель *m* консистенции

C1009 e **consistometer**
d Konsistenzmesser *m*
f consistomètre *m*, sonde *f* de consistance
nl consistentiemeter *m*
r консистометр *m*

C1010 **console** *see* **cantilever**
C1011 e **consolidated immediate shear test**
d Schnellscherversuch *m* mit vorbelasteter Probe
f essai *m* de cisaillement rapide sur échantillon consolidée
nl directe schuifproef *f(m)*
r испытание *n* (*грунта*) на сдвиг при быстром увеличении давления

C1012 e **consolidation**
d Konsolidierung *f*, Verdichtung *f*
f consolidation *f*, compactage *m* (*du sol*)
nl verdichting *f*
r консолидация *f*, уплотнение *n* (*грунта*)

C1013   *e* **consolidation settlement**
    d Verdichtungssetzung *f*
    f tassement *m* dû à la consolidation (du sol)
    nl zetting *f* tengevolge van grondverdichting
    r осадка *f* вследствие уплотнения [консолидации] грунта в основании сооружения

C1014   *e* **consolidation test**
    d Konsolidierungsversuch *m*
    f essai *m* de consolidation
    nl samendrukkingsproef *f(m)*
    r компрессионное испытание *n*

C1015   *e* **consolidation test apparatus, consolidometer**
    d Oedometer   *n*,   Kompressionsgerät *n*
    f œdomètre *m*
    nl samendrukkingsapparaat *n*
    r одометр *m*

C1016   *e* **constant pressure regulator**
    d Konstantdruckregler *m*
    f régulateur *m* à pression constante
    nl regelaar *m* voor constante druk
    r регулятор *m* постоянного давления

C1017   *e* **constant pressure valve**
    d Konstantdruckventil *n*
    f soupape *f* [robinet *m*] à pression constante
    nl ventiel *n* voor constante druk
    r клапан *m* постоянного давления

C1018   *e* **constant volume system**
    d Klimaanlage *f* mit konstantem Luftvolumendurchsatz
    f installation *f* de conditionnement à débit d'air constant
    nl luchtbehandelingssysteem *n* met constant volume
    r система *f* кондиционирования с постоянным расходом воздуха

C1019   *e* **constituents of concrete**
    d Betonbestandteile *m pl*
    f constituants *m pl* du béton
    nl bestanddelen *n pl* van het beton
    r компоненты *m pl* бетонной смеси

C1020   *e* **constraint**
    d Einspannung *f*; Zwang *m*, Zwangsbindung *f*
    f encastrement *m*
    nl samenhang *m*, dwang *m*
    r защемление *n*; налагаемая связь *f*

C1021   *e* **construction**
    d 1. Bau *m*, Bauen *n*, Errichtung *f* 2. Konstruktion *f*; Bauart *f*, Bauweise *f*; Bauausführung *f*
    f 1. construction *f*, exécution *f* des travaux 2. construction *f*; bâtiment *m*; ouvrage *m*; schéma *m* (de fonctionnement)
    nl 1. bouw *m*, bouwwijze *f*; aanleg *m* 2. bouwwerk, constructie *f*

C1022   *r* 1. строительство *n*; возведение *n*, сооружение *n* 2. конструкция *f*; устройство *n*

C1022   **construction adit** *see* **access tunnel**

C1023   *e* **constructional concrete**
    d konstruktiver Beton *m*
    f béton *m* pour ouvrages d'art
    nl constructiebeton *n*
    r конструктивный бетон *m*

C1024   *e* **constructional defect**
    d Baufehler *m*
    f vice *m* [défaut *m*] de construction
    nl constructiefout *f(m)*
    r строительный дефект *m*

C1025   **constructional materials** *see* **building materials**

C1026   *e* **constructional project**
    d 1. Bauobjekt *n*, Bauvorhaben *n* 2. Bauprojekt *n*, Bauentwurf *m*
    f 1. chantier *m* de bâtiment, bâtiment *m* [ouvrage *m*] en stade de construction 2. projet *m* de construction
    nl 1. bouwproject *n* 2. bouwplan *n*
    r 1. строительный объект *m* 2. строительный проект *m*

C1027   *e* **constructional work**
    d Bauarbeiten *f pl*
    f travaux *m pl* de construction [de bâtiment]
    nl bouwwerkzaamheden *f pl*
    r строительные работы *f pl*

C1028   *e* **construction area**
    d konstruktive Fläche *f*
    f surface *f* de construction (*aire sommaire des sections horizontales des murs et des cloisons d'un bâtiment*)
    nl gebouwoppervlak *n*
    r конструктивная площадь *f* (*площадь горизонтальных сечений стен и перегородок здания*)

C1029   *e* **construction budget**
    d Baukostenvoranschlag *m*
    f devis *m* estimatif, pris *m* [coût *m*] de construction (d'un ouvrage)
    nl bouwbegroting *f*
    r (сметная) стоимость *f* строительства

C1030   **construction climatology** *see* **building climatology**

C1031   *e* **construction documents**
    d Bauunterlagen *f pl*, Ausführungsunterlagen *f pl*, Baudokumentation *f*
    f documentation *f* sur la construction; documents *m pl* d'exécution
    nl bestek en tekeningen
    r строительно-техническая документация *f*

C1032   *e* **construction economics**
    d Bauökonomik *f*, Bauwirtschaft *f*

f économie *f* de la construction
nl bouweconomie *f*
r экономика *f* строительства

C1033 *e* **construction engineering**
d Bauwesen *n*; Bautechnik *f*;
Tiefbau *m*
f technique *f* du bâtiment et des
travaux publics; étude *f* de
construction et réalisation des
travaux
nl weg- en waterbouwkunde *f*;
civiele techniek *f*
r строительство *n*; технология *f*
строительного производства;
проектирование *n*, строительство
n и техническая эксплуатация *f*
инженерных сооружений

C1034 *e* **construction equipment**
d Bauausrüstung *f*, Baugeräte *n pl*,
Baumaschinen *f pl*
f équipement *m* de chantier,
matériel *m* de génie civil
nl bouwmachines en -gereedschappen
r строительное оборудование *n*,
строительные машины *f pl*

C1035 *e* **construction expenditure forecast**
d prognostische Baukosten *pl*
f prévision *f* du coût des travaux
(de construction)
nl begroting *f* van bouwkosten,
betalingsplan *n*
r прогноз *m* расходов на
строительство

C1036 **construction fits** *see* **construction
tolerances**

C1037 *e* **construction industry**
d Bauindustrie *f*
f industrie *f* du bâtiment
nl bouw *m*; bouwbedrijf *m*
r строительство *n*; строительная
промышленность *f*

C1038 *e* **construction joint**
d Arbeitsfuge *f*, Baufuge *f*,
Abschnittsfuge *f*
f joint *m* d'éxecution [de reprise, de
bétonnage, de construction]
nl dilatatievoeg *f(m)*
r рабочий [строительный] шов *m*,
шов *m* бетонирования

C1039 *e* **construction loads**
d Baulasten *f pl*,
Montagebelastungen *f pl*
f charge *f* de montage
nl bouwbelastingen *f pl*
r строительные [монтажные]
нагрузки *f pl*

C1040 *e* **construction machinery**
d Baumaschinen *f pl*
f matériel *m* des chantiers de
construction [de génie civil]
nl bouwmachines *f pl*
r строеляные машины *f pl*

C1041 *e* **construction management**
d Bauleitung *f*

f direction *f* de chantier (de
construction), direction *f* des
travaux
nl bouwdirectie *f*
r управление *n* строительством

C1043 *e* **construction methods**
d Bauverfahren *n pl*
f méthodes *f pl* d'exécution des
travaux de construction, pratique *f*
de travaux de construction
nl bouwmethodes *f pl*, bouwwijze *f*
r методы *m pl* строительства,
способы *m pl* производства
строительных работ

C1044 *e* **construction module**
d Baumodul *m*,
Maßordnungseinzelmaß *n*
f module *m* (de construction)
.nl bouwmodulus *m*
r строительный модуль *m*

C1045 **construction operations** *see*
building operations

C1046 *e* **construction paper**
d Baupappe *f*
f papier *m* de construction
nl wegenpapier *n*
r строительный картон *m*

C1047 *e* **construction period**
d Bauzeit *f*, Ausführungszeit *f*
f delai *m* de construction, durée *f*
des travaux de construction
nl bouwtijd *m*
r продолжительность *f*
строительства

C1048 **construction practice** *see*
construction techniques

C1049 *e* **construction recommendations**
d Konstruktionsrichtlinien *f pl*,
Empfehlungen *f pl* für
Baudurchführung
f recommandations *f pl* pour
l'exécution de travaux de
construction
nl richtlijnen *f(m) pl* voor de bouw
r рекомендации *f pl* [указания *n pl*]
по производству строительных
работ

C1050 *e* **construction regulations**
d Bauvorschriften *f pl*,
Baubestimmungen *f pl*
f réglements *m pl* de construction
nl bouwvoorschriften *n pl*,
bouwnormen *f(m) pl*
r строительные нормативы *m pl*

C1051 *e* **construction safety regulations**
d Bausicherheitsvorschriften *f pl*
f prescriptions *f pl* [règles *f pl*] de
sécurité dans les travaux du
bâtiment [de construction]
nl veiligheidsvoorschriften *n pl* voor
de bouw
r правила *n pl* техники
безопасности в строительстве

C1052 *e* **construction schedule**
   d Bauzeitplan *m*, Baufortschrittplan *m*
   f schéma *f* [programme *m*] d'avancement de travaux
   nl werkrooster *n*
   r график *m* производства строительных работ

C1053 *e* **construction sequence**
   d Baufolge *f*, Aufeinanderfolge *f* der Bauarbeiten
   f séquence *f* d'exécution de travaux de construction
   nl bouwvolgorde *f*
   r последовательность *f* выполнения строительных работ [операций]

C1054 *e* **construction site**
   d Baustelle *f*, Bauplatz *m*
   f chantier *m* de construction, terrain *m* à bâtir
   nl bouwplaats *f(m)*, bouwterrein *n*
   r строительная площадка *f*

C1055 *e* **construction specifications**
   d Bauausführungsbestimmungen *f pl*, Baudurchführungsvorschriften *f pl*, technische Bauvorschriften *f pl*
   f cahier *m* des charges applicable aux travaux de construction, cahier *m* des prescriptions techniques générales applicable au chantier de construction
   nl bestek *n*, beschrijving *f* van het werk
   r технические условия *n pl* на производство строительных работ [на строительство объекта]

C1056 *e* **construction standard specification**
   d Baunorm *f*
   f norme *f* de construction, règlement *m* (de travaux) de construction
   nl kwaliteitseisen *pl*
   r строительный стандарт *m*

C1057   **construction steel** *see* **structural steel**

C1058 *e* **construction supervising authority**
   d Bauaufsichtsbehörde *f*, Baupolizei *f*
   f autorité *f* surveillant les constructions [les travaux]
   nl bouw- en woningtoezicht *n*
   r строительный надзор *m* (организация)

C1059 *e* **construction supervision**
   d Bauaufsicht *f*, Bauüberwachung *f*
   f surveillance *f* [contrôle *m*] (technique) des travaux
   nl toezicht *n* op de bouw
   r строительный надзор *m*, технический надзор *m* за строительством

C1060 *e* **construction survey**
   d Schichtlinienplan *m*; topographische Aufnahme *f*;

Anschlüsse *m pl* [Anpassungen *f pl*] des Bauwerkes [des Gebäudes]
   f plan *m* topographique, levé *m* topographique de l'implantation d'un bâtiment
   nl uitzetten *n*
   r топографический план *m*; топографическая съёмка *f*, привязки *f pl* строительного сооружения

C1061 *e* **construction techniques**
   d Arbeitsverfahren *n pl*, Baupraxis *f*
   f techniques *f pl* des travaux, pratique *f* de construction
   nl bouwtechniek *f*
   r приёмы *m pl* [технология *f*] производства строительных работ

C1062 *e* **construction technology**
   d Technologie *f* der Bauproduktion
   f technologie *f* des travaux de construction
   nl bouwtechnologie *f*
   r строительная технология *f*, технология *f* производства строительных работ

C1063 *e* **construction tolerances**
   d Herstellungs- und Einbautoleranzen *f pl*, zulässige Maßabweichungen *f pl*
   f tolérences *f pl* de fabrication et de mise en œuvre des éléments constructifs
   nl fabricage- en uitvoeringstoleranties *f pl*
   r строительные допуски *m pl* (на изготовление и монтаж строительных конструкций)

C1064 *e* **constructiontrain**
   d Bauzug *m*, Bau- und Montagezug *m*
   f convoi *m* [colonne *f*] des engins pour la construction
   nl bouw- en montagetrein *m*
   r строительная колонна *f*; строительный поезд *m*

C1065 *e* **consulting engineer**
   d beratender Ingenieur *m*
   f ingénieur-conseil *m*
   nl raadgevende ingenieur *m*
   r инженер-консультант *m*, эксперт *m*

C1066 *e* **consumer gas piping**
   d Hausgasleitungsnetz *n*
   f canalisation *f* de gaz intérieure, tuyauteries *f pl* d'alimentation en gaz intérieures
   nl gasinstallatie *f* bij de afnemer
   r внутридомовая газовая сеть *f*

C1067 *e* **consumer gas service line**
   d Gasanschlußstrang *m*, Gebäudestrang *m* der Gasleitung
   f branchement *m* d'immeuble (arrivée de gaz au robinet-chef)
   nl gasaansluiting *f* op de woning
   r соединительная ветка *f* газопровода

C1068 e **consumer's supply control**
  d Zähler- und Sicherungstafel *f*
  f tableau *m* d'abonné avec compteur
  nl meter en hoofdafsluiter
  r электрощиток *m* (*со счётчиком и предохранителями*)

C1069 e **consumer's terminals**
  d Anschlußverteilerschalttafel *f*
  f tableau *m* électrique de contrôle [de distribution]
  nl groepenkast *f(m)*
  r электрораспределительный щиток *m* с разводкой к потребителю

C1070 e **consumption curve**
  d Wasserverbrauchskurve *f*
  f courbe *f* de consommation d'eau
  nl verbruikscurve *f*
  r кривая *f* водопотребления

C1071 e **contact aeration**
  d Kontaktbelüftung *f*, Tauchkörperbelüftung *f*
  f aération *f* [aérage *m*] par contact (*des eaux usées*)
  nl contactbeluchting *f* (*riool*water)
  r контактная аэрация *f* (*сточных вод*)

C1072 e **contact aerator**
  d belüfteter Tauchkörper *m*
  f aérateur *m* de contact
  nl contactbeluchter *m*
  r аэротенк *m* с пневматическим аэратором

C1073   contact area *see* **contact surface**

C1074 e **contact bed**
  d Filterkörperfüllung *f*, Einstaufilter *n*, *m*, Füllkörper *m*
  f lit *m* filtrant de contact
  nl beluchtingsfilterbed *n*
  r контактная загрузка *f* (*вихревого реактора*), контактный фильтр *m*

C1075 e **contact filtration**
  d Kontaktfiltration *f*
  f filtration *f* [filtrage ॑*m*] par contact
  nl contactfiltratie *f*
  r контактная фильтрация *f*, фильтрация *f* через адсорбирующий слой

C1076 e **contact flocculator**
  d Kontaktflockulator *m*, Flockungsreaktor *m*
  f bassin *m* de floculation par contact
  nl uitvlokkingsbassin *n*
  r контактный осветлитель *m*

C1077 e **contact pressure**
  d Kontaktdruck *m*, Anpreßdruck *m*, Berührungsdruck *m*
  f pression *f* de contact
  nl contactdruk *m*
  r контактное давление *n*

C1078 e **contact surface**
  d Kontaktfläche *f*, Berührungsfläche *f*
  f surface *f* de contact
  nl contactvlak *n*
  r поверхность *f* контакта, контактная поверхность *f*

C1079 e **containment structure**
  d Reaktorsicherheitshülle *f*
  f caisson *m* de réacteur
  nl reactorafscherming *f*
  r противоаварийная оболочка *f* [внешняя защита *f*] реактора

C1080 e **contamination**
  d Verunreinigung *f*, Verschmutzung *f*
  f contamination *f*, pollution *f*
  nl verontreiniging *f*
  r загрязнение *n*

C1081 e **continental hydrology**
  d Binnengewässerkunde *f*, Hydrologie *f* des Festlandes
  f hydrologie *f* continentale
  nl hydrologie *f* van het vaste land
  r гидрология *f* суши

C1082 ॑e **contingencies**
  d Extraausgaben *f pl*, Unvorhergesehenes *n*, unvorhergesehene Kosten *pl*
  f dépenses *f pl* imprévues [non prévues, supplémentaires]
  nl onvoorziene uitgaven *pl*
  r непредвиденные расходы *m pl*

C1083 e **contingency sum**
  d vorgegebene Kostensumme *f* [veranschlagte Haushaltmittel *n pl*] für Extraausgaben
  f frais *m pl* (*prévus par le devis estimatif*) pour les travaux contingents [imprévus]
  nl post *f(m)* voor de onvoorziene uitgaven
  r ассигнования *n pl* (*по смете*) на непредвиденные работы

C1084 e **continuity**
  d Stetigkeit *f*; Durchlaufwirkung *f*; Kontinuität *f*
  f continuité *f*
  nl doorlopend verband *n*; continuïteit *f*
  r непрерывность *f*; неразрезность *f* (*балки, соединения*); сплошность *f*; неразрывность *f* (*потока*)

C1085 e **continuity bars**
  d durchlaufende Bewehrung *f* [Stahleinlagen *f pl*]
  f barres *f pl* d'armature assurant la continuité (*aux zones d'appui des poutres ou dalles continues en béton armé*)
  nl doorlopende wapening *f*
  r арматурные стержни *m pl* в растянутой зоне над опорами неразрезных балок и плит

C1086 e **continuity equation**
  d Kontinuitätsgleichung *f*
  f équation *f* de continuité
  nl continuïteitsvergelijking *f*
  r уравнение *n* неразрывности

C1087 e **continuity moment**
  d Kontinuitätsmoment n,
    Stützmoment n des
    Durchlaufträgers
  f moment m de continuité
  nl overgangsmoment n
  r момент m неразрезности,
    опорный момент m неразрезной
    балки

C1088 e **continuous batcher**
  d Stetigdosiervorrichtung f
  f doseur m continu
  nl apparaat n voor continue dosering
  r дозатор m непрерывного действия

C1089 e **continuous beam**
  d durchlaufender [kontinuierlicher]
    Balken m [Träger m],
    Durchlaufträger m
  f poutre f continue
  nl balk m op meerdere steunpunten
  r неразрезная балка f

C1090 e **continuous-beam bridge**
  d Durchlaufbalkenbrücke f,
    Durchlaufträgerbrücke f
  f pont m à poutres continues
  nl doorgaande liggerbrug f(m)
  r мост m с неразрезными балками,
    балочный мост m неразрезной
    конструкции

C1091 e **continuous bucket elevator**
  d Reihenbecherwerk n,
    Gurtbecherwerk n, Bandbecherwerk
    n, Gurtelevator m
  f élévateur m (à godets) à courroie
  nl jakobsladder m
  r ковшовый подъёмник m
    непрерывного действия,
    ленточный ковшовый подъёмник m

C1092 e **continuous concreting**
  d kontinuierliches Betonieren n
  f bétonnage m [coulage m de béton]
    continu, mise f en place de béton
    continue
  nl ononderbroken betonstorten n
  r непрерывное бетонирование n

C1093 e **continuous cracking**
  d Durchlaufrißbildung f,
    Durchlaufrisse m pl
  f fissuration f superficielle continue
  nl voortgaande scheurvorming f
  r сплошное трещинообразование n,
    сплошные трещины f pl

C1094 e **continuous-flow aeration tank**
  d längsdurchflossenes
    Belebtschlammbecken n
  f bassin m d'aérage [d'aération] à
    écoulement continu
  nl doorstroom-beluchtingstank m
  r аэротенк-вытеснитель m

C1095 e **continuous-flow settling basin**
  d Absetzbecken n [Klärbecken n]
    mit kontinuierlicher Spülung
  f bassin m de décantation à purge
    continue

  nl doorstroom-bezinkbassin n
  r отстойник m непрерывного
    промыва

C1096 e **continuous footing**
  d Streifenfundament n
  f fondation f continue,
    empattement m continu
  nl strookfundering f
  r ленточный фундамент m

C1097 e **continuous grading**
  d kontinuierliche Kornabstufung f
  f granulométrie f continue
  nl voortdurende meting f van de
    korrelgrootte
  r непрерывная гранулометрия f

C1098 e **continuously graded aggregate**
  d kontinuierlich abgestufter
    Zuschlagstoff m
  f granulat m [agrégat m] à
    granulométrie continue
  nl toeslag m met voortdurend
    gecontroleerde korrelgrootte
  r заполнитель m (подобранного
    гранулометрического состава),
    включающий все фракции

C1099 e **continuously reinforced concrete**
  d Faserbeton m
  f béton m (à armature) en fibres
    d'acier ou de verre
  nl vezelbeton n
  r дисперсно-армированный бетон m
    (со стальным или стеклянным
    волокном), фибробетон m

C1100 e **continuous mixer**
  d Fließmischer m, Stetigmischer m,
    Durchlaufmischer m,
    kontinuierlicher Mischer m
  f bétonnière-malaxeuse f [malaxeur
    m à béton] à production continue
  nl continu-betonmenger m
  r (бетоно)смеситель m непрерывного
    действия

C1101 e **continuous-mix plant**
  d Stetigmischanlage f,
    Durchlaufmischanlage f,
    kontinuierliche Mischanlage f
  f centrale f de malaxage
    [mélangeuse] à production
    continue
  nl continu werkende menginstallatie f
  r смесительная установка f
    непрерывного действия

C1102 e **continuous proportioning plant**
  d kontinuierlich arbeitende
    Dosieranlage f
  f installation f de dosage continu
  nl installatie f voor continue
    dosering
  r дозировочная установка f
    непрерывного действия

C1103 e **continuous slab**
  d Durchlaufplatte f
  f dalle f continue

nl ononderbroken plaat f(m)
r неразрезная плита f

C1104 e **continuous stream**
d perennierender Wasserlauf m
f cours m d'eau permanent
nl permanente waterloop m
r непрерывный поток m;
постоянный [непересыхающий]
водоток m

C1105 e **continuous string**
d durchgehender Wangenträger m
f limon m extérieur continu
nl doorgaande trapboom m
r непрерывная наружная тетива f
по всем маршам лестницы;
непрерывный косоур m

C1106 e **continuous thickener**
d Klärturm m [vertikales
Absetzbecken n] mit
kontinuierlicher Spülung
f décanteur m vertical à purge
continue
nl continu werkend bezinkingstoestel n
r вертикальный отстойник m
непрерывного действия

C1107 e **contour**
d 1. Umriß m, Umrißlinie f
2. Kontur f, Höhenschichtlinie f,
Schichtlinie f
f 1. contour m 2. courbe f de
niveau
nl 1. contour m 2. hoogtelijn f(m)
r 1. контур m, внешнее очертание n
2. горизонталь f

C1107a e **contour bank**
d Bodenschutz- oder
Irrigationsterrasse f, die einer
Höhenlinie folgt
f terrasse f le long d'une courbe de
niveau
nl de hoogtelijnen volgend terras
f (m)
r контурная терраса f
(ирригационная или
почвозащитная)

C1108 e **contour drainage**
d Längsdränung f
f drainage m suivant les courbes de
niveau
nl langsdrainage f
r контурный или продольный
дренаж m

C1109 e **contour paver**
d Profilfertiger m
f finisseur m asphalteur [à
produits noirs], épandeur-
-reprofileur m
nl afwerkmachine f (wegenbouw)
r профилирующий финишер m
(для дорожных покрытий на
органическом вяжущем)

C1110 e **contract**
d Vertrag m, Auftrag m
f contrat m, marché f (des travaux)

nl bouwcontract n
r подрядный договор m, контракт m

C1110a **contract award]** see **award of a
contract**

C1111 e **contract documents**
d Vertragsunterlagen f pl
f documentation f relative à l'offre,
dossier m d'appel d'offres
nl contractdocumenten n pl
r подрядная документация f
(рабочие чертежи, технические
условия, договор)

C1112 e **contracted weir**
d zusammengezogenes [eingeschnürtes] Meßwehr n
f déversoir m à contraction latérale
nl meetstuw m
r измерительный водослив m с
боковым сжатием

C1113 e **contraction**
d 1. Kontraktion f,
Zusammenziehung f,
Querschnittverminderung f;
Einschnürung f 2. Schrumpfung f
f 1. contraction f 2. retrait m
nl 1. insnoering f 2. krimp m
r 1. сжатие n; сужение n;
образование n шейки 2. усадка f

C1114 e **contraction joint**
d Kontraktionsfuge f; Schwindfuge f;
Bewegungsfuge f; Druckfuge f
f joint m de contraction; joint m de
retrait
nl dilatatievoeg f(m)
r температурный шов m; усадочный
шов m; деформационный шов m;
шов m сжатия (в бетонных
конструкциях)

C1115 e **contract manager**
d Bauleiter m, Firmenbauführer m
f maître m de l'œuvre
nl (hoofd)uitvoerder m
r руководитель m [начальник m]
строительства

C1116 e **contractor**
d Bauunternehmer m,
Bauauftragnehmer m, Baufirma f
f entrepreneur m de construction
[des travaux publics]
nl aannemer m
r подрядчик m; подрядная
строительная организация ,
[фирма f]

C1117 e **contract price**
d Auftragwert m, Auftragsumme f
f prix m du marché
nl aannemingssom f(m)
r договорная цена f

C1118 e **contract specifications**
d Bauleistungsbuch n,
Bauleistungsverzeichnis n
f cahier m des charges de marché,
cahier m des charges de contract

nl bouwrapportage f
r технические условия n pl
подрядного договора (*перечень
работ с детальным техническим
описанием*)

C1119 e **control action**
d Regelwirkung f
f action f de réglage
nl regeling f
r регулирующее воздействие n

C1120 e **control cabinet**
d Schaltschrank m
f armoire f de distribution
nl schakelkast f(m)
r распределительный шкаф m

C1121 e **control flume**
d Kontrollkanal m
f canal m jaugeur
nl meetgoot f(m)
r измерительный [контрольный]
лоток m

C1122 e **control gate** see **check gate**

C1123 e **controlled condition**
d Regelgröße f
f condition f contrôlée; grandeur f
réglée
nl bestuurde grootheid f
r регулируемая переменная f;
регулируемый параметр m

C1124 e **controlled pedestrian crossing**
d geregelter Fußgängerüberweg m
f passage m pour piétons régulé
[contrôlé]; passage m clouté
nl beveiligde oversteekplaats f(m)
voor voetgangers
r регулируемый пешеходный
переход m

C1125 e **controlled ventilation**
d kontrollierte Lüftung f
f ventillation f réglable
nl gereguleerde ventilatie f
r организованная вентиляция f

C1126 e **controller**
d Regler m
f dispositif m de réglage,
régulateur m
nl regelaar m, regulateur m
r регулятор m

C1127 e **controlling depth**
d kritische Schiffahrtstiefe f,
zulässige Tauchtiefe f
f profondeur f normative, tirant m
d'eau normatif
nl kritische scheepvaartdiepte f,
toegestane diepgang m
r нормирующая глубина f

C1128 e **control of concrete quality**
d Betongütekontrolle f,
Betongüteüberwachung f
f contrôle m du béton
nl controle f op de betonkwaliteit
r контроль m качества бетона

C1129 e **control point**
d 1. Sollwert m 2. Kontrollpunkt m,
Bezugspunkt m, Festpunkt m;
Vermessungspunkt m,
Höhenmarke f
f 1. valeur f réglée 2. point m de
repère [de contrôle, de référence];
point m géodesique
nl 1. ingestelde waarde f
2. oriëntatiepunt n; vast punt n
r 1. заданное значение n
регулируемой переменной
2. контрольная точка f, точка f
(*на местности*) с известной
отметкой; геодезический пункт
m репер m

C1130 e **control test**
d Kontrollprüfung f
f essai m de contrôle, contre-
-essai m, contre-épreuve f
nl controleproef f(m)
r контрольное испытание n

C1131 e **control valve**
d Regelventil n
f vanne f de réglage
nl regelventiel n, regelafsluiter m
r регулирующий клапан m

C1132 e **convection heating**
d Konvektionsheizung f
f chauffage m par convection
nl verwarming f door convectie
r конвективное отопление n

C1133 e **convection heat transfer**
d Wärmeübertragung f durch
Konvektion
f transfert m de chaleur par
convection
nl warmteoverdracht f(m) door
convectie
r конвективный.перенос m тепла

C1134 e **convective heat exchange**
d Konvektionswärmeaustausch m
f échange m convectif de chaleur,
échange m de chaleur par
convection
nl warmtewisseling f door convectie
r конвективный теплообмен m

C1135 e **conversion**
d 1. Umwandlung f, Verwandlung f;
Umformung f 2. Umbau m
3. Längsschnitt m des Rundholzes
f conversion f, transformation f
nl 1. conversie f, omvorming f
2. ombouwen n
r 1. превращение n; преобразование
n; пластическое формоизменение n
2. перестройка f,
реконструкция f 3. продольная
распиловка f (брёвен)

C1136 e **converted timber**
d Schnittholz n
f bois m débité [équarri]
nl bezaagd hout n
r обрезные пиломатериалы m pl

C1137 *e* convertible attachments
d Austauschgeräte *n pl*,
Umbauwerkzeuge *n pl*,
Anbauvorrichtungen *f pl*
f équipement *m* interchangeable
nl verwisselbaar hulpstuk *n*
r сменное рабочее оборудование
n

C1138 *e* conveyance structure
d Wasserleitungsbauwerk *n*
f ouvrage *m* du transport d'eau
nl kunstwerk *n* voor waterafvoer
r водопроводящее сооружение *n*

C1139 *e* conveyor
d Stetigförderer *m*, Förderer *m*
f convoyeur *m*
nl transportinstallatie *f*
r конвейер *m*

C1140 *e* conveyor belt
d Förderband *n*
f courroie *f* transporteuse [de
transport, de convoyeur]
nl transportband *n*
r конвейерная лента *f*

C1141 *e* conveyor type bucket loader
d Eimerkettenlader *m*,
Becherwerkslader *m*
f chargeur *m* à chaîne de roulement
nl jakobsladder *m*
r многоковшовый погрузчик *m*

C1142 *e* cooler
d 1. Kühler *m* 2. Kältekammer *f*
f 1. refroidisseur *m* 2. chambre *f* du
froid
nl 1. koelapparaat *n* 2. koelkamer *f(m)*
r 1. охладитель *m*; холодильник *m*
2. холодильная камера *f*

C1142a cooling agent *see* cooling medium

C1143 *e* cooling air
d Kühlluft *f*
f air *m* de refroidissement
nl koellucht *f(m)*
r охлаждающий воздух *m*

C1144 *e* cooling and heating unit
d Heiz- und Kühlgerät *n*
f installation *f* de chauffage et de
refroidissement
nl koel- en verwarmingsapparaat *n*
r отопительно-охладительный
агрегат *m*

C1145 *e* cooling coil
d Kühlrohrschlange *f*; Luftkühler *m*
f serpentin *m* refroidisseur;
refroidisseur *m* d'air
nl koelspiraal *f(m)*
r охлаждающий змеевик *m*;
(воздухо)охладитель *m*

C1146 *e* cooling grid
d mehrreihiger Kühler *m*
f grille *f* de refroidissement
nl koelrooster *n*
r многорядный охлаждающий
теплообменник *m*

C1147 *e* cooling load
d Kühllast *f*
f charge *f* frigorifique
nl koelbelasting *f*
r холодильная нагрузка *f*

C1148 *e* cooling medium
d Kühlmittel *n*, Kälteträger *m*
f agent *m* refroidisseur
nl koelmedium *n*
r (вторичный) холодоноситель *m*

C1149 *e* cooling pond
d Kühlteich *m*, Kühlbecken *n*
f étang *m* réfrigérant
nl koelbekken *n*
r охлаждающий бассейн *m*

C1150 *e* cooling power
d Kälteleistung *f*
f capacité *f* frigorifique
nl koelvermogen *n*
r холодопроизводительность *f*

C1151 *e* cooling tower
d Kühlturm *m*, Rückkühlwerk *n*
f tour *f* de réfrigération [de
refroidissement]
nl koeltoren *m*
r градирня *f*

C1152 *e* cooling tower packing
d Kühlturmpackung *f*
f corps *m* de remplissage de tour
de réfrigération
nl koeltorenpakking *f*
r насадка *f* градирни

C1153 *e* cooling unit
d Luftkühler *m*
f bloc *m* refroidisseur
nl koelapparaat *n*
r воздухоохладитель *m*

C1154 *e* cooling water
d Kühlwasser *n*
f eau *f* de refroidissement
nl koelwater *n*
r охлаждающая вода *f*

C1155 *e* coordination drawings
d Fachkoordinierungszeichnungen *f
pl*
f dessins *m pl* [plans *m pl*] de
coordination des travaux
nl overzichtstekeningen *f pl* ter
coördinatie van de onderscheiden
werkzaamheden
r координирующие чертежи *m pl*
(*для координации работ,
выполняемых рабочими
различных специальностей*)

C1156 COP *see* coefficient of performance

C1156a *e* coping
d Mauerabdeckung *f*; Mauerkrone *f*
f couronnement *m* de mur, chaperon
m (de mur)
nl muurafdekking *f*
r верхний ряд *m* кладки; гребень
m стены

C1157 e **corbel**
    d Konsole *f*; Kragarm *m*;
      Auskragung *f*; Wandvorsprung *m*;
      Kragstein *m*; Überlappung *f* der
      Mauerwerksreihen
    f corbeau *m*; saillie *f*; console *f*;
      encorbellement *m*; balèvre *f*
    nl kraagsteen *m*; uitkraging *f* van de
      muur
    r выступ *m*; выпуск *m*; кронштейн
      m; консольный выступ *m* стены;
      консольный камень *m*; напуск *m*
      кирпичной кладки

C1158 e **corbelling**
    d allmählige Überlappung *f* der
      Mauerwerksreihen
    f encorbellement *m* des assises de bri-
      ques
    nl uitkragen *n* van metselwerk ter
      vorming van een console
    r последовательный напуск *m*
      [выступ *m*] рядов кирпичей

C1159 e **corbel piece**
    d Sattelholz *n*
    f sous-longeron *m*, tête *f* du poteau
      en bois
    nl zadelhout *n*, sleutelstuk *n*,
      karbeel *m*
    r подбабок *m*, оголовок *m*
      (*деревянной стойки*)

C1160 e **corded way**
    d abgetreppter Böschungsgang *m*
    f escalier *m* en retraite (*sur le
      talus de terre*)
    nl loopschade *f* aan talud
    r уступная лестница *f* в земляном
      откосе

C1161 e **core**
    d 1. Kern *m* 2. Dammkern *m*,
      Dichtungskern 3. Bohrkern *m*,
      Probekern *m*
    f 1. noyau *m* 2. noyau *m*
      d'étanchéité 3. carotte *f*
    nl 1. kern *f(m)* 2. waterdicht scherm
      n in een waterkering
      3. ongestoord grondmonster *n*
    r 1. ядро *n*; сердечник *m*
      2. (водонепроницаемое) ядро *n*
      (*земляной плотины*) 3. керн *m*

1162 e **core area**
    d Stirnquerschnitt *m*
    f section *f* de face
    nl kopse doorsnede *f*
    r габаритное [лобовое] сечение *n*,
      площадь *f* сечения брутто
      (*воздухораспределительного
      устройства*)

C1163 e **core drill**
    d Kernbohrer *m*
    f foreuse *f* de carottage
    nl kernboor *m*
    r колонковый бур *m*

C1164 e **core former**
    d Hohlraumbildner *m*
    f noyau *m* pour formation des
      évidements (*dans les dalles en
      béton armé*)
    nl sparingmal (*gietwerk*)
    r пустотообразователь *m* (*для
      пустотелых бетонных плит*)

C1165 e **core of column**
    d Stützenkern *m*
    f noyau *m* central du poteau (en
      béton armé)
    nl kolomkern *f(m)*
    r ядро *n* (железобетонной) колонны

C1166 e **core pool**
    d Spülteich *m*, Klärteich *m*,
      Auflandungsteich *m*, Absetzteich *m*
    f étang *m* de boues
    nl bezinkvijver *m*
    r пруд(ок)-отстойник *m*

C1167 e **core sample**
    d Probekern *m*, Bohrkern *m*, Kern
    f carotte *f* découpée
    nl boormonster *n*
    r керн *m* (*образец грунта*)

C1168   **core trench** see **cutoff trench**

C1169 e **core wall**
    d Kernmauer *f*, Dichtungsmauer *f*,
      Kernwand *f*, Dammkern *m*,
      Dichtungskern *m* der Staumauer
    f noyau *m* d'étanchéité
    nl waterdicht scherm *n* in een
      waterkering
    r ядро *n* плотины;
      противофильтрационная диафрагма
      f

C1170 e **core wall dam**
    d Kerndamm *m*, Erddamm *m* mit
      Dichtungskern
    f barrage *m* en terre à noyau
      d'étanchéité
    nl dijk *m* met verdichte kern
    r земляная плотина *f*
      с противофильтрационным ядром

C1171 e **corkboard**
    d Korkplatte *f*
    f plaque *f* de liège
    nl kurkplaat *f(m)*
    r плита *f* из прессованной пробки

C1172 e **cork carpet**
    d Korklinoleum *n*
    f tapis *m* en liège
    nl kurklinoleum *n*, *m*
    r пробочный ковёр *m* (*для
      покрытия полов*)

C1173   **corking** see **cogging**
C1173a e **cork pipe covering**
    d Korkschalenisolierung *f*
    f coquille *f* de liège
    nl kurkschalenisolatie *f*
    r пробковые изоляционные
      скорлупы *f pl* (*для труб*)
C1174   **cork slab** see **corkboard**

C1175 e cork tile
    d Korkplatte *f*, Korkfliese *f*
    f carreau *m* en liège
    nl kurktegel *m*
    r плитка *f* из прессованной пробки
    для полов

C1176   corner bead *see* angle bead

C1177 e corner chisel
    d Winkelmeißel *m*, Winkelbeitel *m*
    f bédane *m* en cornière
    nl driekante guts *f(m)*
    r угловое долото *n*, угловая
    стамеска *f*

C1178 e corner column
    d Eckstütze *f*
    f colonne *f* [poteau *m*] du coin
    nl hoekpilaar *m*
    r угловая колонна *f*

C1179 e corner joint, corner welding joint
    d Eckstoß *m*, Winkelstoß *m*;
    Ecknaht *f*
    f joint *m* (de soudure) d'angle
    nl hoek(las)naad *m*
    r угловое (сварное) соединение *n*;
    угловой (сварной) шов *m*

C1180   cornerlock joint *see* combed joint

C1181   corner post *see* angle post

C1182 e corner trowel
    d Kantenlehre *f*, Eckenputzlehre *f*
    f truelle *f* d'angle [à angle] vif *ou*
    rentrant
    nl hoektroffel *m*
    r угловой полутёрок *m* (*лузговой
    или усёночный*)

C1183 e cornice
    d Gesims *n*
    f corniche *f*
    nl kroonlijst *f(m)*
    r карниз *m*

C1184 e corridor
    d Korridor *m*, Gang *m*, Flur *m*
    f corridor *m*, couloir *m*
    nl gang *m*
    r коридор *m*, проход *m*

C1185 e corrosion control
    d Korrosionsverhütung *f*,
    Korrosionsschutz *m*
    f lutte *f* contre la corrosion
    nl bescherming *f* tegen corrosie
    r предупреждение *n* коррозии,
    борьба *f* с коррозией

C1186 e corrosion-inhibiting admixture
    d Korrosionsschutzmittel *n*,
    Korrosionsinhibitor *m*
    f inhibiteur *m* de corrosion
    nl corrosie-verhinderend middel *n*
    r антикоррозионная добавка *f*,
    ингибитор *m* коррозии

C1187 e corrugated aluminium
    d Wellalu(minium) *n*
    f aluminium *m* ondulé
    nl aluminium golfplaat *f(m)*
    r волнистые алюминиевые листы
    m pl

C1188 e corrugated asbestos
    d Wellasbestzementplatten *f pl*
    f plaque *f* d'amiante-ciment
    ondulée
    nl golfplaat *f(m)* van asbest
    r волнистые асбестоцементные
    листы *m pl*

C1189 e corrugated sheet
    d Wellblech *n*, Welltafel *f*,
    Wellplatte *f*
    f tôle *f* (mince) ondulée
    nl (stalen) golfplaat *f (m)*
    r волнистый (стальной) лист *m*,
    лист *m* волнистого профиля

C1190   corrugated steel fibers *see* crimped
    steel fibers

C1191 e corrugated toothed ring
    d Zahnring-Dübel *m*
    [Alligatordübel *m*, Alligatorring *m*]
    mit Wellstahlverzahnung
    f crampon *m* en couronne denté
    nl golfkram *f(m)*
    r зубчатая волнистая кольцевая
    шпонка *f*

C1192 e corrugations
    d Welligkeit *f* (*Straße*)
    f ondulation *f* (*route*)
    nl golfvorming *f* (*wegdek*)
    r поперечные волны *f pl* и морщины
    f pl (*дефект дорожного
    покрытия*)

C1193 e cost-efficient design
    d wirtschaftlich wirksames
    [rentables] Projekt *n*
    f projet *m* à coût des dépenses
    réduit, projet *m* rentable
    nl economisch ontwerp *n*
    r рентабельный проект *m*;
    экономически эффективное
    конструктивное решение *n*

C1194 e cost estimating
    d Veranschlagung *f*, Kalkulation *f*
    f étude *f* de prix, évaluation *f*,
    élaboration *f* du devis estimatif
    nl calculatie *f*, kostenberekening *f*
    r составление *n* сметы,
    калькуляция *f*

C1195   costimating *see* estimatin

C1196 e costing
    d Kostehkalkulation *f*, Aufstellung
    f der Einheitspreisen
    f estimation *f* du coût (*d'une unité
    de travail*)
    nl kostprijsberekening *f*
    r расчёт *m* стоимости единицы
    работы

C1197 e counter
    d biegsame Strebe *f*;
    Zwischenstrebe *f*, Hilfsstrebe *f*
    f diagonale *f* flexible (*travailant en
    tension*); diagonale *f* secondaire,

contre-diagonale *f* (*d'une poutre à treillis*)
nl langsschoor *m*
r гибкий раскос *m* фермы (*работающий только на растяжение*); вспомогательный раскос *m*

C1198 *e* **counter-arched revetment**
d Breschmauer *f*, entlastete Futtermauer *f*
f mur *m* de soutènement [de revêtement] en arcs multiples
nl steunmuur *m*
r многоарочная подпорная стенка *f*

C1199 *e* **counter-berm**
d Kónterbankett *n*, Gegenbankett *n*
f banquette *f* de pied
nl conterbanket *n*
r контрбанкет *m*

C1200 *e* **counterbracing**
d Kreuzstreben *f pl*
f diagonales *f pl* croisées; contreventements *m pl* croisés
nl windverband *n*, schrankverband *n*
r перекрёстные раскосы *m pl* (*фермы*); перекрёстные связи *f pl*

C1201 *e* **counter ceiling**
d Hängedecke *f*, abgehängte Decke *f*
f plafond *m* suspendu
nl verlaagd plafond *n*
r подвесной потолок *m*

C1202 *e* **counter-current mixer**
d Gegenstrom(beton)mischer *m*
f malaxeur *m* à contre-courant
nl tegenstroom(beton)menger *m*
r противоточный (бетоно)смеситель *m*

C1203 *e* **counter drain**
d Böschungsfußdrän *m*
f drain *m* de pied de talus
nl afwatering *f* langs het talud
r дрена *f*, проложенная вдоль подошвы откоса

C1204 **counter flange** *see* **companion flange**

C1205 *e* **counter flashing** *US*
d Abweiseblech *n*
f contre-solin *m*
nl (lood)slabbe *f(m)*, loodloketten *n pl*
r металлический фартук *m* (*для гидроизоляции швов кладки*)

C1206 *e* **counter floor**
d Blendboden *m*, Blindboden *m*
f plancher *m* brut, planchéichage *m* brut
nl ondervloer *m*
r чёрный пол *m*

C1207 *e* **counterflow**
d Gegenstrom *m*
f contre-courant *m*
nl tegenstroom *m*
r противоток *m*

C1208 *e* **counterflow cooling tower**
d Gegenstromkühlturm *m*
f tour *f* de réfrigération à contre-courant
nl tegenstroom-koeltoren *m*
r противоточная градирня *f*

C1209 *e* **counter flow heat exchanger**
d Gegenstromwärmetauscher *m*
f échangeur *m* de chaleur à contre-courant
nl tegenstroom-warmtewisselaar *m*
r противоточный теплообменник *m*

C1210 *e* **counterfort**
d Mauerpfeiler *m*, Strebepfeiler *m*
f contrefort *m*
nl schoormuur *m*, steunbeer *f(m)*
r контрфорс *m*

C1211 *e* **counterfort dam**
d Pfeilersperre *f*, aufgelöste Staumauer *f*
f barrage *m* à contreforts
nl stuwdam *m* met steunpijlers
r контрфорсная плотина *f*

C1212 *e* **counterforted retaining wall**
d Pfeilerstützmauer *f*, Pfeilerstützwand *f*
f mur *m* de soutènement à contreforts
nl keermuur *m* met steunberen
r контрфорсная подпорная стенка *f*

C1213 *e* **counterfort type powerhouse**
d Pfeilerkrafthaus *n*
f usine-pile *f*, pile-usine *f*, centrale-pile *f*
nl tegen de stuwdam gebouwde waterkrachtcentrale *f*
r здание *n* ГЭС бычкового типа

C1214 **counterpoise bridge** *see* **bascule bridge** *or* **lift bridge**

C1215 **counter reservoir** *see* **equalizing reservoir**

C1216 *e* **countersill**
d Gegenschwelle *f*
f contre-seuil *m*
nl steldorpel *m*
r водобойный порог *m*; водобойная стенка *f*

C1217 *e* **counterweight**
d Gegengewicht *n*
f contrepoids *m*, contrebalance *f*
nl tegengewicht *n*
r противовес *m*

C1218 **counterweight rope** *see* **compensating rope**

C1219 **counterweight shutter** *see* **balanced gate**

C1220 *e* **country road**
d Landstraße *f*
f route *f* rurale [provinciale]
nl provinciale weg *m*
r сельская дорога *f*

C1221 e **couple of forces**
    d Kräftepaar *n*
    f couple *m* des forces
    nl krachtenkoppel *n*
    r пара *f* сил

C1222 e **coupler**
    d Verbindungsmuffe *f* (z. *B. für Stahlrohrgerüst*)
    f raccord *m* (*p. ex. pour les échafaudages tubulaires*); manchon *m* de jonction [de raccordement]
    nl koppeling(sstuk *n*) *f*
    r соединительная деталь *f* (*для трубчатых лесов*); соединительная муфта *f*

C1223 e **couple roof**
    d Sparrendach *n*
    f comble *m* à deux versants sur arbalétriers
    nl zadeldak *n* met dakstoel
    r двускатная крыша *f* с наслонными стропилами

C1224 e **coupling**
    d 1. Gewindemuffe *f*, Rohrverbinder *m* 2. Verbindungsmuffe *f* für Stahlrohrgerüst
    f 1. manchon *m* de raccordement 2. raccord *m*
    nl koppeling *f*
    r 1. нарезная (трубопроводная) муфта *f* 2. зажим *m* трубчатых лесов

C1225 e **course**
    d 1. Mauerlage *f*, Mauerschicht *f*; Blockreihe *f* 2. Wasserlauf *m*, Flußlauf *m*
    f 1. assise *f* de maçonnerie, assise *f* de blocs 2. cours *m* d'eau
    nl 1. gemetselde laag *f(m)* 2. waterloop *m*
    r 1. ряд *m* каменной кладки; курс *m* кладки из массивов 2. водоток *m*, русловой поток *m*

C1226   **coursed ashlar** *US see* **regular--coursed rubble**

C1227 e **coursed blockwork**
    d gerichtetes [geschichtetes] Blockmauerwerk *n*, Schichten-Blockmauerwerk *n*
    f maçonnerie *f* en blocs de béton appareillée [en assises]
    nl in lagen opgetrokken metselwerk *n* van steenblokken
    r правильная рядовая кладка *f* из бетонных массивов

C1228 e **coursed brickwork**
    d geschichtetes Ziegelmauerwerk *n*
    f maçonnerie *f* appareillé [en assise]
    nl laagsgewijs baksteenmetselwerk *n*
    r рядовая кирпичная кладка *f*

C1229 e **coursing joint**
    d 1. Lagerfuge *f*, waagerechte Fuge *f*
    2. Radialfuge *f* des Stein- oder Ziegelbogens
    f 1. joint *m* horizontal de maçonnerie 2. joint *m* radial d'un arc en briques *ou* en pierres
    nl 1. horizontale voeg *f(m)* 2. radiale voeg *f(m)* (in een boog)
    r 1. горизонтальный шов *m* кладки 2. радиальный шов *m* каменной [кирпичной] арки

C1230 e **cove**
    e Eckleiste *f*, Hohlkehle *f*
    f congé *m* (de raccordement)
    nl keel *f(m)*, holle lijst *f(m)*
    r галтель *f*, выкружка *f*

C1231 e **coved ceiling**
    d Decke *f* mit Auskehlung
    f plafond *m* à gorge [à congé de raccordement]
    nl plafond *n* met koof
    r потолок *m*, соединённый со стенами выкружками

C1232 e **cove lighting**
    d indirekte Beleuchtung *f*
    f éclairage *m* en corniche
    nl indirecte verlichting *f*
    r освещение *n* светильниками, скрытыми в карнизе потолка

C1233 e **cover**
    d 1. Betonüberdeckung *f*; Deckbodenschicht *f*, Decklage *f* 2. Deckel *m*, Kappe *f* 3. nutzbare Dach- *oder* Verkleidungsmaterialbreite *f*
    f 1. couche *f* de recouvrement *m*; couche *f* de protection; couche *f* superficielle 2. couvercle *m*; chemise *f*; chapeau *m* 3. largeur *f* utile
    nl 1. buitenlaag *f* (*m*) van het beton 2. kap *m*, bedekking *f*, dek *n* 3. deksel *n* 4. werkende breedte *f*
    r 1. защитный слой *m* (*бетона*); покровный слой *m* (*грунта*) 2. крышка *f*; кожух *m* 3. полезная [кроющая] ширина *f* (*кровельных или облицовочных материалов*)

C1234 e **coverage**
    d 1. überbaubare [überbaute] Fläche *f* 2. Betondeckung *f*, Überdeckung *f* 3. Deckfähigkeit *f*, Deckkraft *f*, Deckvermögen *n*
    f 1. aire *f* [surface *f*] bâtie 2. couche *f* de protection 3. pouvoir *m* couvrant; faculté *f* [capacité *f*] couvrante
    nl 1. bebouwd oppervlak *n* 2. betondekking *f* 3. dekvermogen *n*
    r 1. площадь *f* застройки 2. защитный слой *m* бетона 3. кроющая способность *f*, укрывистость *f* (*краски*)

C1235 e **cover block**
 d Unterlage *f*, Unterlegklötzchen *n*,
 Betonabstandhalter *m*
 f espaceur *m* [distancier *m*] pour
 ferraillage (*assurant l'epaisseur
 spécifiée d'une couche de
 protection*)
 nl afstandhouder *m* (*voor het stellen
 van wapening*)
 r фиксатор *m* арматуры
 (*обеспечивает заданную толщину
 защитного слоя бетона*)

C1236 **covered drainage** *see* **subsurface
 drainage**

C1237 e **covered repair slip**
 d Helling *f m* für Schiffsreparatur
 f cale *f* (couverte) de réparation
 nl overdekte scheepsreparatiehelling *f*
 r судоремонтный эллинг *m*

C1238 e **covered ship-building slip**
 d Schiffbauhelling *f*, *m*
 f cale *f* (couverte) de construction
 nl overdekte scheepshelling *f*
 r судостроительный эллинг *m*

C1239 e **covered slipway**
 d Helling *f*, *m*
 f cale *f* couverte
 nl overdekte scheepshelling *f*
 r эллинг *m*

C1240 e **cover fillet**
 d Abdeckleiste *f*, Fugenleiste *f*,
 Halteleiste *f*
 f couvre-joint *m*, latte *f* de
 recouvrement
 nl aftimmerlat *f(m)*, deklat *f(m)*
 r нащельник *m*, раскладка *f*

C1241 e **cover flashing**
 d senkrechter Fensterablauf *m*
 f bande *f* de solin verticale
 nl (lood)slabbe *f(m)*, loodloket *n*
 r вертикальный отлив *m*

C1242 e **covering layer**
 d Decklage *f*, Deckschicht *f*
 (*Erddamm*)
 f couche *f* de revêtement [de
 recouvrement]
 nl deklaag *f(m)*
 r покрывающий слой *m* (*земляной
 плотины*)

C1243 **covering power** *see* **coverage**

C1244 e **cover meter**
 d Betonanstrichmesser *m*,
 Überdeckungsmesser *m*
 f appareil *m* à mesurer l'épaisseur
 de la couche de protection
 nl laagdiktemeter *m*
 r прибор *m* для измерения толщины
 защитного слоя

C1245 **cover moulding** *see* **cover fillet**

C1246 e **cover plate**
 d 1. Gurtplatte *f*, Gurtlamelle *f*
 2. Decklasche *f*, Deckleiste *f*
 3. Schleppblech *n* (*Brückenfahrbahn*)

f 1. semelle *f* en large plat
 2. couvre-joint *m*, éclisse *f*, tôle
 f [plaque *f*] de recouvrement
 3. tôle *f* [plaque *f*] d'expension du
 tablier de pont
 nl 1. randplaat *f(m)* (*van stalen balk*)
 2. afdekplaat *f(m)*, lat *f(m)*
 3. brugopritplaat *f(m)*
 r 1. поясной лист *m* (*стальной
 балки*) 2. накладка *f*
 3. подвижный лист *m* настила
 (*у швов проезжей части моста*)

C1247 **cover strip** *see* **cover fillet**

C1248 **coving** *see* **cove**

C1249 e **cowl**
 d 1. Schornsteinhaube *f*,
 Schornsteinaufsatz *m*;
 Deflektorhaube *f*, Rohraufsatz *m*
 2. Heberdecke *f*, Heberhaube *f*
 f 1. mitre *f* de cheminée;
 déflecteur *m* 2. hotte *f* (de
 siphon)
 nl 1. schoorsteenkap *f(m)* 2.
 syfondeksel *n*
 r 1. зонт *m*, колпак *m*; дефлектор
 m (*на дымовой или
 вентиляционной трубе*) 2. капор
 m сифона

C1250 e **crab**
 d 1. Laufkatze *f*, Krankatze *f*
 2. tragbare Handwinde *f*
 f 1. chariot *m* porte-palan (*p.ex.
 d'un pont roulant*) 2. treuil *m* à
 bras [manuel]
 nl 1. loopkat *f* 2. handlier *f(m)*
 r 1. грузоподъёмная крановая
 тележка *f* (*напр. мостового крана*)
 2. переносная ручная лебёдка *f*

C1251 e **crack**
 d Riß *m*
 f fissure *f*
 nl scheur *f(m)*
 r трещина *f*

C1252 e **crackage**
 d Gesamtquerschnitt *m* aller
 Undichtigkeiten in den äußeren
 Umfassungskonstruktionen
 f manaque *m* d'étanchéité (*de fenêtre*)
 nl gezamelijke doorsnede *f* van de
 kieren in de omhullende
 constructie
 r суммарное сечение *n* неплотностей
 в притворах дверей и окон
 здания *или* помещения

C1253 e **crack control**
 d Rißbeherrschung *f*
 f prévention *f* de la fissuration
 nl voorkomen *n* van wilde
 scheurvorming
 r предупреждение *n*
 трещинообразования

C1254 e **crack formation**
 d Rißbildung *f*; Rißbeschaffenheit *f*

f fissuration f; tracé m des
fissures

nl scheurvorming f, verloop m van
de scheuren

r образование n трещин,
трещинообразование n; характер
m строения трещин, форма f
трещинообразования

C1255 e cracking
d Rißbildung f
f fissuration f
nl scheuren n
r трещинообразование n,
образование n трещин

C1256 cracking limit state see limit
state of cracking

C1257 e cracking moment
d Rißmoment m
f moment m de fissuration
nl moment n van de scheurvorming
r момент m трещинообразования

C1258 e crack pattern
d Rissebild n
f tracé f des fissures
nl scheurenpatroon n
r форма f трещинообразования

C1259 crack restraint see crack control

C1260 e crack width
d Rißbreite f
f ouverture f des fissures
nl scheurbreedte f
r ширина f раскрытия трещин

C1261 e cradle
d 1. Rohrwiege f, Rohrsattel m
2. Trogwagen m, Stapelschlitten m
3. Hängebühne f, Hängekorb m;
Hängegerüst n 4. Montage-
-Stützrahmen m
f 1. berceau m [support m] pour
tuyau 2. ber m de lancement;
berceau m roulant 3. sellette f,
nacelle f; échafaudage m volant
4. échafaud m de montage
nl 1. zadel m, n (leidingsteun)
2. slede f (scheepshelling)
3. hangsteiger m 4. montagerek n
r 1. седловая опора f
(трубопровода, водовода)
2. тележка f эллинга или слипа
3. подвесная люлька f; висячие
подмости pl 4. монтажная опорная
рама f

C1262 cradle scaffold see cradle 3.

C1263 e cramp
d 1. Schraubzwinge f; Bankhaken m
2. Bauklammer f, Krampe f;
Bankeisen n
f 1. sergent m, serre-joint m;
servante f 2. agrafe f, happe f;
goujon m; ancre m de bâti
dormant
nl 1. lijmtang f(m) klemhaak m
2. dookanker n

r 1. струбцина f; сжим m;
сжимное приспособление n
2. анкер m для скрепления
каменных блоков, пирон m;
закладной анкер m для крепления
дверной коробки к кладке

C1264 crampon, crampoon see nippers

C1265 e crane
d Kran m
f grue f
nl (hijs)kraan f(m)
r (грузоподъёмный) кран m

C1266 e crane air conditioner
d Kran(kabinen)klimagerät n
f conditionneur m de l'air de grue,
climatiseur m de cabine de grue
nl klimaatregelingsapparaat n voor de
kraancabine
r кондиционер m кабины крана,
крановый кондиционер m

C1267 e crane boom
d Kranausleger m
f flèche f de grue
nl giek m, kraanarm m
r крановая стрела f, стрела f крана

C1268 e crane girder
d Kranbahnträger m
f poutre f de roulement (d'un pont
roulant)
nl kraanbaanligger m
r подкрановая балка f

C1269 e crane runway
d Kranbahn f
f chemin m de roulement
nl kraanbaan f(m)
r (под)крановый путь m

C1270 e crane stanchion
d Stützenteil n unter der Kranbahn
f membrure f de poteau supportant la
poutre de roulement
nl kolom f(m) van een bouwkraan
r подкрановая ветвь f колонны

C1271 e crank brace
d Brustbohrer m
f vilebrequin m
nl booromslag m, n
r коловорот m

C1272 e crash barrier
d Leitplanke f
f garde-corps m de sécurité
nl veiligheidshek n
r барьер m безопасности

C1273 crawl channel see crawlway

C1274 e crawler crane
d Gleiskettenkran m, Raupenkran m
f grue f sur chenille(s)
nl rupskraan f(m)
r гусеничный кран m

C1275 e crawler excavator
d Raupenbagger m
f excavateur m [excavatrice f] sur
chenille(s)
nl graafmachine f op rupsbanden
r гусеничный экскаватор m

C1276 e **crawler mounted piling rig**
   d Raupenramme f
   f sonnette f (de battage) sur chenille(s) [chenillée]
   nl heimachine f op rupsbanden
   r сваебойное оборудование n [копёр m] на гусеничном ходу, гусеничный копёр m

C1277 e **crawl space**
   d Kriechraum m, Bekriechungsraum m
   f sous-sol m technique bas
   nl kruipruimte f
   r полупроходное техническое подполье n (ограниченной высоты)

C1278 e **crawlway**
   d bekriechbarer Kanal m
   f galerie f de service [technique] basse
   nl kruipkoker m (leidingkoker)
   r полупроходной канал m

C1279 e **crazing**
   d Haarrisse m pl; Netzrißbildung f
   f fissures f pl capillaires
   nl (vorming f van) haarscheurtjes n pl (beton, verf)
   r волосные трещины f pl (напр. на поверхности бетона); образование n волосных трещин

C1280 e **creep**
   d Kriechen n
   f fluage m
   nl kruip m
   r ползучесть f, крип m, пластическое течение n металла

C1281 e **creep coefficient**
   d Kriechzahl f
   f coefficient m de fluage
   nl kruipcoëfficiënt m
   r коэффициент m ползучести

C1282 e **creep deformation**
   d Kriechverformung f
   f déformation f de fluage
   nl kruipvervorming f
   r деформация f ползучести

C1283 e **creeper crane**
   d Kletter(turmdreh)kran m, Hochhauskletterkran m
   f grue f (à tour) gripmante
   nl meelopende bouwkraan f(m)
   r ползучий кран m (перемещается по крутой наклонной или] вертикальной плоскости)

C1284 e **creep limit**
   d Kriechgrenze f
   f limite f de fluage
   nl kruipgrens f(m)
   r предел m ползучести

C1285 e **creep line**
   d Sickerweg m
   f ligne f de fuite [de saturation]
   nl kruiplijn f(m)
   r путь m [контур m] фильтрации

C1286 e **creep of concrete**
   d Betonkriechen n
   f fluage m du béton
   nl kruip m van het beton
   r ползучесть f бетона

C1287 e **creep rate**
   d Kriechgeschwindigkeit f
   f vitesse f de fluage
   nl kruipsnelheid f
   r скорость f ползучести

C1288 **creep strength** see **creep limit**

C1289 e **creep test**
   d Kriechversuch m, Kriechprüfung f; Dauerstandversuch m
   f essai m de fluage
   nl kruipproef f(m)
   r испытание n на ползучесть; испытание n на длительную прочность при статической нагрузке

C1290 e **crest**
   d 1. Krone f, Dammkrone f, Wehrkrone f; Böschungskante f 2. Hochwasserspitze f 3. Wellenberg m, Wellenkamm m, Wellenscheitel m 4. First m
   f crête f; faîte m; sommet m
   nl 1. kruin f(m) (dijk) 2. hoogste waterstand m 3. golftop m 4. vorst f(m), nok f(m) (dak)
   r 1. гребень m (плотины); порог m (водослива); бровка f (откоса) 2. пик m паводка 3. вершина f [гребень m] волны 4. конёк m (крыши)

C1291 e **crest gate**
   d Kronenverschluß m, Kronenschütz n, Überfallschütz n
   f vanne f de crête [de couronnement]
   nl hoogwaterkering f
   r поверхностный затвор m на гребне плотины

C1292 e **crest level**
   d Kronenhöhe f
   f hauteur f de crête
   nl kruinhoogte f
   r отметка f гребня (плотины)

C1293 e **crib**
   d 1. Balkenrostwerk n; Trägerrost m (im Säulenfuß) 2. Wanderkasten m, Holzpfeiler m 3. Steinkiste f, Steinkasten m, Kribbe f
   f 1. radier m de poutres 2. soutènement m à piles 3. coffrage m en charpente, encoffrement m en charpente
   nl 1. roosterwerk n van balken 2. stapeling f 3. steenkist f(m)
   r 1. балочный ростверк m; балочная клетка f (в основании колонны) 2. костровая клеть f 3. ряж m

C1294 *e* **crib cofferdam**
 d Steinkistenfangdamm *m*,
  Blockfangdamm *m*
 f batardeau-caisson *m*
 nl met stenen opgevulde cofferdam
  m
 r ряжевая перемычка *f*

C1295 *e* **crib dam**
 d Steinkistenwehr *n*,
  Steinkistensperre *f*,
  Steinkastenwehr *n*,
  Steinkastensperre *f*,
  Steinkistendamm *m*,
  Steinkastendamm *m*
 f barrage *m* en charpente
 nl dam *m* van gestapelde
  steenblokken
 r ряжевая дамба *f*; ряжевая
  плотина *f*

C1296 *e* **crib groyne**
 d Steinkistentrennbuhne *f*,
  Steinkistentrenndamm *m*
 f épi *m* [éperon *m*] à coffrages
  remplis d'enrochement
 nl (rivier)krib *f(m)* van gestapelde
  steenblokken
 r ряжевая полузапруда *f* [шпора *f*]

C1297 crib weir *see* crib dam

C1298 *e* **crimped steel fibers**
 d gekröpfte Stahlfasern *f pl*
 f fibres *f pl* en acier ondulées
 nl gebogen staalvezels *f pl*
 r волнистые стальные волокна *n pl*
  (*для дисперсного армирования*
  бетона)

C1299 crippling load *see* critical
  buckling load

C1300 *e* **criss-cross-diagonals**
 d Kreuzstreben *f pl*
 f diagonales *f pl* croisées
 nl kruisschoren *m pl*
 r перекрёстные раскосы *m pl*
  (*фермы*)

C1301 *e* **criterion of buckling**
 d Biegeknickenkriterium *n*,
  Beul(ungs)kriterium *n*,
  Knickmerkmal *n*
 f critère *m* de flambage [de
  flambement]
 nl knikgrens *f(m)*
 r критерий *m* потери устойчивости
  при продольном изгибе

C1302 *e* **critical buckling load**
 d Beullast *f*, Knicklast *f*
 f charge *f* critique de flambage
 nl kritische belasting *f* bij knik
 r критическая нагрузка *f* при
  продольном изгибе

C1303 *e* **critical density**
 d kritische Dichte *f*
 f densité *f* critique
 nl kritische dichtheid *f*
 r критическая плотность *f*

C1304 *e* **critical depth**
 d kritische Tiefe *f*, Grenztiefe *f*
 f profondeur *f* critique
 nl grensdiepte *f*
 r критическая глубина *f*

C1305 *e* **critical discharge**
 d kritischer Abfluß *m*; kritische
  Einheitsergiebigkeit *f*
  [Brunnenspende *f*]
 f débit *m* critique de charriage,
  débit *m* de début d'entraînement;
  débit *m* critique
 nl kritisch debiet *n* (*afstroming*)
 r критический расход *m* (*при*
  котором начинается движение
  наносов); критический удельный
  дебит *m*

C1306 *e* **critical flow**
 d kritische Strömung *f*, kritischer
  Abfluß *m*
 f écoulement *m* critique
 nl kritische stroming *f* (*afstroming*)
 r критическое состояние *n* потока

C1307 *e* **critical head**
 d kritische Druckhöhe *f*
 f hauteur *f* de chute *f* critique
 nl kritische drukhoogte *f*
 r критический напор *m*

C1308 *e* **critical height**
 d kritische Höhe *f*
 f hauteur *f* critique
 nl kritische hoogte *f*
 r критическая высота *f*

C1309 *e* **critical hydraulic gradient**
 d kritische Druckhöhe *f*, kritisches
  hydraulisches Gefälle *n*
 f gradient *m* hydraulique critique,
  pente *f* critique
 nl kritisch verval *n*
 r критический гидравлический
  градиент *m*, критический уклон
  m (*водотока*)

C1310 *e* **critical limit state**
 d kritischer Grenzzustand *m*
 f état *m* limite critique
 nl kritische grenstoestand *m*
 r критическое предельное состояние
  n

C1311 *e* **critical-load design, critical**
  method
 d Verfahren *n* der kritischen Last
 f calcul *m* (de stabilité) à la
  charge critique
 nl methode *f* van kritische belasting
 r расчёт *m* (устойчивости) по
  критическим нагрузкам

C1312 *e* **critical moment**
 d kritisches Moment *n*
 f moment *m* critique
 nl kritisch moment *n*
 r критический момент *m*

C1313 *e* **critical path scheduling**
 d CPM-Planung *f*, Netzwerkplanung *f*

f planning *m* (d'exécution des travaux) par chemin critique
nl netwerkplanning *f*
r планирование *n* (*строительных работ*) методом критического пути, сетевое планирование *n*

C1314 *e* **critical pressure**
d kritischer Druck *m*
f pression *f* critique
nl kritische druk *m*
r критическое давление *n*

C1315 *e* **critical slope**
d 1. Grenzgefälle *n* 2. natürlicher Böschungswinkel *m*, Ruhewinkel *m*
f 1. pente *f* critique 2. talus *m* naturel
nl 1. gevaarlijkste glijvlak *n* 2. natuurlijk talud *n*
r 1. критический уклон *m* водотока 2. угол *m* естественного откоса

C1316 *e* **critical void ratio**
d kritische Porenziffer *f*
f indice *m* de vide critique
nl kritisch poriëngehalte *n*
r критический коэффициент *m* пористости

C1317 *e* **cross**
d Kreuzstück *n*
f croix *f*
nl kruis(stuk) *n*
r крестовина *f* (*трубопровода*)

C1318 *e* **cross arm**
d Querträger *m*, Querriegel *m*, Traverse *f*
f traverse *f* de support
nl dwarsarm *m*
r траверса *f*; поперечина *f*, поперечная балка *f*

C1319 *e* **cross bar slings**
d viersträngige Traverse *f*, Vierstropptraverse *f*
f palonnier *m* de [à] quatre brins
nl gekruiste hijsstrop *m*, *f*
r четырёхветвевая траверса *f*

C1320 *e* **cross beam, crossbeam**
d 1. Holm *m* 2. Querbalken *m*, Querträger *m*
f 1. chapeau *m*, poutre *f* de couronnement 2. poutre *f* transversale
nl 1. sloof *f(m)* 2. kruisbalk *m*, dwarsbalk *m*
r 1. шапочный брус *m* 2. поперечная балка *f*

C1321 *e* **cross bridging** *US*
d Anordnung *f* der Kreuzstackung
f pose *f* des croisillons (*entre poutres de plancher*)
nl dwarskoppeling *f*
r постановка *f* перекрёстных связей между балками перекрытий

C1322 *e* **cross-country crane**
d geländegängiger Kran *m*
f grue *f* tous terrains
nl terreinwaardige kraan *f(m)*
r самоходный кран *m* повышенной проходимости

C1323 *e* **cross cut**
d Hirnschnitt *m*, Querschnitt *m* (*Holzbearbeitung*)
f 1. sciage *m* [coupage *m*] transversal 2. coupe *f* transversal
nl dwarssnede *f* (*houtbewerking*)
r 1. поперечная резка *f* [распиловка *f*] 2. поперечный рез *m* [пропил *m*]

C1324 *e* **cross-cut saw**
d Quersäge *f*
f scie *f* passe-partout, passe-partout *m*
nl zaag *f(m)* met wijde zetting
r пила *f* поперечной резки, поперечная пила *f*

C1325 *e* **cross dike**
d Querdeich *m*, Traverse *f*
f digue *f* transversale
nl dwarsdijk *m*
r поперечная дамба *f*, траверс *m*

C1326 *e* **cross dredging**
d Entegang *m*, Papillionierung *f*
f papillonnage *m*
nl dwars baggeren *n*
r папильонирование *n*

C1327 **cross dyke** *see* **cross dike**

C1328 *e* **cross fall**
d Querneigung *f*
f profil *m* en travers
nl dwarshelling *f*
r поперечный уклон *m*, уклон *m* поперечного профиля (*дороги*)

C1329 *e* **cross-flow**
d Querstrom *m*
f courant *m* croisé
nl dwarsstroom *m*
r перекрёстный ток *m*

C1330 *e* **cross flow fan**
d Querstromventilator *m*, Querstromgebläse *n*
f ventilateur *m* à courants croisés
nl dwarsstroomventilateur *m*
r диаметральный вентилятор *m*

C1331 *e* **cross flow heat exchanger**
d Kreuzstromwärmetauscher *m*
f échangeur *m* de chaleur à courant croisé
nl dwarsstroomwarmtewisselaar *m*
r перекрёстноточный теплообменник *m*

C1332 *e* **cross-garnet**
d Aufsatzband *n*
f penture *f*
nl kruisheng *n*, staartheng *n*
r накладная петля *f* (*для ворот*)

C1333 *e* **cross-head**
   d Fallhöhe *f*
   f chute *f* brute
   nl kruiskop *m*
   r перепад *m* напоров, разность *f*
   уровней бьефов

C1334 *e* **crossing**
   d Straßenkreuzung *f*;
   Bahnübergang *m*;
   Fußgängerübergang *m*
   f croisement *m*, intersection *f* (de
   routes); traversée *f* (de deux voies
   ferrées); passage *m* pour piétons
   nl (spoor)wegkruising *f*
   r пересечение *n* дорог;
   (железнодорожный) переезд *m*;
   пешеходный переход *m*

C1335 *e* **cross joint**
   d Querfuge *f*, Quernaht *f*;
   Stoßfuge *f*, Vertikalnaht *f*
   f joint *m* transversal; joint *m*
   vertical de maçonnerie
   nl stootvoeg *f(m)*; achtervoeg *f(m)*
   r поперечный шов *m*; вертикальный
   шов *m* кирпичной кладки

C1336 *e* **cross-lap joint**
   d gerade Überblattung *f*,
   Kreuzüberblattung *f*
   f assemblage à mi bois formant un X
   nl overkeping *f* (met voorloeven)
   r крестовая врубка *f* в полдерева

C1337 *e* **cross-member**
   d Querträger *m*
   f entretoise *f*, traverse *f*
   nl dwarsdrager *m*
   r поперечина *f*

C1338 *e* **crossover**
   d 1. Überkreuzung *f*,
   Leitungskreuzung *f* 2.
   Abzweigweiche *f*, Bahnanschluß *m*
   3. Straßenüberführung *f*,
   Verkehrsbrücke *f* 4. (wandernde)
   Sandbank *f*, Furt *f*
   f 1. coude *f* de croisement (d'un
   tuyau) 2. jonction *f* ferroviaire 3.
   passage *m* supérieur; viaduc *m*
   4. banc *m*, barre *f*
   nl 1. kruising *f* 2. wissel *m* 3.
   verkeersbrug *f(m)* 4. doorwaadbare
   plaats *f(m)*
   r 1. перекидка *f* [обвод *m*]
   трубопровода 2. железнодорожный
   стрелочный съезд *m* 3.
   путепровод *m* 4. перекат *m*

C1339 *e* **cross reinforcement**
   d Querbewehrung *f*, Querarmierung *f*
   f armatures *f pl* transversales
   nl dwarswapening *f*
   r поперечная арматура *f*

C1340 *e* **cross road**
   d Niveaukreuzung *f*, höhengleiche
   Kreuzung *f*; Rechteckkreuzung *f*
   f croisement *m* [intersection *f*] des
   routes

   nl dwarsweg *m*
   r перекрёсток *m*; пересечение *n*
   дорог (*под прямым углом*)

C1341 *e* **cross section**
   d Querschnitt *m*; Querprofil *n*
   f section *f* transversale
   nl dwarsprofiel *n*
   r поперечное сечение *n*;
   поперечный профиль *m*

C1342 *e* **cross-sectional area**
   d Querschnittsfläche *f*
   f aire *f* de la section transversale
   nl dwarsdoorsnede *f*
   r площадь *f* поперечного сечения

C1343 *e* **cross section of stream**
   d Durchflußprofil *n*,
   Durchflußquerschnitt *m*,
   durchströmter Querschnitt *m*
   f section *f* (transversale) d'un
   cours d'eau
   nl stroomprofiel *n*
   r поперечное сечение *n* потока

C1344 *e* **cross-shaped column**
   d Kreuzstütze *f*
   f colonne *f* [poteau *m*] à section
   cruciforme
   nl kruiskolom *f(m)*
   r колонна *f* крестообразного
   сечения

C1345 *e* **cross ventilation**
   d Querlüftung *f*
   f ventilation *f* transversale
   nl dwarsventilatie *f*
   r поперечная вентиляция *f*

C1346 *e* **cross wall**
   d Querwand *f*, Quermauer *f*
   f mur *m* transversal, paroi *f*
   transversale
   nl dwarsmuur *m*
   r поперечная стена *f*

C1347 *e* **cross-wall construction**
   d (Gebäude-)Konstruktion *f* mit
   tragenden Querwänden
   f bâtiment *m* [construction *f*] à
   parois transversales portantes
   nl constructie *f* met dragende
   dwarsmuren
   r конструкция *f* (*здания*) с
   несущими поперечными стенами

C1348 *e* **cross welt**
   d dem First parallel laufender
   Liegefalz *m* (*Blechdachhaut*)
   f agrafure *f* transversale (*parallèle
   au faîtage*)
   nl dwarsnaad *m* in metalen
   dakbedekking
   r лежачий фальц *m* (*металлической
   кровли*), параллельный коньку

C1349 *e* **cross-wise reinforcement**
   d kreuzweise Bewehrung *f*
   f armatures *f pl* croisées
   nl kruiswapening *f*
   r перекрёстная арматура *f*

C1350 *e* **crow-bar**
    d Brecheisen *n*
    f pince-monseigneur *f*, pied-de-biche *m*
    nl breekijzer *n*
    r лом *m*

C1350a **crowd shovel** *see* **face shovel**

C1351 *e* **crown**
    d 1. Gewölbescheitel *m*,
    Gewölbekrone *f*; Bogenscheitel *m*;
    Rohrscheitel *m* 2. Dammkrone *f*,
    Wehrmauerkrone *f* 3. Scheitel *m*
    des Straßenquerprofils
    4. Kalotte *f* (*Tunnel*)
    f 1. clé *f* de la voûte, sommet *m*
    2. crête *f*, couronnement *m* (*d'un
    barrage*) 3. sommet *m*, profil
    m bombé (*de la route*) 4. voûte *f*
    du tunnel
    nl 1. top *m*, toplijn *f(m)* (*boog,
    gewelf*) 2.,3. kruin *f* (*m*) (*stuw,
    weg*) 4. gewelf *n* (*tunnel*)
    r 1. замок *m* (*свода*); ключ *m*
    (*арки*); шелыга *f* 2. гребень *m*
    (*плотины*) 3. выпуклый
    поперечный профиль *m* (*дороги*)
    4. свод *m* (*туннеля*)

C1352 **crown block** *see* **apex block**

C1353 *e* **crown cover**
    d (Kreissägeblatt-)Schutzhaube *f*
    f capot *m* de protection
    nl beschermkap *f(m)* (*cirkelzaag*)
    r защитный кожух *m* (*напр. над
    диском циркульной пилы*)

C1354 *e* **crown hinge**
    d Scheitelgelenk *n*
    f articulation *f* [rotule *f*] à la clé
    nl topscharnier *n*
    r ключевой шарнир *m*

C1355 *e* **crown level**
    d Scheitelhöhe *f*
    f cote *f* de la crête
    nl kruinhoogte *f*
    r отметка *f* гребня

C1356 *e* **crown of arch**
    d Bogenscheitel *m*
    f clé *f* [clef *f*] de l'arc; clé *f* de
    la voûte
    nl toppunt *n* (*boog*)
    r ключ *m* арки; вершина *f* арки

C1357 *e* **crude sewage**
    d Rohabwasser *n*
    f eaux *f pl* d'égout brutes
    nl onbehandeld afvalwater *n*
    r неочищенные сточные воды *f pl*

C1358 *e* **crude tar**
    d Dickteer *m*, Rohteer *m*
    f goudron *m* brut
    nl ruwe teer *m*, *n*
    r сырой дёготь *m*

C1359 *e* **crumbling, crumbling away**
    d Abbröckeln *n*, Zerbrökeln *n*
    f écaillage *m*, écaillement *m*,
    effritement *m*

    nl afbrokkelen *n*
    r выкрашивание *n*

C1360 *e* **crushed aggregate**
    d Schotter *m*, Splitt *m*
    f agrégat *m* concassé, pierre *f*
    concassée
    nl steenslag *n*
    r щебень *m*

C1361 *e* **crushed gravel**
    d Quetschkies *m*, Brechkies *m*
    f gravier *m* concassé
    nl gebroken grind *n*
    r дроблёный гравий *m*

C1362 **crushed stone** *see* **crushed
    aggregate**

C1363 **crushed stone aggregate** *see*
    crusher-run stone

C1364 *e* **crushed-stone base course,
    crushed-stone bed**
    d Schotterbett *n*, Steinschlagbett *n*
    f lit *m* de pierre concassée
    nl paklaag *f(m)* van gebroken steen
    r щебёночная подготовка *f*

C1365 *e* **crusher**
    d Brecher *m*
    f concasseur *m*, broyeur *m*
    nl steenbreker *m*
    r дробилка *f*

C1366 *e* **crusher dust**
    d Brechmehl *n*
    f poussière *f* de concassage
    nl steenpoeder *n*, *m*
    r каменная пыль *f* (*от
    камнедробилки*)

C1367 *e* **crusher-run stone**
    d ungesiebter gebrochener Stein *m*
    [Schotter *m*, Splitt *m*]
    f pierre *f* concassée non tamisée
    [non criblée]
    nl ongezeefde gebroken steen *m*
    r несортированный щебень *m*

C1368 *e* **crushing**
    d Brechen *n*
    f concassage *m*, broyage *m*
    nl breken *n* van steen
    r дробление *n*

C1369 *e* **crushing-and-grading plant,
    crushing-and-screening plant**
    d Brech- und Siebanlage *f*
    f centrale *f* [station *f*] de concassage-
    -criblage
    nl breek- en zeefinrichting *f*
    r дробильно-сортировочная
    установка *f*

C1370 *e* **crushing strength**
    d Quetschgrenze *f*
    f résistance *f* à l'écrasement [à la
    compression]
    nl druksterkte *f*
    r прочность *f* при раздавливании
    [сжатии]

C1371 *e* **cryogenic insulation**
    d Tieftemperaturisolation *f*

f isolation *f* cryogénique
nl isolatie *f* voor zeer lage
temperaturen
r криогенная изоляция *f*

C1372   **cubage** *see* **cubical content**

C1373 *e* **cube impact strength**
d Würfel-Schlagfestigkeit *f*; an
Würfeln ermittelte
Schlagfestigkeit *f*
f résistance *f* au choc du cube
d'essai [de l'éprouvette cubique]
nl kubusvastheid *f*
r динамическая [ударная]
прочность *f* кубического образца
(*цемента, бетона*)

C1374 *e* **cube mould**
d Würfelform *f*
f moule *m* cubique
nl kubusvorm *m*
r стандартная (металлическая)
форма *f* (*для контрольных
кубических образцов бетона*)

C1375 *e* **cube strength at 28 days**
d Würfel(druck)festigkeit *f* W28
f résistance *f* en compression sur
cube à l'age de 28 jours
nl kubussterkte *f* van het beton na
28 dagen
r кубиковая прочность *f* бетона в
28-суточном возрасте

C1376 *e* **cube strength test, cube test**
d Würfel(druck)probe *f*,
Würfel(druck)versuch *m*
f essai *m* de résistance sur cube
nl kubusdrukproef *f(m)*
r испытание *n* кубического
образца бетона на сжатие

C1377 *e* **cubical content, cubic content**
d umbauter Raum *m*, Kubatur *f*
f cubature *f*, cubage *m*, cube *m*
total (*de bâtiment*)
nl bouwvolume *n*
r строительный объём *m*, кубатура
f (*здания*)

C1378   **cubicle** *see* **booth 1.**

C1379 *e* **cubicle aggregate**
d Zuschlagstoff *m* mit eckigen
[scharfkantigen] Körnern
f granulat *m* [agrégat *m*] (à grain)
anguleux
nl toeslag *m* met scherpe korrels
r заполнитель *m* с угловатыми
зёрнами

C1380 *e* **cultivated soil**
d Kulturboden *m*, bebauter Boden *m*
f sol *m* cultivé
nl cultuurbodem *m*, bebouwde
bodem *m*
r культурная [обработанная]
почва *f*

C1381 *e* **culvert**
d 1. Wasserdurchlaßbauwerk
Durchlaß *m*; Düker *m*

2. Stichkanal *m*, Sohlenkanal *m*
3. Abzug(s)kanal *m* 4. Drumme *f*
f 1. ponceau *m* 2. conduite *f* de
vidange 3. canal *m* de sortie;
canal *m* de décharge 4. goulotte *f*
en planches
nl 1. waterdoorlaat *m*; duiker *m*
2. riool *n* (*sluis*) 3. riool *n*
4. duikersluis *f(m)*
r 1. водопропускное сооружение *n*;
труба *f* под насыпью; водоспуск
m в теле плотины; дюкер *m*
2. выпускной канал *m* шлюзного
водопровода; водопроводная
галерея *f* шлюза 3. сточный
[водоотводной] канал *m*
4. дощатый жёлоб *m*

C1382 *e* **cumulative runoff diagram,
cumulative volume curve**
d Abflußmengensummenlinie *f*
f courbe *f* des débits cumulés
nl cumulatieve afvloeiingskromme *f*
r интегральная кривая *f* стока

C1383 *e* **cunette**
d Wassersammler *m*, Abflußrinne *f*,
Entwässerungsrinne *f*
f cunette *f*; rigole *f*; goulotte *f*
nl greppel *f(m)*; waterloop *m*
r водосборник *m*; водосток *m*;
лоток *m*; жёлоб *m*

C1384 *e* **cupola dam**
d Kuppelstaumauer *f*, Kuppelsperre *f*
f barrage *m* à dôme [à coupole]
l koepelstuwdam *m*
r купольная плотина *f*

C1385   **cup shake** *see* **ring shake**

C1386 *e* **cup-square bolt**
d Vierkant-Holzschraube *f*
f vis *f* à bois carré
nl vierkante houtschroef *f(m)*
r шуруп *m* с квадратной головкой,
глухарь *m*

C1387 *e* **curb**
d Bordstein *m*
f bordure *f* (de trottoir)
nl trottoirband *n*
r бортовой камень *m*

C1388   **cure** *see* **concrete curing**

C1389   **curing** *see* **concrete curing**

C1390 *e* **curing agent**
d Nachbehandlungsmittel *n* für
Beton
f produit *m* de cure
nl nabehandelingsmiddel *n* voor beton
r состав *m* [материал *m*] для ухода
за бетоном; состав *m*,
улучшающий условия
выдерживания бетона

C1391 *e* **curing blanket**
d Abdeckmatte *f*
(*Betonnachbehandlung*)
f matelas *m* [natte *f*] de cure
nl afdekmat *f(m)* (*betonuitharding*)

r защитное покрытие *n* для ухода за
бетоном

C1392 curing compound *see* curing agent

C1393 *e* **curing membrane**
d Nachbehandlungsfilm *m*
f pellicule *f* de cure
nl nabehandelingsfolie *n*
r плёнка *f* для выдерживания
бетона (*наносится методом
распыления плёнкообразующего
состава*)

C1394 *e* **curing period**
d Abbindezeit *f*;
Nachbehandlungszeitraum *m*
f durée *f* de cure; période *m* de
maturation
nl verhardingstijd *m* van beton
r период *m* [время *n*]
выдерживания [созревания]
(*бетона*); продолжительность *f*
ухода за бетоном

C1395 *e* **curing temperature**
d Nachbehandlungstemperatur *f*
f température *f* de cure
nl nabehandelingstemperatuur *f*
r температура *f* выдерживания
(*бетона*)

C1396 *e* **curling-cracking**
d Verwerfung *f* mit Rißbildung
f gauchissement *m* avec fissuration
nl rimpeling *f* met scheurvorming
r коробление *n* с трещинообразова-
нием

C1397 *e* **curling of slabs**
d Verwerfung *f* [Krümmung *f*] der
Betonplatten
f gauchissement *n* des dalles en
béton
nl rimpelen *n* (*betonplaten*)
r искривление *n* [коробление *n*]
бетонных плит

C1398 *e* **current failure**
d Stromausfall *m*
f panne *f* [defaillance *f*]
d'électricité
nl stroomuitval *m*
r перерыв *m* в электроснабжении

C1399 *e* **current meter**
d Meßflügel *m*, Strömungsmesser *m*
f moulinet *m* hydrométrique
nl stromingsmeter *m*
r гидрометрическая вертушка *f*

C1400 *e* **current regime**
d Strömungsregime *n*
f régime *m* du courant
nl stromingsregime *n*
r режим *m* потока

C1401 *e* **current velocity**
d Strömungsgeschwindigkeit *f*,
Stromgeschwindigkeit *f*
f vitesse *f* de courant
nl stroomsnelheid *f*
r скорость *f* течения

C1402 *e* **curtailment of reinforcement**
d Abschneiden *n* der Bewehrungsstäbe
außerhalb der Momentumfläche
f coupe *f* des barres d'armature
au-delà de la courbe des
moments fléchissants
nl wapeningsstaven inkorten *n*
(*buiten de momentenlijn*)
r обрезка *f* арматурных стержней
за пределами эпюры
изгибающих моментов
(*в соответствии с эпюрой
материалов*)

C1403 *e* **curtain grouting**
d Schleierverpressung *f*,
Schleierinjektion *f*,
Schürzen-Vermörtelung *f*
f injection *f* du coulis formant un
rideau [un voile]
nl cementscherminjectie *f*
r создание *n* цементационной
завесы

C1404 *e* **curtain wall**
d nicht tragende Leichtbauaußenwand
f, Vorhangwand *f*
f mur-rideau *m*
nl gordijngevel *m*
r ненесущая (навесная) наружная
стена *f*, стена *f* из лёгких
навесных панелей

C1405 *e* **curved bridge**
d Kurvenbrücke *f*
f pont *m* en courbe
nl boogbrug *f(m)*
r криволинейный мост *m*

C1406 *e* **curve of maximum bending
moments**
d Umhüllende *f* der
Höchstbiegemomente,
Maximalmomentenlinie *f*
f ligne *f* enveloppante des moments
fléchissants maxima
nl omhullende *f* van de maximale
buigmomenten
r кривая *f* [огибающая *f*]
наибольших изгибающих моментов

C1407 *e* **curve plotter**
d Kurvenschreiber *m*
f traceur *m* de courbes
nl schrijfstift *f(m)*
r графопостроитель *m*

C1408 *e* **cushion course**
d Bettung *f*; Bettschicht *f*
f lit *m* de ballaste, couche *f* de
ballast; sous-couche *f*
nl ballast(bed *n*) *m*
r постель *f*; основание *n*;
балластный слой *m*

C1409 *e* **cushioning pool**
d Beruhigungsbecken *n*, Tosbecken *n*;
Wasserpolster *n*
f bassin *m* de tranquillisation
nl golfdempingbassin *n*

r успокоительный бассейн *m*,
водобойный колодец *m*; водяная
подушка *f*

C1410 *e* **custom-built forms, custom-made forms**
    d individuell gefertigte Schalung *f*
    f coffrage *m* non standard [confectionné sur mesure]
    nl op maat vervaardigde mal *m* [bekisting *f*]
    r нетиповая опалубка *f*, опалубка *f* индивидуального изготовления

C1411 *e* **cut**
    d 1. Durchstich *m*; Stichkanal *m*; Vertiefungsrinne *f* 2. Aushub *m*, Abtrag *m*, Einschnitt *m*; Baugrube *f* 3. Abtragmaterial *n* 4. Schnitt *m*
    f 1. percement *m*; coupure *f* de meandre 2. excavation *f*, déblai *m*, fouille *f* 3. déblai *m* 4. coupe *f*, section *f*
    nl 1. ingraving *f* 2. bouwput *m*; doorsnijding *f* 3. baggerspecie *f* 4. doorsnede *f*
    r 1. прокоп *m*; спрямляющий канал *m*; дноуглубительная прорезь *f* 2. выемка *f*; котлован *m* 3. вынутый грунт *m* 4. разрез *m*

C1412 *e* **cut and cover technique**
    d offene Tunnelbauweise *f*
    f construction *f* des tunnels en à **ciel** ouvert [dans une fouille]
    nl open tunnelbouw *m*
    r проходка *f* туннелей открытым способом

C1413 *e* **cut and fill excavation**
    d Abtrag *m* und Auftrag *m*, Massenausgleich *m*, Einschnitt *m* und Damm *m*
    f excavation *f* déblai-remblai (*compensation des terrassements*)
    nl uitgraven en volgend aanvullen
    r метод *m* строительства дорог *или* каналов со сбалансированным объёмом земляных работ

C1414 *e* **cut back asphalt, cut-back bitumen**
    d Verschnittbitumen *n*
    f bitume *m* coupé, bitume *m* cut-back
    nl voorgemengd bitumen *n*
    r разжиженный битум *m*

C1414a **cut drain** *see* **drainage canal**

C1415 *e* **cut member**
    d durchgeschnittenes Element *n*
    f élément *m* sectionnée
    nl theoretische staafdoorsnede *f*
    r разрезанный [рассечённый] элемент *m* (*для расчёта*)

C1416 *e* **cutoff, cut-off**
    d 1. Durchstich *m* 2. Herdmauer *f*, Dichtungssporn *m* 3. Abbrechung *f*, Absperrung *f*; Abschaltung *f*
    f 1. coupure *f*, percement *m* 2. mur *m* étanche de garde, parafouille *f* (*d'une digue*); éperon *m* 3. déconnexion *f*, arrêt *m*; déclanchement *m*, coupure *f*
    nl 1. doorsteek *m* 2. afsluiting *f*
    r 1. прокоп *m*; спрямляющий канал *m* 2. зуб *m* плотины 3. отсечка *f* (*газа, пара*); отключение *n*

C1417 *e* **cutoff collar**
    d Rohrkragen *m*
    f manchon *m* d'étanchéité
    nl afsluitmanchet *f(m)*
    r противофильтрационная манжета *f*, противофильтрационный воротник *m* (*по внешнему периметру трубопровода в теле земляной плотины*)

C1418 *e* **cutoff depth**
    d Gründungstiefe *f* der Dichtungsschürze
    f profondeur *f* de parafouille [d'éperon]
    nl diepte *f* van de dijkvoet
    r глубина *f* заложения зуба плотины (*относительно дна котлована*)

C1419 *e* **cutoff trench**
    d 1. Dichtungsgraben *m*, Verherdung *f* 2. Dichtungssporn *m*, Herdmauer *f*
    f 1. tranchée *f* de la parafouille 2. parafouille *f*
    nl 1. schortgracht *f(m)* 2. schort *f(m)*
    r 1. траншея *f* зуба плотины 2. зуб *m* плотины

C1420 *e* **cutoff valve**
    d 1. Gasmangelsicherung *f* 2. Absperrventil *n*
    f 1. clapet *m* d'arrêt, soupape *f* de détente 2. vanne *f* d'arrêt [de fermeture]
    nl 1. afsluitklep *f(m)* 2. afsluiter *m*
    r 1. отсечный клапан *m*, клапан--отсекатель *m* 2. запорный клапан *m*

C1421 *e* **cutoff wall**
    d Herdmauer *f*, Dichtungsmauer *f*, Dichtungssporn *m*; Fußmauer *f*, Abschlußwand *f*
    f mur *m* de garde; mur *m* de pied; parafouille *f*
    nl dichtingsmuur *m* in de bedding aan de voet van een stuwdam
    r зуб *m* [шпора *f*] плотины; замок *m* гидросооружения; низовой зуб *m* водобоя

C1422 **cut-out valve** *see* **cutoff valve 1.**

C1422ae **cut roof**
    d Terrassendach *n*
    f toiture-terrace *f*, toit-terrace *f*,
    nl terrasdak *n*
    r крыша-терраса *f*

C1423 e **cut stone**
d Haustein *m*, Werkstein *m*
f bloc *m* [parpaing *m*] en pierre
naturelle; pierre *f* carrée [taillée]
nl behouwen steen *m*
r блок *m* из природного камня;
тёсаный камень *m*

C1424 e **cut string**
d Sattelwange *f*
f limon *m* denté [à crémaillère],
crémaillère *f*
nl uitgekeepte trapboom *m*
r наружная тетива *f* со ступенчатыми
вырезами, ступенчатая тетива *f*

C1425 e **cutter-dredger**
d Schneidkopfsaugbagger *m*
f drague *f* à roue à couteaux,
drague *f* suceuse à désagrégateur
nl cutterzuiger *m*
r землесосный снаряд *m*
с фрезерным рыхлителем,
фрезерный землесосный снаряд *m*

C1426 **cutter head** *see* **cutting head**

C1427 **cutter-head suction dredger** *see*
cutter-dredger

C1428 **cutting** *see* **cut 1., 2.**

C1429 e **cutting curb**
d Brunnenkranz *m*, Brunnenschneide *f*
f couteau *m* de caisson
nl beitelkrans *f(m)* (*boorbuis*)
r нижний нож *m* [венец *m*]
опускного колодца

C1430 e **cutting head**
d 1. Schneidrad *n*
(*Tunnelvortriebsmaschine*)
2. Schneidkopf *m* (*Saugbagger*)
f 1. tête *f* coupante 2.
désagrégateur *m* (*d'une drague*)
nl 1. snijkop *m* 2. cutter *m*
(*cutterzuiger*)
r 1. режущая головка *f*,
роторный рабочий орган *m*
(*туннелепроходческой машины*)
2. фрезерный рыхлитель *m*
(*земснаряда*)

C1431 e **cutting iron**
d Hobeleisen *n*
f fer *m* de rabot
nl schaafbeitel *m*
r лезвие *n* рубанка, железко *n*

C1432 e **cutting list**
d 1. Stahlauszug *m*,
Bewehrungsstahlliste *f*,
Schnittliste *f* (*für Bewehrung*)
2. Bauholzliste *f*
f 1. liste *f* de coupage 2. liste *f*
de débitage (*de bois*)
nl 1. wapeningsstaat *m* 2. houtstaat *m*
r 1. спецификация *f* резки арматуры
2. спецификация *f* на
пиломатериалы

C1433 e **cutting-off machine**
d Ablängmaschine *f*

f machine *f* à couper [à cisailler]
nl afkortmachine *f*
r отрезной станок *m*

C1434 e **cutting pliers**
d Schneidzange *f*
f pince *f* coupante [universelle]
nl nijptang *f(m)*, kniptang *f(m)*
r плоскогубцы-кусачки *pl*

C1435 e **cut-water**
d Pfeilerkopf *m*, Pfeilernase *f*
f bec *m* (d'une pile de pont)
nl pijlerkop *m*; golfbreker *m*
r оголовок *m* быка (контрфорса);
ледорез *m*, волнорез *m*;
аванбек *m*

C1436 e **cycle path, cycle track, cycleway**
d Fahrradweg *m*, Rad(fahr)weg *m*
f piste *f* cycliste [cyclable]
nl fietspad *n*
r велосипедная дорожка *f*

C1437 e **cyclic loading**
d Dauerschwingbeanspruchung *f*
f chargement *m* cyclique;
sollicitation *f* cyclique
nl cyclische belasting *f*
r цикличное [циклическое]
нагружение *n*

C1438 e **cyclic load regime**
d zyklische Belastungsweise *f*
f régime *m* des charges cycliques,
régime *m* de la sollicitation
cyclique
nl cyclische belastingswijze *f*
r режим *m* циклических нагрузок

C1439 e **cyclic strains**
d zyklische Verformungen *f pl*
f déformations *f pl* cycliques
nl cyclische vervormingen *f pl*
r циклические деформации *f pl*

C1440 e **cycling**
d Aussetzbetrieb *m*
f fonctionnement *m* périodique
nl intermitterend bedrijf *n*
r периодический режим *m*

C1441 e **cyclone**
d Zyklon(abscheider) *m*
f cyclone *n* dépoussiéreur,
dépoussiéreur *m* à (tube) cyclone
nl cycloon-stofafscheider *m*
r циклон *m*, циклонный
пылеулсвитель *m*

C1442 **cyclone cellar** *see* **storm cellar**

C1443 e **cyclone classifier**
d Naßzyklon *m*,
Aufschwimmklassierer *m*
f (hydro)cyclone *m*
class(ificat)eur, class(ificat)eur *m*
[cyclone *m*] hydraulique
nl hydrocycloon *m*
r гидроциклон *m*,
гидроклассификатор *m*

C1444 e **cyclopean aggregate**
d Zyklopen-Stein *m*, Bruchstein *m*

f moellon m, pierre f brute (de carrière)
nl (grove) breuksteen m
r бутовый камень m (крупностью более 150 мм)

C1445 e cyclopean block
d Zyklopenmassiv n, Zyklopenblock m
f bloc m cyclopéen
nl cyclopenblok m
r циклопический массив m

C1446 e cyclopean concrete
d Zyklopenbeton m, Bruchsteinbeton m
f béton m cyclopéen
nl cyclopenbeton n
r бутобетон m

C1447 e cyclopean masonry, cyclopean rubble masonry
d Zyklopenmauerwerk n
f maçonnerie f cyclopéenne
nl cyclopenverband n (natuursteenmetselwerk)
r циклопическая кладка f

C1448 e cylinder
d 1. Großbohrpfahl m, Mantelpfahl 2. Rundpfahl m 3. zylindrischer Warmwasserbehälter m, Heißwasserspeicher m 4. Gasflasche f
f 1. pieu m caisson 2. pieu m cylindrique 3. chauffe-eau m à accumulation 4. bouteille à gaz
nl 1. holle heipaal m 2. cilinderpaal m 3. met dompelelement verwarmde boiler m 4. gasfles f(m)
r 1. свая-оболочка f 2. круглая свая f 3. ёмкостный водонагреватель m с погружным нагревательным элементом; бак-аккумулятор m перегретой воды 4. газовый баллон m

C1449 cylinder barrel see cylinder shell 2.

C1450 cylinder caisson see open caisson

C1451 e cylinder of concrete
d Betonzylinderprobe f
f éprouvette f cylindrique de béton
nl (beton)proefcilinder m
r цилиндрический бетонный образец m (для испытаний)

C1452 e cylinder penetration test
d Zylinder-Eindring(ungs)versuch m, Zylinder-Drucksondierung f
f essai m de pénétration au cylindre
nl cilinderindringingsproef f(m)
r определение n несущей способности грунта вдавливанием цилиндра

C1453 e cylinder shell
d 1. Zylinderschale f, Zylindermantel m 2. Ringschuß m, Zarge f

f 1. voile m mince cylindrique, coque f cylindrique 2. virole f
nl 1. cilindermantel m 2. deurlijst f(m)
r 1. цилиндрическая оболочка f 2. обечайка f; царга f

C1454 e cylinder strength
d Zylinderfestigkeit f
f résistance f sur cylindre
nl cilindervastheid f
r цилиндрическая прочность f

C1455 e cylinder test
d Betonzylinder-Druckfestigkeitsprüfung f
f essai m de résistance sur eprouvette cylindrique
nl druksterkteproef f(m) op betoncilinder
r определение n предела прочности на сжатие (бетона) испытанием цилиндрических образцов

C1456 cylindrical barrage UK see roller dam

C1457 e cylindrical cofferdam
d Zylinderzellenfangdamm m
f batardeau m en cellules cylindriques, batardeau m cellulaire
nl ronde damwand m
r цилиндрическая ячеистая перемычка f

C1458 cylindrical gate see cylindrical sluice-gate

C1459 cylindrical rolling gate see roller drum gate

C1460 cylindrical shell see cylinder shell

C1461 e cylindrical sluice-gate
d Zylinderschütz n, Zylinderverschluß m
f vanne f cylindrique
nl cilinderschuif f(m)
r вертикальный цилиндрический затвор m

C1462 e cylindrical tank
d zylindrischer Behälter m
f réservoir m cylindrique
nl cilindrische tank m
r цилиндрический резервуар m

D

D1 e dabbing
d Abklopfen n (des Blendsteins zur Erzeugung der punktierten Oberfläche)
f smillage m
nl boucharderen n, punthameren n
r ударная обработка f камня (для получения точечной фактуры)

D2 e daily pondage basin
d Ausgleichbecken n

f bassin *m* de régularisation diurne
nl bassin *n* voor buffering per etmaal
r бассейн *m* суточного
регулирования (ГЭС)

D3 e **dam**
d Talsperre *f*, Damm *m*, Deich *m*
f barrage *m*, levée *f*, digue *f*
nl (stuw)dam *m*
r плотина *f*, дамба *f*

D4 e **dam body**
d Dammkörper *m*, Sperrenkörper *m*
f corps *m* de barrage
nl damlichaam *n*
r тело *n* плотины

D5 e **dam intake**
d Talsperrenentnahme *f*,
Talsperreneinlaß *m*,
Wehrentnahme *f*, Wehreinlauf *m*
f prise *j* à travers le barrage
nl inlaat *m* door de dam
r плотинный водозабор *m*

D6 e **damming**
d Absperrung *f*, Abdämmung *f*;
Stauung *f*
f endiguement *m*, barrage *m*
nl afdamming *f*
r преграждение *n*, перекрытие *n*;
создание *n* подпора

D7 **damming in** *see* **bordering**

D8 e **dam outlet**
d Tief(en)ablaß *m*
f déchargeur *m* [évacuateur *m*] de
fond
nl spui-opening *f*
r глубинный водосброс *m*,
водоспуск *m*

D9 **dampcourse** *see* **damp-proof course**

D10 e **damper**
d 1. Drosselklappe *f*, Luftklappe *f*
2. Dämpfer *m*, Puffer *m*
f 1. registre *m*, vanne *f* d'air
2. damper *m*
nl 1. rookklep *f(m)*, schoorsteenschuif
f(m) 2. demper *m*
r 1. (дроссельный) воздушный
клапан *m*, заслонка *f*
2. успокоитель *m*, амортизатор *m*

D11 **damping capacity**
d Dämpfungsvermögen *n*
f capacité *f* d'amortissement
nl dempingsvermogen *n*
r амортизирующая [демпфирующая]
способность *f*; способность *f*
(*напр. конструкции*) рассеивать
[гасить] динамические усилия

D12 e **damping material**
d Schallschluckstoff *m*
f matériau *m* absorbant le bruit, matériau *m* d'absorption phonique, matériau *m* acoustique
nl geluidabsorberend materiaal *n*
r звукопоглощающий материал *m*

D13 e **damping of concrete**
d Betonbefeuchtung *f*
f humidification *f* du béton
nl bevochtiging *f* van beton
r увлажнение *n* бетона

D14 e **dampness**
d Feuchtigkeit *f*
f humidité *f*
nl vocht *n*, vochtigheid *f*
r влажность *f*, влага *f*

D14a e **damp proof course**
d Sperrschicht *f*, Sperrlage *f*
f couche *f* d'étanchéité
nl dampdichte laag *f(m)*
r гидроизоляционный ковёр *m*
[слой *m*]

D15 e **dampproofing**
d 1. Feuchteisolierung *f*,
Feuchtigkeitsisolierung *f*
2. Grundwasserisolierung *f*
f 1. imperméabilisation *f* à
l'humidité 2. imperméabilisation *f*
à l'eau
nl vochtdicht maken, vochtisolatie *f*
r 1. влагоизоляция *f*, изоляция *f*
против капиллярной влаги
2. гидроизоляция *f* (*против
почвенной влаги и грунтовых вод*)

D16 e **damp-proof membrane**
d Feuchtigkeitsisolierhaut *f*
f membrane *f* d'étanchéite [étanche,
imperméable]
nl vochtisolerende laag *f*
r гидроизоляционная мембрана *f*
[плёнка *f*]

D17 e **dam site**
d Sperrstelle *f*
f emplacement *m* du barrage
nl stuwdam-emplacement *n*
r створ *m* плотины

D18 e **darby**
d Kartätsche *f*, Abziehlatte *f*;
Putzlehre *f*, Putzleiste *f*
f latte *f* de réglage; règle *f* de
plâtrier
nl schuurbord *n*
r правило *n* (*для разравнивания
бетона или штукатурки*);
полутёрок *m*

D20 e **Darcy law**
d Filtergesetz *n* von Darcy
f loi *f* de Darcy
nl wet *n* van Darcy (*filters*)
r закон *m* ламинарной фильтрации,
закон *m* Дарси

D21 e **date of commencement of the work**
d Baubeginndatum *n*
f date *f* de commencement [de début]
des travaux, date *f* de mise
en chantier
nl begindatum *n* van het werk
r дата *f* начала строительных работ
(*юридическая*)

D22 *e* **date of substantial completion**
  d Baufertigstellungstermin *m*
  f date *f* d'achèvement des travaux
  nl opleveringsdatum *n*
  r дата *f* окончания строительных
  работ (*реальная*), дата *f* сдачи
  объекта в эксплуатацию

D23 *e* **datum, datum level**
  d Bezugsfläche *f*, Bezugsebene *f*,
  Bezugshöhe *f*, Nullpunkt *m*,
  Nullmarke *f*
  f plan *m* de référence
  nl referentievlak *n*, nulpunt *n*
  r плоскость *f* нулевых отметок,
  нулевая плоскость *f*

D24 *e* **daubing**
  d 1. Abklopfen *n* des Blendsteins
  (*zur Erzeugung der punktierten
  Oberfläche*) 2. Putzbewurf *m*
  f 1. smillage *m* 2. couche *f* de fond
  (*d'enduit*)
  nl 1. boucharderen *n*, punthameren *n*
  2. ruwe beraping *f*
  r 1. ударная обработка *f* камня (*для
  получения точечной фактуры*)
  2. штукатурный обрызг *m*

D25 *e* **daylight factor**
  d Tageslichtquotient *m*
  f facteur *m* de lumière du jour
  nl daglichtfactor *m*
  r коэффициент *m* естественной
  освещённости

D26 *e* **daylighting**
  d Tagesbeleuchtung *f*, natürliche
  Beleuchtung *f*
  f éclairage *m* diurne [naturel]
  nl daglicht *n*, natuurlijke verlichting *f*
  r освещение *n* естественным светом

D27 *e* **day-supply reservoir**
  d Tagesspeicher *m*
  f bassin *m* journalier
  nl dagreservoir *n*
  r водохранилище *n* суточного
  регулирования

D28 *e* **day-to-day site supervision**
  d laufende Bauaufsicht *f*
  f surveillance *f* technique des travaux
  journalière
  nl dagelijks toezicht *n* op het werk
  r повседневный технический надзор
  m за строительством

D29 *e* **D-cracking**
  d Rißbildung *f* an der
  Betonoberfläche
  f fissuration *f* superficielle [sur la
  surface] de béton
  nl scheurvorming *f* op het
  betonoppervlak
  r трещинообразование *n* на
  поверхности бетона

D31 *e* **dead bolt**
  d Klinke *f*, Schnappverschluß *m*
  f pêne *m* dormant
  nl staartgrendel *m*, kantschuif *f(m)*

  r дверная защёлка *f* (*отпираемая
  ключом или поворотом ручки*)

D32 *e* **dead end**
  d befestigtes Spanngliedende *n*
  f bout *m* mort
  nl dood eind *n* (*leiding*); vast eind *n*
  van voorgespannen wapening
  r закреплённый конец *m*
  напрягаемой арматуры

D33 *e* **dead-end-anchor**
  d Blindanker *m*
  f ancre *f* morte
  nl blind anker *n*
  r глухой анкер *m*; анкер *m* на
  ненатягиваемом конце
  напрягаемой арматуры

D34 *e* **deadleg**
  d Abzweigrohr *n*, Ansatzstück *n*
  f branche *f* de tuyau, tubulure *f*
  nl aftakking *f*
  r отросток *m* (трубопровода)

D35 *e* **dead load**
  d Eigengewicht *n*, ständige Last *f*
  f poids *m* mort, charge *f* permanente
  nl eigen gewicht *n*
  r собственный вес *m* (*конструкции*),
  постоянная нагрузка *f*

D36 *e* **deadman**
  d rückwärtige Verankerung *f*, Ankerblock *m*
  f corps *m* mort, porteau *m* d'ancrage
  nl verankering *f*
  r анкер *m*, заглублённый в грунт,
  «мертвяк» *m*

D37 *e* **dead shore**
  d senkrechte Steife *f*
  f étai *m* [montant *m*, étançon *m*]
  provisoire
  nl lcodrechte stut *m* [stempel *m*]
  r временная [монтажная] стойка *f*

D38 **dead space** *see* **dead zone 2.**

D39 *e* **dead storage**
  d Totraum *m* des Speicherbeckens,
  verlorener Speicherraum *m*
  f réserve *f* inutilisable, volume *m*
  non vidangeable, eau *f* morte
  nl dode berging *f* (*waterreservoir*)
  r мёртвый объём *m* водохранилища

D40 *e* **dead-storage level**
  d Niedrigwasserspiegel *m*,
  Niedrigwasserstand *m*
  f niveau *m* d'eau morte
  nl laagste nuttige waterstand *m*
  r уровень *m* мёртвого объёма

D41 *e* **dead wall**
  d blinde Mauer *f*
  f mur *m* aveugle
  nl blinde muur *m*, akoestisch dcde
  muur *m*
  r глухая стена *f*

D42 *e* **dead water space**
  d Totwassergebiet *n*, Totraum *m*
  eines Wasserquerschnittes

f section f morte de courant
nl gebied n met stilstaand water, dood
gedeelte n van een stroomprofiel
r мёртвое пространство n (в сечении
водотока)

D43 e **dead weight**
d Eigengewicht n, Totgewicht n,
Totlast f
f poids m mort
nl eigen gewicht n
r собственный вес m

D44 e **dead window**
d Scheinfenster n
f fenêtre f dormante [aveugle, feinte],
fausse-fenêtre f
nl blind venster n
r глухое окно n

D45 e **dead zone**
d 1. Unempfindlichkeitszone f,
Unempfindlichkeitsbereich m,
totes Band n 2. Luftstagnationszone
f
f 1. bande f d insensibilité; zone f
morte 2. zone f de stagnation d'air
nl 1. niet-waarneembaar gebied n
2. neutrale zone f
r 1. зона f нечувствительности
(регулятора) 2. зона f застойного
воздуха (в помещении)

D46 e **de-aeration**
d Entlüftung f
f désaération f, évacuation f de
l'air
nl ontluchting f
r деаэрация f

D47 e **de-ashing**
d Entaschung f, Aschenbeseitigung f
f évacuation f [enlèvement m] des
cendres
nl asafvoer m
r золоудаление n

D49 e **debit**
d Ergiebigkeit f, Quellschüttung f
f débit m
nl debiet n
r дебит m

D50 e **debris**
d Trümmer pl, Schutt m;
Bauschutt m; Feststoffe m pl (in
Oberflächengewässern);
Trümmermaterial n, Bruchgestein n
f débris m; détritus m
nl puin n, bouwpuin n; drijvend
materiaal n
r обломки m pl; строительный мусор
m; наносы m pl; обломочные
породы f pl

D51 e **debris dam**
d Geschiebe(stau)sperre f,
Konsolidierungssperre f
f barrage m de sédimentation
nl vuilvanger m
r наносозадерживающая плотина f
[запруда f], запруда-селеуловитель f

D52 e **decay**
d 1. Abnahme f, Verminderung f,
Erschöpfung f 2. Fäulnis f
f 1. épuisement m 2. pourriture f
(du bois)
nl 1. uitputting f (bron) 2. rotting f
(van hout)
r 1. истощение n 2. гниение n
(древесины)

D53 e **deceleration lane**
d Verzögerungsstreifen m
f voie f de décélération [de
ralentissement]
nl uitloopstrook f(m)
r полоса f замедления движения
(дороги)

D54 e **decentering**
d Abrüsten n, Ausrüsten n,
Abbau m (der tragenden Stützen)
der Gewölberüstung
f décintrage m, décintrement m
nl ontkisting f
r раскруживание n, демонтаж m
опорных стоек опалубки

D55 e **decentralized sewerage system**
d dezentralisierte Entwässerung f
f système m d'assainissement
décentralisé
nl gedecentraliseerd rioolstelsel n
r децентрализованная канализация f

D57 e **deck**
d 1. Belag m; Belagplatte f
2. Plattform f, Arbeitsbühne f
3. Bodenplatte f 4. Grundplatte f
f 1. tablier m, platelage m,
planchéiage m; plaque de plancher
2. plate-forme f (de travail) 3.
plaque f de fond 4. plaque f
d'appui
nl 1. (weg)dek n 2. werkvloer m 3.
bodemplaat f(m) 4. grondplaat f(m)
r 1. настил m; плита f настила 2.
платформа f, рабочая площадка f
3. донная плита f 4. опорная плита f

D58 e **deck bridge**
d Deckbrücke f
f pont m à tablier supérieur
nl brug f(m) met hooggelegen rijvloer
r мост m с ездой поверху

D59 e **deck clip**
d Klammer f
f agrafe f de fixation, agrafure f
nl kram f(m)
r клямера (f) (для крепления кровли
к прогонам или обрешётке)

D60 e **deck framing**
d Tragwerk n der Brückenfahrbahn
f charpente f [ossature f] de tablier
(du pont)
nl brugdekconstructie f
r несущие конструкции f pl
проезжей части или ездового
полотна моста

D61 e deck girder
  d Tafelträger m, Fahrbahnträger m
    (Brücke)
  f poutre f transversale du tablier (de
    pont)
  nl dekligger m (brug), langsligger m
  r поперечная балка f верхнего
    строения моста

D62 e decking
  d 1. Belag m 2. Brückentafel f
    3. Oberbau m (Kai)
  f 1. tablier m, platelage; plancher m;
    pose f du tablier 2. tablier m de
    pont 3. superstructure f (de quai)
  nl 1. wegdek n 2. brugdek n
    3. dekvloer m
  r 1. настил m 2. проезжая часть f,
    мостовой настил m 3. верхнее
    строение n (набережной, пирса)

D63 e deck level
  d Höhe f der Brückenbelagplatte [der
    Brückentafeloberkante]
  f niveau m du tablier (de pont)
  nl niveau n van het brugdek
  r отметка f [уровень m] настила

D64   deck spillway US see spillway slab

D65 e deck structure
  d Überbau m, Brückenüberbau m
  f travée f [superstructure f] de pont
  nl brugdekconstructie f
  r верхнее строение n моста

D66 e deck truss
  d Tragwerk n mit obenliegender
    Fahrbahn (Brücke)
  f poutre f à treillis du pont à
    tablier supérieur
  nl vakwerkbrug f(m) met hooggelegen
    rijvloer
  r ферма f моста с ездой поверху

D67   deck type bridge see deck bridge

D68   decohesion see disbonding

D70 e decompression chamber
  d Druckentlastungskammer f
  f chambre f de décompression
  nl decompressietank m
  r декомпрессионная камера f

D71 e decontamination factor
  d Reinigungsgrad m, Reinigungswert m
  f facteur m d'épuration
  nl zuiveringsgraad m
  r коэффициент m очистки

D72 e decoration
  d Dekorieren n
  f décoration f
  nl schilderwerk n, versiering f
  r декоративная отделка f

D73 e decorative block, decorative faced
    block
  d Wandblock m mit Zierverkleidung
  f block m de mur décoratif
    [ornemental], parpaing m de
    décoration

  nl decoratief bekleed gedeelte n van
    een wand
  r стеновой блок m с декоративной
    облицовкой

D74 e decorative glass
  d Zierglas n, Ornamentglas n
  f verre f décoratif [ornemental]
  nl sierglas n
  r декоративное [орнаментальное]
    стекло n

D75 e dedusting
  d Entstaubung f
  f dépoussiérage m
  nl stofvrij maken n
  r обеспыливание n

D77 e dedusting ventilation
  d Entstaubungsventilation f
  f ventilation f de dépoussiérage
  nl ventilatie f ter verwijderingᵥvan stof
  r обеспыливающая вентиляция f

D78 e deep beam
  d wandartiger Träger m,
    Scheibenträger m
  f poutre-cloison f, poutre-paroi f
  nl vollewandligger m
  r балка-стенка f

D79 e deep compaction
  d Tiefenverdichtung f
  f compactage m [serrage m] du sol en
    profondeur
  nl diepteverdichting f
  r глубинное уплотнение n (грунта)

D80 e deep excavation
  d Tiefaushub m; Tiefbaugrube f
  f excavation f profonde; fouille f
    profonde
  nl diepe ontgraving f
  r глубокая разработка f [выемка f]
    грунта; глубокий котлован m

D81 e deep foundation
  d Tiefgründung f, Tieffundation f
  f fondation f profonde
  nl diepe fundering f
  r фундамент m глубокого заложения

D82 e deep gate
  d tiefer Verschluß m
  f vanne f de profondeur
  nl laaggelegen afsluiter m
  r глубинный затвор m

D84 e deep sluice
  d Grundablaß m
  f déchargeoir m [vidange f,
    déchargeur m] de fond
  nl duikersluis f(m)
  r донный водоспуск m

D85 e deep-water berth
  d Tiefwasseranlegestelle f
  f quai m d'accostage [d'amarrage]
    profond
  nl kade f(m) aan diep vaarwater
  r глубоководный причал m

D86 e deep well pump
  d Tiefbrunnenpumpe f

f pompe f de puits profond
nl diepbronpomp f(m)
r глубинный насос m

D87 e **deflected tendon**
d angehobene Litze f
f toron m curviligne
nl gebogen wapeningsbundel m
r криволинейный арматурный пучок m

D88 e **deflection**
d Durchbiegung f
f déflexion f, flexion f, flèche f
nl doorbuiging f
r прогиб m; стрела f прогиба

D89 e **deflection curve**
d Biegelinie f, Biegungslinie f;
elastische Linie f
f ligne f élastique [de flexion]; ligne
f fléchie
nl doorbuigingslijn f(m)
r линия f прогибов; упругая линия f

D90 e **deflectometer**
d Durchbiegungsmesser m
f déflectomètre m
nl buigingsmeter m
r прогибомер m

D91 e **deflector**
d Leitblech n
f déflecteur m
nl leiplaat f(m), deflector m
r отклоняющая
[струненаправляющая] перегородка
f (в воздуховоде)
deflexion... see **deflection...**

D93 e **deformability**
d Verformbarkeit f,
Formänderungsfähigkeit f
f déformabilité f
nl vervormbaarheid f
r деформируемость f, податливость f

D94 e **deformability meter**
d Deformationsmesser m,
Verformungsmesser m
f déformètre m, extensomètre m
nl vervormingsmeter m, tensometer m
r тензометр m

D95 e **deformation**
d Verformung f
f déformation f
nl vervorming f
r деформация f

D96 **deformation capacity** see
deformability

D97 e **deformation per unit of length**
d Dehnung f
f déformation f pour unité de longueur
nl relatieve rek m
r деформация f на единицу длины,
относительная линейная деформация f

D98 e **deformed bars, deformed
reinforcement**
d Betonformstahl m
f barres f pl d'armature crénelées

il профилеерд betonijzer n
r арматура f периодического профиля

D99 e **deformeter** see **deformability meter**

D100 e **defrosting**
d Abtauen n, Entfrosten n
f dégivrage m, dégel m
nl ontdooien n
r размораживание n, оттаивание n

D101 e **degree-day**
d Gradtag m
f degrés-jour m
nl graad-dag m
r градусо-день m

D102 e **degree of compaction**
d Verdichtungsgrad m
f degré m de compactage
nl verdichtingsgraad m
r степень f уплотнения (грунта)

D103 e **degree of consolidation**
d Kompressionsgrad m
f degré m de consolidation
nl verstevigingsgraad m
r степень f естественного уплотнения
[консолидации], степень f сжатия
грунта без бокового расширения
(в одометре)

D104 e **degree of fire resistance**
d Feuerfestigkeitsgrenze f,
Feuerfestigkeitsgrad m
f degré m de résistance au feu
nl mate f(m) van brandwerendheid
r предел m [степень f] огнестойкости

D105 e **degree of freedom**
l Freiheitsgrad m
f degré m de liberté
nl vrijheidsgraad m
r степень f свободы

D106 e **degree of incombustibility**
d Entflammbarkeitsgrad m,
Entzündbarkeitsgrad m
f degré m d'inflammabilité
nl mate f(m) van onbrandbaarheid
r группа f возгораемости (конструкций)

D107 e **degree of saturation**
d Sättigungsgrad m
f pourcentage m de saturation
nl verzadigingsgraad m
r степень f насыщения; коэффициент
m водонасыщения (грунта)

D108 e **Dehottay process**
d Dehottay-Verfahren n
f procédé m Dehottay
nl procédé n volgens Dehottay
r замораживание n водонасыщенного
грунта путём нагнетания
сжиженного углекислого газа

D109 **dehumidification**
d Entfeuchtung f, Trocknung f
f déshumidification f
nl vochtonttrekking f
r осушение n

D110 *e* **dehumidifier**
 d (Luft-)Entfeuchter *m*,
  Entfeuchtungsgerät *n*
 f déshumidificateur *m*
 nl luchtdrogingsapparaat *n*
 r воздухоосушитель *m*

D111 *e* **dehumidifying capacity**
 d Entfeuchtungsleistung *f*
 f capacité *f* de déshumidification
 nl vochtonttrekkend vermogen *n*
 r производительность *f* кондиционера
  по осушению воздуха

D112 **dehydrator** *see* **dryer**

D113 *e* **de-icer, deicing agent**
 d Enteiser *m*
 f dégivreur *m*
 nl ontdooiingsmiddel *n*
 r антиобледенитель *m* (*для*
  дорожных покрытий)

D114 *e* **delamination**
 d Abblättern *n*, Abschuppung,
  Abschälung *f*, Abspaltung *f*
 f exfoliation *f*, écaillement *m*,
  écaillage *m*
 nl afbladdering *f*
 r отслаивание *n*; шелушение *n*

D115 *e* **delay-action detonator**
 d Zeitzünder *m*, Intervallzünder *m*
 f détonateur *m* à action retardée
 nl vertragingsontsteker *m*
 r детонатор *m* замедленного
  действия

D116 *e* **delivery**
 d 1. Fördermenge *f* (*Pumpe,*
  Verdichter) 2. Lieferung *f*
 f 1. débit *m* [refoulement *m*] de
  pompe 2. livraison *f*
 nl 1. capaciteit *f* (*pomp*)
  2. (op)levering *f*
 r 1. подача *f* (*характеристика*
  насоса, компрессора) 2. доставка
  f, поставка *f*

D117 *e* **delivery head**
 d Druckhöhe *f*, Förderhöhe *f*
 f hauteur *f* de refoulement
 nl pershoogte *f*, opvoerhoogte *f*
 r высота *f* нагнетания

D118 *e* **delivery hose**
 d Ablaßschlauch *m*
 f tuyau *m* flexible de refoulement
 nl stortslang *f(m)* (*betonmortel*)
 r нагнетательный шланг *m* (*для*
  подачи раствора, бетона)

D120 **delivery pressure** *see* **head pressure**

D121 *e* **delivery pressure head**
 d 1. geodätische Druckhöhe *f* 2.
  verfügbares Gefälle *n* (*Irrigation*)
 f 1. hauteur *f* géometrique
  [piézométrique] de refoulement
  2. charge *f* [pression *f*] effective
  disponible (*arrosage*)
 nl 1. pershoogte *f* 2. beschikbaar
  verval *n* (*irrigatie*)

 r 1. геометрическая высота *f*
  нагнетания 2. располагаемый напор
  m (*в орошении*)

D123 *e* **demand**
 d Bedarf *m*
 f demande *f*
 nl behoefte *f*, benodigd vermogen *n*
 r требуемая производительность *f*
  (*системы, установки*)

D124 *e* **demand pattern**
 d Bedarfsverlauf *m*
 f diagramme *m* de demande
 nl behoeftepatroon *n*, verloop *m* van
  de behoefte
 r характеристика *f* требуемой
  производительности в зависимости
  от времени

D125 *e* **demand peak**
 d Bedarfsspitze *f*
 f demande *f* de pointe
 nl piekbehoefte *f*, benodigd vermogen
  n
 r максимальное значение *n*
  требуемой производительности

D126 *e* **demi-column**
 d Halbsäule *f*, Wandsäule *f*
 f semi-colonne *f*
 nl halfzuil *f(m)*
 r полуколонна *f*

D127 *e* **demineralization**
 d Entsalzung *f*, Wasserenthärtung *f*
 f déminéralisation *f*, dessalement *m*,
  dessalage *m*
 nl ontzouting *f*
 r деминерализация *f*, обессоливание
  n, опреснение *n*

D128 *e* **demolding**
 d Ausschalen *n*, Entformen *n*
 f démoulage *m*
 nl ontkisting *f*
 r распалубка *f*

D129 *e* **demolition**
 d Abbruch *m*
 f démolition *f*
 nl sloping *f*, sloop *m*, afbreken *n*,
  afbraak *f* (*m*)
 r слом *m*, снос *m*

D130 *e* **demolition hammer**
 d Abbauhammer *m*, Pickhammer *m*
 f marteau-piqueur *m*, marteau *m* de
  démolition
 nl sloophamer *m*
 r отбойный молоток *m*, перфоратор *m*

D131 *e* **demolition work**
 d Abbrucharbeiten *f pl*
 f travaux *m pl* de démolition
 nl sloopwerk *n*, afbraak *f* (*m*)
 r работы *f pl* по сносу зданий

D132 *e* **demountable mast**
 d demontierbarer Mast *m*
 f mât *m* démontable
 nl demonteerbare mast *m*
 r сборно-разборная мачта *f*

D133 e **demountable partition**
d bewegliche [versetzbare] Trennwand *f*
f cloison *f* démontable [amovible, mobile, transformable]
nl verplaatsbare scheidingswand *m*
r разборная [переставляемая, передвижная] перегородка *f*

D134 e **dense concrete**
d Normalbeton *m*, Schwerbeton *m*
f béton *m* dense [à densité élevée]
nl (normaal) beton *n*
r плотный бетон *m*

D135 e **dense-graded aggregate**
d geschlossen abgestuftes Mineralgemisch *n*
f agrégat *m* [granulat *m*] de granulométrie pleine
nl goed gegradeerde toeslag *m*
r смесь *f* плотно подобранного гранулометрического состава; плотно подобранный состав *m* заполнителей (*для бетона*)

D136 e **dense tar surfacing**
d Teerbetondecke *f*, Teerbetonbelag *m*
f couche *f* de surface en goudron **dense**
nl wegdek *n* van teerbeton
r верхний слой *m* дорожного покрытия из плотного дёгтебетона

D137 e **densified impregnated wood**
d getränktes [imprägniertes] Preßvollholz *n*
f bois *m* imprégné à la résine et densifié
nl met kunsthars geïmpregneerd hout *n*
r прессованная пропитанная смолой древесина *f*

D138 e **density control**
d Baustellen-Betondichtekontrolle *f*
f contrôle *m* de densité de béton
nl betondichtheidscontrole *f* (*m*)
r контроль *m* за плотностью бетона

D140 e **dental, dentated sill**
d Zahnschwelle *f*
f seuil *m* denté
nl getande drempel *m*
r зубчатый порог *m* (*водобойного колодца*); пирсовый гаситель *m* (*энергии потока*)

D141 e **deodorization**
d Desodor(is)ierung *f*, Geruchsbeseitigung *f*
f désodorisation *f*
nl deodorisatie *f*
r дезодорация *f*

D142 e **depletion curve**
d Trockenwetterganglinie *f*, Trockenwetterabflußlinie *f*, Rückgangskurve *f*
f courbe *f* de tarissement [d'épuisement]
nl depletielijn *f* (*m*)
r кривая *f* истощения (*стока, расхода*)

D143 e **depolished glass**
d Mattglas *n*
f verre *m* dépoli [mat]
nl matglas *n*
r матовое стекло *n*

D144 e **depositing site**
d Spülfläche *f*, Spülfeld *n*
f plage *f* d'épandage
nl stortplaats *f* (*m*)
r карта *f* намыва

D145 e **deposition of soil**
d Bodenschüttung *f*
f remblayage *m*
nl storten *n* van grond
r отсыпка *f* грунта

D146 e **deposits**
d Ablagerungen *f pl*, Geschiebe *n*, Sedimente *n pl*
f dépôts *m pl*, sédiments *m pl*
nl bezinksels *n pl*, sedimenten *n pl*
r наносы *m pl*, отложения *n pl*

D147 e **depot**
d 1. Lagergebäude *n* 2. Bahnhof *m*, Eisenbahnstation *f*
f 1. dépôt *m*, entrepôt *m*, magasin *m* 2. gare *f*
nl depot *n*; magazijn *n*, opslagplaats *m*
r 1. склад *m*; складское здание *n* 2. вокзал *m*; железнодорожная станция *f*

D148 e **depressed arch**
d gedrückter Bogen *m*
f arc *m* aplati
nl platte boog *m*
r пониженная арка *f*

D149 e **depression curve**
d Sickerlinie *f*, Depressionskurve *f*, Absenkungskurve *f*
f courbe *f* de dépression, ligne *f* d'infiltration
nl depressielijn *f* (*m*)
r кривая *f* депрессии

D150 e **depression head**
d Absenken *n*, Spiegel(ab)senkung *f* (*Grundwasser*)
f hauteur *f* de rabattement (*des eaux souterraines*)
nl verlaging *f* van het grondwaterpeil
r депрессия *f* подземных вод

D151 e **depreter**
d Außenputz *m* mit eingeprägtem Steinschrot
f couche *f* de finition (*d'enduit*) avec morceaux de pierre incorporés
nl buitenbepleistering *f* met ingedrukt split
r наружная штукатурка *f* наборной фактуры (*получаемой вдавливанием в раствор мелкого камня*)

D152 e **depth**
d Tiefe *f*
f profondeur *f*; hauteur *f*; épaisseur *f*

nl diepte *f*; holte *f*; dikte *f*
(*bodemlaag*)
r глубина *f*; высота *f*; толщина *f*

D153 *e* **depth of beam**
d Trägerhöhe *f*
f hauteur *f* de la poutre
nl hoogte *f* van een ligger
r высота *f* балки

D154 *e* **depth of compacted layer, depth of compacted lift**
d Stärke *f* der verdichteten Schicht
f épaisseur *f* de la couche compactée
nl dikte *f* van de verdichte laag
r толщина *f* уплотнённого слоя

D155 *e* **depth of drainage**
d Dränwirkungstiefe *f*
f profondeur *f* de drainage
nl ontwateringsdiepte *f*
r глубина *f* осушения

D156 *e* **depth of frost penetration**
d Frosttiefe *f*
f profondeur *f* de pénétration du gel
nl vorstgrens *f* (*m*)
r глубина *f* промерзания грунта

D157 *e* **depth of runoff**
d Abflußhöhe *f*
f lame *f* [hauteur *f*] d'eau écoulée
nl afvloeiingshoogte *f*
r слой *m* стока

D158 *e* **depth of truss**
d Fachwerkträgerhöhe *f*, Binderhöhe *f*
f hauteur *f* de la ferme
nl spanthoogte *f*, hoogte *f* van een vakwerkligger
r высота *f* фермы

D159 *e* **depth-to span ratio**
d Höhe-Spannweite-Verhältnis *n* (*Balken*)
f rapport *m* hauteur/portée (*d'une poutre*)
nl hoogte-overspanningsverhouding *f* (*balk*)
r отношение *n* высоты к пролёту (*балки*)

D160 *e* **depth velocity curve**
d Vertikalgeschwindigkeitskurve *f*, Geschwindigkeitsdiagramm *n*
f courbe *f* de répartition verticale des vitesses
nl verticale snelheidskromme *f* (*m*), snelheidsdiagram *n*
r годограф *m*, кривая *f* распределения скоростей водотока по вертикали

D161 **derby slicker** *see* **darby**

D162 *e* **derivation**
d Umleitung *f*
f dérivation *f*
nl omleiding *f*
r деривация *f*

D163 *e* **derivation conduit**
d Wasserumleitung *f*
f conduit *m* de dérivation

nl (water)omleidingsleiding *f*
r деривационный водовод *m*

D164 *e* **derivation tunnel**
d Umleitungsstollen *m*
f tunnel *m* [galerie *f*] de dérivation
nl omleidingstunnel *m*
r деривационный туннель *m*

D165 *e* **derrick**
d Derrick(kran) *m*
f derrick *m*, grue *f* derrick
nl dirkkraan *f* (*m*), laadboom *m*
r мачтово-стреловой кран *m*, деррик--кран *m*

D166 *e* **derrick bullwheel**
d Derrick-Wendescheibe *f*
f plate-forme *f* orientable [rotative] du Derrick
nl draaiplaat *f* (*m*) van de dirkkraan
r поворотный круг *m* деррик-крана

D167 *e* **derricking**
d Wippen *n*, Auslegerverstellung *f*
f variation *f* de la portée de flèche (*d'une grue*); levage *m* ou abaissement *m*. de la flèche (*d'une grue*)
nl hijsen *n* met een wipkraan [laadboom]
r изменение *n* вылета стрелы крана; подъём *m* и опускание стрелы крана

D168 *e* **derrick rope**
d Derrickseil *n*, Auslegerseil *n*
f câble *m* de flèche (*d'une grue*)
nl dirkkabel *m*
r стреловой канат *m* (*крана*)

D169 *e* **desalination, desalting**
d Entsalzen *n*, Entsalzung *f*
f dessalage *m*, dessalement *m*
nl ontzouten *n*
r обессоливание *n*, опреснение *n*

D170 *e* **desanding**
d Entsanden *n*, Entsandung *f*
f dessablement *m*, dessablage *m*
nl ontzanden *n*
r удаление *n* песка; очистка *f* от песка

D171 *e* **desiccant**
d Trockenmittel *n*, Trockenmedium *n*
f desséchant *m*
nl droogmiddel *n*
r осушитель *m* (*вещество*)

D172 *e* **design**
d 1. Entwurf *m*, Plan *m*, Projekt *n* 2. Projektieren *n*, Berechnung *f*; Bemessung *f*; Dimensionierung *f* 3. zeichnerische Konstruktion *f*, Ausführung *f*
f 1. projet *m*, plan *m* 2. calcul *m*; étude *f*, élaboration *f* du projet 3. construction *f*; disposition *f* des éléments, conception *f*
nl 1. ontwerp *n*, opzet *m* 2. berekening *f*, dimensionering *f* 3. uitwerking *f*. tekenen *n*

r 1. проект *m* 2. проектирование *n*; расчёт *m*; определение *n* размеров 3. конструкция *f*; конструктивный вариант *m*

D173 *e* **design assumption**
d Bemessungsannahme *f*
f hypothèse *f* de calcul
nl berekeningsaanname *f*|
r расчётное предположение *n*

D174 *e* **design chart**
d Bamessungstafel *f*, Rechentafel *f*; Berechnungstabelle *f*
f abaque *m*, nomogramme *m*; graphique *m* pour le calcul
nl ontwerptabel *f* (*m*); berekeningsdiagram *n*
r расчётный график *m*; расчётная номограмма *f*; расчётная таблица *f*

D175 *e* **design contract**
d Projektierungsvertrag *m*
f contrat *m* [marché *m*] d'étude de projet
nl overeenkomst *f* voor het maken van een ontwerp
r контракт *m* [договор *m*] на проектные работы

D176 *e* **design crest level**
d Entwurfskronenhöhe *f*, Sollhöhe *f* der Krone (*Damm*)
f niveau *m* normal [cote *f* normale] de crête d'un barrage
nl ontwerp-kruinhoogte *f*
r проектная отметка *f* гребня плотины (*после стабилизации осадок*)

D177 *e* **design criteria**
d Entwurfskriterien *n pl*
f critères *m pl* de conception
nl eisen *m pl* voor het ontwerp
r критерии *m pl* проектирования [расчёта]

D178 *e* **design data**
d Entwurfsdaten *pl*, Auslegungsdaten *pl*, Bemessungsangaben *f pl*
f données *f pl* pour le calcul, données *f pl* de base pour l'étude de projet
nl basisgegevens *n pl* voor het ontwerp
r исходные данные *pl* для проектирования

D179 *e* **design decision**
d Entwurfslösung *f*, Entwurfgestaltung *f*
f conception *f* du projet
nl definitief ontwerp *n*
r проектное решение *n*

D181 *e* **design development phase**
d Phase *f* der Entwurfsbearbeitung
f stade *m* [phase *f*] d'élaboration du projet
nl voorbereidingsfase *f*
r стадия *f* разработки рабочего проекта

D182 *e* **design diagram**
d Rechendiagramm *n*, Berechnungsschema *n*

f diagramme *m* [schéma *m*] de **calcul**
nl ontwerpdiagram *n*
r расчётная схема *f*

D183 **design discharge** *see* **design flow**

D184 *e* **design drawings**
d Entwurfszeichnungen *f pl*
f dessins *m pl* d'étude
nl bouwtekeningen *f pl*
r чертежи *m pl* проектно-конструкторских разработок; **эс**кизные чертежи *m pl*

D185 *e* **design feature**
d Konstruktionsmerkmal *n*
f caractéristique *f* [particularité *f*] d'une construction
nl hoofdpunten *m pl* van het ontwerp
r особенность *f* конструкции

D186 *e* **design flexibility**
d Flexibilität *f* [Anpassungsfähigkeit *f*] der Entwurfgestaltung
f flexibilité *f* du projet
nl flexibiliteit *f* van het ontwerp
r гибкость *f* проекта [проектного решения]

D187 *e* **design flood level**
d höchster Hochwasserstand *m*, höchster Wasserstand *m* im Stauraum, rechnerischer Wasserstand *m* für die Hochwasser--Entlastung |
f niveau *m* normal de retenue
nl aangenomen hoogste waterstand *m*
r максимальный подпорный уровень *m* водохранилища

D188 *e* **design flow**
d Berechnungs(wasser)menge *f*, rechnerischer Wasserdurchfluß *m*, Bemessungsdurchfluß *m*
f débit *m* nominal [de projet]
nl aangenomen uitstroming *f*
r расчётный расход *m*

D189 *e* **design formulas**
d Berechnungsformeln *f pl*
f formules *f pl* de calcul
nl berekeningsformules *f*(*m*) *pl*
r расчётные формулы *f pl*

D190 **design fundamentals** *see* **design principles**

D191 *e* **design head**
d errechnete theoretische Fallhöhe *f*
f chute *f* calculée [de calcul]
nl berekende drukhoogte *f*
r расчётный напор *m*

D192 *e* **design heat loss**
d Bemessungswärmeverlust *m*
f pertes *f pl* de chaleur [déperditions *f pl* calorifiques] admises pour le calcul
nl aangenomen peil *n*
r расчётные теплопотери *pl*

D193 *e* **design life**
d bemessene Lebensdauer *f*, Bemessungsliegezeit *f*
f durée *f* de vie [de service] **calculée**

nl berekende levensduur *m*
(*constructie*), standtijd *m*
r проектная долговечность *f*;
проектный срок *m* службы
(*сооружения*)

D194 *e* design limitations
d Begrenzungen *f pl* der technischen
Bauvorschriften
f limitations *f pl* établies par cahier
des charges
nl geldende beperkingen *f pl*
r проектные ограничения *n pl*;
ограничения *n pl*, налагаемые
техническими условиями и нормами
при проектировании

D195 *e* design load
d rechnerische Last *f*, Lastannahme *f*
f charge *f* de calcul
nl aangenomen [berekende] belasting *f*
r расчётная нагрузка *f*

D197 *e* design of concrete mix
d Festlegen *n* des
Mischungsverhältnisses (*Beton*)
f recherche *f* de la composition du
béton; dosage *m* du mélange de béton
nl vaststelling *f* van de mengverhouding
(*beton*)
r проектирование *n или* подбор *m*
состава бетонной смеси

D198 *e* design principles
d Bemessungsgrundlagen *f pl*
f principes *m pl* fondamentaux
[essentiels] de calcule, hypothèses *f pl*
essentielles de calcul
nl grondslagen *m pl* voor het ontwerp
r основные положения *n pl* [принципы
m pl] расчёта

D199 *e* design specifications
d Bemessungsrichtlinien *f pl*,
Entwurfsrichtlinien *f pl*
f règles *f pl* d'établissement des projets,
règlements *m pl* concernant
l'établissement des projets
nl richtlijnen *f (m) pl* voor het ontwerp
r технические условия *n pl* [нормы *f
pl*] на проектирование

D200 *e* design strength
d Sollfestigkeit *f*, rechnerischer
Widerstand *m*
f résistance *f* caractéristique [calculée]
nl berekende sterkte *f*
r расчётное сопротивление *n*

D201 *e* design stress
d Bemessungsspannung *f*
f contrainte *f* de calcul
nl toe te laten spanning *f*
r расчётное напряжение *n*

D202 *e* design temperature map
d Klimakarte *f*
f carte *f* des températures extérieures
de base, carte *f* climatologique
nl klimaatkaart *f (m)*
r карта *f* районирования (страны) по
расчётным температурам

D203 *e* design ultimate load
d rechnerische Grenzbelastung *f*
[Bruchbelastung *f*]
f charge *f* limite de calcul
nl berekende maximum *n* belasting
r расчётная предельная
[разрушающая] нагрузка *f*

D204   design water discharge *see* design flow

D205 *e* desilting
d Entschlammung *f*
f délimonage *m*
nl ontdoen *n* van zwevende
verontreinigingen
r очистка *f* от ила

D206 *e* desilting basin
d Absetzbecken *n*,
Entschlammungsbecken *n*;
Vorspeicher *m*, Verlandebecken *n*
f bassin *m* de décantation; retenue *f*
de dessablement
nl bezinkbekken *n*
r отстойный бассейн *m*;
наносозадерживающее
водохранилище *n*

D207 *e* destructive testing
d zerstörende Prüfung *f*
f essai *m* destructif [de rupture]
nl destructieve beproeving *f*
r испытание *n* на разрушение [на
разрыв]

D208 *e* detachable bit
d lösbare [auswechselbare] Bohrkrone *f*
f couronne *f* de forage rapportée
nl losse boor *m*
r съёмная [сменная] буровая коронка
f

D209 *e* detachable union
d lösbare Rohrverschraubung *f*
f raccord *m* vissé démontable
nl losneembare pijpkoppeling *f*
r разъёмное резьбовое соединение *n*
труб

D210 *e* detached dwelling
d Einzelhaus *n*, alleinstehendes Haus *n*
f maison *f* détachée [isolée]
nl vrijstaande woning *f*
r отдельностоящий жилой дом *m*

D211 *e* detached garage
d Einzelgarage *f*, einzelstehende Garage
f; Garage *f*, durch einen offenen
Gang ans Hauptgebäude
angeschlossen
f garage *m* détaché [isolé]; garage *m*
lié à la maison par passage non
couvert
nl vrijstaande garage *f*
r отдельностоящий гараж *m*; гараж
m, соединённый со зданием
открытым проходом [галереей]

D212 *e* detail
d 1. Detail *n*, Bauteil *n* 2. Teilzeichnung
f
f 1. détail *m* 2. dessin *m* détaillé

nl 1. detail n 2. detailtekening f
r 1. деталь f; узел m (чертежа)
2. деталировочный чертёж m

D213 detail design see detailing

D214 e detail drawing
d Detailplan m, Detailzeichnung f,
Teilzeichnung f
f dessin m détaillé [en detail]
nl detailtekening f
r деталировочный чертёж m

D215 e detailed design stage
d detaillierte Entwurfausarbeitung f,
Etappe f der Ausarbeitung der
Bauzeichnungen
f stade m d'élaboration du projet final;
stade m d'élaboration des dessins
d'exécution
nl stadium n van het gedetailleerd
ontwerpen
r этап m детальной разработки
проекта; этап m разработки
рабочих чертежей

D216 e detailed estimate of construction costs
d detaillierter Baukosten(vor)anschlag
m
f devis m estimatif détaillé
nl gedetailleerde begroting f (van
bouwkosten)
r смета f на строительный объект
(составленная по единичным
расценкам)

D217 e detailing
d Detaillierung f
f élaboration f des dessins détaillés
nl detaillering f
r разработка f деталировочных
чертежей, деталировка f

D218 e detail sheet
d Liste f der Bauelemente
f nomenclature f des détails [des pièces]
de construction
nl stuklijst f (m)
r спецификация f (конструктивных)
деталей

D219 e detensioning
d 1. Ablassen n (Spannkabel)
2. Entspannung f
f 1. détension f, détente f (transmission
au béton de la précontrainte des fils
d'acier) 2. réduction f des contraintes
nl 1. aflaten n (spanning) 2. ontspanning
f
r 1. передача f усилий натяжения
арматуры с упоров на бетон
2. уменьшение n [снижение n]
напряжений

D220 e detention
d Rückhalt m, Retention f
f retention f, accumulation f (des eaux)
nl bufferen n van de afvloeiing
r аккумуляция f стока

D221 detention basin see detention reservoir

D222 e detention dam
d Wassersperrmauer f,
Wassersperrdamm m
f barrage m de retenue
nl afsluitdam m
r водозадерживающая плотина f

D223 e detention reservoir
d Rückhaltebecken n,
Rückhaltespeicher m,
Hochwasserbecken n
f réservoir m de régulation avec
débouchés réglables; réservoir m
tampon
nl bufferbekken n
r задерживающее водохранилище n
(с регулируемыми попусками);
противопаводковое [буферное]
водохранилище n

D224 e detention time
d Aufenthaltszeit f, Durchflußzeit f
f durée f [temps m] de rétention
nl bufferperiode f, n
r продолжительность f пребывания
(напр. сточных вод в очистном
сооружении)

D225 e deterioration
d Verschleiß m; Beschädigung f;
Zerstörung f
f détérioration f
nl achteruitgang m in kwaliteit
r износ m; повреждение n; разрушение
n

D226 e detonating cord
d detonierende Zündschnur f,
Sprengschnur f
f cordeau m détonant
nl slagsnoer n
r детонирующий шнур m

D227 e detonator
d Sprengkapsel f, Zündkapsel f
f détonateur m, capsule f d'amorçage
nl slagspijpje n
r капсюль-детонатор m

D228 e detritor
d Sandfang m
f dessableur m
nl zandvang m
r песколовка f (канализационная)

D229 detritus see debris

D230 detritus chamber see detritor

D231 e development
d 1. Erschließung f 2. Bebauungsplan
m 3. Erweiterung f, Ausbau m
4. Entwicklung f, Ausarbeitung f,
Versuchs- und Konstruktionsarbeiten
f pl 5. Abpumpen n 6. Abwicklung f
f 1. aménagement des terres [du
terrain à bâtir] 2. projet m de
construction d'abitations
3. développment m, réconstruction f
4. élaboration f [étude f] de projet
5. réalisation f, mise f en service
de sources d'eau 6. développement m

nl 1. ontsluiting *f* 2. bebouwingsplan *n*
3. uitbreiding *f* 4. ontwikkeling *f* 5.
openen *n* van een bron 6. uitslag *m*
(*van een gebogen oppervlak*)
r 1. освоение *n* земель; инженерная
подготовка *f* территории к застройке
2. проект *m* застройки 3. развитие *n*;
реконструкция *f* 4. разработка *f*;
опытно-конструкторские работы *f pl*
5. разработка *f* водоисточника
6. развёртка *f* (*поверхности*)

D232 *e* **development bond stress**
d Haftspannung *f* an der Kräfte-
-Übergangszone (*Armierung*)
f contrainte *f* d'adhérence sur la longueur de zone de transmission
nl hechtingsspanning *f* in
het krachten-overgangsgebied
r напряжение *n* сцепления на длине
зоны передачи усилий натяжения
(*напрягаемой арматуры*)

D233 *e* **development length**
d Länge *f* der Kräfte-Übergangszone
f longueur *f* de transmission
nl lengte *f* van krachten-overgangsgebied
r длина *f* зоны передачи
преднапряжения (*от натянутой
арматуры к бетону*)

D234 *e* **development plan**
d Bebauungsplan *m*, Fluchtlinienplan
m, Generalbebauungsplan *m*
f plan *m* [projet *m*] d'aménagement
général, plan *m* d'implantation des
bâtiments à construire
nl algemeen bebouwingsplan *n*
r генеральный план *m* застройки

D235 *e* **development project**
d Ausbauprojekt *n*; Ausbauvorhaben *n*
f projet *m* de développement urbain;
développement *m* fluvial
nl ontwikkelingsproject *n*
r градостроительный проект *m*, проект
m застройки территории; проект
m комплексного освоения водных
ресурсов речного бассейна; проект
m энергетического использования
водотока

D236 *e* **development works**
d Entwicklungsarbeiten *f pl*
f travaux *m pl* d'étude, travail *m* de
recherche
nl ontwikkelingswerk *n*
r проектно-исследовательские
разработки *f pl*

D237 *e* **deviation**
d Abweichung *f* (vom Mittelwert);
Fehler *m*; Lotabweichung *f* der
Bohrlochachse
f déviation *f*, écart *m*; écart *m* de
perpendicularité (*d'un trou de forage*)
nl deviatie *f*, afwijking *f*
r отклонение *n* (от среднего
значения); погрешность *f*;

отклонение *n* оси скважины от
вертикали

D238 *e* **deviation angle**
d Drehwinkel *m*, Schwenkwinkel *m*
(*Straßenbau*)
f angle *m* de déviation
nl hoek *m* van afwijking,
richtingsverandering *f* van een
wegtracé
r угол *m* поворота (*трассы дороги*)

D239 *e* **devil**
d 1. Bauaustrocknungsofen *m*
2. Nagelbrett *n*
f 1. réchauffeur *m* d'air 2. chemin *m*
der fer (*outil de dressage*)
nl 1. luchtverhitter *m* (*voor
gebouwuitdroging*) 2. krabber *m*
r 1. воздушно-отопительный агрегат *m*
для сушки зданий 2. гвоздевая тёрка *f*

D240 *e* **devil float**
d Nagelbrett *n*
f chemin *m* de fer (*outil de dressage*)
nl krabber *m* (*pleisterwerk*)
r гвоздевая тёрка *f* (*для создания
шероховатой фактуры*)

D242 *e* **dewatering**
d Wasserhaltung *f*,
Baugrubenentwässerung *f*;
Dränage *f*, Bodenentwässerung *f*;
Entwässern *n*, Wasserentzug *m*
f épuisement *m*, exhaure *f*; drainage *m*,
assèchement *m*; essorage *m*
nl wateronttrekking *f*, (bron)bemaling *f*
r водоотлив *m*; дренаж *m*, осушение *n*
грунта; обезвоживание *n*

D243 *e* **dewatering coefficient**
d Wasserabgabekapazität *f*
f coefficient *m* de drainage
nl bemalingscapaciteit *f*
r коэффициент *m* дренирования
[водоотдачи]

D245 *e* **dewatering installation**
d Grundwasserabsenkungsanlage *f*;
Entwässerungsanlage *f*,
Wasserhaltungsanlage *f*
f installation *f* de rabatement de la
nappe aquifère; installation *f*
d'épuisement *m* d'eaux
nl bronbemalingsinstallatie *f*;
bemalingsinstallatie *f*
r установка *f* водопонижения;
установка *f* водоотлива

D246 *e* **dewatering of excavation**
d Wasserhaltung *f* (*zur Baugruben-
oder Grabenentwässerung*)
f assèchement *m* d'une fouille
nl bemaling *f* van een bouwput
r водоотлив *m* из котлована *или*
траншеи

D247 *e* **dewatering operation**
d Grundwasserabsenkung *f*;
Wasserhaltung *f*
f rabattement *m* de la nappe aquifère;
épuissement *m* d'eaux

nl bemaling *f*
r водопонижение *n*; водоотлив *m*

D248 *e* **dewpoint temperature**
   d Taupunkttemperatur *f*
   f température *f* du point de rosée
   nl dauwpunt *n*
   r температура *f* точки росы

D249 *e* **diagonal**
   d Diagonale *f*, Schräge *f*,
      Diagonalstab *m*
   f diagonale *f*, barre *f* diagonale
      (*d'une poutre à treillis*)
   nl diagonaal *f* (*m*)
   r раскос *m* (*фермы*)

D250 *e* **diagonal band**
   d Schrägbewehrung *f*,
      Diagonalbewehrung *f*
   f nappe *f* d'armatures diagonales
   nl diagonaalwapening *f*
   r диагональные арматурные стержни
      m pl

D251 *e* **diagonal bond**
   d Stromverband *m*, Festungsverband *m*,
      Fischgrätenverband *m*,
      Kornährenverband *m*
   f appareil *m* diagonal [en arête de
      poisson]
   nl diagonaalverband *n*
   r диагональная перевязка *f*,
      перевязка *f* в ёлку

D252 *e* **diagonal brace**
   d Schrägstab *m*, Diagonalglied *n*;
      Strebe *f*
   f diagonale *f*, moise *f* en écharpe;
      contreventement *m* diagonale
   nl diagonaalschoor *m*; windverband *n*
   r диагональная связь *f*; наклонная
      схватка *f*

D253 *e* **diagonal crack**
   d Schrägriß *m*
   f fissure *f* diagonale [oblique]
   nl diagonaalscheur *f* (*m*)
   r косая [наклонная] трещина *f*

D254 *e* **diagonal cracking**
   d Schrägrißbildung *f*
   f formation *f* des fissures diagonales,
      fissuration *f* en écharpe
   nl vorming *f* van diagonaalscheuren
   r образование *n* косых [наклонных]
      трещин

D255 *e* **diagonal in compression**
   d Druckschräge *f*, Druckdiagonale *f*
   f diagonale *f* [barre *f* diagonale]
      comprimée
   nl drukdiagonaal *f* (*m*)
   r сжатый раскос *m*

D256 *e* **diagonal in tension**
   d Zugdiagonale *f*, gezogene **Schräge**
      f
   f diagonale *f* tendue, barre *f* diagonale
      en traction
   nl getrokken diagonaal *f* (*m*)
   r растянутый раскос *m*

D257 *e* **diagonal sheathing**
   d Diagonalbretterverschalung *f*
   f planchéiage *m* en diagonale,
      revêtement *m* en planches diagonales
   nl diagonale beplanking *f*
   r диагональная дощатая обшивка *f*

D258 *e* **diagonal tensile stress, diagonal
     tension stress**
   d Diagonalzugspannung *f*
   f contrainte *f* de traction sur une
      face(tte) inclinée
   nl diagonale trek *m*
   r главное растягивающее напряжение
      n на наклонной площадке

D259 *e* **diagram**
   d Diagramm *n*; Schaubild *n*
   f diagrame *m*; graphique *m*, schéma *m*
   nl diagram *n*; schema *n*
   r диаграмма *f*; схема *f*

D260 *e* **diagrid floor**
   d Decke *f* aus kreuzweise verlegten
      Diagonalelementen; diagonaler
      Gitterbelag *m*
   f caillebotis *m*; platelage *m* à claire-voie
   nl roosterplafond *n* met diagonaal
      georiënteerde lamellen
   r перекрытие *n* из диагональных
      перекрёстных элементов,
      решётчатый настил *m*

D261 *e* **diametral compression test**
   d Spaltzugprüfung *f*, Spaltzugversuch
      m, Spaltzugprobe *f*
   f essai *m* de compression diamétrale,
      essai *m* brésilien
   nl splijtingsdrukproef *f* (*m*)
   r испытание *n* на раскалывание
      (*бетонного цилиндра, сжимаемого
      вдоль образующей*)

D262 *e* **diamond-head buttress dam**
   d Rhombenkopf-Staumauer *f*,
      Pfeilerstaumauer *f* mit
      diamantkopfähnlichen
      Strebepfeilern
   f barrage *m* à contreforts à tête
      octogonale
   nl keermuur *m* met scherpe rug, beer *m*
   r массивно-контрфорсная плотина *f*
      с полигональными оголовками

D263 *e* **diamond mesh**
   d Rauten(maschen)gewebe *n*
   f toile *f* métallique à mailles en losange
   nl gaas *n* met ruitvormige mazen
   r ромбическая сетка *f*, сетка *f*
      с ромбическими ячейками

D264 *e* **diamond-mesh lath**
   d Streckmetall-Putzträger *m*
   f grillage *m* en métal déployé
   nl diamantgaas-pleisterdrager *m*
   r штукатурная сетка *f* из просечно-
      -вытяжной тонколистовой стали

D265 *e* **diaphragm**
   d 1. Diaphragma *n*, Membrane *f*,
      Trennwand *f*; Versteifungswand *f*,
      steifes Querschott *n* 2. Meßblende *f*,

DIAPHRAGM

Drosselscheibe *f* 3. Dichtungsschirm
m, Dichtungshaut *f*, Dichtungswand
f, Kernmauer *f*
f 1. diaphragme *m*, membrane *f*, paroi
f; diaphragme *m* de rigidité
2. débitmètre *m* à diaphragme,
diaphragme *m* 3. masque *m*
d'étanchéité
nl 1. membraan *n*; tussenschot *n*;
verstijvingsplank *f* (*m*)
2. (meet)diafragma *n* 3. scherm *n*,
vlies *n*, membraan *n*
r 1. диафрагма *f*, мембрана *f*,
перегородка *f*; диафрагма *f*
жёсткости 2. измерительная
диафрагма *f* (*в трубопроводе*) 3.
противофильтрационная диафрагма
f, экран *m* (*плотины*)

D266 *e* **diaphragm pump**
d Membran(e)pumpe *f*, Dia(phragma)-
-Pumpe *f*
f pompe *f* à diaphragme
nl diafragmapomp *f* (*m*), kattekop *m*
r мембранный насос *m*

D267 *e* **diaphragm valve**
d Membranventil *n*
f robinet *f* à membrane
nl diafragmaklep *f* (*m*)
r мембранный клапан *m*

D268 *e* **diaphragm wall**
d Dichtungswand *f*, Kernmauer *f*;
Dichtwand *f*, Schlitzwand *f*
f voile *m* [noyau *m*] d'étanchéité;
écran *m* étanche, mur *m* de fondation
moulé dans le sol
nl kernmuur *m*; waterdicht scherm *n*
r противофильтрационная диафрагма
f, экран *m* (*плотины*); стена *f*
(*фундамента или туннеля неглубокого заложения*) в грунте, стена *f*,
бетонируемая в траншее

D269 *e* **diatomaceous brick**
d Diatomitstein *m*
f brique *f* (de) diatomite
nl diatomeeënsteen *m*
r диатомитовый кирпич *m*

D270 *e* **diatomaceous earth, diatomite**
d Diatomeenerde *f*, Kieselgur *f*
f terre *f* à diatomées, kieselguhr *m*
nl diatomeënaarde *f* (*m*), kiezelgoer *n*
r диатомит *m*, диатомовая земля *f*,
кизельгур *m*

D271 *e* **diatomite filter**
d Diatomitfilter *n*, *m*
f filtre *m* à kieselguhr
nl kiezelgoerfilter *n*, *m*
r диатомитовый фильтр *m*

D272 *e* **dicalcium silicate**
d Dikalziumsilikat *n*
f silicate *f* bicalcique
nl dicalciumsilicaat *n*
r двухкальциевый силикат *m*
(*компонент цемента*)

D273 *e* **diesel pile hammer**
d Dieselhammer *m*
f mouton *m* de battage diesel
nl dieselheiblok *n*
r дизельный молот *m*

D274 *e* **differential pressure**
d Differentialdruck *m*, Differenzdruck *m*
f pression *f* différentielle
nl drukverschil *n*
r дифференциальное давление *n*,
перепад *m* давления

D275 *e* **differential pressure controller**
d Differentialdruckregler *m*,
Differenzdruckregler *m*
f régulateur *m* de pression différentiel
nl differentiaaldrukregelaar *m*
r регулятор *m* дифференциального
давления

D276 *e* **differential settlement**
d relative Setzung *f*,
Setzungsunterschied *m*;
ungleichmäßige Setzungen *f pl* (*z.B.
von Brückenpfeilern*)
f tassement *m* differentiel,
affaissement *m* inégale,
dénivellation *f*
nl ongelijkmatige zetting *f*;
zettingsverschil *n*
r относительная осадка *f*;
неравномерная осадка *f* (*напр. опор
моста*)

D277 *e* **differential surge tank**
d Differentialwasserschloß *n*,
Steigrohrwasserschloß *n*
f cheminée *f* d'équilibre différentielle
nl buffertank *m*
r дифференциальный уравнительный
резервуар *m*

D278 *e* **diffused-air aeration**
d Druckluftverfahren *n* (*Abwasser*)
f aération *f* (*des eaux usées*) par
insufflation d'air diffusé
nl geforceerde beluchting *f*
(*afvalwaterbehandeling*)
r пневматическая аэрация *f* (*сточных
вод*)|

D279 *e* **diffused air device**
d Druckbelüftungseinrichtung *f*
f aérateur *m* à air diffusé
nl beluchtingsinstallatie *f*
(*afvalwaterbehandeling*)
r пневматический аэратор; *m* (*сточных
вод*)

D280 *e* **diffused air supply**
d diffuse Luftführung *f*
[Luftverteilung *f*]
f alimentation *f* [approvisionnement *m*]
en air diffusé
nl gespreide luchttoevoer *m*
r рассеянная подача *f* воздуха

D282 *e* **diffused lighting**
d diffuse Beleuchtung *f*
f éclairage *m* diffusé

nl diffuus licht *n*
r диффузное [рассеянное] освещение *n*

D283 *e* diffuse light luminaire
   d diffusstrahlende Leuchte *f*
   f luminaire *m* à lumière diffusé
   nl armatuur *f* voor indirecte verlichting
   r светильник *m* рассеянного света

D284 *e* diffuser
   d 1. Diffusor *m* 2. Deckenluftverteiler *m*, Anemostat *m* 3. Lichtzerstreuer *m* 4. diffusstrahlende Leuchte *f*, Luftverteiler *m*, Filterplatte *f*, Filtrosplatte *f*
   f 1. diffuseur *m* 2. diffuseur *m* d'air 3. plafon *m* diffusant 4. plaque *f* diffuseuse
   nl 1. diffusor *m* 2. plafonddiffusor *m* (*warme lucht*) 3. armatuur *f* voor diffuus licht 4. filterplaat *f* (*m*)
   r 1. диффузор *m* (*расширяющийся канал*) 2. диффузор *m*, воздухораспределительный плафон *m* 3. светорассеиватель *m* 4. фильтрос *m*

D285 *e* diffuser plate
   d Luftverteiler *m*, Filterplatte *f*, Filtrosplatte *f*
   f plaque *f* diffuseuse, diffuseur *m* d'air
   nl filterplaat *f* (*m*)
   r фильтрос *m*

D286 *e* diffusion block
   d Strahlablenker *m*, Energiebrecher *m* und Auflöser, Schikane *f*
   f déflecteur *m*, bloc *m* de diffusion, chicane *f*
   nl luchtstroomverspreider *m*
   r гаситель-растекатель *m*

D287 *e* digested sludge
   d Faulschlamm *m*, ausgefaulter Sinkstoff *m*
   f boue *f* des égouts digérée
   nl verteerd slib *n* (*afvalwater*)
   r сброженный осадок *m* (*сточных вод*)

D288 digestion tank *see* septic tank

D289 *e* digging
   d Aushubarbeiten *f pl*, Abgrabung *f* von Einschnitten, Aushub *m*
   f creusage *m*, creusement *m*, excavation *f*; travaux *m pl* de terrassement; fouilles *f pl*
   nl graafwerk *n*; baggerwerk *n*
   r землеройные [земляные] работы *f pl*; разработка *f* выемок; выемка *f* грунта

D290 *e* digging bucket
   d Baggerbecher *m*, Grabeimer *m*
   f benne *f* piocheuse
   nl baggerbak *m*, graafemmer *m*
   r землеройный ковш *m*

D291 *e* digging depth
   d Aushubtiefe *f*, Grabtiefe *f*
   f profondeur *f* d'excavation [de fouille, de cavage, de draguage]
   nl baggerdiepte *f*, graafdiepte *f*

r глубина *f* копания [черпания]; глубина *f* выемки

D292 *e* digging face
   d Aushubwand *f*; Ausschachtungswand *f*
   f front *m* d'abattage [de taille, d'attaque]
   nl talud *n* van de ingraving
   r забой *m* карьера

D293 *e* digging ladder
   d Becherleiter *f*; Eimerleiter *f*, Baggerleiter *f*
   f élinde *f* à godets
   nl emmerbagger *m*, excavateur *m*
   r рама *f* цепи многоковшового экскаватора

D294 *e* digging line
   d Grabseil *n*
   f câble *m* tracteur [de halage]
   nl trekdraad *m* (*dragline*)
   r напорный [тяговый] канат *m* экскаватора

D295 *e* digging radius
   d Aushubbereich *m*
   f rayon *m* d'action d'un excavateur [d'une drague]
   nl uitgooilengte *f*
   r радиус *m* копания [черпания]

D296 *e* digging wheel
   d Eimerrad *n*, Becherrad *n*
   f roue *f* à godets
   nl emmerrad *n*
   r ковшовое колесо *n*, ротор *m* (*роторного экскаватора*)

D297 dike *see* dyke

D298 *e* dike breach
   d Damm(ein)bruch *m*, Deich(ein)bruch *m*
   f brèche *f* de digue
   nl dijkbreuk *f* (*m*)
   r прорыв *m* дамбы

D299 *e* dike dam
   d Uferschutzdamm *m*, Deich(damm) *m*, Flußdamm *m*, Absperrdamm *m*
   f digue *f* [jetée *f*] de protection
   nl strekdam *m*
   r берегозащитная [ограждающая] дамба *f*

D300 *e* diking
   d Eindeichung *f*, Bedeichung *f*, Abdeichung *f*
   f endiguement *m*
   nl bedijking *f*
   r обвалование *n*, сооружение *n* дамбы

D301 dilatation *see* expansion

D302 *e* dilatation joint
   d Temperaturdehnungsfuge *f*
   f joint *m* de dilatation
   nl dilatatievoeg *f* (*m*)
   r температурный шов *m*

D303 *e* diluent
   d Verschnittmittel *n*
   f diluant *m*
   nl versnijdingsmiddel *n*
   r разбавитель *m* (*краски*)

D304 *e* dilution
   d 1. Verdünnung *f* 2. Raumlastaufnahme
     f (*durch Zuluft*); Auflösung *f* der
     Emissionen (*in der Luft*)
   f 1. dilution *f* 2. réduction *f* de la
     teneur des agents de pollution nocifs
     en atmosphère
   nl 1. verdunning *f* 2. terugdringen *n*
     van de concentratie
   r 1. разбавление *n* 2. ассимиляция *f*
     вредностей приточным воздухом;
     растворение *n* выделяющихся
     вредностей в окружающей
     атмосфере

D305 *e* dilution ventilation
   d gesamte Belüftung *f*, Belüftung *f* des
     gesamten Raumes
   f ventilation *f* activée à échange d'air
     total
   nl volledige luchtverversing *f*
   r общеобменная приточная вентиляция
     f

D306 *e* dimensional stability
   d Maßhaltigkeit *f*, Maßbeständigkeit *f*
   f stabilité *f* dimensionnelle
   nl maatvastheid *f*
   r способность *f* сохранять размеры,
     постоянство *n* размеров,
     стабильность *f* в отношении размеров

D307 *e* dimensional tolerance
   d Maßtoleranz *f*
   f tolérance *f* dimensionnelle
   nl maattolerantie *f*
   r допуск *m* на размеры

D308 *e* dimensionless unit graph
   d dimensionslose Einheitsganglinie *f*
   f hydrogramme *f* unitaire sans
     dimension
   nl dimensieloze eenheidsgrafiek *f*
   r безразмерный единичный гидрограф
     m

D309 *e* dimension lumber, dimension staff
   d Dimensionsschnittholz *n*
   f bois *m* de charpente de dimension
     nominale
   nl bezaagd hout *n* in handelsmaten
   r строительные пиломатериалы *m pl*
     определённого сортимента

D310 *e* dimension stock
   d Dimensionsstockholz *n*
   f bois *m* de dimensions régulières
   nl onbezaagd hout *n* in standaardmaten
   r пиломатериал *m* крупного сечения

D311 *e* dinas brick *UK*
   d Quarzkalkziegel *m*, Quarzitstein *m*,
     Quarzitziegel *m*
   f brique *f* de Dinas
   nl kiezelzandsteen *m*, kwartszandsteen *m*
   r динасовый кирпич *m*

D312 *e* dinging
   d einschichtiger Außenputz *m*
   f enduit *m* extérieur à une seule couche
   nl grof buitenpleisterwerk *n* in één laag
   r однослойная наружная штукатурка *f*

D313 *e* dining kitchen
   d Eßküche *f*, Wohnküche *f*
   f cuisine *f* de séjour, cuisine-salle *f* à
     manger
   nl eetkeuken *f* (*m*)
   r кухня-столовая *f*

D314 *e* dip
   d 1. Böschungsneigung *f* 2. Einfallen *n*,
     Fallen *n* 3. Durchhang *m*
   f 1. pente *f* (*du terrain, du talus*)
     2. pendage *m* (*d'une strate*) 3. flèche
     (*d'une courbe*)
   nl 1. toename *f* van de helling 2. helling
     f van een bodemlaag 3. doorhang *m*
   r 1. уклон *m* (*земляной поверхности,
     откоса*) 2. падение *n* пласта 3.
     провисание *n*, провес *m*; стрела *f*
     провеса

D315 *e* dipper
   d Baggerlöffel *m*; Grabgefäß *n*
   f godet *m* de pelle mécanique; godet *m*
     de drague
   nl lepelschop *m*; dieplepel *m*
   r ковш *m*; черпак *m*

D316 *e* dipper stick
   d Löffelstiel *m*, Hubarm *m*, Hubgestell *n*
   f bras *m* de godet [de levage, de
     poussée]
   nl steel *m* van de lepelschop
   r рукоять *f* ковша

D317 *e* direct air heating system
   d Außenluftheizung *f*, Frischluftheizung
     f
   f installation *f* de chauffage direct à
     air chaud
   nl luchtverwarming *f* met aanzuiging
     van buitenlucht
   r система *f* прямоточного воздушного
     отопления

D318 *e* direct band
   d Längs- *oder* Querbewehrung *f*
   f nappe *f* d'armature longitudinale *ou*
     transversale
   nl langs- *of* dwarswapening *f* (*bij
     wapening in beide richtingen*)
   r продольные *или* поперечные
     арматурные стержни *m pl*

D320 *e* direct dumping
   d Beschickung *f* von Frischbeton direkt
     in die Schalung
   f déversement *m* [basculage *m*] (*du
     béton frais*) directement dans le
     coffrage
   nl toevoer *m* van het betonmengsel direct
     aan de bekisting
   r подача *f* (*бетонной смеси*)
     непосредственно в опалубку

D321 *e* direct electric curing
   d direkte Elektrowarmbehandlung *f*
   f cure *f* [curing *m*] électrique (*de
     béton*), cure *f* (de béton) par chauffage
     électrique direct
   nl nabehandeling *f* van de mortel door
     directe elektrische verwarming

r прямой электропрогрев *m* (*для
ускорения твердения бетона*)

D322 *e* **direct expansion air cooler**
d Direktverdampfungsluftkühler *m*
f refroidisseur *m* à détente directe
nl luchtkoeler *m* met directe verdamping
r воздухоохладитель *m*
непосредственного расширения
[испарения]

D323 *e* **direct-fired unit heater**
d Feuerlufterhitzer *m*
f réchauffeur *m* d'air à flamme directe
nl luchtverhitter *m* met brander
r огневой калорифер *m*, агрегат *m*
воздушного отопления с непосредственным сжиганием топлива

D324 *e* **direct-flow water heater**
d Durchlauferhitzer *m*,
Durchlaufwasserheizer *m*
f chauffe-eau *m* à écoulement libre
nl warmwaterdoorstroomapparaat *n*
r проточный [контактный]
водонагреватель *m*

D325 *e* **direct heating**
d Heizung *f* mit unmittelbarer
Heizstoffverfeuerung, örtliche
Beheizung *f*; Radiatorenheizung *f*
(*im Gegensatz zur Warmluftheizung*)
f chauffage *m* à flamme directe;
chauffage *m* par radiateurs
nl verwarming *f* onder directe afgifte
van verbrandingswarmte
r отопление *n* с непосредственным
сжиганием топлива, местное
отопление *n*; радиаторное отопление *n*

D326 *e* **direct hot water supply**
d direkte Warmwasserversorgung *f*
f distribution *f* d'eau chaude directe
nl directe warmwatervoorziening *f*
r горячее водоснабжение *n* с
непосредственным нагревом воды в
котельной установке

D327 *e* **direct hot water system**
d Radiatorenwasserheizanlage *f*,
Wasserheizsystem *n* mit Heizkörpern
f système *m* de chauffage à radiateurs
d'eau
nl verwarming *f* door
warmwaterradiatoren
r радиаторная система *f* водяного отопления

D328 *e* **directional fixed grille**
d nicht regulierbares
Lüftungsgitter *n* [Luftverteilgitter *n*]
f grille *f* d'aérage fixe
nl niet-regelbaar luchtrooster *m*, *n*
r нерегулируемая приточная решётка
f

D329 *e* **direct-light luminaire**
d Direktleuchte *f*
f luminaire *m* d'éclairage direct
nl armatuur *f* met directe verlichting
r светильник *m* прямого света

D330 *e* **direct-reading tacheometer**
d Reduktionstachymeter *n*
f tachéomètre *m* à lecture directe
nl direct afleesbare tachometer *m*
r тахеометр *m* прямого отсчёта

D331 *e* **direct stress**
d Normalspannung *f*
f contrainte *f* normale
nl normaalspanning *f*
r нормальное напряжение *n*

D332 *e* **direct system**
d Warmwasserversorgungsanlage *f* mit
Primärenergieverbrauch zur
Warmwasseraufbereitung
f système *m* direct de distribution d'eau
chaude réchauffée par énergie
thermique primaire
nl warmwatervoorzieningssysteem *n* met
gebruik van primaire energie
r система *f* горячего водоснабжения с
нагревом воды за счёт энергии
первичного энергоносителя

D333 disastrous flood *see* catastrophic flood

D334 *e* **disbonding**
d Dekohäsion *f*, Ablösung *f*
f décohésion *f*
nl ontbinding *f*, loslaten *n*
r нарушение *n* сцепления

D335 *e* **discharge**
d 1. Durchfluß *m*, Abfluß *m*;
Ergiebigkeit *f*, Schüttung *f* 2.
Ausfluß *m* 3. Leistung *f*, Fördermenge
f (*Pumpe*) 4. Austrag *m*
f 1. débit *m* 2. écoulement *m*
3. rendement *m* (*de la pompe*)
4. déchargement *m*, décharge *f*
nl 1. waterafvoer *m* 2. uitstroming *f* 3.
pompopbrengst *f* 4. afladen *n*, lossen *n*
r 1. расход *m*; дебит *m* 2. истечение *n*
3. подача *f* (*насоса*) 4. разгрузка *f*;
выпуск *m* (*напр. из смесителя,
дробилки, бункера*)

D336 *e* **discharge area**
d Abflußbereich *m*,
Abflußquerschnittsfläche *f*,
Durchflußquerschnitt *m*
f section *f* mouillée, aire *f* de la
section mouillée
nl doorstromingsprofiel *n*, nat profiel *n*
r площадь *f* живого сечения

D337 *e* **discharge canal**
d Ausflußkanal *m*
f canal *m* de décharge [d'évacuation]
nl afvoerkanaal *n*
r сбросной канал *m*

D338 *e* **discharge capacity**
d 1. Durchlaßvermögen *n*,
Durchlaßkapazität *f*,
Leistungsfähigkeit *f*
2. Förder(leistungs)vermögen *n*,
Fördermenge *f* (*Pumpe*)
f 1. capacité *f* de débit (*d'un canal*)
2. débit *m*, refoulement *m* (*de pompe*)

nl 1. doorlaatvermogen *u*
2. pompvermogen *n*
r 1. пропускная способность *f*
(*водотока, водопроводящего
сооружения*) 2. подача *f* (*насоса*)

D339 *e* discharge carrier
d Überfallrücken *m*;
Überlaufabführungsanlage *f*
f coursier *m*; ouvrage *m* de dérivation
du déversoir
nl rug *m* van de overlaat
r водосливная грань *f*, водоскат *m*;
отводящее сооружение *n* водосброса

D340 *e* discharge chamber
d Regenüberfall *m*
f déversoir *m* d'orage
nl regenoverloop *m*
r ливнесброс *m*

D341 discharge channel *see* tailrace channel

D342 *e* discharge coefficient
d Abflußkoeffizient *m*,
Durchflußkoeffizient *m*
f coefficient *m* de débit
nl afvloeiingscoëfficiënt *m*
r коэффициент *m* расхода

D343 *e* discharge conduit
d Abflußleitung *f*, Ablaßleitung *f*
f conduit *m* d'évacuation
nl afvoerleiding *f*
r спускной водовод *m* [трубопровод *m*]

D344 discharge cross-section *see* discharge area

D345 *e* discharge culvert
d Ablaufkanal *m*, Abflußkanal *m*,
Umlauf *m*
f aqueduc *m* d'évacuation
nl afvoerriool *n*, *f* (*m*)
r водопроводная галерея *f*, галерея *f*
опорожнения (*дока, шлюза*)

D346 *e* discharge curve
d Abfluß(mengen)kurve *f*,
Durchflußkurve *f*,
Pegelschlüsselkurve *f*
f courbe *f* des débits [de tarage]
nl afvoercurve *f*
r кривая *f* расходов (*расход — уровень*)

D347 *e* discharge electrode
d Sprühelektrode *f*
f électrode *f* d'émission
nl ontladingselektrode *f*
r разрядный [коронирующий]
электрод *m* (*электрофильтра*)

D348 *e* discharge gate
d Entleerungsöffnung *f*
f trappe *f* de vidange, porte *f* à trappe
nl schuif *f* (*m*) [klep *f* (*m*)] in de
afvoerleiding
r разгрузочная дверца *f* [заслонка *f*];
разгрузочный затвор *m*

D349 *e* discharge head
d geodätische Druckhöhe *f*, Förderhöhe *f*
f hauteur *f* géométrique de refoulement
nl drukhoogte *f*
r геометрическая высота *f* нагнетания

D350 *e* discharge hydrograph
d Abflußganglinie *f*
f courbe *f* des débits, hydrogramme *f*,
courbe *f* hydrométrique d'écoulement
nl afvloeiingscurve *f*
r гидрограф *m* (стока)

D351 *e* discharge line
d Förderleitung *f*, Druckleitung *f*
f conduit *m* de refoulement
nl drukleiding *f*, persleiding *f*
r линия *f* нагнетания

D352 *e* discharge mass curve
d Abfluß(mengen)-Summenlinie *f*,
Durchfluß(mengen)-Summenlinie *f*,
Durchflußsummen(gang)linie *f*
f courbe *f* des débits cumulés
nl curve *f* van de cumulatieve afvoer
r интегральная кривая *f* стока
[расходов]

D353 *e* discharge modulus
d spezifische Wasserführungsfähigkeit *f*
f coefficient *m* de d'écoulement
nl specifieke afvoercapaciteit *f*
r модуль *m* стока

D354 *e* discharge of solids *US see* sediment
discharge

D355 *e* discharge opening
d Ablauföffnung *f*, Auslauföffnung *f*,
Durchflußöffnung *f*; Wasserauslaß *m*,
Talsperrenablaß *m*
f orifice *f* de décharge
nl doorlaatopening *f*, uitlaat *m* (*dam*)
r водопропускное отверстие *n*

D356 *e* discharge pipe
d 1. Auslauf *m* der
Entwässerungsleitung; Ausflußrohr *n*,
Ablaßrohr *n* 2. Schlammleitung *f*,
Spülleitung *f*
f 1. canal *m* de décharge; tuyau *m* de
décharge 2. conduite *f* de refoulement
nl 1. afvoerbuis *f* (*m*) 2. persleiding *f*
r 1. водосбросный выпуск *m*;
спускная [выпускная] труба *f*
2. пульпопровод *m*

D357 *e* discharge pressure
d Förderdruck *m*
f pression *f* de refoulement
nl persdruk *m*
r давление *n* нагнетания

D358 *e* discharge recorder
d Durchflußmengenschreiber *m*
f enregistreur *m* des débits
nl afvoerregistratietoestel *n*
r самописец *m* расхода воды,
расходомер-самописец *m*

D359 *e* discharge regulation
d Abflußregelung *f*,
Wasserablaufregelung *f*
f régularisation *f* des débits
nl regeling *f* van de waterafvoer
r регулирование *n* стока

D360 *e* discharge rod
d Blitzauffang *m*

f tige *f* de paratonnerre
nl bliksemopvangspits *f* (*m*)
r молниеприёмник *m*

D361 *e* **discharge section**
  d durchströmter Querschnitt *m*,
    Abflußquerschnitt *m*,
    Fließquerschnitt *m*,
    Durchflußquerschnitt *m*
  f section *f* d'écoulement
  nl doorsnede *f* (*m*) van de afvoeropening
  r живое сечение *n* потока

D362 *e* **discharge section line**
  d Abflußquerschnittslinie *f*,
    Abflußmeßstelle *f*
  f section *f* de jaugeage
  nl afvloeiingsmeetpunt *n*
  r гидрометрический створ *m*

D363   **discharge sewer** *see* **subsidiary sewer**

D364 *e* **discharge site**
  d Abflußmeßstelle *f*
  f station *f* de jaugeage
  nl plaats *f* (*m*) van de (meet)overlaat
  r гидрометрическая станция *f*

D365 *e* **discharge spout**
  d offener Wasserauslaß *m*
  f goulotte *f* d'écoulement
  nl open waterlozing *f*
  r открытый водосточный выпуск *m*

D366 *e* **discharge tunnel**
  d Entlastungsstollen *m*, Ringüberlauf
    m; Ablaufstollen *m*, Ableitungsstollen
    m
  f évacuateur *m* souterrain; galerie *f*
    d'évacuation, tunnel *m* de
    dérivation
  nl ontlastriool *n*
  r туннельный водосброс *m*; отводящий
    туннель *m*

D367 *e* **discharge valve**
  d 1. Durchflußregelventil *n*;
    Regulierschieber *m* 2. Druckventil *n*
  f 1. soupape *f* de réglage de débit
    2. vanne *f* de refoulement
  nl doorstroomregelingsklep *f* (*m*)
    2. regelventiel *n*
  r 1. клапан *m* регулирования расхода;
    регулирующая задвижка *f* 2.
    нагнетательный клапан *m* (*насоса*)

D368 *e* **discomfort**
  d Diskomfort *m*
  f inconfort *m*, incommodité *f*
  nl onbehagen *n*
  r дискомфорт *m*

D369 *e* **discontinuity**
  d Diskontinuität *f*, Unstätigkeit *f*,
    Ungänze *f*
  f discontinuité *f*
  nl discontinuiteit *f*
  r разрывность *f*, нарушение *n*
    сплошности

D370 *e* **discontinuous-flow settling basin**
  d Absetzbecken *n* mit periodischer
    Spülung

f décanteur *m* à fonctionnement
  discontinu
nl bassin *n* met periodieke spoeling
r отстойник *m* периодического промыва

D371 *e* **discontinuous grading**
  d unstetige [diskontinuierliche]
    Kornabstufung *f*
  f granulométrie *f* discontinue
  nl onregelmatige korrelgrootteverdeling *f*
  r прерывистая гранулометрия *f*

D372 *e* **disc saw, disk saw**
  d Kreissäge *f*
  f scie *f* circulaire
  nl cirkelzaag *f* (*m*)
  r дисковая пила *f*

D373 *e* **disc-type [disk-type] power float**
  d Kreisglättmaschine *f*
  f machine *f* de lissage à disques
  nl pleisterschuurmachine *f* met ronde
    schijf
  r дисковая затирочная машина *f*

D374 *e* **disinfection of sewage**
  d Abwasserdesinfektion *f*,
    Abwasserentkeimung *f*, Entseuchung *f*
    von Abwässern
  f désinfection *f* [stérilisation *f*] des eaux
    usées
  nl ontsmetting *f* van afvalwater
  r обеззараживание *f* [дезинфекция *f*]
    сточных вод

D375 *e* **disk valve**
  d 1. Keil(platten)schieber *m*
    2. Tellerventil *n*, Scheibenventil *n*,
    Plattenventil *n*
  f 1. robinet-vanne *m* à lunette
    2. soupape *f* à plateau [à disque]
  nl 1. schuifafsluiter *m* 2. schijfklep *f* (*m*)
  r 1. клиновая задвижка *f*
    2. тарельчатый клапан *m*

D376 *e* **dismantling**
  d 1. Abbau *m*, Demontage *f*, Zerlegung
    f, Auseinandernehmen *n* 2. Ausschalen
    n
  f 1. démontage *m* 2. décoffrage *m*
  nl 1. ontmantelen *n*, slechten *n*
    2. ontkisten *n*
  r 1. демонтаж *m*; разборка *f*
    2. распалубка *f*

D377   **dismantling of shuttering** *see*
    dismantling 2.

D378 *e* **dispenser**
  d Verteiler *m*
  f distributeur *m*
  nl verdeler *m*
  r разливочное *или* раздаточное
    (автоматизированное) устройство *n*
    (*на бетонных заводах*)

D380 *e* **dispersing agent**
  d Dispergierungsmittel *n*,
    Dispersionsmittel *n*
  f dispersant *m*, dispersif *m*, agent *m*
    de dispersion
  nl dispersiemiddel *n*

r диспергирующий агент *m*,
диспергатор *m*

D381 *e* displacement
  d 1. Verschiebung *f*, Umlagerung *f*
    2. Wasserverdrängung *f*;
    Werkstoffverdrängung *f*
    3. Fördermenge *f*, Förderung *f*
  f 1. déplacement *m* 2. déplacement *m*
    (du navire), tonnage *m* de
    déplacement 3. débit *m* de la pompe
  nl 1. verplaatsing *f* 2. waterverplaatsing
    f 3. opbrengst *f* (*pomp*), opvoerhoogte *f*
  r 1. перемещение *n* 2. водоизмещение
    n; количество *n* воды, вытесняемое
    плавающим телом 3. подача (*насоса*)

D382 *e* displacement of hydraulic jump
  d Wechselsprungverschiebung *f*,
    Anlaufstrecke *f* des abgedrängten
    Wassersprunges
  f déplacement *m* de ressaut
  nl volume *n* van de watersprong
  r отгон *m* гидравлического прыжка

D383 *e* displacement pile
  d Verdrängungspfahl *m*, Rammpfahl *m*
  f pieu *m* battu
  nl heipaal *m*
  r забивная свая *f*

D384 *e* displacement ventilation
  d Verdrängungslüftung *f*
  f ventilation *f* par déplacement
  nl verdrijvende ventilatie *f*
  r вентиляция *f* вытесняющим
    [псевдоламинарным] потоком
    воздуха

D386 *e* displacer
  d Zyklopenbeton-Stein *m*
  f libage *m*, moellon *m*
  nl grove breuksteen *m*
  r бутовый камень *m* [бут *m*] для
    бутобетона, «изюм» *m*

D387 *e* disposal area
  d 1. Absetzstelle *f* 2. Spülteich *m*,
    Bergeteich *m*; Hydrokippe *f*,
    Spülhalde *f*
  f 1. voirie *f*, décharge *f* 2. terril *m*,
    crassier *m*
  nl 1. stortplaats *f(m)*, vuilnisbelt *m*, *f*
    2. steenberg *m*
  r 1. место *n* свалки, свалка *f* 2.
    хвостохранилище *n*; гидроотвал *m*

D388 *e* disposal well
  d Einlaßbohrung *f*, Schluckbrunnen *m*
  f puits *m* d'évacuation, puits *m* perdu
  nl zinkput *m*, zakput *m*
  r поглощающая скважина *f*

D389   dispose-all *US see* garbage disposer

D390 *e* dissipator
  d Verteiler *m* der Energie
  f dissipateur *m* (d'énergie)
  nl koellichaam *n*, koelplaat *f(m)*
  r гаситель *m* энергии потока

D391 *e* distance between girders
  d Trägerabstand *m*

  f distance *f* entre les poutres
    principales
  nl afstand *m* tussen hoofdbalken
  r расстояние *n* между главными
    балками

D392 *e* distance-measuring device
  d Entfernungsmeßgerät *n*
  f télémètre *m*, distancemètre *m*.
    stadiomètre *m*
  nl afstandsmeter *m*
  r дальномер *m*

D393 *e* distance piece
  d 1. Zwischenstück *n*, Abstandhalter *m*
    2. Betonabstandhalter *m*
  f 1. entretoise *f*, pièce *f* d'écartement
    2. écarteur *m*, séparateur *m*,
    distanceur *m*
  nl 1. afstandstuk *n*, klos *m*
    2. afstandshouder *m*
  r 1. распорка *f* 2. фиксатор *m*
    (арматуры)

D394 *e* distance separation
  d Gebäudeabstand *m*
  f prospect *m* (*distance entre les
    bâtiments*)
  nl tussenruimte *f* tussen gebouwen
  r противопожарный разрыв *m* между
    зданиями

D395 *e* distemper
  d Leimfarbe *f*
  f couleur *f* à la colle
  nl temperaverf *f(m)*
  r клеевая краска *f*

D396   distributary *see* distributing canal

D397 *e* distributed force
  d verteilte Kraft *f*
  f force *f* distribuée [répartie]
  nl verdeelde kracht *f* (*m*)
  r распределённая сила *f*

D398 *e* distributed load
  d Streckenlast *f*, verteilte Last *f*
  f charge *f* distribuée [répartie]
  nl verdeelde belasting *f*
  r распределённая нагрузка *f*

D399 *e* distributing canal
  d Verteil(ungs)kanal *m*,
    Wasserverteiler *m*
  f canal *m* distributeur [de répartition],
    canal *m* secondaire
  nl verdeelkanaal *n*, verdeler *m*
  r распределительный канал *m*,
    распределитель *m*

D400 *e* distributing pipe line
  d Versorgungsrohr *n* (*z.B. von einem
    Wasserbehälter aus*)
  f tuyauterie *f* [conduite *f*] de
    distribution (*d'eau, de gaz*)
  nl distributieleiding *f*, hoofdleiding *f*
  r распределительный *или* разводящий
    трубопровод *m* (*воды, газа*)

D401 *e* distribution-bar reinforcement
  d Verteilungsstabbewehrung *f*
  f barres *f pl* d'armature de répartition

nl verdeelwapening *f*
r распределительная стержневая
арматура *f*

D402 *e* **distribution board**
d Verteilertafel *f*
f tableau *m* [panneau *m*] de
distribution
nl installatiekast *f (m)*
r распределительный щит *m*

D403 *e* **distribution box**
d Verteilungskasten *m*, Verteilerdose *f*,
Anschlußdose *f*
f boîte *f* de distribution
nl aansluitkast *f (m)*, verdeeldoos
f (m)
r распределительный ящик *m*,
распределительная коробка *f*

D404 *e* **distribution cabinet**
d Schaltschrank *m*, Verteilerschrank
m
f armoire *f* de distribution [de
répartition], tableau-armoire *m*
nl meterkast *f (m)*
r распределительный шкаф *m*

D405 **distribution curve** *see* **frequency curve**

D406 *e* **distribution factor**
d Verteilungszahl *f*
(*Momentenausgleichverfahren*)
f coefficient *m* de distribution
(*pour le calcul par la méthode de balancement des moments*)
nl verdelingsfactor *m*
r коэффициент *m* распределения
моментов

D407 *e* **distribution header**
d Verteilerkopf *m*, Verteiler *m*
f collecteur *m* secondaire [de
distribution]
nl verdeelkop *m*
r распределительный коллектор *m*,
«гребёнка» *f*

D408 *e* **distribution network**
d Verteilungsnetz *n*
f réseau *m* de distribution [de
répartition]
nl distributienet *n*
r распределительная [разводящая]
сеть *f*

D409 *e* **distribution of concrete**
d Betonverteilung *f*
f distribution *f* du béton frais [du
mélange de béton]
nl verdeling *f* van betonmortel
r распределение *n* бетонной смеси

D410 *e* **distribution of daylight**
d Tageslichtverteilung *f*
f distribution *f* de lumière diurne
nl verspreiding *f* van het daglicht
r распределение *n* дневного света

D411 *e* **distribution of stresses**
d Spannungsverteilung *f*
f distribution *f* de contraintes

nl spanningsverdeling *f*
r распределение *n* напряжений

D412 *e* **distribution pipe**
d Verteilungsrohr *n*
f tuyau *m* de distribution
nl verdeelleiding *f*
r распределительный трубопровод *m*;
разводка *f* отопительной сети

D413 **distribution pipe line** *see* **distributing
pipe line**

D414 *e* **distribution reinforcement**
d Verteilungsbewehrung *f*
f armature *f* de répartition
nl verdeelwapening *f*
r распределительная арматура *f*
(*железобетона*)

D415 *e* **distribution reservoir**
d Ausgleichbehälter *m*
f réservoir *m* de compensation [de
l'eau potable]
nl reinwaterreservoir *n*;
reinwaterkelder *m*
r распределительный резервуар *m*;
резервуар *m* чистой воды

D416 **distribution steel** *see* **distribution
reinforcement**

D417 *e* **distribution system**
d 1. Bewässerungssystem *n*
(*innerbetriebliche Anlagen
ausgenommen*) 2. Versorgungssystem *n*,
Verteilungsnetz *n*
f 1. réseau *m* d'irrigation 2. réseau *m* de
distribution [de répartition]
nl 1. bevloeiingssysteem *n*
2. distributienet *n*
r оросительная система *f* (*от головного
сооружения до водопотребляющих
хозяйств*) 2. распределительная
система *f*

D418 *e* **distributor**
d 1. Springer *m* (*Abwasserwesen*)
2. Stromverteiler *m* 3. Speiser *m*,
Beschicker *m*, Aufgeber *m*
f 1. arroseur *m* (*du biofiltre*)
2. distributeur *m* de courant
3. alimenteur *m*; distributeur *m*
nl 1. sproeier *m* 2. stroomverdeler *m*
3. toevoerinrichting *f*
r 1. ороситель *m* (*биофильтра*) 2.
токораспределительное устройство *n*
3. питающее устройство *n*;
распределительное устройство *n*

D419 *e* **district**
d Stadtbezirk *m*
f région *f*, district *m*
nl stadswijk *f (m)*
r район *m*

D420 *e* **district boiler house**
d Bezirksheizzentrale *f*
f bâtiment *m* des chaudières régional,
chaufferie *f* régionale
nl wijkketelhuis *n*
r районная котельная *f*

D421 *e* district cooling
  d Fernkälteversorgung *f*, zentrale
    Kälteversorgung *f*
  f système *m* urbain de production et de
    transport de froid
  nl wijkkoelsysteem *n*
  r централизованное [районное]
    холодоснабжение *n*

D422 *e* district heating
  d Fernwärmeversorgung *f*, Fernheizung
    f, Bezirks-Fernheizung *f*
  f chauffage *m* urbain
  nl stadsverwarming *f*
  r районное теплоснабжение *n*

D423 district heat and cold supply, district
    heating and cooling *see* pipe line
    heating and refrigerating

D424 *e* district road
  d Kreisstraße *f*
  f route *f* régionale [communale]
  nl gemeenteweg *m*
  r дорога *f* областного значения

D425 *e* disturbance of equilibrium
  d Gleichgewichtsstörung *f*
  f déséquilibrage *m*, déséquilibre *m*,
    rupture *f* d'équilibre
  nl verstoring *f* van het evenwicht
  r нарушение *f* равновесия

D426 *e* ditch
  d Graben *m*; Rigole *f*, Rinne *f*
  f rigole *f*; fossé *m*
  nl sloot *f* (*m*)
  r канава *f*; траншея; кювет *m*

D427 *e* ditch drainage
  d Grabenentwässerung *f*
  f drainage *m* par tranchées drainantes
    [par fossés ouverts]
  nl open ontwatering *f*
  r траншейный [открытый] дренаж *m*

D428 *e* ditcher
  d Grabenbagger *m*
  f trancheuse *f*, excavateur *m* de
    tranchées, trancheur *m*
  nl slotengraafmachine *f*
  r канавокопатель *m*; траншеекопатель
    m

D429 *e* ditcher plough
  d Grabenpflug *m*
  f charrue *f* fossoyeuse
  nl greppelploeg *m*
  r плужный канавокопатель *m*

D430 ditching shovel *see* ditcher

D431 ditch system *see* open drainage system

D432 *e* diurnal temperature range
  d tägliche Temperaturschwankungen *f pl*
  f variations *f pl* [fluctuations *f pl*]
    de température journalières
  nl dagelijkse gang *m* van de temperatuur
  r суточные колебания *n pl* температуры

D433 *e* diurnal variation
  d Tagesschwankung *f*, tägliche
    Variation *f*
  f variation *f* diurne [journalière]

nl veranderingen *f pl* per dag
  r суточное изменение *n*

D434 *e* diversion
  d 1. Ableitung *f*, Umleitung *f*,
    Derivation *f* 2. Ableitungsgerinne *n*
    3. Umgehungsstraße *f*,
    Umleitungsstraße *f*
  f 1. dérivation *f* des eaux 2. canal *m*
    de dérivation 3. déviation *f*, route *f*
    de déviation
  nl 1. omleiding *f*, aftakking *f* ter
    omleiding 2. omleidingskanaal *n*
    3. wegomleiding *f*
  r 1. отвод *m* воды, деривация *f*
    2. водоотводящий канал *m*
    3. обходная [объездная] дорога *f*

D435 *e* diversion channel
  d Wasserumleitung *f*, Umleitungskanal
    m, Ableitungsgerinne *n*
  f conduite *f* [canal *m*] de dérivation
  nl omleidingskanaal *n*
  r деривационный водовод *m* [канал *m*]

D437 *e* diversion dam
  d Umleit(ungs)sperre *f*, Umleitungswehr
    n, Entnahmesperre *f*, Fassungswehr *n*,
    Einlaßwehr *n*, Stauwehr *n*
  f barrage *m* de dérivation [de
    dénivellation]
  nl afsluitdam *m*
  r водозаборная *или* водоподъёмная
    плотина *f*

D438 diversion gallery *see* diversion tunnel

D439 division gate *see* water divider

D440 *e* diversion passage way
  d Ausbaudurchlaß *m*
  f pertuis *m* de dérivation
  nl afleidingsdoorlaat *m* (*bouw van een
    stuw*)
  r сооружение *n* для пропуска
    строительных расходов (воды)

D441 *e* diversion power plant
  d Umleitungswasserkraftwerk *n*
  f usine *f* de dérivation
  nl waterkrachtcentrale *f* in een aftakking
  r деривационная ГЭС *f*

D442 *e* diversion terrace
  d Kanalisationsterrasse *f*, Dränterrasse *f*
  f terrasse *f* de déviation
  nl afleidingsterras *n*, aflaatbanket *n*
  r водоотводящая терраса *f*

D443 *e* diversion tunnel
  d Umlaufstollen *m*, Umleitungsstollen *m*
  f tunnel *m* de dérivation
  nl omloopriool *n*
  r деривационный туннель *m*

D444 *e* diversion works
  d Umleitungsanlagen *f pl*
  f ouvrages *m pl* de dérivations
  nl omleidingswerken *n pl*
  r деривация *f* (*комплекс сооружений*)

D445 diversity factor *see* load diversity factor

D446 *e* diver's works
  d Taucherarbeiten *f pl*

f travaux *m pl* en plongée
nl onderwaterwerk *n*
r водолазные работы *f pl*

D447 *e* **diverter valve**
d Verteilventil *n*, Stromteiler *m*
f soupape *f* de séparation, diviseur *m* de débit
nl wisselklep *f (m)*
r разделяющий клапан *m*

D448 *e* **divide** *US*
d Wasserscheide *f*
f ligne *f* de partage des eaux
nl waterscheiding *f*
r водораздел *m*

D449 *e* **divided highway** *US*
d geteilte Straße *f*, Straße *f* mit Mittelstreifen
f route *f* à deux voies séparées
nl weg *m* met gescheiden rijbanen
r автомобильная дорога *f* с разделённой проезжей частью (*для встречного движения*)

D450 **divider** *see* **division wall**

D451 **dividing valve** *see* **diverter valve**

D452 *e* **diving bell**
d Taucherglocke *f*
f cloche *f* à plongeur
nl duikerklok *f (m)*
r водолазный колокол *m*

D453 *e* **division gate, divisor**
d Wasser(ver)teiler *m*, Bewässerungsschleuse *f*; Verteilerschütz *n*
f partiteur *m* d'eau; vanne *f* de bifurcation
nl inlaatsluis *f (m)*
r вододелитель *m*, шлюз--распределитель *m*; затвор *m* вододелителя

D454 *e* **division wall**
d Trennungsmauer *f*, Trennungswand *f*, Leitwand *f*
f mur *m* de séparation, paroi *f* de partage, guideau *m*
nl scheidingswand *m*, tussenmuur *m*
r разделительная [направляющая] стенка *f*

D455 **divisor** *see* **division gate**

D456 *e* **dock**
d 1. Hafenbecken *n* 2. Dock *n*
f 1. bassin *m* (*d'un port*) 2. dock *m*
nl 1. havenbassin *n* 2. dok *n*
r 1. портовый бассейн *m* 2. док *m*

D457 *e* **dock gate**
d Docktor *n*
f porte *f* de dock
nl sluisdeur *f(m)*
r ворота *pl* дока

D458 **docking impact** *see* **berthing impact**

D459 *e* **dockyard**
d Schiffswerft *f*
f chantier *m* naval [maritime]

nl scheepswerf *f (m)*
r судостроительная *или* судоремонтная верфь *f*

D460 *e* **dog**
d 1. Bauklammer *f* 2. Anschlag *m* 3. Scharriereisen *n*
f 1. clameau *m*, crampon *m* à deux pointes 2. butée *f*, butoir *m* 3. gradine *f*
nl 1. mastkram *f (m)*, kramplaat *f (m)* 2. aanslag *m* 3. ceseel *m*
r 1. строительная скоба *f* 2. упор *m* 3. троянка *f*, скарпель *f*

D461 *e* **dog iron**
d Befestigungsklammer *f*
f clameau *m*, crampon *m* à deux pointes
nl klamp *m, f*, klemhaak *m*
r скоба *f*

D462 *e* **dogs and chains**
d 1. zweisträngige Anschlagkette *f* mit Lasthaken (*für Steinaufnahme*) 2. Raubvorrichtung *f*, Raubgerät *n*
f 1. appareil *m* d'accrochage à deux chaînes avec crochets terminaux; élingue *f* double en chaîne avec crochets du bout (*pour levage des pièrres tailleés*) 2. appareil *m* pour déboisage
nl 1. kettingstrop *m* met klemhaken 2. duivelsklauw *m*, rooiklauw *m* met ketting
r 1. двухветвевой цепной строп *m* с крюками, захватные крючья *pl* с цепями (*для подъёма каменных блоков*) 2. механизм *m* для извлечения крепи

D463 *e* **dolly**
d 1. Rammjungfer *f*, Rammknecht *m* 2. Gegenhalter *m* (*Nieten*), Nietwippe *f* 3. Fahrgestell *n*
f 1. avant-pieu *m* 2. tas *m* à river *m* 3. chariot *m* à plate-forme basse
nl 1. heimuts *f (m)* 2. tas *m*, tegenhouder (*klinken*) 3. (lage) aanhanger *m*
r 1. подбабок *m* 2. держатель *m* заклёпки 3. тележка *f* с низкой платформой

D464 *e* **dolphin**
d Dalbe *f*, Dalben *m*; Pfahlbündel *n*; Dalbenpfahl *m*, Prallpfahl *m*, Abweisepfahl *m*
f duc *m* d'Albe, dispositif *m* d'amarrage; dophin *m*
nl dukdalf *m*
r швартовный пал *m*; куст *m* свай; отбойная свая *f*

D465 **dolphin berth** *see* **offshore mooring**

D466 *e* **dome**
d Kuppel *f*
f dôme *m*, coupole *f*
nl koepel *m*
r купол *m*

D467 *e* **dome light**
d Lichtkuppel *f*

f lanterneau *n* zénithal en coupole,
lanterneau-dôme *m*
nl lichtkoepel *m*
r зенитный фонарь *m* купольного типа

D468　dome shaped dam *see* cupola dam

D469 *e* dome shell
d Kuppelschale *f*
f voile *m* [coque *m*] en coupole
nl koepelschaal *f* (*m*)
r купольная оболочка *f*

D470 *e* domestic boiler
d Warmwasserbereiter *m*
f chaudière *f* à eau chaude, chauffe-eau
m
nl boiler *m*
r водогрейный котёл *m* бытового
назначения

D471 *e* domestic garbage
d Haus(halts)müll *m*
f ordures *f pl* ménagères
nl huisvuil *n*, vuilnis *n*, huishoudelijk
afval *n*
r бытовой мусор *m*

D472 *e* domestic gas installation
d Gebäudegasinstallation *f*
f installation *f* de gaz d'un immeuble
nl gasinstallatie *f* (*voor een woning*)
r внутридомовый газопровод *m*

D473 *e* domestic heating
d örtliche Beheizung *f*,
Wohnraumheizung *f*, Einzelheizung *f*
f chauffage *m* indépendant [individuel]
nl perceelverwarming *f*
r местное отопление *n*

D474 *e* domestic heating appliance
d lokale Heizungsanlage *f*;
Haushaltswasserheizer *m*
f installation *f* de chauffage individuel
nl verwarmingsinstallatie *f* van het
perceel
r квартирная отопительная *или*
водогрейная установка *f*

D475 *e* domestic hot water
d Gebrauchswarmwasser *n*
f eau *f* chaude de consommation
nl warmwater *n* (*huishoudelijk*)
r горячая вода *f* для бытовых нужд

D476 *e* domestic premises
d Wohnräume *m pl*
f locaux *m pl* d'habitation
nl woningen *f pl*
r жилые помещения *n pl*

D477 *e* domestic sewage
d Haushaltsabwasser *n*, Schmutzwasser
n
f eaux *f pl* usées domestiques
[ménagères]
nl huishoudelijk afvalwater *n*
r бытовые [хозяйственно-фекальные]
сточные воды *f pl*

D478 *e* domestic sewerage system
d Schmutzwasserkanalisation *f*,
Haushaltentwässerungsnetz *n*

f système *m* intérieur d'évacuation des
eaux ménagères
nl huisriolering *f*
r хозяйственно-фекальная бытовая
канализационная сеть *f*

D479 *e* domestic water supply
d Trinkwasserversorgung *f*,
Brauchwasserversorgung *f*
f alimentation *f* en eau potable
nl drinkwatervoorziening *f*
r хозяйственно-питьевое водоснабжение *n*

D480 *e* dominant formative discharge
d laufbildender Abfluß *m*
f débit *m* actif dominant
nl stroombedvormende afvoer *m*
r руслоформирующий сток *m*

D481 *e* donkey
d Doppelwinde *f*
f treuil *m* à deux tambours avec
manœuvre indépendante
nl differentiaalwindas *n*
r двухбарабанная лебёдка *f* с
независимым [раздельным]
управлением барабанами

D482　donut *see* doughnut

D483 *e* dook
d Holzdübel *m*
f taquet *m* de fixation [cheville *f*
tamponnée] en bois
nl houten deuvel *m*, timmermansdeuvel
m
r деревянная пробка *f* в стене (*для*
крепления облицовки)

D484 *e* door
d Tür *f*
f porte *f*
nl deur *f* (*m*)
r дверь *f*

D485　door and frame packaged unit *see* door
assembly 1.

D486 *e* door assembly
d 1. Türelement *n* 2. Ausfüllung *f* der
Türöffnung
f 1. unité *f* complète porte et
encadrement 2. remplissage *m* de
baie de porte
nl 1. kozijn *n* met deur 2. een deur
afhangen
r 1. дверной блок *m* 2. заполнение *n*
дверного проёма

D487 *e* door bolt
d Türbasküle *f*
f verrou *m* de porte
nl deurgrendel *m*
r дверной шпингалет *m*, дверная
задвижка *f*

D488 *e* door catch
d Türklinke *f*
f arrêt *m* de porte, loquet *m*
nl deurklink *f* (*m*)
r дверная щеколда *f*

D489 *e* **door closer**
  d Türschließer *m*
  f ferme-porte *m*
  nl deurensluiter *m*, deurdranger *m*
  r закрыватель *m* двери

D490 *e* **door frame**
  d Türfutter *n*, Türeinfassung *f*,
    Türzarge *f*
  f bâti *m* dormant (*d'une porte*),
    huisserie *f* de porte
  nl deurkozijn *n*
  r дверная коробка *f*

D491 *e* **door furniture**
  d Türbeschläge *m pl*
  f quincaillerie *f* de porte
  nl deurbeslag *n*
  r дверные приборы *m pl* и
    принадлежности *f pl*

D492 *e* **doorjamb**
  d Türpfosten *m*
  f montant *m* du bâti dormant [de
    l'huisserie]
  nl stijl *m* van het deurkozijn
  r вертикальный брус *m* дверной
    коробки, дверной косяк *m*

D493 *e* **door mullion**
  d Türzwischenpfosten *m*
  f meneau *m* [montant *m* central] de la
    porte à deux battants
  nl tussenstijl *m*
  r притворная стойка *f* (*двупольной
    раздельной двери*)

D494 *e* **door opening**
  d Türöffnung *f*; Türöffnungsweite *f*
  f baie *f* [ouverture *f*] de porte;
    largeur *f* d'ouverture de porte
  nl (wijdte *f* van de) deuropening *f*
  r дверной проём *m*; ширина *f* дверного
    проёма по дверной коробке

D495 *e* **door operator**
  d Türbetätiger *m*
  f mécanisme *m* automatique
    d'ouverture de la porte
  nl installatie *f* voor het bedienen van
    deuren
  r автоматический открыватель *m*
    двери

D496 *e* **door plate**
  d Türschild *n*
  f plaque *f* de porte
  nl deurplaat *f* (*m*) (*om het sleutelgat*)
  r дверная табличка *f*

D497 *e* **door rail**
  d Querholz *n*
  f traverse *f* du cadre de porte
  nl deurdorpel *m*, tussenregel *m*
  r горизонтальный брус *m* каркаса
    дверного полотна

D498 *e* **door schedule**
  d Türliste *f*
  f nomenclature *f* des portes
  nl deurenlijst *f* (*m*)
  r спецификация *f* дверей

D499 *e* **door sill**
  d Türschwelle *f*, Türbank *f*
  f seuil *m* de porte
  nl onderdorpel *m*
  r дверной порог *m*, нижний брус *m*
    дверной коробки

D500 *e* **door stile**
  d Türrahmenpfosten *m*
  f montant *m* de porte
  nl stijl *m* van het deurkozijn
  r вертикальный брус *m* каркаса
    дверного полотна

D501 *e* **doorstop, door-stop**
  d Türanschlag *m*
  f butoir *m* de porte
  nl stootdop *m*, deurvastzetter *m*
  r останов *m* двери

D502   **door unit** *see* **door assembly 1.**

D503   **doorway** *see* **door opening**

D506 *e* **dormant window, dormer window**
  d Dachgaupe *f*, Dachgaube *f*
  f lucarne *f*, chien-assis *m*
  nl dakvenster *n*, dakraam *n*
  r слуховое*,* *или* мансардное окно *n*

D507 *e* **dosing chamber, dosing tank**
  d Dosierbehälter *m*,
    Beschickungsbehälter *m*
  f bac *m* doseur [jaugeur], doseur *m*
  nl doseertank *m*
  r дозатор *m*, дозирующий бак *m*

D508 *e* **double-acting door**
  d Pendeltür *f*, Durchschlagtür *f*
  f porte *f* oscillante [va-et-vient]
  nl tochtdeur *f* (*m*)
  r качающаяся дверь *f*; дверь *f*,
    открывающаяся в обе стороны

D509 *e* **double-acting hydraulic jack**
  d doppelwirkende Spannpresse *f*,
    Doppelspannpresse *f*
  f verin *m* hydraulique à double effet
  nl dubbelwerkende hydraulische vijzel
    f (*m*)
  r гидравлический домкрат *m*
    двойного действия

D510 *e* **double-acting pile hammer**
  d Schnellschlagbär *m*, Oberdruck-
    Rammhammer *m*
  f marteau-trépideur *m* [moton *m* de
    sonnette] à double effet
  nl dubbelwerkend heiblok *n*
    (*stoom/diesel*)
  r свайный молот *m* двойного действия

D511 *e* **double angle**
  d Doppelwinkel *m*
  f deux cornières *f pl* dos à dos
  nl dubbelhoekprofiel *n*
  r парные уголки *m pl*

D513   **double box** *see* **double socket**

D514 *e* **double branch pipe**
  d Doppel-T-Stück *n*, Doppelabzweig *m*
  f pièce *f* en croix, embranchement *m*
    double

nl dubbel-T-stuk *n*
r крестовина *f* (*трубопровода*)

D515 **double crossover** *see* **scissors junction**

D516 *e* **double-deck bridge**
d Etagenbrücke *f*, zweistöckige Brücke *f*
f pont *m* à deux niveaux [à deux étages]
nl dubbeldekbrug *f* (*m*)
r двухъярусный мост *m*

D5 7 *e* **double deck screen**
d Zweidecker *m*, Doppeldecksiebmaschine *f*
f crible *m* à deux étages [à deux tamis]
nl zeef *f* (*m*) met dubbele bodem
r двухдечный грохот *m*

D518 *e* **double deck settling basin**
d Emscherbrunnen *m*, Imhoffbrunnen *m*, zweistöckiges Absetzbecken *n*
f décanteur *m* deux étages
nl bezinkvijver *m* met twee verdiepingen
r двухъярусный отстойник *m*

D519 *e* **double-deflection grille**
d Horizontal- und Vertikal-Luftverteilgitter *n*
f grillage *m* de déviation double
nl dubbel luchtverdeelrooster *n*
r двухрядная воздухораспределительная решётка *f* (*с регулированием направления струи в двух плоскостях*)

D520 *e* **double door**
d Doppeltür *f*
f porte *f* double
nl dubbele deur *f* (*m*)
r двупольная дверь *f*

D521 *e* **double-dovetail key**
d Schwalbenschwanzdübel *m*
f clavette *f* à queue d'aronde double
nl dubbele zwaluwstaart *m* (*langshoutverbinding*)
r деревянная шпонка *f* с концами в форме ласточкиного хвоста

D522 *e* **double-drum hoist**
d Doppeltrommelwinde *f*
f treuil *m* à deux tambours
nl differentiaalwindas *f* (*m*)
r двухбарабанная лебёдка *f*

D523 *e* **double-end wrench**
d Doppelschlüssel *m*
f clef *f* [clé *f*] double
nl dubbele moersleutel *m*
r двусторонний гаечный ключ *m*

D524 *e* **double flooring**
d zweischichtiger Fußbodenbelag *m*, Doppelfußbodenbelag *m*
f revêtement *m* de sol double, couvre-sol *m* double
nl vloerbedekking *f* in twee lagen
r двухслойное покрытие *n* пола

D525 *e* **double glazing**
d doppelte Verglasung *f*, Doppelverglasung *f*
f vitrage *m* double

nl dubbele verglazing *f*
r двойное остекление *n*

D526 *e* **double-headed nail**
d Duplexnagel *m*
f clou *m* à deux têtes (*une au dessous de l'autre*)
nl duplexnagel *m* (*bekisting*)
r опалубочный гвоздь *m* с двумя головками (*одна — ниже другой*)

D527 *e* **double house**
d Doppelhaus *n*
f maison *f* double [jumelle]
nl dubbel huis *n*; twee-onder-een-kap
r два дома с общей стеной

D528 *e* **double inlet fan**
d zweiseitig [ansaugender Radiallüfter *m*
f ventilateur *m* à admission double
nl ventilator *m* met tweezijdige aanzuiging
r радиальный вентилятор *m* двустороннего всасывания

D529 *e* **double-intersecting truss**
d Binder *m* mit kreuzweiser Verstrebung
f poutre *f* à treillis en X sans montants
nl ligger *m* [spant *n*] met kruisdiagonalen
r ферма *f* с перекрёстной решёткой

D530 *e* **double-lane lock**
d Zweikammerschleuse *f*, Zwillingsschleuse *f*
f écluse *f* double
nl tweelingssluis *f* (*m*)
r двухниточный [парный] шлюз *m*

D531 *e* **double-leaf bascule bridge**
d zweiarmige Drehklappbrücke *f* [Drehzugbrücke *f*]
f pont *m* basculant double
nl dubbele basculebrug *f* (*m*)
r двукрылый раскрывающийся разводной мост *m*

D532 *e* **double-leaf gate**
d 1. Stemmtorpaar *n*, Stemmtore *n pl*, zweiflügeliges Schleusentor *n* 2. Doppelschütze *f*, Doppelverschluß *m*
f 1. porte *f* de clôture à deux battants 2. vanne *f* à deux corps, vanne *f* double
nl 1. puntdeuren *f* (*m*) *pl* 2. toldeur *f* (*m*) (*sluis*), dubbele schuif *f* (*m*)
r 1. двустворчатые ворота *pl* 2. сдвоенный затвор *m*

D533 *e* **double lock**
d 1. Zweikammerschleuse *f*, Zwillingsschleuse *f* 2. zweitouriges Schluß *n*
f 1. écluse *f* double 2. serrure *f* à deux tours [à double tour]
nl 1. tweelingssluis *f* (*m*) 2. dag- en nachtslot *n*
r 1. двухниточный [парный] шлюз *m* 2. двойной замок *m* (*с двумя позоротами ключа*)

D534 e **double-lock seam, double-lock welt**
    d liegender Doppelfalz *m*
    f double agrafure *f*
    nl dubbele vouwnaad *m*
    r двойной лежачий фальц *m*

D535 e **double pipe condenser**
    d Doppelrohrkondensator *m*
    f condensateur *m* à double tube
    nl tweepijpscondensator *m*
    r двухтрубный конденсатор *m* типа «труба в трубе»

D536 **double pipe heat exchanger** *US see* **pipe-in-pipe**

D537 e **double-pipe heat-supply system**
    d Zweileitersystem *n* der Wärmeversorgung
    f réseau *m* thermique à deux conduits, système *m* de chauffage urbain à deux conduits
    nl centrale verwarming *f* met aparte retourleiding
    r двухтрубная система *f* теплоснабжения

D538 **double pitch roof** *see* **gable roof**

D539 e **double-pole scaffold**
    d Stangengerüst *n* mit zwei Gerüstbaumreihen
    f échafaudage *m* de pied fixe indépendant [à deux rangées de montants]
    nl dubbele bouwsteiger *m*
    r коренные леса *pl* с двумя рядами стоек

D540 e **double post roof truss**
    d doppelt stehender Dachstuhl *m*
    f ferme *f* à deux poinçons
    nl kapspant *n* op twee kolommen
    r двустоечная висячая стропильная ферма *f*

D541 e **double roof**
    d Doppeldach *n*
    f comble *m* [toit *m*] double
    nl gordingdak *n*, gebroken dak *n*
    r крыша *f* с вспомогательными стропилами, уложенными на прогоны

D542 e **double-sheave pulley block**
    d Zweirollenblock *m*
    f moufle *f* à deux poulies
    nl tweeschijfsblok *m*
    r блочная обойма *f* с двумя шкивами

D543 e **double sheetpiling**
    d Doppelspundwand *f*
    f rideau *m* de palplanches double
    nl dubbele damwand *m*
    r двойной шпунтовый ряд *m*

D544 e **double-size pantile**
    d holländische Pfanne *f*, Pfannendachziegel *m*, doppelter Hohlziegel *m*
    f tuile *f* flamande [(cuite) galbée]
    nl Hollandse (dak)pan *f* (*m*)
    r голландская черепица *f*

D545 **double sling** *see* **two-leg sling**

D546 e **double socket**
    d Doppelmuffe *f*
    f manchon *m* à double emboîture [emboîtement]
    nl dubbele sok *f* (*m*)
    r двойной раструб *m*

D547 e **double stack system**
    d Doppelstrangsystem *n*
    f réseau *m* d'évacuation des eaux ménagères (*à l'intérieur du bâtiment*) à chutes et descentes
    nl binnenriolering *f* met twee standleidingen
    r внутренняя канализация *f* с двумя раздельными стояками

D548 e **double standing seam**
    d Doppelstehfalz *m*
    f joint *m* debout double
    nl dubbele staande vouwnaad *m*
    r двойной стоячий фальц *m*

D549 e **double T-beam**
    d Stahlbetondoppelplattenbalken *m*, Doppel-T-Träger *m*, zweistegiger Plattenbalken *m*
    f poutre *f* en T double en béton armé préfabriqué; panneau *m* de mur en T double en béton armé préfabriqué
    nl dubbele-T-balk *m*, H-balk *m*
    r сборная железобетонная балка *f* в форме двойного Т; сборная железобетонная стеновая панель *f* Ш-образного сечения

D550 e **double tenons**
    d Doppelzapfen *m*
    f tenon *m* double
    nl dubbele pen *f* (*m*)
    r двойной шип *m*

D551 e **double-T-fitting** *see* **double branch pipe**

D552 e **double-wall cofferdam**
    d doppelwandiger Fang(e)damm *m*
    f batardeau *m* à deux parois
    nl cofferdam *m*
    r двухрядная перемычка *f*

D553 **double width fan** *see* **double inle fan**

D554 e **double window**
    d Doppelfenster *n*
    f fenêtre *f* double
    nl dubbel venster *n*
    r двойное окно *n*, окно *n* с двумя переплётами

D555 e **doubly prestressed concrete**
    d doppelt vorgespannter [längs- und quervorgespannter] Stahlbeton *m*
    f béton *m* à double précontrainte
    nl dubbel voorgespannen beton *n*
    r бетон *m*, напряжённый в двух (взаимно перпендикулярных) направлениях

D556 e **doubly reinforced concrete**
    d doppelt bewehrter Stahlbeton *m*

f béton *m* armé à barres d'armature
tendues et comprimées
nl dubbel gewapend beton *n*
r железобетон *m* с растянутой и
сжатой арматурой

D557 *e* **doubly symmetrical section**
d zweifach axialsymmetrischer
Querschnitt *m*
f section *f* transversale à deux axes de
symétrie
nl tweezijdig symmetrische doorsnede *f*
r сечение *n* с двумя осями симметрии

D558 *e* **doughnut**
d 1. große Scheibe *f*
2. Unterlegklotz *m* (*aus Beton*)
f 1. rondelle *f* grosse 2. écarteur *m*
[distanceur *m*, séparateur *m*] rond en
béton
nl 1. ring *m*, onderlegplaat *f* (*m*) 2.
afstandhouder *m* voor betonwapening
r 1. шайба *f* крупного размера
2. круглый бетонный фиксатор *m*
арматуры

D559 *e* **dovetail**
d 1. Schwalbenschwanz *m*,
schwalbenschwanzförmige Zinke *f*
2. schwalbenschwanzförmige
Überblattung *f*
f 1. tenon *m* à queue d'aronde
2. assemblage *m* à queue d'aronde,
entaille *f* à queue d'aronde
nl 1. zwaluwstaart *m*
2. zwaluwstaartverbinding *f*
r 1. шип *m* в форме ласточкиного
хвоста 2. врубка *f* [соединение *n*]
с применением шипа в форме
ласточкиного хвоста

D560 **dovetail dowel** *see* **double-dovetail key**
D561 *e* **dowel**
d 1. auskragende Stahleinlage *f*,
Anschlußstab *m* 2. Stift *m*, Bolzen *m*
3. Dübel *m*
f 1. extrémité *f* d'armature en saillie
2. goujon *m*, cheville *f*, goupille *f*
3. pinoche *f*
nl 1. stekeind *n* (*wapening*) 2. stift *f* (*m*),
bout *m* (*klink*) 3. muurplug *f* (*m*)
r 1. арматурный выпуск *m*; стыковая
арматура *f* 2. штырь *m* 3. дюбель
m

D562 **dowel-bar reinforcement** *see* **dowel 1.**
D563 *e* **dowel driver**
d Dübeltreibgerät *n*, Dübelschießgerät *n*
f pistolet *m* de scellement pour goujons
nl boutenschiethamer *m*
r монтажный пистолет *m*

D564 *e* **dowel lubricant**
d Dübelschmiere *f*
f graisse *f* pour goujons (*encastrés dans
les dalles routières*)
nl deuwelsmering *f* (*betonweg*)
r смазка *f* для штырей дорожных плит
(*для предупреждения сцепления
одного конца с бетоном*)

D565 **downcomer** *see* **downpipe**
D566 *e* **down draught**
d 1. Abwärtszug *m*
2. Abwärtsströmung *f*
f 1. tirage *m* renversé 2. tirage *m*
vers le bas
nl 1. naar beneden gerichte
(lucht)trek *m* 2. neerwaartse
luchtstroom *m*
r 1. обратная тяга *f* 2. нисходящий
конвективный поток *m* воздуха

D567 *e* **down-feed system**
d Dampfheizungssystem *n* mit oberer
Verteilung
f installation *f* de chauffage à
distribution haute
nl verwarmingssysteem *n* met
bovenverdeling
r система *f* парового отопления с
верхней разводкой

D568 *e* **downpipe**
d 1. Dachabfallrohr *n*,
Regenablaufrohr *n* 2. Fallrohr *n*,
Falleitung *f*
f 1. tuyau *m* de chute 2. descente *f*,
tuyau *m* de descente
nl 1. hemelwaterpijp *f* (*m*) 2. valpijp *f* (*m*)
r 1. водосточная труба *f* 2. отводящий
стояк *m*

D569 *e* **down service**
d Gravitationsleitung *f*,
Freispiegelleitung *f*
f canalisation *f* d'eau à écoulement
libre
nl gravitatie(water)leiding *f*
r самотёчный водопровод *m*

D570 *e* **downspout** *see* **down pipe 1.**
D571 *e* **downstand beam**
d Unterzug *m*, Unterzugsbalken *m*
f poutre *f* de plancher nervuré
nl moerbalk *m*, onderslagbalk *m*
r балка *f* ребристого перекрытия

D572 *e* **downstream apron**
d Rißberme *f*, Sturzbett *n*, Kolkschutz
m, Schwellen-Verlängerung *f*
f arrière-radier *m*, tapis *m* de
protection
nl stortebed *n*
r рисберма *f*

D573 *e* **downstream control**
d unterwasserseitige Regelung *f*
f régulation *f* par aval
nl benedenpandse regulering *f*
r регулирование *n* по нижнему бьефу

D574 *e* **downstream drains**
d Flachdränung *f*, Filterbett *n*,
Dränpackwerk *n*;
Erdstaumauerentwässerung *f*
f tapis *m* de drainage; drainage *m* du
corps *ou* de la base du barrage en
terre
nl filterbed *n*; drainage *f* van een
damlichaam

r плоский дренаж *m*, дренажный
тюфяк *m*; дренаж *m* тела и
основания земляной плотины

D575 *e* **downstream face**
d Talseite *f*, Luftseite *f*, drucklose
Seite *f*
f parement *m* aval
nl benedenstroomse zijde *f*, front *n*
r низовая грань *f*

D576 *e* **downstream fishway**
d Fischdurchlaß *m*
f passe *f* à poissons
nl vispassage *f*
r рыбоспуск *m*

D577 **downstream floor** *see* **spillway apron**

D578 *e* **downstream level**
d Unterwasserspiegelhöhe *f*,
Unterwasserstand *m*
f niveau *m* aval
nl benedenstrooms waterpeil *n*
r уровень *m* нижнего бьефа

D579 *e* **downstream reach**
d Unterwasserhaltung *f*, Unterwasser *n*
f bief *m* d'aval
nl benedenpand *n*
r нижний бьеф *m*

D580 *e* **downstream slope**
d luftseitige Böschung *f*
f talus *m* [paroi *f*] aval
nl srtoomafwaarts gelegen aftakking *f*
r низовой откос *m*

D582 *e* **downward ventilation**
d Luftführung *f* von oben nach unten
f ventilation *f* du type «de haut en
bas»
nl ventilatie *f* met dalende luchtstroom
r вентиляция *f* по схеме «сверху вниз»

D583 **dozer** *see* **bulldozer**

D584 *e* **dozer shovel**
d Bulldozer-Schaufellader *m*,
Bulldozer-Ladeschaufel *f*
f bull(dozer)-chargeur *m*, tracto-pelle *f*
nl trekker *m* met laadschop
r трактор *m* с фронтальным погрузоч-
ным ковшом, трактор-
-погрузчик *m*

D585 **draft** *US see* **draught**

D586 *e* **draft check damper**
d Zugregler *m*
f registre *m* de tirage
nl trekregelaar *m*
r регулятор *m* тяги

D587 *e* **drafting machine**
d Zeichenmaschine *f*
f machine *f* à dessiner
nl tekenmachine *f*
r кульман *m*

D588 *e* **draft tube** *US*
d Saugrohr *n*, Auslaufrohr *n*,
Saugkrümmer *m*
f tube *m* d'aspiration
nl zuigbuis *f* (*m*)
r отсасывающая труба *f*

D589 *e* **drag**
d 1. Planierschleppe *f*, Abziehschleppe
f 2. Kratzeisen *n* 3. Naßbagger *m*,
Schwimmbagger *m*
4. aerodynamischer Widerstand *m*,
Stirnwiderstand *m*, Kopfwiderstand
m, Luftwiderstand *m*
f 1. rabot *m* tracté 2. racloir *m*
[grattoir *m*] denté, raclette *f* dentée
3. drague *f*, draguer *m* 4. traînée *f*,
résistance *f* frontale (*aérodinamique*)
nl 1. sleepbord *n*, sleepplank *f* (*m*)
2. tandijzer *n* (*pleisterwerk*)
3. sleepzuiger *m*, hopperzuiger *m*
4. luchtweerstand *m*
r 1. дорожный утюг *m*, волокуша *f*,
гладилка *f* 2. зубчатый скребок *m*
(*штукатура*) 3. землесосный снаряд
m, земснаряд *m*; драга *f* 4. аэро-
динамическое сопротивление *n*

D590 *e* **drag cable**
d Zugseil *n*, Förderseil *n*
f câble *m* tracteur [de traction]
nl sleepkabel *m*, trekkabel *m*
r тяговый канат *m*

D591 *e* **drag head**
d Saugkopf *m*, Schleppkopf *m*,
Naßbaggerkopf *m*
f tête *f* dragueuse
nl zuigkop *m*
r грунтозаборное устройство *n*
(*землесосного снаряда*)

D592 *e* **dragline, drag-line**
d Kabelbagger *m*, Schleppschaufel *f*,
Zugseilbagger *m*
f dragline *f*
nl dragline *f* (*m*)
r драглайн *m*

D593 *e* **dragon's tooth**
d Drachenzahn *m*, Schikane *f*,
Bremspfeiler *m*
f dent *f* de tranquillisation, pile *f* de
freinage
nl valbreker *m*
r пирсовый [шашечный] гаситель *m*,
водобойный зуб *m*

D595 *e* **drag spreader, drag spreader box**
d Kastenverteiler *m*, Verteilkasten *m*
f spreader *m* [spreader box *m*] tracté,
épandeuse *f* [gravillonneuse *f*] tractée
nl trekkast *f* (*m*)
r прицепной ящичный
распределитель *m* высевок [щебня]

D596 *e* **drag suction head**
d Saugschleppkopf *m*, Schleppsaugkopf
m
f tête *f* dragueuse suceuse à
désagrégateur [à couteaux]
nl sleepzuigkop *m*
r всасывающая головка *f* с рыхлителем
(*земснаряда*)

D597 *e* **drain**
d 1. Drän *m*; Dränrohr *n*,
Entwässerungsrohr *n* 2. Dränage *f*,

Entwässerung *f*
3. Kanalisationsanschluß *m*,
Hausanschlußabwasserleitung *f*
f 1. drain *m*, tuyau *m* drainant
[draineur, de drainage] 2. drainage *m*
3. branchement *m* d'égout
nl 1. draineerbuis *f* (*m*) 2. ontwatering *f*
3. afvoerleiding *f*
r 1. дрена *f*; дренажная труба *f*
2. сток *m*, дренаж *m*
3. соединительная ветка *f*, домовый
выпуск *m* канализации

D598 *e* drainage
d 1. Dränage *f*, Entwässerung *f*,
Trockenlegung *f* 2. Kanalisation *f*,
Abwasserbeseitigung *f*
f 1. assèchement *m*, drainage *m*
2. assainissement *m*
nl afwatering *f*, drainage *f*
r 1. дренаж *m*, осушение *n*
2. канализация *f*

D599 *e* drainage area
d 1. Entwässerungsfläche *f*;
Abflußgebiet *n*, Einzugsgebiet *n*
2. dränierte Fläche *f*
f 1. aire *f* du bassin hydrographique
[versant] 2. surface *f* de drainage
nl 1. stroomgebied *n* 2. gedraineerd
areaal *n*
r 1. площадь *f* водосбора; водосбор *m*
2. дренированная площадь *f*

D600 *e* drainage basin
d Abflußgebiet *n*, Einzugsgebiet *n*,
Sammelbecken *n*
f bassin *m* hydrographique [versant]
nl stroomgebied *n*, waterwingebied *n*
r водосборный бассейн *m*, водосбор *m*

D601 *e* drainage blanket
d Filterschicht *f*, Drän(ungs)schicht *f*,
Entwässerungsschicht *f*
f couche *f* filtrante, tapis *m* de drainage
nl drainerende laag *f* (*m*)
r дренирующий слой *m*

D602 *e* drainage canal
d Entwässerungskanal *m*
f canal *m* de dessèchement
nl afwateringskanaal *n*
r осушительный канал *m*

D603 drainage coefficient *see* drainage
modulus

D605 *e* drainage ditch
d Abflußgraben *m*, Drängraben *m*,
Sickergraben *m*, Entwässerungsgraben
m, Abstichgraben *m*
f fossé-drain *m*, fossé *m* filtrant
[d'écoulement]
nl afwateringssloot *f* (*m*)
r дренажная [водоотводная] канава *f*

D606 *e* drainage drop
d Gefälleabsturz *m*, Dränabsturz *m*,
Gefällestufe *f*
f chute *f* du réseau de drainage
nl val *m* in het afvloeiingsniveau
r перепад *m* дренажной сети

D607 *e* drainage facilities
d Sielnetz *n*
f réseau *m* de drainage et d'écoulement
nl afwateringsstelsel *n*
r водосточно-дренажная сеть *f*
[система *f*]

D608 *e* drainage fill
d zusätzliche Dränageschicht *f*
(*Straßenbefestigung*);
Entwässerungsteppich *m* (*unter dem
Betonfußboden*)
f couche *f* filtrante [de drainage]
(*sous-couche de chaussée ou de dallage*)
nl (opgebrachte) ontwateringslaag *f* (*m*)
r дополнительный дренирующий слой
m (*дорожной одежды*); дренажная
прослойка *f* (*под бетонным полом*)

D609 *e* drainage flow
d Dränabfluß *m*, Entwässerungsabfluß *m*
f débit *m* [écoulement *m*] de percolation
nl afwatering *f*
r дренажный сток *m*

D610 *e* drainage gallery
d Entwässerungsstollen *m*, Sammler *m*,
Dränagegang *m*, Sickergalerie *f*
f galerie *f* de drainage, collecteur *m*
nl verzamelriool *n*
r дренажная галерея *f*

D611 *e* drainage header
d Binnenvorfluter *m*; Sammeldrän *m*,
Sammler *m*, Auffangdrän *m*
f ouvrage *m* horizontal d'admission
d'eau de nappe dans un drain;
collecteur *m*
nl verzameldraineerleiding *f*
r канал-водоприёмник *m* на
обвалованной территории;
(закрытый) собиратель *m*

D612 *e* drainage layer
d Dränschicht *f*, Entwässerungsschicht *f*
f tapis *m* de drainage, couche *f*
drainante
nl ontwateringslaag *f* (*m*)
r пластовая дрена *f*

D613 *e* drainage main
d Hauptentwässerungskanal *m*
f canal *m* de dessèchement principal
nl hoofdafwateringskanaal *n*
r магистральный осушительный канал
m

D614 *e* drainage mattress
d Dränpackwerk *n*, Filterbett *n*
f tapis *m* de drainage
nl draineermat *f* (*m*)
r дренажный тюфяк *m*

D615 *e* drainage modulus
d Abflußmodul *m*
f débit *m* spécifique de drainage
nl (dagelijks) afwateringsdebiet *n*
r модуль *m* дренажного стока

D616 *e* drainage network
d 1. Entwässerungsnetz *n*, Dränsystem
n 2. Flußnetz *n*, Gewässernetz *n*

f 1. réseau *m* d'assèchement [d'assainissement, d'écoulement, de drainage] 2. réseau *m* fluvial [hydrographique]
nl 1. afwateringsstelsel *n*
2. rivierstelsel *n*
r 1. осушительная [дренажная] сеть *f*
2. речная [гидрографическая] сеть *f*

D617 *e* **drainage outlet**
d Dränageöffnung *f*, Abflußöffnung *f*
f débouché *m* du réseau de drainage
nl wateruitlaat *m*
r дренажный выпуск *m*, водоотводное отверстие *n*

D618 *e* **drainage path**
d Entwässerungsweg *m*
f parcours *m* de drainage
nl afwateringstracé *n*
r трасса *f* дренажа

D619 *e* **drainage pit**
d Sickergrube *f*, Abflußgrube *f*, Entwässerungsschacht *m*
f puisard *m*
nl afwateringsput *m*
r дренажный приямок *m*

D620 *e* **drainage pumping station**
d Schöpfwerk *n*, Entwässerungspumpstation *f*
f station *f* d'assèchement par pompage
nl gemaal *n*
r осушительная насосная станция *f*

D621 *e* **drainage rate**
d Entwässerungsnorm *f*, Grundwasserabsenkungsnorm *f*
f taux *m* d'assèchement
nl afwateringsnorm *f* (*m*)
r норма *f* осушения

D622 **[drainage shaft** *see* **drainage well**

D623 *e* **drainage structures**
d Entwässerungsbauwerke *n pl*
f ouvrages *m pl* de drainage
nl afwateringskunstwerken *n pl*
r дренажные сооружения *n pl*

D624 *e* **drainage system**
d Entwässerungssystem *n*, Dränagesystem *n*, Dränanlage *f*
f système *m* d'assèchement [de drainage]
nl afwateringssysteem *n*
r дренажная [осушительная] система *f*

D625 **drainage trench** *see* **drainage ditch**

D626 *e* **drainage valve**
d Entleerungsventil *n*, Ablaßventil *n*
f soupape *f* de vidange [d'évacuation]
nl aftapklep *f* (*m*)
r спускной клапан *m*

D627 *e* **drainage well**
d Schluckbrunnen *m*; Nadelfilter *n*; Entwässerungsschacht *m*
f puits *m* absorbant; puits *m* de drainage
nl zinkput *m*; draineerput *m*

r дренажный колодец *m*; иглофильтр *m*; дренажная шахта *f*

D628 *e* **drainage works**
d 1. Entwässerungsbauwerke *n pl*
2. Dränagearbeiten *f pl*, Entwässerungsarbeiten *f pl*, Trockenlegungsarbeiten *f pl*
f 1. ouvrages *m pl* de drainage
2. travaux *m pl* de drainage [d'assèchement]
nl 1. afwateringswerken *n pl*
2. droogleggingswerken *n pl*
r 1. дренажные [осушительные] сооружения *n pl* 2. дренажные [осушительные] работы *f pl*

D629 *e* **drain blockage**
d Dränverstopfung *f*
f blocage *m* du drain
nl drainverstopping *f*
r закупорка *f* дрены

D630 *e* **drain cleaner**
d Dränspülmaschine *f*
f cureuse *f* de drain
nl spoelapparaat *n* voor draineerleidingen
r дреноочиститель *m*

D631 *e* **drain depth**
d Dräntiefe *f*
f profondeur *f* du drain
nl draineerdiepte *f*
r глубина *f* заложения дрены

D632 *e* **drained area**
d trockengelegte [entwässerte] Fläche *f*
f surface *f* drainée
nl gedraineerd gebied *n*
r осушенная площадь *f*

D633 *e* **drained ground base**
d dränierte Gründung *f*
f terrain *m* de fondation drainant
nl gedraineerde fundering *f*
r дренируемое грунтовое основание *n*

D634 *e* **drained shear test**
d entwässerter Scherversuch *m*, Langsam-Versuch *m*
f éssai *m* de cisaillement drainé
nl schuifproef *f* (*m*) van een monster gedraineerde grond
r испытание *n* на сопротивление сдвигу дренируемого образца (*грунта*)

D635 *e* **drain filter beds**
d Bodenfilter *n pl* mit Dränage
f lits *m pl* filtrants avec drains
nl bodemfilters *n, m pl* met draineerbuizen
r поля *n pl* фильтрации с дренами

D636 *e* **drain head**
d Dränkopf *n*, Dräneinlauf *m*, Dräneinlaß *m*
f extrémité *f* amont d'un drain, tête *f* du drain
nl draininlaat *m*
r голова *f* дрены

D637 *e* **drain layer**
 d Dränmaschine *f*
 f poseur *m* de tuyaux drainants
 nl draineer(buizenleg)machine *f*
 r дреноукладчик *m*

D638 *e* **drain mouth**
 d Dränausmündung *f*, Dränauslauf *m*
 f bouche *f* de drainage, embouchure *f*
 nl drainuitmonding *f*
 r устье *n* дрены

D639 *e* **drain pipe**
 d Dränrohr *n*, Entwässerungsrohr *n*
 f tuyau *m* de drainage [de dessèchement]
 nl draineerbuis *f* (*m*), afvoerbuis *f* (*m*)
 r дренажная труба *f*

D640 *e* **drain pipeline**
 d Sickerrohrleitung *f*
 f ligne *f* [conduite *f*] de drainage
 nl draineer(pijp)leiding *f*
 r дренажная линия *f*

D641 *e* **drain pocket**
 d Schlammsammler *m*, Schlammfang *m*, Schmutzfänger *m*
 f puisard *m*
 nl slijkvanger *m*
 r грязеуловитель *m*, грязевик *m*

D642 *e* **drain silting**
 d Dränverschlammung *f*
 f envasement *m* du drain
 nl slibafzetting *f* op draineerbuizen
 r заиливание *n* дрены

D644 *e* **drain tile** *US*
 d Dränrohr *n*, Tonrohr *n*
 f tuyau *m* de drainage en terre cuite
 nl keramische draineerbuis *f* (*m*)
 r осушительная дрена *f*; гончарная дренажная труба *f*

D645 **drain trench** *see* **drainage ditch**
D646 *e* **drain valve**
 d Entleerungsventil *n*, Ablaßventil *n*
 f soupape *f* de vidange, robinet *m* à décharge
 nl aftapklep *f* (*m*)
 r дренажный клапан *m*

D647 **drain well** *see* **drainage well**
D648 **draped tendon** *see* **deflected tendon**
D649 *e* **draught** *UK*
 d Luftzug *m*, Zugwirkung *f*
 f écoulement *m* [courant *m*] d'air
 nl trek *m*, tocht *m*
 r тяга *f* (*в трубе*)

D650 *e* **draught gauge**
 d Zugmesser *m*
 f indicateur *m* de tirage
 nl trekmeter *m*
 r тягомер *m*

D651 *e* **draught stabilizer**
 d Zugstabilisator *m*, Zugregler *m*
 f stabilisateur *m* [régulateur *m*] de tirage

 nl trekregelaar *m*
 r стабилизатор *m* [регулятор *m*] тяги

D652 **draught tube** *UK see* **draft tube**
D654 *e* **drawbar pull**
 d Zughaken-PS *f*, Zugkraft *f* am Zughaken
 f effort *m* de traction à la barre
 nl trekkracht *f* (*m*) (aan de trekhaak)
 r тяговое усилие *n* на крюке

D655 *e* **draw-door weir**
 d Schützenwehr *n*
 f barrage-déversoir *m* à hausses
 nl schuivenstuw *m*
 r щитовая плотина *f*, водосливная плотина *f* с подъёмными щитовыми затворами

D656 *e* **draw down, drawdown**
 d Grundwasserabsenkung *f*; Spiegel(ab)senkung *f*; Entnahme *f*
 f rabattement *m* (*de la nappe aquifère*); hauteur *f* de rabattement; abaissement *m* du niveau des eaux
 nl grondwaterdaling *f*
 r депрессия *f* подземных вод; понижение *n* уровня воды; сработка *f* водохранилища

D657 **draw-down cone** *see* **cone of depression**
D658 *e* **drawdown curve**
 d Absenkungskurve *f*, Senkungskurve *f*
 f courbe *f* de dépression, courbe *f* d'abaissement
 nl curve *f* van de verlaging van het grondwaterpeil
 r кривая *f* депрессии (*подземных вод*); кривая *f* спада (*водотока, водоёма*)

D659 *e* **draw-down level**
 d Absenkziel *n*
 f baisse *f* de niveau
 nl gewenst grondwaterpeil *n*
 r горизонт *m* сработки водохранилища

D660 *e* **drawing**
 d Zeichnung *f*
 f dessin *m*, plan *m*
 nl tekening *f*
 r чертёж *m*

D661 *e* **drawing not to scale**
 d nicht maßstäbliche Zeichnung *f*
 f dessin *m* non à l'échelle
 nl tekening *f* niet op schaal
 r чертёж *m* не в масштабе

D662 *e* **draw-off**
 d 1. Entnahme *f*, Spiegelabsenkung *f* 2. Wasserfassung *f*
 f 1. prélèvement *m* d'eau 2. prise *f* d'eau
 nl 1. peilverlaging *f* 2. wateronttrekking *f*
 r 1. сработка *f* водохранилища 2. водозабор *m*

D663 *e* **draw-off cock**
 d Wasserhahn *m*
 f robinet *m* de puisage

nl tapkraan *f* (*m*)
r водоразборный кран *m*

D664 *e* **draw-off point**
d Wasserentnahmestelle *f*,
Wasserzapfstelle *f*
f point *m* de captage
nl wateraftappunt *f* (*m*)
r водоразборная точка *f*

D665 **draw-off tap** *see* **draw-off cock**

D666 **draw-off valve** *see* **drain valve**

D667 *e* **dredge**
d Schwimmbagger *m*, Naßbagger *m*;
Saugbagger *m*, Schwimmsaugbagger
m; Eimerkettennaßbagger *m*
f drague *f*
nl baggermolen *m*
r дноуглубительный снаряд *m*;
землесосный снаряд *m*;
землечерпательный снаряд *m*

D668 *e* **dredge cutter head**
d Schneidkopf *m* des
Schwimmsaugbaggers
f désagrégateur *m* [cutter *m*] d'une
drague suceuse
nl snijkop *m*
r фреза *f* земснаряда

D669 *e* **dredge pump**
d Baggerpumpe *f*
f pompe *f* de dragage
nl grondpomp *f* (*m*)
r грунтовый насос *m*, землесос *m*

D670 **dredge-well** *see* **dredging tube**

D671 *e* **dredge working travel**
d Entegang *m*, Papillionierung *f*
f papillon(n)age *m* de la drague
nl werkslagen *n* van de baggermolen
r папильонирование *n*

D672 *e* **dredging**
d Naßbaggerung *f*, Vertiefen *n*,
Sohlenvertiefungsarbeiten *f pl*
f dragage *m*
nl baggeren *n*, uitbaggeren *n*
r подводная выемка *f* грунта,
землечерпание *n*; дноуглубительные
работы *f pl*, дноуглубление *n*

D673 *e* **dredging ladder**
d Eimerleiter *f*, Baggerleiter *f*
f élinde *f* (à godets)
nl emmerladder *f* (*m*)
r черпаковая рама *f*
(*землечерпательного снаряда*)

D674 *e* **dredging pipe**
d Spülrohr *n*, Druckrohr *n*
f tuyau *m* de refoulement
nl persleiding *f*, persbuis *f* (*m*)
r напорный пульпопровод *m*

D675 **dredging pump** *see* **dredge pump**

D676 *e* **dredging tube**
d Baggerrohr *n*
f tuyau *m* d'aspiration d'une drague
nl zuigbuis *f* (*m*)
r всасывающая труба *f* земснаряда

D677 *e* **drencher head**
d Feuerlöschbrause *f*, Brausenkopf *m*
f pomme *f* d'arrosoir, tête *f*
d'extincteur automatique
nl sprinkler *m* (*brandblusinstallatie*)
r дренчерная головка *f*, дренчер *m*

D678 *e* **drencher system**
d Drencheranlage *f*,
Sprühwasserlöschanlage *f*,
Feuerlöschbrausesystem *n*
f système *m* d'extinction automatique
à pommes d'arrosoir
nl sprinklerinstallatie *f*
r дренчерная система *f*

D679 *e* **drift**
d 1. Abweichung *f* 2. Drift *f* 3.
bleibende Formänderung *f*; elastische
Nachwirkung *f* 4. Bogenschub *m*,
Gewölbeschub *m* 5. Richtstrecke *f*;
Baustollen *m*; Zugangsstollen *m*;
Fensterstollen *m* 6. Steinmeißel *m*
f 1. déviation *f* 2. dérive *f*;
entraînement *m* 3. déformation *f*
résiduelle; retardation *f* élastique 4.
poussée *f* (*d'un arc, d'une voûte*) 5.
galerie *f* amont; galerie *f* d'accès;
galerie *f* latérale 6. poinçon *m*,
tamponnoir *m*, chasse-pointe *m*
nl 1. deviatie *f* 2. drift *f* (*m*) 3.
elastische vervorming *f* 4. boogschuif
f (*m*), gewelfschuif *f* (*m*) 5. mijngang
m 6. steenbeitel *m*
r 1. отклонение *n* 2. снос *m* 3.
остаточная деформация *f*; упругое
последействие *n* 4. распор *m* (*арки,
свода*) 5. передовая штольня *f*,
подходная галерея *f*; боковая
штольня *f* 6. шлямбур *m*

D682 *e* **drift barrier**
d Treibzeugsperre *f*
f barrages *m pl* flottants
nl vanginstallatie *f* voor drijvend vuil
r запань *f*, наплавное сооружение *n*

D686 *e* **drill carriage**
d Bohrunterwagen *m* (*ohne Aufbauten*)
f chariot *m* de forage, jumbo *m*
nl boorwagen *m*
r буровая каретка *f*

D687 *e* **drill core**
d Bohrkern *m*, Kernprobe *f*
f carotte *f*
nl boormonster *n*
r (буровой) керн *m*

D688 **drilled pile** *see* **bored pile**

D689 *e* **drilled well**
d Bohrbrunnen *m*
f puits *m* foré
nl geboorde bron *n*
r буровая водозаборная скважина *f*

D690 *e* **drill feed**
d 1. Bohrervorschubgeschwindigkeit *f*
2. Bohrervorschubeinrichtung *f*
f 1. avance *f* du foret 2. dispositif *m*
d'avance(ment) du foret

nl 1. boorsnelheid *f* 2. booraanzettoestel
n
r 1. скорость *f* подачи бура, скорость *f*
бурения 2. механизм *m* подачи бура

D691 *e* **drill hole**
d Bohrloch *n*
f trou *m* de forage
nl boorgat *n*
r буровая скважина *f*

D692 *e* **drilling**
d Bohrung *f*, Bohrarbeiten *f pl*
f forage *m*, perforation *f* (des roches),
travaux *m pl* de forage
nl boren *n*
r бурение *n*, буровые работы *f pl*

D694 *e* **drilling platform**
d Bohrinsel *f*, Bohrturmbühne *f*
f plate-forme *f* de forage
nl boorplatform *n*
r буровая платформа *f*

D695 *e* **drilling rig**
d Bohranlage *f*
f installation *f* de forage
nl boorinstallatie *f*
r самоходная буровая установка *f*

D697 *e* **drip**
d 1. Wassernase *f*, Unterschneidung *f* 2.
Abzweigrohr *n* mit Kondensatableiter
3. Abtropfwasser *n*
f 1. larmier *m* 2. tubulure *f* de vapeur
épuisée; tubulure *f* de sortie de
condensat 3. eau *m* de condensation
nl 1. waterhol *m* (*muurafdekking*)
2. afvoerbuis *f* (*m*) voor condenswater
3. druipwater *n*
r 1. слезник *m*, слив *m* 2. отвод *m*
трубы (с конденсатоотводчиком) 3.
стекающая капельная влага ,
конденсат *m*

D698 *e* **drip pan, drip tray, drip trough**
d Tropfschale *f*, Tropfwanne *f*,
Tauwasserschale *f*, Tauwasserwanne *f*
f cuvette *f* pour condensat
nl druipbak *m*
r поддон *m* для конденсата,
каплесборник *m*

D699 *e* **driven cast-in-place pile**
d Bohrpfahl *m*, Ortbetonpfahl *m* mit
zurückgewonnenem Füllrohr
f pieu *m* à fourreau temporaire, pieu *m*
Franki
nl in werk gestorte betonpaal *m* met
teruggewonnen mantelbuis
r набивная бетонная свая *f* с
извлекаемой из грунта оболочкой

D700 *e* **driven pile**
d Rammpfahl *m*
f pieu *m* battu
nl heipaal *m*
r забивная свая *f*

D701 *e* **driven shell pile**
d Hülsenrammpfahl *m*
f pieu *m* à tube battu, pieu *m* tubé
nl mantelbuis *f* (*m*) voor betonpaal

r набивная свая *f* с остающейся в
грунте оболочкой

D702 *e* **driven well**
d Rammbrunnen *m*, Schlagbrunnen *m*,
Abessinierbrunnen *m*, Norton-Brunnen
m
f puits *m* foncé
nl nortonput *m*, geslagen bron *f* (*m*)
r забивной колодец *m*

D703 *e* **drive pipe**
d Rammrohr *n*; Mantelrohr *n*,
Vortriebrohr *n*
f tube *f* de sondage; tube *f* fourreau
[de revêtement]
nl boorbuis *f* (*m*); mantelbuis *f* (*m*)
r забивная труба *f* для взятия проб
грунта; обсадная труба *f*

D704 *e* **drive point**
d Brunnenspitze *f*; Rammfilter *n*, *m*,
Einschlagfilter *n*, *m*
f pointe *f* de puits foncé
nl boorpunt *m* (*mantelbuis*); filterpunt *m*
(*bronbuis*)
r башмак *m* обсадной трубы; забивной
фильтр *m* (*нижняя часть забивного
колодца*)

D705 *e* **driveway**
d Auffahrt *f*, Zufahrt *f*
f passage *m* pour voitures; allée *f*
nl toegangsweg *m*; oprijlaan *f* (*m*)
r въезд *m*; подъездная дорога *f*

D706 *e* **driving**
d 1. Vortrieb *m* (*Tunnelbau*)
2. Einrammen *n*, Eintreiben *n*
f 1. creusement *m* (*d'un tunnel*)
2. battage *m*, fonçage *m* (*des pieux*)
nl 1. drijven *n*, boren *n* (*tunnel*) 2. heien *n*
r 1. проходка *f* (*туннеля*) 2. забивка *f*
(*свай, шпунта*)

D707 *e* **driving band**
d Rammring *m*
f frette *f* (*d'une pieu*)
nl paalring *m*, heiband *m*
r бугель *m* (*сваи*)

D708 *e* **driving cap, driving helmet**
d Rammhaube *f*, Rammjungfer *f*
f casque *f* de battage
nl paalmuts *f* (*m*), heimuts *f* (*m*)
r наголовник *m* (*сваи*)

D709 *e* **driving resistance of a pile**
d Verdrängungswiderstand *m*
[Eindringungswiderstand *m*] eines
Rammpfahls
f résistance *f* du sol au battage d'un
pieu
nl grondweerstand *m* van de paal
r сопротивление *n* сваи погружению
[забивке]

D710 *e* **driving test**
d Rammsondieren *n*; Rammversuch *m*,
Rammprobe *f*, Rammprüfung *f*
f sondage *m* de battage; essai *m* de
battage

nl proefheiing *f*
r динамическое зондирование *n* грунта; опытная забивка *f* свай

D712 drop arch *see* depressed arch

D713 *e* **drop-bottom bucket**
d Bodenentleerungskübel *m*
f benne *f* à fond ouvrant
nl onderlossende betonkubel
r бадья *f* с откидным днищем (*для бетонирования*)

D714 drop ceiling *see* suspended ceiling

D715 *e* **dropchute**
d Rutsche *f*, Schurre *f*
f goulotte *f*, goulette *f*
nl stortgoot *f* (*m*)
r спускной жёлоб *m* [лоток *m*]

D716 drop-down curve *see* recession curve

D717 *e* **drop gate**
d Senktor *n*; Fallklappe *f*
f porte *f* descendante; vanne *f* descendante
nl valdeur *f* (*m*) (*sluis*); valluik *n*
r опускные ворота *pl* (*шлюза*); опускной затвор *m* (*гидротехнических сооружений*)

D718 *e* **drop hammer**
d Freifallbär *m*, Fallhammer *m*
f mouton *m* à déclic [à gravité]
nl valhamer *m*, heiblok *n*
r свайный молот *m*, баба *f* (*копра*)

D719 drop heating system *see* overhead heating system

D720 *e* **drop-in beam, drop-in-girder**
d Einhängeträger *m*
f poutre *f* suspendue
nl ingeklemde balk *m*
r подвесная [свободно опёртая] балка *f* (*консольно-балочной системы*)

D721 *e* **drop-inlet dam**
d Damm *m* mit Ringüberlauf
f barrage *m* avec évacuateur à puits
nl dam *m* met ringoverloop
r плотина *f* с шахтным водосбросом

D722 *e* **drop inlet intake**
d Fallschachtfassung *f*, Stollenfassung *f*
f prise *f* d'eau en puits
nl wateronttrekking *f* via tunnels
r шахтный водозабор *m*

D723 *e* **drop-inlet spillway**
d Ringüberlauf *m*
f évacuateur *m* à puits
nl ringoverloop *m*
r шахтный водосброс *m*

D724 *e* **drop manhole**
d Absturz(schacht) *m*
f regard *m* [puits *m* d'accès] de chute
nl valschacht *f* (*m*)
r перепадный колодец *m*

D725 dropped duct *see* duct drop

D726 drop penetration test *see* dynamic penetration test

D727 *e* **drop shaft**
d Senkbrunnen *m*; Fallschacht *m*
f caisson *m* ouvert [perdu]; fût *m* de puits vertical
nl boorbuis *f* (*m*); loodrechte schacht *f* (*m*)
r опускной колодец *m*; вертикальный шахтный ствол *m*

D728 drop siding *see* matched siding

D729 *e* **drop structure**
d Absturzbauwerk *n*, Absturzstufe *f*, Gefällestufe *f*
f ouvrage *m* de chute
nl overstort *n*, *m*
r перепадное сооружение *n*

D730 *e* **drop wall**
d Abfallmauer *f* (*Schleuse*)
f mur *m* de chute (*d'écluse*)
nl stortmuur *m* (*sluis*)
r стенка *f* падения (*шлюза*)

D731 *e* **drop wire**
d Einführungsleitung *f*
f branchement *m* d'abonné
nl huisaansluiting *f*, aansluiting *f* van een gebouw
r ввод *m* провода (*в здание*)

D733 *e* **drowned flow**
d behinderte Strömung *f*
f écoulement *m* noyé
nl verdronken stroming *f*
r движение *n* воды в условиях подпора, течение *n* с глубиной выше критической

D734 drowned hydraulic jump *see* submerged hydraulic jump

D735 drowned weir *see* submerged weir

D737 *e* **drowning ratio**
d Überlaufverhältnis *n*
f taux *m* de submersion
nl overloopverhouding *f*
r коэффициент *m* затопления (*водослива*)

D740 drum gate *US see* sector gate

D741 *e* **drum mixer**
d Freifallmischer *m*, Trommelmischer *m*
f malaxeur *m* [bétonnière *f*] à tambour
nl trommelbetonmenger *m*
r барабанный бетоносмеситель *m*

D742 *e* **drum screen**
d 1. Trommelfilter *n*, *m*
2. Trommelsieb *n*, Wälzsieb *n*, Siebtrommel *f*
f 1. filtre *m* tournant à tambour
2. tambour *m* classeur [cribleur], crible *m* rotatif
nl 1. trommelfilter *n*, *m* 2. zeeftrommel *f* (*m*)
r 1. барабанный фильтр *m*
2. барабанный грохот *m*

D743 *e* **drum weir** *UK*
d Trommel(stau)wehr *n*
f barrage *m* avec vanne à tambour

nl dam *m* met trommelschuif
r плотина *f* с барабанным затвором

D744 *e* **dry air filter**
  d Trockenfilter *n, m,*
    Trockenschicht(luft)filter *n, m*
  f filtre *m* sec à air
  nl droogluchtfilter *n, m*
  r сухой воздушный фильтр *m*

D745 *e* **dry batch**
  d Trockencharge *f,* Trockenmischung *f*
  f gachée *f* sèche
  nl droge mortel *m*
  r сухой замес *m*

D746 *e* **dry-batched aggregate**
  d Trockenbeton *m*
  f mélange *m* de béton sec
  nl droog betonmengsel *n*
  r сухая бетонная смесь *f*

D747 *e* **dry-batch weight**
  d Trockenmischungsgewicht *n*
  f poids *m* de la gachée sèche de béton
  nl droogmengselgewicht *n*
  r масса *f* сухого замеса бетона

D748 *e* **dry concrete**
  d erdfeuchter Beton *m*
  f béton *m* sec, béton *m* de consistance
    de terre humide
  nl aardvochtige betonmortel *m*
  r жёсткий бетон *m*

D750 *e* **dry density**
  d Trocken(roh)dichte *f*
  f densité *f* sèche
  nl droogvolumegewicht *n*
  r плотность *f* в сухом состоянии

D751 *e* **dry density/moisture ratio**
  d Trocken(roh)dichte-Wassergehalt-
    -Verhältnis *n*
  f rapport *m* densité sèche/humidité
  nl verhouding *f* van de bodemdichtheid
    in droge en vochtige toestand
  r отношение *n* средней плотности
    грунта в сухом состоянии к его
    влажности

D752 *e* **dry dock**
  d Trockendock *n*
  f dock *m* de radoub
  nl droogdok *n*
  r сухой док *m*

D753 *e* **dryer**
  d Trockner *m*
  f sécheur *m,* séchoir *m*
  nl droger *m,* droogapparaat *n*
  r сушилка *f*

D754 *e* **drying chamber**
  d Trocknungskammer *f*
  f chambre *f* de séchage
  nl droogkamer *f (m)*
  r сушильная камера *f*

D756 *e* **drying shrinkage**
  d Trocknungsschrumpfen *n;*
    Erhärtungsschwindung *f*
  f retrait *m* au séchage
  nl krimp *m* tengevolge van drogen

---

r усадка *f (бетона)* при высыхании
  или твердении

D757 *e* **drying unit**
  d Trocknungsanlage *f*
  f installation *f* de séchage
  nl droogaggregaat *n*
  r сушильный агрегат *m*

D758 *e* **dry masonry**
  d Steinpackung *f,* Trockenmauerwerk *n*
  f maçonnerie *f* de moellous à sec,
    maçonnerie *f* en pierres sèches
  nl stapelwerk *n* van natuursteen
  r сухая каменная кладка *f*

D759 *e* **dry mix**
  d Trockengemisch *n,* trockene
    Mischung *f*
  f mélange *m* sec
  nl droog mengsel *n*
  r сухая смесь *f*

D760 **dry-mix process** *see* **dry process of
    shotcreting**

D761 *e* **dryness of air**
  d Lufttrockenheit *f*
  f sécheresse *f* de l'air
  nl luchtdroogheid *f*
  r сухость *f* воздуха

D762 *e* **dry pack**
  d steifes [halbtrockenes] Gemisch *n*
  f béton *m* ou mortier *m* de consistance
    demi-sèche
  nl halfdroog mengsel *n*
  r полусухая смесь *f (растворная или
    бетонная)*

D763 *e* **dry-packed concrete**
  d steifes Betongemisch *n*
  f béton *m* de consistence sèche
  nl stampbeton *n*
  r очень плотная бетонная смесь *f*

D764 *e* **dry packing**
  d Einstampfung *f* des steifen
    Betongemisches
  f mise *f* en œuvre du béton de
    consistance très sèche
  nl aanstampen *n* van droge betonmortel
  r укладка *f* очень жёсткой бетонной
    смеси

D765 *e* **dry process**
  d Trockenverfahren *n*
  f voie *f* sèche
  nl droogprocédé *n*
  r сухой способ *m*

D766 *e* **dry process of shotcreting**
  d getrenntes Torkretierverfahren *n*
  f méthode *f* de gunitage par mélange
    à sec
  nl gescheiden torkreteermethode *f*
  r торкретирование *n* при раздельной
    подаче воды и сухой смеси

D767 *e* **dry residue**
  d Trockenrückstand *m,*
    Trockensubstanz *f*
  f résidu *m* sec

nl droge stof *f* (*m*)
r сухой остаток *m*

D768 e dry riser
  d trockene Löschwassersteigleitung *f*
  f colonne *f* montante d'incendie sec
  nl droge stijgbuis *f* (*m*)
  r сухой пожарный стояк *m*

D769 e dry-rodded volume
  d Volumen *n* des stangenverdichteten trockenen Grobzuschlagstoffes
  f volume *m* d'agrégat compacté par piquage à sec
  nl volume *n* van de grove, door porren vedichte, toeslagstof
  r объём *m* уплотнённого штыкованием крупного заполнителя (*для определения средней плотности*)

D770 e dry-rodded weight
  d Gewicht *n* des stangenverdichteten trockenen Grobzuschlagstoffes
  f poids *m* d'agrégat gros compacté par piquage à sec
  nl gewicht *n* van de grove, door porren verdichte, toeslagstof
  r масса *f* уплотнённого штыкованием крупного заполнителя (*для определение средней плотности*)

D771 e dry saturated steam
  d trockener Sattdampf *m*
  f vapeur *f* saturée sèche
  nl droge verzadigde waterdamp *m*
  r сухой насыщенный пар *m*

D772 e dry shake
  d Trockenmörtelgemisch *n*
  f mortier *m* sec
  nl droog mortelmengsel *n*
  r сухая растворная смесь *f*

D773 e dry system of curing
  d Nachbehandlung *f* ohne Feuchthaltung
  f cure *f* du béton sans arrosage
  nl nabehandeling *f* zonder bevochtiging
  r уход *m* за бетоном без увлажнения

D774 e dry technique of construction
  d Trockenbautechnik *f*, Trockenbauweise *f*
  f procédé *m* de construction sec
  nl droge bouwwijze *f*
  r сухой метод *m* отделки стен (*листами сухой штукатурки*)

D775 dry topping *see* dry shake

D776 e dry wall
  d Zwischenwand *f* in Trockenbauweise; Trockenmauer *f*
  f mur *m* revêtu par placoplâtre *ou* par plaques en fibres de bois; perré *m*; mur *m* en pierres sèches
  nl drooggestapelde muur *m*, 'dry wall' wandbekleding *f*
  r внутренняя стена *f*, облицованная сухой штукатуркой *или* фанерой; стена *f* из бута *или* тёсаного камня, сложенная насухо

D777 e dry weather discharge, dry weather flow
  d Trockenwetterabfluß *m*
  f efflux *m* [débit *m*] de temps sec
  nl droogweerafvoer *m*
  r меженный сток *m*

D778 e dry weight
  d Trockengewicht *n*
  f poids *m* sec
  nl drooggewicht *n*
  r сухой вес *m*, вес *m* в сухом состоянии

D779 e dry well
  d 1. Pumpenplatz *m*, Pumpenkammer *f* 2. Kanalisationspumpenschacht *m* 3. Sickerschacht *m*, Sickergrube *f*, Dränschacht *m*, vertikale Dränung *f*
  f 1. superstructure *f* de la station de pompage *f* 2. chambre *f* de pompage des eaux usées 3. puits *m* absorbant de drainage
  nl 1. pompstation *n* 2. rioolpompput *m* 3. loodrechte draineerbuis *f* (*m*)
  r 1. сухая наземная часть *f* здания насосной станции 2. канализационная насосная камера *f* 3. поглощающий [дренажный] колодец *m*; вертикальная дрена *f*

D780 e DSM screen *see* Dutch Staatsmijnen screen

D781 e dual carriageway road
  d Autobahn *f* mit zwei getrennten Fahrbahnen
  f (auto)route *f* à deux chaussées
  nl autoweg *m* met twee gescheiden rijbanen
  r дорога *f* с двумя проезжими частями

D782 e dual duct system
  d Zweikanalsystem *n*, Zweikanalklimaanlage *f*
  f système *m* à deux conduits
  nl twee-leidingssysteem *n* (*klimaatregeling*)
  r двухканальная система *f* (*кондиционирования воздуха*)

D783 e dual-purpose room
  d Doppelzweckraum *m*
  f local *m* à double but
  nl vertrek *n* met tweeledig doel
  r помещение *n* [комната *f*] двойного назначения

D784 e duct
  d Luft(leitungs)kanal *m*; Kabelkanal *m*; Leitungskanal *m*, Sammelkanal *m*; Spannkanal *m*
  f conduite *f* d'air; tube *m* à câble; conduite *f*; gaine *f* pour câble d'armature
  nl luchtkanaal *n*; kabelgoot *f* (*m*); kanaal *n* voor voorspanningswapening
  r воздушный канал *m*, воздуховод *m*; кабельный канал *m*; коммуникационный канал *m*; канал *m* для арматурного пучка

D785 e duct drop
d senkrecht nach unten führende
Luftkanalabzweigung f
f abaissement m (d'un conduit d'air)
nl loodrecht naar beneden gerichte
luchtkanaalaftakking f
r опуск m (воздуховода)

D786 e duct fittings
d Luftleitteile m pl,
Luftkanalformstücke n pl
f raccords m pl de conduit d'air
nl aansluitingen f pl op de luchtkanalen
r фасонные части f pl воздуховода

D788 e ductility
d Duktilität f, Bildsamkeit f,
Verformbarkeit f; Schmiedbarkeit f,
Warmbildsamkeit f
f ductilité f; malléabilité f
nl vervormbaarheid f zonder
sterkteverlies
r пластичность f; тягучесть f; вязкость
f; ковкость f, растяжимость f
(битума)

D790 e duct sizing
d Luftkanalbemessung f,
Luftkanalnetzberechnung f
f calcul m des conduits d'air
nl berekening f van de doorsneden van
luchtkanalen
r расчёт m сечения воздуховодов

D791 duct system see ductwork

D792 e ductube
d Gummischlauchschalung f
f ductube m, tube f souple gonflable
(pour la formation des creux dans la
dalle alvéolée de béton armé)
nl pneumatische holtevormer m
r надувной [пневматический]
каналообразователь m
[пустотообразователь m] (в
железобетонных плитах)

D793 e ductwork
d Kanalsystem n, Luftleitungsnetz n
f réseau m de gaines [de conduits d'air]
nl net n van luchtkanalen
r сеть f воздуховодов

D794 e dug well
d Schachtbrunnen m, Kesselbrunnen m
f puits m creusé [à ciel ouvert]
nl gegraven put m
r копаный или шахтный колодец m

D795 e dummy joint
d Scheinfuge f
f joint m aveugle
nl blinde voeg f (m)
r глухой (неполного профиля)
температурный шов m, ложный шов
m (бетонного покрытия)

D796 e dummy load
d fiktive Belastung f, Scheinbelastung f
f charge f fictive
nl schijnbelasting f
r фиктивная нагрузка f

D797 e dummy unit load
d scheinbare Einheitsbelastung f
f charge f fictive unitaire
nl schijnbare eenheidsbelasting f
r фиктивная единичная нагрузка f

D798 e dummy unit-load method
d Kraftverfahren n (Rahmenberechnung)
f méthode f des forces (calcul des
portiques rigide)
nl krachtenmethode f (berekening van
het skelet)
r метод m сил (для расчёта рам)

D799 e dummy unit moment
d scheinbares Einheitsmoment n
f moment m fictif unitaire
nl schijnbaar eenheidsmoment n
r фиктивный единичный момент m

D800 e dump bank
d Dammkippe f, Halde f, Kavalier m,
Aussatzkippe f
f cavalier m, décharge f, lieu m de
dépôt
nl steenberg m, stortplaats m
r кавальер m, отвал m

D801 e dumper
d Dumper m
f dumper m, tombereau m (automoteur)
nl kipper m, kipwagen m
r самосвал m

D802 e dumping height
d Schütthöhe f, Kipphöhe f
f hauteur f de déchargement
nl kiphoogte f
r высота f разгрузки [опорожнения]

D803 e dumpling
d Bodenmassiv n, Pfeiler m, Feste f
f massif m, stot m
nl bij ontgraving achtergebleven
grondpijler m
r целик m

D804 dumptruck see dumper

D805 e dump valve
d Notablaßventil n
f valve f de vidange de sécurité
nl noodaftapklep f (m)
r клапан m аварийного слива

D806 e duplex
d 1. Maisonette f, doppelstöckige
Wohnung f 2. Zweifamilienhaus n
f 1. maisonnette f, logement m à deux
niveaux 2. maison m à deux
logements [à deux familles]
nl 1. maisonnette f,
tweeverdiepingenhuis n 2. woning f
voor twee gezinnen
r 1. квартира f в двух уровнях
2. двухквартирный дом m

D807 duplex apartment US see duplex 1.

D808 duplex house see duplex 2.

D809 e durability
d Haltbarkeit f, Dauerhaftigkeit f,
Beständigkeit f; Zeitstandfestigkeit f,
Dauerstandfestigkeit f

f durabilité *f*; longévité *f*, résistance *f*
nl duurzaamheid *f*
r стойкость *f*, долговечность *f*;
длительная прочность *f*

D810 *e* **durable material**
   d langlebiges [beständiges] Material *n*
   f matériau *m* durable [résistant]
   nl duurzaam materiaal *n*
   r долговечный [стойкий] материал *m*

D811 *e* **duraluminium**
   d Duraluminium *n*, Dur(al) *n*
   f duraluminium *n*
   nl duraluminium *n*
   r дуралюмин *m*, дюралюминий *m*,
дюраль *m*

D812 *e* **duration curve**
   d Dauerlinie *f*
   f courbe *f* des valeurs classées, courbe *f*
de durée
   nl tijdsduurcurve *f*
   r кривая *f* продолжительности

D813 *e* **duration of mixing**
   d Mischdauer *f*
   f délai *m* [temps *m*] de malaxage
   nl mengtijd *m*
   r время *n* [продолжительность *f*]
перемешивания

D814 *e* **dustbin room**
   d Mülltonnenraum *m*, Müllkübelraum *m*
   f chambre *f* à ordures [pour poubelles]
   nl ruimte *f* voor (het neerzetten van)
vuilnisbakken
   r мусороприёмная камера *f*

D815 *e* **dust control**
   d Staubbekämpfung *f*, Staubfreimachung
f, Entstaubung *f*
   f suppression *f* de la poussière, contrôle
m des poussières, lutte *f* contre les
poussières
   nl stofverwijdering *f*
   r борьба *f* с пылью, обеспыливание *n*

D816 *e* **dustfree environment**
   d staubfreie Umgebung *f*
   f environnement *m* pur [dépoussiéré]
   nl stofvrije omgeving *f*
   r незапылённая окружающая среда *f*;
незапылённое пространство *n*

D817 *e* **dusting**
   d Staubbildung *f*, Stauben *n*
   f poussiérage *m* du béton (*défaut*)
   nl stofvorming *f* bij beton
   r образование *n* пыли;
пыление *n* бетонных поверхностей
(*дефект бетона*)

D818 *e* **dust mask**
   d Staubmaske *f*
   f masque *m* anti-poussière
   nl stofmasker *n*
   r респиратор *m*

D819 *e* **dustpan dredger**
   d Spülbagger *m* [Pumpenbagger *m*] mit
hydraulischer Bodenlösung
   f dragage *f* à désagrégateur hydraulique

nl profielzuiger *m*
r земснаряд *m* с гидравлическим
рыхлителем

D820 *e* **dustproof enclosure**
   d staubdichte Einfassung *f*
   f enceinte *f* étanche aux poussières
   nl stofdichte afsluiting *f*
   r пыленепроницаемое ограждение *n*

D821 *e* **dustproof lighting fitting**
   d staubabgeschützte Leuchte *f*
   f luminaire *m* protégé contre les
poussières
   nl stofdichte verlichtingsarmatuur *f*
   r пылезащищённый светильник *m*

D822 *e* **dust-tightness**
   d Staubundurchlässigkeit *f*,
Staubdichtigkeit *f*
   f étanchéité *f* aux poussières
   nl stofdichtheid *f*
   r пыленепроницаемость *f*

D823 *e* **dusty fraction**
   d staubförmige Fraktion *f*,
Feinstkörnung *f*
   f fraction *f* fine, fines *f pl*
   nl stoffractie *f*
   r пылеватая фракция *f*

D824 *e* **Dutch arch**
   d gemauerter flacher Bogensturz *m*
   f voûte *f* hollandaise
   nl strekse [rechthoekige] boog *m*
   r пологая арочная перемычка *f* из
кирпича

D825   **Dutch brick** *see* **Dutch clinker**

D826 *e* **Dutch bond**
   d 1. holländischer Verband *m*
2. Kreuzverband *m*
   f 1. appareil *m* hollandais
2. appareil *m* l'anglais
   nl kruisverband *n*
   r 1. фламандская перевязка *f*
кирпичной кладки, крестовая кладка
f 2. английская крестовая кладка *f*

D827 *e* **Dutch clinker**
   d holländischer Straßenbauklinker *m*
   f brique *f* hollandaise
   nl kleine gele straatklinker *m*
   r голландский клинкер(ный кирпич) *m*

D828 *e* **Dutch door**
   d holländische Tür *f* (*zweiflügelige Tür
mit zwei vertikal angeordneten Flügeln*)
   f porte *f* normande [à guichet]
   nl onder- en bovendeur *f* (*m*)
   r голландская дверь *f* (*разделённая по
горизонтали на две створки,
открывающиеся раздельно или
вместе*)

D829 *e* **dutchman**
   d 1. Pfropfen *m*, Stöpsel *m*
2. Decklasche *f* 3. Montagemast *m*
   f 1. bouchon *m*, tampon *m* 2. plaque *f*
de recouvrement (*pour cacher le
défaut*); plaque *f* [latte *f*] décorative
3. mat *m* de levage

nl 1. stopstuk *n* 2. vulstuk *n* (*in gapende naad*), dekplaat *f* (*m*) 3. montagemast *m*

r 1. затычка *f*, пробка *f* 2. накладка *f*, скрывающая дефект; декоративная накладка *f* 3. монтажная мачта *f*

D830 e **Dutch mattress**
d Faschinenreisigpackwerk *n*, Faschinenlage *f*, Faschinensenkstück *n*
f matelas *m* de fascines [de fascinages]
nl zinkstuk *n* van rijswerk
r фашинный тюфяк *m*

D831 e **Dutch Staatsmijnen screen**
d Bogensieb *n*
f tamis *m* incurvé, grille *f* courbe
nl boogzeef *f* (*m*)
r криволинейный грохот *m*

D832 e **Dutch tongs**
d Froschklemme *f*
f tendeur-grenouille *m*, pince *f* grenouille
nl draadtang *f* (*m*)
r «лягушка» *f* (*для натягивания проводов*)

D833 e **duty cycle**
d relative Einschaltsdauer *f*
f facteur *m* d'utilisation
nl relatieve inschakelduur *m*
r относительная продолжительность *f* включения (*напр. сварочной машины*)

D834 e **dwarf partition**
d nicht bis zur Geschoßdecke reichende Trennwand *f*
f cloison *f* mi-haute [basse]
nl niet tot het plafond doorlopende scheidingswand *m*
r перегородка *f*, не доходящая до потолка; неполная перегородка *f*

D835 e **dwelling**
d Wohnung *f*; Eigenheim *n*
f habitation *f*, logis *m*, demeure *f*
nl woning *f*
r жилище *n*; индивидуальный жилой дом *m*

D836 e **dwelling house**
d Wohngebäude *n*
f maison *f* d'habitation
nl woonhuis *n*
r жилой дом *m*

D837 e **dwelling unit**
d Wohn(ungs)einheit *f*
f unité *f* de logement
nl wooneenheid *f*
r жилая (планировочная) единица *f*; квартира *f*

D838 **DX-cooler** *US see* **direct expansion air cooler**

D841 e **dyke**
d Eindeichung *f*; Deich *m*, Uferdamm *m*, Sperre *f*; Damm *m*, Aufschüttung *f*
f digue *f*, levée *f*, endiguement *m*

nl dijk *m*, kade *f* (*m*)
r обвалование *n*; дамба *f*, запруда *f*, плотина *f*; насыпь *f*

D842 e **dykeland** *UK*
d Deichland *n*
f terres *f pl* protégées par digues (*contre la mer*)
nl bedijkt land *n*
r земля *f*, отвоёванная у моря; осушенная земля *f*

D843 **dyking** *see* **diking**
dynamical... *see* **dynamic...**

D844 e **dynamic analysis**
d dynamische Berechnung *f*
f calcul *m* dynamique [aux sollicitations dynamiques]
nl dynamische berekening *f*
r динамический расчёт *m*, расчёт *m* на динамическую нагрузку

D847 e **dynamic balance**
d dynamisches Gleichgewicht *n*
f équilibre *m* dynamique
nl dynamisch evenwicht *n*
r динамическое равновесие *n*

D848 e **dynamic buckling**
d dynamisches Knicken *n*
f flambement *m* [flambage *m*] dynamique, flambage *m* dû aux efforts dynamiques
nl dynamische knik *m*
r потеря *f* устойчивости при динамической нагрузке

D849 e **dynamic creep**
d dynamisches Kriechen *n*
f fluage *m* dynamique
nl dynamische kruip *m*
r динамическая ползучесть *f*, ползучесть *f* в условиях переменных нагрузок *или* температур

D850 e **dynamic delivery head**
d dynamische Druckhöhe *f*
f hauteur *f* de charge dynamique totale (de refoulement)
nl dynamische drukhoogte *f*
r полная высота *f* нагнетания

D851 **dynamic equilibrium** *see* **dynamic balance**

D852 e **dynamic head**
d hydrodynamische Druckhöhe *f*
f poussée *f* (hydro)dynamique
nl hydrodynamische drukhoogte *f*
r динамический напор *m*

D853 e **dynamic load, dynamic loading**
d dynamische Belastung *f* [Beanspruchung *f*]
f charge *f* [chargement *m*] dynamique
nl dynamische belasting *f*
r динамическая нагрузка *f*, динамическое нагружение *n*

D854 e **dynamic penetration test**
d Rammsondierung *f*, dynamische Sondierung *f*

f sondage *m* dynamique
nl dynamische sondering *f*
r динамическое зондирование *n*

D855 *e* **dynamic pile-driving resistance**
d dynamischer Rammwiderstand *m* (*Pfahl*)
f résistance *f* dynamique au battage du pieu
nl dynamische heiweerstand *m*
r динамическое сопротивление *n* забивке сваи

D856 *e* **dynamic pile formula**
d dynamische Rammformel *f*
f formule *f* dynamique de battage
nl dynamische heiformule *f* (*m*)
r формула *f* динамического сопротивления сваи при забивке

D857 *e* **dynamic pressure**
d dynamischer Druck *m*
f pression *f* dynamique
nl dynamische druk *m*
r динамическое давление *n*

D858 *e* **dynamic resistance**
d dynamischer Widerstand *m*
f résistance *f* dynamique
nl dynamische weerstand *m*
r динамическое сопротивление *n*

D861 *e* **dynamic similarity**
d dynamische Ähnlichkeit *f*
f similitude *f* dynamique
nl dynamische overeenkomst *f*
r динамическое подобие *n*

D862 **dynamic sounding** *see* **dynamic penetration test**

D863 *e* **dynamic strength**
d dynamische Festigkeit *f*
f résistance *f* dynamique
nl dynamische stevigheid *f*
r динамическая прочность *f*

D864 *e* **dynamic suction head**
d dynamische Saughöhe *f*
f hauteur *f* de charge dynamique d'aspiration
nl dynamische zuighoogte *f*
r полная высота *f* всасывания

D865 *e* **dynamic testing**
d dynamische Prüfung *f*
f essai *m* dynamique
nl dynamische beproeving *f*
r испытание *n* на динамическую нагрузку

D866 **dynamic water-level** *see* **pumping level**

E

E1 *e* **earliest event occurence time**
d frühester Ausführungstermin *m* eines Ereignisses (*Netzplantechnik*)
f date *f* [temps *m*] au plut tôt de l'événement (*planification par chemin critique*)

nl vroegste tijdstip *n* van optreden **van** een gebeurtenis (*netwerkplanning*)
r самый ранний срок *m* наступления события (*сетевое планирование*)

E2 *e* **early finish time**
d frühester Endtermin *m* einer Aktivität *f* (*Netzplantechnik*)
f fin *m* au plus tôt de l'activité (*planification par chemin critique*)
nl vroegste tijdstip *n* van beëindiging van een activiteit (*netwerkplanning*)
r самый ранний срок *m* окончания работы (*сетевое планирование*)

E3 *e* **early morning boost**
d Frühanheizbetrieb *m*
f régime *m* de fonctionnement matinal (*d'une installation de chauffage*)
nl in de ochtend op temperatuur brengen
r режим *m* утреннего разогрева (*отопительной системы*)

E4 *e* **early start time**
d frühester Anfangstermin *m* einer Aktivität (*Netzplantechnik*)
f début *m* au plus tôt de l'activité (*planification par chemin critique*)
nl vroegste tijdstip *n* van begin van een activiteit (*netwerkplanning*)
r самый ранний срок *m* начала работы (*сетевое планирование*)

E5 **early stiffening** *see* **flash set**

E6 *e* **early strength**
d Anfangsfestigkeit *f*, Frühfestigkeit *f*
f résistance *f* initiale (de béton)
nl drukvastheid *f* na één dag
r раннее нарастание *n* прочности

E7 *e* **earth**
d Erde *f*, Grund *m*, Boden *m*
f terre *f*, sol *m*, terrain *m*
nl aarde *f* (*m*), grond *m*, bodem *m*
r земля *f*, грунт *m*

E8 *e* **earth-and-rockfill dam**
d zusammengesetzter Steinschüttdamm *m*, Erd-Fels-Damm *m*
f barrage *m* en terre et enrochements
nl dam *m* met kern van stortsteen
r каменно-земляная плотина *f*

E9 *e* **earth back pressure**
d Bodengegendruck *m*
f contre-pression *f* du sol
nl passieve gronddruk *m*
r отпор *m* грунта

E10 *e* **earth banking**
d Deichdamm *m*
f digue *f* d'encaissement
nl aarden dam *n*, dijk *m*
r дамба *f* обвалования

E11 *e* **earth blanket**
d Erddichtung *f*
f masque *m* d'étanchéité
nl grondafdekking *f*
r грунтовый противофильтрационный экран *m*

E12　*e* **earth borer**
　　d Bohrschnecke *f* [Schnecken-
　　-Erdbohrer *m*] auf LKW montiert
　　f installation *f* de forage automotrice,
　　foreuse *f* automotrice
　　nl op vrachtauto gemonteerde grondboor
　　m
　　r самоходная буровая установка *f*

E13　*e* **earth canal**
　　d unausgekleideter Kanal *m*
　　f canal *m* en terre
　　nl kanaal *n* zonder oeverbekleding
　　r земляной [необлицованный] канал *m*

E15　*e* **earth cofferdam**
　　d Erdfangdamm *m*
　　f batardeau *m* en terre [en sol]
　　nl aarden vangdam *m*
　　r земляная [грунтовая] перемычка *f*

E16　*e* **earth compaction**
　　d Bodenverdichtung *f*
　　f compactage *m* [serrage *m*] de sol
　　nl grondverdichting *f*
　　r уплотнение *n* грунта

E17　*e* **earth consolidation**
　　d Konsolidierung *f*, Eigensetzung *f*,
　　natürliche Bodenverdichtung *f*
　　f consolidation *f* naturelle de sol
　　nl inklinking *f*, zetting *f* van de grond
　　r консолидация *f* [естественное
　　уплотнение *n*] грунта

E18　*e* **earth core**
　　d Erdkern *m*
　　f noyau *m* en terre imperméable
　　nl kern *f* (*m*) van ondoorlatende grond
　　r противофильтрационное ядро *m*
　　(*плотины*)

E19　*e* **earth dam**
　　d Erddamm *m*, Schüttdamm *m*
　　f barrage *m* [digue *f*] en terre
　　nl aarden dam *m*
　　r земляная плотина *f*

E20　*e* **earth drilling rig**
　　d Erdbohrgerät *n*, Bohrschnecke *f*
　　f engine *m* de forage de sol
　　nl grondboorapparaat *n*
　　r шнековая буровая установка *f*

E21　*e* **earth embankment**
　　d Erddamm *m*
　　f remblai *m*
　　nl aarden dam *m*
　　r земляная насыпь *f*

E22　*e* **earth excavation**
　　d Bodenausbaggerung *f*, Bodenaushub *m*
　　f excavation *f* de sol [de terre]
　　nl uitgraving *f*
　　r выемка *f* [экскавация *f*] грунта

E23　　**earth fill** *see* **earth embankment**

E24　*e* **earth-fill cofferdam**
　　d Erdfangdamm *m*, geschütteter
　　Fangdamm *m* [Vordamm *m*]
　　f batardeau *m* en terre
　　nl vangdam *m*, kistdam *m*
　　r земляная (насыпная) перемычка *f*

E25　*e* **earth-fill dam**
　　d geschütteter Erddamm *m*,
　　Schüttdamm *m*
　　f barrage *m* de remblai
　　nl dam *m* van opgebrachte grond
　　r насыпная плотина *f*

E26　*e* **earth-filled pile sheeting cofferdam**
　　d geschütteter Spundwandfang(e)damm
　　m
　　f batardeau *m* à double paroi de
　　palplanches remblayé de terre
　　nl kistdam *m*
　　r двухрядная шпунтовая перемычка *f*
　　с грунтовым заполнением

E27　　**earth filling** *see* **soil filling**

E28　*e* **earth-fill timber dam**
　　d erdgefüllter Holzkastendamm *m*
　　f barrage *m* en (charpente de) bois
　　remblayé de terre
　　nl met grond gevulde houten kistdam *m*
　　r деревоземляная плотина *f*

E29　*e* **earthflow**
　　d 1. Murgang *m*, Schlammstrom *m*
　　2. Hangrutsch *m*, Abgleitung *f*,
　　Fließrutschung *f*
　　f 1. écoulement *m* de sol, avalanche *f*
　　boueuse 2. coulée *f* de boue,
　　glissement *m* de terrain
　　nl 1. modderlawine *f* 2. aardverschuiving *f*
　　r 1. сель *m*, грязекаменный поток *m*
　　2. оплывина *f*, оползень *m*

E30　*e* **earth foundation**
　　d Bodengründung *f*
　　f sol *m* [terrain *m*] de fondation
　　nl fundament *n* op staal
　　r грунтовое основание *n*

E31　*e* **earth grade**
　　d Erdoberflächengefälle *n*
　　f pente *f* de terrain; pente *f* naturelle,
　　talus *m* naturel
　　nl helling *f* van het terrein
　　r уклон *m* земной поверхности;
　　естественный откос *m*

E32　*e* **earthing**
　　d Erdung *f*
　　f mise *f* à la terre
　　nl aarding *f*
　　r заземление *n*

E33　*e* **earthing electrode**
　　d Erdelektrode *f*, Erder *m*
　　f électrode *m* de mise à la terre
　　nl aardelektrode *f*
　　r заземляющий электрод *m*

E34　*e* **earth levee**
　　d Erddeich *m*
　　f levée *f* de terre, remblai *m*
　　nl aarden dijk *m*, zomerkade *f* (*m*)
　　r земляная дамба *f*, земляной вал *m*

E35　*e* **earth mass**
　　d Bodenmasse *f*, Erdmasse *f*
　　f masse *f* de terre [de sol]
　　nl grondmassa *f*
　　r грунтовая масса *f*

E36   *e* **earth mixtures**
    d Erdmischungen *f pl*
    f mélanges *m pl* de sol
    nl grondmengsels *n pl*
    r грунтовые смеси *f pl*

E37   *e* **earthmover**
    d Erdbewegungsmaschine *f*,
      Bodenfördergerät *n*
    f engin *m* de déplacement de terres
    nl grondverzetmachine *f*
    r землеройно-транспортная машина *f*

E38   *e* **earthmoving**
    d Bodenförderung *f*
    f déplacement *m* de terre(s) [de sol]
    nl grondverzet *n*
    r перемещение *n* грунта

E39   **earthmoving equipment** *see* **earth-
    -moving machinery**

E40   **earthmoving gear** *UK see* **earthmover**

E41   *e* **earth-moving machinery, earth-
    -moving plant**
    d Erdbaugeräte *n pl*
    f engins *m pl* de terrassement
    nl materiëel *n* voor grondverzet
    r землеройно-транспортные машины *f
    pl*, землеройно-транспортное
    оборудование *n*

E42   *e* **earth pigment**
    d natürlicher Mineralfarbstoff *m*
    f pigment *m* minéral
    nl natuurlijk mineraal pigment *n*
    r минеральный пигмент *m*

E43   *e* **earth plate**
    d Erdungsplatte *f*
    f plaque *f* de mise à la terre
    nl aardplaat *f (m)*
    r заземляющий электрод *m* в форме
    пластины

E44   *e* **earth pressure**
    d Erddruck *m*
    f poussée *f* des terres; pression *f* sur terre
    nl gronddruk *m*
    r давление *n* грунта; давление *n* на
    грунт

E45   *e* **earth pressure at rest**
    d Ruhedruck *m*
    f pression *f* du sol au repos
    nl rustgronddruk *m*
    r давление *n* грунта в состоянии покоя,
    статическое давление *n* грунта

E46   **earthquake damage** *see* **earthquake-
    -induced failure**

E47   *e* **earthquake forces**
    d Erdbebenkräfte *f pl*, seismische
      Reaktionskräfte *f pl*
    f efforts *m pl* séismiques
    nl aardschokkrachten *f (m) pl*
    r сейсмические силы *f pl* [воздействия
    n pl]

E48   *e* **earthquake-induced failure**
    d Zerstörung *f* durch Erdbeben
    f démolition *f* due à la sollicitation
    d'origine séismique

    nl bezwijken *n* ten gevolge van een
    aardbeving
    r разрушение *n* от землетрясения

E49   *e* **earthquake load**
    d seismische Beanspruchung *f*
    f charge *f* séismique
    nl belasting *f* ten gevolge van
    aardschokken
    r сейсмическая нагрузка *f*

E50   *e* **earthquake period**
    d Erdbebendauer *f*
    f durée *f* de séisme
    nl duur *m* van een aardbeving
    r длительность *f* землетрясения

E51   *e* **earthquake-resistant structure**
    d erdbebensicheres Bauwerk *n*
    f construction *f* résistante aux
    séismes
    nl bouwwerk *n*, bestand tegen
    aardbevingen
    r сейсмостойкое сооружение *n*

E52   *e* **earth ridge**
    d Erdwall *m*
    f bourrelet *m* en terre
    nl aarden wal *m*
    r земляной валик *m (мелиорация)*

E53   *e* **earth-sheltered home**
    d erdbedecktes Haus *n*, Hanghaus *n*
    f bâtiment *m* appuyé contre une
    colline, maison *f* appuyée à la **pente;**
    maison *f* à l'abri de la terre
    nl in een (berg)helling verzonken huis *n*
    r дом *m*, врезанный в холм *или*
    обвалованный грунтом;
    заглублённое здание *n*

E54   *e* **earth slope**
    d Erdböschung *f*
    f talus *m* de terre
    nl grondtalud *n*
    r грунтовый [земляной] откос *m*

E55   *e* **earth terminal**
    d Erdungsklemme *f*
    f borne *f* de mise à la terre
    nl aardklem *f (m)*
    r зажим *m* [вывод *m*] заземления

E56   *e* **earthwork**
    d 1. Erdbau *m*, Erdarbeiten *f pl*
    2. Volumen *n* der Erdarbeiten,
    Erdbaumasse *f* 3. Erdanlage *f*,
    Erdbauwerk *n*
    f 1. travaux *m pl* de terrassement,
    terrassement *m* 2. cube *m* de
    terrassements 3. ouvrage *m* en sol
    nl 1. grondwerk *n* 2. volume van het
    grondverzet
    r 1. земляные работы *f pl*, разработка
    f грунта 2. объём *m* земляных работ
    3. земляное сооружение *n*

E58   **earthwork quantity** *see* **earthwork 2.**

E59   *e* **easement**
    d 1. Übergangskurve *f*, Übergangsbogen
    m 2. Handlaufrundung *f* (*im
    Grundriß*)

f 1. courbe *f* de transition 2. partie *f* courbe d'une main courante
nl 1. overgangsboog *m* 2. afronding *f*, gebogen leuning *f*
r 1. переходная [спрямляющая] кривая *f* (*дороги*) 2. закругление *n* поручней (*в плане*)

E60   easement curve *see* transition curve

E61   *e* easily dug rock
d leichtbearbeitbarer Felsboden *m*, abbauwilliges Felsgestein *n*
f roche *f* tendre facile à excaver
nl gemakkelijk te ontgraven gesteente *n*
r легко разрабатываемая скальная порода *f*

E62   easy bend *see* long radius bend

E63   *e* eave bearer
d Auflagerholz *n*
f coyau *m*
nl spar *m*
r кобылка *f*

E64   *e* eaves
d 1. Traufe *f* 2. Dachrinne *f*
f 1. auvent *m*, queue *f* de vache 2. goutière *f*, chéneau *m*
nl 1. dakoverstek *n* 2. dakrand *m*
r 1. карнизный свес *m* крыши 2. водосточный жёлоб *m*

E65   *e* eaves board
d Stirnbrett *n*, Traufbrett *n*
f planche *f* d'auvent du comble, planche *f* goutière
nl boeiplank *f* (*m*), vogelschroot *m*
r доска *f* карнизного свеса (*на концах стропильных ног*)

E66   *e* eaves channel
d hölzerne Traufrinne *f*
f chéneau *m* en bois
nl houten goot *f* (*m*), mastgoot *f* (*m*)
r карнизный жёлоб *m* (*из досок*)

E67   *e* eaves course
d Traufziegelreihe *f*, untere Dachbelagplattenreihe *f*
f rangée *f* de bordure de tuiles *ou* d'ardoises
nl onderste rij *f* (*m*) pannen; dubbele rij *f* (*m*) leien langs de onderkant
r нижний ряд *m* черепицы *или* кровельных плиток

E68   *e* eaves girder
d Dachsparrenträger *m*
f sablière *f* d'avant-toit
nl eindgording *f*
r подстропильная балка *f*

E69   *e* eaves gutter
d Dachrinne *f*
f chéneau *m*, gouttière *f*
nl dakgoot *m* (*onder het dakoverstek*)
r водосточный [карнизный] жёлоб *m*

E70   *e* eaves tile
d Traufziegel *m*
f tuile *f* cuite d'avant-toit

nl onderste pan *f* (*m*), druippan *f* (*m*)
r краевая черепица *f*

E71   *e* eaves trough *see* eaves gutter

E72   *e* ebb gate
d Binnentor *n*, Ebbetor *n*
f porte *f* intérieure (*d'écluse*)
nl ebdeur *f* (*m*)
r внутренние [отливные] ворота *pl* (*морского шлюза*)

E73   *e* eccentrically loaded column
d außermittig belastete Säule *f*
f poteau *m* sollicité par une charge excentrée
nl excentrisch belaste kolom *f* (*m*)
r внецентренно нагруженная колонна *f*

E74   *e* eccentricity
d Ausmittigkeit *f*, Ausmitte *f*, Exzentrizität *f*
f excentricité *f*
nl excentriciteit *f*
r эксцентриситет *m*

E75   *e* eccentric load
d außermittige [exzentrische] Belastung *f*
f charge *f* excentrée
nl excentrische belasting *f*
r внецентренная нагрузка *f*

E76   *e* eccentric tendon
d außermittiges Bewehrungsbündel *n*
f câble *m* de précontrainte excentré
nl excentrisch geplaatste wapeningsbundel *m*
r внецентренная прядевая *или* пучковая напрягаемая арматура *f*

E77   *e* econocrete *UK*
d minderwertiger Beton *m* mit örtlichen Zuschlagstoffen
f béton *m* «économique» (*à base de granulat élémentaire local*)
nl beton *n* van ter plaatse gewonnen toeslag
r бетон *m* на дешёвых местных заполнителях

E78   *e* economic percentage of steel and concrete
d wirtschaftlich optimaler Bewehrungsanteil *m*
f pourcentage *m* d'armature économique
nl meest economische wapeningspercentage *f*
r экономически оптимальный процент *m* армирования

E79   *e* economic storage
d wirtschaftlich optimales Beckenvolumen *n*
f volume *m* de la retenue économique
nl economisch stuwmeervolume *n*
r экономически оптимальный объём *m* водохранилища

E80   *e* economic yield
d wirtschaftliche Grundwasserentnahmemenge *f*
f débit *m* de soutirage économique

nl economisch verantwoorde
grondwateronttrekking *f*

r экономически допустимый дебит *m*
(*колодца, скважины*)

E81 *e* **economizer cycle**
d energiegünstige Betriebsweise *f* der
Klimaanlage mit größtmöglicher
Außenluftkälteutilisation
f conditonnement *m* de l'air à cycle
économique (*avec utilisation
maximale de l'air extérieur pour le
système de refroidissement*)
nl energiebesparende klimaatregeling *f*
door optimaal gebruik van
buitenlucht voor koeling
r режим *m* системы кондиционирования
воздуха с максимальным
использованием наружного воздуха
для охлаждения

E82 *e* **economy brick**
d gewöhnlicher Bauziegel *m*,
Normalziegel *m*
f brique *f* économique
nl gewone baksteen *m*
r стандартный [«экономичный»]
кирпич *m* [блок *m*]

E83 *e* **economy wall**
d tragende Ziegelwand *f* mit
Pilastern aus keramischen
Formblöcken
f mur *m* en briques «économiques»
nl dragende muur *m* (met pilasters) van
gewone baksteen
r несущая кирпичная стена *f* из
стандартных кирпичей

E84 **edge conditions** *see* **boundary
conditions**

E86 **edge force** *see* **boundary force**

E87 *e* **edge girder**
d Rand(balken)träger *m*, Endträger *m*
f poutre *f* marginal [de rive]
nl rijbalk *m*
r краевая [концевая] балка *f*

E88 *e* **edge joint**
d Parallelstoß *m*
f joint *m* [assemblage *m*] latéral à plat
nl hoekverbinding *f* in de
breedterichting
r шов *m* сплачивания (*досок*)

E89 *e* **edge load**
d Randlast *f*
f charge *f* au bord
nl randbelasting *f*
r краевая нагрузка *f*

E90 *e* **edge moment**
d Randmoment *n*
f moment *m* au bord
nl randmoment *n*
r краевой момент *m*

E91 **edge mouldings** *see* **edge strips**

E92 *e* **edge plate**
d Kantenschutzlasche *f*,
Randeinfassung *f*

f eclisse *f* cornière [d'équerre]
nl hoeklijst *f* (*m*), hoekbescherming *f*
r предохранительная угловая накладка
f

E93 *e* **edger**
d 1. Glätteisen *n* mit umgebördeltem
Rand 2. Parkettenschleifmaschine *f*
für Behandlung der Friese
3. Besäum(kreis)säge *f*
4. Kantenhobelmaschine *f*
f 1. truelle *f* d'angle 2. ponceuse *f*
portative à parquets (*pour ponçage
des frises*) 3. tronçonneuse *f*
circulaire, scie *f* circulaire à
tronçonner 4. raboteuse *f* latérale,
raboteuse-chanfreineuse *f*
nl 1. hoektroffel *f*
2. parketschuurmachine *f* voor de
randen 3. kantrecht(cirkel)zaag *f* (*m*)
4. afschuinmachine *f*
r 1. металлическое усёночное правúло
n, угловая тёрка *f* (*для отделки
кромок бетонных плит*) 2.
паркетно-шлифовальная машина *f*
(*для отделки фризов*) 3. торцовая
обрезная пила *f* 4. кромкострогальный
[кромкофуговальный] станок *f*

E94 *e* **edge shot board**
d Hobelkantbrett *n*
f planche *f* aux bords rabotés
nl plank *f* (*m*) met geschaafde kanten
r доска *f* со строгаными кромками

E95 *e* **edgestone**
d Bordstein *m*, Randstein *m*
f pierre *f* de bordure
nl kantsteen *m*
r бортовой камень *m*

E96 *e* **edge strip of a slab**
d Randstreifen *m* der Platte
f bande *f* de rive d'une dalle
nl randstrip *m* van een plaat
r краевая полоса *f* плиты (*элемент
расчётной схемы*)

E97 *e* **edge strips**
d Halteleisten *f pl*, leistenförmige
Bauelemente *n pl*
f couvres-joints *m pl*, baguettes *f pl*,
lattes *f pl* moulurées
nl hoeklijst *f* (*m*), hoekstrip *m*
r строительные погонажные изделия *n
pl* для оформления углов,
раскладки *f pl*

E98 *e* **edge supported slab**
d randgelagerte [allseitig aufliegende]
Platte *f*
f dalle *f* appuyée sur le périmètre
nl langs de omtrek gesteunde plaat *f* (*m*)
r плита *f*, опёртая по контуру

E99 *e* **edging**
d 1. Kantenbearbeitung *f* (*Beton*)
2. Besäumen *n* (*Holz*)
f 1. finition *f* des bords [des rives]
(*d'une dalle routière en béton*) 2.
coupe *f* des bouts de planches

nl 1. kantenafwerking *f* (*wegdekplaten*)
2. kantrechten *n* (*hout*)
r 1. отделка *f* кромок (*напр. плит, дорожных покрытий*) 2. обрезка *f* (*пиломатериалов*)

E100 edging trowel *see* edger

E101 *e* edifice
d stattliches Bauwerk *n* [Gebäude *n*]
f édifice *m*
nl monumentaal gebouw *n*
r крупное [монументальное] здание *n*

E102 *e* educational building
d Lehrgebäude *n*
f bâtiment *m* d'enseignement [d'instruction]
nl schoolgebouw *n*
r учебное здание *n*

E103 *e* effective area of an orifice
d wirksamer [effektiver] Querschnitt *m*, Wirkungsquerschnitt *m*
f aire *f* effective (*d'une orifice*)
nl effectieve doorstroomopening *f*
r эффективное сечение *n* (*отверстия*)

E104 *e* effective area of concrete
d wirksame Betonquerschnittsfläche *f*
f aire *f* de section effective (*d'une poutre en béton armé*)
nl effectieve betondoorsnede *f*
r расчётная площадь *f* бетона

E105 *e* effective depth
d Nutzhöhe *f*
f hauteur *f* utile [effective]
nl werkzame hoogte *f*
r полезная высота *f* сечения

E106 effective flange width *see* effective width of slab

E107 *e* effective grain diameter, effective grain size
d wirksamer Korndurchmesser *m*
f dimension *f* effective des grains
nl werkelijke korrelgrootte *f*
r эффективный диаметр *m* зёрен

E108 *e* effective head
d nutzbare Fallhöhe *f*
f chute *f* utile
nl nuttige drukhoogte *f*
r полезный напор *m*

E109 *e* effective height
d Knicklänge *f* (*Säule*)
f hauteur *f* de flambage (*d'une colonne*)
nl kniklengte *f* (*kolom*)
r приведённая высота *f* (*стойки, колонны*)

E110 *e* effective lateral restraint
d knickungsverhindernde seitliche Einspannung *f* (*Balken*)
f encastrement *m* latéral effective (*contre flambement lateral d'une poutre fléchie*)
nl dwarsstempeling *f* (*balken*); zijdelingse steun *m* (*kolom*)

r боковое закрепление *n* балок (*против потери устойчивости из плоскости изгиба*)

E111 *e* effective length
d Knicklänge *f*
f longueur *f* de flambage
nl kniklengte *f*
r приведённая длина *f* (*при продольном изгибе*)

E112 *e* effective length factor
d Knicklängenbeiwert *m*
f coefficient *m* de la longueur de flambage
nl slankheidsfactor *m*
r коэффициент *m* приведённой длины

E113 *e* effective prestress
d wirksame Vorspannung *f*
f précontrainte *f* finale [effective]
nl blijvende voorspanning *f*
r установившееся предварительное напряжение *n*

E114 *e* effective prestress in tendon
d wirksame Vorspannung *f* im Bewehrungsbündel
f effort *m* de tension effectif
nl blijvende voorspanning *f* (in wapeningsbundel)
r действующее [установившееся] усилие *n* натяжения арматурного пучка (*после всех потерь*)

E115 *e* effective reinforcement
d Arbeitsbewehrung *f*
f armature *f* principale, ferraillage *m* principal
nl hoofdwapening *f*
r рабочая арматура *f*

E116 *e* effective span
d Stützweite *f*
f portée *f* théorique
nl werkelijke overspanning *f*
r расчётная длина *f* пролёта, расчётный пролёт *m*

E117 *e* effective storage
d wirksames Speichervolumen *n*
f volume *m* utile du réservoir
nl nuttig stuwbekkenvolume *n*
r полезный объём *m* водохранилища

E118 effective stress *see* final stress

E119 *e* effective thickness
d 1. Kleinstmaß *n* des Querschnittes (*zur Bestimmung des Schlankheitsgrades*) 2. reduzierte Stärke *f* (*der Straßendecke*)
f 1. plus petit côté *m* de la section transversale 2. épaisseur *f* [hauteur *f*] réduite (*du corps de chaussée*)
nl 1. kritieke dikte *f* (*bepalend voor de slankheidsfactor*) 2. effectieve wegdekdikte *f*
r 1. наименьший размер *m* (*поперечного сечения*) 2. приведённая толщина *f* (*дорожной одежды*)

E120 *e* **effective velocity of ground water**
   d Grundwassergeschwindigkeit *f*,
     Filtergeschwindigkeit *f* eines
     Grundwasserkörpers
   f vitesse *f* effective de l'eau souterraine
  nl effectieve stroomsnelheid *f*,
     filtratiesnelheid *f* van het
     grondwater
   r эффективная скорость *f* движения
     грунтовых вод, скорость *f*
     фильтрации

E121 *e* **effective width of slab**
   d mitwirkende Plattenbreite *f*
     (*Stahlbeton*)
   f largeur *f* de dalle [de plaque] admise
     pour le calcul
  nl effectieve plaatbredte *f* (*gewapend
     beton*)
   r расчётная ширина *f* плиты

E122 *e* **effect of restraint**
   d Einspannungswirkung *f*
   f effet *m* d'encastrement
  nl invloed *m* van inklemming
   r влияние *n* защемления [заделки]
     (*конструктивного элемента*)

E123 *e* **effects of end conditions**
   d Einfluß *m* der
     Auflagerungsbedingungen
   f effet *m* des conditions d'appui aux
     extrémités (*d'une pièce comprimée*)
  nl invloed *m* van de statische toestand
     van de kolomeinden
   r влияние *n* условий опирания концов
     (*сжатого элемента*)

E124 *e* **efflorescence**
   d Ausblühung *f*, Ausschlag *m*
   f efflorescence *f*
  nl uitbloeiing *f*, uitslag *m*
   r выцветы *m pl*

E125 *e* **effluent**
   d behandeltes Abwasser *n*,
     Klärwasser *n*
   f effluent *m*, eaux *f pl* résiduaires
     traitées
  nl gezuiverd afvalwater *n*
   r очищенная сточная вода *f*

E126 *e* **effluent seepage**
   d Heraussickern *n*, ausfließende
     Versickerung *f*
   f infiltration *f* effluente
  nl infiltratie *f* van het gezuiverde
     afvalwater
   r диффузионное просачивание *n*
     подземных вод на поверхность
     земли, просачивание *n* из глубины,
     фильтрационный отток *m*

E127 *e* **egg-shaped sewer**
   d eiförmige Abwasserleitung *f*
   f égout *m* ovoïde
  nl rioolbuis *f* (*m*) met ei-vormig profiel
   r канализационный коллектор *m*
     яйцевидного профиля

E128 *e* **eggshell gloss**
   d Eierschalenmattglanz *m*
   f lustre *m* du type coquille d'œuf
  nl eiglans *m*
   r приглушённый глянец *m*, глянец *m*
     типа «яичная скорлупа»

E129 *e* **egress**
   d Ausgang *m*; Ausgangsmittel *n*
   f sortie *f*; issue *f*, moyen *m* de sortie
  nl uitgang *m*; uitweg *m*
   r выход *m*; средство *n* выхода
     [эвакуации] (*из здания*)

E130 *e* **ejector**
   d Strahlpumpe *f*, Ejektor *m*
   f éjecteur *m*
  nl éjecteur *m*, straalpomp *f* (*m*)
   r эжектор *m*, отсасывающий
     струйный насос *m*

E131 *e* **elastically supported girder**
   d elastisch aufgelagerter Träger *m*
   f poutre *f* sur appuis élastiques
  nl ligger *m* op verende steunpunten
   r балка *f* на упругих опорах

E132 *e* **elastic aftereffect, elastic after-effect**
   d elastische Nachwirkung *f*,
     Nachwirkungseffekt *m*
   f retardation *f* élastique
  nl elastische nawerking *f*
   r упругое последействие *n*

E133 *e* **elastic behaviour**
   d elastisches Verhalten *n*
   f comportement *m* [tenue *f*] élastique
  nl elastisch gedrag *n*
   r упругая работа *f* (*материала,
     конструкции*)

E134 *e* **elastic buckling**
   d elastisches Knicken *n*
   f flambage *m* [flambement *m*]
     élastique
  nl elastische knik *m*
   r упругий продольный изгиб *m*, упру-
     гое выпучивание *n*

E135 *e* **elastic center**
   d elastischer Mittelpunkt *m*
   f centre *m* élastique
  nl elastisch centrum *n*
   r упругий центр *m*

E136 *e* **elastic constants**
   d elastische Konstanten *f pl*
   f constantes *f pl* élastiques
  nl elastische constanten *f pl*
   r упругие постоянные *f pl*

E137 *e* **elastic curve**
   d elastische Linie *f*, Biegelinie *f*
   f ligne *f* élastique
  nl elastische lijn *f* (*m*), buiglijn *f* (*m*)
   r упругая линия *f*, линия *f* прогибов

E138 *e* **elastic deformation**
   d elastische Verformung *f*, federnde
     Formänderung *f*
   f déformation *f* élastique
  nl elastische vervorming *f*
   r упругая деформация *f*

E139 *e* **elastic design**
   d elastisches Berechnungsverfahren *n*

f calcul *m* élastique
nl elastische berekeningsmethode *f*
r расчёт *m* по допускаемым
напряжениям

E140 *e* **elastic limit**
d Elastizitätsgrenze *f*
f limite *f* élastique
nl elasticiteitsgrens *f* (*m*)
r предел *m* упругости

E141 *e* **elastic loss**
d Vorspannungsverlust *m* durch
elastische Betonzusammenpressung
f perte *f* de la précontrainte due à la
compression élastique du béton
nl voorspanningsverlies *n* tengevolge
van de elastische samendrukking van
het beton
r потери *f pl* преднапряжения от
упругого обжатия бетона

E142 **elastic modulus** *see* **modulus of
elasticity**

E143 *e* **elastic rail spike**
d elastischer Schienennagel *m*
f crampon *m* de rail élastique
nl verende spoorspijker *m*
r упругий костыль *m* (*для крепления
рельсов*)

E144 *e* **elastic range**
d elastischer Verformungsbereich *m*
f domaine *m* élastique
nl elastische zone *f*, elastisch gebied *n*
r упругая область *f*, область *f*
упругих деформаций

E145 *e* **elastic recovery**
d elastische Erholung *f*, Rückfederung *f*
f restitution *f* élastique
nl elastisch herstel *n*, terugvering *f*
r упругое восстановление *n*

E146 *e* **elastic restraint**
d elastische Einspannung *f*
f encastrement *m* élastique
nl elastische inklemming *f*
r упругое защемление *n*

E147 *e* **elastic shortening**
d 1. elastische Verkürzung *f* (*des
Druckgliedes*) 2. elastische
Betonzusammenpressung *f*
f 1. raccourcissement *m* élastique 2.
compression *f* élastique de béton
nl 1. elastische verkorting *f* van een
drukstaaf 2. elastische samendrukking
f (*beton*)
r 1. упругое укорочение *n* (*сжатого
элемента*) 2. упругое обжатие *n*
бетона

E148 *e* **elastic strain**
d elastische Dehnung *f*
f déformation *f* élastique
nl elastische rek *m*
r упругая удельная [относительная]
деформация *f*

E149 **elastic strain range** *see* **elastic range**

E150 *e* **elastic strength**
d Schubproportionalitätsgrenze *f*,
Schubelastizitätsgrenze *f*
f limite *f* d'élasticité au cisaillement
nl elasticiteitsgrens *f* (*m*) bij afschuiving
r предел *m* пропорциональности
[упругости] при сдвиге

E151 *e* **elastic support**
d elastische Unterlage *f*
f appui *m* élastique
nl verende oplegging *f*
r упругая опора *f*

E152 *e* **elastic theory**
d Elastizitätstheorie *f*
f théorie *f* d'élasticité
nl elasticiteitstheorie *f*
r теория *f* упругости

153 *e* **elastic weights method**
d Winkelgewichtsverfahren *n*,
Verfahren *n* der elastischen Lasten
f méthode *f* des poids élastiques
nl elasticiteitsmethode *f*
r метод *m* упругих грузов

E154 *e* **elastomeric bearing**
d Elastomerauflager *n*
f appui *m* [appareil *m* d'appui] en
élastomère
nl oplegging *f* in een elastomeer
r опора *f* [опорная часть *f*] из
эластомера

E155 *e* **elastomeric joint sealant**
d elastomere Fugenvergußmasse *f*
f mastic *m* d'étanchéité à base d'un
élastomère
nl elastomeervoegvulling *f*
r гидроизолирующая мастика *f* на
основе эластомера

E156 *e* **elbow**
d Knierohr *n*, Kniestück *n*, Krümmer *m*
f coude *m*, genou *m* (*d'un tuyau*)
nl knie *f* (*m*); aanzetsteen *m* van een
boog
r колено *n* (*трубопровода*)
electrical... *see* **electric...**

E157 *e* **electric curing**
d Elektrowarmbehandlung *f*
f cure *f* [curing *m*] électrique
nl elektrothermische betonbehandeling *f*
r электропрогрев *m* (*бетона*)

E158 *e* **electrical ground**
d Erdung *f*
f (dispositif *m* de) mise *f* à la terre
nl aarding *f*
r заземление *n*, заземляющее
устройство *n*

E159 *e* **electrically conductive concrete**
d elektrisch leitender Beton *m*
f béton *m* électroconducteur
nl de elektriciteit geleidend beton *n*
r электропроводный [токопроводящий]
бетон *m*

E160 **electrical prestressing** *see*
electrothermal prestressing

E161 *e* **electric analogy method**
  d elektrisches Analogieverfahren *n*
  f méthode *f* d'analogie électrique
    [électrohydrodynamique]
  nl methode *f* naar de analogie met
    elektriciteit
  r метод *m* электрогидродинамической
    аналогии, метод *m* ЭГДА

E162 *e* **electric cable**
  d Elektrokabel *n*, Stromkabel *n*
  f câble *m* électrique
  nl elektriciteitskabel *f* (*m*)
  r электрический кабель *m*,
    электрокабель *m*

E163 *e* **electric calorifier, electric cylinder**
  d Elektroboiler *m*, Elektro-
    -Heißwassererhitzer *m*
  f chauffe-eau *m* électrique
  nl elektrische boiler, *m*, elektrisch
    voorraadwarmwatertoestel *n*
  r электрический водонагреватель *m*

E164 *e* **electric distribution network**
  d elektrisches Verteilungsnetz *n*
  f réseau *m* de distribution électrique
  nl elektriciteits(distributie)net *n*
  r электрическая распределительная
    сеть *f*

E165 *e* **electric drainage**
  d Elektrodränung *f*
  f drainage *m* électrique
  nl zwerfstroomafleiding *f*
  r электродренаж *m*,
    электроосмотическое водопонижение
    n, электроосушение *n*

E166 *e* **electric drill**
  d Drillbohrer *m*, elektrische
    Handbohrmaschine *f*
  f perceuse *f* électrique
  nl elektrische handboormachine *f*
  r электродрель *f*

E167 *e* **electric fixtures** *US*
  d elektrisches Zubehör *n*,
    Elektroarmatur *f*
  f appareils *m pl* électriques
  nl elektrisch armatuur *n*, *f*
  r электротехническая арматура *f*

E168 *e* **electric float**
  d Elektro-Glättmaschine *f*, Elektro-
    -Glättkelle *f*
  f lisseuse *f* électrique
  nl elektrische pleisterschuurmachine *f*,
    elektrische plakspaan *f* (*m*)
  r электрогладилка *f*

E169 **elektric heat accumulator** *see* **electric
    storage heater**

E170 *e* **electric heater**
  d Elektroofen *m*; elektrischer
    Heizkörper *m*
  f appareil *m* de chauffage électrique
    [électrique de chauffage]
  nl elektrische kachel *f* (*m*);
    elektrische radiator *m*
  r электропечь *f*; электроотопительный
    прибор *m*

E171 *e* **electric heating**
  d elektrische Raumheizung *f*,
    Elektroheizung *f*
  f chauffage *m* électrique
  nl elektrische verwarming
  r электроотопление *n*

E172 *e* **electric heating blanket**
  d elektrische Heizmatte *f*
  f couverture *f* électrique de chauffe
  nl elektrische deken *f* (*m*)
  r электронагревательный мат *m*

E173 *e* **electric heating coil**
  d Elektrolufterhitzer *m*
  f calorifère *m* électrique
  nl elektrische luchtverhitter *m*
  r электрокалорифер *m*

E174 *e* **electric heating system**
  d elektrisches Heizungssystem *n*
  f système *m* électrique de chauffage
  nl elektrisch verwarmingssysteem *n*
  r система *f* электроотопления

E175 *e* **electric hoist**
  d Elektro(bau)aufzug [*m*; Elektrotakel
    n
  f élévateur *m* électrique; palan *m*
    électrique
  nl elektrische hijsinrichting *f*;
    elektrische takel *m*, *n*
  r электрический (строительный)
    подъёмник *m*; электроталь *f*

E176 *e* **electric installation**
  d 1. elektrische Installation *f* [Anlage
    2. elektrische Gebäudeausrüstung *f*
  f 1. installation *f* électrique
    2. équipement *m* électrique des bâti-
    ments
  nl elektrische installatie *f*
  r 1. электрическая установка *f*
    2. электрооборудование *n* зданий

E177 *e* **electric lift truck**
  d Elektrohubwagen *m*, Elektrohubkarre
    f
  f chariot *m* élévateur électrique
  nl elektrische heftruck *m*
  r электрокар *m*

E178 *e* **electric nibbler** *US*
  d Elektronibbler *m*, elektrische
    Nibbelschere *f*
  f cisaille *f* vibratoire électrique
  nl elektrische trilschaar *f* (*m*)
  r электровиброножницы *pl*

E179 *e* **electric operator**
  d elektrischer Stellantrieb *m*
  f servo-commande *f* électrique
  nl elektrisch servomechanisme *n*
  r электрический сервомеханизм *m*
    (*напр. для открывания дверей*),
    исполнительный механизм *m*

E180 *e* **electric panel heating**
  d elektrische Plattenheizung *f*
  f chauffage *m* électrique par panneaux
  nl verwarming *f* met elektrische
    paneelradiatoren
  r панельное электроотопление *n*

E181 *e* **electric power pylon**
    d elektrischer Leitungsmast *m*,
    Überlandmast *m*, Freileitungsmast *m*
    f poteau *m* [mât *m*, support *m*] de ligne
    électrique
    nl elektriciteitsmast *m*
    r мачта *f* линии электропередачи

E182 *e* **electric service**
    d 1. provisorische Stromnetze *n pl* und
    elektrische Ausrüstung (*der*
    Baustellen) 2. Stromversorgung *f* (*der*
    Baustelle)
    f 1. canalisations *f pl* et installations *f*
    pl électriques de chantier 2.
    alimentation *f* du chantier en énergie
    électrique
    nl 1. tijdelijke elektriciteitsvoorzieningen
    f pl op de bouwplaats 2.
    stroomvooziening *f* op de bouwplaats
    r 1. временные электрические сети *f pl*
    и оборудование *n* (*на стройплощад-*
    ках) 2. электроснабжение *n* строй-
    площадки

E183 *e* **electric space heater**
    d elektrisches Raumheizgerät *n*,
    elektrischer Raumheizer *m*
    f appareil *m* de chauffage électrique
    nl elektrisch ruimteverwarmingstoestel *n*
    r электроотопительный прибор *m*

E184 *e* **electric storage heater**
    d 1. Elektrospeicherheizungsgerät *n*,
    ESH-Gerät *n* 2.
    Elektrowarmwasseraufbereiter *m*
    f 1. chauffage *m* électrique par
    accumulation 2. chauffe-eau *m*
    électrique à accumulation
    nl 1. accumulerende elektrische kachel
    f (*m*) 2. elektrisch
    voorraadwarmwatertoestel *n*
    r 1. аккумулирующий
    электроотопительный прибор *m*
    2. электрический ёмкостный
    водонагреватель *m*

E185 *e* **electric strike**
    d elektrische Entriegelungseinrichtung *f*
    f dispositif *m* de déblocage électrique
    nl elektrische deuropener *m*
    r электрическое деблокировочное
    устройство *n*

E186 *e* **electric tools**
    d Elektrowerkzeug *n*
    f outillage *m* électrique
    nl elektrisch gereedschap *n*
    r электроинструмент *m*

E187 *e* **electric traction**
    d elektrische Traktion *f*
    f traction *f* électrique
    nl elektrische tractie *f*
    r электрическая тяга *f*, электротяга *f*

E188 **electric trowel** *see* **electric float**

E189 **electric water heater** *see* **electric**
    calorifier

E190 **electric welding** *see* **1. arc welding 2.**
    resistance welding

E191 *e* **electric winch**
    d Elektrowinde *f*
    f treuil *m* électrique
    nl elektrische lier *f* (*m*)
    r электролебёдка *f*

E192 *e* **electrochemical gauging**
    d elektrochemische Hydrometrie *f*
    [Wassermessung *f*]
    f jaugeage *m* électrochimique
    nl elektrochemische hydrometrie *f*
    [watermeting *f*]
    r электрохимическая гидрометрия *f*

E193 *e* **electrode boiler**
    d Elektrodenkessel *m*
    f chaudière *f* à électrodes
    nl elektrodenketel *m*
    r электрокотёл *m*

E194 **electrofilter** *see* **electrostatic air filter**

E195 *e* **electrohydraulic elevator** *US*
    d elektrohydraulischer Aufzug *m*
    f ascenseur *m* électro-hydraulique
    nl elektrohydraulische lift *m*
    r электрогидравлический лифт *m*

E196 *e* **electro-osmosis**
    d 1. Elektro-Osmose *f* 2.
    elektroosmotische Entwässerung *f*
    f 1. électro-osmose *f* 2. rabattement *m*
    de la nappe aquifère par
    électro-osmose
    nl 1. elektro-osmose *f* 2. elektro-
    -osmotische ontwatering *f*
    r 1. электроосмос *m*
    2. электроосмотический способ *m*
    водопонижения

E197 *e* **electroslag welding**
    d Elektroschlacke-Schweißen *n*
    f soudage *m* sous laitier
    (électroconducteur)
    nl elektro-slaklassen *n*
    r электрошлаковая сварка *f*

E198 *e* **electrostatic air filter,**
    electrostatic precipitator
    d Elektrofilter *n*, *m*, elektrostatisches
    Luftfilter *n*, *m*
    f filter *m* d'air électrostatique
    nl elektrostatisch luchtfilter *n*, *m*
    r электростатический воздушный
    фильтр *m*

E199 *e* **electro-thermal curing,**
    electrothermal hardening
    d elektrothermische Härtung *f* (*Beton*)
    f durcissement *m* par chauffage
    électrique (*du béton*)
    nl elektrothermische nabehandeling *r*
    (*beton*)
    r твердение *n* (*бетона*) при
    электронагреве

E200 *e* **electrothermal prestressing, electro-**
    -thermal pre-tensioning
    d elektrothermisches Spannen *n* von
    Spannstahl, Elektrovorspannung *f*
    f précontrainte *f* par chauffage
    électrique

nl elektrothermische voorspanning *f*
(*wapening*)
r электротермическое натяжение *n*,
электротермический способ *m*
предварительного напряжения
(*арматуры*)

E201 *e* **element of construction**
d Bauwerkteil *m*, Gebäudeteil *m*
f élément *m* constructif [de
construction]
nl bouwelement *n*, constructiedeel *n*
r конструктивный элемент *m*

E202 *e* **elephant trunk**
d Betonschüttrohr *n*, Rüssel *m*
f trompe *f* d'éléphant
nl betonstortpijp *f* (*m*), afvoerpijp *f* (*m*)
van een pneumatisch
funderingscaisson
r хобот *m* для подачи бетона

E203 *e* **elevated ditch**
d Kanal *m* im Auftrag
f canal *m* en remblai
nl sloot *f* (*m*) in een ophoging
r канал *m* в насыпи

E204 *e* **elevated flume**
d hochgelegte Rinne *f*
f canal *m* surélevé
nl bovengrondse goot *f* (*m*)
r надземный лоток *m*

E205 **elevated road** *see* **skyway**

E206 *e* **elevated tank**
d Hochbehälter *m*, Hochreservoir *n*
f château *m* d'eau
nl hooggelegen reservoir *n*
r водонапорная башня *f*; водонапорный
бак *m*

E207 *e* **elevating belt conveyor**
d Fahrband *n*, fahrbarer Gurtförderer *m*
f sauterelle *f*, transporteur *m* mobile à
bande, elevateur *m* transporteur
mobile
nl transportband *m*
r передвижной ленточный конвейер *m*

E208 *e* **elevating grader**
d Pflugbagger *m* [Flachbagger-Lader *m*]
mit kreisförmiger Diskuspflugschar,
Grader *m* mit Förderwerk
f niveleuse *f* automotrice [motograder
m] avec chargeur à godet
nl elevatorgrondschaaf *f* (*m*)
r грейдер-элеватор *m*

E209 *e* **elevating scraper**
d Schrapper *m* mit Kettenförderwerk
f scraper *m* avec élévateur
nl zelfladende schraper *m*
r скрепер-элеватор *m*

E210 *e* **elevation**
d 1. Vertikalprojektion *f*, Ansicht *f*,
Aufriß *m* 2. Höhe *f*; Ordinate *f*;
Kote *f*
f 1. projection *f* verticale, élévation *f*
2. cote *f*; repère *m* d'altitude;
hauteur *f*

nl 1. verticale projectie *f*, vooraanzicht *n*
2. (terrein)hoogte *f*, ordinaat *f* (*m*)
3. elevatie *f*, elevatiehoek *m*
r 1. вертикальная проекция *f* 2.
отметка *f*; высота *f*; превышение *n*

E211 *e* **elevation head**
d 1. statische Druckhöhe *f* 2. Ortshöhe *f*
f 1. charge *f*, hauteur *f* de chute
2. altitude *f*, cote *f*
nl 1. statische drukhoogte *f*
2. hoogteligging *f*
r 1. гидростатический [потенциальный,
высотный] напор *m* 2. высота *f*
положения, геометрическая высота *f*,
отметка *f* (*высота над условным
нулевым уровнем*)

E212 *e* **elevator**
d Aufzug *m*; Becher(förder)werk *n*,
Eimerketten-Aufzug *m*,
Aufnahmeelevator *m*
f ascenseur *m*, monte-charge *m*;
élévateur *m* à godets
nl lift *m*; hefwerktuig *n*
r лифт *m*; подъёмник *m*; нория *f*,
ковшовый элеватор *m*

E213 *e* **elevator cage** *US*, **elevator car**
d Fahrstuhl *m*, Fahrkorb *m*
r cabine *f* d'ascenseur
nl liftkooi *f* (*m*)
r кабина *f* лифта

E214 *e* **elevator car safeties**
d Fangvorrichtungen *f pl* des
Fahrkorbes
f arrêt-ascenseur *m*
nl veiligheidsinrichting *f*
[vanginrichting *f*] van een liftkooi
r ловители *m pl* кабины лифта

E215 *e* **elevator dredger**
d Eimerketten-Naßbagger *m*,
Eimerketten-Schwimmbagger *m*
f drague *f* à godets [à chaîne]
nl emmerbaggermolen *m*
r многоковшовый землечерпательный
снаряд *m*

E216 *e* **elevator guide rails**
d Führungsschienen *f pl* des Aufzugs
f guides *m pl* d'ascenseur
nl liftkooigeleiders *m pl*
r направляющие *f pl* лифта

E217 **elevator hoistway** *see* **hoistway**

E218 *e* **elevator landing** *US*
d Aufzugpodest *n*
f palier *m* d'ascenseur
nl liftportaal *n*
r посадочная площадка *f* лифта

E219 *e* **elevator machine room** *US*
d Aufzugmotorraum *m*
f local *m* de machines d'ascenseur
nl liftmachinekamer *f* (*m*)
r машинное отделение *n* лифта

E220 *e* **elevator pit**
d Schachtgrube *f* des Aufzugs
f fosse *f* d'ascensseur

nl liftschachtput *m*, liftschacht *f* (*m*)
r приямок *m* шахты лифта

E221 e elevator shaft
d Aufzugsschacht *m*
f cage *f* [gaine *f*, puits *m*] d'ascenseur
nl liftkoker *m*, liftschacht *f* (*m*)
r шахта *f* лифта

E222 e elevator shaft gates
d Schachttür *f*
f porte *f* de la cage [gaine] d'ascenseur
nl liftschachtdeur *m*
r дверь *f* шахты лифта

E223 e elevator-stairway core
d Gebäudekern *m* mit Aufzug(s)schacht und Treppenhaus
f noyau *m* central (*d'un bâtiment*) avec cage d'escalier et gaines d'ascenseur
nl gebouwkern *f* (*m*) met liftschachten en trappenhuis
r центральное ядро *n* [центральный ствол *m*] здания (*с лифтами и лестничной клеткой*)

E224 elevator well *US see* elevator shaft

E225 e eliminator
d Abscheider *m*
f éliminateur *m*, séparateur *m*
nl afscheider *m*
r сепаратор *m*

E226 e eliminator plate
d Tropfenabscheider *m*, Tropfenfänger *m*
f plaque *f* d'élimination
nl druppelvanger *m*
r каплеуловитель *m*, каплеотделитель *m*

E227 e ellipse of inertia
d Trägheitsellipse *f*
f ellipse *f* d'inertie
nl traagheidsellips *f* (*m*)
r эллипс *m* инерции

E228 e ellipse of stress
d Spannungsellipse *f*
f ellipse *f* des tensions [des contraintes]
nl spanningsellips *f* (*m*)
r эллипс *m* напряжений

E229 e ellipsoid of elasticity
d Elastizitätsellipsoid *n*, Spannungsellipsoid *n*
f ellipsoïde *m* (des efforts) d'élasticité
nl spanningsellipsoïde *f*
r эллипсоид *m* (тензора) напряжений

E230 e elliptical-paraboloid shell
d elliptische Paraboloidschale *f*
f voile *m* mince paraboloïdal elliptique
nl elliptische-paraboloïdeschaal *f* (*m*)
r оболочка *m* в форме эллиптического параболоида

E231 e elongated aggregate
d Zuschlagstoff *m* mit eckiglänglicher Kornform
f granulat *m* [agrégat *m*] à grains allongés

nl toeslag *m* met langwerpige korrels (*beton*)
r заполнитель *m* с зёрнами удлинённой формы

E232 e elongation
d Verlängerung *f*; Bruchdehnung *f*
f dilatation *f* (positive); allongement *m* (de rupture)
nl rek *m*; breukrek *m*
r удлинение *n*; относительное удлинение *n* при разрыве

E233 elongation at rupture *see* breaking elongation

E234 e elongation per unit length
d bezogene [relative] Dehnung *f*
f allongement *m* relatif [unitaire]
nl relatieve rek *m*
r относительное удлинение *n*

E235 e embankment
d 1. Uferschutzwerk *n*, Uferdamm *m*, Sperrdamm *m*, Deich *m*, Damm *m*, Verkehrsdamm *m* 2. Kai *m*, Ufereinfassung *f*
f 1. digue *f*, levée *f*, jetée *f*, remblai *m* 2. quai *m*
nl 1. dijk *m*, kade *f* (*m*) 2. spoordijk *m*, weglichaam *n*
r 1. оградительная дамба *f*; береговой вал *m*; насыпь *f* 2. набережная *f*

E236 e embankment wall
d Damm-Stützmauer *f*
f mur *m* de talus d'un remblai
nl steunmuur *m* van een dam, keermuur *m*
r подпорная стенка *f* насыпи

E237 e embedded column
d Wandsäule *f*, Halbsäule *f*
f colonne *f* engagée, demi-colonne *f*, demi-ceintre *m*
nl halfzuil *f* (*m*)
r колонна *f*, частично встроенная в стену, полуколонна *f*

E238 e embedded heating panel
d Heizwand *f*, Betonheizplatte *f*
f panneau *m* chauffant [dalle *f* chauffante] en béton
nl ingelaten paneelradiator *m*
r встроенная отопительная бетонная панель *f*

E239 e embedment length
d nach der Momentenlinie bestimmte Einbindelänge *f* (*Bewehrungsstab*)
f longueur *f* (*de barre d'armature*) déterminée par la ligne des moments
nl aanhechtingslengte *f*
r длина *f* (*арматурного стержня*), определяемая по эпюре моментов

E240 e emergency exhaust fan
d Notabsauger *m*
f ventilateur *m* aspirant [aspirateur] de secours
nl noodafzuiger *m*
r вентилятор *m* аварийной вытяжки

E241 *e* **emergency exit**
   d Notausgang *m*
   f sortie *f* de secours, issue *f* de fuite
   nl nooduitgang *m*
   r запасный [аварийный] выход *m*

E242 *e* **emergency gallery**
   d Notstollen *m*, Fluchtstollen *m*
   f voie *f* [galerie *f*] de secours
   nl nooduitgang *m* (*balkon*)
   r штольня *f* аварийного выхода, эвакуационная штольня *f*

E243 *e* **emergency gate**
   d Notverschluß *m*
   f vanne *f* de garde [protectrice, de sécurité]
   nl nooddeur *f* (*m*)
   r аварийный затвор *m*

E244 *e* **emergency lighting**
   d Notbeleuchtung *f*
   f éclairage *m* [éclairement *m*, illumination *f*] de secours
   nl noodverlichting *f*
   r аварийное освещение *n*

E245 *e* **emergency outlet**
   d Notablaß *m*
   f déversoir *m* de secours [d'urgence]
   nl nooduitlaat *m*
   r аварийный спуск *m*

E246 *e* **emergency spillway**
   d Notüberlauf *m*, Notentlastungsanlage *f*
   f évacuateur *m* de secours, déversoir *m* de sécurité
   nl noodoverloop *m*, noodaflaat *m*
   r аварийный водоспуск *m* [водосброс] *m*

E247 *e* **emergency ventilation**
   d Notlüftung *f*
   f ventilation *f* de secours
   nl noodventilatie *f*
   r аварийная вентиляция *f*

E248 *e* **emergency water**
   d Feuerlöschwasser *n*
   f eau *f* de lutte contre l'incendie
   nl bluswater *n*
   r противопожарный запас *m* воды

E249 *e* **emergency water valve**
   d Brandschieber *m*
   f vanne *f* de secours
   nl brandkraan *f* (*m*), hydrant
   r пожарная задвижка *f*, задвижка *f* аварийного водоснабжения

E250 *e* **emission**
   d Emission *f*, Abluftauswurf *m*, Stoffauswurf *m*
   f emission *f*, dégagement *m*
   nl emissie *f*, uitstoot *m* van afvoergassen
   r вентиляционный выброс *m*, выброс *m* вещества в атмосферу

E251 *e* **emissivity**
   d Emissionsverhältnis *n*
   f émissivité *f*
   nl emissieverhouding *f*
   r относительная излучающая способность *f* поверхности при данной температуре

E252 *e* **employee identification**
   d Einlaßschein *m*, Kennzeichen *n* (*des Bauarbeiters*)
   f plaque *f* d'identité d'un travailleur (*pour admission au chantier*)
   nl toegangsbewijs *n* (*van bouwmeester*)
   r пропуск-значок *m* (*для прохода на стройку*)

E253 *e* **emptying gate**
   d Ablaßverschluß *m*
   f vanne *f* de vidange
   nl aftapkraan *f* (*m*), aftapventiel *n*
   r затвор *m* водоспуска

E254 *e* **empty weight**
   d Leermasse *f*
   f poids *m* à vide
   nl leeg gewicht *n*
   r масса *f* без груза (*транспортной машины*)

E255 *e* **Emscher tank**
   d Emscherbrunnen *m*, Imhoffbrunnen *m*
   f décanteur *m* à double étage
   nl imhofftank *m*
   r двухъярусный отстойник *m*

E256 *e* **emulsified asphalt**
   d Bitumenemulsion *f*, Kaltasphalt *m*
   f émulsion *f* de bitume (asphaltique), émulsion *f* bitumineuse
   nl asfaltenemulsie *f*
   r битумная эмульсия *f*

E257 *e* **emulsion coating**
   d Emulsionsanstrich *m*, Filmüberzug *m*, Nachbehandlungsfilm *m*
   f couche *f* d'émulsion
   nl beschermende emulsielaag *f* (*m*)
   r эмульсионное покрытие *n*

E258 *e* **emulsion injection**
   d Emulsionsverpressung *f*, Emulsioninjektion *f*
   f stabilisation *f* du sol au bitume (*par injection*)
   nl een emulsie injecteren *n* (ten behoeve van grondstabilisatie)
   r стабилизация *f* грунта нагнетанием битумной эмульсии, битумизация *f* грунта

E259 *e* **emulsion paint**
   d Emulsionsfarbe *f*
   f peinture *f* émulsionnée
   nl emulsieverf *f* (*m*)
   r эмульсионная краска *f*

E260 *e* **emulsion slurry**
   d Straßenbauemulsion *f*
   f émulsion *f* routière
   nl wegenbouwemulsie *f*
   r дорожная эмульсия *f*

E261 *e* **emulsion sprayer**
   d Kaltbindemittelspritzmaschine *f*, Kaltasphaltspritzmaschine *f*

f épandeur *m* d'émulsion
nl emulsiesproeier *m*
r распределитель *m* дорожной
эмульсии

E262 e **enamel**
d 1. Email *n*, Glasur *f* 2. Lackfarbe *f*
f 1. émail *m* 2. peinture *f* au vernis
nl 1. email *n* 2. emailverf *f* (*m*)
r 1. эмаль *f*; глазурь *f* 2. эмалевая
[лаковая] краска *f*

E263 **enameled brick** *see* **glazed brick**

E264 e **enameled tile**
d glasierte Steinzeug(belag)platte *f*
[Steinzeugfliese *f*]
f carreau *m* en grès (cérame) émaillé
nl geglazuurde tegel *m* [dakpan *m*]
r глазурованная облицовочная плитка
f

E265 e **encased sheet pile**
d Kastenspundbohle *f*
f palplanche *f* (en) caisson
nl damplank *f* (*m*)
r коробчатая шпунтовая свая *f*

E266 e **encased structures**
d Stahlrohrkonstruktionen *f pl* aus
betoneingefüllten Röhren
f constructions *f pl* en tubes remplis
de béton
nl constructies *f pl* van met beton
gevulde stalen pijpen
r конструкции *f pl* из металлических
труб, заполненных бетоном

E267 e **encastré moment**
d Einspann(ungs)moment *n*
f moment *m* d'encastrement
nl inklemmingsmoment *n*
r момент *m* защемления [заделки]

E268 e **encaustic tile**
d Kachelplatte *f*
f carreau *m* glacé
nl geglazuurde tegel *m*
r изразцовая плитка *f*, изразец *m*,
кафель *m*

E269 e **enclosed dock**
d Tidebecken *n*, Gezeitenbecken *n*
f bassin *m* de marée
nl sluishaven *f* (*m*)
r приливный бассейн *m* (*приливной*
ГЭС)

E270 **enclosed stairs** *see* **box stairs**

E271 e **enclose pattern**
d Innengestaltung *f*
f aménagement *m* intérieur
nl indeling *f* [plattegrond *m*] van een
gebouw
r внутренняя планировка *f*

E272 e **enclosing sheeting**
d Spundwandumschließung *f*
f rideau *m* de **palplanches**, enceinte *f*
en palphanches
nl damwand *m*
r шпунтовое ограждение *n*, шпунтовая
стенка *f*

E273 e **enclosure**
d 1. Hülle *f*, Umhüllung *f* 2. Kapselung
f (*Lüftung*) 3. Umschließung *f*,
Umzäunung *f* 4. abgeschlossener
Raum *m*
f 1. enveloppe *f*, voile *m* mince, coque *f*
2. hotte *f* de ventilation
[d'aspiration] 3. clôture *f* [enceinte *f*]
de protection 4. local *m* hermétique
nl 1. omhulling *f* 2. ventilatiekap *f* (*m*)
3. omheining *f* 4. omsloten ruimte *f*
r 1. оболочка *f* 2. вентиляционное
укрытие *n* полностью закрытого
типа 3. ограждение *n*
4. герметизированное помещение *n*

E274 e **enclosure wall**
d Fachwerkwand *f*; Einfriedungsmauer
f, Umfassungswand *f*
f mur *m* en pans [à ossature]; mur *m*
de clôture [d'enceinte]
nl vakwerkwand *m*; ringmuur *m*,
omheiningsmuur *m*
r фахверковая стена *f*, стена *f*
заполнения каркаса; наружная
ограждающая стена *f*

E275 e **encroachment line**
d Wasserspiegelrand *m*
f bord *m* d'eau, frontière *f* liquide
nl vloedlijn *f* (*m*)
r урез *m* воды

E276 e **encrustation of pipes**
d Krustenbildung *f* [Verkrustung *f*] von
Rohren
f incrustation *f* dans les tuyauteries
nl afgroeiing *f* in buizen
r инкрустация *f* [зарастание *n*] труб

E277 e **end anchorage**
d Endverankerung *f*
f ancrage *m* d'extrémité
nl eindverankering *f* (*van de spankabels*)
r торцевое анкерное устройство *n*

E278 e **end-bearing pile**
d Säulenpfahl *m*
f pieu *m* résistant à la pointe,
pieu-colonne *m*
nl op stuit berekende paal *m*
r свая-стойка *f*, свая-колонна *f*

E279 e **end block**
d Endblock *m*
f block *m* d'about
nl ankerblok *n* (van de spankabels)
r блок *m* усиления торца
(*предварительно напряжённого*
элемента)

E280 e **end cap**
d Enddeckel *m*
f bouchon *m* d'extrémité
nl eindafsluiting *f*
r заглушка *f*

E281 e **end conditions**
d Auflagerbedingungen *f pl*,
Einspannbedingungen *f pl*

f conditions *f pl* aux extrémités (*pour des pièces comprimées*)
nl eindenvoorwaarden *f pl* (*drukstaaf*)
r условия *n pl* опирания концов (*сжатого элемента*)

E282 *e* **end contraction**
d End-Einschnürung *f*, seitliche Einschnürung *f*
f contraction *f* latérale
nl eindcontractie *f*, insnoering *f* aan het eind
r боковое сжатие *n*

E283 *e* **end distance**
d Abstand *m* zwischen dem Kopfende eines Bauelementes und der Bolzenaugenmitte
f distance *f* au bord
nl afstand *m* tussen het einde van een staaf en het hart van zijn bevestiging
r расстояние *n* от торца [конца] элемента до центра болтового отверстия

E284 *e* **end hook**
d Endhaken *m*
f crochet *m* d'extrémité
nl haak *m* (*wapeningsstaaf*)
r крюк *m* (*арматурного стержня*)

E285 *e* **end joint**
d Endknoten *m*, Stoßverbindung *f*
f assemblage *m* [joint *m*] bout à bout
nl stompe las *f* (*m*), stuiklas *f* (*m*)
r соединение *n* встык

E286 *e* **end lap**
d Überlappungslänge *f*
f longueur *f* de recouvrement
nl overlap *m*
r длина *f* нахлёстки

E287 *e* **end lap joint**
d gerade Ecküberblattung *f*
f entaille *f* d'angle à mi-bois
nl rechte las *f* (*m*) (*hout*)
r угловое соединение *n* в полдерева

E288 **endless saw** *see* **band saw**

E289 *e* **endless sling**
d endloser Stropp *m*
f élingue *f* sans fin
nl strop *m*, *f* zonder eind
r петлевой строп *m*

E290 *e* **end moments**
d Stützmomente *n pl*; Einspann(ungs)momente *n pl*
f moments *m pl* d'appui; moments *m pl* d'encastrement
nl inklemmingsmomenten *n pl*
r опорные *или* концевые моменты *m pl*

E291 *e* **end post**
d Ecksäule *f*, Eckstütze *f*, Schlußstütze *f*
f montant *m* extrême [d'extrémité]
nl eindstijl *m*
r концевая стойка *f*

E292 *e* **end restraint**
d Endeinspannung *f* (*Balken*)

f encastrement *m* d'extrémité (*d'une poutre*)
nl inklemming *f* (*balk*)
r защемление *n* конца (*балки*)

E293 *e* **end section of a tunnel**
d Portalstrecke *f* eines Tunnels
f débouché *m* du tunnel
nl tunnelmond *m*
r устье *n* туннеля

E294 *e* **end sill**
d Gegenschwelle *f*, Endschwelle *f*, Gegensperre *f*, Stoßnase *f*
f contre-barrage *m*, mur *m* déflecteur
nl einddrempel *m* van een woelkom
r водобойная стенка *f*

E295 *e* **end span**
d Endfeld *n*, Außenfeld *n*, Endöffnung *f*, Außenöffnung *f*
f travée *f* externe [terminale, extrême, d'extrémité]
nl eindoverspanning *f*
r крайний пролёт *m*

E296 *e* **end stiffener**
d Endsteife *f*
f nervure *f* de raidissement extrême (*d'une poutre*)
nl eindverstijving *f* (*samengestelde ligger*)
r концевой уголок *m* жёсткости (*балки*)

E297 *e* **end thrust**
d Enddruck *m*
f poussée *f* d'about
nl staafkracht *f* (*m*), kracht *f* (*m*) op het einde van een staaf
r усилие *n*, передаваемое торцом элемента

E298 *e* **endurance limit**
d Ermüdungsgrenze *f*
f limite *f* de fatigue
nl vermoeidheidsgrens *f* (*m*)
r предел *m* выносливости

E299 **endurance strength** *see* **fatigue strength**

E300 **energy dispersion block** *see* **baffle block**

E301 *e* **energy dissipator**
d Energievernichter *m*, Schikane *f*, Stromenergiebrecher *m*
f dissipateur *m* d'énergie, transquillisateur *m*
nl stromingsenergiebreker *m*
r гаситель *m* энергии потока

E302 *e* **energy-efficient building**
d energieökonomisch günstiges Gebäude *n*
f bâtiment *m* à basse consommation d'énergie
nl energiesparend gebouw *n*
r здание *n* с низким энергопотреблением

E303 *e* **energy-efficient precast wall panel**
d energiesparende Fertigbetonwandplatte *f* mit hochwertiger Wärmedämmung

  f panneau *m* de mur préfabriqué à haute résistance thermique
  nl geprefabriceerd wandpaneel *n* met hoge isolatiewaarde
  r сборная стеновая панель *f* с эффективной теплоизоляцией

E305 *e* energy line
  d Energielinie *f*
  f ligne *f* de charge
  nl energielijn *f* (*m*)
  r линия *f* полных напоров, напорная [энергетическая] линия *f*

E306 engaged column *see* attached column

E307 *e* engineering brick
  d Hartbrandstein *m*, Klinkerziegel *m*
  f brique *f* vitrifiée, klinker *m*
  nl klinker *m*
  r клинкер *m*, клинкерный кирпич *m*

E308 *e* engineering geology
  d Ingenieurgeologie *f*, Baugrundgeologie *f*
  f géologie *f* appliquée, géotechnique *f*
  nl toegepaste geologie *f*
  r инженерная геология *f*

E309 *e* engineering requirements
  d technische Forderungen *f pl*
  f prescriptions *f pl* techniques
  nl technische voorwaarden *f pl*
  r технические требования *n pl*

E310 engineering services *see* building services

E311 engineering structures *see* engineering works

E312 *e* engineering survey
  d 1. ingenieurmäßige Erkundungen *f pl* 2. Kontrolle *f* des technischen Zustandes eines Gebäudes *oder* Bauwerkes
  f 1. recherches *f pl* sur les conditions naturelles de site; reconnaissance *f* géotechnique 2. expertise *f* des constructions
  nl 1. bouwtechnisch onderzoek *n* 2. inspectie *f* van de technische toestand van een gebouw
  r 1. инженерные изыскания *n pl* 2. обследование *n* технического состояния (*здания или сооружения*)

E313 *e* engineering works
  d Ingenieurbauten *m pl*, Ingenieurbauwerke *n pl*
  f ouvrages *m pl* d'art
  nl bouwwerken *n pl*, civieltechnische kunstwerken *n pl*
  r инженерные сооружения *n pl*

E314 *e* engineer's chain
  d Meßkette *f*
  f chaîne *f* d'arpenteur
  nl meetketting *f*
  r мерная цепь *f*

E315 *e* English basement
  d Sockelgeschoß *n*
  f sous-sol *m* habitable

  nl souterrain *n*
  r цокольный этаж *m* (*жилого здания*)

E316 *e* English bond
  d Blockverband *m*
  f appareil *m* de maçonnerie anglaise [à l'anglaise]
  nl blokverband *n*, staand verband *n*
  r цепная перевязка *f* [кладка *f*]

E317 *e* English cross bond
  d Kreuzverband *m*
  f appareil *m* croisé [(de maçonnerie) alterné en croix]
  nl kruisverband *n*
  r английская крестовая перевязка *f* [кладка *f*] (*чередование тычковых и ложковых рядов со смещением последних на* ¹/₄ *кирпича*)

E318 *e* English tile
  d Strangbiberschwanz *m*, stranggefertigter Biberschwanz *m*
  f tuile *f* (cuite) plate
  nl Engelse (vlakke dak)pan *f* (*m*)
  r плоская ленточная черепица *f*

E319 enlarged toe pile *see* belled-out pile

E320 *e* enrichment
  d Verzierung *f*
  f ornement *m*
  nl versiering *f*
  r декоративный элемент *m*, украшение *n*, отделка *f*

E321 *e* enrockment
  d Steinschüttung *f*, Steinwurf *m*
  f enrochement *m* (en vrac)
  nl steenbezetting *f*
  r каменная наброска *f*

E322 *e* entrained air
  d Lufteinschluß *m*
  f air *m* entrainé
  nl (bij het gieten) meegekomen luchtbellen *f* (*m*) *pl* (*beton*)
  r воздух *m*, вовлечённый в бетонную смесь (*с помощью добавок*)

E323 *e* entrance
  d 1. Eingang *m* 2. Eingangshalle *f* (*Gebäude*) 3. Mund *m*, Portal *n* (*Tunnel*) 4. Haupt *n* (*Schleuse, Dock*)
  f 1. entrée *f* 2. vestibule *m*; antichambre *f* 3. débouché *m* du tunnel 4. tête *f* (d'écluse, de dock)
  nl 1. ingang *m* 2. vestibule *m* 3. mond *m*, ingang *m* (*tunnel*) 4. hoofd *m* (*sluis*)
  r 1. вход *m* 2. небольшой холл *m или* вестибюль *m*; прихожая *f* 3. устье *n*, портал *m* (*туннеля*) 4. голова *f* (*шлюза, дока*)

E323 *e* entrance foyer
  d Eingangshalle *f*, Empfangshalle *f*
  f hall *m* d'entrée [d'acceuil, de réception]; foyer *m*
  nl ontvangshal *f* (*m*); foyer *m*
  r холл *m*, фойе *n*

E323b entrance head *see* entry head

E323ce **entrance lock**
  d Dockschleuse *f*
  f écluse *f* d'entrée
  nl doksluis *f (m)*
  r входной шлюз *m (дока-бассейна)*

E323d **entrance loss** *see* **entry loss**

E323ee **entrance ramp**
  d Rampe *f*, Auffahrt *f*, geneigte
    Einfahrt *f*
  f rampe *f* d'accès
  nl (hellende) oprit *m*, laadperron *n*
  r наклонный въезд *m*, рампа *f*

E324 *e* **entresol**
  d Halbgeschoß *n*, Zwischengeschoß *n*,
    Mezzanin *n*
  f entresol *m*, mezzanine *f*
  nl tussenverdieping *m (f)*, entresol *m*
  r мезонин *m*

E325 **entry** *see* **entrance**

E326 *e* **entry head**
  d Eintrittsdruckhöhe *f*
  f charge *f* d'entrée
  nl drukhoogte *f (m)* bij bovenhoofd
  r входной напор *m*

E327 *e* **entry loss**
  d Eintrittsverlust *m*
  f perte *f* de charge à l'entrée
  nl schutverliezen *n pl*
  r потеря *f* напора на входе

E328 *e* **envelope**
  d 1. Hülle *f*; Umhüllung *f*; Mantel *m*;
    Umkleidung *f*, Verkleidung *f*
    2. Hüllgebilde *n*, Hüllfläche *f*
  f 1. enveloppe *f*; voile *m*; coque *f*;
    revêtement *m*; capot *m*, chemise *f*
    2. enveloppe *f* imaginaire (*délimitant
    le contour extérieur d'un bâtiment*)
  nl 1. omhulsel *n*; bemanteling *f*
    2. omhullend oppervlak *n* van een
    gebouw
  r 1. оболочка *f*; обшивка *f*; кожух *m*
    2. воображаемая оболочка *f* по
    наружным габаритам сооружения
    (*даёт представление о форме и
    размерах*)

E329 *e* **envelope curve**
  d Umhüllungskurve *f*, Hüllkurve *f*
  f courbe *f* enveloppe
  nl omhullende (kromme *f (m)*) *f*
  r огибающая кривая *f*

E330 *e* **envelope wall**
  d Ummantelungswand *f*
  f enceinte *m* [écran *m*] étanche (*de la
    centrale nucléaire*)
  nl ommantelingsmuur *m*
  r герметизирующая оболочка *f* [стенка
    f] (*АЭС*)

E331 **environmental architecture** *see*
    landscape architecture

E332 *e* **environmental chamber**
  d Klimakammer *f*, Simulierkammer *f*
  f chambre *f* climatique [à climat
    artificiel]

nl klimaatkamer *f (m)*, klimaatkast *f (m)*
  r камера *f* искусственного климата,
    климатическая камера *f*

E333 *e* **environmental design**
  d 1. umweltschutzorientierte
    Projektierung *f* [Gestaltung *f*]
    2. Projektierung *f* der
    raumklimatechnischen Anlagen und
    Einrichtungen
  f 1. élaboration *f* du projet de
    protection de l'environnement
    2. études *f pl* des systèmes de réglage
    de microclimat
  nl 1. milieubewust ontwerpen *n*
    2. ontwerpen *n* van
    luchtbehandelingssystemen
  r 1. природоохранное проектирование
    n, проектирование *n* защиты
    окружающей среды
    2. проектирование *n* систем
    регулирования микроклимата

E334 *e* **environmental hazards**
  d schadhafte Umgebungseinwirkungen *f
    pl*
  f effets *m pl* nocifs de l'environnement
  nl bedreiging *f* van het milieu
  r вредные воздействия *n pl*
    окружающей среды

E335 *e* **environmental system**
  d raumklimatechnische Anlage *f*
  f système *m* de climatisation [de
    microclimat]
  nl klimaatregelingsinstallatie *f*
  r система *f* регулирования
    микроклимата

E336 *e* **environment pollution**
  d Umweltverunreinigung *f*
  f pollution *f* de l'environnement
  nl milieuontreiniging *f*, milieubederf
    n
  r загрязнение *n* окружающей среды

E337 *e* **epoxy adhesive**
  d Epoxydharzkleber *m*
  f adhésif *m* [colle *f*] époxide
  nl epoxyharslijm *m*
  r эпоксидный клей *m*

E338 *e* **epoxy binder**
  d Epoxydharzbindemittel *n*
  f liant *m* époxyde
  nl epoxyharsbindmiddel *n*
  r эпоксидное вяжущее *n*

E339 *e* **epoxy-bitumen material** *US*
  d Epoxydharz *n* mit
    Schwarzbindemittel
  f liant *m* [mastic *m*] à base de bitume
    et époxyde
  nl epoxy-bitumenbindmiddel *n*
  r эпоксидно-битумное вяжущее *n*

E340 *e* **epoxy-coated rebars, epoxy coated
    reinforcing bars**
  d Bewehrungsstäbe *m pl* mit
    Epoxydüberzug
  f barre *f pl* d'armature enrobée par
    époxyde

nl wapeningsstaven *f (m) pl* met
epoxyharsbedekking
r арматура *f* с эпоксидным
покрытием

E341 *e* **epoxy concrete**
d Epoxydbeton *m*
f béton *m* de résine époxyde
nl epoxybeton *n*, beton *n* met
epoxybindmiddel
r эпоксидный полимербетон *m*

E342 *e* **epoxy cure-seal-harden compound**
d Epoxydharz-Nachbehandlungsfilm *m*
mit komplementärer Betonschutz-
funktion
f curing-compound *m* à base d'époxyde
nl epoxy-nabehandelingsmengsel *n*
r эпоксидный состав *m* для
выдерживания свежеуложенного
бетона

E343 *e* **epoxy mortar**
d Epoxydmörtel *m*, Mörtel *m* mit
Epoxydharzbindemittel
f mortier *m* époxyde [à base de résine
époxyde]
nl epoxymortel *m*, mortel *m* met
epoxyharsbindmiddel
r эпоксидный раствор *m*,
полимерраствор *m* на эпоксидной
смоле

E344 *e* **equal friction method**
d Reibungsverlustausgleich-
-Verfahren *n*
f méthode *f* de calcul des conduits
d'air par pertes de frottement égales
nl methode *f* van
wrijvingsverliesvergelijking
r метод *m* эквивалентных длин
(*расчёт сети воздуховодов по
условию равных потерь на трение
на расчётных участках*)

E345 *e* **equalizer**
d Ausgleichbehälter *m*, Ausgleichgefäß
n
f égalisateur *m*
nl vereffeningsinrichting *f*, balans *f (m)*
r уравниватель *m* давления
(*компрессора*)

E347 *e* **equalizing reservoir**
d 1. Ausgleichbecken *n*,
Ausgleichspeicher *m*
2. Ausgleichbehälter *m*
3. Gegenbehälter *m*
f 1., 2. bassin *m* de compensation
3. contre-réservoir *m*, réservoir *m*
d'équilibre
nl 1. vereffeningsbekken *n*
2. balansreservoir *n*
r 1. водохранилище *n*
компенсированного регулирования
2. уравнительный резервуар *m*
3. контррезервуар *m*

E348 *e* **equalizing tank**
d Ausgleichbehälter *m*

f réservoir *m* d'équilibre
nl vereffeningsreservoir *m*
r усреднитель *m*, усреднительный
резервуар *m*

E349 **equalizing valve** *see* **balancing valve**

E350 *e* **equal lay rope**
d Gleichschlagseil *n*
f câble *m* d'acier à câblage parallèle
nl parallelgeslagen kabel *m*
r канат *m* параллельной свивки

E351 *e* **equation of continuity**
d Kontinuitätsgleichung *f*
f équation *f* de continuité
nl continuïteitsvergelijking *f*
r уравнение *n* неразрывности

E352 *e* **equation of the influence line**
d Einflußliniengleichung *f*
f équation *f* de la ligne d'influence
nl vergelijking *f* van de invloedlijn
r уравнение *n* линии влияния

E353 *e* **equilateral roof**
d Satteldach *n* mit Dachneigung von 60°
f comble *m* équilatéral
nl gelijkzijdig zadeldak *n*
r двускатная крыша *f* с уклоном
скатов 60°

E354 *e* **equilibrant**
d Ausgleichskraft *f*
f force *f* équilibrante [d'équilibre]
nl tegenwicht *n*
r уравновешивающая сила *f*

E355 *e* **equilibrium conditions**
d Gleichgewichtsbedingungen *f pl*
f conditions *f pl* d'équilibre
nl evenwichtsvoorwaarden *f pl*
r условия *n pl* равновесия

E356 *e* **equilibrium drawdown**
d Grundwasserabsenkziel *n*
f rabattement *m* d'équilibre
nl maximale grondwaterverlaging *f*
r предельное понижение *n* уровня
грунтовых вод

E357 *e* **equilibrium method**
d Deformationsmethode *f*,
Deformationsverfahren *n*
f méthode *f* des déformations
nl vervormingsmethode *f*
r метод *m* деформаций

E358 *e* **equilibrium of forces**
d Kräftegleichgewicht *n*
f équilibre *m* des forces
nl krachtenevenwicht *n*
r равновесие *n* сил

E359 *e* **equilibrium polygon**
d Gleichgewichtspolygon *n*
f polygone *m* de forces
nl gesloten krachtenveelhoek *m*
r многоугольник *m* сил

E360 *e* **equilibrium pressure**
d Gleichgewichtsdruck *m*
f pression *f* d'équilibre
nl evenwichtsdruk *m*
r равновесное давление *n*

E361 *e* **equilibrium state**
    d Gleichgewichtszustand *m*
    f état *m* d'équilibre
    nl evenwichtstoestand *m*
    r состояние *n* равновесия

E362 *e* **equilibrium temperature**
    d Gleichgewichtstemperatur *f*
    f température *f* d'équilibre
    nl evenwichtstemperatuur *f*
    r равновесная температура *f*

E363 *e* **equipment ground**
    d Erdung *f* des Gehäuses
    f mise *f* à la terre des appareils
    électriques
    nl aarding *f* van een toestel
    r заземление *n* корпуса
    (электрооборудования)

E364 *e* **equipment yard**
    d Bauhof *m*, Gerätepark *m*,
    Maschinenpark *m*
    f parc *m* de matériel [de machines]
    nl machinepark *n*
    r машинный парк *m*

E365 *e* **equivalent diameter**
    d gleichwertiger Durchmesser *m*
    f diamètre *m* équivalent
    nl equivalente diameter *m*
    r эквивалентный диаметр *m* (*трубы,*
    воздуховода)

E366 *e* **equivalent embedment length**
    d äquivalente Einbindungslänge *f*
    f longueur *f* d'adhérence équivalente
    (*d'égale à l'effet du crochet*)
    nl equivalente hechtingslengte *f*
    (*wapeningshaak*)
    r эквивалентная (*концевому крюку*
    арматуры) длина *f* зоны
    сцепления

E367 *e* **equivalent length**
    d äquivalente Länge *f*
    f longueur *f* équivalente
    nl equivalente lengte *f*
    r приведённая [эквивалентная]
    длина *f*

E368 *e* **equivalent load**
    d Vergleichslast *f*, Ersatzlast *f*,
    äquivalente Last *f*
    f charge *f* [surcharge *f*] équivalente
    nl vervangende belasting *f*
    r эквивалентная нагрузка *f*

E369 *e* **equivalent pipe**
    d äquivalentes Rohr *n*
    f tuyau *m* équivalent
    nl gelijkwaardige pijp *f* (*m*)
    r эквивалентная труба *f* (*труба*
    равного гидравлического
    сопротивления при данном
    расходе)

E370 *e* **erecting bill**
    d Montageliste *f*
    f liste *f* des éléments préfabriqués en
    ordre de montage
    nl montagelijst *f* (*m*)
    r монтажная спецификация *f* (*узлов и*
    элементов конструкции в порядке их
    монтажа)

E371 *e* **erecting deck**
    d Montagebühne *f*, Montageplattform *f*
    f plate-forme *f* de montage
    nl montageplatform *n*
    r монтажный помост *m*, монтажная
    площадка *f* [платформа *f*]

E372 *e* **erecting jib**
    d Montageausleger *m*, Aufbauausleger *m*
    f flèche *f* [chèvre *f*] de levage
    nl montageheefboom *m*
    r монтажная стрела *f*

E373 *e* **erecting mast**
    d Montagemast *m*, Aufbaumast *m*
    f mât *m* de montage [de levage]
    nl montagemast *m*
    r монтажная мачта *f*

E374 *e* **erecting tools**
    d Montagewerkzeuge *n pl*
    f outils *m pl* [outillage *m*] d'assemb-
    lage [de montage]
    nl montagewerktuigen *n pl*
    r инструменты *m pl* для сборки и
    монтажа

E375 *e* **erection**
    d Errichtung *f*, Montage *f*
    f montage *m*, assemblage *m*
    nl oprichten *n*, opbouw *m*, plaatsing *f*
    r монтаж *m*

E376   **erection bars** *see* **erection reinforcement**

E377 *e* **erection brace**
    d Montagestrebe *f*; Montageverband *m*;
    Montageschräge *f*
    f entretoise *f* de montage
    nl montageverband *n*; montageschoor *m*
    r монтажная схватка *f или* связь *f*;
    монтажный раскос *m*

E378 *e* **erection by floating**
    d Einschwimmontage *f* (*Brückenbau*)
    f montage *m* à l'aide des engins
    flottants
    nl drijvende montage *f* (*bruggenbouw*)
    r монтаж *m* на плаву (*пролётного*
    строения моста)

E379 *e* **erection by launching**
    d Montage *f* durch Einfahren (*Brücke*)
    f montage *m* par lancement
    nl montage *f* door overrollen (*brugslag*)
    r надвижной монтаж *m* (*моста*)

E381 *e* **erection column**
    d Montagestütze *f*
    f montant *m* [poteau *m*] provisoire de
    montage
    nl montagesteun *m*
    r монтажная опора *f* [стойка *f*]

E382 *e* **erection crane**
    d Montagekran *m*
    f grue *f* pour travaux de montage, grue
    f de chantier
    nl montagekraan *f* (*m*)
    r монтажный кран *m*

E383 *e* erection diagram, erection drawing
  d Montagezeichnung *f*
  f diagramme *m* de montage
  nl montagetekening *f*
  r монтажная схема *f*, монтажный
    чертёж *m*

E384 *e* erection equipment
  d Montageausrüstung *f*
  f équipement *m* des chantiers de
    montage, engins *m* de montage
  nl montage-uitrusting *f*
  r монтажное оборудование *n*

E385 erection joint *see* field joint

E386 *e* erection load, erection loading
  d Montagebelastung *f*, Montagelast *f*
  f charge *f* de montage
  nl montagebelasting *f*
  r монтажная нагрузка *f*

E387 *e* erection of structural steel
  d Montage *f* der Stahlkonstruktionen
  f montage *m* des constructions [de
    charpente] en acier
  nl montage van een staalconstructie
  r монтаж *m* стальных конструкций

E388 *e* erection practice
  d Montage-Arbeitsweisen *f pl*,
    Montageverfahren *n pl*
  f pratiques *f pl* [méthodes *f pl*] de
    montage
  nl montagewerk *n*
  r приёмы *m pl* [методы *m pl*] монтажа

E389 *e* erection procedure
  d Montagevorgang *m*,
    Montageverfahren *n*
  f procédé *m ou* technologie *f* de montage
  nl montagemethode *f*;
    montagevolgorde *f* (*m*)
  r технология *f*, метод *m или* процесс
    m монтажа

E390 *e* erection reinforcement
  d Montagebewehrung *f*,
    Montagestahleinlagen *f pl*
  f armature *f* [ferraillage *m*] de montage
  nl montagewapening *f*
  r монтажная арматура *f*

E391 *e* erection schedule
  d Montageplan *m*
  f programme *m* d'avancement des
    travaux de montage
  nl montagewerkplan *n*, montagerooster
    n
  r график *m* монтажа

E392 *e* erection scheme
  d Montageablaufplan *m*,
    Montageausführungsplan *m*
  f programme *m* d'exécution des travaux
    de montage
  nl montageontwerp *n*, montageplan *n*
  r проект *m* производства монтажных
    работ

E393 *e* erection sequence
  d Montagefolge *f*, Aufbaufolge *f*
  f succession *f* des opérations de

montage, ordre *m* des succession des
    travaux de montage
  nl montagevolgorde *f*
  r последовательность *f* операций
    монтажа, порядок *m* сборки
    конструкций

E394 *e* erection site
  d Montageplatz *m*
  f chantier *m* [aire *f*] de montage
  nl montageplaats *f* (*m*)
  r монтажная площадка *f*

E395 *e* erection stage
  d Montagephase *f*, Montagestadium *n*
  f étape *f* d'exécution des travaux de
    montage, phase *f* de montage
  nl montagestadium *n*
  r стадия *f* [очередь *f*] монтажа

E396 *e* erection stress
  d Montagespannung *f*
  f contrainte *f* due à la charge de
    montage
  nl montagespanning *f*
  r напряжение *n* от монтажной
    нагрузки

E397 *e* erection tower
  d 1. Montagegerüst *n*,
    Aufstellungsgerüst *n* 2. Montageturm
    m, Abspanngerüst *n*
  f 1. échafaudage *m* à tour, tour *f* de
    montage 2. mât *m* (haubané) de
    montage
  nl 1. montagestelling *f* 2. montagemast *m*
  r 1. монтажные леса *pl* башенного
    типа 2. монтажная вышка *f*;
    расчаленная монтажная мачта *f*

E398 *e* erection work
  d Montagearbeiten *f pl*
  f travaux *m pl* de montage
  nl montagewerkzaamheden *f pl*
  r монтажные работы *f pl*

E399 *e* erosion protection works
  d Kolkschutzanlagen *f pl*
  f ouvrages *m pl* de protection contre
    érosion
  nl erosie-bestrijdende maatregelen *m pl*
  r противоэрозионные сооружения *n pl*

E400 *e* escape
  d 1. Entlastungsgerinne *n*,
    Umleitungskanal *m*, Auslaßkanal *m*
    2. Entweichung *f*; Leckverlust *m*
  f 1. ouvrage *m* [canal *m*] de décharge
    2. échappement *m*, fuite *f* (*de gaz*,
    d'eau)
  nl 1. aflaat *m*, spui *n* 2. ontsnappen *n*,
    ontwijken *n*
  r 1. водосброс *m*, отводящее русло *n*
    (*на канале*) 2. утечка *f* (*газа, воды*)

E401 *e* escape canal
  d Abflußkanal *m*, Überlaufkanal *m*
  f canal *m* d'évacuation [de décharge]
  nl aflaatkanaal *n*, spuikanaal *n*
  r водосбросный канал *m*

E402 escape gallery *see* emergency gallery

E403 *e* **escape ladder**
  d Feuerleiter *f*, Fluchtleiter *f*
  f échelle *f* d'incendie
  nl brandladder *m*
  r пожарная лестница *f* (*наружная*)

E404 *e* **escape route**
  d Fluchtweg *m*
  f voie *f* (d'acheminement) de secours;
    sortie *f* de secours
  nl vluchtweg *m*
  r путь *m* эвакуации; запасной выход *m*

E405 *e* **escape stair**
  d Nottreppe *f*
  f escalier *m* de sauvetage
  nl noodtrap *m*
  r пожарная [аварийная] лестница *f*
    (*в здании*)

E406 *e* **espagnolette bolt**
  d Espagnoletteverschluß *m*,
    Treibriegel *m*
  f espagnolette *f*
  nl spanjolet *f* (*m*), espagnolet *f* (*m*)
  r (оконный) шпингалет *m*

E407 *e* **estate**
  d 1. Grundstück *n* 2. Baugrundstück *n*
  f 1. propriété *f* terrienne
    (immobilière) 2. terrain *m* à bâtir
  nl 1. terrein *n* met woningen 2.
    bouwgrond *m*
  r 1. земельный участок *m* 2.
    территория *f* строительства

E408   **etching primer** *see* **wash primer**

E409 *e* **evaporation basin**
  d Verdunstungsbecken *n*
  f bassin *m* d'évaporation
  nl verdampingsbekken *n*
  r испарительный бассейн *m*

E410 *e* **evaporative condenser**
  d Verdunstungsverflüssiger *m*,
    Verdunstungskondensator *m*
  f condenseur *m* à évaporation
  nl verdampingscondensor *m*
  r испарительный конденсатор *m*

E411 *e* **evaporative cooling**
  d Verdunstungskühlung *f*, adiabate
    Kühlung *f*
  f refroidissement *m* par évaporation
  nl verdampingskoeling *f*
  r испарительное [адиабатическое]
    охлаждение *n*

E412 *e* **evaporative heat meter**
  d Verdunstungswärmemesser *m*
  f calorimètre *m* à évaporation
  nl verdampinswarmtemeter *m*
  r испарительный тепломер *m*

E413 *e* **evaporative humidifier**
  d Verdunstungs(luft)befeuchter *m*
  f humidificateur *m* à évaporation
  nl verdampings(lucht)bevochtiger *m*
  r испарительный увлажнитель *m*
    воздуха

E414 *e* **event**
  d Ereignis *n* (*Netzplantechnik*)

  f événement *m* *planification par*
    chemin critique)
  nl gebeurtenis *f*, activiteit *f*
    (*netwerkplanning*)
  r событие *n* (*сетевое планирование*)

E415 *e* **excavated area**
  d Aushubfläche *f*
  f zone *f* de déblai
  nl oppervlak *n* van de uitgraving
  r зона *f* выемки

E416 *e* **excavated material**
  d Aushubmaterial *n*
  f terre *f* excavée, sol *m* excavé
  nl uitgegraven grond *m*
  r извлечённый [вынутый] грунт *m*

E417 *e* **excavated volume**
  d Erdaushubvolumen *n*
  f volume *m* de déblai
  nl volume *n* van de uitgegraven grond
  r объём *m* выемки

E418 *e* **excavating cableway**
  d Kabelkran *m* mit Greifer
  f transporteur *m* à câble pour les
    travaux de terrassement
  nl kabelbaan-graafmachine *f*
  r экскаваторный кабель-кран *m*,
    кабель-кран-экскаватор *m*

E419 *e* **excavating equipment**
  d Erdbaumaschinen *f pl*
  f engins *m pl* de terrassement
  nl graafmachines *f pl*
  r землеройное оборудование *n*

E420 *e* **excavating tools**
  d Erdbaugeräte *n pl*
  f outils *m pl* de terrassement
  nl graafwerktuigen *n pl*
  r инструменты *m pl* для земляных
    работ

E421 *e* **excavation**
  d 1. Aushub *m*, Abtrag *m*, Abbau *m*,
    Ausbaggerung *f*, Aushubarbeiten *f pl*,
    Ausschachtung *f* 2. Baugrube *f*
  f 1. travaux *m pl* d'excavation,
    excavation *f*, fouille *f* 2. fosse *f*
  nl 1. uitgraving *f*, ontgraving *f*
    2. bouwput *m*
  r 1. экскавация *f*, выемка *f*
    [отрывка *f*] грунта 2. котлован *m*,
    выемка *f*

E422 *e* **excavation bottom**
  d Grubensohle *f*
  f fond *m* de fouille [d'excavation]
  nl bodem *m* van de ontgraving
  r дно *n* котлована [выемки]

E423 *e* **excavation depth**
  d Aushubtiefe *f*, Grabtiefe *f*,
    Baugrubentiefe *f*
  f profondeur *f* de fouille
  nl uitgravingsdiepte *f*, diepte *f* van de
    put
  r глубина *f* котлована [выемки]

E424   **excavation works** *see* **excavation 1.**

E425 *e* **excavator**
 d Bagger *m*
 f excavateur *m*, pelle *f* mécanique
 nl graafmachine *f*, excavateur *m*
 r экскаватор *m*

E426 *e* **excess heat**
 d Wärmeüberschuß *m*
 f exès *m* de chaleur, chaleur *f* excessive
 nl warmteoverschot *n*
 r теплоизбытки *m pl*

E427 *e* **excessive air elimination**
 d Betonentlüftung *f*
 f élimination *f* d'air occlus à l'excès
 nl lucht uit betonmortel verdrijven (*door een toeslag*)
 r удаление *n* излишнего воздуха из бетонной смеси

E428 *e* **excessive elastic deformability**
 d übermäßige elastische Verformbarkeit *f*
 f déformabilité *f* élastique excessive
 nl overmatige elastische vervorming *f*
 r чрезмерная упругая деформативность *f*

E429 *e* **excessive stiffness**
 d übermäßige Steifheit *f*
 f rigidité *f* excessive du béton frais
 nl overmatige stijfheid *f*
 r чрезмерная жёсткость *f*

E430 **excess pressure** *see* **surplus pressure**

E431 *e* **exfoliated vermiculite**
 d geblähter Vermikulit *m*
 f vermiculite *f* expansée
 nl geëxpandeerd vermiculiet *n*
 r вспученный вермикулит *m*

E432 *e* **exfoliation**
 d Abschuppung *f*, Abblätterung *f*
 f exfoliation *f*
 nl afbladdering *f*
 r отслаивание *n*, шелушение *n*

E433 *e* **exhaust air**
 d Abluft *f*
 f air *m* évacué
 nl afgevoerde lucht *f* (*m*)
 r вытяжной воздух *m*

E434 **exhaust canopy** *see* **exhaust hood**

E435 *e* **exhaust damper**
 d Abluftklappe *f*, Abluftventil *n*
 f clapet *m* d'exhaustion
 nl uitlaatklep *f* (*m*), schoorsteenklep *f* (*m*)
 r вытяжной клапан *m*

E436 *e* **exhauster, exhaust fan**
 d Absauger *m*, Sauglüfter *m*, Entlüfter *m*, Sauggebläse *n*
 f aspirateur *m*, ventilateur *m* d'aspiration, exhauster *m*
 nl afzuigventilator *m*, exhauster *m*
 r вытяжной вентилятор *m*, эксгаустер *m*

E437 *e* **exhaust flue**
 d Abzugskanal *m*, Fuchs(kanal) *m*
 f conduit *m* [conduite *f*] de fumée

 nl afzuigkanaal *n*, rookkanaal *n*
 r дымовытяжной канал *m*, дымоход *m*

E438 *e* **exhaust funnel**
 d Absaugtrichter *m*, Trichterabzug *m*
 f entonnoir *m* d'échappement
 nl afzuigpijp *f* (*m*), schoorsteenpijp *f* (*m*)
 r вытяжная воронка *f*

E439 *e* **exhaust grille**
 d Entlüftungsgitter *n*, Abluftgitter *n*
 f grille *f* de ventilation
 nl ventilatierooster *n*
 r вытяжная решётка *f*

E440 *e* **exhaust hood**
 d Absaughaube *f*, Abzugshaube *f*
 f hotte *f*
 nl afzuigkap *f* (*m*)
 r вытяжной зонт *m*

E441 *e* **exhaust intake**
 d Lufteinlaß *m*, Abluftdurchlaß *m*
 f bouche *f* d'exhaustion
 nl afzuigpunt *n*
 r вытяжной плафон *m*

E442 *e* **exhaust opening**
 d Entlüftungsöffnung *f*, Abzugsöffnung *f*
 f orifice *m* de soutirage
 nl afzuigopening *f*
 r вытяжное отверстие *n*

E443 *e* **exhaust plant**
 d Absauganlage *f*
 f installation *f* d'aspiration
 nl afzuiginstallatie *f*, ventilatie-inrichting *f*
 r вентиляционно-вытяжной агрегат *m*, вытяжная установка *f*

E444 *e* **exhaust shaft**
 d Entlüftungsschacht *m*
 f puits *m* d'échappement
 nl ventilatieschacht *f* (*m*)
 r вытяжная шахта *f*

E445 *e* **exhaust system**
 d Absauganlage *f*, Entlüftungsanlage *f*
 f système *m* [installation *f*] d'aspiration
 nl afzuiginstallatie *f*, ventilatiesysteem *n*
 r вытяжная система *f*

E446 *e* **exhaust ventilation**
 d Entlüftung *f*
 f ventilation *f* aspirante
 nl geforceerde [kunstmatige] ventilatie *f*
 r (общеобменная) вытяжная вентиляция *f*

E447 *e* **exit**
 d Ausgang *m*
 f sortie *f*, issue *f*
 nl uitgang *m*
 r выход *m*

E448 *e* **exit gradient**
 d Austrittsgefälle *n*
 f gradient *m* de sortie
 nl uitstroomverval *n*
 r выходной градиент *m* (*в конце водобоя*)

E449 *e* **exit loss**
   d Austrittsverluste *m pl*
   f perte *f* à la sortie
   nl uitstroomverlies *n*
   r потеря *f* напора на выходе

E450 *e* **exit opening**
   d Dachausstieg *m*, Dachaussteigluke *f*
   f ouverture *f* [lucarne *f*] d'accès au toit
   nl dakluik *n*
   r люк *m* для выхода на крышу

E451 *e* **exit ramp**
   d Abfahrtrampe *f*, Ausfahrtrampe *f*
   f rampe *f* d'accès, rampe *f* [bretelle *f*] de raccordement
   nl hellende afrit *m*
   r выездной пандус *m*, рампа *f*

E452   **exit route** *see* **escape route**

E453 *e* **expamet**
   d Streckmetall *n* (*Putzgrund*)
   f treillis *m* en métal déployé (*support d'enduit*)
   nl métal déployé, plaatgaas *n*
   r сетка *f* из просечно-вытяжного листового металла (*для штукатурных работ*)

E454 *e* **expanded blast-furnace slag**
   d Hüttenbims *m*, Schaumschlacke *f*, Thermosit *m*
   f laitier *m* expansé
   nl geëxpandeerde hoogovenslak *f* (*m*), thermosiet *m*
   r термозит *m*, шлаковая пемза *f*

E455 *e* **expanded bore pile**
   d Bohrpfahl *m* mit Fußverbreiterung
   f pieu *m* foré à bulbe
   nl boorpaal *m* met verbrede voet
   r буровая свая *f* с уширенным основанием

E456   **expanded cement** *see* **expansive cement**

E457 *e* **expanded clay**
   d Blähtonzuschlag *m*
   f argile *f* expansée [soufflée]
   nl geëxpandeerde klei *f* (*m*), klinkerisoliet *m*
   r керамзит *m*

E458 *e* **expanded column head**
   d Pilzkopf *m* einer Kolonne
   f champignon *m* d'un poteau
   nl kolomkop *m* van een paddestoelvloer
   r грибовидный оголовок *m* колонны

E459   **expanded glass** *see* **foam glass**

E460 *e* **expanded metal**
   d Streckmetall *n*
   f métal *m* déployé
   nl métal déployé, plaatgaas *n*
   r просечно-вытяжной металлический лист *m*

E461 *e* **expanded metal fabric**
   d Streckmetallbewehrung *f*
   f treillis *m* d'armature en métal déployé
   nl plaatgaasmat *f* (*m*) (als wapening)

   r просечно-вытяжная арматурная сетка *f*

E462   **expanded metal lathing** *see* **expamet**

E463 *e* **expanded-metal partition**
   d Streckmetall-Trennwand *f*
   f cloison *f* en métal déployé avec deux couches d'enduit
   nl scheidingswand *m* van aan beide zijden gepleisterd plaatgaas
   r перегородка *f* из просечно-вытяжных стальных листов, оштукатуренных с обеих сторон

E464 *e* **expanded perlite**
   d (auf)geblähter Perlit *m*
   f perlite *f* expansée
   nl geëxpandeerd perliet *n* (*toeslag*)
   r вспученный перлит *m* (*заполнитель*)

E465 *e* **expanded plastics**
   d Schaumstoffe *m pl*
   f mousses *f pl* plastiques
   nl schuimplastic *n*, geëxpandeerde kunststof *f* (*m*)
   r пенопласты *m pl*

E466 *e* **expanded polyurethane**
   d Polyurethan-Hartschaum *m*, Polyurethanschaum *m*, PUR-Schaum *m*
   f polyuréthane *m* expansé, mousse *f* de polyuréthane
   nl schuim-polyurethaan *n*
   r пенополиуретан *m*

E467 *e* **expanded rubber**
   d Schaumgummi *m*
   f caoutchouc-mousse *f*, caoutchouc--éponge *m*
   nl schuimrubber *n*
   r губчатая [микропористая] резина *f*

E468 *e* **expanded shale**
   d Blähschieferton *m*
   f schiste *m* expansé
   nl geëxpandeerde kleischalie *f*
   r вспученный аргиллит *m*

E469 *e* **expanded slate**
   d Blähtonschiefer *m*
   f ardoise *f* expansée
   nl geëxpandeerde leisteen *m*
   r вспученный глинистый сланец *m*

E470   **expanding** *see* **expansion**

E471   **expanding cement** *see* **expansive cement**

E472 *e* **expanding vault**
   d kegeliges Tonnengewölbe *n*
   f voûte *f* conique
   nl puntgewelf *n*
   r конический свод *m*

E473 *e* **expansion**
   d Ausdehnung *f*; Dehnung *f*; Verlängerung *f*; Quellen *n*, Aufblähen *n*, Auftreiben *n*
   f expansion *f*; dilatation *f*; croissance *f*; foisonnement *m*
   nl uitbreiding *f*, uitzetting *f*, expansie *f*; opzwellen *n*, expanderen *n*
   r расширение *n*; удлинение *n*; вспучивание *n*

E474 e expansion bearing
  d eingespanntes bewegliches Auflager n
  f appareil m d'appui mobile
  nl glij-oplegging f
  r защемлённая [защемляющая]
    подвижная опора f

E475 expansion bend see expansion loop

E476 e expansion bolt see expansion-shell
    bolt

E477 expansion cistern see expansion tank

E478 e expansion crack
  d Treibriß n; Wärmedehnungsriß m
  f fissure f d'expansion
  nl scheur f (m) ten gevolge van
    uitzetting
  r усадочная трещина f; трещина f при
    тепловом расширении

E479 e expansion joint
  d 1. Dehn(ungs)fuge f, Trennfuge f,
    Raumfuge f 2. Dehnungsausgleicher m
  f 1. joint m de dilatation
    2. compensateur m (à tuyaux)
  nl 1. dilatatievoeg f (m)
    2. expansiekoppeling f (pijpen)
  r 1. расширительный [температурный]
    шов m 2. компенсатор m
    (трубопровода)

E480 e expansion joint cover
  d Dehnungsfugendeckleiste f
  f couvre-joint m dissimulant un joint
    d'expansion
  nl dekplaat f (m) van de dilatatievoeg
  r накладка f, скрывающая
    температурный шов

E481 e expansion loop
  d Ausgleichsschleife f, Ausgleichbogen
    m
  f boucle f compensatrice [de
    compensation], lyre f de dilatation
  nl expansielus f (m)
  r П-образный или лирообразный
    компенсатор m (трубопровода)

E482 e expansion rollers
  d Rollenauflager n
  f appareil m d'appui à rouleaux
  nl roloplegging f (brug)
  r подвижная катковая опорная часть f
    (моста)

E483 e expansion-shell bolt
  d Spreizhülsenanker m
  f boulon m à fente et coin
  nl keilbout m met huls
  r болт m с разрезной распорной
    гильзой, распорный
    [расширительный] болт m

E484 e expansion strip
  d elastische Fugeneinlage f,
    Fugenstreifen m
  f bande f de dilatation
  nl elastische vulling f van een
    dilatatievoeg
  r эластичная прокладка f в
    температурном шве

E485 e expansion tank
  d 1. Ausdehnungsgefäß n,
    Expansionsbehälter m
    2. Ausgleich(s)behälter m
  f 1. tank m de dilatation, réservoir m
    d'expansion 2. réservoir m
    d'équilibre
  nl 1. expansievat n
    2. vereffeningsreservoir n
  r 1. открытый расширительный бак m
    [сосуд m] 2. регулирующий
    [уравнительный] резервуар m

E486 expansion turbine see turboexpander

E487 e expansion vessel
  d geschlossener Expansionsbehälter m,
    Druckausdehnungsgefäß n
  f vase m couvert d'expansion
  nl gesloten expansievat n
  r закрытый расширительный бак m
    [сосуд m]

E488 e expansive cement
  d Quellzement m, Schwellzement m,
    selbstspannender Zement m
  f ciment m expansif [sans retrait]
  nl expanderend cement n
  r расширяющийся [напрягающий]
    цемент m

E489 e expansive concrete
  d Quell(zement)beton m
  f béton m expansif [expansé, sans
    retrait]
  nl beton n met expanderend cement als
    bindmiddel
  r бетон m на расширяющемся
    цементе

E490 e expansive soil
  d Quellboden m, Schwellboden m
  f sol m expansif
  nl uitleverende grond m
  r набухающий грунт m

E491 e expected life of the structure
  d rechnerische Lebensdauer f des
    Bauwerkes
  f durée f de service calculée d'un
    bâtiment
  nl verwachte levensduur f van een
    gebouw
  r ожидаемый срок m службы
    сооружения

E492 e explanatory note
  d Erläuterungsbericht m,
    Erläuterungen f pl (zum Projekt)
  f note f explicative
  nl toelichting f (b.v. bij een ontwerp)
  r пояснительная записка f (к проекту)

E493 e exploded view
  d Darstellung f in auseinandergezogener
    Anordnung, auseinandergezogene
    Ansicht f
  f vue f détaillée [séparée] des éléments
  nl tekening f van onderdelen in
    gemonteerde volgorde
  r изображение n узла в разобранном
    виде

E494 *e* **explosion-proof luminaire** *US*
  d explosionsgeschützte Leuchte *f*
  f luminaire *m* antidéflagrant
  nl explosiebestendige lamp *f* (*m*)
  r взрывозащищённый светильник *m*

E495 *e* **explosion-proof wiring**
  d elektrische Sprengschutzleitung *f*
  f canalisation *f* électrique résistante à
    l'explosion
  nl explosiebestendige elektrische leiding *f*
  r взрывозащитная электропроводка *f*

E496 *e* **explosive-actuated fastening tool**
  d Bolzenschießgerät *n*
  f pistolet *m* de scellement à
    cartouches explosives
  nl montagepistool *n*, boutenschiethamer *m*
  r строительно-монтажный пистолет *m*

E497 *e* **explosive dust**
  d explosibler Staub *m*
  f poussière *f* explosible
  nl explosief stof *n*
  r взрывоопасная пыль *f*

E498 *e* **explosive limits**
  d Explosionsgrenzen *f pl*
  f limites *m pl* d'explosibilité
  nl explosiegrenzen *f* (*m*) *pl*
  r пределы *m pl* взрывоопасной
    концентрации (*газовоздушной смеси,*
    запылённого воздуха)

E499 *e* **explosive range**
  d Explosionsbereich *m*
  f domaine *m* de concentrations
    explosibles
  nl uitwerkingsdiepte *f* van een
    springlading
  r область *f* [диапазон *m*]
    взрывоопасных концентраций

E500 *e* **exposed aggregate**
  d Sichtbeton *m* mit freigelegten
    Zuschlägen
  f béton *m* à granulats lavés
  nl aan het oppervlak zichtbaar
    toeslagmateriaal *n*
  r бетон *m* с обнажёнными зёрнами
    заполнителя

E501 *e* **exposed aggregate finish**
  d freigelegte Zuschläge *m pl*
    [Zuschlagstoffe *m pl*]
  f crépissage *m* à agregats lavés
  nl vrijgelegde toeslagen *m pl*
    (*behandeling van betonvlak*)
  r фактурная отделка *f* (*бетонной*
    поверхности) с обнажёнными
    зёрнами крупного заполнителя

E501a *e* **exposed concrete**
  d Sichtbeton *m*, Architekturbeton *m*
  f béton *m* à granulats lavés
  nl in het zicht blijvend beton *n*
  r декоративный бетон *m* с обнажёнными
    зёрнами заполнителя

E502 *e* **exposed finish tiles**
  d unglasierte keramische
    Verkleidungsfliesen *f pl*

  f tuiles *f pl* de terra-cotta
  nl ongeglazuurde keramische tegels *m pl*
  r неглазурованные облицовочные
    керамические плитки *f pl*

E503 *e* **exposed intake**
  d freie Flußwasserentnahme *f*,
    Oberflächenwasserfassung *f*
  f prise *f* d'eau ouverte
  nl vrije rivierwateronttrekking *f*
  r открытый речной водозабор *m*

E504 *e* **exposed masonry**
  d unverputztes [rohes] Mauerwerk *n*
  f maçonnerie *f* de pierre sans ravalement
  nl schoon metselwerk *n*
  r неоштукатуренная каменная кладка *f*

E505 *e* **exposed wiring**
  d A.P.-Installation *f*, offen verlegte
    Leitung *f*
  f installation *f* de fils électriques à
    découvert
  nl elektrische leidingen *f pl* in het zicht
  r открытая электропроводка *f*

E506 *e* **exposure**
  d 1. Gebäudeanordnung *f*;
    Raumanordnung *f* 2.
    Windexponierung *f*, Sonnenbelichtung
    f bzw. Lärmexponierung *f* des
    Gebäudes 3. Wirkungsdauer *f*;
    Umgebungseinwirkung *f*
  f 1. orientation *f* (*d'un bâtiment ou*
    d'un local) 2. exposition *f* d'un
    bâtiment (*à l'effet du vent, au*
    rayonnement direct du soleil, au
    bruit) 3. durée *f* d'action des **agents**
    nocifs sur l'homme 4. sollicitation
    f [effet *m*] de l'environnement
  nl 1. oriëntatie *f* van een gebouw 2.
    blootstelling *f* van een gebouw **aan**
    invloeden van buitenaf, bezonning *f*
    3. duur *m* van de blootstelling **aan**
    invloeden van buitenaf
  r 1. ориентация *f* здания *или*
    помещения 2. ветровая,
    инсоляционная *или* шумовая
    экспозиция *f* здания
    3. длительность *f* воздействия
    вредности на человека; воздействие
    n окружающей среды

E507 *e* **exposure of aggregate**
  d Freilegen *n* der Zuschläge
  f mise *f* à nu des grains du granulat
    gros (*sur la surface de béton*)
  nl toeslagkorrels blootleggen *n* (*ter*
    verfraaiing van beton)
  r обнажение *n* зёрен заполнителей
    (*вид отделки бетонных*
    поверхностей)

E508 *e* **express-filter**
  d Hochgeschwindigkeitsfilter *n*, *m*,
    Immediumfilter *n*, *m*
  f filtre-expresse *m*
  nl snelfilter *n*, *m*
  r высокоскоростной фильтр *m*

E509 *e* **expressway**
 d Schnellstraße *f*, kreuzungsfreie
  Autostraße *f*
 f autoroute *f* express
 nl stadsautoweg *m* met ongelijkvloerse
  kruisingen
 r скоростная автомобильная дорога *f*

E510 *e* **extended-life superplasticizer**
 d Superplastifikator *m* mit erhöhter
  Wirkungsdauer
 f superplastifiant *m* à action
  principale de longue durée
 nl sterk plastificerende toeslag *m* met
  verlengde werkingsduur
 r суперпластификатор *m* (*добавка*) с
  длительным эффектом действия

E512 *e* **extensible boom platform**
 d Teleskopausleger-Hebebühne *f*
 f engin *m* élévateur à nacelle
  téléscopique
 nl hoogwerker *m*
 r телескопическая автовышка *f*

E513 *e* **extension device**
 d Verlängerungsvorrichtung *f*
 f dispositif *m* téléscopique
 nl langskoppeling *f* (voor verlenging van
  buissteigerpalen)
 r выдвижное устройство *n*

E514 *e* **extension flush bolt**
 d Türkantriegel *m*
 f verrou *m* à entailler à bascule;
  verrou *m* sur platine
 nl kantschuif *f* (*m*); basculeschuif *f* (*m*)
 r откидная *или* выдвижная дверная
  защёлка *f* [задвижка *f*] (*для
  крепления полотна двери к полу или
  притолоке*)

E515 *e* **extension ladder**
 d Ausziehleiter *f*
 f échelle *f* à coulisse
 nl uitschuifladder *m*
 r выдвижная лестница *f*

E516 *e* **extension trestle ladder**
 d ausziehbare Doppelleiter *f*
 f échelle *f* double télescopique (à
  échelons)
 nl dubbele uitschuifladder *m*
 r выдвижная лестница-стремянка *f*

E518 exterior bridge support *see* abutment

E519 exterior finish *see* outside finish

E520 *e* **exterior hood**
 d offene Absauganlage *f*, Absaughaube *f*
 f hotte *f* extérieure
 nl schoorsteenkap *f* (*m*)
 r открытый отсос *m*

E521 *e* **exterior panel**
 d Außenwandplatte *f*
 f panneau *m* mural [de mur] extérieur
 nl gevelpaneel *n*
 r наружная стеновая панель *f*

E522 *e* **exterior separation**
 d Abstand *m* zwischen Gebäudefassade
  und Straßenachse

 f distance *f* horizontale entre la
  façade d'un bâtiment et la ligne centrale de la rue voisine
 nl afstand *m* uit de as van de straat
 r расстояние *n* от фасада здания до
  оси улицы *или* до оси проезда между
  зданиями

E523 *e* **exterior stair**
 d Außentreppe *f*
 f escalier *m* extérieur
 nl buitentrap *m*
 r наружная лестница *f*

E524 *e* **exterior support**
 d außenseitige Stütze *f*,
  Außenstütze *f*
 f support *m* [poteau *m*] extérieur
 nl schoor *f* (*m*), stut *m* aan de buitenzijde
 r наружная опора *f*

E525 *e* **exterior-type plywood**
 d wasserbeständiges
  Außenverkleidungsfurnier *n*
 f contre-plaqué *m* résistant à l'eau
 nl watervast multiplex *n*
 r водостойкая фанера *f*

E526 *e* **exterior wall**
 d Außenwand *f*
 f mur *m* extérieur
 nl buitenmuur *m*
 r наружная стена *f*

E527 *e* **exterior zone**
 d Außenzone *f*
 f zone *f* extérieur
 nl buitenste zone *f*
 r периферийная зона *f* (*здания*);
  наружная зона *f*

E528 external air *see* outside air

E529 external dormer *see* dormant window

E530 *e* **external downpipe**
 d Dachabfallrohr *n*, Regenfallrohr *n*
 f descente *f* (pluviale), tuyau *m* de
  descente extérieur
 nl regenwaterafvoerpijp *f* (*m*)
 r наружный водосточный стояк *m*

E531 *e* **external prestress**
 d Außenvorspannung *f*
 f précontrainte *f* externe
 nl uitwendige voorspanning *f*
 r предварительное напряжение *n*
  наружными прядями

E532 *e* **external presstressing cable**
 d Außenkabel *n*
 f câble *m* extérieur de précontrainte
 nl uitwendige wapeningskabel *m*
 r наружный напрягаемый арматурный
  канат *m*, наружная арматурная
  прядь *f*

E533 *e* **external protection works**
 d Hafen-Außenwerke *n pl*
 f ouvrages *m pl* de protection extérieurs
 nl in zee doorlopende
  havenbeschermingswerken *n pl*
 r внешние оградительные сооружения
  n pl

E534 *e* **external skin**
  d Außenschale *f*, äußere Schale *f*
  f paroi *f* extérieure
  nl buitenschaal *f* (*m*)
  r внешняя оболочка *f* (ограждающих) конструкций

E535 *e* **external vibrator**
  d Außenrüttler *m*, Schalungsvibrator *m*
  f vibrateur *m* extérieur
  nl triltafel *f* (*m*)
  r наружный вибратор *m*

E536   **external wall** *see* **exterior wall**

E537   **extract air** *see* **exhaust air**

E538 *e* **extract duct**
  d Abluftkanal *m*, Entlüftungskanal *m*
  f gaine *f* d'évacuation d'air
  nl afzuigkanaal *n*
  r вытяжной воздуховод *m*

E539   **extract fan** *see* **exhauster**

E540 *e* **extraction of groundwater**
  d Grundwasserabpumpen *n*
  f pompage *m* des eaux souterraines
  nl onttrekken *n* van grondwater
  r откачка *f* грунтовых вод

E541   **extract opening** *see* **exhaust opening**

E543 *e* **extract ventilation**
  d Absauglüftung *f*
  f ventilation *f* par aspiration
  nl luchtafzuiging *f*
  r вытяжная вентиляция *f* (*местная*)

E544 *e* **extrados**
  d Rücken *m*, Bogenrücken *m*
  f extrados *m*
  nl buitenboog *m* van een gewelf
  r наружная поверхность *f* арки

E545   **extra-rapid-hardening cement** *see* **high-early-strength cement**

E546 *e* **extreme fibre stress**
  d Randspannung *f*
  f contrainte *f* dans la fibre extrême
  nl randspanning *f*
  r краевое напряжение *n*, напряжение *n* в крайнем волокне

E547 *e* **extruded section**
  d Strangpreßprofil *n*
  f profilé *m* extrudé
  nl strengpersprofiel *n*
  r экструдированный [прессованный] профиль *m*

E548 *e* **extrusion chipboard, extrusion particle board**
  d Strangpreßplatte *f*, stranggepreßte Spanplatte *f*
  f panneau *m* de copeaux [de particules] extrudé
  nl geëxtrudeerde spaanplaat *f* (*m*)
  r древесностружечная экструзионная плита *f*

E549 *e* **eye**
  d Öse *f*, Öffnung *f*, Auge *n*
  f œillet *m*; boucle *f*; cosse *f*
  nl oog *n*
  r ушко *n*, проушина *f*; концевая петля *f* (*каната*)

E550 *e* **eyebar**
  d Augenstab *m*
  f barre *f* d'acier avec oeillets aux extrémités
  nl ogenstaaf *f* (*m*)
  r стальной стержень *m* с проушинами на концах

E551 *e* **eyebolt**
  d Aug(en)bolzen *m*, Ringbolzen *m*, Ösenschraube *f*
  f boulon *m* à œillet
  nl oogbout *m*, ringbout
  r болт *m* с проушиной

F

F1 *e* **fabric**
  d 1. Aufbau *m*; Textur *f*; Gefüge *n* 2. Rohbau *m*, Skelett *n* (*Gebäude*) 3. Gewebe *n*
  f 1. structure *f*; texture *f*; construction *f* 2. carcasse *f*, ossature *f*, charpente *f* 3. tissu *m*
  nl 1. constructieve structuur *f* (*gebouw*); structuur *f*; textuur *f* 2. muren *m* *pl*, vloeren *f* (*m*) *pl* en dak *n* van een gebouw, ruwbouw *m* 3. geweven stof *f* (*m*)
  r 1. структура *f*; текстура *f* 2. остов *m* каркас *m* (*здания*) 3. ткань *f*

F2 *e* **fabric air filter**
  d Gewebe(luft)filter *n*, *m*
  f filtre *m* en tissu
  nl weefseluchtfilter *n*, *m*
  r тканевый воздушный фильтр *m*

F3 *e* **fabrication of structural steel**
  d Fertigung *f* der Stahlkonstruktionen
  f fabrication *f* des constructions en acier
  nl vervaardiging *f* van staalconstructies
  r изготовление *n* стальных конструкций

F4 *e* **fabric form**
  d Gewebeschalung *f*
  f coffrage *m* en tissu
  nl bekisting *f* met een weefselbekleding
  r тканевая опалубка *f*

F5 *e* **fabric reinforcement**
  d Mattenbewehrung *f*, Betonstahlmatteneinlage *f*
  f treillis soudé [d'armature] (*pour béton armé*)
  nl wapeningsnet *n*
  r арматурная сетка *f* (*сварная*)

F6 *e* **fabric wrapping**
  d Umwickeln *n* der Wärmeisolierung mit Schutzstoff
  f enveloppe *f* de protection en tissu (*enroulée autour d'une couche d'isolation thermique*)

nl omwikkeling *f* met weefsel
(*pijpisolatie*)
r обёртка *f* теплоизоляции защитной
тканью

F7 *e* **fabridam**
d Schlauchwehr *n*, aufpumpbares
Wehr *n*
f barrage *m* souple (*à récipient en
tissu rempli d'air ou d'eau*)
nl opblaasbare stuw *m*
r мягкая водо- *или*
воздухонаполняемая плотина *f*,
тканевая плотина *f*

F8 *e* **façade**
d Fassade *f*, Vorderansicht *f*
f façade *f* (*d'un bâtiment*)
nl voorgevel *m*
r фасад *m* (*преимущественно: главный*)

F9 *e* **face and bypass damper**
d Frontal- und Umgehungsklappe *f*,
Stirnseiten- und Bypassklappe *f*
f clapet *m* d'entrée d'air jumelé
nl tweevoudige luchtinlaatklep *f* (*m*)
r сдвоенный воздушный клапан *m*

F10 *e* **faced block**
d verkleideter Betonblock *m*
f bloc *m* en béton carrelé, bloc *m*
de parement (en béton)
nl betonsteen *m* met glad oppervlak
r облицованный бетонный блок *m*

F11 *e* **faced plywood**
d Furnierplatte *f* mit aufgeleimten
Außenfurnieren
f contre-plaqué *m* d'ébénisterie
nl gefineerd multiplex *n*
r венированная фанера *f*

F12 *e* **face hammer**
d Handschlageisen *n*
f smille *f* à un manche, martau *m*
à panne
nl vlakhamer *m*
r одноручный закольник *m* (*молоток
для ударной обработки камня*)

F13 *e* **face mix, face mixture**
d Vorsatzmischung *f*, Vorsatzgemisch *n*
f mélange *m* de parement (*pour blocs
de béton*)
nl mortel *m* voor de voorkant
(*betonsteen*)
r смесь *f* для лицевой грани
(*бетонного камня или блока*)

F14 *e* **face opening of exhaust hood**
d Ansaugquerschnitt *m*
f débouché *m* de la hotte de ventilation
[d'aspiration]
nl oppervlak *n* van de afzuigopening
r рабочий проём *m* вытяжного зонта

F15 *e* **face saver**
d Gesichtsschutzschild *m*, Maskenhaube
f
f casque *m* de protection à écran **facial**
nl beschermende bril *m*, masker *n*
r шлем-щиток *m* для защиты лица

F16 *e* **face shield**
d 1. Schutzschirm *m* 2.
Schweißerschutzschild *m*
f 1. panneau *m* [ecran *m*] de protection
2. masque *m* de soudeur
nl 1. beschermend paneel *n* 2. schermkap
f (*m*) (*lassen*)
r 1. защитная панель *f*; защитный
экран *m* 2. (защитный) щиток *m*
сварщика

F17 *e* **face shovel**
d Hochlöffel *m*, Löffelhochbagger *m*
f pelle *f* butte
nl lepelschop *f* (*m*)
r экскаватор *m* с прямой лопатой

F18 *e* **face velocity**
d Geschwindigkeit *f* im
Stirnquerschnitt
f vitesse *f* d'air de face
nl snelheid *f* in frontale dwarssnede
r фасадная скорость *f*, скорость *f*
в лобовом [габаритном] сечении
(*теплообменника, воздухоприёмника*)

F19 *e* **face waling**
d Endspreize *f*, Stirnspreize *f*
f moise *f* de boisage (*au bout de la
tranchée*)
nl eindgording *f*, stut *m* voor wand **van**
bouwput
r горизонтальная пажилина *f* в торце
траншеи (*элемент крепления*)

F20 *e* **face wall**
d 1. Stützwand *f* 2. Blendmauer *f*,
Verblendmauer *f*, Stirnmauer *f*,
Frontwand *f*
f 1. mur *m* de soutènement 2. mur *m*
frontal
nl frontmuur *m*
r 1. подпорная стенка *f* 2. фронтальная
стена *f*

F21 *e* **face work, facing**
d 1. Verkleidung *f*, Auskleidung *f*,
Verblendung *f*; Böschungsabdeckung *f*
2. Verkleidungsarbeiten *f pl*
f 1. parement *m*; revêtement *m*; perré *m*
2. traveax *m pl* de carrelage [de
revêtement]
nl 1. bekleding *f*; taludbekleding *f* 2.
bekleden *n*
r 1. облицовка *f*; одежда *f* откосов
2. облицовочные работы *f pl*

F22 *e* **facing bricks**
d Verblendsteine *m pl*, Sichtmauersteine
m pl
f briques *f pl* de parement
nl verblendsteen *m*, voorwerker *m*
r облицовочные кирпичи *m pl*

F23 *e* **facing wall**
d Betonausbauwandung *f*;
Betonböschungsdecke *f*
f mur *m* de revêtement; mur *m* de face;
mur *m* parafouille
nl bekledingsmuur *m*

r облицовочная стенка *f*; бетонная
одежда *f* (*откоса*)

F24 *e* **factor of safety**
d Sicherheitsbeiwert *m*,
Sicherheitsfaktor *m*
f coefficient *m* de sécurité
nl veiligheidscoëfficiënt *m*
r коэффициент *m* запаса (прочности)

F25 *e* **factory precast unit**
d industrielles Fertigteil *n*,
fabrikmäßig vorgefertigtes Element *n*
f élément *m* [pièce *f*] de construction
préfabriquée
nl geprefabriceerd constructiedeel *n*
r сборный элемент *m* (конструкций)
заводского изготовления

F26 *e* **fading of settlements**
d Ausklingen *n* der Setzungen (*Bauwerk*)
f affaiblissement *m* de tassement
[d'affaissement]
nl afnemen *n* van zettingen (*bouwwerk*)
r затухание *n* осадок (*сооружения*)

F27 **fagot, faggot** *see* **fascine**
F28 *e* **faggoting**
d Faschinenpackwerk *n*
f revêtement *m* de talus en fascines
nl rijspakwerk *n*, bekleding *f* met
fascines [rijshout]
r одежда *f* (*откоса*) из
фашинного тюфяка

F29 **faggot wood** *see* **brush wood**
F30 *e* **faïence, faïence ware**
d Fayenceerzeugnisse *n pl*
f produits *m pl* de construction en
faïence
nl geglazuurde terracottasteen *m*
r фаянсовые строительные изделия *n*
pl

F31 *e* **faïence tile**
d Fayence-Platte *f*, Fayence-Fliese *f*
f carreau *m* de faïence
nl geglazuurde tegel *m*
r фаянсовая керамическая плитка *f*

F32 **failing moment** *see* **failure moment**
F33 *e* **failing stress**
d Bruchspannung *f*
f contrainte *f* de rupture
nl breukspanning *f*
r разрушающее напряжение *n*

F34 *e* **failure criteria**
d Bruchkriterien *n pl*
f critères *m pl* de rupture
nl breukcriteria *n pl*
r критерии *m pl* прочности
[разрушения] (*конструкций*)

F35 *e* **failure hypotheses**
d Festigkeitstheorien *f pl*
f hypothèses *f pl* de rupture
nl plasticiteitstheorie *f*
r теории *f pl* прочности

F36 **failure in buckling** *see* **buckling
breaking**

F37 *e* **failure limit**
d Bruchgrenze *f*
f limite *f* de rupture
nl breukgrens *f* (*m*), vloeigrens *f* (*m*)
r предел *m* прочности

F38 *e* **failure limit state**
d Bruchgrenzzustand *m*
f état-limite *m* de rupture
nl breukgrenstoestand *m*,
bezwijktoestand *m*
r предельное состояние *n* по несущей
способности

F39 *e* **failure load**
d Bruchlast *f*, Einsturzlast *f*
f charge *f* de rupture
nl breukbelasting *f*, bezwijkbelasting *f*
r разрушающая нагрузка *f*

F40 *e* **failure mechanism**
d Bruchmechanismus *m*
f méchanisme *m* de rupture
nl bezwijkmechanisme *n*
r механизм *m* разрушения

F41 *e* **failure moment**
d Bruchmoment *n*
f moment *m* de rupture
nl bezwijkmoment *n*
r разрушающий момент *m*

F42 *e* **failure stage**
d Bruchstadium *n*
f état *m* de rupture
nl bezwijkstadium *n*
r стадия *f* разрушения

F43 *e* **failure strain**
d Bruchdehnung *f*
f déformation *f* de rupture
nl breukrek *m*
r деформация *f* при разрушении
[в момент разрушения]

F44 *e* **failure surface**
d Bruchfläche *f*
f surface *f* de rupture
nl breukvlak *n*
r поверхность *f* разрушения

F45 *e* **failure zone**
d Bruchzone *f*
f zone *f* de rupture
nl breukzone *f*
r зона *f* разрушения

F46 *e* **fair-faced brickwork**
d glattes Sicht(flächen)mauerwerk *n*
f maçonnerie *f* de parement
nl schoon metselwerk *n*
r гладкая лицевая кладка *f*

F47 *e* **fair-faced concrete**
d Sichtbeton *m*, Architekturbeton *m*
f béton *m* architectural [décoratif,
ornemental]
nl schoon uit de kist komend beton *n*
r архитектурно-декоративный бетон *m*

F48 *e* **fairway**
d Fahrwasser *n*, Fahrrinne *f*
voie *f* navigable, chenal *m*, passe *f*

nl vaarwater *n*, vaargeul *f* (*m*)
r фарватер *m*

F49 e **fall block**
d 1. Unterflasche *f* 2. Rammbär *m*,
Fallhammer *m*
f 1. boîte *f* du crochet (d'un palan);
moufle *f* (de poulie) mobile
2. mouton *m* (*de sonnette*)
nl 1. onderste schijvenblok *n* 2. heiblok *n*
r 1. подвижная блочная обойма *f*
полиспаста (*с крюком*) 2. баба *f*
копра

F50 e **falling apron**
d biegsame Böschungsbekleidung *f*
f revêtement *m* souple d'un talus
nl meegevende taludbekleding *f*
r гибкая одежда *f* откоса

F51 e **fall of a lock**
d Schleusenfall *m*, Schleusengefälle *n*
f chute *f* à l'écluse
nl schuthoogte *f*
r падение *n* шлюза, высота *f*
шлюзования

F52 e **false ceiling**
d Blinddecke *f*, Hängedecke *f*
f plafond *m* suspendu
nl verlaagd plafond *n*
r подвесной потолок *m*

F53 e **false joint**
d Scheinfuge *f*
f joint *m* simulé
nl schijnvoeg *f* (*m*), krimpvoeg *f* (*m*)
r декоративный шов *m*;
искусственный шов *m* (*напр.*
в бетоне)

F54 e **false leaders**
d Aufsteckmäkler *m pl*,
Aufstecklaufruten *f pl*
f guides *m pl* [glissières, *f pl*] du
mouton
nl leipalen *m pl* van een heistelling
r направляющие мачты *f pl* копра,
раскреплённые оттяжками

F55 e **false rafter**
d offenes Auflagerholz *n* [Futterholz *n*]
unter dem Dachüberstand
f coyau *m* visible (*sous l'auvent du*
comble)
nl spar *m* onder het overstek
r открытая кобылка *f* под свесом
кровли

F56 **false set** *see* **flash set**
F57 **false tenon** *see* **inserted tenon**
F58 e **false tongue**
d Scheibendübel *m*, Rechteckdübel *m*,
eingestemmter Dübel *m*
f languette *f* rapportée
nl veer *f* (*m*)
r пластинчатый нагель *f*;
пластинчатая шпонка *f*; шкант *m*

F59 e **falsework, false work**
d Schalgerüst *n*, Lehrgerüst *n*,
Rüstung *f*

f étayage *m* de coffrage, éléments *m pl*
porteur du coffrage
nl steigerwerk *n* voor de bekisting
r поддерживающие леса *pl* (опалубки)

F60 e **fan**
d Lüfter *m*, Ventilator *m*, Gebläse *n*
f ventilateur *m*
nl ventilator *m*, aanjager *m*
r вентилятор *m*

F61 e **fan-assisted warm air heating**
d Kraft-Luftheizung *f*
f chauffage *m* à l'air chaud forcé,
chauffage *m* par soufflage d'air chaud
(*a l'aide des ventilateurs*)
nl luchtverwarming *f* met aanjager
r воздушное отопление *n* с нагнетанием
тёплого воздуха (*в отапливаемые*
помещения) вентиляторами

F62 e **fan-assisted warm air heating unit**
d Kraft-Luftheizungsaggregat *n*
f installation *f* de chauffage à l'air
chaud forcé
nl luchtverwarmingsaggregaat *n* met
aanjager
r вентиляторный агрегат *m*
воздушного отопления

F63 e **fan coil unit**
d Luftbehandlungsaggregat *n*
f installation *f* à serpentin ventilé
nl luchtbehandelingsaggregaat *n*
r вентиляторный теплообменник *m*
[доводчик *m*]

F64 e **fan convector**
d Ventilator-Konvektor *m*
f convecteur *m* ventilé
nl ventilatorconvector *m*
r вентиляторный конвектор *m*

F65 e **fan delivery**
d Luftförderstrom *m*, Fördervolumen *n*
des Ventilators
f débit *m* de ventilateur
nl hoeveelheid *f* uitgeblazen lucht
r подача *f* вентилятора

F66 e **fan heater**
d Luftheizgerät *n*, Umlufterhitzer *m*
f chaufferette *f* avec ventilateur
nl ventilatorkachel *f* (*m*)
r вентиляторный [воздушный]
отопительный агрегат *m*

F67 e **fan inlet volume**
d Fördervolumen *n*
f volume *f* aspiré d'un ventilateur
nl hoeveelheid *f* aangezogen lucht
r подача *f* вентилятора по объёму
всасываемого воздуха

F68 e **fan instability**
d aerodynamische Instabilität *f* (des
Betriebsverhaltens) des Lüfters
f instabilité *f* aérodynamique d'un
ventilateur
nl aërodynamische instabiliteit *f* van
een ventilator
r аэродинамическая неустойчивость *f*
режима вентилятора

F69   *e* **fanlight**
    d Türoberlicht *n*
    f imposte *f* (*partie fixe vitrée au-dessus d'une porte*)
    nl bovenlicht *n* (*deur*)
    r фрамуга *f* над дверью

F70   *e* **fan performance curve**
    d Ventilatorkennlinie *f*
    f caractéristique *f* de ventilateur
    nl ventilatorkarakteristiek *f*
    r аэродинамическая характеристика *f* вентилятора

F71   *e* **fan shaft power**
    d Bremsleistung *f* des Lüfters
    f puissance *f* motrice [sur l'arbre] de ventilateur
    nl vermogen *n* op de ventilatoras
    r мощность *f* на валу вентилятора

F72   *e* **fan-type anchorage**
    d Fächerverankerung *f*
    f ancrage *m* en éventail
    nl waaierverankering *f*
    r веерообразная анкеровка *f*

F73   *e* **fan-type cooling tower**
    d Ventilatorkühlturm *m*
    f tour *m* de réfrigération [de refroidissement] à ventilateur
    nl ventilatorkoeltoren *m*
    r вентиляторная градирня *f*

F74   *e* **fan vault**
    d Fächergewölbe *n*, Strahlengewölbe *n*
    f voûte *f* normande [en éventail, à nervures rayonnantes]
    nl waaiergewelf *n*
    r веерный [зонтичный] свод *m*

F75   *e* **farm building**
    d landwirtschaftliches Gebäude *n*
    f bâtiment *m* agricole [de ferme]
    nl boerderij *f*
    r сельскохозяйственное здание *n*

F76   *e* **fascia**
    d 1. Bedachung *f*; Fensterverdachung *f*; Riemchen *n*, Seitenansicht *f* eines Trägers *oder* Balkens 2. Brückenrandbalken *m*
    f 1. bandeau *m* saillant, cordon *m*, filet *m*; moulure *f* 2. poutre *f* de rive (d'un pont), longeron *m* de tablier
    nl 1. timpaan *n*, gevelstrook *f* (*m*) 2. brugrandbalk *m*; boeiboord *n*
    r 1. сандрик *m*; поясок *m*; полка *f*, полочка *f* (*архитектурный облом*) 2. бортовая балка *f* моста

F78   *e* **fascine**
    d Faschine *f*, Drahtsenkwalze *f*
    f fascine *f*
    nl fascine *f*, rijsbos *m*
    r фашина *f*

F79   *e* **fascine dike** *US*, **fascine dyke** *UK*
    d Faschinendamm *m*
    f digue *f* en fascines
    nl rijsdam *m*
    r дамба *f* из фашинной кладки

F80   *e* **fascine groyne**
    d Faschinenbuhne *f*
    f épi *m* en fascines
    nl krib *f* (*m*) van rijswerk
    r фашинная полузапруда *f*

F81   *e* **fascine mattress**
    d Senkstück *n*
    f matelas *m* de fascinages, bloc *m* de fascines
    nl zinkstuk *n*
    r фашинный тюфяк *m*

F82   *e* **fascine revetment**
    d Faschinenbekleidung *f*
    f revêtement *m* (de talus) en fascines
    nl rijsbeslag *n*
    r фашинная одежда *f*

F83   *e* **fascine wood** *see* **brushwood**

F84   *e* **fascine work**
    d Faschinenpackwerk *n*, Packfaschinat *n*
    f fascinage *m*
    nl kribwerk *n*, rijswerk *n*
    r фашинная кладка *f*

F85   *e* **fasteners**
    d Befestigungsmittel *n pl*, Halterungen *f pl*
    f organes *m pl* de fixation, éléments *m pl* d'assemblage, agrafes *f pl*
    nl bevestigingsmiddelen *n pl*
    r крепёжные детали *f pl*, скрепы *f pl*

F86   *e* **fastening**
    d 1. Befestigung *f* 2. Befestigungsmittel *n*, Verbindungselement *n*; lösbare Verbindung *f*
    f 1. fixation *f*; attache *f* 2. pièces *f pl* [éléments *m pl*] d'assemblage; assemblage *m* démontable
    nl 1. bevestiging *f* 2. bevestigingsmiddel *n*
    r 1. крепление *n* 2. средство *n* крепления, соединительный элемент *m*; разъёмное соединение *n*

F87   *e* **fastening devices**
    d Befestigungselemente *n pl*; Befestigungsvorrichtungen *f pl*
    f organes *m pl* de fixation; pièces *f pl* d'assemblage; dispositifs *m pl* de serrage [d'assemblage]
    nl bevestigingsdelen *n pl*
    r крепёжные детали *f pl или* приспособления *n pl*; сжимающие приспособления *n pl*

F88   *e* **fastening tool**
    d Befestigungswerkzeug *n*
    f outils *m* de montage; outillage *m* de monteur
    nl montagegereedschap *n*
    r монтажный инструмент *m*; инструмент *m* для установки креплений

F89   *e* **fast-joint butt, fast-pin-hinge**
    d Einstemmband *n*, Einstellband *n* (*Fenster-bzw. Türbeschlag*)
    f charnière *f* à fiche rivée

nl paumelle f, fits
r врезная петля f (дверная или
оконная) с невынимающейся осью

F90 e **fast setting patching compound**
d schnellhärtende Zusammensetzung f
für Ausbesserungsarbeiten
f mastic m de prise rapide pour
l'exécution du point à temps
nl snelhardend mengsel n voor
wegreparaties
r быстротвердеющий состав m для
ямочного ремонта (дорожных
покрытий)

F91 e **fast-to-light paint**
d lichtechte Farbe f
f peinture f résistant à la lumière
nl lichtbestendige verf f (m)
r светостойкая краска f

F92 e **fat concrete**
d fetter Beton m
f béton m gras [riche]
nl vet beton n
r жирный бетон m

F93 e **fatigue**
d Ermüdung f
f fatigue f
nl vermoeidheid f
r усталость f

F94 e **fatigue failure**
d Dauerbruch m, Ermüdungsbruch m
f cassure f par fatigue, rupture f due à
la fatigue
nl vermoeidheidsbreuk f (m)
r разрушение n от усталости,
усталостное разрушение n

F95 e **fatigue life**
d Lebensdauer f
f durée f de fatigue
nl levensduur m bij vermoeidheid
r циклическая долговечность f (число
циклов переменных напряжений до
разрушения)

F96 e **fatigue limit**
d Dauer(schwing)festigkeit f,
Ermüdungsgrenze f
f limite f de fatigue
nl vermoeidheidsgrens f (m)
r предел m выносливости [усталости]

F97 e **fatigue ratio**
d Dauerfestigkeitsgrenzenverhältnis n
f rapport m résistance à la
fatigue: résistance à la traction
nl vermoeidheidsgrensverhouding f
r отношение n предела выносливости
к пределу прочности при
растяжении

F98 e **fatigue strength**
d Dauerfestigkeit f, Ermüdungsfestigkeit
f, Zeitschwingfestigkeit f
f résistance f à la fatigue
nl vermoeidheidssterkte f
r предел m усталости (при заданном
числе циклов напряжения)

F99 e **fatigue test**
d Ermüdungsversuch m,
Dauer(schwing)versuch m,
Dauerprüfung f
f essai m de fatigue
nl vermoeidheidsduurproef f (m)
r испытание n на усталость

F100 e **fat lime**
d Fettkalk m, Weißkalk m
f chaux f grasse
nl witkalk f (m)
r жирная [белая] известь f

F101 e **fat mortar**
d fetter Mörtel m, Fettmörtel m
f mortier m gras [riche]
nl vetmortel m
r жирный строительный раствор m

F102 e **fattening**
d Verdicken n (Farbe)
f épaississement m (de peinture)
nl indikken n (verf)
r загустение n (краски)

F103 e **faucet**
d Wasserhahn m, Auslaufventil n
f robinet m (d'eau)
nl tapkraan f (m), mofeind n
r водопроводный кран m

F104 e **faulting**
d Stufenbildung f
f dénivellation f des dalles de plancher
[des panneaux de plafond]
nl spatten n (vloerdelen)
r вертикальное смещение n плиты у
шва (дефект потолка)

F105 e **faulty workmanship**
d niedrige [minderwertige]
Arbeitsqualität f
f exécution f défectueuse, malfaçon f
nl slechte uitvoering f
r низкое качество n работ

F106 e **feasibility study**
d Durchführbarkeitsstudie f
f justification f technique et
économique du marché
nl uitvoerbaarheidsonderzoek n,
bruikbaarheidsonderzoek n
r технико-экономическое обоснование
n проекта

F107 e **feather**
d Feder f, Spundfeder f
f languette f; clavette f plate
nl messing f (m) (aan plank); kiel f (m
(voor kloven van natuursteen)
r гребень m, шпунт m; продольный
выступ m

F108 e **feathered board**
d keilartiges Brett n (für vollständig
überlagerte Bretterverkleidung)
f planche f taillé en biseau
nl plank f (m) met messing en groef
r клиновая доска f (для обшивки
внакрой)

F109 *e* **featheredge brick**
 d Keilziegel *m*
 f brique *f* en coin
 nl gewelfsteen *m*
 r клиновой [клинчатый] кирпич *m*

F110 *e* **feather tongue**
 d Federkeil *m*
 f languette *f* à rainure; languette *f* plate
 nl veer *f* (*m*) (*verbinding*)
 r шкант *m*; пластинчатый нагель *m*

F111 *e* **feeder**
 d 1. Aufgabevorrichtung *f*, Beschicker *m*, Speiser *m*, Zubringer *m* 2. Wassergraben *m*, Vorfluter *m*, Speisegraben *m*
 f 1. convoyeur *m* d'alimentation 2. canal *m* adducteur
 nl 1. aanvoerband *m* 2. toevoerkanaal *n*, toevoerleiding *f*
 r 1. питатель *m* 2. водопроводный канал *m*, канал-водоприёмник *m*

F112 *e* **feeder box**
 d Anschlußdose *f*
 f boîte *f* de feeders [d'artère]
 nl lasdoos *f* (*m*)
 r фидерный короб *m*

F113 *e* **feeder canal**
 d Verbindungskanal *m*, Zuleiter *m*; Speisekanal *m*
 f canal *m* de jonction *ou* d'alimentation
 nl irrigatiekanaal *n*; toevoerkanaal *n*
 r обводнительный [водоповодящий] канал *m*; подпитывающий канал *m*

F114 *e* **feeder drain**
 d Sauger *m*, Saugdrän *m*
 f drain *m* de décharge
 nl zuigdrain *m*
 r регулирующая дрена *f*

F115 **feed riser** *see* **flow riser**

F116 *e* **feed screw**
 d Verteilerschnecke *f*, Auflockerungsschnecke *f* (*Schwarzbelageinbaumaschine*)
 f vis *f* transporteuse [doseuse] (*d'un finisseur asphalteur*)
 nl spreidworm *m* (*in asfaltafwerkmachine*)
 r распределительный шнек *m* (*асфальтоукладчика*)

F117 *e* **feeler gauge**
 d Fühllehre *f*
 f calibre *m* d'épaisseur
 nl voelmaat *f* (*m*)
 r щуп *m*; толщиномер *m*

F118 *e* **felt**
 d 1. Filz *m* 2. Rohfilzpappe *f*
 f 1. feutre *m* 2. carton *m* de construction
 nl 1. vilt *n* 2. dakleer *n*
 r 1. войлок *m* 2. строительный картон *m*

F119 *e* **felt base, felted fabric backing**
 d Filzunterlage *f*

 f assise *f* de feutre (*sous un tapis plastique*)
 nl viltlaag *f* (*m*)
 r войлочный подкладочный слой *m* (*напр. линолеума*)

F120 *e* **felt nails**
 d Pappnagel *m pl*
 f clous *m pl* pour carton
 nl asfaltnagels *m pl*
 r толевые гвозди *m pl*

F121 *e* **felt paper**
 d Bitumendachpappe *f*
 f carton *m* goudronné, carton *m* (à couverture) bitumé, papier *m* à toiture
 nl dakvilt *n*, viltpapier *n*, bitumenvilt *n*
 r рубероид *m*, толь *m*

F122 *e* **fence**
 d Zaun *m*
 f clôture *f*, enceinte *f*
 nl schutting *f*; afrastering *f*
 r ограда *f*; изгородь *f*; забор *m*

F123 *e* **fender**
 d Fender *m*
 f pare-chocs *m*; défence *f* (élastique)
 nl wrijfhout *n*; schuurplank *f* (*m*)
 r отбойный амортизатор *m*, отбойное приспособление *n*; кранец *m*

F124 *e* **fender pile**
 d Prallpfahl *m*, Fenderpfahl *m*, Reibepfahl *m*, Abweisepfahl *m*
 f pieu *m* de défence
 nl schutpaal *m*, schuurpaal *m*
 r отбойная свая *f*

F125 *e* **fender post**
 d Dalbe *f*, Poller *m*
 f bollard *m*, borne *f* d'amarrage
 nl dukdalf *m*
 r швартовная тумба *f*, пал *m*

F126 *e* **fenestration**
 d Befensterung *f*
 f fenêtrage *m*
 nl plaatsing *f* van ramen en deuren in een gevel
 r размещение *n* окон по фасаду здания

F127 *e* **Ferrari cement**
 d Ferrarizement *m*
 f ciment *m* Ferrari
 nl ferraricement *n*
 r цемент *m* Феррари (*портландцемент с пониженным глинозёмным модулем*)

F128 *e* **ferrocement**
 d Armozement *m*
 f ferrociment *m*
 nl ferrocement *n*
 r армоцемент *m*, ферроцемент *m*

F129 **ferroconcrete** *see* **reinforced concrete**

F130 *e* **ferrometer**
 d Ferrometer *n*
 f ferromètre *m*

nl ferrometer *m*
r феррометр *m* (*прибор,
определяющий положение
арматуры в бетоне*)

F131 e **fiber balling**
d Faserzusammenballung *f*,
Fiberzusammenballung *f*
(*Störerscheinung bei der Faser- bzw.
Fiberbetonherstellung*)
f formation *f* des boules en fibres
(*pendant le malaxage de béton à fibres*)
nl vezelklontering (*fout in vezelbeton*)
r образование *n* комков волокон (*при
приготовлении фибробетона*)

F132 e **fiberboard** *US*
d Holzfaserplatte, Faserplatte *f*
f panneau *m* en fibre(s) de bois
nl houtvezelplaat *f* (*m*)
r древесноволокнистая плита *f*

F133 e **fiberglass**
d Glasfaser *f*, Faserglas *n*, Fiberglas *n*
f fibre *f* de verre
nl glasvezel *f* (*m*), glasweefsel *n*
r стекловолокно *n*, стеклянное
волокно *n*

F134 e **fiber glass batt insulation**
d Glasfaserisoliermatten *f pl*
f matelas *m* [tapis *m*] isolant en fibre
de verre
nl glaswolisolatiemat *f* (*m*), glaswoldeken
f (*m*)
r изолирующие маты *m pl* из
стекловолокна

F135 e **fiber glass-reinforced plastic form**
d Schalungsform *f* aus
Glasfaserkunststoff
f coffrage *m* en plastique renforcé à la
fibre de verre
nl mal *m* [bekisting *f*] van gewapend
beton
r опалубка *f* из стеклопластика

F136 **fiber reinforced concrete** *see* **fibrous
concrete**

F137 e **fiber stress**
d Faserspannung *f*
f contrainte *f* parallèle aux fibres
nl vezelspanning *f*
r фибровое напряжение *n*,
напряжение *n* вдоль волокон

F138 e **fibrous concrete**
d Fiberbeton *m*
f béton *m* de fibres
nl vezelbeton *n*, polyesterbeton *n*
r фибробетон *m*; бетон *m*,
дисперсно-армированный волокном

F139 e **fibrous glass**
d Glasfasern *f pl*, Fiberglas *n*;
Glaswolle *f*
f fibre *f* de verre; laine *f* de verre
nl glasvezel *n*; glaswol *f* (*m*)
r стекловолокно *n*; стекловата *f*,
стеклянная шерсть *f*

F140 e **fibrous insulation**
d Faser(stoff)isolierung *f*
f isolation *f* [isolement *m*] de fibre
nl isolatie *f* met vezelmateriaal
r волокнистая изоляция *f*

F141 e **fibrous shotcrete**
d Fibertorkretbeton *m*
f gunite *f* [béton *m* projeté] renforcé
aux fibres
nl spuitbeton *n* met glasvezels
r фиброторкретбетон *m*; торкрет-бетон
m, армированный волокнами

F142 e **fictitious force**
d Scheinkraft *f*
f force *f* fictive
nl schijnbare kracht *f* (*m*)
r фиктивная сила *f*

F143 **fictitious load** *see* **dummy load**

F144 e **fiddle block**
d zweifache Rollenflasche *f* mit
komplanaren Rollen
f poulie *f* à violon, violon *m*
nl tweeschijfsblok *n*
r блочная обойма *f* с двумя шкивами
в одной плоскости

F145 e **fidler's gear**
d Hebezeug *n* für Unterwassereinbau
der Betonkörper
f louve *f* à tenailles, griffe *f* de
levage (*pour la pose de blocs massifs*)
nl hijsinrichting *f* voor het laten zakken
van steen- of betonblokken
r грузозахватное устройство *n* для
укладки массивов

F146 e **field bending**
d Biegen *n* der Stahlbewehrung auf der
Baustelle
f pliage *m* [cintrage *m*] des armatures
sur chantier
nl buigen *n* van wapeningsstaven op de
bouwplaats
r заготовка *f* [гибка *f*] арматуры на
стройплощадке

F147 e **field bolt**
d Montagebolzen *m*, Heftbolzen *m*
f boulon *m* d'assemblage [de montage]
nl montagebout *m*
r монтажный болт *m*

F148 e **field book**
d Feldbuch *n*
f carnet *m* de levé topographique
nl veldboek *n*
r журнал *m* полевых (геодезических)
изысканий; журнал *m*
геодезических съёмок

F149 e **field concrete**
d Baustellenbeton *m*
f béton *m* préparé sur chantier
nl op de bouwplaats vervaardigd beton
n
r бетонная смесь *f*, приготовляемая на
стройплощадке

F150 *e* **field cube test**
d Baustellen-Würfelprüfung *f*
f essai *m* au cube de chantier
nl proef *f* (*m*) van betonkubussen op bouwplaats
r испытание *n* контрольных бетонных кубиков на стройплощадке

F151 *e* **field-cured cylinders**
d freiluftgelagerte zylindrische Betonprüfkörper *m pl*
f éprouvettes *f pl* cylindriques de béton conservées à l'air sur chantier
nl in de lucht verharde proefcilinders *m pl*
r контрольные цилиндрические бетонные образцы *m pl*, выдерживаемые в натурных условиях стройплощадки (*для определения срока распалубливания*)

F152 *e* **field drain**
d Landdrän *m*
f drain *m* de dernier ordre
nl vangdrain *m*
r осушительная [полевая] дрена *f*

F153 *e* **field erection**
d Baustellenmontage *f*
f montage *m* sur le chantier
nl montage *f* op de bouwplaats
r монтаж *m* конструкций на строительной площадке

F154 *e* **field joint**
d Montagefuge *f*, Montagestoß *m*
f joint *m* de montage
nl montagevoeg *f* (*m*)
r монтажный шов *m* [стык *m*], монтажное соединение *m*

F155 *e* **field laboratory**
d Baulabor *n*
f laboratoire *m* de chantier
nl laboratorium *n* op de bouwplaats
r полевая строительная лаборатория *f*

F156 *e* **field of load**
d Lastfeld *n*
f champ *m* des charges
nl belastingsveld *n*
r поле *n* нагрузки

F157 *e* **field tile**
d Landdrän *m*
f tuyau *m* de drainage en grès
nl draineerbuis *f* (*m*)
r гончарная полевая дрена *f*, керамическая дренажная трубка *f*

F158 *e* **figured glass**
d Ornamentglas *n*
f verre *m* ornemental
nl gefigureerd glas *n*
r узорчатое стекло *n*

F159 *e* **fill**
d 1. Dammschüttung *f*, Aufschüttung *f* 2. Auffüllung *f*, Ausfüllung *f*, Ausfachung *f* 3. aufgefülltes Gelände *n*, Auftrag *m*; abgelöster Boden *m*; Füllboden *m*, Auftragsboden *m*

f 1. remblai *m* 2. remblayage *m*; comblement *m*; remplissage *m* 3. terril *m*; sol *m* [terre *f*] pour remblai
nl 1. aanaarding *f* 2. opvulling *f* 3. opgebrachte grond *m*
r 1. насыпь *f* 2. засыпка *f*; заполнение *n* 3. отвальный грунт *m*; грунт *m* *или* гравийная смесь *f*, используемые для насыпи

F160 *e* **fill dam**
d Steinerddamm *m*
f barrage *m* en terre et enrochement
nl steendam *m*, zanddam *m*
r каменно-земляная плотина *f*

F161 *e* **filled bitumen**
d künstlicher Asphalt *m*
f mélange *m* de bitume et de poudre calcaire, asphalte *m*
nl asfaltbitumen *n*, mastiek *n*
r битум *m* с минеральным наполнителем, асфальт *m*

F162 *e* **filled soil**
d Auffüllung *f*, Füllboden *m*, Aufschüttboden *m*
f terrain *m* rapporté [remblayé]
nl grondaanvulling *f*, ophoogzand *n*
r насыпной грунт *m*

F163 *e* **filler**
d Füllstoff *m*
f filler *m*, fines *f pl* (*poudre minérale*)
nl vulstof *f* (*m*), vulmiddel *n*
r наполнитель *m*

F164 *e* **filler-joist floor**
d Stahlträgerdecke *f* mit Beton- *oder* Hohlformsteinausfachung
f plancher *m* à poutrelles d'acier avec hourdis de remplissage (*en béton armé ou en terre cuite*)
nl vloer *f* (*m*) van stalen balken met betonopvulling
r перекрытие *n* со стальными прокатными балками с заполнением из бетонных *или* пустотелых керамических блоков

F165 *e* **filler joists**
d Stichbalken *m pl*, Deckenträger *m pl*
f poutrelles *f pl* (en acier) de plancher
nl stalen vloerbalken *m pl*
r прокатные стальные балки *f pl* междуэтажных перекрытий

F166 *e* **fillet**
d 1. Leiste *f*; Kehlleiste *f*, Halteleiste *f*; Deckleiste *f* 2. gerades Riemchen *n* 3. Schweißraupe *f*
f 1. latte *f*, li(s)teau *m*, doucine *f*, bande *f* de recouvrement 2. moulure *f*, bande *f* saillante 3. cordon *m* de soudure
nl 1. keellijst *f* (*m*); (dek)lijst *f* (*m*) 2. vlakke strook *f* (*m*) tussen cannelures (*in een sier- of deklijst*) 3. wouterman *m*, wouterlatje *n*
r 1. рейка *f*, планка *f*; галтель *f*; раскладка *f* (*изделие*); нащельник *m*

2. прямолинейный облом *m*, пояс *m*,
полочка *f* 3. валик *m* (*сварного шва*)

F167 *e* fill-in brickwork
  d Ausfachungsmauerwerk *n*
  f remplissage *m* en briques du pan de
     fer
  nl vulling *f* van metselwerk
  r заполнение *n* каркаса (*каменной
     кладкой*)

F168 *e* filling
  d 1. Schüttung *f*; Aufschüttung *f*,
     Auffüllung *f* 2. Speicherfüllung *f*,
     Unterwassersetzung *f* 3. Füllung *f*,
     Füllmasse *f*, Füllkörper *m*
  f 1. remblayage *m*; bourrage *m* 2. mise *f*
     en eau, remplissage *m* 3. garnissage *m*
     (*d'une colonne*)
  nl 1. opvulling *f*, storting *f* 2. vullen *n*
     3. opvulsel *n*
  r 1. насыпь *f*; отсыпка *f*, засыпка *f*
     2. наполнение *n* (*напр.
     водохранилища*) 3. насадка *f*
     (*контактных аппаратов*)

F169 *e* filling knife
  d Spachtel *m*, *f*
  f spatule *f*
  nl plamuurmes *n*
  r шпатель *m*

F170 *e* filling material
  d Füllmaterial *n*, Ausfüllungsbaustoff *m*
  f matériau *m* de remplissage [de
     garnissage]
  nl vulmateriaal *n*
  r материал *m* заполнения [засыпки]
     (*швов, перекрытий*)

F171 filling up *see* silting

F172 *e* fill insulation
  d Schüttisolierung *f*
  f isolant *m* thermique en vrac
  nl los gestort isolatiemateriaal *n*
  r засыпная изоляция *f*,
     теплоизоляционная засыпка *f*

F173 *e* fillister
  d 1. Kittfalz *m* 2. Kittfalzhobel *m*,
     Grathobel *m*
  f 1. feuillure *f* 2. feuilleret *m*,
     bouvet *m*
  nl 1. glassponning *f* 2. sponningschaaf
     f (*m*), boorschaaf *f* (*m*)
  r 1. фальц *m* (*для остекления*)
     2. фальцгебель *m*, пазник *m*

F174 *e* film type cooling tower
  d Rieselflächenkühler *m*
  f tour *f* de réfrigération [de
     refroidissement] à pellicule,
     refroidisseur *m* à pellicule
  nl koeltoren *m* met watersluierinstroming
  r плёночная градирня *f*

F175 *e* filter
  d Filter *n*, *m*, Filteranlage *f*
  f filtre *m*
  nl filter *n*, *m*
  r фильтр *m*

F176 *e* filter bed
  d 1. Biofilter *n*, *m*, Tropfkörper *m* 2.
     Filterbett *n*, Filterkörper *m* 3.
     Filterpackung *f*, Bodenfilter *n*, *m*,
     Rückhaltefilter *n*, *m*, Filterschicht *f*
  f 1. lit *m* bactérien 2. lit *m* de contact
     3. tapis *m* filtrant, couche *f* filtrante
  nl 1. continufilter *n*, *m*, druppelfilter *n*,
     m 2. filterbed *n* 3. filtreerlaag *f* (*m*);
     natuurlijk grondfilter *n*, *m*
     (*duinwater*)
  r 1. биофильтр *m*, биологический
     фильтр *m* 2. контактная загрузка *f*,
     загрузка *f* фильтра 3. дренажный
     тюфяк *m*; донный фильтр *m*;
     фильтрующий слой *m*

F177 *e* filter drain
  d Steindrän *m*, Rigole *f*,
     Sickergraben *m*, Sauggraben *m*
  f drain *m* français, fossé *m* couvert
  nl draineerbuis *f* (*m*), drainagesleuf *f* (*m*)
  r каменная дрена *f*; дренажная
     канава *f*

F178 *e* filter gallery
  d Sickergalerie *f*, Schlitzfassung *f*,
     Sickerstrang *m*
  f galerie *f* drainante [filtrante, de
     drainage]
  nl filtergalerij *f*
  r фильтрационная галерея *f*

F179 *e* filter pipe
  d Filterrohr *n*
  f tuyau *m* de filtrage [de filtration,
     filtrant]
  nl filterpijp *f* (*m*)
  r фильтровая труба *f* (*колодца*)

F180 *e* filter run
  d Filtergang *m*, Filterbetriebszeit *f*
  f période *f* de filtration
  nl filtercyclus *m*
  r фильтроцикл *m*

F181 *e* filter well
  d Tiefbrunnen *m*, Filterbrunnen *m*;
     Dränagebohrloch *n*
  f puits *m* filtrant
  nl filterbron *f* (*m*), nortonput *m*
  r фильтрационный колодец *m*;
     дренажная скважина *f* с фильтром

F182 *e* filtration
  d Filterung *f*, Filtern *n*, Filtration *f*
  f filtration *f*
  nl filtratie *f*, filtrering *f*
  r фильтрация *f*

F183 *e* filtration beds
  d Filtrationsfelder *n pl*
  f champs *m pl* de filtration
  nl filtratievelden *n pl*, filtratiebed *n*
  r поля *n pl* фильтрации

F184 *e* filtration chamber
  d Filterkammer *f*, Filterraum *m*
  f chambre *f* de filtration
  nl filterkamer *f* (*m*)
  r фильтровальная камера *f*

F185 *e* **filtration pressure**
  d Filterdruck *m*, Sickerdruck *m*
  f pression *f* de filtration
  nl filterdruk *m*
  r фильтрационное давление *n*

F186 *e* **fin**
  d 1. Grat *m* (*Fehler von Betonoberfläche*)
  2. Rippe *f* 3. Grat *m* (*Metall*)
  f 1. bavure *f* du béton; arête *f* sur
  surface de béton 2. ailette *f*
  3. ébarbure *f*, bavure *f*
  nl 1. scherpe kant *f* (*m*) (*betonfout*)
  2. rand *m* 3. baard *m*
  r 1. ребристый выступ *m* цементного
  камня (*дефект бетонных
  поверхностей*) 2. ребро *n* 3. заусенец
  m

F187 *e* **final acceptance**
  d Schlußabnahme *f*
  f réception *f* définitive (*des travaux*)
  nl laatste oplevering *f*
  r окончательная приёмка *f*
  (*строительного объекта*)

F188 *e* **final bending moment diagram**
  d Momentensummenlinie *f*
  f diagramme *m* des moments
  fléchissants définitif
  nl additieve momentenlijn *f* (*m*)
  r суммарная эпюра *f* моментов

F189 *e* **final completion**
  d endgültige Fertigstellung *f*
  f achèvement *m* définitif des travaux
  nl werk *n* voor de laatste oplevering
  r окончательное завершение *n*
  строительных работ

F190 *e* **final finishing**
  d Endbearbeitung *f*, Fertigbehandlung *f*
  f finition *f*
  nl afwerken *n*, klaarmaken *n* voor
  oplevering
  r чистовая [окончательная] отделка *f*

F191 *e* **final payment**
  d Endzahlung *f*
  f solde *m*, dernier paiement *m*
  nl laatste betalingstermijn *m*
  r окончательный расчёт *m*
  (*заказчика с подрядчиком*)

F192 *e* **final position**
  d Entwurfslage *f*, Entwurfsstellung *f*;
  Endlage *f*, Endstellung *f*
  f position *f* finale
  nl definitieve plaats *f* (*m*)
  r проектное положение *n*
  (*конструктивного элемента*);
  конечное положение

F193 *e* **final set, final setting time**
  d Erstarrungsende *n*
  f prise *f* finale, fin *f* de prise
  nl einde *n* van de verharding
  r конец *m* схватывания

F194 *e* **final settling basin**
  d Nachklärbecken *n*
  f bassin *m* de décontation secondaire

  nl laatste bezinkbekken *n*
  r вторичный отстойник *m*

F195 *e* **final strength**
  d Endfestigkeit *f*
  f résistance *f* finale
  nl eindsterkte *f*
  r конечная прочность *f*

F197 **final tank** *see* **final settling basin**

F198 *e* **fine aggregate**
  d Feinzuschlag(stoff) *m*
  f agrégat *m* [granulat *m*] fin
  nl fijne toeslag *m*
  r мелкий заполнитель *m*

F199 *e* **fine cold asphalt**
  d Asphaltkaltbeton *m*
  f béton *m* bitum(in)eux [asphaltique]
  fin à froid
  nl fijnkorrelig koudasfalt *n*
  r мелкозернистый холодный
  асфальтобетон *m*

F200 *e* **fine dust**
  d Feinstaub *m*, Flugstaub *m*
  f poussière *f* fine
  nl fijn stof *n*
  r мелкая пыль *f*

F201 *e* **fine-grained chippings**
  d Feinsplitt *m*
  f gravillon *m* fin
  nl fijn steenslag *n*; split *n*
  r каменная крошка *f*;
  мелкозернистый щебень *m*

F202 *e* **fine grained soils**
  d Feinböden *m pl*
  f sols *m pl* à grain fin
  nl fijnkorrelige grondsoorten *f* (*m*) *pl*
  r мелкозернистые грунты *m pl*

F203 *e* **fine gravel**
  d Feinkies *m*; Grus *m*
  f gravier *m* fin
  nl fijngrind *n*; kiezelzand *n*
  r мелкий гравий *m*; высевки *pl*

F204 *e* **fineness modulus**
  d Feinheitsmodul *m*, Körnungsmodul
  m, Feinheitszahl *f*
  f module *m* de finesse
  nl fijnheidsmodulus *m*
  r модуль *m* крупности

F205 *e* **fines**
  d Feingut *n*, Feinkorn *n*, Feinmaterial *n*
  f agrégat *m* fin; fines *f pl*
  nl zeeffractie *f* met kleinste
  korrelgrootte
  r мелкий заполнитель *m*;
  мелкозернистые фракции *f pl*,
  мелкодисперсный материал *m*

F206 *e* **fine screen**
  d Feinsieb *n*
  f crible *m* [tamis *m*] fin
  nl fijne zeef *f* (*m*)
  r сито *n* с мелкими отверстиями
  (*менее 6,4 мм*)

F207 *e* **finger joint**
  d verzahnte Verbindung *f*

f assemblage *m* en bout à dents (*des planches*)
nl meervoudige slisverbinding *f* [tandverbinding *f*]
r зубчатое соединение *n* (*досок, пластин*)

F208 *e* finish
  d 1. Oberflächengüte *f*, Oberflächenbeschaffenheit *f* 2. Ausbau *m*
  f 1. finition *f* [fini *m*] de la surface 2. travaux *m pl* de finition (*enduit, peinture*)
  nl 1. oppervlaktegesteldheid *f* 2. afwerking *f*, eindbewerking *f*
  r 1. отделка *f* поверхности (*результат*) 2. отделка *f* (*здания*), отделочные работы *f pl*

F209 *e* finish coat
  d 1. Oberputz *m*, Feinputz *m* 2. Deckanstrich *m* 3. Versiegelungsschicht *f*, Absiegelungsschicht *f*, Porenschlußschicht *f* (*Straßenbau*)
  f 1., 2. couche *f* de finition (*enduit, peinture*) 3. couche *f* de fermeture [de scellement]
  nl 1. afwerkpleisterlaag *f* (*m*) 2. laatste verflaag *f* (*m*) 3. afsluitlaag *f* (*m*)
  r 1. накрывка *f*, накрывочный слой *m* (*штукатурки*) 2. верхний отделочный слой *m* (*красочного покрытия*) 3. замыкающий слой *m* (*дорожной одежды*)

F210 *e* finished floor
  d Fußbodenbelag *m*
  f revêtement *m* de sol, couvre-sol *m*
  nl afgewerkte vloer *f* (*m*)
  r чистый пол *m*; чистовой настил *m* пола

F211 *e* finished floor level
  d Fußbodenoberkante *f*
  f cote *f* de niveau du couvre-sol [du revêtement de sol]
  nl niveau *n* van de afgewerkte vloer
  r отметка *f* чистовой поверхности пола

F212   finish flooring *see* finished floor

F213   finish hardware *see* builder's finish hardware

F214 *e* finishing
  d Fertigbearbeitung *f*, Fertigbehandlung *f*, Endbearbeitung *f*
  f finissage *m*, finition *f*; traveaux *f pl* [opérations *f pl*] de finition
  nl afwerking *f*, eindbewerking *f*
  r отделка *f*, отделочные работы *f pl*

F215 *e* finishing belt
  d Glättband *n*, Glättriemen *m* (*Straßenbau*)
  f courroie *f* lisseuse
  nl strijkband *m* (*wegenbouw*)
  r выглаживающее полотнище *n*, выглаживающая лента *f* (*дорожное строительство*)

F216 *e* finishing machine
  d Straßendeckenfertiger *m*
  f finisseur *m*, finisseuse *f*
  nl wegdek-afwerkmachine *f*
  r бетоноотделочная машина, финишер *m*

F217 *e* finishing screed
  d Glättbohle *f*
  f screed *m* flattant, poutre *f* lisseuse
  nl strijkplaat *f* (*m*)
  r выглаживающий брус *m* (*финишера*)

F218 *e* finite-difference method
  d Differenzverfahren *n*, Gitterpunktmethode *f*
  f méthode *f* des différences finies
  nl differentiemethode *f*
  r метод *m* конечных разностей

F219 *e* finite element method
  d Finit-Element-Methode *f*
  f méthode *f* des éléments finis
  nl eindige elementenmethode *f*
  r метод *m* конечных элементов

F220 *e* finned strip heater
  d Lamellenheizkörper *m*
  f radiateur *m* à ailettes
  nl lamelleradiator *m*
  r пластинчатый радиатор *m*

F221 *e* Finsterwalder prestressing method
  d Finsterwalder-Vorspannsystem *n*
  f méthode *f* de Finsterwalder
  nl voorspanmethode *f* volgens Finsterwalder
  r метод *m* преднапряжения пролётных строений моста под действием собственного веса

F222 *e* fire-alarm system
  d Feuermeldeanlage *f*
  f système *m* d'avertissement d'incendie
  nl brandmeldingsinstallatie *f*
  r пожарная сигнализация *f*

F223 *e* fire area
  d Feuerschutzzone *f*
  f zone *f* pare-feu
  nl brandbeschermingsgebied *n*
  r противопожарная зона *f*

F224 *e* fire barrier
  d Brandblende *f*, Feuerschutzabschluß *m*
  f barrière *f* antifeu, coupe-feu *m*
  nl brandwerende constructie *f*
  r противопожарная преграда *f*

F225 *e* firebreak
  d Brandschutz-Gebäudeabstand *m*
  f prospect *m*
  nl brandsingel *m* tussen gebouwen
  r противопожарный разрыв *m* между зданиями

F226 *e* fire brick
  d Schamotteziegel *m*, feuerfester Stein *m*
  f brique *f* à feu
  nl vuurvaste steen *m*, chamottesteen *m*
  r огнеупорный кирпич *m*

F227 *e* **fire cement**
  d feuerbeständiger Zement *m*
  f ciment *m* réfractaire
  nl vuurvaste specie *f*
  r жаростойкий цемент *m*

F228 *e* **fire clay**
  d feuerfester Ton *m*, Schamotte *f*
  f argile *f* réfractaire
  nl vuurvaste klei *f* (*m*), chamotte *f* (*m*)
  r огнеупорная глина *f*

F229 *e* **fire cock**
  d Feuerhahn *m*, Brandhahn *m*
  f bouche *f* d'incendie
  nl brandkraan *f* (*m*)
  r пожарный кран *m*

F230 *e* **fire damper**
  d Feuerschutzklappe *f*
  f registre *m* ignifuge
  nl zelfsluitende brandklep *f* (*m*)
  r пожарный [огнезадерживающий]
    клапан *m*

F231 *e* **fire devil**
  d 1. Bauaustrocknungsofen *m*,
    Kokskorb *m* 2. transportabler
    Erhitzer *m* (*für Straßenbauarbeiten*)
  f brasero *m* de chantier, panier *m*
    à coke 2. réchauffeur *m* mobile pour
    l'outillage routier
  nl 1. bouwdroogkachel *f* (*m*), vuurkorf
    m 2. verplaatsbare verhitter *m*
    (*wegenbouw*)
  r 1. коксовая жаровня *f* для сушки
    помещений 2. передвижной
    разогреватель *m* (*для нагрева
    инструмента при устройстве
    чёрных дорожных покрытий*)

F232 *e* **fire division wall**
  d Brandmauer *f*
  f mur *m* coupe-feu [de feu, pare-feu]
  nl brandmuur *m*
  r брандмауэр *m*

F233 *e* **fire door**
  d Brandschutztür *f*, Feuerschutztür *f*
  f porte *f* coupe-feu [pare-feu]
  nl branddeur *f* (*m*)
  r противопожарная дверь *f*

F234 *e* **fire-duration test**
  d Branddauerprüfung *f*,
    Branddauerversuch *m*
  f essai *m* de feu, essai *m* de
    résistance au feu, essai *m* de
    réaction au feu
  nl test *f* (*m*) op brandbestendigheid
  r определение *n* огнестойкости
    (*материалов*), испытание *n* на
    огнестойкость

F235   **fire endurance** *see* **fire resistance**

F236 *e* **fire escape**
  d Feuer-Fluchtweg *m*
  f sortie *f* *ou* escalier *m* de secours
  nl branduitgang *m*
  r запасной пожарный выход *m*;
    путь *m* эвакуации

F237 *e* **fire extinction basin**
  d Feuerlöschbecken *n*
  f bac *m* [réservoir *m*] d'eau pour la
    lutte contre l'incendie
  nl bluswaterreservoir *n*
  r пожарный водоём *m*

F238 *e* **fire extinguishing equipment, fire-
    -fighting equipment**
  d Feuerlöschgeräte *n pl*,
    Brandbekämpfungsausrüstung *f*
  f équipment *m* d'extinction
  nl (brand)blusmiddelen *n pl*
  r оборудование *n* для пожаротушения

F239 *e* **fire flow**
  d Wasserverbrauch *m* für Feuerlöschen
  f débit *m* d'eau de la canalisation
    d'extinction d'incendie
  nl waterverbruik *n* bij brandblussen
  r расход *m* воды для пожаротушения

F240 *e* **fire-grading of buildings**
  d Brandklasseneinteilung *f* der Gebäude
  f classement *m* des édifices selon leur
    résistance au feu
  nl brandclassificatie *f* van gebouwen
  r классификация *f* зданий по
    огнестойкости

F241 *e* **fire hydrant**
  d Feuerlöschwasserständer *m*
  f bouche *f* d'incendie
  nl brandkraan *f* (*m*), hydrant *m*
  r пожарный гидрант *m*

F242 *e* **fire line**
  d 1. Feuerlöschleitung *f*
    2. Feuerlöschschlauch *m*
  f 1. canalisation *f* d'extinction
    d'incendie 2. tuyau *m* d'incendie
  nl 1. brandleiding *f* 2. brandslang *f* (*m*)
  r 1. пожарный водопровод *m*
    2. пожарный шланг *m* [рукав *m*]

F243 *e* **fire load**
  d Brandlast *f*
  f charge *f* d'incendie
  nl vuurbelasting *f*
  r пожарная [огневая] нагрузка *f*

F244 *e* **fire lobby**
  d Brandabschnitt *m*
  f compartiment *m* [couloir *m*]
    coupe-feu (*d'un édifice*)
  nl brandcompartiment *n*
  r противопожарный отсек *m* (*здания*)

F245 *e* **fire place**
  d Feuerstelle *f*; offener Kamin *m*
  f foyer *m* de cheminée; cheminée *f*
    décor(ative) intérieure
  nl stookplaats *f* (*m*); open haard *m*
  r топка *f* камина *или* печи;
    открытый камин *m*

F246   **fire prevention** *see* **fire protection**

F247 *e* **fireproof construction**
  d feuerbeständige Konstruktion *f*,
    feuerbeständiges Bauwerk *n*
  f construction *f* résistante au feu
  nl brandvrije constructie *f*

r огнестойкая конструкция *f*,
огнестойкое сооружение *n*

F248 *e* **fireproofing**
 d 1. Feuerschutz-Maßnahmen *f pl*
 (*um Bauteile feuerbeständig zu
 machen*) 2. Feuerschutz-Werkstoff
 m
 f 1. ignifugeage *m*, ignifugation *f*
 2. matériau *m* protecteur ignifuge
 nl 1. brandwerend maken
 2. brandwerend materiaal *n*
 r 1. обеспечение *n* [придание *n*]
 огнестойкости 2. огнестойкий
 материал *m*

F249 *e* **fireproofing plaster**
 d Brandschutz(innen)(ver)putz *m*,
 Feuerschutz(innen)(ver)putz *m*
 f enduit *m* pare-feu, plâtre *m*
 résistant au feu
 nl brandwerend pleisterwerk *n*
 r огнестойкая штукатурка *f*
 [обмазка *f*]

F250 *e* **fire protection**
 d Brandschutz *m*, Feuerschutz *m*
 f protection *f* contre l'incendie
 nl brandbeveiliging *f*
 r пожарная охрана *f*;
 противопожарные мероприятия *n pl*;
 пожарная профилактика *f*

F251 *e* **fire-protection equipment**
 d Feuerlöschgeräte *n pl*
 f matériel *m* de protection contre
 l'incendie
 nl (brand)blusmiddelen *n pl*
 r пожарное оборудование *n*

F252 *e* **fire resistance**
 d Feuerbeständigkeit *f*
 f résistance *f* au feu
 nl vuurvastheid *f*, brandwerend
 vermogen *n*
 r огнестойкость *f*

F253 *e* **fire-resistance classification**
 d Brandklasseneinteilung *f*
 f classement *m* d'après le degré de
 résistance au feu
 nl brandclassificatie *f*
 r классификация *f* по пределам
 огнестойкости

F254 *e* **fire-resistance rating**
 d Brandklassenwert *m*
 f degré *m* de résistance au feu
 nl mate *f* (*m*) van brandwerendheid
 r предел *m* огнестойкости (*в часах*)

F255 **fire-resistive construction** *see*
 fireproof construction

F256 *e* **fire-resistive flooring**
 d feuerbeständiger Fußboden *m*
 f plancher *m* ininflammable
 [résistant au feu]
 nl brandwerende vloer *m*
 r невозгораемый пол *m*

F257 **fire-retardant agents** *see* **fire retarding
 materials**

F258 *e* **fire-retardant coating**
 d feuerhemmender Anstrich *m*
 f revêtement *m* [enrobage *m*] résistant
 au feu
 nl brandvertragende bedekking *f*
 r обмазка *f* огнестойкими
 материалами; покрытие *n*
 антипиренами

F259 *e* **fire-retarding component**
 d feuerbeständiges Montageelement *n*
 f composant *m* résistant au feu
 nl brandvertagend element *n*
 r огнестойкий сборный элемент *m*

F260 *e* **fire retarding materials**
 d Feuerhemmstoffe *m pl*
 f produits *m pl* de protection contre le
 feu; antipyrènes *m pl*
 nl brandvertragende middelen *n pl*
 r огнезадерживающие материалы *m pl*
 (*повышающие огнестойкость
 конструкций*); антипирены *m pl*

F261 *e* **fire separation, fire stop**
 d Brandblende *f*, Feuerschutzabschluß
 m; feuerhemmender Bauteil *m*
 f coupe-feu *m*, pare-feu *m*, cloison *f*
 coupe-feu; élément *m* pare-feu
 [coupe-feu]
 nl brandvertragende constructie *f*
 r противопожарная преграда *f*;
 огнезадерживающий элемент *m*
 конструкции (*напр. засыпка
 в полостях*)

F262 *e* **fire ventilation**
 d Brandventilation *f*
 f ventilation *f* de secours en cas de
 l'incendie
 nl rookventilatie *f*
 r пожарная вентиляция *f*

F263 *e* **fire ventilator**
 d Lüfter *m* zum Absaugen von
 Rauchgasen und Dämpfen
 f ventilateur *m* d'aspiration des
 fumées d'incendie
 nl afzuigventilator *m* voor rook en
 gassen
 r пожарный дымовытяжной
 вентилятор *m*

F264 *e* **fire wall**
 d Brandwand *f*, Brandmauer *f*;
 Brandschott *n*
 f mur *m* coupe-feu [pare-feu]
 nl brandmuur *m*
 r брандмауэр *m*; противопожарная
 перегородка *f*

F265 *e* **fire-warning device**
 d Feuermeldeanlage *f*, Brandmelder,
 Feurmelder *m*
 f avertisseur *m* d'incendie
 nl brandwaarschuwingsinstallatie *f*
 r устройство *n* пожарной
 сигнализации; пожарный
 извещатель *m* [сигнализатор *m*]

F266 *e* **fire water supply**
 d Löschwasserversorgung *f*

f réseau *m* de distribution (d'eau)
«incendie»
nl bluswatervoorziening *f*
r пожарное водоснабжение *n*

F267 **firing of refractories** *see* **burning of refractories**

F268 *e* **firm discharge**
d garantierter Mindestabfluß *m*
f débit *m* de garantie [de sécurité]
nl gegarandeerd afvoerminimum *n*
r гарантированный расход *m*

F269 *e* **firm price contract**
d Festpreisvertrag *m*
f contract *m* [marché *m*] à forfai,
marché *m* forfaitaire
nl contract *n* met vaste prijs
r контракт *m* с твёрдой ценой

F270 *e* **firm soils**
d dichte Böden *m pl*
f terrain *m* compact
nl vaste grond *m*
r плотные грунты *m pl*

F271 **firring** *see* **furring**

F272 *e* **first coat**
d Unterputz *m*
f couche *f* de base (d'*enduit*)
nl in het ruw gezette pleisterlaag *f* (*m*),
raaplaag *f* (*m*)
r нижний слой *m* штукатурного
намёта (*при двухслойной
штукатурке*)

F273 *e* **first fixings**
d Holzdübel *m pl*
f tasseaux *m pl* de fixation (*en bois*)
nl houten deuvels *m pl*
r деревянные пробки *f pl* (*для
крепления столярных изделий
к кладке*)

F274 *e* **first floor** *UK*
d erstes Obergeschoß *n*
f premier étage *m*
nl eerste verdieping *f*
r второй этаж *m* (*над первым или
цокольным этажом*)

F275 *e* **first floor** *US*
d Erdgeschoß *n*
f rez-de-chaussée *m*
nl begane grond *m*
r первый этаж *m* (*на уровне земли или
несколько выше*)

F276 *e* **first-stage development**
d erster Bauabschnitt *m*, erste
Ausbaustufe *f*
f première étape *f* d'aménagement
nl eerste ontwikkelingsfase *f*
r первая очередь *f* застройки

F277 *e* **first stage pumping station**
d Pumpstation *f* der ersten Förderstufe
f station *f* de pré-pompage
nl pompstation *n* van eerste produktietrap
r насосная станция *f* первого подъёма

F278 *e* **fish chute**
d Fischgerinne *n*

f passe *f* à poissons
nl visgoot *f* (*m*)
r лотковый рыбоход *m*

F279 **fish elevator** *see* **fish lift**

F280 *e* **fisheyes**
d Blasen *f pl* (*in Putzdeckschicht*)
f soufflures *f pl*
nl blazen *f pl* (*luchtbellen in
pleisterwerk*)
r бугорки *m pl*, «дутики» *m pl*
(*дефект штукатурки*)

F281 *e* **fish ladder**
d Fischtreppe *f*, Fischleiter *f*
f échelle *f* à poissons
nl vistrap *m*
r лестничный [ступенчатый] рыбоход *m*

F282 *e* **fish lift**
d Fischaufzug *m*
f ascenseur *m* à poissons
nl vislift *m*, viselevator *m*
r рыбоподъёмник *m*

F283 *e* **fish lock**
d Fischschleuse *f*
f écluse *f* à poissons
nl vissluis *f* (*m*)
r рыбоходный шлюз *m*

F284 *e* **fish pass**
d Fischpaß *m*, Fischweg *m*,
Fischdurchlaßeinrichtung *f*
f passe *f* à poissons
nl visdoorlaat *m*
r рыбоход *m*, рыбопропускное
устройство *n*

F285 *e* **fishplate**
d Anschlußblech *n*, Stoßlasche *f*
f couvre-joint *m*, plaque *f* de joint,
éclisse *f*, plate-bande *f*
nl lasplaat *f* (*m*), koppelplaat *f* (*m*)
r стыковая накладка *f*

F286 *e* **fish screen**
d Fischsperre *f*, Fischrost *m*,
Fischrechen *m*
f grille *f* à poissons
nl visrooster *n*
r рыбооградитель *m*,
рыбозаградительная решётка *f*

F287 *e* **fishtail bolt**
d Fischschwanzbolzen *m*
f boulon *m* de scellement à bout
fourchu
nl veerbout *m*, bledschroef *f* (*m*)
r анкерный болт *m* с раздвоенным
концом

F288 **fishway** *see* **fish pass**

F289 *e* **fissure drain**
d Schlitzdrän *m*
f drain *m* avec fentes
nl zuigdrain *m* met spleten
r щелевая дрена *f*

F290 *e* **fit**
d 1. Passung *f* 2. Einpassung *f*,
Anpassung *f*; Aufstellung *f*

f 1. ajustement *m*, ajustage *m*
2. mise *f* en place
nl 1. passing *f* 2. passend maken *n*,
aanbrengen *n*
r 1. посадка *f*; пригонка 2. установка *f*
(в проектное положение)

F291 *e* fitch
d 1. kleiner Malerpinsel *m* mit langem
Stiel 2. Stegbrett *n* des
zusammengesetzten Holzstahlträgers
3. Furnier *n*
f 1. pinceau *m* mince à manche longue
2. planche *f* de poutre composite en
bois et acier 3. feuillet *m* de
contre-plaqué
nl 1. varkensharen penseel *n* met lange
steel 2. lijf *n* van een samengestelde
ligger 3. fineer *n*
r 1. тонкая кисть *f* с длинной ручкой
(для окраски труднодоступных
мест) 2. доска *f* деревостальной
составной балки 3. шпон *m*

F292 *e* fitment
d Einbaumöbel *n* pl
f meubles *m* pl encastrés [incorporés]
nl ingebouwde meubels *pl*
r встроенная мебель *f*

F293 *e* fitting
d 1. Rohrverbindungsstück *n*
2. Einpassen *n*, Paßarbeit *f*
3. Montage *f*, Aufbau *m*,
Aufstellen *n*, Aufbauen *n*,
Anbringung *f* 4. Einbaumöbel *n* pl
f 1. raccord *m*, pièce *f* de
raccordement 2. ajustage *m*
3. assemblage *m*, mise *f* en place;
montage *m* 4. meubles *m* pl
incorporés
nl 1. (pijp)fitting *m* 2. inpassen *n*
3. montage *f* 4. inbouwmeubels *pl*
r 1. фитинг *m* 2. пригонка *f* 3. сборка *f*,
монтаж *m* 4. встроенная мебель *f*

F294 *e* fitting-out basin
d Ausrüstungsbecken *n*,
Ausrüstungsbassin *n*
f bassin *m* d'armement [d'achèvement]
nl afbouwhaven *f* (*m*)
r достроечный бассейн *m* (верфи)

F295 *e* fitting-out quay
d Ausrüstungskai *m*
f quai *f* d'achèvement
nl afbouwkade *f* (*m*)
r достроечная набережная *f*
(на верфях)

F296 *e* fittings
d 1. Rohrformstücke *n* pl,
Rohrleitungsarmatur *f*, Fittings *m* pl
2. Kesselarmatur *f* 3. Leuchtkörper *m*
pl, Beleuchtungsarmatur *f*;
Elektroarmatur *f* 4. Fenster- *bzw*.
Türbeschläge *m* pl; Baubeschläge *m*
pl
f 1. raccords *m* pl (*de tuyauterie*);
robinetterie *f* 2. armature *f*

(*p. ex. d'une chaudière*), accessoires *m*
pl 3. appareils *m* pl, appareillage *m*
(*électrique ou d'éclairage*)
4. quincaillerie *f* (*de portes ou de
fenêtres*); ferronnerie *f*
nl installatiebenodigdheden *f* pl,
hulpstukken *n* pl
r 1. фитинги *m* pl, трубопроводная
арматура *f* 2. арматура *f* (*напр.
котла*) 3. осветительные
электрические приборы *m* pl;
электротехническая арматура *f*
4. приборы *m* pl (*оконные или
дверные*); строительные скобяные
изделия *n* pl

F297 *e* fitting-up
d Vormontage *f*
f assemblage *m* préliminaire
[préalable]
nl voormontage *f*
r предварительная сборка *f*
(конструкций)

F298 *e* fitting-up bolt
d Heftbolzen *m*
f boulon *m* de montage
nl hulpbout *m*
r временный [монтажный] болт *m*

F299 *e* fit-up
d 1. Anpassung *f*; Vormontage *f*,
Werkstattmontage *f*
2. Fertigungsschalung *f*,
wiederverwendungsfähige Schalung *f*
f 1. ajustage *m*, assemblage *m*
préalable 2. coffrage *m* préfabriqué
à usage multiple
nl 1. passend maken *n*, proefmontage *f*
2. prefabbekisting *f* voor meermalig
gebruik
r 1. подгонка *f*; контрольная сборка *f*
2. многократно оборачиваемая
опалубка *f* из сборных элементов

F300 *e* fixed beam
d zweiseitig eingespannter Balken *m*
[Träger *m*]
f poutre *f* encastrée aux deux
extrémités
nl tweezijdig ingeklemde balk *m*
r балка *f* с защемлёнными концами

F301 *e* fixed bridge
d feste Brücke *f*
f pont *m* fixe
nl vaste brug *f* (*m*)
r неразводной мост *m*

F302 *e* fixed bridge bearings
d feste Lager *n* pl einer Brücke
f appareils *m* pl d'appui fixe d'une
travée de pont
nl vaste steunpunten *f* (*m*) pl van een
brug
r неподвижные опорные части *f* pl
моста

F303 *e* fixed-edge-slab
d eingespannte Platte *f*
f dalle *f* encastrée

nl ingeklemde plaat *f* (*m*)
r защемлённая (по контуру) плита *f*

F304 *e* **fixed end**
 d eingespanntes Ende *n* (*Balken*)
 f extrémité *f* encastrée (*d'une poutre*)
 nl ingeklemd einde *n* (*balk*)
 r защемлённый конец *m* (*балки*)

F305 *e* **fixed-end arch**
 d eingespannter Bogen *m*
 f arc *m* encastré
 nl ingeklemde boog *m*
 r бесшарнирная арка *f*

F306 **fixed-end beam** *see* **fixed beam**

F307 **fixed-end moment** *see* **fixing moment**

F308 *e* **fixed hinge**
 d unverschiebliches Gelenk *n*
 f articulation *f* fixe
 nl vast scharnier *n*
 r неподвижный шарнир *m*

F309 *e* **fixed joint**
 d steife [starre] Verbindung *f*
 f joint *m* [assemblage *m*] rigide
 nl starre verbinding *f*
 r жёсткое соединение *n*

F310 *e* **fixed ladder**
 d eingebaute Leiter *f*
 f échelle *f* fixe
 nl vaste ladder *m*
 r неподвижная лестница-стремянка *f* (*прикреплённая к конструкции*)

F311 *e* **fixed load**
 d ständige Last *f*, Eigenlast *f*
 f charge *f* permanente [fixe, constante]
 nl vaste belasting *f*
 r постоянная нагрузка *f*

F312 *e* **fixed mooring berth**
 d Anlegeplatz *m* mit Abweisedalben
 f quai *m* d'amarrage à duc d'Albe de guidage
 nl aanlegsteiger *m* met geleidwerken [remmingwerken]
 r причал *m* с отбойными палами

F313 **fixed point** *see* **bench mark**

F314 *e* **fixed retaining wall**
 d starre [steife] Stützmauer *f*
 f mur *m* de soutènement rigide
 nl starre keermuur *m*
 r жёсткая подпорная стенка *f*

F315 *e* **fixed support**
 d eingespannte [gelenklose] Stütze *f*
 f appui *m* [support *m*, poteau *m*] fixe, colonne *f* encastreé
 nl oplegging *f*
 r защемлённая опорная стойка *f* [колонна *f*]

F316 *e* **fixed transom**
 d blinder Oberflügel *m* (*Tür*)
 f imposte *f* fixe (*au-dessus d'une porte*)
 nl vast bovenlicht *n* (*deur*)
 r глухая фрамуга *f* (*над дверью*)

F317 *e* **fixed weir**
 d festes Wehr *n*, blinder Staudamm *m*, überlauflose Staumauer *f*
 f barrage *m* fixe
 nl vaste stuw *m*
 r глухая плотина *f*

F318 *e* **fixed-wheel gate**
 d Rollschütz(e) *n* (*f*), Schiebetor *n*
 f vanne *f* wagon
 nl rolschuif *f* (*m*), roldeur *f* (*m*)
 r плоский колёсный затвор *m*

F319 *e* **fixing**
 d Befestigung *f*
 f fixation *f*, attache *f*
 nl bevestiging *f*
 r (при)крепление *n*, закрепление *n*

F320 *e* **fixing moment**
 d Einspann(ungs)moment *n*
 f moment *m* d'encastrement
 nl inklemmingsmoment *n*
 r момент *m* защемления, изгибающий момент *m*, вызванный защемлением (*конца балки, края плиты*)

F321 *e* **fixing profile, fixing shape, fixing unit**
 d Befestigungsprofil *n*, Halteprofil *n*
 f profilé *m* de fixation
 nl bevestigingsprofiel *n*
 r крепёжный прокатный профиль *m* [элемент *m*] (*уголок, швеллер*)

F322 *e* **fixture**
 d 1. Spannvorrichtung *f* 2. Elektroarmatur *f* 3. Einbauteil *n*
 f 1. dispositif *m* de fixation [de serrage], serre-joint *m* 2. appareil *m* d'éclairage 3. élément *m* (*constructif ou décoratif*) incorporé dans une partie portante de bâtiment
 nl 1. spanklem *f* (*m*) 2. armatuur *f* (*verlichting*) 3. spijkervast bevestigd toebehoren
 r 1. зажимное приспособление *n*; обойма *f*; хомут *m* 2. электрическая осветительная арматура *f* 3. декоративная *или* конструктивная деталь *f*, встроенная в несущую конструкцию здания

F323 *e* **flag**
 d Gehwegplatte *f*, Fußwegplatte *f*; Pflasterplatte *f*
 f dalle *f* [plaque *f*] de pavage
 nl grote trottoirtegel *m*; betonnen putdeksel *n*
 r плита *f* для мощения

F324 *e* **flag paving**
 d Pflasterung *f* mit Stein- *bzw.* Betonplatten
 f pavage *m* en plaques [dalles] de béton
 nl plaveien *n* met platen
 r мощение *n* плитами

F325 **flagstone** *see* **flag**

F326 **flake board** *see* **particle board**

F327 e **flaking**
 d Abblättern n, Abplatzen n (Anstrich)
 f écaillement m, écaillage m,
 exfoliation f
 nl afschilferen n (verf)
 r отслаивание n, отслоение n,
 шелушение (краски)

F328 e **flaky aggregate**
 d plattenförmiger Kornzuschlag(stoff) m
 f granulat m à grains aplatis, agregat
 m lamellaire
 nl vlokkige toeslag m
 r заполнитель m с зёрнами
 пластинчатой формы, лещадный
 заполнитель m

F329 e **flame cleaning**
 d Flammstrahlreinigung f,
 Flammstrahlen n
 f nettoyage m par la flamme
 nl schoonbranden n
 r огневая [газопламенная] очистка f
 (стальных конструкций)

F330 e **flameproofing agent**
 d Flammenschutzmittel n;
 Flammenschutzanstrich m
 f agent m ignifugeant
 nl brandwerend middel n
 r огнезащитное средство n;
 антипирен m

F331 **flameproof lighting fitting** UK see
 explosion-proof luminaire

F332 e **flame-resistant glass**
 d flammwidriges Glas n
 f verre m résistant au feu
 nl brandwerend glas n
 r огнестойкое стекло n

F333 e **flame-retardant partition**
 d feuerhemmende Trennwand f,
 feuerhemmendes Schott n
 f cloison f pare-feu
 nl brandwerende scheidingswand m
 r огнезадерживающая перегородка f
 [переборка f]

F334 e **flange**
 d 1. Flansch m; Flachflansch m
 2. Trägerflansch m
 f 1. bride f 2. aile f, membrure f (d'une
 poutre), semelle f
 nl flens m (van een pijp, ligger of wiel)
 r 1. фланец m 2. полка f (балки)

F335 e **flange angle**
 d Gurtwinkel m
 f cornière f de la membrure (d'une
 poutre rivée)
 nl flenshoekijzer n (geconstrueerde ligger)
 r поясной уголок m (балки)

F336 e **flanged bend**
 d Flanschkrümmer m
 f coude m à bride
 nl flensbocht f (m), gebogen
 flenskoppeling f (pijp)
 r фланцевое колено n

F337 e **flange socket**
 d E-Formstück n, Flanschenmuffenstück
 n
 f bout m de raccord mâle à emboîtement
 nl rechte flenskoppeling f, flensstuk n
 r патрубок m «фланец — раструб»

F338 e **flange spigot**
 d F-Formstück n
 f bout m de raccord mâle à bride
 nl pijpeinde n met flens
 r патрубок m «фланец — гладкий
 конец»

F339 e **flange splice**
 d Stoßverbindung f der Gurtelemente
 f joint m des semelles (d'une poutre
 composée)
 nl flensverbinding f (ligger)
 r стык m поясных элементов (балки)

F340 e **flank**
 d Bankett n (Straße), Fahrbahnrand m
 f bord m de chaussée
 nl zijkant m van de weg, wegberm m
 r наружная кромка f проезжей части
 дороги

F341 e **flanking transmission of sound**
 d Flankenübertragung f,
 Nebenwegübertragung f
 f transmission f latérale (du son)
 nl flankerende geluidoverdracht f (m)
 r боковая передача f шума (через
 звуковые мостики)

F342 e **flank wall**
 d Giebelmauer f, Giebelwand f
 f mur m latéral [(de) pignon]
 nl zijgevel m
 r боковая [торцевая] стена f (здания)

F343 e **flap gate**
 d Klappenverschluß m
 f vanne f à clapet
 nl klepstuw m
 r клапанный затвор m

F344 e **flap valve**
 d Klappenventil n, Rückschlagklappe f
 f clapet m de retenue
 nl terugslagklep f (m)
 r обратный клапан m

F345 e **flap weir**
 d bewegliches Wehr n
 f barrage-déversoir m démontable
 nl klepstuw m
 r разборная плотина f

F346 e **flared column head**
 d Pilzkopf m
 f chapiteau m de colonne évasé
 [élargi] vers le haut, tête f de
 poteau en champignon
 nl kolomkop m van paddestoelvloer
 r уширенная кверху капитель f
 колонны (для опирания безбалочных
 перекрытий)

F347 e **flared support**
 d Gegenkonussäule f
 f support m pyramidal ou conique

évasé vers le haut
nl trechtervormig verbrede kolom *f* (*m*)
r расширяющаяся кверху опора *f*
из наклонных стоек

F348 **flashboards** *see* **stop logs**

F349 *e* **flash coat**
d dünne Torkretbeton-Deckschicht *f*
f couche *f* de finition de gunite
nl fijne deklaag *f* (*m*) van spuitbeton
r тонкий слой *m* торкретбетона (*для
исправления мелких дефектов
бетонных поверхностей*)

F350 *e* **flashing**
d 1. Dichtungsblech *n*, Abdeckblech *n*
2. Wassersperre *f*
f 1. chaperon *m*, noquet *m* 2. garniture *f*
d'étanchéite
nl 1. loodslabbe *f* (*m*), loodloketten *n pl*
2. waterdichte afsluiting *f* tussen
twee vlakken
r 1. слив *m*, фартук *m* (*элементы
кровли*) 2. гидроизолирующая
прокладка *f* (*в швах*)

F351 *e* **flash set**
d vorzeitiges Abbinden *n*; falsche
Abbindung *f*
f prise *f* momentanée [fausse]
nl snel verharden *n*
r мгновенное схватывание *n* (*бетона,
раствора*); ложное схватывание *n*

F352 *e* **flat**
d 1. Wohnung *f* 2. Flachstahl *m*
f 1. appartement *m*, logement *m*
2. fer *m* plat
nl 1. etagewoning *f* 2. bandstaal *n*
r 1. квартира *f* 2. полосовая сталь *f*

F353 *e* **flat arch**
d scheitrechter [gerader] Bogen *m*
f arc *m* déprimé; linteau *m*
nl strekse boog *m*, hanekam *m*
r плоская арка *f*; перемычка *f*

F354 *e* **flat-chord truss**
d Parallel(fachwerk)träger *m*,
Parallelgurtbinder *m*
f poutre *f* à treillis à membrures
parallèles
nl parallelligger *m*
r ферма *f* с параллельными поясами

F355 *e* **flat course**
d Läuferschicht *f*
f assise *f* de panneresses
nl strekkenlaag *f* (*m*), strekse laag *f* (*m*)
r ложковый ряд *m* (*кладки*)

F356 *e* **flat drainage**
d Flachdränung *f*
f tapis *m* de drainage
nl vlakke drainage *f*
r плоский дренаж *m*

F357 *e* **flat hinge**
d Flachgelenk *n*
f articulation *f* [charnière *f*] plate
nl vlak scharnier *n*
r плоский шарнир *m*

F358 *e* **flat jack**
d Freyssinet-Kapselpresse *f*
f vérin *m* plat
nl platte vijzel *f* (*m*)
r плоский домкрат *m*

F359 *e* **flat joint**
d bündige Fuge *f*
f joint *m* plein [plat]
nl platvoeg *f* (*m*)
r шов *m* вподрезку (*каменной кладки*)

F360 *e* **flat-joint jointed pointing**
d geschlossene Vollfuge *f*
f joint *m* plein jointoyé
nl aangestreken platvolvoeg *f* (*m*)
r шов *m* вподрезку с расшивкой
(*каменной кладки*)

F361 **flat-joint pointing** *see* **flat pointing**

F362 *e* **flat-oval duct**
d flach-ovaler Kanal *m*
f gaine *f* plate-ovale
nl vlak-ovale luchtleiding *f*
r воздуховод *m* плоскоовального
сечения

F363 *e* **flat paint**
d Schleiflackfarbe *f*
f peinture *f* mate
nl matte verf *f* (*m*)
r матовая окраска *f*

F364 *e* **flat paintbrush**
d Flachpinsel *m*
f pinceau *m* plat
nl platte verniskwast *m*
r флейц *m*, плоская кисть *f*

F365 *e* **flat pointing**
d Ausfugen *n* der Mauerfugen mit der
Kelle
f jointoiement *m* effectué à la truelle
nl platvolvoegen *n*
r подрезка *f* швов (*кирпичной кладки*)
кельмой (*простейший метод
расшивки*)

F366 *e* **flat roof**
d Flachdach *n*
f toit *m* plat, toit-terrasse *m*,
toiture-terrasse *f*
nl plat dak *n*
r плоская крыша *f*; плоское покрытие *n*

F367 *e* **flat rope**
d Flachseil *n*, Bandseil *n*
f câble *m* plat
nl vlak kabeltouw *n*
r плоский (стальной) канат *m*

F368 *e* **flat seam**
d liegender Falz *m*
f agrafure *f* horizontale
nl platte vouwnaad *m*
r лежачий фальц *m* (*металлической
кровли*)

F369 *e* **flat skylight**
d flache Firstlaterne *f* [Scheitellaterne *f*]
f lanterneau *m* (de toiture) plat
nl platte daklantaarn *f* (*m*)
r плоский зенитный фонарь *m*

F370 *e* flat slab
   d Pilzdeckenplatte *f*
   f dalle *f* plane en béton armé (*avec armature à deux directions*);
   dalle *f* de plancher champignon
  nl balkloze vloer *m*, paddestoelvloer *m*
   r плоская железобетонная плита *f* (*армированная в двух направлениях*);
   плита *f* безбалочного перекрытия

F371 *e* flat-slab buttress dam
   d Plattenpfeiler(stau)mauer *f*, Pfeiler-Plattensperre *f*, Ambursen--Damm *m*, verstrebte Stauwand *f*
   f barrage *m* à contreforts à dalle plane
  nl vlakke stuwdam *m* met steunberen
   r контрфорсная плотина *f* с плоскими перекрытиями

F372 *e* flat slab construction
   d Pilzdecke *f*
   f plancher *m* champignon
  nl paddestoelvloer *m*
   r безбалочное (железобетонное) перекрытие *n*

F373 flat-slab deck dam *see* flat-slab buttress dam

F374 *e* flat shell
   d krumme [flache] Schale *f*
   f voile *m* mince à faible courbure
  nl flauw gebogen schaal *f*
   r плоская оболочка *f*, оболочка *f* малой кривизны

F375 *e* flat slide valve
   d Flachschieber *m*
   f soupape *f* à tiroir plat
  nl vlakke schuif *f* (*m*)
   r плоская задвижка *f*

F376 flat wall brush *see* flat paintbrush

F377 *e* flatwork
   d flache [ebene] Konstruktionen *f pl*
   f ouvrages *m pl* plats en béton
  nl vlakke betonconstructies *f pl*
   r плоскостные бетонные конструкции *f pl*

F378 *e* flemish bond
   d holländischer Verband *m*
   f appareil *m* flamand
  nl vlaams verband *n*
   r фламандская перевязка *f* (*кирпичной кладки*)

F379 *e* flexibility
   d Biegsamkeit *f*
   f flexibilité *f*, souplesse *f*
  nl buigzaamheid *f*, flexibiliteit *f*
   r гибкость *f*

F380 flexible carpet *see* flexible pavement

F381 flexible connection *see* flexible joint

F382 *e* flexible connector
   d Elastikschlauch-Einsatzstück *n*
   f manchon *m* (de raccord) flexible, raccord *m* flexible
  nl flexibele verbinding *f* (*koppeling*)
   r гибкая вставка *f*

F383 *e* flexible dolphin
   d elastische Dückdalbe *f*, elastischer Dückdalben *m*
   f poteau *m* d'amarrage flexible
  nl veerkrachtige dukdalf *m*
   r гибкий пал *m*

F384 *e* flexible dropchute
   d biegsame [anpassungsfähige] Betonrutsche *f*
   f goulotte *f* à béton flexible
  nl flexibele betongoot *f* (*m*)
   r гибкий лоток *m* (*для подачи бетонной смеси*)

F385 *e* flexible filling blade
   d biegsamer Stahlspachtel *m*
   f spatule *f* (*de peintre*) flexible
  nl buigzaam plamuurmes *n*
   r гибкий стальной шпатель *m*

F386 *e* flexible foam, flexible foamed plastic
   d (Kunststoff-)Weichschaum *m*, Weichkunststoffschaum *m*, Weichschaum(kunst)stoff *m*
   f mousse *f* (plastique) souple
  nl zachte schuimkunststof *f* (*m*)
   r мягкий пенопласт *m*

F387 *e* flexible foundation
   d elastische Gründung *f*
   f fondation *f* flexible
  nl elastische fundering *f*
   r гибкий фундамент *m*

F388 *e* flexible insulation
   d Mattenisolierung *f*
   f isolant *m* flexible, matelas *m* [tapis *m*] isolant flexible
  nl flexibele pijpisolatie *f*
   r гибкая теплоизоляция *f*, теплоизоляционные маты *m pl*

F389 *e* flexible joint
   d flexibler Anschluß *m*
   f assemblage *m* [raccord *m*] flexible
  nl flexibele verbinding *f*
   r гибкое соединение *n*

F390 *e* flexible metal roofing
   d biegsames Stahlblechdach *n*, Stahlblechdachhaut *f*
   f couverture *f* [toiture *f*] en tôle (d'acier), couverture *f* [toiture *f*] métallique
  nl dakbedekking *f* van metalen (golf)platen
   r металлическая кровля *f* (*из тонколистового металла*)

F391 *e* flexible mounting
   d elastische Stütze *f*
   f assise *f* élastique
  nl elastische oplegging *f*
   r упругое амортизирующее опорное устройство *n*

F392 *e* flexible pavement
   d schmiegsame Decke *f*, Schwarzdecke *f*, Schwarzbelag *m*
   f revêtement *m* routier souple, chaussée *f* souple

nl veerkrachtig wegdek *n*
r нежёсткое дорожное покрытие *n*, нежёсткая дорожная одежда *f*

F393 *e* **flexible retaining wall**
 d elastische Stützmauer *f* [Stützwand *f*]
 f mur *m* de soutènement flexible
 nl flexibele keerwand *m*
 r гибкая подпорная стенка *f*

F394 **flexion** *see* **bending**

F395 *e* **flexural bending**
 d Biegeknickung *f*
 f flambage *m* [flambement *m*] par flexion
 nl buigingsknik *m*
 r потеря *f* устойчивости при изгибе

F396 *e* **flexural bond**
 d Biege-Haftfestigkeit *f*
 f contrainte *f* d'adhésion en flexion
 nl aanhechtingsspanning *f* bij buiging
 r напряжение *n* сцепления при изгибе

F397 *e* **flexural cracking**
 d Biegerißbildung *f*
 f fissures *f pl* par flexion; fissuration *f* due à la flexion
 nl scheurvorming *f* bij buiging
 r трещинообразование *n* при изгибе; трещины *f pl* от изгиба

F398 *e* **flexural formula**
 d Biegegleichung *f*
 f formule *f* de flexion
 nl buigvergelijking *f*
 r уравнение *n* изгиба, формула *f* для определения напряжений при изгибе

F399 *e* **flexural member**
 d Biegestab *m*
 f élément *m* fléchi, pièce *f* fléchie
 nl op buiging belast constructiedeel *n*
 r изгибаемый элемент *m*, элемент *m* (конструкции), работающий на изгиб

F400 *e* **flexural rigidity, flexural stiffness**
 d Biegesteifigkeit *f*, Biegungswiderstand *m* (*Balken*)
 f rigidité *f* (*de la poutre*) à la flexion
 nl buigstijfheid *f* (*balk*)
 r жёсткость *f* (*балки*)

F401 *e* **flexural strength**
 d Biegefestigkeit *f*
 f résistance *f* à la flexion
 nl buigsterkte *f*
 r предел *m* прочности [временное сопротивление *n*] при изгибе

F402 *e* **flexural stress**
 d Biegespannung *f*
 f contrainte *f* [tension *f*] de flexion
 nl buigspanning *f*
 r напряжение *n* при изгибе

F403 *e* **flexure**
 d Durchbiegung *f*; Biegung *f*
 f flexion *f*, fléchissement *m*
 nl doorbuiging *f*; buiging *f*
 r изгиб *m*, изгибание *n*; прогиб *m*, выгиб *m*

F404 *e* **flexure curve**
 d Durchbiegungskurve *f*, Biegungslinie *f*, elastische Linie *f*
 f ligne *f* élastique
 nl elastische lijn *f* (*m*)
 r кривая *f* прогибов, упругая линия *f*

F405 *e* **flight conveyor**
 d Kratzkettenförderer *m*, Mitnehmerförderer *m*
 f convoyeur *m* à plateaux [à palettes]
 nl schrapertransporteur *m*
 r пластинчатый конвейер *m*

F406 *e* **flight of locks**
 d Mehrkammerschleuse *f*, Kuppelschleuse *f*, Schleusentreppe *f*
 f écluses *f pl* étagées; échelle *f* d'écluses
 nl trapsluis *f* (*m*)
 r многокамерный шлюз *m*; каскад *m* шлюзов, шлюзовая лестница *f*

F407 *e* **flight of stairs**
 d Treppenlauf *m*
 f volée *f* d'escalier
 nl trap *m*
 r лестничный марш *m*

F408 *e* **flight sewer**
 d Gefälleabsturz *m*, Stufenabsturz *m*
 f chute *f* en gradins (*de la ligne d'égout*)
 nl getrapt riool *n*
 r ступенчатый перепад *m* (*канализационной линии*)

F409 *e* **flight slab**
 d Treppenlaufplatte *f*
 f dalle *f* de volée (d'escalier)
 nl trapplaat *f* (*m*)
 r сборный лестничный марш *m*, плита *f* лестничного марша

F410 *e* **flitch**
 d 1. unbearbeitetes Rundholz *n* 2. Furnier *n* 3. unbesäumter Balken *m*, unbesäumtes Kantholz *n*; Schwarte *f* 4. Bohle *f* für einen zusammengesetzten Träger
 f 1. grume *f* d'œuvre 2. pli *m* (*de contre-plaqué*) 3. dosse *f*, dosseau *m* 4. planche *f* rabotée pour poutre composée
 nl 1. onbewerkt rondhout *n* 2. fineerplaat *f* (*m*) 3. beslagen balk *m* zwaarder dan 10×30 cm², boskant *m* 4. plank *f* (*m*) voor een samengestelde draagbalk
 r 1. бревно-заготовка *n* 2. шпон *m* 3. брус *m* с обзолами, горбыль *m* 4. доска *f* для составной балки

F411 *e* **flitch beam, flitch girder**
 d zusammengesetzter Holzstahlträger *m*
 f poutre *f* en planches de bois et en plaques métalliques serrées par boulons
 nl uit twee houten balken en een stalen plaat samengestelde balk *m*

r деревостальная составная балка *f*
(*из досок и стальных полос,
стянутых болтами*)

F412 *e* **float**
  d 1. Reibebrett *n*, Glättkelle *f*
  2. Maurerkelle *f* 3. Schwimmer *m*
  f 1. lisseuse *f* 2. truelle *f* 3. flotteur *m*
  nl 1. plakspaan *f* (*m*), pleisterspaan *f* (*m*)
  2. troffel *m* 3. vlotter *m*
  r 1. тёрка *f*, гладилка *f* (*для затирки
  поверхности*) 2. мастерок *m*
  3. поплавок *m*

F413 *e* **float-actuated gate**
  d Schwimmerschütze *f*
  f vanne *f* [porte *f*] d'écluse actionnée
  par flotteur
  nl vlotterafsluiter *m*
  r затвор *m* с поплавковой камерой,
  поплавковый затвор *m*

F414 *e* **floated coat**
  d Ziehschicht *f*, Glättschicht *f*
  (*Putzentechnik*)
  f couche *f* d'enduit lissée à la truelle
  nl blauw pleister *n*
  r слой *m* штукатурки с поверхностной
  затиркой

F415 *e* **float finish**
  d Reibeputz *m*
  f finition *f* de surface (*p. ex en béton*)
  par lissage à la truelle
  nl blauw pleister *n* (*betonafwerking*)
  r отделка *f* (*поверхности*) гладилкой

F416 *e* **float gauge**
  d Schwimmerpegel *m*, Schwimmermesser
  m, Wasserstandanzeiger *m* mit
  Schwimmer
  f limnimètre *m* à flotteur
  nl vlotterpeil *n*
  r гидрометрический поплавок *m*;
  поплавковый уровнемер *m*

F417 *e* **float glass**
  d Floatglas *n*
  f verre *m* flotté
  nl spiegelglas *n*, floatglas *n*
  r полированное листовое стекло *n*,
  флатовое стекло *n*

F418 *e* **floating**
  d 1. Verreiben *n*, Verputzen *n*,
  Glattstreichen *n* 2. zweite
  Unterputzschicht *f* 3. Einschwimmen
  n 4. Streifenbildung *f*
  f 1. lissage *m*, aplanissement *m* (*p. ex.
  du béton*) 2. couche *f* de fond (*d'enduit*)
  3. flotaison *m*, flottage *m*, transport *m*
  par flottage 4. apparition *f* de
  rayures *ou* de taches colorées (*défaut
  de peinture*)
  nl 1. gladpleisteren *n* 2. tweede ruwe
  pleisterlaag *f* (*m*) 3. invaren *n*
  4. strook- *of* vlekvorming *f*
  r 1. затирка *f* поверхности 2. грунт *m*
  (*второй слой трёхслойной
  штукатурки*) 3. подача *f* (*конструкции*) на плаву

4. образование *n* полос *или* пятен
(*дефект окрашенной поверхности*)

F419 *e* **floating boom**
  d Floßrechen *m*, schwimmender
  Holzfang *m*; Schwimmsperre *f*
  f barrage *m* pour le bois flottable,
  estacade *f* flottante
  nl drijfhoutversperring *f*
  r запань *f*; бон *m*

F420 *e* **floating breakwater**
  d schwimmender Wellenbrecher *m*
  f brise-lames *m* flottant
  nl drijvende golfbreker *m*
  r плавучий волнолом *m*

F421 *e* **floating caisson**
  d 1. Schwimmkasten *m* 2.
  Schwimmponton *m*, Abschlußponton
  m, Schwimmtor *n*
  f 1. caisson *m* flottant 2. bateau-porte
  m
  nl 1. drijvende caisson *m*, caissondeur
  f (*m*) 2. schipdeur *f* (*m*)
  r 1. массив-гигант *m*, плавучий кессон
  m 2. плавучий затвор *m*, батопорт *m*

F422 *e* **floating control**
  d astatische Regelung *f*
  f réglage *m* astatique
  nl astatische regeling *f*
  r астатическое регулирование *n*

F423 *e* **floating crane**
  d Schwimmkran *m*
  f grue *f* flottante
  nl drijvende kraan *f* (*m*)
  r плавучий кран *m*

F424 *e* **floating dock**
  d Schwimmdock *n*
  f dock *m* flottant, cale *f* flottante
  nl drijvend dok *n*
  r плавучий док *m*

F425 *e* **floating dredger**
  d schwimmender Saugbagger *m*,
  Schwimmbagger *m*, Naßbagger *m*
  f drague *f* flottante
  nl drijvende baggermolen *m*
  r плавучий земснаряд *m*

F426 *e* **floating earth**
  d Triebsand *m*, Fließsand *m*
  f terrain *m* coulant [mouvant]
  nl drijfzand *n*
  r плывун *m*

F427 *e* **floating floor**
  d schwimmender Estrich *m*
  f plancher *m* flottant
  nl gegoten vloer *m*
  r плавающий пол *m* (*на сплошных или
  отдельных упругих прокладках*)

F428 *e* **floating foundation**
  d schwimmende Fundation *f*,
  schwimmender Gründungskörper *m*
  f fondation *f* flottante
  nl drijvende fundering *f*
  r плавающий фундамент *m*

F429 **floating harbour** *see* **floating breakwater**

F430 *e* **floating method**
  d Einschwimmverfahren *n*
  f méthode *f* de flotaison
  nl methode *f* van drijvende montage
  r наплавной метод *m* строительства
  подводных туннелей

F431 **floating pile** *see* **friction pile**

F432 *e* **floating pile-driver**
  d Schwimmramme *f*, Rammschiff *n*
  f sonnette *f* flottante
  nl drijvende heistelling *f*
  r плавучий копёр *m*

F433 *e* **floating pile foundation**
  d schwimmende [schwebende]
  Pfahlgründung *f*
  f fondation *f* sur pieux flottants
  nl op kleef geheide paalfundering *f*
  r висячий свайный фундамент *m*

F434 *e* **floating pipeline**
  d schwimmende Spülleitung *f*
  f conduit *m* [pipeline *m*] flottant
  nl drijvende pijpleiding *f*
  r плавучий трубопровод *m* (*напр.*
  пульпопровод)

F435 *e* **floating pumping station**
  d schwimmende Pumpstation *f*,
  schwimmendes Pumpwerk *n*,
  Schwimm-Pumpanlage *f*
  f station *f* [installation *f*] de pompage
  flottante
  nl drijvend pompstation *n*
  r плавучая насосная станция *f*

F436 *e* **floating rigs**
  d Schwimmgeräte *n pl*
  f matériel *m* flottant, installations *f pl*
  flottantes
  nl drijvend materiëel *n*
  r плавсредства *n pl*, суда *n pl*
  технического флота

F437 *e* **floating rule**
  d Richtscheit *n*, Abgleichlatte *f*,
  Kartätsche *f*, Abziehplatte *f*
  f règle *f* égaliseuse [surfaceuse, de
  finition]
  nl rei *f* (*m*)
  r правило *n*

F438 *e* **floating slab**
  d schwimmende Fußbodenplatte *f*
  f dalle *f* flottante
  nl zwemmende vloer *m*
  r «плавающая» бетонная плита *f* (*пола*)

F439 *e* **floating strainer**
  d schwimmende Einlaßvorrichtung *f*
  f filtre *m* à tamis [à crépine] flottant
  nl drijvend inlaatfilter *n*, *m*
  r плавучий фильтр *m* (*оголовок
  речного водозаборного устройства*)

F440 **float level indicator** *see* **float gauge**

F441 *e* **float scaffold**
  d Hängerüstung *f*, Hängegerüst *n*
  f échafaudage *m* volant
  nl hangsteiger *m*
  r подвесные подмости *pl*

F442 *e* **float steam trap**
  d Schwimmerabscheider *m*,
  Schwimmerableiter *m*
  f purgeur *m* de vapeur à flotteur
  nl vlottercondenspot *m*
  r поплавковый конденсатоотводчик *m*

F443 *e* **float stone**
  d Reib(e)stein *m*
  f bloc *m* aplanissoir
  nl vlaksteen *m*
  r камень *m* для черновой затирки
  (*бетонных поверхностей*)

F444 *e* **float switch**
  d Schwimmerschalter *m*
  f interrupteur *m* à flotteur
  nl vlotterschakelaar *m*
  r поплавковое реле *n* уровня,
  поплавковый выключатель *m*

F445 *e* **float time**
  d Ereignisspiel *n*
  f réserve *f* de temps
  nl speling *f* in de tijd (*netwerkplanning*)
  r резервное время *n* (*сетевое
  планирование*)

F446 *e* **float valve**
  d Schwimmerventil *n*
  f régleur *m* [clapet *m*] à flotteur
  nl vlotterkraan *m*
  r поплавковый клапан *m*

F447 *e* **flocculation tank**
  d Flockungsbecken *n*
  f bassin *m* de floculation
  nl uitvlokkingsbak *m*
  r камера *f* хлопьеобразования

F448 *e* **flood coat**
  d Bitumendecklage *f*
  f revêtement *m* d'étanchéité de bitume
  nl bitumineuze slijtlaag *f* (*m*)
  r накрывочный слой *m* битума

F449 **flood control canal** *see* **floodway**

F450 *e* **flood control dam**
  d Hochwasserdamm *m*,
  Hochwasserdeich *m*,
  Hochwasserauffangsperre *f*
  f digue *f* contre les inondations, digue *f*
  de défense contre les crues
  nl hoogwaterkering *f*, stormvloedkering *f*
  r противопаводковая дамба *f*

F451 *e* **flood-control works**
  d Hochwasserschutzwerke *n pl*
  f ouvrages *m pl* de contrôle des crues
  nl kunstwerken *n pl* voor de beheersing
  van hoge waterstanden
  r сооружения *n pl* для регулирования
  паводков

F452 *e* **flooded dike**
  d überströmbarer Deich *m*
  f digue *f* submersible
  nl overlaat *m*
  r затопляемая дамба *f*

F453 *e* **flooded roof**
  d wasserbeaufschlagte Dachhaut *f*,
  überflutetes Dach *n*

17*

259

f couverture *f* [toiture *f*] recouverte
d'une couche d'eau
nl blankstaand dak *n*
r заливаемая [затопляемая] кровля *f*

F454 *e* **flood gate**
d 1. Hochwasserüberlauf *m*
2. Hochwasser-Verschluß *m*
3. Flutschleuse *f*, Fluttor *n*
f 1. évacuateur *m* de crue 2. vanne *f*
de protection contre les crues
3. écluse *f* de marée
nl 1. hoogwateraflaat *m* 2. vloedkering *f*
3. getijdesluis *f* (*m*)
r 1. паводковый водосброс *m*
[водослив *m*] 2. противопаводковый
затвор *m* 3. приливный шлюз *m*;
приливные ворота *pl* морского
шлюза

F455 *e* **flood hydrograph**
d Hochwasserganglinie *f*
f courbe *f* des débits d'une crue,
hydrogramme *m* de la crue
nl hoogwaterafvoercurve *f*
r гидрограф *m* паводка

F456 *e* **flooding dock**
d Flutbecken *n*
f dock *m* coulé
nl droogdok *n*
r наливной док *m*

F457 **flooding zone** *see* **flood zone**

F458 *e* **flood light**
d Flutlichtstrahler *m*, Lichtfluter *m*
f projecteur *m* pour illumination
nl floodlight *n*
r прожектор *m* заливающего света

F459 *e* **flood mark**
d Hochwassermarke *f*, Ebbemarke *f*
f marque *f* des hautes eaux
nl floedmerk *n*, hoogwaterlijn *f* (*m*)
r метка *f* высоких вод

F460 *e* **flood peak**
d Hochwasserspitze *f*, höchstes
Hochwasser *n*, HHW, höchster
Hochwasserstand *m*; höchste
Hochwasserdurchflußmenge *f*
f niveau *m* maximum d'une crue;
débit *m* de crue maximum
nl hoogste hoogwaterpeil *n*
r пик *m* [наивысший уровень *m*]
паводка; максимальный расход *m*
паводка

F461 *e* **floodplain**
d Auenebene *f*, Flußaue *f*, Talaue *f*,
Hochflutebene *f*, Hochwasserbett *n*
f plaine *f* inondable [d'inondation]
nl uiterwaard *f* (*m*), winterbedding *f* (*m*)
r пойма *f*

F462 *e* **flood proofing, flood protecting**
d Hochwasserschutz *m*
f défense *f* [protection *f*] contre les
inondations
nl bescherming *f* tegen overstroming

r защита *f* от затопления,
противопаводковая защита *f*

F463 *e* **flood protection works**
d Hochwasserschutzwerke *n pl*
f ouvrages *m pl* de protection contre
les crues
nl tegen overstroming beschermende
kunstwerken *n pl*
r противопаводковые сооружения *n pl*

F464 *e* **flood routing**
d 1. Hochwasserberechnung *f*
2. Hochwasserführung *f*
f 1. calcul *m* de la propagation des
crues 2. évacuation *f* des crues
nl 1. berekening *f* van het
hoogwaterverloop 2. beheersing *f*
van de hoogwaterafvoer
r 1. расчёт *m* трансформации паводка
2. пропуск *m* паводка

F465 *e* **flood spillway**
d Hochwasserentlastungsanlage *f*
f évacuateur *m* de crue
nl hoogwateraflaatkanaal *n*
r паводковый водосброс *m*

F466 **floodwall** *see* **flood control dam**

F467 *e* **floodwater retarding dam**
d Verzögerungsdamm *m*,
Hochwasserschutzbauwerk *n*
f barrage *m* protecteur contre les
inondations
nl vloedgolfvertragingsstuw *m*
r паводкозадерживающая плотина *f*
(с нерегулируемым водосливом)

F468 *e* **floodway**
d Hochwasserkanal *m*, Flutweg *m*,
Hochwasserbett *n*
f voie *f* [terrain *m*] d'évacuation des
crues
nl hoogwaterafvoerkanaal *n*
r паводковый водосброс *m*, паводочное
русло *n*

F469 *e* **flood zone**
d Überflutungsfläche *f*,
Überschwemmungszone *f*,
Inundationsgebiet *n*, Flutungsgebiet *n*
f zone *f* d'inondation, terrain *m*
submergé
nl overstomingsgebied *n*
r зона *f* затопления

F470 *e* **floor**
d 1. Fußboden *m* 2. Decke *f*
3. Geschoß *n*
f 1. couvre-sol *m*, revêtement *m* de sol
2. plancher *m* 3. étage *m*
nl 1. vloer *m* 2. rijdek *n* 3. verdieping *f*
r 1. пол *m* 2 перекрытие *n* 3. этаж *m*

F471 *e* **floor-and-wall tiling work**
d Fliesen- und Plattenarbeiten *f pl*
f travaux *m pl* de carrelage
nl tegelzetten *n*
r облицовка *f* плитками (*стен и пола*)

F472 *e* **floor area**
d Deckenfläche *f*

f surface *f* de plancher, aire *f* de parquet
nl vloeroppervlak *n*
r общая площадь *f* всех этажей здания (*по внутренним габаритам
этажей*)

F473 *e* **floor batten**
d Fußbodenbalken *m* (*auf der
Betonbettung verlegt*)
f lambourde *f*
nl op betonbed gelegde vloerbalk *m*
r лага *f* пола (*на бетонной
подготовке*)

F474 *e* **floor beam**
d 1. Deckenbalken *m*, Deckenträger *m*
2. Querbalken *m* [Querträger *m*] der
Brückenfahrbahn
3. (Treppen-)Absatzbalken(träger) *m*
f 1. poutre *f* de plancher; solive *f*
2. poutrelle *f* de tablier de pont
3. poutre *f* de palier
nl 1. vloerbint *n* 2. dwarsdrager *m* van
een brugdek 3. bordesdraagbalk *m*
r 1. балка *f* перекрытия 2. поперечная
балка *f* проезжей части моста
3. балка *f* лестничной площадки

F475 *e* **floor board**
d Fußbodenbrett *n*
f planche *f* de sol
nl vloerplank *f* (*m*)
r доска *f* для чистого пола

F476 *e* **floor boarding**
d Holzfußbodenbelag *m*
f planchéiage *m*, couvre-sol *m* en bois,
revêtement *m* de sol en bois
nl houten vloer *m*
r настил *m* дощатых полов, покрытие
n полов из досок

F477 *e* **floor box**
d Bodenverteilerdose *f*
f boîte *f* de distribution de parquet
nl tafelcontactdoos *f* (*m*)
r электрораспределительная коробка *f*,
скрытая в полу

F478 *e* **floor brick**
d Fußbodenziegel *m*; Deckenziegel *m*
f brique *f* de revêtement de sol;
brique *f* de planchers [d'entrevous]
nl plavuis *m*
r клинкер *m* [облицовочный кирпич *m*]
для пола; кирпич *m* для заполнения
часторебристых перекрытий

F479 *e* **floor clamp**
d Dielenzwinge *f*
f serre-joint *m* pour la pose des planches
de sol
nl vloer(aan)drijver *m*
r сжим *m* для сплачивания половых
досок

F480 *e* **floor clearance**
d Türspaltweite *f*
f hauteur *f* libre au-dessous de la porte
nl deurspleet *f* (*m*)
r зазор *m* между полом и дверью

F481 *e* **floorcloth**
d Fußbodenbelaggewebe *n*
f revêtement *m* de sol textile (*en
rouleau*)
nl vloerbedekking(smateriaal *n*) *f*
r тканевый рулонный материал *m*
для покрытия полов

F482 **floor component** *see* **floor unit**

F483 *e* **floor cover, floor covering**
d Fußbodenbelag *m*
f revêtement *m* du plancher
nl vloerbedekking *f*
r верхний (*отделочный*) слой *m*
покрытия пола

F484 *e* **floor covering work**
d Fußbodenbelagarbeiten *f pl*,
Fußbodenverlegearbeiten *f pl*
f travaux *m pl* [pose *f*] de revêtement
de sol
nl vloerbedekking leggen
r устройство *n* [работы *f pl* по
устройству] полов

F485 **floor cramp** *see* **floor clamp**

F486 *e* **floor diffuser**
d Fußbodendiffusor *m*,
Fußbodenluftverteiler *m*
f diffuseur *m* de plancher, grillage *m*
arrangé à plancher
nl vloerrooster *n* (*luchttoevoer*)
r напольный диффузор *m*,
напольный (встроенный в пол)
воздухораспределитель *m*

F487 *e* **floor dog** *see* **floor clamp**

F488 *e* **floor duct**
d Bodenkanal *m*, Bodenschlitz *m*
Unterflurkanal *m*
f conduit *m* sous-plancher
nl kanaal *n* onder de vloer
r подпольный канал *m*

F489 *e* **floor fill**
d Deckenfüllstofflage *f*
f isolation *f* de plancher en vrac
nl ondertapijt *n*, isolerende
egalisatielaag *f* (*m*)
r слой *m* засыпки в перекрытии

F490 **floor finish** *see* **floor cover**

F491 *e* **floor framing**
d Fußbodenstützkonstruktion *f*
f charpente *f* de plancher
nl vloerconstructie *f*
r опорная конструкция *f* пола
(*система балок и лаг*)

F492 *e* **floor guide**
d Führungsnut *f* im Fußboden
f rainure-guide *f* de parquet (*pour
portes coulissantes*)
nl geleiding *f* in de vloer (*schuifdeur*)
r направляющий паз *m* в полу (*для
раздвижных дверей*)

F493 *e* **floor gully**
d Bodenablauf *m*
f orifice *m* d'écoulement des eaux

nl schrobputje *n*
r трап *m* (*для стока жидкости*)

F494 *e* **floor hardener**
d Fußbodenhärtemittel *n*
f durcisseur *m* de surface de
(couvre-sol en) béton
nl harde toeslag *m* voor de slijtlaag
(*betonvloer*)
r материал *m* для поверхностного
упрочнения монолитного бетона

F495 *e* **floor heating**
d Fußboden(strahlungs)heizung *f*,
FB-Heizung *f*
f chauffage *m* par le plancher
nl vloerverwarming *f*
r система *f* отопления с обогреваемым
полом

F496 **floor height** *see* **storey height**

F497 *e* **floor hole**
d Deckenöffnung *f*
f trou *m* dans le plancher
nl doorvoering *f* door de vloer
r отверстие *n* в перекрытии

F498 *e* **flooring**
d Fußbodenbelag *m*
f revêtement *m* de sol, couvre-sol *m*
nl dekvloer *m*; vloeren *n*
r покрытие *n* [настил *m*] пола

F499 *e* **flooring nail**
d Dielen-Baunagel *m*
f clou *m* à planche [pour planches de sol]
nl vloerspijker *m*
r гвоздь *m* для деревянных полов
(*специальный фасонный гвоздь*)

F500 **floor joist** *see* **floor beam 1.**

F501 *e* **floor opening**
d Deckendurchbruch *m*,
Bodendurchbruch *m*
f ouverture *f* dans le plancher
nl vloeropening *f*
r проём *m* в перекрытии (*для пропуска
людей и грузов*)

F502 *e* **floor operated crane**
d flurgesteuerter Kran *m*
f grue *f* commandée à distance
nl beneden bestuurde kraan *f* (*m*)
r кран *m*, телеуправляемый с пола

F503 *e* **floor paint**
d Fußboden(anstrich)farbe *f*
f peinture *f* à plancher
nl vloerverf *f* (*m*)
r краска *f* для пола

F504 *e* **floor panel**
d Deckenbelagplatte *f*
f panneau *m* de plancher [de revêtement
de sol]
nl vloerplaat *f* (*m*)
r панель *f* настила пола

F505 *e* **floor plate**
d Bodenblech *n*
f 1. tôle [plaque *f*] de plancher
2. plaque *f* d'acier encastrée dans le

plancher (*pour la fixation de
d'équipement*)
nl staalplaat *f* (*m*) (*staalplaat-
-betonvloer*)
r 1. плита *f* металлического настила
2. стальная пластина *f* [плита *f*]
с прорезями в полу (*для крепления
оборудования*)

F506 *e* **floor plug, floor receptacle**
d Fußboden-Steckdose *f*
f prise *f* de courant de parquet
nl vloercontactdoos *f* (*m*)
r электророзетка *f* в полу

F507 *e* **floor slab**
d 1. Deckenplatte *f* 2. Bodenplatte *f*
f 1. dalle *f* de plancher 2. dalle *f* de
plancher en sous-sol
nl vloerplaat *f* (*m*)
r 1. плита *f* [панель *f*] перекрытия
2. плита *f* пола подвала

F508 *e* **floor sleeve**
d Fußbodenhülse *f*, Rohrhülse *f* in der
Geschoßdecke
f manchon *m* encastré dans le plancher
(*pour faire passer la canalisation*)
nl mantelbuis *f* (*m*) (*bij doorvoer van
elektrische leidingen*)
r трубная гильза *f* в перекрытии (*для
пропуска трубопроводов*)

F509 **floor space** *see* **floor area**

F510 *e* **floor system**
d Deckenkonstruktion *f*,
Deckensystem *n*
f système *m* [charpente *f*] de plancher
nl vloersysteem *n*
r система *f* [конструкция *f*]
перекрытий

F511 *e* **floor tile**
d Bodenplatte *f*, Bodenfliese *f*
f carreau *m* céramique pour sols
nl vloertegel *m*
r керамическая плитка *f* для полов

F512 *e* **floor-to-ceiling window**
d raumhohes Fenster *n*
f fenêtre *f* à toute hauteur de la pièce
nl kamerhoog venster *n*, glaspui *f* (*m*)
r окно *n* высотой от пола до потолка

F513 *e* **floor unit**
d Deckenelement *n*
f élément *m* du plancher
nl vloerelement *n*
r элемент *m* конструкции перекрытия

F514 **floor warming** *see* **floor heating**

F515 *e* **floor with precast beams placed close
together**
d Balkendecke *f* mit dicht verlegten
Balken
f plancher *m* en poutres préfabriquées
posées bord à bord
nl betonbalkenvloer *m*
r перекрытие *n* из уложенных рядом
сборных (железобетонных) балок
[из железобетонных настилов]

F516 *e* **flour limestone**
   d Kalk(stein)mehl *n*,
   Kalk(stein)pulver *n*
   f farine *f* calcaire [de calcaire, de chaux]
   nl kalksteenmeel *n*
   r молотый известняк *m*, известковая
   мука *f*

F517 *e* **flow**
   d Ausbreitmaß *n* (*Betongemisch*)
   f étalement *m* (du béton frais)
   nl zetmaat *f* (*m*) (*betonmortel*)
   r расплыв *m* (*бетонной смеси*)

F518 *e* **flow area**
   d 1. Durchgangsquerschnitt *m*,
   durchströmter [freier, lichter]
   Querschnitt *m* 2. Abflußquerschnitt *m*
   f 1. section *f* de passage 2. section *f*
   d'écoulement
   nl 1. doorstromingsprofiel *n*
   2. stroomprofiel *n*
   r 1. проходное сечение *n* 2. живое
   сечение *n* потока

F519 *e* **flow coefficient**
   d Durchflußzahl *f*; Abflußbeiwert *m*
   f coefficient *m* de débit; coefficient *m*
   d'écoulement
   nl afvoercoëfficiënt *m*
   r коэффициент *m* расхода;
   коэффициент *m* стока

F520 *e* **flow cone**
   d Fließkegel *m*
   f cône *m* d'écoulement
   nl vloeikegel *m*
   r конус *m* для измерения подвижности
   цементного раствора

F521 *e* **flow controller**
   d Durchflußmengenregler *m*
   f régulateur *m* de débit
   nl stroomregelaar *m*
   r регулятор *m* расхода

F522 *e* **flow-duration curve**
   d Abfluß(mengen)dauerlinie *f*
   f courbe *f* des débits classés
   nl afvoergrafiek *f*
   r кривая *f* обеспеченности (*стока,
   расходов*)

F523 *e* **flow factor**
   d Fließfähigkeit *f* des Einpreßmörtels
   f coefficient *m* de fluidité d'un coulis
   d'injection
   nl vloeibaarheidsgetal *n* (*injectiemortel*)
   r показатель *m* текучести
   [подвижности] инъекционного
   раствора

F524 *e* **flowing concrete**
   d flüssiges Betongemisch *n*
   f béton *m* fluide [de fluidité]
   nl vloeibare betonmortel *m*
   r разжиженная бетонная смесь *f*
   (*с суперпластифицирующими
   добавками*)

F525 *e* **flowing well**
   d frei ausfließende [selbstausfließende]
   Bohrung *f*
   f puits *m* jaillissant
   nl welput *m*
   r самоизливающийся колодец *m*

F526 *e* **flow mass**
   d Abflußsumme *f*
   f volume *m* (d'eau) écoulé
   nl afgevoerde watermassa *f*
   r объём *m* стока

F527 *e* **flow mass curve**
   d Abfluß(mengen)-Summenlinie *f*,
   Durchfluß(mengen)-Summenlinie *f*
   f courbe *f* des débits [des apports]
   cumulés
   nl gecumuleerde afvoergrafiek *f*
   r интегральная кривая *f* стока

F528 *e* **flow meter**
   d Durchflußmesser *m*,
   Durchflußmeßgerät *n*
   f débitmètre *m*, fluxmètre *m*
   nl verbruikmeter *m* (*water, gas*)
   r расходомер *m*

F529 *e* **flow net**
   d Stromliniennetz *n*, Strömungsnetz *n*,
   hydrodynamisches Netz *n*,
   Sickernetz *n*
   f réseau *m* d'écoulement, réseau *m* de
   lignes de courant
   nl patroon *f* (*m*) van ondergrondse
   afstroming
   r гидродинамическая сетка *f*

F530 *e* **flow rate**
   d Durchfluß(menge) *m* (*f*)
   f débit *m*
   nl afvoer *m*, debiet *n*
   r расход *m* (*газа, жидкости*)

F531   **flow separation** *see* **breakaway**

F532 *e* **flow table**
   d Ausbreittisch *m*
   f table *m* d'étalement
   nl uittrektafel *f* (*m*)
   r вибрационный столик *m* (*для измерения расплыва бетонной или растворной смеси*)

F533 *e* **flow test of concrete**
   d Fließprobe *f* [Fließprüfung *f*] von
   Beton(gemisch)
   f essai *m* d'étalement du béton
   nl vloeiproef *f* (*m*) van beton
   r испытание *n* бетона [бетонной смеси]
   на расплыв

F534 *e* **flow trough**
   d Gefällebetonrutsche *f*
   f goulotte *f* inclinée (*pour distribution
   et coulage du béton par gravité*)
   nl beton(mortel)goot *f* (*m*), stortgoot
   f (*m*) voor mortel
   r наклонный лоток *m* (*для подачи
   бетонной смеси самотёком*)

F535 *e* **flow-through system**
   d Durchlaufsystem *n*
   f système *m* à écoulement direct
   nl stortgootsysteem *n*

r проточная [прямоточная] система *f*
(*отопления, вентиляции*)

F536 *e* **flow type calorifier**
  d Durchlaufgaswasserheizer *m*
  f chauffe-eau *m* instantané à gaz
  nl warmwaterdoorstroomtoestel *n*
  r проточный газовый водонагреватель
  m, газовая колонка *f*

F537 *e* **flow variator**
  d Luftmengenstromregler *m*
  f variateur *m* de la circulation d'air
  nl luchtstroomregelaar *m*
  r автоматический регулятор *m* расхода
  воздуха

F538 **fluctuating load** *see* **alternating load**

F539 *e* **fluctuating stresses**
  d Schwellspannungen *f pl*
  f contraintes *f pl* [tensions *f pl*]
  variables; contraintes *f pl* répétées
  ou ondulées
  nl wisselende spanningen *f pl*
  r переменные напряжения *n pl*;
  цикличные напряжения *n pl* (*одного
  знака*)

F540 *e* **flue**
  d Fuchs *m*, Rauchkanal *m*,
  Schornsteinzug *m*, Abzugschacht *m*
  f conduit *m* de fumée
  nl rookkanaal *n*, schoorsteenkanaal *n*
  r дымоход *m*, дымовой канал *m*,
  газоход *m*

F541 *e* **flue gas, flue gases**
  d Rauchgas(e) *n* (*pl*), Abgas(e) *n* (*pl*)
  f gaz *m* de fumée
  nl rookgas(sen) *n* (*pl*)
  r дымовые газы *m pl*

F542 *e* **flue liner**
  d Rauchkanalausfütterung *f*
  f revêtement *m* [garniture *f*] réfractaire
  de cheminée
  nl inzetpijp *f* (*m*) in schoorsteen
  r футеровка *f* дымохода [газохода] *или*
  дымовой трубы

F543 *e* **fluid concrete**
  d flüssiger Beton *m*
  f béton *m* fluide [coulé, coulant]
  nl vloeibaar beton *n*
  r литой бетон *m*

F544 *e* **fluid-filled column**
  d wassergefüllte Hohlstütze *f*
  [Kastenstütze *f*, Rohrstütze *f*]
  f poteau *m* creux en acier rempli d'eau
  (*pour augmenter la résistance au feu*)
  nl met water gevulde stalen kolom *f* (*m*)
  r пустотелая стальная колонна *f*,
  наполненная водой (*для повышения
  огнестойкости при пожаре*)

F545 *e* **fluidifier**
  d Verflüssigungsmittel *n* (*Beton*)
  f fluidifant *m*, additif *m* fluidifant
  (*du béton*)
  nl toeslag *m* ter verhoging van de
  vloeibaarheid (*betonmortel*)

r разжижитель *m*,
суперпластифицирующая добавка *f*
(*бетонной смеси*)

F546 *e* **fluidity of the concrete**
  d Fließfähigkeit *f* des Betongemisches
  f fluidité *f* de béton
  nl vloeibaarheid *f* [zetmaat *f* (*m*)] van
  betonmortel
  r подвижность *f* [текучесть *f*] бетонной
  смеси

F547 *e* **flume**
  d Rinne *f*; Wasserbrücke *f*,
  Überleitung *f*
  f canal *m* sur appuis; aqueduc *m*,
  pont-aqueduc *m*
  nl betonstortgoot *f* (*m*), goot *f* (*m*);
  aquaduct *n*, waterleiding *f*
  r лоток *m*; акведук *m*, лотковый
  водовод *m*

F548 *e* **fluorescent paint**
  d Leuchtfarbe *f*
  f peinture *f* lumineuse
  nl lichtgevende verf *f* (*m*)
  r светящаяся [люминесцентная]
  краска *f*

F549 *e* **flush**
  d 1. Spülauftrag *m*, Spülauffüllung *f*,
  Spülaufschüttung *f* 2. Spülversatz *m*
  3. Spülen *n* mit Hochdruckstrahl
  4. Wildbettabgabe *f*
  f 1. remblayage *m* hydraulique
  2. remblai *m* hydraulique [à l'eau]
  3. chasse *f* d'eau, nettoyage *m* à la
  lance 4. lâchure *m*
  nl 1. opspuiten *n* 2. opgespoten grond *m*
  3. doorspoelen *n*, doorspuiten *n*
  4. spuien *n*
  r 1. намыв *m* (*грунта в тело
  гидросооружений*) 2. гидравлическая
  закладка *f* 3. промывка *f* высоко-
  напорной струёй 4. попуск *m* воды

F550 **flush curb** *see* **flush kerb**

F551 *e* **flush door**
  d Tür *f* mit glattem Blatt
  f porte *f* plane
  nl vlakke deur *f* (*m*)
  r гладкая щитовая дверь *f*

F552 *e* **flushing**
  d 1. Spülen *n* 2. Ebenen *n*,
  Egalisieren *n*, Nivellieren *n*
  3. Abbrechen *n*
  f 1. chasse *f* (d'eau), curage *m*
  2. arasement *m*, égalisation *f*,
  nivellement *m* 3. cassure *f*
  nl 1. doorspoelen *n* (*riool*) 2. gelijkmaken *n*
  3. breuk *m* (*steen*)
  r 1. промывка *f* 2. выравнивание *n*
  3. откол *m* (*камня*); излом *m*

F553 **flushing asphalt** *see* **bleeding asphalt**

F554 **flushing basin** *see* **scouring sluice**

F555 *e* **flushing canal**
  d Spülkanal *m*

f canal *m* de purge [purgeur]
nl spuikanaal *n*
r промывной канал *m*

F556 e **flushing cistern**
d Spülkasten *m*
f réservoir *m* de chasse
nl stortbak *m*
r смывной бачок *m*

F557 e **flushing gallery**
d Spülgalerie *f*, Spülstollen *m*
f galerie *f* de chasse
nl spoeltunnel *m*
r промывная галерея *f*

F558 e **flushing manhole**
d Spülschacht *m*
f regard *m* de chasse
nl spoelschacht *f* (*m*)
r промывной канализационный колодец
m

F559 e **flushing tank**
d Spülbehälter *m*
f réservoir *m* de chasse
nl spoeltank *m*
r промывной бак *m* (*для промывки*
канализационного выпуска)

F560 **flush joint** *see* **flat joint**

F561 e **flush kerb**
d Tiefbord(stein) *m*
f pierre *f* de bordure à ras de chaussée
nl verzonken trottoirband *n*
r бортовой [бордюрный] камень *m*
заподлицо с дорожным покрытием
(*по наружным кромкам*)

F562 e **flute**
d Riefe *f*, Kannelur(e) *f* (*Säule*)
f cannelure *f* (*d'une colonne*)
nl verticale groef *f* (*m*), cannelure *f* (*m*)
(*zuil*)
r каннелюра *f* (*колонны*)

F563 e **fluxed bitumen**
d gefluxtes Bitumen *n*,
Verschnittbitumen *n*
f bitume *m* fluidifié [fluxé]
nl verdund bitumen *n*
r разжиженный нефтяной битум *m*
(*на нелетучем разбавителе*)

F564 e **flux oil**
d Fluxöl *n*; Flußmittel *n*; Stellöl *n*
(*zum Einstellen der Viskosität*),
Verschnittöl *n*
f huile *f* fluxante [de fluxage]
nl fluxolie *f* (*asfaltbitumen*)
r разбавитель *m*; гудрон *m* [масло *n*]
для размягчения битума

F565 e **fly ash**
d Flugasche *f*
f cendre *f* volante
nl vliegas *n*
r летучая зола *f*, зола-унос *f*

F566 e **fly-ash concrete**
d Flugaschebeton *m*
f béton *m* de cendre volante

nl vliegasbeton *n*
r золобетон *m*

F567 **flying buttress** *see* **arc-boutant**

F568 e **flying form**
d zusammensetzbare Deckenschalung *f*
f coffrage-outil *m* pour planchers
(*dit «volant»*)
nl monteerbare bekisting *f*
r сборный блок *m* переставной
опалубки перекрытий
(*устанавливаемый краном*)

F569 **flying fox** *see* **cableway**

F570 e **flying scaffold**
d hängendes Gerüst *n*
f échafaudage *m* volant [suspendu]
nl vliegende steiger *m*, steeksteiger *m*
r подвесные подмости *f pl*

F571 e **flyover**
d 1. Überführung *f* 2. Kreuzungsbauwerk
n
f 1. passage *m* superieur; viaduc *m*
2. croisement *m* à niveaux
différents, ouvrage *m* de croisement
dénivelé
nl fly-over
r 1. мост-путепровод *m* 2. пересечение
n автодорог в разных уровнях

F572 **flyover junction** *see* **flyover 2.**

F573 e **fly rafter**
d ornamentales Windbrett *n*
f bordure *f* de pignon ornementale
nl spar *m* van dakoverstek
r декоративная фронтонная
доска *f*

F574 **foamed blast-furnace slag** *see*
expanded blast-furnace slag

F575 e **foamed concrete**
d Schaumbeton *m*
f béton-mousse *m*
nl cellenbeton *n*
r пенобетон *m*

F576 **foamed glass** *see* **foam glass**

F577 e **foamed-in-place insulation, foamed-**
-in-situ plastics
d an-ort-eingeschäumte Wärmedämmung
f
f mousse *f* isolante projetée, isolant *m*
en mousse plastique préparé sur
chantier
nl in het werk aangebrachte
schuimisolatie *f*
r теплоизоляция *f* из пенопласта,
приготовляемого на месте укладки

F578 e **foamed plastics**
d Schaumstoff *m*
f mousse *f* plastique
nl schuimplastic *n*
r пенопласт *m*

F579 e **foamed slag**
d Schaumschlacke *f*, Hüttenbims *m*
f ponce *f* de laitier, laitier *m* expansé
nl geëxpandeerde hoogovenslak *f*

r термозит *m*, вспученный шлак *m*, шлаковая пемза *f*

F580 *e* **foamed slag aggregate**
  d Schaumschlacke-Zuschlag *m*
  f granulat *m* en ponce de laitier
  nl schuimslak-toeslag *m*
  r лёгкий заполнитель *m* из термозита

F581 *e* **foam-entraining admixture**
  d oberflächenaktiver Mikroschaumbildner *m*
  f générateur *m* de mousse à cellules microscopiques
  nl schuimvormende toevoeging *f*
  r микропенообразующая полимерная добавка *f*

F582 *e* **foam-filled block**
  d Betonstein *m* mit schaumstoffausgefüllten schlitzförmigen Hohlräumen
  f bloc *m* creux en béton rempli par mousse plastique
  nl holle betonsteen *m* met schuimvulling
  r бетонный блок *m* с заполнением пустот пенопластом

F583 *e* **foam-forming admixture**
  d Schaummittel *n*, Schaumbildner *m*
  f générateur *m* de mousse
  nl schuimmiddel *n*
  r пенообразующая добавка *f*

F584 *e* **foam glass**
  d Schaumglas *n*
  f verre *m* moussé
  nl schuimglas *n*
  r пеностекло *n*

F585 *e* **foaming**
  d Aufschäumen *n*, Schaumbildung *f*
  f moussage *m*
  nl schuimvorming *f*
  r вспенивание *n*, пенообразование *n*

F586 *e* **foaming agent**
  d Schaummittel *n*
  f agent *m* mossant, écumant *m*
  nl schuimmiddel *n*
  r пенообразователь *m*

F587 *e* **foam silicate**
  d Schaumsilikat *n*
  f mousse *f* silico-calcaire [de silicate]
  nl schuimsilicaat *n*
  r пеносиликат *m*

F588 *e* **foam-silicate concrete**
  d Schaumsilikatbeton *m*
  f béton *m* mousse silico-calcaire
  nl schuimsilicaatbeton *n*
  r пеносиликатобетон *m*

F589 *e* **fog curing**
  d 1. Betonfeuchtlagerung *f* 2. Sprühbefeuchtung *f*, Anfeuchtung *f* durch Wasservernebelung
  f 1. curing *m* du béton dans l'atmosphère humide 2. humidification *f* à pulvérisation
  nl 1. nabehandeling *f* in vochtige atmosfeer 2. vochtig houden door verneveling van water
  r 1. выдерживание *n* (бетона) во влажной среде 2. увлажнение *n* тонкораспылённой водой

F590 *e* **fogging**
  d Erzeugung *f* der Feuchtluftatmosphäre, Wasservernebelung *f*, Nebelbildung *f*
  f brumisage *m*
  nl produceren *n* van nevel
  r создание *n* влажностно-воздушной среды, туманообразование *n*

F591 **fog room** *see* **moist room**

F592 *e* **fog seal**
  d versprühbare Bitumenemulsion *f* (*zur Absiegelung der Straßendecke*)
  f émulsion *f* routière appliquée par pulvérisation
  nl sproeilaag *f* (*m*) van bitumenemulsie
  r дорожная битумная эмульсия *f*, наносимая методом распыления

F593 *e* **fog spraying**
  d Wasservernebelung *f*
  f pulverisation *f* de brouillard
  nl waterverneveling *f*
  r распыление *n* воды в виде тумана (*метод ухода за бетоном*)

F594 *e* **foil**
  d Folie *f*
  f feuille *f* (mince) métallique
  nl foelie *n*
  r фольга *f*

F595 *e* **fold**
  d Falte *f*
  f pli *m*
  nl vouw *f* (*m*)
  r складка *f*

F596 *e* **folded-plate concrete roof**
  d Faltwerkbetondach *n*
  f comble-voûte *m* prismatique en béton
  nl vouw-schaalbetondak *n*
  r складчатое бетонное покрытие *n*

F597 *e* **folded plate construction**
  d Faltwerk *n*, Faltkonstruktion *f*
  f structure *f* plissée prismatique
  nl vouwschaal-dakconstructie *f*
  r складчатая конструкция *f*

F598 *e* **folded plates**
  d Faltwerk *n*
  f dalle *f* (de couverture) prismatique
  nl vouwschalen *f* (*m*) *pl*
  r складки *f pl* (складчатого) покрытия

F599 *e* **folded-plate theory**
  d Faltwerkstheorie *f*
  f théorie *f* des structures plissées
  nl vouwschaaltheorie *f*
  r теория *f* складок

F600 *e* **folding door**
  d Falttür *f*, Harmonikatür *f*
  f porte *f* accordéon [pliante]

nl vouwdeur *f (m)*
r складчатая дверь *f*

F601 *e* **folding partition**
d Falttrennwand *f*
f cloison *f* en accordéon
nl vouwwand *m*
r складчатая раздвижная
перегородка *f*

F602 *e* **follower**
d Rammjungfer *f*, Schlagjungfer *f*
(*Pfahl*)
f avant-pieu *m* (*en bois*)
nl oplanger *m*, opzetter *m* (*paal*)
r деревянный подбабок *m* (*сваи*)

F603 *e* **follower-ring valve**
d Absperrschieber *m*,
Regulierschieber *m*, Keilschieber *m*
f robinet-vanne *m* à lunette
nl regelafsluiter *m*
r клиновая задвижка *f*

F604 **fondu** *see* **high-alumina cement**

F605 *e* **foot bearing**
d Stützlager *n*, Spurlager *n*, Fußlager *n*
f butée *f*, pied *m* d'appui pivotant
nl taatspot *m*, vloerveer *f (m)*
r упорный подшипник *m*, опорная
поворотная пята *f* (*мачты*)

F606 *e* **foot block**
d 1. Sohlblock *m*, Lagerkörper *m*
2. Stützbalken *m* 3. Sohlenquader *m*
f 1. bloc *m* [massif *m*] de fondation
2. semelle *f*, sole *f* 3. blocks *m pl*
d'appui, posés au fond d'écluse
nl 1. poer *f (m)* 2. funderingsblok *n*,
steunbalk *m* 3. grondbalk *m*, neut
f (m)
r 1. опорный блок *m*, опорная
подушка *f* 2. опорный лежень *m*
3. штучный камень *m*
(*закладываемый в днище шлюза под
шандорным пазом*)

F607 *e* **footbridge**
d Fußgängerbrücke *f*
f passerelle *f*
nl voetbrug *f (m)*, loopbrug *f (m)*
r пешеходный мостик *m*

F608 *e* **footing**
d Fundament *n*, Gründungskörper *m*;
Sockel *m*; Untergrund *m*;
Gründung *f*; Treppenarmbalken *m*
f fondation *f*; base *f*; socle; embase *f*;
empattement *m*; semelle *f*; assise *f*;
patin *m*
nl fundering *f*; voetmuur *m*; voeting *f*
r фундамент *m*; нижняя
уширенная часть *f* фундамента *или*
стены; основание *n*; подкосоурная
балка *f*

F609 *e* **footing beam**
d Bundbalken *m*, Spannbalken *m*,
Binderbalken *m*
f tirant *m*, entrait *m*
nl trekplaat *f (m)*

r затяжка *f* стропильной фермы
(*у нижних концов стропильных ног*)

F610 *e* **footing course**
d untere Mauerwerksschicht *f*
f assise *f* inférieure de maçonnerie
nl vrijlaag *f (m)* (*gemetselde fundering*)
r нижний ряд *m* кладки

F611 *e* **footing excavation**
d Fundamentgrubenaushub *m*,
Fundamentgrabenausbaggerung *f*
f excavation *f* des tranchées pour la
fondation
nl funderingssleuf *f (m)*
r отрывка *f* котлована *или* траншеи
под фундамент

F612 *e* **footing stop**
d Holzeinlage *f* (*in Betonierfuge*)
f joue *f* provisoire (*au joint de reprise*)
nl houten tussenlaag *f (m)* (*in betonvoeg*)
r временная деревянная прокладка *f*
(*в рабочем шве бетонируемой
конструкции*)

F613 *e* **footpath**
d 1. Gehweg *m* 2. Gehsteig *m*, Trottoir *n*
3. Laufsteg *m*, Laufbrücke *f*
f 1. voie *f* pour piétons 2. trottoir *m*
3. passerelle *f*
nl 1. voetpad *n* 2. trottoir *n* 3. voetbrug
f (m)
r 1. пешеходная дорожка *m*
2. тротуар *m* 3. переходной мостик *m*

F614 *e* **footplate**
d Grundbalken *m*, Sohlbalken *m*
f semelle *f* de la paroi à pans de
bois
nl voetplaat *f (m)*; steekbalk *m*,
balksleutel *m*
r лежень *m* [брус *m*] нижней обвязки
(*деревянной рамы*)

F615 *e* **foot screws**
d Fußschrauben *f pl*
f vis *m pl* d'ajustage [de réglage]
nl stelschoeven *f (m) pl*
r установочные винты *m pl*
(*геодезического прибора*)

F616 *e* **foot valve**
d Fußventil *n*, Bodenventil *n*
f soupape *f* [clapet *m*] de pied
nl voetventiel *n*
r нижний [приёмный] клапан *m*
(*обратный клапан в нижнем конце
всасывающей трубы*)

F617 *e* **footway**
d Gehweg *m*, Bürgersteig *m*; Fußweg *m*,
Gehsteig *m*; Fahrort *m*, Fahrung *f*;
Kontrollgang *m*
f trottoir *m*; voie *f* pour piétons;
galerie *f* de circulation, boyeau *m*;
galerie *f* de révision [de contrôle]
nl trottoir *n*; voetpad *n*; schouwpad *n*;
controlegang *m*
r тротуар *m*; пешеходная дорожка;
ходок *m*; потерна *f*

F618 *e* **force**
   d Kraft *f*
   f force *f*, effort *m*
   nl kracht *m*
   r сила *f*, усилие *n*

F619 *e* **force application**
   d Kraftangriff *m*, Kraftanbringung *f*
   f application *f* de force
   nl aanbrengen *n* [uitoefenen *n*] van een kracht
   r приложение *n* силы

F620 *e* **forced circulation**
   d erzwungene Zirkulation *f*, Zwangsumlauf *m*, Zwangsumwälzung *f*
   f circulation *f* forcée
   nl gedwongen circulatie *f*
   r принудительная циркуляция *f*

F621 *e* **forced convection**
   d erzwungene Konvektion *f*
   f convection *f* forcée
   nl gedwongen convectie *f*
   r вынужденная конвекция *f*

F622 *e* **forced draft** *US*
   d Druckzug *m*
   f tirage *m* forcé
   nl geforceerde trek *m*
   r дутьё *n*, принудительная тяга *f*

F623 *e* **forced-draft cooling tower**
   d zwangsbelüfteter Kühlturm *m*
   f refroidisseur *m* d'eau à ventilation par refoulement
   nl koeltoren *m* met geforceerde trek
   r градирня *f* с принудительной циркуляцией воздуха

F624 *e* **forced ventilation**
   d Zwangslüftung *f*
   f ventilation *f* forcée
   nl kunstmatige ventilatie *f*
   r принудительная вентиляция *f*

F625 *e* **force main**
   d Hauptdruckleitung *f*
   f conduite *f* de refoulement
   nl hoofdpersleiding *f*
   r напорная магистраль *f*

F626 *e* **force-polygon pole**
   d Pol *m* eines Kräftepolygons
   f pôle *m* de polygone des forces
   nl pool *m* van een krachtenveelhoek
   r полюс *m* силового многоугольника

F627 *e* **force resolution**
   d Kräftezerlegung *f*
   f décomposition *f* de la force
   nl ontbinding *f* van krachten
   r разложение *n* силы

F628 *e* **force system**
   d Kräftesystem *n*
   f système *m* des forces
   nl krachtenstelsel *n*
   r система *f* сил

F629 *e* **force triangle**
   d Kräftedreieck *n*

   f triangle *m* des forces
   nl krachtendriehoek *m*
   r силовой треугольник *m*

F630 *e* **forebay**
   d 1. Vorbecken *n*, Druckbecken *n* 2. Werkgraben *m* 3. Oberwasser *n*
   f 1. bassin *m* de mise en charge; avant-port *m* 2. canal *m* d'amenée usinière 3. eau *f* d'amont
   nl 1. reservoir *n* (*begin persleiding*) 2. toevoerkanaal *n* 3. bovenpand *n*
   r 1. водорегулирующий бассейн *m*; аванкамера *f*; напорный бассейн *m* 2. подводящий канал *m* (*к турбине*) 3. верхний бьеф *m*

F631 *e* **forecourt**
   d 1. Vorgarten *m* 2. Vorhof *m*
   f 1. terrain *m* devant la maison 2. avant-cour *f*
   nl 1. voortuin *m* 2. voorplaats *f* (*m*)
   r 1. площадка *f* перед зданием (*между фасадом и улицей*) 2. передний двор *m*

F632 *e* **foregallery**
   d Vorstollen *m*, Richtstollen *m*
   f galerie *f* d'avancement
   nl voorgalerij *f*
   r передовая штольня *f*

F633 *e* **foreland**
   d Vorland *n*
   f avant-terrain *m*
   nl voorland *n*, landtong *f* (*m*)
   r территория *f* перед ограждающей дамбой (*не защищённая от затопления*)

F634 *e* **foreman**
   d Vorarbeiter *m*, Brigadier *m*
   f chef *m* d'équipe, contremaître *m*
   nl voorman *m*, ploegbaas *m*
   r бригадир *m*

F635 *e* **fork grip**
   d Klammergabel *f* (*Gabelstapler*); Krangabel *f*
   f pince *f* à fourche
   nl gaffelklauw *m*, *f*, stapelvork *f* (*m*)
   r вилочный захват *m*

F636 *e* **fork junction**
   d Gabelkreuzung *f*, Straßengabelung *f*
   f jonction *f* de deux routes en bifurcation
   nl wegsplitsing *f*, t-kruising *f*
   r вилообразное примыкание *n* автодорог

F637 *e* **fork-lift truck**
   d Gabelstapler *m*, Hubstapler *m*
   f chargeur *m* à fourches
   nl vorkheftruk *m*
   r вилочный погрузчик *m*

F638 *e* **form**
   d Schalung *f*; Schalungsform *f*
   f coffrage *m*; moule *m*
   nl bekisting *f*; mal *m*
   r опалубка *f*; форма *f*

F639 *e* **form anchor**
  d Anker *m* für Befestigung der Schalung zum erhärteten Beton
  f ancre *m* supportant le coffrage (*attaché au béton durci*)
  nl anker *n* ter bevestiging van de bekisting aan het verharde beton
  r анкер *m* для закрепления опалубки (*к ранее уложенному и затвердевшему бетону*)

F640 *e* **form assembly**
  d Schalungsblock *m*
  f unité *f* préfabriquée de coffrage, coffrage-outil *m*
  nl systeembekisting *f*
  r сборный опалубочный блок *m*

F641 *e* **formation**
  d Erdplanum *n*, Bodenplanum *n*; Bettungskoffer *m*
  f plate-forme *f* (*partie supérieure de l'infrastructure routière*)
  nl oppervlak *n* van de aardebaan (*wegbouw*)
  r проектная поверхность *f* земляного полотна; дно *n* корыта (*для дорожной одежды*)

F642 *e* **formation level**
  d Erdplanumhöhe *f*
  f profil *m* du fond de déblai
  nl peil *n* van de aardebaan
  r проектный уровень *m* [профиль *m*] земляной выемки

F643 *e* **form coating**
  d Schalungsschmierung *f*, Schalungsbefeuchtung *f*
  f enduisage *m*, humidification *f* (*de la surface intérieure du coffrage*)
  nl insmeren *n* van bekisting
  r обмазка *f* [увлажнение *n*] опалубки (*внутренней поверхности*)

F644 *e* **formed steel construction**
  d Profilblechbelagplatte *f* mit Obergurtblech
  f tôle *f* profilée avec semelle supérieure
  nl vloerconstructie *f* van geprofileerde staalplaat
  r стальной профилированный настил *m* с верхним поясным листом

F645   **formers** *see* **form liners**

F646 *e* **form hanger**
  d Schalungsaufhängung *f*
  f suspente *f* [tige *f* de suspension] de coffrage
  nl hanger *m* voor bekisting
  r опалубочная подвеска *f*

F647 *e* **forming-and-reinforcing system**
  d Schalungs-Bewehrungsblock *m*
  f unité *f* préfabriqué de coffrage avec l'armature
  nl bekistings- en wapeningseenheid *f*
  r арматурно-опалубочный блок *m*

F648 *e* **form insulation**
  d Schalungswärmedämmung *f*

f isolation *f* thermique extérieure du coffrage
  nl warmte-isolatie *f* van bekisting
  r наружная теплоизоляция *f* опалубки

F650 *e* **form pressure**
  d Schalungsdruck *m* des Betons, Frischbetondruck *m*
  f pression *f* sur le coffrage (*du béton frais*)
  nl betondruk *m* op bekisting
  r давление *n* бетонной смеси на стенки опалубки

F651 *e* **form release agent**
  d Entschalungsmittel *n*, Betonschalmittel *n*, Schalungsfett *n*, Entschalungsöl *n*
  f huile *f* de coffrage
  nl ontkistingsmiddel *n*
  r опалубочная смазка *f*

F652 *e* **form removal**
  d Ausschalung *f*
  f décoffrage *m*
  nl ontkisten *n*
  r распалубливание *n*, распалубка *f*

F653 *e* **form scabbing**
  d Außenlunker *m pl* (*in Beton*)
  f formation *f* de cavités (*sur la surface de béton*) due à l'adhérence du coffrage au béton
  nl kistmoeten *f* (*m*) *pl*
  r дефекты *m pl* [раковины *f pl*] (*на бетонной поверхности*), вызванные прилипанием опалубки

F654 *e* **form spacer**
  d Bewehrungshalterung *f*
  f écarteur *m* de barres de l'armature
  nl centerpen *f* (*m*)
  r фиксатор *m* арматуры

F655   **form spreader** *see* **spreader**

F656 *e* **form tie**
  d Schalungsanker *m*
  f tirant *m* de coffrage
  nl bekistingsklamp *m*
  r стяжка *f* элементов опалубки

F657 *e* **formwork** *US*
  d 1. Schalung *f*, Verschalung *f*, Schalgerüst *n* 2. Schalarbeit *f*
  f 1. coffrage *m* 2. traveau *m pl* de coffrage
  nl 1. bekisting *f*; bekistingsconstructies *f pl* 2. bekistingswerk *n*
  r 1. опалубка *f*; опалубочные конструкции *f pl* 2. опалубочные работы *f pl*

F658 *e* **formwork contractor**
  d Schalungsmontagebetrieb *m*
  f entrepreneur *m* en coffrage
  nl aannemer *m* voor het bekistingswerk
  r специализированная строительная организация *f*, монтирующая опалубку

F659 *e* **formwork drawings**
  d Schalungszeichnungen *f pl*

f dessins m pl de coffrage
nl bekistingstekeningen f pl
r чертежи m pl опалубки

F660 formwork oil see form release agent

F661 e formwork plywood
d Schalungsfurnier n,
Schalungssperrholz n
f contreplaqué m pour coffrage
nl bekistingstriplex n, betonplex n
r опалубочная фанера f, фанера f для
(устройства) опалубки

F662 e formwork safety
d Schalungssicherheit f
f sécurité f de coffrage
nl bekistingsveiligheid f
r безопасность f [надёжность f]
(работы) опалубки

F663 e formwork specifications
d Schalungsrichtlinien f pl
f cahier m de charge pour conception et
montage des coffrages
nl beschrijving f van de bekisting
r технические условия n pl на
проектирование и возведение
опалубочных конструкций

F664 formwork tie see form tie

F665 e foul sewer, foul water drain
d Schmutzwasser(ab)leitung f
f conduite f pour eaux résiduaires,
canalisation f d'eau usée
nl afvalwaterleiding f, vuilwaterriool n
r трубопровод m бытовой канализации

F666 e foundation
d Fundation f, Fundament n, Gründung
f
f fondation f, base f, assise f
nl fundering f, fundament n
r фундамент m, искусственное
основание n

F667 e foundation area of a dyke
d Maifeld n (Bodenstreifen, auf dem ein
Deich geschüttet ist)
f terrain m de fondation de le digue
nl zool f (m) van een dijk
r основание n дамбы

F668 e fondation base
d 1. Fundamentsohle f
2. Einzelfundament n
f 1. assise f de fondation 2. fondation f
individuelle
nl 1. funderingsgrondslag m
2. afzonderlijke fundering f
r 1. подошва f фундамента
2. отдельный или столбчатый
фундамент m

F669 e foundation beam
d Fundamentbalken m,
Fundamentträger m; Randträger m
f poutre f de fondation
nl funderingssloof f (m)
r фундаментная балка f; рандбалка f

F670 e foundation bed
d Gründungssohle f

f sol m de fondation; assise f de
fondation
nl funderingsbodem m
r основание n (грунтовое); подошва f
фундамента

F671 e foundation block
d Fundamentblock m
f bloc m de fondation
nl funderingsblok n
r фундаментный блок m

F672 e foundation bolt
d Ankerbolzen m, Verankerungsschraube
f, Fundamentschraube f,
Steinschraube f
f boulon m de fondation [d'ancrage]
nl funderingsbout m, ankerbout m
r фундаментный [анкерный] болт m

F673 e foundation damp proofing course
d Fundamentisolierung f
f étanchement m de la fondation
nl vochtisolatielaag f (m) in de fundering
r гидроизоляция f фундамента

F674 e foundation depth
d Gründungstiefe f, Fundationstiefe f
f profondeur f de fondation
nl funderingsdiepte f
r глубина f заложения фундамента

F675 e foundation engineering
d Grundbau m, Grundbautechnik f,
Gründungsbau m, Gründungswesen n
f technique f des fondations
nl funderingstechniek f
r расчёт m и проектирование n
фундаментов; фундаментостроение n

F676 e foundation pad
d Fundamentauflager n
f semelle f de fondation
nl funderingsbasis f
r подушка f фундамента

F677 e foundation plan
d Fundamentplan m,
Fundamentzeichnung f
f plan m de fondation
nl funderingsplan n, funderingstekening f
r план m фундаментов

F678 e foundation slab
d Fundamentplatte f
f dalle f de fondation
nl funderingsstrook f (m)
r фундаментная плита f

F679 e foundation soil
d Baugrundboden m, Gründungsboden m
f sol m [terrain m] de fondation
nl ondergrond m, grondslag m
r грунт m основания; грунтовое
основание n

F680 e foundation wall
d Grundmauer f, Fundamentmauer f
f mur m de fondation
nl funderingsmuur m
r фундаментная стена f

F681 e four-leg sling
d viersträngiger Stropp m

f élingue *f* à quatre brins
nl viersprongstrop *m*, *f*
r четырёхветвевой строп *m*

F682 *e* **four pipe system**
   d Vierrohrklimaanlage *f*
   f système *m* de climatisation à quatre conduites
   nl vierpijps klimaatregelingsinstallatie *f*
   r четырёхтрубная система *f* кондиционирования воздуха

F683 *e* **four-way reinforcement**
   d kreuzweise angeordnete Plattenbewehrung *f* in vier Richtungen, Vierwegbewehrung *f* der Platten
   f ferraillage *m* de dalle en quatre directions
   nl kruiswapening *f* van een plaat
   r перекрёстное армирование *n* плиты в четырёх направлениях (*параллельно двум сторонам и двум диагоналям*)

F684 *e* **fraction**
   d Kornklasse *f*, Korngrößenbereich *m*
   f fraction *f*
   nl fractie *f*, korrelgrootte *f*
   r фракция *f*, класс *m* крупности

F685 **fractional size analyses** *see* **particle-size analysis**

F686 *e* **fracture initiation**
   d Anfangserscheinungen *f pl* [Anfangsmerkmale *n pl*] des Bruchs
   f indices *m pl* initiaux de rupture
   nl beginverschijnselen *n pl* bij breuk
   r начальные признаки *m pl* разрушения

F687 **fracture load** *see* **breaking load**

F688 *e* **fracture mechanics**
   d Bruchmechanik *f*
   f mécanique *f* de rupture
   nl breukmechanica *f*
   r механика *f* разрушения

F689 *e* **fracture mechanism**
   d Bruchmechanismus *m*
   f mécanisme *m* de rupture
   nl breukmechanisme *n*
   r механизм *m* разрушения

F690 *e* **fracture propagation**
   d Bruchfortpflanzung *f*
   f propagation *f* de rupture
   nl voortschrijdende breuk *m*
   r распространение *n* [развитие *n*] разрушения

F691 **fracture strength** *see* **breaking strength**

F692 *e* **fracture test**
   d Bruchlastversuch *m*
   f essai *m* de rupture
   nl breukproef *f* (*m*)
   r испытание *n* на разрыв

F693 *e* **frame**
   d Rahmen *m*; Fachwerk *n*; Skelett *n*
   f portique *m*; cadre *m*, bâti *m*; charpente *f*

nl raam *n*; vakwerk *n*; skelet *n*
r рама *f*; каркас *m*

F694 *e* **framed breakwater**
   d durchgehender Wellenbrecher *m*
   f brise-lames *m* à claire voie
   nl havendam *m* met betonskelet
   r сквозной волнолом *m*

F695 *e* **frame construction**
   d 1. Rahmenkonstruktion *f*
     2. Holzgerüstbau *m*, Holzskelettkonstruktion *f*
   f 1. construction *f* en portique
     2. ouvrage *m* à charpente en bois
   nl 1. portaalconstructie *f*
      2. skeletbouw *m*
   r 1. рамная конструкция *f*
     2. конструкция *f* с деревянным каркасом

F696 *e* **framed building**
   d Rahmenskelettbau *m*, Skelettbau *m*
   f bâtiment *m* à ossature [à charpente]
   nl raamgebouw *n*, skeletgebouw *n*
   r каркасное [рамное] здание *n*, здание *n* рамного типа

F697 *e* **framed dam**
   d Gerüst-Sperre *f*, Rahmensperre *f*; bewegliches Wehr *n*
   f barrage *m* en charpente; barrage *m* mobile (*de cours d'eau*)
   nl schotbalkkering *f*
   r рамная плотина *f*; разборчатая [разборная] плотина *f*

F698 *e* **framed door**
   d Rahmentür *f*
   f porte *f* encadrée (*par montants et traverses*)
   nl spiegeldeur *f* (*m*), opgeklampte deur *f* (*m*)
   r дверь *f* рамной конструкции

F700 *e* **framed partition**
   d Gerippetrennwand *f*
   f cloison *f* à ossature
   nl scheidingswand *m* van stijl- en regelwerk
   r каркасная перегородка *f*

F701 *e* **frame structures**
   d Rahmenkonstruktionen *f pl*, Rahmentragwerk *n*
   f constructions *f pl* à cadres, charpentes *f pl*
   nl skeletconstructies *f pl*, gebinten *n pl*
   r рамные конструкции *f pl*

F702 *e* **frame-type bridge**
   d Rahmenbrücke *f*
   f pont-portique *m*
   nl vaste brug *f* (*m*), oeverbrug *f* (*m*)
   r рамный мост *m*, мост *m* рамного типа

F703 **frame weir** *see* **framed dam**

F704 *e* **frame wood square**
   d Baugerüstfeld *n*
   f cadre *m* d'échafaudage (*en bois*)
   nl steigerveld *n*
   r конверт *m* строительных лесов

F705 e framework
 d Rahmenwerk n
 f ossature f, charpante f
 nl skelet n, gebint n
 r рамная конструкция f, каркас m

F706 e framing
 d 1. Holzgerippe n, Holzskelett n;
 Holztragwerk n 2. Umrahmung f,
 Einrahmung f, Einfassung f
 f 1. charpente f [ossature f] en bois,
 bâti m 2. chambranle m,
 encadrement m (de la baie de porte)
 nl 1. houtskelet n 2. kozijn n
 r 1. деревянный каркас m, остов m;
 деревянная несущая конструкция f
 (крыши) 2. обрамление n проёма
 (дверного)

F707 e Franki pile
 d Frankipfahl m
 f pieu m Franki
 nl Franki-paal m
 r набивная свая f с уширенным
 основанием, свая f Франки

F708 frazil ice see slush ice

F709 e freeboard
 d Freibord n
 f revanche f
 nl vrijboord n
 r запас m гребня; надводный борт m
 (гидросооружения)

F710 e free float time
 d freie Zeitreserve f
 f flottement m de l'activité libre
 nl vrije tijdreserve f (m)
 r свободный резерв m времени

F711 e free-flow gallery, free-flow tunnel
 d Freispiegelstollen m, Freilaufstollen m
 f tunnel m [galerie f] à écoulement
 libre
 nl vrij stromende tunnel m (riool)
 r безнапорный туннель m

F712 e free-flow weir
 d freier [nichteingestauter,
 vollkommener] Überfall m
 f déversoir m dénoyé
 nl gestreken stuw m
 r незатопленный водослив m

F713 e free groundwater
 d freies [nicht gespanntes]
 Grundwasser n
 f nappe f libre
 nl niet-overspannen grondwater n
 r безнапорные подземные воды f pl

F714 e free-hand drawing, free-hand sketch
 d Freihandskizze f
 f dessin m à main levée, esquisse f,
 croquis m
 nl schets(tekening) f
 r эскиз m, набросок m

F715 e freely movable bearing
 d freibewegliches Auflager n
 f appui m mobile [glissant], appareil m
 d'appui glissant

 nl vrij beweegbaar steunpunt n
 r подвижная опора f [опорная часть f]

F716 e freely supported structure
 d beweglich gelagerte Konstruktion f
 f construction f à appui libre
 nl vrij opgelegde constructie f
 r свободно опёртая конструкция f

F717 e free nappe weir
 d Sturzwehr n
 f barrage-déversoir m à nappe libre
 nl vrij overstroomde stuw m
 r водосбросная плотина f (со свободно
 падающей струёй)

F718 e free roller gate
 d Stoneyschütze f
 f vanne f wagons [Stoney]
 nl rolschuif f (m), roldeur f (m)
 r плоский катковый затвор m

F719 e freestanding tower structure
 d freistehendes Turmbauwerk n
 f tour f non haubanée, mât m
 autoportant [non haubané]
 nl vrijstaand torengebouw n
 r свободностоящее башенное
 сооружение n

F720 e free support
 d freie Auflagerung f, freies Auflager n
 f appui m libre
 nl vrije oplegging f
 r свободное опирание n

F721 free-surface tunnel see free-flow gallery

F722 e free water level
 d freier Wasserspiegel m, freie
 Oberfläche f
 f surface f libre de l'eau
 nl vrij afvloeiend water n
 r уровень m свободной водной
 поверхности

F723 e freeway
 d Autobahn f, Schnell(verkehrs)straße f
 f autoroute f, autostrade f
 nl autosnelweg m
 r автострада f, автомагистраль f

F724 e free weir
 d Wehr n mit freiem [vollkommenem]
 Überfall, freies Wehr n, echtes
 Überström-Wehr n,
 nichteingestauter Überlauf m
 f déversoir m libre [dénoyé]
 nl vrij overstroomde stuw m
 r незатопленный водослив m

F725 e freeze-thaw durability tests
 d Gefrier-Auftau-Prüfung f,
 Frostprobe f
 f essai m de résistance au gel
 nl vorst-dooiproef f (m)
 r определение n морозостойкости

F726 e freeze up
 d Einfrieren n; Verstopfung f infolge
 Eispfropfenbildung (Rohrleitungen)
 f congélation f, prise f (d'un cours
 d'eau); obturation f par le gel,
 blocage m par suite de glace

nl bevriezing f (*pijpleiding*)
r замерзание n воды; образование n
  ледяных пробок (*в трубопроводе,
  теплообменнике*)

F727 e freezing
  d Einfrieren n (z.B. des Bodens)
  f congélation f
  nl bevriezen n (*ook van grond*)
  r замораживание n (*напр. грунта*);
    замерзание n

F728 e freezing and thawing cycle
  d Frost-Tau-Wechsel m, Gefrier- und
    Auftauzyklus m
  f cycle m du gel et du dégel
  nl vorst-dooicyclus m
  r цикл m замораживания и оттаивания

F729 e freezing stresses
  d Frostspannungen f pl,
    Einfrierspannungen f pl
  f contraintes f pl dues au gel
  nl spanning f tengevolge van bevriezing
  r напряжения n pl, возникающие при
    замораживании [замерзании]

F730 freight berth see cargo berth

F731 e freight elevator US
  d Materialaufzug m, Lastenaufzug m
  f monte-charge m
  nl goederenlift m
  r грузовой лифт m [подъёмник m]

F732 freight terminal see cargo berth

F733 French arch see Dutch arch

F735 e French drainage
  d Faschinendränage f
  f drainage m fasciné
  nl drainage f over rijshout
  r фашинный дренаж m

F736 e frequency curve
  d Häufigkeitskurve f
  f courbe f de fréquences [de
    distribution de fréquence]
  nl frequentiekromme f (m)
  r кривая f повторяемости
    [обеспеченности], кривая f
    распределения частот

F737 e frequency diagram
  d Häufigkeitsdiagramm n
  f histogramme m
  nl frequentiediagram n
  r гистограмма f

F738 e fresh concrete
  d Frischbeton m
  f béton m frais
  nl vers beton n
  r свежеприготовленная бетонная смесь
    f

F739 e freshly-placed concrete
  d frisch eingebrachtes Betongemisch n,
    eingebrachter Frischbeton m
  f béton f frais mis en place
  nl versgestort beton n
  r свежеуложённая бетонная смесь f,
    свежеуложенный бетон m

F740 e fretting
  d Abbröckeln n, Schlaglochbildung f
  f arrachement m (*défaut d'un enduit
    superficiel*)
  nl ontstaan n van gaten in het wegdek
    (*slijtage*)
  r отрыв m зёрен щебня *или* гравия от
    дорожного покрытия, образование n
    выбоин

F741 e friction
  d Reibung f
  f frottement m, friction f
  nl wrijving f
  r трение n

F742 e frictional prestress loss
  d reibungsbedingter Spannkraftverlust
    m
  f perte f de précontrainte par frottement
  nl voorspanningsverlies n door wrijving
  r потеря f предварительного
    напряжения вследствие трения

F743 e frictional resistance
  d Reibungswiderstand m
  f résistance f du frottement
  nl wrijvingsweerstand m
  r сопротивление n трения,
    фрикционное сопротивление n

F744 e frictional soil
  d nichtbindiger [kohäsionsloser] Boden m
  f sol m faiblement cohérent, terre f
    meuble
  nl onsamenhangende grond m
  r несвязный [сыпучий] грунт m

F745 e friction factor
  d Reibungsbeiwert m, Reibungsziffer f,
    Reibungszahl f
  f coefficient m de frottement
  nl wrijvingscoëfficiënt m
  r коэффициент m трения

F746 friction grip bolt UK see high-strength
    bolt

F747 e friction head, friction loss
  d Reibungs(druck)verlust m;
    Reibungs(widerstands)höhe f
  f perte f de charge par frottement;
    perte f de précontrainte par
    frottement
  nl weerstandsdrukverlies n
  r потеря f напора на трение, потеря f
    напора по длине; потеря f
    предварительного напряжения на
    трение (*о стенки канала*)

F748 e friction pile
  d Schwebepfahl m, Reibungspfahl m
  f pieu m flottant
  nl op kleef geheide paal m
  r висячая свая f

F749 e friction slope
  d 1. Reibungsgefälle n
    2. Reibungswinkel m
  f 1. perte f de charge due au frottement
    par unité de longeur 2. angle m de
    frottement

nl 1. hydraulische gradiënt *m*
2. wrijvingshoek *m*
r 1. гидравлический уклон *m* 2. угол *m* трения

F750　fringe conditions *see* boundary conditions

F751　*e* fringe water
d Kapillarwasser *n*, Porensaugwasser *n*
f eau *f* de la frange capillaire
nl capillair gebonden water *n*, poriënwater *n*
r вода *f* капиллярной зоны, капиллярная влага *f*

F752　*e* frog rammer
d Frosch-Ramme *f*, Explosionsramme *f*
f dame-grenouill *f*, dame *f* sauteuse [à explosion]
nl grondstamper *m*, grondverdichter *m*
r (ручная) механизированная трамбовка *f*, трамбовка *f* взрывного действия

F753　*e* frontage line
d Gebäudeflucht *f*, Bauflucht *f*; Fluchtlinie *f*, Bau(flucht)linie *f*
f ligne *f* de délimitation d'un chantier *ou* d'une zone habitée du côté de la rue; ligne *f* de façade [d'alignement], alignement *m*
nl rooilijn *f* (*m*)
r граница *f* домовладения *или* стройплощадки со стороны улицы; красная линия *f* застройки

F754　*e* frontage road
d Begleiterstraße *f*
f route *f* cotière [de côte]
nl parallelweg *m*, ventweg *m*
r дорога *f* вдоль канала, реки *или* железной дороги

F755　*e* front door
d Eingangstür *f*; Wohnungseingang *m*
f porte *f* d'entrée
nl voordeur *f* (*m*)
r входная дверь *f*

F756　*e* front-end attachments, front-end equipment
d Front-Anbaugeräte *n* *pl*, Front-Zusatzwerkzeuge *n* *pl*, Front-Umbauvorrichtungen *f* *pl*
f équipement *m* porté frontal
nl aan de voorkant aantekoppelen hulpstukken *n* *pl*
r передненавесное оборудование *n*, передненавесные орудия *n* *pl*

F757　*e* front-end loader
d Frontschaufellader *m*, Frontschaufler *m*, Ladeschaufel *f*
f chargeur *m* à benne [à godet], traxcavateur *m*
nl laadschop *f* (*m*)
r фронтальный одноковшовый погрузчик *m*

F758　*e* front-entrance intake
d frontale Wasserfassung *f* [Wasserentnahme *f*]

f prise *f* d'eau frontale
nl frontale wateronttrekking *f*
r фронтальный водозабор *m*

F759　*e* front view
d Vorderansicht *f*
f vue *f* de face, élévation *f*
nl vooraanzicht *n*
r вид *m* спереди

F760　*e* frost blanket
d Frostschutzschicht *f*
f couche *f* anti-gel
nl vorstwerende laag *f* (*m*)
r морозозащитный слой *m*

F761　*e* frost damage
d Frostschaden *m*
f dommage *m* [dégât *m*] causé par le gel
nl vorstschade *f*
r повреждение *n* от замерзания [от мороза]

F762　*e* frost depth
d Frosteindringtiefe *f*
f profondeur *m* de congélation des sols
nl vorstdiepte *f*
r глубина *f* промерзания (грунта)

F763　*e* frost effects
d Frostwirkung *f*
f effet *m* du gel
nl vorstinwerking *f*
r воздействие *n* мороза

F764　*e* frost heave, frost heaving
d Frosthebung *f*, Frostaufbruch *m*, Frostauftreibung *f*
f gonflement *m* [soulèvement *m*] (*du sol de la chaussée*) dû au gel
nl uitzetten *n* van de grond door vorst
r вспучивание *n* мёрзлого грунта, морозная пучина *f*

F765　*e* frost lens
d Eislinse *f*, Frostlinse *f*
f lentille *f* de glace
nl ijslens *f* (*m*)
r ледяная линза *f*

F766　*e* frost line
d Frostgrenze *f*
f ligne *f* de pénétration du gel
nl vorstgrens *f* (*m*) (*in de bodem*)
r предельная глубина *f* промерзания (грунта)

F767　frostproof course *see* frost blanket

F768　frost protection layer *see* frost blanket

F769　*e* frost resistance
d Frostbeständigkeit *f*
f résistance *f* au gel
nl weerstand *m* tegen vorst
r морозостойкость *f*

F770　*e* frost-resisting admixture
d Frostschutzzusatz *m*
f agent *m* améliorant la résistance au gel
nl vorstwerende toeslag *m*
r добавка *f*, повышающая морозостойкость

F771 *e* **frost thermostat**
   d Frost(schutz)thermostat *m*
   f thermostat *m* de gel
   nl vorstthermostaat *m*
   r морозозащитный термостат *m*,
   реле *n* температуры обмерзания

F772 *e* **frost thrust**
   d Frostschub *m*
   f poussée *f* du sol congélé
   nl uitzettingskracht *f* (*m*) bij bevriezing
   r давление *n* замёрзшего грунта

F773 *e* **frost valve**
   d Frostschutzventil *n*
   f soupape *f* automatique antigel
   nl vorstafsluiter *m*
   r автоматический морозозащитный
   клапан *m*

F774 *e* **frost zone**
   d Frostzone *f*
   f zone *f* de gel
   nl vorstzone *f*
   r зона *f* промерзания

F775 *e* **froth breaker**
   d Schaumbrecher *m*
   f abat-mousse *m*, antimousse *m*,
   brise-mousse *m*
   nl schuimbreker *m*
   r пеногаситель *m*

F776 *e* **froth breaking**
   d Schaumbekämpfung *f*, Entschäumung
   f (*im Lüftungsbecken*)
   f abattage *m* de mousse
   nl ontschuiming *f*
   r пеногашение *n* (*в аэротенке*)

F777 *e* **frozen ground**
   d gefrorener Boden *m*
   f sol *m* gelé, terre *f* gelée
   nl bevroren grond *m*
   r мёрзлый грунт *m*

F778 *e* **full-cell process**
   d Volltränkverfahren *n*
   f traitement *m* antiseptique du bois par
   le vide et pression
   nl impregneren *n* onder druk, volle
   bereiding *f* (*houtconservering*)
   r антисептирование *n* (*древесины*) под
   давлением с предварительным
   вакуумированием

F779 *e* **full-depth asphalt pavement**
   d Asphaltbeton-Straßendecke *f*
   f corps *m* de chaussée en béton
   asphaltique
   nl volledigde asfaltbetonweg *m*
   r асфальтобетонная дорожная одежда *f*

F780 *e* **full-face tunneling**
   d Vollausbruch *m*
   f percement *m* d'un tunnel à pleine
   section
   nl uitbreken *n* over de gehele
   doorsnede (*tunnelbouw*)
   r проходка *f* туннеля полным профилем

F781 *e* **Fuller's curve**
   d Fullerkurve *f*

   f courbe *f* granulométrique [de Fuller]
   nl Fuller-kromme *f* (*m*)
   r кривая *f* оптимального
   гранулометрического состава
   заполнителей, кривая *f* Фуллера

F782 *e* **full slewing crane**
   d Vollschwenkkran *m*, Volldrehkran *m*
   f grue *f* à rotation totale
   nl volledig draaibare kraan *f* (*m*)
   r полноповоротный грузоподъёмный
   кран *m*

F783 *e* **full restraint**
   d Volleinspannung *f*
   f encastrement *m* complet
   nl volledige inklemming *f*
   r полное защемление *n*

F784 *e* **full-scale load test**
   d Belastungsprüfung *f* im Maßstab 1:1,
   Vollastprüfung *f*
   f essai *m* à pleine charge
   nl beproeving *f* met volle belasting
   r испытание *n* полной расчётной
   нагрузкой

F785 *e* **full size drawing**
   d Zeichnung *f* im Maßstab 1:1 [in
   natürlicher Größe]
   f dessin *m* à vrai grandeur [à la **cote**
   exacte]
   nl tekening *f* in ware grootte
   r чертёж *f* в натуральную величину

F786 *e* **full splice**
   d gleichfeste Verbindung *f*;
   gleichfeste Spleißung *f*
   f épissure *f* d'égale résistance
   nl lange splits *f* (*m*)
   r равнопрочное соединение *n*;
   равнопрочное сплетение *n* (*троса*)

F787 *e* **full supply level**
   d Wasserstand *m* bei Vollfüllung,
   höchster geregelter Wasserstand *m*
   (*Wasserspeicher*);
   Vollbedarfswasserstand *m* (*Kanal*)
   f niveau *m* maximum normal de
   retenue; niveau *m* maximum normal
   nl waterpeil *n* bij volledige vulling;
   normaal stuwpeil *m*
   r нормальный подпорный уровень *m*
   воды в водохранилище;
   максимальный нормальный уровень
   m воды в канале

F788 *e* **full-tide cofferdam**
   d Volltidefangdamm *m*
   f bâtardeau *m* de haute marée
   nl hoogwatervangdam *m*
   r перемычка *f*, обеспечивающая
   защиту от воды во все периоды
   прилива

F789   **full trailer** *see* **trailer**

F790 *e* **full-width paver**
   d Brückenfertiger *m*
   f finisseur *m* pour surfaçage du
   tablier de pont à une seule passe
   nl asfaltafwerkmachine *f*

r дорожная отделочная машина *f* с шириной захвата, равной полной ширине проезжей части (*моста*)

F791 *e* **full-width weir**
d unbeengter Wasserüberlauf *m*
f déversoir *m* sans contraction latérale
nl stuw *m* over de volle breedte
r водослив *m* без бокового сжатия потока

F792 *e* **fully adjustable air diffuser**
d Diffusor *m* mit Luftmengen- und Richtungsregelung
f diffuseur *m* d'air avec réglage de débit et des directions de jets d'air
nl instelbare diffuseur *m*
r диффузор *m* с регулированием расхода воздуха и направления воздушных струй

F793 *e* **fully fixed member**
d voll eingespanntes Element *n* [Glied *n*, Bauteil *n*]
f élément *m* encastrée à deux extrémités
nl tweezijdig ingeklemd element *n*
r элемент *m* с защемлёнными концами

F794 *e* **fully penetrating well**
d vollkommener [voll wirksamer] Brunnen *m*
f puits *m* complet
nl optimale boorbron *f* (*m*)
r совершенная скважина *f*, совершенный колодец *m*

F795 *e* **fume cupboard, fume hood**
d Digestorium *n*, Abzugsschrank *m*
f hotte *f* de tirage
nl afzuigkast *f* (*m*), zuurkast *f* (*m*)
r вытяжной шкаф *m*

F796 *e* **funicular polygon**
d Seilzug *m*, Seileck *n*, Seilpolygon *n*
f polygone *m* funiculaire
nl stangenveelhoek *m*
r верёвочный многоугольник *m*

F797 *e* **funicular railway**
d Personen-Seilschwebebahn *f*
f funiculaire *m*
nl kabelbaan *f* (*m*)
r фуникулёр *m*, подвесная железная дорога *f*

F798 **funnel** *see* **flue**

F799 *e* **furnace**
d 1. Ofen *m*, Industrieofen *m* 2. Feuerung *f*, Feuerraum *m*
f 1. poêle *m* 2. foyer *m*
nl 1. oven *m* 2. verbrandingsruimte *f*
r 1. печь *f* 2. топка *f*, топочная камера *f*

F800 *e* **furnace clinker**
d Kesselschlacke *f*
f mâchefer *m*
nl slakken *f* (*m*) *pl*
r котельный шлак *m*

F801 *e* **furnace clinker concrete**
d Schlackenbeton *m*
f béton *m* de mâchefer

nl slakkenbeton *n*
r шлакобетон *m*

F802 *e* **furring**
d 1. Richtleiste *f*, Putzleiste *f*, Lehrstreifen *m* 2. Ausgleichen *n* der Wandinnenseite nach Richtpunkten 3. Unterkonstruktion *f* (*Putzträger*), Unterfutterung *f* 4. Verbretterung *f*
f 1. cuieillie *f*, cueillé *f* 2. aplanissement *m* [égalisation *f*] de l'enduit aux cueillées 3. lattes *f pl* à plâtre 4. revêtement *m* en bois; planchéiage *m*
nl 1. richtlijst *f* (*m*) 2. binnenmuur vlakken *n* op richtpunten 3. pleisterwerk *n* op latten 4. beschieting *f*
r 1. маячная рейка *f*, маяк *m* 2. выравнивание *n* внутренней поверхности стены по маякам 3. основание *n* под штукатурку (*дранка, обрешётка*) 4. обшивка *f* досками

F803 *e* **furring brick**
d Tonhohlziegel *m* [Lehmhohlziegel *m*] zur Innenwandverkleidung
f brique *f* creuse de parement intérieur
nl holle baksteen *m* voor wandbekleding
r облицовочный пустотелый кирпич *m*

F804 *e* **furring tile**
d keramischer Rippen-Formstein *m* zur Innewandverkleidung
f bloc *m* creux en ceramique de parement intérieur
nl bekledingssteen *m* voor buitenmuren (*isolatie*)
r керамический блок *m* для внутренней облицовки стен

F805 *e* **fuse**
d 1. elektrische Sicherung *f* 2. Zündschnur *f*
f 1. coupe-circuit *m* fusible 2. cordeau *m* porte-feu
nl 1. zekering *f* 2. vuurkoord *n*
r 1. плавкий предохранитель *m* 2. огнепроводный шнур *m*

G

G1 *e* **gabion**
d 1. Drahtschotterkasten *m*, Drahtschotterbehälter *m* 2. Klein--Zellenfangdamm *m*
f gabion *m*
nl 1. tenen mand *f* (*m*) met aarde 2. kribwerk *n*
r 1. габион *m* 2. ячейка *f* из металлического шпунта

G2 *e* **gabion dam**
d Wehr *n* aus Drahtschotterbehältern
f digue *f* en gabions

nl dam *m* van kribwerk
r габионная плотина *f*

G3 e **gable**
d Giebel *m*
f pignon *m*, gable *m*
nl driehoekige topgevel *m*, geveltop *m*
r фронтон *m*, щипец *m*

G4 e **gable board**
d Windbrett *n*, Windfeder *f*
f bordure *f* de pignon
nl windveer *f* (*m*) (*dakrand*)
r фризовая доска *f*, обрамляющая
фронтон

G5 e **gable roof**
d Satteldach *n*
f toiture *f* à double pente, comble *m*
à 2 versants
nl zadeldak *n*
r двускатная [щипцовая] крыша *f*

G6 e **gable wall**
d Giebelwand *f*
f mur *m* pignon
nl gevelmuur *m*
r фронтонная *или* щипцовая стена *f*

G7 e **gable window**
d Giebelfenster *n*
f fenêtre *f* de fronton; fenêtre *f* en
forme de fronton, fenêtre *f*
triangulaire
nl zolderraam *n*, gevelvenster *n*
r фронтонное окно *n*, окно *n*
в фронтоне; треугольное окно *n*

G8 **gage** *US see* **gauge**

G9 e **gain**
d Loch *n*; Nest *n*; Nut *f*; Aussparung *f*;
Ausschnitt *m*
f entaille *f*; encoche *f*; rainure *f*
nl nest *n*; uitsparing *f*
r гнездо *n*; паз *m*; вырез *n*
(*в деревянных элементах*)

G10 e **gallery**
d 1. Galerie *f*; Brunnengalerie *f*,
Sickergalerie *f*; Kontrollstollen *m*,
Kontrollgang *m*; Stollen *m*
2. Veranda *f* 3. Balkon *m*
f 1. galerie *f*; galerie *f* de captage;
galerie *f* de visite; galerie *f* d'accès
2. veranda *f* 3. balcon *m*
nl 1. galerij *f*, omloop *m*; controlebuis
f (*m*) (*tunnel*) 2. veranda *f* 3. balkon *m*
r 1. галерея *f*; водосборная галерея *f*;
потерна *f*; штольня *f* 2. веранда *f*
3. балкон *m*

G11 e **gallery appartment house**
d Außenganghaus *n*, Laubenganghaus *n*
f maison *f* d'appartements à balcons
d'accès (*à chaque niveau*)
nl galerijflat *m*
r жилой дом *m* галерейного типа
(*с входами в квартиры через общий
балкон*)

G12 e **gallery canal**
d Stollenkanal *m*, Tunnelkanal *m*

f tunnel *m* pour canal
nl tunnelkanaal *n*
r судоходный туннель *m*, туннельный
канал *m*

G13 e **gallery concreting train**
d fahrbare Stollenbetonieranlage *f*,
Stollenbetonierzug *m*, Misch- und
Einbauzug *m* für Tunnelauskleidung
f train *m* à béton [de bétonnage]
nl betontrein *m* (voor tunnelbouw)
r бетонопоезд *m*, передвижной
комплексный агрегат *m* для
бетонирования обделки туннелей

G14 e **gallery machine**
d Stollenbohrgerät *n*
f excavateur *m* de tunnel du type à
foreuse
nl tunnelboorinstallatie *f*
r туннелепроходческий комбайн *m*
с буровым исполнительным органом

G15 e **gallery portal**
d Stollenportal *n*, Tunnelportal *n*
f entrée *f* de la galerie, tête *f*
[débouché *m*] du tunnel
nl tunnelportaal *n*
r портал *m* туннеля

G16 e **galvanized fabric**
d verzinktes Stahldrahtgewebe *n*
f treillis *m* en acier galvanisé
nl gegalvaniseerd draadnet *n*
r оцинкованная стальная сетка *f*

G17 e **galvanized steel**
d verzinkter Stahl *m*
f acier *m* galvanisé
nl gegalvaniseerd staal *n*
r оцинкованная сталь *f*

G18 e **gambrel roof**
d Walmdach *n*
f toit *m* [comble *m*] en croupe
nl gebroken dak *n*, mansardedak *n*
r вальмовая крыша *f*

G19 e **ganged form**
d vergrößerte Schalungsplatte *f*
f gros panneau *m* de coffrage (*assemblé
de petits éléments*)
nl systeembekisting *f*
r укрупнённая опалубочная панель *f*,
укрупнённый опалубочный щит *m*

G20 e **ganger**
d Polier *m*
f chef *m* d'équipe
nl ploegbaas *m*
r бригадир *m* (*рабочей бригады*)

G21 e **gang mould**
d Reihenform *f*
f moule *m* multiple
nl meervoudige betongietmal *m*,
mallenreeks *f* (*m*)
r батарейная форма *f* (*для
изготовления сборных
железобетонных элементов*)

G22 e **gang of wells**
d Brunnenreihe *f*, Brunnengalerie *f*

f batterie *f* de puits
nl bronnengroep *f* (*m*)
r группа *f* колодцев

G23　*e* **gang saw**
d Bundgatter *n*
f scie *f* mécanique à plusieurs lames,
scie *f* multiple (*à roches*)
nl raamzaag *f* (*m*)
r пильная рама *f* (*для распиловки камня*)

G24　*e* **gangway** *US*
d 1. Arbeitsbrücke *f*, Arbeitsbühne *f*;
Laufsteg *m* 2. Bedienungssteg *m*,
Bedienungsbrücke *f* (*am Stauwehr*)
f 1. plate-forme *f* de travail
2. passerelle *f* [pont *m*] de service
nl 1. werkplatform *n*; bordes *n*
2. loopbrug *f* (*m*)
r 1. рабочая платформа *f*; подмости *pl*
(*у строящегося здания*)
2. служебный мостик *m*

G25　*e* **gantry**
d 1. Gerüst *n* 2. Arbeitsbühne *f*
3. Krangerüst *n*, Portal *n*
f 1. chevalet *m* 2. plate-forme *f* pour
le matériel 3. portique *m* de la grue,
grue *f* à portique
nl 1. stelling *f*, (bouw)steiger *m*
2. werkbrug *f* (*m*) 3. portaal *n* (*kraan*)
r 1. опорная конструкция *f* (*рама, каркас*) рабочей платформы
2. рабочая платформа *f* для
строительного оборудования
3. портал *m* крана

G26　*e* **gantry crane**
d Portalkran *m*; Bockkran *m*,
Gerüstkran *m*
f grue *f* à portique; grue-chevalet *f*
nl portaalkraan *f* (*m*); loopkraan *f* (*m*)
r портальный *или* козловой кран *m*

G27　*e* **gap-filling glue**
d Klebekitt *m*, Kitt *m* zum Ausfüllen
größerer Klebefugen
f colle *f* pour remplissage des joints
nl vullende lijm *m*, voegenkit *f* (*m*)
r клей *f* для заполнения швов

G28　*e* **gap-graded aggregate**
d unstetig abgestufter Zuschlagstoff *m*
f granulat *m* [agrégat *m*] à granulométrie
discontinue
nl onregelmatig gegradeerde toeslag *m*
r заполнитель *m*, не содержащий одной
или нескольких фракций

G29　**garage door** *see* **overhead door**

G30　*e* **garbage disposer**
d Müllschlucker *m*, Müllabwurfschacht
m
f gaine *f* à ordures, vide-ordures *m*
nl vuilstortkoker *m*
r мусоропровод *m*

G31　*e* **garden apartment**
d Erdgeschoßwohnung *f* mit Ausgang
in den Garten; zwei- bis
dreigeschossiges Wohnhaus *n* mit
gemeinsamem Garten
f appartement *m* de rez-de-chaussée
avec la sortie dans le jardinet privé;
maison *f* à deux *ou* trois étages avec
le jardin commun
nl parterre-woning *f* met tuin,
etagewoning *f* met gemeenschappelijke
tuin
r квартира *f* на первом этаже с выходом
в сад; жилой дом *m* высотой
2—3 этажа с общим садом

G32　*e* **garden tile**
d Baukeramik *f* für Grünanlagen und
Fußwege
f produits *m pl* céramiques pour les
jardins *ou* allées
nl wandelpadtegel *m*, tuintegel *m*
r строительная керамика *f* для садов
и дорожек

G33　*e* **garnet paper**
d feinkörniges Schmirgelpapier *n*
f papier *m* d'émeri, papier-émeri *m*
nl fijn schuurpapier *n*
r тонкая наждачная бумага *f*

G34　*e* **garret**
d Dachboden *m*; Dachkammer *f*,
Dachstube *f*
f grenier *m*; mansarde *f*; local *m*
habité sous le comble
nl zolderverdieping *f*; zolderkamer *f* (*m*)
r чердак *m*; жилое чердачное
помещение *n*, жилой чердак *m*

G35　*e* **gas analyser**
d Gasprüfgerät *n*, Gasprüfer *m*
f analyseur *m* de gaz
nl gasanalysator *m*
r газоанализатор *m*

G36　*e* **gas appliance**
d Gasgerät *n*, Haushaltgasgerät *n*
f appareil *m* à gaz, appareil *m*
ménager utilisant le gaz
nl gasapparaat *n*, gastoestel *n*
r газовый прибор *m*

G37　*e* **gas-ash silicate concrete**
d Gas-Aschen-Silikatbeton *m*
f béton *m* silico-calcaire au gaz à base
de cendre
nl gashoudend vliegas-silicaatbeton *n*
r газозолосиликатный бетон *m*

G38　*e* **gas balance**
d Gasbilanz *f*
f bilan *m* de gaz
nl gasboiler *m*
r газовый баланс *m* (*вентиляция*)

G39　*e* **gas boiler**
d Gaskessel *m*, gasgefeuerter
Heizkessel *m*
f chaudière *f* à gaz
nl gasketel *m*
r газовый котёл *m*

G40　*e* **gas circulator**
d Umlaufgaswasserheizer *m*, Gas-
-Durchlauf-Wasserheizer *m*

f chauff-eau *m* à gaz [réchauffeur *m* d'eau] à circulation
nl gasgeiser *m*
r циркуляционный [проточный] газовый водонагреватель *m*

G41 *e* **gas-compression plant**
d Gasdruckerhöhungsstation *f*
f centrale *f* de compresseurs de gaz
nl gascompressie-installatie *f*
r газокомпрессорная станция *f*

G42 *e* **gas concrete**
d Gasbeton *m*
f béton-gaz *m*, béton *m* gazeux [au gaz]
nl gasbesmetting *f*
r газобетон *m*

G43 *e* **gas contamination of premises**
d Gasanreicherung *f* des Raumluftes
f pollution *f* des locaux par le gaz
nl gascontaminatie *f* van ruimten
r загазованность *f* помещений

G44 *e* **gas-control automatic equipment**
d Gasreglerautomatik *f*
f système *m* [appareillage *m*] automatique de réglage des réseaux de gaz
nl automatische gasdrukregeling *f*
r газовая автоматика *f* регулирования; газорегуляторное автоматическое устройство *n*

G45 *e* **gas control unit**
d Gasregleranlage *f*
f poste *m* de réglage de gaz
nl gasregelingsinstallatie *f*
r газорегуляторная установка *f*

G46 *e* **gas cutting**
d Gasschneiden *n*
f coupage *m* aux gaz
nl autogeen snijden *n*
r газовая резка *f*

G47 *e* **gas detector**
d Gasspürgerät *n*
f détecteur *m* (de fuites) de gaz
nl opsporingstoestel *n* voor gas(lekken)
r индикатор *m* утечки газа

G48 *e* **gas distribution network**
d Gasrohrnetz *n*
f réseau *m* de distribution de gaz
nl gasdistributienet *n*
r газовая распределительная сеть *f*

G49 *e* **gas-distribution plant**
d Gasverteilerstation *f*, Gasreglerstation *f*
f centre *m* de distribution de gaz
nl gasontfangsstation *n*
r газораспределительная станция *f*

G50 *e* **gas emission**
d Gasausscheidungen *f pl*, Gasemission *f*
f libération *f* [dégagement *m*] de gaz
nl gasemissies *f pl*
r газовыделения *n pl*

G51 *e* **gaseous air pollutants**
d gasförmige Luftverunreinigungen *f pl*

f agents *m pl* de pollution gazeux, impuretés *f pl* gazeuses
nl luchtverontreinigingen *f pl* in gasfase
r газообразные загрязнения *n pl* воздуха

G52 *e* **gas filter**
d Gasfilter *n*, *m*
f filtre *m* de gaz
nl gasfilter *n*, *m*
r газовый фильтр *m*

G53 *e* **gas-fired heating**
d Gasheizung *f*
f chauffage *m* au gaz
nl gasverwarming *f*
r газовое отопление *n*

G54 *e* **gas-fired stove**
d Gas(heiz)ofen *m*, Gasraumheizer *m*
f four *m* [cuisinière *f*] à gaz
nl gaskachel *f* (*m*)
r бытовая газовая печь *f*

G55 *e* **gas-fired unit heater**
d Gasraumheizer *m*, Gaslufterhitzer *m*, Gasluftheizgerät *n*
f réchauffeur *m* d'air à gaz
nl met gas gestookte luchtverhitter *m*
r газовый воздухонагреватель *m*

G56 *e* **gas fittings**
d Gasanschlußarmaturen *f pl*
f robinetterie *f* pour gaz
nl appendages voor gasinstallaties
r газопроводная арматура *f*

G57 *e* **gas flue**
d Rauchgaskanal *m*, Gaszug *m*, Rauchgasabzug *m*, Fuchs *m*
f carneau *m*, renard *m*, tuyau *m* de cheminée
nl gaskanaal *n*
r газоход *m*, газовый канал *m*

G58 *e* **gas-forming admixture**
d Gasbildner *m*, Gasentwickler *m*, gaserzeugender Zusatz *m*
f générateur *m* de gaz
nl gasvormende toevoeging *f*
r газообразующая добавка *f*

G60 *e* **gas grid**
d Gasversorgungsnetz *n*
f réseau *m* de tuyaux de gaz, tuyauterie *f* pour gaz
nl gasdistributienet *n*, gasleidingennet *n*
r сеть *f* газоснабжения

G61 *e* **gas heating system**
d Gasheizsystem *n*
f système *m* de chauffage à gaz
nl gasverwarmingsinstallatie *f*
r система *f* газового отопления

G62 *e* **gas-holder, gasholder**
d Gasometer *n*, Gasbehälter *m*
f gazomètre *m*
nl gashouder *m*
r газгольдер *m*

G64 *e* **gasholder station**
d Gastanklager *n*
f station *f* [groupe *m*] de gazomètres

nl gashouderemplacement n
r газгольдерная станция f

G65 e gas indicator
d 1. Gasanzeiger m, Gasindikator m
2. Gasdruckmesser m
f 1. indicateur m de débit de gaz
2. manomètre m à gaz
nl 1. gasindicator m 2. gasmanometer m
r 1. газоиндикатор m, индикаторная трубка f 2. газовый манометр m

G66 e gas installation
d innere Gasleitung f, Gasinstallation f
f installation f de gaz, conduit m de gaz intérieur
nl gasinstallatie f bij de afnemer
r внутренний газопровод m

G67 e gasket
d Dichtungsring m, Flanschdichtung f
f garniture f (de joint), joint m
nl pakking f, pakkingring m
r уплотняющая прокладка f, уплотняющее кольцо n; уплотняющий шнур m (вокруг двери)

G68 e gaskin
d Hanfstrick m, Hanfriste f
f toron m en (fibres de) chanvre
nl hennepstreng f (m)
r пеньковая прядь f

G69 gas main see gas transmission pipeline

G70 e gas metering station
d Gasmeßstation f
f poste m de compteurs de gaz
nl gasmeetstation n
r газоизмерительная станция f

G71 e gasoline interceptor US
d Benzinabscheider m
f collecteur m d'essence
nl benzine-opvanginrichting f
r бензоуловитель m

G72 e gas permeability
d Gasdurchlässigkeit f
f perméabilité f aux gaz
nl doorlatendheid f voor gas
r газопроницаемость f

G73 e gas pipe
d Gasrohr n
f tuyau m de gaz, conduite f à gaz
nl gaspijp f (m)
r газовая труба f

G74 e gas pipeline
d Gas(versorgungs)leitung f
f canalisation f de gaz
nl gasleiding f
r газопровод m

G75 e gas pipeline construction rig
d Gasleitungsbaugerät n
f train m d'engins pour la construction des gazoducs
nl constructiemateriëel n voor gasleidingen
r комплексный агрегат m для сооружения газопроводов

G76 e gas pliers
d Gasrohrzange f
f pince f de gaz
nl gastang f (m), buizentang f (m)
r газовые клещи pl

G77 e gas pressure drop indicator
d Gasdruckabfallmelder m
f indicateur m de chute de pression pour gaz
nl drukdalingsindicator m
r сигнализатор m падения давления газа

G78 e gas-pressure reducer
d Gasreduzierventil n, Gasdruckminderregler m
f détendeur m de gaz
nl reduceerventiel n
r газовый редуктор m

G79 e gas pressure regulator
d Gasdruckregler m
f régulateur m de pression de gaz
nl drukregelaar m
r регулятор m давления газа

G80 e gas purification
d Gasreinigung f
f épuration f de gaz
nl gasreiniging f
r очистка f газа

G81 e gas range
d Gasherd m
f fourneau m [cuisinière f] à gaz
nl gasfornuis n
r газовая плита f

G82 e gas-range burner
d Kochbrenner m
f brûleur m de la cuisinière à gaz
nl gaspit f (m)
r конфорочная горелка f

G83 gas regulating governor see gas pressure regulator

G84 e gas relay
d Gasdrucküberwachung f
f relais m à gaz
nl gasdrukrelais n
r газовое реле n, реле n давления газа

G85 e gas riser
d Gassteigrohr n
f conduite f montante de gaz, colonne f à gaz
nl verticale gasleiding f
r газовый стояк m

G86 e gas safety automation
d Gaszündsicherung f
f équipement m automatique de sécurité pour installations de gaz, système m automatique de sécurité des réseaux de gaz
nl automatische waakvlambeveiliging f
r газовая автоматика f безопасности

G87 e gas service pipe
d Gasanschlußleitung f
f branchement m de gaz

nl gas-dienstleiding f
r газовый ввод m

G88 e gas supply
d Gasversorgung f
f alimentation f en gaz
nl gaslevering f
r газоснабжение n

G89 e gas transmission pipeline
d Gasfernleitung f, Ferngasleitung f
f conduite f principale de gaz
nl gastransportleiding f
r магистральный газопровод m

G90 gas water heater see geyser

G91 e gas welding
d Gasschweißen n
f soudage m aux gaz
nl autogeen lassen n
r газовая сварка f

G92 e gate
d 1. Tor n 2. Schütz n, Schütze f,
Verschluß m
f 1. porte f de clôture [d'accès]
2. vanne f (d'écluse)
nl 1. poort f (m), sluisdeur f (m)
2. klep f (m), sluiting f
r 1. ворота pl 2. (гидротехнический)
затвор m

G93 e gate chamber
d Torkammer f; Verschlußkammer f
f chambre f des portes; chambre f des
vannes
nl deurkas f (m) (sluis)
r шкафная камера f (в голове шлюза);
камера f затворов

G94 e gate groove
d Schütznische f
f rainure f de la vanne
nl schuifsponning f
r паз m затвора

G95 e gate operator
d Torschließer m
f appareil m électro-mécanique
d'ouverture et de fermeture de la
porte d'accès
nl automatische deurbediening f
r электромеханический открыватель-
-закрыватель m ворот

G96 e gate pier
d Torpfeiler m
f pilier m de porte de clôture
nl poortpilaar m
r воротный столб m (из камня, бетона)

G97 e gate posts
d Torpfosten m pl
f montants m pl [poteaux m pl] de porte
de clôture
nl stijl m van een poortdeur
r стойки f pl ворот (деревянные)

G98 e gates crusher
d Kegelbrecher m, Kegelmühle f
f concasseur m giratoire [à cône]
nl kegelsteenbreker m
r конусная (камне)дробилка f

G99 e gate valve
d Absperrschieber m, Plattenschieber m
f robinet-vanne m, vanne f à glissière
nl schuifafsluiter f (m)
r плоская задвижка f, шибер m

G100 e gateway
d Halbschleuse f, Dockschleuse f
f écluse f simple
nl doksluis f (m)
r полушлюз m

G101 gathering ring see master link

G102 e gauge
d 1. Blechdicke f; Drahtdurchmesser m,
Drahtdicke f 2. Wanddicke f
3. Skale f, Maßstab m 4. Lehre f,
Schablone f 5. Meßgerät n
6. Pegellatte f, Pegel m,
Wasserstandanzeiger m 7. Spurweite f,
Gleisspur f
f 1. calibre m, jauge f 2. épaisseur f
de la paroi 3. échelle f 4. gabarit m
5. appareil m de mesure, mesureur m
6. échelle f hydrométrique,
limnimètre m, échelle f d'étiage
7. écartement m de la voie,
entrerails m
nl 1. kaliber(maat) f (m) 2. wanddikte f
3. schaal f (m) 4. (teken)maal n;
5. meetapparaat n 6. peil n,
peilstok m 7. spoorwijdte f
r 1. калибр (листовой стали,
проволоки) 2. толщина f стенки
3. шкала f 4. шаблон m, лекало n
5. измерительный прибор m (напр.
уровнемер, манометр, плювиометр)
6. водомерная рейка f, водомерный
пост m 7. ширина f колеи

G103 e gauge box
d Abmeßkiste f, Zuteilkiste f,
Dosierkiste f
f boîte f de mesure [de dosage]
nl meetkist f (m)
r мерный ящик m

G104 e gauged arch
d Bogen m, Bogensturz m
f arceau m, arc m, linteau m cintré en
pierres
nl uit boogstenen gemetselde boog m
r арка f, арочная перемычка f

G105 e gauge datum
d Pegelnull f, Pegelnullpunkt m
f altitude f du zero de l'échelle
nl normaal peil n
r нуль m графика водомерного поста

G106 e gauged brick
d Formziegel m, Keilziegel m
f brique f en coin
nl boogsteen m, gewelfsteen m
r фасонный [клиновой] кирпич m
(для кладки арок)

G107 e gauged mortar
d Kalk-Zement-Mörtel m
f mortier m chaux-ciment [de chaux
et de ciment]

nl metselmortel *m*, kalk-cementmortel *m*
r известково-цементный раствор *m*

G108 *e* **gauged skim coat**
d Kalk-Gips-Oberputz *m*, Kalk-Gips-
-Putzdeckschicht *f*
f couche *f* de finition d'enduit en
mortier de chaux et de plâtre
nl gipskalkpleister *n*
r тонкий отделочно-накрывочный слой
m известково-гипсовой штукатурки
(*затираемый до получения гладкой*
плотной поверхности)

G109 **gauged staff** *see* **gauging plaster**

G110 *e* **gauged work**
d 1. Keilziegelmauerwerk *n*,
Gewölbeziegelmauerwerk *n*,
Formziegelmauerwerk *n*
2. Stuckarbeiten *f pl* mit Stuckplastik
f 1. maçonnerie *f* de briques radiales
ou en coin 2. travaux *m pl* d'enduit
avec la fixation des moulures
nl 1. metselwerk *n* van radiaalstenen
2. stucwerk *n* aan lijsten en
ornamenten
r 1. кладка *f* из клинового
[лекального] кирпича
2. штукатурные работы *f pl*
с установкой лепных изделий

G111 **gauge pile** *see* **guide pile**

G112 *e* **gauge pressure**
d manometrischer Druck *m*;
Überdruck *m*
f pression *f* manométrique;
surpréssion *f*
nl manometrische druk *m*; overdruk *m*
r манометрическое давление *n*;
избыточное давление *n*

G113 *e* **gauge rod**
d Schichtenlehre *f* (*Maurer*)
f règle *f* de maçon (*pour régler les*
assises des briques)
nl lagenstok *m*
r порядовка *f* (*каменщика*)

G114 *e* **gauging**
d 1. Zumessen *n*, Dosieren *n*
2. Durchmischung *f* von Hand
3. Hydrometrie *f*
f 1. dosage *m* 2. malaxage *m* à la main
3. hydrométrie *f*
nl 1. dosering *f* 2. menging *f* met de
hand 3. hydrometrie *f*
r 1. дозирование *n* 2. перемешивание *n*
(*составляющих смеси*) вручную
3. гидрометрия *f*

G115 *e* **gauging board**
d Mischplatte *f*, Mischbühne *f*
f planche *f* [plate-forme *f*] de gâchage
nl mengplateau *n*
r боёк *m* (*платформа для*
приготовления бетонной или
штукатурной смеси вручную)

G116 *e* **gauging plaster**
d Oberputzgips *m*

f enduit *m* de finition; mortier *m* de
plâtre et de chaux (*pour l'enduit de*
finition)
nl afwerkpleister *n*
r известково-гипсовый штукатурный
раствор *m* для накрывки
[накрывочного слоя]; накрывочный
слой *m* штукатурки

G117 *e* **gauging section, gauging site**
d Abflußmeßstelle *f*
f section *f* de jaugeage
nl afvloeiingsmeetpunt *n*
r замыкающий створ *m*

G118 *e* **gauging station**
d 1. Pegelmeßstation *f*,
Abflußmeßstelle *f* 2. Pegel *m*,
Wasserstandanzeiger *m*
f 1. station *f* de jaugeage 2. limnimètre *m*
nl 1. meetstation *n* 2. peilstation *n*
r 1. гидрометрическая станция *f*
2. водомерный пост *m*

G119 *e* **gauging trowel**
d Verputzkelle *f*
f truelle *f*
nl pleistertroffel *m*
r кельма *f*, мастерок *m*, гладилка *f*

G120 *e* **gauging water**
d Mischwasser *n*, Anmach(e)wasser *n*.
f eau *f* de gâchage
nl aanmaakwater *n*
r вода *f* затворения

G121 *e* **gauze wire cloth**
d Drahtgewebe *n*
f grillage *m* [toile *f*] métallique,
treillis *m* (en fil) métallique
nl metaalgaas *n*, draadgaas *n*
r проволочная сетка *f*

G122 *e* **G-cramp**
d Schraubzwinge *f*, Zwinge *f*
f serre-joint *m*, presse *f* à vis, sergent *m*
nl klemschroef *f* (*m*), ketelklem *f* (*m*)
r струбцина *f*

G123 *e* **gear pump**
d Zahnradpumpe *f*
f pompe *f* à engrenages
nl tandradpomp *f* (*m*)
r шестерённый насос *m*

G124 **gemel** *see* **jimmer**

G125 *e* **gemel window**
d Zwillingsfenster *n*
f fenêtre *f* géminée [jumelée]
nl dubbel venster *n*
r парное [спаренное] (*по фасаду*)
окно *n*

G126 *e* **general arrangement drawing**
d Übersichtszeichnung *f*, Gesamtansicht
f
f plan *m* d'ensemble [d'arrangement],
vue *f* générale [d'ensemble]
nl situatietekening *f*
r чертёж *m* общего расположения,
компоновочный чертёж *m*; общий
вид *m* (*на чертеже*)

G127 *e* **general buckling**
   d allgemeiner Stabilitätsverlust *m*
   f flambage *m* [flambement *m*] général
  nl geheel bezwijken *n*
   r общая потеря *f* устойчивости

G128 *e* **general contract**
   d Generalbauvertrag *m*
   f marché *m* de travaux général
  nl algemeen bouwcontract *n*
   r генеральный подрядный договор *m*

G129 *e* **general contractor**
   d Generalauftragnehmer *r.,*
     Hauptunternehmer *m*
   f entrepreneur *m* de contruction
     principal, maître *m* de l'œuvre
  nl hoofdaannemer *m*
   r генеральный подрядчик *m*

G130 *e* **general design considerations**
   d allgemeine Entwurfsbestimmungen *f* pl
   f dispositions *f* pl générales rélatives
     à l'étude de projets
  nl algemene ontwerpdoelstellingen *f* pl
   r общие положения *n'pl* проектирования

G131 *e* **general drawings**
   d Übersichtszeichnungen *f* pl
   f dessins *m* pl d'ensemble (*d'un bâtiment*)
  nl overzichtstekeningen *f* pl
   r чертежи *m* pl общего вида (*сооружения*)

G132 *e* **general exhaust ventilation**
   d Entlüftung *f*, Unterdrucklüftung *f*
   f ventilation *f* aspirante [par aspiration] générale
  nl centrale afzuiging *f*
   r общеобменная вытяжная вентиляция *f*

G133 *e* **general foreman**
   d Polier *m*, Oberpolier *m*;
     Bauleiter *m*
   f chef *m* de chantier
  nl uitvoerder *m*
   r начальник *m* стройплощадки;
     старший производитель *m* работ

G134 *e* **generalized force**
   d verallgemeinerte [generalisierte] Kraft *f*
   f force *f* généraliseé
  nl gegeneraliseerde kracht *f* (*m*)
   r обобщённая сила *m*

G135 *e* **general layout**
   d Generalplan *m*
   f plan *m* général [d'ensemble]
  nl algehele opzet *m*
   r генеральный план *m*

G136 *e* **general lighting**
   d Allgemeinbeleuchtung *f*
   f éclairage *m* général
  nl algemene verlichting *f*
   r общее освещение *n*

G137 *e* **general principles of the design**
   d allgemeine Berechnungsgrundlagen *f* pl;
     allgemeine Entwurfsgrundlagen *f* pl
   f principes *m* pl généraux de calcul;
     principes *m* pl généraux d'étude de projet
  nl grondslagen *m* pl voor het ontwerp
   r основные расчётные положения *n* pl;
     основные принципы *m* pl проектирования

G138 *e* **general purpose Portland cement**
   d Normalportlandzement *m*
   f ciment *m* portland pour usage courant
  nl gewone portlandcement *n*
   r портландцемент *m* общего назначения, рядовой портландцемент *m*

G139 *e* **general requirements**
   d allgemeine Forderungen *f* pl [Bestimmungen *f* pl] der technischen Bauvorschriften
   f prescriptions *f* pl [dispositions *f* pl] générales (*du cahier des charges*)
  nl algemene voorwaarden pl
   r общие требования *n* pl [положения *n* pl] технических условий на строительство

G140 *e* **geodetic dome**
   d geodätische Kuppel *f*
   f dôme *m* [coupole *f*] géodésiquе
  nl geodetische koepel *m*
   r геодезический купол *m*

G141   **geodetic construction** *see* **stressed-skin construction**

G142 *e* **geodetic head, geodetic pressure head**
   d geodätische Druckhöhe *f* [Förderhöhe *f*]
   f hauteur *f* de chute géométrique
  nl geodetische drukhoogte
   r геометрический напор *m*

G143 *e* **geodetic suction head**
   d geodätische [geometrische] Saughöhe *f*
   f hauteur *f* d'aspiration géométrique
  nl geodetische zuighoogte *f*
   r геометрическая высота *f* всасывания

G144 *e* **geodetic survey**
   d geodätische Aufhahme *f*
   f levé *m* géodésique
  nl landmeting *f*, opmeting *f*
   r геодезическая съёмка *f*

G145 *e* **geodimeter**
   d Geodimeter *n*
   f géodimètre *m*
  nl geodimeter *m*
   r геодиметр *m* (*электронно-оптический дальномер*)

G146 *e* **geological section**
   d geologisches Profil *n*
   f coupe *f* géologique
  nl geologisch profiel *n*
   r геологический профиль *m* [разрез *m*]

G147 *e* **geometric properties of sections**
   d geometrische Querschnittwerte *m* pl
   f caractéristiques *f* pl géométriques des sections [des profilés]

nl geometrische waarden *f* (*m*) *pl* van de
dwarsdoorsnede
r геометрические характеристики *f* *pl*
сечений

G148 *e* **geometry of specimens**
   d Probengeometrie *f*, Prüflingsgeometrie
f
   f géométrie *f* [caractéristiques *f* *pl*
géométriques] des échantillons
   nl geometrische waarden *f* (*m*) *pl* van de
monsters
   r геометрические характеристики *f* *pl*
образцов

G149 *e* **geometry of the structure**
   d Bauwerksgeometrie *f*
   f géométrie *f* de l'ouvrage, schéma *m*
[diagramme *f*] géométrique de
l'ouvrage
   nl vorm *m* en afmetingen *f* *pl* van het
gebouw
   r геометрическая схема *f* сооружения

G150 *e* **geophysical prospecting**
   d geophysikalische Schürfung *f*
   f prospection *f* géophysique
   nl geophysische opsporing *f*
   r геофизические методы *m* *pl*
исследования грунтов, геофизическая
разведка *f*

G151 *e* **geotechnical fabric**
   d geotechnisches Kunststoff-Gitter *n*
(*zur Bodenverfestigung im Straßenbau*)
   f tissu *m* [treillis *m*] géotechnique
(en plastique) (*utilisé pour le
renforcement du sol*)
   nl kunststofweefsel *n* voor wegfundering
   r геотехническая (пластмассовая)
сетка *f* (*для укрепления
поверхностного слоя грунта*)

G152 *e* **geotechnical processes**
   d geotechnische Verfahren *n* *pl*
   f processus *m* *pl* géotechniques
   nl geotechnische processen *n* *pl*
   r геотехнические процессы *m* *pl*,
процессы *m* *pl*, изменяющие
свойства грунтов

G153 *e* **geothermal energy**
   d geothermische Energie *f*
   f énergie *f* géothermique
   nl geothermische energie *f*
   r геотермальная энергия *f*

G154 *e* **gesso**
   d Spachtelmasse *f*, Porenfüller *m*
   f mastic *m* bouche-pores *m*, apprêt *m*
(*pour la peinture décorative*)
   nl plamuur *m*, *n*
   r шпаклёвка *f*, шпатлёвка *f* (*для
декоративной окраски*)

G155 *e* **geyser**
   d Gas-Wasserbereiter *m*, Gasbadeofen *m*
   f chauffe-eau *m* à gaz
   nl (gas)geiser *m*
   r газовый водонагреватель *m*,
газовая водогрейная колонка *f*

G156 *e* **giant**
   d Druckstrahlbagger *m*,
Hydromonitor *m*, Spritzkanone *f*,
Wasserwerfer *m*
   f monitor *m* hydraulique
   nl waterkanon *n*
   r гидромонитор *m*

G157 *e* **gib**
   d Nachstelleiste *f*, Stellkeil *m*,
Fixierkeil *m*
   f contre-clavette *f*, réglette *f* de guidage
   nl stelwig *f* (*m*)
   r стальная зажимная планка *f*,
зажимной клин *m*

G158   **gilled pipe** *see* **ribbed pipe**

G159 *e* **Gillmore needle**
   d Gillmore-Nadel *f*,
Normengewichtsnadel *f*
(*Abbindezeitbestimmung*)
   f aiguille *f* de Gillmore (*pour la
détermination du temps de prise des
liants hydraulisques*)
   nl Gillmore-naald *f* (*m*) (*ter bepaling van
de verhardingstijd*)
   r игла *f* Гилмора (*для определения
сроков схватывания вяжущих*)

G160 *e* **gilsonite**
   d Gilsonitasphalt *m*, Uintait *m*
   f gilsonite *f*
   nl gilsoniet *n* (*een natuurlijk soort
asfaltbitumen*)
   r гильсонит *m* (*асфальтовая порода*)

G161 *e* **gimlet**
   d Holzbohrer *m*
   f foret *m* à bois, vrille *f*, queue-de-
-cochon *f*
   nl fretboor *m*
   r бурав *m*

G162 *e* **gin block**
   d Baurolle *f*
   f poulie *f* simple
   nl bouwrol *f* (*m*)
   r простой блок *m* с одним шкивом

G163 *e* **gin pole derrick**
   d Trossen-Derrick(kran) *m* ohne
Ausleger, Hebezeug-Mast *m*
   f mât *m* de levage
   nl hijsmast *m*
   r монтажная мачта *f*

G164 *e* **gin pole type concrete spouting plant**
   d Betongießmast *m*
   f tour *f* distributrice de béton
   nl betongietmast *m*
   r бетонолитная мачта *f*

G165 *e* **girder**
   d Hauptträger *m*, Unterzug *m*, Riegel *m*;
mehrteiliger Träger *m*; Binder *m*
   f poutre *f* maîtresse [principale],
longrine *f*; poutre *f* composée
[composite] de grande hauteur;
poutre *f* à treillis
   nl draagbalk *m*, onderslagsbalk *m*, ligger
m; samengestelde ligger *m*; bint *n*

r главная балка *f*, прогон *m*, ригель *m*;
крупная составная балка *f*; ферма *f*
с параллельными поясами

G166 *e* **girder bridge**
  d Trägerbrücke *f*
  f pont *m* à poutres
  nl liggerbrug *f* (*m*)
  r балочный мост *m*

G167 *e* **girder casing**
  d Schutzummantelung *f* eines
  Stahlträgers
  f gaine *f* de protection (*d'une poutre
en acier*)
  nl liggerbekleding *f*
  r обделка *f* стальной балки (*в форме
короба*)

G168 *e* **girder grillage**
  d Trägerrost *m*, Trägerkreuzwerk *n*
  f grillage *m* de poutres
  nl vakwerk *n*
  r балочный ростверк *m*, балочная
решётка *f*

G169 *e* **girderless floor construction**
  d Pilzdecke *f*
  f plancher *m* champignon
  nl balkloze vloer *m*, paddestoelvloer *m*
  r безбалочное перекрытие *n*

G170 *e* **girder post**
  d Stütze *f* des Trägers [des Balkens]
  f poteau *m* supportant la poutre
  nl kolom *f* (*m*)
  r стойка *f*, поддерживающая балку

G171 *e* **girder stiffener**
  d Versteifung *f* des Trägersteges
  f raidisseur *m* d'âme (*d'une poutre*)
  nl verstijving *f* van een lijf (*van een
balk*)
  r элемент *m* жёсткости стенки балки

G172 *e* **girder web**
  d Trägersteg *m*
  f âme *f* de la poutre
  nl lijf *n* van een ligger
  r стенка *f* балки

G173 *e* **girt, girth**
  d 1. Horizontalverband *m*
  2. Fachwerkriegel *m*
  3. Sattelschwelle *f*, Brustschwelle *f*
  f 1. ceinture *f* de renforcement
(*d'un bâtiment*); barre *f* horizontale
de contreventements 2. traverse *f* de
pan de fer 3. traverse *f* haute de bâti
en bois
  nl 1. horizontale verbinding *f*
2. bovenrand *m* van een vakwerkligger
3. dwarsbalk *m* van een houtskelet
  r 1. горизонтальный пояс *m* здания
(*напр. антисейсмический*);
горизонтальный элемент *m*
ветровых связей 2. ригель *m*
фахверка 3. верхняя обвязка *f*
деревянного каркаса

G174 *e* **girth weld**
  d Umfangsschweißnaht *f*

  f soudure *f* circulaire
  nl ringlas *f* (*m*)
  r кольцевой (сварной) шов *m*

G175 *e* **gland**
  d Stopfbuchsendeckel *m*;
Stopfbuchsenbrille *f*
  f chapeau *m* [tampon *m*] de boîte
à bourrage
  nl drukstuk *n* van pakking
  r крышка *f* *или* нажимная втулка *f*
сальника

G176 *e* **gland cock**
  d Packhahn *m*
  f robinet *m* avec presse-étoupe
  nl stopkraan *f* (*m*)
  r сальниковый кран *m*

G177 *e* **glass block, glass brick**
  d Glasbaustein *m*, Glasziegel *m*
  f brique *f* de verre
  nl glazen bouwsteen *m*
  r стеклянный блок *m*, стеклоблок *m*

G178 *e* **glass cloth**
  d Glasfasergewebe *n*
  f tissu *m* de verre
  nl glasweefsel *n*
  r стеклоткань *f*

G179 *e* **glass concrete constructions**
  d Glasbetonkonstruktionen *f pl*
  f béton *m* (armé) translucide
  nl glasbetonconstructies *f pl*
  r стеклобетонные конструкции *f pl*

G180 *e* **glass domed roof light**
  d Glaslichtkuppel *f*
  f coupole *f* d'eclairage en verre, dôme
m en verre (*pour éclairage zénithal*)
  nl lichtkoepel *m* van glas
  r стеклянный световой купол *m*

G181   **glass door** *see* **glazed door**

G182   **glass fabric** *see* **glass cloth**

G183 *e* **glass-façade building**
  d Glasfassadengebäude *n*
  f édifice *m* à façade entièrement vitrée
  nl gebouw *n* met gordijngevel
[vliesgevel]
  r здание *n* со стеклянным фасадом

G184 *e* **glass fiber** *US*
  d Glasfaser *f*
  f fibre *m* de verre
  nl glasvezel *f* (*m*)
  r стекловолокно *n*

G185 *e* **glass fiber mat**
  d Glas(faser)matte *f*
  f matelas *m* en fibre de verre
  nl glasvezelmat *f* (*m*)
  r стекломат *m*

G186 *e* **glass fiber reinforced concrete**
  d glasfaserverstärkter Beton *m*,
Glasfaserbeton *m*
  f béton *m* de fibres de verre, béton *m*
renforcé par fibres de verre
  nl glasvezelbeton *n*
  r стеклофибробетон *m*, бетон *m*,
армированный стекловолокном

G187 *e* **glass fiber reinforced plastics**
    d glasfaserverstärkter Kunststoff *m*
    f plastique *m* renforcé [armé] par la fibre de verre
    nl met glasvezel gewapende kunststof *f* (*m*)
    r стеклопластик *m*

G188 *e* **glass-fiber strand**
    d Glasfaserlitze *f*, Glasfaserbündel *n*
    f toron *n* en fibres de verre
    nl streng *f* (*m*) glasvezels
    r прядь *f* из стекловолокна (*для армирования бетона*)

G189 *e* **glass fiber wrapping material**
    d Glas(faser)vlies *n*
    f tissu *m* [toile *f*] en fibres de verre
    nl glasvlies *f*
    r стеклохолст *m*

G190   **glass fibre** *UK see* **glass fiber**
G191   **glasshouse** *see* **greenhouse**

G192 *e* **glass paper**
    d Glaspapier *n*, Schmirgelpapier *n*
    f papier *m* de verre
    nl schuurpapier *n*
    r тонкая наждачная бумага *f*

G193 *e* **glass reinforced waterproof paper**
    d glasfaserverstärktes Baupapier *n*
    f papier *m* imperméable renforcé par fibres de verre
    nl met glasvezel versterkt waterdicht papier *n*
    r водоизолирующая бумага *f*, армированная стекловолокном

G194   **glass silk** *see* **glass wool**

G195 *e* **glass slates**
    d Dachglasstein *m*
    f tuile *f* de toiture en verre
    nl glazen dakpan *f* (*m*)
    r стеклянные кровельные плитки *f pl*, стеклянная черепица *f*

G196   **glass stop** *see* **glazing bead**
G197   **glass tiles** *see* **glass slates**

G198 *e* **glass tissue**
    d Glasfilz *m*
    f tissu *m* de verre
    nl glasvilt *n*
    r стекловойлок *m*

G199 *e* **glass wool**
    d Glaswolle *f*
    f laine *f* de verre
    nl glaswol *f* (*m*)
    r стеклянная вата *f*

G200 *e* **glass wool roll**
    d Glaswollbahn *f*
    f laine *f* de verre en rouleau
    nl glaswoldeken *f* (*m*)
    r рулонный стекловойлок *m*

G201 *e* **glaze**
    d Glasur *f*
    f glaçure *f*
    nl glazuur *n*
    r глазурь *f*

G202 *e* **glaze coat**
    d 1. Bitumendeckschicht *f* (*weiche Dachhaut*) 2. Glasur *f*
    f 1. couche *f* d'étanchéité en bitume (*de la toiture*) 2. couche *f* de la peinture supérieure transparente
    nl 1. verdichte bitumineuze dakbedekking *f* 2. glazuurlaag *f* (*m*)
    r 1. верхний накрывочный слой *m* битума (*мягкой кровли*) 2. верхний полупрозрачный окрасочный слой *m*

G203 *e* **glazed brick**
    d Glasurziegel *m*
    f brique *f* émaillée [vitrifiée]
    nl geglazuurde baksteen *m*
    r глазурованный кирпич *m*

G204 *e* **glazed clayware pipe**
    d Steinzeugrohr *n*
    f tuyau *m* en grès cérame
    nl gresbuis *f* (*m*)
    r керамическая [гончарная] труба *f*

G205 *e* **glazed door**
    d verglaste Tür *f*; Fenstertür *f*, französisches Fenster *n*
    f porte *f* vitrée; porte *f* française
    nl glazen deur *f* (*m*)
    r остеклённая дверь *f*; дверь-окно *f* (*с выходом на террасу или балкон*), французское окно *n*

G206 *e* **glazed interior tile**
    d glasierte Porzellanfliese *f*
    f tuile *f* vitrifiée pour revêtement intérieur
    nl geglazuurde wandtegel *m*
    r глазурованная плитка *f* для внутренней облицовки

G207 *e* **glazed work**
    d Mauerwerk *n* mit Einsatz von Glasurziegel
    f maçonnerie *f* avec revêtement en briques vitrifiées
    nl met geglazuurde steen bekleed metselwerk *n*
    r каменная кладка *f* с применением глазурованного кирпича

G208 *e* **glazier's point**
    d Fensterstift *m*
    f pointe *f* de vitrier
    nl raampen *f* (*m*)
    r шпилька *f* для крепления стекол (*в деревянной раме*)

G209 *e* **glazier's putty**
    d Glaserkitt *m*
    f mastic *m* de vitrier [à vitres], lut *m* de vitrier [à vitres]
    nl stopverf *f* (*m*)
    r замазка *f* для остекления

G210 *e* **glazing**
    d 1. Verglasen *n*, Verglasung *f* 2. Glasieren *n*
    f 1. pose *f* des vitres, vitrage *m* 2. glaçure *f*
    nl 1. met glas bezetten 2. glazuren *r*

r 1. остекление *n* 2. глазурование *n*, покрытие *n* глазурью

G211 *e* **glazing bar**
  d Fenstersprosse *f*
  f petit bois *m*
  nl glasroede *f* (*m*)
  r горбылёк *m* (*оконного переплёта*)

G212 *e* **glazing bead**
  d Falzleiste *f*
  f parclose *m*, parclause *f*
  nl glaslat *f* (*m*)
  r штапик *m*

G213 *e* **glazing clip**
  d Verglasungsklammer *f*
  f agrafe *f* à vitrage
  nl glasveer *f* (*m*), glaspen *f* (*m*)
  r клямера *f* (*для остекления*)

G214 **glazing fillet** *see* **glazing bead**

G215 *e* **glazing knife**
  d Glasermesser *n*
  f couteau *m* de vitrier
  nl stopmes *n*
  r нож *m* для остекления, нож *m* стекольщика

G216 *e* **glazing moulding**
  d Falzkehlleiste *f*, Falzleiste *f* mit Kehlung
  f parclause *f* [parclose *f*] moulurée
  nl geprofileerde glaslat *f* (*m*)
  r штапик *m* с калёвкой

G217 *e* **glazing work**
  d Glasarbeiten *f pl*, Verglasungsarbeiten *f pl*
  f travaux *m pl* de vitrage
  nl glaszetten *n*
  r стекольные работы *f pl*

G218 *e* **gliding window**
  d Schiebefenster *n*
  f fenêtre *f* coulissante
  nl schuifraam *n*
  r раздвижное окно *n*

G219 *e* **globe valve**
  d 1. Kugelventil *n* 2. Kugelschieber *m*
  f 1. soupape *f* à boulet 2. vanne *f* sphérique
  nl 1. kogelventiel *n* 2. bolafsluiter *m*
  r 1. вентиль *m* 2. шаровой затвор *m*

G220 **glory-hole spillway** *see* **shaft spillway**

G221 *e* **gloss**
  d Glanz *m*
  f brillant *m*, lustre *m*
  nl glans *m*
  r глянец *m*

G223 *e* **glued connection**
  d Leimverbindung *f*
  f assemblage *m* collé
  nl lijmverbinding *f*
  r клеевое соединение *n*

G224 *e* **glued-laminated timber, glued--laminated wood**
  d 1. Elemente *n pl* von verleimten

Schichtkonstruktionen 2. Holzplatte *f*, **Holztafel** *f*
  f 1. bois *m* lamellé-collé 2. dalle *f* collée en planches posées à l'arête
  nl 1. gelamineerd hout *n* 2. multiplex *n*
  r 1. элементы *m pl* многослойных деревянных клеевых конструкций 2. деревоплита *f*

G225 *e* **glued structures**
  d geklebte [geleimte] Konstruktionen *f pl*
  f constructions *f pl* collées
  nl gelijmde constructies *f pl*
  r клеёные конструкции *f pl*, конструкции *f pl* с клеевыми соединениями

G226 **glulam** *see* **glued laminated timber**

G227 *e* **glulam beam**
  d geleimter Balken *m* [Träger *m*], Schichtholzbalken *m*, Schichtholzträger *m*
  f poutre *f* collée (en planches)
  nl gelamineerde houten ligger *m*
  r дощато-клеёная балка *f*

G228 *e* **glulam column**
  d lamellenverleimte Holzstütze *f*
  f poteau *m* collé (en planches)
  nl gelamineerde houten kolom *f* (*m*)
  r дощатая клеёная стойка *f*

G229 *e* **going**
  d 1. Auftritt *m*, Trittbrett *n*; Stufenbreite *f* 2. Grundrißlänge *f* der Treppe
  f 1. giron *m* 2. portée *f* horizontale de la volée d'escalier
  nl 1. aantrede *f* (*m*) 2. horizontale lengte *f* van een trap
  r 1. проступь *f*; ширина *f* ступени 2. заложение *n* лестничного марша

G230 **going of the flight** *see* **going** 2.

G231 *e* **gold foil, gold leaf**
  d Blattgold *n*
  f or *m* mussif; feuille *f* d'or
  nl bladgoud *n*
  r сусальное золото *n*, сусаль *f*

G232 *e* **goliath crane**
  d Schwerlast-Bockkran *m*
  f grue *f* géante [titan]
  nl kraan *f* (*m*) voor zeer zware lasten
  r козловой кран *m* большой грузоподъёмности (*свыше 50 m*)

G233 *e* **goods lift** *UK see* **freight elevator**

G234 *e* **goods shed**
  d Warenschuppen *m*
  f remise *f*; entrepôt *m*
  nl goederenloods *f* (*m*)
  r материальный склад *m*

G235 *e* **go-out**
  d Abfallschleuse *f*
  f écluse *f* de décharge (*dans la digue de garde*)
  nl uitwateringssluis *f* (*m*)
  r водосбросный шлюз *m* в дамбе обвалования

G237 *e* **gorge dam**
  d Schluchtenstaumauer *f*
  f barrage *m* [digue *f*] de gorge
  nl stuwdam *m* in een kloof
  r плотина *f*, перегораживающая
    ущелье

G238 *e* **gouge**
  d Drechslerröhre *f*
  f gouge *f*
  nl guts *f* (*m*), holle beitel *m*
  r полукруглая желобчатая стамеска *f*

G240 *e* **grab**
  d Greifer *m*
  f benne *f* preneuse [à griffe]
  nl grijper *m*
  r грейферный ковш *m*

G241 *e* **grab dredge, grab dredger**
  d Greifer-Naßbagger *m*, Greifer-
    -Schwimmbagger *m*
  f drague *f* à benne preneuse
  nl baggermachine *f* met grijper
  r грейферный земснаряд *m*

G242 *e* **gradall**
  d Hydrauliktieflöffelbagger *m* mit
    Teleskopausleger; Grabenbagger *m*
    mit großer Ausladung des Auslegers
  f pelle *f* retro hydraulique à flèche
    télescopique
  nl slotengraafmachine *f* met telescooparm
  r гидравлическая обратная лопата *f*
    с телескопической стрелой;
    длинностреловой траншеекопатель *m*

G243   gradation *see* **particle size distribution**

G244 *e* **gradation of the fire safety**
  d Einteilung *f* (*der Konstruktionen und
    Baustoffen*) nach der
    Entflammbarkeitsklassen
  f classement *m* (*des constructions et des
    matériaux*) selon leur degré
    d'inflammabilité
  nl indeling *f* van bouwmaterialen naar
    hun brandveiligheid
  r классификация *f* по возгораемости
    (*строительных конструкций
    и материалов*)

G245 *e* **grade**
  d 1. geneigter Straßenabschnitt *m*;
    Gefälle *n*; Steigung *f*, Anstieg *m*;
    Neigung *f* 2. Geländekote *f*,
    Bodenniveau *n*; tatsächliche Höhe *f*
    3. Sorte *f*, Güteklasse *f* 4. Grad *m*
  f 1. rampe *f*, pente *f*, inclinaison *m*,
    montée *f ou* descente *f* 2. niveau *m*
    naturel (*des terres*) 3. qualité *f*;
    grade *m*, catégorie, classe *f*
    4. grade *m*
  nl 1. helling *f* 2. terreinpeil *n* 3. soort
    f (*m*); kwaliteit *f*
  r 1. наклонный участок *m* дороги;
    спуск *m*, подъём *m*; уклон *m*;
    наклон *m* 2. отметка *f* земли; чёрная
    отметка *f* 3. сорт *m*; качество *n*;
    категория *f* (*материала*) 4. градус *m*

G246 *e* **grade beam**
  d Fundamentbalken *m*
  f poutre *f* de fondation [de rive]
  nl fundamentbalk *m*
  r фундаментная балка *f*

G247 *e* **graded aggregate**
  d abgestufter Zuschlagstoff *m*
  f granulat *m* [agrégat *m*] à granulométrie
    établie, granulat *m* tamisé
  nl gegradeerde toeslag *m*
  r заполнитель *m* подобранного
    гранулометрического состава

G248 *e* **grade depressions**
  d Absackungsstellen *f pl* der Straßendecke
  f dépressions *f pl* du revêtement routier
  nl verzakkingen *f pl* in het wegdek
  r места *n pl* просадок дорожного
    покрытия

G249 *e* **graded filter**
  d abgestuftes Filter *n*, *m*
  f filtre *m* inversé
  nl retourfilter *n*, *m*
  r обратный фильтр *m*

G250 *e* **graded samples**
  d nach Korngröße getrennte Proben *f
    pl*
  f échantillons *m pl* de granulat séparés
    selon leur granulométrie
  nl gefractioneerde monsters *n pl* van de
    toeslag
  r фракционированные образцы *m pl*
    (*заполнителя*)

G251 *e* **grade line**
  d Nivellementzug *m*; abgesteckte
    Linienführung *f*
  f cheminement *m* de nivellement
  nl beloop *m*; hellinglijn *f* (*m*)
  r нивелирный ход *m*; вынесенный на
    местность участок *m* трассы

G252 *e* **grade of concrete**
  d Betongüte *f*
  f classe *f* de qualité [de résistance] de
    béton
  nl betonkwaliteit *f*
  r марка *f* бетона

G253   **grade of cement** *see* **brand of cement**

G254 *e* **grader**
  d Planiergerät *n*, Straßenhobel *m*
  f niveleuse *f*, grader *m*
  nl grondschaaf *f* (*m*), grader *m*
  r грейдер *m*

G255 *e* **grade-separated fork junction**
  d Gabelkreuzung *f* in getrennten
    Ebenen
  f jonction *f* de routes en Y à niveaux
    différents
  nl ongelijkvloerse wegsplitsing *f*
  r Y-образное примыкание *n* дорог в
    разных уровнях

G256 *e* **grade-separation junction**
  d plankreuzungsfreie
    [überschneidungsfreie] Kreuzung *f*,
    Kreuzung *f* auf verschiedenen Ebenen

f croisement *m* des routes dénivelé
[à niveaux différents]
nl ongelijkvloerse wegkruising *f*
r пересечение *n* дорог в разных
уровнях

G257 *e* **grade-stabilising structure**
d Gefällestabilisierungsbauwerk *n*
(*Querdamm, Sohlabsturz*)
f ouvrage *m* de stabilisation de pente
nl terrasmuur *m*, terrasseringswerk *n*
r сооружение *n* (*плотина, перепад*)
для стабилизации продольного
уклона и предотвращения
регрессивной эрозии русла

G258 *e* **grade stake**
d Höhenpflock *m*
f piquet *m* de nivellement
nl piketpaal *m*, piket *n*
r нивелировочный колышек *m*,
колышек *m* с отметкой

G259 *e* **grade strip**
d Richtlatte *f*
f latte *f* de niveau
nl richtlat *f* (*m*)
r маячная рейка *f* [планка *f*]
(*фиксирующая уровень укладки
бетона*)

G260 *e* **gradient**
d Gradiente *f*
f gradient *m*; rampe *f*
nl helling *f*, gradiënt *m*, montée *f*
r уклон *m*, градиент *m*; наклонный
участок *m* дороги

G261 *e* **grading**
d 1. Kornzusammensetzung *f*, Kornabstufung *f* 2. Kornanalyse *f*, Siebanalyse *f* 3. Klassierung *f*, Korngrößentren-
nung *f* 4. Profilieren *n*, Planieren *n*,
Planierarbeiten *f pl*
f 1. granulométrie *f*, composition *f*
granulométrique 2. analyse *f*
granulométrique 3. tamisage *m*
4. profilage *m*, mise *f* au profil du sol,
nivellement *n*, régalage *m*
nl 1. korrelgrootteverdeling *f*
2. korrel(grootte)analyse *f*
3. korrelsortering *f*
r 1. гранулометрический [зерновой]
состав *m* 2. определение *n*
гранулометрического состава,
ситовый анализ *m* 3. классификация *f*
[фракционирование *n*] по
крупности зерна 4. профилирование
n, планировка *f* (*грунта*)

G262 *e* **grading curve**
d Siebkurve *f*, Sieblinie *f*,
Körnungskurve *f*
f courbe *f* granulométrique
nl zeefkromme *f* (*m*)
r кривая *f* гранулометрического
состава

G263 *e* **grading plan**
d Entwurf *m* [Zeichnung *f*] der
Geländeausgleichung

f projet *m* de mise en forme du terrain,
plan *m* de terrassement de finition,
plan *m* d'aplanissement du terrain
nl terrasseringsplan *n*,
nivelleringsplan *n*
r проект *m* [чертёж *m*] вертикальной
планировки (*участка*)

G264 *e* **grading timber**
d Bauholz- *bzw.* Schnittholzsortieren *n*
f classement *m* du bois
nl sorteren *n* van bezaagd hout
r сортировка *f* лесо- и пиломатериала
по качеству [по сортам]

G265 *e* **gradual application of the prestressing
force**
d stufenweises Anbringen *n* der
Vorspannkraft
f application *f* graduelle de l'effort de
précontrainte
nl de voorspanning geleidelijk
aanbrengen *n*
r постепенное [поэтапное] приложение
n усилий преднапряжения

G266 *e* **gradual settlement**
d allmähliche Setzung *f*
f affaissement *m* [tassement *m*] graduel
nl geleidelijke zetting *f*
r постепенная осадка *f*

G267 *e* **gradual stress increase**
d allmähliche Spannungszunahme *f*
f accroissement *m* de tension graduel
nl geleidelijke spanningstoename *f* (*m*)
r постепенный рост *m* [возрастание *n*]
напряжений

G268 *e* **grafting tool**
d Stechspaten *m*
f bêche *f*, pioche *f*
nl spade *f* (*m*)
r заступ *m*, узкая штыковая лопата *f*
(*для разработки плотных грунтов*)

G269 *e* **grain size**
d Korngröße *f*
f dimension *f* des grains
nl korrelgrootte *f*
r размер *m* зёрен

G270 *e* **grain-size classification**
d Einteilung *f* der Böden nach
Korngrößenverteilung
f classification *f* de sols d'après la
grosseur des grains
nl bodemclassificatie *f* naar korrelgrootte
r классификация *f* грунтов по
крупности зёрен

G271 *e* **grain size distribution**
d Korngrößenverteilung *f*,
Kornzusammensetzung *f*
f granulométrie *f*, granularité *f*
nl korrelgrootteverdeling *f*
r гранулометрический [зерновой]
состав *m*

G272 **grain-size distribution curve** *see*
grading curve

G274 *e* **granolithic concrete**
   d Hartbeton *m*
   f béton *m* dur
   nl hard beton *n*
   r износостойкий бетон *m*

G275 *e* **granolithic flooring**
   d Hartbetonfußbodenbelag *m*
   f couvre-sol *m* [revêtement *m* de sol] en
   béton dur
   nl cementvloer *f* (*m*) met slijtlaag
   r бесшовный пол *m* из износостойкого
   бетона

G276 *e* **granular-bed filter**
   d Schüttschichtfilter *n*, *m*
   f filtre *m* à gravier fin
   nl korrelfilter *n*, *m*
   r зернистый фильтр *m*

G277 *e* **granular-fill insulation, granular
insulation**
   d Wärmeisolierung *f* aus körnigem
   Dämmstoff; körniger Dämmstoff *m*
   f isolant *m* thermique granulé
   nl warmte-isolatie *f* met losse korrels
   r теплоизоляция *f* из гранулированного
   материала; гранулированный
   теплоизоляционный материал *m*

G278 *e* **granular materials**
   d 1. körnige Baustoffe *m pl* 2. körnige
   Böden *m pl*
   f 1. matériaux *m pl* granuleux 2. sols *m*
   pl granuleux, terre *f* granuleuse
   nl bouwmaterialen *n pl* in korrelvorm
   r 1. зернистые материалы *m pl*
   2. зернистые грунты *m pl*

G279   granular soils *see* granular materials 2.

G280 *e* **granulated blast-furnace slag**
   d gekörnte Hochofenschlacke *f*
   f laitier *m* de haut fourneau granulé
   nl gegranuleerde hoogovenslak *m*, *n*
   r гранулированный доменный шлак *m*

G281 *e* **granulated cork**
   d granulierter Kork *m*
   f liège *m* granulaire [granulé, en grain]
   nl kurkkorrels *pl*
   r пробковая крошка *f*

G282 *e* **granulator**
   d Granulator *m*
   f granulateur *m*, concasseur *m*
   nl granulator *m*, verkruimelaar *m*
   r гранулятор *m*, дробилка *f*

G283 *e* **graphical analysis**
   d graphische Berechnung *f*
   f calcul *m* graphique
   nl grafische berekening *f*
   r графический расчёт *m*

G284 *e* **graphical construction**
   d zeichnerisches Verfahren *n*
   f construction *f* graphique
   nl grafische constructie *f*
   r графическое построение *n*

G285 *e* **graphic statics**
   d graphische Statik *f*, Graphostatik *f*
   f statique *f* graphique, graphostatique *f*

   nl grafostatica *f*
   r графическая статика *f*, графостатика
   f

G286 *e* **graphite paint**
   d Graphitfarbe *f*
   f couleur *f* graphitée
   nl grafietverf *f* (*m*)
   r графитовая краска *f* (*для
   металлоконструкций*)

G287 *e* **grate**
   d Gitterrost *m*; Abdeckgitter *n*
   f grille *f*; grillage *n*
   nl staafrooster *n*; rooster *n*
   r колосниковая *или* напольная
   решётка *f*

G288 *e* **grate inlet**
   d Rosteinlauf *m*
   f bouche *f* d'égout des eaux pluviales à
   grille
   nl straatkolk *m* met rooster
   r уличный дождеприёмник *m*
   с решёткой [с решётчатой крышкой]

G289 *e* **grate tamper**
   d gitterförmiger | Stampfer *m*
   f dame *f* à grille
   nl roosterstamper *m*
   r решётчатая трамбовка *f* (*для
   придания бетону шероховатой
   фактуры*)

G291 *e* **gravel**
   d Kies *m*, Geröll *n*
   f gravier *m*
   nl grind *n*, kiezel *n*
   r гравий *m*

G292 *e* **gravel board**
   d Sockelbrett *n*
   f sablière *f* inférieure enterrée
   nl funderingsmuur *m*
   r замятина *f* (*нижняя доска забора на
   уровне грунта*); цокольная доска *f*
   (*двери*)

G293 *e* **gravel filter well**
   d Kiesschüttungsbrunnen *m*,
   Kiesfilterbrunnen *m*
   f puits *m* avec filtre en gravier
   nl put *m* met grindfilter
   r колодец *m* с гравийным фильтром

G294 *e* **gravel fraction**
   d Kieskörnung *f*, Kiesanteil *m*
   f fraction *f* de gravier
   nl grindfractie *f*
   r гравийная фракция *f*

G295 *e* **gravel pack, gravel packing**
   d Kiesschüttung *f*, Kiesmantel *m*,
   Kiesfilter *n*, Kiespackung *f*
   f filtre *m* à gravier
   nl grindmantel *m*
   r гравийный экран *m* [фильтр *m*]
   (*колодца*)

G296   gravel plank *see* gravel board

G297 *e* **gravel pocket**
   d Kiesnest *n*, Steinnest *n*
   f nid *m* de cailloux

nl grindnest *n* (*in beton*)
r гравийное гнездо *n* (*в бетоне*)

G298 *e* **gravel pump**
d Kiespumpe *f*
f pompe *f* à gravier
nl grindpomp *f* (*m*)
r центробежный насос *m* для подачи
гравийной смеси в потоке [струе]
воды

G299 *e* **gravel screen**
d Kiessortieranlage *f*
f crible *m* à gravier
nl grindzeef *f* (*m*)
r гравиесортировка *f*

G300 *e* **gravel trap**
d Kiesfang *m*
f piège *m* à gravier
nl grindvang *m*
r гравиеловка *f*

G301 *e* **gravel washer, gravel washing plant**
d Kieswaschanlage *f*, Kieswäsche *f*
f laveur *m* à gravier
nl grindwasserij *f*
r гравиемойка *f*

G302 *e* **gravimetric air filter test**
d gravimetrische Luftfilterprüfung *f*
f essai *m* gravimétrique du filtre à
air
nl gravimetrische lichtfilterproef *m*
r гравиметрический контроль *m*
загрязнённости воздушного фильтра

G303 *e* **graving dock**
d Trockendock *n*
f dock *m* [cale *f*] de radoub, cale *f* sèche
nl droogdok *n*
r сухой док *m*

G304 *e* **gravitational dust collector**
d Schwerkraftabscheider *m*
f collecteur *m* [aspirateur *m*] à
poussières gravitaire
nl gravitatie-stofafscheider *m*
r гравитационный пылеуловитель *m*

G305 **gravitational water** *see* gravity water 1.

G306 *e* **gravity air circulation**
d natürliche Luftkonvektion *f*
f circulation *f* naturelle de l'air
nl natuurlijke luchtcirculatie *f*
r естественная конвекция *f*

G307 *e* **gravity-arch dam**
d Bogengewichtsmauer *f*,
Schwergewichts-Bogenmauer *f*,
massive Bogenmauer *f*
f barrage-poids *m* voûte [courbe],
barrage-gravité *m* voûte
nl stabiele gebogen stuwdam *m*
r арочно-гравитационная плотина *f*

G308 *e* **gravity circulation**
d Schwerkraftzirkulation *f*,
Schwerkraftumlauf *m*, natürliche
Zirkulation *f*
f circulation *f* par gravité
nl natuurlijke circulatie *f*
r гравитационная циркуляция *f*

G309 *e* **gravity classifier**
d Schwerkraftklassierer *m*
f classeur *m* [classificateur *m*]
gravitaire [par gravité]
nl gravitatie-scheider *m*
r гравитационный классификатор *m*

G310 *e* **gravity dam**
d Schwergewichtsstaumauer *f*
f barrage-poids *m*, barrage *m* de gravité
nl zware [stabiele] stuwdam *m*
r гравитационная плотина *f*

G311 *e* **gravity drainage**
d Entwässerung *f* im freien Gefälle
f assèchement *m* [drainage *m*] par
gravité
nl afwatering *f* met natuurlijk verloop
r самотёчное осушение *n*

G312 *e* **gravity feed**
d Schwerkraftförderung *f*,
Freifallbeschickung *f*
f alimentation *f* par gravité
nl toevoer *m* door zwaartekracht
r гравитационная подача *f*,
гравитационное питание *n*

G313 *e* **gravity feeder**
d Schwergewichtsaufgeber *m*
f alimenteur *m* par gravité
nl toevoer *m* met vrij verval
r гравитационный питатель *m*

G314 *e* **gravity filter**
d offenes [druckloses] Sandfilter *n*, *m*
f filtre *m* gravitaire
nl open zandfilter *n*, *m*
r гравитационный [безнапорный
песчаный] фильтр *m*

G315 *e* **gravity-flow heating system**
d Schwerkraft(-Warm)wasserheizung *f*
f installation *f* de chauffage central à
circulation par gravité
nl centrale verwarmingsinstallatie *f* met
natuurlijke circulatie
r система *f* отопления с естественной
циркуляцией

G316 **gravity groundwater** *see* gravity water 1.

G317 *e* **gravity hammer**
d Freifallramme *f*
f marteau *m* à chute libre
nl vrijvallende smeedhamer *m*
r свайный молот *m* свободного падения,
копёр *m* со свободно падающей бабой

G318 *e* **gravity main**
d Schwergewichtsflußleitung *f*
f conduite *f* à écoulement gravitaire
nl waterleiding *f* met vrij verval
r самотёчная водопроводная
магистраль *f*

G319 *e* **gravity quay wall**
d Schwergewichtskaimauer *f*
f mur *m* de quai type gravité
nl zware kademuur *m*
r гравитационная набережная *f*

G320 **gravity roof ventilator** *see* roof
ventilator 2.

G321　gravity scheme *see* **gravity water-supply system**

G322 *e* **gravity sewerage system**
　　d Schwergewichtsflußanlage *f*
　　f système *m* de conduites d'assainissement à écoulement gravitaire
　　nl rioolstelsel *n* met vrij verval
　　r самотёчная система *f* канализации

G323 *e* **gravity tipping skip**
　　d Schwergewicht-Kippmulde *f*
　　f benne *f* chargeuse à basculement par gravité
　　nl zelfkippende laadbak *m*
　　r самоопрокидывающийся загрузочный ковш *m*

G324 *e* **gravity vent**
　　d Belüftungsöffnung *f*
　　f orifice *m* [trou *m*] d'aérage
　　nl ontluchtingsopening *f*
　　r аэрационное отверстие *n*, аэрационный проём *m*

G325 *e* **gravity water**
　　d 1. Gravitationswasser *n* 2. Wasser *n* in Freispiegelwasserleitung
　　f 1. eau *f* de gravité 2. eau *f* dans le système d'adduction gravitaire
　　nl 1. gravitatiewater *n* 2. water *n* in een open waterleiding
　　r 1. гравитационная вода *f* 2. вода *f* в самотёчной системе водоснабжения

G326 *e* **gravity water-supply system**
　　d Freispiegelwasserleitung *f*,
　　f système *m* d'adduction d'eau gravitaire
　　nl waterleiding(stelsel *n*) *f* met natuurlijke druk
　　r самотёчная система *f* водоснабжения

G327 *e* **grease air filter**
　　d Fettabscheider *m*, Fettfilter *n*, *m*, Ölnebelabscheider *m*
　　f filtre-capteur *m* de graisse (*au dessus de la cuisinière à gaz*)
　　nl vetfilter *n*, *m* (*boven fornuis*)
　　r надплитный фильтр-жироуловитель *m*

G328　grease catcher *see* **grease interceptor**

G329 *e* **grease ice**
　　d Eisbrei *m*
　　f glace *f* pâteuse, sorbet *m*
　　nl papachtige ijslaag *f* (*m*), pap-ijs *n*
　　r ледяное сало *n*, шуга *f*

G330 *e* **grease interceptor, grease trap**
　　d Fettfang *m*, Fettabscheider *m*
　　f dégraisseur *m*, siphon *m* de depôt de graisses
　　nl vetvanger *m*, vetafscheider *m*
　　r жироловка *f*, жироуловитель *m*

G331 *e* **greenbelt**
　　d Grüngürtel *m*, Grünring *m*
　　f ceinture *f* [zone *f*] verte [de verdure]
　　nl groene zone *f*
　　r зелёный пояс *m* (*вокруг города*)

G332 *e* **green concrete**
　　d Frischbeton *m*
　　f béton *m* frais; béton *m* jeune
　　nl vers gestort beton *n*
　　r свежеуложенный бетон *m*; схватившийся, но не затвердевший бетон *m*

G333 *e* **greenhouse**
　　d Gewächshaus *n*; Treibhaus *n*
　　f serre *f*, forcerie *f*; bâche *f*
　　nl broeikas *f* (*m*); warenhuis *n*
　　r оранжерея *f*; парник *m*

G334 *e* **grid**
　　d 1. Raster *m*, Systemliniennetz *n* 2. Rostwerk *n*, Rost *m* 3. Rost *m*; Rechen *m*
　　f 1. quadrillage *m*, réseau *m* (*de référence ou modulaire*) 2. grillage *m*, treillis *m* (*de fondation*) 3. grille *f*
　　nl 1. raster *n*, coördinatiestelsel *n* 2. roosterwerk *n* 3. rooster *n*
　　r 1. сетка *f* (*напр. координационная, модульная*) 2. ростверк *m* (*из перекрёстных балок*) 3. решётка *f*

G335 *e* **grid connection**
　　d Stromanschluß *m*
　　f entree *f* de câble *ou* de ligne électrique (*dans un bâtiment*)
　　nl netaansluiting *f*, huisaansluiting *f*
　　r электрический ввод *m*, присоединение *n* к внешней сети

G336 *e* **grid failure**
　　d Stromausfall *m*, Netzausfall *m*
　　f panne *f* de courant [d'électricité]
　　nl netstoring *f*
　　r нарушение *n* электроснабжения

G337 *e* **grid floor, grid floor cover**
　　d Rostfußbodenbelag *m*; Rostfußboden *m*
　　f platelage *m* en métal déployé; platelage *m* en treillis métallique [en fers plats]
　　nl roostervloer *f* (*m*)
　　r решётчатый металлический настил *m*; металлический решётчатый пол *m*

G338 *e* **grid foundation**
　　d gitterförmige Fundamentplatte *f* [Gründungsplatte *f*]; Stahlbetonrost *m*
　　f fondation-dalle *f* en treillis, grillage *m* en béton armé
　　nl roosterfundering *f*
　　r фундамент *m* в виде решётчатой плиты; железобетонный ростверк *m*

G339 *e* **grid gas**
　　d Ferngas *n*
　　f gaz *m* transporté par gazoducs
　　nl gaslevering *f* over grote afstand
　　r газ *m* дальнего транспорта

G340 *e* **gridiron distribution system**
　　d Wasserversorgungsnetz *n* mit einseitigem Anschluß der Abzweigleitungen zur Hauptleitung

f système *m* d'adduction d'eau à embranchements unilatérals
nl waterleidingssysteem *n* met eenzijdig aan het distributienet aangesloten dienstleidingen
r сеть *f* водоснабжения с односторонним ответвлением распределительных линий от магистрали

G341 *e* **grid line**
d Rasterlinie *f*
f ligne *f* du (quadrillage) modulaire
nl rasterlijn *f* (*m*), modulaire lijn *f* (*m*)
r модульная линия *f*, линия *f* модульной сетки

G342 *e* **grid network**
d Verbundnetz *n*
f réseau *m* énergétique interconnecté, interconnexion *f*
nl koppelnet *n*, maasnet *n*
r объединённая энергосистема *f*

G343 **grid plan** *see* **reference grid**

G344 *e* **grid roller**
d Gitterwalze *f*
f rouleau *m* à grille
nl traliewals *f* (*m*)
r сетчатый дорожный каток *m*, каток *m* с сетчатыми вальцами

G345 *e* **grid structures**
d Raumfachwerke *n pl*
f structures *f pl* [constructions *f pl*] spatiales réticulées [tridimensionnelles à treillis]
nl ruimtelijke vakwerken *n pl*
r решётчатые пространственные конструкции *f pl*

G346 *e* **grillage**
d 1. Gründungsrost *m*, Kreuzwerk *n*; Schwellenfeld *n* 2. Gitter *n*
f 1. radier *m* 2. grille *f*; réseau *m*; treillis *m*
nl 1. roosterwerk *n* 2. traliewerk *n*
r 1. ростверк *m*; штвальная клетка *f* 2. решётка *f*; сетьа *f*

G347 **grillage flooring** *see* **grid floor**

G348 *e* **grillage foundation**
d Fundamentrost *m*, Gründungsrost *m*
f fondation *f* en grillage
nl roosterwerkfundering *f*
r ростверковый фундамент *m*, ростверковое основание *n*

G349 *e* **grille**
d 1. Gitter *n* (*z.B. Ziergitter*) 2. Gründungsrost *m* 3. Lüftungsgitter *n*, Luftverteilgitter *n*
f 1. grille *f*, treillis *m*, réseau *m* 2. radier *m* 3. grille *f* à air
nl 1. traliewerk *n* 2. roosterwerk *n* 3. luchtrooster *n*
r 1. решётка *f* 2. ростверк *m* 3. нерегулируемая вентиляционная [воздухораспределительная] решётка *f*

G350 *e* **grinder-finisher**
d Betondeckenschleifgerät *n*, Fußbodenpoliermaschine *f*
f polisseuse *f* [polissoir *m*] de sol, machine *f* à polir les dallages
nl vloerschuurmachine *f*
r шлифовально-отделочная машина *f*

G351 *e* **grinding**
d 1. Zermahlen *n*, Zerkleinern *n* 2. Schleifen *n* 3. Schärfen *n*
f 1. mouture *f*, broyage *m*, désagrégation *f* 2. meulage *m*; polissage *m* 3. affûtage *m* (*d'un outil*)
nl 1. malen *n* 2. slijpen *n* 3. scherpen *n*
r 1. измельчение *n*, помол *m*; дробление *n* 2. шлифовка *f* 3. (за)точка *f* (*инструмента*)

G352 *e* **grinding wheel**
d Schleifscheibe *f*
f disque *m* abrasif
nl slijpsteen *f*
r абразивный диск *m*

G353 *e* **grip**
d 1. Blechpaketdicke *f* (*bei Bolzen- oder Nietverbindung der Bleche*) 2. nutzbare Länge *f* des Bolzenschaftes 3. Kraftschluß *m* [Reibschluß *m*] zwischen den Elementen der schubsicheren Hochfestbolzenverbindung 4. Haftung *f*, Haftkraft *f* (*von Fahrzeugrädern von Straßenoberfläche*) 5. Klemme *f*; Greifer *m*; Seilklemme *f* 6. Backen *m pl* (*Schraubstock*) 7. Griff *m*
f 1. épaisseur *f* de serrage, épaisseur *f* totale des pièces assemblées par boulons *ou* rivets 2. longueur *f* du boulon en prise, longueur *f* de serrage 3. force *f* de frottement entre éléments assemblés par boulons à haute résistance 4. adhérence *f* des roues 5. pince *f*; douille *f* de serrage de câble 6. mâchoires *f pl* d'un étau 7. poignée *f* (*d'un outil*)
nl 1. pakketdikte *f* (*klinknagelverbinding*) 2. werkende lengte *f* van een bout 3. wrijvingskracht *f* tussen de platen bij een boutenverbinding 4. grip *m* (*op de weg*) 5. kabelklem *f* (*m*) 6. bek *m* (*bankschroef*) 7. handgreep *m*
r 1. толщина *f* пакета листов (*стягиваемых болтами или заклёпками*) 2. рабочая длина *f* [длина *f* рабочей части] болта 3. сила *f* трения между контактными поверхностями элементов сдвигоустойчивого соединения на высокопрочных болтах 4. сцепление *n* (*автомобильных колёс с поверхностью дороги*) 5. захват *m*; канатный сжим *m*; зажим *m* 6. губки *f pl* (*тисков*) 7. ручка *f* (*инструмента*)

G354 *e* **grip length**
  d 1. Einbindelänge *f*, Verbundlänge *f*
  2. Haftlänge *f* (*eines Bolzens*)
  f 1. longueur *f* d'adhérence 2. longueur
  f du boulon en prise, longueur *f* de
  serrage du boulon
  nl 1. aanhechtingslengte *f* (*wapening*)
  2. werkende lengte *f* van een bout
  r 1. длина *f* анкеровки арматуры
  2. зажимная длина *f* болта

G355 *e* **grit**
  d Grieß *m*, grober Sand *m*;
  Kleingestein *n*, Feinsplitt *m*
  f sable *m* gros; déchets *m pl* pierreux
  [de carrière de pierre]
  nl scherp zand *m*; split *n*
  r крупный песок *m*, гравий *m*;
  каменная мелочь *f*

G356 *e* **grit arrestor**
  d Flugaschenfänger *m*,
  Aschenabscheider *m*; Sandfang *m*
  f séparateur *m* [collecteur *m*] de
  cendres; dessableur *m*
  nl vliegasvanger *m*
  r золоуловитель *m*; песколовка *f*

G357 *e* **grit basin, grit chamber**
  d Sandfang *m*, Sandkammer *f*; Kiesfang
  m
  f dessableur *m*; piège *m* à gravier
  nl zandvang *m*
  r песколовка *f*; гравиеуловитель *m*,
  отстойник *m* для гравия

G358 *e* **grit gulley**
  d Straßenablauf *m* mit Schlammfang
  f bouche *f* d'égout à puisard de
  décantation
  nl straatkolk *m*, *f* met zandvanger
  r дождеприёмник *m* [водосточный
  колодец *m*] с отстойником

G359 *e* **gritter**
  d Splittstreuer *m*, Splittstreumaschine *f*;
  Sandstreuer *m*
  f gravillonneuse *f* mécanique;
  (r)épandeur *m* [(r)épandeuse *f*] de sable
  nl splitstrooier *m*
  r распределитель *m* высевок [щебня];
  пескоразбрасыватель *m*

G360 **gritting material** *see* **grit**
G361 *e* **groin** US
  d 1. Buhne *f*, Höfter *m* 2. Kreuzgurt *m*
  f 1. épi *m* 2. arc *m* de voûte croisée
  nl 1. krib *f* (*m*) 2. ribbe *f* (*m*)
  (*kruisgewelf*)
  r 1. полузапруда *f*, шпора *f*, буна *f*
  2. ребро *n* крестового свода

G362 *e* **groined arch**
  d Kreuzbogen *m*
  f arc *m* de voûte croisée
  nl kruisboog *m*
  r ребро *n* крестового свода

G363 *e* **groined slab**
  d Kassettendecke *f*, kreuzweise
  gerippte Decke *f*

f plancher *m* [plafond *m*] (à) caisson(s),
  plafond *m* alvéolaire
  nl caissonzoldering *f*, cassettenplafond *n*
  r кессонное перекрытие *n*

G364 *e* **groined vault**
  d Kreuzgewölbe *n*
  f voûte *f* croisée
  nl kruisgewelf *n*
  r крестовый свод *m*

G365 *e* **groining** US
  d Buhnenbau *m*
  f régularisation *f* d'un cours d'eau par
  épis
  nl riviernormalisering *f* met behulp van
  kribben
  r регулирование *n* реки с помощью
  полузапруд

G366 **groin vault** *see* **groined vault**
G367 *e* **groove**
  d 1. Hohlkehle *f*; Rille *f*; Nut(e) *f*
  2. Hohlkehle *f*, Hohlstreifen *m*
  3. Seilrille *f* 4. Ausfalzung *f*,
  Viertel *n*
  f 1. rainure *f*, encoche *f* 2. cannelure *f*
  (*d'une colonne*) 3. gorge *f* (*d'une
  poulie*) 4. feuillure *f* (*p.ex. dans un
  montant de porte*)
  nl 1. sleuf *f* (*m*) 2. cannelure *f* 3. groef
  f (*m*) op een poelie 4. sponning *f*
  (*deurkozijn*)
  r 1. выточка *f*; канавка *f*; паз *m*
  2. каннелюра *f* (*колонны*) 3. ручей *m*
  (*блока*) 4. четверть *f* (*напр.
  в стойке дверной коробки*)

G368 *e* **grooved drum**
  d Rillentrommel *f*
  f tambour *m* rainuré
  nl gegroefde trommel *f* (*m*) (*lier*)
  r барабан *m* (*лебёдки*) с канавками
  для каната

G369 *e* **grooved slab**
  d genutete Wärmedämmplatte *f*
  f dalle *f* d'isolation thermique à
  rainures
  nl gegroefde isolatieplaat *f* (*m*)
  r теплоизоляционная плита *f*
  с прорезями

G370 *e* **groove joint**
  d 1. Scheinfuge *f* 2. Stumpfnaht *f*
  f 1. joint à rainure *f* 2. soudure *f* en bout
  nl 1. schijnvoeg *f* (*m*) 2. stompe naad *m*
  r 1. пазовый шов *m*, шов *m* в форме
  канавки 2. стыковой (сварной) шов *m*

G371 *e* **groover**
  d Nutenmeißel *m*
  f rabot *m* à rainurer, bouvet *m*
  nl sponningschaaf *f* (*m*), boorschaaf *f* (*m*)
  r пазник *m*

G372 *e* **groove seam**
  d liegender Falz *m*
  f agrafure *f*
  nl liggende vouw *f* (*m*)
  r лежачий фальц *m*

G373 *e* **grooving**
 d 1. Falzen *n*, Falzformen *n*, Spunden *n*
 2. Nutung *f*; Einschneiden *n*,
 Einschnitt *m*
 f 1. agrafage *m* 2. bouvetage *m*,
 formation *f* de rainures
 nl 1. vouwen *n* 2. een sponning schaven
 r 1. фальцевание *n* 2. шпунтование *n*,
 выборка *f* пазов

G374 *e* **gross calorific value**
 d oberer Heizwert *m*
 f valeur *f* calorifique maximale [brute]
 nl ketelcapaciteit *f* (*bovenwaarde*)
 r высшая теплота *f* сгорания

G375 *e* **gross duty of water**
 d gesamte Wasserabgabe *f*
 f tâche *f* brute de l'eau d'irrigation
 nl bruto-waterafgifte *f*
 r валовая норма *f* воды, норма *f*
 воды брутто

G376 *e* **gross section**
 d Bruttoquerschnitt *m*
 f section *f* brute
 nl bruto-doorsnede *f* (*m*)
 r сечение *n* брутто

G377 *e* **gross/floor area**
 d Bruttodeckenfläche *f*,
 Bruttogeschoßfläche *f*
 f aire *f* [surface *f*] brute de plancher
 [de parquet]
 nl bruto-vloeroppervlak *n* (*van een
 gebouw*)
 r общая площадь *f* этажа здания (*по
 внутреннему периметру стен*)

G378 *e* **gross storage capacity**
 d Gesamtfassungsvermögen *n*, Brutto-
 -Fassungsvermögen *n* des Staubeckens,
 Bruttostauraum *m*
 f capacité *f* totale du réservoir, **gros**
 volume *m* de retenue
 nl totale waterberging *f* van een stuwmeer
 r полный объём *m* водохранилища,
 ёмкость *f* водохранилища брутто

G379 *e* **ground**
 d 1. Grund *m*, Boden *m*; Terrain *n*;
 Bauplatz *m* 2. Holzpfropfen *m*,
 Holzdübel *m* 3. Putzlatte *f*,
 Putzstreifen *m* 4. Erd(ungs)leiter *f*,
 Erdleitung *f*
 f 1. sol *m*; terre *f*; terrain *m*; localité *f*;
 chantier *m* 2. taquet *m* [brique *f*] de
 bois 3. cueillie *f*, cueillée *f*
 4. conducteur *m* de terre; fil de (**mise**
 à la) terre
 nl 1. grond *m* 2. spijkerklos *m* in
 metselwerk 3. pleisterlat *f* (*m*)
 4. aardleiding *f*
 r 1. грунт *m*; местность *f*; площадка *f*
 2. пробка *f* (*в кладке*)
 3. штукатурный маяк *m*
 4. заземляющая [нулевая] линия *f*
 электросети; заземляющий провод *m*

G380 *e* **ground base**
 d Bodengründung *f*

 f sol *m* de fondation, couche *f* de
 fondation en terre
 nl funderingsgrondslag *m*
 r грунтовое основание *n*

G381 *e* **ground beam**
 d 1. Fundamentbalken *m*
 2. Grundbalken *m*, Grundschwelle *f*
 f 1. poutre *f* de fondation; poutre *f*
 ceinture 2. semelle *f* de charpente
 en bois
 nl 1. funderingsbalk *m* 2. drempel *m*
 r 1. фундаментная балка *f*, рандбалка
 f 2. нижняя обвязка *f* каркаса;
 лежень *m*

G382 *e* **ground brush**
 d Rundpinsel *m*
 f pinceau *m* rond, brosse *f* ronde
 nl verfkwast *m*, rond penseel *n*
 r ручник *m* (*кисть*)

G383 *e* **ground cable**
 d 1. Erdungskabel *n* 2. Erdkabel *n*,
 Untergrundkabel *n*
 f 1. câble *m* de mise à la terre 2. câble
 m souterrain [enterré]
 nl 1. aardingskabel *m* 2. ingegraven
 kabel *m*
 r 1. заземляющий кабель *m*
 2. подземный кабель *m*

G384 *e* **ground coat**
 d Grundanstrich *m*, Grundierung *f*
 f sous-couche *f*, couche *f* de fond
 nl grondverf(laag) *f* (*m*)
 r грунт *m*, грунтовочный слой *m*
 (*окраски*)

G385 *e* **ground conditions**
 d Bodenbeschaffenheit *f*,
 Bodenverhältnisse *n pl*
 f conditions *f pl* du sol
 nl bodemgesteldheid *f*
 r грунтовые условия *n pl*

G386 *e* **ground-controlled crane**
 d kabinenloser Kran *m*
 f grue *f* commandée à distance, grue *f*
 télécommandée
 nl van de grond af bediende kraan *f* (*m*)
 r подъёмный кран *m* с дистанционным
 управлением

G387 *e* **ground floor**
 d Erdgeschoß *n*
 f rez-de-chaussée *m*; sous-sol *m* situé au
 niveau du socle
 nl begane grond *m*
 r первый *или* цокольный этаж *m*

G388 *e* **ground ice**
 d Grundeis *n*, Bodeneis *n*
 f glace *f* de fond
 nl bodemijs *n*
 r донный лёд *m*

G389 *e* **grounding**
 d Erdung *f*, Erdleitung *f*
 f mise *f* à la terre
 nl aarding *f*
 r заземление *n*

G390 *e* **grounding conductor**
 d Erdelektrode *f*, Erder *m*
 f conducteur *m* [fil *m*] de terre
 nl aardelektrode *f*
 r заземлитель *m*

G391 *e* **grounding system**
 d Erdungsanlage *f*
 f système *m* de mise à la terre
 nl aardingsstelsel *n*
 r система *f* заземления

G392 **grounding wire** *see* **grounding conductor**

G393 *e* **ground investigation**
 d Bodenerkundung *f*,
 Bodenuntersuchung *f*
 f reconnaissance *f* des terrains [des sols]
 nl bodemonderzoek *m*
 r разведка *f* грунтов

G394 *e* **ground level, ground line**
 d Erdgleiche *f*, Geländeoberfläche *f*
 f niveau *m* du terrain [du sol]
 nl maaiveld *n*
 r уровень *m* земли

G395 *e* **ground oil storage tank**
 d Übertage(-Erd)öltank *m*
 f réservoir *m* terrestre de stockage du pétrol
 nl bovengrondse olietank *m*
 r наземный нефтерезервуар *m*

G396 *e* **ground plan**
 d Grundriß *m* des Erdgeschosses;
 Grundriß *m* des Gebäudes auf der Nullhöhe
 f plan *m* de rez-de-chaussée
 nl plattegrond *m* van de benedenverdieping
 r план *m* первого этажа здания; план *m* здания на нулевой отметке

G397 *e* **ground pressure**
 d 1. Bodenpressung *f*,
 Untergrundpressung *f* 2. Erddruck *m*,
 Bodendruck *m* 3. Gebirgsdruck *m*
 f 1. pression *f* sur le sol 2. poussée *f* des terres [du sol] 3. pression *f* de terrain, poussée *f* de roches
 nl 1. gronddruk *m* op de ondergrond 2. gronddruk *m* door de omringende grond
 r 1. давление *n* на грунт 2. давление *n* грунта 3. горное давление *n*

G398 *e* **ground relief**
 d Geländerelief *n*
 f relief *m* topographique; relief *m* du terrain
 nl reliëf *n* van het terrein
 r рельеф *m* местности *или* площадки

G399 *e* **ground research**
 d Baugrundforschung *f*
 f étude *f* des sols (*pour le chantier de construction*)
 nl bodemonderzoek *n*
 r исследование *n* грунтов

G400 *e* **ground shaping**
 d Profilieren *n* des Untergrundes
 f profilage *m* du sol [du terrain]
 nl onderprofiel brengen van het terrein
 r профилирование *n* грунта

G401 *e* **ground sill**
 d 1. Grundbuhne *f* 2. Grundschwelle *f*,
 Sohlschwelle *f*, Sohlriegel *m*,
 Grundwehr *n* 3. Grundbalken *m*,
 Grundschwelle *f*
 f 1. épi *m* de fond 2. seuil *m* à la base 3. semelle *f* de charpente en bois
 nl 1. grondkrib *f* (*m*) 2. kesp *f*, sloof *f* (*m*) 3. grondbalk *m*
 r 1. донная полузапруда *f* 2. донный порог *m* 3. нижняя обвязка *f* каркаса; лежень *m*

G402 *e* **ground stabilization**
 d Baugrundverfestigung *f*
 f stabilisation *f* [consolidation *f*] du sol
 nl grondstabilisatie *f*
 r стабилизация *f* [укрепление *n*] грунта

G403 *e* **ground stability**
 d Grundbruchsicherheit *f*
 f stabilité *f* du sol
 nl stabiliteit *f* van de grond
 r устойчивость *f* грунта

G404 *e* **ground surface**
 d Geländeoberfläche *f*
 f surface *f* du terrain
 nl maaiveld *n*
 r поверхность *f* земли

G405 *e* **groundwater**
 d Grundwasser *n*
 f eau *f* souterraine, nappe *f* phréatique
 nl grondwater *n*
 r грунтовые [подземные] воды *f pl*

G40? *e* **ground water control**
 d Regelung *f* der Grundwasserverhältnisse
 f régularisation *f* des sources [des eaux] souterraines, régularisation *f* de la nappe phréatique
 nl grondwaterbeheersing *f*
 r регулирование *n* режима грунтовых вод

G408 *e* **ground water dam**
 d Grundwasserstau *m*
 unterirdische Talsperre *f*
 f barrage *m* souterrain
 nl grondwaterstuw *m*
 r плотина *f* подземных вод ·

G409 *e* **groundwater equation**
 d Grundwassergleichung *f*
 f équation *f* du bilan d'une nappe souterraine
 nl grondwatervergelijking *f*
 r уравнение *n* баланса подземных вод

G410 *e* **ground-water exploration**
 d Grundwassererschließung *f*
 f recherches *f pl* hydrogéologiques
 nl grondwateronderzoek *n*
 r гидрогеологические изыскания *n pl*

G411 *e* **groundwater extraction**
d Grundwasserentnahme *f*
f captage *m* de l'eau souterraine,
captage *m* dans la nappe phréatique
nl grondwateronttrekking *f*
r каптаж *m* грунтовых вод

G412 **groundwater increment** *see*
groundwater recharge

G413 *e* **groundwater inventory**
d Grundwasserbestandsaufnahme *f*
f bilan *m* des eaux souterraines
nl grondwaterinventarisatie *f*
r баланс *m* подземных вод

G414 **groundwater lowering** *see*
groundwater recession 2.

G415 *e* **groundwater management**
d Grundwasserwirtschaft *f*
f aménagement *m* des eaux
souterraines, aménagement *m* de la
nappe phréatique [aquifère]
nl grondwaterhuishouding *f*
r система *f* мероприятий по
сохранению и рациональному
использованию подземных водных
ресурсов

G416 *e* **groundwater recession**
d 1. Absinken *n* des
Grundwasserspiegels
2. Grundwasserabsenkung *f*;
Wasserhaltung *f*
f épuisement *m* [rabattement *m*] de la
nappe aquifère
nl daling *f* van het grondwater(peil)
r 1. истощение *n* [общее понижение *n*
уровня] подземных вод в бассейне
2. водопонижение *n*; водоотлив *m*

G417 *e* **groundwater recharge**
d Grundwasseranreicherung *f*,
Wiederauffüllung *f* des Grundwassers
f alimentation *f* d'une nappe souterraine,
emmagasinement *m* des eaux
souterraines
nl grondwateraanvulling *f*
r пополнение *n* подземного бассейна,
магазинирование *n* подземных вод

G418 *e* **groundwater runoff**
d Grundwasserabfluß *m*
f écoulement *m* souterrain
nl afvloeiing *f* van het grondwater
r сток *m* грунтовых вод, подземный
сток *m*

G419 *e* **groundwater table**
d Grundwasserspiegel *m*,
Grundwasserstand *m*
f surface *f* de la nappe phréatique,
niveau *m* phréatique
nl grondwaterstand *m*
r зеркало *n* [свободная поверхность *f*,
уровень *m*] подземных вод

G420 **groundwater tapping** *see* **captation**

G421 *e* **grout curtain**
d Einpreß-Dichtwand *f*,
Injektionsschürze *f*,
Dichtungsschleier *m*,
Injektionsschleier *m*,
Mörtelschleier *m*, Verpreßmörtelzone *f*
f rideau *m* [voile *m*] d'injection,
coupure *f* étanche injectée, écran *m*
d'étanchéité
nl dichtingsscherm *n*, injectiescherm *n*
r противофильтрационная
[цементационная, инъекционная]
завеса *f* [диафрагма *f*]

G422 *e* **grouted-aggregate concrete**
d Skelettbeton *m*, Schlämmbeton *m*,
Prepaktbeton *m*, Ausgußbeton *m*
f béton *m* prépact [prépakt]
nl injectiebeton *n*
r бетон *m* раздельной укладки

G423 *e* **grouted bolt**
d zementierter Anker *m* [Ankerbolzen *m*]
f boulon *m* de scellement [d'ancrage]
nl ankerbout *m*
r фундаментный болт *m*,
замоноличиваемый после
крепления конструкции

G424 **grouted concrete** *see* **colloidal concrete**

G425 **grouted cut-off wall** *see* **grout curtain**

G426 *e* **grouted joint**
d Vergußfuge *f*
f joint *m* rempli par coulis, joint *m*
scellé
nl te cementeren voeg *f* (*m*)
r шов *m*, заливаемый цементным
раствором

G427 *e* **grouted macadam**
d Tränkmakadam *m*, *n*, getränkte
Schotterdecke *f*
f macadam-ciment *m*; macadam-
-mortier *m*, macadam *m* bitumineux
[au bitume]
nl doordrenkte macadam *n*
r гравийное *или* щебёночное
(дорожное) покрытие, пропитанное
битумом *или* жидким цементным
раствором

G428 *e* **grouted masonry**
d Mauerwerk *n* aus Betonhohlblöcken
mit Mörtel-Hohlraumausfüllung
f maçonnerie *f* en blocs creux remplis
par mortier *ou* coulis
nl metselwerk *m* van holle betonsteen
met mortelvulling
r кладка *f* из пустотелых бетонных
блоков с заполнением пустот
раствором

G429 *e* **grouted tendon**
d verpreßtes [ausgepreßtes] Spannglied *n*
f câble *m* (d'armature) injecté
nl gecementeerde voorspankabel *m*
r напрягаемый арматурный элемент
m, уложенный в канал
заполняемый раствором

G430 *e* **grout hole**
d Mörteleinpreßloch *n*,
Mörtelinjektionsloch *n*

f trou m [orifice f] d'injection
nl cementatiegat n
r цементационная скважина f, канал m (в толще железобетонной конструкции) для инъецирования раствора

G431 e grouting cup
d Injizierglocke f
f cloche f d'injection
nl injectieklok f (m)
r воронка f для инъецирования раствора

G432 e grouting equipment
d Einpreßmörtelanlage f, Injektionsmörtelanlage f
f dispositif m d'injection de coulis
nl gereedschap n voor mortelinjectie
r оборудование n для нагнетания [инъецирования] раствора

G433 e grouting gallery
d Injektionsgang m, Einpreßstollen m, Verpreßstollen m
f galerie f d'injection
nl cementatiegalerij f
r инъекционная [цементационная] галерея f

G434 grouting hole see grout hole

G435 e grouting hose
d Einpreßschlauch m, Injektionsschlauch m
f tuyau m d'injection
nl cementinjectieslang f (m)
r шланг m для подачи [нагнетания] раствора

G436 e grouting installation
d Injektions(mörtel)anlage f, Verpreß(mörtel)anlage f, Auspreß(mörtel)anlage f
f installation f d'injection de coulis ou de mortier
nl installatie f voor mortelinjectering
r установка f для нагнетания (раствора), инъекционный аппарат m

G437 e grouting lance
d Handverpreßrohr n, Handinjektionsrohr n, Injektionsstutzen m
f tube m d'injection
nl mortelinjectielans f (m)
r инъекционная трубка f для нагнетания раствора ручным насосом

G438 e grouting machine
d Zementmörtel-Einpreßgerät n, Druckluft-Injektor m
f machine f d'injection
nl installatie f voor het injecteren van cementmortel
r установка f для нагнетания цементного раствора

G439 e grouting mortar
d Einpreßmörtel m, Injektionsmörtel m, Auspreßgut n, Verpreßgut n, Injektionsgut n, Injiziergut n

f coulis m d'injection
nl injectiemortel m
r инъекционный раствор m

G440 grout injection see cementation

G441 e grout mixer and pump
d Einpreßmörtelmisch- und -pumpanlage f
f mélangeuse-pompe f à mortier
nl gietmortelmeng- en -pompinstallatie f
r агрегат m растворомешалка--растворонасос

G442 e grout
d dünnflüssiger Mörtel m; Verpreßmörtel m, Injektionsmörtel m
f coulis m; mortier m d'injection
nl dunvloeibare mortel m; gietspecie f
r жидкий строительный раствор m; цементационный раствор m

G443 e grout pump
d Injektionspumpe f, Injizierpumpe f, Verpreßpumpe f; Zementeinspritzapparat n, Zementinjektor m
f injecteur m de coulis
nl injectiepomp f (m)
r растворонасос m для инъекционных работ

G444 e grout remixing
d Mörtelnachmischung f
f remalaxage m de coulis [de mortier]
nl herhaalde mortelmenging f
r повторное перемешивание n раствора

G446 groyne UK see groin

G447 e grubbing machine
d Rodemaschine f, Roder m
f arrache-souche m, arracheuse f, essoucheur m
nl rooimachine f, ripper m
r корчеватель m (машина)

G448 e grubbing tool
d Rodewerkzeug n
f outil m à essoucher [pour essochement]
nl rooigereedschapp n, stobbelichter m
r корчевальное орудие n, орудие n [инструмент m] для корчевания

G449 e grubbing up
d Rodung f; Geländeräumung f
f arrachage m [extraction f] des souches, essouchage m; débroussaillement m
nl van begroeiing ontdoen (bouwterrein)
r корчевание n (пней); расчистка f земельного участка от кустарника

G450 e grub saw
d Stein(hand)säge f
f scie f à main pour sciage de roche tendre (tuf, marbre)
nl handzaag f (m) voor marmer ножовка f для распиловки мягкого камня (мрамора, туфа), камнерезная пила-ножовка f

G451 *e* **guaranteed discharge**
  d niedrigster Wasserdurchfluß *m*
  f débit *m* garanti
  nl gegarandeerd debiet *n*, gegarandeerde
  waterafvoer *m*
  r гарантированный расход *m*

G452 *e* **guaranteed strength**
  d garantierte [gewährleistete]
  Festigkeit *f*
  f résistance *f* garantie
  nl gegarandeerde vastheid *f*
  r гарантируемая прочность *f*

G453 *e* **guard board**
  d Schutzbrett *n*
  f plinthe *f*
  nl kantplank *f* (*m*) (*aan steiger*)
  r защитная бортовая доска *f*
  (*у настила строительных лесов*)

G454 *e* **guard boom**
  d Flößrechen *m*, Fangrechen *m*
  f barrage *m* pour le bois flotté, radeau
  m de protection
  nl vangbalk *m* voor drijfhout
  r запань *f*

G455 **guard gate** *see* **repair gate**

G456 *e* **guard lock**
  d Schutzschleuse *f*, Sperrschleuse *f*
  f écluse *f* de sûreté
  nl zeesluis *f* (*m*)
  r заградительный шлюз *m*

G457 **guard post** *see* **bollard**

G458 *e* **guard-rail, guardrail**
  d Geländer *n*; Schutzgeländer *n*,
  Leitplanke *f*
  f garde-fou *m*, garde-corps *m*; barrière *f*
  nl leuning *f*, balustrade *f*; reling *f* (*m*)
  r перила *pl*; ограждение *n*

G459 **guard shield** *see* **protective shield**

G460 *e* **guard wall**
  d 1. Leitdalben *m* 2. Brüstungsmauer *f*
  f 1. estacade *f* de guidage 2. parapet *m*
  nl 1. remmingwerk *n* 2. borstwering *f*
  r 1. направляющий пал *m*
  2. парапетная стенка *f*

G461 *e* **guidance sign**
  d Leitzeichen *n*
  f signal *m* d'orientation [de direction]
  nl (weg)wijzer *m*, richtingbord *n*
  r указатель *m*, указательный знак *m*

G462 *e* **guide**
  d 1. Führung *f*, Führungsstück *n*
  2. Mäkler *m*, Laufrute *f*, Läuferrute *f*
  f 1. guide *m*, pièce *f* de guidage
  2. guide *m* de mouton [de battage],
  montant *m* de sonnette
  nl 1. geleider *m* 2. leipaal *m*, leider *m*
  (*heistelling*)
  r 1. направляющая *f* 2. стрела *f* копра

G463 *e* **guide peg**
  d Markierpflöckchen *n*
  f piquet *m* de contrôle en bois, jalon *m*
  en bois avec cote de nivellement

nl piket *n*
  r контрольный колышек *m* (*для
  перенесения проектной точки на
  местность*)

G464 *e* **guide pile**
  d Richtpfahl *m*, Führungspfahl *m*
  f pieu *m* de direction
  nl schoorpaal *m*, verankeringspaal *m*
  r маячная [направляющая] свая *f*

G465 *e* **guide rail**
  d Führungsschiene *f*
  f guide *m*, rail *m* de guidage
  nl dwangrail *f* (*m*), leirail *f* (*m*)
  r направляющая *f*

G467 *e* **guide roller**
  d Führungsrolle *f*, Leitrolle *f*,
  Führungswalze *f*
  f galet *m* conducteur [de guidage, de
  direction]
  nl leirol *f* (*m*)
  r направляющий ролик *m*

G468 *e* **guide specifications**
  d bautechnische Richtlinien *f pl*
  f cahier *m* des charges type, devis *m*
  descriptif type
  nl bouwtechnische richtlijnen *f* (*m*) *pl*
  r типовые технические условия *n pl*

G469 *e* **guide strip**
  d Richtleiste *f*
  f cueillie *f*
  nl richtlijst *f* (*m*)
  r маяк *m* (*для выравнивания
  поверхностей*)

G470 *e* **guide vane axial fan**
  d Axialventilator *m* mit Leitschaufeln
  f ventilateur *m* axial à aubes
  directrices
  nl axiaalventilator *m* met leischoepen
  r осевой вентилятор *m*
  с направляющим аппаратом

G471 *e* **guide wall**
  d Leitwand *f*
  f mur *m* guideau [de guidage]
  nl leimuur *m*
  r направляющая стенка *f*

G472 *e* **guiding dolphin**
  d Führungsdalbe *f*, Leitdalbe *f*
  f estacade *f* de guidage
  nl geleidewerken *n pl*
  r направляющий пал *m*

G473 *e* **guiding edge strip**
  d Leitstreifen *m* (*Fahrbahn*)
  f bande *f* du bord de chaussée
  nl kantstreep *f* (*m*) (*rijbaan*)
  r направляющая линия *f* кромки
  (*проезжей части*)

G474 *e* **guillotine shear**
  d Guillotineschere *f*
  f cisaille *f* à guillotine
  nl guillotineschaar *f* (*m*)
  r гильотинные ножницы *pl*

G475 *e* **gullet**
  d Spülwasser-Ablaufrinne *f*

f rigole f de purge
nl afvoergoot f (m), eerste smalle
ingraving f tot de volle diepte
r лоток m для сброса промывной
воды

G476 e **gulley**
d Ablauf m, Einlauf m, Rinne f,
Gully m, Regen(wasser)ablauf m,
Regen(wasser)einlauf m,
Rinneneinlaß m
f puits m d'entrée, bouche f d'égout,
déversoir m d'orage, rigole f
nl regenwaterafloop m, goot f (m)
r дождеприёмник m, водоприёмный
[ливневый] колодец m, ливнеспуск m,
ливневой водосток m

G477 e **gulley sucker**
d Schlammsauger m, Reinigungsgerät n
(*Abwasserwesen*)
f engin m de curage (*des bouches d'égout*)
nl slibzuiger m (*riool*)
r илосос m (*автоцистерна для
откачки водоприёмных колодцев*)

G478 **gully** *see* **gulley**

G479 e **gun**
d Zementkanone f, Zementspritze f;
Druckluftauspressungsgerät n
f cément-gun m, guniteuse f;
injecteur m de ciment
nl cementspuit f (m), mortelinjecteur m
r цемент-пушка f; пневмонагнетатель
m для подачи бетонной смеси

G480 e **gunite**
d Torkretbeton m, Spritzbeton m,
Gunit n
f gunite f, béton m projeté
nl spuitbeton n (*droog mengsel van zand
en cement*)
r торкрет-бетон m, шприцбетон m

G481 e **gunite work, guniting**
d Torkretieren n, Torkret-Verfahren n
f gunitage m
nl torkreteren n, betonspuiten n
r торкретирование n

G482 **guniting machine** *see* **gun**

G484 e **gusset**
d Knotenblech n
f gousset m (*d'assemblage*), plaque f de
jonction [d'assemblage]
nl koppelplaat f (m), knoopplaat f (m)
r фасонка f

G485 e **gutter**
d Dachrinne f, Wasserabflußrohr n;
Abflußgraben m, Entwässerungsrinne
f, Rinnstein m, Bordrinne f, Gosse f
f chéneau m, gouttière f, fossé m, fosse f
d'égout, caniveau m
nl dakgoot f (m); waterafvoerpijp f (m);
greppel f (m)
r водосточный жёлоб m, водосточная
труба f; сточная канава f, кювет m,
ливневый лоток m

G486 e **guy**
d Abspannung f, Abspannseil n,
Ankerseil n
f hauban m, câble m de haubanage
nl tui n, scheerlijn f (m), spandraad m
r оттяжка f, расчалка f, ванта f

G487 e **guy anchor**
d Trossenanker m
f ancre f de hauban
nl grondanker m, vast punt n voor tui
r анкер m оттяжки [ванты]

G488 e **guy derrick**
d Trossen-Derrick(kran) m, Seil-
-Derrick(kran) m
f derrick m haubané
nl getuide mastkraan f (m),
kabelderrik-kraan f (m)
r вантовый деррик-кран m, вантовый
кран m

G489 e **guying**
d Verankern n mit Drahtseilen,
Abfangen n mit Trossen,
Seilverspannung f
f haubanage m
nl tuien n
r крепление n анкерными оттяжками,
расчаливание n

G490 **guy rope** *see* **guy**

G491 e **gypsum**
d Gips m
f gypse m, plâtre m
nl gips n
r гипс m

G492 e **gypsum backing boards**
d Unterleg-Gipsputzplatten f pl (*bei
zweilagiger Verkleidung*)
f couche f de fond en placoplâtre
nl gipsplaten f (m) als ondergrond
r выравнивающие подкладочные листы
m pl сухой штукатурки (*при
двухслойной облицовке*)

G493 e **gypsum block**
d Gipsbaustein m, Gipsblock m
f parpaing m [bloc m] de plâtre
nl gipsblok n
r гипсобетонный камень m [блок m],
гипсоблок m

G494 e **gypsum board**
d Gipskartonplatte f, Gipsputzplatte f,
Trockenputzplatte f
f placoplâtre m, planche f murale au
plâtre
nl gipsplaat f (m)
r лист m сухой штукатурки

G495 e **gypsum cement**
d Estrichgips m, Anhydrit m
f plâtre m cuit à haute temperature; plâtre
m anhydre
nl vloergips n, estrichgips n
r ангидритовый цемент m, эстрихгипс
m

G496 e **gypsum concrete**
d Gipsbeton m

f béton *m* de plâtre [de gypse]
nl gipsbeton *n*
r гипсобетон *m*, гипсовый бетон *m*

G497 *e* **gypsum fiber concrete**
d Gipsfaserbeton *m*
f béton *m* fibreux de plâtre
nl gipsvezelbeton *n*
r гипсоволокнистый бетон *m*

G498 *e* **gypsum lath**
d Gipsbautafel *f*
f sappor *m* d'enduit en placoplâtre
nl gipsplaat *f* (*m*) als basis voor
   pleisterwerk
r гипсовый лист *m* (*основание под
   штукатурку*)

G499 *e* **gypsum-lath nail**
d Rohrnagel *m*, Lattennagel *m*
f clou *m* à latte, clou *m* pour planches
   de plâtre
nl plafondnagel *m*
r штукатурный гвоздь *m*

G500 *e* **gypsum-lime mortar**
d Gipskalkmörtel *m*
f mortier *m* de plâtre et de chaux
nl gipskalkmortel *m*
r гипсоизвестковый раствор *m*

G501 *e* **gypsum neat plaster**
d reiner Baugips *m* [Stuckgips *m*,
   Formgips *m*]
f plâtre *m* de construction [de gypse
   pur]
nl stukadoorgips *n*
r чистый строительный гипс *m* (*без
   добавок*)

G502 *e* **gypsum panel**
d Gipsputztafel *f*
f panneau *m* de plâtre; placoplâtre *m*
nl gips-pleisterpaneel *n*; gipsplaat *f* (*m*)
r гипсобетонная панель *f*; гипсовая
   сухая штукатурка *f*

G503 *e* **gypsum plank**
d Gipsdiele *f*
f planche *f* [plaque *f*] de plâtre
nl gipsstrook *f* (*m*)
r гипсовая доска *f* [плита *f*]

G504 *e* **gypsum plaster**
d 1. Baugips *m* 2. Gipsputz *m*
f 1. plâtre *m* de construction; plâtre *m*
   de gypse 2. enduit *m* de plâtre
nl 1. bouwgips *n* 2. gipspleister *n*
r 1. строительный гипс *m*; алебастр *m*
   2. гипсовая штукатурка *f*

G505 *e* **gypsum plasterboard**
d Gipskartonplatte *f*, Gipsputzplatte *f*
f placoplâtre *m*
nl gipsplaat *f* (*m*)
r гипсовая сухая штукатурка *f*, лист *m*
   сухой штукатурки

G506 *e* **gypsum sheathing**
d Trockenputzverkleidung *f*,
   Verkleidung *f* aus Gipsputzplatten
f revêtement *m* de fond en placoplâtres
nl bekleding *f* gipsplaten

r обшивка *f* из листов сухой
   штукатурки (*осноэание для
   облицовки стен*)

G507 **gypsum tile** *see* **gypsum block**

G508 *e* **gypsum wallboard**
d Gipswand(bau)platte *f*
f panneau *m* mural de revêtement en
   plâtre
nl gipswandplaat *f* (*m*)
r гипсовая стеновая плита *f*

G509 *e* **gyratory (cone) crusher**
d Kegelbrecher *m*, Kreiselbrecher *m*
f concasseur *m* à cône [giratoire]
nl kegel-steenbreker *m*
r конусная дробилка *f*

H

H1 *e* **habitable room**
d Wohnzimmer *n*, Wohnraum *m*
f pièce *f* habitable, chambre *f*
nl bewoonbare kamer *f* (*m*), beschikbare
   woonruimte *f*
r жилая комната *f*, жилое помещение *n*

H2 *e* **hacking**
d 1. Aufrauhen *n* (*zum Verputzen*)
   2. unregelmäßiges
   Schichtenmauerwerk *n*
f 1. hachure *f*, rayures *f* *pl* (*entailles
   pratiquées sur une surface*)
   2. maçonnerie *f* de briques en assises
   saillantes *ou* en retrait
nl 1. ruwen *n* (*vóór bepleistering*)
   2. onregelmatig metselwerk *n* van
   natuursteen
r 1. насечка *f* (*напр. бетонной
   поэерхности*) 2. кирпичная кладка *f*
   с напуском *или* отступом рядов

H3 *e* **hacking knife**
d Entkittungsmesser *n*
f couteau *m* à démastiquer
nl hakmes *n* (voor het verwijderen van
   stopverf)
r нож *m* для очистки оконных фальцев
   (*напр. от старой замазки*)

H4 *e* **hacksaw**
d Bügelsäge *f*
f scie *f* à main pour métaux
nl beugelzaag *f* (*m*), beugelzaagmachine *f*
r бугельная пила *f* (*для металла*)

H5 *e* **haft**
d Griff *m*
f manche *m*, poignée *f* (*d'un outil*)
nl handgreep *m*, greep *m*, heft *n*
r ручка *f* (*инструмента*)

H6 *e* **hair crack**
d Haarriß *m*
f fissure *f* capillaire
nl haarscheur *f* (*m*)
r волосная трещина *f*, волосовина *f*

H7   *e* haired cement mortar *UK*
    d Haarzementmörtel *m*
    f mortier *m* [enduit *m*] de ciment avec
      cheveux
    nl haarcementmortel *n*
    r цементный раствор *m* с добавкой
      волоса

H8   *e* half bat, half brick
    d halber Ziegel *m*, Halbziegel *m*,
      Zweiquartier *m*
    f demi-brique *f*
    nl halve steen *m*
    r половняк *m* (*кирпич*), полкирпича *f*

H9   *e* half-brick wall
    d halbziegelstarke Wand *f*,
      Halbsteinwand *f*
    f mur *m* demi-brique
    nl halfsteensmuur *m*
    r кирпичная стена *f* толщиной
      в полкирпича

H10 *e* half column
    d Halbsäule *f*
    f demi-poteau *m*, demi-colonne *f*
    nl halfzuil *f* (*m*)
    r полуколонна *f*

H11 *e* half-glass door
    d halbverglaste Tür *f*, Tür *f* mit
      Glasfüllung im Oberteil
    f porte *f* demi-vitrée [vitrée dans la
      partie supérieure]
    nl deur *f* (*m*) met glasbezetting in de
      bovenhelft
    r панельная дверь *f*, застеклённая
      в верхней части

H12 *e* half-landing
    d Zwischenpodest *m* (*n*), Zwischenabsatz
      m
    f palier *m* intermédiaire
    nl trapbordes *n*
    r промежуточная лестничная
      площадка *f*

H13 *e* half-lap joint
    d Überblattung *f*, Blockverband *m*
    f assemblage *m* [entaille *f*] à
      demi-bois, entablure *f*
    nl rechte (lip)las *f* (*m*), halfhoutse
      lipverbinding *f*
    r соединение *n* [врубка *f*] в полдерева

H14 *e* half-lattice girder
    d Warrenträger *m*
    f poutre *f* à treillis en V
    nl vakwerkligger *m* met horizontale
      randen en diagonale staven
    r прямоугольная ферма *f*
      с треугольной решёткой, ферма *f*
      с параллельными поясами и
      треугольной решёткой

H15 *e* half-mortise hinge
    d Aufstemm-Auflege-Türband *n*
      (*Aufstemmteil seitens des Türpfostens*)
    f charnière *f* [paumelle *f*] demi-entaillée
    nl geheng *n*
    r полунакладная дверная петля *f*

H16   halfpace *see* half-landing
H17 *e* halfpace stair
    d zweiläufige Treppe *f*
    f escalier *m* à deux volées
    nl bordestrap *m*
    r двухмаршевая лестница *f*

H18   half-space landing *see* half landing
H19 *e* halfspan loading
    d Halbfeldbelastung *f*
    f (sur)charge *f* répartie sur la demi-
      -portée [demi-poutre]
    nl verdeelde belasting *f* op halve
      overspanning
    r нагрузка *f*, распределённая на
      половине длины пролёта

H20 *e* half storey
    d Dachgeschoß *n*, Dachraum *m*
    f étage *m* des combles
    nl vliering *f*, zolderverdieping *f*
    r чердачный этаж *f*, чердачное
      помещение *n*

H21 *e* half-surface hinge
    d Auflege-Aufstemm-Türband *n*
      (*Aufstemmteil seitens des Türflügels*)
    f charnière *f* [paumelle *f*] demi-
      -encloisonnée
    nl half-ingelaten deurhengsel *n*
    r полуврезная дверная петля *f*

H22 *e* half-tide cofferdam
    d Halbtidefangdamm *m*
    f batardeau *m* de mi-marée
    nl halftij-vangdam *m*
    r низкая перемычка *f*, не защищающая
      берег от полной воды

H23 *e* half tiled wall
    d Wandfliesensockel *m*
    f socle *m* carrelé
    nl tegellambrizering *f*
    r цокольная часть *f* стены
      облицованная плиткой

H24 *e* half-track
    d Halb(gleis-)kettenschlepper *m*,
      Halbraupenschlepper *m*
    f tracteur *m* à demi-chenille
      [à semi-chenille]
    nl halfrupstrekker *m*
    r полугусеничный трактор *m* или
      тягач *m*, трактор *m* на
      полугусеничном ходу

H25 *e* hallway
    d Hausgang *m*, Hausflur *m*
    f couloir *m*, corridor *m*
    nl gang *m*
    r коридор *m*; проход *m* внутри
      здания

H26 *e* halogen leak detector
    d Halogenprüfgerät *n*
    f détecteur *m* de fuites à halogène
    nl halogeen-lekdetector *m*
    r галогенный течеискатель *m*

H27   halved joint *see* half-lap joint
H28 *e* hammer
    d Hammer *m*

f marteau *m*
nl hamer *m*
r молоток *m*

H29 *e* **hammer beam**
d Sattelholz *n*, Unterbalken *m*,
Stützkonsole *f*
f blochet *m*
nl steekbalk *m*, balksleutel *m*
r опорная консоль *f* стропильной **ноги**
(*вместо затяжки*)

H30 **hammer dressed ashlar** *see* **common ashlar**

H31 *e* **hammer-dressed stone**
d behauener Stein *m*, Werkstein *m*
f mœllon *m* smillé [taillé]
nl behakte steen *m*
r околотый [тёсаный] камень *m*

H32 *e* **hammer drill**
d Bohrhammer *m*
f marteau *m* perforateur
nl boorhamer *m*, klopboor *m*
r бурильный молоток *m*

H33 *e* **hammer grab**
d Bohrgreifer *m*
f benne *f* preneuse de forage [de puits];
grappin *m* à fonçage de puits
nl grijperboor *m*
r грейферный ковш *m* ударного
действия (*для отрывки шурфов или
колодцев*)

H34 *e* **hammer head crane, hammerhead crane**
d Hammerkopfkran *m*
f grue *f* à fléchette avec contrepoids
nl torenkraan *f* (*m*)
r стреловой кран *m* с гуськом,
оснащённым противовесом

H35 *e* **hammering**
d 1. Pfahlrammen *n* 2. Schmieden *n*
f 1. battage *m* [enfoncement *m*] des
pieux 2. martelage *m*
nl 1. heien *n* 2. smeden *n*
r 1. забивка *f* свай 2. ковка *f*

H36 *e* **hand brace**
d Handbohrmaschine *f*
f perceuse *f* à main
nl omslagboor *m*, borstboor *m*
r (ручная) дрель *f*.

H37 *e* **hand clamp**
d Schraubzwinge *f*
f presse *f* à main, serre-joint *m*;
mâchoire *f* de serrage
nl lijmknecht *m*
r ручной винтовой зажим *m*,
струбцина *f*

H38 *e* **hand distributor**
d Handspritzgerät *n*,
Handspritzmaschine *f*
f goudronneuse *f* à main
nl handspuit(machine) *f*
r ручной гудронатор *m*

H39 **hand drill** *see* **hand brace**

H40 *e* **hand finisher**
d 1. Abgleichlatte *f* 2. Handfertiger *m*

f 1. barre *f* [règle *f*] égaliseuse, barre *f*
lisseuse 2. règle *f* vibrante
nl 1. rei *f* (*m*) 2. handtrilplaat *f* (*m*)
r 1. правило *n* 2. ручная виброрейка *f*

H41 *e* **hand float**
d Reibebrett *n*, Glättkelle *f*
f taloche *f*, lisseuse *f*
nl plakspaan *f* (*m*), raapbord *n*
r ручная тёрка *f*, гладилка *f*

H42 *e* **hand held compactor**
d handgeführtes Stampfgerät *n*
f compacteur *m* commandé à la main,
dame *f* [dameuse *f*] mécanisée
commandée à main
nl met de hand geleide trilmachine *f*
r трамбовка *f* с ручным управлением

H43 *e* **hand-held roller**
d handgeführte Walze *f*
f rouleau *m* compacteur à main
nl handroller *m*
r ручной дорожный каток *m*

H44 *e* **handle**
d Griff *m*
f poignée *f*, manche *f*; levier *m*
nl handgreep *m*, greep *m*, hendel **n,**
m
r ручка *f*, рукоятка *f*; рукоять *f*

H45 *e* **handling**
d 1. Behandlung *f*, Bearbeitung *f*
2. Förderung *f* 3. Handhabung *f*,
Bedienung *f*
f 1. traitement *m*; usinage *m*;
façonnage *m* 2. transport *m*,
manutention *f* 3. manipulation *f*,
manœuvre *f*
nl 1. behandeling *f* 2. vervoer *n*;
3. hanteren *n*
r 1. обработка *f* 2. транспортирование
n, перемещение *n* 3. манипулирование
n; обслуживание *n*

H46 *e* **handling equipment**
d Hebe- und Fördermittel *n pl*
f matériel *m* de manutention et de
levage
nl hijs- en transportmiddelen *n pl*
r подъёмно-транспортное оборудование
n

H47 *e* **handling reinforcement**
d Montagebewehrung *f*
f armature *f* de montage
nl transportwapening *f*
r монтажная арматура *f*

H48 *e* **handling stresses**
d Transportbeanspruchung *f*,
Transportspannungen *f pl*
f contraintes *f pl* dues à la
manutention et au montage
nl montage- en vervoerspanningen *f pl*
r напряжения *n pl*, возникающие при
транспортировании и монтаже

H49 *e* **hand operated bar cutter**
d Betonstahlhandschneider *m*,
Betonstahlschere *f*

f tronçonneuse *f* à main, cisailles *f pl*
à barres à main
nl betonstaalschaar *f (m)*,
betonstaalknipmachine *f*
r станок *m* ручного действия для
резки арматуры

H50 *e* **handrail**
d Handlauf *m*
f main-courante *f*, barre *f* d'appui,
lisse *f*
nl leuning *f*, trapleuning *f*
r поручни *m pl*, перила *pl*

H51 *e* **hand rammer**
d Handramme *f*
f dame *f*, fouloir *m* manuel [à main]
nl handstamper *m*
r ручная трамбовка *f*

H52 *e* **handsaw**
d Handsäge *f*
f scie *f* courte [à main]
nl handzaag *f (m)*
r ножовка *f*

H53 *e* **hand sprayer**
d Handspritzgerät *n*,
Handspritzmaschine *f*
f pulvérisateur *m* à main; pistolet *m* à
peindre
nl handspuit(toestel *n*) *f (m)*
r ручной распылитель *m*, удочка *f*
(*малярная*)

H54 *e* **hand tool**
d Handwerkzeug *n*
f outil *m* à main
nl handgereedschap *n*
r ручной инструмент *m*

H55 *e* **handwork**
d Handarbeit *f*, manuelle Arbeit *f*
f travail *m* à la main, ouvrage *m*
manuel
nl handwerk *n*
r работа *f*, выполняемая вручную;
немеханизированная работа *f*

H56 *e* **hanger**
d Hänger *m*, Aufhänger *m*
f suspente *f*, barre *f* [tige *f*] de
suspension; bride *f* [étrier *m*] de
suspension
nl hanger *m*; gootbeugel *m*,
ophangconsole *f*
r подвеска *f* (*элемент крепления*);
опорный хомут *m* (*для подвешивания
балок*)

H57 *e* **hanging**
d Aufhängung *f* (*z.B. Tür- bzw.
Fensterflügel*); Befestigung *f* der
Wandverkleidung
f accrochage *m* (*d'une porte ou d'une
fenêtre*); pose *f* d'un revêtement mural
nl afhangen *n* (*deuren en ramen*);
ophangen *n*
r навеска *f* (*дверных полотен или
оконных створок*); крепление *n*
стеновой облицовки

H59 *e* **hanging gutter**
d Hängerinne *f*
f gouttière *f* pendante
nl mastgoot *f (m)*, bakgoot *f (m)*
r подвесной водосточный жёлоб *m*
(*на кровле*)

H60 *e* **hanging leaders**
d Hängemäkler *m pl*, Hänge-
-Läuferruten *f pl*, Hänge-Laufruten *f
pl*
f guides *m pl* de mouton suspendus à
la flèche d'une grue
nl (aan een kraangiek) hangende leiders
m pl van een heistelling
r направляющие *f pl* копра,
подвешиваемые к стреле крана

H61 **hanging post** *see* **gate post**

H62 *e* **hanging stile**
d Hangpfosten *m* des Türrahmens
f montant *m* [battant *m*] d'assemblage
(*d'une porte*)
nl deurpost *m* (*hangzijde*), kozijnstijl *m*
(*hangzijde*)
r навешиваемый вертикальный брус *m*
дверной рамы

H63 **hanging water** *see* **suspended water**

H64 *e* **harbour**
d Hafen *m*
f port *m*
nl zeehaven *f (m)*
r порт *m*, гавань *f*

H65 *e* **harbour basin**
d Hafenbecken *n*
f bassin *m* de port
nl havenbekken *n*
r портовый бассейн *m*

H66 *e* **harbour of refuge**
d Schutzhafen *m*, Nothafen *m*
f port *m* de refuge [d'abri]
nl vluchthaven *f (m)*
r порт-убежище *m*, отстойная гавань *f*

H67 *e* **hard asphalt**
d Hartbitumen *n*, Hartasphalt *m*
f bitume *m* dur
nl hard asfalt *n*
r твёрдый битум *m*

H68 *e* **hardboard**
d Holzfaserplatte *f*, Faserplatte *f*
f plaque *f* [dalle *f*] en fibres de bois,
plaque *f* fibreuse en bois
nl hardbord *n*
r древесноволокнистая плита *f*

H69 *e* **hard brick**
d Hart(brand)ziegel *m*
f clinker *m*, brique *f* de pavage
nl straatklinker *m*
r клинкер *m* (*кирпич*)

H70 *e* **hard compact soils**
d dichte tragfähige Böden *m pl*
f terrain *m* résistant, sols *m pl*
cohérents [compacts]
nl vaste grondslag *m*
r плотные [устойчивые] грунты *m pl*

H71   hard-burnt plaster *see* gypsum cement

H72 *e* **hardcore**
     d 1. Grobschlag *m* 2. Packlage *f*,
        Schüttpacklage *f*
     f 1. matériau *m* granulaire dur
        2. enrochement *m*, remblai *m*
        rocheux
     nl 1. gestort puin *n* 2. paklaag *f* (*m*)
     r 1. твёрдый кусковой [зернистый]
        материал *m* 2. основание *n* из
        крупного колотого камня, каменная
        наброска *f*

H73 *e* **hardcore bed**
     d verdichtete Schüttlage *f*,
        Kiespackung *f*
     f fondation *f* d'une route en granulat
        concassé
     nl verdichte spreidlaag *f* (*m*)
     r утрамбованное щебёночное *или*
        гравийное основание *n* дорожного
        покрытия

H74 *e* **hard-drawn wire reinforcement**
     d hartgezogene Drahtbewehrung *f*
     f armatures *f pl* en fil d'acier étiré à
        froid
     nl wapening *f* van koudgetrokken draad
     r проволочная арматура *f*,
        упрочнённая вытяжкой

H75 *e* **hardened cement paste**
     d Zementstein *m*, erhärteter
        Zementleim *m*
     f pâte *f* de ciment durcie
     nl verharde cementbrij *m*
     r цементный камень *m*;
        затвердевшее цементное тесто *n*

H76 *e* **hardened concrete**
     d erhärteter Beton *m*, Festbeton *m*
     f béton *m* solidifié [durci]
     nl verhard beton *n*
     r затвердевший бетон *m*

H77 *e* **hardened verge**
     d befestigter Randstreifen *m*
     f bande *f* stabilisée de l'accotement
     nl verharde berm *m*
     r укреплённая часть *f* обочины

H78 *e* **hardener**
     d Härter *m*, Härtemittel *n*,
        Härtungsmittel *n*
     f agent *m* chimique provoquant
        l'accroissement du résistance de
        béton à l'usure; durciseur *m*
        (*de peinture*), agent *m* de durcissement
        (*de peinture*)
     nl verharder *m*; verhardingsmiddel *n*
     r химическая добавка *f*,
        повышающая износостойкость
        бетона; отвердитель *m* (*краски*)

H79 *e* **hardening accelerating admixture,
        hardening accelerator**
     d Erhärtungsbeschleuniger *m*
     f accélérateur *m* de durcissement
     nl verhardingsversneller *m*
     r добавка *f*, ускоряющая твердение
        (*напр. бетонной смеси*)

H80 *e* **hard finish**
     d harte Putzdeckschicht *f*
     f couche *f* de finition dense (*de l'enduit
        au plâtre*)
     nl hard pleister *n*
     r плотный накрывочный слой *m*
        штукатурки

H81 *e* **hardness of water**
     d Wasserhärte *f*
     f dureté *f* de l'eau
     nl hardheid *f* van water
     r жёсткость *f* воды

H82 *e* **hard pan, hardpan**
     d Hartschicht *f*, Ortstein *m*;
        tragfähige Schicht *f*, tragfähiger
        Untergrund *m*
     f terrain *m* résistant, terre *f* compacte
     nl vaste steengrond *m*; harde
        draagkrachtige laag *f* (*m*)
     r твёрдый (подпочвенный) слой *m*
        (*трудный для разработки*);
        сцементированная прослойка *f*
        грунта, ортштейн *m*; грунтовое
        основание *n*, способное нести
        нагрузку *n*

H83 *e* **hard parking-place**
     d befestigte Fahrzeugparkfläche *f*
     f aire *f* de parcage [de stationnement]
        avec revêtement rigide
     nl parkeerplaats *f* (*m*) met verharde
        bedekking
     r автомобильная стоянка *f* с твёрдым
        покрытием

H84 *e* **hard pavement**
     d Hartbelag *m*, Hartdecke *f*
     f chaussée *f* rigide
     nl hard wegdek *n*
     r твёрдое (дорожное) покрытие *n*

H85 *e* **hard plaster**
     d schnellhärtender Baugips *m*,
        Formgips *m*, Stuckgips *m*,
        Hartformgips *m*, Formenhartgips *m*
     f plâtre *m* à mouler [de moulage] dur,
        plâtre *m* à prise rapide
     nl hard pleister *n*
     r быстротвердеющий строительный
        гипс *m*, формовочный гипс *m*

H86 *e* **hard rock**
     d massiger Fels *m*, Felsgründung *f*
     f rocher *m*, roc *m*; terrain *m* de
        fondation rocheux
     nl vaste rots *f* (*m*), rotsbodem *m*
     r скала *f*; скальное основание *n*

H87 *e* **hard shoulder**
     d befestigte Standspur *f*
     f accotement *m* stabilisé [traité]; zone *f*
        de stationnement
     nl verharde berm *m*; vluchtstrook *f* (*m*)
     r краевая полоса *f* (*обочины*)
        с одеждой; полоса *f* стоянки

H88 *e* **hard surfacing**
     d 1. Hartbelag *m*, Hartdecke *f*
        2. Auftragsschweißen *n*
        3. Oberflächenhärten *n*

f 1. chaussée *f* rigide 2. rechargement *m* dur 3. trempe *f* superficielle
nl 1. aanbrengen *n* van een slijtvaste laag 2. oplassen *n* 3. oppervlakteharden *n*
r 1. твёрдое дорожное покрытие *n* 2. твёрдая наплавка *f* 3. поверхностная закалка *f*

H89 *e* **hard-top**
d befestigte Straße *f*
f route *f* dure [à revêtement dur]
nl met hard wegdek; verhard (*wegdek*)
r дорога *f* с твёрдым покрытием

H90 *e* **hard-troweled surface**
d geglättete Zementmörteloberfläche *f*, mit Zementmörtel überzogener Beton *m*
f surface *f* de béton lissée à la truelle avec ciment
nl met cementpleister afgewerkt betonoppervlak *n*
r железнённая бетонная поверхность *f*

H91 *e* **hardware**
d Beschläge *m pl*, Kleineisenzeug *n*
f articles *m pl* de quincaillerie; ferronrie *f*; serrurerie *f* (*du bâtiment*)
nl hang- en sluitwerk en ijzerwaren *f* (*m*) *pl*
r строительные металлические изделия *n pl* [детали *f pl*]

H92 *e* **hardware mounting machine**
d Beschlag-Einlaßmaschine *f*
f machine *f* à encastrer *ou* entailler les articles de quincallerie
nl machine *f* voor het inkrozen van deur- en vensterbeslag
r машина *f* для врезки дверных и оконных приборов

H93 *e* **hardwood**
d Hartholz *n*, Laubholz *n*
f bois *m* dur
nl hardhout *n*
r твёрдая древесина *f*, древесина *f* твёрдых пород

H94 *e* **Hardy Cross method**
d Momentausgleichverfahren *n* nach Cross, Cross-Verfahren *n*
f méthode *f* de Cross [de la répartition des moments]
nl momentenvereffeningsmethode *f* volgens Cross (*skeletberekening*)
r метод *m* распределения моментов, метод *m* Кросса (*расчёт рам*)

H95 **harped tendon** *see* **deflected tendon**

H96 *e* **harsh mix**
d erdfeuchte [steife] Betonmischung *f*
f béton *m* sec, béton *m* de consistance de terre humide
nl aardvochtige betonmortel *m*
r жёсткая бетонная смесь *f*

H97 *e* **hasp**
d Schließband *n*, Kettel *m*, Überwurf *m*
f moraillon *m*

nl overval *m* (*in sluiting met een hangslot*)
r накладка *f* с пробоем (*для висячего замка*)

H98 **hasp-and-staple** *see* **hook-and-eye fastener**

H99 *e* **hatch**
d 1. Klappe *f*, Luke *f*, Durchreiche *f* 2. Schraffur *f*, Schraffierung *f*
f 1. trappe *f*, trou *m* d'homme 2. hachure *f* (*sur un dessin*)
nl 1. luik *n* 2. arcering *f*
r 1. люк *m*, небольшой закрываемый проём *m* (*напр. в полу*) 2. штриховка *f* (*на чертеже*)

H100 *e* **hatchet**
d Handbeil *n*
f hachette *f*, hachereau *m*, hache *f* à marteau
nl handbijl *f* (*m*)
r молоток-кирочка *m*

H101 *e* **hatchway**
d Dachausssteigluke *f*
f trappe *f* d'accès à la toiture
nl mangat *n*, dakluik *n*
r люк *m* в кровле (*для выхода на крышу*)

H102 *e* **haul**
d Förderweite *f*, Förderstrecke *f*, Transportweite *f*
f distance *f* de transport (*p.ex. de sol*); parcours *m*
nl materiaaltransport *n* uitgedrukt in tonkilometers
r расстояние *n* перевозки (*напр. грунта*)

H103 *e* **haulage, hauling**
d Förderung *f*, Transport *m*
f transport *m*, traction *f*, halage *m*
nl vervoer *n*, transport *n*
r перемещение *n*, транспортирование *n*; тяга *f*

H106 *e* **haul road**
d Baustellenzufahrt *f*, Anfahrtstraße *f*, Anfuhrweg *m*; Erdförderweg *m*, Erdbeförderungsweg *m*
f chemin *m* [route *f*] d'accès au chantier, chemin *m* de service; chemin *m* pour le transport des terres
nl toevoerweg *m*
r подъездная дорога *f* к стройплощадке; дорога *f* для транспортирования грунта

H107 *e* **haunch**
d 1. Halbbogen *m*, Bogenschenkel *m* 2. aus der Deckenebene vorspringendes Balkenunterteil *n* 3. Voute *f*, Schräge *f* 4. Randstreifen *m*, Bankett *n*
f 1. demi-arc *m* 2. partie *f* inférieure de la poutre (*en saillie au-dessous du plancher*) 3. rein *m*, aisselle *f*, esselle *f* (*d'une voûte*), gousset *m*, voûtain *m*

(*d'une poutre*) 4. bande *f* [voie *f*]
extérieure de chaussée, accotement *n*
nl 1. boogschinkel *m* 2. onder het
plafond zichtbaar gedeelte *n* van een
vloerbalk 3. onderste booghelft *f* (*m*)
r 1. полуарка *f* 2. нижняя часть *f*
балки (*выступающая ниже плиты
перекрытия*) 3. вут *m* 4. внешняя
полоса *f* проезжей части дороги

H108 *e* **haunch board**
 d Seitentafel *f* der Balkenschalung
 f panneau *n* de joue (*de coffrage*)
 nl zijvlak *n* van een balkbekisting
 r боковой [вертикальный] щит *m*
 опалубочного короба балки

H109 *e* **haunched beam**
 d Voutenbalken *m*, Voutenträger *m*
 f poutre *f* à voûtains [à aisselles]
 nl gewelfde ligger *m*
 r балка *f* с вутами

H111 *e* **hawk**
 d Fugmörtelbrett *n*
 f taloche *f*
 nl spaarbord *n*
 r сокол *m* (*штукатурный*)

H112 *e* **haydite**
 d Blähschiefer-Zusatzstoff *m*
 f granulat *m* [agrégat *m*] en schiste
 expansé
 nl geëxpandeerde leisteen *m*
 r лёгкий заполнитель *m* из
 вспученных сланцев, шунгизит *m*

H113 *e* **hazardous gas emissions**
 d Gasemissionen *f pl*
 f dégagements *m pl* de gaz nocif,
 agents *m pl* de pollution
 atmosphérique nocifs gazeux
 nl schadelijke gasemissie *f*
 r вредные газовые выбросы *m pl*

H114 *e* **hazardous materials**
 d entflammbare [explosionsfähige]
 Stoffe *m pl*
 f matériaux *m pl* inflammable
 [déflagrants, explosibles]
 nl gevaarlijke stoffen *f pl*
 r огнеопасные [взрывоопасные]
 материалы *m pl*

H115 *e* **hazardous substances**
 d Schadstoffe *m pl*
 f substances *f pl* nocives
 nl schadelijke stoffen *f* (*m*) *pl*
 r вредные вещества *n pl*

H116 *e* **haze**
 d trübe Flecken *m pl* (*auf der
 Farbschicht*)
 f taches *f pl* mates (*défaut de la
 peinture*)
 nl troebeling *f*, troebele vlekken *f* (*m*) *pl*
 (*op verflaag*)
 r жухлые пятна *n pl* (*на окрасочном
 слое*)

H117 *e* **H-beam**
 d Breitflansch-Träger *m*
 f poutre *f* à larges ailes

nl breedflensbalk *m* (*I-balk m*)
r широкополочный двутавр *m*

H118 *e* **head**
 d 1. Druckhöhe *f*, Förderhöhe *f* 2. Haupt
 n (*z.B. Dockhaupt, Schleusenhaupt*) 3.
 Fallhöhe *f*, Gefälle *n*,
 Niveauunterschied *m* 4. Dachziegel *m*
 am äußeren Dachrand 5. Rähm *m*,
 Rahmholz *n* 6. Schienenkopf *m*;
 Nietkopf *m*, Nagelkopf *m*;
 Kopfstück *n*; Stutzenkopf *m*
 f 1. chute *f*, hauteur *f* de chute, charge
 f; hauteur *f* de refoulement 2. tête *f*
 d'un ouvrage hydraulique (*p. ex. d'une
 d'écluse*) 3. hauteur *f* de chute;
 différence *f* de niveaux 4. tuile *f* de
 bordure [d'extrémité] 5. traverse *f*
 haute de bâti, poutrelle *f* de tête
 6. tête *f* (*de rivet, de clou, de poteau*);
 champignon *m*, boudin (*de rail*);
 tête *f* de poteau
 nl 1. drukhoogte *f* 2. hoofd *n* (*sluis*)
 3. valhoogte *f* 4. onderste rij *f* (*m*)
 dakpannen 5. latei *f* (*m*), bovendorpel
 m 6. kop *m*
 r 1. напор *m*; высота *f* нагнетания
 2. головная часть *f* гидросооружения
 (*напр. шлюза*) 3. высота *f* падения,
 перепад *m*; разность *f* уровней
 4. концевая черепица *f* («*половинки*»)
 5. верхний [обвязочный] брус *m*
 6. головка *f* (*напр. рельса,
 заклёпки*); шляпка *f* (*напр. гвоздя*);
 оголовок *m* (*напр. стойки*)

H119 *e* **head across the weir**
 d Stauhöhe *f*, Wehrgefälle *n*
 f hauteur *f* de retenue
 nl stuwhoogte *f*
 r напор *m* водоподпорного сооружения

H120 *e* **head bay**
 d Vorbecken *n*, Vorkammer *f*
 f avant-port *m*, bief *m* d'amont
 nl bovenhoofd *n* (*sluis*), bovenpand *n*
 r аванкамера *f* шлюза

H121 *e* **head-capacity curve**
 d Förderhöhe-Förderleistungs-Kurve *f*,
 Pumpenkennlinie *f*
 f courbe *f* caractéristique de la pompe
 (*rapport «chargedébit»*)
 nl pompkarakteristiek *f*
 r характеристика *f* насоса (*кривая
 «напор — подача»*)

H122 *e* **head casing**
 d obere Verkleidungsleiste *f*
 f chambranle *m* supérieur
 nl bovendorpel *m* (*deurkozijn*)
 r верхний наличник *m* (*оконного или
 дверного проёма*)

H123 *e* **head-control gate**
 d Einlaßschleuse *f*, Einlaßregler *m*,
 Kopfregler *m*, Ableitungskopf *m*
 f régulateur *m* de prise d'eau
 nl regelbare schuif *f* (*m*) (*in sluisdeur*)
 r головной регулятор *m*

H124 *e* **header**
   d 1. Sammelrohr *n*; Sammler *m*,
   Sammelkanal *m* 2. Hauptstrang *m*,
   Hauptleitung *f*, Hauptrohr *n*
   3. Binder(stein) *m*, Strecker *m*
   4. Sturz *m*; Fachwerkriegel *m*;
   Unterzug *m*, Dachsparrenträger *m*,
   Verteilerbalken *m*,
   Verteilungsträger *m* 5. Kopfbalken
   m, Jochschwelle *f*
   f 1. collecteur *m* 2. conduite *f*
   maîtresse [principale], grand
   collecteur *m* 3. brique *f* en boutisse,
   boutisse *f* 4. linteau *m*, traverse *f*
   de pan; solive *f* supportant des
   fermes 5. chapiteau *m* de pilotis
   nl 1. verzamelriool *n* 2. hoofdleiding *f*,
   3. kop *m* (*van een baksteen*)
   4. muurplaat *f* (*m*) 5. kesp *f* (*m*)
   r 1. коллектор *m*, гребёнка *f*;
   собиратель *m* 2. магистральный
   трубопровод *m*, магистраль *f*
   3. тычковый кирпич *m*, тычок *m*
   (*кирпичной кладки*) 4. балка-
   -перемычка *f*, ригель *m* фахверка;
   подстропильная [распределительная]
   балка *f* 5. насадка *f* свайного ряда,
   шапочный брус *m*

H125 *e* **header bond**
   d Binderverband *m*, Kopfverband *m*
   f appareil *m* en boutisses
   nl koppenverband *n* (*metselwerk*)
   r тычковая перевязка *f*
   (*кирпичной кладки*)

H126 *e* **header course**
   d Binderschicht *f*, Streckerschicht *f*
   f assise *f* de boutisse
   nl koppenlaag *f* (*m*) (*metselwerk*)
   r тычковый ряд *m* (*кладки*)

H127 *e* **header joist**
   d Wechselholz *n*, Wechselbalken *m*
   f chevêtre *f*
   nl raveelbalk *m*
   r междубалочный ригель *m*
   (балочного перекрытия)

H128 *e* **header pipe**
   d Hauptrohr *n*; Kopfrohr *n*,
   Abflußleitung *f*
   (*Nadelfilteranlage*)
   f collecteur *m* (*p. ex. de distribution*),
   conduite *f* collectrice; collecteur *m*
   d'une station de pointes filtrantes
   nl hoofdpijp *f* (*m*)
   r коллектор (*сборный или
   распределительный*); коллектор *m*
   иглофильтровой установки

H129 *e* **headgate**
   d Einlaßverschluß *m*; Einlaufbauwerk
   n; Hauptschütz *n*; Obertor *n*, oberes
   Schleusentor *n*
   f vanne *f* de tête; vanne *f* de prise;
   porte *f* d'amont
   nl deur *f* (*m*) in bovenhoofd (*sluis*)
   r головной затвор *m*; головное

сооружение *n* (*канала, водовода*);
головные [верхние] ворота *pl* (*шлюза*)

H130 *e* **heading**
   d 1. Dränagestollen *m*,
   Entwässerungsstollen *m*,
   Sickergalerie *f*, Filtergalerie *f*,
   Dränagegalerie *f* 2. Baustollen *m*
   3. Angriff *m* (*Bergbau*)
   4. Streckenvortrieb *m*, Auffahren *n*
   f 1. galerie *f* de drainage, tunnel *m*
   d'évacuation des eaux 2. galerie *f*
   de dérivation provisoire 3. cycle *m*
   de creusement 4. creusement *m*
   [percement *m*] des souterrains
   [des galeries]
   nl 1. waterafvoergalerij *f* 2. bouwtunnel
   m 3. richting *f* 4. boren *n* van een
   tunnel
   r 1. дренажная штольня *f* [галерея *f*]
   2. строительный туннель *m*
   3. проходческий цикл *m*
   4. проходка *f* горизонтальных
   выработок

H131 **heading course** *see* **header course**
H132 *e* **heading joint**
   d Stoßverbindung *f* (*Holzbau*);
   Stoßfuge *f* (*Mauerwerk*)
   f assemblage *m* par aboutement [en
   bout], assemblage *m* bord à bord;
   joint *m* vertical entre pièrres de
   maçonnerie (*dans une assise*)
   nl stuiklas *f* (*m*) (*hout*); stootvoeg *f* (*m*)
   (*metselwerk*)
   r стыковое [торцовое] соединение *n*
   (*деревянных элементов*);
   вертикальный шов *m* (*каменной
   кладки*)

H133 *e* **heading-up**
   d Stau *m*, Stauung *f*
   f remous *m*
   nl stuwing *f*
   r подпор *m*

H134 *e* **head jamb**
   d oberes Türfutter *n*, Türriegel *m*
   f traverse *f* d'huisserie
   nl bovendorpel *m* (*deurkozijn*)
   r верхний брус *m* дверной коробки

H135 **head joint** *see* **heading joint**
H136 *e* **head loss**
   d Druckhöhenverlust *m*,
   Strömungsenergieverlust *m*,
   Förderdruckverlust *m*
   f perte *f* de charge
   nl drukverlies *n*
   r потеря *f* напора

H137 *e* **head loss across the filter**
   d Druckabfall *m* [Druck(höhen)verlust
   m] am Filter
   f perte *f* de charge dans le filtre
   nl drukverlies *n* door filter
   r потеря *f* напора на фильтре

H138 *e* **head loss coefficient**
   d Druckverlustfaktor *m*,
   Druckverlustbeiwert *m*

f coefficient *m* de perte de charge
nl drukverliesfactor *m*
r коэффициент *m* потери напора [давления]

H139 *e* head loss due to friction
    d Reibungsgefälle *n*, Reibungsverlust *m*
    f perte *f* de charge due à la friction
    nl wrijvingsdrukverlies *n*
    r потери *f pl* напора на трение

H140   head of drain *see* drain head

H141 *e* head of groyne
    d Buhnenkopf *m*
    f tête *f* d'épi
    nl hoofd *n* van een krib
    r голова *f* полузапруды [шпоры, буны]

H143 *e* head piece
    d 1. Aufsatz *m*, Kappholz *n* 2. Rähm *m*
    f 1. tête *f* (*p.ex. d'un pilotis*)
    2. chapeau *m* (*d'une cloison en bois*)
    nl 1. kesp *f* (*m*) 2. bovenregel *m*
    r 1. насадка *f* (*напр. деревянной опоры моста*) 2. верхняя обвязка *f* (*деревянной перегородки*)

H144   head plate *see* wall plate

H145 *e* head pressure
    d Förderdruck *m*
    f pression *f* de refoulement
    nl stuwdruk *m*
    r давление *n* нагнетания

H146 *e* head race
    d Ober(-Trieb)wasserkanal *m*, Obergraben *m*, Zubringerkanal *m*, Werkkanal *m*, Zuleitung *f*, Werkgraben *m*
    f canal *m* [bief *m*] d'amenée, amenée *f*, canal *m* d'alimentation
    nl watertoevoerkanaal *n*
    r подводящий канал *m* ГЭС (*от регулирующего бассейна к турбине*)

H147   head regulator *see* head-control gate

H148   head reservoir *see* head tank

H149 *e* headroom
    d lichte Höhe *f* des Raumes, lichte Raumhöhe *f*
    f hauteur *f* libre sous plafond
    nl vrije hoogte *f* binnenwerks
    r высота *f* помещения (в свету)

H150 *e* head tank
    d Druckbehälter *m*
    f réservoir *m* à pression
    nl waterdrukvat *n*
    r водонапорный бак *m*

H151 *e* head tower
    d Kabelkran-Maschinenturm *m*, Antriebsturm *m*
    f support *m* [pylône *m*] à installation motrice (*d'un téléphérique*)
    nl machinentoren *m* (*kabelkraan*)
    r машинная башня *f* кабельного крана

H152 *e* head wall
    d 1. Stirnmauer *f*, Kopfwand *f* 2. Abschlußmauer *f* 3. Flügelmauer *f*

f 1. mur *m* amont 2. bajoyer *m*, mur *m* bajoyer 3. talus *m* revêtu de la culée de pont
nl 1. buitenwand *m* van het bovenhoofd (*sluis*) 2. kolkmuur *m* (*sluis*) 3. bekleed talud *n* [vleugelmuur *m*] van landhoofd
r 1. входной оголовок *m*; откосная стенка *f* водопропускной трубы *или* тоннеля 2. торцовая стенка *f* дока, замыкающая стенка *f* 3. облицованный откос *m* берегового устоя моста

H153 *e* head water, headwater
    d Oberwasser *n*
    f bief *m* amont
    nl bovenwater *n*
    r верхний бьеф *m*

H154   headwater level *see* banked-up water level

H155   headway *see* headroom

H156 *e* headworks
    d 1. Einlaufbauwerk *n*, Einlaßbauwerk *n*, Entnahmebauwerk *n* 2. Regulierungsbauwerk *n*
    f 1. aménagement *m* hydraulique de tête 2. ouvrages *m pl* de régularisation
    nl 1. bovenstrooms gedeelte *n* van een waterkrachtinstallatie 2. normalisatiewerken *n pl* in een rivier
    r 1. головные сооружения *n pl*, головной узел *m*; водозаборные сооружения *n pl* 2. регуляционные сооружения *n pl*

H158 *e* health protection
    d Arbeitshygiene *f*, Gesundheitsschutzmaßnahmen *f pl* (*im Bauwesen*)
    f hygiène *f* professionnelle; sécurité *f* et hygiène *f* sur le chantier
    nl bedrijfsgeneeskundige gezondheidszorg *f* (*m*)
    r гигиена *f* труда; меры *f pl* по защите [охране] здоровья

H159 *e* heaping of waters by wind
    d Windstau *m*
    f relèvement *m* du plan d'eau sous l'effet du vent
    nl opstuwing *f* door de wind, opwaaiing *f*
    r ветровой нагон *m*

H160 *e* hearting
    d Hintermauerung *f*, Mauerhinterfüllung *f*
    f remplissage *m*, rembourrage *m*
    nl binnenwerk *n* van een muur; metselen *n* van de achterwerkers
    r забутка *f* (*кирпичной стены*)

H161 *e* hearting concrete
    d Kernbeton *m*
    f béton *m* de remplissage
    nl betonvulling *f*

    r бетон *m* внутренней зоны; бетон *m* заполнения (*напр. пустотелой стены*)

H162 *e* **heart shake**
    d Herzriß *m*, Kernriß *m*
    f gerçure *f* radiale [de cœur] (*dans le bois*)
    nl hartscheur *f* (*m*) (*hout*)
    r радиальная трещина *f*, начинающаяся от сердцевины (*дерева*)

H163 *e* **heat-absorbing glass**
    d wärmeabsorbierendes Glas *n*, Wärmeschutzglas *n*
    f verre *f* athermane [absorbant l'infrarouge]
    nl warmteabsorberend glas *n*
    r теплопоглощающее стекло *n*

H164 *e* **heat absorption**
    d Wärmeaufnahme *f*, Wärmeabsorption *f*
    f absorption *f* de chaleur
    nl warmteabsorptie *f*
    r теплопоглощение *n*, теплоусвоение *n*

H165 *e* **heat accumulation**
    d Wärmespeicherung *f*
    f accumulation *f* de chaleur
    nl accumulatie *f* van warmte
    r аккумуляция *f* тепла

H166 *e* **heat-affected zone**
    d wärmebeeinflußte Zone *f*
    f zone *f* affectée par la chaleur, zone *f* d'influence thermique
    nl door de warmte beïnvloede zone *f* (*lassen*)
    r зона *f* термического влияния (*при сварке*)

H167   **heat-and-power supply** *see* **combined heat and power generation**

H168 *e* **heat balance**
    d Wärmebilanz *f*, Wärmehaushalt *m*
    f bilan *m* thermique
    nl warmtebalans *f* (*m*)
    r тепловой баланс *m*

H169 *e* **heat bridge**
    d Wärmebrücke *f*
    f pont *m* de chaleur
    nl warmtebrug *f* (*m*)
    r тепловой мостик *m*

H170 *e* **heat capacity**
    d 1. Wärmeleistung *f* 2. Wärmespeicherkapazität *f* 3. spezifische Wärmeaufnahmefähigkeit *f*
    f 1. puissance *f* [débit, pouvoir *m*] calorifique, rendement *m* thermique 2. capacité *f* calorifique 3. chaleur *f* spécifique
    nl 1. calorisch vermogen *n* 2. warmte-opslagcapaciteit *f* 3. soortelijke warmte *f*
    r 1. теплопроизводительность *f* 2. теплоаккумулирующая ёмкость *f* 3. удельная теплоёмкость *f*

H171 *e* **heat comfort**
    d Wärmebehaglichkeit *f*, Wärmekomfort *m*
    f confort *m* thermique
    nl behaaglijkheid *f* tengevolge van de verwarming
    r тепловой комфорт *m*

H172 *e* **heat consumption**
    d Wärmeverbrauch *m*
    f consommation *f* de chaleur
    nl warmteverbruik *n*
    r теплопотребление *n*

H173 *e* **heat cure period**
    d Warmbehandlungsdauer *f*
    f durée *f* d'étuvage par vapeur, durée *f* de cure [de traitement] à vapeur
    nl warmtebehandelingsduur *m* (*beton*)
    r длительность *f* цикла пропаривания [гидротермальной обработки]

H174 *e* **heat demand**
    d Wärmebedarf *m*
    f charge *f* de chaleur [thermique, calorifuge]
    nl warmtebehoefte *f*
    r требуемое количество *n* тепла, тепловая нагрузка *f*

H175 *e* **heat dissipation**
    d Wärmestreuung *f*
    f dissipation *f* de chaleur
    nl warmteverlies *n*
    r рассеяние *n* тепла

H177 *e* **heat emission**
    d Wärmeabgabe *f*, Wärme(ab)strahlung *f*
    f émission *f* de chaleur
    nl warmteafgifte *f*
    r теплоотдача *f*, тепловыделение *n*, теплоизлучение *n*

H178 *e* **heat engineering**
    d Wärmetechnik *f*
    f technique *f* de la chaleur
    nl warmtetechniek *f*
    r теплотехника *f*

H179   **heater** *see* **heating appliance**

H180 *e* **heater planer**
    d Heiß-Planiermaschine *f*
    f aplanisseuse *f* [aplanisseur *m*] à plaque chauffante (*pour les chaussées en matériaux enrobés*)
    nl hitte-planeermachine *f*
    r ремонтер *m* (*дорожная машина*)

H182 *e* **heat exchanger**
    d Wärme(aus)tauscher *m*, Wärmeübertrager *m*
    f échangeur *m* thermique [de chaleur]
    nl warmtewisselaar *m*
    r теплообменник *m*

H183 *e* **heat exchange surface**
    d Wärmeübertragungsfläche *f*
    f surface *f* d'échange de chaleur
    nl warmte-overdrachtsoppervlak *n*
    r поверхность *f* теплообмена

H184 *e* **heat flow meter**
  d Wärmestrommesser *m*
  f fluxmètre *m* thermique
  nl warmtestromingsmeter *m*
  r прибор *m* для измерения теплового потока

H185 *e* **heat gain**
  d Wärmegewinn *m*, Wärmeanfall *m*, Wärmeaufnahme *f*
  f gain *m* de chaleur
  nl warmte-opname *f*
  r теплопоступление *n*

H186 *e* **heat generator**
  d Wärmeerzeuger *m*
  f générateur *m* de chaleur
  nl warmtegenerator *m*
  r теплогенератор *m*

H187 *e* **heating appliance**
  d Heizungsanlage *f*, Heizeinrichtung *f*; Heizgerät *n*, Heizkörper *m*
  f installation *f* de chauffage
  nl verwarmingstoestel *n*
  r отопительная установка *f*, отопительный агрегат *m*; отопительный прибор *m*

H188 *e* **heating boiler**
  d Heizkessel *m*
  f chaudière *f* de chauffage
  nl verwarmingsketel *m*
  r отопительный котёл *m*

H189 *e* **heating cable**
  d Heizkabel *n*
  f câble *m* de chauffage électrique
  nl elektrische verwarmingskabel *m*
  r греющий электрокабель *m*

H190 *e* **heating capacity**
  d Heizfähigkeit *f*, Wärmeabgabe *f*
  f capacité *f* de chauffage, émission *f* de chaleur
  nl verwarmingscapaciteit *f*
  r теплоотдача *f* (*отопительного прибора*)

H191 *e* **heating circuit**
  d Heizkreis *m*
  f circuit *m* de chauffage
  nl verwarmingsringleiding *f*
  r циркуляционное кольцо *n* системы отопления

H192 *e* **heating coil**
  d Heizschlange *f*; Lufterhitzer *m*
  f serpentin *m* de chauffage; réchauffeur *m* d'air, calorifère *m*
  nl verwarmingsspiraal *f* (*m*) (*luchtverwarming*)
  r обогревающий змеевик *m*; воздухонагреватель *m*, калорифер *m*

H193 *e* **heating demand**
  d Heizlast *f*
  f bésoin *m* en chaleur
  nl warmtebehoefte *f* (*van een gebouw*)
  r отопительная нагрузка *f* (*здания*)

H194 *e* **heating duct**
  d Heizungskanal *m*
  f galerie *f* visitable *ou* caniveau *m* de la canalisation de chauffage urbain
  nl verwarmingskanaal *n*
  r канал *m* теплотрассы

H195 **heating installation** *see* **heating system**

H196 *e* **heating line**
  d Heizleitung *f*
  f canalisation *f* de distribution de chaleur
  nl verwarmingsleiding *f*
  r теплопровод *m*

H197 **heating load** *see* **heat load**

H198 *e* **heating panel**
  d Heizplatte *f*, Heiztafel *f*
  f panneau *m* radiant [rayonnant]
  nl paneelradiator *m*
  r отопительная панель *f*

H199 **heating period** *see* **heating season**

H200 *e* **heating riser**
  d Heizstrang *m*
  f colonne *f* verticale de chauffage centrale
  nl stijgleiding *f* (*in een centrale--verwarmingssysteem*)
  r отопительный стояк *m*

H201 *e* **heating season**
  d Heizungsperiode *f*, Heizsaison *f*
  f période *f* de chauffage
  nl stookseizoen *n*
  r отопительный период *m*

H202 *e* **heating surface**
  d Heizfläche *f*
  f surface *f* de chauffe
  nl verwarmend oppervlak *n*
  r поверхность *f* нагрева

H203 *e* **heating system**
  d Heizsystem *n*, Heiz(ungs)anlage *f*
  f installation *f* de chauffe
  nl verwarmingssysteem *n*
  r система *f* отопления

H204 *e* **heating tape**
  d Heizband *n*
  f ruban *m* chauffant
  nl platte verwarmingskabel *m*
  r ленточный электронагревательный элемент *m* (*для обогрева трубопроводов*)

H205 **heating unit** *see* **heating appliance**

H206 *e* **heating-up time**
  d Aufheizperiode *f*, Anheizdauer *f*
  f temps *m* de préchauffage
  nl opwarmtijd *m*
  r время *f* разогрева отопительной системы

H207 *e* **heating water**
  d Heizwasser *n*, Kreislaufwasser *n*
  f eau *f* chaude de circulation
  nl heetwater *n*
  r сетевая вода *f* (*в отопительной сети*)

H208 *e* **heating water pump**
  d Heizwasserpumpe *f*

f pompe f à eau chaude de circulation
nl circulatiepomp f (m)
r насос m сетевой воды

H209 e heat input
d Wärmezufuhr f, Wärmezufluß m,
Wärmeeinfall m
f apport m [amenée f] de chaleur
nl warmtetoevoer m
r теплоприток m

H210 e heat-insulating concrete
d Wärmedämmbeton m
f béton m pour isolation thermique,
béton m isolant
nl warmte-isolerend beton n
r теплоизоляционный бетон m

H211 e heat insulation
d Wärmedämmung f
f isolation f thermique;
calorifugeage m; isolement m
thermique
nl warmte-isolatie f
r теплоизоляция f

H212 e heat insulation work
d Wärmeisolierungsarbeiten f pl
f travaux m pl d'isolation thermique
nl warmte-isolatiewerk n
r теплоизоляционные работы f pl

H213 e heat jacket
d Heizmantel m
f chemise f de réchauffage
nl verwarmingsmantel m
r обогревающая рубашка f

H214 e heat load
d Wärmelast f; Heizlast f
f charge f de chaleur [thermique,
calorifique]; besoin m de chaleur
nl warmtebehoefte f
r тепловая нагрузка; отопительная
нагрузка f

H215 e heat loss
d Wärmeverluste m pl
f pertes f pl thermiques [de chaleur]
nl warmteverliezen n pl
r теплопотери f pl

H216 e heat meter
d Wärme(mengen)messer m,
Wärmezähler m
f compteur m (de la quantité) de
chaleur
nl warmtemeter m
r тепломер m

H217 e heat of hydration
d Hydratationswärme f
f chaleur f d'hydratation
nl hydratatiewarmte f
r теплота f гидратации

H218 e heat output
d Wärmeabgabe f, Wärmeleistung f
f puissance f calorifique
nl warmteproductie f
r теплоотдача f;
теплопроизводительность f,
тепловая мощность f

H219 e heat pipe
d Wärmerohr n
f tube m [tuyau m] calorifique [de
chauffe, de réchauffage]
nl verwarmingspijp f (m)
r тепловая труба f

H220 e heat pipeline
d Fernheizleitung f, Fernheiztrasse f
f tuyauterie f de chauffage
nl stadsverwarmingsleiding f
r теплотрасса f

H221 e heat pollution
d thermische Umweltbelastung f
f pollution f thermique
nl thermische milieubelasting f
r термальное загрязнение n среды

H222 e heat pump
d Wärmepumpe f
f pompe f de chaleur
nl warmtepomp f
r тепловой насос m

H223 e heat radiation
d Wärmestrahlung f
f thermorayonnance f, rayonnement m
de chaleur, radiation f thermique
nl warmtestraling f
r теплоизлучение n, тепловая радиация
f

H224 heat reclaim, heat reclamation see
heat recovery

H225 e heat reclaim unit
d Wärmerückgewinnungsanlage f,
Abwärmeverwertungsanlage f
f récupérateur m de chaleur (perdue);
économiseur m
nl warmte-recuperator m
r утилизатор m тепла,
теплоутилизационная установка f

H226 e heat recovery
d Wärmeregenerierung f,
Wärmerückgewinnung f
f récupération f de la chaleur
nl warmte-recuperatie f
r регенерация f [утилизация f] тепла

H227 heat regenerator see heat reclaim unit

H229 e heat release
d Wärmeabgabe f
f émission f calorifique [de chaleur],
dégagement m de chaleur
nl warmteafgifte f
r теплоотдача f, выделение n тепла

H230 heat requirement see heat demand

H231 e heat resistance
d Hitzebeständigkeit f
f résistance f aux actions de hautes
températures, résistance f
chauffante; réfractairité f
nl chittebestendigheid f
r жаростойкость f

H232 e heat-resistant concrete
d hitzebeständiger Beton m
f béton m réfractaire

nl hittebestendig beton n
r жаростойкий бетон m

H234 e heat source
d Wärmequelle f
f source f calorifique
nl warmtebron f (m)
r источник m тепла [тепловыделений]

H235 e heat storage
d 1. Wärmespeicher m 2.
Wärmespeicherung f
f 1. accumulateur m de chaleur
2. accumulation f de chaleur
nl 1. warmteaccumulator m 2.
accumulatie f van warmte
r 1. теплоаккумулятор m
2. аккумуляция f тепла

H236 e heat supply
d Wärmezufuhr f; Wärmeversorgung f
f alimentation f en chaleur; apport m
thermique [de chaleur]
nl warmtetoevoer m; verwarming f
r подвод m тепла; теплоснабжение n

H237 e heat supply main
d Fernheizleitung f
f ligne f de chauffage urbain
nl hoofdverwarmingsleiding f
r магистральный теплопровод m

H238 e heat supply source
d Wärmeversorgungsquelle f
f source f d'alimentation en énergie
thermique
nl verwarmingsbron f (m)
r источник m теплоснабжения

H239 e heat supply system
d Fernwärmeversorgungssystem n,
Fernheizungsanlage f
f système m d'alimentation en énergie
thermique
nl wijkverwarmingssysteem n
r система f теплоснабжения

H240 heat transfer see heat transmission

H241 e heat transfer area
d Wärmeaustauschfläche f
f surface f d'échange thermique
[de chaleur]
nl warmte-overdragend oppervlak n
r поверхность f теплообмена

H242 e heat transfer fluid US, heat transfer
medium
d Wärmeübertragungsmedium n
f milieu m d'échange calorifique
nl warmtetransportmiddel n
r теплопередающая среда f;
теплохладоноситель m

H243 e heat transmission
d Wärmeübertragung f
f transmission f de chaleur
nl warmte-overdracht f (m)
r теплопередача f

H244 e heat transmission by conductivity
d Transmissionswärmeaustausch m
f transmission f de la chaleur par
conductibilité

nl warmteoverdracht f (m) door geleiding
r кондуктивная теплопередача f

H245 e heat transmission by convection
d Konvektionswärmeaustausch m
f transmission f de la chaleur par
convection
nl warmte-overdracht f (m) door
convectie
r конвективная теплопередача f

H246 e heat transmission coefficient
d Wärmeübergangszahl f,
Wärmedurchgangszahl f, K-wert m
f coefficient m de transmission
[de transfer] de chaleur
nl warmtedoorlaatcoëfficiënt m
r коэффициент m теплопередачи

H247 e heat treatment
d Wärmebehandlung f
f traitement m thermique
nl warmtebehandeling f, thermische
behandeling f
r тепловая [термическая] обработка f,
термообработка f

H248 heat utilization see heat recovery

H249 heat waste see heat loss

H250 e heat wheel
d Wärmerad n, Regenerativ-
-Energieübertrager m
f échangeur m de chaleur tournant
nl roterende warmtewisselaar m
r вращающийся (регенеративный)
теплообменник m

H251 heave lift see heaving

H252 e heave-off hinge
d Aufsatzband n
f charnière f à fiche mobile, paumelle f
nl paumelle f, fits
r разъёмная дверная петля f

H253 e heaving
d Hebung f
f soulèvement m, gonflement m (du sol)
nl opkomen n van de grond
r пучение n, выпор m (грунта)

H254 heaving soil see expansive soil

H255 e heavy abutment
d massives Uferwiderlager n
f culée f massive
nl massief landhoofd n
r массивная береговая опора f (моста)

H256 e heavy-bodied paint
d zähflüssige Farbe f
f peinture f à viscosité élevée
nl lijvige verf f (m)
r вязкая краска f, краска f вязко-
-текучей консистенции

H257 e heavy concrete
d Schwerstbeton m
f béton m très lourd
nl beton n met zware (metaal)toeslag
r особо тяжёлый бетон m

H258 e heavy-duty scaffold
d hochbelastbares Gerüst n

    f échafaudage *m* d'une grande portance
  nl zware steiger *m*
   r леса *pl* [подмости *pl*] большой несущей способности

H259 *e* **heavy-edge reinforcement**
   d Bewehrungsmatte *f* mit verstärkten Randstäben (*Betonfahrbahndecke*)
   f armature *f* en treillis avec barres extérieures grosses
  nl wapeningsnet *n* met verzwaarde randstaven
   r арматурная сетка *f* с усиленными краевыми стержнями

H260 *e* **heavy equipment transporter**
   d Schwerlasttransporter *m*
   f camion *m* gros porteur
  nl vrachtauto *m* voor zwaar vervoer
   r автомобиль-тяжеловоз *m*; транспортное средство *n* для перевозки тяжеловесных грузов

H261 *e* **heavy grading**
   d umfangreiche Erdarbeiten *f pl*
   f terrassements *m pl* de cubage considérable
  nl omvangrijk grondverzet *n* (*egalisatie*)
   r земляные работы *f pl* значительного объёма

H262 *e* **heavy soil**
   d schwerer Boden *m*
   f sol *m* compact (à grains fins)
  nl zware grond *m*
   r плотный (мелкозернистый) грунт *m*

H263 *e* **heavyweight aggregate**
   d Schwerzuschlag(stoff) *m*
   f granulat *m* [agrégat *m*] lourd
  nl zware toeslag *m*
   r тяжёлый заполнитель *m* (*из плотных скальных пород или стали*)

H264   **heavyweight concrete** *see* **heavy concrete**

H265 *e* **heck**
   d 1. zweiteilige Tür *f* mit vertikal angeordneten Flügeln 2. Tür *f* mit einer Öffnung *oder* einem Gitter im oberen Türfüllungsbereich 3. Gittertor *n*
   f 1. porte *f* sectionnée horizontalement en deux battants 2. porte *f* à panneau supérieur à clairevoie *ou* grillé 3. porte *f* cochère grillée
  nl 1. onder- en bovendeur *f* (*m*) 2. deur *f* (*m*) met een (getralied) raampje in het bovenste paneel 3. traliedeur *f* (*m*)
   r 1. дверь *f*, разделённая по горизонтали на две створки 2. дверь *f* с верхней незаполненной *или* решётчатой панелью 3. решётчатые ворота *pl*

H266 *e* **hedge**
   d Hecke *f*
   f haie *f* (vive)
  nl heg *f* (*m*), haag *f*
   r зелёная изгородь *f*

H267 *e* **heel**
   d 1. Ferse *f*, oberwasserseitiger Fuß *m*, Druckbank *f* 2. Auslegeranlenkpunkt *m*; Stützenfuß *m*
   f 1. pied *m* amont 2. pied *m* (*de flèche, de poteau*)
  nl 1. voet *m* aan het bovenhoofd van een sluis 2. ondereinde *n* van een stijl
   r 1. подошва *f* верхового откоса плотины 2. опорная пята *f* (*стрелы, стойки*)

H269 *e* **heightening**
   d Aufhöhung *f*; Aufstockung *f*
   f surélévation *f* (*p. ex. du barrage*); exhaussement *f* (*p. ex. d'une maison*)
  nl ophoging *f* (*dijk*); verhoging *f* (*gebouw*)
   r наращивание *n* (*плотины, дамбы*); надстраивание *n*, надстройка *f* (*здания*)

H270 *e* **height of arc**
   d Pfeilhöhe *f*, Stich *m* (*Bogen*)
   f flèche *f* de l'arc
  nl porring *f*, pijl *m* (*gewelf*)
   r стрела *f* [высота *f* подъёма] арки

H271 *e* **height of embankment**
   d Dammhöhe *f*
   f hauteur *f* de remblai
  nl dijkhoogte *f*
   r высота *f* насыпи

H272 *e* **height of face**
   d Abbauhöhe *f*, Orthöhe *f*
   f hauteur *f* du front de taille [d'attaque]
  nl fronthoogte *f* (*steengroeve*)
   r высота *f* забоя (*в карьере*)

H273 *e* **height of lands**
   d Wasserscheide(linie) *f*
   f partage *m* des eaux, ligne *f* de partage des eaux
  nl waterscheiding *f*
   r водораздел *m*; линия *f* водораздела

H274 *e* **height of water**
   d Wasserstand *m*, Wasserhöhe *f*; Druckhöhe *f* [Förderhöhe *f*] des Wassers
   f niveau *m* [hauteur *m*] de l'eau
  nl waterstand *m*, drukhoogte *f*
   r уровень *m* [высота *f* подъёма] воды

H275 *e* **height under hook**
   d Haken(hub)höhe *f*
   f hauteur *f* de levage du crochet, course *f* haute du crochet
  nl hefhoogte *f* van een kraan
   r высота *f* подъёма крюка

H276   **held water** *see* **capillary water**

H277 *e* **heliarc welding**
   d Heliarcschweißen *n*
   f soudage *m* (à l'arc) en atmosphère d'hélium
  nl heliumbooglassen *n*
   r электродуговая сварка *f* в защитной среде гелия

H278   **helical binding** *see* **helical reinforcement**

H279 e helical cage
    d Spiralbewehrungskorb *m*
    f cage *f* d'armature [de ferraillage] en spirale
    nl spiraalwapeningskorf *m*
    r спиральный арматурный каркас *m*

H280 e helical conveyor
    d Rohrschneckenförderer *m*
    f vis *f* [hélice *f*] transporteuse, transporteur *m* [convoyeur *m*] à vis
    nl wormtransporteur *m*
    r шнек *m*, винтовой конвейер *m*

H281 e helical reinforcement
    d Spiralbewehrung *f*
    f armature *f* hélicoïdale [spirale]
    nl spiraalwapening *f*
    r спиральная арматура *f*

H282 e helical strake
    d wendelartige Flachstahlrippe *f*
    f arête *f* hélicoïdale en bande d'acier
    nl spiraalvormige bandstaalribbe *f* (*m*)
    r спиральное ребро *n* из полосовой стали

H284 e hemihydrate plaster
    d Gips-Halbhydrat *n*, Halbhydratbinder *m*
    f plâtre *m* de construction semi--hydraté
    nl gips *n*, pleister *n*
    r полуводный [полугидратный] строительный гипс *m*

H285 e hemp
    d Hanf *m*
    f chanvre *m*
    nl hennep *m*
    r пенька *f*

H286 herkinghead *see* hipped gable

H287 e herring-bone strutting
    d Kreuzaussteifung *f*
    f croisillons *m pl*, entretoisement *m*
    nl kruisverstijving *f*
    r система *f* перекрёстных связей

H288 e herringbone drain system
    d Fischgrätensystem *n*, Fischgrätendränage *f*
    f système *m* en arête de poisson
    nl drainage *f* in visgraatpatroon
    r система *f* дренажа с расположением дрен «ёлочкой»

H289 e hessian
    d Sackleinen *n*, Hessian *n*
    f toile *f* à sac
    nl zakkengoed *n*
    r мешочная ткань *f*

H290 hick joint *see* flat joint

H291 e hidden wiring
    d verdeckt verlegte Energieleitung *f*, Unterputzinstallationsleitung *f*
    f canalisation *f* électrique dissimulée, fils *m pl* conducteurs installés dans les murs
    nl elektrische leiding *f* uit zicht
    r скрытая электропроводка *f*

H292 hide glue *see* animal glue

H293 e hiding power
    d Deckkraft *f*, Deckvermögen *n*, Deckfähigkeit *f*
    f pouvoir *m* couvrant (*de peinture*)
    nl dekkend vermogen *n* (*verf*)
    r кроющая способность *f* (*краски*)

H294 e high air
    d Druckluft *f*
    f air *m* comprimé
    nl perslucht *f* (*m*)
    r сжатый воздух *m*

H295 e high-alumina cement
    d Tonerde(schmelz)zement *m*, Aluminozement *m*
    f ciment *m* alumineux [fondu]
    nl aluminiumcement *n*
    r глинозёмистый цемент *m*

H296 high-bond bars *see* deformed bars

H297 high capacity trickling filter *see* high rate trickling filter

H298 high-density concrete *see* heavy concrete

H299 e high-density plywood
    d Preßsperrholz *n*, verdichtetes Sperrholz *n*
    f contre-plaqué *m* à haute densité
    nl multiplex *n* van hoge persing
    r высокопрочная фанера *f*, фанера *f* высокой плотности и прочности

H300 high-discharge mixer *see* inclined-axis mixer

H301 e high-early-strength cement
    d frühhochfester Zement *m*, Schnellerhärter *m*
    f ciment *m* à durcissement rapide
    nl snelverhardend cement *n*
    r быстротвердеющий цемент *m*; цемент *m*, обеспечивающий быстрое нарастание прочности бетона

H302 e high-early-strength concrete
    d frühtragfester Beton *m*
    f béton *m* à durcissement rapide
    nl snelverhardend beton *n*
    r быстротвердеющий бетон *m*

H303 e high efficiency air filter
    d absolutes Filter *n*, HEPA-Filter *n*, Hochleistungsfilter *n*
    f filtre *m* absolu
    nl volledig werkend luchtfilter *n*, *m*
    r абсолютный фильтр *m*

H304 e higher critical velocity
    d obere kritische Geschwindigkeit *f*
    f vitesse *f* critique supérieure (*du cours d'eau*)
    nl bovenste kritische snelheid *f*
    r верхняя критическая скорость *f* (*потока*)

H305 e higher high water
    d hohes Tidehochwasser *n*
    f pleine-mer *f* supérieure

nl springvloed *m*, hoog springtij *n*
r наивысший уровень *m* полной воды
H306 *e* **higher low water**
d hohes Tideniedrigwasser *n*
f basse-mar *f* supérieure
nl laag springtij *n*
r наивысший уровень *m* малой воды
H307 *e* **highest tailwater level**
d höchster Unterwasserstand *m*
f niveau *m* maxi d'eau aval
nl hoogste waterstand *m* beneden de sluis
r наивысший уровень *m* нижнего бьефа
H308 *e* **highest water storage level**
d Stauziel *n*, maximaler Stauspiegel *m*, Höchststau *m*
f retenue *f* maximale
nl hoogste stuwpeil *n*
r максимальный подпорный уровень *m*
H309 *e* **high-frequency finisher**
d Hochfrequenzfertiger *m*
f finisseur *m* vibrant de haute fréquence
nl afwerkmachine *f* met hoogfrequente trilbalk
r бетоноукладчик *m* с высокой частотой колебаний уплотняющего бруса
H310 *e* **high-frequency vibrator**
d Hochfrequenzvibrator *m*
f vibrateur *m* de haute fréquence
nl hoogfrequente vibrator *m*
r высокочастотный вибратор *m*
H311 *e* **high-frequency vibratory compactor**
d Hochfrequenz-Vibrationsverdichter *m*
f compacteur *m* à plaque vibrante de haute fréquence
nl trilwals *f* (*m*) met hoge frequentie
r высокочастотный виброуплотнитель *m*
H312 *e* **high-head dam**
d Hochdruckstaudamm *m*
f barrage *m* à haute chute
nl stuwdam *m* met grote stuwhoogte
r высоконапорная плотина *f*
H313 *e* **high-head development, high-head water power plant**
d Hochdruck-Wasserkraftwerk *n*
f usine *f* à haute chute
nl hoge-drukwaterkrachtcentrale *f* (*m*)
r высоконапорная ГЭС *f*
H314 *e* **high-joint pointing**
d Ausfugen *n* [Fugenausfüllung *f*] mit Überhöhung
f jointoiement *m* saillant [en saillie]
nl voegen *n* in knipwerk
r выпуклая расшивка *f* швов (кладки)
H315 *e* **high-level bridge**
d Hochbrücke *f*
f pont *m* au-dessus d'un fleuve navigable
nl brug *f* (*m*) met grote doorvaarthoogte
r высоководный мост *m*

H316 *e* **high level cistern**
d Hochspülkasten *m*
f réservoir *m* de chasse haut
nl hoog opgehangen stortbak *m*
r высокорасположенный смывной бачок *m*
H317 *e* **high-lift grouting**
d Wandhohlraumausfüllung *f* (*mit Mörtel*) in geschoßhohen Lagen
f remplissage *m* des murs creux (*avec du coulis*) par étages
nl mortelinjectie *f* in holle muren over de volle etagehoogte
r заполнение *n* (*раствором*) пустотелой стены на высоту этажа
H318 *e* **high localized stresses**
d Kerbspannungen *f pl* (*örtliche Spannungskonzentration*)
f contraintes *f pl* locales à hautes concentrations
nl spanningsconcentratie *f*
r напряжения *n pl* высокой концентрации
H319 *e* **high-melting point asphalt**
d hochschmelzendes Bitumen *n*, Hartbitumen *n*
f bitume *m* à haut point de fusion
nl bitumen *n* met hoog smeltpunt
r битум *m* с высокой температурой размягчения
H320 *e* **high modulus polymer fibers**
d hochelastische Polymerfibern *f pl*
f fibres *f pl* en polymères à haut module d'élasticité
nl sterk elastische polymeervezels *f* (*m*) *pl*
r высокоупругие полимерные волокна *n pl*
H321 *e* **high pressure air-conditioning system**
d Hochdruckklimaanlage *f*
f système *m* de climatisation [de conditionnement d'air] à haute pression
nl klimaatregelingsinstallatie *f* met hoge druk
r система *f* кондиционирования воздуха высокого давления
H322 *e* **high-pressure gas pipeline**
d Hochdruckgasleitung *f*
f conduite *f* à gaz à haute pression
nl gasleiding *f* met hoge druk
r газопровод *m* высокого давления
H323 *e* **high-pressure line**
d Hochdruckleitung *f*
f conduite *f* [canalisation *f*, ligne *f*] haute pression
nl hoge-drukleiding *f*
r трубопровод *m* высокого давления
H324 **high pressure steam curing** *see* **autoclaving**
H325 *e* **high pressure terminal unit**
d Hochdruckklimagerät *n*
f unité *f* de climatisation terminale, dispositif *m* de conditionnement d'air terminal

nl luchtbehandelingstoestel *n* met **hoge druk**

r доводчик *m* высокого давления

H326 *e* **high-production loader**
    d Hochleistungslader *m*
    f chargeur *m* à grand rendement
    nl laadmachine *f* met groot vermogen
    r высокопроизводительный погрузчик *m*

H327 *e* **high-rate biofilter**
    d Hochleistungs-Tropfkörper *m*
    f lit *m* bactérien à haute puissance [à haut rendement]
    nl snelwerkend bacteriologisch filter *n, m*
    r высоконагружаемый биофильтр *m*

H328 *e* **high-rate sand filter**
    d Hochleistungssandfilter *n*
    f filtre *m* rapide à sable
    nl snelwerkend zandfilter *n, m*
    r скорый песчаный фильтр *m*

H329 *e* **high-rate trickling filter**
    d Spültropfkörper *m*
    f lit *m* bactérien percolateur à haute puissance
    nl snelwerkend druppelfilter *n, m*
    r высоконагружаемый капельный биофильтр *m*

H330 *e* **high-rise block, high-rise building**
    d Hochhaus *n*
    f immeuble *m* élevé, bâtiment *m* à étages multiples
    nl torengebouw *n*
    r высотное [многоэтажное] здание *n*

H331 *e* **high-rise erection**
    d Montage *f* der Hochhäuser und turmartiger Bauwerke
    f montage *m* du bâtiment de grande hauteur
    nl hoogbouw *m*
    r монтаж *m* высотных сооружений [многоэтажных зданий]

H332 *e* **high-slump concrete**
    d Weichbeton *m*
    f béton *m* plastique, béton *m* (en) consistance plastique
    nl plastisch beton *n*
    r пластичный бетон *m*

H333 *e* **high-speed passenger elevator** *US*
    d Schnellfahrstuhl *m*, Schnell--Personenaufzug *m*
    f ascenseur *m* rapide [express]
    nl snelle personenlift *m*
    r высокоскоростной пассажирский лифт *m*

H334 *e* **high strength bolt** *US*
    d hochfest vorgespannte Schraube *f*
    f boulon *m* à haute résistance
    nl bout *m* van hoogwaardig staal
    r высокопрочный болт *m*

H335 *e* **high-strength concrete**
    d hochfester Beton *m*
    f béton *m* à haute résistance
    nl hoogwaardig beton *n*

    r высокопрочный [высокомарочный] бетон *m*

H336 **high strength friction grip bolt** *UK see* **high strength bolt**

H337 **high strength reinforcement** *see* **high tensile reinforcement**

H339 *e* **high-temperature-resisting refractory concrete**
    d hochfeuerfester hitzebeständiger Beton *m*
    f béton *m* réfractaire à haute réfractarité
    nl hoogwaardig hittebestendig beton *n*
    r высокоогнеупорный жаростойкий бетон *m*

H340 *e* **high-temperature water**
    d Heißwasser *n*
    f eau *f* surchauffée
    nl heet water *n*, oververhit water *n*
    r высокотемпературная [перегретая] вода *f* (*выше 100 °C*)

H341 **high-tensile bolt** *see* **high strength bolt**

H342 *e* **high tensile reinforcement**
    d hochzugfeste Bewehrung *f*
    f armature *f* d'acier à haute résistance
    nl wapening *f* met grote trekvastheid
    r арматура *f* из высокопрочной стали

H343 *e* **high tensile steel**
    d hochzugfester Stahl *m*
    f acier *m* à haute résistance [de haute tension]
    nl staal *n* met grote trekvastheid
    r высокопрочная сталь *f*

H344 **high-tension bolt** *see* **high strength bolt**

H345 *e* **high tied static crane**
    d gebäudeverankerter Turmkran *m*
    f grue *f* à mât fixe de grande hauteur (*attaché au bâtiment*)
    nl aan een gebouw verankerde torenkraan *f* (*m*)
    r высокий приставной башенный кран *m*

H346 *e* **high-velocity air-conditioning system**
    d Hochgeschwindigkeitsklimaanlage *f*
    f système *m* de conditionnement d'air à haute vitesse
    nl luchtbehandelingssysteem *n* met grote luchtsnelheid
    r система *f* кондиционирования с высокими скоростями воздуха

H347 *e* **high water**
    d 1. Hochwasser *n*; Hochflut *f* 2. Oberhaltung *f*, Oberwasser *n*
    f 1. hautes eaux *f pl*, marée *f* montante; crue *f* 2. eau *f* d'amont, bief *m* supérieur
    nl 1. hoogwater *n* 2. vloed *m*
    r 1. полная [высокая] вода *f*; паводок *m* 2. верхний бьеф *m*

H348 *e* **high-water level**
    d Hochwasserstand *m*
    f niveau *m* des eaux hautes

nl bovenpand *n*, hoogwaterpeil *n*,
hoogwaterstand *m*
r уровень *m* высоких вод [полной воды]

H349 *e* **high waterline**
d Tidehochwasserlinie *f*
f laisse *f* de haute mer
nl vloedlijn *f* (*m*)
r линия *f* прилива полной воды

H350 **high-water mark** *see* **flood mark**

H351 *e* **highway**
d Fernverkehrsstraße *f*, Landstraße *f*
der ersten Ordnung *f*
f voie *f* routière, route *f*
nl snelweg *m*
r автомобильная дорога *f* общего
назначения [пользования]

H352 *e* **highway bridge**
d Straßenbrücke *f*
f pont-route *m*
nl verkeersbrug *f* (*m*)
r автодорожный мост *m*

H353 *e* **highway bridge loads**
d Berechnungslast *f*
[Transportbelastung *f*] der
Straßenbrücke
f charge *f* sur le pont-route, charge *f*
pour le calcul des ponts-routes
nl lastenstelsel *n* voor de berekening van
een verkeersbrug
r расчётные нагрузки *f pl* для автодо-
рожных мостов

H354 *e* **highway construction**
d Straßenbau *m*
f construction *f* routière [de routes]
nl wegenbouw *m*
r дорожное строительство *n*

H355 *e* **highway crane**
d straßenfahrbarer Kran *m*
f grue *f* automotrice [sur camion]; grue
f sur roues [sur pneu(matique)s]
nl mobiele kraan *f* (*m*), wielkraan *f* (*m*)
r самоходный кран *m* на
автомобильном *или* пневмоколёсном
ходу

H356 *e* **highway engineering**
d Straßentechnik *f*, Straßenbau *m*
f technique *f* [industrie *f*] routière;
construction *f* de routes
nl wegbouwkunde *f*
r проектирование *n* и строительство *n*
автомобильных дорог, дорожное
строительство *n*

H357 *e* **highway grinder**
d Straßenschleifmaschine *f*
f machine *f* à polir le revêtement routier
en béton, polisseuse *f* [polissoir *m*]
de dallages
nl wegenslijpmachine *f*
r дорожно-шлифовальная машина *f*

H359 **highway junction** *see* **road intersection**

H360 **highway planing machine** *see* **heater planer**

H361 *e* **highway project**
d 1. Straßenbauobjekt *n*,
Straßenbauvorhaben *n*
2. Straßenbauprojekt *n*
f 1. chantier *m* routier [de route]
2. projet *m* de route, projet *m* de
construction d'une route
nl wegenbouwproject *n*
r 1. дорожно-строительный объект *m*,
объект *m* дорожного строительства
2. проект *m* строительства дороги

H362 *e* **highway-rail bridge**
d Straßen- und Eisenbahnbrücke *f*
f pont *m* mixte [route-rail]
nl gecombineerde spoor- en
verkeersbrug *f* (*m*)
r совмещённый мост *m*
(*автомобильно-железнодорожный*)

H363 *e* **highway structures**
d Kunstbauwerke *n pl*, Kunstbauten *pl*
einer Straße
f ouvrages d'art *m pl* routiers
nl wegenbouwtechnische kunstwerken *n*
pl
r дорожные сооружения *n pl*,
сооружения *n pl* на дорогах

H364 *e* **highway trailer**
d Straßenanhänger *m*
f remorque *f* sur roues [sur pneus]
nl aanhangwagen *m*, trailer *m*
r автомобильный прицеп *m*

H365 *e* **highway transition curve**
d Straßenübergangsbogen *m*
f courbe *f* de transition [de
raccordement] du tracé autoroutier
nl overgangsboog *m* (*in een weg*)
r дорожная переходная кривая *f*

H366 *e* **highway tunnel**
d Straßentunnel *m*
f tunnel *m* (auto)routier, tunnel-route *m*
nl verkeerstunnel *m*
r автодорожный туннель *m*

H367 *e* **high-workability concrete**
d flüssiges Betongemisch *n*, Gießbeton
m
f béton *m* frais à haute ouvrabilité;
béton *m* frais fluide
nl vloeibare betonmortel *m*
r бетонная смесь *f* с высоким
показателем удобоукладываемости;
подвижная *или* литая бетонная
смесь *f*

H368 *e* **high yield strength reinforcement**
d Bewehrung *f* aus Stahl mit hoher
Streckgrenze
f armature *f* d'acier à haute limite
élastique
nl wapening *f* van hoogwaardig staal
r арматура *f* из стали с высоким
пределом текучести

H369 *e* **hillside cut and fill**
d Anschnitt *m*
f déblai *m* et remblai *m* (*sur le flanc
de côteau*)

nl ingraving *f* en aanvulling *f* (*op een helling*)
r полувыемка-полунасыпь *f* (*на склоне холма*)

H370 **hillside ditch** *see* **interception channel**
H371 *e* **hinge**
  d 1. Scharnier *n*, Gelenk *n* 2. Türangel *f*
  f 1. articulation *f*, charnière *f*, rotule *m* 2. charnière *f*, paumelle *f*
  nl 1. scharnier *n* 2. deurhengsel *n*
  r 1. шарнир *m* 2. дверная петля *f*

H372 *e* **hinged bearing**
  d Kipplager *n*
  f appui *m* articulé
  nl scharnierende oplegging *f*
  r шарнирная опора *f* [опорная часть *f*]

H373 *e* **hinged gates**
  d Flügeltor *n*
  f porte *f* de clôture à charnière
  nl draaideur *f* (*m*), draaiende sluisdeur *f* (*m*)
  r распашные ворота *pl*

H374 *e* **hinged girder**
  d Gelenkträger *m*
  f poutre *f* [ferme *f*] articulée
  nl Gerberligger *m*
  r шарнирная балка *f* [ферма *f*]

H375 *e* **hinged joint**
  d Gelenkstoß *m*, Gelenkverbindung *f*, Gelenkknoten *m*
  f articulation *f*, joint *m* à charnière
  nl scharnierpunt *n*, scharnier *n*
  r шарнирное соединение *n*, шарнирный узел *m*

H376 *e* **hinged-leaf gate** US
  d Klappenschütz *n*, Klappenschütze *f*
  f vanne *f* à tablier articulé
  nl scharnierende afsluiter *m*, klepafsluiter *m*
  r поворотный щитовой затвор *m*, клапанный затвор *m*

H377 *e* **hinged pier**
  d Pendelpfeiler *m*
  f support *m* pendulaire [à pendule], appui *m* oscillant
  nl pendeloplegging *f*
  r качающаяся опора *f*

H378 *e* **hinged structures**
  d gelenkige Konstruktionen *f pl*
  f structures *f* articulées, charpentes *f pl* à nœuds articulés
  nl scharnierende constructies *f pl*
  r шарнирные конструкции *f pl*

H379 *e* **hinge jamb**
  d Türpfosten *m* [Schenkel *m*] mit angehängtem Türflügel
  f poteau *m* d'assemblage d'une huisserie de porte (*sur lequel on attache les paumelles*)
  nl deurpost *m* (*hangzijde*), kozijnstijl *m* (*hangzijde*)

r навесной косяк *m* [вертикальный брус *m*] дверной рамы (*к которому крепят дверное полотно*)
H380 *e* **hingeless frame**
  d gelenkloser Rahmen *m*
  f portique *f* [charpente *f*] encastrée à nœuds rigides
  nl skelet *n* [portaal *n*] met stijve knooppunten
  r бесшарнирная рама *f*

H381 **hinge stile** *see* **hanging stile**
H382 *e* **hip**
  d 1. Grat *m* (*Walmdach*) 2. Walmdachverband *m*; Ecksparren *m*, Gratsparren *m* 3. oberer Randknoten *m* des Trapezbrückenträgers
  f 1. chevron *m* d'arête 2. arbalétriers *m pl* de croupe; chevron *m* de croupe 3. nœud *m* supérieur extrême d'une poutre à treillis trapézoïdale de pont
  nl 1. hoekkeper *m*, dakrib *f* (*m*) 2. hoekkeper(spant) *m* (*n*), graatspar *m* 3. bovenste knooppunt *n* van een trapeziumligger
  r 1. ребро *m* вальмовой крыши 2. вальмовые стропила *pl*, укосная стропильная нога *f* 3. крайний верхний узел *m* мостовой трапецеидальной фермы

H383 *e* **hip-and-valley roof**
  d mehrseitig geneigtes Dach *n*
  f comble *m* à plusieurs versants (*à orientations différentes*)
  nl dak *n* met hoek- en kilkepers
  r многоскатная крыша *f*; крыша *f* сложной формы (*с несколькими скатами и разжелобками*)

H384 *e* **hip jack**
  d Gratschifter *m*, Walmschifter *m*
  f chevron *m* de croupe
  nl dakspar *m* van het eindschild
  r нарожник *m* вальмы

H385 *e* **hipped end**
  d Walm *m*
  f croupe *f*
  nl eindschild *n*
  r вальма *f*

H386 *e* **hipped gable**
  d Krüppelwalm *m*, Schopfwalm *m*
  f demi-croupe *f*
  nl wolfeind *n*
  r полувальма *f*

H387 *e* **hipped-gable roof**
  d Krüppelwalmdach *n*, Schopfwalmdach *n*
  f comble *m* [toit *m*] en demi-croupe
  nl wolfsdak *n*
  r полувальмовая крыша *f*

H388 *e* **hipped roof**
  d Walmdach *n*
  f comble *m* [toit *m*] en croupe, comble *m* [toit *m*] à quatre versants

   nl schilddak *n*
   r вальмовая [четырёхскатная] крыша *f*

H389 *e* **hip rafter**
   d Gratbalken *m*
   f chevron *m* de rive
   nl hoekkeper *m*, graatspar *f* (*m*)
   r накосная [диагональная] стропильная нога *f*

H390   **hip roof** *see* **hipped roof**

H391 *e* **hip tile**
   d Gratstein *m*, Walmstein *m*
   f tuile *f* de croupe, tuile *f* arrêtière [faîtière]
   nl keperpan *f* (*m*)
   r коньковая черепица *f* (*для покрытия рёбер кровли*)

H392 *e* **hit and miss damper, hit and miss ventilator**
   d Schlitzschieber *m*
   f grille *f* d'aération réglable (*en deux tôles glissantes à rainures*)
   nl regelbaar luchtrooster *n*, *m*
   r воздушный клапан *m*, состоящий из двух листов (*неподвижного и скользящего*) с идентичными прорезями; регулируемая вентиляционная решётка *f*

H394 *e* **hog**
   d Aufwölbung *f*, Überhöhung *f*
   f contre-flèche *f*
   nl opbuiging *f*, welving *f*, uitwijking *f* (*muur*)
   r выгиб *m*, искривление *n*

H395 *e* **hoggin, hogging**
   d Fußwegkies *m*, tonhaltiger Kiessand *m*
   f mélange *m* gravier-argileux
   nl grind-en-kleimengsel *n* ter verharding van paden
   r природная гравийно-глинистая смесь *f*

H396 *e* **hogging bending**
   d Aufwölbung *f*
   f flexion *f* due aux moments fléchissants négatifs
   nl opbuigen *n*
   r изгиб *m*, вызванный отрицательными моментами

H397 *e* **hogging moment**
   d Aufwölbungsmoment *n*
   f moment *m* fléchissant [de flexion] négatif
   nl negatief buigend moment *n*
   r отрицательный изгибающий момент *m*

H398 *e* **hogging of beam**
   d Aufwölbung *f* des Balkens
   f contre-flèche *f* d'une poutre
   nl opbuiging *f* van een balk
   r выгиб *m* балки

H399 *e* **hoist**
   d 1. Aufzug *m* 2. Winde *f*
   f 1. élévateur *m*, monte-charge *m* 2. treuil *m*

   nl 1. takel *m*, *n*, hijsinrichting *f*, 2. windas *n*, lier *f* (*m*)
   r 1. подъёмник *m* 2. лебёдка *f*

H400 *e* **hoisting and conveying cableway**
   d Kabelkran *m*
   f blondin *m*, grue *f* aérienne, grue *f* à câbles; téléphérique *m*, transporteur *m* à câbles
   nl kabelkraan *f* (*m*)
   r кабельный кран *m*, кабель-кран *m*

H401 *e* **hoisting machine**
   d Hebezeug *n*
   f engin *m* de levage, mécanisme *m* élévateur
   nl hijswerktuig *n*, kraan *f* (*m*)
   r подъёмная машина *f*, подъёмный механизм *m*

H402 *e* **hoisting rope**
   d Hubseil *n*, Lastseil *n*, Aufzugseil *n*, Förderseil *n*
   f câble *m* de monte-charge [de levage]
   nl hijskabel *m*
   r грузовой [грузоподъёмный] канат *m*

H403 *e* **hoist limit device**
   d Hubbegrenzer *m*
   f limiteur *m* de course haute (du crochet), limiteur *m* de course du crochet en montée
   nl hijsbegrenzer *m*
   r ограничитель *m* подъёма (*груза*)

H404 *e* **hoist tower**
   d Aufzugturm *m*
   f tour *f* élévatrice
   nl hijstoren *m*
   r башенный подъёмник *m*

H405 *e* **hoistway**
   d Aufzugsschacht *m*, Fahrstuhlschacht *m*
   f gaine *f* [cage *f*] d'ascenseur
   nl liftschacht *f* (*m*)
   r шахта *f* лифта

H406 *e* **hoistway door**
   d Aufzugsschachttür *f*
   f porte *f* de cage [de gaine] d'ascenseur
   nl schachtdeur *f* (*m*) (*lift*)
   r дверь *f* шахты лифта

H407 *e* **holdfast**
   d 1. Ankerblock *m*, Verankerungsblock *m* 2. Bankschraubstock *m* 3. Zwinge *f*; Klammer *f*
   f 1. massif *m* d'ancrage (en béton); pilier *m* d'ancrage 2. valet *m* d'établi 3. serre-joints *m*, servante *f*, sergent *m*
   nl 1. verankering *f* 2. bankschroef *f* (*m*) 3. houvast *n*
   r 1. якорь *m* (*инвентарный*) для оттяжек *или* расчалок 2. тиски *pl* столярного верстака 3. струбцина *f*, сжим *m*

H408   **holding-down bolt** *see* **anchor bolt**

H410 *e* **holding rope**
   d Halteseil *n*

f câble *m* porteur [de retenue]
nl torntouw *n*
r удерживающий [несущий] канат *m*

H411 *e* **hold-over system**
d Wärmespeichersystem *n*
f système *m* accumulateur de chaleur
nl warmte-accumulerend systeem *n*
r теплоаккумулирующая система *f* отопления

H412 *e* **hollow block**
d Hohlblockstein *m*
f bloc *m* [corps *m*, parpaing *m*] creux
nl holle bouwsteen *m*
r пустотелый блок *m*

H413 *e* **hollow block floor**
d Hohlblockdecke *f*, Hohlsteindecke *f*, Hourdidecke *f*, Hohlkörperdecke *f*
f plancher *m* à hourdis en céramique
nl balkenvloer *m* met holle baksteen als vulelement
r балочное перекрытие *n* с заполнением межбалочного пространства пустотелыми керамическими блоками

H414 *e* **hollow brick**
d Lochziegel *m*
f brique *f* perforée; brique *f* creuse
nl holle baksteen *m*
r дырчатый кирпич *m*; пустотелый кирпич *m*

H415 *e* **hollow clay block**
d Lochziegelplatte *f*
f hourdis *m* en céramique; élément *m* de remplissage en terre cuite
nl element *n* van holle baksteen
r пустотелый керамический блок *m*

H416 *e* **hollowcore slab**
d Deckenhohlplatte *f*, Hohldeckenplatte *f*
f dalle *f* creuse [alvéolée, vide]
nl holle plaatvloer *f* (*m*)
r многопустотная железобетонная плита *f*

H417 **hollow dam** *US see* **cavity dam**

H418 *e* **hollow glass block**
d hohler Glasbaustein *m*, Hohlglas(bau)stein *m*
f brique *f* de verre creuse, bloc *m* en verre creux
nl holle glastegel *m*
r полый стеклоблок *m*

H419 *e* **hollow pile**
d Hohlpfahl *m*
f pieux *m* creux
nl holle paal *m*
r пустотелая свая *f*

H420 **hollow pot** *see* **hollow block**

H421 *e* **hollow preformed pile**
d Hohlfertigpfahl *m*
f pieu *m* préfabriqué creux
nl holle prefabpaal *m*
r пустотелая забивная свая *f*

H422 *e* **hollow shell pile**
d Mantelpfahl *m*
f pieu *m* foré de grand diamètre, pieu-caisson *m*
nl mantelpaal *m*
r свая-оболочка *f*

H423 *e* **hollow tile**
d Hohlblockstein *m*
f hourdis *m* (*en céramique ou en béton*)
nl holle tegel *m*
r пустотелый (*бетонный или керамический*) блок *m* (*для заполнения перекрытий*)

H424 *e* **hollow-tile floor slab**
d Füllkörperdeckenplatte *f*
f dalle *f* de plancher en béton armé avec hourdis en céramique
nl vloer *f*(*m*) van betonnen balken met opvulling van keramische blokken
r железобетонная плита *f* перекрытия с вкладышами из керамических блоков

H425 **hollow wall** *see* **cavity wall**

H426 *e* **hollow walling**
d zweischaliges Mauerwerk *n*
f mur *m* à double parois, mur *m* creux
nl spouwmuur *m*
r пустотелая стена *f*

H427 *e* **home of high insulation standard**
d Gebäude *n* [Haus *n*] mit verbesserter Wärmedämmung
f maison *f* avec haut degré d'isolation thermique
nl goed geïsoleerd huis *n*
r здание *n* [дом *m*] с эффективной теплоизоляцией

H428 *e* **homogeneous earth dam**
d homogener Erd(stau)damm *m*
f barrage *m* en sol [en terre] homogène
nl homogene aardedam *m*
r земляная плотина *f* из однородного грунта

H429 *e* **honeycomb**
d Lunker *m*
f alvéole *m*, cavité *f* (*dans le béton*); soufflure *f* (*dans le métal*)
nl luchtbellen *pl* (*beton*); gietgallen *pl* (*metaal*)
r раковина *f* (*дефект бетона, металла*)

H430 *e* **honeycombed area**
d Zone *f* der wabenförmigen Rißbildung (*Beton*)
f zone *f* défectueuse (*de béton*) avec cavités
nl gebied *n* met honingraatvormige scheuren
r дефектный участок *m* (*бетонной поверхности*) с раковинами

H431 *e* **honeycombing**
d wabenförmige Rißbildung *f* (*z.B. in Beton*); innere Rißbildung *f* (*im Holz*)

  f formation *f* de cavités sur la surface
   de béton; formation *f* de fentes
   intérieures dans le bois
  nl optreden *n* van haarscheuren (*beton*);
   hartscheuren (*hout*)
  r образование *n* раковин и пустот;
   щербатость *f* (*дефект бетонной*
   поверхности); образование *n*
   внутренних трещин (*в древесине*)

H432 *e* **honeycomb structure**
  d 1. Wabenstruktur *f*
   2. Wabenkonstruktion *f*
  f 1. structure *f* en nid d'abeilles
   2. construction *f* [élément *m* de
   construction] en nid d'abeilles
  nl 1. honingraatstructuur *f*
   2. honingraatconstructie *f*
  r 1. сотовая структура *f* 2. сотовая
   [ячеистая] конструкция *f*

H433 **honeycomb wall** *see* **shelter wall**

H434 *e* **hood**
  d Haube *f*, Abzughaube *f*,
   Entlüftungshaube *f*, Absaughaube *f*;
   Abzugsschrank *m*; Lüftungsabdeckung;
   f, Erfassungseinrichtung *f*
  f hotte *f*; capuchon *m*; chapeau *m* de
   cardinal
  nl afzuigkap *f* (*m*); ventilatorrozet *n*
  r вытяжной зонт *m*; вытяжной шкаф
   m; вентиляционное укрытие *n*

H435 *e* **hooded roof ventilator**
  d Dachventilator *m* [Dachlüfter *m*] mit
   Deflektorhaube
  f ventilateur *m* de toit à déflecteur
  nl dakventilator *m* met deflector
  r крышный вентилятор *m*
   с дефлектором

H436 *e* **hood inlet opening**
  d Eintrittsöffnung *f* der Absaughaube
  f orifice *m* d'admission de la hotte
  nl inlaatopening *f* van een afzuigkap
  r входное отверстие *n* вытяжного
   зонта

H437 *e* **hook**
  d Haken *m*
  f crochet *m* porte-charge [de levage];
   crochet *m* de l'armature
  nl haak *m*; haak *m* van een
   wapeningsstaaf
  r 1. грузоподъёмный крюк *m* 2. крюк
   m на конце арматурного стержня

H438 *e* **hook-and-eye fastener**
  d Haken *m* und Öse *f*
  f crochet *m* et piton *m*
  nl overval *m* (*deur*), haakje *n* (*op een*
   deur)
  r запорный крюк *m* с пробоем

H439 *e* **hook bolt**
  d Hakenbolzen *m*
  f boulon *m* à crochet
  nl schroefhaak *m*
  r болт *m* с крюком на конце

H440 *e* **hooked bar**
  d Bewehrungsstab *m* mit Endhaken

  f barre *f* d'armature avec crochets aux
   extrémités
  nl wapeningstaaf *f* (*m*) met haken
  r арматурный стержень *m* с крюками
   на концах

H441 *e* **hooked scraper**
  d Anhängeschürfwagen *m*
  f scraper *m* tracté, décapeuse *f* tractée
  nl getrokken sc(h)raper *m*
  r прицепной скрепер *m*

H442 *e* **hook gage** *US* *see* **micromanometer**
H442a *e* **hook gage** *US*, **hook gauge** *UK*
  d Hakenstechpegel
  f indicateur *m* limnimétrique[de niveau]
   à pointe recourbée
  nl haakvormig steckpeil *n*
  r крючковая (игольчатая) водомерная
   рейка *f*

H443 **hook height** *see* **height under hook**

H445 *e* **hooping, hoop reinforcement**
  d Ringbewehrung *f*
  f frettage *m*
  nl ringwapening *f*
  r кольцевая арматура *f*

H446 *e* **hoop stress**
  d Ringspannung *f*; Ringzugkraft *f*
  f tension *f* circonférentielle; effort *m*
   de traction dans l'anneau *ou* virole
  nl tangentiale trekspanning *f*
  r тангенциальное напряжение *n*

H447 *e* **hopper**
  d 1. Trichter *m* 2. Kippwagen *m*
  f 1. trémie *f* 2. berline *f* culbuteuse
   [à renversement latéral]
  nl 1. trechter *m* 2. onderlosser *m*
  r 1. бункер *m*; приёмная воронка *f*
   2. вагонетка *f* с опрокидным кузовом

H448 *e* **hopper dredger**
  d Schachtpumpenbagger *m*,
   Hopperbagger *m*
  f drague *f* suceuse autoporteuse, drague
   f aspiratrice porteuse
  nl hopperzuiger *m*, zelflossende
   zandzuiger *m*
  r самоотвозящий дноуглубительный
   снаряд *m*

H449 *e* **hopper frame**
  d Fensterrahmen *m* mit nach innen
   aufgehendem Kippflügel
  f chassis *m* (de fenêtre) à vantail
   supérieur basculant
  nl raamkozijn *n* met valraam
  r оконная рама *f* с откидывающейся
   внутрь верхней фрамугой [створкой]

H450 *e* **hopper light, hopper vent, hopper**
   ventilator
  d nach innen aufgehender Kippflügel
   m
  f vantail *m* [battant *m*] de fenêtre
   basculant [à bascule]
  nl valraam *n*
  r фрамуга *f* [створка *f*] окна,
   откидывающаяся внутрь

H451 *e* **horizontal alignment**
 d Straßenzug *m*, Linienführung *f* im Grundriß (*Straßenbau*)
 f tracé *m* de la route
 nl tracé *n* van een weg, aslijn *f* (*m*) van de weg
 r трасса *f* дороги

H452 *e* **horizontal-axis mixer**
 d Betonmischer *m* mit horizontal rotierender Trommel
 f malaxeur *m* [mélangeur *m*] à arbre [à axe] horizontal
 nl betonmolen *m* met horizontaal draaiende trommel
 r бетоносмеситель *m* с барабаном, вращающимся вокруг горизонтальной оси

H453 *e* **horizontal boiler**
 d Horizontal-Kessel *m*, liegender Dampfkessel *m*
 f chaudière *f* horizontale
 nl liggende stoomketel *m*
 r горизонтальный котёл *m*

H454 *e* **horizontal braces**
 d Horizontalverband *m*
 f système *m* d'entretoisements [de contreventements] horizontaux
 nl horizontaal windverband *n*
 r горизонтальные связи *f pl*

H455 *e* **horizontal cell tile**
 d keramischer Langlochstein *m*, Hourdi *n*
 f bloc *m* [parpaing *m*] en céramique à alvéoles horizontaux
 nl baksteen *m* met horizontale holten
 r керамический блок *m* с горизонтальными пустотами

H456 *e* **horizontal curve**
 d Kurve *f* im Grundriß (S*traßenbau*)
 f courbe *f* (*du tracé de la route*) en plan
 nl bocht *f* (*m*) in horizontale projectie (*wegenbouw*)
 r горизонтальная кривая *f* (*дороги*)

H457 *e* **horizontal drain**
 d horizontaler Dränagegang *m*, horizontales Dränagerohr *n*
 f drain *m* horizontal
 nl horizontale drain *m*
 r горизонтальная дрена *f*

H458 *e* **horizontal drainage blanket**
 d Flächenfilter *n*, *m*, Entwässerungsfilter *n*, *m*, horizontaler Entwässerungsteppich *m*
 f tapis *m* filtrant [de drainage]
 nl horizontale filterlaag *f* (*m*), filterbed *n*
 r дренажный тюфяк *m*, плоский тюфячный дренаж *m* (*земляных плотин*)

H459 *e* **horizontal filter well**
 d Horizontalfilterbrunnen *m*
 f prise *f* d'eau rayonnante; puits *m* rayonnant horizontal; drain *m* rayonnant
 nl bron *f* (*m*) met horizontaal filter
 r лучевой водозабор *m*; горизонтальная фильтрационная скважина *f*; лучевая дрена *f*

H460 *e* **horizontal flow sedimentation basin**
 d längsdurchflossenes [horizontal durchflossenes] Absetzbecken *n*, liegender Klärbehälter *m*, Langbecken *n*
 f décanteur *m* horizontal
 nl horizontale bezinkbak *m*
 r горизонтальный отстойник *m*

H461 *e* **horizontal panel**
 d liegende Wandplatte *f* (*mit horizontal angeordneter größerer Seite*)
 f panneau *m* (mural) horizontal (*dont le grand côté est horizontal*)
 nl paneel *n* met de langste zijde horizontaal gesteld
 r прямоугольная стеновая панель *f* с большей стороной, расположенной горизонтально

H462 *e* **horizontal restraint**
 d waagerechte Einspannung *f*
 f encastrement *m* horizontal (*contre les déplacements horizontaux*)
 nl inklemming *f* tegen horizontale verplaatsing
 r защемление *n* против горизонтального смещения

H463 **horizontal screen well** *see* **horizontal filter well**

H464 *e* **horizontal sheeting**
 d 1. Horizontalschalung *f* 2. horizontaler Holzausbau *m*
 f 1. revêtement *m* en planches posées horizontalement 2. planches *f pl* de blindage horizontales (*d'une tranchée*)
 nl 1. liggende beschieting *f* 2. horizontale bekleding *f* van een gegraven sleuf
 r 1. горизонтальная дощатая обшивка *f* 2. горизонтальное крепление *n* (*стенок траншей*)

H465 *e* **horizontal-shaft mixer**
 d Betonmischer *m* mit horizontaler Rührflügelwelle
 f malaxeur *m* [mélangeur *m*] à un ou plusieurs arbres horizontaux à palettes
 nl betonmolen *m* met horizontaal draaiend roerwerk
 r бетоносмеситель *m* с одним *или* несколькими горизонтальными лопастными валами

H466 *e* **horizontal shore**
 d waagerechter Schalungsunterzug *m* (*beim Bau der Ortbetondecke*)
 f traverse *f* sous platelage (*du coffrage des planchers à poutres armées*)
 nl gestelde (horizontale) bekisting *f*
 r опорный прогон *m* опалубочной конструкции (*железобетонного перекрытия*)

H467 *e* **horizontal sliding door**
   d Horizontalschiebetür *f*
   f porte *f* à coulisse [coulissante,
     glissante]
   nl schuifdeur *f* (*m*)
   r раздвижная [откатная, скользящая]
     дверь *f*

H468 *e* **horizontal sliding window**
   d Horizontalschiebefenster *n*
   f fenêtre *f* coulissante
   nl opzij openschuivend venster *n*
   r раздвижное окно *n*

H469 *e* **horizontal slipforming**
   d Ziehen *n* (*Gleitschalungsbau*)
   f bétonnage *m* avec coffrage mobile
     [glissant horizontalement]
   nl glijdende bekisting *f* (*betonweg*)
   r бетонирование *n* с применением
     катучей опалубки

H470 *e* **horizontal tie**
   d Zugband *n*; Horizontalverband *m*
   f tirant *m* horizontal; contrevent *m*
     horizontal, entretoise *f* horizontale
   nl trekband *m*, horizontale trekstaaf *f* (*m*)
   r горизонтальная затяжка *f*;
     горизонтальная связь *f*

H471 *e* **horn**
   d auskragendes Hirnende *n* (*z.B. des*
     Firstbalkens); Vorsprung *m*,
     Überstand *m*, Auskragung *f*
   f about *m* (*p.ex. de faîtière*); bout *m* en
     saillie (*p.ex. de traverse du bâti*)
   nl uitstek *n*, neus *m*
   r выступающий торец *m* (*напр.*
     конькового бруса); выступ *m* (*напр.*
     ригеля обвязки)

H472 *e* **horse**
   d 1. Gerüstbock *m* 2. Treppenwange *f*
   f 1. chevalet *m*, chèvre *f*, tréteau *m*
     2. limon *m* d'escalier
   nl 1. zaagbok *m* 2. trapboom *m*
   r 1. козлы *pl* 2. тетива *f* лестницы

H473 *e* **horse mould**
   d verschiebbares Schablonenbrett *n*,
     Ziehlatte *f* (*fürs Ziehen der Gesimse*)
   f calibre *m* mobile (*pour exécution des*
     moulures de corniche)
   nl mal *m* voor het maken van
     (kroon)lijsten
   r передвижной шаблон *m* для тяги
     карнизов

H474 *e* **horse scaffold**
   d Bockgerüst *n*, Schrägengerüst *n*
   f échafaudage *m* sur trétaux
   nl schraagsteiger *m*
   r подмости *pl* на козлах, козловые
     подмости *pl*

H475 *e* **horseshoe curve**
   d Hufeisenkurve *f*
   f courbe *f* en forme de fer à cheval
     (*p.ex. d'un tracé de route*)
   nl hoefijzerbocht *f* (*m*)
   r овальная кривая *f* (*дороги*)

H476 *e* **horsing-up**
   d Ziehen *n* der Gesimse
   f exécution *f* de moulures de corniche
     à l'aide de calibres
   nl maken *n* van (kroon)lijsten
   r вытягивание *n* [тяга *f*] штукатурных
     карнизов

H477 *e* **hose**
   d Schlauch *m*, Schlange *f*
   f tuyau *m* souple [flexible], boyau *m*
     d'arrosage, manche *f* (d'incendie)
   nl slang *f* (*m*), brandslang *f* (*m*)
   r шланг *m*, (пожарный) рукав *m*

H478   **hose connection** *see* **hose coupling**

H479 *e* **hose coupling**
   d Schlauchkupplung *f*
   f raccord *m* mâle ou femelle de boyau
   nl slangkoppeling *f*
   r шланговое соединение *n*,
     соединительная муфта *f* для шлангов

H480 *e* **hose-stream test**
   d Druckstrahlprüfung *f*
     (*der Baukonstruktionen*)
   f essai *m* à l'action dynamique du jet
     d'eau
   nl waterstraalproef *f* (*m*)
     (*van bouwconstructies*)
   r испытание *n* (*строительных*
     конструкций) на динамическое
     воздействие водяной струи

H481 *e* **hot aggregates**
   d erwärmte Zuschlagstoffe *m pl*
   f agrégats *m pl* chauffés
   nl verwarmde toeslagen *m pl*
   r подогретые заполнители *m pl*

H482 *e* **hot concrete**
   d Heißbeton *m*
   f béton *m* frais chauffé par vapeur
     (*dans le tambour du malaxeur*)
   nl verwarmde betonmortel *m*
   r горячий бетон *m*, бетонная смесь *f*,
     подогреваемая паром
     (*в бетоносмесителе*)

H483 *e* **hot face**
   d Feuerseite *f*
   f parement *m* exposé au feu (*d'un*
     produit réfractaire)
   nl vuurzijde *f* (*m*)
   r поверхность *f* (*огнеупорного*
     изделия), обращённая к огню

H484 *e* **hot-face insulation**
   d feuerseitige Wärmeisolierung *f*
   f isolant *m* calorifuge appliqué sur la
     surface chaude (*exposée à l'action de la*
     chaleur)
   nl warmte-isolatie *f* aan de vuurzijde
   r теплоизоляция *f* на стороне высокой
     температуры

H485 *e* **hot feed**
   d Warmwasser-Speiseleitung *f*
   f canalisation *f* d'eau chaude
   nl tapleiding *f* voor warm water
   r горячий водопровод *m*

H486 *e* **hot-laid mixtures**
  d Heißeinbaugemische *n pl*,
  Heißeinbaumischungen *f pl*
  f enrobés *m pl* [mélanges *m pl*
  bitumineux] posés au chaud
  nl warm te verwerken bitumina *pl*
  r смеси *f pl*, укладываемые и
  уплотняемые в горячем состоянии

H487 *e* **hot mix, hot-mix asphalt**
  d Heißasphaltbeton *m*
  f asphalte *m* chaud, béton *m*
  asphaltique à chaud, enrobé *m*
  à chaud
  nl warm asfaltbeton *n*
  r горячий асфальтобетон *m*

H488 *e* **hot-mix asphalt plant**
  d Walzasphalt(misch)anlage *f*,
  Heißmischanlage *f*
  f poste *m* [installation *f*, centrale *f*]
  d'enrobage à chaud
  nl warm-asfaltinstallatie *f*
  r установка *f* для приготовления
  горячей асфальтобетонной смеси

H489 **hot plant mix** *see* **hot mix**

H490 *e* **hot-rolled wire**
  d warmgewalzter Draht *m*
  f fil *m* laminé à chaud
  nl warm gewalste draad *m*
  r горячекатаная проволока

H491 *e* **hot setting adhesive**
  d warmabbindender Klebstoff *m*
  f colle *f* thermodurcissable [durcissant
  à température élevée]
  nl warmverhardende lijm *m*
  r термореактивный клей *m*, клей *m*
  горячего отверждения

H492 *e* **hot spraying**
  d Heißspritzen *n*
  f pulvérisation *f* à chaud
  nl warm sproeien *n*
  r горячее распыление *n*

H493 *e* **hot water**
  d Warmwasser *n*; Heißwasser *n*
  f eau *f* chaude; eau *f* surchauffée
  nl warm water *n*
  r 1. горячая вода *f* (*до 100 °C*)
  2. высокотемпературная [перегретая]
  вода *f* (*выше 100 °C*)

H494 *e* **hot water accumulator**
  d Warmwasserspeicher *m*
  f accumulateur *m* à eau chaude
  nl warmwatervoorraadtank *m*
  r бак-аккумулятор *m* горячей воды

H495 *e* **hot water appliance**
  d Heißwassergerät *n*
  f chauffe-eau *m*
  nl warmwatertoestel *n*
  r водонагреватель *m*

H496 *e* **hot water boiler**
  d Heißwasserheizungskessel *m*, Boiler *m*
  f chaudière *f* de chauffage à l'eau
  surchauffée

  nl boiler *m*, warmwatervoorradtoestel *n*
  r водогрейный котёл *m*

H497 *e* **hot-water circulation system**
  d Warmwasserzirkulationsnetz *n*
  f réseau *m* (de distribution) d'eau
  chaude
  nl warmwatercirculatiesysteem *n*
  r циркуляционная система *f* горячего
  водоснабжения

H498 *e* **hot water circulator**
  d Heißwasserumwälzpumpe *f*
  f pompe *f* de circulation à eau chaude
  nl warmwatercirculatiepomp *f* (*m*)
  r циркуляционный насос *m* горячей
  воды

H499 *e* **hot water cylinder**
  d Gebrauchswasserspeicher *m*,
  Warmwasserbehälter *m*
  f réservoir *m* accumulateur cylindrique
  d'eau chaude, bac *m* à eau chaude
  nl warmwaterreservoir *n*
  r бак *m* горячей воды, цилиндрический
  бак-теплоаккумулятор *m* (*в местной
  системе горячего водоснабжения*)

H500 *e* **hot water heating system**
  d Heißwasserheizungssystem *n*,
  Hochdruckheißwasserheizung *f*
  f chauffage *m* à l'eau surchauffée
  nl verwarmingssysteem *n* met oververhit
  water
  r высокотемпературная система *f*
  (водяного) отопления, система *f*
  отопления высокого давления

H501 *e* **hot water storage heater**
  d Warmwasserspeichergerät *n*
  f chauffe-eau *m* à accumulation
  nl warmwatervoorraadtoestel *n*, boiler
  m
  r ёмкостный водонагреватель *m*

H502 *e* **hot water supply**
  d Warmwasserversorgung *f*
  f production *f* et distribution *f* de l'eau
  chaude
  nl warmwatervoorziening *f*
  r горячее водоснабжение *n*

H503 *e* **hot water system**
  d Wasserheizungssystem *n*;
  Warmwasserversorgungssystem *n*,
  Warmwasserleitungsnetz *n*
  f installation *f* [système *m*] de chauffage
  par l'eau chaude; système *m* de
  production et de distribution de l'eau
  chaude, réseau *m* de distribution de
  l'eau chaude
  nl warmwaterinstallatie *f*
  r система *f* водяного отопления;
  система *f* [сеть *f*] горячего
  водоснабжения

H504 *e* **hot water tank**
  d Heißwasserbehälter *m*
  f réservoir *m* accumulateur d'eau
  chaude (*non cylindrique*)
  nl warmwaterreservoir *n*

r замкнутый бак-аккумулятор *m* горячей воды (*нецилиндрической формы*)

H505 *e* **hot well**
  d Kondensatsammelrohr *n*
  f collecteur *m* de condensat
  nl condensaatverzamelpijp *f* (*m*), condensleiding *f*
  r конденсатосборник *m*, коллектор *m* конденсата

H506 *e* **hot wire anemometer**
  d Hitzdrahtanemometer *n*
  f anémomètre *m* à fil chaud
  nl thermo-elektrische windkrachtmeter *m*
  r электротермоанемометр *m*

H507 *e* **hot work**
  d Warmbetriebsarbeiten *f pl*
  f travaux *m pl* liés à la chauffe *ou* à l'utilisation des flammes nues
  nl bewerkingen *f pl* met gebruik van warmte
  r работы *f pl*, связанные с нагревом *или* применением пламени

H508 *e* **house** *UK*
  d Wohnhaus *n*, Einfamilienhaus *n*
  f maison *f* (d'habitation) à une famille
  nl woonhuis *n*
  r односемейный жилой дом *m*

H509 *e* **housebuilding industry**
  d Montagebauweise *f* mit fabrikmäßig hergestellten Fertigteilen
  f construction *f* industrialisée des habitations [des immeubles] d'habitation
  nl industriële woningbouw *m*
  r промышленное [заводское] домостроение *n*

H510 *e* **house connection**
  d 1. Hausanschluß *m* 2. Endstrang *m*
  f 1. branchement *m* d'eau général 2. branchement *m* d'égout
  nl 1. huisaansluiting *f* 2. perceelaansluiting *f*
  r 1. домовый ввод *m* (*водопровода*) 2. домовый выпуск *m* (*канализации*)

H511 house drain *see* house connection 2.

H512 *e* **house drainage, household sewage**
  d häusliches Abwasser *n*
  f eaux *f pl* domestiques [ménagères]
  nl huishoudelijk afvalwater *n*
  r хозяйственно-бытовые сточные воды *f pl*

H513 *e* **housekeeping** *US*
  d Aufräumung *f*, Säuberung *f*
  f nettoyage *m* d'emplacements de travail, de voies de passage, d'aires de montage (*sur le chantier*)
  nl de bouwplaats opruimen
  r очистка *f* рабочего места, проходов, зоны монтажа (*на стройплощадке*)

H514 *e* **house plumbing**
  d Wasserleitungs- und Entwässerungsnetz *n* des Gebäudes, Hausinstallation *f*, Hauswasserleitung *f* und -entwässerung *f*
  f réseau *m* de tuyautérie domestique; conduites *f pl* domestiques
  nl riool- en drinkwaterleidingen *f pl* van het perceel
  r водопроводно-канализационная сеть *f* здания

H516 *e* **house sewerage**
  d Hausentwässerung *f*, Gebäudeentwässerung *f*
  f réseau *m* d'évacuation des eaux ménagères
  nl binnenriolering *f*
  r домовая канализация *f*

H517 *e* **house sewer connection**
  d Endstrang *m*
  f branchement *m* d'égout
  nl huisaansluitleiding *f* (*riool*)
  r соединительная ветка *f* канализации, домовый выпуск *m*

H518 *e* **housing colony**
  d Wohnsiedlung *f*
  f localité *f*; bourg *m*
  nl (nieuwe) woonwijk *f* (*m*)
  r посёлок *m*

H519 *e* **housing construction**
  d Wohn(ungs)bau *m*
  f construction *f* des habitations [des logements]
  nl woningbouw *m*
  r жилищное строительство *n*

H520 *e* **housing density**
  d Behausungsdichte *f*, Baudichte *f*
  f densité *f* de construction
  nl bebouwingsdichtheid *f*
  r плотность *f* застройки жилого района

H521 *e* **housing estate**
  d 1. Wohnsiedlung *f*, Wohngebiet *n* 2. Gebäudegrundstück *n*
  f 1. unité *f* d'habitation [de voisinage]; ilôt *m* urbain 2. domaine *m* résidentiel
  nl 1. woonwijk *f* (*m*) 2. bouwperceel *n*
  r 1. жилой микрорайон *m*; жилая застройка *f* 2. домовладение *n*

H522 *e* **housing project**
  d 1. Bebauungsplan *m* 2. Wohnungsbauvorhaben *n*
  f 1. projet *m* de construction d'habitations 2. chantier *m* de construction des logements
  nl woningbouwproject *n*
  r 1. проект *m* застройки 2. объект *m* жилищного строительства

H523 *e* **housing scheme**
  d Wohnsiedlungsplan *m*
  f 1. projet *m* de construction [d'aménagement] d'une unité de voisinage [d'un ilôt urbain] 2. unité *f* de voisinage aménagée, ilôt *m* urbain aménagé
  nl (plan *n* voor een) wooneenheid *f*

r 1. проект *m* застройки
(*микрорайона*) 2. застраиваемый
микрорайон *m*

H524　**hovercraft construction vehicles** *see*
air cushion construction vehicles

H525 *e* **Howe truss**
d Howescher Träger *m*, Howe-Träger *m*
f poutre *f* Howe
nl Howe-vakwerkligger *m*
r ферма *f* Гау, ферма *f* со сжатыми
раскосами и растянутыми стойками

H526 *e* **H-pile**
d H-Pfahl *m*, Stahlträgerpfahl *m*
f pieu *m* en H
nl kolom *f* (*m*) in I-profiel
r стальная свая *f* двутаврового
профиля

H527 *e* **Huckbolt** *US*
d Huckbolzen *m*
f boulon *m* Huckbolt
nl huckbout *m*
r хакболт *m pl*

H528　**humidification efficiency** *see*
saturation efficiency

H529 *e* **humidification load**
d Befeuchtungslast *f*
f charge *f* d'humidification
nl bevochtigingslast *f* (*m*)
r нагрузка *f* по увлажнению воздуха

H530 *e* **humidifier**
d Befeuchter *m*, Befeuchtungseinrichtung
f
f humidificateur *m* (d'air)
nl luchtbevochtiger *m*
r увлажнитель *m* воздуха

H531 *e* **humidifying capacity**
d Befeuchtungsleistung *f*
f capacité *f* d'humidification
nl bevochtigend vermogen *n*
r производительность *f* по увлажнению
(*кондиционирование воздуха*)

H532 *e* **humidistat, humidity controller**
d Feuchtigkeitsregler *m*, Hygrostat *m*
f humidistat *m*, hygrostat *m*
nl automatische luchtvochtigheidsrege-
laar *m*
r гигростат *m*, регулятор *m*
влажности

H534　**humidity ratio** *US see* **specific
humidity**

H535 *e* **humus**
d Humus *m*
f 1. humus *m* terre *f* végétale
2. matières *f pl* en suspension
décantées dans le décanteur secondaire
nl humus *m*
r 1. гумус *m*, растительный слой *m*
2. осадок *m* из вторичного
отстойника (*после биофильтров*)

H536 *e* **humus tank**
d Nachklärbecken *n*
f décanteur *m* secondaire, bassin *m* de
décantation secondaire

nl nabezinktank *m*, compostbed *n*
r вторичный отстойник *m* очистной
станции с биофильтрами,
осветлитель-перегниватель *m*

H537 *e* **hung floor construction**
d Geschoßhängedecke *f*
f plancher *m* suspendu [supporté par
suspentes]
nl hangende verdiepingsvloer *f* (*m*)
r подвесное междуэтажное
перекрытие *n*

H538 *e* **hung scaffold**
d Hängegerüst *n*, Hängerüstung *f*
f échafaudage *m* volant
nl vliegende steiger *m*
r подвесные подмости *pl*

H539 *e* **hung-span beam**
d Gelenkträger *m*, Gerberträger *m*
f poutre *f* articulée, poutre *f*
(système) Gerber
nl Gerberligger *m*
r многопролётная консольно-
-подвесная балка *f*, балка *f* Гербера

H540 *e* **hunting**
d Pendelschwingung *f*, Pendelung *f*
f pulsation *f*
nl slingeren *n*
r рыскание *n* (*неустановившиеся
колебания регулируемой величины*)

H541 *e* **hurdle work**
d Flechtwerk *n*, Weidengeflecht *n*;
Schlickzaun *m*
f clayonnage *m*; épi *m* à clayonnage
nl vlechtwerk *n* van rijshout
r плетнёвое крепление *n*;
берегозащитный плетень *m*;
решётчатые [свайно-плетнёвые]
наносозадерживающие заграждения
n pl

H542 *e* **hurricane barrier, hurricane dam,
hurricane wall**
d Sperrwerk *n* zum Schutz einer
Flußmündung vor Sturmfluten
f barrière *f* de protection contre le
relèvement d'eau
nl stormvloedkering *f*
r заградительное сооружение *n* для
защиты устья от штормовых нагонов

H544 *e* **hybrid system**
d hybride Wärmepumpenanlage *f*
f installation *f* de chauffage hybride
(*à pompe de chaleur*)
nl warmtepompinstallatie *f*
r теплонасосная система *f* с двумя
взаимодополняющими источниками
тепловой энергии

H545 *e* **hybrid beam**
d zusammengesetzter Balken *m* (*aus
verschiedenartigen Stählen*)
f poutre *f* hybride
nl samengestelde ligger *m*
r составная балка *f* (*из сталей разной
прочности*)

H546 *e* **hydrant**
  d (Feuer-)Löschwasserständer *m*
  f hydrante *m*, bouche *f* d'incendie;
  poteau *m* d'incendie
  nl hydrant *f*, brandkraan *f* (*m*)
  r (пожарный) гидрант *m*

H547 *e* **hydrated cement**
  d hydratisierter Zement *m*
  f ciment *m* hydraté
  nl gehydrateerd cement *n*
  r гидратированный цемент *m*;
  цементное тесто *n*

H548 *e* **hydrated lime**
  d hydraulischer Kalk *m*
  f chaux *f* éteinte [hydratée]
  nl gebluste kalk *m*
  r гашёная [гидратированная] известь *f*

H549 *e* **hydration shrinkage**
  d Schrumpfung *f* infolge Hydratation
  f retrait *m* dû à l'hydratation
  nl hydratatiekrimp *m*
  r усадка *f* при гидратации

H550 *e* **hydraulic additive**
  d hydraulischer Zusatz(stoff) *m*
  f adjuvant *m* hydraulique
  nl hydraulische toeslag *m*
  r гидравлическая добавка *f*

H551 *e* **hydraulically operated gate**
  d hydraulisch betätigter Wasserschluß
  m
  f vanne *f* à manœuvre hydraulique
  nl hydraulisch bediende afsluiter *m*
  r затвор *m* гидравлического действия

H552 **hydraulic binders** *see* **hydraulic
  cementing materials**

H553 *e* **hydraulic breakwater**
  d hydraulischer Wellenbrecher *m*
  f brise-lames *m* hydraulique
  nl hydraulische golfbreker *m*
  r гидравлический волнолом *m*

H554 *e* **hydraulic cement**
  d 1. hydraulischer Zement *m*
  2. hydraulisches Bindemittel *n*
  f 1. ciment *m* hydraulique 2. liant *m*
  hydraulique
  nl 1. hydraulisch cement *n*, *m*
  2. hydraulisch bindmiddel *n*
  r 1. гидравлический цемент *m* (*цемент
  с гидравлическими добавками для
  подводных работ*) 2. гидравлическое
  вяжущее *n*

H555 *e* **hydraulic cementing materials,
  hydraulic cementitious materials**
  d hydraulische Bindemittel *n pl*
  [Binder *n pl*]
  f liants *m pl* hydrauliques
  nl hydraulische bindmiddelen *n pl*
  r гидравлические вяжущие материалы
  m pl

H556 *e* **hydraulic classifier**
  d hydraulischer Klassifikator *m*,
  Naßklassierer *m*
  f classificateur *m* hydraulique

  nl hydraulische klasseerinrichting *f*
  r гидроклассификатор *m*

H557 *e* **hydraulic concrete**
  d hydraulischer Beton *m*
  f béton *m* hydraulique
  nl hydraulisch beton *n*
  r гидротехнический бетон *m*

H558 *e* **hydraulic conductivity**
  d hydraulisches Leitungsvermögen *n*,
  Durchlässigkeitskoeffizient *m*
  f conductivité *f* hydraulique,
  coefficient *m* de conductivité
  hydraulique, coefficient *m* de Darcy,
  perméabilité *f*
  nl hydraulische doorlaatcoëfficiënt *m*
  r коэффициент *m* водопроницаемости,
  гидравлическая проводимость *f*,
  коэффициент *m* фильтрации

H559 *e* **hydraulic cyclone**
  d hydraulischer Zyklon *m*,
  Hydrozyklon *m*
  f cyclon *m* hydraulique
  nl hydrocycloon *m*
  r гидроциклон *m*

H560 **hydraulic damper** *see* **hydraulic shock
  absorber**

H561 *e* **hydraulic dredger**
  d Schwimmsaugbagger *m*,
  Naßsaugbagger *m*,
  Pumpen(naß)bagger *m*
  f drague *f* suceuse flottante
  nl snijkopzuiger *m*, cutterzuiger *m*
  r землесосный снаряд *m*

H562 *e* **hydraulic earth-fill cofferdam**
  d aufgespülter Vordamm *m*
  f batardeau *m* en remblai hydraulique
  nl opgespoten dam *m*
  r намывная перемычка *f*

H563 *e* **hydraulic ejector, hydraulic elevator**
  d 1. Hydroelevator *m*;
  Pumpenschachtrohr *n* 2. hydraulischer
  [hochgedrückter] Aufzug *m*;
  hydraulischer Heber *m*
  f 1. pompe *f* à jet, éjecteur *m*
  (hydraulique) 2. ascenseur *m*
  hydraulique; élévateur *m* hydraulique
  nl 1. hydraulische elevator *m*
  2. hydraulische lift *m*
  r 1. гидроэлеватор *m* 2. гидравлический
  лифт *m* [подъёмник *m*]

H564 *e* **hydraulic engineering**
  d 1. Hydrotechnik *f* 2. Wasserbau *m*,
  Wasserbauwesen *n*
  f 1. hydrotechnique *f* 2. travaux *m pl*
  hydrauliques
  nl 1. hydraulica *f* 2. waterbouwkunde *f*
  r 1. гидротехника *f* 2. гидротехническое
  строительство *n*

H565 *e* **hydraulic erosion dredger**
  d Spülbagger *m*
  f drague *f* suceuse à désagrégateur
  hydraulique
  nl profielzuiger *m*

r землесосный снаряд *m*
с гидравлическим рыхлителем

H566 *e* **hydraulic excavation**
d Spülaushub *m*
f remblayage *m* [remblaiement *m*]
hydraulique
nl hydraulisch graafwerk *n*
r гидромеханизированная разработка *f*
грунта; гидравлическая экскавация *f*

H567 *e* **hydraulic excavator**
d Hydrobagger *m*, Hydraulikbagger *m*
f pelle *f* hydraulique
nl hydraulische graafmachine *f*
r гидравлический экскаватор *m*

H568 *e* **hydraulic fill**
d 1. Aufspülung *f*, eingespülte
Auffüllung *f*, Spülkippe *f*
2. Spülversatz *m*
f 1. remblai *m* (à transport) hydraulique
2. remblayage *m* [remblaiement *m*]
hydraulique
nl 1. opgespoten terrein *n* 2. opspuiten *n*
r 1. намывная насыпь *f*; гидроотвал *m*
2. гидрозакладка *f*

H569 *e* **hydraulic fill dam**
d gespülter Erddamm *m*,
Spül(stau)damm *m*
f barrage *m* en terre remblayée
hydrauliquement
nl opgespoten dam *m*
r намывная плотина *f*

H570 *e* **hydraulic fill pipeline**
d Spülleitung *f*, Schlammleitung *f*,
Pulpenleitung *f*
f pulpeduc *m*, conduite *f* pour
remblayage hydraulique
nl persleiding *f* voor het opspuiten
r пульпопровод *m*

H571 *e* **hydraulic friction**
d Reibungsgefälle *n*, hydraulische
Reibung *f*
f perte *f* de charge due au frottement
par unité de longeur, frottement *m*
hydraulique
nl hydraulische wrijving *f*
r гидравлическсе сопротивление *n*
трения

H572 **hydraulic friction factor** *see* **resistance coefficient**

H573 *e* **hydraulic grade line** *US*
d Drucklinie *f*, piezometrische Linie *f*;
Wasserspiegellinie *f*
f ligne *f* piézométrique, ligne *f* d'eau
nl hydraulische gradiënt *m*
r пьезометрическая линия *f*, линия
гидравлического уклона *f*; линия *f*
свободной поверхности открытого
водотока

H574 *e* **hydraulic gradient**
d Druckgefälle *n*, Energieliniengefälle *n*,
hydraulisches Gefälle *n*,
Wasserspiegelgefälle *n*
f gradient *m* hydraulique

nl hydraulische gradiënt *m*, verval *n*
(*van een rivier*)
r гидравлический градиент *m*
[уклон *m*]; уклон *m* свободной
поверхности в открытом русле

H575 **hydraulic hammer** *see* **hydraulic ram 1.**

H576 *e* **hydraulic investigations**
d wasserbauliche Forschungsarbeiten *f pl*
f reconnaissance *f* hydraulique
nl waterbouwkundig onderzoek *n*
r гидротехнические изыскания *n pl*

H577 *e* **hydraulicity**
d Hydraulizität *f*
f hydraulicité *f*
nl hydraulisch vermogen *n* (*het vermogen
om onder water te verharden*)
r гидравлическая активность *f*

H578 *e* **hydraulic jack**
d hydraulischer Hebebock *m*
f vérin *m* hydraulique
nl hydraulische vijzel *m*
r гидравлический домкрат *m*

H579 *e* **hydraulic jump**
d Wassersprung *m*, Wechselsprung *m*,
hydraulischer Sprung *m*,
Fließwechsel *m*
f ressaut *m* hydraulique
[d'exhaussement]
nl watersprong *m*, hydraulische sprong *m*
r гидравлический прыжок *m*

H580 *e* **hydraulicking**
d 1. Aufspülung *f*, Spülung *f*,
Spülverfahren *n* 2. Spülabbau *m*
f 1. remblayage *m* hydraulique
2. méthode *m* hydraulique
nl 1. opspuitmethode *f* 2. hydraulische
ontgraving *f*
r 1. намыв *m*; метод *m* намыва
2. гидравлическая
[гидромеханизированная]
разработка *f*

H581 *e* **hydraulic lime**
d hydraulischer Kalk *m*
f chaux *f* hydraulique
nl hydraulische kalk *m*
r гидравлическая известь *f*

H582 *e* **hydraulic mean depth**
d hydraulischer Radius *m*,
Profilhalbmesser *m*, hydraulische
Querschnittstiefe *f*
f rayon *m* hydraulique
nl hydraulische straal *m*
r гидравлический радиус *m*

H583 **hydraulic mining** *see* **hydraulicking 2.**

H584 *e* **hydraulic modelling**
d Strömungsmodellierung *f*
f simulation *f* hydraulique, étude *f* sur
modèles hydrauliques
nl waterloopkundig modelonderzoek *n*
r гидравлическое моделирование *n*

H585 *e* **hydraulic monitor**
d Hydromonitor *m*,

Wasserspritzkanone *f*,
Druckstrahlbagger *m*
f monitor *m* hydraulique, lance *f*
à eau
nl hydraulische spuit *f* (*m*)
r гидромонитор *m*

H586 *e* **hydraulic mortar**
d hydraulischer Mörtel *m*,
Wassermörtel *m*
f mortier *m* hydraulique
nl hydraulische mortel *m*
r гидравлический раствор *m* (*для
подводных работ*)

H587 *e* **hydraulic permeability**
d Wasserdurchlässigkeit *f*
f perméabilité *f* (à l'eau)
nl waterdoorlatendheid *f*
r гидравлическая проницаемость *f*,
водопроницаемость *f*

H588 *e* **hydraulic pile driver**
d Spülpfahlabsenker *m*
f enfonçeur *m* de palplanches
hydraulique
nl centrale spuitlans *f* (*m*) (*betonpalen*)
r гидравлический сваепогружатель *m*

H589 *e* **hydraulic platform**
d selbstfahrender Hydraulik-Heber *m*;
Hydraulik-Hebebühne *f*,
Hydraulik-Arbeitsbühne *f*
f engin *m* élévateur à nacelle (à bras
hydraulique)
nl hydraulisch bediend hefplatform *n*,
hoogwerker *m*
r гидравлический (автомобильный)
подъёмник *m*; гидравлическая
подъёмная платформа *f*, рабочая
площадка *f* с гидравлическим
подъёмником

H590 *e* **hydraulic pressure test**
d Abdrücken *n*, Abdrückprüfung *f*,
Innendruckprobe *f*,
Wasserabpreßversuch *m*,
hydraulische Druckprobe *f*
f épreuve *f* hydraulique
nl waterdrukproef *f* (*m*)
r гидравлическое испытание *n*,
гидравлическая опрессовка *f*
(*трубопровода*)

H591 **hydraulic radius** *se* **hydraulic mean
depth**

H592 *e* **hydraulic ram**
d 1. hydraulischer Widder *m*,
Stoßheber *m* 2. Hydraulikzylinder *m*
f 1. bélier *m* hydraulique 2. cylindre *m*
hydraulique
nl 1. hydraulische plunjer *m*
2. hydraulische cilinder *m*
r 1. гидравлический таран *m*
2. гидроцилиндр *m*

H593 *e* **hydraulic scaffolding**
d Hydraulik-Gerüst *n*, Hydro-Rüstung *f*
f échafaudage *m* télescopique à vérins
hydrauliques
nl hydraulisch bediende stelling *f*

r выдвижные телескопические
подмости *pl* с гидродомкратами

H594 *e* **hydraulic seal**
d 1. hydraulische Dichtung *f*
2. Geruchverschluß *m*
f 1. joint *m* hydraulique 2. vanne *f*
à commande hydraulique
nl 1. hydraulische dichting *f*
2. waterslot *n*
r 1. гидравлическое уплотнение *n*
2. гидравлический затвор *m*

H595 *e* **hydraulic shock absorber**
d 1. Flüssigkeitsdämpfer *m*,
hydraulischer Stoßdämpfer *m*
2. Stoßdruckdämpfer *m*
f 1. amortisseur *m* hydraulique
2. dissipateur *m* d'énergie hydraulique
nl 1. hydraulische schokdemper *m*
2. waterslagdemper *m*
r 1. гидравлический амортизатор *m*
2. гаситель *m* гидравлического удара

H596 *e* **hydraulic slope**
d hydraulisches Gefälle *n*,
Energieliniengefälle *n*
f pente *f* de la ligne d'eau
nl hydraulische gradiënt *m*
r гидравлический уклон *m*

H597 *e* **hydraulic soil transportation** *US*
d hydraulische Erdstofförderung *f*,
Spül(boden)förderung *f*
f transport *m* hydraulique du sol
nl grondtransport *n* via spoelgoten
r гидравлический транспорт *m* грунта

H598 *e* **hydraulic splitter**
d hydraulisches Steinspaltgerät *n*
f coin *m* hydraulique (*pour fendre des
pierres*)
nl hydraulische steenklover *m*
r гидравлический клин *m* (*для
раскалывания камня*)

H599 *e* **hydraulic spoil bank**
d Hydrokippe *f*
f dépôt *m* de terres [terril *m*]
à transport hydraulique
nl mijnsteenberg *m*
r гидроотвал *m*

H600 **hydraulic spraying** *see* **airless spraying**

H601 *e* **hydraulic stripping**
d Spülung *f* von Abraum, hydraulische
Abraumbeseitigung *f*
f découverte *f* hydraulique
nl geforceerde afspoeling *f* van deklagen
r гидравлическая вскрыша *f*

H602 **hydraulic structures** *see* **hydraulic
works**

H603 *e* **hydraulic tunnel**
d Wasserleitungsstollen *m*,
Wasserversorgungsstollen *m*
f tunnel *m* hydraulique
nl waterleidingtunnel *m*
r гидротехнический туннель *m*

H604 *e* **hydraulic works**
d Wasserbauten *m* *pl*

f ouvrages *m pl* hydrotechniques,
ouvrages *m pl* d'équipment
hydraulique
nl hydrotechnische werken *n pl*
r гидротехнические сооружения *n pl*

H605 hydrocarbon black *see* carbon black

H606 *e* hydrodynamic bulging
d hydrodynamischer Grundbruch *m*
f soulèvement *m* de terrain dû aux
actions hydrodynamiques
nl welving *f* van de bodem door
grondwaterbewegingen
r гидродинамическое выпирание *n*
грунта

H607 *e* hydrodynamic pressure
d hydrodynamischer Druck *m*
f pression *f* hydrodynamique
nl hydrodynamische druk *m*
r гидродинамическое давление *n*

H608 hydroelectric complex *see* multiple-use
development

H609 *e* hydroelectric dam
d Wasserkraft-Talsperre *f*
f barrage *m* hydroélectrique
nl stuwdam *m* met waterkrachtcentrale
r плотина *f* ГЭС, станционная
плотина *f*

H610 *e* hydroelectric development
d Kraftstufe *f*, Wasserkraftanlage *f*
f aménagement *m* hydroélectrique
nl ontwikkeling *f* van (het gebruik van)
waterkracht
r энергетический гидроузел *m*

H611 *e* hydroelectric power station
d Wasserkraftwerk *n*
f usine *f* hydroélectrique
nl waterkrachtcentrale *f* (*m*)
r гидроэлектрическая станция *f*,
гидроэлектростанция *f*, ГЭС

H612 *e* hydroelectric scheme
d 1. Wasserkraftprojekt *n*
2. Wasserkraftwerksystem *n*
f 1. projet *m* d'aménagement
hydraulique 2. système *m* hydraulique
[d'ensemble énergétique]
nl 1. hydro-elektrisch project *n*
2. hydro-elektrisch systeem *n*
r 1. проект *m* гидроэнергетического
узла 2. гидроэнергетическая
система *f*

H613 hydrofiller *see* bituminated filler

H614 *e* hydrograph
d Ganglinie *f*
f hydrogramme *m*
nl hydrogram *n*
r гидрограф *m*

H615 *e* hydrographic survey
d hydrografische Aufnahme *f*
f levé *m* hydrographique
nl hydrografische opmeting *f*
r гидрографическая съѐмка *f*

H616 *e* hydrological forecast
d hydrologische Vorhersage *f*

[Prognose *f*], Abfluß-Vorhersage *f*
f prévision *f* hydrologique
nl hydrologische prognose *f*
r гидрологический прогноз *m*

H617 *e* hydrological regime
d hydrologisches Regime *n*
f régime *m* hydrologique
nl hydrologisch regime *n*
r гидрологический режим *m*

H618 *e* hydrologic study
d hydrologische Erkundungen *f pl*
f études *f pl* hydrologiques
nl hydrologisch onderzoek *n*
r гидрологические изыскания *n pl*

H619 *e* hydro-mechanisation
d Hydromechanisierung *f*,
Spülverfahren *n*
f méthode *f* hydraulique, remblayage *m*
hydraulique
nl hydromechanisatie *f*
r гидромеханизация *f*

H620 *e* hydrometric network
d Abflußmeßnetz *n* (der
Abflußmeßstellen)
f réseau *m* hydrométrique
nl meetnet *n* afvloeiingsmeetpunten
r гидрометрическая сеть *f*

H621 *e* hydrometric section
d Wassermeßprofil *n*
f ligne *f* de la section de jaugeage
nl watermeetprofiel *n*
r гидрометрический створ *m*

H622 *e* hydromodule
d Hydromodul *n*
f hydromodule *m*, module *m*
d'arrosage [d'irrigation]
nl hydromodulus *m*
r удельный оросительный расход *m*,
гидромодуль *m*

H623 *e* hydrophobic cement
d wasserabstoßender [wasserabweisender,
hydrophober] Zement *m*
f ciment *m* hydrophobe
nl waterafstotend cement *n*
r гидрофобный цемент *m*

H624 *e* hydropower engineering
d Hydroenergetik *f*
f industrie *f* hydroélectrique
nl hydro-elektrische techniek *f*
r гидроэнергетика *f*

H625 *e* hydro power plant
d hydraulische Kraftanlage *f*;
Wasserkraftwerk *n*
f équipement *m* [installation *f*]
hydroélectrique
nl waterkrachtcentrale *f* (*m*)
r гидроэнергетическая установка *f*;
гидроэлектростанция *f*

H626 *e* hydrostatic load
d hydrostatische Belastung *f*,
Wasserdruck *m*
f charge *f* hydrostatique

nl hydrostatische belasting *f*
r гидростатическая нагрузка *f*

H627 *e* **hydrostatic pressure**
    d hydrostatischer Druck *m*,
    Ruhewasserdruck *m*
    f pression *f* hydrostatique
    nl hydrostatische druk *m*
    r гидростатическое давление *n*

H628 *e* **hydrostatic pressure ratio**
    d Koeffizient *m* des aktiven Erddrucks
    f coefficient *m* de poussée active des
    terres
    nl coëfficiënt *m* van actieve gronddruk
    r коэффициент *m* активного
    давления грунта

H629 **hydrostatic test** *see* **hydraulic pressure
test**

H630 *e* **hygrometer**
    d 1. Hygrometer *n*,
    Feuchtigkeitsmesser *m*
    2. Psychrometer *n*
    f 1. hygromètre *m* 2. psychromètre *m*
    nl 1. hygrometer *m* 2. psychrometer *m*
    r 1. гигрометр *m* 2. психрометр *m*

H631 *e* **hygroscopic moisture**
    d hygroskopisches Wasser *n*
    f humidité *f* [eau *f*] hygroscopique
    nl hygroscopisch gebonden water *n*
    r гигроскопическая влага *f*

H632 **hygrostat** *see* **humidistat**

H633 *e* **hyperbolic-paraboloid shell**
    d HP-Schale *f*, hyperbolische
    Paraboloidschale *f*
    f coque *f* [voile *m* mince] en
    paraboloïde hyperbolique
    nl hyperbolische paraboloïdeschaal *f* (*m*)
    r гипар *m*, оболочка *f* в форме
    гиперболического параболоида

H634 *e* **hyperbolic shell**
    d hyperbolische Schale *f*
    f coque *f* hyperbolique
    nl hyperbolische schaal *f* (*m*)
    r гиперболическая оболочка *f*

H635 *e* **hyper-critical flow**
    d Schießen *n*, schießender Abfluß *m*,
    reißende Strömung *f*, Sturmströmung *f*
    f écoulement *m* torrentiel [jaillissant]
    nl schieten *n*, schietende *f* stroming *f*
    r сверхкритическое течение *n*,
    бурное состояние *n* потока

H636 **hyperstatic structure** *see* **statically
indeterminate structure**

I

I1 *e* **I-beam**
    d Doppel-T-Träger *m*, I-Träger *m*
    f poutre *f* [poutrelle *f*, profilé *m*] en I
    nl I-balk *m*
    r двутавровая балка *f*, двутавр *m*

I2 *e* **ice apron, ice breaker**
    d Pfeilerkopf *m*, Eisbrecherpfeiler *m*,
    Eisschollenbrecher *m*
    f brise-glace *m*
    nl ijsbok *m*, ijsbreker *m*
    r ледорез *m*

I6 *e* **ice chute**
    d Eisablaß *m*, Eisklappe *f*
    f passe *f* de sorbet
    nl aflaat *m* voor drijfijs
    r ледосброс *m*, шугосброс *m*

I7 *e* **ice concrete**
    d Eisbeton *m*, Schmelzbeton *m*
    f béton *m* à la glace
    nl ijsbeton *n*
    r ледяной бетон *m* (*ячеистый бетон*,
    *в котором образование пор
    обеспечивается введением в смесь льда*)

I8 *e* **ice dam**
    d Eisbarre *f*, Eisstau *m*, Eisversetzung *f*
    f embâcle *m*
    nl ijsdam *m*
    r ледяной затор *m*

I9 **ice drift** *see* **ice run**

I10 *e* **ice gate**
    d Eisablaß *m*, Eisklappe *f*
    f ouvrage *m* de contrôle des glaces
    nl doorlaat *m* voor drijfijs
    r ледосбросное сооружение *n*

I13 *e* **ice load, ice loading**
    d Eisbelastung *f*
    f charge *f* de glace
    nl ijsbelasting *f*
    r ледовая нагрузка *f*

I14 **ice pass** *see* **slush ice chute**

I15 *e* **ice pressure**
    d Eisdruck *m*
    f pression *f* de glace
    nl ijsdruk *m*
    r давление *n* льда

I16 *e* **ice regime**
    d Eisverhältnisse *n pl*
    f régime *m* de glace
    nl ijsomstandigheiden *f pl*
    r ледовый режим *m*

I17 *e* **ice run**
    d Eisgang [*m*, Eistrieb *m*, Eistreiben
    n
    f dérive *f* de la glace, débâcle *f*
    nl ijsgang *m*, kruien *n* van het ijs
    r ледоход *m*

I19 *e* **i,d-diagram**
    d I,d-Diagramm *n*, Mollier-Diagramm *n*
    f diagramme *m* psychrométrique
    nl diagram *n* van Mollier,
    psychrometrisch diagram *n*
    r i,d-диаграмма *f*, психрометрическая
    диаграмма *f*

I21 *e* **imaginary hinge**
    d imaginäres [fiktives] Gelenk *n*
    f articulation *f* fictive [imaginaire]
    nl fictief scharnier *n*
    r фиктивный шарнир *m*

122   *e* **Imhoff tank**
   d Imhoffbrunnen *m*
   f bassin *m* de décantation à deux étages
   nl imhofftank *m*, voorbezinktank *m*
   r двухъярусный отстойник *m*

123   **immediate deformation** *see* **instantaneous deformation**

124   **immediate runoff** *see* **surface runoff**

125   *e* **immersed tunnel**
   d in offener Baugrube hergestellter Tunnel *m*
   f tunnel *m* immergé
   nl in open bouwput gemaakte tunnel *m*, tunnel *m* van afgezonken elementen (*open bouwput methode*)
   r туннель *m*, построенный методом открытой проходки

126   *e* **immersion heater**
   d Taucherhitzer *m*
   f chauffe-eau *m* électrique à élément chauffant immergé
   nl warmwatertoestel *n* met ondergedompelde verwarmingselementen, elektrische boiler *m*
   r водонагреватель *m* с погружным электронагревательным элементом

127   *e* **immersion needle**
   d Vibrationsnadel *f*, Rüttelnadel *f*, Schwing(ungs)nadel *f*
   f aiguille *f* vibrante
   nl trilnaald *f* (*m*)
   r виброигла *f*

128   **immersion vibrator** *see* **internal vibrator**

129   *e* **impact bending strength**
   d Schlagbiegefestigkeit *f*
   f résistance *f* au choc en flexion, résistance *f* à la flexion au choc
   nl slagbuigsterkte *f*
   r ударная прочность *f* [сопротивление *n* удару] при изгибе

130   *e* **impact compacter**
   d Schlagverdichter *m*
   f compacteur *m* [dameur *m*, dameuse *f*] à percussion
   nl slagverdichter *m*
   r ударный уплотнитель *m*

131   *e* **impact crusher**
   d Prallbrecher *m*
   f concasseur *m* percuteur [à percussion]
   nl slagbreker *m*
   r ударно-отражательная дробилка *f*

132   *e* **impact damage**
   d Stoßbeschädigung *f*
   f dommage *m* dû au choc
   nl beschadiging *f* door stoten
   r повреждение *n* от удара

133   *e* **impact-driven pile**
   d Rammpfahl *m*
   f pieu *m* battu

   nl heipaal *m*
   r забивная свая *f*

134   *e* **impact electric wrench**
   d Elektroschlagschrauber *m*
   f clé *f* [clef *f*] électrique à choc [à percussion]
   nl elektrische moersleutel *m*
   r электрический гайковёрт *m* ударного действия

135   *e* **impact factor**
   d Stoßfaktor *m*, Stoßzahl *f*, Stoßwert *m*
   f coefficient *m* de majoration dynamique
   nl stootcoëfficiënt *m*
   r динамический коэффициент *m*, коэффициент *m* динамического воздействия

137   *e* **impact load**
   d Schlagbeanspruchung *f*, Stoßlast *f*, Stoßbelastung *f*
   f charge *f* de choc
   nl stootbelasting *f*
   r ударная нагрузка *f*

138   *e* **impact loss**
   d Stoßverlust *m*
   f perte *f* de charge due au coup de bélier
   nl energieverlies *n* bij een slag van een heiblok
   r динамические потери *f pl*, потери *f pl* напора при гидравлическом ударе

139   *e* **impact noise**
   d Stoßschall *m*, Trittschall *m*
   f bruit *m* d'impact
   nl contactgeluid *n*
   r ударный шум *m*

140   *e* **impact of a ship**
   d Schiffstoß *m*, Auflaufen *n* des Schiffes
   f choc *m* de navire
   nl oploop *m* van een schip
   r навал *m* судна

141   *e* **impactor**
   d Prallmühle *f*
   f moulin *m* à percussion
   nl slagmolen *m*
   r ударно-отражательная мельница *f*

142   *e* **impact pile driver**
   d Vibrationsramme *f*
   f enfonceur *m* de palplanches à percussion [à vibrations]
   nl blok *n* voor trillend heien
   r вибромолот *m*

143   *e* **impact pressure**
   d Geschwindigkeitsdruck *m*, Staudruck *m*
   f pression *f* dynamique
   nl dynamische belasting *f*, stuwdruk *m*
   r скоростное давление *n*, скоростной напор *m*

144   **impact resistance** *see* **impact strength**

145   *e* **impact screen**
   d Stoßsieb *n*, Prallsieb *n*
   f crible *m* à percussion
   nl schudzeef *f* (*m*)

r вибрационный грохот *m*,
виброгрохот *m*

146 e **impact strength**
d 1. Kerbschlagfestigkeit *f*
2. Schlagfestigkeit *f*
f 1. résilience *f* 2. résistance *f*
dynamique [au choc]
nl 1. kerfslagwaarde *f* 2. schokvastheid *f*,
slagweerstand *m*
r 1. ударная вязкость *f* 2. ударная
прочность *f*, сопротивление *n* удару

147 e **impact stresses**
d dynamische Spannungen *f pl*,
Schlagbeanspruchung *f*,
Stoßbeanspruchung *f*
f contraintes *f pl* dues au choc
[sollicitation dynamique]
nl stootbelastingen *f pl*
r напряжения *n pl* от удара
[ударной нагрузки]

148 e **impact test**
d 1. Kerbschlagversuch *m*
2. Schlagversuch *m*, Schlagprüfung *f*
f 1. essai *m* de résilience 2. épreuve *f*
au choc
nl 1. kerfslagproef *f* (*m*) 2. slagproef *f* (*m*)
r 1. определение *n* ударной вязкости
2. испытание *n* на удар

149 e **impact wrench**
d Schlagschrauber *m*, Schraubmaschine *f*
f clé *f* [clef *f*] à choc [à percussion]
nl pneumatische moersleutel *m*
r ударный гайковёрт *m*

151 e **impeller pump**
d Kreiselpumpe *f*
f pompe *f* centrifuge
nl centrifugaalpomp *f* (*m*)
r центробежный насос *m*

152 e **impending slough**
d maximal zulässige Fließfähigkeit *f*
(*Torkretbeton*)
f fluidité *f* limite admissible du béton
projeté
nl hoogst toelaatbare vloeibaarheid *f* van
spuitbeton
r предельная подвижность *f*
[текучесть *f*] (*торкретбетона*)

153 e **imperfect frame**
d geometrisch veränderliches Stabsystem
n [Stabwerk *n*]
f système *m* à treillis incomplet,
construction *f* à ossature déformable
nl geometrisch [meetkundig,
onvolkomen] veranderlijk vakwerk *n*
r геометрически изменяемая
стержневая система *f*

155 e **imperfect well**
d unvollkommener Brunnen *m*
f puits *m* incomplet
nl onvolkomen bron *f* (*m*) [wel *f* (*m*)]
r несовершенный колодец *m*

157 e **impermeability factor**
d 1. Abflußkoeffizient *m*
2. Undurchlässigkeitsbeiwert *m*

f 1. coefficient *m* d'écoulement 2. **taux**
m d'imperméabilité à l'eau
nl 1. afvloeiingscoëfficiënt *m* 2. factor *m*
van de ondoorlatendheid
r 1. коэффициент *m* стока
2. коэффициент *m*
водонепроницаемости

158 e **impermeable rock**
d wasserundurchlässiges Gestein *n*
f roche *f* imperméable à l'eau
nl waterdicht [ondoorlatend] gesteente *n*
r водонепроницаемая горная порода *f*

159 e **impervious blanket**
d Dichtungsvorlage *f*, Dichtungsteppich
m, Dichtungsschürze *f*; Außendichtung
f, Dichtungsschicht *f*
f revêtement *m* d'étanchéité, cuvelage
m; masque *m* d'étanchéité, écran *m*
nl ondoorlatend scherm *n* van de stuw
r водонепроницаемое покрытие *n*;
понур *m*; противофильтрационный
экран *m*, уплотняющий слой *m*
(*на откосе земляной плотины*)

160 e **impervious core**
d Dichtungskern *m*, Dammkern *m*,
undurchlässiger Kern *m*
f noyau *m* d'étanchéité [imperméable]
nl dichte kern *f* (*m*) (*aardedam*)
r противофильтрационное ядро *n*
(*земляной плотины*)

161 e **impervious course**
d Sperrschicht *f*, Abdichtungsschicht *f*,
Sperrlage *f*
f couche *f* de scellement [de fermeture]
nl ondoorlatende laag *f* (*m*),
afdichtingslaag *f* (*m*)
r изоляционный [уплотняющий] слой *m*

162 e **impervious diaphragm**
d Dicht(ungs)wand *f*, Sperrwand *f*,
Dichtungsschirm *m*
f voile *m* imperméable, diaphragme *m*
d'étanchéité
nl afdichtingsscherm *n*
r диафрагма *f* (*плотины*)

162a **impervious layer** *see* **impervious
course**

163 e **impervious soils**
d wasserundurchlässige Böden *m pl*
f sols *m pl* [terrains *m pl*] imperméables
nl niet-doorlatende gronden *m pl*
r водонепроницаемые грунты *m pl*

164 e **impingement air filter**
d ölbenetztes Metallzellenfilter *n*, *m*,
Ölluftfilter *n*, *m*
f filtre-crépine *m* huilé [ensimé] à air
nl oliebadluchtfilter *n*, *m*
r воздушный фильтр *m* ударно-
-вязкостного типа; фильтр *m*
с замасливаемой металлической
сеткой высокой пылеёмкости

165 e **implementation of program**
d Planerfüllung *f*
f exécution *f* [mise *f* en œuvre] d'un
programme de construction

nl realisatie *f* van het plan
r выполнение *n* [реализация *f*] плана (*работ*)

166 *e* **imposed load**
d Nutzlast *f*, eingeleitete Last *f*
f charge *f* appliquée
nl aangebrachte belasting *f*
r приложенная нагрузка *f* (*любая нагрузка, кроме собственного веса конструкции*)

167 *e* **imposition of the prestress**
d Vorspannkräfteeintragung *f*
f mise *f* en tension, application *f* des efforts de précontrainte
nl aanbrengen *n* van de voorspanning
r приложение *n* (усилий) преднапряжения

168 *e* **impost**
d Bogenanfänger *m*, Bogenansatz *m*; Gewölbeanfang *m*, Kämpfer *m*; Kämpferholz *n*, Fensterkämpfer *m*, Mittelstück *n*
f imposte *f*
nl impost *m*, aanzet(steen) *m*; kalf *n*, tussendorpel *m*
r импост *m*, пята *f* арки *или* свода; импост *m* окна

169 *e* **impounded water**
d Stauwasser *n*
f eau *f* accumulée
nl stuwwater *n*
r подпёртый бьеф *m*

170 *e* **impounding dam**
d Speichersperre *f*, Sperrdamm *m*, Staudamm *m*
f barrage *m* de retenue
nl stuwdam *m*
r водоудерживающая плотина *f*, запруда *f*

171 *e* **impounding reservoir**
d Staubecken *n*, Speicherbecken *n*, Talsperrenbecken *n*, Stausee *m*; Wassersammler *m*, Vorratsbehälter *m*
f bassin *m* de retenue, réservoir *m* d'accumulation, lac *m* de barrage
nl stuwbekken *n*; stuwmeer *n*; voorraadsreservoir *n*
r водохранилище *n* для водоснабжения; резервуар-накопитель *m*; водосборник *m*

172 *e* **impoundment**
d Sammelbecken *n*, Speicherbecken *n*
f retenue *f*, lac *m* de retenue
nl verzamelbekken *n*, voorraadsbassin *n*
r водоём *m* (*созданный путём накопления воды*)

174 *e* **improved alignment**
d verbesserte Trasse *f*
f tracé *m* amélioré [perfectionné]
nl verbeterd tracé *n*
r улучшенная трасса *f*

175 *e* **improved land**
d erschlossenes Baugelände *n*
f terrain *m* aménagé
nl bouwrijp terrein *n*
r благоустроенная территория *f*

176 *e* **improved wood**
d veredeltes Schnittholz *n*, vergütetes Holz *n*
f bois *m* amélioré
nl verbeterd bezaagd hout *n*
r улучшенный пиломатериал *m*

177 *e* **impulse steam trap**
d dynamischer Kondensatableiter *m*
f purgeur *m* de vapeur actionné par impulsion
nl dynamische condenspot *m*
r импульсный [динамический] конденсатоотводчик *m*

178 *e* **impurities**
d Verunreinigungen *f pl*; Beimengungen *f pl*
f impuretés *f pl*
nl verontreinigingen *f pl*, vervuilingen *f pl*; bijmengels *n pl*
r загрязнения *n pl*; примеси *f pl*

179 *e* **inaccuracies of fabrication**
d Fertigungsungenauigkeiten *f pl*
f inexactitude *f* de fabrication [de confection]
nl fabricage-afwijkingen *f pl*
r неточности *f pl* изготовления

180 *e* **in-built furniture**
d Einbaumöbel *n pl*
f meubles *m pl* incorporés
nl inbouwmeubel *n*
r встроенная мебель *f*

181 *e* **incineration**
d Müllverbrennung *f*, Abfallverbrennung *f*
f crémation *f* [incinération *f*] des ordures
nl vuilverbranding *f*
r мусоросжигание *n*

182 *e* **incinerator**
d Müllverbrennungsofen *m*, Müllverbrennungsanlage *f*
f usine *f* d'incinération
nl (vuil)verbrandingsoven *m*
r мусоросжигательная установка

183 *e* **inclination**
d 1. Längsneigung *f* 2. Neigungswinkel *m*
f 1. pente *f*, inclinaison *f*; dévers *m*; déclivité *f* 2. angle *m* d'inclinaison
nl 1. (lengte)helling *f*; inclinatie *f* 2. hellingshoek *m*
r 1. продольный уклон *m*; наклон *m* 2. угол *m* наклона

184 *e* **inclined-axis mixer**
d Kipptrommel(-Transport)betonmischer *m*
f camion *m* malaxeur [mélangeur] à tambour basculant (*pour la décharge*)
nl truckmixer *m*
r автобетоносмеситель *m* с опрокидным барабаном (*в сторону выгрузки смеси*)

335

I85 e inclined cableway
    d Schräg-Luftseilbahn *f*
    f téléphérique *m* [téléférique *m*,
    transporteur *m* à cable] gravitaire
    nl hellende kabelbaan *f (m)*
    r наклонная канатная подвесная
    дорога *f*

I86 e inclined drop
    d Schußrinne *f*, Schnellflußstrecke *f*
    f rapide *m*, chute *f*
    nl stroomversnelling *f*
    r быстроток *m*

I87 e inclined end post
    d Kopfstrebe *f*, Endstrebe *f*
    f diagonale *f* extrême de la poutre à
    treillis
    nl eindschoor *m*, einddiagonaal *f (m)*
    r концевой раскос *m* фермы

I88 e inclined gallery
    d Schrägstollen *m*
    f galerie *f* inclinée
    nl hellende galerij *f*
    r наклонная галерея *f*

I89 e inclined intake
    d Schrägfassung *f*
    f prise *f* d'eau en profondeur à puits
    filtrants inclinés
    nl wateronttrekking *f* door hellende
    inlaten
    r лучевой водозабор *m*
    с наклонными скважинами

I90 e inclined links
    d Schrägbügel *m pl*
    f étriers *m* inclinés
    nl schuine schakels *m pl*,
    diagonaalbeugels *m pl*
    r наклонные арматурные хомуты *m pl*

I91 e inclined plane
    d schiefe Ebene *f*
    f plan *m* incliné, rampe *f*
    nl hellend vlak *n*
    r наклонная плоскость *f*

I92 e inclinometer
    d Neigungsmesser *m*
    f inclinomètre *m*, boussole *f*
    d'inclinaison
    nl hellingmeter *m*
    r инклинометр *m*

I93 e inclosure wall
    d nichttragende Außenwand *f* in
    Skelettbauweise
    f mur-rideau *m*, mur *m* non-portant du
    bâtiment en ossature
    nl niet-dragende muur *m* in een
    skeletconstructie
    r ограждающая стена *f* каркасного
    здания

I94 e incomplete well
    d nichtvollständiger [partieller]
    Brunnen *m*
    f puits *m* incomplet
    nl onvolledige bron *m*
    r несовершенный колодец *m*;
    несовершенная скважина *f*

I95 e increaser
    d Erweiterungsstück *n*, Anschluß *m* für
    Gußeisenrohr an Steinzeugrohr
    f raccord *m* fonte-grès [grès-fonte]
    nl aanpassingsstuk *n* voor gietijzer op
    gres
    r раструбный переход *m* для
    присоединения чугунной
    канализационной трубы
    к керамической

I96 e increments of settlement
    d Setzungszunahme *f*
    f accroissements *m pl* d'affaissement,
    tassement *m* [affaissement *m*]
    progressif
    nl toename *f* van de zetting
    r рост *m* осадок сооружения
    (*во времени*)

I97 e incrustation
    d 1. Ansatz *m*, Belag *m*
    2. Kesselstein *m* 3. Inkrustation *f*,
    Verkrustung *f*
    f 1. dépôt *m* dur (*sur le béton*,
    maçonnerie) 2. entartrage *m*, tartre *m*
    3. incrustation *f*
    nl 1. harde aanslag *m* 2. ketelsteen *m*
    3. incrustatie *f*
    r 1. твёрдое наслоение *n* (*на бетоне*,
    каменной кладке) 2. накипь *f*
    3. инкрустация *f*

I98 e indented bars
    d Nockenstäbe *m pl*
    f barres *f pl* d'armature crénelées
    [déformées]
    nl geprofileerde wapeningsstaven *f (m)*
    pl
    r арматурные стержни *m pl*
    периодического профиля,
    холодносплющенная стержневая
    арматура *f*

I99 e indented wire
    d profilierter Draht *m*, Profildraht *m*
    f fil *m* à empreintes [façonné, profilé]
    nl profieldraad *m*
    r холодносплющенная арматурная
    проволока *f*

I100 e indenting roller
    d Riffelwalze *f*
    f cylindre *m* à billes striées
    nl ribbelwals *f (m)*
    r дорожный каток *m* с рифлёными
    вальцами

I101 e independent boiler
    d freistehender Heizkessel *m*, Kessel *m*
    mit eingebauter Feuerung
    f chaudière *f* indépendante
    nl mobiele ketel *m* met ingebouwde
    vuurhaard
    r передвижной котельный агрегат *m*
    со встроенной топкой

I102 e independent insulated hot water heater
    d Heißwasserspeicher *m*
    f chauffe-eau *m* calorifugé
    nl vrijstaande boiler *m*

r теплоизолированный ёмкостный
   водонагреватель *m*

I104   **independent-pole scaffolding** *see*
      double-pole scaffold

I105 *e* **index of liquidity**
    d Fließ(fähigkeits)index *m*, Fließzahl *f*
    f indice *m* de liquidité
   nl viscositeitsindex *m*
    r показатель *m* [индекс *m*] текучести
      (*грунта*)

I106 *e* **index of plasticity**
    d Plastizitätsindex *m*, Bildsamkeitszahl
      f, Bildsamkeit *f*
    f indice *m* de plasticité
   nl plasticiteitsindex *m*
    r показатель *m* [число *n*] пластичности
      (*грунта*)

I107 *e* **index properties**
    d Klassifizierungseigenschaften *f pl*
    f propriétés *f pl* caractéristiques
   nl kenmerkende eigenschappen *f pl*
    r классификационные свойства *n pl*
      [признаки *m pl*]

I108 *e* **indirect cylinder**
    d Wärmeaustauscher *m* mit Speicherung
      für Warmwasserbereitung
    f réservoir *m* réchauffeur à serpentin
   nl warmtewisselaar *m* met
      voorraadsreservoir
    r ёмкостный водонагреватель *m*
      (в местной системе горячего
      водоснабжения) (*водяной бак
      с греющим змеевиком*)

I109 *e* **indirect evaporation air conditioner**
    d Oberflächenverdunstungs-
      -Klimaanlage *f*
    f climatiseur *m* à évaporation indirecte
   nl klimaatregelingsinstallatie *f* met
      indirecte verdampingskoeling
    r кондиционер *m* косвенного
      испарительного охлаждения

I110 *e* **indirect expansion system**
    d Klimaanlage *f* mit indirekter
      Kühlung
    f système *m* frigorifique indirect
      [à détente indirecte]
   nl koelsysteem *n* met indirecte expansie
    r система *f* кондиционирования
      воздуха с косвенным испарительным
      охлаждением

I111 *e* **indirect heating**
    d indirekte Heizung *f*
    f chauffage *m* indirect
   nl indirecte verwarming *f*
    r косвенный нагрев *m*

I112 *e* **indirect hot water supply**
    d Warmwasserversorgung *f* mit
      indirektem Anschluß an die
      Wärmequelle
    f production *f* et distribution *f* d'eau
      chaude à réchauffage indirect
   nl indirect systeem *n* voor
      warmwatervoorziening

    r система *f* горячего водоснабжения
      с присоединением к источнику тепла
      по косвенной схеме (*через
      теплообменник*)

I113 *e* **individual base**
    d Sockelfundament *n*, Einzelfundament
      n, Sockel(fundamentkörper) *m*,
      Fundamentsockel *m*
    f fondation *f* isolée [individuelle,
      détachée], chandelle *f*
   nl poer *f* (*m*), fundering *f* van een kolom
    r отдельный столбчатый фундамент *m*,
      фундаментный стул *m*

I115 *e* **indoor climate**
    d Raumklima *n*, Innenklima *n*
    f climat *m* intérieur
   nl binnenklimaat *n*
    r микроклимат *m* помещений

I116 *e* **indoor lighting**
    d Innenbeleuchtung *f*
    f éclairage *m* intérieur
   nl binnenverlichting *f*
    r внутреннее освещение *n*

I117 *e* **indoor piping**
    d Innenrohrleitungen *f pl*
    f canalisations *f pl* intérieures
   nl binnenleidingen *f pl* (*gas en water*)
    r внутренние трубопроводы *m pl*

I118 *e* **indoor powerhouse, indoor power plant**
    d geschlossenes Wasserkraftwerk *n*,
      Wasserkraftwerk *n* mit Hochbau
    f usine *f* du type barrage
   nl eigen stroomvoorziening *f*
    r ГЭС *f* закрытого типа

I119 *e* **indoor swimming pool**
    d Hallen(schwimm)bad *n*
    f piscine *f* couverte
   nl overdekt zwembad *n*, binnenbad *n*
    r крытый плавательный бассейн *m*

I120 *e* **indoor temperature**
    d Raumtemperatur *f*, Innentemperatur *f*
    f température *f* intérieure [interne]
   nl kamertemperatuur *f*;
      binnentemperatuur *f*
    r комнатная температура *f*,
      температура *f* помещения

I121 *e* **induced-draft cooling tower**
    d Kühlturm *m* mit Sauglüfter
    f tour *f* de refroidissement à
      ventilation par aspiration
    l koeltoren *m* met afzuiging
    r градирня *f* со всасывающим
      вентилятором

I122 *e* **induced draught**
    d 1. künstlicher Zug *m* 2. Saugzug *m*
    f 1. tirage *m* mécanique, appel *m* d'air
      2. courant *m* d'air aspiré par un
      ventilateur
   nl 1. kunstmatige ventilatie *f*
      2. afzuiging *f* met ventilator
    r 1. искусственная тяга *f* 2. поток *m*
      воздуха на стороне всасывания
      вентилятора

I123 e **induced draught fan**
d Saugzugventilator *m*
f aspirateur *m*, ventilateur *m* aspirant,
exhausteur *m*
nl afzuiger *m*, exhauster *m*
r дымосос *m*, эксгаустер *m*

I126 e **induction air conditioning system**
d Induktionsklimaanlage *f*
f système *m* de climatisation à groupe
d'induction
nl inductie-
-luchtbehandelingsinstallatie *f*
r эжекционная система *f*
кондиционирования воздуха

I127 e **induction flowmeter**
d Induktionszähler *m*
f débitmètre *m* à induction
nl inductie-watermeter *m*
r индукционный расходомер *m*

I128 e **induction ratio**
d Ansaugverhältnis *n*,
Induktionsverhältnis *n*
f rapport *m* d'induction
nl inductie-verhouding *f*
r коэффициент *m* эжекции

I129 e **induction unit**
d Induktionsklimagerät *n*, Düsenkon-
vektor *m*, Klimakonvektor *m*
f climatiseur *m* terminal d'induction
nl inductie(klimaatregelings)toestel *n*
r эжекционный доводчик *m*

I130 e **induction welding**
d Induktionsschweißen *n*
f soudage *m* par induction
nl inductielassen *n*
r индукционная сварка *f m*

I131 e **industrial air conditioning**
d Industrieklimatisierung *f*
f conditionnement *m* de l'air industriel
nl industriële klimaatregeling *f*
r промышленное кондиционирование *n*
воздуха

I132 e **industrial building**
d Industriegebäude *n*
f bâtiment *m* industriel
nl fabrieksgebouw *n*
r промышленное здание *n*

I133 e **industrial construction**
d Industriebau *m*
f construction *f* industrielle
[de bâtiments industriels]
nl industriële bouw *m*
r промышленное строительство *n*

I134 e **industrial development area**
d Industrieansiedlungszone *f*
f zone *f* de développement industriel
nl industriegebied *n* (in ontwikkeling)
r зона *f* промышленной застройки

I136 e **industrial housing**
d industrialisierter Wohnungsbau *m*
f construction *f* de bâtiments
industrialisée
nl fabriekswoningen *f pl*

r индустриализованное жилищное
строительство *n*

I137 e **industrialization of construction
industry**
d Industrialisierung *f* des Bauwesens
f industrialisation *f* de construction
nl industrialisatie *f* van de bouw
r индустриализация *f* строительства

I138 e **industrialized building**
d 1. Gebäude *n* aus fabrikmäßig
hergestellten Bauteilen, industriel
hergestelltes Fertighaus *n*
2. industrielles [industrialisiertes]
Bauen *n*, Montagebau *m*;
industrielle Fertighausherstellung *f*
f 1. bâtiment *m* préfabriqué
2. construction *f* industrialisée
nl 1. systeembouw *n* 2. industriëel
bouwen *n*
r 1. сборное здание *n* (из элементов)
заводского изготовления
2. индустриализованное
строительство *n* зданий;
промышленное изготовление *n*
[поточный монтаж *m*] сборных зданий

I139 e **industrialized building systems**
d industrielle Baukonstruktionsweisen *f*
pl [Baukonstruktionsmethoden *f pl*];
Konstruktionssysteme *n pl* der
Gebäude in Vollmontagebauweise
f systèmes *m pl* constructifs des
bâtiments préfabriqués
[industrialisés]
nl industriële bouwsystemen *n pl*
r конструктивные системы *f pl*
сборных зданий заводского
изготовления

I140 e **industrialized production of houses**
d industrielle Fertighausherstellung *f*
f production *f* industrielle des
bâtiments
nl industriële woningbouw *m*
r индустриальное производство *n*
сборных домов

I141 e **industrial noise**
d Industrielärm *m*
f bruit *m* de travail
nl industrielawaai *n*
r промышленный шум *m*

I142 e **industrial sewage, industrial waste
water**
d Industrieabwasser *n*
f eaux *f pl* usées industrielles
nl industriëel afvalwater *n*
r промышленные сточные воды *f pl*

I143 e **industrial water**
d Industriewasser *n*, Betriebswasser *n*
f eau *f* d'usage industriel
nl bedrijfswater *n*, industriewater *n*
r техническая вода

I144 e **inelastic behaviour**
d unelastische [plastische] Arbeit *f*
f comportement *m* [tenu *f*] non
élastique

   nl niet-elastisch gedrag *n*
   r неупругая [пластическая] работа *f*
   (конструкции)

I145 *e* **inelastic buckling**
   d Knickung *f* [Längsbiegung *f*] im
   plastischen Bereich, plastische
   [unelastische] Knickung *f*
   f flambage *m* plastique [inélastique],
   flambage *m* dans le domaine plastique
   nl knik *m*
   r продольный изгиб *m* в неупругой
   стадии (*за пределами упругости*)

I146 *e* **inelastic deformation**
   d nichtelastische Verformung *f*
   [Formänderung *f*]
   f déformation *f* inélastique
   nl niet-plastische vormverandering *f*
   r неупругая [пластическая]
   деформация *f*

I147 *e* **inelastic range**
   d nichtelastischer [plastischer] Bereich
   m, Plastizitätsgebiet *n*
   f domaine *m* inélastique [plastique]
   nl niet-elastische gebied *n*
   r неупругая [пластическая] область *f*

I148 **inelastic strain** *see* **inelastic deformation**

I150 *e* **inertia block**
   d träger Fundamentblock *m*
   (*schwingungsdämpfende Fundation*)
   f block *m* massif de fondation sur
   assise antivibratile
   nl trillingsvrije fundatie *f*
   r инерционный фундаментный блок *m*
   на виброизолирующем основании
   (*для машин*)

I151 *e* **inertia effects**
   d Beharrungswirkungen *f pl*
   f effets *m pl* [action *f*] des forces
   d'inertie
   nl traagheidsverschijselen *n pl*
   r инерционные воздействия *n pl*;
   проявление *n* инерционных сил

I152 *e* **infilled wall**
   d Fachwerkwand *f* mit Ausfüllung
   f pan *m* (*de fer ou de bois*) bardé, mur *m*
   à pans avec remplissage, paroi *f* à
   pans avec bardage
   nl vakwerkmuur *m* met vulling
   (van metselwerk)
   r каркасная [фахверковая] стена *f*
   с заполнением

I153 *e* **infiller panel**
   d Ausfachungstafel, Ausfüllungstafel *f*
   f panneau *m* de remplissage
   nl wandpaneel *n*
   r панель *f* заполнения каркаса
   [стенового заполнения]

I154 *e* **infilling**
   d Ausfüllung *f*, Ausfachung *f*
   f remplissage *m*, matériau *m* de
   remplissage, hourdis *m*; matériau *m*
   en vrac
   nl vulling *f*, vulwand *m*

   r заполнение *n* (*каркаса, фахверка,
   межбалочного пространства*);
   засыпка *f*

I155 *e* **infilling masonry**
   d Ausfachungsmauerwerk *n*,
   Mauerwerkausfachung *f*
   f remplissage *m* de pan [d'ossature]
   avec briques
   nl vulling *f* van metselwerk
   r кирпичное заполнение *n* фахверка

I156 **infilling panel** *see* **infiller panel**

I157 *e* **infiltration area**
   d Abgabebereich *m*, Infiltrationsbereich
   m, Sickerungsbereich *m* (*Brunnen,
   Bohrloch*)
   f surface *f* de captage effectif d'un
   puits
   nl infiltratiegebied *n*; zuigbereik *n*
   van een pomp
   r водоприёмная поверхность *f*
   (*колодца, скважины*)

I158 *e* **infiltration coefficient**
   d Versickerungskoeffizient *m*
   f coefficient *m* d'infiltration
   nl infiltratiecoëfficiënt *m*
   r коэффициент *m* инфильтрации

I159 *e* **infiltration gallery**
   d Sickerstollen *m*
   f galerie *f* filtrante
   nl drainagetunnel *m*, drainagegalerij *f*
   r дренажная галерея *f*

I160 *e* **infiltration head**
   d Sickerdruck *m*, Filtrationsdruck *m*
   f charge *f* d'écoulement en milieu
   poreux, charge *f* d'infiltration
   [de filtration]
   nl infiltratiedruk *m*
   r фильтрационный напор *m*

I161 *e* **infiltration heat loss**
   d Infiltrationswärmeverluste *m pl*
   f pertes *f pl* [déperdition *f*] de chaleur
   par infiltration
   nl warmteverlies *n* door infiltratie
   r инфильтрационные теплопотери *f pl*

I162 *e* **infiltration routing**
   d Infiltrationsberechnung *f*
   f détermination *f* de l'infiltration
   nl berekening *f* van de infiltratie
   r расчёт *m* инфильтрации

I164 *e* **inflatable gasket**
   d aufblasbare Dichtungseinlage *f*,
   Druckluftpackung *f*
   f garniture *f* de joint gonflable
   nl opblaasbare pakking *f* [afdichting *f*]
   r пневматическая уплотняющая
   прокладка *f*, пневматический
   сальник *m*

I165 *e* **inflatable structures**
   d pneumatische Konstruktionen *f pl*,
   Tragluft(hallen)konstruktionen *f pl*
   f ouvrages *m pl* [constructions *f pl*]
   gonflables
   nl opblaasbare gebouwen *m pl*

22*

r пневматические (воздухоопорные)
сооружения *n pl* [конструкции *f pl*]

1166 *e* **inflatable void formers**
d Druckluft-Hohlraumbildner *m*,
pneumatischer [aufblasbarer]
Hohlraumbildner *m*, aufblasbarer
Gummischlauch *m* für Kanäle im
Beton
f noyau *m* gonflable (*pour la
formation de vides*)
nl pneumatische sparingmakers *m pl*
r пневматические надувные
пустотообразователи *m pl* (*для
бетонных плит*)

1167 *e* **inflatable weir**
d Schlauchwehr *n*, pneumatisches
[aufpumpbares] Wehr *n*
f barrage *m* gonflable
nl opblaasbare stuw *m*
r мягкая водо- *или*
воздухонаполняемая плотина *f*

1168 *e* **inflation system**
d Aufblassystem *n* für
Tragluftkonstruktionen
f équipement *m* [installation *f*] de
gonflement
nl opblaassysteem *m*
r система *f* подачи воздуха
в пневматические сооружения

1169 **inflected arch** *see* **inverted arch**

1170 *e* **inflow**
d Zufluß *m*, Wasserzufuhr *f*,
Zuflußmenge *f*
f arrivée *f* [venue *f*] d'eau
nl instroming *f*, instroomvolume *n*
r приток *m* воды (*в водоприёмник*),
водоприток *m*

1171 *e* **influence area**
d Einflußfeld *n*, Einflußfläche *f*
f surface *f* [aire *f*] d'influence
nl invloedsgebied *n*
r площадь *f* влияния, инфлюэнтная
площадь *f*

1172 *e* **influence basin**
d Absenkungsbereich *m*
f zone *f* d'appel
nl gebied *n* waarbinnen de
grondwaterstand word verlaagd
r зона *f* влияния скважины

1173 *e* **influence line**
d Einflußlinie *f*
f ligne *f* d'influence
nl invloedslijn *f* (*m*)
r линия *f* влияния

1174 *e* **influence line for reaction**
d Einflußlinie *f* der Auflagerkraft
f ligne *f* d'influence de la réaction
d'appui
nl invloedslijn *f* (*m*) van de oplegkracht
r линия *f* влияния опорной реакции

1175 *e* **influent**
d zufließendes Abwasser *n*
f effluents *m pl*

nl aangevoerd afvalwater *n*
r сточная вода *f*, поступающая
на очистку

1176 *e* **influent seepage**
d einfließende Versickerung *f*;
Grundwasserzutritt *m*
f infiltration *f* affluente
nl verzinking *f* van effluent;
infiltratie *f* van gezuiverd afvalwater
r фильтрационный приток *m*;
инфильтрационное просачивание *n*

1177 *e* **infrared drying**
d Infrarottrocknung *f*
f séchage *m* par radiations infrarouge
nl droging *f* met infrarood
r высушивание *n* [просушка *f*]
с помощью инфракрасного излучения

1178 *e* **infra-red element**
d Infrarot(heiz)strahler *m*
f élément *m* chauffant par radiation
infrarouge
nl infraroodverwarmingselement *n*
r инфракрасный нагреватель *m*

1179 *e* **infra-red heating**
d Infrarotheizung *f*
f chauffage *m* aux rayons infrarouges
nl infraroodverwarming *f*
r инфракрасное отопление *n*

1180 *e* **infra-red pavement heater**
d Infrarot-Deckenheizgerät *n*
f réchauffeur *f* de revêtement en
asphalte à radiation infrarouge
nl infraroodstraler *m* voor
wegdekverwarming
r дорожный асфальторазогреватель *m*
беспламенного типа (*с нагревом
инфракрасным излучением*)

1181 *e* **infra-red radiator**
d Infrarotstrahler *m*
f radiateur *m* à rayonnement infra-
-rouge
nl infraroodstraler *m*
r инфракрасный излучатель *m*

1182 *e* **ingress of ground water**
d Grundwasserandrang *m*,
Grundwassereinbruch *m*,
Grundwasserzutritt *m*
f afflux *m* d'eau souterraine
nl binnendringen *n* van grondwater
r напор *m* [приток *m*] грунтовых вод

1183 **inherent moisture of aggregates** *see*
water contained in aggregates

1184 *e* **inherent settlement**
d Eigensetzung *f*
f affaissement *m* [tassement *m*] propre
[inhérant] (*c.-à-d. dû à la
consolidation du sol sous le support
sans prendre en considération le
tassement des supports adjacents*)
nl zakking *f* door eigen gewicht
r осадка *f* опоры *или* фундамента под
нагрузкой (*исключая влияние
осадки соседних опор*)

1185 *e* **inhibiting pigment**
d Korrosionsschutzpigment *n*
f pigment *m* anticorrosif
nl corrosie-werend pigment *n*
r антикоррозионный пигмент *m*

1186 *e* **initial drying shrinkage**
d 1. Anfangsschwindung *f*
[Anfangsschwund *m*] bei
Austrocknung,
Anfangstrockenschwindung *f*
2. Anfangstrockenschwundmaß *n*
f 1. retrait *m* de séchage initial
2. taux *m* de retrait dû au séchage
initial
nl 1. beginkrimp *m* bij droging
2. beginkrimp-factor *m*
r 1. начальная усадка *f* при высыхании
2. коэффициент *m* начальной усадки
при высыхании

1187 *e* **initial prestress**
d Anfangsvorspannung *f*, anfängliche
Vorspannung *f*
f tension *f* [prétension *f*] initiale
nl initiële voorspanning *f*
r начальное преднапряжение *n*

1188 *e* **initial rate of absorption**
d Anfangswasseraufnahmefähigkeit *f*
f taux *m* initial d'absorption
nl initiëel absorptievermogen *n*
r начальная водопоглощающая
способность *f*

1189 *e* **initial set**
d Abbindebeginn *m*, Erstarrungsbeginn
m
f prise *f* initiale
nl beginbinding *f*
r начальное схватывание *n*

1190 *e* **initial setting time**
d Abbindezeit *f*, Erstarrungsdauer *f*
f temps *m* de prise initiale
nl begin *m* van de binding
r начало *n* схватывания

1191 *e* **initial stress**
d Anfangsspannung *f*
f contrainte *f* initiale
nl aanvangsspanning *f*
r начальное напряжение *n*

1193 *e* **injected foundation**
d Auspreßgründung *f*,
Injektionsgründung *f*,
Injiziergründung *f*
f fondation *f* stabilisée par cimentage
à projection
nl door injectie gestabiliseerde
funderingsgrondslag *m*
r основание *n*, укреплённое
цементацией

1194 *e* **injecting grout**
d Einspritzmörtel *m*, Einpreßmörtel *m*,
Verpreßmörtel *m*, Injiziermörtel *m*
f coulis *m* injecté; coulis *m* pour
injection
nl injectiemortel *m*

r инъецируемый раствор *m*; раствор *m*
для инъецирования

1196 **injection gallery** *see* **grouting gallery**

1197 *e* **injection hole**
d Injektionsloch *n*, Einpreßloch *n*,
Auspreßöffnung *f*
f trou *m* d'injection
nl injectieopening *f*
r инъекционная скважина *f*, отверстие
n для нагнетания цементного
раствора

1198 *e* **injection lance**
d Verpreßlanze *f*, Einpreßlanze *f*,
Injizierlanze *f*
f buse *f* d'injection
nl injectielans *f* (*m*)
r насадка *f* для нагнетания
цементного раствора

1199 **injection mortar** *see* **injecting grout**

1200 *e* **injection of grout**
d Zementmörtelinjektion *f*,
Zementmörteleinpressung *f*
f injection *f* de coulis
nl injectie *f* van cementmortel
r нагнетание *n* [инъецирование *n*]
цементного раствора

1201 **injection well** *see* **inverted well**

1202 *e* **injector**
d Injektor *m*, Strahlpumpe *f*
f injecteur *m*, pompe *f* à jet
nl injector *m*, straalpomp *f* (*m*)
r инжектор *m*, струйный насос *m*

1203 *e* **inland harbour**
d Binnenhafen *m*, Flußhafen *m*
f port *m* fluvial [en rivière]
nl binnenhaven *f* (*m*)
r речной порт *m*

1204 *e* **inland waterway**
d Binnenwasserstraße *f*,
Binnenschiffahrtsstraße *f*
f voie *f* navigable [fluviale]
nl waterweg *m* voor de binnenvaart
r внутренний водный путь *m*

1205 *e* **inlet**
d 1. Einlaßbauwerk *n*, Einlaufbauwerk
n 2. Dräneinlaß *m*, Dräneinlauf *m*,
Dränkopf *m* 3. Bucht *f* 4. Einlauf *m*,
Einlaßöffnung *f* 5. Belüftungsventil *n*
f 1. entrée *f*; prise *f* d'eau 2. extremité *f*
amont d'un drain 3. baie *f* 4. orifice *m*
d'admission; entrée *f*; arrivée *f* d'air
5. bouche *f* d'air; clapet *m*
d'admission d'air
nl 1. inlaatwerk *n* 2. begin *n* van een
aanvoerleiding 3. zeearm *m*
4. inlaatopening *f*
r 1. головное сооружение *n* (*водовода*,
канала) 2. голова *f* дрены 3. залив *m*
4. впускное [входное] отверстие *n*;
ввод *m* 5. аэрационный клапан *m*
(*на трубопроводе*)

1206 *e* **inlet grating**
d Einlaufrost *m*

f grille *f* d'entrée
nl inlaatrooster *m, n*
r водоприёмная решётка *f*

1208 *e* **inlet opening**
d Einlauföffnung *f*
f orifice *f* d'entrée [d'admission]
nl inlaatopening *f*
r входное [впускное] отверстие *n*

1209 *e* **inlet sill**
d Einlaufschwelle *f*, Oberdrempel *m*
f seuil *m* [radier *m*] d'écluse supérieur
nl bovendrempel *m* (*sluis*)
r верхний порог *m* (*шлюза*)

1210 **inlet structure** *see* **intake structure**

1212 *e* **inlet well**
d 1. Einlaufschacht *m* 2. Einpreß-bohrung *f*
f 1. puits *m* collecteur 2. trou *m* d'injection
nl 1. welput *m* 2. injectieboring *f*
r 1. водоприёмный колодец *m* 2. инъекционная [цементационная] скважина *f*

1213 *e* **inner bench, inner berm**
d Binnenbankett *n*, Binnenberme *f*
f berme *f* intérieure
nl binnenbanket *n*, binnenberm *m*
r банкет *m* [берма *f*] на внутреннем откосе дамбы

1214 *e* **inner face**
d Innenfläche *f*, Innenseite *f*
f surface *f* intérieure, face *f* interne
nl binnenvlak *n*
r внутренняя поверхность *f* [грань *f*] (*напр. стены*)

1215 **inner forces** *see* **internal forces**

1216 *e* **inner slope**
d Binnenböschung *f*
f talus *m* interne
nl binnentalud *n*
r внутренний откос *m* (*дамбы*)

1217 **inner span** *see* **interior span**

1218 *e* **innings**
d 1. Deichland *n*, eingedeichtes Land *n* 2. Polder *m*
f 1. terrains *m pl* desséchés 2. polder *m*
nl 1. bedijkt land *n* 2. polder *n, m*
r 1. угодья *n pl* на осушенном дне водоёма *или* болота 2. польдер *m*

1219 *c* **in-place slump test**
d Setzprobe *f* des Frischbetons an Ort und Stelle
f essai *m* de consistance du béton jeune
nl zetproef *f* (*m*) op de bouwplaats
r определение *n* подвижности свежеуложенного бетона

1220 **input ventilation** *see* **supply ventilation**

1221 *e* **in-rush of water**
d Wassereinbruch *m*
f venue *f* d'eau
nl waterinbraak *f* (*m*)
r прорыв *m* воды

1223 *e* **insertion connection**
d dichtende Steckverbindung *f*
f raccord *m* intercalé [inséré] (*d'une conduite d'air*)
nl steekmofverbinding *f*
r вставное соединение *n* (*воздуховодов*)

1224 *e* **inserts**
d Stahleinlagen *f pl*; Einsatzstücke *n pl*
f pièces *f pl* (d'ancrage) incorporées
nl ingezette delen *n pl*
r закладные детали *f pl*

1225 **inside air** *see* **room air**

1226 *e* **inside casing**
d innere Blendleiste *f*
f chambranle *m* intérieur (*de fenêtre ou de porte*)
nl kantstuk *n*; dagstuk *n*
r внутренний наличник *m* (*окна, двери*)

1227 *e* **inside embankment toe**
d wasserseitiger Dammfuß *m*
f pied *m* amont (d'un barrage)
nl teen *m* (*dijk*)
r подошва *f* верхового откоса (*плотины*)

1229 *e* **in-situ concrete**
d Ortbeton *m*
f béton *m* coulé en [sur] place
nl in het werk gestort beton *n*, monolietbeton *n*
r монолитный бетон *m*

1230 *e* **in-situ concrete pile**
d Betonortpfahl *m*, Ortbetonpfahl *m*
f pieu *m* moulé dans le sol
nl in de grond gevormde betonpaal *m*
r набивная бетонная свая *f*

1231 *e* **in-situ construction**
d Ortbetonbau *m*
f construction *f* en béton coulé sur place
nl constructie *f* met in het werk gestort beton
r строительство *n* сооружений из монолитного бетона

1233 *e* **insolation**
d Sonneneinstrahlung *f*, Insolation *f*, Besonnung *f*
f insolation *f*, ensoleillement *m*
nl insolatie *f*, bezonning *f*
r инсоляция *f*, инсоляционные теплопоступления *n pl*

1234 *e* **inspection**
d 1. Bauaufsicht *f*, Bauüberwachung *f* 2. Arbeitsaufmaß *n*
f 1. contrôle *m* [surveillance *f*] (technique) des travaux 2. metré *m* de contrôle
nl 1. (bouw)inspectie *f*, bouwopzicht *n* 2. maatcontrole *f*
r 1. контроль *m* [надзор *m*] за качеством работ 2. контрольный обмер *m*

1235 *e* **inspection cap**
d Inspektionskappe *f*, Revisionskappe *f*, Reinigungskappe *f*

f regard *m* de visite
nl inspectiedeksel *n*
r смотровой [очистной] лючок *m*
(*на трубопроводе*)

1236 e **inspection certificate**
d Protokoll *n* über eine technische
Kontrolle
f certificat *m* d'inspection
[de vérification]
nl inspectiecertificaat *n*
r акт *m* технического осмотра

1237 **inspection chamber** *see* **inspection manhole**

1238 e **inspection cock**
d Probeentnahmehahn *m*
f robinet *m* d'échantillonnage
nl controlekraan *f* (*m*)
r кран *m* для отбора проб (воды)

1239 e **inspection cover**
d Revisionsdeckel *m*
f couvercle *m* d'inspection
nl inspectiedeksel *n* (van een put)
r ревизионная крышка *f* (колодца)

1240 e **inspection door**
d Schauöffnung *f*, Schauloch *n*,
Einsteigöffnung *f*
f trou *m* de regard [d'homme]
nl inspectieluik *n*, mangat *n*
r смотровой люк *m*

1241 e **inspection eye**
d Reinigungsstück *n*,
Reinigungsöffnung *f*
f regard *m*, bouchon *m* de visite
nl controleopening *f*; ontstoppingsstuk *n*
r прочистка *f* (*деталь трубопровода*);
очистное отверстие *n*, ревизия *f*

1242 e **inspection gallery**
d Kontrollgang *m*, Kontrollstollen *m*,
Beobachtungsstollen *m*
f galerie *f* de visite [de contrôle]
nl inspectiegang *m*
r потерна *f*, смотровая галерея *f*
(*плотины*)

1243 **inspection hole** *see* **inspection door**

1244 **inspection junction** *see* **cleanout**

1245 e **inspection list**
d Nacharbeitsliste *f*,
Abschlußleistungsverzeichnis *n*,
Verzeichnis *n* der Abschlußarbeiten
f liste *f* de contrôle (*liste des travaux*
de construction à exécuter)
nl inspectieregister *m*, lijst *f* (*m*) van de
nog uit te voeren werkzaamheden voor
de oplevering
r контрольный перечень *m* работ
(*подлежащих выполнению или*
исправлению подрядчиком)

246 e **inspection manhole**
d Kontrollschacht *m*, Einsteigschacht *m*,
Revisionsschacht *m*, Mannloch *n*
f puits *m* [régard *m*, bouche *f*] d'accès,
trou *m* de visite

nl inspectieschacht *f* (*m*), controleschacht
f (*m*), mangat *n*
r смотровой колодец *m*

1247 e **inspector**
d Aufsichtsbeamte *m*
f surveillant *m* [inspecteur *m*] des
travaux
nl bouwopzichter *m*, bouwinspecteur *m*
r представитель *m* технадзора
за строительством

1248 e **instability**
d Instabilität *f*, Labilität *f*,
Unbeständigkeit *f*
f instabilité *f*
nl instabiliteit *f*
r неустойчивость *f*

1249 e **installation**
d 1. Installation *f*, Anlage *f* 2. innere
Leitungsnetze *n pl* 3. technische
Gebäudeausrüstung *f* 4. Einbau *m* der
Ausrüstung
f 1. installation *f* 2. canalisations *f pl*
intérieures 3. équipement *m*
technique des bâtiments 4. montage *m*
nl installatie *f*
r 1. установка *f* 2. внутренние
инженерные сети *f pl* 3. инженерное
оборудование *n* (зданий) 4. монтаж *m*
(инженерного оборудования)

1250 e **installation drawing**
d Einbauzeichnung *f*,
Montagezeichnung *f*
f dessin *m* de montage
nl installatietekening *f*
r монтажный чертёж *m*

1251 e **installed capacity, installed power**
d installierte Leistung *f*
f puissance *f* installée
nl geïnstalleerd vermogen *n*
r установленная мощность *f*

1252 e **instantaneous center of rotation**
d Momentanzentrum *n*
f centre *m* instantané de rotation
nl momentaan draaiingsmiddelpunt *n*
r мгновенный центр *m* вращения

1253 e **instantaneous deformation**
d momentane Verformung *f*
f déformation *f* instantanée
nl momentane vervorming *f*
r мгновенная деформация *f*

1254 e **instantaneous elevator safeties**
d Fangvorrichtungen *f pl* für sofortige
Bremsung
f frein *m* à machoires à action
instantanée (*pour ascenseur ou*
monte-charge)
nl liftbeveiliging *f*
r ловители *m pl* мгновенного действия
(*для лифта, подъёмника*)

1255 e **instantaneous heat gain**
d Momentanwärmeeinfall *m*,
augenblicklicher Wärmeeinfall *n*
f gain *m* de chaleur instantané

    nl onmiddelijke warmte-opname *f*
    r концентрированные
    тепловыделения *n pl*

1256 *e* **instantaneous water heater**
    d Durchlauferhitzer *m*
    f chauffe-eau *m* instantané
    nl geiser *m*
    r проточный водонагреватель *m*

1257 *e* **instant lock**
    d automatisches Schloß *n*
    f serrure *f* automatique
    nl automatisch slot *n*
    r самозапирающийся замок *m*

1258 *e* **instrument panel**
    d Armaturenbrett *n*, Schalttafel *f*,
    Bedienungstafel *f*
    f tableau *m* de commande
    nl bedieningspaneel *n*
    r приборный щит *m*, щит *m* [панель *f*]
    управления

1259 *e* **insulating board**
    d Isolierpappe *f*; Dämmplatte *f*
    f panneau *m* isolant
    nl isolatieplaat *f* (*m*)
    r листовой теплоизоляционный
    материал *m* (*из древесного или*
    минерального волокна);
    теплоизоляционная плита *f*

1260 *e* **insulating brick**
    d Dämmstein *m*, Isolierstein *m*,
    Porenziegel *m*
    f brique *f* réfractaire isolante
    nl holle baksteen *m*
    r огнеупорный теплоизоляционный
    кирпич *m*

1261 *e* **insulating concrete**
    d Leichtdämmbeton *m*
    f béton *m* isoléger [léger isolant]
    nl warmteïsolerend beton *n*
    r лёгкий теплоизоляционный бетон *m*

1262 *e* **insulating form board**
    d Wärmedämm-Schalungsplatte *f*
    f panneau *m* de coffrage calorifuge
    (*perdu*)
    nl warmteïsolerend materiaal *n* voor
    bekistingen
    r теплоизоляционная (несъёмная)
    опалубочная плита *f*

1263 *e* **insulating glass unit**
    d Glasverbundscheibe *f*,
    Doppelscheibenglas *n*
    f thermopane *m*
    nl dubbel-glas ruit *f* (*m*)
    r стеклопакет *m*

1264 *e* **insulating jacket**
    d Wärmedämmantel *m*
    f enveloppe *f* [chemise *f*] calorifuge
    nl warmteïsolerende mantel *m*
    r теплоизоляционный капот *m*
    [кожух *m*] (*по размерам изолируемого*
    сосуда)

1265 *e* **insulating layer**
    d Dämmschicht *f*; Isolierschicht *f*

    f couche *f* isolante; couche *f* de
    scellement [de fermeture, de cure]
    nl isolatielaag *f* (*m*)
    r изоляционный слой *m*; изолирующая
    прослойка *f* (*дорожного покрытия*)

1266 *e* **insulating refractory**
    d Isolierstein *m*
    f réfractaires *m pl* isolants, produits *m*
    pl réfractaires isolants
    nl vuurvaste isolatiesteen *m*
    r изоляционный огнеупор *m*

1267 *e* **insulating value**
    d Dämmwert *m*, Isolierwert *m*
    f pouvoir *m* isolant
    nl warmtedoorgangscoëfficiënt *m*
    r коэффициент *m* изоляции

1268 *e* **insulating work**
    d Isolier(ungs)arbeiten *f pl*,
    Dämmarbeiten *f pl*
    f travaux *m pl* d'isolation [d'isolement]
    nl isolatiewerk *n*
    r изоляционные работы *f pl*

1269 *e* **insulation materials**
    d Isolierstoffe *m pl*, Dämmstoffe *m pl*
    f matériaux *m pl* d'isolation
    [d'isolement]
    nl isolatiematerialen *n pl*
    r изоляционные материалы *m pl*

1270 *e* **insulation quilt**
    d Dämmatte *f*
    f matelas *m* isolant cousu, natte *f*
    isolante cousue
    nl isolatiedeken *f* (*m*)
    r теплоизоляционный прошивной
    мат *m*

1271 *e* **insulation sleeve**
    d Isolierrohr *n*
    f manchon *m* calorifuge
    nl isolatiekous *f* (*m*)
    r теплоизолирующий стакан *m*,
    теплоизолирующая гильза *f* (*для*
    пропуска труб и кабелей через
    ограждения)

1272 *e* **insulator string**
    d Isolatorkette *f*
    f chaîne *f* [chapelet *m*] d'isolateurs
    nl keten *f* (*m*) van isolatoren
    r гирлянда *f* изоляторов

1273 *e* **intake**
    d 1. Entnahmebauwerk *n*,
    Wasserentnahme *f* 2.
    Außenluftentnahme *f* 3. Einlauf *m*,
    Einlauföffnung *f*, Ansaugöffnung *f*
    f 1. prise *f* d'eau 2. prise *f* d'air
    3. bouche *f* d'air, orifice *m*
    d'aspiration
    nl 1. wateronttrekking *f* 2. luchtinlaat *m*
    3. inlaat *m*
    r 1. водозабор *m* 2. воздухозабор *m*
    3. заборное [всасывающее]
    отверстие *n*

1274 *e* **intake basin**
    d Einlaufbecken *n*

f bassin *m* de réception
nl verzamelbekken *n*
r бассейн *m* водоприёмника

1275 e **intake channel**
d Einlaufkanal *m*
f canal *m* adducteur [d'amenée]
nl inlaatkanaal *n*
r подводящий канал *m*

1276 e **intake gallery**
d Einlaufstollen *m*
f galerie *f* d'amenée [de prise (d'eau)]
nl wateronttrekkingsgalerij *f*
r водозаборная галерея *f*, туннель *m* водозабора

1277 e **intake heading**
d Einlaufbauwerk *n*, Einlaßbauwerk *n*; Entnahmebauwerk *n*
f ouvrage *m* de prise d'eau
nl inlaatwerk *n*
r головное сооружение *n* гидроузла; водозаборное сооружение *n*

1280 e **intake pipe**
d Einlaufrohr *n*
f tuyau *m* de prise [d'admission]
nl inlaatbuis *f* (*m*)
r впускная [входная] труба *f*

1281 e **intake screen**
d Einlaufrechen *m*
f grille *f* d'entrée
nl inlaatrooster *m*, *n*
r входная [водоприёмная] решётка *f*

1282 e **intake structure**
d Wasserfassung *f*, Wasserentnahme *f*, Entnahmebauwerk *n*, Einlauf *m*, Einlaufbauwerk *n*
f prise *f* d'eau, ouvrage *m* d'entrée
nl inlaatwerk *n*
r водозаборное [водоприёмное] сооружение *n*

1283 **intake system** see **intake works**

1284 e **intake tower**
d Entnahmeturm *m*, Einlaufturm *m*, Turmfassung *f*, Turmeinlaß *m*
f tour *f* de prise d'eau
nl inlaattoren *m*
r башенный водозабор *m*

1285 **intake tunnel** US see **intake gallery**

1287 e **intake works**
d 1. Entnahmebauwerk *n*, Einlaufbauwerk *n*, Einlauf *m*, Einlaßbauwerk *n*, Einlaß *m*
2. Zuleitungsumleitung *f*, Obergraben *m*
f 1. ouvrages *m pl* de prise d'eau
2. ouvrages *m pl* de dérivation
nl 1. inlaatwerk *n* 2. toevoerleiding *f*
r 1. водоприёмные сооружения *n pl*
2. подводящая деривация *f*

1288 e **integral coloring**
d Betoneinfärbung *f*, Betondurchfärbung *f*
f coloration *f* [teinture *f*] du béton dans la masse

nl door-en-door kleuren *n* van beton
r окраска *f* бетона (*введением красителя в смесь*)

1289 e **integral control**
d Integralregelung *f*, I-Regelung *f*
f réglage *m* par intégration [intégral]
nl integrale regeling *f*
r регулирование *n* по интегралу, интегральное регулирование *n*

1291 e **integral waterproofing**
d Hydrophobierung *f* des Betons
f imperméabilisation *f* du béton par incorporation de l'additif hydrofuge
nl waterdichtmaken *n* van beton
r гидрофобизация *f* бетона (*введением добавки*)

1292 e **integrated ceiling**
d Kombinationsdecke (*Hängedecke mit eingebauten Leitungsnetzen*)
f plafond *m* suspendu avec canalisations incorporées
nl verlaagd systeemplafond *n* met ruimte voor leidingen
r подвесной потолок *m* со встроенными инженерными сетями

1293 e **integrated distribution floor system**
d Deckensystem *n* mit eingebauten Leitungsnetzen
f plancher *m* avec canalisations incorporées
nl vloersysteem *n* met ingebouwde leidingen
r перекрытие *n* со встроенными инженерными сетями

1294 e **integrated flow curve**
d Abfluß(mengen)-Summenlinie *f*
f courbe *f* des débits cumulés
nl totale-afvoerkromme *f*
r интегральная кривая *f* стока

1295 e **integrated power grid**
d Verbundnetz *n*
f système *m* unifié de réseaux électriques
nl koppelnet *n*
r объединённая энергосистема *f*

1296 e **integrated river basin development**
d wasserwirtschaftliche Großplanung *f*
f aménagement *m* d'ensemble d'un bassin fluvial
nl ontwikkelingsplan *n* voor een rivierbekken
r комплексный проект *m* водохозяйственного использования речного бассейна

1297 e **integrating flow meter**
d Wasserzähler *m*, integrierender Durchflußmesser *m*
f compteur-totalisateur *m* d'eau
nl volumestroommeter *m*
r водомер-интегратор *m*

1298 e **integrating heat meter**
d Wärmezähler *m*
f compteur-totalisateur *m* de chaleur
nl warmte(verbruik)meter *m*
r тепломер-интегратор *m*

I299 *e* **integration**
  d Anordnung *f*, Raumgestaltung *f*,
    Komposition *f*
  f arrangement *m* [organisation *f*] des
    composants
  nl ruimtelijke integratie *f*
  r компоновка *f*

I300 *e* **intensity of load**
  d Laststärke *f*
  f intensité *f* de la (sur)charge
  nl belastingsintensiteit *f*
  r интенсивность *f* нагрузки

I301 *e* **interbasin diversion**
  d Abflußüberleitung *f*,
    Wasserüberleitung *f*
  f dérivation *f* du débit
    [du cours d'eau]
  nl waterafleiding *f* naar ander stuwmeer
  r переброска *f* стока

I302   intercepting *see* interception
I303   intercepting drain *see* catch water
    drain

I304 *e* **intercepting sewer**
  d Mischwassersammler *m*,
    Hauptabwasserleitung *f*
  f collecteur *m* d'interception, émissaire
    m d'évacuation
  nl hoofdriool *n*
  r коллектор *m* общесплавной
    канализации

I305 *e* **intercepting trap**
  d Wasserverschluß *m*
  f siphon *m* à garde d'eau, siphon *m*
    d'égout
  nl waterslot *n*
  r водяной затвор *m*

I306 *e* **interception**
  d Abfangen *n* der Quelle;
    Grundwasserabfangung *f* durch
    Dräne
  f captage *m*, interception *f*
  nl aantappen *n* (*bronwater*)
  r каптаж *m* (источника); перехват *m*
    (грунтовой воды дренами)

I307 *e* **interception channel, interception ditch**
  d Auffanggraben *m*, Sammelgraben *m*,
    Saumgraben *m*, Hintergraben *m*,
    Hanggraben *m*, Bergabflußgraben *m*
  f fossé *m* de crête [de garde,
    d'interception]
  nl opvangsloot *f* (*m*), opvanggreppel *f* (*m*)
  r нагорная канава *f*, перехватывающий
    канал *m*

I308 *e* **interceptor**
  d Abfangleitung *f*;
    Hauptabwasserleitung *f*,
    Hauptsammler *m*; Sammeldrän *m*,
    Abfangsammler *m*; Geruchverschluß
    m, Siphon *m*
  f collecteur *m* abducteur; collecteur *m*
    d'égout principal; drain *m* collecteur;
    joint *m* hydraulique, syphon *m*
    d'égout

  nl rioolput *m*; stankafsluiter *m* (in
    huisaansluitleiding)
  r отводящий коллектор *m*; главный
    канализационный коллектор *m*;
    дрена-собиратель *f*; гидравлический
    затвор *m*, сифон *m*

I309 *e* **interceptor manhole**
  d Einsteigschacht *m* mit
    Wasserverschluß
  f bouche *f* d'accès [d'égout] à siphon
  nl rioolputmangat *n*
  r колодец *m* (*канализационной сети*)
    с водяным затвором

I310 *e* **interchange**
  d Zufahrtstelle *f*, Anschlußstelle *f*,
    Kreuzungsanlage *f*
  f croisement *m* à niveaux différants,
    ouvrage *m* de croisement dénivelé
  nl ongelijkvloerse wegkruising *f*
  r транспортная развязка *f*
    [пересечение *n* автомагистралей]
    в разных уровнях

I311 *e* **interchangeable equipment**
  d Austauschgerät *n*, Umbauwerkzeug *n*,
    Zusatzvorrichtung *f*
  f équipement *m* porté interchangeable
  nl uitwisselbaar materieel *n*
  r сменное навесное рабочее
    оборудование *n*

I312 *e* **intercupola, interdome**
  d Doppelkuppelzwischenraum *m*;
    Kuppelschalen-Zwischenraum *m*
  f espace *m* entre coupole extérieure et
    coupole intérieure; espace *m* entre
    coque extérieure et coque intérieure
    de la coupole double [du dôme
    double]
  nl ruimte *f* tussen de binnen- en
    buitenkoepel
  r пространство *n* между наружным и
    внутренним куполом; пространство *n*
    между наружной и внутренней
    оболочками купола

I313 *e* **interfenestration**
  d Fensterpfeilerbreite *f*
  f intervalle *f* entre deux fenêtres
    adjacentes
  nl penant *n*
  r расстояние *n* между окнами
    по фасаду здания

I314 *e* **interference between wells**
  d Wechselwirkung *f* der Brunnen
  f interférence *f* des puits
  nl wisselwerking *f* tussen putten
  r взаимодействие *n* колодцев
    [скважин]

I315 *e* **interference settlement**
  d Setzung *f* infolge gegenseitiger
    Beeinflussung benachbarter
    Gründungen
  f affaissement *m* [tassement *m*] du
    support dû à l'effet de chargement
    des supports adjacents

  nl zakking *f* onder invloed van
   aangrenzende fundamenten
  r осадка *f* (опоры) под влиянием
   нагрузки на смежные опоры

1316 *e* **interior architecture**
  d Innenarchitektur *f*
  f architecture *f* des intérieurs
  nl binnenhuisarchitectuur *f*
  r архитектура *f* интерьеров

1317 **interior bridge support** *see* **pier**

1318 **interior casing** *see* **inside casing**

1319 **interior climate** *see* **indoor climate**

1320 *e* **interior finish**
  d Innenausbau *m* [Innenausstattung *f*]
   des Gebäudes
  f décoration *f* des intérieurs
  nl interieurafwerking *f*
  r внутренняя отделка *f* (здания)

1321 *e* **interior glazed window**
  d innenseitig verglastes Außenfenster *n*
  f fenêtre *f* de façade vitrée à l'intérieur
  nl venster *n* met binnenraam
  r наружное окно *n*, остеклённое
   изнутри

1322 *e* **interior span**
  d Mittelfeld *n*, Mittelspanne *f*,
   Mittelöffnung *f*, Innenfeld *n*
  f travée *f* [portée *f*] intérieure
   [intermédiaire]
  nl binnenoverspanning *f*
  r промежуточный [средний] пролёт *m*;
   промежуточное пролётное строение *n*

1323 *e* **interior support**
  d Innenstütze *f*, Mittelstütze *f*
  f support *m* intérieur, appui *m*
   intermédiaire
  nl binnenkolom *f* (*m*)
  r внутренняя *или* промежуточная
   опора *f*

1324 **interior temperature** *see* **indoor temperature**

1325 *e* **interior-type plywood**
  d Zierfurnier *n*, Ausstattungsfurnier *n*
  f contreplaqué *m* pour la menuiserie
   intérieure
  nl multiplexplaten *f* (*m*) *pl* voor
   binnenwerk
  r фанера *f* для внутренней отделки

1326 *e* **interior wall**
  d Innenwand *f*
  f mur *m* intérieur, paroi *f* intérieure
  nl binnenmuur *m*
  r внутренняя стена *f*

1327 *e* **interior works**
  d Ausbau *m*, Ausbauarbeiten *f pl*
  f travaux *m pl* de finition des intérieurs
  nl afbouw *m*, afwerking *f*
  r отделка *f* (здания), отделочные
   работы *f pl*

1328 *e* **interior zone**
  d Innenzone *f*
  f zone *f* intérieure

  nl binnenzone *f*
  r внутренняя зона *f* (здания)

1329 *e* **interjoist**
  d Balkenabstand *m*
  f entrevous *m*
  nl balkafstand *m*
  r расстояние *n* между балками

1330 *e* **interlock**
  d Verriegelung *f*, Sperrung *f*;
   Schloßverbindung *f*
  f verrouillage *m*; enclenchement *m*
  nl koppeling *f*; slotverbinding *f*
  r блокировка *f*; запирание *n*;
   замковое соединение *n*

1331 *e* **interlocking concrete blocks**
  d ineinandergreifende Betonsteine *m pl*
   [Betonblöcke *m pl*]
  f blocs *m pl* en béton (avec joints)
   enclanchés
  nl in elkaar grijpende betonstenen *m pl*,
   profielkeien *f pl*
  r бетонные блоки *m pl* с замковым
   соединением

1332 *e* **interlocking piles**
  d Spundpfähle *m pl*
  f palplanches *f pl*
  nl stalen damplanken *f* (*m*) *pl*
  r шпунтовые сваи *f pl*

1333 *e* **interlocking tile**
  d Falzziegel *m*, Dachfalzstein *m*
  f tuile *f* en terre cuite à emboîtement
   [emboîtée]
  nl dakpan *f* (*m*) met sluiting
  r пазовая черепица *f*

1334 *e* **intermediate gate**
  d Zwischentor *n* (*Schleuse*)
  f porte *f* d'écluse intermédiaire
  nl tussendeur *f* (*m*) (*sluis*)
  r промежуточные ворота *pl* (шлюза)

1335 *e* **intermediate gate post**
  d Zwischenpfosten *m*
  f montant *m* intermédiare] de la porte
   de clôture
  nl tussenstijl *m*
  r средняя стойка *f* ворот

1336 *e* **intermediate product**
  d Zwischenprodukt *n*
  f produit *m* intermédiaire
  nl tussenproduct *n*
  r промежуточный продукт *m*
   (*обогащения*)

1337 **intermediate rafter** *see* **common rafter**

1338 *e* **intermediate settling basin**
  d Zwischenklärbecken *n*
  f décanteur *m* intermédiaire
  nl tussengelegen bezinkbekken *n*
  r промежуточный отстойник *m*

1339 *e* **intermediate span bridge**
  d mittelgroße Brücke *f*
  f pont *m* de la portée moyenne
  nl middelgrote brug *f* (*m*)
  r мост *m* со средней длиной пролёта
   (*до 45—50 м*)

1340 *e* **intermediate support**
   d Zwischenstütze *f*
   f support *m* [appui *m*] intermédiaire
   nl tussensteunpunt *n*
   r промежуточная опора *f*

1341 *e* **intermittent filter**
   d intermittierendes Filter *n*
   f filtre *m* à action discontinue [intermittente]
   nl intermitterend filter *n*, *m*
   r фильтр *m* периодического действия

1342 *e* **intermittent freezing**
   d Frostwechsel *m*
   f cycles *m pl* intermittents gel-dégel
   nl intermitterend bevriezen *n*
   r переменное замораживание *n* и оттаивание *n*

1343 *e* **intermittent heating**
   d unterbrochener Heizbetrieb *m*, Aussetzheizung *f*
   f chauffage *m* intermittent, marche *f* intermittente de chauffage
   nl intermitterende verwarming *f*
   r периодическое отопление *n*

1344 *e* **intermittent operation**
   d Aussetzbetrieb *m*
   f fonctionnement *m* intermittent, marche *f* intermittente
   nl intermitterend bedrijf *n*
   r периодический режим *m*

1345 *e* **intermittent sand filter**
   d absatzweise arbeitendes Sandfilter *n*
   f filtre *m* en sable à fonctionnement discontinu [intermittent]
   nl intermitterend zandfilter *n*, *m*
   r песчаный фильтр *m* периодического действия

1346   **intermittent type settling basin** *see* **discontinuous-flow settling basin**

1347 *e* **internal conditions**
   d Raumklimabedingungen *f pl*
   f confort *m* des intérieurs
   nl binnen(huis)klimaat *n*
   r внутренние условия *n pl* помещения; микроклимат *m* помещения

1348 *e* **internal downpipe**
   d innenliegendes Regenfallrohr *n*
   f tuyau *m* de descente intérieur
   nl hemelwaterafvoerpijp *f (m)* binnen het gebouw
   r внутренний водосточный стояк *m*, внутренний водосток *m*

1350 *e* **internal forces**
   d innere Kräfte *f pl*
   f forces *f pl* intérieures
   nl inwendige krachten *f (m) pl*
   r внутренние силы *f pl* [усилия *n pl*]

1351 *e* **internal friction**
   d innere Reibung *f*, Eigenreibung *f*
   f frottement *m* interne
   nl inwendige wrijving *f*
   r внутреннее трение *n*

1352 *e* **internal glazing**
   d Innenverglasung *f*
   f vitrage *m* intérieur; vitrage *m* des cloisons et des impostes intérieures
   nl glasbezetting *f* binnenhuis
   r внутреннее остекление *n*; остекление *n* внутренних перегородок и фрамуг

1353   **internal leaf** *see* **internal skin**

1354 *e* **internal moment**
   d Moment *n* der inneren Kräfte [des inneren Kräftepaares]
   f moment *m* intérieur [des forces intérieures, du couple intérieur]
   nl moment *n* van inwendige krachten
   r момент *m* внутренних сил [внутренней пары сил]

1355 *e* **internal pipework**
   d innere Rohrleitungen *f pl*
   f canalisations *f pl* intérieures
   nl binnenleidingen *f (m) pl*
   r внутренние трубопроводы *m pl* (здания)

1356 *e* **internal-quality brick**
   d Ziegel *m* für verdeckte Maurerarbeiten
   f brique *f* pour maçonnerie dissimulée
   nl baksteen *m* van mindere kwaliteit, achterwerkers
   r кирпич *m* для скрытых каменных конструкций

1357 *e* **internal radial pressure**
   d radialer Innendruck *m*
   f pression *f* interne radiale
   nl radiale druk *m*
   r внутреннее радиальное давление *n*

1358 *e* **internal skin**
   d innere Wandschale *f*, Innenschale *f* (*Hohlmauer*)
   f paroi *f* intérieure (*d'un mur creux*)
   nl binnenspouwblad *n* (*muur*); binnenwand *m* (*lichtkoepel*)
   r внутренняя продольная стенка *f* облегчённой кладки

1359   **internal span** *see* **interior span**

1360 *e* **internal vibrator**
   d Innenrüttler *m*, Innenvibrator *m*
   f vibr(at)eur *m* interne, pervibr(at)eur *m*
   nl inwendige trilinrichting *f*, trilnaald *f (m)*
   r внутренний [глубинный] вибратор *m*

1361 *e* **internal wiring**
   d innere Elektroinstallation *f*, Innenverdrahtung *f*
   f canalisation *f* électrique intérieure
   nl huisinstallatie *f* (*electr.*)
   r внутренняя электропроводка *f*

1362 *e* **intersection**
   d Kreuzung *f*; Straßenknoten *m*; Schnittpunkt *m*
   f croisement *m* [croisée *f*, intersection *f*] de routes, carrefour *m*; point *m* d'intersection
   nl (weg)kruising *f*; kruispunt *f (m)*

r пересечение *n*; перекрёсток *m*;
точка *f* пересечения

1363 e **intersection angle**
d 1. Drehwinkel *m*, Schwenkwinkel *m* (*Straßenbau*) 2. Schnittwinkel *m*
f 1. angle *m* de virage 2. angle *m* d'intersection
nl 1. draa ingshoek *m* (*weg*), richtingverandering *f* (*weg*) 2. snijdingshoek *m*
r 1. угол *m* поворота (*дороги*) 2. угол *m* пересечения

1364 e **intersection point**
d 1. Tangentenschnittpunkt *m* (*Straßenbau*) 2. Schnittpunkt *m*
f 1. point *m* d'intersection des tangentes 2. point *m* d'intersection
nl 1. tangentensnijpunt *n* (*snijpunt van de rechte wegassen bij een bocht*) 2. snijpunt *n*
r 1. вершина *f* угла поворота (*горизонтальной дорожной кривой*) 2. точка *f* пересечения

1365 e **interstices**
d Hohlräume *m pl*; Zwischenräume *m pl*; Risse *f pl*, Spalten *f pl*, Fugen *f pl*; Poren *f pl*
f interstices *m pl*; alvéoles *f pl*
nl holten *f pl*; tussenruimten *f pl*; poriën *f pl*
r пустоты *f pl*, полости *f pl*; зазоры *m pl*; трещины *f pl*; щели *f pl*; поры *f pl*

1366 e **intrados**
d Bogenleibung *f*, innere Bogenfläche *f*
f intrados *m*
nl binnenwelfvlak *n*
r внутренняя поверхность *f* арки [свода]

1367 e **intrapermafrost water**
d Intrapermafrostwasser *n*
f eau *f* dans la couche de pergélisol
nl grondwater *n* binnen de permafrostzone
r межмерзлотная вода *f*

1368 e **inundation**
d Überschwemmung *f*, Überflutung *f*, Inundation *f*
f inondation *f*
nl onderwaterzetten *n*
r затопление *n*, наводнение *n*

1369 e **inundation canal**
d Flutkanal *m*, Entlastungskanal *m*
f canal *m* alimenté au fil de l'eau
nl overlaatkanaal *n*
r лиманный канал *m* (*питание зависит от уровня воды в реке*)

1370 **inundation zone** see **flood zone**

1372 e **invert**
d Sohle *f* (*als umgekehrtes Gewölbe*)
f radier *m*
nl vloer *m* (*sluis*); bodem *m* (*riool*)
r лоток *m*

1373 e **inverted arch**
d Gegenbogen *m*, Konterbogen *m*

f arc *m* renversé
nl omgekeerde boog *m*
r обратная арка *f*

1374 e **inverted asphalt emulsion**
d umgekehrte Bitumenemulsion *f*
f émulsion *f* de bitume inversée
nl inverte bitumen-emulsie *f*
r битумная эмульсия *f* (с битумом в качестве дисперсной среды)

1375 e **inverted capacity**
d Sickerbrunnen-Kapazität *f*
f capacité *f* d'un puits absorbant
nl opnamevermogen *n* (van een put)
r поглощающая способность *f* (колодца)

1376 **inverted filter** see **loaded filter**

1377 e **inverted flat roof**
d umgekehrtes Flachdach *n*
f toit-terrasse *f* inversée (*toit dont la couche calorifuge est posée au-dessus du revêtement d'étanchéité*)
nl plat dak *n* met warmteïsolatie aan de bovenzijde
r обратная крыша *f* (*крыша, у которой теплоизолирующий слой уложен поверх гидроизолирующего ковра*)

1378 e **inverted siphon**
d Düker *m*, umgekehrter Heber *m*
f siphon *m* inversé
nl duiker *m*
r дюкер *m*

1379 e **inverted T-shaped footing**
d abgetrepptes [umgekehrtes T-förmiges] Fundament *n*, Stufenfundament *n*
f fondation *f* de béton en T inversé
nl strokenfundering *f* met verstijvingsrib
r ступенчатый фундамент *m*, бетонный фундамент *m* в форме перевёрнутого **T**

1380 e **inverted U-shaped precast concrete unit**
d umgekehrtes U-förmiges Betonfertigteil *n*
f composant *m* [élément *m* préfabriqué] de béton un U inversé
nl prefabbetonelement *n* in omgekeerd U-profiel
r П-образный сборный железобетонный элемент *m*

1381 e **inverted well**
d umgekehrter Brunnen *m*, Schluckbrunnen *m*, Sickerbrunnen *m*, Einleitungsbrunnen *m*
f puits *m* absorbant [d'alimentation]
nl zinkput *m*
r поглощающий [подпитывающий] колодец *m*

1383 e **ionite filter**
d Ionenaustauscheranlage *f*
f filtre *m* à ionite

nl ionenfilter *n, m*
r ионитовый фильтр *m*

1384 *e* **ionization**
d Ionisation *f,* Ionisierung *f*
f ionisation *f*
nl ionisatie *f*
r ионизация *f (воздуха)*; ионирование *n (воды)*

1385 *e* **iris damper**
d Iris-Luftklappe *f,* Lamellenklappe *f*
f registre *m* d'air du type iris
nl lamellenklep *f (m)*
r лепестковый воздушный клапан *m,* клапан *m* типа «ирис»

1386 *e* **Irish bridge**
d überschwemmbare Kleinbrücke *f;* befestigte Furt *f;* offener Steindrän *m;* Entwässerungsgerinne *n* quer zur Straße verlegt
f pont *m* submercible; route-cassis *f* surélevée; drain *m* ouvert à pierres; rigole *f (pour deriver des eaux pluviales en travers de la route)*
nl overstroombare brug *f (m);* verharde doorwaadbare plaats *f (m);* stenen afvoergoot *f (m)* over de straat
r затопляемый мост *m;* мощёный брод *m;* открытая каменная дрена *f;* дорожный лоток *m (для пропуска воды поперёк дороги)*

1387 *e* **iron-alumina ratio**
d Tonerdemodul *m*
f module *m* alumino-ferrique
nl ijzer-aluminiumverhouding *f,* ijzermodulus *m*
r глинозёмный модуль *m*

1388 *e* **iron cement**
d **1.** Eisenerzzement *m,* Eisenportlandzement *m,* EPZ **2.** Eisenkitt *m,* Metallkitt *m,* Graphitzement *m*
f **1.** ciment *m* de fer **2.** mastic *m* de fer [à limaille], lut *m* de fer
nl **1.** ijzercement *n* **2.** ijzerkit *f (m)*
r **1.** железистый [рудный] цемент *m* **2.** замазка *f* с добавкой железных опилок

1389 **iron-ore cement** *see* **iron cement 1.**

1390 *e* **iron paving**
d Fußbodenbelag *m* aus Gußeisenplatten
f pavage *m* [revêtement *m*] en plaques de fonte
nl wegdek *n* van gietijzeren platen
r покрытие *n* из чугунных плит *(дорожное или аэродромное)*

1391 *e* **iron Portland cement**
d Eisenportlandzement *m*
f ciment *m* portland de fer
nl ijzerportlandcement *n*
r шлакопортландцемент *m (до 30% шлака)*

1392 *e* **irregular aggregate**
d Zuschlagstoff *m* mit unregelmäßiger Kornform

f granulat *m* [agrégat *m*] à grains irréguliers
nl toeslagstoffen *f (m) pl* met onregelmatige korrelvorm
r заполнитель *m* с зёрнами различной формы

1393 *e* **irrevocable water consumption**
d Wasserverbrauch *m* ohne Rückleitung
f consommation *f* d'eau irrévocable
nl onherroepelijk waterverbruik *m*
r безвозвратное водопотребление *n*

1396 *e* **irrigation canal**
d Bewässerungskanal *m,* Bewässerungsgraben *m*
f canal *m* d'irrigation
nl bevloeiingskanaal *n,* irrigatiekanaal *n*
r оросительный канал *m*

1400 **irrigation ditch** *see* **irrigation canal**

1401 **irrigation module** *see* **hydromodule**

1402 *e* **irrigation network**
d Bewässerungsnetz *n*
f réseau *m* d'irrigation
nl bevloeiingssysteem *n,* irrigatiestelsel *n*
r оросительная сеть *f* [система *f*]

1403 *e* **irrigation requirement**
d Bewässerungswasserbedarf *m*
f besoins *m pl* en eau d'irrigation
nl waterbehoefte *f* voor irrigatiedoeleinden
r потребность *f* в воде для орошения, оросительная норма *f* нетто

1404 *e* **irrigation works**
d Bewässerungsbauwerke *n pl*
f ouvrages *m pl* d'irrigation
nl bevloeiingswerken *n pl*
r оросительные сооружения *n pl*

1405 *e* **I-section**
d **1.** Doppel-T-Querschnitt *m* **2.** Doppel-T-Träger *m,* I-Träger *m*
f profilé *m* en I; poutre *f* en double T [en I]
nl **1.** T-profiel *n* **2.** T-balk *m*
r **1.** двутавровое сечение *n* **2.** двутавровая балка *f*

1406 **isolated foundation** *see* **pad foundation**

1407 *e* **isolated pier**
d freistehender Pfeiler *m*
f poteau *m* isolé [individuel, détaché], colonne *f* isolée
nl vrijstaande pijler *m*
r свободностоящая опора *f* [колонна *f*]

1408 *e* **isolating valve**
d **1.** Gasdruckmangelsicherung *f* **2.** Rohrbruchwächter *m,* Selbstschlußventil *n*
f **1.** dispositif *m* protecteur contre les pertes de pression de gaz **2.** vanne *f* de sectionnement
nl **1.** druk-uitvalbeveiliging *f* **2.** zelfsluitend ventiel *n*

r 1. отсечный клапан *m*, клапан-
-отсекатель *m* (*газовый*)
2. аварийный [изолирующий]
клапан *m*

I409 *e* **isolation joint**
d Trenn(ungs)fuge *f*; Dehnungsfuge *f*
f joint *m* de sectionnement; joint *m* de
dilatation
nl dilatatievoeg *f* (*m*)
r разделительный шов *m*;
температурный шов *m*

I410 **isolation mounting** *see* **antivibration
mounting**

I411 *e* **isolator**
d 1. Isolator *m* 2. Trennschalter *m*
f 1. isolateur *m* 2. sectionneur *m*
nl 1. isolator *m*
2. scheidingsschakelaar *m*
r 1. изолятор *m* 2. рубильник *m*

I412 *e* **isometric view**
d isometrische Ansicht *f* [Darstellung *f*],
Isometrie *f*
f représentation *f* isométrique
nl isometrisch aanzicht *n*
[perspectief *n*]
r изометрическая проекция *f*,
изометрия *f*

I414 **italian roof** *see* **hipped roof**
I415 *e* **item**
d Position *f*, Gegenstand *m* (*z.B. im
Leistungsverzeichnis, in der Stückliste*)
f item *m*, article *m*
nl afzonderlijk artikel *n* (*b.v. op
stuklijst*)
r отдельное изделие *n*, деталь *f*
(*в спецификации*)

J

J1 *e* **jack**
d 1. Hebebock *m*, Schraubenwinde *f*,
Winde *f* 2. Steckdose *f* 3. Stützbock *m*
f 1. vérin *m*, cric *m* 2. fiche *f* femelle,
conjoncteur *m* 3. étai *m*, chandelle *f*,
étançon *m*, montant *m*
nl 1. vijzel *f* (*m*) 2. steekdoos *f* (*m*)
3. stut *m*; zuil *f* (*m*)
r 1. домкрат *m* 2. штепсельная
розетка *f* 3. подпорка *f*, стойка *f*
(*лесов, опалубки*)

J2 *e* **jack arch**
d scheitelrechter Bogen *m*,
Geradbogen *m*; Plattbogen *m*
f arc *m* droit, arc *m* (en) plate-bande;
voutain *m* de plancher
nl hanekam *m*, strekse boog *m*;
rechthoekige boog *m*
r плоская арка *f*, арочная перемычка *f*;
пологий свод *m* (*между балками
перекрытия*)

J3 *e* **jackhammer**
d leichter Bohrhammer *m*
f marteau *m* de mine; marteau *m*
piqueur
nl pneumatische boorhamer *m*
r бурильный *или* отбойный молоток *m*

J4 *e* **jacking**
d 1. Einpressen *n* (*Pfahl, Spundbohle*)
2. Anspannung *f* (*Spannbeton*)
f 1. enfoncement *m* (*des pieux*) par
vérin 2. mise *f* en tension (de
l'armature) de précontrainte
nl 1. inpersing *f* (*heipalen*) 2. inspanning
f (*wapening*)
r 1. задавливание *n*, вдавливание *n*
(*свай*) 2. натяжение *n* (*напрягаемой
арматуры*)

J6 *e* **jacking device**
d Spannvorrichtung *f*
f dispositif *f* de précontrainte
nl spaninrichting *f*
r устройство *n* для натяжения
(*арматуры*)

J7 *e* **jacking end of a bed**
d Spannende *n* eines Spannbettes
f extrémité *f* de tension du banc de
précontrainte
nl spaneinde *n* van een spanstand
r натяжной конец *m* стенда для
преднапряжения арматуры

J8 *e* **jacking force**
d 1. kontrollierte Spannkraft *f*
2. Hebekraft *f*
f 1. effort *m* de précontrainte contrôlé
2. force *f* de levage du coffrage
glissant (crée par vérins)
nl controleerde spankracht *f* (*m*)
(*wapening*)
r 1. контролируемое усилие *n*
натяжения (*арматуры*) 2. усилие *n*
подъёма (*скользящей опалубки*)

J9 *e* **jacking plate**
d Eindrückplatte *f*
f plaque *f* d'appui [de support] d'un
vérin
nl steunplaat *f* (*m*) (voor dommekracht)
r плита *f* для опирания домкрата

J11 *e* **jackknife drilling mast**
d Klappbohrmast *m*
f tour *m* de forage [de sondage] pliante
nl opklapbare boortoren *m*
r складывающаяся буровая вышка *f*

J12 *e* **jack pair**
d Doppelheber *m*, Heberpaar *n*,
Doppelspannpresse *f*
f vérins *m pl* (hydrauliques) jumelés
nl tweelingvijzel *f* (*m*)
r спаренный (гидро)домкрат *m*

J13 *e* **jack pile**
d eingedrückter Fertigpfahl *m*
f pieu *m* enfoncé par vérin
nl weggedrukte heipaal *m*
r свая *f*, вдавливаемая [погружаемая]
домкратом

J14 *e* **jack plane**
d Langhobel *m*, Rauhhobel *m*,
Schrupphobel *m*
f riflard *m*
nl voorloper *m*
r шерхебель *m*

J15 *e* **jack rafter**
d Walmschifter *m*, Schiftsparren *m*
f empanon *m*, chevron *m* de croupe
nl spar *m*
r нарожник *m*, короткая стропильная
нога *f*

J16 *e* **jack rib**
d niedrige Aussteifungsrippe *f*
[Kuppelrippe *f*, Gewölberippe *f*]
f nervure *f* courte (*d'une coupole ou
d'une voûte*)
nl steunribbe *f* (*m*) (*koepel, gewelf*)
r короткое ребро *n* (*купола, свода*)

J17 *e* **jack roll**
d Handwinde *f*, Förderhaspel *f*
f treuil *m* à bras [à main, manuel]
nl handlier *f* (*m*)
r ручной ворот *m*

J18 *e* **jack screw, jackscrew**
d Hebespindel *f*, Schraubenwinde *f*
f vérin *m* à vis
nl vijzel *f* (*m*), dommekracht *f* (*m*)
r винтовой домкрат *m*

J19 *e* **jack shore**
d ausfahrbare Stütze *f* (*mit Hebebock*)
f étai *m* [étançon *m*, montant *m*]
extensible
nl uitschuifbare stempel *m* met vijzel
r выдвижная опорная стойка *f*

J20 *e* **jack truss**
d Walmdachstuhl *m*
f ferme *f* de toit [de comble] secondaire,
ferme *f* de la croupe
nl schilddakstoel *m*
r вальмовая [вспомогательная]
стропильная ферма *f*

J21 **jack-up rig** *see* **mobile work platform 2.**

J22 *e* **jamb**
d Pfosten *m*; Gewände *n*; Leibung *f*
f montant *m* [poteau *m*] du bâti
dormant (*d'une porte ou d'une
croisée*); ébrasement *m*
nl kozijnstijl *f* (*m*); post *m*
r косяк *m* [вертикальный брус *m*]
оконной *или* дверной коробки;
оконный *или* дверной откос *m*

J23 *e* **jamb anchor**
d Befestigungsanker *m* (*für Türe,
Fenster*)
f patte *f* à scellement (*pour fixer bâti
dormant à la maçonnerie*)
nl kozijnanker *n*
r дверная *или* оконная закрепа *f* (*для
крепления к каменной кладке*)

J25 *e* **jamb lining**
d Gewändeauskleidung *f*; Türfutter *n*
f chambranle *m* (de fenêtre *ou* de porte)

nl dagbekleding *f* (*venster*); dagstuk *n*
(*deur*)
r оконный *или* дверной наличник *m*

J26 *e* **jaw breaker, jaw crusher**
d Backenbrecher *m*
f concasseur *m* à mâchoires
nl kaakbreker *m*
r щековая дробилка *f*

J27 **jedding axe** *see* **cavil**

J28 **jemmy** *see* **jimmy**

J29 *e* **jenny**
d Dampfstrahlgebläse *n*,
Dampfstrahlapparat *n*
f dispositif *m* de nettoyage à vapeur,
nettoyeur *m* au jet de vapeur
nl stoomstraalblazer *m*
r пароструйная машина *f* (*для
очистки поверхностей*)

J30 *e* **jesting beam**
d Zierbalken *m*
f poutre *f* ornementale
nl sierbalk *m*
r декоративная балка *f*

J32 *e* **jet-action trencher**
d Unterwasser-Düsengrabenzieher *m*
f excavateur *m* de tranchées
hydraulique; trancheuse *f*
hydraulique (*pour travaux sous l'eau*)
nl hydraulische slotengraafmachine *f*
r гидравлический траншеекопатель *m*
(*для подводных работ*)

J34 **jetcrete** *see* **gunite**

J35 *e* **jet disperser**
d Strahlverzehrer *m*, Strahlvernichter *m*
f dissipateur *m* d'énergie des jets d'eau
nl straalverspreider *m*
r струегаситель *m*

J38 *e* **jet piercer**
d Flammstrahlbohrgerät *n*
f installation *f* de forage thermique
nl installatie *f* voor thermisch boren
r установка *f* для термического
бурения

J39 *e* **jet piercing lance**
d Thermobohrer *m*
f lance *f* de thermoforage
nl thermische lans *f* (*m*)
r термобур *m*

J40 *e* **jet pump**
d Strahlpumpe *f*, Ejektor *m*
f pompe *f* à jet, éjecteur *m*
nl straawerper *m*
r струйный насос *m*, эжектор *m*

J43 *e* **jetted pile**
d Spülpfahl *m*, eingespülter Pfahl *m*
f pieu *m* lancé
nl ingespoten paal *m*
r подмывная свая *f* (*погружаемая
с подмывом*)

J44 *e* **jetting**
d 1. Einspülen *n* 2. hydraulisches
Bohren *n*; Preßluftbohren *n*

f 1. enfoncement *m* hydraulique
2. forage *m* à jet d'eau; forage *m*
pneumatique
nl 1. inspuiten *n* met waterstraal
2. boren *n* met water- *of* luchtstraal
r 1. гидравлическое погружение *n*
2. гидравлическое *или*
пневматическое бурение *n*

J45 jetting gear *see* giant

J46 *e* jetting pump
d Einspülpumpe *f*, Spülpumpe *f*
f refouleur *m*, pompe *f* à déblais
nl grondpomp *f* (*m*)
r грунтовой насос *m*, пульпонасос *m*

J47 *e* jetting tip
d Spülspitze *f*
f embout *m* de pointe filtrante
nl spoeltop *m* (*naaldfilter*)
r наконечник *m* иглофильтра

J48 *e* jetty
d 1. Regelungsbauwerk *n* (*Buhne*, S*porn*,
Höfter) 2. Anlegebrücke *f*,
Landungsbrücke *f*, Pier *m* 3. Mole *f*,
Hafendamm *m*, Wellenbrecher *m*
4. Vorbau *m*
f 1. ouvrage *m* de réglage (*epi*, *éperon*,
épi de branchage) 2. quai *m*
d'amarrage 3. jetée *f*, digue *f*;
brise-lames *m* 4. avant-corps *m*; parti *f*
saillante d'un bâtiment (*p. ex fenêtre
en ancorbellement*)
nl 1. regelbouwwerk *n* (*krib*)
2. aanlegsteiger *m* 3. havenhoofd *n*,
pier *m* 4. uitbouw *m* aan voorzijde
r 1. регуляционное сооружение *n*
(*буна, шпора, полузапруда*)
2. причальная эстакада *f*;
причальное сооружение *n*; пирс *m*
3. мол *m*; волнолом *m* 4.
выступающая часть *f* здания (*напр.
эркер*)

J49 jetty cylinder *see* screw pile

J50 *e* jib
d 1. Ausleger *m* 2. Auslegeransatz *m*,
Spitzenausleger *m*
f 1. flèche *f* (*d'une grue*) 2. fléchette *f*,
flèche *f* supplémentaire [auxiliaire]
nl 1. kraanarm *m*, giek *m* 2. laadboom
m, schalk *m*
r 1. стрела *f* (*крана, экскаватора*)
2. гусёк *m* (*стрелы крана*)

J51 jib boom *see* jib 2.

J52 *e* jib crab *UK*
d Auslegerlastkatze *f*,
Ausleger(lauf)katze *f*
f chariot *m* de flèche
nl loopkat *f* (*m*)
r грузовая тележка *f* крана

J53 *e* jib crane
d Auslegerkran *m*, Schwenkkran *m*
f grue *f* à la flèche
nl giekkraan *f* (*m*), laadmast *m*
r стреловой кран *m*

J54 *e* jib door
d Geheimtür *f*, Tapetentür *f*
f porte *f* dérobée [au ras de mur]
nl niet-zichtbare deur *f* (*m*)
r скрытая дверь *f*, дверь *f* заподлицо
со стеной

J55 *e* jig
d Lehre *f*; Montagelehrgerüst *n*;
Haltevorrichtung *f*,
Spannvorrichtung *f*
f gabarit *m*, dispositif *m* d'assemblage
[de montage]; dispositif *m* de serrage;
guide-outil *m*
nl montagesteiger *m*, boorstandaard *m*
r кондуктор *m*; сборочно-монтажное
приспособление *n*; держатель *m*;
зажимное приспособление *n*

J56 jigger saw *see* jigsaw

J57 jigging chute *see* shaker conveyor

J58 *e* jigsaw
d Laubsäge *f*, Rahmenspaltsäge *f*
f scie *f* sauteuse [à chantourner,
à découper]
nl figuurzaag *f* (*m*)
r лобзик *m*

J59 *e* jim crow, jim-crow
d Schienenbiegeeinrichtung *f*
f appareil *m* manuel à cintrer les rails
nl buigapparaat *n* voor rails, koevoet *m*
r рельсогибочный станок *m* ручного
действия

J60 *e* jimmer
d Gegenstücke *n pl*, Doppelelemente
n pl
f pièces *f pl* jumelées; paire *f* de pièces
de construction semblables
nl gepaard constructie-element *n*
r парные элементы *m pl* конструкции,
парные строительные детали *f pl*;
ответные детали *f pl* (*напр. фланцы*)

J61 *e* jimmy
d kleine Brechstange *f*
f pince(-monseigneur) *m*, levier *m* de fer
nl sloopbeitel *m*
r короткий лом *m*, ломик *m*

J62 jinnie wheel *see* gin block

J63 *e* jitterbug
d Gitterstampfer *m*
f dame *f* [dameuse *f*, pilonneuse *f*]
à grille
nl roosterstamper *m*
r решётчатая трамбовка *f*

J64 *e* job
d 1. Bauobjekt *n* 2. Gewerbe *n*
f 1. ouvrage *m* (à bâtir) 2. travail *m*;
tâche *f*; corps *m* d'état
nl 1. karwei *n* 2. baan *f* (*functie*)
r 1. объект *m* строительства 2. вид *m*
строительных работ; выполняемая
работа *f*

J65 *e* job captain
d Chefingenieur *m* eines Projektes
f ingénieur *m* chef d'étude

nl ingenieur *f* belast met de leiding van een bouwproject, projectingenieur *m*
r главный инженер *m* проекта; руководитель *m* проекта

J66 *e* **job cleanup**
d Baustellen(auf)räumung *f*
f nettoyage *m* de chantier de construction
nl opruimen *n* van de bouwplaats
r очистка *f* строительной площадки

J68 *e* **job-mix concrete**
d Baustellenbeton *m*
f béton *m* préparé sur le chantier, béton *m* de chantier
nl op de bouwplaats vervaardigd beton *n*
r бетон *m*, приготовляемый на строительной площадке

J69 *e* **job-mix formula**
d Baustellen-Mischungsformel *f*
f composition *f* du béton spécifiée
nl mengverhouding *f* voor op de bouwplaats vervaardigde mortel, samenstelling *f* van het betonmengsel
r заданный [проектный] состав *m* (бетонной) смеси

J70 *e* **job practices**
d technologische Methoden *f pl*, Arbeitsverfahren *n pl*
f techniques *f pl* des travaux
nl in de praktijk gebruikelijke werkmethoden *f pl*
r технологические приёмы *m pl* [технология *f*] производства работ

J71 **job site** *see* **building site**

J72 *e* **job superintendant**
d Oberbauleiter *m*
f chef *m* de chantier; conducteur *m* de travaux
nl hoofdopzichter *m*
r начальник *m* строительного участка; старший производитель *m* работ

J73 *e* **job workshop**
d Baustellenwerkstatt *f*
f atelier *m* de chantier [de réparation sur chantier]
nl bij het werk behorende werkplaats *f (m)*
r приобъектная мастерская *f*

J74 *e* **joggle**
d 1. Nut *f* und Feder *f*; Verschränkung *f*, Verzahnung *f*; Absatz *m* im Betonierungsabschnitt 2. Zapfen *m*, Zapfenverbindung *f*, Verzapfung *f*
f 1. encoche *f ou* partie *f* saillante dans des pièces à assembler; harpe *f*, redan *m* 2. adent *m*, goujon *m*, joint *m* à adent
nl 1. kraag *m*, borst *f (m)* 2. dook(anker) *n*, korte pen *f (m)* voor blinde pen- en gatverbinding
r 1. ответный выступ *m или* паз *m* в соединяемых элементах (*напр. в стеновых блоках*); штраба *f*, уступ *m*

в блоке бетонирования 2. шип *m*, шиповое соединение *n*

J75 *e* **joggle joint**
d 1. Steinverklammerung *f* 2. Verbindung *f* mit Feder und Nut, gespundete Verbindung *f*
f 1. assemblage *m* à gradins 2. assemblage *m* à ténon [à rainure et languette], joint *m* à adent
nl 1. dookverbinding *f* 2. verbinding *f* met pen en gat [met messing en groef]
r 1. соединение *n* с ответными выступами и пазами (*напр. стеновых блоков*) 2. соединение *n* в гнездо и шип [в шпунт и гребень]

J77 *e* **joiner's chisel**
d Meißel *m*, Stemmeisen *n*
f bédane *m*, bec-d'âne *m*, ciseau *m* à bois
nl steekbeitel *m*
r стамеска *f*

J78 **joiner's gauge** *see* **marking gauge**

J79 *e* **joiner's glue**
d Tischlerleim *m*
f colle *f* de menuiserie
nl houtlijm *m*
r столярный клей *m*

J80 *e* **joinery**
d 1. Tischlerarbeit *f* 2. Tischlererzeugnisse *n pl*
f 1. travaux *m pl* de menuiserie 2. menuiserie *f*
nl 1. schrijnwerk *n* 2. timmerwerk *n*
r 1. столярные работы *f pl* 2. столярные изделия *n pl*

J81 **joinery work** *see* **joinery 1.**

J82 *e* **joining means**
d Verbindungsmittel *n pl*
f pièces *f pl* d'assemblage [de raccord]; raccords *m pl*
nl verbindingsmiddelen *n pl*
r соединительные элементы *m pl*; средства *n pl* соединения

J83 *e* **joint**
d 1. Fuge *f*, Stoßfuge *f*; Stoß *m*; Verbindung *f* 2. Rahmenknoten *m*
f 1. joint *m*; jointure *f*; assemblage *m* 2. nœud *m* (*d'un treillis, d'une ossature*)
nl 1. naad *m*, voeg *m* 2. knooppunt *n* (*skelet*)
r 1. шов *m*; стык *m*; соединение *n* 2. узел *m* (*фермы, рамы*)

J84 **joint angle** *see* **connecting angle**

J85 **joint box** *see* **junction box**

J86 *e* **joint cement**
d Vergußmasse *f*, Füllmasse *f*
f masse *f* de jointoiement [de scellement, de remplissage]
nl vulmiddel *n*, voegvulmassa *f*
r уплотняющая масса *f*, заливочная мастика *f*

J87 *e* **joint cutter**
d Fugenschleifgerät *n*
f coupe-joint *m*, découpeuse *f* de joints;
machine *f* à découper les joints
nl asfaltbetonzaagmachine *f* voor het
snijden van krimpvoegen
r инструмент *m* или машина для
нарезки швов (*в бетонном
дорожном покрытии*)

J88 *e* **joint efficiency**
d Nahtwertigkeit *f*,
Verschwächungsbeiwert *m*
f efficacité *f* du joint
nl relatieve sterkte *f* van een lasnaad
r показатель *m* равнопрочности
сварного шва (*в процентах*);
коэффициент *m* ослабления сечения
(*в сварном соединении*)

J89 *e* **jointer**
d 1. Fugeneinbaugerät *n* 2. Fugeisen *n*,
Fugenkelle *f* 3. Fügehobel *m*,
Langhobel *m*, Rauhbank *f*
4. Verbindungsbügel *m*, Schäkel *m*,
Befestigungsbügel *m*,
Befestigungsklammer *f* 5.
Elektromeißel *m*, Druckluftmeißel *m*
f 1. coupe-joint *m* manuel
2. jointoyeur *m* 3. rabot *m* plat,
varlope *f* 4. happe *f* en fer plat
5. machine *f* à mortaiser
nl 1. voegspijker *m* 2. zwaard *n* 3. sabel
m, voorloper *m* 4. verbindingsklem
f (*m*) 5. langgatboor(machine) *m* (*f*)
r 1. инструмент *m* для ручной нарезки
швов в свежем бетоне 2. расшивка *f*
(*инструмент*) 3. фуганок *m*
4. скоба *f* из полосовой стали
5. долбёжник *m*

J90 **jointer plane** *see* **jointer 3.**

J91 *e* **joint filler**
d 1. Fugenvergußmasse *f*, Fugenkitt *m*
2. Fugeneinlage *f*, Fugenstreifen *m*
f 1. pâte *f* de joints, produit *m*
d'étanchéité, matériau *m* de
remplissage 2. garniture *f* waterstop
[d'étanchéité] profilée
nl 1. voegmortel *m* 2. voegstrip *m*
r 1. герметизирующая паста *f*,
герметик *m* 2. профильная
эластичная герметизирующая
прокладка *f*

J93 *e* **joint grouting**
d Fugenvermörtelung *f*, Fugenverguß *m*
f remplissage *m* [scellement *m*] des
joints
nl vullen *n* van voegen
r цементация *f* [замоноличивание *n*]
швов

J94 *e* **jointing**
d 1. Ausfugen *n*, Verfugen *n*,
Fugenverschluß *m* 2. Abrichten *n*,
Ausfügen *n* 3. Pflasterfugenausfüllung
f

f 1. jointoiement *m*, garnissage *m* de
joints 2. rabotage *m* de finition
3. remplissage *m* des joints du pavage
nl 1. voegen *n* (*metselwerk*)
2. vijnschaven *n*, stropen *n*
3. voegvulling *f* (*wegdek*)
r 1. расшивка *f* швов (*в кладке*)
2. чистовая строжка *f*, фугование *n*
3. заполнение *n* швов (*брусчатой
мостовой*)

J95 *e* **jointing compound**
d Dicht(ungs)masse *f*
f mélange *m* étanche [d'étanchéité],
masse *f* d'étanchéité
nl vulmassa *f* (*m*)
r герметизирующий состав *m* (*для
стыков труб*)

J96 *e* **jointing medium**
d Dichtungskitt *m*, Dichtungsmittel *n*
f produit *m* d'étanchéité
nl vulmiddel *n*
r герметик *m*

J97 **jointing tool** *see* **jointer 1.**

J98 *e* **jointless flooring**
d Estrich *m*
f revêtement *m* de sol coulé
nl naadloze vloer *f* (*m*) (*bv magnesiet*)
r бесшовный пол *m*

J99 *e* **joint load**
d Knotenlast *f*, Knotenbelastung *f*
f charge *f* appliquée au nœud (*d'une
poutre à treillis*)
nl krachten *f* (*m*) *pl* in een knooppunt
r узловая нагрузка *f* (*фермы*)

J100 *e* **joint movement**
d Öffnen *n* und Schließen *n* der
Dehnungsfuge
f mouvement *m* du joint de dilatation
nl beweging *f* in een dilatatievoeg
r раскрытие *n* и закрытие *n*
температурного шва

J101 *e* **joint rigidity**
d Knotensteifigkeit *f*
f rigidité *f* de nœud (*d'une ossature*)
nl stijfheid *f* van een knooppunt (*skelet*)
r жёсткость *f* узла (*рамы*)

J102 *e* **joint ring**
d Dichtungsring *m*, O-Ring *m*
f bague *f* de garniture, rondelle *f* de
joint
nl dichtingsring *m*
r кольцевая уплотняющая прокладка *f*

J103 *e* **joint sawing**
d Einschleifen *n* von Fugen
f sciage *m* des joints (*dans le
revêtement routier en béton*)
nl krimpvoegen zagen (*wegdek*)
r нарезка *f* швов (*в дорожном
покрытии*)

J104 **joint sealant** *see* **joint filler**

J105 *e* **joint sealing compound**
d Fugenvergußmasse *f*,
Abdichtungsmastix *m*

 f masse *f* [mélange *m*] d'étanchéité
nl voegvulmateriaal *n*
 r состав *m* [смесь *f*] для
гидроизоляции швов

J106 *e* **joint sealing machine**
 d Fugenvergußgerät *n*
 f machine *f* à sceller les joints
nl voegenvulmachine *f*
 r машина *f* для герметизации
[заполнения] швов

J107 *e* **joint strip**
 d Fugeneinlage *f*; Fugenstreifen *m*
 f bande *f* de recouvrement, couvre-
-joint *m*; garniture *f* de joint profilée
nl deklat *f* (*m*); architraaf *f* (*m*)
 r стыковая рейка *f*; профильный
уплотняющий элемент *m* стыковых
соединений

J108 *e* **joint tape**
 d Klebeband *n* für Fugenabdichtung
 f bande *f* adhésive [d'étanchéité],
ruban *m* de joint
nl plakband *n* voor voegafdichting
(*gipsplaat*)
 r лента *f* для заклейки швов (*напр.
между листами сухой штукатурки*)

J109 *e* **joist**
 d Deckenbalken *m*, Deckenträger *m*
 f poutre *f*, lambourde *f*, poutrelle *f*,
solive *f*
nl bint *n*, draagbalk *m*
 r балка *f* (*перекрытия, потолка,
покрытия*)

J110 *e* **joisted floor**
 d 1. Balkendecke *f*, Trägerdecke *f*
2. Fußboden *m* auf Lagerhölzern
 f 1. plancher *m* à poutrelles 2. couvre-sol
m en madriers posés sur solives
nl 1. vloer *f* (*m*) op houten balklaag
2. houten zoldering *f*
 r 1. балочное перекрытие *n* 2. пол *m*,
укладываемый по лагам [балкам]

J111 *e* **jolting table**
 d Schocktisch *m*, Rütteltisch *m*
 f table *f* à secousses
nl schoktafel *f* (*m*), triltafel *f* (*m*)
 r вибрационный стол *m*, вибростол *m*

J112 *e* **jolting vibrator**
 d Schockrüttler *m*, Schockvibrator *m*
 f vibrateur *m* à choc [à secousse]
nl schokvibrator *m*
 r встряхивающий вибратор *m*

J113 *e* **Joosten process**
 d Joosten-Verfahren *n*
 f procédé *m* de Joosten
nl Joosten-procédé *n* (*grondverbetering
door injectie*)
 r способ *m* Джустена
(*последовательное нагнетание
хлорида кальция и силиката натрия
при закреплении грунта*)

J114 *e* **jubilee wagon**
 d Feldbahn-Muldenkipper *m*

 f wagonnet *m* basculant [basculeur]
nl kiplorrie *f*
 r опрокидная вагонетка *f*

J115 *e* **jumbo**
 d 1. Bohrwagen *m* 2. fahrbares
Tunnelgerüst *n*
 f 1. jumbo *m*, chariot *m* de forage
2. chariot *m* de coffrage mobile
nl 1. boorwagen *m* 2. mobiele
tunnelbekisting *f*
 r 1. буровая каретка *f* 2. тележка *f*
тоннельной опалубки

J116 *e* **jump**
 d Fundamentabsatz *m*
 f gradin *m* de fondation
nl versnijding *f* van een fundament
 r уступ *m* [ступень *f*] в фундаменте
(*каменном или бетонном*)

J117 *e* **jump distance**
 d Sprungweite *f*
 f longueur *f* de ressaut hydraulique
nl sprongwijdte *f*, grootte *f* van de
hydraulische sprong
 r длина *f* отгона гидравлического
прыжка

J118 *e* **jumpforming**
 d Klettern *n* (*Kletterschalungsbau*)
 f déplacement *m* du coffrage grimpant
nl maken *n* van een klimbekisting
[klimkist]
 r перемещение *n* подъёмно-
-переставной опалубки

J119 *e* **jump forms, jumping formwork**
 d Kletterschalung *f*
 f coffrage *m* grimpant
nl klimbekisting *f*, klimkist *f* (*m*)
 r подъёмно-переставная опалубка *f*

J120 *e* **junction**
 d 1. Einmündung *f*; Straßenkreuzung *f*
2. Stoß *m*, Anschluß *m* (*von
Wasserströmen*) 3. Rohrverbindung *f*
4. Abzweigung *f*
 f 1. bifurcation *f* (*de routes*); jonction *f*
(*de routes, de deux rues*) 2. confluent *m*
(*de cours d'eau*) 3. raccordement *m*,
abouchement *m*, raccord *m*
4. embrachement *m*
nl 1. wegkruising *f* 2. samenloop *m* van
waterstromen 3. verbinding *f* (*pijpen*)
4. spruitstuk *n*, T-stuk *n*
 r 1. примыкание *n* (*дороги*); пересечение
n (*дорог, улиц*) 2. слияние *n*
(*водотоков*) 3. соединение *n* труб
4. ответвление *n* (*трубопровода*)

J121 *e* **junction box**
 d Anschlußkasten *m*, Verbindungsdose *f*,
Abzweigdose *f*, Verteilerkasten *m*
 f boîte *f* de jonction [de connexion,
de branchement]
nl aansluitkast *f* (*m*), lasdoos *f* (*m*)
 r соединительная [ответвительная]
коробка *f* (*электросети*)

J122 *e* **junction canal**
 d Verbindungskanal *m*

f canal *m* de communication
[de jonction, de raccordement]
nl verbindingskanaal *n*
r соединительный канал *m*

J123 *e* **junction plate**
d Knotenblech *n*
f gousset *m*
nl schetsplaat *f* (*m*), knoopplaat *f* (*m*)
r фасонка *f*, косынка *f*

J124 *e* **junction point**
d Verbindungspunkt *m*
f point *m* de raccordement [de jonction]
nl splitsingspunt *f* (*m*), kruispunt *n*
r точка *f* сопряжения кривой поворота
и переходной кривой (*дороги*)

J125 *e* **jute rope**
d Jutefaserseil *n*, Jutefasertau *n*
f câble *m* [corde *f*] en jute
nl jutetouw *n*
r джутовый канат *m*

J126 *e* **jut window**
d hervorstehendes [auskragendes]
Fenster *n*
f fenêtre *f* en saillie [saillante]
nl erker *m*
r эркерное окно *n*

K

K1 *e* **Kangaroo tower crane**
d Kletter(-Turmdreh)kran *m*,
Hochbaukletterkran *m*
f grue *f* à tour grimpante du type
«Kangourou»
nl kangoeroe-torenkraan *f* (*m*)
r самоподъёмный башенный кран *m*
типа «Кенгуру»

K2 *e* **keel**
d Rotocker *m*, Rötel *m*
f ocre *m* rouge
nl okerrood *n*
r красная охра *f*

K3 *e* **keel arch**
d Kielbogen *m*, persischer Bogen *m*
f arc *m* en accolade
nl kielboog *m*
r килевидная арка *f*

K4 **Keene's cement** *see* **gypsum cement**

K5 *e* **keeper**
d Halteeinrichtung *f*, Halter *m*
f arrêtoir *m* [arrêt *m*] automatique
nl automatische grendel *m*
r (автоматический) останов *m* (*ворот*,
дверей)

K6 *e* **Kelly ball**
d Halbkugel(-Betonprüf)gerät *n* von
Kelly
f bille *f* de Kelly
nl hardheidstester *m* volgens Kelly
(*beton*)

r прибор *m* Келли для определения
консистенции бетонной смеси

K7 *e* **Kelly ball test**
d Frischbetonsteifeprüfung *f* nach Kelly
f essai *m* à bille de Kelly
nl betonhardheidsproef *f* (*m*) volgens
Kelly
r определение *n* консистенции
бетонной смеси методом Келли
(*вдавливанием металлического
плунжера с полусферическим концом*)

K8 *e* **kentledge**
d Widerlager *n*, Gegengewicht *n*
f contrepoids *m*
nl contragewicht *n*, tegengewicht *n*
r контргруз *m*, противовес *m*

K9 **kerb** *see* **curb**

K10 *e* **kerf**
d Sägeschnitt *m*, Kerbe *f*, Einschnitt *m*
f trait *m* [coup *m*] de scie, coupure *f*
nl kerf *f* (*m*), inkeping *f*
r пропил *m*, надрез *m*

K11 *e* **kern area**
d Querschnittskern *m*
f noyau *m* central (*d'une section*)
nl kern *f* (*m*) van de doorsnede
r ядро *n* сечения

K12 *e* **kevel**
d 1. Schlageisen *n* 2. Holzknagge *f*
(*zwischen Holzstützen*)
f 1. laie *f*, laye *f* 2. bloc *m* de bois
(*réliant les deux montants*)
nl 1. steenhouwershamer *m*
2. belegklamp *m*, *f*
r 1. закольник *m*, зубатка *f*
2. бобышка *f* (*связывающая
деревянные стойки*)

K13 *e* **key**
d 1. Dübel *m*, Keil *m* 2. Querkeil *m*,
Vorstecker *m*, Splint *m* 3. Keilbohle *f*
4. Spritzputz *m* 5. Adhäsionsvermögen
n 6. Schlußstein *m* 7. Nut *f*
8. Schlüssel *m* 9. Bogenscheitel *m*
f 1. clavette *f* 2. goupille *f*, cheville *f*
3. madrier *m* en coin *m* (*utilisé pour le
procédé de plancheiage*) 4. couche *f* de
fond (*d'enduit*) 5. pouvoir *m*
[capacité *f*] d'adhésion 6. claveau *m*
[voussoir *m*] de la tête (de l'arc)
7. rainure *f* (*pour languette*) 8. clé *f*
[clef *f*] de serrure 9. clé *f* [clef *f*]
d'un arc
nl 1. deuvel *m* 2. spie *f* (*m*) 3.
vloeraandrijver *m* 4. spuitmortel *m*
5. hechtvermogen *n*; goed hechtende
ondergrond *m* 6. sluitsteen *m* 7. groef
f (*m*) 8. sleutel *m* (*slot*) 9. boogsteen *m*
r 1. шпонка *f* 2. чека *f* 3. клиновая
доска *f* (*для сплачивания досок
дощатых полов*) 4. обрызг *m*
(*штукатурного намёта*)
5. адгезионная способность *f*
6. замковый камень *m* 7. паз *m*

(для гребня) **8.** ключ *m* (*замка*)
9. ключ *m* арки

K14 key block *see* apex block

K15 *e* key bolt
d Splintbolzen *m*, Keilbolzen *m*
f boulon *m* à clavette
nl spiebout *m*
r болт *m* с чекой

K16 *e* key brick
d 1. Keilziegel *m*, Gewölbeziegel *m*
2. Schlußstein *m*
f 1. brique *f* en coin 2. claveau *m*
[voussoir *m*] de clé
nl 1. gewelfsteen *m* 2. sluitsteen *m*
r 1. клиновой кирпич *m* 2. ключевой
[замковый] камень *m*

K17 *e* keyed beam
d verdübelter Balken *m*
f poutre *f* composée en bois à clavettes
nl samengestelde balk *m* met deuvels
r составная деревянная балка *f* на
шпонках

K18 *e* keyed joint
d Federkeilverbindung *f*, Verdübelung *f*
f assemblage *m* par clavettes *ou* organes
d'assemblage
nl verbinding *f* met deuvels
r шпоночное соединение *n*,
соединение *n* на шпонках

K19 keyed pointing *see* key joint pointing

K20 *e* key for plaster
d Putzunterlage *f*
f support *m* pour enduit
nl ondergrond *m* voor pleister
r основание *n* под штукатурку

K21 *e* key hole saw, keyhole saw
d Stichsäge *f*, Spitzsäge *f*, Lochsäge *f*
f scie *f* d'entrée [à guichet]
nl schrobzaag *f* (*m*), sleutelgatzaag *f* (*m*)
r ножовка *f* с очень узким полотном

K22 *e* keying in
d Anschlußmauerverband *m*
f liaison *f* entre une maçonnerie
inachevée et maçonnerie ultérieure,
joint *m* de reprise (*d'un mur de
maçonnerie*)
nl dwarsverband *n* in gemetselde muur
r соединение *n* [перевязка *f*] новой
кирпичной стены со старой

K23 *e* key joint pointing
d Hohlfugenverschluß *m*
f jointoiement *m* rainuré
nl voegen met verdiepte voeg
r вогнутая расшивка *f* швов (*кладки*)

K24 *e* key plan
d Übersichtsplan *m*
f plan *m* de disposition, plan-repère *m*
nl situatietekening *f*
r схематический чертёж *m*
расположения основных
конструктивных элементов здания,
план *m* с отметками

K25 *e* key plate
d Schlüssellochdeckel *m*,
Schlüsselschild *m*
f entrée *f* de serrure
nl sleutelplaatje *n*
r личина *f*, планка *f* замка

K26 keystone *see* apex block

K27 key-valve *see* wedge valve

K28 *e* kick-atomizing pile hammer
d Dieselbär *m* mit Schlagzerstäubung,
SZ-Dieselbär *m*, DELMAG-Bär *m*
f mouton *m* diesel à pulvérisation par
choc, mouton *m* DELMAG
nl diesel-heiblok *n* met slagverstuiver
r трубчатый свайный дизель-молот
m

K29 *e* kickout
d plötzliches Nachlassen *n* [plötzlicher
Einsturz *m*] der Graben- *bzw.*
Baugrubenaussteifung
f affaissement *m* *ou* effondrement *m*
subit des éléments du blindage des
fouilles
nl plotseling bezwijken *n* van het
stutwerk van een bouwput
r внезапное ослабление *n* или
обрушение *n* элементов крепления
(*траншей, котлованов*)

K30 *e* kickplate
d Stoßplatte *f*, Trittplatte *f*
f plaque *f* à pieds [de poussée]
nl schopplaat *f* (*m*) (*deur*)
r предохранительная пластинка *f* для
дверей

K31 *e* kid
d Strauchwerkbündel *n*,
Buschwerkbündel *n*
f fascine *f* en verges [en branches]
nl takkenbos *n*, fascine *f*
r хворостяная фашина *f*

K32 *e* kidding
d Packwerk *n*, Faschinenpackwerk *n*
f pose *f* des fascines
nl rijswerk *n*
r тюфячная [фашинная] кладка *f*

K33 kieselguhr *see* diatomite

K34 *e* kieselguhr brick
d Diatomitstein *m*, Schamotteziegel *m*
f brique *f* (de) diatomite [de kieselguhr]
nl chamottesteen *m*
r диатомовый [шамотный] кирпич *m*

K35 kiln brick *see* refractory brick

K36 *e* kinematic viscosity
d kinematische Viskosität *f*
f viscosité *f* cinématique
nl kinematische viscositeit *f*
r кинематическая вязкость *f*

K37 *e* kinetic head
d Geschwindigkeitshöhe *f*
f pression *f* dynamique
nl stuwdruk *m*
r скоростной напор *m*

K38 *e* **king bolt**
    d Hängestange *f*
    f aiguille *f* pendante
    nl stalen middenstijl *m*
    r центральная металлическая подвеска
    f [тяж *m*] стропильной фермы

K39 *e* **king closer**
    d Ziegel *m* mit einer abgeschrägten Ecke
    f brique *f* biaise [chanfreinée]
    nl baksteen *m* met één afgeschuinde
    hoek, drieklezoor
    r кирпич *m* со скошенным углом
    (*наполовину ширины тычка*)

K40 *e* **king pile**
    d Richtpfahl *m*
    f pieu *m* d'alignement [de direction]
    nl schoorpaal *m*
    r маячная [направляющая] свая *f*

K41 *e* **king post**
    d Hängesäule *f*
    f poinçon *m* (*de comble*)
    nl koningsstijl *m*, hanger *m*
    r (центральная жёсткая) подвеска *f*
    стропильной фермы, бабка *f*
    (висячих стропил)

K42 *e* **king-post joint**
    d 1. Kopfbauwerk *n* einer hölzernen
    Schleuse, Hauptfachbaum *m*
    2. Anschlußknoten *m* zwischen
    Hängesäule und Dachbalken
    f 1. tête *f* d'écluse en bois 2. nœud *m*
    de jonction entre poinçon et entrait
    nl 1. houten sluishoofd *n* 2. verbinding *f*
    tussen hanger en trekplaat
    r 1. головной [королёвый] узел *m*
    (*деревянного шлюза*) 2. узел *m*
    крепления центральной подвески
    [бабки] стропил к затяжке

K43 *e* **king-post truss**
    d einsäuliges Hängewerk *n*, einfacher
    Hängebock *m*
    f ferme *f* simple
    nl hangspant *n*
    r висячие стропила *pl* с одной
    подвеской [бабкой]

K44   king rod *see* king bolt

K45 *e* **kink**
    d Kink *f*
    f coque *f* [tortillement *m*] d'un câble
    métallique
    nl kink *f* (*m*) (*kabel*)
    r перегиб *m* (*троса*)

K46 *e* **kitchen building block unit**
    d Küchenzelle *f*
    f bloc-cuisine *m*
    nl keukenblok *n*, bouwelement *n* van
    systeemkeuken
    r сборный пространственный блок-
    -кухня *m*

K47 *e* **kitchen-dining room**
    d Eßküche *f*, Wohnküche *f*
    f salle *f* à manger-cuisine, cuisine *f* de
    séjour [d'habitation]
    nl eetkeuken *f* (*m*)
    r кухня-столовая *f*

K48 *e* **kite winder**
    d Wendelstufe *f*, Winkelstufe *f*
    f marche *f* balancée [dansante, gironnée]
    triangulaire
    nl schuine trede *f* (*m*)
    r забежная ступень *f* треугольной
    формы

K49 *e* **Kjellmann-Franki machine**
    d Dränmaschine *f*
    f machine *f* pour pose des tuyaux de
    drainage
    nl draineermachine *f*
    r дреноукладчик *m*

K50 *e* **knapping hammer**
    d Vorschlaghammer *m*; Bossierhammer
    m
    f massette *f*; têtu *m* à pointe
    nl moker *m*
    r кувалда *f*; закольник *m* (*с плоским
    и заострённым бойками*)

K51 *e* **knapsack duster, knapsack sprayer**
    d Tornister-Spritzgerät *n*
    f pulvérisateur *m* à havresac [à dos]
    nl rugspuit *f* (*m*)
    r ранцевый распылитель *m*

K52 *e* **knee, knee bend**
    d Knierohr *n*, Kniestück *n*
    f coude *m* d'équerre [en équerre]
    nl kniestuk *n*
    r колено *n* (*трубопровода*)

K53 *e* **knee board**
    d Knieschutztafel *f*
    f planche *f* [panneau *m*] d'appui pour
    les genoux (*à l'usage des ouvriers
    travaillant à genoux*)
    nl knieplank *f* (*m*)
    r подколенный опорный щит *m* (*для
    работы при отделке незатвердевших
    бетонных полов*)

K54 *e* **knee brace**
    d Kopfband *n*
    f bracon *m*
    nl hoekverstijving *f*; bintstijl *m*
    r угловой подкос *m или* раскос *m*;
    угловая связь *f*

K55 *e* **knee piece, knee rafter**
    d Ecksparren *m*, Gratsparren *m*,
    Dachschifter *m*
    f chevron *m* du toit [du comble] en
    mansarde
    nl hoekspar *f* (*m*)
    r стропильная нога *f* (*мансардной
    крыши, шатровых стропил*)

K56 *e* **knee roof**
    d Mansarddach *n*
    f toit *m* [comble *m*] en mansarde
    nl mansardedak *n*
    r мансардная крыша *f*

K57 *e* **knife-edge loading**
    d Schneidenlast *f*, Streckenlast *f*
    (*Berechnung der Brücken*)

 f charge *f* équivalente uniformément répartie (*calcul des ponts*)
 nl lijnbelasting *f*
 r равномерно распределённая эквивалентная погонная нагрузка *f* (*расчёт мостов*)

K58 *e* **knife switch**
 d Messerschalter *m*
 f iterrupteur *m* à lame, commutateur *m* à couteau
 nl messchakelaar *m*
 r рубильник *m*

K59 *e* **knob**
 d 1. Türknauf *m* 2. Betätigungsknopf *m*, Bedienknopf *m* 3. Auswölbung *f*, Überhöhung *f*
 f 1. bouton *m* poignée de porte 2. bouton *m* de réglage 3. bosse *f* (*sur surface*)
 nl 1. deurknop *m* 2. knop *m* (*schakelaar*) 3. knobbel *m*
 r 1. дверная ручка-кнопка *f* 2. кнопка *f* управления 3. выпуклость *f* (*на поверхности*)

K60 *e* **knobbing, knobbling**
 d Behauen *n*
 f smillage *m*, dressage *m*, démaigrissement *m* (*d'une pierre*)
 nl behakken *n* (*steen*)
 r околка *f* (*природного камня*)

K61 *e* **knob latch**
 d drehbarer Türknauf *m* (*mit Schnappverschluß*)
 f loquet *m* à bouton
 nl loopslot *m* met knop
 r дверная фалевая ручка-кнопка *f*

K62 *e* **knocked down building components**
 d Fertigbauteile *n pl* in auseinandergenommenem [zerlegtem] Zustand beliefert
 f composants *m pl* préfabriqués livrés au chantier par éléments
 nl in gedemonteerde toestand geleverde samengestelde bouwelementen *n pl*
 r сборные строительные конструкции *f pl*, поставляемые отдельными элементами

K63 *e* **knocked-down frame**
 d auseinandergenommene Türzarge *f*
 f bâti *m* dormant démontable
 nl uiteengenomen kozijn *f* (*m*)
 r разбираемая дверная коробка *f* (*поставляемая в разобранном виде*)

K64 knot *see* **system point**

K65 *e* **knuckle bend**
 d Rohrwinkel *m*, starkgekrümmtes Bogenrohr *n*
 f coude *m* [raccord *m*] courbé raide
 nl haakse bocht *f* (*m*), knie *f* (*m*)
 r крутоизогнутый отвод *m*

K66 *e* **kraft, kraft paper**
 d Kraftpapier *n*
 f papier *m* kraft

 nl kraftpapier *n*
 r крафт-бумага *f*

K67 *e* **K-truss, K-type truss**
 d K-Fachwerk *n*
 f poutre *f* à treillis en K
 nl K-vakwerk *n*
 r ферма *f* с полураскосной решёткой

L

L1 *e* **labeled door**
 d feuerbeständige [feuerfeste] Tür *f*
 f porte *f* coupe-feu avec la résistance au feu spécifiée
 nl branddeur *f* (*m*)
 r огнестойкая [противопожарная] дверь *f*

L2 *e* **labour saving devices, labour saving tools**
 d Kleinmechanisierungsmittel *n pl*, arbeitssparende Vorrichtungen *f pl*
 f dispositifs *m pl* assurant l'économie de main-d'œuvre
 nl arbeidsbesparende hulpmiddelen *n pl*
 r средства *n pl* малой механизации, приспособления *n pl*, повышающие производительность труда

L3 *e* **laced column**
 d zusammengesetzte Stütze *f*, Gitterstütze *f*
 f poteau *m* composé (*assemblé par treillis ou par des plats*)
 nl samengestelde kolom *f* (*m*)
 r составная колонна *f* (*с соединительной решёткой или планками*)

L4 *e* **laced valley**
 d Dachkehle *f* mit Stückbelag
 f noue *f* revêtue par petits éléments (*tuiles, ardaises*)
 nl dakkeel *f* (*m*) met leibedekking
 r ендова *f*, покрытая штучным кровельным материалом (*черепицей, плитками*)

L5 *e* **lacing**
 d 1. Gitter *n*, Verbindungslaschen *f pl* Querbleche *n pl*, Vergitterung *f* (*Säule*) 2. Verbände *m pl*, Verstrebung *f* (*Gerüst*) 3. Fugendeckleisten *f pl* (*Grabenverschalung*) 4. Zwischenschicht *f* von Ziegeln *oder* Blöcken (*im Bruchsteinmauerwerk*) 5. Verteilerbewehrung *f*
 f 1. treillis *m*, plats *m pl* (*pour poteau composé en acier*) 2. entretoisement *m*, contreventement 3. planches *f pl* de recouvrement de blindage 4. assise *f* de liaison en briques (dans la maçonnerie en moellons) 5. ferraillage *m* [armature *f*] de répartition
 nl 1. verbindingsstaven *f* (*m*) *pl* 2. windverband *n* 3. bekledingsplanken

f (*m*) *pl* 4. muur *m* van natuursteen in
onregelmatig verband
5. verdeelwapening *f*
r 1. соединительная решётка *f*,
соединительные планки *f pl* (*ветвей
колонны*) 2. связи *f pl* (*между
стойками лесов*) 3. нащельники *m pl*
(*дощатой обшивки траншей*)
4. связующий ряд *m* кирпичей *или*
блоков (*в бутовой кладке*)
5. распределительная арматура *f*

L6   **lacing course** *see* **lacing 4.**

L7   **lacing wire** *see* **tie-wire**

L8   *e* **lack of homogeneity**
d Inhomogenität *f* (*des Gemisches*)
f inhomogénéité *f*, manque *m*
d'homogénéité (*du mélange*)
nl gebrek *n* aan homogeniteit
r неоднородность *f* (*смеси*)

L9   *e* **lack of parallelisme**
d Lageabweichung *f* von der
Parallelität
f non-parallélisme *m*
nl onevenwijdigheid *f*
r непараллельность *f*, перекос *m*

L10   *e* **lacquer**
d Öllack *m*, Lackfarbe *f*, Farblack *m*
f vernis *m* à l'huile
nl vernis *m*, lak *n*, *m* (op oliebasis)
r масляный лак *m*

L11   *e* **ladder**
d 1. Unterleiter *f*, Eimerleiter *f*
2. Steigleiter *f*
f 1. élinde *f* 2. échelle *f*
nl 1. emmerladder *f* (*m*) 2. ladder *f* (*m*)
r 1. ковшовая [черпаковая] рама *f*
(*землечерпательного снаряда,
цепного канавокопателя*)
2. лестница-стремянка *f*;
приставная лестница *f*

L12   *e* **ladder cleats**
d Sprossen *f pl*
f échelons *m pl* (*d'une échelle*)
nl sporten *f* (*m*) *pl* (*ladder*)
r поперечины *f pl* лестницы

L13   *e* **ladder ditcher**
d Eimerkettengrabenbagger *m*
f excavateur *m* de tranchée(s) à élinde
nl slotengraafmachine *f* met emmerladder
r траншеекопатель *m* с ковшовой цепью

L14   *e* **ladder jack scaffold, ladder scaffolding**
d Leitergerüst *n*
f échafaudage *m* sur [à] échelles
nl laddersteiger *m*
r лестничные леса *pl*

L15   *e* **ladder stand**
d Stehleiter *f*
f échelle *f* double (à marches),
escabeau *m*
nl trapleer *n*
r (нераздвижная) лестница-стремянка *f*

L16   **ladder trencher** *see* **ladder ditcher**

L17   *e* **Lafarge cement**
d Lafargezement *m* (*Kalk-Gips-Zement
mit Marmorpulverzusatz*)
f ciment *m* de Lafarge (*liant à base de
chaux et plâtre avec addition de poudre
de marbre*)
nl Lafarge-cement *n* (*kalk-gipscement met
bijslag van marmerpoeder*)
r цемент Лафарга *m* (*известково-
-гипсовое вяжущее с добавлением
молотого мрамора*)

L18   *e* **lag bolt**
d Vierkantholzschraube *f*
f tire-fond *m*
nl houtdraadbout *m* met vierkante kop,
tire-fond *n*
r крупный шуруп *m* с квадратной
головкой

L19   *e* **lagging**
d 1. Wärmedämmung *f* (*Rohrleitungen,
Behälter*) 2. Wärmedämmstoff *m*;
Schalldämmstoff *m* 3. wärmedämmende
Ausbauschalen *f pl*
4. Schalungsoberfläche *f*,
Bretterschalung *f*
5. Bohlenabsteifung *f* (*der Gräben*)
f 1. revêtement *m* calorifuge *m*,
enveloppe *f* isolante 2. calorifuge *m*;
isolant *m* calorifuge [thermique,
isotherme] *ou* acoustique [phonique]
3. coquille *f* d'isolation thermique
4. surface *f* de contact d'un coffrage,
joue *f* de coffrage 5. blindage *m* (de la
fouille) en planches
nl 1. isolatie *f* 2. isolerend materiaal *n*
3. warmteïsolerende bekisting *f*
4. bekistingoppervlak *n* 5. houten
bekleding *f* van een ingraving
r 1. теплоизоляция *f* (*труб,
резервуаров*) 2. тепло- или
звукоизоляционный материал *m*
3. теплоизоляционные скорлупы *f pl*
4. рабочая поверхность *f*, дощатая
обшивка *f* (*опалубки*) 5. дощатое
крепление *n* (*траншей*)

L20   *e* **lags**
d wärmedämmende Ausbauplatten *f pl*
(*für zylindrische Behälter*)
f panneaux *m* verticaux d'isolation
thermique (*de réservoirs cylindriques*)
nl warmteïsolerende siloplaten *f pl*
r вертикальные теплоизолирующие
панели *f pl* (*цилиндрических
резервуаров*)

L21   **lag screw** *see* **lag bolt**

L22   *e* **laid-on-edge course**
d Hochkantschicht *f*
f assise *f* de champ
nl rollaag *f* (*m*)
r ряд *m* кирпичей, уложенных на ребро
(*в кладке*)

L23   *e* **laid-on moulding**
d leistenförmige Auflageelemente *n pl*
f baguettes *f pl*, moulures *f pl* (*en bois*)

nl opgelegde versiering *f*
r накладные погонажные изделия *n pl*

L24 *e* laitance
d Zementmilch *f*, Feinschlämme *f*
f lait *m* de ciment
nl cementbrij *m*, cementmelk *f* (*m*)
r цементное молоко *n*

L25 *e* lake asphalt
d Seeasphalt *m*
f bitume *m* Trinidad [de lac]
nl asfaltbitumen *n*
r тринидадский битум *m*

L26 *e* lakes
d gedeckte Farbstoffe *m pl*, Lackfarben
f pl, Farblacke *m pl*, organische
Pigmente *n pl*
f pigments *m pl* organiques
nl lakverven *pl*
r органические пигменты *m pl*
(*яркого оттенка*)

L27 *e* lamella
d Gewölbepfeiler *m* (*Zollbaulamellendach*)
f lamelle *f*
nl gewelfspijler *m*
r косяк *m* кружально-сетчатого свода
(*стандартный элемент*)

L28 *e* lamella roof
d Zollbaulamellendach *n*
f voûte *f* en lamelles
nl lamellengewelf *n*
r кружально-сетчатый свод *m*

L29 *e* laminate
d Schichtstofferzeugnis *n*;
Schicht(preß)stoff *m*,
Schichtkunststoff *m*
f stratifié *m*, matériau *m* stratifié;
matière *f* stratificé [lamellée]
nl gelaagd bouwmateriaal *n*
r слоистое изделие *n*; многослойный
материал *m*

L30 *e* laminated arch
d geleimter [geklebter] Bohlenbogen *m*
f arc *m* lammellé(-collé)
nl boog *m* van gelijmde planken
r дощатая клеёная арка *f*

L31 *e* laminated beam
d geklebter Holzträger *m*
f poutre *f* lamellée
nl gelamineerde houten balk *m*
r дощатая клеёная балка *f*

L32 *e* laminated glass
d Verbundglas *n*
f verre *m* de sécurité
nl gelaagd glas *n*
r многослойное [безопасное,
безосколочное] стекло *n*

L33 laminated joint *see* finger joint

L34 *e* laminated wood
d Schichtholz *n*, verleimte Bretter *n pl*
f bois *m* lamellé(-collé)
nl gelamineerd hout *n*
r клеёная древесина *f*

L35 *e* lamp post, lamp standard
d Lichtmast *m*, Beleuchtungsmast *m*
f poteau *m* d'éclairage
nl lantaarnpaal *m*
r осветительный столб *m*, опора *f*
уличного светильника

L36 lanced arch *see* acute arch

L37 land accretion *see* land reclamation 2.

L39 *e* land-clearing rake
d Wurzelrechen *m*, Wurzelharke *f*
f débroussailleuse *f*, débroussailleur
m
nl wroeter *m*
r кусторез *m*

L40 *e* land development project
d Meliorationsvorhaben *n*
f projet *m* d'amélioration
nl ruilverkavelingsplan *n*
r проект *m* мелиорации

L41 *e* land drain
d Landdrän *m*
f drain *m* enterré
nl drainageleiding *f*
r осушительная дрена *f*

L42 *e* land drainage
d Meliorationen *f pl*, Landentwässerung
f, Landtrockenlegung *f*
f drainage *m* des terres [des terrains]
nl drainage *f*
r осушение *n* земель, мелиорация *f*

L43 *e* land fill
d Geländeauffüllung *f*,
Geländeaufschüttung *f*
f remblaiement *m* [remblayage *m*,
comblement *m*] de terrain
nl ophoging *f* van het terrein
r отсыпка *f* грунта

L44 *e* land filtration
d Bodenfiltration *f*
f filtration *f* [filtrage *m*] par le sol
nl bodemfiltratie *f*
r почвенная очистка *f* сточных вод
(*на полях фильтрации*)

L45 *e* landing
d 1. Treppenabsatz *m*, Podest *n*
2. Landungsbrücke *f*, Anlegestelle *f*
f 1. palier *m* 2. quai *m* (d'accostage)
nl 1. trapportaal *n* 2. aanlegsteiger *m*
r 1. лестничная площадка *f*
2. причал *m*, пристань *f*

L46 *e* landing stage
d Landungsbrücke *f*, Anlegestelle *f*
f quai *m*, appontement *m*
nl aanlegplaats *f* (*m*)
r причал *m*, пристань *f*

L47 *e* land pier
d Uferpfeiler *m*, Landpfeiler *m*
f culée *f*, pile-culée *f*, pile *f* de culée
nl landpijler *m*
r береговой устой *m*, береговая опора
f

L48 *e* land reclamation
d 1. Melioration *f* 2. Neulandgewinnung

f 1. amelioration *f* 2. aménagement *m*
du littorale
nl landwinning *f*
r 1. мелиорация *f* 2. расширение *n*
и освоение *n* прибрежной полосы

L49 *e* **land reclamation by enclosure**
d Landgewinnung *f* durch Eindeichung
f assèchement *m* des polders par la
création des levées
nl landwinning *f* door indijken
r создание *n* польдеров обвалованием

L50 *e* **land reclamation by filling**
d Landgewinnung *f* durch Aufschüttung
f assèchement *m* des terrains par
remblaiement
nl landwinning *f* door ophoging
r расширение *n* суши отсыпкой грунта
на мелководье

L51 *e* **landscape architecture**
d Landschaftsarchitektur *f*
f architecture *f* paysagiste
[de paysage]
nl tuin- en landschapsarchitectuur *f*
r ландшафтная [садово-парковая]
архитектура *f*

L52 *e* **landscaped central reservation**
d mittlerer Grünstreifen *m*
f terre-plein *m* central boisé, bande *f*
médiane boisée
nl begroeide middenberm *m*
r центральная полоса *f* зелёных
насаждений (*на дороге*)

L53 *e* **landscaping**
d Landschaftsgestaltung *f*;
Freiflächenplanung *f*; Begrünung *f*
f aménagement *m* de paysages
[de parterres, de jardins paysagers];
aménagement *m* des espaces verts;
plantation *f*
nl groenvoorziening *f*
r проектирование *n* ландшафта;
создание *n* искусственного
ландшафта; озеленение *n*

L54 *e* **landslide**
d Absetzung *f*, Rutschung *f*,
Erdrutsch *m*
f glissement *m* de terrain, éboulement *m*
nl aardverschuiving *f*
r оползень *m*

L55 *e* **landslide control**
d Erdrutschsicherungsmaßnahmen *f pl*
f protection *f* contre les glissements de
terrain
nl maatregelen *m pl* tegen
aardverschuivingen
r противооползневые мероприятия *n pl*

L56 *e* **landslide protection works**
d Rutschsicherungsbauwerke *n pl*
f ouvrages *m pl* de protection contre
l'éboulement des terres
nl werken *n pl* ter bescherming tegen
aardverschuivingen
r противооползневые сооружения *n pl*

L57 *e* **land tie**
d Verankerungsstab *m*, Ankerstab *m*
f tirant *m* d'ancrage
nl ankerstaaf *f* (*m*)
r анкер *m*, анкерная тяга *f*

L58 *e* **land tile**
d perforiertes Tondränagerohr *n*
f tuyau *m* de drainage en terre cuite
perforé
nl geperforeerde draineerbuis *f* (*m*),
kannebuis *f* (*m*)
r гончарная дренажная труба *f*
с перфорацией

L60 *e* **lane**
d 1. Verkehrsstreifen *m* 2. schmaler
Heckenweg *m*; Gasse *f*
f 1. voie *f* [bande *f*] de circulation
2. allée *f*, passage *m*
nl 1. verkeersstrook *f* (*m*) 2. pad *n*,
steeg *f* (*m*)
r 1. полоса *f* движения, дорожная
полоса *f* 2. узкая аллея *f*; проход *m*;
проезд *m*

L61 *e* **Lang lay**
d Albertschlag *m*, Gleichschlag *m*,
Parallelschlag *m*, Längsschlag *m*
f câblage *m* Lang à droite, câblage *m*
parallèle, torsion *f* Lang
nl Langslag *m*, gelijkslag *m* (*kabel*)
r параллельная свивка *f* (*стального
каната*), свивка *f* Лэнга

L62 *e* **lantern**
d 1. Oberlicht *n*, Dachreiter *m*,
Dachlaterne *f* 2. Straßenleuchte *f*,
Laterne *f*
f 1. lanterneau *m* 2. luminaire *m* des
rues
nl 1. lichtkap *f* (*m*) 2. lantaarn *f* (*m*)
r 1. фонарь *m* (крыши) 2. светильник *m*
уличного освещения

L63 **lantern-light** *see* **lantern 1.**

L64 *e* **lanyard**
d Schutzgurthalteseil *n*
f corde *f* d'assurance (*d'une ceinture de
securité*)
nl veiligheidsgordel *m*
r строп *m* предохранительного пояса

L65 *e* **lap**
d Überdeckung *f*, Überlappung *f*
f chevauchement *m*, chevauchure *f*
nl overlapping *f*
r нахлёстка *f*

L66 *e* **lap joint**
d Überlapp(ungs)stoß *m*,
Überlappverbindung *f*
f joint *m* de recouvrement
nl rechte liplas *f* (*m*)
r соединение *n* внахлёстку

L67 *e* **lap length**
d Überlappungslänge *f*
f longueur *f* de recouvrement
nl overlap *f* (*m*)
r длина *f* нахлёстки

L68   **lapping** *see* **lap joint**

L69 *e* **lap siding**
    d Stülpschalung *f*
    f parement *m* à recouvrement
    nl potdekselen *n*
    r горизонтальная дощатая обшивка *f*
      внакрой

L70   **lap splice** *see* **lap joint**

L71 *e* **lap weld**
    d Überlapp(schweiß)naht *f*
    f soudure *f* à recouvrement
    nl lasnaad *m* met overlap
    r (сварной) шов *m* внахлёстку

L72 *e* **large block masonry**
    d Großblockmauerwerk *n*
    f maçonnerie *f* en grands blocs
    nl metselwerk *n* van grote blokken
    r крупноблочная кладка *f*

L73 *e* **large job**
    d Großbauvorhaben *n*; Großbaustelle *f*
    f chantier *m* à grande échelle; gros
      chantier *m*
    nl groot bouwwerk *n*
    r крупный объект *m* строительства;
      крупная строительная площадка *f*

L74 *e* **large panel structures**
    d Großplattenkonstruktionen *f pl*
    f bâtiments *m pl ou* constructions *f pl*
      en grands panneaux préfabriqués
    nl bouwen *m pl* met grote prefab-
      -elementen
    r крупнопанельные конструкции *f pl*
      или сооружения *n pl*

L75 *e* **larmier**
    d Traufenüberstand *m*; Wassernase *f*,
      Ablauf *m*
    f auvent *m*; queue *f* de vache; larmier *m*
    nl waterlijst *f (m)*; lekdorpel *m*
    r свес *m* карниза; слезник *m*, отлив *m*

L76 *e* **larry**
    d Weichmacherspaten *m*
    f bouloir *m*, croc à chaux, doloire *f*
    nl kalkhouw *f (m)*
    r мотыга-скребок *f (для*
      перелопачивания раствора)

L77 *e* **Larssen section**
    d Larssen-Profil *n*, Larssenbohle *f*
    f palplanche *f* Larssen
    nl Larssenprofiel *n*
    r профиль *m* [шпунтовая свая *f*
      системы] Ларсена

L78 *e* **laser level**
    d Lasernivellier *m*
    f niveau *m* à laser
    nl laser-waterpasinstrument *n*
    r лазерный нивелир *m*

L79 *e* **lashing**
    d Rödelung *f*, Würgeverbindung *f*
    f ligature *f* avec fil de fer (*de*
      l'armature)
    nl binding *f* (*wapening*)
    r связывание *n*, перевязка *f*
      (*арматуры*)

L79a *e* **latch**
    d Sperrklinke *f*
    f pêne *m* battant [demi-tour réversible]
    nl klink *f (m)*, snapslot *n*
    r дверная пружинная защёлка *f*

L80 *e* **lateral buckling**
    d (seitliche) Ausknickung *f*,
      Knickung *f*
    f flambage *m* (latéral), flambement *m*
    nl knik *m*
    r боковое выпучивание *n*, потеря *f*
      устойчивости при продольном изгибе

L81 *e* **lateral canal**
    d Seitenkanal *m*, Lateralkanal *m*,
      Nebenkanal *m*
    f canal *m* latéral de navigation
    nl lateraal kanaal *n*
    r боковой канал *m* (*проложенный*
      параллельно естественному руслу
      реки)

L82 *e* **lateral contraction**
    d Querkontraktion *f*,
      Querzusammenziehung *f*
    f contraction *f* latérale [transversale]
    nl dwarscontractie *f*
    r поперечное сжатие *n*

L83 *e* **lateral earth pressure**
    d horizontaler Erddruck *m*
    f pression *f* latérale du sol
    nl actieve gronddruk *m*
    r горизонтальное [боковое] давление *n*
      грунта

L84   **lateral exhaust** *see* **lateral hood**

L85   **lateral flexure** *see* **buckling**

L86 *e* **lateral-flow spillway**
    d seitlicher Überlauf *m*, Entlastung ·*f*
      mit seitlichem Überlauf
    f évacuateur *m* latéral avec canal
      d'évacuation parallèle au déversoir
    nl overlaat *m*
    r водосброс *m* с боковым отводом
      воды

L87 *e* **lateral force**
    d Seitenkraft *f*, Horizontalkraft *f*
    f force *f* latérale [horizontale]
    nl zijwaartse kracht *f (m)*
    r боковая [горизонтальная] сила *f*

L87a *e* **lateral force design**
    d Berechnung *f* auf horizontale
      seismische Belastungen
    f calcul *m* aux sollicitations séismiques
      latérales [horizontales]
    nl berekening *f* van horizontale
      seismische belasting
    r расчёт *m* на действие
      горизонтальных сейсмических сил

L88 *e* **lateral hood**
    d Seitenhaube *f*, Seitenabsaugung *f*,
      Seitenabzug *m*
    f hotte *f* latérale
    nl inrichting *f* voor zijdelings afzuigen
    r бортовой отсос *m*

L88a e **lateral load**
 d 1. Windbelastung f 2. horizontale
   seismische Belastung f 3.
   Querbelastung f; seitliche Belastung f,
   Seitenlast f, Seitenschub m
 f 1. charge f due au vent 2.
   sollicitations f pl séismiques latérales
   [horizontales] 3. charge f
   transversale; charge f latérale
 nl zijdelingse belasting f (wind;
   seismische krachten)
 r 1. ветровая нагрузка f 2.
   горизонтальное сейсмическое
   воздействие n 3. поперечная или
   боковая нагрузка f

L89 e **lateral reinforcement**
 d Querbewehrung f; Ringbewehrung f;
   Spiralbewehrung f, Bügelbewehrung f
 f armature f transversale, ferraillage m
   [acier m] transversal; frette f en
   hélice [hélicoïdale]; fréttage m
 nl dwarswapening f; spiraalwapening f
 r поперечная арматура f; кольцевая
   арматура f (колонн, свай);
   спиральная арматура f

L94 e **lateral restraint**
 d Seiteneinspannung f,
   Seitenbefestigung f
 f encastrement m latéral
 nl zijdelingse inklemming f
 r боковое защемление n
   [закрепление n]

L95 e **lateral sewer**
 d Abwassersammler m der 3. Ordnung
 f égout m latéral
 nl lateraal riool n
 r канализационный коллектор m
   третьего порядка (для приёма стоков
   из домовых выпусков)

L96 e **lateral soil load**
 d Belastung f durch horizontalen
   Erddruck
 f charge f [sollicitation f] due à la
   poussée latérale du sol
 nl actieve gronddruk m
 r нагрузка f от бокового давления
   грунта

L97 **lateral storage** see **bank storage**

L98 e **lateral strain**
 d Seitendehnung f, Querdehnung f
 f déformation f lataérle [transversale]
 nl dwarsvervorming f
 r поперечная деформация f

L99 e **lateral support**
 d seitliches Auflager n, Seitenstütze f
 f appui m [support m] latéral
 nl stempel m
 r горизонтальная [боковая] опора f;
   горизонтальный опорный стержень m

L100 **lateral sway** see **sidesway**

L100a e **lateral-torsional buckling**
 d Biegedrehknicken n
 f flambage m [flambement m] général

par compression et torsion
 nl wringingsknik m
 r потеря f устойчивости при
   продольном изгибе и кручении;
   выпучивание n от продольного
   изгиба и кручения

L101 e **latest finish date**
 d spätester Endtermin m einer
   Aktivität
 f date f [fin f] au plus tard
 nl laatste tijdstip m van beëindiging
 r самый поздний срок m окончания
   работ

L102 e **latest start date**
 d spätester Anfangstermin m einer
   Aktivität
 f date f [début m] au plus tôt
 nl laatste begintijdstip m van een
   activiteit
 r самый поздний срок m начала работ

L103 e **latex-modified concrete**
 d Beton m mit Latexzusatz
 f béton m à liant mixte à base de
   ciment et de latex
 nl beton n met latextoeslag
 r полимерцементный бетон m на
   латексе

L104 e **lath**
 d Latte f
 f latte f; lame f (de jalousie); volige f
 nl lat f (m); tengel m; ondergrond m
   voor pleisterwerk
 r планка f; рейка f; дранка f; дрань f;
   штукатурная сетка f

L105 e **lath hammer**
 d Lattenhammer m
 f hachotte f
 nl stukadoorshamer m
 r штукатурный молоток m

L106 e **lathing**
 d 1. Putzträger m; Latten f pl,
   Reißlatten f pl 2. Annageln n von
   Latten
 f 1. support m d'enduit; lattes f pl,
   lattis m 2. lattage m
 nl 1. pleisterdrager m 2. aanbrengen n
   van steengaas
 r 1. основание n под штукатурку;
   дранка f; дрань f 2. крепление n
   драни или штукатурной сетки

L107 **lath laid-and-set** see **two-coat work 1.**

L108 **lath, plaster, and set** see **two-coat work 1.**

L110 e **lattice**
 d Gitter n; Fachwerk n; Ausfachung f
 f treillis m
 nl rooster m, n; traliewerk n
 r решётка f; решётка f фермы;
   соединительная решётка f
   составной колонны

L111 e **lattice beam, lattice girder**
 d Fachwerkbinder m, Gitterträger m
 f poutre f à treillis

nl vakwerkspant *n*, tralieligger *m*
r сквозная [решётчатая] балка *f*,
   балочная ферма *f*

L112 e **lattice retaining wall**
   d aufgelöste Stützmauer *f*
   f mur *m* de soutènement à treillis
   nl roosterkeermuur *m*
   r решётчатая [сквозная] подпорная
     стенка *f*

L113 e **lattice truss**
   d Rautenfachwerkträger *m* mit
     Parallelgurten
   f poutre *f* à treillis en X
   nl tralieligger *m*
   r ферма *f* с параллельными поясами
     и перекрёстной решёткой

L114 e **lattice work**
   d Fachwerk *n*
   f structure *f* à résille, construction *f* en
     treillis
   nl vakwerk *n*, traliewerk *n*
   r решётчатая [стержневая]
     конструкция *f*

L115 e **launching truss**
   d Vorbauschnabel *m*
   f poutre *f* de lancement
   nl lanceerbalk *m*, neus *m* (*mil. brugslag*)
   r ферма *f* для надвижки пролётных
     строений моста (*при монтаже*),
     аванбек *m*

L116 e **lavatory**
   d 1. Waschbecken *n* 2. Waschraum *m*;
     Spülklosettraum *m*, WC
   f 1. lavabo *m* 2. cabinet *m* de toilette
     [d'aisances]
   nl 1. wasbak *m* [2. wasvertrek *n*; toilet
     n
   r 1. умывальник *m* (*санитарный
     прибор*) 2. умывальная [туалетная]
     комната *f*, туалет *m*, уборная *f*

L117 e **law of continuity**
   d Kontinuitätssatz *m*
   f loi *f* [équation *f*] de continuité
   nl wet *f* (*m*) van de continuïteit
   r закон *m* неразрывности

L118 e **lay**
   d Schlag *m* (*Drahtseil*)
   f câblage *m*
   nl slag *m* (*touw*)
   r свивка *f* (*каната*)

L119 e **lay bar**
   d horizontale Fenstersprosse *f*
   f petit bois *m* horizontal
   nl horizontale glasroede *f* (*m*)
   r горизонтальный горбылёк *m*
     оконного переплёта

L120 e **lay-by**
   d Rastplatz *m*, Parkplatz *m*
   f aire *f* de repos; halte *f* simple (*le long
     d'une route*)
   nl parkeerplaats *f* (*m*)
   r площадка *f* отдыха
     (*у автомобильной дороги*)

L121 e **layer**
   d Lage *f*, Schicht *f*
   f couche *f*; assise *f* (*de briques*)
   nl laag *f* (*m*) (*metselwerk*); mortellaag *f* (*m*)
   r слой *m*, ряд *m* (*кирпичной кладки*)

L122 e **laying**
   d 1. Einbringen *n*, Aufbringen *n*
     (*Beton*) 2. Verlegen *n*, Verlegung *f*
     (*Rohrleitungen*) 3. Trassenabsteckung
     f 4. Unterputz *m* (*Grundschicht des
     gewöhnlichen Putzes*)
   f 1., 2. pose *f*, mise *f* en œuvre [en place]
     3. piquetage *m* du tracé 4. couche *f*
     de fond (*d'enduit*)
   nl 1. storten *n* 2. aanleggen (*leidingen*)
     uitzetten *n* van het tracé 4. eerste
     laag *f* (*m*) (*pleisterwerk*)
   r 1. укладка *f* (*бетона*) 2. прокладка *f*
     (*труб*) 3. разбивка *f* (*трассы*)
     4. первый слой *m* двухслойной
     штукатурки

L123 e **laying and finishing machine**
   d Straßendeckenfertiger *m*
   f finisseur *m* [finisseuse *f*] de revêtement
     routier en béton
   nl betonverwerkingsmachine *f*
   r бетоноукладочная и отделочная
     машина *f*

L124 e **laying course**
   d Sandunterbettung *f*,
     Sandbettungsschicht *f*
   f forme *f* en sable
   nl aardebaan *f* (*m*), zandbed *n*
   r песчаная постель *f* (*мостовых*)

L125 e **laying out**
   d Absteckung *f*, Linienführung *f*;
     Markierung *f*
   f traçage *m*, tracé *m*, piquetage *m*;
     marquage *m*
   nl uitzetten *n*, afpalen *n*; aftekening *f*
     (*materiaal*)
   r трассирование *f*, разбивка *f*;
     разметка *f*

L126 e **laylight**
   d lichtdurchlässige Deckenplatte *f*,
     Scheitellaterne *f*, Firstlaterne *f*
   f panneau-plafond *m* lumineux
   nl plafondvenster *n*, bovenlicht *n*
   r потолочная (застеклённая) световая
     панель *f*, потолочный зенитный
     фонарь *m*

L127 e **layout, lay-out**
   d 1. Übersichtsplan *m*,
     Übersichtszeichnung *f*, Generalplan
     m; Lageplan *m*; Kompositionsplan *m*,
     Anordnungsplan *m*, Aufstellungsplan
     m 2. Trassierung *f*, Linienführung *f*,
     Streckenfestlegung *f*
   f 1. disposition *f* d'ensemble, plan *m*
     de masse; plan *m* de situation; schéma
     m de disposition 2. traçage *m*, tracé *m*
   nl 1. situatietekening *f* 2. uitzetten *n*
   r 1. генеральный план *m*; ситуационный
     план *m*; компоновочный план *m*,

схема *f* расположения
2. трассирование *n*

L128 *e* **L-beam**
d Winkelbalken *m*, Winkelträger *m*
f poutre *f* de section en L
nl hoekprofiel *n*
r балка *f* Г-образного поперечного сечения

L129 *e* **L-column**
d L-Säule *f*; einhüftiger Rahmen *m*
f colonne *f* en L, poteau *m* à console
nl L-kolom *f* (*m*)
r Г-образная колонна *f*

L131 *e* **leader**
d 1. Dachabfallrohr *n*
2. Warmluftkanal *m* (*Luftheizung*)
f 1. tuyau *m* de descente 2. conduit *m* [conduite *f*] à air chaud
nl 1. regenpijp *f* (*m*) 2. warme-
-luchtkanaal *n*
r 1. водосточная труба *f* 2. воздуховод *m* тёплого воздуха

L132 **leader drain** *see* **main drain**

L133 *e* **leaders**
d Mäkler *m*, Laufrute *f*, Läuferrute *f*
f guides *m pl* [guidage *m*] de sonnette
nl leipalen *m pl* (*heistelling*)
r направляющие *f pl* (рамы) копра

L134 **leading jetty** *see* **guiding dolphin**

L135 *e* **lead sheath**
d Blei(kabel)mantel *m*
f gaine *f* de plomb
nl loden kabelmantel *m*
r свинцовая оболочка *f* (кабеля)

L136 *e* **leaf**
d 1. Türflügel *m*; Torflügel *m* 2. Flügel *m* der Klappbrücke, Brückenflügel *m* 3. Schale *f* (*Hohlmauerwerk*) 4. Tafel *f*, Stauklappe *f* (*Klappenwehr*) 5. Luftklappenblatt *n*, Blatt *n*
f 1. battant *m* [ventail *m*] de porte 2. travée *f* basculante de pont 3. paroi *f* de mur creux 4. clapet *m* de vanne 5. lame *f* de registre d'air
nl 1. deurvleugel *m* 2. klep *f* (*m*) (*ophaalbrug*) 3. blad *n* (*spouwmuur*) 4. schuif *f* (*m*) (*stuw*) 5. blad *n* van een luchtklep
r 1. полотно *n* (*двери*); полотнище *n* (*ворот*) 2. крыло *n* (*раскрывающегося моста*) 3. параллельная стенка *f* (*колодцевой кладки*) 4. щит *m*, клапан *m* (*гидротехнического затвора*) 5. створка *f*, перо *n*, лепесток *m* (*воздушного клапана*)

L137 *e* **leaf dam**
d Klappenwehr *n*
f barrage *m* à clapet
nl klepstuw *m*
r плотина *f* с клапанным затвором

L138 *e* **leafing aluminium**
d Aluminiumfolie *f*

f feuille *f* (mince) en aluminium
nl aluminiumfolie *f* (m)
r алюминиевая фольга *f*

L139 **leaf-type damper** *see* **iris damper**

L140 *e* **leakage**
d Leckage *f*, Sickerverlust *m*; Leckstelle *f*
f fuite *f*
nl lekkage *f*
r утечка *f*; течь *f*

L141 *e* **leakage coefficient**
d Leckverlustfaktor *m*, Leckrate *f*
f coefficient *m* de fuite
nl coëfficiënt *m* van lekverliezen
r коэффициент *m* утечки

L142 *e* **leakage loss**
d Leck(wasser)verlust *m*, Sickerverlust *m*, Filtrationsverlust *m*
f perte *f* par infiltration
nl lekverlies *n*
r фильтрационные потери *f pl* воды

L143 *e* **leakage test**
d 1. Dichtigkeitsprüfung *f* 2. Bestimmung *f* der Grundwasserinfiltration
f 1. essai *m* [épreuve *f*] d'étanchéité 2. mesure *f* d'infiltration des eaux souterraines
nl 1. dichtheidsproef *f* (m) 2. bepaling *f* van de grondwaterinfiltratie
r 1. испытание *n* на герметичность 2. измерение *n* инфильтрации грунтовых вод

L144 **leakage water** *see* **percolating water**

L145 *e* **leak detector**
d Lecksuchgerät *n*, Lecksucher *m*
f détecteur *m* de fuites
nl lekdetector *m*
r течеискатель *m*, детектор *m* течей

L146 *e* **lean concrete**
d Magerbeton *m*, Sparbeton *m*
f béton *m* maigre
nl mager [schraal] beton *n*
r тощий бетон *m*

L147 *e* **lean-to**
d Anbau *m* mit Pultdach
f appentis *m*
nl afdak *n*
r пристройка *f* с односкатной крышей (*небольшая*)

L148 *e* **lean-to-roof**
d freitragendes Pultdach *n*
f comble *m* [toit *m*] en appentis
nl (vrijdragend) afdak *n*
r наслонная односкатная крыша *f* (*пристройки*)

L149 *e* **leaping weir**
d Regenüberfall *m*
f déversoir *m* intercepteur
nl overlaat *m*
r ливнеспуск *m*

L150 *e* **least moment of inertia**
d Kleinstträgheitsmoment *n*

f moment *m* d'inertie [quadratique]
minimum
nl kleinste traagheidsmoment *n*
r наименьший момент *m* инерции
(сечения)

L151 *e* **leave-in-place form**
d verlorene Schalung *f*
f coffrage *m* perdu
nl verloren bekisting *f*
r несъёмная опалубка *f*

L152 *e* **leaving air**
d Abluft *f*, Fortluft *f*
f air *m* évacué
nl afgevoerde lucht *f* (*m*)
r уходящий [вытяжной] воздух *m*

L153 *e* **ledge**
d 1. Vorsprung *m*, Ansatz *m* 2. Gesims
n, Sims *m* 3. Querleiste *f*, Gurtholz *n*
4. Berme *f* 5. Querholz *n*
f 1. saillie *f* 2. bandeau *f* 3. moise *f*
4. berme *f* 5. traverse *f* médiane
d'encadrement de porte
nl 1. uitstek *n* 2. kroonlijst *f* (*m*)
3. klamp *m*, *f* 4. berm *m* 5. tussendorpel *m*
r 1. выступ *m* 2. поясок *m* 3. схватка *f*
4. берма *f* 5. средний горизонтальный
брус *m* обвязки дверного полотна

L154 *e* **ledged-and-braced door**
d Brettertür *f* mit Bug
f porte *f* en planche à écharpes
nl opgeklampte deur *f* (*m*) met
schoorklampen
r дощатая дверь *f* на планках
с диагональными схватками

L155 *e* **ledged door**
d Lattentür *f*, aufgedoppelte
Rahmentür *f*
f porte *f* en planches, porte-pleine *f*
nl opgeklampte deur *f* (*m*)
r дощатая дверь *f* на планках

L156 *e* **ledger**
d 1. Gurtholz *n*, Zangenholz *n*
(*Tafelschalung*) 2. Riegel *m*,
Rahmenholz *n*; Rödelbalken *m*
3. Längsstange *f*, Streichstange *f*
(*Holzgerüst*) 4. Hirnholzlatte *f*
f 1. lisse *f* des panneaux de joue (*d'un
coffrage*) 2. traverse *f* d'une charpente
en bois 3. longeron *m*, moise *f*
(*élément d'échafaudage*) 4. tasseau *m*,
lambourde *f*
nl 1. koppelhout *n* (*bekisting*)
2. dwarsbalk *m* 3. korteling *f*, bulsing
f (*steiger*) 4. klamp *m*, *f*, stempel *m*
r 1. горизонтальное ребро *n*, схватка *f*
(*щитовой опалубки*) 2. ригель *m*,
обвязка *f* (*деревянного каркаса*);
пажилина *f* 3. ригель *m* лесов
4. черепной брусок *m*

L157 *e* **leadger stringer**
d Gurtholz *n* des Ständergerüstes
f longeron *m* d'échafaudage
nl scheerhout *n*, aanbinder *m*

r горизонтальная схватка *f* стоечных
лесов

L158 **ledger strip** *see* **ledger** 4.

L159 *e* **left-hand lay**
d Linksschlag *m* (*Seil*)
f câblage *m* à gauche
nl links geslagen (*touw*)
r левая свивка *f* (*каната*)

L160 **left-in-place formwork** *see* **leave-in-
-place form**

L161 *e* **leg of tower**
d Tragfuß *m* des Turmes
f pied *m* d'une tour
nl voet *m* van een toren
r опорная нога *f* башни; нижняя
опорная секция *f* ствола башни

L162 *e* **lengthening**
d Verlängerung *f*
f allongement *m*, rallongement *m*
nl verlenging *f*
r удлинение *n*; наращивание *n*

L163 *e* **lengthening joint**
d Längsverband *m*
f joint *m* de rallongement
nl verbinding *f* in lengterichting
r соединение *n* [шов *m*] наращивания

L164 *e* **lengthening structural timber**
d Verlängerung *f* des Schnittholzes
f rallongement *m* des éléments de
charpente en bois
nl verlenging *f* van bezaagd hout
r наращивание *n* элементов деревянных
конструкций

L165 *e* **length of cantilever**
d Kragarmlänge *f*
f portée *f* de console [de porte-à-faut]
nl lengte *f* van de uitkraging
r вылет *m* консоли

L166 *e* **length of haul**
d Förderweite *f*, Transportweite *f*
f longueur *f* de parcours [de transport]
nl transportafstand *m*
r дальность *f* перевозки
[транспортировки]

L167 *e* **length of hydraulic jump**
d Sprungweite *f*
f longueur *f* de ressaut hydraulique
nl hoogte *f* van de watersprong
r длина *f* отгона гидравлического
прыжка

L168 *e* **length of lay**
d Schlaglänge *f* (*Seil*)
f longueur *f* de câblage
nl slaglengte *f* (*touw*)
r длина *f* свивки (*каната*)

L169 *e* **length of restraint**
d Einspannlänge *f*
f longueur *f* d'encastrement
nl inklemmingslengte *f*
r глубина *f* заделки, длина *f*
защемления

L170 *e* **let-in brace**
d eingeschnittenes Diagonalzangenholz *n*

f entretoise *f* entaillée
nl ingelaten schoor *m*
r врезная диагональная схватка *f*

L171 **letting of a contract** *see* **award of a contract**

L172 *e* **levee**
d Leitdeich *m*, Leitdamm *m*;
Hochwasserdeich *m*, Uferdamm *m*;
Eindeichung *f*; Strandwall *m*,
Schluffbank *f*
f digue *f*, levée *f*; cordon *m* littoral
[de plage]; endiguement *m*
nl stranddam *m*
r направляющая дамба *f*;
ограждающая [берегозащитная]
дамба *f*; обвалование *n*;
естественный береговой вал *m*

L173 *e* **levee breach**
d Deicheinbruch *m*, Dammeinbruch *m*
f rupture *f* d'une digue
nl dijkbreuk *f* (*m*)
r прорыв *m* дамбы

L174 *e* **leveed area**
d eingedeichtes Gelände *n*, Deichland *n*
f terrain *m* endigué
nl bedijkt gebied *n*
r обвалованная территория *f*

L175 *e* **level**
d 1. Höhenkote *f*, Höhe *f*, Niveau *n*
2. Nivellierinstrument *n*;
Wasserwaage *f*
f 1. cote *f* de niveau 2. niveau *m*
(*instrument*)
nl 1. peil *n*, niveau *n*
2. waterpas(toestel) *n*
r 1. отметка *f*, уровень *m* 2. нивелир *m*;
уровень *m* (*инструмент*)

L176 *e* **level controller**
d Niveau-Regler *m*
f régulateur *m* de niveau
nl niveau-regelaar *m*
r регулятор *m* уровня

L177 *e* **leveler** *US*
d Flachbagger *m*, Planiergerät *n*
f niveleuse *f* de route
nl grondschaaf *f* (*m*), grader *m*
r землеройно-планировочная машина *f*

L178 *e* **level gauge, level indicator**
d Füllstandanzeiger *m*,
Niveauanzeiger *m*
f indicateur *m* du niveau
nl peilglas *n*
r указатель *m* уровня, уровнемер *m*

L179 *e* **leveling**
d 1. Ausgleichen *n*, Abgleichen *n*,
Planierung *f* 2. Höhenmessung *f*
f 1. nivellement *m*, aplanissement *m*,
régalage *m*, régalement *m*, arasement
m 2. nivellement *m*
nl 1. egaliseren *n* 2. waterpassen *n*
r 1. выравнивание *n*, планировка *f*
2. нивелирование *n*

L180 **leveling beam** *see* **screed board**

L181 *e* **leveling board**
d Abziehlatte *f*, Kardätsche *f*
f planche *f* lisseuse [de finition]
nl rij *f* (*m*)
r правило *n*, разравнивающая рейка *f*

L182 *e* **leveling concrete**
d Ausgleichbeton *m*
f béton *m* d'egalisation
nl uitvlakbeton *n*
r выравнивающий слой *m* бетона

L183 *e* **leveling course**
d Ausgleichlage *f*, Ausgleichschicht *f*
f couche *f* d'arasement [d'aplanissage]
nl uitvlaklaag *f* (*m*)
r выравнивающий слой *m*

L184 *e* **leveling operations**
d Planieren *n*, Planier(ungs)arbeiten *f pl*
f travaux *m pl* de nivellement
nl egalisatiewerk *n*
r планировочные *или*
нивелировочные работы *f pl*

L185 *e* **leveling rod, leveling staff**
d Nivellierlatte *f*
f mire *f*
nl nivelleerlat *f* (*m*)
r нивелирная рейка *f*

L187 **leveling work** *see* **leveling operations**

L188 **levelling...** *see* **leveling...**

L189 *e* **level-luffing crane**
d Wippdrehkran *m*, Wippschwenkkran *m*
f grue *f* tournante [orientable,
pivotante] à flèche relevable
nl wipkraan *m*
r кран *m* с горизонтальным
перемещением груза при изменении
вылета стрелы

L190 *e* **level mark**
d Niveauhöhe *f*, Spiegelhöhe *f*
f cote *f* de niveau
nl hoogtemerk *n*
r отметка *f* уровня

L191 *e* **level of illumination**
d Lichtniveau *n*
f éclairement *m*, niveau *m* d'éclairement
nl verlichtingssterkte *f*
r уровень *m* освещённости

L192 *e* **level of noise**
d Lärmpegel *m*
f niveau *m* de bruits
nl geluidspeil *n*
r уровень *m* шума

L193 *e* **level recorder**
d Schreibpegel *m*, Limnigraph *n*
f limnigraphe *m*
nl zelfregistrerende peilschaal *f* (*m*)
r самописец *m* уровня воды,
лимниграф *m*

L194 *e* **lever arm**
d Hebelarm *m*; Arm *m* des inneren
Kräftepaares
f bras *m* (*de levier, du couple de forces*)
nl arm *m* van een koppel
r плечо *m* (*рычага, внутренней пары*)

L195 *e* **lever jack**
   d Hebelwinde *f*, Lastwinde *f*
   f cric *m* [verin *m*] à levier
  nl krik *m*
   r рычажный домкрат *m*

L196   **lever shears** *see* **alligator shears**

L197 *e* **Lévy facing**
   d Schutzverkleidung *f* nach Lévy
   f masque *m* Lévy
  nl beschermende bekleding *f* volgens
     Lévy
   r противофильтрационная маска *f*
     Леви

L198 *e* **lewis**
   d Steinwolf *m*, Keilklaue *f*
   f louve *f*
  nl wolf *m*, wigklauw *m*, *f*
   r захват *m* «волчья лапа» (*для*
     подъёма каменных блоков)

L199 *e* **lewis bolt**
   d Ankerschraube *f*, Steinschraube *f*
   f boulon *m* de fondation à bout barbelé
     [d'ancrage barbelé]
  nl funderingsbout *m*, ankerbout *m*
   r анкерный [фундаментный] болт *m*
     (с утолщённым коническим
     зазубренным стержнем)

L200 *e* **L-head**
   d L-förmiger Stutzenaufsatz *m*
   f tête *f* de poteau [de montant] en L
  nl L-vormige stutkop *m*
   r Г-образный оголовок *m* стойки

L201 *e* **L-headed jetty**
   d L-förmige Kaizunge *f*
   f jetée *f* à tête en L
  nl L-vormige aanlegkade *f* (*m*)
   r Г-образный причал *m*

L202   **lid** *see* **blind flange**

L203 *e* **life**
   d Liegezeit *f*, Lebensdauer *f* (*Geräte,*
     Werkzeuge); Standzeit *f* (*Werkzeuge,*
     Ventile); Betriebszeit *f*
   f durée *f* de vie [de service]; longévité *f*
  nl levensduur *m*; standtijd *m*
   r срок *m* службы [эксплуатации]
     (*оборудования*); стойкость *f*
     (*инструмента*); долговечность *f*,
     технический ресурс *m*

L204 *e* **lifeline**
   d Sicherheitsseil *n*
   f corde *f* [câble *m*] d'assurance
  nl veiligheidstouw *n*
   r строп *m* предохранительного пояса

L205 *e* **life of reservoir**
   d rechnungsmäßige Lebensdauer *f* des
     Wasserspeichers
   f durée *f* [période *f*] d'exploitation
     calculée d'une retenue
  nl berekende levensduur *m* van een
     stuwbekken
   r расчётный срок *m* эксплуатации
     водохранилища

L206 *e* **life test**
   d Lebensdauerprüfung *f*
   f essai *m* de durée (de vie)
  nl duurproef *f* (*m*)
   r испытание *n* на долговечность

L207   **lifetime** *see* **life**

L208 *e* **lift**
   d 1. Heben *n* (*einer Last*) 2. Aufzug *m*,
     Fahrstuhl *m*, Lift *m*
     3. Schiffshebewerk *n*, Trog *m* 4.
     Einbauschicht *f* (*Beton*) 5. Hubhöhe *f*
     (*Wasserhebung*) 6. Hub *m*
     (*Schleusungshöhe*) 7. Hublast *f*
     8. Hebung *f*, Auftreibung *f*
     (*Bodenmechanik*)
   f 1. élévation *f* (d'un poids), levage *m*
     2. ascenseur *m*, monte-charge *m*
     3. ascenseur *m* à bateaux 4. hauteur *f*
     d'une coulée de béton, levée *f* de
     béton 5. hauteur *f* d'élévation d'eau
     6. chute *f* à l'écluse 7. charge *f* de
     levage (à lever) 8. soulèvement *m*
     de sol
  nl 1. (op)heffen *n* 2. lift *m* 3.
     scheepslift *m* 4. stortlaag *f* (*m*)
     5. hefhoogte *f* 6. schuthoogte *f* heflast
     f (*m*) 8. bodemverheffing *f*
   r 1. подъём *m* (*груза*) 2. лифт *m*,
     подъёмник *m* 3. судоподъёмник *m*
     4. высота *f* слоя бетона, уложенного
     за смену 5. высота *f* подъёма воды
     (*из реки, канала*) 6. высота *f*
     шлюзования 7. монтажный груз *m*
     8. выпучивание *n* [выпор *m*] грунта

L209 *e* **lift and force pump**
   d Saug- und Druckpumpe *f*
   f pompe *f* aspirante et foulante
  nl zuigperspomp *f* (*m*)
   r насос *m*, работающий на всасывание
     и нагнетание одновременно

L210 *e* **lift bridge**
   d Hubbrücke *f*
   f pont-levis *m*
  nl hefbrug *f* (*m*)
   r вертикально-подъёмный мост *m*

L212 *e* **lift-car door** *UK*
   d Aufzugtür *f*
   f porte *f* de la cabine d'ascenseur
  nl deur *f* (*m*) van een liftcabine
   r дверь *f* кабины лифта

L213 *e* **lift gate**
   d 1. Hubtor *n* 2. Hubschütz(e) *n* (*f*)
   f 1. porte *f* levante 2. vanne *f* levante
     plane [plate]
  nl 1. hefdeur *f* (*m*) (*sluis*) 2. schuif *f* (*m*)
   r 1. подъёмные шлюзные ворота *pl*
     2. подъёмный щитовой затвор *m*

L214 *e* **lifting appliance**
   d Hebevorrichtung *f*
   f appareil *m* de levage
  nl hefinrichting *f*
   r грузоподъёмное устройство *n*

L215 *e* **lifting beam**
   d Hubtraverse *f*

f palonnier *m*, palonneau *m*, entretoise *f*
nl hijstraverse *f*
r подъёмная траверса *f*

L216 *e* **lifting block**
    d Flaschenzug *m*
    f poulie *f* (de levage), moufle *f*, cayorne *f*
    nl schijvenblok *n*, blok *n*
    r грузоподъёмный блок *m*, грузовая блочная обойма *f*

L217   **lifting capacity** *see* **load capacity**

L218 *e* **lifting eye**
    d Aufnahmeöse *f*, Huböse *f*
    f œil *m* [œilleton *m*] de levage, manille *f*
    nl hijsoog *n*
    r подъёмная серьга *f*, проушина *f*

L219   **lifting gate** *see* **lift gate 2.**

L220 *e* **lifting gear**
    d 1. Hebezeug *n*, Anschlagmittel *n pl* 2. Hubwerk *n*
    f 1. accessoires *m pl* de levage 2. appareil *m* de levage
    nl 1. hijsmiddelen *n pl* 2. hefinrichting *f*
    r 1. подъёмные такелажные устройства *n pl* 2. подъёмный механизм *m*

L221 *e* **lifting hook**
    d Lasthaken *m*, Hubhaken *m*, Transporthaken *m*
    f crochet *m* de levage
    nl hijshaak *m*
    r грузовой [грузоподъёмный] крюк *m*

L222 *e* **lifting inserts**
    d Montagestahleinlagen *f pl*
    f pièces *f pl* [garnitures *f pl*] d'ancrage noyées dans le béton
    nl ingestorte hijsogen *f (m) pl*
    r закладные монтажные детали *f pl*

L223   **lifting lug** *see* **lifting eye**

L224 *e* **lifting magnet**
    d Last(hebe)magnet *m*, Hubmagnet *m*
    f aimant *m* [électro-aimant *m*] de levage
    nl hijsmagneet *m*
    r магнитный грузоподъёмный захват *m*

L225 *e* **lifting platform**
    d Hebebühne *f*, Hubplattform *f*
    f plate-forme *f* élévatrice [de levage], pont *m* élévateur
    nl hefplatform *n*
    r подъёмная платформа *f*

L226 *e* **lifting tackles**
    d 1. Hebezug *m*, Flaschenzug *m* 2. Anschlagmittel *n pl*, Gehänge *n*, Hebezeug *n*
    f 1. palan *m*, moufle *f* 2. accessoirs⁻ *m pl* de levage
    nl 1. takel *m*, *n* 2. hijsmiddelen *n pl*
    r 1. полиспаст *m*; таль *f* 2. такелажные устройства *n pl*

L227 *e* **lift joint**
    d Betonierfuge *f*, Arbeitsfuge *f*

f joint *m* de reprise [de construction]
nl stortnaad *m* (beton)
r рабочий шов *m* между слоями бетона (*укладываемыми за смену*)

L228 *e* **lift latch**
    d Klinke *f*
    f loquet *m*
    nl klink *f*
    r щеколда *f*

L229 *e* **lift line**
    d 1. sichtbare Betonschichtabgrenzungslinie *f* (*Betonfehler*) 2. Aufzugseil *n*
    f 1. ligne *f* visible de partage des levées de béton consécutives (*défaut*) 2. câble *m* de levage d'ascenseur
    nl 1. zichtbare stortnaad (*betonfout*) 2. hijskabel *m*
    r 1. видимая линия *f* раздела укладываемых слоёв (*дефект бетона*) 2. подъёмный канат *m* лифта

L230 *e* **lift machine room**
    d Aufzugmotorraum *m*
    f salle *f* de moteurs [de machines] d'un ascenseur
    nl liftmachinekamer *f (m)*
    r машинное отделение *n* лифта

L231 *e* **lift of a lock**
    d Schleusenfallhöhe *f*, Schleusengefälle *n*
    f chute *f* à l'écluse
    nl verval *n* van een sluis
    r напор *m* шлюза

L232 *e* **lift platform**
    d Aufzugsplattform *f*
    f plate-forme *f* [plateau] de la cabine d'ascenseur; plate-forme *f* [plateau] de monte-charge
    nl liftplatform *n*
    r платформа *f* кабины лифта *или* подъёмника

L233   **lift pump** *see* **suction pump**

L234 *e* **lifts**
    d Gerüstetagen *f pl*
    f plates-formes *f pl* de travail des échafaudages à plusieurs étages
    nl steigervloeren *f (m) pl*
    r ярусы *m pl* строительных лесов

L235   **lift shaft** *UK see* **elevator shaft**

L236 *e* **lift slab**
    d 1. Hubplattenbauweise *f*, Hubplattenverfahren *n*, Geschoßhebeverfahren *n* 2. Hubplatte *f*, Hub-Betondecke *f*
    f 1. méthode *f* de levage de planchers (*en béton armé*) coulés au sol 2. plancher *m* en béton armé coulé au sol et levé en position finale
    nl 1. vloerenhefsysteem *n*, lift-slabsysteem *n* 2. op begane grond gestorte verdiepingsvloer *f (m)*
    r 1. метод *m* подъёма перекрытий 2. поднимаемое железобетонное перекрытие *n*

L237 lift-slab concrete floor *see* lift slab 2.

L238 lift-slab method *see* lift slab 1.

L239 *e* lift span
  d Hubfeld *n*, Huböffnung *f*
  f travée *f* levante (*d'un pont*)
  nl overspanning *f* van een hefbrug
  r подъёмное пролётное строение *n*
    (*моста*)

L240 *e* lift tower
  d 1. Aufzugsturm *m*, Förderturm *m*
    2. Aufzugsschacht *m*,
    Fahrstuhlschacht *m*
  f 1. tour *f* élévatrice 2. gaine *f*
    d'ascenseur
  nl 1. heftoren *m* 2. liftschacht *f* (*m*)
  r 1. башенный подъёмник *m* 2. шахта *f*
    лифта

L241 *e* lift truck
  d Hubstapler *m*,
    Hebetransportstapler *m*,
    Hubkarren *m*
  f chariot *m* élévateur
  nl heftruck *m*
  r автопогрузчик *m* с вертикальной
    рамой

L242 lift well *UK see* elevator shaft

L243 *e* light aggregate concrete
  d Leichtbeton *m*
  f béton *m* de granulats [d'agrégats]
    légers
  nl lichtbeton *n*
  r лёгкий бетон *m*, бетон *m* на лёгких
    заполнителях

L244 light aperture *see* light opening

L245 *e* light-diffusing grate
  d Blendschutzgitter *n*
  f grille *f* de diffusion des rayons
    lumineux
  nl lichtverstrooiiend rooster *n*, *m*
  r светорассеивающая решётка *f*

L246 *e* light-duty scaffold
  d Leichtgerüst *n*
  f échafaudage *m* légers
  nl lichte steiger *m*
  r лёгкие леса *pl или* подмости *pl*

L247 *e* light-gauge cold-formed steel
    structural member
  d dünnwandiges kaltverformtes
    Stahlbauelement *n*
  f élément *m* de construction en profilé
    d'acier plié à froid
  nl koudgevormd plaatstalen
    bouwelement *n*
  r тонкостенный холодногнутый
    стальной конструктивный элемент *m*

L248 *e* light-gauge steel construction
  d dünnwandige Stahlkonstruktion *f* aus
    kaltverformten Profilelementen
  f construction *f* métallique en profilés
    pliés à froid
  nl constructie *f* in plaatstalen profielen
  r стальная конструкция *f* из
    холодногнутых профилей

L249 *e* light-gauge steel framing
  d Stahlskelettbau *m* aus dünnwandigen
    kaltverformten Elementen
  f charpente *f* en profilés d'acier légers
    pliés à froid
  nl staalskelet *n* van plaatstalen
    elementen
  r стальной каркас *m* из тонкостенных
    холодногнутых профилей

L250 *e* lighting fitting *UK*, lighting fixture
  d 1. Beleuchtungskörper-Armatur *f*
    2. Leuchte *f*; Lampenfassung *f*
  f 1. appareil *m* d'éclairage
    2. luminaire *m*; douille *f* de lampe
  nl 1. verlichtingsarmatuur *n*, *f*
    2. lichtpunt *n*, fitting *m*
  r 1. осветительная арматура *f*
    2. светильник *m*; патрон *m*
    электролампы

L251 *e* lighting mains
  d Lichtleitung *f*
  f canalisation *f* d'éclairage
  nl lichtleiding *f*
  r осветительная проводка *f*

L252 lighting mast *see* lamp post

L253 lighting standard *UK see* lamp post

L254 *e* lightning arrestor
  d Blitzableiter *m*
  f parafoudre *m*, paratonnerre *m*
  nl bliksemafleider *m*
  r молниеотвод *m*

L255 *e* lightning protection
  d Blitzschutz *m*
  f protection *f* contre la foudre
  nl bliksembeveiliging *f*
  r молниезащита *f*

L256 *e* lightning rod
  d Blitzauffang *m*, Blitzableiter *m*
  f tige *f* de paratonnerre
  nl bliksemopvangstang *f* (*m*)
  r молниеприёмник *m*

L257 *e* lightning switch
  d Blitzschutzschalter *m*
  f commutateur *m* de mise à la terre
  nl beveiligingsschakelaar *m* voor
    blikseminslag
  r грозовой переключатель *m*

L258 *e* light noncombustible constructions
  d unbrennbare Leichtkonstruktionen *f pl*
  f constructions *f pl* légères non
    inflammables
  nl lichte en onbrandbare constructies *f pl*
  r лёгкие невозгораемые конструкции *f*
    pl

L259 *e* light opening
  d Lichtöffnung *f*
  f baie *f* d'admission de lumière
  nl lichtopening *f*
  r световой проём *m*

L260 light railway *UK see* narrow-gauge
    railway

L261 *e* light ventilation area
  d Lichtschacht *m*

f cour *f* anglaise
nl lichtkoker *m*
r приямок *m* у окна подвального
помещения

L262 *e* **lightweight aggregate**
d Leichtbetonzuschlag(stoff) *m*
f agrégat *m* [granulat *m*] léger
nl lichte toeslag *m* (*beton*)
r лёгкий заполнитель *m*

L263 *e* **lightweight concrete**
d Leichtbeton *m*
f béton *m* léger
nl licht beton *n*
r лёгкий бетон *m*

L264 *e* **lightweight insulating concrete**
d Leichtdämmbeton *m*
f béton *m* isoléger [isolant léger, léger
isolant]
nl warmteïsolerend licht beton *n*
r лёгкий теплоизоляционный бетон
m

L265 **lightweight metal section** *see*
lightweight section

L266 *e* **lightweight plaster**
d Leichtputz *m*
f enduit *m* léger
nl licht pleister *n*
r штукатурка *f* с лёгкими
наполнителями

L267 *e* **lightweight section**
d Leichtprofil *n*
f profilé *m* léger
nl licht profiel *n*
r тонкостенный профиль *m*

L268 *e* **lightweight structural concrete**
d Konstruktionsleichtbeton *m*
f béton *m* léger porteur [structurel]
nl licht beton *n* voor constructies
r лёгкий бетон *m* для несущих
конструкций, лёгкий
конструктивный бетон *m*

L269 *e* **lightweight structures**
d Leichtkonstruktionen *f pl*,
Leichtbauten *m pl*
f structures *f pl* [constructions *f pl*]
légères, ouvrages *m pl* légers
nl lichte constructies *f pl*
r лёгкие конструкции *f pl*
[сооружения *n pl*]

L270 *e* **like-grained concrete**
d Einkornbeton *m*
f béton *m* de granulat monogranulaire
nl monogranulair beton *n*
r бетон *m* с однофракционным
заполнителем

L271 *e* **lime**
d Kalk *m*
f chaux *f*
nl kalk *m*
r известь *f*

L272 *e* **lime agitator**
d Kalkrührwerk *n*, Kalkmischer *m*
f agitateur *m* pour chaux

nl kalkroerwerk *n*
r известемешалка *f*

L273 *e* **lime cement**
d Kalkkitt *m*; Kalkzement *m*
f liant *m* à chaux; mortier *m* de chaux
nl kalkbindmiddel *n*
r известковое вяжущее *n*;
известковый раствор *m*

L274 *e* **lime cement mortar**
d Kalkzementmörtel *m*
f mortier *m* de ciment et de chaux
nl kalkcementmortel *m*
r известково-цементный раствор *m*

L275 **lime clay** *see* **calcareous clay**

L276 *e* **lime concrete**
d Kalkbeton *m*
f béton *m* à base de chaux vive [incuite]
moulue
nl kalkbeton *n*
r бетон *m* на молотой извести-кипелке

L277 *e* **lime efflorescence**
d Kalkausschlag *m*, Kalkausblühung *f*
f efflorescence *f* de chaux
nl kalkuitslag *m*
r известковые выцветы *m pl*

L278 *e* **lime mortar**
d Kalkmörtel *m*
f mortier *m* de chaux
nl kalkmortel *m*
r известковый раствор *m*

L279 **lime paste** *see* **lime putty**

L280 *e* **lime pit**
d Kalk(lösch)grube *f*, Löschgrube *f*
f fosse *f* à chaux
nl kalkput *m*
r известковая [творильная] яма *f*,
яма *f* для гашения извести

L281 *e* **lime plaster**
d Kalkputz *m*
f enduit *m* de chaux
nl kalkpleister *n*
r известковая штукатурка *f*

L282 *e* **lime process**
d Kalkverfahren *n*, Kalkklärung *f*
f traitement *m* à la chaux
nl behandeling *f* met kalk
r известкование *n*

L283 *e* **lime putty**
d Kalkbrei *m*, Kalkteig *m*
f pâte *f* de chaux
nl kalkbrij *m*
r известковое тесто *n*

L284 *e* **lime-sandstone**
d Kalksandstein *m*
f grès *m* calcaire
nl kalkzandsteen *n*, *m*
r известковый песчаник *m*

L285 *e* **lime slaker**
d Kalklöschtrommel *f*,
Trommellöscher *m*
f tambour *m* extincteur [extincteur *m*]
de chaux

nl kalkblustrommel *f* (*m*)
r известегасильный аппарат *m*,
известегасилка *f*

L286 *e* **lime-soda process**
d Kalk-Soda-Verfahren *n*
f adoucissement *m* d'eau calcosodique
nl kalk-sodamethode *f* voor ontharding
van water
r известково-содовое умягчение *n*,
обработка *f* воды известью
с кальцинированной содой

L287 *e* **lime stabilization**
d Kalkstabilisierung *f*
f stabilisation *f* des sols par chaux,
traitement *m* des sols à la chaux
nl grondstabilisatie *f* met kalk
r укрепление *n* [стабилизация *f*]
известью, известкование *f* (*грунта*)

L288 *e* **limestone**
d Kalkstein *m*
f calcaire *m*, pierre *f* calcaire [à chaux]
nl kalksteen *n*, *m*
r известняк *m*

L289 *e* **lime-water**
d Kalkwasser *n*
f eau *f* calcique [de chaux]
nl kalkwater *n*
r известковая вода *f*

L290 **liming** *see* **lime process**

L291 **limit crack growth** *see* **limit state of**
cracking

L292 **limit design** *see* **limit state design**

L293 **limiting discharge** *see* **critical discharge**

L294 *e* **limiting gradient**
d höchstzulässige Gradiente *f*
f rampe *f* [pente *f*, inclinaison *f*] limite
nl maximaal toelaatbare helling *f*
r предельный [руководящий] уклон *m*

L295 *e* **limiting velocity**
d kritische Geschwindigkeit *f*,
Grenzgeschwindigkeit *f*
f vitesse *f* limite
nl kritische snelheid *f*
r критическая скорость *f*

L296 **limit load** *see* **load limit**

L297 **limit-load design** *see* **ultimate load**
analysis

L298 *e* **limit of backwater**
d Staugrenze *f*
f limite *f* du remous
nl stuwgrens *f* (*m*)
r граница *f* подпора

L299 *e* **limit of proportionality**
d Proportionalitätsgrenze *f*, P-Grenze *f*
f limite *f* de proportionnalité
nl evenredigheidsgrens *f* (*m*)
r предел *m* пропорциональности

L300 *e* **limit state**
d Grenzzustand *m*
f état *m* limite
nl grenstoestand *m*
r предельное состояние *n*

L301 *e* **limit state design**
d Berechnung *f* nach Grenzzuständen
f calcul *m* à l'état limite
nl berekening *f* naar grenstoestand
r расчёт *m* по предельному состоянию

L302 *e* **limit state of cracking**
d Grenzzustand *m* der Rißbildung
f état *m* limite de fissuration
nl grenstoestand *m* volgensscheurvorming
r предельное состояние *n* по
трещинообразованию

L303 *e* **limit state of crack width**
d Grenzzustand *m* der Rißbreite
f état *m* limite de l'ouverture des
fissures
nl grenstoestand *m* volgens scheurbreedte
r предельное состояние *n* по
раскрытию трещин

L304 *e* **limit-stop**
d Endanschlag *m*
f arrêt *m* de fin de course
nl eindaanslag *m*
r останов-ограничитель *m*

L305 *e* **limit switch**
d End(aus)schalter *m*,
Begrenzungsschalter *m*,
Grenzschalter *m*
f interrupteur *m* de fin de course
nl eindschakelaar *m*
r предельный выключатель *m*

L306 **limnigraph** *see* **level recorder**

L308 *e* **Limpet asbestos**
d Spritzasbest *m*
f asbeste *m* projeté
nl spuitasbest *n*
r асбестовый огнестойкий раствор *m*,
наносимый распылением
[набрызгом]

L309 *e* **linear air terminal device**
d Gitterband *n*, Schlitzdiffusor *m*,
Schlitzverteiler *m*, Linearauslaß *m*
f diffuseur *m* à fentes
nl spleetvormige luchtverdeler *m*
r щелевой воздухораспределитель *m*

L310 *e* **linear elastic analysis**
d Berechnung *f* nach dem Hookschen
Gesetz, Bemessung *f* nach der
Elastizitätstheorie
f calcul *m* élastique [selon la loi de
Hooke]
nl berekening *f* volgens de
elasticiteitstheorie
r расчёт *m* по закону Гука
(*основанный на линейно-упругой
зависимости между напряжениями и
деформациями*)

L312 *e* **linear elasticity**
d lineare Elastizität *f*
f élasticité *f* linéaire
nl lineaire elasticiteit *f*
r линейная упругость *f*

L313 *e* **linear load**
d Streckenlast *f*, Linienlast *f*

f charge *f* linéaire
nl lineaire belasting *f*
r линейная [линейно-распределённая] нагрузка *f*

L314 *e* **linear prestressing**
d lineare Vorspannung *f*
f précontrainte *f* linéaire [axiale]
nl lineaire voorspanning *f*
r линейное преднапряжение *n*, преднапряжение *n* линейных конструктивных элементов

L315 *e* **linear strain**
d lineare Verformung *f* [Deformation *f*], Längenänderung *f*
f déformation *f* linéaire
nl lineaire vervorming *f*
r линейная деформация *f*

L316 *e* **linear town**
d Bandstadt *f*, lineare Stadt *f*
f ville *f* linéaire
nl lintbebouwing *f*
r линейный город *m*, город *m* линейной планировки

L317 *e* **lined ditch**
d verkleideter Bewässerungsgraben *m* [Bewässerungskanal *m*]
f canal *m* d'irrigation aux talus revêtus
nl bekleed bevloeiingskanaal *n*
r оросительный канал *m* с противофильтрационной облицовкой откосов

L318 *e* **line diffuser**
d Schlitzgitter *n*
f diffuseur *m* à fentes
nl spleetrooster *m*, *n*
r щелевой диффузор *m*

L319 *e* **line network**
d Leitungsnetz *n*
f réseau *m* de lignes [des canalisations]
nl leidingnet *n*
r сеть *f* коммуникаций

L320 *e* **line of creep**
d Sickerweg *m*, Kriechweg *m*, Kriechlinie *f*
f ligne *f* de fuite [de cheminement]
nl insijpelingskromme *f* (*m*)
r путь *m* фильтрации (*по подземному контуру сооружения*)

L321 *e* **line of seepage**
d Sickerlinie *f*
f ligne *f* de saturation
nl kromme *f* van het phreatisch oppervlak
r линия *f* депрессии

L322 *e* **line of thrust**
d Stützlinie *f*, Druckkurve *f*
f axe *m* de centres de poussée
nl druklijn *f* (*m*)
r кривая *f* давления (*в арке*)

L323 **line of total head** *see* **energy line**

L324 *e* **line of traffic**
d Kolonne *f* von Verkehrsmitteln

f colonne *f* [file *f*] de véhicules
nl convooi *n* vrachtwagens
r колонна *f* транспортных средств, транспортная колонна *f*

L325 *e* **lining**
d Auskleidung *f*, Bekleidung *f*, Verkleidung *f*; Mauerung *f* (*Tunnel*); Kanalauskleidung *f*; Einschaltung *f*, Austäfelung *f*; Ausfutterung *f*; Verbau *m*, Verschalung *f* (*Graben, Baugrube*)
f revêtement *m* (*d'un tunnel, d'un canal*); cuvelage *m* (*d'un puits*); bardage *m*, recouvrement *m*; garniture *f* (réfractaire); blindage *m*
nl bekleding *f*; mantel *m*; paneelwerk *n*; beschot *m*
r облицовка *f*; обделка *f* (*туннеля*); одежда *f* (*канала*); обшивка *f*; футеровка *f*; крепление *n* (*траншеи, котлована*)

L326 *e* **lining brick**
d Verkleidungsziegel *m*; Futterziegel *m*, Futterstein *m*; Hintermauerungsziegel *m*
f brique *f* de revêtement; brique *f* de remplissage
nl bekledingssteen *m*; achterwerker *m*
r облицовочный [футеровочный] кирпич *m* [блок *m*]; кирпич *m* для забутки

L327 *e* **link**
d Kettenglied *n*; Verbindungsstange *f*, Kuppelstange *f*, Bindung *f*
f maillon *m*, maille *f*, chaînon *m* (*d'une chaîne*); tige *f* d'assemblage
nl schalm *m*; verbindingsstang *f* (*m*)
r звено *n* (*цепи*); соединительная штанга *f*, соединительный тяж *m*

L328 *e* **links**
d Bügelbewehrung *f*
f étriers *m pl* (*pour barres d'armatures*)
nl beugelwapening *f*
r арматурные хомуты *m pl*

L329 *e* **linoleum**
d Linoleum *n*
f linoléum *m*
nl linoleum *n*, *m*
r линолеум *m*

L340 *e* **linseed oil**
d Leinöl *n*
f huile *f* de lin
nl lijnolie *f* (*m*)
r льняное масло *n*; олифа *f*

L341 *e* **lintel, lintol**
d Sturz *m*, Oberschwelle *f*
f linteau *m*, sommier *m*; linceau *m*
nl latei *f* (*m*) (*venster, deur*)
r перемычка *f* (*оконная, дверная*)

L342 *e* **lintel beam**
d Sturzbalken *m*
f linteau *m* préfabriqué
nl latei *f* (*m*)
r балочная перемычка *f*

L343 *e* lip
   d 1. Schwelle *f* (*im Wasserbau*)
      2. Verteilerstein *m*, Lippe *f* (*des Schußwehrs*)
   f 1. seuil *m* 2. bec *m* (*d'un barrage*)
   nl 1. drempel *m* 2. verdelersteen *m*
   r 1. порог *m* (*гидросооружений*)
      2. струерасширительная консоль *f* (*плотины*)

L344 *e* liquefied natural gas
   d flüssiges [verflüssigtes] Erdgas *n*
   f gaz *m* naturel liquéfié
   nl vloeibaar aardgas *n*
   r сжиженный природный газ *m*

L345 *e* liquid accelerator
   d flüssiger Erhärtungsbeschleuniger *m*
   f accélérateur *m* de durcissement liquide
   nl vloeibare verhardingsversneller *m*
   r жидкий ускоритель *m* твердения

L346 *e* liquid asphalt
   d Verschnittbitumen *n*
   f bitume *m* liquide [liquéfié]
   nl vloeibaar bitumen *n*
   r жидкий [разжиженный] битум *m*

L347 *e* liquid concrete admixture
   d flüssiges Betonzusatzmittel *n*
   f additif *m* [adjuvant *m*] liquide du béton
   nl vloeibare bijslag *m* voor het beton
   r жидкая добавка *f* к бетону

L348 *e* liquid limit
   d Fließgrenze *f*
   f limite *f* de liquidité
   nl verwekingsgrens *f* (*m*)
   r предел *m* текучести (*грунта*)

L349 *e* liquid line
   d Flüssigkeitsleitung *f*
   f conduite *f* de réfrigérant liquide
   nl vloeistofleiding *f*
   r трубопровод *m* жидкого хладоагента

L350 *e* listening methods
   d akustische Lecksuchverfahren *n pl*
   f méthodes *f pl* acoustiques de détection des fuites
   nl zoeken *n* van lekken op het gehoor
   r акустические методы *m pl* обнаружения течей

L351 *e* list of bid items and quantities
   d Leistungsverzeichnis *n*, Baupreisbuch *n*
   f bordereau *m* de prix, devis *m* estimatif (relatif à l'offre)
   nl lijst *f* (*m*) van eenheidsprijzen en verrekenbare hoeveelheden
   r смета *f* на строительство объекта, представленная подрядчиком (*на торгах*)

L352 *e* litter bin
   d Mülltonne *f*
   f boîte *f* aux ordures, poubelle *f*
   nl vuilnisbak *m*
   r мусороуборочный бачок *m*

L353 *e* live load
   d 1. Verkehrslast *f* (*Wind- und Schneelast nicht miteingerechnet*)
      2. Nutzlast *f*, Gebrauchslast *f*
   f 1. charge *f* mobile 2. charge *f* utile
   nl nuttige belasting *f*
   r 1. временная нагрузка *f* (*моста*)
      2. полезная нагрузка *f*

L354 *e* live steam
   d Frischdampf *m*
   f vapeur *f* vive
   nl verse stoom *m*
   r острый пар *m*

L355 *e* live storage
   d Nutzinhalt *m*, nutzbarer Speicherinhalt *m* [Speicherraum *m*]
   f capacité *f* utile du réservoir, réserve *f* d'eau utile
   nl nuttige inhoud *m*, effectieve voorraad *n*
   r полезный объём *m* водохранилища

L356 *e* living area
   d Wohnfläche *f*
   f surface *f* habitable [d'habitation]
   nl woonoppervlakte *f*
   r жилая площадь *f*

L357 *e* living room
   d Wohnzimmer *n*
   f chambre *f* [pièce *f*, salle *f*] de sejour
   nl woonkamer *f* (*m*)
   r жилая комната *f*, комната *f* дневного пребывания

L358 *e* load
   d 1. Last *f*, Belastung *f* 2. Schuttlast *f*, Geschiebelast *f*
   f 1. charge *f*, surcharge *f* 2. débit *m* solide
   nl last *f* (*m*), belasting *f*
   r 1. нагрузка *f* 2. транспортируемые наносы *m pl*, расход *m* наносов

L359 *e* load application
   d Belasten *n*, Lastaufbringung *f*, Lastangriff *m*, Lasteintragung *f*
   f application *f* de charge, chargement *m*
   nl belasten *n*
   r нагружение *n*, приложение *n* нагрузки

L360 load-bearing capacity *see* load capacity

L361 *e* load-bearing cross wall
   d tragende Querwand *f*
   f mur *m* transversal [en traverse]
   nl dragende binnenmuur *m*
   r поперечная несущая стена *f*

L362 *e* load bearing panel
   d tragende Wandplatte *f*
   f panneau *m* mural portant [porteur]
   nl dragend muurgedeelte *m*
   r несущая (стеновая) панель *f*

L363 *e* load bearing partition
   d tragende Trennwand *f*
   f paroi *f* [cloison *f*] portante
   nl dragende tussenmuur *m* [scheidingswand *m*]
   r несущая перегородка *f*

L364 *e* **load bearing stiffener**
d Tragrippe *f*, lasttragende
Versteifungsrippe *f*
f nervure *f* vertical raidisseuse
nl dragende verstijvingsribbe *f (m)*
r несущее вертикальное ребро *n*
жёсткости

L365 *e* **load-bearing structural insulating
materials**
d tragfähige Dämmstoffe *m pl*
f matériaux *m pl* isolants de bâtiment
nl dragende en isolerende
bouwmaterialen *n pl*
r несущие конструктивно-
-изоляционные материалы *m pl*

L366 *e* **load-bearing wall**
d (last)tragende Wand *f*
f mur *m* porteur
nl dragende muur *m*
r несущая стена *f*

L367 *e* **load capacity, load carrying capacity**
d 1. Tragfähigkeit *f*; Hubfähigkeit *f*;
Ladeinhalt *m*, Laderaum(inhalt) *m*
2. Transportvermögen *n* des Flusses
f 1. capacité *f* portante; capacité *f* de
charge(ment) 2. capacité *f* de
transport de sédiments, puissance *f*
de transport de fleuve
nl 1. draagvermogen *n*; laadvermogen *n*
2. transportvermogen *n* van een rivier
r 1. несущая способность *f*;
грузоподъёмность *f*;
грузовместимость *f*
2. наносотранспортирующая
способность *f* (*водотока*)

L368 *e* **load cell**
d Druckdose *f*, Belastungsmeßdose *f*
f cellule *f* dynamométrique
nl meetdoos *f (m)*
r динамометрический датчик *m*

L369 *e* **load cell weighing equipment**
d Dosierwaage *f* mit Belastungsmeßdose
f balance *f* doseuse et enregistreuse
nl doseerbalans *f (m)*
r дозировочные весы *pl* с
динамометрическим датчиком

L370 *e* **load combination**
d Lastenkombination *f*
f combinaison *f* de charges
nl samenstel *n* van belastingen
r сочетание *n* нагрузок

L371 **load control** *see* **load handling**

L372 *e* **load cycle**
d Lastspiel *n*, Lastwechsel *m*
f cycle *m* de chargement
nl afwisselende belasting *f*
r цикл *m* нагружения [нагрузки]

L373 **load-deformation curve** *see* **load-strain
curve**

L374 *e* **load distribution**
d Lastverteilung *f*
f distribution *f* des charges

nl krachtverdeling *f*, spreiding *f* van de
belasting
r распределение *n* нагрузки

L375 *e* **load diversity factor**
d Gleichzeitigkeitsfaktor *m*
f facteur *m* de diversité
nl gelijktijdigheidsfactor *m*
r коэффициент *m* разновременности
нагрузок

L376 *e* **load due to wind pressure**
d Winddrucklast *f*
f charge *f* due à la poussée du vent
nl winddruk *m*
r ветровая нагрузка *f*

L377 *e* **loaded area**
d Lastfläche *f*, Belastungsfläche *f*,
Lastfeld *n*
f surface *f* de chargement
nl belast oppervlak *n*
r площадь *f* нагружения [действия
нагрузки]; поле *n* нагрузки

L378 *e* **loaded axle**
d belastete Achse *f*
f essieu *m* chargé [sollicité]
nl belaste as *f (m)*
r нагруженная ось *f*

L379 *e* **loaded chord**
d Traggurt *m*, Lastgurt *m*
f membrure *f* chargée
nl belaste rand *m* (*ligger*)
r нагруженный пояс *m*

L380 *e* **loaded filter**
d Rückhaltefilter *n, m*, umgekehrtes
Filter *n, m*, Entlastungsfilter *n, m*
f filtre *m* inversé
nl invertfilter *n, m*
r обратный фильтр *m*

L381 *e* **load equivalent**
d Lastäquivalent *n*
f charge *f* équivalente
nl equivalente belasting *f*
r эквивалентная нагрузка *f*

L382 *e* **loader**
d 1. Auflader *m*, Lademaschine *f*
2. Beschickungsvorrichtung *f*
f chargeuse *f*, chargeur *m*
nl 1. laadinrichting *f* 2. aanvoerinstallatie
f
r 1. погрузчик *m* 2. загрузочное
устройство *n*; питатель *m*

L383 *e* **load factor**
d Lastfaktor *m*, Belastungsfaktor *m*;
Sicherheitsbeiwert *m*
f facteur *m* de charge; facteur *m* de
sécurité de la rupture
nl belastingsfactor *m*
r коэффициент *m* нагрузки; коэффициент *m* запаса прочности

L384 *e* **load factor design**
d Traglastverfahren *n*, *n*-freies
Bemessungsverfahren *n*
(*Stahlbetonberechnung*)
f calcul *m* à la rupture

nl draaglastmethode f
r расчёт m по разрушающим нагрузкам

L385 e **load handling**
d Lastenbewegung f, Lastenumschlag m
f manutention f de matériaux
nl behandeling f en transport n van lading
r перемещение n [транспортирование n] грузов, перегрузочные операции f pl

L386 e **load hoist**
d Ladewinde f
f treuil m de chargement [de flèche]
nl laadwindas n (kraan)
r грузовая лебёдка f (крана)

L387 e **load hoisting mechanism**
d Hubwerk n
f engin m [mécanisme m] de levage
nl hijsinrichting f
r грузоподъёмный механизм m

L388 e **load increment**
d Belastungsstufe f, Laststufe f
f accroissement m de la charge
nl toename f van de last
r приращение n нагрузки

L389 e **load indicating bolt**
d lastanzeigender Hochspannungsbolzen m
f boulon m à haute résistance spécial (à indicateur de couple)
nl momentaanwijzende bout m
r высокопрочный болт m, контролирующий [лимитирующий] усилие натяжения

L390 e **load indicating washer**
d lastanzeigende Unterlegscheibe f
f rondelle f limitant la force ou le couple de serrage
nl momentaanwijzende onderlegring m
r шайба f, регистрирующая усилие натяжения высокопрочного болта

L391 e **loading**
d 1. Belastung f 2. Beladen n
f chargement m
nl 1. belasting f 2. laden n
r 1. нагрузка f, нагружение n 2. погрузка f

L392 e **loading bin**
d Fertiggut-Silo m, Verladesilo m, Ladebunker m
f trémie f [silo m] de chargement
nl silo m voor stortgoed
r силос m для конечного продукта, погрузочный силос m, погрузочный бункер m

L393 e **loading bridge**
d Brückenkran m; Ladebrücke f; Ladebühne f
f pont m roulant; pont m de chargement; estacade f de chargement
nl laadbrug f (m)
r мостовой кран m; перегрузочный мост m; погрузочная эстакада f

L394 e **loading capacity**
d Tragfähigkeit f
f capacité f portante; charge f utile
nl draagvermogen n; laadvermogen n
r несущая способность f; грузоподъёмность f

L395 **loading combination** see **load combination**

L396 e **loading conditions**
d Belastungsbedingungen f pl
f conditions f pl de chargement
nl belastingsvoorwaarden f pl
r условия n pl нагружения

L397 e **loading gauge**
d Lichtraumprofil n
f gabarit m de chargement [de passage, d'encombrement]
nl laadmal m, ladingprofiel n
r габарит m подвижного состава

L398 e **loading history**
d Belastungsgeschichte f
f histoire f de chargement
nl laadrapport n
r картина f изменения нагрузки во времени, предыстория f нагружения, предшествующее нагружение n

L399 e **loading hopper**
d Ladetrichter m
f trémie f de chargement
nl laadtrechter m, laadbunker m
r загрузочный бункер m

L400 e **loading platform**
d Verladebühne f, Ladebühne f, Laderampe f
f plate-forme f [estacade f] de chargement
nl laadperron n
r погрузочная [перегрузочная] платформа f, погрузочная эстакада f

L401 e **loading shovel**
d Schaufellader m, Hochlader m, Ladeschaufel f
f chargeur m [chargeuse f] à godet [à benne]
nl laadschop f (m)
r одноковшовый погрузчик m

L402 e **loading technique**
d Lasteintragungsverfahren n
f méthode f de chargement [d'application de la charge]
nl belastingsmethode f
r приём m [способ m] нагружения, способ m приложения нагрузки

L403 e **loading test**
d Belastungsprüfung f
f essai m de charge
nl belastingsproef f (m)
r испытание n нагружением

L404 e **load limit**
d Lastgrenze f, Grenzbelastung f
f charge f limite

nl toelaatbare [maximale] belasting *f*
r предельная нагрузка *f*

L405 *e* **load line**
    d Lastlinie *f*, Belastungslinie *f*
    f ligne *f* de charge
    nl lastlijn *f* (*m*) (*mechanica*); laadlijn *f* (*m*) (*schip*)
    r эпюра *f* нагрузки

L406 *e* **load-line of crane**
    d Lastseil *n*, Hubseil *n*
    f câble *m* de levage du poids
    nl hijskabel *m*, hijsdraad *m*
    r грузовой канат *m* крана

L407 *e* **load moment**
    d Lastmoment *n*
    f moment *m* de charge
    nl belastingsmoment *n*
    r грузовой момент *m*; момент *m* от нагрузок

L408 *e* **load pattern**
    d Lastverlauf *m*, Lastgang *m*
    f régime *m* de charge(ment)
    nl belastingspatroon *n*
    r режим *m* нагружения [нагрузки]

L409 *e* **load plane**
    d Belastungsebene *f*
    f plan *m* d'action des forces extérieures [des charges]
    nl belastingsvlak *n*
    r плоскость *f* действия внешних сил [нагрузки]

L410 *e* **load rating**
    d Lastwert *m*
    f capacité *f* portante [de charge] nominale
    nl nominaal draagvermogen *n*, toelaatbare belasting *f*
    r номинальная несущая способность *f*

L411 *e* **load reset**
    d lastabhängige Regelung *f* (*Klimatechnik*)
    f ajustage *m* par charge (*du système de climatisation*)
    nl lastafhankelijke regeling *f* (*luchtbehandeling*)
    r регулирование *n* по нагрузкам (*кондиционирование воздуха*)

L412 *e* **loads equidistant from midspan**
    d von der Feldmitte gleich entfernte Lasten *f pl*
    f charges *f pl* équidistant du milieu de la poutre
    nl symmetrische belasting *f* van een overspanning
    r нагрузки *f pl*, равноотстоящие от середины пролёта

L413 *e* **load-settlement graph**
    d Lastsetzungslinie *f*
    f courbe *f* charge-tassement
    nl grafiek *f* van transport en afzetting
    r кривая *f* зависимости осадки от нагрузки

L414 *e* **load-strain curve**
    d Last-Dehnungskurve *f*
    f courbe *f* charges-déformations
    nl last-vervormingskromme *f*, spanning-rekdiagram *n*, trekdiagram *n*
    r диаграмма *f* зависимости деформации от нагрузки

L415 *e* **load-supporting ability of ground**
    d Tragfähigkeit *f* des Bodens
    f capacité *f* portante [de charge] du sol
    nl draagvermogen *n* van de grond
    r несущая способность *f* грунта

L416 *e* **load surface**
    d Belastungsfläche *f*, Lastfläche *f*
    f surface *f* chargée [de charge]
    nl belast oppervlak *n*
    r площадь *f* распределения нагрузки

L417 *e* **load testing of structures**
    d Bauwerkprüfung *f* durch Probebelastung
    f essai *m* des constructions par charge(ment)
    nl proefbelasting *f* van een constructie
    r испытание *n* сооружений нагрузкой [нагружением]

L418 *e* **load transfer, load transmission**
    d Lastübertragung *f*
    f transmission *f* de charge
    nl lastoverbrenging *f*
    r передача *f* нагрузки

L419 *e* **loam**
    d Lehm *m*
    f limon *m*, sol *m* argileux
    nl leem *n*, *m*
    r суглинок *m*

L420 *e* **lobby**
    d 1. Vorhalle *f*, Diele *f*, Vestibül *n* 2. Foyer *n* 3. Stallung *f*
    f 1. entrée *f*, antichambre *f*, vestibule *m* 2. hall *m* d'entrée, foyer *m* 3. petit parc *m* de bétail
    nl 1. portaal *n* 2. lobby *m* 3. stalling *f*
    r 1. прихожая *f*, передняя *f*; вестибюль *m* (*небольшой*) 2. фойе *n* 3. загон *n* для скота (*небольшой*)

L421 *e* **local bond stress**
    d örtliche Verbundspannung *f* [Haftspannung *f*]
    f tension *f* [contrainte *f*] d'adhérence locale
    nl plaatselijke hechtspanning *f*
    r местное напряжение *n* сцепления

L422 *e* **local buckling**
    d örtliches Knicken *n*
    f flambage *m* [flambement *m*] local
    nl plaatselijk plooien *n*
    r местное выпучивание *n*, местная потеря *f* устойчивости

L423 *e* **local building regulations**
    d örtliche Hochbauvorschriften *f pl* [Hochbaubestimmungen *f pl*]
    f règlement *m* de construction local, lois *f pl* sur la construction locales

nl plaatselijke bouwverordeningen *f pl*
r местные строительные правила *n pl*

L424 *e* **local exhaust**
d örtliche Absaugung *f*
[Absaugeinrichtung *f*]
f échappement *m* local
nl plaatselijke afzuiging *f*
r местный отсос *m*

L425 *e* **local exhaust ventilation**
d örtliche Absauglüftung *f*
f ventilation *f* aspirante locale
nl plaatselijke kunstmatige ventilatie *f*
r местная вытяжная вентиляция *f*

L426 *e* **local failure**
d örtlicher Bruch *m*
f rupture *f* locale
nl plaatselijke breuk *f (m)*
r местное разрушение *n*

L427 *e* **local heating**
d Einzelheizung *f*, örtliche Beheizung
f
f chauffage *m* individuel
nl afzonderlijke verwarming *f*
r местное отопление *m*; местный
обогрев *m*

L428 *e* **local lighting**
d örtliche Beleuchtung *f*,
Arbeitsplatzbeleuchtung *f*
f éclairage *m* local
nl plaatselijke verlichting *f*
r местное освещение *n*, освещение *n*
рабочего места

L429 *e* **local losses**
d örtliche Verluste *m pl* [Verlusthöhe *f*],
Einzeldruckverluste *m pl*,
Druckverluste *m pl* durch
Einzelwiderstände
f pertes *f pl* de charge locales
nl plaatselijke drukverliezen *n pl*
(*water, gas*)
r местные потери *f pl* (*давления,
напора*), потери *f pl* на местных
сопротивлениях

L430 *e* **local materials**
d örtliche Baumaterialien *n pl*
f matériaux *m pl* de construction
obtenus des sources locales
nl ter plaatse winbare
bouwmaterialen *n pl*
r местные строительные материалы *m pl*

L431 *e* **local resistance**
d Einzelwiderstand *m*, örtlicher
Widerstand *m*
f résistance *f* locale
nl plaatselijke weerstand *m*
r местное сопротивление *n*

L432 *e* **local sewerage system**
d Kleinkläranlage *f*
f réseau *m* d'évacuation des eaux
usées local, système *m* d'égouts local
nl plaatselijke rioleringssysteem *n*
r местная [малая] канализация *f*

L434 *e* **local water supply**
d Eigenwasserversorgung *f*, örtliche
Wasserversorgung *f*
f alimentation *f* individuelle en eau,
approvisionnement *m* individuel en
eau, distribution *f* d'eau individuelle
nl eigen watervoorziening *f*
r местное водоснабжение *n*

L435 *e* **location**
d 1. Trassierung *f*, Linienführung *f*,
Streckenfestlegung *f* 2. örtliche Lage
f, Standort *m*
f 1. traçage *m*, piquetage *m*
[implantation *f*] de tracé 2. site *m*,
emplacement *m*
nl 1. uitzetten *n* van het werk [tracé]
2. locatie *f* (van een werk)
r 1. трассирование *n* 2. местоположение
n

L436 *e* **location line**
d Bautrasse *f*, Trasse *f*,
Streckenverlauf *m*, Linienführung *f*
f tracé *m* (*d'une route, d'une
canalisation*)
nl tracé *n*
r трасса *f* (*дороги, трубопровода*)

L437 *e* **location of reinforcement**
d Bewehrungsverteilung *f*, Anordnung *f*
der Bewehrung
f disposition *f* des armatures
nl plaatsing *f* [verdeling *f*] van de
wapening
r расположение *n* арматуры

L348 **location plan** *see* **site plan**

L439 *e* **lock**
d 1. Schleuse *f* 2. Schloß *n*, Verschluß *m*
f 1. écluse *f* 2. surrure *f*, verrou *m*
nl 1. sluis *f (m)* 2. sluiting *f*, slot *n*
r 1. шлюз *m* 2. замок *m*

L440 *e* **lockage**
d 1. Schiffsschleusung *f*,
Durchschleusen *n*
2. Durchschleusen *n*, Schleusen *n*
3. Schleusenfüllung *f*
f 1. éclusage *m* 2. sassement *m*
3. remplissage *m* d'une écluse
nl 1. sluiswerken 2. schutten *n*
3. schutwater *n*
r 1. шлюзование *n* судов
2. шлюзование *n*, пропуск *m* людей
и грузов через шлюз (*напр.
кессона*) 3. наполнение *n* камеры
шлюза

L441 *e* **lockage water**
d Schleusenwasser *n*
f éclusée *f*
nl schutwater *n*
r сливная призма *f* шлюза

L442 **lock bay** *see* **lock chamber**

L443 *e* **lock bolt**
d Sicherungsschraube *f*,
Verriegelungsbolzen *m*
f boulon-rivet *m*, écrou-rivet *m*

nl borgbout *m*, borgschroef *f* (*m*)
r болт-заклёпка *m*

L444 *e* **lock chamber**
d Schleusenkammer *f*
f sas *m* [chambre *f*] d'écluse
nl sluiskamer *f* (*m*), schutkolk *f* (*m*)
r камера *f* шлюза

L445 *e* **locked-coil rope**
d patentverschlossenes Seil *n*
f câble *m* (en fils) d'acier fermé
nl gesloten staalkabel *m*
r закрытый проволочный канат *m*

L446 *e* **lock-filling system**
d Schleusenfüllanlage *f*
f système *m* de remplissage de l'écluse
nl spui-installatie *f* (*schutkolk*)
r система *f* наполнения шлюза

L447 **lock flight** *see* **flight of locks**

L448 *e* **lock former**
d Blechfalzmaschine *f*
f machine *f* à former les agrafures
nl blikvouwmachine *f*
r фальцегибочный станок *m*

L449 *e* **lock gate**
d Schleusentor *n*
f porte *f* d'écluse
nl sluisdeur(en) *f* (*m*) (*pl*) (*in één sluishoofd*)
r ворота *pl* шлюза

L450 *e* **lock gates, lock head**
d Schleusenhaupt *n*
f tête *f* d'écluse
nl sluishoofd *n*
r голова *f* шлюза

L451 **lock joint** *see* **lock seam**

L452 *e* **locknut**
d Gegenmutter *f*, Kontermutter *f*, Sicherungsmutter *f*
f contre-écrou *m*, écrou *m* de blocage
nl borgmoer *f* (*m*), contramoer *f* (*m*)
r контргайка *f*, стопорная гайка *f*

L453 *e* **lock paddle**
d Umlaufverschluß *m* (*Schleuse*)
f étanche *f*
nl schuif *f* (*m*), rinket *n* (*sluisdeur*)
r затвор *m* шлюза

L454 *e* **lock seam**
d Liegefalz *m*
f agrafure *f*
nl liggende vouw *f* (*m*)
r лежачий фальц *m*

L455 *e* **lockset**
d Schließbeschlag *m*, Verschlußbeschlag *m*
f appareil *m* de fermeture (*de porte ou de fenêtre*), verrous *m* pl
nl sluitwerk *n* (*venster en deuren*)
r запорный прибор *m* (*для окон и дверей*)

L456 *e* **lock-shield valve**
d doppelt einstellbares Heizkörperventil *n*, Drosselventil *n* mit Abdeckkappe, Behördenventil *n*
f robinet *m* à double réglage

nl regelventiel *n*, radiatorkraan *f* (*m*) met regelschroef
r кран *m* [клапан *m*] двойной регулировки

L457 *e* **lock sill**
d Drempel *m*
f seuil *m* à redan
nl sluisdrempel *m*, slagdrempel *m*
r порог *m* шлюза

L458 *e* **lock stile**
d Türrahmenpfosten *m* mit Einsteckschloß
f montant *m* de porte avec la serrure
nl deurstijl *m* met insteekslot
r вертикальный брус *m* обвязки двери, в который врезается замок

L459 *e* **lock-up**
d Lückenbau *m*
f bâtiment *m* construit dans l'intervalle entre deux autres bâtiments
nl gapingbouw *m*
r здание *n*, возводимое в разрыве между другими строениями

L460 *e* **lock up stage**
d Abschlußetappe *f* des Bauens
f étape *f* [phase *f*] finale de construction
nl voltooiingsfase *f* van het werk
r завершающая стадия *f* строительства

L461 *e* **lock-weir network**
d Ableitungsnetz *n* für Überlaufwasser
f réseau *m* de colature
nl afvoerstelsel *n* van een overlaat
r водосбросная сеть *f*

L462 *e* **locomotive crane**
d Lokomotivkran *m*
f grue *f* (automotrice) sur rails
nl op rails rijdende kraan *f* (*m*)
r железнодорожный кран *m*

L463 *e* **loess**
d Löß *m*
f loess *m*
nl loess *f* (*m*)
r лёсс *m*, лёссовый грунт *m*

L464 *e* **loft**
d Dachboden *m* (*als Speicherraum*), Rumpelkammer *f*
f grenier *m*
nl vliering *f* (*m*)
r складское чердачное помещение *n*

L465 *e* **loft building**
d Handels- *bzw.* Produktionsgebäude *n*
f bâtiment *m* d'atelier *ou* de commerce
nl handels- *of* industriegebouw *m*
r торговое *или* производственное здание *n*

L466 *e* **log boom**
d Anlegeponton *m*, Bon *m*, Schwimmsperre *f*, Bonensperre *f*
f barrière *f* flottante

nl drijvende versperring f
r бон m

L468 e log chute
d Floßgasse f, Floßdurchlaß m, Trift f,
Holzpaß m, Floßrinne f
f passe f de flottage
nl doorlaat m voor houtvlotten
r бревноспуск m, лесосплавный лоток
m

L469 e log crib
d Steinkasten m, Steinkiste f
f encoffrement m en charpente de bois
nl kistdam m
r ряж m (из брёвен)

L470 e loggia
d 1. Loggia f 2. Hauslaube f
3. Arkadengalerie f
f 1. loggia f 2. arcade f 3. galerie f
ouverte avec arcades
nl 1. loggia f (m) 2. arcade f
r 1. лоджия f 2. аркада f 3. открытая
галерея f с аркадой, аркадная
галерея f

L471 e log mean temperature difference
d mittlere logarithmische
Temperaturdifferenz f
f différence f moyenne logarithmique de
températures
nl gemiddeld logarithmisch
temperatuurverschil n
r среднелогарифмическая разность f
температур

L472 e log sluice
d Flößereianlage f, Holzflößanlage f
f ouvrage m de flottage
nl plaats f (m) waar houtvlotten worden
samengesteld
r лесосплавное сооружение n

L473 log-way see log chute

L474 e long column
d schlanke Säule f [Stütze f]
f poteau m long (le cas du flambage)
nl slanke kolom f (m)
r гибкая колонна f [стойка f]
(работающая на продольный изгиб)

L475 long dolly see follower

L476 e long duration load
d Dauerbelastung f
f charge f de longue durée
nl langdurige belasting f
r долговременная [длительная]
нагрузка f

L478 e longitudinal bar
d Längsstab m
f barre f longitudinale
nl langsstaaf f (m)
r продольный стержень m

L479 e longitudinal center joint
d Mittellängsfuge f
f joint m longitudinale central
nl overlangse middenvoeg f (m)
r срединный продольный шов m

L480 e longitudinal compression
d Längsdruck m
f compression f longitudinale
nl samendrukking f in de lengte
r продольное сжатие n

L481 e longitudinal dike, longitudinal
embankment
d Längsdamm m, Längswerk n,
Leitwerk n, Richtwerk n,
Streichwerk n, Parallelwerk n
f digue f longitudinale
nl dijk m
r продольная [направляющая] дамба f

L482 longitudinal finishing machine see
bullfloat finishing machine

L483 e longitudinal force
d Längskraft f
f force f longitudinale
nl langskracht f (m)
r продольная сила f

L484 e longitudinal grade
d Längsgefälle n, Gradiente f,
Längsneigung f
f pente f [rampe f, inclinaison f]
longitudinale
nl langshelling f
r продольный уклон m

L485 e longitudinal joint
d Längsnaht f; Lagerfuge f
f joint m longitudinal; joint m
d'appareil de briques en long
nl langsverbinding f, lintvoeg f (m)
r продольное соединение n,
продольный шов m; ложковый шов
m (кладки)

L486 e longitudinal prestressing
d Längsvorspannung f
f précontrainte f [mise f en tension]
axiale [longitudinale]
nl voorspanning f in lengterichting
r продольное преднапряжение n

L487 e longitudinal profile
d Langsprofil n, Längsschnitt m
f profil m longitudinal
nl langsprofiel n
r продольный профиль m

L488 e longitudinal rigidity, longitudinal
stiffness
d Längssteifigkeit f, Längsstarrheit f
f rigidité f longitudinale
nl langsstijfheid f
r продольная жёсткость f

L489 e longitudinal strain
d Längsdehnung f
f déformation f longitudinale [axiale]
nl langsrek m, vervorming f in de lengte
r продольная деформация f

L490 e longitudinal tensile stress, longitudinal
tension stress
d Längszugspannung f
f contrainte f [tension f] de traction
longitudinale
nl langstrekspanning f

r продольное растягивающее
напряжение *n*

L491 *e* **long radius bend**
d sanfter Bogen *m* (*Rohrformstück*)
f coude *m* aplati
nl flauwe bocht *f* (*m*)
r отвод *m* с нормальным радиусом
закругления

L492 *e* **longscrew**
d Gewindestutzen *m*,
Überwurfmutterverbindung *f*
f tubulure *f* de jonction [de
raccordement] filetée
nl schroefmof *f* (*m*), draadsok *f* (*m*)
r сгон *m*, патрубок *m* с резьбой

L493 *e* **long-span bridge**
d Brücke *f* mit großer Spannweite
f pont *m* de grande portée
nl brug *f* (*m*) met grote overspanning
r большепролётный мост *m*

L494 *e* **long-term strength**
d Dauerstandfestigkeit *f*,
Zeitstandfestigkeit *f*
f résistance *f* de grande durée
nl sterkte *f* op lange termijn
r долговременная [длительная]
прочность *f*

L495 *e* **loop anchorage**
d Schlaufenverankerung *f*
f ancrage *m* bouclé
nl lusverankering *f*
r петлевая анкеровка *f*

L496 *e* **loose core**
d entmischter Grobzuschlag(stoff) *m*
f granulat *m* [agrégat *m*] ségrégé
nl ontmengde grove toeslag *m*
r сегрегированный заполнитель *m*,
крупный заполнитель *m*,
отслоившийся от бетонной смеси

L497 *e* **loose fill**
d 1. lose Schüttung *f* 2. Wärmedämmung
f mit Schüttstoffen
f 1. remblayage *m* en terre meuble
2. isolant *m* thermique en vrac
nl 1. losse storting *f* 2. los gestort
isolatiemateriaal *n*
r 1. засыпка *f* несвязным грунтом *или*
зернистым материалом 2. засыпная
теплоизоляция *f*

L498 *e* **loose-fill insulation**
d Schüttisolierung *f*, Wärmedämmung *f*
mit Schüttstoffen
f isolant *m* thermique en vrac
nl warmte-isolatie *f* met los gestort
materiaal
r засыпная теплоизоляция *f*

L499 *e* **loose ground**
d lockerer Boden *m*
f terre *f* meuble, sol *m* non cohérent
nl losse grond *m*
r несвязный [сыпучий] грунт *m*

L500 **loose insulation** *see* **loose-fill
insulation**

L501 *e* **loose-joint butt, loose-joint hinge**
d Aufsatzband *n* mit festeingebauter
Achse
f charnière *f* pivotant sur gond,
paumelle *f*
nl paumelle *f*, fits
r разъёмная дверная петля *f*
с невынимаемой осью

L502 *e* **loose lintel**
d freiaufgelagerter Fertigteilsturz *m*
f linteau *m* [sommier *m*] préfabriqué
nl prefab latei *f* (*m*)
r сборная перемычка *f*, свободно
уложенная поверх проёма

L503 *e* **loose-pin hinge**
d Aufsatzband *n* mit herausziehbarer
Achse
f charnière *f* à broche amovible
nl gewoon scharnier *n*
r разъёмная дверная петля *f* с
вынимаемой осью

L504 *e* **loose rock dam**
d trocken gepackter
Bruchsteinmauerwerksdamm *m*,
Trockenbruchsteinmauerwerksdamm
m, Steinpackungsdamm *m*
f barrage *m* en maçonnerie ordinaire
à sec
nl steendam *m*, dam *m* van stortsteen
r плотина *f* из сухой бутовой кладки

L505 **loose soil** *see* **loose ground**

L506 **lorry mounted mixer** *UK see* **mixer-
-type truck**

L507 *e* **loss coefficient**
d Druckverlustfaktor *m*,
Verlustkoeffizient *m*
f coefficient *m* de perte de charge
nl drukverliesfactor *m*
r коэффициент *m* потерь напора
[давления]

L508 *e* **loss of bond strength**
d Haftfestigkeitsverlust *m*
f perte *f* d'adhérence (*de l'armature au
béton*)
nl vermindering *f* van de
aanhechtingsspanning (*betonwapening*)
r нарушение *n* сцепления (*напр.
арматуры с бетоном*)

L509 *e* **loss of head**
d Druck(höhen)verlust *m*,
Verlusthöhe *f*, Fallhöhenverlust *m*,
Förderdruckverlust *m*
f perte *f* de charge
nl drukverlies *n*
r потеря *f* напора [давления]

L510 *e* **loss of prestress**
d Vorspannverlust *m*,
Spannkraftverlust *m*
f perte *f* de précontrainte
nl verlies *n* aan voorspanning
r потеря *f* предварительного
напряжения

L511 e **loss of prestress due to friction**
  d reibungsbedingter Spannkraftverlust *m*
  f perte *f* de précontrainte due à la friction
  nl verlies *n* aan voorspanning ten gevolge van wrijving
  r потеря *f* преднапряжения от трения

L512 e **loss of prestress due to shrinkage of the concrete**
  d schwundbedingter Spannkraftverlust *m*
  f perte *f* de précontrainte due au retrait de béton
  nl verlies *n* aan voorspanning door krimp van het beton
  r потеря *f* преднапряжения от усадки бетона

L513 **loss of prestressing** *see* **loss of prestress**

L514 e **loss of workability**
  d Schüttfähigkeitsverlust *m*
  f perte *f* [diminution *f*] d'ouvrabilité [de workabilité]
  nl vermindering *f* van de verwerkbaarheid
  r потеря *f* удобоукладываемости

L515 e **lot** *US*
  d 1. Grundstück *n* 2. Baugrundstück *n* 3. Bauplatz *m*
  f 1. lot *m* [parcelle *f*] de terrain 2. terrain *m* à bâtir 3. chantier *m*
  nl 1. perceel *n* 2. bouwterrein *n*
  r 1. земельный участок *m* 2. строительный участок *m*; участок *m* застройки 3. строительная площадка *f*

L516 e **lot line**
  d Grundstücksgrenze *f*
  f limites *f pl* d'une parcelle de terrain
  nl perceelgrens *f* (*m*)
  r граница *f* земельного участка

L517 e **loudness level**
  d Schallpegel *m*, Lautstärkepegel *m*
  f niveau *m* (d'intensité) sonore
  nl geluidniveau *n*
  r уровень *m* громкости

L518 e **louvers, louvres**
  d 1. Jalousie *f*, Gliederklappe *f* 2. Jalousiegitter *n*, Jalousieklappe *f*
  f 1. jalousie *f*; persienne *f* 2. volet *m* d'aérage
  nl 1. jaloezie *f*, louvre-luik *n* 2. ventilatiekap *f* (*m*)
  r 1. жалюзи *pl*, жалюзийная штора *f* 2. жалюзийный [многостворчатый] воздушный клапан *m*

L519 e **low-alkali cement**
  d alkaliarmer Zement *m*
  f ciment *m* à faible teneur en alcali
  nl alkali-arm cement *n*
  r цемент *m* с низким содержанием щелочных компонентов

L521 e **low block, low building**
  d Flachgebäude *n*
  f bâtiment *m* bas
  nl laag gebouw *n*
  r малоэтажное здание *n*

L522 e **low cost homes**
  d billige Standardhäuser *n pl*
  f maisons *m pl* collectives standardisées à bon marché
  nl goedkope standaardhuizen *n pl*
  r стандартные дома *m pl* низкой стоимости

L523 e **low cost housing**
  d Massenwohnungsbau *m*, Bebauung *f* mit kostengünstigen Wohnhäusern
  f construction *f* en grande série des maisons collectives à bon marché
  nl volkswoningbouw(complex) *m* (*n*)
  r массовое дешёвое жилищное строительство *n*, застройка *f* жилыми домами низкой стоимости

L524 e **low-density concrete**
  d Leichtbeton *m*
  f béton *m* léger
  nl licht beton *n*
  r бетон *m* малой плотности, лёгкий бетон *m*

L525 e **lower chord**
  d Untergurt *m* (*Balken*)
  f membrure *f* inférieure
  nl rand *m* (*ligger*)
  r нижний пояс *m* (*балки, фермы*)

L526 e **lower-chord panel points**
  d Untergurtknoten *m*
  f nœuds *m pl* de la mambrure inférieure (*d'une poutre à treillis*)
  nl knooppunten *n pl* in de onderrand (*vakwerkligger*)
  r узлы *m pl* [узловые точки *f pl*] нижнего пояса фермы

L527 e **lower critical velocity**
  d untere kritische Geschwindigkeit *f*
  f vitesse *f* critique inférieure
  nl laagste kritische snelheid *f*
  r нижняя критическая скорость *f*

L528 e **lower explosive limit**
  d untere Explosionsgrenze *f*
  f limite *f* inférieure d'explosibilité
  nl laagste explosiegrens *f* (*m*)
  r нижний предел *m* взрывоопасной концентрации (*пыли, паровоздушной смеси*)

L529 e **lower gate**
  d Untertor *n* (*Schleuse*)
  f porte *f* aval
  nl deur(en) *f* (*m*) (*pl*) in het benedenhoofd (*sluis*)
  r нижние ворота *pl* (*шлюза*)

L530 e **lower layer**
  d Unterlage *f*, Unterschicht *f* (*Fußboden*)
  f sous-couche *f* (*du revêtement de sol*)
  nl onderste laag *f* (*m*) (*vloer*)
  r подстилающий слой *m*, подоснова *f* (*покрытия пола*)

L531 e lower lock head
  d Schleusenunterhaupt n
  f tête f d'aval
  nl benedenhoofd n (sluis)
  r нижняя голова f шлюза

L532 e lower reservoir, lower storage basin
  d Unterbecken n, Tiefbecken n
  f bassin m inférieur
  nl benedenbekken n
  r нижний бассейн m
  (гидроаккумулирующей
  электростанции); низовой бассейн m
  (приливной электростанции)

L533 e lower yield point
  d untere Streckgrenze f [Fließgrenze f]
  f point m d'écoulement inférieur
  nl onderste vloeigrens f (m)
  r нижний предел m текучести

L534 e lowest water level
  d niedrigster Wasserstand m
  f niveau m d'eau minimum
  nl laagste waterstand m
  r наинизший уровень m воды

L535 e low-flow channel
  d Niedrigwasserbett n
  f lit m mineur [d'étiage]
  nl zomerbed n
  r меженное русло n

L536 e low-frequency vibration
  d Niederfrequenzrütteln n
  f vibrations m pl à basse fréquence
  nl laagfrequent trillen n
  r низкочастотное вибрирование n
  [виброуплотнение n]

L537 e low grade concrete
  d Beton m niederer Güte; Magerbeton m
  f béton m de qualité inférieure;
  béton m maigre
  nl mager beton n
  r бетон m низких марок; тощий бетон m

L538 e low-head dam
  d Niederdruckstaudamm m,
  Niederdruckwehr n
  f barrage m à faible hauteur
  nl stuwdam m met geringe hoogte
  r низконапорная плотина f

L539 e low-head hydroelectric power station,
  low-head water power plant
  d Niederdruck-Wasserkraftwerk n
  f usine f à basse chute
  nl lagedruk-waterkrachtcentrale f
  r низконапорная ГЭС f

L540 e low-heat cement, low-heat-of-
  -hydratation cement
  d Zement m mit geringer
  Hydra(ta)tionswärme
  f ciment m à faible chaleur
  d'hydratation
  nl cement n met lage hydratatiewaarde
  r низкотермичный цемент m

L542 e low-level bridge
  d überflutbare Brücke f, Tauchbrücke
  f, Tiefbrücke f
  f pont m à bas niveau
  nl brug f (m) met overspoelbaar rijdek
  r низководный мост m

L543 e low-level cistern
  d Tiefspülbecken n
  f réservoir m de chasse bas
  nl laaggeplaatst closetreservoir n
  r низкорасположенный смывной
  бачок m

L544 e low level mixing plant
  d Horizontalmischanlage f
  f poste m d'enrobage à disposition
  horizontale
  nl horizontale menginstallatie f
  r смесительная установка f
  горизонтальной компоновки

L545 e low level sluiceway
  d Grundablaß m
  f évacuateur m de fond
  nl duikersluis f (m)
  r донный водоспуск m

L546 e low pressure air conditioning system
  d Niederdruckklimaanlage f
  f système m de conditionnement d'air
  à basse pression
  nl lagedruk-luchtbehandelingstoestel n
  r система f кондиционирования воздуха
  низкого давления

L548 e low pressure hot water system
  d Warmwasserheizungssystem n
  f installation f de chauffage par l'eau
  chaude (à) basse pression
  nl verwarmingssysteem n met water
  onder lage druk
  r система f водяного отопления
  низкого давления

L549 e low-pressure induction unit
  d Niederdruck-Düsenkonvektor m
  f appareil m de climatisation terminal
  à thermoconvection de basse pression
  nl lagedruk-straalconvector m
  r эжекционный доводчик m низкого
  давления

L551 e low-pressure wet gasholder
  d Niederdruckgasbehälter m,
  Niederdruckgasometer n
  f gasomètre m [réservoir m à gaz] à
  basse pression
  nl lagedruk-gashouder m
  r газгольдер m низкого давления

L552 e low rise buildings
  d weniggeschossige Gebäude n pl
  f bâtiments m pl non élevés [bas]
  nl gebouwen n pl met slechts enkele
  verdiepingen
  r малоэтажные здания n pl

L553 e low velocity system
  d Niederdruckklimaanlage f
  f système m d'air à basse vitesse
  nl luchtbehandelingstoestel n met lage
  luchtsnelheid
  r система f кондиционирования с
  низкими скоростями воздуха

L555 *e* **low-water discharge, low-water flow**
  d Niedrigwasserabfluß *m*,
    Niedrigwasserdurchfluß *m*,
    Trockenwetterabfluß *m*
  f écoulement *m* [débit *m*] d'étiage
  nl laagwater-afvoer *m*
  r меженный сток *m* [расход *m*]

L556 *e* **low-water stage**
  d Niedrigwasserstand *m*
  f niveau *m* d'étiage
  nl laagwaterstand *m*
  r меженный уровень *m*, уровень *m*
    низких вод

L557 *e* **low workability concrete**
  d Betongemisch *n* mit geringem
    Ausbreitmaß
  f béton *m* de basse ouvrabilité
  nl te droge, en daarom slecht
    verwerkbare betonmortel *m*
  r бетонная смесь *f* с низким
    показателем удобоукладываемости,
    малоподвижная бетонная смесь *f*

L558 *e* **L-shore**
  d L-köpfige Montagestütze *f*, L-Stütze *f*,
    Winkelstütze *f*
  f était *m* [montant *m*] à tête en L
  nl montagesteun *m* van hoekprofiel
  r опорная *или* монтажная стойка *f*
    с Г-образным оголовком

L559 *e* **lucarne**
  d liegendes Dachfenster *n*
  f lucarne *f*
  nl dakkapel *f* met het raam gedeeltelijk
    in de gevel
  r низкое слуховое окно *n*

L560 *e* **luffing**
  d Wippen *n*, Auslegerverstellen *n*
  f variation *f* de la portée (*d'une flèche*
    de grue)
  nl de giekhoek veranderen
  r изменение *n* вылета [подъём *m* и
    опускание *n*] стрелы крана

L561 *e* **luffing cableway mast**
  d Pendelturm *m*
  f mât *m* [pylône *m*] incliné réglable
    d'une grue à câble
  nl pendeltoren *m*
  r регулируемая наклонная мачта *f*
    кабель-крана

L562 *e* **luffing crane**
  d Wippkran *m*, Einziehkran *m*
  f grue *f* à flèche relevable
  nl giekkraan *f* (*m*)
  r кран *m* с подъёмной стрелой

L563 *e* **lumber** *US*
  d Bauholz *n*, Nutzholz *n*
  f bois *m* de construction [de charpente]
  nl bezaagd hout *n*
  r пиломатериал *m*

L564 *e* **lumber shrinkage**
  d Trockenschwund *m* [Eintrocknen *n*]
    des Holzes
  f retrait *m* du bois

  nl krimp *m* bij bezaagd hout
  r усыхание *n* [усушка *f*] древесины

L565 *e* **luminaire**
  d 1. Leuchte *f* 2. Lüftungsleuchte *f*,
    ventilierte Leuchte *f*
  f 1. luminaire *m*, appareil *m*
    d'éclairage 2. panneau-plafond *m*
    d'aérage et d'éclairage
  nl 1. verlichtingsarmatuur *f*
    2. verlichtingspaneel *n* met
    ventilatierooster
  r 1. светильник *m* 2. вентиляционно-
    -осветительный плафон *m*,
    светильник-воздухораспределитель *m*

L566 *e* **luminescent lamp**
  d Leuchtstofflampe *f*, Fluoreszenzlampe
    f
  f lampe *f* fluorescente
  nl fluorescentielamp *f* (*m*), TL-buis *f* (*m*)
  r люминесцентный светильник *m*

L568 *e* **luminous emittance**
  d spezifische Lichtausstrahlung *f*
  f émittance *f* lumineuse
  nl lichtsterkte *f*
  r светимость *f*

L569 *e* **luminious paint** *see* **fluorescent paint**

L570 *e* **lump quick lime**
  d Stückenkalk *m*
  f chaux *m* en pierre [en morceaux]
  nl stukken ongebluste kalk *m*
  r комовая (негашёная) известь *f*

L571 **lump-sum contract** *see* **firm price**
   contract

L572 *e* **lute**
  d 1. Betonabstreicher *m* 2. Richtscheit *n*
    3. Abdichtungskitt *m*, Isoliermastik *m* ,
    Fugenabdichtungsmasse *f*
  f 1. racloir *m* 2. règle *f* égaliseuse
    3. mastic *m* d'étanchéité; mastic *m*
    pour (garnissage) joints
  nl 1. betonrei *f* (*m*) 2. richtlat *f* (*m*)
    3. kit *m*
  r 1. скребок *m* для разравнивания
    бетона 2. правило *n* каменщика
    3. герметизирующая [изолирующая]
    мастика *f*

M

M1 *e* **macadam**
  d 1. Makadamdecke *f*, Makadam *m*
    2. Schotter *m* für Makadamdecke
  f 1. macadam *m*, empierrement *m*
    (*couche de chaussée*) 2. macadam *m*
    (*pierres cassées*)
  nl macadam *n*
  r 1. щебёночное *или* гравийное
    дорожное покрытие *n*; дорожное
    покрытие типа «макадам» 2. щебень
    m или гравий *m* для покрытия типа
    «макадам»

M2   e macadam aggregate
     d grobkörniges Gestein n für
       Makadamdecken
     f agrégat m, macadam m (pierres
       cassées)
     nl breuksteen m voor macadamweg
     r зернистый каменный материал m
       (щебень, гравий) для дорожных
       покрытий

M3   e macadam base, macadam foundation
     d Schotterunterbau m,
       Schottertragschicht f,
       Steinschlagtragschicht f (Straßenbau)
     f couche f de base en macadam [en
       agrégat]
     nl steenslagfundering f (wegenbouw)
     r щебёночное основание n дорожной
       одежды

M4   machine application see mechanical
     application

M5   e machine cavern
     d Maschinenkaverne f,
       Kraftwerkkaverne f
     f caverne f de la centrale
     nl ondergrondse machinehal f (m)
       (waterkrachtcentrale)
     r подземный машинный зал m ГЭС

M6   e machinery and equipment yard
     d Betriebshof m
     f parc m mécanique des chantiers
     nl (bouw)machinepark n
     r машинный парк m строительной
       организации

M7   e machinery foundation
     d Maschinenfundament n,
       Maschinengründungskörper m
     f fondation f de machine
     nl fundatie f van een machine
     r фундамент m под машинное
       оборудование

M8   e machinery room
     d Maschinenhalle f; Maschinenraum m
     f halle f des machines; compartiment m
       des machines
     nl machinekamer f (m)
     r машинный зал m; машинное
       помещение n [отделение n]

M9   e machinery trailer
     d Maschinen-Tiefladeanhänger m
     f remorque f porte-engins de travaux
       publics
     nl dieplader m
     r низкорамный прицеп m для
       транспортирования строительных
       машин

M10 e machine troweling
     d maschinelles Verreiben n
       [Glattstreichen n]
     f lissage m mécanique (des surfaces de
       béton)
     nl machinaal afrijen n (betonweg)
     r механическая затирка f (бетонных
       поверхностей)

M11 e made ground, made-up ground
     d Aufschüttung f, Auftrag m,
       aufgefülltes [aufgetragenes] Gelände
       n; Dammschicht f, Dammlage f
     f terrain m remblayé; couche f de
       remblai
     nl aanvulling f; opgebrachte grond m
     r насыпной грунт m; насыпной слой m
       грунта

M13 e magnesia brick
     d Magnesitstein m, Magnesitziegel m,
     f brique f de magnésie
     nl magnesietsteen m
     r магнезитовый огнеупорный кирпич m

M14 e magnesia cement
     d Magnesitzement m, Magnesitbinder m
     f ciment m magnésien [de magnésie]
     nl magnesietcement n
     r магнезиальный цемент m,
       магнезиальное вяжущее n

M15 e magnesia insulation
     d Magnesiadämmstoff m
     f calorifuge m à base de magnésie
     nl warmteïsolerend materiaal n op basis
       van magnesia
     r магнезиальный теплоизоляционный
       материал m

M16 magnesia refractory see magnesia brick
M17 e magnesite flooring
     d Magnesitfußboden m,
       Steinholzfußboden m
     f revêtement m de sol à magnésie
     nl magnesietvloer m, asbestgranietvloer
       m
     r ксилолитовый пол m, бесшовный пол
       m на магнезиальном вяжущем
       magnetical... see magnetic...

M18 e magnetic crack detection
     d magnetische Rißprüfung f
     f détection f des fissurations par
       l'examen magnétique
     nl magnetisch onderzoek n op scheuren
     r магнитная дефектоскопия f

M19 e magnetic separator
     d Magnetabscheider m
     f séparateur m magnétique
     nl magnetische scheider m
     r магнитный сепаратор m

M20 e main
     d Hauptleitung f; Hauptstrang m
     f conduite f principale [maîtresse];
       conduite f de distribution
     nl hoofdleiding f; distributieleiding f
     r магистральный трубопровод m;
       распределительная магистраль f

M21 e main air duct
     d Hauptluftkanal m
     f conduit m d'air principal
     nl hoofdluchtkanaal n
     r магистральный воздуховод m

M22 e main bar
     d Hauptbewehrungsstab m,
       Grundarmierungsstab m

  f barre f principale
nl hoofdwapeningsstaaf f (m)
 r стержень m рабочей арматуры

M23 e main beam
  d Hauptbalken m, Hauptträger m
  f poutre f maîtresse
 nl moerbalk m
  r главная балка f

M24 e main bridge members
  d Hauptelemente n pl des
    Brückentragwerkes
  f ensemble m des éléments porteurs du
    tablier de pont
 nl hoofdelementen n pl van de
    brugconstructie
  r основные несущие конструкции f pl
    пролётного строения моста

M25 e main canal
  d Hauptkanal m
  f canal m principal
 nl hoofdkanaal n
  r магистральный канал m

M26 main contractor see general contractor
M27 e main dike
  d Schardeich m, Hauptdeich m
  f digue f principale
 nl schaardijk m
  r передовой вал m, главная дамба f

M28 e main drain
  d Hauptsammler m, Hauptdrän m,
    Sammeldrän m
  f collecteur m principal
 nl hoofdafvoerleiding f
  r главный магистральный коллектор m

M29 main dyke see main dike
M30 main girder see main beam
M31 main rafter
  d Hauptsparren m
  f arbalétrier m
 nl spantbeen n
  r главная стропильная нога f

M32 e main reinforcement
  d Grundbewehrung f,
    Hauptarmierung f
  f armature f principale
 nl hoofdwapening f
  r рабочая арматура f

M33 e main runner
  d Hauptunterzug m (einer Hängedecke)
  f poutre f principale du plafond
    suspendu
 nl draagprofiel n (van verlaagd plafond)
  r главный несущий прогон m
    подвесного потолка
    (поддерживаемый подвесками)

M34 e mains distribution box
  d Hauptverteilkasten m
  f boîte f de distribution des lignes
    électriques principales, coffret m de
    branchement [de connexion] des
    câbles principaux
 nl hoofdverdeelkast f (m)

  r магистральная разветвительная
    коробка f
M35 e mains failure
  d Netzausfall m
  f panne f de courant
 nl netstoring f
  r нарушение n электроснабжения.
    выпадение n сети

M36 e mains pressure differential
  d Netzdruckunterschied m
  f différence f de pressions dans les
    artères d'alimentation et de retour
    (de chauffage urbain)
 nl aanvoer- en retourleiding f
    (wijkverwarming)
  r разность f давлений в подающей
    и обратной магистралях теплосети

M37 mains subway see service duct
M38 e mains supply
  d Netzstromversorgung f,
    Netzanschluß m
  f alimentation f secteur [par ligne
    électrique]
 nl netstroomvoorziening f, netaansluiting
    f
  r электроснабжение n от сети, сетевое
    питание n

M39 e mains water
  d Leitungswasser n, Versorgungswasser
    n
  f eau f de la ville [de robinet]
 nl leidingwater n
  r водопроводная вода f

M40 e maintenance
  d Unterhaltung f
  f entretien m, maintien m,
    maintenance f
 nl onderhoud n
  r техническое обслуживание n, уход m

M41 e maintenance hangar
  d Werft f (auf einem Luftfahrtgelände)
  f hangar m de montage [d'entretien]
 nl onderhoudshangar m
  r ангар-мастерские m

M42 e maintenance patching materials
  d Straßenbaumaterialien n pl für
    Schlaglochbeseitigung [für
    Flickarbeiten]
  f matériaux m routier pour point à
    temps
 nl materialen n pl voor wegreparaties
  r материалы m pl для ямочного
    ремонта (асфальтобетонных
    дорожных покрытий)

M43 e main tie
  d Dachbinderbalken m
  f entrait m principal (d'une ferme de
    comble)
 nl trekbalk m, trekplaat f (m)
  r стропильная затяжка f (на уровне
    опор)

M44 e main truss
  d Hauptträger m

f ferme *f* principale [maîtresse],
poutre *f* à treillis principale
nl hoofdligger *m*
r главная ферма *f*

M45 *e* **main valve**
d Hauptschieber *m*
f vanne *f* principale
nl hoofdafsluiter *m*, hoofdkraan *f (m)*
r главная задвижка *f (водоснабжение)*

M46 **maisonette** *see* **duplex 1.**

M47 *e* **major axes of stress ellipse**
d Hauptachsen *f pl* der Spannungsellipse
f axes *m pl* principaux d'ellipse des contraintes
nl hoofdassen *f pl* van de spanningsellips
r главные оси *f pl* эллипса напряжений

M48 *e* **make-up air**
d Zusatzluft *f*
f air *m* additionnel [supplémentaire]
nl toegevoerde lucht *f (m)*
r компенсационный воздух *m (приточный воздух, возмещающий вытяжку из помещения)*

M49 *e* **make up pump**
d Füllpumpe *f*
f pompe *f* d'alimentation
nl voedingspomp *f (m)*
r подпиточный насос *m*

M50 **make-up reservoir** *see* **equalizing reservoir**

M51 *e* **make-up water**
d Zusatzwasser *n*, Auffüllwasser *n*
f eau *f* d'appoint [de compensation]
nl voedingswater *n*
r подпиточная вода *f*

M52 *e* **male**
d Innenteil *m (Steck- oder Gewindeverbindung)*
f partie *f* [pièce *f*] mâle *(d'un raccordement)*
nl mannelijk deel *n* van een steekverbinding, vaareinde *n*
r охватываемая часть *f* соединения *(напр. раструбного или резьбового)*

M53 **mall** *see* **maul**

M54 *e* **mallet**
d Schlägel *m*, Schlegel *m*, Holzhammer *m*; Gummihammer *m*
f maillet *m* de charpentier; maillet *m* en bois *ou* en caoutchouc
nl sleg *f*, houten hamer *m*
r киянка *f*; деревянный *или* резиновый молоток *m*

M55 **mammoth pump** *see* **air lift**

M56 *e* **mandrel**
d Rohrkern *m*; Dorn *m*; Biegedorn *m*
f mandrin *m*
nl pijpopruimer *m*; doorn *m*
r сердечник *m (тонкостенной обсадной трубы или трубчатой сваи);* оправка *f*

M57 *e* **manhole**
d 1. Mannloch *n*, Einsteigloch *n* 2. Einsteigschacht *m*, Kontrollschacht *m*, Revisionsschacht *m*; Bodenschacht *m*
f 1. trou *m* d'homme 2. puits *m* de visite [de surveillance, d'accès]
nl 1. mangat *n* 2. controleschacht *f (m)*
r 1. люк-лаз *m* 2. смотровой колодец *m*

M58 *e* **manhole cover**
d Schachtdeckel *m*, Schachtabdeckung *f*, Mannlochdeckel *m*
f couvercle *m* [plaque *f*] de regard [de visite]
nl deksel *n* van de controleschacht
r крышка *f* смотрового колодца

M59 *e* **manhole rings**
d Brunnenringe *m pl*, Schachtringe *m pl*
f anneaux *m pl* en béton pour puits
nl putringen *m pl*
r бетонные кольца *n pl* для колодцев

M60 *e* **manhole step irons**
d Mannlochsteigeisen *n pl*
f barreaux *m* d'échelle; échelons *m pl* en fer *(scellés dans la paroi d'un puits de visite)*
nl klimbeugels *m pl (in een schacht)*
r ходовые скобы *f pl (для спуска в смотровой колодец)*

M61 *e* **manifold**
d Sammler *m*, Sammelrohr *n*, Sammelleitung *f*, Kollektor *m*, Verteiler *m*
f collecteur *m*, distributeur *m* d'admission
nl verzamelpijp *f (m)*, spruitstuk *n*
r коллектор *m*, гребёнка *f*

M62 *e* **Manilla rope**
d Manila(-Hanf)seil *n*
f câble *m* en chanvre de Manille
nl manillatouw *n*
r канат *m* из манильской пеньки, манильский канат *m*

M63 *e* **man-lock**
d Personen(-Druckluft)schleuse *f*
f sas *m* de personnel
nl luchtsluis *f (m) (in caisson)*
r шлюз *m* для людей *(в кессоне)*

M64 **man-made construction materials** *see* **manufactured construction materials**

M65 *e* **man-made stone**
d künstlicher Stein *m*
f pierre *f* artificielle
nl kunststeen *m*
r искусственный камень *m*

M66 **manometric pressure** *see* **gauge pressure**

M67 *e* **mansard roof**
d Mansardendach *n*, gebrochenes Dach *n*
f comble *m* [toit *m*] en mansarde, comble *m* brisé
nl mansardedak *n*, gebroken dak *n*
r мансардная крыша *f*

M68 *e* mantelpiece, mantelshelf
d Kaminsims *m, n*
f manteau *m* de cheminée; linteau *m* de cheminée
nl schoorsteenmantel *m*
r обрамление *n* [облицовка *f*] камина; каминная полочка *f*

M69 *e* manual batcher
d Handdosierer *m*, Handdosiereinrichtung *f*
f doseur *m* manuel, dispositif *m* de dosage manuel
nl handdoseerapparaat *n*
r ручной дозатор *m*, дозатор *m* ручного действия

M70 *e* manual cycling control
d Handsteuerung *f* der Taktabläufe *bzw.* der zyklischen Vorgänge
f commande *f* manuelle des opérations cycliques
nl regeling *f* met de hand van cyclische processen
r ручное управление *n* цикличными операциями

M71 *e* manual damper
d Handluftklappe *f*
f registre *m* réglé à main
nl handbediende trekregelaar *m*
r ручной воздушный клапан *m*

M72 *e* manually propelled mobile scaffold
d handgesteuertes Fahrgerüst *n*
f échaffaudage *m* roulant (*déplacé à la main*)
nl met handkracht verplaatsbare stelling *f*
r катучие подмости *pl*

M73 *e* manual proportioning control
d handgesteuerte mechanische Dosierung *f*
f dosage *m* mécanique à commande manuelle
nl met de hand geregelde mechanische dosering *f*
r механическое дозирование *n* с ручным управлением

M74 *e* manufactured construction materials
d Kunstbaustoffe *m pl*, Industriebaustoffe *m pl*
f matériaux *m pl* de construction artificiels
nl kunstmatig vervaardigde bouwmaterialen *n pl*
r искусственные строительные материалы *m pl*

M75 *e* manufactured sand
d Brechsand *m*, Steinsand *m*
f sable *m* artificiel [concassé, de concassage]
nl breukzand *n* (*fijne steenslag*)
r искусственный песок *m* (*получаемый дроблением камня, гравия или шлака*)

M76 *e* map cracking *US*
d Netzrißbildung *f*, Maronage *f*

f fissures *f pl* en réseau
nl polygoon-krimpscheuren *f pl*
r сеть *f* усадочных трещин

M77 *e* marbling
d Marmorierung *f*
f marbrure *f*
nl marmeren *n* (*schilderwerk*)
r имитация *f* [окраска *f*] под мрамор

M78 *e* margin
d 1. Toleranz *f*, Spielraum *m*, zulässige Grenze *f* 2. Randstreifen *m* 3. Rand *m*, Saum *m* 4. Einfassung *f*, Umrandung *f*
f 1. écart *m*, tolérance *f* 2. bande *f* latérale 3. marge *f* 4. encadrement *m* (*p. ex d'un panneau*)
nl 1. tolerantie *f* 2. onverharde berm *m* 3. zoom *m* 4. omranding *f*
r 1. допуск *m*, допустимый предел *m* 2. краевая полоса *f* (*дороги*) 3. край *m*, кайма *f* 4. обрамление *n* (*панели*)

M79 *e* marginal concrete strip finisher
d Randstreifenfertiger *m*
f finisseur *m* [finisseuse *f*] de bandes latérales (*de chausée*)
nl betonverwerkingsmachine *f* voor de kantstrook
r бетоноотделочная машина *f* [финишер *m*] для краевой полосы проезжей части

M80 *e* marginal strip
d 1. Randstreifen *m* 2. Friesbrett *n*
f 1. bordure *f* de chaussée 2. planche *f* d'encadrement (*du sol en parquet*)
nl 1. kantstrook *f* (*m*) 2. rand *m* (*parketvloer*)
r 1. кромка *f* (*проезжей части дороги*) 2. фризовая доска *f* (*пола*)

M81 *e* margin of safety
d Sicherheitsfaktor *m*, Sicherheitszuschlag *m*, Sicherheitstoleranz *f*, Gestaltfestigkeit *f*
f marge *f* [coefficient *m*] de sécurité
nl veiligheidsmarge *f* (*m*)
r коэффициент *m* безопасности, надёжность *f*, запас надёжности, запас *m* прочности

M82 *e* margin of stability
d Stabilitätsreserve *f*
f marge *f* [facteur *m*] de stabilité
nl stabiliteitsreserve *f* (*m*)
r запас *m* устойчивости

M83 *e* marine construction
d Seebau *m*
f construction *f* maritime, travaux *m pl* maritimes
nl maritieme bouw *m*
r морское строительство *n*

M84 *e* marine dredger
d Seebagger *m*
f drague *f* marine [de haute mer]
nl zeebaggermolen *m*
r морской земснаряд *m*

M85 *e* marine glue
  d Marineleim *m*
  f colle *f* marine
  nl watervaste lijm *m*
  r водостойкий клей *m* (*на смолах и каучуке*)

M86 *e* marine paint
  d Schiffs(anstrich)farbe *f*
  f painture *f* marine
  nl scheepsverf *f (m)*
  r краска *f*, стойкая против воздействия морской воды и солнца

M87 *e* marine structures, marine works
  d Seebauten *pl*, Meeresbauten *pl*
  f ouvrages *m pl* maritimes
  nl maritieme werken *n pl*
  r морские сооружения *n pl*

M88 *e* maritime canal *see* seaway canal

M89 *e* marked face
  d Sichtfläche *f* (*Schnittholz*)
  f face *f* rabotée de planches
  nl geschaafde zijde *f (m)* (*plank*)
  r лицевая сторона *f* (*пиломатериала*)

M90 *e* marker post
  d Leitpfosten *m*
  f borne *f* de repérage; borne *f* de jalonnement
  nl bermreflector *m*
  r грунтовый репер *m*, реперный столб *m*; межевой столб *m*

M91 *e* marking gauge
  d Streichmaß *n*
  f trusquin *m*
  nl kruishout *n*, ritshout *n*
  r рейсмус *m*

M92 *e* marl
  d Mergel *m*
  f marne *f*
  nl mergel *m*
  r мергель *m*

M93 *e* marquee
  d Vordach *n*, Schrägdach *n* (*über Tür*)
  f marquise *f*; abat-vent *m*
  nl luifel *f (m)* (*boven ingang*)
  r навес *m* (*над входом в здание, над окном*)

M94 *e* marquetry
  d Holzeinlegearbeit *f*
  f marquetrie *f*
  nl inlegwerk *n*; intarsia *f*
  r инкрустация *f или* мозаика *f* по дереву

M95 *e* marsh
  d Marsch *f*; Moor *n*, Sumpf *m*, Morast *m*
  f marais *m*, marécage *m*
  nl moerasland *n*, moeras *n*
  r марши *m pl*; болото *n*

M96 *e* marshy ground
  d Bruchmoor *n*, Fenn *n*, Fehn *n*
  f sol *m* [terrain *m*] marécageux
  nl moerasgrond *m*
  r топь *f*, торфяник *m*, торфяное болото *n*

M97 *e* mash hammer
  d Vorschlaghammer *m*, Steinmetzhammer *m*
  f massete *f*
  nl moker *m*
  r кувалда *f* для ударной обработки камня, камнетёсная киянка *f*

M98 *e* mashroom valve
  d Ventilschütz *n*
  f vanne *f* conique
  nl kegelafsluiter *m*
  r конусный затвор *m*

M99 *e* masonry
  d Mauerwerk *n*
  f maçonnerie *f*
  nl metselwerk *n*
  r каменная кладка *f*

M100 *e* masonry anchor
  d Maueranker *m*, Mauerschlauder *f*
  f patte *f* de scellement (*assurant la liaison huisserie-maçonnerie*)
  nl muuranker *n*
  r закрепа *f* (*для крепления дверной металлической коробки к кладке*)

M101 *e* masonry block
  d Wandblock *m*, Mauerstein *m*
  f bloc *m* [corps *m*] de maçonnerie, parpaing *m*
  nl blok *n* (*natuursteen*); bouwblok *n* (*beton*)
  r стеновой блок *m* [камень *m*]

M102 *e* masonry bridge
  d Mauerwerkbrücke *f*
  f pont *m* en maçonnerie
  nl stenen brug *f (m)*
  r каменный мост *m*

M103 *e* masonry cement
  d Mauermörtelzement *m*, Mauerwerkzement *m*, Putz- und Mauerbinder *m*, PM-Binder *m*
  f ciment *m* à maçonner
  nl metselcement *n*
  r цемент *m* для кладочных растворов

M104 *e* masonry drill
  d Mauerbohrer *m*, Steinbohrer *m*
  f ciseau *m* à pierre [du maçon]
  nl steenboor *m*
  r шлямбур *m*

M105 *e* masonry mortar
  d Mauermörtel *m*, Fugenmörtel *m*
  f mortier *m* de maçonnerie
  nl metselspecie *f*
  r кладочный раствор *m*

M106 *e* masonry nail
  d Mauernagel *m*
  f clou *m* à maçonnerie
  nl stalen nagel *m*
  r гвоздь *m* для забивки в кладку *или* в бетон

M107 *e* masonry plate
  d Unterlegplatte *f*, Auflagerplatte *f*
  f plaque *f* d'assise de poutre, plaque *f* mural

nl gemetselde oplegging *f* voor een balk
r подферменная плита *f*, опорная подкладка *f* под балку

M108 *e* **masonry veneer**
  d Steinverkleidung *f*, Ziegelverkleidung *f*
  f revêtement *m* de maçonnerie *ou* de briques non appareillé avec le mur
  nl bekleding *f* met metselwerk
  r прислонная каменная [кирпичная] облицовка *f* (*не связанная со стенной перевязкой*)

M109 *e* **mason's adjustable suspension scaffold**
  d selbstverstellbares Maurerhängegerüst *n*
  f échafaudage *m* de maçon volant réglable
  nl verstelbare vliegende metselsteiger *m*
  r подвесные самоподъёмные подмости *pl* для каменщика

M110 **mason's ax** *see* **axhammer**

M111 *e* **mason's hammer**
  d Maurerhammer *m*
  f marteau *m* de maçon
  nl moker *m*, vuist *f* (*m*)
  r молоток-кирочка *m* каменщика

M112 *e* **mason's scaffold**
  d Maurergerüst *n*
  f échafaud *m* [échafaudage *m*] de maçon
  nl metselsteiger *m*, dubbele stalen bouwsteiger *m*
  r леса *pl* [подмости *pl*] для каменной кладки

M113 *e* **mass concrete**
  d 1. Massenbeton *m*, Massivbeton *m* 2. Kernbeton *m*, Hauptbeton *m*
  f 1. béton *m* de masse 2. béton *m* pour le noyau de la digue
  nl 1. ongewapend beton *n* 2. kernbeton *n*
  r 1. массовый бетон *m* (*для гидросооружений*) 2. бетон *m* внутренней части плотины

M114 *e* **mass curve**
  d 1. Mengenkurve *f*, Summen(gang)linie *f*, Integralkurve *f* 2. Kurve *f* des Erdmassenausgleichs, Massenlinie *f*, Massenprofil *n*
  f 1. courbe *f* integrale [de masse, des valeurs cumulées] 2. épure *f* de terrassement [du mouvement des terres]
  nl 1. hoeveelheids(integraal)kromme *f* (*m*) 2. profiel *n* van het massanivellement
  r 1. интегральная кривая *f* (*напр. стока*) 2. кривая *f* объёмов земляных работ, диаграмма *f* перемещения земляных масс

M115 **mass diagram** *see* **discharge mass curve**

M116 *e* **mass force**
  d Massenkraft *f*
  f force *f* volumique
  nl massa *f* (*als kracht*)
  r массовая [объёмная] сила *f*

M117 *e* **mass foundation**
  d Massivfundament *n*
  f fondation *f* massive (antivibratile)
  nl fundering *f* met grote massa
  r массивный фундамент *m*

M118 **mass-haule curve** *see* **mass curve 2.**

M119 *e* **massive concrete structures**
  d Massenbetonbauten *m pl*
  f ouvrages *m pl* massifs en béton
  nl monoliet-betonconstructies *f pl*
  r массивные бетонные сооружения *n pl*

M120 *e* **massive head buttress dam**
  d Massivkopf-Staumauer *f*, Pfeilergewichtsstaumauer *f*
  f barrage *m* à contreforts à tête massive
  nl pijlerstuwdam *m* met massieve bovenrand
  r массивно-контрфорсная плотина *f*, контрфорсная плотина *f* с массивными оголовками

M121 *e* **mass-produced structural units**
  d Massenfertigungsbauelemente *n pl*
  f éléments *m pl* de construction préfabriqués mis en série
  nl in massaproductie vervaardigde bouwelementen *n pl*
  r сборные элементы *m pl* конструкций массового изготовления

M122 *e* **mass runoff**
  d Abflußsumme *f*
  f volume *m* d'eau écoulée, débit *m* total
  nl afvoer *m*, debiet *n*
  r объём *m* стока

M123 *e* **masstone**
  d natürliche Farbe *f*, Naturfarbe *f*
  f couleur *f* franche, teinte *f* mère (*d'un pigment, d'une peinture*)
  nl natuurlijke kleur *f* (*m*)
  r натуральный цвет *m* пигмента *или* краски (*без разбавления и смешения*)

M124 *e* **mast**
  d Mast *m*; Stütze *f*; Pfosten *m*
  f mât *m*; poteau *m*; pylône *m*
  nl mast *m*; paal *m*
  r мачта *f*; опора *f*; столб *m*

M125 *e* **mast arm**
  d Mastausleger *m* (*Lichtmast*)
  f bras *m* du poteau d'éclairage
  nl mastconsole *f* (*m*) (*lichtmast*)
  r консоль *f* [кронштейн *m*] осветительного столба

M126 *e* **mast cap spider**
  d Mastkopf *m* (*Trossenderrickkran*)
  f croisillon *m* de derrick
  nl mastkop *m*
  r паук *m* (*вантовых деррик-кранов*)

M127 *e* **master plan**
  d Hauptbebauungsplan *m*, Generalbebauungsplan *m*
  f plan *m* directeur (*d'aménagement*)

nl structuurplan *n* (*bouwproject*)
r генеральный план *m* застройки

M128 *e* **mastic**
d Mastix *m*; Kitt *m*
f mastic *m*
nl mastiek *m*, *n*; kit *f* (*m*)
r мастика *f*; замазка *f*

M129 *e* **mastic asphalt**
d 1. Asphaltmastix *m* 2. Gußasphalt *m*, Streichasphalt *m*, Mastixasphalt *m*
f 1. mastic *m* d'asphalte 2. asphalte *m* coulé, mortier *m* (d')asphalte
nl 1. asfaltmastiek *m*, *n* 2. gietasfalt *n*
r 1. асфальтовая [битумная] мастика *f* 2. литой асфальт *m*

M130 *e* **mastic compound**
d 1. Latex *m*, Latexmischung *f* 2. Isoliermasse *f*
f 1. latex *m*, mélange *m* de latex 2. pâte *f* isolante, compound *m* [composé *m*] isolant
nl 1. latex *m*, *n* 2. isolatiemastiek *m*, *n*
r 1. латекс *m*, латексная смесь *f* 2. изоляционная мастика *f*

M131 *e* **mastic cooker**
d Gußasphaltkocher *m*, Asphalt-Schmelzkessel *m*
f fondoir-citerne *m* à bitume
nl mastiekketel *m*
r битумный котёл-цистерна *m*

M132 *e* **mast with arm** *US*
d Auslegerbeleuchtungsmast *m*
f poteau *m* d'éclairage à bras [à console]
nl lichtmast *m* met console
r осветительная мачта *f* с консолью

M133 *e* **mat**
d 1. Wärmedämmatte *f* 2. Filz *m*, Faserfilz *m* 3. Bewehrungsmatte *f* 4. Teppichbelag *m*, Verschleißschicht *f* (*Straßenbau*) 5. Sauberkeitsschicht *f*, Unterbeton *m* 6. Packwerk *n*, Senkstück *n*, Faschinenfloß *n*, Buschmatratze *f* 7. Rostwerk *n* 8. Einzelfundament *n*
f 1. tapis *m* isolant 2. feutre *m* 3. treillis *m* d'armature 4. couche *f* d'usure [de roulement] 5. chape *f* de béton 6. natte *f* de fascines 7. radier *m* 8. fondation *f* détachée [isolé]
nl 1. mat *f* (*m*) 2. vilt *n* 3. wapeningsnet *n* 4. deklaag *f* (*m*) 5. werkvloer *m* 6. matras *f* (*m*) (*zinkstuk*) 7. roosterwerk *n* 8. afzonderlijke fundering *f*
r 1. теплоизоляционный мат *m* [тюфяк *m*] 2. войлок *m* 3. арматурная сетка *f* 4. слой *m* износа, слой *m* поверхностной обработки (*дорожного покрытия*) 5. бетонная подготовка *f* 6. фашинный тюфяк *m* 7. ростверк *m* 8. отдельный фундамент *m*

M134 *e* **matchboard**
d Profilbrett *n*

f planche *f* profilé [bouvetée, embrevée]
nl plank *f* (*m*) met messing en groef
r шпунтовая доска *f*

M135 *e* **matchboarding**
d Profilbretterbeschlag *m*
f lambris *m*, lambrissage *m*
nl vloer *m* van geschaafde en geploegde delen
r декоративная дощатая обшивка *f*

M136 *e* **matched siding**
d Horizontalschalung *f* mit gespundeten *oder* Falzverbindungen
f bardage *m* en planches horizontales avec joints en feuillure
nl aaneenpassende zijbekisting *f*
r горизонтальная дощатая обшивка *f* в четверть *или* в шпунт

M137 *e* **material agressive to concrete**
d Betonschädling *m*
f matière *f* nocive pour le béton
nl schade toebrengend materiaal *n* (*beton*)
r материал *m*, вызывающий коррозию бетона

M138 *e* **material behavior**
d Materialverhalten *n*
f comportement *m* du matériau
nl materiaalgedrag *n*
r поведение *n* [работа *f*] материала

M139 *e* **material debris chute**
d Bauschuttrutsche *f*
f gaine *f* d'évacuation de débris, gaine *f* à débris
nl puinstortgoot *f* (*m*)
r короб *m* для спуска строительного мусора

M141 *e* **material handling**
d Fördern *n* und Heben *n*; Materialförderung *f*, Materialumschlag *m*
f travaux *m pl* de manutention; manutention *f* de matériaux
nl materiaaltransport en -opslag
r подъёмно-транспортные работы *f pl*; транспортирование *n* материалов

M142 *e* **material handling bridge**
d Verladebrücke *f*, Förderbrücke *f*
f pont *m* de manutention
nl laadbrug *f* (*m*)
r мост-перегружатель *m*

M143 *e* **material handling system**
d Transportsystem *n*
f système *m* de manutention [de transport] de matériaux
nl bedrijfstransportsysteem *n*
r система *f* подачи [доставки] материалов

M144 *e* **material hoist**
d Last(en)aufzug *m*
f monte-matériaux *m*, monte-charge *m*
nl goederenhijsinrichting *f*
r грузовой подъёмник *m*

M145 *e* **material hose**
  d Förderschlauch *m*
  f tuyau *m* souple à matériaux liquéfiés
  nl transportslang *f* (*m*)
  r шланг *m* для подачи разжиженных материалов

M146 *e* **material platform hoist**
  d Plattform(bau)aufzug *m*
  f monte-charge *m* à tour [à mât]
  nl bouwlift *m*, goederenlift *m*
  r башенный [мачтовый] подъёмник *m*

M147 *e* **material retained on sieve**
  d Siebrückstand *m*
  f refus *m* au tamis
  nl zeefrest *m*
  r остаток *m* на сите, надрешётный продукт *m*

M148 *e* **materials**
  d 1. Materialien *n pl*, Werkstoffe *m pl*
     2. Böden *m pl*
  f 1. matériaux *m pl* 2. sols *m pl*, terres *f pl*
  nl (bouw)materialen *n pl*
  r 1. материалы *m pl* 2. грунты *m pl*

M149 *e* **materials cage**
  d Hebebühne *f*
  f plateau *m* de monte-charge
  nl liftkooi *f* (*m*) (*bouwlift*)
  r платформа *f* мачтового *или* рамного подъёмника

M150 *e* **materials lock**
  d Materialschleuse *f*
  f sas *m* de matériaux [pour les matériaux]
  nl materiaalsluis *f* (*m*) (*caisson*)
  r грузовой шлюз *m* (*кессона*)

M151 *e* **materials testing laboratory**
  d Materialprüfungsanstalt *f*
  f laboratoire *m* d'essais des matériaux
  nl laboratorium *n* voor materiaalonderzoek
  r лаборатория *f* по испытанию материалов

M152 *e* **mat foundation**
  d Plattenfundament *n*, Plattengründung *f*
  f fondation-dalle *f*
  nl plaatfundering *f*
  r плитный фундамент *m*

M153 *e* **mat reinforcement bender**
  d Biegemaschine *f* für Bewehrungsmatten
  f cintreuse *f* [coudeuse *f*, plieuse *f*] pour treillis d'armature
  nl buigmachine *f* (*wapeningsstaven*)
  r машина *f* для гнутья арматурных сеток

M154 *e* **matrix**
  d Bindemittel *n*; Zementstein *m*
  f pâte *f* de ciment; pâte *f* de ciment durci
  nl bindmiddel *n* (*mortel*)
  r растворная часть *f* (*бетона*);

цементное тесто *n* (*строительного раствора*); цементный камень *m* (*бетона, раствора*)

M155 *e* **matted glass, matte-surface glass**
  d Mattglas *n*
  f verre *m* mat
  nl matglas *n*
  r матовое стекло *n*

M156 *e* **mattock**
  d Breithaue *f*, Platthacke *f*
  f pic *m* à tranche, pioche *f*, pioche-hache *f*
  nl hak *f* (*m*)
  r кирка-мотыга *f*

M157 *e* **matt paint**
  d Matt(anstrich)farbe *f*
  f peinture *f* mate
  nl matverf *f* (*m*)
  r матовая краска *f*

M158 *e* **mattress**
  d 1. Betongrundplatte *f*
     2. Faschinensenkstück *n*, Faschinenlage *f*, Faschinenmatratze *f*, Faschinenfloß *n* 3. Isoliermatte *f*, Wärmedämmatte *f*
  f 1. fondation-dalle *f* pour machine
     2. natte *f* [matelas *m*] en fascines
     3. matelas *m* isolant [d'isolation thermique]
  nl 1. betonfundatieplaat *f* (*m*)
     2. rijszinkstuk *n* 3. isolatiedeken *f* (*m*)
  r 1. плитный бетонный фундамент *m*
     2. фашинный тюфяк *m* 3. теплоизоляционный тюфяк *m* [мат *m*]

M159 *e* **mattress revetment**
  d Packwerk *n*
  f revêtement *m* en fascines, fascinage *m*
  nl rijspakwerk *n*
  r тюфячная одежда *f* (*откоса*)

M160 *e* **matured cement**
  d abgelagerter Zement *m*
  f ciment *m* stocké
  nl langdurig bewaard cement *n*
  r лежалый цемент *m*

M161 *e* **maturing**
  d Alterung *f*; Reifen *n*
  f maturation *f*
  nl rijpen *n*
  r старение *n*; твердение *n* (*напр. раствора*); вызревание *n* (*бетона*)

M162 *e* **maturing of concrete**
  d 1. Betonfestigkeitszunahme *f*, Ansteigen *n* der Betonfestigkeit
     2. Nachbehandlung *f* des Betons
  f 1. maturation *f* du béton 2. cure *f*, curing *m*, traitement *m* après prise (*de béton*)
  nl 1. verharding *f* van beton
     2. nabehandeling *f* (*beton*)
  r 1. набор *m* бетоном прочности, вызревание *n* бетона 2. выдерживание *n* бетона; уход *m* за бетоном

M163 *e* **maul**
d 1. schwerer Holzhammer *m*
2. Handstampfer *m*
f 1. maillet *m* 2. dame *f* (en bois),
demoiselle *f*
nl 1. (zware) houten hamer *m*
2.handstamper *m*
r 1. большая киянка *f*
2. деревянная ручная трамбовка *f*

M164 *e* **maximum allowable concentration**
d maximale Arbeitsplatz-
-Konzentration *f*, MAK-Wert *m*
f concentration *f* maximale admissible
nl hoogst toelaatbare concentratie *f*
r предельно допустимая концентрация
f (*напр. примеси в воздухе
рабочей зоны*), ПДК

M165 *e* **maximum allowable emission**
d Emissionsgrenzwert *m*, maximal
zulässige Emission (s-Konzen-
tration) *f*, MEK-Wert *m*
f émission *f* maximale admissible,
dégagement *m* maximal admissible
nl hoogst toelaatbare uitstoot *m*
r предельно допустимый выброс *m*

M166 *e* **maximum capacity of well**
d Brunnenhöchstergiebigkeit *f*,
maximale Brunnenergiebigkeit *f*
f débit *m* maximum d'un puits,
capacité *f* d'un puits
nl hoogste opbrengst *f* van de bron
r максимальный дебит *m* колодца
[скважины]

M167 *e* **maximum density of soil**
d maximale Bodendichte *f*
f densité *f* maximale du sol compacté
nl grootst mogelijke dichtheid *f* van
grond
r максимальная плотность *f*
сложения уплотнённого грунта

M168 *e* **maximum dry density**
d Größtwert *m* der Trockenrohdichte
f densité *f* sèche maximale
nl grootst mogelijke dichtheid *f* in droge
toestand
r максимальная плотность *f*
сложения грунта в сухом состоянии

M169 *e* **maximum flood discharge**
d höchster Hochwasserdurchfluß *m*
f débit *m* de crue maximal
nl grootste afvoer *m*
r максимальный паводковый расход *m*

M170 *e* **maximum gradient**
d Höchstlängsneigung *f*,
Höchstgefälle *n*
f inclinaison *f* [pente *f*] maximale
nl maximale langshelling *f*, maximaal
verval *m*
r максимальный (продольный) уклон
m

M171 **maximum load design** *see* **load factor
design**

M172 *e* **maximum rated load**
d Höchstnennlast *f*

f charge *f* prévue maximale
nl volle belasting *f*
r максимальная номинальная
[нормативная] нагрузка *f*

M173 *e* **maximum safe load**
d zulässige Grenzlast *f*,
höchstzulässige Belastung *f*
f charge *f* de sécurite
nl hoogst toelaatbare belasting *f*
r предельная [максимально
допустимая] нагрузка *f*

M174 *e* **maximum simultaneous demand**
d Spitzenwärmebedarf *m*
f besoin *m* de chaleur maximal
nl piekbelasting *f*
r пиковая тепловая нагрузка *f*
(*системы*)

M175 *e* **maximum size of aggregate**
d Größtkorn(maß) *n*
f grosseur *m* maximal de grains de
granulat
nl maximale korrelgrootte *f*
r максимальный размер *m*
[максимальная крупность *f*] зёрен

M176 *e* **meager lime**
d Magerkalk *m*
f chaux *f* maigre
nl magere kalk *m*
r тощая известь *f*

M178 *e* **mean cycle stress**
d Mittelspannung *f*
(*Wechselbeanspruchung*)
f contrainte *f* moyenne du cycle
nl gemiddelde spanning *f* bij wisselende
belasting
r среднее напряжение *n* цикла

M179 *e* **mean depth**
d 1. mittlere Querschnittstiefe *f*
2. hydraulischer Radius *m*,
Profilradius *m*
f 1. profondeur *f* moyenne 2. rayon *m*
hydraulique
nl gemiddelde diepte *f* (van het nat
profiel)
r 1. средняя глубина *f* живого сечения
2. гидравлический радиус *m*

M180 *e* **mean radiant temperature**
d mittlere Strahlungstemperatur *f*
f température *f* moyenne radiante
nl gemiddelde stralingstemperatuur *f*
r средняя радиационная температура
f

M181 *e* **means of conveyance**
d Transportmittel *n pl*
f moyens *m pl* de transport
nl transportmiddelen *n pl*
r транспортные средства *n pl*

M182 *e* **means of egress**
d Fluchtwege *m pl*
f moyens *m pl* de sortie
nl vluchtweg *m*
r пути *m pl* выхода [эвакуации]
(*из здания*)

M183 *e* **means of slinging**
 d Anschlagmittel *n pl*
 f accessoires *m pl* de levage
 nl hulpmiddelen *n pl* voor het aanslaan
  van een last
 r такелажные приспособления *n pl*

M184 **means of transportation** *see* **means of conveyance**

M186 *e* **mean temperature difference**
 d mittlere Temperaturdifferenz *f*
 f différence *f* moyenne de températures
 nl gemiddeld temperatuurverschil *n*
 r средняя разность *f* температур

M187 *e* **mean velocity of flow**
 d mittlere Querschnittsgeschwindigkeit
  f [Strömungsgeschwindigkeit *f*]
 f vitesse *f* moyenne d'un cours d'eau
 nl gemiddelde stroomsnelheid *f*
 r средняя скорость *f* потока

M188 *e* **measured drawing**
 d Vermessungszeichnung *f*
 f dessin *m* exécuté d'après le métré
  (*du bâtiment bâti*)
 nl revisietekening *f* (na opmeting)
 r натурный чертёж *m*
  (*существующего здания*)

M189 *e* **measuring chain**
 d Meßkette *f*
 f chaîne *f* d'arpentage
 nl meetketting *f*
 r мерная цепь *f* (*для геодезических работ*)

M190 *e* **measuring flume**
 d Kanalmesser *m*, Durchflußmeßgerinne *n*
 f canal *m* jaugeur
 nl meetgoot *f* (*m*)
 r гидрометрический лоток *m*

M191 **measuring frame** *see* **batch box**

M192 *e* **measuring section**
 d Meßquerschnitt *m*, Abflußmeßstelle *f*
 f section *f* de jaugeage
 nl meetprofiel *n*
 r гидрометрический створ *m*

M193 *e* **measuring tape**
 d Bandmaß *n*, Meßband *n*
 f ruban *m* à mesurer
 nl meetband *m*, meetlint *n*
 r рулетка *f*, измерительная лента *f*

M194 *e* **measuring weir**
 d Meßwehr *n*, Meßüberfall *m*,
  hydrometrischer Überlauf *m*
 f déversoir *m* de mesure
 nl meetstuw *m*
 r измерительный [гидрометрический]
  водослив *m*

M195 *e* **measuring worm conveyor**
 d Dosier(ungs)schnecke *f*,
  Zuteilschnecke *f*, Zumeßschnecke *f*
 f convoyeur *m* à vis doseur
 nl doserende wormtransporteur *m*
 r шнековый дозатор *m*

M196 *e* **mechanical aeration**
 d mechanische Belüftung *f* (*Abwasser*)
 f aération *f* [aérage *m*] mécanique
  (*des eaux usées*)
 nl beluchting *f* met roerwerk (*afvalwater*)
 r механическая аэрация *f* (*сточных вод*)

M197 *e* **mechanical analysis**
 d kombinierte Sieb- und
  Schlämmanalyse *f*
 f analyse *f* granulométrique par
  tamisage et sédimentation
 nl zeef- en slibanalyse *n*
 r комбинированный ситовый и
  седиментационный анализ *m*
  гранулометрического состава
  (*грунта, заполнителей*)

M198 *e* **mechanical application**
 d maschinelles Aufbringen *n* (*Farbe, Putz*)
 f application *f* (*d'une peinture, d'un enduit*) mécanique
 nl machinaal opbrengen *n* (*verf, pleisterwerk*)
 r нанесение *n* (*краски, штукатурки*)
  механическим способом

M199 *e* **mechanical area**
 d Installationsraum *m*
 f locaux *m pl* (*dans un bâtiment*) pour
  équipement mécanique et installations
  sanitaires
 nl ruimte *f* voor technische installaties
 r зона *f* размещения инженерного
  оборудования (*в здании*)

M200 *e* **mechanical bond**
 d mechanische Haftung *f*, mechanischer
  Verbund *m*
 f adhérance *f* [cohésion *f*] mécanique
 nl mechanische hechting *f*
 r механическое сцепление *n*

M201 *e* **mechanical classification**
 d mechanische Korntrennung *f*
 f classement *m* granulométrique (*des matériaux granulaires*)
 nl mechanische korrelsscheiding *f*
  [classificatie *f*]
 r механическая классификация *f*
  (*зернистого материала*)

M202 *e* **mechanical cooling**
 d mechanische [maschinelle]
  Luftkühlung *f*
 f réfrigération *f* mécanique (*d'air*)
 nl machinale luchtkoeling *f*
 r машинное охлаждение *n* (воздуха)

M203 *e* **mechanical core**
 d 1. Installationskern *m*, Betriebskern *m*
  (*eines Gebäudes*)
  2. Sanitärkern *m*, Installationskern *m*
  (*einer Wohnungseinheit*)
  3. vorgefertigtes
  [Rohrbündel *n* (*ggf. mit
  Anschlußleitungen*) 4.
  Installationselement *n*,
  Montagegruppe *f*
  (*Elektroinstallationen*)

1. noyau *m* technique [de service]
(*d'un bâtiment*) 2. bloc-eau *m*, bloc *m*
sanitaire 3. paquet *m* de tuyaux
[de tubes], faisceau *m* tubulaire
[de tuyaux] 4. tronçon *m* préfabriqué
d'une colonne montante (*d'une
installation électrique intérieure*]
nl 1. installatieschacht *f* (*m*) 2. natte-
cel *f* (*m*) 3. geprefabriceerde leiding-
bundel *m* 4. montagegroep *f*
(*elektrische installaties*)
r 1. техническое ядро *n* здания
2. санитарно-технический узел *m*,
санузел *m* 3. сборный
узел *m* трубопроводов; пакет *m*
труб 4. сборный блок *m*
электрооборудования зданий

M204 *e* **mechanical core wall**
d Betriebskernwand *f*,
Installationswand *f*
f panneau *m* mural à tubes
en castrés
nl wandelement *n* met leidingen
r санитарно-техническая панель *f*

M205 *e* **mechanical coupling link**
d Verbindungsglied *n*, Schäkel *m*
(*Kette, Stropp*)
f maille *f* [maillon *m*] de
raccordement [d'attache]
nl verbindingsschakel *m*, *f* (*ketting*)
r соединительная скоба *f* (*цепи,
стропа*)

M206 **mechanical draft cooling tower** *see*
fan-type cooling tower

M207 *e* **mechanical dust collector**
d mechanischer Staubabscheider *m*
f dépoussiéreur *m* mécanique
nl mechanische stofscheider *m*
[stofafscheider *m*]
r механический пылеуловитель *m*

M208 *e* **mechanical filter**
d mechanisches Filter *n*, *m*
f filtre *m* mécanique
nl mechanisch filter *n*, *m*
r механический фильтр *m*

M209 *e* **mechanical operator**
d mechanischer Antrieb *m* für Öffnen
und Schließen der Fenster
f appareil *m* d'ouverture et de
fermeture (*d'une fenêtre*)
nl aandrijving *f* voor het mechanisch
openen en sluiten van een venster
r механический открыватель *m*,
механизм *m* открывания и
закрывания (окон)

M210 *e* **mechanical plant**
d mechanische Ausrüstung *f* der
Baustelle, Baugeräte *n pl*
f équipement *m* des chantiers
nl bouwmachines *f pl*
r механизированное оборудование *n*
стройплощадок, строительные
машины *f pl*

M211 **mechanical purification of sewage** *see*
mechanical treatment of sewage

M212 *e* **mechanical rake**
d Rechenreiniger *m*; Schlammkratzer
m
f dégrilleur *m*; appareil *m* racleur
tournant (*pour les boues*)
nl mechanische (slib)krabber *m*
r механический скребок *m* (для
очистки решётки); илоскрёб *m*

M213 *e* **mechanical rammer**
d Fallramme *f*
f dame *f* [dameur *m*, dameuse *f*]
mécanique
nl mechanische stamper *m*
r механическая трамбовка *f*,
механический уплотнитель *m*;
трамбующая машина *f*

M214 *e* **mechanical refrigerating system**
d Kälteanlage *f* mit Maschinenkühlung
f système *m* de réfrigération mécanique
nl mechanisch koelsysteem *n*
r система *f* машинного охлаждения

M215 **mechanical refrigeration** *see* **chilling**

M216 *e* **mechanical shear connectors**
d mechanische Schubverbinder *m pl*
f connecteurs *m pl* de liaison fer-béton
(*organes d'ancrage reliant poutres en
acier avec dalles de béton*)
nl mechanische schuifverbinders *m pl*
(*koppeling van vloer aan balken bij
gewapend-betonconstructies*)
r детали *f pl*, обеспечивающие
совместную работу стальных балок
и железобетонных плит
(*в сталежелезобетонных
конструкциях*)

M217 *e* **mechanical strength characteristics**
d mechanische Festigkeitseigenschaften
f pl
f caractéristiques *f pl* [propriétés *f pl*]
(mécaniques) de résistance
nl mechanische sterkte-eigenschappen *f pl*
r прочностные свойства *n pl*

M218 *e* **mechanical testing**
d mechanische Prüfung *f*
f essais *m pl* [épreuves *f pl*] mécaniques
nl mechanische beproeving *f*
r механические испытания *n pl*,
испытание *n* механических свойств

M219 *e* **mechanical tooling of concrete**
d mechanische Betonbearbeitung *f*
f traitement *m* [façonnage *m*] mécanique
du béton
nl mechanische betonbewerking *f*
r механическая обработка *f* бетона

M220 *e* **mechanical treatment of sewage**
d mechanische Abwasserklärung *f*
[Abwasserreinigung *f*]
f traitement *m* mécanique de l'effluent
nl mechanische afvalwaterzuivering *f*
r механическая очистка *f* сточных
вод

M221 e **mechanical trowel**
   d Estrichglättmaschine f
   f lisseuse f [lisseur m] mécanique
  nl afstrijkingsmachine f (betonvloer)
   r затирочная лопастная машина f
     (для бетона)

M222 e **mechanical ventilation**
   d mechanische [erzwungene] Lüftung f,
     Zwangslüftung f
   f ventilation f mécanique
  nl mechanische ventilatie f
   r механическая [принудительная]
     вентиляция f

M223 e **median**
   d (nichtbefahrbarer) Mittelstreifen m
     (Straße)
   f terre-plein m axial [central], bande f
     médiane
  nl middenberm m
   r разделительная полоса f (дороги)

M224 e **median barrier**
   d Mittelstreifen-Schutzbarriere f
   f barrière f de protection [de sécurité]
     sur la bande médiane
  nl middenbermbeveiliging f
   r защитное ограждение n на
     разделительной дорожной полосе

M225 e **medical lock**
   d Krankenschleuse f
   f sas m médical de recompression
  nl luchtsluis f (m) in recompressieruimte
   r медицинская воздушная шлюзовая
     камера f

M226 e **medium-duty scaffold**
   d mittelschweres Gerüst n
   f échafaudage m d'une portance moyenne
  nl middelzware steiger m
   r леса pl [подмости pl] средней
     несущей способности

M228 e **meeting rail**
   d Fensterkreuz n
   f traverse f d'imposte
  nl wisseldorpel m (schuifraam)
   r средник m оконного переплёта

M229 e **meeting stile**
   d Schlagleiste f (Fenster, Tür)
   f montant m de battement
  nl sluitstijl m (venster, deur)
   r притворный брус m (окна, двойной
     двери)

M230   melioration see **land reclamation**

M231 e **member**
   d 1. Bauglied n, Konstruktionsteil n, m
     2. Gitterstab m, Füllstab m
   f 1. élément m constructif [de
     construction], pièce f de charpente,
     composant m (de construction) 2. barre
     f de treillis
  nl 1. onderdeel n, constructie-element n
     2. staaf f (m) van een vakwerk
   r 1. строительный элемент m, элемент
     m конструкции 2. стержень m
     [элемент m заполнения] решётки

M232 e **membrane**
   d Membran(e) f; Film m;
     Wasserabdichtungsfolie f
   f membrane f; pellicule f; membrane f
     étanche [d'étanchéité]
  nl membraan n, f (m); vlies n
   r мембрана f; плёнка f; плёночная
     гидроизоляция f

M233 e **membrane concrete curing**
   d Betonnachbehandlung f mit
     Abdichtungsmitteln
   f cure f [curing m] de béton par
     pulvérisation de la composition
     filmogène
  nl nabehandeling f door afdekking
     (beton)
   r выдерживание n бетона путём
     нанесения плёнкообразующих
     составов [эмульсий]

M234 e **membrane filter**
   d Membran(e)filter n, m
   f filtre m à membrane
  nl membraanfilter n, m
   r мембранный фильтр m

M235 e **membrane fireproofing**
   d Feuerschutzmembrane f, Brandblende
     f (bewehrter Feuerschutzputz)
   f couche f protectrice [de protection]
     ignifuge, enduit m protecteur [de
     protection] ignifuge
  nl brandwerende laag f (m)
   r огнезащитный слой m (из
     огнестойкой штукатурки и сетки)

M236 e **membrane-forming type bond braker**
   d filmbildendes Haftablösemittel n
   f film m [pellicule f] de protection contre
     l'adhérence (p.ex. du béton frais au
     coffrage)
  nl filmvormend losmiddel n (ontkisten
     van beton)
   r плёнкообразующий состав m,
     предупреждающий сцепление (напр.
     опалубки с бетоном)

M237 e **membrane theory**
   d Membran(e)theorie f
   f théorie f de la membrane
  nl membraantheorie f
   r безмоментная теория f (расчёт
     оболочек)

M238 e **membrane waterproofing**
   d 1. Membranisolierung f gegen Wasser,
     Wassersperre f aus Isolierpappe
     2. Sperranstrich m, Dichtungsanstrich
     m
   f 1. revêtement m d'étanchéité en feutre
     bitumé 2. revêtement m
     d'étanchéité bitumé [en asphalte]
     coulé
  nl 1. waterdichtmaken n met een vlies
     (b.v. bitumenkit) 2. aanbrengen n van
     een waterdichte laag (b.v. gietasfalt)
   r 1. обклеечная гидроизоляция f
     2. обмазочная гидроизоляция f
     (в несколько слоёв битума)

M239 *e* **meridian stress**
  d Meridianspannung *f*
  f contrainte *f* méridienne
  nl meridiaanspanning *f*
  r меридиональное напряжение *n*
  (*в купольной оболочке*)

M240 *e* **mesh**
  d 1., 2. Masche *f* 3. Bewehrungsmatte *f*,
  Netzbewehrung *f*
  f 1. maille *f* (de tamis) 2. mesh *m*,
  nombre *m* de mailles (de tamis) par
  un pouce 3. toile *f* métallique,
  treillis *m* d'armature
  nl 1. maas *f* (*m*) 2. zeefmaat *f* (*m*)
  (*aantal openingen per cm²*)
  3. plaatgaas *n*, wapeningsnet *n*
  r 1. ячейка *f*, отверстие *n* (*в сите, сетке,*
  решётке) 2. меш *m*, число *n*
  отверстий на дюйм 3. сетка *f*;
  арматурная сетка *f*

M241 *e* **mesh laying jumbo**
  d Bewehrungsmattenverleger *m*,
  Verlegegerät *n* für Bewehrungsmatten
  f machine *f* de pose des treillis
  d'armature
  nl machine *f* voor het leggen van
  wapeningsmatten
  r машина *f* для укладки арматурных
  сеток

M242 *e* **mesh-reinforced shotcrete**
  d netzbewehrter [netzarmierter]
  Spritzbeton *m*
  f gunite *f* renforcée par l'armature en
  treillis, gunite *f* renforcée [armée] par
  treillis d'armature
  nl spuitbeton *n* met wapeningsnet
  r торкрет-бетон *m*, усиленный
  арматурной сеткой

M243 *e* **mesh reinforcement**
  d Betonstahlgewebe *n*,
  Bewehrungsmatte *f*, Armierungsnetz
  n, Streckmetallbewehrung *f*,
  Streckmetalleinlagen *f pl*
  f armature *f* en treillis, treillis *m*
  d'armature; armature *f* en acier
  déployé
  nl wapeningsmat *f* van plaatmat
  r арматурная сетка *f*, арматура *f* из
  просечно-вытяжной листовой стали

M244 *e* **mesh series**
  d Maschenweitenfolge *f*, Maschensatz *m*
  f jeu *m* de tamis
  nl stel *n* (van) zeven
  r комплект *m* сит

M245 *e* **mesh size**
  d Lochweite *f*; Gewebenummer *f*
  (*Siebtechnik*)
  f ouverture *f* des mailles du tamis;
  numéro *m* de toile
  nl maaswijdte *f*; zeefnummer *n*
  r размер *m* ячейки сита; номер *m*
  сита

M246 *e* **metal**
  d 1. Metall *n* 2. Straßenbauschotter *m*

  f 1. métal *m* 2. empierrement *m*,
  ballast *m* (*d'une voie ferrée, d'une*
  route)
  nl 1. metaal *n* 2. steenslag *n*
  r 1. металл *m* 2. щебень *m* для
  дорожно-строительных работ

M247   **metal-clad cable** *see* **armoured cable**

M248 *e* **metal-clad fire door**
  d Feuerschutztür *f* [Brandschutztür *f*]
  mit Blechüberzug
  f porte *f* pare-feu revêtue de métal
  nl branddeur *f* (*m*) met metaalbekleding
  r противопожарная дверь *f*,
  обитая листовым металлом

M249 *e* **metal cleading**
  d Metallhaut *f*, Feinblechumhüllung *f*
  f enveloppe *f* métallique (*du calorifuge*)
  nl bekleding *f* van metaalplaat
  r металлическая защитная оболочка *f*
  (*теплоизоляция*)

M250 *e* **metal curtain wall**
  d Vorhangwandplatte *f* mit
  Metallgerippe
  f mur-rideau *m* à ossature métallique
  nl gordijngevel *m*, metalen vliesgevel *m*
  r стеновая навесная панель *f*
  с металлическим каркасом

M251 *e* **metal floor decking**
  d Metallbelagplatte *f*
  f tôles *f pl* ondulées [façonnées,
  profilées] pour plancher
  nl metalen vloer *m*
  r профилированный металлический
  настил *m* перекрытия

M252 *e* **metal grating**
  d Gitterrostabdeckung *f*
  f caillebotis *m* métallique
  nl metalen roostervloer *m*
  r металлический решётчатый настил *m*,
  металлическая решётка *f*

M253 *e* **metal lathing**
  d Metallgewebe *n*, Putzdrahtgeflecht *n*
  f treillis *m* métallique pour enduit(s),
  grillage *m* à poules
  nl plaatgaas *n*, steengaas *n* (*als drager*
  van pleisterwerk)
  r металлическая сетка *f* под
  штукатурку

M254   **metalled road** *UK see* **macadam**

M255 *e* **metallic sprayed coating**
  d gespritzter Metallüberzug *m*
  f revêtement *m* métallisé (*appliqué*
  au pistolet)
  nl gespoten metaallaag *f* (*m*)
  r напылённое металлопокрытие *n*

M257   **metallurgical cement** *see* **supersulphated**
  cement !

M258 *e* **metal mesh fabric**
  d Drahtnetz *n*, Drahtgewebe *n*
  f toile *f* métallique, treillis *m* en fil
  métallique
  nl fijnmazig metaalgaas *n*
  r металлическая проволочная сетка *f*

M259 e **metal runner**
  d Metalleinrahmung *f*
    [Metalleinfassung *f*] eines
    Trennwand-Holzgerüstes
  f chapeau *m* métallique du bâti de
    cloison
  nl metalen omranding *f* van een houten
    wandelement
  r металлическая обвязка *f* каркаса
    перегородки

M260 e **metal sash putty**
  d Stahlfensterkitt *m*
  f mastic *m* à vitres pour châssis
    métallique
  nl stopverf *f* (*m*) voor stalen ramen
  r замазка *f* для остекления
    металлических оконных переплётов

M261 e **metal sheet roof covering**
  d Blechdacheindeckung *f*,
    Metalldachbelag *m*
  f couverture *f* [toiture *f*] en tôle
  nl dakbedekking *f* van metaalplaat
  r металлическая кровля *f*

M262 e **metal stud**
  d Matallprofilstütze *f* (*Trennwände*,
    V*orhangplatten*)
  f montant *m* [poteau *m*] métallique de
    cloison *ou* de mur-rideau
  nl metalen stijl *m* (*tussenwand, gevelplaat*)
  r металлическая стойка *f* (*перегородок,
    навесных панелей*)

M263 e **metal valley**
  d Dachkehle *f* mit Metalleindeckung
  f noue *f* (à revêtement d'étanchéité)
    métallique
  nl met metaal beklede kilgoot *f* (*m*)
  r ендова *f* с металлическим покрытием

M264 e **metal water stop**
  d Metalldichtungsstreifen *m*
  f garniture *f* métallique (*calfeutrement
    des joints dans les ouvrages
    hydrauliques*)
  nl metalen afdichtingsband *m*
  r уплотняющая металлическая полоса
    f [шпонка *f*] (*в швах гидросооружений*)

M265 e **meter pit**
  d Zählerkammer *f*
  f fosse *f* [puits *m*] à compteur de débit
  nl meterput *m*
  r камера *f* [приямок *m*] для
    расходомера (*на подземном
    трубопроводе*)

M266 e **methane tank**
  d Schlammfaulanlage *f*, Faulkammer *f*,
    Faulbecken *n*
  f digesteur *m* de boues
  nl rottingskamer *f* (*m*)
  r метантенк *m*

M267 e **methane-tank charge dose**
  d Dosis *f* für Faulraumbelastung
  f dose *f* de chargement d'un digesteur
    de boues

  nl beladingsdosis *n* voor een
    rottingskamer
  r доза *f* загрузки метантенка

M268 e **method of elastic weights**
  d Verfahren *n* der elastischen Lasten
  f méthode *f* des poids élastiques
  nl methode *f* van elastische lasten
  r метод *m* упругих грузов

M269 e **method of joints, method of joint
    isolation**
  d Knotenschnittverfahren *n*,
    Knotenpunktverfahren *n*
  f méthode *f* des nœuds
  nl knooppuntenmethode *f*
  r метод *m* вырезания узлов
    (*определение усилий в стержнях
    фермы*)

M270 e **method of least work, method of
    minimum strain energy**
  d Minimalprinzip *n* für die Spannungen,
    Castiglianosches Prinzip *n*,
    Castigliano-Prinzip *n*
  f principe *m* de Castigliano, principe *m*
    du travail minimum [du moindre
    travail]
  nl Castigliano-beginsel *n*
  r принцип *m* наименьшей работы,
    вторая теорема *f* Кастильяно

M271 **method of redundant reactions** *see*
    moment area method

M272 e **method of sections**
  d Schnittverfahren *n*
  f méthode *f* des sections [de Ritter]
  nl snedemethode *f*
  r метод *m* сечений (*определение
    усилий в стержнях фермы*)

M273 e **method of the substitute redundant
    members**
  d Verfahren *n* des Ersatzes der
    Verbindungskräfte
  f méthode *f* de substition des liaisons
    surabondantes
  nl substitutiemethode *f*
  r метод *m* замены связей

M274 e **method of zero moment points**
  d Momentennullpunktsverfahren *n*
  f méthode *f* des nœuds à moment zero
  nl methode *f* van de momentennulpunten
  r метод *m* моментных нулевых точек

M275 **mezzanine** *see* **entresol**

M276 e **mezzanine floor**
  d Zwischengeschoß *n*, Halbgeschoß *n*
  f mezzanine *f*
  nl tussenverdieping *f*, entresol *m*
  r промежуточный этаж *m*, полуэтаж
    m, антресоль *f*

M277 e **micro-crack**
  d Feinstriß *m*, Mikroriß *m*
  f microfissure *f*
  nl haarscheur *f* (*m*)
  r микротрещина *f*, волосная трещина *f*

M278 e **micropolitan**
  d Ortschaft *f*; Städtchen *n*; Siedlung *f*

f petite agglomération f
nl stadje n; dorp n
r населённый пункт m; небольшой
город m; посёлок m

M279 e **micro-strainer**
d Sieb-Mikrofilter n, m
f microfiltre m à crépine
nl zeefmicrofilter n, m
r сетчатый микрофильтр m

M280 e **middle strip**
d Mittelstreifen m (*Plattenberechnung*)
f bande f médiane (*calcul des plaques*)
nl middenstrook f (m) (*betonberekening*)
r средняя расчётная полоса f
(*расчёт балочных плит*)

M281 e **middle surface**
d Mittelfläche f
f surface f moyenne
nl middenvlak n
r срединная поверхность f (*оболочки*)

M282 e **middle third**
d mittleres Drittel n
f tiers m central
nl middelste derde deel n
r средняя треть f сечения (*стены,*
колонны)

M283 e **midget construction crane**
d Bau(stellen)kleinkran m
f grue f de construction [de travaux
publics] légère
nl lichte bouwkraan f (m)
r лёгкий строительный кран m

M284 e **midspan**
d Feldmitte f
f milieu m de travée (*d'une poutre*)
nl midden n van een overspanning
r середина f [средняя точка f] пролёта

M285 e **midspan deflection**
d Durchbiegung f in der Feldmitte
f flèche f au milieu de travée
nl doorbuiging f in het midden van een
overspanning
r прогиб m в середине [средней точке]
пролёта

M286 e **mild steel reinforcement**
d schlaffe Bewehrung f, nichtgespannte
Armierung f
f armature f en acier doux
nl wapening f van normaal staal
r арматура f из низкоуглеродистой
стали

M287 e **milk of lime**
d Kalkmilch f
f lait m [blanc m] de chaux, badigeon m
à la chaux
nl kalkmelk f (m)
r известковое молоко n

M288 e **mill**
d 1. Farbmühle f 2. Mühle f
f 1. broyeur-malaxeur m à peinture
2. moulin m
nl molen m
r 1. краскотёрка f 2. мельница f

M289 e **millboard**
d Zellstoffpappe f
f gros carton m
nl cellulosekarton n
r толстый картон m

M290 e **milled joint**
d Werkstoß m
f assemblage m exécuté en atelier
(*constructions métalliques*)
nl werkverbinding f (*metaalconstructies*)
r заводское соединение n
(*металлоконструкций*)

M291 e **mineral dust**
d mineralisches Pulver n, Steinmehl n;
Mineralfüller m, Mineralfüllstoff m
f farine f minérale; poudre f minérale,
fines f pl de roche
nl minerale vulstof f (m)
r минеральный порошок m;
минеральный наполнитель m

M292 e **mineral fibers**
d Mineralfasern f pl
f fibres f pl minérales
nl minerale vezels f (m) pl
r минеральные волокна n pl

M293 e **mineral fiber tiles**
d Mineralfaser-Schalldämmsteine m pl
f carreaux m pl insonorisants en fibres
minérales
nl akoestische tegels m pl van minerale
vezels
r минерально-волокнистые
акустические плитки f pl

M294 e **mineral-filled asphalt**
d Bitumenbindemittel n mit
Mineralfüllstoff
f bitume m avec filler, liant m
bitumineux avec poudre minérale
nl bitumen n met minerale vulstof
r асфальтовое вяжущее n (*смесь*
битума с минеральным порошком)

M295 e **mineral filler**
d Mineralfüller m, Mineralfüllstoff m
f filler m, fines f pl, poudre f minéral
fine
nl minerale vulstof f (m)
r минеральный наполнитель m

M296 e **mineralogical composition of**
aggregates
d mineralogische Zusammensetzung f
der Zuschläge
f composition f minérale des granulats
[des agrégats]
nl mineralogische samenstelling f van
de toeslagen
r минералогический состав m
заполнителей

M297 **mineral pigment** *see* **earth pigment**

M298 e **mineral-surfaced roofing**
d weiche Dachhaut f mit
mineralbestreuter Bitumendeckschicht,
Kiesschütt(dach)eindeckung f
f couverture f recouverte au gravier
enrobé

nl bitumineuze dakbedekking *f* met minerale deklaag
r мягкая кровля *f* с посыпкой высевками по битумному слою

M299 *e* **mineral wool**
d Mineralwolle *f*
f laine *f* minérale
nl steenwol *f* (*m*)
r минеральная вата *f* [шерсть *f*]

M300 *e* **minimum air requirement**
d Mindestluftbedarf *m*, Mindestluftrate *f*
f bésoin *m* en air minimum [minimal]
nl minimaal luchtverbruik *n*
r минимальная потребность *f* в свежем воздухе, минимальный нормативный расход *m* наружного воздуха

M301 **minimum reliable yield** *see* **safe yield**

M302 **minimum water storage elevation** *see* **draw-down level**

M303 *e* **mining shovel**
d Steinbruchbagger *m*, Abbaubagger *m*
f pelle *f* de mineur [de carrière]
nl lepelschop *f* (*m*)
r карьерный экскаватор *m*, экскаватор *m* для производства горных работ

M304 **minus size of a screen** *see* **undersize**

M305 *e* **miser**
d Hand-Schneckenbohrer *m*
f cuiller *m* à tarière manuelle, tarrière *f* à main
nl (hand)lepelboor *f* (*m*) (*grondboor*)
r ручная буровая желонка *f* (*для исследования грунтов*)

M306 *e* **misfire**
d Fehlschuß *m*, Fehlzündung *f*, Versagen *n*
f raté *f* (de sautage)
nl weigering *f* (*springlading, ontsteking*)
r отказ *m* заряда (*при взрывных работах*)

M307 *e* **mission tile**
d Hohlziegel *m*
f tuile *f* creuse [en auge]
nl trogvormige tegel *m*, goottegel *m*
r желобчатая черепица *f*

M308 *e* **mist-spraying**
d Versprühung *f*, Zerstäubung *f*
f pulvérisation *f* fine
nl verstuiven *n*
r тонкое распыление *n*

M309 *e* **miter box**
d Gehrungs(schneid)lade *f*
f boîte *f* à coupe [à onglets]
nl versteklade *f*, verstekbak *m*
r стусло *n*

M310 *e* **miter cut**
d Gehrungsschnitt *m*
f coupe *f* biaise, trait *m* de scie biais
nl verstek *n*
r рез *m* [пропил *m*] под углом

M311 *e* **miter gate** *US*
d Stemmtorpaar *n*, Stemmtore *n pl*, zweiflügeliges Schleusentor *n*
f porte *f* busquée double
nl puntdeuren *pl*
r двустворчатые шлюзные ворота *pl*

M312 *e* **miter joint**
d Gehrungsstoß *m*
f assemblage *m* à onglet
nl verstekverbinding *f*
r соединение *n* в ус

M313 **miter sill** *see* **lock sill**

M314 *e* **miter square**
d Gehrlade *f*, Gehrungswinkel *m*; Stellwinkel *m*, Schmiege *f*, Gehrmaß *n*
f équerre *f* à onglet, fausse-équerre *f*, sauterelle *f*
nl verstekhaak *m*
r ерунок *m*; малка *f* (*столярный инструмент*)
mitre... *see* **miter...**

M316 **mitred joint** *see* **miter joint**

M317 *e* **mix**
d 1. Mischung *f* 2. Mischungsverhältnis *n*
f 1. mélange *m* 2. composition *f* de mélange
nl 1. mengsel *n* 2. mengverhouding *f*
r 1. смесь *f* 2. состав *m* смеси

M318 *e* **mix control**
d Überwachung *f* [Kontrolle *f*] des Mischvorgangs; Regelung *f* der Mischungszusammensetzung
f contrôle *m* de la confection [préparation] de béton; contrôle *m* de la composition de béton
nl regeling *f* van de mengverhouding
r контроль *m* за приготовлением (бетонной) смеси; регулирование *n* состава (бетонной) смеси

M319 *e* **mix design**
d Bestimmen *n* der Betonzusammensetzung; Festlegen *n* des Mischungsverhältnisses
f recherche *f* de la composition de béton; dosage *m* (de constituants) de béton
nl bepalen *n* van de betonsamenstelling
r проектирование *n* (бетонной) смеси; подбор *m* состава (бетонной) смеси

M320 *e* **mixed bituminous macadam**
d bituminöser Mischmakadam *m*, *n*
f macadame *m* bitumineux [enrobé de bitume]
nl met bitumen gemengd macadam *n*
r чёрное щебёночное покрытие *n*, выполненное способом смешения в установке

M321 *e* **mixed-in-place method**
d An-Ort-Mischverfahren *n*; Bodenmischverfahren *n*
f procédé *m* [méthode *f*] (de) mix-in--place, stabilisation *f* sur place
nl oppervlaktebehandeling *f* met ter plaatse vervaardigd mengsel

r строительство *n* дорожных покрытий по способу смешения на дороге

M322 *e* **mixed-in-place road mix**
   d An-Ort-Mischbelag *m*
   f mélange *m* (*sol-ciment, sol-bitume, gravier-bitume*) confectionné sur place [in situ]
   nl gemengde wegverharding *f*
   r слой *m* дорожного покрытия, устроенного смешением на дороге

M323   **mixed-in-place surface treatment** *see* **mixed-in-place method**

M324 *e* **mixer**
   d Mischer *m*, Mischmaschine *f*
   f mélangeur *m*, malaxeur *m*
   nl mengmachine *f*, molen *m*
   r смеситель *m*; бетоносмеситель *m*

M325 *e* **mixer drum**
   d Mischtrommel *f*
   f tambour *m* mélangeur
   nl mengtrommel *f* (*m*)
   r смесительный барабан *m*, барабан *m* бетоносмесителя

M326 *e* **mixer efficiency**
   d Funktionstüchtigkeit *f* des Betonmischers, Misch(ungs)wirkungsgrad *m*
   f productivité *f* [effet *m* utile, rendement *m*] d'un mélangeur
   nl doelmatigheid *f* van een betonmolen
   r работоспособность *f* [продуктивность *f*] бетоносмесителя (*способность выдавать однородную смесь в течение заданного периода*)

M327 *e* **mixer skip**
   d Aufzugskasten *m*, Kippkübel *m*, Beschicker *m*
   f benne *f* chargeuse (*de bétonnière*)
   nl vulbak *m* van een betonmolen
   r опрокидной загрузочный ковш *m* (*бетоносмесителя*)

M328 *e* **mixer trestle**
   d Mischerbock *m*
   f plate-forme *f* de travail du malaxeur
   nl bok *m* van de betonmolen
   r эстакада *f* смесителя

M329   **mixer truck** *see* **truck mixer**

M330 *e* **mixer-type truck** *US*
   d Transportbetonmischer *m*, Betontransportwagen *m* mit aufgebautem Rührwerk
   f camion *m* mélangeur pour béton, camion *m* (avec) malaxeur
   nl truckmixer *m*
   r автобетоносмеситель *m*

M331   **mix formulation** *see* **mix proportions**

M332 *e* **mixing box, mixing chamber**
   d Mischkammer *f*; Mischkasten *m*
   f chambre *f* de mélange; boîte *f* de mélange
   nl mengbak *m*
   r смесительная камера *f*; воздухосмеситель *m*

M333 *e* **mixing cycle**
   d 1. Mischspiel *n* 2. Mischspieldauer *f*
   f 1. cycle *m* de malaxage 2. temps *m* [durée *f*] du cycle de malaxage
   nl (duur *m* van de) mengcyclus
   r 1. цикл *m* работы бетоносмесителя 2. продолжительность *f* цикла работы бетоносмесителя

M334 *e* **mixing damper**
   d Mischklappe *f*
   f registre *m* de mélange
   nl mengklep *f* (*m*)
   r смесительный клапан *m* (*воздушный*)

M335   **mixing drum** *see* **mixer drum**

M336 *e* **mixing of concrete**
   d Betonmischen *n*
   f malaxage *m* du béton
   nl beton mengen
   r приготовление *n* бетонной смеси

M337 *e* **mixing-placing train**
   d Misch- und Betonierzug *m*, Misch- und Einbauzug *m*
   f train *m* de bétonnage
   nl betontrein *m* (*tunnelbouw*)
   r комплексный агрегат *m* для приготовления и укладки бетонной смеси

M338 *e* **mixing plant**
   d 1. Betonmischanlage *f* 2. bituminöse Mischanlage *f*
   f 1. central *f* à béton 2. poste *f* d'enrobage
   nl 1. betoncentrale *f* 2. asfaltinstallatie *f*
   r 1. бетоносмесительная установка *f* 2. асфальтосмесительная установка *f*

M339 *e* **mixing ratio**
   d Mischungsverhältnis *n*, Mischungszahl *f*
   f rapport *m* de mélange (*conditionnement de l'air*)
   nl mengverhouding *f*
   r коэффициент *m* смешения (*кондиционирование воздуха*)

M340 *e* **mix ingredients**
   d Mischungsbestandteile *m pl*
   f composants *m pl* [éléments *m pl* constitutifs] d'un mélange (de béton)
   nl bestanddelen *n pl* van het mengsel
   r составляющие *n pl* [компоненты *m pl*] (бетонной) смеси

M341 *e* **mixing screw**
   d Mischschnecke *f*
   f vis *f* mélangeuse
   nl mengworm *m*
   r смесительный шнек *m*

M342 *e* **mixing speed**
   d Mischungsgeschwindigkeit *f*
   f vitesse *f* de malaxage
   nl mengsnelheid *f*
   r скорость *f* перемешивания

M343 *e* **mixing temperature**
   d Mischungstemperatur *f*

f température f de malaxage
nl mengtemperatuur f
r температура f смешения

M344 e **mixing time**
d Mischdauer f, Mischzeit f
f durée f [temps m] de malaxage
nl mengtijd m
r продолжительность f [время n]
перемешивания

M345 e **mixing valve**
d Mischventil n
f vanne f à mélange
nl mengklep f (m)
r смесительный клапан m (для жидких
сред)

M346 e **mixing water**
d Anmach(e)wasser n, Mischwasser n
f eau f de gâchage
nl aanmaakwater n
r вода f затворения (смеси)

M347 **mix-in-place** see **mixed-in-place method**

M348 e **mix-in-place machine, mix-in-place travel plant**
d Aufnahmemischer m,
Aufnehmermischer m,
Bodenvermörtelungsmaschine f mit
hochliegendem Zwangsmischer
f malaxeur m «travel-plant»,
enrobeur-automoteur-autochargeur m
nl stabilisatiemachine f,
stabilisatietrein m
r передвижная асфальтосмесительная
[грунтосмесительная] установка f

M349 e **mix proportions**
d Mischungsformel f
f composition f d'un mélange
nl mengformule f (m)
r состав m смеси

M350 **mixture** see **mix**

M351 e **mobile bituminous mixing plant**
d fahrbare bituminöse Mischanlage f
f groupe m d'enrobage mobil
nl mobiele asfaltinstallatie f
r передвижной
асфальтобетоносмеситель m

M352 e **mobile crane**
d Mobilkran m
f grue f automotrice [mobile],
mobilgrue f
nl mobiele kraan f (m)
r самоходный стреловой кран m

M353 **mobile field office** see **mobile site office**

M354 e **mobile form**
d bewegliche Schalung f, Rollschalung
f, Wanderschalung f
f coffrage m roulant [mobile]
nl verplaatsbare [verrijdbare] bekisting f
r передвижная [катучая] опалубка f

M355 e **mobile hoist**
d straßenfahrbarer Plattform-
-Bauaufzug m

f monte-charge m mobile [roulant]
nl mobiele bouwlift m
r передвижной подъёмник m

M356 **mobile jib crane** see **mobile crane**

M357 e **mobile job crane**
d leichter hydraulischer Laufkran m
f chariot-grue m hydraulique
nl mobiele kraan f (m)
r лёгкий катучий гидравлический
кран m

M358 e **mobile scaffold tower**
d fahrbarer Gerüstturm m
f échafaudage m à tour roulante
[mobile]
nl verrijdbare steiger m
r передвижные подмости pl башенного
типа

M359 e **mobile site office**
d Bürowagen m
f bureau m de chantier mobile
nl directiekeet f (m)
r передвижная контора f прорабского
участка, передвижная строительная
контора f

M360 e **mobile space heater**
d fahrbares Warmluftgebläse n
f chauffrette f mobile, réchauffeur m
d'air mobile
nl mobiele luchtverhitter m
r передвижной воздухонагреватель m
(для временного отопления или
сушки)

M361 e **mobile tower crane**
d nichtgleisgebundener
Turmdrehkran m
f grue f à tour [à pylône] mobile
nl mobiele torenkraan f (m)
r передвижной башенный кран m

M362 e **mobile work platform**
d 1. transportabler Turm m;
Fahrgerüst n, fahrbares Gerüst n
2. Hubinsel f
f 1. plate-forme f de travail roulante
2. plate-forme f de travail
autoélévatrice flottante
nl 1. mobiele steiger 2. hoogwerker m
r 1. передвижная вышка f;
передвижные подмости pl
2. плавучая самоподъёмная
платформа f (для морских работ)

M363 e **model testing**
d Modellprüfung f, Modellversuch m
f essai m sur modèle
nl modelbeproeving f
r испытание n на моделях, модельное
испытание n

M364 e **mode of buckling**
d Knickform f
f forme f de flambage [de flambement]
nl knikvorm f (m)
r форма f потери устойчивости

M365 e **mode of failure**
d Bruchart f. Bruchform f

f forme *f* [caractère *m*] de rupture
nl wijze *f* van bezwijken
r характер *m* разрушения

M366 *e* **moderate exposure**
   d gemäßigte atmosphärische Einwirkung *f*, gemäßigter Witterungseinfluß *m*
   f sollicitation *f* atmosphérique [climatique] modérée
   nl gematigde blootstelling *f* aan weersinvloeden
   r умеренное воздействие *n* условий среды

M367 *e* **modification**
   d Abänderung *f*
   f modification *f* (du projet)
   nl wijziging *f*, modificatie *f*
   r письменное дополнение *n* к проектно-сметной документации, распоряжение *n* об изменениях в проекте

M368 *e* **modified I beam**
   d Doppel-T-Träger *m* mit herausstehenden Bewehrungsbügeln im Obergurt
   f poutre *f* préfabriquée en double T modifiée (*à étriers en saillie sur la membrure supérieure pour la liaison avec une dalle coulée sur place*)
   nl prefab betonbalk *m* in I-profiel met uitstekende luswapening in de bovenflens
   r сборная железобетонная балка *f* с выпусками хомутов из верхнего пояса

M369 *e* **modular brick**
   d Modulziegel *m*, Modulstein *m*
   f brique *f* modulaire
   nl modulaire steen *m*
   r модульный кирпич *m*

M370 *e* **modular building unit**
   d Modulbauelement *n*, Modulbaukörper *m*
   f élément *m* [unité *f*] de construction modulaire
   nl modulair bouwelement *n*
   r модульный блок *m* [строительный элемент *m*]

M371 *e* **modular construction**
   d 1. Modulbau *m* 2. Modulkonstruktion *f*
   f 1. construction *f* modulaire 2. construction *f* modulée
   nl modulaire constructie *f*
   r 1. строительство *n*, основанное на модульной системе 2. конструкция *f*, основанная на модуле

M372 *e* **modular coordination**
   d Maßordnung *f*
   f coordination *f* modulaire
   nl modulaire coördinatie *f*
   r модульная координация *f* (*размеров*)

M373 *e* **modular design**
   d Baukastenbemessungsverfahren *n*, Baukastensystem *n*

f élaboration *f* du projet à base d'un module
nl modulair ontwerp *n*
r модульное проектирование *n*, проектирование *n* на базе модуля

M374 *e* **modular ratio**
   d Verhältnis *n* der Elastizitätsmodule
   f rapport *m* des modules d'élasticité, coefficient *m* d'équivalence
   nl verhouding *f* der elasticiteitsmoduli van beton en staal
   r отношение *n* модулей упругости (*арматурной стали и бетона*)

M375 *e* **modular ratio method**
   d n-Verfahren *n*
   f méthode *f* élastique du calcul de béton armé (*par le rapport des modules d'élasticité du béton et de l'acier dit le coefficient d'équivalence*)
   nl n-methode *f* (*toelaatbare spanningen in gewapend beton*)
   r метод *m* расчёта железобетонных конструкций по допускаемым напряжениям (*по отношению модулей упругости стали и бетона*)

M376 *e* **modular size**
   d Modulgröße *f*, Modulmaß *n*
   f dimension *f* [mesure *f*] modulaire
   nl modulaire maat *f* (*m*)
   r модульный размер *m*

M377 *e* **modular system**
   d System *n* der Maßordnung im Bauwesen
   f système *m* modulé en construction
   nl modulair (maat)systeem *n*
   r модульная система *f* в строительстве

M378 *e* **modular unit**
   d 1. Modulbaustein *m* 2. Modulbauelement *n*, Modulbaukörper *m*
   f 1. brique *f* ou bloc *m* mural modulaire 2. unité *f* de construction modulaire
   nl 1. modulaire bouwsteen *m* 2. modulair bouwelement *n*
   r 1. кирпич *m* или стеновой блок *m*, размеры которого кратны установленному модулю 2. модульный строительный элемент *m*

M380 *e* **module**
   d 1. Modul *m* 2. Wassermeßvorrichtung *f* 3. Modul *m*, Bauwerksabschnitt *m*
   f 1. module *m* 2. prise *f* d'eau du type module, module *m* 3. module *m*, compartiment *m*
   nl 1. moduul *m*, modulus *m* 2. watermeetinrichting *f* 3. compartiment *n*
   r 1. модуль *m* (*строительный, планировочный*) 2. модуль(ный водовыпуск) *m* 3. модуль *m*, отсек *m*

M381 *e* **modulus of creep**
   d Kriechmodul *m*
   f module *m* de fluage

nl kruipmodulus *m*
r модуль *m* ползучести

M382 *e* **modulus of deformation**
   d Deformationsmodul *m*,
     Verformungswert *m*
   f module *m* de déformation
   nl vormveranderingsmodulus *m*
   r модуль *m* деформации

M383 *e* **modulus of elasticity**
   d Elastizitätsmodul *m*, E-Modul *m*
   f module *m* d'élasticité
   nl elasticiteitsmodulus *m*
   r модуль *m* упругости

M384   **modulus of flow** *see* **specific discharge**

M385   **modulus of foundation support** *see*
    modulus of subgrade reaction

M386 *e* **modulus of resilience**
   d spezifische Formänderungsarbeit *f* bis
     zur Elastizitätsgrenze, Resilienz *f*
   f module *m* du travail de déformation
     élastique
   nl elastische vormveranderingsarbeid *m*
   r удельная работа *f* деформации

M387 *e* **modulus of rigidity**
   d Schubmodul *m*
   f module *m* de glissement [d'élasticité
     transversal]
   nl stijfheidsfactor *m*
   r модуль *m* сдвига, модуль *m*
     упругости при сдвиге

M388 *e* **modulus of rupture**
   d 1. Bruchmodul *m*,
     Querbiegefestigkeit *f*
     2. Verdrehungsfestigkeit *f*,
     Torsionsfestigkeit *f*
   f 1. module *m* de rupture par flexion
     2. module *m* de rupture par torsion
   nl 1. buigvastheid *f* 2. torsievastheid *f*
   r 1. предел *m* прочности на растяжение
     при изгибе 2. предел *m* прочности при
     кручении

M389   **modulus of section** *see* **section modulus**

M390 *e* **modulus of subgrade reaction**
   d Bettungsziffer *f*, Bettungszahl *f*
   f module *m* de réaction [de résistance]
     du sol, module *m* de réaction de
     fondation
   nl beddingsconstante *f* (*m*)
   r коэффициент *m* постели, модуль *m*
     упругости основания

M391 *e* **moire fringe method**
   d Moiréstreifenverfahren *n*,
     Moirémethode *f*
   f méthode *f* de moirage
   nl moirémethode *f*
   r метод *m* муаровых полос
     (*исследование напряжений*)

M392 *e* **moist room**
   d Feuchtkammer *f*
   f chambre *f* humide
   nl vochtige kamer *f* (*m*)
   r влажная камера *f* (*для хранения*
     образцов)

M393 *e* **moisture barrier**
   d Feucht(igkeits)sperre *f*, Dampfsperre *f*
   f membrane *f* d'étanchéité
   nl waterkerende laag *f* (*m*)
   r гидроизолирующая прокладка *f*,
     гидроизолирующий слой *m*

M400 *e* **moisture gradient**
   d Feuchtigkeitsgradient *m*
   f gradient *m* d'humidité
   nl vochtigheidsgradiënt *m*
   r влажностный градиент *m*

M401 *e* **moisture migration, moisture**
    movement
   d Feuchtigkeitswanderung *f*
   f mouvement *m* [migration *f*]
     d'humidité
   nl verplaatsing *f* van het vocht
   r миграция *f* влаги

M402 *e* **moisture-proof luminaire**
   d feuchtigkeitsgeschützte [wasserdichte]
     Leuchte *f*
   f luminaire *m* étanche à l'immersion
   nl waterdicht ornament *n*
   r влагозащищённый светильник *m*

M403 *e* **moisture-resistant insulating material**
   d feuchtigkeitsbeständiger
     Wärmedämmstoff *m*
   f matériau *m* isolant résistant à
     l'humidité
   nl vochtbestendig isolatiemateriaal *n*
   r влагостойкий теплоизоляционный
     материал *m*

M404   **moisture seal** *see* **moisture barrier**

M405 *e* **moisture sensing probe**
   d Feuchtigkeitssonde *f*
   f détecteur *m* [capteur *m*] d'humidité
   nl vochtigheidsvoeler *m*
   r гигрометрический зонд *m*, датчик *m*
     влажности

M406 *e* **moisture tons** *US*
   d latente Wärmebelastung *f*
   f charge *f* frigorifique latente (*en tonnes*
     de froid)
   nl latente warmtebelasting *f*
   r холодильная нагрузка *f* по скрытому
     теплу (*в тоннах холода*)

M407 *e* **mold**
   d 1. Form *f* 2. Setzbecher *m*
     (*Ausbreitversuch*)
   f 1. moule *m* 2. moule *m* tronconique
     (*pour l'essai d'affaissement du béton*)
   nl vorm *f* (*m*) (*reproductie*)
   r 1. форма *f* 2. конус *m* для определения
     консистенции бетонной смеси

M408 *e* **mold board**
   d Schild *m* (*Schürfschlepper*);
     Hobelmesser *n* (*Straßenhobel*)
   f lame *f* de travail, bouclier *m*
     (*d'un bulldozer*)
   nl dozerblad *n*, gradermes *n*
   r отвал *m* (*бульдозера, грейдера*)

M409 *e* **molded brick**
   d Formziegel *m*

    f brique *f* profilé [de forme, moulurée]
    nl geprofileerde steen *m*
    r фасонный декоративный кирпич *m*

M410 *e* **molded gutter**
    d Traufrinne *f* mit profilierter
    Sichtfläche
    f chéneau *m* mouluré
    nl goot *f* (*m*) met geprofileerd
    buitenboeiboord
    r карнизный водосточный жёлоб *m*
    с фигурной лицевой стенкой

M411 *e* **molded insulation**
    d wärmedämmende Ausbauplatten *f pl*
    [Segmente *n pl*]
    f coquilles *f pl* d'isolement thermique
    nl voorgevormde isolatieplaten *f pl*
    r теплоизоляционные скорлупы *f pl*
    [сегменты *m pl*]

M412 *e* **molded plywood**
    d Sperrbiegeholz *n*
    f contre-plaqué *m* cintré
    nl gevormd multiplex *n*
    r гнутая фанера *f*

M413 *e* **moldings**
    d 1. Riemchen *n pl*, Kymas *n pl*
    2. leistenförmige Architekturdetails *n*
    pl (*z.B. Halte-, Fuß-, Kehlleisten,*
    Hohlkehlen) 3. dekorative Formgüsse
    m pl
    f 1. moulures *f pl* 2. baguettes *f pl*
    3. pièces *f pl* moulées décoratives
    nl 1. gevelversieringselementen *n pl*
    2. lijstwerk *n* 3. van te voren gegoten
    ornamenten *pl*
    r 1. архитектурные обломы *m pl*
    2. погонажные архитектурные
    профильные детали *f pl* (*раскладки,*
    плинтусы, калевки, галтели)
    3. декоративные отливки *f pl*

M414 *e* **mold oil**
    d Entschalungsöl *n*, Schalöl *n*
    f graisse *f* de coffrage
    nl bekistingsolie *f* (*m*)
    r опалубочная смазка *f*, масло *n* для
    смазки форм

M415 **mold release** *see* **release agent**

M416 *e* **molds reuseability**
    d Wiederverwendbarkeit *f* der Formen
    f réemploi *m* [réutilisation *f*] du coffrage
    nl herhaald gebruik *n* van de bekisting
    r оборачиваемость *f* форм

M417 *e* **mole**
    d 1. Mole *f*, Hafendamm *m*
    2. Wellenbrecher *m*
    f 1. jetée *f* 2. brise-lames *m*
    nl 1. havendam *m*, pier *m*
    2. golfbreker *m*
    r 1. мол *m* 2. волнолом *m*

M419 *e* **mole drain**
    d Maulwurfdrän *m*, rohrloser Drän *m*
    f drain *m* taupe [de charrue-taupe]
    nl moldrain *m*
    r кротовая дрена *f*

M420 *e* **mole drainage**
    d Maulwurfdränung *f*, rohrlose
    Dränung *f*, Fräsrinnendränung *f,*
    Schlitzdränung *f*
    f drainage-taupe *m*
    nl moldrainage *f*
    r кротовый дренаж *m*

M421 *e* **mole plough**
    d Maulwurfpflug *m*
    f charrue *f* à taupe
    nl molploeg *m*, *f*
    r кротовый плуг *m*

M422 *e* **moler brick**
    d Diatomitstein *m*
    f brique *f* (de) diatomite
    nl kiezelgoersteen *m*
    r диатомовый кирпич *m*

M423 *e* **moling**
    d Fräsrillendränherstellung *f,*
    Schlitzdränherstellung *f*
    f aménagement *m* de drains de
    charrue-taupe
    nl uitvoering *f* van moldrains
    r устройство *n* кротовых дрен

M424 **Mollier diagram** *see* **i,d-diagram**

M425 *e* **moment area**
    d Moment(en)fläche *f*
    f aire *f* des moments
    nl momentenvlak *n*
    r площадь *f* эпюры моментов,
    моментная площадь *f*

M426 *e* **moment area method**
    d Kraftmethode *f,*
    Kraft(größen)verfahren *n*
    f méthode *f* des forces
    nl momentenvlakmethode *f*
    r метод *m* сил

M427 **moment at fixed end** *see* **encastré
moment**

M428 *e* **moment at support**
    d Stützenmoment *n*
    f moment *m* au support [d'appui,
    à l'appui]
    nl steunpuntmoment *n*
    r опорный момент *m*

M429 *e* **moment buckling**
    d Momentenknickung *f*
    f flambage *m* [flambement *m*] par
    flexion
    nl knik *m*
    r продольный изгиб *m* под действием
    моментной нагрузки

M430 *e* **moment-curvature law**
    d Zusammenhang *m* zwischen Moment
    und Krümmung
    f loi *f* moment-courbure
    nl verband *n* tussen moment en
    doorbuiging
    r зависимость *f* между изгибающим
    моментом и кривизной

M431 **moment diagram** *see* **bending moment
diagram**

M432 *e* **moment distribution method**
 d Momentenausgleichverfahren *n*
 f méthode *f* de distribution des moments
 nl momentenvereffeningsmethode *f*
 r метод *m* распределения моментов

M433 *e* **moment equation**
 d Momentengleichung *f*
 f équation *f* de moments
 nl momentenvergelijking *f*
 r уравнение *n* моментов

M434 *e* **moment influence line**
 d Momenteneinflußlinie *f*
 f ligne *f* d'influence de moments
 nl momenteninvloedslijn *f* (*m*)
 r линия *f* влияния моментов

M435 *e* **moment of couple**
 d Moment *n* des Kräftepaares
 f moment *m* d'un couple
 nl moment *n* van een koppel
 r момент *m* пары сил

M436 *e* **moment of inertia**
 d Trägheitsmoment *n*
 f moment *m* d'inertie
 nl traagheidsmoment *n*
 r момент *m* инерции

M437 *e* **moment of resistance**
 d Widerstandsmoment *n*, Moment *n* der inneren Kräfte [des inneren Kräftepaares]
 f moment *m* résistant [de résistance, des forces intérieures]
 nl weerstandsmoment *n*
 r момент *m* внутренних сил [внутренней пары сил]

M438 *e* **moment of span**
 d Feldmoment *n*
 f moment *m* de travée
 nl veldmoment *n*
 r момент *m* в пролёте

M439 *e* **moment reinforcement**
 d Bewehrung *f* gegen negative Biegemomente, Biegebewehrung *f*, Biegearmierung *f*
 f armature *f* [ferraillage *m*] de flexion
 nl wapening *f* tegen negatieve buigende momenten
 r арматура *f*, воспринимающая моментные нагрузки

M440 *e* **moment-resisting space frame**
 d momentbeanspruchtes Raumfachwerk *n*
 f charpente *f* [ossature *f*] tridimensionnelle rigide [à nœuds rigides]
 nl stijf ruimtelijk vakwerk *n*
 r жёсткая пространственная рама *f*

M441 *e* **moment resulting from sidesway**
 d durch seitliche Verschiebung bedingtes Moment *n*
 f moment *m* dû au déplacement latéral
 nl moment *n* tengevolge van horizontale verplaatsing
 r момент *m*, вызванный боковым смещением рамы

M442 *e* **moment splice**
 d momentbeanspruchte Verbindung *f*
 f assemblage *m* rigide (*travaillant en flexion*)
 nl stijve verbinding *f*
 r соединение *n*, воспринимающее момент(ную нагрузку)

M443 *e* **moment zero point**
 d Momentnullpunkt *m*
 f point *m* de moment nul
 nl momentennulpunt *n*
 r точка *f* нулевого момента

M444 *e* **monitor**
 d 1. Belichtungs- und Belüftungslaterne *f* 2. Wasserkanone *f*, Hydromonitor *m*, Druckstrahlbagger *m*
 f 1. lanterneau *m* d'éclairage et d'aération 2. monitor *m* (hydraulique)
 nl 1. licht- en ventilatielantaarn *f* (*m*) 2. waterkanon *n*, monitor *m*
 r 1. фонарная надстройка *f* над кровлей, светоаэрационный фонарь *m* 2. гидромонитор *m*

M445 *e* **monitoring**
 d Überwachung *f*, Kontrolle *f*
 f surveillance *f*, contrôle *m*
 nl controle *f*
 r диспетчерский контроль *m* (*в централизованной системе управления инженерным оборудованием здания*)

M446 *e* **monitor roof**
 d Oberlichtdach *n*, Laternendach *n*
 f comble *m* à lanterneau
 nl dak *n* met lantaarn
 r крыша *f* со световым фонарём, фонарное покрытие *n*

M447 *e* **monkey**
 d 1. Kleinmastenkran *m*, Kleinderrick *m* 2. Freifallhammer *m*, Freifallbär *m*
 f 1. mât *m* de levage léger 2. mouton *m* à chute libre
 nl 1. lichte laadmast *m* 2. vrijvallend heiblok *n*, valblok *n*
 r 1. лёгкая грузоподъёмная мачта *f*; лёгкий деррик-кран *m* 2. свободно падающий свайный молот *m*, (свободно) падающая баба *f* (*копра*)

M448 *e* **monkey wrench**
 d Universalschraubenschlüssel *m*
 f clé *f* à machoires mobiles
 nl verstelbare moersleutel *m*
 r разводной гаечный ключ *m*

M449 *e* **monocable**
 d Einseilschwebebahn *f*
 f blondin *m* à un cable porteur
 nl kabelbaan *f* (*m*) (aan één kabel)
 r одноканатная подвесная дорога *f*

M450 *e* **monolith**
 d 1. Schwimmkasten *m*, gestrandeter Senkkasten *m* 2. Betonstaumauer--Abschnitt *m* zwischen den vertikalen Trennfugen

f 1. caisson *m* flottant échoué, caisson *m* à fond 2. section *f* de bétonnage d'un barrage
nl funderingscaisson *m* (van beton *of* steen)
r 1. пустотелый массив-гигант *m* с перегородками 2. секция *f* *или* блок *m* бетонирования плотины

M451 *e* **monolithic concrete**
d Monolithbeton *m*, Ortbeton *m*
f béton *m* coulé sur [en] place
nl monolietbeton *n*
r монолитный бетон *m*

M452 *e* **monolithic slab and foundation wall**
d monolithisch betonierte Grundmauer *f* mit Kellerdeckenplatte
f dalle *f* et mur *m* de sous-sol coulés simultanément
nl gewapend-betonnen plaat *f* (*m*) met aangestorte funderingsmuren
r надподвальная плита *f* перекрытия и фундамент *m*, бетонируемые совместно

M453 *e* **monolithic terrazzo**
d monolithischer Terrazzo *m*
f revêtement *m* de sol en terrazzo
nl monoliet-terrazzovloer *m*
r террациевое покрытие *n* пола

M454 *e* **monolithic topping**
d verschleißfeste Betondeckschicht *f*
f couche *f* supérieure [de surface] en béton
nl slijtvaste cementvloer *m*
r износостойкий покровный слой *m* бетона

M455 *e* **monopitch roof**
d Pultdach *n*, Halbdach *n*, einhängiges Dach *n*
f toit *m* [comble *m*] à un égout [en appentis]
nl lessenaardak *n*
r односкатная крыша *f*, односкатное верхнее покрытие *n*

M456 *e* **monorail**
d Tragschiene *f*, Laufschiene *f*, Einschienenbahn *f*
f monorail *m*
nl luchtspoor *n*, zweefbaan *f* (*m*) (*met ophangstaaf*)
r монорельсовая дорога *f*

M457 *e* **monorail system**
d Einschienen-Hängebahn *f*, Einschienen-Schwebebahn *f*
f monorail *m* suspendu
nl monorailspoor *n*
r однорельсовая подвесная дорога *f*

M458 *e* **monorail transporter**
d Einschienen-Baubahn *f*
f convoyeur *m* à monorail suspendu
nl monorailtransporteur *m*
r монорельсовый конвейер *m* [транспортёр *m*]

M459 *e* **monotower crane**
d ortsfester Turmdrehkran *m*

f grue *f* monotour [à tour] fixe
nl vaste torenkraan *f* (*m*)
r стационарный полноповоротный башенный кран *m*

M460 *e* **monument**
d Vermessungspunkt *m*
f borne *f* (*d'arpentage*)
nl meetpunt *n*, vast punt (*triangulatie*)
r межевой знак *m* [столб *m*]

M461 **mooring** *see* **berthing facilities**

M462 *e* **mooring accessories, mooring appurtenances**
d Anlegevorrichtungen *f pl*
f dispositifs *m pl* d'amarrage
nl outillage *f* voor afmeren
r швартовные [причальные] устройства *n pl*

M463 *e* **mooring dolphin**
d Festmachedalbe *f*
f poteau *m* d'amarrage, duc *m* d'Albe
nl meerpaal *m*
r швартовный пал *m*

M464 *e* **mooring pile**
d Prellpfahl *m*
f pieu *m* de défense
nl stootpaal *m*
r отбойная свая *f*

M465 **mooring post** *see* **bollard**

M466 *e* **mooring ring**
d Schiffshaltering *m*, Schiffshaltekreuz *n*
f organeau *m*
nl tuiring *m*
r рым *m*, причальное кольцо *n*

M467 **mop plate** *see* **kickplate**

M468 *e* **morning glory spillway** *see* **shaft spillway**

M469 *e* **mortar**
d Mörtel *m*; Bindemittel *n*
f mortier *m*
nl mortel *m*; specie *f*
r (строительный) раствор *m*; растворная часть *f* бетона

M470 *e* **mortar additive**
d Mörtelzusatz *m*
f additif *m* [adjuvant *m*] du mortier (de ciment)
nl toevoeging *f* aan de mortel
r добавка *f* к (цементному) раствору

M471 *e* **mortar base**
d Mörtelunterlage *f*
f couche *f* d'arasement [de nivellement, d'aplanissement] de mortier
nl mortelonderlaag *f* (*m*)
r подстилающий [выравнивающий] слой *m* раствора

M472 *e* **mortar bed**
d 1. Mörtelbett *n*, Mörtelschicht *f* 2. Kalklöschkasten *m*
f 1. lit *m* de mortier 2. auge *f*
nl 1. mortelbed *n* 2. kalkbak *m*
r 1. растворная постель *f* 2. творильный ящик *m*, творило *n*

M473 *e* **mortarboard**
 d 1. Fugmörtelbrett *n* 2. Mischplatte *f* für Mörtel, Mischbühne *f*
 f 1. bouclier *m*; taloche *f* à mortier 2. petite plate-forme *f* pour la préparation des mortiers
 nl 1. speciebord *n* 2. plaat *f* (*m*) voor het mengen van de mortel
 r 1. сокол *m* 2. боёк *m* [небольшой помост *m*] для приготовления смесей (*напр. мастик, растворов*)

M474 **mortar box** *see* **mortar bed 2.**

M475 *e* **mortar cube**
 d Mörtelwürfel *m*
 f cube *m* d'essai en mortier
 nl mortelkubus *m*
 r контрольный кубик *m* для испытания строительного раствора

M476 *e* **mortar fraction**
 d Mörtelingrediens *n* des Betons, Bindemittel *n*
 f mortier *m* du béton frais
 nl mortelfractie *f* van het beton
 r растворная часть *f* бетона

M477 *e* **mortar gun**
 d Mörtelspritzmaschine *f*, Mörtelkanone *f*
 f projeteur *m* de mortier
 nl mortelspuit(inrichting) *f*
 r растворомёт *m*

M478 **mortar mill** *see* **mortar mixing machine**

M479 **mortar mix** *see* **mortar**

M480 *e* **mortar mixing machine**
 d Mörtelmischer *m*, Mörtelmischmaschine *f*
 f malaxeur *m* [malaxeuse *f*] à mortier
 nl betonmolen *m*
 r растворосмеситель *m*

M481 *e* **mortar plasticizer**
 d Mörtelweichmacher *m*
 f plastifiant *m* du mortier
 nl plastificerend middel *n* voor mortel
 r пластификатор *m* [пластифицирующая добавка *f*] раствора

M482 *e* **mortar pump**
 d Mörtelpumpe *f*
 f pompe *f* à mortier
 nl mortelpomp *f* (*m*)
 r растворонасос *m*

M483 *e* **mortar sand**
 d Mörtelsand *m*
 f sable *m* à mortier
 nl mortelzand *n*
 r песок *m* для раствора

M484 *e* **mortar spreader**
 d Mörtelverteiler *m*
 f épandeur *m* [épandeuse *f*] de mortier
 nl mortelverspreider *m*
 r распределитель *m* раствора

M485 **mortar streaks** *see* **streaks of mortar**

M486 *e* **mortice** *UK*, **mortise**
 d Ankerloch *n*, Nut *f*, Schlitz *m*
 f mortaise *f*
 nl gat *n* (*voor een pen, tap, stift*)
 r гнездо *n*, паз *m*

M487 *e* **mortise-and-tenon joint**
 d Zapfenverbindung *f*, Zapfenfuge *f*, Verzapfung *f*
 f assemblage *m* à tenons et mortaises [à enture et goujon]
 nl pen-en-gatverbinding *f*
 r соединение *n* в шип

M488 *e* **mortise chisel**
 d Stemmeisen *n*
 f bédane *m*, bec *m* d'âne
 nl hakbeitel *m*, vermoorbeitel *m*
 r долото *n*; стамеска *f*

M489 *e* **mortise gauge**
 d Streichmaß *n* für Tischler- und Zimmererarbeiten
 f trusquin *m* pour mortaises
 nl kruishout *n*, ritshout *n*
 r рейсмус *m* (*для разметки шипов или гнёзд*)

M490 **mortise joint** *see* **mortise-and-tenon joint**

M491 *e* **mortise lock**
 d Einsteckschloß *n*, Einlaßschloß *n*, Blindschloß *n*
 f serrure *f* encastrée [mortaisée, lardée]
 nl insteekslot *n*
 r врезной замок *m*

M492 *e* **mortise machine**
 d Stemmaschine *f*
 f mortaiseuse *f*, machine *f* à mortaiser
 nl langgatboormachine *f*
 r долбёжник *m*

M493 *e* **mortise pin**
 d Dübelnagel *m*
 f cheville *f* (*d'un assemblage à tenon et mortaise*)
 nl toognagel *m*
 r нагель *m* [шкант *m*] для закрепления шипа в гнезде

M494 *e* **mortising**
 d Ausstemmen *n*, Einstemmen *n*, Einschlitzen *n*, Verzapfen *n*, Zapfenlochen *n*
 f mortaisage *m*
 nl gaten steken
 r долбление *n* гнезд; выборка *f* пазов

M495 *e* **mosaic floor, mosaic flooring**
 d Mosaikfußboden(belag) *m*
 f revêtement *m* de sol [couvre-sol *m*] mosaïque
 nl mozaïekvloer *m*
 r мозаичный пол *m*

M496 *e* **mosaic tile**
 d Mosaikplatte *f*, Mosaikfliese *f*
 f carreau *m* (en) mosaïque
 nl mozaïektegel *m*
 r мозаичная плитка *f*

M497 *e* **motor-bug**
   d Motorjapaner *m*
   f benne *f* automotrice, motobrouette *f*
   nl motorjapanner *m*
   r моторизованная тележка *f* [тачка *f*]

M498 *e* **motor damper**
   d Motorklappe *f*
   f régistre *m* d'air mécanisé, clapet *m*
   d'entrée d'air à commande
   nl klep *f* (*m*) met servomotor
   r приводной воздушный клапан *m*

M499 *e* **motor-generator set**
   d Motor-Generator-Aggregat *n*,
   Motorgenerator *m*
   f groupe *m* moteur-générateur
   nl motorgenerator *m*
   r мотор-генераторная силовая
   установка *f*, двигатель-генератор *m*

M500 *e* **motor grader**
   d Motor-Straßenhobel *m*, Motor-
   -Straßenplanierer *m*
   f motograder *m*, niveleuse *f*
   automotrice
   nl motorgrader *m*, wegschaaf *f* (*m*)
   r самоходный грейдер *m*, автогрейдер
   m

M501 *e* **motor-in-head vibrator**
   d Innenrüttler *m* mit Einbaumotor,
   Rüttellanze *f*, Nadelvibrator *m*
   [Nadelrüttler *m*] mit eingebautem
   Motor
   f vibr(at)eur *m* [pervibrateur *m*] à
   aiguille à moteur incorporé, aiguille *f*
   vibrante à moteur incorporé
   nl trilnaald *f* (*m*) met ingebouwde
   motor
   r вибробулава *f*, погружной вибратор
   m (для бетона) со встроенным
   двигателем в наконечнике

M502 *e* **motorized solar control blinds**
   d mechanisch betätigte Sonnenschutz-
   -Jalousie *f*
   f jalousie *f* à lames mobiles à
   commande (*pouvant régulariser
   l'admission de la lumière solaire*)
   nl mechanisch bediende zonwering *f*
   r механические жалюзи *pl*,
   регулирующие поступление
   солнечного света

M503 *e* **motor operated valve**
   d Ventil *n* mit Stellantrieb,
   mechanisch betätigtes Ventil *n*
   f soupape *f* à commande mécanique
   nl mechanisch bediend ventiel *n*
   r клапан *m* с механическим приводом

M504 *e* **motor scraper**
   d Autoschrapper *m*, Motor-Schürfzug *m*
   f motorscraper *m*, décapeuse *f*
   automotrice
   nl motorschraper *m*
   r самоходный скрепер *m*

M505 *e* **motorway** *UK*
   d Autobahn *f*

   f autoroute *f*
   nl autosnelweg *m*
   r автомобильная дорога *f*
   с пересечениями в разных уровнях;
   автострада *f*

M506 *e* **mottled discoloration**
   d Fleckigkeit *f*; Fleckenbildung *f*
   infolge Entfärbung (*Betonfehler*)
   f tacheture *f*; formation *f* de taches
   décolorées
   nl vlekkerigheid *f*; pleksgewijze
   ontkleuring *f* (*betonfout*)
   r пятнистость *f*; образование *n*
   обесцвеченных пятен (*дефект
   бетона*)

M507 *e* **mottled surface**
   d gesprenkelte Oberfläche *f* (*Fehler*)
   f surface *f* à souillures (*défaut de
   surface*)
   nl gespikkeld oppervlak *n*
   r испещренная поверхность *f*,
   поверхность *f*, покрытая крапинками
   (*дефект*)

M508   **mottle effect** *see* **mottled discoloration** 2.
   mould... *see* **mold...**

M509 *e* **mound breakwater**
   d abgeböschte Meeresschutzanlage *f*,
   wellenbrechende Seeschutzanlage *f*,
   Steinschüttungs-Wellenbrecher *m*
   f ouvrage *m* de défense [brise-lames *m*]
   à talus incliné
   nl golfbreker *m* van stortsteen
   r внешнее оградительное сооружение
   n откосного профиля, откосное
   оградительное сооружение *n*

M510 *e* **mouthpiece**
   d Ansatzstutzen *m*, Ansatzstück *n*
   f ajutage *m*, embouchure *f*
   nl mondstuk *n*
   r гидравлический насадок *m*

M511 *e* **movable bridge**
   d bewegliche Brücke *f*
   f pont *m* mobile
   nl beweegbare brug *f* (*m*)
   r разводной мост *m*

M512   **movable dam** *see* **movable weir**

M513 *e* **movable distributor**
   d Fahrsprenger *m*, Wandersprenger *m*
   f distributeur *m* d'eau mobile
   nl sproeiwagen *m*
   r подвижной водораспределитель *m*
   (*на биофильтре*)

M514 *e* **movable joint**
   d bewegliche Verbindung *f*
   f joint *m* mobile [glissant]
   nl beweeglijke verbinding *f*
   r подвижное [податливое] соединение

M515 *e* **movable partition**
   d bewegliche [versetzbare] Trennwand *f*
   f cloison *m* amovible [mobile]
   nl wegschuifbare scheidingswand *m*
   r передвижная [переставная]
   перегородка *f*

M516 *e* **movable rocker bearing**
 d bewegliches Bolzenkipplager *n*
 f appareil *m* d'appui mobile aux balanciers
 nl beweeglijke oplegging *f* (*pendeloplegging*)
 r подвижная шарнирная опора *f*

M517 *e* **movable scaffolding**
 d bewegliches Gerüst *n*
 f échafaudage *m* roulant
 nl verplaatsbare steiger *m*
 r передвижные [катучие] подмости *pl*

M518 *e* **movable span**
 d bewegliche Öffnung *f*, bewegliches Feld *n* (*Brücke*)
 f travée *f* mobile (*d'un pont*)
 nl beweegbare overspanning *f*
 r разводной пролёт *m* (*моста*)

M519 *e* **movable tangential bearing**
 d bewegliches Tangentiallager *n*
 f appareil *m* d'appui mobile tangentiel [à plaques de glissement]
 nl beweegbaar tangentiëellager *n*
 r подвижная тангенциальная опора *f*

M520 *e* **movable weir**
 d bewegliches [absenkbares] Wehr *n*, mehrteilige Staumauer *f*
 f barrage *m* mobile [à bouchures mobiles]
 nl stuw *m* met beweegbare keringen
 r разборная [разборчатая] плотина *f*

M521 **moving forms** *see* **mobile form**

M522 *e* **moving load**
 d Wanderlast *f*, bewegliche Last *f*
 f charge *f* mobile
 nl rollende [mobiele] last *f* (*m*)
 r подвижная нагрузка *f*

M523 *e* **moving ramp**
 d steigendes Fußgängerlaufband *n*
 f rampe *f* pour piétons roulante, plan *m* incliné roulant
 nl vlakke roltrap *f* (*m*)
 r плоский эскалатор *m* (*без ступеней*); движущийся пандус *m*

M525 *e* **moving walk**
 d Fußgängerlaufband *n*, Fußgängerrollband *n*, rollender Gehsteig *m*, Roll(geh)steig *m*
 f troittoir *m* roulant
 nl rollend frottoir *n*
 r движущийся [подвижной] тротуар *m*

M526 **mucker** *see* **muck loader**

M527 *e* **mucker belt**
 d Schuttergurt *m*
 f convoyeur *m* de taille (d'abattage) [d'extraction]
 nl puintransportband *m*
 r породоуборочный конвейер *m*

M528 *e* **mucking**
 d 1. Schuttern *n*, Fortschaffen *n* der Ausbruchmassen *n*
 2. Kontrolle *f* der Entwurfslage der Bewehrung (*beim Betonieren*)

 f 1. extraction *f* d'abattage 2. contrôle *m* [vérification *f*] de disposition admise de l'armature pendant le bétonnage
 nl 1. afvoer *m* van breuksteen 2. controle *f* op de plaatsing van de wapening
 r 1. уборка *f* породы; выемка *f* грунта (*строительство туннелей*) 2. контроль *m* за проектным положением арматуры (*при бетонировании*)

M529 *e* **muck loader**
 d Haufwerklader *m*, Haufwerklademaschine *f*, Schuttermaschine *f*
 f chargeuse *f* d'abattage, pelle-chargeuse *f* de mine
 nl laadmachine *f* voor uitgebroken steen
 r породопогрузочная машина *f*

M530 **muck-lock** *see* **materials lock**

M531 *e* **mud**
 d 1. Bohrspülung *f* 2. Schlick *m*, schlickiger Ton *m*; Schlamm *m*
 f 1. boue *f* de forage 2. boue *f*, boue *f* schlammeuse [de l'argile]
 nl 1. boorsuspensie *f* 2. modder *m*
 r 1. буровой раствор *m* 2. ил *m*, иловатая глина *f*; шлам *m*

M532 *e* **mud auger**
 d Schappe *f*, Schappenbohrer *m*
 f tarière *f* ouverte [à cuiller]
 nl grondboor *f* (*m*)
 r буровая ложка *f*, желонка *f*

M533 *e* **mud box**
 d Schlammfang *m*, Schlammsammler *m*
 f puisard *m*
 nl slibvanger *m*
 r грязевик *m* (*на трубопроводе*)

M534 *e* **mud-capping**
 d Auflegerschießen *n*, Schießen *n* mit aufgelegter Ladung
 f tir *m* [sautage *m*] de gros blocs par charges superficielles
 nl springen *n* met opgestopte lading
 r подрывание *n* (*валунов*) поверхностным зарядом

M535 **mud filling** *US see* **colmatage**

M537 *e* **mud-jacking**
 d Mud-Jack-Verfahren *n*
 f égalisation *f* [aplanissement *m*] des dalles routiers par injection du coulis sous leur surface d'appui (à travers des trous forés)
 nl verzakte betonplaten weer stellen door dunne specië via geboorde gaten te injecteren
 r выравнивание *n* просевшего дорожного покрытия нагнетанием под бетонные плиты раствора через просверленные отверстия

M538 **mud mat** *see* **mud slab**

M539 *e* **mud outlet**
 d Schlammablaßstutzen *m*

f tubulure *f* d'évacuation de boue
nl slijkuitlaatbuis *f* (*m*)
r грязеспускной патрубок *m*

M541 *e* **mudsill**
d Grundbalken *m*, Grundsohle *f*,
Schwelle *f*; unterer Kranz *m*
f sablière *f* de boue (*reposant sur le sol*)
nl drempel *m*
r лежень *m*; опорная подкладка *f*
[опорный брус *m*] под стойками;
нижний венец *m*

M542 *e* **mud slab**
d Betongrundplatte *f* im feuchten
Schlickboden
f dalle *f* de fondation reposant sur le
terrain meuble et humide
nl funderingsplaat *f* (*m*) van gewapend
beton op slappe grond
r бетонное основание *n или* бетонная
подушка *f*, укладываемые в мягких
влажных грунтах

M543 *e* **mud soil**
d Schlammboden *m*
f sol *m* [terrain *m*] limoneux
nl moddergrond *m*
r илистый грунт *m*

M544 *e* **mud sump, mud trap**
d Schmutzfänger *m*, Schmutzfangkorb *m*,
Schlammfang *m*, Schlammfänger *m*,
Schlammkasten *m*
f collecteur *m* de boue, puisard *m*
nl slibvanger *m*
r грязевик *m*, грязеуловитель *m*

M545 *e* **mullion**
d 1. leichte Stahlstütze *f*
2. Fachwerkstütze *f*
3. Mittelpfosten *m* 4. Kämpfer *m*,
Anschlagleiste *f*
f 1. support *m* léger en acier 2. montant
m de pan de fer 3. meneau *m*
4. montant *m* central [de battement]
nl 1. lichte stalen steun *m* 2. verticaal
f (*m*) in vakwerk 3. middenstijl *m*
4. sluitstijl *m*
r 1. лёгкая стальная опора *f*
2. стойка *f* фахверка 3. средник *m*
4. импост *m*, притворный брус *m*

M546 *e* **mulseal**
d Bitumenemulsion *f*, Asphaltemulsion *f*
f bitume *m* émulsifié, émulsion *f* de
bitume
nl bitumenemulsie *f*
r эмульгированный битум *m*,
битумная эмульсия *f*

M547 *e* **multi-arch dam**
d Gewölbereihen(-Pfeiler)staumauer *f*,
Pfeilergewölbestaumauer *f*,
Bogenreihenstaumauer *f*,
Vielfachbogen-Mauer *f*
f barrage *m* à voûtes multiples
nl pijlerstuwdam *m* met steunbogen
r многоарочная плотина *f*

M548 **multiblade damper** *see* **multiple leaf
damper**

M549 *e* **multi-branch fitting**
d Abzweigstück *n* (*mit mehr als einer
Abzweigung*)
f raccord *m* à plusieurs branches,
nourrice *f*
nl verdeelstuk *n*
r разветвление *n* (*фитинг*)

M550 *e* **multu-bucket dredger**
d Eimerketten-Naßbagger *m*
f drague *f* à chaîne à godets
nl emmerbaggermolen *m*
r многочерпаковый земснаряд *m*

M551 *e* **multi-bucket excavator**
d Eimer(trocken)bagger *m*
f excavateur *m* à chaîne à godets
nl excavateur *m*
r многоковшовый экскаватор *m*

M552 *e* **multi-cell battery mould**
d Formbatterie *f*
f moule *m* multiple
nl mallenbatterij *f*
r батарейная [групповая] форма *f* (*для
сборных железобетонных
конструкций*)

M553 *e* **multi-cell dust collector**
d Zellenluftfilter *n*, *m*,
Mehrzellenentstauber *m*
f dépoussiéreur *m* à cellules multiples
nl cellenluchtfilter *n*, *m*
r ячейковый пылеуловитель *m*

M554 *e* **multicolour finish**
d Mehrfarbenanstrich *m*, mehrfarbige
Farbgebung *f*
f peinturage *m* [peinture *f*] multicolore
nl schilderwerk *n* in meer kleuren
r разноцветная окраска *f*

M555 *e* **multi-compartment building**
d Mehrfamilienhaus *n*
f bâtiment *m* collectif [d'appartements]
nl appartementengebouw *n*, flatgebouw
n; meergezinswoning *f*
r многоквартирный дом *m*,
многоквартирное жилое здание *n*

M556 *e* **multi-compartment settling basin**
d Mehrkammerabsetzbecken *n*,
Mehrkammerkläranlage *f*
f bassin *m* de décantation à plusieurs
compartiments
nl gecompartimenteerd bezinkbekken *n*
r многокамерный отстойник *m*

M557 *e* **multi-cored brick**
d Lochziegel *m*
f brique *f* perforée
nl geperforeerde baksteen *m*
r многодырчатый кирпич *m*

M558 *e* **multideck-screen**
d Mehrdecker *m*, Mehrdeckersieb *n*
f crible *m* à étages multiples
nl etagezeef *f* (*m*)
r многодечный грохот *m*

M559 *e* **multi-degree system**
d System *n* mit mehreren
Freiheitsgraden

413

f système *m* à plusieurs degrés de
liberté
nl systeem *n* met veel vrijheidsgraden
r система *f* со многими степенями
свободы

M560 *e* **multielement member**
d zusammengesetztes vorgespanntes
Bauelement *n*
f pièce *f* [poutre *f*] composée d'éléments
préfabriqués (*juxtaposés et
postcontraints*)
nl samengesteld (voorgespannen)
bouwelement *n*
r составной предварительно
напряжённый элемент *m* из
отдельных блоков (*с натяжением
арматуры на торцы крайних блоков*)

M561 *e* **multielement prestressing**
d Vorspannen *n* des zusammengesetzten
Bauelements
f postcontrainte *f* des pièces [des
poutres] composées d'éléments
préfabriqués
nl samengestelde bouwelementen
voorspannen
r предварительное напряжение *n*
конструкций, состоящих из
отдельных сборных блоков
(*соединяемых напрягаемой
арматурой*)

M562 *e* **multifolding door**
d Harmonikatür *f*, Falttür *f*
f porte *f* en accordéon à vantaux
multiples
nl vouwdeur *f* (*m*)
r подвесная складывающаяся дверь *f*
(*из нескольких полотен*)

M563 *e* **multi-lane roadway**
d mehrspurige Fahrbahn *f*
f chaussée *f* à plusieurs voies
nl weg *m* met meer rijstroken,
meerbaansweg *m*
r многополосная проезжая часть *f*
дороги

M564 *e* **multi-layer consolidation**
d schichtenweise Bodenverdichtung *f*
f consolidation *f* du sol par couches
nl laagsgewijze verdichting *f*
r послойное уплотнение *n* грунта

M565 *e* **multilayer insulation**
d mehrschichtige Wärmedämmung *f*
[Isolation *f*]
f isolation *f* [isolement *m*] thermique
multicouche
nl meervoudige warmteïsolatie *f*
r многослойная (тепло)изоляция *f*

M566 **multileaf damper** *see* **multiple leaf
damper**

M567 *e* **multi-legged sling**
d mehrsträngiger Stropp *m*,
Vielseilanhängemittel *n*
f élingue *f* à plusieurs brins
nl strop *m* met meer dan één lus
r многоветвевой строп *m*

M568 *e* **multi-level guyed tower**
d mehrfach abgespannter Mast *m*
f mât *m* à plusieurs niveaux de haubans
nl op meer dan één niveau getuide mast
m
r мачта *f* с несколькими ярусами
оттяжек

M569 *e* **multi-level junction**
d Kreuzung *f* in mehreren Ebenen
f croisement *m* à plusieurs niveaux,
ouvrage *m* de croisement dénivelé
nl kruising *f* op verschillende niveaus
r пересечение *n* дорог в разных
уровнях, сложная транспортная
развязка *f*

M570 *e* **multipass aeration tank, multiple
aeration tank**
d Mehrgangbelebungsanlage *f*
f bassin *m* d'aération [d'aérage] à
plusieurs couloirs
nl meervoudig oxydatiebed *n*
r многокоридорный аэротенк *m*

M571 *e* **multiple-arch bridge**
d Mehrbogenbrücke *f*
f pont *m* en arcs multiples
nl bogenbrug *f* (*m*)
r многопролётный арочный мост *m*

M572 *e* **multiple-dome dam**
d Pfeilerkuppel-Staumauer *f*,
Kuppelreihenstaumauer *f*,
Vielfach-Kuppelsperre *f*
f barrage *m* à dômes multiples
nl stuwdam *m* met steungewelven
r многокупольная плотина *f*

M573 **multiple dwelling building** *see*
multi-compartment building

M574 *e* **multiple-flue chimney**
d mehrzügiger Kamin *m*
f cheminée *f* à plusieurs carneaux
nl schoorsteen *m* met meerdere
rookkanalen
r многоствольная дымовая труба *f*

M575 *e* **multiple glass**
d Verbundglas *n*, Mehrscheibenglas *n*
f verre *m* multiple (isolant)
nl isolatieglas *n*
r многослойное стекло *n*

M576 *e* **multiple glazing**
d Verglasung *f* mit Verbundscheiben,
Vielfachverglasung *f*,
Mehrfacheinglasung *f*;
Verbundscheibe *f*,
Thermoscheibe *f*
f vitrage *m* multiple
nl beglazing *f* met isolatieglas
r многослойное остекление *n*;
стеклопакет *m*

M577 *e* **multiple leaf damper**
d Jalousieklappe *f*, Gliederklappe *f*
f registre *m* à persiennes
nl jaloezieklep *f* (*m*)
r многостворчатый [жалюзийный]
воздушный клапан *m*

M578 *e* **multiple-lift construction**
  d mehrschichtige Einbauweise *f*
  (*Straßenbau*)
  f corps *m* de chaussée à plusieurs
  couches
  nl wegconstructie *f* in meerdere lagen
  r многослойная конструкция *f*
  (*дорожного покрытия*)
M579 **multiple louver** *see* **multiple leaf
  damper**
M580 *e* **multiple-pipe inverted siphon**
  d Mehrfachdüker *m*
  f siphon *m* renversé à plusieurs tuyaux
  nl meervoudige duiker *m*
  r многотрубный дюкер *m*
M581 *e* **multiple purpose project**
  d 1. Mehrzweckausbauprojekt *n*,
  Mehrzweckausbauvorhaben *n*,
  Verbundprojekt *n* 2. Vielzweckbau-
  werk *n*, Mehrzweckausbau *m*
  f 1. projet *m* à buts multiples
  2. aménagement *m* hydraulique à buts
  multiples
  nl bouwproject *n* met verschillende
  bestemmingen
  r 1. многоцелевой проект *m*
  2. гидротехнический комплекс *m*
  многоцелевого назначения
M582 *e* **multiple-purpose reservoir**
  d Mehrzweckbecken *n*,
  Mehrzweck(wasser)speicher *m*
  f réservoir *m* à plusieurs usages
  [à usage multiple]
  nl stuwbekken *n* voor veschillende
  doeleinden
  r водохранилище *n* многоцелевого
  назначения
M583 *e* **multiple-row heating coil**
  d mehrreihiger Lufterhitzer *m*
  f serpentin *m* réchauffeur [de
  réchauffage] à spires multiples
  nl luchtverwarmingstoestel *n* met
  meerdere spiralen
  r многорядный калорифер *m*
M584 *e* **multiple sedimentation tank**
  d Mehrkammerabsetzbecken *n*
  f bassin *m* de décantation à plusieurs
  compartiments
  nl meervoudige sedimentatietank *m*
  r многокамерный отстойник *m*
M585 *e* **multiple span structure**
  d Mehrfeldbauwerk *n*
  f ouvrage *m* d'art à travées multiples
  nl constructie *f* met meerdere
  overspanningen
  r многопролётное сооружение *n*
M586 *e* **multiple surface treatement**
  d mehrfache Oberflächenbehandlung *f*
  (*Straßenbau*)
  f traitement *m* superficiel multicouche
  nl oppervlaktebehandeling *f* in meerdere
  lagen
  r многослойная поверхностная
  обработка *f* (*дорожного покрытия*)

M587 *e* **multiple-use development**
  d hydrotechnische Anlage *f*
  f aménagement *m* hydroélectrique
  [à buts multiples]
  nl energieproject *n* met meervoudige
  bestemming
  r гидроузел *m*
M588 *e* **multiple-web systems**
  d mehrteilige Fachwerke *n pl*
  [Gitterwerke *n pl*]
  f systèmes *m pl* de treillis multiples
  nl bouwsystemen *n pl* met meervoudige
  vakwerken
  r сложные [комбинированные] системы
  f pl решёток ферм
M589 *e* **multi-point heater**
  d Warmwasserbereiter *m* mit mehreren
  Anzapfstellen
  f chauffe-eau *m* [chauffe-bain *m*] à
  plusieurs prises d'eau
  nl warmwatertoestel *n* met verscheidene
  tappunten
  r водонагреватель *m*, питающий
  несколько водоразборных точек
M591 *e* **multi-purpose coating plant**
  d Makadam- und Asphalt-
  -Mischanlage *f*
  f poste *m* d'enrobage universel
  [à plusieurs usages]
  (*pour préparation des enrobés
  hydrocarbonés différents*)
  nl macadam- en asfaltinstallatie *f*
  r универсальная смесительная
  установка *f* для приготовления
  разных смесей на органическом
  вяжущем (*для дорожных покрытий*)
M592 **multi-purpose scheme** *see* **multiple-
  -purpose project**
M593 **multi-rubber-tire roller** *US see*
  multi-wheel roller
M594 *e* **multistage construction**
  d abgestuftes Bauvorgehen *n*,
  stufenweiser Ausbau *m*
  f réalisation *f* de construction par
  étapes
  nl uitvoering *f* (van het werk) **in fasen**
  r поэтапное строительство *n*
M595 *e* **multi-stage fan**
  d Mehrstufenventilator *m*,
  Vielstufengebläse *n*
  f ventilateur *m* multi-étagé
  nl meertrapsventilator *m*
  r многоступенчатый вентилятор *m*
M596 *e* **multistage gas-supply system**
  d druckgestuftes Gasversorgungssystem *n*
  f système *m* d'approvisionnement en
  gaz à plusieurs étages
  nl getrapt gasdistributiesysteem *n*
  r многоступенчатая система *f*
  газоснабжения
M597 *e* **multistage stressing**
  d Vorspannung *f* in mehreren Stufen
  f précontrainte *f* à plusieurs étapes

nl in fasen opgewekte voorspanning *f*
r многоступенчатое
предварительное напряжение *n*
(*железобетонных конструкций*)

M598 *e* **multi-storey building**
d Etagenhaus *n*, Geschoßhaus *n*,
Stockwerkbau *m*
f immeuble *m* à plusieurs étages
nl gebouw *n* met meer dan één
verdieping
r многоэтажное здание *n*

M599 *e* **multiuse building**
d Mehrzweckgebäude *n*
f bâtiment *m* à usage multiples
nl gebouw *n* met meervoudige
bestemming
r здание *n* многоцелевого
назначения, многофункциональное
здание *n*

M600 **multiway junction** *see* **multilevel
junction**

M601 *e* **multi-wheel roller**
d pneumatische Vielradwalze *f*,
Gummireifenvielfachwalze *f*
f compacteur *m* [rouleau *m*, cylindre *m*]
à pneus multiples
nl bandenwals *f* (*m*)
r многоколёсный пневмокаток *m*

M602 *e* **multi-zone system**
d Mehrzonenklimaanlage *f*
f installation *f* de conditionnement
d'air à zones multiples
nl klimaatregelingssysteem *n* met
meerdere zones
r многозональная система *f*
кондиционирования воздуха

M603 *e* **municipal engineering**
d 1. Städtebau *m* 2. Stadtwirtschaft *f*,
städtischer Tiefbau *m*
f 1. urbanisme *m*, technique *f* de
l'aménagement des villes, planification
f urbaine 2. exploitation *f* et
construction *f* des équipements
collectifs [des équipements
d'infrastructure]
nl openbare werken *n* *pl*
r 1. градостроительство *n*
2. городское хозяйство *n*;
проектирование *n*, строительство *n*
и эксплуатация *f* городских
инженерных сетей и сооружений

M604 *e* **municipal facilities**
d Ingenieurbauwerke *n* *pl* der
Stadtwirtschaft,
Stadtversorgungs- und
Entsorgungsnetze *n* *pl*
f équipements *m* *pl* collectifs;
services *m* *pl* communautaires d'une
ville
nl gemeentelijke diensten *m* *pl*
r инженерные сооружения *n* *pl*
[сети *f* *pl*] городского хозяйства

M605 *e* **municipal sewage**
d städtisches Abwasser *n*

f effluents *m* *pl* urbains, eaux *f* *pl*
usées urbaines
nl gemeentelijk afvalwater *n*
r городские сточные воды *f* *pl*

M606 *e* **municipal sewers**
d Stadtentwässerungsnetz *n*
f réseau *m* d'égouts urbain
[d'assainissement des villes,
d'évacuation des eaux usées
urbaines]
nl gemeentelijk rioleringsstelsel *n*
r городская канализационная сеть *f*

M607 *e* **municipal treatment plant**
d städtische Abwasserreinigungsanlage *f*
f installation *f* urbaine d'épuration
[de traitement] des eaux usées
nl gemeentelijke
rioolwaterzuiveringsinstallatie *f*
r городская станция *f* очистки сточных
вод, очистная станция *f*

M608 *e* **muntin**
d Fenstersprosse *f*
f petit-bois *m*
nl middenstijl *m* van een deur
r горбылёк *m* (*оконного переплёта*)

M609 *e* **mushroom construction**
d Pilzdecke *f*
f plancher *m* champignon
nl paddestoelvloer *m*
r безбалочное (железобетонное)
перекрытие *n*

M610 *e* **mushroom-head column**
d Pilzdeckenstütze *f*
f poteau *m* à champignon [supportant
le plancher champignon]
nl kolom *f* (*m*) van een paddestoelvloer
r опорная колонна *f* безбалочного
перекрытия

M611 *e* **mushroom slab**
d Pilzdeckenplatte *f*
f dalle *f* de plancher champignon
nl paddestoelvloer *m*
r плита *f* безбалочного перекрытия

M612 **mushroom slab construction** *see*
mushroom construction

N

N1 *e* **nail**
d Nagel *m*
f clou *m*, pointe *f*
nl nagel *m*, spijker *m*
r гвоздь *m*

N2 *e* **nailable concrete**
d Nagelbeton *m*, nagelbarer Beton *m*
f béton *m* clouable
nl spijkerbaar beton *n*
r гвоздимый бетон *m*

N3 *e* **nail claw**
d Nagelzieher *m*

    f arrache-clou *m*, tire-clou *m*
nl spijkertrekker *m*
  r гвоздодёр *m*

N4 *e* **nailed beam**
   d genagelter Balken *m* [Träger *m*]
   f poutre *f* clouée
 nl gespijkerde balk *m*
  r гвоздевая балка *f*

N5 *e* **nailed connection, nailed joint**
   d Nagelverbindung *f*
   f assemblage *m* cloué
 nl genagelde verbinding *f*
  r гвоздевое соединение *n*

N6 *e* **nailed roof truss**
   d Nageldachbinder *m*
   f ferme *f* de comble [de toit] à
    assemblages par clous
 nl gespijkerd dakspant *n*
  r гвоздевая дощатая стропильная
    ферма *f*

N7   nailer *see* **nailing strip**
N8   nail float *see* **devil float**
N9 *e* **nailing**
   d Annageln *n*; Nageln *n*
   f clouage *m*; clouement *m*
 nl nagelen *n*, spijkeren *n*
  r прибивка *f* гвоздями; забивка *f*
    [забивание *n*] гвоздей

N10   nailing block *see* **wood brick**
N11   nailing concrete *see* **nailable concrete**
N12 *e* **nailing marker**
   d Nagellehre *f*, Nagelschablone *f*
   f gabarit *m* de clouage
 nl nagelsjabloon *f* (*m*)
  r шаблон *m* для забивки гвоздей

N13 *e* **nailing strip**
   d eingebaute nagelbare Leiste *f*
   f bande *f* [tringle *f*] de clouage
 nl ingestorte spijkerstrook *f* (*m*)
  r деревянная планка *f*, вделанная
    в поверхность жёсткого элемента
    для забивки гвоздей

N14 *e* **nail plate**
   d Nagel-Knotenblech *n*, Nagel-
    -Anschlußblech *n*
   f plate-bande *f* [plaque *f* métallique]
    clouée
 nl nagelplaat *f* (*m*)
  r стальная накладка *f* [фасонка *f*],
    прибиваемая гвоздями

N15 *e* **nail punch, nail set**
   d Nageltreiber *m*
   f chasse-clou *m*, chasse-pointe *m*
 nl drevel *m*, doorslag *m*
  r добойник *m* (*для утапливания
    шляпок гвоздей*)

N16 *e* **nail stake**
   d Stahlrohr-Absteckpflock *m*
   f piquet *m* de nivellement tubulaire
    [en tube]
 nl jalon *m*
  r инвентарный стальной трубчатый
    колышек *m* (*для геодезических работ*)

N17 *e* **naked flooring**
   d Blendboden *m*, Blindboden *m*,
    Rohboden *m*, Rohdecke *f*
   f sol *m* brut, plancher *m* (porteur) brut,
    sous-sol *m*
 nl kale [onafgewerkte] vloer *m*
  r чёрный пол *m*

N18 *e* **name block**
   d Schriftfeld *n*, Grundschriftfeld *n*
   f cartouche *m*, titre *m*
 nl naam *m* van de tekening, rechter
    onderhoek *m*
  r штамп *m* (*чертежа*)

N19 *e* **names of parts**
   d Stückliste *f*
   f nomenclature *f* des pièces
 nl stuklijst *f* (*m*)
  r спецификация *f* деталей

N20 *e* **Naples yellow**
   d Neapelgelb *n*
   f jaune *m* de Naples
 nl Napels geel *n*
  r сурьмяный жёлтый пигмент *m*

N21 *e* **nappe**
   d freier Überfallstrahl *m*,
    Schuß(wasser)strahl *m*
   f nappe *f* libre
 nl vrije overloop *m*, waterfilm *m*
  r свободная плоская струя *f* воды,
    слой *m* воды, переливающийся через
    порог *или* гребень гидросооружения

N22 *e* **nappe-shaped crest profile**
   d normalkroniger Überfall *m*
   f déversoir *m* de Creager
 nl vlakke overloop *m*
  r водослив *m* практического профиля

N23 *e* **narrow gauge**
   d Schmalspur *f*
   f voie *f* étroite
 nl smalspoor *n*
  r узкая колея *f*

N24 *e* **narrow-gauge railway**
   d Schmalspurbahn *f*
   f chemin *m* de fer à voie étroite
 nl smalspoor(baan) *n* (*f* (*m*))
  r узкоколейная железная дорога *f*

N25 *e* **natatorium**
   d Schwimmbad *n*
   f piscine *f*, bassin *m* de natation
 nl zwembassin *n*
  r плавательный бассейн *m*

N26   native asphalt *US see* **natural asphalt**
N27 *e* **natural adhesive**
   d natürlicher Kleber *m*
   f adhésif *m* naturel
 nl natuurlijk kleefmiddel *n*,
    natuurlijke lijm *m*
  r клей *m* из природных веществ

N28 *e* **natural asphalt**
   d Naturasphalt *m*, natürliches
    Bitumen-Mineralgemisch *n*;
    Naturbitumen *n*
   f asphalte *m* [bitume *m*] natif [naturel]

nl natuurlijk asfalt *n*
r природный асфальт *m* или битум *m*

N29 *e* **natural bed**
d Wildbett *n*
f lit *m* naturel
nl natuurlijke bedding *f*
r естественное русло *n*

N30 *e* **natural building sand**
d Naturbausand *m*
f sable *m* naturel pour travaux de construction
nl natuurzand *n* (*bouwmateriaal*)
r природный песок *m* для строительства

N31 *e* **natural cement**
d Naturzement *m*; Mischzement *m*; hydraulischer Kalk *m*
f ciment *m* naturel; ciment *m* mélangé [mixte]
nl natuurcement *n*
r натуральный портландцемент *m*, цемент *m* из природного мергеля; смешанный цемент *m* (*с гидравлическими добавками*); гидравлическая известь *f* (*получаемая обжигом ниже предела спекания*)

N32 *e* **natural circulation**
d natürlicher Umlauf *m*, natürliche [freie] Zirkulation *f*
f circulation *f* naturelle
nl natuurlijke circulatie *f*
r естественная циркуляция *f*

N33 **natural clear varnish** *see* **natural finish**

N35 *e* **natural concrete aggregate**
d natürlicher Betonzuschlag(stoff) *m*
f granulat *m* [agrégat *m*] naturel de béton
nl natuurlijke betontoeslag *m*
r природный заполнитель *m* для бетона

N36 *e* **natural convection**
d natürliche [freie] Konvektion *f*
f convection *f* naturelle
nl (natuurlijke) convectie *f*
r естественная конвекция *f*

N37 *e* **natural convector**
d Konvektor *m* mit natürlichem Luftumlauf, stiller Konvektor *m*
f convecteur *m* à circulation naturelle
nl convector *m* met natuurlijke luchtcirculatie
r конвектор *m* с естественной циркуляцией воздуха

N38 *e* **natural depth**
d natürliche Wassertiefe *f*
f profondeur *f* naturelle
nl natuurlijke diepte *f*
r бытовая глубина *f*

N39 *e* **natural draft cooling tower**
d selbstsaugender Kühlturm *m*, Kühlturm *m* mit freier Luftbewegung
f tour *f* de réfrigération [de refroidissement] à tirage naturel

nl koeltoren *m* met natuurlijke trek
r градирня *f* с естественной тягой

N40 *e* **natural drainage**
d natürliche Dränung *f* [Entwässerung *f*]
f drainage *m* naturel [libre]
nl natuurlijke drainage *f*
r естественный дренаж *m*

N41 *e* **natural draught**
d natürlicher Zug *m*
f tirage *m* naturel
nl natuurlijke trek *m*
r естественная тяга *f*

N42 *e* **natural escape**
d Überlaufkanal *m* ohne Auskleidung
f canal *m* latéral de décharge en terre
nl overloopkanaal *n* zonder oeverbekleding
r боковой водосбросный канал *m* без облицовки

N43 *e* **natural features**
d spezifische Merkmale *n pl* [Besonderheiten *f pl*] der natürlichen Umgebung
f particularités *f pl* naturelles
nl natuurlijke eigenschappen *f pl*
r природные особенности *f pl*

N44 *e* **natural finish**
d Naturton *m* (*Holzoberfläche*)
f fini *m* naturel de surface de bois
nl natuurlijk oppervlak *n*
r прозрачная отделка *f* (*древесины*)

N45 *e* **natural flow**
d Flußdarbietung *f*, natürlicher Abfluß *m*, Abfluß vor der Regelung, unausgeglichener Abflußgang *m*
f écoulement *m* [débit *m*] naturel
nl natuurlijke afvloeiing *f*
r незарегулированный [естественный, бытовой] сток *m*

N46 *e* **natural foundation**
d tragfähiger Boden *m*, Bodengründung *f*
f fondation *f* naturelle, sol *m* de fondation naturel
nl fundering *f* op staal
r естественное [грунтовое] основание *n*

N47 *e* **natural ground**
d Erdreich *n*, Untergrund *m*; gewachsener Boden *m*
f sol *m* de fondation naturel; terrain *m* [sol *m*] naturel
nl natuurlijke grondslag *m*, ongeroerde grond *m*
r грунт *m* в основании сооружения, грунтовый массив *m*; естественный грунт *m*

N48 *e* **natural lighting**
d Tageslichtbeleuchtung *f*, natürliche Beleuchtung *f*
f éclairage *m* naturel
nl verlichting *f* met daglicht
r естественное освещение *n*

N49 *e* **natural mineral materials**
d natürliche Mineralbaustoffe *m pl*

f matériaux *m pl* minéraux naturels
nl natuurlijke minerale bouwstoffen
f (m) pl
r природные неорганические
материалы *m pl*

N50 *e* **natural pigment** *see* **earth pigment**

N53 *e* **natural rock asphalt**
d Naturasphalt *m*, natürliche
Bitumen-Mineralmischung *f*
f asphalte *m* en roche naturel, roche *f*
bitumineuse (naturelle)
nl bitumineus gesteente *n*
r асфальтовая [битумная] порода *f*

N54 *e* **natural sett**
d Naturpflasterstein *m*
f pavé *m* naturel [de pierre naturelle]
nl keisteen *n*, *m*, straatkei *m*
r природный камень *m* для мощения,
брусчатка *f*

N55 *e* **natural slope**
d 1. natürliche Böschung *f*
2. Ruhewinkel *m*, natürlicher
Böschungswinkel *m*
f 1. talus *m* naturel 2. angle *m* de
talus naturel
nl 1. natuurlijke helling *f*
2. natuurlijk talud *n*
r 1. естественный откос *m* 2. угол *m*
естественного откоса

N56 **natural soil** *see* **undisturbed soil**

N57 *e* **natural stone**
d Naturstein *m*
f pierre *f* naturelle
nl natuursteen *n*
r природный (строительный) камень *m*,
природные каменные материалы *m pl*

N58 *e* **natural stone veneer**
d Natursteinverblendung *f*
f perré *m*, revêtement *m*
en pierre de taille
nl bekleding *f* met natuursteen
r облицовка *f* естественным камнем

N59 *e* **natural stonework**
d Naturstein-Mauerwerk *n*
f maçonnerie *f* en pierres naturelles
nl metselwerk *n* van natuursteen
r кладка *f* из естественного камня

N60 *e* **natural ventilation**
d natürliche [freie] Lüftung *f*
f ventilation *f* naturelle
nl natuurlijke ventilatie *f*
r естественная вентиляция *f*

N61 *e* **natural water level**
d natürlicher Wasserstand *m*,
Wasserspiegel *m* vor der Regelung
f niveau *m* naturel, plan *m* d'eau
primitif
nl natuurlijk waterpeil *n*
r бытовой уровень *m*

N62 *e* **Navier's hypothesis**
d zweidimensionale Theorie *f*
(*Ebenbleiben des Querschnittes*)
f hypothèse *f* [théorie *f*] de Navier

nl tweedimensie-theorie *f*, hypothese *f*
van Navier
r гипотеза *f* плоских сечений, гипотеза
f Навье

N63 *e* **navigable canal**
d Schiffahrtskanal *m*
f chenal *m* [canal *m*] navigable
nl bevaarbaar kanaal *n*
r судоходный канал *m*

N64 *e* **navigation facilities**
d Schiffahrtsanlagen *f pl*
f ouvrages *m pl* permettant la
navigation
nl scheepvaartvoorzieningen *f pl*
r судоходные сооружения *n pl*

N65 *e* **navigation lock**
d Schiff(ahrt)sschleuse *f*
f écluse *f* de navigation
nl schutsluis *f (m)*
r судоходный шлюз *m*

N66 *e* **navigation pass**
d Schiffahrtsbauwerk *n*
f ouvrage *m* de franchissement
nl doorvaart *f (m)*
r судопропускное сооружение *n*

N67 *e* **nearside lane**
d Bordstreifen *m*, Befestigungsstreifen *m*
(*Autobahn*)
f bande *f* latérale (*d'une chaussée*)
nl buitenste rijstrook *f (m)*
r краевая [укрепительная] полоса *f*
(*дороги*)

N68 *e* **neat cement**
d 1. reiner Zementmörtel *m*,
Zementpaste *f* 2. nichthydratisiertes
hydraulisches Bindemittel *n*
f 1. pâte *f* de ciment pure 2. ciment *m*
non hydraté
nl 1. cementmortel *m* zonder zand
2. hydraulisch bindmiddel *n*
r 1. цементное тесто *n*
2. гидравлическое вяжущее *n*
в дегидратированном состоянии

N69 *e* **neat cement grout, neat cement paste**
d 1. Zementpaste *f* (*ohne Zuschläge*)
2. Zementstein *m*
f 1. pâte *f* de ciment pure (*sans
adjuvents*) 2. pâte *f* de ciment durcie,
mortier *m* de béton durci
nl 1. cementbrij *m* zonder toeslag
2. cementsteen *n*, *m*
r 1. цементное тесто *n* (*без добавок*)
2. цементный камень *m*

N70 *e* **neat line**
d 1. Grenze *f* zwischen Tage- und
Untertagebau 2. Baufluchtlinie *f*
3. Gebäudeflucht *f*
f 1. ligne *f* délimitant un souterrain
2. alignement *m* des bâtiment
3. ligne *f* [alignement *m*] de façade
nl 1. grens *f (m)* tussen fundering en
schoon werk 2. rooilijn *f (m)* 3. lijn
f (m) van de voorgevel

r 1. граница *f* горных выработок
2. красная *линия f* застройки
3. наружная граничная *линия f* здания, *линия f* лицевого фасада

N72 e **necking**
d Einschnürung *f*
f striction *f* (progressive)
nl insnoering *f*
r образование *n* шейки, сужение *n* поперечного сечения (*растягиваемого образца*)

N73 e **needle**
d 1. Nadel *f* 2. Abfangträger *m* 3. Tragarm *m* für Auslegergerüst
f 1. aiguille *f* 2. poutre *f* de reprise en sous-œuvre 3. poutre *f* en porte-à-faux supportant la plate-forme d'un échafaudage
nl 1. naald *f* (*m*) (*stuw*) 2. opvangbalk *m*, stempelbalk *m* 3. console(juk) *f* (*n*) van een steeksteiger
r 1. спица *f* (*затвора, плотины*) 2. балка *f* для опирания стены (*при усилении фундамента*) 3. палец *m*, пропущенный сквозь стену (*для выпускных лесов*)

N74 e **needle apparatus**
d Nadelgerät *n*
f appareil *m* à aiguille
nl penetrometer *m*
r пенетрометр *m*

N75 e **needle beam**
d 1. Abfangträger *m* 2. Nadellehne *f*
f 1. poutre *f* de reprise en sous-œuvre 2. poutre *f* de support supérieure de la vanne à pointeau
nl 1. stempelbalk *m*, dwarsdrager *m* (*brugdek*) 2. draagbalk *m* van een naaldstuw
r 1. балка *f* для опирания стены (*при усилении фундамента*) 2. верхний упорный прогон *m* спицевого затвора

N76 e **needle beam scaffold**
d Auslegergerüst *n*, fliegendes Gerüst *n*
f échafaudage *m* en porte-à-faux
nl steeksteiger *m*
r выпускные леса *pl* (*на пальцах, пропускаемых через проёмы в стене*)

N77 e **needle gate**
d Nadelverschluß *m*
f vanne *f* [robinet *m*] à pointeau
nl naaldafsluiting *f*
r спицевой затвор *m*

N78 e **needle shoring**
d 1. Unterfangen *n*, Unterfangung *f* (*mittels Abfangträger*) 2. Abfangträger *m pl* zur Abstützung der Wand
f 1. reprise *f* en sous-œuvre par poutres transversales 2. système *m* de poutres de reprises en sous-œuvre
nl 1. stutten *n* met een stempelbalk 2. stempeling *f* voor het stutten van een muur

N79 e **needle valve**
d 1. Nadelventil *n* 2. Ringschieber *m*
f 1. soupape *f* à pointeau 2. vanne *f* à pointeau
nl 1. naaldklep *f* (*m*) 2. ringschuif *f* (*m*)
r 1. игольчатый клапан *m* 2. игольчатый затвор *m*

N80 e **needle vibrator**
d Nadelvibrator *m*, Nadelrüttler *m*
f vibrateur *m* à aiguille, aiguille *f* vibrante
nl trilnaald *f* (*m*)
r вибробулава *f*, виброигла *f*

N81 e **needle weir**
d Nadelwehr *n*
f barrage *m* à aiguilles
nl naaldstuw *m*
r спицевая плотина *f*

N82 e **needling**
d 1. Abfangträger *m pl* zur Abstützung einer Wand (*beim Unterfangen*) 2. Durchbrechen *n*, Durchschlagen *n* (*im Mauerwerk oder Felsgestein zum Einsetzen der Tragarme*)
f 1. système *m* de poutres de support (*pour réprise en sous-œuvre des fondations*) 2. percement *m* du mur (*pour la pose des poutres de support*)
nl 1. methode *f* om het gewicht van de muur op te vangen tijdens reparatie aan de fundering 2. maken van openingen in een muur ter bevestiging van de steigerjukken
r 1. система *f* балок для опирания стены (*при усилении фундамента*) 2. пробивка *f* отверстий в стене (*для установки пальцев*)

N83 e **negative friction**
d negative Mantelreibung *f*
f friction *f* négative
nl negatieve wrijving *f*
r отрицательное поверхностное трение *n*

N84 e **negative moment**
d negatives Moment *n*
f moment *m* négatif
nl negatief moment *n*
r отрицательный момент *m*

N85 e **negative reinforcement**
d Bewehrung *f* im Bereich negativer Momente
f armature *f* soumise à l'action des moments fléchissants négatifs
nl wapening *f* tegen negatieve buigende momenten
r рабочая арматура *f*, воспринимающая усилия от отрицательных моментов

N86 *e* **negative pressure**
   d negativer Druck *m*
   f dépression *f*, contre-pression *f*
   nl negatieve druk *m*
   r разрежение *n*

N87   **negative source** *see* **sink 2., 3.**

N88   **negative well** *see* **absorbing well**

N89 *e* **neighbourhood unit**
   d Nachbarschaftseinheit *f*
   f unité *f* de voisinage
   nl buurteenheid *f*
   r микрорайон *m*

N90 *e* **neoprene paint**
   d Neopren(anstrich)farbe *f*
   f peinture *f* à base de néoprène
   nl neopreenverf *f (m)*
   r неопреновая краска *f*, окрасочный состав *m* на основе неопрена

N91 *e* **nerve, nervure**
   d Gewölberippe *f*
   f nervure *f*
   nl rib *f (m)* van een gewelf
   r нервюра *f*

N92 *e* **nest of sieves**
   d Prüfsiebsatz *m*
   f jeu *m* de tamis
   nl stel *m* zeven
   r набор *m* контрольных сит (*для ситового анализа*)

N93 *e* **net**
   d Netz *n*, Netzwerk *n*; Drahtgewebe *n*
   f filet *m*, grillage *m*, treillis *m*; toile *f* métallique
   nl net *n*, netwerk *n*
   r сетка *f*; металлическая ткань *f*

N94 *e* **net cut**
   d positive Erdmassenbilanz *f*
   f bilan *m* déblai-remblai positif
   nl grondoverschot *n*, overschot *n* van het massanivellement
   r положительный баланс *m* земляных масс; излишек *m* вынутого грунта (*после отсыпки всех насыпей*)

N95 *e* **net duty of water** *US*
   d Nettoabgabe *f*
   f tâche *f* nette de l'eau d'irrigation
   nl netto-waterafgifte *f*
   r норма *f* воды нетто, норма *f* воды для хозяйства

N96 *e* **net fill**
   d negative Erdmassenbilanz *f*
   f bilan *m* déblai-remblai négatif
   nl grondtekort *n*, tekort *n* bij het massanivellement
   r отрицательный баланс *m* земляных масс, нехватка *f* грунта (*для засыпки выемок*)

N97 *e* **net floor area**
   d reine Bebauungsfläche *f*
   f surface *f* bâti nette
   nl netto bebouwd oppervlak *n*

   r площадь *f* (застройки) здания нетто (*за вычетом конструктивной площади*)

N98   **net line** *see* **neat line**

N99 *e* **net positive suction head**
   d 1. zulässige Hohlsogreserve *f*
       2. geodätische Saughöhe *f*
   f 1. réserve *f* de cavitation admissible
       2. hauteur *f* d'aspiration d'une pompe
   nl 1. toelaatbare onderdruk *m*
       2. geodetische zuighoogte *f (pomp)*
   r 1. допускаемый кавитационный запас *m* 2. геометрическая высота *f* всасывания (*насоса*)

N100 *e* **net room area**
   d reine Zimmerfläche *f*, Nettoraumfläche *f*
   f aire *f* de la surface d'une pièce, superficie *f* de pièce nette
   nl netto vloeroppervlak *n* van een vertrek
   r площадь *f* комнаты

N101 *e* **net sectional area**
   d Nettoquerschnittsfläche *f*
   f section *f* [aire *f* de la section] nette
   nl netto oppervlakte *n* van de dwarsdoorsnede
   r площадь *f* сечения нетто

N102 *e* **net settlement**
   d reine Pfahlsetzung *f* (*nach der Wegnahme von Versuchslast*)
   f tassement *m* [affaissement *m*] d'un pieu net (*après l'enlèvement de la charge d'essai*)
   nl netto zakking *f* van de paal
   r чистая осадка *f* сваи (*после удаления пробной нагрузки*)

N103 *e* **network**
   d 1. Leitungsnetz *n* 2. Netzdiagramm *n*
   f 1. réseau *m*, canalisation *f* 2. graphe *m* de programme par chemin critique
   nl 1. netwerk *n* 2. netwerkdiagram *n*
   r 1. инженерная сеть *f* 2. сетевой график *m*

N104 *e* **network analysis, network planning**
   d Netzplanung *f*
   f planification *f* [planning *m*] par chemin critique
   nl netwerkplanning *f*
   r сетевое планирование *n*

N105 *e* **neutral axis**
   d neutrale Achse *f*, Nullinie *f*
   f axe *m* neutre
   nl neutrale lijn *f (m)*
   r нейтральная ось *f*, нулевая линия *f*

N106 *e* **neutral equilibrium**
   d indifferentes Gleichgewicht *n*
   f équilibre *m* indifférent
   nl indifferent evenwicht *n*
   r безразличное равновесие *n*

N107 *e* **neutralisation tank**
   d Neutralisationsbecken *n*, Neutralisationsreaktor *m*

f bassin *m* de neutralisation,
neutralisateur *m* (*des eaux usées industrielles*)
nl neutralisatiereactor *m*
r нейтрализатор *m* (*промышленных сточных вод*)

N108 e **neutral plane**
d neutrale Ebene *f* [Fläche *f*, Schicht *f*], Nullschicht *f*
f plan *m* neutre, surface *f* des fibres neutres
nl neutraal vlak *n*
r нейтральная плоскость *f*

N109 e **neutral pressure**
d neutraler Druck *m*, neutrale Druckspannung *f* [Flächenlast *f*] (*Bodenmechanik*)
f pression *f* neutre
nl neutrale korreldruk *m* [gronddruk] (*grondmechanica*)
r нейтральное [пороговое] давление *n*

N111 e **neutral surface**
d neutrale Fläche *f*
f surface *f* neutre
nl neutraal oppervlak *n*
r нейтральная поверхность *f* (*оболочки*)

N112 e **neutron shield**
d Neutronenschild *m*, Neutronenabschirmung *f*
f écran *m* protecteur contre les neutrons [la radiation neutronique]
nl neutronenscherm *n*
r экран *m* для защиты от нейтронов

N113 e **newel**
d 1. Treppenspindel *f* 2. Antritts- *bzw.* Austrittspfosten *m*
f 1. noyau *m* d'escalier à vis 2. poteau *m* d'arrivée
nl 1. spil *f* (*m*) (*van een spiltrap*) 2. trappaal *m*, hoofdbaluster *m*
r 1. центральная стойка *f* (*винтовой лестницы*) 2. концевая опорная стойка *f* (*лестничных перил*)

N114 **newel-post** *see* **newel** 2.

N115 e **newel stair**
d Spindeltreppe *f*
f escalier *m* à vis [en hélice]
nl spiltrap *m*
r винтовая [витая] лестница *f*

N116 e **newly-placed concrete**
d frisch eingebauter Beton *m*
f béton *m* frais
nl vers gestort beton *n*
r свежеуложенный бетон *m*

N117 e **nibbed tile**
d Dachziegel *m* mit Aufhängenasen
f tuile *f* à emboîtement (simple ou double)
nl dakpan *f* (*m*) met nok [neus]
r черепица *f* с шипами (*для зацепления за обрешётку*)

N118 e **nib guide**
d Simsabziehplatte *f*

f calibre *m* pour moulures d'un corniche
nl mal *m* voor het trekken van kroonlijsten
r правило *n* [шаблон *m*] для тяги карнизов

N119 e **niche**
d Nische *f*
f niche *f*
nl nis *f*(*m*)
r ниша *f*

N120 **nick** *US see* **notch** 1.

N121 e **nick-break test** *US*
d Kerbbruchversuch *m*
f essai *m* de rupture sur l'entaille
nl kerfslagbuigproef *f*(*m*)
r испытание *n* на излом образца с надрезом

N122 e **nidged ashlar, nigged ashlar**
d Naturbaustein *m* mit bearbeiteter Sichtfläche, außenflächiger Naturbaustein *m*
f pierre *f* de taille à face façonnée
nl natuursteen *m* met gebouchardeerd oppervlak
r офактуренный природный камень *m*

N123 e **nigging**
d Bouchardieren *n*, Aufspitzen *n*, Stocken *n* (*Stein*)
f bouchardage *m* des pierres de taille
nl boucharderen *n* (*natuursteen*)
r ударная обработка *f* (*бучардой*) (*камня*)

N124 e **night setback**
d Nachtabsenkung *f*
f réduction *f* de la température pour les heures de nuit; mise *f* (*d'un climatiseur*) en position pour régime thermique de nuit
nl aanpassen *n* van de regeling voor de nacht
r понижение *n* температуры на ночной период; перенастройка *f* (*регуляторов*) на пониженную температуру помещения *в* (*ночной период*)

N125 e **nippers**
d 1. Vorschneider *m* 2. Steingreifzange *f*
f 1. pinces *f pl* coupantes 2. pinces *f pl* d'accrochage
nl 1. nijptang *f*(*m*) 2. steentang *f*(*m*) (*voor het hanteren van steenblokken*)
r 1. кусачки *pl* (*для проволоки*) 2. грузоподъёмный захват *m* (*для каменных блоков*)

N126 e **nipple**
d Nippel *m*
f manchon *m* de tuyau (à *filet mâle*)
nl nippel *m*
r ниппель *m*

N127 e **Nissen hut**
d zerlegbares Bogenbauwerk *n* (*Bauart Nissen*) aus Wellblech
f hutte *f* de Nissen (*bâtiment en voûte préfabriqué de tôles métallique ondulées cintrées*)

nl nissenhut, geheel uit gebogen golfplaten opgetrokken loods
r сборное сооружение *n* типа свода, собираемое из гнутых листов волнистой стали

N128 *e* **no-bond prestressing, no-bond stretching, no-bond tensioning**
d Vorspannung *f* ohne Verbund
f précontrainte *f* sans adhérence
nl voorspanning *f* zonder aanhechting
r натяжение *n* арматуры на бетон без (обеспечения) последующего сцепления

N129 *e* **nodal forces**
d Knotenbelastung *f*, Knotenkräfte *f pl*
f forces *f pl* nodales, charge *f* nodale
nl knooppuntbelasting *f*, krachten *f(m) pl* in een knooppunt
r узловая нагрузка *f*, силы *f pl*, приложенные в узлах фермы

N130 *e* **node**
d 1. Knoten(punkt) *m*
2. Verzweigungspunkt *m*
f 1. nœud *m* (*p. ex d'une poutre à treillis*) 2. jonction *f* des conducteurs
nl 1. knooppunt *n* 2. vertakkingspunt *n*
r 1. узел *m* 2. место *n* соединения электрических проводов

N131 *e* **no-fines concrete**
d entfeinter Beton *m*, Grobporenbeton *m*
f béton *m* sans granulats [agrégats] fins
nl beton *n* met grote poriën, beton *n* zonder fijne toeslag
r беспесчаный [крупнопористый] бетон *m*

N132 *e* **no-flow shutoff head**
d Schließdruck *m* (*Pumpe*)
f pression *f* de fermeture
nl sluitdruk *m* (*pomp*)
r напор *m* (*насоса*) при перекрытии линии нагнетания

N133 *e* **nog**
d 1. nagelbarer Holzblock *m* (*im Mauerwerk*) 2. Spreizstück *n*, Querholz *n*, Holzriegel *m*
f 1. taquet *m* [cheville *f*] en bois 2. entretoise *f* d'une charpente en bois
nl 1. spijkerklos *m* (*in metselwerk*) 2. regel *m* (*in houten geraamte*)
r 1. деревянная пробка *f* (*в каменной кладке*) 2. распорка *f* (*деревянного каркаса*)

N134 *e* **nogging**
d 1. Queraussteifung *f*
2. Ziegelausfachung *f*
f 1. traverse *f* 2. remplissage *m* de la carcasse murale par briques
nl 1. dwarsverband *n* 2. vullen *n* van vakwerk met metselwerk
r 1. поперечная связь *f* 2. кирпичное заполнение *n* фахверка

N135 *e* **nogging piece**
d Querholz *n*, Spreizstück *n*
f entretoise *f* [étrésillon *f*] d'une ossature de bois
nl regel *m* in houten raamwerk
r распорка *f* деревянного каркаса

N136 *e* **no-hinged frame**
d gelenkloser Rahmen *m*
f portique *m* encastré (sans articulation), charpente *f* [ossature *f*] sans articulation
nl scharnierloos skelet *n* [portaal *n*]
r бесшарнирная рама *f*

N137 **noise abatement** *see* **noise control**

N138 *e* **noise abatement wall**
d Schluckwand *f*, Schallschirm *m*
f écran *m* acoustique [d'insonorisation phonique]
nl geluidwerende muur *m*, geluidwal *m*
r стенка *f*, экранирующая звук; акустический экран *m*

N139 *e* **noise absorption factor**
d Schalldämpfungsfaktor *m*, Schallabsorptionsfaktor *m*, Schallschluckgrad *m*
f pouvoir *m* d'absorption du son, coefficient *m* d'absorption acoustique
nl geluidabsorptiefactor *m*
r коэффициент *m* поглощения шума [звукопоглощения]

N140 *e* **noise attenuation**
d Geräuschdämpfung *f*
f atténuation *f* [affaiblissement *m*] du bruit
nl geluiddemping *f*
r ослабление *n* шума

N141 *e* **noise barrier**
d Lärmsperre *f*
f écran *m* [barrière *f*] d'insonorisation phonique
nl lawaaibarrière *f(m)*
r звукоизолирующая преграда *f*

N142 *e* **noise control**
d Lärmbekämpfung *f*, Lärmminderung *f*
f contrôle *m* du bruit, lutte *f* contre le bruit
nl lawaaibestrijding *f*
r борьба *f* с шумом, снижение *n* уровня шума

N143 **noise criteria curves** *see* **noise rating curves**

N144 **noise damper** *see* **silencer**

N145 *e* **noise level**
d Geräuschpegel *m*
f niveau *m* sonore [de bruit]
nl geluidniveau *n*
r уровень *m* шума

N146 *e* **noise meter**
d Lautstärkemesser *m*, Geräuschmeßgerät *n*
f mesureur *m* de niveau de bruit
nl geluidsterktemeter *m*
r шумомер *m*

N147 *e* **noise nuisance**
   d Lärmbelästigung *f*, Lärmstörung *f*
   f ennui *m* [malaise *m*] causé par le
     bruit
  nl geluidhinder *m*, lawaai-overlast *m*
   r шумовые помехи *f pl*, раздражающий
     шум *m*

N148 *e* **noise rating**
   d Lärmbewertung *f*
   f estimation *f* de bruit
  nl bepaling *f* van de geluidwaarde
     volgens frequentiekarakteristieken
   r оценка *f* шумности по частотным
     характеристикам

N149 *e* **noise rating curves**
   d Schallpegel-Frequenz-Diagramm *n* mit
     Linien gleicher Lautstärke;
     Lärmbewertungskurven *f pl*
   f courbes *f pl* d'estimation de bruit
  nl frequentiekarakteristieken *f pl* voor
     de bepaling van de geluidwaarde
   r частотные характеристики *f pl* для
     оценки шума, нормированные
     кривые *f pl* шумности

N150 *e* **noise reduction coefficient**
   d Schallschluckzahl *f*
   f coefficient *m* d'absorbtion phonique
  nl geluiddempingscoëfficiënt *m*
   r коэффициент *m* звукопоглощения

N151 *e* **nominal bore, nominal diameter**
   d Nennweite *f*
   f diamètre *m* nominal
  nl nominale diameter *m*
   r номинальный диаметр *m*

N152 *e* **nominal dimension**
   d Nennmaß *n*, Nennabmessung *f*
   f dimension *f* [cote *f*] nominale
  nl nominale maat *f(m)*
   r номинальный размер *m*

N153 *e* **nominal horsepower**
   d Nennleistung *f*
   f puissance *f* nominale
  nl nominaal vermogen *n*
   r номинальная мощность *f*

N154 *e* **nominal maximum size of aggregate**
   d Nenn-Höchstkorngröße *f*
   f calibre *m* (de grain) nominal maximal
  nl nominale bovengrens *f(m)* voor de
     korrelgrootte
   r номинальный максимальный размер
     m зёрен

N155 *e* **nominal mix**
   d Nenn-Mischungsverhältnis *n*
   f composition *f* admise du béton
  nl nominale mengverhouding *f*
   r заданный состав *m* бетонной смеси

N156 *e* **nominal size**
   d Nenngröße *f*
   f grosseur *m* nominale
  nl nominale grootte *f*
   r типоразмер *m*; номинальный размер *m*

N157 *e* **nonagitating unit**
   d Betontransportwagen *m*

   f camion *m* à benne (basculante) sans
     agitateur
  nl betontransportwagen *m* zonder menger
   r автобетоновоз *m* без побудительного
     перемешивающего устройства

N158 **nonbearing partition** *see* **non-load-
-bearing partition**

N159 *e* **nonbearing wall**
   d unbelastete [selbsttragende] Wand *f*
   f mur *m* non portant [non porteur]
  nl niet-dragende wand *m*
   r самонесущая стена *f*

N160 *e* **non-bonded joint**
   d Knirschfuge *f*, Trockenfuge *f*
   f joint *m* (de maçonnerie) sec
  nl droge voeg *f(m)*
   r сухой шов *m*, шов *m* сухой каменной
     кладки

N161 *e* **nonbreakable glass**
   d unzerbrechliches Glas *n*
   f verre *m* de sécurité [de sûreté], verre
     m incassable
  nl onbreekbaar glas *n*
   r небьющееся стекло *n*

N162 *e* **non-carbonate hardness**
   d Nichtkarbonathärte *f*, bleibende
     Härte *f*
   f dureté *f* non carbonatée [permanente]
  nl hardheid *f* van het water onder
     uitsluiting van carbonaationen
   r некарбонатная [постоянная]
     жёсткость *f* (*воды*)

N163 *e* **non-changeover system**
   d nichtumkehrbares System *n*
   f système *m* d'air climatisé
     non-inversible
  nl niet-omkeerbaar systeem *n*
   r эжекционная система *f*
     кондиционирования воздуха
     с качественным регулированием по
     первичному воздуху, система *f* без
     сезонного переключения режимов
     работы

N164 *e* **non-clogging filter**
   d verstopfungsfreies Filter *n*, *m*
   f filtre *m* incolmatable (*non
     susceptible de se boucher*)
  nl verstoppingsvrij filter *n*, *m*
   r незабивающийся фильтр *m*

N165 *e* **non-clogging screen**
   d verstopfungsfreies Sieb *n*
   f crible *m* [tamis *m*] incolmatable
  nl verstoppingsvrije zeef *f(m)*
   r незабивающийся грохот *m*

N166 *e* **non-cohesive soil**
   d nichtbindiger [kohäsionsloser]
     Boden *m*
   f sol *m* non cohérent, terre *f* meuble
  nl onsamenhangende grond *m*
   r несвязный [сыпучий] грунт *m*

N167 *e* **noncombustible construction**
   d nichtbrennbare Konstruktion *f*

f construction f incombustible
(*qui ne brûle pas ou très mal*)
nl brandwerende constructie f
r невозгораемая конструкция f

N168 e **noncombustible materials**
d nichtbrennbare [nichtentzündbare]
Baustoffe m pl
f matériaux m pl ininflammables
[incombustibles]
nl brandwerende materialen n pl
r невозгораемые материалы m pl

N169 e **non-concussive tap**
d wasserschlagfreier Wasserhahn m
f robinet m antibélier
nl waterslagvrije kraan f(m)
r водоразборный кран m, не
создающий гидравлических ударов

N170 e **nonconforming work**
d Bauarbeit f, die den Anforderungen
des Projekts *oder* der
Baubestimmungen nicht entspricht
f travail m exécuté non conformément
aux dispositions du devis descriptif
nl van het bestek afwijkende uitvoering
f
r работа f, не отвечающая требованиям
проекта *или* технических условий

N171 e **noncreeping material**
d kriechfestes Material n
f matériau m non susceptible de fluage
nl kruipvast materiaal n
r материал m, мало подверженный
ползучести

N172 e **nondestructive testing**
d zerstörungsfreie Werkstoffprüfungen
f pl
f essais m pl non destructifs
nl niet-destructieve materiaalbeproeving
f
r неразрушающие [адеструктивные]
испытания n pl

N174 e **nondomestic building**
d unbewohnbares Gebäude n
f bâtiment m non résidentiel
nl niet voor bewoning bestemd gebouw
n
r нежилое здание n

N175 e **nondomestic premises**
d nicht für Wohnzwecke gebaute
Räume m pl
f locaux m pl non à usages
d'habitation
nl niet voor bewoning bestemde
ruimten f pl
r нежилые помещения n pl

N176 **nonelastic deformation** *see* **inelastic deformation**

N177 e **nonelastic range**
d unelastischer [plastischer] Bereich m,
Plastizitätsgebiet n
f domaine m plastique [inélastique],
zone f plastique
nl niet-elastisch [plastisch] gebied n

r область f пластичности, область f
неупругих [пластических]
деформаций

N178 e **non-elevating road mixer** *see* **mix-in-place machine**

N179 e **nonhomogeneity of materials**
d Inhomgenität f der Materialien
f inhomogénéité f [nonhomogénéité f,
hétérogénéité f] des matériaux
nl gebrek n aan homogeniteit f van de
materialen
r неоднородность f материалов

N180 e **nonhomogeneous state of stress**
d nichthomogener Spannungszustand m
f état m de contrainte non uniforme
nl niet-homogene spanningstoestand m
r неоднородное напряжённое
состояние n

N181 e **nonhydraulic lime**
d Luftkalk m
f chaux f aérienne [(durcissant) à l'air]
nl niet-hydraulische kalk m, luchtkalk
m
r воздушная известь f

N182 e **non-hydraulic mortar**
d Kalksandmörtel m, Luft(kalk)mörtel
m
f mortier m aérien [à chaux aérienne]
nl niet-hydraulische kalkmortel m,
luchtmortel m
r раствор m воздушного твердения
[на воздушной извести],
известково-песчаный раствор m

N183 e **nonlinear distribution of stresses**
d nichtlineare Spannungsverteilung f
f distribution f de contraintes non-
-linéaire
nl niet-lineare spanningsverdeling f
r нелинейное распределение n
напряжений

N184 e **nonlinear elastic behavour**
d nichtlineares elastisches Verhalten n
f comportement m élastique non-
-linéaire
nl niet-lineair elastisch gedrag n
r нелинейно упругая работа f

N185 e **nonlinear elasticity**
d nichtlineare Elastizität f
f élasticité f non-linéaire
nl niet-lineaire elasticiteit f
r нелинейная упругость f

N186 e **nonlinear plastic theory**
d genaue [nichtlineare] plastische
Theorie f
f théorie f de plasticité non-linéaire
nl theorie f van de niet-lineaire
plasticiteit
r нелинейная теория f пластичности

N187 e **non-load-bearing partition**
d nichttragende Trennwand f
f paroi f non portante
nl niet-dragende tussenmuur m
r ненесущая перегородка f

N189 *e* **nonoverflow section**
   d überfallfreie Strecke *f* (*Damm*)
   f section *f* insubmersible (*du barrage*)
   nl damsectie *f* zonder overstort
   r глухая секция *f* (*плотины*)

N190 *e* **nonparallel-chord truss**
   d Träger *m* mit nicht parallelen Gurten
   f poutre *f* à treillis avec membrures non-parallèles
   nl ligger *m* met niet-parallele randen
   r ферма *f* с непараллельными поясами

N191 *e* **non-pressure drain**
   d Freispiegeldrän *m*
   f drain *m* à écoulement libre
   nl drain *m* [buis *f(m)*] met vrije afvoer
   r безнапорная дрена *f*

N192 *e* **non-pressure pipe**
   d Freispiegelrohr *n*
   f tuyau *m* pour eau sans pression
   nl buis *f(m)* met vrije afvoer
   r безнапорная труба *f*

N193 *e* **nonprestressed reinforcement**
   d schlaffe [nichtgespannte] Bewehrung *f* [Armierung *f*]
   f armature *f* ordinaire [non précontrainte]
   nl niet-voorgespannen wapening *f*
   r ненапрягаемая арматура *f*

N194 *e* **non-regulated discharge**
   d natürlicher Durchfluß *m*
   f débit *m* naturel
   nl ongestuwde afvoer *m*
   r бытовой расход *m*

N195 **nonresidential building** *see* **nondomestic building**

N196 *e* **non-return valve**
   d Rückschlagventil *n*
   f clapet *m* de retenue
   nl terugslagklep *f(m)*
   r обратный клапан *m*

N197 *e* **nonrigid carriageway**
   d Schwarzdeckenfahrbahn *f*
   f chaussée *f* souple
   nl rijbaan *f(m)* met zwart wegdek
   r проезжая часть *f* дороги с нежёстким покрытием

N198 *e* **nonrigid pavement**
   d nichtstarre [schmiegsame] Straßendecke *f*
   f revêtement *m* routier souple, corps *m* de chaussée souple
   nl niet-star wegdek *n*
   r нежёсткое дорожное покрытие *n*

N199 *e* **non-rotating rope**
   d drallfreies Seil *n*
   f câble *m* antigiratoire
   nl draaivrij touw *n*
   r нераскручивающийся канат *m*

N200 *e* **non-scouring velocity**
   d kritische Sohlengeschwindigkeit *f*, maximal zulässige Geschwindigkeit *f*, die noch nicht zur Erosion führt
   f vitesse *f* limite d'érosion

   nl grootste stroomsnelheid waarbij nog geen erosie optreedt
   r неразмывающая скорость *f*

N201 *e* **nonshrink concrete**
   d schwindfreier [raumbeständiger] Beton *m*
   f béton *m* expansif [sans retrait]
   nl krimpvrij beton *n*
   r безусадочный бетон *m*

N202 *e* **non-shrinking cement**
   d schwindfreier [raumbeständiger] Zement *m*
   f ciment *m* sans retrait
   nl krimpvrij cement *n*
   r безусадочный цемент *m*

N203 *e* **non-silting velocity**
   d Mindestgeschwindigkeit *f*, die noch nicht zur Verlandung führt
   f vitesse *f* limite de dépôt [d'envasement]
   nl kleinste stroomsnelheid waarbij nog geen sedimentatie plaatsvindt
   r незаиливающая скорость *f*

N204 *e* **nonsimultaneous prestressing**
   d ungleichzeitiges Spannen *n* (*Bewehrung*)
   f précontrainte *f* non simultanée
   nl niet gelijktijdig aanbrengen van de voorspanning
   r неодновременное [поэтапное] натяжение *n* (*арматуры на бетон*)

N205 *e* **nonsiphon trap**
   d Ablauf *m* [Traps *m*] ohne Geruchverschluß
   f collecteur *m* [puisard *m*] sans siphon
   nl stankafsluiter *m* zonder sifon
   r трап *m* без гидравлического затвора

N206 *e* **non-skid carpet, non-skid surface**
   d Gleitschutzteppich *m*
   f tapis *m* antidérapant, couche *f* de roulement antidérapante
   nl stroeve vloerbedekking *f*
   r нескользкий слой *m* износа, нескользкое дорожное покрытие *n*

N207 *e* **non-skid surfacing**
   d Rauhbelag *m*, Rauhdecke *f*
   f revêtement *m* antidérapant
   nl stroef oppervlak *n*
   r шероховатое [нескользящее] покрытие *n* (*дороги, пола*)

N208 *e* **non-slewing crane**
   d Kran *m* mit nicht drehbarem [schwenkbarem] Ausleger
   f grue *f* à flèche fixe (*non tournante*)
   nl kraan *f(m)* met niet-zwenkbare arm
   r кран *m* с неповоротной стрелой

N209 *e* **non-slip floor**
   d rutschfester [gleitsicherer, griffiger] Bodenbelag *m*
   f revêtement *m* de sol antidérapant [non glissant]
   nl slipvrije vloer *f(m)*
   r нескользкий [шероховатый] пол *m*

N210 *e* **nonstaining mortar**
    d alkaliarmer Putzmörtel *m*
    f mortier *m* à faible teneur en alcali
    nl alkali-arme pleistermortel *m*
    r штукатурный раствор *m* с низким
    содержанием щелочных компонентов
    (*не дающий выцветов*)

N211 *e* **non-standart component**
    d nichttypisiertes Bauteil *n*
    f élément *m* de construction non
    standardisé
    nl niet-gestandaardiseerd constructie-
    -element *n*
    r нетиповой элемент *m* конструкции

N213 *e* **nonsticky soil**
    d nichtanhaftender [nichtklebriger]
    Boden *m*
    f sol *m* non gluant
    nl niet-klevende grond *m*
    r неприлипающий грунт *m*

N214 *e* **non-storage calorifier**
    d Durchlauf-Wasserheizer *m*,
    Durchlauferhitzer *m*
    f chauffe-eau *m* à écoulement libre
    nl warmwater-doorstroomtoestel *n*
    r проточный водонагреватель *m*

N215 *e* **non-structural component**
    d nichttragendes Bauteil *n*
    f élément *m* de construction non portant
    nl niet-dragend constructie-element *n*
    r ненесущий конструктивный элемент
    m

N216 *e* **non-tilting drum mixer**
    d Umkehr(beton)mischer *m*,
    Freifallmischer *m* mit Umkehraustrag
    f mélangeur *m* à tambour non basculant,
    mélangeur *m* [malaxeur *m*] avec cuve
    non basculante
    nl betonmolen *m* met niet-kippende
    trommel
    r бетоносмеситель *m* с неопрокидным
    барабаном

N217 *e* **non-toxic tar**
    d unschädlicher [giftfreier]
    Straßenteer *m*
    f goudron *m* raffiné [non toxique]
    nl gifvrije teer *m*
    r нетоксичный дорожный дёготь *m*

N218 *e* **non-volatile matter**
    d nichtflüchtiger Stoff *m*
    f constituant *m* [composant *m*] non
    volatil; agent *m* non volatil
    nl niet-vluchtige stof *f(m)*
    r нелетучий компонент *m* (*краски*);
    нелетучее вещество *n*

N219 *e* **nonvolatile vehicle**
    d nichtflüchtiger Lösemittelanteil *m*
    f véhicule *m* [solvant *m*, dissolvant *m*]
    non volatil
    nl niet-vluchtig oplosmiddel *n*
    r нелетучий растворитель *m*

N220 *e* **noria**
    d Schöpfwerk *n*

    f noria *f*
    nl noria *f(m)*, kettingpomp *f(m)*
    r нория *f*, черпаковый подъёмник *m*

N222 *e* **normal annual runoff**
    d mittlerer jährlicher Abfluß *m*,
    Abflußnorm *f*
    f écoulement *m* annuel moyen
    nl gemiddelde jaarlijkse afvloeiing *f*
    r средний многолетний сток *m*, норма
    f стока

N223 *e* **normal bend**
    d 90-Grad-Bogen *m*
    f coude *m* à 90°
    nl normale bocht *f(m)*
    r отвод *m* под углом 90°

N224 **normal concrete** *see* **normal-weight
concrete**

N225 *e* **normal consistency**
    d Normalsteife *f*, Normaldichte *f*
    f consistance *f* [fluidité *f*] normale
    [normalisée]
    nl normale consistentie *f*
    r нормальная консистенция *f*
    [густота *f*]

N226 *e* **normal force**
    d Normalkraft *f*
    f force *f* normale, effort *m* normal
    nl normaalkracht *f(m)*
    r нормальная сила *f*

N227 *e* **normal force diagram**
    d Längskraftlinie *f*
    f épure *f* [diagramme *m*] des forces
    normales
    nl diagram *n* van normaalkrachten
    r эпюра *f* продольных сил

N228 *e* **normal operating level**
    d Normalspiegel *m*, normales Stauziel *n*
    f niveau *m* (de la retenue) normal
    nl normaal peil *n*, normaal stuwniveau *n*
    r нормальный подпорный уровень *m*

N229 *e* **normal section**
    d Normalschnitt *m*
    f section *f* normale
    nl dwarsdoorsnede *f(m)*
    r нормальное сечение *n*

N230 *e* **normal stress**
    d Normalspannung *f*
    f contrainte *f* [tension *f*] normale
    nl normaalspanning *f*
    r нормальное напряжение *n*

N231 *e* **normal-weight concrete**
    d Normalbeton *m*, Schwerbeton *m*
    f béton *m* normal [lourd]
    nl normaal beton *n*
    r тяжёлый [обычный] бетон *m*

N232 *e* **north-light glazing**
    d Shedverglasung *f*
    f vitrage *m* de la toiture en shed
    nl sheddakbeglazing *f*
    r остекление *n* шедового покрытия

N233 *e* **north-light gutter**
    d Sägedach-Kehlrinne *f*, Sheddach-
    -Kehlrinne *f*

f noue *f* du comble (à) shed
nl sheddakgoot *f(m)*, zaagdakgoot *f(m)*
r ендова *f* шедового покрытия

N234 *e* **north-light roof**
  d Sheddach *n*, Säge(zahn)dach *n*
  f comble *m* [toit *m*] (à) shed, shed *m*,
    toit *m* en dents de scie
  nl sheddak *n*
  r шедовая крыша *f*, шедовое покрытие
    n

N235 *e* **north-light shell roof**
  d Säge(zahn)schalendach *n*,
    Shedschalendach *n*
  f toit-voûte *m* [comble-voûte *m*] à shed
  nl cilinderschaaldak *n*
  r шедовое покрытие *n* с
    криволинейными скатами

N236 **Norton tube well** *see* **driven well**

N237 **nose** *US see* **nosing 1.**

N238 *e* **nose of groyne**
  d Buhnenkopf *m*
  f tête *f* [extrémité *f*] d'épi
  nl kop *m* van de krib
  r голова *f* буны

N239 *e* **nosing**
  d 1. Winkeleckleiste *f*,
    Kantenschutzleiste *f* (*Treppen*)
    2. Ausladung *f*, Kante *f* 3. Wulst *m*
    (*Trittstufe*)
  f 1. éclisse *f* cornière 2. arête *f* 3. nez *m*
    de marche (*d'escalier*)
  nl 1. trapneus *m*, hoekijzer *n* 2. ronde
    rand *m*, overstek *n* 3. wel *f(m)*
    (*traptrede*)
  r 1. угловая [защитная] накладка *f*
    (*кромки ступеней*) 2. выступ *m*,
    кромка *f* 3. валик *m* (*ступени*)

N240 *e* **no-slump concrete**
  d erdfeuchter Beton *m*
  f béton *m* sec
  nl aardvochtig beton *n*
  r жёсткая бетонная смесь *f*

N241 *e* **notch**
  d 1. Kerbe *f*, Einschnitt *m* 2. Ausschnitt
    m im Meßwehr
  f 1. entaille *f*, encoche *f* 2. échancrure *f*
  nl 1. kerf *f(m)* 2. uitsnijding *f*
    (*meetschot*)
  r 1. надрез *m*, вырез *m* 2. отверстие *n*
    измерительного водослива

N242 **notch drop** *see* **notch fall**

N243 **notched bars** *see* **deformed bars**

N244 *e* **notched bar test**
  d Kerbschlagversuch *m*
  f essai *m* sur éprouvette entaillée
  nl kerfslagproef *f(m)*
  r определение *n* ударной вязкости

N245 *e* **notched specimen**
  d Kerb(schlag)probe *f*,
    Kerbschlagprobestab *m*
  f éprouvette *f* entaillée, éprouvette *f*
    Charpy

nl kerfslagproefmonster *n*
r образец *m* с надрезом (*для
    определения ударной вязкости*)

N246 **notched weir** *see* **measuring weir**

N247 *e* **notch effect**
  d Kerbwirkung *f*
  f effet *m* d'entaille
  nl kerfeffect *n*
  r влияние *n* надреза (*эффект
    концентрации напряжений*)

N248 *e* **notch fall**
  d Spaltenabsturzbauwerk *n*, Absturz *m*
    mit Kammschwelle
  f chute *f* à déversoir
  nl drempeloverloop *m*
  r щелевой перепад *m*

N249 **notch impact strength** *see* **impact
strength 1.**

N250 *e* **notching**
  d 1. stufenförmiger Aushub *m*
    2. Ausklinken *n*, Abflanschen *n*
    3. Überschneidung *f*; Einzapfen *n*
  f 1. excavation *f* par gradins 2.
    entaille *f* dans la semelle d'une
    poutre 3. entaille *f* d'assemblage
    (*dans une pièce de bois*)
  nl 1. trapsgewijze ontgraving *f*
    2. inkerving *f* 3. inkeping *f*
  r 1. ступенчатая экскавация *f*
    [разработка *f*] 2. срез *m* полки
    (*балки*) 3. врубка *f* (*деревянные
    конструкции*); поперечная вязка *f*
    (*деревянных элементов*)

N251 *e* **notching machine**
  d Ausklinkmaschine *f*,
    Abflanschmaschine *f*
  f machine *f* à encocher
  nl knabbelschaar *f(m)*
  r вырубной пресс *m*

N252 *e* **notch sensitivity**
  d Kerbempfindlichkeit *f*
  f sensibilité *f* à l'effet d'entaille
  nl kerfgevoeligheid *f*
  r чувствительность *f* к концентрации
    напряжений при надрезе

N253 **notch shock test** *see* **nick-break test**

N254 **notch toughness** *see* **impact strength 1.**

N255 *e* **not-water-carriage toilet facility**
  d Trockenabort *m*
  f installation *f* sanitaire isolée (*non
    attachée à l'égout public*)
  nl closet *n* zonder waterspoeling,
    droogcloset *n*
  r уборная *f*, не подключённая к
    канализации (*с химической
    обработкой фекалиев в выгребных
    резервуарах*)

N256 *e* **novelty flooring**
  d Fußbodenbelag *m* nach einem neuen
    Muster
  f revêtement *m* de sol avec un dessin
    ornamental hors série
  nl vloerbedekking *f* met afwijkend dessin

r покрытие *n* пола с необычным
[нестандартным] рисунком

N257 **novelty siding** *see* **matched siding**

N258 *e* **noxious fumes**
   d schädliche Abgase *n pl* und
     Verdunstungen *f pl*
   f fumées *f pl* nocives; vapeurs *f pl*
     et gaz *m pl* nocifs
   nl schadelijke dampen *n pl*
   r вредные дымы *m pl*; вредные пары
     m pl и газы *m pl*

N259 *e* **nozzle**
   d Düse *f*; Mundstück *n*; Zerstäuber *m*
   f tuyère *f*; ajutage *m*; embouchure
     f
   nl mondstuk *n*, straalpijp *f(m)*;
     spuitmond *m*
   r сопло *n*; гидравлический насадок
     m; форсунка *f*

N260 *e* **nozzle meter**
   d Venturirohr *n*, Venturimeter *n*
   f compteur *m* Venturi
   nl venturimeter *m*
   r расходомер *m* Вентури

N262 **N.T. mixer** *see* **non-tilting drum mixer**

N263 *e* **N-truss**
   d Pratt-Träger *m*
   f poutre *f* à treillis en N
   nl N-spant *n*
   r ферма *f* с параллельными поясами
     и нисходящими раскосами

N264 *e* **nuclear energy structures**
   d Bauwerke *n pl* einer
     Atomenergieanlage
   f ouvrages *m pl* [équipements *m pl*]
     d'énergie nucléaire
   nl gebouwen *n pl* van een kerncentrale
   r строительные сооружения *n pl*
     атомной электростанции

N265 *e* **nucleus**
   d Stadtkern *m*
   f centre *m* de ville
   nl stadscentrum *n*
   r городской центр *m*, центр *m* города

N266 *e* **nut**
   d Mutter *f*, Schraubenmutter *f*
   f écrou *m*
   nl moer *f(m)*
   r гайка *f*

N267 *e* **nut anchorage**
   d Verankerung *f* mit Mutter
     (*Spannbeton*)
   f ancrage *m* à écrou
   nl verankering *f* met moeren
   r анкеровка *f* (*напрягаемых*
     арматурных стержней) с помощью
     гайки

N268 *e* **nut runner**
   d Schlagschrauber *m*, Drehschrauber
     m
   f clé *f* [clef *f*] à choc
   nl pneumatische moersleutel *m*
   r гайковёрт *m*

N269 *e* **nut socket driver**
   d Muttereinschraubmaschine *f* mit
     auswechselbaren Aufsatzköpfen
   f clé *f* [clef *f*] à choc avec douilles
     amovibles
   nl automatische moeraanzetter *m*
   r гайковёрт *m* со сменными головками

N270 *e* **nut torque**
   d Anzugsmoment *n*
   f moment *m* de rotation appliqué
     à l'écrou
   nl aantrekmoment *n* (*moer*)
   r момент *m* закручивания гайки

N271 *e* **N-year flood**
   d Hochwasser *n* mit Jährlichkeit 1/N
   f crue *f* de probabilité 1/N
   nl N-jaarlijks hoogwater *n*
   r паводок *m* повторяемостью раз в N лет

N272 *e* **nylon rope**
   d Nylontrosse *f*
   f câble *m* en nylon
   nl nylontouw *n*
   r нейлоновый [найлоновый] канат *m*

O

O1   *e* **oakum**
   d Werg *n*
   f étoupe *f*; filasse *f*
   nl werk *n* (*touwwerk*)
   r пакля *f*; пенька *f*

O2   *e* **oblique butt joint**
   d schräge Überblattung *f* [Blattung *f*],
     schräges Blatt *n*
   f sifflet *m* simple, entaille *f* à mi-bois
     en sifflet
   nl schuine liplas *f(m)*
   r косое лобовое примыкание *n*, косой
     прируб *m*

O3   *e* **oblique projection**
   d schiefe Parallelprojektion *f* auf
     lotrechte Bildebene
   f projection *f* oblique
   nl scheve projectie *f*
   r косоугольная проекция *f*

O4   *e* **oblique scarf**
   d französisches Blatt *n*, schräges
     Hakenblatt *n*
   f assemblage *m* à trait de Jupiter
   nl schuine haaklas *f(m)*
   r врубка *f* косым зубом

O5   *e* **oblique section**
   d Schrägschnitt *m*
   f section *f* [coupe *f*] oblique
   nl schuine doorsnede *f*
   r косой разрез *m*, косое сечение *n*

O6   *e* **obscure glass**
   d Mattglas *n*, Milchglas *n*
   f verre *m* op**acifié** [mat]
   nl ondoorzichtig **glas** *n*
   r глушёное **стекло** *n*

O7 e **observation of the work**
    d Autorenkontrolle *f*, Bauübersicht *f*
      durch den Autor eines Projektes
    f contrôle *m* technique des travaux par
      auteur du projet
    nl inspectie *f* door de ontwerper
    r авторский надзор *m*

O8 **observation point** *see* **vertical**

O9 e **observation well**
    d Beobachtungsbrunnen *m*
    f puits *m* d'observation [témoin, de
      surveillance]
    nl peilbuis *f(m)*, controleput *m*
    r наблюдательная скважина *f*,
      наблюдательный колодец *m*

O10 e **obstruction light**
    d Hindernisfeuer *n*
    f feu *m* d'obstacles
    nl obstakelverlichting *f*
    r сигнальный огонь *m* (*на высотных
      сооружениях*)

O11 e **occupancy**
    d Besetzungsgrad *m*, Belegung *f*
    f nombre *m* d'habitans par bâtiment,
      capacité *f* de bâtiment
    nl bezettingsgraad *m*
    r населённость *f* (*расчётное число
      людей в здании или помещении*)

O12 e **occupancy load**
    d 1. rechnerische Belegung *f*
      [Besetzung *f*] (*eines ˌRaumes*)
      2. Menschenlast *f*
    f 1. capacité *f* (*p.ex. d'une salle de
      cinéma, d'un ascenseur*) 2. charge *f*
      (*mécanique, thermique*) due à l'activité
      des hommes à l'intérieur les locaux
    nl 1. berekende bezetting *f* 2. belasting
      tengevolge van het gebruik
    r 1. вместимость *f* по количеству
      людей (*кинозала, лифта*) 2. нагрузка
      f от людей (*механическая, тепловая*)
      в помещении

O13 e **occupancy rate**
    d Belegungsdichte *f*
    f capacité *f* de logement, densité *f*
      residentielle d'un logement
    nl bezettingsgraad *m*
    r плотность *f* заселения (*напр.
      квартиры*)

O14 e **occupants**
    d Benutzer *m pl*
    f usagers *m pl*
    nl gebruikers *m pl*, bewoners *m pl*
    r пользователи *m pl*

O15 e **occupant space requirements**
    d Wohnflächennorm *f*
    f surface *f* nécessaire par habitant
    nl woonruimte *f*
    r норма *f* площади на одного человека

O16 e **occupied zone**
    d Aufenthaltszone *f*, Aufenthaltsbereich
      m

    f zone *f* d'occupation
    nl verblijfszone *f*
    r зона *f* обслуживания; зона *f*
      пребывания людей

O17 e **ocher, ochre**
    d Ocker *m*
    f ocre *f*
    nl oker *m*
    r охра *f*

O18 e **ocrated concrete**
    d Okratbeton *m*
    f béton *m* ocraté
    nl met fluorsilicaatverbinding
      bewerkt beton *n*
    r бетон *m*, обработанный
      четырёхфтористым кремнием

O19 e **ocrate process, ocrating**
    d Okratieren *n*
    f ocratation *f* (*procédé de traitement du
      béton par du tétrafluorure de silicium*)
    nl bewerking *f* van beton met fluorsilicaat
    r обработка *f* бетона четырёхфтористым
      кремнием

O20 e **odor control**
    d Desodorierung *f*
    f désodorisation *f*, enlèvement *m* de
      l'odeur
    nl stankwering *f*
    r дезодорация *f*

O21 e **odor suppression**
    d Geruchsbeseitigung *f*
    f suppression *f* de l'odeur
    nl desodoratie *f*, stankverdrijving *f*
    r устранение *n* запахов
      odour... *see* **odor...**

O23 e **oedometric test**
    d Oedometerversuch *m*
    f essai *m* oedométrique
    nl odometrische proef *f(m)*
    r одометрическое испытание *n*,
      испытание *n* в одометре

O24 e **off-center load**
    d ausmittige [exzentrische] Last *f*
    f charge *f* excentrée
    nl excentrische belasting *f*
    r внецентренная нагрузка *f*

O25 **offer** *see* **bid**

O26 e **off-formwork concreting**
    d schalungslose Betonierung *f*
    f bétonnage *m* sans coffrage
    nl betonstorten *n* zonder bekisting
    r безопалубочное бетонирование *n*

O27 e **off-grade size**
    d Ausfallkörnung *f*
    f grains *m pl* déclassés
    nl niet-bruikbare korrelgrootte *f*
    r зёрна *n pl*, не удовлетворяющие
      заданному гранулометрическому
      составу

O28 e **off-ground hauler**
    d Muldenkipper *m*, Mulden-
      -Erdbaufahrzeug *n*, Groß-Förderwagen
      m

f camion *m* «hors route» pour le transport de sol [de terres]
nl zware kipauto *m* voor grondvervoer, (zware) zandauto *m*
r грунтовоз *m* для работы в условиях бездорожья

O29 *e* **office building**
d Bürogebäude *n*
f édifice *m* [immeuble *m*] à bureaux
nl kantoorgebouw *n*
r административное [конторское] здание *n*

O30 **off-loading ramp** *see* **unloading ramp**

O31 *e* **off-peak electric heating**
d Nachtspeicherheizung *f*
f chauffage *m* électrique hors-pointe
nl verwarming *f* op nachtstroom
r внепиковое электроотопление *n*

O32 *e* **off-peak load**
d Belastung *f* außerhalb der Spitzenzeit
f charge *f* hors-pointe
nl belasting *f* buiten de piekuren
r внепиковая нагрузка *f*

O33 *e* **off-peak period**
d Nichtspitzenzeit *f*, belastungsspitzenfreie Zeitspanne *f*
f période *f* normale [sans charge de pointe, hors-pointe]
nl tijdvak *n* buiten de piekuren
r период *m* внепиковой нагрузки

O34 *e* **offset**
d 1. seitliche Verschiebung *f* 2. Sprungmaß *n* 3. Absatzstück *n*, Sprungrohr *n* 4. Kröpfung *f* 5. Absatz *m* 6. Mauerabsatz *m* 7. bleibende Regelabweichung *f*, Versetzung *f*
f 1. déplacement latéral 2. longueur *f* [grandeur *f*] de déplacement latéral 3. double coude *m*, siphon *m* 4. rebord *m* (*de tige*) 5. berme *f*, gradin *m*, banquette *f* 6. retrait *m* d'un mur, ressaut *m* 7. décentrage *m*; écart *m* de réglage, décalage *m*
nl 1. zijdelingse verschuiving *f* 2. verschuivingsmaat *f(m)* 3. dubbele bocht *f(m)* 4. kniebocht *f(m)* 5. banket *n* 6. versnijding *f* van de fundering 7. justeerfout *m*
r 1. боковое смещение *n* 2. размер *m* смещения 3. отступ *m* (*соединительная деталь*) 4. отгиб *m*, выгиб *m* (*стержня, профильного элемента*) 5. уступ *m*; берма *f* 6. обрез *m* стены (*в месте изменения толщины*) 7. рассогласование *n*, ошибка *f* регулирования, отклонение *n* от заданного значения

O35 *e* **offset bend**
d Abbiegung *f* der Bewehrung
f relevé *m* d'un rond à béton
nl verbuiging *f* in de wapening
r отгиб *m* арматуры

O36 *e* **offset block**
d Beton-Eckbaustein *m*, Beton--Formbaustein *m*, Betonformblock *m*
f bloc *m* [parpaing *m*, corps *m*] profilé *ou* d'angle en béton
nl betonsteen *m* van afwijkende vorm, geprofileerde betonsteen *m*
r фасонный *или* угловой бетонный камень *m* [блок *m*]

O37 *e* **offset canal**
d Stichkanal *m*, Scheitelkanal *m*
f canal *m* à travers une ligne de partage des eaux
nl dwarskanaal *n*
r канал-прокоп *m*

O38 *e* **off-set joints**
d versetzte Fugen *f pl*
f joints *m pl* en quinconce
nl verspringende voegen *f(m) pl*
r швы *m pl* вразбежку

O39 *e* **offset yield strength**
d definierte Fließgrenze *f*
f limite *f* de fluage conventionnelle
nl elasticiteitsgrens *f(m)*
r условный предел *m* текучести

O40 **offshore boring island** *see* **drilling platform**

O41 *e* **offshore mooring**
d Reedenankerplatz *m*
f amarrage *m* en mer
nl op de rede voor anker gaan
r рейдовый причал *m*

O42 *e* **off-shore structures**
d küstenferne Seebauten *pl*
f ouvrages *m pl* maritimes [à la mer, en mer]
nl offshore constructies *f pl*
r глубоководные морские сооружения *n pl*

O43 **off-shuttering concreting** *see* **off--formwork concreting**

O44 *e* **offstream storage**
d Pumpenspeicherbecken *n*
f bassin *m* haut
nl kunstmatig gevuld voorraadbekken *n*
r наливное водохранилище *n*

O45 *e* **off-the-site work**
d Arbeiten *f pl*, die außerhalb der Baustelle ausgeführt werden
f travaux *m pl* exécutés hors chantier
nl werkzaamheden *f pl* buiten de bouwplaats
r работы *f pl*, производимые за пределами строительной площадки

O46 *e* **off-white**
d schwachpigmentiertes Weiß *n*
f blanc *m* légèrement teinté
nl gebroken wit *n* (*verf*)
r белила *pl* с небольшим добавлением пигмента

O47 *e* **ogee**
d 1. Karnies *n*, Sims *m*, *n* 2. Helmdach *n*

f 1. doucine *f*, talon *m* 2. toit *m* en dos
d'âne
nl 1. keellijst *f(m)*, ojief *n*, talon *m*
2. helmdak *n*
r 1. гусёк *m*, каблучок *m*
2. килевидная крыша *f*

O48 *e* ogee arch
d Spitzbogen *m*
f arc *m* en accolade
nl ogiefboog *m*
r килевидная арка *f*

O49 *e* ogee curve
d 1. S-Kurve *f* 2. Dammschulter *f*
3. Ausrundung *f* der Überfallstaumauer
f 1. courbe *f* en S 2. talus *m* en S
3. évacuateur *m* [déversoir *m*] en
doucine
nl 1. S-vormige lijn *f(m)* 2. S-vormig talud
n 3. afronding *f* van een overlaat
r 1. S-образная кривая *f* 2. откос *m*
насыпи S-образного очертания
3. водосливная грань *f* (*плотины*)
практического профиля

O50 *e* ogee spillway, ogee weir
d Freiflut *f*, Freifluter *m*,
Oberflächenüberlauf *m*;
normalkroniger Überfall *m*
f déversoir *m* superficiel, évacuateur *m*
aérien [de surface]; déversoir *m* de
Creager [en doucine]
nl overloop *m* met normale kruin
r поверхностный водосброс *m*; водослив
m практического профиля

O51 *e* Ohio cofferdam
d zweireihiger Spundwandfangdamm *m*
f batardeau *m* à double paroi de
palplanches
nl kistdam *m*
r двухрядная шпунтовая перемычка *f*

O52 *e* oil-bath air filter
d Ölbadluftfilter *n*, *m*
f épurateur *m* d'air à bain d'huile
nl oliebad-luchtfilter *n*, *m*
r воздушный масляный самоочищающийся фильтр *m* !

O53 *e* oil buffer
d Öldämpfer *m*
f amortisseur *m* hydraulique (à l'huile)
nl oliebad-luchtfilter *n*, *m*, oliedemper *m*
r гидравлический [масляный]
амортизатор *m*

O54 *e* oil-fired heater
d Ölluftheizer *m*
f réchauffeur *m* d'air à l'huile
nl oliegestookte luchtverhitter *m*
r воздухонагреватель *m* [калорифер
m], работающий на жидком топливе

O55 *e* oil fired heating system
d Ölheizung *f*
f installation *f* de chauffage à l'huile
nl oliegestookte verwarmingsinstallatie
f, oliestook *m*
r система *f* отопления, работающая на
жидком топливе

O56 *e* oil paint
d Ölfarbe *f*
f peinture *f* [couleur *f*] à l'huile
nl olieverf *f(m)*
r масляная краска *f*

O57 *e* oil pipeline
d Rohölrohrleitung *f*
f oléoduc *m*
nl oliepijpleiding *f*
r нефтепровод *m*

O58 *e* oil stain
d Ölbeize *f*
f mordant *m* à l'huile
nl oliebeits *m*, *n*
r масляная протрава *f*

O59 *e* oil trap
d Ölabscheider *m*
f 1. récupérateur *m* [capteur *m*] d'huile
2. piège *m* d'huile, attrape-pétrole *m*
nl olieafscheider *m*
r 1. маслоуловитель *m*,
маслоулавливатель *m* 2. нефтеловушка *f*

O60 oil-washed air cleaner *see* oil-bath air
filter

O61 *e* oil-well cement
d Erdölzement *m*, Ölbohrzement *m*,
Tiefbohrzement *m*
f ciment *m* pour puits de pétrole
[pour forages]
nl cement *n* voor aardolie voerende
boorputten
r тампонажный цемент *m*

O62 *e* Omnia concrete floor
d Omniadecke *f*,
Verbundgitterträgerdecke *f*
f plancher *m* Omnia
nl Omnia-vloer *f(m)*
r сборно-монолитное железобетонное
перекрытие *n* типа «Омниа»
(*перекрытие из сборных
железобетонных пластин,
выполняющих роль несъёмной
опалубки, пустотелых бетонных
блоков заполнения и слоя
монолитного бетона*)

O63 *e* Omnia concrete plank
d Omnia-Vollbetonplatte *f*
f planche *f* Omnia (*en béton armé
léger préfabriqué*)
nl Omnia-betonplaat *f(m)*
(*breedplaatvloer*)
r сборная железобетонная пластина *f*

O64 *e* Omnia trimmer concrete plank
d schmale Omnia-Wechselvollbeton-
platte *f*
f planche *f* d'encadrement Omnia (*en
béton armé léger préfabriqué*)
nl smalle Omnia-betonplaat *f(m)*
(*breedplankvloer*)
r сборная железобетонная пластина *f*,
обрамляющая проём в перекрытии
(*выполняет роль несъёмной
опалубки проёма*)

O65 e once-through water-supply system
d Durchlaufwasserversorgung *f*
f système *m* [réseau *m*] de
distribution d'eau à passage direct
nl directe watervoorziening *f*
r прямоточная система *f*
водоснабжения

O66 e one-brick wall
d ein-Stein-starke Mauer *f*
f mur *m* d'une brique d'épaisseur
nl eensteens muur *m*, steens muur *m*
r стена *f* толщиной в один кирпич

O67 e one-centered arch
d Kreisbogen *m*
f arc *m* en plein cintre
nl rondboog *m*
r круговая арка *f*

O68 e one-coat work
d einlagiger Putz *m*
f enduit *m* à une couche
nl beraping *f*
r однослойная штукатурка *f*

O69 e one-component material
d Ein-Komponenten-Material *n*
f matériau *m* à un composant [à un
constituant]
nl een-componentmateriaal *n*
r однокомпонентный материал *m*

O70 e one-degree system
d System *n* mit einem einzigen
Freiheitsgrad
f système *m* à un degré de liberté
nl systeem *n* met één vrijheidsgraad
r система *f* с одной степенью свободы

O72 one-family dwelling *see* house

O73 e one-hand block, one-hand tile
d Einhandblock *m*, Einhandblockstein *m*
f bloc *m* mural à poser d'une main
nl met één hand op te pakken bouwblok
n
r мелкий стеновой блок *m* (*для
ручной кладки*)

O74 e one-hinged arch
d Eingelenkbogen *m*
f arc *m* à une articulation
nl boog *m* met één scharnier
r одношарнирная арка *f*

O75 e one-off design
d Sonderentwurf *m*
f projet *m* individuel
nl afzonderlijk ontwerp *n*
r индивидуальный проект *m*

O76 e one-off houses
d provisorische Gebäude *n pl*,
Behelfsbauten *pl*
f maisons *m pl* provisoires
[temporaires]
nl semipermanente gebouwen *n pl*
r дома-времянки *m pl*

O77 e one-pass aeration tank
d Eingangbelebungsanlage *f*
f aérateur *m* [bassin *m* d'aération] à un
seul couloir

nl enkelvoudig oxydatiebed *n*
r однокоридорный аэротенк *m*

O78 e one-pipe circulation
d Einrohr(-Strom)kreis *m*
f circulation *f* [anneau *m* de circulation]
à un seul tube [à un tuyau]
nl een-pijpcirculatie *f*
r однотрубное кольцо *n* циркуляции

O79 e one pipe heating system
d 1. Einrohrheizungssystem *n*
2. Einrohr-Wärmeversorgungssystem
n, Einleiter-Wärmeversorgungssystem
n
f 1. chauffage *m* central à un tuyau
2. chauffage *m* urbain à un seul
conduit
nl enkelpijps verwarmingssysteem *n*
r 1. однотрубная система *f* отопления
2. однотрубная система *f* теплоснабжения

O80 e one-pipe loop system
d Einrohrwasserheizungssystem *n* mit
Kurzschlußstrecken
f chauffage *m* central à un tuyau bouclé
nl enkelpijpsysteem *n* met
kortsluitleidingen
r (насосная) однотрубная система *f*
отопления с замыкающими участками

O81 e one-pipe system
d 1. Einrohrsystem *n*, Einleitersystem
n (*Wärmeversorgung*)
2. Einrohrhausentwässerung *f*,
Mischsystem *n* der Gebäudeentwässerung
f 1. installation *f* de chauffage (*à eau
chaude*) à un tuyau 2. réseau *m*
d'évacuation intérieure aux chutes
uniques
nl 1. enkelpijpsysteem *n* (*verwarming*)
2. gemengd rioolwatersysteem *n*
r 1. однотрубная система *f* (*напр.
отопления*) 2. общесплавная
система *f* домовой канализации

O82 e one-sided connection
d einseitige Verbindung *f*, einseitiger
Anschluß *m*
f assemblage *m* [joint *m*,
branchement *m*] unilateral
nl eenzijdige aansluiting *f*
r одностороннее (при)соединение *n*

O83 e one-side welding
d einseitiges Schweißen *n*
f soudage *m* unilateral, soudure *f*
simple
nl eenzijdig lassen *n*
r односторонняя сварка *f*

O84 e one-span two-hinged frame
d einfeldriger Zweigelenkrahmen *m*
f portique *m* à deux articulations
nl tweescharnierportaal *n*
r однопролётная двухшарнирная рама
f

O85 e one-storey house *US*
d Flachhaus *n*

f maison *m* à un seul étage
nl bungalow *m* (*huis met alle vertrekken op één niveau*)
r одноэтажный дом *m*

O86 e **one-way slab**
d einsinnig bewehrte Platte *f*
f dalle *f* en béton armé à l'armature disposée dans une seule direction
nl in slechts één richting gewapende plaat *f* (*m*)
r плита *f*, армированная в одном направлении

O87 e **one-way street**
d Einbahnstraße *f*
f rue *f* à un sens [à sens unique]
nl straat *f*(*m*) met eenrichtingsverkeer
r улица *f* с односторонним движением

O88 e **one-way system**
d Einweg-Bewehrungssystem *n*, einsinnige Bewehrungsanordnung *f*
f système *m* de ferraillage dans une seule direction
nl wapeningssysteem *n* in één richting
r система *f* армирования в одном направлении

O89 **one-way valve** *see* **non-return valve**

O90 **on-job laboratory** *see* **field laboratory**

O91 e **on-off valve**
d Zweipunktregelventil *n*
f soupape *f* de réglage à deux positions
nl tweepunts regelventiel *n*
r двухпозиционный регулирующий клапан *m*

O92 e **onramp**
d Auffahrt *f*, Einfahrtrampe *f*
f rampe *f* de raccordement [d'accès]
nl oprit *m* (*autoweg*)
r въезд *m*, въездная аппарель *f* (*на автостраде*)

O93 e **on-site connection**
d Baustellenverbindung *f*; Baustellenstoß *m*
f joint *m* de montage; assemblage *m* sur le chantier
nl montageverbinding *f*
r монтажное соединение *n*; монтажный стык *m*

O94 e **on-the-job service unit**
d Werkstattwagen *m*
f camion-atelier *m*
nl werkplaatswagen *m*
r передвижная мастерская *f*

O95 e **opal glass**
d Opalglas *n*
f verre *m* opale
nl opaalglas *n*
r молочное стекло *n*

O96 e **opaque ceramic-glazed tile**
d glasierte Fliese *f* mit undurchsichtiger Farbglasur
f tuile *f* cuite à glaçure colorée opaque
nl keramische tegel *m* met ondoorzichtig glazuur

r керамическая плитка *f* с непрозрачной цветной глазурью

O97 e **open-air intake works**
d drucklose Umleitung *f*
f dérivation *f* [prise *f* d'eau] à air libre
nl ongestuwde afleiding *f*
r безнапорная деривация *f*, безнапорный водозабор *m*

O98 e **open-air plant**
d Freiluftanlage *f*, Außenanlage *f*
f installation *f* extérieure
nl installatie *f* in de open lucht
r открытая (технологическая) установка *f*

O99 e **open-air water power plant**
d offenes Wasserkraftwerk *n*, Wasserkraftwerk *n* in Freiluftbauweise, Freiluft--Wasserkraftanlage *f*
f usine *f* du type «outdoor»
nl open waterkrachtcentrale *f*
r ГЭС *f* открытого типа

O100 e **open boarding**
d offene Dachbelattung *f*
f planchéiage *m* à claire-voie
nl dak *n* zonder dakbeschot
r сквозная обрешётка *f* кровли

O101 e **open caisson**
d Senkbrunnen *m*, offener Senkkasten *m*
f caisson *m* ouvert
nl open caisson *m*
r опускной колодец *m*

O102 e **open caisson foundation**
d Brunnengründung *f*, Senkbrunnen-fundamentkörper *m*
f fondation *f* sur caissons ouverts
nl caissonfundering *f*
r фундамент *m* на опускных колодцах

O104 **open chamber needle valve** *see* **mashroom valve**

O105 e **open channel**
d offenes Gerinne *n*
f canal *m* découvert [à ciel ouvert]
nl open geul *f*(*m*) [waterloop *m*]
r открытое русло *n*, открытый водовод *m*

O106 e **open-channel flow**
d Freispiegel-Strömung *f*, Strömung *f* in offenem Wasserlauf
f écoulement *m* à surface libre
nl vrije stroming *f*
r поток *m* со свободной поверхностью, движение *n* воды в открытом русле, безнапорное движение *n*

O107 e **open-channel hydraulics**
d Hydraulik *f* offener Gerinne, Flußhydraulik *f*
f hydraulique *f* fluviale
nl hydraulica *f* van de vrije stroming
r гидравлика *f* руслового потока, русловая гидравлика *f*

O108 e **open-conduit drop**
d offener Absturz *m*

f chute *f* d'eau ouverte
nl open overstort *n*
r открытый перепад *m*

O109 *e* **open cornice**
d offenes Gesims *n*
f corniche *f* ouverte
nl open kroonlijst *f(m)*
r открытый карниз *m*

O110 *e* **open cut**
d offene Baugrube *f*, Graben *m*
f fouille *f* à ciel ouvert
[*nl* open bouwput *m*
r открытая выемка *f*

O111 *e* **open-cut tunneling**
d Herstellung *f* eines Tunnels in
offener Baugrube
f exécution *f* des tunnels à ciel ouvert,
exécution des tunnels dans une fouille
nl tunnelbouw *m* in open bouwput
r строительство *n* туннеля открытым
способом

O112 **open deck car park** *see* **open parking
structure**

O113 *e* **open distribution**
d offenverlegte Verteilung *f*
f distribution *f* ouverte
nl net *n* van open leidingen
r открытая разводка *f*

O114 **open-ditch drainage** *see* **open drainage**

O115 *e* **open drain**
d offener Drängraben *m*
f drain *m* (à ciel) ouvert
nl open afvoerleiding *f*
r открытая дрена *f*

O116 *e* **open drainage**
d Oberflächenentwässerung *f*
f drainage *m* à ciel ouvert [par fossés]
nl afwatering *f* door open leidingen
r открытый дренаж *m*

O117 *e* **open floor**
d offene Balkendecke *f*
f plancher *m* ouvert (sans plafond)
nl balkenzoldering *f*
r балочное перекрытие *n* с открытыми
балками, открытый потолок *m*

O118 **open eaves** *see* **open cornice**

O119 *e* **open-end well**
d Rohrbrunnen *m* ohne Bodenfilter
f puits *m* à fond ouvert
nl pompbuis *f(m)* zonder bodemfilter
r колодец *m* с донным питанием

O120 *e* **open expansion tank heating system**
d offenes Heizungssystem *n*
f installation *f* de chauffage ouvert
nl verwarmingssysteem *n* met open
expansievat
r открытая система *f* отопления

O121 *e* **open floor duct**
d offener Bodenkanal *m*
f canal *m* dans le plancher à dalles de
couverture amovibles

nl open leidingkoker *m*
r канал *m* в полу со съёмными плитами

O122 *e* **open-frame girder**
d Vierendeelträger *m*, Rahmenträger
m
f poutre *f* Vierendeel
nl Vierendeelligger *m*
r безраскосная ферма *f*, ферма *f*
Виренделя

O123 *e* **open-graded aggregate**
d offen abgestuftes Mineralgemisch *n*
f aggrégat *m* [granulat *m*] sans fractions
fines
nl aggregaat *n* zonder fijne fracties
r каменный зернистый материал *m*
[заполнитель *m*] без мелких
и пылевидных фракций

O124 *e* **open hood**
d offene Absaughaube *f* [Absauganlage
f]
f hotte *f* ouverte
nl open afzuigkap *f(m)*
r открытый отсос *m*

O125 *e* **opening**
d 1. Öffnung *f* 2. Einschnitt *m*
3. Aufklaffen *n* 4. Aufschwenken *n*,
Aufdrehen *n* (*Schwenkbrücke*)
5. Brückenfeld *n*, Brückenöffnung *f*
f 1. ouverture *f*, trou *m*, orifice *f*,
baie *f* 2. fouille *f*, tranchée *f* (*pour
tuyau*) 3. ouverture *f* de la soudure
(*en V, en X*) 4. pivotement *m* de la
travée tournante d'un pont mobile
5. ouverture *f* [travée *f*] d'un pont
nl 1. opening *f* 2. sleuf *f(m)* 3. openstaan
n (*naad*) 4. openen *n* (*brug*)
5. doorvaartwijdte *f* (*brug*)
r 1. отверстие *n*; проём *m* 2. прорезь *f*,
траншея *f* (*для трубопровода*)
3. раскрытие *n* (*шва*) 4. разводка *f*
(*поворотного моста*) 5. пролёт *m*
(*моста*)

O126 *e* **opening light**
d oberer Belüftungsflügel *m*,
Fensterlüftungsflügel *m*;
bewegliches Fenster *n*
f imposte *f* mobile ouvrante, vasistas
m; chassis *m* ouvrant
nl beweegbaar raam *n*
r аэрационная фрамуга *f*;
открывающаяся створка *f* окна

O127 *e* **opening of bids**
d Eröffnung *f* der Ausschreibung
f ouverture *f* des offres
nl openstellen *n* van de inschrijving
r открытие *n* торгов на подряд

O129 *e* **open jetty**
d durchbrochene Anlegebrücke *f*
[Landungsbrücke *f*], Pier *m*
[Anlegekai *m*] in durchbrochener
Bauweise
f quai *m* sur pieux avec plate-forme,
jetée *f* à claire-voie
nl aanlegsteiger *m*

r причальная эстакада *f*, набережная-
-эстакада *f*, сквозной пирс *m*

O130 *e* **open joint**
  d offene Fuge *f*; Stumpfstoß *m*
  f joint *m* ouvert; joint *m* [assemblage
    m] en bout ouvert
  nl open stootnaad *m*
  r открытый шов *m*; стыковое
    соединение *n* с зазором

O131 *e* **open mortise**
  d Zapfenloch *n*
  f mortaise *f* passante
  nl open gat *n* (*pen-en-gatverbinding*)
  r прорезной паз *m*

O132 *e* **open-newel stair**
  d Wendeltreppe *f* mi offener Spindel;
    mehrarmige [mehrläufige] Treppe
    mit Zentralöffnung
  f escalier *m* à vis sans noyau; escalier
    m à noyau ouvert [creux, évidé]
  nl rechte steektrap *m* met schalmgat,
    bordestrap *m*
  r витая лестница *f* без центральной
    стойки (*с опиранием ступеней на
    стену*); многомаршевая лестница *f*
    с центральным колодцем между
    маршами

O133 *e* **open-pan mixer**
  d Trogmischer *m*
  f bétonnière-malaxeuse *f*, malaxeur *m*
    [malaxeuse *f*] à béton à axes
    horizontaux (à palettes)
  nl trogmenger *m*
  r бетоносмеситель *m* с вращающимися
    горизонтальными лопастными
    валами

O134 *e* **open parking structure**
  d offenes Parkhaus *n*
  f garage-parc *m* [garage *m* de parcage]
    ouvert
  nl open parkeergarage *f*
  r открытый гараж *m* (*без боковых стен*)

O135 *e* **open-pier construction**
  d durchbrochene [aufgelöste]
    Bauwefise *f* (*Wasserbau*)
  f construction *f* à claire-voie
  nl doorbroken constructie *f*
  r сквозная (свойная) *f* конструкция

O136 *e* **open piping**
  d offene Rohrleitung *f*
  f réseau *m* de tuyauteries [de
    conduites] extérieur
  nl terreinleiding *f* (*verwarming*)
  r открытый трубопровод *m*

O137 *e* **open pit**
  d 1. Tagebau *m*, Grube *f*
    2. Schachtbrunnen *m*
  f 1. carrière *f* à ciel ouvert 2. puits *m*
    à ciel ouvert (*d'un diamètre
    supérieur à 0,9 m*)
  nl 1. groeve *f* 2. schachtput *m*
  r 1. открытая разработка *f*, карьер *m*
    2. шахтный колодец *m* (*диаметром
    свыше 0,9 м*)

O138 *e* **open plan**
  d offene Grundrißplanung *f*,
    Großraumplanung *f*
  f plan *m* ouvert (*d'un bâtiment*)
  nl structuurplan *n*
  r открытая планировка *f* (*здания*)

O139 *e* **open return steam heating system**
  d Niederdruck-Dampfheizungsanlage *f*,
    offene Dampfheizung *f*
  f installation *f* de chauffage à la
    vapeur à basse pression
  nl lagedruk-stoomverwarmingssysteem *n*
    met open condensafvoer
  r система *f* парового отопления
    низкого давления, разомкнутая
    система *f* парового отопления

O140 *e* **open-riser stair**
  d Leitertreppe *f*
  f escalier *m* à contre-marches à claire-
    -voie
  nl open trap *m*
  r лестница *f* с открытыми
    подступёнками

O142 *e* **open roof**
  d Dach *n* ohne Dachboden, Warmdach
  f toit *m* ouvert (*sans plafond*)
  nl kap *f(m)* zonder plafond, vide *f*
  r бесчердачная крыша *f*, крыша *f*
    без потолка

O143 *e* **open section**
  d offener Querschnitt *m*
  f section *f* ouverte
  nl open doorsnede *f*
  r открытое сечение *n*

O144 *e* **open sheathing, open sheeting**
  d gitterförmige Bretterverschalung *f*
    (*Grabensicherung*)
  f blindage *m* [coffrage *m*] ajouré
    d'une tranchée
  nl wandbekleding *f* met roosters
    (*bouwput*)
  r сквозная дощатая обшивка *f*
    крепления траншей

O145 *e* **open slating**
  d Verlegen *n* der Dachbelagplatten mit
    offenen Fugen in waagerechten
    Reihen
  f disposition *f* des ardoises à claire-
    -voie
  nl dakbedekking *f* met leien met open
    voegen in horizontale rijen
  r расположение *f* кровельных плиток
    с зазорами в горизонтальных рядах

O146 *e* **open space**
  d offener Raum *m*
  f espace *f* libre
  nl open ruimte *f*
  r открытое архитектурное простран-
    ство *n*

O147 *e* **open spillway**
  d Freistrahlüberlauf *m*, freier
    Überlauf *m*
  f déversoir *m* à l'air libre

nl open overloop *m*
r открытый водослив *m*

O148 *e* **open stairway**
d Freitreppe *f*
f escalier *m* intérieur dégagé
nl vrijstaande open trap *m*
r открытая
лестница *f*

O149 *e* **open steam heating system** *see* **open return steam heating system**

O150 *e* **open storage area, open storage ground**
d offener Lagerplatz *m*, Freilagerplatz *m*
f aire *f* de stockage ouverte
nl open opslagplaats *f(m)*
r открытая складская площадка *f*

O151 *e* **open storm drainage**
d Oberflächenentwässerung *f* durch offene Gräben
f système *m* d'évacuation des eaux de pluie ouvert
nl hemelwaterafvoer *m* langs open goten
r открытый водоотвод *m* (*ливневых вод*)

O152 *e* **open-tank treatment**
d Trogtränkung *f*
f traitement *m* de bois dans le réservoir ouvert (*sans pression*)
nl verduurzamen *n* in open bad
r антисептирование *n* [пропитка *f*] (*древесины*) в открытом резервуаре (*без давления*)

O153 *e* **open tendering**
d öffentliche Ausschreibung *f*
f appel *m* d'offres public
nl openbare aanbesteding *f*
r открытые торги *m pl* (*на заключение строительного контракта*)

O154 *e* **open-textured wearing course**
d offene Verschleißschicht *f*
f couche *f* d'usure [de roulement] en enrobé ouvert
nl open slijtlaag *f(m)*
r слой *m* износа из пористого асфальтобетона

O155 *e* **open-timber floor**
d Decke *f* mit sichtbaren Holzbalken
f plancher *m* à poutres de bois ouvert (*sans plafond*)
nl balkenzoldering *f*
r перекрытие *n* с открытыми деревянными балками

O156 **open timbering** *see* **open sheathing**

O157 **open-timber roof** *see* **open roof**

O158 *e* **open-top mixer**
d Schaufelmischer *m*
f malaxeur *m* à tambour fixe ouvert avec axe vertical à palettes; malaxeur *m* à auge avec axe horizontal à palettes
nl schoepenmenger *m*
r лопастной смеситель *m* (*бетоносмеситель с вертикальным барабаном или растворосмеситель с горизонтальным лотком*)

O159 *e* **open traverse**
d offener Polygonzug *m*
f cheminement *m* (d'angles) non fermé; polygone *m* ouvert
nl niet-gesloten landmeting *f*
r незамкнутый (полигональный) ход *m*

O160 *e* **open type exhaust canopy**
d offene Saughaube *f*
f hotte *f* d'évacuation ouverte
nl open afzuigkap *f(m)*
r открытый вытяжной зонт *m*

O161 *e* **open valley**
d offene Kehle *f*
f noue *f* ouverte métallique
nl open kilgoot *f(m)*
r ендова *f*, покрытая кровельной сталью

O162 *e* **open vent**
d offenes Entlüftungsrohr *n*
f tuyau *m* d'aération ouvert
nl open ventilatiepijp *f(m)*
r открытая вентиляционная труба *f*

O163 *e* **open web**
d offener Steg *m*
f âme *f* en treillis; âme *f* (*d'une poutre*) ajourée [évidée]
nl geperforeerd lijf *n* (*balk*)
r решётчатая *или* перфорированная стенка *f* (*балки*)

O164 **open-web girder** *see* **lattice beam**

O165 *e* **open-wed steel joist**
d Stahlgitterpfette *f*
f panne *f* en treillis; poutrelle *f* à âme évidée [ajourée]
nl lichte stalen tralieligger *m*
r лёгкий стальной решётчатый прогон *m*; прутковый прогон *m*

O166 *e* **open well**
d Schachtbrunnen *m*
f puits *m* à ciel ouvert
nl open put *m*
r шахтный колодец *m*

O167 *e* **open well stair**
d Wendeltreppe *f* mit offener Spindel
f escalier *m* à vis sans noyau
nl wendeltrap *m*
r винтовая лестница *f* с центральной стойкой

O168 *e* **open wharf**
d durchbrochene Anlegebrücke *f* [Landungsbrücke *f*], durchbrochener Landesteg *m*
f quai *m* sur pieux avec plate-forme
nl landingsbrug *f(m)*
r сквозное причальное сооружение *n*

O169 *e* **open wiring**
d Überputzinstallationsleitung *f*, offen verlegte Energieleitung *f*
f installation *f* (de fils) à découvert, circuit *m* ouvert
nl leidingen *f pl* in het zicht
r открытая (электрическая) проводка *f*

O170 **openwork jetty** *see* **open wharf**

O171 *e* operable partition
    d ausziehbare [verschiebbare]
      Großplattentrennwand *f*
    f cloison *f* amovible [en panneaux
      amovibles]
    nl wegschuifbare paneelwand *m*
    r раздвижная панельная перегородка *f*

O172 *e* operable transom
    d oberer Belüftungsflügel *m*
    f imposte *f* tournante [mobile]
    nl beweegbaar bovenlicht *n*
    r открывающаяся фрамуга *f* над
      дверью

O173 operable wall *see* operable partition

O174 *e* operating bridge
    d Bedienungsbrücke *f*, Bedienungssteg *m*
    f pont *m* de service
    nl bedienbare brug *f(m)*
    r служебный мостик *m*

O175 *e* operating width
    d Arbeitsbreite *f*, Einbaubreite *f*
      (*Straßenbaumaschinen*)
    f largeur *f* de passage [de passe]
    nl werkbreedte *f*
    r ширина *f* захвата (*дорожной*
      машины); ширина *f* прохода (*катка*)

O176 *e* operational range
    d Arbeitsbereich *m*, Schwenkradius *m*
      (*Kran, Bagger*)
    f portée *f*, rayon *m* d'action (*d'une*
      grue, d'un excavateur)
    nl reikwijdte *f* (*kraan, graafmachine*)
    r зона *f* действия; радиус *m*
      действия (*крана, экскаватора*)

O177 *e* operational waste
    d Betriebsverlust(e) *m* (*pl*) (*Wasser,*
      Strom)
    f gaspillage *m* (*d'eau, d'énergie*) au
      cours de distribution
    nl bedrijfsverliezen *n pl* (*water, stroom*)
    r потери *f pl* в распределительной
      сети (*воды, электроэнергии*)

O178 *e* operation drawing
    d Werkstattzeichnung *f*, Flußdiagramm
      n, Ablaufschema *n*
    f schéma *m* de fonctionnement; plan *m*
      de disposition d'équipement
    nl werktekening *f*; uitvoeringsschema *n*
    r технологический чертёж *m*;
      технологическая схема *f*

O179 *e* operation schedule
    d Betriebsablaufplan *m*,
      Betriebsdiagramm *n*
    f programme *m* de marche des travaux
    nl werkplan *n*
    r технологический график *m*

O180 *e* operative, operator
    d 1. Arbeiter *m* 2. Maschinist *m*,
      Bedienungsmann *m*
    f 1. ouvrier *m*, travailleur *m*
      2. opérateur *m*, conducteur *m*,
      mécanicien *m*
    nl 1. arbeider *m* 2. machinist *m*

    r 1. рабочий *m* 2. оператор *m*,
      механик *m*

O181 *e* opposite blade damper
    d gegenläufige Jalousieklappe *f*
    f régistre *m* d'air à lames opposées
    nl jaloezie-terugslagklep *f(m)*
    r встречноствopчатый воздушный
      клапан *m*

O182 *e* optical sag
    d optischer Durchhang *f*, optische
      Durchbiegung *f*
    f courbure *f* optique [visuelle]
    nl zichtbare doorbuiging *f*
    r оптический [зрительный] прогиб *m*

O183 *e* optical square
    d Winkelspiegel *m*
    f équerre *f*
    nl hoekspiegel *m*
    r экер *m*

O184 *e* opus quadratum
    d regelmäßiges Schichtenmauerwerk *n*
    f maçonnerie *f* appareillée [en assises
      réglées]
    nl regelmatig steenverband *n*
      (*natuursteen*)
    r регулярная кладка *f* из каменных
      блоков, каменная кладка *f*
      регулярными рядами

O185 *e* orange peel
    d Apfelsinenschaleneffekt *m*,
      apfelsinenschalenartige Oberfläche *f*
      (*Fehler*)
    f rugosité *f* superficielle d'une couche
      de peinture (*défaut*); coulures *f pl*
      sur carraux de faïence (*défaut*)
    nl sinaasappelhuid *f(m)* (*verfspuiten*)
    r шероховатость *f* окрасочного слоя
      (*дефект*); мелкие наплывы *m pl или*
      крапинки *f pl* (*дефект на фаянсовых*
      плитках)

O186 *e* orange-peel bucket
    d Apfelsinenschalengreifer *m*,
      Polygreifer *m*
    f benne *f* preneuse à mâchoires
      multiples curvilignes
    nl poliepgrijper *m*
    r многочелюстный грейфер *m* с
      криволинейными челюстными
      створками

O187 *e* orbital sander
    d orbitale Handschleifmaschine *f*
    f ponceuse *f* à disque à mouvement
      orbital
    nl aan de omtrek werkende
      handschuurmachine *f*
    r орбитальная ручная шлифовальная
      машина *f*

O188 *e* ordinary construction
    d Gebäude *n* mit unbrennbaren
      Wänden, Holzdecken und
      Holzdachkonstruktion
    f bâtiment *m* traditionnel *ou*
      ordinaire (*avec murs en maçonnerie,*
      planchers et comble en bois)

nl traditionele bouwwijze *f*
r здание *n* с невозгораемыми стенами,
деревянными перекрытиями
и крышей

O190 *e* **ordinary portland cement**
d gewöhnlicher Portlandzement *m*,
Normalportlandzement *m*
f ciment *m* portland pour usage courant
nl normaal portlandcement *n*
r обычный портландцемент *m*

O191 *e* **ordinary structural concrete**
d tragender Schwerbeton *m*
f béton *m* normal [lourd, ordinaire
dense]
nl normaal gewapend beton *n*
r тяжёлый конструктивный бетон *m*

O192 *e* **ordnance bench mark**
d Höhenkote *f* (*im System des
staatlichen Festpunktnetzes*);
Festpunkt *m* mit absoluter Höhenkote
f cote *f* de niveau (*du canevas
géodésique national*); repère *m*
d'altitude absolute
nl hoogtemerk *n* van rijksdriehoeksnet
r высотная отметка *f* (*в системе
общенациональной опорной
геодезической сети*); репер *m* с
абсолютной отметкой

O193 *e* **or-equal specification**
d Abänderungs-Klausel *f* (*Bestimmung
der technischen Bedingungen, die dem
Auftragnehmer das Recht zuspricht,
die Abänderungen vorzuschlagen*)
f cahier *m* des charges permettant au
contracteur d'apporter au projet, en
cours d'exécution, des modifications
nl wijzigingsclausule *f(m)*
r технические условия *n pl*,
позволяющие подрядчику предлагать
изменения

O194 *e* **organic binder**
d organisches Bindemittel *n*
f liant *m* organique
nl organisch bindmiddel *n*
r органическое вяжущее *n*

O195 *e* **organic content**
d Gehalt *m* an organischen Stoffen
[Beimengungen]
f teneur *m* en matières organiques
nl gehalte *n* aan organische stoffen
r содержание *n* органических веществ

O196 *e* **organic impurity**
d organische Verunreinigung *f*
f impuretés *f pl* organiques
nl organische verontreiniging *f*
r органическая примесь *f*

O197 *e* **organic loading**
d Raumbelastung *f* der Kläranlagen
f charge *f* organique
nl organische belasting *f*
(*zuiveringsinstallatie*)
r нагрузка *f* (очистного сооружения)
по органическим загрязнениям

O198 *e* **organic silt**
d organischer Schluff *m*
f silt *m* [limon *m*] organique
nl organisch slib *n*
r пылевидный грунт *m* с
органическими включениями

O199 *e* **organic test**
d Ätznatronprobe *f*, Ätznatronprüfung *f*
f essai *m* organique
nl organische-stofbepaling *f*
r определение *n* содержания
органических примесей в грунте

O200 *e* **organic wastes**
d organisch verschmutztes Abwasser *n*
f eaux *f pl* usées avec impuretés
organiques
nl afvalwater *n* met organische
bestanddelen
r сточные воды *f pl* с органическими
загрязнениями

O201 *e* **oriel window**
d auskragender Erker *m*
f fenêtre *f* en encorbellement [en
saillie]
nl uitgekraagde erker *m*
r консольный эркер *m*

O202 *e* **orientation of building**
d Gebäudeanordnung *f*,
Fassadenanordnung *f*
f orientation *f* (*d'un bâtiment*)
nl oriëntatie *f* van een gebouw
r ориентация *f* (фасадов) здания

O203 *e* **orifice**
d 1. Öffnung *f* 2. Meßblende *f*,
Normblende *f*
f 1. orifice *m* 2. diaphragme *m*
circulaire sensible
nl 1. opening *f* 2. gecalibreerde opening *f*
r 1. отверстие *n* 2. кольцевая
измерительная диафрагма *f*
(*в трубопроводе*)

O204 *e* **orifice meter**
d Meßblende *f*, Meßblenden-
-Durchflußmesser *m*
f débitmètre *m* à diaphragme
nl stuwschijfmeter *m*
r измерительная диафрагма *f*

O205 *e* **orifice plate**
d Drosselscheibe *f*, Drosselblende *f*
f diaphragme *m* d'un débitmètre
nl stuwschijf *f(m)*
r дроссельная шайба *f* [диафрагма *f*]

O206 *e* **orifice tube test**
d Beweglichkeitsprüfung *f* des
Betongemisches
(*Beweglichkeit wird als Funktion der
Ausflußdauer bestimmt*)
f essai *m* de consistance [de fluidité]
du béton à l'aide d'un tronçon de
tube (*on mesure le temps d'écoulement
du béton frais hors de ce tronçon*)
nl consistentiebepaling *f* van
betonmortel (*consistentie als functie
van de uitlooptijd*)

r определение *n* подвижности
бетонной смеси (*путём измерения
времени вытекания из отрезка
трубы*)

O207 *e* original cross-sectional area
   d Ausgangsquerschnittsfläche *f*
   f aire *f* de la section initiale
  nl oppervlak *n* van de oorspronkelijke
     dwarsdoorsnede
   r начальная площадь *f* поперечного
     сечения

O208 *e* origin of force
   d Kraftangriffspunkt *m*
   f point *m* d'application de la force
  nl aangrijpingspunt *f(m)* van een kracht
   r точка *m* приложения силы

O209 *e* O-ring
   d Vollgummiring *m*, O-Ring *m*
   f anneau *m* d'étanchéite
  nl afsluitring *m*, O-ring *m*
   r уплотняющее кольцо *n* круглого
     сечения

O210 *e* ornamental concrete
   d Zierbeton *m*, Dekorativbeton *m*,
     Sichtbeton *m*
   f béton *m* décoratif [architectural]
  nl sierbeton *n*
   r декоративный бетон *m*

O211 *e* ornamental concrete finishing
   d Zierbetonverkleidung *f*
   f revêtement *m* décoratif en béton
     architectural
  nl afwerking *f* met sierbeton
   r отделка *f* декоративным бетоном

O212 **orthogonal anisotropic plate** *see*
     orthotropic plate

O213 *e* orthogonal projection, orthography
   d orthogonale Projektion *f*
   f projection *f* orthogonale
  nl orthogonale [rechte] projectie *f*
   r ортогональная проекция *f*

O214 *e* orthotropic plate
   d orthotrope Platte *f*, orthotroper
     Brückenüberbau *m*,
     Stahlblechfahrbahn *f*, mitwirkende
     Stahlfahrbahn *f* (*Brücke*)
   f plaque *f* orthotrope
  nl meewerkende stalen rijvloer *f(m)*
   r ортотропная плита *f* (*проезжей
     части моста*)

O215 *e* orthotropic structures
   d orthotrope Konstruktionen *f pl*
   f constructions *f pl* orthotropes
  nl meewerende [orthotrope]k
     constructies *f pl*
   r ортотропные конструкции *f pl*

O216 *e* oscillating machinery
   d Rüttelmaschinen *f pl*, Vibrations-
     maschinen *f pl*
   f engins *m pl* vibrants, machines *f pl*
     vibrantes [vibratoires]
  nl trilapparatuur *f*
   r вибрационные машины *f pl*

O217 *e* oscillating screen
   d Schwingsieb *n*
   f crible *m* oscillant
  nl trilzeef *f(m)*
   r качающийся грохот *m*;
     виброгрохот *m*

O218 *e* oscillation conveyor
   d Rüttelförderer *m*, Schwingförderer *m*,
     Vibrationsförderer *m*
   f convoyeur *m* oscillant
  nl triltransporteur *m*
   r качающийся конвейер *m*

O219 *e* ossature
   d Gerippe *n*, Skelett *n*, Gerüst *n*
   f ossature *f*, charpente *f*, carcasse *f*
  nl geraamte *n*, skelet *n*
   r каркас *m*

O220 *e* outbuilding
   d Außengebäude *n*
   f appentis *m*; bâtiment *m* extérieur;
     bâtiment *m* auxiliaire
  nl bijgebouw *n*
   r вспомогательное здание *n*;
     надворное строение *n*

O221 **outdoor air** *see* **outside air**

O222 **outdoor air inlet** *see* **outside air
     opening**

O223 *e* outdoor exposure test
   d Alterungsversuch *m* durch
     Freibewitterung
   f essai *m* des matériaux à l'action des
     intempéries
  nl verouderingsproef *f(m)* door
     blootstellen aan weersinvloeden
   r испытание *n* (*материала*) на
     старение выдерживанием на
     открытом воздухе

O224 **outdoor installation** *see* **open-air**

O225 **outdoor powerhouse** *see* **open-air water
     power plant**

O226 **outdoor power station** *see* **open-air
     water power plant**

O227 **outdoor temperature** *see* **outside
     temperature**

O228 *e* outer court
   d Vorhof *m*
   f cour *f* extérieure
  nl voorhof *m*
   r передний двор *m* (*открытый со
     стороны подъезда к зданию*)

O229 *e* outer fiber
   d Außenfaser *f* (*z.B. im Träger*)
   f fibre *f* extrême [extérieure, externe]
     (*p. ex. d'une poutre*)
  nl buitenste vezel *f(m)* (*b.v. balk*)
   r крайнее волокно *n* (*напр. балки*)

O230 *e* outer forces
   d äußere Kräfte *f pl*
   f forces *f pl* extérieures
  nl uitwendige krachten *f(m) pl*
   r внешние силы *f pl*

O231 **outer gate** *see* **lower gate**

O232 *e* **outer harbor**
   d Außenhafen *m*
   f avant-port *m*
   nl buitenhaven *f(m)*, voorhaven *f(m)*
   r аванпорт *m*

O233   **outer lining** *see* **outside architrave**

O234   **outermost fiber** *see* **outer fiber**

O235 *e* **outer slope**
   d Außenböschung *f*
   f talus *m* extérieur
   nl buitentalud *n*
   r низовой откос *m* (*плотины*);
   внешний откос *m* (*дамбы*)

O236 *e* **outer string**
   d Außenwange *f* (*Treppe*)
   f limon *m* extérieur
   nl vrije trapboom *m*
   r наружный косоур *m*, наружная
   тетива *f* (*лестницы*)

O237 *e* **outer support**
   d Endstütze *f*, Randstütze *f*,
   Endauflager *n*, Randauflagerung *f*
   f support *m* [appui *m*] extérieur
   nl randsteun *m*, eindoplegging *f*
   r наружная [крайняя] опора *f*

O238 *e* **outfall**
   d Auslaß *m*, Ausmündung *f*
   f débouché *m*, décharge *f* (*d'un égout*)
   nl uitloop *m*
   r выпуск *m* (*канализационного
   коллектора*)

O239   **outfall channel** *see* **discharge channel**

O240 *e* **outfall drain**
   d 1. natürlicher Vorfluter *m*,
   Dränagewassersammelgerinne *n*
   2. Sammeldrän *m*, Abfangsammler *m*
   f 1. drain *m* de décharge 2. drain *m*
   collecteur
   nl 1. afwateringssloot *f(m)*
   2. verzamelleiding *f*
   r 1. водоприёмник *m* (*для воды из
   осушительной системы*);
   дренирующий водоток *m* 2. дрена-
   -собиратель *f*, коллектор *m*

O241 *e* **outfall sewer**
   d Ableitungssammler *m*
   f égout *m* de décharge, collecteur *m*
   d'évacuation
   nl hoofdriool *n*, afvoerriool *n*
   r выводной [отводящий]
   (канализационный) коллектор *m*

O242 *e* **outfall structure**
   d Auslaufbauwerk *n*, Auslaßbauwerk *n*,
   Betriebsauslaß *m*
   f ouvrages *m pl* de décharge
   nl uitlaatwerk *n*
   r выпускное сооружение *n*, выпуск *m*,
   эксплуатационный водоспуск *m*

O243 *e* **outfit**
   d Ausrüstung *f*, Bestückung *f*,
   Ausstattung *f*
   f appareillage *m*, appareils *m pl*,
   équipement *m*, outillage *m*

   nl uitrusting *f*, outillage *f*
   r снаряжение *n*, оснащение *n*,
   оборудование *n*, технологическая
   оснастка *f*

O244 *e* **outflow channel**
   d Auslaufgerinne *n*
   f rigole *f* de déversement
   [d'écoulement], canal *m* de décharge
   nl uitloopgoot *f(m)*
   r выпускной жёлоб *m*, сточный лоток
   m

O245 *e* **outhouse**
   d 1. Hofklosett *n* 2. Viehstall *m*
   3. Schuppen *m*
   f 1. latrinnes *f pl* 2. remise *f* 3. grange
   f
   nl 1. buitenprivaat *n* 2. veestal *m*
   3. loods *f*
   r 1. надворная уборная *f* 2. хлев *m*
   3. сарай *m*

O246 *e* **outlet**
   d 1. Auslaßbauwerk *n*, Auslaufbauwerk
   n, Wasserauslaß *m*, Wasserauslauf *m*,
   Wasserabfluß *m*; Auslauf *m*,
   Dränauslaß *m*, Regenauslaß *m*
   2. Auslauföffnung *f*, Ablaßöffnung *f*,
   Austrittsöffnung *f*; Betriebsauslaß *m*,
   Ablauföffnung *f* 3. Wasserableitung *f*;
   Ablaufgerinne *n*, Ablaufkanal *m*
   4. Abflußmeßstelle *f* 5. Steckdose *f*
   f 1. débouché *m*; conduite *f* de vidange
   2. point *m* de rejet; orifice *m* de
   décharge 3. conduite *f*
   de vidange; canal *m* d'évacuation
   4. exutoire *m* 5. prise *f* de courant
   nl 1. uitlaatwerk *n*, uitstroomopening *f*
   2. aflaatwerk *n*, overloopbuis *f(m)*
   3. afleidingskanaal *n* 4. hydromet-
   risch meetstation *n* 5. wandcontactdoos
   f(m)
   r 1. выпускное сооружение *n*,
   (водо)выпуск *m*, водоспуск *m*;
   канализационный выпуск *m*, устье
   n (*дрены, ливнеотвода*) 2. выпускное
   [спускное] отверстие *n* 3. водоотвод
   m; водосбросное русло *n*
   4. замыкающий створ *m*
   5. штепсельная розетка *f*

O247 *e* **outlet box**
   d Verteilkasten *m*, Abzweigkasten *m*
   f boîte *f* de sortie
   nl verdeelkast *f(m)*, afsluitkast *f(m)*
   r выходная коробка *f* (*электросети*)

O248 *e* **outlet channel**
   d Auslaufgerinne *n*, Auslaufkanal *m*
   f canal *m* de dérivation
   nl afleidingskanaal *n*
   r отводящий канал *m*

O250 *e* **outlet gate**
   d Ablaßverschluß *m*
   f vanne *f* de vidange
   nl uitwateringssluis *f(m)*
   r затвор *m* водоспуска

O251   **outlet opening** *see* **discharge openin**

O252 *e* **outlet pipe**
　　d 1. Rohrüberlauf *m*, Rohrüberfall *m*;
　　Rohrsiel *n* 2. Ablaßrohr *n*,
　　Saugrohr *n* 3. Entnahmerohr *n*
　　f 1. vidange *f* tubulaire 2. tuyau *m*
　　de décharge 3. tuyau *m* de prise d'eau
　　nl 1. overloopbuis *f(m)* 2. uitloopbuis
　　f(m) 3. inlaatpijp *f(m)*
　　r 1. трубчатый водоспуск *m*
　　2. отводящая [отсасывающая]
　　труба *f* 3. водозаборная труба *f*

O253 **outlet section** *see* **point of concentration**

O254 *e* **outlet structure**
　　d Auslaufbauwerk *n*, Auslaßbauwerk *n*,
　　Betriebsauslaß *m*;
　　Ausmündungsbauwerk *n*
　　f ouvrage *m* d'évacuation; ouvrage *m*
　　de vidange
　　nl uitlaatwerk *n*; uitmondingswerk *n*
　　r выпускное [сбросное] сооружение *n*;
　　устьевое сооружение *n* (*на выходе*
　　дренажной сети)

O255 **outlet tunnel** *see* **discharge tunnel**

O256 *e* **outlet valve**
　　d Ablaßschieber *m*, Entleerungsschieber
　　m, Auslaßschieber *m*
　　f soupape *f* d'échappement [de
　　décharge]
　　nl uitlaatklep *f(m)*, uitlaatschuif *f(m)*
　　r спускная задвижка *f*, задвижка *f*
　　опорожнения

O257 *e* **outlet works**
　　d Auslaßbauwerke *n pl*
　　f ouvrages *m pl* de vidange
　　nl uitlaatwerken *n pl*
　　r водовыпускные сооружения *n pl*

O258 *e* **outline drawing**
　　d Umrißzeichnung *f*
　　f plan *m* d'encombrement
　　nl schetsontwerp *n*, globaal plan *n*
　　r габаритный чертёж *m*

O259 *e* **outline specification**
　　d allgemeine technische
　　Bauvorschriften *f pl* [Bedingungen *f*
　　pl]
　　f cahier *m* des charges [de clauses]
　　générales
　　nl algemene technische
　　bouwvoorschriften *n pl*
　　r общие технические условия *n pl*
　　на строительство

O260 **outmost fiber** *see* **outer fiber**

O261 *e* **out-port**
　　d Vorkammer *f*; Vorhafen *m*
　　f bassin *m* d'attente; avant-port *m*
　　nl buitenhaven *f(m)*; voorhaven *f(m)*
　　r аванкамера *f*; аванпорт *m*

O262 *e* **outrigger**
　　d Abstützarm *m*, ausfahrende
　　Seitenstütze *f*, Hilfsstütze *f* (*Kran*)
　　f stabilisateur *m* (*d'une grue*); porte-
　　en-dehors *m*

　　nl stempel(arm) *m* (*mobiele kraan*);
　　console *f* van een steegsteiger
　　r аутриггер *m*, выносная опора *f*
　　(*крана*)

O263 *e* **outrigger base**
　　d Stützfuß *m*, Abstützplatte *f* (*Kran*)
　　f base *f* [pied *m*] du stabilisateur
　　(*d'une grue*)
　　nl bodemplaat *f(m)* van de stempel
　　r плита *f* выносной опоры (*крана*)

O264 *e* **outrigger jack**
　　d Stützpresse *f* (*Kran*)
　　f verin *m* [cric *m*] du stabilisateur
　　(*d'une grue*)
　　nl stempelpomp *f(m)*
　　r домкрат *m* выносной опоры (*крана*)

O265 *e* **outrigger scaffold**
　　d Kragstützengerüst *n*
　　f échafaudages *m pl* sur consoles
　　nl consolestelling *f*
　　r консольные леса *pl*

O266 *e* **outrigger shore**
　　d ausziehbare Seitenstütze *f*
　　f console *f* provisoire, support *m* en
　　porte-à-faux provisoire
　　nl afneembare stempel *m*
　　r временная консольная опора *f*,
　　временный кронштейн *m*

O267 *e* **outside air**
　　d Außenluft *f*
　　f air *m* extérieur
　　nl buitenlucht *f(m)*
　　r наружный воздух *m*

O268 *e* **outside air opening**
　　d Außenlufterfasser *m*,
　　Außenluftdurchlaß *m*,
　　Außenluftansaugöffnung *f*
　　f ouverture *f* d'air extérieur
　　nl aanzuigopening *f* voor buitenlucht
　　r наружный воздухозабор *m*
　　(*устройство*)

O269 *e* **outside architrave, outside casing**
　　d äußere Verkleidungsleiste *f* (*Fenster*)
　　f chambranle *m* extérieur (*d'une*
　　fenêtre)
　　nl buitenomraming *f* (*venster*)
　　r наружный наличник *m*, наружное
　　обрамление *n* (*окна*)

O270 **outside corner moulding** *see* **angle bead**

O271 *e* **outside door**
　　d Außentür *f*
　　f porte *f* extérieure
　　nl buitendeur *f(m)*
　　r наружная дверь *f*

O272 *e* **outside film coefficient**
　　d äußere Wärmeübergangszahl *f*
　　f coefficient *m* de transmission de
　　chaleur superficielle
　　nl warmtedoorgangscoëfficiënt *m*
　　r внешний коэффициент *m* теплоотдачи,
　　коэффициент *m* теплоотдачи
　　наружной поверхности

O273 e outside finish
 d Außenputz *m*, Außenverkleidung *f*
   (*Gebäude*)
 f fini *m* extérieur, finition *f* (*d'un
   bâtiment*) extérieure
 nl buitenafwerking *f* (*gebouw*)
 r наружная отделка *f* (*здания*)

O274 e outside foundation line
 d äußere Fundamentumrißlinie *f*
 f ligne *f* de (mur de) fondation
   extérieure
 nl uitwendige funderingsomtrek *m*
 r наружная граничная линия *f*
   фундамента [фундаментной стены]

O275 e outside glazing
 d Außenverglasung *f*
 f vitrage *m* extérieure [de façade]
 nl buitenbeglazing *f*
 r остекление *n*, производимое с фасада
   здания, наружное остекление *n*

O276 e outside scaffolding
 d Außenrüstung *f*, Außenbaugerüst *n*
 f échafaudages *m pl* extérieurs
 nl buitensteiger *m*
 r наружные леса *pl*

O277 e outside slope
 d 1. luftseitige Böschung *f* (*Talsperre*)
   2. Außenböschung *f*, Außenseite *f*
   (*Deiche*)
 f talus *m* aval; talus *m* extérieur
 nl buitenvlak *n* (*stuwdam*); buitentalud *n*
   (*dijk*)
 r низовой откос *m* (*водохранилищной
   плотины*); внешний откос *m*
   (*ограждающей дамбы*)

O278 e outside studding plate
 d 1. Rahmenoberteil *m*, Rahmholz *n*
   2. Rahmenunterteil *m*, Schwelle *f*
 f traverse *f* d'une ossature de bois
   (*semelle basse ou sablière haute*)
 nl bovendorpel *m*; onderdorpel *m* (*raam*)
 r обвязочный брус *m* (*нижний или
   верхний*) деревянного каркаса

O279 e outside temperature
 d Außen(luft)temperatur *f*
 f température *f* extérieure [externe]
 nl buitentemperatuur *f*
 r наружная температура *f*, температура
   f наружного воздуха

O280 e outsize
 d Fehlkorn *n*
 f grains *m pl* déclassé(s), fractions *f pl*
   granulométriques rejetées, refus *m* de
   crible, rejet *m* au tamis
 nl van de voorgeschreven maat afwijkende korrel *m*
 r некондиционные фракции *f pl*
   (*зернистого материала,
   сортируемого по крупности*)

O281 e outward bulging
 d Ausbeulen *n*, Ausbauchen *n*
 f soulèvement *m*, bombement *m*,
   voilement *m*, surélévation *f*

 nl uitpuilen *n*, opbollen *n*
 r выпучивание *n*

O282 e ovals
 d Marmorgrus *m* (*mit glatten
   abgerundeten Körnern*)
 f grains *m pl* de marbre (*d'une forme
   ovale*)
 nl marmerslag *m* met gladde ovale
   korrels
 r мраморная крошка *f* (*гладкой
   овальной формы*)

O283 e ovendry wood
 d ofengetrocknetes Holz *n*
 f bois *m* séché en tuve, bois *m* de
   séchage artificiel
 nl in de oven gedroogd hout *n*
 r древесина *f* [лесоматериал *m*]
   печной сушки

O284 e overall coefficient of heat transfer
 d Wärmedurchgangszahl *f*, k-Wert *m*
 f coefficient *m* global de transmission
   thermique
 nl totale warmtedoorgangscoëfficiënt *m*
 r коэффициент *m* теплопроводности

O285 e overall depth of section
 d Gesamtquerschnittshöhe *f*
 f hauteur *f* total de la section
 nl hoogte *f* van de volle doorsnede
 r полная высота *f* сечения

O286 e overall dimensions
 d Außenmaße *n pl*, Außenabmessungen
   f pl, Gesamtabmessungen *f pl*
 f dimensions *f pl* hors tout, encombrement *m*
 nl buitenwerkse maten *f(m) pl*
 r наружные [габаритные] размеры *m
   pl*

O288 e overall housing
 d Wärmeschutzeinhausung *f* (*auf
   Baustellen*)
 f abri *m* chauffant (*pour le bétonnage
   en hiver*)
 nl beschuttende omhulling *f* (*vers beton*)
 r тепляк *m*

O289 e overall length
 d Gesamtlänge *f*
 f longueur *f* hors tout [d'encombrement]
 nl volle lengte *f*
 r габаритная длина *f*

O290 e overbridge
 d Straßenüberführung *f*
 f pont *m* supérieur
 nl overbrugging *f*, viaduct *n* (*over een
   weg in uitgraving*)
 r путепровод *m* над автомагистралью

O291 e overburden
 d Abraum *m*
 f terrain *m* de recouvrement
 nl af te voeren bovengrond *m*
 r вскрышной слой *m* грунта,
   вскрышная порода *f*

O292 e overburden excavator
 d Abraumbagger *m*

f excavateur *m* [excavatrice *f*]
à morts-terrains, excavateur *m* de
deblais
nl graafmachine *f*, scraper *m*
r вскрышной экскаватор *m*

O293 *e* **overburden pressure**
d Überlagerungsdruck *m* (*Tunnel*)
f pression *f* de surcharge [de terrain
de recouvrement]
nl druk *m* van de deklaag (*tunnel*)
r давление *n* пород кровли (*туннеля*)

O294 *e* **overburden stripping**
d Abraumbaggerung *f*, Abräumen *n*
f travaux *m pl* de déblaiement
nl verwijderen *n* van de begroeiing en
de bovengrond
r вскрышные работы *f pl*

O295 *e* **overdamming**
d Überstau *m*
f remous *m* excessif [d'excès]
nl overstuwing *f*
r переподпор *m*

O296 *e* **overdesign**
d Überdimensionierung *f*
f calcul *m* des constructions à base
d'une marge de sécurité élevée
nl overdimensionering *f*
r проектирование *n* [расчёт *m*]
конструкций с повышенным запасом
прочности

O297 *e* **overdevelopment**
d Entnahmeüberschreitung *f*
f superexploitation *f*
nl uitputting *f*
r истощение *n* подземного бассейна

O298 *e* **overdosage**
d Überdosierung *f*
f surdosage *m*
nl overdosering *f*
r превышение *n* установленной
нормы расхода (*напр. цемента*) при
дозировке, повышенная дозировка *f*

O299 *e* **overdraft**
d Überpumpen *n*
f pompage *m* excessif
nl overmatig pompen *n*
r откачка *f* сверх гарантированного
дебита (*колодца, скважины*)

O300 *e* **overfall**
d 1. Überfall *m*, Überlauf *m*
2. Überfallwehr *n*, Überlaufwehr *n*
f 1. déversoir *m*, évacuateur *m*
2. barrage-déversoir *m*
nl 1. overloop *m* 2. overloopstuw· *m*,
overval *m*
r 1. водослив *m*, водосброс *m*
2. водосливная плотина *f*

O301 *e* **overfall dam**
d Überfallsperre *f*, Überfallwehr *n*,
Überfallstaumauer *f*, Sturzwehr *n*,
Schußwehr *n*
f barrage-déversoir *m*
nl stuwdam *m* met overstort [overlaat]

r водосливная [водосбросная]
плотина *f*
O302 **overfall gap** *see* **spillway opening**
O303 **overfall weir** *see* **overflow weir**
O304 *e* **overflow**
d 1. Überlauf *m*, Ablauf *m*, Abfluß *m*;
Überlaufrinne *f* 2. Überlaufrohr *n*,
Überlaufvorrichtung *f*,
Überlaufsicherung *f*,
Überfüllsicherung *f*
f 1. déversement *m*, débordement *m*
2. égout *m* [conduite *f*] de trop-plein,
trop-plein *m*
nl 1. afvloeiing *f*; overloopgoot *f(m)*
2. overlooppijp *f(m)*
r 1. перелив *m*, перепуск *m*, слив *m*;
водосток *m* 2. переливная труба *f*;
переливное устройство *n*

O305 *e* **overflow buttress**
d Überfallpfeiler *m*
f éperon *m* [contrefort *m*] du barrage-
-déversoir
nl overlooppijler *m*
r контрфорс *m* водосливной части
плотины

O307 *e* **overflow chute**
d Hochwasserentlastungskanal *m*,
Überlaufkanal *m*, Überfallkanal *m*
f déversoir *m*, canal *m* d'écoulement
[d'évacuation]
nl overlaat *m*, afleidingskanaal *n*
r водосливный [водосбросный] канал *m*

O308 **overflow dam** *see* **overfall dam**
O309 *e* **overflow height**
d Überfallhöhe *f*
f chute *f* à crête déversante [de
déversoir]
nl overloophoogte *f*
r напор *m* на гребне водослива [на
водосливе]

O310 *e* **overflow pipe**
d Überlaufrohr *n*
f trop-plein *m*, tuyau *m* d'écoulement
[de trop-plein]
nl overlooppijp *f(m)*
r переливная труба *f*

O311 **overflow section** *see* **spillway section**
O312 *e* **overflow spillway**
d Überlaufstrecke *f*; Freiflut *f*;
Oberflächenüberlauf *m*
f déversoir *m* de superficie, évacuateur
m aérien de surface
nl overlaat *m* met afvloeiing over het
terrein
r водосливная секция *f* бетонной
плотины; поверхностный водослив
m; водослив *m* со свободным
падением струи

O313 **overflow stand** *see* **overflow pipe**
O314 *e* **overflow surface**
d Überfallrücken *m*
f surface *f* déversante

nl kruin *f(m)* van de overlaat
r водосливная поверхность *f*,
водоскат *m*

O315 *e* **overflow tower**
d Überlaufturm *m*
f tour *f* déversante
nl overlooptoren *m*
r башенный водосброс *m*

O316 *e* **overflow valve**
d Schlabberventil *n*, Überströmventil *n*
f soupape *f* de bypass, vanne *f* de dérivation
nl omloopklep *f(m)*
r перепускной клапан *m*

O317 **overflow water** *see* **tail water**

O318 *e* **overflow weir**
d 1. Überfallwehr *n*, Überlaufwehr *n* 2. Wehrkrone *f*
f 1. barrage *m* déversoir (à basse chute) 2. seuil *m* du déversoir
nl 1. overloopstuw *m* 2. stuwdrempel *m*
r 1. низконапорная водосливная плотина *f* 2. водосливный порог *m*

O319 *e* **overhand work**
d Außenwanderrichtung *f* mit Benutzung von Innengerüst
f travaux *m pl* de maçonnerie exécutes à l'aide des échafaudages intérieurs fixes
nl vanaf een binnenstelling een buitenmuur metselen *n*
r кирпичная кладка *f* наружных стен с внутренних лесов *или* подмостей

O320 *e* **overhang**
d 1. Überhang *m*, Überstand *m* 2. Ausladung *f*, Auskragung *f* 3. Überhang *m*
f 1. auvent *m* 2. porte-à-faux (*d'une poutre*); portée, volée (*d'un porte-à-faux*) 3. saillie *f*, partie *f* saillante (*d'un bâtiment*)
nl 1. overhangend gedeelte *n* 2. overstek *n* 3. uitbouw *m*
r 1. свес *m* 2. консоль *f* (*балки*); вылет *m* (*консоли*) 3. выступ *m*, консольная часть *f* (*здания*)

O321 *e* **overhead clearance**
d lichte Höhe *f*
f hauteur *f* libre
nl vrije hoogte *f*
r верхний габарит *m* приближения конструкций

O322 *e* **overhead crane**
d Laufkran *m*
f pont *m* roulant
nl bovenloopkraan *f(m)*
r мостовой кран *m*

O323 *e* **overhead crane girder**
d Laufkranträger *m*
f poutre *f* du pont roulant
nl loopkraandrager *m*
r балка *f* мостового крана

O324 *e* **overhead distribution**
d Oberleitungsverteilernetz *n*
f ligne *f* électrique aérienne
nl distributienet *n* met bovengrondse leidingen
r воздушная (электрическая) сеть *f*

O325 *e* **overhead door**
d Klapptür *f*
f porte *f* basculante
nl kanteldeur *f(m)*
r подъёмно-поворотная дверь *f*

O326 *e* **overhead heating system**
d Heizungssystem *n* mit oberer Verteilung
f chauffage *m* à eau chaude par gravité
nl verwarmingssysteem *n* met bovenverdeling
r система *f* отопления с верхней разводкой

O327 **overhead irrigation** *see* **spray irrigation**

O328 *e* **overhead line**
d 1. Oberleitung *f* 2. Freileitung *f*
f 1. ligne *f* de contact 2. ligne *f* électrique aérienne
nl 1. bovenleiding *f* 2. luchtleiding *f*, bovengrondse leiding *f*
r 1. контактный провод *m* 2. воздушная (электрическая) линия *f*

O329 *e* **overhead line mast**
d Oberleitungsmast *m*
f mât *m* [poteau *m*, pylône *m*] d'une ligne électrique
nl elektriciteitspaal *m*, hoogspanningsmast *m*
r мачта *f* электрической сети

O330 *e* **overhead monorail crane**
d Einschienen-Laufkran *m*
f palan *m* électrique
nl loopkat *f(m)*
r тельфер *m*

O331 *e* **overhead pipeline**
d Hochrohrleitung *f*, oberirdische Rohrleitung *f*, Überkopfrohrleitung *f*
f pipe-line *m* aérien, tuyautérie *f* [canalisation *f*, conduite *f*] aérienne
nl bovengrondse pijpleiding *f*
r надземный [воздушный] трубопровод *m*

O333 **overhead-travel(l)ing crane** *see* **overhead crane**

O334 *e* **overhead water-storage tank**
d Hochbehälter *m*, Wasserdruckbehälter *m*
f chateau *m* d'eau
nl watertoren *m*
r наземный напорный резервуар *m* для воды

O335 *e* **overhead welding**
d Überkopfschweißen *n*, Zwölf-Uhr--Schweißen *n*
f soudage *m* au plafond
nl boven het hoofd lassen *n*

r потолочная сварка *f*, сварка *f*
в потолочном положении

O336 *e* overhead wire *see* overhead line

O337 *e* overhead work
d Arbeit *f* in der Überkopfhöhenlage
f travaux *m pl* effectués à une hauteur
nl werken *n* boven hoofdhoogte
r работа *f* на высоте

O338 *e* overhung door
d Hänge-Schiebetür *f*
f porte *f* suspendue coulissante
[à coulisse]
nl hangende schuifdeur *f(m)*
r подвесная раздвижная дверь *f*

O339 *e* overland flow
d Überland-Abfluß *m*, flächenhafter
[oberirdischer] Abfluß *m*
f ruissellement *m* de surface
nl afvloeiing *f* over het land,
overstroming *f*
r поверхностный сток *m*

O340 *e* overlap
d 1. Überlappung *f* (*Schweißen*)
2. Überdeckung *f* (*Dachstein*)
f 1. recouvrement *m* (*d'un joint soudé*)
2. chevauchement *m*, imbrication *f*
(*des ardoises*)
nl 1. overlapping *f* (*lassen*) 2. overlap *m*
(*leien*)
r 1. нахлёстка *f* (*при сварке*)
2. перекрытие *n* (*черепицы*)

O341 *e* overlapping astragal
d Anschlagleiste *f*, Fugendeckleiste *f*
f battement *m* mouluré rapporté
nl deklijst *f(m)*, deklat *f(m)*
r накладная притворная планка *f*,
притворный нащельник *m*,
притворная накладка *f*

O342 *e* overlapping construction schedule
d vereinigter Arbeitsablaufplan *m*
f programme *m* d'avancement des
travaux pour divers corps d'état,
planning *m* général d'avancement des
travaux par corps d'état
nl gezamenlijk werkplan *n*
r совмещённый график *m* производства
строительных работ

O343 *e* overlay
d 1. Überzug *m*, Ausgleichschicht *f*,
Abgleichschicht *f*,
2. Verschleißschicht *f*, Deckschicht *f*,
Über(zug)decke *f*
f 1. chape *f* de réglage [de nivellement]
(*en béton ou mortier*) 2. couche *f*
d'usure [de roulement]
nl 1. afwerklaag *f(m)*, deklaag *f(m)*
(*beton*) 2. slijtlaag *f(m)* (*wegdek*)
r 1. бетонная *или* растворная стяжка *f*
2. слой *m* износа (*дорожного
покрытия*)

O345 *e* overlay paper
d Abdeckpapier *n* (*Nachbehandlung des
Betons*)
f papier *m* pour cure du béton

nl afdekpapier *n* (*nabehandeling van
beton*)
r бумага *f* [картон *m*] для укрытия
выдерживаемого бетона

O346 *e* overlay pavement
d Über(zug)decke *f*
f chaussé *f* avec couche d'usure
nl deklaag *f(m)*, wegdek *n* met slijtlaag
r дорожное покрытие *n*, подвергнутое
поверхностной обработке;
дорожное покрытие *n* со слоем износа

O347 *e* overlength
d Zugabe *f*, erforderliche Überlänge *f*
(*Bewehrungsstäbe*)
f surlongueur *f*
nl overlengte (*wapeningsstaven*)
r припуск *m* [запас *m*] длины
(*арматурных стержней*)

O348 *e* overload
d Überlastung *f*
f surcharge *f*
nl overbelasting *f*
r перегрузка *f*

O349 *e* overloader
d Überkopf-Lader *m*, Wurfschaufellader
m, Rückwärtslader *m*
f chargeur *m* basculeur
nl laadschop *f(m)*
r перекидной одноковшовый погрузчик
m

O350 *e* overload limiter
d Überlastsicherung *f*
f limiteur *m* de surcharge
nl belastingsbegrenzer *m*,
overbelastingsveiligheid *f*
r ограничитель *m* грузоподъёмности

O351 *e* overload protection
d Überlast(ungs)schutz *m*
f protection *f* contre les surcharges
nl beveiliging *f* tegen overbelasting
r защита *f* от перегрузок

O352 *e* overlying ground
d Überdeckung *f* (*Tunnelbau*)
f ciel *m* de galerie
nl gronddekking *f* (*tunnelbouw*)
r кровля *f* (*туннеля*)

O353 *e* over-mortared concrete mix
d Betongemisch *n* [Betonmischung *f*]
mit Mörtelüberschuß
f béton *m* frais à haut teneur [dosage]
en mortier
nl te rijke betonmortel *m*
r бетонная смесь *f* с большим
содержанием растворной части

O354 *e* overpass
d Überführung *f*
f passage *m* supérieur
nl viaduct *n*, ongelijkvloerse kruising *f*
r путепровод *m* над дорогой

O335 *e* overpass for pedestrians
d Fußgängersteg *m*,
Fußgängerüberführung *f*,
Fußwegübergang *m*

f passage *m* supérieur pour piétons,
pont *m* pour piétons
nl voetbrug *f(m)*
r надземный переход *m*,
пешеходный мостик *m*

O356 *e* **overpressure protection**
d Überdruckschutz *m*
f protection *f* contre les surpressions
nl overdrukbeveiliging *f*
r защита *f* от избыточного давления

O357 *e* **overpressure release valve**
d Überdruckentlastungsventil *n*
f soupape *f* de sûreté, détendeur *m* de
pression
nl overdrukventiel *n*
r предохранительный клапан *m*

O358 *e* **oversailing course**
d auskragende Ziegelschicht *f*
f assise *f* de briques saillantes;
bandeau *m* saillant (en briques)
nl overkragende steenlaag *f(m)*
r выступающий ряд *m* каменной
кладки; поясок *m*

O359 *e* **oversanded mix**
d Gemisch *n* [Mischung *f*] mit
Sandüberschuß
f mélange *m* (*p. ex. de béton*) à grand
teneur en sable
nl mortel *m* met overmaat aan zand
r смесь *f* (*напр. бетонная*) с избыточным
содержанием песка

O360 *e* **oversite concrete**
d Betonbettung *f*, Betonunterschicht *f*,
Betonausgleich(s)schicht *f*
f lit *m* de béton, couche *f* d'arasement
en béton, sous-couche *f* en béton
nl werkvloer *f(m)* van beton
r бетонная подготовка *f*

O361 *e* **oversize**
d 1. Überkorn *n*, Siebüberlauf *m*
2. Sandaustrag *m*
f 1. refus *m* (*d'un crible*) 2. surverse *f*
nl 1. zeefrest *f(m)* 2. zanduitvoer *m*
r 1. надрешётный продукт *m*
2. сгущённый продукт *m*
(*гидроклассификатора*)

O362 *e* **overslabbing**
d 1. Betonüberzugeinbau *m*
2. Betonplattenüberzug *m*
f 1. dallage *m* en béton (*constitués d'une
sous-couche et d'une couche d'usure en
béton*) 2. chaussée *f* rigide en dalles
de béton
nl 1. betonneren *n* van een wegdek
2. wegdek *n* van betonplaten
r 1. бетонирование *n* дорожного
покрытия 2. дорожное покрытие *n*
из бетонных плит

O363 *e* **overtaking lane**
d Überholspur *f*
f voie *f* de dépassement
nl inhaalstrook *f(m)*
r полоса *f* обгона

O364 *e* **overstressing, overstretching**
d Überspannen *n* (*Spannbeton*)
f surtension *f* (*d'un câble de
précontrainte*)
nl overspannen *n* (*voorgespannen beton*)
r перетяжка *f*, перенапряжение *n*
(*напрягаемой арматуры*)

O365 *e* **overturning moment**
d Kippmoment *n*
f moment *m* renversant [de
renversement]
nl kantelmoment *n*
r опрокидывающий момент *m*

O366 *e* **overvibration**
d Übervibrieren *n*
f survibration *f* (*d'un béton frais*)
nl overmatig trillen *n*
r чрезмерное вибрирование *n*,
перевибрирование *n* (*бетонной смеси*)

O367 *e* **owner**
d Bauauftraggeber *m*, Auftraggeber *m*,
Bauherr *m*
f maître *m* de l'ouvrage
nl bouwheer *m*, principaal *m*
r заказчик *m*, застройщик *m*

O368 *e* **owner-architect agreement**
d Auftraggeber-Architekt-Vereinbarung
f
f contrat *m* de service entre le maître
de l'ouvrage et l'architecte
nl overeenkomst *f* tussen opdrachtgever
en architect
r соглашение *n* заказчика с
архитектором

O369 *e* **owner-contractor agreement**
d Bau(übernahme)vertrag *m*
f marché *m* de travaux (de
construction)
nl bouwcontract *n*,
aannemingsovereenkomst *f*
r подрядный договор *m*, договор *m*
[контракт *m*] заказчика с
подрядчиком

O370 *e* **owner's inspector**
d Bauaufseher *m* des Auftraggebers
f inspecteur *m* des travaux (*de la part
du maître de l'œuvre*)
nl bouwkundig opzichter *m*,
toezicht-houdende vertegenwoordiger
m van de opdrachtgever
r представитель *m* [инспектор *m*]
технадзора от заказчика

O371 *e* **oxidation capacity**
d Abbauleistung *f*,
Oxydationsleistungsfähigkeit *f*
(*Kläranlage*)
f puissance *f* oxydable
nl oxydatiecapaciteit *f*
(*zuiveringsinstallatie*)
r окислительная мощность *f*
(*очистного сооружения*)

O372 *e* **oxidation column**
d Turm-Tropfkörper *m*

f tour *f* d'oxydation, lit *m* bactérien
(du type) tour
nl druipkolom *f(m)*
(*zuiveringsinstallatie*)
r башенный биофильтр *m*

O373 *e* oxidation ditch
d Oxydationsgraben *m*, Belebungs-
graben *m*, Umlaufbecken *n*
f fosse *f* [fossé *m*] d'oxydation
nl oxydatiesloot *f(m)*, Pasveersloot *f(m)*
r окислительная траншея *f*,
циркуляционный окислительный
канал *m*

O374 *e* oxidation lagoon, oxidation pond
d Oxydationsteich *m*, Abwasserteich *m*
f étang *m* biologique
nl oxydatievijver *m*
r окислительный [биологический]
пруд *m*

O375 oxidized asphalt, oxidized asphaltic
bitumen *see* blown asphalt

O376 *e* oxidized linseed oil
d oxidiertes Leinöl *n*, Linoxyn *n*
f linoxine *f*, linoxyne *f*, huile *f* de lin
oxydée
nl geoxydeerde lijnolie *f(m)*
r линоксин *m*, оксидированное
льняное масло *n*

O377 oxidizing recirculation channel *see*
oxidation ditch

O378 *e* oxyacetylene cutting torch
d Azetylensauerstoffschneidbrenner *m*
f chalumeau *m* découpeur [à découper]
oxyacétylénique
nl acetyleen-snijbrander *m*
r газовый [кислородно-ацетиленовый]
резак *m*

O379 *e* oxychloride cement
d Magnesialzement *m*
f ciment *m* magnésien, ciment *m* Sorel
nl Sorelcement *n*
r магнезиальный цемент *m*, цемент *m*
Сореля

O380 *e* oxygenator
d Belüftungseinrichtung *f* (*Abwasser*)
f aérateur *m* des eaux usées
nl aëratie-inrichting *f* (*afvalwater*)
r аэратор *m* (*сточных вод*)

O382 *e* oxygen deficit
d Sauerstoffmangel *m*,
Sauerstofffehlbetrag *m* (*in Wasser*)
f insuffisance *f* d'oxygène
nl tekort *n* aan zuurstof in het water
r дефицит *m* кислорода (*в воде*)

O383 *e* oxytank
d Oxydationsbecken *n*
f bassin *m* d'aération des eaux usées
nl oxydatiebed *n*
r окситенк *m*

O384 *e* ozonation, ozone treatment
d Ozonierung *f*, Ozonisierung *f*
f ozonisation *f*

nl ozonisatie *f*
r озонирование *n*

O385 *e* ozonizing chamber
d Ozoni(si)erungsbehälter *m*
f chambre *f* d'ozonisation
nl ozonisatiekamer *f(m)*
r камера *f* озонирования

P

P1 pace *see* landing

P2 *e* packaged air conditioner
d anschlußfertiges Klimagerät *n*,
Kompaktklimagerät *n*
f conditionneur *m* de l'air indépendant,
climatiseur *m* compact
nl gebruiksgereed luchtbehandelingstoestel *n*
r агрегатированный автономный
кондиционер *m*

P3 *e* packaged boiler
d transportabler Kessel *m*
f générateur *m* de vapeur transportable
nl transportabele ketel *m*
r транспортабельный котельный
агрегат *m*

P4 *e* packaged building program
d Komplexbauprogramm *n*
f programme *m* de construction
complexe
nl alomvattend bouwprogramma *n*
r комплексная программа *f*
строительства

P5 *e* packaged concrete
d vorgepackter Beton *m*
f béton *m* sec, mélange *m* à sec,
gâchée *f* sèche
nl droog voorgemengde betonmortel *m*
r сухая бетонная смесь *f*
(*поставляемая в упаковках*)

P6 *e* package dealer
d Hauptunternehmer *m* für Projektierung,
Bau- und Montagearbeiten
f organisation *f* autorisée par le
contrat d'exécuter le calcul
et réalisation des ouvrages
nl hoofdaannemer *m* voor het ontwerp
en de uitvoering
r фирма *f или* организация *f*,
ответственная по договору за
проектирование и строительство
сооружения

P7 *e* packaged water chiller
d Kompaktwasserkühler *m*,
Kompaktwasserkühlaggregat *m*,
Kompaktkaltwassersatz *m*
f refroidisseur *m* compact de l'eau
nl compacte waterkoeler *m*
r агрегатированный проточный
водоохладитель *m*

P8 *e* packed heat insulation
d Wärmedämmung *f* mit Füllstoffen
f isolation *f* thermique en bourrage
nl warmte-isolatie *f* met vulstoffen
r набивная теплоизоляция *f*

P9 *e* packed joint
d abgedichtete Fuge *f*
f joint *m* bourré
nl afgedichte voeg *f(m)*
r уплотнённый шов *m*

P10 *e* packing
d 1. Packung *f*, Verdichtung *f*
2. kleinstückiges Gestein *n* zur
Hohlraumausfüllung im
Feldsteinmauerwerk
f 1. étoupage *m*, bourrage *m*,
garniture *f*, garnissage *m*,
étanchement *m* 2. blocage *m*, pierre *f*
de blocage, blocaille *f*
nl 1. pakking *f*; dichting *f* 2. stukjes
n pl natuursteen voor het opvullen
van openingen
r 1. набивка *f*; уплотнение *n* 2. мелкий
камень *m* для заполнения пустот
в бутовой кладке

P11 *e* packless valve
d stopfbuchsenloses Ventil *n*
f robinet *m* sans garniture d'étanchéité
nl ventiel *n* zonder pakkingbus
r бессальниковый вентиль *m*

P12 pad *see* padstone

P13 *e* paddle
d 1. Umlaufverschluß *m* (*Schleuse*)
2. Schaufel *f*
f 1. vannette *f* (*de porte d'écluse*)
2. palette *f*, pale *f*, aube *f*, ailette *f*
nl 1. omlooprioolafsluiter *m* 2. schoep
f(m)
r 1. затвор *m* (*водопроводной
галереи шлюза*) 2. лопасть *f*

P14 *e* paddle aerator
d Paddelradbelüfter *m*,
Oberflächenbelüfter *m* (*Abwasser*)
f aérateur *m* à aubes [à palettes]
nl oppervlaktebeluchter *m* (*afvalwater*)
r лопастной поверхностный аэратор *m*
(*сточных вод*)

P15 *e* paddle hole
d Umlaufkanalöffnung *f*,
Umgehungskanalöffnung *f* (*Schleuse*)
f orifice *m* de la galerie d'écoulement
d'une écluse
nl opening *f* in sluisdeur, rinket *n*;
omloopriool *n*
r впускное *или* выпускное отверстие *n*
(*водопроводной галереи шлюза*)

P16 *e* paddle loader
d Seitengriffbandlader *m*
f chargeur *m* à râcloirs
nl zijgreepbandlader *m*
r ленточный погрузчик *m* с
подгребными лопастями

P17 paddle mixer *see* open-top mixer

P18 *e* pad foundation
d Einzelfundament *n*, Einzelgründung *f*
f fondation *f* isolée [solitaire]
nl afzonderlijk fundament *n*
r столбчатый фундамент *m*

P19 *e* padlock
d Vorhängeschloß *n*
f cadenas *m*
nl hangslot *n*
r висячий замок *m*

P20 *e* padstone
d Auflagerstein *m*, Auflagerquader *m*
Trägerauflager *m*
f dalle *f* [plaque *f*] d'appui
nl draagsteen *m*
r подферменник *m*, опорная плита *f*
для балок

P21 *e* pail
d Eimer *m*; Kübel *m*
f seau *m*; baille *f*
nl emmer *m*
r ведро *n*; бадья *f*

P22 *e* paint
d Farbe *f*
f peinture *f*, couleur *f*
nl verf *f(m)*
r краска *f*

P23 *e* paint base
d Farbuntergrund *m*; Haftgrund *m*
f base *f* de peinture; liant *m* pour
couleurs
nl hechtlaag *f(m)* voor de verf
r основа *f или* связующее *n* краски

P24 *e* paint-brush
d Malerpinsel *m*
f pinceau *m*, brasse *f* de peintre
nl penseel *n*, verfkwast *m*
r малярная кисть *f*

P25 *e* painters' hand tools
d Handwerkzeug *n* für Malerarbeiten
f outils *m pl* manuels [à main] de
peintre
nl handgereedschap *n* voor schilderwerk
r ручной малярный инструмент *m*

P26 *e* painting work
d Malerarbeiten *f pl*
f travaux *m pl* de peinture
nl schilderwerk *n*
r малярные работы *f pl*

P27 *e* paint kettle
d Malereimer *m*
f cuve *f* à peinture
nl verfemmer *m*, verfpot *m*
r бачок *m* для краски

P28 *e* paint peeling
d Abblättern *n* [Abschuppung *f*] der
Farbschicht
f écaillage *m* de la peinture
nl afbladderen *n* van een verflaag
r шелушение *n* окрасочного слоя

P29 paint pot *see* paint kettle

P30 *e* paint roller
d Auftragwalze *f*

f rouleau *m* à peindre
nl verfroller *m*
r валик *m* для окраски

P31 *e* **paint spray gun**
d Farbspritzpistole *f*
f pistolet *m* à peindre [de peinture]
nl verfspuit *m*, spuitpistool *n*
r краскораспылительный пистолет *m*,
пистолет-краскораспылитель *m*

P32 *e* **paint spraying**
d Farbspritzen *n*
f peinture *f* par pulvérisation [au
pistolet]
nl verfspuiten *n*
r аэрографическая окраска *f*, окраска
f распылением

P33 *e* **paint spraying machine**
d Farbzerstäubungsmaschine *f*
f installation *f* de la peinture par
pulverisation
nl verfspuitinstallatie *f*
r краскораспылительная установка *f*

P34 *e* **pair of doors**
d aufgedoppelte Tür *f*, Doppeltür *f*
f porte *f* double
nl dubbele deur *f(m)*
r двойная дверь *f*

P35 *e* **palette knife**
d Spa(ch)tel *m*
f amassette *f*, couteau-palette *m*,
spatule *f*
nl tempermes *n*, spatel *f(m)*, paletmes *n*
r шпатель *m*, мастехин *m*

P36 *e* **pale**
d 1. Pflock *m*, Bodenpfahl *m*;
Zaunlatte *f* 2. Stakete *f*, Lattenzaun
m, Palisade *f*
f 1. pieu *m* (*de clôture*);
planchette *f* 2. clôture *f* en pieux
ou en planchettes; palissade *f*
nl 1. piket *n*, paaltje *n*, schuttingpaal *m*
2. palissade *f*
r 1. кол *m*; штакетина *f* 2. частокол *m*;
штакетник *m*

P37 *e* **paling**
d Vertikalschalung *f*
f bardage *m* [revêtement *m*] en planches
verticales
nl verticale beplanking *f* (*beschieting*)
r вертикальная дощатая обшивка *f*

P38 *e* **palisade**
d Palisade *f*, Stakete *f*, Lattenzaun *m*
f palissade *f*
nl palissade *f*, omheining *f*
r частокол *m*; ограда *f*

P39 *e* **pallet**
d 1. Palette *f*, Stapelplatte *f*
2. Spa(ch)tel *m* 3. Mauerdübel *m*
f 1. palette *f* (*de manutention*)
2. spatule *f*, amassette *f*, couteau-
-palette 3. tringle *f* de clouage,
taquet *m*
nl 1. pallet *n* 2. spatel *f* 3. houten strip
m (*in muurvoeg*)

r 1. поддон *m* 2. шпатель *m*
3. деревянная пробка *f*
(*закладываемая в кирпичную стену*)

P40 *e* **palletized load**
d Palettengut *n*
f poids *m* [matériau *m*] palettisé
nl lading *f* op pallets
r груз *m* на поддоне

P41 *e* **pan**
d 1. Mauerlatte *f* 2. Feld *n* der
hölzernen Fachwerkwand (*zwischen
zwei Wandstielen*) 3. muldenförmiges
Element *n* der Stahl- *oder*
Plaststoffschalung (*für
Kassettenplatten*) 4. Türbandloch *n*
f 1. sablière *f*, panne *f* sablière 2. pan
m [panneau *m*] entre deux tournisses
adjacentes 3. coffrage *m* STA-KA
4. encoche *f*, entaille *f*
nl 1. muurplaat *f(m)* 2. veld *n* van
metsel- *of* pleisterwerk (*tussen twee
stijlen*) 3. trogvormig bekistingelement
n 4. grendelgat *n*
r 1. мауэрлат *m* 2. часть *f* стены
между стойками деревянного
фахверка 3. элемент *m* опалубки
в форме чаши (*для кессонных
конструкций*) 4. гнездо *n* для
(дверной) петли

P42 *e* **panel**
d Feld *n*; Tafel *f*, Platte *f*;
Vertäfelungsplatte *f*
f panneau *m*; lambris *m*; plaque *f*
nl veld *n*; plaat *f(m)*; paneel *n*
r панель *f*; облицовочная панель *f*;
плита *f*

P43 *e* **panel air system**
d Paneelklimasystem *n*
f système *m* de conditionnement d'air
à panneaux radiants
nl paneelluchtbehandelingssysteem *n*
r система *f* кондиционирования
воздуха с радиационными панелями

P44 *e* **panel construction**
d 1. Plattenkonstruktion *f* 2.
Plattenbauweise *f*
f construction *f* en panneaux
préfabriqués
nl 1. paneelconstructie *f* 2. bouwsysteem
n met prefab panelen
r 1. панельная конструкция *f*
2. панельное строительство *n*

P45 *e* **panel cooler**
d Paneelkühler *m*
f panneau *m* refroidissant
nl paneelkoeler *m*
r панель *f* радиационного охлаждения

P46 *e* **panel cooling**
d Paneelkühlung *f*, Plattenkühlung *f*
f refroidissement *m* par panneaux
nl paneelkoeling *f*
r панельное охлаждение *n*

P47 *e* **panel door**
d Stemmtür *f*

f porte *f* en lambris à cadres
nl paneeldeur *f(m)*
r филёнчатая дверь *f*

P48 *e* **panel heating**
d Paneelstrahlungsheizung *f*,
Plattenheizung *f*
f chauffage *m* par panneaux rayonnants
nl paneelverwarming *f*
r панельное [лучистое] отопление *n*

P49 *e* **paneling**
d Vertäfelung *f*, Getäfel *n*
f lambrissage *m*, revêtement *m* en
panneaux de bois [de lambris]
nl bekleden *n*; lambrizering *f*
r отделка *f или* облицовка *f* панелями

P50 *e* **panelled ceiling**
d 1. Kassettendecke *f* 2. getäfelte
Decke *f*
f 1. plafon *m* alvéolaire [à caisson]
2. plafon *m* en panneaux
nl 1. cassettenplafond *n*
2. paneelzoldering *f*
r кессонный *или* панельный
подшивной потолок *m*

P51 *e* **panel load**
d Knotenlast *f*
f charge *f* (appliquée) au nœud
nl knooppuntbelasting *f*
r узловая нагрузка *f*

P52 *e* **panel node, panel point**
d Knoten *m*
f nœud *m* d'une ferme
nl knooppunt *n*
r узел *m* (*фермы*)

P53 *e* **panel radiator**
d Plattenheizkörper *m*
f panneau *m* radiant
nl paneelradiator *m*
r панельный [плоский штампованный]
радиатор *m*

P54 *e* **panel strip**
d Fugendeckleiste *f*
f couvre-joint *m*
nl deklat *f(m)*
r раскладка *f*, нащельник *m* (*для
панелей*)

P55 *e* **panel wall**
d Plattenwand *f*
f mur *m* en panneaux, mur-rideau *m*
nl paneelwand *n*
r панельная стена *f*

P56 *e* **pane of glass**
d Glasscheibe *f*; Flachglas *n*,
Fensterglas *n*
f carreau *m*, vitre *f*, panneau *m* de
verre
nl ruit *m*
r оконное [листовое] стекло *n*

P57 *e* **pan floor**
d Kassettendecke *f*
f plancher *m* à caisson
nl cassettenplafond *n*
r кессонное перекрытие *n*

P58 *e* **pan humidifier**
d Befeuchter *m* mit Wanne
f humidificateur *m* à bac
nl bevochtigingsapparaat *n* met reservoir
r испарительный увлажнитель *m*
с поддоном

P59 **pan mixer** *see* **open-top mixer**

P60 **pan socket** *see* **sanitary socket**

P61 *e* **pantile**
d holländische Pfanne *f*, Hohlpfanne *f*,
Dachpfanne *f*
f tuile *f* flamande
nl holle dakpan *f(m)*
r S-образная ленточная черепица *f*

P62 **pan vibrator** *see* **slab vibrator**

P63 *e* **paper-backed lath**
d Reißlattenplatte *f* mit
Papierunterlage
f lattis *m* à enduire sur le support en
papier
nl pleisterplaat *f(m)* met papieronderlaag
r драночный щит *m* на бумажной
основе

P64 *e* **paper-board**
d Baupappe *f*
f carton *m* de construction
nl bouwkarton *n*
r строительный картон *m*

P65 *e* **paper filter**
d Papierfilter *n*, *m*
f filtre *m* à papier
nl papierfilter *n*, *m*
r бумажный фильтр *m*

P66 *e* **paper form**
d Baupappe-Schalung *f*
f coffrage *m* en papier de construction
nl bekisting *f* van bouwkarton
r опалубка *f* из вощёного
строительного картона

P67 *e* **parabolic shell**
d parabolische Schale *f*
f coque *f* [voile *m* mince] parabolique
nl parabolische schaal *f(m)*
r параболическая оболочка *f*

P68 *e* **parallel blade damper**
d parallel(laufend)e Jalousieklappe *f*
f registre *m* d'air à lames parallèles
nl jaloezieklep *f(m)* :
r параллельностворчатый воздушный
клапан *m*

P69 *e* **parallel chord truss**
d Parallelfachwerkbinder *m*,
parallelgurtiger Fachwerkbinder *m*
f poutre *f* à treillis à membrures
parallèles
nl parallelligger *m*
r ферма *f* с параллельными поясами

P70 *e* **parallel flow heat exchanger**
d Gleichstromwärmetauscher *m*
f échangeur *m* à courants parallèles
nl warmteuitwisselaar *m* in
evenwijdige stroom
r прямоточный теплообменник *m*

P71   parallel lay rope *see* equal lay rope

P72 *e* **parallelogram of forces**
   d Kräfteparallelogramm *n*
   f parallélogramme *m* des forces
   nl parallelogram *n* van krachten
   r параллелограмм *m* сил

P73 *e* **parallel swivel bench vice**
   d Parallel-Bankschraubstock *m*
   f étau *m* d'établi à mors parallèles
   nl franse voortang *f(m)* met
   parallelgeleiding
   r верстачные слесарные тиски *pl*
   с параллельными губками

P74 *e* **parallel stair**
   d zweiläufige Treppe *f*
   f escalier *m* à deux volées
   nl bordestrap *m*
   r двухмаршевая лестница *f*

P75 *e* **parallel wire unit**
   d Spannkabel *n* mit parallel
   angeordneten Drähten
   f élément · *m* de précontrainte en fils
   ou en torons parallèles
   nl voorspanningseenheid *f* met
   evenwijdige kabels
   r пучок *m* параллельных проволок
   или прядей (*для преднапряжения
   бетона*)

P76 *e* **parapet**
   d 1. Brüstung *f*; Geländer *n*
   2. Brüstungsmauer *f* 3. Strandmauer
   f, Küstenschutzmauer *f*
   f 1. parapet *m*; garde-fou *m*, garde-
   -corps *m* 2. mur *m* de parapet,
   parapet *m* 3. brise-lames *m*
   nl 1. leuning *f* 2. borstwering *f*
   3. golfbreker *m*
   r 1. парапет *m*; перила *pl*
   2. парапетная стена *f*
   3. волноотбойная стенка *f*

P77   parapet wall *see* parapet 2., 3.

P78   parent matrix *see* matrix

P79 *e* **parget**
   d 1. Zierputz *m* 2. Innenverputz *m*
   eines Schornsteins 3. Zementmörtel-
   -Sperranstrich *m* 4. Rapputz *m*
   f 1. plâtre *m* décoratif, crépi *m*
   décoratif [ornemental] 2. fourrure *f*
   de cheminée 3. enduit *m*
   d'imperméabilisation en mortier de
   ciment, cuvelage *m* en mortier de
   ciment 4. couche *f* d'enduit à
   gravillon incorporé
   nl 1. sierpleister *n* 2.
   binnenbepleistering *f* van een
   schoorsteen 3. waterdicht
   cementpleister *n* 4. vertinning *f*
   r 1. орнаментальная штукатурка *f*
   2. внутренняя футеровка *f*
   [облицовка *f*] дымохода
   3. обмазочная гидроизоляция *f* из
   цементного раствора 4. штукатурный
   намёт *m* с мелким гравием без
   затирки

P80 *e* **parging**
   d 1. Schornsteinreinigung *f*,
   Rauchabzugreinigung *f* 2. Sichtputz
   m 3. Zementmörtel-Sperranstrich *m*
   f 1. lissage *m* (*de la fourrure de
   cheminée*) 2. crépi *m* décoratif 3. couche
   f d'étanchéité en mortier de ciment
   nl 1. gladpleisteren *n* van de
   binnenkant van de schoorsteen
   2. sierpleister *n* 3. waterdicht
   cementpleister *n*
   r 1. швабровка *f* (*дымохода*)
   2. декоративная штукатурка *f*
   3. гидроизолирующий слой *m* из
   цементной штукатурки

P81 *e* **paring chisel**
   d Breitstahl *m*, Flachstahl *m*
   f ciseau *m* long
   nl steekbeitel *m*
   r долото *n* для ручной затёски *или*
   долбления

P82   parker's cement *see* Roman cement

P83 *e* **parker truss**
   d Fachwerkträger *m* mit
   Polygonalobergurt
   f poutre *f* à treillis Parker, poutre
   à treillis à membrure supérieure
   polygonale
   nl vakwerkligger *m* met polygonale
   bovenrand
   r ферма *f* с полигональным верхним
   поясом

P84 *e* **parking apron**
   d Abstellfläche *f*, Abstellplatz *m*
   f aire *f* de parcage [de stationnement]
   nl parkeerplaats *f(m)*
   r место *n* стоянки автомобилей

P85 *e* **parking area, parking lot**
   d Parkplatz *m*
   f zone *f* [parc *m*] de stationnement
   nl parkeerterrein *n*
   r стоянка *f* [место *n* открытой
   стоянки] автомобилей

P86 *e* **parking tower**
   d Parkturm *m*, Autosilo *n*, *m*
   f garage-tour *m*, garage *m* de grande
   hauteur, garage *m* en élévation
   nl parkeertoren *m*
   r гараж *m* башенного типа

P87 *e* **parquet floor**
   d Stabparkettfußboden *m*
   f parquet *m*
   nl parketvloer *f(m)*
   r паркетный пол *m*

P88 *e* **partial air-conditioning**
   d Teilklimatisierung *f*
   f conditionnement *m* de l'air partiel
   nl gedeeltelijke klimaatregeling *f*
   r неполное кондиционирование *n*
   воздуха; кондиционирование *n*
   воздуха в отдельных помещениях
   или зонах здания

P89 *e* **partial cover plate**
   d verkürzte Gurtplatte *f*

f semelle *f* raccourcie (*d'une poutre à âme pleine*)
nl verkorte flens *f(m)* (*van een vollewandligger*)
r укороченный поясной лист *m* (*стальной составной балки*)

P90 **partial discharge** *see* **specific discharge**

P91 *e* **partially drowned weir**
d unvollkommener Überfall *m*
f déversoir *m* partiellement noyé
nl onvolkomen overlaat *m*
r подтопленный водослив *m*

P92 *e* **partially penetrating well**
d unvollkommener [teilweise wirksamer] Brunnen *m*
f puits *m* incomplet
nl onvolkomen put *m*
r несовершенная скважина *f*; несовершенный колодец *m*

P93 *e* **partially-separate system**
d teilweise Trennentwässerung *f*, teilweises Trennverfahren *n*, Teiltrennsystem *n*
f système *m* séparatif partiel
nl gedeeltelijk gescheiden systeem *n*
r полураздельная система *f* канализации

P94 *e* **partial prestressing**
d teilweise [beschränkte] Vorspannung *f*
f précontrainte *f* partielle
nl gedeeltelijke voorspanning *f*
r частичное предварительное напряжение *n*

P95 *e* **partial safety factors**
d partielle Sicherheitsbeiwerte *m pl* (*für die Berechnung nach dem Grenzzustand*)
f coefficients *m pl* de sécurité partiels (*employés pour le calcul aux états limites*)
nl beperkte veiligheidscoëfficiënten *m pl*
r коэффициенты *m pl*, учитываемые при расчёте конструкций по предельному состоянию

P96 *e* **particle board**
d Spanplatte *f*
f panneau *m* en particules [en copeaux] de bois
nl spaanplaat *f(m)*
r древесностружечная плита *f*

P97 *e* **particle shape**
d Kornform *f*, Korngestalt *f*
f forme *f* [configuration *f*] des grains d'un granulat [d'un agrégat]
nl korrelvorm *m*
r форма *f* частиц [зёрен]

P98 *e* **particle size**
d Korngröße *f*, Teilchengröße *f*
f dimension *f* [grosseur *f*] de grains
nl korrelgrootte *f*, deeltjesgrootte *f*
r размер *m* [крупность *f*] частиц [зёрен]

P99 *e* **particle-size analysis**
d Kornanalyse *f*
f analyse *f* granulométrique
nl korrelbepaling *f*
r гранулометрический [ситовый] анализ *m*

P100 **particle size distribution** *see* **grading 1**

P101 *e* **particle size distribution curve**
d Siebkurve *f*; Kornverteilungskurve *f*
f courbe *f* de tamissage; courbe *f* granulométrique
nl zeefkromme *f*
r кривая *f* ситового анализа; кривая *f* гранулометрического состава

P102 **parting agent** *see* **release agent**

P103 *e* **partition**
d Trennwand *f*
f paroi *f*, cloison *f*
nl scheidingswand *m*, tussenmuur *m*
r перегородка *f*

P104 *e* **partition block**
d Trennwandstein *m*, Trennwandblock *m*
f bloc *m* [parpaing *m*, corps *m*] pour cloisons
nl bouwblok *n* voor tussenmuren
r блок *m* для кладки внутренних самонесущих перегородок

P105 *e* **partition tile**
d Keramikhohlstein *m* für Trennwände
f bloc *m* creux en terre-cuite pour cloisons
nl holle baksteen *m* voor scheidingswanden
r пустотелый керамический блок *m* для внутренних перегородок

P106 **partition wall** *see* **partition**

P107 **part of a building** *see* **building part**

P108 *e* **part-swing shovel**
d nicht vollschwenkbarer Löffelbagger *m*
f pelle *f* mécanique à rotation partielle
nl zwenkbare lepelschop *f(m)* met beperkte zwenkhoek
r неполноповоротная механическая лопата *f*

P109 *e* **party parapet**
d Parapett *n* [Brüstung *f*] der gemeinschaftlichen Giebelmauer
f parapet *m* de mur mitoyen
nl borstwering *f* van gemeenschappelijke gevelmuur
r парапет *m* общей стены двух зданий

P110 *e* **party wall**
d Brandmauer *f*, gemeinschaftliche Giebelmauer *f*; Grenzmauer *f*; Wohnungstrennwand *f*
f mur *m* mitoyen
nl scheidingsmuur *m*
r общая стена *f* двух зданий; межевая стена *f*, межквартирная раздельная стена *f*

P111 *e* passage
   d Durchgang *m*; Flur *m*
   f passage *m*; corridor *m*, couloir *m*
   nl doorgang *m*; corridor *m*
   r проход *m*; коридор *m*

P112 *e* passageway
   d 1. Wasserdurchgang *m*,
   Feuchtemigrationsweg *m*
   2. Durchgang *m*; Flur *m*
   f 1. passage *m* d'eau 2. corridor *m*
   nl 1. waterdoorgang *m*   2. corridor *m*,
   gang *m*
   r 1. проход *m* (*для влаги*)
   2. коридор *m*, проход

P113 *e* passenger elevator *US*, passenger lift
   UK
   d Personenaufzug *m*, Fahrstuhl *m*
   f ascenseur *m* de personnes [pour
   passagers]
   nl personenlift *m*
   r пассажирский лифт *m*

P114 *e* passing place
   d Ausweichstelle *f*, Überholstelle *f*
   f point *m* d'évitement
   nl uitwijkplaats *f(m)*
   r местное уширение *n* проезжей части
   дороги (*для разъезда или обгона*)

P115 *e* passive earth pressure, passive
   resistance of ground
   d Erdwiderstand *m*, passiver Erddruck *m*
   f poussée *f* passive des terres, pression
   f de butée
   nl passieve gronddruk *m*
   r отпор *m* [пассивное давление *n*]
   грунта

P116 *e* paste
   d Paste *f*, Kleister *m*, Brei *m*
   f pâte *f*
   nl pasta *m*, deeg *n*
   r паста *f*; тесто *n* (*напр. цементное*)

P117 *e* paste content
   d Zementbreianteil *m* (*im Beton*)
   f teneur *m* en pâte de ciment (*dans le
   béton*)
   nl gehalte *n* aan cementbrij
   r содержание *n* цементного теста
   (*в бетоне*)

P118 *e* paste paint
   d Farbpaste *f*
   f peinture *f* en pâte
   nl verfpasta *m*
   r (готовая) густотёртая краска *f*

P119   paste volume *see* paste content

P120 *e* pat
   d Zementkuchen *m*
   f gâteau *m* de pâte de ciment
   (*echantillon d'essai normalisé*)
   nl cementkoek *m*
   r лепёшка *f* из цементного теста (*для
   испытания*)

P121 *e* patch
   d 1. Flickstelle *f* 2. Flickmasse *f*,
   Reparaturmasse *f*

   f 1. réparation *f* localisée de chaussée
   2. granulat *m* enrobé pour point
   à temps
   nl 1. plaatselijke wegdekreparatie *f*
   2. reparatiemassa *f*
   r 1. место *n* ямочного ремонта
   (*дорожного покрытия*) 2. состав *m*
   для заделки повреждений (*напр.
   в дорожном покрытии*)

P122 *e* patcher
   d Flickmaschine *f*, Schwarzdecken-
   -Instandsetzungsmaschine *f*
   f point *m* à temps (*petite répandeuse
   pour réparations localisées de
   chaussée*)
   nl wegdekherstelmachine *f*
   r машина *f* для ямочного ремонта
   (*дорожных покрытий*)

P123 *e* patching
   d 1. Flickarbeit *f*, Lochausfüllung *f*
   2. Straßenausbessern *n*, Flicken *n*,
   Schlaglochbeseitigung *f*,
   Betonausbesserungsarbeiten *f pl*
   f 1. replâtrage *m*; bouchage *m* (*de
   cavités*) 2. point *m* à temps
   nl 1. vullen *n* van gaten in de weg
   2. plaatselijk wegherstel *n*
   r 1. заделка *f* отверстий и раковин
   2. ямочный ремонт *m* (*дорожного
   покрытия*)

P124 *e* patching compound, patching mix
   d Flickmasse *f*, Reparaturmasse *f*
   f mélange *m* d'un liant avec gravillon
   pour le point à temps
   nl reparatiemassa *f* (*wegdek*)
   r смесь *f* для ямочного ремонта
   (*дорожных покрытий*)

P125 *e* patent glazing
   d kittloses Verglasen *n*
   f vitrage *m* [pose *f* des vitres] sans
   mastic (*avec parcloses*)
   nl kitloos glas zetten *n*
   r остекление *n* без замазки (*на
   штапиках или прокладках*)

P126 *e* paternoster
   d 1. Umlaufaufzug *m*, Paternoster-
   Aufzug *m* 2. Eimerketten-
   -Schwimmbagger *m*
   f 1. ascenseur-patrenôtre *m*, ascenseur
   m à marche continue 2. drague *f*
   (à chaîne) à godets
   nl 1. paternosterlift *m*
   2. emmerbaggermolen *m*
   r 1. лифт *m* непрерывного действия,
   патерностер *m* 2. многоковшовый
   землечерпательный
   [дноуглубительный] снаряд *m*

P127 *e* path of seepage
   d Sickerweg *m*, Kriechweg *m*
   f ligne *f* [cheminement *m*] d'infiltration
   nl filtratieweg *m*
   r путь *m* фильтрации

P128 *e* patio
   d Lichthof *m*, Innenhof *m*

f patio *m*
nl binnenplaats *f*
r патио *m*, внутренний дворик *m*

P129 *e* **patten**
d 1. Säulenfuß *m*, Basis *f* 2.
Fundamentplatte *f*, Grundplatte *f*
3. Schwelle *f*
f 1. pied *m* de colonne 2. dalle *f* de
fondation 3. semelle *f*, seuil *m*
nl 1. basement *n*, kolomvoet *m*, zuilvoet
f(m) 2. fundatieplaat *f(m)* 3. drempel
m
r 1. база *f* колонны 2. фундаментная
плита *f* 3. основание *n*, порог *m*

P130 *e* **pattern**
d 1. Schablone *f*, Modell *n* 2. Muster *n*,
Probestück *n* 3. Abdruck *m* 4. Dessin
n, Vorbild *n*
f 1. gabarit *m*; modèle *m* 2. échantillon
m (*d'un produit*) 3. moule *m*, forme *f*
4. dessin *m*; ornement *m*
nl 1. sjabloon *n*, model *n* 2. monster *n*
3. afdruk *m* 4. dessin *n*
r 1. шаблон *m*; модель *f*; трафарет *m*
2. образец *m* 3. слепок *m* 4. узор *m*,
рисунок *m*

P131 *e* **patterned finish**
d gemusterte Fakturierung *f*
f enduit *m* en bossage (*avec dessin
répété réguliérement*)
nl reliëfafwerking *f*
r фактурная [рельефная] отделка *f*
(*с соблюдением установленного
рисунка*)

P132 *e* **paumelle**
d Auflageband *n* (mit herausnehmbarer
Achse) (*Tür, Fenster*)
f paumelle *f*; charnirère *f*
nl paumelle *f*
(*deur, venster*)
r накладная (*дверная или оконная*)
петля *f* (*с вынимающейся осью*)

P133 *e* **paved median**
d gefestigter Mittelstreifen *m*
f terre-plein *m* central pavé [revêtu]
nl verharde middenstrook *f(m)*
r мощёная разделительная полоса *f*;
разделительная полоса *f* с покрытием

P134 *e* **paved shoulders**
d bedeckte Schulter *f pl*, befestigte
Standspuren *f pl*
f accotements *m pl* pavés [revêtus]
nl verharde bermen *m pl*
r мощёные обочины *f pl*, обочины *f pl*
с искусственным покрытием

P135 *e* **pavement**
d Belag *m*, Decke *f*; Fahrbahnbefesti-
gung *f*, Straßenpflaster *n*
f revêtement *m*; pavé *m*
nl wegverharding *f*; bestrating *f*
r покрытие *n*; мостовая *f*

P136 *e* **pavement base**
d Straßenunterbau *m*
f couche *f* de base (*de chaussée*)

nl fundering *f* van het wegdek
r основание *n* дорожного покрытия

P137 **pavement breaker** *see* **paving breaker**

P138 *e* **pavement concrete**
d Straßen(decken)beton *m*
f béton *m* routier [de chaussée]
nl wegenbeton *n*
r бетон *m* для дорог [дорожных
покрытий]

P139 *e* **pavement design**
d 1. Deckenkonstruktion *f*, Belag-
-Konstruktion *f* 2. Deckenbemessung
f, Bemessung von Straßenbelägen
f 1. corps *m* de chaussée, ensemble *m*
de couches de chaussées 2. calcul *m*
(de corps) de chaussée
nl 1. wegconstructie *f* 2. ontwerpen *n*
van een wegverharding
r 1. конструкция *f* дорожной одежды
2. проектирование *n* дорожной
одежды

P140 *e* **pavement-marking machine**
d Straßenmarkierungsmaschine *f*,
Strichziehgerät *n*
f machine *f* pour marquage des
chaussées
nl wegmarkeringsmachine *f*
r маркировщик *m*, машина *f* для
маркировки дорожных покрытий

P141 *e* **pavement overlay**
d Verschleißschicht *f*
f couche *f* de roulement [d'usure]
nl slijtlaag *f(m)*
r слой *m* износа (*дорожного
покрытия*)

P142 **pavement paint striper** *see* **pavement-
-marking machine**

P143 *e* **pavement saw**
d Fugenschleifgerät *n*
f scie *f* à béton
nl zaagmachine *f* voor betonwegen
r самоходная пила *f* для нарезки
швов (*в бетонном покрытии*)

P144 **pavement slab** *see* **paving slab**

P145 **pavement structure** *see* **pavement
design 1.**

P146 *e* **pavement undersealing**
d Unterpressen *n* (*Betonstraßendecke*),
Unterfüllen *n* von hohlliegenden
Betonfahrbahnplatten mit Bitumen
f injection *f* de bitume pour relevage
des dalles affaissées
nl inpersen *n* van bitumen (onder een
verzakte betonplaat)
r инъецирование *n* [нагнетание *n*]
битума под осевшие дорожные
бетонные плиты (*для заполнения
пустот и подъёма плит*)

P147 *e* **paver**
d 1. Straßenbetoniermaschine *f* 2.
Pflasterer *m*, Steinsetzer *m*

f 1. motopaveur *m*, finisseur *m*
[finisseuse *f*] d'autoroute 2. paveur
m, dalleur *m*
nl 1. betonverwerkingsmachine *f*
2. straatmaker *m*
r 1. дорожный асфальтоукладчик *m*
2. мостильщик *m*; мостовщик *m*

P148 *e* **paver's hammer**
d Zurichtehammer *m*, Steinsetzerhammer *m*
f marteau *m* de paveur
nl straatmakershamer *m*
r молоток *m* мостовщика

P149 **pavestone** *see* **pavior** 1.

P150 *e* **pavilion**
d 1. Vorbau *m* 2. Seitenflügel *m*
3. Pavillon *m*; Gartenhäuschen *n*,
Laube *f*
f 1. partie *f* saillante d'un bâtiment
2. annexe *m* 3. pavillon *m*; kiosque *m*
nl 1. voorbouw *m* 2. vleugel *m*
3. paviljoen *n*; prieel *n*, tuinhuisje *n*
r 1. выступающая часть *f* здания
2. флигель *m* 3. павильон *m*;
беседка *f*

P151 *e* **pavilion roof**
d Pavillondach *n*
f comble *m* en pavillon
nl paviljoendak *n*
r пирамидальная [полигональная]
крыша *f*

P152 *e* **paving**
d 1. Straßenpflaster *n*, Steinvorlage *f*
2. Straßendecke *f*, Straßenbelag *m*,
Straßenbefestigung *f* 3. Pflastern *n*,
Deckeneinbau *m*
f 1. pavé *m*, pavage *m* 2. revêtement *m*
routier; dallage *m* 3. pavage *m* des
rues
nl 1. plaveisel *n* 2. verharding *f*
3. bestraten *n*
r 1. мостовая *f*; отмостка *f* 2.
дорожное покрытие *n* 3. мощение *n*
улиц

P153 *e* **paving asphalt**
d Straßenbaubitumen *n*
f bitume *m* routier, ciment *m*
asphaltique
nl wegenasfalt *n*
r дорожный битум *m*

P154 *e* **paving blocks**
d Pflasterstein *m*, Pflasterwürfel *m*,
Pflasterklotz *m*
f blocs *m pl* [dalles *f pl*] de pavage
nl straatklinkers *m pl*, betonklinkerkeien
m pl
r брусчатка *f*

P155 *e* **paving breaker**
d Staßenaufbrechhammer *m*,
Aufreißhammer *m*
f brise-béton *m*
nl pneumatische breekhamer *m*
r бетонолом *m*

P156 *e* **paving brick**
d Pflasterziegel *m*, Straßenbauklinker
m
f brique *f* de pavage
nl straatklinker *m*
r клинкер *m* для мостовых

P157 *e* **paving flag**
d Gehwegplatte *f*, Gehsteigplatte *f*,
Fußwegplatte *f*
f dalle *f* [plaque *f*] de pavage
nl trottoirtegel *m*
r плита *f* для мощения тротуаров

P158 *e* **paving slab**
d Pflasterplatte *f*
f dalle *f* routière [de chaussée]
nl betonplaat *f(m)* (*weg*)
r плита *f* для мощения дорог и
улиц

P159 **paving stone** *see* **pavior** 1.

P160 *e* **paving train**
d Deckeneinbauzug *m*
f train *m* de bétonnage (*de la route*)
nl betontrein *m*
r дорожно-строительный отряд *m*,
дорожно-строительная колонна *f*

P161 *e* **pavior**
d 1. Straßenklinker *m*; Pflasterstein *m*
2. Stampfer *m*, Stößel *m*, Ramme *f*,
Rammklotz *m* 3. Pflasterer *m*,
Steinsetzer *m*
f 1. brique *f* de pavage 2. dame *f*,
dameuse *f*, dameur *m* 3. paveur *m*,
dalleur *m*
nl 1. straatsteen *m* 2. stamper *m*
3. straatmaker *m*
r 1. дорожный клинкер *m*; камень *m*
для мощения 2. трамбовка *f*
3. мостильщик *m*; мостовщик *m*

P162 **pavior's hammer** *see* **paver's hammer**

P163 *e* **pea gravel**
d Perlkies *m*, Erbskies *m*
f gravier *m* de grosseur de pois
nl parelgrind *m*
r мелкий окатанный гравий *m*

P164 *e* **pea gravel grout**
d dünnflüssiger Baumörtel *m* mit
Perlkieszusatz
f mortier *m* liquide avec
petit gravillon, coulis *m* avec **petit**
gravillon
nl vloeibare mortel *m* met parelgrind
r жидкий строительный раствор *m*
с добавлением мелкого гравия

P165 **peak arch** *see* **pointed arch**

P166 *e* **peak demand**
d Spitzenbedarf *m*; Spitzenbelastung *f*
f demande *f* de pointe; charge *f* de
pointe
nl piekbehoefte *f*; piekbelasting *f*
r пиковая потребность *f*; пиковая
нагрузка *f*

P167 *e* **peak discharge**
d Spitzenabfluß *m*

   f débit *m* de crue maximum
  nl piekafvoer *m*
   r максимальный [пиковый] расход *m*
     паводка

P168 *e* **peak load**
   d Belastungsspitze *f*, Spitzenbelastung *f*
   f pointe *f* de charge, charge *f*
     maximale [de pointe]
  nl piekbelasting *f*
   r пиковая нагрузка *f*

P169 *e* **peak stress**
   d Spitzenspannung *f*, Spannungsspitze *f*
   f pointe *f* de contraintes, contrainte *f*
     de pointe
  nl piekspanning *f*
   r пиковое напряжение *n*, пик *m*
     напряжения (в точках концентрации
     напряжений)

P170 *e* **peat soil, peaty soil**
   d Torfboden *m*, torfhaltiger Boden *m*
   f sol *m* tourbeux
  nl veengrond *m*
   r торфяной грунт *m*

P171 *e* **pebble**
   d Kiesel(stein) *m*
   f galet *m*, caillou *m*
  nl kiezelstenen *m pl*, grind *n*
   r (окатанный крупный) гравий *m*,
     галька *f*

P172 *e* **pebble clarifier**
   d Kies (schüttungs) filter *n*, *m*
   f filtre *m* en gravier
  nl grindfilter *n*, *m*
   r гравийный фильтр *m*

P173 *e* **pebble dash**
   d Steinputz *m*
   f enduit *m* avec éclats de pièrre *ou*
     gravillon
  nl grindpleister *n*
   r штукатурка *f* с каменной крошкой
     или гравием

P174 *e* **pedestal**
   d Säulenfuß *m*, Sockel *m*, Postament *n*,
     Standsockel *m*
   f socle *m* de colonne [de poteau]
  nl voetstuk *n*, sokkel *m*, postament *n*
   r подколонник *m*, цоколь *m* колонны

P175 *e* **pedestal foot**
   d Klumpfuß *m*, Sprengfuß *m*
   f bulbe *m* du pieu
  nl paalvoet *m*
   r камуфлетная пята *f*
     (*набивной сваи*)

P176 *e* **pedestal footing**
   d Einzelfundament *n*, Sockelfundament *n*
   f fondation *f* (isolée) à [sur] pilier
  nl poer *m*
   r столбчатый фундамент *m* (*колонны*)

P178 *e* **pedestrian bridge**
   d Fußgängerbrücke *f*
   f passerelle *f*, pont *m* pour piétons
  nl voetgangersbrug *f(m)*
   r пешеходный мост(ик) *m*

P179 *e* **pedestrian guard rail**
   d Fußgänger-Schutzgeländer *n*
   f barrière *f* (de sécurité); barrage *m* **de**
     route
  nl voetgangersrailing *f*
   r перильное ограждение *n*
     (*тротуара или на разделительной*
     полосе дороги)

P180 *e* **pedestrian protection**
   d Fußgängerschutzmaßnahmen *f pl*
   f moyens *m pl* de protection des piétons
     (*autour des chantiers urbains*)
  nl maatregelen *m pl* voor voetgangers
   r средства *n pl* защиты пешеходов
     (*в зоне городских стройплощадок*)

P181 *e* **pedestrian railing**
   d Brückengeländer *n*
   f garde-corps *m* [garde-fou *m*, parapet
     m] de pont (*pour piétons*)
  nl brugleuning *f* voor voetgangers
   r перильное ограждение *n* моста (*для
     пешеходов*)

P182 *e* **pedestrian subway**
   d Fußgängertunnel *m*
   f passage *m* souterrain [inférieur]
  nl voetgangerstunnel *m*
   r пешеходный туннель *m*

P183 *e* **pediment**
   d Ziergiebel *m*
   f fronton *m*
  nl gevelveld *n*, timpaan *n*
   r фронтон *m*, щипец *m*

P184 *e* **peeling**
   d Abschälen *n*, Abblättern *n*
   f écaillage *m* (*de la peinture*)
  nl afschilferen *n* (*verflaag*)
   r отслоения *n pl*, отлупы *m pl*
     (*красочной плёнки*)

P185 *e* **peg**
   d 1. Pflock *m* 2. Stift *m*, eingebohrter
     [eingestemmter] Dübel *m*
   f 1. piquet *m*, fiche *f* 2. cheville *f*,
     goujon *m*
  nl 1. piket *n*, paaltje *n* 2. houten pen *f(m)*
   r 1. колышек *m* 2. штифт *m*,
     шкант *m*

P186 *e* **pegging out**
   d Abpfählen *n*, Abstecken *n*, Abpflocken
     n
   f jalonnement *m*, bornage *m*; tracement
     m, traçage *m*
  nl afbakenen *n*, uitzetten *n*
   r установка *f* межевых знаков [вех];
     трассирование *n*

P187 *e* **pelleted mineral wool**
   d granulierte Mineralwolle *f*
   f laine *f* minérale en pelotes [en
     boulettes]
  nl gegranuleerde mineraalwol *f(m)*
   r гранулированная минеральная вата *f*

P188 *e* **penalty clause**
   d Strafeklausel *f*, Verzugsstrafeklausel *f*
   f clause *f* (contractuelle) de pénalité
     de retard

nl strafclausule *f*

r статья *f* (*подрядного договора*), устанавливающая штраф за невыполнение строительства в срок

P189 *e* **pencil rod**
d Stabmaterial *n*
f rond *m* métallique
nl dun staafmateriaal *n*
r металлический стержень *m* [пруток *m*]

P190 *e* **pendentive dome**
d Hänge(zwickel)kuppel *f* Eckzwickelkuppel *f*, Pendentifkuppel *f*
f coupole *f* [dôme *m*] à [sur] pendentifs
nl koepel *m* met hangbogen
r парусный купол *m*

P191 *e* **pendentives**
d Gewölbezwickel *m pl*, Hängebögen *m pl*, Hängezwickel *m pl*
f pendentifs *m pl* (*d'une coupole*)
nl gewelfboog *m*, hangboog *m*
r паруса *m pl* (*свода, купола*)

P192 *e* **pendulum bearing**
d Pendellager *n*
f appuis *m* [support *m*] pendulaire
nl pendeloplegging *f*
r качающаяся [маятниковая] опора *f*

P193 *e* **pendulum saw**
d Pendelsäge *f*
f scie *f* à pendule
nl slingerzaag *f(m)*
r маятниковая пила *f*

P194 *e* **penetrability**
d Durchdringbarkeit *f*, Durchlässigkeit *f*
f pénétrabilité *f*
nl permeabiliteit *f*
r проницаемость *f*

P195 *e* **penetration per blow**
d Pfahlabteufung *f* per Schlag
f pénétration *f* [enfoncement *m*] d'un pieu résultant d'un seul coup
nl zakking *f* van de paal per slag
r погружение *n* сваи под действием одного удара

P196 *e* **penetration probe**
d Durchdringungssonde *f*
f sonde *f* de pénétration
nl indringingssonde *f* (*beton*)
r зонд *m* для определения механической пробиваемости (*бетона*)

P197 *e* **penetration resistance**
d Eindring(ungs)widerstand *m*
f résistance *f* à la pénétration
nl indringingsweerstand *m*
r сопротивление *n* внедрению зонда в грунт (*при динамическом зондировании*)

P198 *e* **penetration test**
d Eindringungsversuch *m*
f essai *m* de pénétration
nl indringingsproef *f(m)*

r испытание *n* на пенетрацию [на проникание]

P199 *e* **penetration treatment**
d Oberflächendurchtränkung *f*, Halbtränkung *f* (*Straßendecke*)
f (semi-)pénétration *f* (*traitement de chaussée*)
nl oppervlakte-impregnering *f*
r поверхностная пропитка *f* (*дорожного покрытия*)

P200 *e* **penning gate**
d Schleusenhubtor *n*, Schleusenhebetor *n*
f porte *f* levante
nl hefdeur *f(m)* (*sluis*)
r подъёмные ворота *pl* шлюза

P201 *e* **penstock**
d Druckleitung *f*, Triebwasserleitung *f*, Turbinendruckleitung *f*
f conduite *f* forcée, canal *m* d'amenée
nl turbinewaterleiding *f*
r напорный трубопровод *m* ГЭС, турбинный водовод *m*

P203 *e* **penstock dam gallery**
d Einlaufbauwerkstollen *m*
f gallerie *f* d'amenée à la centrale, gallerie *f* forcée
nl inlaattunnel *m*
r туннель *m* водоприёмного сооружения ГЭС

P204 *e* **penstock pipe**
d Panzerrohr *n* (*Triebwasserleitung*)
f conduite *f* forcée [d'amenée] blindée
nl pantserpijp *f(m)* (*turbinewaterleiding*)
r бронированный напорный водовод *m* ГЭС

P205 *e* **penthouse**
d 1. Dachaufbau *m*, Dachaufsatz *m* 2. Anbau *m* mit Schleppdach 3. Schutzdach *n*, Wetterdach *n*
f 1.surélévation *f* de comble; appartement *m* sur toit 2. appentis *m* 3. auvent *m*
nl 1. verhoogd dak *n* 2. koepel *m* 3. afdak *n*, luifel *f*(*m*)
r 1. надстройка *f* над крышей 2. малая пристройка *f* с односкатной крышей 3. навес *m*, зонт *m* над входом

P206 *e* **pentroof**
d Pultdach *n*, Schleppdach *n*
f auvent *m*, appentis *m*
nl luifel *m*
r односкатная крыша *f*; пристройка *f*, примыкающая к скату крыши главного здания

P207 *e* **percentage by volume**
d Raumanteil *m*
f pourcentage *m* en volume
nl volumepercentage *n*
r процентное содержание *n* по объёму

P208 *e* **percentage by weight**
d Gewichtsanteil *m*
f pourcentage *m* en poids

nl gewichtspercentage *n*
r процентное содержание *n* по массе

P209 *e* **percentage reinforcement**
d Bewehrungsanteil *m*,
Bewehrungsprozentsaatz *m*
f pourcentege *m* d'armature [de ferraillage]
nl wapeningspercentage *n*
r процент *m* армирования

P210 *e* **percent compaction**
d prozentualer Verdichtungsgrad *m*
f degré *m* de compactage du sol (*en pourcentage*)
nl percentuele verdichtingsgraad *m*
r степень *f* уплотнённости (грунта) в процентах

P211 *e* **percent fines**
d 1. prozentualer Feinstkorngehalt *m* (*im Zuschlagstoff, Korngröße weniger als 75 mkm*) 2. prozentualer Sandgehalt *m* (*im Beton, in Bezug auf die Gesamtzuschlagmenge*)
f 1. pourcentage *m* de fines (*contenues dans un granulat*) 2. teneur *m* en sable (*d'un béton*)
nl 1. percentage *n* fijnkorrelige toeslag 2. percentage *n* zand (*begrepen op de totale toeslag*)
r 1. процентное содержание *n* пылевидных частиц (*в заполнителе*) 2. процентное содержание *n* песка (*в бетоне*)

P212 **percolating filter** *see* **bacteria bed**

P213 *e* **percolating water**
d Sickerwasser *n*
f eau *f* de cheminement
nl zakwater *n*
r просачивающаяся вода *f*; вода *f* зоны свободной фильтрации

P214 **percolation** *see* **seepage**

P215 **percolation beds**
d Filtrationsfelder *n pl*, Filterbett *n*
f lit *m* filtrant
nl vloeivelden *n pl*, filtratiebedden *n pl*
r поля *n pl* фильтрации

P216 *e* **percolation gauge**
d Versickerungsmesser *m*, Lysimeter *n*
f lysimètre *m*
nl infiltratiemeter *m*
r инфильтрометр *m*, лизиметр *m*

P217 *e* **percussion drill hammer**
d Schlagbohrhammer *m*
f perforateur *m* [perforatrice *f*] à percussion
nl klopboormachine *f*
r ударный перфоратор *m*, бурильный молоток *m*

P218 *e* **percussion drilling**
d Schlagbohrung *f*
f forage *m* [sondage *m*] à percussion
nl kloppend boren *n*
r ударное бурение *n*

P219 *e* **percussion penetration method**
d Rammsondierung *f*
f procédé *m* [méthode *f*] de reconnaissance par sondage(s) à percussion
nl slagsondering *f*
r динамическое зондирование *n*

P220 *e* **perennial stream**
d ganzjähriger [perennierender] Wasserlauf *m*
f cours *m* d'eau pérenne
nl bestendige waterloop *m*
r постоянный водоток *m*

P221 *e* **perfect jump**
d idealer Wechselsprung *m*
f ressaut *m* hydraulique parfait
nl volledige watersprong *m*
r совершенный гидравлический прыжок *m*

P222 *e* **perfect well**
d vollkommener Brunnen *m*
f puits *m* filtrant parfait [complet]
nl volkomen bron *m*
r совершенный колодец *m*, совершенная скважина *f*

P223 *e* **perforated breakwater**
d Wellenbrecher *m* in aufgelöster Bauweise
f brise-lames *m* à claire-voie
nl golfbreker *m* met openingen
r волнолом *m* сквозной конструкции

P224 *e* **perforated brick**
d Viellochziegel *m*
f brique *f* perforée [à trous]
nl geperforeerde steen *m*
r дырчатый кирпич *m*

P225 *e* **perforated-casing well**
d Schlitzrohrbrunnen *m*, Mantelrohrbrunnen *m*
f puits *m* avec tubage perforé, puits *m* à barbacanes
nl boorput *m* met geperforeerde bronbuis
r колодец *m* с перфорированной трубой

P226 *e* **perforated ceiling board**
d perforierte Hängedeckenplatte *f*
f plaque *f* [dalle *f*] de plafond perforée
nl geperforeerde plafondplaat *f(m)*
r перфорированная панель *f* подвесного потолка

P227 *e* **perforated plate**
d perforierte Platte *f*, Lochplatte *f*, Lochtafel *f*, durchlochte Tafel *f*
f panneau *m* perforé; tôle *f* perforée
nl geperforeerde plaat *f(m)*
r перфорированная панель *f*; перфорированный лист *m*

P228 *e* **performance analysis**
d Analyse *f* der Nutzungseigenschaften eines Bauwerkes
f analyse *f* de comportement (d'un ouvrage) en service
nl doelmatigheidsanalyse *f*
r анализ *m* эксплуатационных качеств сооружения

P229 *e* **performance characteristic**
    d Leistungskennwert *m*,
    Arbeitskenngröße *f*; Betriebskennlinie *f*
    f caractéristiques *f pl* de fonctionnement [de service]
    nl arbeidskarakteristiek *f*
    r эксплуатационная [рабочая] характеристика *f*

P230 *e* **performance test**
    d Leistungsprüfung *f*, Funktionsprüfung *f*, Leistungsprobe *f*
    f essai *m* de foctionnement [de service]
    nl prestatieonderzoek *n*
    r проверка *f* [испытание *n*] эксплуатационных качеств

P231 *e* **perimeter beam**
    d Randbalken *m*, Umfassungsbalken *m*
    f poutre *f* marginale
    nl randbalk *m*
    r наружная [краевая] балка *f*

P232 *e* **perimeter grouting**
    d Perimetralinjektion *f*, Perimetraleinpressung *f*, **Grundstückumfang-Bodenzementierung** *f*
    f stabilisation *f* du sol (au ciment) le long du périmètre d'un terrain
    nl grondstabilisatie *f* door injectie langs de omtrek van een bouwterrein
    r цементация *f* грунта по периметру участка

P233 *e* **perimeter heating**
    d Perimetralheizung *f*
    f chauffage *m* périmètral
    nl verwarming *f* langs de buitenkant
    r периметральное отопление *n*

P234 *e* **perimeter wall**
    d Umfassungswand *f*
    f mur *m* extérieur [périphérique] d'un bâtiment
    nl buitenmuur *m*
    r наружная стена *f* здания

P235 *e* **perimeter zone**
    d Perimetralzone *f*
    f zone *f* périmètrale
    nl omtrekzone *f*
    r периферийная зона *f*

P236 *e* **peripheral stress**
    d Umfangsspannung *f*
    f effort *m* circonférenciel
    nl omtrekskracht *f*
    r окружное напряжение *n*

P237 *e* **perlite plaster**
    d Perlitputz *m*; Gipsperlitmörtel *m*
    f enduit *m* en perlite; mortier *m* de plâtre et de perlite
    nl perlietpleister *m*; gipsperlietmortel *m*
    r перлитовая обмазка *f*; гипсо-перлитовый раствор *m*

P238 *e* **permafrost**
    d Permafrost *m*, Dauerfrostboden *m*
    f congélation *f* perpétuelle [éternelle, permanente], pergélisol *m*

    nl permafrost *m*
    r вечная мерзлота *f*

P239 *e* **permafrost table**
    d obere Grenze *f* des Dauerfrostbodens
    f nappe *f* supérieure de pergélisol [de congélation perpétuelle]
    nl bovengrens *f(m)* van de permafrost
    r верхняя граница *f* вечной мерзлоты

P240 *e* **permanent deformation**
    d bleibende Verformung *f*
    f déformation *f* permanente [plastique]
    nl blijvende vormverandering *f*
    r остаточная [пластическая] деформация *f*

P241 *e* **permanent hardness**
    d bleibende [permanente] Härte *f*, Dauerhärte *f* (*Wasser*)
    f dureté *f* permanente [non carbonatée]
    nl blijvende hardheid *f* (*water*)
    r постоянная [некарбонатная] жёсткость *f* (воды)

P242 *e* **permanent load**
    d Dauerlast *f*, bleibende [ständige] Last *f*, Stetiglast *f*
    f charge *f* permanente
    nl permanente belasting *f*
    r постоянная [статическая] нагрузка *f*

P243 *e* **permanent position**
    d Endlage *f*, Entwurfsstellung *f*
    f position *f* final
    nl definitieve plaats *f(m)* van een constructie-element
    r проектное положение *n* (*элемента конструкции*)

P244   **permanent set** *see* **permanent deformation**

P245 *e* **permanent shuttering**
    d verlorene Schalung *f*
    f coffrage *m* perdu
    nl blijvende bekisting *f*
    r конструктивная опалубка *f* (*остающаяся в теле сооружения*)

P246   **permanent stream** *see* **perennial stream**

P247 *e* **permanent structure**
    d Dauerbauwerk *n*
    f ouvrage *m* permanent, structure *f* permanente
    nl duurzaam bouwwerk *n*, permanent gebouw *n*
    r постоянное сооружение *n*

P248 *e* **permeability**
    d Durchlässigkeit *f*
    f perméabilité *f*
    nl doorlatendheid *f*
    r проницаемость *f*; водопроницаемость *f* (*напр. грунта*)

P249 *e* **permeability apparatus**
    d Durchlässigkeitsmesser *m*
    f appareil *m* de perméabilité
    nl doordringbaarheidsmeter *m*
    r аппарат *m* для определения водопроницаемости

P250   **permeability coefficient** *see* **coefficient of permeability**

P251 *e* **permeability of soil**
  d Bodendurchlässigkeit *f*
  f perméabilité *f* des sols
  nl doorlatendheid *f* van de grond
  r водопроницаемость *f* грунта

P252 *e* **permeability-reducing admixture**
  d durchlässigkeitsmindernder Zusatz *m*
  f réducteur *m* de perméabilité (*adjuvant*)
  nl toevoeging *f* ter vermindering van de doorlatendheid
  r добавка *f*, снижающая водопроницаемость

P253 *e* **permeable groyne**
  d durchlässige Trennbuhne *f*
  f épi *m* perméable
  nl doorlatende krib *f(m)*
  r полузапруда *f* сквозной конструкции

P254 *e* **permeameter**
  d Durchlässigkeitsmesser *m*
  f perméamètre *m*
  nl doordringbaarheidsmeter *m*
  r пермеаметр *m*, фильтрометр *m*, фильтрационный прибор *m*

P255 *e* **permissible deflection**
  d zulässige Durchbiegung *f*
  f déflexion *f* admissible
  nl toelaatbare doorbuiging *f*
  r допускаемый прогиб *m*

P256   **permissible load** *see* **allowable load**

P257   **permissible stress** *see* **allowable stress**

P258 *e* **permissible stress method**
  d Berechnung *f* nach zulässigen Spannungen
  f méthode *f* des contraintes [des tensions] admissibles
  nl berekening *f* naar toelaatbare spanning
  r метод *m* расчёта по допускаемым напряжениям

P259 *e* **permissible velocity**
  d zulässige Geschwindigkeit *f*, Grenzgeschwindigkeit *f*
  f vitesse *f* admissible
  nl toelaatbare snelheid *f*
  r допускаемая скорость *f*; неразмывающая скорость *f*

P260 *e* **personnel job hoist**
  d Mannschaftsaufzug *m*
  f ascenseur *m* pour le personnel du chantier, ascenseur *m* de chantier
  nl personenlift *m* (*op het werk*)
  r подъёмник *m* для рабочих

P261 *e* **perspective drawing**
  d Perspektivzeichnung *f*
  f dessin *m* en perspective
  nl perspectieftekening *f*
  r чертёж *m* в перспективе

P262 *e* **pervious blanket**
  d Dränschicht *f*
  f tapis *m* de drainage
  nl doorlatende laag *f(m)*
  r пластовая дрена *f*

P263 *e* **pessimistic time estimate**
  d längstmögliche Aktivitätsdauer *f*, pessimistische Zeit *f* (*bei PERT*)
  f estimation *f* pessimiste de la durée, durée *f* maximale (*d'une opération*)
  nl maximale tijdsduur *m* (*netwerkplanning*)
  r пессимистическая оценка *f* продолжительности работы (*сетевое планирование*)

P264 *e* **pet cock**
  d Luftauslaßhahn *m*; Kondenswasserhahn *m*; Regulierhahn *m*
  f robinet *m* d'air; robinet *m* de purge; robinet *m* de contrôle
  nl condenskraan *f(m)*; controlekraan *f(m)*; ontluchtingskraan *f(m)*
  r воздуховыпускной кран *m*; кран *m* для удаления конденсата; регулировочный кран *m*

P265 *e* **petroleum asphalt, petroleum bitumen**
  d Erdölbitumen *n*
  f bitume *m* de pétrol
  nl aardoliebitumen *n*
  r нефтяной битум *m*

P266 *e* **petroleum tar**
  d Asphaltteer *m*, Goudron *m*
  f goudron *m*
  nl asfaltolie *m*
  r гудрон *m*

P267 *e* **petrol interceptor, petrol trap**
  d Benzinabscheider *m*
  f piège *m* de pétrole
  nl benzine-afscheider *m*
  r бензиноуловитель *m*, нефтеловушка *f*

P268 *e* **photo-elastic test**
  d optische Spannungsprüfung *f*
  f essai *m* photoélastique
  nl optisch spanningsonderzoek *n*
  r оптический метод *m* исследования напряжений, метод *m* фотоупругости

P270 *e* **phreatic line**
  d 1. Sickerlinie *f*, Kriechlinie *f* 2. Absenkungskurve *f*, Depressionskurve *f*
  f 1. ligne *f* de fuite 2. courbe *f* de dépression
  nl 1. freatische lijn *f(m)* 2. depressiekromme *f*
  r 1. контур *m* фильтрации 2. кривая *f* депрессии

P271   **phreatic surface** *see* **groundwater table**

P272 *e* **phreatic water**
  d ungespanntes [freies] Grundwasser *n*
  f nappe *f* phréatique
  nl grondwater *n*
  r безнапорные подземные воды *f pl*

P273 *e* **pick**
  d Pickel *m*, Picke *f*, Hacke *f*
  f pioche *f*, pic *m*

nl pikhouweel *n* (met punt en punt)
r кирка *f*

P274 *e* pickaxe
d Kreuzhacke *f*, Kreuzpickel *m*
f pic *m* à tranche, pioche *f*
nl pikhouweel *n* (met punt en bijl)
r киркомотыга *f*

P275 pick-dressed ashlar *see* common ashlar

P276 *e* pick dressing
d Absplittern *n* des Natursteines
f taille *f* pointée
nl afsplinteren *n* van natuursteen
r окалывание *n* [обтёска *f*] природного камня

P278 *e* pier
d 1. Pfeiler *m* 2. Mole *f*, Hafendamm *m*, Wellenbrecher *m*; Pier *m* 3. Ziegelwandpfeiler *m* 4. Ziegelsäule *f* 5. Fensterpfeiler *m*, Fensterschaft *m*
f 1. pile *f*, pilier *m* 2. jetée *f*, épi *m* digue; brise-lames *m* 3. pilastre *m*, colonne *f* adossée 4. colonne *f* [poteau *m*] en briques 5. trumeau *m*, entre-fenêtre *m*
nl 1. pijler *m* 2. havendam *m* 3. steunbeer *m* 4. gemetselde kolom *f(m)* 5. penant *n*
r 1. бык *m*, промежуточная опора (*плотины, моста*) 2. мол *m*; волнолом *m*; пирс *m* 3. пилястра *f* (*кирпичной стены*) 4. кирпичная колонна *f* 5. простенок *m*

P279 *e* pier cap
d Pfeilerkopf *m*
f tête *f* d'une pile
nl pijlerkop *m*
r массивный оголовок *m* (*быка, контрфорса*)

P280 pier foundation *see* caisson foundation

P281 *e* pier head power station
d Pfeiler-Wasserkraftwerk *n*
f usine *f* pile
nl pijler-waterkrachtcentrale *f(m)*
r бычковая гидроэлектростанция *f*

P282 *e* piezometer
d Piezometer *n*
f piézomètre *m*
nl piëzometer *m*
r пьезометр *m*; пьезометрическая трубка *f*

P283 *e* piezometer tube
d Piezometerrohr *n*, Beobachtungsrohr *n*, Standrohr *n*; Beobachtungsbrunnen *m*
f tube *m* piézométrique
nl peilbuis *f(m)*
r пьезометрическая трубка *f* (*забиваемая в грунт*); наблюдательный колодец *m*; контрольная скважина *f*

P284 *e* piezometric gradient
d piezometrisches Gefälle *n*, Druckliniengefälle *n*, spezifischer Druckverlust *m*
f gradient *m* piézométrique
nl piëzometrische gradiënt *m*
r пьезометрический градиент *m*

P285 *e* piezometric head
d Standrohrspiegelhöhe *f*, Druckhöhe *f*, piezometrischer Druck *m*
f pression *f* [charge *f*, hauteur *f*] piézométrique
nl hydrostatische druk *m*
r пьезометрический [гидростатический] напор *m*

P286 *e* piezometric level
d Piezometerstand *m*, Beobachtungsrohrspiegelhöhe *f*
f niveau *m* piézométrique
nl drukniveaulijn *f(m)*
r пьезометрический уровень *m*

P287 piezometric slope *see* piezometric gradient

P288 *e* pigment
d Pigment *n*
f pigment *m*
nl pigment *n*
r пигмент *m*

P289 *e* pilaster
d Pilaster *m*, Wandpfeiler *m*, eingebundener Pfeiler *m*
f pilastre *m*, colonne *f* adossée
nl pilaster *m*, muurpijler *m*
r пилястра *f*

P290 *e* pilaster face
d Vorderseite *f* [Sichtfläche *f*] des Pilasters
f face *f* frontale de pilastre, nu *m* de pilastre
nl voorkant *m* van de pilaster
r лицевая грань *f* пилястры

P291 *e* pilaster side
d Seitenfläche *f* des Pilasters
f face *f* latérale de pilastre
nl zijvlak *n* van de pilaster
r боковая грань *f* пилястры

P292 *e* pile
d 1. Pfahl *m* 2. Stapel *m*; Haufen *m*
f 1. pieu *m*, pilot *m*, pilotis *m* 2. pile *f*, tas *m*
nl 1. heipaal *m* 2. stapel *m*; hoop *m*
r 1. свая *f* 2. штабель *m*; куча *f*, груда *f*

P293 *e* pile after-driving
d Nachrammung *f*
f battage *m* supplémentaire (*d'un pieu*)
nl naheien *n*
r добивка *f* сваи

P294 *e* pile-and-cribwork foundation
d Pfahl-Kastengründung *f*
f fondation *f* sur pilotis et encoffrement en charpente
nl paal-en-kistfundering *f*
r свайно-ряжевый фундамент *m*

P295 e **pile and fascine revetment**
d Pfahl-Faschinenpackung *f*
f revêtement *m* en pilotis et fascines
nl beschoeiing *f* van palen en rijshout
r свайно-фашинная одежда *f*

P296 **pile band** see **pile ferrule**

P297 e **pile bent**
d Pfahljoch *n*
f palée *f* d'appui [sur pilotis]
nl paaljuk *n*
r плоская свайная опора *f*

P298 e **pile casing**
d Pfahlrohr *n*, Mantelrohr *n*
f fourreau *m*, tubage *m* (*d'un pieu coulé*)
nl mantelbuis *f(m)*
r обсадная труба *f* (*набивной сваи*)

P299 e **pile cluster**
d Pfahlbündel *n*, Pfahlgruppe *f*,
Pfahlpoller *m*
f groupe *f* [bouquet *m*] de pieux, duc *m*
d'Albe
nl dukdalf *m*
r куст *m* свай

P300 e **pile dolphin**
d Pfahlpoller *m*, Dalbe(n) *f(m)*
f duc *m* d'Albe, poteau *m* d'amarrage
nl dukdalf *m*
r свайный пал *m*

P301 e **piled quay**
d Pfahlrostkaimauer *f*
f quai *m* sur pieux
nl kade *f(m)* met damwand
r свайная набережная *f*

P302 **pile drawer** see **pile extractor**

P303 e **pile driver**
d Schlagramme *f*
f sonnette *f*
nl heimachine *f*
r свайный копёр *m*

P304 e **pile driving**
d Pfahlrammen *n*
f battage *m* de pieux
nl heien *n*
r забивка *f* свай

P305 e **pile driving by vibration**
d Vibrationsrammen *n*
f vibrofonçage *m* (de pieux)
nl vibratieheien *n*
r вибропогружение *n* (свай)

P306 e **pile-driving equipment**
d Pfahlrammausrüstung *f*
f matériel *m* de battage (de pieux)
nl hei-uitrusting *f*
r сваебойное оборудование *n*

P307 e **pile-driving formula**
d Rammformel *f*
f formule *f* de battage
nl heiformule *f*
r формула *f* несущей способности сваи

P308 e **pile driving frame**
d Rammgerüst *n*
f charpente *f* de la sonnette (de battage)

nl heistelling *f*
r станина *f* копра

P309 **pile-driving hammer** see **pile hammer**

P310 **pile-driving plant** see **pile driving rig**

P311 e **pile driving pontoon**
d Schwimmramme *f*, Rammschiff *n*,
Rammponton *m*
f sonnette *f* flottante
nl heiponton *m*
r плавучий копёр *m*

P312 e **pile driving rig**
d Rammanlage *f*
f installation *f* de battage
nl hei-installatie *f*
r сваебойная установка *f*

P313 **pile extension** see **pile splicing**

P314 e **pile extractor**
d Pfahl(aus)zieher *m*
f arracheur *m* de pieux, arrache-pieu *m*
nl paaltrekker *m*
r сваеизвлекатель *m*,
сваевыдёргиватель *m*

P315 e **pile ferrule**
d Pfahlring *m*
f anneau *m* de pile [de pieu]
nl paalring *m*
r бугель *m* сваи

P316 **pile follower** see **driving cap**

P317 e **pile footing**
d Einzel-Pfahlfundament *n*
f fondation *f* isolée sur pieux [sur
pilotis]
nl funderingsblok *n* op palen
r отдельный свайный фундамент *m*

P318 e **pile foundation**
d Pfahlgründung *f*
f fondation *f* sur pieux [sur pilotis]
nl paalfundering *f*
r свайный фундамент *m*

P319 e **pile grillage**
d Pfahlrost *m*, Pfahlwerk *n*
f grillage *m* en pieux
nl paalroosterwerk *n*
r свайный ростверк *m*

P320 **pile grillage foundation** see **pile
foundation**

P321 **pile group** see **pile cluster**

P322 e **pile hammer**
d Rammhammer *m*, Bär *m*
f marteau *m* de battage, mouton *m*
nl heiblok *n*
r свайный [сваебойный] молот *m*

P323 e **pile head**
d Pfahlkopf *m*
f tête *f* de pieu
nl paalkop *m*
r оголовок *m* сваи

P324 e **pile helmet**
d Rammhaube *f*, Schlaghaube *f*
f casque *m* de pieu
nl paalmuts *f(m)*
r наголовник *m* сваи

P325   pile hoop *see* pile ferrule
P326 *e* pile jacking
   d Erpressen *n* des Pfahles,
      Pfahleinpressen *n*
   f enfoncement *m* de pieu (par vérin)
   nl palen te grond in drukken
   r задавливание *n* сваи (домкратом)
P327 *e* pile jetting
   d Einspülen *n* eines Pfahles
   f lançage *m*, lancement *m* (d'un pieu)
   nl palen inspuiten
   r погружение *n* сваи с подмывом
P328 *e* pile load capacity
   d Tragfähigkeit *f* des Pfahles
   f capacité *f* portante d'un pieu, pouvoir
      m porteur d'un pieu
   nl draagvermogen *n* van de paal
   r несущая способность *f* сваи
P329 *e* pile pedestal
   d Pfahlzopf *m*
   f base *f* de pieu en bulbe, bulbe *m* de
      pieu
   nl verzwaarde paalvoet *m*
   r комель *m* [уширение *n*] набивной
      сваи
P330 *e* pile pier
   d Pfahlrostpfeiler *m*
   f pilot *m*, palée *f*
   nl paaljuk *n*
   r свайная опора *f*
P332 *e* pile positioning
   d Aufstellen *n* eines Pfahles
   f mise *f* en fiche du pieu
   nl een paal stellen *n*
   r установка *f* сваи
P333   pile ring *see* pile hoop
P334 *e* pile row
   d Pfahlreihe *f*
   f palée *f*, file *f* [rangée *f*] de pieux
   nl palenrij *f(m)*
   r свайный ряд *m*
P335 *e* pile screwing
   d Einschrauben *n* von Pfählen
   f vissage *m* de pieux, mise *f* en place
      (des pieux) par vissage
   nl palen inschroeven *n*
   r завинчивание *n* свай
P336 *e* pile shoe
   d Pfahlschuh *m*, Rammspitze *f*
   f sabot *m* de pieu
   nl paalschoen *m*
   r башмак *m* сваи
P337   pile-sinking *see* pile driving
P338 *e* pile splicing
   d Aufjungfern *n* von Pfählen
   f aboutement *m* [rallongement *m*] du
      pieu
   nl **palen oplangen**
   r **наращивание** *n* **сваи**
P339   pile stockage *see* pile row
P340 *e* pile tip, pile toe
   d Pfahlspitze *f*

   f pointe *f* de pieu
   nl paalpunt *f*
   r остриё *n* сваи
P341   pile top *see* pile head
P342 *e* pilework
   d Pfahlwerk *n*, Pfahlkonstruktion *f*
   f pilotage *m*, pilotis *m*, ouvrage *m* sur
      pilotis
   nl paalfundering *f*
   r свайная конструкция *f*
P343 *e* piling
   d 1. Spundwand *f*, Pfahlreihe *f*
      2. Pfahlgründung *f* 3. Pfahlrammen *n*,
      Rammen *n* 4. Stapelung *f*
      5. Haldenschüttung *f*
   f 1. palée *f*, palplanches *f pl*, pilotis *m*
      2. fondation *f* sur pieux 3. battage *m*
      des pieux 4. empilage *m*, empilement
      m 5. remblayage *m*, remblaiement *m*
   nl 1. damwand *m* 2. paalfundering *f*
      3. heien *n* 4. stapeling *f* 5. storten *n*
      (*op een hoop*)
   r 1. шпунтовый *или* свайный ряд *m*
      2. свайный фундамент *m* 3. забивка *f*
      свай; погружение *n* свай 4.
      штабелирование *n* 5. отсыпка *f*
      в отвал (*грунта, породы*)
P344 *e* pillar
   d Pfeiler *m*; Stütze *f*; Pfosten *m*;
      Stiel *m*, Säule *f*; Pylone *f*
   f colonne *f*; poteau *m*; pilier *m*;
      montant *m*
   nl pilaar *m*, kolom *f(m)*; pyloon *m*
   r стойка *f*; опора *f*; столб *m*;
      колонна *f*; пилон *m*
P345 *e* pillar crane
   d Turmkran *m*; Mastenkran *m*;
      Schwenkmast *m*
   f grue *f* à colonne; grue *f* à pylône;
      grue *f* à potence
   nl torenkraan *m*; mastkraan *m*
   r кран *m* башенного типа; мачтовый
      кран *m*; кран-укосина *m*
P346 *e* pillar drain
   d lotrechter Steindrän *m*
   f drain *m* vertical en pierres
   nl grindpaal *m*, zandpaal *m*
   r вертикальная каменная дрена *f*
P347 *e* pilot channel, pilot cut
   d Rösche *f*, Aufschlußaushub *m*
   f tranchée *f* pilote
   nl leigeul (*m*)
   r пионерная траншея *f* [прорезь *f*]
P348 *e* pilot flame
   d Zündflamme *f*
   f torche *f* d'allumage
   nl waakvlam *f(m)*
   r запальный факел *m*
P349   pilot heading *see* pilot tunnel
P350   pilot trench *see* pilot channel
P351 *e* pilot tunnel
   d Richtstollen *m*, Vortriebstollen *m*
   f galerie *f* pilote [de direction]

nl richttunnel *m*
r направляющая [передовая, головная] штольня *f*

P352 *e* **pilot valve**
d Steuerventil *n*
f robinet *m* pilote
nl regelventiel *n*
r управляющий клапан *m*

P353 *e* **pimple**
d Pickel *m*
f soufflure *f*, bombement *m*
nl puist *f* (*op beton*)
r бугорок *m*, вздутие *n* (*на бетоне*)

P354 *e* **pin**
d Stift *m*, Bolzen *m*
f 1. cheville *f*, goupille *f*, goujon *m*, boulon *m* 2. axe *m*, pivot *m*; gond *m*, tourillon *m*
nl pen *f(m)*; bout *m*; stift *f(m)*
r шпилька *f*; болт *m*; штырь *m* (*дверной петли*)

P355 *e* **pinch bar**
d Brechstange *f*
f pince-monseigneur *f*
nl breekijzer *n*
r лом *m* с лапой (*для распалубливания*)

P356 *e* **pin connection**
d Steckbolzenverbindung *f*; Gelenkverbindung *f*
f assemblage *m* par goujon [goujonné]; articulation *f* à tourillon
nl scharnierverbinding *f*
r одноболтовое *или* шарнирное соединение *n*

P357 *e* **pin-ended column**
d gelenkig angeschlossene Stütze *f*
f colonne *f* [poteau *m*] à pied articulé
nl scharnierend opgelegde kolom *f(m)*
r шарнирно опёртая колонна *f*

P358 *e* **pinholes**
d 1. Blasen *f pl*, Lunker *m pl* (*Betonoberflächenfehler*) 2. Bolzenaugen *n pl*
f 1. piqûres *f pl* (*défauts de surface de béton*) 2. trous *m pl* de boulon
nl 1. poriën *f pl* (*fout van betonoppervlak*) 2. bouten *n pl*
r 1. поры *f pl*; небольшие раковины *f pl* (*дефект бетонных поверхностей*) 2. болтовые отверстия *n pl*

P359 **pin joint** *see* **pin connection**

P360 *e* **pin plug**
d Stiftstecker *m*
f fiche *f* de contact [de connexion], prise *f* de courant, contact *m* à fiche
nl penstekker *m*
r электрическая [штепсельная] вилка *f*

P361 *e* **pioneer road**
d provisorische Baustraße *f*
f route *f* provisoire (*d'accès au chantier*)
nl weg *m* voor bouwverkeer
r временная дорога *f*

P362 *e* **pipe**
d Rohr *n*
f tube *m*, tuyau *m*
nl pijp *f(m)*, buis *f(m)*
r труба *f*

P363 *e* **pipe anchor**
d Rohr-Festpunkt *m*
f appui *m* fixe [rigide] d'un conduit
nl buis(grond)anker *n*
r (неподвижная) анкерная опора *f* трубопровода

P364 *e* **pipe bend**
d Bogen *m*, Krümmer *m*
f coude *m* d'un tuyau
nl bocht *f*, elleboog *m*
r отвод *m*

P365 *e* **pipe bender, pipe bending machine**
d Rohrbiegemaschine *f*; Rohrbieger *m*
f machine *f* à cintrer les tuyaux, cintreuse *f* [coudeuse *f*] à tuyaux, plieuse *f* à tube
nl pijpenbuigmachine *f*; pijpenbuiger *m*
r трубогибочный станок *m*; трубогиб *m*

P366 *e* **pipe box**
d wärmedämmender [mit Schüttstoff ausgefüllter] Rohrleitungskasten *m*
f enveloppe *f* calorifuge pour tuyaux, chemise *f* [enveloppe *f*] de tuyau avec isolation thermique en vrac
nl geïsoleerde leidingkoker *m*
r трубоизоляционный короб *m* с засыпной теплоизоляцией

P367 **pipe bridge** *see* **pipeline bridge**

P368 *e* **pipe clamp**
d Rohrhalter *m*, Rohrschelle *f*
f collier *m* de retenue, collier *m* fixant tube, carcan *m*
nl pijpbeugel *m*
r трубодержатель *m*, хомут *m* для труб

P369 *e* **pipe cleaning**
d Rohrreinigung *f*
f curage *m* des tuyaux
nl pijpreiniging *f*
r прочистка *f* труб

P370 *e* **pipe column**
d Rohrstütze *f*
f poteau *m* [colonne *f*] tubulaire
nl kolom *f(m)* van (stalen) pijp
r трубчатая колонна *f или* стойка *f*

P371 *e* **pipe culvert**
d Rohrdurchlaß *m*
f ponceau *m* de route, dalot *m*
nl duiker *m* (*onder aardedam*)
r дорожная труба *f* (*под насыпью*)

P372 *e* **pipe cutter**
d Rohrtrnngerät *n*, Rohrschneider *m*
f coupe-tuyau *m*, coupe-tube *m*, coupeur *m* [coupeuse *f*] de tuyaux
nl pijpafsnijder *m*
r труборез *m*

P373 *e* **pipe-cutting machine**
   d Rohrschneidemaschine *f*
   f machine *f* à tronçonner les **tubes**
   nl pijpen-afsnijmachine *f*
   r трубоотрезной станок *m*

P374 *e* **pipe drain**
   d Rohrdrän *m*
   f drain *m* en tuyau(terie)
   nl afvoerbuis *f(m)*
   r трубчатая дрена *f*

P375 *e* **pipe drainage**
   d Rohrdränung *f*, Dränrohrentwässerung *f*
   f drainage *m* tubulaire
   nl buizendrainage *f*
   r трубчатый дренаж *m*

P376 *e* **pipe elbow**
   d Kniestück *n*, Rohrknie *n*
   f coude *m*, genou *m* (*d'un tuyau*)
   nl knie *m* (*pijp*)
   r колено *n* (*трубопровода*), отвод *m*

P377 *e* **pipe fitting**
   d Rohrformstück *n*, Fitting *n*
   f pièce *f* de raccordement
   nl fitting *m* (*voor pijp*)
   r фасонная часть *f* трубопровода

P378 *e* **pipe flow**
   d Rohrströmung *f*
   f écoulement *m* dans un tube
   nl stroming *f* in een pijp
   r течение *n* [поток *m*] в трубопроводе

P379   **pipe-forcing system** *see* **pipe pusher**

P380 *e* **pipe for hydraulicking**
   d Spülrohr *n*
   f conduite *f* [canalisation *f*, tuyaterie *f*] de refoulement hydraulique
   nl opspuitbuis *f(m)*
   r намывной пульпопровод *m*

P381 *e* **pipe guide**
   d Gleitführungslager *n*, Gleitschelle *f*, bewegliche Rohrunterstützung *f*
   f appui *m* à glissement (*d'un conduit*)
   nl schuifoplegging *f* (*pijp*)
   r скользящая опора *f* (*трубопровода*)

P382 *e* **pipe hanger**
   d Rohrschelle *f*, Rohrbügel *m*, Rohrsattel *m*, Rohraufhänger *m*
   f étrier *m* de suspension, suspente *f* pour tuyau, tige *f* de suspension pour tuyau
   nl pijpzadel *n*, pijphanger *m*
   r подвеска *f* для труб

P383 *e* **pipe hydraulics**
   d Rohrhydraulik *f*
   f hydraulique *f* de conduites
   nl hydraulica *f* van pijpleidingen
   r гидравлика *f* трубопроводов

P384 *e* **pipe-in-pipe**
   d Doppelrohrwärmeaustauscher *m*
   f échangeur *m* thermique de tube double
   nl dubbele-pijpwarmtewisselaar *m*
   r теплообменник *m* типа «труба в трубе»

P385 *e* **pipe insulation**
   d Rohrisolierung *f*, Rohrummantelung *f*
   f isolation *f* [isolement *m*] d'un tuyau
   nl pijpisolatie *f*
   r изоляция *f* труб

P386 *e* **pipe-jacking**
   d Rohrdurchstoßen *n*, Rohrdurchdrücken *n*
   f pose *f* de tuyau par percement [perçage, enfoncement]
   nl een pijp door een grondlichaam drukken
   r продавливание *n* труб (через насыпь)

P387   **pipe-jacking system** *see* **pipe pusher**

P388 *e* **pipe joint**
   d Rohrverbindung *f*
   f raccord *m* de tuyau; raccordement *m* [assemblage *m*] de tuyauterie
   nl pijpverbinding *f*
   r соединение *n* труб

P389 *e* **pipe jointing**
   d Stoßen *n* der Rohre
   f assemblage *m* de tubes bout à bout
   nl pijpen koppelen
   r стыкование *n* труб

P390 *e* **pipe jointing compound**
   d Vergußmasse *f*, Füllmasse *f*
   f mastic *m* [pâte *f*] pour raccords de tuyaux
   nl vulmassa *f(m)*
   r герметизирующий состав *m* (для стыков труб)

P391   **pipe lagging** *see* **pipe insulation**

P392 *e* **pipelayer**
   d Rohrleger *m*, Rohrverlegekran *m*
   f poseur *m* de tuyaux [de canalisations]
   nl pijplegger *m*
   r трубоукладчик *m*

P393 *e* **pipe laying**
   d Rohrverlegung *f*
   f pose *f* de tuyaux [de canalisations]
   nl pijpen leggen
   r укладка *f* [прокладка *f*] трубопроводов

P394 *e* **pipeless drain**
   d rohrloser Drän *m*
   f drain *m* sans tuyaux
   nl pijploze drainage *f*
   r беструбная дрена *f*

P395 *e* **pipe lifting tongs**
   d Rohrhebezange *f*
   f pince *f* de levage à tubes
   nl pijpheftang *f(m)*
   r монтажный захват *m* (для подъёма труб)

P396 *e* **pipeline**
   d Rohrleitung *f*, Rohrstrang *m*
   f tuyauterie *f*, canalisation *f*, conduite *f*, conduit *m*
   nl pijpleiding *f*
   r трубопровод *m*

P397 *e* **pipeline bridge**
  d Rohr(leitungs)brücke *f*
  f pont *m* pour canalisations; traversée *f*
    aérienne
  nl pijpleidingsbrug *f(m)*
  r мост-трубопровод *m*;
    трубопроводный переход *m*

P398 *e* **pipeline excavation**
  d Rohrgrabenaushub *m*
  f creusement *m* [excavation *f*] de
    tranchées de canalisation [de
    tuyauterie]
  nl de pijpleiding ingraven
  r отрывка *f* трубопроводных траншей

P399 *e* **pipeline gradient**
  d Leitungsgefälle *n*
  f pente *f* d'une conduite
  nl helling *f* van een pijpleiding
  r уклон *m* трубопровода

P400 *e* **pipe line heating and refrigerating**
  d Fernwärme- und Kälteversorgung *f*
  f canalisation *f* urbaine d'approvision-
    nement en chaleur et en froid
  nl afstandverwarming en -koeling *f*
  r централизованное теплохолодоснаб-
    жение *n*

P401 **pipeline laying** *see* **pipe laying**

P402 *e* **pipe line refrigerating**
  d Fernkälteversorgung *f*, zentrale
    Kälteversorgung *f*
  f canalisation *f* urbaine d'alimentation
    en froid
  nl centrale koeling *f*
  r централизованное холодоснабжение *n*

P403 **pipeline section** *see* **pipe run**

P404 *e* **pipeline transport**
  d Rohrleitungstransport *m*
  f transport *m* en conduite [par tuyau]
  nl pijpleidingtransport *n*
  r трубопроводный транспорт *m*

P405 *e* **pipelining machine**
  d Rohrauskleidungsmaschine *f*,
    Rohrisoliermaschine *f*
  f machine *f* à revêtir [à enrober] (les
    tuyaux), enrobeuse *f* mécanique
  nl pijpommantelingsmachine *f*
  r трубоизолировочная машина *f*

P406 *e* **pipe losses**
  d Leitungsverlust *m*
  f pertes *f pl* (*d'énergie, de chaleur*) dans
    la ligne (de la tuyauterie)
  nl leidingverliezen (*energie, warmte*)
  r потери *f pl* энергии потоком среды
    в трубопроводе; потери *f pl* тепла
    с поверхности трубы

P407 *e* **pipemaking machine**
  d Betonrohrfertiger *m*,
    Betonrohrmaschine *f*
  f machine *f* pour la fabrication des
    tuyaux en béton
  nl betonbuizenmachine *f*
  r машина *f* для изготовления
    бетонных труб

P408 *e* **pipe nipple**
  d Nabe *f*, Verbindungsnippel *m*,
    Rohrnippel *m*
  f manchon *m* de tuyau à fil extérieur
  nl pijpnippel *m*
  r трубный ниппель *m*

P409 *e* **pipe outlet**
  d Rohrauslaß *m*
  f prise *f* d'eau à buse
  nl buisuitlaat *m*
  r трубчатый водовыпуск *m*
    (*распределительного канала
    оросительной сети*)

P410 **pipe pier** *see* **pipe anchor**

P411 *e* **pipe pile**
  d Rohrpfahl *m*
  f pieu *m* tubulaire
  nl buispaal *m*
  r трубчатая свая *f*

P412 *e* **pipe pusher**
  d Rohrdurchstoßgerät *n*,
    Rohrdrückgerät *n*, Horizontal-
    -Preßanlage *f*
  f pousse-tube *m*
  nl installatie *f* vor het horizontaal
    drukken van pijpen
  r установка *f* для продавливания
    трубопроводов (через насыпь)

P413 **pipe pushing** *see* **pipe-jacking**

P414 *e* **pipe reamer**
  d Rohrerweiterer *m*
  f fraise *f* à tuyau, fraise *f* conique
    mâle
  nl pijpuitruimer *m*, uitzettang *f(m)*
    (*loden pijp*)
  r трубораширитель *m*

P415 *e* **pipe rigging**
  d Einbindung *f*, Verflanschung *f*
  f tuyauterie *f* [ensemble *m* des tuyaux]
    d'une installation industrielle
  nl installatie *f* van pijpleidingen
  r трубопроводная обвязка *f*

P416 *e* **pipe run**
  d Rohrstrecke *f*, Rohrstrang *m*
  f tronçon *m* de la conduite
  nl pijplengte *f*, strang *m*
  r звено *n* (трубы); участок *m*, нитка *f*
    (*трубопровода*)

P417 *e* **pipe saddle**
  d Rohrsattel *m*
  f selle *f* d'appui [de repos] pour tuyau
  nl pijpzadel *n*
  r седлообразная опора *f* трубопровода

P418 **pipe section** *see* **pipe run**

P419 *e* **pipe sections**
  d 1. wärmedämmende Zylinderschalen
    f pl, Dämmschalen *f pl*
    2. Rohrstränge *m pl*
  f 1. coquilles *f pl* d'isolement
    (thermique) 2. tronçons *m pl* de
    tuyauterie
  nl 1. warmte-isolerende cilinderschalen
    f(m) pl 2. pijplengtes *f pl*

r 1. цилиндрические теплоизоляционные скорлупы *f pl* 2. участки *m pl* трубопровода

P420 *e* **pipe shell**
d Bandage *f*, Rohrschelle *f*
f bandage *m* (*d'un tuyau*)
nl isolatieschaal *f(m)*
r бандаж *m* (*трубопровода, воздуховода*)

P421 *e* **pipe size**
d Nennweite *f* (*Rohr*)
f diamètre *m* nominal d'un tuyau
nl pijpdiameter *m*
r номинальный диаметр *m*, условный проход *m* (*трубопровода*)

P422 *e* **pipe sizing**
d Rohrleitungsberechnung *f*, Rohrbemessung *f*
f calcul *m* de tuyauterie
nl berekening *f* van leidingen
r расчёт *m* трубопроводов

P423 **pipe socket** *see* **bell end**

P424 *e* **pipe support**
d Rohrunterstützung *f*, Rohrfestpunkt *m*, Rohrsattel *m*, Rohrtraverse *f*
f support *m* [appui *m*] d'un conduit
nl pijpoplegging *f*
r опора *f* трубопровода

P425 *e* **pipe systems**
d Rohrsysteme *n pl*
f tuyauterie *f*, canalisations *f pl*, réseau *m* de conduites [de tuyauteries]
nl buizennet *n*
r трубопроводные системы *f pl*

P426 *e* **pipe thread sealing tape**
d Rohrgewindedichtungsband *n*
f ruban *m* d'étanchéité pour filet des tuyaux
nl pijpdraadpakking *f*
r лента *f* для уплотнения резьбовых соединений

P427 *e* **pipe tongs**
d Rohrzange *f*
f pince *f* à tuyaux, serre-tube *m*
nl pijpentang *f(m)*
r трубные клещи *pl*

P428 *e* **pipe tracer**
d Heizband *n*
f ruban *m* chauffant, bande *f* chauffante (*pour la protection des tuyaux contre le gel*)
nl verhitterband *m*
r электронагревательная лента *f* (*для защиты трубопроводов от замерзания*)

P429 *e* **pipe transport of concrete**
d Rohrförderung *f* von Beton
f transport *m* de béton (frais) en conduite [par tuyau]
nl betonmorteltransport *n* door pijp
r трубопроводный транспорт *m* бетонной смеси

P430 *e* **pipe union**
d Rohrkupp(e)lung *f*, Rohrverschraubung *f*
f raccord *m* de tuyau fileté
nl pijpkoppeling *f*, schroefverbinding *f*
r резьбовая соединительная муфта *f*

P431 *e* **pipe vice**
d Rohrschraubstock *m*
f étau *m* à tuyaux
nl bankschroef *f(m)* voor pijpen
r тиски *pl* для труб

P432 *e* **pipeway**
d Rohr(leitungs)kanal *m*
f souterrain *m* pour tuyauterie
nl leidingkoker *m*
r сборный коллектор *m*, подземный коммуникационный канал *m*

P433 *e* **pipe welding machine**
d Rohrschweißmaschine *f*
f soudeuse *f* à tubes, machine *f* à souder pour tubes
nl pijplastoestel *n*
r трубосварочная машина *f*

P434 **pipework** *see* **piping** 1.

P435 *e* **piping**
d 1. Verrohrung *f*, Rohrnetz *n* 1. hydraulischer Grundbruch *m*, innere Erosion *f*
f 1. tuyauterie *f*, canalisation *f*; pose *f* des tuyaux 2. renard *m*, érosion *f* interne
nl 1. leidingstelsel *n*, buizennet *n* 2. ondergrondse erosie *f*
r 1. сеть *f* трубопроводов; укладка *f* трубопроводов 2. внутренняя эрозия *f*, суффозия *f*

P436 *e* **piping diagram**
d Montageplan *m* [Schema *n*] der Rohrleitungen
f schéma *m* des tuyaux [de la tuyauterie]
nl installatieplan *n* voor leidingen
r схема *f* трубопроводов

P437 *e* **pisé**
d Piseebau *m*, Pisébau *m*, Erdstampfbau *m*
f pisé *m*
nl pisébouw *m*
r глинобитная постройка *f*

P438 **piston pump** *see* **reciprocating pump**

P439 *e* **piston sampler**
d Kolbenprobenehmer *m*, Kolbenbohrer *m*
f sonde *f* à piston
nl zuigersondeertoestel *n*
r поршневой грунтонос *m*, поршневая желонка *f* (*для отбора образцов грунта*)

P440 *e* **pit**
d 1. Baugrube *f* 2. Vertiefung *f*, Grube *f* 3. Schürfung *f*, Schürfgrube *f* 4. Steinbruch *m*, Grube *f*
f 1. fouille *f*, excavation *f*, fossé *m* 2. puisard *m* 3. puits *m* (de recherche) 4. carrière *f*

nl 1. kuil *m* 2. groeve *f*, bouwput *m*
3. proefput *m* 4. steengroeve *f*
r 1. котлован *m* 2. приямок *m*
3. шурф *m* 4. карьер *m*

P441 *e* **pitch**
d 1. Teerpech *n*, Teer *m*, Pech *n*
2. Gefälle *n*, Neigung *f*, Steigung *f*
3. Abdachungswinkel *m*
4. Gewindesteigung *f*
f 1. brai *m* 2. angle *m* d'inclinaison
(*vers l'horizon*) 3. pente *f* du toit
4. pas *m* (*p. ex. d'un filetage*)
nl 1. pek *n* 2. helling *f* (*talud*) 3.
dakhelling *f* 4. spoed *m* (*schroef*)
r 1. пек *m*, каменноугольная смола *f*
2. угол *m* наклона 3. уклон *m* ската
(*крыши*) 4. шаг *m* (*напр. резьбы*)

P442 *e* **pitch-bitumen binder**
d Teer-Bitumen-Gemisch *n*,
Bitumenteer *m*
f liant *m* à base de brai de houille et
de bitume, mélange *m* bitume-brai de
houille
nl teerasfalt *n*
r пекобитумное вяжущее *n*

P443 *e* **pitch chain**
d kalibrierte Kette *f*
f chaîn *f* à maillons calibrés
nl gecalibreerde ketting *m*, *f*
r калиброванная звеньевая цепь *f*

P445 *e* **pitched roof**
d geneigtes Dach *n*, Schrägdach *n*
f comble *m* [toit *m*] en pente
nl schuin dak *n*
r скатная [наклонная] крыша *f*,
крыша *f* со скатами

P446 *e* **pitched work**
d Steinvorlage *f*, handversetztes
Steindeckwerk *n*, Steinpackung *f*
f perré *m*, pavage *m*
nl steenbekleding *f*, steenbezetting *f*
r отмостка *f*, мощение *n*

P447 **pitcher tee** *see* **angle tee**

P448 *e* **pitching**
d 1. Uferbefestigung *f*, Ufervorbau *m*,
Böschungssicherung *f* 2.
Setzpacklage *f*, Steinvorlage *f*,
Steindeckwerk *n*, Pflasterung *f*,
Steinpackung *f* 3. Aufstellen *n*,
Errichten *n*, Instellungbringen *n*
f 1. protection *f* des rives 2. perré *m*,
pavage *m* 3. installation *f* [pose *f*] en
position finale
nl 1. oeverbekleding *f*; steenbezetting *f*
(*dijk of krib*) 2. paklaag *f(m)*
3. opstellen *n*
r 1. берегоукрепление *n*,
берегоукрепительные работы *f pl*,
укрепление *n* откосов 2. каменная
отмостка *f*; мощение *n* 3. установка *f*
в проектное положение

P449 *e* **pitching stone**
d Setzpacklagestein *m*, Vorlagestein *m*
f moellon *m* [pierre *f*] du hérisson

nl zetsteen *m* voor paklaag van een weg
r булыжный камень *m* для мощения

P449a *e* **pitch mastic**
d Pechmastix *m*
f mastic *m* de goudron
nl teermastiek *m*
r дёгтевая мастика *f*

P450 *e* **pitch pocket, pitch streak**
d Harztasche *f*
f fente *f* résinifère (*défaut du bois*)
nl harskanaal *n* (*hout*)
r засмолок *m* (*дефект древесины*)

P451 **pit-run gravel** *see* **as-dug gravel**

P452 *e* **pitting**
d Schürfen *n*, Schürfung *f*
f exécution *f* de fouilles de recherche
nl proefboring *f*
r шурфование *n*

P453 *e* **pivot**
d 1. Spurzapfen *m*, Fußgelenk *n*,
Kämpfergelenk *n*, Kupplungsbolzen
m, Zapfen *m* 2. Drehachse *f*
f 1. pivot *m*, tourillon *m*, fusée *f*
2. axe *m* de rotation
nl 1. draaipunt *n*, taats *m* 2. draaiingsas
f(m)
r 1. пятовый шарнир *m*; шкворень *m*;
цапфа *f* 2. ось *f* вращения

P454 **pivot-leaf gate** *see* **tilting gate**

P455 **placeability** *see* **workability**

P456 *e* **placement of concrete**
d Betoneinbau *m*, Betoneinbringung *f*,
Betonieren *n*
f mise *f* en place [en œuvre] du béton
nl betonstorten *n*
r укладка *f* бетона

P457 *e* **placing boom**
d Betonierausleger *m*
f flèche *f* distributrice articulée (pour
le bétonnage à la pompe)
nl giek *m* met stortgoten
r шарнирно-сочленённая стрела *f* (для
укладки бетонной смеси бетононасо-
сами)

P458 **placing of concrete** *see* **placement of
concrete**

P459 *e* **placing sequence**
d Einbaufolge *f* (*z. B. Betoneinbaufolge*)
f succession *f* des opérations de mise
en place [en œuvre] (du béton frais)
nl volgorde *f* bij het plaatsen en
betonstorten
r последовательность *f* укладки
(бетона)

P460 *e* **placing train**
d Betonierzug *m*
f convoi *m* de machines de bétonnage,
train *m* de bétonnage [à beton]
nl betontrein *m*
r передвижной комплексный агрегат
m для бетонирования, бетонный
поезд *m*

P461 *e* **plain bars**
　　d glatte Bewehrungsstähle *m pl*
　　f armatures *f pl* [barres *f pl*
　　d'armature] lisses
　　nl glad wapeningsstaal *n*
　　r гладкие (арматурные) стержни *m pl*

P462 *e* **plain concrete**
　　d unbewehrter Beton *m*
　　f béton *m* simple; béton *m* faiblement
　　armé
　　nl ongewapend beton *n*
　　r неармированный *или*
　　слабоармированный бетон *m*

P463 　　plain girder *see* solid-web girder

P464 *e* **plan**
　　d 1. Plan *m*; Grundriß *m*; Entwurf *m*
　　2. Horizontalprojektion *f*
　　f 1. plan *m*; dessin *m*; projet *m*
　　2. projection *f* en plan
　　nl 1. plan *n*, ontwerp *n* 2. horizontale
　　projectie *f*
　　r 1. план *m*; чертёж *m*; проект *m*
　　2. горизонтальная проекция *f*

P465 *e* **plane**
　　d 1. Ebene *f* 2. Hobel *m* 3. axialer
　　Längsschnitt *m* (*Säule*)
　　f 1. plan *m* 2. rabot *m* 3. section *f*
　　longitudinale d'une colonne
　　nl 1. plat vlak *n* 2. schaaf *f(m)*
　　3. langsdoorsnede *f* (*zuil*)
　　r 1. плоскость *f* 2. рубанок *m*
　　3. продольный осевой разрез *m*
　　(*колонны*)

P466 *e* **plane cross section**
　　d ebener Querschnitt *m*
　　f section *f* transversale plane
　　nl vlakke dwarsdoorsnede *f*
　　r плоское поперечное сечение *n*

P467 *e* **plane frame**
　　d ebener Rahmen *m*
　　f ossature *f* [charpente *f*] plane
　　nl plat raam *n*
　　r плоская рама *f*

P468 *e* **plane gate**
　　d Tafelschütze *f*, Schieberverschluß *m*
　　f vanne *f* plate [à glissières]
　　nl vlakke schuif *f(m)*
　　r плоский затвор *m*

P469 *e* **plane of bending**
　　d Biegungsebene *f*, Biegeebene *f*
　　f plan *m* de flexion
　　nl buigingsvlak *n*
　　r плоскость *f* изгиба

P470 *e* **plane of failure**
　　d Bruchfläche *f*, Bruchebene *f*
　　f plan *m* de cassure, surface *f* de
　　rupture
　　nl breukvlak *n*
　　r плоскость *f* *или* поверхность *f*
　　разрушения

P471 *e* **plane of section**
　　d Querschnittsebene *f*
　　f plan *m* de section

nl vlak *n* van de dwarsdoorsnede
　　r плоскость *f* сечения

P472 *e* **plane of weakness**
　　d Sollbruchstelle *f*
　　f section *f* affaiblie
　　nl verzwakte doorsnede *f*
　　r ослабленное сечение *n*

P473 *e* **plane-sections hypothesis**
　　d Navier'sche Hypothese *f*
　　f hypothèse *m* [théorie *f*] de Navier
　　nl hypothese *f* van Navier
　　r гипотеза *f* плоских сечений

P474 *e* **plane stress analysis**
　　d ebenes Problem *n*, Analysis *f* des
　　ebenen Spannungszustandes
　　f problème *m* plan
　　nl analyse *f* van vlakspanningstoestand
　　r исследование *n* плоского
　　напряжённого состояния, плоская
　　задача *f*

P475 *e* **plank**
　　d Bohle *f*
　　f madrier *m*, planche *f* épaisse, ais *m*
　　nl plank *f(m)*
　　r брус *m*; толстая доска *f*, пластина *f*

P476 *e* **planking and strutting**
　　d Holzabsteifung *f* der Gräben
　　f blindage *m* [étaiement *m*] des parois
　　de fouilles, boisage *m*
　　nl stutten *n* van de wanden van een
　　ingraving
　　r деревянное крепление *n* траншей

P477 *e* **planned-stage construction**
　　d Bauen *n* nach dem Stufenbauplan
　　f construction *f* par étapes
　　nl gefaseerde bouw *m*
　　r поэтапное строительство *n*,
　　строительство *n* в несколько
　　очередей

P478 *e* **planning**
　　d Planung *f*
　　f planification *f*
　　nl een plan opstellen, plannen *n*
　　r планирование *n*

P479 *e* **planning and design**
　　d Projektierung *f*
　　f conception *f* et élaboration *f* du projet
　　nl ontwerpen *n*
　　r проектирование *n*

P480 *e* **planning of houses**
　　d Innenraumgestaltung *f*, Raum- und
　　Umrißlösung *f* der Gebäude
　　f aménagement *m* intérieur
　　nl ontwerpen *n* van de indeling van
　　huizen
　　r (внутренняя) планировка *f* зданий

P481 *e* **planning permission**
　　d Baugenehmigung *f*
　　f permis *m* de construire
　　nl bouwvergunning *f*
　　r разрешение *n* на строительство

P482 *e* **planning standard specification**
　　d Planungsnormen *f pl*

f normes *f pl* concernant l'aménagement intérieur
nl ontwerpnormen *f(m) pl*
r планировочные нормативы *m pl*

P483 *e* **plant and process ventilation**
d Betriebsventilation *f*, industrielle Ventilation *f*
f ventilation *f* d'installations industrielles
nl bedrijfsventilatie *f*
r промышленная вентиляция *f*

P484 *e* **planting**
d Begrünung *f*, Grünanpflanzung *f*
f plantation *f* (*d'arbres*)
nl beplanting *f*
r озеленение *n*

P485 *e* **plant-mix base**
d Mischanlagenunterbau *m* der Straßendecke
f fondation *f* [couche *f* de fondation] (*d'une route*) en enrobé
nl wegfundering *f* van asfaltbeton
r основание *n* дорожного покрытия из чёрного щебня

P486 *e* **plant-mixed concrete**
d Handelsfrischbeton *m*, Lieferbeton *m*
f béton *m* pré-mélangé, béton *m* de centrale, béton *m* prêt à l'emploi
nl transportbeton *n*
r бетонная смесь *f* заводского изготовления, товарная бетонная смесь *f*

P487 *e* **plant-mixed surface treatment**
d Einbringung *f* der Mischanlagendecke von Asphaltbeton (*auf den Unterbau oder die bestehende Straßendecke*)
f mise *f* en œuvre de la couche d'usure en béton asphaltique
nl deklaag *f(m)* van asfaltbeton
r укладка *f* поверхностного слоя асфальтобетона

P488 *e* **plant room**
d Maschinenraum *m*, Maschinenhalle *f*
f halle *f* [salle *f*] des machines
nl machinehal *f*
r машинный зал *m*

P489 *e* **plan view**
d Draufsicht *f*
f vue *f* en plan
nl bovenaanzicht *n*
r вид *m* в плане

P490 *e* **plaque diffuser**
d Platten-Luftverteiler *m*
f diffuseur *m* d'air de plafon en forme d'une plaque
nl platen-luchtverdeler *m*
r дисковый потолочный воздухораспределитель *m*

P491 *e* **plaster**
d 1. Baugips *m*, Stuckgips *m* 2. Gipsputz *m*, Stuck *m*
f 1. plâtre *m* 2. enduit *m* au plâtre

nl 1. bouwgips *n* 2. pleister *n*
r 1. строительный гипс *m* 2. штукатурка *f* (*внутренняя*)

P492 *e* **plaster base**
d Putzträger *m*
f support *m* [fond *m*] pour enduit (au plâtre)
nl draagconstructie *f* voor pleisterwerk
r основание *n* под штукатурку

P493 *e* **plasterboard**
d Gips(bau)platte *f*, Gipskartonplatte *f*, Trockenputz *m*
f panneau *m* (en enduit) de plâtre
nl gipsplaat *f(m)*
r сухая штукатурка *f*

P494 *e* **plastered wall**
d verputzte Wand *f*
f mur *m* avec enduit au plâtre
nl bepleisterde wand *m*
r оштукатуренная стена *f*

P495 *e* **plasterer's float**
d Reib(e)brett *n*
f planchette *f* à régaler, lisseuse *f* du plâtrier
nl plakspaan *m*
r штукатурная тёрка *f*

P496 *e* **plasterer's trowel**
d Spaten *m*, Putzerkelle *f*
f truelle *f* de plâtrier
nl pleistertroffel *m*
r штукатурная лопатка *f*, кельма *f*, мастерок *m*

P497 *e* **plastering**
d 1. Verputzen *n*, Putzarbeiten *f pl* 2. Innenputz *m*
f 1. plâtrage *m*, plâtrerie *f*, travaux *m pl* de plâtrerie 2. enduit *m*, couche *f* d'enduit
nl 1. pleisteren *n* 2. pleisterwerk *n*; stucwerk *n*
r 1. штукатурные работы *f pl* 2. штукатурка *f*; штукатурный намёт *m*

P498 **plastering machine** *see* **plaster--throwing machine**

P499 *e* **plaster mould**
d Gipsform *f*
f moule *m* en plâtre
nl gipsvorm *m*
r гипсовая форма *f*

P500 *e* **plaster of Paris**
d gebrannter Gips *m*, Stuckgips *m*, Formgips *m*
f plâtre *m* de Paris [de moulage]
nl gebrande gips *n*
r обожжённый чистый гипс *m*

P501 *e* **plaster rendering**
d Verputz *m*
f enduit *m* de plâtre, crépi *m*
nl pleisterberaping *f*
r штукатурный намёт *m*

P502 *e* **plaster skim coat**
d Feinputz *m*

f enduit *m* fin [de finition], couche *f* (d'enduit) de finition
nl afwerkpleisterlaag *f(m)*
r чистый накрывочный слой *m* штукатурки

P503 *e* **plaster-throwing machine**
d Putzwerfer *m*, Putz-Spritz-Apparat *m*, Putzmaschine *f*, Verputzanlage *f*
f projeteur *m* de mortier d'enduit
nl spuitpleisterinstallatie *f*
r штукатурная машина *f*, машина *f* для нанесения штукатурного раствора

P504 *e* **plaster work**
d 1. Verputzarbeit *f* 2. Gipsarbeit *f*, Stuckarbeit *f*
f plâtrerie *f*, travaux *m pl* de plâtrerie
nl pleisterwerk *n*; stucwerk *n*
r штукатурные *или* лепные работы *f pl*

P505 *e* **plastic aggregates**
d Kunststoffzuschläge *m pl*
f granulats *m pl* en (matière) plastique
nl kunststoftoeslagen *m pl*
r пластмассовые заполнители *m pl*

P506 *e* **plastic behaviour**
d plastisches Verhalten *n*
f comportement *m* [tenue *f*] plastique (*d'une construction, d'un matériau*)
nl plastisch gedrag *n*
r пластическая работа *f* (*конструкции, материала*)

P507 *e* **plastic clay**
d verformbarer Ton *m*
f argile *f* plastique
nl plastische klei *f(m)*, pijpaarde *f*
r пластичная глина *f*

P509 *e* **plastic concrete**
d plastisches Betongemisch *n*
f béton *m* plastique
nl plastische betonspecie *f*
r пластичная бетонная смесь *f*

P510 *e* **plastic consistency**
d plastische Konsistenz *f*
f consistance *f* plastique
nl plastische consistentie *f*
r пластичная консистенция *f*

P511 *e* **plastic cracking**
d plastische Rißbildung *f*
f fissuration *f* plastique
nl plastische scheurvorming *f*
r пластическое трещинообразование *n*

P512 *e* **plastic deformation**
d plastische Verformung *f*
f déformation *f* plastique
nl plastische vormverandering *f*
r пластическая деформация *f*

P513 *e* **plastic design**
d plastisches Berechnungsverfahren *n*
f calcul *m* d'après la méthode plastique, calcul *m* selon la théorie de plasticité
nl plasticiteitsberekeningsmethode *f*
r расчёт *m* по теории пластичности; расчёт *m* с учётом пластических деформаций

P514 *e* **plastic equilibrium**
d plastisches Gleichgewicht *n*
f équilibre *m* plastique
nl plastisch evenwicht *n*
r пластическое равновесие *n*

P515 *e* **plastic flow**
d plastisches Fließen *n*; Kriechen *n*
f écoulement *m* plastique; fluage *m*
nl plastisch vloeien *n*; kruip *m*
r пластическое течение *n*; ползучесть *f*

P516 *e* **plastic hinge**
d plastisches Gelenk *n*, Fließgelenk *n*
f articulation *f* plastique
nl plastisch scharnier *n*
r пластический шарнир *m*

P517 *e* **plasticiser**
d Betonverflüssiger *m*, Plastifizierungsmittel *n*, Weichmacher *m*
f plastifiant *m*
nl weekmaker *m*, plastificeermiddel *n*
r пластифицирующая добавка *f*

P518 *e* **plasticity**
d Plastizität *f*
f plasticité *f*
nl plasticiteit *f*
r пластичность *f*

P519 *e* **plasticity index**
d Plastizitätsindex *m*
f indice *m* de plasticité
nl plasticiteitsgetal *n*
r число *n* пластичности

P520 *e* **plasticity modulus**
d Plastizitätsmodul *m*, Modul *m* der plastischen Formänderung
f module *m* de plasticité
nl plasticiteitsmodulus *n*
r модуль *m* пластичности

P521 *e* **plasticized cement**
d Zement *m* mit Plastifikatorzusatz
f ciment *m* plastifié
nl cement *n* met plastificeermiddel
r пластифицированный цемент *m*

P522 **plasticizer** *see* **plasticiser**

P523 *e* **plasticizing**
d Plastifizierung *f*
f plastification *f*
nl plastificering *f*; toevoeging *f* van een plastificeermiddel
r придание *n* свойств пластичности; пластификация *f*, введение *n* пластифицирующих добавок

P524 *e* **plasticizing mineral powder**
d plastifizierendes Mineralpulver *n*
f poudre *f* minérale plastifiante
nl plastificerend mineraalpoeder *n*, *m*
r пластифицирующий минеральный порошок *m*

P525 *e* **plastic limit**
d Ausrollgrenze *f*
f limite *f* de plasticité
nl plasticiteitsgrens *f(m)*
r предел *m* пластичности

P526 *e* **plastic moment**
   d plastisches Moment *n*
   f moment *m* plastique
  nl plastisch moment *n*
   r пластический момент *m*, момент *m*, вызывающий образование пластического шарнира

P527 *e* **plastic mortar**
   d plastischer [bildsamer] Baumörtel *m*
   f mortier *m* plastique
  nl plastische mortel *m*
   r пластичный строительный раствор *m*

P528 *e* **plastic pipe**
   d Plastikrohr *n*
   f tube *f* en plastique
  nl kunststofbuis *f(m)*
   r пластмассовая труба *f*

P529 *e* **plastic range**
   d plastischer [unelastischer] Bereich *m*, plastische Zone *f*
   f domaine *m* plastique
  nl vloeigebied *n*
   r область *f* пластических деформаций, пластическая область *f*

P530 *e* **plastics coating**
   d Plastüberzug *m*, Kunststoffbeschichtung *f*
   f revêtement *m* [enrobage *m*] en matière plastique
  nl kunststofbekleding *f*
   r пластмассовое покрытие *n*

P532   **plastics finish** *see* **plastics coating**

P533 *e* **plastic shrinkage**
   d plastische Schwindung *f*, plastisches Schwinden *n*
   f retrait *m* plastique
  nl plastische krimp *m*
   r пластическая усадка *f*

P534 *e* **plastic shrinkage cracking**
   d Rißbildung *f* infolge des plastischen Schwindens
   f fissuration *f* due au retrait plastique
  nl scheurvorming *f* tengevolge van plastische krimp
   r трещинообразование *n*, вызванное пластической усадкой

P535 *e* **plastic soil**
   d bildsamer [plastischer] Boden *m*
   f sol *m* plastique
  nl plastische bodem *m*
   r пластичный грунт *m*

P536 *e* **plastic state**
   d bildsamer [plastischer] Zustand *m*
   f état *m* plastique
  nl plastische toestand *n*
   r пластическое состояние *n*

P537   **plastic strain** *see* **plastic deformation**

P538 *e* **plastification**
   d Plastifizierung *f* des Querschnittes (*Baustatik*); Plastizitätssteigerung *f*
   f plastification *f* de la section
  nl vloeien *n* van een dwarsdoorsnede (*statica*)
   r образование *n* пластического шарнира; нарастание *n* пластических деформаций (*в сечении балки*), распространение *n* фибровой текучести внутрь сечения балки

P539 *e* **platband**
   d 1. Bandrippe *f* 2. Anfänger *m*, Kämpfer *m* 3. Sturzbalken *m*, Tür- *bzw.* Fenstersturz *m*
   f 1. bandeau *m* saillant (*d'un mur*) 2. imposte *f* (*pierre en saillie sous le coussinet d'un arc*) 3. linteau *m*, plate-bande *f* de baie
  nl 1. muurlijst *f* 2. impost *f*, kapiteel *m* 3. bovendorpel *m*
   r 1. поясок *m* (*на стене*) 2. импост *m* (*опора для пяты арки*) 3. перемычка *f* (*над оконным или дверным проёмом*)

P540 *e* **plate**
   d 1. Platte *f*, Tafel *f* 2. Blatt *n*, Scheibe *f*, Blech *n*
   f 1. plaque *f* 2. tôle *f*
  nl 1. metalen plaat *f(m)* 2. plaat *f(m)* (*materiaal*)
   r 1. пластина *f*; пластинка *f*; плита *f* 2. лист *m*; листовой материал *m*

P541 *e* **plate bending**
   d Biegung *f* der Platte
   f flexion *f* de la plaque mince
  nl buiging *f* van een plaat
   r изгиб *m* пластины [пластинки]

P542 *e* **plate electrostatic precipitator**
   d Plattenelektrofilter *n*, *m*
   f filtre *m* électrostatique à lames
  nl elektrostatisch plaatfilter *n*, *m*
   r пластинчатый электрофильтр *m*

P543 *e* **plate girder**
   d Blechträger *m*, Vollwandbinder *m*
   f poutre *f* composée (en acier)
  nl vollewandligger *m*
   r составная стальная балка *f*

P544 *e* **plate-girder bridge**
   d Blechträgerbrücke *f*
   f pont *m* à poutres composées
  nl vollewandliggerbrug *f(m)*
   r мост *m* с составными стальными балками

P545 *e* **plate saddle hanger**
   d Sattelband-Aufhängung *f*
   f selle *f* [étrier *m*] de suspension en plat d'acier (*pour la suspension du coffrage*)
  nl zadelband-ophanging *f*
   r охватывающая подвеска *f* (*из полосовой стали*) для поддержания опалубки

P546 *e* **plate vibrator**
   d Plattenvibrator *m*, Plattenrüttler *m*
   f vibr(at)eur à plaque, plaque *f* vibrante
  nl trilplaat *f(m)*, triltafel *f(m)*
   r виброплита *f*

P547 e **platform**
   d 1. Gerüst n, Arbeitsbühne f, Podest
      m; Holzbelag m (auf Erdreich)
      2. kletternder Montagekorb m
      3. Bahnsteig m 4. Bodenplatte f
      5. Wartungsbühne f, Laufbühne f
      6. Bodenplanum n, Erdplanum n
   f 1. plate-forme f de travail
      2. nacelle f; plate-forme f de levage
      3. quai m d'une gare 4. dalle f de
      fond; dalle f de fondation 5. plate-
      -forme f de service; passerelle f
      6. plate-forme f (d'une route)
  nl 1. platform n 2. hoogwerker m
      3. perron n 4. bodemplaat f(m) 5. bor-
      des n 6. bovenvlak n aardebaan
   r 1. помост m; (деревянный) настил m
      (для повышения несущей
      способности грунта) 2. подъёмная
      люлька f 3. перрон m, пассажирская
      платформа f 4. донная (фундаментная)
      плита f 5. площадка f обслуживания,
      ходовая площадка f 6. земляное
      полотно n (дороги)

P548 e **platform hoist**
   d Plattform(bau)aufzug m
   f monte-matériaux à plateau
  nl bouwlift m
   r рамный или башенный подъёмник m

P549 e **platform ladder**
   d Stehleiter f mit Arbeitsbühne
   f échelle f double à plate-forme de
     travail
  nl bordesladder f(m)
   r лестница-стремянка f с рабочей
     площадкой

P550 e **plenum chamber**
   d Druckluftkammer f; Druckraum m,
     Überdruckkammer f,
     Luftverteilkammer f; Luftmischraum
     m
   f chambre f de répartition d'air;
     chambre f de mélange d'air
  nl verdeelkamer f(m); luchtmengkamer
     f(m)
   r камера f статического давления;
     воздухораспределительная камера f;
     смесительная камера f

P551 e **plenum heating**
   d Überdruckluftheizung f
   f chauffage m à air chaud refoulé
  nl overdrukluchtverwarming f
   r воздушное отопление n (с подачей
     воздуха в помещения по
     воздуховодам)

P552 e **plenum space**
   d Luftraum m, Druckraum m
   f espace m de plenum
  nl drukruimte f
   r замкнутое воздушное пространство
     n

P553 e **plenum system**
   d Überdrucklüftung f,
     Überdrucksystem n

   f ventilation f par refoulement,
     système m à surpression
  nl overdrukventilatie f
   r система f приточной вентиляции,
     принудительная вентиляция f

P554 e **plinth**
   d 1. Fußleiste f, Scheuerleiste f
      2. Sockel m, Mauersockel m
      3. Plinthe f, Säulenfuß m
   f 1. plinthe f, antebois m 2. socle m,
     mur m de soubassement 3. pied m
     de colonne
  nl 1. plint f(m) 2. muurplint f(m)
      3. sokkel m
   r 1. плинтус m 2. цоколь m (здания);
     цокольная часть f (стены)
     3. подколонная плита f;
     подколонник m

P555 e **plinth wiring**
   d Fußleisten-Elektroinstallation f
   f installation f de fils dissimulés sous
     la plinthe, canalisation f électrique
     dissimulée sous la plinthe
  nl elektrische installatie f in plintkokers
   r плинтусная электропроводка f

P556 e **plotter**
   d Kurvenschreiber m, Plotter m
   f traceur m (de courbes)
  nl plotter m
   r графопостроитель m

P557 e **plow**
   d 1. Pflug m, Straßenpflug m
      2. Schneepflug m 3. Kehlhobel m,
     Nuthobel m
   f 1. charrue f 2. chasse-neige m,
     charrue f à neige 3. rabot m à rainure
  nl 1. ploeg m (weg) 2. sneeuwploeg m
      3. boorschaaf f(m), sponningschaaf
     f(m)
   r 1. плуг m (дорожный)
     2. снегоочиститель m 3. пазник m,
     шпунтгебель m

P558 e **plug**
   d 1. Stopfen m 2. Bodenfilter n, m
     (Rückhaltefilter am Brunnenboden),
     Brunnenfilter n, m
   f 1. bouchon m, obturateur m, tampon
     m 2. filtre m de fond
  nl 1. plug m 2. bronfilter n, m
   r 1. пробка f (трубопровода)
     2. донный фильтр m (обратный
     фильтр в основании колодца)

P559 e **plug cock**
   d 1. Kükenhahn m 2. Gasleitungshahn m
   f 1. robinet m à clé [à boisseau]
      2. robinet m (de) gaz
  nl 1. plugkraan f(m) 2. gaskraan f(m)
   r 1. пробковый [стержневой] кран m
     2. газовый кран m

P560 e **plugging leaks**
   d Leckstellenabdichtung f
   f élimination f des fuites
  nl dichten n van lekken
   r устранение n течей

P561 *e* **plug hole**
   d Ausguß *m*
   f bonde *f*, trou *m* d'écoulement
   nl aftapopening *f*
   r сливное отверстие *n* (*санитарного прибора*)

P562 *e* **plug-in connection**
   d Steckeranschluß *m*, Steckverbindung *f*, Steck(kontakt)kupplung *f*
   f accouplement *m* [raccordement *m*] à fiches
   nl stekkerverbinding *f*
   r вставное соединение *n*, штепсельный [штеккерный, штыковой] разъём *m*

P563 *e* **plug socket**
   d Steckdose *f*
   f prise *f* de courant (murale), prise *f* femelle
   nl stopcontact *n*; wandcontactdoos *f(m)*
   r штепсельная розетка *f*; штепсельное гнездо *n* для кабеля

P564   **plug tap** *see* **plug cock**
P565   **plug valve** *see* **plug cock 1.**
P566 *e* **plum**
   d Gesteinblock *m* (*im Massenbeton*)
   f grosses pierres *m pl* pour béton
   nl ruwe steenblokken *n pl* (*in massief beton*)
   r «изюм» *m*, бутовый камень *m* (*в бетоне массивных сооружений*)

P567 *e* **plumb bob**
   d Lot *n*, Senkblei *n*
   f plomb *m*, fil *m* à plomb
   nl schietlood *n*
   r отвес *m*

P568 *e* **plumber's dope**
   d Gewinde-Dichtungskitt *m*
   f mastic *m* à enduire les raccords vissés
   nl fitterskit *f(m)*
   r состав *m* для уплотнения резьбовых соединений труб

P569 *e* **plumbing**
   d 1. Klempner- und Installationsarbeiten *f pl* 2. sanitäre Installation *f*; innere Rohrleitungen *f pl* 3. Abloten *n*, Absenken *n*
   f 1. travaux *m pl* de plomberie, plomberie *f* 2. tuyauterie *f* [canalisation *f*] de distribution d'eau 3. plombage *m*
   nl 1. loodgieterswerk *n* 2. water-, gas- en afvoerleidingen *f pl* 3. solderen *n*
   r 1. слесарно-водопроводные работы *f pl*; санитарно-технические работы *f pl* 2. санитарно-техническое оборудование *n* здания; внутренние трубопроводы *m pl* 3. установка *f* [выверка *f*] по отвесу

P570 *e* **plumbing fixtures**
   d Wasserleitungsarmatur *f*
   f robinetterie *f* eau

   nl kranen *m pl* en afsluiters *m pl*
   r водопроводная арматура *f*

P571 *e* **plumbing unit**
   d Installationsblock *m*, Installationswand *f*
   f bloc-sanitaire *m*, bloc-eau *m*, bloc-bain *m*, bloc-toilette *m*
   nl sanitair bouwelement *n*
   r санитарно-технический блок *m*

P572   **plumbing work** *see* **plumbing 1.**
P573 *e* **plumb line**
   d Lotschnur *f*
   f fil *m* à plomb
   nl schietlood *n*
   r шнур *m* отвеса, отвес *m*

P574 *e* **plunger pump**
   d Plungerpumpe *f*, Tauchkolbenpumpe *f*, Verdrängerpumpe *f*
   f pompe *f* à piston plongeur
   nl plunjerpomp *f(m)*
   r плунжерный насос *m*

P575   **plus size of a screen** *see* **screen oversize**
P576 *e* **ply**
   d 1. Lage *f* 2. Seillitze *f* 3. Dicke *f* eines Elementes der Gewindeverbindung
   f 1. pli *m*, couche *f* 2. brin *m*, toron *m* (*d'un câble*) 3. épaisseur *f* d'un élément d'assemblage à boulons
   nl 1. laag *f(m)* 2. kabelstreng *f(m)* 3. dikte *f* van een element van een boutverbinding
   r 1. слой *m* 2. прядь *f* (*каната*) 3. толщина *f* одного элемента болтового соединения

P577 *e* **plywood**
   d Sperrholz *n*
   f contre-plaqué *f*
   nl multiplex *n*
   r фанера *f*

P578 *e* **plywood concrete forms**
   d Sperrholzschalung *f*
   f coffrage *m* en contre-plaqué
   nl multiplexbekisting *f*
   r фанерная опалубка *f*

P579 *e* **pneumatic aeration**
   d Druckluftbelüftung *f* (*Abwasser*)
   f aération *f* pneumatique (*des eaux usées*)
   nl beluchting *f* met perslucht (*afvalwater*)
   r пневматическая аэрация *f* (*сточных вод*)

P580   **pneumatically applied mortar** *see* **shotcrete**
P581 *e* **pneumatically operated concrete placer**
   d Druckluft-Betonförderer *m*
   f transporteur *m* [convoyeur *m*] pneumatique de béton
   nl perslucht-betonstorter *m*
   r пневмобетоноукладчик *m*

P582 *e* **pneumatically operated discharge gate**
   d pneumatisch betriebener Absperrschieber *m* (*Bunker*)

f trappe *f* de vidange à commande
pneumatique
nl pneumatisch bediende uitlaatschuif
f(m)
r пневмоуправляемая разгрузочная
дверца *f*, пневматический затвор *m*
(*бункера*)

P583 *e* **pneumatically placed concrete**
d Torkretbeton *m*, Spritzbeton *m*
f béton *m* projaté
nl spuitbeton *n*
r торкрет-бетон *m*, шприцбетон *m*

P584 **pneumatic breakwater** *see* **bubblle
breakwater**

P585 *e* **pneumatic caisson**
d Caisson *m*, Druckluftsenkkasten *m*
f caisson *m* à air comprimé
nl caisson *m*
r кессон *m*

P586 **pneumatic concrete placer** *see*
**pneumatically operated concrete
placer**

P587 *e* **pneumatic conveying**
d pneumatische Förderung *f*,
Druckluftförderung *f*
f transport *m* pneumatique
nl pneumatisch transport *n*
r пневматический транспорт *m*,
транспортирование *n* с помощью
пневматических средств

P588 *e* **pneumatic conveying system**
d Druckluftförderanlage *f*
f système *m* de transport pneumatique
nl pneumatisch transportsysteem *n*
r установка *f* пневматического
транспорта

P589 *e* **pneumatic drill**
d Drucklufthammer *m*
f perforatrice *f* pneumatique
nl pneumatische boor *f(m)*
r пневмоперфоратор *m*

P591 *e* **pneumatic feed**
d pneumatische Materialzuführung *f*
[Beschickung *f*, Speisung *f*]
f alimentation *f* pneumatique [par jet
d'air]
nl pneumatische voeding *m*
r пневматическая подача *f*
(материалов)

P592 **pneumatic hammer** *see* **air hammer**

P593 *e* **pneumatic hoist**
d Druckluftaufzug *m*, pneumatischer
Aufzug *m*
f élévateur *m* [treuil *m*] pneumatique
nl pneumatisch hefwerktuig *n*
r пневматический подъёмник *m*

P594 *e* **pneumatic pile hammer**
d Druckluftbär *m*, Preßluftbär *m*
f mouton *m* pneumatique [à air
comprimé]
nl pneumatisch heiblok *n*
r пневматический свайный молот *m*

P595 *e* **pneumatic shield driving**
d Druckluftschildvortrieb *m*
f percement *m* [attaque *f*] au bouclier
à l'air comprimé
nl een tunnelschild pneumatisch
voortschuiven
r пневмощитовая проходка *f*

P596 *e* **pneumatic structure**
d pneumatische Konstruktion *f*
f bâtiment *m* [ouvrage *m*] pneumatique
nl opblaasbaar gebouw *n*
r пневматическое сооружение *n*

P597 *e* **pneumatic test**
d Luftdruckprobe *f*, Luftdruckversuch
m, Luftdruckprüfung *f*
f essai *m* [épreuve *f*] pneumatique, essai
m d'étanchéité par air comprimé
nl afpersen *n*, gasdichtheidscontrole *f*
r пневматическое испытание *n* (*на
герметичность*)

P598 *e* **pneumatic tool**
d Preßluftwerkzeug *n*
f outil *m* pneumatique
nl pneumatisch gereedschap *n*
r пневматический инструмент *m*

P599 **pneumatic transport installation** *see*
pneumatic conveying system

P600 *e* **pneumatic-tyred roller**
d Pneuwalze *f*, Luftreifenwalze *f*
f cylindre *m* [rouleau *m*] à pneus
[à pneumatiques]
nl bandenwals *f(m)*
r дорожный каток *m* на пневматических
шинах

P601 **pneumatic winch** *see* **air winch**

P602 *e* **pocket**
d 1. Aussparung *f*; Tasche *f*, Sack *m*
2. Lunker *m*
f 1. poche *f*, cavité *f*, vide *m*, alvéole *f*
2. retassure *f*, soufflure *f*
nl 1. uitsparing *f*, holte *f* 2. luchtbel *f*
r 1. гнездо *n*; выемка *f*; углубление *n*
2. раковина *f*

P603 *e* **pocket filter**
d Taschenfilter *n*, *m*
f filtre *f* à manche
nl zakfilter *n*, *m*
r карманный [плоскорукавный]
фильтр *m*

P605 *e* **point block**
d Punkt(hoch)haus *n*, Punktgebäude *n*
f bâtiment *m* [bâtiment-tour *m*]
concentré, maison-tour *f* d'habitation
détachée
nl torengebouw *n*
r многоквартирное здание *n* точечной
застройки; отдельностоящее
многоквартирное здание *n*

P606 *e* **pointed arch**
d Spitzbogen *m*
f arc *m* ogival [en ogive]
nl spitsboog *m*
r остроконечная арка *f*

P607   **point load** *see* **concentrated load**

P608 *e* **point of application**
d Angriffspunkt *m*
f point *m* d'application
nl aangrijpingspunt *n*
r точка *f* приложения (*силы*)

P609 *e* **point of concentration**
d Abschlußprofil *n*
f exutoire *m* d'un bassin versant
nl afsluitprofiel *n*, uitloop *m* van een bassin
r замыкающий створ *m*

P610 *e* **point of contraflexure, point of inflection**
d Wendepunkt *m*
f point *m* d'inflexion
nl buigpunt *n*
r точка *f* перегиба

P611 *e* **point of support**
d Stützpunkt *m*, Auflagerpunkt *m*
f point *m* d'appui
nl steunpunt *n*
r точка *f* опоры

P612 *e* **point of zero moment**
d Momentennullpunkt *m*
f point *m* de moment nul
nl momentensteunpunt *n*
r точка *f* нулевого момента

P613 *e* **point source of air pollution**
d punktförmige Emissionsquelle *f*
f source *f* ponctuelle de pollution atmosphérique
nl puntvormige emissiebron *f(m)*
r точечный источник *m* выбросов

P614 *e* **Poisson's ratio**
d Querdehnungszahl *f*
f coefficient *m* de contraction latérale, coefficient *m* de Poisson
nl contractiecoëfficiënt *m* van Poisson
r коэффициент *m* Пуассона, коэффициент *m* поперечного сжатия

P615   **poker vibrator** *see* **internal vibrator**

P616 *e* **polar moment of inertia**
d polares Trägheitsmoment *n*
f moment *m* d'inertie polaire
nl polair traagheidsmoment *n*
r полярный момент *m* инерции

P617 *e* **polder**
d Polder *m*, Ko(o)g *m*
f polder *m*
nl polder *m*
r польдер *m*

P618 *e* **pole**
d 1. Mast *m*; Pfosten *m* 2. Gerüstpfosten *m* 3. Fluchtstange *f*, Absteckpfahl *m*
f 1. mât *m*; poteau *m*, pilier *m* 2. montant *m* d'échafaudage 3. perche *f*; jalon *m*
nl 1. paal *m* 2. steigerpaal *m* 3. stang *f(m)*
r 1. мачта *f*; столб *m* 2. стойка *f* (*лесов*) 3. веха *f*

P619   **pole shore** *see* **post shore**

P620 *e* **polish, polishing**
d Polieren *n*
f polissage *m*
nl polijsten *n*
r полировка *f*

P621 *e* **pollutant**
d Verunreinigungsstoff *m*, Verunreinigung *f*, Pollutant *m*, verunreinigende Beimengung *f*
f agent *m* [substance *f*] de pollution
nl verontreiniging *f*
r нетоксичное загрязняющее вещество *n*, загрязняющая примесь *f*

P622 *e* **pollutant effluents**
d verunreinigte [verseuchte] Industrieabwässer *n pl*
f eaux *f pl* usées des industries [industrielles]
nl verontreinigd industrieël afvalwater *n*
r загрязнённые (промышленные) сточные воды *f pl*

P623 *e* **pollutant emission**
d Schadstoffemission *f*
f emission *f* des agents de pollution
nl uitstoot *m* van schadelijke stoffen
r выделение *n* [выброс *m*] вредностей [вредных веществ]

P624 *e* **pollution load**
d Abwasserlast *f*
f charge *f* (en matière) organique polluante (*dans l'effluent urbain*)
nl afvalwaterbelasting *f*
r (удельная) нагрузка *f* сточных вод (*на водоприёмник или водоток*)

P625 *e* **polyethylene coated craft paper**
d polyäthylenbeschichtetes Kraftpapier *n*
f papier *m* kraft protégé par pellicule étanche de polyéthylène
nl wegenpapier *n* met polyethyleenbekleding
r крафт-бумага *f* с гидроизолирующей полиэтиленовой плёнкой

P626 *e* **polygon of forces**
d Kräftepolygon *n*
f polygone *m* des forces
nl krachtenveelhoek *f(m)*
r многоугольник *m* сил, силовой многоугольник *m*

P627 *e* **polymer-cement concrete**
d Beton *m* mit Plastzusatz, Plastzusatzbeton *m*
f béton *m* à liant mixte
nl betonmortel *m* met toevoeging van een polymeerbindmiddel
r полимерцементный бетон *m*

P628 *e* **polymer concrete**
d Polymer(isations)beton *m*
f béton *m* de résines
nl polymeerbeton *n*
r полимербетон *m*

P629 *e* **polymer-modified cement mortar**
d Polymer(isations)zementmörtel *m*, Zementmörtel *m* mit Plastzusatz

      f mortier *m* à liant mixte
    nl mortel *m* met polymeertoevoeging
     r полимерцементный раствор *m*

P630   **polymer-modified concrete** *see* **polymer-
-cement concrete**

P631 *e* **polymer-modified glass-fiber
reinforced concrete**
    d glasfaserbewehrter Plastzusatzbeton *m*
    f béton *m* à liant mixte renforcé [armé]
par fibres de verre
    nl polymeerbeton *n* met
glasvezelwapening
    r полимерцементный бетон *m*,
армированный стекловолокном,
стеклофибpoлимерцементный
бетон *m*

P632 *e* **polymer mortar**
    d Polymer(isations)mörtel *m*
    f mortier *m* de résines
    nl polymeermortel *m*
    r полимерраствор *m*

P633 *e* **polystyrene board**
    d Styropor-Tafel *f*
    f plaque *f* [dalle *f*] en polystyrène
expansé
    nl polystyreenplaat *f(m)*
    r плита *f* из пенополистирола
(*теплоизоляция*)

P634 *e* **polystyrene flakes**
    d Styropor-Flocken *pl*
    f polystyrène *m* en flocons
    nl polystyreenvlokken *f pl*
    r пенополистирол *m* в хлопьях
(*теплоизоляционный материал*)

P635 *e* **polystyrene foam balls**
    d Styroporkügelchen *n pl* (*Zuzatzstoff*)
    f granulats *m pl* en (forme de) billes
de polystyrène expansé
    nl polystyreenkorrels *m pl*
    r заполнитель *m* в форме шариков из
пенополистирола

P636 *e* **polystyrene foam concrete**
    d Leichtbeton *m* mit Styroporzusatz
    f béton *m* léger à base de granulats en
polystyrène expansé
    nl lichtbeton *n* met polystyreentoeslag
    r лёгкий бетон *m* на заполнителях из
пеностирола

P637 *e* **polystyrene sheet**
    d Styropor-Platte *f*
    f plaque *f* en polystyrène expansé
    nl polystyreenplaat *f(m)* (*warmte-isolatie*)
    r листовой пенополистирол *m*
(*теплоизоляция*)

P638 *e* **polythene sheeting**
    d Polyäthylenplatte *f*
    f feuille *f* [plaque *f*] en polyéthylène
    nl polyetheenfoelie *f(m)*
    r полиэтиленовый листовой материал
m

P639 *e* **polyvinyl acetate concrete**
    d PVA-Beton *m*
    f béton *m* de résine polyvinylacétate

    nl PVA-beton *n*, polyvinylacetaatbeton
n
    r поливинилацетатный полимербетон
m

P640 *e* **pond**
    d 1. Teich *m*, Weiher *m*, Becken *n*
2. Kanalhaltung *f* 3. Spülfläche *f*,
Spülfeld *n*
    f 1. étang *m*; bassin *m*, réservoir *m*
2. bief *m* de canal 3. plage *f* d'épandage
    nl 1. vijver *m*, bekken *n* 2. kanaalpand *n*
3. vloeiveld *n*
    r 1. пруд *m*; бассейн *m*, водоём *m*
2. промежуточный бьеф *m* канала
3. карта *f* намыва

P641 *e* **pondage method**
    d Wasserhaltungsverfahren *n*
    f méthode *m* du tronçon de canal isolé
    nl schutten *n* met gebruik van
spaarbekkens
    r метод *m* глухих отсеков (*измерение
потерь на фильтрацию из канала*)

P642 *e* **ponding**
    d 1. Aufstauung *f*, Anstauung *f*,
Stauung *f* 2. Bluten *n*, Ausschwitzen
(*Schwarzdecken*) 3. Teichverfahren *n*,
Einsumpfen *n* (*Betonnachbehandlung*)
4. Überflutung *f* von Flachdach mit
Kühlwasserschicht
    f 1. endiguement *m* 2. ressuage *m*
(*rémontée du liant hydrocarbone vers
la surface de revêtement*) 3. inondation
f de la surface bétonnée (*pour assurer
la maturation de béton*) 4. inondation
f de la toiture-terrasse (*accidentelle
ou artificielle pour la protection
contre le soleil*)
    nl 1. opvoeren *n* van het waterpeil
2. zweten *n* (*bitumineus wegdek*)
3. onder water zetten *n* van het
betonoppervlak (*nabehandeling*)
4. onder water zetten *n* van plat dak
    r (*koeling*)
1. запруживание *n*, подпружива-
ние *n* 2. выпотевание *n* (*напр.
чёрного вяжущего на поверхность
дорожного покрытия*) 3. затопление
n водой забетонированной
горизонтальной поверхности (*для
обеспечения вызревания бетона*)
4. затопление *n* водой плоской крыши (*для защиты от чрезмерного
нагрева солнцем*)

P643 *e* **pontoon crane**
    d Schwimmkran *m*, Kranschiff *n*,
Pontonkran *m*
    f grue-ponton *f*, grue *f* flottante
    nl drijvende kraan *f(m)*
    r плавучий кран *m*

P644 *e* **pool**
    d 1. Becken *n*, Weiher *m*, Teich *m*
2. Wasserhaltung *f*
    f 1. piscine *f*, réservoir *m*, bassin *m*
2. bief *m*

nl 1. waterbassin *n* 2. deel *n* van het bekken in de nabijheid van de stuw
r 1. водоём *m*, бассейн *m* 2. бьеф *m*

P645 e **pool fishway**
d Tümpelfischpaß *m*, Wildpaß *m*, Teichfischweg *m*
f passe *f* à poissons à bassins successifs
nl visdoorgang *m*
r прудковый рыбоход *m*

P646 e **poor bearing stratum**
d schlecht tragende Bodenschicht *f*
f terrain *m* de mauvaise qualité; couche *f* de sol faible
nl weinig draagkrachtige ondergrond *m*
r слабый грунт *m*, прослойка *f* слабого грунта

P647 e **pop-corn concrete**
d entfeinter Beton *m*
f béton *m* poreux [caverneux, à pores]
nl poriënbeton *n*
r крупнопористый бетон *m*

P648 e **population equivalent**
d Einwohnergleichwert *m*, EGW
f population *f* équivalente
nl inwonersequivalent *n* (*riolering*)
r эквивалентное число *n* жителей (*канализация*)

P649 e **pores**
d Poren *f pl*, Hohlräume *m pl*
f pores *f pl*
nl poriën *f pl*
r поры *f pl*, пустоты *f pl*

P650 **pore volume** *see* **voids volume**

P651 e **porosity**
d Porosität *f*, Hohlraumgehalt *m*
f porosité *f*
nl poreusheid *f*
r пористость *f*

P652 e **porous diffuser plate**
d Filtrosplatte *f*
f plaque *f* de diffusion poreuse
nl poreuze diffusorplaat *f(m)*
r фильтрос *m*

P653 **porous fill** *see* **drainage fill**

P654 e **port**
d 1. Hafen *m* 2. Durchgangsöffnung *f*, Duchtrittsöffnung *f*
f 1. port *m* 2. orifice *m* de passage
nl 1. zeehaven *f(m)* 2. doorlaatopening *f*, toegang *m*
r 1. порт *m* 2. проходное отверстие *n*

P655 e **portable grinder**
d Handschleifmaschine *f*
f meuleuse *f* portative
nl handslijpmachine *f*
r ручная шлифовальная машина *f*

P656 e **portable structure**
d transportables Bauwerk *n*
f ouvrage *m* transportable
nl transportabel bouwwerk *n*
r транспортабельное сооружение *n*

P657 e **portable tank**
de tragbare *oder* befahrbare Zisterne *f*
f réservoir *m* [citerne *f*] transportable
nl verplaatsbare tank *m*
r переносная *или* передвижная цистерна *f*

P658 e **portable welder**
d transportabler Schweißapparat *m*
f soudeuse *f* [machine *f* à souder] transportable
nl verplaatsbaar lasapparaat *n*
r передвижной [переносный] сварочный аппарат *m*

P659 e **portal**
d Portal *n* (*z.B. Tunnelportal*); Portalrahmen *m*
f portail *m*; portique *m*
nl portaal *n*
r портал *m* (*тоннеля*); поперечная концевая портальная ферма *f* (*между главными фермами моста*)

P660 e **portal crane**
d Portalkran *m*
f grue *f* sur [à] portique
nl portaalkraan *f(m)*
r портальный кран *m*

P661 e **portal frame**
d Portalrahmen *m*
f portique *m*, cadre *m* portail
nl portaal *n*
r портальная рама *f*

P662 e **portal framed building**
d Gebäude *n* mit tragenden Portalrahmen
f bâtiment *m* à ossature en portiques
nl gebouw *n* met dragende portalen
r одноэтажное здание *n* с несущими портальными рамами

P663 e **portal structures**
d Portalrahmenbauwerke *n pl*
f ouvrage *m* à charpente à portiques; structures *f pl* à portiques
nl portaalconstructies *f pl*
r конструкции *f pl* портального типа *или* с портальными рамами

P664 e **Portland blast-furnace cement**
d Hochofen(-Portland)zement *m*
f ciment *m* portland au laitier, ciment *m* de laitier au clinker
nl hoogovencement *n*
r шлакопортландцемент *m*

P665 e **Portland cement**
d Portlandzement *m*
f ciment *m* portland
nl portlandcement *n*
r портландцемент *m*

P666 e **Portland cement clinker**
d Portlandzementklinker *m*
f clinker *m* (de ciment) portland
nl portlandcementklinker *m*
r портландцементный клинкер *m*

P667 e **Portland cement concrete**
d Portlandzementbeton *m*

479

f béton m portland
nl portlandcementbeton n
r бетон m на портландцементе

P668 e **Portland-pozzolan cement**
d Portlandzement m mit
Puzzolanzuschlag
f ciment m portland à la pouzzolane,
ciment m portland artificiel
pouzzolanique
nl portlandcement n met
puzzolaantoeslag
r пуццолановый портландцемент m

P669 **Portland slag cement** see
Portland blast-furnace cement

P670 e **position head**
d statische Druckhöhe f, hydrostatischer
Druck m
f charge f [poussée f] hydrostatique
nl hydrostatische druk m,
potentiaaldrukhoogte f
r гидростатический напор m

P671 e **positioning of beams**
d Balkenverlegung f; Balkenanordnung f
f pose f [mise f en place, positionnement
m] des poutres
nl leggen n van balken
r укладка f [размещение n] балок

P672 e **positioning of tendons**
d Verlegung f der Bewehrungsbündel
f positionnement m des câbles
d'armature [des tendons]
nl aanbrengen n van de wapeningsbundels
r укладка f [размещение n]
арматурных пучков

P673 **position of the reinforcement** see
arrangement of reinforcement

P674 e **positive bending moment**
d positives Biegemoment n
f moment m fléchissant positif
nl positief buigend moment n
r положительный﹐изгибающий момент m

P675 e **positive confining bed**
d Grundwasserdeckschicht f
f couche f encaissante positive
nl ondoorlatende laag f(m) boven het
grondwater
r кровля f напорного водоносного
пласта

P676 e **positive displacement meter**
d Verdrängungszähler m
f débitmètre m volumétrique
nl opbrengstmeter m
r расходомер m объёмного типа

P677 e **positive displacement pump**
d Verdrängerpumpe f,
Verdrängungspumpe f
f pompe f volumétrique
nl persluchtpomp f(m), mammoetpomp
f(m)
r объёмный насос m

P678 e **positive reinforcement**
d Bewehrung f gegen positive
Biegemomente

f armature f positive, armature f
soumise aux efforts dus aux moments
fléchissants positifs
nl wapening f tegen positieve buigende
momenten
r арматура f (в балке), воспринимающая усилия от положительных
изгибающих моментов

P679 e **post**
d Pfosten m; Ständer m, Stempel m;
Stütze f, Säule f
f poteau m; montant m; était m
nl paal m; stut m; zuil f(m)
r столб m; стойка f; подпорка f;
колонна f

P680 e **post-and-beam structure**
d Ständerbau m
f construction f [ossature f] à poteaux
et à poutres
nl skeletbouw m
r стоечно-балочный каркас m

P681 e **post-buckling strength**
d überkritische Knickfestigkeit f
f résistance f au flambement au-delà
d'une contrainte critique, résistance f
au flambement postcritique
nl kniksterkte f
r несущая способность f при
продольном изгибе в новой форме
упругого равновесия

P682 e **post-cracking strength**
d Festigkeit f im Zustand nach
Rißbildung
f résistance f après la fissuration
[après la formation des fissures]
nl sterkte f na scheurvorming
r прочность f после
трещинообразования

P683 e **post footing**
d Fundament n aus Einzelstützen,
Pfeilerfundament n
f fondation f de pilier
nl fundering f van een kolom
r столбчатый фундамент m

P684 e **posts and timbers**
d Kantholz n (mit Querschnitt von
125 × 135 mm und mehr)
f bois m carrée (d'une section
125 × 125 mm et plus)
nl kantrecht hout n (met doorsnede van
125 mm × 125 mm en opwaarts)
r деревянные брусья m pl (сечением
125 × 125 мм и более)

P685 e **post shore**
d tragende Stütze f, Gerüstbaum m
f était m [montant m] de coffrage ou
d'échafaudages
nl bekistingsstut m, steigerpaal m
r стойка f опорной конструкции
опалубки или лесов

P686 e **post-tensioned construction**
d vorgespannte Konstruktion f mit
nachträglichem Verbund
f construction f en béton postcontraint

nl nagespannen betonconstructie *f*
r предварительно напряжённая
железобетонная конструкция *f*
с натяжением арматуры на бетон

P687 *e* post-tensioned steel
d Spannbewehrung *f* mit
nachträglichem Verbund
f armature *f* (de précontrainte) mise
en tension sur le béton durci
nl nagespannen wapening *f*
r арматура *f*, натягиваемая на бетон

P688 *e* post-tensioning
d Vorspannung *f* mit nachträglichem
Verbund
f postcontrainte *f*, post-tension *f*
nl naspannen *n*
r натяжение *n* (арматуры) на бетон,
напряжение *n* бетона методом
натяжения арматуры на бетон

P689 *e* post-tensioning field log
d Bau(tage)buch *n* [Baujournal *n*] für
Vorspannen
f carnet *m* de travaux de mise en tension
[en précontrainte]
nl bouwdagboek *m* voor het aanbrengen
van de voorspanning
r полевой журнал *m* производства
работ по натяжению арматуры на
бетон

P690 *e* post-tension unit
d Spannbetonmontageelement *n*,
Spannbetonfertigteil *n*, *m*
f composant *m* postcontraint
nl na te spannen prefabbetonelement *n*
r предварительно напряжённый сборный железобетонный элемент *m* (с
натяжением арматуры на затвердевший бетон)

P691 *e* potable water supply
d Trinkwasserversorgung *f*,
Gebrauchswasserversorgung *f*
f distribution *f* d'eau potable,
alimentation *f* en eau potable
nl drinkwatervoorziening *f*
r питьевое *или* хозяйственно-
-бытовое водоснабжение *n*

P692 potential head *see* position head
P693 pot floor *see* hollow block floor
P694 *e* pothole
d 1. Lunker *m* (*in Beton*) 2. Schlagloch *n*
f 1. cavité *f* (*dans le béton*) 2. nid *m*
de poule (*dans une chaussée*)
nl 1. krimpholte *f* (*beton*) 2. gat *n*
(*wegdek*)
r 1. раковина *f* (*в бетоне*) 2. выбоина *f*
(*в дорожном покрытии*)

P695 *e* pot life
d Topfzeit *f*, Standzeit *f*,
Gebrauchsdauer *f*
f durée *f* de conservation en pot
nl houdbaarheid *f*
r срок *m* годности материала (при
хранении в контейнере)

P696 *e* pot sink
d Spülbecken *n*, Spültisch *m*
f évier *m* de cuisine
nl spoelbak *m*
r кухонная мойка *f*

P697 *e* pouring in lifts
d schichtweises [lagenweises]
Betonieren *n*
f bétonnage *m* [mise *f* en place du béton]
par couches
nl betonstorten *n* in lagen
r послойное бетонирование *n*

P698 *e* pour point
d 1. Stockpunkt *m* 2. Fließpunkt *m*
f 1. température *f* de coulée 2. point *m*
de solidification
nl 1. vloeipunt *n* 2. stolpunt *n*
r 1. температура *f* текучести
2. температура *f* застывания
[затвердевания]

P699 *e* powder-actuated fastening tool
d Schußgerät *n*, Bolzensetzwerkzeug *n*
f appareil *m* de scellement
à cartouches explosives
nl boutenschiethamer *m*
r строительный пистолет *m*

P700 *e* powdered asphalt
d Asphaltmehl *n*, Naturasphaltrohmehl
n
f bitume *m* en poudre
nl poederbitumen *n*
r порошковый [молотый] твёрдый
битум *m*

P701 *e* power
d Leistung *f*
f puissance *f*
nl 1. vermogen *n* 2. arbeidsvermogen *n*
r 1. мощность *f* 2. производительность
f

P702 power bender *see* bar bender
P703 *e* power buggy
d Motorjapaner *m*
f chariot *m* automoteur à benne
nl motorjapanner *m*
r самоходная кузовная тележка *f*;
самоходная тачка *f*

P704 power canal, power channel *see*
diversion channel
P705 power conduit *see* penstock
P706 *e* power comsumption
d Leistungsaufnahme *f*; Strombedarf *m*,
Energiebedarf *m*
f puissance *f* absorbée
nl energieverbruik *n*
r расход *m* мощности [энергии];
энергопотребление *n*

P707 *e* power dam
d Talsperre *f* zur Energieerzeugung,
Sperrstelle *f* mit Kraftwerk
f barrage *m* usine [hydroélectrique]
nl stuwdam *m* voor energiewinning
r станционная плотина *f*

P708 *e* **power float**
   d Betonglättmaschine *f*,
      Abreibmaschine *f*
   f lisseuse *f* [lisseur *m*] mécanique
   nl mechanische plakspaan *f(m)*
   r затирочная машина *f*, механическая
      тёрка *f* (для бетонных полов)

P709   **power grinder** *see* **portable grinder**

P710 *e* **power house**
   d Krafthaus *n*, Maschinenhaus *n*,
      Turbinenhaus *n*
   f bâtiment *m* d'usine hydraulique,
      bâtiment *m* des machines
   nl elektrische centrale *f*
   r здание *n* гидроэлектростанции

P711 *e* **power-operated bush hammer**
   d kraftbetriebener Steinschlägel *m*
      Kraftsteinhammer *m*
   f boucharde *f* mécanique
   nl mechanische bouchardhamer *m*
   r механизированная бучарда *f*

P712 *e* **power rammer**
   d Kraftramme *f*
   f dameuse *f* [dameur *m*] mécanique
   nl motorstamper *m*
   r механизированно-ручная трамбовка *f*

P713 *e* **power shovel**
   d Kraftschaufel *m*; Löffelbagger *m*,
      Hochlöffelbagger *m*
   f pelle *f* mécanique, excavateur *m*
   nl lepelbagger *m*, excavateur *m*
   r механическая лопата *f*; экскаватор
      m

P714 *e* **power supply source**
   d Stromquelle *f*
   f source *f* d'énergie
   nl stroombron *f(m)*
   r источник *m* энергоснабжения

P715 *e* **power take-off**
   d 1. Kraftabnahme *f* 2. Zapfwelle *f*,
      Zapfgetriebe *n*
   f 1. prise *f* de force [de puissance, de
      mouvement] 2. arbre *m* de prise de
      force [de mouvement]
   nl 1. krachtafnemer *m* 2. aftakas *f(m)*
   r 1. отбор *m* мощности 2. вал *m*
      отбора мощности

P716 *e* **power trowel**
   d Estrichglättmaschine *f*
   f lisseur *m* [lisseuse *f*] mécanique de
      béton
   nl mechanische betonplakspaan *f(m)*
   r машина *f* для затирки бетонных
      покрытий

P717 *e* **pozzolan**
   d Puzzolane *f* hydraulischer Zuschlag *m*
   f pouzzolane *m*
   nl puzzolaan *n*
   r пуццолан *m*

P718 *e* **pozzolan cement**
   d Puzzolanzement *m*
   f ciment *m* pouzzolanique, ciment *m*
      à [de] pouzzolane

   nl puzzolaancement *n*
   r пуццолановый цемент *m*

P719 *e* **pozzolanic aggregate**
   d Puzzolanzuschlag *m*
   f granulat *m* [agrégat *m*] pouzzolanique
   nl puzzolaanhoudende toeslag *m*
   r пуццолановый заполнитель *m*

P720 *e* **pozzolanic materials**
   d Puzzolanen *f pl*, Puzzolanzusätze *m pl*
   f matériaux *m pl* pouzzolaniques,
      pouzzolanes *m pl*
   nl puzzolanen *n pl*
   r пуццоланы *m pl*, пуццолановые
      добавки *f pl*

P721   **Pratt truss** *see* **N-truss**

P722 *e* **preaerator**
   d Vorbelüftungsbecken *n*,
      Vorbelüfter *m*
   f préaérateur *m*
   nl voorbeluchter *m*
   r преаэратор *m*

P723 *e* **pre-assembled member**
   d vorgefertigtes Element *n*
      (*einer Stahlkonstruktion*)
   f élément *m* constructif pré-assemblé
   nl prefab-element *n* (*staalconstructies*)
   r сборный элемент *m* (*металлических
      конструкций*)

P724 *e* **precast concrete**
   d Fertig(teil)beton *m*; vorgefertigter
      Stahlbeton *m*
   f béton *m* préfabriqué
   nl prefab-beton *n*
   r сборный бетон *m или* железобетон

P725 *e* **precast concrete building**
   d Gebäude *n* aus Stahlbetonfertigteilen
   f bâtiment *m* en béton armé préfabriqué
   nl gebouw *n* met prefab-betonconstructie
   r здание *n* из сборного железобетона

P726 *e* **precast concrete cladding panel**
   d vorgefertigte Betonverkleidungsplatte
      f
   f panneau *m* de revêtement préfabriqué
      en béton
   nl prefab-betonbekledingsplaat *f(m)*
   r сборная бетонная облицовочная
      панель *f*

P727 *e* **precast concrete connection details**
   d Verbinder *m pl* der Betonfertigteile
      [Stahlbetonfertigteile]
   f pièces *f pl* d'assemblage des éléments
      préfabriqués en béton (armé)
   nl verbindingsdelen *n pl* van prefab-
      -betonelementen
   r соединительные детали *f pl* сборных
      железобетонных элементов

P728 *e* **precast concrete construction**
   d 1. Bauwerk *n* [Gebäude *n*] aus
      Betonfertigteilen 2. Beton-Montage-
      bau *m*, Beton-Fertigteilbau *m*
   f 1. ouvrage *m* [bâtiment *m*] en béton
      armé préfabriqué *f* 2. construction *f*
      en béton armé préfabriqué

nl 1. prefab-betonconstructie *f*
2. montagebouw *m*
r 1. сборное железобетонное
сооружение *n* [здание *n*]
2. сборное строительство *n* (из
бетона)

P729 *e* **precast concrete industry**
d industrielle Fertigung *f* der
Stahlbetonelemente
f production *f* industrielle du béton
armé préfabriqué
nl prefab-betonindustrie *f*
r промышленное производство *n*
сборных бетонных изделий

P730 *e* **precast concrete shaft ring**
d vorgefertigter Schachtring *m*
f anneau *m* de cuvelage en béton
préfabriqué
nl betonnen putring *m*
r сборный элемент *m* кольцевой крепи
(*колодца, шахты*)

P731 *e* **precast constructions**
d Fertigteil(stahl)betoлkonstruktionen
f pl
f constructions *f pl* de béton (armé)
préfabriqué
nl prefab-betonconstructies *f pl*
r сборные (железо)бетонные
конструкции *f pl*

P732 *e* **precast factory**
d Fertigteilwerk *n*
f usine *f* de préfabrication
nl fabriek *f* van betonproducten
r завод *m* сборных (железобетонных)
конструкций [изделий]

P733 *e* **precasting**
d Vorfertigung *f* (*Betonbau*)
f préfabrication *f* d'éléments de
construction en béton (armé)
nl productie *f* van betonelementen
r изготовление *n* сборных
(железо)бетонных изделий

P734 *e* **precasting machine**
d Betonsteinmaschine *f*,
Betonsteinpresse *f*
f machine *f* de préfabrication
nl betonsteenmachine *f*,
betonblokmachine *f*
r машина *f* для изготовления сборных
бетонных блоков

P735 *e* **precasting yard**
d Betonierplatz *m*, Vorfertigungsplatz *m*
f chantier *m* de préfabrication [de
bétonnage] des éléments de
construction, aire *f* de bétonnage
nl plaats *f(m)* voor het storten van
betonnen prefab-onderdele
r полигон *m* [площадка *f*] для
изготовления сборных (железо)-
бетонных изделий

P736 *e* **precast member**
d Betonfertigteil *m, n*
f élément *m* (de construction)
préfabriqué

nl geprefabriceerd constructie-element *n*,
prefabonderdeel *n*
r сборный (железо)бетонный элемент
m

P737 *e* **precast nailable concrete plank**
d nagelbare Leichtbetonbohle *f*
f planche *f* (préfabriquée) en béton léger
clouable
nl lichte nagelbare betonplank *f(m)*
r легкобетонная (сборная) гвоздимая
доска *f*

P738 *e* **precast pile**
d Betonfertigpfahl *m*
f pieu *m* préfabriqué
nl geprefabriceerde betonpaal *m*
r сборная (железобетонная) свая *f*

P739 *e* **precast prestressed concrete unit**
d Spannbeton-Fertigteil *m, n*,
Spannbeton-Montageteil *m, n*
f élément *m* de construction préfabriqué
en béton précontraint
nl geprefabriceerd en voorgespannen
betonelement *n*
r сборный предварительно
напряжённый железобетонный
элемент *m*

P740 **precast unit** *see* **precast member**

P741 *e* **precipitation**
d 1. Niederschlag *m* 2. Fällung *f*
f 1. précipitations *f pl* atmosphériques
2. précipitation *f* (*des particules fines*)
nl 1. neerslag *m* 2. neerslaan *n*,
afzetting *f*
r 1. атмосферные осадки *m pl*
2. осаждение *n* (*твёрдых частиц*)

P742 *e* **precipitator**
d 1. elektrostatischer Staubabscheider
m, Abscheider *m* 2. Ausflockungsmittel
n, Ausfällungsmittel *n*
f 1. précipitateur *m* électrostatique des
poussières 2. précipitant *m*, agent *m*
de coagulation
nl 1. elektrostatische stofscheider *m*
2. uitvlokkingsmiddel *n*
r 1. электростатический пылеуловитель *m* 2. коагулянт *m*, флокулянт *m*

P743 *e* **precision air conditioning**
d Präzisionsklimatisierung *f*
f conditionnement *m* de l'air de
précision
nl precisieluchtbehandeling *f*
r прецизионное кондиционирование *n*
воздуха

P745 *e* **prefabricated building**
d Gebäude *n* aus Fertigteilen,
Fertigteilbau *m*
f bâtiment *m* préfabriqué
nl gebouw *n* van geprefabriceerde
constructie-elementen
r сборное здание *n*, здание *n* из
сборных конструкций

P746 **prefabricated building unit** *see* **precast member**

P748   prefabricated heat insulation *see* preformed insulation

P749 *e* prefabricated housing
    d Fertigteilwohnungsbau *m,* Montagewohnungsbau *m*
    f construction *f* domiciliaire [de logements] en éléments préfabriqués, construction *f* de logements industrialisée
    nl montage-woningbouw *m*
    r индустриализованное жилищное строительство *n*

P750 *e* prefabricated structural element
    d tragendes Fertigteil *n, m*
    f élément *m* porteur préfabriqué
    nl geprefabriceerd dragend constructie-element *n*
    r сборный несущий конструктивный элемент *m*, несущий элемент *m* конструкции

P751 *e* prefabricated unit
    d Fertigteil *n, m*
    f élément *m* préfabriqué
    nl prefab-onderdeel *n*
    r сборный элемент *m*

P752 *e* prefabricated window-wall unit
    d Montage-Fensterwand *f*
    f bloc-fenêtre *m* [panneau-fenêtre *m*] préfabriqué
    nl montage-vensterwand *m*
    r сборная панель *f* с окном [с оконным проёмом]

P753 *e* prefabrication
    d 1. Vorfertigung *f* 2. Fertigteilbauweise *f*
    f 1. préfabrication *f* 2. construction *f* en éléments préfabriqués
    nl 1. prefabriceren *n* 2. bouwen *n* met prefab-onderdelen
    r 1. предварительное изготовление *n* сборных элементов 2. сборное строительство *n*

P754 *e* prefiring
    d vorläufiger Brandversuch *m*, vorläufige Feuerprüfung *f* (*des hitzebeständigen Betons*)
    f préchauffage *m* du béton réfractaire (*type d'essai préliminaire*)
    nl voorlopige vuurproef *f(m)* (*hittebestendig beton*)
    r предварительные испытания *n pl* нагревом (*жаростойкого бетона*)

P755 *e* preflex girder
    d Preflex-Träger *m*
    f poutre *f* préflexe
    nl preflexligger *m*
    r балка *f* «префлекс» (*стальная широкополочная балка с обетонированным нижним поясом, преднапряжение которой создано предварительным выгибом*)

P756 *e* preformed asphalt joint filler
    d bitumengetränkte[r] Fugeneinlage *f* [Fugenstreifen *m*]

    f garniture *f* de joint profilée en asphalte
    nl in bitumen gedrenkte voegstrip *m*
    r профильная герметизирующая прокладка *f* из битумно-волокнистой смеси [композиции]

P757 *e* preformed foam
    d Sonderschaum *m*, Spezial-Schaum *m*
    f mousse *f* préparée dans la batteuse
    nl speciaal schuim *n*
    r пена *f*, образованная в пеновзбивателе

P758 *e* preformed insulation
    d vorgeformte Wärmedämmung *f*, Wärmedämmung *f* aus Fertigteilen
    f isolant *m* préformé [profilé]
    nl voorgevormd warmte-isolatiemateriaal *n*
    r сборная теплоизоляция *f*

P759 *e* preformed pile
    d Fertigpfahl *m*, Rammpfahl *m*
    f pieu *m* préfabriqué
    nl geprefabriceerde heipaal *m*
    r готовая (забивная) свая *f*

P760 *e* preheat coil, preheater
    d Vorerhitzer *m*, Luftvorwärmer *m*
    f préchauffeur *m*, serpentin *m* de préchauffage
    nl luchtvoorverwarmer *m*
    r калорифер *m* предподогрева

P761 *e* preheating
    d Luftvorwärmung *f*; Anheizung *f*
    f préchauffage *m*
    nl voorverwarmen *n*
    r предподогрев *m*, первый подогрев *m* (*воздуха*); разогрев *m* (*системы отопления*)

P762 *e* preliminary clarification tank
    d Vorklärbecken *n*
    f bassin *m* clarificateur [de clarification] primaire, décanteur *m* primaire
    nl voorbezinktank *m*
    r первичный отстойник *m*

P763 *e* preliminary cube test
    d Eignungs-Würfelprüfung *f*
    f essai *m* préliminaire sur cube
    nl kubusproef *f(m)* (*drukvastheid*)
    r предварительное определение *n* прочности (бетона)

P764 *e* preliminary design
    d Vorentwurf *m*
    f avant-projet *m*
    nl voorontwerp *n*
    r проектное задание *n*

P765 *e* preliminary drawings
    d Vorprojekt *n*
    f dessins *m pl* préliminaires
    nl schetsontwerp *n*
    r эскизный проект *m*

P766 *e* preliminary specifications
    d vorläufige technische Bauvorschriften *f pl*

f devis *m* descriptif préliminaire
nl voorlopige technische bouwvoorwaarden *f pl*
r предварительные технические условия *n pl* на строительство объекта

P768 *e* **preliminary treatment**
d Vorbehandlung *f* (*Abwasser*)
f épuration *f* préalable (*des eaux usées*); traitement *m* préliminaire
nl voorbehandeling *f* (*afvalwater*)
r предварительная очистка *f* (*сточных вод*); предварительная обработка *f*

P768a **premature stiffening** *see* **flash set**
P769 **pre-mixed waterbound macadam** *see* **wet-mix macadam**
P770 *e* **prepackaged concrete**
d Sackbeton *m*
f béton *m* ensaché [en sacs]
nl in zakken verpakte betonmortel *m*
r бетон *m* в мешках

P771 *e* **prepacked concrete, preplaced aggregate concrete**
d Prepaktbeton *m*, vorgepackter Beton *m*, Skelettbeton *m*
f béton *m* prépack [prépakt]
nl injectiebeton *n* waarvan eerst de toeslag word gestort
r бетон *m* раздельной укладки с предварительной укладкой заполнителя

P772 *e* **pre-post-tensioned concrete**
d vor- und nachgespannter Beton *m*
f béton *m* précontraint par prétension et post-tension
nl voor- en nagespannen beton *n*
r преднапряжённый бетон *m* с арматурой, натянутой частично на упоры и частично на бетон

P773 *e* **pre-post tensioning**
d Vor- und Nachspannung *f* (*Bewehrung*)
f précontrainte *f* par prétension et post--tension
nl voor- en naspanning *f*
r натяжение *n* (*арматуры*) частично на упоры и частично на бетон

P774 *e* **prescribed mix**
d vorgeschriebene Betonzusammensetzung *f*
f composition *f* de béton spécifiée
nl voorgeschreven mortelsamenstelling *f*
r бетонная смесь *f* заданного состава

P775 *e* **preshrunk concrete**
d vorgemischter Transportbeton *m*
f béton *m* prémélangé partiellement en centrale (*avant le transport par la bétonnière motorisée*); béton *m* prémélangé partiellement 1—3 heures avant la mise en place
nl voorgemengd transportbeton *n*
r бетон *m*, частично перемешанный на заводе (*перед загрузкой в*

автобетоносмеситель) *или за* 1—3 ч до его укладки (*для уменьшения усадки*)

P776 *e* **pressed bend**
d Preßbogen *m*
f coude *m* embouti
nl geperste bocht *f(m)*
r штампованный отвод *m*

P777 *e* **pressed sheet steel**
d Preßblech *n*
f tôle *f* ondulée en acier
nl gegolfd plaatstaal *n*
r волнистая [гофрированная] сталь *f*

P778 *e* **pressostat**
d Niederdruckwächter *m*
f pressostat *m*
nl pressostaat *m*
r прессостат *m*, реле *n* низкого давления

P779 *e* **pressure**
d 1. Druck *m*, Pressung *f* 2. Spannung *f*
f 1. pression *f*; compression *f*; poussée *f* 2. tension *f*, voltage *m*
nl 1. druk *m* 2. (elektrische) spanning *f*
r 1. давление *n*; сжатие *n*; напор *m* 2. (электрическое) напряжение *n*

P780 *e* **pressure against the ends**
d Vorspannkraftangriff *m* (*an die Balkenköpfe*)
f pression *f* sur les abouts (*d'une poutre précontrainte*)
nl druk *m* op de kopeinden van een voorgespannen balk
r давление *n* на торцы (*преднапрягаемой балки*)

P781 *e* **pressure balancing**
d Druckabgleich *m*, Druckausgleich *m*
f compensation *f* [égalisation *f*] de pressions
nl vereffenen *n* van de druk
r увязка *f* давлений (*в узлах сети воздуховодов, трубопроводов*); уравнивание *n* давлений

P782 *e* **pressure conduit**
d Druckwasserleitung *f*, Druckrohrleitung *f*
f conduite *f* forcée [en charge]
nl waterpersleiding *f*
r напорный водовод *m* [трубопровод *m*]

P783 **pressure controller** *see* **pressure regulator**
P784 *e* **pressure control valve**
d Druckregelventil *n*
f soupape *f* de régulation [de contrôle] de la pression
nl drukregelventiel *n*
r клапан *m* регулирования давления

P785 *e* **pressure distribution**
d Druckverteilung *f*
f distribution *f* de pression
nl drukverdeling *f*
r распределение *n* давления

P786 pressure equalizing *see* **pressure balancing**

P787 *e* pzessure filter
 d Filterpresse *f*, Druckfilter *n*, *m*, geschlossenes Schnellfilter *n*, *m*
 f filtre-presse *m*
 nl persfilter *n*, *m*
 r фильтр-пресс *m*, напорный фильтр *m*

P788 *e* pressure filtration
 d Druckfiltration *f*
 f filtration *f* [filtrage *m*] sous pression
 nl filtratie *f* onder druk
 r напорная фильтрация *f*

P789 pressure grouting *see* **cementation**

P790 *e* pressure gun
 d Druckpistole *f*
 f pistolet *m* á air (comprimé)
 nl drukpistool *n*
 r нагнетательный пистолет *m*, шприц *m* (*для нагнетания уплотняющих и гидроизолирующих мастик*)

P791 *e* pressure head
 d Druckhöhe *f*
 f hauteur *f* piézométrique [de charge]
 nl drukhoogte *f*
 r высота *f* напора, статический напор *m*

P792 *e* pressure intake works
 d Druckumleitung *f*
 f dérivation *f* en charge
 nl inlaatwerk *n*
 r напорная деривация *f*

P793 *e* pressure line
 d 1. Drucklinie *f* 2. Druckleitung *f*
 f 1. ligne *f* de pressions 2. ligne *f* en charge, conduite *f* de pression
 nl 1. druklijn *f(m)* 2. drukleiding *f*
 r 1. линия *f* давления (*напр. в арке*) 2. напорный трубопровод *m*

P794 *e* pressure loss factor
 d Druckverlustfaktor *m*, Druckverlustbeiwert *m*
 f facteur *m* de perte de pression
 nl drukverliesfactor *m*
 r коэффициент *m* потерь давления

P795 *e* pressure pile
 d Preßbeton(bohr)pfahl *m*
 f pieu *m* foré à air comprimé
 nl in de grond gedrukte betonnen mantelpaal *m* [boorpaal *m*]
 r (буро)набивная свая *f* (*с уплотнением бетона сжатым воздухом*)

P796 *e* pressure pipeline
 d Druckrohr *n*; Druck(rohr)leitung *f*
 f tayau *m* de [sous] pression; conduite *f* forcée [en pression]
 nl persleiding *f*
 r нагнетательная труба *f*; напорный трубопровод *m*

P797 *e* pressure regulator
 d Druckregler *m*
 f régulateur *m* de pression

 nl drukregelaar *m*
 r регулятор *m* давления

P798 *e* pressure-relief valve
 d Sicherheitsventil *n*, Entlastungsventil *n*, Überdruckventil *n*
 f soupape *f* de sûreté [de décharge]
 nl ontlastklep *f(m)*
 r предохранительный клапан *m*, клапан *m* сброса давления

P799 *e* pressure surge
 d Druckstoß *m*, Wasserschlag *m*
 f bélier *m* (hydraulique)
 nl waterslag *m*
 r гидравлический удар *m*

P800 *e* pressure switch
 d Druckschalter *m*, Druckwächter *m*
 f contacteur *m* manométrique
 nl drukschakelaar *m*
 r реле *n* давления

P801 pressure tank lorry *see* **bulk asphaltic--bitumen distributor**

P802 *e* pressure tap
 d Druckrohr *n*, Drucksonde *f*; Druckentnahmebohrung *f*, Zapfstelle *f* zur Druckprüfung
 f robinet *m* de manomètre; robinet *m* pour air comprimé
 nl persluchtkraan *m*
 r приёмник *m* давления; штуцер *m* для отбора давления, отвод *m* к манометру

P803 pressure test *see* **pneumatic test**

P804 *e* pressure-tight joint
 d druckdichte Verbindung *f*
 f joint *m* [raccord *m*] étanche [imperméable]
 nl water- en luchtdichte verbinding *f*
 r герметичное соединение *n*; герметичный шов *m*

P805 *e* pressure tunnel
 d Druckstollen *m*
 f galerie *f* en charge
 nl druktunnel *m*
 r напорный туннель *m*

P806 *e* pressure vessel
 d Druckgefäß *n*, Druckbehälter *m*
 f réservoir *m* à pression, autoclave *m*
 nl drukvat *n*
 r сосуд *m* под давлением, напорный сосуд *m*

P807 pressure void ratio curve *see* **compression curve**

P808 *e* pressure zone
 d Druckzone *f*; Druckbereich *m*
 f zone *f* à pression; zone *f* en compression; zone *f* comprimeé
 nl drukzone *f*
 r зона *f* давления; зона *f* сжатия; сжатая зона *f* (*напр. сечения балки*)

P809 *e* pressurization
 d 1. Erzeugung *f* des Luftüberdrucks, Luftüberdruckhaltung *f* (*in Räumen*,

im Gebäude); Druckhaltung *f* (*in geschlossenem Rohrnetz*)
2. Hermetisierung *f*, Abdichtung *f*
3. Aufladung *f*, Lufteinblasen *n*
f pressurisation *f*; mise *f* sous pression
nl opbouwen *n* van een overdruk; onderhouden *n* van een overdruk; lucht inblazen
r 1. создание *n* [поддержание *n*] подпора воздуха (*в помещениях, зданий*); поддержание *n* давления (*в замкнутой сети трубопроводов*) 2. герметизация *f* 3. наддув *m*

P810 *e* pressurization system
d Druckhalteanlage *f*, Druckdiktieranlage *f*; Überdrucksystem *n*
f système *m* de pressurisation
nl overdruksysteem *n*
r средства *n pl* поддержания давления (*в замкнутой системе охлаждения или водяного отопления*); вентиляционная система *f*, создающая подпор воздуха в здании

P811 *e* pressurization unit
d Druckerhöhungsanlage *f*, Druckhalteanlage *f*
f appareil *m* de pressurisation
nl drukverhogingsinstallatie *f*
r гидрофор *m*, агрегат *m* для поддержания давления в системе

P812 *e* pressurized expansion tank
d Druckausdehnungsgefäß *n*
f réservoir *m* d'expansion fermé
nl (druk)expansievat *n*
r закрытый расширительный бак *m*

P813 *e* pressurized hot water system
d geschlossenes Heißwasserheizungssystem *n*
f installation *f* de chauffage par l'eau chaude à haute température
nl gesloten heetwaterverwarmingssysteem *n*
r высокотемпературная система *f* водяного отопления

P814 *e* prestressed brickwork
d vorgespanntes Mauerwerk *n*
f maçonnerie *f* précontrainte
nl voorgespannen metselwerk *n*
r предварительно напряжённая каменная кладка *f*

P815 *e* prestressed concrete
d Spannbeton *m*
f béton *m* précontraint
nl voorgespannen beton *n*
r предварительно напряжённый (железо)бетон *m*

P816 *e* prestressed concrete beam
d Spannbetonträger *m*
f poutre *f* en béton précontraint
nl voorgespannen betonbalk *m*
r предварительно напряжённая железобетонная балка *f*

P817 *e* prestressed concrete dam
d Spannbeton-Staumauer *f*, vorgespannte Staumauer *f*
f barrage *m* en béton précontraint
nl stuwmuur *m* van voorgespannen beton
r предварительно напряжённая бетонная плотина *f*

P818 *e* prestressed concrete pile
d Spannbetonpfahl *m*
f pieu *m* en béton précontraint
nl voorgespannen-betonpaal *m*
r предварительно напряжённая железобетонная свая *f*

P819 *e* prestressed constructions
d vorgespannte Konstruktionen *f pl* [Bauwerke *n pl*]
f ouvrages *m pl* précontraints, constructions *f pl* précontraintes
nl voorgespannen constructies *f pl*
r предварительно напряжённые конструкции *f pl*

P820 *e* prestressed double tee
d vorgespannte Doppel-T-Platte *f*
f dalle *f* en T double de béton précontraint
nl voorgespannen dubbele-T-plaat *f(m)*
r предварительно напряжённая железобетонная панель *f* в форме двойного T

P821 *e* prestressing
d Vorspannen *n*, Vorspannung *f*
f précontrainte *f*, prétension *f*
nl voorspanning *f*
r предварительное напряжение *n*

P822 prestressing bed *see* **pretensioning bed**

P823 *e* prestressing by winding
d Bewehrungsvorspannung *f* durch Aufwickeln
f précontrainte *f* par enroulement
nl ringvoorspanning *f*
r предварительное напряжение *n* методом навивки спиральной арматуры

P824 *e* prestressing force
d Vorspannkraft *f*
f force *f* de précontrainte
nl voorspankracht *f(m)*
r усилие *n* натяжения (*арматуры*); усилие *n* предварительного напряжения (*конструкций*); усилие *n* обжатия (*бетона*)

P825 *e* prestressing force after transfer
d wirksame Spannkraft *f* (*auf Beton übertragene Vorspannkräfte*)
f effort *m* de précontrainte du béton après la transmission (*de la tension des aciers au béton*)
nl effectieve spankracht *f(m)*
r значение *n* преднапряжения бетона после передачи ему усилий натяжения арматуры, передаточное преднапряжение *n*

P826 *e* **prestressing force in the tendon**
   d kontrollierte Spannkraft *f* im Bewehrungsbündel
   f effort *m* de précontrainte dans le câble d'armature
   nl gekontroleerde spankracht *f(m)* in wapeningsbundel
   r контролируемое усилие *n* натяжения пучка

P827 *e* **prestressing steel**
   d Spannstahl *m*, Spannbewehrung *f*
   f acier *m* de précontrainte
   nl voorspanstaal *n*
   r напрягаемая арматура *f*

P828 *e* **prestressing techniques**
   d Vorspannverfahren *n pl*
   f pratique *f* courante de réalisation des précontraintes
   nl voorspanmethoden *f pl*
   r технологические приёмы *m pl* [способы *m pl*] предварительного напряжения

P829 *e* **prestressing tendon**
   d 1. Spannglied *n* 2. Bewehrungsbündel *n*
   f 1. acier *m* mis en tension 2. câble *m* de précontrainte
   nl 1. voorspanstaaf *f(m)* 2. wapeningsbundel *m*
   r 1. напрягаемая арматура *f* 2. арматурный пучок *m*

P830 *e* **pretensioned construction**
   d vorgespannte Konstruktion *f* mit sofortigem Verbund
   f construction *f* précontrainte (*par câbles tendus sur le béton durci*)
   nl voorgespannen constructie *f* met onmiddellijke aanhechting
   r предварительно напряжённая конструкция *f* с натяжением арматуры на упоры

P831 *e* **pretensioned steel**
   d Spannglied *n* mit sofortigem Verbund
   f acier *m* mis en tension avant le durcissement du béton
   nl voorgespannen wapening *f* (*voorgerekt*)
   r арматура *f*, натягиваемая на упоры

P832 *e* **pretensioning**
   d Vorspannung *f* mit sofortigem Verbund
   f précontrainte *f* par fils adhérents
   nl voorspanning *f* met voorgerekt staal
   r натяжение *n* (*арматуры*) на упоры

P833 *e* **pretensioning bed, pretensioning bench**
   d Spannbett *n*, Spannbahn *f*
   f banc *m* de précontrainte [de mise en tension]
   nl spanbank *m*
   r натяжной стенд *m*, стенд *m* для преднапряжения сборных железобетонных конструкций

P834 *e* **preventive maintenance**
   d vorbeugende Instandhaltung *f*
   f entretien *m* préventif
   nl preventief onderhoud *n*
   r предупредительный ремонт *m*; эксплуатационное обслуживание *n*

P835 *e* **prick punch**
   d Handkörner *m*, Punktiereisen *n*
   f pointeau *m* (de traçage)
   nl kornagel *m*, centerpunt *m*
   r кернер *m*

P836 *c* **primary air**
   d Primärluft *f*
   f air *m* primaire
   nl primaire lucht *f(m)*
   r первичный воздух *m*

P837 *e* **primary circuit**
   d Primärkreislauf *m*
   f circuit *m* primaire
   nl primaire stroomkring *f(m)*
   r первичный контур *m* циркуляции; контур *m* первичного теплоносителя

P839 *e* **primary crusher**
   d Vorbrecher *m*
   f concasseur *m* [broyeur *m*] primaire
   nl primaire steenbreker *m*
   r дробилка *f* первичного измельчения

P840 *e* **primary filter**
   d Primärfilter *n*, *m*, Vorfilter *n*, *m*
   f filtre *m* primaire
   nl primair filter *n*, *m*, voorfilter *n*, *m*
   r предфильтр *m*, фильтр *m* первой ступени

P841 *e* **primary fluid**
   d Primärwärmeträger *m*; Primärkälteträger *m*
   f fluide *m* primaire
   nl primair warmtemedium *n*; primaire koelvloeistof *m*
   r первичный теплохладоноситель *m*

P842 *e* **primary member**
   d Hauptbauelement *n*, Grundelement *n* (*der Konstruktion*)
   f élément *m* principal
   nl hoofdbouwelement *n*
   r главный элемент *m* (*конструкции*)

P843 *e* **primary-secondary system**
   d Primär-Sekundär-System *n*
   f système *m* primaire-secondaire
   nl primair-secundairsysteem *n*
   r система *f* с кольцами первичной и вторичной циркуляции теплоносителя

P844 *e* **primary sedimentation**
   d Vorklärung *f*
   f décantation *f* [sédimentation *f*] primaire
   nl voorbezinking *f*
   r первичное отстаивание *n*

P845 *e* **primary sludge**
   d Frischschlamm *m* (*untersetzter Schlamm im Absetzbecken*)
   f boues *f pl* d'égout primaire
   nl bezonken slijk *n* (*afvalwater*)
   r сырой осадок *m* (*сточных вод*)

P846 e **primary treatment of wastes**
d mechanische Abwasserklärung *f*
f traitement *m* [épuration *f*] des eaux
d'égout primaire
nl mechanisch zuiveren *n* van
afvalwater
r первичная очистка *f* сточных вод

P847 e **primary water**
d Klärwasser *n*, geklärtes Wasser *n*
f eau *f* clarifiée
nl gezuiverd water *n*
r осветлённая вода *f*

P848 e **prime coat**
d Grundierung *f*, Grund(ier)schicht *f*,
Voranstrich *m*
f première couche *f* (de peinture), couche
f d'impression [de fond]
nl grondlaag *f(m)*
r грунтовочный слой *m*

P849 e **primer**
d 1. Grundiermittel *n*,
Voranstrichmittel *n*; Grund(ier)-
schicht *f*, Voranstrich *m* 2. Schlagpa-
trone *f*, Bohrpatrone *f*,
f 1. peinture *f* (pour couche) de fond;
couche *f* (de peinture) d'impression
2. amorce *f*
nl 1. grondlaag *f(m)*, grondverf *f(m)*
2. ontsteker *m*
r 1. грунтовка *f*, грунтовочное
покрытие *n*, грунтовочный слой *m*
2. инициирующий заряд *m*
[патрон *m*]

P850 e **priming**
d 1. Erstfüllung *f* (*eines Kanals*,
Wasserspeichers) 2. Pumpenfüllung *f*;
Anspringen *n* des Hebers 3.
Grundierung *f*, Voranstreichen *n*;
Grundanstrich *m* 4. Auftragen *n* der
Kleb(e)schicht (*Straßendecke*)
f 1. mise *f* en eau 2. amorçage *m*
3. peinture *f* d'impression 4. répandage
m superficiel (du bitume)
nl 1. eerste vulling *f* (*kanaal*) 2. net
vloeistof vullen *n* (*pomp*) 3. in de
grondverf zetten, gronden *n* 4. de
hechtlaag aanbrengen
r 1. первое заполнение *n* водой
(*канала, водохранилища*) 2. заливка
f насоса (*перед пуском*); зарядка *f*
сифона 3. огрунтовка *f*;
грунтовочное покрытие *n*
4. первичная обработка *f* битумом
(*дорожного покрытия*)

P851 **priming coat** *see* **prime coat**

P852 e **principal**
d Hauptsparren *m*; Binder *m*,
Dachträger *m*
f arbalétrier *m*; ferme *f* de comble [de
toit]
nl kapspant *n*
r главная стропильная нога *f*;
стропильная ферма *f*

P853 e **principal axes**
d Hauptachsen *f pl*
f axes *m pl* principals
nl hoofdassen *f(m) pl*
r главные оси *f pl*

P854 e **principal curvature**
d Hauptkrümmung *f*
f courbure *f* principale
nl hoofdkromming *f*
r главная кривизна *f*

P855 e **principal design sections**
d rechnerische Hauptschnitte *m pl*
f sections *f pl* critiques principales
(*d'une plaque*)
nl voor de berekening belangrijke
doorsneden *f pl* (*van een plaat*)
r главные расчётные сечения *n pl*
(*плиты*)

P856 **principal rafter** *see* **principal**

P857 e **principal reinforcement**
d Hauptarmierung *f*, Hauptbewehrung *f*
f armature *f* principale, ferraillage *m*
principal
nl hoofdwapening *f*
r (расчётная) рабочая арматура *f*

P858 e **principal road**
d Hauptstraße *f*
f route *f* principale
nl hoofdweg *m*
r главная дорога *f*

P859 e **principal stresses**
d Hauptspannungen *f pl*
f contraintes *f pl* principales
nl normaalspanningen *f pl*
r главные напряжения *n pl*

P860 e **principal tensile stress**
d Hauptzugspannung *f*
f contrainte *f* principale de traction
nl normaalrekspanning *f*
r главное растягивающее напряжение
n

P861 e **principle of least work**
d Prinzip *n* der kleinsten
Formänderungsarbeit
f principe *m* de travail minimum
nl beginsel *n* van de kleinste
vormveranderingsarbeid
r принцип *m* наименьшей работы

P862 e **principle of superposition**
d Superpositionsprinzip *n*,
Überlagerungsprinzip *n*
f principe *m* de superposition
nl superpositieprincipe *n*
r принцип *m* независимости действия
сил, принцип *m* наложения
[суперпозиции]

P863 e **principle of virtual displacements**
d Prinzip *n* der virtuellen Verschiebung
[Arbeit]
f principe *m* des déplacements virtuels,
théorème *m* du travail virtuel
nl beginsel *n* der virtuele verplaatsingen
r принцип *m* возможных перемещений

P864 *e* print
  d Lichtpause *f*
  f bleu *m*, tirage *m* bleu
  nl lichtdruk *m*
  r светокопия *f* (*чертежа*), синька *f*

P865 *e* prior loading
  d vorherige Belastung *f*
  f chargement *m* précédent [antécédent]
  nl voorafgaande belasting *f*
  r предшествующее нагружение *n*

P866 *e* prismatic beam
  d Rechteckbalken *m* mit gleichbleibendem Querschnitt; prismatischer Balken *m* [Träger *m*]
  f poutre *f* prismatique; poutre *f* de la section rectangulaire
  nl balk *m* met rechthoekige doorsnede
  r призматическая балка *f*; балка *f* прямоугольного сечения

P867 *e* probabilistic design
  d statistisches Verfahren *n*, statistische Methode *f*
  f méthode *f* statistique (*de calcul*)
  nl statistische berekeningsmethode *f*
  r статистический метод *m* (*расчёта*)

P868 *e* probability curve
  d Wahrscheinlichkeitsverteilung *f* (*z. B. der hydrologischen Größen*); Wahrscheinlichkeitskurve *f*
  f courbe *f* de probabilité
  nl waarschijnlijkheidskromme *f*
  r кривая *f* обеспеченности; кривая *f* распределения; вероятностная кривая *f*

P869 *e* probe
  d Sonde *f*; Meßfühler *m*
  f sonde *f*
  nl sonde *f*; voeler *m*
  r зонд *m*; щуп *m*

P870 *e* process heating
  d Prozeßheizung *f*, Industrieheizung *f*
  f chauffage *m* industriel
  nl procesverwarming *f*, industriële verwarming *f*
  r промышленное отопление *n*

P871 *e* process wastes
  d 1. Industrieabfälle *m pl* 2. Industrieabwasser *n*, Betriebsabwässer *n pl*
  f 1. résidus *m pl* industriels; déchets de fabrication 2. eaux *f pl* industrielles [des idustries], eaux *f pl* résiduaires
  nl 1. industriële afvalstoffen *m pl* 2. industrieel afvalwater *n*
  r 1. промышленные отходы *m pl* 2. промышленные сточные воды *f pl*

P872 *e* process water
  d Betriebswasser *n*, technisches Wasser *n*
  f eau *f* industrielle
  nl bedrijfswater *n*, proceswater *n*
  r вода *f* для технологических нужд

P873 *e* production bay
  d Produktionshalle *f*
  f hall *m* de production
  nl productiehal *f(m)*
  r цех *m*; производственный пролёт *m*

P874 *e* production building
  d Industriegebäude *n*
  f bâtiment *m* industriel
  nl fabrieksgebouw *n*
  r производственное здание *n*

P875 *e* production drawings
  d Arbeits(ausführungs)zeichnungen *f pl*, Bauzeichnungen *f pl*, Werk(statt)zeichnungen *f pl*
  f dessins *m pl* d'exécution
  nl werktekeningen *f pl*
  r рабочие (строительные) чертежи *m pl*

P876 *e* production of concrete
  d Zubereitung *f* des Betongemisches
  f confection *f* [fabrication *f*, préparation *f*] de béton
  nl bereiden *n* van een betonmortel
  r приготовление *n* бетонной смеси

P877 *e* profile
  d 1. Walzprofil *n* 2. Straßenlängsprofil *n* 3. geologisches Profil *n*; Bodenprofil *n*
  f 1. profilé *m* laminé 2. profil *m* en long de la route 3. profil *m* géologique
  nl 1. walsprofiel *n* 2. langsprofiel *n* van een weg 3. bodemprofiel *n*
  r 1. прокатный профиль *m* (*изделие*) 2. продольный профиль *m* дороги 3. геологический профиль *m*; почвенный профиль *m*

P878 *e* programmed controller
  d Programmregler *m*, Zeitplanregler *m*
  f régulateur à programme
  nl geprogrammeerde regelaar *m*
  r программный регулятор *m*

P879 *e* programming of costruction
  d Bauplanung *f*; Aufstellung *f* der Bauzeitpläne
  f élaboration *f* d'un programme de construction
  nl opstellen *n* van het bouwprogramma
  r планирование *n* строительных работ; составление *n* графиков производства строительных работ

P880 *e* progress payment
  d Abschlagszahlung *f*
  f décomte *m* provisoire (*des travaux*)
  nl termijnbetaling *f*
  r промежуточная оплата *f* (*заказчиком*) выполненного объёма работ

P881 *e* project
  d 1. Projekt *n* 2. Bauvorhaben *n*
  f 1. projet *m* 2. chantier *m* de construction; ensemble *m* d'un chantier
  nl 1. project *n* 2. bouwwerk *n*, werk *n*
  r 1. проект *m* 2. строительный объект *m*

P882 *e* **project appraisal**
d Bewertung *f* [Begutachtung *f*] des Projektes
f expertise *f* du projet
nl goedkeuring *f* van het ontwerp
r оценка *f* [экспертиза *f*] проекта

P883 *e* **project documentation**
d Projektunterlagen *f pl*, Entwurfsunterlagen *f pl*
f documents *m pl* [dossier *m*] d'exécution [de réalisation] d'un projet
nl bestek *n*
r проектная документация *f*

P884 *e* **project management**
d Bauleitung *f* bei hauptkontraktmässiger Bauorganisation
f gestion *f* de chantier de construction basée sur le choix d'un entrepreneur général
nl projectleiding *f*
r управление *n* строительством на основе генподряда

P885 *e* **projection**
d 1. Vorsprung *m* 2. Projektion *f*
f 1. saillie *f* 2. projection *f*
nl 1. kraging *f* 2. projectie *f*
r 1. выступ *m* 2. проекция *f*

P886 *e* **projection plastwork**
d Spritzputz *m*, Spritzmörtel *m*
f enduit *m* [mortier *m* de plâtre] projeté
nl spuitpleisterwerk *n*
r штукатурка *f*, наносимая методом распыления

P887 *e* **projection welding**
d Buckelschweißen *n*, Dellenschweißen *n*
f soudage *m* par projection
nl reliëflassen *n*
r рельефная сварка *f*

P888 *e* **project planning**
d Bauplanung *f*
f planification *f* (d'un projet), organisation *f* du chantier
nl opstellen *n* van een bouwplan
r разработка *f* проекта организации строительства

P889 *e* **project specifications**
d objektgebundene technische Bauvorschriften *f pl*
f cahier *m* des charges pour la construction (d'un ouvrage), devis *m* descriptif de travaux à exécuter
nl technische voorschriften *n pl* voor het werk
r технические задания *n pl* на строительство объекта

P890 *e* **project supervision**
d Projektüberwachung *f*
f surveillance *f* [contrôle *m*] technique des travaux
nl toezicht *n* op het werk
r технический надзор *m* за строительством

P891 *e* **project under construction**
d im Bau befindliches Objekt *n*
f ouvrage *m* en phase [en stade] de construction
nl werk *n* in uitvoering
r строящийся [сооружаемый] объект *m*

P892 *e* **proof load**
d Prüflast *f*, Versuchslast *f*
f charge *f* d'essai
nl proefbelasting *f*
r пробная [испытательная, контрольная] нагрузка *f*

P893 *e* **proof stress**
d 1. definierte Fließgrenze *f* [Dehngrenze *f*] 2. Prüf(ungs)spannung *f*
f 1. limite *f* conventionnelle d'élasticité 2. tension *f* d'essai [d'épreuve]
nl 1. elasticiteitsgrens *f(m)* 2. proefspanning *f*
r 1. условный предел *m* текучести 2. испытательное напряжение *n*

P894 *e* **proof stress at 0.2 per cent set**
d 0,2-Dehngrenze *f*
f limite *f* conventionnelle d'élasticité (correspondant à un allongement permanent de 0.2%)
nl elasticiteitsgrens *f(m)* bij 0,2% blijvende rek
r условный предел *m* текучести (при 0,2% остаточной деформации)

P895 **prop** *see* **post**

P896 *e* **propeller fan**
d Propellerlüfter *m*, Propellerventilator *m*
f ventilateur *m* à hélice
nl propellerventilator *m*
r пропеллерный [лопастной] вентилятор *m*

P897 **propeller pump** *see* **axial-flow pump**

P898 *e* **properties of materials**
d Werkstoffeigenschaften *f pl*
f propriétés *f pl* de matériaux
nl materiaaleigenschappen *f(m) pl*
r свойства *n pl* материалов

P899 *e* **proportional balancing**
d verhältnisgleicher Mengenstromausgleich *m* (*Verfahren zur Einregulierung der Luftkanalnetze*)
f équilibrage *m* proportionnel (*des conduits de ventilation*)
nl instellen *n* van de juiste verhouding der luchtstromen in een luchtbehandelingssysteem
r пропорциональная наладка *f* (*вентиляционных сетей*)

P900 *e* **proportional control**
d Proportionalregelung *f*, P-Regelung *f*
f régulation *f* proportionelle
nl proportionele regeling *f*
r пропорциональное регулирование *n*

P901 *e* **proportional elastic limit**
d Proportionalitätsgrenze *f*
f limite *f* de proportionnalité
nl evenredigheidsgrens *f(m)*
r предел *m* пропорциональности

P902 *e* **proportionality of stress to strain**
d Spannungs-Dehnungs-
-Proportionalität *f*
f proportionnalité *f* des contraintes
aux déformations, proportion *f* entre
les contraintes et (les) déformations
nl evenredigheid *f* van spanning en rek
r пропорциональность *f*
[пропорциональная зависимость *f*]
между напряжениями и
деформациями

P904 *e* **proportioning**
d Festlegen *n* des
Mischungsverhältnisses; Dosierung *f*,
Zuteilung *f*, Zumessung *f*
f dosage *m* (*d'un mélange de béton*)
nl bepalen *n* van de
betonmortelsamenstelling
r проектирование *n* состава смеси;
дозировка *f* составляющих
(*бетонной или растворной смеси*)

P905 *e* **proposal form**
d Angebotsform *f*
f modèle *m* de soumission
nl inschrijfbiljet *n*
r форма *f* заявки (на участие в торгах)

P906 *e* **propped cantilever beam** *US*
d einseitig eingespannter und einseitig
freiaufliegender Träger *m*
f poutre *f* encastrée à une extrémité
et appuyée à l'autre
nl eenzijdig ingeklemde en aan de
andere zijde vrij opgelegde ligger *m*
r балка *f* с одним защемлённым
и другим шарнирно опёртым концом

P907 *e* **prospecting**
d Erkundung *f*, Schürfung *f*,
Forschung *f*
f prospection *f*, recherche *f*, exploration *f*
nl terreinonderzoek *n*, exploratie *f*
r разведка *f*, изыскания *n pl*

P908 *e* **protection by dykes**
d Kehren *n*
f protection *f* par digues
nl bedijken *n*
r обвалование *n* низменных
участков (*для защиты от
наводнений*)

P909 *e* **protection embankment**
d Schutzdamm *m*, Schutzdeich *m*,
Hochwasserdeich *m*, Stromdeich *m*
f rempart *m*, remblai *m*, digue *f* de
protection
nl bandijk *m*, schaardijk *m*, dijk *m*
r ограждающий вал *m*

P910 *e* **protection works**
d Hafenaußenwerke *n pl*,
Meeresschutzanlagen *f pl*,
Seeschutzanlagen *f pl*

f ouvrages *m pl* extérieurs de protection
des ports
nl beschermingswerken *n pl* (*haven*)
r внешние оградительные сооружения
n pl (порта)

P911 *e* **protective coating**
d Schutzüberzug *m*, Schutzschicht *f*
f couche *f* protectrice, revêtement *m*
de protection
nl beschermende laag *f(m)*
r защитное покрытие *n*

P912 *e* **protective concrete layer**
d Schutzbetonschicht *f*,
Betonschutzschicht *f*
f couche *f* protectrice en béton
nl beschermende laag *f(m)* van beton
r защитный слой *m* бетона

P913 *e* **protective cowl**
d Wetterschutzhaube *f*, Regenhaube *f*
(*Schornstein*)
f abat-vent *m* [mitre *f*] de cheminée
nl schoorsteenkap *f(m)*
r защитный колпак *m* (*дымовой трубы*)

P914 **protective dike** *see* **dike dam**

P915 *e* **protective guard**
d Schutzgeländer *n*, Schutzbarriere *f*;
Schutzgitter *n*
f treillis *m* [grillage *m*] de protection
nl veiligheidsleuning *f*
r защитное ограждение *n*;
предохранительная решётка *f*

P916 **protective harbour structure** *see*
protection works

P917 *e* **protective helmet**
d Schutzhelm *m*
f casque *m* protecteur [de protection]
nl veiligheidshelm *m*
r защитный шлем *m*, защитная каска *f*

P918 *e* **protective screening**
d Schirm *m*, Sicherheitsnetz *n*
f filet *m* de sécurité
nl veiligheidsnet *n*
r предохранительная (металлическая)
сетка *f* (*для улавливания тяжёлых
падающих предметов*)

P919 *e* **protective shield**
d Schutzschild *m*
f écran *m* protecteur
nl veiligheidsscherm *n*
r оградительный щиток *m*,
защитный экран *m*

P920 *e* **protective site clothing**
d Arbeitsschutzbekleidung *f* für
Baustelle
f vêtement *m* de protection pour
chantier
nl beschermende werkkleding *f*
r спецодежда *f* для строителей

P921 *e* **protruding bar**
d herausstehendes Bewehrungseisen *n*,
Anschlußstab *m*
f barre *f* saillante

nl uitstekende wapeningsstaaf *f(m)*, stek *m*
r выступающий (арматурный) стержень *m*; выпуск *m* арматуры

P922 *e* **psychrometric chart**
d Psychrometerdiagramm *n*
f diagramme *m* psychrométrique
nl psychrometrisch diagram *n*
r психрометрическая диаграмма *f*

P923 *e* **public building**
d öffentliches Gebäude *n*
f édifice *m* [immeuble *m*] public
nl openbaar gebouw *n*
r общественное здание *n*

P924 *e* **public roads**
d öffentliche Verkehrsstraßen *f pl*
f réseau *m* de routes publiques
nl openbare wegen *m pl*
r система *f* общественных дорог

P925 *e* **public service building**
d Dienstleistungsgebäude *n*
f bâtiment *m* de services publics
nl gebouw *n* voor openbare diensten
r здание *n* учреждений бытового обслуживания

P926 *e* **public transport system**
d öffentliches Massenverkehrssystem *n*
f système *m* de transport public
nl openbaar-vervoersysteem *n*
r система *f* общественного транспорта

P927 *e* **public utilities**
d 1. Versorgungs- und Entsorgungsanlagen *f pl* 2. Stadtwerke *n pl*, öffentliche Versorgungsbetriebe *m pl*
f 1. utilités *f pl*, réseaux *m pl* urbains 2. entreprises *f pl* de services publics
nl 1. openbare voorziening *f* 2. openbare nutsbedrijven *n pl*
r 1. санитарно-технические и энергетические сооружения *n pl* 2. коммунальные предприятия *n pl*

P928 *e* **public waterwork system**
d Stadtwasserleitung *f*
f réseau *m* de distribution d'eau urbain, canalisation *f* d'eau urbaine
nl gemeentelijke waterleiding *f*
r городской водопровод *m*

P929 *e* **public works**
d 1. öffentliche Bauarbeiten *f pl* 2. öffentliche Bauten *pl*
f 1. travaux *m pl* publics 2. ouvrages *m pl* de travaux publics
nl 1. werk-uitvoering *f* in opdracht van de overheid 2. openbare gebouwen *n pl* en installaties *f pl*
r 1. работы *f pl* по строительству общественных зданий и сооружений 2. общественные здания *n pl* и сооружения *n pl*

P930 *e* **public works truck**
d Muldenkipper(-LKW) *m* für Baustellen

f camion-benne *m* travaux publics
nl kipwagen *m* voor bouwplaatsen
r автомобиль-самосвал *m* для дорожных или строительных работ

P931 *e* **puddle clay**
d Lehmschlag *m*, Tonschlag *m*, Puddle *m*, Lettenschlag *m*
f glaise *f*, mélange *m* argileux
nl vulklei *m*
r глинобитная смесь *f*

P932 **pug** *see* **puddle clay**

P933 *e* **pugging**
d 1. gekneteter Lehm *m*, Lehmschlag *m* 2. Auffüllung *f* (*beim Einschub von Holzbalkendecken*) 3. Ausfüllung *f* 4. Weichkneten *n* (*Lehm*)
f 1. mélange *m* argileux; glaise *f* damée; argile *f* pétrie battue 2. enduisage *m* en argile *f* (*de faux plancher*) 3. hourdis *m*, hourdage *m* 4. battage *m* de l'argile
nl 1. leemmortel *m* 2. geluiddempende vulling *f* tussen houten vloerbalken 3. pleisterlaag *f(m)* van leem 4. kneden *n* (*klei of leem*)
r 1. мятая [трамбованная] глина *f* 2. смазка *f* деревянного наката (*звуко- и теплоизолирующая*) 3. заполнение *n* (*перекрытий или каркасных перегородок*) 4. разминание *n* глины

P934 *e* **pugmill**
d 1. Knetmischer *m*, Kneter *m* 2. Zwangsmischer *m* mit vertikalem Rührwerk
f 1. broyeur *m* à meules (pour glaise) 2. malaxeur *m* d'un poste d'enrobage, malaxeur *m* d'une centrale de malaxage
nl 1. kleikneder *m* 2. speciemolen *m*
r 1. глиномялка *f* 2. мешалка *f* асфальтобетоносмесителя

P935 *e* **pull**
d Zugkraft *f*; Zug *m*; Spannung *f*
f effort *m* de traction; effort *f* de tirage; tension *f*; tirage *f*
nl trekkracht *f(m)*; trekspanning *f*
r сила *f* тяги; тяга *f*; натяжение *n*; растяжение *n*

P936 *e* **pull-in**
d Abstellplatz *m*
f zone *f* de stationnement (*aménagée sur un accotement*)
nl parkeerstrook *f(m)* langs een weg
r место *n* (для) стоянки автомобилей (*на обочине дороги*)

P937 *e* **Pullman-kitchen**
d Pullman-Küche *f* (*kleine Einbau-Küche*)
f cuisine *f* installeé
nl kleine ingebouwde keuken *f(m)*
r небольшая кухня *f*, встроенная в нишу (*в квартире*)

P938 *e* **pullout method**
  d Ausziehversuch *m*, Ausziehprüfung *f*
  f arrachement *m* d'une barre
  d'armature (*méthode d'essai*)
  nl uittrekproef *f(m)* (*wapening*)
  r вырывание *n* [выдёргивание *n*]
  стального прутка из бетона (*метод
  испытания*)

P939 *e* **pulp**
  d Pulpe *f*
  f pulpe *f*, pâte *f*
  nl pulp *f(m)*, brij *m*
  r пульпа *f*

P940 *e* **pulsating stress**
  d Schwellbeanspruchung *f*
  f contrainte *f* pulsatoire
  nl pulserende belasting *f*
  r пульсирующее [импульсное]
  напряжение *n*

P941 **pulverator** *see* **pulverizer 2.**

P942 *e* **pulverizer**
  d 1. Zerstäuber *m*, Sprühdüse *f*
  2. Prallmühle *f*
  f 1. pulvérisateur *m* (de l'eau)
  2. broyeur *m*; atomiseur *m*
  nl 1. verstuiver *m* 2. poedermolen *m*,
  vergruizer *m*
  r 1. (водо)распылитель *m* 2. ударно-
  -отражательная мельница *f*

P943 *e* **pulverizing mixer**
  d Bodenmischer *m*, Bodenmischmaschine
  f; Mehrwellen-Bodenzwangsmischer *m*
  f fraise *f* routière; pulvimixer *m*,
  tritureuse *f*
  nl stabilisatiemachine *f*
  r дорожная фреза *f*; грунтосмеситель
  m, грунтосмесительная машина *f*

P944 *e* **pulverizing nozzle**
  d Zerstäubungsdüse *f*
  f buse *f* à pulvériser
  nl verstuivingsmondstuk *n*
  r распылительное сопло *n*

P945 *e* **pumice aggregate**
  d Bims(beton)zuschlag(stoff) *m*
  f agrégat *m* [granulat *m*] de ponce
  nl bimstoeslag *m*
  r пемзовый заполнитель *m* [щебень *m*]

P946 *e* **pumice concrete**
  d Bimsbeton *m*
  f béton *m* (de) ponce
  nl bimsbeton *n*
  r пемзобетон *m*

P947 *e* **pump**
  d Pumpe *f*
  f pompe *f*
  nl pomp *f(m)*
  r насос *m*

P948 *e* **pumpability of concrete**
  d Betonpumpfähigkeit *f*,
  Betonpumpbarkeit *f*
  f pompabilité *f* de béton
  nl pompbaarheid *f* van betonmortel

  r способность *f* бетонной смеси
  к транспортированию бетононасосом

P949 *e* **pumpage**
  d 1. Pumpmenge *f* 2. Fördermenge *f*
  f 1. volume *m* de liquide pompé
  2. débit *m* de liquide pompé
  nl 1. verpompte hoeveelheid *f*
  2. opbrengst *f* van een pomp
  r 1. объём *m* откачки 2. подача *f*
  насоса

P950 *e* **pumpcrete**
  d Pump(kret)beton *m*
  f béton *m* pompé
  nl verpompt beton *n*
  r бетонная смесь *f*, подаваемая
  насосом

P951 *e* **pumpcrete pipe**
  d Betonförderrohr *n*
  f conduit *m* [tuyau *m*] à béton pompé
  nl betonpompbuis *f(m)*
  r бетоновод *m*

P952 *e* **pump curve**
  d Pumpenkennlinie *f*
  f caractéristique *f* de la pompe
  nl pompkarakteristiek *f*
  r характеристика *f* насоса (*кривая
  «давление — подача»*)

P953 *e* **pump delivery head**
  d Förderdruck *m*, Pumpendruck *m*
  f chasse *f* de la pompe
  nl opvoerhoogte *f*
  r напор *m* насоса

P954 *e* **pumpdown**
  d Abpumpen *n*
  f pompage *m* de décharge
  nl afpompen *n*
  r откачка *f*

P955 **pumped concrete** *see* **pumpcrete**

P956 *e* **pumped storage**
  d 1. Pumpenspeicherbecken *n*,
  Oberbecken *n* eines
  Pumpenspeicherwerkes
  2. Pumpenspeicherung *f*,
  hydraulische Energiespeicherung *f*
  f 1. bassin *m* accumulateur [haut]
  2. accumulation *f* par pompage
  nl 1. spaarbekken *n* 2. opgepompte
  watervoorraad *m*
  r 1. аккумулирующий бассейн *m*;
  наливное водохранилище *n*
  2. насосное аккумулирование *n*

P957 *e* **pumped-storage hydropower plant**
  d Speicher(wasser)kraftwerk *n*,
  Pumpspeicherwerk *n*
  f usine *f* de pompage, centrale *f* de
  pompage [d'accumulation]
  nl waterkrachtcentrale *f(m)* met
  spaarbekken
  r гидроаккумулирующая
  электростанция *f*

P958 *e* **pumped water line**
  d Druckwasserleitung *f*

f conduite *f* [tuyauterie *f*] d'eau sous pression
nl persleiding *f* voor water
r напорный водопровод *m*

P959 **pump house** *see* **pumping plant**

P960 *e* **pumping**
d 1. Abpumpen *n*; Umpumpen *n*
2. Betonplattenpumpen *n*
f 1. pompage *m* 2. pumping *m*
nl 1. bemaling *f* 2. betonpompen *n*
r 1. откачка *f*; перекачка *f*
2. фонтанирование *n* жидкого грунта (*через швы и трещины бетонного дорожного покрытия при наезде автомобилей*)

P961 **pumping boom** *see* **placing boom**

P962 *e* **pumping drainage**
d künstliche Entwässerung *f*
f desséchement *m* mécanique
nl kunstmatige ontwatering *f* [drainage *f*]
r машинное [механическое] осушение *n*

P963 *e* **pumping level**
d beeinflußter Grundwasserstand *m*
f nappe *f* phréatique dynamique, niveau *m* dynamique
nl grondwaterstand *m* in een waterwingebied
r динамический уровень *m* подземных вод

P964 *e* **pumping main**
d Pumpleitung *f*
f canalisation *f* de pompage
nl hoofdwaterleiding *f*, transportleiding *f*
r насосная линия *f* водопровода

P965 *e* **pumping of concrete**
d Betonpumpen *n*; Einbringen *n* von Pumpbeton
f pompage *m* de béton; bétonnage *m* à la pompe
nl betonpompen *n*
r подача *f* бетонной смеси бетононасосом; бетонирование *n* с применением бетононасоса

P966 *e* **pumping plant, pumping station**
d Pumpwerk *n*, Pumpstation *f*, Hebeanlage *f*
f station *f* de pompage
nl pompstation *n*
r насосная станция *f*

P967 *e* **pumping test**
d Pumpversuch *m*, Pumpprobe *f*
f pompage *m* expérimental, épuisement *m* d'essai, essai *m* de pompage
nl pompproef *f(m)*, proefbemaling *f*
r опытная откачка *f*

P968 *e* **pumping through pipes**
d Förderung *f* in Rohrleitungen
f pompage *m* à travers les conduits
nl waterafvoer *m* door leidingen
r подача *f* [перекачка *f*] по трубам, трубопроводный транспорт *m* (*жидкости*)

P969 **pump sump** *see* **pump well**

P970 *e* **pump unit**
d Pumpensatz *m*
f motopompe *f*; groupe *m* pompe-moteur
nl pompaggregaat *n*
r насосный агрегат *m*

P971 *e* **pump well**
d Pumpensumpf *m*
f puisard *m* d'une pompe
nl boorbron *m*, pompput *m*
r приямок *m* [зумпф *m*] насоса

P972 *e* **punching shear**
d Eindruck-Scherspannung *f*
f contrainte *f* de cisaillement au percement
nl druk-schuifspanning *f*
r напряжение *n* (среза) при продавливании

P973 *e* **punner**
d Handramme *f*
f dame *f*
nl stamper *m*
r ручная трамбовка *f*

P974 *e* **punning**
d 1. Stampfen *n* 2. Stocherverdichtung *f*, Stochern *n* (*Beton*)
f 1. damage *m* à main 2. piquage *m* (du béton)
nl 1. stampen *n* 2. porren *n* (*beton*)
r 1. трамбование *n* ручной трамбовкой 2. штыкование *n*, уплотнение *n* (*бетона*)

P975 *e* **pure bending**
d reine Biegung *f*
f flexion *f* pure [simple]
nl zuivere buiging *f*
r чистый изгиб *m*

P976 *e* **pure shear**
d reiner Schub *m*, reine Scherung *f*
f cisaillement *m* pur [simple]
nl zuivere afschuiving *f*
r чистый сдвиг *m* [срез *m*]

P977 *e* **purge**
d Abblasen *n*, Reinigung *f*, Spülung *f*
f purge *f*
nl leegblazen *n*, doorspoelen *n*
r продувка *f*

P978 *e* **purger, purge valve**
d Abblaseventil *n*, Entleerungsventil *n*, Ablaßeinrichtung *f*
f robinet *m* de purge
nl spuiklep *f(m)*, aftapklep *f(m)*
r продувочный клапан *m*

P979 **purging** *see* **purge**

P980 *e* **purification**
d Reinigung *f* (*Wasser*)
f purification *f* (d'eau)
nl zuivering *f* (*water*)
r очистка *f* (*воды*)

P981 *e* **purification plant**
d Wasseraufbereitungsanlage *f*

f installation *f* de traitement [de préparation] d'eau
nl waterzuiveringsinstallatie *f*
r станция *f* водоподготовки [водоочистки]

P982 *e* purlin
d Pfette *f*
f panne *f*
nl gording *f*
r прогон *m* (крыши); обрешётина *f*

P983 *e* pusher
d Schubfahrzeug *n*, Schubmaschine *f*, Schubraupentraktor *m*
f (tracteur-)pousseur *m*, pousseur-
-dozer *m*
nl dozer *m*
r трактор-толкач *m*

P984 *e* putlogs
d Querstangen *f pl*, Querriegel *m pl*
f taquets *m pl*
nl kortelingen *f pl* (*steigerwerk*)
r пальцы *m pl*, консоли *f pl* (*выпускных лесов*)

P985 *e* putty
d Kitt *m*, Dichtungsmasse *f*
f mastic *m*, enduit *m*, lut *m*
nl stopverf *f(m)*, vulmassa *f*
r уплотняющая масса *f*, герметик *m*

P986 putty knife *see* glazing knife

P987 *e* pycnometer test
d pyknometrische Dichtebestimmung *f*
f essai *m* pycnométrique
nl pyknometrische dichtheidsbepaling *f*
r пикнометрический метод *m* определения плотности

P988 *e* pylon tower
d Pylone *f*, Pylon *m* (*Brücke*)
f pylône *m*, tour *f* de pont
nl pyloon *m* (*pijler van hangbrug*)
r пилон *m* (*моста*)

Q

Q1 *e* quad, quadrangle
d 1. Hofviereck *n* 2. quadratische Gebäudeanordnung *f*
f 1. quadrilatère *m*, cour(ette) *f* intérieure (*entourée par bâtiments*), carré *m* 2. bâtiments *m pl* arrangés en carré
nl 1. binnenplaats *m* 2. gebouwen *n pl* opgesteld in een vierkant
r 1. внутренний двор *m или* газон *m*, окружённый зданиями 2. здания *n pl*, расположенные четырёхугольником

Q2 *e* quadrel
d 1. quadratische Fliese *f* 2. quadratische *oder* rhombische Glasscheibe *f*
f 1. carreau *m* carré 2. verre *f* en forme d'un carré *ou* d'un losange

nl 1. vierkante tegel *m* 2. vierkante *of* ruitvormige glasruit *f(m)*
r 1. квадратная плитка *f* 2. квадратное *или* ромбовидное стекло *n*

Q3 *e* quadripartite vault
d Kreuzrippengewölbe *n*
f voute *f* sur croisée d'ogives
nl kruisgewelf *n*
r крестовый свод *m* на нервюрах

Q4 *e* quake-proof structure
d erdbebensicheres Bauwerk *n*
f ouvrage *m* résistant au tremblements de terre
nl tegen seismische schokken bestendig bouwwerk *n*
r сейсмостойкое сооружение *n*

Q5 *e* qualification test
d Eignungsprüfung *f*
f essai *m* [épreuve *f*] de qualification
nl vakbekwaamheidsproef *f(m)*
r квалификационное испытание *n* (*напр. сварщика*)

Q6 *e* quality assurance
d Qualitätssicherungsmaßnahmen *f pl*
f garantie *f* de qualité d'exécution des travaux
nl kwaliteitsgarantie *f*
r меры *f pl* по обеспечению установленного уровня качества строительных работ

Q7 *e* quality controlled concrete
d gütegesicherter Beton *m*
f béton *m* à qualité contrôlée
nl beton *n* van gegarandeerde kwaliteit
r бетон *m*, контролируемый по составу и качеству (*марке*)

Q8 *e* quality specifications
d Gütevorschriften *f pl*
f spécifications *f pl* techniques pour la qualité, normes *f pl* qualitatives
nl kwaliteitseisen *m pl*
r требования *n pl* к качеству

Q9 *e* quality standards for water
d Wassergütenormen *f pl*
f normes *f pl* relatives à la qualité de l'eau
nl normen *f(m) pl* voor de waterkwaliteit
r стандарты *m pl* на качество воды

Q10 *e* quantity surveying
d 1. Aufstellen *n* der Leistungsverzeichnisse, Massenberechnung *f* 2. Arbeitsaufmaß *n*, Ausmaß *n*
f 1. établissement *m* de métrés [de l'état quantatif] 2. métré *m*   Л
nl 1. opnemen *n* van het verrichte werk 2. opmeten *van* de in het werk gebrachte hoeveelheden
r 1. составление *n* ведомостей и подсчёт *m* объёмов строительных работ (*для смет*) 2. обмер *m* выполненных строительных работ

Q11 *e* **quantity surveyor**
d Aufmaßnehmer *m*, Massenberechner *m*, Kalkulator *m*
f métreur *m*
nl functionaris *m* die de verwerkte hoeveelheden vaststelt
r инженер-сметчик *m*

Q12 **quarrel** *see* quadrel 2.

Q13 *e* **quarry**
d 1. Steinbruch *m* 2. offener Einschnitt *m*
f 1. carrière *f* 2. ouvrage *m* à ciel ouvert
nl 1. steengroeve *f(m)* 2. ontginning *f* in dagbouw
r 1. карьер *m* 2. открытая горная выработка *f*

Q14 *e* **quarry bench**
d Strosse *f*
f gradin *m* d'une carrière
nl vloer *m* [trap *m*] van een steengroeve
r уступ *m* карьера

Q15 *e* **quarry face**
d Abbauwand *f*, Baggerstoß *m*
f front *m* de taille
nl inbreekwand *m*
r забой *m* карьера, экскаваторный забой *m*

Q16 *e* **quarry fines**
d feiner Grubensteinsplitt *m*
f grenaille *f* (de pierre), fines *m* *pl*, fraction *f* fine
nl split *n*
r карьерная [каменная] мелочь *f*

Q17 *e* **quarrying**
d Abbau *m*, Abbauen *n*; Tagebau *m*
f abattage *m* (des pierres) en carrière
nl ontginnen *n* van een steengroeve
r разработка *f* карьера, добыча *f* камня в карьере

Q18 *e* **quarrying old pavements**
d Gewinnung *f* von Zuschlagstoffen aus alten Betonstraßendecken
f désagrégation *f* de vieux revêtements routiers en béton
nl opgebroken betonwegdek tot betontoeslag verwerken
r разработка *f* [дробление *n*] старых бетонных дорожных покрытий (*метод получения заполнителей*)

Q19 *e* **quarry run, quarry stone**
d Bruchstein *m*
f moellon *m*, pierre *f* de carrière
nl breuksteen *m*
r карьерный [рваный] камень *m*, бут *m*

Q20 *e* **quarrystone bond**
d unregelmäßiges Mauerwerk *n*
f appareil *m* de moellons bruts
nl metselwerk *n* van natuursteen in onregelmatig verband
r нерегулярная перевязка *f* каменной кладки, бутовая кладка *f*

Q21 *e* **quarry tile**
d 1. Bodenplatte *f*, Natursteinplatte *f*, Fußbodenplatte *f* 2. Gehwegplatte *f*
f 1. carreau *m* de carrière [de pavage] 2. dalle *f* de trottoir en pierre
nl 1. natuursteen plavuis *m* 2. natuurstenen trottoirtegel *m*
r 1. каменная плитка *f* для пола 2. каменная плита *f* тротуара

Q22 *e* **quarry waste**
d Abfallsteine *m* *pl*, Steinbruchabfall *m*
f rebuts *m* *pl* de carrière [de ballastière]
nl afvalsteen *n*
r карьерные отходы *m* *pl*

Q23 *e* **quarter**
d 1. senkrechtes Rahmenholz *n* einer Fachwerktrennwand 2. quadratische Platte *f* [Tafel *f*]
f 1. montant *m* [poteau *m*] en bois d'une cloison en charpente 2. panneau *m* carré
nl 1. stijl *m* in een houten wand 2. vierkant paneel *n*
r 1. деревянная стойка *f* каркасной перегородки 2. квадратная панель *f*

Q24 *e* **quarter closer, quarter closure**
d Quartier *m*, Quartierstück *n*, Viertelstück *n*, Viertelziegel *m*, Viertelstein *m*, Riemchen *n*
f quart *m* (*forme de brique*), mulet *m*, mulot *m*
nl klezoor *m*
r четверть *f* кирпича, четвертной кирпич *m* (*в кладке*)

Q25 *e* **quarter red partition**
d Trennwand *f* aus quadratischen Platten
f cloison *f* de [en] panneaux carrés
nl scheidingswand *m* bestaand uit vierkantige platen
r перегородка *f* из квадратных панелей [плит]

Q26 *e* **quartering**
d 1. Vierteilung *f* 2. Probenahmeverfahren *n* mit Vierteilung der Probe 3. senkrechtes Rahmenholz *n* der Fachwerk(trenn)wand
f 1. division *f* [partage *m*] en quatre parties 2. quartation *f* (*méthode de préparation des spécimens par le partage successif d'une portion en quatre parties*) 3. montant *m* de cloison
nl 1. in vieren delen 2. vierendeling *f* (*methode van monstername*)
r 1. деление *n* на четыре части, квартование *n* 2. методика *f* отбора проб (*напр. бетонной смеси*) делением взятой порции на четыре части с последующим перемешиванием двух из них и т. д. 3. стойки *f* *pl* каркасной перегородки [стены]

Q27 *e* **quarter landing** *UK*, **quarterpace,**
    quarterpace landing *US*
  d Viertelpodest *n*
  f quart-palier *m* (*palier m intermédiaire*
    entre deux volées perpendiculaires)
  nl trapbordes *n* waarin de stijglijn 90°
    van richting verandert
  r промежуточная лестничная
    площадка *f* (*между*
    перпендикулярными маршами)

Q28 *e* **quarter pegs**
  d Viertelpunkt-Pflöcke *m pl* am
    Straßenquerprofil
  f piquets *m pl* [fiches *f pl* en bois]
    du profil en travers (*d'une route*)
  nl vierkante piketten *n pl* waarmee het
    dwarsprofiel van een weg wordt
    aangegeven
  r пикеты *m pl* (*колышки*) на
    поперечном профиле дороги (*на*
    четверти ширины между бровкой
    и гребнем)

Q29 **quarterspace landing** *see* **quarter
    landing**

Q30 *e* **quarter timber**
  d Viertelholz *n*
  f bois *m* de refend, quart *m* de rond
  nl vierkant bezaagd hout *n*
  r четвертина *f* (*брус*)

Q31 *e* **quartz aggregate**
  d Quarzzuschlag(stoff) *m*
  f granulat *m* [agrégat *m*] de quartz
  nl kwartstoeslag *m*
  r кварцевый заполнитель *m*

Q32 *e* **quartz glass**
  d Quarzglas *n*
  f verre *m* de quartz
  nl kwartsglas *n*
  r кварцевое [силикатное] стекло *n*

Q33 *e* **quartz powder**
  d Quarzmehl *n*
  f farine *f* de quartz
  nl kwartspoeder *n*
  r кварцевая пыль *f* (*наполнитель*)

Q34 **quatation** *see* **tender**

Q35 *e* **quay**
  d Kai *m*
  f quai *m*
  nl kade *f(m)*
  r набережная *f*; причал *m*; пристань *f*

Q36 *e* **quay pier**
  d Kaipfeiler *m*
  f pile *f* [support *m*] de quai
  nl kadepijler *m*
  r контрфорс *m* набережной стенки

Q37 *e* **quay stair**
  d Kaitreppe *f*
  f escalier *m* de quai
  nl kadetrap *m*
  r лестничный сход *m* набережной

Q38 *e* **quay wall**
  d Kaimauer *f*, Kaje *f*, Uferwand *f*
  f mur *m* de quai

  nl kademuur *m*
  r набережная [причальная] стенка *f*

Q39 **queen bolt** *see* **queen rod**

Q40 *e* **queen closer**
  d englischer Halbziegel *m*
  f demi-brique *f* anglaise
  nl klisklezoor *m*
  r продольная половинка *f* кирпича

Q41 *e* **queen post**
  d 1. Sprengwerkspfosten *m*
    2. Hängesäule *f*
  f 1. poinçon *m* de brancard 2. faux
    poinçon (*de comble*)
  nl extra hanger *m* bij een hangspant met
    een overspanning van meer dan
    10 m
  r 1. стойка *f* шпренгеля 2. подвеска *f*
    стропил

Q42 *e* **queen-post truss**
  d doppelter Hängebock *m*, doppeltes
    Hängewerk *n*
  f ferme *f* de comble [de toit] à deux
    poinçons
  nl hangspant *n* met twee hangers
  r деревянная стропильная ферма *f*
    с двумя вертикальными подвесками
    [стойками]; висячая ферма *f*
    с двумя подвесками

Q43 *e* **queen rod**
  d Hängestab *m*
  f aiguille *f* pendante (*d'une ferme de*
    comble)
  nl hangstaaf *f(m)*
  r металлическая подвеска *f*
    (*стропильной фермы*)

Q44 **query-random blockwork** *US see*
    random blocks

Q45 *e* **quetta bond**
  d bewehrtes Schachtmauerwerk *n*
  f maçonnerie *f* (de brique) creuse armée
  nl gewapend metselwerk *n*, werk van
    holle baksteen
  r армированная колодезная кирпичная
    кладка *f*

Q46 *e* **quick-acting clamp**
  d Moment(schraub)zwinge *f*,
    Schnellschraubknecht *m*
  f serre-joint *m* rapide
  nl snelwerkende klemschroef *f(m)*
  r быстродействующий винтовой зажим
    m

Q47 *e* **quick-acting coupling**
  d Schnellverbindung *f*,
    Rohrschnellkupplung *f*
  f raccord *m* instantané [rapide]
  nl snap-mofverbinding *f*
  r быстродействующее соединение *n*
    (*труб*)

Q48 *e* **quick-acting gate**
  d Schnellverschluß *m*
  f vanne *f* à fermeture rapide
  nl snelwerkende afsluiter *m*
  r быстродействующий затвор *m*

Q49 e quick-action water heater
d Warmwasserschnellerhitzer *m*
f chauffe-eau *m* instantané
nl warmwater-doorstroomtoestel *n*
r быстродействующий водонагреватель
m

Q50 e quick clay
d Quickton *m*, Fließton *m*
f argile *f* fluide [mouvante]
nl vloeibare klei *f(m)*
r плывунная глина *f*

Q51 e quick-closing valve
d Schnellschlußventil *n*,
Schnellschlußschieber *m*
f valve *f* à fermeture rapide
nl snelwerkende klep *f(m)*
r быстрозакрывающийся клапан *m*,
быстрозакрывающаяся задвижка *f*

Q52 e quick-drying paints
d schnelltrocknende Farbstoffe *m pl*
f peinture *f* siccative [à séchage rapide]
nl sneldrogende verven *f(m) pl*
r быстросохнущие красочные составы
m pl

Q53 quick-hardening lime *see* hydraulic
lime

Q54 e quicklime
d Branntkalk *m*, gebrannter Kalk *m*,
Brennkalk *m*
f chaux *f* vive
nl ongebluste kalk *m*
r негашёная известь *f*; известь-
-кипелка *f*

Q55 quick-release coupling *see* quick-acting
coupling

Q56 e quicksand
d Fließsand *m*, Schwimmsand *m*,
Triebsand *m*, Quicksand *m*
f sables *m pl* boulants [mouvants]
nl drijfzand *n*, loopzand *n*
r песок-плывун *m*

Q57 quick set *see* flash set

Q58 e quick setting admixture
d Schnellabbindemittel *n*,
Abbindebeschleuniger *m*
f accélérateur *m* [agent *m* accélérateur]
de prise
nl verhardingsversneller *m*
r добавка-ускоритель *f* схватывания

Q59 e quick-setting cement
d schnellabbindender Zement *m*
f ciment *m* à prise rapide [prompte]
nl snelwerkend cement *n*
r быстросхватывающийся цемент *m*

Q60 e quick shear test
d Schnellscherversuch *m*
f essai *m* de cisaillement rapide
nl snelschuifproef *f(m)*
r испытание *n* на быстрый сдвиг

Q61 e quick-slacking lime
d schnellöschender Kalk *m*
f chaux *f* à extinction rapide

nl snelblussende kalk *m*
r быстрогасящаяся известь *f*

Q62 e quick test
d Kurzzeitversuch *m*, Schnellversuch *m*
f essai *m* rapide
nl snelle proef *f(m)*
r ускоренное испытание *n*

Q63 e quiescent load
d ruhende Last *f*, Eigenlast *f*
f charge *f* fixe [statique]
nl rustende belasting *f*
r неподвижная нагрузка *f*

Q64 e quiet piling rig
d geräuscharme Pfahlramme *f*
f sonnette *f* «calme» (*avec niveau de
bruit bas*)
nl geluidarme heistelling *f*
r «спокойный» свайный копёр *m*
(*обеспечивающий низкий уровень
шума*)

Q65 e quilt, quilt insulation
d Steppmatte *f*
f isolant *m* en laize, laize *f* d'isolant
matelassé
nl doorgestikte mat *m* (*warmte-isolatie*)
r прошивной теплоизоляционный мат
m

Q66 e quoin
d 1. Eckstein *m*, Eckziegel *m*
2. Gebäudeecke *f*; Maueraußenecke *f*
f 1. pierre *f* d'angle 2. angle *m* d'un
bâtiment; angle *m* d'un mur
nl 1. hoeksteen *m* 2. buitenhoek *m* van
een gebouw
r 1. кордонный угловой камень *m*,
угловой кирпич *m* (*для усиления
углов кладки*) 2. угол *m* здания *или*
стены

Q67 e quoin bonding
d Eckverband *m*
f appareil *m* d'angle en besace
nl hoekverband *m*
r каменная кладка *f* углов здания
с чередованием ложковых и
тычковых камней, угловая перевязка
f

Q68 e quoin brick
d Eckziegel *m*
f brique *f* d'angle
nl hoeksteen *m*
r угловой кирпич *m* (*в кладке*)

Q69 e quoin header
d Eck-Binderstein *m*
f boutisse *f* d'angle
nl hoeksteen *m* met de kop in voorgevel
(en de strek in de zijmuur)
r угловой тычковый камень *m*
(*в кладке*)

Q71 e Quonset hut
d Quonset-Baracke *f* (*zerlegbares
Gebäude aus bogenförmigen
Wellblechsektionen*)
f hutte *f* Quonset

nl Quonset-hut *f*
r сборное здание *n* из волнистых
стальных листов (*в форме свода*)

Q72 *e* **quotation**
d Preisangebot *n*
f prix *m*, cotation *f* (*de matériaux,
de travaux*)
nl prijsopgave *f*, inschrijving *f*
r цена *f*, предложенная подрядчиком
или поставщиком

R

R1 *e* **race**
d 1. Stromschnelle *f*, Flußschnelle *f*
2. Schußrinne *f*, Schußgerinne *n*
3. Werkgraben *m*, Werkkanal *m*
f 1. rapide *m* 2. rapide *m*,
déchargeur *m* 3. canal *m* d'amenée;
canal *m* de fuite
nl 1. stroomversnelling *f* 2. stroomgoot
f(m) 3. werkleiding *f* (*krachtcentrale*)
r 1. стремнина *f*, быстроток *m*
2. быстроток *m* (*сооружение*)
3. канал *m* ГЭС (*подводящий или
отводящий*)

R2 *e* **raceway**
d Kabelkanal *m*, Leitungskanal *m*
f gaine *f* [conduit *m*] à câbles
nl leidingkoker *m*
r канал *m* для электрических
проводов *или* кабелей (*внутри
зданий*)

R3 *e* **rack**
d 1. Stabrechen *m* 2. Zahnstange *f*
3. Gestell *n*
f 1. grille *f* de protection; grille *f*
à barreaux 2. crémaillère *f* 3. casier
m, rayonnage *m*
nl 1. staafrooster *n* 2. tandheugel
f(m) 3. stelling *f*, rek *n*
r 1. сороудерживающая решётка *f*;
колосниковая решётка *f* 2.
зубчатая рейка *f* 3. стеллаж *m*

R4 *e* **rack-and-pinion jack, rack
building hoist**
d Zahnstangenaufzug *m* (*für Baustellen*)
f cric *m* à crémaillère
nl kelderwind(e) *f(m)*
r реечный домкрат *m*; строительный
подъёмник *m* с реечным приводом

R5 *e* **racking course**
d Verfüllschicht *f*
f couche *f* d'égalisation en granulat
enrobé
nl spreilaag *f(m)* van gebitumineerd
steenslag
r гравийный *или* щебёночный
выравнивающий слой *m*,
пропитанный чёрным вяжущим

R6 *e* **radial brick**
d radialer Querwölber *m*

f brique *f* radiale
nl radiaalsteen *m*, putsteen *m*
r клинчатый кирпич *m* (*для арок*)

R7 *e* **radial cone bottom**
d radial-konischer Behälterboden *m*
f fond *m* radial conique
nl kegelvormige bodem *m*
r радиально-коническое днище *n*
(*резервуара*)

R8 *e* **radial blade fan**
d Radiallüfter *m* mit geraden Schaufeln
f ventilateur *m* centrifuge à palettes
radiales
nl ventilator *m* met rechte schoepen
r радиальный вентилятор *m*
с плоскими радиальными лопатками

R9 *e* **radial fan**
d Radialventilator *m*, Radiallüfter *m*
f ventilateur *m* radial
nl radiaalventilator *m*
r радиальный вентилятор *m*

R10 *e* **radial flow settlement tank**
d radialdurchflossenes Absetzbecken *n*
f décanteur *m* radial
nl radiaal bezinkbassin *n*
r радиальный отстойник *m*

R11 *e* **radial gate**
d Segmentschütz *n*, Segmentverschluß
m, Segmentwehr *n*
f vanne *f* (à) segment
nl segmentschuif *f(m)*
r сегментный затвор *m*

R12 *e* **radial road**
d Radialstraße *f*
f route *f* radiale
nl radiale straatweg *m*
r радиальная дорога *f*

R13 *e* **radial sett paving**
d Kleinpflaster *n* in Kreisbogenform
f pavage *m* mosaïque
nl bestrating *f* in waaierverband
r мощение *n* дороги дугами
(*круговыми рядами*)

R14 *e* **radial stresses**
d Radialspannungen *f pl*
f contraintes *f pl* radiales
nl radiaalspanningen *f pl*
r радиальные напряжения *n pl*

R15 *e* **radial travelling cableway**
d Schwenkkabelkran *m*, radial
verfahrbarer Kabelkran *m*
f blondin *m* à déplacement radial,
blondin *m* du type à point fixe et
piste circulaire
nl zwenkkabelkraan *m*
r радиальный кабель-кран *m* (*с одной
неподвижной и одной
перемещающейся башней*)

R16 *e* **radial well**
r Horizontalfilterbrunnen *m*,
Schrägfilterbrunnen *m*
f puits *m* à drains rayonnants
nl radiale bron *m*

r лучевой водозабор *m*; центральный
колодец *m* лучевого водозабора

R17 *e* **radiant cooling**
d Strahlungskühlung *f*
f refroidissement *m* à radiation
nl afkoeling *f* door straling
r радиационное охлаждение *n*

R18 *e* **radiant heat**
d Strahlungswärme *f*
f chaleur *f* rayonnante
nl stralingswarmte *f*
r лучистая теплота *f*

R19 *e* **radiant heater**
d Strahlungsheizgerät *n*
f dispositif *m* de chauffage rayonnant
[radiant], radiateur *m*
nl radiator *m*
r прибор *m* радиационного отопления,
радиатор *m*

R20 *e* **radiant heat exchange**
d Strahlungswärmeaustausch *m*
f échange *m* de chaleur rayonnante
[par rayonnement]
nl warmte-uitwisseling *f* door straling
r лучистый теплообмен *m*

R21 *e* **radiant heating**
d Strahlungsheizung *f*
f chauffage *m* rayonnant [par
rayonnement]
nl verwarming *f* door middel van straling
r радиационное отопление *n*

R22 *e* **radiant heat transfer**
d Wärmeübertragung *f* durch Strahlung
f transport *m* de chaleur à rayonnement
nl warmteoverdracht *f* door straling
r теплопередача *f* излучением

R23 *e* **radiant panel**
d Strahl(ungsheiz)platte *f*;
Strahlungskühlplatte *f*
f panneau *m* rayonnant
nl radiator *m*; koelplaat *f(m)*
r отопительная [теплоизлучающая]
панель *f*; панель *f* лучистого
нагрева *или* охлаждения

R24 *e* **radiant strip**
d Bandstrahler *m*
f bande *f* de chauffe rayonnante,
dispositif *m* de chauffage rayonnant
à bande
nl verwarmingsband *f*
r ленточный излучатель *m*

R25 *e* **radiating bridge**
d Fächerbrücke *f*
f pont *m* en éventail
nl waaierbrug *f(m)*
r веерообразный мост *m*

R26 *e* **radiation shield concrete**
d Abschirmbeton *m*, Strahlenschutzbeton
m
f béton *m* de protection (*contre le
rayonnement radioactif*)
nl beton *n* voor bescherming tegen
radioactieve straling

r экранирующий бетон *m* (*для
защиты от облучения*)

R27 *e* **radiation shield wall**
d Abschirmmauer *f*, Abschirmwand *f*
f mur-écran *m* protecteur (*pour la
protection contre les radiations
radioactives*)
nl schermmuur *m* (voor bescherming)
tegen radioactieve straling
r экранирующая стена *f* (*для
защиты от облучения*)

R28 *e* **radiator**
d Heizkörper *m*, Radiator *m*
f corps *m* de chauffe, radiateur *m*
nl radiator *m*
r радиатор *m*

R29 *e* **radiator grouping**
d Gruppieren *n* der Heizkörperglieder
f assemblage *m* [groupement *m*] des
radiateurs
nl plaatsing *f* van radiatoren
r группировка *f* радиаторов

R30 *e* **radiator valve**
d Heizkörperventil *n*
f soupape *f* de corps de chauffe, soupape
f de radiateur
nl radiatorkraan *m*
r радиаторный клапан *m*

R31 *e* **radiator vent cock**
d Heizkörperentlüftungshahn *m*,
Heizkörperluftauslaßventil *n*
f robinet *m* de désaération du radiateur
nl ontluchtingsventiel *n* van radiator
r воздуховыпускной кран *m*
радиатора

R32 *e* **radius of curvature**
d Krümmungshalbmesser *m*
f rayon *m* de courbure
nl kromtestraal *m*
r радиус *m* кривизны

R33 *e* **radius of gyration**
d Trägheitsradius *m*
f rayon *m* de giration
nl traagheidsmoment *n*
r радиус *m* инерции (*сечения*)

R34 *e* **radius of influence**
d Radius *m* der Absenkung
f rayon *m* d'appel
nl straal *m* van het beïnvloede gebied
r радиус *m* влияния (*колодца,
скважины*)

R35 *e* **raft**
d 1. Rost *m*, Rostwerk *n* 2. Grundplatte
f, Fundamentplatte *f*
f 1. radier *m* 2. semelle *f*
nl 1. rooster(werk) *n* 2. funderingsplaat
f(m)
r 1. ростверк *m* 2. опорная плита *f*

R36 *e* **rafter**
d Sparren *m*
f chevron *m*, arbalétrier *m*
nl (dak)spar *m*
r стропильная нога *f*, стропило *n*

R37 *e* raft foundation
d Plattengründung *f*
f fondation *f* sur radier, radier *m* de fondation
nl plaatfundering *f*
r плитный фундамент *m*

R38 *e* rafting canal
d Floßgasse *f*, Floßkanal *m*
f canal *m* de flottage
nl vlotgoot *f(m)*, vlotkanaal *n*
r плотоход *m*, лесосплавный канал *m*

R39 *e* rag-bolt
d Verankerungsbolzen *m*, Steinschraube *f*, Klauenschraube *f*
f boulon *m* d'ancrage [de scellement]
nl dookbout *m*
r заершённый анкерный болт *m*, болт *m* с расширенным концом

R40 *e* rail
d 1. Schiene *f* 2. Geländer *n* 3. Grundbalken *m*; Fußbodenlagerholz *n*; Fußschwelle *f*, Sohlschwelle *f* 4. Fenstersturz *m*; Türsturz *m*
f 1. rail *m* 2. parapet *m*; garde-fou *m*, garde-corps *m*; main *f* courante 3. traverse *f*; entretoise *f*; longeron *m* 4. linteau *m*, plate-bande *f* de baie
nl 1. spoorstaaf *m* 2. leuning *f* 3. regel *m*, draagbalk *m* 4. latei *f(m)*
r 1. рельс *m* 2. перильное ограждение *n*; поручень *m* 3. лежень *m*; лага *f*; опорный брус *m* 4. оконная *или* дверная перемычка *f*

R41 *e* rail clamp
d 1. Gleisklemme *f* 2. Schienenzange *f*
f 1. serre-frein *m*, garde-frein *m* (*pour grue sur rails*) 2. pince *f* à rails
nl 1. railklem *f(m)* (*bouwkraan*) 2. railtang *f(m)*
r 1. противоугон *m*, рельсовый зажим *m* (*для крепления крана к рельсам*) 2. клещи *pl* для переноски рельсов

R42 *e* railings
d Geländer *n*
f garde-corps *m*, balustrade *f*
nl leuning *f*
r перила *pl*; перильные ограждения *n pl*

R43 *e* rail ladder
d Hängeleiter *f*, angehängte Leiter *f*
f échelle *f* extérieure suspendue
nl hangsteiger *m*
r навесная наружная лестница *f*

R44 *e* rail steel reinforcement
d steife Stahlschienen-Bewehrung *f*
f armature *f* rigide en rails
nl wapening *f* van spoorstaven
r жёсткая арматура *f* из стальных рельсов

R45 *e* rail track
d Gleis *n*
f voie *f* ferrée
nl spoorbaan *f(m)*
r рельсовый путь *m*

R46 *e* railway bridge
d Eisenbahnbrücke *f*
f pont-rail *m*
nl spoorbrug *f(m)*
r железнодорожный мост *m*

R47 *e* railway bridge loads
d Eisenbahnbrückenbelastung *f*
f charge *f* roulante sur pont-rail
nl belasting *f* op een spoorbrug
r нагрузка *f* на железнодорожный мост

R48 *e* railway gauge
d Spurweite *f*
f écartement *m* de voie, entre-rail *m*, largeur *f* de voie
nl spoorbreedte *f*
r ширина *f* железнодорожной колеи

R49 *e* railway pile drivers
d schienengebundene Ramme *f*
f sonette *f* automotrice sur rails
nl spoorwegheimachine *f*
r железнодорожный свайный копёр *m*

R50 *e* rainer
d Regner *m*
f asperseur *m*, arroseur *m*, installation *f* d'aspersion
nl beregeningsinstallatie *f*
r дождевальная установка *f*

R51 *e* rainfall
d Regenfall *m*; Regenguß *m*
f précipitations *f pl*; averse *f* atmosphérique
nl regenval *m*
r атмосферные осадки *pl*; ливень *m*

R52 *e* rainfall discharge
d Regenwasserspende *f*
f ruisellement *m* des eaux de pluie [pluviales]
nl regenwaterafvoer *m*
r ливневый сток *m*

R53 rainfall runoff *see* storm runoff
R54 rain-gun *see* rainer

R55 *e* rainwater
d Regenwasser *n*
f eau *f* de pluie, eaux *f pl* pluviales
nl regenwater *n*
r дождевые [ливневые] сточные воды *f pl*

R56 *e* rainwater head
d Regenwasserfang *m*, Rinnenkasten *m*
f cuvette *f* de chéneau
nl vergaarbak *m*, kiezelbak *m*
r водосточная воронка *f*

R57 *e* rainwater inlet
d Regeneinlauf *m*, Regenablauf *m*, Rinneneinlaß *m*, Gully *n*
f puits *m* d'entrée, puisard *m*
nl straatkolk *m*, regenput *m*
r дождеприёмник *m*

R58 *e* rainwater installation
d Regenwasser-Ableitung *f*

f canalisations *f pl* d'eaux pluviales
nl regenwaterafvoer *m*
r водостоки *m pl* здания

R59 *e* **rainwater pipe**
d Regenfallrohr *n*
f tuyau *m* de descente (pluviale)
nl hemelwaterafvoerpijp *f(m)*
r водосточный стояк *m*

R60 *e* **rainwater shoe**
d Dachabfallrohrauslaß *m*
f pied *m* de descente pluviale
nl voet *m* van de regenpijp
r выпуск *m* водосточной трубы

R61 *e* **raised flooring**
d doppelter Fußboden *m*, Doppeldecke *f*
f plancher *m* double
nl dubbele vloer *m*, dubbeling *f*
r фальшпол *m*

R62 *e* **raising device**
d Hebegerät *n*, Hebezeug *n*, Hubgerät *n*, Hubzeug *n*
f dispositif *m* [engin *m*, appareil *m*] de levage
nl hefwerktuig *n*
r подъёмное устройство *n* [приспособление *n*], подъёмный механизм *m*

R63 *e* **raising of a dam**
d Staudammerhöhung *f*
f surélévation *f* du barrage
nl ophogen *n* van een stuwdam
r наращивание *n* плотины

R64 *e* **rake**
d 1. Neigung *f* 2. Kratze *f*, Mörtelhaue *f*
f 1. inclinaison *f*, fruit *m* 2. racleur *m*, raclette *f*, rateau *n*
nl 1. helling *f* 2. krabber *m*; kalkhouw *m*; hark *f*
r 1. наклон *m* к вертикали (*напр.* *сваи*) 2. скребок *m*

R65 *e* **rake classifier**
d Schrapperklassifikator *m*
f class(ificat)eur *m* à raclettes
nl krabberclassificator *m*
r скребковый классификатор *m*

R66 *e* **raked joint**
d ausgekratzte Fuge *f*
f joint *m* ouvert
nl uitgekrabde voeg *f(m)*
r незаполненный шов *m*; пустошовка *f*

R67 *e* **raking bond**
d Stromverband *m*, Fischgrätenverband *m*, Kornährenverband *m*
f appareil *m* à assises de boutisses obliques
nl diagonaalverband *n* (*metselwerk*)
r перевязка *f* (*кладки*) с косым расположением тычковых рядов

R68 **raking pile** *see* **battered pile**

R69 *e* **rakings**
d Rechengut *n*
f matières *f pl* détenues

nl op het rooster achtergebleven grove bestanddelen *n pl*
r отбросы *m pl* с решётки

R70 **ram filter** *see* **drive point**

R71 *e* **ram lift**
d hochgedrückter Lift *m*
f ascenseur *m* à piston
nl hydraulische lift *m*
r выжимной лифт *m*

R72 *e* **rammed-earth construction**
d Lehmstampfbau *m*
f bâtiment *m* en terre compactée
nl opbouw *m* met verdichte grond
r сооружение *n* из уплотнённого связного грунта

R73 *e* **ramming**
d Abrammen *n*, Feststampfen *n*
f damage *m*, pilonnage *m*
nl stampen *n*
r трамбование *n*, уплотнение *n*

R74 *e* **ramp**
d 1. Auffahrtrampe *f* 2. Schußrinne *f*
f 1. rampe *f* 2. coursier *m*, rapide *m*
nl 1. hellende oprit *m* 2. schietgoot *f(m)*
r 1. аппарель *f*, пандус *m*, въезд *m*, съезд *m* (*с дороги*) 2. быстроток *m*

R75 *e* **random blocks**
d halb in Verband verlegte Blöcke *m pl* (*Hafenbau*)
f maçonnerie *f* de blocs massifs en rangeés irrégulières
nl in onregelmatig verband gemetselde natuursteen
r кладка *f* из массивов с частично упорядоченной перевязкой

R77 *e* **random sampling**
d Stichprobenentnahme *f*
f échantillonnage *m* au hasard
nl monsterneming *f* door steekproeven
r случайный [произвольный] отбор *m* образцов [проб]

R78 *e* **random test**
d Stichprobe *f*, Stichprüfung *f*
f essai *m* au hasard
nl steekproef *f(m)*
r выборочное испытание *n*

R79 *e* **range**
d 1. Flucht *f* 2. Reichweite *f*, Aktionsradius *m* 3. Wertbereich *m*
f 1. alignement *m*, site *m* 2. portée *f*, étendue *f* 3. gamme *f*; bande *f*; amplitude *f*
nl 1. rooilijn *f(m)* 2. reikwijdte *f* 3. amplitude *f*, bereik *m*
r 1. створ *m* 2. радиус *m* действия 3. амплитуда *f*, диапазон *m*

R80 *e* **range finder**
d Entfernungsmeßgerät *n*
f télémètre *m*
nl afstandmeter *m*
r дальномер *m*

R81 **range masonry** *see* **regular-coursed rubble**

R82 *e* **range of rolling temperature**
 d Abwalztemperaturbereich *m*
 f limites *f pl* des températures de
  cylindrage (*du béton bitumineux*)
 nl grenswaarde *f* van de
  verwerkingstemperatuur van het
  asfalt
 r предельные значения *n pl* температур
  укатки асфальтобетонных
  дорожных покрытий

R83 *e* **range of tide**
 d Flutgröße *f*, Flutintervall *n*,
  Tidehub *m*
 f amplitude *f* de la marée, marnage *m*
  de marée
 nl verschil *n* tussen eb en vloed
 r амплитуда *f* прилива

R84 **ranger** *see* **wale**

R85 **rapid** *see* **inclined drop**

R86 *e* **rapid-curing asphalt, rapid-curing**
  cutback
 d schnellabbindendes
  Verschnittbitumen *n*
 f bitume *m* fluidifié à durcissement
  rapide [à séchage rapide]
 nl sneldrogend vloeibaar bitumen *n*
 r быстровысыхающий разжиженный
  дорожный битум *m*

R87 *e* **rapid design method**
 d Schnellbemessung *f*,
  Schnelldimensionierung *f*
 f méthode *f* simplifiée de calcul
 nl vereenvoudigde berekening *f*
  [rekenwijze *f*]
 r упрощённый расчёт *m*

R88 *e* **rapid erection**
 d Schnellmontage *f*
 f montage *m* rapide
 nl snelle montage *f*
 r скоростной монтаж *m*

R89 *e* **rapid filter**
 d Schnellfilter *n*, *m*
 f filtre *m* rapide
 nl snelfilter *n*, *m*
 r скорый фильтр *m*

R90 *e* **rapid-hardening cement**
 d schnell(er)härtender [frühhochfester]
  Zement *m*, Schnellerhärter *m*
 f ciment *m* à durcissement rapide
 nl snelwerkend cement *n*, *m*
 r быстротвердеющий цемент *m*

R91 *e* **rapid-setting cement**
 d Schnellbinder *m*, schnellabbindender
  Zement *m*
 f ciment *m* à prise rapide
 nl snelwerkend cement *n*, *m*
 r быстросхватывающийся цементный
  раствор *m*

R92 *e* **rate action control**
 d Vorhalteregelúng *f*
 f réglage *m* à action dérivée, régulation
  f par dérivation

 nl aangepaste regeling *f*
 r регулирование *n* по производной

R93 *e* **rated capacity**
 d 1. Nenntragfähigkeit *f*
  2. Nennfassungsvermögen *n*
  3. Nennleistung *f*
 f 1. capacité *f* portante nominale
  2. capacité *f* nominale (*volume*)
  3. capacité *f* de rendement nominale
 nl 1. nominaal hefvermogen *n*
  2. nominale capaciteit *f* 3. nominaal
  vermogen *n*
 r 1. номинальная грузоподъёмность *f*
  2. номинальная вместимость *f*
  3. номинальная производительность
  f; номинальная мощность *f*

R94 *e* **rate of air circulation**
 d Luftumlaufzahl *f*, Luftwechselzahl *f*
 f taux *m* de circulation d'air
 nl luchtcirculatiegetal *n*
 r кратность *f* воздухообмена

R95 *e* **rate of application**
 d 1. Gießnorm *f*, Bewässerungsgabe *f*
  2. Verbrauchsnorm *f*,
  Anstrichstoffverbrauchsnorm *f*
 f 1. norme *f* [dose *f*] d'arrosage
  2. dosage *m* d'application (*p. ex. de*
  peinture)
 nl 1. sproeinorm *f(m)* 2. verbruiksnorm
  f(m)
 r 1. поливная норма *f* 2. норма *f*
  расхода (*напр. краски*)

R96 *e* **rate of concrete placement**
 d Steiggeschwindigkeit *f* (*beim*
  Betonieren aufgehender Bauwerke)
 f norme *f* de la mise en œuvre du
  béton
 nl tempo *n* van het betonstorten
 r норма *f* [темп *m*] укладки бетона
  (*измеряемая высотой слоя*)

R97 **rate of construction** *see* **rate of progress**
 of the construction work

R98 *e* **rate of flow**
 d Durchfluß *m*
 f débit *m*
 nl stroomsterkte *f*, debiet *n*
 r расход *m* (*жидкости, газа*)

R99 *e* **rate of foundation settlement**
 d Setzungsgeschwindigkeit *f* des
  Fundaments
 f vitesse *f* de tassement des
  fondations, taux *m* d'affaissement des
  fondations
 nl snelheid *f* van de zetting van de
  fundering
 r скорость *f* осадки фундамента

R100 *e* **rate of hardening**
 d Erhärtungsgeschwindigkeit *f*
 f vitesse *f* [taux *m*] de durcissement
 nl verhardingssnelheid *f*
 r скорость *f* твердения

R101 *e* **rate-of-hardening parameters**
 d Einflußgrößen *f pl* der
  Erhärtungsgeschwindigkeit

f paramètres *m pl* gouvernant la vitesse de durcissement
nl factoren *m pl* van invloed op de verhardingssnelheid
r параметры *m pl*, определяющие скорость твердения

R102 *e* **rate of infiltration**
d Infiltrationsgeschwindigkeit *f*, Sickergeschwindigkeit *f*
f taux *m* [vitesse *f*] d'infiltration
nl infiltratiesnelheid *f*
r интенсивность *f* [скорость *f*] инфильтрации

R103 *e* **rate of load application, rate of loading**
d Belastungsgeschwindigkeit *f*
f taux *m* de chargement
nl stijgtempo *n* van de belasting
r скорость *f* нагружения; скорость *f* возрастания нагрузки

R104 **rate of performance, rate of pour** *see* **rate of concrete placement**

R105 *e* **rate of progress of the construction work**
d Bautempo *n*, Arbeitstempo *n*
f vitesse *f* d'avancement des travaux de construction
nl bouwtempo *n*
r темп *m* строительства; темп *m* производства строительных работ

R106 **rate of runoff** *see* **specific discharge**

R107 *e* **rate of sediment delivery**
d Feststoffabtrag *m*
f débit *m* solide spécifique
nl afvoer *m* van de vaste stoffen
r модуль *m* твёрдого стока

R108 *e* **rate of setting**
d Abbindegeschwindigkeit *f*; Erstarrungsgeschwindigkeit *f*
f vitesse *f* [rapidité *f*] de prise
nl verhardingssnelheid *f*
r скорость *f* или время *n* схватывания

R109 *e* **rate of spread**
d spezifische Auftragsmenge *f*, Streumaterialverbrauch *m* je Flächeneinheit der Straßendecke
f dosage *f* de répandage de matériaux granuleux
nl splitverbruik *n*
r расход *m* высевок *или* щебня (*на единицу площади дорожного покрытия*)

R110 *e* **rate of strain**
d Dehnungsgeschwindigkeit *f*
f vitesse *f* de déformation
nl vervormingssnelheid *f*
r скорость *f* деформации

R111 *e* **rate of strength gain**
d Geschwindigkeit *f* der Festigkeitszunahme
f vitesse *f* d'accroissement de la résistance
nl snelheid *f* van de sterktetoename
r скорость *f* нарастания прочности

R112 *e* **rate of water demand**
d Wasserbedarfswert *m*, Wasserverbrauchswert *m*
f taux *m* de consommation d'eau
nl waterverbruiksnorm *f*
r норма *f* водопотребления

R113 *e* **rate of water loss**
d Wasserverlustgeschwindigkeit *f*; Verdunstungsgeschwindigkeit *f*
f taux *m* de perte(s) d'eau
nl mate *f(m)* van waterverlies
r скорость *f* потери влаги (*напр. бетоном*); скорость *f* испарения влаги

R114 *e* **rate of work**
d Arbeitsnorm *f*, Leistungsnorm *f*
f norme *f* de production
nl produktienorm *f(m)*
r норма *f* выработки

R115 *e* **rating**
d 1. Nennleistung *f* 2. Nennversuch *m*, Nennprobe *f*, Nennprüfung *f* 3. Eichung *f*
f 1. puissance *f* nominale 2. évaluation *f* de la puissance nominale (*par essai*) 3. étalonnage *m*
nl 1. nominale produktiecapaciteit *f* 2. bepalen *n* van de produktiecapaciteit 3. ijking *f*
r 1. номинальная [паспортная] производительность *f* 2. определение *n* номинальной производительности (*испытанием в стандартных условиях*) 3. тарирование *n*

R116 *e* **rating curve**
d Abflußkurve *f*, Eichkurve *f*, Pegelschlüsselkurve *f*, Durchflußkurve *f*, Ablauflinie *f*
f courbe *f* de tarage [de relation hauteur-débit, des débits]
nl afvoerkromme *f(m)*
r кривая *f* расходов (*уровень — расход*)

R117 **rating flume**
d Meßkanal *m*
f canal *m* jaugeur
nl meetkanaal *n*
r тарировочный лоток *m*

R118 *e* **rating of lifting gear**
d Festlegung *f* der Nenntragfähigkeit der Anschlagmittel
f détermination *f* de la charge nominale d'utilisation d'un dispositif de levage
nl vaststellen *n* van het nominaal hefvermogen van hefwerktuigen
r определение *n* номинальной грузоподъёмности такелажных устройств

R119 *e* **ratiometer**
d Gerät *n* zur Bestimmung des W/Z-Faktors
f appareil *m* électrique pour mesurer le rapport eau-ciment
nl toestel *n* voor het bepalen van de water-cementfactor

r (электрический) прибор *m* для
определения водоцементного
отношения

R120 **ratio method of balancing** *see*
proportionate balancing

R121 *e* **raveling**
  d 1. Schotterdeckenversackung *f*
  2. Abbröckeln *n*
  f 1. affaissement *m* du macadam
  2. arrachement *m* des grains de
  granulat du revêtement routier
  nl 1. onstaan *n* van gaten in een grindweg
  2. afbrokkelen *n*
  r 1. просадка *f* щебёночного дорожного
  покрытия *или* основания 2. отрыв *m*
  зёрен заполнителей от дорожного
  покрытия

R122 *e* **raw mix**
  d Rohmischung *f*, Rohgemisch *n*
  f mélange *m* de matières premières
  nl grondstoffenmengsel *n*
  r сырьевая смесь *f* (*напр. цементная*)

R123 *e* **raw sewage**
  d ungereinigtes Abwasser *n*,
  Rohabwasser *n*
  f eaux *f pl* d'égout brutes
  nl ongezuiverd afvalwater *n*
  r неочищенная сточная жидкость *f*;
  неочищенные сточные воды *f pl*

R124 **raw sludge** *see* **primary sludge**

R125 *e* **raw water**
  d Rohabwasser *n*; Rohwasser *n*
  f eau *f* brute
  nl ongezuiverd water *n*
  r необработанная сточная вода *f*;
  неочищенная (водопроводная) вода
  f

R126 *e* **Raymond pile**
  d Ortbetonpfahl *m* mit
  zurückgewonnenem Füllrohr
  f pieu *m* Raymond
  nl in de grond gevormde betonpaal *m*
  met teruggewonnen mantelbuis
  r набивная свая *f*, бетонируемая
  в извлекаемой обсадной трубе

R127 *e* **reach**
  d 1. Haltung *f*, Stauhaltung *f*
  2. Kanalhaltung *f* 3. Reichweite *f*,
  Aktionsradius *m* 4. (Ausleger-)-
  Ausladung *f*
  f 1. bief *m* 2. bief *m* de canal
  3. portée *f*; rayon *m* d'action
  4. volée *f* [portée *f*] de la grue
  nl 1. stuwpand *n* 2. kanaalpand *n*
  3. draagwijdte *f* 4. vlucht *m* (*kraan*),
  reikwijdte *f* (*kraan*)
  r 1. подпёртый бьеф *m* 2.
  промежуточный бьеф *m* канала
  3. радиус *m* действия 4. вылет *m*
  стрелы (*крана*)

R128 *e* **reaction**
  d Reaktionskraft *f*, Gegenwirkung *f*
  f réaction *f*

  nl tegendruk *m*, reactie *f*,
  reactiekracht *f(m)*
  r реакция *f*, реактивная сила *f*;
  противодействие *n*

R129 *e* **reaction couple**
  d gegenwirkendes Kräftepaar *n*;
  Gegenmoment *n*
  f couple *m* des forces de réaction
  nl tegengesteld moment *n*
  r противодействующая пара *f* сил;
  реактивный момент *m*

R130 *e* **reaction force**
  d Reaktionskraft *f*
  f force *f* de réaction
  nl reactiekracht *f(m)*
  r сила *f* реакции, реакция *f*

R131 *e* **reaction of the support**
  d Auflagerreaktion *f*
  f réaction *f* d'appui, action *f* de
  contact au appui
  nl steunpuntreactie *f*
  r опорная реакция *f*

R132 *e* **reactive aggregates**
  d reaktionsfähige [reaktive]
  Zuschlagstoffe *m pl*
  f agrégats *m pl* réactifs
  nl reactieve toeslagen *m*
  r реакционноспособные заполнители
  m pl (*способные реагировать
  с цементом*)

R133 *e* **reactive silica material**
  d reaktionsfähiger [reaktiver]
  Silikastoff *m*
  f matière *f* silicieuse réactive
  nl reactief silicaatmateriaal *n*
  r реакционноспособный
  кремнезёмистый материал *m*

R134 *e* **reactor building**
  d Reaktorgebäude *n*, Reaktorhalle *f*,
  Reaktorhaus *n*
  f bâtiment *m* de pile
  nl reactorgebouw *n*
  r здание *n* для реактора, реакторный
  зал *m*

R135 *e* **reactor containment structure**
  d Einschlußbauwerk *n* [Container *m*]
  einer Reaktoranlage,
  Reaktorsicherheitshülle *f*
  f enceinte *m* étanche de la centrale
  nucléaire
  nl afschermingsconstructie *f* van een
  reactor
  r оболочка *f* [внешняя защита *f*]
  реактора

R136 *e* **reactor shell**
  d Containerschale *f*
  f enveloppe *f* de protection de la **pile**
  [du réacteur]
  nl omhulsel *n* van een reactor
  r оболочка *f* реактора

R137 *e* **reactor shielding**
  d Strahlenschutz *m*,
  Strahlenabschirmung *f*

f protection *f* biologique
nl absorberend scherm *n*
r биологическая [радиационная] защита *f* (*реактора*)

R138 *e* **reactor vessel**
d Druckbehälter *m*, Reaktorbehälter *m*
f caisson *f* [cuve *f*] de réacteur
nl reactorvat *n*
r корпус *m* реактора

R139 *e* **ready-made partition**
d Fertig-Trennwand *f*
f paroi *f* préfabriquée
nl prefab scheidingswand *m*
r сборная перегородка *f*

R140 *e* **ready-mix concrete, ready mixed concrete**
d Transportbeton *m*, Lieferbeton *m*, Fertigbeton *m*
f béton *m* prêt à l'emploi, béton *m* pré-mélangé [préfabriqué] en centrale
nl transportbeton *n*
r товарный бетон *m*

R141 *e* **ready-mixed concrete lorry**
d Transportbetonmischer *m*
f camion-malaxeur *m*
nl truckmixer *m*
r автобетоносмеситель *m*

R142 *e* **ready-mixed concrete plant**
d Transportbetonwerk *n*, Betonlieferwerk *n*
f centrale *f* de béton préparé
nl betonmortelcentrale *f*
r завод *m* товарного бетона; бетонный завод *m*

R143 *e* **ready-mixed joint compound**
d werkgemischte Vergußmasse *f*
f mastic *m* préparé pour joints
nl gebruiksklare voegvulling *f*
r готовый состав *m* для заделки швов в конструкциях

R144 *e* **ready-mixed mortar**
d Fertigmörtel *m*
f mortier *m* préparé d'avance
nl aangevoerde mortel *m*
r готовый строительный раствор *m*; раствор *m* централизованного приготовления (*на растворном узле*)

R145 *e* **ready-mixed plaster**
d Fertigputz *m*
f mortier *m* d'enduit sec préparé d'avance
nl droogklaar pleister *n*
r готовая сухая смесь *f* для приготовления штукатурного раствора

R146 *e* **ready-mix paints**
d streichfertige Farben *f pl*
f peinture *f* prête à l'emploi
nl gebruiksklare verven *f(m) pl*
r краски *f pl*, готовые к употреблению

R147 *e* **reagent mixer**
d Reagenzmischanlage *f*, Rührwerk *n* für Reagenzmittel

f mélangeur *m* à agent moussant
nl reagensmenger *m*
r смеситель *m* пенообразующего реагента

R148 *e* **real load**
d wirkliche Last *f*; einwirkende Last *f*, Angriffslast *f*
f charge *f* réelle; charge *f* sollicitante
nl werkende belasting *f*
r действительная *или* действующая нагрузка *f*

R149 **real size** *see* **actual size**

R150 **rear apron** *see* **downstream apron**

R151 **rear arch** *see* **arrière-voussure**

R152 *e* **rear-dump lorry** *UK*, **rear-dump truck** *US*
d Rückwärtskipper *m*, Hinterkipper *m*
f camion *m* basculant [basculeur], camion-benne *m* à basculement sur arrière
nl achterwaarts kiepende kipauto *m*
r автосамосвал *m* с задней разгрузкой

R153 *e* **rebar**
d Bewehrungsstab *m*, Armierungsstab *m*
f barre *f* d'armature [de ferraillage]
nl wapeningsstaaf *f(m)*
r арматурный стержень *m*

R154 *e* **rebar mat**
d Betonstahlmatte *f*, Bewehrungsmatte *f*
f résille *f*, quadrillage *m*, armatures *f pl* croisées, treillis *m* pour béton armé
nl wapeningsnet *n*
r арматурная сетка *f*

R155 *e* **rebar spacing**
d Abstand *m* der Bewehrungsstäbe
f espacement *m* des barres d'armatures, écartement *m* des barres de ferraillage
nl afstand *m* tussen wapeningsstaven
r шаг *m* арматурных стержней

R156 *e* **rebar tie encased in concrete**
d Stahlbetonzugbalken *m*
f tirant *m* (d'arc) en barres de ferraillage enrobées dans le béton
nl gewapend betonnen trekbalk *m*
r затяжка *f* из арматурных стержней, омоноличенных бетоном

R157 *e* **rebate**
d Falz *m* (*Holzbearbeitung*)
f rainure *f*, feuillure *f*
nl sponning *f*
r шпунт *m*, четверть *f*, фальц *m* (*в деревянных конструкциях*)

R158 *e* **rebated joint**
d Falzverbindung *f*
f joint *m* feuillé, assemblage *m* à feuillure, jointure *f* à recouvrement
nl liplas *f*
r соединение *n* в четверть

R159 *e* **rebated weatherboarding**
d Stülpschalung *f* mit Wechselfalz

f revêtement *m* en planches assemblées
à feuillure
nl beplanking *f* met sponningdelen
r обшивка *f* досками в четверть

R160 *e* **rebated wooden flooring**
d gefalzte [halbgespundete] Dielung *f*
f planchéiage *m* demi-bouveté
[à feuillure]
nl vloer *m* van sponningdelen
r дощатый пол *m* [настил *m*] в четверть

R161 *e* **rebound**
d 1. Rückprall *m*; Rückstoß *m*
2. elastische Rückbildung *f*
f 1. rebondissement *m*; ricochet *m*
2. reprise *f* élastique
nl 1. terugstoot *m* 2. elastisch herstel *n*
na vervorming
r 1. отскакивание *n*; отдача *f*
2. упругое восстановление *n*

R162 *e* **rebound deflection**
d elastische Durchbiegung *f*
f reprise *f* élastique de la flèche
nl elastische buiging *f*
r упругое восстановление *n* прогиба

R163 *e* **rebound of pile**
d Pfahlrücksprung *m* (*nach*
Einrammung bzw. Entlastung)
f rebondissement *m* du pieu
nl elastische rijzing *f* na ontlasting
r упругий подъём *m* сваи

R164 *e* **receiver**
d Aufnehmer *m*, Aufnahmegefäß *n*,
Druckausgleicher *m*
f récipient *m* (*d'air comprimé*)
nl luchtketel *m*
r ресивер *m* (*сжатого воздуха*)

R165 *e* **recess**
d Aussparung *f*; Rücksprung *m*;
Absatz *m*
f évidement *m*; niche *f*; cavité *f*;
encoche *f*, entaille *f*; retraite *f*
nl uitsparing *f*; nis *f(m)*; holte *f*
r проём *m*; паз *m*; ниша *f*; углубление *n*

R166 *e* **recessed lighting fitting**
d Einbauleuchte *f*
f luminaire *m* encastré
nl ingebouwd verlichtingsarmatuur *n*
r встроенный светильник *m*

R167 *e* **recession curve**
d Senkungskurve *f*
f courbe *f* de décrue
nl dalingskromme *f(m)*
r кривая *f* спада

R168 *e* **recharge of aquifer**
d Wiederauffüllung *f* des Grundwassers,
Grundwasseranreicherung *f*,
Grundwasserauffüllung *f*
f alimentation *f* d'une nappe souterraine, emmagasinement *m* des eaux
souterraines
nl aanvulling *f* van het grondwaterreser-
voir

r подпитка *f* водоносного горизонта,
пополнение *n* запасов подземных
вод

R169 *e* **recharge well**
d Einleitungsbrunnen *m*,
Anreicherungsbrunnen *m*,
Verteilbrunnen *m*
f puits *m* d'alimentation
nl voedingsbron *m*
r подпитывающий [распределитель-
ный] колодец *m*

R170 *e* **reciprocal deflection relationship**
d Satz *m* von der Gegenseitigkeit der
Verschiebung
f théorème *m* de réciprocité des
déflexions, principe *m* de Maxwell
nl theorema *n* van de wederkerigheid
van verplaatsingen
r теорема *f* о взаимности перемещений

R171 *e* **reciprocal force polygon**
d reziproker Kräfteplan *m*
f diagramme *m* de Cremona
nl reciproke krachtenveelhoek *m*
r диаграмма *f* Кремоны —
Максвелла

R172 *e* **reciprocal value**
d reziproker Wert *m*, Kehrwert *m*
f valeur *f* réciproque
nl reciproke waarde *f*
r обратная величина *f*

R173 *e* **reciprocating compressor**
d Kolbenkompressor *m*,
Kolbenverdichter *m*
f compresseur *m* à piston
nl zuigercompressor *m*
r поршневой компрессор *m*

R174 *e* **reciprocating pump**
d Kolbenpumpe *f*
f pompe *f* à plongeur [à piston]
nl zuigerpomp *f(m)*
r поршневой насос *m*; плунжерный
насос *m*

R175 **recirculated air** *see* **recirculation air**
R176 *e* **recirculated water**
d Umlaufwasser *n*
f eau *f* de récirculation
nl circulerend water *n*
r циркулирующая вода *f*

R177 *e* **recirculating air fan**
d Ventilator *m* für Luftumwälzung
f ventilateur *m* de circulation d'air
nl ventilator *m* voor luchtcirculatie
r вентилятор *m*, обеспечивающий
циркуляцию воздуха

R178 *e* **recirculation**
d Rezirkulation *f*, Rückführung *f*,
Umluftbetrieb *m*
f récirculation *f* (*d'air*)
nl hercirculatie *f*
r рециркуляция *f* (*воздуха*)

R179 *e* **recirculation air**
d Umluft *f*

f air *m* recyclé [récirculé]
nl opnieuw in circulatie gebrachte lucht *f(m)*
r рециркуляционный воздух *m*

R180 *e* **recirculation duct**
  d Umluftkanal *m*
  f conduit *m* de récirculation de l'air
  nl recirculatiekanaal *n*
  r рециркуляционный воздуховод *m*, канал *m* рециркуляционного воздуха

R181 *e* **recirculation ratio**
  d Rücklaufverhältnis *n*
  f coefficient *m* de récirculation (des eaux usées)
  nl recirculatiefactor *m*
  r коэффициент *m* рециркуляции сточных вод (*в очистных сооружениях*)

R182 **reclaimed energy** *see* **recovered energy**

R183 *e* **reclaimed heat**
  d regenerierte Wärme *f*
  f chaleur *f* recyclée
  nl geregenereerde warmte *f*
  r регенерированное тепло *n*

R184 *e* **reclaimed waste water**
  d rückgewonnenes gereinigtes Abwasser *n*
  f eaux *f pl* usées récupérées
  nl teruggewonnen afvalwater *n*
  r регенерированные сточные воды *f pl*

R185 *e* **reclamation**
  d Melioration *f*
  f mise *f* en valeur (*des terres*)
  nl landaanwinning *f*; terugwinning van bruikbare stoffen
  r мелиорация *f* земель

R186 *e* **reconnaissance**
  d Erkundung *f*
  f reconnaissance *f*
  nl verkenning *f*
  r инженерная разведка *f*, инженерные изыскания *n pl*

R187 **reconstructed stone** *see* **cast stone**

R188 *e* **reconstruction**
  d Wiederherstellung *f*, Wiederaufbau *m*
  f reconstruction *f*
  nl reconstructie *f*
  r реконструкция *f*

R189 *e* **recool system**
  d Klimaanlage *f* mit Nachkühlung
  f installation *f* de climatisation avec refroidisseurs locaux [zonals]
  nl luchtbehandelingsinstallatie *f* met nakoeling
  r система *f* кондиционирования воздуха с зональными доохладителями

R190 *e* **recorded settlement**
  d überwachte [registrierte] Setzung *f*
  f tassement *m* contrôlé [enregistré], affaissement *m* contrôlé
  nl geregistreerde zetting *f*
  r контролируемая [регистрируемая] осадка *f*

R191 **record flood** *US see* **catastrophic flood**

R192 *e* **recovered energy**
  d rückgewonnene Energie *f*
  f énergie *f* récupérée
  nl teruggewonnen energie *f*
  r регенерируемая энергия *f*

R193 **recovered heat** *see* **reclaimed heat**

R194 *e* **recovery factor**
  d Rückgewinnungsfaktor *m*
  f facteur *m* de régénération
  nl terugwinfactor *m*
  r коэффициент *m* регенерации (*тепла, холода*); коэффициент *m* полезного действия регенерационного теплообменника; степень *f* регенерации

R195 **recovery pegs** *see* **reference pegs**

R196 *e* **recovery time**
  d Aufheizzeit *f* der Speichermasse
  f durée *f* de la mise à température (*du fluide caloporteur*)
  nl hersteltijd *m*
  r время *n* нагрева (*теплоаккумулирующей массы отопительного прибора*) до рабочей температуры

R197 *e* **recreation area**
  d Erholungsgebiet *n*
  f zone *f* de repos [de récréation]
  nl recreatiegebied *n*
  r зона *f* отдыха, рекреационная зона *f*

R198 *e* **rectangular beam**
  d Rechteckträger *m*
  f poutre *f* rectangulaire
  nl rechthoekige balk *m*
  r балка *f* прямоугольного сечения

R199 **rectangular channel** *see* **box channel**

R200 *e* **rectangular tube**
  d Vierkantrohr *n*
  f tube *f* rectangulaire, profilé *m* tubulaire rectangulaire
  nl rechthoekige pijp *f(m)*
  r труба *f* прямоугольного сечения, короб *m*

R201 *e* **rectilinear building**
  d langgestrecktes Gebäude *n*
  f bâtiment *m* rectiligne
  nl rechthoekig gebouw *n*
  r прямолинейное здание *n*

R202 *e* **recurrence interval**
  d Wiederholungs-Intervall *n*, Wiederkehrintervall *n*
  f intervalle *m* de récurrence
  nl herhalingsinterval *n*
  r интервал *m* возврата [повторения, повторяемости]

R203 *e* **recycled concrete**
  d wiederverwendeter Altbeton *m*
  f béton *m* recyclé [réutilisé] (*p.ex. pour la préparation des agrégats*)
  nl wedergebruikt beton *n*

r повторно используемый бетон *m*
(*напр. для приготовления
заполнителей*)

R204 recycle sludge *see* **return sludge**

R205 *e* **recycling**
d 1. Wiederverwendung *f*
2. Rückführung *f*
f 1. récyclage *m* [réutilisation *f*] des
matériaux 2. récirculation *f*
nl 1. wedergebruiken *n* van materialen
2. hercirculatie *f*
r 1. повторное использование *n*
материалов (*после вторичной
переработки*) 2. рециркуляция *f*

R206 *e* **redistribution of forces**
d Kräfteumlagerung *f*, Änderung *f* des
Kräftespiels
f redistribution *f* des forces
nl herverdelen *n* van krachten
r перераспределение *n* сил

R207 *e* **redistribution of moments**
d Momentenumlagerung *f*,
Momentenausgleich *m*
f redistribution *f* des moments
nl herverdelen *n* van momenten
r перераспределение *n* моментов

R208 *e* **red label goods**
d leichtentzündliche Materialien *n pl*
f produits *m pl* [matière *f*] à étiquette
rouge (*pouvant pendre feu à la
température au dessous de 10° C*)
nl licht ontvlambare materialen *n pl*
r особо легковоспламеняющиеся
материалы *m pl* (*хранятся в
упаковках с красным ярлыком,
могут воспламеняться при
температуре ниже 10° С*)

R209 *e* **red lead**
d Bleimennige *f*
f minium *m* de plomb
nl loodmenie *f*, rode menie *f*
r свинцовый сурик *m*

R210 *e* **red pigment**
d rote Mineralfarbe *f*
f rouge *m* de Saturne, pigment *m* rouge
nl rode kleurstof *f(m)*
r красный пигмент *m*

R211 *e* **reduced level**
d Bezugshöhe *f*
f cote *f* [niveau *m*] de référence
nl hoogte *f* ten opzichte van een
aangenomen peil
r относительная [условная] отметка *f*

R212 *e* **reducer**
d 1. Übergangsstück *n* 2.
Untersetzungsgetriebe *n*, Reduktor *m*
3. Verdünnungsmittel *n*
4. Reduzierventil *n*
f 1. raccord *m* [manchon *m*] de réduction
2. réducteur *m* 3. diluant *m*
4. réducteur *m* de pression
nl 1. verloopstuk *n*
2. reductieoverbrenging *f*
3. verdunner *m* 4. reduceerventiel *n*

r 1. переходная деталь *f* 2. редуктор *m*
3. разбавитель *m* 4. редукционный
клапан *m*

R213 *e* **reducing fitting**
d Reduzierstück *n*, Übergangsrohr *n*
f raccord *m* réducteur [de réduction]
nl verloopstuk *n*
r переходной фитинг *m*,
соединительный элемент *m*
(*трубопровода*)

R214 *e* **reducing tee**
d reduzierendes T-Stück *n*,
Übergangs-T-Stück *n*
f raccord *m* en T réducteur
nl T-stuk *n* met verloop
r переходной [неравнопроходной]
тройник *m*

R215 *e* **reducing valve**
d Reduzierventil *n*, Minderventil *n*
f clapet *m* ralantisseur, soupape *f* de
réduction, réducteur *m* de pression
nl reduceerventiel *n*
r редукционный клапан *m*

R216 *e* **reduction of bleeding**
d Verminderung *f* des
Betongemischblutens
f réduction *f* de ressuage du béton
nl vermindering *f* van waterafscheiding
uit de betonmortel
r уменьшение *n* водоотделения
бетонной смеси

R217 *e* **redundancy**
d statische Unbestimmtheit *f*
f indétermination *f* statique,
hyperstaticité *f*
nl statische onbepaaldheid *f* door
overtollige elementen
r статическая неопределимость *f*

R218 *e* **redundants**
d überschüssige Bindungen *f pl*
f liaisons *f pl* surabondantes
nl overtollige elementen *n pl*
r лишние связи *f pl* (*создающие
статическую неопределимость*)

R219 *e* **redundant support**
d überzählige Stütze *f*
f appui *m* [support *m*] surabondant
nl overmatige ondersteuning *f*
r лишняя опора *f* (*придающая
системе статическую
неопределимость*)

R220 *e* **reel**
d Spule *f*; Trommel *f*; Drahthaspel *f*
f bobine *f*; rouleau *m*
nl spoel *f(m)*; haspel *m*
r катушка *f*; барабан *m*

R221 *e* **reeving**
d Seileinlauf *m*, Seilaufwickelung *f*
f enroulement *m* de câble
nl inscheren *n* (*kabel*)
r система *f* запасовки каната

R222 *e* **reference grid**
d Rasternetz *n*

f réseau *m* modulaire [de référence]
nl coördinatenstelsel *n*, raster *n*
r модульная сетка *f*

R223 *e* **reference line**
d Bezugslinie *f*
f ligne *f* modulaire [de référence]
nl meetlijn *f(m)*
r координационная модульная линия *f*

R224 **reference mark** *see* **reference point**

R225 *e* **reference pegs**
d Markierpflöckchen *n pl*,
Nivellierpflöckchen *n pl*
f piquets *m pl* de référence
nl meetpiketten *n pl*
r реперные [нивелирные] колышки
m pl

R226 *e* **reference plane**
d Libellenebene *f*, Bezugsebene *f*
f plan *m* de référence
nl referentievlak *n*
r уровенная [базовая] плоскость *f*

R227 *e* **reference point**
d Bezugspunkt *m*
f point *m* de référence, repère *m*
nl meetpunt *n*, vast punt *n*
r репер *m*; точка *f* отсчёта

R228 **refilling** *see* **backfill**

R229 *e* **refined lake asphalt**
d reines natürliches Bitumen *n*, reiner
Seeasphalt *m*
f bitume *m* naturel pur [épuré]; bitume
m de Trinidad
nl geraffineerd natuurlijk bitumen *n*
r чистый природный битум *m*;
тринидадский битум *m*

R230 *e* **refined tar**
d gereinigter Teer *m*
f goudron *m* raffiné
nl geraffineerde teer *m*, *n*
r очищенный дёготь *m*

R231 *e* **refit, refitting**
d Altbaumodernisierung *f*,
Neugestaltung *f* der Installationen
f rénovation *f* de l'équipement technique
de vieux bâtiments
nl vernieuwing *f* van de installaties
van een oud gebouw
r реконструкция *f* инженерного
оборудования старых зданий

R232 **refitting quay** *see* **repair quay**

R233 *e* **reflected-light luminaire**
d Indirektleuchte *f*
f luminaire *m* d'éclairage indirect
nl armatuur *n* voor indirecte belichting
r светильник *m* отражённого света

R234 *e* **reflecting concrete**
d Leuchtbeton *m*
f béton *m* réfléchissant
nl reflecterend beton *n*
r светящийся [светоотражающий]
бетон *m* (*дорожное строительство*)

R235 *e* **reflective insulant**
d reflektierendes Isoliermaterial *n*
f isolant *m* réfléchissant
nl reflecterend isolatiemateriaal *n*
r отражательный теплоизоляционный
материал *m*

R236 *e* **reflective insulation**
d reflektierende Wärmeisolierung *f*
[Wärmedämmung *f*]
f isolation *f* thermique réfléchissante
nl terugkaatsende warmte-isolatie *f*
r отражательная теплоизоляция *f*

R237 **reflux valve** *see* **check valve**

R238 *e* **refractories**
d feuerfeste Erzeugnisse *n pl*
f réfractaires *m pl*, produits *m pl*
réfractaires
nl vuurvaste materialen *n pl*
r огнеупорные материалы *m pl*,
огнеупоры *m pl*

R239 *e* **refractoriness**
d Feuerfestigkeit *f*
f réfractairité *f*
nl vuurvastheid *f*
r огнеупорность *f*

R240 *e* **refractory brick**
d feuerfester Stein *m* [Ziegel *m*]
f brique *f* réfractaire
nl vuurvaste steen *m*
r огнеупорный кирпич *m*

R241 *e* **refractory cement**
d feuerbeständiger Zement *m*
f ciment *m* réfractaire
nl vuurvast cement *n*
r огнеупорный [жаростойкий] цемент *m*

R242 *e* **refractory concrete**
d feuerfester Beton *m*
f béton *m* réfractaire
nl vuurvast [hittebestendig] beton *n*
r огнеупорный [жаростойкий] бетон *m*

R243 *e* **refractory insulating brick**
d feuerfester Isolierstein *m*
f brique *f* réfractaire isolante
nl vuurvaste isolatiesteen *m*
r огнеупорный теплоизоляционный
кирпич *m* (*для температур выше
1100° C*)

R244 *e* **refractory insulating concrete**
d feuerfester Dämmbeton *m*
f béton *m* réfractaire isolant
nl vuurvast [hittebestendig]
isolatiebeton *n*
r жаростойкий теплоизоляционный
бетон *m*

R245 *e* **refractory lining**
d Schamottebekleidung *f*,
Schamottefutter *n*
f revêtement *m* [garniture *f*]
réfractaire
nl vuurvaste bekleding *f*
r шамотная [огнеупорная] футеровка *f*

R246 *e* **refrigerant**
d Kältemittel *n*

f réfrigérant *m*, fluide *m* frigorigène
nl koelmiddel *n*
r хладоагент *m*, первичный
  хладоноситель *m*

R247 refrigerating capacity *see* cooling power

R248 *e* refrigerating load
d Kälteverbrauch *m*; Kühllast *f*
f charge *f* frigoritique [de
  refroidissement]
nl koelbelasting *f*
r расход *m* холода; холодильная
  нагрузка *f*

R249 refrigerating medium *see* refrigerant

R250 *e* refrigerating plant
d Kälteanlage *f*
f installation *f* frigorifique [de
  réfrigération]
nl koelinstallatie *f*
r холодильная установка *f*

R251 *e* refrigeration
d mechanische [maschinelle] Kühlung *f*
f réfrigération *f* [refroidissement *m*]
  mécanique
nl mechanische afkoeling *f*
r машинное *или* искусственное
  охлаждение *n*

R252 *e* refrigeration compressor
d Kälte(mittel)kompressor *m*
f compresseur *m* frigorifique
nl koelcompressor *m*
r холодильный компрессор *m*

R253 *e* refuge island
d Fußgänger-Schutzinsel *f*
f refuge *m* (pour piétons)
nl vluchtheuvel *m*
r островок *m* безопасности

R254 *e* refusal
d Standfestigkeit *f* (des Pfahls)
f refus *m* (d'un pieu) au battage
nl stuit *m* (heipaal)
r отказ *m* (сваи)

R255 *e* refuse
d Abfall(stoff) *m*; Müll *m*
f déchets *m pl*, détritus *m*, gravats
  m pl; ordures *f pl*
nl afval *m*; vuilnis *n*
r отходы *m pl*; мусор *m*

R256 *e* refuse disposal
d Müllbeseitigung *f*, Abfallbeseitigung *f*
f décharge *f* publique, évacuation *f* des
  ordures
nl verzamelen *n* en afvoeren *n* van het
  vuilnis
r удаление *n* мусора и отбросов

R257 *e* refuse disposal works
d Abfallbeseitigungsanlage *f*
f usine *f* de traitement des déchets
  urbains, usine *f* d'incinération
nl vuilverwerkingsbedrijf *n*
r установка *f* для уничтожения мусора

R258 *e* refuse incinerator
d Abfallverbrennungsofen *m*

f incinérateur *m* de déchets, installation
  f d'incinération des déchets
nl vuilverbrandingsinstallatie *f*
r мусоросжигательная печь *f*

R259 *e* refuse processing plant
d Abfallverwertungsanlage *f*
f installation *f* de traitement des
  déchets urbains
nl vuilverwerkinginstallatie *f*
r завод *m* или установка *f* по
  переработке городского мусора

R260 *e* refuse tip
d Deponie *f*, Müllabladeplatz *m*
f voirie *f*, décharge *f* publique
nl vuilnisbelt *m*, *f*
r свалка *f* мусора

R261 *e* regaining of workability
d Rückgewinn *m* der Einbringbarkeit
f rétablissement *m* de l'ouvrabilité
  (*p. ex. du béton*)
nl verbeteren *n* van de verwerkbaarheid
r восстановление *n*
  удобоукладываемости

R262 regenerated heat *see* reclaimed heat

R263 *e* regenerated water
d Filterwasser *n*, filtriertes Wasser *n*;
  Umlaufwasser *n*, Rücklaufwasser *n*
f eau *f* filtrée; eaux *f pl* recyclées [de
  restitution]
nl teruggewonnen water *n*
r фильтрованная вода *f*; возвратные
  воды *f pl*

R264 *e* regenerative air heater
d Regenerativlufterhitzer *m*
f réchauffeur *m* d'air de régénération
nl regeneratieve luchtverhitter *m*
r регенеративный воздухоподогреватель
  m

R265 *e* regenerative turbine pump
d Pumpenturbine *f*, umkehrbare
  Turbine *f*
f turbine-pompe *f*, turbine *f* en pompe
nl omkeerbare turbine *f*
r обратимая гидротурбина *f*, насос-
  -турбина *m*

R266 *e* regime
d stabiler Laufzustand *m*
f régime *m* établi [stable]
nl stabiele toestand *m*
r режим *m* устойчивого русла

R267 *e* regional planning
d Raumplanung *f*, Regionalplanung *f*
f planification *f* régionale
nl regionale planning *f*
r районная планировка *f*

R268 *e* region of stress concentration
d Bereich *m* der Spannungskonzentration, Kerbwirkungsbereich *m*
f zone *f* [domaine *m*] de la
  concentration des contraintes
nl zone *f* van spanningsconcentratie
r область *f* концентрации
  напряжений

R 269 *e* **register**
　d Luftungsgitter *n* mit
　　Mengeneinstellung; Essenschieber *m*,
　　Schornsteinschieber *m*
　f registre *m* [bouche *f*] d'air à lames
　　[à persiennes]; registre *m* de tirage,
　　grille *f* à registre
　nl schoorsteenschuif *f(m)*; register *n*
　r вентиляционная решётка *f*
　　с многоствочатым клапаном;
　　заслонка *f* [задвижка *f*] дымохода

R 270 *e* **regular concrete sand**
　d Betonsand *m*
　f sable *m* à béton [utilisé dans les
　　bétons]
　nl zand *n* voor betonmortel
　r песок *m* для бетонов

R 271 *e* **regular-coursed rubble**
　d regelmäßiges Schichtenmauerwerk *n*
　f maçonnerie *f* de moellons en rangées
　　régulières
　nl metselwerk *n* van natuursteen in
　　regelmatig verband
　r рядовая бутовая кладка *f*, бутовая
　　кладка *f* равномерными рядами

R 272 *e* **regular gypsum wallboard**
　d Normal-Gipskartonbauplatte *f*
　f panneau *m* de plâtre normal(isé),
　　placoplâtre *m* normal(isé)
　nl normale gipsplaat *f(m)*
　r стандартный лист *m* гипсовой сухой
　　штукатурки

R 273 *e* **regulated discharge**
　d regulierter Durchfluß *m*
　f débit *m* aménagé
　nl gereguleerde uitstroming *f*
　r зарегулированный расход *m*

R 274 *e* **regulated flow**
　d regulierter [ausgeglichener] Abfluß
　　m
　f débit *m* aménagé, débit *m* [écoulement
　　m] réglé
　nl gereguleerde afvoer *m*
　r зарегулированный сток *m*

R 275 *e* **regulated stream**
　d regulierter Fluß *m*
　f débit *m* réglé d'un cours d'eau
　nl genormaliseerde rivier *f(m)*
　r зарегулированный водоток *m*

R 276 *e* **regulating course**
　d Ausgleichschicht *f*, Ausgleichdecke *f*
　f couche *f* de profilage
　nl vereffeningslaag *f(m)*
　r выравнивающий слой *m*
　　(*наносимый на дорожное*
　　покрытие при ремонте)

R 277 *e* **regulating damper**
　d Regelklappe *f*
　f registre *m* de tirage
　nl regelklep *f(m)*
　r регулирующий воздушный клапан *m*

R 278 *e* **regulating storage capacity**
　d Normalstauraum *m*

　f emmagasinement *m* de **régulation**,
　　capacité *f* d'emmagasinage de
　　régulation
　nl stuwcapaciteit *f* ten behoeve van
　　regularisatie
　r мобильный объём *m* водохранилища

R 279 *e* **regulating structure**
　d Regelungsbauwerk *n*, Flußbauwerk
　　n
　f ouvrage *m* régulateur [de réglage]
　nl normalisatiewerk *n*
　r регуляционное сооружение *n*

R 280 *e* **regulating tank**
　d Gegenbehälter *m*
　f réservoir *m* de régulation
　nl compensatietank *m*
　r контррезервуар *m*

R 281 *e* **regulating valve**
　d Regulierschieber *m*; Regelventil *n*
　f valve *f* [clapet *m*] de réglage, soupape *f*
　　régulatrice
　nl regelschuif *f(m)*; regelklep *f(m)*
　r регулирующая задвижка *f*;
　　регулирующий клапан *m*

R 282 *e* **regulation**
　d Regelung *f*, Regulierung *f*
　f réglage *f*, régulation *f*
　nl regeling *f*
　r регулирование *n*

R 283 *e* **regulation of streams**
　d Flußregelung *f*, Abflußregelung *f*
　f régularisation *f* des fleuves [des débits]
　nl afvoerregulering *f*, normalisering *f*
　　van rivieren
　r регулирование *n* рек [речного стока]

R 284 *e* **regulations**
　d Vorschriften *f pl*
　f règlements *m pl*; prescriptions *f pl*
　nl voorschriften *n pl*
　r нормы *f pl*; правила *n pl*; указания
　　n pl; инструкции *f pl*

R 285 *e* **regulation works**
　d Regelungsbauwerk *n*
　f ouvrage *m* régulateur [de réglage]
　nl normalisatiewerk *n*
　r регуляционное сооружение *n*

R 286 *e* **regulator storage**
　d Ausgleichsbecken *n*
　f réservoir *m* régulateur [de réglage]
　nl vereffeningsbekken *n*
　r регулирующее водохранилище *n*

R 287 *e* **rehabilitation**
　d Rekonstruktion *f*, Umgestaltung *f*;
　　Sanierung *f*, Erneuerung *f*;
　　Wiederherstellung *f*
　f réhabilitation *f*; rénovation *f*
　nl renovatie *f*
　r реконструкция *f*; благоустройство *n*;
　　санация *f* (*памятников*);
　　восстановление *n*

R 288 *e* **reheat**
　d Nachheizung *f*
　f réchauffage *m*

nl opnieuw verwarmen *n*,
naverwarmen *n*
r второй подогрев *m*; зональный
[местный] подогрев *m*

R289 *e* **reheat coil, reheater**
d Nachheizgerät *n*, örtlicher
Nachwärmer *n*
f réchauffeur *m* d'air
nl naverwarmingstoestel *n*
r доводчик-подогреватель *m*

R290 *e* **reheat induction unit**
d Induktionsklimagerät *n* mit
Primärluftnachwärmung
f réchauffeur *m* d'air primaire à
induction
nl luchtbehandelingstoestel *n* met
luchtnaverwarming van de
aangevoerde lucht
r эжекционный доводчик *m*
с подогревом первичного воздуха

R291 *e* **reheat system**
d Klimaanlage *f* mit Nachheizung
f système *m* de conditionnement d'air
à réchauffage
nl luchtbehandelingsinstallatie *f* met
naverwarming
r система *f* кондиционирования
воздуха со вторым подогревом *или*
с зональными подогревателями

R292 **reheat unit** *see* **reheat coil**

R293 *e* **reinforced brickwork**
d bewehrtes Ziegelmauerwerk *n*,
armierte Ziegelkonstruktion *f*
f maçonnerie *f* en briques armée
[renforcée]
nl gewapend metselwerk *n*
r армированная кирпичная кладка *f*,
армокаменная конструкция *f*

R294 *e* **reinforced cement mortar**
d Stahlmörtel *m*, stahlspannverstärkter
verschleißfester Beton *m*
f mortier *m* de ciment avec raclures
métalliques
nl betonmortel *m* met staalgruis
r стальбетон *m*

R295 *e* **reinforced concrete**
d Stahlbeton *m*
f béton *m* armé
nl gewapend beton *n*
r железобетон *m*

R296 *e* **reinforced concrete beam**
d Stahlbetonträger *m*
f poutre *f* en béton armé
nl gewapend-betonbalk *m*
r железобетонная балка *f*

R297 *e* **reinforced concrete bridge**
d Stahlbetonbrücke *f*
f pont *m* en béton armé
nl gewapend-betonbrug *f(m)*
r железобетонный мост *m*

R298 *e* **reinforced concrete column**
d Stahlbetonstütze *f*, Stahlbetonsäule *f*
f colonne *f* [poteau *m*] en béton armé

nl gewapend betonnen kolom *f(m)*
r железобетонная колонна *f*

R299 *e* **reinforced concrete floor**
d Stahlbetondecke *f*
f plancher *m* en béton armé
nl vloer *m* van gewapend beton
r железобетонное перекрытие *n*

R300 *e* **reinforced concrete hollow slab**
d Stahlbetonhohlplatte *f*
f hourdis *m* creux [dalle *f* alvéolaire]
en béton armé
nl holle vloerplaat *f(m)* van gewapend
beton
r пустотелая железобетонная панель
f [плита *f*] перекрытия

R301 **reinforced concrete mat** *see* **raft
foundation**

R302 *e* **reinforced masonry**
d bewehrtes Mauerwerk *n*
f maçonnerie *f* armée [renforcée]
nl gewapend metselwerk *n*
r армированная каменная кладка *f*;
армокаменная конструкция *f*

R303 *e* **reinforced plaster**
d verstärkter Putz *m* (*z. B. mit
Glasfaser*)
f enduit *m* en plâtre armé
nl gewapend pleisterwerk *n*
r армированная штукатурка *f*

R304 *e* **reinforced plastic**
d verstärkter Kunststoff *m*
f plastique *m* renforcé [armé], matière *f*
plastique armée
nl gewapende kunststof *f(m)*
r армированный пластик *m*,
армированная пластмасса *f*

R305 *e* **reinforcement**
d 1. Bewehrung *f*, Armierung *f*
2. Verstärkung *f*
f 1. armatures *f pl*, fers *m pl*, aciers
m pl 2. renforcement *m*
nl 1. wapening *f* 2. versterking *f*
r 1. арматура *f* 2. усиление *n*

R306 *e* **reinforcement assembly shop**
d Biegerei *f*
f atelier *m* d'armature [de ferraillage]
nl werkplaats *f(m)* voor het buigen van
de wapeningsstaven
r арматурный цех *m*

R307 **reinforcement bar** *see* **rebar**

R308 **reinforcement cage** *see* **cage of
reinforcement**

R309 *e* **reinforcement cutting shears**
d Rundstahlschere *f*
f cisaille *f* pour ronds de béton [pour
acier d'armature]
nl betonijzerschaar *f(m)*
r ножницы *pl* для резки арматуры,
арматурные ножницы *pl*

R310 *e* **reinforcement displacement**
d Bewehrungsverschiebung *f*
f déplacement *m* de l'armature (*par
rapport à la position prévue*)

nl verschuiving *f* van de wapening
r смещение *n* арматуры (*от проектного положения*)

R311 *e* reinforcement percentage
d Bewehrungsprozentsatz *m*
f pourcentage *m* d'armatures
nl wapeningspercentage *n*
r процент *m* армирования

R312 *e* reinforcement placement
d Verlegung *f* [Einbau *m*] der Bewehrung
f mise *f* en place des armatures
nl stellen *n* van de wapening
r укладка *f* арматуры

R313 *e* reinforcement ratio
d Bewehrungsverhältnis *n*, Armierungsverhältnis *n*, Bewehrungsanteil *m*
f rapport *m* d'armatures [de ferraillage]
nl wapeningspercentage *n*
r коэффициент *m* армирования

R314 reinforcement shop *see* reinforcement assembly shop

R315 *e* reinforcement stirrup
d Bügel *m*
f étrier *m*
nl wapeningsbeugel *m*
r арматурный хомут *m*

R316 *e* reinforcement stresses
d Zugspannung *f* in der Bewehrung
f contraintes *f pl* dans les barres d'armature tendues
nl trekspanning *f* in de wapening
r напряжение *n* в растянутой арматуре

R317 *e* reinforcement yard
d zentraler Betonstahlverarbeitungsplatz *m*
f atelier *m* de façonnage des armatures
nl centrale werkplaats *f(m)* voor betonstaal
r арматурный двор *m*

R318 *e* reinforcing
d 1. Bewehren *n*, Armieren *n* 2. Verstärkung *f* 3. Bewehrung *f*, Armierung *f*
f 1. mise *f* en place de ferraillage 2. renforcement (*d'une construction*) 3. ferraillage *m*, armatures *f pl*
nl 1. wapenen *n*, plaatsen *n* van de wapening 2. versterking' *f* 3. wapening *f*
r 1. армирование *n* 2. усиление *n* 3. арматура *f*

R319 *e* reinforcing angle
d Versteifungswinkel *m*
f cornière *f* de raidissement [de renfort, raidisseuse]
nl versterkingshoekstaal *n*
r уголок *m* жёсткости

R320 reinforcing bar *see* rebar

R321 *e* reinforcing bar bending schedule
d Biegeliste *f*

f nomenclature *f* des aciers [de barres à béton]
nl buigstaat *m*
r спецификация *f* арматуры

R322 *e* reinforcing beam
d Versteifungsträger *m*
f poutre *f* de rigidité [de raidissement, de renfort]
nl verstijvingsbalk *m*
r балка *f* жёсткости

R323 *e* reinforcing fabric, reinforcing mesh
d Bewehrungsnetz *n*, Bewehrungsmatte *f*; Stahlgewebeeinlage *f*
f treillis *m* soudé [d'armature]; toile *f* à fils d'acier pour la protection des talus
nl wapeningsmat *f(m)*
r сварная арматурная сетка *f*; стальная сетка *f* для укрепления откосов

R324 ? reinforcing rib
d Aussteifungsrippe *f*
f nervure *f* de raidissement [de renforcement]
nl verstijvingsrib *f(m)*
r ребро *n* жёсткости

R325 *e* reinforcing ring
d Verstärkungsring *m* (*Schornstein*)
f frette *f* de renforcement de la cheminée
nl schoorsteenband *m*
r армирующее кольцо *n* (*ствола дымовой трубы*)

R326 *e* reinforcing steel
d Bewehrungsstahl *m*, Betonstahl *m*
f acier *m* d'armature [à béton]
nl wapeningsstaal *n*, betonstaal *n*
r арматурная сталь *f*

R327 *e* reinforcing work
d Bewehrungsarbeiten *f pl*, Armierungsarbeiten *f pl*
f travaux *m pl* de ferraillage
nl betonstaalvlechten *n*
r арматурные работы *f pl*

R328 *e* rejects
d 1. Siebrückstand *m*, Siebüberlauf *m* 2. Abgänge *m pl*, Abfall *m* 3. Ausschuß *m* (*Aufbereitung*)
f 1. rejet *m* au tamis, refus *m* de tamisage 2. schistes *m pl* [déchets *m pl*] de lavage 3. rejets *m pl*
nl 1. zeefrest *f(m)* 2. afval *m*; wasstenen *m pl* 3. uitschot *n*
r 1. остаток *m* на сите, надрешётный продукт *m* 2. хвосты *m pl* обогащения 3. некондиционный продукт *m* (*обогащения заполнителей*)

R329 *e* relative compaction
d relative Verdichtung *f*
f compacité *f* (du sol) relative
nl relatieve verdichting *f*
r относительная плотность *f* сложения грунта (*процентное отношение плотности грунта*

в естественном залегании к его
максимальной плотности)

R330 e relative density
d Dichtezahl *f*, Dichteverhältnis *n*,
relative Dichte *f*
f densité *f* relative
nl relatieve dichtheid *f*
r относительная плотность *f*

R331 e relative humidity
d relative Luftfeuchtigkeit *f*
f humidité *f* relative
nl relatieve vochtigheid *f*
r относительная влажность *f*
(воздуха)

R332 e relative roughness
d relative Rauhigkeit *f*
f rugosité *f* relative
nl relatieve ruwheid *f*
r относительная шероховатость *f*

R333 e relative stability
d relative Abwasserstabilität *f*
f stabilité *f* relative (des eaux
usées)
nl relatieve stabiliteit *f* (van de
lozing van afvalwater)
r относительная стабильность *f*
(сточных вод)

R334 e relaxation
d Relaxation *f*, Entspannung *f*,
Spannungsfreimachen *n*
f relaxation *f*
nl relaxatie *f*
r релаксация *f*

R335 e relaxation time
d Relaxationszeit *f*
f temps *m* [délai *m*] de relaxation
nl relaxatietijd *m*
r время *n* релаксации

R336 e release
d Ablassen *n*, Regulierungsabgabe *f*,
Zuschußwasser *n*
f lâchure *f*, éclusée *f*
nl spuien *n*
r попуск *m*

R337 e release agent
d Entschalungsmittel *n*
f huile *f* de coffrage [de décoffrage]
nl bekistingsolie *m*
r опалубочная смазка *f*; состав *m*,
предупреждающий сцепление
бетона с опалубкой

R338 e release emulsion
d Entschalungsemulsion *f*
f émulsion *f* de coffrage [de décoffrage]
nl ontkistingsemulsie *f*
r эмульсия *f* для смазывания опалубки

R339 e release valve
d Entlastungsventil *n*
f soupape *f* d'échappement
nl ontlastklep *f(m)*
r выпускной клапан *m*

R340 e reliability
d Zuverlässigkeit *f*, Betriebssicherheit *f*
f fiabilité *f*; sûreté *f* de fonctionnement
nl betrouwbaarheid *f*
r надёжность *f*; безотказность *f*

R341 e relief damper
d Sicherheitsklappe *f*
f clapet *m* de sûreté
nl veiligheidsklep *f(m)*
r предохранительный (воздушный)
клапан *m*

R342 e relief drain
d Sauger *m*
f drain *m* de décharge
nl regelende drain *m*
r регулирующая дрена *f*

R343 relief pit *see* relief shaft

R344 e relief road
d Entlastungsstraße *f*
f route *f* [voie *f*] urbaine de dégagement
nl ontlastende weg *m*
r вспомогательная городская дорога *f*
для разгрузки транспортных
потоков, разгрузочная дорога *f*

R345 e relief sewer
d Entlastungsleitung *f*
f égout *m* secondaire
nl ontlastende leiding *f*
r запасный (канализационный)
коллектор *m*

R346 e relief shaft
d Entlastungsschacht *m*,
Entlastungsbrunnen *m*
f puits *m* de décharge
nl ontlastende schacht *f(m)*
r разгружающая шахта *f*

R347 e relief valve
d Entlastungsventil *n*,
Überdruckventil *n*, Sicherheitsventil *n*
f soupape *f* de sûreté, clapet *m* d'excès
de pression
nl overdrukventiel *n*
r клапан *m* сброса давления,
предохранительный клапан *m*

R348 e relief vent
d Entlüftungsrohr *n*
f tube *f* de ventilation secondaire
(*d'un appareil sanitaire*)
nl ontspanningsleiding *f* (*riolering*)
r вентиляционная труба *f*,
присоединённая к отводной трубе
санитарного прибора

R349 e relief well
d Dränschacht *m*, Abzapfbrunnen *m*,
Entlastungsbrunnen *m*,
Entlastungsschacht *m*,
Entspannungsbrunnen *m*
f puits *m* de drainage ascendant [de
décharge, de décompression]
nl ontspanningsbron *m*
r дренажный [разгружающий]
колодец *m*, разгружающая скважина
f

R350 e **relieving arch**
 d Entlastungsbogen *m*
 f arc *m* de décharge
 nl ontlastende boog *m*
 r разгрузочная арка *f*

R351 e **relieving wall**
 d Entlastungswand *f*
 f mur *m* de décharge
 nl ontlastende muur *m*
 r разгрузочная стенка *f*

R352 e **reloading**
 d Nachbelastung *f*
 f rechargement *m*
 nl herbelasten *n*
 r повторное нагружение *n*

R353 e **reloading curve**
 d Nachbelastungskurve *f*
 f courbe *f* de rechargement
 nl herbelastingskromme *f(m)*
 r кривая *f* повторного нагружения

R354 e **relocatable building**
 d verfahrbares Gebäude *n*
 f bâtiment *m* déplaçable
 nl verplaatsbaar gebouw *n*
 r передвигаемое [перемещаемое]
 здание *n*

R355 **relocatable partition** see **demountable partition**

R356 e **relocation of building**
 d Standortveränderung *f* von Gebäude
 f déplacement *m* [translation *f*] d'un édifice
 nl verplaatsen *n* van een gebouw
 r перемещение *n* [передвижка *f*] здания

R357 e **remoldability**
 d Einformbarkeit *f* (*des Betongemisches*)
 f capacité *f* du remplissage (*des bétons*), maniabilité *f*, ouvrabilité *f* (*du béton*)
 nl verwerkbaarheid *f* van de betonmortel
 r удобоформуемость *f* (*бетонной смеси*)

R358 e **remolding test**
 d Steifeprüfung *f*
 f essai *m* Powers [Vebe]
 nl plasticiteitsproef *f(m)*
 r определение *n* жёсткости
 (*бетонной смеси*)

R359 e **remote condenser**
 d Splitkondensator *m*
 f condenseur *m* séparé
 nl afzonderlijke condensator *m*
 r вынесенный конденсатор *m*

R360 e **render, float and set**
 d dreilagiger Putz *m* auf Putzträgergewebe
 f enduit *m* au plâtre à trois couches
 nl pleisterwerk *n* in drie lagen
 r трёхслойная штукатурка *f*

R361 e **rendering**
 d 1. Außenputz *m* 2. Unterputz *m*
 f 1. enduit *m* extérieur 2. sous-enduit *m*, couche *f* de fond
 nl 1. buitenpleisterwerk *n* 2. primaire raaplaag *f(m)*
 r 1. наружная штукатурка *f* 2. обрызг *m* (*первый слой штукатурного намёта*)

R362 e **repair gate**
 d Reparaturverschluß *m*
 f vanne *f* de dépannage [de garde, d'isolement]
 nl reparatie-afsluiter *m*
 r ремонтный затвор *m*

R363 e **repair materials**
 d Reparaturwerkstoffe *m* pl, Reparaturmaterialien *n* pl
 f matériaux *m* pl de réparation
 nl reparatiematerialen *n* pl
 r материалы *m* pl для ремонтных работ

R364 e **repair parts**
 d Ersatzteile *m, n* pl
 f pièces *f* pl de rechange
 nl verwisselstukken *n* pl
 r запасные части *f* pl

R365 e **repair quay**
 d Reparaturkai *m*, Ausbesserungskai *m*
 f quai *m* de réparations
 nl reparatiekade *f(m)*
 r ремонтная набережная *f*

R366 e **repeated cycles of freezing and thawing**
 d Frost-Tau-Wechsel *m*, Gefrier-Auftau-Folge *f*
 f cycles *m* pl répétés du gel et du dégel
 nl afwisselend vriezen *n* en dooien *n*
 r повторные циклы *m* pl замораживания и оттаивания

R367 e **repeated load**
 d wiederholte Belastung *f*
 f charge *f* répétée
 nl wisselende belasting *f*
 r повторная нагрузка *f*

R368 e **repeated stress cycles**
 d Lastspiele *n* pl bei Schwellast, Schwellbeanspruchung *f*
 f cycles *m* pl de contraintes répétées
 nl herhaalde spanningscyclussen *m* pl
 r циклы *m* pl повторных знакопостоянных напряжений

R369 e **repeated stresses**
 d Schwellspannungen *f* pl
 f contraintes *f* pl répétées [ondulées]
 nl wisselende spanningen *f* pl
 r повторные (знакопостоянные) напряжения *n* pl

R370 e **repelling groin** *US*, **repelling groyne** *UK*
 d Schöpfbuhne *f*, Wasserleitdamm *m*, Abweiser *m*
 f épi *m* de dérivation
 nl leidam *m*, strekdam *m*
 r струеотбойная [струенаправляющая] дамба *f*

R371 e **repetitive housing**
 d Serienwohnungsbau *m*

 f construction *f* de bâtiments
d'habitation en série; construction *f*
de maisons d'habitation d'après les
projets type
 nl seriewoningbouw *m*
 r строительство *n* жилых домов по
типовым проектам; серийное
[массовое] жилищное строительство *n*

R372 *e* **repetitive loading**
 d mehrmalige Belastung *f*
 f sollicitation *f* alternée [périodique,
répétée], sollicitation *f* par des
forces variables dans le temps
 nl wisselende belasting *f*
 r многократное нагружение *n*

R373 **repetitive stresses** *see* **repeated stresses**

R374 *e* **replacement filter**
 d Wegwerffilter *n, m*
 f filtre *m* à un seul emploi
 nl wegwerpfilter *n, m*, filter *n, m* voor
eenmalig gebruik
 r фильтр *m* однократного применения

R375 **reposting** *see* **reshoring**

R376 *e* **reproducible methods**
 d reproduzierbare Verfahren *n pl*
 f méthodes *f pl* reproductibles
 nl reproduceerbare methoden *f pl*
 r воспроизводимые методы *m pl*
(*напр. испытаний*)

R377 *e* **reproducible test results**
 d reproduzierbare Versuchsergebnisse
n pl
 f résultats *m pl* d'essai reproductibles
 nl reproduceerbare testresultaten *n pl*
 r воспроизводимые результаты *m pl*
испытаний

R378 *e* **reproductibility**
 d Reproduzierbarkeit *f*
 f reproductibilité *f*
 nl reproduceerbaarheid *f*
 r воспроизводимость *f*

R379 *e* **required air quantity**
 d erforderliche Luftmenge *f*
 f quantité *f* d'air prévue, débit *m* d'air
spécifié
 nl benodigde hoeveelheid *f* lucht
 r требуемый расход *m* воздуха

R380 *e* **required design load**
 d vorgeschriebene Rechnungslast *f*
 f charge *f* de calcul spécifiée, charge *f*
de projet
 nl voorgeschreven belasting *f*, waarop
een constructie moet worden berekend
 r заданная расчётная нагрузка *f*

R381 *e* **research program**
 d Forschungsprogramm *n*
 f programme *m* de recherches
 nl onderzoekprogramma *n*
 r программа *f* исследований

R382 *e* **reservoir**
 d Staubecken *n*, Speicher *m*, Stausee *m*,
Talsperre *f*; Behälter *m*; Becken *n*;
Pumpspeicher *m*

 f retenue *f*; réservoir *m*; bassin *m*
d'accumulation; bassin *m* de retenu
 nl reservoir *n*, spaarbekken *n*
 r водохранилище *n*; резервуар *m*;
бассейн *m*; гидроаккумулятор *m*

R383 *e* **reservoir basin**
 d 1. Staubeckenmulde *f* 2. Einzugsgebiet
n des Staubeckens
 f 1. cuvette *f* de la retenue 2. bassin *m*
versant de la retenue
 nl 1. kom *f(m)* van het stuwbekken
2. watervanggebied *n* van een
stuwbekken
 r 1. чаша *f* водохранилища
2. водосбор *m* водохранилища

R384 *e* **reservoir capacity**
 d Staubeckeninhalt *m*,
Staubeckenvolumen *n*
 f capacité *f* [volume *m* du bassin] de
retenue
 nl capaciteit *f* van het stuwbekken
 r объём *m* водохранилища

R385 *e* **reservoir emptying**
 d Entleeren *n*, Ablassen *n*,
Stauraumentleerung *f*
 f vidange *f* (*d'un bassin de retenue*)
 nl het stuwbekken laten leeglopen
 r опорожнение *n* водохранилища

R386 *e* **reservoir inflow**
 d Staubeckenzufluß *m*
 f venues *f pl* d'eau
 nl instroming *f* van een stuwbekken
 r пополнение *n* водохранилища

R387 *e* **reservoir life**
 d Berechnungsnutzungsdauer *f* des
Staubeckens
 f vie *f* utile de reservoir, durée *f* de
vie d'un bassin de retenue
 nl levensduur *f* van het stuwbekken
 r расчётный срок *m* эксплуатации
водохранилища

R388 *e* **reservoir routing**
 d Stauraumberechnung *f*
 f calcul *m* de la propagation des crues
en présence d'un réservoir
 nl projecteren *n* [berekenen *n*] van een
stuwbekken
 r расчёт *m* водохранилища

R389 **reservoir sedimentation** *see* **sanding-up**

R390 *e* **resetting of forms**
 d Versetzen *n* der Schalung
 f déplacement *m* des coffrages
 nl opnieuw plaatsen *n* van de bekisting
 r перестановка *f* опалубки

R391 *e* **resettling**
 d Nachklären *n*, Nachklärung *f*
 f décantation *f* secondaire
 nl nabezinking *f*
 r вторичное отстаивание *n*

R392 *e* **reshoring**
 d provisorische Abstützung *f* der
entschaleten Betonkonstruktionen

f étayage *m* [étaiement *m*] du béton
décoffré
nl tijdelijk stutten *n* van een
betonconstructie na het ontkisten
r усиление *n* стойками распалубленных
бетонных конструкций

R393 *e* residential allotment
d Siedlungsbaugebiet *n*
f lotissement *m* résidentiel
nl bouwgrond *m* voor woningbouw
r земельные участки *m pl*,
выделенные для жилищного
строительства

R394 residential area *see* residential district

R395 *e* residential buildings
d Wohnbauten *m pl*
f immeubles *m pl* d'habitations
[résidentiels]
nl woongebouwen *n pl*
r жилые здания *n pl*

R396 *e* residential community
d Residenzeinheit *f*, Wohnkomplex *m*
f ensemble *m* résidentiel
nl woongemeenschap *m*, woonwijk
f(m)
r группа *f* жилых домов; жилой
массив *m*

R397 *e* residential density
d Bevölkerungsdichte *f* eines
Wohnbezirks
f densité *f* résidentielle
nl bevolkingsdichtheid *f* van een
woonwijk
r плотность *f* населения в жилом
районе, плотность *f* заселения
жилого района

R398 *e* residential district
d Wohnviertel *n*, Wohngebiet *n*
f quartier *m* résidentiel
nl woonwijk *f(m)*
r жилой квартал *m* (*города*)

R399 *e* residential neighbourhood
d Nachbarschaftseinheit *f*,
Wohnkomplex *m*
f unité *f* de voisinage
nl woonbuurt *m*
r микрорайон *m*

R400 *e* residential unit
d Wohn(ungs)einheit *f*
f unité *f* de logement
nl wooneenheid *f*
r жилая единица *f*, условная квартира *f*

R401 *e* residential zone
d Wohngebietsfläche *f*
f zone *f* résidentielle
nl woongebied *n*
r селитебная [жилая] зона *f*

R402 *e* resident engineer
d Ingenieur *m* der Bauaufsicht (*seitens
des Auftraggebers*)
f ingénieur *m* en résidence
nl ingenieur *m* belast met het toezicht

r инспектор *m* технадзора на
строительной площадке
(*представитель заказчика*)

R403 *e* residual deflection
d bleibender Durchhang *m*, plastische
Durchbiegung *f*
f flèche *f* résiduelle
nl blijvende doorbuiging *f*
r остаточный прогиб *m*

R404 *e* residual deformation, residual strain
d Restverformung *f*, bleibende
Verformung *f*, plastische
Deformation *f*
f déformation *f* résiduelle
nl blijvende vervorming *f*
r остаточная деформация *f*

R405 *e* residual stresses
d Restspannungen *f pl*
f contraintes *f pl* résiduelles
nl blijvende spanningen *f pl*
r остаточные напряжения *n pl*

R406 *e* resilience
d 1. Formänderungsenergie *f*,
elastische Energie *f* 2. elastiche
Rückbildung *f* 3. Nachgiebigkeit *f*
f résilience *f*
nl 1. vervormingsarbeid *m* 2. elastisch
herstel *n* 3. veerkracht *f(m)*
r 1. удельная работа *f* деформации
2. упругое восстановление *n*
3. упругая податливость *f*

R407 *e* resilient flooring
d elastischer Fußbodenbelag *m*
f revêtement *m* de sol élastique, couvre-
-sol *m* élastique
nl zwevende vloer *m*
r эластичное покрытие *n* пола

R408 *e* resilient joint
d elastische Verbindung *f*
f joint *m* élastique
nl elastische verbinding *f*
r упругое соединение *n*

R409 *e* resilient support
d elastische Stütze *f*
f appui *m* [support *m*] élastique
nl verende oplegging *f*
r упруго-податливая опора *f*

R410 resin-bonded chipboard *see* chipboard

R411 *e* resin-bonded plywood
d kunstharzgebundenes Sperrholz *n*
f contre-plaqué *m* collé par résine
synthétique
nl met kunsthars gelijmd multiplex *n*
r фанера *f*, проклеенная синтетической
смолой

R412 resin concrete *see* polymer concrete

R413 *e* resin emulsion paint
d Emulsions-Kunstharzfarbe *f*
f peinture-émulsion *f* à base de résines
nl kunsthars-emulsieverf *f(m)*
r эмульсионная краска *f* на основе
смол

R414 *e* **resin gun**
   d Kunstharzspritzpistole *f*
   f pistolet *m* à calfeutrer les joints avec
     des résines
   nl kunsthars-spuitpistool *n*
   r шприц-пистолет *m* для заполнения
     швов синтетической смолой

R415   **resin mortar** *see* **polymer mortar**

R416 *e* **resinous waterproof adhesive**
   d wasserbeständiges Bindemittel *n* auf
     Kunstharzbasis
   f adhésif *m* à base de résine résistant
     à l'eau
   nl watervaste lijm *m* op kunstharsbasis
   r водостойкий клей *m* на основе смол

R417 *e* **resistance**
   d Widerstand *m*
   f résistance *f*
   nl weerstand *m*
   r сопротивление *n*, сопротивляемость
     f

R418 *e* **resistance coefficient**
   d Strömungswiderstandsbeiwert *m*,
     hydraulische Widerstandszahl *f*
   f coefficient *m* de Darcy [de résistance
     hydraulique, de perte de charge
     linéaire]
   nl hydraulische wrijvingscoëfficiënt *m*
   r коэффициент *m* гидравлического
     сопротивления в трубах,
     коэффициент *m* Дарси

R419 *e* **resistance strain gauge**
   d Dehnmeßstreifen *m*
   f extensomètre *m* [jauge *f*] à résistance
   nl rekstrookje *n*
   r тензометр *m*, тензодатчик *m*
     сопротивления, проволочный
     тензодатчик *m*

R420 *e* **resistance to cracking**
   d Rißfestigkeit *f*
   f résistance *f* à la fissuration
   nl bestendigheid *f* tegen scheurvorming
   r трещиностойкость *f*

R421 *e* **resistance to frost**
   d Frostwiderstand *m*,
     Frostbeständigkeit *f*
   f résistance *f* au gel
   nl vorstbestendigheid *f*
   r морозостойкость *f*

R422 *e* **resistance welding**
   d Widerstandsschweißen *n*
   f soudage *m* [soudure *f*] par résistance
   nl weerstandslassen *n*
   r сварка *f* сопротивлением, контактная
     сварка *f*

R423 *e* **resisting moment**
   d Moment *n* der inneren Kräfte,
     Widerstandsmoment *n*
   f moment *m* résistant [de résistance]
   nl weerstandsmoment *n*
   r момент *m* внутренних сил,
     момент *m* сопротивления

R424 *e* **resolution of forces**
   d Zerlegung *f* der Kräfte,
     Kräftezerlegung *f*
   f résolution *f* des forces
   nl ontbinding *f* van krachten
   r разложение *n* сил

R425 *e* **resonance screen**
   d Resonanz(-Schwing)sieb *n*
   f crible *m* à résonance, pulsocrible *m*
   nl trilzeef *f(m)*
   r резонансный (вибро)грохот *m*

R426 *e* **resonance table vibrator**
   d Resonanztisch *m*
   f table *f* vibreuse à résonance
   nl resonantietafel *f(m)*
   r резонансная виброплощадка *f*

R427 *e* **resonant pile driver**
   d Resonanz-Pfahlramme *f*
   f sonnette *f* [mouton *m*] à résonance
   nl reesonantie-heiblok *n*
   r резонансный вибропогружатель *m*
     свай

R428 *e* **respirator**
   d Staubmaske *f*, Atemschutzgerät *n*
   f respirateur *m*, masque *f* antipoussière
   nl stofmasker *n*
   r респиратор *m*

R429 *e* **restoring of topsoil**
   d Wiederherstellung *f* des Mutterbodens
     [der Bodenkrume]
   f restauration *f* [restitution *f*] de la
     couche de terre végétale
   nl terugstorten *n* van de bovengrond
   r восстановление *n* растительного слоя

R430 *e* **restrained beam**
   d eingespannter Balken *m*
   f poutre *f* encastrée
   nl ingeklemde balk *m*
   r балка *f* с защемлёнными
     неподвижными опорами [концами]

R431 *e* **restrained slab**
   d eingespannte Platte *f*
   f dalle *f* encastrée
   nl ingeklemde plaat *f(m)*
   r защемлённая плита *f*

R432 *e* **restrained support**
   d eingespannte Stütze *f*
   f appui *m* [support *m*] fixe encastré
   nl ingeklemde stijl *m* [kolom *f(m)*]
   r неподвижная защемлённая опора *f*

R433 *e* **restraint**
   d Einspannung *f*
   f encastrement *m*
   nl inklemming *f*
   r защемление *n*; заделка *f*

R434 *e* **restraint conditions**
   d Einspannbedingungen *f pl*
   f conditions *f pl* d'encastrement aux
     appuis
   nl inklemmingsvoorwaarden *f pl*
   r условия *n pl* наложения связей;
     условия *n pl* защемления опор

R435 *e* **restraint forces**
  d Einspannkräfte *f pl*
  f forces *f pl* d'encastrement
  nl inklemmingskrachten *f(m) pl*
  r усилия *n pl* защемления; усилия
    n pl, вызванные наложением связей

R436 *e* **restraint moment**
  d Einspannmoment *n*
  f moment *m* d'encastrement
  nl inklemmingsmoment *n*
  r момент *m* защемления [заделки]

R437 *e* **restraint of concrete**
  d Betondehnungsbehinderung *f*
  f restriction *f* des déformations libres
    du béton
  nl opsluiten *n* van het beton
  r стеснение *n* [ограничение *n*]
    свободного деформирования бетона

R438 *e* **restraint of displacement**
  d Verschiebungs(be)hinderung *f*
  f restriction *f* des déplacements
  nl verhinderen *n* van verschuiving
  r ограничение *n* [стеснение *n*]
    перемещений

R439 *e* **restricted area**
  d Schutzgebiet *n*, Gefahrenzone *f*
  f zone *f* interdite
  nl verboden gebied *n*
  r запретная зона *f*

R440 *e* **restricted tender**
  d beschränkte Ausschreibung *f*
  f appel *m* d'offres restreint
  nl onderhandse aanbesteding *f*
  r торги *m pl* (*на заключение*
    строительного контракта) с
    ограниченным числом участников

R441 *e* **resultant of forces**
  d Resultierende *f*, Resultante *f* (*der*
    Kräfte)
  f résultante *f*, force *f* résultante
  nl resulterende kracht *m*, resultante *f*
  r равнодействующая *f* сил

R442 *e* **retaining crib wall**
  d Steinkastenstützmauer *f*
  f mur *m* de soutènement à encoffrements
  nl beschoeiing *f*, keerwand *m* van stijlen
    en planken
  r ряжевая подпорная стенка *f*

R443 *e* **retaining dike** *US*
  d Abschlußdamm *m* der Spülfläche
    [des Spülfeldes]
  f digue *f* de garde [de protection]
  nl wal *m* van het vloeiveld
  r дамба *f* обвалования (*карты намыва*)

R444 *e* **retaining wall**
  d Böschungsmauer *f* (*vor*
    aufgeschüttetem Boden)
  f mur *m* de soutènement
  nl keermuur *m*
  r подпорная стенка *f* откоса

R445 *e* **retaining wing**
  d Böschungsflügel *m* (*des Widerlagers*),
    Flügelmauer *f*
  f mur *m* en aile [en retour]
  nl vleugelmuur *m*
  r откосное крыло *n* (*берегового устоя*)

R446 *e* **retardation**
  d Verzögerung *f*
  f retardement *m*, retardation *f*
  nl vertraging *f*
  r замедление *n*

R447 *e* **retarder, retarding admixture**
  d Abbindeverzögerer *m*,
    Erstarrungsverzögerer *m*
  f retard(at)eur *m* [adjuvant-retard *m*] de
    prise
  nl verhardingsvertrager *m*
  r замедлитель *m* схватывания

R448 *e* **retarding reservoir**
  d Verzögerungsbecken *n*,
    Rückhaltebecken *n*,
    Rückhaltespeicher *m*,
    Hochwasserbecken *n*, Pufferbecken *n*,
    Ausgleichbecken *n*
  f bassin *m* de régulation à débit non
    réglable, réservoir *m* tampon
  nl bufferreservoir *n*
  r паводкоаккумулирующее
    водохранилище *n* с нерегулируемым
    выпуском воды, буферное
    водохранилище *n*

R449 *e* **retempering of concrete**
  d Wiederaufbereitung *f* von Beton
  f remalaxage *m* du béton
  nl namenging *f* van betonmortel
  r повторное перемешивание *n*
    бетонной смеси

R450 *e* **retention**
  d Rücklage *f*, Wasserrückhaltung *f*
  f rétention *f*
  nl retentie *f*
  r влагозадержание *n*

R451 *e* **retention basin**
  d Rückhaltebecken *n*
  f bassin *m* récepteur
  nl retentiebekken *n*
  r накопительный резервуар *m*

R452 *e* **reticulated masonry**
  d Ziermauerwerk *n* mit geneigten
    Binderschichten
  f maçonnerie *f* ornementale en boutisses
    inclinées
  nl metselwerk *n* in netverband
  r орнаментальная кирпичная кладка *f*
    с наклонным расположением
    тычковых рядов

R453   **retrofitting** *US see* **refit**

R454 *e* **return**
  d 1. Rückführung *f* 2. Rücklauf *m*,
    Rückleitung *f*
  f 1. retour *m* de (fluide) caloporteur
    2. conduite *f* de retour
  nl 1. terugloop *m* 2. retourleiding *f*
  r 1. возврат *m* 2. обратная линия *f*
    (*трубопровода*)

R455 *e* return air duct
   d Rückluftkanal *m*
   f conduite *f* d'air de retour
   nl retourluchtkanaal *n*
   r воздуховод *m* рециркуляционного воздуха (*подвергаемого вторичной обработке*)

R456 *e* return air fan
   d Umluftventilator *m*
   f ventilateur *m* d'air de retour
   nl retourluchtventilator *m*
   r вентилятор *m* рециркуляционного воздуха

R457 *e* return air system
   d Umluftsystem *n*, Rückluftanlage *f*
   f système *m* d'air de retour
   nl retourluchtsysteem *n*
   r рециркуляционная система *f*

R458 *e* return flow
   d Wasserrückfluß *m*, Rückversickerung *f*
   f écoulement *m* restitué
   nl terugvloeiing *f*
   r возвратный сток *m*

R459 *e* return period
   d Wiederholungsperiode *f*
   f période *f* de récurrence
   nl herhalingsperiode *f*
   r период *m* повторяемости

R460 *e* return pipeline
   d Rücklaufleitung *f*, Rücklauf *m*
   f canalisation *f* [tuyau *m*] de retour
   nl retourleiding *f*
   r обратный трубопровод *m*

R461 *e* return riser
   d Fallstrang *m*, Rücklaufstrang *m*
   f tuyau *m* vertical de retour
   nl valstrang *m*
   r обратный стояк *m*

R462 *e* return sludge
   d Rücklaufschlamm *m*, Impfschlamm *m*, Rücknahmeschlamm *m*
   f boue *f* activée de retour
   nl teruggevoerd slib *n*
   r возвратный активный ил *m*

R463 *e* return tapping
   d Anschlußöffnung *f* [Anschlußstutzen *m*] für Rücklaufrohr
   f branchement *m* de retour
   nl aansluitopening *f* voor retourleiding
   r присоединительный патрубок *m* для обратной [отводящей] трубы

R464 *e* return temperature
   d Rücklauftemperatur *f*
   f température *f* de retour
   nl retourwatertemperatuur *f*
   r температура *f* обратного теплоносителя (*напр. воды*)

R465 *e* return temperature limiter
   d Rückwassertemperaturbegrenzer *m*
   f limiteur *m* de température de retour
   nl begrenzer *m* van de retourwatertemperatuur

   r ограничитель *m* температуры обратной воды

R466 *e* return water
   d Umlaufwasser *n*, Rücklaufwasser *n*
   f eaux *f pl* de restitution recyclées
   nl retourwater *n*
   r возвратные воды *f pl*, обратная вода *f*

R467 *e* reusability, reuse
   d Wiederverwendungsmöglichkeit *f*, Wiederverwendbarkeit *f* (*z. B. der Schalung*)
   f réemploi *m* (*p.ex. de coffrages*)
   nl wedergebruik *n* (*b.v. van bekisting*)
   r оборачиваемость *f* (*напр. опалубки*), возможность *f* многократного использования

R468 *e* reveal
   d Leibung *f*
   f embrasement *m*
   nl negge *f(m)*
   r откос *m* проёма (*дверного или оконного*)

R469 *e* reverberation
   d 1. Nachhall *m* 2. Wassertstoß *m*, Wasserschlag *m*, Druckstoß *m*
   f 1. réverbération *f* 2. coup *m* (de) bélier, marteau *m* d'eau
   nl 1. nagalm *m* 2. waterslag *m*
   r 1. реверберация *f* 2. гидравлический удар *m*

R470 *e* reverse-acting thermostat
   d Umkehrthermostat *m*
   f thermostat *m* à action inverse
   nl omkeerthermostaat *m*
   r термостат *m* обратного действия

R471 *e* reverse curve
   d Gegenbogen *m*, Gegenkurve *f*, Gegenkrümmung *f*
   f courbe *f* contraire, contre-courbe *f*
   nl tegengestelde bocht *f(m)* (*wegenbouw*)
   r обратная кривая *f* (*дороги*)

R472 *e* reverse cycle heating
   d Heizung *f* durch Prozeßumkehrung
   f échauffement *m* par cycle inversé
   nl verwarming *f* door omkering van het proces
   r отопление *n* реверсивным циклом

R473 *e* reversed filter
   d Entlastungsfilter *n*, *m*, Rückhaltefilter *n*, *m*, umgekehrtes Filter *n*, *m*
   f filtre *m* inversé
   nl omgekeerd filter *n*, *m*
   r обратный фильтр *m*

R474   reverse gradient *see* adverse slope

R475 *e* reversible fan
   d umschaltbarer [reversierbarer] Lüfter *m*
   f ventilateur *m* réversible
   nl omkeerbare ventilator *m*
   r реверсивный вентилятор *m*

R476 reversible window *see* vertical
pivoted window

R477 *e* reversing valve
d Umsteuerventil *n*, Umstellventil *n*
f robinet *m* de renversement
nl omkeerklep *f(m)*
r реверсивный клапан *m*

R478 *e* revetment
d Verkleidung *f*; Auskleidung *f*;
Decke *f*; Belag *m*; Befestigung *f*
f revêtement *f*
nl beschoeiing *f*; grondkering *f*
r облицовка *f* (*напр. стен*); одежда *f*,
покрытие *n* (*откосов, дорог, пола*)

R479 *e* revetment wall
d Futtermauer *f*; Verkleidungsmauer *f*
f mur *m* de revêtement; épaulement *m*
nl keermuur *m*
r подпорная стенка *f* для защиты
берегового откоса; облицовочная
стенка *f* для защиты уступа скальной
породы

R480 *e* revibration
d Nach(ein)rüttelung *f*
f révibration *f*
nl natrillen *n* (*van betonmortel*)
r повторное вибрирование *n* (*бетона*)

R481 revolving-blade mixer *see* open-top
mixer

R482 *e* revolving door
d Drehtür *f*
f porte *f* tournante [tambour]
nl draaideur *f(m)*
r вращающаяся дверь *f*

R483 *e* revolving drier, revolving dryer
d Trommeltrockner *m*
f tambour *m* sécheur, sécheur *m*
à tambour
nl trommeldroger *m*
r барабанная сушилка *f*

R484 *e* revolving paddle finisher
d Betonbahnautomat *m*
f finisseur *m* à béton à palettes
rotatives
nl verdichtingsmachine *f* (*betonweg*)
r бетоноотделочный агрегат *m*
[финишер *m*] с вибратором
и вращающимся разравнивателем

R485 *e* revolving screen
d 1. Wälzfilter *n*, *m* 2. Siebtrommel *f*,
Trommelsieb *n*
f 1. filtre *m* à tambour 2. crible *m*
rotatif, trommel-classeur *m*
nl 1. trommelfilter *n*, *m* 2. trommelzeef
f(m)
r 1. вращающийся фильтр *m*
2. вращающийся грохот *m*

R486 *e* revolving shovel
d Schwenk-Löffelbagger *m*, Dreh-
-Löffelbagger *m*
f pelle *f* (mécanique) à rotation totale
nl draaibare lepelschop *f(m)*
r (полно)поворотный экскаватор *m*

R487 *e* rewater
d Umlaufwasser *n*, Zirkulationswasser
n, Kreislaufwasser *n*
f eau *f* de (ré)circulation [récyclée]
nl circulerend water *n*
r оборотная [циркуляционная] вода *f*

R488 *e* rib
d Rippe *f*; Steife *f*; Längsträger *m*;
Steg *m* des Stahlbeton-T-Trägers
f nervure *f*; renfort *m*; poutre *f*
longitudinalle (de pont); âme *f* d'une
poutre de béton armé en T
nl rib *f(m)*; verstijvingsrib *f(m)*;
langsligger *m*; lijf *n* van T-balk
r ребро *n*; ребро *n* жёсткости;
продольная балка *f* (*моста*); стенка *f*
(*железобетонной тавровой балки*)

R488a ribbed floor *see* ribbed-slab floor

R489 *e* ribbed pipe
d Rippenrohr *n*, Lamellenrohr *n*
f tuyau *m* nervuré, tube *m* nervuré
[à ailettes]
nl ribbenbuis *f(m)*
r ребристая [оребрённая] труба *f*

R490 *e* ribbed reinforcing bars
d Betonrippenstahl *m*, quergerippter
Betonformstahl *m*
f barre *f* d'armature crénelée
nl geribd betonstaal *n*
r ребристая арматурная сталь *f*,
арматура *f* периодического профиля

R491 *e* ribbed slab
d Rippenplatte *f*
f dalle *f* nervurée
nl ribbenplaat *f(m)*
r ребристая плита *f*

R492 *e* ribbed-slab floor
d Rippendecke *f*
f plancher *m* à nervures
nl ribbenvloer *m*
r ребристое перекрытие *n*

R493 *e* ribbon development
d Bandbebauung *f*
f développement *m* urbain en bande
nl lintbebouwing *f*
r ленточная застройка *f*

R494 *e* ribbon loading
d gleichzeitige Dosierung *f* und
Beschickung *f* aller Ingredienzien des
Betons
f chargement *m* [dosage *m*] simultané
nl gelijktijdige dosering *f*
(*mortelbereiding*)
r одновременное дозирование *n* (*всех
компонентов бетонной смеси*)

R495 ribbon window *see* window band

R496 *e* rich concrete
d zementreicher Beton *m*
f béton *m* riche
nl betonmortel *m* met hoog cementgehalte
r бетон *m* с высоким содержанием
цемента, жирный бетон *m*, жирная
бетонная смесь *f*

R497　rich lime *see* **fat lime**

R498 *e* **rich mixture**
　　d fette Mischung *f*, fettes Gemisch *n*
　　f mélange *m* gras [riche]
　　nl rijk mengsel *n*
　　r жирная смесь *f*

R499 *e* **rideability**
　　d Fahreigenschaften *f pl* (*Straße*)
　　f qualité *f* de roulement [roulante]
　　nl berijdbaarheid *f* (*straat*)
　　r ездовые качества *n pl* (*дороги*)

R501 *e* **ridge**
　　d 1. First *m*, Förste *f* 2. Grat *m*
　　　3. Kamm *m*
　　f 1. faîte *m*, faîtage *m* 2. arêtier *m*
　　　3. crête *f*, seuil *m*
　　nl 1. nok *f(m)* (*dak*) 2. hoekkeper *m*
　　　3. kruin *f(m)* (*dijk*)
　　r 1. конёк *m* (*крыши*) 2. ребро *n*
　　　(*крыши*) 3. гребень *m* (*напр. дамбы*)

R502 *e* **ridge board**
　　d Firstbrett *n*
　　f planche *f* de faîtage
　　nl ruiter *m*
　　r коньковая доска *f*

R503 *e* **ridge gusset plate**
　　d First-Stoßlasche *f*
　　f ferrure *f* de faîtage, gousset *m*
　　　d'assemblage faîtière
　　nl hechtplaat *f(m)*, spijkerplaat *f(m)*
　　　(*nokgording*)
　　r коньковая соединительная накладка *f*

R504 *e* **ridge purlin**
　　d Firstpfette *f*
　　f panne *f* faîtière
　　nl nokgording *f*
　　r коньковый прогон *m*

R505 *e* **ridge tile**
　　d Firstziegel *m*, Firststein *m*,
　　　Walmstein *m*
　　f tuile *f* faîtière
　　nl nokvorst *m*, vorstpan *f(m)*
　　r коньковая черепица *f*

R506 *e* **rig**
　　d 1. Vorrichtung *f*; Anlage *f*;
　　　Betakelung *f*; Ausrüstung *f*
　　　2. Anhängeausrüstung *f*
　　　3. Bohrausrüstung *f*; Bohrmaschine *f*,
　　　Bohranlage *f* 4. Pfahlramme *f*
　　f 1. équipement *m*; outillage *m*
　　　2. équipement *m* porté; outil *m* porté
　　　(*p.ex. d'un excavateur*) 3. sondeuse *f*;
　　　appareil *m* de forage 4. sonnette *f*
　　nl 1. uitrusting *f* 2. hulpstuk *n*
　　　3. boormachine *f* 4. hei *f(m)*,
　　　heistelling *f*
　　r 1. установка *f*; оснастка *f*;
　　　оборудование *n* 2. навесное
　　　оборудование *n* (*напр. экскаватора*)
　　　3. буровой станок *m*, буровая
　　　установка *f* 4. копёр *m*

R507 *e* **right-of-way**
　　d Wegerecht *n*, Freistreifen *m*
　　　(*Straßenbau*)

　　f emprise *f*
　　nl voorrang *f(m)*
　　r полоса *f* отвода

R508 *e* **rigid frame**
　　d Steifrahmen *m*, starrer **Rahmen** *m*
　　f portique *m* rigide; charpente *f*
　　　[assature *f*] rigide
　　nl stijf skelet *n* [portaal *n*]
　　r рама *f*; рамная конструкция *f*

R509 *e* **rigid insulation**
　　d steife Wärmedämmung *f*,
　　　Plattenisolierung *f*
　　f isolation *f* rigide
　　nl stijve warmte-isolatie *f*
　　r жёсткая теплоизоляция *f*

R510 *e* **rigidity**
　　d Steifigkeit *f*
　　f rigidité *f*
　　nl stijfheid *f*
　　r жёсткость *f*

R511 *e* **rigid joint**
　　d starre Verbindung *f*; Steifknoten *m*
　　f joint *m* rigide
　　nl stijve verbinding *f*
　　r жёсткое соединение *n*; жёсткий узел
　　　m

R512 *e* **rigid pavement**
　　d starre Decke *f*, starrer Belag *m*,
　　　Betonstraßendecke *f*,
　　　Betonstraßenbelag *m*
　　f corps *m* de chaussée rigide, revêtement
　　　m routier rigide
　　nl star wegdek *n*
　　r жёсткая дорожная одежда *f*,
　　　жёсткое покрытие *n*

R513 *e* **rigid plastic structures**
　　d starrplastische Konstruktionen *f pl*
　　f constructions *f pl* plasto-rigides
　　　[à nœuds plasto-rigides]
　　nl stijf-elastische constructies *f pl*
　　r жёсткопластичные конструкции *f pl*,
　　　конструкции *f pl* с жёстко-
　　　пластичными соединениями
　　　[узлами]

R514 *e* **rigid-plastic theory**
　　d starrplastische Theorie *f*
　　f théorie *f* du calcul des constructions
　　　plasto-rigides
　　nl theorie *f* van de stijf-elastische
　　　arbeid van constructiedelen
　　r теория *f* (*расчёта*), основанная на
　　　предположении жёсткопластичной
　　　работы конструкции

R515 *e* **rigid sheets**
　　d Hartplatten *f pl*; gesickte Bleche
　　　n pl
　　f tôle *f* mince rigide; tôle *f* mince
　　　avec éléments de rigidité
　　nl stijve platen *f(m) pl*
　　r жёсткие листы *m pl*; листовые
　　　материалы *m pl* с элементами
　　　жёсткости

R516 *e* **rigid structures**
　　d starre Konstruktionen *f pl*

f constructions *f pl* rigides
[à nœuds rigides]
nl stijve constructies *f pl*
r жёсткие конструкции *f pl*;
конструкции *f pl* с жёсткими
соединительными узлами

R517 *e* **rigging**
d 1. Anbauausrüstung *f* (*der
Baumaschinen*), Anbaugeräte *n pl*
2. Tauwerk *n*, Betakelung *f*,
Seilgehänge *n* 3. Seilführung *f*
4. Kleinmechanisierungsmittel *n pl*
5. Anschlagen *n* 6. Einbindung *f*,
Verflanschung *f*
f 1. équipement *m* porté 2. câbles *m pl*
et cordages utilisés sur des appareils
de levage 3. enroulement *m* d'un
câble sur les poulies 4. outillage *m*
accessoire, accessoires *m pl*
5. élingage *m* 6. tuyautérie *f*,
manifold *m*
nl 1. hulpuitrusting *f* van bouwmachine
2. takelage *f* 3. kabelvoering *f*
4. mechanisch handgereedschap *n*
5. aanslaan *n* 6. installatie *f*
r 1. навесное оборудование *n*
(*строительных машин*) 2. такелаж
m, канатно-блочная оснастка *f*
3. запасовка *f* канатов 4. средства
n pl малой механизации
5. строповка *f* 6. трубопроводная
обвязка *f*

R518 *e* **ring beam**
d Ringträger *m*, Ringbalken *m*
f poutre *f* annulaire
nl ringbalk *m*
r кольцевая балка *f*

R519 *e* **ring bolt**
d Haltekreuz *n*, Haltering *m*
f boulon *m* à œil, anneau *m* à fiche
nl ringbout *m*
r рым-болт *m*

R520 *e* **ring dam**
d Ringdeich *m*
f barrage *m* circulaire, digue *f*
annulaire
nl ringdijk *m*
r кольцевая дамба *f*

R521 *e* **ring gate**
d Ringverschluß *m*
f vanne *f* annulaire
nl ringvormige afsluiter *m*
r кольцевой затвор *m*

R522 **ring girder** *see* **ring beam**

R523 **ring levee** *see* **ring dam**

R524 *e* **ring main**
d Ringleitung *f*
f conduite *f* circulaire
nl ringleiding *f*
r кольцевой трубопровод *m*,
кольцевая (трубопроводная)
магистраль *f*

R525 *e* **ring road**
d Ringstraße *f*

f route *f* de ceinture
nl rondweg *m*
r кольцевая дорога *f*

R526 *e* **ring seal joint**
d O-Ring-Muffenverbindung *f*
f raccord *m* à emboîtement à bague
d'étanchéité
nl snap-mofverbinding *f*
r раструбное соединение *n*
с упругим уплотняющим кольцом

R527 *e* **ring shake**
d Ringriß *m*, Ringschäle *f*
f gerçure *f* annulaire
nl losringigheid *f* (*hout*)
r кольцевая трещина *f*, кольцевой
отлуп *m*

R528 *e* **ring shear apparatus**
d Ringschergerät *n*
f appareil *m* de cisaillement à
couronne cylindrique [par rotation,
rotatif]
nl gatenzaag *f(m)*
r кольцевой аппарат *m* для
определения сопротивления сдвигу

R529 *e* **ring water main**
d Ringwasserleitung *f*
f conduite *f* de distribution d'eau
annulaire
nl ringwaterleiding *f*
r кольцевая водопроводная
магистраль *f*

R530 *e* **rinse type air humidifier**
d berieselter Kapillarzellen-
-Luftbefeuchter *m*, Berieselungs-
-Luftbefeuchter *m*
f humidificateur *m* d'air capillaire
superficielle
nl luchtbevochtiger *m* met besproeide
capillaire cellen
r воздухоувлажнитель *m* с
орошаемым капиллярным слоем,
поверхностный воздухоувлажнитель
m

R531 *e* **rinse water**
d Spülwasser *n*, Waschwasser *n*
f eau *f* de lavage
nl spoelwater *n*
r промывочная [отмывочная] вода *f*

R532 *e* **rinsing tub**
d Spülbad *n*, Spülbecken *n*
f évier *m*
nl spoelbad *n*
r (кухонная) мойка *f*

R533 *e* **ripper**
d Aufreißer *m*, Bodenauflockerer *m*
f ripper *m*, défonceuse *f* (portée),
scarificateur *m* (porté)
nl ripper *m*, opbreektanden *m pl*
r рыхлитель *m*, кирковщик *m*

R534 *e* **ripping**
d Aufreißen *n*, Aufbruch *m*, Aufbrechen
n
f scarification *f*, défonçage *m*

nl opbreken *n* van de bodem
r рыхление *n*, киркование *n*

R535 *e* **ripping timber**
d Längssägen *n*, Längsschnitt *m*
f sciage *m* longitudinal du bois
nl in langsrichting zagen *n*, schulpen *n*
r продольная распиловка *f* (*лесоматериала*)

R536 *e* **riprap, rip-rap**
d Steinschüttung *f*; Steinwurf *m*; Steinbestürzung *f*; Steinschüttvorlage *f*; Steinschüttdeckwerk *n*, Steinschüttbelag *m*
f enrochement *m* (de protection); remblai *m* rocheux
nl steenstorting *f*
r каменная отсыпка *f* [наброска *f*]; каменно-набросная береговая одежда *f*; каменно-набросная одежда *f* откоса

R537 *e* **riprap cofferdam**
d Steinschüttungsvordamm *m*, geschütteter Steinfang(e)damm *m*, Felsschüttungsfang(e)damm *m*
f batardeau *m* en enrochement(s)
nl dam *m* van stortsteen
r каменно-набросная перемычка *f*

R538 *e* **rise**
d 1. Dachsteigungshöhe *f*, Pfeilhöhe *f*, Bogenstich *m* 2. Treppenlaufhöhe *f* 3. Setzstufe *f*, Futterstufe *f*, Stoßtritt *m* 4. Lastenhub *m*
f 1. montée *f* de comble; flèche *f* d'arc 2. hauteur *f* de volée 3. contremarche *f* 4. élévation *f* (*de charge, de poids*)
nl 1. pijl *m* (van een boog) 2. hoogte *f* van een trap 3. optrede *f* 4. hijsen *n* (*van goederen*)
r 1. высота *f* подъёма (*крыши*); стрела *f* (*арки*) 2. высота *f* лестничного марша 3. подступёнок *m* 4. подъём *m* (груза)

R539 *e* **rise of a truss**
d Fachwerk(träger)shöhe *f*
f hauteur *m* [montée *f*] de ferme
nl hoogte *f* van een vakwerkligger
r высота *f* фермы

R540 *e* **riser**
d 1. Falleitung *f*, Fallrohr *n*; Steigleitung *f*, Steigrohr *n* 2. Setzstufe *f*, Futterstufe *f*, Stoßtritt *m*
f 1. tube *m* [tuyau *m*] ascendant [montant], tuyau *m* de montée, colonne *f* montante 2. contremarche *f*
nl 1. stijgpijp *f*(*m*) 2. stootbord *n* (*trap*)
r 1. стояк *m* 2. подступёнок *m*

R541 *e* **riser duct**
d Lüftungsschacht *m*, vertikaler Luftleitkanal *m*
f conduite *f* d'air ascendante
nl ventilatieschacht *f*(*m*)
r вентиляционный стояк *m*

R542 *e* **rising arch**
d ansteigender [abfallender, einhüftiger, einschenkliger] Bogen *m*
f arc *m* rampant
nl klimmende boog *m*
r ползучая арка *f*

R543 **rising heating system** *see* **up-feed heating system**

R544 *e* **rising main**
d Vorlaufstrang *m*, Steigstrang *m*, Steigleitung *f*
f tuyau *m* vertical [ascendant]
nl aanvoerstijgleiding *f*
r подающий стояк *m* (*водопроводный, газовый*)

R545 *e* **river-bank dike**
d Fluß(ufer)damm *m*, Flußdeich *m*
f digue *f* de rivière [fluviale]
nl rivierdijk *m*
r речная береговая дамба *f*

R546 *e* **river-bank water intake**
d Uferentnahme *f*, Ufereinlaßbauwerk *n*
f prise *f* d'eau en rivière
nl inlaatwerk *n* [watervang *m*] aan de rivier
r береговой водозабор *m*

R547 *e* **river bed evolution**
d Flußbettprozeß *m*
f évolution *f* du lit de fleuve
nl evolutie *f* van het rivierbed
r русловый процесс *m*

R548 *e* **river capture**
d Flußenthauptung *f*, Flußanzapfung *f*
f capture *f* d'un fleuve [d'une rivière], décapitation *f*
nl stroomonthoofding *f*
r перехват *m* реки

R549 *e* **river closure**
d Flußbettabriegelung *f*, Flußsperrung *f*
f coupure *f* du cours d'eau, coupure *f* fluviale
nl afsluiting *f* van het rivierbed
r перекрытие *n* русла реки

R550 *e* **river channel**
d Entlastungsrinne *f*, Entlastungskanal *m*
f lit *m* de décharge
nl ontlastingskanaal *n*
r разгружающее русло *n*

R551 *e* **river control**
d Flußregelung *f*, Flußregulierung *f*
f régularisation *f* des fleuves [d'un cours d'eau]
nl riviernormalisatie *f*
r регулирование *n* рек (*комплекс технических мероприятий*); регулирование *n* водотока

R552 *e* **river development**
d Stromausbau *m*
f aménagement *m* fluvial [du lit]
nl rivierverbetering *f*

r освоение *n* [использование *n*]
водотока

R553 *e* river diversion
   d Wasserumlenkung *f*
   f dérivation *f* [déviation *f*] du fleuve
   nl omleiden *n* [afleiden *n*] van een rivier
   r отвод *m* реки

R554 *e* river diversion tunnel
   d Umleitungsstollen *m*
   f galerie *f* de fuite [de restitution];
   galerie *f* d'amenée, tunnel *m* de
   dérivation
   nl omleidingstunnel *m*
   r отводящий (строительный) туннель
   m; деривационный туннель *m*

R555 *e* river engineering
   d Strombau *m*
   f régularisation *f* des cours d'eau [des
   fleuves]
   nl normalisatie *f* van een rivier
   r регулирование *n* рек (техническая
   дисциплина); строительство *n*
   регуляционных сооружений

R556 *e* river flow
   d Durchfluß *m*, Durchfluß(wasser)menge
   f
   f débit *m* de fleuve [de cours d'eau]
   nl afvoer *m*
   r расход *m* водотока

R557 *e* river gravel
   d Flußkies *m*
   f gravier *m* de rivière
   nl riviergrind *n*
   r речной гравий *m*

R558 *e* river intake
   d Flußwasserentnahme *f*,
   Stromwasserfassung *f*; freie
   Wasserfassung *f*, sperrenlose
   Wasserentnahme *f*
   f prise *f* d'eau en rivière; prise *f* directe
   [d'eau sans barrage]
   nl wateronttrekking *f* uit een rivier
   r речной водозабор *m*; бесплотинный
   водозабор *m*

R559 *e* river levee
   d Flußdeich *m*, Flußdamm *m*
   f digue *f* de protection d'un fleuve
   nl rivierdijk *m*, bandijk *m*
   r речная дамба *f*

R560 *e* river outlet
   d 1. Betriebsauslaß *m*, Auslaufbauwerk
   n 2. Wiedereinleitungsbauwerk *n*
   3. Stromauslauf *m*, Flußauslauf *m*
   f 1. ouvrage *m* de décharge 2. ouvrage *m*
   de dérivation 3. ouvrage *m*
   d'écoulement des eaux usées;
   débouché *m*
   nl 1. uitlaatwerk *n* voor bedrijfswater
   2. invoerwerk *n* bij een omleiding
   3. stroomuitmonding *f*
   r 1. эксплуатационный водоспуск *m*,
   выпускное сооружение *n* (*ГЭС*)
   2. отводящее сооружение *n* (*ГЭС*)
   3. речной выпуск *m* (*канализация*)

R561 *e* river power plant
   d Fluß(bett)-Wasserkraftwerk *n*,
   Stromkraftwerk *n*
   f centrale *f* en rivière
   nl rivierbed-waterkrachtcentrale *f*
   r русловая гидроэлектростанция *f*

R562   river regulation *see* river control

R563 *e* river runoff
   d Abfluß *m*
   f écoulement *m* fluvial
   nl afwatering *f*
   r речной сток *m*

R565 *e* river sand
   d Flußsand *m*
   f sable *f* de rivière
   nl rivierzand *n*
   r речной песок *m*

R566 *e* river system
   d Flußsystem *n*
   f système *m* fluvial
   nl rivierstelsel *n*
   r речная система *f*

R567   river training *see* river control

R568 *e* rivet
   d Niet *m*
   f rivet *m*
   nl klinknagel *m*
   r заклёпка *f*

R569 *e* riveting hammer
   d Niethammer *m*, Nietrevolver *m*
   f marteau-riveur *m*
   nl klinkhamer *m*
   r заклёпочник *m*, клепальный молоток
   m

R570 *e* road
   d Straße *f*
   f route *f*, chemin *m*
   nl straat *f(m)*, weg *m*
   r дорога *f*

R571 *e* roadbase
   d Straßenunterbau *m*
   f couche *f* de base [de fondation] (*du
   corps de chaussée*)
   nl fundering *f* van een weg
   r основание *n* дорожной одежды

R572 *e* roadbed
   d 1. Straßenbett *n*, Straßenkoffer *m*
   2. Gleisschotterbett *n*
   f 1. plate-forme *f*; forme *f* (*de la route*)
   2. ballast *m*
   nl 1. aardebaan *f* 2. ballastbed *n*
   r 1. земляное полотно *n* дороги;
   корыто *n* дороги 2. железнодорожный
   балласт *m*

R573 *e* road breaker
   d Betonbrecher *m*,
   Betonaufbruchhammer *m*
   f brise-béton *m*, marteau *m* brise-béton
   nl betonbreekhamer *m*
   r бетонолом *m*

R574 *e* road bridge
   d Straßenbrücke *f*
   f pont-route *m*

nl verkeersbrug *f m*)
r автодорожный мост *m*

R575   road building *see* road construction

R576 *e* road-building machinery
   d Straßenbaumaschinen *f pl*
   f engins *m pl* de travaux de voirie routière; matériel *m* routier [pour la construction des routes]
   nl wegenbouwmachines *f pl*
   r дорожно-строительные машины *f pl*

R577   road burner *see* road heater

R578 *e* road concrete
   d Straßenbeton *m*
   f béton *m* routier
   nl wegenbeton *n*
   r бетон *m* для дорожных покрытий, дорожный бетон *m*

R579 *e* road construction
   d Straßenbau *m*
   f construction *f* des routes
   nl wegenbouw *m*, wegaanleg *m*
   r строительство *n* дорог, (авто)дорожное строительство *n*

R580 *e* road embankment
   d Straßendamm *m*
   f remblai *m* routier
   nl weg *m* in ophoging
   r дорожная насыпь *f*

R581   road excavation *see* roadway excavation

R582 *e* road forms
   d Betondeckenschalung *f*, Straßenbauschalung *f*, Seitenschalung *f*
   f coffrage *m* latéral (de chaussée en béton)
   nl bekisting *f* voor betonweg
   r бортовая опалубка *f* бетонного дорожного покрытия

R583 *e* road grader
   d Straßenhobel *m*, Straßenplanierer *m*
   f niveleuse *f*, grader *m*
   nl wegschaaf *f(m)*, grader *m*
   r грейдер *m*, планировщик *m*

R584 *e* road grinder
   d Straßenschleifmaschine *f*
   f polisseuse *f* [polissoir *m*] de revêtement routier (*en béton*), machine *f* à polir le revêtement routier (*en béton*)
   nl stratenslijpmachine *f*
   r дорожно-шлифовальная машина *f*

R585 *e* road grooving machine
   d Straßenaufrauhmaschine *f*, Straßenfräse *f*
   f pulvimixer *m*, tritureuse *f*
   nl wegfrees *f(m)*
   r дорожная фреза *f*

R586 *e* road gully
   d Straßenablauf *m*
   f avaloir *m* de chaussée
   nl (straat)goot *f(m)*
   r уличный водосток *m*, ливнеспуск *m*

R587 *e* road heater
   d Straßendeckenerhitzer *m*
   f machine *f* chauffante [de chauff(ag)e] (*pour revêtement d'asphalte*)
   nl wegdekverhitter *m*
   r разогреватель *m* асфальтобетонных дорожных покрытий, асфальторазогреватель *m*

R588 *e* road intersection, road junction
   d Straßenkreuzung *f*
   f croisement *m* (*de routes*), intersection *f*, échangeur *m*
   nl wegkruising *f*
   r транспортный перекрёсток *m*; развилка *f* [узел *m*] дорог; пересечение *n* дорог (*в одном уровне*)

R589 *e* road kerb
   d Bordstein *m*
   f bordure *f*
   nl kantopsluiting *f*, trottoirband *m*
   r бордюрный камень *m*

R590 *e* road kettle
   d Straßenkocher *m*
   f chaudière *f* à bitume [de fusion pour bitume], fondoir *m* pour goudron et bitume
   nl asfaltketel *m*
   r битумный котёл *m*

R591 *e* road marking
   d Fahrbahnmarkierung *f*
   f lignes *f pl* [marquage *m* de lignes] de signalisation, marque *f* de la chaussée
   nl wegmarkering *f*
   r линии *f pl* безопасности, разметка *f* проезжей части дороги

R592   road-marking machine *see* pavement--marking machine

R593 *e* road materials
   d Straßenbaustoffe *m pl*
   f matériaux *m pl* routiers
   nl wegenbouwmaterialen *n pl*
   r дорожно-строительные материалы *m pl*

R594 *e* road mesh
   d Straßenbewehrungsmatte *f*
   f treillis *m* soudé pour revêtement routier en béton armé
   nl wegennet *n*
   r стальная сетка *f* для армирования дорожных покрытий

R595   road mixer *see* soil stabilizing machine

R596 *e* road mix surface
   d Baumischbelag *m*
   f revêtement *m* routier stabilisé sur place
   nl slijtlaag *f(m)*
   r щебёночное *или* гравийное покрытие *n*, выполненное способом смешения на дороге

R597 *e* road network
   d Straßennetz *n*

f réseau *m* routier
nl wegennet *n*
r дорожная сеть *f*

R598 e **road oil**
d Straßenbaubitumen *n*
f road-oil *m*, bitume *m* routier
nl wegenbitumen *n*
r дорожный битум *m*

R599 e **road-rail bridge**
d Straßen- und Eisenbahnbrücke *f*
f pont *m* route-rail
nl verkeers- en spoorbrug *f(m)*
r совмещённый мост *m* (*для автомобильного и железнодорожного транспорта*)

R600 e **road-railway tunnel**
d Straßen- und (Eisen-)Bahntunnel *m*
f tunnel *m* route-rail [rail-route]
nl verkeers- en spoorwegtunnel *m*
r совмещённый транспортный туннель *m*

R601 e **road roller**
d Straßenwalze *f*
f cylindre *m* [rouleau *m*] routier
nl wals *f(m)*
r дорожный каток *m*

R602 e **roadside**
d 1. Freistreifen *m*, Geländestreifen *m* neben der Fahrbahn
2. Trenn(ungs)streifen *m*, Mittelstreifen *m*
f 1. bord *m* [accotement *m*] de la route
2. terre-plein *m* central
nl 1. vluchtstrook *f* 2. middenstrook *f*
r 1. краевая полоса *f*, обочина *f*
2. разделительная полоса *f*

R603 e **road side development**
d Straßeneinrichtungen *f pl*, Verkehrsanlagen *f pl* neben der **Straße**
f aménagement *m* de la route
nl voorzieningen *f pl* langs de weg
r дорожные устройства *n pl*, обстановка *f* пути

R604 e **roadside ditch**
d Straßengraben *m*
f fossé *f* (de route)
nl bermsloot *f*
r придорожная канава *f*, кювет *m*

R605 e **roadside vegetation**
d Straßenbepflanzung *f*
f forêt *f* de protection au bord de la route
nl wegbeplanting *f*
r зелёная защита *f* дорог

R606 e **road slab**
d Betonfeld *n*, Betondeckenplatte *f*
f dalle *f* routière
nl wegdekbetonplaat *f(m)*
r дорожная плита *f*

R607 e **road structure**
d Straßenkörper *m*; Aufbau *m* der Straßenbefestigung
f corps *m* de chaussée

nl wegconstructie *f*
r конструкция *f* дорожной одежды

R608 e **road surfacing**
d Straßenbelag *m*, Straßendecke *f*
f revêtement *m* routier, couche *f* de surface (de chaussée)
nl wegverharding *f*
r дорожное покрытие *n*

R609 e **road sweeper-collector**
d Selbstaufnahme-Kehrmaschine *f*
f balayeuse-ramasseuse *f*
nl straatveegauto *m*
r подметально-уборочная машина *f*

R610 e **road tar**
d Straßenteer *m*
f goudron *m* routier [pour route]
nl wegenteer *m*, *n*
r дорожный дёготь *m*

R611 e **road tar emulsion**
d Straßenteeremulsion *f*
f émulsion *f* routière de goudron
nl wegenteeremulsie *f*
r дорожная дёгтевая эмульсия *f*

R612 e **road tar-type penetration macadam**
d Teertränkemakadam *m*, *n*
f macadam *m* au goudron par pénétration
nl teermacadam *n*
r щебёночное покрытие *n*, пропитанное дёгтем

R613 e **road transition curve**
d Straßenübergangsbogen *m*
f courbe *f* de transition [de raccordement]
nl overgangsboog *m*
r дорожная переходная кривая *f*

R614 e **road tunnel**
d Straßentunnel *m*
f tunnel *m* routier [automobile]
nl verkeerstunnel *m*
r автодорожный туннель *m*

R615 e **road vibrating and finishing machine**
d Rüttelbohlenfertiger *m*
f finisseur *m* vibrant, finisseuse *f* **vibrante**
nl asfaltafwerkmachine *f*
r бетоноотделочная машина *f* [финишер *m*] с вибрационным брусом

R616 e **roadway**
d Fahrbahn *f*
f chaussée *f*
nl rijbaan *f(m)*
r проезжая часть *f* дороги

R617 e **roadway delineation**
d Auftragen *n* von Sicherheitsstreifen auf die Straßendecke
f marquage *m* de lignes de signalisation
nl aanbrengen *n* van de strepen op het wegdek
r нанесение *n* линий безопасности на дорожное покрытие

R618 *e* **roadway excavation**
  d Straßenbau-Erdarbeiten *f pl*
  f terrassements *m pl* routiers
  nl grondwerk *n* bij wegenbouw
  r производство *n* земляных работ при строительстве дорог

R619 *e* **roadway lighting**
  d Straßenbeleuchtung *f*
  f éclairage *m* de routes [de voies publiques]
  nl wegverlichting *f*
  r освещение *n* проезжей части (дороги)

R620 *e* **rock anchor**
  d Fels(-Zug)anker *m*, Ausbauanker *m*, Gebirgsanker *m*
  f boulon *m* d'ancrage ancré dans la roche, boulon *m* expansible
  nl rotsanker *n*
  r анкер *m* штанговой крепи

R621 *e* **rock asphalt**
  d Asphaltgestein *n*, Bergasphalt *m*
  f roche *f* asphaltique, pierre *f* d'asphalte
  nl gebitumineerd steenslag *n*
  r битумная горная порода *f*

R622 *e* **rock asphalt pavement**
  d Bitumengestein-Straßendecke *f* in Tränkmakadambauweise
  f revêtement *m* routier en roche asphaltique
  nl wegdek *n* van gebitumineerd steenslag
  r дорожное покрытие *n* из битумных горных пород (*с последующей обработкой битумом или гудроном*)

R623 *e* **rock bolting, rock bolts**
  d Ankerbolzen *m pl*, Gestängeschachtausbau *m*
  f boulonnage *m* de roches, soutènement *m* par boulons d'ancrage
  nl verankering *f* in de rots
  r штанговая крепь *f*

R624 *e* **rock boring**
  d Bohrloch *n*, Schußloch *n*, Schießloch *n*
  f trou *m* (de mine)
  nl boorgat *n*
  r буровая скважина *f*; шпур *m*

R625 *e* **rock classification**
  d Einteilung *f* der Gesteine, Gesteinsklassifikation *f*
  f classification *f* de roches
  nl gesteenteclassificatie *f*
  r классификация *f* горных пород

R626 *e* **rock cut**
  d Felseinschnitt *m*
  f déblai *m* rocheux [en roches]
  nl insnijding *f* in de rots
  r скальная выемка *f*

R627 *e* **rock dash**
  d Steinputz *m*
  f enduit *m* en cailloux intercalés, enduit *m* au gravier intercalé
  nl grof buitenpleister *n* met siersteenslag

  r наружная штукатурка *f* с наборной фактурой (*с втапливанием в слои раствора мелких камней*)

R628 *e* **rock drill**
  d Bohrhammer *m*
  f marteau *m* perforateur
  nl boorhamer *m*
  r перфоратор *m*, бурильный молоток *m*

R629 *e* **rocker**
  d Wippe *f*, Wiege *f*
  f bascule *f*, balancier *m*
  nl tuimelaar *m*
  r балансир *m*

R630 *e* **rocker bar bearing**
  d Schwingenlager *n*
  f appui *m* à biellette
  nl slingerstijloplegging *f*
  r балансирная опора *f*

R631 *e* **rocker bearing**
  d Kipp(auf)lager *n*
  f appui *m* oscillant [à bascule]
  nl pendeloplegging *f*
  r качающаяся опора *f*

R632 *e* **rocker-dump hand-cart**
  d Japaner *m*, Kippkarren *m*
  f charrete *f* à bras à bascule
  nl stortkar *f(m)*, kipkar *f(m)*
  r ручная тележка *f* с опрокидывателем

R633 *e* **rock excavation**
  d Felsbaggerung *f*, Felsaushub *m*
  f excavation *f* de roches [de rocs]
  nl uitgraving *f* in de rots
  r скальная выемка *f*, скальные работы *f pl*, разработка *f* скальных грунтов

R634 **rock-faced brick** *see* **ashlar 1.**

R635 *e* **rock fill, rockfill**
  d Steinschüttung *f*, Steinwurf *m*, Steindeckwerk *n*
  f enrochement *m*, remblai *m* rocheux, perré *m*
  nl aanvulling *f* met stortsteen
  r каменная наброска *f*; каменная отсыпка *f* (*в основание или в тело сооружения*)

R636 *e* **rock-fill cofferdam**
  d Steinfang(e)damm *m*, Felsschüttungsfang(e)damm *m*
  f batardeau *m* en enrochements
  nl damwand *m* met steenstorting
  r каменно-набросная перемычка *f*

R637 *e* **rockfill dam**
  d Steinschüttdamm *m*, Steingerüstdamm *m*, Felsschüttungsstaudamm *m*, Steinfülldamm *m*, Steinwurfdamm *m*, Steinsperre *f*
  f barrage *m* en enrochements, barrage *m* mixte en enrochements et en terre
  nl dam *m* met kern van stortsteen

r каменно-набросная плотина f;
каменно-земляная плотина f
(с объемом камня свыше 50%)

R638 e **rock-fill foundation**
d Steinschüttung f (als offene
Gründung für Kaimauern und
Molen)
f fondation f en enrochements
nl fundering f op steenstorting
r каменно-набросный фундамент m

R639 e **rock-fill revetment**
d Steinschüttbelag m
f revêtement m en enrochements
nl oeverbekleding f met stortsteen
r каменно-набросная одежда f

R640 e **rock-fill timber crib**
d Steinkiste f, Steinkasten m
f coffrage m en bois rempli
d'enrochement
nl steenkist f(m)
r ряж m с каменной наброской

R641 e **rock flour**
d Steinmehl n
f poudre f de roche [de pierre]
nl steenpoeder n
r каменная мука f

R642 e **rock foundation**
d Gründungsfels m, Felsgründung f
f fondation f en roche(s)
nl rotsfundament n
r скальное основание n

R643 e **rock gallery**
d Felsstollen m
f galerie f en roches
nl rotstunnel m
r штольня f в скальной породе

R644 e **rocking ball bearing**
d Kugelkipplager n
f appui m à rotule oscillant, appareil
m d'appui à bascule
nl kogelkiplager n
r шаровая качающаяся опора f

R645 **rock lath** see **gypsum lath**

R646 e **rock pocket**
d Kiesnest n, Steinnest n
f nid m de cailloux
nl grindnest n
r гравийное гнездо n (дефект
бетонных поверхностей)

R647 e **rock tunnel**
d Felstunnel m
f tunnel m en roches
nl rotstunnel m
r туннель m в скальной породе

R648 e **rock wool**
d Mineralwolle f, Steinwolle f
f laine f minérale [de rocaille, de roche]
nl steenwol f(m)
r минеральная вата f

R649 e **rod**
d Stab m, Gitterstab m;
Armierungsstab m
f tige f; barre f, rond m

nl staaf f(m)
r стержень m, элемент m
стержневой системы; арматурный
стержень m

R650 e **rod bender**
d Betonstahlbieger m
f cintreuse f [coudeuse f, plieuse f]
à ronds de béton
nl betonstaalbuiger m
r станок m для гибки арматурных
стержней, гибочный станок m

R652 e **rodding**
d 1. Stochern n, Stocherverdichtung f
2. Stocherreinigung f der
Rohrleitungen
f 1. piquage m (du béton)
2. débouchement m (des drains, des
tuyaux) par piquage
nl 1. porren n (van betonmortel)
2. doorsteken n van afvoerleidingen
r 1. уплотнение n бетонной смеси
штыкованием 2. прочистка f дрен
или канализационных труб
гибкими стержнями

R653 e **rod element**
d Stabheizkörper m
f élément m de chauffage électrique
à barre [à tige]
nl staafvormig verwarmingselement n
r стержневой электронагреватель m

R654 e **rod shears**
d Betonstahlschere f
f cisaille f à bras (pour rond de béton)
nl betonstaalschaar f(m)
r (ручные) ножницы pl для резки
арматурных стержней

R655 e **rod spacers**
d Bewehrungshalterungen f pl
f espaceur m [distancieur m] pour
ferraillage en barres rondes
nl afstandhouders m pl voor
wapeningsstaven
r фиксаторы m pl стержневой
арматуры

R656 e **rod test**
d Stabeinschlagverfahren n der
Bodenprüfung
f essai m de sol par battage de tige
nl grondonderzoek n met visiteerijzer
r испытание n грунта забивкой
стального стержня

R657 e **rolled asphalt**
d Walzasphalt m
f asphalte m roulé
nl gewalst asfaltbeton n
r укатанный [уплотнённый]
асфальтобетон m

R658 e **rolled concrete**
d Walzbeton m
f béton m laminé
nl gewalst beton n
r прокатный бетон m

R659 *e* **rolled earth fill**
    d abgewalzte Erdschüttung *f*;
       Walzdamm *m*
    f remblai *m* (de terre) cylindré; remblai
       m serré par cylindrage
    nl afgewalste grondaanvulling *f*
       [ophoging *f*]
    r укатанный насыпной грунт *m*;
       укатанная насыпь *f*

R660 *e* **rolled-steel beam, rolled steel joist**
    d Walz-Stahlträger *m*
    f poutre *f* laminée en acier, poutrelle *f*
       laminée
    nl stalen walsprofiel *n*
    r стальная прокатная балка *f*,
       стальной прокатный балочный
       профиль *m*

R661 *e* **rolled wire**
    d Walzdraht *m*, gewalzter Draht *m*
    f fil *m* laminé
    nl gewalste draad *m*
    r катаная проволока *f*, катанка *f*

R662 *e* **roller**
    d 1. Straßenwalze *f* 2. Walze *f*, Rolle *f*;
       Rad *n*
    f 1. rouleau *m* (de cylindrage),
       cylindre *m* 2. rouleau *m*; galet *m*;
       cylindre *m*; roue *f*
    nl 1. wegwals *f(m)* 2. rol *f(m)*
    r 1. дорожный каток *m* 2. валец *m*;
       барабан *m*; цилиндр *m*; ролик *m*;
       колесо *f*

R663 *e* **roller balance bridge, roller bascule bridge**
    d Rollklappbrücke *f*, Schaukelbrücke *f*,
       Abrollbrücke *f*
    f pont *m* mobile à bascule
    nl rolbrug *f(m)*
    r откатно-раскрывающийся мост *m*

R664 *e* **roller bearing**
    d Walzenlager *n*, Rollenlager *n*
    f appui *m* à rouleaux, palier *m*
       à rouleaux
    nl rollager *n*
    r катковая опора *f*, катковая опорная
       часть *f*

R665 *e* **roller blind**
    d Rolladen *m*
    f store *m* vénitien, jalousie *f* vénitienne,
       volet *m* roulant
    nl oprolbare zonwering *f*
    r роликовые шторы *f pl*

R666 *e* **roller bridge**
    d Rollbrücke *f*, Schiebebrücke *f*
    f pont *m* roulant [glissant]
    nl rolbrug *f(m)*
    r откатной мост *m*

R667 *e* **roller compacted concrete**
    d walzverdichteter Beton *m*
    f béton *m* compacté par rouleau
       [cylindre] vibratoire
    nl met trilwals verdicht beton *n*
    r бетон *m*, уплотняемый дорожным
       виброкатком

R668 *e* **roller conveyor**
    d Rollenförderer *m*
    f convoyeur *m* à rouleaux
    nl rollenbaan *f(m)*
    r роликовый [катковый] конвейер *m*

R669 *e* **roller cover**
    d austauschbare Gummihülle *f* der
       Auftragwalze
    f couvre-rouleau *m* [chemise *f* du
       rouleau] de peinturage
    nl afneembaar overtrek *n* van de
       verfroller
    r сменяемая оболочка *f* малярного
       валика

R670 *e* **roller dam** *US*
    d Walzenwehr *n*
    f barrage *m* à vannes roulantes
    nl stuw *m* met segmentschuiven
    r плотина *f* с вальцовым затвором

R671 *e* **roller drum gate, roller gate**
    d Walzenverschluß *m*, Walzenwehr *n*,
       Rollschütz *n*
    f vanne *f* rotative [roulante, à rouleau,
       à cylindre]
    nl segmentschuif *f(m)*
    r вальцовый затвор *m*

R672 *e* **roller gate shield**
    d Walzenschnabel *m*
    f bec *m* du rouleau
    nl schild *n* van de segmentschuif
    r козырёк *m* [щиток *m*] вальцового
       затвора

R673 *e* **roller grinding mill, roller mill**
    d Walzenmühle *f*; Walzenfarbmühle *f*,
       Walzenfarbreibwerk *n*
    f moulin *m* [broyeur *m*] à cylindres,
       broyeuse *f* à rouleaux
    nl wals(verf)molen *m*
    r вальцовая мельница *f* или
       краскотёрка *f*

R674 *e* **roller-mounted leaf gate**
    d Rollschütze *f* mit endloser Rollenkette
    f vanne *f* wagon avec chaîne sans fin
       à rouleaux
    nl vlakke wielsluiting *f* met eindloze
       rolketting
    r плоский колёсный затвор *m*
       с бесконечной роликовой цепью

R675 *e* **roller passage**
    d Walzgang *m*
    f passage *m* [passe *f*] de rouleau
    nl roldeur *m* met rondgaande ketting
    r проход *m* дорожного катка

R676 *e* **roller screeding**
    d Betonausgleichen *n* mit Walzen
    f aplanissement *m* du béton par
       rouleaux
    nl vlakken *n* door te walsen
    r разравнивание *n* бетонной смеси
       валками

R677   **roller shield** *see* **roller gate shield**

R678 *e* **roller shutters**
    d Rolladen *m*

f volets *m pl* roulants
nl oprolbare zonwering *f*
r роликовые ставни *m pl* [шторы *f pl*]

R679 *e* **roller table**
d Rollentisch *m*
f table *f* à rouleaux
nl rollenbaan *f(m)*
r рольганг *m*

R680 *e* **roller-type shear**
d Rollenschere *f*
f cisaille *f* à galets
nl rollenschaar *f(m)*
r роликовые ножницы *pl*

R681 *e* **rolling**
d 1. Walzverdichtung *f* 2. Walzen *n*
f 1. cylindrage *m*, roulage *m*,
roulement *m* 2. laminage *m*
nl 1. afwalsen *n (grond)* 2. walsen *n*
r 1. укатка *f (грунта, дорожного
покрытия)* 2. вальцевание *n*;
прокатка *f*

R682 *e* **rolling bearing**
d Wälzlager *n*
f appui *m* à rouleaux [à galets]
nl walskussenblok *n*
r катковая опора *f*

R683 *e* **rolling contact joint**
d Wälzgelenk *n*
f articulation *f* à roulement
nl walsscharnier *n*
r шарнир *m* с трением качения

R684 *e* **rolling door**
d Rolltür *f*, Rolltor *n*
f porte *f* glissante roulante
nl roldeur *f(m)*
r откатная дверь *f*; откатные
ворота *pl*

R685 **rolling drawbridge** *see* **roller bridge**

R686 *e* **rolling gate**
d 1. Rollschütz(e) *n (f)* 2. Schiebetor *n*,
Rollponton *m*
f 1. vanne *f* wagon 2. porte *f* roulante
nl rolschuif *f(m)*
r 1. плоский колёсный затвор *m*
2. откатные ворота *pl (шлюза)*

R687 *e* **rolling lift bridge**
d Wälzklappbrücke *f*
f pont-levis *m* roulant
ni rolbasculebrug *f(m)*
r откатно-раскрывающийся мост *m*

R688 *e* **rolling load**
d rollende Last *f*, Wanderlast *f*
f charge *f* roulante [mobile]
nl rollende last *f(m)*
r подвижная нагрузка *f (от
колёсных транспортных средств)*

R689 *e* **rolling requirements**
d technische Anforderungen *f pl* für
Walzen *(Verdichtung)*
f prescriptions *f pl* d'exécution
concernant le roulage [cylindrage]
nl eisen *f pl* voor het walsen
r требования *n pl* к укатке

R690 *e* **rolling scaffold**
d Rollgerüst *n*, Rollrüstung *f*
f échafaudage *m* roulant
nl rolsteiger *m*
r катучие леса *pl* [подмости *pl*]

R691 *e* **rollway**
d Überlaufstrecke *f*, Überfallstrecke *f*
f partie *f* déversante d'un barrage
nl overlaat *n*
r водослив *m*, водосливная секция *f*
плотины

R692 *e* **Roman cement**
d Romanzement *m*, Romankalk *m*
f ciment *m* romain
nl romaankalk *m*, limitkalk *m*
r романцемент *m*

R693 **rone** *see* **eaves gutter**

R694 *e* **roof**
d 1. Dach *n* 2. First *m*, Hangendes *n*
f 1. toit *m*, toiture *f*, comble *m*
2. toit *m (de la galerie)*
nl 1. dak *n* 2. hangende zoldering *f*
r 1. крыша *f* 2. кровля *f* выработки

R695 *e* **roof aerial**
d Dachantenne *f*
f antenne *f* sur toit
nl dakantenne *f(m)*
r крышная антенна *f*

R696 *e* **roof battens**
d Dachlattung *f*, Dachlatten *f pl*
f voligeage *m* [lattes *f pl*] de toiture
[de toit, de comble]
nl panlatten *f pl*
r обрешётка *f* крыши [кровли]

R697 *e* **roof beam**
d Dachbalken *m*, Dachträger *m*
f poutre *f* de comble, entrait *m*
nl dakbalk *m*
r стропильная затяжка *f*; балка *f*
чердачного перекрытия,
подстропильный брус *m*; лежень *m*

R698 *e* **roof bearer**
d Dechpfette *f*
f poutre *f* de toiture, panne *f*
nl dwarsbalk *m*
r прогон *m* покрытия

R699 **roof cladding** *see* **roof covering**

R700 *e* **roof clay tile**
d Dachziegel *m*, Tondachstein *m*
f tuile *f* cuite
nl dakpan *f(m)*
r кровельная (глиняная) черепица *f*

R701 **roof coating** *see* **roof covering**

R702 *e* **roof component**
d Dachbauteil *m, n*, Dachelement *n*
f élément *m* [composant *m*] de comble
[de toit]
nl dakelement *n*
r элемент *m* конструкции крыши

R703 *e* **roof covering**
d Dach(ein)deckung *f*, Deckung *f*,
Bedachung *f*, Dachhaut *f*,
Dachbelag *m*

f couverture f, toiture f
nl dakbedekking f
r кровельное покрытие n

R704 e roof crane
d Dachkran m
f grue f (montée) sur toit
nl dakkraan f(m)
r крышный кран m (на крыше
складских зданий)

R705 e roof deck
d Dachdecke f
f dalle f de toiture
nl dakbeschot n
r настил m кровли

R706 e roof drain
d Dachablauf m, Dachgully m
f cuvette f de chéneau, trémie f de toit
nl dakafvoer m
r водосточная воронка f

R707 e roof drainage
d Dachentwässerung f
f descentes f pl pluviales incorporées;
système m d'évacuation des eaux
pluviales d'une couverture
nl afwatering f van het dak
r внутренняя ливневая канализация f;
водоотвод m с кровли

R708 e roofer's crane
d Dachdeckerkran m
f grue f pour les travaux de couverture
nl dakdekkerskraan f(m)
r кран m для обслуживания
кровельных работ

R709 e roofer's nail
d Dachdeckernagel m
f clou m de couverture [de toiture]
nl daknagel m, asfaltnagel m
r кровельный гвоздь m

R710 e roof extract unit
d Dachabsauger m
f ventilateur m (d'extraction) de toit
nl dakafzuiger m
r крышный вытяжной вентилятор m

R711 e roof fan
d Dachlüfter m, Dachventilator m
f ventilateur m de toit
nl dakventilator m
r крышный вентилятор m

R712 e roof form
d Dachform f, Dachgestalt f
f forme f de toit [de comble]
nl dakvorm m
r конструктивная форма f крыши
[покрытия]

R713 e roof frame, roof framing
d Dachtragwerk n, Dachstuhl m,
Dachrahmen m
f charpente f de toit [de comble]
nl kapconstructie f
r несущие конструкции f pl покрытия
[крыши]

R714 e roof gate
d Dachverschluß m,
Doppelklappenverschluß m
f vanne f levante à deux corps, vanne f
toit
nl valdeur f(m)
r крышевидный затвор m

R715 roof hatch see hatchway

R716 roofing see roof covering

R717 e roofing application
d Aufbringung f der weichen Dachhaut
f application f de revêtement
d'étanchéité en feutre bitumé
nl opbrengen n van dakvilt
r устройство n мягкой кровли;
укладка f мягкого кровельного
покрытия

R718 e roofing asphalt US
d Dachbitumen n
f bitume m de couverture [de toiture]
nl bitumineuze dakbedekking f
r кровельный битум m

R719 roofing battens see roof battens

R720 e roofing bracket
d Dachkonsole f
f console f de toit pour échafaudage
suspendu ou volant
nl dakconsole f(m)
r консоль f на крыше (напр. для
подвесных подмостей)

R721 e roofing felt
d Dachpappe f; Teerpappe f
f carton m bitumé [asphalté], feutre m
bitumé [bitumeux, asphalté]; carton m
goudronné
nl dakvilt n
r рубероид m; кровельный картон m;
толь m

R722 e roofing materials
d Dach(eindeckungs)stoffe m pl,
Bedachungsstoffe m pl
f matériaux m pl de couverture [de
toiture]
nl dakbedekkingsmaterialen n pl
r кровельные материалы m pl

R723 e roofing slate
d Dachschiefer m
f ardoise f de couverture [pour
toitures]
nl leislag n
r кровельный сланец m или шифер m

R724 e roofing tile
d Dachziegel m, Dachstein m
f tuile f de toiture
nl dakpan f(m)
r кровельная черепица f

R725 e roofing work
d Dach(ein)deckungsarbeiten f pl
f travaux m pl de couverture
nl dakdekkerswerk n
r кровельные работы f pl

R726 e roof insulating slab
d Dachisolierplatte f

f dalle *f* isolante de toiture, plaque *f* isolante de couverture
nl dakisolatieplaat *f(m)*
r изоляционная плита *f* покрытия

R727 *e* **roof insulation**
d Dachwärmeisolierung *f*
f isolation *f* thermique de toiture [de couverture]
nl dakisolatie *f*
r теплоизоляционный слой *m* кровли; теплоизоляция *f* кровли

R728 *e* **roof jib**
d Dachausleger *m*
f potence *f* de [sur] toit
nl dakzwenkmast *m*
r кран-укосина *m* для обслуживания кровельных работ

R729 **roof lathing** *see* **roof battens**

R730 *e* **roof light**
d Dachoberlicht *n*; Dachfenster *n*
f lanterneau *m* (de toit); fenêtre de toit, lucarne *f*
nl zoldervenster *n*, dakraam *n*
r фонарь *m* верхнего света; слуховое окно *n*

R731 *e* **roof load, roof loading**
d Dachlast *f*, Dachbelastung *f*
f charge *f* de toiture
nl dakbelasting *f*
r нагрузка *f* на покрытие

R732 *e* **roof membrane**
d Dachhaut *f*
f revêtement *m* d'étanchéité (de toiture)
nl dakhuid *f(m)*
r гидроизолирующий ковёр *m* кровли, водоизолирующий слой *m* кровли

R733 **roof outlet** *see* **roof drain**

R734 *e* **roof overhang**
d Dachüberstand *m*, Dachvorsprung *m*
f auvent *m*, queue *f* de vache
nl dakoverstek *n*
r свес *m* крыши

R735 *e* **roof panel**
d Dachplatte *f*, Dachtafel *f*
f dalle *f* de couverture [de toiture]; plaque *f* de couverture [de toiture]
nl dakplaat *f(m)*
r кровельная плита *f* [панель *f*]

R736 *e* **roof pitch**
d Dachneigung *f*, Dachgefälle *n*
f pente *f* de couverture [de toit]
nl dakhelling *f*
r уклон *m* ската кровли

R737 *e* **roof ridge**
d Dachfirst *m*, First *m*
f faîte *m*, faîtage *m*
nl nok *f(m)*, (dak)vorst *f(m)*
r конёк *m* крыши

R738 *e* **roof sheating**
d Belattung *f*
f voligeage *m* (*de couverture*)
nl dakbeschot *n*

r сплошная обрешётка *f* (кровли) из досок

R739 *e* **roof shingle**
d Dachschindel *f*
f bardeau *m* (de bois) de couverture
nl shingle *f*
r кровельный гонт *m*

R740 *e* **roof slab**
d Dachplatte *f*, Bedachungsplatte *f*
f dalle *f* de toiture
nl dekplaat *f(m)*
r плита *f* покрытия

R741 **roof slate** *see* **roofing slate**

R742 *e* **roof space**
d Dachraum *m*
f comble *m*, grenier *m*
nl zolderruimte *f*
r чердак *m*, чердачное помещение *n*

R743 *e* **roof stanchion**
d oberster Stützenschuß *m*
f lame *f* du poteau composé supportant le comble
nl standvink *m*
r шатровая [надкрановая] ветвь *f* колонны

R744 *e* **roof structure**
d Dachkonstruktion *f*; Dachverband *m*
f charpente *f* de toit [de comble]
nl kapconstructie *f*
r конструкция *f* крыши; стропильная конструкция *f*

R745 **roof tile** *see* **roofing tile**

R746 *e* **roof top air conditioner**
d Dachklimazentrale *f*, Dachklimagerät *n*
f conditionneur *m* de l'air de toit
nl op het dak aangebrachte airconditioning-installatie *f*
r крышный кондиционер *m*

R747 *e* **roof-top heliport**
d Hubschrauber-Dachflugplatz *m*
f héliport *m* sur toit
nl plat dak *n*, ingericht als landingsplaats voor hefschroefvliegtuigen
r посадочная площадка *f* на крыше для вертолётов

R748 *e* **roof truss**
d Dachbinder *m*, Binder *m*
f ferme *f* de toit
nl (kap)spant *n*
r стропильная ферма *f*

R749 *e* **roof valley**
d Kehle *f*, Dachkehlung *f*, Einkehle *f*
f noue *f*, no(u)let *m*, goulet *m*
nl kil *f(m)*, kilkeper *m*
r ендова *f*, разжелобок *m*

R750 *e* **roof ventilator**
d 1. Dachlüftungsaggregat *n*; Dachlüfter *m* 2. Deflektor *m*, Lüftungsaufsatz *m*
f 1. groupe *m* [batterie *f*] de ventilateurs de toit; ventilateur *m* de toit 2. aspirateur *m* statique de toit

nl 1. afzuiginrichting *f* op het dak
    2. ventilatiekap *f(m)*
r 1. крышный вентиляционный
    агрегат *m*; крышный вентилятор *m*
    2. дефлектор *m* (*вытяжное
    устройство на крыше*)

R751 roof weir *see* roof gate

R752 *e* roof window
  d Dachfenster *n*
  f lucarne *f*
  nl dakraam *n*
  r слуховое окно *n*

R753 *e* room
  d Raum *m*; Zimmer *n*
  f chambre *f*, local *m*; pièce *f*
  nl lokaal *n*; kamer *f(m)*
  r помещение *n*; комната *f*

R754 *e* room acoustics
  d Raumakustik *f*, Bauakustik *f*
  f acoustique *f* de salle
  nl akoestiek *f*
  r архитектурная акустика *f*;
  акустика *f* помещений

R755 *e* room air
  d Raumluft *f*
  f air *m* de local [ambiant]
  nl binnenlucht *f(m)*
  r внутренний воздух *m*, воздух *m*
  помещения

R756 *e* room air-conditioner
  d Raumklimagerät *n*
  f conditionneur *m* de l'air de chambre,
  climatiseur *m* de pièce [local]
  nl klimaatregelingstoestel *n*
  r комнатный *или* местный кондиционер
  m

R757 *e* room air cooler
  d Raumluftkühler *m*
  f refroidisseur *m* d'air local [de
  chambre]
  nl koeleenheid *f* voor kamerlucht
  r местный охладитель *m*,
  доохладитель *m* воздуха

R758 *e* room area
  d Zimmerfläche *f*, Raumfläche *f*
  f aire *f* (de la surface) d'une pièce,
  surface *f* d'une pièce
  nl kameroppervlak *n*
  r площадь *f* комнаты

R759 *e* room constant
  d Raumkonstante *f*
  f constante *f* du local
  nl kamerconstante *f(m)*
  r постоянная *f* помещения
  (*строительная акустика*)

R760 room-cooling unit *see* room air cooler

R761 *e* room spray-type humidifier
  d Düsenbefeuchter *m* der Raumluft
  f humidificateur *m* local à
  pulvérisation
  nl straalbuisluchtbevochtiger *m*
  r туманообразующий комнатный
  доувлажнитель *m* воздуха

R762 *e* room thermostat
  d Raumthermostat *m*
  f thermostat *m* d'intérieur
  nl kamerthermostaat *m*
  r комнатный термостат *m*

R763 *e* room unit
  d Raumblock *m*, Raumzelle *f*
  f bloc-pièce *m*, cellule *f* préfabriquée
  nl prefab-woningelement *n*
  r сборный блок *m* на комнату, блок-
  -труба *m*

R764 *e* room ventilation
  d Raumlüftung *f*
  f ventilation *f* des locaux
  nl kamerventilatie *f*
  r вентиляция *f* помещений

R765 *e* rootdozer
  d Planier-Gleiskettengerät *n* für
  Stubbenrodearbeiten
  f dessoucheur *m* [dessoucheuse *f*,
  arrache-souche *m*] sur bulldozer
  nl op bulldozer gemonteerde
  stobbenrooier *m*
  r гусеничный корчеватель *m*;
  бульдозер *m* с корчевателем

R766 rooter
  d Anhängeaufreißer *m*
  f scarificateur *m*, rooter *m*, défonceuse *f*
  [piocheuse *f*] tractée
  nl opbreektanden *m pl*
  r прицепной кирковщик *m*

R767 *e* root rake
  d Wurzelrechen *m*, Wurzelharke *f*
  f arrache-souche *m*, dessoucheur *m*,
  dessoucheuse *f*
  nl wroeter *m*
  r корчеватель *m*

R768 *e* rope
  d Seil *n*
  f câble *m*
  nl touw *n*, kabel *m*
  r канат *m*, трос *m*

R769 *e* rope block
  d Seilflasche *f*, Block *m*
  f poulie *f* à câble
  nl blok *n*
  r (канатный) блок *m*, шкив *m*

R770 *e* rope hoist
  d Seilwinde *f*
  f treuil *m* à câble
  nl windas *f(m)*, lier *f(m)*
  r канатная лебёдка *f*

R771 rope sag *see* cable sag

R772 *e* rope-suspended cantiliver(ed) roof
  d Kabelkragdach *n*, Kabelauslegerdach
  n, Seilkragdach *n*, Seilauslegerdach *n*
  f comble *m* [toit *m*] suspendu en
  encorbellement [en porte-à-faux]
  nl kabelkraagdak *n*
  r вантовое висячее покрытие *n*
  консольного типа

R773 ropeway *see* cableway

R774 rope winch *see* rope hoist

R775 *e* **rops**
 d Schutzschirm *m*, Schutzdach *n* über
  der Fahrerkabine (*der Baumaschine*)
 f ossature *f* de protection sur cabine
  de conducteur (*d'un engin de
  chantier*)
 nl veiligheidsraam *n* (*boven cabine van
  bouwmachines*)
 r защитный козырёк *m* (*над
  кабиной водителя*)

R776 *e* **rotary and percussion boring**
 d Schlag-Drehbohren *n*, Dreh-
  -Schlagboren *n*
 f forage *m* roto-percutant, perforation *f*
  roto-percutante
 nl stotend en draaiend boren *n*
 r ударно-вращательное бурение *n*

R777 *e* **rotary bucket excavator**
 d Schaufelradbagger *m*
 f excavateur *m* à roue à godets [à roue-
  -pelle], roue-pelle *f*, excavateur *m*
  rotatif
 nl schepradgraafmachine *f*
 r роторный экскаватор *m*

R778 *e* **rotary compressor**
 d Rotationskompressor *m*
 f compresseur *m* (à piston) rotatif
 nl rotatiecompressor *m*
 r ротационный компрессор *m*

R779 *e* **rotary core drilling**
 d Rotary-Kernbohrung *f*
 f carottage *m* au rotary
 nl monstername *n* met roterende boor
 r вращательное колонковое бурение *n*

R780 *e* **rotary distributor**
 d Drehsprenger *m*
 f distributeur *m* tournant
 nl draaiende sproeier *m*
 r реактивный водораспределитель *m*
  [ороситель *m*] (*на биофильтре*)

R781 *e* **rotary drill hammer**
 d Drehbohrhammer *m*
 f marteau-perforateur *m*,
  perforateur *m*, perforatrice *f*
 nl boorhamer *m*
 r бурильный молоток *m*, перфоратор
  m

R782 *e* **rotary drilling**
 d Rotarybohren *n*, Drehbohren *n*
 f forage *m* [perforation *f*] par rotation,
  forage *m* rotatif, perforation *f* rotative
 nl draaiend boren *n*
 r вращательное бурение *n*

R783 *e* **rotary heat exchanger**
 d Wärmerad *n*
 f échangeur *m* thermique rotatif
 nl draaiende warmtewisselaar *m*
 r вращающийся теплообменник *m*

R784 *e* **rotary intersection**
 d Kreisverkehrsplatz *m*,
  Kreiselkreuzung *f*
 f croisement *m* à circulation **giratoire,**
  rond-point *m*

 nl verkeersplein *n*, rotonde *f*,
  verkeersrotonde *f*
 r кольцевая транспортная развязка *f*,
  кольцевое пересечение *n*
  (автомобильных) дорог (*в одном
  уровне*)

R785 *e* **rotary kiln**
 d Drehofen *m*
 f four *m* tournant [rotatif]
 nl draaioven *m*
 r вращающаяся печь *f*

R786 *e* **rotary screen**
 d Trommelsieb *n*
 f crible *m* à tambour; tambour-cribleur
  m
 nl trommelzeef *f(m)*
 r барабанный грохот *m*;
  барабанный фильтр *m*

R787 *e* **rotary shear**
 d Kreisschere *f*
 f cisaille *f* circulaire [à disques],
  cisailles *f pl* rotatives
 nl cirkelschaar *f(m)*
 r дисковые ножницы *pl*

R788 *e* **rotary tower crane**
 d Turmdrehkran *m*
 f grue *f* à tour pivotante [tournante]
 nl torendraaikraan *f(m)*
 r башенный поворотный кран *m*

R789 *e* **rotary washing screen**
 d Siebwaschtrommel *f*
 f crible *m* laveur rotatif
 nl zeefwastrommel *f(m)*
 r промывной барабанный грохот *m*

R790 *e* **rotating boom**
 d Schwenkausleger *m*, Schwenkarm *m*
 f flèche *f* pivotante [tournante,
  orientable]
 nl zwenkbare giek *m*
 r поворотная стрела *f*

R791 *e* **rotating crane**
 d Schwenkkran *m*, Drehkran *m*
 f grue *f* pivotante [tournante], grue *f*
  à flèche pivotante
 nl draaikraan *f(m)*
 r поворотный кран *m*, кран *m*
  с поворотной стрелой

R792 *e* **rotating drum water strainer**
 d Trommel(drehwasser)filter *n*, *m*
 f tamis *m* d'eau à tambour rotatif
 nl trommelfilter *n*, *m*
 r барабанный фильтр *m*

R793 **rotating furnace** *see* **rotary kiln'**

R794 *e* **rotating mast crane**
 d Drehmastkran *m*
 f grue *f* à mât tournant
 nl draaimastkraan *f(m)*
 r мачтовый кран *m* с поворотной
  стрелой

R795 *e* **rotational slide**
 d drehkörperartiges Gleiten *n*,
  kreisförmiges Rutschen *n*
 f glissement *m* rotative

nl draaiend schuiven *n*
r круговое скольжение *n*,
скольжение *n* по дуге окружности

R796 *e* **rough concrete**
d Rauhbeton *m*, schalungsrauher
Beton *m*
f béton *m* brut (de décoffrage)
nl ruw beton *n*
r неотделанный бетон *m*, бетон *m*
после распалубки

R797 *e* **roughness**
d Rauheit *f*, Rauhigkeit *f*
f rugosité *f*
nl ruwheid *f*
r шероховатость *f*

R798 *e* **roughness coefficient**
d Rauhigkeitsbeiwert *m*
f coefficient *m* de rugosité
nl ruwheidscoëfficiënt *m*
r коэффициент *m* шероховатости

R799 *e* **roughometer**
d Rauhigkeitsmesser *m*;
Straßenunebenheitsmesser *m*
f rugosimètre *m*; viagraphe *m*,
profilographe *m*
nl ruwheidsmeter *m* (*voor het wegdek*)
r профилометр *m*; измеритель *m*
шероховатости дорожного покрытия

R800 *e* **rough terrain**
d durchschnittenes Gelände *n*
f terrain *m* accidenté
nl moeilijk begaanbaar terrein *n*
r неровная [пересечённая] местность *f*

R801 *e* **rough timber boarding**
d Fehlboden *m*, Rohboden *m*
f entrevous *m*, sol *m* brut, sous-sol *m*
nl ruwe houten vloer *m*
r чёрный пол *m*

R802 **roundabout** *see* **rotary intersection**

R803 *e* **round aggregate**
d runder Zuschlag(stoff) *m*, Rollkies *m*,
Rundkies *m*
f agrégat *m* roulé
nl toeslag *m* met ronde korrels
r заполнитель *m* с окатанными
зёрнами; окатанный гравий *m*

R804 *e* **round bar**
d Rundstab *m*
f barre *f* ronde, rond *m*
nl ronde staaf *f(m)*
r круглый стержень *m*

R805 *e* **round bars**
d Rundstahl *m*
f ronds *m pl* en acier
nl rondstaal *n*
r круглая сталь *f*

R806 *e* **round-head buttress dam**
d Rundkopfstaumauer *f* von Noetzli,
Pfeilerkopfsperre *f*,
Pfeilerkopfstaumauer *f*,
Pilzkopfstaumauer *f*
f barrage *m* à contreforts à tête ronde
nl pijlerstuwdam *m* met ronde bovenrand

r массивно-контрфорсная плотина *f*
с округлёнными оголовками

R807 *e* **round-hole screen**
d Rundlochsieb *n*
f crible *m* à trous ronds [circulaire]
nl zeef *f(m)* met ronde gaten
r сито *n* с круглыми отверстиями

R808 *e* **route**
d Trasse *f*
f tracé *m*
nl tracé *n*
r трасса *f*

R809 *e* **row construction, row construction
method**
d Reihenbau(weise) *m(f)*
f construction *f* en rangées
nl lintbebouwing *f*
r линейная застройка *f*, застройка *f*
рядами

R810 *e* **rowlock**
d 1. Ziegel *m* auf hochkant verlegt
2. hochkantige Einzel-Bogenschicht *f*
des mehrschichtigen gemauerten
Bogensturzes
f 1. brique *f* posée de champ 2. rangée *f*
de briques posées en arceau
nl rollaag *f(m)*
r 1. кирпич *m*, поставленный на ребро
2. раздельный концентрический
ряд *m* многослойной кирпичной арки
[перемычки]

R811 *e* **rowlock cavity wall**
d hochkantiges Schachtmauerwerk *n*
f maçonnerie *f* creuse en briques posées
de champ
nl spouwmuur *m* waarvan het binnenblad
is gemetseld met rollagen
r колодезная кирпичная кладка *f* из
кирпичей, поставленных на ребро

R812 *e* **row of piles**
d Pfahlreihe *f*, Pfahlwand *f*
f rangée *f* de pieux
nl palenrij *f(m)*
r свайный ряд *m*

R813 *e* **row of wells**
d Brunnengalerie *f*, Filterbrunnenreihe *f*
f galerie *f* de drainage avec une rangée
de puits filtrants; prise *f* d'eau avec
une rangée de puits filtrants
nl bronnenserie *f*
r дренажная галерея *f* с рядом
колодцев; водозабор *m* с рядом
колодцев

R814 *e* **rubbed finish**
d Schleifbearbeitung *f* einer
Betonoberfläche
f finition *f* de surface de béton par
abrasifs, traitement *m* du béton par
abrasifs
nl geslepen oppervlak *n* van beton
r абразивная обработка *f* [отделка *f*]
бетонной поверхности

R815 *e* **rubber bearing**
   d Gummilager *n*
   f appareil *m* d'appui en caoutchouc;
     appui *m* en caoutchouc
   nl rubber oplegging *f*
   r резиновая опорная часть *f*;
     резиновая опора *f*

R816 *e* **rubber-bitumen sealing compound**
   d Gummi-Bitumen-Vergußmasse *f*
   f composition *f* [mastic *m*] de scellement
     à base de caoutchouc et de bitume
   nl rubber-bitumenvulmassa *f(m)*
   r резино-битумная уплотняющая
     мастика *f* (*для заливки швов*)

R817 *e* **rubber elephant trunk**
   d Betonschüttrohr *n*, Schlauchansatz *m*
     am Betonkübel zum Schütten in engen
     Schalungen
   f trompe *f* [goulotte *f*] d'éléphant en
     caoutchouc (pour la mise en œuvre
     du béton)
   nl rubberslang *f(m)* om beton in nauwe
     bekisting te kunnen storten
   r резиновый хобот *m* для спуска
     бетонной смеси

R818 *e* **rubber-emulsion paint**
   d Latexemulsionsfarbe *f*
   f peinture *f* émulsionnée [peinture-
     -émulsion *f*] à base de caoutchouc
   nl rubber-emulsieverf *f(m)*
   r каучуковая [латексная]
     эмульсионная краска *f*

R819 *e* **rubber-faced steel plate**
   d gummibeschichtete Stahlplatte *f*
   f tôle *f* [plaque *f*] d'acier avec
     revêtement en caoutchouc
   nl met rubber beklede staalplaat *f(m)*
   r гуммированная стальная плита *f*;
     гуммированная листовая сталь *f*

R820 *e* **rubber floor covering, rubber flooring**
   d Gummifußbodenbelag *m*
   f revêtement *m* de sol en caoutchouc,
     couvre-sol *m* en caoutchouc
   nl rubber dekvloer *m*
   r резиновое покрытие *n* пола

R821 *e* **rubber gasket**
   d Gummidichtung *f*
   f joint *m* [garniture *f*] en caoutchouc
   nl rubber pakking *f*
   r резиновое уплотнение *n*

R822 *e* **rubber glazing channel**
   d Gummiprofil *n* zur Verglasung
   f garniture *f* profilée à vitrage en
     caoutchouc
   nl gasketprofiel *n*
   r резиновый профильный элемент *m*
     для монтажа остекления; фасонная
     резиновая прокладка *f* для
     остекления

R823 *e* **rubberized asphalt** *US*
   d Bitumen-Gummi-Gemisch *n*
   f asphalte *m* caoutchouté
   nl bitumen-rubbermengsel *n*
   r дорожный битум *m* с добавкой
     синтетического каучука

R824 *e* **rubber mallet**
   d Gummihammer *m*
   f maillet *m* en caoutchouc; marteau *m*
     en caoutchouc
   nl rubber hamer *m*
   r резиновая киянка *f*; резиновый
     молоток *m*

R825 *e* **rubber modified bitumen**
   d Kautschukmilch-Bitumen-
     -Vergußmasse *f*, Bitumen-Gummi-
     -Gemisch *n*, Bitumen-Gummi-
     -Mischung *f*
   f bitume-caoutchouc *m*, mélange *m*
     bitume-caoutchouc
   nl rubber-bitumenmastiek *f*
   r битумно-резиновая композиция *f*
     [смесь *f*]

R826 *e* **rubber-mounted crane**
   d Fahrzeugkran *m*
   f grue *f* (automotrice) sur pneus
   nl wielkraan *f(m)*
   r пневмоколёсный кран *m*,
     автокран *m*

R827 *e* **rubber-mounted crusher**
   d Autosteinbrecher *m*, Fahrbrecher *m*
   f concasseur *m* [broyeur *m*] mobile sur
     pneus
   nl zelfrijdende betonbreker *m*
   r передвижная дробилка *f* на
     пневмоколёсном ходу

R828 *e* **rubber pad**
   d Gummiunterlage *f*, Gummikissen *n*
   f plaque *f* en caoutchouc
   nl rubberonderlaag *f(m)*, rubberkussen *n*
   r резиновая подкладка *f* [подушка *f*]

R829 *e* **rubber screen cloth**
   d Gummisiebgewebe *n*
   f toile-caoutchouc *f* pour le tamis;
     tamis *m* en toile-caoutchouc
   nl rubber zeefdoek *n*
   r резиновая сетка *f*; резиновая
     сетчатая ткань *f*

R830 *e* **rubber tape**
   d Gummiisolierband *n*
   f ruban *m* de toile isolante, chatterton
     m
   nl rubber isolatieband *n*
   r резиновая изоляционная лента *f*

R831 *e* **rubber tile**
   d Gummiplatte *f*, Gummifliese *f*
   f carreau *m* (en) caoutchouc
   nl rubbertegel *m*
   r резиновая плитка *f* (*для пола*)

R832 *e* **rubber-tired roller**
   d Gummiradwalze *f*, luftbereifte Walze
     f, Pneuwalze *f*
   f cylindre *m* [compacteur *m*, rouleau *m*]
     à pneus
   nl bandenwegwals *f(m)*
   r дорожный каток *m* на пневматических
     шинах

R833 *e* **rubber-tired scraper**
  d gummibereifter Radschrapper *m*
  f scraper *m* [décapeuse *f*] à pneus
  nl scraper *m* op banden
  r скрепер *m* на пневматическом ходу

R834 *e* **rubber-wheeled hydraulic excavator**
  d gummibereifter Hydraulikbagger *m* [Hydrobagger *m*]
  f excavateur *m* [pelle *f* mécanique] hydraulique sur pneu(matique)s
  nl hydraulische graafmachine *f* op luchtbanden
  r пневмоколёсный гидравлический экскаватор *m*

R835 *e* **rubbish**
  d 1. Gemüll *n*, Bauschutt *m*, Müll *m*, Abfall *m* 2. Sinkgut *n* (*bei Schwimm-Sink-Scheidung*)
  f 1. gravats *m pl*, gravois *m pl*, abat(t)is *m* 2. sédiments *m pl*
  nl 1. puin *n* vuilnis *n*, *f* 2. bezinksel *n*
  r 1. строительный мусор *m*; отходы *m pl* 2. осевший продукт *m* (*при обогащении в тяжелых средах*)

R836 *e* **rubbish grinder**
  d Müllzerkleinerer *m*, Müllwolf *m*
  f broyeur *m* d'ordures ménagères
  nl afvalvermaler *m*
  r мусородробилка *f*

R837 *e* **rubble**
  d 1. Rollkies *m*, Felsgerölle *n*, Gerölle *n* 2. Bruchsteine *m pl* 3. Bauschutt *m*
  f 1. gros gravier *m* roulé, galet *m*, cailloux *m pl* 2. moellon *m*, pierre *f* brute (de carriere) 3. abat(t)is *m*, gravats *m pl*, gravois *m pl*
  nl 1. rolstenen *m pl* 2. breuksteen *m* 3. puin *n*
  r 1. окатанный гравий *m*, галька *f*; валуны *m pl* 2. карьерный бутовый [рваный] камень *m* 3. строительный мусор *m*

R838 *e* **rubble chute**
  d Schuttrutsche *f*
  f goulotte *f* [goulette *f*] de vidage
  nl puinstortgoot *f(m)*
  r спускной жёлоб *m* [лоток *m*] (для строительного мусора)

R839 *e* **rubble concrete**
  d Zyklopenbeton *m*, Bruchsteinbeton *m*
  f béton *m* cyclopéen [de pierraille]
  nl cyclopenbeton *n*
  r бутобетон *m*

R840 *e* **rubble cushion**
  d Bruchsteinpolster *n*, *m*
  f lit *m* [matelas *m*] de moellons bruts
  nl bed *n* van breuksteen
  r постель *f* из рваного камня

R841 *e* **rubble drain**
  d Steindrän *m*, Steinrigole *f*, Sickergraben *m*, Sauggraben *m*
  f drain *m* en pierres
  nl steendrain *m*
  r дрена *f* из крупного камня

R842 *e* **rubble masonry**
  d Bruchsteinmauerwerk *n*
  f maçonnerie *f* en moellons
  nl metselwerk *n* van breuksteen
  r бутовая кладка *f*

R843 *e* **rubble-mound breakwater**
  d Steinschüttungs-Wellenbrecher *m*
  f brise-lames *m* en enrochement
  nl stortsteen-golvbreker *m*
  r волнолом *m* из каменной наброски

R844 *e* **rubble stone**
  d Bruchstein *m*
  f moellon *m* (brut)
  nl breuksteen *m*
  r бутовый камень *m*

R845 *e* **ruberoid roofing**
  d Ruberoiddacheindeckung *f*
  f couverture *f* [toiture *f*] en feutre bitumé [en carton bitumé]
  nl ruberoid dakbedekking *f*
  r рубероидное кровельное покрытие *n*

R846 **rugged terrain** *see* **rough terrain**

R847 **rugosity** *see* **roughness**

R848 *e* **rule**
  d 1. Vorschrift *f* 2. Lineal *n*, Maßstab *m*
  f 1. règlement *m*, prescription *f* 2. règle *f*
  nl 1. voorschrift *n* 2. lineaal *m*
  r 1. технический норматив *m*; указание *n*; норма *f*; правило *n*; предписание *n* 2. линейка *f*

R849 *e* **ruling gradient**
  d höchstzulässige Straßenneigung
  f déclivité *f* limite admissible
  nl hoogst toelaatbare helling *f*
  r предельно допустимый уклон *m*

R850 *e* **run**
  d 1. Anlage *f*, Gründung *f* 2. Auflagestufe *f*, Trittstufe *f* 3. Abschnitt *m*, Rohrstrang *m* 4. Rohrleitungsschema *n*, Rohranordnung *f* 5. Läufer *m pl*, Gardinen *f pl*, Nasen *f pl*
  f 1. portée *f* horizontale (*d'une volée d'une pente de comble*) 2. pas *m* de marche, giron *m* 3. tronçon *m* de tuyauterie. schéma *m* [disposition *f*] de tuyauterie 5. coulure *f*
  nl 1. vlucht *f(m)* (*horizontale afstand*) 2. aantrede *f* 3. recht gedeelte *n* van een leiding 4. leidingschema *n* 5. loper *m*
  r 1. заложение *n* (*лестничного марша, ската крыши*) 2. проступь *f* 3. отрезок *m*; секция *f* (*трубопровода*) 4. общая схема *f* трубопроводов (*в здании*) 5. потёки *m pl* (*краски*)

R851 *e* **rungs**
  d Leitersprossen *f pl*
  f barreaux *m pl*, échelons *m pl* (*d'une échelle*)
  nl sporten *m pl* (*ladder*)
  r ступеньки *f pl*, перекладины *f pl* (*лестницы*)

R852 *e* **running sand**
  d Triebsand *m*, Schwimmsand *m*,
    Fließsand *m*
  f sable *m* mouvant
  nl drijfzand *n*
  r плывунный песчаный грунт *m*,
    плывун *m*

R853 *e* **runoff**
  d Abfluß *m*
  f écoulement *m*, ruissellement *m*,
    débit *m*
  nl afvoer *m*
  r сток *m*

R854 **run-of-bank gravel** *see* **bank-run gravel**

R855 *e* **runoff coefficient**
  d Abflußverhältnis *n*
  f coefficient *m* de ruisselement
    [d'écoulement]
  nl afvoercoëfficiënt *m*
  r коэффициент *m* стока

R857 *e* **run-out**
  d Abzweigstück *n*, Ansatzstück *n*
  f tubulure *f*
  nl aftakkende pijp *f(m)*; T-stuk *n*
  r отросток *m* (*трубопровода*);
    патрубок *m*

R858 *e* **runway**
  d 1. Durchgang *m*; Durchfahrt *f*
    2. Übergangsbrücke *f* 3. Rollbahn *f*,
    Piste *f* 4. Steg *m*, Schubkarrensteg *m*
  f 1. passage *m* 2. passerelle *f* 3. piste *f*
    d'envol et d'atterrissage 4. planche *f*
    de roulement (*pour brouettes*)
  nl 1. doorvaart *f(m)* 2. doorgang *m*
    3. startbaan *f(m)* 4. loopplank *f(m)*
  r 1. проход *m*; проезд *m* 2. переходной
    мостик *m* 3. взлётно-посадочная
    полоса *f* 4. катальный ход *m*

R859 *e* **runway beam**
  d Kranbahnträger *m*
  f poutre-roulante *f*
  nl kraanbaanligger *m*
  r кран-балка *f*

R860 *e* **rupture**
  d 1. Bruch *m* 2. Deich(ein)bruch *m*,
    Damm(ein)bruch *m*
  f 1. rupture *f* 2. brèche *f*, rupture *f*
    (de digue)
  nl 1. breuk *f(m)* 2. dijkbreuk *f(m)*
  r 1. разрушение *n*, разрыв *m*
    2. прорыв *m* (*плотины*)

R861 **rupture cross-section** *see* **breaking
cross-section**

R863 **rupture load** *see* **breaking load**

R864 **rupture modulus** *see* **modulus of rupture**

R865 **rupture strength** *see* **modulus of
rupture**

R866 *e* **rupture test**
  d Bruchprüfung *f*, Bruchprobe *f*,
    Bruchversuch *m*; Biegefestigkeitsprobe
    f
  f essai *m* de rupture; essai *m* de
    résistance à la flexion

  nl breekproef *f(m)*
  r испытание *n* на разрыв;
    определение *n* прочности (*балки*) при
    изгибе

R867 **rushing flow** *see* **hyper-critical flow**

R868 *e* **rush mat**
  d Binsenmatte *f*
  f natte *f* de jonc
  nl biezenmat *f(m)*
  r камышовый мат *m*

R869 *e* **rustic quoin**
  d Rustikaquader *m*, Bossenstein *m*
  f pierre *f* d'angle rustique
  nl rustiek steenblok *n*
  r рустованный камень *m*

R870 *e* **rustic work**
  d Bossenmauerwerk *n*
  f maçonnerie *f* rustiquée
  nl rustiek metselwerk *n*
  r рустованная кладка *f* (*из тёсаного
    природного камня*)

R871 *e* **rust mark**
  d Rostfleck *m*
  f tache *f* de rouill
  nl roestvlek *f(m)*
  r след *m* ржавчины, ржавое пятно *n*
    (*дефект отделки*)

R872 *e* **rut**
  d Spur *f*
  f ornière *f* (d'un chemin)
  nl spoor *n* (*in wegdek*)
  r колея *f*; борозда *f*

R873 **ruts** *see* **channels**

S

S1   *e* **sack rubbing, sack scrubbing**
  d Glattstreichen *n*, Gleitstrich *m* (*von
    Betonoberflächen mit Sackgewebe*)
  f lissage *m* [surfaçage *m*] de béton par
    toile à sacs
  nl betonoppervlak *n* met zakkenlinnen
    gladstrijken
  r затирка *f* [отделка *f*] мешковиной
    (*бетонных поверхностей*)

S2   *e* **saddle**
  d 1. Schwelle *f* 2. Umlenklager *n*
    3. Rohrsattel *m* 4. sattelförmige
    Dacherhöhung *f* 5. Anbohrschelle *f*
    6. Flansch-Glockenmuffe *f*
  f 1. seuil *m*, semelle *f*, sabot *m*
    2. selle *f* de repos (*pour câlles*)
    3. selle *f* d'appui (*d'un tuyau*)
    4. saillie *f* sur toiture en forme d'une
    selle 5. sellette *f* 6. manchon *m*
    à emboiture et à bride
  nl 1. drempel *n* 2. stut *n* 3. zadel *n*
    4. zadelvormig dakoverstek *n*
    5. zadelklem *f(m)* 6. flenskoppelstuk *n*
  r 1. порог *m*; башмак *m*
    2. седловидная опорная часть *f* (*для*

несущего каната) **3.** седловидная
опора *f* (*трубопровода*)
4. седлообразное возвышение *n*
кровли (*у мест примыкания
к трубе, слуховому окну*)
5. седёлка *f* **6.** раструбно-фланцевая
муфта *f*

S4 *e* **saddle pipe support**
   d Rohrsattel *m*
   f selle *f* d'appui (*d'un tuyau*)
   nl zadel *n*
   r седловидная опора *f* трубопровода

S5 *e* **Sadgrove maturity figure**
   d Betonfestigkeitszunahme-Kennwert *m*
     nach Sadgrove
   f indice *f* de maturation (de béton)
     d'après Sadgrove
   nl vastheidswaarde *f* tijdens verharding
     volgens Sadgrove
   r показатель *m* вызревания
     (*нарастания прочности*) бетона по
     Сэдгреву

S6 *e* **safe leg load**
   d zulässige Gerüstbaumbelastung *f*
   f charge *f* de sécurité sur un montant
     d'échafaudage
   nl toelaatbare belasting *f* op een
     steigerpaal
   r безопасная нагрузка *f* на стойку
     лесов

S7 *e* **safety against cracking**
   d Rißsicherheit *f*
   f sécurité *f* à [contre] la fissuration
   nl scheurveiligheid *f*
   r надёжность *f* по
     трещинообразованию

S8 *e* **safety against rupture**
   d Bruchsicherheit *f*
   f sécurité *f* à [contre] la rupture
   nl breukveiligheid *f*
   r запас *m* прочности на
     разрыв

S9 *e* **safety barrier**
   d Fußgänger-Schutzgeländer *n*
   f barrière *f* de protection [de sécurité],
     garde-corps *m* de protection
   nl leuning *f*
   r защитное [предохранительное]
     ограждение *n*

S10 *e* **safety belt**
   d Sicherheitsgurt *m*
   f ceinture *f* de sécurité
   nl veiligheidsgordel *m*
   r предохранительный пояс *m*

S11 *e* **safety curtain**
   d Brandschutzvorhang *m*
   f rideau *m* coupe-feu
   nl brandwerend gordijn *n*, brandscherm
     n
   r противопожарный занавес *m*

S12 *e* **safety factor**
   d Sicherheitsfaktor *m*
   f facteur *m* de sécurité

   nl veiligheidsfactor *m*
   r запас *m* прочности; коэффициент *m*
     запаса (прочности)

S13 **safety fence** *see* **safety barrier**

S14 *e* **safety gate**
   d Sicherheitstor *n* (*Schleuse*)
   f porte *f* de garde (*d'une écluse*)
   nl vloeddeur *f(m)* (*van een sluis*)
   r предохранительные ворота *pl*
     (*шлюза*)

S15 *e* **safety goggles**
   d Schutzbrille *f*
   f lunettes *f pl* protectrices [de
     protection]
   nl veiligheidsbril *m*
   r предохранительные [защитные] очки
     pl

S16 **safety helmet** *see* **protective helmet**

S17 *e* **safety hook**
   d Sicherheitshaken *m*
   f crochet *m* (de levage) à mousqueton
   nl borghaak *m*, veiligheidshaak *m*
   r грузоподъёмный крюк *m*
     с карабином, безопасный крюк *m*

S18 *e* **safety precautions**
   d Sicherheitsmaßnahmen *f pl*,
     Arbeitsschutzmaßnahmen *f pl*
   f mesures *f pl* de sécurité
   nl veiligheidsmaatregelen *m pl*
   r меры *f pl* по технике безопасности

S19 *e* **safety valve**
   d Sicherheitsventil *n*,
     Entlastungsventil *n*
   f soupape *f* de sûreté
   nl veiligheidsklep *f(m)*
   r предохранительный клапан *m*

S20 **safe velocity** *see* **permissible velocity**

S21 *e* **safe working conditions**
   d sichere Arbeitsbedingungen *f pl*
   f conditions *f pl* de sécurité du travail
   nl veilige arbeidsomstandigheden *f pl*
   r безопасные условия *n pl* производства
     работ

S22 *e* **safe working pressure**
   d höchstzulässiger Arbeitsdruck *m*
   f pression *f* de service admissible
   nl hoogst toelaatbare werkdruk *m*
   r предельно допустимое рабочее
     давление *n*

S23 *e* **safe yield**
   d Wasserdarbietung *f*; Wasserdargebot
     n
   f débit *m* de sécurité; ressources *f pl*
     disponibles en eau, quantité *f* d'eau
     disponible
   nl gegarandeerd debiet *n*
   r гарантированный дебит *m*,
     водообеспеченность *f*

S24 *e* **sag**
   d **1.** Durchhang *m*, Durchbiegung *f*,
     Pfeilhöhe *f* **2.** Senkung *f*, Sackung *f*
     3. Wanne *f*, Konkavität *f*

f 1. flèche *f* 2. tassement *m*,
affaissement *m* 3. partie *f* concave
(*d'une route*)
nl 1. pijl(hoogte) *n* (*f*) van de
doorhanging 2. scheefzakking *f*
3. dal *n* in een weg
r 1. стрела *f* прогиба *или* провеса
2. оседание *n* 3. вогнутость *f* (*дороги*)

S25 *e* **sagging**
d 1. Senkung *f*, Setzung *f*, Absackung *f*
2. Farbschmitzen *f pl*,
Farbstreifen *m pl*, Farbnasen *f pl*,
Läufer *m pl*
f 1. affaissement *m* 2. coulure *f*
nl 1. zakking *f* inklinking *f*
2. uitzakken *n* van de verf
r 1. осадка *f*, просадка *f* (*грунта, шва*)
2. потёки *pl* (*краски*)

S26 *e* **sag pipe**
d Düker *m*, Rohrdüker *m*
f siphon *m* inversé [renversé],
aqueduc-siphon *m*
nl duiker *m*
r дюкер *m*

S27 *e* **salinity, salt content**
d Salzgehalt *m*, Salzhaltigkeit *f*
f salinité *f*, teneur *m* en sel
nl zoutgehalte *n*
r солёность *f*, минерализация *f* воды,
содержание *n* минеральных солей

S28 *e* **sample**
d Probe *f*
f échantillon *m*
nl monster *n*
r образец *m*, проба *f*

S29 *e* **sampler**
d 1. Probeentnahmeapparat *m*,
Probeentnahmestanze *f*
2. Probenehmer *m*
f 1. carottier *m* 2. échantillonneuse *f*
nl 1. bemonsteringstoestel *n* 2. grondboor
f(m)
r 1. грунтонос *m* 2. пробоотборник *m*

S30 *e* **sampling**
d 1. Probenahme *f*
2. Stichprobenerhebung *f*,
Stichprobenprüfung *f*
f 1. échantillonage *m*, prélèvement *m*
2. examen *m* arbitraire
nl 1. monstername *f* 2. steekproef *f*
r 1. отбор *m* [взятие *n*] проб [образцов]
2. выборочное исследование *n*;
выборочный контроль *m*

S31 *e* **sampling disturbance**
d Entnahmestörung *f*
f remaniement *m* de la prise
d'échantillon
nl verstoring *f* van de bodemstructuur
bij monstername
r нарушение *n* структуры грунта при
отборе образцов

S32 *e* **sampling spoon**
d Probelöffel *m*, Schappe *f*

f cuillère *f* pour prélèvement
d'échantillon
nl lepelboor *f(m)*
r буровая ложка *f* для отбора
образцов; ложечный бур *m*;
желонка *f*

S33 **sampling tube** *see* **sampler**

S34 *e* **sand**
d Sand *m*
f sable *m*
nl zand *n*
r песок *m*; песчаная фракция *f*
(*заполнителя*)

S35 *e* **sand-and-gravel filter**
d Sand-Kies-Filter *n*, *m*
f filtre *m* de sable et de gravier
nl zand-grindfilter *n*
r песчано-гравийный фильтр *m*

S36 *e* **sand-and-gravel washer**
d Sand- und Kieswaschmaschine *f*
f laveur *m* [laveuse *f*] de sable et de
gravier
nl zand- en grindwasmachine *f*
r пескогравиемойка *f*

S37 *e* **sand asphalt**
d Sandasphalt *m*
f asphalte *m* sablé
nl zandasfalt *n*
r песчаный асфальт *m*

S38 *e* **sand bed**
d Sandbett *n*
f lit *m* de sable; charge *f* filtrante,
couche *f* filtrante
nl zandbed *n*
r песчаная подушка *f* [постель *f*];
песчаная загрузка *f* (*фильтра*)

S39 *e* **sand blanket**
d Sandbett *n*; Sandunterlage *f*
f lit *m* de sable; couche *f* protectrice
de sable
nl zandafdeklaag *f(m)*
r песчаная постель *f* [подушка *f*];
песчаный защитный слой *n*
(*укладываемый по глинистому дну
каналов, водохранилищ*)

S40 *e* **sandblaster**
d Sandstrahlgebläse *n*
f sableuse *f* pour décapage, appareil *m*
à jet de sable
nl zandstraalapparaat *n*
r пескоструйный аппарат *m*,
аппарат *m* для пескоструйной
обработки

S41 *e* **sandblasting**
d Sandstrahlen *n*, Sandstrahlreinigung *f*
f sablage *m*, nettoyage *m* [décapage *m*]
au jet de sable
nl zandstralen *n*
r пескоструйная очистка *f*
[обработка *f*]

S42 *e* **sand boil**
d Sandquelle *f*
f cratère *m* de renard

nl zandbron *f(m)*
r грифон *m*

S43 **sand box** *see* **sand jack**

S44 *e* **sand-coarse aggregate ratio**
d Feinzuschlag-Grobzuschlag-
-Verhältnis *n*
f rapport *m* sable-agrégat
nl zand-toeslagverhouding *f* (*beton*)
r отношение *n* песок — заполнитель
(*в бетоне*)

S45 **sand cushion** *see* **sand bed**

S46 *e* **sand drain**
d 1. Sanddrän *m*, Sandbrunnen *m*
2. Sickerdohle *f*, Sickerrinne *f*
f 1. drain *m* vertical en sable 2. rigole *f*
dans le sol
nl 1. zandpaal *m* 2. draineersleuf *f(m)*
r 1. вертикальная песчаная дрена *f*
2. подземный сточный лоток *m*

S47 *e* **sand factor**
d Sandanteil *m* (*der Betonmischung*)
f pourcentage *m* de sable (*dans le
béton*); teneur *f* en sable
nl zandpercentage *n*
r процент *m* содержания песка
(*в бетонной смеси*)

S48 *e* **sand fill**
d Sandaufschüttung *f*, Sand(auf)füllung
f
f remblai *m* de sable, remblaiment *m*
en sable; ballast *m* de sable
nl zandaanvulling *f*, zandbed *n*
r песчаная отсыпка *f*; песчаная
насыпь *f*; песчаный балласт *m*

S49 **sand filling** *US* *see* **sanding-up**

S50 *e* **sand filter**
d Sandfilter *n*, *m*
f filtre *m* à sable [de sable]
nl zandfilter *n*, *m*
r песчаный фильтр *m*

S51 *e* **sand fraction**
d Sandanteil *m*, Sandfraktion *f*
f fraction *f* fine, fines *m pl*
nl zandfractie *f*
r песчаная фракция *f*

S52 *e* **sand gradation**
d Sand-Kornabstufung *f*; Abstimmung *f*
der Korngrößenverhältnisse für
Feinzuschlag
f granulométrie *f* [granularité *f*] du
sable
nl bepalen *n* van de vereiste
korrelgrootte van betonzand
r гранулометрический состав *m*
песка; подбор *m* гранулометрического
состава песка

S53 *e* **sand grout**
d Zement-Sand-Schlämme *f*, Zement-
-Sand-Mörtel *m*
f coulis *m* de ciment à sable fin
nl zand-cementmortel *m*
r цементно-песчаный раствор *m*

S54 *e* **sanding-up**
d Versandung *f*
f ensablement *m*
nl bezanden *n*
r занос *m* песком; пескование *n*

S55 *e* **sand jack**
d Sandtopf *m* (*Gerüst*)
f boîte *f* à sable, sablier *m*
nl zandkist *f(m)*
r песочный домкрат *m* (*ящик
с песком для опирания стоек
опалубки*)

S56 **sand-lime brick** *see* **calcium-silicate
brick**

S57 *e* **sand lines**
d Sandstreifen *m pl*, Sandschmitzen *f
pl*
f lignes *f pl* [bandes *f pl*] de sable
(*défaut de surface de béton*)
nl zandstrepen *f(m) pl* (*op het betonoppervlak*)
r прожилки *f pl* песка (*дефект бетонной поверхности*)

S58 *e* **sand pile**
d Sandpfahl *m*
f pieu *m* de sable
nl zandpaal *m*
r песчаная набивная свая *f*

S59 *e* **sand pocket**
d Sandnest *n*
f nid *m* [poche *f*] de sable
nl zandnest *n*
r песчаный карман *m*

S60 *e* **sand screen**
d 1. Siebfilter *n*, *m*, Netzfilter *n*, *m*
2., 3. Sandsieb *n*
f 1. crépine *f* 2. écran *m* à
sédiments 3. crible *m* [claie *f*] à
sable
nl 1. zuigkorf *f(m)* 2. zandzeef *f(m)*
r 1. сетчатый фильтр *m* 2. песчаный
экран *m* (*наносозадерживающее устройство в канале перед
водоприёмным регулятором*) 3. сито
n [сетка *f*] для просеивания песка

S61 *e* **sandstone**
d Sandstein *m*
f grès *m*
nl zandsteen *n*
r песчаник *m*

S62 **sand streaks** *see* **sand lines**

S63 *e* **sand trap**
d Entsander *m*, Sandfang *m*
f dessableur *m*, séparateur *m* de
sable
nl zandvang *m*
r канализационная песколовка *f*;
колодец-отстойник *m*

S64 *e* **sand washer**
d Sandwaschmaschine *f*,
Sandwaschanlage *f*
f laveur *m* [laveuse *f*] de sable,
appareil *m* de lavage du sable

nl zandwasmachine *f*
r пескомойка *f*

S65 *e* sandwich *UK*
d Sperre *f*, Sperrschicht *f*
f couche *f* imperméable [hydrofuge]
mise entre deux couches de béton
nl waterdichte tussenlaag *f pl* (*beton*)
r водонепроницаемая прослойка *f*
между двумя слоями бетона

S66 *e* sandwich construction
d Sandwichkonstruktion *f*,
mehrschichtige Verbundkonstruktion
f
f construction *f* composée
multicouche, construction *f*
sandwich
nl gelaagde constructie *f*, sandwich-
-constructie *f*
r многослойная составная
конструкция *f*

S67 *e* sandwich panel, sandwich slab
d Sandwichplatte *f*
f panneau *m* [plaque *f*] sandwich,
panneau *m* multicouche
nl sandwich-paneel *n*
r многослойная панель *f*

S68 *e* sandwich wall
d mehrschichtige Wand *f*
f mur *m* multicouche [sandwich]
nl gelaagde muur *m*
r многослойная стена *f*

S69 *e* sanitary appliance
d Sanitärgerät *n*
f appareil *m* sanitaire
nl sanitair toestel *n*
r санитарный прибор *m*

S70 *e* sanitary engineering
d Sanitärtechnik *f*
f technique *f* sanitaire
nl gezondheidstechniek *f*
r санитарная техника *f*

S71 *e* sanitary faience
d Baufayence *f*
f faïence *f* sanitaire
nl sanitair porselein *n*
r санитарно-технический фаянс *m*

S72 *e* sanitary fitments
d Sanitärgeräte *n pl*, sanitärtechnische
Ausrüstung *f*
f équipements *m pl* sanitaires
nl sanitaire uitrusting *f*
r санитарные приборы *m pl*;
санитарно-техническое оборудование
n

S73 *e* sanitary fittings, sanitary fixtures
d Sanitärarmatur *f*
f appareils *m pl* de robinetterie
sanitaires
nl sanitair *n*
r санитарная [санитарно-
техническая] арматура *f*

S74 *e* sanitary installation
d 1. Schmutzwasser-Abführung *f*
2. sanitäre Einrichtung *f*
f 1. canalisation *f* [réseau *m*] d'évacuation *f* intérieure 2. équipement *m*
sanitaire; installation *f*
sanitaire
nl 1. binnenriolering *f* 2. sanitaire
installatie *f*
r 1. внутренняя канализация *f*
2. санитарно-техническое
оборудование *n*

S75 *e* sanitary protection zone
d Sanitärschutzgebiet *n*,
Wasserschutzgebiet *n*
f zone *f* sanitaire
nl waterwingebied *n*
r зона *f* санитарной охраны

S76 *e* sanitary sewage
d Schmutzwasser *n*
f eaux *f pl* usée domestiques, eaux *f*
pl usées ménagères
nl huishoudelijk afvalwater *n*
r бытовые сточные воды *f pl*

S77 *e* sanitary sewers
d Ortskanalisation *f*
f réseaux *m pl* d'assainissement
urbains
nl plaatselijk rioolstelsel *n*
r канализация *f* населённых мест

S78 *e* sanitary stoneware
d Sanitärsteinzeug *n*, Sanitärsteingut
n
f grès *m* cérame sanitaire, faïence *f*
sanitaire
nl sanitair porselein en keramiek
r санитарные керамические изделия
n pl

S79 *e* sanitary works
d Sanitärarbeiten *f pl*
f travaux *m pl* d'assainissement
nl loodgieterswerk *n*
r санитарно-технические работы *f pl*

S80 *e* sash *US*
d Fensterflügel *m*, Flügel *m*
f vantail *m* [battant *m*] de fenêtre
nl venstervleugel *m*, raam *n*
r оконная створка *f*

S81 *e* saturated soil
d wassergesättigter Boden *m*
f sol *m* saturé
nl met water verzadigde grond *m*
r водонасыщенный грунт *m*

S82 *e* saturated vapour
d gesättigter Dampf *m*
f vapeur *f* saturée
nl verzadigde damp *m*
r насыщенный пар *m*

S83 *e* saturation
d Sättigung *f*
f saturation *f*
nl verzadiging *f*
r насыщение *n*

S84   saturation curve *see* saturation line

S85 *e* **saturation deficit**
    d Sättigungsdefizit *n*, Untersättigung
    f, Sättigungsfehlbetrag *m*
    f deficit *m* de saturation
    nl onvoldoende vochtigheid *f*,
    vochtgebrek *n*
    r дефицит *m* влажности (*воздуха*)

S86 *e* **saturation efficiency**
    d Befeuchtungswirkungsgrad *m*,
    Sprühkammergütegrad *m*
    f rendement *m* d'humidification
    nl doelmatigheid *f* van de
    luchtbevochtiging
    r эффективность *f* увлажнения
    воздуха (*в камере орошения*)

S87 *e* **saturation line**
    d **1.** Sickerlinie *f*, obere
    Begrenzungslinie *f* der
    Sickerströmung; Absenkungskurve *f*,
    Depressionskurve *f* **2.**
    Sättigungslinie *f*
    f ligne *f* de saturation
    nl verzadigingslijn *f(m)* (*bovengrens*
    van het zakwater)
    r линия *f* насыщения (*верхняя*
    граница фильтрационного потока
    через земляную плотину); кривая *f*
    депрессии; зеркало *n* подземных
    вод; кривая *f* насыщения (*на*
    диаграмме состояния воздуха)

S88 *e* **saturation pressure**
    d Sättigungs(dampf)druck *m*
    f pression *f* de vapeur saturée
    nl verzadigingsdruk *m*
    r давление *n* насыщения, давление *n*
    насыщенного пара

S89 *e* **saturation ratio**
    d Sättigungsverhältnis *n*
    f degré *m* de saturation
    nl verzadigingsgraad *m*
    r степень *f* насыщения

S90 *e* **saturation zone**
    d Sättigungszone *f*
    f zone *f* de saturation
    nl verzadigingszone *f*
    r зона *f* насыщения

S91 *e* **sausage dam**
    d Wurstdamm *m*, Faschinendamm *m*,
    Grundschwelle *f oder* Damm *m* aus
    Drahtschottwalzen
    f digue *f* saucisse, digue *f* en pierres
    retenues par un grillage
    nl rijsdam *m*, fascinedam *m*
    r дамба *f или* донный порог *m* из
    цилиндрических габионов

S92 *e* **saw**
    d Säge *f*
    f scie *f*
    nl zaag *f*
    r пила *f*

S93 *e* **saw cutting**
    d Einschleifen *n* (*Fugen in*
    Betonstraßendecken)
    f sciage *m* (*de joints de chaussée*)
    nl krimpvoegen zagen in betonweg
    r пропил *m*, нарезка *f* швов (*в*
    бетоне)

S94 *e* **sawdust concrete**
    d Holzspanbeton *m*, Sägespanbeton *m*
    f béton *m* à sciure de bois [à
    copeaux de bois]
    nl houtgraniet *n*
    r арболит *m* (*на заполнителях из*
    древесных опилок или
    стружки)

S95 *e* **sawed joint, sawn joint**
    d eingeschliffene Fuge *f*
    f joint *m* scié
    nl gezaagde voeg *f(m)*, krimpvoeg
    (*betonweg*)
    r пропиленный шов *m* (*в бетоне*)

S96 *e* **sawtooth roof**
    d Sheddach *n*, Sägezahndach *n*
    f comble *m* [toit *m*] (à) shed, comble
    m en dents de scie
    nl sheddak *n*, zaagdak *n*
    r шедовая крыша *f*; шедовое
    покрытие *n*

S97 *e* **scab**
    d Holzlasche *f*, hölzerne Decklasche *f*
    f couvre-joint *m* en planche, planche
    f de raccordement
    nl houten schetsplaat *f(m)*
    r соединительная (дощатая)
    накладка *f*

S98 *e* **scaffold**
    d Gerüst *n*, Rüstung *f*, Baugerüst *n*
    f échafaud(age) *m*
    nl steiger *m*
    r (строительные) леса *pl*; подмости
    pl

S99 *e* **scaffold bridge**
    d Estakade *f*, Gerüstbrücke *f*,
    Bockbrücke *f*
    f estacade *f*, route *f* surélevée
    nl schraagsteiger *m*, brugsteiger *m*
    r эстакада *f*

S100 *e* **scaffolding bearer**
    d Gerüstausleger *m*; Gerüstbalken *m*
    f boulin *m*
    nl korteling *f*; steigerbalk *m*
    r палец *m*, поперечина *f*
    (*строительных лесов*)

S101 *e* **scaffolding platform**
    d Arbeitsbühne *f*
    f plate-forme *f* (de travail) sur
    l'échafaudage
    nl werkvloer *m*
    r рабочий помост *m*; подмости *pl*

S102 *e* **scaffold squares**
    d rechteckige Gerüstbelagtafeln *f pl*
    f cadres *m pl* entretoisés de
    l'échafaudage

nl rechthoekige steigerelementen
r конверты *m pl* рабочих подмостей

S103 **scaffold trestle** *see* **scaffold bridge**

S104 *e* **scale depositing**
d Absetzen *n* von Niederschlag, Kesselsteinbildung *f*
f entartrage *m*
nl vorming *f* van ketelsteen
r накипеобразование *n*, отложение *n* накипи

S105 *e* **scale model**
d maßstabgerechtes Modell *n*
f modèle *m* à l'échelle
nl schaalmodel *n*
r масштабная модель *f*

S107 *e* **scale-resistant concrete**
d abblätternsicherer Beton *m*
f béton *m* résistant à exfoliation [à écaillement]
nl weerbestendig beton *n*
r бетон *m*, стойкий к отслаиванию

S108 *e* **scale trap**
d Schmutzfänger *m*
f séparateur *m* de boue
nl vuilvanger *m*
r грязеуловитель *m*

S109 *e* **scaling**
d Abschälen *n*, Abschuppen *n*, Abblättern *n*
f écaillage *m*, écaillement *m*, exfoliation *f* (*de béton*)
nl afschilferen *n* (*beton*)
r отслаивание *n*, шелушение *n* (*бетона*)

S110 *e* **scalper**
d Grobsieb *n*, Trennungssieb *n*
f crible-scalpeur *m* (*pour pré--criblage*)
nl grove zeef *f(m)*
r грохот *m или* сито *n* для отсева крупных фракций

S111 *e* **scalping**
d Vorabsiebung *f*, Grobsiebung *f*
f criblage *m* (de) gros, scalpage *m*
nl grove zeving *f* vooraf
r предварительное грохочение *n*; сортировка *f* просеиванием

S112 **scalpings** *see* **oversize**

S113 *e* **scarf joint**
d 1. schräges Hakenblatt *n* 2. Stumpfstoß *m* unter 45° (*Schweißung*)
f 1. joint *m* [assemblage *m*] à trait de jupiter [à mi-bois] 2. soudure *f* biaise
nl 1. rechte (lip)las *f(m)* 2. las *f(m)* met schuine overlap
r 1. соединение *n* косым замком (*врубка*) 2. соединение *n* внапуск (*сварка*)

S114 *e* **scarifier**
d Aufreißer *m*

f scarificateur *m*, piocheuse *f*
nl wegopbreker *m*
r кирковщик *m*

S115 *e* **scheduled erection time**
d zeitplangemäße Montagefrist *f*, vorgesehener Montageabschlußtermin *m*
f délai *m* (d'exécution) de montage prévu par le programme de travaux
nl montagetijdsduur *m* volgens het werkplan
r продолжительность *f* монтажа по графику [по плану]; запланированный срок *m* окончания монтажных работ

S116 *e* **schematic**
d Schema *n*; Plan *m*
f schéma *m*; plan *m*
nl schema *n*; plan *n*
r схема *f*; план *m*

S117 *e* **scheme**
d 1. Projekt *n* 2. Schema *n*
f 1. projet *m* 2. schéma *m*
nl 1. plan *n*, ontwerp *n* 2. schema *n*, schets *m*
r 1. проект *m* 2. схема *f*

S118 *e* **scissors junction**
d schräge Straßenkreuzung *f*
f croisement *m* (de routes) oblique
nl schuine wegkruising *f*
r пересечение *n* (автомобильных) дорог под острым углом

S119 *e* **scoop**
d 1. Absaughaube *f* 2. Meßtrichter *m* 3. Schaufel *f* 4. Baggerlöffel *m* 5. Kübel *m* 6. Schöpfbecher *m*
f 1. hotte *f*, manteau *m* de cheminée 2. cône *m* [entonnoir *m*] de mesure 3. pelle *f* à main 4. godet *m* de pelle mécanique 5. benne *f* 6. drague *f*
nl 1. afzuigkap *f* 2. meettrechter *m* 3. spade *f* 4. graaflepel *m* 5. kipwagen *m* 6. schepbak *m*
r 1. вытяжной зонт *m* 2. измерительная воронка *f*, измерительный конус *m* 3. совковая лопата *f* 4. ковш *m* экскаватора 5. бадья *f* 6. черпак *m*

S120 *e* **scour**
d 1. Wegspülung *f*, Hinterspülung *f*, Unterspülen *n*; Auskolkung *f*, Flußbetterosion *f* 2. Betonerosion *f*
f 1. affouillement *m* 2. érosion *f* du béton
nl 1. onderspoeling *f* 2. betonerosie *f*
r 1. размыв *m*, подмыв *m*, русловая эрозия *f* 2. эрозия *f* бетона

S121 *e* **scoured hole**
d Auswaschungstrichter *m*, Kolkvertiefung *f*
f fosse *f* d'affouillement
nl kolkgat *n*
r воронка *f* размыва

S123 *e* **scouring gallery**
  d Spülstollen *m*
  f galerie *f* de chasse
  nl spoelgalerij *f*
  r промывная галерея *f*

S124 *e* **scouring sluice**
  d 1. Spülschleuse *f* 2. Leerschuß *m*;
  Grundablaß *m*, Geschiebeauslaß *m*
  f 1. écluse *f* de chasse [de fuite]
  2. déversoir *m* de réglage; vidange *m*
  de fond
  nl 1. spuisluis *f(m)* 2. spui-opening *f*
  r 1. промывной шлюз *m* 2. холостой
  водосброс *m*; донный водоспуск *m*

S125 *e* **scour protection**
  d Kolkschutz *m*
  f protection *f* contre affouillement
  nl bescherming *f* tegen uitschuring
  r противоэрозионные мероприятия
  n pl; защита *f* от эрозии;
  берегоукрепление *n*

S126 *e* **scour valve**
  d Spülventil *n* (*der Spülleitung*)
  f vanne *f* de purge
  nl spuikraan *m*
  r промывной клапан *m*
  (*пульпопровода*)

S127 *e* **scow sucker**
  d Schutensauger *m*
  f drague *f* (suceuse auto)porteuse
  nl hopperzuiger *m*
  r самоотвозный земснаряд *m*

S128 *e* **scraper**
  d 1. Schrapper *m*; Seilschaufler *m*;
  Seilschrapperkasten *m*
  2. Abstreifer *m*, Nahtausreiber *m*,
  Kratzeisen *n*, Kratzer *m*,
  Schabhammer *m* 3. Schlammkratzer
  m, Ausräumer *m*
  f 1. décapeuse *f*, scraper *m* 2.
  raclette *f*, racloir *m* 3. racleur *m*
  nl 1. scraper *m* 2. krasnaald *m*
  3. schraapijzer *n*, krabber *m*
  r 1. скрепер *m*; скрепер-волокуша
  m; ковш *m* скрепера 2. скребок *m*
  (*напр. для прочистки швов*);
  скребок-гладилка *m* 3. иловый
  скребок *m*, илоскрёб *m*

S129 *e* **scraper conveyor**
  d Kratzer *m*, Kratz(band)förderer *m*
  f convoyeur *m* racleur [à raclettes]
  nl jakobsladder *m*
  r скребковый конвейер *m*

S130 **scraper knife** *see* **stripping knife**

S131 *e* **scraping straightedge**
  d Glätte *f*, Glättleiste *f*
  f latte *f* de réglage, règle *f* de
  finition, poutre *f* lisseuse
  nl rei *f(m)*
  r рейка-гладилка *f*

S132 *e* **scratch coat**
  d Unterputz *m*
  f couche *f* de fond (*d'enduit*)

  nl beraping *f* (*onderste pleisterlaag*)
  r обрызг *m* (*первый слой
  штукатурного намёта трёхслойной
  штукатурки*)

S133 *e* **screed**
  d 1. Richtleiste *f* 2. Abziehvorrichtung
  f 3. Estrich(fußboden) *m* 4.
  Putzleiste *f*, Putzlatte *f*
  f 1. latte *f* de réglage, règle *f*
  d'égalisation [à niveler, à araser]
  2. barre *f* niveleuse [à niveler]
  3. chape *f* de réglage (*en béton ou en
  mortier*) 4. cueillie *f*, cueillée *f*
  nl 1. geleider *m* 2. afreibalk *m*
  3. estrich *m* 4. mal *m* voor pleisterwerk
  r 1. правило *n* 2. разравнивающий
  брус *m* 3. стяжка *f* (*бетонная или
  цементная*) 4. (штукатурный)
  маяк *m*, маячная рейка *f*

S134 *e* **screed board**
  d Glättbohle *f*, Abziehbohle *f*,
  Abgleichbohle *f*, Abstreichbohle *f*
  f latte *f* de réglage, règle *f*
  d'égalisation
  nl afreibalk *m*, rei *f(m)*
  r рейка *f* [шаблон *m*] для
  разравнивания бетонной смеси

S135 *e* **screeding**
  d Abziehen *n*
  f réglage *m*, égalisation *f*,
  aplanissement *m* (*du béton frais*)
  nl afreien *n* (*beton*)
  r разравнивание *n* (*бетона или
  раствора рейкой или правилом*)

S136 **screeding beam** *see* **screed board**

S137 *e* **screed of coarse stuff**
  d Putzleiste *f* (aus Mörtel)
  f cueillie *f* [cueillée *f*] de mortier
  nl pleisterstrook *f(m)* voor
  binnenpleister
  r штукатурная маячная полоса *f*

S138 *e* **screen**
  d 1. Netzfilter *n*, *m*, Siebfilter *n*, *m*
  2. Auffangrechen *m* 3. Siebanlage *f*,
  Siebmaschine *f* 4. Dichtungsschürze
  f, Dichtungsteppich *m*,
  Außenhautdichtung *f*, wasserseitige
  Außendichtung *f*
  f 1. crépine *f* 2. grille *f* 3. tamis *m*,
  crible *m* 4. masque *m* [écran *m*]
  d'étanchéité (*d'un barrage*)
  nl 1. filter *n* 2. opvangrooster *n* 3.
  zeef(machine) *f* 4. dichtingsscherm *n*
  r 1. сетчатый фильтр *m* 2.
  сороудерживающая решётка *f* 3.
  грохот *m* 4. противофильтрационный экран *m* плотины

S139 **screen analysis** *see* **sieve analysis**

S140 **screened well** *see* **strainer well**

S141 *e* **screen filter**
  d Siebwasserfilter *n*, *m*;
  Trommelsiebfilter *n*, *m*

f crépine *f*; filtre *m* à tambour
nl filterwand *m*, filterkous *f(m)*
r сетчатый водяной фильтр *m*;
барабанный фильтр *m*

S142 *e* **screening**
   d 1. Klassierungssiebung *f*,
      Absiebung *f*, Siebung *f*,
      Siebklassierung *f*, Sortierung *f* 2.
      Durchseihen *n*, Durchsieben *n*
   f 1. criblage *m*, tamisage *m*
      2. filtrage *m*, passage *m* au filtre
   nl 1. classificeren *n* door zeven,
      sorteren *n* 2. filtreren *n*
   r 1. сортировка *f*; грохочение *n*,
      просеивание *n*, ситовая классифика-
      ция *f* 2. процеживание *n* (*через*
      решётку, сито)

S143 *e* **screening and washing plant**
   d Sieb- und Waschanlage *f*
   f centrale *f* [installation *f*] de
      criblage-lavage
   nl classificeer -en wasinstallatie *f*
   r сортировочно-моечная установка *f*

S144 *e* **screenings**
   d Rechengut *n*, Rechenrückstände *m*
      pl, Siebrückstände *m pl*
   f criblures *f pl*, fraction *f* passante,
      déchets *m pl* de carrière
   nl zeefrest *m*
   r отсев *m*, высевки *f pl*

S145 *e* **screening tower**
   d Siebturm *m*
   f tour *f* de criblage
   nl zeeftoren *m*
   r башенная сортировочная установка
      f

S146 *e* **screening washer**
   d Waschsieb *n*
   f crible *m* laveur
   nl waszeef *f(m)*
   r промывочный грохот *m*

S147 *e* **screen oversize, screen reject**
   d Siebüberlauf *m*, Siebrückstand *m*,
      Überkorn *n*
   f refus *m* de crible
   nl zeefrest *m* (*die op de grofste zeef*
      achterblijft)
   r надрешётный продукт *m*, остаток *m*
      на сите

S148 *e* **screen size**
   d Siebweite *f*, Siebgröße *f*
   f dimension *f* de maille de tamis
   nl zeefmaat *n*
   r размер *m* отверстий сита

S149   **screen wall** *see* **shelter wall**

S150 *e* **screen water filter**
   d Siebwasserfilter *n*, *m*, Netzfilter *n*,
      m
   f crépine *f*, filtre *m* en treillis
      métallique (*pour l'eau*)
   nl filterscherm *n*, filterkous *f(m)*
   r сетчатый водяной фильтр *m*

S151 *e* **screw**
   d Schraube
   f vis *f*
   nl schroef *f*
   r винт *m*

S152 *e* **screw anchor**
   d Schraubanker *m*; Schraubenklemme *f*
   f ancre *f* à vis
   nl schroefanker *n*
   r винтовой анкер *m*; винтовой
      зажим *m*

S153 *e* **screw compressor**
   d Schraubenkompressor *m*,
      Schraubenverdichter *m*
   f compresseur *m* rotatif à vis,
      compresseur *m* à hélice
   nl schroefcompressor *m*
   r винтовой компрессор *m*

S154 *e* **screw-down stop valve**
   d Absperrventil *n*
   f robinet *m* d'arrêt à soupape
   nl stopkraan *m*
   r запорный вентиль *m*

S155 *e* **screwed fittings**
   d Gewindefittings *n pl*, Schraubfit-
      tings *n pl*
   f raccords *m pl* (de tuyauterie) à vis
   nl schroeffittingen *m pl* en
      -appendages
   r резьбовые соединительные детали
      f pl (*трубопровода*)

S156 *e* **screwed joint**
   d Schraubenverbindung *f*
   f joint *m* vissé [à vis]
   nl schroefverbinding *f*
   r резьбовое соединение *n*

S157 *e* **screwed joint connector**
   d Muffenverbinder *m*, Schraubmuffe *f*
   f manchon *m* (de raccord) vissé
   nl schroefmof *f*, nippel *m*
   r соединительная муфта *f* с резьбой

S158 *e* **screw jack**
   d Schraubenwinde *f*
   f vérin *m* à vis
   nl schroefvijzel *f(m)*
   r винтовой домкрат *m*

S159 *e* **screw pile**
   d Schraubenpfahl *m*
   f pieu *m* hélicoïdal [à vis], pilot *m*
      à vis
   nl schroefpaal *m*
   r винтовая свая *f*

S160 *e* **scriber**
   d Reißnadel *f*
   f pointe *m* à tracer
   nl afschrijfnaald *f*
   r разметчик *m* (*инструмент*),
      чертилка *f*

S162 *e* **scroll tank**
   d Bundstahlbehälter *m*
   f réservoir *m* en rouleaux
   nl houder *m* voor gerold bandstaal

r стальной резервуар *m*,
монтируемый из рулонированных
конструкций, рулонированный
резервуар *m*

S163 *e* **scrubber**
d Waschturm *m*, Rieselturm *m*,
Skrubber *m*, Wäscher *m*
f laveur *m*, épurateur *m*
nl wastoren *m*, scrubber *m*
r скруббер *m*

S164 *e* **S-curve**
d S-Kurve *f*
f courbe *f* en S
nl S-curve *f*
r кривая *f* суммирования, S-кривая *f*

S165 *e* **sea defence works**
d Uferschutzwerke *n pl*,
Küstenschutzwerke *n pl*,
Wellenschutzanlagen *f pl*
f ouvrages *m pl* côtiers [de défense
de littoral]
nl zeewering *f*
r морские берегозащитные
сооружения *n pl*

S166 *e* **sea dike**
d Seedeich *m*, Meeresdeich *m*,
Seeuferdamm *m*
f digue *f* maritime [en mer]
nl zeedijk *m*
r морская береговая дамба *f*

S167 *e* **sea dredging**
d Seebaggerung *f*
f dragage *m* en mer
nl baggerwerken *n pl* op zee
r морские дноуглубительные
работы *f pl*

S168 *e* **sealant**
d Dichtungsmittel *n*,
Dichtungskitt *m*
f produit *m* d'étanchéité
nl pakking *f*, pakkingkit *m*
r герметик *m*

S169 *e* **sealant strip**
d Dichtungsband *n*
f bande *f* extrudée d'étanchéité,
bande-garniture *f* étanche
[antifuite]
nl pakkingstrook *f(m)*
r эластичная герметизирующая
прокладка *f*

S170 *e* **sealed expansion vessel**
d Membranausdehnungsgefäß *n*
f vase *m* d'expansion à membrane
élastique
nl membraan-expansievat *n*
r замкнутый расширительный сосуд
m с гибкой мембраной

S172 **sea levee** *see* **sea dike**

S173 **sea-level canal** *see* **seaway canal**

S174 *e* **sealing**
d Abdichtung *f*
f étanchement *m*

nl dichting *f*
r уплотнение *n*, герметизация *f*

S175 *e* **sealing coat**
d Sperrschicht *f*, Absiegelungsschicht
f, Verschlußschicht *f*,
Abdichtungslage *f* (*Straßendecke*)
f couche *f* imperméable [de
scellement] (*d'une route*)
nl dichtingslaag *f(m)*, afsluitlaag *f(m)*
r водоизолирующая прослойка *f*,
верхний водоизолирующий слой *m*
(*в конструкции дорожной одежды*)

S176 *e* **sealing compound**
d Dichtungsmittel *n*,
Abdichtungsmastix *m*
f mastic *m* d'étanchéité, composé *m*
hermétique, lut *m*
nl dichtingsmastiek *m*
r уплотняющая мастика *f*, герметик
m, герметизирующий состав *m*

S177 *e* **sealing gasket**
d abdichtende Zwischenlage *f*
f joint *m* [garniture *f*] d'étanchéité
nl pakking *f*
r уплотняющая [герметизирующая]
прокладка *f*

S178 *e* **sea lock**
d Seeschleuse *f*
f écluse *f* maritime
nl zeesluis *f*
r морской шлюз *m*

S179 *e* **seam**
d 1. Naht *f*, Fuge *f* 2. Falz *m*,
Falzverbindung *f* 3. Flöz *m*,
Schicht *f*, Lage *f*
f 1. joint *m* 2. agrafe *f*, agrafage *m*
(*des tôles*) 3. couche *f*
nl 1. naad *m* 2. felsrand *m*,
felsverbinding *f* 3. laag *f* (*erts*)
r 1. шов *m* 2. фальц *m*, фальцевое
соединение *n* 3. пласт *m*, слой *m*

S180 *e* **seamless floor finishing**
d fugenloser Fußbodenbelag *m*,
Estrich *m*
f revêtement *m* de sol sans joints
nl naadloze dekvloer *m*
r бесшовное покрытие *n* пола

S181 *e* **seam welding**
d Rollennahtschweißen *n*
f soudage *m* au galet
nl naadlassen *n*
r шовная контактная сварка *f*

S182 *e* **seat angle**
d Stützwinkel *m*
f cornière-support *f*, tasseau *m* en
cornière (*soudé sur poteau*)
nl steun *m* van hoekstaal
r монтажная полочка *f*, монтажный
поддерживающий уголок *m*

S183 *e* **sea wall**
d Strandmauer *f*, Küstenschutzmauer *f*
f brise-mer *m*, brise-lames *m*, digue *f*
maritime

nl zeewering *f*
r волноотбойная стенка *f*

S184 *e* seawater concrete
d Meereswasserbeton *m*,
Seewasserbeton *m*, Beton *m* für
Seebauten
f béton *m* à la mer [pour travaux
maritimes]
nl zeewaterbestendig beton *n*
r бетон *m*, устойчивый против
воздействия морской воды

S185 *e* seawater corrosion
d Seewasserkorrosion *f*
f corrosion *f* par l'eau de mer
nl corrosie *f* door zeewater
r коррозия *f* в морской воде

S186 *e* seaway canal
d Seekanal *m*
f canal *m* maritime
nl zeekanaal *n*
r морской канал *m*

S187 *e* secant modulus of elasticity
d Sekante-Elastizitätsmodul *m*
f module *m* (d'élasticité) sécant
nl secans-elasticiteitsmodulus *m*
r секущий модуль *m* упругости
(*определяемый углом наклона
секущей*)

S188 *e* secondary air
d Sekundärluft *f*
f air *m* secondaire
nl secundaire lucht *f*
r вторичный воздух *m*

S189 *e* secondary beam
d Querträger *m*, Zwischenträger *m*
f poutre *f* secondaire
nl secundaire langsligger *m*
r второстепенная балка *f*

S190 *e* secondary buckling
d örtliches Ausbeulen *n*
[Ausknicken *n*]
f flambage *m* [flambement *m*]
secondaire
nl secundaire knik *f(m)*
r местное выпучивание *n*, местная
потеря *f* устойчивости

S191 *e* secondary circuit
d Sekundärkreislauf *m*
f circuit *m* secondaire
nl secundaire kringloop *m*
r вторичный контур *m* циркуляции;
контур *m* вторичного теплоносителя

S192 *e* secondary circulation
d Sekundärzirkulation *f*
f circulation *f* secondaire
nl secundaire circulatie *f*
r вторичная циркуляция *f*;
циркуляция *f* в контуре
вторичного теплоносителя

S193 secondary clarifier *see* secondary
sedimentation tank

S194 *e* secondary consolidation

d sekundäre Verdichtung *f*,
Nachverdichtung *f*
f consolidation *f* secondaire
nl secundaire zetting *f*
r вторичная консолидация *f*
(*грунта*)

S195 *e* secondary crusher
d Nachbrecher *m*, Zweitbrecher *m*
f concasseur *m* secondaire
[intermédiaire]
nl nabreker *m*
r дробилка *f* вторичного *или*
среднего дробления

S196 *e* secondary filter
d Nachfilter *n*, *m*
f filtre *m* secondaire
nl secundair filter *n*, *m*
r фильтр *m* второй ступени

S197 *e* secondary heating surface
d Sekundärheizfläche *f*
f surface *f* de chauffe secondaire
nl secundair verwarmingsvlak *n*
r вторичная поверхность *f* нагрева

S198 *e* secondary main
d Verteil(ungs)leitung *f*; Zuleitung *f*
f conduite *f* secondaire
nl verdeelleiding *f*; toevoerleiding *f*
r распределительный *или* питающий
трубопровод *m*

S199 *e* secondary member
d Konstruktionshilfselement *n*
f élément *m* de construction
secondaire
nl secundair constructie-element *n*
r второстепенный элемент *m*
конструкции

S200 *e* secondary moment
d Nebenmoment *n*, sekundäres
Moment *n*, Zusatzmoment *n*,
Umlagerungsmoment *n*
f moment *m* secondaire
nl secundair moment *n*
r дополнительный [вторичный]
момент *m*

S201 *e* secondary refrigerant
d Kälteträger *m*, Sekundärkältemittel
n
f réfrigérant *m* secondaire, fluide *m*
frigorigène secondaire
nl secundair koelmiddel *n*
r вторичный хладоноситель *m*

S202 *e* secondary reinforcement
d Zusatzbewehrung *f*
f armature *f* [ferraillage *m*]
secondaire
nl hulpwapening *f*
r вспомогательная [вторичная]
арматура *f*

S203 *e* secondary sedimentation tank
d Nachklärbecken *n*
f décanteur *m* secondaire
nl nabezinkingstank *m*
r вторичный отстойник *m*

S204 e **secondary stresses**
   d Sekundärspannungen f pl,
      Zusatzspannungen f pl
   f contraintes f pl secondaires
   nl secundaire spanningen f pl
   r вторичные [дополнительные]
      напряжения n pl

S205 e **secondary treatment**
   d Sekundärreinigung f
      [Nachbehandlung f] von Abwasser
   f traitement m des eaux usées
      secondaire
   nl nabehandeling f van afvalwater
   r вторичная очистка f сточных вод

S206 e **secondary unit**
   d Sekundärwärmetauscher m
   f échangeur m thermique secondaire
   nl secundaire warmtewisselaar m
   r вторичный теплообменник m

S207 **second moment of area** see **moment of inertia**

S208 e **section**
   d 1. Schnitt m 2. Bauprofil n,
      Walzprofil n; Profilstahl m
      3. Schuß m, Rohrstrang m
   f 1. section f, coupe f, profil m 2.
      profilé m (métallique) 3. section f,
      tronçon m (d'une tuyauterie)
   nl 1. doorsnede f 2. walsprofiel n
      3. strang f(m), leidingsectie f
   r 1. сечение n, разрез m (на
      чертеже) 2. прокатный профиль
      m, сортовая сталь f 3. звено n,
      нитка f, участок m (трубопровода)

S209 e **sectional area**
   d Querschnittsfläche f
   f aire f de la section transversale,
      surface f sectionnée
   nl oppervlak n van de dwarsdoorsnede
   r площадь f (поперечного) сечения

S210 e **sectional boiler**
   d Gliederkessel m
   f chaudière f sectionnée
   nl gelede ketel m
   r секционный горизонтально-
      -водотрубный котёл m

S211 e **sectional properties**
   d geometrische Eigenschaften f pl
      der Profilelemente
   f caractéristiques f pl géométriques
      des profilés [des sections]
   nl meetkundige eigenschappen van
      walsprofielen
   r геометрические характеристики f pl
      сечений

S212 e **sectional retaining wall**
   d aufgelöste [gegliederte] Stützmauer f
   f mur m de soutènement sectionné
      [en sections]
   nl gelede keermuur m
   r расчленённая подпорная стенка f

S213 e **sectional steel**
   d Profilstahl m

   f acier m profilé
   nl profielstaal n
   r сортовая сталь f, сортовой прокат
      m

S214 e **section modulus**
   d Widerstandsmoment n
   f module m de résistance [d'inertie]
   nl weerstandsmoment n
   r момент m сопротивления сечения

S215 e **sector gate**
   d Sektorverschluß m
   f vanne f (à) secteur
   nl segmentschuif f(m)
   r секторный затвор m

S216 e **securing attachments**
   d Befestigungselemente n pl
   f pièces f pl de fixation
   nl bevestigingsmiddelen n pl
   r анкера m pl или детали f pl для
      крепления (напр. облицовки)

S217 e **sediment**
   d Absatz m, abgesetzte Stoffe m pl,
      Sediment n, Ablagerung f
   f dépôt m, sédiment m, résidu m
   nl sediment n, bezinksel n
   r осадок m; нанос m

S218 e **sedimentation**
   d 1. Auflandung f, Anschwemmung f,
      natürliche Kolmation f,
      Verlandung f, Verschlickung f
      2. Absetzen n, Abschlämmen n,
      Aufschlämmen n 3. Absetzklärung f,
      mechanische Klärung f von
      Abwasser
   f 1. alluvionnement m; envasement m
      2. sédimentation f 3. décantation f,
      clarification f mécanique (des eaux
      d'égout)
   nl 1. aanslibbing f 2. afzetting f
      3. neerslag n
   r 1. отложение n наносов, заиление n
      2. осаждение n, седиментация f,
      отстаивание n 3. осветление n,
      механическая очистка f сточных
      вод

S219 e **sedimentation analysis**
   d Schlämmanalyse f, Absetzprobe f
   f analyse f par sédimentation
   nl sedimentatieanalyse f
   r анализ m отмучиванием,
      седиментационный анализ m

S220 **sedimentation basin** see **settling basin**

S221 **sedimentation compartment** see **settling chamber**

S222 e **sedimentation plant**
   d Absetzanlage f
   f installation f [bassin m] de
      décantation
   nl bezinkinstallatie f
   r (вертикальный) отстойник m
      водоочистной станции

S223 e **sedimentation pond**
    d Klärteich *m*
    f bassin *m* de clarification,
      clarificateur *m*
    nl bezinkbassin *n*
    r отстойный [осветлительный]
      бассейн *m*, прудок-отстойник *m*

S224 **sedimentation tank** *see* **settling
basin**

S225 e **sediment-carrying capacity**
    d Transportvermögen *n* des Flusses
    f capacité *f* de transport de sédiments
    nl transportvermogen *n* van een
      rivier
    r наносотранспортирующая
      способность *f* (*водотока*)

S226 e **sediment charge**
    d spezifische Feststofführung *f*
    f débit *m* de charge en suspension,
      débit *m* solide par unité de largeur
      (de lit)
    nl belasting *f* met vaste stoffen
    r относительный расход *m* наносов
      (*отношение объёма или массы
      транспортируемых наносов
      к сечению водотока*)

S227 e **sediment concentration**
    d Schwebstoffbelastung *f*; spezifischer
      Schwebstoffgehalt *m*
    f concentration *f* des sédiments,
      teneur *f* en éléments fins;
      turbidité *f* d'eau
    nl gehalte *f* aan vaste stoffen
    r концентрация *f* наносов; мутность
      f воды

S228 **sediment deposition** *see* **sedimentation**

S229 e **sediment discharge**
    d Feststofführung *f*,
      Feststofftransport *m*,
      Feststoffabtrag *m*,
      Schwemmlandabfluß *m*,
      Geschiebemenge *f*
    f débit *m* solide
    nl vaste-stoffenafvoer *m*
    r сток [расход *m*] наносов, твёрдый
      сток *m* [расход *m*]

S230 e **sediment diverting gallery**
    d Geschiebeumleitstollen *m*
    f galerie *f* d'interception des
      sédiments
    nl sedimentafleidingstunnel *m*
    r наносоперехватывающая галерея *f*

S231 e **sediment exclusion works**
    d Schwemmstoffleitwerke *n pl*
    f ouvrages *m pl* pour l'élimination
      des sédiments
    nl sedimentscheidingswerken *n pl*
    r наносоперехватывающие сооруже-
      ния *n pl*

S232 **sediment runoff** *see* **sediment
discharge**

S233 e **sediments**
    d Feststoffe *m pl*, Schwemmstoffe *m
      pl*, Ablagerungen *f pl*
    f sédiments *m pl*
    nl sedimenten *n pl*, afzettingen *f pl*
    r наносы *m pl*, отложения *n pl*

S234 e **sediment sampler**
    d Geschiebefänger *m*
    f turbidisonde *f*
    nl sedimentvanger *m*
    r батометр *m* для взятия проб
      наносов

S235 e **Seduct heater**
    d Außenwandgasheizer *m*, Se-duct-
      -Anlage *f*
    f appareil *m* de chauffage à gaz du
      type Seduct (*avec circuit gaz-air
      isolé*)
    nl gevelkachel *m*
    r газовый отопительный прибор *m*
      закрытого типа (*с изолированным
      газовоздушным трактом*)

S236 e **seepage**
    d Durchsickern *n*, Einsickern *n*,
      Sickerung *f*; Durchströmung *f*,
      Infiltration *f*
    f suintement *m*, filtration *f*;
      infiltration *f*
    nl infiltratie *f*
    r просачивание *n*, фильтрация *f*;
      инфильтрация *f*

S237 e **seepage face**
    d Sickerfläche *f*
    f surface *f* d'infiltration
    nl infiltratievlak *n*
    r площадь *f* просачивания

S238 e **seepage failure**
    d hydraulischer Grundbruch *m*
    f rupture *f* due au mouvement de
      l'eau
    nl aardverschuiving *f* door zakwater
    r обрушение *n* грунта вследствие
      просачивания, гидродинамическое
      выпирание *n* грунта

S239 e **seepage flow**
    d Sickerwasserströmung *f*
    f courant *m* de filtration
    nl zakwaterstroom *f*
    r фильтрационный поток *m*

S240 e **seepage force**
    d Sickerdruck *m*, Sickerkraft *f*
    f force *f* due à l'écoulement en milieu
      poreux
    nl infiltratiedruk *m*
    r фильтрационное [гидродинамиче-
      ское] давление *n*, фильтрационные
      силы *f*

S241 **seepage head** *see* **infiltration head**

S242 **seepage intensity coeffic∙ent** *see*
coefficient of permeability

S243 e **seepage loss**
    d Filtrationsverlust *m*, Sickerverlust *m*
    f fuite *f*, pertes *f pl* par infiltration

nl infiltratieverlies *n*
r потери *f pl* на фильтрацию

S244 *e* **seepage pit**
   d Sickergrube *f*, Dränageschacht *m*,
     Sickerschacht *m*
   f puisard *m*
   nl draineerschacht *m*, zinkput *m*
   r дренажный приямок *m*; дренажная
     [фильтрационная] шахта *f*

S245 *e* **seepage pressure**
   d Sickerströmungsdruck *m*
   f pression *f* de filtration
   nl infiltratiedruk *m*
   r фильтрационное давление *n*

S246 *e* **seepage trench**
   d Sickergraben *m*,
     Entwässerungsgraben *m*
   f tranchée *f* drainante
   nl afwateringssloot *f(m)*
   r фильтрационная [дренажная]
     траншея *f*

S247 *e* **seepage velocity**
   d Filtrationsgeschwindigkeit *f*,
     Sickergeschwindigkeit *f*
   f vitesse *f* de filtration
   nl infiltratiesnelheid *f*
   r скорость *f* фильтрации

S248 *e* **seepage water drainage**
   d Sickerwasserableitung *f*
   f drainage *m* des eaux d'infiltration
   nl afvoer *m* van zakwater
   r отвод *m* фильтрационных вод

S249 *e* **segment**
   d 1. Segment *n* 2. Tübbing *m*
   f 1. segment *m* 2. cuvelage *m*
   nl schachtring *m*
   r 1. сегмент *m* 2. тюбинг *m*

S250 *e* **segmental bridge**
   d Kastenträgerbrücke *f* aus
     vorgefertigten Stahlbetonteilen
   f pont *m* à tronçons
   nl boogbrug *f(m)*
   r мост *m* из сборных коробчатых
     блоков

S251 *e* **segmental member**
   d zusammengesetztes
     Spannbetonmontageelement *n*
   f élément *m* (*de construction*)
     précontraint préfabriqué en
     tronçons
   nl samengesteld voorgespannen
     betonelement *n*
   r составной преднапряжённый
     элемент *m* (*собранный из
     нескольких частей с натяжением
     сквозных арматурных пучков*)

S252 *e* **segmental precast pile**
   d zusammengesetzter
     Stahlbetonpfahl *m*
   f pieu *m* préfabriqué segmentaire
   nl prefab schakelpaal *f(m)* van
     gewapend beton
   r составная железобетонная свая *f*

S253 *e* **segmental shell**
   d Segmentschale *f*
   f coque *f* en éléments préfabriqués
   nl samengestelde boog *m*
   r оболочка *f* из сборных элементов

S254   **segmental sluice gate** *see* **radial gate**

S255 *e* **segment placer**
   d Tübbingeinbaumaschine *f*
   f machine *f* pour la pose mécanique
     du cuvelage
   nl machine *f* voor het plaatsen van
     schachtringen
   r машина *f* для установки тюбинго-
     вой крепи

S256 *e* **segment ring**
   d Tübbingring *m*
   f anneau *m* de cuvelage
   nl tubingring *m*
   r кольцо *n* тюбинговой крепи

S257 *e* **segment type cellular cofferdam**
   d Segmentzellenfangdamm *m*
   f batardeau *m* cellulaire
   nl segmentcellenvangdam *m*
   r сегментная ячеистая перемычка *f*

S258 *e* **segregation**
   d Entmischung *f*
   f ségrégation *f*
   nl ontmenging *f*
   r расслоение *n*, расслаивание *n*

S259 *e* **seismic analysis**
   d Berechnung *f* auf
     Erdbebensicherheit
   f calcul *m* aux sollicitations
     séismiques, calcul *m* aux séisms
     [aux efforts séismiques]
   nl berekening *f* op aardbevingsveilig-
     heid
   r расчёт *m* на сейсмостойкость

S260 *e* **seismic building code**
   d Baunormen *f pl* und
     Baubestimmungen *f pl* für
     erdbebengefährdete Gebiete
   f réglements *m pl* antiséismiques [de
     construction pour les régions
     séismiques]
   nl bouwvoorschriften *n pl* voor door
     aardbevingen bedreigde gebieden
   r строительные нормы *f pl* для
     сейсмических районов

S261 *e* **seismic damage**
   d Erdbebenbeschädigungen *f pl*
   f endommagement *m* des bâtiments
     dû aux séismes
   nl schade *f* tengevolge van
     aardbevingen
   r повреждение *n* от землетрясения

S262 *e* **seismic design**
   d Erdbebenbemessung *f*
   f calcul *m* aux efforts séismiques
   nl berekening *f* op
     aardbevingskrachten
   r расчёт *m* на сейсмические
     нагрузки

S263 *e* **seismic forces**
   d Erdbebenkräfte *f pl*
   f efforts *m pl* s(é)ismiques
  nl seismische krachten *f(m) pl*
   r сейсмические воздействия *n pl*

S264 *e* **seismic load**
   d Erdbebenlast *f*, seismische
     Beanspruchung *f*
   f sollicitation *f* d'origine s(é)ismique
  nl seismische belasting *f*
   r сейсмическая нагрузка *f*

S265 *e* **seismic prospecting**
   d seismische Prospektion *f*
   f prospection *f* [recherche *f*,
     exploration *f*] séismique
  nl seismisch bodemonderzoek *n*
   r сейсмическая разведка *f*, разведка
     f сейсмическими методами

S266 *e* **seismic region**
   d Erdbebengebiet *n*
   f région *f* [zone *f*] séismique
  nl aardbevingsgebied *n*
   r сейсмический район *m*

S267 *e* **seismic resistance**
   d Erdbeben-Widerstandsfähigkeit *f*,
     Erdbebensicherheit *f*
   f résistance *f* aux séismes [aux
     efforts séismiques]
  nl weerstand *m* tegen aardbevingen
   r сейсмостойкость *f*

S268 *e* **selective heating**
   d selektive Heizung *f*
   f chauffage *m* sélectif
  nl selectieve verwarming *f*
   r селективное центральное
     отопление *n*

S269 *e* **self-acting movable flood dam**
   d Hochwasserschutzwehr *n* mit
     automatischer Drehklappe
   f barrage *m* mobile à clapet
  nl automatisch werkende
     hoogwaterkering *f*
   r водосливная плотина *f* с
     автоматическим клапанным
     затвором

S270 *e* **self-cleaning air filter**
   d selbstreinigendes Luftfilter *n, m*
   f filtre *m* à air à nettoyage
     automatique
  nl zelfreinigend luchtfilter *n, m*
   r самоочищающийся воздушный
     фильтр *m*

S271 *e* **self-cleaning velocity**
   d Mindestgeschwindigkeit *f*, die noch
     nicht zur Verlandung führt
   f vitesse *f* d'autocurage
  nl laagste stroomsnelheid *f* waarbij
     nog geen afzetting plaatsvindt
   r незаиливающая скорость *f* (*в*
     трубопроводах и каналах)

S272 *e* **self-cleansing gradient**
   d erforderliche Mindestgradiente *f*
   f déclivité *f* [pente *f*] autocurage

  nl kleinste verval *n* waarbij nog
     geen afzetting plaatsvindt
   r минимальный уклон *m*
     (*трубопровода, канала*) по
     условию незаиления

S273 *e* **self-closing damper**
   d Rückschlagklappe *f*
   f registre *m* à fermeture automatique,
     clapet *m* à clôture automatique
  nl terugslagklep *f(m)*
   r обратный воздушный клапан *m*

S274   **self-closing device** *see* **automatic**
     closing device

S275 *e* **self-compensation**
   d Selbstkompensation *f*
   f autocompensation *f*
  nl zelfcompensatie *f*
   r самокомпенсация *f* (*трубопровода*)

S276 *e* **self-contained air-conditioner**
   d anschlußfertiges [autonomes]
     Klimagerät *n*
   f conditionneur *m* d'air indépendant,
     climatiseur *m* individuel
  nl zelfstandig luchtbehandelingstoestel
     n
   r автономный кондиционер *m*

S277 *e* **self-erecting crane**
   d Selbstmontagekran *m*,
     selbstaufrichtender Kran *m*
   f grue *f* automontante [à auto-
     -montage]
  nl zelfmontagekraan *f(m)*
   r самомонтирующийся кран *m*

S278 *e* **self-ignition, self-inflammation**
   d Selbstentzündung *f*
   f auto-allumage *m*, auto-inflammation *f*
  nl zelfontbranding *f*
   r самовозгорание *n*,
     самовоспламенение *n*

S279 *e* **self-leveling level**
   d selbsthorizontierendes Nivellier *n*
   f niveau *m* automatique [auto-
     -ajustable]
  nl zellinstellende vizierlijn *f(m)* van
     een waterpasinstrument
   r самоустанавливающийся нивелир
     m, нивелир-автомат *m*

S280 *e* **self-lifting forms**
   d selbstverstellende Kletterschalung *f*
   f coffrage *m* auto-relevable
  nl zelflossende bekisting *f*
   r самоподъёмная переставная
     опалубка *f*

S281 *e* **self-limiting fan**
   d Lüfter *m* mit Überlastungsschutz
   f ventilateur *m* à limiteur
     automatique
  nl ventilator *m* met belastingsbegrenzing
   r вентилятор *m* с защитой от
     перегрузки

S282 *e* **self-priming siphon**
   d selbstansaugender Heber *m*

f siphon *m* [coupe-air *m*] à
amorçage automatique
nl zelfaanzuigende hevel *m*
r самозаряжающийся сифон *m*

S283 *e* **self-propelled dredger**
d Baggerschiff *n*
f drague *f* automotrice
nl baggermolen *m* met eigen
voortstuwing
r самоходный земснаряд *m*

S284 *e* **self-propelled scraper**
d Motorschürfkübelwagen *m*
f motoscraper *m*, décapeuse *f*
automotrice
nl motorscraper *m*
r самоходный скрепер *m*

S285 *e* **self-purification**
d Selbstreinigung *f*
f autocurage *m*
nl zelfreiniging *f*
r самоочищение *n*

S286 **self-raising forms** *see* **self-lifting
forms**

S287 *e* **self-regulating barrage**
d automatisches Wehr *n*
f barrage *m* à vanne autoréglable
nl zelfregelende stuw *m*
r низконапорная плотина *f* с
автоматическим затвором

S288 *e* **self-stressing concrete**
d selbstspannender Beton *m*;
Quellbeton *m*, Expansivbeton *m*
f béton *m* auto(pré)contraint
nl expansiebeton *n*
r самонапрягающий *или*
расширяющийся бетон *m*

S289 *e* **self-tapping screw**
d selbstschneidende Blechschraube *f*,
Blechtreibschraube *f*
f vis *f* à autotaraudage [d'autovissage]
nl zelftappende schroef *f(m)*
r самонарезающий винт *m*

S290 *e* **semi-circular cofferdam cell**
d Segmentzelle *f*
f cellule *f* semi-circulaire
nl segmentcel *f(m)*
r сегментная ячейка *f* (*шпунтовой
ячеистой перемычки*)

S291 *e* **semi-low trailer**
d Sattel-Tiefladeanhänger *m*, Sattel-
Tieflader *m*
f sémi-remorque *f* surbaissée
nl zadeldieplader *m*
r низкорамный полуприцеп *m*

S292 *e* **semi-outdoor-type power plant**
d Halbfreiluft-Kraftwerk *n*,
halboffenes Wasserkraftwerk *n*
f usine *f* hydraulique du type «semi-
outdoor», usine *f* hydraulique du
type à tout bas
nl half-open waterkrachtcentrale *f*
r ГЭС *f* полуоткрытого типа

S293 *e* **semi-pervious blinding coat**
d halbdurchlässige Randzonenlage *f*
f couche *f* de propriété semi-
perméable
nl halfdoorlatende slijtlaag *f(m)*
r переходный слой *m*
полупроницаемого материала (*на
низовом откосе земляной плотины*)

S294 *e* **semiportal crane**
d Halbportalkran *m*, Halbtorkran *m*
f grue *f* à demi-portique
nl halfportaalkraan *f(m)*
r полупортальный кран *m*

S295 *e* **semi-rigid joint**
d nachgiebige Verbindung *f*
f nœud *m* semi-rigide, joint *m* (de
raccord) semi-rigide
nl halfstarre verbinding *f*
r упруго-податливое [полужёсткое]
соединение *n* (*трубопровода*)

S296 *e* **semisubmerged jump**
d überstauter Wechselsprung *m*
f ressaut *m* semi-noyé
nl overstuwde watersprong *m*
r подтопленный гидравлический
прыжок *m*

S297 *e* **semitrailer**
d Halbanhänger *m*, Auflieger *m*,
Sattelanhänger *m*
f semi-remorque *f*
nl oplegger *m*
r полуприцеп *m*

S298 *e* **sensible cooling**
d Kühlung *f* durch Abführung von fühlbarer Wärme
f refroidissement *m* á chaleur
sensible
nl merkbare koeling *f*
r охлаждение *n* отводом явного
тепла

S299 *e* **sensible heat**
d fühlbare Wärme *f*
f chaleur *f* sensible
nl voelbare warmte *f*
r явное тепло *n*

S300 **sensible heat factor** *see* **sensible heat
ratio**

S301 *e* **sensible heat gain**
d Zunahme *f* an fühlbarer Wärme,
fühlbarer Wärmeanfall *m*
f gain *m* de la chaleur sensible
nl voelbare warmtetoename *f*
r явные теплопоступления *n pl*

S302 *e* **sensible heat load**
d fühlbare Wärmelast *f*
f besoin *m* de froid correspondant à
la chaleur sensible
nl voelbare warmtebelasting *f*
r нагрузка *f* по явному теплу

S303 *e* **sensible heat ratio**
d sensibler Wärmefaktor *m*
f coefficient *m* de chaleur sensible

nl coëfficiënt *m* van voelbare warmte
r доля *f* явного тепла

S304 *e* **separate system**
d Trennsystem *n*, Trennkanalisation *f*, Trennverfahren *n*
f système *m* séparatif d'assainissement
nl gescheiden rioolstelsel *n*
r раздельная система *f* канализации

S305 *e* **separator**
d Abscheider *m*; Windsichter *m*
f séparateur *m*, purgeur *m*; séparateur *m* d'air
nl afscheider *m*
r сепаратор *m*, очиститель *m*; воздушный сепаратор *m*; элутриатор *m*

S306 *e* **septic tank**
d Faulbecken *n*, Faulkammer *f*
f digesteur *m* (de boues), fosse *f* septique
nl septic tank *m*
r септик *m*

S307 *e* **sequence of prestressing**
d Ablauffolge *f* des Vorspannens
f séquence *f* de précontrainte [de mise en tension]
nl volgorde *f* van het voorspannen
r последовательность *f* операций предварительного напряжения

S308 *e* **sequence-stressing loss**
d Vorspannungsverlust *m* beim ungleichzeitigen Spannen der Bewehrungsbündel
f perte *f* de précontrainte due à la séquence de la mise en tension (*des câbles d'armatures*)
nl voorspanningsverlies *n* tengevolge van samendrukking van het beton
r потеря *f* преднапряжения при неодновременном натяжении прядей

S309 **series loop system** *see* **single-pipe heating system**

S310 *e* **serviceability criteria**
d Brauchbarkeitskriterien *n pl*
f critères *m pl* d'aptitude (*d'un ouvrage*) au fonctionnement normal
nl criteria *n pl* van de bedrijfsbruikbaarheid
r критерии *m pl* эксплуатационной пригодности сооружения

S311 *e* **serviceability limit state**
d Grenzbrauchbarkeitszustand *m*
f état *m* limite de service normal
nl grenstoestand *m* van de bruikbaarheid
r предельное состояние *n* по непригодности к нормальной эксплуатации

S312 *e* **serviceability of structures**
d Brauchbarkeit *f* [Funktionstüchtigkeit *f*] der Bauwerke

f fiabilité *f* et longévité *f* des ouvrages en service
nl bruikbaarheid *f* van bouwwerken
r эксплуатационная надёжность *f* сооружений

S313 **service basin** *see* **service reservoir**

S314 *e* **service behaviour**
d Betriebsverhalten *n*
f comportement *m* (du bâtiment) aux charges de service
nl gedrag *n* onder bedrijfsomstandigheden
r поведение *n* (*конструкции, элемента*) при эксплуатации

S315 *e* **service bridge**
d Dienstbrücke *f*
f pont *m* de service
nl dienstbrug *f(m)*
r служебный [смотровой] мостик *m*

S316 *e* **service conditions**
d Betriebsbedingungen *f pl*, Betriebsverhältnisse *n pl*
f conditions *f pl* de service [de travail]
nl bedrijfsomstandigheden *f pl*, bedrijfsvoorwaarden *f pl*
r условия *n pl* эксплуатации, рабочие условия *n pl*

S317 *e* **service district**
d Druckzone *f*, Versorgungszone *f* (*Wasserversorgung*)
f zone *f* de pression
nl voorzieningsgebied *n* (*waterleiding*)
r зона *f* давления (*в зонном водопроводе или зонированной системе водоснабжения*)

S318 *e* **service duct**
d Leitungskanal *m*, Sammelkanal *m*
f gaine *f* de service, caniveau *m*
nl leidingkoker *m*
r коммуникационный канал *m*, канал *m* для инженерных сетей

S319 *e* **service gate**
d Hauptverschluß *m*
f vanne *f* principale
nl hoofdschuif *f(m)*
r основной [рабочий] затвор *m*

S320 *e* **service load**
d Gebrauchslast *f*, Nutzlast *f*, Betriebslast *f*; Regellast *f*, Rechnungslast *f*
f charge *f* de service; charge *f* de service établie pour le calcul à l'état limite
nl belasting *f* tengevolge van het gebruik waarvoor het gebouw is ontworpen
r эксплуатационная нагрузка *f*; нормативная нагрузка *f* для расчёта по предельному состоянию

S321 *e* **service main**
d Verteilerleitung *f*; Anschlußleitung *f*
f ligne *f* principale de distribution

d'eau; canalisation *f* de raccord;
branchement *m* d'eau
nl distributieleiding *f*, hoofdleiding *f*
r распределительная магистраль *f*
соединительный трубопровод *m*

S322 *e* **service pipe**
d Anschlußstrang *m*; Versorgungsrohr
n
f branchement *m* d'eau général;
conduite *f* distributrice d'eau,
canalisation *f* de distribution d'eau
nl aansluitleiding *f*
r водопроводный ввод *m*;
распределительный трубопровод *m*

S323 *e* **service reservoir**
d Nutz- und Trinkwasserspeicherbek-
ken *n*; Ausgleichbehälter *m*
f réservoir *m* à eau potable;
réservoir *m* de compensation
nl compensatiereservoir *n*
r распределительный резервуар *m*,
резервуар *m* чистой воды;
напорно-регулирующая ёмкость *f*

S324 *e* **service riser**
d Gasstandrohr *n*
f conduite *f* montante (de gaz)
nl opgaande gasleiding *f*
r газовый стояк *m*

S325 *e* **service road**
d Anliegerstraße *f*; Zufahrtsstraße *f*,
Zubringerstraße *f*
f route *f* d'accès; route *f* de service
nl toegangsweg *m*, ventweg *m*
r подъездная *или* служебная дорога *f*

S326 *e* **services**
d Hausleitungen *f pl*,
Hausinnenleitungsnetz *n*
f services *m pl* utilitaires (pour une
maison)
nl binnenleidingen *f pl*
r внутренние инженерные сети *f pl*

S327 *e* **services below street level**
d unterirdische Leitungen *f pl* (*in
einem Stadtgebiet*)
f services *m pl* urbaines souterrains
nl ondergrondse leidingen *f pl* (*in een
stadswijk*)
r городские подземные коммуникации
f pl

S328 *e* **service tank**
d Tagestank *m*, Verbrauchsbehälter *m*;
Durchflußbehälter *m*
f réservoir *m* auxiliaire; réservoir *m*
de déversement
nl verbruikstank *m*
r вспомогательный *или* расходный
резервуар *m*

S329 *e* **service valve**
d Wartungsventil *n*
f valve *f* d'abonné, robinet *m*
d'immeuble
nl hoofdkraan *f(m)*
r изолирующий вентиль *m*, вентиль
m обслуживания

S330 **service walkway** *see* **service bridge**

S331 *e* **service water outlet**
d Betriebs(wasser)auslaß *m*
f vidange *f* (d'eau) de service
nl bedrijfswateruitlaat *m*
r рабочий водовыпуск *m*

S332 *e* **service-water storage tank**
d Warmwasserbehälter *m*,
Gebrauchswasserspeicher *m*
f réservoir *m* d'eau de consommation
nl drinkwaterreservoir *n*
r бак-аккумулятор *m* горячей воды
(*в системе горячего водоснабжения*)

S333 *e* **set**
d 1. Abbinden *n*, Erstarrung *f*
2. bleibende Formänderung *f*
[Verformung *f*] 3. Satz *m*
4. Setzung *f* 5. Aggregat *n*
6. Versatzstück *n*, Doppelbogen *m*
f 1. prise *f* 2. déformation *f*
résiduelle 3. jeu *m*, lot *m*,
assortiment *m* 4. affaissement *m*,
tassement *m* 5. groupe *m*,
installation *f* 6. double coude *m*
nl 1. verharden *n* 2. blijvende
vormverandering 3. stel *n*
4. zetting *f* 5. aggregaat *n*
6. dubbele bocht *f(m)*
r 1. схватывание *n* 2. остаточная
деформация *f* 3. комплект *m*;
набор *m* 4. осадка *f* 5. агрегат *m*;
установка *f* 6. двойной отвод *m*
отступ *m*, «утка» *f*

S334 *e* **set control**
d Abbinderegelung *f*
f régulation *f* de prise
nl regeling *f* van het verhardingsproces
r регулирование *n* схватывания

S335 *e* **set per blow**
d Eindringtiefe *f* pro Schlag
f enfoncement *m* du pieu par un seul
coup de mouton
nl zakking *f* per slag (*paal*)
r глубина *f* погружения сваи под
действием одного удара

S336 *e* **set point adjuster**
d Sollwerteinsteller *m*, Sollwertgeber *m*
f dispositif *m* de changement de la
valeur de consigne
nl instelknop *m*
r задатчик *m* (*в системах
регулирования*)

S337 **set-retarding admixture** *see* **setting
retarder**

S338 *e* **sett**
d 1. Rammjungfer *f* 2. Pflasterstein *m*
f 1. avant-pieu *m* 2. caillou *m*
nl 1. opzetter *m*, oplanger *m*
2. straatsteen *m*, straatkei *m*
r 1. подбабок *m* 2. булыжник *m*

S339 *e* **setting**
d 1. Abbinden *n*, Erstarrung *f*
2. Aufstellung *f*, Anordnung *f*

3. Markierung *f* 4. Deckschicht *f*,
Oberputz *m* 5. Einstellen *n*,
Einstellung *f*
f 1. prise *f* 2. pose *f* 3. traçage *m*
4. couche *f* de finition 5. réglage *m*
nl 1. verharding *f* 2. plaatsing *f* 3.
aanduiden *n* 4. deklaag *f(m)* van de
bepleistering *f* 5. instelling *f*,
regeling *f*
r 1. схватывание *n* 2. установка *f*,
размещение *n* 3. разметка *f*
4. отделочный слой *m*, накрывка *f*
5. регулировка *f*

S340 *e* **setting accelerator**
d Abbindebeschleuniger *m*
f accélérateur *m* de prise
nl verhardingsversneller *m* (*beton*)
r ускоритель *m* схватывания
(*добавка*)

S341 **setting out** *see* **pegging out**

S342 *e* **setting rate**
d Abbindegeschwindigkeit *f*
f rapidité *f* de prise
nl verhardingssnelheid *f*
r скорость *f* схватывания

S343 *e* **setting retarder**
d Abbindeverzögerer *m*
f retardeur *m* de prise
nl verhardingsvertrager *m* (*beton*)
r замедлитель *m* схватывания
(*добавка*)

S344 *e* **setting shrinkage**
d Erstarrungsschwindung *f*,
Abbindungsschwindung *f*
f retrait *m* de prise
nl krimp *m* (*beton*)
r пластическая усадка *f* (*бетона
перед концом схватывания*)

S345 *e* **setting time**
d Abbindezeit *f*, Erstarrungszeit *f*
f temps *m* [durée *f*] de prise
nl verhardingstijd *m*
r время *n* схватывания

S346 *e* **settled sewage**
d geklärtes Abwasser *n*
f eaux *f pl* d'égout clarifiées,
eaux *f pl* usées clarifiées
nl gezuiverd afvalwater *n*
r осветлённая сточная вода *f*

S347 *e* **settlement**
d 1. Setzung *f*, Senkung *f*, Absacken *n*
2. Absetzklärung *f* 3. Sediment *n*,
Ablagerung *f*
f 1. tassement *m*, affaissement *m*
2. clarification *f* des eaux usées
3. sédiment *m* de précipitation
nl 1. zakking *f* 2. zuivering *f* van
afvalwater 3. bezinksel *n*
r 1. осадка *f* 2. осветление *n*
[механическая очистка *f*] сточных
вод 3. осадок *m*

S348 *e* **settlement joint**
d Setz(ungs)fuge *f*, Bewegungsfuge *f*

f joint *m* de tassement
nl dilatatievoeg *f(m)*
r осадочный шов *m*

S349 *e* **settlement stresses**
d Setzungsspannungen *f pl*
f contrainte *f* due au tassement
nl zettingsspanningen *f pl*
r напряжения *n pl* от осадки

S350 *e* **settlement tank**
d Absetzbecken *n*
f bassin *m* de décantation
nl bezinktank *m*
r отстойник *m*, осветлительный
бассейн *m*

S351 *e* **settling**
d Absetzen *n*, Abschlämmen *n*,
Aufschlämmen *n*
f décantation *f*
nl afzetting *f*
r осаждение *n*

S352 *e* **settling basin**
d Absetzbecken *n*, Ablagerbecken *n*,
Klärteich *m*
f décanteur *m*, bassin *m* de
clarification [de décantation]
nl bezinkvijver *m*
r отстойник *m*, отстойный бассейн *m*

S353 *e* **settling chamber**
d 1. Absetzraum *m*, Klärraum *m*
2. Staub(ablagerungs)kammer *f*
f 1. chambre *f* de décantation
2. chambre *f* à poussières
nl 1. bezinkruimte *f* 2. stofkamer *f(m)*
r 1. осадочная камера *f* отстойника
2. пылевая [пылеосадочная]
камера *f*

S354 *e* **settling electrode**
d Niederschlagselektrode *f*
f électrode *f* collectrice
nl afzettingselektrode *f*
r осаждающий электрод *m*
(*электрофильтра*)

S355 **settling pond** *see* **sedimentation pond**

S356 **settling rate** *see* **settling velocity**

S357 *e* **settling time**
d Absetzzeit *f*
f temps *m* de sédimentation
nl bezinktijd *m*
r продолжительность *f* осаждения

S358 *e* **settling velocity**
d Abschlämmgeschwindigkeit *f*,
Sinkgeschwindigkeit *f*
f vitesse *f* de décantation
nl bezinksnelheid *f*
r скорость *f* осаждения [оседания]

S359 *e* **settling well**
d Absetzgrube *f*, Absetzbrunnen *m*,
Absetzschacht *m*
f puits *m* de clarification
nl zinkput *m*, zakput *m*
r осадочный [отстойный] колодец *m*,
колодец-отстойник *m*

S360 *e* **set-up**
 d **1.** Einrichten *n* der Baustelle
 2. Aufstellpunkt *m* (*Vermessung*)
 3. Anlage *f*, Maschinen-Aggregat
 n, Maschinensatz *m*
 f **1.** préparation *f* du terrain de
 chantier **2.** stationnement *m* d'un
 appareil géodésique **3.** installation *f*.
 groupe *m*
 nl **1.** inrichten *n* van de bouwplaats
 2. opstellen *n* (*van een instrument*)
 3. installatie *f*
 r **1.** подготовка *f* строительной
 площадки 2. стоянка *f*
 (*геодезического прибора*)
 3. установка *f*; агрегат *m*

S361 *e* **severe exposure**
 d harte Bewitterung *f*
 f exposition *f* aux intempéries
 sévères
 nl **1.** blootstelling *f* aan strenge
 weersinvloeden **2.** ongunstige
 opstelling *f* van een gebouw
 r воздействие *n* суровых условий
 внешней среды

S362 *e* **sewage**
 d Abwasser *n*, Abwässer *n pl*
 f eaux *f pl* usées [d'égouts,
 résiduaires]
 nl rioolwater *n*, afvalwater *n*
 r сточные воды *f pl*

S363 *e* **sewage aeration**
 d Abwasserbelüftung *f*
 f aération *f* des eaux usées [d'égout]
 nl aëratie *f* van afvalwater
 r аэрация *f* сточных вод

S364 *e* **sewage clarification**
 d Abwasserklärung *f*
 f clarification *f* des égouts
 nl zuivering *f* van afvalwater
 r осветление *n* сточных вод

S365 *e* **sewage disposal**
 d **1.** Abwasserbeseitigung *f*,
 Abwasserableitung *f*,
 Abwasserabführung *f*
 2. Abwasserreinigung *f*
 f **1.** dérivation *f* des eaux d'égout
 2. épuration *f* des eaux résiduaires
 [usées]
 nl **1.** afvoeren *n* van afvalwater
 2. zuivering *f* van afvalwater
 r **1.** водоотведение *n* 2. очистка *f*
 сточных вод

S367 *e* **sewage distribution chamber**
 d Abwasserverteilerkammer *f*
 f chambre *f* de distribution des eaux
 usées [d'égouts]
 nl rioolwaterverdeelkamer *f(m)*
 r распределительная канализационная
 камера *f*

S368 *e* **sewage effluent**
 d gereinigtes [behandeltes] Abwasser *n*
 f eaux *f pl* usées épurées [traitées]

 nl behandeld afvalwater *n*
 r очищенные сточные воды *f pl*

S369 *e* **sewage farming, sewage farms**
 d Rieselfelder *n pl*,
 Rieselfelderwirtschaft *f*,
 Rieselgüter *n pl*
 f terrains *m pl* d'épandage
 nl vloeivelden *n pl*
 r сельскохозяйственные поля *n pl*
 орошения

S370 *e* **sewage flow rate**
 d Abwasserbelastung *f*,
 Abwasseranfall *m*, spezifische
 Abwassermenge *f*
 f débit *m* des eaux usées
 nl hoeveelheid *f* afgevoerd afvalwater
 r норма *f* водоотведения

S371 **sewage gas** *see* **sludge gas**

S372 *e* **sewage inlet chamber**
 d Anschlußschacht *m* der
 Kanalisation
 f chambre *f* de réception des eaux
 usées
 nl aansluitput *m*
 r канализационная приёмная камера

S373 *e* **sewage intake basin**
 d Abwasservorfluter *m*
 f bassin *m* de réception [de prise]
 des eaux usées
 nl afvalwaterverzamelbassin *n*
 r приёмник *m* сточных вод

S374 *e* **sewage pond**
 d Abwasserteich *m*
 f bassin *m* d'activation (des eaux
 usées)
 nl afvalwatervijver *m*
 r биологический пруд *m*

S375 *e* **sewage pump**
 d Abwasserpumpe *f*
 f pompe *f* à eau d'égout
 nl rioolpomp *f(m)*
 r канализационный насос *m*

S376 *e* **sewage pumping station**
 d Abwasserpumpwerk *n*,
 Abwasserhebeanlage *f*
 f station *f* de pompage des eaux
 usées
 nl rioolgemaal *n*
 r канализационная насосная станция
 f

S377 *e* **sewage purification, sewage
 treatment**
 d Abwasserbehandlung *f*
 f traitement *m* [épuration *f*] des eaux
 usées
 nl rioolwaterzuivering *f*
 r обработка *f* [очистка *f*] сточных
 вод

S379 *e* **sewage treatment works**
 d Abwasserbehandlungs- und
 -beseitigungsanlage *f*, Kläranlage *f*
 f ouvrages *m pl* d'épuration des
 eaux usées

nl rioolwaterzuiveringsinstallatie *f*
r канализационные очистные
сооружения *n pl*; станция *f*
очистки сточных вод

S380 *e* **sewer**
d 1. Abwassersammler *m*,
Abwasserleitung *f*,
Abwasserkanal *m*
2. Entwässerungsstollen *m*
f 1. collecteur *m* des eaux usées
2. canal *m* d'assèchement
nl 1. riool *n* 2. afwateringstunnel *m*
r 1. канализационный коллектор *m*;
наружный канализационный
трубопровод *m* 2. дренажная
галерея *f* (*у подошвы верхового
откоса земляной плотины*)

S381 *e* **sewerage**
d Kanalisation *f*, Kanalisierung *f*,
Ortsentwässerung *f*
f assainissement *m*, canalisation *f*;
réseau *m* de canalisation
nl riolering *f*, rioolstelsel *n*
r канализация *f*; система *f*
канализации

S382 *e* **sewer appurtenances**
d Einbauten *m pl* im Abwassernetz
f ouvrages *m pl* annexes des réseaux
d'évacuation
nl bouwwerken *n pl* in het rioolstelsel
r сооружения *n pl* на
канализационной сети

S383 *e* **sewer connection**
d Anschlußkanal *m*,
Grundstückanschlußleitung *f*,
Kanalisationsanschluß *m*
f branchement *m* d'égout
nl huisaansluitleiding *f*
r соединительная ветка *f*
канализации

S384 *e* **sewer flusher**
d Kanalspüler *m*
f appareil *m* [dispositif *m*] de
chasse des égouts
nl spui-inrichting *f* voor het riool
r каналопромыватель *m*

S385 *e* **sewer network**
d Kanalisationsnetz *n*, Abwassernetz *n*
f réseau *m* [système *m*] d'égouts
nl rioolstelsel *n*, riolering *f*
r канализационная сеть *f*

S386 *e* **sewer outfall**
d Abwasserauslaß *m*,
Kanalisationsauslauf *m*;
Auslaufbauwerk *n*
f débouché *m* [décharge *f*] de l'égout;
émissaire *m* d'évacuation,
déversoir *m* (d'eau) d'égout
nl riooluitlaat(werk) *m(n)*
r канализационный выпуск *m*;
выпускное сооружение *n*

S387 *e* **sewer pipe**
d Abwasserrohr *n*
f conduite *f* [tuyau *m*] d'égout

nl rioolbuis *f(m)*
r канализационная труба *f*

S388 *e* **sewer tunnel**
d Abwasserstollen *m*
f galerie *f* [tunnel *m*] d'évacuation
des éaux usées
nl riooltunnel *m*
r канализационный туннель *m*

S389 *e* **sewer zone**
d Kanalisationszone *f*.
Entwässerungsgebiet *n*
f zone *f* d'assainissement
nl gerioleerd gebied *n*
r бассейн *m* [зона *f*] канализования

S390 *e* **shade factor**
d Verschattungskoeffizient *m*
f facteur *m* d'ombrage
nl schaduwfactor *m*
r коэффициент *m* затенения

S391 *e* **shaft**
d 1. Schacht *m*; Brunnen *m*
2. Müllschlucker *m*
f 1. puits *m* 2. vide-ordures *m*
nl 1. schacht *f(m)* 2. vuilniskoker *m*
r 1. шахта *f*; колодец *m*
2. мусоропровод *m*

S392 *e* **shaft-and-culvert spillway**
d Rohrablaß *m*, Schacht- und
Rohrüberfall *m*
f évacuateur *m* à puits combiné
avec un aqueduc souterrain
nl schacht- en buisoverlaat *m*
r башенный трубчатый водосброс *m*

S393 *e* **shaft-and-tunnel spillway**
d Ringüberlauf *m*
f évacuateur *m* à tour combiné avec
une galerie souterraine
nl schacht- en tunneloverloop *m*
r башенный туннельный водосброс *m*

S394 *e* **shaft lock**
d Schachtschleuse *f*
f écluse *f* à sas du type puits
nl duikersluis *f(m)*
r шахтный шлюз *m*

S395 *e* **shaft spillway**
d Schachtüberfall *m*,
Schachtüberlauf *m*, Trichter-
-Hochwasserentlastungseinlauf *m*
f déversoir *m* [évacuateur *m*] en puits
nl aflaatschacht *f(m)*
r шахтный водосброс *m*

S396 *e* **shaker**
d Klopfwerk *n*, Rütteleinrichtung *f*;
Rütteltisch *m*
f agitateur *m*, secoueur *m* (*p.ex. de
filtre à poussière*); table *f* [plate-
-forme *f*] vibrante ; banc *m* de
vibration
nl schudtoestel *n*, triltafel *f(m)*
r встряхивающее устройство *n*;
виброплощадка *f*; вибростенд *m*

S397 *e* **shaker conveyor**
d Schüttelrutsche *f*

f couloir *m* à secousses; transporteur *m* à secousses
nl schudgoot *f(m)*
r виброжёлоб *m*; вибрационный конвейер *m*

S398 *e* **shaker filter**
d Staubfilter *n*, *m* mit Klopfwerk
f filtre *m* à secousses
nl stoffilter *n*, *m* met schudinrichting
r фильтр *m* с регенерацией механическим встряхиванием

S399 *e* **shaking screen**
d Schüttelsieb *n*
f crible *m* à secousse
nl schudzeef *f(m)*
r качающийся грохот *m*

S400 *e* **shale**
d Schiefer *m*
f schiste *m*
nl leisteen *n*, lei *m*
r сланец *m* (*порода*)

S401 *e* **shallow**
d Untiefe *f*; wandernde Sandbank *f*, Furt *f*, Flußübergang *m*
f banc *m*, maigre *m*; seuil *m*
nl zandbank *f(m)*; ondiepte *f*
r мель *f*; перекат *m*

S402 *e* **shallow cut digging**
d Flachbaggern *n*
f terrassement *m* en couches minces
nl afgraven *n* in dunne lagen
r разработка *f* грунта тонкими слоями

S403 *e* **shallow foundation**
d Flachgründung *f*, Flachfundament *n*
f fondation *f* superficielle [de surface]
nl fundering *f* op staal
r фундамент *m* неглубокого [мелкого] заложения

S404 *e* **shallow lift**
d dünne Schicht *f* [Lage *f*]
f couche *f* mince
nl dunne laag *f(m)*
r тонкий слой *m*

S405 *e* **shallow manhole**
d Mannloch *n* mit unveränderlichem Querschnitt
f puits *m* d'accès [de visite] superficiel
nl ondiepe inspectieput *m*
r мелкий смотровой колодец *m*

S406 *e* **shallow well**
d Flachbrunnen *m*; nicht vollständiger Brunnen *m*
f puits *m* de faible profondeur; puits *m* incomplet
nl ondiepe bron *f(m)*
r мелкий колодец *m* (*глубиной до 30 м*); несовершенный колодец *m*

S407 *e* **shape factor**
d Formfaktor *m*
f facteur *m* de forme

nl vormfactor *m*
r коэффициент *m* формы; аэродинамический коэффициент *m* (*для определения ветровой нагрузки*)

S408 *e* **shaping**
d Profilieren *n*
f profilage *m*
nl profilering *f*
r профилирование *n* (*напр. откоса*)

S409 *e* **sharpcrested weir**
d scharfkantiger Überfall *m*
f déversoir *m* en mince paroi
nl overloopstuw *m* met scherpe rand
r тонкостенный водослив *m*, водослив *m* с тонкой стенкой

S410 *e* **sharp sand**
d scharfkantiger Sand *m*
f sable *m* à grains anguleux
nl scherp zand *n*
r крупный песок *m* с остроугольными зёрнами

S411 *e* **shear**
d 1. Querkraft *f* 2. Abscheren *n*, Schub *m*, Scherung *f*
f 1. effort *m* tranchant, force *f* transversale 2. cisaillement *m*
nl 1. dwarskracht *f(m)* 2. afschuiving *f*
r 1. поперечная сила *f* 2. сдвиг *m*, срез *m*, скалывание *n*

S412 *e* **shear angle**
d Scherwinkel *m*
f angle *m* de cisaillement
nl afschuivingshoek *m*
r угол *m* сдвига

S413 *e* **shear center**
d Biegungsmittelpunkt *m*, Schubmittelpunkt *m*
f centre *m* de flexion [de cisaillement]
nl middelpunt *n* van de kromming
r центр *m* изгиба

S414 *e* **shear connectors**
d Dollen *f pl*, Holzdübel *m pl*; Schubverbinder *m pl*, Schubdübel *m pl*
f organes *m pl* d'assemblage (*dans la construction en bois*); pièces *f pl* d'attache travaillant en cisaillement; pattes *f pl* assurant l'adhérence du béton au fer (*dans une solive mixte fer-béton*)
nl deuvels *m pl*
r шпонки *f pl* (*в деревянных конструкциях*); детали *f pl*, предупреждающие сдвиг *или* обеспечивающие совместную работу бетона и стали в объединённых конструкциях

S415 *e* **shear diagram**
d Schubdiagramm *n*, Querkraftdiagramm *n*
f diagramme *m* desefforts tranchants
nl diagram *n* van de schuifkrachten

r эпюра *f* поперечных сил

S416 *e* **shear failure**
d Schubbruch *m*; Grundbruch *m*
f rupture *f* par cisaillement;
rupture *f* de sol due au
cisaillement (*à la base d'un ouvrage*)
nl grondverschuiving *f*; afschuivings-
breuk *m*
r разрушение *n* от сдвига [среза,
скалывания]; разрушение *n* грунта
(*в основании сооружения*)

S417 *e* **shear force**
d Querkraft *f*, Scherkraft *f*
f effort *m* tranchant
nl schuifkracht *f(m)*
r поперечная сила *f*

S418 **shear force diagram** *see* **shear diagram**

S419 *e* **shearhead**
d Säulenkopf *m*, Stützenaufsatz *m*
f chapiteau *m* [tête *f*] de colonne du
plancher-champignon
nl kolomkop *m*
r оголовок *m* [капитель *f*] колонны,
поддерживающей безбалочное
перекрытие

S420 *e* **shearing**
d Abscheren *n*, Schub *m*,
Scherung *f*
f cisaillement *m*
nl afschuiving *f*
r срез *m*, сдвиг *m*, скалывание *n*

S421 *e* **shearing area**
d Scherfläche *f*
f surface *f* de cisaillement
nl schuifvlak *n*
r площадь *f* сдвига [среза]

S422 **shearing failure** *see* **shear failure**

S423 **shearing force** *see* **shear force**

S424 **shearing modulus** *see* **modulus of rigidity**

S425 *e* **shearing strain**
d Gleitung *f*, Schiebung *f*,
Schubverformung *f*
f déformation *f* de cisaillement
nl vervorming *f* bij afschuiving
r деформация *f* сдвига [среза]

S426 *e* **spearleg derrik**
d Zweibaum *m*, Zweibock *m*
f chèvre *f* de levage à deux pieds,
derrick *m* à deux tirants rigides
nl schrank *m*
r двуногий деррик-кран *m*

S427 *e* **shear plate**
d runde Flaschenscheibe *f*, runder
Holzverbinder *m*
f anneau *m* non denté (*organe
d'assemblage*)
nl ringdeuvel *m* (*houtverbinding*)
r гладкая кольцевая шпонка *f*

S428 *e* **shear reinforcement**
d Schubbewehrung *f*
f armature *f* de cisaillement
nl schuifwapening *f*
r арматура *f*, работающая на срез

S429 *e* **shears**
d Schere *f*
f cisailles *f pl*, ciseaux *m pl*
nl schaar *f(m)*
r ножницы *pl*

S430 **shear slide** *see* **slip**

S431 *e* **shear strength**
d Schubfestigkeit *f*, Schubwiderstand
m
f résistance *f* au cisaillement
nl schuifsterkte *f*
r прочность *f* на срез [на сдвиг, на
скалывание]

S432 *e* **shear stress, shear tension**
d Schubspannung *f*, Scherspannung *f*
f contrainte *f* de cisaillement
nl schuifspanning *f*
r напряжение *n* среза [сдвига,
скалывания], срезывающее
напряжение *n*

S433 *e* **shearwall**
d Aussteifungswand *f*, Windscheibe *f*,
Binderscheibe *f*
f paroi *f* raidisseuse [de raidissement,
de renfort]
nl stabiliteitswand *m*, verstijvingswand
m
r стена *f* [перегородка *f*] жёсткости,
стенка-диафрагма *f*

S434 *e* **sheath**
d Scheide *f*, Kabelhülle *f*,
Hüllrohr *n*
f gaine *f*
nl kabelmantel *m*, mantelbuis *f(m)*
r трубчатая оболочка *f*

S435 *e* **sheathing**
d 1. Verschalung *f*, Dachschalung *f*;
Verkleidung *f*; Absteifung *f*
2. Schalungsoberfläche *f*
f 1. bardage *m*; voligeage *m*; planches
f pl ou panneaux *m pl* de blindage
2. surface *f* de contact de coffrage
nl 1. beschieting *f*, bekleding *f*
2. bekistingsoppervlak *n*
r 1. обшивка *f*; сплошная
обрешётка *f* (кровли); элементы
m pl крепления стенок траншей
(*доски, щиты*) 2. рабочая
поверхность *f* опалубки

S436 **sheathing screw** *see* **self-tapping screw**

S437 *e* **she bolt**
d Spannschraube *f* mit stirnseitiger
Gewindepfanne
f boulon *m* de serrage femelle,
boulon *m* (de serrage) à encoche
femelle filetée
nl klemschroef *f(m)*

r стяжной болт *m* с нарезным
гнездом в торце

S438 *e* sheepfoot roller
d Schaffußwalze *f*
f cylindre *m* [rouleau *m*] à pieds-de-
-mouton
nl schapenpootwals *f(m)*
r кулачковый (дорожный) каток *m*

S439 *e* sheet
d 1. Platte *f* 2. Schußstrahl *m*,
Überfallstrahl *m*
f 1. tôle *f*, feuille *f* 2. nappe *f*
nl 1. plaat *f(m)* 2. vrije vlakke
straal *m*, *f*
r 1. лист *m* 2. свободная плоская
струя *f*

S440 *e* sheet asphalt
d Sandasphalt *m*
f sheet-asphalt *m*, micro-béton *m*
nl zandasfalt *n*
r песчаный асфальтобетон *m*; асфальтовый раствор *m*

S441 *e* sheet backing coat
d Spannschloß *n*, Ausgleichsschicht *f*
aus Platten
f couche *f* de nivellement en
panneaux minces
nl tussenvloer *m* van platen
r сборная стяжка *f*, выравнивающий
слой *m* из тонких панелей

S442 *e* sheet flow
d flächenhafter Abfluß *m*,
Oberflächenwasserabfluß *m*
f écoulement *m* en nappe
nl glooiingsafvloeiing *f*
r склоновый сток *m*

S443 *e* sheet glass
d Flachglas *n*, Tafelglas *n*,
Fensterscheibe *f*
f verre *m* plat, vitre *f* de fenêtre
nl getrokken glas *n*, vensterglas *n*
r листовое оконное стекло *n*

S444 *e* sheeting
d 1. Verbau *m*, Verkleidung *f*
(*Graben, Baugrube*)
2. Wandbekleidung *f*; dichte
Dachbelattung *f*, Dachschalung *f*
3. Schalbohlen *f pl*,
Schalungstafeln *f pl* 4. Spundwand *f*
5. Bleche *n pl*
f 1. planches *f pl* de blindage
2. bardage *m*; voligeage *m* jointif
3. planches *f pl* de coffrage 4. palée *f*
5. tôle *m*, feuille *f*
nl 1. beschoeiing *f* (van ingraving) 2.
dakbeschot *n* 3. bekistingsplanken *f*
pl 4. damwand *m* 5. metaalplaat
f (m)
r 1. обшивка *f* стенок траншей 2.
обшивка *f* стен; сплошная
обрешётка *f* кровли 3. доски *f pl*
опалубки 4. шпунтовая стенка *f*
5. листовой материал *m*

S445 sheeting cofferdam *see* steel
sheetpile cofferdam

S446 *e* sheeting driver
d Spundwandramme *f*
f sonnette *f* pour battage des
palplanches
nl heistelling *f* voor damwand
r копёр *m* для забивки шпунта

S447 *e* sheeting rail
d 1. Querholz *n* (*als Unterlage für
Verkleidung*) 2. Fachwerkriegel *m*
f 1. barre *f* transversale 2. traverse *f*
de pan (*de bois ou de fer*)
nl 1. regel *m* voor bekleding 2. regel
m (in stijl- en regelwerk)
r 1. поперечина *f* для крепления
обшивки 2. ригель *m* фахверка

S448 *e* sheet-metal waterstop
d Blechdichtungsstreifen *m*
f membrane *f* d'étanchéité en tôle
mince
nl afdichtingsstrook *f(m) pl* van
metaal
r металлическая противофильтрационная диафрагма *f* (*уплотняющий
элемент шва в массивном
гидросооружении*)

S449 *e* sheet panel
d Wärmedämmtafel *f*
f plaque *f* isothermique [calorifuge]
nl warmte-isolerend plaatmateriaal *n*
r листовой теплоизоляционный
материал *m*

S450 *e* sheet pile
d Spundbohle *f*
f palplanche *f*
nl damplank *f(m)*
r шпунтовая свая *f*, шпунтина *f*;
шпунтовый профиль *m*

S451 *e* sheet pile bulkhead
d Spundwand-Kofferdamm *m*,
Spundwandfangdamm *m*; Bollwerk *n*
f batardeau *m* de palplanches; mur *m*
de soutènement en palplanches
nl keerwand *m* [beschoeiing *f*] van
damplanken
r шпунтовая перемычка *f*; больверк
m

S452 *e* sheetpile cell
d Spundwandzelle *f*
f cellule *f* [gabion *m*] de palplanches
nl gesloten (ronde) damwand *m*
r шпунтовая ячейка *f*

S453 *e* sheet pile cut-off
d Dichtungsspundwand *f*
f parafouille *f* étanche en palplanches
nl damwandafsluiting *f*
r шпунтовая диафрагма *f*;
шпунтовый замок *m* плотины

S454 *e* sheet piles
d Stahlspundbohle *f*
f palplanche *f* métallique

nl damplank *f(m)*
r металлический шпунт *m*

S455 *e* **sheet pile screen**
d Dichtungsspundwand *f*
f rideau *m* de palplanches; écran *m*
de palplanches
nl strijkdam *m*
r ограждающая шпунтовая стенка *f*;
шпунтовая диафрагма *f*

S456 *e* **sheet piling**
d Spund(bohlen)wand *f*;
Verspundung *f*, Ausspundung *f*
f rideau *m* de palplanches; mur *m*
parafouille, parafouille *f*; mur *m*
de fosse
nl damwand *m*, strijkdam *m*
r шпунтовый ряд *m*; шпунтовая
стенка *f*; шпунтовое крепление *n*
[ограждение *n*]

S457 *e* **sheet-piling cofferdam**
d Spundwandfangdamm *m*
f batardeau *m* [enceinte *f*] de
palplanches
nl keerwand *m* van damplanken
r шпунтовая перемычка *f*

S458 *e* **sheet steel**
d Stahl(fein)blech *n*
f tôle *f* d'acier
nl plaatstaal *n*
r тонколистовая сталь *f*

S459 shelf life *see* pot life

S460 shelf retaining wall *see*
cantilevered retaining wall

S461 *e* **shell**
d 1. Rohbau *m* 2. Schale *f*
f 1. carcasse *f*, cage *f* (*d'une*
bâtisse) 2. coque *f*, enveloppe *f*,
voile *m*
nl 1. ruwbouw *m* 2. schaal *f(m)*
r 1. здание *n* без отделки; остов *m*
здания 2. оболочка *f*

S462 *e* **shell and coil condenser**
d Schlangenrohrkesselverflüssiger *m*
f condenseur *m* calandre et serpentins
nl spiraalcondensor *m*
r кожухозмеевиковый конденсатор *m*

S463 *e* **shell and tube heat exchanger**
d Rohrbündelwärmeübertrager *m*
f échangeur *m* thermique de faisceau
tubulaire
nl pijpenbundelwarmtewisselaar *m*
r кожухотрубный теплообменник *m*

S464 *e* **shell constructions**
d Schalentragwerke *n pl*
f constructions *f pl* en coque,
enveloppes *f pl*, voiles *m pl*,
coques *f pl*
nl schaalconstructies *f pl*
r оболочечные конструкции *f pl*,
оболочки *f pl*

S465 *e* **shell of revolution**
d drehsymmetrische Schale *f*,
Rotationsschale *f*
f enveloppe *f* de révolution
nl omwentelingsschaal *f*
r оболочка *f* вращения

S466 *e* **shell pile**
d Mantelpfahl *m*, Hülsen(rohr)pfahl *m*
f pieu *m* à fouerreau perdu [coffre]
nl mantelpaal *m*
r набивная свая *f* с оболочкой

S467 *e* **shell roof**
d Schalendach *n*
f voile *m* (mince)
nl schaaldak *n*
r покрытие *n* в виде
(свода-)оболочки

S468 *e* **shell type boiler**
d Flamm(en)rohrkessel *m*
f chaudière *f* à tube-foyer
nl vlampijpketel *f(m)*
r жаротрубный котёл *m*

S469 *e* **shell vault**
d Schalengewölbe *n*
f voûte *f* en voile mince [en coque]
nl schaalgewelf *n*
r свод-оболочка *m*

S470 *e* **shelter**
d 1. Schutzraum *m*, Schutzbunker *m*
2. Schuppen *m*
f 1. abri *m* 2. auvent *m*
nl 1. schuilkelder *m* 2. afdak *n*
r 1. убежище *n*, укрытие *n* 2. сарай
m; навес *m*

S471 *e* **shelter wall**
d 1. Schutzmauer *f* (*Hafenbau*)
2. Reaktor-Strahlenschutzanlage *f*
f 1. mur *m* de protection 2. mur *m*
de protection biologique
nl 1. schermmuur *m* 2. stralen-
-afschermingsmuur *m*
r 1. парапетная [волноотбойная]
стенка *f* 2. стенка *f* для защиты
от излучения, защитная оболочка
f реактора

S472 *e* **shield**
d Vortriebsschild *m*
f bouclier *m*
nl schild *n*
r проходческий щит *m*

S473 *e* **shielding concrete**
d Strahlenschutzbeton *m*,
Abschirmbeton *m*
f béton *m* de la protection contre le
rayonnement radioactif
nl stralingwerend beton *n*
r бетон *m*, обеспечивающий
биологическую защиту (*от*
радиации)

S474 *e* **shield tunneling**
d Schildbauweise *f*, Schildvortrieb *m*
f percement *m* [creusement *m*] (de
tunnel) au bouclier

nl schildbouw *m* van tunnels
r щитовая проходка *f* туннелей

S475 *e* shift
   d 1. Kurvenversetzung *f*
     2. Arbeitsschicht *f*
     3. Wandaufteilung *f* durch
     Vertikalnähte 4. Verschiebung *f*,
     Sprung *m*
   f 1. déplacement *m* de virage
     2. équipe *f*, poste *m* de travail
     3. appareil *m* alterné des joints
     verticaux 4. faille *f*
  nl 1. oeververplaatsing *f* 2. werkploeg
     f(m) 3. verspringen van de
     stootvoegen 4. breuk (*aardoppervlak*)
   r 1. сдвижка *f* (дорожной) кривой
     2. рабочая смена *f* 3. разрезка *f*
     стены вертикальными швами
     кладки 4. сброс *m* (*геологический*)

S476 *e* shim
   d Einlage *f*, Zwischenlegscheibe *f*
   f cale *f*
  nl vulstuk *n*
   r подкладка *f*

S477 *e* shingle
   d 1. Gerölle *n*, sandloser Kies *m*,
     Rundkies *m* 2. Schindel *f*
   f 1. galet *m* 2. bardeau *m*
  nl 1. grof grind *n* 2. dakspaan *f(m)*,
     shingle
   r 1. галька *f*; гравий *m* (*крупный*)
     2. гонт *m*

S478 *e* ship channel
   d Fahrrinne *f*
   f voie *f* navigable, chenal *m*
  nl vaargeul *f(m)*
   r фарватер *m*

S479 *e* shiplap
   d Falzverbindung *f*, Falzung *f* (*Holz*)
   f jointure *f* à recouvrement [à
     feuillure]
  nl verbinding *f* tussen sponningplanken
   r соединение *n* [сплачивание *n*] в
     четверть

S480 *e* ship lift
   d Schiffshebewerk *n*
   f ascenseur *m* de bateaux
  nl scheepslift *m*
   r судоподъёмник *m*

S481   shipping lock *see* navigation lock
S482   ship scaffold *see* float scaffold

S483 *e* shock hazard
   d Berührungsgefahr *f*
   f risque *f* d'électrocution
  nl gevaarlijk aanraken
   r опасность *f* поражения электрото-
     ком

S484 *e* shock load
   d Stoßbelastung *f*
   f charge *f* de choc
  nl stoorbelasting *f*
   r ударная нагрузка *f*

S485 *e* shock loss
   d Stoßverlust *m*
   f saut *m* de pression
  nl stoorverlies *n*
   r потери *f pl* в скачке давления; поте-
     ри *f pl* в результате резкого
     изменения режима потока

S487 *e* shooting flow
   d schießende Strömung *f*, Schießen *n*
   f écoulement *m* torrentiel
  nl schietende stroom *m*
   r бурное движение *n* потока

S488 *e* shore
   d 1. Stütze *f*, Strebe *f*
     2. Grubenabstützung *f*,
     Grubenabsteifung *f* 3. Gestade *n*,
     Strand *m*, Küste *f*
   f 1. étrésillon *m*, étai *m*
     2. soutènement *m* 3. côte *f*, bord *m*
  nl 1. schoor *m* 2. bekleding *f* van een
     ingraving 3. kust *f(m)*
   r 1. стойка *f*; подкос *m*; подпорка *f*
     2. горная крепь *f* 3. берег *m*

S489 *e* shore erosion
   d Ufer-Erosion *f*
   f érosion *f* littorale
  nl oevererosie *f*
   r береговая эрозия *f*

S490 *e* shore intake
   d Uferentnahmebauwerk *n*,
     Ufereinlaßbauwerk *n*
   f prise *f* d'eau latérale (en galerie)
  nl watervang *m* aan de oever
   r береговой водозабор *m*

S491 *e* shore protection
   d Uferschutz *m*, Uferbefestigung *f*
   f défense *f* de côte [du littorale],
     protection *f* des côtes
  nl kustverdediging *f*
   r берегозащитные мероприятия *n pl*

S492 *e* shore spillway
   d Uferüberlaufanlage *f*
   f déversoir *m* rivrain
  nl overlaat *m*
   r береговой водосброс *m*

S493 *e* shoring
   d 1. Absteifung, Aussteifung *f*
     2. Einrüstung *f* 3. Abstützung *f*
   f 1. blindage *m*, étaiement *m* des
     parois d'une tranchée   2. montage
     m d'échafaudage 3. étayage *m*,
     étaiement *m*, étançonnage *m*,
     chevalement *m*
  nl 1. stutten *n* 2. steigermontage *f*
     3. schoren *n*
   r 1. крепление *n* котлована
     [траншеи] 2. установка *f* лесов
     или подмостей 3. укрепление *n*
     конструкции стойками

S494 *e* short column
   d unelastische Säule *f* (*mit*
     Schlankheit kleiner als 15)

f poteau *m* court (*avec élancement moins de 15*)
nl stijve kolom *f(m)* (*slankheid minder dan 15*)
r жёсткая [негибкая] колонна *f* (*с гибкостью менее 15*)

S495 *e* **short duration load**
d Kurzzeitlast *f*
f sollicitation *f* de courte durée
nl kortstondige belasting
r кратковременная нагрузка *f*

S496 *e* **short span bridge**
d Kleinbrücke *f*, kurzgespannte Brücke *f* (*Spannweite bis 18 m*)
f pont *m* à travée courte, pont *m* de courte travée
nl brug *f(m)* met kleine overspanning
r мост *m* небольшого пролёта (*до 18 м*)

S497 *e* **short-term elastic modulus**
d kurzzeitiger Elastizitätsmodul *m*
f module *m* d'élasticité de courte durée
nl elasticiteitsmodulus *m* bij kortstondige belasting
r кратковременный модуль *m* упругости

S498 **short term load** *see* **short duration load**

S499 *e* **short-term strength**
d kurzzeitige Festigkeit *f*
f résistance *f* de courte durée
nl sterkte *f* bij kortstondige belasting
r кратковременная прочность *f*

S500 *e* **short-term tests**
d Kurzzeitprüfungen *f pl*
f essais *m pl* de courte durée
nl kortdurende proeven *f(m) pl*
r кратковременные испытания *n pl*

S501 *e* **shotcrete**
d Spritzbeton *m*, Torkretbeton *m*
f béton *m* projeté, gunite *f*
nl spuitbeton *n*
r торкрет-бетон *m*

S502 *e* **shotcrete lining**
d Spritzbetonabdeckung *f*
f revêtement *m* en gunite [en béton projeté]
nl bekleding *f* met spuitbeton
r торкрет-бетонная обделка *f или* облицовка *f*

S503 *e* **shot hole**
d Schußloch *n*
f trou *m* de mine
nl boorgat *n*
r шпур *m*

S504 *e* **shoulder**
d 1. Seitenraum *m*, Schulter *f* 2. Absatz *m* an Betonoberfläche (*Fehler*)
f 1. accotement *m* 2. épaulement *m* sur la surface de béton (*défaut*)

nl 1. wegberm *m* 2. uitsteeksel *n* (*fout op betonoppervlak*)
r 1. обочина *f* (*дороги*) 2. уступ *m или* выступ *m* на бетонной поверхности (*дефект*)

S505 *e* **shovel**
d 1. Schaufel *f*, Schippe *f* 2. Standbagger *m*, Hochbagger *m*
f 1. pelle *f* 2. pelle *f* mécanique
nl 1. schop *f(m)* 2. lepelschop *f*
r 1. лопата *f* 2. механическая лопата *f*, одноковшовый экскаватор *m*

S506 *e* **shovel attachment**
d Baggerhochlöffel *m* (*Anhängeausrüstung*)
f pelle *f* butte (*équipement porté*)
nl graaflepel *m* (*hulpstuk*)
r прямая лопата *f* (*навесное оборудование*)

S507 *e* **shovel dozer**
d Bulldozer-Ladeschaufeleinrichtung *f*
f chargeur-niveleur *m*, chargeur *m* avec décapeur
nl dozer-laadschop *f*
r фронтальный погрузчик-планировщик *m* на базе трактора (*выполняет разработку, погрузку, перемещение и планировку грунта*)

S508 *e* **shovel stick**
d Löffelstiel *m*, Auslegerstiel *m*
f bras *m* de godet (*d'excavateur*); bras *m* de levage
nl steel *m* van lepelschop
r рукоять *f* ковша экскаватора

S509 *e* **shower head**
d Brausekopf *m*
f pomme *f* de douche
nl douchekop *m*, broes *f(m)*
r душевая головка *f*

S510 *e* **shrinkage**
d Schrumpfen *n*, Schwinden *n*, Schwindung *f*, Schwund *m*
f retrait *m*, contraction *f*
nl krimp *f(m)*
r усадка *f*

S511 *e* **shrinkage coefficient**
d Volumenminderungsfaktor *m*
f coefficient *m* [facteur *m*] de retrait
nl krimpcoëfficiënt *m*
r коэффициент *m* усадки

S512 *e* **shrinkage compensating concrete**
d schwindfreier Beton *m*
f béton *m* sans retrait
nl krimpvrij beton *n*
r безусадочный бетон *m*

S513 *e* **shrinkage crack**
d Schwindriß *m*
f fissure *f* de retrait
nl krimpscheur *f*
r усадочная трещина *f*

S514 *e* **shrinkage cracking**
d Schwindrißbildung *f*

f fissuration f de retrait
nl krimpscheurvorming f
r усадочное трещинообразование n

S515 shrinkage factor see shrinkage coefficient

S516 e shrinkage joint
d Schwindfuge f, Schrumpfungsfuge f
f joint m de retrait
nl krimpvoeg f(m)
r усадочный шов m

S517 e shrinkage limit
d Schwindgrenze f, Schrumpfgrenze f
f limite f de retrait
nl krimpgrens f(m)
r предел m усадки

S518 e shrinkage loss
d Spannungsverlust m infolge Schwindens von Beton
f perte f de précontrainte due au retrait de béton
nl voorspanningsverlies n door krimp
r потеря f предварительного напряжения от усадки бетона

S519 e shrinkage reinforcement
d Schwindbewehrung f
f armature f de retrait
nl krimpwapening f
r (противо)усадочная арматура f

S520 e shrinkage stress
d Schwindspannung f
f contrainte f de retrait
nl krimpspanning f
r усадочное напряжение n

S521 e shrink-mixed concrete
d nachgemischter Beton m
f béton m préparé partiellement en usine et partiellement en bétonnière motorisé
nl nagemengde betonmortel m
r товарный бетон m, приготовленный частично на заводе и частично в автобетоносмесителе

S522 e shrunk-on ring
d Schrumpfring m
f frette f posée à chaud
nl krimpring m
r термоусадочный бандаж m, термоусадочная манжета f

S523 e shuga
d Eisbrei m, Sulzeis n
f bouillie f de glace, shuga f
nl smeltijs n, ijsbrij m
r шуга f

S524 shunt valve see bypass valve

S525 e shut-off damper
d Absperrklappe f, Absperrschieber m
f soupape f d'arrêt à air, volet m [clapet m] d'air
nl afsluitschuif f(m)
r запорный воздушный клапан m, заслонка f

S526 e shut-off rotary valve
d Kugelabschluß-Schieber m

f vanne f à obturateur sphérique
nl kogelafsluiter m
r шаровой затвор m

S527 shut-off valve see stop valve

S528 e shutter
d 1. Laden m 2. Klappenverschluß m, Stauklappe f 3. Gelenkklappe f
f 1. volet m, contrevent m 2. hausse f 3. registre m d'air à volets
nl 1. blind n 2. klep f(m) van een stuw 3. jaloezieklep f(m)
r 1. ставня f, ставень m 2. (клапанный) затвор m 3. шторный воздушный клапан m (из шарнирно-сочленённых элементов)

S529 shuttering see formwork

S530 e shuttering work
d Schalarbeit f
f travaux m pl de coffrage
nl bekistingswerk n
r опалубочные работы f pl

S531 shutter vibrator see external vibrator

S532 e shutter weir
d 1. Klappenwehr n 2. Tafelwehr n, Schützenwehr n
f 1. barrage m à clapet 2. barrage m à hausses
nl 1. klepstuw m 2. stuw m met schuiven
r 1. плотина f с клапанным затвором 2. водосливная плотина f со щитами на гребне

S533 e sideboom pipe-layer
d Seitenausleger-Rohrverleger m
f poseur m de tuyaux à flèche latérale
nl pijpenlegger m met zijconsole
r трубоукладчик m с боковой стрелой

S534 e side-channel spillway
d Überlaufgraben m
f déversoir m lateral avec canal d'évacuation
nl overlaat m met afvoerkanaal
r водосброс m с боковым отводом воды, траншейный водосброс m; береговой водосливный канал m

S535 e side contraction
d seitliche Einschnürung f
f contraction f latérale
nl zijdelingse samendrukking, insnoering f
r боковое сжатие n

S536 e side formwork
d Straßenbauschalung f, Seitenschalung f
f coffrage-rail m
nl zijbekisting f van betonweg
r рельс-форма f, боковая опалубка f (бетонного покрытия дороги)

S537 *e* **sidehill cut**
  d Hangeinschnitt *m*
  f déblai *m* dans le flanc de côteau
  nl ingraving *f* in een helling
  r выемка *f* в косогоре

S538 *e* **side intake**
  d seitliche Wasserentnahme *f*
  f prise *f* d'eau latérale
  nl zijdelingse wateronttrekking
  r боковой водозабор *m*

S539 *e* **side overfall**
  d Rand(kanal)überlauf *m*
  f déversoir *m* latéral
  nl overlaat *m*
  r боковой водослив *m*

S540 *e* **side pond**
  d Sparbecken *n*
  f bassin *m* économiseur d'eau
  nl spaarbekken *n*
  r сберегательный бассейн *m* (*шлюза*)

S541 *e* **side-rolling ladder**
  d fahrbare Anlegeleiter *f*
  f échelle *f* roulante [coulissante] sur
    rail supérieur
  nl rijdende montageladder *f(m)*
  r наклонная приставная катучая
    лестница *f*

S542 *e* **side service spillway**
  d seitlicher Betriebsüberlauf *m*
  f évacuateur *m* latéral
  nl overlaat *m*
  r боковой рабочий водосброс *m*

S543 *e* **side slope**
  d Böschungsneigung *f*
  f pente *f* du talus
  nl hellingshoek *m*, talud *n*
  r уклон *m* откоса

S544 *e* **sidesway**
  d Horizontalverschiebung *f*
  f déplacement *m* horizontal [latéral]
  nl horizontale verschuiving *f*
  r боковое смещение *n*

S545 *e* **side underflow**
  d Umläufigkeit *f*
  f courant *m* de fond latéral
  nl achterloopsheid *f*
  r фильтрация *f* в обход сооружения

S546 *e* **side view**
  d Seitenansicht *f*, Seitenriß *m*
  f vue *f* latérale
  nl zijaanzicht *n*
  r вид *m* сбоку

S547 *e* **sidewalk**
  d Bürgersteig *m*, Gehweg *m*
  f trottoir *m*
  nl trottoir *n*
  r тротуар *m*

S548 *e* **side wall**
  d 1. Ulm(e) *f* 2. Wehrwange *f*
  f 1. mur *m* latéral 2. culée *f* de barrage
  nl 1. zijwand *m* 2. landhoofd *n*,
    schermwand *m*
  r 1. откосная стенка *f* 2. береговой
    устой *m* плотины

S549 *e* **side weir**
  d Streichwehr *n*, Randüberfall *m*,
    Überlaufgraben-Einlaßschwelle *f*
  f déversoir *m* latéral, seuil *m* du déver-
    soir latéral
  nl stuw *m* met zijdelingse overlaat
  r боковой водослив *m*, порог *m* тран-
    шейного водосброса

S550 *e* **sidings**
  d Stülpschalungsbrett *n*
  f bardage *m* de finition (*en*
    planches ou en plaques)
  nl eindbekisting *f*
  r боковая обшивка *f* (*досками или*
    листовым материалом)

S551 *e* **sieve**
  d Sieb *n*; Siebmaschine *f*
  f tamis *m*; crible *f*
  nl zeef(machine) *m* (*f*)
  r сито *n*; грохот *m*

S552 *e* **sieve analysis**
  d Siebanalyse *f*
  f analyse *m* de tamisage
    [granulométrique par tamisage]
  nl zeefanalyse *f*
  r ситовый анализ *m*

S553 *e* **sieve number**
  d Sieb(gewebe)nummer *f*
  f numéro *m* de toile (*du tamis*)
  nl zeefnummer *n*
  r номер *m* сита

S554 *e* **sieve residue**
  d Siebrückstand *m*
  f refus *m* de crible [de tamisage]
  nl zeefrest *m*
  r остаток *m* на сите

S555 *e* **sieve size**
  d lichte Maschenweite *f*, Siebgröße *f*
  f ouverture *f* de maille
  nl zeefmaat *f(m)*
  r номинальный размер *m*
    отверстий сита

S556 *e* **sight glass**
  d Schauglas *n*, verglastes Schauloch *n*
  f verre *m* de regard
  nl kijkglas *n*
  r смотровое стекло *n*

S557 *e* **sighting**
  d Einvisieren *n*, Visieren *n*
  f visée *f*
  nl zichten *n*
  r визирование *n*

S558 *e* **signal-controlled intersection,
    signal street intersection**
  d signalgesteuerter Knoten *m*
  f carrefour *m* réglementé par signals,
    croisement *m* à circulation
    commandé
  nl kruispunt *n* met verkeerslichten
  r регулируемый перекрёсток *m*

S560 *e* **silencer**
  d Schalldämpfer *m*
  f amortisseur *m* de bruit

nl geluiddemper *m*
r шумоглушитель *m*,
звукопоглотитель *m*

S561 *e* silica
d Siliziumdioxid *n*
f silice *m*, oxide *m* de silicium
nl siliciumdioxyde *n*
r кремнезём *m*, окись *f* кремния

S562 *e* silica brick
d Quarzitstein *m*, Silikastein *m*
f brique *f* de silice [de Dinas]
nl 1. kalkkwartssteen *n*
2. leemkwartssteen *n*
r динас, динасовый кирпич *m*

S563 *e* silica flour, silica powder
d Kieselerde-Pulver *n*
f farine *f* [poudre *f*] de silice
nl kiezelaardepoeder *n, m*
r тонкомолотый порошок *m*
кремнезёма

S565 *e* silicate concrete
d Silikatbeton *m*, Kalksandbeton *m*
f béton *m* silico-calcaire
nl silicaatbeton *n*, kalkzandbeton *n*
r силикатный [известково-
песчаный] бетон *m*

S566 *e* silicate cotton
d Hüttenwolle *f*
f laine *f* de laitier
nl steenwol *f(m)*
r шлаковата *f*, шлаковая вата *f*

S567 *e* silicated surfacing
d Silikatdecke *f*, Silikatmakadam *m*,
n, Wasserglasbelag *m*
f revêtement *m* routier en sol
silicaté, macadame *m* silicaté
nl silicaatdek *n*
r (дорожное) покрытие *n* из
силикатизированного грунта

S568 *e* silicate of soda
d Natronwasserglas *n*
f siliicate *m* de soude
nl natronwaterglas *n*
r натриевое жидкое стекло *n* (*для
укрепления грунтов*)

S569 *e* silicate paint
d Silikatfarbe *f*
f peinture *f* aux silicates
nl silicaatverf *f(m)*
r силикатная краска *f*

S570 *e* silicatization, silicification
d Silikatisierung *f*
f silicatisation *f* (*p.ex. du sol*),
silicification *f*
nl silicatisering *f*
r силикатизация *f* (*напр. грунта*)

S571 *e* sill
d 1. Überlaufschwelle *f*,
Überfallschwelle *f*, Überlaufkante *f*;
Grundschwelle *f*, Sohlschwelle *f*
Grundwehr *n* 2. Grundbalken *m*,
Sohlbalken *m*, Grundsohle *f*;
Schwelle *f* 3. Sohlbankriegel *m*,

Schwellholz *n*, Schwellbohle *f*,
Schwellbrett *n* 4. Fensterablauf *m*
f 1. seuil *m* déversant [du
déversoir, de l'évacuateur] ; seuil *m*
à la base 2. semelle *f*, seuil *m*,
semelle *f* basse (*d'un pan de bois*)
3. pièce *f* d'appuis 4. jet *m* d'eau
nl 1. overloopdrempel *m*
2. bodemdrempel *m* 3. onderdorpel *m*
r 1. водосливный *или* донный порог *m*
2. лежень *m*; опорный брус *m*
деревянного каркаса 3. нижний
брус *m* оконной рамы *или*
дверной коробки 4. отлив *m*
оконной рамы

S572 *e* silo
d Silo *m*
f silo *m*
nl silo *m*
r силос *m*; бункер *m*

S573 *e* silo effect
d Brückenbildung *f*
f effet *m* de silo
nl brugvorming *f* (*silo*)
r зависание *n* (*сыпучего материала*)

S574 *e* silt
d Sinkstoffe *m pl*; Schluff *m*,
Schlamm *m*; feinster Staubsand *m*
f vase *f*, boue *f*; limon *m*
nl slib *n*, slijk *n*, modder *m*
r ил *m*; пылевидные фракции *f pl*
[частицы *f pl*]

S575 siltation *see* colmatage
S576 *e* silt basin
d 1. Dränschacht *m* mit Sand- und
Schlammfang, Absetzbrunnen *m* 2.
Vorspeicher *m*, Verlandebecken *n*
f 1. dessableur *m* 2. retenue *f* pour
charges en suspension
nl 1. draineerschacht *f(m)* met zand-
en slikvanger 2. bezinkbekken *n*
r 1. колодец-отстойник *m*
2. наносозадерживающее
водохранилище *n*

S577 *e* silt content
d 1. Schluffgehalt *m*,
Feinstkornanteil *m*
2. Schwebstoffanteil *m* 3.
Schwebstofführung *f* 4. Trübung *f*
f 1. teneur *f* [proportion *f*] de silt
[de limon] 2. teneur *f* en
suspension 3. débit *m* solide
4. turbidité *f*
nl gehalte *n* aan vaste stoffen
r 1. содержание *n* пылевидных
фракций 2. содержание *n*
[концентрация *f*] взвесей
[взвешенных частиц] 3. расход *m*
[сток *m*] взвешенных наносов
4. мутность *f*

S578 *e* silting
d Verlandung *f*, Verschlämmung *f*,
Verschlickung *f*, Auflandung *f*,
Anschwemmung *f*

f envasement *m*, remblaiment *m*,
alluvionnement *m*
nl aanslibbing *f*, dichtslibbing *f*
r заиление *n*

S579   **silting up** *see* **colmatage**

S580 *e* **silty soil**
d Schluffboden *m*
f sol *m* limoneux
nl modderige grond *m*
r илистый [иловатый] грунт *m*

S581 *e* **similarity criterion**
d Ähnlichkeits(kenn)zahl *f*
f critère *m* de similitude
nl vergelijkingscriterium *n*
r критерий *m* подобия

S582 *e* **similarity test**
d Ähnlichkeitsversuch *m*
f essai *m* sur modèles de
l'hydraulique
nl vergelijkend onderzoek *n*
r испытание *n* гидравлически
подобной модели (*гидросооружения*)

S583 *e* **simple beam**
d einfach (auf)gelagerter Träger *m*
f poutre *f* simplement appuyée,
poutre *f* sur appuis simples
nl vrij opgelegde balk *m*
r простая [свободно опёртая,
однопролётная] балка *f*

S584 *e* **simple bending**
d reine [einfache] Biegung *f*
f flexion *f* simple
nl zuivere buiging *f*
r чистый изгиб *m*

S585 *e* **simple support**
d einfache Auflagerung *f*,
Freilagerung *f*
f appui *m* simple
nl vrije oplegging *f*
r простая *или* шарнирная опора *f*

S586 *e* **simple torsion, simple twist**
d einfache Torsion *f*
f torsion *f* simple
nl zuivere torsie *f*
r простое [чистое] кручение *n*

S587 *e* **simplified calculation**
d Näherungsrechnung *f*
f calcul *m* simplifié
nl verkorte berekening *f*, benadering *f*
r упрощённый расчёт *m*

S588 *e* **simply supported span**
d freigelagerter [freiaufliegender]
Brückenüberbau *m*
f travée *f* (du pont) simplement
appuyée
nl vrij opgelegde overspanning
r свободно опёртое пролётное
строение *n*

S589 *e* **simultaneous loading**
d gleichzeitiges Belasten *n*
f chargement *m* simultané,
application *f* de charges simultanée
nl simultane belasting *f*

r одновременное нагружение *n*;
одновременно действующие
нагрузки *f pl*

S590 *e* **single-acting pile hammer**
d einfach wirkender Rammhammer *m*
f marteau-pilon *m* [mouton *m* de
battage] à simple effet
nl enkelwerkend heiblok *n*
r свайный молот *m* простого
действия

S591 *e* **single-arch beam**
d Einbogenbrücke *f*
f pont *m* (arqué) à un seul arc
nl boogligger *m*
r мост *m* с одним арочным
пролётным строением

S592 *e* **single duct system**
d Einkanalklimaanlage *f*
f système *m* de climatisation à un
seul conduit
nl eenkanaals   klimaatinstallatie *f*
r одноканальная система *f*
кондиционирования воздуха

S593 *e* **single glazing**
d Einfachverglasen *n*
f vitrage *m* simple
nl enkele beglazing *f*
r одинарное остекление *n*

S594 *e* **single-hinged arch**
d Eingelenkbogen *m*
f arc *m* à une seule articulation
nl één-scharnierboog *m*
r одношарнирная арка *f*

S595 *e* **single inlet fan**
d einseitig ansaugender Radiallüfter *m*
f ventilateur *m* à une seule
admission
nl radiaalventilator *m* met eenzijdige
aanzuiging
r радиальный вентилятор *m*
одностороннего всасывания

S596 *e* **single-lane roadway**
d einspurige Fahrbahn *f*
f chaussée *f* á une seule voie
nl rijbaan *f* (*m*) met één rijstrook
r однополосная проезжая часть *f*

S597 *e* **single layer grid**
d einschichtige Netzwerkschale *f*
f structure *f* spatiale à résille
nl enkelvoudige vakwerkschaal *f(m)*
r однопоясная сетчатая оболочка *f*

S598 *e* **single-leaf bascule bridge**
d einarmige Baskülebrücke *f*
f pont *m* basculant (à *une travée
basculante*)
nl enkele basculebrug *f(m)*
r однокрылый разводной
раскрывающийся мост *m*

S599 *e* **single leaf damper**
d einflügelige Drehklappe *f*
f registre *m* d'air
nl luchtklep *f(m)* met één blad

r одностворчатый [однопёрьевой] воздушный клапан *m*

S600 *e* **single-leaf door**
d einflügelige Tür *f*, Einflügeltür *f*
f porte *f* à un seul battant [vantail]
nl deur *f(m)* met één blad
r одностворчатая [однопольная] дверь *f*

S601 *e* **single-level road junction**
d höhengleicher Knotenpunkt *m*
f croisement *m* à niveau
nl niveaukruising *f*
r пересечение *n* дорог в одном уровне

S602 *e* **single-lift lock, single lock**
d Einkammerschleuse *f*
f écluse *f* à sas simple
nl sluis *f(m)* met één kolk, enkelvoudige sluis *f(m)*
r однокамерный шлюз *m*

S602a *e* **single-line piping layout**
d Ein-Strang-Trassierung *f*
f schéma *m* de disposion de la tuyautérie unifilaire
nl leidingschema *n* voor éénpijpsysteem
r однолинейная схема *f* трубопроводов

S603 *e* **single oblique junction**
d einfacher schräger Abzweig *m*
f branchement *m* oblique simple
nl T-stuk *n* met schuine spruit
r косой тройник *m*

S604 *e* **single-pipe heating system**
d Einrohrheizung *f*, Einrohrheizungsanlage *f*
f chauffage *m* à un tuyau
nl éénpijp-verwarmingssysteem *n*
r однотрубная система *f* отопления

S605 *e* **single-pipe heat-supply system**
d Einleitersystem *n* der Wärmeversorgung
f système *m* d'alimentation en chaleur à un seul conduit
nl éénpijpsysteem voor warmwatervoorziening
r однотрубная система *f* теплоснабжения

S607 *e* **single pitch roof**
d freitragendes Pultdach *n*
f toiture *f* [comble *m*] à une seule pente, appentis *m*, toit *m* en appentis
nl lessenaardak *n*
r односкатная *или* наслонная крыша *f*

S608 *e* **single-point adjustable suspension scaffold**
d Hängegerüst *n* mit Einpunktaufhängung
f échafaudage *m* volant suspendu en un seul point

nl verstelbare hangsteiger *m* met éénpuntsophanging
r подъёмные подмости *pl*, подвешенные на одном канате

S609 *e* **single-pole scaffold**
d Ständergerüst *n* mit einer Ständerreihe
f échafaudage *m* de pied (à une rangée de montants)
nl enkele bouwsteiger *m*
r одностоечные леса *pl*, леса *pl* с одним рядом стоек

S610 *e* **single-purpose road**
d Einzweckstraße *f*
f route *f* pour une seule classe de véhicules
nl straat *f(m)* voor één bepaald type voertuig
r дорога *f* для движения определённого вида транспорта

S611 *e* **single-shaft mixer**
d Einwellenmischer *m*
f malaxeur *m* à un axe horizontal à palettes
nl éénassige betonmenger *m*
r бетоносмеситель *m* с одним горизонтальным лопастным валом

S612 *e* **single sheetpiling**
d einfache Spundwand *f*
f paroi *f* à une seule rangée de palplanches
nl enkele damwand *m*
r одинарный шпунтовый ряд *m*

S613 *e* **single-sized aggregate**
d Einkornzuschlag(stoff) *m*
f granulat *m* en une seule gamme de grains
nl toeslag met gelijke korrelgrootte
r заполнитель *m* из зёрен одной фракции

S614 *e* **single-span bridge**
d Ein-Feld-Brücke *f*, Brücke *f* mit einer Öffnung
f pont *m* à une travée
nl oeverbrug *f(m)*
r однопролётный мост *m*

S615 *e* **single-storey heating system**
d Etagenheizung *f*, Stockwerkheizung *f*
f chauffage *m* d'étage
nl etageverwarming *f*
r поэтажное отопление *n*

S616 *e* **single-wall cofferdam**
d einreihiger Fangdamm *m*
f batardeau *m* à simple paroi
nl vangdam *m*
r однорядная перемычка *f*

S617 *e* **single-way heating coil**
d Einweglufterhitzer *m*, einzügiger [eingängiger] Lufterhitzer *m*
f serpentin *m* de chauffage à une seule voie
nl enkelgangs luchtverhitter *m*
r одноходовой калорифер *m*

S618 *e* sink
    d 1. Sumpf *m* 2. Senke *f*, negative
    Quelle *f* 3. Wärmesenke *f*
    4. Auswaschungstrichter *m*
    5. Spülbecken *n*, Spüle *f*, Spültisch *m*
    f 1. puisard *m* 2. source *f* négative
    3. puits *m* de chaleur 4. fosse *f*
    d'affouillement 5. évier *m*
    nl 1. put *m* 2. zinkput *m*
    3. koellichaam *n* 4. straatkolk *m*
    5. gootsteen *m*
    r 1. зумпф *m*, приямок *m* 2.
    отрицательный источник *m*, сток *m*
    3. теплоприёмник *m*, тепловой сток
    m 4. воронка *f* размыва 5. мойка
    f; раковина *f*

S620 *e* sinking
    d 1. Abteufen *n* 2. Versinken *n*,
    Niederbringen *n*
    f 1. fonçage *m* 2. enfoncement *m*,
    fonçage *m*, immersion *f*,
    affaissement *m*
    nl 1. delven *n* 2. laten zakken
    r 1. проходка *f* вертикальных
    выработок 2. погружение *n*

S621 *e* sinking into walls
    d Unterputzverlegung *f*
    f pose *f* sous enduit [dans le mur]
    nl montage *f* uit zicht
    r скрытая прокладка *f*
    (*внутренних проводок*); укладка *f*
    инженерных сетей в стене

S623 sink unit *see* sink 5.

S624 *e* sintered expanded clay
    d Sinter-Blähton *m*
    f keramsite *f*, gravillon *m* en argile
    expansé
    nl klinkerisoliet *n*, gesinterde
    geëxpandeerde klei *m*
    r керамзит *m*

S625 *e* sintered expanded slag
    d Sinterblähhochofenschlacke *f*
    f laitier *m* de haut fourneau expansé
    nl geëxpandeerde hoogovenslak *f(m)*
    r вспученный доменный шлак *m*

S626 *e* siphon
    d Heber *m*
    f siphon *m*
    nl hevel *m*, sifon *m*
    r сифон *m*

S627 *e* siphonage
    d Heberwirkung *f*
    f siphonage *m*
    nl hevelwerking *f*
    r сифонирование *n*

S628 siphon duct *see* barrel

S629 *e* siphon hood
    d Saughaube *f*, Heberhaube *f*,
    Heberdecke *f*
    f hotte *f* d'aspiration
    nl hevelkap *f(m)*
    r капор *m* сифона

S630 siphonic action *see* siphonage

S631 siphon outlet *see* siphon-type outlet

S632 *e* siphon spillway
    d Siphonüberlauf *m*, Siphonüberfall *m*,
    Heberüberfall *m*, Saugüberfall *m*
    f évacuateur *m* [déversoir *m*] à
    siphon
    nl hevelaflaat *m*, heveloverloop *m*
    r сифонный водосброс *m*

S633 *e* siphon-type outlet
    d Saugauslaß *m*, Heberauslaß *m*
    f déversoir *m* à siphon
    nl wateruitlaat *m* met sifon
    r сифонный водовыпуск *m*

S634 *e* site
    d Bauplatz *m*, Baugrundstück *n*;
    Baustelle *f*; Standort *m*,
    örtliche Lage *f*
    f chantier *m* de construction; terrain
    m à bâtir
    nl bouwterrein *n*; locatie *f*
    r территория *f* строительства;
    строительная площадка *f*; место *n*
    размещения строительного объекта

S635 *e* site assembly
    d Zusammenbau *m* auf der
    Baustelle
    f assemblage *m* sur le chantier
    nl montage *f* op de bouwplaats
    r укрупнительная сборка *f* на
    строительной площадке

S636 site bolt *see* field bolt

S637 *e* site casting yard
    d Baustellenfertigungsplatz *m*
    f aire *f* de préfabrication sur le
    chantier de construction
    nl betonwerf *f(m)* op de bouwplaats
    r приобъектный полигон *m*
    железобетонных изделий

S638 *e* site equipment
    d Baustellenausrüstung *f*
    f équipement *m* de chantier
    nl uitrusting van de bouwplaats
    r строительные машины *f pl* и
    оборудование *n* для
    строительных площадок

S639 *e* site investigation
    d Baugrunduntersuchung *f*
    f recherches *f pl* sur le chantier de
    construction; prospections *f pl*
    géotechniques ; reconaissance *f* du
    terrain
    nl onderzoek *n* van het bouwterrein
    r инженерно-геологические
    изыскания *n pl*

S640 site-marking *see* pegging out

S641 *e* site-mixed concrete
    d Baustellenbeton *m*
    f béton *m* préparé sur le chantier
    nl op de bouwplaats vervaardigd
    beton *n*
    r бетон *m*, приготовленный на
    стройплощадке

S642 *e* **site plan**
d Lageplan *m*, Baustellenlageplan *m*
f plan *m* de situation
nl situatietekening *f*
r ситуационный план *m*
стройплощадки

S643 *e* **site planning**
d Planung *f* von Bebauungsgebiet
f plan *m* d'occupation des sols
nl bebouwingsplan *n*
r планировка *f* района застройки

S644 *e* **site precasting, site prefabrication**
d Baustellenvorfertigung *f*
f préfabrication *f* sur chantier
nl prefabricatie *f* op de bouwplaats
r построечное изготовление *n*
сборных элементов строительных
конструкций

S646 *e* **site preparation**
d Baureifmachung *f* des Geländes
f préparation *f* de terrain
nl bouwplaatsvoorbereiding *f*
r расчистка *f* строительной площадки;
подготовительные работы *f pl*
на стройплощадке

S647 *e* **site work**
d Baustellenarbeit *f*
f travaux *m pl* sur chantier
nl werk *m pl* op de bouwplaats
r (строительные) работы *f pl* на
стройплощадке

S648 *e* **site workshop**
d Bau(stellen)werkstatt *f*
f atelier *m* (de réparation) sur le
chantier de construction
nl werkplaats *f(m)* op het bouwterrein
r припостроечная [приобъектная]
мастерская *f*

S649 *e* **siting of houses**
d Anschluß *m* der Gebäude;
Anpassung *f* der Gebäude
f implantation *f* des bâtiments
nl situeren *n* van gebouwen
r посадка *f* зданий на местности;
привязка *f* зданий

S650 *e* **size**
d 1. Nennweite *f* 2. Baumaß *n*,
Nenngröße *f* 3. Leim *m*
f 1. diamètre *m* 2. grandeur *f*
nominale 3. colle *f*
nl maat *f(m)*, afmeting *f*
r 1. номинальный диаметр *m*
2. номинальный размер *m*,
типоразмер *m* 3. клей *m*

S651 *e* **size bracket, size fraction**
d Kornklasse *f*, Korngrößenbereich *m*,
Fraktion *f*
f fraction *f* granulométrique
nl korrelfractie *f*
r гранулометрическая фракция *f*

S652 *e* **sizing**
d 1. Leimen *n*, Schlichten *n*;
Grundieren *n* 2. Dimensionierung *f*
3. Klassieren *n*
f 1. collage *m*, apprêtage *m*, apprêt *m*
2. calibrage *m*, mise *f* à la côte
3. classification *f* granulométrique
nl 1. gronden *n* 2. dimensionering *f*
3. classificeren *n* door te zeven,
sorteren *n*
r 1. проклеивание *n*, приклеивание
n; грунтовка *f* 2. подбор *m*
оборудования *или* элементов
конструкций по типоразмеру;
расчёт *m* сечения элементов
3. классификация *f*, сортировка *f*
(*по крупности зёрен*)

S653 *e* **sizing characteristic**
d Körnungskennlinie *f*,
Kornklassenbild *n*
f courbe *f* granulométrique
nl zeefkromme *f*
r кривая *f* гранулометрического
состава

S654 *e* **skeleton construction**
d Skelettbau *m*, Rahmenbau *m*
f construction *f* à ossature [à
charpente]
nl skeletbouw *m*
r каркасная *или* рамная
конструкция *f*

S655 *e* **skeleton structure**
d Skelettbau *m*, Skelettbauwerk *n*
f ouvrage *m* à ossature [à charpente]
nl skeletbouw *m*
r каркасное сооружение *n*

S656 *e* **skewback**
d Bogenkämpfer *m*, Kämpferstein *m*,
Anfänger *m*
f naissance *f* [sommier *m*] de l'arc
nl aanzetlaag *f(m)*
r пята *f* арки; пятовый камень *m*

S656a *e* **skew bridge**
d schiefe Brücke *f*
f pont *m* biais
nl scheve brug *f(m)*
r косой мост *m*

S657 *e* **skid resistance**
d Griffigkeit *f*, Gleitsicherheit *f*,
Rutschsicherheit *f*
f résistance *f* (*d'un revêtement
routier*) au dérapage
nl stroefheid *f* (*van wegdek*)
r сопротивление *n* (*дорожного
покрытия*) заносу (*автомобилей*)

S658 *e* **ski jump drop**
d Sprungschanzen-Überfall *m*,
Sprungschanzen-Überlauf *m*
f chute *f* en porte à faux
nl console-overloop *m*
r консольный перепад *m*

S659 *e* **ski jump spillway**
  d Sprungschanze *f*, Skisprung-
      -Überlauf *m*
  f déversoir *m* en saut de ski
  nl springschans-overloop *m*
  r рассеивающий трамплин *m*,
      трамплинный водослив *m*

S660 *e* **skimmer**
  d 1. Abstreicher *m* 2. Kratzbagger *m*,
      Planierbagger *m*, Flachlöffelbagger
      m
  f 1. grille *f* de retenue (*pour les
      corps flottants*) 2. pelle *f* (équipée
      en) niveleuse
  nl 1. afstrijker *m* 2. vlaklepelschop
      f(m)
  r 1. скребок *m* (*для удаления
      всплывших веществ с
      поверхности воды*) 2. экскаватор-
      -струг *m*, экскаватор-планировщик
      m, скиммер *m*

S661 *e* **skimming**
  d 1. Abziehen *n*, Abschürfen *n*,
      Flachbaggerung *f* 2. Abheben *n*
      (*der Schwimmschicht im Klärbecken*)
  f 1. nivellement *m* [aplanissement *m*]
      préliminaire 2. écumage *m*,
      dégraissage *m* des eaux d'égout
  nl 1. egaliseren *n* 2. afschuimen *n*
  r 1. предварительная планировка *f*
      (*участка*) 2. слив *m* (*пены*)

S662 *e* **skip**
  d 1. Förderkübel *m* 2. Kippwagen *m*
      3. Kübelaufzug *m*
  f 1. benne *f*; godet *m* 2. skip *m*
      [wagonnet *m*] basculant [basculeur]
      3. skip *m* élévateur
  nl 1. container *m*, bak *m* 2. kipwagen
      m 3. liftkooi *f(m)*
  r 1. бадья *f*; ковш *m*; скип *m*
      2. опрокидная вагонетка *f*
      3. скиповый подъёмник *m*

S663 *e* **skip-hoist**
  d 1. Beschickungswerk *n*
      2. Kübel(bau)aufzug *m*
  f 1. benne *f* de chargement (*de la
      bétonnière*) 2. élévateur *m* à skip,
      skip *m* élévateur
  nl 1. invoerinstallatie *f* voor de
      betonmolen 2. bouwlift *m* met
      kipbak
  r 1. загрузочное устройство *n*
      (*бетоносмесителя*)
      2. строительный скиповый
      подъёмник *m* (*для подачи
      бетонной смеси*)

S664 *e* **skip joist system floor**
  d Stahlbetonrippendecke *f* mit
      vergrößertem Rippenabstand
  f plancher *m* à nervures espacées à
      grande distance (*jusqu'à 1,8 m*)
  nl gewapend betonnen ribbenvloer *m*
      met een afstand tot 1.8 m tussen
      de ribben

  r ребристое железобетонное
      перекрытие *n* с увеличенными
      (*до 1,8 м*) расстояниями между
      рёбрами

S665 **skirting board heating, skirting
      heating** *see* **baseboard heating**

S666 *e* **skullcracker**
  d Zertrümmerungskugel *f*
  f boulet *m* [poire *f*] de démolition
  nl sloopkogel *m*
  r баба *f*, шар *m* (*подвешиваемые к
      стреле крана для разрушения
      зданий*)

S667 *e* **skylight**
  d Oberlicht *n*
  f lucarne *f* (à lunette)
  nl dakraam *n*, lantaarn *m*
  r зенитный фонарь *m*

S668 *e* **skylight dome**
  d Lichtkuppel *f*
  f dôme *m* pour éclairage zénithal,
      coupole *f* d'éclairage [éclairante]
  nl lichtkoepel *f(m)*
  r купольный зенитный фонарь *m*

S669 *e* **skyway** *US*
  d Hochstraße *f*
  f route *f* surélevée
  nl hooggelegen straat *f(m)*
  r эстакадная дорога *f*

S670 *e* **slab**
  d Platte *f*
  f dalle *f*, plaque *f*
  nl plaat *f(m)*, vloer *m*
  r плита *f*

S671 *e* **slab bridge**
  d Plattenbrücke *f*
  f pont-dalle *m*
  nl plaatbrug *f(m)*
  r плитный мост *m*

S672 *e* **slab floor**
  d Plattendecke *f*
  f plancher-dalle *m*, plancher *m* en
      dalle
  nl plaatvloer *m*
  r перекрытие *n* из плит

S673 *e* **slab foundation**
  d Plattenfundament *n*
  f fondation-dalle *f*, fondation *f* sur
      semelle plate
  nl funderingsplaat *f(m)*
  r плитный фундамент *m*

S674 *e* **slab insulant**
  d Plattenisoliermaterial *n*,
      Plattendämmstoff *m*
  f isolant *f* en plaque
  nl isolatieplaat *f(m)*
  r плитный теплоизоляционный
      материал *m*

S675 *e* **slab insulation**
  d Wärmeisolierung *f* aus Platten
  f dalle *f* isothermique [calorifuge],
      plaque *f* d'isolation [d'isolement]

thermique, isolation f [isolement m]
thermique en dalles calorifuges
nl warmte-isolatie f van platen
r плитная теплоизоляция f

S676 e **slab jacking**
d Aufwinden n [Hochbocken n] der
abgesunkenen Straßenplatten
f nivellement m [nivelage m] des
dalles routières par vérins
nl verzakte wegplaten lichten
r выравнивание n бетонных
дорожных плит домкратами

S677 e **slab-stringer bridge**
d Brücke f in Verbundbauweise,
Verbundbrücke f
f pont m mixte acier-béton (avec
poutres en acier et platelage en
béton solidarisés)
nl brug f(m) van samengestelde
constructie
r мост m объединённой конструкции
(с главными балками из стали
и железобетонными плитами
настила)

S678 **slab strip** see **middle strip**

S679 e **slab vibrator**
d Plattenvibrator m, Plattenrüttler m
f plaque f vibrante, vibr(at)eur m à
plaque [à plateau]
nl triltafel f(m)
r площадочный вибратор m,
виброплощадка f

S681 e **slack-line excavator**
d Schlaffseil-Kabelbagger m
f éxcavateur m à câble equipé en
dragline
nl dragline m
r канатно-скребковый экскаватор m

S682 e **slack water**
d strömungsloses Wasser n
f eau f dormante
nl gestuwd water n
r стоячая вода f

S683 e **slag**
d Schlacke f
f laitier m, scorie f, mâchefer m
nl slak f(m)
r шлак m

S684 e **slag cement**
d Hüttenzement m, Schlackenzement m
f ciment m de laitier
nl slakkencement n
r шлаковый цемент m

S685 **slag wool** see **silicate cotton**

S685a e **slaked lime**
d Löschkalk m, gelöschter Kalk m
f chaux f éteinte [hydratée];
chaux f éteinte en poudre
nl gebluste kalk m
r гашёная известь f; известь-
-пушонка f

S686 e **slanting construction**
d sprengdruckwellensicheres
Bauwerk n
f construction f résistante à l'onde
explosive
nl tegen een schokgolf bestendig
gebouw n
r сооружение n, стойкое к
воздействию взрывной волны

S687 e **slat bucket**
d Gitter-Baggerlöffel m für Aushub von
wassergesättigtem klebrigem Boden
f benne f pour éxcavation des sols
adhésifs (d'une construction
évidée)
nl bak m met sleuven voor het
uitgraven van kleverige, met water
verzadigde grond
r ковш m (решетчатой
конструкции) для разработки
липких водонасыщенных грунтов

S688 e **slate**
d 1. Dachschiefer m
2. Schieferplatte(n) f (pl)
f ardoise f
nl 1. lei f(m) 2. daklei f(m)
r 1. сланец m 2. (асбо)шифер m;
(асбо)шиферные плитки f pl

S689 e **sleeper**
d Bahnschwelle f, Gleisschwelle f
f traverse f
nl dwarsligger m, biel f(m)
r шпала f

S690 e **sleeve**
d 1. Hülse f, Hülsrohr n
2. Manschette f 3. Muffe f ohne
Gewinde, Überschiebmuffe f
f 1. douille f, manchon m
2. manchon m souple, fourreau m
3. douille f mobile, manchon m
coulissant
nl 1. huls f(m) 2. manchet f(m)
3. mof f(m) zonder schroefdraad,
inschuifsok f(m)
r 1. гильза f, втулка f 2. манжета f
3. надвижная муфта f

S691 **sleeve joint** see **socket joint**

S692 e **sleeve socket**
d Rohrmuffe f
f manchon m d'un tuyau
nl buismof f(m)
r трубопроводная муфта f

S693 e **sleeving of bars**
d Muffenverbindung f der
Bewehrungsstäbe
f aboutement m des ronds à béton
par manchons
nl verbinding f van wapeningsstaven
door moffen
r соединение n арматурных стержней
муфтами

S694 e **slender beam**
d schlanker Balken m

f poutre *f* élancée [flexible]
nl slanke balk *m*
r гибкая балка *f*

S695 *e* **slender column**
d schlanke Säule *f* [Stütze *f*]
f poteau *m* élancé, colonne *f* élancée
nl slanke kolom *f(m)*
r гибкая колонна *f*; колонна *f*, работающая на продольный изгиб

S696 *e* **slenderness**
d Schlankheit *f*, Schlankheitsverhältnis *n*
f élencement *m*
nl slankheid *f (kolom)*
r гибкость *f*, коэффициент *m* гибкости (*сжатого элемента*)

S697 *e* **slewing**
d Drehen *n*
f rotation *f*, pivotement *m* (*autour d'un axe vertical*)
nl draaiing *f*
r вращение *n* (*вокруг вертикальной оси*)

S698 **slewing crane** *see* **revolving crane**

S699 *e* **sliced blockwork**
d geneigte Blocklagen *f pl*, Schrägverband *m*
f maçonnerie *f* de gros blocs en assises inclinées
nl metselwerk *n* met schuine lagen
r кладка *f* из массивов, выполненная наклонными рядами

S700 *e* **sliced blockwork breakwater**
d Schrägverband-Wellenbrecher *m*
f brise-lames *m* en gros blocs posés en assises inclinées
nl met schuine blokkenlagen gemetselde havendam *m* [pier *m*]
r волнолом *m* из массивов, уложенных наклонными рядами

S700a **slide** *see* **landslide**

S701 *e* **slide damper**
d Schieber *m*
f registre *m* à guillotine
nl schuif *f(m)*
r заслонка *f*; задвижка

S702 **slide gate** *see* **sliding gate**

S703 *e* **slide plane**
d Gleitebene *f*
f plan *m* de glissement
nl glijvlak *n*
r поверхность *f* [плоскость *f*] скольжения

S704 **slide valve** *see* **gate valve**

S705 *e* **sliding agent**
d Gleitmittel *n*
f graisse *f* pour câble d'armature
nl glijmiddel *n*
r смазка *f* (*арматурного пучка*)

S706 **sliding caisson** *see* **caisson gate**

S707 **sliding damper** *see* **blast gate**

S708 **sliding formwork** *see* **slipform**

S709 *e* **sliding gate**
d 1. Gleitschütz *n*, Falle *f*, Tafelschütz(e) *n* (*f*)
2. Schleusen-Schiebetor *n*
f 1. vanne *f* à glissières [à glissement] 2. porte *f* roulante d'écluse
nl 1. schuif *f(m)* 2. roldeur *f(m)*
r 1. щитовой скользящий плоский затвор *m* 2. откатные ворота *pl* шлюза

S710 *e* **sliding joint**
d Gleitfuge *f*, reibungsbehinderte Fuge *f*; Gleitverbindung *f*
f joint *m* glissant [à coulisse]
nl glijverbinding *f*
r скользящее соединение *n* (*допускающее взаимный сдвиг соединяемых элементов*)

S711 *e* **sliding-panel weir**
d Gleitschützenwehr *n*, Fallenwehr *n*
f barrage *m* à hausses, barrage *m* à pertuis
nl schuivenstuw *m*
r щитовая плотина *f*, плотина *f* со скользящими плоскими затворами

S712 *e* **sliding sash**
d Schiebe(fenster)flügel *m*
f vantail *m* [battant *m*] coulissant [glissant]
nl schuifraam *n*
r раздвижная створка *f* (*окна*)

S713 **sliding sluice gate** *see* **sliding gate**

S714 **sliding valve** *see* **gate valve**

S715 *e* **sliding wedge**
d Gleitkeil *m*
f prisme *m* de poussée [de rupture]
nl glijwig *f(m)*; glijspie *f(m)*
r призма *f* обрушения; клин *m* скольжения

S716 *e* **sliding-wedge method**
d Gleitkeilverfahren *n*
f méthode *f* de prisme de rupture, méthode *f* de surfaces de glissement
nl glijwigmethode (*berekening van steunwanden*)
r метод *m* клина скольжения (*расчёт подпорных стенок*)

S717 *e* **slime** US
d Schlamm *m*, Schlammüberlauf *m* (*Aufbereitung*)
f vase *f*, boue *f* (*de classificateur*)
nl modder *m*, slijkoverloop *m*
r шлам *m*, слив *m* (*классификатора*)

S718 *e* **slime coating**
d Schleimdecke *f*, biologischer Rasen *m*
f couche *f* visqueuse biofiltrante
nl slijklaag *f(m)* (*biologisch filter*)
r биологическая плёнка *f* (*биофильтра*)

S719 *e* sling
  d Lasthebeschlinge *f*; Stropp *m*
  f élingue *f*
  nl strop *m*; hijsstrop *m*
  r строп *m*; (грузоподъёмная) петля *f*

S720 *e* slip
  d 1. Aufschleppe *f*, Slip *m*;
  Helling *f*, Helgen *m*; Stapel *m*
  2. Erdrutsch *m* 3. Gleiten *n*;
  Rutschen *n*; Schlupf *m* 4. Spließ
  m, Fugenblech *n* 5. Brei *m*
  6. Schlicker *m*, feiner Emailbrei *m*
  f 1. slip *m*, cale *f* (de construction)
  2. éboulement *m* 3. glissement *m*
  4. couvre-joint *m* 5. pâte *f*
  6. barbotine *f*
  nl 1. stapel *m* (*scheepsbouw*)
  2. aardverschuiving *f* 3. glijden
  n; 4. spijkerstrip *m* 5. brij *f(m)*
  6. emailbrij *f(m)*
  r 1. слип *m*; эллинг *m*; стапель *m*
  2. оползень *m* 3. скольжение *n*;
  проскальзывание *n*; сдвиг *m*;
  4. нащельник *m* (*для кровельных
  покрытий*) 5. тесто *n*,
  кашицеобразная масса *f*
  6. шликер *m*

S722   slip dock *see* building berth

S723 *e* slip factor
  d Schlupfzahl *f*, Gleitungskoeffizient
  m
  f coefficient *m* de glissement
  nl glijdingsmodulus *m*
  r коэффициент *m* скольжения

S724 *e* slipform
  d Gleitschalung *f*
  f coffrage *m* glissant
  nl glijdende bekisting *f*
  r скользящая [подвижная] опалубка *f*

S725 *e* slipforming machine, slip-form
  paver
  d Gleit(schalungs)fertiger *m*
  f bétonneuse *f* [bétonnière *f*] mobile
  à coffrage glissant
  nl betonverwerkingsmachine *f* met
  glijdende bekisting
  r бетоноотделочная машина *f*
  [бетоноукладчик *m*] со
  скользящими формами

S726 *e* slip joint
  d Gleitfuge *f*, dichtende
  Steckverbindung *f*
  f joint *m* coulissant
  nl mofverbinding *f*
  r надвижное соединение *n*
  (*воздуховодов, труб*)

S727 *e* slip resistance
  d Gleitwiderstand *m*
  f résistance *f* au glissement (*des
  aciers enrobés dans le béton*)
  nl glijweerstand *m*
  r сопротивление *n* скольжению
  [проскальзыванию] (*арматурных
  стержней в бетоне*)

S728 *e* slip road
  d Anschlußrampe *f*,
  Verbindungsrampe *f* (*Autobahn*)
  f rampe *f* de raccordement [d'accès]
  nl oprit *m*
  r съезд *m*; въезд *m*

S729 *e* slip surface
  d Gleitbahn *f*, Gleitfläche *f*
  f surface *f* de glissement
  nl glijvlak *n*
  r поверхность *f* скольжения

S730 *e* slipway
  d Aufschleppe *f*, Schlipp *m*, Slip *m*;
  Helling *f*, Helgen *m*; Stapel *m*
  f slip *m*, cale *f* de halage [de
  lancement]
  nl scheepshelling *f*
  r слип *m*, эллинг *m*, стапель *m*

S731 *e* slope
  d 1. Neigung *f*, Gefälle *n* 2. Hang *m*,
  Böschung *f*
  f 1. pente *f*, inclinaison *f* 2. talus *m*
  nl 1. helling *f* 2. talud *n*, glooiing *f*
  r 1. наклон *m*, уклон *m*
  2. откос *m*

S732 *e* slope concrete paver
  d Böschungsbetoniermaschine *f*
  f finisseur *m* de revêtement en béton
  de talus
  nl machine *f* voor het betonneren van
  hellingen
  r машина *f* для бетонирования
  откосов

S733 *e* slope deflection method
  d Drehwinkelverfahren *n*
  f méthode *f* des déformations
  (angulaires), méthode *f* des
  déplacements
  nl draaihoekmethode *f*
  r метод *m* угловых деформаций;
  метод *m* перемещений

S734   sloped pile *see* batter pile

S735 *e* slope drainage
  d Böschungsentwässerung *f*
  f drainage *m* de talus
  nl afwatering *f* van hellingen
  r дренаж *m* откосов

S736 *e* slope failure
  d Böschungsrutschung *f*,
  Hangrutschung *f*
  f rupture *f* de talus
  nl kalven *n* van talud
  r оползание *n* откоса

S737 *e* slope filter
  d Böschungsfilter *n*, *m*,
  Schrägfilter *n*, *m*
  f filtre *m* de parement, couche *f* de
  drainage amont
  nl hellingsfilter *n*, *m*
  r откосный фильтр *m*

S738 *e* slope gauge
  d Gefällepegel *m*; Schrägpegel *m*,
  Böschungspegel *m*

f limnimètre m pour mesurer la
pente de la ligne d'eau;
limnimètre m incliné
nl verhanglijn f(m)
r уклонная или наклонная
водомерная рейка f

S739 slope gradient see **slope inclination**

S740 e slope grading
d Böschungsplanierung f
f aplanissement m des talus
nl taludmaken n
r планировка f откосов

S741 e slope inclination
d Böschungsneigung f,
Hangneigung f
f inclinaison f de talus
nl helling(shoek) f(m)
r крутизна f откоса [склона]

S742 slope leveling see **slope grading**

S743 e slope line
d Böschungsprofil n
f profil m de talus
nl hellingsprofiel n
r профиль m откоса

S744 e slope protection
d Böschungssicherung f,
Böschungsschutz m,
Böschungsbefestigung f
f protection f des talus [de la partie
inclinée]
nl taludverdediging f
r крепление n откосов

S745 sloper see **slope trimming machine**

S746 e slope revetment
d Böschungsverkleidung f
f revêtement m de talus
nl taludbekleding f
r одежда f откоса

S747 e slope roller
d Böschungswalze f
f rouleau m [cylindre m] pour talus
nl wals f(m) voor taludafwerking
r каток m для откосов

S748 e slope scraper
d Böschungsschrapper m
f scraper m [décapeuse f] pour talus
nl taludscraper m
r скрепер m для откосов

S749 e slope stability
d Böschungsstandfestigkeit f
f stabilité f de talus
nl stabiliteit f van het talud
r устойчивость f откоса

S750 e slope trimming
d Böschungsprofilierung f,
Abböschen n, Böschen n
f talutage m, mise f en talus,
dressage m de talus
nl taludmaken n
r профилирование n откосов

S751 e slope trimming machine
d Böschungsprofiliermaschine f,
Böschungsplaniermaschine f
f taluteuse f
nl taludafwerkingsmachine f, dozer m
voor het effenen van hellingen
r планировщик m откосов

S752 e sloping bond
d Schrägverband m (bei geneigten
Blocklagen)
f appareillage m de maçonnerie en
gros blocs inclinés
nl diagonaalverband n (met schuine
blokkenlagen)
r перевязка f массивов при их
укладке наклонными рядами
(во внешнем оградительном
сооружении)

S753 e sloping section
d Schrägschnitt m
f section f oblique
nl schuine doorsnede f
r наклонное [косое] сечение n,
сечение n под углом

S754 e sloping top chord
d geneigter Obergurt m, Schrägober-
gurt m (Träger)
f membrure f supérieure inclinée
nl hellende bovenrand m (ligger)
r наклонный верхний пояс m
(балки)

S755 e slot
d Nut f; Schlitz m; Spalte f
f rainure f; mortaise f; caniveau m
nl sponning f(m); gleuf f
r паз m; вырез m; щель f; прорезь f

**S756 e slot air terminal device, slot
diffuser**
d schlitzförmiger Luftverteiler m,
Schlitzgitter n, Schlitzdiffusor m;
Schlitzöffnung f
f diffuseur m d'air à fentes; sortie f
fendue
nl spleetdiffusor m
r щелевой воздухораспределитель m;
щелевидное приточное отверстие n

S757 e slotted breakwater
d durchbrochener Wellenbrecher m
f brise-lames m à claire-voie
nl pier f(m) met doorstromingsopenin-
gen
r волнолом m сквозной конструкции

S758 slotted outlet, slot-type air diffuser
see **slot air terminal device**

S759 slow-hardening cement see **slow-
-setting cement**

S760 e slow sand filter
d Langsamsandfilter n, m
f filtre m lent au sable
nl langzaam zandfilter n, m
r медленный песчаный фильтр m

S761 e slow-setting cement
d langsam abbindender Zement m

f ciment *m* à prise lente
nl langzaam verhardend cement *n*, *m*
r медленносхватывающийся цемент *m*

S762 *e* sludge
   d Schlamm *m*
   f boue *f*
   nl slib *n*, slijk *n*
   r шлам *m*, ил *m*, осадок *m*

S763 *e* sludge age
   d Schlammalter *n*
   f âge *m* de la boue activée
   nl slijkleeftijd *m*
   r возраст *m* активного ила

S764 sludge collector *see* sludge scraper

S765 *e* sludge dewatering
   d Schlamm(aus)trocknung *f*,
     Schlammentwässerung *f*
   f dessiccation *f* [séchage *m*] des boues
   nl slijkdroging *f*
   r обезвоживание *n* осадка сточных вод

S766 sludge digester *see* septic tank

S767 *e* sludge digestion
   d Schlamm(aus)faulung *f*
   f digestion *f* [fermentation *f*] des boues
   nl slijkgisting *f*
   r сбраживание *n* осадка сточных вод

S768 *e* sludge digestion plant
   d Ausfaulanlage *f*
   f installation *f* de digestion des boues
   nl slijkgistingsinstallatie *f*
   r септик *m*, метантенк *m*

S769 sludge draw-off pump *see* sludge pump

S770 sludge drying *see* sludge dewatering

S771 *e* sludge drying bed
   d Faulschlamm-Trockenbeet *n*,
     Trocknungsbeet *n*
   f lit *m* de séchage
   nl slijkdroogbed *n*
   r иловая площадка *f*

S772 *e* sludge gas
   d Biogas *n*, Faulgas *n*, Klärgas *n*
   f gaz *m* des boues
   nl biogas *n*, rottingsgas *n*
   r биологический газ *m*, биогаз *m*

S773 *e* sludge index
   d Schlammindex *m*
   f indice *m* de la boue
   nl slijkindex *m*
   r иловый индекс *m*

S774 *e* sludge lagoon
   d Schlammteich *m*, Schlammsammler *m*, Schlammfänger *m*
   f lagune *f* des boues
   nl sterfput *m*
   r иловый пруд *m*, илонакопитель *m*, илоуловитель *m*

S775 *e* sludge liquor
   d Schlammwasser *n*

f eau *f* des boues
nl modderwater *n*
r иловая вода *f*

S776 *e* sludge pipe
   d Schlammrohr *n*
   f conduit *m* des boues
   nl slijkbuis *f(m)*
   r иловая труба *f*; илопровод *m*

S777 *e* sludge pump
   d Schlammpumpe *f*
   f pompe *f* à boue
   nl slijkpomp *f(m)*
   r иловый [шламовый] насос *m*; пульпонасос *m*

S778 *e* sludge return
   d Schlammrückführung *f*
   f retour *m* des boues
   nl slijkterugvoer *m*
   r рециркуляция *f* активного ила

S779 *e* sludge scraper
   d Schlammkratzer *m*,
     Schlamm(aus)räumer *m*
   f racleur *m* (tournant), racloirs *m pl* montés sur une chaîne (*pour ramasser les boues*)
   nl slijkkrabber *m*
   r илоскрёб *m*

S780 sludge tank *see* sludge lagoon

S781 *e* sludge thickener
   d Schlammverdichter *m*,
     Schlammeindicker *m*
   f compacteur *m* des boues
   nl slijkverdichter *m*
   r илоуплотнитель *m*; сгуститель *m* шлама

S782 *e* sludge valve
   d Schlamm(abfluß)ventil *n*
   f soupape *f* de boues
   nl spuiklep *f(m)*
   r спускной клапан *m* (*резервуара*)

S783 *e* sluice
   d 1. Halbschleuse *f*, Dockschleuse *f*; Siel *n*, Deichschleuse *f*, Flutschleuse *f*, Ablaufschleuse *f*; Spülschleuse *f* 2. Einlaßschleuse *f* 3. Spülöffnung *f*; Überfallöffnung *f*, Wehröffnung *f* 4. Verschluß *m*; Absperrglied *n*
   f 1. écluse *f* simple *m*; écluse *f* de décharge; écluse *f* de chasse [de fuite] 2. prise *f* d'eau à vanne glissante, écluse *f* d'écoulement 3. orifice *f* de chasse; orifice *f* de purge 4. organe *f* de fermeture
   nl 1. keersluis *f(m)*, uitwateringssluis *f(m)* 2. inlaatsluis *f(m)* 3. spuikanaal *n* 4. schut *n*, sluisdeur *f(m)*
   r 1. полушлюз *m*; водосбросный шлюз *m*; промывной шлюз *m* 2. шлюз-регулятор *m* (*орошение*) 3. промывное *или* водосливное отверстие *n* 4. затвор *m*; запорный орган *m*

S784 *e* sluice chamber
d Schleusenkammer *f*
f sas *m* d'écluse
nl schutkolk *m*, *f*; sluiskolk *m*
r камера *f* шлюза

S785 *e* sluice gate
d 1. Spülschütze *f*, Spülverschluß *m* 2.
Tiefenverschluß *m* 3. Schleusentor *n*
f 1. vanne *f* de chasse 2. vanne *f* de
fond 3. porte *f* d'écluse
nl 1. spuischuif *f(m)* 2. rinket *n*
3. sluisdeur *f(m)*
r 1. промывной затвор *m* 2.
глубинный затвор *m* 3. шлюзные
ворота *pl*

S786 *e* sluice-gate chamber
d Torkammer *f*, Tornische *f*
f chambre *f* des portes
nl deurkas *f(m)* van de sluisdeur
r шкафная ниша *f* [часть *f*]
шлюзных ворот

S787 *e* sluice valve
d 1. Schieberschütz(e) *n* (*f*),
Abzugschieber *m*
2. Überlaufventil *n*
3. Abortdruckspüler *m* 4. Schieber *m*
f 1. vanne *f* (*d'écluse*) 2. vanne *f* de
déviation 3. vanne *f* de chasse
4. vanne *f*, robinet-vanne *m*, valve *f*
nl 1. schuif *f(m)*, rinket *n*
2. overloopventiel *n*
3. drukspoelkraan *f(m)* (*watercloset*)
4. klep *f(m)*
r 1. шлюзный затвор *m* 2.
перепускной клапан *m* 3. смывной
клапан *m* 4. задвижка *f*; шибер *m*
(*номинальным диаметром от
50 мм*)

S788 *e* sluiceway
d Grundablaß *m*, Ablaßöffnung *f*
f évacuateur *m* de fond
nl spuiopening *f*; sluizenkanaal *n*
r глубинный [донный] водоспуск *m*

S789 sluiceway channel *see* flushing canal

S790 *e* sluice weir
d Schützenwehr *n*, Tafelwehr *n*
f barrage *m* à hausses
nl klepstuw *m*
r щитовая плотина *f*

S791 *e* slump
d 1. Einsackung *f*, Absackung *f*,
Senkung *f* 2. Ausbreitmaß *n*,
Konussenkung *f* (*Beton*)
f 1. affaissement *m*, tassement *m*
(*des terres*) 2. affaissement *m* au
[de] cône
nl 1. inzinking *f* 2. zetting *f*,
zetmaat *f(m)*
r 1. осадка *f*; просадка *f* 2. осадка *f*
конуса (*бетон*)

S792 *e* slump cone
d Setzbecher *m*

f cône *m* d'affaissement [d'Abrams]
(*pour l'essai d'affaissement du
béton*)
nl standaardkegel *m*
r стандартный конус *m* (*для
определения подвижности
бетонной смеси*)

S793 *e* slump meter
d Betonkonsistometer *n*
f consistomètre *m* à béton, appareil
m de mesure de la consistance des
bétons
nl consistentiemeter *m*
r измеритель *m* подвижности
[жёсткости] бетонной смеси

S794 *e* slump test
d Setzversuch *m*,
Ausbreit(ungs)versuch *m*
f essai *m* d'affaissement du béton
nl zakproef *f(m)* (*betonmortel*)
r определение *n* подвижности
бетонной смеси (*по осадке конуса*)

S795 *e* slurry
d Wasser-Erdstoff-Gemisch *n*,
Fördergemisch *n*, Pulpe *f*;
Schlämme *f*; Zementbrei *m*;
Suspension *f*
f pulpe *f*; pâte *f* de ciment;
suspension *f*
nl brij *f* van water en grondspecie;
dikspoeling *f*
r гидросмесь *f*, пульпа *f*; жидкое
цементное тесто *n*; шлам *m*;
суспензия *f*

S796 *e* slurry pipeline
d Schlammleitung *f*
f tuyau *m* d'extraction des boues
nl persleiding *f* voor zand of
baggerspecie
r шламопровод *m*, пульпопровод *m*

S797 slurry pump *see* sludge pump

S798 *e* slurry seal
d Bitumen-Abdichtungsmastix *m*
f coulis *m* bitumineux, slurry seal *m*
nl bitumenmastiek *m*
r эмульгированная гидроизолирующая
битумная мастика *f* (*для
дорожных покрытий*)

S799 *e* slurry sump
d Schlammabsetzbecken *n*
f puisard *m* de décantation
nl bezinkput *m*
r шламовый отстойник *m*

S800 slurry thickener *see* sludge thickener

S801 *e* slurry wall
d Schlitzwand *f*
f paroi *f* en béton moulée dans le sol
nl perskade *f(m)*
r стена *f* в грунте

S802 *e* slush ice
d Matscheis *n*, Breieis *n*, Eisschweb
m, Eisbrei *m*, Sorbett *n*

f bouillie f de glace, shuga f,
  sorbet m
nl fijn drijfijs n
r снежно-ледяная каша f, шуга f

S803 e slush ice chute
  d Sorbettablaß m
  f déversoir m pour le sorbet, passe f
    [pertuits m] de sorbet
  nl aflaat m voor drijfijs
  r шугосброс m

S804 e slush ice drift
  d Eisbreitreiben n, Eisbreigang m
  f passage m de sorbet
  nl ijsgang m
  r шугоход m

S805 slush pump see solids-handling
  pump

S806 e small bore heating system
  d Kleinrohrheizung f
  f installation f de chauffage à tubes
    en cuivre de petit diamètre
  nl verwarmingsinstallatie met dunne
    leidingen
  r система f водяного отопления
    с (медными) трубами малого
    диаметра

S807 e small tools
  d Kleinwerkzeuge n pl
  f petit outillage m
  nl handgereedschap n
  r мелкий ручной инструмент m

S808 e smoke box
  d Rauchkasten m, Rauchkammer f,
    Rauchbüchse f
  f boîte f à fumée
  nl rookkast f(m)
  r дымовая коробка f, камера f
    дымовых газов

S809 e smoke control system
  d Rauchschutzanlage f,
    Entrauchungsanlage f
  f système m d'échappement [de
    tirage] des fumées
  nl rookafzuigsysteem n
  r дымозащитная или дымовытяжная
    система f

S810 e smoke damper
  d Rauchklappe f
  f registre m de tirage des fumées
  nl rookklep f(m)
  r дымов(ыпускн)ой клапан m

S811 e smoke density
  d optische Rauchdichte f
  f densité f optique de fumée
  nl optische rookdichtheid f
  r оптическая плотность f дыма

S812 e smoke emission
  d Rauch(gas)emission f
  f émission f [dégagement m] des
    fumées
  nl rookuitstoot m
  r дымовыделения n pl

S813 e smoke protection damper
  d Rauchschutzklappe f
  f registre m anti-fumée, coupe-fumée
    f, pare-fumée m
  nl rookbeschermingsschuif f(m)
  r дымозадерживающий клапан m

S814 e smoke test
  d Rauchprobe f
  f épreuve f aux fumées
  nl rookproef f(m)
  r (за)дымление n (метод
    визуализации воздушных потоков
    и контроля эффективности
    систем воздухораспределения)

S815 e smoke tube
  d Rauchrohr n
  f tube m à fumée
  nl rookpijp f(m)
  r дымогарная труба f

S816 e smooth blasting
  d Sprengen n des Sollprofils vom
    Untertagebau
  f sautage m de réglage
  nl egaliseren n met springstof
  r метод m взрывной проходки
    подземных выработок с точным
    соблюдением проектного профиля

S817 e smoothing
  d Abglätten n, Glätten n
  f lissage m, égalisation f
  nl gladmaken n, egaliseren n
  r сглаживание n; заглаживание n;
    выравнивание n

S818 e smoothing beam finisher
  d Glättbohlenfertiger m
  f finisseur m [finisseuse f] à barre
    lisseuse,
  nl verdichtingsmachine f met afreibalk
  r финишер m с выглаживающим
    брусом

S820 e snap lock
  d 1. Schnappverbindung f,
    Schnappfalz m, Nockenfalz m
    2. Schnappschloß n
  f 1. assemblage m par agrafage
    [agrafures] 2. serrure f à ressort
  nl 1. snapverbinding f, snapsluiting f
    2. snapslot n, valslot n
  r 1. фальцевое соединение n на
    защёлках 2. замок m с засовом-
    -защёлкой

S821 e snap tie
  d Rödelung f, Würg(e)verbindung f
    (für Befestigung der Schaltafeln)
  f torsade f [tendeur m] en fils
    d'acier (pour la fixation des
    panneaux de coffrage)
  nl snelverbinding f voor de
    bevestiging van bekistingselementen
  r стяжка f [скрутка f] для
    крепления опалубочных щитов

S822 e snow fence
  d Schnee(schutz)zaun m

f pare-neige *m*, pare-à-neige *m*
nl sneeuwscherm *n*
r снегозащитное ограждение *n* (*на дорогах*)

S823 *e* snow load
d Schneelast *f*
f charge *f* de neige [due à la neige]
nl sneeuwbelasting *f*
r снеговая нагрузка *f*

S824 *e* snubber
d Pulsationsdämpfer *m*, Schwingungsdämpfer *m*, Vibrationsdämpfer *m*
f amortisseur *m* (d'oscillations)
nl schokdemper *m*, trillingdemper *m*
r амортизатор *m*, демпфер *m*, виброизолятор *m*

S825 *e* soakaway
d Sickerschacht *m*, Sickergrube *f*
f puits *m* absorbant [de drainage], fosse *f* de drainage
nl zinkput *m*, zakput *m*
r поглощающий *m* [дренажный, инфильтрационный] колодец *m*

S826 *e* soap test
d Seifenblasentest *m*
f contrôle *m* par bulbes de savon
nl zeepbellentest *m* (*bepaling van waterhardheid; zoeken van lekken in gasleiding*)
r испытание *n* соединений труб на герметичность методом обмыливания; определение *n* жёсткости воды с использованием стандартного раствора мыла

S827 *e* socket
d 1. Glockenmuffe *f* 2. Muffe *f*; Hülse *f*, Nabe *f* 3. Steckdose *f*, Sockel *m*
f 1. emboîtement *m*, emboîture *f* 2. manchon *m*, douille *f* 3. prise *f* de courant
nl 1. buismof *f(m)* 2. sok *f*; huls *f(m)* 3. contactdoos *f(n)*
r 1. раструб *m* 2. трубопроводная муфта *f*; втулка *f* 3. штепсельная розетка *f*; штепсельное гнездо *n*; штепсельный разъём *m*

S828 socket-and-spigot joint *see* spigot--and-socket connection

S829 *e* socket joint
d Muffenverbindung *f*
f joint *m* à manchon
nl mofverbinding *f*
r муфтовое соединение *n*

S830 *e* socket pipe
d Muffenrohr *n*
f tuyau *m* emboîté
nl mofpijp *f(m)*
r раструбная труба *f*

S831 *e* soffit
d untere Leibungsfläche *f*, innere Bogenfläche *f*, Bogenleibung *f*, Untersicht *f*

f soffite *f*, plafond *m*, intrados *m*
nl binnenwelving *f*
r нижняя поверхность *f* (*свода, потолка, балки*), софит *m*

S832 *e* softboard
d leichte wärmeisolierende Faserplatte *f*
f panneau *m* calorifuge léger (*en fibre de bois ou en fibre minéral*)
nl zachtboard *n*
r лёгкая теплоизоляционная плита *f* (*из древесного или минерального волокна*)

S833 *e* soft soils
d weiche Böden *m pl*
f sols *m pl* mous
nl zachte grondsoort *f(m)*
r мягкие [слабые] грунты *m pl*

S834 *e* soft water
d Weichwasser *n*
f eau *f* douce
nl zacht water *n*
r умягчённая [мягкая] вода *f*

S835 *e* softwood
d Weichholz *n*, Nadelholz *n*
f bois *m* tendre
nl naaldhout *n*
r мягкая древесина *f*, древесина *f* хвойных пород

S836 *e* soil
d 1. Boden *m*, Grund *m* 2. Baugrund *m*, Gründungsboden *m* 3. Hausabwasser *n*, häusliche Abwässer *n pl*
f 1. sol *m*, terre *f*, terrain *m* 2. sol *m* de fondation 3. effluants *m pl*, eaux *f pl* ménagères
nl 1. bodem *m*; grond *m* 2. bouwgrond *m* 3. huishoudelijk afvalwater *n*
r 1. почва *f*; грунт *m* 2. грунтовое основание *n*; грунт *m* основания 3. сточная жидкость *f*; бытовые сточные воды *f pl*

S837 *e* soil analysis
d Bodenanalyse *f*
f analyse *f* du sol
nl bodemanalyse *f*
r лабораторный анализ *m* [исследование *n*] грунта

S838 *e* soil appliance
d Sanitärgerät *n* (*Fäkalabwasser--Abflußstelle*)
f récepteur *m* des eaux usées (*appareil sanitaire*)
nl closetpot *m*
r приёмник *m* фекальных стоков (*санитарный прибор*)

S839 *e* soil auger
d Bohrschnecke *f*, Spiralbohrer *m* zur Bodenuntersuchung
f tarière *f* de sondage [à cuiller]
nl grondboor *f(m)* (*voor monstername*)
r ложечный бур *m*, желонка *f* (*для отбора проб грунта*)

S840 *e* soil bulging
d Bodenhebung *f*
f soulèvement *m*, gonflement *m*
(*de sol*)
nl bodemverheffing *f*
r выпор *m* [выпучивание *n*] грунта

S841 *e* soil-cement
d Erdbeton *m*, Bodenbeton *m*
f sol-ciment *m*
nl gestabiliseerd zand *n*
r грунтобетон *m*, грунтоцементная смесь *f*

S842 *e* soil cementation
d Bodenzementierung *f*, Bodenstabilisierung *f*
f cimentation *f* [cimentage *m*] du sol
nl grondstabilisatie *f*
r цементация *f* грунта

S843 *e* soil-cement base
d zementverfestigte Unterbausohle *f*
f fondation *f* [couche *f* de fondation] en sol-ciment ; sous-couche *f* de sol-ciment
nl fundering *f* van gestabiliseerd zand
r цементированное грунтовое основание *n* (*дорожной одежды*)

S844 *e* soil classification
d Bodenklassifikation *f*, Bodenklassifizierung *f*
f classification *f* de sols
nl bodemclassificatie *f*
r классификация *f* грунтов

S845 *e* soil compacting machine
d Bodenverdichter *m*
f engin *m* de compactage [de serrage] de sol, compacteur *m* de sol
nl grondverdichtingsmachine *f*
r грунтоуплотняющая машина *f*

S846 *e* soil compaction
d (künstliche) Bodenverdichtung *f*
f compactage *m* de sol
nl grondverdichting *f*
r уплотнение *n* грунта

S847 *e* soil conditions
d Bodenverhältnisse *n pl*, Bodenzustand *m*
f conditions *f pl* de sol [de terrain]
nl bodemgesteldheid *f*
r грунтовые условия *n pl*

S848 *e* soil consolidation
d Bodenverfestigung *f*
f consolidation *f* (*d'un sol*)
nl grondverdichting *f*
r консолидация *f* [уплотнение *n*] грунта

S849 *e* soil filling
d Bodenschüttung *f*
f remblayage *m*, remblaiment *m*
nl ophoging *f*
r отсыпка *f* грунта

S850 soil grouting *see* soil cementation

S851 *e* soil investigation
d Bodenuntersuchung *f*
f étude *f* du sol
nl bodemonderzoek *n*
r исследование *n* грунта [грунтов]

S852 *e* soil mechanics
d Bodenmechanik *f*, Baugrundmechanik *f*
f mécanique *f* des sols [des terrains]
nl grondmechanica *f*
r механика *f* грунтов

S853 *e* soil moisture
d Bodenfeuchtigkeit *f*
f humidité *f* de sol
nl bodemvochtigheid *f*
r почвенная влага *f*

S854 *e* soil pipe
d Abwasserfallrohr *n*, Abortwasserrohr *n*
f tuyau *m* d'égout [de descente]
nl standleiding *f*
r канализационный стояк *m*, канализационная труба *f*

S855 *e* soil pressure
d 1. Erddruck *m*, Bodendruck *m* 2. Bodenpressung *f*, Untergrundpressung *f*
f 1. pression *f* du sol [du terrain] 2. pression *f* sur sol
nl gronddruk *m*
r давление *n* грунта *или* на грунт

S856 *e* soil profile
d 1. Bodenprofil *n* 2. geologisches Profil *n*
f 1. profil *m* du sol [du terrain] 2. profil *m* [coupe *f*] géologique
nl bodemprofiel *n*
r 1. профиль *m* грунтового основания 2. геологический разрез *m*

S857 *e* soil resistance
d Bodenwiderstand *m*
f résistance *f* du sol
nl grondweerstand *m*
r сопротивление *n* грунта

S858 *e* soil skeleton
d Bodenskelett *n*, Bodengerüst *n*
f squelette *m* de sol
nl grondstructuur *f*, bodemstructuur *f*
r скелет *m* грунта

S859 *e* soil stabilization
d Bodenverfestigung *f*, Baugrundverfestigung *f*, Bodenstabilisierung *f*, Bodenvermörtelung *f*
f stabilisation *f* du sol
nl grondstabilisatie *f*
r укрепление *n* [закрепление *n*, стабилизация *f*] грунта

S860 *e* soil stabilizing machine
d Bodenvermörtelungsmaschine *f*
f stabilisateur *m* de sols
nl stabilisatiemachine *f*

r машина *f* для стабилизации
[укрепления] грунтов

S861 *e* **soil stack installation**
d Abwasserinstallation *f* (*eines Gebäudes*)
f réseau *m* d'évacuation à l'intérieur du bâtiment, évacuation *f* intérieure
nl binnenriolering *f*
r канализация *f* здания, внутренняя канализация *f*

S862 *e* **soil support value**
d Bodentragwert *m*
f capacité *f* portante du sol, pouvoir *m* porteur du sol
nl draagvermogen *n* van de grond
r несущая способность *f* грунта

S863 *e* **soil survey**
d geologische Suche *f* [Prospektion *f*], geologisches Aufsuchen *n*
f recherche *f* [exploration *f*, prospection *f*] géologique
nl bodemkartering *f*
r инженерно-геологические изыскания *n pl*

S864 *e* **soil system**
d Schmutzwasserleitungssystem *n*
f conduites *m pl* d'eaux usées
nl binnenriolering *f*
r система *f* хозяйственно-фекальной канализации

S865 *e* **soil water**
d Bodenwasser *n*
f eaux *f pl* de sous-sol
nl grondwater *n*
r почвенные воды *f pl*

S866 *e* **sol-air temperature**
d solare Lufttemperatur *f*, Sonnenlufttemperatur *f*
f température *f* conventionnelle du local compte tenu l'effet thermique du soleil
nl luchttemperatuur *f* in de zon
r условная температура *f* помещения с учётом солнечных теплопоступлений

S867 *e* **solar absorption coefficient**
d Absorptionskoeffizient *m* der Sonnenstrahlung
f coefficient *m* d'absorption de la radiation solaire
nl absorptiecoëfficiënt *m* van zonnestraling
r коэффициент *m* поглощения солнечной радиации

S868 *e* **solar building**
d Sonnenhaus *n*
f maison *f* [immeuble *m*] à chauffage solaire [par soleil]
nl gebouw *n* met verwarming door zonne-energie
r здание *n* с системой солнечного отопления

S869 *e* **solar cell roofing material**
d Dachmaterial *n* mit eingebauten Sonnenbatterien
f matériau *m* de couverture avec piles solaires encastrées
nl dakmateriaal *n* met ingebouwde zonnecellen
r кровельный материал *m* со встроенными солнечными термоэлементами [батареями]

S870 *e* **solar collector**
d Sonnen(wärme)kollektor *m*, Solarkollektor *m*
f collecteur *m* solaire
nl zonnecollector *m*
r коллектор *m* солнечной энергии, солнечный коллектор *m*

S871 *e* **solar collector roof panels**
d Dachtafeln *f pl* mit eingebauten Sonnenwärmekollektoren
f dalles *f pl* [panneaux *m pl*] de toiture avec collecteurs solaires
nl dakzonnepanelen *n pl*
r кровельные панели *f pl* с солнечными коллекторами

S872 *e* **solar-electric roof**
d Dach *n* mit eingebauten Sonnenbatterien
f comble *m* [toit *m*] avec piles solaires incorporées [encastrées]
nl dak *n* met ingebouwde zonnecellen
r крыша *f* со встроенными термоэлементами

S873 *e* **solar exposure**
d Ausgesetztsein *n* der Sonnenbestrahlung, Insolationslage *f*
f ensoleillement *n*, insolation *f*, exposion *f* au soleil (*des bâtiments*)
nl bezonning *f* (*van een gebouw*)
r инсоляционная экспозиция *f* (*здания, помещения*)

S874 *e* **solar heat gain**
d Sonnen(strahlungs)wärmeeinfall *m*
f gain *m* de chaleur solaire
nl solaire warmtetoevoer *m*
r солнечные теплопоступления *n pl*

S875 *e* **solar heating system**
d Solarheizanlage *f*
f installation *f* de chauffage solaire
nl zonneverwarmingssysteem *n*
r система *f* солнечного отопления

S876 *e* **solar plant**
d Solaranlage *f*
f installation *f* [centrale *f*] solaire
nl verwarmingsinstallatie *f* op zonne--energie
r гелиотермическая установка *f*

S877 *e* **solar protection devices**
d Sonnenschutzvorrichtungen *f pl*
f dispositif *m* de protection contre le rayonnement solaire
nl zonweringssystemen *n pl*

r солнцезащитные устройства *n pl*

S878 *e* solar reflective curing membrane
d Sonnenschutz-Nachbehandlungsfilm *m*
f membrane *f* étanche réfléchissante pour la cure de béton
nl zonlichtreflecterende folie *f* voor de nabehandeling van beton
r солнцеотражающая плёнка *f* для защиты бетона

S879 *e* solar water heater
d Solar-Warmwasserbereitungsanlage *f*
f chauffe-eau *m* solaire
nl zonneboiler *m*
r солнечный водонагреватель *m*

S880 *e* soldier
d 1. Absteifung *f*, Stütze *f*
2. hochkant gestellter [aufrechtstehender] Ziegel *m*
3. «Mönch» *m* (*Wasserauslaß*)
f 1. étai *m* (vertical) de blindage
2. brique *f* de champ 3. «moine» *m* (*puits de décharge*)
nl 1. stut *m*, stijl *m* 2. rechtop gestelde baksteen *n*, rollaag *f(m)* 3. wateruitlaat *m*
r 1. стойка *f* крепления траншеи 2. кирпич *m*, уложенный вертикально 3. «монах» *m* (*водосбросный колодец*)

S881 *e* solepiece
d Sohlbalken *m*, Grundschwelle *f*; Unterlage *f*
f sablière *f*; semelle *f*
nl onderregel *m*
r лежень *m*; упорная подкладка *f* (*подкоса, стойки*)

S882 *e* soleplate
d Sohlbalken *m*; Grundsohle *f*; Sohlschwelle *f*; Sohlplatte *f*
f semelle *f*; sablière *f*; plaque *f* d'appui
nl grondplaat *f(m)*, sloof *f(m)*
r лежень *m*; нижний опорный брус *m*; опорная подкладка *f*

S883 *e* solid bitumen
d Hartbitumen *n*
f bitume *m* dur
nl hard bitumen *n*
r твёрдый битум *m*

S884 *e* solid floor
d Betondecke *f*, Massivdecke *f*; Beton(fuß)bodenbelag *m*
f plancher *m* en béton (armé), plafond *m* en dalle pleine; couvre--sol *m* en béton
nl vloer *m* van houtblokjes op betonnen ondervloer
r монолитное (бетонное) перекрытие *n*; бетонный пол *m*

S885 *e* solid foundation
d Massivgründung *f*, massives Fundament *n*
f fondation *f* massif
nl betonfundering *f*
r массивный фундамент *m*

S886 *e* solid fuel heating system
d Festbrennstoffheizung *f*
f installation *f* de chauffage à combustible solide
nl verwarmingsinstallatie *f* voor vaste brandstof
r система *f* отопления, работающая на твёрдом топливе

S887 *e* solid panel floor
d Stahlbetonvollplattendecke *f*; Fußboden *m* aus Massivbetonplatten
f plancher *m* en dalles pleines préfabriquées (en béton armé); couvre-sol *m* en panneaux préfabriqués (en béton)
nl gewapend-betonplaten *f(m) pl*
r перекрытие *n* или пол *m* из сплошных [беспустотных] сборных железобетонных плит

S888 *e* solid pile
d Voll(querschnitt)pfahl *m*, Massivpfahl *m*
f pieu *m* pleine [massif]
nl massieve betonpaal *f(m)*
r свая *f* сплошного сечения, массивная свая *f*

S889 *e* solid retaining wall
d massive Stützmauer *f*, Schwergewichtsstützmauer *f*
f mur *m* de soutènement massif, mur *m* (de soutènement) poids
nl massieve steunmuur *m*
r массивная [гравитационная] подпорная стенка *f*

S890 *e* solids-handling pump
d Schlammpumpe *f*, Dickstoffpumpe *f*
f pompe *f* dragueuse [à déblais]
nl zandpomp *f(m)*
r грунтовый насос *m*, землесос *m*

S891 *e* solid slab
d Vollplatte *f*, Massivplatte *f*
f dalle *f* pleine
nl massieve plaat *f(m)*
r плита *f* сплошного сечения, сплошная плита *f*

S892 *e* solid wall
d massive Wand *f*, Vollwand *f*
f mur *m* plein
nl massieve muur *m*
r массивная стена *f*

S893 *e* solid-web girder
d Vollwandträger *m*
f poutre *f* à âme pleine
nl vollewandligger *m*
r балка *f* со сплошной стенкой

S894 *e* solid wharf
d Anlegemauer *f*
f quai *m* d'accostage
nl aanlegkade *f(m)*
r причальная стенка *f*

S895 *e* solum
   d bebaute Fläche *f*
   f surface *f* bâtie
   nl bebouwd oppervlak *n*
   r площадь *f* застройки

S896 *e* solvent-based paint remover
   d flüssiges [lösendes] Abbeizmittel *n*
    [Beizmittel *n*]
   f décapant *m* pour peinture à base
    d'un solvant
   nl vloeibaar afbijtmiddel *n* (*ten grond-
    slag aan verdunningsmiddel*)
   r жидкий состав *m* для удаления
    старого лакокрасочного покрытия
    (*на основе растворителя*)

S897 *e* soot
   d Ruß *m*
   f noir *m* de fumée
   nl roet *n*
   r сажа *f*

S898 *e* soot blower
   d Rußblasevorrichtung *f*, Rußbläser *m*
   f souffleuse *f* (pour la purge par air
    des chaudières)
   nl roetblazer *m*
   r обдувочный аппарат *m* (для
    очистки котла от сажи)

S899 *e* Sorel cement
   d Sorelzement *m*
   f ciment *m* de Sorel
   nl Sorel-cement *n*
   r цемент *m* Сореля, магнезиальное
    вяжущее *n*

S900   sound absorber *see* silencer

S901 *e* sound-absorbing paint
   d Akustikfarbe *f*, Schallschluckfarbe *f*,
    Schallabsorptionsfarbe *f*
   f peinture *f* acoustique [absorbant le
    bruit], peinture *f* d'absorption
    phonique [d'insonorisation]
   nl geluidabsorberende verf *f(m)*
   r звукопоглощающая краска *f*

S902 *e* sound absorption
   d 1. Schallabsorption *f* 2.
    Schallabsorptionsvermögen *n*
   f 1. absorption *f* du son [phonique],
    amortissement *m* acoustique 2.
    capacité *f* d'insonorisation,
    pouvoir *m* d'absorption des sons
   nl 1. geluidabsorptie *f* 2.
    geluidabsorberend vermogen *n*
   r 1. звукопоглощение *n*; глушение *n*
    звука [шума] 2. звукопоглощающая
    способность *f* (*материала*)

S903 *e* sound absorption coefficient
   d Schallabsorptionsgrad *m*,
    Schallschluckgrad *m*
   f pouvoir *m* d'absorption du son,
    coefficient *m* d'absorption
    acoustique
   nl geluidabsorptiecoëfficiënt *m*
   r коэффициент *m* звукопоглощения

S904 *e* sound attenuation
   d Schalldämpfung *f*
   f atténuation *f* du son
   nl geluiddemping *f*
   r затухание *n* звука; поглощение *n*
    звука

S905 *e* sound barrier
   d Schallsperre *f*, Geräuschsperre *f*
   f barrière *f* contre le bruit
   nl geluidbarrière *f*
   r звукоизолирующий экран *m*,
    звукозащитный барьер *m*

S906 *e* sound boarding
   d Balkeneinschub *m*,
    Deckeneinschub *m*, Fehlboden *m*,
    Streifboden *m*
   f planchéiage *m* (en bois) brut,
    plancher *m* brut
   nl tussenvloer *m* waarop de
    geluidisolatie wordt aangebracht
   r накат *m* чёрного пола

S907 *e* sound control
   d Lärmbekämpfung *f*
   f lutte *f* contre le bruit
   nl lawaaibestrijding *f*
   r борьба *f* с шумом

S908 *e* sound-deadening coating
   d Antidröhnbelag *m*
   f revêtement *m* d'insonorisation
    [d'absorption phonique]
   nl geluidabsorberende deklaag *f(m)*
   r звукопоглощающее покрытие *n*

S909 *e* sounding
   d 1. Sondierung *f* 2. Peilung *f*
   f 1. sondage *m* 2. sondage *m*
    [mesurage *m*] de la profondeur
   nl 1. sondering *f* 2. peiling *f*
   r 1. зондирование *n* (*грунта*)
    2. промер *m* глубин

S910 *e* sound insulating panel
   d Schallschluckplatte *f*,
    Akustikplatte *f*
   f panneau *m* isolant acoustique
    [phonique]
   nl geluidisolerend paneel *n*
   r звукоизолирующая панель *f*

S911 *e* sound insulation
   d Schalldämmung *f*
   f isolation *f* acoustique [phonique,
    sonore], insonorisation *f*
   nl geluidisolatie *f*
   r звукоизоляция *f*

S912 *e* sound insulation materials
   d Schalldämmstoffe *m pl*
   f isolants *m pl* phoniques
    [acoustiques]
   nl geluidisolerende materialen *n pl*
   r звукоизоляционные материалы
    m pl

S913   sound isolation *US see* sound
    insulation

S914 *e* soundness
   d 1. Kontinuität *f* 2. Mängelfreiheit *f*,
     Lunkerfreiheit *f*
     3. Raumbeständigkeit *f* (*Zement*)
   f 1. solidité *f*, continuité *f*
     2. absence *f* de défaux intérieurs
     3. stabilité *f* [constance *f*] de volume
     (*lors de prise du ciment*)
  nl 1. onveranderlijkheid *f*
     2. afwezigheid *f* van gebreken
     3. volumebestendigheid *f* (*bij het*
     verharden van cement)
   r 1. сплошность *f* (*твёрдого тела*),
     неизменяемость 2. отсутствие *n*
     внутренних дефектов
     3. нерасширяемость при схватывании
     (*цемента*)

S915 *e* sound power
   d Schalleistung *f*
   f puissance *f* sonore
  nl geluidvermogen *n*
   r акустическая мощность *f*

S916 *e* sound pressure
   d Schalldruck *m*
   f pression *f* sonore [acoustique]
  nl geluiddruk *m*
   r звуковое давление *n*

S917 *e* soundproofing enclosure
   d Lärmschutzkapsel *f*, schalldämmende
     Kapselung *f*
   f clôture *f* insonorisante [amortissant
     le bruit]
  nl geluidwerende omsluiting *f*
   r звукоизолирующее ограждение *n*

S918 *e* soundproofing varnish
   d Antidröhnlack *m*
   f vernis *m* d'insonorisation
  nl geluidabsorberende lak *n*, *m*
   r звукопоглощающий лак *m*

S919 *e* sound reflection coefficient
   d Schallreflexionsgrad *m*
   f coefficient *m* de réflexion sonore
  nl geluidreflectiecoëfficiënt *m*
   r коэффициент *m* звукоотражения

S920 *e* sound spectrum
   d Schallspektrum *n*
   f spectre *m* sonore
  nl geluidspectrum *n*
   r звуковой спектр *m*

S921 *e* sound transmission coefficient
   d Schallübertragungskoeffizient *m*
   f coefficient *m* de transmission du son
  nl geluidoverdrachtcoëfficiënt *m*
   r коэффициент *m* звукопередачи *или*
     звукопроницаемости

S922 *e* sound trap
   d Rohrschalldämpfer *m*
   f amortisseur *m* de bruit tubulaire
     (*pour conduits d'air*)
  nl geluiddemper *m* in luchtkanaal
   r трубчатый шумоглушитель *m* (*для*
     воздуховодов)

S923 *e* space air
   d Raumluft *f*
   f air *m* de locaux
  nl binnenlucht *f* van een ruimte
   r воздух *m* помещения, внутренний
     воздух *m*

S924 *e* space air distribution
   d Luftführung *f* im Raum
   f distribution *f* d'air dans l'espace
     clos
  nl luchtverdeling *f* in een ruimte
   r воздухораспределение *n*
     в помещении; схема *f*
     вентилирования помещения

S925 *e* space enclosing structure
   d raumumschließendes Bauwerk *n*
   f enveloppe *f* extérieure du bâtiment
  nl omsluitende constructie *f*
   r ограждающие конструкции *f pl*;
     наружная оболочка *f* сооружения

S926 *e* space flow
   d Luftführung *f* im Raum;
     Raumströmungen *f pl*
   f distribution *f* d'air dans un local;
     courants *m pl* d'air dans l'espace
     clos
  nl luchtstromen *m pl* in een ruimte
   r воздухораспределение *n*
     в помещении; воздушные потоки
     m pl в помещении

S927 *e* space frame, space frame work
   d räumliches Stabwerk *n* [Fachwerk
     n], räumliche Gitterkonstruktion *f*;
     räumliches Rahmensystem *n*
   f ossature *f* (rigide)spatiale; ossature
     f spatiale à cadre
  nl ruimtevakwerk *n* (*constructie*)
   r пространственная стержневая
     конструкция *f*; пространственная
     рамная конструкция *f* [система *f*]

S928 *e* space heater
   d Raumheizer *m*, Raumheizgerät *n*
   f chaufferette *f* (d'espace), appareil
     m de chauffage individuel
  nl verwarmingstoestel *n*
   r местный отопительный агрегат *m*

S929 *e* space heating
   d Raumheizung *f*
   f chauffage *m* des locaux
  nl ruimteverwarming *f*
   r отопление *n* помещений

S930 *e* space microclimate
   d Mikroklima *n* der Räume,
     Raumklima *n*
   f microclimat *m* des locaux
  nl microklimaat *n*
   r микроклимат *m* помещений

S931 *e* spacer
   d Abstandhalter *m*
   f 1. entretoise *f* 2. espaceur *m* (de
     coffrage), écarteur *m*, séparateur *m*,
     distanceur *m* (*pour travaux de*
     bétonnage)

nl 1. afstandstuk *n* 2. afstandhouder *m*
(*wapening*)
r 1. распорка *f* 2. фиксатор *m*
(*арматуры*)

S932 *e* **space structures**
d Raumkonstruktionen *f pl*,
Raumtragwerke *n pl*
f structures *f pl* [ossatures *f pl*]
spatiales
nl overspanningsconstructies *f pl*,
ruimtevakwerken *n pl*
r пространственные сооружения *n pl*
[конструкции *f pl*]

S933 *e* **spacing of buildings**
d Gebäudeabstand *m*; Baulücke *f*
f prospect *m*
nl afstand *m* tussen gebouwen
r разрыв *m* между зданиями

S934 *e* **spacing of reinforcement**
d Stababstand *m*, Stabanordnung *f*
f espacement *m* [écartement *m*]
des ronds à béton
nl afstand *m* tussen wapeningsstaven
r шаг *m* арматурных стержней,
расстояние *n* между арматурными
стержнями

S935 *e* **spading**
d Stochern *n* (*Betongemisch*)
f piquage *m* (*du béton*) par la pelle
nl porren *n* (*betonmortel*)
r штыкование лопатой (*бетонной
смеси*)

S936 *e* **spall**
d Splitter *m*
f éclat *m*, débris, fragment *m*
nl scherf *f(m)*, splinter *m*
r осколок *m*, отколотый кусок *m*

S937 *e* **spalling**
d Abschälen *n*, Abplatzung *f* (*Beton*)
f écaillage *m*, écaillement *m*,
exfoliation *f*
nl afschilferen *n*, barsten *n* (*beton*)
r отслаивание *n*, откалывание *n*,
выкрашивание *n* (*бетона*)

S938 *e* **span**
d Spannweite *f*; Stützweite *f*;
Brückenfeld *n*
f travée *f* ; portée *f* ; entre-
-colonnement *m*
nl overspanning *f*; spanwijdte *f*
r пролёт *m*; шаг *m* колонн [опор],
расстояние *n* между опорами;
пролётное строение *n* моста

S939 *e* **spandrel**
d 1. Fensterbrüstung *f*,
Brüstungsmauer *f*
2. Bogenzwickel *m*, Zwickel *m*
f 1. allège *f* 2. pendentif *m*; rein *m*
de la voûte
nl 1. borstwering *f* van een venster
2. ruimte *f* tussen geboorte en
onderkant van een boog
r 1. подоконная стенка *f* 2. парус *m*
или пазуха *f* свода

S940 *e* **spandrel beam**
d Fachwerkriegel *m*; Fassadenträger *m*,
Randbalken *m*
f poutre *f* [traverse *f*] de pan ;
poutre *f* de façade (*entre deux
colonnes*)
nl regel *m* in vakwerkmuur, draagbalk
m
r ригель *m* фахверка; балка *f*
между наружными колоннами
(*поддерживающая стену или
перекрытие*), рандбалка *f*

S941 *e* **span pipe line**
d Rohrfreileitung *f*
f conduit *m* aérien [posé à ciel
ouvert]
nl bovengrondse pijpleiding *f*
r надземный трубопровод *m*

S942 *e* **span structure**
d Brückentragwerk *n*,
Brückenüberbau *m*, Brückenfeld *n*
f travée *f* de pont, superstructure *f*
du pont
nl overspanning *f*; brugconstructie
f
r пролётное строение *n* моста

S943 *e* **spare parts, spares**
d Ersatzteile *n pl*
f pièces *f pl* de rechange
nl reservedelen *n pl*
r запасные части *f pl*

S944 *e* **sparge pipe**
d Rieselrohr *n*; Durchblaserohr *n*
f tube *m* de pulvérisation; tube *m*
de purge [de soufflage]
nl 1. sproeipijp *f(m)* 2. afblaaspijp
f(m)
r оросительная *или* продувочная
труба *f*

S945 *e* **spark arrester**
d Funkenfänger *m*, Funkenlöscher *m*
f extincteur *m* d'étincelles
nl vonkenvanger *m*
r искрогаситель *m*

S946 *e* **spark proof fan, spark resistant fan**
d funkensicherer Ventilator *m*;
explosionsgeschützter Ventilator *m*,
Ex-Ventilator *m*
f ventilateur *m* antiétincelle ;
ventilateur *m* antidéflagrant
nl vonkvrije ventilator *m*;
explosieveilige ventilator *m*
r искрозащищённый вентилятор *m*;
вентилятор *m* во взрывобезопасном
исполнении

S947 *e* **special road**
d Sonderstraße *f*
f (auto)route *f* spéciale (*pour
certaines classes de véhicules*)
nl speciale autoweg *m*
r (автомобильная) дорога *f* для
транспортных средств определённых
классов

S948 *e* specifications
 d technische Bedingungen *f pl*;
 technische Beschreibung *f*,
 Bauleistungsverzeichnis *n*
 f prescriptions *f pl* techniques;
 cahier *m* des charges
 nl technische voorwaarden *n pl*;
 bestek *n*
 r технические условия *n pl*;
 техническое описание *n*
 (*строительного объекта*)

S949 *e* specifications and conditions of
 contract
 d Bauleistungsverzeichnis *n* und
 Vertragsbedingungen *f pl*
 f cahier *m* des charges et conditions
 f pl d'un contrat
 [d'un marché de travaux]
 nl beschrijving *f* van het werk en de
 voorwaarden *f pl*
 r условия *n pl* и особые положения
 n pl договора [контракта]

S950 *e* specifications writing
 d Ausarbeitung *f* des
 Bauleistungsverzeichnisses
 f élaboration *f* du cahier des charges
 nl uitwerking *f* van het bestek
 r разработка *f* [составление *n*]
 технических условий

S951 *e* specific deformation
 d spezifische Deformation *f*
 f déformation *f* spécifique
 nl specifieke vormverandering *f*
 r удельная деформация *f*

S952 *e* specific discharge
 d 1. Abflußspende *f* 2. spezifische
 Transportmenge *f*
 f 1. coefficient *m* d'écoulement 2.
 débit *m* spécifique d'entraînement
 nl 1. debiet *n*, afvoer *m* 2.
 soortelijke transporthoeveelheid *f*
 r 1. модуль *m* стока 2. удельный
 транспортирующий расход *m*

S954 *e* specific heat
 d spezifische Wärme *f*
 f chaleur *f* spécifique
 nl soortelijke warmte *f*
 r удельная теплоёмкость *f*

S955 *e* specific heat ratio
 d Adiabatenexponent *m*
 f exposant *m* adiabatique
 nl specifieke-warmtegetal *n*
 r показатель *m* адиабаты

S956 *e* specific humidity
 d spezifische Feuchtigkeit *f*
 f humidité *f* spécifique
 nl soortelijke vochtigheid *f*
 r удельная влажность *f*

S957 *e* specific solid discharge
 d Feststoffabtrag *m*
 f débit *m* solide spécifique, débit *m*
 spécifique de charriage [de vase]
 nl vaste-stoffentransport *n* per volume-
 -eenheid

 r модуль *m* твёрдого стока

S958 *e* specific speed
 d spezifische Geschwindigkeit *f*
 [Drehzahl *f*],
 Schnelläufigkeitsfaktor *m*
 f vitesse *f* spécifique
 nl toerental *n* van een bepaald type
 pomp *of* ventilator
 r быстроходность *f* (*лопаточной
 машины*)

S959 *e* specific surface
 d spezifische Oberfläche *f*
 f surface *f* spécifique
 nl korreloppervlak *n* per volume-
 -eenheid
 r удельная поверхность *f*

S960 *e* specific volume
 d spezifisches Volumen *n*
 f volume *f* spécifique
 nl soortelijk volume *n*
 r удельный объём *m*

S961 *e* specific yield
 d 1. spezifischer Grundwasserertrag *m*
 2. spezifische Ergiebigkeit *f*
 f 1. coefficient *m* de drainage
 2. débit *m* spécifique
 nl 1. specifieke grondwaterafvoer *m*
 2. specifieke opbrengst *f* van een
 bron
 r 1. коэффициент *m* водоотдачи;
 коэффициент *m* дренирования
 грунта 2. удельный дебит *m*
 скважины

S962 *e* specified method of erection
 d vorgeschriebenes Montageverfahren *n*
 f méthode *f* de montage spécifiée
 nl voorgeschreven montagewijze *f(m)*
 r предписанный метод *m* монтажа

S963 *e* specified requirements
 d normative Forderungen *f pl*
 f prescriptions *f pl* spécifiées
 nl gespecificeerde eisen *f pl*
 r требования *n pl* технических
 условий

S964 *e* specified works cube strength
 d normative Würfelfestigkeit *f*
 f résistance *f* sur cube spécifiée
 nl voorgeschreven [vereiste]
 kubussterkte *f* van beton op het
 werk
 r заданная [проектная] кубиковая
 прочность *f* бетона

S965 *e* speedway
 d Schnell(verkehrs)straße *f*, Autobahn *f*
 f autostrade *f*, autoroute *f*
 nl snelverkeersweg *m*, autosnelweg *m*
 r автодорога *f* (для) скоростного
 движения; автострада *f*

S966 *e* speedy drying
 d Schnelltrocknung *f*
 f séchage *m* accéléré
 nl snelle [versnelde] droging *f*
 r ускоренная сушка *f*

S967 *e* speedy erection system
  d Schnellmontageverfahren *n*
  f méthode *f* de montage (*des bâtiments*) accéléré
  nl snelmontagemethode *f*
  r метод *m* скоростного монтажа (*зданий*)

S968 *e* spherical gate
  d Kugelschieber *m*
  f vanne *f* sphérique
  nl kogelschuif *f(m)*
  r шаровой затвор *m*

S969 *e* spherical shell
  d Kugelschale *f*
  f coque-sphère *f*
  nl kogelschaal *f(m)*
  r сферическая оболочка *f*

S970 *e* spherical tank
  d Kugelbehälter *m*, Kugeltank *m*
  f réservoir *m* sphérique
  nl kogeltank *m*, kogelreservoir *n*
  r сферический [шаровой] резервуар *m*

S971 *e* spigot
  d glattes Ende *n*, Spitzende *n*, Einsteckende *n*
  f bout *m* mâle
  nl glad einde *n* van een buis
  r гладкий конец *m* (*трубы*)

S972 *e* spigot and socket connection, spigot-and-socket joint
  d Glockenmuffenverbindung *f*
  f joint *m* d'emboîtement [à cloche]
  nl klokvormige mofverbinding *f*
  r раструбное соединение *n*

S974 *e* spill, spillover
  d Überströmung *f*, Überfall *m*
  f déversement *m*
  nl overloop *m*, overlaat *m*
  r сброс *m* воды

S975 *e* spillway
  d Überlauf *m*, Überfall *m*, Hochwasserentlastungsanlage *f*, Entlastungsbauwerk *n*
  f déversoir *m*, évacuateur *m*, trop--plein *m*
  nl ontlastingskanaal *n*
  r водосброс *m*, водослив *m*

S976 *e* spillway apron
  d Sturzbett *n*, Sturzboden *m*, Schußboden *m*, Tosbecken *n*; Kolkschutz *m*, Rißberme *f*, Ablaufboden *m*
  f radier *m*, arrière-radier *m*
  nl overlaatbekleding *f*
  r водобой *m*; рисберма *f*

S977 *e* spillway bucket
  d Sprungnase *f*, Überfallsprung *m*
  f auge *f* de pied de barrage déversoir
  nl overloopsprong *m*
  r водосливный носок *m* плотины

S978 *e* spillway capacity
  d Leistungsfähigkeit *f*

  [Durchlaßkapazität *f*] der Entlastungsanlage
  f capacité *f* de décharge d'un évacuateur
  nl overlaatcapaciteit *f*
  r пропускная способность *f* водосброса

S979 *e* spillway crest
  d Überlaufkante *f*, Überlaufkrone *f*, Überlaufschwelle *f*, Überfallschwelle *f*
  f seuil *m* du déversoir [déversant]
  nl drempel *m* van het aflaatwerk
  r водосливный порог *m*, гребень *m* водослива

S980   spillway dam *see* overfall dam

S981 *e* spillway face *US*
  d Überfallrücken *m*
  f coursier *m*
  nl kruin *f(m)* van de overlaat
  r водосливная грань *f*, водоскат *m*

S982   spillway gallery *see* spillway tunnel

S983 *e* spillway gate
  d Kronenschütz *n*, Überfallschütz *n*, Kronenverschluß *m*, Überlaufverschluß *m*
  f vanne *f* de couronnement [de crête, de déversoir]
  nl aflaatklep *f*
  r поверхностный затвор *m* на гребне плотины

S984 *e* spillway opening
  d Überfallöffnung *f*, Durchflußöffnung *f*
  f orifice *m* de vidange
  nl aflaatopening *f*
  r водосливное отверстие *n*

S985 *e* spillway section
  d Überfallstrecke *f*, Überlaufstrecke *f*
  f partie *f* déversante d'un barrage
  nl overlaatgedeelte *n*
  r водосливная секция *f* плотины

S986 *e* spillway slab
  d Überfalldecke *f*
  f dalle *f* déversante
  nl overloopplaat *f(m)*
  r водосливная плита *f*

S987 *e* spillway structure
  d Überlaufbauwerk *n*, Entlastungsanlage *f*
  f déversoir *m*, ouvrage *m* de décharge [d'évacuation, de chasse]
  nl aflaatwerk *n*
  r водосбросное сооружение *n*

S988 *e* spillway tunnel
  d Stollenablaß *m*, Überlaufstollen *m*, Hochwasserentlastungsstollen *m*
  f galerie *f* d'évacuation, galerie *f* amont sous pression
  nl aflaattunnel *m*
  r туннельный водопуск *m*, водосбросный туннель *m*

S989 *e* spillway weir *see* overflow weir

S990 e spine wall
  d Mittellängswand f
  f mur m de refend porteur
    longitudinal
  nl dragende binnenmuur m
  r внутренняя продольная
    несущая стена f

S991 e spinning disk air washer
  d Scheibensprühanlage f
  f chambre f d'humidification à
    pulvérisateur d'eau centrifuge
    [rotatif]
  nl luchtbevochtiger m met
    schijfsproeier
  r увлажнительная камера f
    с центробежным водораспылителем

S992 e spinning disk humidifier
  d Schleuderscheibenbefeuchter m,
    Sprühscheibenzerstäuber m,
    Scheibensprühaggregat n
  f humidificateur m à disque (rotatif)
  nl roterende schijfbevochtiger m
  r вращающийся [центробежный]
    дисковый увлажнитель m

S993 e spiral chamber
  d Spiralkammer f
  f bâche f spirale
  nl spiraalkamer f(m) (turbine)
  r спиральная камера f

S994 e spiral classifier
  d Spiralklassierer m, Sandschnecke f,
    Schneckenwäsche f
  f classificateur m [classeur m]
    hydraulique spiral [hélicoïdal]
  nl wormclassificator m
  r спиральный гидроклассификатор m

S995 spiral duct see spiral-wound air
    duct

S996 e spirally reinforced column
  d spiralbewehrte Säule f
  f colonne f frettée, colonne f armée
    par frettes hélicoïdales
  nl zuil f(m) met spiraalwapening
  r колонна f со спиральной
    арматурой

S997 e spiral reinforcement
  d Spiralbewehrung f
  f armature f spirale [hélicoïdale],
    frettes f pl hélicoïdales
  nl spiraalwapening f
  r спиральная арматура f

S998 e spiral stair
  d Wendeltreppe f
  f escalier m à vis [en hélice, en
    colimaçon]
  nl spiltrap f
  r спиральная [винтовая] лестница f

S999 e spiral-wound air duct
  d Wendelfalzrohr n,
    Wickelfalzrohr n, Spiralfalzrohr n
  f conduit m d'air enroulé en hélice
  nl in een spiraal gevouwen
    luchtleiding f

  r спирально-навивной воздуховод m

S1000 e spirit level
  d Wasserwaage f
  f niveau m (à bulle d'air)
  nl buisniveau n
  r (строительный) уровень m

S1001 e spirit leveling
  d Nivellieren n
  f nivellement m
  nl inspelen n van het buisniveau
  r нивелирование n, нивелировка f

S1002 e splay
  d 1. Fahrbahnverbreiterung f 2.
    Sichtweite f (bei Straßenkreuzung)
    3. Abkantung f, Abschrägung f
  f 1. élargissement m des chaussées
    2. zone f de la libre visibilité
    (au croisement) 3. chanfrein m,
    coupe f en biseau, biseau m
  nl 1. verbreding f van de rijbaan 2.
    gedeelte n met onbeperkt uitzicht
    bij een wegkruising 3. schuine
    kant m
  r 1. уширение n проезжей части
    автомагистрали 2. зона f
    свободной видимости в месте
    пересечения дорог 3. скос m

S1003 e splay branch
  d Schrägabzweigung f, einfacher
    Schrägabzweig m
  f culotte f [embranchement m]
    oblique
  nl spruitstuk n
  r косой тройник m

S1004 e splayed scarf, splay scarf
  d schräger Überblattungsstoß m
  f entaille f à mi-bois oblique
  nl schuine liplas f(m)
  r косой замок m (тип врубки)

S1005 e splice
  d 1. Stoß m, Spleiß m, Überlappung
    f 2. Überlappungslänge f
  f 1. épissure f de câbles; ligature f
    (de fils) ; rallongement m ;
    assemblage m [joint m] à
    recouvrement 2. longueur f de
    recouvrement (p.ex. des ronds à
    béton)
  nl 1. rechte liplas f(m) 2. laslengte f
    van wapeningsstaven
  r 1. сросток m (канатов); место n
    сращивания (напр. проволок);
    наращивание n; соединение n
    внахлёстку 2. длина f нахлёстки
    (напр. арматурных стержней)

S1006 e splice plate
  d Stoßblech n, Stoßlasche f
  f couvre-joint m
  nl stalen strip m
  r стыковая накладка f
    (воспринимает продольные,
    поперечные силы и изгибающие
    моменты)

S1007 *e* **splicing of steel wire ropes**
 d Stahlseilverspleißung *f*
 f épissure *f* des câbles en fils
   d'acier
 nl splitsen *n* van staalkabels
 r сращивание *n* стальных канатов

S1008 *e* **splicing sleeve**
 d Verbindungsmuffe *f*
 f manchon *m* de raccordement
 nl kabel(verbindings)mof *f(m)*
 r соединительная муфта *f*

S1009 **split block** *see* **split-face block**

S1010 *e* **split collar**
 d geteilte Überschiebemuffe *f*
 f collier *m* [manchon *m*] fendu
 nl gespleten schuifmof *f(m)*
 r надвижная разрезная муфта *f*

S1011 *e* **split-face block**
 d Rustikabetonblock *m*
 f bloc *m* de béton à face fracturée
   [fissurée]
 nl splitbetonblok *n*
 r бетонный блок *m* с рустованной
   лицевой гранью

S1012 *e* **split-level house**
 d Halbgeschoßhaus *n*
 f maison *m* à niveaux multiples
 nl huis *n* met tussenverdiepingen
 r здание *n* с разными отметками
   перекрытий в смежных секциях

S1013 *e* **split room air-conditioner**
 d Raumklimagerät *n* in
   Splitausführung, Splitklimagerät *n*
 f conditionneur *m* [climatiseur *m*]
   autonome du type split
 nl afzonderlijke
   kamerluchtbehandelingsinstallatie *f*
 r раздельный комнатный
   кондиционер *m*, автономный
   комнатный кондиционер *m* с
   выносным компрессорно-
   -конденсаторным агрегатом

S1014 *e* **split system**
 d Splitklimaanlage *f*, Klimaanlage *f*
   in Splitausführung
 f système *m* (de conditionnement
   d'air) du type split
 nl splitklimaatinstallatie *f*
 r раздельная установка *f*
   кондиционирования воздуха

S1015 *e* **splitter**
 d Luftverteileinrichtung *f*,
   Leitblech *n*
 f répartiteur *m* d'air
 nl luchtverdeelinrichting *f*
 r рассекатель *m*, направляющая
   перегородка *f* (в воздуховоде)

S1016 *e* **splitter damper**
 d 1. Wechselklappe *f*,
   Luftstromteiler *m* 2.
   Leitschaufelverteiler *m*,
   Luftverteileinrichtung *f*

 f 1. clapet *m* distributeur à air
   2. diviseur *m* de courant d'air
 nl 1. luchtstroomverdeler *m*
   2. leischoepverdeler *m*
 r 1. перекидной воздушный клапан
   m; распределительный
   воздушный клапан *m*
   2. рассекатель *m* (в воздуховоде)

S1017 *e* **splitter-type muffler**
 d Zellenschalldämpfer *m*
 f amortisseur *m* de bruit du type
   cellulaire
 nl cellengeluiddemper *m*
 r ячейковый [сотовый]
   шумоглушитель *m*

S1018 *e* **splitter wall**
 d Trennungswand *f*
 f mur *m* de séparation, paroi *f* de
   partage
 nl scheidingsmuur *m*
 r раздельная стенка *f*

S1019 *e* **splitting tensile strength**
 d Spaltzugfestigkeit *f*
 f résistance *f* à la compression
   diamétrale, résistance *f* à la
   rupture par fendage
 nl splijttrekvastheid *f*
 r прочность *f* на раскалывание

S1020 **splitting tensile test** *see* **diametral
   compression test**

S1021 *e* **spoil**
 d Aushub(boden) *m*,
   Aussatzmaterial *n*
 f déblai *m*
 nl grondoverschot *n*
 r отвальный грунт *m*, пустая
   порода *f*

S1022 *e* **spoil area**
 d Spülhalde *f*; Spülteich *m*
 f bassin *m* à déchets de lavage
 nl stortplaats *f(m)*
 r хвостохранилище *n*

S1023 *e* **spoil bank, spoil dump**
 d Halde *f*, Aussatzkippe *f*
 f terril *m*
 nl gronddepot *n*
 r отвал *m*

S1024 *e* **sponge rubber float**
 d Schwammgummi-Glättkelle *f*
 f truelle *f* à lame en caoutchouc
   spongieux
 nl schuimrubberplakspaan *f(m)*
 r тёрка *f* или гладилка *f*
   с полотном из губчатой резины

S1025 *e* **sports building**
 d Sportgebäude *n*; Sportbauwerk *n*
 f bâtiment *m* sportif
 nl sportgebouw *n*
 r спортивное здание *n*; спортивное
   сооружение *n*

S1026 *e* **spot air cooling**
 d lokale [örtliche] Kühlung *f* mit
   Frischluftstrahl, Luftbeduschung *f*

f refroidissement *m* localisé
nl plaatselijke luchtkoeling *f*
r местное охлаждение *n*; воздушное
душирование *n*

S1027 *e* **spot cooling installation**
d Frischluftdusche *f*
f douche *f* d'air frais,
refroidisseur *m* à douche d'air
nl gericht inblazen *n* van koele lucht
r воздушный душ *m*

S1028 *e* **spot cooling jet**
d Kühlluftstrahl *m*,
Frischluftstrahl *m*
f jet *m* d'air frais
nl koelluchtstraal *m*, *f*
r душирующая воздушная струя *f*

S1029 *e* **spot heating**
d lokale [örtliche] Erwärmung *f*,
Platzheizung *f*
f chauffage *m* localisé
nl plaatselijke verwarming *f*
r местный обогрев *m*, обогрев *m*
рабочего места

S1030 *e* **spot welding**
d Punktschweißung *f*
f soudage *m* par points
nl puntlassen *n*
r точечная сварка *f*

S1031 *e* **spout**
d Ausgußrinne *f*, Auslauf *m*,
Auslaufrutsche *f*; Ablaufrohr *n*,
Entleerungsleitung *f*;
Regenwasserfallrohr *n*, Fallrohr *n*
f goulotte *f* de vidange, tuyau *m*
de décharge; descente *f* pluviale
nl 1. goot *f(m)*, afvoerpijp *f(m)*
2. regenpijp *f(m)*
r спускной лоток *m* [жёлоб *m*];
рештак *m*; спускная *или*
водосточная труба *f*

S1032 *e* **spouting plant**
d Beton-Gießrinnenanlage *f*
f centrale *f* distributrice de béton
à goulottes
nl installatie *f* voor het maken van
betongoten
r лотковая бетонолитная установка
f

S1033 *e* **spray**
d 1. Zerstäubung *f* 2. Zerstäuber *m*,
Vernebler *m*, Düse *f* 3.
Verspritzstrahl *m*, Sprühstrahl *m*;
Sprühnebel *m*
f 1. pulvérisation *f*, atomisation *f*
2. pulvérisateur *m*, atomiseur *m* ;
pistolet *m* à air 3. poussière *f*
d'eau *ou* de l'huile
nl 1. verstuiving *f* 2. verstuiver *m*,
verneveler *m* 3. sproeistraal *m*;
nevel *m*
r 1. распыление *n* 2. распылитель
m; форсунка *f* 3. факел *m*
распыла; туман *m* (*водяной*,
масляный)

S1034 *e* **spray air washer**
d Düsenkammer *f*, Sprühkammer *f*
f chambre *f* de pulvérisation
nl sproeierkamer *f (m)*
r форсуночная камера *f* орошения

S1035 *e* **spray booth**
d Spritzkabine *f*
f cabine *f* de peinture au pistolet,
chambre *f* pour peinture par jet
nl verfspuitcabine *f*
r окрасочная кабина *f* (*для
окраски распылением*)

S1036 **spray chamber** *see* **spray air
washer**

S1037 *e* **spray cooling**
d Rieselkühlung *f*
f refroidissement *m* à pulvérisation
nl koeling *f* door besproeien
r охлаждение *n* разбрызгиванием
[орошением]

S1038 *e* **spray cooling tower** *see* **spray
filled cooling tower**

S1039 *e* **sprayed coil**
d berieselte Batterie *f*, Rieselkühler
m
f serpentin *m* arrosée ; batterie *f*
arrosée
nl besproeide koelspiraal *f(m)*
r орошаемый теплообменник *m*

S1040 *e* **sprayed insulation**
d Spritzisolierung *f*
f isolant *m* appliqué par
pulvérisation
nl gespoten isolatie *f*
r теплоизоляция *f*, наносимая
набрызгом

S1041 **sprayed mortar** *see* **shotcrete**

S1042 *e* **sprayed-on membrane**
d aufgespritzter Film *m*
f membrane *f* (d'etanchéité)
appliquée par pulvérisation
nl opgespoten film *m*
r плёнка *f*, образованная
распылением

S1043 *e* **sprayed-on-method of curing**
d Aufsprüh-Verfahren *n* der
Nachbehandlung von Beton
f cure *f* de béton par pulvérisation
des filmogènes
nl nabehandeling *f* van beton door
opspuiten van een film
r метод *m* ухода за бетоном путём
распыления плёнкообразующих
веществ

S1044 *e* **sprayed-on plaster**
d Spritzputz *m*, Torkretputz *m*
f enduit *m* deposé par
pulvérisation, enduit *m* projeté
nl spuitpleister *n*
r торкретная штукатурка *f*;
штукатурка *f*, наносимая методом
набрызга

S1045 e **sprayed roof**
d wasserbespülte Dachhaut *f*
f comble *m* [toit *m*] pulvérisé
nl besproeid dak *n*
r орошаемая кровля *f*

S1046 e **spray filled cooling tower**
d Sprühkühlturm *m*
f tour *f* réfrigérante à pulvérisation
nl sproeikoeltoren *m*
r градирня *f* с орошаемой
насадкой

S1047 e **spraygrip**
d kiesgefüllte Bitumen-Epoxyharz-
-Emulsion *f*
f mélange *m* d'émulsion bitume-
-époxyde et de gravier fin
(*pour point à temps*)
nl mengsel *n* van grind en een
emulsie van bitumen-epoxyhars
r смесь *f* битумно-эпоксидной
эмульсии с гравием (*для
ямочного ремонта дороги*)

S1048 e **spray gun**
d 1. Zementspritze *f*,
Torkretiergerät *n*
2. Farbspritzpistole *f*
f 1. guniteuse *f* ; injecteur *m* du
ciment 2. pistolet *m* de peinture
nl 1. cementspuit *f(m)*
2. verfspuitpistool *n*
r 1. цемент-пушка *f* 2. пистолет-
-краскораспылитель *m*

S1049 **spray humidifier** *see* **spray type
humidifier**

S1051 e **spraying rate**
d Verbrauch *m* des verspritzten
Materials, Spritzaufbringungsrate *f*
f dosage *m* [débit *m*] d'un
matériau pulverisé
nl verbruik *n* van spuitmateriaal
r расход *m* распыляемого
материала

S1052 e **spray irrigation**
d Beregnung *f*
f arrosage *m* en pluie, aspersion *f*
nl beregening *f*
r дождевание *n*

S1053 e **spray painting**
d Farbspritzen *n*
f peinture *f* au pistolet [par
pulvérisation]
nl verfspuiten *n*
r окраска *f* распылением

S1054 e **spray pond**
d Spritzteich *m*, Kühlbecken *n* für
Zirkulationswasser
f bassin *m* refroidisseur d'eau par
aspersion, étang *m* de pulvérisation
nl sproeivijver *m*
r брызгальный бассейн *m*

S1055 e **spray tap**
d Perlatorauslauf *m*

f robinet *m* de puisage avec
pulvérisateur
nl kraan *f(m)* met straalbreker
r излив *m* с аэратором

S1056 e **spray-type air cooler**
d Sprühkühler *m*, Düsenluftkühler *m*
f refroidiseur *m* de pulvérisation,
groupe *m* refroidisseur à
pulvérisation
nl sproeier-luchtkoeler *m*
r форсуночный воздухоохладитель *m*

S1056a e **spray-type air dehumidifier**
d Sprühentfeuchter *m*,
Düsentrockner *m*
f déhumidificateur *m* de
pulvérisation
nl sproei-ontvochtiger *m*
r орошаемый воздухоосушитель *m*

S1057 e **spray-type air washer**
d Luftwäscher *m*, Berieselungskammer
f
f chambre *f* d'arrosage
nl sproeikamer *f(m)*, luchtwasser *m*
r камера *f* орошения

S1058 e **spray-type heat exchanger**
d berieselter Wärmeaustauscher *m*
f échangeur *m* thermique arrosé
nl bevloeide warmte-uitwisselaar *m*
r орошаемый теплообменник *m*

S1059 e **spray type humidifier**
d Düsenbefeuchter *m*
f humidificateur *m* à pulvérisation
nl sproeibevochtiger *m*
r распылительный воздухоувлажнитель *m*

S1060 **spray washer** *see* **spray-type air
washer**

S1061 e **spread**
d 1. Strahlausbreitung *f*
2. Aufbringungsrate *f*
3. Baumaschinenpark *m*,
Gerätebestand *m* 4. Baukolonne *f*
f 1. divergence *f* 2. norme *f* de
consommation des matériaux de
couverture 3. parc *m* d'engins de
chantier 4. convoi *m* [colonne *f*]
de machines
nl 1. straalverbreding *f* 2.
verbruiksnorm *f(m)* van
spuitmateriaal 3. machinepark *n*
van bouwmachines 4. colonne *f*
van bouwmachines
r 1. расширение *n* приточной струи
2. норма *f* расхода материала
покрытия (*на единицу площади*)
3. парк *m* строительных машин
(*подрядчика*) 4. строительная
колонна *f* (*напр. дорожных
машин*)

S1062 e **spread anchorage**
d fächerförmige Verankerung *f*
f ancrage *m* en éventaille
nl waaiervormige verankering *f*

r распределённая [веерная]
анкеровка *f*

S1063 *e* **spreader**
d 1. Betonverteiler *m* 2.
Schüttgutverteiler *m* 3. Spreize *f*
f 1. distributeur *m* de béton 2.
gravillonneuse *f*, camion-
gravillonneur *m* 3. étrésillon *m* ;
entretoise *f* ; moise *f*
nl 1. betonverdeelapparaat *n* 2.
grindstrooimachine *f* 3. schoor *m*
r 1. распределитель *m* бетона 2.
распределитель *m* щебня *или*
гравия 3. распорка *f*; схватка *f*

S1064 *e* **spreader bar**
d Traverse *f*, Ausgleichgehänge *n*
f traverse *f* de répartition
nl afstandhouder *m*
r траверса *f* (*грузозахватное
приспособление*)

S1065 *e* **spreader beam**
d Verteilerbalken *m*,
Verteilungsträger *m*
f poutre *f* de répartition (*des
charges concentrées*)
nl verdeelbalk *m*
r распределительная балка *f* (*для
распределения сосредоточенной
нагрузки*)

S1066 **spreader finisher** *see* **bituminous
road surfacing finisher**

S1067 *e* **spread footing**
d Flächenfundament *n*; Fundament *n*
mit verbreitertem Auflagerteil
f fondation *f* continue; fondation *f*
distributrice (*élargie vers le bas*)
nl fundering *f* met verbrede voet
r сплошной фундамент *m*;
уширенный (книзу) фундамент *m*

S1068 *e* **spreading**
d Berieselung *f*; Begießung *f*;
Verteilung *f* (*Beton*); Auftrag *m*,
Aufbringen *n* (*Anstrichstoffe*)
f épandage *m* ; distribution *f* ;
répartition *f*
nl spreiden *n* ; opbrengen *n*
r полив *m*; распределение *n*
(*бетонной смеси*); нанесение *n*
(*краски*)

S1069 *e* **spreading basin**
d Verteilerbauwerk *n*
(*Abwasserwesen*)
f bassin *m* de répartition
nl verdeelbassin *n* van de riolering
r канализационный
распределительный бассейн *m*

S1070 *e* **spreading rate**
d Deckkraft *f*, Deckfähigkeit *f*
(*Anstrichstoffe*)
f capacité *f* [faculté *f*] couvrante,
pouvoir *m* couvrant
nl dekkend vermogen *n* (*verf*)
r кроющая способность *f*,
укрывистость *f* (*окраски*)

S1071 *e* **spread of concrete**
d Ausbreitung *f*, Ausbreitmaß *n*
(*Betonprüfung*)
f étalement *m* de béton frais
nl schudmaat *f*(*m*) (*betonmortel*)
r расплыв *m* бетонной смеси

S1072 *e* **spring flood**
d Frühjahrshochwasser *n*, Hochflut *f*
f eaux *f pl* hautes, crue *f*
nl springvloed *f*(*m*)
r весенний паводок *m*, половодье *n*

S1073 *e* **springing**
d 1. Kämpfer *m*, Kämpferpunkt *m*,
Bogenansatz *m* 2. vorgekesseltes
Loch *n*
f 1. naissance *f* (*de l'arc ou de la
voûte*) 2. fond *m* du trou élargi
par camouflet
nl 1. geboorte *f* 2. met een
springlading gevormde holte *f* in
de grond
r 1. пята *f* (*арки, свода*) 2.
камуфлетное уширение *n* в забое
скважины

S1074 *e* **springing line**
d Kämpferlinie *f*
f ligne *f* de naissance
nl geboortelijn *f*(*m*)
r линия *f* пят (*арки, свода*)

S1075 *e* **sprinkler**
d Sprinkler *m*
f sprinkler *m*
nl sprinkler *m*
r спринклер *m*, автоматическая
водораспыляющая насадка *f*

S1076 *e* **sprinkler head**
d Sprinklerdüse *f*
f tête *f* d'extinction [d'extincteur]
automatique, sprinkler *m*
nl sprinklerkop *m*
r спринклерная головка *f*

S1077 *e* **sprinkler system**
d 1. Strahlregneranlage *f*
2. Berieselungsanlage *f*,
Sprinkleranlage *f*
f 1. système *m* d'arrosage par
aspersion 2. installation *f* (par)
sprinkler
nl 1. beregeningsinstallatie *f*
2. sprinklersysteem *n*
r 1. дождевальная система *f*
2. спринклерная система *f*

S1078 *e* **sprocket, sprocket piece**
d Sturmlatte *f*, Windrispe *f*,
Strebeschwarte *f*
f coyau *m*
nl opzetwig *f*(*m*) op dakspar ten
behoeve van de dakrand
r кобылка *f* (*стропил*)

S1080 *e* **spud**
d 1. Pickelhacke *f* 2. Verzinkung *f*,
Zapfen *m*
3. Überwurfmutter(verbindung) *f*
4. Zugmesser *n* 5. Schlagbohrer *m*

6. Haltepfahl *m*
f 1. pelle-pioche *f* 2. tenon *m*
3. tubulure *f* de raccordement
4. plane *f* 5. barre *f* à mine
6. pieu *m* de fixation
nl 1. pikhak *f(m)* 2. tap *m*
3. aansluitstuk *n* 4. trekmes *n*
5. klopboor *f(m)* 6. ankerpaal *m*
r 1. киркопата *f* 2. шип *m*
(*в столярных изделиях*)
3. соединительный патрубок *m*,
сгон *m* 4. скобель *m* 5. ударный
бур *m* 6. папильонажная свая *f*

S1081 *e* **spud vibrator** *US*
d Innenrüttler *m*, Innenvibrator *m*
f vibr(at)eur *m* interne,
previbr(at)eur *m*
nl naaldtriller *m*
r глубинный вибратор *m*

S1082 *e* **spun concrete**
d Schleuderbeton *m*
f béton *m* centrifugé
nl slingerbeton *n*
r центрифугированный бетон *m*

S1083 **spurdike** *see* **spur dyke**

S1084 *e* **spur duct**
d 1. Stich(luft)kanal *m*
2. Abzweig(luft)kanal *m*
f 1. conduit *m* d'aérage à cul-de-sac
2. embranchement *m* de conduit
d'aérage principal
nl 1. doodlopend luchtkanaal *n*
2. aftakking *f* van luchtkanaal
r 1. тупиковый воздуховод *m*
2. ответвление *n* магистрального
воздуховода

S1085 *e* **spur dyke**
d Buhne *f*, Krippe *f*, Höfter *m*;
Querdeich *m*, Flügeldeich *m*,
Sporn *m*
f épi *m*, éperon *m*; digue *f*
transversale
nl krib *f(m)*
r буна *f*, шпора *f*, струенаправляющая полузапруда *f*;
поперечная дамба *f*

S1086 *e* **square bridge**
d gerade Brücke *f*
f pont *m* droit
nl rechte brug *f(m)*
r прямой мост *m* (*пересекающий
водоток под прямым углом*)

S1087 *e* **square key**
d Vierkantdübel *m*
f clavette *f* rectangulaire
nl wig *f(m)*, keg *f(m)*
r квадратная шпонка *f*

S1088 *e* **square-sawn timber**
d Kantholz *n*
f bois *m* carré; basting *m*, madrier
m
nl bezaagd hout *n*
r брус *m* (*пиломатериал*)

S1089 **SSV** *see* **soil support value**

S1090 *e* **stabbing**
d Befestigung *f* der
Wärmeisoliermatten mit Stiften
f fixation *f* [attachement *m*] des
matelas isolants par goujons
nl bevestiging *f* van warmte-
-isolerende matten met stiften
r крепление *n* теплоизоляционных
матов шпильками

S1091 **stabilised soil** *see* **stabilized soil**

S1092 **stabilising** *UK see* **stabilizing** *US*

S1093 *e* **stability**
d Standsicherheit *f*; Stabilität *f*
f stabilité *f*
nl stabiliteit *f*
r устойчивость *f*; стабильность *f*

S1094 *e* **stability factor, stability number**
d Standsicherheitsfaktor *m*,
Standsicherheitszahl *f*
f coefficient *m* de stabilité
nl stabiliteitsfactor *m*
r коэффициент *m* устойчивости

S1095 *e* **stability of slope**
d Standfestigkeit *f* der Böschung,
Böschungsstandsicherheit *f*
f stabilité *f* du talus
nl stabiliteit *f* van een talud
r устойчивость *f* откоса

S1096 *e* **stabilization basin**
d Stabilisierungsteich *m*
f bassin *m* de stabilisation
nl stabilisatievijver *m*
r стабилизационный [биологический]
пруд *m*

S1097 *e* **stabilization of effluent**
d Stabilisierung *f* des Abwassers
f stabilisation *f* des effluents
nl stabilisatie *f* van afvalwater
r стабилизация *f* сточных вод

S1098 *e* **stabilization train**
d Vermörtelungsbauzug *m*
f convoi *m* [colonne *f*] d'engins pour
stabilisation de sol
nl stabilisatietrein *m*
r колонна *f* машин, выполняющих
стабилизацию грунта (*дорожное
строительство*)

S1099 *e* **stabilized settlement**
d abklingende Setzung *f*
f tassement *m* [affaissement *m*]
stabilisé
nl gestabiliseerde zetting *f*
r стабилизированная осадка *f*

S1100 *e* **stabilized soil**
d stabilisierter [verfestigter] Boden *m*
f sol *m* stabilisé
nl gestabiliseerde grond *m*
r стабилизированный грунт *m*,
укреплённый грунт *m*

S1101 *e* **stabilizer**
   d 1. Verfestigungsmittel *n*
     2. Stabilisator *m* 3. Abstützarm
     m, ausfahrbare Stütze *f*
   f 1. produit *m* [agent *m*] stabilisant
     (du sol) 2. émulsifiant *m*,
     émulsificateur *m* 3. stabilisateur
     m (*de grue mobile*)
   nl 1. stabilisatiemiddel *n* 2.
     emulsiestabilisator *m* 3. stempel *m*
   r 1. стабилизирующий материал *m*
     (*для укрепления грунтов*)
     2. стабилизатор *m* эмульсии
     3. выносная опора *f*

S1102 *e* **stabilizing** *US*
   d Stabilisieren *n*, Verfestigung *f*
   f stabilisation *f*, consolidation *f*
   nl verdichten *n*, stabilisatie *f*
   r стабилизация *f*, упрочнение *n*

S1103 *e* **stable design**
   d standfeste [standsichere, stabile]
     Konstruktion *f*
   f construction *f* stable
   nl stabiele constructie *f*
   r устойчивая конструкция *f*

S1104 *e* **stable framework**
   d stabiles Stabwerk *n*, stabile
     Rahmenkonstruktion *f*
   f charpente *f* stable
   nl vormvast [stijf] skelet *n*
   r устойчивая стержневая [рамная]
     конструкция *f*

S1105   **stack effect** *see* **chimney effect**

S1106 *e* **stacker**
   d 1. Bandabsetzer *m*, Absetzanlage *f*
     2. Stapelkarre *f*, Stapelförderer *m*,
     Hubstapler *m*
   f 1. empileur *m*, empileuse *f* à
     bande 2. gerbeur *m*, gerbeuse *f*,
     grue *f* gerbeuse
   nl stapelinstallatie *f*; stapelwagen *m*;
     heftruck *m*
   r 1. экскаватор-отвалообразователь
     m; ленточный отвалообразователь
     m 2. штабелеукладчик *m*

S1107 *e* **stacker crane, stacking crane**
   d Stapelkran *m*
   f chariot *m* élévateur, gerbeur *m*,
     empileur *m*, chariot-stockeur *m*
   nl stapelkraan *f*(*m*)
   r кран-штабелёр *m*

S1108 *e* **stack vent**
   d Luftrohr *n*
   f colonne *f* de ventilation
   nl ontspanningsleiding *f* van de
     binnenriolering
   r вентиляционный стояк *m*
     канализации

S1109 *e* **staff gauge**
   d Pegel *m*, Wasserstandanzeiger *m*
   f échelle *f* d'étiage, limnimètre *m*
   nl peilschaal *f*(*m*)
   r реечный водомерный пост *m*;
     водомерная рейка *f*

S1110 *e* **stage**
   d 1. Wasserhöhe *f*, Wasserstand *m*,
     Wasserniveau *n* 2. Stadium *n*,
     Phase *f*, Stufe *f* 3. Theaterbühne *f*
     4. Gerüstboden *m*, Arbeitsbühne *f*
   f 1. hauteur *f* d'eau [limnimétrique],
     niveau *m* d'eau 2. étape *f*,
     phase *f* 3. scène *f* [plateau *m*]
     d'un théâtre 4. échafaud *m*
   nl 1. waterpeil *n* 2. stadium *n*,
     trap *m* 3. tonel *n* 4. steiger *m*
   r 1. отметка *f* водной поверхности,
     уровень *m* воды 2. стадия *f*,
     этап *m* 3. сцена *f* (театра)
     4. помост *m*, рабочий настил *m*,
     подмости *pl*

S1111 *e* **stage-discharge relation**
   d Abflußkurve *f*
   f courbe *f* des débits jaugés
   nl verband *n* tussen afvoer en peil
   r кривая *f* расхода (*для водотока*)

S1112 *e* **stage of construction**
   d Ausführungsphase *f*, Baustadium *n*;
     Ausbaustufe *f*, Bauabschnitt *m*
   f phase *f* de construction; état *m*
     d'avancement de la construction
   nl bouwstadium *n*, uitvoeringsfase *f*
   r этап *m* [стадия *f*] строительства;
     очередь *f* строительства

S1113 *e* **staggered joints**
   d 1. versetzte Fugen *f pl*
     2. Läuferverband *m*,
     Halbsteinverband *m*
   f 1. joints *m pl* en quinconce
     2. appareil *m* à demi-briques
     (*en travers ou en long*)
   nl 1. verspringende voegen *f*(*m*) *pl*
     2. halfsteenverband *n*
   r 1. швы *m pl* вразбежку,
     шахматное расположение *n* швов
     2. перевязка *f* в полкирпича
     (*тычковая или ложковая*)

S1114 *e* **staging**
   d 1. Gerüst *n*, Arbeitsbrücke *f*;
     Arbeitsbühne *f* 2. Anlegebrücke *f*,
     Landungsbrücke *f*, Landungssteg
     m
   f 1. échafaudage *m*; plate-forme *f*
     de travail 2. estacade *f*
     d'accostage, quai *m* d'amarrage
   nl 1. werkplatform *n* 2. aanlegsteiger
     m
   r 1. подмости *pl*, леса *pl*; рабочая
     платформа *f* 2. причальная
     эстакада *f*

S1115 *e* **stagnant air**
   d stagnierende Luft *f*
   f air *m* stagnant
   nl verbruikte lucht *f*(*m*)
   r застойный воздух *m*

S1116 *e* **stagnant zone, stagnation zone**
  d Stauungszone *f* ; Stagnationszone *f*
    Trägheitsbereich *m*
  f zone *f* de stagnation (*d'air, d'eau*)
  nl zone *f* van afnemende
    stroomsnelheid (*lucht, water*)
  r застойная зона *f*; зона *f*
    застойного воздуха; мёртвая зона *f*
    водотока

S1117 *e* **staircase**
  d 1. Treppenhaus *n* 2. Treppe *f*
  f 1. cage *f* d'escalier 2. escalier *m*
  nl 1. trappenhuis *n* 2. trap *m*
  r 1. лестничная клетка *f*
    2. лестница *f*

S1118 *e* **stair platform**
  d Treppenpodest *n*, *m*
  f palier *m*
  nl trapportaal *n*
  r лестничная площадка *f*

S1120 *e* **stair railing**
  d Treppengeländer *n*
  f balustrade *f* d'escalier; garde-
    -corps *m* [garde-fou *m*] d'escalier
  nl balustrade *f*, trapleuning *f*
  r перила *pl* лестницы

S1121   **stairs** *see* **stairway**

S1122 *e* **stairwalking hand truck**
  d Stechkarre *f* zum Transport über
    Treppen
  f cabrouet *m* [chariot *m*] pouvant
    grimper l'escalier
  nl steekwagen *n* voor het vervoer van
    goederen op trappen
  r ручная тележка *f* для перевозки
    грузов по лестницам

S1123 *e* **stairway**
  d Treppe *f*, Treppenanlage *f*;
    Treppenlauf *m*, Treppenarm *m*
  f escalier *m*; volée *f* d'escalier *m*,
    rampe *f*
  nl trap *m*
  r лестница *f*; лестничный марш *m*

S1124 *e* **stair well**
  d Treppenauge *n*, Treppenloch *n*
  f puits *m* d'escalier
  nl schalmgat *n*
  r лестничная шахта *f*

S1125 *e* **staking, staking out**
  d Abmarkung *f*, Absteckung *f*,
    Trassierung *f*
  f piquetage *m*
  nl afbakenen *n*, uitzetten *n*
  r трассировка *f*, разбивка *f* трассы

S1126 *e* **stall urinal**
  d PP-Becken *n*, Wandurinal *n*,
    Wandschüssel *f*
  f urinoir *m* de mur
  nl wandurinoir *n*
  r настенный писсуар *m*

S1127 *e* **stamped concrete**
  d Beton *m* mit eingepreßtem
    Ornament

  f béton *m* (à dessin) estampé
  nl beton *n* met ingeperst ornament
  r бетон *m* с вдавленным рисунком
    (*полученным вдавливанием
    решётчатых штампов*)

S1128 *e* **stamping tool**
  d Einprägestempel *m*
  f outil *m* d'estampage (*pour
    estamper les joints faux concaves
    dans le béton frais du revêtement
    routier*)
  nl ponsgereedschap *n*
  r штамп *m* для выдавливания
    ложных швов в бетонном
    покрытии (*для фактурной
    отделки*)

S1129 *e* **stanchion**
  d Stahlsäule *f*, Stahlstütze *f*;
    Ständer *m*; Strebe *f*; Pfosten *m*
  f poteau *m*; montant *m*; étai *m*;
    colonne *f*; appui *m*
  nl zuil *f(m)*, staander *m*; staaf *f(m)*
  r стальная колонна *f* или стойка *f*;
    подпорка *f*; столб *m*

S1130 *e* **stanchion base**
  d Stahlstütze-Basis *f*; Säulenbasis *f*
  f base *f* de poteau; pied *m* de
    colonne
  nl zuilvoet *m*
  r база *f* или башмак *m* стальной
    колонны [стойки]

S1131 *e* **stanchion cap**
  d Stahlstützenaufsatz *m*; Säulenkopf
    m
  f tête *f* de poteau [de colonne]
  nl kop *n* van een zuil
  r оголовок *m* стальной колонны
    [стойки]

S1132 *e* **stanchion casing**
  d Stahlstützen-Betonummantelung *f*
  f revêtement *m* en béton de
    poteaux en acier
  nl betonmantel *f* om een stalen
    kolom
  r бетонная облицовка *f* стальной
    колонны

S1133 *e* **standard air**
  d Standardluft *f*
  f air *m* normal
  nl normale lucht *f(m)*
  r стандартный воздух *m*

S1134 *e* **standard atmosphere**
  d Standardatmosphäre *f*,
    Normalatmosphäre *f*
  f atmosphère *f* normal
  nl normale atmosfeer *f(m)*
  r стандартная атмосфера *f*

S1135 *e* **standard cube of concrete**
  d Normenbetonwürfel *m*
  f cube *m* de béton normalisé
  nl standaard proefkubus *m* (*beton*)
  r стандартный бетонный кубик *m*

S1136 *e* **standard curing procedure**
    d 1. Normennachbehandlung *f* (*der Betonwürfel*) 2. gewöhnliche Nachbehandlungsweise *f* (*Beton*)
    f 1. procédé *m* normalisé de traitement des cubes de béton 2. pratique *f* courante de cure de béton
    nl 1. gestandaardiseerde nabehandeling *f* van proefkubussen 2. gebruikelijke nabehandelingsmethode *f* van beton
    r 1. стандартная методика *f* выдерживания бетонных образцов 2. обычная практика *f* ухода за бетоном

S1137 *e* **standard design**
    d 1. Typenprojekt *n*, Typenentwurf *m* 2. Normalausführung *f*, Einheitsbauweise *f*
    f 1. projet *m* type [typifié] 2. construction *f* type unifié
    nl 1. standaardontwerp 2. standaardconstructie *f*
    r 1. типовой проект *m* 2. стандартная конструкция *f*

S1138 *e* **standard hook**
    d Normenhaken *m*
    f crochet *m* normal (*au bout d'une barre d'armature*)
    nl standaardhaak *n* (*wapening*)
    r стандартный крюк *m* (*на конце арматурного стержня*)

S1139 *e* **standardized building**
    d Typengebäude *n*
    f bâtiment *m* type
    nl standaardgebouw *n*
    r типовое здание *n*

S1140 *e* **standardized structural element**
    d typisierter Fertigteil *m*
    f élément-type *m* préfabriqué, unité-type *f* préfabriquée
    nl standaardbouwelement *n*
    r типовой сборный элемент *m*

S1141 *e* **standard of construction**
    d Qualitätsniveau *n* des Bauens; technisches Niveau *n* des Bauens
    f standard *m* de construction; niveau *m* technique de construction; niveau *m* de qualité de construction
    nl kwaliteitsniveau *n* van het bouwen
    r уровень *m* качества строительства; технический уровень строительства

S1142 *e* **standard of quality**
    d Qualitätsniveau *n*
    f standard *m* de qualité; niveau *m* de qualité
    nl kwaliteitsniveau *n*
    r уровень *m* качества

S1143   **standard pile** *see* **guide pile**

S1144 *e* **standard plan**
    d 1. Typenentwurf *m* 2. Typenzeichnung *f*
    f 1. projet *m* type 2. dessins *m pl* du projet type
    nl 1. standaardontwerp *n* 2. tekening *f* van het standaardmodel
    r 1. типовой проект *m* 2. типовой чертёж *m*

S1145 *e* **standard railing**
    d Inventarschutzsperren *f pl*, Inventarschutzgeländer *n pl*
    f garde-corps *m* [garde-fou *m*] (de protection) standardisé
    nl standaardleuningen *f pl*
    r инвентарные защитные ограждения *n pl*

S1146 *e* **standard sand**
    d Normensand *m*
    f sable *m* normal(isé)
    nl standaardzand *n*
    r стандартный песок *m* (*для приготовления образцов при испытании цементов*)

S1147 *e* **standards of acceptance**
    d Abnahmevorschriften *f pl*, Abnahmebestimmungen *f pl*
    f normes *f pl* de réception (*p.ex. de matériaux, de produits*)
    nl afnamevoorschriften *n pl*
    r правила *n pl* приёмки (*напр. материалов, изделий*)

S1148 *e* **standard specifications**
    d Ausführungsbestimmungen *f pl*, Standard-Bauleistungsverzeichnis *n*
    f cahier *m* des charges-type
    nl standaardbepalingen *f pl* voor de uitvoering
    r типовые технические условия *n pl*

S1149 *e* **standard truck**
    d Normalkraftwagen *m*
    f camion-type *m*, convoi-type *m* (*pour le calcul des ponts*)
    nl standaardvoertuig *n*
    r стандартный (расчётный) автомобиль *m*

S1150 *e* **standard truck loading**
    d normative Fahrzeugbelastung *f* (*auf die Brücke*)
    f convoi-type *m* (*surcharge sur le pont*)
    nl standaardlaststelsel *n* voor een verkeersbrug
    r нормативная автомобильная нагрузка *f* на мост (*в форме колонны грузовых автомобилей*)

S1151 *e* **stand-by-pump**
    d Zusatzpumpe *f*, Reservepumpe *f*
    f pompe *f* de secours
    nl reservepomp *f(m)*
    r резервный насос *m*

S1152 *e* **standing pier**
    d Zwischenpfeiler *m*
    f pile *f* de pont

nl stroompijler *m*
r промежуточная опора *f* моста

S1153 *e* **standing-water level**
d Ruhe(wasser)spiegel *m*
f niveau *m* piézométrique [hydrostatique]
nl statische waterstand *m*
r статический [установившийся] уровень *m* воды

S1154 **standing wave** *see* **clapotis**

S1155 *e* **standpipe**
d 1. Wasserdrucksäule *f* 2. Ventilbrunnen *m*, Zapfständer *m*, Speisesäule *f*; Hydrant *m*
3. Steigrohr *n*, Steigleitung *f*
4. Piezometerrohr *n*
f 1. cheminée *f* d'eau, chateau *m* d'eau 2. bouche *f* à eau; hydrant *m*, hydrante *f* 3. colonne *f* montante 4. tube *m* piézométrique
nl 1. watertoren *m* 2. standpijp *f(m)*; hydrant *m* 3. stijgleiding *f*
4. open standpijp *f(m)* ter begrenzing van de druk in het systeem
r 1. водонапорная колонна *f*
2. водоразборная колонка *f*; гидрант *m* 3. стояк *m*
4. пьезометрическая трубка *f*

S1156 *e* **stank**
d kleiner tongedichteter Holzfangdamm *m*
f petit batardeau *m* en bois et en argile
nl kleine klei-vangdam *m*
r малая деревянная перемычка *f*, уплотнённая глиной

S1157 **starter bar** *see* **stub bar**

S1158 *e* **starting**
d Inbetriebnahme *f*
f mise *f* en fonctionnement
nl inbedrijfstelling *f*
r пуск *m*, ввод *m* в эксплуатацию

S1159 *e* **starting torque**
d Anlauf(dreh)moment *n*, Anlaß(dreh)moment *n*
f moment *m* de démarrage
nl aanloopkoppel *n*
r пусковой момент *m*

S1160 *e* **state of strain**
d verformter Zustand *m*, Formänderungszustand *m*
f état *m* de déformation
nl vervormde toestand *m*
r деформированное состояние *n*

S1161 *e* **state of stress**
d Spannungszustand *m*
f état *m* de contrainte
nl spanningstoestand *m*
r напряжённое состояние *n*

S1161a **statical...** *see* **statically...**, **static...**

S1162 *e* **statically determinate structure**
d statisch bestimmte Konstruktion *f*

f construction *f* isostatique
nl statisch bepaalde constructie *f*
r статически определимая конструкция *f*

S1163 *e* **statically indeterminate structure**
d statisch unbestimmte Konstruktion *f*
f construction *f* hyperstatique
nl statisch onbepaalde constructie *f*
r статически неопределимая конструкция *f*

S1164 *e* **statical moment of area**
d statisches Querschnittsmoment *n*
f moment *m* statique d'une surface
nl statisch moment *n* van de doorsnede
r статический момент *m* площади

S1165 *e* **static analysis**
d statische Berechnung *f*
f calcul *m* statique
nl statische berekening *f*
r статический расчёт *m*

S1166 *e* **static compaction**
d statische Verdichtung *f*
f compactage *m* [serrage *m*] statique
nl statische verdichting *f*
r статическое уплотнение *n* (*грунта*)

S1167 *e* **static delivery head**
d statische Förderhöhe *f* (*Pumpe*)
f pression *f* statique, chasse *f* (*de la pompe*)
nl opvoerhoogte (*pomp*)
r статический напор *m* (*насоса*)

S1168 *e* **static fan duty**
d statische Ventilatorleistung *f* (*Förderleistung des Ventilators bei angenommenem statischem Förderdruck*)
f débit *m* [rendement *m*] statique du ventilateur
nl stuwdruk *m* van een ventilator
r подача *f* вентилятора, развивающего требуемое статическое давление

S1169 *e* **static fan efficiency**
d statischer Ventilator- -Wirkungsgrad *m*
f rendement *m* statique de ventilateur
nl statisch rendement *n* van een ventilator
r коэффициент *m* полезного действия вентилятора по статическому давлению

S1170 *e* **static head**
d statische Druckhöhe *f*
f hauteur *f* de charge statique
nl statische drukhoogte *f*
r (гидро)статический напор *m*

S1172 *e* **static load**
d statische Last *f*
f charge *f* statique
nl statische belasting *f*
r статическая нагрузка *f*

S1173 *e* static load test
 d statischer Pfahlbelastungsversuch *m*
 f essai *m* (*de portance de pieux*)
 par mise en charge statique;
 essai *m* par charge statique
 nl proefbelasting *f* van een paal
 r определение *n* несущей
 способности сваи статическим
 нагружением

S1174 static mixer *see* vortex generator

S1175 *e* static pressure
 d statischer Druck *m*
 f pression *f* statique
 nl statische druk *m*
 r статическое давление *n*

S1176 *e* static regain
 d statischer Rückgewinn *m*
 f remise *f* en pression statique
 nl herstel *n* van statische druk
 r восстановление *n* статического
 давления

S1177 *e* statics
 d Statik *f*
 f statique *f*
 nl statica *f*
 r статика *f*

S1178 *e* static suction head
 d statische [geodätische] Saughöhe *f*
 f hauteur *f* géométrique d'aspiration
 nl statische zuighoogte *f*
 r геометрическая высота *f*
 всасывания (*насоса*)

S1179 static water level *see* standing-
 -water level

S1180 *e* staunching piece
 d Dichtungsglied *n*
 f étanchement *m* du joint vertical
 de barrage
 nl vulstuk *n* voor de verticale
 voegen in een stuwmuur
 r противофильтрационное
 уплотнение *n* вертикального
 осадочного шва плотины

S1181 *e* staunching plate
 d Dichtungsblech *n*
 f plaque *f* [diaphragme *m*]
 d'étanchéité
 nl dichtingsplaat *f*(*m*)
 r противофильтрационная мембрана
 f (*в швах гидросооружений*)

S1182 *e* staunching rod
 d Dichtungsstab *m*.
 Dichtungsabschluß *m*,
 Dichtungsglied *n*
 f lame *f* d'étanchéité
 nl voegafdichting *f*
 r противофильтрационная шпонка *f*
 (*гидросооружений*)

S1183 *e* stave
 d 1. Daube *f* 2. Leitersprosse *f*,
 Sprosse *f*
 f 1. douve *f* (en bois) 2. barreau *m*
 [échelon *m*] d'échelle

nl 1. duig *f*(*m*) 2. sport *f*
 (*v.e. ladder*)
 r 1. клёпка *f* (*бочарная*)
 2. перекладина *f* (*переносной*)
 лестницы

S1184 *e* stay
 d 1. Abspannseil *n*, Verspannung *f*
 2. Strebe *f*, Spreize *f*,
 Abstandhalter *m*
 f 1. hauban *m* 2. étai *m*, contre-
 -fiche *f*, entretoise *f*
 nl 1. tui *m*, trekstang *f*(*m*) 2.
 schoor *m*, stut *m* 3. afstandhouder
 m (*wapening*)
 r 1. оттяжка *f*; ванта *f* 2. подкос
 m, распорка *f*

S1185 stayed-cable bridge *see* cable-
 -stayed bridge

S1186 *e* stay-in-place form, stay-in-place
 formwork
 d Dauerschalung *f*, verlorene
 Schalung *f*
 f coffrage *m* perdu
 nl verloren bekisting *f*
 r конструктивная опалубка *f*
 (*остающаяся в теле
 бетонируемого сооружения*)

S1187 *e* stay pile
 d Ankerpfahl *m*,
 Verankerungspfahl *m*
 f pieu *m* pour ancrage des haubans;
 pieu *m* d'ancrage
 nl ankerpaal *m*
 r свая *f* для анкеровки оттяжек
 или расчалок; анкерная свая
 f шпунтовой стенки

S1188 *e* steady filtration
 d beständige Filtration *f*
 f (in)filtration *f* permanente
 nl permanente infiltratie *f*
 r установившаяся фильтрация *j*

S1189 *e* steady water level
 d Beharrungsspiegel *m*
 f niveau *m* dynamique constant
 nl permanente inertiewaterstand *m*
 r установившийся уровень *m* воды
 (*в скважине при откачке
 с постоянным дебитом*)

S1190 *e* steam boiler
 d Dampfkessel *m*
 f chaudière *f* à vapeur
 nl stoomketel *m*
 r паровой котёл *m*, парогенератор
 m

S1191 *e* steam converter
 d Dampfumformer *m*
 f transformateur *m* de vapeur
 nl stoomomvormer *m*
 r паропреобразователь *m*

S1192 *e* steam curing
 d Dampfbehandlung *f*
 f traitement *m* à la vapeur, étuvage
 m
 nl stoombehandeling *f*

r пропаривание *n*,
паровлажностная обработка *f*

S1193 steam curing at high pressure *see*
autoclaving

S1194 *e* steam-curing cycle
d Dampfbehandlungsspiel *n*
f cycle *m* de traitement à la
vapeur
nl cyclus *m* van de stoombehandeling
r цикл *m* пропаривания

S1195 *e* steam curing room
d Wärmebehandlungskammer *f*,
Dampfbehandlungsraum *m*
f chambre *f* d'étuvage
nl stoombehandelingsruimte *f*
r пропарочная камера *f*

S1196 *e* steam emission
d Dampfentwicklung *f*
f dégagement *m* de vapeur d'eau
nl stoomontwikkeling *f*
r паровыделения *n pl*

S1197 *e* steam heating coil
d Dampflufterhitzer *m*
f serpentine *m* chauffé à la vapeur,
serpentine *m* à vapeur
nl stoomverwarmingstoestel *n*
r паровой калорифер *m*

S1198 *e* steam heating system
d Dampfheizanlage *f*
f système *m* de chauffage à vapeur
nl stoomverwarmingsinstallatie *f*
r система *f* парового отопления

S1199 *e* steam humidification
d Dampfluftbefeuchtung *f*
f humidification *f* à vapeur
nl luchtbevochtiging *f* met stoom
r увлажнение *n* воздуха паром

S1200 *e* steam humidifier
d Dampfluftbefeuchter *m*
f humidificateur *m* à vapeur
nl stoombevochtiger *m*
r паровой воздухоувлажнитель *m*

S1201 *e* steam injector
d Dampfstrahlpumpe *f*
f pompe *f* à jet de vapeur
nl stoominjector *m*
r пароструйный насос *m*, эжектор *m*

S1202 *e* steam jacket
d Dampfmantel *m*, Heizmantel *m*
f chemise *f* [enveloppe *f*] de vapeur
nl stoommantel *m*
r паровая рубашка *f*

S1203 *e* steam pile hammer
d Dampfbär *m*, Dampfhammer *m*
f mouton *m* à vapeur
nl stoomhei *m*
r паровой свайный молот *m*

S1204 *e* steam pipeline
d Dampfleitung *f*
f conduite *f* de vapeur, tuyau *m*
à vapeur
nl stoomleiding *f*
r паропровод *m*

S1205 *e* steam trap
d Kondensatableiter *m*,
Kondensatabscheider *m*,
Kondenstopf *m*
f séparateur *m* de condensat
nl condenspot *m*
r конденсатоотводчик *m*

S1206 *e* steam water heater
d dampfbeheizter
Warmwasserbehälter *m*
f chauffe-eau *m* à vapeur
nl stoomboiler *m*
r паровой водонагреватель *m*

S1207 *e* steel and concrete structures
d Verbundbauwerke *n pl* Stahl/
Beton, Bauwerke *n pl* aus Stahl
und Beton
f constructions *f pl* mixtes (*en
acier et béton*)
nl constructies *f pl* [bouwwerken *n pl*]
van staal en beton
r сталежелезобетонные [составные]
конструкции *f pl* (*из стали и
бетона*)

S1208 *e* steel bridge
d Stahlbrücke *f*
f pont *m* en acier
nl stalen brug *f(m)*
r стальной мост *m*

S1209 *e* steel casing
d Bohrrohr *n*, Futterrohr *n*
f tube *m* de revêtement [fourreau]
nl boorbuis *f(m)*
r обсадная труба *f*

S1210 *e* steel corrosion damage of concrete
d Beschädigung *f* von Beton durch
Stahlkorrosion
f endommagement *m* du béton armé
causé par la corrosion d'armature
nl beschadiging *f* van beton door
corrosie van wapeningsstaal
r повреждение *n* железобетона,
вызванное коррозией арматуры

S1211 *e* steel-crossing construction joint
d verdübelte Fuge *f*
f joint *m* de dilitation goujonné
(*entre deux dalles de béton
voisines*)
nl expansievoeg *f(m)* met stalen
deuvels
r шов *m* расширения со стальными
штырями (*в бетонном
дорожном покрытии*)

S1212 *e* steel dowel pin
d Stahlnagel *m*, Stahlstift *m*
f boulon *n*, goujon *m*, cheville *f*,
goupille *f*
nl stalen deuvel *m*
r стальной нагель *m*

S1213 *e* steel-fiber reinforced concrete,
steel fibrous concrete
d Stahlfiberbeton *m*,
Stahlfaserbeton *m*

f béton *m* à fibres d'acier, béton *m* renforcé par fibres d'acier
nl beton *m* met staalvezelwapening
r бетон *m*, армированный стальными волокнами; сталефибробетон *m*

S1214 *e* steel fibrous shotcrete
d Stahlfaser-Torkretbeton *m*, Stahlfiber-Spritzbeton *m*
f béton *m* projeté renforcé par fibres d'acier, gunite *f* à fibres d'acier
nl spuitbeton *n* met staalvezelwapening
r торкретбетон *m*, армированный стальными волокнами; сталефиброторкретбетон *m*

S1215 *e* steelfixer's pliers
d Zange *f* für Bewehrung
f pinces *f pl* de ferrailleur
nl wapeningstang *f(m)*
r арматурные клещи *pl*

S1216 *e* steel formwork
d Stahl(blech)schalung *f*
f coffrage *m* en acier
nl stalen bekisting *f*
r стальная опалубка *f*

S1217 *e* steel grid
d 1. Stahlgitter *n* 2. Stahlgitterbelag *m*
f 1. grille *f* d'acier 2. platelage *m* de roulement en métal deployé
nl 1. stalen rooster *m*, *n* 2. stalen roosterdek *n* (*brug*)
r 1. стальная решётка *f* 2. стальной решётчатый настил *m* (*моста*)

S1218 *e* steel grid decking, steel grid floor
d Stahlgitterfußboden *m*
f caillebotis *m* en grille d'acier
nl stalen roostervloer *m*
r стальной решётчатый настил *m* [пол *m*]

S1219 *e* steel landing mats
d Flugplatz-Stahlmatten *f pl*
f plaques *f pl* perforées d'envol grille d'acier (*pour le revêtement de routes en terre ou pistes d'envol*)
nl stalen wegenmat *f(m)* voor versterking van wegen en rolbanen
r стальные маты *m pl* для укрепления грунтовых покрытий (*грунтовых дорог, взлётно-посадочных полос*)

S1220 *e* steel mesh reinforcement
d Bewehrungsmatte *f*, Betonstahlmatte *f*
f armature *f* en treillis d'acier
nl stalen wapeningsnet *n*
r стальная арматурная сетка *f*

S1221 *e* steel pan forms
d Stahl-Kassetten-Deckenschalung *f*

f coffrage *m* STA-KA (*en tôle d'acier pour la construction de planchers alvéolaires en béton armé*)
nl vloerbekisting *f* van stalen cassettenplaat
r стальная опалубка *f* для часторебристых *или* кессонных перекрытий

S1222 *e* steel plate
d Stahlblech *n*; Stahlplatte *f*
f tôle forte d'acier; plaque *f* en acier
nl staalplaat *f(m)*, plaatstaal *n*
r толстый стальной лист *m*; стальная плита *f*

S1223 *e* steel sheet
d 1. Profilblech *n* 2. Feinblech *n*
f 1. tôle *f* profilée 2. tôle *f* mince (d'acier)
nl 1. geprofileerde staalplaat 2. bladstaal *n*
r 1. профилированный (холодногнутый) стальной настил *m* 2. тонкий стальной лист *m*

S1224 *e* steel sheetpile cofferdam
d Spundwandfangdamm *m*
f batardeau *m* en palplanches d'acier
nl stalen damwand *m*
r стальная шпунтовая перемычка *f*

S1225 *e* steel sheet piling
d Stahlspundwand *f*
f rideau *m* de palplanches en acier
nl stalen damwand *m*
r стальной шпунтовый ряд *m*; стальная шпунтовая стенка *f*

S1226 *e* steel shoe
d Stahlschuh *m*
f sabot *m* de pieu en acier
nl stalen paalschoen *m*
r стальной свайный башмак *m*

S1227 *e* steel-wheel roller
d Glattmantelwalze *f*, statische Walze *f*
f rouleau *m* [cylindre *m*] (à jante) lisse
nl wals *f(m)* met stalen rollen
r дорожный каток *m* с гладкими вальцами

S1228 *e* steel wire
d Stahldraht *m*
f fil *m* d'acier
nl staaldraad *m*
r стальная проволока *f*

S1229 *e* steel wire rope
d Stahldrahtseil *n*
f câble *m* (en fils) d'acier
nl staal(draad)kabel *m*
r стальной канат *m*, трос *m*

S1230 *e* steelwork erection
d Montage *f* der Stahlkonstruktionen

f montage *m* de la charpente en
   acier
nl montage *f* van staalconstructies
r монтаж *m* стальных конструкций

S1231 *e* **steening**
   d Brunnenausmauerung *f*
   f revêtement *m* de puits en
      maçonnerie
   nl bekleden *n* van een schacht
   r возведение *n* каменной крепи
      колодца

S1232 *e* **steep arch**
   d Stelzbogen *m*, gestelzter Bogen *m*
   f arc *m* exhaussé
   nl verhoogde boog *m*
   r крутая [возвышенная] арка *f*

S1233 *e* **steepness of slope**
   d Böschungssteilheit *f*
   f pente *f* de talus
   nl hellingshoek *f*, talud *n*
   r крутизна *f* откоса

S1234 *e* **stem**
   d 1. Säulenschaft *m* 2. lotrechter
      Teil *m* der Betonstützmauer,
      Steg *m* 3. Ventilspindel *f*
      4. Nietschaft *m*
   f 1. fût *m* (*d'une colonne*) 2. partie *f*
      verticale d'un mur de
      soutènement (*angulaire ou à*
      console) 3. tige *f* de clapet 4. tige *f*
      de rivet
   nl 1. schacht *f(m)* van een zuil 2.
      vleugelmuur *m* 3. as *f(m)*
      [spindel *n*] van een ventiel 4.
      klinknagelsteel *m*
   r 1. ствол *m*, тело *n* (*колонны*) 2.
      вертикальная часть *f*
      (*железобетонной подпорной*
      стенки) 3. шпиндель *m* (*вентиля*)
      4. стержень *m* (*заклёпки*)

S1235 *e* **step**
   d 1. Treppenstufe *f* 2. Galerie *f*
      (*in der Wand des Trockendocks*) 3.
      Arbeitsgang *m*; Etappe *f*
   f 1. marche *f* (*d'un escalier*) 2.
      redan *m* d'une cale de radoub 3.
      opération *f* de travail; phase *f* de
      travaux
   nl 1. trede (*m*) 2. versnijding *f* van
      de muur van een droogdok 3.
      (werk)fase *f*
   r 1. ступень *f* (*лестницы*) 2. уступ
      m в стенке сухого дока (*для*
      скуловых опор) 3. рабочая
      операция *f*; этап *m*

S1236 *e* **step iron**
   d Klettereisen *n*, Steigeisen *n*
   f échelon *m* métallique (*dans la*
      paroi de la cheminée ou puits)
   nl klimijzer *n*
   r ступень *f*, скоба *f* (*в стенке*
      колодца или дымовой трубы)

S1237 *e* **stepladder**
   d Stufenleiter *f*

f escabeau *m*, échelle *f* double
nl trapleer *f(m)*
r лестница-стремянка *f*

S1238 *e* **stepped diffuser**
   d abgetreppter mehrkegeliger
      Deckenluftverteiler *m*
   f diffuseur *m* d'air de plafon
      multicône
   nl getrapte luchtverdeler *m*
      [diffusor *m*]
   r многоконусный потолочный
      диффузор *m*

S1239 *e* **stepped drop**
   d Stufenabsturz *m*
   f chutes *f pl* en cascade [en gradins]
   nl cascade *m*
   r ступенчатый перепад *m*

S1240 *e* **stepped footing, stepped
      foundation**
   d abgetrepptes Fundament *n*,
      abgetreppter Gründungskörper *m*
   f fondation *f* en gradins
   nl getrapte [versneden] fundering *f*
   r фундамент *m* с уступами,
      ступенчатый фундамент *m*

S1241 *e* **stepping**
   d Abtreppung *f*
   f excavation *f* à gradins
   nl getrapte ingraving *f*
   r выемка *f* уступным забоем

S1242 *e* **sterilisation of water**
   d Entkeimung *f* [Sterilisation *f*] von
      Wasser
   f désinfection *f* [stérilisation *f*,
      décontamination *f*] de l'eau
   nl watersterilisatie *f*
   r обеззараживание *n* воды

S1243 *e* **sticky cement**
   d abgelagerter [anbackender,
      brückenbildender] Zement *m*
      (*im Silo*)
   f ciment *m* agglutiné (*résultat de*
      stockage)
   nl klonterend cement *n*, *m*
   r слёживающийся [слипшийся]
      цемент *m* (*на складах*);
      зависающий цемент *m* (*в*
      силосах)

S1244 *e* **sticky material**
   d 1. klebriges Material *n* 2.
      klebriger [anhaftender] Boden *m*
   f 1. matériau *m* adhésif 2. terrain
      m collant [gluant]
   nl 1. kleverig materiaal *n* 2. klevende
      grond *m*
   r липкий материал *m* или грунт *m*

S1245 *e* **stiff clay**
   d dichter [fester] Ton *m*
   f argile *f* dense
   nl vaste [zware] klei *f(m)*
   r плотная глина *f*

S1246 *e* **stiff concrete**
   d Trockenbeton *m*, steifer Beton *m*

f béton *m* sec
nl aardvochtig beton *n*
r жёсткий бетон *m*

S1247 *e* **stiffener**
d Versteifungsbauteil *m*,
Versteifungselement *n*
f élément *m* de raidissement
nl verstijvingselement *n*
r элемент *m* (*уголок*, *ребро*)
жёсткости

S1248 *e* **stiffening girder**
d Versteifungsträger *m*,
Aussteifungsträger *m*
f poutre *f* de rigidité [de
raidissement, de renforcement]
nl verstijvingsligger *m*
r балка *f* жёсткости

S1249 **stiffening member** *see* **stiffener**

S1250 *e* **stiffening rib**
d Versteifungsrippe *f*,
Verstärkungsrippe *f*
f nervure *f* raidisseuse [de
raidissement]
nl verstijvingsrib *f(m)*
r ребро *n* жёсткости

S1251 *e* **stiff-leg derrick**
d Streben-Derrick *m*, Bock-
-Derrick *m*
f derrick *m* à jambes de force
nl geschoorde laadmast *m*
r жестконогий деррик-кран *m*

S1252 *e* **stiffness**
d Steifigkeit *f*, Starrheit *f*
f rigidité *f*, raidissement *m*,
raidissage *m*
nl stijfheid *f*
r жёсткость *f*

S1253 *e* **stiffness factor**
d relative Steifigkeit *f*
f rigidité *f* de la barre relative
(*rapportée à l'unité de longueur
de la barre*)
nl stijfheidsfactor *m*
r относительная жёсткость *f*

S1254 *e* **stiffness of structure**
d Bauwerkssteifigkeit *f*;
Gebäudesteifigkeit *f*
f rigidité *f* de la charpente
nl stijfheid *f* van de constructie
r жёсткость *f* сооружения *или*
конструкции

S1255 **stiffness ratio** *see* **stiffness factor**

S1256 **stilling basin** *UK see* **stilling pool**

S1257 *e* **stilling basin sill**
d Tosbeckenschwelle *f*
f seuil *m* du bassin d'amortissement
nl woelkomdrempel *m*
r водобойный порог *m*

S1258 *e* **stilling pool** *US*
d Tosbecken *n*, Sturzbecken *n*,
Beruhigungsbecken *n*, Energie-
-Umwandlungsanlage *f*

f bassin *m* d'amortissement,
dissipateur *m* d'énergie
nl woelkom *f(m)*
r успокоительный бассейн *m*,
водобойный колодец *m*, камера *f*
гашения энергии потока

S1259 *e* **stilling well**
d Brunnen *m* mit
Beruhigungskammer *oder* mit
Beruhigungsrohr
f puits *m* de tranquillisation
nl put *m* met woelruimte
r успокоительный колодец *m*

S1260 *e* **stirrup**
d Bügel *m*
f étrier *m*, collier *m*
nl beugel *m* (*wapening*)
r (арматурный) хомут *m*; скоба *f*

S1261 *e* **stirrup spacing**
d Bügelabstand *m*
f écartement *m* [espacement *m*] des
étriers
nl beugelafstand *m*
r шаг *m* хомутов

S1262 **stockage area** *see* **storage area**

S1263 *e* **stoker**
d Feuerrostbeschicker *m*
f chargeur *m* mécanique (*du
combustible solide*)
nl mechanische stookinrichting *f*
r устройство *n* для механической
подачи твёрдого топлива

S1264 *e* **stone breaker**
d Schotterbrecher *m*, Steinbrecher *m*
f concasseur *m* de pierres
nl steenbreker *m*
r дробилка *f*

S1265 *e* **stone chippings, stone chips**
d Splitt *m*
f grenaille *f*, gravillon *m* concassé
nl split *n*
r каменная мелочь *f*, каменные
высевки *pl*, мелкий щебень *m*

S1266 *e* **stone drain**
d Steindrän *m*, Steinrigole *f*
f drain *m* à pierres sèches
nl steendrain *m*
r каменная дрена *f*

S1267 *e* **stone facing**
d Natursteinverblendung *f*,
Natursteinauskleidung *f*,
Steinverblendung *f*
f revêtement *m* en pierres taillées,
perré *m*
nl steenbekleding *f*
r каменная облицовка *f*

S1268 *e* **stonemesh mattress**
d Senkwurst *f*
f enveloppe *f* de grillage remplie de
pierres
nl steenwiep *f(m)* van metaalgaas
r каменно-сетчатый тюфяк *m*

S1269 *e* **stone paving**

d Abpflasterung *f*
f pavage *m* en pierre
nl keibestrating *f*
r каменная отмостка *f*

S1270 *e* **stone pitching**
d Steinvorlage *f*, handversetztes Steindeckwerk *n*
f perré *m*; perré *m* protecteur [de protection]
nl steenbezetting *f*
r каменная облицовка *f*; крепление *n* верхового откоса отмосткой

S1271 *e* **stone riprap**
d Steinschüttung *f*, Steinwurf *m*
f enrochement *m*, remblai *m* rocheux
nl steenbestorting
r каменная наброска *f*

S1272 *e* **stone sand**
d Steinbruchsand *m*
f sable *m* naturel de carrière, grenaille *f*
nl steengruis *n*
r каменная мелочь *f* из карьера

S1273 *e* **stone slab revetment**
d Steinplattenbelag *m* (*an Böschungen*), Steinplattenverkleidung *f*
f revêtement *m* en dalles de pierre
nl taludbekleding *f* van natuursteen
r одежда *f* откоса из каменных плит

S1274 *e* **stone waste**
d Steinschutt *m*
f déchets *m pl* de carrière
nl steenafval *n* uit een steengroeve
r каменная мелочь *f*; отходы *m pl* карьера

S1275 *e* **Stoney gate**
d Stoney-Schütz *n*, Stoney-Schütze *f*
f vanne *f* Stoney
nl Stoney-schuif *f(m)*
r плоский катковый затвор *m*

S1276 *e* **stoning**
d 1. Oberflächenbearbeitung *f* mit Schleifkörper
2. Kolkschutzschotterschicht *f*
f 1. abrasion *f*, ponçage *m*
2. enrochement *m*
nl stortebedden en ontvangbedden in de sluiskolk
r 1. обработка *f* поверхности (абразивным) камнем
2. каменная отсыпка *f* (*в зоне колебаний уровня воды*)

S1277 *e* **stop cock**
d Absperrhahn *m*
f soupape *f* d'arrêt
nl stopkraan *f(m)*
r запорный кран *m*

S1278 *e* **stop end**
d Sperreinlage *f* in der Arbeitsfuge
f plaque *f* [joue *f*] mise au joint de reprise

nl afsluiting *f* [plaat *f(m)*] in de betonvoeg
r ограничительная прокладка *f* в рабочем шве бетонируемой конструкции

S1279 *e* **stop limit switch**
d Fahrt-End(aus)schalter *m*
f interrupteur *m* de fin de course (*d'un ascenseur*)
nl eind-schakelaar *m* (*lift*)
r концевой выключатель *m* (*для остановки лифта на этажах*)

S1280 *e* **stoplog groove**
d Dammnut *f*, Dammfalz *m*, Dammbalkenschlitz *m*
f rainure *f* à poutrelles
nl schotbalksponning *f(m)*
r шандорный паз *m*

S1281 *e* **stop-logs**
d Staubalken *m pl*, Dammbalken *m pl*, Staubohlen *f pl*
f vanne *f* à poutrelles; aiguilles *f pl* d'écluse
nl schotbalken *m pl*
r шандоры *pl*

S1282 *e* **stopping knife**
d Stahlspachtel *m*, *f*
f lissoir *m*, lisseuse *f*; spatule *f*
nl stopmes *n*, plamuurmes *n*
r металлический шпатель *m*; нож *m* для заделки трещин

S1283 **stop-planks** *see* **stop-logs**
S1284 *e* **stop valve**
d Absperrventil *n*
f soupape *f* [vanne *f*] d'arrêt
nl afsluitklep *f(m)*
r запорный клапан *m* [вентиль *m*]; затвор *m*

S1285 *e* **storage**
d 1. Lager *n*; Speicher *m*
2. Speicherung *f*; Vorrat *m*
3. (Wasser-)Speicher *m*, Stausee *m*, Speicherbecken *n*
f 1. dépôt *m*, magasin *m*
2. emmagasinement *m* 3. bassin *m*, réservoir *m* à eau
nl 1. opslagplaats *f(m)* 2. opslagvoorraad *m* 3. spaarbekken *n*
r 1. склад *m*; хранилище *n*
2. накопление *n*; запас *m*
3. водохранилище *n*

S1286 *e* **storage area**
d 1. Lagerfläche *f* 2. Staubereich *m*
f 1. aire *f* de stockage 2. zone *f* de remous
nl 1. opslagterrein *n* 2. stuwzone *f*
r 1. площадка *f* для складирования и хранения (*материалов, конструкций*) 2. зона *f* подпора

S1287 *e* **storage calorifier**
d Heißwasserspeicher *m*
f chauffe-eau *m* à accumulation
nl boiler *m*
r ёмкостный водонагреватель *m*

S1288 e **storage capacity**
d Fassungsvermögen n,
Speicherkapazität f,
Speicherfähigkeit f, Speicherraum
m, Stauraum m
f capacité f de la retenue, tranche f
utile du réservoir
nl capaciteit f van het spaarbekken
r полезная ёмкость f водохранилища

S1289 e **storage chamber**
d Sparkammer f
f chambre f d'épargne (d'une
écluse)
nl spaarkom m (sluis)
r сберегательная камера f
(шлюза)

S1290 e **storage coefficient**
d 1. Rückhaltefaktor m,
Speicherungskoeffizient m eines
Wasserleiters, Koeffizient m der
Wasserabgabe 2. Nutzkoeffizient m
der Speicherkapazität;
Koeffizient m der
Ausgleichwirkung des Speichers
f 1. coefficient m d'emmagasinement
d'une nappe captive;
coefficient m de rendement d'eau
2. taux m d'emmagasinement,
coefficient m de capacité de
retenue; coefficient m de débit
dérivé
nl 1. stuwcoëfficiënt m 2. nuttig
rendement n van het
spaarbekken
r 1. коэффициент m
водовместимости
(подземного бассейна);
коэффициент m водоотдачи
2. коэффициент m полезной
ёмкости водохранилища;
коэффициент m регулирования
стока

S1291 e **storage curve**
d Speicherinhaltskurve f
f courbe f hauteur-volume
nl curve f van de stuwmeerinhoud
r кривая f запаса [объёмов]
(водохранилища)

S1292 e **storage cycle**
d Speicher-Zyklus m,
Speicherperiode f
f cycle m d'emmagasinement
nl stuwcyclus m, stuwperiode f
r цикл m накопления; период m
заполнения (водохранилища)

S1293 e **storage dam**
d Speichersperre f, Speicherdamm m
f barrage m d'accumulation [de
retenue], barrage-réservoir m
nl dam m van het spaarbekken
r водохранилищная плотина f

S1294 e **storage heater**
d Speicherofen m

f appareil m de chauffage à
accumulation
nl warmwatervoorraadtoestel n,
boiler m
r электроаккумуляционная печь f

S1295 e **storage heating**
d Speicherheizung f
f chauffage m par batteries
[accumulateurs] électriques
nl verwarming f met in accu's
opgeslagen energie
r электроаккумуляционное
отопление n

S1296 **storage level** US see **banked-up
water level**

S1297 e **storage racks**
d Lagergestelle n pl
f rayonnage m [casier m] de
stockage
nl opslagstellingen f pl
r складские стеллажи m pl;
стеллажи m pl для хранения

S1298 **storage ratio** see **storage
coefficient 2.**

S1299 e **storage reservoir**
d 1. Speicherbecken n,
Wasserspeicher m
2. Vorratsbehälter m
f 1. retenue f, bassin m de
retenue 2. réservoir m à eau
potable
nl 1. spaarbekken n
2. voorraadreservoir n
r 1. водохранилище n
2. резервуар m чистой воды
(водоснабжение)

S1300 **storage water heater** see **storage
calorifier**

S1301 e **storey**
d Geschoß n, Stockwerk n, Etage f
f étage m
nl etage f, verdieping f
r этаж m

S1302 e **storey height**
d Geschoßhöhe f
f hauteur f d'étage
nl verdiepingshoogte f
r высота f этажа (от пола до
пола)

S1303 e **storey-height standpipe**
d geschoßhohes Steigrohrelement n,
Geschoßanschlußrohr n
f conduite f hauteur d'étage
nl stijgbuis f(m) ter hoogte van een
verdieping
r этажестояк m

S1304 e **storey-height wall panel**
d geschoßhohe Wandtafel f
f panneau m de la hauteur d'un
étage
nl wandpaneel n ter hoogte van een
verdieping
r стеновая панель f высотой в один
этаж

S1305 *e* **storm cellar**
 d Schutzkeller *m*
 f cave-refuge *f* (*contre les orages*)
 nl schuilkelder *m* (*voor tornado's*)
 r подвал-убежище *m* (*для укрытия
 от ураганов*)

S1306 **storm drain** *see* **storm sewer**

S1307 *e* **storm-overflow sewer**
 d Regenauslaß *m* der
 Mischkanalisation,
 Regen(wasser)überlauf *m*,
 Regenüberfall *m*
 f évacuateur *m* [déversoir *m*]
 d'orage, émissaire *m* pluvial
 nl regenuitlaat *m* van het gemengde
 rioolstelsel, regenwateroverloop *m*
 r ливнеспуск *m*, ливнесброс *m*

S1308 *e* **storm runoff**
 d Regen(wasser)abfluß *m*
 f ruissellement *m* de la pluie,
 écoulement *m* d'averse
 nl regenwaterafvoer *f(m)* (*debiet*)
 r ливневый [дождевой] сток *m*

S1309 *e* **storm sewer**
 d Regenwassersammler *m*,
 Regenwasserkanal *m*,
 Regenwasserleitung *f*,
 Flutdrän *m*
 f égout *m* pluvial, canalisation *f*
 d'eaux pluviales
 nl regenwaterriool *n*
 r коллектор *m* ливневой канализации, дождевой водосток *m*,
 ливнеотводный канал *m* [водовод
 m]

S1310 *e* **storm sewers**
 d Regenwasserableitung *f*
 f égouts *m pl* pluvials
 nl regenwaterafvoer *f*
 r ливневая канализация *f*,
 водостоки *m pl*

S1311 *e* **storm surge**
 d Sturmflut *f*, Sturmtide *f*
 f marée *f* de tempête, onde *f* de
 crue
 nl stormvloed *m*, stormtij *n*
 r штормовой нагон *m*

S1312 *e* **storm wall**
 d Brustmauer *f*, Brustwehr *f*,
 Brüstung *f*, Strandmauer *f*,
 Parapett *n*
 f mur *m* de garde
 nl stormvloedkering *f*
 r парапетная [волноотбойная]
 стенка *f*

S1313 *e* **storm water**
 d Regenwasser *n*; Regenabwasser *n*
 f eau *f* pluviale [de pluie]; eaux *f*
 pl pluviales
 nl regenwater *n*
 r ливневая вода *f*; ливневые
 сточные воды *f pl*

S1314 *e* **storm-water drainage**

S1315 *e* **storm-water inlet**
 d Regenüberlauf *m*,
 Regenüberfall *m*; Regeneinlauf
 m
 f déversoir *m* d'orage [d'eau de
 pluie] ; bouche *f* des eaux
 pluviales
 nl regenoverloop *m*; rioolput *m*
 r ливнеспуск *m*; дождеприёмник
 m

S1316 *e* **storm water outfall**
 d Regenwasserauslaßbauwerk *n*,
 Regenauslaß *m*, Regenauslauf *m*
 f débouché *m* [ouvrage *m* de
 décharge] pluvial
 nl regenwateraflaatwerk *n*
 r ливнеотвод *m*

S1317 **storm water overflow** *see* **storm-
 -overflow sewer**

S1318 *e* **storm water tank**
 d Regenwasserbecken *n*
 f bassin *m* d'eau de pluie
 nl regenwaterbassin *n*
 r отстойник *m* для дождевых вод

S1319 *e* **storm window**
 d äußerer Fensterrahmen *m*
 f chassis *m* de fenêtre extérieur
 nl buitenste raam *n*
 r наружная рама *f* окна

S1320 *e* **stove heating**
 d Ofenheizung *f*
 f chauffage *m* par poêles
 nl kachelverwarming *f*
 r печное отопление *n*

S1321 *e* **straight blade impeller**
 d Radial-Kreiselrad *n*, Schaufelrad
 n mit geraden Schaufeln
 f roue *f* motrice radiale
 nl scheprad *n*, waaier *m* met rechte
 schoepen
 r радиальное рабочее колесо *n*,
 рабочее колесо *n*
 (*вентилятора, турбины*)
 с радиальными лопатками

S1322 *e* **straight bridge**
 d gerade Brücke *f*
 f pont *m* droit
 nl rechte brug *f(m)*
 r прямолинейный мост *m*

S1323 *e* **straightedge**
 d Abziehlatte *f*; Kardätsche *f*
 f latte *f* de réglage, règle *f*
 d'égalisation
 nl richtlat *f(m)*; rijlat *f(m)*
 r разравнивающая рейка *f*;
 правило *n*

S1324 *e* straight reinforcement bars
  d gerade Eisen *n pl*
  [Bewehrungsstäbe *m pl*]
  f barres *f pl* d'armature droites
  nl rechte wapeningsstaven *f(m) pl*
  r прямые арматурные стержни *m pl*

S1325 straight-run bitumen *see*
  petroleum asphalt

S1326 *e* straight-through gas-fired water
  heater
  d Gasdurchlauferhitzer *m*
  f chauffe-eau *m* à écoulement
  direct à gaz
  nl gasgeiser *m*
  r проточный газовый
  водонагреватель *m*

S1327 *e* straight way cock
  d Durchgangshahn *m*, Zweiwegehahn
  f robinet *m* droit
  nl rechte afsluiter *m*
  r проходной кран *m*

S1328 *e* straight-way valve
  d Durchgangsventil *n*,
  Durchlaufventil *n*,
  Durchflußventil *n*
  f valve *f* de passage
  nl recht ventiel *n*
  r проходной клапан *m*

S1329 *e* straight web, straight web
  sheetpile
  d Flachspundbohle *f*, Flachprofil *n*
  f palplanche *f* à âme plate
  nl damplank *f(m)* met vlak lijf
  r плоская шпунтовая свая *f*;
  плоский стальной шпунт *m*

S1330 *e* strain
  d Verformung *f*, Formänderung *f*
  f déformation *f*
  nl vervofming *f*, vormverandering *f*
  r деформация *f*

S1331 *e* strain capacity
  d Verformbarkeit *f*
  f déformabilité *f*
  nl vervormbaarheid *f*
  r деформационная способность *f*,
  деформируемость *f*; податливость
  f (конструкции)

S1332 *e* strain energy
  d Verformungsenergie *f*,
  Formänderungsenergie *f*;
  Verformungsarbeit *f*,
  Formänderungsarbeit *f*
  f énergie *f* de déformation; travail
  m de déformation
  nl vervormingsenergie *f*
  r энергия *f* деформации; работа *f*
  деформации

S1333 *e* strain-energy equation
  d Verformungsenergiegleichung *f*
  f équation *f* de l'énergie [du
  travail] de déformation
  nl vergelijking *f* van de
  vervormingsenergie

r уравнение *n* работы деформации

S1334 *e* strainer well
  d Brunnen *m* mit Filterkorb [mit
  Filterrohr]
  f puits *m* crépine
  nl bron *f(m)* met filter
  r колодец *m* с фильтром

S1335 *e* strain hardening
  d Verfestigung *f*, Kaltverfestigung *f*
  f consolidation *f*; durcissement *m*
  mécanique
  nl koudharden *n*
  r упрочнение *n (при деформации)*;
  наклёп *m*

S1336 *e* strain-hardening range
  d Verfestigungsbereich *m*
  f zone *f* durcie [de durcissement
  mécanique]; zone d'écrouissage
  nl verstijvingszone *f*
  r зона *f* [область *f*] упрочнения
  или наклёпа

S1337 *e* strand
  d Litze *f*
  f toron *m*; câble *m* d'armature
  nl streng *f(m) (kabel, wapening)*
  r прядь *f* каната; арматурная
  прядь *f*

S1338 stranded caisson *see* box caisson

S1339 *e* strand grip
  d Bewehrungsbündel-Ankerklemme *f*
  f dispositif *m* d'ancrage (d'un
  câble d'armature)
  nl ankerklem *f(m)* voor
  wapeningsstrengen
  r зажим *m* для анкеровки
  арматурной пряди

S1340 *e* strand-laid rope
  d Litzenseil *n*
  f câble *m* d'acier à torons
  nl gevlochten staalkabel *f(m)*
  r многопрядный стальной канат *m*

S1341 *e* S-trap
  d S-Traps *m*,
  Röhrengeruchsverschluß *m*
  f siphon *m* (d'évier) en S
  nl zwanehals *m*, sifon *m*
  r двухоборотный сифон *m*

S1342 *e* strategic planning of public works
  d Perspektivplanung *f* der
  öffentlichen Bauarbeiten
  f planification *f* perspective [à long
  terme] de la construction des
  ouvrages de génie civil
  nl bouwplan *n* voor de lange termijn
  van openbare werken
  r перспективное планирование *n*
  строительства общественных
  сооружений

S1343 *e* stratification
  d Schichtung *f*
  f stratification *f*; ségrégation *f*
  nl stratificatie *f*

r стратификация f (воздуха,
бетона)

S1344 e strawboard
d Strohplatte f
f panneau m en paille, mat(elas) m
de paille
nl stro-plaat f(m)
r соломит m (плиты из
прессованной соломы с
картонной облицовкой)

S1345 e streaks of mortar
d Mörtelstreifen m pl
f protubérances f pl dans le béton
(correspondant à des jours entre
les planches du coffrage)
nl mortelstrepen f(m) pl
r гребни m pl [наплывы m pl]
цементного камня (в местах
сопряжения опалубочных досок)

S1347 e stream centre line
d Strommitte f
f axe m du cours d'eau
nl as f(m) van de rivier
r осевая линия f потока

S1348 e stream crossing
d Flußdurchquerung f,
Flußüberquerung f
f traversée f de cours d'eau [de
rivière]
nl rivierovergang m
r переход m через реку
(мостовой, трубопроводный,
кабельный)

S1349 e stream cross-section
d Wasserquerschnitt m
f section f transversale d'un
course d'eau
nl nat profiel n
r водное сечение n

S1350 e streamflow
d 1. Abflußwassermenge f,
Durchflußwassermenge f,
Flußabfluß m 2. Flußströmung f
f 1. écoulement m [débit m]
fluvial 2. courant m d'une
rivière
nl 1. afvoer m van een rivier
2. stroming f van een rivier
r 1. речной сток m 2. русловый
поток m

S1351 e stream flow measuring station
d Abflußmeßstelle f
f station f de jaugeage, poste f
hydrologique
nl afvoermeetplaats f(m)
r гидрологический пост m

S1352 e stream gauging
d Strommessung f,
Wassermengenmessung f,
Abflußmessung f
f jaugeage m d'un cours d'eau
nl debietmeting f
r измерение n расходов (водотока)

S1352a stream-gauging network see
hydrometric network

S1353 e stream hydraulics
d Flußhydraulik f
f hydraulique f fluviale
nl rivierhydraulica f
r речная гидравлика f

S1354 e streaming flow
d Strömen n, ruhiges Fließen n
f écoulement m tranquille
[laminaire]
nl stromen n (rustige stroming)
r докритический поток m;
спокойное течение n

S1355 e stream intake
d Flußwasserfassung f
f prise f d'eau en rivière
nl watervang m aan een rivier
r речной водозабор m

S1357 e stream profile
d Flußlängenschnitt m
f profil m longitudinal du lit de
rivière [de fleuve]
nl lengteprofiel n van de rivier
r продольный профиль m реки

S1358 e stream routing
d Wasserlaufberechnung f
f calcul m de la propagation des
crues
nl berekening f van het
stroomverloop
r расчёт m речного стока

S1359 e stream velocity
d Strömungsgeschwindigkeit f
f vitesse f de courant
nl stroomsnelheid f
r скорость f потока

S1360 e street
d Straße f, Stadtstraße f
f rue f; voie f urbaine
nl straat f(m), weg m
r улица f, городская дорога f

S1361 e street crossing
d Stadtstraßenkreuzung f
f croisement m de rues
nl straatkruising f
r уличный перекрёсток m

S1362 e street inlet
d Regeneinlaß m, Regeneinlauf m
f bouche f d'égout [de caniveau]
nl straatkolk m
r дождеприёмник m

S1363 e strength
d 1. Festigkeit f 2. Konzentration f
von Verschmutzungen
in Abwasser
f 1. résistance f 2. concentration f
des agents de pollution dans les
eaux d'égout
nl 1. vastheid f, sterkte f
2. concentratie f van vervuilingen
in afvalwater

r 1. прочность f 2. концентрация f
загрязняющих примесей
в сточных водах

S1364 e strength analysis
d Festigkeitsberechnung f
f calcul m de résistance
nl sterkteberekening f
r расчёт m прочности [на
прочность]

S1365 e strength criterium
d Festigkeitskriterium n,
Festigkeitsbedingung f
f critère m de résistance
nl sterktecriterium n
r критерий m прочности

S1367 e strength developing
d Festigkeitszunahme f
f accroissement m de résistance
nl vastheidstoename f
r нарастание n прочности

S1368 e strength grade
d Güteklasse f (nach Festigkeit)
f qualité f (d'un matériau ou d'un
produit) définie par sa
résistance à la rupture
nl kwaliteitsklasse f naar sterkte
r марка f материала или изделия
(по прочности)

S1369 e strength limit
d Festigkeitsgrenze f
f limite f de résistance
nl grenswaarde f van de vastheid
r предел m прочности

S1370 e strength-maturity factor
d Koeffizient m des
Festigkeitszuwachses mit der
Zeit
f facteur m de maturation du
béton (caractérisant
l'accroissement de résistance de
béton dans le temps)
nl vastheidstoenamecoëfficiënt m van
beton
r коэффициент m,
характеризующий нарастание
прочности бетона (в
зависимости от времени)

S1371 e strength of concrete at twenty-
-eight days
d 28-tägige Betonfestigkeit f
f résistance f du béton à l'âge de
28 jours
nl drukvastheid f na 28 dagen
r 28-суточная прочность f бетона

S1372 e strength of materials
d Festigkeitslehre f
f résistance f des matériaux
nl sterkteleer f(m)
r сопротивление n материалов

S1373 e strength test
d Festigkeitsprüfung f
f essai m de résistance
nl vastheidsproef f(m)

r определение n прочности,
испытание n на прочность

S1374 e strength under sustained load
d Dauerstandfestigkeit f
f résistance f sous charge de
longue durée
nl sterkte f bij ononderbroken
belasting
r прочность f при длительной
нагрузке, длительная
прочность f

S1375 e stress
d Spannung f, Beanspruchung f
f contrainte f, tension f
nl spanning f, belasting f
r напряжение n

S1376 e stress analysis
d Spannungsberechnung f
f calcul m des contraintes
nl spanningsberekening f
r расчёт m напряжений

S1377 e stress circle, stress circle of Mohr
d (Mohrscher) Spannungskreis m,
Mohr-Kreis m
f cercle m de contraintes, cercle m
de Mohr
nl spanningscirkel m van Mohr
r круг m напряжений (Мора)

S1378 e stress components
d Spannungskomponenten f pl
f composantes f pl de contrainte
nl spanningscomponenten m pl
r составляющие f pl [компоненты
f pl] напряжения

S1379 e stress concentration
d Spannungskonzentration f
f concentration f de contraintes
nl spanningsconcentratie f
r концентрация f напряжений

S1380 e stress corrosion
d Spannungskorrosion f
f corrosion f sous tension
nl spanningscorrosie f
r коррозия f под напряжением

S1381 e stress distribution
d Spannungsverteilung f
f distribution f des contraintes
nl spanningsverdeling f
r распределение n напряжений

S1382 e stress due to prestress
d Spannung f durch Vorspannung
f contrainte f due à la précontrainte
nl drukspanning f in beton
tengevolge van voorspanning
r усилие n (в бетоне), вызванное
предварительным напряжением

S1383 e stressed member
d beanspruchtes Element n
f élément m (de construction)
sollicité [sous charge]
nl belast (constructie-)element n
r напряжённый [нагруженный]
элемент m (конструкции)

S1384 *e* **stressed-skin construction**
d Schalenbauweise *f*, beanspruchte
[gespannte] Schalenkonstruktion *f*
f structure *f* géodésique
nl schaalconstructie *f*;
constructie *f*
met dragende huid
r пространственная стержневая
конструкция *f* с работающей
ограждающей оболочкой
[обшивкой] (*напр.*
геодезический купол)

S1385 *e* **stressed-skin roof deck**
d beanspruchte [gespannte]
Membranüberdachung *f*
f toiture *f* [couverture *f*]
géodésique
nl schaaldak *n*
r кровля *f*, работающая совместно
с опорными конструкциями
(*фермами и прогонами*)

S1386 *e* **stresses beyond the elastic limit**
d Spannungen *f pl*, die
Elastizitätsgrenze überschreiten
f contraintes *f pl* au delà de la
limite élastique
nl spanningen *f pl* boven de
elasticiteitsgrens
r напряжения *n pl* за пределом
упругости

S1387 *e* **stresses in the elastic range**
d Spannungen *f pl*, die
Elastizitätsgrenze nicht
überschreiten
f contraintes *f pl* dans le domaine
élastique
nl spanningen *f pl* binnen het
elasticiteitsbereik
r напряжения *n pl* в упругой
области

S1388 *e* **stresses in the plastic range**
d Spannungen *f pl* im
Plastizitätsbereich
f contraintes *f pl* dans le domaine
plastique
nl spanningen *f(m) pl* in het elastische
gebied
r напряжения *n pl* в
пластической области

S1389 **stresses within the elastic limit** *see*
stresses in the elastic range

S1390 *e* **stress function**
d (Airysche) Spannungsfunktion *f*
f fonction *f* de contrainte [d'Airy]
nl spanningsfunctie *f*, functie *f* van
Airy
r функция *f* напряжений

S1391 *e* **stressing abutments**
d Spannblöcke *m pl* (*zum Vorspannen
der Bewehrung*)
f massifs *m pl* d'ancrage (*pour mise
en tension de l'armature de
précontrainte*)

nl spanblokken *n pl* (*voor het
voorspannen van de wapening*)
r натяжные упоры *m pl* [массивы
m pl]

S1392 *e* **stressing device**
d Spannvorrichtung *f*
f dispositif *m* de tension
nl voorspanapparaat *n*
r натяжное устройство *n*

S1393 *e* **stressing equipment**
d Bewehrungsspannvorrichtung *f*
f équipement *m* de mise en tension
nl voorspanuitrusting *f*
r оборудование *n* для
преднапряжения

S1394 *e* **stressing jack**
d Bewehrungsspannpresse *f*
f vérin *m* de mise en tension
nl vijzel *n* voor het spannen van de
wapening
r домкрат *m* для натяжения
арматуры

S1395 *e* **stressing operations**
d Bewehrungsspannvorgänge *m pl*;
Spannverfahren *n*
f opérations *f pl* de précontrainte
[de mise en tension]
nl aanbrengen *n* van de voorspanning;
voorspanprocedure *f*
r операции *f pl* преднапряжения
конструкции; создание *n*
предварительного напряжения

S1396 *e* **stress jump**
d Spannungssprung *m*
f saut *m* de contrainte
nl spanningssprong *m*
r скачок *m* напряжений

S1397 *e* **stress loss**
d Spannungsverlust *m*
f perte *f* de contraintes
nl spanningsverlies *n*
r потеря *f* напряжения

S1398 *e* **stress raiser**
d Spannungskonzentrator *m*
f concentrateur *m* de contraintes
nl spanningsversterker *m*
r концентратор *m* напряжений

S1399 *e* **stress range**
d Spannungsbereich *m*
f zone *f* des contraintes ; étendue *f*
[amplitude *f*] des contraintes
nl spanningsbereik *n*, spanningszone *f*
r зона *f* [область *f*] напряжений;
диапазон *m* [амплитуда *f*]
напряжений

S1400 *e* **stress reduction factor**
d Knickbeiwert *m*, Knickzahl *f*,
Spannungsverminderungsfaktor *m*
f coefficient *m* de (calcul au)
flambage [flambement]
nl knikcoëfficiënt *m*

r коэффициент *m* продольного
изгиба; коэффициент *m*
уменьшения допускаемых
напряжений *(при продольном
изгибе)*

S1401 *e* **stress relaxation**
d Spannungsrelaxation *f*
f relaxation *f* des contraintes
nl spanningsvermindering *f*
r релаксация *f* напряжений;
ослабление *n* напряжений

S1402 *e* **stress relieving**
d Spannungsfreimachen *n,*
Entspannung *f*
f suppression *f* des contraintes
nl spanningsvrij maken *n*
r снятие *n* (остаточных)
напряжений

S1403 **stress state** *see* **state of stress**
S1404 *e* **stress-strain curve**
d Spannungs-Dehnungs-Diagramm *n*
f courbe *f* [diagramme *m*]
contrainte-déformation
nl spanning-rekdiagram *n*
r кривая *f* [диаграмма *f*]
зависимости напряжений от
деформаций

S1405 *e* **stretcher**
d Läufer *m* (*Mauerwerk*)
f panneresse *f*; brique *f* en
panneresse
nl strekse steen *m*
r ложок *m* (*кирпича*); кирпич *m,*
уложенный ложком

S1406 *e* **stretcher bond**
d Läuferverband *m,*
Halbsteinverband *m*
f appareil *m* à demi-brique en
long
nl halfsteens verband *n* (*muur*),
loopverband *n* (*vloer*)
r ложковая перевязка *f*
(*кирпичной кладки*)

S1407 *e* **stretching**
d Dehnung *f*, Streckung *f*,
Anspannung *f*; Verlängerung *f*
f allongement *m*, élongation *f* ;
dilatation *f*
nl rek *f*; verlenging *f*
r растяжение *n*, вытягивание *n*;
удлинение *n*

S1408 **strike-off screed** *see* **screed board**
S1409 *e* **striking**
d Ausrüstung *f*; Ausschalung *f,*
Entformung *f*
f décintrage *m*, décintrement *m*;
décoffrage *m*
nl lossen *n* van formeel; ontkisten *n*
r раскружаливание *n*; распалубка *f*

S1410 *e* **striking period**
d Ausschalfrist *f*, Ausschalzeit *f*
f délai *m* de décoffrage
nl ontkistingstermijn *m*

r срок *m* распалубливания
S1411 *e* **string**
d 1. Schnur *f*; Seil *n* 2. Treppenwange
f
f 1. cordon *m*, corde *f* ; lacet *m*
2. limon *m*
nl 1. snoer *n*; koord *n* 2. trapboom *m*
r 1. шнур *m*; верёвка *f* 2. тетива *f,*
косоур *m* (*лестницы*)

S1412 *e* **stringer**
d 1. Längsträger *m* 2. Treppenwange
f 3. Rödelträger *m,*
Rödelbalken *m*, Reitel *m*
f 1. longeron *m* ; poutre *f*
longitudinale 2. limon *m*
3. longrine *f* ; tirant *m*
nl 1. langsligger *m* 2. trapboom *m*
3. trekbalk *m*
r 1. продольная балка *f* 2. косоур
m, тетива *f* 3. пажилина *f,*
продольный брус *m*

S1413 **string polygon** *see* **funicular
polygon**

S1414 *e* **strip city**
d Bandstadt *f*, lineare Stadt *f*
f ville *f* ruban [en ruban, linéaire]
nl stad *f* met lintbebouwing
r город *m* ленточной застройки

S1415 *e* **strip-cutting shears**
d Längsschnittschere *f*
f cisaille *f* pour coupe dans le sens
de la longueur
nl schaar *f(m)* voor langsknippen
r ножницы *pl* продольной резки

S1416 **strip footing** *see* **continuous
footing**

S1417 *e* **strip foundation**
d Streifenfundament *n,*
Gründungsstreifen *m*
f fondation *f* continue [sur semelle]
nl strokenfundering *f*
r ленточный фундамент *m*

S1418 *e* **strip heater**
d 1. Heizband *n*
2. Lamellenheizkörper *m*
f 1. dispositif *m* de chauffage à
bande 2. radiateur *m* à lammes
[à lamelles]
nl 1. bandverwarmer *m*
2. lamellenradiateur *m*
r 1. полосовой нагревательный
элемент *m*, ленточный излучатель
m 2. пластинчатый радиатор *m*

S1419 *e* **strip heating**
d 1. Heizband-Heizung
2. Lamellenheizung *f,*
Grillheizung *f*
f 1. installation *f* de chauffage par
rubans chauffants 2. chauffage *m*
par radiateurs à ailettes [à
lamelles]
nl verwarmingssysteem *n* met
lamellenradiatoren

r система *f* отопления
с ленточными излучателями

S1420 *e* **stripping**
d 1. Abräumung *f*,
Abraumbetrieb *m*; Abheben *n*
von Rasen und Humus
2. Ausschalung *f*, Ausrüstung *f*
f 1. découverte *f*; dégazonnement *m*
2. décoffrage *m*
nl 1. blootleggen *n* 2. ontkisten *n*
r 1. вскрыша *f*, вскрышные работы
f pl снятие *n* растительного слоя
2. распалубка *f*

S1421 *e* **stripping knife**
d Abstreifer *m*
f couteau *m* à raser; racloir *m*,
raclette *f*
nl handkrabber *m*, trekmes *n*,
haalmes *n*
r ручной скребок *m* (*малярный*
инструмент)

S1422 *e* **stripping shovel**
d Abraumbagger *m*
f excavateur *m* [excavatrice *f*]
morts-terrains [de déblais],
excavateur *m* pour découverte
nl laadschop *f(m)* voor het
schoonmaken van het bouwterrein
r вскрышной экскаватор *m*

S1423 *e* **strongback**
d Traverse *f*
f traverse *f* [poutre *f*] de renfort
[de renforcement]
nl dwarsbalk *m*
r траверса *f* [балка *f*],
усиливающая монтируемый
элемент конструкции

S1424 *e* **structural adequacy**
d Entsprechen *n* den bautechnischen
[konstruktiven] Anforderungen
f conformité *f* aux règlements de
construction
nl voldoen *n* aan de gestelde
constructieve eisen
r соответствие *n* строительным *или*
конструктивным требованиям

S1425 *e* **structural application**
d bautechnische Anwendung *f*,
Verwendung *f* im
Bau(wesen)
f emploi *m* [utilisation *f*] à
l'effet de construction
nl toepassing *f* in de bouw
r применение *n* [использование *n*]
в строительных целях

S1426 *e* **structural bearings**
d Auflagerteile *m pl*
f appareils *m pl* d'appui, appuis
m pl
nl opleggingen *f pl*
r опорные части *f pl*

S1427 *e* **structural behaviour**
d Verhalten *n* der Konstruktion

f comportement *m* [tenue *f*] d'un
ouvrage
nl gedrag *n* van de bouwconstructie
r работа *f* строительной
конструкции *или* сооружения (*под*
нагрузкой)

S1428 *e* **structural component**
d Bauelement *n*
f composant *m* [élément *m*] de
construction
nl constructie-element *n*
r элемент *m* конструкции

S1429 *e* **structural concept**
d konstruktive Lösung *f*
f conception *f* [solution *f*]
constructive
nl constructief concept *n*
r конструктивное решение *n*

S1430 *e* **structural concrete**
d Konstruktionsbeton *m*,
Bauwerkbeton *m*
f béton *m* pour ouvrages d'art
nl beton *n* voor constructiedoeleinde n
r конструктивный бетон *m*

S1431 *e* **structural connections**
d Verbindungen *f pl* der
Baukonstruktionen
f joints *m pl* [assemblages *m pl*]
des éléments constructifs
nl verbindingsmiddelen *n pl* voor
bouwconstructies
r соединения *n pl* [узлы *m pl*]
строительных конструкций

S1432 *e* **structural design**
d 1. konstruktive Ausbildung *f*,
konstruktiver Entwurf *m*
2. statische Berechnung *f* mit
Bemessung [Dimensionierung]
f 1. élaboration *f* du projet de
construction [de génie civil]
2. méthode *f* de calcul des
constructions
nl 1. constructieve vormgeving *f*
2. statische berekening *f*
r 1. строительное проектирование
n 2. статический расчёт *m*;
конструирование *n*, расчёт *m*
(*конструкций*)

S1433 *e* **structural drawings**
d Bauzeichnungen *f pl*
f dessins *m pl* de bâtiment
nl bouwtekeningen *f pl*
r строительные чертежи *m pl*;
строительная часть *f* проекта

S1434 *e* **structural dynamics**
d Bau(werks)dynamik *f*
f dynamique *f* des constructions
[des ouvrages]
nl bouwdynamica *f*
r динамика *f* сооружений

S1435 *e* **structural element**
d Bauteil *m*, Bauelement *n*
f élément *m* [composant *m*] de
construction [constructif]

nl bouwelement *n*, constructie-
-element *n*
r конструктивный элемент *m*,
элемент *m* конструкции;
строительная деталь *f*

S1436 *e* **structural engineering**
d konstruktiver Ingenieurbau *m*,
Ingenieurhochbau *m*
f étude *f* de constructions, étude *f*
de projet
nl ontwerpen *n* en uitvoeren *n* van
gebouwen en werken
r проектирование *n* зданий
и сооружений

S1437 *e* **structural failure**
d Versagen *n* der Baukonstruktionen
f endommagement *m* des édifices
[des ouvrages]; rupture *f* des
constructions
nl bezwijken *n* van een constructie
r повреждение *n* строительных
конструкций [сооружений]

S1438 *e* **structural fire protection**
d bauseitige Feuerschutzmaßnahmen
f pl
f mesures *f pl* constructives de
protection (*des édifices*) contre le
feu
nl maatregelen *pl* voor de
brandwerendheid
r конструктивные меры *f pl*
противопожарной защиты
(*зданий*)

S1439 *e* **structural glass** *US*
d Glasbauelemente *n pl*
f produits *m pl* verriers pour la
construction
nl glazen bouwelementen *n pl*
r строительные изделия *n pl* из
стекла

S1441 *e* **structural integrity**
d Integrität *f* der
Baukonstruktionen; Kontinuität
f der Baukonstruktionen
f intégrité *f* d'une construction;
continuité *f* d'une construction
nl integriteit *f* van een
bouwconstructie; continuïteit *f*
van een bouwconstructie
r конструктивная целостность *f*;
сплошность *f* конструкции

S1442 *e* **structural lightweight concrete**
d Leichtkonstruktionsbeton *m*,
Leichtbauwerkbeton *m*
f béton *m* léger pour ouvrages d'art
nl lichtbeton *n* voor
bouwconstructies
r конструктивный лёгкий бетон *m*

S1443 *e* **structural materials**
d Baustoffe *m pl*
f matériaux *m pl* de construction
nl bouwstoffen *f(m) pl*,
bouwmaterialen *n pl*
r конструкционные материалы *m pl*

S1444 *e* **structural mechanics**
d Baumechanik *f*, Baustatik *f*
f statique *f* de constructions
nl toegepaste mechanica *f*, statica *f*
r строительная механика *f*; статика
f сооружений

S1445 **structural member** *see* **structural element**

S1446 *e* **structural performance**
d Fertigungsqualität *f* der
Baukonstruktionen
f qualité *f* d'exécution d'un ouvrage
nl uitvoeringskwaliteit *f* van de
constructie
r качество *n* выполнения
конструкции

S1447 *e* **structural performance of
pavement**
d Nutzungseigenschaften *f pl*
[Fahreigenschaften *f pl*] der
Straßendecke
f caractéristiques *f pl* techniques de
la route
nl technische eigenschappen *f pl*,
hoedanigheid *f* van de verharding
r конструктивные [ездовые]
качества *n pl* дорожного
покрытия

S1448 *e* **structural requirements**
d konstruktive Forderungen *f pl*
f prescriptions *f pl* techniques;
cahier *m* des clauses techniques
nl constructieve eisen *pl*
r конструктивные требования *n pl*

S1449 *e* **structural safety**
d Bauwerkssicherheit *f*
f sécurité *f* de service [de
comportement] des ouvrages
nl veiligheid *f* van de constructie
r безопасность *f* [надёжность *f*]
сооружения [конструкции]

S1450 *e* **structural slab**
d biegungsbeanspruchte Platte *f*
f dalle *f* travaillant en flexion
nl vrijdragende plaat *f(m)* [vloer
f(m)]
r конструктивная плита *f*; плита *f*,
работающая на изгиб

S1451 *e* **structural sound insulation**
d Körperschalldämmung *f*
f insonorisation *f* [isophonie *f*]
des constructions
nl contactgeluidinstallatie *f*
r изоляция *f* структурного шума
изоляция *f* от шума,
распространяющегося по
конструкциям

S1452 *e* **structural-spatial concept**
d konstruktive und räumlich-
-gestalterische Lösung *f*
f aménagement *m* des ouvrages en
espace
nl ruimtelijk constructief concept *n*

r объёмно-конструктивное решение n

S1453 e **structural steel**
    d 1. Baustahl m
      2. Stahltragwerke n pl,
      Stahlkonstruktionen f pl,
      Stahlbauten m pl
    f 1. acier m de construction
      2. constructions f pl [charpentes f
      pl] en acier
    nl 1. constructiestaal n
      2. staalconstructies f pl
    r 1. строительная сталь f;
      конструкционная сталь f
      2. стальные строительные
      конструкции f pl, стальные
      сооружения n pl

S1454 e **structural steel erection**
    d Stahlbau m
    f montage m des constructions [des
      charpentes] en acier
    nl constructie f
    r монтаж m стальных конструкций

S1455   **structural steelwork** see **structural steel 2.**

S1456 e **structural strength**
    d Gestaltfestigkeit f
    f résistance f dépendant de la
      forme constructive d'un
      composant
    nl sterkte f van de constructie
    r прочность f, обусловленная
      формой элемента

S1457 e **structural system**
    d konstruktive Durchbildung f,
      Bausystem n
    f système m constructif [porteur d'un
      bâtiment]
    nl aard m van de constructie,
      constructiewijze f
    r система f строительных
      конструкций зданий
      [сооружений]

S1458 e **structural theory**
    d Baukonstruktionslehre f
    f théorie f de calcul des structures
    nl constructieleer f(m)
    r теория f расчёта сооружений

S1459 e **structural timber**
    d Konstruktionsholz n,
      Bau(nutz)holz n
    f bois m de construction [de
      charpente]
    nl bouwhout n, timmerhout n
    r строительный лесоматериал m

S1460   **structural use** see **structural application**

S1461 e **structure**
    d 1. Bauwerk n, Tragwerk n,
      Konstruktion f 2. Aufbau m,
      Gefüge n, Struktur f
    f 1. ouvrage m; structure f,
      construction f 2. structure f

nl 1. gebouw n 2. structuur f,
    gebouw m, constructie f
r 1. сооружение n; конструкция f
    2. структура f, строение n

S1462 e **structure beyond repair**
    d reparaturunfähiges Bauwerk n
    f ouvrage m hors d'état d'être
      réparé
    nl onherstelbare constructie f,
      onherstelbaar gebouw n
    r конструкция f, не подлежащая
      восстановлению (ввиду
      чрезмерного износа)

S1463 e **structure heat gain**
    d Transmissionswärmeeinfall m
    f gain m de chaleur dû à l'échange
      à travers des parois
    nl warmtebelasting f van een gebouw
      door zonnestraling
    r трансмиссионные теплопоступления
      n pl (обусловленные
      теплопередачей через
      ограждения)

S1464 e **structure plan**
    d Bauwerkplan m
    f plan m d'ouvrage
    nl plan n van een bouwwerk
    r план m сооружения

S1465 e **structure types**
    d Bauwerkstype m pl
    f structures f pl types; ouvrages m
      pl [bâtiments m pl] types
    nl constructiesoorten n pl, soorten
      n pl van gebouwen
    r типы m pl сооружений [зданий]

S1466 e **structure under construction**
    d Bauwerk n im Bau
    f ouvrage m en cours des travaux
      [en état de construction]
    nl werk n in uitvoering
    r сооружение n в стадии
      строительства

S1467 e **strut**
    d 1. Druckglied n 2. Quersteife f,
      Sprieße f; Kopfband n,
      Strebeband n, Schrägsteife f;
      Strebe f, Stütze f, Stempel m,
      Bolzen m
    f 1. élément m comprimé
      2. étrésillon m, entretoise f,
      contre-fiche f; béquille f;
      diagonale f; était m, montant m
    nl 1. drukstaaf f(m), kolom f(m),
      stijl m 2. schoor m; stut m;
      stempel m; diagonaal m
    r 1. сжатый элемент m 2. распорка
      f; поперечная связь f; подкос m
      (у вершины стойки); раскос m;
      стойка f

S1468 e **strut-framed beam**
    d Sprengwerkbalken m,
      Sprengwerkträger m

f poutre f sous-bandée
nl geschoorde balk m
r шпренгельная балка f

S1469 e strutted sheet piling
d abgespreizte Verspundung f
f palée f entretoisée
nl strijkdam m
r свайно-распорное крепление n

S1470 e strutting
d Verstrebung f, Verspreizung f
f entretoisement m, contreventement m
nl schoring f
r система f раскосов [подкосов, распорок]; система f связей

S1471 e stub bar
d Anschlußstab m, Anschlußstahl m
f barre f [fer m, armature f] en attente
nl wapeningsstek m
r арматурный выпуск m

S1472 e stub tenon
d einfacher Zapfen m
f tenon m droit
nl blinde pen f(m)
r простой шип m

S1473 e stucco
d 1. Außenglatt(ver)putz m
2. Baugips m, Stuckgips m
f 1. crépi m 2. stuc m
nl 1. gladde buitenbepleistering f
2. stukadoorgips n
r 1. наружная штукатурка f 2. отделочный [формовочный] гипс m

S1474 e stucco work
d 1. Stuckwerk n;
Stuck(gips)arbeiten f pl
2. Stuckgipserzeugnis n
f 1. stucage m; ouvrage m de stuc
2. produit m de stuc
nl 1. stukadoorwerk n 2. stucwerk n
r 1. наружные штукатурные работы f pl с гладкой отделкой поверхности 2. изделие n из отделочного гипса

S1475 e stud
d 1. Bundsäule f
2. Stiftschraube f,
Schraubenbolzen m;
Dübelbolzen m
f 1. montant m (d'une ossature de mur) 2. prisonnier m; goujon m (pour les appareils de scellement à cartouches)
nl 1. stijl m van een houten wand
2. draadeind n
r 1. стойка f (деревянного) каркаса; стойка f фахверка
2. шпилька f; дюбель m (забиваемый строительно--монтажными пистолетами), дюбель-гвоздь m

S1476 e stud gun

d Bolzenschießgerät n,
Bolzensetzer m
f pistolet m de scellement (à cartouches explosives)
nl boutenschiethamer m
r строительно-монтажный пистолет m для забивки дюбель-гвоздей

S1477 e studio apartment
d Ein-Raum-Wohnung f, Ein--Zimmer-Wohnung f
f appartement m [logement m] à une (seule) pièce
nl éénkamerappartement n
r однокомнатная квартира f

S1478 e stud partition
d Fachwerktrennwand f
f cloison f en treillis [à ossature]
nl nouten scheidingswand m, wand m van stijl- en regelwerk
r каркасная перегородка f

S1479 e stud welding gun
d Bolzenanschweißpistole f
f pistolet m pour soudage des goujons
nl boutenlastang f(m)
r пистолет m для приварки штырей [шпилек]

S1480 e stuffing box
d Stopfbuchse f
f presse-étoupe m
nl pakkingbus f(m)
r сальник m; набивочная камера f сальника

S1481 e stuffing-box packing
d Stopfbuchsendichtung f
f garniture f de presse-étoupe
nl opsluiting f van pakkingbus
r набивочный материал m; набивка f сальника

S1482 e stumper
d Stubbenrodegerät n, Stubbenroder m
f arrache-souche m, dessoucheuse f
nl stobbenrooimachine f
r корчеватель m пней, корчевальная машина f

S1483 e subaqueous pipeline
d Unterwasserrohrleitung f
f pipe-line m sous-marin, conduite f sous-marine
nl onderzeese pijpleiding f
r подводный трубопровод m

S1484 e subaqueous tunnel
d Unterwassertunnel m
f tunnel m subaquatique [sous--marin]
nl onderwatertunnel m
r подводный туннель m

S1485 e subaqueous tunneling
d Unterwassertunnelbau m
f percement m du tunnel sous--marin
nl onderwater-tunnelbouw m

r строительство *n* подводных туннелей

S1486 *e* **sub-atmospheric heating system**
d Vakuumdampfheizung *f*, Differential-Vakuumheizung *f*
f système *m* [installation *f*] de chauffage à vapeur sous vide
nl vacuüm-stoomverwarmingssysteem *n*
r вакуум-паровая система *f* отопления

S1487 *e* **subbase**
d untere Tragschicht *f*; Sauberkeitsschicht *f*; Unterbau *m* der Straßendecke
f couche *f* de base (*du corps de chaussée*); couche *f* de fondation
nl fundering *f* van de weg
r нижний слой *f* основания дорожного покрытия; однослойное основание *n* дорожного покрытия

S1488 *e* **sub-basement**
d Tiefkellergeschoß *n*, zweites unterirdisches Geschoß *n*
f deuxième sous-sol *m*
nl tweede kelderverdieping *f*
r второй подвальный этаж *m*

S1489 *e* **sub-contractor**
d Nachauftragnehmer *m*, Subauftragnehmer *m*
f sous-traitant *m*
nl onderaannemer *m*
r субподрядчик *m*

S1490 *e* **sub-critical flow**
d Fließen *n*, subkritische Strömung *f*, strömender Fließzustand *m*
f écoulement *m* fluvial
nl subkritische stroming *f*
r спокойное течение *n*, докритический режим *m* потока

S1491 *e* **subdiagonal**
d Halbdiagonale *f*, Halbstrebe *f*
f diagonale *f* secondaire (*d'une treillis en K*)
nl korte vakwerkstaaf *f(m)*
r полураскос *m* (*решётки фермы*)

S1493 *e* **subgrade**
d Unterbettungsschicht *f*, verdichtete Decklage *f* des Erdplanums
f couche *f* supérieure de la plate-forme (*routière*); couche *f* de fondation
nl aardebaan *f(m)*
r подстилающий слой *m*; верхний уплотнённый слой *m* земляного полотна

S1494 **submain sewer** *see* **branch sewer**

S1495 *e* **submerged breakwater**
d überfluteter [überspülter] Wellenbrecher *m*
f brise-mer *m* immergé
nl verzonken golfbreker *m*

r подводный волнолом *m*

S1496 *e* **submerged coil evaporator**
d überfluteter Verdampfer *m*
f évaporateur *m* inondé [noyé]
nl ondergedompelde verdamper *m*
r испаритель *m* затопленного типа

S1497 *e* **submerged float**
d 1. Peillatte *f*, Stabschwimmer *m* 2. Tauchschwimmer *m*
f 1. perche *f* de sondage [de jaugeage] 2. flotteur *m* submergé [de fond]
nl 1. peillat *f(m)* 2. diepvlotter *m*
r 1. гидрометрический шест *m* 2. глубинный поплавок *m*

S1498 *e* **submerged hydraulic jump**
d eingestauter [rückgestauter] Wechselsprung *m*
f ressaut *m* noyé
nl verzonken hydraulische sprong *m*
r затопленный гидравлический прыжок *m*

S1499 *e* **submerged intake**
d Unterwasserfassung *f*, Unterwassereinlauf *m*, Unterwassereinlaß *m*
f prise *f* d'eau noyée [en profondeur]
nl diepwaterwinning *f*, diepwaterbeheersing *f*, diepe watervang *m*, inlaat *m* in diep water
r глубинный водозабор *m*

S1500 **submerged overfall** *see* **submerged weir**

S1501 **submerged roof** *see* **flooded roof**

S1502 *e* **submerged water power plant**
d Wehrschwellenkraftwerk *n*, Wasserkraftwerk *n* in überströmter Bauweise
f usine-barrage *m* déversoir
nl waterkrachtcentrale *f(m)* in de voet van de stuwdam
r встроенная водосливная гидроэлектростанция *f*

S1503 *e* **submerged weir**
d Stauschwelle *f*, Grundwehr *n*, untergetauchtes Wehr *n*, eingestauter Überfall *m*
f déversoir *m* noyé
nl overstromende stuw *m*
r затопленный водослив *m*

S1504 *e* **submergence**
d Überströmung *f*
f submersion *f*
nl overstroming *f*, onderdompeling *f*
r затопление *n*, подтопление *n* (*водослива*)

S1505 *e* **submergence ratio**
d Überlaufverhältnis *n*
f taux *m* de submersion
nl overloopverhouding *f*

r коэффициент m затопления;
относительное затопление n
водослива

S1506 e **submersible pump**
d Unterwasserpumpe f, Tauchpumpe f
f pompe f immergée [submersible]
nl onderwaterpomp $f(m)$
r погружной насос m

S1507 submersion see submergence

S1508 e **sub-pavement tunnel**
d Unterpflastertunnel m
f tunnel m sous la rue
nl tunnel m onder de rijweg
r туннель m под проезжей частью
улицы

S1509 e **subsidence**
d Setzung f, Senkung f; Absetzung f,
Ausfällung f
f affaissement m (du sol);
abaissement m (du terrain);
tassement m, dénivellement m
(d'un ouvrage)
nl zetting f, inklinking f; afkalving
f, inkalving f, inzinking f
r просадка f (грунта); оседание n
(сооружения); осаждение n

S1510 e **subsidiary sewer**
d Nebensammler m
f collecteur m secondaire
nl ontlastend riool n, nevenriool n
r вспомогательный канализационный
коллектор m

S1511 e **sub-size**
d Siebdurchfall m, Siebfeines n
f criblures f pl, tamisat m, passées
f pl, passants m pl, fraction f
passante
nl doorval m (zeven)
r подрешётный продукт m

S1512 e **subsoil**
d 1. Mutterbodenschicht f,
Erdreich n 2. Baugrundboden m,
natürliche Bodengründung f
f 1. sous-sol m 2. couche f de
fondation naturelle
nl 1. ondergrond n 2. vaste
grondslag m
r 1. подпочвенный слой m грунта
2. естественное грунтовое
основание n

S1513 e **subsoil drain**
d verdeckter Drän m
f drain m récouvert [enterré]
nl buizendrain m
r закрытая дрена f

S1514 e **substitution of members**
d Stabvertauschung f
f changement m des barres
nl vervangen n van staven
(berekening)
r замена f стержней (приём
расчёта)

S1516 e **substructure**

d Unterbau m
f infrastructure f; partie f
souterraine d'un ouvrage; couche f
de fondation (d'une autoroute)
nl onderbouw m
r подземная часть f (сооружения);
нижнее строение n (пути);
основание n (дорожной одежды)

S1517 e **subsurface current**
d Unterstrom m,
Unterwasserströmung f
f courant m de profondeur
nl onderstroom m
r подповерхностное течение n

S1518 **subsurface drain** see **blind drain**

S1519 e **subsurface drainage**
d geschlossene Dränung f
f drainage m couvert [souterrain]
nl gesloten drainage f
r закрытый дренаж m

S1520 **subsurface erosion** see **washout**

S1521 **subsurface structures** see
underground structures

S1522 e **subsurface water**
d unterirdisches Wasser n
[Gewässer n]
f eaux f pl souterraines
nl grondwater n
r подземные [грунтовые] воды f pl

S1523 e **subterranean backwater zone**
d Grundwasserstaubereich m
f zone f de retenue hydrostatique
des eaux souterraines
nl grondwaterstuwzone f
r зона f подпора грунтовых вод

S1524 e **subterranean crossing**
d Unterquerung f (Straßen,
Rohrleitungen)
f croisement m [ouvrage m de
croisement] souterrain
nl ondergrondse kruising f (van
wegen of pijpleidingen)
r подземное пересечение n (дорог,
трубопроводов)

S1525 e **subway**
d 1. Fußgängertunnel m,
Gehwegtunnel m 2. U-Bahn f
3. begehbarer Kanal m
f 1. passage m souterrain
2. métro(politain) m
3. souterrain m pour tuyauterie
et câbles
nl 1. voetgangerstunnel m 2. metro
n, ondergrondse (spoorweg) m
3. leidingtunnel m
r 1. подземный пешеходный переход
m 2. метрополитен m, метро n
3. проходной канал m (для
подземных коммуникаций)

S1526 e **succession of blows**
d Schlagfolge f
f succession f des coups (de
mouton)

nl serie *f* slagen, tocht *m*
r серия *f* ударов (*свайного молота*)

S1527 *e* **suction**
d Ansaugen *n*; Sog *m*
f aspiration *f*
nl zuiging *f*
r всасывание *n*; подсос *m*

S1528 *e* **suction-cutter apparatus**
d Schneidsaugkopf *m*, Wühl-Kopf-Fräser *m*
f appareil *m* désagrégateur de la tête dragueuse
nl snijzuigkop *m*
r фрезерный рыхлитель *m* (*грунтозаборного устройства землесосного снаряда*)

S1529 *e* **suction-cutter dredger**
d Schneidkopfsaugbagger *m*
f drague *f* suceuse [aspiratrice] à désagrégateur
nl snijkopzuiger *m*
r землесосный снаряд *m* с фрезерным рыхлителем

S1530 *e* **suction dredger**
d 1. Saugbagger *m*, Pumpenbagger *m* 2. Baggerpumpe *f*
f 1. drague *f* suceuse 2. pompe *f* à déblais
nl 1. zandzuiger *m*, baggerzuiger *m* 2. grondpomp *f(m)*
r 1. землесосный снаряд *m* 2. грунтовый насос *m* (*землесосного снаряда*)

S1531 *e* **suction fan**
d Sauglüfter *m*
f ventilateur *m* aspirant [d'aspiration], aspirateur *m*
nl zuigventilator *m*
r вытяжной [всасывающий] вентилятор *m*

S1532 *e* **suction head**
d 1. Saughöhe *f* 2. Saug(e)kopf *m*, Schleppkopf *m*
f 1. hauteur *f* d'aspiration 2. tête *f* dragueuse
nl 1. zuighoogte *f* 2. zuigkop *m*, sleepkop *m*
r 1. (геометрическая) высота *f* всасывания 2. грунтозаборное устройство *n* (*землесосного снаряда*)

S1533 *e* **suction header**
d Saugsammelleitung *f*
f collecteur *m* d'aspiration
nl zuigleiding *f*
r всасывающий сборный коллектор *m*

S1534 *e* **suction lift**
d Saughöhe *f*
f hauteur *f* d'aspiration
nl zuighoogte *f*
r высота *f* всасывания (*насоса*)

S1535 *e* **suction line**
d Ansaugleitung *f*
f conduite *f* d'aspiration
nl zuigleiding *f*
r всасывающий трубопровод *m*; линия *f* всасывания

S1536 *e* **suction overfall**
d Saugüberfall *m*, Vakuumüberlauf *m*
f déversoir *m* aéré [à vide]
nl zuigoverloop *m*
r вакуумный водослив *m*

S1537 *e* **suction panel**
d Sauglochplatte *f*
f panneau *m* aspirant [d'aspiration]
nl zuigpaneel *n*
r всасывающая панель *f*

S1538 *e* **suction pressure**
d Saugdruck *m*
f pression *f* d'aspiration
nl zuigdruk *m*
r давление *n* всасывания

S1539 *e* **suction pump**
d Ansaugpumpe *f*; Baggerpumpe *f*
f pompe *f* suceuse [d'aspiration]
nl zuigpomp *f(m)*; grondpomp *f(m)*
r всасывающий [откачивающий] насос *m*; багерный насос *m*; землесос *m*

S1540 *e* **suction valve**
d Saugventil *n*
f soupape *f* d'aspiration
nl zuigklep *f(m)*
r всасывающий клапан *m*

S1541 **suction well** *see* **absorbing well**

S1542 *e* **sudden drawdown**
d plötzliche Wasserspiegelsenkung *f*
f décrue *f* subite [rapide]
nl plotselinge daling *f* van de waterstand
r быстрый спад *m* (*уровня воды*)

S1543 **suffosion** *see* **washout**

S1544 *e* **sulfate attack**
d Schwefelkorrosion *f*, Sulfatangriff *m*
f corrosion *f* sulfurée
nl zwavelcorrosie *f*, sulfaatcorrosie
r сульфатная агрессия *f* (*разрушение бетона*)

S1545 *e* **sulfate-resisting cement**
d sulfatbeständiger Zement *m*
f ciment *m* sursulfaté
nl sulfaatbestendig cement *n*
r сульфатостойкий цемент *m*

S1546 *e* **sulfur asphalt**
d Schwefel-Bitumen-Gemisch *n*
f asphalte *m* à soufre
nl zwavelasfalt *n*
r смесь *f* битума с серой (*вяжущий дорожный материал*)

S1547 *e* **sulfur-asphalt mix**
d Schwefelasphaltbeton *m*
f béton *m* asphaltique à soufre

nl mengsel *n* van bitumen en zwavel
r серный асфальтобетон *m*
(*с жёлтой серой в качестве
дополнительного вяжущего*)

S1548 e sulfur cement
d Schwefelzement *m*
f liant *m* à base de soufre, ciment *m*
à soufre
nl zwavelcement *n*
r вяжущее *n* на базе жёлтой серы

S1549 e sulfur concrete
d Schwefelbeton *m*
f béton *m* à soufre
nl zwavelbeton *n*
r серный бетон *m*, сульфур-бетон *m*
(*из жёлтой серы и заполнителей*)

S1550 e sulfur modified concrete
d Beton *m* mit Schwefelzusatz
f béton *n* à base de ciment et de
soufre
nl beton *n* met zwaveltoeslag
r бетон *m* на сернисто-
-цементном вяжущем
sulphate... *see* sulfate...

S1552 e summary of reinforcement
d Stahlauszug *m*, Bewehrungsstahlliste
f
f nomenclature *f* d'armatures
nl specificatie *f* van de wapening
r спецификация *f* арматуры

S1553 e summer air conditioning
d Sommerklimatisierung *f*
f conditionnement *m* de l'air d'été
nl zomerklimaatregeling *f*
r летнее кондиционирование *n*
воздуха

S1554 e summer dike
d Sommerdeich *m*, Überlaufdeich *m*
f digue *f* d'été
nl zomerdijk *m*
r сезонная летняя дамба *f*

S1555 e summer mode
d Sommerbetrieb *m*
f mode *m* de fonctionnement prévu
pour été (*d'un système de
climatisation*)
nl zomerbedrijf *n*
r летний режим *m* (*системы
кондиционирования воздуха*)

S1556 e summer polder
d Sommerpolder *m*
f polder *m* d'été
nl zomerpolder *m*
r летний польдер *m*

S1557 e summit
d Kuppe *f* (*Längenschnitt
einer Straße*); Kuppenstrecke *f*
f profil *m* bombé; sommet *m* du
profil en long (de la route)
nl rug *m* van een weg, hoogste punt
n in het langsprofiel van een weg
r выпуклый перелом *m*

продольного профиля **дороги**;
участок *m* дороги на выпуклой
кривой

S1558 e summit canal
d Scheitelkanal *m*
f canal *m* liant deux rivières
séparées par une ligne de
partage des eaux
nl kanaal *n* door een waterscheiding
r канал-прокоп *m* (*через
водораздел*)

S1559 e summit reach
d Scheitelhaltung *f*
f bief *m* de partage
nl waterscheidingspand *n*, kruinpand *n*
r водораздельный бьеф *m*

S1560 e sump pipe
d Grundablaßrohr *n*
f tuyau *m* de vidange
nl bodemaflaatpijp *f(m)*
r сливная труба *f* (*резервуара*)

S1561 e sump pit, sumpwell
d Pumpengrube *f*, Sumpf *m*;
Absetzbrunnen *m*,
Sammelbrunnen *m*
f puisard *m*; puits *m* de vidange
nl pompput *m*, pompgat *n*
r приямок *m*, зумпф *m* (*насосной
станции*); отстойный
[водосборный] колодец *m*

S1562 e sunbreaker
d Sonnenblende *f*
f pare-soleil *m*, brise-soleil *m*
nl zonwering *f*
r внешнее солнцезащитное
[затеняющее] устройство *n*
(*навес, козырёк, рёбра*)

S1563 e sun effect *US*
d 1. Sonnenwärmeeinfall *m* 2.
Insolation *f*, Sonneneinstrahlung *f*
f 1. gain *m* de chaleur solaire
2. insolation *f*
nl 1. zonnewarmtebelasting *f*
2. zoninstraling *f*
r 1. солнечные теплопоступления *n*
pl 2. инсоляция *f*

S1564 e sunk fascine
d Steinsenkfaschine *f*, Senkwurst *f*
f fascine *f* immergée
nl zinkfascine *f*
r тяжёлая фашина *f*

S1565 e sun protection glass
d Sonnenschutzglas *n*
f verre *m* réfléchissant
nl zonwerend glas *n*
r солнцезащитное стекло *n*

S1566 super-critical flow *see* hyper-
-critical flow

S1567 e superelevation
d Kurve *f*, Überhöhung *f*
f virage *m*
nl verkanting *f*

r вираж *m* (*на дороге*)

S1568 e **superheat**
d Überhitzungswärme *f*
f surchauffe *f*
nl oververhitting *f*
r перегрев *m*

S1569 e **superheated steam**
d überhitzter Dampf *m*
f vapeur *f* surchauffée
nl oververhitte stoom *m*
r перегретый пар *m*

S1570 e **superheater**
d Überhitzer *m*
f surchauffeur *m* (de vapeur)
nl stoomoverhitter *m*
r пароперегреватель *m*

S1571 e **superimposed load**
d 1. vorübergehende Last *f*
2. Auflast *f*
f 1. charge *f* [surcharge *f*] utile,
charge *f* imposée [passagère]
2. lest *m*, surcharge *f*
nl 1. veranderlijke nuttige
belasting *f* 2. tijdelijke
belasting *f*
r 1. временная нагрузка *f*
2. пригрузка *f*

S1572 **superplasticized concrete** *see*
flowing concrete

S1573 e **superplasticizer, superplasticizing**
admixture
d Superplastifikator *m*
f superplastifiant *m*
nl superplastificeermiddel *n*
r суперпластифицирующая **добавка**
f; разжижитель *m* (*бетонной*
смеси)

S1574 e **superstructure**
d 1. Oberbau *m*, Überbau *m*;
Aufbau *m* 2. Brückenüberbau *m*
3. Oberwagen *m*
f 1. superstructure *f*
2. superstructure *f* de pont
3. superstructure *f* pivotante
(*d'une grue à flèche*)
nl 1. bovenbouw *m*; opbouw *m*
2. bovenbouw *m* van een brug
3. bovenbouw *m* (*draaikraan,*
graafmachine)
r 1. наземная часть *f* здания *или*
сооружения 2. пролётное
строение *n* моста 3. верхняя
поворотная часть *f* (*крана,*
экскаватора)

S1575 e **supersulphated cement**
d Sulfat-Hüttenzement *m*
f ciment *m* métallurgique
sursulfaté
nl sulfaat-hoogovencement *n*
r сульфатно-шлаковый цемент *m*

S1576 e **supervision**
d Aufsicht *f*; Überwachung *f*
f surveillance *f*; inspection *f*,
contrôle *m*

nl toezicht *n*, inspectie *f*
r надзор *m*; контроль *m*

S1577 e **supplementary effect of the**
admixture
d Nebenwirkung *f* des Zusatzes
f effet *m* [action *f*] secondaire de
l'adjuvant
nl nevenwerking *f* van de toevoeging
r дополнительный эффект *m*
добавки

S1578 e **supply air**
d Zuluft *f*
f air *m* d'approvisionnement
nl luchttoevoer *m*
r приточный воздух *m*

S1579 e **supply air grille**
d Lufteintrittsgitter *n*,
Zuluftgitter *n*
f grillage *m* d'approvisionnement
en air
nl luchtinlaatrooster *m*, *n*
r приточная решётка *f*

S1580 e **supply and exhaust ventilation**
d Be- und Entlüftung *f*,
Verbundlüftung *f*, ausgeglichene
Lüftung *f*
f ventilation *f* par soufflage et par
aspiration
nl toevoer- en afzuigventilatie *f*
r приточно-вытяжная вентиляция *f*

S1581 e **supply channel**
d Hauptbewässerungsgraben *m*,
Verteilungskanal *m*
f canal *m* principal d'irrigation,
canal *m* d'alimentation
nl hoofdbevloeiingskanaal *m*,
irrigatiekanaal *m*
r магистральный оросительный
канал *m*, распределитель *m*

S1582 e **supply ditch**
d Verteil(er)graben *m*
f canal *m* d'irrigation
nl verdeelsloot *f*(*m*)
r оросительный канал *m*,
ороситель *m*

S1583 e **supply duct**
d Zuluftkanal *m*
f gaine *f* d'approvisionnement
nl luchttoevoerkanaal *n*
r приточный воздуховод *m*

S1584 e **supply jet**
d Zuluftstrahl *m*
f jet *m* d'insufflation
nl toestromende luchtstraal *m*, *f*
r приточная струя *f*

S1585 **supply line** *see* **flow line**

S1586 e **supply opening**
d Zuluftdurchlaß *m*, Zuluftöffnung
f, Lufteintrittsöffnung *f*,
Lufteinlaß *m*
f orifice *m* d'admission d'air
nl luchtinlaat *m*
r приточное отверстие *n*

623

ˢS1587 *e* **supply pipe**
   d 1. Zuleitungsrohr *n*,
      Zuführungsrohr *n*
      2. Grundstückstrang *m*,
      Anschlußleitung *f*
   f 1. tuyau *m* d'amenée [d'admission]
      2. branchement *m* général
   nl 1. aanvoerbuis *f(m)*, aanvoerpijp
      f(m) 2. aansluitleiding *f*,
      dienstleiding *f*
   r 1. подающий трубопровод *m*;
      подводящая труба *f* 2. домовый
      ввод *m* (*часть*
      присоединительного
      трубопровода, контролируемая
      водопотребителем)

ˢS1588 *e* **supply pipe from source**
   d Rohwasserzuflußrohr *n*
   f tuyau *m* d'alimentation en eau
      brute
   nl aanvoerleiding *f* van ongezuiverd
      water
   r водозаборная труба *f*

ˢS1589 *e* **supply register**
   d Luftregler *m*, Auslaß *m* mit
      verstellbarer Jalousieklappe
   f registre *m* d'air réglable
   nl regelbaar luchtinlaatrooster *n*, *m*
   r регулируемая приточная
      решётка *f*

ˢS1590 *e* **supply tunnel**
   d Zuleitungsstollen *m*,
      Zulaufstollen *m*,
      Triebwasserstollen *m*
   f galerie *f* d'amenée
   nl wateraanvoertunnel *m*
   r подводящий туннель *m*

S1591 *e* **supply ventilation**
   d Belüftung *f*
   f ventilation *f* soufflante [par
      refoulement, par soufflage]
   nl ventilatie *f*, luchtverversing *f*
   r приточная вентиляция *f*

S1592 *e* **support**
   d Auflager *n*
   f appui *m*, support *m*
   nl oplegging *f*
   r опора *f*

S1593 *e* **support conditions**
   d Aulagerbedingungen *f pl*
   f conditions *f pl* aux appuis
      [d'appui]
   nl opleggingsvoorwaarden *f pl*
   r условия *n pl* опирания

S1594   **supported member** *see* **supporting member**

S1595 *e* **supporting layer**
   d Filterstützschicht *f*
   f couche *f* de support
   nl draagkrachtige laag *f(m)*
   r поддерживающий слой *m*
      (*фильтра*)

S1596 *e* **supporting member**

   d tragendes Element *n*, Tragelement
      n
   f élément *m* (de construction)
      supportant [de support]
   nl dragend constructie-element *n*
   r поддерживающий
      (конструктивный) элемент *m*

S1597 *e* **support moment**
   d Stützmoment *n*, Auflagermoment *n*
   f moment *m* d'appui
   nl steunpuntmoment *n*, oplegmoment
      n
   r опорный момент *m*, момент *m*
      на опоре

S1598 *e* **support reaction**
   d Auflagerreaktion *f*,
      Auflagergegendruck *m*
   f réaction *f* d'appui,
      réaction *f* aux appuis
   nl oplegreactie *f*
   r опорная реакция *f*

S1599 *e* **support settlement**
   d Stützensenkung *f*
   f tassement *m* du support [de
      l'appui]
   nl zakking *f* van de oplegging
   r осадка *f* опор(ы)

S1600 *e* **suppressed weir**
   d Wehr *n* ohne Seiteneinengung
   f déversoir *m* sans contraction
      latérale
   nl overlaat *m* zonder zijvernauwing
   r измерительный водослив *m* без боко-
      вого сжатия

S1601 *e* **surcharge**
   d Auflast *f*
   f recharge *f*, lest *m*
   nl extra belasting *f*, overladen *n*
   r пригрузка *f*

S1602 *e* **surcharged reservoir level**
   d Höchststauspiegel *m*,
      Schutzstauziel *n*
   f niveau *m* maximum de la
      retenue, niveau *m* de la retenue
      forcé maximum
   nl hoogste stuwpeil *n*
   r форсированный подпорный
      уровень *m*

S1603 *e* **surcharged wall**
   d Stützwand *f* mit Auflast
   f mur *m* de soutènement à lest
      supplémentaire
   nl keermuur belast met ophoging
      boven het bovenvlak van de muur
   r подпорная стенка *f* с пригрузкой

S1604 *e* **surcharge storage**
   d Hochwasserschutzraum *m*,
      Reserveraum *m*, Sicherheitsraum
      m
   f capacité *f* de surcharge, réserve *f*
      de retenue pour le débit de crue
   nl reservecapaciteit *f* van een
      stuwbekken voor hoogwater

r аварийная ёмкость *f* [резервный противопаводочный объём *m*] водохранилища

S1605 *e* surface active agent
d oberflächenaktiver Stoff *m*
f agent *m* tensio-actif
nl oppervlakte-actieve stof *f(m)*
r поверхностно-активное вещество *n*

S1606 *e* surface aerator
d Oberflächenbelüfter *m*
f aérateur *m* mécanique (superficie*l*)
nl oppervlaktebeluchter *m*
r механический (поверхностный) аэратор *m* (*сточных вод*)

S1607 *e* surface blemishes
d Oberflächenfehler *m pl*
f défauts *m pl* de surface
nl oppervlaktegebreken *n pl*
r поверхностные дефекты *m pl* [пороки *m pl*]

S1608 *e* surface compaction
d Oberflächenverdichtung *f*
f compactage *m* [serrage *m*] (de sol) superficiel
nl oppervlakteverdichting *f*
r поверхностное уплотнение *n*

S1609 *e* surface concrete membrane
d wasserseitige Betondichtung *f*
f membrane *f* étanche en béton, écran *m* d'étanchéité en béton (*du barrage*)
nl dichtingsscherm *n* van beton
r бетонный противофильтрационный экран *m* (*плотины*)

S1610 *e* surface cooler
d Rieselkühler *m*
f refroidisseur *m* superficiel
nl oppervlaktekoeler *m*
r поверхностный охладитель *m*

S1611 *e* surface course
d Verschleißschicht *f*
f couche *f* d'usure [de roulement]
nl slijtlaag *f(m)*
r слой *m* износа (*дорожной одежды*)

S1612 *e* surface drain
d Tagwasserdrän *m*, Oberflächenwasserdrän *m*
f drain *m* à ciel ouvert
nl oppervlaktedrain *m*
r поверхностная дрена *f*

S1613 *e* surface drainage
d 1. Oberflächenentwässerung *f* 2. Oberflächenwasserableitung *f*
f 1. drainage *m* de surface 2. dérivation *f* des eaux de surface
nl 1. oppervlakte-drainage *f* 2. afvoeren *n* van het oppervlaktewater
r 1. открытый поверхностный дренаж *m* 2. отвод *m* поверхностных вод

S1614 *e* surface-dry aggregate

d oberflächentrockener Zuschlagstoff *m*
f granulat *m* sec à l'extérieur
nl oppervlakkig droge toeslag *m* (*beton*)
r поверхностно-сухой заполнитель *m* (*для бетона*)

S1615 *e* surfaced timber
d gehobeltes Bauholz *n* [Schnittholz *n*]
f bois *m* raboté
nl geschaafd hout *n*
r строганый пиломатериал *n*

S1616 *e* surface film conductance
d Wärmeübergangszahl *f*
f coefficient *m* de transmission thermique de surface [de transmission superficielle]
nl warmteoverdrachtscoëfficiënt *m*
r коэффициент *m* теплоотдачи

S1617 *e* surface finish from formwork
d schalungsbedingte Beschaffenheit *f* der Beton-Sichtfläche
f aspect *m* de la surface d'un béton brut de décoffrage
nl betonoppervlaktestructuur *f* ontstaan door bekisting
r фактура *f* бетонных поверхностей, созданная опалубкой

S1618 *e* surface heat transfer resistance
d äußerer Wärmeleitwiderstand *m*, Wärmedurchlaßwiderstand *m*
f résistance *f* thermique extérieure
nl warmtedoorgangsweerstand *m* van de oppervlaktelaag
r внешнее термическое сопротивление *n*

S1619 *e* surface loading
d spezifische Oberflächenbelastung *f* des Filters
f charge *f* spécifique [unitaire] sur le filtre
nl specifieke oppervlaktebelasting *f* van een filter
r удельная нагрузка *f* на фильтр

S1620 *e* surface pipeline
d oberirdische Rohrleitung *f*, Landrohrleitung *f*
f tuyauterie *f* [conduite *f*] terrestre
nl bovengrondse pijpleiding *f*
r наземный трубопровод *m*

S1621 surface power house, surface power station *see* open-air water power plant

S1623 *e* surface quality
d Oberflächenbeschaffenheit *f*
f qualité *f* de la surface
nl oppervlaktekwaliteit *f*, oppervlaktegesteldheid *f*
r качество *n* поверхности

S1624 *e* surface retarder
d Oberflächen-Abbindeverzögerer *m*

f retard(at)eur *m* (de prise)
superficiel
nl uitwendige verhardingsvertrager *m*
r поверхностный плёночный
замедлитель *m* схватывания

S1625 *e* **surface retarder paper**
d abbindeverzögerndes
Kraftpapier *n* (*auf die
Oberfläche einwirkend*)
f papier *m* kraft retardeur de prise
nl kraftpapier *n* ter vertraging van
de verharding (*buitenste
betonlaag*)
r крафт-бумага *f* для замедления
схватывания (*бетона*)

S1626 *e* **surface roughness**
d Oberflächenrauhigkeit *f*
f rugosité *f* de surface
[superficielle]
nl ruwheid *f* van het oppervlak
r неровность *f* или
шероховатость *f* поверхности

S1627 *e* **surface runoff**
d Oberflächenabfluß *m*,
oberirdischer Abfluß *m*
f ruissellement *m*, écoulement *m*
superficiel
nl bovengrondse afvloeiing *f*
r поверхностный сток *m*

S1628 *e* **surface slope**
d Wasserspiegelgefälle *n*
f pente *f* superficielle
nl oppervlaktehelling *f*
r уклон *m* водной поверхности,
поверхностный уклон *m*
(*напр. водотока*)

S1629 *e* **surface structures**
d oberirdische Bauwerke *n pl*
f ouvrages *m pl* terrestres
nl bovengrondse bouwwerken *n pl*
r наземные сооружения *n pl*

S1631 *e* **surface temperature**
d Oberflächentemperatur *f*
f température *f* de surface
nl oppervlaktetemperatuur *f*
r поверхностная температура *f*

S1632 *e* **surface texture**
d Oberflächenstruktur *f*
f texture *f* superficielle
nl oppervlaktestructuur *f*
r текстура *f* поверхности

S1633 *e* **surface treatment**
d Oberflächenbehandlung *f*
f traitement *m* superficiel
nl oppervlaktebehandeling *f*
r поверхностная обработка *f*

S1634 *e* **surface vibrator**
d Oberflächenrüttler *m*
f vibrateur *m* de surface
nl oppervlaktevibrator *m*
r поверхностный вибратор *m*

S1635 *e* **surface voids**
d Oberflächenporen *f pl*,

Außenlunker *m pl*
f cavités *f pl* superficielles
nl holtes *n pl* in het oppervlak
r поверхностные раковины *f pl*

S1636 *e* **surface water**
d 1. Oberflächenwasser *n*,
Oberflächengewässer *n*,
Tagwasser *n* 2. oberirdischer
Abfluß *m*
f 1. eaux *f pl* à ciel ouvert [de
surface, pluviales] 2. écoulement *m*
superficiel, ruissellement *m*
nl 1. hemelwater *n*
2. bovengrondse afwatering *f*
r 1. поверхностные воды *f pl*
2. поверхностный сток *m*

S1637 *e* **surface water heater**
d Oberflächenwasserheizer *m*
f chauffe-eau *m* superficiel
nl oppervlak-waterverwarmer *m*
r поверхностный водонагреватель *m*

S1638 *e* **surfacing**
d 1. Flächenbearbeiten *n*
2. Straßendecke *f*, Belag *m*
f 1. surfaçage *m*, finition *f* de la
surface 2. couche *f* de surface
[de roulement]
nl 1. oppervlakteafwerking *f*
2. wegdek *n*
r 1. обработка *f* [отделка *f*]
поверхности 2. покрытие *n*
дорожной одежды

S1639 **surfacing train** *see* **paving train**

S1640 *e* **surge**
d 1. Schwall *m*, fortschreitende
Welle *f*, Translationswelle *f*,
Versetzungswelle *f*,
Hochwasserschwall *m*
2. Wasserschlag *m*, Wasserstoß
m, Druckstoß *m*
f 1. onde *f* de translation; onde *f*
progressive; onde *f* de
déplacement; flot *m* de crue
2. coup *m* de bélier
nl 1. vloedgolf *f*(*m*) 2. waterslag *m*
r 1. волна *f* перемещения; волна *f*
попуска; паводочная волна *f*
2. гидравлический удар *m*

S1641 **surge chamber** *see* **surge tank**

S1642 *e* **surge shaft**
d Schachtwasserschloß *n*,
Schwallschacht *m*
f puits *m* antibélier
nl vereffeningsschacht *f*(*m*),
dempingschacht *f*(*m*)
r уравнительная шахта *f*

S1643 *e* **surge tank**
d Ausgleichskammer *f*;
Schwallraum *m*; Wasserschloß *n*;
Beruhigungskammer *f*
f bassin *m* de compensation
[d'équilibre]; cheminée *f*

d'équilibre; chambre *f*
d'équilibre
nl waterslagtank *m*,
waterslagdemper *m*
r уравнительный резервуар *m*;
уравнительная камера *f*

S1644 *e* **surge tower**
d Turmwasserschloß *n*,
Schwallturm *m*
f tour *f* antibélier
nl dempingstoren *m*
r уравнительная башня *f*

S1645 *e* **surplus pressure**
d Überdruck *m*
f pression *f* excédente, surpression *f*
nl overdruk *m*
r избыточное давление *n*

S1647 **surrounding embankment** *see*
retaining dike

S1648 *e* **surveying by laser**
d Laseraufnahme *f*, Aufnahme *f* mit
Lasergeräten
f levé *m* géodésique à laser
nl landmeten *n* met gebruik van
laserstralen
r геодезическая съёмка *f*
с применением лазерных приборов

S1649 *e* **susceptibility to corrosion**
d Korrodierbarkeit *f*
f susceptibilité *f* à la corrosion,
corrodabilité *f*
nl gevoeligheid *f* voor corrosie
r подверженность *f* коррозии,
склонность *f* к коррозии

S1650 *e* **suspended ceiling**
d Hängedecke *f*, abgehängte Decke *f*,
Unterhangdecke *f*
f plafond *m* suspendu
nl zwevend [verlaagd] plafond *n*
r подвесной потолок *m*

S1651 *e* **suspended floor**
d freitragende Geschoßdecke *f*
f plancher *m* suspendu (sans
supports intermédiaires)
nl zwevende vloer *m*
r перекрытие *n* без промежуточных
опор

S1652 *e* **suspended formwork**
d Hängeschalung *f*
f coffrage *m* suspendu
nl hangende bekisting *f*,
hangbekisting *f*
r подвесная опалубка *f*

S1653 *e* **suspended roof**
d Hängedach *n*
f toit *m* [comble *m*] suspendu
nl hangdak *n*
r висячее покрытие *f*

S1654 *e* **suspended sediment load**
d Schwebstoff fracht *f*
f débit *m* solide en suspension
nl lading *f* door zwevende stoffen
r расход *m* взвешенных наносов

S1655 *e* **suspended solids**
d Schwebstoffe *m pl*
f matières *f pl* solides еп
suspension
nl zwevende vaste stoffen *pl*
r взвешенные вещества *n pl*;
взвешенные наносы *m pl*

S1656 *e* **suspended span**
d Einhängefeld *n*
f travée *f* (*du pont*) à poutre
cantilever
nl tussenoverspanning *f* van een
gerberliggerbrug [cantileverbrug]
r консольно-балочный пролёт *m*
(*моста*)

S1657 *e* **suspended-span beam**
d Einhängeträger *m*
f poutre *f* centrale suspendue (*du
pont à poutre cantilever*)
nl tussenstuk *n* van een gerberligger
r подвесная балка *f* консольно-
-балочного пролёта (*моста*)

S1658 *e* **suspended water**
d schwebendes Grundwasserstockwerk
n
f eau *f* suspendue [pendante]
nl pendulair water *n*
r подвешенная вода *f*,
верховодка *f*

S1659 *e* **suspension bridge**
d Hängebrücke *f*
f pont *m* suspendu
nl hangbrug *f*(*m*)
r висячий мост *m*

S1660 *e* **suspension cable**
d Hängekabel *n*, Hängedrahtseil *n*
f câble *m* suspendu
nl hangkabel *m*
r висячий кабель *m*; ванта *f*;
подвешенный канат *m*

S1661 *e* **suspension structural system**
d Hängebausystem *n*,
Hängekonstruktion *f*,
Hängewerk *n*
f système *m* suspendu,
construction *f* suspendue
nl hangconstructie *f*
r висячая система *f* [конструкция *f*]

S1662 *e* **sustained load**
d Dauerlast *f*, Dauerbelastung *f*
f charge *f* [sollicitation *f*] de longue
durée
nl blijvende belasting *f*
r длительная [долговременная]
нагрузка *f*

S1663 *e* **swamp area, swamp land**
d Sumpfgebiet *n*, sumpfiges
Gelände *n*
f terrain *m* marécageux
nl moerassig terrein *n*,
moerasgebied *n*
r заболоченная территория *f*

S1664 *e* **swamp shooting method**

d Moorsprengung f,
Schüttsprengverfahren n
f compactage m de terrains
marécageux par explosions [par
tirs]
nl grondverdichtingsmethode f
gebaseerd op explosies, explosieve
verdichting
r взрывной метод m уплотнения
заболоченных грунтов (для
возведения сооружений на
заболоченной территории)

S1665 e swampy soil
d Morastboden m, Sumpfboden m
f sol m marécageux
nl moerasgrond m
r болотистый [заболоченный]
грунт m

S1666 e swan-neck boom US, swan-neck
jib UK
d Schwanenhalsausleger m
f flèche f en col de cygne
nl zwanenhalsgiek f(m)
r стрела f (грузоподъёмного
крана) с гуськом

S1667 e swaybraces
d Schlingverbände m pl,
Querverbände m pl
f contreventement(s) m (pl),
entretoisement(s) m(pl)
nl windverband m
r ветровые связи f pl; связи f pl
жёсткости

S1668 e swaying due to wind
d Schlingern n
f déplacement m (élastique)
horizontal dû au vent
nl slingeren n (door wind)
r горизонтальное (упругое)
перемещение n [сдвиг m]
(конструкции) от ветра

S1669 e sweating
d 1. Bluten n,
Wasserabsonderung f,
Wasserabscheidung f (Beton)
2. Schwitzen n,
Schwitzwasserbildung f
f 1. ressuage m (du béton)
2. transpiration f, condensation f
nl 1. waterafscheiding f (van beton)
2. zweten n
r 1. водоотделение n (бетона)
2. отпотевание n

S1670 e sweeping motion
d Auftragen n des Putzmörtels mit
einem Aufziehbrett von unten
nach oben
f application f d'enduit par
bouclier [taloche] (de bas en haut)
nl zwaaiende beweging f van het
schuurbord bij het maken van
(fijn)schuurwerk

r нанесение n штукатурного раствора
(соколом) снизу вверх (рабочий
приём)

S1671 sweep tee see swept tee

S1672 e swell
d 1. Wellenschlag m,
Wellengang m, Dünung f
2. Schwall m 3. Schwellung f,
Schwellen n (Boden)
f 1. houle f 2. houle f franche,
flots {m pl 3. gonflement m,
foisonnement m (de sol)
nl 1. golfslag m 2. vloedgolf f (m)
3. zwelling f (bodem)
r 1. накат m ветровых волн,
волнение n; зыбь f 2. волна f
попуска 2. набухание n,
пучение n (грунта)

S1673 e swell control
d Quellschutzmaßnahmen f pl
f mesures f pl de protection contre
le foisonnement [le gonflement]
de sol
nl beschermende maatregelen f pl
tegen zwelling van de grond
r противопучинные мероприятия n pl

S1674 e swelling
d Schwellen n, Bodenhebung f,
Aufquellen n (des Bodens)
f gonflement m, foisonnement m
nl zwelling f; uitlevering f;
verdikking f in het midden van
een kolom
r разбухание n, набухание n;
вспучивание n, пучение n

S1675 e swelling capability of soil
d Schwellvermögen n des Bodens
f susceptibilté f des sols [terres]
au gonflement [au foisonnement]
nl zwelvermogen n van de bodem
r пучинистость f грунта

S1676 e swell soil
d Quellboden m, Schwellboden m
f sol m gonflable
nl zwelgrond m
r пучинистый [набухающий] грунт m

S1677 e swept tee
d Bogenabzweig m
f raccord m courbé
nl stroom-T-stuk n
r отвод m криволинейного
[плавного] очертания

S1678 e swing
d 1. Schwenkbewegung f
2. Ausschlag m, Schwingung f
f 1. pivotement m, rotation f
(autour d'un axe vertical)
2. envergure f; amplitude f
nl 1. zwenking f 2. uitslag m,
slingeren n
r 1. поворот m вокруг вертикальной
оси 2. размах m, амплитуда f

S1679 e swing bridge
d Drehbrücke f
f pont m pivotant
nl draaibrug f(m)
r разводной поворотный
(в горизонтальной плоскости)
мост m

S1680 e swing door
d Schwingflügeltür f, Pendeltür f,
Drehtür f
f porte f oscillante
nl doordraaiende deur f(m),
tochtdeur f(m)
r дверь f, открывающаяся в обе
стороны

S1681 swinging gate see miter gate

S1682 swing saw see pendulum saw

S1683 e swirl diffuser
d Luftwirbler m,
Drall(düsen-Luft)auslaß m
f diffuseur m à tourbillonnement
nl wervelluchtverdeler m
r вихревой воздухораспределитель m

S1684 e switchboard !
d Schalttafel f, Schaltbre;t n
f tableau m de distribution
nl schakelbord n
r распределительный щит m

S1685 e switchbox
d Schaltkasten m
f boîte f de distribution
nl schakelkast m
r распределительный ящик m,
распределительная коробка f

S1687 e symmetrical load
d symmetrische Belastung f
f charge f symétrique
nl symmetrische belasting f
r симметричная нагрузка f

S1688 e synthetic materials
d synthetische Materialien n pl
[Baustoffe m pl]
f matériaux m pl synthétiques
nl synthetische materialen n pl
r синтетические материалы m pl

S1689 e system adjustment
d Nachstellen n [Einstellung f]
eines Systems
f ajustage m du système
nl afregeling f van het systeem
r регулировка f [настройка f]
системы

S1690 e system balance
d 1. Mengenstrom-, Druck- oder
Geschwindigkeitsausgleich m im
Leitungsnetz 2. beständige
Strömungsverhältnisse n pl 3.
Gleichgewicht n des
(mechanischen) Systems
f 1. équilibrage m du système (de
canalisation) 2. régime m
équilibré (de fonctionnement du

système) 3. état m d'équilibre
(d'un système mécanique)
nl 1. uitbalanceren n van een
leidingsysteem op druk en
snelheid 2. regelmatige stroming
3. evenwicht n van een
mechanisch systeem
r 1. увязка f расходов, давлений
или скоростей в узлах сети
(трубопроводов, воздуховодов)
2. устойчивый (гидравлический
или аэродинамический) режим m
системы 3. равновесие n
(механической) системы

S1691 system centre see system point

S1692 e system characteristic
d Rohrnetzkennlinie f,
Leitungsnetzkennlinie f
f caractéristiques f pl du réseau
(de tuyauterie)
nl systeemkenmerken n pl,
karakteristiek f van een
leidingstelsel
r характеристика f сети
(трубопроводов или
воздуховодов)

S1693 e system head curve
d Pumpsystemkennlinie f,
Förderstrom-Druck-Kurve f
f courbe f caractéristique du
système de pompage
nl pompkarakteristiek f
r характеристика f насосной
системы (расход в функции
напора)

S1694 e system in equilibrium
d Gleichgewichtssystem n
f système m équilibré [en
équilibre]
nl systeem n in evenwicht
r равновесная система f

S1695 e system instability
d Instabilität f des Systems
f instabilité f du système
nl instabiliteit f [onbalans f(m)]
van een systeem
r неустойчивость f системы

S1696 e system of forces
d Kräftegruppe f
f système m des forces
nl krachtenstelsel n
r система f сил

S1697 e system point
d Knotenpunkt m
f nœud m (du système triangulè)
nl knooppunt n
r узел m (стержневой системы)

S1698 systems building see
industrialized building

S1699 e system surging
d Druckpulsation f im
Leitungssystem

f contre-foulement *m* de l'eau (*dans une conduite*)
nl waterslag *m* [drukstoot *m*] in leidingstelsel
r пульсации *f pl* давления в системе

S1700 *e* **system water**
d Heizwasser *n*, Kreislaufheißwasser *n*
f eau *f* chaude de circulation
nl ketelwater *n*
r сетевая вода *f* (*в системе отопления*)

T

T1 *e* **table of quantities**
d Massenauszug *m*
f tableau *m* des quantités d'ouvrages à exécuter, liste *f* des travaux à exécuter
nl materiaalboek *n*, staat *m* van de verrekenbare hoeveelheden
r ведомость *f* объёмов работ

T2 *e* **table saw**
d Besäumkreissäge *f*
f scie *f* circulaire de menuiserie à table
nl tafel-cirkelzaagmachine *f*
r круглопильный обрезной станок *m*

T3 *e* **table vibrator**
d Rütteltisch *m*
f table *f* vibrante
nl triltafel *f(m)*
r вибрационный стол *m*, вибростол *m*

T4 *e* **tack coat**
d Bindeschicht *f*
f couche *f* d'accrochage
nl hechtlaag *f(m)*
r слой *m* сцепления, связующая прослойка *f*

T5 *e* **tackle**
d Flaschenzug *m*; Seilgehänge *n*, Seilgeschirr *n*
f palan *m*; moufle *f*
nl takel *m, n*, katrol *f(m)*
r полиспаст *m*; канатная оснастка *f*

T6 *e* **tail bay**
d untere Wasserberuhigungskammer *f*
f chambre *f* de tête aval
nl benedenhoofd *n* van een sluis
r камера *f* нижних ворот шлюза

T7 *e* **tail beam, tailbeam**
d Stichbalken *m*
f solive *f* de remplissage
nl staartbalk *m* [kreupele balk *m*] in een raveling
r укороченная деревянная балка *f* (*у проёма в перекрытии*)

T8 *e* **tailgate**
d 1. Unterwassernotverschluß *m* 2. Untertor *n*
f 1. vanne *f* de vidange de secours 2. portes *f pl* d'aval
nl 1. wateruitlaat *m* 2. benedendeur *f(m)*, ebdeur *f(m)*
r 1. глубинный аварийный затвор *m* 2. нижние ворота *pl* шлюза

T9 *e* **tailings**
d 1. unhältiges Gut *n*, Abgänge *m pl*, Berge *pl* 2. Siebrückstand *m*
f 1. déchets *m pl* de lavage [de traitement] 2. refus *m* de crible [sur tamis]
nl 1. wasstenen *m pl* 2. overloop *m*, zeefrest *f(m)*
r 1. отходы *m pl*, хвосты *m pl* обогащения 2. остаток *m* на сите [грохоте]

T10 **tail joist** *see* **tail beam**

T11 *e* **tail pond**
d Unterbecken *n*, Tiefbecken *n*
f bief *m* aval
nl benedenbekken *n* van een spaarbekkencentrale
r нижний бьеф *m*

T12 *e* **tail race channel**
d Untergraben *m*, Unter(-Trieb)-wasserkanal *m*, Wasserableitungstrakt *m*
f canal *m* d'aval [de restitution]
nl afvoerkanaal *n*
r водотводящий канал *m* [тракт *m*], отводящее русло *n*

T13 *e* **tail water**
d 1. untere Haltung *f*, Unterwasser *n* 2. Überlaufwasser *n*
f 1. eau *f* [bief *m*] d'aval 2. eau *f* supplémentaire
nl 1. onderwater *n* 2. overloopwater *n*
r 1. нижний бьеф *m* 2. сбросные воды *f pl* (*гидросооружения*)

T14 **taintor gate** *see* **radial gate**

T16 *e* **tall structures**
d hohe Bauwerke *n pl*
f ouvrages *m pl* de grande hauteur
nl hoge bouwwerken *n pl*
r высотные сооружения *n pl*

T17 *e* **tamped concrete**
d Stampfbeton *m*
f béton *m* damé
nl stampbeton *n*
r трамбованный бетон *m*

T18 *e* **tamper**
d 1. Stampfbär *m*, Fallbär *m*; Rüttelstampfmaschine *f* 2. Schaffußwalze *f* 3. Bohlenstampfer *m*, Stampfbohle *f*
f 1. pilonneuse *f*, dame *f*, dameuse *f* 2. rouleau *m* tournepieds [à pieds de mouton] 3. poutre *f* [planche *f*] vibrante

nl 1. stamper *m* 2. schapenpoot *f*
3. stampbalk *m*
r 1. вибрационная трамбовка *f*
2. кулачковый каток *m*
3. трамбующий брус *m*

T19 *e* **tamper-crane rammer**
d Freifall-Kranstampfer *m*,
Baggerstampfer *m*,
Fallplattenkran *m*
f dame-grue *f* à chute libre
nl vrije-valkraanstamper *m*
r кран-экскаватор *m* с
трамбующей плитой свободного
падения

T20 *e* **tamper finisher**
d Schlagbohlenfertiger *m*,
Stampfbohlenfertiger *m*
f finisseur *m* [finisher *m*] à lame
dameuse vibrante
nl asfalt-afwerkmachine met
stampbalk
r финишер *m* с трамбующим брусом

T21 *e* **tamping**
d 1. Stampfverdichtung *f*
2. Besatz *m*, Verdämmung *f*
f 1. pilonnage *m*, damage *m*,
compactage *m* 2. bourrage *m*
nl 1. stampen *n*, verdichten *n*,
porren *n* 2. opstoppen *n* (van een
boorgatlading)
r 1. трамбование *n*, уплотнение *n*
2. забойка *f* (*шпура, скважины*)

T22 **tamping beam finisher** *see* **tamper
finisher**

T23 *e* **tamping rod**
d Stocherstange *f*
f tige *f* de damage [de pilonnage]
nl porstaaf *f(m)*
r уплотняющий стержень *m*

T24 *e* **tamping roller**
d Anhängewalze *f*
f cylindre *m* [rouleau *m*] tracté
nl trilwals *f(m)*
r прицепной дорожный каток *m*

T25 *e* **tandem roller**
d zweitrommelige Selbstfahrwalze *f*,
Tandemwalze *f*
f cylindre *m* [rouleau *m*] tandem,
tandem *m*
nl tandemwals *f(m)*
r двухвальцовый дорожный каток *m*,
каток-тандем *m*

T26 *e* **tang**
d Schaft *m*
f queue *f*
nl angel *m*, arend *m*, doorn *m*
r хвостовик *m*

T27 **tangential flow fan** *see* **cross flow
fan**

T28 *e* **tangent modulus of elasticity**
d Tangentialmodul *m*
f module *m* d'élasticité tangentiel
nl tangentiële elasticiteitsmodulus *m*

r касательный [тангенциальный]
модуль *m* упругости

T29 *e* **tangent point**
d Krümmungsanfangspunkt *m*;
Krümmungsendpunkt *m*,
Tangentenpunkt *m*
f point *m* de tangence [de tengente]
nl raakpunt *n*, beginpunt *n* of
eindpunt *n* van een bocht
r точка *f* начала *или* конца
горизонтальной кривой (*дороги*)

T30 **tank sprayer** *see* **binder distributor**

T31 *e* **tank winding machine**
d Behälterwickelmaschine *f*
f machine *f* enrouleuse
nl machine *f* voor het ontwikkelen
van wapeningsstaal op een trommel
r машина *f* для навивки арматуры
на резервуар

T32 *e* **tap**
d Wasserhahn *m*
f robinet *m*
nl waterkraan *f(m)*
r водоразборный кран *m*

T33 **tap cock** *see* **plug cock**

T34 *e* **tapered haunch**
d Voute *f*
f voûtin *m*
nl voûte *f*, schuin aflopende
bekisting *f*
r вут *m*

T35 *e* **taper pipe**
d Übergangsstück *n*, Reduzierstück *n*
f tuyau *m* de réduction, réducteur *m*
nl verloopstuk *n*
r переход *m* трубопровода

T36 *e* **tapers**
d verengende Brunnenringe *m pl*
f anneaux *m pl* côniques en béton
nl tapse put *m*
r суживающиеся бетонные кольца
n pl (*колодца*)

T37 *e* **tapping screw**
d selbstschneidende Schraube *f*
f vis *f* taraud
nl zelftappende schroef *f* (*m*)
r самонарезающий винт *m*

T38 **tap water** *see* **city water**

T39 *e* **tar-bitumen binder**
d Teer-Bitumen-Bindemittel *n*
f liant *m* à base de goudron et
bitume
nl teer-bitumenbindmiddel *n*
r дёгтебитумное вяжущее *n*

T40 *e* **tar emulsion**
d Teeremulsion *f*
f émulsion *f* de goudron
nl teeremulsie *f*
r дёгтевая эмульсия *f*

T41 *e* **tar felt**
d Teerpappe *f*
f carton *m* goudronné (pour toiture)

nl teervilt *n*
r кровельный толь *m*

T42 *e* **tarmacadam**
d Teermakadam *m*
f macadam *m* au goudron, macadam-
-goudron *m*, goudron-macadam *m*
nl teermacadam *n*, met teer gebonden
steenslagverharding *f*
r щебёночное дорожное покрытие *n*,
обработанное дёгтем

T43 **tar paper** *see* **tar felt**

T44 *e* **tarpaulin**
d Persenning *f*, geteertes Segeltuch *n*,
Plane *f*
f bâche *f*
nl dekkleed *n*, geteerd zeildoek *n*, *m*
r брезент *m*

T45 *e* **tar paving**
d Teerdecke *f*, Teerbelag *m*
f revêtement *m* routier goudronné
nl teerwegdek *n*
r дёгтевое дорожное покрытие *n*

T46 *e* **tarring**
d Teeren *n*, Teerung *f*
f goudronnage *m*
nl teren *n*
r гудронирование *n*, обмазка *f*
гудроном

T47 *e* **T-beam**
d T-Profil-Träger *m*
f poutre *f* en T
nl T-profiel *n*; T-balk *m*
r тавровая балка *f*

T48 *e* **team**
d Mannschaft *f*, Brigade *f*
f équipe *f*
nl ploeg *f*, team *n*
r бригада *f*

T49 *e* **technical supervision**
d technische Überwachung *f*,
Bauüberwachung *f*
f contrôle *m* [surveillance *f*]
technique des travaux de
construction
nl bouwtoezicht *n*
r технический надзор *m* (за
строительством)

T50 *e* **tectonics**
d Architektonik *f*, Tektonik *f*
f (archi)tectonique *f*
nl (archi)tektoniek *f* (*bouwkunst*)
r (архи)тектоника *f*

T51 *e* **tee**
d T-Stück *n*
f té *m*, raccordement *m* en T
nl T-stuk *n*
r тройник *m*

T52 *e* **tee hinge**
d T-Scharnier *n*
f té *m* à charnière, penture *f*
[charnière *f*] en T
nl kruisheng *m*
r Т-образная петля *f*

T53 *e* **telescopic crane**
d Teleskopkran *m*
f grue *f* à flèche télescopique
nl teleskopische kraan *f(m)*
r телескопический кран *m*, кран *m*
с телескопической стрелой

T54 *e* **telescopic gate**
d Teleskopschütze *f*,
Teleskopenverschluß *m*
f vanne *f* télescopique
nl teleskoopsluiter *m*
r телескопический затвор *m*

T55 *e* **telescopic metal joist**
d teleskopischer [ausziehbarer]
Stahlträger *m*
f poutre *f* télescopique en acier
nl teleskopische stalenligger *m*
r телескопическая [выдвижная]
стальная балка *f*

T56 *e* **telescoping excavator**
d Teleskopbagger *m*
f excavateur *m* à flèche télescopique
nl graafmachine *f* met telescopische
arm
r экскаватор *m* с телескопической
стрелой

T57 *e* **telescoping formwork**
d Ausziehschalung *f*,
Teleskopschalung *f*
f coffrage *m* télescopique
nl uittrekbare bekisting *f*
r телескопическая [выдвижная]
опалубка *f*

T58 *e* **telescoping shoring column**
d Teleskop-Einrüststütze *f*
f montant *m* télescopique
d'échafaudage
nl verstelbare [uitschuifbare] stempel *m*
r телескопическая стойка *f* лесов

T59 *e* **telpher**
d Einschienen-Flaschenzug *m*,
Elektro-Flaschenzug *m*,
Einschienenlaufkatze *f*
f palan *m* monorail [électrique]
nl monorailloopkat *f(m)*
r (электро)тельфер *m*

T60 *e* **temperature rebars, temperature
reinforcement**
d Temperaturbewehrung *f*
f armature *f* destinée à supporter les
contraintes de température
nl krimpwapening *f*
r арматура *f*, воспринимающая
температурные напряжения

T61 *e* **temperature shrinkage**
d Wärmeschwindung *f*
f retrait *m* thermique
nl krimp *m* bij lage temperatuur
r температурная усадка *f*

T62 *e* **temperature stresses**
d Temperaturspannungen *f pl*
f contraintes *f pl* thermiques

nl spanningen *f pl* door
temperatuurwisseling
r температурные напряжения *n pl*

T63 *e* tempered glass *US*
d gehärtetes Glas *n*
f verre *m* trempé
nl gehard [voorgespannen] glas *n*
r закалённое стекло *n*

T64 *e* tempered hardboard
d gehärtete [vergütete] Faserplatte *f*
f panneau *m* de particules de bois
à haute densité
nl hardboard *n* met hoge persing
r высокоплотная древесностружечная
панель *f* [плита *f*]

T65 *e* template, templet
d Profillehre *f*
f calibre *m*, gabarit *m*, modèle *m*,
cerce *f*
nl mal *m*, sjabloon *f(m)*; draagsteen *m*
r шаблон *m*

T66 *e* temporary encroachments
d Bauplatzeinplankung *f*
f clôture *f* des chantiers
nl bouwplaatsomheining *f*
r временные ограждения *n pl*
(*стройплощадок*)

T67 *e* temporary services
d provisorische Leitungsnetze *n pl*
f réseaux *m pl* de canalisations
provisoires
nl tijdelijke bouwplaatsaansluitingen
f pl
r временные инженерные сети *f pl*

T68 *e* temporary structure
d Behelfsbauwerk *n*; Hilfsbauwerk *n*
f ouvrage *m* provisoire; ouvrage *m*
auxiliaire
nl tijdelijk bouwwerk *n*;
hulpbouwwerk *n*
r временное *или* вспомогательное
сооружение *n*

T69 *e* tender
d 1. Angebot *n* 2. Kostenanschlag *m*
f 1. offre *f* 2. devis *m* estimatif
[évaluation]
nl 1. aanbieding *f*, inschrijving *f*
2. begroting *f* van een aannemer
r 1. заявка *f* на выполнение подряда
2. смета *f* (*подрядчика на торгах*)

T70 *e* tender conditions
d Ausschreibungsbedingungen *f pl*
f conditions *f pl* d'adjudication
nl aanbestedingsvoorwaarden *f pl*
r условия *n pl* проведения торгов

T71 *e* tenderer
d Bieter *m*, Bewerber *m*,
anbietende Firma *f*
f soumissionaire *m*
nl inschrijver *m*
r фирма-участник *f* торгов (на
заключение подрядного договора)

T72 *e* tendon
d Spannglied *n*, Vorspannglied *n*
f élément *m* [câble *m*] de
précontrainte
nl voorspanwapening *f*
r напрягаемый элемент *m*
(*арматурный*)

T73 *e* tendon profile
d Spanngliedverlauf *m*
f tracé *m* de l'armature de
précontrainte
nl beloop *m* van de voorspanwapening
r очертание *n* напрягаемой арматуры

T74 tenement-house *see* apartment block

T75 .tenon
d Zapfen *m*
f tenon *m*
nl pen *f(m)*, tap *m*
r шип *m*

T76 *e* tenon saw
d Fuchsschwanz *m*
f scie *f* à faire la rainure
nl kapzaag *f(m)*, verstekzaag *f(m)*
r наградка *f* (*ручная пила*)

T77 *e* tensile forces
d Zugkräfte *f pl*
f forces *f pl* de traction
nl trekkrachten *m pl*
r растягивающие силы *f pl*
[усилия *n pl*]

T78 *e* tensile reinforcement
d Zugbewehrung *f*
f armature *f* tendue
nl trekwapening *f*
r растянутая арматура *f*

T79 *e* tensile splitting strength
d Spaltzugfestigkeit *f*
f résistance *f* à la traction par
fendage
nl splijtingstreksterkte *f*
r предел *m* прочности при
раскалывании

T80 *e* tensile strength
d Zugfestigkeit *f*
f résistance *f* à la traction
nl treksterkte *f*, trekvastheid *f*
r предел *m* прочности при растяжении, временное сопротивление *n*
разрыву

T81 *e* tensile stress
d Zugspannung *f*
f contrainte *f* [tension *f*] de
traction
nl trekspanning *f*
r растягивающее напряжение *n*

T82 *e* tensile test
d Zugprüfung *f*, Zugprobe *f*
Zugversuch *m*
f essai *m* à la traction
nl trekproef *f(m)*
r испытание *n* на растяжение

T83 *e* tensile zone
d Zugzone *f*

f zone f tendue
nl trekzone f
r зона f растяжения
[растягивающих напряжений]

T84 e tension
d Zug m; Zugkraft f; Spannen n
f traction f, effort m de traction ;
tension f ; mise f en tension
nl trekkracht m; spannen n
r растяжение n; растягивающее
усилие n; напряжение n;
натяжение n

T85 e tensioning apparatus, tensioning
equipment
d Spannvorrichtungen f pl
(Spannbeton)
f équipement m [matériel m] de
mise en tension
nl spanapparatuur m (wapening)
r оборудование n для натяжения
арматуры

T86 e tensioning jack
d Spannpresse f
f verin m de mise en tension
nl spanvijzel f(m)
r домкрат m для натяжения
арматуры

T87 e tension member
d Zugglied n, Zugstab m
f élément m tendu, pièce f tendue
nl trekstaaf f(m)
r растянутый элемент m
(конструкции)

T88 e tension pile
d Zugpfahl m
f pieu m tendu
nl trekpaal m
r свая f, работающая на растяжение

T89 tension stress see tensile stress

T90 tentative specifications see
preliminary specifications

T91 e tent roof
d Zeltdach n
f 1. toit m [comble m] en pavillon
2. toit m [comble m] flexible en
textile artificiel
nl tentdak n
r 1. шатровое покрытие n,
шатровая крыша f 2. гибкое
тентовое покрытие n

T92 e terminal reheat system
d Klimaanlage f mit Nachwärmern
f système m de climatisation avec
appareils de réchauffage terminaux
nl luchtbehandelingstoestel n met
naverhitters
r система f кондиционирования
воздуха с доводчиками-
-подогревателями

T94 e terminal unit
d 1. Nachbehandlungsgerät n,
Endaggregat n
2. Luftverteilungseinrichtung f

f 1. unité f terminale, appareil m
[climatiseur m] terminal 2.
dispositif m de distribution d'air
nl 1. nabehandelingstoestel n
2. luchtverdeelinrichting f
r 1. доводчик m; неавтономный
местный кондиционер m
2. воздухораспределительное
устройство n

T95 e terrace
d 1. Terrasse f 2. Häuserzeile f,
Reihenbebauung f, Zeilenbebauung
f 1. terrasse f 2. rangée f de
maisons ; implantation f de
maisons linéaire
nl 1. terras n 2. rijbebouwing f
r 1. терраса f 2. ленточная
застройка f; ряд m
примыкающих друг к другу
домов

T96 e terraced houses
d Reihenhäuser n pl, Zeilenhäuser
n pl
f maisons f pl en rangée
nl naast elkaar langs een straat staande
huizen n pl
r дома m pl рядовой застройки

T97 e terrace roof
d Terrassendach n
f toit-terrasse m, toit m en
terrasse
nl terrasdak n
r плоская крыша f

T98 e terrazzo, terrazzo concrete
d Terrazzobeton m,
Marmor(zuschlagstoff)beton m
f terrazzo m, granito m
nl kunstgraniet n, terrazzo n
r бетон m типа «террaццо»
(с заполнителем из цветной
каменной крошки)

T99 e terrazzo slab
d Terrazzoplatte f
f dalle f en terrazzo
nl terrazzoplaat f(m)
r бетонно-мозаичная плита f

T100 e territory planning
d Landesplanung f, Raumordnung f,
Gebietsplanung f
f planification f territoriale,
aménagement m des territoires
nl ruimtelijk streekplan n
r планировка f и застройка f
территорий

T101 e tertiary beam
d Tertiärbalken m
f poutre f tertiaire
nl tertiaire balk m
r продольная вспомогательная балка
f, балка f настила (в системе
балочной клетки)

T102 e test beam
d Prüfbalken m
f poutre f d'essai

nl proefbalk *m*
r испытательная балочка *f*

T103 *e* **test cube**
   d Probewürfel *m*
   f cube *m* d'essai
   nl proefkubus *m* (*beton*)
   r испытательный (бетонный) кубик *m*

T104 *e* **test cylinder**
   d Prüfzylinder *m*
   f cylindre *m* d'essai
   nl proefcilinder *m*
   r цилиндрический образец *m* для
     испытания (*бетона*)

T105 *e* **tested capacity**
   d durch Versuch ermittelte
     Brunnenschüttung *f*
   f débit *m* d'essai en régime
   nl gemeten opbrengst *f* van een bron
   r испытательный дебит *m*
     (*колодца, скважины*)

T106   **test pit** *US see* **trial pit**

T107 *e* **test pressure**
   d Prüfdruck *m*, Probedruck *m*
   f pression *f* d'épreuve [d'essai]
   nl proefdruk *m*
   r испытательное давление *n*

T108 *e* **tetrapod**
   d Tetrapod *n*, Vierfußblock *m*
   f tétrapode *m*
   nl tetrapode *f*
   r тетрапод *m*

T109 *e* **texturing**
   d Riffelung *f*, Riffeln *n*
   f surfaçage *m*, traitement *m* de
     finition superficiel
   nl aanbrengen *n* van
     oppervlaktestructuur
   r придание *n* заданной фактуры,
     офактуривание *n* поверхности

T110 *e* **T-head**
   d 1. T-förmiger Stützenaufsatz *m*
     [Stützenkopf *m*], T-Kopfstück *n* 2.
     zweischenklige Knagge *f*
   f 1. tête *f* du poteau préfabriqué en
     T 2. tête *f* de poteau en forme
     d'une traverse à brancons
   nl 1. T-vormige kolomkop *m* 2.
     kolomkop *m* bestaande uit een
     dwarsbalk met twee flenzen
   r 1. Т-образный оголовок *m*
     (*колонны*) 2. подбабок *m*
     (*опорной стойки*)

T111 *e* **T-headed buttress dam**
   d Pfeilersperre *f* mit T-förmigen
     Pfeilerköpfen
   f barrage *m* à contreforts à tête en T
   nl contrefortdam *m* met T-vormige
     pijlerkoppen
   r контрфорсная плотина *f*
     с Т-образными оголовками

T112 *e* **T-headed jetty**
   d T-förmiger Anlegekai *m*
   f quai *m* en T

nl T-vormige aanlegkade *f*(*m*)
r Т-образный причал *m*

T113 *e* **T-head pier**
   d T-förmiger Pier *m*
   f jetée *f* en T
   nl T-vormige pier *m*
   r Т-образный пирс *m*

T114 *e* **theory of plastic behaviour, theory**
     of plasticity
   d Plastizitätstheorie *f*,
     Bildsamkeitstheorie *f*
   f théorie *f* plastique [de plasticité]
   nl plasticiteitstheorie *f*
   r теория *f* пластичности

T115 *e* **theory of plates**
   d Theorie *f* der dünnen Platten
   f théorie *f* (du calcul) des plaques
     minces
   nl theorie *f* van de dunne platen
   r теория *f* тонких пластин(ок)

T116 *e* **theory of shells**
   d Schalentheorie *f*
   f théorie *f* (du calcul) des voiles
     minces
   nl theorie *f* van schalen
   r теория *f* оболочек

T117 *e* **theory of structures**
   d Bauwerkstheorie *f*; Baustatik *f*,
     Baumechanik *f*
   f théorie *f* des structures
   nl constructieleer *f*
   r теория *f* сооружений; строительная
     механика *f*

T118 *e* **thermal balance**
   d Wärmebilanz *f*
   f bilan *m* thermique [de chaleur]
   nl warmtebalans *f*(*m*)
   r тепловой баланс *m*

T119 *e* **thermal barrier**
   d Wärme(dämm)sperre *f*,
     Thermosperre *f*
   f barrière *f* thermique
   nl warmte-isolatielaag *f*(*m*)
   r теплоизолирующая прослойка *f*

T120 *e* **thermal comfort**
   d thermische Behaglichkeit *f*, Behag-
     lichkeitsgefühl *n*
   f confort *m* thermique
   nl thermische behaaglijkheid *f*
   r тепловой комфорт *m*

T121 *e* **thermal conductance**
   d Wärmeübergangszahl *f* infolge
     Wärmeleitung
   f conductance *f* thermique
   nl warmtedoorgangscoëfficiënt *m*,
     warmtegeleidingsvermogen *n*
   r коэффициент *m* теплопередачи за
     счет теплопроводности

T122 *e* **thermal deformation**
   d thermische Verformung *f*,
     Wärmeformänderung *f*
   f déformation *f* thermique

nl thermische vervorming *f*
r температурная деформация *f*

T123 *e* thermal foil
    d wärmeisolierende Folie *f*
    f feuille *f* d'isolation thermique
    nl warmte-isolerende folie *f*
    r теплоизоляционная
    [теплоотражающая] фольга *f*
    [плёнка *f*]

T124 *e* thermal insulation
    d Wärmedämmung *f*,
    Wärmeisolierung *f*
    f isolation *f* thermique
    nl warmte-isolatie *f*
    r теплоизоляция *f*

T125 *e* thermal insulation slab
    d Wärmedämmplatte *f*
    f plaque *f* [dalle *f*] isothermique
    nl warmte-isolatieplaat *f(m)*
    r теплоизоляционная плита *f*

T126 *e* thermal insulation material
    d Wärmedämmstoff *m*
    f matériau *m* thermique
    nl warmte-isolatiemateriaal *n*
    r теплоизоляционный материал *m*

T127 *e* thermally insulated cladding panel
    d wärmeisolierte Verkleidungsplatte *f*
    f panneau *m* d'isolation thermique
    nl warmte-isolerend bekledingspaneel *n*
    r теплоизоляционная ограждающая
    панель *f*

T128 *e* thermal stability of heating system
    d Wärmestabilität *f*
    [Wärmebeständigkeit *f*] einer
    Heizungsanlage
    f stabilité *f* thermique du système
    chauffant
    nl stabiliteit *f* van een
    verwarmingssysteem
    r теплоустойчивость *f* системы
    отопления

T129 *e* thermal storage heating system
    d elektrische Speicherheizung *f*,
    Elektrospeicherheizung *f*
    f installation *f* de chauffage
    électrique à accumulation
    nl verwarmingssysteem *n* met
    energieopslag in accu's
    r аккумуляционное электроотопление
    n

T130 *e* thermal stress
    d Wärmespannung *f*, thermische
    Spannung *f*
    f contrainte *f* [tension *f*] thermique
    nl warmtespanning *f*
    r теплонапряжённость *f*,
    теплонапряжение *n*

T131 *e* thermal stresses
    d Temperaturbeanspruchungen *f pl*,
    Temperaturspannungen *f pl*
    f contraintes *f pl* thermiques
    nl temperatuurspanningen *f pl*
    r температурные напряжения *n pl*

T132 *e* thermal wheel
    d Wärmerad *n*, Regenerativ-
    -Energieübertrager *m*
    f échangeur *m* de chaleur tournant
    nl warmte(wisselaar)wiel *n*
    r вращающийся теплообменник *m*

T133 *e* thermodynamic heating system
    d Wärmepumpenheizung *f*
    f chauffage *m* thermodynamique
    nl warmtepompverwarming *f*
    r теплонасосная система *f*
    отопления

T135 *e* thermoplastics
    d Thermoplaste *m pl*
    f thermoplastiques *m pl*, matières
    f pl thermoplastiques
    nl thermoplastische kunststoffen *f(m)*
    pl
    r термопластичные пластмассы *f pl*

T136 *e* thermostatic control
    d thermostatische Regelung *f*
    f commande *f* thermostatique
    nl thermostatische regeling *f*
    r термостатическое регулирование *n*

T137 *e* thermostatic mixing valve
    d thermostatisches Mischventil *n*
    f vanne *f* thermostatique à mélange
    nl thermostatische mengklep *f(m)*
    r термостатический смеситель *m*

T138 *e* thick arch dam
    d Bogengewichtsstaumauer *f*
    f barrage *m* (à) poids-voûte
    nl zware boogvormige stuwdam *m*
    r гравитационно-арочная плотина *f*

T139 *e* thimble
    d 1. Kausche *f*, Kausse *f* (*Seil*)
    2. Überschiebemuffe *f*; Mantelhülse *f*
    f 1. cosse *f* (*de câble*) 2. manchon *m*
    droit; bandage *m*
    nl 1. kous *f(m)* (*kabel*) 2. nisbus *f(m)*
    r 1. коуш *m* 2. надвижная муфта *f*;
    бандаж *m* (*в стенке
    дымовой трубы*)

T140 *e* thin arch dam
    d Schalenstaumauer *f*
    f barrage *m* en voûte mince
    nl schaalstuwmuur *m*
    r тонкая арочная плотина *f*

T141   T-hinge *see* tee hinge

T142 *e* thinner
    d Verdünner *m*, Verdünnungsmittel *n*
    f diluant *m*
    nl verdunningsmiddel *n*, verdunner *m*
    r разбавитель *m*

T143 *e* thin-plate weir
    d Plattenwehr *n* mit scharfkantiger
    Krone, scharfkantiges Wehr *n*
    f déversoir *m* en mince paroi
    nl plaatstuw *m*
    r водослив *m* с тонкой стенкой

T144 *e* thin-shell precast units
    d dünnschalige [dünnwandige]
    Stahlbetonfertigteile *m pl*

f composants *m pl* (en béton armé) à paroi mince
nl dunwandige prefabbetonelementen *n pl*
r тонкостенные сборные железобетонные элементы *m pl*

T145 *e* thin-shell structures
d dünnwandige Schalen(konstruktionen) *f pl*
f structures *f pl* en voiles minces
nl schaalconstructies *f pl*
r тонкостенные оболочки *f pl* (*конструкции*)

T146 *e* thin walled steel structure
d dünnwandige Stahlkonstruktion *f*
f construction *f* en profilés d'acier à paroi mince
nl dunwandige staalconstructie *f*
r сооружение *n* из стальных тонкостенных элементов

T147 *e* thixotropic liquid method
d Schlitzbauweise *f*
f méthode *f* par liquide thixotropique
nl sleufbouwmethode *f*
r экранирование *n* зданий от вибраций методом отрывки траншей, заполняемых тиксотропной жидкостью

T148 *e* thixotropic suspension
d thixotrope Suspension *f*
f suspension *f* thixotropique
nl thixotropische suspensie *f*
r тиксотропная суспензия *f*

T149 *e* threaded anchorage
d Gewindeverankerung *f*
f ancre *m* fileté, tige *f* d'ancrage filetée
nl schroefanker *n*
r анкер *m* [анкерное устройство *n*] с резьбой

T150 *e* threaded inserts
d Gewindestahleinlagen *f pl*
f manchons *m pl* taraudés noyés dans le béton
nl schroefhulzen *f pl* (*in beton*)
r закладные детали *f pl* с внутренней резьбой

T151 *e* three-axle self-propelled scraper
d dreiachsiger Motorschrapper *m*
f motorscraper *m* à trois essieux
nl drie-assige zelfrijdende scraper *m*
r трёхосный самоходный скрепер *m*

T152 *e* three compartment bin
d Dreiwabensilo *m*
f trémie *f* à trois compartiments
nl silo *m* met drie afdelingen
r бункер *m* с тремя отделениями

T153 *e* three-dimensional analysis
d Berechnung *f* von Raumsystemen [von dreidimensionalen Tragwerken]
f calcul *m* des systèmes tridimensionnels
nl berekening *f* van driedimensionale systemen

r расчёт *m* пространственных систем

T154 *e* three dimensional module
d Raumblock *m*, Raumzelle *f*
f bloc-pièce *m*, cellule *f* préfabriquée
nl driedimensionaal gebouwmoduul *m*
r объёмный блок *m* (*здания*)

T155 *e* three-dimensional module house building
d Montagebau *m* in Raumzellenbauweise
f construction *f* préfabriquée en cellules [en blocs-pièces]
nl montagebouw *m* van prefab blokken
r сборное строительство *n* домэв из объёмных блоков

T156 *e* three-hinged arch
d Dreigelenkbogen *m*
f arc *m* à trois articulations
nl driescharnierboog *m*
r трёхшарнирная арка *f*

T157 *e* three-hinged arch bridge
d Dreigelenkbogenbrücke *f*
f pont *m* en arc à trois articulations
nl driescharnierbrug *f(m)*
r трёхшарнирный арочный мост *m*

T158 *e* three hinged frame
d Dreigelenkrahmen *m*
f portique *f* à trois articulations
nl driescharnierportaal *n*
r трёхшарнирная рама *f*

T159 *e* three-level grade separation
d dreistöckige Kreuzung *f*
f croisement *m* à trois niveaux
nl wegkruising *f* op drie niveaus
r транспортная развязка *f* в трёх уровнях

T160 *e* three part line tackle
d Dreistrang-Flaschenzug *m*
f palan *m* triple [à trois courants]
nl takel *m* met drie parten, drieloper *m*
r трёхветвевой полиспаст *m*

T161 three pined frame *see* three hinged frame

T162 *e* three-pipe heat-supply system
d Dreileitersystem *n* der Fernheizung, Dreirohrsystem *n* der Wärmeversorgung
f système *m* de chauffage urbain à trois conduites
nl driepijps afstandsverwarmingssysteem *n*
r трёхтрубная система *f* теплоснабжения

T163 *e* three-quarter bat, three-quarter brick
d Dreiquartier *m*, Dreiviertelziegel *m*
f trois quarts *m pl* de brique

nl drieklezoor *m*
r трёхчетвертной кирпич *m*

T164   **three-roll tandem roller** *see* **three--axle tandem roller**

T165 *e* **three-way cock**
d Dreiwegehahn *m*
f robinet *m* à trois voies
nl driewegkraan *m*
r трёхходовой кран *m*

T166 *e* **three-wheel roller**
d Dreirad(-Straßen)walze *f*
f cylindre *m* [rouleau *m*] trijante
nl driewielwals *f*
r трёхвальцовый двухосный (дорожный) каток *m*

T167 *e* **thrift lock**
d Sparschleuse *f*, Speicherschleuse *f*
f écluse *f* avec bassins d'épargne
nl sluis *f(m)* met spaarkom, spaarsluis *f(m)*
r шлюз *m* со сберегательными бассейнами

T168 *e* **throat**
d 1. Wasserschenkel *m*, Ablauf *m* 2. Wassernase *f* 3. Querschnittverringerung *f* in Schornsteinbasis 4. Kehle *f*; Einschnürung *f* 5. Nahthöhe *f*, Nahtdicke *f*
f 1. jet *m* d'eau, réverseau *m* 2. larmier *m* 3. gorge *f* d'une cheminée 4. collet *m* 5. épaisseur *f* du cordon de soudure
nl 1. waterhol *n*, druiphol *n* 2. waterneus *m*, druiplijst *f* 3. vernauwing *f* in de schoorsteenboezem 4. hals *m* 5. lasnaaddikte *f*
r 1. отлив *m* 2. капельник *m*, слезник *m* 3. сужение *n* в основании дымохода 4. горловина *f*; шейка *f* 5. толщина *f* сварного шва

T169   **throttling orifice** *see* **orifice plate**

T170 *e* **through bolt**
d Durchsteckschraube *f*
f boulon *m* traversant [passant]
nl doorgaande bout *m*
r сквозной болт *m*

T171 *e* **through bridge**
d Trogbrücke *f*
f pont *m* à tablier inférieur
nl brug *f(m)* met laaggelegen rijvloer
r мост *m* с ездой понизу

T172 *e* **through tenon**
d Schlitzzapfen *m*
f tenon *m* passant
nl doorgaande pen *f(m)*
r сквозной шип *m*

T173 *e* **through truss**
d Tragwerk *n* [Träger *m*] mit untenliegender Fahrbahn

f poutre *f* à treillis d'un pont à tablier inférieur
nl hoofdligger *m* met laaggelegen rijvloer
r ферма *f* пролётного строения моста с ездой понизу

T174   **throw** *see* **air throw**

T175 *e* **throw-away filter**
d Wegwerffilter *n, m*
f filtre *m* à jeter après l'usage
nl wegwerpfilter *n, m*
r воздушный фильтр *m* однократного применения

T176 *e* **thrust**
d 1. Schub *m*, Horizontalschubkomponente *f* 2. Druckkraft *f* 3. Verwerfung *f*
f 1. poussée *f* 2. butée *f*, pression 3. faille *f*
nl 1. schuif *f(m)* 2. drukkracht *m*
r 1. распор *m* 2. сжимающее усилие *n*; напор *m* (*механический*), отпор *m* грунта 3. сброс *m*, взброс *m* (*геологический*)

T177 *e* **thrust borer**
d Rohrdurchstoßgerät *n*, Rohrdrückgerät *n*, Horizontal--Preßanlage *f*
f presse *f* pour percer des trous dans un massif de sol
nl drukboormachine *f* (*voor het horizontaal onder een weg of dijk doordrukken van buizen*)
r пресс *m* для продавливания отверстий в грунте насыпи

T178 *e* **tidal marsh**
d Gezeitenmarsch *f*
f marais *m* maritime
nl kwelder *m*, gors *f*, schor *f*
r марши *m pl*

T179 *e* **tidal mud flat**
d Watten *n pl*
f wadden *m*, watten *m*
nl wadden *n pl*
r ватты *pl*

T180 *e* **tidal power plant**
d Gezeitenkraftwerk *n*, Flutkraftwerk *n*
f usine *f* marémotrice
nl getijcentrale *f*
r приливная электростанция *f*

T181 *e* **tide gate**
d Gezeitentor *n*, Siel *n*
f clapet *m* à marée
nl getijsluis *f(m)* (*uitwatering*)
r приливный затвор *m*

T182 *e* **tide lock**
d Flutschleuse *f*
f écluse *f* à marée
nl getijsluis *f(m)* (*schutsluis*)
r приливный шлюз *m*

T183 *e* **tie**
d 1. Zugglied *n*, Zugband *n*
2. Schienenschwelle *f*
3. Verbindung *f* 4. Verrödelung *f*
5. Zugstab *m*
f 1. tirant *m* 2. traverse *f*
3. ligature *f* 4. tirant *m* en fil
d'acier (*pour coffrage*) 5. poinçon
m, aiguielle *f* pendante (*d'une
ferme*) ; montant *m* du treillis
nl 1. trekstang *f(m)*, anker *n*
2. dwarsligger *m*, biel *m*
3. verbinden *n* 4. spandraad *m*
5. trekstaaf *f(m)*
r 1. элемент *m*, работающий на
растяжение; затяжка *f* 2. шпала *f*
3. связывание *n*, сращивание *n*
4. проволочная стяжка *f*
5. подвеска *f*, стойка *f* (*решётки
фермы*)

T184 *e* **tie bar**
d Fugenanker *m*, Ankereisen *n*,
Zuganker *m*, Ankerbolzen *m*
f entrait *m*, tirant *m*, barre *f*
d'entretoisement
nl verbindingsstaaf *f(m)*
r стяжка *f*, соединительный штырь *m*

T185 *e* **tied column**
d querversteifte Säule *f*
f poteau *m* entretoisé
nl geschoorde [getuide] kolom *f(m)*
r колонна *f*, раскреплённая связями

T186 *e* **tied retaining wall**
d verankerte Stützwand *f*
f mur *m* de soutènement ancré
nl verankerde keermuur *m*
r заанкерованная подпорная
стенка *f*

T187 *e* **tie beam**
d 1. Spannbalken *m*, Spannriegel *m*
2. Pfahlholm *m*
f 1. entrait *m* de ferme, tirant *m*
2. chapeau *m* [traverse *f*] de pilotis
nl 1. trekplaat *f(m)*, trekbalk *m* 2.
sloof *f(m)*, kesp *f*
r 1. затяжка *f* стропил 2. свайная
насадка *f*

T188 *e* **tie laying machine**
d Schwellenverlegemaschine *f*
f machine *f* à poser les traverses
nl bielslegger *m*
r шпалоукладочная машина *f*,
шпалоукладчик *m*

T189 *e* **tier**
d Stockwerk *n*; Etage *f*
f étage *m*
nl laag *f(m)*; verdieping *f*
r ярус *m*; этаж *m*

T190 *e* **tier of blocks**
d Blockreihe *f*
f assise *f* de blocs
nl blokkenlaag *f(m)*
r ряд *m* кладки из массивов; ряд *m*
блоков

T191 tiers *see* **lifts**

T192 *e* **ties**
d Verbände *m pl*
f entretoises *f pl*, contrevents *m pl*,
entretoisement *m*,
contreventement *m*
nl verband *n*
r связи *f pl*, система *f* связей

T193 *e* **tie tamper**
d Gleisstopfer *m*,
Schwellenstopfer *m*,
Schotterstopfer *m*
f appareil *m* [machine *f*] pour le
bourrage des traverses
nl stopmachine (*dwarsliggers*)
r шпалоподбивочная машина *f*,
шпалоподбойка *f*

T194 *e* **tie-wire**
d Rödeldraht *m*, Bindedraht *m*
f fil *m* de ligature
nl wurgdraad *m*, binddraad *m*
r вязальная проволока *f*, проволока
f для вязки (арматуры)

T195 *e* **tight working space**
d eingeschränkter Arbeitsraum *m*
f espace *f* confinée, emplacement *m*
de travail confiné
nl beperkte werkruimte *f*
r ограниченное [стеснённое] рабочее
пространство *n*

T196 *e* **tile**
d 1. Fliese *f* 2. Keramikhohlbock *m*,
Keramikhohlstein *m* 3. Tonrohr *n*
4. Dachziegel *m*
f 1. carreau *m* (en) céramique
2. bloc *m* mural en terre-cuite
3. tuyau *m* en grès 4. tuile *f*
nl 1. tegel *m* 2. hol keramisch blok
3. kannebuis *f(m)*, gresbuis *f(m)*
4. dakpan *f(m)*
r 1. керамическая облицовочная
плитка *f* 2. керамический
пустотелый блок *m* 3. керамическая
труба *f* 4. кровельная черепица *f*

T197 *e* **tile adhesive**
d Fliesenkleber *m*, Belagplattenkleber
m
f adhésif *m* pour carreaux
nl tegellijm *m*
r клеящий состав *m* [мастика *f*] для
крепления плиток

T198 *e* **tile cutting machine**
d Fliesentrennmaschine *f*,
Fliesenschneider *m*
f machine *f* à tailler les carreaux
nl tegelsnijmachine *m*
r устройство *n* [машина *f*] для
резки керамических плиток

T199 *e* **tiled finish**
d Fliesenverkleidung *f*
f revêtement *m* en carreaux
nl tegelbekleding *f*
r облицовка *f* плитками

T200 *e* tile drain
   d Röhrendrän *m*, Tonrohrdrän *m*
   f drain *m* en poterie
   nl buizendrain *m*
   r гончарная дрена *f*

T201 *e* tile drainage *US*
   d Tonrohrdränung *f*,
    Dränrohrentwässerung *f*
   f drainage *m* en poterie
   nl drainage *f* met kannebuizen
   r гончарный дренаж *m*

T202   tile fixing *see* floor-and-wall tiling
   work

T203   tile mastic *see* tile adhesive

T204 *e* tiler work
   d Fliesenarbeit *f*,
    Fliesenlegerarbeiten *f pl*
   f travaux *m pl* de carrelage
   nl tegelwerk *n*, tegelzetten *n*
   r облицовочные работы *f pl*

T205 *e* tile trim units
   d Formfliesen *f pl* und keramische
    Formelemente *n pl*
   f pièces *f pl* façonnées pour
    carrelage en céramique
   nl geprofileerde [gevormde] tegels *m pl*
   r фасонные керамические плитки *f pl*
    и детали *f pl*

T207 *e* tiling
   d 1. Verkleidung *f* mit Fliesen 2.
    Verlegen *n* von Dachziegeln
    3. Ziegeldach *n*, Ziegeleindeckung *f*
   f 1. pose *f* des carreaux, carrelage *m*
    2. pose *f* des tuiles (*sur un*
    comble) 3. couverture *f* en tuiles
   nl 1. betegelen *n* 2. leggen *n* van
    dakpannen 3. pannendak *n*
   r 1. облицовка *f* плитками 2.
    укладка *f* черепицы 3.
    черепичная кровля *f*

T207ₐ tiltdozer *see* tilting dozer

T208 *e* tilting concrete mixer
   d Kipptrommel-Betonmischer *m*
   f bétonnière *f* avec tambour
    basculant [basculeur]
   nl betonmolen *m* met kiptrommel
   r бетоносмеситель *m* с опрокидным
    барабаном

T209 *e* tilting dozer
   d Schürfschlepper *m* mit Kippschild
   f tiltdozer *m*
   nl bulldozer *m* met kantelbaar blad
   r бульдозер *m* с наклоняющимся
    (*в вертикальной плоскости*)
    отвалом

T210 *e* tilting drum
   d Kipptrommel *f*
   f tambour *m* basculant
   nl kiptrommel *f(m)*
   r опрокидной барабан *m*
    (*бетоносмесителя*)

T211 *e* tilting force
   d Kippkraft *f*

   f effort *m* de renversement
   nl kantelingskracht *f(m)*
   r опрокидывающая сила *f*

T212 *e* tilting gate
   d Pendelklappe *f*, Klappenverschluß
    m, (Gleichgewichts-)Klappschütz *n*
   f vanne *f* à clapet [à hausse], clapet
    m équilibré
   nl pendelklep *f(m)*
   r автоматический (уравновешенный)
    клапанный затвор *m*

T213   tilting mixer *see* tilting concrete
   mixer

T214 *e* tiltster
   d selbstentladender Tieflader *m*
   f remorque *f* surbaissée pour transport
    d'engins
   nl dieplader *m*
   r прицеп *m* с низкой
    наклоняющейся платформой

T215 *e* tilt-up method
   d Aufkipp-Bauweise *f*,
    Richtaufbauweise *f*
   f méthode *f* (de montage) tilt-up
   nl opklap-montagemethode *f*
   r метод *m* монтажа «тилт-ап»
    (*с поворотом сборных элементов*
    конструкций вокруг нижней
    опорной кромки)

T216 *e* timber
   d Bauholz *n*
   f bois *m* de construction [de
    charpente]; bois *m* débité; bois *m*
    équarri; madriers *m pl* pour boisage
   nl bouwhout *n*; timmerhout *n*;
    ongezaagd *of* bezaagd hout *n*
   r строительный лесо- *или*
    пиломатериал *m*; деревянные
    брусья *m pl*

T217 *e* timber boarding
   d Bretterbelag *m*
   f entrevous *m*; planchéiage *m*,
    bardage *m* en planches,
    voligeage *m*
   nl plankenzoldering *f*; beplanking *f*,
    beschieting *f*
   r накат *m*; дощатая обшивка *f*

T218 *e* timber connectors
   d Holzverbinder *m pl*
   f organes *m pl* d'assemblage
   nl houtverbindingsmiddelen *n pl*
   r соединительные детали *f pl*
    элементов деревянных
    конструкций

T219 *e* timber floor
   d Holzfußboden *m*
   f revêtement *m* du sol en bois
   nl houten vloer *m*
   r деревянный пол *m*; деревянный
    [дощатый] настил *m*

T220 *e* timbering
   d Ausbauen *n*, Ausbau *m*,
    Zimmerung *f*, Aussteifung *f*

 f boisage *m*, blindage *m* des fouilles
 nl bekleding *f* met hout van een
  ingraving
 r установка *f* крепи, крепление *n*
  стенок траншей

T221 *e* **timber pile**
 d Holzpfahl *m*
 f pieu *m* en bois
 nl houten paal *m*
 r деревянная свая *f*

T223 *e* **timber sheet piling**
 d Holzbohlenwand *f*
 f palplanches *f pl* [palée *f*] en bois
 nl houten beschoeiing *f*
 r деревянная шпунтовая стенка *f*

T224 *e* **time-dependent analysis**
 d Dauerstandfestigkeitsberechnung *f*;
  Berechnung *f* auf Kriechen
 f calcul *m* aux sollicitations
  dépendant du temps; calcul *m* de
  fluage
 nl krimp- en kruipberekening *f*
 r расчёт *m* с учётом факторов,
  зависящих от времени; расчёт *m*
  на ползучесть

T225 *e* **time-dependent deformation,**
  time-dependent strain
 d Kriechverformung *f*
 f déformation *f* dépendant du temps;
  fluage *m*
 nl tijdsafhankelijke vervorming *f*;
  kruip *f*
 r деформация *f*, зависящая от
  времени; ползучесть *f*

T227 *e* **time of setting**
 d Abbindezeit *f*, Erstarrungszeit *f*
 f temps *m* [durée *f*] de prise
 nl bindtijd *m*, verhardingstijd *m*
 r время *n* схватывания

T228 *e* **time schedule**
 d Bauzeitplan *m*
 f programme *m* d'avancement des
  travaux
 nl werkplan *n*
 r график *m* строительства,
  календарный график *m*
  производства работ

T229 *e* **time-settlement graph**
 d Zeit-Setzungs-Kurve *f*
 f diagramme *m* de tassements sous
  charges en fonction du temps
 nl zakkingsgrafiek *m*
 r кривая *f* «осадка — время»

T230 *e* **tingle**
 d 1. Klammer *f*, Haftblech *n* 2.
  Richtziegel *m*
 f 1. tringle *f* 2. brique *f* d'alignement
  (*pour la fixation du cordon de*
  maçon)
 nl 1. klang *f*, bled *n* 2. richtsteen *m*
 r 1. клямера *f* 2. маячный кирпич *m*
  (*для фиксации шнура-причалки*)

T231   tip *see* **spoil bank**

T232 *e* **tipping lorry**
 d Kipper *m*, Kipplastkraftwagen *m*,
  Kipp-LKW *m*
 f camion *m* basculant [basculeur]
 nl kipauto *m*
 r автомобиль-самосвал *m*

T233 *e* **title-block**
 d Grundschriftfeld *n*
 f cartouche *m* d'inscription
 nl rechteronderhoek *f(m)*
 r штамп *m* (чертежа)

T234 *e* **T-joist**
 d T-Träger *m*
 f poutre *f* en T
 nl T-balk *m*
 r тавровая балка *f*

T235 *e* **T-junction**
 d T-Verzweigung *f*
 f bifurcation *f* [jonction *f*] en T
  (de deux routes)
 nl kruising *f*, splitsing *f*
 r Т-образное примыкание *n*
  (автомобильных) дорог,
  примыкание *n* дорог под прямым
  углом

T236 *e* **toe basin**
 d Tosbecken *n*, Sturzbecken *n*,
  Stoßbecken *n*, Beruhigungsbecken *n*
 f bassin *m* d'amortissement
 nl woelkom *f*
 r водобойный бассейн *m*

T237 *e* **toeboard**
 d Randbrett *n*, Randeinfassung *f*,
  Schutzbarriere *f*
 f plinthe *f*
 nl kantplank *m*
 r бортовая доска *f*, ограждающий
  борт *m* (*настила лесов*)

T238   toe ditch *see* toe trench

T239 *e* **toe drain, toe filter**
 d Fußfilterkörper *m*, Sickerkörper *m*,
  Zehenentwässerung *f*
 f drain *m* de pied
 nl draineerprisma *n*, voetfilterlaag
  f (m)
 r дренажная призма *f* (*плотины*)

T240 *e* **toe of the dam**
 d 1. Dammfuß *m*, Böschungsfuß *m*
  2. Steinbankett *m*, Bankett *n*
 f 1. pied *m* [base *f*] du barrage
  2. banquette *f* de pied; drain *m*
  de pied (*d'un barrage*)
 nl 1. teen *n* van de dam
  2. steenbestorting *f*, banket *n*
 r 1. подошва *f* откоса плотины
  2. банкет *m*; дренажная призма
  f (*плотины*)

T241 *e* **toe-shooting**
 d Moorsprengung *f*,
  Schüttsprengverfahren *n*
 f enfoncement *m* du marais au
  moyen de charges d'explosifs
 nl explosieve grondverdichting *f*

r взрывной метод *m* уплотнения
основания под сооружение
(*на заболоченной территории*)

T242 *e* **toe trench**
　　d Böschungsfußgraben *m*
　　f tranchée *f* de pied d'un talus
　　nl greppel *f(m)* aan de voet van een dijk
　　r траншея *f* у подножия откоса

T243 *e* **toe trench method**
　　d Böschungsfußgraben-Verfahren *n*
　　f méthode *f* de la tranchée de pied
　　nl drainagegreppelmethode *f*
　　r метод *m* траншеи у подножия
откоса (*метод укрепления откоса*)

T244 **toe wall** *see* **cutoff wall**

T245 **toggle joint** *see* **tongue-and-groove joint**

T246 *e* **toilet unit**
　　d 1. Klosettbecken *n* mit Spülkasten
2. Toilettenraum *m*
　　f 1. cuvette *f* à réservoir de chasse
2. cabinet *m* de W.C., water-closet
m
　　nl 1. closetcombinatie *f* 2. toilet *n*,
WC *m*
　　r 1. унитаз *m* со смывным бачком
2. уборная *f*, туалет *m*
(*помещение*)

T247 *e* **tongue**
　　d Feder *f*
　　f languette *f*, tenon *m*
　　nl veer *f*, messing *f(m)*
　　r гребень *m*, шип *m*
(*в соединениях*)

T248 *e* **tongue-and-groove joint**
　　d gespundete [genutete] Fuge *f*
　　f assemblage *m* à rainure et
languette
　　nl messing-en-groefverbinding *f*
　　r шпунтовый шов *m*,
шпунтовое соединение *n*

T249 *e* **toothing course**
　　d Vertikalverzahnung *f*
　　f arrachement *m*
　　nl staande tand *m* (*metselwerk*)
　　r вертикальная штраба *f*

T250 *e* **top**
　　d 1. Oberkante *f* 2. Scheitel *m* 3.
Verschleißschicht *f*
　　f 1. face *f* supérieure 2 faîte *m*,
haut *m*, tête *f* 3. couche *f* d'usure
[de roulement]
　　nl 1. bovenkant *m* 2. kruin *f(m)*
(*van boog*) 3. slijtlaag *f(m)*
(*wegdek*)
　　r 1. верхняя грань *f* 2. шелыга *f*,
вершина *f* 3. слой *m* износа

T251 *e* **top bars**
　　d Druck(bewehrungs)stäbe *m pl*
　　f barres *f pl* supérieures
　　nl bovenwapening *f*
　　r верхние (арматурные) стержни
m pl

T252 **top beam** *see* **collar beam**

T253 *e* **top edge of foundation**
　　d Fundamentoberkante *f*
　　f chape *f* d'arasement de fondation
　　nl bovenrand *m* van de fundering
　　r верхний обрез *m* фундамента

T254 *e* **top fiber**
　　d obere Faser *f*
　　f fibre *f* supérieure
　　nl bovenste vezel *f(m)*
　　r верхнее волокно *n* (*сечения балки*)

T255 *e* **top flange**
　　d Obergurt *m*
　　f aile *f* supérieure
　　nl bovenflens *m*, bovenrand *m*
　　r верхняя полка *f* (*балки*)

T256 *e* **top form**
　　d Obenschalung *f*
　　f coffrage *m* du comble
　　nl bovenbekisting *f*
　　r опалубка *f* бетонного покрытия
здания

T257 *e* **top frame member**
　　d Rahmenriegel *m*
　　f poutre *f* supérieure d'un portique
　　nl skeletbalk *m*, portaalbalk *m*,
bovenregel *m*
　　r ригель *m* рамы

T258 *e* **top hinged window** *see* **awning window**

T259 *e* **top level**
　　d 1. Oberkante *f* 2. Stauhöhe *f*,
Stauziel *n*
　　f 1. cote *f* [niveau *m*] de la face
supérieure 2. niveau *m* de retenue
　　nl 1. bovenwaterpeil *n* 2. stuwhoogte *f*
　　r 1. отметка *f* [уровень *m*] верхней
поверхности *или* грани 2. высота
f подпора, проектный подпор *m*

T260 *e* **topping**
　　d 1. Decklage *f*, Betondeckschicht *f*,
Verschleißschicht *f*
(*Betonfußboden, Straßendecke*)
2. Zementestrich *m*,
Zementdeckschicht *f*
(*Dachbelag*) 3. Terrazzobeton *m*
4. Betondeckschicht *f* auf
Profilstahlbelag
　　f 1. chape *f* en béton (*de couche
d'usure*) 2. chape *f* de
nivellement en mortier (*sous le
carralage ou revêtement
d'étanchéité du toit*) 3. mortier *m*
pour revêtement de sol en terrazzo
4. couche *f* de béton posée sur
tôles en profilé de plancher
　　nl 1. bovenlaag *f(m)* van het
betondek 2. uitvullaag *f(m)* van
beton op een plat dak 3.
betonmortel *m* voor een
terrazzovloer 4. betonlaag *f(m)*
op stalenvloer
　　1. верхний слой *m* бетонного
покрытия *n* (*пола, дороги*) 2.

цементная стяжка *f* для сборного
[штучного] покрытия (*пола,
крыши*) 3. бетон *m* типа «террацио»
4. слой *m* бетона на стальном
профилированном настиле

T261 *e* **topping lift**
   d Auslegerseilflasche *f*,
     Auslegerkopfrolle *f* (*Kran*)
   f moufle *m* de flèche (*d'une grue*)
  nl hijsschijf *f(m)* in de giektop van
     een kraan
   r стреловой полиспаст *m* (*крана*)

T262 *e* **topping lift cable**
   d Auslegerseil *n* (*Kran*)
   f câble *m* de flèche (*d'une grue*)
  nl hijsdraad *m* (*kraan*)
   r стреловой канат *m* (*крана*)

T263 *e* **topping of dike**
   d Erhöhen *n* des Deiches
   f hausse *f* [surélévation *f*] de la
     crête du barrage
  nl ophogen *n* van een dijk
   r наращивание *n* дамбы

T264 *e* **top rail**
   d Rahmenoberteil *m* (*Tür, Fenster*)
   f traverse *f* supérieure (*d'un bâti,
     d'une croisée*)
  nl bovendorpel *m* (*deur, venster*)
   r верхний брус *m* (*дверного полотна,
     оконного переплёта*)

T265 *e* **top reinforcement**
   d obere Bewehrung *f*
   f armature *f* supérieure
  nl bovenwapening *f*
   r верхняя арматура *f*

T266 *e* **top-running crane beam**
   d obenlaufender Kranbalken *m*
   f pont *m* roulant monorail [léger]
  nl kraanbalk *m* van een loopkraan
   r кран-балка *f*

T267 *e* **top soil**
   d Mutterboden *m*
   f sol *m* végétal
  nl bovengrond *m*
   r почвенный [растительный] слой
     m, дёрн *m*

T268 *e* **topsoil excavation**
   d Mutterbodenberäumung *f*
   f dégazonnement *m*, déblaiement *m*
     [déblayage *m*] du sol végétal
  nl afgraven *n* van de bovengrond
   r снятие *n* растительного слоя
     [дёрна]

T269 *e* **top surface of the beam**
   d Balkenoberkante *f*, Trägeroberkante *f*
   f face *f* supérieure de la poutre
  nl bovenkant *m* van een balk
   r верхняя грань *f* балки

T270 *e* **topwater**
   d schwebendes Grundwasserstockwerk *n*
   f nappe *f* suspendue temporaire
  nl grondwater *n* boven het freatisch vlak
   r верховодка *f*

T271 *e* **torquemeter**
   d Drehmomentmesser *m*
   f torsiomètre *m*
  nl draaimomentmeter *m*,
     torsiemeter *m*
   r вращательный динамометр *m*

T272 *e* **torque wrench**
   d Drehmomentschlüssel *m*
   f clé *f* dynamométrique
  nl momentensleutel *m*, torsiesleutel *m*
   r динамометрический гайковёрт
     m

T273 *e* **Torshear bolt**
   d hochfester Gewindebolzen *m* mit
     Ringnut
   f boulon *m* à haute résistance du
     type Torshear
  nl bout *m* van hoogwaardig staal met
     ringgroef
   r высокопрочный болт *m* с
     кольцевой выточкой

T274 *e* **torsion**
   d Torsion *f*, Verdrehung *f*,
     Drillung *f*
   f torsion *f*
  nl verdraaiing *f*, wringing *f*, torsie *f*
   r кручение *n*

T275 *e* **torsional cracking**
   d verwerfungsbedingte Rißbildung *f*
   f fissuration *f* par torsion
  nl scheurvorming *f* tengevolge van
     torsie
   r трещинообразование *n*,
     вызванное короблением

T276 *e* **torsional load**
   d Torsionslast *f*,
     Torsionsbeanspruchung *f*
   f charge *f* [sollicitation *f*] de torsion
  nl torsiekracht *f(m)*
   r крутящая [скручивающая]
     нагрузка *f*

T278 *e* **torsional moment**
   d Torsionsmoment *n*,
     Drill(ungs)moment *n*,
     Verdrehmoment *n*
   f moment *m* de torsion
  nl torsiemoment *n*
   r крутящий момент *m*

T279 *e* **torsional stiffness**
   d Torsionssteifigkeit *f*,
     Verdreh(ungs)steifigkeit *f*
   f rigidité *f* de torsion
  nl torsiestijfheid *f*
   r крутильная жёсткость *f*,
     жёсткость *f* при кручении

T280 *e* **torsional stress**
   d Torsionsspannung *f*,
     Verdrehungsspannung *f*
   f contrainte *f* de torsion
  nl torsiespanning *f*
   r напряжение *n* при кручении

T281 *e* **torsion reinforcement**
   d Drillbewehrung *f*, Torsionsarmierung *f*

f armature f de torsion
nl torsiewapening f
r арматура f, воспринимающая
усилия кручения

T282 e torsion strength
d Torsionsfestigkeit f
f résistance f à la torsion
nl torsiesterkte f
r предел m прочности при кручении

T283 e total energy
d 1. gesamte Energie f 2. autonome
Einschienen-Energieversorgung f
des Gebäudes
f 1. énergie f totale 2. utilisation f
d'une seule espèce de porteur de la
chaleur
nl 1. volledige energie f 2. gebruik n
van één energiedrager in de eigen
energievoorziening van een gebouw
r 1. полная энергия f 2. применение
n одного первичного энергоносителя
в системе автономного
энергоснабжения здания

T284 e total energy system
d 1. autonomes
Einschienen-
-Energieversorgungssystem n
2. Klimaanlage f mit
autonomer Energieversorgung
f 1. système m autonome
d'approvisionnement en énergie
basé sur un seul caloporteur
2. installation f de climatisation
avec un système autonome
d'approvisionnement en énergie
nl 1. eigen energievoorzieningssysteem n
met één primaire energiedrager
2. luchtbehandelingssysteem n met
eigen energievoorziening
r 1. автономная система f
энергоснабжения (здания или
группы зданий) с одним
первичным энергоносителем
2. система f кондиционирования
воздуха с автономным
энергоснабжением

T285 total head line see energy line

T287 e total storage capacity
d Gesamtspeicherinhalt m,
Stauraum m, Speichervolumen n
f capacité f totale du réservoir,
réserve f totale
nl stuwruimte f
r полный объём m водохранилища

T288 e total strain energy
d Deformationsenergie f,
Formänderungsenergie f,
Spannungsenergie f, Formenergie f
f énergie f (totale) de déformation
nl totale energie f
r общая [суммарная] энергия f
упругой деформации

T289 e total stress
d Gesamtspannung f

f effort m total
nl totaalspanning f
r суммарное напряжение n

T290 e toughened glass UK see tempered glass

T291 e toughness
d 1. Zähigkeit f; Kerbschlagzähigkeit
f 2. Steifigkeit f;
Widerstandsfähigkeit f
3. Sprödbruchunempfindlichkeit f
f 1. ductilité f; résilience f
2. ténacité f, dureté f, résistance f
3. résistance f à la cassure fragile
nl 1. taaiheid f; kerfslagtaaiheid f
2. stijfheid f
r 1. вязкость f; ударная вязкость f
2. жёсткость f; прочность f 3.
сопротивление n хрупкому излому

T292 e towed roller
d Schleppwalze f, Anhängewalze f
f cylindre m [rouleau m] tracté
nl getrokken wals f(m)
r прицепной (дорожный) каток m

T293 e towed scraper
d Anhängeschürfkübelwagen m
f scraper m tracté, décapeuse f
tractée
nl getrokken schraper m
r прицепной скрепер m

T294 e towel rail
d Handtuchtrockner m;
Handtuchstange f
f séchoir m à serviettes; porte-
-serviettes m
nl handdoekhanger m
r полотенцесушитель m; вешалка f
для полотенец

T295 e tower biofilter
d Turmtropfkörper m
f filtre m bactérien à tour
nl biologisch torenfilter n, m
r башенный биофильтр m

T296 e tower concrete spouting plant
d Betongießturm m, Ausleger-
-Gießturm m
f tour f distributrice de béton (à
flèche)
nl betongiettoren m met spier
r бетонолитная башня f со стрелой

T297 e tower crane
d Turmdrehkran m
f grue f à tour, grue f monotour,
grue f sur pylône
nl torendraaikraan f(m)
r башенный поворотный кран m

T298 e tower hoist
d Turmgerüst(bau)aufzug m
f monte-charge m de chantier (à cage)
nl torensteiger-bouwkraan m,
bouwlift m
r башенный подъёмник m

T299 e tower intake
d Turmfassung f, Turmeinlauf m,
Turmentnahme f

    f prise-d'eau *m* à tour
  nl toreninloop *m*
   r башенный водозабор *m*

T300 *e* **tower loader**
   d Frontschaufellader *m* mit
     Hubrahmen
   f chargeur *m* élévateur à mât
     frontal avec benne de chargement
  nl schop *f(m)* met universele hefmast
   r фронтальный одноковшовый
     погрузчик *m* с грузоподъёмной
     мачтой [башней]

T301 *e* **town drainage main**
   d Abwassersammelkanal *m*
   f collecteur *m* de drainage urbain
  nl stadsregenwaterafvoerleiding *f*
   r городской водосточный коллектор
     m

T302 *e* **town planning**
   d Stadtplanung *f*, Städteplanung *f*;
     städtebauliche Planung *f*
   f planification *f* urbaine;
     urbanisme *m*
  nl stadsplanning *f*; stedelijk
     structuurplan *n*
   r планировка *f* городов;
     градостроительное проектирование *n*

T303 *e* **towpath**
   d Treidelpfad *m*
   f chemin *m* de halage
  nl jaagpad *n*
   r бечевник *m* (*дорога, проложенная
     вдоль канала или реки для
     службы навигации*)

T304 **T-piece** *see* tee

T305 *e* **tracing**
   d 1. Trassierung *f*, Linienführung *f*,
     Abstecken *n* 2. Rohrheizung *f*,
     Begleitheizung *f* 3. Pause *f*
   f 1. traçage *m*, tracé *m*
     2. réchauffage *m* des tuyaux
     3. calque *f*, dessin *m* sur calque;
     croquis *m*
  nl 1. traceren *n* 2. pijpverwarming *f*
     3. calque *f(m)*
   r 1. трассирование *n*, провешивание *n*
     2. обогрев *m* трубопроводов
     3. калька *f*, чертёж *m* на кальке

T306 *e* **track adjustment**
   d Streckenausrichtung *f*,
     Gleisrichtung *f*
   f ajustement *m* [ajustage *m*, réglage
     m] de la voie ferrée; ajustage *m*
     des railles
  nl het spoor richten
   r регулировка *f* рельсового пути;
     рихтовка *f* пути

T307 *e* **tracked vehicle**
   d Gleiskettenfahrzeug *n*,
     Raupenfahrzeug *n*
   f véhicule *m* chenillé [à chenilles]
  nl rupsvoertuig *n*
   r гусеничная транспортная машина *f*

T308 *e* **track gauge**
   d Gleisspur *f*, Spurweite *f*
   f gabarit *m* (d'écartement) de voie,
     écartement *m* de voie
  nl spoorbreedte *f*, spoorwijdte *f*
   r ширина *f* колеи

T309 *e* **tracking**
   d Spurbildung *f*
   f ornièrage *m*
  nl spoorvorming *f* in het wegdek
   r образование *n* колеи (*дефект
     дорожного покрытия*)

T310 *e* **track laying crane**
   d Gleisbau-Baggerkran *m*
   f grue *f* pour la pose de la voie
     ferrée
  nl spoorwegkraan *m*
   r кран-путеукладчик *m*

T311 **track tamper** *UK see* tie tamper

T312 *e* **traction lift, traction sheave**
    elevator *US*
   d Treibscheibenaufzug *m*
   f ascensseur *m* commandé par un
     treuil (à adhérence)
  nl tractielift *m*
   r лифт *m* с канатоведущим шкивом

T313 *e* **tractor-drawn scraper**
   d traktorgezogener Schrapper *m*
   f scraper *m* tracté (par tracteur),
     décapeuse *f* tracté
  nl getrokken scraper *m*
   r скрепер *m* с тракторной тягой

T314 *e* **tractor scraper**
   d selbstfahrender Schürfwagen *m*
   f motorscraper *m*, décapeuse *f*
     automotrice
  nl zelfrijdende scraper *m*
   r самоходный скрепер *m*

T315 **tractor shovel** *see* loading shovel

T316 *e* **trade effluent, trade wastes**
   d Industrieabwasser *n*,
     Betriebsabwasser *n*
   f eaux *f pl* usées industrielles
  nl bedrijfsafvalwater *n*
   r производственные [промышленные]
     сточные воды *f pl*

T317 *e* **traffic capacity**
   d maximale *bzw.* rechnerische
     Durchlaßfähigkeit *f*
     [Durchlaßkapazität *f*] (*einer Straße*)
   f débit *m* d'une route, capacité *f* du
     trafic (*maximale ou calculée*)
  nl verkeerscapaciteit *f*
   r (максимальная *или* расчётная)
     пропускная способность *f*
     (*дороги*)

T318 *e* **traffic-carrying deck**
   d verkehrsbeanspruchte
     Brückendeckplatte *f*
   f tablier *m* (*d'un pont*) soummis à
     l'action de trafic
  nl brugdek *n*
   r настил *m* проезжей части моста

      (*воспринимающий нагрузки*
      от транспортных средств)

T319 *e* **traffic cone**
    d Lübecker Hütchen *n*
    f cône *m* de signalisation (*pour la*
      protection des lieux de travail sur
      chaussées)
   nl verkeerskegel *m* (*markering van*
      wegwerken)
    r сигнальный [оградительный]
      конус *m* (*для ограждения*
      рабочих мест на дороге)

T320 *e* **traffic decking**
    d Fahrbahnplatte *f* (*Brücke*)
    f dalle *f* de tablier
   nl rijvloer *m* (*brug*)
    r плита *f* проезжей части (*моста*)

T321   **traffic deck surfacing** *see* **topping**

T322 *e* **traffic island**
    d Verkehrsinsel *f*
    f ilôt *m* de circulation
   nl vluchtheuvel *m*
    r островок *m* безопасности (*на*
      дороге)

T323 *e* **traffic lane**
    d Fahrspur *f*, Fahrstreifen *m*
    f voie *f* de circulation
   nl rijstrook *f(m)*
    r полоса *f* движения (*на дороге*)

T324 *e* **traffic line**
    d Verkehrslinie *f*, Markierungsstrich
      m, Farbstreifen *m*
      (*Straßenmarkierung*)
    f ligne *f* de sécurité
   nl rijstrookmarkeringsstreep *f(m)*
    r линия *f* разметки (*на дороге*)

T325 *e* **traffic load**
    d Verkehrslast *f*
    f charge *f* roulante [routière]
   nl verkeersbelasting *f*
    r нагрузка *f* от транспортных
      средств *или* от подвижного состава

T326 *e* **traffic paint**
    d Straßenmarkierungsfarbe *f*
    f peinture *f* pour marquage (*de*
      lignes de signalisation)
   nl wegenverf *f*
    r красочный состав *m* [краска *f*]
      для дорожной разметки

T327 *e* **traffic railing**
    d Straßenschutz *n*, Schutzgeländer *n*
    f garde-corps *m* [garde-fou *m*] de
      sécurité ; glissière *f* de sûrete [de
      sécurité]
   nl vangrail *m*
    r защитные [перильные] ограждения
      n pl (*на дорогах*)

T328 *e* **traffic structure**
    d Verkehrsbauwerk *n*
    f ouvrage *m* autoroutier [d'art de
      trafic]
   nl verkeerskunstwerk *n*
    r транспортное сооружение *n*

T329 *e* **traffic tunnel**
    d Verkehrstunnel *m*
    f tunnel *m* de trafic
   nl verkeerstunnel *m*
    r транспортный туннель *m*

T330 *e* **trailed vibrating roller**
    d Anhänge-Vibrationswalze *f*,
      Schlepp-Rüttelwalze *f*
    f rouleau *m* [cylindre *m*] vibrant
      tracté
   nl getrokken trilwals *f(m)*
    r прицепной виброкаток *m*

T331 *e* **trailer**
    d Anhänger *m*, Nachläufer *m*,
      Schleppfahrzeug *n*
    f remorque *f*
   nl aanhangwagen *m*
    r прицеп *m*

T332   **training embankment** *see*
    longitudinal dyke

T333   **training structures** *see* **training**
    works

T334 *e* **training wall**
    d 1. Leitmauer *f*, Leitdamm *m*,
      Leitwerk *n* 2. Längsdamm *m*,
      Längswerk *n*, Parallelwerk *n*
    f 1. guideau *m*, mur *m* guide-eau
      2. digue *f* longitudinale
   nl 1. leidam *m* 2. langsdijk *m*,
      schaardijk *m*
    r 1. (струе)направляющая стенка *f*
      2. продольная дамба *f*

T335 *e* **training works**
    d Einschränkungsbauwerke *n pl*,
      Regelungsbauwerke *n pl*
    f ouvrages *m pl* de régularisation
   nl regularisatiewerken *n pl*
    r руслорегулирующие
      [водостеснительные] сооружения
      n pl

T336   **trans-divide diversion** *see*
    interbasin diversion

T337 *e* **transfer**
    d Vorspannkraftübertragung *f* auf
      Beton
    f transmission *f* au béton de la
      précontrainte de l'armatur
   nl overdracht *f(m)* van de
      voorspanningskracht naar het
      beton
    r передача *f* усилия предваритель-
      ного напряжения на бетон

T338 *e* **transfer bond**
    d Haftverbund *m*
    f adhérence *f* assurant la
      transmission au béton des efforts
      de précontrainte de l'armature
   nl aanhechting *f*
    r силы *f pl* сцепления,
      обеспечивающие напряжение
      бетона

T339 *e* **transfer girder**
    d Unterfangriegel *m*

f poutre *f* de répartion
nl draagbalk *m* voor de hiet tot de
fundering doorgaande kolommen
r ригель *m* каркаса высотного
здания, поддерживающий
промежуточные колонны
(*не доходящие до фундамента*)

T341 *e* **transfer strength**
d Übertragungsfestigkeit *f*
(*Spannbeton*)
f résistance *f* du béton au moment
de la détente
nl transfersterkte *f*
r передаточная прочность *f*
(*минимальная прочность бетона,
допускающая его преднапряжение*)

T342 *e* **transformed composite cross-
-section, transformed section**
d reduzierter Querschnitt *m*
j section *f* composite transformée
nl gereduceerde doorsnede *f*
r приведённое сечение *n*

T343 *e* **transit**
d Theodolit *m*
f théodolite *m*
nl omkeren van de kijker van een
theodoliet
r теодолит *m*

T344 *e* **transition curve**
d Übergangskurve *f*, Übergangsbogen
m
f courbe *f* de raccordement
nl overgangsboog *m* (*weg*)
r переходная кривая *f* (*дороги*)

T345 *e* **transit-mix concrete**
d Transportmischbeton *m*
f béton *m* préparé et transporté
dans le camion malaxeur
nl transportbeton *n*
r бетон *m*, приготовляемый
и доставляемый автобетоносмесителем

T346 **transit mixer** *see* **truck mixer**

T347 *e* **transit mixing**
d Transportmischung *f* (*Beton*)
f préparation *f* de béton dans le
camion malaxeur
nl mengen *n* van een betonmortel
tijdens transport
r приготовление *n* бетонной смеси
в автобетоносмесителе

T348 *e* **transit pipeline**
d 1. Fernwärmeleitung *f*
2. Hauptleitung *f*,
Fernrohrleitung *f*
f 1. canalisation *f* de transport de
chaleur urbaine, artère *f* de
chauffage 2. pipe-line *m*,
canalisation *f* [ligne *f*]
principale, oléoduc *m*
nl 1. afstandverwarmingsgeleiding *f*
2. hoofdpijpleiding *f*
r 1. тепломагистраль *f*, трубопровод
m дальнего теплоснабжения

2. транзитный [магистральный]
трубопровод *m*

T349 *e* **translucent concrete**
d Glasbeton *m*
f béton *m* armé translucide
nl prefab-betonpaneel *n* met
glastegels
r сборная железобетонная панель *f*
с остеклением

T350 *e* **transmissibility coefficient**
d Ableitungsfaktor *m* nach Darcy,
Durchlässigkeitsfaktor *m*
f coefficient *m* de Darcy [de
filtration]
nl afleidingsfactor *m* van Darcy,
doorlaatcoëfficiënt *m*
r коэффициент *m* фильтрации

T351 **transmission coefficient** *see*
coefficient of permeability

T352 *e* **transmission zone**
d Transmissionszone *f*,
Verankerungszone *f* (*Spannbeton*)
f zone *f* de transmission (*de la
tension*)
nl overdrachtszone *f*, verankeringszone *f*
r зона *f* передачи (*усилий*); зона *f*
анкеровки

T353 *e* **trans-mountain water diversion**
d Überleitung *f*, Beileitung *f*
f captage *m* [dérivation *f*] du débit
nl afleiding *f* over de waterscheiding
r переброска *f* стока

T354 *e* **trans-mountain water diversion
gallery**
d Überleitungsstollen *m*
f galerie *f* de captage en travers de
la ligne de partage
nl afleidingstunnel *m*
r туннель *m* для переброски стока

T355 *e* **transom**
d 1. Kämpfer *m*, Losholz *n*
2. oberer Belüftungsflügel *m*,
Fensterbelüftungsklappe *f*
3. Jochbalken *m*
f 1. traverse *f* d'imposte, meneau *m*
horizontal 2. imposte *f* 3. chapeau
m, traverse *f*
nl 1. tussendorpel *m*, kalf *n* 2.
bovenlicht *n* 3. junkbalk *m*, kesp *m*
r 1. средник *m*, горизонтальный импост
m (*оконной коробки*) 2. фрамуга *f*
3. шапочный брус *m*

T356 *e* **transom light**
d oberer Belichtungsflügel *m*
f imposte *f* vitrée
nl bovenlicht *n*
r застеклённая фрамуга *f*

T357 *e* **transporter crane**
d Umladekran *m*, Überladekran *m*
f portique *m* roulant
nl overladerkraan *f*(*m*), rijdende
kraan *m*
r кран-перегружатель *m*

T358 *e* **transporting cableway**
   d Drahtseilschwebebahn *f*,
     Seilhängebahn *f*
   f câble *m* aérien de transport
   nl kabelbaan *f*
   r подвесная канатная дорога *f*

T359 *e* **transverse beam**
   d Querbalken *m*
   f poutre *f* transversale
   nl dwarsbalk *m*
   r поперечная балка *f*

T360 *e* **transverse bending**
   d Querbiegen *n*
   f flexion *f* transversale
   nl dwarsbuiging *f*
   r поперечный изгиб *m*

T361 *e* **transverse contraction**
   d Querkontraktion *f*, Einschnürung *f*
   f contraction *f* transversale [latérale]
   nl dwarscontractie *f*, insnoering *f*
   r поперечное [боковое] сжатие *n*

T362 *e* **transverse cracks**
   d Querrisse *m pl*
   f fissures *f pl* transversales
   nl dwarsscheuren *f(m) pl*
   r поперечные трещины *f pl*

T363 *e* **transverse drain**
   d Sammler *m*, Sammelkanal *m*,
     Binnenvorfluter *m*
   f drain *m* transversal
   nl verzamelkanaal *n*, verzameldrain *m*
   r дрена-собиратель *f* (*в*
     осушительной системе)

T364 *e* **transverse dyke**
   d Querdeich *m*, Querdamm *m*,
     Buhne *f*, Krippe *f*
   f traverse *f*, épi *m* de liaison
   nl dwarsdijk *m*, dwarsdam *m*, krib *f(m)*
   r траверс *m*; полузапруда *f*,
     буна *f*

T365 **transverse fiouw fan** *see* **cross flow
fan**

T366 *e* **transverse joint**
   d Querfuge *f*
   f joint *m* transversal
   nl dwarsvoeg *f(m)*; dwarsverbinding *f*
   r поперечный шов *m*; поперечное
     соединение *n*

T367 *e* **transverse load**
   d Querlast *f*, Querbelastung *f*
   f charge *f* transversale
   nl zijdelingse belasting *f*
   r поперечная нагрузка *f*

T368 *e* **transverse member**
   d Traverse *f*; Querglied *n*, Querholz *n*,
     Querriegel *m*
   f traverse *f*; entretoise *f*
   nl dwarshout *n*, tussenregel *m*
   r поперечина *f*

T369 *e* **transverse prestress**
   d Quervorspannung *f*
   f précontrainte *f* transversale

   nl dwarse voorspanning *f*
   r поперечное преднапряжение *n*

T370 *e* **transverse reinforcement**
   d Querbewehrung *f*, Querarmierung *f*
   f armature *f* transversale
   nl dwarswapening *f*
   r поперечная арматура *f*

T371 **transverse strength** *see* **flexural
strength**

T372 *e* **trap**
   d 1. Geruchverschluß *m*, Traps *m*
     2. Geruchverschluß *m* mit
     Revisionsöffnung 3. Abscheider *m*
   f 1. joint *m* hydraulique, fermeture *f*
     d'eau, siphon *m*, coupe-air *m*
     2. siphon *m* regard, regard *m* à
     siphon, boite *f* de nettoyage à
     siphon 3. collecteur *m*,
     séparateur *m*
   nl 1. waterslot *n* 2. stankafsluiter *m*
     met ontstoppingsopening 3.
     opvanginrichting *f*, afscheider *m*
   r 1. гидравлический [водяной]
     затвор *m*, сифон *m* (*сантехника*)
     2. затвор-ревизия *m*
     (*санитарных приборов*)
     3. ловушка *f*, сепаратор *m*,
     отделитель *m*

T373 *e* **trap gate**
   d Klapptor *n* (*Schleuse*)
   f porte *f* rabattante (*d'écluse*)
   nl klepdeur *f(m)* (*sluis*)
   r клапанные ворота *pl* (*шлюза*)

T374 *e* **trapped gully**
   d Wasserablauf *m* mit Geruchverschluß
   f caniveau *m* de rue à siphon
   nl waterafvoer *m* met waterslot
   r уличный водосток *m* с водяным
     затвором

T375 *e* **trash can**
   d Müllsammelgefäß *n*
   f récepteur *m* d'ordures
   nl afvalemmer *m*, vuilnisemmer *m*
   r мусоросборник *m*

T376 *e* **trash chute**
   d 1. Müllschlucker *m* 2.
     Müllabwurfschacht *m*
   f 1. gaine *f* à ordures, vide-ordures *m*
     2. gaine *f* [goulotte *f*] à gravats
   nl vuilniskoker *m*
   r 1. мусоропровод *m* 2. короб *m*
     для спуска строительного мусора

T377 *e* **trash rack**
   d Auffangrechen *m*, Einlaufrechen *m*
   f grille *f* (d'arrêt)
   nl vuilopvangrooster *n*
   r сороудерживающая решётка *f*

T378 **traveled way** *see* **traffic lane**

T379 *e* **traveler**
   d 1. Schalwagen *m* 2. Brückenkran *m*,
     Laufkran *m*
   f 1. coffrage *m* roulant [mobile] en
     portique 2. pont *m* roulant

nl 1. П-vormige sectie *f* van een
rolbekisting 2. brugkraan *m*
r 1. П-образная секция *f* катучей
опалубки 2. мостовой кран *m*

T380 *e* **traveling cable**
d Schleppkabel *n*
f câble *m* électrique traîné (*p. ex.,
pour alimenter une grue à tour*)
nl elektrische kabel *m* voor een
rijdende machine
r электрокабель *m*, питающий
передвижную строительную
машину (*напр. башенный кран*)

T381 *e* **traveling cradle**
d 1. Hängerüstung *f*,
Hänge(bau)gerüst *n* 2. Hängekorb *m*
f 1. échafaudage *m* volant
2. selette *f*
nl 1. hangsteiger *m* 2. hangkorf *m*
r 1. висячие леса *pl* [подмости *pl*]
2. подвесная люлька *f*

T382 *e* **traveling crane**
d 1. Brückenkran *m*, Laufkran *m*
2. beweglicher Turmkran *m*
f 1. pont *m* roulant 2. grue *f* à tour
mobile
nl 1. loopkraan *f(m)* 2. rijdende
torenkaan *m*
r 1. мостовой кран *m*
2. передвижной башенный кран *m*

T383 *e* **traveling formwork**
d Wanderschalung *f*
f coffrage *m* roulant
nl rolbekisting *f*
r катучая опалубка *f*

T384 *e* **traveling spray booth**
d bewegliche [verschiebbare]
Spritzkabine *f*
f cabine *f* de peinture (*par
pulverisation*) mobile [roulante]
nl verplaatsbare verfspuitcabine *f*
r накатное вентиляционное
укрытие *n*, накатная кабина *f*
(*для окраски распылением*)

T385 *e* **traveling stage**
d Fahr-Arbeitsplattform *f*,
Fahrbühne *f*
f échafaudage *m pl* roulant ; plate-
-forme *f* de travail roulante
nl rolsteiger *m*, mobiel montageplatform
n
r передвижные подмости *pl*;
передвижная рабочая площадка *f*

travelling... *see* **traveling...**

T386 *e* **traverse**
d 1.Querriegel *m*, Querträger *m*,
Traverse *f* 2. Dübel *m*
3. Querdamm *m*, Querdeich *m*,
Schutzdamm *m* 4. Polygonzug *m*,
Theodolitzug *m* 5. Verteilung *f* der
Geschwindigkeit *oder* des Drucks
im Stromquerschnitt (*z.B. im
Luftkanal*)
f 1. traverse *f*, poutre *f* [solive *f*]

transversale 2. clavette *f* 3. digue *f*
transversale, duit *m* 4. traversée *f*
[cheminement *m*] d'angle
5. distribution *f* des vitesses *ou* des
pressions dans la section
transversale
nl 1. dwarsbalk *m* 2. deuvel *m*
3. dwarsdijk *m* 4. slag *m* van een
landmeting 5. snelheids- *of*
drukverdeling *f* in stroomdoorsnede
r 1. поперечина *f*, поперечная
балка *f*, траверса *f* 2. шпонка *r*
3. траверс *m*, поперечная дамба *f*
4. полигонометрический
[теодолитный] ход *m*
5. распределение *n* скорости *или*
давления в поперечном сечении
потока (*напр. в воздуховоде*)

T387 *e* **traversing**
d 1. Polygonierung *f* 2. Fahren *n*
(*Kran, Laufkatze*)
3. Geschwindigkeits- und
Druckmessungen *f pl* in bestimmten
Punkten des Luftkanalquerschnittes
f 1. cheminement *m* d'angle
2. translation *f* latérale (*p. ex d'un
chariot de pont roulant*)
3. mesurage *m* de vitesses et de
pressions d'air (*dans la section
transversale d'une conduite de
ventilation*)
nl 1. landmeting *f* 2. dwarsbeweging *f*
(*loopkat*) 3. snelheids- en
drukmetingen *f pl* in bepaalde
punten van een luchtleiding
r 1. полигонометрическая съёмка *f*
2. рабочее перемещение *n* (*напр.
подъёмного крана, тележки
тельфера*) 3. траверсирование *n*
(*измерение скоростей и давлений
в определённых точках поперечного
сечения воздуховода*)

T388 *e* **traversing slipway**
d Queraufschleppe *f*, Querschlipp *m*,
Querslip *m*
f cale *f* de lancement transversale
nl dwarsslip *m*
r поперечный слип *m*

T389 *e* **tread**
d 1. Auftritt *m*, Trittstufe *f*
2. Sprosse *f* 3. Spurweite *f*
(*Raupenfahrwerk*)
f 1. giron *m* 2. échelon *m*
d'échelle 3. écartement *m* de
chenilles
nl 1. traptrede *f(m)* 2. sport *f*
3. spoorbreedte *f*
r 1. проступь *f* 2. перекладина *f*
лестницы-стремянки 3. ширина *f*
колеи (*гусеничного хода*)

T390 *e* **tread width**
d Stufenbreite *f*
f largeur *f* de giron
nl breedte *f* van de trede
r ширина *f* ступени [проступи] *f*

T391 *e* treated joint
   d behandelte [wasserabgedichtete]
     Fuge *f*
   f joint *m* scellé [calfeutré]
   nl waterdicht afgesloten voeg *f(m)*
     [naad *m*]
   r герметизированный
     [гидроизолированный] шов *m*

T392 *e* treated pile
   d imprägnierter Holzpfahl *m*
   f pieu *m* en bois imprégné
   nl geïmpregneerde houten paal *m*
   r антисептированная деревянная
     свая *f*

T393 *e* treating load
   d Abwasserbelastung *f*
   f charge *f* sur les ouvrages
     d'épuration des eaux usées
   nl afvalwaterbelasting *f*
   r нагрузка *f* на очистные сооружения

T394 *e* treatment facilities, treatment
     plant, treatment works
   d Abwasserbehandlungsanlage *f*,
     Abwasserreinigungsanlage *f*;
     Kläranlage *f*
   f ouvrages *m* d'assainissement;
     station *f* d'épuration
   nl afvalwaterzuiveringsinstallatie *f*
   r очистные сооружения *n pl*;
     станция *f* очистки сточных вод

T395 *e* tree and existing plant protection
   d Baum- und Pflanzenschutz *m*
     (*auf Baustellen*)
   f protection *f* des arbres (*sur le
     chantier*)
   nl sparen *n* van bomen en planten
   r сохранение *n* деревьев и растений
     (*на стройплощадке*)

T396 *e* tree and shrub planting
   d Baum- und Strauchbepflanzung *f*,
     Grünanpflanzung *f*, Begrünung *f*
   f plantation *f* des arbres et des
     buissons
   nl groenverzorging *f*
   r озеленение *n*, посадка *f* деревьев
     и кустов

T397 *e* tremie
   d Kontraktor(beton)rohr *n*,
     Trichterrohr *n*, Betonversenkrohr *n*
   f trémie *f*, tube *m* de bétonnage,
     goulotte *f* (*pour le bétonnage sous l'eau*)
   nl trechterpijp *f(m)* (*voor het
     betonstorten onder water*)
   r бетонопровод *m*, труба *f*
     с воронкой (*для подводного
     бетонирования*), хобот *m*

T398 *e* tremie concrete
   d Kontraktorbeton *m*,
     Unterwassergußbeton *m*
   f béton *m* frais mis en place par
     goulotte
   nl onderwaterbeton *n*

   r бетон *m* подводной укладки (*по
     методу вертикального
     перемещения трубы*)

T399 *e* tremie concreting
   d Kontraktorverfahren *n*, Betonieren
     n mit senkrechten Schüttrohren
   f méthode *f* de bétonnage par
     goulotte
   nl contractormethode *f*, betonstorten *n*
     met loodrechte stortpijpen
   r подводное бетонирование *n*
     методом вертикального
     перемещения трубы

T400 *e* trench
   d Graben *m*; Grube; *f*
   f tranchée *f*; fossé *m*
   nl sleuf *f(m)*; bouwput *m*
   r траншея *f*, канава *f*; ров *m*;
     котлован *m*

T401 *e* trench braces
   d bewegliche Quersteifen *f pl*
     [Grabensteifen *f pl*]
   f étais *m pl* de blindage
   nl bekledingsschoren *m pl*, stutten
     m pl
   r распорки *f pl* траншейных крепле-
     ний

T402   trench digger *see* trencher

T403 *e* trench drain
   d Graben-Entwässerung *f*,
     Grabendrän *m*
   f tranchée *f* drainante
   nl drainage *f* van de bouwput
   r траншейный дренаж *m*

T404   trench drainage *see* ditch drainage

T405 *e* trencher, trench excavator,
     trenching machine
   d Grabenbagger *m*
   f excavateur *m* de tranchée, pelle *f*
     de tranchées, trancheuse *f*
   nl sleuvengraver *m*
   r траншеекопатель *m*,
     канавокопатель *m*

T406 *e* trench jack
   d Spreizwinde *f* (*für Absteifung der
     Gräben*)
   f vérin *f* pour étayage de tranchée
   nl schoorvijzel *m*
   r траншейный домкрат *m*

T407 *e* trenchless pipelayer
   d grabenfreie Dränmaschine *f*,
     Maulwurfdränmaschine *f*
   f poseur *m* de tuyaux [de
     canalisations] sans creusement de
     tranchées
   nl drainagemachine *f* (*molploegsysteem*)
   r бестраншейный дреноукладчик *m*
     [трубоукладчик *m*]

T408 *e* trench plough
   d Grabenpflug *m*
   f charrue *f* fossoyeuse [cureuse de
     fossés], fossoyeuse *f*
   nl greppelploeg *f*, *m*

r траншейный плуг *m*, плужный канавокопатель *m*

T409 *e* trench shield
d bewegliche Kanaldiele *f*
f blindage *m* des fouilles mobile [roulant]
nl verplaatsbaar schild *n* voor een ingraving
r передвижное траншейное крепление *n*

T410 *e* trench shoring system
d Grabenverbau *m*
f système *m* d'étayage [de blindage] des fouilles
nl bekledings- en stutsysteem *n* voor ingravingen
r траншейные крепления *n pl*, крепления *n pl* стенок траншей

T411 *e* trench water intake
d Grabenwasserentnahme *f*
f prise *f* d'eau en tranchée
nl wateronttrekking *f* uit een greppel
r траншейный водозабор *m*

T412 *e* trestle
d 1. Estakade *f*, Gerüstbrücke *f*
2. Gerüst *n*
f 1. pont *m* de chevalets, estacade *f*
2. échafaudages *m pl* sur trétaux
nl 1. schraag *m*, schraagbrug *f(m)*
2. schraagsteiger *m*
r 1. эстакада *f* 2. подмости *pl*

T413 *e* trestle bridge
d Bockbrücke *f*
f pont *m* sur chevalets [sur tréteaux]; estacade *f*
nl schraagbrug *f(m)*
r мост-эстакада *m*; эстакада *f* на кóзлах *(напр. для бетонных работ)*

T414 *e* trestle concreting
d Betonieren *n* von einem Gerüst aus
f bétonnage *m* à l'aide d'estacade
nl betonstorten *n* vanaf de brug
r бетонирование *n* с эстакад(ы)

T415 *e* trestle hydraulicking
d Anspülen *n* mit Spülgerüst
f remblayage *m* hydraulique à l'aide des estacades
nl opspuiten *n* met gebruik van schragen
r эстакадный намыв *m*

T416 *e* trestle ladder
d Doppelleiter *f*, Klappleiter *f*
f échelle *f* double
nl trapleer *f*, dubbele ladder *f(m)*
r лестница-стремянка *f*

T417 *e* trial batch, trial concrete mix
d Probecharge *f*, Versuchscharge *f*
f gâchée *f* d'essai [d'épreuve]
nl proefcharge *f(m)* van betonmortel
r пробный замес *m*]

T418 *e* triangular weir
d Dreieckswehr *n*, Wehr *n* mit Dreiecksprofil
f déversoir *m* (à profil) triangulaire
nl driehoekige stuw *m*
r треугольный водослив *m*

T419 *e* triaxial compression test
d triaxialer Druckversuch *m*, triaxiale Druckprobe *f*
f essai *m* triaxial
nl drieassige drukproef *f(m)*
r испытание *n* на трёхосное сжатие

T420 *e* triaxial prestress
d dreiachsige Vorspannung *f*
f précontrainte *f* triaxiale
nl triaxiale voorspanning *f*
r трёхосное преднапряжение *n*

T421 *e* triaxial stress state
d dreiachsiger [räumlicher] Spannungszustand *m*
f état *m* de contrainte [de tension] à trois dimensions, état *m* triple de tension
nl triaxiale spanningstoestand *m*
r трёхосное напряжённое состояние *n*

T422 triaxial test *see* triaxial compression test

T423 *e* trickling filter
d berieselter Tropfkörper *m*
f lit *m* bactérien percolateur
nl druppelfilter *n*, *m*
r капельный биофильтр *m*

T424 trief cement *see* Portland blast-furnace cement

T425 *e* trim
d Einfassung *f*, Umrahmen *n*
f encadrement *m*
nl afwerking *f* (*om een deur- of vensteropening*)
r обкладка *f*, обрамление *n*, отделка *f*

T426 trimmer joist *see* tail beam

T427 *e* trimming
d 1. Profilgebung *f*, Profilieren *n* 2. Abbinden *n*, Abbund *m* (*Zimmerei*) 3. Einfassen *n* von Öffnungen, Umrahmung *f* 4. Besäumung *f*
f 1. profilage *m* [mise *f* au profil] de finition (*de la plate-forme d'une route*) 2. assemblage *m* des élément d'une charpente en bois 3. encadrement *m* d'une baie 4. taille *f* [coupe *f*] des bords
nl 1. profilering *f* (*wegenbouw*) 2. samenstellen *n* van een houtconstructie 3. aftimmeren *n*
r 1. чистовая планировка *f* насыпей и выемок (*в дорожном строительстве*) 2. вязка *f* [изготовление *n*] деревянных конструкций 3. обрамление *n* проёма (*оконного, дверного*) 4. обрезка *f* кромок

T428 *e* **trim paint**
  d Alkydharz-Ölfarbe *f*
  f peinture *f* à base de l'huile et de
    résine alkyd
  nl oliealkydverf *f(m)*
  r масляно-алкидная краска *f*

T429 *e* **triple mould**
  d dreifache Schalungsform *f*
  f moule *f* triple, triple moule *f*
  nl drievoudige mal *m* (*voor
    betonproducten*)
  r тройная форма *f* (*для
    железобетонных изделий*)

T430 *e* **tripod rig**
  d 1. Dreifuß-Bohrgerät *n*
    2. Dreifußramme *f*
  f 1. sondeuse *f* tripode 2. sonnette *f*
    tripode
  nl 1. driepoot-sondeerapparaat *n*
    2. driepoot-heistelling *f*
  r 1. треножный буровой станок *m*
    2. треножный копёр *m*

T431 *e* **troffer**
  d Lichtband *n*
  f chemin *m* lumineux encastré
  nl ingebouwde lichtstrook *f(m)*
  r встроенная световая полоса *f*
    (*в потолке, стене*), троффер *m*

T432 *e* **trolley**
  d Laufkatze *f*
  f trolley *m*, chariot *m*
  nl loopkat *f(m)*
  r тележка *f*

T433 *e* **trolley ladder**
  d angehängte Rolleiter *f*
  f échelle *f* suspendue roulante (à
    chariots)
  nl verrijdbare hangladder *f(m)*
  r подвесная роликовая лестница *f*

T434 *e* **trough lighting fitting**
  d Leuchtwanne *f*, Lichtwanne *f*
  f bac *m* lumineux
  nl lichtbak *m*, trogarmatuur *n*
  r многоламповый софит *m*

T435 **trough mixer** *see* **open-top mixer**

T436 *e* **trowel**
  d Kelle *f*
  f truelle *f*
  nl troffel *m*
  r лопатка *f*, кельма *f*, мастерок *m*

T437 *e* **trowelability**
  d Glättbarkeit *f* (*Beton*)
  f aptitude *f* (*du béton frais mis en
    place*) à lissage à la truelle
  nl gladstrijkbaarheid *f* (*beton*)
  r способность *f* (*бетона*) к затирке

T438 *e* **trowel finish**
  d Kellenglattstrich *m*
  f lissage *m* de finition à la truelle
  nl afwerken *n* van het betonoppervlak
  r чистовая затирка *f* (*бетона
    ручным инструментом*)

T439 *e* **troweling**
  d Glätten *n* (*Betonoberfläche*)
  f lissage *m* du béton
  nl afstrijken *n* (met troffel *of*
    mechanische plakspaan)
  r затирка *f* (*бетона*)

T440 *e* **troweling machine**
  d Glättmaschine *f*
  f lisseur *m* [lisseuse *f*] mécanique
  nl mechanische plakspaan *f(m)*
  r затирочная машина *f*

T441 *e* **truck**
  d 1. Lastkraftwagen *m*, LKW *m*,
    Lastfahrzeug *n* 2. Fahrgestell *n*
  f 1. camion *m*, camion *m*
    automobile 2. châssis *m* (roulant),
    bogie *f*, train *m*
  nl 1. vrachtauto *m* 2. wielstel *n*
  r 1. грузовой автомобиль *m*
    2. тележка *f*; колёсная ходовая
    часть *f*, шасси *n*

T442 *e* **truck crane**
  d Lastkraftwagenkran *m*,
    Aufbaukran *m*, Autokran *m*,
    Mobilkran *m*
  f grue *f* sur pneus [sur camion]
  nl wielkraan *m*, mobiele kraan *m*
  r пневмоколёсный кран *m*

T443 *e* **truck loader**
  d luftbereifter Schaufellader *m*
  f chargeur *m* (sur pneus)
  nl wiellaadschop *m*
  r колёсный (одноковшовый) погрузчик
    m

T444 **truck-mixed concrete** *see* **transit-mix
  concrete**

T445 *e* **truck mixer**
  d Beton-Transportmischer *m*,
    Liefermischer *m*, Mischerfahrzeug *n*
  f camion *m* (avec) malaxeur
  nl truckmixer *m*
  r автобетоносмеситель *m*

T446 **truck mounted crane** *see* **truck
  crane**

T447 **truck mounted distributor** *US see*
  bulk asphaltic-bitumen distributor

T448 *e* **truck-mounted placing boom**
  d zusammenlegbarer Aufbau-
    -Betonierarm *m*
  f flèche *f* distributrice de béton à
    sections articulées sur camion
  nl op vrachtauto gemonteerde giek *m*
    voor het storten van gietbeton
  r шарнирно-сочленённая стрела *f*
    для подачи бетона,
    установленная на грузовом
    автомобиле

T449 *e* **trumpet junction**
  d Trompetenkreuzung *f*
  f croisement *m* en trompette
  nl trompetkruising *f*

r пересечение *n* автомагистралей
в разных уровнях с петлей
поворота на 270°

T450 *e* **trunk duct**
d Hauptkanal *m*, Hauptluftstrang *m*
f conduit *m* [canal *m*] d'air
principal
nl hoofdluchtleiding *f*
r магистральный воздуховод *m*

T451 *e* **trunk main**
d Haupt(verbindungs)leitung *f*
f conduite *f* maîtresse [principale],
canalisation *f* principale
nl hoofdwaterleiding *f*
r магистральный водовод *m*

T452 *e* **trunk road**
d Überlandstraße *f*
f grande route *f*, autoroute *f*
nl hoofdweg *m*
r магистральная автомобильная
дорога *f*

T453 *e* **trunk sewer**
d Hauptabwasserleitung *f*,
Hauptsammler *m*
f égout *m* collecteur, grand
collecteur *m*
nl hoofdriool *n*
r магистральный канализационный
коллектор *m*

T454 *e* **truss**
d Binder *m*, Fachwerk *n*
f ferme *f*, poutre *f* à treillis
nl ligger *m*, vakwerkligger *m*,
spant *n*
r ферма *f*

T455 *e* **truss bars**
d aufgebogene [abgebogene]
Bewehrungsstäbe *m pl*
f barres *f pl* relevées
nl opgebogen wapeningsstaven *f(m) pl*
r отогнутые арматурные стержни
m pl

T456 *e* **trussed beam**
d 1. Balkenträger *m*,
Parallelgurtbinder *m*,
Fachwerkträger *m* mit
parallelen Gurten 2. umgekehrter
Hängebock *m*, unterspannter
Balken *m*, Sprengwerkträger *m*
f 1. poutre à treillis, poutre-ferme *f*
à membrures parallèles 2. poutre *f*
sous-bandée
nl 1. parallelligger *m*, vakwerkligger *m*
2. geschoorde balk *m*, schoorwerk *n*
r балочная ферма *f*, ферма *f*
с параллельными поясами
2. шпренгельная балка *f*

T457 *e* **trussed joist**
d Pfette *f* mit Rundsttahlausfachung,
R-Pfette *f*
f panne *f* métallique à treillis en
barre ronde (*pliée en zigzag*)
nl vakwerkligger *m* van rond staal
r прутковый прогон *m*

T458 *e* **trussed partition**
d Fachwerkwand *f* [Riegelwand *f*,
tragende Gerippentrennwand *f*]
aus Holz
f cloison *f* en ossature [en charpente]
portante
nl scheidingswand *m* van stijl- en
regelwerk
r несущая каркасная перегородка *f*

T459 *e* **trussed system**
d Gitterwerk *n*, Fachwerk *n*
f système *m* réticulé [à treillis]
nl vakwerk *n*
r решётчатая система *f*
[конструкция *f*]

T460 *e* **truss web**
d Trägerfachwerk *n*,
Trägerausfachung *f*
f treillis *m* d'une ferme
nl lijf *n* van een ligger
r решётка *f* фермы

T461 *e* **T-shaped pier**
d T-förmiger Brückenpfeiler *m*
f pile *f* de pont en T
nl T-vormige brugpijler *m*
r T-образная мостовая опора *f*

T462 *e* **T-shore**
d Stütze *f* mit T-Aufsatz
f montant *m* avec tête en T
nl schoor *m* met T-vormige kop
r стойка *f* с Т-образным оголовком

T463 *e* **tub**
d 1. Badewanne *f* 2. Faß *n*; Kübel *m*
f 1. baignoire *f* 2. cuve *f*; cuvette *f*
nl 1. badkuip *f(m)* 2. vat *n*; kubel *m*
r 1. ванна *f* 2. бак *m*; бадья *f*

T464 *e* **tubbing**
d Tübbingausbau *m*
f tubage *m*, cuvelage *m* à segments
nl mijnschacht- *of* galerijbekleding *f*
met tubingen
r тюбинговая крепь *f*

T465 *e* **tubbing ring**
d Tunnelring *m*
f anneau *m* de cuvelage
nl tubingring *m*
r кольцо *n* тюбингов

T466 *e* **tube**
d 1. Rohr *n*, Röhre *f* 2.
Tunnelröhre *f*, U-Bahn-Rohrtunnel *m*
f 1. tube *m*, tuyau *m* 2. tunnel *m* de
métropolitain
nl 1. pijp *f(m)*, buis *f(m)* 2.
tunnelbuis *f(m)* (*van een ondergrondse
spoorweg*)
r 1. труба *f*; канал *m* замкнутого
профиля 2. тюбинговый туннель
m; туннель *m* метрополитена

T467 *e* **tube-and-coupler scaffold**
d Stahlrohrgerüst *n* mit
einheitlichen Klemmkupplungen
f échafaudages *m pl* tubulaires avec
raccords serrés par boulons

nl stalen steigerwerk *n* met
klemkoppelingen
r трубчатые леса *pl* с крепёжными
хомутами

T468 *e* **tube-and-coupler shoring**
d Stahlrohr-Schalungsgerüst *n* mit
einheitlichen Klemmkupplungen
f étayage *m* de coffrage en tubes
nl bekistingsdraagconstructie *f* van
stalen steigerwerk
r трубчатые опорные конструкции
f pl опалубки

T469 *e* **tubeaxial fan**
d Röhrenaxiallüfter *m*
f ventilateur *m* tubulaire axial
nl buisventilateur *m*
r осевой вентилятор *m* в канальном
исполнении

T470 *e* **tube drainage**
d Rohrdränage *f*, Röhrendränung *f*,
Dränrohrentwässerung *f*,
Dränrohrwerk *n*
f drainage *m* tubulaire [en tuyaux]
nl buisdrainage *f*
r трубчатый дренаж *m*

T471 *e* **tube electrostatic precipitator**
d Röhrenelektrofilter *n*, *m*
f filtre *m* électrostatique en tube
nl buizenelektrofilter *n*, *m*
r трубчатый электрофильтр *m*

T472 *e* **tube expander**
d Rohraufwalzgerät *n*
f extendeur *m* (des tubes), mandrin *m*
à évaser les tubes
nl opruimer *m*
r вальцовка *f*, труборасширитель *m*

T473 *e* **tube pile**
d Rohrpfahl *m*
f pieu *m* tubulaire
nl buispaal *m*
r трубчатая свая *f*

T474 *e* **tube plate**
d Rohrboden *m*
f plaque *f* de tubes
nl pijpenplaat *f(m)*
r трубная доска *f*

T475 *e* **tube segment**
d Tunnelsegment *n*
f segment *m* de cuvelage (*d'un
tunnel*)
nl tubingringsegment *n*,
tunnelsegment *n*
r сегмент *m* тюбингового кольца

T476 *e* **tube trench**
d Tunnelrinne *f*
f tranchée *f* [fouille *f*] pour tunnel
nl leidingsleuf *f*
r выемка *f* для туннеля (*при
открытой проходке*)

T477 **tube well** *see* **tubular well**

T478 **tub mixer** *see* **open-top mixer**

T479 *e* **tubular design**
d 1. Hochhauskonstruktion *f* mit

Aussteifungskern (*im Unterbau
eingespannt*) 2. Rohrkonstruktion *f*
f 1. maison-tour *f* [bâtiment-tour *m*]
à noyau central porteur (*encastré
dans la fondation*) 2. construction *f*
tubulaire [en tubes]
nl 1. torenbouw *m* met stijve kern (*in
fundering ingeklemd*)
2. buisconstructie *f*
r 1. конструкция *f* высотных зданий
с несущим стволом [ядром]
(*защемлённом в фундаменте*)
2. трубчатая конструкция *f*

T480 *e* **tubular leg**
d Rohrfuß *m*
f pied *m* tubulaire
nl buispoot *m*
r трубчатая опора *f*

T481 *e* **tubular pile**
d Rohrpfahl *m*; Mantelpfahl *m*
f pieu *m* creux; pieu-voile *m*
nl mantelbuispaal *m*
r трубчатая свая *f*; свая-оболочка *f*

T482 *e* **tubular scaffolding**
d Rohrgerüst *n*, Rohrrüstung *f*
f échafaudage *m* tubulaire
nl stalen steigerwerk *n*
r трубчатые леса *pl*

T483 **tubular structure** *see* **tubular
design 2.**

T484 *e* **tubular welded frame scaffold**
d Baugerüst *n* mit geschweißtem
Rohrrahmenstützwerk
f échafaudage *m* sur portiques
tubulaires [soudés en tubes]
nl stalen raamsteiger *m*,
torkretsteiger *m*
r леса *pl* на сварных рамных
трубчатых опорах

T485 *e* **tubular well**
d Rohrbrunnen *m*
f puits *m* tubulaire
nl buisbron *f(m)*, boorbron *m*
r трубчатый колодец *m*

T486 *e* **tuck pointing**
d 1. Fugenschneiden *n* und
-ausfüllen *n* 2. vorspringende
Fugenbildung *f* (*Mauerwerk*)
f 1. jointoyage *m* au mortier *ou* au
mastic 2. jointoiement *m*
[jointoyage *m*] saillant
nl 1. snijden *n* en opvullen *n* van
voegen (*met kit of mastiek*)
2. voegvulling *f*
r 1. разрезка *f* и заполнение *n*
швов раствором *или* мастикой
2. выпуклая расшивка *f* швов
(*кладки*)

T487 *e* **tufted carpet**
d Florteppich *m*
f tapis *m* à poil
nl tapijt *n* met pool (*vloerbedekking*)
r ворсовый ковёр *m*

T488 *e* **tunnel**
  d Tunnel *m*; Stollen *m*
  f tunnel *m*; galerie *f*; souterrain *m*
  nl tunnel *m*; galerij *f*
  r туннель *m*; галерея *f*; штольня *f*,
    горизонтальная выработка *f*

T489 *e* **tunnel boring machine**
  d Tunnelbohrmaschine *f*
  f machine *f* de creusement (de
    galeries), machine *f* à creuser
  nl tunnelboormachine *f*
  r буровая туннелепроходческая
    машина *f*

T490 *e* **tunnel concreting train**
  d Tunnelbetonieranlage *f*,
    Betonierzug *m*
  f train *m* de bétonnage
  nl tunnelbetontrein *m*
  r бетонопоезд *m*, передвижной
    комплексный агрегат *m* для
    бетонирования обделки туннеля

T491 *e* **tunnel digging and lining machine**
  d Tunnelvortriebs- und
    -auskleidungsmaschine *f*
  f machine *f* à creuser et revêtir les
    tunnels
  nl tunnelgraaf- en bekledingsmachine *f*
  r туннелепроходческий комбайн *m*

T492   **tunnel form** *see* **tunneling formwork**

T493 *e* **tunneling**
  d 1. Tunnelbau *m* 2. Zerklüftung *f*
    (*der Erddämme*)
  f 1. creusement *m* [percement *m*]
    des tunnels 2. développement *m* des
    fissures dans les barrages en terre
  nl 1. tunnelbouw *m* 2. scheurvorming *f*
    in droge dijken
  r 1. проходка *f* туннелей
    [горизонтальных выработок]
    2. развитие *n* трещин (*в плотинах*)

T494 *e* **tunneling formwork**
  d Tunnelschalung *f*, räumlich
    umsetzbare Schalung *f* aus
    Portalsektionen
  f coffrage-tunnel *m*
  nl tunnelbekisting *f*
  r объёмно-переставная опалубка *f*
    из П-образных секций

T495   **tunneling machine** *see* **tunnel
    boring machine**

T497   **tunnel liner** *see* **tunnel lining
    machine**

T498 *e* **tunnel lining**
  d Tunnelauskleidung *f*
  f revêtement *m* du tunnel
  nl tunnelbekleding *f*
  r обделка *f* [облицовка *f*] туннеля

T499 *e* **tunnel lining machine**
  d Tunnelauskleidungsmaschine *f*
  f machine *f* pour revêtement de
    galeries
  nl tunnelbekledingsmachine *f*
  r агрегат *m* для облицовки туннелей

T500 *e* **tunnel portal**
  d Tunnelportal *n*
  f tête *f* [entrée *f*] de tunnel
  nl tunnelingang *m*
  r портал *m* туннеля

T501 *e* **tunnel shield**
  d Tunnelschild *m*, Vortriebsschild *m*
  f bouclier *m* d'avancement
  nl tunnelschild *n*
  r проходческий щит *m*

T502 *e* **tunnel-type tail race**
  d Unter-Triebwasserstollen *m*
  f galerie *f* de dérivation, canal *m*
    d'évacuation [de fuite]
  nl uitstroomtunnel *m*, laaggelegen
    toevoertunnel *m*
  r отводящий туннель *m* ГЭС

T503   **tunnel work** *see* **tunneling**

T504 *e* **turbidimeter**
  d 1. Trübungsmesser *m*
    2. Feinheitsprüfgerät *n*
  f 1. turbidimètre *m* 2. appareil *m* à
    mesurer la finesse de mouture de
    ciment
  nl 1. troebelheidsmeter *m* 2. toestel *n*
    voor controle van de
    cementkorrelgrootte
  r 1. турбидиметр *m*, прибор *m* для
    измерения мутности (воды) 2.
    прибор *m* для контроля тонкости
    помола (цемента)

T505 *e* **turbidimeter fineness**
  d bestimmte Mahlfeinheit *f* nach
    der Trübungsmessungsmethode
  f finesse *f* de mouture (de ciment)
    turbidimétrique
  nl nefelemetrische fijnheid *f* van
    gemalen cementklinker
  r турбидиметрическая тонкость
    f помола цемента

T506 *e* **turbine flow meter**
  d Turbinen-Durchflußmesser *m*
  f débimètre *m* à turbine
  nl turbinedoorstromingsmeter *m*
  r турбинный расходомер *m*

T507 *e* **turbine mixer**
  d Turbomischer *m*
  f turbomélangeur *m*, turbomalaxeur *m*
  nl turbomenger *m*
  r турбосмеситель *m*, роторный
    смеситель *m*

T508 *e* **turbine pit**
  d Turbinenkammer *f*,
    Turbinengrube *f*
  f puits *m* de turbine, chambre *f* de
    turbine
  nl turbineschacht *f(m)*
  r турбинная камера *f*

T509 *e* **turbine pump**
  d 1. Turbinen-Pumpe *f*, Turbopumpe *f*
    2. Pumpturbine *f*
  f 1. turbopompe *f* 2. turbine-pompe *f*

nl 1. turbopomp *f(m)*
2. turbinepomp *f(m)*
r 1. турбонасос *m* 2. насос-турбина *f*, обратимая гидротурбина *f*

T510 *e* **turfing**
d 1. Berasen *n*, Rasenverkleidung *f* 2. Rasendecke *f*
f 1. gazonnement *m* 2. tapis *m* vert [de gazon]
nl 1. aanleggen *n* van een grasveld, beleggen *n* met zoden 2. grasveld *n*, gazon *n*
r 1. одернение *n*, обкладывание *n* дерном 2. дерновый покров *m*

T511 **turn bridge** *see* **swing bridge**

T512 *e* **turnbuckle**
d Spannmutter *f*, Spannschloß *n*
f écrou *m* de tendeur, tendeur *m* à vis, manchon *m* de serrage
nl spanschroef *f(m)*, draadspanner *m*
r стяжная муфта *f*, талреп *m*

T513 *e* **turning basin**
d Wendebecken *n*
f bassin *m* de virement
nl zwaaikom *f(m)*, kanaal *n*
r разворотный бассейн *m*

T514 *e* **turn-key type building**
d schlüsselfertiges Gebäude *n*
f bâtiment *m* clé sur porte, maison *f* [bâtiment *m*] du type clé [clef] en mains
nl gebouw *m*, gereed voor gebruik
r здание *n*, построенное «под ключ»

T515 *e* **turn-key type of contract**
d Entwurf- und Bauauftrag *m*
f contrat *m* [marché *m*] du type clé sur porte
nl ontwerp- en bouwopdracht *f(m)*
r контракт *m* на проектирование и строительство объекта «под ключ»

T515a *e* **turn-of-nut method**
d Spannen *n* der Hochspannungsbolzen durch regelbares Mutterdrehen
f méthode *f* de turn-of-nut, méthode *f* de tension des boulons à haute résistance par rotation de l'écrou à un angle prévu
nl aandraaien *n* van de moer van voorspanbouten over een bepaalde hoek
r метод *m* натяжения высокопрочных болтов поворотом гайки на заданный угол

T516 **turn-out wall** *US see* **wing wall**

T517 *e* **turnover**
d Fluktuation *f* der Arbeitskräfte
f rotation *f* du personnel
nl personeelsaflossing *f*
r текучесть *f* рабочей силы

T518 *e* **twenty eight day cube strength**
d 28-Tage-Würfelfestigkeit *f*

f résistance *f* à la compression (*de béton*) après 28 jours
nl kubussterkte *f* van beton na 28 dagen
r кубиковая прочность *f* бетона в возрасте 28 суток

T519 *e* **turn-batch mixing drum**
d Zweikammer-Mischtrommel *f*
f tonneau *m* malaxeur [mélangeur] à deux compartiments
nl tweekamer-mengtrommel *f(m)*
r двухкамерный смесительный барабан *m*

T520 *e* **twin-box girder**
d Doppelkastenträger *m*
f poutre *f* caisson jumelée
nl dubbele kokerligger *m*
r двухпустотный коробчатый настил *m*

T521 *e* **twin elbow**
d Doppelkrümmer *m*
f coude *m* double
nl dubbelkniestuk *n*, sprongstuk *n*
r двойное колено *n*

T522 *e* **twin gates**
d Doppelschütz(e) *n(f)*, Doppelverschluß *m*
f vanne *f* double [à deux corps]
nl dubbele afsluiter *m*
r сдвоенный затвор *m*

T523 **twin lock** *see* **double-lane lock**

T524 *e* **twin tunnel**
d Zwillingstunnel *m*, Doppelrohrtunnel *m*
f tunnel *m* jumelé
nl tweepijpstunnel *m*
r двуствольный [двухпролётный] туннель *m*

T525 *e* **twin-shaft paddle mixer, twin shaft pan mixer**
d Doppelwellen-Schaufelmischer *m*
f malaxeur *m* [malaxeuse *f*] à béton à deux axes avec palettes
nl schoepenbetonmolen *m*
r двухвальный бетоносмеситель *m*

T526 *e* **twin-twisted bars, twin-twisted reinforcement**
d Istegstahl *m*
f armature *f* [ferraillage *m*, aciers *m pl*] Isteg
nl Isteg wapeningsstaal *n*
r витая прядь *f* из двух арматурных стержней

T527 *e* **twin-webbed T-beam**
d zweistegiger Plattenbalken *m*
f dalle *f* en T double
nl T-balk *m* met twee lijven
r балка-плита *f* с двумя стенками, железобетонная плита *f* [панель *f*] в форме двойного Т

T528 *e* **twist**
d 1. Verdrehung *f*, Verwindung *f*, Torsion *f* 2. Verwerfung *f*,

Querauskippung *f* 3. aerodynamische
Verwindung *f*, Drallbewegung *f*
4. Drehimpuls *m*, Drall *m*
f 1. torsade *f*, torsion *f* 2. dévers *m*
(*d'un madrier*); gondolage *m* (*d'une
tôle*) 3. tourbillon *m*, mouvement *m*
tourbillonnaire 4. moment *m*
cinétique
nl 1. verdraaiing *f*, wringing *f*,
torsie *f* 2. scheluw [krom] trekken,
verbuigen *n* 3. turbulentie *f*,
werveling *f*
r 1. скручивание *n* 2. перекос *m*
поперечного сечения (*бруса*,
балки) 3. закручивание *n*
(*потока*), вихревое движение *n*
4. момент *m* количества движения

T529 *e* **twisted-steel fabric**
d Bewehrungsmatte *f* aus Drillstahl
f treillis *m* soudé en barres
torsadées
nl gelast wapeningsnet *n* van
getwiste draden [staven]
r сварная арматурная сетка *f* из
витых стержней

T530 *e* **twisting**
d Verwinden *n*, Verdrehen *n*
f torsade *f*, torsadage *n*
nl torderen *n*, ineendraaien *n*
r скручивание *n*

T531 *e* **two-axle scraper**
d Vierradschrapper *m*
f scraper *m* [décapeuse *f*] à deux
essieux
nl tweeassige scraper *m*
r двухосный скрепер *m*

T532 **two-chamber lock** *see* **double lock**

T533 *e* **two-chamber surge tank**
d Zweikammer-Wasserschloß *n*
f cheminée *f* d'équilibre à deux sas
nl vereffeningsreservoir *n*
[waterslagtank *m*] met twee kamers
r двухкамерный уравнительный
резервуар *m*

T534 *e* **two-coat paint**
d zweischichtiges Anstrichsystem *n*
f peinture *f* à deux couches
nl verfwerk *n* in twee lagen
r двухслойная окраска *f*

T535 **two-coat plastering, two-coat
plaster work** *see* **two-coat work** 1.

T536 *e* **two-coat work**
d 1. zweilagiges [zweischichtiges]
Verputzen *n* 2. zweischichtiges
Anstrichsystem *n*, zweischichtige
Asphaltbetondecke *f*
f 1. enduit *m* (en plâtre) à deux
couches 2. peinture *f* à deux
couches 3. revêtement *m* routier au
bitume à deux couches
nl 1. pleisterwerk *n* in twee lagen
2. verfwerk *n* in twee lagen
3. asfaltwegdek *n* in twee lagen

r 1. двухслойная штукатурка *f*
2. двухслойная окраска *f*
3. двухслойное асфальтобетонное
покрытие *n*

T537 *e* **two-family house**
d Zweifamilienhaus *n*
f maison *f* à deux familles
nl huis *n* voor twee gezinnen
r двухквартирный дом *m*

T538 *e* **two-hinged arch**
d Zweigelenkbogen *m*
f arc *m* à deux articulations
nl tweescharnierboog *m*
r двухшарнирная арка *f*

T539 *e* **two-leg sling**
d zweisträngiger Stropp *m*
f élingue *f* à deux brins
nl tweestrengsstrop *m*, *f*
r двухветвевой строп *m*

T540 *e* **two-part line tackle**
d zweifach geschertes Seil *n*
f palan *m* à deux brins [courants]
nl derdehandstakel *m*, tweeloper *m*
r двухветвевой полиспаст *m*

T541 *e* **two-pipe system**
d Zweirohrsystem *n*
f installation *f* [système *m*] à deux
tuyaux
nl tweepijpssysteem *n*
r двухтрубная система *f*

T542 *e* **two-point suspension scaffold**
d Hängegerüst *n* mit
Zweipunktaufhängung
f échafaudage *m* volant suspendu à
deux points
nl hangsteiger *m* met tweepunts-
ophanging
r подвесные подмости *pl* с
подвесками в двух точках

T543 *e* **two point workability test**
d Einbringbarkeitsprüfung *f* nach
zwei Parametern
f essai *m* de workabilité basé sur
deux paramètres
nl verwerkbaarheidscontrole *f* op
twee punten (*betonmortel*)
r определение *n* удобоукладываемости
(*напр. бетонной смеси*) по двум
параметрам

T544 *e* **two-span bridge**
d Zweifeldbrücke *f*
f pont *m* à deux travées
nl brug *f(m)* met twee overspanningen
r двухпролётный мост *m*

T545 *e* **two-stage curing**
d zweistufige Warmbehandlung *f*
(*Beton*)
f traitement *m* de béton à deux
phases
nl warmtebehandeling *f* in twee fasen
(*beton*)
r двухэтапная тепловлажностная
обработка *f* (*бетона*)

T546 e **two-way reinforcement**
 d kreuzweise Bewehrung *f*
 f armature *f* [ferraillage *m*] à deux
 directions, armature *f* croisée
 nl kruiswapening *f*; plaatwapening *f*
 in twee richtingen
 r армирование *n* в двух
 направлениях; перекрёстная
 арматура *f*

T547 **two-way ribbed slab** *see* **beam** [and
 girder slab

T548 e **two-way slab**
 d zweiseitig gespannte Platte *f*
 f dalle *f* à armatures croisées
 nl plaat *f(m)* gewapend in twee
 richtingen
 r плита *f*, армированная в двух
 направлениях

T549 **two-way system** *see* **two-way**
 reinforcement

T550 **two-way Y** *see* **wye fitting**

T551 e **tying wire**
 d Rödeldraht *m*, Bindedraht *m*
 f fil *m* d'acier pour ligature
 nl binddraad *m*
 r вязальная проволока *f*, проволока
 f для вязки арматуры

T552 e **typical apartment building**
 d Typenwohngebäude *n*
 f bâtiment *m* résidentiel standardisé,
 bâtiment *m* d'habitation type
 [typifié]
 nl standaard-appartementengebouw *n*,
 standaard-flatgebouw *n*
 r типовое жилое здание *n*

T553 e **typical construction**
 d Typenkonstruktion *f*
 f construction *f* standardisée
 nl standaardconstructie *f*
 r типовая конструкция *f*

U

U1 e **U-bend**
 d Doppelkrümmer *m*, U-Rohr *n*
 f coude *m* (d'un tuyau) en U,
 coude *m* double
 nl zwanehals *m*, stankafsluiter *m*
 r двойной отвод *m*

U2 e **U-bolt**
 d Schäkel *m*; Hakenbügel *m*
 f agrafe *f* filetée, bride *f* à écrous,
 étrier *m* à vis
 nl staaldraadklem *f*; haakbeugel *m*
 r болт-скоба *m*, U-образный болт *m*,
 серьга *f*; анкерный хомут *m*
 (для зацепления шандор)

U4 e **ultimate bearing capacity**
 d Grenztragfähigkeit *f*,
 Grenztragvermögen *n*,

Grenztragwiderstand *m* (*Baugrund*)
 f résistance *f* limite (*d'un terrain*)
 nl grensdraagvermogen *n*
 (*bouwgrond*)
 r предельная несущая способность *f*
 (*грунта*)

U5 e **ultimate bending moment**
 d Bruchbiegemoment *n*
 f moment *m* fléchissant ultime [de
 rupture]
 nl breukbuigmoment *n*
 r разрушающий [предельный]
 изгибающий момент *m*

U6 e **ultimate compressive strength**
 d Bruchdruckfestigkeit *f*
 f résistance *f* à la compression
 nl breukdrukvastheid *f*
 r предел *m* прочности при сжатии,
 временное сопротивление *n*
 сжатию

U7 e **ultimate design resisting moment**
 d Bruchmoment *n* des inneren
 Kräftepaares
 f couple *m* intérieur ultime [de
 rupture]
 nl voor de berekening aangehouden
 grootste weerstandsmoment *n*
 r предельный момент *m* внутренней
 пары

U8 e **ultimate factor of safety**
 d Grenzsicherheitsfaktor *m*
 f coefficient *m* de sécurité par
 rapport à la charge de rupture
 nl grensveiligheidscoëfficiënt *m*
 r коэффициент *m* запаса прочности
 (*по отношению к пределу
 прочности*)

U9 e **ultimate limit state**
 d Bruchgrenzzustand *m*, Bruchgrenze *f*
 f état *m* limite ultime (*relatif à la
 rupture ou au flambement de la
 construction*)
 nl breukgrens *f(m)* (*toestand*)
 r предельное состояние *n* по
 прочности *или* устойчивости

U10 e **ultimate load**
 d Bruchlast *f*, Grenzlast *f*,
 Bruchwiderstand *m*
 f charge *f* ultime [critique, de
 rupture]
 nl breukbelasting *f*
 r разрушающая [предельная,
 критическая] нагрузка *f*

U11 e **ultimate load analysis, ultimate
 load design**
 d Traglastverfahren *n*; n-freies
 Berechnungsverfahren *n*
 (*Stahlbeton*)
 f calcul *m* à la rupture
 nl berekeningsmethode *f* uitgaande
 van de breuktoestand
 r расчёт *m* по разрушающим
 нагрузкам

U12 *e* ultimate loading
  d Bruchbelastung *f*
  f sollicitation *f* de rupture
  nl breukbelasting *f*
  r предельное нагружение *n*

U13 ultimate resistance *see* ultimate strength

U14 *e* ultimate shear stress
  d Bruchscherspannung *f*,
    Bruchschubspannung *f*
  f résistance *f* au cisaillement
  nl breukschuifspanning *f*
  r предел *m* прочности [временное сопротивление *n*] при сдвиге [срезе]

U15 *e* ultimate shrinkage
  d Endschwindung *f*
  f retrait *m* total [global]
  nl eindkrimp *m*
  r предельная [конечная] усадка *f*

U16 *e* ultimate strength
  d Bruchfestigkeit *f*
  f résistance *f* à la rupture
  nl breuksterkte *f*, breukvastheid *f*
  r предел *m* прочности, временное сопротивление *n*

U18 *e* ultimate stress
  d Bruchspannung *f*
  f contrainte *f* critique; contrainte *f* de rupture
  nl breukvastheid *f*, breukspanning *f*
  r критическое напряжение *n*; временное сопротивление *n*

U19 *e* ultimate tensile strength
  d Bruchzugfestigkeit *f*
  f résistance *f* à la traction
  nl trekvastheid *f*
  r предел *m* прочности [временное сопротивление *n*] при растяжении

U20 *e* ultra-high-strength concrete
  d ultrahochfester Beton *m*
  f béton *m* à très haute résistance
  nl beton *n* met zeer grote drukvastheid
  r сверхвысокопрочный бетон *m* (*с пределом прочности* 98,1 МПа *и выше*)

U22 *e* ultrasonic pulse velocity method
  d Ultraschallimpuls-Prüfverfahren *n*
  f méthode *f* de contrôle (*du béton*) par ondes ultra-soniques impulsives
  nl onderzoek *n* met ultrasone pulsen
  r ультразвуковой (импульсный) метод *m* (*определения прочностных характеристик бетона*)

U25 *e* umbrella roof
  d Regenschirmschalendach *n*
  f comble *m* [toit *m*] en parapluie; comble *m* en pavillon
  nl parapludak *n*
  r зонтичное покрытие *n*; зонтичная крыша *f*

U26 *e* unbalanced load
  d unsymmetrische Belastung *f*
  f charge *f* non equilibrée [non pondérée]
  nl onsymmetrische belasting *f*
  r неуравновешенная нагрузка *f*

U27 *e* unbalanced moment
  d unausgeglichenes Moment *n*
  f moment *m* non équilibré
  nl niet-vereffend moment *n*
  r неуравновешенный момент *m*

U28 *e* unbonded member
  d Spannbetonteil *m* mit gleitunbehinderten Spanngliedern
  f pièce *f* en béton postcontraint sans adhérence de l'armature
  nl betonelement *n* met voorspanstaven zonder aanhechting
  r преднапряжённый железобетонный элемент *m* с натяжением арматуры на бетон без сцепления

U29 *e* unbonded posttensioning
  d Vorspannung *f* ohne Verbund
  f procédé *m* de post-tension *f* sans adhérence de l'armature
  nl naspanning *f* zonder aanhechting
  r натяжение *n* арматуры на бетон без последующего сцепления её

U30 *e* unbonded tendon
  d gleitunbehindertes Spannglied *n*
  f câble *m* *pl* de précontrainte non adhérant
  nl voorspanstaaf *f(m)* zonder aanhechting
  r напрягаемая арматура *f*, не сцеплённая с бетоном

U31 *e* unbraced length
  d freie Länge *f*, lichte Weite *f*
  f longueur *f* libre
  nl overspanning *f*, vrije lengte *f*
  r свободная [нераскреплённая] длина *f*, длина *f* элемента между точками закрепления

U32 *e* uncased pile
  d Ortpfahl *m* ohne Mantelrohr
  f pieu *m* coulé sur place sans fourreau
  nl in de grond gevormde paal *m* zonder mantelbuis
  r набивная свая *f* без обсадной трубы

U33 *e* unconfined compression strength
  d Druckfestigkeit *f* bei unbehinderter Seitenausdehnung
  f résistance *f* à la compression sans contrainte latérale
  nl drukvastheid *f* (*van grond*) bij onbelemmerde zijvervorming
  r сопротивление *n* сжатию (*грунта*) в условиях одноосного напряжённого состояния

U34 *e* unconfined groundwater
  d freies Grundwasser *n*
  f nappe *f* (souterraine à surface) libre

42*

nl vrij [niet-gespannen] grondwater *n*
r безнапорные подземные воды *f pl*

U35 *e* **uncontrolled crest spillway**
d unverschlossener [offener] Überlauf *m*
f évacuateur *m* sans organes de réglage, évacuateur *m* non commandé
nl niet-regelbare overloop *m*
r нерегулируемый водослив *m*, водосброс *m* без затворов на гребне

U36 *e* **uncontrolled pedestrian crossing**
d ungeschützter Fußgängerüberweg *m*
f passage *m* pour piétons sans régulation du traffic
nl niet-beveiligde voetgangersoversteekplaats *f(m)*
r нерегулируемый пешеходный переход *m*

U37 *e* **uncontrolled ventilation**
d ungeordnete Lüftung *f*, Schwerkraftlüftung *f*
f ventilation *f* non régulée
nl niet-regelbare ventilatie *f*
r неорганизованная вентиляция *f*

U38 *e* **uncoursed rubble masonry**
d ungeschichtetes Bruchsteinmauerwerk *n*
f maçonnerie *f* de moellons bruts non appareillée [en rangées irrégulières]
nl breuksteenmuurwerk *n* zonder gelijke rijen
r нерегулярная бутовая кладка *f*

U39 *e* **underbed**
d Mörtelbett *n*
f sous-couche *f* (*du dallage en ciment*)
nl breuksteenmetselwerk *n* in onregelmatig verband
r нижний слой *m* пола (*цементного, террацевого*)

U40 *e* **underbridge**
d Unterführung *f*
f passage *m* supérieur, pont *m* au--dessus d'une autoroute
nl onderdoorgang *m*
r путепровод *m* на трассе дороги, проходящей по насыпи (*для пропуска под ним поперечной дороги*)

U41 *e* **undercoat**
d Unterputz *m*, Rauhwerk *n*
f couche *f* d'impression [d'apprêt, primaire de fond], enduit *m* d'impression
nl eerste pleisterlaag *f(m)*
r нижний слой *m* штукатурки, первый намёт *m*

U42 *e* **undercreep**
d Viehdurchlaß *m* (*unter einem Verkehrsweg*)
f passage *m* (inférieur) à bétail

nl veetunnel *m* (*onder een verkeersweg*)
r скотопрогонный туннель *m* (*под дорогой*)

U44 *e* **undercrossing**
d Unterquerung *f*, Unterführung *f*
f passage *m* inférieur
nl onderdoorgang *m*, tunnel *m* (*onder een spoorlijn*)
r подземный переход *m*, туннель *m* под дорогой

U45 *e* **undercutting**
d 1. Unterspülung *f*, Auskolkung *f*, Auswaschung *f* 2. Einbrandkerbe *f*
f 1. affouillement *m*, délavage *m* 2. caniveau *m* de soudure
nl 1. onderspoeling *f*, kalving *f* 2. inbrandkerf *f(m)* (*lassen*)
r 1. подмыв *m*, размыв *m* (*берега*) 2. подрез *m* (*сварного шва*)

U46 *e* **underdrainage**
d Untergrundentwässerung *f*, Bodenentwässerung *f*, Tiefenentwässerung *f*
f drainage *m* de fond
nl ondergrondse drainage *f*
r дренаж *m* основания, глубинный дренаж *m*

U47 *e* **underflooding**
d Grundwasseranstieg *m*, Überflutung *f* von Senken infolge Grundwasseranstiegs
f submersion *f* partielle
nl grondwaterstijging *f*
r подтопление *m*

U48 *e* **underfloor duct**
d Unterbodenkanal *m*
f conduite *f* posée sous revêtement de sol, caniveau *m* sous couvre-sol
nl ondergronds kanaal *n*
r подпольный канал *m*

U49 *e* **underfloor heating**
d Fußboden(strahlungs)heizung *f*
f chauffage *m* par le plancher [par panneaux rayonnants posés sous couvre-sol]
nl vloerverwarming *f*
r система *f* напольного панельного отопления

U50 *e* **underfloor heating cable**
d Heizkabel *n* einer Elektrofußbodenheizung
f câble *m* chauffant [de chauffage] posé en sous-couche de plancher
nl verwarmingskabel *m* van elektrische vloerverwarming
r греющий кабель *m* напольного отопления

U51 *e* **underfloor raceway**
d Kabelkanal *m* unter dem Fußboden, Unterboden-Kabelkanal *m*
f caniveau *m* pour câbles électriques sous couvre-sol
nl leidingkoker *m* onder de vloer

r подпольный канал *m* для
электросетей

U52 *e* underflow
d 1. Strom *m* unter dem Flußbett
2. Unterstrom *m*,
Unterwasserströmung *f*
3. Unterläufigkeit *f* 4. Durchlauf *m*
Durchgang *m*
f 1. courant *m* sous le lit du fleuve
2. courant *m* de profondeur
3. courant *m* d'infiltration sous
l'ouvrage 4. passant *m*, fraction *f*
passante
nl 1. stroming *f* onder de rivierbedding
2. onderstroom *m*
3. onderloopsheid *f* 4. kwel *f*
r 1. подрусловой поток *m*
2. подповерхностное течение *n*
3. фильтрационный поток *m*
под сооружением *или* фундаментом
4. подситовой [подрешётный]
продукт *m*

U53 *e* underground cable
d Erdkabel *n*, unterirdisches Kabel *n*
f câble *m* souterrain
nl ondergrondse kabel *m*
r подземный кабель *m*

U54 *e* underground construction work
d unterirdische Bauarbeiten *f pl*
f travaux *m pl* [ouvrages *m pl*]
souterrains
nl bouwwerk *n* onder het maaiveld
r подземное строительство *n*,
подземные работы *f pl*

U55 *e* underground garage
d Tiefgarage *f*, unterirdische Garage *f*,
Kellergarage *f*
f garage *m* sous-sol [souterrain]
nl ondergrondse garage *f*
r подземный гараж *m*

U56 *e* underground gas storage reservoir
d unterirdischer Gasspeicherbehälter *m*
f stockage *m* [réservoir *m* de
stockage] souterrain de gaz
nl ondergrondse gashouder *m*
r подземное газохранилище *n*

U57 *e* underground hydroelectric power
station
d Kavernen-Wasserkraftwerk *n*,
Kavernen-Krafthaus *n*
f usine *f* hydroélectrique souterraine
nl ondergrondse waterkrachtcentrale *f*
r ГЭС *f* подземного типа

U58 *e* underground laying
d Erdverlegung *f*
f pose *f* [mise *f* en place] des lignes
enterrées
nl ondergrondse aanleg *m* (*pijpen*,
kabels)
r подземная бесканальная прокладка
f (*труб*, *кабелей*)

U59 *e* underground pipeline
d unterirdische Rohrleitung *f*

f pipe-ligne *m* enterré, conduite *f*
enterrée [souterraine]
nl ondergrondse pijpleiding *f*
r подземный трубопровод *m*

U60 *e* underground powerhouse,
underground power plant
d Hohlbaukrafthaus *n*,
unterirdisches Krafthaus *n*,
Untergrundturbinenhaus *n*
f centrale *f* hydroélectrique
souterraine, bâtiment *m* de
turbines souterrain
nl ondergrondse waterkrachtcentrale *f*
r подземное здание *n* ГЭС

U61 *e* underground railway
d U-Bahn *f*, Untergrundbahn *f*
f métropolitain *m*, métro *m*;
chemin *m* de fer (en) souterrain
nl metro *m*; ondergrondse (spoorweg) *m*
r метрополитен *m*; подземная
железная дорога *f*

U62 *e* underground run-off
d Dränabfluß *m*, Entwässerungsabfluß
m
f débit *m* [écoulement *m*] de
percolation
nl drainafvoer *m*
r дренажный сток *m*

U63 *e* underground service
d unterirdische Versorgungsleitungen
f pl
f réseaux *m pl* divers souterrains
nl ondergronds leidingnet
r подземные коммуникации *f pl*

U64 *e* underground shelter
d unterirdischer Bunker *m*
f abri *m* souterrain
nl onderaardse schuilplaats *f*(*m*)
r подземное убежище *n* [укрытие *n*]

U65 *e* underground structures
d unterirdische Bauwerke *n pl*,
Untergrund-Bauwerke *n pl*
f ouvrages *m pl* souterrains
nl ondergrondse gebouwen *n pl*
r подземные сооружения *n pl*

U66 *e* underground tank
d Erdbehälter *m*, Erdtank *m*,
unterirdischer Behälter *m*
f réservoir *m* souterrain, citerne *f*
[cuve *f*] enterrée
nl ondergronds reservoir *n*
r подземный резервуар *m*

U67 underground water *see* subsurface
water

U68 *e* underground water head
d Grundwassersäule *f*,
Grundwasserdruck *m*
f charge *f* des eaux souterraines
nl grondwaterdruk *m*
r напор *m* подземных вод

U69 underground works *see* underground
structures

U70 *e* under inlet
d Schlucker *m*
f ouvrage *m* d'admission d'eau de
nappe dans un drain
nl kolk *m*
r поглотитель *m*

U71 undermining *see* undercutting 1.

U72 *e* underpass
d Unterführung *f*
f passage *m* inférieur
nl onderdoorgang *m*
r путепровод *m* (*под дорогой*),
туннель *m* на пересечении дорог

U73 *e* underpinning
d Unterfangen *n*, Unterfangung *f*
f reprise *f* en sous-œuvre, étayage *m*
d'une muraille
nl stutten *n*, stempelen *n*
r усиление *n* фундамента с
устройством временного
опирания стены

U74 *e* underplanting
d Unterbestückung *f* (*der Baustelle
mit den Mechanisierungsmitteln*)
f mécanisation *f* d'un chantier
insuffisante, approvisionnement *m*
d'un chantier insuffisant
nl onvoldoende uitrusting *f* van de
bouwplaats met machines en
gereedschap
r недостаточное оснащение *n*
строительной площадки средствами
механизации

U75 *e* under-river shield
d Schild *m* für Tunnelbau unter
einem Strom
f bouclier *m* pour creusement des
tunnels sous-fluviaux
nl tunnelschild *n* voor tunnelbouw
onder water
r щит *m* для проходки туннеля
под рекой

U76 *e* under-river tunnel
d Stromtunnel *m*
f tunnel *m* sous-fluvial
nl riviertunnel *m*
r туннель *m* под рекой

U77 *e* under-running stacker crane
d Einschienen-Stapelkran *m*
f pont *m* gerbeur suspendu
nl hangende stapelkraan *f(m)*
r стеллажный подвесной кран *m*;
монорельсовый кран-штабелёр *m*
(*подвесного типа*)

U78 *e* undersanded concrete
d Beton *m* mit zu wenig Feinzuschlägen
f béton *m* pauvre en sable
nl betonspecie *f* met te weinig zand
r бетонная смесь *f* с недостаточным
содержанием песка

U79 *e* undersize
d Siebdurchfall *m*, Unterkorn *n*;
Schlammüberlauf *m*

f fraction *f* passante, passant *m*
tamisat *m*, produit *m* traversé;
débordement *m* du classeur,
surverse *f*
nl doorval *m* (*zeef*); sliboverloop *m*
r подрешётный продукт *m*
(*грохота*); слив *m*, верхний
продукт *m* (*гидроклассификатора*)

U80 underwashing *see* undercutting 1.

U81 *e* underwater apron
d Flutbett *n*, Wehrboden *m*; Vorboden *m*,
Dichtungsteppich *m*; Sturzbett *n*;
Kolkschutz *m*, Ablaufboden *m*
f arrière-radier *m*, radier *m*
nl ontvangbed *n*, stortebed *n*
r флютбет *m*, водобой *m*, рисберма *f*

U82 *e* underwater concreting
d Unterwasserbetonieren *n*
f bétonnage *m* [coulage *m* du béton
sous l'eau
nl betonstorten *n* onder water
r подводное бетонирование *n*

U83 *e* underwater digging, underwater
excavation
d Unterwasseraushub *m*,
Unterwasserabbau *m*
f excavation *f* [cavage *m*] sous l'eau
nl ontgraving *f* onder water
r подводная выемка *f* [разработка *f*]
грунта

U84 *e* underwater rockfill
d Unterwasser(stein)schüttung *f*
f enrochement *m* sous l'eau,
remblai *m* rocheux sous l'eau
nl onderwater-steenbestorting *f*
r подводная каменная наброска *f*

U85 *e* underwater structures
d Unterwasserbauwerke *n pl*
f ouvrages *m pl* sous-marins
[subaquatiques]
nl onderwater-bouwwerken *n pl*
r подводные сооружения *n pl*

U86 *e* underwater tube
d Unterwasser-Tunnelröhre *f*
f tunnel *m* subaquatique (*en
tronçons préfabriqués de section
ronde*)
nl onderwater-tunnelbuis *f(m)*
r подводный туннель *m* (из
сборных секций круглого сечения)

U87 undistorted ground *see* undisturbed
soil

U88 *e* undisturbed sample
d ungestörte Probe *f*
f échantillon *m* intact [non remanié]
nl ongeroerd grondmonster *n*
r образец *m* грунта с ненарушенной
структурой

U89 *e* undisturbed soil
d gewachsener [natürlicher] Boden *m*
f sol *m* [terrain *m*] naturel [intact]
nl ongeroerde grond *m*
r грунт *m* ненарушенной структуры,

грунт *m* в условиях
естественного залегания,
естественный грунт *m*

U90 *e* **undrained shear strength**
d undrainierte Scherfestigkeit *f* ohne
Entwässerung
f résistance *f* au cisaillement à
teneur en eau constante
nl schuifweerstand *m* zonder
ontwatering
r сопротивление *n* сдвигу в условиях
отсутствия дренажа

U91 *e* **undressed timber** *US*
d ungehobeltes Schnittholz *n*
f bois *m* débité non raboté
nl ongeschaafd gezaagd hout *n*
r нестроганый пиломатериал *m*

U92 *e* **unframed door**
d rahmenloses Türblatt *n*
f porte *f* en planches sans
encadrement
nl deur *f(m)* zonder kozijn
r дверное полотно *n* без каркаса

U93 *e* **ungauged lime plaster**
d Kalkputz *m* (*ohne Gips*)
f enduit *m* en chaux (*sans plâtre*)
nl kalkpleister *n*
r известковая штукатурка *f* (*без
гипса*)

U94 *e* **unglazed tile**
d unglasierte Fliese *f*
f carreau *m* non vitrifié [non glacé]
nl ongeglazuurde tegel *m*
r неглазурованная плитка *f*

U95 *e* **uniform grading**
d gleichförmige Kornzusammensetzung
f
f granulométrie *f* [granularité *f*]
uniforme
nl gelijkvormige korrelsamenstelling *f*
r однородный гранулометрический
состав *m*

U97 *e* **uniformly distributed load**
d gleichmäßig verteilte Last *f*
f charge *f* uniformement répartie
nl gelijkmatig verdeelde belasting *f*
r равномерно распределённая
нагрузка *f*

U98 *e* **uniform settlement**
d gleichmäßige Setzung *f*
f affaissement *m* [tassement *m*]
uniforme
nl gelijkmatige zetting *f*
r равномерная осадка *f*

U99 *e* **uniform stress**
d ebener Spannungszustand *m*
f contrainte *f* [tension *f*] uniforme
nl gelijkmatige spanningsverdeling *f*
r плоское напряжённое состояние *n*

U100 *e* **union**
d Stutzen *m*; Verschraubung *f*;
Überwurfmutter *f*

f raccord *m*; raccordement *m*;
manchon *m* de raccord
nl aansluitstuk *n*; koppelstuk *n*
r патрубок *m*, штуцер *m*; резьбовое
соединение *n* (*труб*); сгон *m*

U101 *e* **union flange**
d Bundflansch *m*, Anschlußflansch *m*,
Verbindungsflansch *m*
f bride *f* de raccordement
nl koppelflens *f(m)*
r присоединительный фланец *m*

U102 *e* **unit**
d 1. Einheit *f* 2. Satz *m* 3.
Bauelement *n*
f 1. unité *f* 2. ensemble *m*, groupe *m*
3. élément *m* (de construction)
nl 1. eenheid *f* 2. samenstel *n* 3.
bouwelement *n*
r 1. единица *f* 2. комплект *m* 3.
элемент *m* (конструкции)

U103 *e* **unit air conditioner**
d Klimakonvektor *m*, Düsen-
-Konvektor *m*
f conditionneur *m* de l'air local,
climatiseur *m* local
nl klein luchtbehandelingstoestel *n*
местный автономный кондиционер
m малой мощности

U104 *e* **unit air heater**
d Deckenluftheizgerät *n*
f rechauffeur *m* d'air de plafond
nl plafondluchtverwarmingstoestel *n*
r потолочный рециркуляционный
агрегат *m* воздушного отопления

U105 *e* **unit area discharge**
d Abflußspende *f*, spezifischer
Abfluß *m*
f débit *m* par unité d'air
nl afvloeiingsdebiet *n* per
oppervlakte-eenheid
r модуль *m* стока

U106 *e* **unitary air conditioner**
d Klimageräteeinheit *f*,
(anschlußfertiges)
Kompaktklimagerät *n*
f conditionneur *m* de l'air unitaire
[compact, indépendant]
nl airconditioning-eenheid *f*
r агрегатированный автономный
кондиционер *m*

U107 **unit construction** *see* **modular
construction**

U108 *e* **unit cooler**
d Luftkühlgerät *n*, Raumkühlgerät *n*
f group *m* refroidisseur,
refroidisseur *m* d'air
nl kamerkoelapparaat *n*
r вентиляционно-охладительный
агрегат *m*

U109 *e* **unit deformation**
d bezogene Verformung *f*
f déformation *f* unitaire [relative]
nl relatieve vervorming *f*

r относительная [удельная]
деформация *f*

U110 *e* **unit flow**
d elementares Hochwasser *n*
f crue *f* unitaire
nl gemiddeld hoogwater *n*
r единичный [элементарный] паводок
m

U112 *e* **unit generalized load**
d verallgemeinerte Einheitslast *f*
f charge *f* [force *f*] unitaire
généralisée
nl algemene standaardbelasting *f*
r обобщённая единичная нагрузка *f*

U113 *e* **unit heater**
d Luftheizgerät *n*
f calorifère *m* à air chaud
nl luchtverwarmingstoestel *n*
r калорифер *m*

U115 *e* **unit lateral strain**
d relative Querverformung *f*
f déformation *f* latérale unitaire
nl relatieve dwarsvervorming *f*
r относительная [удельная]
поперечная деформация *f*

U116 *e* **unit load**
d Einheitslast *f*
f charge *f* unitaire
nl eenheidslast *f(m)*
r единичная нагрузка *f*

U118 *e* **unit of area**
d Flächeneinheit *f*; Einheitsfläche *f*
f unité *f* de surface
nl oppervlakte-eenheid *f*
r единичная площадь *f*;
элементарная площадка *f*,
единица *f* площади

U119 *e* **unit pressure**
d spezifischer Preßdruck *m*
f pression *f* unitaire
nl specifieke persdruk *m*
r удельное давление *n*

U120 *e* **unit-producing yard**
d Feldfabrik *f*, Vorfertigungsplatz *m*
f chantier *m* [aire *f*] de préfabrication
chantier *m* de bétonnage des
éléments préfabriqués
nl werf *f(m)* bij de bouwplaats voor
het vervaardigen van betonelementen
r полигон *m* для производства
сборных железобетонных изделий

U121 *e* **unit storm**
d Einheitsregen *m*
f pluie *f* [averse *f*] unitaire
nl eenheidsregen *m*
r единичный [элементарный] ливень
m

U122 *e* **unit strain** *US*
d relative Dehnung *f*
f déformation *f* unitaire [spécifique]
nl relatieve vervorming *f*
r относительная [удельная]
деформация *f*

U123 *e* **unit water content**
d erforderliche Anmachewassermenge *f*
pro Kubikmeter Beton;
Anmachewassermenge *f*, nach
Wasser-Zement-Faktor bestimmt
f eau *f* de gâchage par unité de
volume du béton frais; eau *f* de
gâchage détermineé par le rapport
eau-ciment
nl **vereiste hoeveelheid *f* aanmaakwater;**
hoeveelheid *f* aanmaakwater naar
water-cementfactor
r расход *m* воды затворения (*на
единицу объема бетонной смеси*),
удельная водопотребность *f*
бетонной смеси; расход *m* воды
затворения, определяемый
водоцементным отношением

U124 *e* **universal base excavator**
d **Universal-Grundbagger *m***
f excavateur *m* de base universel
nl universele graafmachine *f*
r универсальный (одноковшовый)
экскаватор *m*

U125 *e* **universal bed**
d Universal-Spannbett *n*
f banc *m* de précontrainte universel
nl universele spanbank *f(m)*
r универсальный натяжной стенд *m*

U126 *e* **universal piling plant**
d Universal-Ramme *f*
f sonnette *f* universelle
nl universele heistelling *f*
r универсальный копёр *m*

U127 *e* **universal plane**
d Universalhobel *m*
f rabot *m* universel (*à fers
interchangeables*)
nl universele schaaf *f(m)*
r универсальный ручной рубанок *m*
(*со сменными лезвиями*)

U128 **unlined canal** *see* **earth canal**

U129 *e* **unloaded chord**
d unbelasteter Gurt *m*
f membrure *f* non sollicitée
nl onbelaste rand *m*
r ненагруженный пояс *m*

U130 *e* **unloader**
d 1. Druckausgleich(s)leitung *f* 2.
Ablader *m*, Entladeeinrichtung *f*
f 1. by-pass *m* de décharge 2.
dispositif *m* de décharge(ment)
nl 1. drukvereffeningsleiding *f* 2.
aflaadinrichting *f*
r 1. уравнитель *m* давления,
разгрузочное устройство *n*
(*компрессора*) 2. разгрузочное
устройство *n*

U131 *e* **unloading ramp**
d Abladerampe *f*
f rampe *f* de déchargement
nl laadperron *m*
r разгрузочная платформа *f*
[аппарель *f*]

U132   **unregulated flow** *see* **natural flow**

U133   **unreinforced concrete** *see* **plain concrete**

U134 *e* **unrestrained member**
- *d* gelenkig gelagertes Element *n*
- *f* élément *m* non encastré [aux appuis articulés]
- *nl* niet-ingeklemd scharnierend opgelegd element *n*
- *r* шарнирно опёртый элемент *m*

U135 *e* **unset concrete**
- *d* Frischbeton *m*, unabgebundener Beton *m*
- *f* béton *m* frais [jeune]
- *nl* vers beton *n*, onverhard beton *n*
- *r* свежеуложенный [несхватившийся] бетон *m*

U136 *e* **unslaked and ground quicklime**
- *d* ungelöschter gemahlener Kalk *m*, gemahlener Branntkalk *m*, Mahlkalk *m*
- *f* chaux *f* vive broyé
- *nl* ongebluste gemalen kalk *m*
- *r* негашёная молотая известь *f*, известь-кипелка *f*

U137 *e* **unsound plaster**
- *d* blasenbildender Putzmörtel *m*
- *f* enduit *m* au mortier insuffisamment hydraté
- *nl* blaasvormende stukadoormortel *m*
- *r* штукатурный раствор *m* с неполностью гидратированными частицами (*образующими «дутики»*)

U138 *e* **unsound wood**
- *d* verfaultes [stockiges] Holz *n*, Faulholz *n*
- *f* bois *m* échauffé [pourri]
- *nl* hout *n* met gebreken, aangetast hout *n*
- *r* лесоматериал *m* с гнилью

U140 *e* **unstable soil**
- *d* nachgiebiger Boden *m*
- *f* sol *m* [terrain *m*] instable
- *nl* onstabiele bodem *m*
- *r* неустойчивый грунт *m*

U141 *e* **unsteady filtration**
- *d* unbeständige Filtration *f*
- *f* filtration *f* non permanente
- *nl* onregelmatige filtratie *f*
- *r* неустановившаяся фильтрация *f*

U142 *e* **unstiffened edge slab**
- *d* randgelagerte [allseitig aufliegende] Platte *f*
- *f* dalle *f* aux bords posés librement [aux bords non encastrés]
- *nl* alzijdig opgelegde, niet--ingeklemde balk *m*
- *r* плита *f*, свободно опёртая по контуру, плита *f* с незащемлёнными краями

U143 *e* **unstiffened member**
- *d* unversteiftes Element *n*

- *f* élément *m* de construction sans raidisseurs
- *nl* onverstijfd element *n*
- *r* элемент *m* без уголков *или* рёбер жёсткости

U144 *e* **unsymmetrical bending**
- *d* schiefe Biegung *f*
- *f* flexion *f* gauche [composée, déviée]
- *nl* torsiebuiging *f*
- *r* косой [сложный] изгиб *m*

U145 *e* **untensioned reinforcement**
- *d* schlaffe Bewehrung *f*, nichtgespannte Stahleinlagen *f pl*
- *f* armature *f* non prétendue [non précontrainte]
- *nl* niet-gespannen wapening *f*
- *r* ненапрягаемая арматура *f*

U147 *e* **unusual structure**
- *d* nichtgetyptes Bauwerk *n*
- *f* ouvrage *m* individuel
- *nl* niet-standaard bouwwerk *n*
- *r* нетиповое сооружение *n*

U149   **unwrought timber** *UK see* **undressed timber**

U151 *e* **up-feed heating system**
- *d* Heizungssystem *n* mit unterer Verteilung
- *f* chauffage *m* central ascendante
- *nl* verwarmingssysteem *n* met onderverdeling
- *r* система *f* отопления с нижней разводкой

U152 *e* **uphand welding**
- *d* Aufwärtsschweißen *n*
- *f* soudage *m* ascendante
- *nl* opwaarts lassen *n*
- *r* сварка *f* вертикальных швов снизу вверх

U153 *e* **upheaval**
- *d* Auftreiben *n*, Quellen *n*, Bodenhebung *f*; Aufblähung *f*
- *f* soulèvement *m* de chaussée; gonflement *m* [soulèvement *m*] du sol
- *nl* oprijzen *n* [spatten *n*] van de weg; opheffing *f* van de bodem
- *r* вздутие *n* (*дорожного покрытия*); пучение *n* (*грунта*)

U154 *e* **uphill grade**
- *d* Steigung *f*
- *f* contre-pente *f*, montée *f* de la route
- *nl* helling *f* omhoog, stijging *f*
- *r* обратный уклон *m*, подъём *m* (*дороги*)

U155 *e* **uplift**
- *d* 1. Hebung *f* 2. Auftrieb *m*, Sohlenwasserdruck *m*
- *f* 1. soulèvement *m*, élévation *f* (*du terrain*) 2. sous-pression *n*
- *nl* 1. verheffing *f* 2. opwaartse waterdruk *m*
- *r* 1. выпор *m* грунта фильтрационным

потоком 2. взвешивающее давление
n, противодавление *n*

U156 *e* **uplift pressure** *US*
  d Auftrieb *m*, Sohlenwasserdruck *m*
  f pression *f* de filtration, sous-
   -pression *f*
  nl opwaartse waterdruk *m*
  r взвешивающее давление *n*,
   противодавление *n*

U157 *e* **up-lighting**
  d Oberlichtbeleuchtung *f*
  f éclairage *m* supérieur [de plafond]
  nl plafondverlichting *f*
  r система *f* верхнего освещения

U158 *e* **upper chord**
  d Obergurt *m*
  f membrure *f* supérieure
  nl bovenrand *m* (*ligger*)
  r верхний пояс *m* (*балки*)

U159 *e* **upper gate**
  d Obertor *n* (*Schleuse*)
  f porte *f* de tête (*d'une écluse*)
  nl vloeddeuren *f pl* (*sluis*), deuren in
   het bovenhoofd
  r верхние ворота *pl* (*шлюза*)

U160 *e* **upper gates**
  d Oberhaupt *n* (*Schleuse*);
   Binnenhaupt *n* (*Seeschleuse*)
  f tête *f* amont
  nl bovenhoofd *n* van een sluis
  r верхняя голова *f* (*речного шлюза*);
   внутренняя голова *f* (*морского*
   шлюза)

U161 *e* **upper load block**
  d Lasthubrolle *f*
  f mouffle *f* du haut de palan (*d'une*
   grue), poulie *f* de tête de flèche
  nl hijsschijf *f*(*m*)
  r грузовой блок *m* (*крана*)

U162   **upper lock head** *see* **upper gates**

U163 *e* **upper pool**
  d Hochbecken *n*, Oberbecken *n*
  f bassin *m* supérieur
  nl bovenbekken *n*
  r верхний бассейн *m*

U164   **upper reach** *see* **headwater**

U165 *e* **upper reinforcement**
  d obere Bewehrung *f*
  f armature *f* supérieure
  nl bovenwapening *f*
  r арматура *f* сжатой зоны, верхний
   слой *m* арматуры

U166 *e* **upper yield point**
  d obere Streckgrenze *f* [Fließgrenze *f*]
  f point *m* [limite *f*] d'écoulement
   supérieur(e)
  nl bovenste vloeigrens *f*(*m*)
  r верхний предел *m* текучести

U167 *e* **upright**
  d Vertikale *f*, Lotrechte *f*,
   Vertikalstab *m*; Pfosten *m*,
   Ständer *m*; Grabenausbaustütze *f*,
   Brustholz *n*

  f élément *m* vertical (*d'une*
   construction), montant *m*, barre *f*
   verticale; poteau *m*, pilier;
   étrésillon *m* (des blindages)
  nl verticaal constructie-onderdeel *n*;
   staander *m*, stijl *m*;
   bekledingspaal *m*
  r вертикальный элемент *m*
   конструкции; стойка *f*; стойка *f*
   крепления (*траншеи*)

U168 *e* **up stairs**
  d hinaufführende Treppe *f*
  f escalier *m* servant seulement pour
   l'ascension
  nl omhoogvoerende trap *m*
  r лестница *f* только для подъёма
   (*напр. в школах*)

U169 *e* **upstand**
  d Aufkantung *f*
  f bande *f* de solin
  nl opstaande rand *m*
  r отгиб *m* (кровельного покрытия)
   вверх на вертикальную плоскость
   (*напр. на парапетную стенку*)

U171 *e* **upstanding beams slab**
  d Stahlbetonrippenplatte *f* mit
   aufwärts gerichteten Rippen
  f dalle *f* en béton armé avec
   nervures saillant vers le haut
  nl vloer *m* van gewapend beton met
   opwaarts gerichte ribben
  r ребристая железобетонная плита *f*
   с выступающими вверх рёбрами

U172 *e* **upstream apron, upstream bottom**
  d oberwasserseitige Vorlage *f*,
   Dichtungsschürze *f*, Vorboden *m*,
   Bodenteppich *m*
  f avant-radier *m*
  nl bovenstroomse bodembescherming *f*
  r понур *m*

U173 *e* **upstream cofferdam**
  d Vorsperre *f*
  f batardeau *m* amont
  nl bovenstroomse keerdam *m* (*stuw*)
  r верховая перемычка *f*

U174 *e* **upstream control**
  d oberwasserseitige Regelung *f*
  f régulation *f* par amont
  nl regeling *f* in het bovenpand
  r регулирование *n* по верхнему
   бьефу

U175 *e* **upstream engineering**
  d Oberlauf-Erosionsschutzmaßnahmen
   f pl
  f protection *f* contre l'érosion
  nl maatregelen *m pl* tegen erosie in de
   bovenloop
  r инженерные мероприятия *n pl*
   по борьбе с эрозией водосбора,
   антиэрозионные мероприятия *n pl*
   на водосборе

U176 *e* **upstream face**
  d Wasserseite *f*
  f parement *m* amont

nl bovenstroomse zijde *f*
r верховая [напорная] грань *f*

U177 upstream floor *see* **upstream apron**

U178 *e* **upstream level**
d Oberwasserspiegelhöhe *f*,
Oberwasserstand *m*
f niveau *m* amont
nl bovenwaterpeil *n*
r уровень *m* верхнего бьефа

U180 *e* **upstream reach**
d Oberwasserhaltung *f*, Oberwasser *n*
f bief *m* amont
nl bovenpand *n*
r верхний бьеф *m*

U181 *e* **upstream slope**
d Oberwasserböschung *f*,
wasserseitige Böschung *f*
f talus *m* [paroi *f*] amont
nl boventalud *n*
r верховой откос *m*

U182 upthrust *see* **buoyancy**

U183 upturn *see* **upstand**

U184 *e* **upward-acting force, upward force**
d Auftriebskraft *f*
f force *f* ascensionnelle
nl opwaartse kracht *f(m)*
r сила *f*, направленная вверх

U185 *e* **urban area**
d Stadtgebiet *n*
f territoire *m* urbain
nl stadsgebied *n*
r территория *f* города

U186 *e* **urban centre**
d Stadtzentrum *n*
f centre *m* urbain, centre-ville *m*
nl stadscentrum *n*
r центр *m* города, городской центр *m*

U187 *e* **urban communications**
d städtische Kommunikationen *f pl*
f réseaux *m pl* urbains, voiries *f pl*
et réseaux *m pl* divers (v.r.d.);
viabilité *f*
nl stedelijke verbindingen *f pl*
r городские коммуникации *f pl*

U188 *e* **urban development**
d Stadtentwicklung *f*
f développement *m* urbain
nl stadsontwikkeling *f*
r развитие *n* города; застройка *f*
города

U189 *e* **urban freeway**
d Stadtautobahn *f*
f autostrade *f* [autoroute *f*] urbaine
nl stads-autoweg *m*
r городская дорога *f* скоростного
движения

U190 urban planning *see* **town planning**

U191 *e* **urban renewal**
d Stadtumbau *m*, Stadterneuerung *f*
f rénovation *f* [renouvellement *m*]
urbaine, réorganisation *f* de vieux
quartiers (*d'une ville*)

nl stadsvernieuwing *f*, renovatie *f*
r обновление *n* города

U192 *e* **urinal**
d Pissoir *n*, Urinalbecken *n*,
Pißbecken *n*
f urinoir *m*
nl urinoir *n*
r писсуар *m*

U193 *e* **usable floor area**
d Nutzfläche *f*
f surface *f* (habitable) utile
nl bruikbaar vloeroppervlak *n*
r полезная площадь *f*

U195 *e* **use district**
d Sonderbebauungsbereich *m*
f quartier *m* de ville de
développement spécial
nl stadswijk *f(m)* met bijzondere
bestemming
r городская зона *f* специальной
застройки

U197 useful capacity *see* **live storage**

U198 *e* **U-stirrup**
d offener Bügel *m*, «Haarnadel» *f*
f étrier *m* en U
nl U-vormige beugel *m* (*wapening*)
r U-образный арматурный хомут *m*

U199 utensil sink *see* **pot sink**

U200 *e* **utilities**
d 1. Versorgungsnetze *n pl*
2. Elektroenergie *f* und durch
Versorgungsnetze zu verteilende
Energieträger *m pl* (*z.B. Gas,
Warmwasser*) 3. kommunale
Einrichtungen *f pl*
[Dienstleistungen *f pl*]
f 1. réseaux *m pl* urbains,
canalisations *f pl* urbaines 2.
utilités *f pl* (*c.-à-d. l'eau, le gaz,
l'électricité*) 3. services *m pl* publics
nl 1. voorzieningsnetten *n pl* 2.
technische overheidsvoorzieningen
f pl 3. gemeentelijke diensten *m pl*
r 1. инженерные сети *f pl* 2.
коммунальные услуги *f* (*напр. газ,
горячая вода*), поставляемые
потребителям централизованно
3. коммунальные предприятия *n pl*
обслуживания

U201 *e* **utility core**
d 1. Gebäudekern *m* Hauskern *m*
2. Installationskern *m*,
Installationszelle *f*
f 1. noyau *m* (central) d'un
bâtiment à étages multiples
2. bloc-eau *m*
nl 1. centraal gedeelte *n* van een gebouw
2. installatiekern *f(m)*
r 1. центральное ядро *n*, ствол *m*
(*высотного здания*) 2.
санитарно-техническая кабина *f*

U202 *e* **utility elements**
d Versorgungselemente *n pl*

f éléments *m pl* des canalisations
intérieures incorporés
nl in prefab bouwelementen
opgenomen voorzieningen voor
leidingen
r закладные трубы *f pl* в сборных
(железобетонных) конструкциях

U203 *e* **utility gallery**
d Versorgungsstollen *m*
f souterrain *m* urbain (*pour câbles
et tuyauterie*)
nl leidingtunnel *m*
r туннель *m* для городских подземных
коммуникаций

U204 *e* **utility gas**
d Haushaltgas *n*
f gaz *m* domestique [urbain, de
ville]
nl gas *n* voor huishoudelijk gebruik
r бытовой газ *m*

U206 *e* **utility piping**
d Versorgungsrohrleitungen *f pl*
f tuyauterie *f* urbaine enterrée
nl distributieleidingnetten *n pl*
r трубопроводы *m pl* подземного
городского хозяйства, подземные
коммуникации *f pl*

U207 *e* **utility pole**
d Leitungsmast *m*
f poteau *m* d'éclairage *ou*
téléphonique
nl telefoonpaal *m*
r столб *m* для телефонных *или*
осветительных воздушных сетей

U208 *e* **utility room**
d 1. Vorratskammer *f*;
Rumpelkammer *f* 2. Zählerraum *m*
f 1. chambre *f* de débarras 2. local
m pour compteurs (*électriques, de
gaz, de chaleur etc.*)
nl vooraadkamer *f(m)*; opslagplaats
f(m) 2. ruimte *f* voor meters en
tellers
r 1. кладовая *f*; чулан *m*
2. помещение *n* для счётчиков

U209 *e* **utility shaft**
d Betriebsschacht *m* (*im Hochhaus*)
f gaine *f* pour canalisations
intérieures
nl leidingschacht *f(m)* (*in hoog
gebouw*)
r шахта *f* для инженерных
коммуникаций

U210 *e* **utility structures**
d Tiefbauanlagen *f pl*
f ouvrages *m pl* urbains
souterrains
nl ondergrondse voorzieningen
[constructies] *f pl*
r подземные сооружения *n pl*
городского хозяйства

U211 *e* **utility tractor**
d Allzwecktraktor *m*
f tracteur *m* de base

nl kleine universele trekker *m*
r базовый трактор *m*

U212 *e* **utility window**
d serienmäßig hergestellter
Stahlfensterrahmen *m*
f chassis *m* (de fenêtre) en acier
fabriqué en série
nl standaard metalen raam *n*
r стальная оконная рама *f*
серийного производства

U213 *e* **U-type expansion bend**
d U-förmiger Ausgleichsbogen *m*
f coude *m* compensateur en U
nl U-vormige expansiebocht *f(m)*
r П-образный [дуговой] компенсатор
m

V

V1 *e* **vacuum concrete**
d Vakuumbeton *m*
f béton *m* traité à vide
nl vacuümbeton *n*
r вакуумированный бетон *m*

V2 *e* **vacuum core**
d Vakuumkern *m*
f noyau *m* à vide
nl vacuümkern *f(m)*
r вакуумирующий
пустотообразователь *m*

V3 *e* **vacuum dewatering plant**
d Vakuumentwässerungsbecherwerk *n*
f installation *f* d'essorage
[d'égouttage] au vide
nl vacuümontwateringsinstallatie *f*
r вакуумная обезвоживающая
установка *f*

V4 *e* **vacuum filter**
d Vakuumfilter *n, m*
f filtre *m* à vide
nl vacuümfilter *n, m*
r вакуум-фильтр *m*

V5 *e* **vacuum filtration**
d Vakuumfilterung *f*,
Vakuumfiltration *f*
f filtrage *m* par le vide
nl vacuümfiltrering *f*
r вакуум-фильтрование *n*

V6 **vacuum heating system** *see* **sub-
-atmospheric heating system**

V7 *e* **vacuum lifting equipment**
d Vakuumhebevorrichtungen *f pl*
f accéssoires *m pl* de levage à vide
nl vacuümhefinstallaties *f pl*
r вакуумные грузозахватные
устройства *n pl*

V8 *e* **vacuum mat**
d Saugmatte *f*
f tapis *m* d'essorage au vide
nl zuigmat *f(m)*
r мат *m* для вакуумирования

V9　*e* **vacuum method wellpoint**
　d Vakuumbrunnen *m*, Filterbrunnen
　　m nach Vakuumverfahren
　f pointe *f* filtrante
　nl vilterbron *f(m)* volgens
　　vacuümmethode
　r иглофильтр *m*

V10　*e* **vacuum mold, vacuum mould**
　d Vakuumform *f*, Saugform *f*
　f moule *m* à vide
　nl vacuümmal *m*
　r вакуум-форма *f*

V11　*e* **vacuum pad**
　d Vakuumlasthaftgerät *n*
　f pince *f* à vide
　nl zuignap *f*
　r вакуумный грузозахватный
　　присос *m*

V12　*e* **vacuum suction head**
　d vakuummetrische Saughöhe *f*
　f hauteur *f* d'aspiration à vide
　nl zuighoogte *f*
　r вакуумметрическая высота *f*
　　всасывания

V13　*e* **vadose water** *US*
　d 1. Gravitationswasser *n*
　　2. schwebendes Grundwasserstock
　　werk *n*
　f 1. eau *f* vadose [de gravité]|
　　2. nappe *f* suspendue temporaire
　nl 1. zakwater *n* 2. grondwater *n*
　　boven het freatisch vlak
　r 1. гравитационная вода *f*
　　2. верховодка *f*

V14　*e* **valley**
　d Kehle *f*
　f noue *f*
　nl kilkeper *m*, kil *f*
　r ендова *f*, разжелобок *m*

V15　*e* **valley crossing**
　d Taldurchquerung *f*,
　　Talüberführung *f*, Talübergang
　　m
　f traversée *f* au-dessus d'une vallée
　nl dal-oversteek *m*
　r путепровод *m* [виадук *m*] над
　　долиной

V16　*e* **valley flashing**
　d Kehlblech *n*
　f noquet *m*
　nl zinken kilgoot *f(m)*
　r металлическое покрытие *n* ендовы
　　[разжелобка]

V17　*e* **valley gutter**
　d Kehlrinne *f*
　f noue *f* ronde [en chéneau]
　nl kilgoot *f (m)*
　r лоток *m* разжелобка; разжелобок *m*

V18　*e* **valley jack**
　d Kehlschifter *m*
　f empanon *m* de noue [à noulet]
　nl kort kilspant *n*

　r нарожник *m* (*короткая
　　стропильная нога*) ската
　　разжелобка

V19　*e* **valley rafter**
　d Kehlsparren *m*
　f arrêtier *m* [chevron *m*] de noue
　nl kilkeperbalk *m*
　r стропильная нога *f* разжелобка

V20　*e* **valley roof**
　d 1. Schmetterlingsdach *n* 2.
　　mehrseitig geneigtes Kehlendach *n*
　f 1. toit *m* [comble *m*] en V 2. toit
　　m à plusieurs versants (*formant les
　　noues*)
　nl 1. V-vormig zadeldak *n* 2. dak *n*
　　met kilkepers
　r 1. двускатная крыша *f* с
　　обращёнными внутрь скатами,
　　V-образная крыша *f*
　　2. многоскатная крыша *f* с
　　разжелобками

V22　*e* **valve**
　d Ventil *n*; Schieber *m*
　f valve *f*; soupape *f*; clapet *m*
　nl klep *f(m)*; afsluiter *m*; ventiel *n*
　r клапан *m*; шибер *m*; задвижка *f*

V23　*e* **valve bag**
　d Ventilsack *m*
　f sac *m* à valve
　nl ventielzak *m*
　r мешок *m* с клапаном, клапанный
　　мешок *m* (*напр. для цемента*)

V24　*e* **vandal proof equipment**
　d vandalsichere Ausrüstung *f*
　f équipement *m* [appareillage *m*] à
　　l'épreuve des vandales
　nl solide gereedschap *n*
　r оборудование *n*, защищённое от
　　порчи при актах вандализма

V25　*e* **vaned outlet**
　d Gliederklappenaustritt *m*,
　　Jalousieauslaß *m*
　f bouche *f* grillagée à réglage
　nl schoepenrooster *n*
　r регулируемая приточная решётка *f*

V26　*e* **vanishing point**
　d Fluchtpunkt *m*
　f centre *m* de projection [de
　　perspective rayonnée]
　nl verdwijnpunt *n*
　r центр *m* перспективы, точка *f* схода

V27　*e* **vapor lock**
　d Dampfsack *m*, Dampfpfropfen *m*
　f bouchon *m* de vapeur
　nl gasslot *n*, verstopping *f*
　　tengevolge van een gasbel
　r паровая пробка *f*

V28　*e* **variable air volume system**
　d Klimaanlage *f* mit
　　Luftstromregelung [mit variablem
　　Förderstrom]
　f installation *f* de conditionnement
　　d'air à volume variable

nl luchtbehandelingsinstallatie *f* met
regelbare luchtstroom
r система *f* кондиционирования
с переменным расходом воздуха

V29　*e* **variable concentrated load**
d ortsveränderliche Einzellast *f*
f charge *f* concentrée mobile
nl puntlast *f(m)* op wisselende
plaatsen
r подвижная сосредоточенная
нагрузка *f*

V30　*e* **varnish**
d Lack *m*, Firnis *m*
f vernis *m*, laque *f*
nl lak *n*, vernis *n*, *m*
r лак *m*

V31　*e* **varnish paint**
d Lackfarbe *f*
f peinture *f* à vernis
nl lakverf *f(m)*
r лаковая краска *f*

V32　*e* **vault**
d Gewölbe *n*
f voûte *f*
nl gewelf *n*
r свод *m*

V33　*e* **vault abatment**
d Gewölbewiderlager *n*
f soutènement *m* [support *m*] de la
voûte
nl schoormuur *m* (*gewelf*)
r опора *f* свода

V34　*e* **vault bay**
d Gewölbe(dach)feld *n*
f travée *f* du toit-voûte [du comble-
-voûte]
nl gewelfveld *n*, travee *f*,
gewelfkluis *m*
r пролёт *m* [отсек *m*] сводчатого
покрытия, звено *n* [поле *n*] свода

V35　*e* **vaulting tile**
d keramischer Hohlformstein *m* (*für
Gewölbemauerwerk*)
f bloc *m* creux de voûte en terre-
-cuites, bloc *m* façonné creux en
céramique (*pour la construction
des voûtes*)
nl gebakken holle steen *m* voor
gewelven, gewelfsteen *m*
r фасонный пустотелый
керамический блок *m* (*для кладки
сводов*)

V36　*e* **vault rib**
d Gewölberippe *f*
f nervure *f* [arête *f*] d'une voûte
nl gewelfribbe *f(m)*
r арочное ребро *n* свода

V37　*e* **vault thrust**
d Gewölbeschub *m*
f poussée *f* de la voûte
nl horizontale oplegkracht *f(m)* van
een gewelf
r распор *m* свода

V38　**VAV-system** *see* **variable air
volume system**

V39　*e* **V-beam sheeting**
d Stahlbelag *m* mit V-förmigem
Profil
f tôle *f* d'acier profilée en V
nl staalplaat *f(m)* met V-vormig
profiel
r стальной профилированный настил
m V-образного сечения

V40　*e* **V-brick**
d Lochziegel *m*
f brique *f* perforée
nl holle steen *m*
r дырчатый кирпич *m*

V41　*e* **Vebe apparatus**
d Vebe-Konsistenzmesser *m*
f appareil *m* (de) Vebe
nl Vebe-consistentiemeter *m*
r прибор *m* Вебе (*для определения
жёсткости бетонной смеси*)

V43　*e* **vehicle**
d 1. Fahrzeug *n* 2. Verdünnungsmittel
n, Verschnittmittel *n*;
Lösungsmittel *n* 3. Farbbindemittel
n; Trägersubstanz *f*, Trägerstoff *m*
f 1. véhicule *m* 2. diluant *m*,
fluidifiant *m* ; solvant *m* 3. liant *m*
pour couleurs
nl 1. voertuig *n* 2.
transporterend medium *n*
r 1. самоходное транспортное
средство *n*; автомобиль *m* 2.
разжижитель *m*, разбавитель *m*;
растворитель *m* (*для краски*) 3.
связующее *n* для краски

V44　*e* **vehicular lane**
d Fahrspur *f*, Verkehrsstreifen *m*
f voie *f* [bande *f*] de circulation
nl rijstrook *f(m)*
r полоса *f* движения

V45　*e* **vehicular live load**
d Fahrzeuglast *f*
f convoi *m* type
nl belasting *f* van een voertuig
r автомобильная нагрузка *f* (*на
мост*)

V46　*e* **vehicular tunnel**
d Fahrzeugtunnel *m*
f tunnel *m* pour véhicules, tunnel-
-route *m*
nl verkeerstunnel *m*
r автодорожный туннель *m*

V47　*e* **velocity head**
d Geschwindigkeitshöhe *f*,
Geschwindigkeitsgefälle *n*
f charge *f* cinétique [due à le vitesse]
nl kinetische-drukhoogte *f*
r скоростной напор *m*

V48　*e* **velocity of approach**
d 1. Anströmgeschwindigkeit *f*
2. Annäherungsgeschwindigkeit *f*
f vitesse *f* d'approche (*de l'eau*)

nl 1. toestromensnelheid *f*
2. snelheid *f* van het toestromende
water
r 1. скорость *f* набегающего потока,
скорость *f* подхода 2. средняя
скорость *f* над порогом
измерительного водослива

V49 *e* **velocity of retreat**
d Geschwindigkeitsabfall *m* unterhalb
eines Abflußhindernisses
f vitesse *f* de départ [de sortie]
nl stroomsnelheid *f* van het afvloeiende
water
r падение *n* средней скорости
потока за препятствием

V50 *e* **velocity pressure**
d Geschwindigkeitsdruck *m*
f pression *f* dynamique [de vitesse]
nl stuwdruk *m*
r скоростное давление *n*

V51 *e* **velocity profile**
d Geschwindigkeitsprofil *n*
f profil *m* des vitesses
nl snelheidsprofiel *n*
r эпюра *f* [профиль *m*] скоростей

V52 *e* **vena contracta**
d Strahleinschnürung *f*, verengter
Flüssigkeitsstrahl *m*;
Einschnürungsquerschnitt *m*
f section *f* contractée [étranglée]; jet
m rétréci
nl straalcontractie *f*
r сжатое сечение *n*; местное сужение
n потока

V53 *e* **veneer**
d 1. Furnier *n* 2. nicht lasttragende
Mauerverkleidung *f*
[Ziegelverkleidung *f*]
f 1. pli *m* de contre-plaqué 2.
revêtement *m* mural de briques
nl 1. fineer *n* 2. niet-dragende
muurbekleding *f*
r 1. шпон *m* (фанеры)
2. прислонная кирпичная
облицовка *f* стены

V54 *e* **veneer base**
d Gipsstreifentafel *f* (*Putzträger*)
f sous-couche *f* en lattes de plâtre
(*support pour enduit*)
nl pleisterdrager *m*, ondergrond *m*
voor pleisterwerk
r щит *m* из гипсовых полос
(*основание под штукатурку*)

V55 *e* **veneered construction**
d Skelettbau *m* mit Außenverkleidung,
verkleidete Skelettkonstruktion *f*
f charpente *f* avec bardage
extérieur en tôles
nl skeletconstructie *f* waarvan de in
het zicht blijvende delen zijn
bekleed
r каркасная конструкция *f*
с наружной облицовкой
[обшивкой]

V56 *e* **veneered door**
d furnierte Tür *f*
f porte *f* contre-plaqué
nl gefineerde deur *f(m)*
r офанерованная дверь *f*

V57 *e* **veneered plywood**
d Zierfurnier *n*
f contre-plaqué *m* décoratif
[ornemental, de décoration]
nl multiplex *n* met fineerbekleding
r декоративная фанера *f*

V58 *e* **veneered wall**
d Wand *f* mit loser Ziegelverkleidung
f mur *m* à parement [avec
revêtement extérieur de briques]
nl muur *m* met steenbekleding
r облицованная стена *f*
(*с прислонной облицовкой*)

V59 *e* **veneer plaster**
d Fertigdeckmörtel *m*
f mortier *m* d'enduit de finition
prêt à l'emploi
nl gebruiksgereed afwerkpleister *n*
r готовый накрывочный
штукатурный раствор *m*

V60 *e* **veneer tie, veneer wall tie**
d Mauerverkleidungsanker *m*
f attache *f* [ancre *f*] murale (*pour la
fixation du revêtement au mur*)
nl muurbekledingsanker *n*
r анкер *m* [закрепа *f*] для
крепления наружной облицовки

V61 *e* **Venetian blind**
d Jalousie *f*
f jalousie *f*, store *m* vénitien
nl jaloezie *f*
r венецианская штора *f* (*из
тонких полосок*)

V62 *e* **Venetian light**
d Venezianisches Fenster *n*
f fenêtre *f* vénitienne
nl venitiaans venster *n*
r венецианское [трёхстворчатое]
окно *n*

V63 *e* **Venetian mosaic**
d Terrazzo *m*
f mosaïque *f* vénitienne, terrazzo *m*
nl terrazzo *n*
r террацо *m*

V64 *e* **vent**
d Entlüftungsloch *n*,
Lüftungsöffnung *f*; Entlüftungsrohr
n; Entlüftungskanal *m*, Luftabzug *m*
f orifice *m* d'aérage; trou *m* de
ventilation; conduit *m* de
ventilation
nl ventilatieopening *f*;
ventilatiekanaal *n*
r вентиляционное отверстие *n*;
вентиляционная труба *f*;
вентиляционный канал *m*

V65 *e* **vented form**
d ventilierte [entlüftete] Schalung *f*

f coffrage *m* aéré [d'aérage]
nl geventileerde bekisting *f*
r проветриваемая опалубка *f*

V66 *e* **vented gas heater**
d Außenwandgas(raum)heizer *m*
f réchauffeur *m* d'air à gaz du type isolé
nl gevelgaskachel *m*
r газовый воздухонагреватель *m* закрытого типа

V67 *e* **ventilated casing**
d Staubschutzummantelung *f*
f cage *f* [chemise *f*] aérée
nl geventileerde mantel *m*
r защитно-обеспыливающий кожух *m*

V68 *e* **ventilated light fitting**
d Lüftungsleuchte *f*, ventilierte Leuchte *f*
f plafonnier *m* avec orifice de ventilation
nl plafonnier met ventilatieopeningen
r вентиляционно-осветительный плафон *m*

V69 *e* **ventilating chamber**
d Lüfterkammer *f*, Lüftungsstation *f*
f chambre *f* de ventilation [d'aérage, d'aération]
nl ventilatiekamer *f(m)*
r вентиляционная камера *f*

V70 **ventilating installation** *see* **ventilation plant**

V71 *e* **ventilating pipe**
d Lüftungsrohr *n*
f tuyau *m* [conduit *m*] de ventilation
nl ventilatiepijp *f(m)*
r вентиляционная труба *f*

V72 *e* **ventilating shaft**
d Lüftungsschacht *m*, Luftschacht *m*
f gaine *f* de ventilation, puits *m* d'aérage
nl ventilatieschacht *f(m)*
r вентиляционная шахта *f*

V73 *e* **ventilating skylight**
d Lüftungslaterne *f*, Lüftungsoberlicht *n*
f lucarne *f* d'aérage [d'aération, de ventilation]
nl ventilatielantaarn *f(m)*
r аэрационный фонарь *m*

V74 *e* **ventilation**
d Ventilation *f*, Lüftung *f*
f ventilation *f*
nl ventilatie *f*
r вентиляция *f*

V75 *e* **ventilation aperture**
d Ventilationsöffnung *f*, Belüftungsöffnung *f*
f orifice *m* d'aérage, baie *f* d'aération
nl ventilatieopening *f*
r вентиляционный проём *m*

V76 *e* **ventilation brick**
d Entlüftungsziegel *m*
f brique *f* de ventilation
nl ventilatiesteen *m*
r вентилируемый кирпич *m* (с отверстиями)

V77 *e* **ventilation cavity**
d Lüftungsschlitz *m*
f vide *m* d'aérage
nl ventilatieholte *f*
r вентиляционная полость *f* (в строительных конструкциях)

V78 *e* **ventilation duct**
d Lüftungskanal *m*
f conduit *m* [canal *m*, gaine *f*] de ventilation
nl ventilatiekanaal *n*
r вентиляционный воздуховод *m*

V80 *e* **ventilation heat gain**
d Ventilationswärmeeinfall *m*
f gain *m* en chaleur par ventilation
nl warmtewinst *f* tengevolge van ventilatie
r вентиляционные теплопоступления *n pl*

V81 *e* **ventilation heat loss**
d Ventilationswärmeverlust *m*
f perte *f* de chaleur par ventilation
nl warmteverliezen *n pl* tengevolge van ventilatie
r вентиляционные теплопотери *f pl*

V82 *e* **ventilation hood**
d Entlüftungshaube *f*
f hotte *f* de ventilation
nl ventilatiekap *f(m)*
r вентиляционный колпак *m*, вытяжной зонт *m*

V83 *e* **ventilation load**
d Lüftungsheizlast *f*, Lüftungswärmebedarf *m*
f charge *f* calorifique [thermique] de ventilation
nl warmteverlies *n* tengevolge van ventilatie
r вентиляционная тепловая нагрузка *f*

V84 *e* **ventilation opening**
d Lüftungsöffnung *f*
f trou *m* [bouche *f*] d'aération [d'aérage]
nl ventilatieopening *f*
r вентиляционное отверстие *n*

V85 *e* **ventilation plant**
d Lüftungsanlage *f*, lufttechnische Anlage *f*
f installation *f* de ventilation [d'aérage, d'aération]
nl ventilatiesysteem *n*
r вентиляционная установка *f*

V86 *e* **ventilation plant room**
d Lüfterraum *m*
f salle *f* des ventilateurs
nl ventilatie-machinekamer *f(m)*
r вентиляторная *f* (помещение)

V87 e ventilation rate
d Luftwechselzahl f
f taux m de ventilation
nl luchtverversingsgraad m
r кратность f воздухообмена

V88 e ventilation system
d Lüftungssystem n
f système m de ventilation
nl ventilatiesysteem n
r система f вентиляции

V89 e ventilator
d 1. Lüftungsöffnung f, Ent- und
Belüftungsöffnung f 2. Lüfter m,
Ventilator m
f 1. trou m d'aération, ouverture f
de ventilation 2. ventilateur m,
aérateur m
nl 1. luchtopening f 2. ventilator m
r 1. аэрационное отверстие n
2. вентилятор m

V90 e ventlight
d Lüftungsflügel m; oberer
Fensterflügel m
f imposte f ouvrante (au-dessus d'une
croisée); vasistas m
nl beweegbaar bovenraam n
r (открывающаяся оконная) фрамуга
f; форточка f

V91 e vent pipe
d Entlüftungsrohr n
f tuyau m de ventilation; colonne f
de ventilation secondaire
nl ventilatiepijp f(m)
r вентиляционная [вытяжная]
труба f (в канализации);
воздушная труба f (котельных
и водогрейных установок)

V92 vent sash see ventlight

V93 vent shaft see vent pipe

V94 e vent stack
d Entlüftungsstrang m
f évent m, colonne f d'évent [de
ventilation]
nl ontspanningsleiding f,
ontluchtingsleiding f
r вентиляционный стояк m (в
системе канализации)

V95 e venturi meter
d Venturimesser m, Venturimeter n
f débitmètre m Venturi
nl venturimeter m
r расходомер m Вентури

V96 e vent valve see bleeder valve

V97 e verge
d 1. Bankett n, Außenstreifen m
2. Rand m, Außenkante f
3. Ortgang m; giebelseitige
Dachtraufe f, giebelseitiger
Dachvorsprung m 4. Schaft m
Rumpf m, Stamm m (Säule)
f 1. accotement m 2. bord m, arête
f, rive f 3. planche f de rive;
avant-toit m d'un pignon, saillie f

de rive 4. fût m de la colonne
nl 1. berm m 2. buitenkant m
3. dakrand m (buiten de gevel)
4. zuilschacht f(m)
r 1. обочина f 2. кромка f, бровка f
3. ветровая фронтонная доска f;
свес m кровли над фронтоном
4. ствол m колонны

V98 e verge board
d Windbrett n, Windfeder f
f soffite m de l'avant toit
nl windveer f(m)
r ветровая доска f

V99 e verge cutter
d Abbordgerät n,
Bordkantenschneidgerät n,
Straßenbordfräs- und
-räummaschine f
f coupe-gazon m, tranche-gazon m
nl bermmaaier m
r газонокосилка f

V100 e verge fillet
d Einschalung f des Dachüberstandes
(auf der Giebelseite)
f voligeage m [revêtement m en
planches] de la saillie de rive
(d'une toiture)
nl beplanking f van de dakrand
(boven de gevel)
r дощатая обшивка f свеса (над
щипцом или фронтоном)

V101 e verge purlin
d Randpfette f, Saumpfette f
f latte f de l'avant toit
nl randgording f
r обрешётина f у свеса крыши

V102 e vermiculated work
d wurmstichförmige Gestaltung f der
Wandoberfläche
f finition f décorative ayant l'aspect
d'une surface vermoulue
nl pleisterwerk n met wormvormige
sleuven, granol m
r декоративная отделка f стен
с прожилками и канавками (в виде
червоточины)

V103 e vermiculite aggregate
d Vermikulitzuschlagstoff m
f agrégat m [granulat m] de
vermiculite
nl vermiculiettoeslag f(m)
r лёгкий заполнитель m из
вспученного вермикулита

V104 e vermiculite concrete
d Vermikulitbeton m
f béton m (de) vermiculite
nl vermiculietbeton n
r вермикулитобетон m]

V105 e vermiculite plaster
d Vermikulitputz m
f enduit m de vermiculite
nl vermiculietpleister n
r вермикулитовая штукатурка f

V106 *e* **versed sine**
   d Bogenpfeil *m*, Bogenstich *m*,
     Bogenhöhe *f*
   f flèche *f* de l'arc
   nl booghoogte *f*, porring *f*, pijl *m*
   r стрела *f* подъёма арки

V107 *e* **vertex**
   d Bogenscheitel *m*
   f clef *f* [clé *f*] de l'arc *ou* de voûte
   nl toppunt *n*, boogkruin *f(m)*
   r ключ *m* арки *или* свода

V108 *e* **vertical**
   d 1. Meßlotrechte *f* 2. Vertikalstab *m*
     3. Vertikale *f*
   f 1. verticale *f* de mesure 2. élément
     m de construction vertical
     3. verticale *f*
   nl 1. loodlijn *f(m)* 2. loodrecht
     constructie-element *n* 3. verticale
     lijn *f(m)*
   r 1. гидрометрическая вертикаль *f*
     2. вертикальный конструктивный
     элемент *m* 3. вертикаль *f*

V109 *e* **vertical alignment**
   d Linienführung *f* im Abriß,
     Längsprofil *n*
   f profil *m* en long
   nl langsprofiel *n* (*weg*)
   r продольный профиль *m*

V110 *e* **vertical bar**
   d Fensterpfosten *m*
   f meneau *m* vertical
   nl vensterstijl *m*, middenstijl *m*
   r вертикальный импост *m* окна

V111 *e* **vertical boom type trencher**
   d Grabenbagger *m* mit vertikaler
     Eimerleiter
   f trancheuse *f* avec chaîne à godets
     verticale
   nl slootgraafmachine *f* mit verticale
     graafketting
   r траншеекопатель *m* с
     вертикальной рамой

V112 *e* **vertical bucket elevator**
   d Senkrechtbecherwerk *n*,
     Senkrechtelevator *m*
   f élévateur *m* à godets verticale
   nl verticale jacobsladder *m*
   r ковшовый элеватор *m*
     вертикального подъёма

V113 *e* **vertical exit**
   d Vertikalausgang *m*
   f moyen *m* de sortie verticale
   nl verticale uitgang *f*
   r вертикальный выход *m* [путь *m*
     эвакуации]

V114 *e* **vertical hitch**
   d einsträngiger Stropp *m*
   f brin *m* de suspension
   nl eenstrengige strop *m*
   r одноветвевой (вертикальный)
     строп *m*

V115 *e* **vertical lift bridge**
   d Hubbrücke *f*
   f pont *m* levant
   nl hefbrug *f(m)*
   r разводной вертикально-
     -подъёмный мост *m*

V116 *e* **vertical lift gate**
   d Hubtor *n*, Hebetor *n*;
     Hubschütz(e) *n* (*f*)
   f porte *f* levante (d'*écluse*); vanne *f*
     levante
   nl hefdeur *f(m)*
   r подъёмные ворота *pl* (*шлюза*);
     подъёмный затвор *m*

V117 *e* **vertical lift weir**
   d Schützenwehr *n*, Fallenwehr *n*
   f barrage *m* à hausses
   nl stuw *m* met verticale schuiven
   r плотина *f* с плоскими затворами,
     щитовая плотина *f*

V118 *e* **vertical links**
   d lotrechte Bewehrungsbügel *m pl*
   f étriers *m pl* verticaux
   nl verticale beugels *m pl*
     (*wapening*)
   r вертикальные арматурные хомуты
     m pl

V119 *e* **vertical pivoting window**
   d Wendeflügelfenster *n*
   f fenêtre *f* pivotante
   nl taatsraam *n*
   r окно *n*, вращающееся вокруг
     вертикальной оси

V120 **vertical meeting rail** *see* **meeting**
   stile

V121 **vertical sash** *see* **vertical sliding**
   window

V122 **vertical shoring** *see* **underpinning**

V123 *e* **vertical sliding window**
   d Senkrechtschiebefenster *n*,
     Vertikal-Schiebe(fenster)flügel *m*
   f fenêtre *f* à guillotine
   nl verticaal schuifvenster *n*
   r опускное окно *n*

V124 *e* **vertical strip door**
   d Lattentür *f* mit waagerechten
     Latten
   f porte *f* en planches verticales
   nl opgeklampte deur *f(m)*
   r дверь *f* из вертикальных досок на
     планках

V125 *e* **vertical ties**
   d Vertikalverband *m*
   f contreventements *m pl* verticaux
   nl verticale trekstaven *f(m) pl*
   r вертикальные связи *f pl*

V126 *e* **vertical tiling**
   d Wandfliesenbelag *m*
   f carrelage *m* de mur
   nl wandtegelbekleding *f*
   r облицовка *f* стен плитками

V127 *e* **vertical tray conveyors**
   d Paternosteraufzug *m*, Schaukeleleva-
     tor *m*, Schaukelförderer *m*
   f convoyeur *m* vertical à balancelles,
     élévateur-transporteur *m*
  nl paternosterlift *m*
   r вертикальный полочный *или* люлеч-
     ный элеватор *m*

V128 *e* **vertical web member**
   d Vertikalstab *m*, Lotrechte *f*, Hänge-
     pfosten *m*; Vertikalsteife *f*, Verstei-
     fungsrippe *f* (*Trägersteg*)
   f montant *m* du treillis; élément *m*
     vertical de raidissement (*d'une poutre*)
  nl verticale staaf *f* (*m*); stijl *f* (*m*)
   r стойка *f*, подвеска *f* (*решётки фер-
     мы*); вертикальный элемент *m* жёст-
     кости (*стенки балки*)

V129 *e* **vestibule**
   d Vorraum *m*; Vestibül *n*
   f antichambre *f*; vestibule *m*
  nl vestibule *m*, hal *f* (*m*)
   r вестибюль *m*; холл *m*

V130 *e* **vest-pocket park**
   d Kleingrünanlage *f*
   f petit jardin *m* public de ville
  nl klein plantsoen *n*
   r небольшой городской парк *m*, сквер *m*

V131 *e* **viaduct**
   d 1. Viadukt *m* 2. Talbrücke *f*
   f 1. viaduc *m* 2. pont *m* au-dessus d'une
     vallée (*à travées multiples*)
  nl 1. viaduct *m*, *n* 2. brug *f* (*m*) over
     een dal
   r 1. виадук *m* 2. многопролётный мост
     m через долину

V132 **vibrating** *see* **vibration**

V133 *e* **vibrating beam**
   d Rüttelbohle *f*, Vibrations(glätt)bohle *f*
   f poutre *f* [règle *f*] vibrante
  nl trilbalk *m*
   r виброрейка *f*, вибробрус *m*

V134 *e* **vibrating circular pipe-line**
   d Schwingförderrohr *n*, Rohrförderer *m*,
     Vibrationsrohr *n*
   f transporteur *m* [convoyeur *m*] à tuyau
     vibrant
  nl triltransportpijp *f* (*m*)
   r вибрационный транспортирующий
     трубопровод *m*

V135 *e* **vibrating chute**
   d Vibrationsschurre *f*; Schwingförder-
     rinne *f*
   f couloir *m* vibrant; gouttière *f* [gou-
     lotte *f*] vibrante
  nl trilgoot *f* (*m*)
   r качающийся рештак *m*; виброжёлоб *m*

V136 *e* **vibrating conveyor**
   d Schwingförderer *m*
   f vibro-transporteur *m*
  nl triltransporteur *m*, trilgoot *f* (*m*)
   r вибрационный конвейер *m*

V137 **vibrating float** *see* **vibratory bullfloat**

V138 *e* **vibrating grate**
   d Rüttelrost *m*, Vibrationsrost *m*
   f vibro-crible *m*, crible *m* vibrant
  nl trilrooster *n*
   r виброгрохот *m* с колосниковой ре-
     шёткой

V139 *e* **vibrating mold, vibrating mould**
   d Rüttelform *f*, Rüttelschalung *f*, Vi-
     brationsform *f*
   f moule *m* vibrant
  nl trilmal *m*
   r виброформа *f*

V140 *e* **vibrating pebble mill**
   d Schwingsteinmühle *f*, Schwingkugel-
     mühle *f*
   f moulin *m* vibratoire à boulets [à billes]
  nl trilsteenmolen *m*
   r вибрационная шаровая мельница *f*

V143 *e* **vibrating plate**
   d Rüttelplatte *f*, Schwingplatte *f*
   f plateau *m* vibrant, plaque *f* vibrante
  nl trilplaat *f* (*m*)
   r виброплощадка *f*, виброплита *f*

V144 **vibrating roller** *see* **vibratory roller**

V145 **vibrating screed** *see* **vibrating beam**

V146 *e* **vibrating screen**
   d Schüttelsieb *n*, Vibrationssieb *n*, Vi-
     brationssiebmaschine *f*
   f tamis *m* vibrant
  nl trilzeef *f* (*m*)
   r виброгрохот *m*, вибросито *n*

V147 *e* **vibrating stressing bed**
   d Vibrations-Spannbett *n*
   f banc *m* [lit *m*] de mise en tension
     à vibration
  nl trilspanbank *m*
   r вибронатяжной стенд *m*

V148 *e* **vibrating table**
   d Rütteltisch *m*, Schwingtisch *m*
   f table *f* vibrante
  nl triltafel *f* (*m*)
   r вибрационный стол *m*, вибростол *m*

V149 **vibrating tamper** *see* **vibratory compactor**

V150 *e* **vibration**
   d 1. Schwingung *f*, Vibration *f* 2. Ver-
     dichten *n* durch Vibration (*Beton*)
   f vibration *f*
  nl 1. trilling *f*, vibratie *f* 2. verdichten *n*
     door de trillen (*beton*)
   r 1. вибрация *f* 2. вибрирование *n*
     (*уплотнение бетона*)

V151 *e* **vibration control, vibration isolation**
   d Schwingungsdämpfung *f*
   f absorption *f* [amortissement *m*] de
     vibration
  nl trillingsdemping *f*
   r виброизоляция *f*

V152 *e* **vibration limit**
   d Rüttelgrenze *f*
   f temps *m* [durée *f*] limite de vibration
     (*du béton*)

nl maximale trillingsduur m
r предельно допустимая продолжительность f вибрирования (бетона)

V153 e **vibrator**
d Vibrator m, Rüttler m
f vibr(at)eur m
nl vibrator m, triller m
r вибратор m, вибрационный уплотнитель m

V154 **vibrator beam** see **vibrating beam**

V155 e **vibrator cylinder**
d Flaschenrüttler m
f pervibrateur m à aiguille, aiguille f vibrante
nl trilnaald m
r вибробулава f

V156 e **vibratory bullfloat**
d Vibrationsglätter m
f vibrolisseuse f, vibrolisseur m
nl tril-plakspaan f (m)
r вибрационная гладилка f [затирочная машина f]

V157 **vibratory compacting** see **vibration 2.**

V158 e **vibratory compactor**
d Vibrationsverdichter m
f compacteur m vibrant, vibro-pilonneuse f, pilonneuse f vibrante, vibro-dameuse f
nl trilmachine f, trilslede f, trilstamper m
r вибрационная уплотняющая [трамбующая] машина f

V159 e **vibratory pile driving**
d Pfahleinrütteln n, Pfahlabsenkung f durch Vibration
f enfoncement m [fonçage m] de pieux par vibrations
nl palen intrillen
r вибропогружение n свай

V160 e **vibratory pile hammer**
d Vibrations(ramm)bär m, Vibrations-(ramm)hammer m
f vibro-marteau m, vibromouton m, vibro-fonceur m
nl trilblok n
r вибромолот m, вибропогружатель m свай

V161 e **vibratory roller**
d Vibrationswalze f, Schwingwalze f, Rüttelwalze f
f rouleau m vibrant [à vibration], cylindre m vibrant
nl trilwals f (m)
r вибрационный дорожный каток m

V162 **vibratory screed** see **vibrating beam**

V163 **vibratory screen** see **vibrating screen**

V164 e **vibratory spiral elevator**
d Wendelschwingrinne f
f élévateur m à vis sans fin vibratoire
nl trilwormtransporteur m
r червячный [винтовой] виброжёлоб m

V165 **vibro-driver** see **vibratory pile hammer**

V166 e **vibroflotation soil-compaction**
d Rütteldruckverfahren n
f compactage m par vibro-flottation, vibro-flottation f
nl diepteverdichting f door trilling
r гидровибрационное уплотнение n, уплотнение n грунтов с помощью гидровибраторов

V167 e **vibroflot machine**
d Auflastrüttler m, Hydrovibrator m, Tiefenrüttler m für nichtbindige wassergesättigte Böden
f vibrateur m hydraulique
nl trilmachine f voor diepteverdichting
r гидровибратор m (для уплотнения грунтов)

V168 **vibromold** see **vibrating mold**

V169 e **vibro pile**
d Vibropfahl m
f pieu m enfoncé par vibro-fonceur
nl vibropaal m
r свая f, погружаемая вибрацией

V170 **vibropiling** see **vibratory pile driving**

V171 e **vibrotrunk**
d Rüttelschwenkrohr n
f trompe f [goulotte f] vibrante
nl trillende schommelpijp f (m)
r виброхобот m

V172 e **vice**
d Schraubstock m
f étau m
nl bankschroef f (m)
r тиски pl

V173 e **Vierendeel girder**
d Vierendeelträger m
f poutre f Vierendeel
nl Vierendeelligger m
r безраскосная ферма f, балка f Виренделя

V174 e **vinyl-asbestostiles**
d Vinylasbest(-Fußbodenbelag)platten f pl
f carreau m thermoplastique vinyl amiante
nl asbestvinylvloertegels m pl
r асбестосмоляные плитки f pl для полов (на базе поливинилхлоридной смолы)

V175 e **vinyl tiles**
d PVC-Fliesen f pl, PVC-Platten f pl
f carreau m (en) P.V.C. [chlorure de polyvinyle]
nl vinyltegels m pl
r поливинилхлоридные плитки f pl (для полов)

V177 e **viscoelastic strain**
d viskoelastische Verformung f
f déformation f viscoélastique
nl visco-elastische vervorming f
r вязкоупругая деформация f

V178 e **viscosity**
d Viskosität f, Zähigkeit f
f viscosité f

nl viscositeit *f*
r вязкость *f*

V179 *e* **viscous air filter**
    d Viskositätsfilter *n*, *m*, Naßluftfilter *n*,
      m, ölbenetztes Luftfilter *n*, *m*
    f filtre *m* à air visqueux
    nl oliebad-luchtfilter *n*, *m*
    r вязкостный воздушный фильтр *m*

V180   **visionproof glass** *see* **obscure glass**

V181   **visual concrete** *see* **exposed concrete**

V182 *e* **vitrified brick**
    d 1. Klinker *m*, Straßenbauziegel *m*
      2. glasierter Ziegel *m*
    f 1. brique *f* de pavage 2. brique *f*
      vitrifiée [émaillée]
    nl 1. verglaasde steen *m* 2. geglazuurde
      steen *m*
    r 1. клинкер *m*, клинкерный кирпич *m*
      2. глазурованный кирпич *m*

V183 *e* **vitrified ceramic drain**
    d Keramikrohrdränung *f*
    f tuyau *m* céramique de drainage
    nl gresbuis *f (m)*
    r керамическая дрена *f*

V184 *e* **V-notch weir**
    d Dreiecksüberfall *m*
    f déversoir *m* en V [en forme triangu-
      laire]
    nl stuw *m* met driehoek-overlaat
    r водослив *m* треугольного профиля

V185   **V-joint** *see* **V-shaped joint**

V186 *e* **void-cement ratio**
    d Hohlraum-Zement-Verhältnis *n*
    f rapport *m* vides/ciment
    nl verhouding *f* cement-holle ruimten
    r отношение *n* объёма пор (*в бетоне,
      растворе*) к объёму цемента

V187 *e* **void filler**
    d porenschließender Sperrstoff *m*, Po-
      renfüller *m*, Porenfüllstoff *m*
    f bouche-pores *m*
    nl poriënvuller *m*
    r шпаклёвка *f*, шпатлёвка *f* (*для за-
      делки пор*)

V188 *e* **void formers**
    d Hohlraumbildner *m pl*
    f noyaux *m pl* incorporés dans le moule
    nl poriënvormers *m pl*
    r пустотообразователи *m pl*

V189 *e* **void pockets**
    d Lunker *m pl*
    f cavités *f pl*, cavernes *f pl*
    nl holtes *f pl*
    r раковины *f pl*

V190 *e* **voids**
    d Hohlräume *m pl*; Poren *f pl*
    f vides *m pl*; pores *m pl*
    nl holtes *f pl*, holle ruimten *f pl*; poriën
      f pl
    r пустоты *f pl*; поры *f pl*

V191 *e* **voids content**
    d Porengehalt *m*
    f teneur *m* en vides

nl volume *n* van de holle ruimten
r пористость *f*

V192 *e* **voids ratio**
    d Porenziffer *f*
    f rapport *m* volumétrique des vides
    nl poriënfactor *m*
    r коэффициент *m* пустотности

V193 *e* **voids volume**
    d Porenvolumen *n*, Porenraum *m*
    f volume *m* des vides
    nl poriënvolume *n*
    r объём *m* пор

V194 *e* **volume batching**
    d Volumendosierung *f*, Raumteilmi-
      schung *f*, Raumzumessung *f*
    f dosage *m* volumétrique
    nl volumedosering *f*
    r дозирование *n* по объёму

V195 *e* **volume curve**
    d Volumenlinie *f*, Inhaltslinie *f (Tal-
      sperre)*
    f courbe *f* hauteur-volume [de jaugeage
      d'un réservoir]
    nl inhoudscurve *f*
    r кривая *f* объёмов (*водохранилища*)

V196   **volume of runoff** *see* **mass runoff**

V197 *e* **volume ratio**
    d Mischungsverhältnis *n* nach Raumin-
      halt
    f rapport *m* volumétrique des consti-
      tuants d'un mélange
    nl volume-verhouding *f*, samenstelling *f*
      in volumedelen
    r объёмное соотношение *n* компонентов
      смеси

V198 *e* **volumetric batch plant**
    d Raumteildosieranlage *f*
    f doseur *m* volumétrique
    nl volumedoseerinrichting *f*
    r установка *f* объёмного дозирования,
      объёмный дозатор *m*

V200 *e* **vortex air diffuser**
    d Drall(düsen-Luft)auslaß *m*, Luftwirb-
      ler *m*
    f diffuseur *m* d'air à tourbillonnement
    nl wervelende luchtverdeler *m*
    r вихревой воздухораспределитель *m*

V201 *e* **vortex air supply panel**
    d Luftverteilungswirbelplatte *f*
    f panneau *m* de distribution d'air à
      tourbillonnement
    nl verwellucht-inlaatrooster *n*
    r вихревая воздухораспределительная
      панель *f*

V203 *e* **vortex generator**
    d Drallerzeuger *m*, Drallkörper *m*,
      Dralleinsatz *m*
    f générateur *m* de tourbillons
    nl wervelvormer *m*
    r закручиватель *m* (воздушного) по-
      тока

V204 *e* **vortex reactor**
    d Wirbelreaktor *m*

    f réacteur *m* de tourbillons
  nl wervelreactor *m*
   r вихревой реактор *m*

V205 *e* **vortex tube**
   d Wirbelrohr *n*
   f tube *m* vortex
  nl wervelpijp *f* (*m*)
   r вихревая труба *f*

V206 *e* **voussoir**
   d keilförmiger Stein *m*
   f voussoir *m*
  nl radiaalsteen *m*, sluitsteen *m*
   r клинчатый замковый камень *m*, камень *m* свода

V207 *e* **V-shaped joint**
   d V-förmige Ausfugung *f* (*Mauerwerk*)
   f joint *m* creux en V (*de la maçonnerie*)
  nl V-vormige voeg *f* (*m*) (*metselwerk*)
   r шов *m* (*кладки*) с остроугольной расшивкой

W

W1 *e* **waffle slab**
   d Kassettenplatte *f*
   f dalle *f* [plaque *f*] caisson(s), dalle *f* alvéolaire
  nl cassettenvloer *m*
   r железобетонная плита *f* с перекрёстными рёбрами, кессонная плита *f* перекрытия

W2 *e* **Wagner fineness**
   d Mahlfeinheit *f* nach Wagner, Wagnersche Mahlfeinheit *f*
   f finesse *f* de broyage [de mouture] d'après Wagner
  nl maalfijnheid *f* naar Wagner
   r тонкость *f* помола (*цемента*) по Вагнеру

W3 *e* **wagon ceiling**
   d gewölbte Decke *f*
   f plafond *m* voûté
  nl gewelfd plafond *n*
   r сводчатый потолок *m*

W4 *e* **wagon drill**
   d Bohrwagen *m*
   f jumbo *m* pneumatique, foreuse *f* mobile
  nl mobiele grondboorinstallatie *f*
   r буровая тележка *f* для пневматического бурения

W5 *e* **wainscot**
   d Lamberie *f*, Lambris *m*, Sokkeltäfelung *f*
   f boiserie *f* murale, lambris *m*
  nl houten wandbekleding *f*, lambrizering *f*
   r панельная обшивка *f* (*стен, перегородок*)

W6 *e* **waisted specimen**
   d achterförmiger Probekörper *m*

   f éprouvette *f* [échantillon *m*] de mortier en forme d'un huit
  nl proefstuk *n* in de vorm van een 8 ter bepaling van de trekvastheid
   r образец *m* в форме восьмёрки, образец-восьмёрка *m* (*для испытания на растяжение*)

W7 *e* **wale** *US*, **waling** *UK*
   d Ankergurt *m*, Holm *m*
   f moise *f*, traverse *f*
  nl gording *f* van een grondkerende wand
   r горизонтальная схватка *f*, горизонтальный брус *m*, соединяющий вертикальные стойки, насадка *f*

W8 *e* **walked-on roof covering**
   d begehbare Dacheindeckung *f*
   f toiture *f* permettant le déplacement des ouvriers avec équipement léger (*p. ex. avec brouettes*)
  nl begaanbaar dak *n*
   r кровля *f*, допускающая хождение по ней рабочих с лёгким оборудованием (*напр. с тачками*)

W9   **walker** *see* **walking excavator**

W10 *e* **walking driver**
   d Schreitramme *f*
   f sonnette *f* marcheuse
  nl trilmachine *f*, trilslede *f* (*grondverdichtingsmachine*)
   r шагающий копёр *m*

W11 *e* **walking excavator**
   d Schreitbagger *m*
   f walker-dragline *m*, excavateur *m* marcheur, dragline *m* marcheur [promneur], dragline *f* marcheuse [promneuse]
  nl stappende graafmachine *f*
   r шагающий экскаватор *m*

W12 *e* **walking line**
   d Lauflinie *f*, Ganglinie *f*, Teilungslinie *f* (*Treppe*)
   f ligne *f* de marche [de foulée, de giron]
  nl looplijn *f* (*m*) (*trap*), stijglijn *f* (*m*)
   r средняя линия *f* [ось *f*] лестничного марша

W13 *e* **walking spud**
   d Schreitpfahl *m*, linker Pfahl *m* (*Naßbagger*)
   f pieu *m* de papillonage
  nl beweegbare paal *m* (*baggermolen*)
   r папильонажная свая *f* (*земснаряда*)

W14   **walkover** *see* **pedestrian bridge**

W15 *e* **walk plank**
   d Laufbohle *f*, Karrbohle *f*
   f planche *f* pour le mouvement des brouettes
  nl loopplank *f* (*m*)
   r катальная доска *f*

W16 *e* **walks**
   d Brettersteg *m* mit Querleisten
   f planche *f* à tasseaux
  nl vonder *n*, vlonder *n*
   r сходни *pl*

W17 *e* **walk-up**
d Gebäude *n* ohne Aufzug
f bâtiment *m* sans ascenceurs
nl gebouw *n* zonder lift
r здание *n* без лифта

W18 *e* **walkway**
d 1. Laufsteg *m* 2. Parksteig *m*, Parkweg *m* 3. Fußgängerübergang *m*
f 1. passerelle(s) *f* (*pl*), planche *f* à tasseaux (*sur toit*) 2. allée *f* de parc 3. passage *m* pour piétons
nl 1. vonder *n*, vlonder *n* 2. wandelweg *m* 3. voetbrug *f* (*m*)
r 1. переходный мостик *m*; мостки *pl* 2. дорожка *f* (*в парке*) 3. пешеходный переход *m*

W19 *e* **wall**
d 1. Mauer *f*, Wand *f* 2. Abschirmung *f*; Schutzwand *f*; Schleier *m*
f 1. mur *m*, muraille *f*, paroi *f* 2. écran *m*
nl 1. muur *m*, wand *m* 2. scherm *n*, kade *f* (*m*), wal *m*
r 1. стена *f* 2. экран *m*; защитная перегородка *f*

W20 *e* **wall air conditioner**
d Wandklimagerät *n*
f conditionneur *m* d'air du type mural
nl aan de wand te bevestigen airconditioning-installatie *f*
r настенный кондиционер *m*

W21 *e* **wall anchor**
d Balkenanker *m*
f ancre *m* mural [de mur]
nl muuranker *n*, balkanker *n*
r анкер *m* для крепления балки к стене

W22 *e* **wall base**
d Wandsockel *m*
f socle *m*, soubassement *m*, mur *m* de soubassement
nl funderingsmuur *m*
r цоколь *m* стены

W23 **wall-bearing partition** *see* **load bearing partition**

W24 *e* **wallboard**
d 1. Gipskartonplatte *f*, Trockenputz *m* 2. Wandelement *n*, Wand(bau)platte *f*
f 1. plâcoplatre *m* 2. panneau *m* mural [de mur]
nl 1. pleisterplaat *f* (*m*) 2. wandplaat *f* (*m*)
r 1. листовая сухая штукатурка *f* 2. стеновая панель *f*

W26 *e* **wall bracket**
d Wandhalterung *f*
f console *f* fixée au mur
nl wandconsole *f*
r стеновая консоль *f*, стеновой кронштейн *m*

W27 *e* **wall-breakwater**
d Mauer-Wellenbrecher *m*
f brise-lames *m* à parois verticales

nl havenmuur *m*, golfbreker *m* met verticale zijvlakken
r волнолом *m* вертикального профиля

W28 *e* **wall channeler**
d Mauerfräse *f*
f machine *f* pour entailler les rainures dans les murs
nl muurvrees *f* (*m*)
r ручная машина *f* для фрезерования пазов в стене

W29 *e* **wall chase**
d Mauernut *f*
f rainure *f* dans un mur en maçonnerie
nl sleuf *f* (*m*) in een muur
r паз *m* [жёлоб *m*] в стене

W30 *e* **wall column**
d Wandsäule *f*, Halbsäule *f*, eingebundene Säule *f*
f colonne *f* adossée; colonne *f* engagée [fasciculée]
nl halfzuil *f* (*m*)
r колонна *f*, полностью *или* частично заделанная в стену, полуколонна *f*

W31 *e* **wall component**
d Wand(bau)element *n*, Wandfertig-(bau)element *n*
f composant *m* de paroi, élément *m* préfabriqué de paroi
nl wandelement *n*
r стеновой сборный элемент *m*, сборный элемент *m* стеновой конструкции

W32 **wall covering** *see* **wall facing**

W33 *e* **wall crane**
d Konsolkran *m*, Wandkran *m*
f grue *f* murale
nl muurkraan *f* (*m*)
r настенный консольный поворотный кран *m*

W34 *e* **wall crossing**
d Durchführung *f*, Wanddurchgang *m*
f traversée *f* d'une canalisation à travers d'un mur
nl muurdoorvoering *f*
r узел *m* пересечения стены (*трубопроводом, воздуховодом*)

W35 *e* **waller's hammer**
d Abputzhammer *m*, Maurerhammer *m*
f marteau *m* de maçon
nl kaphamer *m*
r кирочка *f* [молоточек *m*] каменщика

W36 *e* **wall facing**
d Wandverkleidung *f*, Wandbekleidung *f*
f revêtement *m* mural, parement *m* (d'un mur)
nl muurbekleding *f*
r облицовка *f* стены

W37 *e* **wall footing**
d Wandfundament *n*, Mauergründungskörper *m*
f fondation *f* de mur [de paroi]
nl voetmuur *m*, fundering *f*
r фундамент *m* стены

W38 *e* **wall form**
 d Wandschalung *f*
 f coffrage *m* mural, coffrage-paroi *m*
 nl wandbekisting *f*
 r стеновая опалубка *f*, опалубка *f* дл бетонирования стен

W39 *e* **wall handrail**
 d Handlauf *m* an der Treppenraumwand
 f main-courante *f* (*d'un escalier*) murale
 nl muurleuning *f*
 r лестничный поручень *m*, прикрепляемый к стене

W40 *e* **wall heating panel**
 d Heizwandplatte *f*
 f panneau *m* de chaufage de mur, panneau *m* chauffant mural
 nl wandradiator *m*, plaatradiator *m*
 r стеновая отопительная панель *f*

W41 *e* **wall-hung water closet**
 d wandangehängter Abort *m*
 f water-closet *m* mural
 nl wandcloset *n*
 r настенный унитаз *m*

W42 *e* **walling**
 d 1. Wände *f pl*, Wandung *f* 2. Wanderrichtung *f* 3. Wandbaumaterialien *n pl*
 f 1. murs *m pl*, parois *f pl* (*d'un bâtiment*) 2. murage *m*, construction *f* de murs (*en maçonnerie ou en briques*) 3. matériaux *m pl* pour la construction des murs
 nl 1. muren *m pl* en wanden *m pl* 2. bouwen *n* van muren 3. bouwmaterialen *n pl* voor muren
 r 1. стены *f pl*, стеновые конструкции *f pl* 2. возведение *n* стен 3. стеновые материалы *m pl*

W43 *e* **wall in trench**
 d Schlitzwand *f*
 f paroi *f* moulée dans le sol
 nl diepwand *m*
 r стена *f* в грунте

W44 *e* **wall iron**
 d Traggabel *f*, Rohrschelle *f* für Befestigung der Regenfallrohre an der Außenwand
 f collet *m* de la descente pluviale
 nl muurbeugel *m* (voor regenpijp)
 r ухват *m* для крепления водосточных труб

W45 *e* **wall line**
 d Mauerflucht *f*
 f alignement *m* de façade (*d'un mur*)
 nl rooilijn *f* (*m*)
 r линия *f* по наружной грани стены (*в плане*), красная линия *f*

W46 *e* **wall mounted heater**
 d Wandluftheizgerät *n*
 f calorifière *m* à air chaud mural
 nl wandluchtverhitter *m*
 r пристенный агрегат *m* воздушного отопления

W47 *e* **wall opening**
 d Wanddurchbruch *m*, Maueröffnung *f*
 f baie *f* [ouverture *f*] de mur
 nl muuropening *f*
 r стеновой проём *m*

W48 *e* **wall outlet**
 d Wand-Einbausteckdose *f*
 f prise *f* de courant mural
 nl ingebouwde wandcontactdoos *f* (*m*)
 r стеновая электророзетка *f* (*заподлицо со стеной*)

W49 *e* **wall panel**
 d Wandscheibe *f*, Wandplatte *f*; Vertäfelung *f*, Wandtafel *f*
 f panneau *m* mural [de mur]; panneau *m* mural de remplissage [de revêtement], mur-rideau *m*
 nl wandpaneel *n*
 r стеновая панель *f*; стеновая облицовочная панель *f*

W50 *e* **wall paper**
 d Tapete *f*, Papiertapete *f*
 f papier-tenture *m*, papier *m* peint [de tenture]
 nl behangsel(papier) *n*
 r обои *pl*

W51 *e* **wallpapering work**
 d Tapezierarbeit *f*
 f collage *m* [encollage *m*] de papier-tenture
 nl behangen *n*
 r оклейка *f* обоями

W52 *e* **wall plate**
 d 1. Mauerlatte *f* 2. Streichbalken *m* 3. Fußpfette *f*, Traufpfette *f*
 f 1. sablière *f* 2. lambourde *f*, solive *f* 3. panne *f* inférieure [sablière]
 nl 1. muurplaat *f* (*m*) 2. vloerrib *f* (*m*) 3. voetgording *f*
 r 1. мауэрлат *m* 2. брус *m*, прилегающий к стене *или* лежащий на выступе стены 3. карнизный прогон *m*, прогон *m* у свеса кровли

W53 *e* **wall-pointing machine**
 d Mauerfuggerät *n*
 f machine *f* à jointoyer [à lisser les joints], jointoyeur *m* mécanique
 nl voegmachine *f*
 r механизированный инструмент *m* для расшивки швов

W54 *e* **wall run**
 d provisorische Maueröffnung *f*
 f ouverture *f* [baie *f*] provisoire dans le mur
 nl tijdelijke opening *f* in de muur
 r временный проём *m* в стене

W55 *e* **walls**
 d 1. Wände *f pl*, Wandung *f* 2. Grabenwandung *f*
 f 1. murs *m pl*, parois *f pl* 2. parois *f pl* verticales d'une tranchée
 nl 1. muren *m pl* 2. bekleding *f* [stutwerk *n*] van de wanden van een ingraving

    r 1. стены *f pl* (*здания*) 2. вертикаль-
ные стенки *f pl* выемки [траншеи]

W56 *e* **wall skin**
    d 1. Wanddichtung *f* 2. Wandhaut *f*
    f 1. cuvelage *m* de sous-sol, revêtement
    m d'étanchéité de la cave 2. bardage *m*
    nl 1. waterafdichting *f* van kelderwand
    2. muurbekleding *f*
    r 1. гидроизоляция *f* стен подвала
    2. стеновая обшивка *f*

W57 *e* **wall string**
    d Wandwange *f*
    f limon *m* mural [de mur], contreli-
mon *m*
    nl muurboom *m*
    r косоур *m или* тетива *f* у стены

W58 *e* **wall tap**
    d Wandhahn *m*
    f robinet *m* d'eau scellé dans le mur
    nl tapkraan *f (m)*
    r настенный водоразборный кран *m*

W59 *e* **wall tie**
    d Wandverbinder *m*
    f ancre *m* mural [de mur], attache *f*
murale
    nl muuranker *n*
    r стеновой анкер *m*, закрепа *f* для
крепления стеновой облицовки

W60 *e* **wall tile**
    d Wandfliese *f*, Wandplatte *f*
    f carreau *m* mural [de mur]
    nl wandtegel *m*
    r стеновая керамическая облицовоч-
ная плитка *f*

W61   **wall unit** *see* **wall component**

W62 *e* **wall urinal**
    d Wandurinal *n*, PP-Becken *n*, Urinal-
becken *n*
    f urinoir *m* appliqué
    nl wandurinoir *m*
    r настенный писсуар *m*

W63 *e* **wall-wash luminaire**
    d Wand-Einbauleuchte *f*
    f luminaire *m* [appareil *m* d'éclairage]
encastré dans la paroi
    nl wand-inbouwlamp *f (m)*
    r светильник *m*, встроенный в стену

W64 *e* **wane, waney edge**
    d Fehlkante *f*, Schalkante *f*
    f flache *f*
    nl wankant *f (m)*
    r обзол *m*

W65 *e* **warehouse set**
    d Ablagerung *f* (*Zement*)
    f prise *f* partielle du ciment en dépôt
    nl gedeeltelijke verharding *f* van cement
tijdens de opslag
    r частичное схватывание *n* цемента,
хранящегося на складе

W66 *e* **warm air curtain**
    d Warmluftschleier *m*
    f rideau *m* d'air chaud

    nl warme-luchtgordijn *n*, *f (m)*
    r воздушно-тепловая завеса *f*

W67 *e* **warm air duct**
    d Warmluftkanal *m*, Warmluftleitung *f*
    f gaine *f* d'air chaud
    nl warme-luchtkanaal *n*
    r воздуховод *m* тёплого воздуха

W68 *e* **warm air heating**
    d Warmluftheizung *f*
    f chauffage *m* par air chaud
    nl luchtverwarming *f*
    r воздушное отопление *n*

W69 *e* **warm air heating unit**
    d Luftheizgerät *n*
    f installation *f* de chauffage par air
chaud, générateur *m* d'air chaud
    nl luchtverwarmingsapparaat *n*
    r агрегат *m* воздушного отопления

W71 *e* **warm water method**
    d abgekürzte Festigkeitsprüfung *f* des
Betons (*nach 24-stündiger Warmwas-
serbehandlung*)
    f méthode *f* accélérée d'essai de béton
(*après la cure des eprouvettes dans
l'eau chaude pendant 24 h.*)
    nl bepaling *f* van de drukvastheid van
beton na een 24-uurs-warmwaterbe-
handeling
    r ускоренное определение *n* прочности
бетона (*после суточного выдерживания в тёплой воде*)

W72 *e* **warning system**
    d Alarmanlage *f*
    f systéme *m* d'alarme
    nl alarmsysteem *n*
    r система *f* аварийной сигнализации

W73 *e* **warp**
    d Verwerfung *f*, Verziehen *n*
    f gauchissement *m*
    nl scheluwtrekken *n*
    r коробление *n* (*древесины*)

W74   **warping stresses** *see* **temperature stresses**

W75 *e* **Warren girder, Warren truss**
    d Warrenträger *m*
    f poutre *f* à triangulation Warren
    nl Warren-ligger *m*
    r ферма *f* с параллельными поясами
и треугольной решёткой

W76 *e* **wash basin**
    d Waschbecken *n*, Waschtrog *m*, Wasch-
kübel *m*
    f lavabo *m*
    nl wasbak *m*, wastafel *f (m)*
    r умывальник *m*

W77   **washboarding** *see* **waves**

W78 *e* **wash fountain**
    d Waschbrunnen *m*
    f lavabo *m* collectif
    nl gemeenschappelijke wasbak *m*
    r круглый умывальник *m* группового
пользования

W79   **washing** *see* **scour**

W80   *e* **washing and screening plant**
    d Wasch- und Siebanlage *f*
    f installation *f* de criblage et de lavage
    nl was- en zeefinstallatie *f*
    r моечно-сортировочная установка *f*

W81   *e* **washing-out** *see* **washout**

W82   *e* **washing screen**
    d Waschsieb *n*
    f crible *m* laveur
    nl waszeef *f* (*m*)
    r промывочный грохот *m*

W83   *e* **washing screw**
    d Sandschnecke *f*, Waschschnecke *f*,
      Spiral(naß)klassierer *m*, Schrauben-
      klassierer *m*
    f class(ificat)eur *m* [trieur *m*] à spirale
    nl wormclassificeerder *m*
    r спиральный классификатор *m*

W84   *e* **washing trough**
    d Reihenwaschanlage *f*
    f lavabo-auge *m*
    nl wastrog *f* (*m*)
    r многоместный рядный умывальник
      m

W85   *e* **washout**
    d 1. Auswaschung *f*, Unterwaschung *f*,
      Kolk *m*, Auskolkung *f*, Suffosion *f*,
      innere Erosion *f* 2. Ufereinsturz *m*
      3. Reinigungsöffnung *f*; Spülauslaß *m*
    f 1. suffosion *f*, érosion *f* interne
      2. effondrement *m* 3. regard *m* de net-
      toyage; débouché *m* de chasse
    nl 1. uitspoelen *n* 2. ontstoppingsope-
      ning *f* 3. verzakking *f*
    r 1. размыв *m*, суффозия *f*, глубинная
      эрозия *f* 2. обрушение *n* от размыва
      3. прочистной люк *m* (*трубопрово-*
      да); промывное отверстие *n*, про-
      мывной водовыпуск *m*

W86   *e* **washout valve**
    d Spülschieber *m*
    f vanne *f* de dévasement
    nl aftapventiel *n*
    r затвор *m* промывного шлюза

W87   *e* **wash primer**
    d Haftgrund *m*
    f wash primer *m*
    nl grondverf *f* (*m*) voor metaal
    r протравная грунтовка *f*, грунтовка *f*
      с протравой

W88   *e* **wash-water trough**
    d Spülwasserrinne *f*
    f auge *f* de lavage
    nl afvoergoot *f* (*m*)
    r промывной жёлоб *m*

W89   *e* **waste**
    d 1. Ablauf *m* (*konstruktives Element*)
      2. Spülicht *n* 3. Abfall(stoff) *m*; Bau-
      schutt *m* 4. überschüssiger Aushub-
      (boden) *m* 5. Wasserverlust *m* (*im*
      Verteilungsnetz)
    f 1. vidange *m*, bonde *f* 2. eaux *f pl*
      ménagères [domestiques] 3. déchets

m pl, rési**dus** *m pl*; gravats *m pl*,
gravois *m pl*, détritus *m* 4. déblais
m pl [matériau *m* excavé] en excé-
dant 5. pertes *f pl* d'eau
    nl 1. afwatering *f* 2. huishoudelijk afval-
      water *n* 3. rommel *m* 4. overtollige
      grond *m* 5. waterverliezen *n pl*
    r 1. слив *m* (*санитарного прибора*)
      2. бытовые сточные воды *f pl* 3. от-
      ходы *m pl*; строительный мусор *m*
      4. неиспользованный грунт *m* 5. по-
      тери *f pl* воды *в сети водоснабжения*)

W90   *e* **waste appliances**
    d Sanitärgeräte *n pl* für Küche und Bad
    f appreils *m pl* sanitaires pour la cui-
      sine et salle de bain
    nl garnituur *n* voor de afvoer van huis-
      houdelijk afvalwater
    r приёмники *m pl* нефекальных быто-
      вых стоков

W91   **waste bank** *see* **spoil bank**

W92   **waste canal** *see* **floodway**

W93   *e* **waste-disposal unit, waste-food grinder**
    d Abfallzerkleinerer *m*, Hausmüllzer-
      kleinerer *m*
    f broyeur *m* d'ordures domestiques [mé-
      nagères], broyeur *m* évier
    nl afvalvermaler *m*
    r мусородробилка *f*

W94   *e* **waste pipe**
    d Bad- und Küchenabwasserrohr *n*
    f descente *f* [tuyau *m* de descente]
      d'eaux ménagères
    nl aansluitleiding *f*
    r канализационный стояк *m* для не-
      фекальных бытовых стоков

W96   *e* **waste sluice gate**
    d Freilaufschütz(e) *n* (*f*), Leerlauf-
      schütz(e) *n* (*f*)
    f vanne *f* compensatrice
    nl vrijloopschuif *f* (*m*)
    r затвор *m* холостого сброса

W97   **waste stack** *see* **waste pipe**

W98   *e* **waste trap**
    d Geruchverschluß *m*
    f siphon *m*, coupe-air *m*
    nl stankafsluiter *m*, waterslot *n*
    r сифон *m*, водяной затвор *m*

W99   *e* **waste water**
    d Abwasser *n*; gebrauchtes Wasser *n*;
      Ablaufwasser *n*
    f eau *f* résiduaire; eau *f* d'échappement;
      eaux *f pl* pluviales
    nl rioolwater *n*
    r отработанная вода *f*; сточные воды
      f pl

W100   *e* **waste water outlet**
    d Leerschuß *m* des Wasserkraftwerkes
    f déversoir *m* du déchargeur [de sécu-
      rité]
    nl waterlozing *f* van de waterkrachtcen-
      trale
    r холостой водосброс *m* ГЭС

W101 *e* **waste water spillway**
 d Entlastungsüberfall *m*
 f déversoir *m* de réglage
 nl aflaatkanaal *n*, overlaat *m*
 r холостой водослив *m*

W102 *e* **wasteway**
 d 1. Hochwasserüberlauf *m*, Hochwasserentlastungsanlage *f* 2. Entlastungsrinne *f*
 f 1. déversoir *m* 2. canal *m* d'évacuation
 nl 1. overlaat *m* 2. afleidingskanaal *n*
 r 1. водосброс *m*, водослив *m* 2. сбросное [отводящее] русло *n*

W103 *e* **wasting**
 d rohes Behauen *n* des Steinblockes
 f dégrossissage *m* des blocs en pierre
 nl kloven *n* van ruwe natuursteen
 r оспицовка *f* [приближённая ударная обработка *f*] камня

W104 *e* **watchman services**
 d Wachdienst *m*
 f service *f* de garde des chantiers
 nl terreinbewaking *f*
 r служба *f* охраны объектов строительства *или* стройплощадок

W105 *e* **water after-treatment**
 d Feuchthalten *n*, Anfeuchten *n*, Naßhalten *n* (*Betonnachbehandlung*)
 f traitement *m* (ultérieur) du béton à l'eau, humidification *f* du béton
 nl nathouden *n* [vochtighouden *n*] van beton
 r увлажнение *n* бетона (*способ ухода за бетоном*)

W106 *e* **water-air heating system**
 d Wasserluftheizung *f*
 f installation *f* de chauffage à eau et air chaud
 nl water-luchtverwarmingssysteem *n* met water als warmtemedium
 r водовоздушная система *f* отопления

W107 *e* **water-bearing ground**
 d 1. wassergesättigter Boden *m* 2. Grundwasserleiter *m*
 f 1. sol *m* aquifère [saturé], terre *f* saturée 2. nappe *f* aquifère
 nl 1. met water verzadigde grond *m* 2. grondwatervoerende grondlaag *f* (*m*)
 r 1. водонасыщенный грунт *m* 2. водоносный горизонт *m*

W108 *e* **water bearing rock**
 d wassergesättigtes Felsgestein *n*
 f roche *f* aquifère
 nl met water verzadigde rots *f* (*m*)
 r водонасыщенная скальная порода *f*

W109 *e* **water binder ratio**
 d Wasser-Bindemittel-Faktor *m*
 f rapport *m* eau-liant
 nl water-cementfactor *m*
 r водовяжущее отношение *n*

W110 *e* **water-bloom**
 d Wasserblüte *f*
 f floraison *f* [fleuraison *f*] de l'eau
 nl algengroei *f*
 r цветение *n* воды

W111 *e* **water body**
 d Wasserkörper *m*, Gewässer *n*
 f pièce *f* d'eau
 nl watermassa *f*, waterberging *f*
 r водный объект *m*; водоём *m*; водная масса *f*

W112 *e* **water booster system**
 d Druckerhöhungsanlage *f*
 f installation *f* de surpression, surpresseur *m*, pompe-relais *f*, pompe *f* de renfort
 nl drukverhogingsinstallatie *f*
 r установка *f* повышения давления (*во внутренней водопроводной сети*)

W113 *e* **water-borne dumper**
 d schwimmender Kippentlader *m* (*für den Wasserbau*), Steinschütter *m*
 f chaland *m* basculeur [à clapets] (*à matériaux rocheux*)
 nl onderlosser *m*
 r судно *n* для отсыпки камня в воду

W114 *e* **water-borne equipment**
 d schwimmende Bauausrüstung *f* [Baugeräte *n pl*]
 f matériel *m* [équipement *m*] flottant
 nl drijvend bouwgereedschap *n*
 r оборудование *n* для работы на плаву, плавучее оборудование *n*, плавсредства *n pl*

W116 *e* **water boundary**
 d Uferlinie *f*, Wasser(spiegel)rand *m*
 f bord *m* d'eau
 nl oeverlijn *f* (*m*), waterrand *m*
 r урез *m* воды

W117 *e* **water-bound macadam**
 d wassergebundene Schotterdecke *f*, wassergebundener Makadam *m*
 f macadam *m* [empierrement *m*] à l'eau
 nl macadam *n*, watergebonden macadam *n*
 r водосвязное щебёночное покрытие *n*

W118 *e* **water cadastre**
 d Wasserkataster *m*, *n*, Wassergrundbuch *n*
 f cadastre *f* hydraulique
 nl waterkadaster *n*
 r водный кадастр *m*

W119 **water capacity** *see* **water-holding capacity**

W120 *e* **water-carrying capacity**
 d Leistungsfähigkeit *f*, Durchlaßkapazität *f*, Durchlaßfähigkeit *f*
 f capacité *f* de transport [d'écoulement] de l'eau
 nl afvoercapaciteit *f*
 r водопропускная способность *f*

W121 *e* **water-cement ratio**
 d Wasser-Zementfaktor *m*
 f rapport *m* eau-ciment
 nl water-cementfactor *m*
 r водоцементное отношение *n*

W122 *e* **water channel**
  d 1. Wassernase *f* 2. Wasserkanal *m*
  f 1. larmier *m*, goutte *f* d'eau, chasse-
  -goutte *m* 2. conduit *m* d'eau
  nl 1. waterbord *n*, waterhol *n* 2. water-
  kanaal *n*
  r 1. слезник *m* 2. гидролоток *m*; гид-
  роканал *m*

W123 *e* **water closet**
  d Abort *m*, Wasserklosett *n*
  f water-closet *m*, cabinet *m* d'aisances
  nl watercloset *n*
  r уборная *f*, туалет *m*

W124 *e* **water collector**
  d Wassersammelanlage *f*, Wassersammler *m*
  f collecteur *m* d'eau
  nl verzamelbekken *n*
  r водосборное сооружение *n*, водосборник *m*

W125 *e* **water conduit**
  d Zubringerleitung *f*, Wasserleitung *f*
  f conduite *f* [ligne *f*] d'eau
  nl waterleiding *f*
  r водовод *m*

W126 *e* **water consumption**
  d Wasserverbrauch *m*
  f consommation *f* d'eau
  nl waterverbruik *n*
  r водопотребление *n*

W127 **water consumption rate** *see* **water-demand rate**

W128 *e* **water contained in aggregates**
  d Eigenfeuchtigkeit *f* der Betonzuschläge
  f humidité *f* propre des granulats
  nl watergehalte *n* van de toeslagen
  (*beton*)
  r природная влажность *f* заполнителей
  (бетона)

W129 *e* **water control**
  d Wasser-Abflußregelung *f*
  f aménagement *m* de l'écoulement
  nl beheersing *n* van de waterafvoer
  r регулирование *n* водного режима
  и водного баланса

W130 *e* **water cooled room air-conditioner**
  d wassergekühltes Raumklimagerät *n*
  f conditionneur *m* de l'air de chambre
  à l'eau, climatiseur *m* de chambre
  avec refroidissement par eau
  nl klimaatregelingstoestel *n* met water-
  koeling
  r комнатный кондиционер *m* с водяным охлаждением

W131 *e* **watercourse**
  d Wasserlauf *m*; Kanal *m*
  f cours *m* d'eau
  nl waterloop *m*; kanaal *n*
  r водоток *m*; русло *n*; канал *m*

W132 *e* **water creep**
  d Durchsickerung *f* unter der Bauwerk-
  sohle; Umsickerung *f* des Bauwerkes

  f infiltration *f* de l'eau sous la semelle
  ou en détour d'un bâtiment
  nl optrekken *n* van water
  r фильтрация *f* воды в основании *или*
  в обход сооружения

W133 *e* **water cushion**
  d Wasserpolster *n*, *m*
  f matelas *m* d'eau
  nl waterkussen *n*
  r водобойный колодец *m*; водяная подушка *f*

W134 *e* **water demand**
  d Wasserbedarf *m*
  f besoins *m pl* en eau
  nl waterbehoefte *f*
  r потребность *f* в воде; расход *m* воды

W135 *e* **water-demand rate**
  d spezifischer Wasserbedarf *m*,
  Wasserverbrauchswert *m*
  f besoins *m pl* en eau par habitant par
  jour
  nl gemiddeld waterverbruik *n* per per-
  soon per dag
  r норма *f* водопотребления

W136 **water-development works** *see* **hydraulic
  works**

W137 *e* **water disposal**
  d Wasserableitung *f*
  f évacuation *f* des eaux
  nl rioolwaterafvoer *m*
  r водоотвод *m*, водоотведение *n*

W138 *e* **water distribution installation**
  d Wasserverteilungsanlage *f*
  f installation *f* de distribution d'eau
  nl waterleidingbedrijf *n*
  r система *f* водораспределения

W139 *e* **water diversion gallery**
  d Wasserumleitungsstollen *m*
  f galerie *f* d'évacuation d'eau, galerie *f*
  de fuite
  nl waterafleidingstunnel *m*
  r деривационная галерея *f*, деривационный туннель *m*

W140 *e* **water-diverting structure**
  d Wasserableitungsbauwerk *n*
  f ouvrage *m* de dérivation
  nl afleidingswerk *n*
  r деривационное сооружение *n*

W142 **water economy** *see* **water management**

W143 **water edge** *see* **water boundary**

W144 **water engineering** *see* **hydraulic engineering**

W145 *e* **water extraction**
  d Wassergewinnung *f*
  f prise *f* d'eau
  nl waterwinning *f*
  r водозабор *m* (*из источника водоснабжения*)

W146 *e* **water face**
  d wasserseitige Böschung *f*, Oberwasserböschung *f*
  f talus *m* amont

nl bovenstroomse zijde *f* van een stuwdam
r верховой откос *m* дамбы

W147 *e* **water-filled roof**
d wasserbeschichtetes Dach *n*, Wannen-
dach *n*, Wasserdach *n*
f toit *m* plat à accumulation d'eau,
couverture *f* recouverte d'une couche
d'eau
nl bevloeid [blankstaand] dak *n*
r водонаполненная [заливная] кровля
f

W148 *e* **water-filled steel framed structure**
d wassergefülltes Stahlrahmentragwerk
n
f charpente *f* métallique irriguée
nl met water gevuld staalskelet *n*
r водонаполненный рамный стальной
каркас *m* коробчатого сечения (*для
повышения огнестойкости*)

W149 *e* **water filter**
d Wasserfilter *n*, *m*
f filtre *m* à eau
nl waterfilter *n*, *m*
r фильтр *m* для очистки воды

W150 *e* **water filtration plant**
d Wasserfilteranlage *f*
f station *f* de filtration [de filtrage]
d'eau
nl waterfiltreerinstallatie *f*
r станция *f* водоочистки; фильтровальная станция *f*

W151 *e* **water fittings**
d Wasserarmaturen *f pl*
f appareils *m pl* de robinetterie
nl fittingen *f pl* voor waterleidingen
r водопроводно-канализационная арматура *f*

W152 **water flow** *see* **discharge**

W153 *e* **water flow meter**
d Wassermesser *m*
f compteur *m* d'eau
nl watermeter *m*
r водомер *m*, расходомер *m* воды

W154 *e* **waterfront**
d 1. Wasserfront *f* 2. Wasserfassade *f*
f front *m* d'eau
nl waterfront *n*
r 1. причальный [напорный] фронт *m*;
береговая линия *f*, урез *m* воды
2. фасад *m*, обращённый к воде

W155 **water gain** *see* **bleeding**

W156 *e* **water gallery**
d Wasserstollen *m*
f galerie *f* d'eau
nl waterlozingstunnel *m*, watergang *m*
r гидротехнический туннель *m*

W157 *e* **water glass**
d 1. Wasserglas *n* 2. Wasserstandglas *n*,
Wasserstandrohr *n*
f 1. verre *f* soluble 2. tube *f* de niveau
d'eau, tube *f* de verre d'indicateur
de niveau d'eau

nl 1. waterglas *n* 2. peilbuis *m*
r 1. жидкое стекло *n* 2. водомерное
стекло *n*, водомерная трубка *f*

W158 *e* **water glass paint**
d Silikatfarbe *f*, Wasserglasfarbe *f*,
S-Farbe *f*
f peinture *f* aux silicates
nl silikaatverf *f*, waterglasverf *f*
r силикатная краска *f*, краска *f* на
жидком стекле

W159 *e* **water hammer**
d Wasserschlag *m*, Wasserstoß *m*
f coup *m* de bélier
nl waterslag *m*
r гидравлический удар *m*

W160 *e* **water hardness**
d Wasserhärte *f*
f dureté *f* de l'eau
nl waterhardheid *f*
r жёсткость *f* воды

W161 *e* **water heater**
d Warmwasserbereiter *m*
f chauffe-eau *m*, réchauffeur *m* d'eau
nl warmwatertoestel *n*
r водонагреватель *m*, водоподогреватель *m*

W162 *e* **water-holding capacity**
d Wasserhaltekapazität *f*
f capacité *f* [pouvoir *m*] de rétention
d'eau
nl maximale wateropnamecapaciteit *f*
r влагоёмкость *f*, водоудерживающая
способность *f* (*напр. грунта*)

W163 **water inflow** *see* **inflow**

W164 **water inlet** *US see* **water intake**

W165 **water installation** *see* **water service 1.**

W166 *e* **water intake**
d Wasserfassung *f*, Wasserentnahme *f*;
Einlaufbauwerk *n*
f ouvrage *m* d'entrée [de prise d'eau];
prise *f* d'eau
nl inlaatwerk *n*; watervang *m*, water-
inlaat *m*
r водозабор(ное сооружение *n*) *m*; водоприёмник *m*

W167 *e* **water intake sill**
d Entnahmeschwelle *f*
f seuil *m* de prise
nl waterinlaatgrens *f* (*m*)
r порог *m* водозабора

W168 *e* **water-intake tunnel**
d Wassereinlaufstollen *m*, Wasserfas-
sungsstollen *m*
f tunnel *m* de prise d'eau
nl waterinlaattunnel *m*
r водозаборный туннель *m*

W169 **water intake works** *see* **intake structure**

W170 **water-jet driving** *see* **water jetting 2.**

W171 *e* **water-jet pump**
d Wasserstrahlpumpe *f*
f pompe *m* à jet d'eau
nl waterstraalpomp *f* (*m*)

r водоструйный насос *m*, гидроэлеватор *m*

W172 *e* **water jetting**
 d 1. Spülabbau *m* 2. Einspülen *n* (*Pfahl, Spundbohle*)
 f 1. excavation *f* par jets d'eau 2. fonçage *m* de pieux à la lance [au jet d'eau]
 nl 1. spuiten *n* (*met water*) 2. inspuiten *n* (*paal, damplank*)
 r 1. гидромониторная разработка *f* грунта 2. погружение *n* (*свай, шпунта*) подмывом

W173 *e* **water level**
 d 1. Wasserhöhe *f*, Wasserstand *m*, Wasserspiegel *m* 2. Wasserwaage *f*
 f 1. niveau *m* [plan *m*] d'eau 2. niveau *m* (*instrument*)
 nl 1. waterstand *m* 2. waterpas *n* (*instrument*)
 r 1. уровень *m* воды, водное зеркало *n* 2. уровень *m* (*инструмент*)

W174 **water level control dam** *see* **barrage**

W175 *e* **water level width**
 d Wasserspiegelbreite *f*
 f largeur *f* au plan d'eau
 nl breedte *f* over de waterspiegel
 r ширина *f* водотока по урезу воды

W176 **water lime** *see* **hydraulic lime**

W177 *e* **water line**
 d Wasserleitung *f*
 f conduite *f* [conduit *m*] d'eau
 nl waterleiding *f*
 r водовод *m*, водопровод *m*

W178 *e* **water line network**
 d Wasserleitungsnetz *n*
 f réseau *m* de distribution d'eau
 nl waterleidingnet *n*
 r водопроводная сеть *f*

W179 *e* **waterlogged soils**
 d wasserführende Böden *m pl*
 f sol *m* aquifère
 nl waterhoudende gronden *m pl*
 r водонасыщенные [водоносные] грунты *m pl*

W180 **water-loss shrinkage** *see* **drying schrinkage**

W181 *e* **water main**
 d Wasserversorgungsleitung *f*, Hauptwasserleitung *f*
 f conduite *f* principale d'eau
 nl hoofdleidingnet *n*
 r водопроводная магистраль *f*

W182 *e* **water management**
 d Wasserwirtschaft *f*
 f économie *f* hydraulique, gestion *f* des eaux
 nl waterhuishouding *f*
 r водное хозяйство *n*; использование *n* вод

W183 *e* **water meter**
 d Wasserzähler *m*, Wassermengenmeßgerät *n*, Wasseruhr *f*
 f compteur *m* hydraulique [d'eau]
 nl watermeter *m*
 r водомер *m*; интегрирующий расходомер *m*

W184 *e* **water meter unit**
 d Wasserzählanlage *f*, Wasserzählergarnitur *f*
 f bloc *m* des compteurs hydrauliques
 nl groep *f* (*m*) watermeters (in een woningcomplex)
 r водомерный узел *m*

W185 **water need** *see* **water demand**

W186 **water object** *see* **water body**

W187 *e* **water of hydration**
 d Hydratationswasser *n*
 f eau *f* d'hydratation
 nl kristalwater *n*
 r вода *f* гидратации

W188 *e* **water outlet**
 d Wasserablaß *m*, Wasserauslauf *m*, Wasserauslaß *m*, Siel *n*
 f sortie *f* d'eau
 nl waterlozing *f*, verlaat *n* (*sluis*)
 r водоспуск *m*, водовыпуск *m*

W189 *e* **water paint**
 d Wasserfarbe *f*
 f peinture *f* à l'eau
 nl waterverf *f*
 r водная краска *f*

W191 *e* **water pipe**
 d Wasserleitungsrohr *n*
 f tuyau *m* [conduite *f*] d'eau
 nl waterleidingbuis *f*(*m*)
 r водопроводная труба *f*

W192 *e* **water post**
 d Wasserentnahmesäule *f*, Zapfständer *m*, Ventilbrunnen *m*
 f borne-fontaine *f*, borne *f* fontaine
 nl drinkwaterfontein *f*
 r водоразборная колонка *f*

W194 *e* **water power dam**
 d Wasserkraft-Talsperre *f*
 f barrage-usine *f*
 nl stuwdam *m* met waterkrachtcentrale
 r станционная плотина *f*

W195 **water power development** *see* **hydroelectric development**

W196 **water power scheme** *see* **hydroelectric scheme 1.**

W197 *e* **water pressure**
 d Wasserdruck *m*
 f pression *f* d'eau
 nl waterdruk *m*
 r давление *n* воды

W198 *e* **waterproof cement**
 d wasserundurchlässiger [hydrophober] Zement *m*
 f ciment *m* à adjuvent hydrophobisant
 nl cement *n*, *m* voor waterdicht werk
 r цемент *m* с водоотталкивающей добавкой, водоотталкивающий [гидрофобный] цемент *m*

W199 *e* **waterproofing**
    d Wasserabdichtung *f*, Wassersperre *f*
    f étanchéité *f*, étanchement *m*, hydrofugation *f*
    nl waterafdichtmaken *n*
    r гидроизоляция *f*

W200 *e* **waterproofing compound**
    d Sperrzusatz(mittel *n*) *m*
    f hydrofuge *m* en pâte, pâte *f* hydrofuge, mastic *m* d'étanchéité
    nl dichtingsmengsel *n*
    r гидроизоляционный обмазочный состав *m*

W201 *e* **waterproofing course**
    d druckwasserhaltende Isolierschicht *f*, Sperrschicht *f*
    f couche *f* d'étanchéité
    nl waterkerende laag *f* (*m*)
    r гидроизоляционный [противофильтрационный] слой *m*

W202 *e* **waterproofing emulsion**
    d Sperremulsion *f*
    f émulsion *f* hydrophobisante [hydrofuge]
    nl waterafstotende emulsie *f*
    r гидрофобизирующая эмульсия *f*

W203 *e* **waterproofing felt**
    d wassersperrende Pappe *f*; wassersperrendes Papier *n*
    f feutre *m* pour étanchéité [isolant]; carton *m* d'imperméabilisation [d'étanchéité, étanche]; papier *m* pour cure du béton
    nl waterafstotend vilt *n*; waterafstotend papier *n*
    r гидроизоляционный строительный картон *m*; водоизолирующая крафт-бумага *f*

W204 *e* **waterproofing materials**
    d Sperrstoffe *m pl*, Wassersperrmittel *n pl*
    f matériaux *m pl* hydrofuges, hydrofuge *m* (de masse)
    nl dichtingsmiddelen *n pl*, middelen *n pl* voor het waterdichtmaken
    r гидроизоляционные материалы *m pl*

W205   **waterproofing paper** *see* **waterproofing felt**

W206 *e* **waterproofing work**
    d Dichtungsarbeit *f*
    f traveaux *m pl* d'étanchement [d'étanchéisation]
    nl waterdichtmaken *n*
    r гидроизоляционные работы *f pl*

W207 *e* **waterproof membrane**
    d Sperrschicht *f*; Bitumen-Dichtungsschicht *f*
    f membrane *f* d'étanchéité [imperméable]; couche *f* d'étanchéité de bitume
    nl waterdicht membraan *n*, waterafstotende bitumenlaag *f* (*m*)
    r гидроизоляционная мембрана *f*; битумный изолирующий слой *m*

W208 *e* **waterproof paint**
    d wasserisolierende Farbe *f*
    f peinture *f* hydrofuge [d'étanchéité]
    nl waterafstotende verf *f* (*m*)
    r водоизолирующая краска *f*

W209 *e* **waterproof plaster**
    d Sperrputz *m*
    f enduit *m* imperméable [hydrofuge,. étanche]
    nl waterdicht pleister *n*
    r гидроизоляционная штукатурка *f*

W210 *e* **water protection zone**
    d Wasserschutzgebiet *n*
    f zone *f* de la protection sanitaire
    nl waterwingebied *n*
    r водоохранная зона *f*

W212 *e* **water putty**
    d Dichtungskitt *m*
    f mastic *m* d'étanchéité [d'étanchement], pâte *f* hydrofuge, hydrofuge *m* en pâte
    nl dichtingskit *f* (*m*)
    r гидроизоляционная мастика *f*, водоизолирующая замазка *f*

W213 *e* **water-reducing admixture, water-reducing agent**
    d wasseranspruchmindernder Zusatz *m*
    f réducteur *m* d'eau (*adjuvant*)
    nl waterverminderende toeslag *m*
    r добавка *f*, уменьшающая водопотребность

W214 *e* **water regime**
    d Wasserregime *n*, Wasserhaushalt *m*
    f régime *m* hydraulique
    nl waterhuishouding *f*
    r водный режим *m*

W215 *e* **water removal**
    d Wasserhaltung *f*
    f évacuation *f* des eaux
    nl waterlozing *f*, bemaling *f*
    r водоотлив *m*

W216 *e* **water repellent cement**
    d hydrophober Zement *m*
    f ciment *m* ressuant
    nl waterafstotend cement *n*, *m*
    r гидрофобный цемент *m*

W217 *e* **water repellent concrete**
    d Sperrbeton *m*
    f beton *m* étanche [hydrofuge, imperméable]
    nl waterafstotend beton *n*
    r водонепроницаемый бетон *m*

W218 *e* **water-repellent facing**
    d wasserabweisende Verkleidung *f*
    f parement *m* [revêtement *m*] imperméable [étanche] à l'eau
    nl waterafstotende bekleding *f*
    r водонепроницаемая облицовка *f*

W219 *e* **water-repelling admixture**
    d hydrophobisierender Zusatz *m*
    f agent *m* répulsif à l'eau
    nl waterafstotende toevoeging *f*

r водоотталкивающая [гидрофобизирующая] добавка *f*

W220 **water requirement** *see* **water demand**

W221 *e* **water resistance**
d Wasserbeständigkeit *f*
f résistance *f* à l'eau
nl watervastheid *f*
r водостойкость *f*

W222 *e* **water-retaining admixture**
d wasserbindender Zusatz *m*
f agent *m* rétenteur d'eau
nl waterbindende toevoeging *f*
r водоудерживающая добавка *f*

W223 *e* **water-retaining capacity**
d spezifische Retention *f*, Wasserhaltevermögen *n*
f capacité *f* [pouvoir *m*] de rétention (d'eau)
nl waterbindend vermogen *n*
r водоудерживающая способность *f*, удельная водовместимость *f* (*грунта*)

W224 *e* **water retaining structure, water retaining works, water retention works**
d Stauanlage *f*, Stauwerk *n*; Wasserschutzbauwerk *n*
f ouvrage *m* de retenue; ouvrage *m* de protection [de défense]
nl stuwwerk *n*, stuw *m*; waterkering *f*
r водоподпорное *или* водозащитное сооружение *n*

W226 *e* **water scheme**
d Wasserbauprojekt *n*
f projet *m* hydraulique
nl waterbouwkundig project *n*
r проект *m* водохозяйственного комплекса *или* гидроузла

W227 *e* **water service**
d 1. Wasserleitungsanlage *f*, Wasserversorgungssystem *n* 2. Wasserversorgung *f*; Wasserversorgungsdienst *m*
f 1. conduite *f* [conduit *m*, ligne *f*] d'eau; canalisation *f* d'eau 2. distribution *f* d'eau, alimentation *f* [approvisionnement *m*] en eau
nl 1. waterleiding *f* 2. waterleidingbedrijf *n*
r 1. водопровод *m*, водовод *m* 2. водоснабжение *n*; служба *f* водоснабжения

W228 *e* **water service pipe**
d Hausanschluß *m* an die Wasserleitung *f*, Hausanschlußleitung *f*
f branchement *m* d'eau général
nl dienstleiding *f*
r ввод *m* водопровода

W229 *e* **watershed**
d 1. Wasserscheide *f* 2. Abflußgebiet *n*, Einzugsgebiet *n*
f 1. ligne *f* de partage des eaux 2. bassin *m* versant
nl 1. waterscheiding *f* 2. waterwingebied *n*

r 1. водораздел *m* 2. водосборный бассейн *m*

W230 **watershed divide** *US see* **watershed 1.**

W231 *e* **watershed management**
d Bewirtschaftung *f* eines Einzugsgebiets
f exploitation *f* d'un bassin versant
nl beheer *n* van het waterwingebied
r система *f* мероприятий, регулирующих сток с водосбора

W232 *e* **water slope** *see* **water face**

W233 *e* **water softening plant**
d Wasserenthärtungsanlage *f*
f installation *f* d'adoucissement (de l'eau)
nl wateronthardingsinstallatie *f*
r станция *f* водоумягчения

W234 *e* **watersphere**
d Kugelbehälter *m* für Wasser
f réservoir *m* sphérique à eau
nl watertoren *m* met bolvormige reservoirs
r шаровой водонапорный резервуар *m*

W235 *e* **water storage tank**
d Wasserbehälter *m*
f bac *m* [réservoir *m*] à eau
nl waterreservoir *n*
r водяной бак *m*

W236 *e* **water strainer**
d Siebwasserfilter *n*, *m*
f crépine *f* de succion
nl waterfilter *n*, *m*
r водяной сетчатый фильтр *m*

W237 *e* **water supply**
d Wasserversorgung *f*
f approvisionnement *m* [alimentation *f*] en eau
nl watervoorziening *f*
r водоснабжение *n*

W238 *e* **water supply canal**
d Bewässerungskanal *m*
f canal *m* adducteur
nl watertoevoerkanaal *n*
r обводнительный канал *m*

W239 *e* **water-supply facilities**
d Wasserversorgungsanlagen *f pl*
f installation *f* d'approvisionnement d'eau
nl installaties *f pl* voor de watervoorziening
r сооружения *n pl* и технические средства *n pl* водоснабжения

W240 *e* **water supply fittings**
d Wasserarmaturen *f pl*
f robinetterie *f* eau
nl appendages *f pl*
r водопроводная арматура *f*

W241 *e* **water supply network**
d Außenwasserleitungsnetz *n*, Wasserversorgungsnetz *n*
f réseau *m* d'alimentation [de distribution] d'eau

nl waterleidingsnet *n*
r наружная водопроводная сеть *f*

W242 *e* **water-supply point**
d Wasserzapfstelle *f*, Wasserentnahme-stelle *f*
f poste *m* d'eau
nl watertappunt *n*
r водоразборная точка *f*

W243 *e* **water-supply reservoir**
d Wasserversorgungsspeicher *m*
f retenue *f* (de barrage) destinée à l'alimentation en eau
nl watervoorzieningsreservoir *n*
r водохранилище *n* для водоснабжения

W244 *e* **water-supply scheme, water-supply system**
d Wasserversorgungssystem *n*
f système *m* de distribution d'eau
nl watervoorzieningssysteem *n*
r система *f* водоснабжения

W246 *e* **water-supply tunnel**
d Wasserleitungstunnel *m*
f galerie *f* d'alimentation en eau
nl waterleidingtunnel *m*
r водопроводный туннель *m*

W247 *e* **water-table**
d 1. Grundwasserspiegel *m* 2. Wasser-schenkel *m*
f 1. surface *f* de la nappe aquifère, niveau *m* hydrostatique, nappe *f* d'eau souterraine 2. rejéteau *m* (*d'une fondation*), chanfrein *m* du socle
nl 1. grondwaterspiegel *m* 2. waterneus *m*
r 1. зеркало *n* грунтовых вод 2. сливная плита *f*, поясок *m* (цоколя) с отливом

W249 **watertight barrier** *see* **core wall**

W250 *e* **watertight concrete**
d wasserdichter Beton *m*
f béton *m* étanche
nl waterdicht beton *n*
r водонепроницаемый бетон *m*

W251 *e* **watertight core**
d Kerndichtung *f*, Dichtungskern *m* (*Staudamm*)
f noyau *m* d'étanchéite (*d'un barrage*)
nl waterdichte kern *f* (*m*) (*stuwdam*)
r противофильтрационное ядро *n* (*плотины*)

W252 **watertight diaphragm** *US see* **impervious diaphragm**

W253 *e* **watertight facing arch**
d Abdichtungsgewölbe *n* (*Staudamm*)
f revêtement *m* étanche en voûte (*d'un barrage-voûte*)
nl waterdichte steunboog *m* (*stuwdam*)
r противофильтрационная арочная облицовка *f* (*плотины*)

W254 *e* **water-to-air ratio**
d Wasser-Luft-Zahl *f*, Sprüh-Koeffizient *m*, Berieselungszahl *f*

f rapport *m* eau-air (*conditionnement de l'air*)
nl water-luchtverhouding *f*
r коэффициент *m* орошения (*кондиционирование воздуха*)

W255 *e* **water-to-air system**
d Wasser-Luft-System *n*, Wasser-Luft--Klimaanlage *f*
f installation *f* de climatisation de l'eau à air
nl water-luchtsysteem *n*
r водовоздушная система *f* (*кондиционирования воздуха*)

W256 *e* **water-to-earth ratio**
d Wasser-Erdaushub-Volumenverhältnis *n*
f rapport *m* eau-deblai
nl water-grondverhouding *f*
r отношение *n* полезного объёма водохранилища к объёму земляных работ

W257 *e* **water-to-water heater**
d Wasser-Wasser-Wärmeübertrager *m*, wasserbeheizter Warmwasserbereiter *m*
f chaufe-eau *m* fonctionnant par circulation d'eau chaude
nl met water verwarmde boiler *m*
r водоводяной водонагреватель *m*

W258 *e* **water-to-water system**
d Wasser-Wasser-Wärmepumpe *f*, Wasser-Wasser-System *n*
f installation *f* de refroidissement *ou* de chauffage de l'eau à eau
nl water-naar-waterwarmtepomp *f* (*m*)
r водоводяная система *f* (*охлаждения или отопления — охлаждения*)

W259 *e* **water tower**
d Wasserturm *m*
f château *m* d'eau
nl watertoren *m*
r водонапорная башня *f*

W260 *e* **water treatment plant, water treatment works**
d 1. Wasseraufbereitungsanlage *f* 2. Kläranlage *f*
f 1. station *f* de traitement des eaux 2. installation *f* de purification des eaux
nl waterzuiveringsinstallatie *f*
r 1. станция *f* водоподготовки 2. водоочистная станция *f* (*водоснабжения*)

W261 *e* **waterway**
d 1. Wasserweg *m*, Wasserstraße *f* 2. Wasserlauf *m* 3. Überfallöffnung *f*, Wehröffnung *f* 4. Durchflußquerschnitt *m*, freier Querschnitt *m*
f 1. voie *f* navigable 2. cours *m* d'eau 3. pertuis *m* 4. section *f* mouillés [d'écoulement]
nl 1. waterweg *m* 2. waterloop *m*, watergang *m* 3. stuwopening *f* 4. nat profiel *n*

r 1. водный путь *m* 2. водоток *m* 3. водосливное отверстие *n* 4. живое сечение *n* (водотока)

W262 *e* **water well**
d Brunnen *m*
f puits *m* d'eau
nl bron *f(m)*, wel *f(m)*, welput *m*
r колодец *m*, скважина *f*

W263 *e* **water-wheel**
d Wasserrad *n*
f roue *f* hydraulique
nl waterrad *n*; scheprad *n*
r водоподъёмное [водяное] колесо *n*

W264 *e* **water works**
d Wasserwerke *n pl*
f ouvrages *m pl* d'adduction et de distribution d'eau
nl waterleiding *f*, waterleidingbedrijf *n*
r водопроводные сооружения *n pl*

W265 *e* **water yield**
d Ergiebigkeit *f* eines Einzugsgebietes
f apport *m* d'eau
nl opbrengst *f* van het waterwingebied
r водоотдача *f* (*общий сток с водосборного бассейна через поверхностные русла и водоносные слои*)

W266 **wattage draw** *see* **power consumption**

W267 *e* **wave erosion**
d Brandungserosion *f*, Wellenerosion *f*
f érosion *f* par les vagues
nl golfslag-erosie *f*
r волновая эрозия *f*

W268 *e* **wave force**
d Wellenkraft *f*
f effort *m* de vagues
nl golfenergie *f*
r энергия *f* волн

W269 *e* **wave generator**
d Wellenerzeuger *m*, Wellenerreger *m*
f appareil *m* à produire la houle
nl golfslaggenerator *m*
r волнопродуктор *m* (*для гидравлических лабораторных исследований*)

W270 *e* **wave impact**
d Wellenstoß *m*
f choc *m* de la houle [de l'onde, de la vague]
nl golfslag *m*
r удар *m* волны

W271 *e* **wave load**
d Wellenbelastung *f*
f charge *f* d'ondes
nl golfbelasting *f*
r волновая нагрузка *f*

W272 *e* **wave pressure**
d Wellendruck *m*
f pression *f* d'ondes
nl golfdruk *m*
r волновое давление *n*

W273 *e* **waves**
d Welligkeit *f* (*Straßendecke*)
f vagues *f pl* (*déformation ondulée de la chaussée*)

nl golving *f* (*wegdek*)
r волнистость *f* (*дорожного покрытия*)

W274 **wave wall** *see* **storm wall**

W275 *e* **wax emulsion**
d Paraffinemulsion *f*
f émulsion *f* à la cire
nl paraffine-emulsie *f*
r парафиновая эмульсия *f*

W276 *e* **WC-and-bathroom unit**
d Bad-Abort-Raumzelle *f*
f bloc-eau *m*
nl natte cel *f(m)* met toilet- en badinstallatie
r объёмный блок *m* санузла

W277 **WC bowl** *see* **WC pan**

W278 *e* **WC flushing cistern**
d Spülkasten *m*
f réservoir *m* de chasse, chasse *f* d'eau
nl stortbak *m*, closetreservoir *n*
r смывной бачок *m*

W279 *e* **WC pan**
d Toilettenschüssel *f*
f cuvette *f* de W.C., bol *m* de toilettes, cuvette *f* à chasse
nl closetschotel *m*
r чаша *f* унитаза

W280 *e* **WC suite**
d komplette WC-Anlage *f*
f installation *f* de W.C.
nl komplete WC-installatie *f*
r укомплектованное оборудование *n* санузла

W281 *e* **wearing course**
d 1. Verschleißschicht *f*, Nutzschicht *f* 2. Deckschicht *f*, Laufschicht *f*, Gehschicht *f*
f 1. couche *f* d'usure 2. couvre-sol *m* résistant à l'usure
nl 1. slijtlaag *f(m)* (*wegdek*) 2. dekvloer *m*
r 1. слой *m* износа (*дорожного покрытия*) 2. износостойкое покрытие *n* (*пола*)

W282 *e* **weather-boarding**
d Stülpschalung *f*
f planches *f pl* à recouvrement, revêtement *m* en planches à recouvrement
nl potdekseling *f*
r обшивка *f* досками внакрой

W283 *e* **weather groove**
d Wassernase *f*, Unterschneidung *f*
f larmier *m*
nl waterhol *n*
r слезник *m*

W284 *e* **weather resistance**
d Wetterbeständigkeit *f*, Wetterfestigkeit *f*
f résistance *f* aux intempéries
nl weerbestendigheid *f*
r атмосферостойкость *f*

W285 **weather strip** *see* **air lock 3.**

W286 *e* **weather stripping**
d Fugenabdichtung *f*; Dichtungsstreifen *m pl*

f calfeutrement *m* des joints; garnitures *f pl* pour joints waterstop; bourrelets *m pl*
nl voegdichting *f*
r уплотнение *n* швов (*в строительных конструкциях*); уплотнительные элементы *m pl*

W287 *e* **web**
d Steg *m*, Stegblech *n* (*Balken*)
f âme *f* (*d'une poutre*)
nl lijf *f(m)* (*balk*)
r стенка *f* (*балки*)

W288 **web bars** *see* **shear reinforcement**

W289 *e* **web member**
d Fachwerkstab *m*, Füllungsstab *m*
f élément *m* du treillis
nl vakwerkelement *n*
r элемент *m* решётки фермы

W290 *e* **web plate**
d Stegblech *n*
f âme *f* en tôle [pleine]
nl lijfplaat *f(m)*
r сплошная стенка *f* (*составной металлической балки*)

W291 **web reinforcement** *see* **shear reinforcement**

W292 *e* **web splice**
d Laschenstoßverbindung *f* im Steg des zusammengesetzten Stahlträgers
f joint *m* bout à bout de l'âme d'une poutre en tôles
nl stuiklas *f(m)* in het lijf van een samengestelde ligger
r стыковое соединение *n* листов стенки металлической составной балки

W293 *e* **web stiffener**
d Stegsteife *f*
f élément *m* de raidissement, raidisseur *m* de l'âme (*d'une poutre*)
nl lijfversterking *f*
r элемент *m* жёсткости стенки балки

W294 *e* **wedge**
d Keil *m*
f coin *m*, clavette *f*
nl spie *f(m)*, wig *f(m)*
r клин *m*

W295 *e* **wedge anchor**
d Keilverankerung *f*
f clavette *f* d'ancrage
nl muuranker *n* met keilbout
r клиновое анкерное устройство *n*

W296 *e* **wedge anchorage**
d Keilverankerung *f*
f ancrage *m* par clavettes
nl verankering *f* met keilbout
r анкеровка *f* клиньями (*арматуры*)

W297 *e* **wedge of failure**
d Bruchprisma *n*, Gleitkeil *m*
f prisme *m* de rupture [de glissement]
nl afschuivend grondprisma *n*
r призма *f* обрушения [скольжения]

W298 *e* **wedge valve**
d Keilschieber *m*

f robinet-vanne *m* à coins
nl tapse kraanafsluiter *m*
r клиновая задвижка *f*

W299 *e* **wedging**
d Verkeilung *f*
f coinçage *m*, calage *m*, clavetage *m*
nl opwiggen *n*
r закрепление *n* клиньями, расклинивание *n*

W300 **weightbatcher** *see* **weigh batch plant**

W301 *e* **weigh-batching**
d Gewichtsdosierung *f*, Gewichtszumessung *f*
f dosage *m* en poids
nl gewichtsdosering *f*
r дозировка *f* по весу, весовая дозировка *f*

W302 *e* **weigh batch plant**
d Wiegedosieranlage *f*, Wiege-Zumeßanlage *f*, Wiege-Zuteilanlage *f*
f installation *f* de dosage en poids
nl gewichtsdoseerapparaat *n*
r весовой дозатор *m*

W303 **weighed filter** *see* **loaded filter**

W304 *e* **weigh-hopper**
d Wiegebehälter *m*, Wiege(dosier)gefäß *n*
f trémie *f* de pesage
nl weegbak *m*, weegtrechter *m*
r весовой дозаторный бункер *m*

W305 *e* **weighing apparatus**
d Wägevorrichtung *f*, Wiegegerät *n*
f balance *f*
nl weegapparaat *n*
r взвешивающее устройство *n*, весы *pl*

W306 *e* **weight ratio**
d Gewichtsverhältnis *n*
f proportion *f* en poids [pondérale]
nl gewichtsverhouding *f*
r весовое соотношение *n*

W307 *e* **weir**
d 1. Wehr *n*, Stauwehr *n*, Schußwehr *n* 2. Wehrüberfall *m*, Wehrüberlauf *m*
f 1. barrage-déversoir *m* 2. déversoir *m*
nl 1. weer *n*, keerdam *m* 2. weer *m* met overlaat
r 1. водоподъёмная [водосливная] плотина *f* 2. водослив *m*

W309 *e* **weld-crete**
d Betonklebstoff *m*, Betonkleber *m*
f colle *f* pour béton, adhésif *m* à béton
nl hechtmiddel *n* voor betonelementen
r клеевой [адгезионный] состав *m* для бетона

W310 *e* **welded reinforcement**
d geschweißte Bewehrung *f*
f armature *f* soudée
nl gelaste wapening *f*
r сварная арматура *f* (*сетка, каркас*)

W311 *e* **welded steel base**
d geschweißter Stahlstützenfuß *m*
f pied *m* soudé
nl gelaste stalen voet *m*

44*

r сварная стальная база *f* [башмак *m*]
колонны

W312 *e* **welded structures**
d geschweißte Konstruktionen *f pl*
f constructions *f pl* soudées
nl lasconstructies *f pl*
r сварные конструкции *f pl*

W313 *e* **welded wire fabric**
d punktgeschweißte Bewehrungsmatte *f*
[Armierungsmatte *f*]
f treillis *m* soudé
nl gepuntlast wapeningsnet *n*
r сварная арматурная сетка *f*

W314 *e* **welding goggles**
d Schweißbrille *f*, Schutzbrille *f*
f lunettes *f pl* de protection [de secu-
rité] (*pour les soudeurs*)
nl lasbril *m*
r защитные очки *pl* (*сварщика, рез-
чика*)

W315 *e* **welding helmet**
d Schweißerschutzhelm *m*, Schweißer-
schutzhaube *f*
f casque *f* [masque *f*] de soudeur
nl lashelm *m*
r защитный шлем *m* сварщика

W316 *e* **weldment**
d Schweißteil *n*; Schweißkonstruktion *f*
f élément *m* de construction soudé
nl gelast constructie-element *n*
r сварной элемент *m*; сварная конструкция *f*

W317 *e* **well**
d 1. Brunnen *m* 2. Bohrloch *n*, Bohrung *f*
3. Schacht *m* (z. B. *Aufzugsschacht*)
f 1. puits *m* 2. trou *m* foré [de forage]
3. gaine *f*; cage *f* (*d'ascenseur*)
nl 1. bron *f(m)*, wel *f(m)* 2. boorput *m*
3. lichtschacht *m*; liftschacht *m*
r 1. колодец *m* 2. буровая скважина *f*
3. шахта *f* (*напр. лифта*)

W318 *e* **well-burned brick**
d Hart(brand)ziegel *m*
f brique *f* dure [bien cuite]
nl hardgrauw *m*
r хорошо обожжённый кирпич *m*

W319 *e* **well capacity**
d Brunnenleistung *f*
f capacité *f* d'un puits, débit *m* de puits
nl capaciteit *f* van een bron
r производительность *f* колодца

W320 *e* **well cementing**
d Bohrlochzementierung *f*, Auszemen-
tieren *n* des Bohrlochs
f cimentation *f* de forage
nl cementatering *f* van het boorgat
r цементация *f* скважины

W321 *e* **well field**
d Brunnenreihe *f*, Brunnenkette *f*
f champ *m* de puits
nl bronnenrij *f*
r поле *n* колодцев, группа *f* колодцев
[скважин]

W322 *e* **well foundation**
d Brunnengründung *f*
f fondation *f* par fonçage de puits, fondation *f* sur puits
nl puttenfundering *f*
r фундамент *m* в форме опускных
колодцев

W323 *e* **well head**
d Mundloch *n*, Tagkranz *m*; Bohrloch-
kopf *m*, Brunnenkopf *m*
f tête *f* de forage
nl curve *f* van de grondwaterstand
r устье *n* скважины; оголовок *m* скважины *или* трубчатого колодца

W324 *e* **well interference**
d gegenseitige Beeinflussung *f* von
Brunnen
f interférence *f* des puits
nl wisselwerking *f* van de bronnen
r взаимодействие *n* колодцев

W325 *e* **wellpoint**
d 1. Nadelfilter *n, m* 2. Rohrbrunnenfil-
ter *n, m*, Rammfilter *n, m*
f 1. puits *m* filtrant 2. pointe *f* de puits
foncé
nl 1. naaldfilter *n, m* 2. bron *m*
r 1. иглофильтр *m* 2. фильтр *m* трубчатого забивного колодца, забивной
фильтр *m*

W326 *e* **wellpoint system**
d Vakuumbrunnen-Anlage *f*
f installation *f* de puits filtrants [de
pointes filtrantes]
nl bronbemaling *f*
r иглофильтровая установка *f*, система *f* водопонижения с помощью игло-
фильтров

W327 *e* **well pump**
d Brunnenpumpe *f*
f pompe *f* à puits
nl bronpomp *f(m)*
r скважинный насос *m*

W328 *e* **well screen**
d Brunnenfilter *n, m*
f crépine *f* de puits
nl bronfilter *n, m*
r фильтр *m* трубчатого колодца *или*
скважины

W329 **wet cell humidifier** *see* **capillary cell
washer**

W330 **wet collector** *see* **scrubber**

W331 *e* **wet consistency**
d Frischbetonsteife *f*, Frischbetonkon-
sistenz *f*
f consistance *f* [fluidité *f*] du béton
frais
nl consistentie *f* van verse betonmortel
r консистенция *f* свежеуложенной бетонной смеси

W332 *e* **wet cut**
d Naßaushub *m*
f fouille *f* dans l'eau

nl uitgraving *f* onder water
r подводная выемка *f*

W333 *e* **wet dock**
    d Hafenbecken *n*; Flutbecken *n*
    f bassin *m* de port, darse *f*; bassin *m* à flot
    nl getijdehaven *m*, havenbekken *n*; drijvend dok *n*
    r открытый портовый бассейн *m*; док-бассейн *m*, наливной док *m*

W334   **well drain** *see* **absorbing well**

W335 *e* **wet landslide**
    d Murgang *m*, Mure *f*
    f coulée *f* de boue, avalanche *f* boueuse [de boue]
    nl slibstroom *m*
    r селевой поток *m*

W336   **wet-mix macadam** *see* **waterbound macadam**

W337 *e* **wet-mix shotcrete**
    d Spritzbeton *m*, der fertig angemacht durch die Schläuche gefördert wird
    f gunite *f* préparée d'avance
    nl vloeibaar spuitbeton *n*
    r готовая мокрая торкретбетонная смесь *f*

W338 *e* **wet screed**
    d Putzlatte *f*
    f cueillée *f* [cueillie *f*] en mortier (de plâtre)
    nl pleisterlat *f(m)* van mortel
    r штукатурный маяк *m* из раствора

W339 *e* **wet screening**
    d Absieben *n* von Frischbeton; Naßsiebung *f*
    f criblage *m* de béton frais; criblage *m* humide
    nl nat zeven *n* (*betonspecie*)
    r грохочение *n* свежеприготовленной бетонной смеси; мокрое грохочение *n*

W340 *e* **wet sump**
    d Pumpensumpf *m*
    f puisard *m* (*d'une pompe*)
    nl pompput *m*
    r зумпф *m* [приямок *m*] насоса

W341 *e* **wettest stable consistency**
    d maximal zulässige Bildsamkeit *f* [Fließfähigkeit *f*] (*Betongemisch, Frischmörtel*)
    f fluidité *f* limite (*dépassée, la matière perd sa cette limite stabilité structurale*)
    nl maximaal toelaatbare vloeibaarheid *f* (*mortel*)
    r предельная подвижность *f* пластичной бетонной *или* растворной смеси, при которой ещё сохраняется их структурная прочность

W342 *e* **wetting agent**
    d 1. Benetzungsmittel *n*, Netzmittel *n* 2. Betonverflüssiger *m*, Weichmacher *m*
    f 1. additif *m* mouillant, agent *m* de mouillage 2. plastifiant *m*, plastificateur *m*
    nl 1. bevochtigingsmiddel *n* 2. plastificeermiddel *n* (*beton*)
    r 1. смачивающая добавка *f* 2. пластификатор *m* бетона

W343 *e* **wet well**
    d Pumpwerksumpf *m*
    f 1. partie *f* souterraine de la salle de pompes 2. puisard *m*, cuvette *f* (*d'une installation de pompage*)
    nl 1. pompenkelder *m* 2. pompput *m*
    r 1. подземная часть *f* здания насосной станции 2. зумпф *m* [прямок *m*] насоса

W346 *e* **wharf**
    d 1. Kaianlage *f*, Kai *m*; Landungssteg *m*, Landungsbrücke *f*, Anlegebrücke *f* 2. Anlegestelle *f* 3. Werft *f*
    f 1. ouvrage *m* d'accostage; quai *m* 2. point *m* d'amarrage 3. chantier *m* naval
    nl 1. kade *f(m)*; aanlegsteiger *m* 2. aanlegplaats *f(m)* 3. werf *f(m)*
    r 1. причальное сооружение *n*; пристань *f* 2. якорное место *n* 3. верфь *f*

W347 *e* **wheelabrating**
    d Schrotstrahlen *n*
    f grenaillage *m*
    nl staalstralen *n*
    r дробеструйная обработка *f*

W348   **wheel-barrow** *see* **barrow**

W349 *e* **wheel excavator**
    d 1. Schaufelradbagger *m* 2. luftbereifter Bagger *m*, Mobilbagger *m*
    f 1. roue-pelle *f*, excavateur *m* rotatif, excavatrice *f* rotative, excavateur *m* à roue à godets 2. excavateur *m* sur pneus
    nl 1. schepradgraafmachine *f* 2. graafmachine *f* op wielen
    r 1. роторный экскаватор *m* 2. пневмоколёсный экскаватор *m*

W350 *e* **wheel loader**
    d Radlader *m*
    f chargeur *m* [chargeuse *f*] sur roues
    nl wiellaadschop *f(m)*
    r колёсный (одноковшовый) погрузчик *m*

W351 *e* **wheel-mounted crane**
    d luftbereifter Kran *m*
    f grue *f* automatrice sur roues pneu(matique)s, grue *f* sur pneu(matique)s
    nl wielkraan *f(m)*
    r пневмоколёсный кран *m*

W353 *e* **wheel type trencher**
    d Schaufelrad-Grabenbagger *m*, Kreiseimerleiter-Grabenbagger *m*
    f trancheuse *f* [excavateur *m* de tranchées] à roue (à godets)
    nl scheprad-sleuvengraafmachine *f*
    r роторный канавокопатель *m*

W354   **whip line** *see* **auxiliary hoist**

W356 *e* **white cement**
   d Weißzement *m*
   f ciment *m* blanc
   nl wit cement *n*
   r белый цемент *m*

W357 *e* **white lead**
   d Bleiweiß *n*
   f céruse *f*, blanc *m* de plomb
   nl loodwit *n*
   r свинцовые белила *pl*

W358 *e* **whitening, whitewash**
   d Kalkmilch *f*, Tünche *f*, Weiße *f*
   f badigeon *m* à la chaux, lait *m* [blanc *m*] de la chaux
   nl witkalk *m*
   r известковое молоко *n*, известковая побелка *f*

W359 *e* **whole-brick wall**
   d ganzziegeldicke Wand *f*
   f mur *m* (d')une brique d'épaisseur
   nl steensmuur *m*
   r стена *f* (толщиной) в один кирпич

W360 **whole-tide cofferdam** *see* **full-tide cofferdam**

W361 *e* **wicket**
   d 1. Stautafel *f*, Stauklappe *f* 2. Torschütz(e) *n* (*f*), Klappschütz(e) *n* (*f*), Absperrschieber *m* 3. Obergurt *m* 4. Spülschütz(e) *n* (*f*) 5. Torpforte *f*
   f 1. vanne *f* pivotante [à pivot] 2. vanne *f* de dévasement 3. face *f* supérieure d'une vanne secteur 4. vanne *f* de chasse 5. guichet *m* (*dans la porte de clôture*)
   nl 1. stuwklep *f*(*m*) 2. rinket *n* 3. klinketdeur *f*(*m*), loopdeur *f*(*m*) 4. spuischuif *f*(*m*) 5. hekje *n*
   r 1. поворотный щитовой затвор *m* 2. клинкетный затвор *m*, клинкет *m* 3. верховое полотнище *n* секторного затвора 4. промывной затвор *m*, затвор *m* промывного отверстия 5. калитка *f* (*в воротах*)

W362 *e* **wicket dam**
   d Klappenwehr *n*
   f barrage *m* à vannes pivotantes
   nl klepstuw *m*
   r разборная плотина *f* с поворотными щитовыми затворами

W363 *e* **widener**
   d Verbreiterer *m*
   f machine *f* pour élargir les chaussées
   nl verbredingsmachine *f* (*wegenbouw*)
   r машина *f* для расширения дорожного полотна

W364 *e* **widening of bridge**
   d Brückenverbreiterung *f*
   f élargissement *m* de la travée de pont
   nl verbreden *n* van een brug
   r расширение *n* пролётного строения моста

W365 **wild flooding** *see* **basin irrigation**

W366 *e* **wimble**
   d Bohrwinde *f*
   f vilebrequin *m*
   nl krukboor *f*(*m*), borstboor *f*(*m*)
   r коловорот *m*

W367 *e* **winch**
   d Winde *f*
   f treuil *m*
   nl windas *n*
   r лебёдка *f*

W368 **wind beam** *see* **collar beam**

W369 **wind brace** *see* **wind bracing 1.**

W370 *e* **wind bracing**
   d 1. Windstrebe *f* 2. Windverband *m*, Windversteifung *f*, Windverspannung *f*
   f 1. contrevent *m*, entretoise *f* 2. contreventement *m*, entretoisement *m*
   nl 1. windschoor *m* 2. windverband *n*
   r 1. ветровая связь *f* 2. система *f* ветровых связей

W371 *e* **windbreak**
   d 1. Waldschutzstreifen *m*, Waldschutzgürtel *m* 2. Windschutzschirm *m*, Windschutztafel *f*
   f 1. bande *f* forestière de protection 2. écran *m* protecteur contre le vent, barrière *f* contre le vent
   nl 1. beschermende bosgordel *m* 2. windscherm *n*
   r 1. защитные лесонасаждения *n pl*, защитная лесополоса *f* 2. ветрозащитный экран *m* [щит *m*], ограждение *n* для защиты от ветра

W372 *e* **wind effect**
   d Windeffekt *m*
   f effet *m* du vent
   nl windeffect *n*
   r ветровое воздействие *n*

W373 *e* **winder**
   d Wendelstufe *f*, Winkelstufe *f*
   f marche *f* dansante [balancée, gironnée]
   nl hoektrede *f*(*m*), bochttrede *f*(*m*)
   r забежная ступень *f*

W374 *e* **wind girder**
   d Windträger *m*
   f poutre *f* de contreventement
   nl windbalk *m*
   r ветровая балка *f* [ферма *f*]

W375 *e* **winding machine**
   d Wickelmaschine *f*
   f machine *f* à enrouler [d'enroulement] les fils d'acier
   nl wikkelmachine *f* (*voor het wikkelen van de wapening*)
   r навивочная машина *f* (*для навивки арматуры*)

W376 *e* **wind load**
   d Windlast *f*
   f charge *f* de vent [due au vent]
   nl windbelasting *f*
   r ветровая нагрузка *f*

W377 *e* **window**
   d Fenster *n*
   f fenêtre *f*
   nl venster *n*, raam *n*
   r окно *n*

W378 *e* **window air conditioner**
   d Fensterklimagerät *n*
   f conditionneur *m* d'air type fenêtre, climatiseur *m* de fenêtre
   nl in venster in te bouwen airconditioning-apparaat *n*
   r оконный кондиционер *m*

W379 **window and frame packaged unit** *see* **window unit**

W380 *e* **window band**
   d Fensterband *n*, Lichtband *n*
   f bande *f* vitrée, fenêtre *f* gisante [en longueur], fenêtres *f pl* continues
   nl raamstrook *m*, glasstrook *m*
   r ленточное окно *n*

W381 *e* **window bar**
   d 1. Fenstersprosse *f* 2. Mittelstück *n*, Mönch *m* 3. Fugendeckleiste *f* 4. Fensterbasküleverschluß *m*
   f 1. petit-bois *m* 2. traverse *f* d'imposte 3. battement *m*, bande *f* couvre-joint 4. espagnolette *f*
   nl 1. vensterroede *f(m)* 2. tussendorpel *m* 3. naald *m*, deklijst *f(m)* 4. spanjolet *f(m)*
   r 1. оконный горбылёк *m* 2. средник *m* 3. уплотняющий нащельник *m* 4. (оконный) шпингалет *m*

W382 **window bay** *see* **bay window**

W383 *e* **window board**
   d Fensterbrett *n*
   f tablette *f* de fenêtre, planche *f* d'appui
   nl vensterbank *f(m)*
   r подоконная доска *f*

W384 **window casing** *see* **window frame**

W385 *e* **window catch**
   d Fensterschließe *f*, Fensterbasküleverschluß *m*
   f loquet *m*, loqueteau *m*, cliquet *m* d'arrêt
   nl raamboompje *n*
   r оконная задвижка *f*, (оконный) шпингалет *m*

W386 **window efficiency ratio** *see* **daylight factor**

W387 *e* **window fittings**
   d Fensterbeschläge *m pl*
   f quincaillerie *f* de fenêtre
   nl vensterbeslag *n*
   r оконные приборы *m pl*

W388 *e* **window frame**
   d Fensterrahmen *m*
   f châssis *m* de fenêtre
   nl raamkozijn *n*
   r оконная рама *f*

W389 *e* **window glass**
   d Fensterglas *n*
   f verre *m* à vitres
   nl vensterglas *n*
   r оконное стекло *n*

W390 **window hardware** *see* **window fittings**

W391 *e* **window head**
   d Fensterriegel *m*
   f traverse *f* haute (*du bâtis dormant ou de la croisée*)
   nl latei *f*
   r верхний брус *m* оконной рамы *или* коробки

W392 *e* **windowless building**
   d fensterloses Gebäude *n*
   f bâtiment *m* sans fenêtres
   nl vensterloos gebouw *n*
   r здание *n* без окон

W393 *e* **window lintel**
   d Fenstersturz *m*
   f linteau *m* de fenêtre
   nl latei *f*
   r оконная перемычка *f*

W394 *e* **window opening**
   d Fensteröffnung *f*
   f baie *f* [ouverture *f*] de fenêtre
   nl vensteropening *f*
   r оконный проём *m*

W395 *e* **window panel**
   d Wandplatte *f* mit Fensteröffnung, Fensterplatte *f*
   f panneau *m* fenêtre [baie]
   nl wandpaneel *n* met venster
   r стеновая панель *f* с окном

W396 *e* **window pier**
   d Fensterpfeiler *m*
   f trumeau *m*
   nl penant *n*, muurdam *m*
   r межоконный простенок *m*

W397 *e* **window reveal**
   d äußere Fensterleibung *f*
   f tableau *m* (de baie), embrasement *m* extérieur de la fenêtre
   nl negge *f(m)*
   r наружный откос *m* окна

W398 *e* **window schedule**
   d Stückliste *f* der Fensterelemente (*für das im Bau befindliche Gebäude*)
   f nomenclature *f* des fenêtres
   nl lijst *f(m)* van ramen
   r спецификация *f* окон [оконных блоков]

W399 *e* **window sill**
   d Fensterbank *f*, Sohlbank *f*
   f pièce *f* d'appui du bâti dormant d'une croisée
   nl raamdorpel *m*, lekdorpel *m*
   r нижний брус *m* оконной коробки

W400 *e* **window space**
   d gesamte [verglaste] Lichtöffnungsfläche *f*
   f surface *f* vitrée
   nl totaal glasoppervlak *n*
   r общая площадь *f* застеклённых оконных проёмов

W401 **window stool** *see* **window board**

W402 *e* **window unit**
d Fensterelement *n*
f bloc-fenêtre *m*
nl samengesteld kozijn *n*, buitenlicht-kozijn *n* met ramen
r оконный блок *m*

W403 *e* **window wall**
d Fensterwand *f*
f paroi *f* de verre [en verre]
nl (gevel)pui *f(m)*
r застеклённая навесная стена *f*

W404 *e* **wind pressure**
d Winddruck *m*
f pression *f* dynamique du vent
nl winddruk *m*, stuwdruk *m*
r давление *n* [скоростной напор *m*] ветра

W405 *e* **wind pressure coefficient**
d Winddruckbeiwert *m*
f coefficient *m* aérodynamique
nl stuwdrukcoëfficiënt *m*
r аэродинамический коэффициент *m* (ветрового давления)

W406 *e* **wind-protection screen**
d Windschutztafel *f*
f pare-brise *m*, écran *m* de protection contre le vent
nl windscherm *n*
r ветрозащитный щит *m*

W407 **wind-raised level** *see* **wind setup**

W408 *e* **windrow**
d Längsreihe *f*, Streifhaufen *m*, Boden-schwaden *m*
f cordon *n* de terre [de sol], levée *f* de terre
nl wal *m*
r вал *m* насыпного грунта

W409 *e* **windrowing**
d Längsreihenverteilung *f*
f endiguement *m*, formation *f* de levées de terre
nl vormen *n* [opwerpen *n*] van wallen
r обвалование *n* грунта, отсыпка *f* грунта валами

W411 *e* **wind setup**
d Windstau *m*
f relèvement *m* du plan d'eau sous l'effet du vent
nl opwaaiing *f*, opstuwing *f*
r ветровой нагон *m*

W412 *e* **wind speed alarm**
d Windmelder *m*
f avertisseur *m* de vitesse dangéreuse du vent
nl stormwaarschuwing *f*
r сигнализатор *m* опасной скорости ветра

W413 **wind stop** *see* **air lock 3.**

W414 **wind tide** *see* **wind setup**

W415 *e* **wind turbine fan**
d Windturbinenventilator *m*
f ventilateur *m* à turbine éolienne

nl windturbineventilator *m*
r ветротурбинный вентилятор *m*

W416 *e* **wind uplift**
d Windsog *m*
f sous-pression *f* du vent
nl windzuiging *f*
r отрицательное давление *n* ветра, ветровой отсос *m*

W417 *e* **windward side**
d Windseite *f*
f côté *m* du vent
nl windzijde *f*
r наветренная сторона *f*

W418 *e* **wind wave**
d Windwelle *f*
f vague *f* d'origine éolienne
nl windgolf *f(m)* (*water*)
r ветровая волна *f*

W419 *e* **wind-wave flume**
d Windwellen-Modellgerinne *n*
f petit canal *m* pour la génération [simulation] de houle
nl goot *f(m)* voor het opwekken van windgolven
r лоток *m* для гидравлического моделирования ветровых волн

W420 *e* **wing**
d 1. Flügel *m*, Seitenflügel *m* 2. Drehflügel *m* (*Dreh- oder Karusselltür*)
f 1. aile *f* 2. vantail *m* [battant *m*] de la porte tournante
nl 1. vleugel *m* (*gebouw*) 2. deurvleugel *m*
r 1. крыло *n* (*здания*), флюгель *m* 2. створка *f* (*вращающейся двери*)

W421 **wing dam** *US see* **groin**

W422 *e* **wing pile**
d konischer Pfahl *m*
f pieu *m* avec tête élargie
nl paal *m* met verdikte kop
r бетонная свая *f* с уширенной верхней частью

W423 *e* **wing wall**
d Wehrwange *f*, Flügelmauer *f*, Böschungsflügel *m*
f mur *m* en aile
nl schermwand *f*
r открылок *m* (*плотины, берегового устоя моста*); крыло *n* (*подпорной стенки*)

W424 *e* **winter building techniques**
d Winterbauverfahren *n pl*
f technique *f* des travaux d'hiver
nl technieken *f pl* voor het bouwen bij lage temperaturen
r методы *m pl* [приёмы *m pl*] выполнения зимних работ

W425 *e* **winter mode**
d Winterbetrieb *m*
f régime *m* de travail [de marche, de fonctionnement] du système en hiver
nl winterbedrijf *n*
r зимний режим *m*

W426 *e* **wire**
 d 1. Draht *m* 2. elektrischer Leiter *m*
 f 1. fil *m* métallique 2. conducteur *m* (*électrique*)
 nl 1. draad *m* 2. elektrische leiding *f*
 r 1. проволока *f* 2. электрический провод *m*

W427 *e* **wire anchorage**
 d Verankerung *f* der Drähte (*im Spann-beton-Bauelement*)
 f ancrage *m* des fils de précontrainte
 nl verankering *f* van de voorspandraden
 r анкеровка *f* (арматурных) проволок

W428 *e* **wire brushing**
 d Reinigen *n* mit Drahtbürste
 f brossage *m*, nettoyage *m* par brosse métallique
 nl reiniging *f* met draadborstel
 r очистка *f* проволочной щёткой

W429 *c* **wire cable**
 d Stahldrahtseil *n*
 f câble *m* en fils d'acier
 nl staalkabel *m*
 r стальной канат *m* [трос *m*]

W431 *e* **wire dam**
 d Deich *m* in Drahtschotterbauweise
 f barrage *m* avec enveloppe en treillage métallique
 nl dijk *m* van in gaas gepakte stortsteen
 r габионная дамба *f*

W432 *e* **wire glass**
 d Drahtglas *n*
 f verre *m* armé [de sécurité]
 nl draadglas *n*
 r армированное [безопасное, безоско-лочное] стекло *n*

W433 *e* **wire lath**
 d Rabitz(draht)geflecht *n*
 f grillage *m* d'enduit [à poules], treillis *m* métallique pour enduit, treillis *m* Rabitz
 nl rabitzmetaaldrager *m* voor pleister-werk
 r сетка *f* Рабица, сетка *f* для штукатур-ных работ

W434 *e* **wireless tower**
 d Antennenturm *m*, Antennenmast *m*
 f tour *f* [mât *m*] d'antenne, pylône *m* support d'antenne
 nl antennemast *m*
 r антенная башня *f* [мачта *f*]

W435 *e* **wire rope sling**
 d Stropp *m*, Anschlagseil *n*, Seilge-hänge *n*
 f élingue *f* en câble d'acier
 nl strop *m*, *f*
 r строп *m* (из стального каната)

W436 *e* **wire tie**
 d 1. Drahtspannschloß *n* 2. Drahtanker *m*
 f 1. tirant *m* en fils d'acier 2. ancre *m* en fil d'acier
 nl 1. spandraad *m* 2. draadanker *n*
 r 1. проволочная стяжка *f* 2. прово-лочный анкер *m*

W437 **wire winding machine** *see* **winding machine**

W438 *e* **withe**
 d Trennmauer *f* der Schornsteinlängs-züge
 f paroi *f* entre carneaux des fumées adjacents
 nl scheidingswand *m* tussen schoor-steenkanalen
 r перегородка *f* в дымовых каналах и трубах

W439 *e* **wobble coefficient**
 d Reibungskoeffizient *m* bei Spann-gliedern mit gekrümmtem Verlauf
 f coefficient *m* de frottement dû aux petites ondulations aléatoires des gaines (*pour câbles de précontrainte*)
 nl wrijvingscoëfficiënt *m* van de span-elementen tengevolge van toevallige afwijkingen van het kanaaltracé
 r коэффициент *m* трения напрягаемой арматуры о стенки канала *или* обо-лочки, вызванного их случайной волнистостью (*отклонением от про-ектного положения*)

W440 *e* **wobble effect**
 d Reibung *f* im Spannkanal infolge zufälliger Welligkeit des Spannkanals
 f frottement *m* dû aux petites ondula-tions aléatoires des gaines (*pour câbles de précontrainte*)
 nl wobble-effect *n* (*wrijving tengevolge van kleine afwijkingen in de sparingen*)
 r трение *n* от волнистости арматурных каналов (*при отклонении от проект-ного положения*)

W441 *e* **wobbly wheel roller**
 d Gummiradwalze *f* mit flatternden Rädern
 f rouleau *m* [compacteur *m*, cylindre *m*] à pneus dandinants
 nl waggelwielwals *f*
 r дорожный каток *m* с «виляющими» пневматическими шинами

W442 *e* **wood**
 d Holz *n*
 f bois *m*
 nl hout *n*; bos *n*
 r дерево *n*; древесина *f*; лес *m*; лесо-материал *m*

W443 *e* **wood block**
 d 1. Parkettäfelchen *n*, Parkettstab *m* 2. Holzpflasterklotz *m*
 f 1. lame *f* de parquet 2. pavé *m* de bois
 nl 1. parketblokje *n* 2. straatblok *n* van hout
 r 1. паркетная клёпка *f* [дощечка *f*] 2. деревянный блок *m* [шашка *f*] для торцовой мостовой

W444 *e* **wood brick**
 d nagelbarer Holzblock *m*
 f bloc *m* de clouage

nl spijkerklos *m*
r пробка *f* (*закладываемая в кладку*); деревянный кирпич *m*

W445 *e* **wood-cement concrete**
d Holzspänebeton *m*, Holzzementbeton *m*
f béton *m* (de) bois et ciment
nl houtvezelbeton *n*
r арболит *m*, опилкобетон *m*

W446 *e* **wood chipboard**
d Holzspanplatte *f*
f panneau *m* de particules [copeaux] de bois
nl spaanplaat *f(m)*
r древесностружечная плита *f*

W447 *e* **wooden turned baluster**
d Holzbaluster *m*
f balustre *m* façonné en bois
nl getorste [gedraaide] houten baluster *m*
r деревянная фигурная [точёная] балясина *f*

W448 *e* **wooden weir**
d Blockwandwehr *n*
f barrage *m* en charpente de bois
nl houten stuw *m*
r бревенчатая ряжевая плотина *f*

W449 *e* **wood fiber concrete**
d Holzfaserbeton *m*
f béton *m* (de) fibres de bois
nl houtvezelbeton *n*
r деревобетон *m*, бетон *m* с древесно-волокнистым заполнителем

W450 *e* **wood-fibered plaster**
d Holzfasermörtel *m*
f enduit *m* de plâtre avec fibres de bois
nl houtvezelpleister *n*
r штукатурный раствор *m* с древесным волокном

W451 *e* **wood-fibered insulation**
d Holzfaserdämmung *f*
f isolation *f* (thermique) en fibre de bois
nl houtvezelwarmte-isolatie *f*
r теплоизоляция *f* из древесноволокнистых материалов

W452 *e* **wood-fiber slab, wood fibre board**
d Holzfaserplatte *f*
f dalle *f* [plaque *f*] en fibre(s) de bois, dalle *f* fibreuse de bois
nl (hout)vezelplaat *f(m)*
r древесноволокнистая плита *f*

W453 *e* **wood float**
d Holzspachtel *m*, *f*, Reibebrett *n*, Streichbrett *n*, Maurerhobel *m*
f taloche *f* en bois
nl schuurbord *n*
r деревянная тёрка *f* [гладилка *f*]

W454 *e* **wood flooring**
d Dielenfußboden *m*, Fußbodendielung *f*
f sol *m* en planche, revêtement *m* de sol en planches
nl houten vloer *m*
r дощатый пол *m*

W455 *e* **wood flour**
d Holzmehl *n*
f farine *f* de bois
nl houtmeel *n*
r древесная мука *f*

W456 *e* **wood-frame construction**
d Holzfachwerkbau *m*
f charpente *f* en bois
nl houtskelet *n*, houten vakwerk *n*
r деревянная каркасная конструкция *f*

W457 *e* **wood laminated arch**
d Schichtholzbogen *m*, geleimter Bogen *m*
f arc *m* en lame(lle)s [(en bois) lamellé](-collé)
nl gelijmde houten boog *m*
r дощатая многослойная клеёная арка *f*

W458 *e* **wood lath**
d Holzlatte *f*, Spaltlatte *f*, Lattenputzträger *m*
f latte *f* en bois
nl tengel *m*, tengelwerk *n*
r дранка *f*, драночная рейка *f*

W459 *e* **wood moulding**
d Kehlleiste *f*, Holzprofillatte *f*
f latte *f* moulurée en bois, baguette *f*, listeau *m* [listel *m*] décoratif de bois
nl geprofileerde houten lijst *f(m)*
r деревянная профилированная раскладка *f* [рейка *f*]

W460 *e* **wood paving**
d Holzpflaster *n*
f pavage *m* en bois
nl plaveisel *n* van hout
r торцовая мостовая *f*; дорожное покрытие *n* из деревянных блоков [шашек]

W461 *e* **wood-stave pipe**
d Holzdaubenrohr *n*
f tuyau *m* en douves de bois
nl buis *f(m)* van houten duigen
r труба *f* из деревянной клёпки

W462 *e* **wood wool**
d Holzwolle *f*
f laine *f* de bois
nl houtwol *f(m)*
r древесная шерсть *f*

W463 *e* **wood wool cement slab**
d Holzwollebetonplatte *f*
f dalle *f* en ciment et laine de bois, dalle *f* en béton léger à base de ciment et laine de bois
nl houtwolcement *n*
r плита *f* из арболита

W464 *e* **wood wool concrete**
d Holzwollebeton *m*
f béton *m* léger à base de laine de bois
nl houtwolbeton *n*
r арболит *m*

W465 *e* **wood wool slab**
d Holzwolleplatte *f*

f dalle *f* [plaque *f*] isolante en laine de bois
nl houtwolplaat *f(m)*
r плита *f* из древесной шерсти, древошерстная плита *f*

W466 *e* **woodwork**
   d 1. Holzartikel *m pl*, Holzbauelemente *n pl* 2. Tischlerarbeiten *f pl*, Holzarbeiten *f pl*
   f 1. boiserie *f*, menuiserie *f*; constructions *f pl* en bois charpenterie 2. travaux *m pl* de menuiserie (en bâtiment)
   nl 1. houtwerk *n* 2. timmerwerk *n*; houtbewerking *f*
   r 1. деревянные строительные изделия *n pl*, деревянные элементы *m pl* конструкций 2. столярные работы *f pl*

W467 *e* **woodworker's vise**
   d Hobelbankschraubstock *m*
   f étau *m* d'établi de menuisier
   nl timmermansbankschroef *f(m)*
   r тиски *pl* на столярном верстаке

W468 *e* **woodworking machinery**
   d Holzbearbeitungsmaschinen *f pl*
   f machines *f pl* à bois
   nl houtbewerkingsmachines *f pl*
   r деревообрабатывающие станки *m pl*

W469 *e* **workability**
   d Verarbeitbarkeit *f*
   f ouvrabilité *f*, maniabilité *f*, workabilité *f*
   nl bewerkbaarheid *f*
   r удобообрабатываемость *f*, удобоукладываемость *f* (*смеси*)

W470 **workability agent** *see* **wetting agent 2.**

W471 *e* **workable weather**
   d baugünstiges Wetter *n*, baustellengerechte [baufreundliche] Wetterbedingungen *f pl*
   f temps *m* ouvrable
   nl werkbaar weer *n*
   r погода *f*, благоприятная для производства работ

W472 *e* **«work as executed» drawing**
   d Revisionszeichnung *f*, Abnahmezeichnung *f*
   f dessin *m* de réception
   nl afnametekening *f*
   r чертёж *m* сооружения по данным натурного обмера

W473 *e* **work bench**
   d Arbeitsbank *f*, Werkbank *f*, Werktisch *m*
   f étable *m*, banc *m* d'ajusteur, table *f* à ouvrage
   nl werkbank *f(m)*
   r верстак *m*

W474 *e* **working area**
   d Arbeitsfläche *f*
   f aire *f* de travaux [de travail]
   nl werkterrein *n*, werkzone *f*
   r рабочая площадка *f* [зона *f*]

W475 *e* **working chamber**
   d Arbeitskammer *f*
   f chambre *f* de travail
   nl werkkamer *f(m)*
   r рабочая камера *f*

W476 **working deck** *see* **working platform**

W477 *e* **working design**
   d Ausführungsentwurf *m*
   f projet *m* d'exécution
   nl uitvoeringsontwerp *n*
   r рабочий проект *m*

W478 *e* **working drawings**
   d Ausführungszeichnungen *f pl*, Bauzeichnungen *f pl*
   f dessins *m pl* d'exécution
   nl werktekeningen *f pl*
   r рабочие чертежи *m pl*

W479 *e* **working face**
   d 1. Vortriebstelle *f*, Ort *n*; Ortsbrust *f* 2. Straßenbaustelle *f* 3. Arbeitsfläche *f*
   f 1. front *m*; front *m* d'attaque 2. chantier *m* de route [routier] 3. surface *f* travaillante
   nl 1. inbreekplaats *m* (*tunnelbouw*), front *n* 2. wegenwerk *n* in uitvoering 3. werkvlak *n*
   r 1. забой *m*; грудь *f* забоя 2. фронт *m* дорожно-строительных работ 3. рабочая поверхность *f*

W480 *e* **working load**
   d 1. Gebrauchslast *f*, Nutzlast *f*, Betriebsbeanspruchung *f* 2. Traglast *f*
   f 1. charge *f* utile [de travail] 2. charge *f* maximale d'utilisation (*d'un appareil de levage*)
   nl 1. nuttige belasting *f*; veranderlijke belasting *f* 2. hijslast *f*
   r 1. полезная нагрузка *f* 2. грузоподъёмность *f* (*крана*)

W481 *e* **working load limit**
   d maximal zulässige Belastung *f*, Grenzlast *f*
   f 1. charge *f* de sécurité 2. charge *f* maximale d'utilisation
   nl 1. maximaal toelaatbare belasting *f* 2. maximum hijslast *f*, hefvermogen *n*
   r 1. предельная допускаемая нагрузка *f* 2. максимальная грузоподъёмность *f*

W482 **work order** *see* **notice to proceed**

W483 *e* **working place**
   d Arbeitsplatz *m*, Arbeitsstelle *f*
   f emplacement *m* [poste *f*, lieu *m*] de travail
   nl plaats *f(m)* van tewerkstelling
   r рабочее место *n*

W484 *e* **working platform**
   d Arbeitsplattform *f*, Arbeitsbühne *f*
   f plate-forme *f* de travail; échafaud *m*
   nl werkvloer *m*
   r рабочая платформа *f*, рабочий помост *m*; леса *pl*, подмости *pl*

W485 *e* **working platform of a scaffold**
d Gerüstbelag *m*
f plate-forme *f* d'un échafaudage
nl steigervloer *m*
r рабочий настил *m* лесов [подмостей]

W486 *e* **working pressure**
d Betriebsdruck *m*
f pression *f* de travail [de service, d'utilisation]
nl werkdruk *m*
r рабочее давление *n*

W487 *e* **working range**
d Arbeitsbereich *m*
f zone *f* d'action
nl vlucht *f(m)* (*van een kraan*)
r зона *f* действия, рабочая зона *f* (*крана, экскаватора*); рабочий диапазон *m*

W488 **working space** *see* **working area**

W489 *e* **working spud**
d Arbeitspfahl *m*, rechter Pfahl *m* (*Naßbagger*)
f pieu *m* d'ancrage (*d'une drague*)
nl werkpaal *m* (*baggermolen*)
r напорная свая *f* (*земснаряда*)

W490 *e* **working stress**
d zulässige Spannung *f*; Gebrauchsspannung *f*
f contrainte *f* admissible; tension *f* de travail
nl toelaatbare spanning *f* (*vloeigrens gedeeld door veiligheidscoëfficiënt*)
r допускаемое напряжение *n*; действующее [рабочее] напряжение *n*

W491 *e* **working-stress design**
d Gebrauchslastverfahren *n*, Bemessungsverfahren *n* mit zulässigen Spannungen
f méthode *f* des contraintes [tensions] admissibles
nl berekening *f* gebaseerd op toelaatbare spanningen
r метод *m* расчёта по допускаемым напряжениям

W492 **working temperature** *see* **operating temperature**

W493 *e* **working width**
d Arbeitsbreite *f*, Einbaubreite *f*
f largeur *f* de travail
nl werkbreedte *f*
r рабочая ширина *f*, ширина *f* захвата

W494 *e* **working zone**
d Arbeitszone *f*
f zone *f* de travail, aire *f* des travaux [de travail]
nl werkzone *f*
r рабочая зона *f*

W495 *e* **work in progress**
d laufende Bauarbeiten *f pl*
f travaux *m pl* en cours (d'exécution)
nl werk *n* in uitvoering

r строительные работы *f pl* в стадии выполнения, выполняемые строительные работы *f pl*

W496 *e* **work level**
d Arbeitsbühne *f*, Arbeitsebene *f*
f aire *f* de travail élevée, plate-forme *f* de travail
nl werkplatform *n*
r рабочая площадка *f* [платформа *f*] (*выше уровня пола или земли*)

W497 *e* **workmanship**
d 1. Ausführung *f*, Fertigungsgüte *f*, Herstellungsgüte *f*; Arbeitsgüte *f* 2. Fachkönnen *n*
f 1. qualité *f* de travail [d'exécution] 2. qualification *f* professionnelle
nl 1. afwerking *f*, uitvoering *f* 2. vakbekwaamheid *f*
r 1. качество *n* исполнения [изготовления]; качество *n* работ 2. квалификация *f*, мастерство *n*

W498 *e* **workmen's shelter**
d Baubude *f*, Leutebude *f*
f abri *m* à l'usage [l'intention] des travailleurs
nl personeelsverblijf *n*, schaftlokaal *n*
r бытовая времянка *f* для рабочих

W499 *e* **work progress**
d Baufortschritt *m*, Baufortgang *m*
f avancement *m* des travaux
nl voortgang *m* [vordering *f*] van de werkzaamheden
r ход *m* строительства

W500 *e* **works**
d 1. Bauwerke *n pl*, Bauten *pl* 2. Bauarbeiten *f pl*
f 1. ouvrages *m pl* 2. travaux *m pl* de construction
nl 1. bouwwerken *n pl* 2. bouwen *n*, werk *n*
r 1. сооружения *n pl* 2. строительные работы *f pl*

W501 *e* **works beam test**
d Baustellen-Balkenprüfung *f*
f épreuve *f* de poutres (pour l'essai de béton) sur le chantier
nl beproeving *f* van balken op de bouwplaats
r испытание *n* стандартных бетонных балок на стройплощадке

W502 *e* **works cubes**
d Probewürfel *m pl*, Versuchswürfel *m pl*, Prüfwürfel *m pl*
f cubes *m pl* [éprouvettes *f pl* cubiques] d'essais en béton
nl proefkubussen *m pl*
r бетонные кубики *m pl* (*изготовляемые на стройплощадке*)

W503 *e* **works cube strength**
d Würfeldruckfestigkeit *f*
f résistance *f* sur cube
nl kubussterkte *f* (*beton*)
r действительная кубиковая прочность *f* (бетона)

W505 *e* **workshop**
  d Werkstatt *f*
  f atelier *m*
  nl werkplaats *f(m)*
  r мастерская *f*

W506   **worm conveyor** *see* **screw conveyor**

W507 *e* **worm elevator**
  d Senkrecht-Förderschnecke *f*, Senk-recht-Scheckenförderer *m*
  f élévateur *m* hélicoïdal [á helike]
  nl wormelevator *m*
  r шняковый элеватор *m*, грузоподъемный шнек *m*

W508 *e* **worm's-eye view**
  d Froschperspektive *f*
  f perspective *f* militaire
  nl perspectif *f(m)* met oogpunt op het maaiveld
  r лягушечья перспектива *f*

X

X1 *e* **X-braces**
  d Kreuzverband *m*, Kreuzverstrebung *f*
  f entretoises *f pl* croisées [en X]
  nl kruisverband *n*
  r перекрёстные связи *f pl*

X2   **XPM** *see* **expanded metal**

X3 *e* **X-ray examination, X-ray test**
  d Röntgenprüfung *f*
  f contrôle *m* aux rayons X
  nl röntgenonderzoek *n*
  r рентгенографический *или* рентгеноскопический контроль *m*, рентгенодефектоскопия *f*

X4 *e* **xylolite**
  d Steinholz *n*
  f xylolithe *f*, béton *m* magnésien
  nl asbestgraniet *n*, magnesiet *n*
  r ксилолит *m*

Y

Y1 *e* **yard**
  d 1. Hof *m* 2. Betonierplatz *m*; Lagerfläche *f* 3. Werft *f*; Stapelplatz *m*
  f 1. cour *f* 2. chantier *m* expérimental [d'essai]; terrain *m* de stockage 3. chantier *m* de construction naval [maritime]
  nl 1. binnenplaats *f(m)*, erf *n* 2. werf *f(m)*, buitenopslag(plaats) *m (f(m))* 3. scheepswerf *f(m)*
  r 1. двор *m (здания)* 2. полигон *m*; площадка *f (напр. для складирования материалов)* 3. верфь *f*; стапельное место *n*

Y3 *e* **yard drainage**
  d Grundstückentwässerung *f*
  f réseau *m* de branchements d'égout
  nl afwatering *f* van het perceel
  r дворовая канализация *f*

Y4 *e* **yard gulley**
  d Hofablauf *m*, Hofeinlauf *m*
  f rigole *f* d'écoulement [de déversement] de cour *(d'une maison)*
  nl afvoer *m* [afvoergoot *f(m)*] van de binnenplaats
  r дворовый водосток *m*

Y5 *e* **yard lumber**
  d Schnittholz *n*
  f bois *m* de construction debité
  nl bezaagd hout *n*
  r строительный пиломатериал *m*

Y6 *e* **yard trap**
  d Hofablauf-Geruchverschluß *m*
  f siphon *m* de cour
  nl afvoerput *m* met stankscherm voor de binnenplaats
  r водяной затвор *m* дворового водостока

Y7 *e* **Y-branch**
  d Abzweigstück *n*, Abzweigrohr *n*, schräger Abzweig *m*
  f embranchement *m* simple, raccord *m* à culotte [en Y]
  nl enkel spruitstuk *n*
  r косой тройник *m*

Y8 *e* **year-round air-conditioning**
  d ganzjährige Klimatisierung *f*
  f conditionnement *m* de l'air toute l'année
  nl airconditioning *f* tijdends het gehele jaar
  r круглогодичное кондиционирование *n* воздуха

Y9 *e* **yellow ocher, yellow ochre**
  d Gelbocker *m*, Berggelb *n*
  f ocre *m* jaune
  nl gele oker *m*, berggeel *n*
  r жёлтая охра *f (пигмент)*

Y12 *e* **yield, yielding**
  d 1. plastisches Fließen *n*, Fließverhalten *n* 2. Nachgiebigkeit *f* 3. Ertrag *m*, Ausbringen *n*
  f 1. écoulement *m* [fluage *m*] plastique 2. déformabilité *f* 3. rendement *m (de la production)*
  nl 1. vloeien *n* 2. doorbuigen *n* 3. produktie *f*, opbrengst *f*
  r 1. пластическое течение *n*; текучесть *f* 2. податливость *f* 3. выход *m (продукции)*

Y13 *e* **yielding ground**
  d nachgiebiger Boden *m*
  f terrain *m* instable
  nl slappe grond *m*
  r неустойчивый грунт *m*

Y14 *e* **yielding support**
  d elastische Stütze *f*

f appui *m* élastique
nl elastische ondersteuning *f* [oplegging *f*]
r упругая опора *f*

Y15 *e* **yield point**
d Streckgrenze *f*, Quetschgrenze *f*, Fließgrenze *f*
f point *m* d'écoulement; limite *f* élastique [d'élasticité]
nl vloeigrens *f(m)*
r предел *m* текучести, начальное напряжение *n* текучести

Y16 *e* **yield range**
d Fließbereich *m*
f domaine *f* d'écoulement
nl plastisch gebied *n*
r область *f* текучести

Y17 *e* **yield strength**
d definierte Fließgrenze *f*
f limite *f* convetionnelle d'écoulement [d'élasticité]
nl onderste vloeigrens *f(m)*
r условный| предел *m* текучести

Y18 *e* **yield stress**
d untere Streckgrenze *f*, Fließspannung *f*
f contrainte *f* d'écoulement
nl onderste vloeigrens *f(m)*
r нижний предел *m* текучести; напряжение *n* текучести

Y19 *e* **Y-junction**
d schiefe Straßeneinmündung *f* [Straßengabelung *f*]
f bifurcation *f* routière [de routes], jonction *f* de deux routes en Y
nl wegsplitsing *f*
r У-образное примыкание *n* автомобильных дорог

Y20 *e* **yoke**
d 1. Bügel *m*, Befestigungsstrebe *f* 2. Horizontalelement *n* des Blendrahmens 3. Traverse *f*
f 1. collier *m*, étrier 2. traverse *f* de bâti dormant 3. traverse *f*, barre *f* transversale
nl 1. koppelhout *n* 2. dorpel *m* van een deur *of* raam 3. dwarsbalk *m*, jukbalk *m*
r 1. хомут *m*, бугель *m* 2. горизонтальный брус *m* дверной *или* оконной коробки 3. траверса *f*

Y21 *e* **Young's modulus of elasticity**
d Elastizitätsmodul *m*, E-Modul *m*
f module *m* [coefficient *m*] d'élasticité, module *m* de Young
nl elasticiteitsmodulus *m*, E-modulus *m*
r модуль *m* Юнга, модуль *m* упругости (*при осевом растяжении - сжатии*)

Z

Z1 *e* **zebra crossing**
d Zebrastreifen *m* (*Verkehrsüberweg*)
f passage *m* avec bandes zébrées [de peinture]
nl oversteekplaats *f(m)* voor voetgangers, zebra *m*
r пешеходный переход *m* типа «зебра»; нерегулируемый пешеходный переход *m*

Z2 *e* **zeolite water softening plant**
d Zeolithwasserenthärtungsanlage *f*
f station *f* d'adoucissement de l'eau par zéolithe
nl installatie *f* voor waterontharding met gebruik van zeoliet
r станция *f* цеолитового водоумягчения

Z3 *e* **zero end of survey**
d Nullpflock *m* (*Straßentrasse*)
f piquet *m* terminal [d'extrémité] (*du tracé de la route*)
nl nulpiket *f(m)* (*wegtracé*)
r нулевой пикет *m* (*дорожной трассы*)

Z4 **zero line** *see* **neutral axis**

Z5 *e* **zero moment**
d Nullmoment *n*
f moment *m* nul
nl momentennulpunt *n*
r нулевой момент *m*

Z6 *e* **zero-moment method**
d Nullmomentenmethode *f*
f méthode *f* des moments nuls [zéro]
nl momentennulpuntmethode *f*
r метод *m* нулевых моментов

Z7 **zero of gauge** *see* **gauge datum**

Z8 *e* **zero slump concrete**
d Steifbetongemisch *n* mit Nullkegelsetzmaß
f béton *m* frais ferme [à l'affaissement au cône nul]
nl stijve betonmortel *m* met zetmaat nul, stampbeton *n*
r жёсткая бетонная смесь *f* с нулевой осадкой конуса

Z9 *e* **zinc white**
d Zinkweiß *n*
f blanc *m* de zinc [de neige], oxyde *m* de zinc
nl zinkwit *n*
r цинковые белила *pl*

Z10 *e* **zonal ventilation**
d Zonenlüftung *f*
f ventilation *f* [aération *f*] zonale
nl zone-ventilatie *f*
r зональная вентиляция *f*

DEUTSCH

Beispiele für die alphabetische Anordnung:

Balken
~ auf elastischer Bettung
Balken‖abstand
~anker
Balkenanschluß
~ an die Säule [Stütze]

zulässige Belastung
~ ~ des Pfahles
~ Bodenpressung
zulässiger Auflagerdruck
~ Geräuschpegel
Zulaufstollen

Die Tilde (~) ersetzt das sich wiederholende Wort sowie den durch den senkrechten Doppelstrich (‖) getrennten Teil des Wortes. ·
Der Schrägstrich (/) verweist auf die Veränderung der Wortstellung in einer Wortverbindung:
Kräfte/ sich kreuzende = sich kreuzende Kräfte
Einzählung von Hinweisungen auf jeweiligen englischen Wortbegriff:
Absaughaube E440, 520; H434 = Absaughaube E440, E520, H434

A

Abänderung M367
Abänderungs-Klausel O193
Abbau D376, E421, Q17
Abbau‖bagger M303
~hammer D130
~höhe H272
~leistung O371
~wand Q15
abbauwilliges Felsgestein E61
Abbiegung der Bewehrung O35
Abbinde‖beginn I189
~beschleuniger A44, 46; C854, Q58, S340
~geschwindigkeit R108, S342
~n S333, 339, T427
~regelung S334
~verzögerer R447, S343
abbindeverzögerndes Kraftpapier S1625
Abbinde‖zeit C1394, I190, S345, T227
~zeitbeschleuniger A44
Abbindungsschwindung S344
Abblasen B474, P927
Abblättern D114, F327, P184, S109
~ der Farbschicht P28
abblätternsicherer Beton S107
Abbordgerät V99
Abböschen S750
Abbrechung C1416
Abbröckeln C1359, F740, R121
Abbruch D129
Abbrucharbeiten D131

Abbund T427
Abbürsten B701
Abdachungswinkel P441
Abdämmung D6
Abdeck‖blech F350
~gitter G287
~leiste C1240
~matte C1391
~papier O346
Abdeichung D300
abdichtende Zwischenlage S177
Abdichtung P809, S174
Abdichtungs‖gewölbe W253
~lage S175
~mastix J105, S176
~schicht I61
Abdruck P130
Abdrücken H590
Abdrückprüfung H590
Abessinierbrunnen D702
Abfahrtrampe E451
Abfall R255, 328, 835, W89
Abfall‖beseitigung R256
~beseitigungsanlage R257
Abfälle A1
Abfall‖mauer D730
~schleuse G235
~steine Q22
~stoff R255, W89
~verbrennung I81
~verbrennungsofen R258
~verwertungsanlage R259
~zerkleinerer W93
Abfangdrän C687
Abfangen der Quelle I306
~ mit Trossen G489

Abfang‖graben C200
~leitung I308
~sammler C200, 201, I308, O240
~träger N73, 75
Abfasung C363
Abflanschen N250
Abflanschmaschine N251
Abfluchtung A320
Abfluß D335, O304, R563, 853
Abfluß‖beiwert F519
~bereich D336
~dauerlinie F522
~fläche C193
~ganglinie D350
~gebiet C193, D599, 600, W229
~graben D605, G485
~grube D619
~höhe D157
~kanal D345, E401
~koeffizient D342, I57
~kurve D346, R116, S1111
~leitung D343, H128
~mengendauerlinie F522
~mengen-Summenlinie C1382, D352, F527, I294
~meßstelle D362, 364; G117, 118; M192, O246, S1351
~messung S1352
~modul D615
~norm N222
~öffnung D617
~querschnitt A604, D361, F518
~regelung D359, R283
~rinne C1383
~spende S952, U105
~summe F526, M122

gerades Riemchen F166
Geradschild B796
Geräte‖bestand S1061
~park E364
~vorhaltung C742
geräuscharme Pfahlramme Q64
Gerberträger H539
gerichtetes Blockmauerwerk C1227
Gerinne C369
Gerinnenhydraulik C375
Gerippe O219
Gerippetrennwand F700
Geröll G291
Gerölle R837, S477
Geruchverschluß H594, I308, T372, W98
~ mit Revisionsöffnung T372
Gerüst G25, O219, P574, S98, 1114; T412
Gerüst‖ausleger S100
~balken S100
~baum P685
~belag W485
~bock H472
~boden S1110
~brücke S99, T412
~etagen L234
~kran G26
~pfosten P618
~-Sperre F697
Gesamt‖abmessungen O286
~ansicht G126
~baukosten B729
Geschäftshaus C738
geschichtetes Blockmauerwerk C1227
~ Ziegelmauerwerk C1228
Geschiebe B281, D146
Geschiebe‖auslaß S124
~durchfluß B281
~falle B283
~fang B283
~fänger S234
~fracht B281
~last B281, L358
~menge S229
~mergel B588
~sperre D51
~stausperre D51
~umleitstollen S230
geschlossene Dränung B441, S1519
~ Vollfuge F360
geschlossener Absturz C563
~ Polygonzug C571
geschlossenes Dampfheizungssystem C572
~ Heißwasserheizungssystem P813
~ Kräftepolygon C567
~ Wasserkraftwerk I118
Geschoß F470 S1301
Geschoß‖anschlußrohr S1303
~haus M598
~hängedecke H537
~hebeverfahren L236
~höhe S1302
geschoß‖hohe Wandtafel S1304
~hohes Steigrohrelement S1303
geschütteter Erddamm [Fang-damm] E24
~ Spundwandfang(e)damm E26
~ Steinfangedamm R537
~ Vordamm E24
geschweißte Bewehrung W310
~ Konstruktionen W312
geschweißter Stahlstützenfuß W311
Geschw ndigkeits‖diagramm D160
~druck I43, V50
~gefälle V47
~höhe K37, V47
~profil V51
~zahl C646
gesickte Bleche R515
Gesims B105, C1183, L153
gespannte Membranüberdachung S1385
~ Schalenkonstruktion S1384
gespülter Erddamm H569
gespundete Fuge T248

~ Verbindung J75
Gestade S488
Gestaltfestigkeit M81, S1456
Gestängeschachtausbau R623
Gestell R3
gestelzter Bogen S1232
gestrandeter Senkkasten B603, M450
Gestrüppharke B703
Getäfel P49
getäfelte Decke P50
geteilte Straße D449
getränkte Schotterdecke G427
getränktes Preßvollholz D137
gewachsener Boden N47, U89
Gewächshaus G333
Gewände J22
Gewändeauskleidung J25
Gewässer W111
Gewässernetz D616
Gewichts‖dosierung W301
~teile P208
~verhältnis W306
~zumessung W301
Gewinde‖-Dichtungskitt P568
~stahleinlagen T150
~stutzen L492
~verankerung T149
Gewölbe V32
Gewölbe‖anfang I68
~bildung A573, B683
~dachfeld V34
~feld C766, V34
~krone C1351
~pfeiler L27
~reihen-Pfeilerstaumauer M547
~reihenstaumauer M547
~rippe V36
~scheitel C1351
~schub D679, V37
~stützmauer A570
~widerlager V33
~wirkung A573
~ziegel K16
~-Ziegelmauerwerk G110
~zwickel P191
gewölbte Decke W3
~Mauer A567
gezahnte Maurerschippe C617
gezahntes Kratzeisen C617
Gezeiten‖becken C60, E269
~kraftwerk T180
~marsch T178
~tor T181
gezogene Schräge D256
Giebel G3
Giebel‖fenster G7
~ mauer H342
giebel‖seitige Dachtraufe V97
~seitiger Dachvorsprung V97
Giebelwand F342, G6
Gieß‖beton H367
~norm R95
giftfreier Straßenteer N217
Gillmore-Nadel G159
Gilsonitasphalt G160
Gips G491
Gips‖arbeit P504
~bauplatte P493
~baustein G493
~bautafel G498
~beton G496
~block G493
~diele G503
~faserbeton G497
~form P499
~-Halbhydrat H284
~kalkmörtel G500
~kartonplatte G494, 505; P493, W24
~perlitmörtel P237
~platte P493
~putz G504, P491
~putzplatte G494, 502, 505,
~streifentafel V54
~wand(bau)platte G508
Gitter G346, 349; L5, 110
Gitter‖band L309

~belag D260
gitterförmige Bretterverschalung O144
~ Fundamentplatte G338
gitterförmiger Stampfer G289
Gitter‖konstruktionen B617
~punktmethode F218
~rost G287
~rostabdeckung M252
~stab M231, R649
~stampfer J63
~stütze L3
~tor H265
~träger L111
~walze G343
~werk T459
Glas‖arbeiten G217
~bauelemente S1439
~baustein G177
~beton T349
~betonkonstruktionen G179
Glaser‖kitt G209
~messer G215
Glas‖faser G133, 184
~faserbeton G186
glasfaserbewehrter Plastzusatzbe-ton P631
Glasfaser‖bündel G188
~gewebe G178
~isoliermatten G134
~litze G188
~matte G185
Glasfasern F139
glasfaserverstärkter Beton G186
Glasfaser‖vlies G189
~zement B526
Glas‖fassadengebäude G183
~filz G198
glasierte P rzellanfliese G206
~ Steinzeug(belag)platte [Stein-zeugfliese] E264
glasierter Ziegel V182
Glas‖lichtkuppel G180
~matte G185
~papier G192
~scheibe P56
Glasur E262, G201, 202
Glasur‖-Steingutrohrschneiden B7
~ziegel G203
Glas‖verbundscheibe I263
~vlies G189
~wollbahn G200
~wolle F139, G199
Glanz G221
Glätt‖band F215
~barkeit T437
~bohle F217, S134
~bohlenfertiger S818
Glätte S131
glatte Bewehrungsstähle P461
Glätteisen mit umgebördeltem Rand E93
Glätten S817, T439
glattes Ende S971
~ Sicht(flächen)mauerwerk F46
Glätt‖kelle F412, H41
~leiste S131
Glattmantelwalze S1227
Glätt‖maschine T440
~riemen F215
~schicht F414
Glattstreichen F418, S1
gleichfeste Bewehrung B78
~ Spleißung F786
~ Verbindung F786
gleichförmige Kornzusammenset-zung U95
Gleichgewichts‖bedingungen E355
~druck B80, E360
~-Klappschütz T212
~polygon E359
~störung D425
~system S1694
~zustand E361
gleichmäßig verteilte Last U97
gleichmäßige Setzung U98
Gleich‖schlag L61
~schlagseil E350

I

imaginäres Gelenk I21
ımhoffbrunnen D518, E255, I22
Immediumfilter E508
Impfschlamm R462
indifferentes Gleichgewicht N106
indirekte Beleuchtung C1232
~ Heizung I111
~ Spülung B39
Indirektleuchte R233
Induktions‖verhältnis I128
~zähler I127
Industrie‖abfälle P871
~abwasser I142, P871, T316
~ansiedlungszone I134
~bau I133
~baustoffe M74
~gebäude I132, P874
~heizung P870
~klimatisierung I131
~lärm I141
Infiltration S236
Infiltrations‖berechnung I162
~bereich I157
~geschwindigkeit R102
~wärmeverluste I161
Infrarot‖-Deckenheizgerät I180
~heizstrahler I178
~heizung I179
~strahler I178, 181
~trocknung I177
Ingenieur‖bau C481
~bauklinker C554
~bauten E313
Ingenieurbauwerke E313
~ der Stadtwirtschaft M604
Ingenieur‖geologie E308
~hochbau S1436
ingenieur‖mäßige Erkundungen
    E312
~technische Ausstattung [Er-
    schließung] des Baugeländes A394
~technischer Ausbau A394
Inhaltslinie A606, V195
Inhomogenität L8
Injektions‖anlage G436
~gang G433
~gründung I193
~gut G439
~loch I197
~mörtel G439, 442
~mörtelanlage G432, 436
~pumpe G443
~schlauch G435
~schleier G421
~schürze G421
~stutzen G437
Injektor I202
Injizier‖glocke G431
~gründung I193
~gut G439
~lanze I198
~mörtel I194
~pumpe G443
Inkrustation I97
Innen‖architektur I316
~ausbau [~ausstattung] des Ge-
    bäudes I320
~beleuchtung I116
~bogen A624
~druckprobe H590
~feld I322
~fenster B565
~fläche I214
~gestaltung E271
~gewölbe A624
~hof A596, P128
~klima I115
~rohrleitungen I117
~luftaufstau B760
~putz P497
~raumgestaltung P480
Innenrüttler I360, S1081
~ mit Einbaumotor M501
Innen‖schale I358
~seite I214
innenseitig verglastes Außenfen-
    ster I321
Innen‖stütze I323
~teil M52

~temperatur I120
~verdrahtung I361
~verglasung I352
~vibrator I360, S1081
~wand I326
~zone I328
innere Blendleiste I226
~ Bogenfläche I366, S831
~ Dachentwässerung C986
~ Elektroinstallation I361
~ Erosion P435, W85
~ Gasleitung G66
~ Kräfte I350
~ Leitungsnetze I249
~ Reibung I351
~ Rißbildung H431
~ Rohrleitungen I355, P569
~ Wandschale I358
innerer Reibungsbeiwert C629
~ Reibungswinkel A461
~ Verankerungskonus A431
Insolation I233, S1563
Insolationslage S873
Inspektionskappe I235
Instabilität I248
Installation I249
Installations‖block P571
~element M203
~kern M203, U201
~raum M199
~wand M204, P571
~zelle U201
installierte Leistung I251
Integrität der Baukonstruktionen
    S1441
Intervallzünder D115
Intrapermafrostwasser I367
Inundation I368
Inundationsgebiet F469
Inventarschutz‖geländer S1145
~sperren S1145
Isolator I411
Isolatorkette I272
Isolier‖arbeiten I268
~massa M130
~mastik L572
~matte B200, 414, 416; M158
~pappe I259
~platte B200
~rohr I271
~schicht I265
~stein I260 266
~stoffe I269
Isolierungsarbeiten I268
Isolierwert I267
Istbauzeit A95
Istegstahl T526
Iterationsverfahren A404
I-Träger I1, 405

J

Jagdzapfen A39
jährlicher Abfluß A487
Jolousie A2, B447, L518, V61
Jalousie‖auslaß V25
~gitter L518
~klappe L518, M577
Japaner R632
Jochbalken T355
Joosten-Verfahren J113
Jumbo-Bohrwagen C149

K

Kabel C4
Kabel‖bagger C22 D592
~kanal C11, 987; D784, R2
~klammer C6
~klemme C6
~kran C21, H400
~kran-Maschinenturm H151
~muffe C12

~rohr C987
~säge C18
~schutzrohr C15
~verlegemaschine C14
Kabine C1, 2; B544
kabinenloser Kran G386
Kachelplatte E268
Kai E235, Q35, W346
Kai‖anlage B337, 338; W346
~mauer B788, Q38
~pfeiler Q36
~treppe Q37
Kaje Q38
Kalfaterhammer C217
Kalfatern C216
Kalk L271
Kalk‖ausblühung L277
~ausschlag L277
~beton L276
~brei L283
~-Gips-Oberputz G108
~-Gips-Putzdeckschicht G108
~grube L280
~härte C38
~kitt L273
~klärung L282
~löschgrube L280
~löschkasten M472
~löschtrommel L285
~mehl F516
~milch M287, W358
~mischer L272
~mörtel L278
~pulver F516
~putz L281, U93
~rührwerk L272
~sandbeton S565
~sandmörtel N182
~sandstein C41, L284
~schutzanlagen E399
~stabilisierung L287
~-Soda-Verfahren L286
~stein L288
~steinmehl F516
~steinpulver F516
~teig L283
~ton C33
~verfahren L282
~wasser L289
~zement L273
~zementmörtel C273, 782; G107,
    L274
Kalotte C1351
Kalt‖asphalt E256
~asphaltspritzmaschine E261
~bindemittelspritzmaschine E261
Kälteanlage R250
kalt‖einbaufähiger Asphaltbeton
    C662
~eingebaute Asphaltbetondecke
    C663
Kälte‖kammer C1142
~kompressor R252
~leistung C1150
~mittel R246
~quelle C664
~system C667
~träger C1148, S201
~verbrauch R248
~versorgung C666
Kaltverfestigung C672, S1335
kaltverformter Betonstahl C671
Kaltverformung C672
Kalziumchlorid C37
Kamin C424
Kamin‖drän C429
~entwässerung C429
~formstein C426
~sims M68
~stein C426
~vorsprung C427
Kamm R501
Kammer C360
Kämpfer A36, I68, M545 **S1078,**
    T355
Kämpfer‖gelenk P453
~holz I68
~linie S1074

kleine Brechstange J61
Kleineisenzeug H91
Klein‖gestein G355
~grünanlage V130
~mastenkran M447
~mechanisierungsmittel L2, R517
~-Planierschlepper C53
~räumer C53
~rohrheizung S806
~werkzeuge S807
~-Zellenfangdamm G1
Kleister P116
Klemme G353
Klemmvorrichtung C485
Klempner- und Installationsarbeiten P569
Kletter‖eisen S1236
~kran C1283, K1
Klettern J118
kletternder Montagekorb P547
Kletter‖schalung C550, J119
~turmdrehkran C1283, K1
Klimaaggregat A194, 200
Klimaanlage in Splitausführung S1014
~ mit autonomer Energieversorgung T284
~ mit indirekter Kühlung I110
~mit konstantem Luftvolumen C1018
~ mit Luftstromregelung V28
~ mit Nachheizung R291
~ mit Nachkühlung R189
~ mit Nachwärmern T92
~ mit variablem Förderstrom V28
Klima‖gebiet C547
~gerät A194
~geräteeinheit U106
~kammer C548, E332
~karte D202
~konvektor U103
~last A197
~raum C548
~schrank C3a
~station A199
~technik A195
klimatisierter Raum C982
Klimatisierung A195
klimatologische Daten C549
Klima‖unterzentrale A199
~zentrale A198, C304
~zone C547
Klinke D31, L228
Klinker C552; V182
Klinker‖pflaster B662
~ziegel E307
Klinometer C555
Klopfwerk S396
Klosett C574
Klosettbecken mit Spülkasten T246
Klump‖fuß P175
~fußpfahl B293, 552
Klotz B55
Knagge B260, C540
Kneter P934
Knetmischer P934
Knick B313, 718, 719
Knick‖ausleger A634
~beiwert B722, S1400
~berechnung B720
~festigkeit B727
~figur B723
~form M364
Knicklänge B724, E109, 111
~ der Säule C704
Knick‖längenbeiwert E112
~last B725, C1302
~merkmal C1301
~spannung B728
Knickung B718, 719; L80
Knick‖w derstand B727
~zahl B722, S1400
Knie‖rohr E156, K52
~schutztafel K53
~stück B313, E156, K52, P376
Knirschfuge N160
Knoten N131, P52
Knoten‖belastung J99, N129

~blech G484, J123
~kräfte N129
~ ast J99, P51
~punkt N130, S1697
~punktverfahren M269
~schnittverfahren M269
~steifigkeit J101
Kochbrenner G82
Kog P617
Kohäsion C653
kohäsionsloser Boden C654, F744, N166
Kohlen‖dioxid-Feuerlöscher C134
~faden C135
~wasserstoffbindemittel B395
Kokskorb F231
Kolk W85
Kolk‖schutz D572, S125 976
~schutzschotterschicht S1276
~vertiefung S121
Kollaudierung A47
Kollektor M61
Kolloidalbeton C690
kolloidaler Zementmörtel C691
Kolloidmörtelmischer C692
Kolmatage C560, 693
Komfort C730
Komfort‖bedingungen C733
~kühlung C734
~luftkühlanlage C735
~zone C737
Kompakt‖kaltwassersatz P7
~klimagerät P2, U106
~wasserkühler P7
Komplexbauprogramm P4
Komposition C796, 782; I299
Kompositionsplan L127
Kompression C822
Kompressions‖gerät C1015
~grad D103
~versuch C834
~welle C836
Kompressor A193, C843
Konditionierung C983
konjugierte Tiefen A365
konkave Mauerfuge C847
Konkavität S24
Konoidschale C1004
Konsole B619, C1157
Konsolegerät B620
Konsolidierung C1012, E17
Konsolidierungs‖beiwert C620
~sperre D51
~versuch C1014
Konsol‖kran W33
~stützmauer C104
~träger B256
Konstruktion C1021, S1461
Konstruktions‖beton C1430
~hilfselement S199
~holz C139, S1459
~leichtbeton L268
~merkmal D185
~richtlinien C1049
~teil M231
Kontakt‖belüftung C1071
~druck C1077
~filtration C1075
~fläche C1078
Konter‖bankett C1199
~bogen I373
~mittel L452
kontinuierliche Kornabstufung C1097
~ Mischanlage C1101
kontinuierlicher Balken C1089
~ Mischer C1100
~ Träger C1089
kontinuierliches Betonieren C1092
Kontinuität C1084, S914
~ der Baukonstruktionen S1441
Kontinuitäts‖gleichung C1086, E351
~moment C1087
~satz L117
Kontraktion C1113
Kontraktions‖fuge C1114
~koeffizient C621

Kontraktor‖beton T398
~(beton)rohr T397
~verfahren T399
Kontrollabsturz C400
Kontrolle C399, M445
Kontrollgang F617, G10, I242
kontrollierte Lüftung C1125
~ Spannkraft J8
~ ~ im Bewehrungsbündel P826
Kontroll‖kanal C1121
~prüfung C1130
~punkt C1129
~schacht M57, I246
~stollen G10, I242
Kontur C1107
Konvektions‖heizung C1132
~wärmeaustausch C1134, H245
konzentrierte Kraft C850
Koog P617
Kopf‖balken H124
~band A442, K54, S1467
~regler H123
~rohr H128
~stein C610
~stück H118
~strebe I87
~verband H125
~wand H152
~widerstand D589
Koppelschleuse C352
Korbbogen C802
Kork‖fliese C1175
~linoleum C1172
~platte C1171, 1175
~schalenisolierung C1173a
Kornabstufung G261
~ der Zuschläge A167
Korn‖ährenverband D251, R67
~analyse G261, P99
~form P97
~gestalt P97
~größe G269, P98
Korngrößen‖bereich F684, S651
~trennung S1400
~verteilung G271
körnige Baustoffe G278
~ Böden D278
körniger Dämmstoff G277
Korn‖klasse F684, S651
~klassenbild S653
Körnungs‖kennlinie S653
~kurve G262
~modul F204
Korn‖verteilungskurve P101
~zusammensetzung G261, 271
Körper B489
Körperschalldämmung S1451
Korridor C1184
Korrosions‖inhibitor C1186
~schutz C1185
~schutzanstrich A499
~schutzmittel C1186
~schutzpigment I185
~verhütung C1185
Kostenkalkulation C1196
Kote E210
Kraft F618
Kraft‖abnahme P715
~angriff F619
~angriffspunkt O208
Kräfte‖dreieck F629
~gleichgewicht E358
~gruppe S1196
~paar C1221
~parallelogramm P72
~polygon P626
~system P628
~umlagerung R206
~zerlegung F627, R424
Kraft‖größenverfahren M426
~haus P710
~komponente C785
~-Luftheizung F61
~-Luftheizungsaggregat F62
~methode A600, M426
~papier K66
~ramme P712
~schaufel P713

Papier‖filter P65
~tapete W50
Papillionierung C1326, D671
Pappe C141
Pappnagel F120
Papputz P79
parabolische Schale P67
Parallel‖fachwerkbinder P69
~fachwerkträger F354
~gurtbinder F354, T456
parallel(laufend)e Jalousieklappe P68
Parallel‖projektion A853
~schlag L61
~stoß E88
~träger F354
~werk L481, T334
Parapett S1312
Parkettäfelchen W443
Parkettafelfußboden B454
Parkett‖asphalt A687
~-Klebemasse A687
~stab W443
Park‖platz C143, L120, P85
~steig W18
~turm A800, P86
~weg W18
Passage A608
Paßbreite F293
Passung F290
Paste P116
Paternoster-Aufzug P126, V127
Pause T305
Pavillon P150
Pavillondach P151
Pech P441
Pech‖ausbringen B434
~mastix P449a
Pegel B312, G102, 118; S1109
Pegel‖latte G102
~meßstation G118
~null(punkt) G105
Peillatte S1497
Peilung S909
Pendel‖klappe T212
~lager P192
~pfeiler H377
~säge P193
~schwingung H540
~tür D508, S1680
~turm L561
Pendelung H540
Pendentifkuppel P190
perforierte Hängedeckenplatte P226
~ Platte P227
Perimetral‖heizung P233
~einpressung P232
~zone P235
Perlatorauslauf S1055
Perlitputz P237
Perlkies P163
Permafrost P238
Persenning T44
persischer Bogen K3
Personen‖aufzug P113
~-Druckluftschleuse M63
~schleuse M63
~-Seilschwebebahn F797
Pfahl P292
Pfahl‖absenkung durch Vibration V159
~abteufung per Schlag P195
~auszieher P314
~bündel D464, P299
~einpressen P326
~einrütteln V159
~-Faschinenpackung P295
~gruppe P299
~gründung P318, 343
~holm T187
~joch P297
~-Kastengründung P294
~konstruktion P342
~kopf P323
~poller P299, 300
~rammausrüstung P306
~ramme R506

~rammen H35, P304, 343
~reihe P334, 343; R812
~rohr P298
~rost P319
~rostkaimauer P301
~rostpfeiler P330
~rücksprung R163
~schuh P336
~spitze P340
~tragfähigkeit A345
~wand R812
~werk P319, 342
~zieher P314
~zopf P329
Pfannendachziegel D544
Pfeiler B831, D803, P278, 344
Pfeiler‖fundament P683
~gewichtsstaumauer M120
~gewölbestaumauer M547
~kopf B834, C1435, I2, P279
~kopfsperre R806
~kopfstaumauer R806
~krafthaus B835, C1213
~kuppel-Staumauer M572
~nase C1435
~-Plattensperre F371
~sperre B833, C1211
~staumauer B833
~stützmauer C1212
~stützwand C1212
~-Wasserkraftwerk B836, P281
Pfeilhöhe H270, R538, S24
Pfette B257, P982
Pflasterer P147, P161
Pflaster‖fugenausfüllung J94
~klotz P154
Pflastern P152
Pflaster‖platte F323, P158
~stein 154, 161; S338
Pflasterung P448
Pflaster‖würfel P154
~ziegel P156
Pflock P36, 185
Pflug P557
Pfosten C700 J22, M124, P344, 618, 679; S1129, U167
Pfriem(en) A823
Pfropfen D829
Picke P273
Pickel P273, 353
Pickelhacke S1080
Pickhammer D130
Pier J48, P278
~ in durchbrochener Bauweise O129
Piezometer P282
Piezometer‖rohr P283, S1155
~stand P286
piezometrische Linie H573
piezometrischer Druck P285
piezometrisches Gefälle P284
Pigment C697, P288
Pilaster P289
Pilz‖decke F372, G169, M609
~deckenplatte F370, M611
~deckenstütze M610
Pilzkopf C707, F346
Pilzkopfstaumauer R806
Pinsel B698
Pinselanstrich B701
Pisébau P437
Piseebau P437
Pißbecken U192
Pissoir U192
Pissoirbecken B599
Piste R858
Plan A620, D172, P464, S116
Plane C107, T44
plangleiche Kreuzung A729
Planier‖arbeiten G261, L184
~bagger S660
~gerät G254, L177
~raupe B796
~schild B412
~schleppe D589
~schneckenfräse A749
Planierung L179
Planierungsarbeiten L184

Planung P478
Planungsnormen P482
Plastifizierung P523
Plastifizierungsmittel P517
Plastikrohr P528
plastische Arbeit I144
~ Deformation R404
~ Durchbiegung R403
~ Knickung I145
~ Konsistenz P510
~ Rißbildung P511
~ Schwindung P533
~ Verformung P512
~ Zone P529
plastischer Baumörtel P527
~ Bereich N177, P529, I147
~ Boden P535
~ Zustand P536
plastisches Berechnungsverfahren C676
~ Betongemisch P509
~ Fließen P515, Y12
~ Gelenk P516
~ Gleichgewicht P514
~ Moment P526
~ Schwinden P533
Plastizität P518
Plastizitäts‖gebiet I147, N177
~index I106, P519
~modul P520
~theorie T114
Plast‖überzug P530
~zusatzbeton P627
Plattbogen J2
Platte P42, 540; S439, 670
Platten‖bandspeiser A544
~bauweise P44
~brücke S671
~dämmstoff S674
~decke S672
~elektrofilter P542
plattenförmiger Kornzuschlag-(stoff) F328
Platten‖fundament B809, M152, S673
~gründung B809, M152, R37
~heizkörper P53
~heizung P48
~isoliermaterial B484, S674
~isolierung R509
~konstruktion P44
~kühlung P46
~-Luftverteiler P490
Plattenpfeiler(stau)mauer F371
Platten‖rüttler P546, S679
~schieber G99
~ventil D375
~vibrator P546, S679
~wand P55
Plattform D57
Plattform(bau)aufzug M146, P548
Platthacke M156
Plattierungsschicht C482
Platz A596
Platzheizung S1029
Plinthe B158, P554
Plotter P556
PM-Binder M103
pneumatische Beschickung P591
~ Förderung P587
~ Glättkelle A235
~ Konstruktion(en) I165, P596
~ Winde A314
pneumatischer Hohlraumbildner I166
~ Aufzug A245, P593
~ Betonförderer A269
~ Heber C811
~ Spatenhammer A216
~ Wellenbrecher A181, B708
pneumatisches Wehr I167
~ Werkzeug A300
pneumatisch-hydraulischer Hebebock A249
Pneuwalze P600, R832
Podest L45, P547
polares Trägheitsmoment P616
Polder I218, P617

ZUSATZ

FRANÇAIS

Les termes sont disposés dans l'ordre alphabétique. Dans la famille de mots le terme principal est toujours mis à la première place et est remplacé par le tilde (~). Par exemple:

bâtiment
~ à charpente
~ annexe
~ à ossature
~ bas
~ d'atelier
~ de ferme
~ des machines
~ de turbines
~-tour concentré
~ type
bâtiments agricoles
~ bas
~ non élevés

La combinaison contenant des lettres précédant des chiffres C92, 1136; T216

il faut comprendre comme
C92, C1136; T216

A

abaissement D785, S1509
~ du niveau D656
abaque C396, D174
abatis R835, 837
abat-mousse F775
abattage de mousse F776
~ (des pierres) en carrière Q17
abattement A1
abattis R835, 837
abat-vent A2, M93
~ de cheminée P913
abducteur d'égout principal J308
abouchement J120
about A39, H471
aboutement A36
~ des ronds à béton par manchons S693
~ du pieu P338
abrasifs A18
abrasion A9, S1276
abri C1, S470
~ à l'usage [l'intention] des travailleurs W498
~ chauffant O288
~ du conducteur de travaux C2
~ souterrain U64
absence de défaux intérieurs S914
absorbant l'infrarouge H163
absorption de chaleur H164
~ de vibration V151
~ du son S902

~ phonique S902
absorptivité acoustique A73
accélérateur A46
~ de durcissement C854, H79
~ de durcissement liquide L345
~ de prise A44, Q58, S340
accès A50, 536
accessoires A56, F296, R517
~ appartenances A542
~ de chaudière B496
~ de levage L220, 226
~ de levage à vide V7
accotement H107, V97, S504
~ de la route R602
~ stabilisé H87
~ traité H87
accotements pavés P134
~ revêtus P134
accouplement à fiches P562
accrochage H57
accroissement de la charge L388
~ de pression dans un conduit B35
~ de résistance S1367
~ de tension graduel G267
accroissements d'affaissement J96
accumulateur à eau chaude H494
~ de chaleur H235
accumulation D220
~ de chaleur H165, 235
~ par pompage P956
achèvement C780
~ définitif des travaux F189
acier allié A350
~ à béton R326

~ à haute résistance H343
~ coulé C184
~ d'armature R326
~ de construction S1453
~ de haute tension H343
~ de précontrainte P827
~ fondu C184
~ galvanisé G17
~ mis en tension P829
~ moulé C184
~ plaqué C484
~ profilé S213
~ transversal L89
aciers F305
~ Isteg T526
acoustique architectural A576
~ de salle R754
action composite C786
~ de contact à l'appui R131
~ de réglage C1119
~ des forces d'inertie I151
~ secondaire de l'adjuvant S1577
additif de ciment C255
~ du béton C855
~ du mortier (de ciment) M470
~ fluidifiant F545
~ liquide du béton L347
~ mouillant W342
addition A101, 123
adent J74
adhérence B514, C653
~ mécanique M200
~ des roues G353

bande médiane boisée L52
~ saillante F166
~ stabilisée de l'accotement H77
~ vitrée W380
bandes de sable S57
bandeau B105; 303; L153
~ saillant F76, P539
~ saillant (en briques) O358
banquette B127, O34
~ de pied C1199, T240
barbotine B205, S720
bardage C482, L325, S435, 441;
W56
~ de finition S550
~ en amiante-ciment A658
~ en bois B483
~ en planches T217
~ en planches jointées en biais
B344
~ en planches verticales P37
bardeau S477
~ de bois de couverture R739
~ de couverture R739
barème C396
barrage D3, 6
~ à aiguilles N81
~ à bouchures mobiles M520
~ à clapet L137, S532
~ à contreforts B833, C1211
~ à contreforts à dalle plane
F371
~ à contreforts à tête en T T111
~ à contreforts à tête massive
M120
~ à contreforts à tête ronde C100,
R806
~ à contreforts à tête octogonale
D262
~ à coupole C1384
~ à dôme C1384
~ à dômes multiples M572
~ à faible hauteur L538
~ à (fermettes et) aiguilles C677
~ à hausses S532, 711, 790; V117
~ à haute chute H312
~ à pertuis S711
~ à poids-voûte T138
~ à vanne autoréglable S287
~ à vannes pivotantes W362
~ à vannes roulantes R670
~ avec évacuateur à puits D721
~ avec enveloppe en treillage
métallique W431
~ avec vanne à tambour D743
~ à voûtes multiples M547
~ cellulaire C228, 247
~ circulaire R520
~ d'accumulation S1293
~ de dénivellation B139, D437
~ de dérivation D437
~ de garde B139
~ de gorge G237
~ de gravité G310
~ de remblai E25
~ de retenue D222, I70, S1293
~ de route P179
~ de sédimentation D51
~ déversant C533
~-déversoir C533, O300, 301, 318;
W307
~-déversoir à basse chute O318
~-déversoir à hausses D655
~-déversoir à nappe libre F717
~-déversoir démontable F345
~ en béton C892
~ en béton précontraint P817
~ en bois remblayé de terre E28
~ en charpente C1295, F697
~ en charpente de bois W448
~ en charpente de bois remblayé
de terre E28
~ en enrochements R637
~ en maçonnerie ordinaire à sec
L504
~ en rivière B139
~ en sol homogène H428
~ en terre E19

~ en terre à noyau d'étanchéité
C1170
~ en terre et enrochement F160
~ en terre et enrochements E8
~ en terre homogène H428
~ en terre remblayée hydrauliquement H569
~ en toit B270
~ en voûte mince T140
~ fixe F317
~ gonflable I167
~-gravité voûte G307
~ hydroélectrique H609, P707
~ mixte en enrochements et en
terre R637
~ mobile F697, M520
~ mobile à clapet S269
~-poids G310
~-poids à voûte A572
~-poids courbe G307
~-poids voûte G307, T138
~ pour le bois de flot G454
~ pour le bois flottable F419
~ protecteur contre les inondations F467
~-réservoir S1293
~ souple F7
~ souterrain G408
~-usine B456, P707, W194
~-voûte A567
barre B128, C1338, R649
~ à béton tordue C668
~ à mine S1080
~ ancrée A434
~ à niveler S133
~ d'acier avec œillets aux extrémités E550
~ d'ancrage A434
~ d'appui B259, H50
~ d'armature B128, R153
~ d'armature avec crochets aux
extrémités H440
~ d'armature crénelée R490
~ d'armature enrobée par époxyde E340
~ de ferraillage R153
~ d'entretoisement T184
~ de suspension H56
~ de treillis M231
~ diagonale D249
~ diagonale comprimée D255
~ diagonale en traction D256
~ égaliseuse H40
~ en attente S1471
~ horizontale de contreventements
G173
~ lisseuse H40
~ longitudinale L478
~ niveleuse S133
~ principale M22
~ relevée B334
~ ronde R804
~ saillante P921
~ transversale S447, Y20
~ verticale U167
barreau d'échelle S1183
barreaux R851
~ d'échelle M60
barres crénelées I98
~ déformées I98
~ d'armature assurant la continuité C1085
~ d'armature crénelées D98
~ d'armature de répartition D401
~ d'armature droites S1324
~ d'armature lisses P461
~ de suspension B244
~ relevées T455
~ supérieures T251
barricade B145
barrière P179
~ antifeu F224
~ contre le bruit A82, **S905**
~ contre le vent W371
~ de protection S9
~ de protection contre le relèvement d'eau H542

~ de protection sur la bande médiane M224
~ de sécurité P179, S9
~ de sécurité sur la bande médiane
M224
~ d'insonorisation phonique N141
~ flottante L466
~ protection G458
~ thermique T119
basculage directement dans le coffrage D320
bascule R629
base B170, F608, 666
~ de peinture P23
~ de pieu en bulbe P329
~ de poteau C701, S1130
~ du barrage T240
~ du stabilisateur O263
basse de colonne C701
~-mer supérieure H306
bassin B182, C766, D456, P640,
644; S1285
~ accumulateur P956
~ à déchets de lavage S1022
~ à flot W333
~ clarificateur primaire P762
~ d'accumulation R382
~ d'achèvement F294
~ d'activation (des eaux usées)
S374
~ d'aérage à écoulement continu
C1094
~ d'aérage à plusieurs couloirs
M570
~ d'aération à écoulement continu C1094
~ d'aération à plusieurs couloirs
M570
~ d'aération à un seul couloir O77
~ d'aération des eaux usées O383
~ d'alimentation d'une formation
aquifère C194
~ d'amortissement A30, S1256,
T236
~ d'armement F294
~ d'arrosage C399
~ d'attente O261
~ d'eau de pluie S1318
~ de boues activées A136
~ de clarification S233, 352
~ de clarification primaire P762
~ de coagulation C590
~ de compensation B86, C774,
E347, S1643
~ de décantation D206, S222,
350, 352
~ de floculation F447
~ de marée E269
~ de mise en charge F630
~ de natation N25
~ de neutralisation N107
~ de port B183, H65, W333
~ de prise des eaux usées S373
~ d'équilibre S1643
~ de réception I274
~ de réception des eaux usées S373
~ de régulation à débit non
réglable R448
~ de régularisation diurne D2
~ de répartition S1069
~ de retenue I71, R382, S1299
~ de stabilisation S1096
~ de submersion C399
~ de tranquillisation A30, C1409
~ d'évaporation E409
~ de virement T513
~ économiseur d'eau S540
~ haut O44, P956
~ hydrographique D600
~ inférieur L532
~ intérieur A548
~ journalier D27
~ portuaire B183
~ récepteur R451
~ refroidisseur d'eau par aspersion S1054
~ supérieur U163
~ versant C193, D600, W229

bride de raccordement U101
~ de suspension H56
~ feinte B443
brillant G221
brin P576
~ de suspension V114
brique B653
~ aérée A182
~ à feu F226
~ antiacide A72
~ à trous P224
~ biaise A444, K39
~ bien cuite B490, W318
~ brute A125
~ chanfreinée A444, K39
~ creuse H414
~ creuse de parement intérieur F803
~ crue A125, C609
~ cuite B66
~ d'aérage A182
~ d'alignement T230
~ d'alumine réfractaire A388
~ d'ancrage B520
~ d'angle Q68
~ d'arc A562
~ d'argile ordinaire C746
~ de bois G379
~ de champ B800, S880
~ de clinker C554
~ de diatomite D269, K34, M422
~ de Dinas D311, S562
~ de forme M409
~ de kieselguhr K34
~ de liaison B520
~ de maçonnerie B742
~ de magnésie M13
~ d'entrevous F478
~ de pavage C4554, H69, P156, 161; V182
~ de planchers F478
~ de remplissage B23, L326
~ de revêtement L326
~ de revêtement de sol F478
~ de silice C41, S562
~ de terre cuite ordinaire C746
~ de ventilation A182, V76
~ de verre G177
~ de verre creuse H418
~ de voûte A562
~ deux quartiers B187
~ diatomite D269, K34, M422
~ dure W318
~ économique E82
~ émaillée G203, V182
~ en béton C870
~ en boutisse H124
~ en coin F109, G106, K16
~ en panneresse S1405
~ hollandaise D827
~ modulaire M369
~ moulurée M409
~ murale modulaire M378
~ perforée H414, M557, P224, V40
~ posée de champ R810
~ pour cheminées C433
~ pour maçonnerie dissimulée I356
~ profilée M409
~ radiale R6
~ réfractaire R240
~ réfractaire isolante R243, I260
~ siliceuse C41
~ vitrifiée E307, G203, V182
briques de parement F22
~ de remplissage B35, 37
briqueterie B668
briqueteur B658
briquette B686
brise-avalanche A811
brise-béton C869, P155, R573
brise-glace I12
brise-lames B642, J48, M417, P76, 278; S183
~ à batardeau cellulaire C252
~ à claire-voie F694, P223, S757
~ à parois verticales W27
~ en enrochement R843

~ flottant F420
~ hydraulique H553
~ pneumatique A181, B708
brise-mer S183
~ à batardeau cellulaire C252
~ immergé S1495
brise-mousse F775
brise-soleil S1562
brossage B701, W428
brosse B698
brouette B149
broyage C1368, G351
broyeur B630, C1365, P942
~ à boulets B100
~ à ciment C265
~ à cylindres R673
~ à meules P934
~ à meules pour glaise P934
~ d'ordures domestiques W93
~ d'ordures ménagères R836, W93
~ évier W93
~-malaxeur à peinture M288
~ mobile sur pneus R827
~ primaire P839
broyeuse à rouleaux R673
bruit aérien A180
~ ambiant A390
~ de travail I141
bruit d'imact I39
brûleur B816
~ de la cuisinière à gaz G82
brumisage F590
bulbe B783
~ du pieu P175, 329
bulldozer B796
~-chargeur D584
~ de petites dimensions C53
bulle B448, 796
~ chargeur D584
bureau de chantier mobile M359
buse à pulvériser P944
~ d'injection I198
butée D460, F605, T176
butoir D460
~ de porte D501
by-pass B842, 844
~ de décharge U130

C

cabestan C125
cabine B544, C1, 2
~ d ascenseur E213
~ de peinture au pistolet S1035
~ de peinture mobile [roulante] T384
cabinet d'aisances C574, I116, W123
~ de toilette L116
~ de W.C. T246
câblage L118
~ à gauche L159
~ Lang à droite L61
~ parallèle L61
câble C4, 986; R768
~ à âme composé de six torons C13
~ aérien de transport T358
~ antigiratoire N199
~ armé A614
~ chauffant posé en sous-couche de plancher U50
~ d'acier S1229
~ d'acier à câblage parallèle E350
~ d'acier à torons S1340
~ d'acier fermé L445
~ d'armature S1337
~ d'armature de précontrainte C4
~ d'armature injecté G429
~ d'assurance L204
~ de chauffage électrique H189
~ de chauffage posé en sous-couche de plancher U50
~ de contrepoids C776
~ de fermeture C578
~ de flèche D168, T262
~ de halage D294

~ de haubanage G486
~ de levage H402
~ de levage d'ascenseur L229
~ de levage du poids L406
~ de mise à la terre G383
~ de monte-charge H402
~ de précontrainte P829, T72
~ de précontrainte excentré E76
~ de précontrainte non adhérant U30
~ de retenue H410
~ de traction D590
~ électrique E162
~ électrique armé A614
~ électrique traîné T380
~ en chanvre de Manille M62
~ en fibres de nylon tressées B622
~ en fils d'acier S1229, W429
~ en fils d'acier fermé L445
~ en jute J125
~ en nylon N272
~ enterré G383
~ extérieur de précontrainte E532
~ grue C21
~ injecté C429
~ plat F367
~ porteur H410
~ souterrain G383, U53
~ suspendu S1660
~ tracteur D294, 590
cadastre hydraulique W118
cadenas P19
cadre F693
~ A A149
~ contreventé B616
~ d'échafaudage F704
~ en A A149
~ portail P661
cage S461
~ aérée V67
~ d'armature C25
~ d'armature en spirale H279
~ d'ascenseur E221, H405
~ de ferraillage en spirale H279
cahier des charges S948
~ des charges de contrat C1118
~ des charges de marché C1118
~ des charges du fournisseur C739
~ des charges et conditions d'un contrat S939
~ des charges et conditions d'un marché de travaux S949
~ des charges générales O259
~ des charges pour la construction (d'un ouvrage) P889
~ des charges type G468, S1148
~ des clauses générales O259
~ des clauses techniques S1448
caillebotis D260
~ en grille d'acier S1218
~ métallique M252
caillou C610, P171, S338
cailloux R837
caisson C648
~ à air comprimé P585
~ à fond M450
~ de réacteur C1079, R138
~ du réacteur nucléaire B148
~ en béton C873
~ flottant B603, F421
~ flottant échoué M450
~ géant B603
~ ouvert C26, D727, O101
~ ouvert en béton C873
~ perdu D727
~ pneumatique C26
calage W299
calcaire L288
~ argileux A609
calcul D172
~ à la charge critique C1311
~ à la rupture L384, U11
~ à l'état limite L301
~ approximatif A541
~ aux efforts séismiques S259, 262
~ aux séismes S259

maçonnerie de brique creuse armée Q45
~ de briques B660
~ de gros blocs en assises inclinées S699
~ de moellons à sec D758
~ de moellons bruts non appareillée U38
~ de moellons en rangées régulières R271
~ de parement F46
~ de pierre sans ravalement E504
~ de pierre(s) taillé(s) A674
~ de remplissage B9, 22, 35, 37
~ en assise C1228
~ en assises réglées O184
~ en blocs de béton C867
~ en blocs de béton appareillée C1227
~ en briques B660
~ en briques à demi-sec B654
~ en briques armée R293
~ en briques renforcée R293
~ en galets C612
~ en grands blocs L72
~ en moellons C612, R842
~ en pierres brutes C612
~ en pierres de taille A674
~ en pierres naturelles N59
~ en pierres sèches D758
~ en pierres taillées A671
~ massive en moellons B464
~ ornementale en boutisses inclinées R452
~ précontrainte P814
~ renforcée R302
~ rustique R870
madrier B225, C92, P475, S1088
~ en coin K13
madriers pour boisage T216
magasin D147, S1285
maigre S401
maille L327, M240
~ d'attache M205
~ de raccordement M205
~ de tamis M240
maillet B290, M163
~ de charpentier M54
~ en caoutchouc M54, R824
maillon L327
~ d'attache M205
~ de raccordement M205
main courante H50, R40
maintenance M40
maintien M40
maison à chauffage par soleil S868
~ à chauffage solaire S868
~ à deux familles D806, T537
~ à deux logements D806
~ à l'abri de la terre E53
~ à niveaux multiples S1012
~ appuyée à la pente E53
~ à une famille H508
~ à un seul étage O85
~ avec haut degré d'isolation thermique H427
~ d'appartements à balcons d'accès G11
~ détachée D210
~ d'habitation D836
~ d'habitation à une famille H508
~ double D527
~ du type clé[clef] en mains T514
~ en rondins (de bois) C613
~ isolée D210
~ jumelle D527
~ tour à noyau central porteur T479
~ tour d'habitation détachée P605
maisonnette D806
maisons dos à dos B34
~ collectives standardisées à bon marché L522
~ en rangée T96
~ provisoires O76
~ temporaires O76
mîatre de l'ouvrage O367

~ de l'œuvre C1115, **G129**
~ d'ouvrage C546
malaise causé par le bruit N147
malaxage B440, G114
~ à la main G114
~ des granulats A163
~ du béton M336
malaxeur A175, M324
~ à ailettes vibratoires C911
~ à arbre horizontal H452
~ à auge avec axe horizontal à palettes O158
~ à axe horizontal H452
~ à béton C910
~ à béton à axes horizontaux (à palettes) O133
~ à béton à deux axes avec palettes T525
~ à béton à production continue C1100
~ à contre-courant C1202
~ à mortier M480
~ à tambour D741
~ à un axe horizontal à palettes S611
~ avec cuve non basculante N216
~ discontinu B196
~ d'une centrale de malaxage P934
~ d'un poste d'enrobage P934
~ pour mortier Colgrout C692
malaxeuse à ailettes vibratoires C911
~ à béton C910
~ à béton à axes horizontaux (à palettes) O133
~ à béton à deux axes avec palettes T525
~ à mortier M480
~ discontinue B196
malfaçon F105
malléabilité D788
manche B58, H5, 44, 477
~ armée A618
~ d'incendie H477
manchette C679
manchon C679, S690, 827
~ à double emboîtement D546
~ à double emboîture D546
~ à emboîture et à bride S2
~ coulissant S690
~ d'assemblage C679
~ de jonction C1222
~ de raccord U100
~ de raccordement C1222, 1224; S1008
~ de raccord(ement) à brides **C682**
~ de raccord flexible F382
~ de raccord vissé S157
~ de réduction R212
~ de serrage B718, T512
~ de tuyau N126
~ de tuyau à fil extérieur P408
~ d'un tuyau S692
~ encastré dans le plancher F508
~ fendu S1010
~ flexible F382
~ souple S690
manchons taraudés **noyés dans le** béton T150
~ vissés S157
mandrin M56
~ à évaser les tubes T472
maniabilité R357, W469
manifold R517
manille C545, L218
manipulation H45
manœuvre H45
manomètre à gaz G65
manque d'homogénéité L8
mansarde G34
manteau de cheminée **M68, S119**
manutention H45, L385
~ de matériaux M141
marais M95
~ maritime T173
marbre artificiel A644
marbruere M77
marche C11103, 1235

~ à forfait F269
~ balancée C743, W373
~ balancée triangulaire K48
~ dansante C743, W373
~ dansante triangulaire K48
~ de travaux O369
~ de travaux de construction O369
~ d'étude de projet D175
~ du type clé sur porte T515
~ forfaitaire F269
~ gironnée C743, W373
~ gironnée triangulaire K48
~ intermittente I344
~ intermittente de chauffage I343
marécage M95
marée de tempête S1311
~ montante H347
marge M78
~ de sécurité M81
~ de stabilité M82
marnage de marée R83
marne M92
marquage L125
~ de lignes de signalisation R591, 617
marque de la chaussée R591
~ des hautes eaux F459
marqueterie M94
marquise M93
marteau H28
~ à air comprimé A240
~ à chute libre G317
~ à mater C217
~ à panne F12
~ brise-béton B630, **C869, R573**
~ d'eau R469
~ de battage P322
~ de charpentier C145
~ de démolition D130
~ de maçon A829, B656, **M111,** W35
~ de mine J3
~ de paveur P148
~ en caoutchouc R824
~ perforateur H32, R628, 781
~ perforateur à colonne C703
~ pilon à simple effet S590
~ piqueur J3, D130
~ pneumatique A210
~-riveur R569
~-trépideur à double effet **D510**
martelage H35
mascaret B549
masque B647
~ anti-poussière D818, R428
~ de soudeur F16, W315
~ d'étanchéité D265, I59, S138, E11
~ étanche d'argile C507
~ Lévy L197
masse de jointoiement J86
~ de remplissage J86
~ de scellement J86
~ de sol E35
~ d'étanchéité J95, 105
~ de terre E35
massette K50, M97
massif D803
~ creux en béton C873
~ d'ancrage H407, S1391
~ d'ancrage en béton H407
~ de fondation F606
mastic M128, P985
~ à base de bitume et époxyde E339
~ à enduire les raccords **vissés** P568
~ à limaille I388
~ à vitres pour châssis métallique M260
~ bitum(in)eux A701, B404
~ bitum(in)eux pour remplissag de joints A698
~ bouche-pores G154
~ d'asphalte A701, M129
~ d'asphalte pour remplissage **de** joints A698
~ de fer I38

PIÈCES

puits d'accès I246, M57
~ d'accès de chute D724
~ d'accès superficiel S405
~ d'aérage A255, V72
~ d'alimentation I381, R169
~ d'ascenseur E221
~ d'eau W262
~ de captage central C321
~ de chaleur S618
~ d'échappement E444
~ de clarification S359
~ de décharge B433, R346, 349
~ de décompression B433, F349
~ de drainage D627, 779; R349, S825
~ de faible profondeur S406
~ d'entrée G476, R57
~ de recherche P440
~ d'escalier S1124
~ de surveillance M57, O9
~ de tranquillisation S1259
~ de turbine T508
~ d'évacuation D388
~ de visite M57
~ de visite superficiel S405
~ d'observation O9
~ filtrant B433, F181, W325
~ filtrant complet P222
~ filtrant parfait P222
~ foncé D702
~ foré B553, D689
~ incomplet 155, 94; P92, S406
~ jaillissant F525
~ rayonnant horizontal H459
~ perdu D388
~ témoin O9
~ tubé C161
~ tubulaire T485
~ tubulaire à diamètre variable C807
pulpe P939, S795
pulpeduc H570
pulsation H540
pulsocrible R425
pulvérisateur P942, S1033
~ à dos K51
~ à havresac K51
~ à main H53
~ de l'eau P942
pulvérisation A735, S1033
~ à chaud H492
~ airless A258
~ de brouillard F593
~ fine M308
~ sans air A258
pulvimixer P943, R585
pumping P960
purge B468, P977
~ de la chaudière B474
purgeur S305
~ d'air A763, B432
~ de vapeur actionné par impulsion I77
purification P980
pylône M124, P988
~ à installation motrice H151
~ incliné réglable d'une grue à câble L561
~ support d'antenne W434

Q

quadrillage R154, G334
quadrilatère Q1
quai E235, L45, 46; Q35, W346
~ d'accostage B789, S45, 894
~ d'accostage de chargement C142
~ d'accostage profond D85
~ d'achèvement B740, F295
~ d'amarrage B337, J48, S1144
~ d'amarrage à duc d'Albe de guidage F312
~ d'amarrage profond D85
~ de chargement C142
~ de réparations R365
~ d'une gare P547

~ en T T112
~ sur pieux P301
~ sur pieux avec plate-forme O129, 168
qualification professionnelle W497
qualité G245
~ de ciment B628
~ définie par sa résistance à la rupture S1368
~ de la surface S1623
~ de roulement R499
~ de travail W497
~ d'exécution W497
~ d'exécution d'un ouvrage S1446
~ roulante R499
quantité d'air prévue R379
~ d'eau disponible S23
quart Q24
~ de rond Q30
~ palier Q27
quartation Q26
quartier B451
~ de ville de développement spécial U195
~ résidentiel R398
queue T26
~-de-cochon G161
~ de vache E64, L75, R734
quincaillerie F296
~ de bâtiment B734
~ de fenêtre W387
~ de porte D491
~ de portes ou de fenêtres F296
quotient d'écoulement R855

R

rabattement D656
~ de la nappe aquifère D247, G416
~ d'équilibre D356
rabot P465
~ à moulures B224
~ à rainurer B50, G371, P557
~ à repasser J89
~ cintré C769
~ plat J89
~ tracté D589
~ universel U127
rabotage de finition J94
raboteuse latérale E93
raccord B623, C1222, 1224; F293, J120, U100
~ à culotte Y7
~ à emboîtement à bague d'étanchéité R526
~ à plusieurs branches M549
~ courbé S1677
~ courbé raide K65
~ de réduction R212, 213
~ de tuyau P388
~ de tuyau fileté P430
~ de tuyaux capillaires C117
~ en T à trois brides A333
~ en Y Y7
~ étanche P804
~ flexible F382, 389
~ fonte-grès I95
~ grès-fonte I95
~ imperméable P804
~ inséré I223
~ instantané Q47
~ intercalé I223
~ mâle ou femelle de boyau H479
~ rapide Q47
~ réducteur R213
~ vissé démontable D209
raccordement C997, J120, U100
~ à fiches P562
~ de tuyauterie P388
~ en T T51
raccourcissement élastique E147
raclette B205, R64, S128, 1421
~ à dents C617
~ dentée D589
racleur R64, S128

~ (tournant) S779
racloir B205, L572, S128, 1421
~ à dents C617
~ denté D589
racloirs montés sur une chaîn S779
radeau de protection F419, G454
radiateur R19, 28
~ à ailettes F220
~ à lamelles S1418
~ à lammes S1418
~ à plinthe B164
~ à rayonnement infra-rouge I181
~ de chauffage ventilé B418
~ de fonte C182
radiation thermique H223
radier A543, B569, G346, 349; I372, M133, R35, S976, U81
~ d'écluse supérieur I209
~ de fondation R37
~ de poutres C1293
raidissage S1252
raidissement S1252
raidisseur d'âme G171
rail R40
~ de guidage G465
rainure C397, G9, 367; K13, R157, S755
~ à poutrelles S1280
~ dans un mur en maçonnerie W29
~ de la vanne G94
~-guide de parquet F492
rallongement L162, S1005
~ des éléments de charpente en bois L164
~ du pieu P338
rampe E323e, G245, 260; I91, R74, S1123
~ d'accès A58, E451, O92, S728
~ de déchargement U131
~ de raccordement E451, O92, S728
~ limite L294
~ longitudinale L484
~ pour piétons roulante M523
rangée de bordure de tuiles ou d'ardoises E67
~ de briques posées en arceau R810
~ de maisons T95
~ de pieux P334, R812
rapide B74, I86, R1
~ de prise R108, S342
rapidité d'action O176, R127
rapport âge-résistance A155
~ agrégat-ciment A164
~ ciment-eau C286
~ d'allongement A679
~ d'armatures R313
~ de circulation C474
~ de ferraillage R313
~ de la largeur à la hauteur A679
~ de mélange M339
~ densité sèche/humidité D751
~ des modules d'élasticité M374
~ d'induction I128
~ eau-air W254
~ eau-ciment W121
~ eau-déblai W256
~ eau-liant W109
~ entre l'âge et la résistance A155
~ hauteur/portée D159
~ sable-agrégat S44
~ vides/ciment V186
~ volumétrique des vides V192
raté M306
~ de sautage M306
rateau R64
ratoteuse-chanfreineuse E93
rayon d'action O176, R127
~ d'appel R34
~ de courbure R32
~ de giration R33
~ hydraulique H582, M179
rayonnage R3
~ de stockage S1297
rayonnement de chaleur H223
rayures H2

NEDERLANDS

Bij de alfabetisering zijn per trefwoord de uitdrukkingen gescheiden van de samengestelde termen, met dien verstande, dat eerst de uitdrukkingen in alfabetische volgorde zijn opgenomen, en daarna de samengestelde termen:

bitumen
bitumen met asbesttoeslag
bitumen met minerale vulstof
bitumenaangietmiddel
bitumenvulstof

Indien het eerste gedeelte van een trefwoord ook deel uitmaakt van direct daarop volgende trefwoorden, dan wordt dit gedeelte daar vervangen door een tilde (~). Het te vervangen gedeelte wordt begrensd hetzij door een spatie, hetzij door het scheidingsteken (‖):

bitumen
~ met asbesttoeslag
~ met minerale vulstof
bitumen ‖ aangietmiddel
~ verf

Bij uitdrukkingen, waarvan niet het eerste woord als trefwoord is opgenomen, is dit aangegeven door middel van een schuine streep (/):

betonpaal/ in het werk gestorte: in het werk gestorte betonpaal

Waar verwezen wordt naar meer dan één artikel met dezelfde lettercode, is bij d tweede en volgende verwijzingen deze lettercode weggelaten:

aanjager B470, 541; aanjager B470, B541

A

aambeeld A513
aanaarding B9, F159
aanbesteding C58
aanbestedingsvoorwaarden T70
aanbeveling van de architect A575
aanbieding T69
aanbinder L157
aanbouw A101, 438
aanbrengen F290
~/ dakvilt A529
~/ de hechtlaag P850
~/ de voorspanning geleidelijk G265
~ van de voorspanning I67, S1395
~ van de wapeningsbundels P672
~ van een slijtvaste laag H88
~ van een waterdichte laag M238
~ van oppervlaktestructuur T109
~ van slijtlaag B445
~ van spreidlaag B445
~ van steengaas L106
~ van strepen op het wegdek R617
aanbrug A539
aandrijving voor het mechanisch

openen en sluiten van een venster M209
aanduiden S339
aaneenpassende zijbekisting M136
aangebouwde garage A741
aangebrachte belasting A534, I66
aangenomen belasting D159
~ hoogste waterstand D187
~ peil D192
~ uitstroming D188
aangepaste regeling R92
aangestreken platvolvoeg F360
aangestampte kleilaag C516
aangetast hout U138
aangevoerd afvalwater I175
aangevoerde mortel R144
aangewalste ophoging R659
aangrijpen van een kracht F619
aangrijpings‖moment A535
~punt P608
~punt van een kracht O208
aangroeiing in buizen E276
aanhangwagen H364, T331
aanhanger D463
aanhechting I338
aanhechtings‖lengte E239, G354
~spanning bij buiging F396
aanjager B470, 541; F60

aanleg C1021
~ van de fundering B580
aanlegfaciliteiten B338
aanleggen L122
~ van een grasveld T510
aanleg‖kade B789, S894
~plaats B337, L46, W346
aanlegsteiger J48, L45, O129, S1114, W346
~ met remmingwerken [geleidewerken] F312
aanloopkoppel S1159
aanmaakwater G120, M346
aanname voor de berekening C46
aannemer C1116
~ voor het bekistingswerk F658
aannemings‖overeenkomst O369
~som C1117
aanpassen van de regeling voor de nacht N124
aanpassingsstuk voor gietijzer op gres I95
aanslaan R517
aanslag B221, D460
aanslag(dorpel) van sluisdeur C399
aanslibben C693
aanslibbing S218, 578
aansluiting A36

~ achter een keermuur B22
grond‖afdekking E11
~anker A410, G487
~balk F606, G401
grondboor A748, M532, S29, 839
~/ op vrachtauto gemonteerde E12
grond‖boorapparaat E20
~dekking O352
~depot S1023
gronddruk S855
~ door de omringende grond G397
~ op de ondergrond G397
gronden P850, S652
grond‖ijs A428
~kering R478
~krib B279, G401
grondlaag P848, 849
~ voor de pleisterlaag B697
grond‖massa E35
~mechanica S852
~mengsels E36
~muur B170
~onderzoek met visiteerijzer R656
~overschot N94
~pijler/ bij ontgraving achtergebleven D803
~plaat B173, D57, J9, S882
~pomp B669, J46, S1530, 1539
~schaaf G254, L177
~slag F679
~slagen voor het ontwerp D198,
G137
grondstabilisatie G402, S842, 859
~ door injectie langs de omtrek
van een bouwterrein P232
~ met bitumen A711
~ met kalk L287
grond‖stamper F752
~stoffenmengsel R122
~structuur S858
~talud E54
~tekort N96
~transport via spoelgoten H597
~verdichter B10, F752
~verdichting E16, S846, 848
grondverdichtings‖machine S845
~methode/ gebaseerd op explosies
S1664
grondverf G384, P849
~ voor metaal W87
grond‖verf(laag) G384
~verschot S1021
~verschuiving S416
~verzet E38
~verzetmachine E37
~vlak B569
~volume in verdichte toestand
C754
grondwater G405, P272, S865, 1522
~ binnen de permafrostzone I367
~ boven het freatisch vlak T270,
V13
grondwater‖aanvulling G417
~beheersing G407
~daling D656
~druk U68
~huishouding G415
~inventarisatie G413
~onderzoek G410
~onttrekking G411
~spiegel W247
grondwaterstand G419
~ in een waterwingebied P963
grondwater‖stijging U47
~stroming B167
~stuw G408
~stuwzone S1523
~vergelijking G409
grondwatervoerende grondlaag
W107
~ laag A553
grondweerstand S857
~ van de paal D709
grondwerk E56
~ bij wegenbouw R618
groofbuitenpleister met sier-steen-
slag R627
groot bouwwerk L73

grootste afvoer M169
grote ketel C214
~ trottoirtegel F323
grove bestanddelen/ op het rooster
achtergebleven R69
~ breuksteen C1444, D386
~ toeslag C594
~ zeef S110
~ zeving voorat S111
guillotineschaar G474
gunning A48
gunning (bij aanbesteding) A822
guts G238

H

haag H266
haak E284, H437
~ van een wapeningsstaaf H437
haakbeugel U2
haakje H438
haakse bocht K65
~ las A38
haalmes S1421
haar‖cementmortel H7
~scheur H6, M277
~scheurtjes C1279
hak M156
hak‖beitel M488
~mes H3
hal A496, V129
halfdoorlatende slijtlaag S293
halfdroog mengsel D762
~ metselwerk B654
halfhoutse lipverbinding H13
half-ingelaten deurhengsel H21
half-open waterkrachtcentrale S292
half‖portaalkraan S294
~rupstrekker H24
halfstarre verbinding S295
halfsteens verband S1406
halfsteensmuur H9
halfsteensverband S1113
~ met om de vijf lagen een koppenlaag C745
halftij-vangdam H22
halfzuil A740, D126, E237, H10,
W30
halogeen-lekdetector H26
hals T168
halve steen B187, H8
hamer H28
handbediende trekregelaar M71
hand‖bijl H100
~doekhanger T294
~doseerapparaat M69
handelsgebouw C738, L465
handgereedschap H54, S807
~ voor schilderwerk P25
hand‖greep G353, H5, 44
~krabber S1421
~lepelboor M305
~lier C1250, J17
~roller H43
~schuurmachine/ aan de omtrek
werkende O187
~slijpmachine P655
~snijbank voor betonstaal B135
~spuit(machine) H38
~spuit(toestel) H53
~stamper H51, M163
~takel C350
~trilplaat H40
~werk H55
handzaag H52
~ voor marmer G450
hane‖balk C680
~kam F353, J1
hang spant K43
hang‖boog P191
~brug C20, S1659
~constructie S1661
~dak S1653
hangende bekisting S1652
~ leiders van een heistelling H60
~ schuifdeur O338

~ stapelkraan U77
~ verdiepingsvloer H537
~ zoldering R694
hang- en sluitwerk en ijzerwaren
H91
hanger H56, K41
~ voor bekisting F646
hang‖kabel S1660
~korf T381
~slot P19
~spant met twee hangers Q42
~staaf Q43
hangsteiger B485, C1261, F441,
R43, T381
~ met tweepunts-ophanging T542
hanteren H45
hard asfalt H67
~ beton D274
~ bitumen S883
~ pleister H80, 85
~ wegdek H84
~ wegdek/ met H89
hardbord H68
~ met hoge persing T64
harde aanslag I97
~ draagkrachtige laag H82
~ kleihoudende gesteentes A610
~ toeslag voor de slijtlaag
F494
hardgrauw W318
hardheid van water H81
hardheidstester volgens Kelly K6
hardhout H93
hark R64
harskanaal P450
hart‖lijn C290
~scheur C399, H162
~scheuren H431
haspel R220
haven‖bassin D456
~bekken A548, B182, 183; H65,
W333
havenbeschermingswerken/ in zee
doorlopende E533
havendam B642, M417, P278
~ met betonskelet F694
~/ met schuine blokkenlagen gemetselde S700
haven‖hoofd J48
~muur W27
~voorzieningen B338
H-balk D549
hechtend pleister B527
hechtende kit B526
hechting A111, B514, 528, 530
~ van cement aan de grove vulstof
B531
~ van het cement aan het toeslagmateriaal C256
hechtings‖spanning in het krachten-overgangsgebied D238
~verbeterend middel B525
~vlak B515
hechtlaag T4
~ voor asfaltbetonweg A683
~ voor de verf P23
hechtmiddel A105
~ voor betonelementen W399
hechtplaat R503
hechtspanning B532
~ in verankeringsbereik van de
wapening A411
hecht‖sterkte B530
~vermogen K13
~vlak B533
heen- en-weerbuigproef A367
heet water H340
hefbegrenzer van de giek B540
hefboomschaar A336
~ voor betonstaal B135
hef‖brug L210, V115
~deur L213, P200, V116
heffen L208
hefhoogte L208
~ van een kraan H275
hef‖inrichting L214, 220
~last L208
~platform L225

∼/ spanningsvrij S1402
∼/ stofvrij D75
∼/ talud B124
∼/ uitsparingen in beton B609
∼ van een klimbekisting [klim-kist] J118
∼ van (kroon)lijsten H476
∼/ vochtdicht D15
mal C170, F638, G102, T65
∼/ op maat vervaardigde D1410
∼ van gewapend polyester F135
∼ voor het maken van betonbal-ken B245
∼ voor het maken van (kroon)-lijsten H473
∼ voor het trekken van kroonlijs-ten N118
∼ voor pleisterwerk S133
malen G351
mallen‖batterij M552
∼reeks G21
mammoetpomp P677
manchet C679, S690
mangat A54, H101, I240, 246; M57
manillatouw M62
mannelijk deel van een steekver-binding M52
manometrische druk G112
mano-vacuümmeter C803
mansardedak G18, K56, M67
mantel L325
mantelbuis C15, 167, 987; D703, F508, P298, S434
∼ voor betonpaal D701
∼ voor de spankabel C11
mantel‖buispaal T481
∼paal H422, S466
mantelplaat/ in de grond gedrukte betonnen P795
markies A824
maritieme bouw M83
∼ werken M87
marmeren M77
marmerslag met gladde ovale kor-rels O282
masker F15
massa M116
massanivellement B72
massief landhoofd H255
massieve betonplaat S888
∼ muur S892
∼ plaat S891
∼ steunmuur S889
mast M124
∼ op meer dan één niveau getuide M568
mast‖console M125
∼goot E66, H59
mastiek F161, M128
mastiekketel M131
mast‖kop M126
∼kraan P345
∼kram D460
mat M133
mate van brandwerendheid D104, F255
∼ van circulatie C474
∼ van onbrandbaarheid D106
∼ van waterverlies R113
materiaal/ met cement gestabili-seerd C283
materiaal‖boek T1
∼eigenschappen P898
∼gedrag M138
∼sluis M150
∼transport en -opslag M141
materialen M148
∼/ licht ontvlambare R208
∼ voor werkreparatie M42
materieel C75
∼ voor grondverzet E41
matglas D143, M155
matras M133
matverf M157
maximaal toelaatbare belasting W481
∼ ∼ helling L294
∼ ∼ scheurvorming L291

∼ ∼ vloeibaarheid W341
maximaal verval M170
maximale belasting L404
∼ grondwaterverlaging E356
∼ korrelgrootte M175
∼ langshelling M170
∼ tijdsduur P263
∼ trillingsduur V152
∼ wateropnamecapaciteit W162
maximum hijslast W481
mechanisch bediend ventiel M503
∼ bediende zonwering M502
∼ gereedschap R517
∼ koelsysteem M214
∼ ilter M208
∼ zuiveren van afvalwater P846
mechanische afkoeling R251
∼ afvalwaterzuivering M220
∼ beproeving M218
∼ betonbewerking M219
∼ betonplakspaan P716
∼ bouchardhamer P711
∼ classificatie M201
∼ korrelscheiding M201
∼ krabber M212
∼-luchtkoeling M202
∼ hechting M200
∼ plakspaan P708, T440
∼ schuifverbinders M216
∼ slibkrabber M212
∼ stamper M213
∼ sterkte-eigenschappen M217
∼ stof(af)scheider M207
∼ stookinrichting S1263
∼ ventilatie M222
meegevende taludbekleding F50
meelopende bouwkraan C1283
meerbaanweg M563
meer‖gezinswoning M555
∼paal B509, M463
∼stoel B292
∼traps compressor C801
∼trapsventilator M595
meervoudig oxydatiebed M570
meervoudige betongietmal G21
∼ dampwerende laag B781
∼ duiker M580
∼ sedimentatietank M584
∼ slisverbinding C710, F207
∼ tandverbinding F207
∼ warmteïsolatie M565
meest economische wapeningsper-centage E78
meet‖apparaat G102
∼band B109, M193
∼diafragma D265
∼doos L368
∼goot C1121, M190
∼kanaal R117
∼kast B189
∼kist G103
∼ketting B109, C343, E314, M189
∼kundige eigenschappen van walsprofielen S211
∼lijn R223
∼lint M193
∼net van afvloeiingsmeetpunten H620
∼piketten R225
∼profiel M192
∼punt M460, R227
∼schot van Cipoletti C455
∼station G118
∼stuw C1112, M194
∼trechter S119
meewerkende constructies O215
∼ stalen rijvloer O214
membraan D265, M232
membraan‖-expansievat S170
∼filter M234
∼theorie M237
meng‖bak M332
∼cyclus M333
mengen B440
∼/ beton M336
∼ van de toeslagmaterialen A163
∼ van een betonmortel tijdens transport T347

menger voor colloïdale specie C692
mengformule M349
menging met de hand G114
meng‖klep M334, 345
∼klepcombinatie C714
∼machine M324
∼plateau G115
∼pomp B439
mengsel C799, M317
∼ van bitumen en zwavel S1547
∼ van grind en een emulsie van bitumen-epoxydhars S1047
meng‖snelheid M342
∼tijd D813, M344
∼temperatuur M343
∼trommel A174, M325
mengverhouding M317, 339
∼ voor op de bouwplaats vervaar-digde mortel J69
mengworm M341
mergel M92
meridiaanspanning M239
merkbare koeling S298
merkteken B312
messchakelaar K58
messing F107, T247
messing-en-groefverbinding T248
metaal M246
metaal‖gaas G121
∼kit I388
∼plaat S444
métal déployé E453, 460
metalen afdichtingsband M264
∼ gebouw A339
∼ omranding van een houten wandelement M259
∼ plaat P540
∼ roostervloer M252
∼ stijl M262
∼ vliesgevel M250
∼ vloer M251
meter en hoofdafsluiter C1068
meter‖kast D404
∼put M265
methode naar de analogie met elek-triciteit E161
∼ van de momentennulpunten M274
∼ van drijvende montage F430
∼ van elastische lasten M268
∼ van kritische belasting C1311
∼ van wrijvingsverliesvergelijking E344
metselaar B658
metselaarshamer A829, B656
metselblok voor achterwerk B36
metselcement M103
metselen/ muren B665
∼/ vanaf een binnenstelling een buitenmuur O319
∼ van de achterwerkers H160
metsel‖mortel G107
∼specie M105
metselsteen B742
∼ voor opgaand metselwerk B490
metselsteiger M112
metselverband B514
∼ met verspringende voegen B635
metselwerk B660, 666; M99
∼ in natuursteen A671, 674
∼ in netverband R452
∼/ met geglazuurde steen bekleed G207
∼ met schuine lagen S699
∼ onder het maaiveld B300
∼/ trapsgewijs verlopend B205
∼ van betonsteen C867
∼ van breuksteen R842
∼ van grote blokken L72
∼ van holle betonsteen met mor-telvulling G428
∼ van natuursteen N59
∼ van natuursteen in onregelma-tig verband Q20
∼ van natuursteen in regelmatig verband R271
∼ van radiaalsteen G110

~erosie S489
~lijn W116
~stabilisatie door begroeiing B376
~verplaatsing S475
offshore constructies O42
ogenstaaf E550
ogiefboog O48
ojief O47
oker O17
okerrood K2
olie‖afscheider O59
~alkydverf T428
~bad-luchtfilter I64, O52
~beits O58
~demper O52
oliegestookte luchtverhitter O54
~ verwarmingsinstallatie O55
olie‖-luchtfilter V179
~pijpleiding O57
~stook O55
~verf O56
ombouwen C1135
omgekeerd filter R473
omgekeerde boog I373
omgevings‖geluid A390
~temperatuur A391
~voorwaarden A389
omheining E273, P38
omhoog U154
omhoogvoerende trap U168
omhullend oppervlak (van een gebouw) E328
omhullende kromme E329
~ van de maximale buigmomenten C1406
omhulling E273
~ van de (uranium)brandstofstaven C64
omhulsel C482, E328
~ van een reactor R136
omkeerbare turbine R265
~ ventilator R475
omkeer‖klep R477
~spoeling B39
~thermostaat R470
omkeren van de kijker aan een theodoliet T343
omleiden van een rivier R553
omleiding D162, 434
omleidings‖kanaal D434, 435
~leiding D163
~tunnel D164, R554
~werken D444
omloop G10
omloop‖klep O316
~koeling C471
~leiding B842, 845
~riool D443, P15
~rioolafsluiter P13
~ventiel B846
ommantelings‖muur E330
~plaat C483
ommuring in baksteen B665
Omnia-‖betonplaat O63
~vloer O62
omranding B574, M78
omschakelsysteem C368
omslagboor H36
omsloten ruimte E273
omsluitende constructie S925
~ damwand B606
omtrek‖skracht P236
~spanning C476
~zone van een gebouw P235
omvangrijk grondverzet H261
omvorming C1135
omwentelingsschaal S465
omwikkeling met weefsel F6
onafgewerkte vloer N17
onbalans van een systeem S1695
onbehagen D368
onbehandeld afvalwater C1357
onbelaste rand U129
~ wapening A805
onbewerkt rondhout F410
onbezaagd hout in standaardmaten D310

onbreekbaar glas N161
onderaardse schuilplaats U64
onder‖aannemer S1489
~bouw S1516
onderdeel M231
~/ in het beton opgenomen C174
onder‖dompeling S1504
~doorgang U40, 44, 72
~dorpel B583, 585; D499, O278, S571
~druk B28
~einde van een stijl H267
~- en bovendeur D828, H265
~gedompelde verdamper S1496
ondergrond F679, S1512
~/ goed hechtende K13
~ voor pleister K20
~ voor pleister- of verflaag B165
~ voor pleisterwerk B1, K13, L104, V54
onder‖gronds kanaal U48
~ leidingnet U63
~ reservoir B301, U66
ondergrondse (spoorweg) S1525, U61
ondergrondse aanleg U58
~ constructies U210
~ drainage U46
~ draineringsleiding B441
~ erosie P435
~ garage U55
~ gashouder U56
~ gebouwen U65
~ kabel U53
~ kruising S1524
~ leidingen S327
~ machinehal M5
~ pijpleiding U59
~ spoorweg S1525, U61
~ voorzieningen U210
~ waterkrachtcentrale U57, 60
onder‖handse aanbesteding R440
~houd M40
~houden van een overdruk P809
~houdshangar M41
~invoer B578
onderkant van een balk B231
~ van een leuningbalk B645
onderlaag B35, 166, 275
~ van een fineerbekleding B1
onder‖legplaat B265, D558
~loopsheid U52
~lossende betonkubel D713
~losser H447, W113
~rand (van een vakwerk) B570, C448
~regel S881
~slagbalk B257, D571, G165
~spoeling C224, S120, U45
onderste booghelft H107
~ laag L530
~ laag van een bitumineuze dakbedekking B176
~ laag van een gemetselde muur B166
~ pan E70
~ rij dakpannen H118
~ rij pannen E67
~ schijvenblok F49
~ vezel B574
~ vloeigrens L533, Y17, 18
onder‖stroom S1517, U52
~tapijt F489
~uitlaat B573
~verdeling B582
~vlak B569
~vloer C1206
~water T13
onderwater‖beton T398
~bouwwerken U85
~pomp S1506
~-steenbestorting U84
~tunnel S1484
~-tunnelbouw S1485
~-tunnelbuis U86
~werk D446
~zetten I368
onderzeese pijpleiding S1483
onderzijde B569

onderzoek met ultrasone pulsen U22
~ van het bouwterrein S639
onderzoekprogramma R381
ondiepe bron S406
~ inspectieput S405
ondiepte S401
ondoorlatend gesteente I58
~ scherm van de stuw I59
ondoorlatende laag I61
~ laag boven het grondwater P675
ondoorzichtig glas O6
oneffenheid van de bodem C380
onevenwijdigheid L9
ongebluste gemalen kalk U136
~ kalk Q54
ongeglazuurde keramische tegels E502
~ tegel U94
ongelijkmatige korrelgrootteverdeling D371
~ zetting D276
ongelijkvloerse kruising O354
~ wegkruising G256, I310
~ wegsplitsing G255
ongeroerde grond N47, U89
ongeschaafd gezaagd hout U91
ongesorteerd grind A668
ongestoord grondmonster C1161
ongestuwde afleiding O97
~ afvoer N194
ongewapend beton M113, P462
ongezaagd hout T216
ongezeefd grind A338
ongezeefde gebroken steen C1367
ongezuiverd afvalwater R123
~ water R125
ongunstige opstelling S361
onherroepelijk waterverbruik I393
onherstelbaar gebouw S1462
onherstelbare constructie S1462
onmiddellijke warmte-opname I255
ononderbroken plaat C1103
~ betonstorten C1092
onregelmatig gegradeerde toeslag G28
~ metselwerk van natuursteen H2
onregelmatige filtratie U141
onsamenhangende grond F744, N166
onstabiele bodem U140
onsymmetrische belasting U26
ontbinding D334
~ van krachten F627, R424
ontdoen/ de bouwplaats van begroeiing C531
~ van zwevende verontreinigingen D205
~ van begroeiing G449
ontdooien D100
ontdooiingsmiddel D113
ontgavingsgebied B566
ontginnen van een steengroeve Q17
ontginning in dagbouw Q13
ontgraving E421
~ onder water U83
ontgravings‖diepte E423
~gebied B564
ontkisten D376, F652, S1409, 1420
ontkisting D54, 128
ontkistings‖emulsie R338
~middel F651
~tengel W514
~termijn S1410
ontkleuring B429
ontladingselektrode D347
ontlastend riool S1510
ontlastende boog R350
~ leiding R345
~ muur R351
~ schacht R346
~ weg R344
ontlastings‖boog B3
~kanaal R550, S975

ontlast‖kanaal C385
~klep P798, R339
~riool D366
ontluchting D46
ontluchtings‖leiding V94
~opening G324
ontluchtingsventiel A305, B432, 475; P264
~ van radiator R31
ontmantelen D376
ontmengde grove toeslag L496
ontmengen van betonmortel C942
ontmenging S258
ontschuiming F776
ontsluiting D231
ontsmetting van afvalwater D374
ontsnappen E400
ontspanning D219
ontspanningsbron R349
ontspanningsleiding R348, V94
~ van de binnenriolering S1108
ontspanningsopening W85
ontstaan van gaten in een grindweg R121
~ van gaten in het wegdek F740
ontsteker B426, P849
ontstekingstoestel B427
ontstoppingsstuk I241
onttrekken van grondwater E540
ontvangbed U81
ontvangsthal F323a
ontwatering D597
ontwaterings‖diepte D155
~laag D608, 612
ontwerp D172, P464, S117
ontwerpdiagram D182
ontwerpen P479
~ en uitvoeren van gebouwen en werken S1436
~ van de indeling van huizen P480
~ van een wegverharding P139
~ van luchtbehandelingssystemen E333
ontwerp- en bouwopdracht T515
ontwerp‖-kruinhoogte D176
~normen P482
~tabel D174
ontwijken E400
ontwikkeling D231
~ van (het gebruik van) waterkracht H610
ontwikkelingsplan voor een rivierbekken I296
ontwikkelings‖project D235
~werk D236
ontzanden D170
ontzouten D169
ontzouting D127
onveilige zone B417
onveranderlijkheid S914
onverhard beton U135
onverharde berm M78
onverstijfd element U143
onvoldoende uitrusting van de bouwplaats met machines en gereedschap U74
~ vochtigheid S85
onvolkomen bron I55
~ overlaat P91
~ put P92
~ vakwerk I53
onvolledige bron 194
onvoorziene uitgaven C1082
oog E549
oogbout E551
opaalglas O95
opblaasbaar gebouw A288, P596
opblaasbare afdichting I164
~ gebouwen I165
~ pakking I164
~ stuw F7, I167
opblaassysteem I168
opblazen B424
opbollen O281
opbouw E375, S1574
~ met verdichte grond R72
opbouwen van een overdruk P809

opbreektanden R533, 766
opbreken B641
~ van de bodem R534
opbrengen S1068
~/ een laag C608
~ met lijm en spijkers A110
~ van dakvilt R717
opbrengmethoden A530
opbrengst D381, V12
~ van een pomp P949
~ van het waterwingebied W265
opbrengstmeter P676
opbuigen H396
opbuiging H394
~ van een balk H398
open afvoerleiding O115
~ afzuigkap O124, 160
~ bouwput O110
~ caisson O101
~ doorsnede O143
~ gat O131
~ haard F245
~ kilgoot O161
~ kroonlijst O109
~ leidingkoker O121
~ ontwatering D427
~ opslagplaats O150
~ overloop O147
~ overstort O108
~ parkeergarage O134
~ put O166
~ ruimte O146
~ slijtlaag O154
~ standpijp ter begrenzing van de druk in het systeem S1155
~ stootnaad O130
~ trap O140
~ tunnelbouw C1412
~ ventilatiepijp O162
~ waterkrachtcentrale O99
~ waterloop O105
~ waterlozing D365
~ zandfilter G314
openbaar gebouw P923
~ vervoersysteem P926
openbare aanbesteding O153
~ gebouwen en installaties P929
~ nutsbedrijven P927
~ voorziening P927
~ wegen P924
~ werken M603
openen (brug) O125
~ van een bron D231
opening A518, O125, 203
~ in sluisdeur P15
openslaand venster C166
openstaan O125
openstellen van de inschrijving O127
opgaande gasleiding S324
opgebogen (wapenings)staaf B334
~ wapeningsstaven T455
opgebrachte grond F159, M11
~ laag C605
~ ontwateringslaag D608
opgehoogde weg in laag terrein C219
opgeklampte deur B203, F698, L155, V124
~ ~ met schoorklampen L154
opgelegde versiering L23
opgepompte watervoorraad P956
opgespoten dam H562, 564
~ film S1042
~ grond F549
~ terrein H568
ophang‖band C557
~console H56
ophangen H57
opheffen L208
opheffing van de bodem U153
ophogen van een dijk T263
~ van een stuwdam R63
ophoging B113, C60, E56, H269, S849
~ met overmaat voor de zetting A120
~ van het terrein L43

opkomen van de grond H253
opklapbare boortoren J11
opklap-montagemethode T215
oplangen/ palen P338
oplanger F602, S338
oplassen H88
oplegdruk D266, 267
oplegger S297
oplegging B258, S1592
~ in een elastomeer E154
opleggingen S1426
opleggings‖deel B268
~voorwaarden S1593
opleg‖moment S1597
~plaat B265
~reactie S1598
~stoel B260
oplevering A47, D116
opleveringsdatum D22
oploop van een schip I40
opmeten van de in het werk gebrachte hoeveelheden Q10
opmeting G144
opnamevermogen (van een put) I375
opnemen van het verrichte werk Q10
oppervlak van aardebaan F641
~ van de afzuigopening F14
~ van de dwarsdoorsnede S209
~ van de oorspronkelijke dwarsdoorsnede O207
~ van de uitgraving E415
oppervlakkig droge toeslag S1614
oppervlakte A596
~ met artistieke afwerking A587
oppervlakte-actieve stof S1605
oppervlakteafwerking S1638
oppervlaktebehandeling S1633
~ in meerdere lagen M586
~ met geëmulgeerd bitumen A694
~ met ter plaatse vervaardigd mengsel M321
oppervlakte‖berekening C49
~beluchter P14, S1606
~drain S1612
~drainage S1613
~eenheid U118
~gebreken S1607
~gesteldheid F208, S1623
~harden H88
~helling S1628
~impregnering P199
~koeler S1610
~kwaliteit S1623
~moment A599
~momentenberekening A600
~structuur S1632
~temperatuur S1631
oppervlakteverdichting S1608
~ door injecteren B415
oppervlakte‖verhouding van een monster A605
~vibrator S1634
oppervlak-waterverwarmer S1637
oprichten E375
~ van het skelet C138
oprijlaan A817, D705
oprijzen van de weg U153
oprit A58, E323e, O92, S728
oprolbare zonwering R665, 678
opruimen (van) de bouwplaats H513, J66
opruimer T472
opruimingswerk C527
opschuifbare bekisting C550
opslag in waterwingebied C195
opslag‖plaats D147, S1285, U208
~stellingen S1297
~terrein S1286
~voorraad S1285
opsluiten van het beton R437
opsluiting van pakkingbus S1481
opsporingstoestel voor gas (lekken) G47
opspuitbuis P380
opspuiten F549, H568

РУССКИЙ

Примеры алфавитного расположения терминов:

балка
~ жесткости
~ ~ контрфорса
~ коробчатого сечения
~, несущая подвесной потолок
~ чердачного перекрытия

балка-||перемычка
~стенка
балкон

бетонная дорога
~ облицовка
бетонное покрытие
~ сооружение
бетонный блок
~ куб(ик)

весов||ая дозировка
~ое соотношение

Тильда (~) заменяет повторяющееся слово или неизменяемую часть слова, которая отделяется вертикальными параллельными линиями (||).

Косая черта (/) указывает на обратный порядок слов в терминологическом словосочетании:

болт/ U-образный = U-образный болт
Буквенно-цифровые сочетания типа
C12, 137
следует читать:
C12, C137

А

абразивн||ая обработка [отделка] бетонной поверхности R814
~ые материалы A18
абразивный диск G352
~ износ A8
~ круг A14
абразивы A18
абсолютная влажность A23
~ деформация A20
абсолютное давление A24
~ перемещение A21
абсолютный фильтр A22, H303
абсорбирующая опалубка A34
аванбек C1435, L115
аванкамера F630, O261
~ шлюза H120
аванпорт O232, 261
аварийная вентиляция E247

~ ёмкость водохранилища S1604
~ лестница E405
аварийное освещение E244
аварийно-сигнальн||ая система A317
~ое устройство A317
аварийный водосброс E246
~ водоспуск E246
~ выход E241
~ ~ через балкон B89
~ затвор E243
~ клапан I408
~ спуск E245
авеню A817
автобетоновоз A172
автобетоносмеситель A172, M330, R141, T445
~ с опрокидным барабаном S84
авто||грейдер M500
~гудронатор B362, 405

автодорога (для) скоростного движения S965
автодорожное строительство R579
автодорожный мост H352, R574
~ туннель H336, R614, V46
автоклавн||ая обработка A755
~ый ячеистый бетон A753
авто||кран R826, A798
~магистраль A631, F723
автомат для резки кирпича-сырца A770
автоматизированн||ая котельная установка A766
~ое устройство для наклейки герметизирующих [уплотняющих] лент A792
автоматическая водораспыляющая насадка S1075
~ кислородная резка A785
~ противопожарная дверь A777

гидравлическое погружение J44
~ сопротивление трения H571
~ уплотнение H594
гидрант H546, S1155
гидратированн‖ая известь H548
~ый цемент H547
гидро‖аккумулирующая электростанция P957
~аккумулятор R382
~вибратор V167
~вибрационное уплотнение V166
~геологические изыскания G410
гидрограф D350, H614
~ паводка F455
~ стока D350
гидрографическая сеть D616
~ съёмка H615
гидродинамическая сетка F529
гидродинамическое выпирание грунта H606, S238
~ давление H607, S240
гидро‖закладка H568
~изол A652, 663
~изолированный шов T391
гидроизолирующая прокладка F350, M393
гидроизолирующие ограждения C482
гидроизолирующий воротник C679
~ ковёр кровли R732
~ слой M393
гидроизоляционная мастика W212
~ мембрана D16, W207
~ плёнка D16
~ штукатурка W209
гидроизоляционные материалы W204
~ работы W206
гидроизоляционный ковёр D14
~ обмазочный состав W200
~ слой D14, W201
~ строительный картон W203
гидроизоляция D15, W199
~ стен подвала W56
~ фундамента F673
гидро‖канал W122
~классификатор C1443, H556
гидрологические изыскания H618
гидрологический пост S1351
~ прогноз H616
~ режим H617
гидрология суши C1081
гидролоток W122
гидрометрическая вертикаль V108
~ вертушка C1399
~ сеть H620
~ станция D364, G118
гидрометрический водослив M194
~ лоток M190
~ поплавок F416
~ створ D362, H621, M192
~ шест S1497
гидро‖метрия G114
~механизация H619
гидромеханизированная разработка H580
~ ~ грунта H566
гидро‖модуль H622
~монитор G156, H585, M444
~ниторная разработка грунта W172
~окись кальция C39
~отвал D387, H568, 599
~смесь S795
гидростатическ‖ая нагрузка H626
~ий напор E211, P285, 670; S1170
~ое давление H627
гидротехника H564
гидротехнические изыскания H576
гидротехнический бетон H557
~ затвор G92
~ туннель H603, W156

гидротехническое строительство H564
гидроузел M587
гидрофобизация бетона I291
гидрофобизирующая добавка W219
~ эмульсия W202
гидрофобный цемент H623, W198, 216
гидро‖фор P811
~циклон H559
~цилиндр H592
~элеватор H563, W171
гидроэлектростанция H611, H625
~ совмещённого типа B456
гильотинные ножницы G474
гильсонит G160
гипар H633
гиперболическая оболочка H634
гипотеза Навье N62
~ плоских сечений N62, P473
гипс G491
гипсобетон G496
гипсобетонная панель G502
гипсобетонный блок [камень] G493
гипсоблок G493
гипсовая доска G503
~ плита G503
~ стеновая плита G508
~ сухая штукатурка G502, 505
~ форма P499
~ штукатурка G504
гипсоволокнистый бетон G497
гипсовый бетон G496
~ лист G498
гипсо-известковый раствор G500
гипсо-перлитовый раствор P237
гирлянда изоляторов I272
гистограмма F737
главная балка B257, G165, M23
~ дорога P858
~ задвижка M45
~ кривизна P854
~ мостовая ферма B682
~ стропильная нога M31, P852
~ ферма M44
главное стропило B1
главные напряжения P859
~ оси P853
~ эллипса напряжений M47
~ расчётные сечения P855
главный инженер проекта J65
гладилка D589, F412, G119, H41
гладкая кольцевая шпонка S427
~ лицевая кладка F46
~ щитовая дверь F551
гладкие (арматурные) стержни P461
глазурованная облицовочная плитка E264
глазурованный кирпич G203, V182
глазурь E262, G201
глина C506
глинизация A612, C517
глинистый грунт C506
~ известняк A609
глинобетон C508
глинобетонная смесь C516
глинобитная постройка P437
~ смесь P931
глинозём A372
глинозёмистый огнеупорный кирпич A388
~ цемент A387, H295
глинозёмный модуль H387
глиномялка P934
глиняная кровельная черепица C518
глиняный противофильтрационный экран C507
~ раствор C514, 516
глубина D152
~ выемки D291
~ ~ котлована E423
~ ~ дрены D631
~ ~ фундамента F674

~ котлована E423
~ промерзания F762
~ ~ грунта D156, F762
~ черпания D291
глубинн‖ая эрозия W85
~ое уплотнение D79
глубинный аварийный затвор T8
~ вибратор I360, S1081
~ водозабор S1499
~ водосброс D8
~ водоспуск S788
~ дренаж U46
~ затвор B575, D82, S785
~ насос B557, D86
~ поплавок S1497
глубокая выемка [разработка] грунта D80
глубокий котлован D80
глубоководн‖ые сооружения O42
~ый причал D85
глухарь C1386
глухая плотина F317
~ секция N189
~ стена D41
~ фрамуга F316
глухое окно D44
~ сопряжение балок B253
глухой анкер D33
~ температурный шов D795
~ фланец B443
глушёное стекло O6
глянец G221
гнездо A518, G9, M486, P602
~ для (дверной) петли P41
гнутая фанера F412
гнутьё B315
Г-образная колонна L129
Г-образный оголовок стойки L200
~ причал L201
годограф D160
голландская дверь D828
~ черепица D544
голландский клинкер(ный кирпич) D827
голова E323
~ буны H141, N238
~ дрены D636, I205
~ канала D71
~ полузапруды H141
~ шлюза L450
~ шпоры H141
головка H118
головная часть (гидросооружения) H118
~ штольня P351
головное сооружение C71, H129, I205
~ ~ гидроузла I277
головной затвор H129
~ регулятор H123
~ узел C71, H156, K42
головные ворота H129
~ сооружение H156
гонт S477
гончарная дрена C332, 518; T200
~ дренажная труба D644
~ полевая дрена F157
гончарный дренаж T201
горбатик C769
горбылёк G211, M608
горбыль F410
горелка B816
горизонт сработки водохранилища D659
горизонталь C1107
горизонтальная выработка T488
~ дрена H457
~ затяжка H470
~ опора L99
~ проекция P464
~ связь H470
~ сила L87
~ схватка W7
~ стоечные лесов L157
горловина T168
горн‖ая крепь S488
~ое давление G397
город ленточной застройки S1414

крюк E 284
ксилолит X4
ксилолитовый пол M17
кубатура C1377
~ здания A589
~ земляных работ B118
кувалда K50
кулачковый (дорожный) каток S438, T18
кульман D587
культурная почва C1380
купол D466
купольная оболочка D469
~ плотина C1384
купольный зенитный фонарь S668
куратор C544
курс кладки из массивов C1225
кусачки N125
куст свай D464, P299
кусторез L39
кухня-столовая D313, K47
кухонная мойка P696, R532
~ раковина B182
куча P292
кювет D426, G485, R604

Л

лабораторный анализ грунта S837
лавинозадерживающее сооружение A811
лавинозащитная галерея A812
~ стена A810
лавинозащитные сооружения A815
лавиноотбойная дамба A814
лага R40
~ пола F473
лазерный нивелир L78
лак V30
лаковая краска E262, V31
ландшафтная архитектура L51
латекс M130
латексная смесь M130
~ эмульсионная краска R818
лебёдка H399, W367
левая свивка L159
лёгкие конструкции L269
~ леса L246
~ подмости L246
~ сооружение L269
лёгкий бетон L243, 263, 524
~ деррик-кран M447
~ заполнитель L262
~ строительный кран M283
~ теплоизоляционный бетон I261, L264
ледов‖ая нагрузка I13
~ый режим I16
ледо‖рез C1435, I2
~сброс I6
ледосбросное сооружение I10
ледоход I17
ледян‖ая линза F765
~ое сало G329
ледяной бетон I7
~ затор I8
лежалый цемент M160
лежачий фальц F368, G372, L354
лежень G381, 401; M541, R40, 697; S571, 881, 882
~ нижней обвязки F614
лезвие B412
~ рубанка C1431
лекало G102
лента для заклейки швов J108
~ для уплотнения резьбовых соединений P426
ленточная застройка R493, T95
~ пила B111
ленточное окно W380
~ сито B112
ленточный водяной фильтр B309
~ излучатель R24, S1418
~ ковшовый подъёмник C1091
~ конвейер B302

~ отвалообразователь S1106
~ погрузчик B305
~ фильтр B110
~ фундамент S1417
~ электронагревательный элемент H204
лепестковый воздушный клапан I385
лепесток L136
лепёшка C32
~ из цементного теста P120
лепные работы P504
лес W442
леса S98, 1114
~ большой несущей способности H258
~ для каменной кладки M112
~ с одним рядом стоек S609
~ средней несущей способности M226
лесоматериал W442
лесосплавное сооружение L472
лесосплавный канал R38
~ лоток L468
лёсс L463
лёссовый грунт L463
лестница S1117, 1123
~, опирающаяся на стеновые косоуры [на стены] B612
~ с открытыми подступёнками O140
~ только для подъёма U168
лестница-стремянка L11, 15; S1237, T416
~ с рабочей площадкой P549
лестничная клетка S1117
~ площадка L45, S1118
~ шахта S1124
лестничные леса L14
лестничный марш F407, S1123
~ поручень, прикрепляемый к стене W39
~ рыбоход F281
~ сход набережной Q37
летнее кондиционирование воздуха S1553
летний польдер S1556
~ режим S1555
летучая зола F565
лещадный заполнитель F328
ливень R51
ливневая вода S1313
~ канализация S1310, 1314
ливневые сточные воды R55, S1313
ливневый водосток G476
~ колодец G476
~ лоток G485
~ сток R52, S1308
ливнеотвод S1316
ливнеотводный водовод S1309
~ канал S1309
ливне‖сброс D340, S1307
~спуск G476, L149, R586, S1307, 1315
лизиметр P216
лиманный канал I369
лимниграф L193
линейка R848
линейная деформация L315
~ застройка R809
~ нагрузка L313
~ упругость L312
линейное преднапряжение L314
линейно-распределённая нагрузка L313
линейный город L316
линзовый компенсатор B298
линии безопасности R591
линия визирования C689
~ влияния I173
~ ~ моментов M434
~ ~ опорной реакции I174
~ ~ опорных реакций A321
~ водораздела H273
~ всасывания S1535
~ гидравлического уклона H573
~ давления P793
~ депрессии L321

~ застройки B752
~ лицевого фасада N70
~ модульной сетки G341
~ нагнетания D351
~ насыщения S87
~ по наружной грани стены W45
~ прогибов D89, E137
~ пят S1074
~ разметки T324
линоксин O376
линолеум L329
липкий грунт S1244
~ материал S1244
лирообразный компенсатор E481
лист P540, S439
~ волнистого профиля C1189
~ сухой штукатурки G494, 505
листовая сухая штукатурка W24
листовое оконное стекло S443
~ стекло P56
~ ~ с фаской B343
листовой материал P540, S444
~ пенополистирол P637
~ теплоизоляционный материал I259, S449
листовые материалы с элементами жёсткости R515
литая бетонная смесь H367
~ сталь C184
литой асфальт M129
~ бетон F543
лифт E212, L208
~ непрерывного действия P126
~ с автоматически открывающимися дверьми A775
~ с канатоведущим шкивом T312
лифтовая шахта без выходов на промежуточных этажах B444
лицевая грань B1
~ ~ пилястры P290
~ сторона M89
личина K25
лишн‖ие связи R218
~яя опора R219
лобзик J58
ловители кабины лифта E214
~ мгновенного действия I254
ловушка T372
~ донных наносов B283
ловчие дрены C201
~ каналы C201
лоджия L470
ложе B273
ложечный бур A748, S32, 839
ложковая перевязка S1406
ложковый ряд F355
~ шов L485
ложн‖ое схватывание F351
~ый шов D795
ложок S1405
лом L485
~ с лапой P355
ломик J61
лопастной вентилятор P896
~ поверхностный аэратор P14
~ смеситель O158
лопасть B412, P13
лопата S505
лопатка B51, 412, T436
лотковая бетонолитная установка S1032
лотковый водовод F547
~ рыбоход F278
лоток C451, 1383; F547, I1372, R838
~ для строительного мусора R838
~ разжелобка V17
лузговое правило A628
лучевая дрена H459
лучевой водозабор H459, R16
~ ~ с наклонными скважинами I89
люк H99
~ в кровле H101
~ для выхода на крышу E450
люк-лаз M57
люлечный элеватор V127

О

проволочный анкер W436
прогиб D88, F403
~ балки B235
~ в середине [в средней точке] пролёта M285
прогибомер D90
прогноз расходов на строительство C1035
прогон B225, 257; G165, P982
~ крыши P982
~ покрытия R698
~ у свеса кровли W52
программа исследований R381
продавливание труб (через насыпь) P386
продолжительность монтажа по графику [по плану] S115
~ перемешивания D813, M344
~ работы A93
~ строительства C1047
~ цикла работы бетоносмесителя M333
продольная балка R488, 1412
~ вспомогательная балка T101
~ дамба L481, T334
~ деформация A831, L489
~ жёсткость L488
~ кривизна B598
~ половинка кирпича Q40
~ распиловка C1135, R535
~ ~ брёвен C1135
~ сила L483
продольное коробление B598
~ преднапряжение A838, L486
~ растягивающее напряжение L490
~ сжатие L480
~ соединение L485
продольно-поперечный изгиб C718
продольные арматурные стержни D318
продольный брус S1412
~ выступ F107
~ дренаж C1108
~ изгиб B718, 719
~ осевой разрез P465
~ профиль L487, V109
~ ~ дороги P877
~ ~ реки S1357
~ разрез A840
~ стержень L478
~ уклон I83, L484
~ шов L485
продувка B468, P977
продутый битум B472
проезд A332, 608; L60, R858
проезжая часть C151, D62
~ ~ дороги R616
~ ~ ~ с нежёстким покрытием N197
~ ~ с бетонным покрытием C875
проект D172, P464, 881; S117
проектирование D172, M603, P479
~ бетонной смеси M319
~ дорожки P139
~ защиты окружающей среды E333
~ зданий B749
~ зданий и сооружений S1436
~ и строительство автомобильных дорог H356
~ ландшафта L53
~ на базе модуля M373
~ систем регулирования микроклимата E333
~ смеси M319
~ состава бетонной смеси D197
~ состава смеси P904
~ строительных конструкций A583
проектная документация P883
~ долговечность D193
~ кубиковая прочность бетона S964
~ отметка гребня плотины D176
~ поверхность земляного полотна F641

проектное задание P764
~ положение F192, P243
~ решение D179
проектно-исследовательские разработки D236
проектные ограничения D194
проектный подпор T259
~ профиль земляной выемки F642
~ состав смеси J69
~ срок службы D193
проём A518, O125, R165
~ в перекрытии F501
прожектор заливающего света F458
прожилки песка S57
прозрачная отделка N44
производственн||ые C109, P701
~ое здание L465, P874
~ые сточные воды T316
~ый пролёт P873
производство взрывных работ B424
прокатный бетон R658
~ профиль P877, S208
прокладка L122
~ трубопроводов P393
прокол C1411, 1416
пролёт B215, O125, S938
~ в свету C534
~ свода C766
~ сводчатого покрытия V34
пролётное строение моста S938, 942, 1574
промежуточная лестничная площадка H12, Q227
~ опора I323, 340; P278
~ ~ моста B679, S1152
~ тетива C150
промежуточн||ое пролётное строение I322
~ые ворота (шлюза) I334
промежуточный бьеф канала P640, R127
~ пролёт I322
~ этаж M276
промер глубин S909
~ длин C356
промывка F552
промывн||ая галерея F557, S123
~ое отверстие B474, S783, W85
промывной бак F559
~ барабанный грохот R789
~ водовыпуск W85
~ жёлоб W88
~ затвор S785, W361
~ канал F555
~ канализационный колодец F558
~ клапан S126
~ шлюз S124, 783
промывочн||ая вода R531
~ый грохот S146, W82
промышленное домостроение H509
~ здание I132
~ изготовление сборных зданий I138
~ кондиционирование воздуха I131
~ отопление P870
~ производство сборных бетонных изделий P729
~ строительство I133
промышленные отходы P871
~ сточные воды I142, P871, T316
промышленный шум I141
пропаривание S1192
пропарочная камера S1195
пропил K10, S93
~ под углом M310
пропуск людей и грузов через шлюз L440
~ паводка F464
пропуск-значок E252
пропускная способность C109, 155; D338, T317
проран C583

прорезь O125, S755
прорыв R860
~ воды I221
~ дамбы D298, L173
просадка S25, 791, 1509
~ грунта S1509
просачивание B434, S236
~ из глубины E126
просеивание S142
прослойка слабого грунта P646
проспект A817
простенок P278
пространственная панельная конструкция B613
~ рамная конструкция S927
~ ~ система S927
~ стержневая конструкция S927
пространственные конструкции S932
~ сооружения S932
пространственный арматурный каркас C25
прострелка (скважины, шпура) C361
проступь G229, R850, T389, 390
противо||вес C1217, K8
~вспенивающая добавка A502
~гнилостная краска A503
~давление B28, U155, 156
~действие R128
~действующая пара сил R129
~лавинные сооружения A815
~морозная добавка A504
~обвальная галерея A812
~противооползневые мероприятия L55
~ сооружения L56
противопаводковая дамба F450
~ защита F462
противопаводков||ое водохранилище D223
~ые сооружения F463
~ый затвор F454
противопенная добавка A502
противопожарная дверь F233, L1
~ зона C766, F223
~ перегородка F264
~ преграда F224, 261
противопожарные мероприятия F250
противопожарный занавес S11
~ запас воды E248
~ отсек C766, F244
~ разрыв между зданиями D394, F225
противопучинные мероприятия S1673
противоток C1207
противоточная градирня C1208
противоточный (бетоно)смеситель C1202
~ теплообменник C1209
противоугон R41
противоусадочная арматура S519
противофильтрационная арочная облицовка W253
~ диафрагма C1169, D265, 268; G421
~ завеса G421
~ манжета C1417
~ маска Леви L197
~ мембрана S1181
~ перемычка A507
~ шпонка S1182
противофильтрационное ядро E18, I60, W251
противофильтрационный воротник C1417
~ слой W201
~ экран I59
~ ~ плотины S138
противоэрозионные мероприятия S125
~ сооружения E399
проточная система F535
проточный водонагреватель D324, 1256, N214
протравная грунтовка W87

соединительный уголок C998
~ фланец U101
~ штырь T184
~ элемент C1003, F86, R213
создание влажностно-воздушной среды F590
~ искусственного ландшафта L53
~ подпора D6
~ ~ воздуха P809
~ предварительного напряжения S1395
~ цементационной завесы C1403
сокол H111, M473
солёность S27
солнечный водонагреватель S879
~ коллектор S870
солнцезащитн‖ое стекло S1565
~ые устройства S877
солнцеотражающая плёнка для защиты бетона S878
~ теплоизоляция из алюминиевой фольги A383
соломит S1344
сооружаемый объект P891
сооружение B738, C1021, S1461
~ в стадии строительства S1466
~ дамбы D300
~ из бетона C884
~ из стальных тонкостенных элементов T146
сооружения W500
~ для регулирования паводков F451
~ и технические средства водоснабжения W239
~ на дорогах H363
~ на канализационной сети S382
соотношение сторон сечения A679
сопло N259
сопротивление R417
~ внедрению зонда в грунт P197
~ грунта S857
~ заносу S657
~ истиранию A11
~ материалов S1372
~ проскальзыванию S727
~ сваи погружению [забивке] D709
~ скольжению S727
~ трения F743
~ удару I46
~ при изгибе I29
~ хрупкому излому T291
сопротивляемость R417
сопряжение A36, C997
~ балок B252
~ ~ с колоннами B254
сопряжённые глубины A365
сороудерживающая решётка C598, R3, S138, T374
сорт G245
сортировка S142, 652
сортировочно-моечная установка S143
сортов‖ая сталь S208, 213
~ой прокат S213
сосредоточение C852
сосредоточенная нагрузка C851
~ сила C850
состав C782, 796, 799
~ бетона C881
~ для гидроизоляции швов J105
~ для заделки повреждений P121
~ для уплотнения резьбовых соединений труб P568
~ для ухода за бетоном C1390
~ смеси M317, 349
составление ведомостей и подсчёт объёмов строительных работ Q10
~ графиков производства строительных работ P879
~ сметы C1194
~ ~ по стоимости единицы площади здания A598
~ технических условий S950

составляющая C784
~ силы C785
составляющие бетонной смеси M340
~ напряжения S 378
~ смеси M340
составная балка B779, C800, H545
~ деревянная балка C800
~ ~ ~ на шпонках K17
~ железобетонная свая S252
~ колонна C790, L3
~ конструкция C791
~ стальная балка P543
составной единичный гидрограф C795
~ конструктивный элемент C793
~ преднапряжённый элемент S251
~ элемент C793
~ в ~ деревянных конструкций B782
состояние равновесия E361
сотовая конструкция H432
~ структура H432
сотовый шумоглушитель C253, S1017
софит S831
сохранение деревьев и растений T395
сочетание нагрузок L370
спаренн‖ое окно G125
~ый (гидро)домкрат J12
спецификация арматуры B138, R321, S1552
~ дверей D498
~ деталей D218, N19
~ конструктивных деталей D218
~ на пиломатериалы C1432
~ окон(ных) блоков) W398
~ резки арматуры C1432
спиральная арматура H281, L89, S997
~ камера S993
~ лестница S998
спирально-навивной воздуховод S999
спиральный арматурный каркас H279
~ гидроклассификатор S994
~ классификатор W83
спица N73
спицев‖ая плотина C677, N81
~ой затвор N77
сплачивание в четверть S479
сплошная обрешётка из досок R738
~ ~ кровли C562, S444
~ ~ ~ из досок R738
~ плита S891
~ стенка W290
сплошн‖ое трещинообразование C1093
~ой фундамент B809, S1067
сплошность C1084, S914
~ конструкции S1441
сплошные трещины C1093
спокойное течение S1354, 1490
спортивное здание S1025
~ сооружение S1025
способ нагружения L402
~ приложения нагрузки L402
способность к затирке T437
~ сохранять размеры D306
спринклер S1075
спринклерная головка S1076
~ система S1077
спрямляющ‖ая кривая E59
~ий канал C1411, 1416
спуск G245
~ воды B474
спускная задвижка O256
~ труба D356, S1031
спускное отверстие O246
спускной водовод D343
~ жёлоб C451, D715, R838, S1031
~ для строительного мусора R838
~ клапан B475, D626, S782
~ лоток D715, S1031

~ трубопровод D343
сработка водохранилища D656, 662
сращивание T183
~ стальных канатов S1007
средин‖ная поверхность M281
~ый продольный шов L479
среднегодовые осадки A818
среднее напряжение цикла M178
средний многолетний сток N222
~ пролёт I322
~ расход водотока A820
средник M545, T355, W381
~ оконного переплёта M228
средняя глубина живого сечения M179
~ линия лестничного марша W12
~ опора C301
~ плотность B787
~ расчётная полоса M280
~ стойка ворот I335
средства защиты пешеходов P180
~ малой механизации L2, R517
средство A154
~ крепления F86
срез S411, 420
~ под углом C363
~ полки N250
срезывающее напряжение S432
срок годности материала (при хранении в контейнере) P695
~ распалубливания S1410
~ службы L203
~ эксплуатации L203
сросток S1005
сруб C613
срыв потока B629
стабилизатор тяги D651
~ эмульсии S1101
стабилизационный пруд S1096
стабилизация S1102
~ грунта G402, S859
~ извести L287
~ сточных вод S1097
стабилизированн‖ая осадка S1099
~ый грунт S1100
стабилизирующий материал S1101
стабильность S1093
~ окраски [цвета] C698
ставень S528
ставни B447
ставня S528
стадия S1110
~ монтажа E395
~ разработки рабочего проекта D181
~ разрушения F42
~ строительства S1112
стаканный фундаментный блок C706
сталежелезобетонный мост C788
сталефибробетон S1213
сталефибротокретбетон S1214
стальбетон R294
стальная арматурная сетка S1220
~ зажимная планка G157
~ колонна S1129
~ опалубка S1216
~ ~ для кессонных перекрытий S1221
~ ~ для часторебристых перекрытий S1221
~ пластина с прорезями в полу F505
~ плита S1222
~ ~ с прорезями в полу F505
~ проволока S1228
~ прокатная балка R660
~ решётка S1217
~ свая двутаврового профиля H526
~ сетка для армирования дорожных покрытий R594
~ ~ для укрепления откосов R323
~ стойка S1129
~ шпунтовая перемычка S1224

тупиковый воздуховод S1084
турбидиметр T504
турбидиметрическая тонкость помола (цемента) T505
тычковая перевязка H125
тычковый кирпич B520, H124
~ ряд H126
~ ~ кладки B516
тычок H124
тюбинг S249
тюбингов‖ая крепь T464
~ый туннель T466
тюфячная кладка К32
~ одежда M159
тяга D649, H103, P935
~ в дымовой трубе C430
~ штукатурных карнизов H476
тягомер D650
тягучесть D788
тяж стропильной фермы К38
тяжёлая фашина S1564
тяжёлый бетон N231
~ заполнитель H263
~ конструктивный бетон O191

У

убежище S470
убежный уступ каменной кладки B205
уборка породы M528
~ строительных объектов или площадок C527
уборная C574, L116, T246, W123
увлажнение бетона D13, W105
~ опалубки F643
угловая врубка в стропильной ноге B382
~ деформация A475
~ капитель A446
~ колонна C1178
~ накладка N239
~ перевязка Q67
~ пилястра A495
~ ручная сверлильная машина A442
~ связь A442, К54
~ стамеска C1177
~ стойка A469
~ стропильная нога A470
~ схватка A442
~ тёрка E93
угловое долото C1177
~ перемещение A455
~ правило A452
~ соединение A455, C1179
~ ~ в полдерева E287
~ ~ прямыми шипами C710
угловой бетонный блок [камень] O36
~ вентиль A473
~ кирпич Q66, 68
~ контрфорс A445
~ кронштейн A443
~ подкос К54
~ полутёрок A452, C1182
~ раскос К54
~ тычковый камень Q69
~ шов C1179
углубление P602, R165
угол A440
~ естественного откоса A462, C1315, N55
~ здания Q66
~ наклона I83, P441
~ ~ стрелы B534
~ обрушения A459
~ откоса A467
~ пересечения I363
~ сдвига A465, S412
~ скольжения A466
~ стены Q66
~ трения A460, F749

уголковая накладка A441, 448
~ сталь A454
уголковый профиль A440, 471
уголок A440, 471
~ жёсткости R319
угольный фильтр A86
удар B468
~ волны W270
ударная вязкость I46, T291
~ нагрузка I37, S484
~ обработка (бучардой) N123
~ ~ камня D1, 24
~ прочность I46
~ ~ кубического образца C1373
~ ~ при изгибе I29
ударно-вращательное бурение R776
ударное бурение P218
ударно-канатн‖ое бурение C10
~ый буровой станок C9
ударно-отражательная дробилка I31
~ мельница I41, P942
ударный бур S1080
~ гайковёрт I49
~ перфоратор P217
~ уплотнитель I30
удельная влажность S956
~ водовместимость W223
~ водопотребность бетонной смеси U123
~ деформация S951, U109, 122
~ поперечная деформация U115
~ работа деформации M386, R406
удлинение E232, 473; L162, S1407
~ при разрыве B634
удобо‖обрабатываемость W469
~укладываемость W469
~формуемость R357
узел A724, D212, J83, N130, P52, S1697
~ дорог R588
~ пересечения стены W34
узкая аллея L60
~ колея N23
~ штыковая лопата G268
узкоколейная железная дорога N24
узловая нагрузка J99, N129, P51
узловые точки [узлы] нижнего пояса фермы L526
узлы строительных конструкций S1431
узор P130
узорчатое стекло F158
укатанная насыпь R659
укатанный асфальтобетон R657
~ насыпной грунт R659
укатка R681
укладка L122
~ арматуры R312
~ балок P671
~ бетона P456
~ инженерных сетей в стене S621
~ трубопроводов P393, 435
~ черепицы T207
уклон D314, G245, 260; S731
~ откоса B205
~ ската P441
~ ~ кровли R736
~ трубопровода P399
уклонная водомерная рейка S738
уклономер C555
укороченн‖ая деревянная балка T7
~ый поясной лист P89
украшение E320
укрепительная полоса N67
укрепление грунта G402, S859
~ известью L287
~ конструкции стойками S493
~ откосов P448
~ плетняком B702, 707
укреплённая часть обочины H77
укрупнённ‖ая опалубочная панель G19
~ый опалубочный щит G19
укрупнительная сборка на строи-

тельной площадке S635
укрывистость C1234, S1070
укрытие S470
улица S1360
~ с односторонним движением O87
уличный водосток R586
~дождеприёмник с решёткой [с решётчатой крышкой] G288
~ перекрёсток S1361
умывальная комната L116
умывальник B182, L116, W76
унитаз со смывным бачком T246
уплотнение C758, 1012; P10, 974; R73, S174, T21
~ бетонной смеси штыкованием R652
~ грунта E16, S846, 848
~ катками C759
~ швов W286
уплотнённый асфальтобетон R657
~ шов P9
уплотнитель бетонной смеси C880
~ насыпи B10
уплотнительн‖ые элементы W286
~ый валик A441
уплотняемость C752
уплотняющая масса J86, P985
~ мастика S176
~ для заливки швов в бетонных конструкциях C904
~ прокладка B221, G67, S177
уплотняющее кольцо G67
уплотняющий битумный состав B385
~ брусок B221
~ нащельник W381
~ слой I59, 61
~ состав B275
~ шнур G67
упор D460
упрочнение S1102, 1335
упрощённый расчёт R87, S587
упругая деформация E138
~ область E144
~ опора E151, Y14
~ работа E133
~ удельная деформация E148
упругий костыль E143
~ подъём сваи R163
~ центр E135
упругое амортизирующее опорное устройство F391
~ восстановление E145, R161, 406
~ прогиба R162
~ выпучивание E134
~ защемление E146
~ обжатие бетона E147
~ последействие D681, E132
~ соединение E137
уравни(ва)тель давления E345, U130
уравнительная башня S1644
~ камера S1643
~ шахта S1642
уравнительный бассейн B87, E347
~ резервуар E485, S1643
уравновешенная тара B71
уравновешенный затвор B75
~ момент B77
уравновешивание B82
~ напряжений B85
уравновешивающая сила E354
урез воды E275, W116, 154
уровень L175, S1000, W173
~ верхнего бьефа U178
~ воды H274, S1110, W173
~ высоких вод H348
~ громкости шума L517
~ настила D63
~ нижнего бьефа D578
~ низких вод L556
~ освещённости L191
~ подземных вод G419
~ полной воды H348
~ свободной водной поверхности F722

931

фракционированные образцы G250
фракция F684
фрамуга B565, T355, V90
~ над дверью F69
~ окна, откидывающаяся внутрь H450
французское окно G205
фреза земснаряда D668
фрезерный землесосный снаряд C1425
~ рыхлитель C1430
фриз B547
фризовая доска M80
~ ~, обрамляющая фронтон G4
фронт дорожно-строительных работ W479
фронтальная стена F20
фронтальный водозабор F758
~ одноковшовый погрузчик F757
~ погрузчик-планировщик на тракторе S507
фронтон A743, G3, P183
фронтонн||ая стена G6
~ое окно G7
фуганок J89
фугование J94
фундамент B158, 170; F608, 666
~ в виде решётчатой плиты G338
~ в форме опускных колодцев W322
~ глубокого заложения D81
~ из железобетонных свай C923
~ из массивов B455
~ из массивов-гигантов C29
~ из опускных колодцев C29
~ колонны C705
~ мелкого заложения S403
~ на опускных колодцах O102
~ неглубокого заложения S403
~ под машинное оборудование M7
~ стены W37
~ с уступами S1240
фундаментная балка F669, G381
~ плита F678, P129
~ стена F680
фундаментный блок F671, 672
~ болт A420, L199
~ стул I113
фундаментостроение F675
функикулёр F797
футеровка L325
~ газохода F542
~ дымовой трубы F542
~ дымохода F542
футеровочный блок [кирпич] L326

Х

хакболт H527
хворост B705
хворостяная выстилка B704
~ одежда B704
~ фашина K31
хворостяной тюфяк B706
хвостовик T26
хвостохранилище D387, S1022
хвосты обогащения R328, T9
химическая водоподготовка C415
~ гидрометрия C410
~ обработка воды C415
~ потребность в кислороде C412
химическое осаждение C413
~ укрепление грунта C411
хлев O245
хлоратор C445
хлораторная установка C444
хлорирование C443
хлоркаучуковые краски C442
~ окрасочные составы C442
хобот T397
~ для подачи бетона E202
ход строительства W499
ходов||ая площадка P547

~ые скобы M60
ходок F617
хозяйственно-бытов||ое водоснабжение P691
~ые сточные воды H512
хозяйственно-питьев||ое водоснабжение D479
~ой водопровод C669
хозяйственно-фекальн||ая канализационная сеть D478
~ые сточные воды D477
холл V129, E323a
холодильная камера C1142
~ установка R250
холодильник C1142
холодный асфальтобетон C662
холостой водосброс S124
~ ~ ГЭС W100
~ водослив W101
хомут B361, 718; C485, 556, 557; F322, S1260, Y20
~ для труб P368
хорда C448
хранилище S1285
хрупк||ий разрыв B688
~ое разрушение B688
хрупкость B690

Ц

цапфа P453
царга C1453
цвет C695
цветение воды W110
цветной цемент C696
цветостойкость C698
целик D803
цельнометаллическое здание A339
цельносварная стальная конструкция A358
цемент C254
~ для кладочных растворов M103
~ из природного мергеля N31
~ Лафарга L17
~ навалом N269
~ насыпью B785
~ с водоотталкивающей добавкой W198
~ с воздухововлекающей добавкой A227
~ Сореля O379, S899
~ Феррари F127
цементационная галерея G433
~ диафрагма G421
~ завеса G421
~ скважина G430, I212
цементационный раствор G442
цементация C257
~ грунта S842
~ ~ по периметру участка P232
~ скважины W320
~ швов J93
цементная краска C276
~ мельница C265
~ стяжка C279
~ черепица C903, 952
цементно-водное отношение C286
цементное молоко C277, L24
~ тесто C277, H547, M154, N68, 69
цементно-||известковый раствор C782
~песчаный раствор S53
~щебёночное (дорожное) покрытие C260
цементные минералы C274
цементный бункер C259, 280
~ завод C288
~ камень H75, M154, N69
~ клинкер C261
~ раствор C275
~ ~ с добавкой волоса H7
~ сырьевой шлам C266
~ шлам C281

цементовоз B786
цемент-пушка C268, 899; G479, S1048
центр вращения C296
~ города N265, U186
~ давления C295
~ изгиба S413
~ кривизны C292
~ масс C293
~ перспективы V26
централизованная система горячего водоснабжения C314
центральная вентиляционная установка C306
~ воздухоприготовительная установка C306
~ котельная C307, 311
~ опора C301
~ ось C330
~ полоса зелёных насаждений L52
~ разделительная полоса C320
~ система C308
~ ~ воздушного отопления C322
~ ~ кондиционирования воздуха C317
~ стойка C301, N113
~ установка кондиционирования воздуха C304
~ холодильная станция C319
центральное отопление C309
~ ядро E223, U201
центральный воздухоувлажнитель C313
~ колодец лучевого водозабора R16
~ кондиционер A198, C303
~ момент инерции C316
~ пролёт C300
~ ствол здания B747, E223
~ тепловой пункт C312
центрирование A320
центрифугированн||ая железобетонная труба C327
~ый бетон C326, S1082
цепная кладка C916
~ ~ контактная подвеска C203
~ линия C202
~ перевязка E316
~ пила C354
~ рычажная лебёдка C350
цепной долбёжник C351
~ строп C355
~ трубный ключ C357
цепь C343
цех D971
~ укрупнительной сборки конструкций A725
цикл автоклавной обработки A756
~ дозировки B194
~ замораживания и оттаивания F728
~ нагружения L372
~ нагрузки L372
~ пропаривания S1194
~ работы бетоносмесителя M333
циклическ||ие деформации C1439
~ое нагружение C1437
~ичн||ое нагружение C1437
~ые напряжения F539
~ый бетоносмеситель B196
циклон C1441
циклопическ||ая кладка C1447
~ий массив C1445
цилиндр R662
цилиндрическая оболочка B142, C1453
~ прочность C1454
~ ячеистая перемычка C1457
цилиндрические теплоизоляционные скорлупы C1459
цилиндрический бак-теплоаккумулятор H499
~ бетонный образец C1451
~ каменный блок A726
~ образец для испытания T104
~ резервуар C1462
~ свод A491, B142, 144

933

Ч

Ш